Metals Handbook®
Ninth Edition

Volume 5
Surface Cleaning, Finishing, and Coating

Prepared under the direction of
the ASM Handbook Committee

Planned, prepared, organized and reviewed by the ASM Surface Treating and Coating Division Council

Coordinator

William G. Wood

William H. Cubberly, Director of Publications
Robert L. Stedfeld, Assistant Director of Reference Publications
Kathleen Mills, Managing Editor
Lennard G. Kruger, Technical Editor
Bonnie Sanders, Production Editor
Laura Borges Monroe, Copy Editor

AMERICAN SOCIETY FOR METALS
METALS PARK, OHIO 44073

First printing, October, 1982

Metals Handbook is a collective effort involving thousands of technical specialists. It brings together in one book a wealth of information from world-wide sources to help scientists, engineers and technicians solve current and long-range problems.

Great care is taken in the compilation and production of this volume, but it should be made clear that no warranties, express or implied, are given in connection with the accuracy or completeness of this publication, and no responsibility can be taken for any claims that may arise.

Nothing contained in the Metals Handbook shall be construed as a grant of any right of manufacture, sale, use or reproduction, in connection with any method, process, apparatus, product, composition or system, whether or not covered by letters patent, copyright or trademark, and nothing contained in the Metals Handbook shall be construed as a defense against any alleged infringement of letters patent, copyright or trademark, or as a defense against any liability for such infringement.

Comments, criticisms and suggestions are invited, and should be forwarded to the American Society for Metals.

Library of Congress Cataloging in Publication Data

American Society for Metals

Surface cleaning, finishing, and coating.

(Metals Handbook; 9th ed., v. 5)
Includes bibliographical references.
1. Metals—Finishing. 2. Protective coatings.
I. Wood, William G. II. American Society for Metals. Handbook Committee.
III. ASM Surface Treating and Coating Division Council. IV. Series.
TA459.A5 9th ed. vol. 5 [TS213] 669s [671.7] 82-13844 ISBN 0-87170-011-5

SAN 204-7586
Printed in the United States of America

Foreword

As technology has grown more complex and sophisticated, the operating and environmental demands on metals have steadily grown more stringent. Temperatures in turbines and other energy systems have soared. The range of corrosive environments that metals must withstand has broadened. And the complexity of conditions that metals must now handle — combined conditions of stress, corrosion, temperature — puts new demands on metals and the finishing and coating systems used to protect them.

So it is not surprising that, since the publication of the last Metals Handbook on this subject 18 years ago, the technology of cleaning, finishing, and coating has expanded greatly. This growth is reflected in the present volume, both in the range of information covered and in the depth of treatment.

In the 8th Edition, only half of Volume 2 was needed to present the subject adequately; a completely separate volume has been required for the 9th Edition. And this Volume 5 is double the size of the 362 pages devoted to cleaning and finishing in the 8th Edition.

Continuing the unique process for ensuring quality and completeness of the Metals Handbook series, every article in this book has been prepared and reviewed by technical experts from industry and academia. Nearly 250 specialists contributed their knowledge and practical experience to the writing and editing of this comprehensive record of cleaning, coating, and finishing technology.

For their efforts, a tribute is due the Handbook Committee, the Surface Treating and Coating Division Council, the editorial staff of the Handbook, and the many individual contributors and reviewers from industrial companies, research organizations, government establishments, and educational institutions. For all of us who will learn from and be helped by their efforts, we extend our gratitude and appreciation.

David Krashes
President

Allan Ray Putnam
Managing Director

The Ninth Edition of Metals Handbook
is dedicated to the memory of
TAYLOR LYMAN, A. B. (Eng.), S. M., Ph. D.
(1917-1973)
Editor, Metals Handbook 1945-1973

Preface

Surface cleaning, finishing, and coating technology is normally included with one or more disciplines in the technical literature produced by the American Society for Metals. With the Ninth Edition of the *Metals Handbook*, the explosive growth of new information in the field necessitates the publication of a separate volume.

Complete quality control programs, essential for maintaining both the excellence of performance and the durability required of modern manufactured products, require the most effective cleaning systems available. The basic principles of cleaning are relatively unchanged, but new techniques are constantly being developed to expand capabilities and ensure product quality. Environmental regulations are also a major concern in cleaning technology. Both areas are given top priority in this book.

Finishing technology in the aircraft engine, hydraulic, and bearing industries has always been considered a primary functional operation because of the human safety factors involved. In contrast, many manufactured products traditionally have been finished only to meet aesthetic or simple appearance requirements. This situation is rapidly changing as competitors in the world market strive to make their assemblies reliable and useful for maximum periods of time. Finishing processes are constantly being improved to accommodate high productivity and quality requirements. Each of the finishing operations described in Volume 5 incorporates the latest technology available at the time of printing and is expected to meet today's stringent requirements in the materials field.

Coating of materials either before or after forming is one of the most rapidly growing areas of scientific development in this century. The need for materials and processes to improve wear, corrosion resistance, conductivity, absorption, and formability is a major challenge. Lowering frictional resistance, reducing vehicle weight, and increasing the lifetime of materials used in rapidly changing environments are other demands being met with modern coatings. The coatings described in Volume 5 reflect this new technology.

The contributors to this Volume of the *Metals Handbook* have undertaken their work with the express purpose of producing a publication that will meet the needs of industry and the people who bring about progress. You should find this to be true.

William G. Wood
Coordinator
Metals Handbook Activity
Surface Treating and
Coating Division

Policy on Units of Measure

By a resolution of its Board of Trustees, the American Society for Metals has adopted the practice of publishing data in both metric and customary U.S. units of measure. In preparing this Handbook, the editors have attempted to present data primarily in metric units based on Système Internationale d'Unités (SI), with secondary mention of the corresponding values in customary U.S. units. The decision to use SI as the primary system of units was based on the aforementioned resolution of the Board of Trustees, the widespread use of metric units throughout the world, and the expectation that the use of metric units in the United States will increase substantially during the anticipated lifetime of this Handbook.

For the most part, numerical engineering data in the text and in tables are presented in SI-based units with the customary U.S. equivalents in parentheses (text) or adjoining columns (tables). For example, pressure, stress and strength are shown in both SI units, which are pascals (Pa) with a suitable prefix (see the description of SI at the back of the volume), and in customary U.S. units, which are pounds per square inch (psi). To save space, large values of psi have been changed to kips per square inch (ksi), where one kip equals 1000 pounds. Some strictly scientific data are presented in SI units only.

To clarify some illustrations that depict machine parts described in the text, only one set of dimensions is presented on artwork. References in the accompanying text to dimensions in the illustrations are presented in both SI-based and customary U.S. units.

On graphs and charts, grids correspond to SI-based units, which appear along the left and bottom edges; where appropriate, corresponding customary U.S. units appear along the top and right edges. Some previously published charts, particularly histograms depicting statistical distribution of values of mechanical properties, could not be redrawn because of the absence of the original data points; these have been reproduced in their original forms, with SI equivalents on the top and right edges.

Data pertaining to a specification published by a specification-writing group may be given in only the units used in that specification or in dual units, depending on the nature of the data. For example, the typical yield strength of aluminum sheet made to a specification written in customary U.S. units would be presented in dual units, but the thickness ranges listed in that specification might be presented only in inches.

Data obtained according to specified test methods for which the specification implies a particular system of units are presented in the units of that system. Wherever feasible, equivalent units are also presented.

Conversions and rounding have been done in accordance with ASTM Standard E380, with careful attention to the number of significant digits in the original data. For example, an annealing temperature of 1575 °F contains three significant digits (and possibly only two), because few commercial heat treatment systems can control the temperature of an entire load of parts within a spread of 10 °F. In this instance, the equivalent temperature would be given as 860 °C, or perhaps 850 °C depending on the degree of accuracy meant to be conveyed in the conversion; the exact conversion to 857.22 °C would not be appropriate. In many instances (especially in tables and data compilations), temperature values in °C and °F are alternatives rather than conversions.

The policy on units of measure in this Handbook contains several exceptions to strict conformance to ASTM E380; in each instance, the exception has been made to improve the clarity of the Handbook. Three examples of such exceptions are the use of "L" rather than "l" as the abbreviation for litre, reporting temperature in °C rather than K and reporting stress intensity in $MPa\sqrt{m}$ rather than $MNm^{-3/2}$.

SI practice requires that only one virgule (diagonal) appear in units formed by combination of several basic units. Therefore, all of the units preceding the virgule are in the numerator and all units following the virgule are in the denominator of the expression (and no parentheses are required to prevent ambiguity).

Handbook Committee, Officers and Trustees

Members of the ASM Handbook Committee (1981-1982)

Officers and Trustees of the American Society for Metals

Previous Chairmen of ASM Handbook Committee

R. S. Archer
(1940-1942) (Member, 1937-1942)

L. B. Case
(1931-1933) (Member, 1927-1933)

E. O. Dixon
(1952-1954) (Member, 1947-1955)

R. L. Dowdell
(1938-1939) (Member, 1935-1939)

J. P. Gill
(1937) (Member, 1934-1937)

J. D. Graham
(1965-1968) (Member, 1961-1970)

J. F. Harper
(1923-1926) (Member, 1923-1926)

C. H. Herty, Jr.
(1934-1936) (Member, 1930-1936)

J. B. Johnson
(1948-1951) (Member, 1944-1951)

R. W. E. Leiter
(1962-1963)
(Member, 1955-1958, 1960-1964)

G. V. Luerssen
(1943-1947) (Member, 1942-1947)

Gunvant N. Maniar
(1978-1980) (Member, 1974-1980)

W. J. Merten
(1927-1930) (Member, 1923-1933)

N. E. Promisel
(1955-1961) (Member, 1954-1963)

G. J. Shubat
(1972-1975) (Member, 1966-1975)

W. A. Stadtler
(1968-1972) (Member, 1962-1972)

Raymond Ward
(1976-1978) (Member, 1972-1978)

Martin G. H. Wells
(1980-1982) (Member, 1976-1982)

D. J. Wright
(1964-1965) (Member, 1959-1967)

ASM Surface Treating and Coating Division Council

D. Cameron Perry
Chairman
Armco Inc. (retired)

Howard S. Bender
Senior Research Chemist
Polymers Department
General Motors Corp.

Raymond J. Donahue
Director, Advanced Materials
Mercury Marine

William D. Fields
Vice President
ELNIC, Inc.

Herman L. Gewanter
Manager, Technical Service
Pfizer, Inc.

Arthur D. Godding
Manager, Technical Service & Equipment Engineering
Heatbath Corp.

Robert D. Grear
Finishing Consultant
International Harvester Co.

J. Bernard Hignett
Vice President
Harper Co.

William Hintalla
Manager of Quality Control
McLough Steel Corp.

Robert F. Hochman
Professor and Associate Director for Metallurgy
Georgia Institute of Technology

Ronald J. Joniec
General Manager
Harry Miller Corp.

Farrell M. Kilbane
Senior Staff Physicist
Armco Inc.

John B. Kittredge
Consultant

Joseph F. Loersch
Supervisor, Metal Finishing
Pratt & Whitney Aircraft Group

William J. Schumacher
Senior Staff Engineer
Armco Inc.

Stephen R. Shatynski
Professor
Rensselaer Polytechnic Institute

Daryl E. Tonini
Manager, Technical Services
American Hot Dip Galvanizers Association

William G. Wood
Vice President
Kolene Corp.

Reed Yount
General Electric Co.

Donald R. Zaremski
Research Specialist
Allegheny Ludlum Steel Corp.

Stanley E. Zwick
Plant Metallurgist
Caterpillar Tractor Co.

Author Committees

Metal Cleaning

George Shepard
Chairman
Research Engineer
Republic Steel Corp.

Richard W. Clement
Sales Manager
Industrial Equipment Division
Detrex Chemical Industries Inc.

Michael P. D'Angelo
Experimental Product Manager
MacDermid Inc.

H. L. Gewanter
Manager, Technical Service
Chemical Division
Pfizer, Inc.

Harry Hicks
Senior Research Engineer
Boeing Aerospace Co.

Robert M. Hudson
Research Consultant
Research Laboratory
U. S. Steel Corp.

Ronald J. Joniec
General Manager
Harry Miller Corp.

Raymond M. Leliaert
Technical Manager
Materials Cleaning Systems Division
Wheelabrator-Frye Inc.

Kenneth Lindblad
Specialist Engineer
Boeing Aerospace Co.

Thomas F. McCardle
Supervisor–Sales
Kolene Corp.

Homer O. Meserve
Specialist Engineer
Boeing Aerospace Co.

Donald P. Murphy
Senior Research Chemist
Parker Surface Treatments
Occidental Chemical Corp.

Thomas P. O'Keeffe
Industrial Hygienist
Boeing Aerospace Co.

John Pilznienski
Senior Research Chemist
Kolene Corp.

John H. Rains
Applications Chemist
Ethyl Corp.

William J. Roberts
Senior Sales Engineer
Chemical Group
GAF Corp.

Stephen R. Shatynski
Professor
Rensselaer Polytechnic Institute

Robert Simmons
Research Supervisor
Dow Chemical Co.

Bob Srinivasan
Research Manager
Diversey Wyandotte Corp.

Kenneth S. Surprenant
Research Leader
Dow Chemical Co.

Neil Weightman
Sales Manager
Comco, Inc.

James D. Winchester
Technical Service Representative
Chemical Division
Pfizer Inc.

Michael M. Woelfel
Technical Services Manager – Impact Products
Potters Industries, Inc.

William G. Wood
Vice President
Technology/Research and Development
Kolene Corp.

Mechanical Finishing

J. Bernard Hignett
Chairman
Vice President
The Harper Co.

Thomas C. Andrew
Executive Vice President
The Harper Co.

James Belanger
President
Belanger Inc.

John Coffield
President
The Harper Co.

William Downing
Process Engineer
Lea Manufacturing Co.

Ernest J. Duwell
Corporate Scientist
3M Co.

Robert W. Johannesen
President
Rodeco Co.

John B. Kittredge
Consultant

Clifford S. Mehelich
Senior Vice President
Metal Improvement Co., Inc.

Alfred F. Scheider
Vice President
Research & Development
Osborn Manufacturing Corp.

Gary A. Snell
Sales Representative
Wisconsin Porcelain Co.

Edward H. Tulinski
Regional Manager
The Harper Co.

L. Van Kuiken
President
Progressive Blasting Systems

Michael M. Woelfel
Technical Services Manager–Impact Products
Potters Industries, Inc.

Plating and Electropolishing

Myron E. Browning
Co-Chairman
Project Manager, Materials and Producibility
Rocketdyne
Division of Rockwell International

Donald Gagas
Co-Chairman
Materials and Process Engineer
Centralab

Roger E. Marce
Co-Chairman
Manager, Technical Services
Zinc Institute Inc.

Donald W. Baudrand
Vice President
New Market Development
Allied-Kelite
Division of the Richardson Co.

Craig J. Brown
Vice President and General Manager
Eco-Tec Ltd.

Hyman Chessin
Senior Research Associate
Plating Division
M & T Chemicals Inc.

Paul E. Davis
Manager
Tin Research Institute, Inc.

Marshall J. Deitsch
Vice President
Vanguard Pacific, Inc.

Ronald N. Duncan
Director, Research & Engineering
ELNIC, Inc.

Everett H. Fernald, Jr.
President
Induplate, Inc.

William D. Fields
Vice President
ELNIC, Inc.

Herbert Geduld
Board Chairman
Columbia Chemical Corp.

Louis Gianelos
Manager, Central Technical Staff
Surface Technology Department
Industrial Finishing Group
Harshaw Chemical Co.

Juan Hajdu
Vice President, Technology
Enthone Inc.

Russell A. Henry, Jr.
President
Wear-Cote International, Inc.

Alexander Korbelak
Consultant
Smith Precious Metals Co.

Douglas W. Maitland
Applications Engineer
Vanguard Pacific, Inc.

Warren McMullen
Commercial Development Manager
M & T Chemicals Inc.

Fred I. Nobel
Executive Vice President
LeaRonal Inc.

Art O'Cone
Marketing Manager
3M Co.

Vincent Paneccasio
Supervisor, Organic Research
Enthone Inc.

Nicholas J. Spiliotis
Supervisor, Technical Service
Allied Chemical Co.

Henry Strow
Oxyphen Products Co.

Lavern M. Weisenberger
Product Development Manager
Allied-Kelite Division
WITCO Chemical Corp.

Louis S. Winter
Vice President
Hydrite Chemical Co.

Joseph R. Zickgraf
Technical Services Director
ELNIC, Inc.

Metallic Coating Processes Other Than Plating

Daryl E. Tonini
Chairman
Manager, Technical Services
American Hot Dip Galvanizers Association, Inc.

Robert Baboian
Senior Fellow
Texas Instruments Inc.

Edward A. Barsditis
Senior Manufacturing Engineer, Metals Joining
Westinghouse Electric Corp.

Edwin S. Bartlett
Senior Researcher
Battelle Columbus Laboratories

Harry A. Beale
Associate Section Manager
Battelle Columbus Laboratories

Serge Belisle
Research Engineer
Noranda Research Centre

John M. Blocher, Jr.
CVD Consultant

James H. Clare
Principal Research Scientist
Battelle Columbus Laboratories

Daryl E. Crawmer
Research Scientist
Battelle Columbus Laboratories

T. W. Fisher
Supervisor, Metallic Coatings Development
Homer Research Laboratories
Bethlehem Steel Corp.

Daniel Gittings
Chief Research Engineer – Coated Sheets
U. S. Steel Research Laboratory

Hart F. Graff
Principal Research Engineer
Armco Inc.

Robert F. Hochman
Associate Director for Metallurgy
Georgia Institute of Technology

Farrell M. Kilbane
Senior Staff Physicist
Armco Inc.

Donald M. Mattox
Supervisor, Surface Metallurgy Division
Sandia National Laboratories

Daniel J. Maykuth
Manager
Tin Research Institute, Inc.

Carl F. Mietzner
Manager
Corrosion and Coatings Research Section
Bethlehem Steel Corp.

Wolf-Dieter Münz
Manager
Research & Development Department for Coating
Leybold-Heraeus GmbH

Richard A. Nickola
Supervising Research Engineer
Inland Steel Co.

David C. Pearce
Senior Research Metallurgist
ASARCO Inc.

David V. Rigney
Manager, Advanced Programs
General Electric Co.

D. J. Schardein
Director, Finishing Technology Section
Reynolds Metals Co.

John A. Thornton
Vice President
Research and Development
Telic Co.

H. E. Townsend
Supervisor, Corrosion and Surface Research
Homer Research Laboratories
Bethlehem Steel Corp.

Nonmetallic Coating Processes

Dean M. Berger
Corrosion Specialist
Gilbert/Commonwealth Companies

Clifton G. Bergeron
Professor and Head
Department of Ceramic Engineering
University of Illinois at Urbana

G. Thomas Cavanaugh
Manager of Finishing Engineering
Jenn-Air Corp.

John E. Cox
Special Projects Manager
O. Hommel Co.

James W. Davis
Group Leader
Amchem Products, Inc.

James W. Elliott
Vice President, Manufacturing
Porcelain Industries, Inc.

Richard A. Eppler
Scientist, Research & Development
Pemco Products
Mobay Chemical Corp.

Wayne L. Gasper
Chief Process Engineer
Maytag Co.

M. B. Gibbs
Assistant Superintendent
Technical Service Department
Inland Steel Co.

Albert L. Gugeler
Manager, Frit Quality Control
Ferro Corp.

Howard H. Hovey
Chief, Standardization Branch
Packaging Division
Department of the Army

Albert O. Hungerford
Chemical and Coatings Specialist
Butler Manufacturing Co.

Roy C. Kissler
Program Manager, Environment & Energy
General Electric Co.

Howard G. Lasser
Materials Engineer
Materials Research Consultants

Robert K. Laird
Ceramic Engineer
American Standard, Inc.

Daniel H. Luehrs
Ceramic Engineer
Clyde Division
Whirlpool Corp.

James I. Maurer
Director, Technology
Parker Division
Occidental Chemical Corp.

Dennis E. McCloskey
Production Manager
White Consolidated Industries — Mansfield

John C. Oliver
Executive Vice President
Porcelain Enamel Institute, Inc.

James F. Quigley
Manager, Porcelain Enamel Coatings and Frit Operations
Ferro Corp.

Melvin H. Sandler
Consulting Services for Protective Coatings

Donald R. Sauder
Division Finishing Manager
Tappan Co.

Albert J. Schmidt
Director of Research
American Porcelain Enamel Co.

Howard F. Smalley
Technical Service Coordinator
Pemco Products
Mobay Chemical Corp.

L. N. Smith
Technical Director
Porcelain Metals Corp.

Thomas L. Stalter
Manager, Porcelain Enamel Development
Pemco Products
Mobay Chemical Corp.

Larry L. Steele
Research Metallurgist
Armco Inc.

Lester E. Steinbrecher
Director, Research & Development
Amchem Products, Inc.

Carl G. Strobach
Director, Advanced Engineering and Technical Services
Rheem Water Heating Division
City Investing Co.

James D. Sullivan
Manager, Ceramic Research Laboratory
A. O. Smith Corp.

Thomas A. Taylor
Consultant, Materials Development
Engineering Products Division
Union Carbide Corp.

Donald A. Toland
Associate Research Consultant
U. S. Steel Corp.

Donald B. Tolly
Senior Application Engineer — Ceramics
General Electric Co.

Carl M. Varga
General Sales Manager
Man-Gill Chemical Co.

James K. White
Chemical Process Engineer
Caterpillar Tractor Co.

Daniel R. Yearick
Senior Manufacturing Engineer
Caloric Corp.

Floyd I. Young
Principal Project Engineer
Motor Wheel Corp.

Cleaning and Finishing of Stainless Steel and Heat-Resisting Alloys

Joseph A. Douthett
Senior Staff Engineer
Armco Inc.

Robert R. Gaugh
Senior Staff Engineer
Armco Inc.

Leonard Rozenberg
Research Project Engineer
Wyman-Gordon Co.

Bob Srinivasan
Research Manager
Diversey Wyandotte Corp.

J. D. VanDevender
Chemist
Huntington Alloys, Inc.

Cleaning and Finishing of Nonferrous Metals

Keith Ball
Manufacturing Engineer Manager
Brooks & Perkins

Howard G. Lasser
Materials Engineer
Materials Research Consultants

Donald J. Levy
Senior Staff Scientist
Lockheed Palo Alto Research Laboratory

Roger E. Marce
Manager, Technical Services
Zinc Institute Inc.

Allan W. Morris
Technical Specialist
McDonnell Aircraft Co.

Robert M. Paine
Senior Chemist
Brush Wellman, Inc.

John F. Pashak
Project Leader
Dow Chemical Co.

D. J. Schardein
Director, Finishing Technology Section
Reynolds Metals Co.

Bob Srinivasan
Research Manager
Diversey Wyandotte Corp.

J. D. VanDevender
Chemist
Huntington Alloys, Inc.

Chun T. Wang
Senior Research Metallurgist
Teledyne Wah Chang Albany

R. Terrence Webster
Principal Metallurgical Engineer
Teledyne Wah Chang Albany

Douglas H. Wilson
Manager, Customer Technical Services
RMI Co.

William G. Wood
Vice President
Technology/Research and Development
Kolene Corp.

Walter G. Zelley
Technical Manager
Surface Technology Division
Alcoa Laboratories

Contents

Metal Cleaning

Selection of Cleaning Process

By the ASM Committee on Selection of Cleaning Process*

CLEANING PROCESSES used for removing soils and contaminants are varied and can be compared for different levels of effectiveness. Major emphasis is placed on the cleaning of ferrous metals, but the methods discussed are often applicable to the cleaning of nonferrous metals.

The processing procedures, equipment requirements, effects of variables, and safety precautions that are applicable to individual cleaning processes are dealt with in separate articles in this Volume on vapor degreasing, solvent cleaning, acid cleaning, emulsion cleaning, alkaline cleaning, salt bath descaling, pickling, abrasive blast cleaning, and barrel finishing. A brief overview of EPA, OSHA, and NFPA regulations as they affect cleaning operations is also included in this article.

The cleaning of nonferrous and highly alloyed ferrous metals is considered in detail in the articles on specific processes enumerated above, as well as in articles in this Volume concerning the cleaning and finishing of specific metals, such as aluminum, copper alloys, heat-resisting alloys, magnesium, zinc, nickel, lead, titanium, and stainless steel.

Selection of Cleaning Process

In selecting a metal cleaning process, many factors must be considered, including: (*a*) the identification and characterization of the soil to be removed; (*b*) identification of the substrate to be cleaned and the importance of the condition of the surface or structure to the ultimate use of the part; (*c*) degree of cleanness required; (*d*) capabilities of the available facilities; (*e*) impact of the process on the environment; (*f*) overall cost of the process; and (*g*) subsequent operations to be applied, such as phosphating, plating, and painting. Very few factors in these analyses can be accurately quantified, which results in a subjective analysis. Frequently several sequences of operations may be chosen which all produce the desired end result. As in most industrial operations, the tendency is to provide as much flexibility and versatility in a facility as the available budget will allow. The size and shape of the largest predicted workpiece is generally used to establish the cleaning procedure, equipment sizes, and handling techniques involved.

Because of the variety of cleaning materials available and the process step possibilities, the selection of a cleaning procedure depends greatly on the degree of cleanliness required and subsequent operations to be performed. Abrasive blasting produces the lowest degree of cleanliness. Solvent, solvent vapor degrease, emulsion soak, alkaline soak, alkaline electroclean, alkaline plus acid cleaning, and finally ultrasonics each progressively produce a cleaner surface. In addition to these conventional methods, very exotic and highly technical procedures exist which have been developed in the electronics and space efforts to produce clean surfaces far above the normal requirements for industrial use.

Cleaning Media. An understanding of the mechanics of the cleaning action is helpful in selecting a process. Solvent cleaning, as the name implies, is the dissolution of the contaminant by a liquid, such as organic solvents and chlorinated hydrocarbons; for example, trichloroethylene, methylene chloride, toluene, or benzene. The mechanics are accomplished through swabbing, tank immersion, spray or solid stream flushing, or condensation of the vapor phase, as is found in vapor degreasing. Temperature elevation accelerates the activity. One major drawback to solvent

*Homer O. Meserve, *Chairman*, Specialist Engineer, Boeing Aerospace Co.; Harry Hicks, Senior Research Engineer, Boeing Aerospace Co.; Kenneth Lindblad, Specialist Engineer, Boeing Aerospace Co.; Thomas P. O'Keeffe, Industrial Hygienist, Boeing Aerospace Co.; Bob Srinivasan, Research Manager, Diversey Wyandotte Corp.

cleaning is the possibility of leaving some residues on the surface, often requiring additional cleaning.

Emulsion cleaning depends on the physical action of emulsification, in which discrete particles of contaminant are suspended in the cleaning medium and then separated from the surface to be cleaned. Emulsion cleaners can be water or water solvent-based solutions; for example, emulsions of hydrocarbon solvents such as kerosine and water containing an emulsifiable surfactant. To maintain stable emulsions, coupling agents such as oleic acid are added.

Alkaline cleaning is the mainstay of industrial cleaning and may employ both physical and chemical actions. These cleaners contain combinations of ingredients such as surfactants, sequestering agents, saponifiers, emulsifiers, and chelators, as well as various forms of stabilizers and extenders. Except for saponifiers, these ingredients are physically active and operate by reducing surface or interfacial tension, by formation of emulsions and by suspension or flotation of insoluble particles. Solid particles on the surface are generally assumed to be electrically attracted to the surface. During the cleaning process, these particles are surrounded by wetting agents to neutralize the electrical charge and are floated away, held in solution suspension indefinitely, or eventually are settled out as a sludge in the cleaning tank. Saponification is a chemical reaction involving hydrolysis of an ester to produce a soap.

Electrolytic cleaning is a modification of alkaline cleaning in which an electrical current is imposed on the part to produce vigorous gassing on the surface to promote the release of soils. Electrocleaning can be either anodic or cathodic cleaning. Anodic cleaning is also called "reverse cleaning", and cathodic cleaning is called "direct cleaning". The release of oxygen gas under anodic cleaning or hydrogen gas under cathodic cleaning in the form of tiny bubbles from the work surface greatly facilitates lifting and removing surface soils.

Abrasive cleaning uses small sharp particles propelled by an air stream or water jet to impinge on the surface, removing contaminants by the resulting impact force. A wide variety of abrasive media in many sizes is available to meet specific needs. Abrasive cleaning is often preferred for removing heavy scale and paint, especially on large, otherwise inaccessible areas. Abrasive cleaning is also frequently the only allowable cleaning method for steels sensitive to hydrogen embrittlement. This method of cleaning is also used to prepare metals, such as stainless steel and titanium, for painting to produce a mechanical lock for adhesion because conversion coatings cannot be applied easily to these metals.

Acid cleaning is used more often in conjunction with other steps than by itself. Acids have the ability to dissolve oxides, which are usually insoluble in other solutions. Straight mineral acids, such as hydrochloric, sulfuric, and nitric acids, are used for most acid cleaning, but organic acids, such as citric, oxalic, acetic, tartaric, and gluconic acids, occupy an important place in acid cleaning because of their chelating capability.

Molten salt cleaning is not commonly used, although it is very effective for removing many soils, especially paints and heavy scale. The very high operating temperature and high facility cost discourage widespread use. Ultrasonic cleaning uses sound waves passed at very high frequency through liquid cleaners, which could be alkaline, acid, or even organic solvents. The passage of ultrasonic waves through the liquid media creates tiny gas bubbles, which offer vigorous scrubbing action on the parts being cleaned. This action brings about the most efficient of all the cleaning possible. Because of high capital investment required for ultrasonic units, the process is often cost prohibitive and limited to applications requiring extreme cleanliness.

Substrate Considerations. The selection of a cleaning process must be based on the substrate being cleaned as well as the soil to be removed. Metals such as aluminum and magnesium require special consideration because of their sensitivity to attack by chemicals. Aluminum is dissolved rapidly by both alkalis and acids. Magnesium is resistant to alkaline solutions with a high pH value, but is attacked by many acids. Copper is merely stained by alkalis, yet severely attacked by oxidizing acids (such as nitric acid) and only slightly by others. Zinc and cadmium are attacked by both acids and alkalis. Steels are highly resistant to alkalis and attacked by essentially all acidic material. Corrosion-resistant steels, also referred to as stainless steels, have a high resistance to both acids and alkalis, but the degree of resistance depends on the alloying elements. Titanium and zirconium have come into common use because of their excellent chemical resistance. These two metals are highly resistant to both alkalis and acids with the exception of acid fluorides which attack them rapidly and severely.

Types of soil may be broadly classified into six groups: (*a*) pigmented drawing compounds, (*b*) unpigmented oil and grease, (*c*) chips and cutting fluids, (*d*) polishing and buffing compounds, (*e*) rust and scale, and (*f*) miscellaneous surface contaminants, such as lapping compounds and residue from magnetic particle inspection. These six types of soil are dealt with separately in the order listed.

Removal of Pigmented Drawing Compounds

All pigmented drawing lubricants are difficult to remove from metal parts. Consequently, many plants review all aspects of press forming operations to avoid the use of pigmented compounds. Pigmented compounds most commonly used contain one or more of the following substances: whiting, lithopone, mica, zinc oxide, bentonite, flour, graphite, white lead (which is highly toxic), molybdenum disulfide, and soaplike materials. Some of these substances are more difficult to remove than others. Because of their chemical inertness to acid and alkali used in the cleaners and tight adherence to metal surfaces, graphite, white lead, molybdenum disulfide, and soaps are the most difficult to solubilize and remove.

Certain variables in the drawing operation may further complicate the removal of drawing lubricants. For example, as drawing pressures are increased, the resulting higher temperatures increase the adherence of the compounds to the extent that some manual scrubbing is often an essential part of the subsequent cleaning operation. Elapsed time between the drawing and cleaning operations is also a significant factor; the longer the compound is allowed to dry, the more difficult it becomes to remove.

Table 1 indicates cleaning processes typically selected for removing pigmented compounds from drawn and stamped parts such as Parts 1 through 6 in Fig. 1.

Emulsion cleaning is one of the most effective methods for removing pigmented compounds, because it relies on mechanical wetting and floating the contaminant away from the sur-

Table 1 Metal cleaning processes

Processes are listed in order of decreasing preference

Type of production	In-process cleaning	Preparation for painting	Preparation for phosphating	Preparation for plating
Removal of pigmented drawing compounds(a)				
Occasional or intermittent	Hot emulsion hand slush, spray emulsion in single stage, vapor slush degrease(b)	Boiling alkaline, blow off, hand wipe Vapor slush degrease, hand wipe Acid clean(c)	Hot emulsion hand slush, spray emulsion in single stage, hot rinse, hand wipe	Hot alkaline soak, hot rinse (hand wipe, if possible), electrolytic alkaline, cold water rinse
Continuous high production	Conveyorized spray emulsion washer	Alkaline soak, hot rinse, alkaline spray, hot rinse	Alkaline or acid(d) soak, hot rinse, alkaline or acid(d) spray, hot rinse	Hot emulsion or alkaline soak, hot rinse, electrolytic alkaline, hot rinse
Removal of unpigmented oil and grease				
Occasional or intermittent	Solvent wipe Emulsion dip or spray Vapor degrease Cold solvent dip Alkaline dip, rinse, dry or dip in rust preventive)	Solvent wipe Vapor degrease or phosphoric acid clean(d)	Solvent wipe Emulsion dip or spray, rinse Vapor degrease	Solvent wipe Emulsion soak, barrel rinse, electrolytic alkaline rinse, hydrochloric acid dip, rinse
Continuous high production	Automatic vapor degrease Emulsion, tumble, spray, rinse, dry	Automatic vapor degrease	Emulsion power spray, rinse Vapor degrease Acid clean(c)	Automatic vapor degrease, electrolytic alkaline rinse, hydrochloric acid dip, rinse(e)
Removal of chips and cutting fluid				
Occasional or intermittent	Solvent wipe Alkaline dip and emulsion surfactant Stoddard solvent or trichlorethylene Steam	Solvent wipe Alkaline dip and emulsion surfactant Solvent or vapor	Solvent wipe Alkaline dip and emulsion surfactant(f) Solvent or vapor	Solvent wipe Alkaline dip, rinse, electrolytic alkaline(g), rinse, acid dip, rinse(h)
Continuous high production	Alkaline (dip or spray) and emulsion surfactant	Alkaline (dip or spray) and emulsion surfactant	Alkaline (dip or spray) and emulsion surfactant	Alkaline soak, rinse, electrolytic alkaline(g), rinse, acid dip and rinse(h)
Removal of polishing and buffing compounds				
Occasional or intermittent	Seldom required	Solvent wipe Surfactant alkaline (agitated soak), rinse Emulsion soak, rinse	Solvent wipe Surfactant alkaline (agitated soak), rinse Emulsion soak, rinse	Solvent wipe Surfactant alkaline (agitated soak), rinse, electroclean(j)
Continuous high production	Seldom required	Surfactant alkaline spray, spray rinse Agitated soak or spray, rinse(k)	Surfactant alkaline spray, spray rinse Emulsion spray, rinse	Surfactant alkaline soak and spray, alkaline soak, spray and rinse, electrolytic alkaline(j), rinse, mild acid pickle, rinse

(a) For complete removal of pigment, parts should be cleaned immediately after the forming operation, and all rinses should be spray where practical. (b) Used only when pigment residue can be tolerated in subsequent operations. (c) Phosphoric acid cleaner-coaters are often sprayed on the parts to clean the surface and leave a thin phosphate coating. (d) Phosphoric acid for cleaning and iron phosphating. Proprietary products for high- and low-temperature application are available. (e) Some plating processes may require additional cleaning dips. (f) Neutral emulsion or solvent should be used before manganese phosphating. (g) Reverse-current cleaning may be necessary to remove chips from parts having deep recesses. (h) For cyanide plating, acid dip and water rinse are followed by alkaline and water rinses. (j) Other preferences: stable or diphase emulsion spray or soak, rinse, alkaline spray or soak, rinse, electroclean; or solvent presoak, alkaline soak or spray, electroclean. (k) Third preference: emulsion spray rinse

face, rather than chemical action which would be completely ineffective on such inert materials. However, emulsions alone will not do a complete cleaning job, particularly when graphite or molybdenum disulfide is the contaminant. Emulsion cleaning is an effective method of removing pigment because emulsion cleaners contain organic solvents and surfactants, which can dissolve the binders, such as stearates, present in the compounds.

Diphase or multiphase emulsions, having a concentration of 1 to 10% in water and used in a power spray washer, yield the best results in removing pigmented compounds. The usual spray time is 30 to 60 s; emulsion temperatures may range from 54 to 77 °C (130 to 170 °F), depending on the flash point of the cleaner. In continuous cleaning, two adjacent spray zones or a hot water (60 to 66 °C or 140 to 150 °F) rinse stage located between the two cleaner spraying zones is common practice.

Cleaning with an emulsifiable solvent, a combination of solvent and emul-

Fig. 1 Parts that can be cleaned through various processes

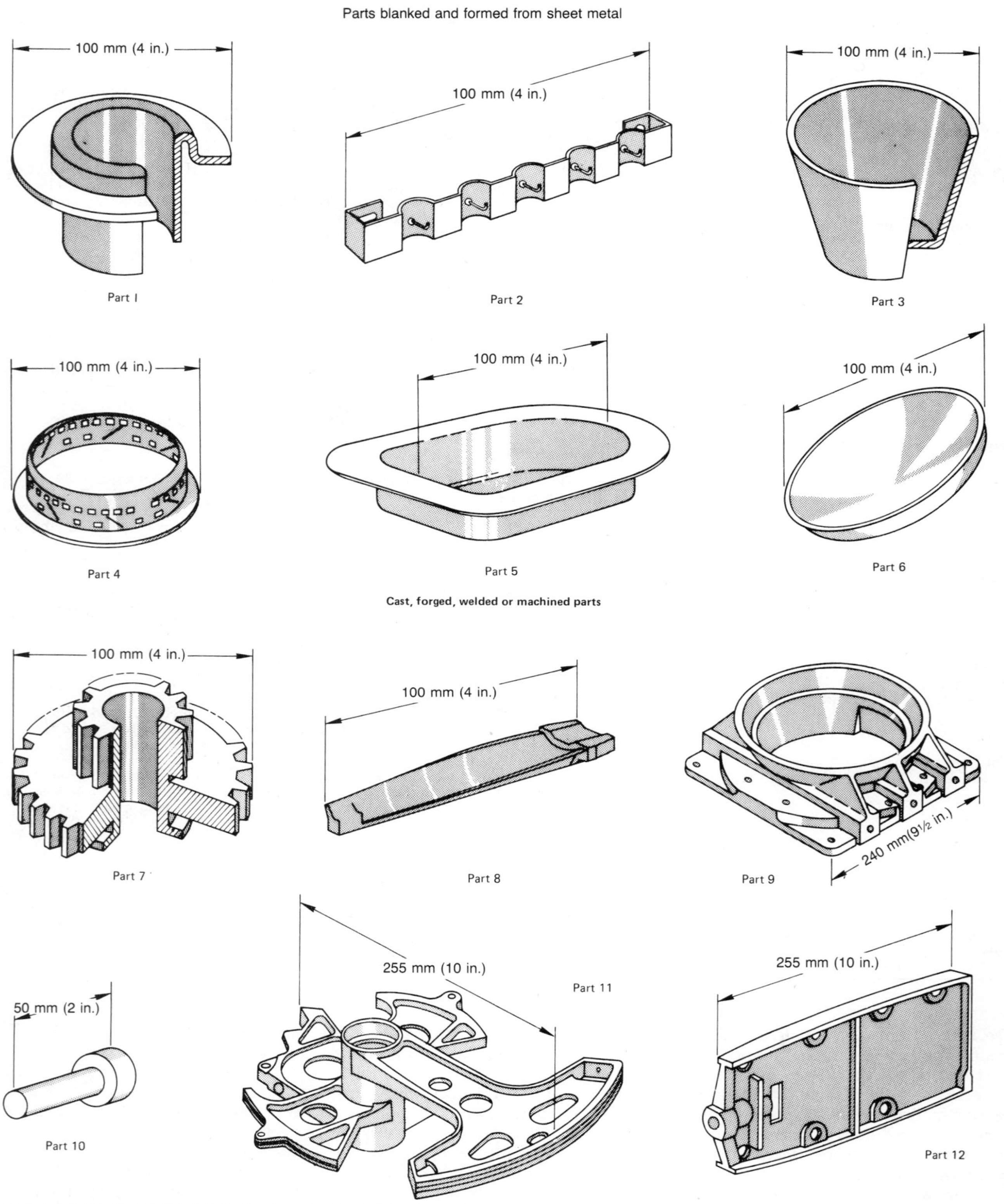

Table 2 Alkaline cleaning cycle for removing pigmented drawing compounds

Process sequence	Concentration g/L	Concentration oz/gal	Time, min	Temperature °C	Temperature °F	Anode current A/dm²	Anode current A/ft²	Remarks
Alkaline soak clean								
Barrel(a)	65 to 90	9 to 12	3 to 5	Boiling	Boiling	...	...	...
Rack(b)	65 to 90	9 to 12	3 to 5	Boiling	Boiling	...	...	...
Hot water rinse, immersion, and spray								
Barrel(a)	...	...	3(c)	43	110	...	...	Spray jet if barrel is open type
Rack(b)	...	...	2(c)	43	110	...	...	Spray rinse, immerse, and spray rinse
Electrolytic alkaline clean								
Barrel(a)	50 to 65	7 to 9	2	82 to 99	180 to 210	4 to 6	40 to 60	...
Rack(b)	65 to 90	9 to 12	2	82 to 99	180 to 210	4 to 6	40 to 60	...
Hot water rinse, immersion, and spray(d)								
Barrel(a)	...	...	3(c)	43	110	...	...	Spray jet if barrel is open type
Rack(b)	...	...	2(c)	43	110	...	...	Spray rinse, immerse, and spray rinse
Cold water rinse, immersion, and spray(e)								
Barrel(a)	...	...	2(c)	...	...	...	...	Spray jet if barrel is open type
Rack(b)	...	...	1(c)	...	...	...	...	Spray rinse, immerse, and spray rinse

(a) Rotate during entire cycle. (b) Agitate arm of rack, if possible. (c) Immersion time. (d) Maintain overflow at approximately 8 L/min (2 gal/min). (e) Clean in cold running water

sion cleaning, is an effective technique for removing pigmented compounds. Emulsifiable solvents may either be used full strength or be diluted with hydrocarbon solvent, 10 parts to 1 to 4 parts of emulsifiable solvent. Workpieces with heavy deposits of pigmented compound are soaked in this solution, or the solution is slushed or swabbed onto heavily contaminated areas. After thorough contact has been made between the solvent and the soil, workpieces are rinsed in hot water, preferably by pressure spray. Emulsification loosens the soil and permits it to be flushed away. Additional cleaning, if required, is usually done by either a conventional emulsion or an alkaline cleaning cycle.

Most emulsion cleaners can be safely used to remove soil from any metal. However, a few highly alkaline emulsion cleaners with pH higher than 10 must be used with caution in cleaning aluminum or zinc because of chemical attack. Low alkaline pH (8 to 9) emulsion cleaners, safe on zinc and aluminum, are available. Emulsion cleaners with a pH above 11 should not be used on magnesium alloys.

Alkaline cleaning, when used exclusively, is only marginally effective in removing pigmented compounds. Success depends mainly on the type of pigmented compounds present and the extent to which they have been allowed to dry. If the compounds are the more difficult types, such as graphite or white lead, and have been allowed to harden, hand slushing and manual brushing will be required for removing all traces of the pigment. Hot alkaline scale conditioning solutions can be used to remove graphite and molybdenum disulfide pigmented hot forming and heat treating protective coatings. The use of ultrasonics in alkaline cleaning is also highly effective in removing tough pigmented drawing compounds.

The softer pigmented compounds can usually be removed by alkaline immersion and spray cycles (Table 1). The degree of cleanness obtained depends largely on thorough mechanical agitation in tanks or barrels, or strong impingement if a spray is used. A minimum spray pressure of 0.10 MPa (15 psi) is recommended.

Parts such as 1 to 6 in Fig. 1 can be cleaned effectively by immersion or immersion and spray when the parts are no larger than about 508 mm (20 in.) across. Larger parts of this type can be cleaned more effectively by spraying.

Operating conditions and the sequence of processes for a typical alkaline cleaning cycle are listed in Table 2. This cycle has removed pigmented compounds effectively from a wide variety of stampings and drawn parts. Energy saving low-temperature solventized-alkaline cleaners are available for soak cleaning. Similarly low-temperature electrocleaners also are effectively employed in industry, operating at 27 to 49 °C (80 to 120 °F).

Electrolytic alkaline cleaning is seldom used as a sole method for the removal of pigmented compounds. Although the generation of gas at the workpiece surface provides a scrubbing action that aids in removal of the pigment, the cleaner becomes contaminated so rapidly that its use is impractical except for final cleaning before plating (Table 1).

Copper alloys, aluminum, lead, tin, and zinc are susceptible to attack by uninhibited alkaline cleaners (pH 10 to 14). Inhibited alkaline cleaners (pH below 10), which have reduced rates of reaction, are available for cleaning these metals. These contain silicates and borates.

Acid Cleaning. Acid cleaners, composed of detergents, liquid glycol ether, and phosphoric acid have proved effective in removing pigmented compounds from engine parts, such as sheet rocker covers and oil pans, even after the pigments have dried. These acid compounds, mixed with water and used in a power spray, are capable of cleaning such parts without hand scrubbing.

A power spray cycle used by one plant is given in Table 3. A light blowoff follows the rinsing cycle. Parts with recesses should be rotated to allow complete drainage. This cleaning procedure suitably prepares parts for painting, but for parts to be plated, the acid cleaning cycle is conventionally followed by electrolytic cleaning which is usually alkaline, but sometimes done with sulfuric or hydrochloric acid. Phosphoric acid cleaners will not etch steel, although they may cause some discoloration.

Table 3 Power spray acid cleaning for removing pigmented compounds

Steel parts cleaned by this method are suitable for painting, but electrolytic cleaning normally follow if parts are to be electroplated; solventized, phosphoric acid-based, low-temperature (27 to 49 °C or 80 to 120 °F) products are successfully used for power spray cleaning

Cycle	Phosphoric acid g/L	Phosphoric acid oz/gal	Solution temperature °C	Solution temperature °F	Cycle time, min
Wash	15-19	2-2.5	74-79	165-175	3-4
Rinse	4-7.5	0.5-1	74-79	165-175	1-1.5

Aluminum and aluminum alloys are susceptible to some etching in phosphoric acid cleaners. Chromic acid or sodium dichromate with either nitric or sulfuric acid is used to deoxidize aluminum alloys. Nonchromated deoxidizers are preferred environmentally. Ferric sulfate and ferric nitrate are used in place of hexavalent chromium. However, nonchromated deoxidizers tend to produce smut on the workpiece, especially 2000 and 7000 series alloys, when the deoxidizer etch rate is maintained (normally with fluoride) above 0.003 μm/side per hour (0.1 μin./side per hour). For more information on removing smut from aluminum, see the article on cleaning and finishing of aluminum in this Volume.

Vapor degreasing is of limited value in removing pigmented compounds. The solvent vapor will usually remove soluble portions of the soil, leaving a residue of dry pigment that may be even more difficult to remove by other cleaning processes. However, modifications of vapor degreasing, such as slushing, spraying, ultrasonic, or combinations of these, can be utilized for 100% removal of the easier-to-clean pigments, such as whiting, zinc oxide, or mica.

The latter practice is often used for occasional or intermittent cleaning (Table 1). However, when difficult-to-clean pigments such as graphite or molybdenum disulfide are present, it is unlikely that slush or spray degreasing will remove 100% of the soil.

Vapor degreasing of titanium should be limited to detailed parts and should not be used on welded assemblies, which will see later temperatures in excess of 290 °C (550 °F). Subsequent pickling in nitric-fluoride etchants may relieve this concern.

Solvent cleaning, because of its relatively high cost, lack of effectiveness, rapid contamination, and health and fire hazards, is seldom recommended for removing pigmented compounds, except for occasional preliminary or rough cleaning before other methods. For example, parts are sometimes soaked in solvents such as kerosine or mineral spirits immediately following the drawing operation to loosen and remove some of the soil, but the principal effect of this operation is to condition parts for easier cleaning by more suitable methods, such as emulsion or alkaline cleaning.

Removal of Unpigmented Oil and Grease

Common shop oils and greases, such as unpigmented drawing lubricants, rust-preventive oils, and quenching and lubricating oils, can be effectively removed by several different cleaners. Selection of the cleaning process depends on production flow as well as on the required degree of cleanness, available equipment, and cost. For example, steel parts in a clean and dry condition will rust within a few hours in a humid atmosphere. Thus, parts that are thoroughly clean and dry must go to the next operation immediately, be placed in hold tanks, or be treated with rust preventives or water displacing oils. If rust preventives are used, the parts will probably require another cleaning before further processing. Accordingly, a cleaner that leaves a temporary rust-preventive film might be preferred.

Table 1 lists cleaning methods frequently used for removing oils and greases from the 12 types of parts in Fig. 1. Similar parts that are four or five times as large would be cleaned in the same manner, except for methods of handling. Variation in shape among the 12 parts will affect racking and handling techniques.

Advantages and disadvantages of the cleaners shown in Table 1, as well as other methods for removing common unpigmented oils and greases, are discussed in the following paragraphs.

Emulsion Cleaning. Emulsion cleaners, although fundamentally faster but less thorough than alkaline cleaners, are widely used for intermittent or occasional cleaning, because they leave a film that protects the steel against rust. Emulsion cleaners are most widely used for inprocess cleaning, preparation for phosphating, and precleaning for subsequent alkaline cleaning before plating (Table 1).

Vapor degreasing is an effective and widely used method for removing a wide variety of oils and greases. It develops a reproducible cleanliness because the degreasing fluid is distilled and filtered. Vapor degreasing is used as a preliminary step before alkaline and acid cleaning as well as to remove corrosion protective oils, greases, and forming lubricants before heat treating.

Vapor degreasing has proved especially effective for removing soluble soil from crevices, such as rolled or welded seams that may permanently entrap other cleaners. Vapor degreasing is particularly well adapted for cleaning oil-impregnated parts, such as bearings, and for removing solvent-soluble soils from the interiors of storage tanks.

Solvent cleaning may be used to remove the common oils and greases from metal parts. Methods vary from static immersion to multistage washing. Eight methods of solvent cleaning listed in increasing order of their effectiveness are as follows:

- Static immersion
- Immersion with agitation of parts
- Immersion with agitation of both the solvent and the parts
- Immersion with scrubbing
- Pressure spraying in a spray booth
- Immersion scrubbing, followed by spraying
- Multistage washing
- Hand application with wiper

A number of solvents and their properties are found in the articles on vapor degreasing and solvent cleaning in this Volume. Solvent cleaning is most widely used as a preliminary or conditioning cleaner to decrease both the time required in and contamination of the final cleaner.

Shape of the part influences the cycle and method selected. For example, parts that will nest or entrap fluids (Parts 3 and 6 in Fig. 1) are cleaned by dipping in a high-flash naphtha, Stoddard solvent, or chlorinated hydrocarbon for 5 to 30 s at room temperature. Time depends on the type and amount of soil. Parts that are easily bent or otherwise damaged, such as Part 2 in Fig. 1, are sprayed for 30 s to 2 min at room temperature. Complex parts, such as Part 9 in Fig. 1, are soaked at room temperature for 1 to 10 min.

Acid Cleaning. Acid cleaners such as the phosphoric acid-ethylene glycol monobutyl ether type are efficient in the removal of oil and grease. Also, they remove light blushing rust and form a thin film of phosphate that provides temporary protection against rusting and functions as a suitable base for paint (Table 1).

Acid cleaners are usually used in a power spray washer. The cycle shown for removing pigmented compounds in Table 3 also removes unpigmented compounds.

Although acid cleaners are comparatively high in cost, they are often used on large ferrous components, such as truck cabs, before painting. Acid cleaners will etch aluminum and other nonferrous metals.

Alkaline Cleaning. Alkaline cleaners are efficient and economical for removing oil and grease and are capable of cleaning to a no-water-break surface. They remove oil and grease by saponification or emulsification, or both. The types that saponify only are quickly exhausted.

Mineral, lard, and synthetic unpigmented drawing compounds are easily removed by alkaline cleaners. Silicones, paraffin, and sulfurized, chlorinated, oxidized, or carbonized oils are difficult, but can be removed by alkaline cleaners.

Alkaline cleaners will etch aluminum and other nonferrous metal parts unless inhibitors are used, and aqueous solutions of alkaline cleaners cannot be tolerated on some parts or assemblies. On assemblies comprised of dissimilar metals, the presence of alkaline solution in crevices may result in galvanic corrosion, and even a trace of alkali will contaminate paint and phosphate coating systems; therefore, rinsing must be extremely thorough.

Electrolytic alkaline cleaning is effective as a final cleaning process for removing oil and grease from machined surfaces when extreme cleanness is required. It is almost always used for final cleaning of steel parts before electroplating, such as precision parts (fitted to ±0.0076 mm or ±0.0003 in.) in refrigeration and air conditioning equipment. Electrolytic alkaline cleaning provided a cleanness of 0.0005 g/10 parts on the small plate assembly (Part 13) in Fig. 2, and of 0.003 g/10 parts on the 165-mm (6.5-in.) diam part (Part 14). This degree of cleanness was obtained by using a conveyor system and the following cycle:

1. Soak in alkali, 45 to 60 g/L (6 to 8 oz/gal) at 77 to 88 °C (170 to 190 °F) for 1 to 2 min. Energy saving, solventized-alkaline low-temperature soak cleaners, suitable for ferrous and nonferrous metals are available. Similarly, low-temperature electrocleaners are also used. Both operate at 27 to 49 °C (80 to 120 °F)
2. Alkaline clean with reverse current, using current density of 5 A/dm^2 (50 A/ft^2), same time, concentration, and temperature as in step 1. Avoid making the part cathodic when cleaning high-strength steels or titanium to avoid hydrogen embrittlement
3. Rinse in cold water containing chromic acid for rust prevention
4. Rinse in cold water containing ammonia
5. Rinse in hot water containing 0.1% sodium nitrate
6. Dry in hot air
7. Place parts in solvent emulsion prior to manganese phosphate coating

Removal of Chips and Cutting Fluids from Steel Parts

Cutting and grinding fluids used for machining may be classified into three groups, as follows:

- Plain or sulfurized mineral and fatty oils (or combinations of the two), chlorinated mineral oils, and sulfurized chlorinated mineral oils
- Conventional or heavy-duty soluble oils with sulfur or other compounds added and soluble grinding oils with wetting agents
- Chemical cutting fluids, which are water-soluble and generally act as cleaners. They contain soaps, amines, sodium salts of sulfonated fatty alcohols, alkyl aromatic sodium salts of sulfonates, or other types of soluble addition agents

Usually, all three types of fluids are easily removed, and the chips fall away during cleaning, unless the chips or part become magnetic. Plain boiling water is often suitable for removing these soils, and in some plants, mild detergents are added to the water to increase its effectiveness. Steam is widely used for in-process cleaning, especially for large components. Table 1 indicates cleaning processes typically used for removing cutting fluids to meet specific production requirements.

Fig. 2 Parts for refrigerators or air conditioners

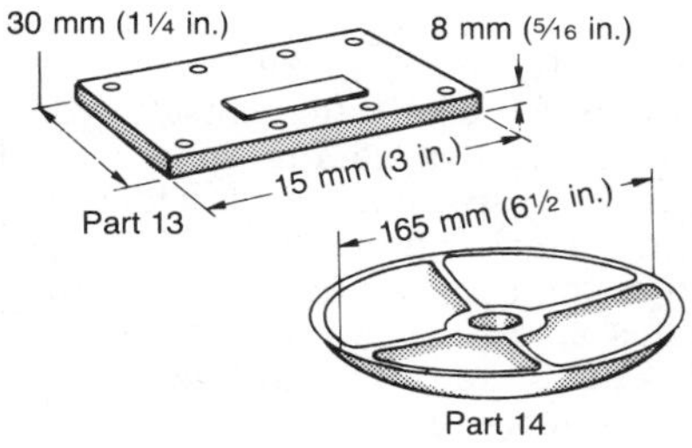

Electrolytic alkaline cleaning is required for attaining a high degree of cleanness

Emulsion cleaning is an effective and relatively inexpensive means of removing all three types of cutting fluids. Attendant fire hazard is not great if operating temperatures are at least 8 to 11 °C (15 to 20 °F) below the flash temperature of the hydrocarbon used. Parts may be cleaned by either dipping or spraying. Many parts are immersed and then sprayed, particularly parts with complex configurations, such as Part 9 in Fig. 1.

It has often proved economical to remove a major portion of the soil by alkaline cleaning first and then to use an emulsion surfactant, an emulsion containing surface-activating agent. This sequence prevents the possible contamination of painting or phosphating systems with alkaline solution.

Most emulsion cleaners can be safely used for removing these soils from nonferrous metals. Only the emulsions having pH values higher than 10 are unsafe for cleaning nonferrous metals.

Alkaline Cleaners. Alkaline cleaners are effective for removing all three types of cutting and grinding fluids. Alkaline cleaning is usually the least expensive process and is capable of delivering parts that are clean enough to be phosphate coated or painted. Inhibited alkaline cleaners are required for removing cutting and grinding fluids from aluminum and zinc and their alloys.

Electrolytic alkaline cleaning, which invariably follows conventional alkaline cleaning for parts that are to be plated, is also recommended for removing cutting fluids when extra cleanness is required. For example, Parts 7 and 9 in Fig. 1 would be cleaned electrolytically before scaleless heat treating.

Vapor degreasing will remove cutting fluids of the first group easily and completely, but fluids of the second and third groups may not be completely removed and are likely to cause deterioration of the solvent. Water contained in these soluble fluids causes hydrolysis of the degreasing solvent and produces hydrochloric acid, which will damage steel and other metals. Vapor degreasing solvents have inhibitors to reduce corrosion by stabilizing the pH. A potential fire hazard exists when water or moisture and aluminum chips are allowed to accumulate in a vapor degreaser.

If vapor degreasing is used to remove water-containing soils, perchlorethylene may be the preferred solvent because its higher boiling point (120 °C or 250 °F) causes most of the water to be driven off as vapor. However, prolonged immersion at 120 °C (250 °F) may also affect the heat treated condition of some aluminum alloys. Used exclusively, the vapor phase will not remove chips or other solid particles. Therefore, combination cycles, such as warm liquid and vapor, are ordinarily used. An air blowoff also aids in removing chips.

Solvent cleaning by soaking (with or without agitation), hand wiping, or spraying is frequently used for removing chips and cutting fluids. Solvents preferentially remove cutting fluids of the first group. Solvent cleaning is commonly used for cleaning between machining operations, to facilitate inspection or fixturing.

Acid Cleaning. Phosphoric or chromic acid cleaners used in a power spray or soak cleaning when followed by pressure spray rinsing are effective in removing most types of cutting fluids. However, they are expensive and are seldom used for routine cleaning. In some applications, acid cleaners have been used because they also remove light rust from ferrous metals and oxide and scale from aluminum alloys.

Removal of Polishing and Buffing Compounds

Polishing and buffing compounds are difficult to remove because the soil they deposit is composed of burned-on grease, metallic soaps, waxes, and vehicles that are contaminated with fine particles of metal and abrasive. Consequently, cleaning requirements should be considered when selecting polishing and buffing compounds. Compounds used for obtaining buffed and polished finishes may be classified by cleaning requirements:

- *Liquids*: mineral oils and oil-in-water emulsions or animal and vegetable oils with abrasives
- *Semisolids*: oil based, containing abrasives and emulsions, or water based, containing abrasives and dispersing agents
- *Solids*: greases containing stearic acid, hydrogenated fatty acid, tallow, hydrogenated glycerides, petroleum waxes, and combinations that produce either saponifiable or unsaponifiable materials, in addition to abrasives

Table 1 lists preferred and alternate methods for removing polishing and buffing compounds from sheet metal parts. However, some modification may be required for complete removal of all classes of these soils. Characteristics of polishing compounds and their effects on cleaning for the three broad classifications of soil are described in the following paragraphs.

Liquid compositions are oil based and flow readily, leaving a thin film of oil that contains particles of metal and abrasive on the work. Under extreme heat and pressure, some oils polymerize and form a glaze that is difficult to remove.

Mineral oils are usually unsaponifiable and are not readily removable by conventional alkaline cleaners. Solvent wiping, alkaline, or emulsion cleaning, using surfactant cleaners containing surface-activating agents, are more effective in removing residues from mineral oils.

Most animal and vegetable oils can be saponified at a slow rate. These oils are insoluble in water, but can be removed by soaking or spraying in hot alkaline solutions (82 °C or 180 °F). Spraying is preferred because it removes adhering particles more effectively. Surfactants are suitable also, but their higher cost cannot always be justified.

Semisolid compounds are mixtures of liquid binders and abrasives that contain emulsifying or dispersing agents to keep the abrasive in suspension. When subjected to heat and pressure, these compounds usually form a heavy soil on the surface and may cake and fill in depressions and corners. Such compounds vary from unsaponifiable to completely saponifiable. Hand wiping with solvent or emulsion cleaner is effective in removing these compounds. Impingement from power washers usually removes most of the soil, regardless of the cleaner used. If power washers are not available, soak in agitated solutions containing surfactants, followed by a thorough rinsing, for satisfactory results.

Solid Compounds. The oil phases of solid compounds are easily removed, but the remaining residues cling tenaciously to metal surfaces and must be dislodged by scrubbing action. Power washers are the most effective. Most agitated surfactant cleaners are also effective, but the agitation must be strong enough to dislodge the soils.

Removal Methods

Solvent cleaning is effective for precleaning, but is more costly than alkaline or emulsion methods. Cleaning with chlorinated solvents in a mechanical degreaser or brushing or spraying with petroleum solvents quickly removes most of the gross soil after buffing or polishing.

Emulsion Cleaning. Emulsion cleaners containing one part of emulsion concentrate to 50 to 100 parts of water, and operated at 54 to 60 °C (130 to 140 °F) are effective for removing mineral oils and other unsaponifiable oils from polished work. For effectively removing semisolid compounds, the temperature must be raised to 66 to 71 °C (150 to 160 °F) and the concentration increased to one part concentrate to 10 to 20 parts water. Agitation helps dislodge soil from corners or grooves. Table 4 describes cleaning cycles for removing polishing and buffing compounds. Thickened emulsion cleaners may be applied with an airless spray pump. Allow 5 to 10 min dwelling time before cold water rinsing. Emulsion cleaners applied manually at ambient temperature are suitable for many applications, especially for buffed aluminum parts.

Removing solid soils or those containing grit requires the use of higher temperature (71 to 82 °C or 160 to 180 °F) and increased concentration (one part concentrate to ten parts water). If the soil is heavy, caked, or impacted in corners, a spray washer is required, and the proper ratio of concentrate to water is between 1 to 20 and 1 to 50 (Table 4).

All emulsion methods must be followed by a thorough water spray rinse. The cleaner will loosen and remove most of the soil, but only a strong water spray can remove the remainder. Warm

Table 4 Emulsion cleaning cycles for removing polishing and buffing compounds

All workpieces were rinsed using water spray

Type of compound	Temperature °C	°F	Time, min	Concentration, emulsion to water	Agitation
Oil	66-71	150-160	3-5	1:10-20	Soak
Semisolid	54-60	130-140	3-5	1:50-100	Solution movement
Solid	71-82	160-180	1½	1:20-50	Spray wash

Note: All emulsion cleaned parts should be subsequently cleaned by alkaline soaking and electrolytic alkaline cleaning before electroplating

water is preferred, but cold water can be used.

In spray equipment, concentration must be controlled to avoid foaming or breaking the emulsion. When soil removal requires a critical concentration, a foam depressant may be added to the cleaner. Polishing compounds containing soap or soap-forming material will cause excessive foaming during agitation, which may reduce the efficiency of the cleaner and the washer. The performance of emulsion cleaners can sometimes be improved by using them in conjunction with alkaline solutions, particularly in spray washers. Alkaline cleaning compounds at a concentration of about 4 g/L (½ oz/gal) may be used, but the surface being cleaned will still have an oily film after rinsing.

Although the preceding information is applicable primarily to ferrous metal parts, it can be applied also to brass and to zinc-based die castings. The following is a cycle that proved successful for removing polishing and buffing soil from zinc-based die castings in high-volume production:

- Preclean by soaking for 4 min in diphase cleaner, using kerosine as the solvent; temperature, 71 °C (160 °F); concentration, 1 to 50; plus a 75-mm (3-in.) layer of kerosine. Parts are sprayed with solution as they are being withdrawn from the tank
- Fog spray rinse
- Alkaline spray cleaner, 7.5 g/L (1 oz/gal), 71 °C (160 °F), for 1½ min
- Alkaline soak cleaner, 30 to 45 g/L (4 to 6 oz/gal), 71 °C (160 °F), for 4 min
- Spray rinse
- Transfer to automatic plating machine or electrolytic alkaline cleaning

Alkaline cleaning, or one of its modifications, is an effective and usually the least expensive method for removing soils left by polishing and buffing. Mineral oils and other saponifiable oils are difficult to remove by soak cleaning. Oil that floats to the surface redeposits on the work unless the bath is continually skimmed. Agitation of the bath to minimize oil float and proper rinsing of parts as they are withdrawn from the tank minimizes the retention of oil by cleaned parts.

Removing liquid or solid compounds that contain abrasives requires agitation. Most soak cleaners foam excessively if agitated sufficiently to dislodge hardened soil from recesses or pockets. A mildly agitated surfactant cleaner, followed by a strong water spray, can loosen these soils (Table 1).

Operating conditions for soak, spray, and electrolytic alkaline cleaning methods for removing polishing and buffing compounds are listed in Table 5. When the soil is charged with abrasive, alkaline cleaners must be renewed more frequently to prevent the accumulation of dirt that will clog screens and nozzles.

Electrolytic alkaline cleaning provides a high level of agitation close to the work surface because of the gas generated and is an effective method for removing polishing and buffing residues. Electrocleaners can be easily contaminated by polishing and buffing compounds as well as steel particles which may be attracted to the work and cause surface roughness during plating. Precleaning is necessary. Parts on which mineral oil has been used as a polishing compound should always be precleaned before being electrocleaned. Use of both heavy duty alkaline soak cleaners and electrocleaners is often necessary to provide a water-break-free surface necessary for good plating quality and adhesion. The presence of large amounts of animal or vegetable oils or fatty acids and abrasives in the polishing and buffing compounds will react with free caustic and form soaps in the electrocleaner and shorten its life.

Acid Cleaning. Acid cleaners are chemically limited in their ability to remove polishing and buffing compounds. Soaps and other acid-hydrolyzable materials present in these compounds are decomposed by acid cleaners into insoluble materials, which precludes the use of acid cleaners in most instances.

Acid cleaners can be used alone for the more easily removed polishing and buffing compounds, such as fresh and unpolymerized liquids. In these applications, the acid cleaner must be used at the maximum operating temperature recommended for the specific cleaner in conjunction with the maximum agitation obtainable by spraying or scrubbing.

Acid cleaners may be desirable for removing acid-insensitive soils in special instances such as the following two examples: (*a*) where slight surface attack (short of pickling) is needed for dislodging particles or smut, and (*b*) in conjunction with alkaline or alkaline emulsion cleaners, when successive reversal of pH proves to be advantageous. A light pickle in dilute hydrochloric, hydrofluoric, or sulfuric acid may be added to the cleaning sequence to remove fine metal particles, tarnish, or light scale to activate the surface for electroplating.

Removal of Rust and Scale

The seven basic methods used for removing rust and scale from ferrous mill products, forgings, castings, and fabricated metal parts are:

- Abrasive blasting (dry or wet)
- Tumbling (dry or wet)
- Brushing
- Acid pickling
- Salt bath descaling
- Alkaline descaling
- Acid cleaning

The most important considerations in selecting one of the above methods are:

- Thickness of rust or scale
- Composition of metal
- Condition of metal (product form or heat treatment)
- Allowable metal loss
- Surface finish tolerances
- Shape and size of workpieces
- Production requirements
- Available equipment
- Cost
- Freedom from hydrogen embrittlement

Combinations of two or more of the available processes are frequently used to advantage.

Abrasive blast cleaning is widely used for removing all classes of scale and rust from ferrous mill products, forgings, castings, weldments, and heat treated parts. Depending on the finish requirements, blasting may be the sole means of scale removal, or it may be

Table 5 Alkaline cleaning for removing polishing and buffing compounds

Soak and spray cleaning are followed by electrolytic cleaning if parts are to be electroplated; electrolytic cleaning is usually preceded by soak or spray cleaning

Method of cleaning	Concentration g/L	oz/gal	Temperature °C	°F	Time, min
Soak(a)	30-90	4-12	82-100	180-212	3-5
Spray(b)	4-15	½-2	71-82	160-180	1-2
Electrolytic(c)	30-90	4-12	82-93	180-200	1-3

Note: Use great care in cleaning brass and zinc die cast, because these materials are easily attacked at high concentration, temperature, and current density of alkaline cleaners. Anodic cleaning is best, using a concentration of 30 to 45 g/L (4 to 6 oz/gal) at a temperature of 54 to 66 °C (130 to 150 °F) at a current density of 3 to 5 A/dm^2 (30 to 50 A/ft^2)
(a) For removing light oils, semisolid compounds, and solid compounds if not impacted or burned on work; must be followed by a strong spray rinse. (b) For removing light mineral oils, semisolids, and solids if impacted or caked on work; followed by a rinse. (c) For removing light oil films and semisolids. Solids are difficult to remove, especially if combined with grit or metal particles

used to remove the major portion of scale, with pickling employed to remove the remainder. Glass bead cleaning (blasting) is used for cleaning threaded or precision parts, high strength steel, titanium, and stainless steel.

Tumbling is often the least expensive process for removing rust and scale from metal parts. Size and shape of parts are the primary limitations of the process. Tumbling in dry abrasives (deburring compounds) is effective for removing rust and scale from small parts of simple shape, such as Part 10 in Fig. 1. However, parts of complex shape with deep recesses and other irregularities cannot be descaled uniformly by tumbling and may require several hours of tumbling if that method is used. Adding descaling compounds rather than deburring compounds often decreases the required tumbling time by 75%.

Brushing is the least used method of descaling parts, although it is satisfactory for removing light rust or loosely adhering scale. It is better suited for workpieces formed from tubing than for castings or forgings.

Pickling in hot, strong solutions of sulfamic, phosphoric, sulfuric, or hydrochloric acid is used for complete removal of scale from mill products and fabricated parts. However, pickling is declining in use as a single treatment for scale removal. With increasing frequency, pickling, at acid concentrations of about 3% and at temperatures of 60 °C (140 °F) or lower, is being used as a supplementary treatment following abrasive blasting or salt bath descaling. Use of deoxidizing aluminum alloys in room temperature chromic-nitric-sulfuric acid solutions to remove heat treat scale is common practice.

Electrolytic pickling, although more expensive than conventional pickling, can remove scale twice as fast and may prove economical where the time is limited. In an automatic plating installation, electrolytic pickling removes light scale and oxides during the time allowed in the pickling cycle and eliminates a preliminary pickling operation. For this purpose, a solution of 30% hydrochloric acid is used at 55 °C (130 °F) and 3 to 6 V for 2 to 3 min. Cathodic current is used.

Sulfuric acid formulas also are used electrolytically. A cycle for removing light scale from spot welded parts is a solution of 10% sulfuric acid at 82 °C (180 °F) and 3 to 6 V for 5 to 20 s.

The main objection to electrolytic pickling is high cost. In addition to the requirement for more elaborate equipment, all workpieces must be racked.

Salt bath descaling is an effective means of removing or conditioning scale on carbon, alloy, stainless and tool steels, heat-resisting alloys, copper alloys, nickel alloys, titanium, and refractory metals. Several types of salt baths either reduce or oxidize the scale. Various baths operate within a temperature range of 400 to 525 °C (750 to 975 °F).

Except in the descaling of pure molybdenum, molten salt baths are seldom used alone for scale removal. Usually, salt bath descaling and quenching are followed by acid pickling as a final step in removing the last of the scale. The supplementary pickling is done with more dilute acids at lower temperatures and for shorter times than are used in conventional pickling. A solution of 3% sulfuric acid at a maximum temperature of about 60 °C (140 °F) is commonly used for pickling after salt bath descaling. Other acids are used at comparable concentrations. Metal loss and the danger of acid embrittlement are negligible in this type of pickling.

Alkaline descaling or alkaline derusting is used to remove rust, light scale, and carbon smut from carbon, alloy, and stainless steels and from heat-resisting alloys. Alkaline descaling is more costly and slower in its action than acid pickling of ferrous alloys, but no metal is lost using the alkaline method, because chemical action stops when the rust or scale is removed. Alkaline descaling also allows complete freedom from hydrogen embrittlement. Alkaline etch cleaning of aluminum alloys is less expensive than acid pickling solutions for descaling, removing shot peen residue, removing smeared metal prior to penetrant inspection, chemical deburring, and decorative finishing of nonclad surfaces.

A number of proprietary compounds is available. They are composed mainly of sodium hydroxide (60% or more) but also contain chelating agents.

Immersion baths are usually operated from room temperature to 71 °C (160 °F), but can be used at 93 to 99 °C (200 to 210 °F) with concentrations of about 0.9 kg (2 lb) of compound to 4 L (1 gal) of water. Required immersion time depends on the thickness of the rust or scale.

The rate of removal of oxide can be greatly increased by the use of current in the bath, either continuous direct or periodically reversed. In one instance, an electrolyzed bath descaled steel parts in 1½ min, as compared to 15 min for a nonelectrolytic bath doing the same job. However, parts must be racked for electrolytic descaling, increasing cost because of the additional equipment, increased power requirement, and decreased bath capacity.

The addition of about 0.5 kg (1 lb) of sodium cyanide per 4 L (1 gal) of water increases the effectiveness of electrolyzed baths. However, when cyanide is used, the bath temperature should be kept below 54 °C (130 °F) to prevent excessive decomposition of the cyanide. One manufacturer descales heat treated aircraft parts in an alkaline descaling bath, using direct current and cyanide additions. Another manufacturer descales similar work in an alkaline bath operated at 82 to 93 °C (180 to 200 °F) with a lower concentration of descaling compound, 60 to 90 g/L (8 to 12 oz/gal), and no cyanide. The latter bath is operated at a current density of 2 to 20 A/dm^2 (20 to 200 A/ft^2) and with periodic current reversal (55 s anodic, followed by 5 s cathodic). Alkaline permanganate baths are also used for descaling. Proprietary products available are used at about 120 g/L (1 lb/gal), 82 to 93 °C (180 to 200 °F), 30 min or longer, depending on scale thickness and condition.

Despite the high cost of alkaline descaling baths, they can be economical. Because alkaline descaling baths are compounded for detergency as well as derusting, chemical cleaning and derusting are accomplished simultaneously. Paint, resin, varnish, oil, grease, and carbon smut are removed along with rust and scale. Thus, in a single operation, work is prepared for phosphating, painting, or electroplating. If parts are to be plated, the cost of electrolytic descaling may be comparable to that of the nonelectrolytic process, because in either case workpieces must be racked before final cleaning and plating. An electrolytic descaling bath may serve as the final cleaner.

Alkaline descalers are used for applications on critical parts such as turbine blades for jet engines where risk of hydrogen embrittlement, loss of metal, or etched surfaces cannot be tolerated. Alkaline descaling may also be chosen for parts made of high-carbon steel or cast iron, because acid pickling will leave smut deposits on these metals. Because of the time required, alkaline descaling is seldom used for removing heavy scale from forgings.

Acid Cleaning. Acid cleaners more dilute than acid pickling solutions are effective for removing light, blushing rust, such as the rust that forms on ferrous metal parts in storage under conditions of high humidity or short-time exposure to rain. Acid deoxidizing solutions specifically designed for use on aluminum remove oxides and should be used before electroplating or chemical coating. Various organic acid based solutions, such as citric acid, are used to remove rust from stainless steels, including the 400 series and the precipitation hardening steels.

The following examples illustrate the considerations that influence the choice of process for removing rust and scale. Additional criteria for selection of process are included in Table 6, which compares advantages and disadvantages of abrasive blast cleaning, pickling, and salt bath descaling.

Table 6 Advantages and disadvantages of the three principal processes for removing scale and rust from steel parts

Advantages	Disadvantages
Abrasive blast cleaning	
A variety of equipment and abrasives is available	Some of the metal will be abraded from workpieces, especially from corners
Does not interfere with properties established by heat treatment	May alter dimensions of machined parts or damage corners
Size of workpiece is limited only by available equipment	If sufficiently drastic to remove scale, process may cause more surface etching or roughness than can be tolerated
A wide variety of shapes can be blasted	Complex configurations will not receive equal blasting on all surfaces without special handling, which may be too costly
All metals can be safely blasted	
Adaptable to either intermittent low or continuous high production	
Pickling	
Formulations can be adjusted to meet individual requirements in removing scale from various ferrous and nonferrous alloys	Potential source of hydrogen embrittlement in some metals, such as carbon and alloy steels of high carbon content, especially if these materials have been heat treated to high strength levels
Equipment required is simple and relatively inexpensive	Up to 3% of the metal may be lost in pickling—particularly significant for the more costly metals such as stainless steels or heat-resisting alloys
Materials are relatively low in cost, and process control usually is not difficult	Fume control and disposal of spent acids are major problems
Adaptable to products of virtually any size or shape	Process is likely to deposit smut on cast iron
Installations can be adapted to either low or high, intermittent or continuous production	Excessive pitting may occur in the pickling of cast steels and irons
Temperatures used will not affect properties of heat treated steel	
Salt bath descaling	
Reduction or oxidation of the scale is almost instantaneous after workpieces reach bath temperature	Not economical for intermittent production, because high operating temperatures necessitate special heating and handling equipment, and because the bath must be kept molten between production runs
No loss of metal and no danger of hydrogen embrittlement	The required water quenching may cause cracking or excessive warping of complex workpieces
Preliminary cleaning is unnecessary unless there is so much oil on the work that a fire hazard is involved as workpieces enter the bath	The process is not suitable for metals (such as some grades of stainless steel) that precipitation harden at the temperature of the salt bath
Different metals can be descaled in the same bath	Operating temperature of the bath can cause carbide precipitation in unstabilized stainless steels
Workpieces of complex shape can be processed, although special handling may be required to obtain complete removal of salt	Properties of heat treated workpieces may be impaired if their tempering temperature is below that of the salt bath
Processing temperature may provide useful stress relieving	Subsequent acid cleaning is usually required to neutralize remaining salts, complete the descaling, and brighten the finished product
For some heat-resisting and refractory metals, molten salt is the only satisfactory method	
Will not damage sensitized stainless steels, whereas acid pickling would be harmful	

Example 1. Barrel or vibratory tumbling is probably the most economical method for removing scale or rust from steel parts like Part 10 in Fig. 1, if they are no larger than about 50 to 75 mm (2 to 3 in.). For similar but larger parts, abrasive blasting is usually a better choice.

However, if such parts are close to finished dimensions and these dimensions are critical, a nonabrasive method of cleaning should be chosen. If parts are made of low-carbon steel and are not heat treated, pickling in inhibited hydrochloric or sulfuric acid is satisfactory and less expensive, and hydrogen embrittlement is not a factor. However, if such parts are made of high-carbon (or carburized) steel and are heat treated, acid pickling would be hazardous and alkaline descaling would be preferred.

Example 2. The gear illustrated as Part 7 in Fig. 1 is made of 8620 steel, carburized, and hardened to about 56 to 58 HRC. Although the part is processed in a controlled atmosphere, a descaling operation is required. Abrasive blasting with fine steel grit or chilled iron shot (SAE G40 or S170) proved the most economical method for cleaning large tonnages of such parts used in the manufacture of trucks, tractors, and similar vehicles. Acid pickling was precluded because of hydrogen embrittlement, and descaling in molten salt was unsuitable because of the softening effect of the high-temperature bath.

Conventional abrasive blasting may deleteriously affect the dimensions of precision gears or pinions. In these special applications, alkaline descaling or wet blasting with a fine abrasive, such as glass beads, under carefully controlled conditions, is indicated.

Example 3. The turbine blade shown as Part 8 in Fig. 1 is made of type 403 stainless steel. If such parts are made in continuous production, molten salt bath descaling would be the preferred cleaning method. If production is intermittent, the molten salt method would be too costly, and alkaline descaling would be more practical. Abrasive blasting is unsuitable for this application because of close dimensional requirements; pickling cannot be used because of metal loss and the risk of hydrogen embrittlement.

Example 4. Scale resulting from welding of the low-carbon steel component shown as Part 12 in Fig. 1 could be removed satisfactorily and economically by either abrasive blasting or acid pickling. Because the part is phosphated and painted, surfaces are not critical. Acid pickling would probably be preferred, because it would make more uniform contact with all areas without the need for special handling. Even if a large quantity of parts were to be cleaned, salt bath descaling would not be used, because the water quench from about 425 °C (800 °F) would cause excessive warpage. The cost of alkaline descaling in an aqueous solution would not be justified for this class of work.

Example 5. Normally, abrasive blasting would be the preferred method for removing rust and scale from a rough ferrous metal casting like Part 11 in Fig. 1. Chilled iron shot or steel abrasives are usually the most economical abrasives for this purpose.

Pickling is seldom used for descaling castings, such as cast iron, because smut is deposited and must be removed by another cleaning operation. Severe pitting is also likely to result.

Salt baths have been successfully used for descaling ferrous castings, but there is danger of cracking and excessive distortion for configurations such as Part 11.

Removal of Residues from Magnetic Particle and Fluorescent Penetrant Inspection

Successful removal of the iron oxide particles deposited on ferrous parts during magnetic particle inspection requires complete demagnetization of the part. After demagnetization, emulsion cleaning is an effective and practical means of removing both the iron oxide residues and oil. Fluorescent pigments used for similar inspection of aluminum parts can be removed with hot alkaline cleaners.

For low-to-moderate production, an efficient procedure consists of immersing parts in a light, undiluted, oil-based emulsion cleaner at room temperature or slightly above. Parts are then drained to remove excess cleaner and rinsed in water, using either agitation or forced spray at room temperature or slightly above. For higher-volume production, power washers are successful. Parts can be handled singly or in baskets or carriers.

Parts with complex configurations such as Part 9 in Fig. 1, fine threads, or serrations are difficult to clean thoroughly. As-cast or as-forged surfaces also cause the magnetic oxide particles to cling tenaciously. However, immersion in a cleaning emulsion with sufficient agitation or the use of a power washer, with properly placed nozzles and with suitable handling equipment, will clean almost any part. All oxide particles must be removed before the part is dried, or hand wiping or brushing will be required.

A type of emulsion cleaner that incorporates a rust preventive is usually preferred, because it provides protection until the next operation is performed. If rust-preventive films are objectionable in the next operation, they can be removed easily with alkaline cleaners.

Special Procedures for the Removal of Grinding, Honing, and Lapping Compounds

Residues remaining on parts after honing or grinding are usually mixtures of metallic and abrasive particles with oil-based or water-based cutting fluids. Thus, the methods recommended earlier in this article for the removal of chips and cutting fluids are applicable also for the removal of grinding residues in a majority of instances.

Lapped parts are usually more difficult to clean than honed or ground parts. Lapping residues are composed of extremely fine particles of various abrasives, minute metal particles, semi-solid greases and oils, and some graphite. Even if graphite is not a part of the original lapping compound, it accumulates from wear of cast iron laps. Allowing compounds to dry increases cleaning difficulty. In many instances, methods used for removing polishing and buffing compounds are applicable also for removing lapping compounds. However, parts that are precision ground, honed, or lapped present special cleaning problems because: (*a*) such parts are commonly used in precision machinery, and consequently the degree of cleanness required is higher than for most commercial work; (*b*) they are frequently intricate in design (an example is Part 15 in Fig. 3); and (*c*) they are commonly susceptible to damage and frequently require special handling.

Numerous modifications of conventional processes are used for obtaining the required high degree of cleanness without damage to expensive and delicate parts, such as those used in instruments or fuel injection equipment. One modification of solvent cleaning uses small batch washing machines, which often resemble home laundry equipment. Mineral solvents, such as high-flash naphthas, are usually used. Parts are placed in racks. Agitation, by mechanical movement of the racks or of the solvent, dislodges and floats away the last trace of solids resulting from grinding or lapping. The solvent is constantly filtered and recirculated. Complete removal of solid particles may require an hour or more of constant agitation. The presence of dissolved oil in the solvent will leave a slight oil film on the parts. This is frequently desirable, but can be removed easily by vapor degreasing before assembly. Figure 3 shows a precision part that is successfully cleaned with such a procedure.

Spraying with an atomized solvent, such as filtered high-flash naphtha, in specially designed equipment has also been used successfully for removing the final trace of oil film from precision instrument parts.

Similar procedures are often used for final cleaning of intricate relays just before they are sealed in an assembly. A small spray gun, using filtered solvent and filtered air, is directed at the critical areas. The spray is picked up by a suction vent to eliminate fire hazard.

Room-Temperature Cleaning

Room-temperature or cold cleaners are aqueous solutions for removing soil

Table 7 Examples of application of room temperature alkaline cleaners

Part	Surface from last operation	Relative amount of original soil on part	Residual soil on part after cleaning, mg
Aluminum alloy piston	Ground	Heavy, up to 0.75 g per part	0.1
Pinion gear, ferrous	Ground and lapped	Heavy	0.1
Ring gear, ferrous	Ground and lapped	Heavy	0.1
Engine heads, ferrous	Fully machined	Heavy	1.5
Engine intake manifold, ferrous	Fully machined	Heavy	10
Carburetor throttle body, ferrous	Fully machined	Medium to heavy	0.5
Automatic transmission pump, ferrous	Fully machined	Medium to heavy	6

without the aid of heat other than that resulting from pumping and circulating the solution or being transferred from the surrounding atmosphere. The operating range of such cleaners is usually from 21 to 46 °C (70 to 115 °F). For additional information, see the article on alkaline cleaning in this Volume.

Cold alkaline cleaners, such as the silicate types (orthosilicate or tetrasodium pyrophosphate), are chiefly used for cleaning where heat is not available, where heated solutions are not permitted, or when heating the parts above about 46 °C (115 °F) is not desirable. In a cold process for iron phosphating, for example, parts that have been cleaned in a heated solution and have not cooled sufficiently before entering the phosphate solution will yield an unacceptable phosphate coat. In some applications, an unheated cleaning solution is preferred in order to facilitate the checking of part dimensions at room temperature without the delay involved in cooling the parts after cleaning. This procedure is used for cam shafts, honed cylinder walls, and valve-guide holes in engine heads. One automotive plant utilizes a cold alkaline cleaner for removing soil from engine blocks in a power washer at the rate of 300 per hour. In another application, carburetor parts are cleaned at a rate of 600 to 700 per hour. Table 7 provides several detailed examples of the application of cold alkaline cleaners.

Cold cleaners may also reduce costs by using simpler equipment, eliminating the expense of energy for heating, and reducing maintenance requirements.

Cold acid cleaners, such as monosodium phosphate containing a detergent, are also available. Their chief use is for cleaning immediately before iron phosphating, where the advantage of a lower pH is significant. These acid cleaners have a pH of about 6 and thus impart a surface compatible with the iron phosphate bath, which has a pH of 4.5 to 5.5.

Some proprietary products now offer simultaneous cleaning and iron phosphating at room or low temperatures. In a few other isolated applications, cold acid cleaners perform satisfactorily, but in most instances heated solutions are much more efficient.

Ultrasonic Cleaning

Ultrasonic energy can be used in conjunction with several types of cleaners, but it is most commonly applied to chlorinated hydrocarbon solvents, water, and water with surfactants. Ultrasonic cleaning, however, is more expensive than other methods, because of higher initial cost of equipment and higher maintenance cost, and consequently the use of this process is largely restricted to applications in which other methods have proved inadequate. Areas of application in which ultrasonic methods have proved advantageous are as follows:

- Removal of tightly adhering or embedded particles from solid surfaces
- Removal of fine particles from powder-metallurgy parts
- Cleaning of small precision parts, such as those for cameras, watches, or microscopes
- Cleaning of parts made of precious metals
- Cleaning of parts with complex configurations, when extreme cleanness is required
- Cleaning of parts for hermetically sealed units

Fig. 3 Part for fuel control mechanism

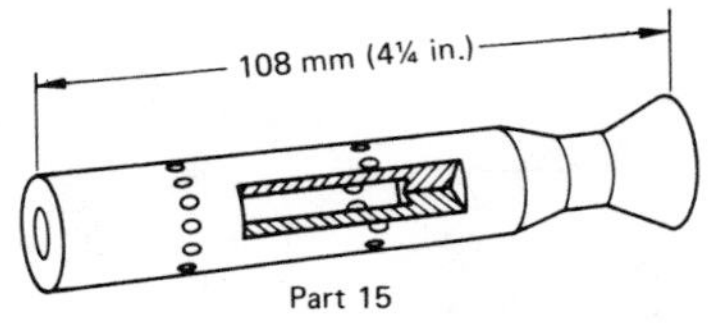

Requires special modification of solvent cleaning for the removal of grinding and lapping compounds

- Cleaning of printed circuit cards and electronic assemblies

Despite the high cost of ultrasonic cleaning, it has proved economical for applications that would otherwise require hand operations.

Part size is a limitation, although no definite limits have been established. The commercial use of ultrasonic cleaning has been limited principally to small parts. The process is used as a final cleaner only, after most of the soil is removed by another method. Ultrasonic cleaning, in some cases, has resulted in fatigue failure of parts. Proper racking and isolation from tank wall will often solve this problem.

Surface Preparation for Phosphate Coating

Because the chemical reaction that results in the deposit of a phosphate coating depends entirely on good contact between the phosphating solution and the surface of the metal being treated, parts should always be sufficiently clean to permit the phosphating solution to wet the surface uniformly. Soil that is not removed can act as a mechanical barrier to the phosphating solution, retarding the rate of coating, interfering with the bonding of the crystals to the metal, or, at worst, completely preventing solution contact. Some soils can be coated with the phosphate crystals, but adherence of the coating will be poor, and this will in turn affect the ability of a subsequent paint film to remain continuous or unbroken in service.

Soils such as cutting oils, drawing compounds, coolants, and rust inhibitors can react with the basis metal and form a film that substantially changes the nature of the coating. Precautions must be taken to avoid carryover of clean-

Fig. 4 Flow charts and operating conditions of processes for preparing steels for electroplating

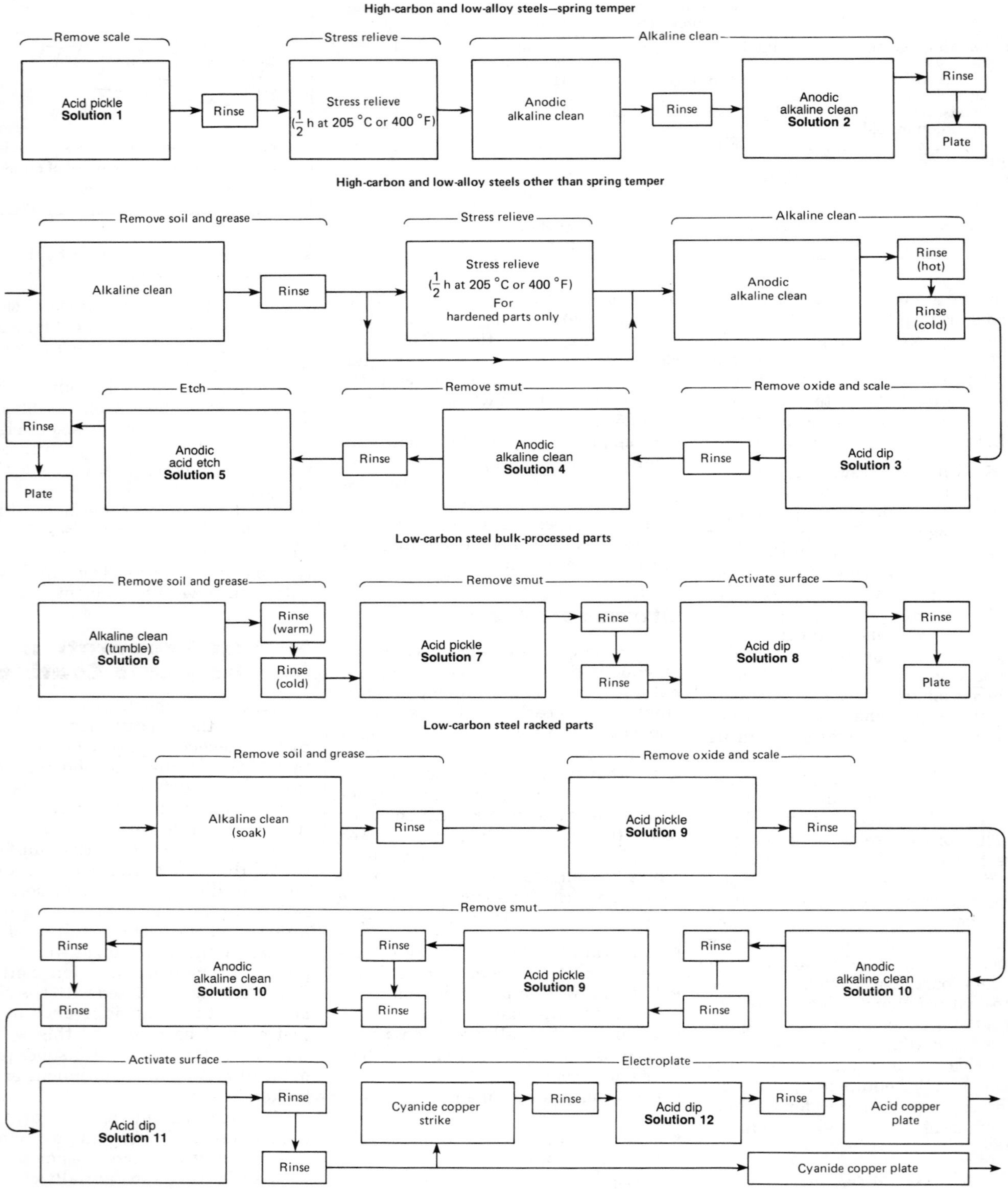

See table on following page for operating conditions

Table to Fig. 4 (operating conditions)

Solution No.	Type of solution	Composition	Amount	Operating temperature °C	°F	Cycle time, s
High-carbon and low-alloy steels, spring temper						
1	Acid pickle	HCl HNO_3	20-80 vol % 1-5 vol %	Room temperature		(a)
2	Anodic alkaline cleaner(b)	NaCN	20-45 g/L (3-6 oz/gal)	49-54	120-130	30-60
High-carbon and low-alloy steels other than spring temper						
3	Acid dip	HCl	1-10 vol %	Room temperature		(a)
4	Anodic alkaline cleaner(b)	NaCN	20-45 g/L (3-6 oz/gal)	Room temperature		30-60
5	Anodic acid etch(c)	H_2SO_4	250-1005 g/L (33.5-134 oz/gal)	30 max	86 max	60 max
Low-carbon steel bulk-processed parts						
6	Alkaline cleaner(d)	Alkali	30-60 g/L (4-8 oz/gal)	82-99	180-210	60-120
7	Acid pickle	HCl	25-85 vol %	Room temperature		5-15
8	Acid dip	H_2SO_4	4-10 vol %	Room temperature		5-15
Low-carbon steel racked parts(e)						
9	Acid pickle	HCl	25-85 vol %	Room temperature		(a)
10	Anodic alkaline cleaner(f)	Alkali	60-120 g/L (8-16 oz/gal)	93-99	200-210	60-120
11	Acid dip	H_2SO_4	4-10 vol %	Room temperature		5-15
12	Acid dip	H_2SO_4	1 vol %	Room temperature		5-10

(a) Minimum time for removal of scale. (b) Current density, 1.5 to 2.0 A/dm^2 (15 to 20 A/ft^2). (c) Current density, 1.50 A/dm^2 (150 A/ft^2). (d) Tumble, without current. (e) Cycles for copper plating included in chart are applicable to all steels here, except that for high-carbon and low-alloy steels, a cyanide copper strike precedes cyanide copper plating. (f) Current density, 5.0 to 10.0 A/dm^2 (50 to 100 A/ft^2)

ing materials into phosphating tanks. This is particularly true for alkaline cleaners, which can neutralize the acid phosphating solutions, rendering them useless. Additional information can be found in the article on phosphate coating in this Volume.

Surface Preparation for Painting

Surface preparation has a direct effect on the performance of paint films. The best paint available will fail prematurely if applied to a contaminated or improperly prepared surface. The surface also will influence the final appearance of the paint film. Surface irregularities may not be hidden by the paint, but may instead be reflected as apparent irregularities of the paint film. The principal surface contaminants that are deleterious to the performance of paint films include oil, grease, dirt, weld spatter, alkaline residues, rust, mill scale, water, and salts such as chlorides and sulfides.

Mechanical and chemical cleaning operations may be used in combination to meet a rigid requirement of surface cleanliness. For example, on scale-bearing steel intended for an application involving exposure to chemical environments, complete removal of all oil, grease, rust, mill scale, and any other surface contaminants is mandatory. Nonferrous alloys such as aluminum require chemical conversion pretreatment plus chromated primers for maximum life and corrosion protection.

Further discussion can be found in the article on painting in this Volume.

Surface Preparation for Electroplating

Preparation for plating is one of the most critical of all cleaning operations, because maximum adhesion of the plated coating to the substrate is the major requirement for quality work. Maximum adhesion depends on both the elimination of surface contaminants in order to induce a metallurgical bond whenever possible and the generation of a completely active surface to initiate plating on all areas. In addition to pickling or other descaling operations, adequate cleaning requires multistage cycles, usually comprised of the following steps: (*a*) precleaning with a solvent to remove most of the soil; (*b*) intermediate cleaning with alkaline cleaners; (*c*) electrocleaning to remove the last traces of solids and other contaminants that are especially adherent; (*d*) acid treatment and surface conditioning to remove light oxide films formed during previous cleaning processes and to microetch the surface; and (*e*) electrolytic (anodic) desmutting to remove any smut formed during acid pickling of heat treated high-carbon steel parts. Low-carbon steels do not require this desmutting step. Anodic electrocleaning also offers oxidation or conditioning of scale. The oxidized or softened scale is easily removed in subsequent acid pickling. The types of cleaning usually employed in the above steps are:

- *Precleaning*: cold solvent, vapor degreasing, emulsifiable solvent, solvent emulsion spray, or alkaline spray with or without solvent emulsion
- *Intermediate alkaline cleaning*: soak cleaning with 30 to 90 g/L (4 to 12 oz/gal) of cleaner at 82 °C (180 °F) to boiling, spray cleaning with 4 to 15 g/L (0.5 to 2 oz/gal) at 66 to 82 °C (150 to 180 °F), and barrel cleaning with 7.5 to 45 g/L (1 to 6 oz/gal) at temperatures below 82 °C (180 °F)
- *Electrocleaning*: cathodic, anodic, or periodic-reverse
- *Acid treatment*: practice is highly specific for the metal being processed
- *Anodic desmutting*: necessary to remove carbon smut

ASTM recommended practices for cleaning various metals prior to plating are given below:

A380	Descaling and cleaning of stainless steel surfaces
B183	Preparation of low-carbon steel for electroplating
B242	Preparation of high-carbon steel for electroplating
B252	Preparation of zinc-based die castings for electroplating
B253	Preparation of and electroplating on aluminum alloys
B254	Preparation of and electroplating on stainless steel
B281	Preparation of copper and copper-based alloys for electroplating
B319	Preparation of lead and lead alloys for electroplating
B480	Preparation of magnesium and magnesium alloys for electroplating
B322	Cleaning metals before electroplating

Process sequences and operating details in surface preparation for electroplating are presented in articles in this Volume on cadmium plating, finishing of stainless steel, finishing of aluminum alloys, finishing of copper alloys, finishing of magnesium alloys, and finishing of titanium alloys. The procedures used for preparing the surfaces of high-carbon and low-alloy steels, low-carbon steel, and zinc-base die castings are discussed below.

Steels may be cleaned and otherwise prepared for electroplating according to the procedures outlined by the flow charts and tabulated data in Fig. 4. The preparation of low-carbon steel for electroplating consists essentially of cleaning to remove oil and caked-on grease, pickling to remove scale and oxide films, cleaning to remove smut left on the surface, and reactivating the surface for plating.

Plating on low-carbon steels represents the bulk of industrial plating. The steps generally used before plating low-carbon steels are the following:

- Vapor degrease, if necessary
- Alkaline soak clean
- Water rinse
- Descale, if necessary
- Water rinse
- Alkaline electroclean
- Water rinse
- Acid activate
- Water rinse
- Plate, as required

These steps are a general guideline and should not be construed as firm recommendations. The actual required cycle would depend on extent of grease and oil contamination, type of scale, and facilities available for the plating operation. Some of the options available to the plater are described below:

- Emulsion cleaning may be used in place of vapor degreasing. In this case, additional water rinsing is required
- Anodic electrocleaning is preferred over cathodic cleaning which can cause smut on parts because of plating of polar soils in the cleaner. Electrocleaners are generally used at 60 to 75 g/L (8 to 10 oz/gal) and at 8.0 to 10.0 A/dm^2 (80 to 100 A/ft^2). Temperature will depend on the type of cleaner. Low temperature cleaners operate at 27 to 49 °C (80 to 120 °F), high temperature cleaners operate at 82 to 93 °C (180 to 200 °F)
- If parts are not excessively dirty, soak cleaning can be used instead of electrocleaning. Specially compounded alkaline cleaners are sometimes used to remove slight amounts of oxides. Elevated temperatures are recommended for all alkaline cleaning
- Alkaline cleaners are difficult to rinse. Carryover of residues can produce staining, skip plating, or loss of adhesion. Warm water is recommended in the first rinse along with good agitation. Two or more countercurrent (cascade) rinses are highly desirable both from the standpoint of good rinsing and conservation of water. If both alkaline soak cleaning and alkaline electrocleaning are used, the two cleaning steps should be separated with a thorough rinse
- Plating is initiated on an active surface. A wide variety of activators is available, and most are acidic in nature. Hydrochloric, sulfuric, or fluoboric acids are commonly used
- Water rinse after activation is critical to avoid contaminating the sensitive plating solution. Countercurrent rinsing with two or more rinse tanks is desirable
- High-carbon and low-alloy steels are susceptible to hydrogen embrittlement
- Proprietary inhibited acid pickles are available for the effective removal of scale and rust with reduced danger of hydrogen embrittlement and base metal attack

Unless the acids used contain inhibiting agents, the acid treatments for surface preparation must be very mild and of short duration. If electrolysis is necessary, it should be used with anodic current. This is especially significant for spring-temper parts and parts that have been case hardened. Mechanical methods of descaling can often eliminate the need for pickling.

During the anodic etch, a high acid content, low solution temperature, and high current density will minimize smut formation. Carryover of water into the anodic etching solution should be held to a minimum, and long transfer times after the anodic etch should be avoided.

Cold rolled steel that has been subjected to deep drawing and certain prepickled hot rolled steels with glazed brownish-colored surfaces may be exceedingly difficult to clean. For these materials, a solution of 25 to 85 vol % hydrochloric acid and 1 to 5 vol % nitric acid has proved effective.

Paint Stripping

Infrequently, parts have to be stripped and repainted. Possibly there is a problem with appearance; the wrong paint or color may have been used. Tools, fixtures, and automatic spray line fixtures must be periodically cleaned of old paint buildup as well. Some paints are easier to strip than others, while some paint stripping methods are incompatible with some metals. A hot alkaline cleaning bath, which is a part of a metal process line, should not be used as a paint stripping tank. Even if the cleaning bath works, the bath quality would be degraded and uncontrolled impurities introduced. Paint cannot be effectively removed from a soiled part, so any part should first be cleaned. Table 8 compares various stripping methods and lists appropriate financial considerations. Selection of strippers is summarized in Table 9. In paint stripping, two processes are widely used, hot stripping and cold stripping.

Table 8 Methods of stripping paint

Method	Facility	Cost
Immersion	One or more tanks, water rinse capability required	Slow removal rate, low labor, costly facility, disposal cost
Spray or brush-on	Area, ventilation, rinse capability required	Slow removal rate, higher labor, lesser cost facility, disposal cost
Abrasive	Sand or shot blast facility	Slow removal, high labor, may use existing facility, disposal cost
Molten salt	Specialized facility for steel only	Rapid removal rate, costly facility, low labor, very efficient, lower disposal cost, fume collection required

Hot stripping uses high caustic level and high temperatures. Alkaline paint strippers contain caustic soda, sodium gluconate, phenols, or cresols. The bath is used at 80 to 95 °C (180 to 200 °F). Depending on the type of paint and coating thickness, stripping can be done in 30 min to 6 to 8 h. Hot stripping is slow, but economical and environmentally safe. Hot alkaline paint strippers will attack brass, zinc, and aluminum. These strippers are safe for steel and copper.

Cold Stripping. As the name indicates, paint stripping is done without any heating. The stripping bath consists of powerful organic solvents, such as methylene chloride; also organic acids, such as phenols or cresols. Many of the organic solvent strippers available in the market contain two layers. The heavier bottom layer is the organic solvent layer, in which the actual paint stripping takes place. The lighter top layer is the aqueous layer which prevents the evaporation of the highly volatile organic solvents from the bottom layer.

Cold solvent stripping, when applicable, is fast. The process, however, is

Table 9 Selection of strippers

Type of organic finish to be removed	Approved metal substrates	Means of application	Approved strippers and methods	Operating temperature °C	°F	Remarks
Epoxy primer, epoxies, polyurethanes ...	All(a)	Spray or brush on	Proprietary phenolic chromated methylene chloride	10-38(b)	50-100(b)	Good ventilation and protective clothing. Must be approved for high-strength steels
All others	Steel	Immersion	Low viscosity(c)	10-38(b)	50-100(b)	Good ventilation and protective clothing
	All(a)	Spray or brush on	High viscosity(c)	10-38(b)	50-100(b)	Must be approved for high-strength steels
All..............	Steel(d)	Immersion	Proprietary molten salt	As specified by vendor		2-5 min follow with water quench and rinse. Smoke and fume control required
Primers, wax, overspray, and temporary coatings.......	All	Wipe or squirt on	Butyl cellosolve methyl isobutyl ketone, ethyl alcohol xylene, toluene	Room temperature(e)		Xylene and toluene are normally only effective on waxes and some temporary coatings
All except epoxy based	All	Immersion	Caustic stripper	10-38(b)	50-100(b)	Water base, 10-12 pH
Epoxy...........	All	Dry abrasive blast	MIL-G-5634 Type III	Room temperature		Adjust pressure to part fragility
	Aluminum	Immersion	Chromic acid solution, 360-480 g/L (3-4 lb/gal) Chromic acid plus nitric acid solution	74 ± 3	165 ± 5	Maximum allowable immersion time is 15 min. Water rinse parts as soon as possible on removal from solution CrO_3 360-480 g/L (3-4 lb/gal) HNO_3 5% total volume
All..............	Aluminum	Immersion	Nitric acid solution 50-78% HNO_3	34 ± 6	110 ± 10	Maximum allowable immersion time, 20 min

Note: Heavy metals plus stripping chemicals require appropriate means of disposal to meet EPA regulations

(a) Except steel heat treated above 1500 kPa (220 psi). (b) Optimum temperature range: 18 to 29 °C (65 to 85 °F). (c) Proprietary: phenolic, chromated, methylene, chloride. (d) Except heat treated steel. (e) Do not exceed 32 °C (90 °F)

very expensive, and waste disposal could be a problem. Unlike hot strippers, the organic cold strippers can be used on all base metals such as steel, copper, aluminum, brass, and zinc.

Glass Bead Cleaning

Glass bead cleaning is a low energy, nonpolluting method for use with both small and delicate parts as well as large turbines and engines. Glass bead air systems equal or surpass the finish quality provided by liquid abrasive slurry. Other benefits include no measurable amount of metal removed from close tolerance surfaces (fine threaded screws) and noncontamination of work surfaces with wide range of bead sizes (170 to 400+ grit). Glass bead cleaning has been successfully applied to a wide diversity of uses such as: (*a*) preparation of surfaces for painting, plating, brazing, welding, bonding; (*b*) finishing of castings; (*c*) production of matte finish on metal, glass, and plastics for decorative purposes; (*d*) reclamation of tools such as files and saws; (*e*) stripping of paint; and (*f*) removal of solder from electrical assemblies.

Air pressures recommended for this procedure range from 70 to 415 kPa (10 to 60 psi). An angle of 40 to 60° for nozzle to work direction should be used to minimize bounce back and reduce bead consumption because of breakage. The selection of bead size should be based on the smallest particle that will give the desired surface. This provides the maximum number of impacts per pound. Working distances of 100 to 200 mm (4 to 8 in.) from nozzle to work will provide greatest impact (velocity) with the best pattern.

Pollution Control and Resource Recovery

The increasing cost of waste disposal has a greater impact on process cost and should be considered in selecting cleaning processes. Treatment of waste within the plant should be considered to reduce cost, reduce liability, permit reuse of the raw material, and improve process control. A good example of closed loop recycling is the distillation purification of vapor degreasing solvent. The Federal EPA has established compliance guidelines, but state and local regulations are often more stringent. For more information on waste disposal, see the article on disposal of plating wastes in this Volume.

Safety

In the use of any metal cleaning process, there are possible safety, health, and fire hazards which need to be considered. The degree of hazard is dependent upon such factors as the specific materials and chemicals involved, the duration of employee exposure, and the specific operating procedures.

Information is presented in Table 10 on the types of hazards which may be associated with each cleaning process and the general control measures which would be used for each hazard.

The Occupational Safety and Health Administration has established in its General Industry Standards (29 CFR 1910) regulations pertaining to a variety of safety and health hazards. Those sections of the standards which may

Table 10 Safety and health hazards

Cleaning process	Hazard/air contaminant	Control measures	OSHA/NFPA references
Abrasive blasting	Silica dust/total dust exposures	Local exhaust ventilation	(29 CFR)
		Respiratory protection	1910.94(a)
		Goggles or face shield	1910.95
	Noise exposures	Noise exposures	1910.133
		Hearing protective devices	1910.134
	Skin abrasion	Leather protection garments	1910.1000
			Table Z-3
Acid cleaning	Acid gas or mist exposure	Local exhaust ventilation	1910.94(L)
		Respiratory protection	1910.133
		Goggles or face shield	1910.134
	Skin contact	Impervious gloves and garments	1910.1000
			Table Z-1
Alkaline cleaning	Alkaline mist exposure	Local exhaust ventilation	1910.94(d)
		Respiratory protection	1910.133
		Goggles or face shield	1910.134
	Skin contact	Impervious gloves and garments	1910.1000
			Table Z-1
Emulsion cleaning	Petroleum or chlorinated hydrocarbons	Local exhaust ventilation	1910.94(d)
		Respiratory protection	1910.132
	Alkaline mist exposures	Local exhaust ventilation	1910.133
			1910.134
			1910.1000
			Tables Z-1, Z-2
Emulsion cleaning	Alkaline mist exposures	Respiratory protection	
		Goggles or face shield	
	Skin contact	Impervious gloves and garments	
Pickling	Acid gas or mist exposures	Local exhaust ventilation	1910.94(d)
		Respiratory protection	1910.133
		Goggles or face shield	1910.134
	Skin contact	Impervious gloves and garments	1910.1000
			Table A
Salt bath descaling	Burns	Heat resistant gloves and garments	1910.132
		Face shield	1910.133
	Toxic gasses	Local exhaust ventilation	1910.134
		Respiratory protection	1910.1000
			Table Z-1
	Fire/explosion	Proper facility design, construction, maintenance	NFPA 86C, Chapter 11
		Proper controls for tank	
		Proper work procedures	
Solvent cleaning	Petroleum or chlorinated hydrocarbon exposure	Local exhaust ventilation	1910.94(d)
			1910.132
			1910.133
		Respiratory protection	1910.134
			1910.1000
	Skin contact	Impervious gloves and garments	Tables Z-1, Z-2
Tumbling	Noise exposure	Noise enclosure for equipment	1910.95
		Hearing protective devices	
Vapor degreasing	Chlorinated hydrocarbon exposure	Provide condenser cooling system and appropriate thermostats	1910.94(d)
		Operate to minimize dragout	
		Install local exhaust ventilation	
	Solvent decomposition products	Eliminate hot surfaces above 400 °C (750 °F) in the vicinity	
		Eliminate sources of ultraviolet radiation in the vicinity	
		Monitor solvent for acid buildups to prevent exothermic decomposition	

apply to each cleaning process are referenced in Table 10. Because of the unusual fire hazard associated with salt bath descaling, an applicable chapter of the NFPA standards has also been referenced.

Tests for Cleanliness

The final evaluation of the effectiveness of a cleaning process should come from a performance test. Eight well-known methods of determining the degree of cleanness of the work surface are discussed below.

Water-break test is a simple test, widely used in industry. It consists of dipping the work into clean water to reveal a break in the water film in the soiled area. However, because the test depends on the thickness of the applied water film, a factor which cannot be controlled, false results can be obtained because of bridging of residues. A mild acid dip before testing for water break has been found advantageous.

Nielson method requires that ten soiled panels be processed individually to determine the time required for each to be cleaned. Panels are checked by the water-break test and then by the acid copper test. In the acid copper test, the ferrous panel is immersed in a copper sulfate solution (typical composition, 140 g (5 oz) of copper sulfate and 30 cm^3 (1 fluid oz) of sulfuric acid per gallon of water). On clean surface areas, copper will be deposited by chemical activity, forming a strongly adherent, semibright coating that is free of spots.

An average of the times required to clean the ten panels is taken as a measure of the effectiveness of the cleaning solution.

Atomizer Test. In the atomizer test, panels are cleaned, acid dipped, dried, placed in a vertical position, and sprayed with an atomizer containing a blue dye solution. Just before the droplets begin to run, the spray is stopped and the panel is placed in a horizontal position. Heat is applied to freeze the pattern. The cleaning index is the percentage of the total area that appears clean. This is determined by placing a grid over the panel, estimating the cleaning for several random squares and then averaging for the reported value. The atomizer test is 10 to 30 times as sensitive as the water-break test.

Fluorescent method requires soiling with a fluorescent oil, cleaning, and inspecting under ultraviolet light. It is very slow and is less sensitive than the water-break and atomizer tests.

Weight of residual soil is also an evaluation of cleanness. The cleaned panel is washed with ether, the washings are evaporated, and the residue is then weighed. A modified method is to clean, dry, and weigh the test panel, then soil, clean, dry and reweigh it. The increase in weight represents the amount of residual soil present.

Wiping method is a qualitative test. A panel is coated with pigmented soil, cleaned, and then wiped with a white cloth or paper. The presence of soil on the cloth or paper indicates poor cleaning.

In the residual pattern method, cleaned panels are dried at 49 °C (120 °F) for 20 min. After drying, the presence of a stained area indicates residual soil and incomplete cleaning.

Radioisotope tracer technique requires that radioactive atoms be mixed with the soil. Panels are coated uniformly with the soil, and their radioactivity is determined. The panels are then subjected to various cleaning cycles, after which their radioactivity is again determined. The cleaning ability of each of the various cycles can be evaluated by the amount of radioactivity remaining on the panels. This is the most sensitive test; however, dealing with radioactive materials requires an AEC license, trained personnel, and special types of equipment.

Alkaline Cleaning

By Donald P. Murphy
Senior Research Chemist
Occidental Chemicals
Parker Surface Treatments

ALKALINE CLEANING is used to remove soils from the surface of metals. Soils removed through alkaline cleaning include: oil, grease, waxy solids, metallic particles, dust, carbon particles, or silica. Alkaline cleaners are applied by immersion or spray, and the metals are cleaned by emulsification, dispersion, saponification, or combinations of these mechanisms. The cleaning step is usually followed by a water rinse and a subsequent operation such as conversion coating or electroplating.

Method of Application

Immersion Cleaning. When an alkaline cleaner is used by immersion, the parts to be cleaned are placed in the cleaning solution, which allows the cleaning solution to come in contact with the entire surface of the part. After the alkaline cleaner has affected the soil on the part, the soil is removed from the metal surface by convection currents in the solution which are created by heating units or through some mechanical means of solution movement.

Electrocleaning is a specialized form of alkaline cleaning in which electrodes are placed in a cleaning solution. Direct current is passed through the solution, and the part to be cleaned is made the anode and the electrode is the cathode. The cleaning process is enhanced by the scrubbing action of the oxygen which evolves at the anode. Additional information on this process can be found in the article on electrolytic cleaning in this Volume.

Relatively expensive electrolytic cleaners known as derusters are used extensively, especially by electroplaters. Derusters are used with periodic reverse cycles, in which both anodic and cathodic cleaning are used in the operation, as well as for nonelectrolytic immersion cleaning.

Derusters are more effective than standard cleaners. When used with periodic reverse, these cleaners remove a small amount of metal from the work being cleaned, leaving the work with a fresh, clean surface. The cycle must end with the workpiece as the anode to prevent a layer of unbonded metal from being deposited on the workpiece. In a periodic reverse cycle, the surface is washed by hydrogen when the workpiece is the cathode. Because electrolysis produces twice as much hydrogen as oxygen, there is twice as much washing action when the part is the cathode. The drawback to derusters is that the efficient ones contain cyanide. This means added cost for the treatment of rinse waters and for the disposal of the solutions.

Other variations of immersion cleaning include (*a*) barrel cleaning, in which small parts are agitated inside a barrel which rotates in the cleaner; (*b*) immersion cleaning, where parts are moved by either a rotary screw conveyor or a moving conveyor chain; (*c*) immersion cleaning where solution is agitated by either a recirculating pump, mechanical mixer, or ultrasonic sound waves; or (*d*) cleaning with the aid of external forces, such as brushes or squeegees.

Spray cleaning is accomplished by pumping the cleaning solution from a reservoir through a large pipe, or header, to a series of smaller pipes off of the header, called risers, and finally out of spray nozzles on the riser onto the part to be cleaned (Fig. 1). Spray pressure can vary from as low as 14 kPa (2 psi) to as much as 1380 kPa (200 psi) or more. Pressures usually range from 70 to 210 kPa (10 to 30 psi). In general, the higher the spray pressure, the more mechanical help is provided in removing soil from the metal surface. This is especially true with small electrically charged particles such as dust, carbon smut, and silica. Spray cleaners are prepared with low foaming detergents, which are not usually as effective as those found in immersion cleaners. Thus, the impingement of a spray cleaner plays a vital role in the removal of soil.

While spray cleaning is effective on most parts, certain parts, such as the interior of an automobile tail pipe, have soiled areas inaccessible to the sprayed cleaning solution. In these instances, an immersion cleaning is more effective because all surfaces of the part can be

Fig. 1 Equipment for spray cleaning operation

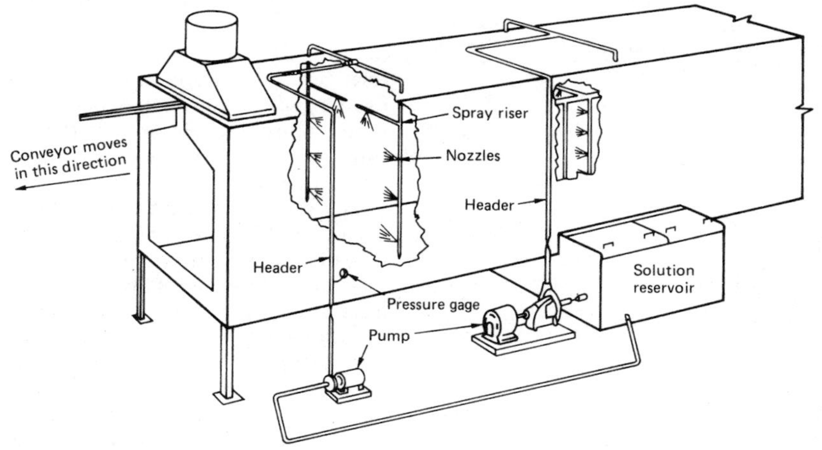

in contact with cleaning solution. Variations of spray cleaning include steam cleaning, where the cleaning solution is injected into a high pressure steam, and flow cleaning, where the cleaner is flooded onto the part at relatively low pressure from a large diameter opening.

Cleaning Mechanisms

Cleaning is accomplished by one or more of three major mechanisms: (*a*) saponification, (*b*) emulsification, or (*c*) dispersion. The three mechanisms can operate independently or in combination with each other. The saponification mechanism is limited to soils containing fats or other compounds that will chemically react with alkaline salts. The emulsification and dispersion mechanisms are effective on almost any liquid organic soil that is insoluble in water. The following is a brief description of these mechanisms.

Saponification. In saponification, fatty compounds, both animal and vegetable, react with the alkaline salts in an alkaline cleaner to form a water-soluble soap. Fatty soils are removed by dissolving them into the cleaning solution.

Emulsification joins together two mutually insoluble liquids such as oil and water. This is made possible by the detergent, more properly called a surface active agent or surfactant, which has a water soluble chemical grouping on one end and an oil soluble chemical grouping on the other end. The surfactant acts as a connector to keep the two insoluble liquids together as though they were one unit. In a cleaner dissolved in water, the water soluble end of the detergent has been solubilized in the water with the oil soluble end still needing a media where it will be soluble. Upon encountering the oil on a part, the oil soluble end is immediately solubilized, and because there is more water than oil, the water phase completely surrounds the oil. Most emulsions are cloudy or milky in appearance, although it is possible to form what appears to be a clear emulsion.

Environmental limitations in many areas, especially fish breeding areas, minimize the use of emulsifiers. Emulsifiers tie up oils and minute amounts of metals in the oils and prevent their removal during the treatment of solutions and rinse waters before discharging them into rivers and streams.

Dispersion. In dispersion, the surfactant acts to lower the surface tension of the cleaner at the metal surface, allowing the cleaner to cover the metal uniformly. The interfacial tension of the cleaner is lowered, permitting the cleaner to penetrate the oil film and break it into smaller units. As a result, the oil is dispersed into small droplets, which are undercut by the film of cleaner spreading across the metal surface. These droplets lose their attraction to the metal, are released, and because they are lighter than the cleaner solution, float to the cleaner surface where they reassemble into a semicontinuous film.

Rinsing

A good water rinse is essential for good cleaning. The water rinse may be either hot, warm, or cold, but it should be kept relatively clean. In order to ensure good rinsing, the water rinse should contain no more than 3% of the concentration of the cleaner solution. If the cleaner is at 30 g/L (4 oz/gal), the rinse water should contain no more than 0.9 g/L (0.12 oz/gal). In some instances, the water rinse removes as much as 50% of the soil. The cleaner conditions the soil, and the water rinse removes it. However, in most cases the water rinse is used to remove the cleaner solution and perhaps a minor amount of soil. Water rinsing can usually be done by using either immersion or spray, with spray being the most effective in most instances. Warm is usually the best temperature for a water rinse. Cold rinses do not have as much ability to remove the alkaline cleaner film, and the soils and hot rinses tend to promote formation of an oxide film such as rust.

Alkaline Cleaner Composition

Alkaline cleaners are comprised basically of three major types of components: (*a*) the builders, which make up the largest portion of the cleaner, (*b*) organic or inorganic additives which promote better cleaning or act to affect the metal surface in some way, and (*c*) the surfactants.

Builders are the alkaline salts in an alkaline cleaner. They are usually a blend selected from the following groups: alkali metal orthophosphates, alkali metal condensed phosphates, alkali metal hydroxides, alkali metal silicates, alkali metal carbonates, alkali metal bicarbonates, and alkali metal borates. A blend of two or more of these builders is used for cleaning performance, physical properties of the dry blend, and economics (Table 1). The alkali metal is usually sodium.

Phosphates serve a multiple function in the cleaner. They act to soften the water, eliminating the flocculent precipitate caused by calcium, magnesium, and iron. They act as a soil dispersant, a source of alkalinity, and as a buffer, which prevents large changes in the level of alkalinity. Some common phosphates used are trisodium phosphate, disodium phosphate, tetrasodium pyrophosphate, and sodium tripolyphosphate known as tripoly. Phosphates also have a moderate detersive value. Like emulsifiers, phosphates are subject to environmental regulations which limit

Table 1 Alkaline cleaning formulas for various metals

	Formula, wt %							
	Aluminum		Steel			Zinc		
Constituent	Immersion	Spray	Immersion	Spray	Electro-cleaning	Immersion	Spray	Electro-cleaning
Sodium hydroxide	...	...	38	50	55	...	...	16
Sodium carbonate(a)	55	18	36	17	18	10	20	50
Sodium metasilicate(b)	37	...	12	...	10	15	10	...
Sodium metasilicate(c)	...	60	...	...	10	...	...	32
Tetrasodium pyrophosphate	...	20	9	20	6	20	65	...
Sodium tripolyphosphate	...	...	...	...	...	50	...	...
Trisodium phosphate	...	...	...	10	...	...	...	...
Fatty acid esters	1	...	3	0.6	...	...	...	1
Ethoxylated alkylphenol	...	...	2	0.2	...	...	...	...
Ethoxylated alcohol	...	2	...	2	...	...	5	...
Sodium alkylbenzene sulfonate	5	...	...	...	...	5	...	1
Naphthalene sulfonate	2	...	...	0.2	1	...	...	...

(a) Dense. (b) Anhydrous. (c) Hydrate

the amount of phosphorus allowed in discharged water.

Silicates are also multifunctional. They provide alkalinity, keep soil suspended, provide some detergency, and act as an inhibitor, protecting metals such as aluminum and zinc from attack by other alkaline salts. Commonly used silicates include sodium metasilicate and sodium orthosilicate. Silicates are difficult to rinse and may cause trouble in subsequent plating operations if they are not completely removed during rinsing.

Carbonates are a cheap source of alkalinity. They act as a buffer and an absorbing media for the liquid components of the cleaner. Hydroxides are another relatively inexpensive source of strong alkalinity. Borates are somewhat like silicates in that they provide some detergency, act as a buffer, and provide some metal protection.

Additives are either organic or inorganic compounds that provide additional cleaning or surface modification. Chemical compounds such as glycols, glycol ethers, chelating agents, and polyvalent metal salts could be considered additives. For environmental reasons, chelating agents are replacing phosphates in some cleaning formulas. These additives can soften water and complex or tie up metal ions. Some widely used chelating agents include: sodium gluconate, sodium citrate, tetrasodium ethylenediamine tetraacetate (EDTA), trisodium nitrilotriacetate (NTA), and triethanolamine.

Surfactants are organic compounds which provide detergency, emulsification, and wetting in an alkaline cleaner. There are four major types of surfactants: (*a*) anionic, (*b*) cationic, (*c*) nonionic, and (*d*) amphoteric. An anionic surfactant is one in which the largest part of the molecule is the anion, or the negatively charged ion. Conversely, in the cationic type the cation, or positively charged ion, is the larger molecule. The amphoteric type is pH dependent, being anionic in an alkaline media and cationic in an acidic media. In all three, the surfactant ionizes in water, separating into negative ions and positive ions which function independently of each other. A nonionic surfactant, as its name implies, does not ionize, but rather remains as an intact molecule. An example of each surfactant is listed below:

- Anionic: sodium alkylbenzene sulfonate
- Nonionic: ethoxylated long chain alcohol
- Cationic: quaternary ammonium chloride
- Amphoteric: alkyl substituted imidazoline

In alkaline spray cleaners, the nonionic type of surfactant is used almost exclusively because, in general, it is the only type that can be obtained in a low foaming form and still provide good detergency. The immersion cleaners can use any of the four types because foam is not a problem. However, the surfactants used are usually of the anionic or nonionic type.

The amphoteric type is seldom used because it becomes anionic in an alkaline media and cationic in an acidic media. Since these types are already available at a lower cost, there is generally no advantage to using an amphoteric. The cationic type is good for certain special applications as an emulsifier, but generally is not as good a detergent as the anionic or nonionic types. Additionally, the cationics are frequently substantive to metal, that is they form a film on the metal, thus defeating the intended purpose of providing a clean metal surface.

Operating Conditions

The various types of alkaline cleaners are applied differently, using different equipment and operating parameters. The following examples indicate typical operating conditions for several types of cleaners and include information on concentration of the cleaner, the temperature at which it is used, the time normally required to clean the metal, and certain specialty data such as current density for electrocleaners and spray pressures for spray cleaners.

Immersion cleaners

Amount . . . . 30 to 90 g/L (4 to 12 oz/gal)
Temperature . . . . 71 to 99 °C (160 to 210 °F)
Processing time . . . . 1 to 5 min

Electrocleaners

Amount . . . . 60 to 120 g/L (8 to 16 oz/gal)
Temperature . . . . 71 to 93 °C (160 to 200 °F)
Processing time . . . . ½ to 3 min
Current . . . . 6 to 9 V
Current density . . . . . . 6 to 15 A/dm^2 (60 to 150 A/ft^2)

Spray cleaners

Amount . . . . 1.9 to 22.5 g/L (0.25 to 3 oz/gal)
Temperature . . . . 60 to 82 °C (140 to 180 °F)

Spray pressure....... 35 to 210 kPa (5 to 30 psi)
Processing time.........½ to 3 min

The recent discovery of certain select blends of surfactants, the addition of other organic compounds, and the invention of new surfactant technology has allowed the temperature of cleaners to be reduced to as low as 27 °C (81 °F) in spray cleaners and 49 °C (120 °F) in immersion cleaners without sacrificing performance. In most cases, the concentration of the cleaner does not have to be increased, and a considerable savings results from the decrease in energy demand. For example, the cost of heating a solution by steam to 88 °C (190 °F) is about three times that of heating that solution to 49 °C (120 °F). Ongoing research in this area is being directed at further reducing the temperature necessary to provide top quality cleaning of metal surfaces in both spray and immersion applications.

Testing and Control of Cleaners

An alkaline cleaner loses strength through use and dilution caused by replacing lost cleaning solution with water. For this reason, a method of determining not only what the strength of the cleaner is at any given time, but also what it should be to provide the best performance is necessary. A chemical procedure known as an acid-base titration is most commonly used. In this procedure, a known amount of alkaline cleaner is placed in a container, and an acid of precise concentration (titrating solution) is measured into the cleaner until a certain pH is achieved. Certain organic compounds known as indicators change color at specific pH levels, and this color change is used to identify when the desired pH level is achieved in the titration. By determining the amount of acid titrating solution required to achieve the desired pH level in an alkaline cleaner of known volume and concentration, a factor can be developed which will allow the calculation of the concentration of any solution of the same cleaner by multiplying the factor by the number of millilitres of titrating solution required to achieve the desired pH.

The following example shows acid-base titration used for testing alkaline cleaner. A 5-mL sample of cleaner is tested with 1*N* acid used for titrating. (1*N* refers to a specific concentration of acid.) The indicator is phenolphthalein, which changes from red to colorless at a pH of about 8.7. When the 5-mL sample of a precisely measured 10 g/L (1.3 oz/gal) concentration of cleaner is titrated, 5 mL of acid is needed to turn the indicator from red to colorless. 5 mL equals 10 g/L (1.3 oz/gal) or each 1 mL equals 2 g/L (0.27 oz/gal). The factor is 2.0 and the equation for calculating concentration in grams per litre would be:

$$\text{Concentration} = 2 \times \text{millilitres of acid}$$

The procedure outlined above is a method for determining concentration based on the amount of free alkalinity. Another check on the condition of the cleaner can be done by using another indicator called methyl orange, which changes from yellow to orange at about 3.9 pH. This measures the amount of total alkalinity, which in many cleaners is in the ratio of 1.2 to 1, total alkalinity to free alkalinity. This ratio is different for each cleaner and should be determined for each cleaner. A rule of thumb sometimes used is dumping the cleaner and preparing a new solution when the ratio doubles. This comes about because the total alkalinity constantly increases with additions of cleaner while the free alkalinity is held fairly constant. As a result of this, total alkalinity titration is a direct measure of the amount of contamination in the cleaner because the ratio of total alkalinity to free alkalinity increases only when cleaner is added to replace free alkalinity lost in reacting with the soil. Dragout of solution on parts does not cause an increase in the ratio because both free and total alkalinity are lost in proportion to their concentration in the solution.

Because of the blend of builders used, some alkaline cleaners do not have any free alkalinity, and they must be controlled by total alkalinity titrations. These cleaners are usually specialized, and the pH of the cleaner solution is intentionally maintained under nine to maintain stability of one or more of the cleaner's components, keep the cleaner from attacking the metal substrate, or for some other special purpose. Total alkalinity titration is performed exactly like the free alkalinity titration, except the indicator is methyl orange (3.9 pH) instead of phenolphthalein (8.7 pH). The only drawback to this method of control is the inability to measure the degree of soil contamination of the cleaner by a titration procedure.

Equipment for Alkaline Cleaners

All equipment for alkaline cleaners may be constructed of low-carbon steel, although 300 series stainless steel is often used for steam coils and pumps which are exposed to other corrosive media such as water and air. The simplest cleaning line is immersion, where the equipment consists of a tank, a source of heat, such as gas, electricity, steam, or heat exchanger, and an exhaust system to draw off the steam being generated by the hot cleaner. For a spray system, a pump, piping, and a spray-zone bonnet must be added. For electrocleaning, a rectifier to supply direct current electricity and electrodes must be added. Periodically, usually one to three times a year, alkaline cleaner tanks and spray equipment must be drained and cleaned with an inhibited acid solution to remove accumulated hard water salt deposits which build up on the heating coils, tank walls, and inside the spray piping. These salts cause reduced spray pressures and inefficient heating. If not removed, these salts could permanently damage the equipment.

Electrolytic Cleaning

By Michael P. D'Angelo
Experimental Product Manager
MacDermid, Inc.

ELECTROLYTIC CLEANING, or electrocleaning, is the process by which a workpiece is made anodic or cathodic in a specially formulated alkaline cleaning solution. Between 3 and 12 V direct current are applied to yield current densities ranging from 1 to 15 A/dm^2 (10 to 150 A/ft^2) of work area. An electrocleaning operation usually follows an alkaline soak and serves two purposes: (*a*) it removes any residual soils that may have been left behind by the soak cleaner, and (*b*) the electrocleaner activates the metal surface, that is, it eliminates the passive condition but only when cleaned in the cathodic mode. After electrocleaning, parts are usually pickled in mineral acids or solutions of proprietary acid salts, which neutralize any residual alkali film, remove oxides and smuts, and further activate the workpiece for subsequent electroplating. The quality and adherence of the final finish of an electroplated piece depends on the pretreatment process, and the alkaline electrocleaner is an integral part of that process.

Although electrolytic cleaning is sometimes used in other metal finishing processes, such as phosphating, chromating, and painting, it is almost always used with electroplating, where a greater degree of cleanliness is required. When selecting an electrocleaner, there are two major considerations: (*a*) the substrate to be cleaned, and (*b*) the type of plating solution to be used.

Metal substrates that are commonly electrocleaned include steels, copper and its alloys, nickel and its alloys, and zinc die castings. Electrocleaning is performed in rack, barrel, and continuous strip operations.

Alkaline cyanide-plating baths are not as critical to cleaning as noncyanide baths, and cleaner selection is easier. It is not uncommon to see a mid- or high-cyanide zinc bath with fair amounts of oil and grease floating on top. An acid chloride zinc or nickel-plating solution, on the other hand, does not tolerate soiled work or oil drag-in, so cleaner efficiency is important. Also, some compounds used in formulating cleaners, such as silicates and gluconates, may be detrimental if dragged into some plating baths.

Anodic Electrocleaning

Also referred to as reverse current cleaning, anodic electrocleaning is most commonly used on ferrous metals. Figure 1 illustrates a workpiece being cleaned anodically and gives the accompanying reaction. Because the workpiece is actually the anode (positive), free electrons are discharged by the hydroxyl ions to the metal, resulting in the liberation of gaseous oxygen. The oxygen bubbles liberated at the workpiece create a scrubbing action, which actually blasts dirt particles off the workpiece. These bubbles rise to the top of the cleaner, increasing solution agitation and continually bringing fresh cleaning solution to the work.

Anodic electrocleaning is desirable in the preplate cycle because the positive charge creates an oxidizing condition at the metal interface. This may remove metallic oxides and smuts, but also prevents the deposition of other positively charged metallic ions, which otherwise may result in a detrimental film on the workpiece that usually is undetectable until after electroplating. It also prevents hydrogen embrittlement during cleaning, which is important for high-strength steels.

Nonferrous metals, such as zinc die castings, copper, and brass, may be successfully cleaned anodically; however, prolonged times, high temperatures, and high current densities, must be avoided to prevent etching, pitting, and dezincification of brass. Because these nonferrous metals may be oxidized because of oxygen liberation, specially formulated cleaning compounds containing inhibiting additives are used.

Cathodic Electrocleaning

Direct current, or cathodic electrocleaning, employs a negative charge on the part. Insoluble steel or nickel-plated steel anodes are used. The cleaning action obtained is somewhat better than that of anodic cleaning because

Fig. 1 Anodic electro-cleaning

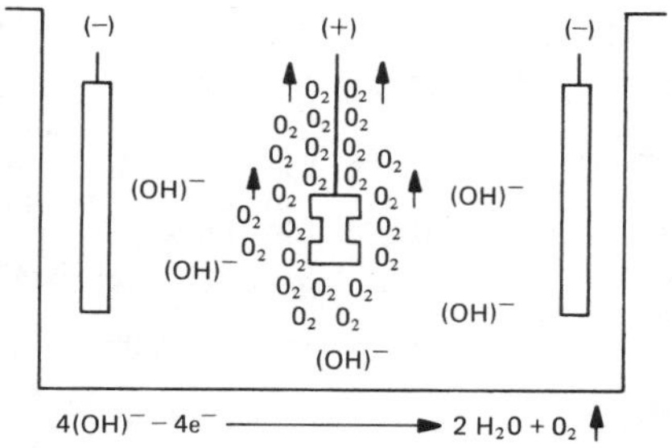

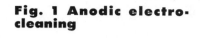

Four electrons are discharged by four hydroxyl (OH)⁻ ions at the anode, or workpiece, to liberate one molecule of oxygen (O_2)

Fig. 2 Cathodic electro-cleaning

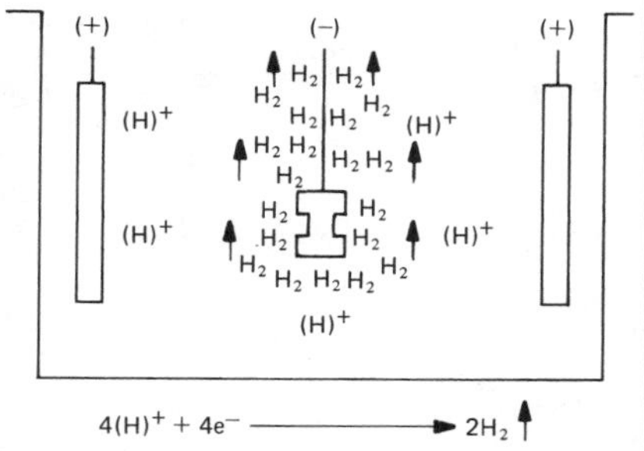

Reaction of electrons with positively charged hydrogen ions results in liberation of hydrogen gas

hydrogen gas is evolved at the cathode (workpiece), and at a given current density, twice the volume of hydrogen is liberated at the cathode than oxygen at the anode. The added mechanical scrubbing action and solution agitation, coupled with the fact that the negatively charged workpiece repels negatively charged particles of soil, are the major advantages of cathodic cleaning. Figure 2 illustrates cathodic electrocleaning. The electrons are supplied to the cathode where they combine with positive hydrogen ions to liberate hydrogen gas.

A disadvantage of cathodic electrocleaning is that the negatively charged workpiece attracts positively charged species, including metallic ions such as Cu^+ and Zn^{+2}, and other species such as soaps and colloids. These particles may plate out onto the workpiece, producing a loosely adherent film that can lead to subsequent electroplating problems. These problems may include irregular plate, blistering, poor adhesion, roughness, staining, skip plating, and/or cloudy deposits. For this reason, cathodic cleaning usually is not used as the final step in the preplate cycle.

Hexavalent chromium (Cr^{+6}) is another common impurity in electrocleaners, but it is more detrimental in cathodic applications. Chromium-bearing solution from chrome plating baths, chromates, and etchants becomes entrapped in blind holes and seams on the workpiece, as well as in cracks and crevices in plastic rack coating materials. The solution can be leached out in the hot alkaline cleaners. On many automatic return-type platers, the cleaners are directly opposite chromium-bearing solutions and mists of hexavalent chrome may be in the air and can contaminate the electrodes. This phenomenon is referred to as chrome blowover. Hexavalent chromium causes the metal to become passive and resist electroplating. It usually can be reduced to the harmless trivalent state using sugars or proprietary reducing agents, but the best control is preventive maintenance. Racks should be repaired and replaced as needed, and plastic dividers between process tanks are effective in preventing chrome mist from contaminating electrocleaners. Also, a more frequent dump schedule is needed with cathodic cleaners to prevent buildup of impurities.

Another disadvantage of cathodic cleaning is hydrogen embrittlement. Certain steels may become embrittled due to the occlusion of gaseous hydrogen. Parts adversely affected by hydrogen embrittlement, such as spring steel and steels with hardness values exceeding 40 HRC, should be baked at 205 °C (400 °F) for at least 1 h, immediately after plating, to drive off the hydrogen gas. Loud, annoying hydrogen explosions often result from hydrogen entrapped in foam on the surface of the cleaner, which can be ignited by sparks caused when racks make and break contact. These may be more common where cycles are used with no rinse between soak cleaners and electrocleaners. The more heavily wetted soak cleaners are dragged into the electrocleaner, producing heavy foam blankets that trap hydrogen. Proprietary defoaming agents usually are effective in reducing the foam blanket.

Cathodic electrocleaning plays an important role in cleaning certain alloys, such as nickel, that are easily passivated if cleaned anodically. Buffed nonferrous metals can be subjected to direct current for longer periods than are safe with reverse current. This is because the negative charge on the metal part represses the tendency of a nonferrous metal to dissolve in an alkaline cleaning solution and also because the presence of hydrogen protects the nonferrous surface from the tarnishing effect of oxygen.

Periodic Reverse Electrocleaning

Periodic reverse (PR) electrocleaning is a combination of the two aforementioned procedures, whereby the workpiece is subjected to a number of cycles of anodic and cathodic electrocleaning. Cleaning solutions that are high in sodium hydroxide content and contain fair amounts of metal chelating agents usually are incorporated in PR cleaning. The most effective cycles are those in which the reverse (anodic) time is equal to, or greater than, the direct (cathodic) time. A typical 10/10 PR cycle is employed on many ferrous metals, that is 10 s of anodic direct current, followed by 10 s of cathodic direct current. In 1 min, the part is subjected to three 10-s cycles of anodic and three 10-s cycles of cathodic current. The workpiece should always be removed during the anodic cycle, preferably toward the end, to prevent deposition of loose metallic smuts.

The major advantage of PR cleaning is better cleaning and removal of oxides, scale, and rust. The cleaning mechanisms used are those of alkaline soil removal combined with oxidation and reduction, and strong metal chelation. Periodic reverse electrocleaning has, in some situations, allowed acid-free cleaning cycles. This is beneficial because it further reduces hydrogen embrittlement and can eliminate bleedout problems associated with parts that have blind holes and seams. Unlike acid pickles, PR cleaners remove rust and oxides from steel without etching or smut formation.

Table 1 Compositions and operating conditions for electrocleaner baths

Substrate	Formula No.	Bath composition, wt %: Sodium hydroxide (NaOH)	Sodium carbonate (Na_2Co_3)	Sodium metasilicate ($NaSiO_3$)	Sodium tripolyphosphate ($Na_5P_3O_{10}$) (a)	Surfactants (anionic)	Use concentration g/L	oz/gal	Current density A/dm^2	A/ft	Temperature °C	°F	Common polarity
Steel	1	50	24	20	5	1	75-120	10-16	5-15	50-150	77-93	170-200	Anodic or cathodic
Zinc	2	20	28	40	10	2	30-60	4-8	1-5	10-50	54-65	130-150	Cathodic
Copper	3	40	23	25	10	2	45-75	6-10	1-5	10-50	71-82	160-180	Cathodic
Brass	4	25	33	30	10	2	30-60	4-8	1-5	10-50	65-82	150-160	Anodic or cathodic

(a) Part or all of the phosphate may be substituted by chelator

Cleaner Formulation

Soil removal in electrocleaning is accomplished by the same mechanisms described in the article on alkaline cleaning in this Volume. These methods include wetting, emulsification, saponification, peptization, dispersion, flocculation, water softening, and chelation. Electrocleaning adds the beneficial effects of mechanical scrubbing action and solution agitation caused by gas evolution at the metal-solution interface. Desirable characteristics of alkaline electrocleaners include:

- Ability to wet metal surfaces
- High alkalinity
- High conductivity
- Low foaming
- Minimal attack of substrate
- Free rinsing
- Ability to soften water (prevent insoluble soap formation)
- Good buffering (neutralize acid drag-in)
- Nonclumping and free flowing in powder form

Alkaline electrocleaners are formulated from a carefully selected combination of alkalis, wetting agents, soaps, and chelators. The presence of amounts of certain compounds must be adjusted depending on the substrate to be cleaned. Most metal finishers utilize proprietary cleaning compounds formulated by suppliers who give careful consideration to soil types and difficulty of removal, as well as substrate attack. Some typical electrocleaner bath compositions and operating conditions are given in Table 1.

Sodium hydroxide, or caustic soda, is one of the most important ingredients in alkaline electrocleaners. It offers higher alkalinity per pound than any other alkali and is the most conductive. Among the disadvantages of sodium hydroxide are poor soil removal, especially on the nonsaponifiable soils, and difficulty in rinsing. Lesser amounts of sodium hydroxide are incorporated in nonferrous metal cleaner formulations to inhibit substrate attack.

Carbonates, and sodium carbonate (soda ash), are low-cost sources of alkalinity and offer good buffering, but their major purpose in dry mix cleaners is as a carrier for liquid surfactants, which yield a clump-free, free-flowing powder. High carbonate contents are used in nonferrous metal cleaner formulations because they buffer well at lower pH levels.

Silicates are usually present in most heavy-duty alkaline electrocleaners because they offer good emulsifying characteristics and act as pH buffers. The most common silicate used is sodium metasilicate.

Phosphates are good water softeners and aid in sequestering and chelation. The most common phosphate used is sodium tripolyphosphate. Phosphates also aid in buffering, add alkalinity, and improve ability to rinse. Part or all of the phosphate may be replaced with organic chelating agents to boost detergency. A disadvantage of phosphates is increased algae growth (eutropication) in streams and lakes, which is becoming an increasing ecological problem. Today, however, many phosphate-free cleaners are available that are as effective in cleaning as their phosphated counterparts. Phosphates are replaced by chelators or other water softeners which convert the otherwise insoluble calcium or magnesium salts into free-rinsing precipitates or sequester them into solution as soluble complexes.

Surfactants, which are usually either monionic or anionic in charge, provide the required wetting and dispersing properties at the operating pH of the cleaner. Care must be taken in formulating to limit their concentration because they may become insoluble and float to the top as scum if cleaner concentration is high. In electrocleaners, however, lower levels of surfactants are generally used to reduce foam levels. Exceptional detergency usually is not required because the parts already have been soak cleaned, and the majority of the heavy soils have been removed.

Surfactants used in electrocleaners are commonly of the monionic detergent type in conjunction with an anionic coupling agent. A modern trend is toward biodegradable surfactants. These are destroyed by bacteria present in waste water treatment and sewage systems. Surfactants are usually biodegradable if they are composed of straight-chain organic aliphatics. Organic compounds composed of branched chains are usually difficult to decompose. Surfactants decrease surface tension (improve wetting) and improve detergency.

A variety of electrocleaner additives for specific problems are available. These include fume suppressors, chelating agents, and chrome reducers.

Process and Cleaning Cycles

Process Cycles. The selection of an electrocleaner, its use concentration, temperature, time, and polarity depend on the nature of the oil, the substrate, and the final plating process. Some typical cycles are given below. As described, anodic electrocleaning is used most widely on ferrous metals, whereas cathodic current is employed on difficult cleaning jobs and on nonferrous metals. Periodic reverse current is used on heavily scaled work and for acid-free cycles to reduce hydrogen embrittlement.

Cleaning Cycles. When selecting a cleaning cycle for parts to be electroplated, consideration must be given to the substrate, soils, and surface treatment (the final finish). A properly designed pretreatment cycle removes soils and activates the work quickly, but sufficiently, to perform the final operation with minimal substrate attack. Different substrates will require different pretreatment processing. But, it is also true that the same substrate, with the same soil, may not require cleaning as critical for one plating process as it might be for another.

Some typical cleaning cycles are given below. These include a variety of substrates, soils, and final surface treatments.

Steel stampings

To remove light stamping oils (mineral or sulfurized) for a final cyanide zinc plate (barrel) finish, the following cleaning cycle is suggested:

- Alkaline soak clean in 75 to 90 g/L (10 to 12 oz/gal) of solution at 77 to 82 °C (170 to 180 °F) for 4 min
- Alkaline electroclean in 90 g/L (12 oz/gal) of Formula 1 (Table 1) at 77 °C (170 °F) with 6 to 12 V anodic current, 10 to 15 A/dm^2 (100 to 150 A/ft^2) for 1½ min
- Two 30-s cold water rinses with counterflow current
- Acid pickle in 30 to 40 vol % hydrochloric acid at 27 °C (80 °F) for 2 min
- Two 30-s cold water rinses with counterflow current
- Cyanide zinc plate

Steel wire fabrications

To remove mineral oils, drawing lubricants, weld scale, light rust for a final bright nickel plate (rack) finish, the following cleaning cycle is suggested:

- Alkaline soak clean in 75 to 90 g/L (10 to 12 oz/gal) of solution at 77 to 82 °C (170 to 180 °F) for 5 min
- Alkaline electroclean in 90 g/L (12 oz/gal) of Formula 1 (Table 1) with chelator at 82 °C (180 °F) with 3 to 5 V, 8 to 15 A/dm^2 (80 to 150 A/ft^2) cathodic current for 2 min
- Two 30-s cold water rinses with counterflow current
- Acid pickle in 60 to 120 g/L (8 to 16 oz/gal) of proprietary acid salt at 27 to 49 °C (80 to 120 °F) for 3 min (may be used with cathodic current)
- Two 30-s cold water rinses with counterflow current
- Alkaline electroclean in 90 g/L (12 oz/gal) of Formula 1 (Table 1) at 82 °C (180 °F) with 3 to 5 V anodic current for 1½ min
- Two 30-s cold water rinses with counterflow current
- Acid dip in 10 to 20 vol % sulfuric acid at 27 to 49 °C (80 to 120 °F) for 2 min
- 30-s cold water rinse
- Bright nickel plate

Brass and zinc die castings

To remove buffing compounds for a final finish of cyanide copper/nickel plate (rack), the following cleaning cycle is suggested:

- Alkaline soak clean in 75 g/L (10 oz/gal) of low caustic solution at 77 °C (170 °F) for 3 min
- Ultrasonic clean
- Warm water rinse
- Spray clean in 7.5 to 30 g/L (1 to 4 oz/gal) of solution at 60 °C (140 °F) for 2 min
- Cold water rinse
- Alkaline electroclean in 60 g/L (8 oz/gal) of Formula 2 or 4 (Table 1) at 65 °C (150 °F) with 3 to 5 V cathodic current for 2 min
- Two 30-s cold water rinses with counterflow current
- Acid dip in 45 to 75 g/L (6 to 10 oz/gal) of proprietary acid salt at 27 °C (80 °F) for 1 min
- Two 30-s cold water rinses with counterflow current
- Cyanide copper strike

Steel fasteners (nuts and bolts)

To remove heavy heat-treat scale, carbonaceous soils for a final finish of cadmium plate (acid-free cycle), the following cleaning cycle is recommended:

- Alkaline soak clean in 75 g/L (10 oz/gal) of solution at 82 to 88 °C (180 to 190 °F) for 4 min
- Alkaline electroclean in 120 g/L (16 oz/gal) of heavily chelated solution of Formula 1 at 85 to 90 °C (185 to 195 °F) with 6 to 8 V current 10 A/dm^2 (100 A/ft^2) periodic reverse, 10/4 cycle) for 10 min
- Two 1-min cold water rinses with counterflow current
- Cyanide cadmium plate

Solution Control

Maintenance and control of electrocleaners are simple and require only a minimal amount of time. Important control factors and methods are:

Factor	Controlled by:
Cleaner concentration	Titration
Temperature	Thermostat
Voltage	Voltmeter
Current density	Ammeter and total work area
Time	Programming on automatics or workers on hand lines
Cleaner replacement*	Dump schedules

*Highly dependent on amount of dirt and soil entering cleaner

Usually, the electrocleaner concentration is determined by titration for total alkalinity, and multiplication by a factor based on the alkalinity of the cleaner. This simple, direct titration should be run prior to each work shift. Alkaline test kits are available which allow the titration to be performed on-line during production. More elaborate chemical analysis, such as free alkalinity, surface tension, pH, and percentage of soil load, may be performed; however, for most applications, the total alkalinity titration is sufficient. Electrocleaner temperatures should be maintained within a tolerance of ±3 °C (±5 °F), and the electrocleaner should be operated at the highest temperature recommended by the supplier that the substrate can withstand. In general, the activity of a cleaner doubles for each 11 °C (20 °F) increase in temperature; the higher cleaner temperature lowers the electrical resistance, permitting higher current densities. Voltages and current densities are controlled by reliable metering devices and accurate measurements of work surface area.

A loss of brightness due to residual soils left on the substrate, blistering or poor adhesion, staining, irregular patterns, or other blemishes in the final finish are indications that a cleaning solution has reached its intended service limits. The cleaner should be dumped prior to this point; a few hours of rejected work caused by these conditions may offset the cost of a new bath make-up. Most metal finishers base dump schedules on production time or by the square footage of processed work. The service life of an electrocleaner varies from operation to operation depending on the type and degree of soiling, as well as the efficiency of the soak cleaner and type of work being processed; therefore, dump schedules can only be accurately determined by experienced personnel. Recordkeeping is very important for all chemical processes used in the finishing line. One person should be assigned the responsibility of collecting and tabulating all pertinent data on a daily basis in an organized chart form for permanent record.

Cleanliness of Parts

It is difficult to determine if a metal part has been sufficiently cleaned to be subsequently electroplated. However, the effects of plating a soiled part are easily discernible as poor adhesion or a multitude of other problems. Aside from adhesion testing, little quantitative data relating to cleaning can be obtained from the finished part. Usually, tests performed on the finished product are destructive tests that may be performed to evaluate the efficiency of soil removal. In practice, most of these tests are too elaborate and time consuming for the plater.

Water-Break Test. This is the most widely used test for a clean surface throughout the industry. A clean metal surface will sustain an unbroken film of water. The metal part is cleaned, rinsed, acid dipped, and rinsed thoroughly. A part that sustains a water-break film after a 30-s drainage period can almost always be successfully electroplated.

Copper Displacement Test. In this test, the cleaned specimen (ferrous metal) is immersed in a solution of 15 g/L (1.7 oz/gal) copper sulfate with 0.9 g/L (0.1 oz/gal) H_2SO_4 at room temperature for 20 s. An immersion copper deposit forms on all clean areas because the copper replaces the iron galvanically. Soiled or passive areas remain bare, or the copper may be removed easily by wiping lightly with a soft cloth.

Another on-line test that may be used on steel parts consists of removing the part from the plating line following the acid pickle, rinsing thoroughly, and allowing it to stand for about 10 min exposed to the atmosphere. A uniform, light flash rust should cover the entire part. If the part does not rust in some areas, soils are presumed to be present in those areas.

Pumice can be used to scrub clean one part thoroughly. This part is then processed through the cleaning line together with an unscrubbed part. After plating, the two parts are examined visually or bend tested for adhesion to determine if the unscrubbed part is comparable to the cleaned, scrubbed part. By visual inspection, the plater should evaluate the unscrubbed piece for brightness. It should be as bright as the scrubbed parts, with no patterns or stains in the deposit. The deposit should be free of blisters, pits, and roughness and should be completely covered, even in low-current density areas. Any visually discernable difference between the two parts should be further investigated. The plater can assume that the piece scrubbed with pumice is sufficiently cleaned.

Fig. 3 Electrolytic alkaline cleaning tank

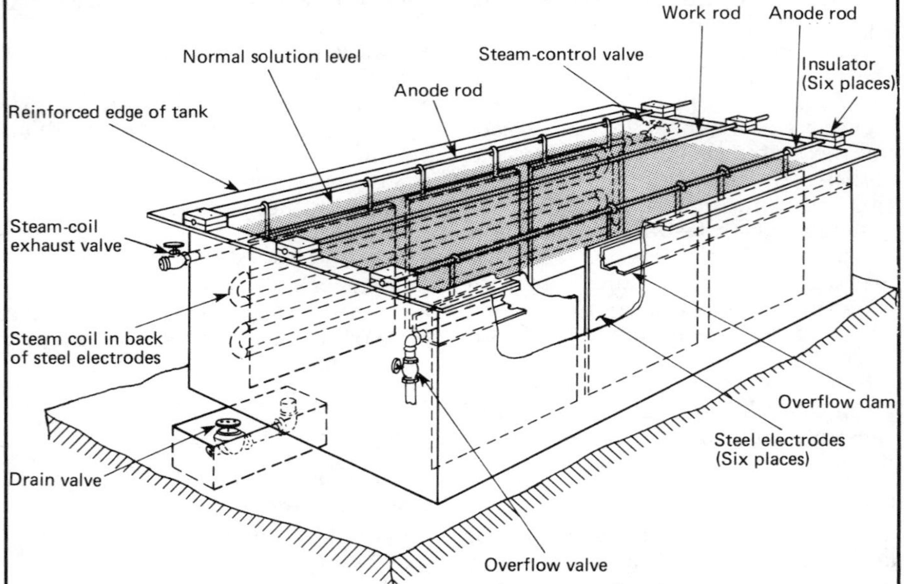

Porcelain insulators are attached to bottom edges of electrodes to prevent contact with the sides of the tank

Once a cleaning problem is identified, the source of the problem must be found by process of elimination. It may be necessary to make up fresh cleaning solution in small containers off-line and eliminate each possibility one step at a time. However, before this is done, the plater should check the electrocleaner in the following order:

- Electrical contact (gassing originating from the workpiece or anode)
- Polarity
- Cleaner temperature
- Voltage
- Concentration
- Current density

When designing a new plating line or cleaning a new part with a different type of soil, the plater should utilize the resources of the metal finishing supplier by submitting information, parts, and samples of the soil, if possible, for analysis. Suppliers can analyze soil, simulate production conditions, and make recommendations to eliminate much guesswork on the part of the plater.

Equipment Requirements

The basic equipment requirements for alkaline electrocleaning are simple: a steel or rubber-lined tank equipped with heating elements, anodes, and accompanying bus work connected to some direct current power supply. More elaborately designed equipment is usually employed to add safety and efficiency to the operation. Figure 3 illustrates a typical electrocleaning tank and its accessories.

Tanks should be wide enough to allow a safe distance (152 mm or 6 in.) between the workpiece and the electrodes to prevent the formation of a heavy foam blanket and reduce the possibility of short circuits. Busbars should be located high enough above the solution level to ensure that normal foam blankets do not reach them.

Table 2 Approximate current-carrying capacity of copper busbars

For distance not to exceed 9.1 m (30 ft)

Current, A	Size, mm	Size, in.	Diameter, mm	Diameter, in.	Weight, kg/m	Weight, lb/ft
Flat copper busbar						
125	3.175 by 25.4	0.125 by 1	...	...	0.72	0.48
250	3.175 by 50.8	0.125 by 2	...	...	1.46	0.98
250	6.35 by 25.4	0.25 by 1	...	...	1.45	0.97
500	6.35 by 50.8	0.25 by 2	...	...	2.88	1.93
1000	6.35 by 101.6	0.25 by 4	...	...	5.75	3.86
Round copper rod						
50	...	...	6.35	0.25	0.30	0.2 approx
75	...	...	7.87	0.31	0.45	0.3
110	...	...	9.525	0.375	0.633	0.425
196	...	...	12.70	0.50	1.126	0.756
300	...	...	15.875	0.625	1.76	1.18
500	...	...	19.05	0.75	2.53	1.7
600	...	...	22.225	0.875	3.44	2.31
800	...	...	25.40	1.0	4.50	3.02
1000	...	...	28.575	1.125	5.71	3.83
1250	...	...	31.75	1.25	7.03	4.72
1500	...	...	34.925	1.375	8.52	5.72

Electrodes/Busbars. Steel electrodes should be plated with a heavy nickel plate to extend service life and facilitate cleaning. Contact is made to busbars with steel hooks. Copper is recommended as the material of construction of busbars because of its excellent current-carrying capacity. Copper bars should be protected with a heavy nickel plate to prevent cleaner contamination. Porcelain or plastic insulators should be attached to the ends of the busbars to prevent contact with the sides of the tank. Table 2 lists the approximate current-carrying capacity of various sizes of copper busbars.

Power Supplies. Electrocleaner tanks should be powered by reliable rectifiers. Regulated and unregulated power supplies are available that provide filtered direct current with good reliability. A regulated power supply may be of the silicon-controlled or saturable reactor type. These power supplies maintain constant current or voltage under varied loads and usually are utilized in large operations. The more common unregulated types include the tap switch and manual powerstat. Both air- and liquid-cooled rectifiers are available. The major advantage of the more expensive liquid-cooled type is that the unit is completely sealed, and rectifier components are not exposed to the atmosphere, which in a plating shop may be very corrosive. As a general rule, a rectifier should be operated at a minimum of 50% of its maximum rate output current to provide low ripple direct current. High ripple can also mean that the rectifier is deficient. In PR cleaning, the power supply and buswork would be connected in the same manner, with an automatic (timer) switch mechanism connected in series in the line between the power supply and tank.

Solution Heating Equipment. The electrocleaner may be heated by immersed steel steam coils or electric immersion heaters. These should be placed behind the electrodes on the operating side of the tank so that convection currents occur toward the rear of the tank. Holes of 51- to 76-mm (2- to 3-in.) diam may be drilled in the electrodes near the heating coils to further increase solution agitation. The tank heater should be capable of bringing the solution to operating temperature within 30 to 60 min.

Tanks. Electrocleaner tanks may be fabricated from hot rolled, low-carbon steel of minimum thickness of 6.4 mm (0.25 in.). Large tanks, 1.8 m (6 ft) or more deep, require thicker walls and horizontal braces to support the sides. Tanks should be lined with suitable rubber linings that are rated to withstand the normal operating temperatures. Linings protect the steel tank and eliminate any stray currents. The tank should be equipped with an overflow weir to skim oils. A valve in the weir is used to prevent solution loss when large parts are processed. The tank should be pitched slightly toward a drain valve to facilitate drainage, and a spring-loaded water inlet valve is recommended to eliminate the possibility of tank overflow when the filling tank is unattended.

The most effective ventilation for electrocleaner tanks is the push-pull type in which fresh air is forced over the operating side of the tank to the back side, where it is exhausted by hoods. The air flow across the tank should be at least 0.38 m^3/s per square metre (75 ft^3/min per square foot) of cleaner surface area.

Safety

Each year, many serious accidents including chemical burns, electrical shock, eye injuries, and respiratory problems occur, which may have been avoided if correct safety measures were taken. Some important safety tips regarding alkaline electrocleaners include:

- Never add dry powdered cleaner to a hot electrocleaner tank; dry cleaner should be predissolved in water with adequate mixing to avoid violent solution eruption. It should be dissolved at such a rate that undissolved cleaner does not accumulate at the bottom of the mixing tank
- Safety clothing should be worn whenever dry cleaner is being handled: (*a*) dust mask or respirator, (*b*) rubber apron, (*c*) rubber gloves, and (*d*) safety glasses with side shields
- When making up fresh cleaner, use warm water (49 °C or 120 °F) rather than cold
- Keep matches and open flames away from electrocleaner tanks because hydrogen gas is flammable; avoid hydrogen explosions by keeping busbars and racks clean to maintain good electrical contact
- If splattered with hot alkaline cleaning solution, rinse skin thoroughly with cold water, wash with vinegar and rinse again; contact physician
- Always turn off the rectifier when performing maintenance on an electrocleaning tank

SELECTED REFERENCES

- P.N. Burkard, Preparation for Plating, *Modern Electroplating,* 3rd ed., New York: John Wiley & Sons, 1973, p 571-589
- Frey, D.A. Swalheim, Cleaning and Pickling for Electroplating, *AES Illustrated Lecture,* 1970
- W.P. Innes, Metal Cleaning, *Metal Finishing Guidebook and Directory,* Hackensack: Metals and Plastics Publications, 1980, p 118-143

- Lux, Linford, Alkaline Cleaning of Metals, *Electroplating Engineering Handbook,* 3rd ed., New York: Van Nostrand Reinhold Publishing, 1971, p 153-174
- Blum, Hogaboom, *Principles of Electroplating and Electroforming,* 1949, p 200-211
- Lowenheim, *Electroplating,* New York: McGraw-Hill, 1978, p 67-92
- Linford, Saubestre, *Plating,* Vol 37, 1950, p 1265; Vol 38 (No. 60), 1951, p 158
- J. A. Murphy, Ed., *Surface Preparation and Finishes for Metals,* New York: 1971
- Pollack, Wesphal, *Metal Degreasing and Cleaning,* Great Britain: Robert Drapes, 1963

Emulsion Cleaning

By the ASM Committee on Emulsion Cleaning*

EMULSION CLEANING is a process for removing heavy soils from the surfaces of metals and nonmetals by using organic solvents dispersed in an aqueous medium aided by an emulsifying agent. Depending on the solvent used, cleaning is done at temperatures from 10 °C (50 °F) to 60 to 82 °C (140 to 180 °F).

An emulsion system contains two mutually insoluble or nearly insoluble phases, one of which is dispersed in the other in the form of globules. One phase is usually a hydrocarbon and the other is water. The dispersed phase is distributed as globules in the liquid, continuous phase.

The stability of emulsion cleaners depends on the properties of emulsifying agents that are capable of causing oil and water to mix and form a more stable system. Because oil and water do not mix, an oil-in-water dispersion that does not contain an emulsifying agent, or dispersant, requires constant mechanical agitation to prevent the oil and water from separating into two layers. Emulsifying agents can be placed in two categories: (*a*) those that promote the formation of solvent-in-water emulsions with water constituting the continuous phase, and (*b*) those that form water-in-solvent emulsions, in which water is the dispersed phase.

Emulsion cleaners are broadly classified into four groups on the basis of stability. Because stability is a relative term, definitions can overlap. A stable, stable single phase, or permanent emulsion is one in which the discontinuous phase is dispersed throughout the continuous phase. This requires no more agitation to maintain a uniform dispersion than that provided by thermal gradients and the motion of the work being cleaned.

An unstable single phase emulsion has a uniformly dispersed phase that tends to separate and form a solvent layer. Solvents with specific gravity of less than 1.0 form a top layer, and those with a specific gravity greater than 1.0 form a bottom layer. These cleaners require moderate to considerable agitation to maintain complete dispersion.

A diphase, multiphase, or floating layer emulsion cleaner forms two layers in the cleaning tank and is used in this separated condition. Work is immersed through the solvent-rich surface layer into the water-rich lower layer, permitting both cleaning phases to come in contact with the surfaces to be cleaned. When used in a spray system, a diphase cleaner resembles an unstable single phase cleaner, because the solvent and water phases are mixed in the pumping action.

An emulsifiable-solvent system is one in which the as-received, undiluted solvent is applied to the surface to be cleaned by hand or by use of a dip tank and is followed by a water rinse that emulsifies and removes the solvent and soil.

Composition of Emulsion Cleaners

Stable, unstable, diphase, and other emulsion cleaners cover a wide range of solvent and emulsifier compositions. The solvent is generally of petroleum origin and may be heterocyclic (M-pyrol), naphthenic, aromatic, and of hydrocarbon nature (kerosine). Solvents are available with boiling points of 60 to 260 °C (140 to 500 °F) and flash points ranging from room temperature to above 93 °C (200 °F). Because the solubility factor increases as the molecular weight of the solvent approaches that of water, low-to-medium molecular weight solvents are usually more effective in removing soils; however, fire hazards and evaporation losses increase as boiling and flash points decrease.

Emulsifiers include: (*a*) nonionic polyethers and high molecular weight sodium or amine soaps of hydrocarbon sulfonates, (*b*) amine salts of alkyl aryl sulfonates (anionic), (*c*) fatty acid esters of polyglycerides, (*d*) glycerols, or (*e*) polyalcohols. Cationic ethoxylated long-chain amines and their salts are also used in emulsions.

Emulsifiers must have some solubility in the solvent phase. When solubility is low, it can be increased by adding a coupling agent (hydrotrope), such as a higher molecular weight alcohol, ester, or ether. These additives are soluble in oil and water.

*William J. Roberts, Senior Sales Engineer, Chemical Group, GAF Corporation; Bob Srinivasan, Research Manager, Diversey Wyandotte Corp.

Concentration ranges of emulsion cleaners are 2 to 5% of concentrate for spray applications and 4 to 10% for soak applications. Floating layer diphase systems usually require about a 50-mm (2-in.) layer of solvent over a sufficient depth of water to permit the workpiece to be submerged. Water-in-solvent emulsions are operated at higher concentrations, ranging from 15 to 25%.

The compositions of emulsion cleaners for spray and soak operations are similar; however, soaps or wetting agents used in spray operations must have low-foaming characteristics. Compositions and operating temperature ranges for emulsion concentrations are given in Table 1. Product clarity and emulsion stability depend on pH, specifically related to the composition of the product.

Table 1 Compositions and operating temperatures for emulsion concentrates

Maximum safe temperature depends on the flash point of the hydrocarbon (petroleum) solvent used as the major component

Component	Composition, parts by volume Stable(a)	Unstable(b)	Diphase(c)
Petroleum solvent(d)	250-300	350-400	250-300
Soaps(e)	10-15	15-25	None
Petroleum (or mahogany) sulfonates(f)	10-15	None	1-5
Nonionic surface-active agents(g)	5-10	None	1-5
Glycols, glycol ethers(h)	1-5	1-5	1-5
Aromatics(j)	5-10	25-50	5-10
Water(k)	5-10	None	None

(a) Operating temperature range: 4 to 66 °C (40 to 150 °F). (b) Operating temperature range: 4 to 66 °C (40 to 150 °F). (c) Operating temperature range: 10 to 82 °C (50 to 180 °F). (d) Two frequently used solvents are deodorized kerosine and mineral seal oil. (e) Most soaps are based on rosin or other short-chain fatty acids, saponified with organic amines or potassium hydroxide. (f) Low molecular weight petroleum sulfonates (mahogany sulfonates) are used for good emulsification plus some rust protection. High molecular weight sulfonates, with or without alkaline-earth sulfonates, offer good rust inhibition and fair emulsification. (g) Increased content improves stability in hard water, but increases cost. (h) Glycols and glycol ethers are used in amounts necessary to act as couplers in stable and unstable emulsions. These agents are frequently used with diphase and detergent cleaners to provide special cosolvency of unique or unusual soils. (j) Aromatic solvents are used to provide cosolvency for special or unique soils. They also serve to inhibit odor-causing or rancidifying bacteria. (k) Water or fatty acids, or both, are used to adjust the clarity and the stability of emulsion concentrate, particularly those which are stable or unstable

Selecting an Emulsion System

Emulsion systems are used to remove heavy oily soils and are usually followed by an alkaline cleaning system to remove the last traces of contaminants completely. An emulsion cleaner is used more to remove gross soils than to produce a clean, water-break-free surface. Factors that influence the selection of stable, unstable, or diphase emulsion systems include: (*a*) type of soil to be removed, (*b*) size and quantity of work, (*c*) need for rust protection, (*d*) water condition, (*e*) cleaning sequence (especially if emulsion cleaning is preceded by alkaline cleaning), and (*f*) cost. Production applications for the principal emulsion cleaners and pertinent operating data are given in Table 2.

Stable emulsion cleaners are the most economical of the three emulsion cleaners. They are practical for removing light shop soils, especially in applications where in-plant rust protection is required. These cleaners contain hydrocarbon solvents such as kerosine, which can dissolve and clean light soils. Two to three weeks of rust protection can be expected for ferrous metal parts cleaned by a properly constituted stable cleaner. Such a cleaner maintains an emulsion with water for many hours, requiring a minimum amount of agitation.

A 2% stable emulsion spray rinse often follows alkaline cleaning. This procedure has provided rust protection for as long as 3 to 4 weeks in storage areas where humidity is not excessive and unusual changes in temperature are not encountered.

Although 77 °C (170 °F) is the recommended maximum operating temperature, stable emulsions can be operated safely at temperatures up to 82 °C (180 °F). The higher temperatures, sometimes advantageous when rapid drying of the work is desired, increase evaporation rates and may cause polymerization of emulsion and the formation of a varnishlike film that is difficult to remove from work. When large quantities of parts are cleaned in a continuous production flow in automatic spray washers, stable emulsion cleaners are preferred because of their lower initial cost and ease of maintenance.

Stable emulsion cleaners do have disadvantages. Their efficiency is low in removing hydrocarbon soils if more than 10% of the soil has a solidification temperature within 6 °C (10 °F) of the temperature of the emulsion. In hard water, stable emulsions form insoluble precipitates that may plug drains and increase maintenance.

Unstable emulsion cleaners, although higher in cost than stable emulsions, perform more efficiently in removing heavy shop soils, such as oil-based rust preventives and lubricants used in stamping and extruding. The hydrocarbon fraction of unstable emulsion cleaners makes more intimate contact with the work surface, permitting greater action of the solvent on soil. Unstable emulsions are also successful in hard waters that cause stable emulsions to break down.

Unstable emulsions, as well as the equipment required for using them, are less costly than diphase emulsion systems. However, the cleaning power of these emulsions approaches that of diphase systems, and they are widely used for the removal of heavy hydrocarbon soils. Phosphates may be added to hard waters to increase the efficiency of unstable emulsions.

The concentration of an unstable emulsion is not easily determined. However, operation above or below the preferred concentration range lowers cleaning efficiency and causes excessive cleaner consumption. The operating temperature of an unstable emulsion is critical and must not exceed 71 °C (160 °F). The usual range is 63 to 68 °C (145 to 155 °F).

When cleaning in unstable emulsions follows alkaline cleaning, the alkali carryover forms soap and causes excessive foaming. Therefore, stable emulsion cleaners are preferred when these cleaning methods are used in sequence.

Diphase emulsion cleaners are used for removing the most difficult hydrocarbon soils, such as lapping compounds, buffing compounds, and oxidized oils. They can provide a higher degree of cleanness than can be obtained with stable or unstable emulsions. The flash points of diphase emulsion cleaners cover a wide range, permitting operating temperatures up to 82 °C (180 °F). The monomolecular layer of oil that remains after diphase cleaning provides good rust protection.

In diphase cleaning, the solvent in the bottom phase is very powerful and a 100% concentrated product. It is not

Table 2 Production applications of emulsion cleaning
Data represent practices reported by a number of plants

Part	Soils removed	Cleaning cycles	Cleaning time, min	Subsequent operations
Stable emulsion, dip cleaning				
Cast iron parts and machined parts	Machining oil, chips	Alkaline clean, emulsion clean	1	Storage
Stable emulsion, spray cleaning				
Aluminum and brass carburetor parts	Machining oil, shop dirt	Clean, blow off(a)	1	Assembly, storage
Aluminum and brass	Dirt, machining oil	Clean, blow off	2	Assembly, storage
Aluminum cabinets	Machining oil, chips	Clean(b)	1	Assembly, storage
Aluminum housing (automatic transmission)	Alkali	Alkaline clean, emulsion clean(c)	1	Assembly, storage
Automobile wheel assembly, 0.103 m^2 (160 $in.^2$)	Drawing compound, chips	Clean, no rinse	1	Assembly, storage
Brass valves	Machining oil	Clean, blow off	2	Assembly, storage
Cast iron motor blocks	Machining oil, chips	Clean, no rinse	2	Assembly, storage
Cast iron motor heads	Machining oil, chips	Clean, no rinse	1	Assembly, storage
Retainer plate, 0.01 m^2 (16 $in.^2$)	Shop dirt, drawing compound	Clean	1	Assembly, storage
Steel rings, 100-mm (4-in.) diam	Machining oil	Clean, no rinse	1	Assembly, storage
Steel sinks	Drawing compound, oil	Clean	4	Alkaline soak, then enamel
Tractor parts	Machining oil, dirt	Clean, blow off	1	Wash, then paint
Valves (steel and brass)	Machining oil	Clean, blow off	1	Assembly, storage
Washing machine tubs	Drawing compound	Clean, no rinse	3	Alkaline soak, then paint
Unstable emulsion, spray cleaning				
Brake assembly, 0.01 m^2 (20 $in.^2$)	Shop dirt, chips	Clean, no rinse	1	Assembly, storage
Brake plates, 200-mm (8-in.) diam	Machining oil, chips	Clean, no rinse	1	Assembly, storage
Brake cases, 100 by 100 mm (4 by 4 in.)	Drawing compound	Clean, blow off	2	Assembly, storage
Diphase emulsion, dip cleaning				
Brass or zinc die castings	Buffing dirt	Soak, spray, electro clean, acid pickle	4	Wash, then plate

(a) Emulsion does not plug holes of the needle valves and does not interfere with subsequent gaging operations. (b) Emulsion does not spot or dull aluminum. (c) Emulsion furnishes lubricity for interlocking gear parts

an emulsion with water. Hence, these cleaners provide better cleaning than regular emulsion cleaners.

Diphase cleaners are most frequently used in dip tanks. However, with specially designed equipment or the addition of emulsifiers to retard separation into solvent and water layers, these cleaners can be used in recirculating spray washers. Diphase cleaners also have disadvantages which include:

- The cleaners are adversely affected by hard water, and preconditioning the water with phosphates is unsuccessful
- Vaporization of hydrocarbon layers requires more ventilation than is needed for stable and unstable cleaners to avoid fire and health hazards
- No easy tests are available to determine cleaner concentration
- Diphase cleaners cost more than stable or unstable cleaners

Emulsion Cleaning Cycles

Emulsion cleaning cycles vary, depending on the type of soil. Light shop soils require less cleaning time and emulsion concentration. Hard soils, such as embedded drawing or buffing compounds, require higher cleaner concentration and more cleaning time.

The immersion and spray cleaning cycles used in one manufacturing plant are presented in Table 3. Although cycle times vary depending on the cleaning method and on whether parts are easy or difficult to clean, the sequence of processing methods remains the same. A definition of easy-to-clean and difficult-to-clean parts, based on the type of soils to be removed, is also given in the table. Additional data relating to the operation of immersion and spray systems are given in Table 4.

Processing Variables

Agitation. In order of increasing effectiveness, agitation is provided by pumping, mechanical stirring, or air pressure stirring. When air pressure is used, maximum effect is obtained by admitting the air through small holes in a pipe located directly beneath the workpiece. Spray systems require a minimum pressure of 100 kPa (15 psi).

Temperature. Although significant, the effect of bath temperature is less important in emulsion cleaning than it is in alkaline or detergent cleaning. The dispersed oil phase in an emulsion cleaner dissolves oil-based soils, and its efficiency is seldom affected by temperature. Exceptions do exist, however. For example, an increase in temperature is helpful in removing high-melting greases, buffing compounds, or waxes. When bath temperature is below the melting point of the soil, heat softens the soil and facilitates its removal.

The temperature to which an emulsion cleaner can be heated safely depends on the flash point of the hydrocarbon. For safety, bath temperature

Table 3 Cycles for immersion and spray emulsion cleaning

Process sequence	Cycle time, min: Easy cleaning(a) Immersion(c)	Easy cleaning(a) Spray(d)	Difficult cleaning(b) Immersion(e)	Difficult cleaning(b) Spray(f)
Clean(g)	2-4	½-1	4-10	1-2½
Rinse(h)	¼-½	¼-½	¼-½	¼-½
Rinse(j)	¼-1	¼-1	¼-1	¼-1
Air dry(k)	½-2	¼-2	½-2	¼-2

(a) Removing cutting oils and chips from machined surfaces, shop dirt and oil from sheet metals, and drawing compounds from automotive trim. (b) Removing embedded buffing compounds, impregnated carbonized oils from cast iron motor blocks, and quenching oil from heat treated forgings. (c) Concentration of cleaner, 1.5 to 6 vol %. (d) Concentration, 0.6 to 1.5 vol %. (e) Concentration, 3 to 9 vol %. (f) Concentration, 0.75 to 1.5 vol %. (g) 10 to 82 °C (50 to 180 °F). (h) Unheated rinse. (j) 54 to 71 °C (130 to 160 °F). (k) 10 to 71 °C (50 to 160 °F)

Table 4 Operating conditions for emulsion cleaners

Classification of cleaner	Concentration, %	Operating temperature °C	°F	Time, min
Immersion systems				
General-purpose	5-15	10-71	50-160	2-8
Unstable single-phase(a)	10	21	70	1-10
Kerosine-based(b)	15-25(b)	21	70	2-10
Diphase, heavy-duty	15-25	21-54	70-130	2-10
Emulsifiable solvent	100	21-60	70-140	½-2
Spray systems				
General-purpose	1-5	10-71	50-160	½-3
General-purpose	2-5	10-77	50-170	½-3
Light cleaning	1-2	10-71	50-160	1-3
Diphase	2-5	10-77	50-170	½-3

(a) Requires vigorous agitation. (b) Water-in-solvent emulsion, 15 to 25% water in kerosine

should be maintained at least 8 to 11 °C (15 to 20 °F) below the flash point. Working temperatures for emulsion cleaners are indicated in Tables 1, 3, and 4. To limit volatilization losses, spray cleaners should be used at temperatures 6 to 11 °C (10 to 20 °F) lower than cleaners for immersion systems.

Duration of exposure to the emulsion, whether in soaking or spraying, usually ranges from ½ to 5 min, although some difficult-to-remove soils may require up to 10 min exposure. Longer exposure times are seldom practical. For cleaning extremely difficult workpieces, a combination of soak and spray systems is more effective than an extended soaking time.

Concentration. The ranges of emulsion concentration suggested in Tables 3 and 4 indicate that concentration is not critical. Although the capacity for dissolving soil is increased as emulsion concentration increases, solubilizing rates are not similarly affected by increases in concentration. Different types of soil react differently, as indicated in Fig. 1.

The number of washing stages prepared with a recommended concentration should be increased, rather than increasing concentration and using fewer washing stages. Increasing the number of washing stages helps prevent the cleaner from becoming saturated with soil; this extends bath life, while providing efficient cleaning. If increasing the number of washing stages is not feasible, more frequent replacement of emulsion baths is recommended. Cleanness of the emulsion is important, and the bath should be discarded when (*a*) smut or other deposits appear on workpieces, or (*b*) the cleaning action is noticeably less effective.

Rinsing parts after emulsion cleaning is an important adjunct to the cleaning operation, except in applications where rinsing is omitted to gain maximum temporary rust prevention. Rinsing is most effectively accomplished in two stages: (*a*) an initial rinse to remove most of the soil, and (*b*) a final rinse to ensure cleanness and preheat parts for drying.

Efficiency of the rinse depends on the following factors: (*a*) cleanness of the water, (*b*) water temperature, (*c*) impinging action (of a spray) or agitation (in tanks), and (*d*) quantity of water used. Hot water is seldom used for the initial rinse, because it is more economical to use a larger volume of cold water than a smaller volume of hot water. Water at 66 to 93 °C (150 to 200 °F) is frequently used for final rinsing, because it preheats the work and facilitates drying. However, hot water is not desired for final rinsing in all applications. For example, a portion of the hydrocarbon phase is often left on the work to prevent rust. Hot rinsing may result in this protective coat being removed.

Analysis of the more stable emulsion cleaners can be made at the tank. However, distillation techniques are used for the unstable and diphase cleaners, requiring analysis in a laboratory.

To obtain a good representative sampling, samples should be taken from various locations in immersion tanks with a glass tube. In a spray installation, samples should be taken from the jets after the washer has been in operation for some time, because soluble oils become more emulsified as spraying continues. One simple and rapid method of analysis is the following:

- Place approximately 90 to 95 mL of emulsion in a 100-mL glass stoppered graduated cylinder
- Measure and record the actual amount of sample
- Cautiously add 5 mL of sulfuric acid
- Place stopper on graduate and shake until emulsion begins to break
- Allow emulsion to cool to 21 °C (70 °F) and separate completely
- Measure and record the amount of separated emulsifiable material or oil

The volume of soluble oil divided by the volume of original sample times 100 is the volume percentage of oil in the emulsion.

Process Selection

The size and shape of a part are the main considerations that influence the choice between immersion and spray methods.

Spray Cleaning. Although spray cleaning has been traditionally used with large rackable parts, applications have been expanded. Large stampings, flat sheet metal components, and simple castings are parts that can be spray cleaned efficiently and economically, but spray cleaning of small parts processed through rotary washers is increasingly common for in-process operations.

Spray methods are also used for large components, such as truck cabs, for which tanks of required size are not available. The use of emulsifiable sol-

Fig. 1 Approximate relationship of time and concentration for emulsion cleaners used to remove two different soils

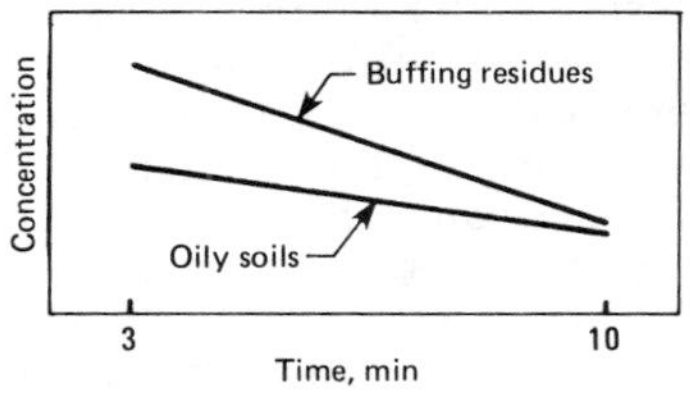

vents is often more practical in such applications, because these solvents can be sprayed on the work in a concentrated form and then rinsed with a power spray.

Immersion cleaning is preferred for small parts that must be placed in baskets. Tank agitation helps to remove soil from grooves or deep recesses. Combination cycles of immersion followed by spray washing and power spray rinsing are often used for difficult-to-clean parts. Another practical way to clean difficult parts is to take the solution from the immersion tank and spray it on the parts as they emerge from the tank. The sprayed solution drains back into the tank. Unless followed by another cleaning method, some parts are not recommended for any type of emulsion cleaning because entrapped emulsion would impair subsequent finishing operations. Examples include:

- Tubular parts used in making furniture
- Sandcore brass castings, such as plumbing fixtures (the rough interior retains emulsion)
- Parts with rolled edges, such as bicycle rims
- Parts with lapped and spot welded sections

Vapor degreasing may be used after emulsion cleaning to remove entrapped emulsion.

Equipment for Immersion Systems

Tanks for cleaning solution should be constructed of hot rolled steel. Depending on tank capacity, steel gage requirements are as follows:

- Up to 380 L (100 gal), 12 gage
- Up to 1890 L (500 gal), 10 gage
- Over 1890 L (500 gal), 7 gage

All seams should be welded on the inside and outside. Channel or angle iron reinforcements should be welded wherever they are required for strength or rigidity. Because of weight, tanks of 1140-L (300-gal) capacity or above must be provided with a solid footing.

A shield with two rows of holes along the top should be fastened over the steam coils to increase agitation of the cleaning solution. The holes permit the violent discharge of solution from the section enclosed by the agitation shield. As the solution passes through the holes, a corresponding amount of solution is drawn underneath the bottom edge, rises, and, in turn, is discharged. Thus, a constant surging of the solution in the tank is ensured.

A support for the agitation shield is welded along the shield's entire length. The length is welded to the front of the tank, and the slanting ends are welded against the sides of the tank. If this support is drilled and tapped on about 100-mm (4-in.) centers, the agitation shield is easily secured to the support by means of 8-mm (5/16-in.) machine screws. For the agitation shield to operate properly, it must fit tightly against each end of the tank, ensuring that the only passage for the surging solution is through the holes or perforations in the top of the shield. Support and shield should be made of steel thick enough not to be dented or damaged by the parts being cleaned.

The immersion tank should contain a leakproof overflow dam consisting of a length of 38 or 50-mm (1½- or 2-in.) angle iron, welded full length to the back of the tank. Other forms of structural steel may also be used for the overflow dam. After the tank has been fitted into its final position, the top edge of the dam must be perfectly level to allow the solution to overflow uniformly at all points.

The heating coil should be made up as a single complete unit. Unions are used to make the heating coil easy to remove for repairs and for cleaning.

Rinsing. The immersion tank should be followed in sequence by a spray-rinse tank with two series of nozzles from which high-pressure water can be discharged. The spray-rinse tank is made as a weldment with 16-gage (or heavier) black iron sheet and a 64-mm (2½-in.) iron pipe-sized drain opening at the bottom.

The size of the tank, the number of nozzles in each set, and the number of sets depend on the size and shape of the articles to be rinsed. For some applications, it is advisable to have more spray nozzles on one side, as for example, in cleaning parts with deep recesses. The nozzles are fitted to a 25-mm (1-in.) water line that extends around the inner walls of the tank. This line connects with the top of a 25-mm (1-in.) pipe riser which, in turn, connects with the water supply. A 25-mm (1-in.) foot or pedal valve is located at the floor level at a point convenient for the operator to use while lowering parts. The lower set of rinse nozzles should be staggered so that the nozzle tips are not directly below those of the upper set.

Maintenance. The following is a maintenance schedule for emulsion immersion systems:

Daily

- Check temperature and instrumentation
- Check level of emulsion in tank
- Check operation of agitation devices, pumps, stirrers, or compressed air
- Check exhaust system

Weekly

- Drain emulsion tank
- Flush and clean tank, making certain that any sludge and other soil contaminants are removed
- Check operation of agitation devices

Weekly operations may be scheduled to coincide with changing to fresh emulsion.

Equipment for Spray Systems

Spray cleaning units are divided into zones for cleaning, rinsing, and forced-air drying, and each zone must be designed to retain workpieces for a time sufficient to perform the necessary function. The production capacity of a spray unit depends not only on unit size and part size, but also on retention time and part arrangement or spacing. If the cleaning chamber of a spray unit is large enough to contain eight parts, and the minimum retention time for effective cleaning is 2 min, the capacity of this zone of the unit is 240 parts per hour.

The over-all efficiency of a spray cleaning process depends on the mechanical energy delivered by the spray against the work as well as on the effectiveness of the emulsion cleaner used. Because mechanical energy greatly facilitates cleaning performance, the design and operation of the spray cleaner and the arrangement of the parts in the cleaning chamber should provide maximum practicable hydraulic energy to remove the dirt mechanically from the part, rather than rely entirely on chemical cleaning action by the emulsion.

The design and construction of spray cleaning units depend on the specific operation for which they are to be used. All spray units require some supplementary equipment, such as reservoir tanks, piping systems, housings, conveyors, pumps, sources of heat, and facilities for forced-air drying.

Reservoir tanks should have a volume at least three times the capacity of the pump, expressed in terms of volume per minute to allow ample time for solids and foam to settle. The base of the tank should be sloped like the hull of a ship and provided with a drain valve, and the tank should be equipped with a sump.

The pump intake should be at least 100 mm (4 in.) above the bottom of the tank and should be equipped with a double screen to prevent the intake of sediment and chips. In handling unstable emulsions, pump intakes should be located at interfaces of oil and water. In some applications, more than one intake is necessary. The reservoir tank is usually constructed of low-carbon steel and is double-welded at the joints. The thickness of the steel depends on the size of the equipment, but it should not be less than 10 gage. Tanks should be covered with a metallic grill to allow solution drainage and prevent any parts from dropping accidentally from the rack into the cleaning solution.

Piping Systems. For effective spray cleaning, nozzle pressure should be at least 100 kPa (15 psi), to provide adequate mechanical action at the surface of the workpiece. The nozzles should be readily accessible and removable for cleaning. To prevent overspraying, end nozzles in the cleaning and rinsing chambers should be deflected inward approximately 30°. All nozzles should be staggered to ensure complete coverage of the workpieces.

Housings may be made of low-carbon steel and should contain access doors at convenient locations to permit removal, adjustment, and cleaning of all risers and nozzles, as well as the removal of any material that may have fallen from the conveyor.

Conveyors. The use of a variable speed conveyor should be considered in the initial installation to permit some latitude in the retention time of parts in the cleaning cycles.

Heating. Steam is widely used as a source of heat in spray cleaning units. Gas immersion burners are not recommended, because they present a fire hazard. The capacity of the steam coils or plates should be sufficient to heat the solution to operating temperature in 30 min to 1 h.

Air Drying. Forced air used to dry parts after cleaning and rinsing may be kept at room temperature or heated. Heated air has three advantages: (*a*) drying is hastened; (*b*) floor space is conserved; and (*c*) less air is required for the same number of parts.

Maintenance. The following is a maintenance schedule for emulsion spray systems:

Daily

- Check temperature and instrumentation
- Check levels of emulsion and rinse water
- Check operation of pumps and motors
- Check operation of exhaust system
- Check filters and screening devices
- Check overflow for proper drainage
- Check water leveling devices
- Check any drain-off area for accumulated solvent, a potential fire hazard

Weekly

- Drain emulsion and rinse water tanks
- Flush and clean tanks, making certain that any sludge and other soil contaminants are removed
- Clean filters thoroughly
- Check piping, pump intakes, and spray nozzles; clean and adjust as necessary
- Check and lubricate pumps and motors
- Add water in tanks to pump intake level; pump water to check spray pattern. Make certain that nozzles do not cross spray

Equipment Location

To reduce handling and temporary storage, emulsion cleaning systems are usually located near the production line. Some other factors that must be considered in the location of an emulsion cleaning system are (*a*) potential fire hazards, (*b*) ventilation, and (*c*) accessibility of steam and water.

Fire Hazard. Emulsion cleaners containing Stoddard solvent, kerosine, or similar petroleum derivatives have relatively low flash points and are a potential fire hazard. When these cleaners are used, the system should be kept away from heat treating, welding, and similar operations. If the cleaner is heated, steam must be used as the heating medium. Any welding for repair and maintenance of tanks or structures should be done only after covering the tank with a lid. Failure to do this could pose a potential fire hazard.

Ventilation. The rate of ventilation required for health and safety depends largely on cleaner characteristics, operating temperature, and the efficiency of the baffles on cleaning machines. The removal rate of volatiles varies from as little as zero to as much as 7.1 m^3/min (250 ft^3/min) for each square foot of cleaner surface area. Because both the area of air-liquid interface and the velocity are greater, spray washers require more ventilation than immersion systems.

Advantages and Limitations of Emulsion Cleaning

Emulsion cleaning is an effective means of removing a wide variety of soils from metal parts, including: (*a*) pigmented drawing lubricants, (*b*) unpigmented oils and greases, (*c*) cutting fluids, (*d*) polishing and buffing residues, and (*e*) residues resulting from magnetic particle inspection.

Emulsion cleaning is used when rapid superficial cleaning is required and when some protection by a light residual film of oil is desired. Emulsion cleaning is usually less costly than solvent cleaning with 100% chlorinated solvent, because relatively small concentrations of expensive solvents and high concentrations of water are used. Hazards from fire and toxic fumes are not great if temperatures less than the maximum recommended for the specific cleaner are used.

Emulsion cleaners are usually safe to use with most metals, provided that the pH of the cleaner is kept between 7 and 10. If the cleaner initially has a pH higher than 10 or if alkali contamination has brought it above this value, the cleaner should not be used with aluminum, zinc, or other soft metals. Emulsion cleaners can be formulated to be compatible with most water conditions, and the size of workpiece is limited only by available equipment.

Rust Prevention. Emulsion cleaners leave a thin film of oil on the work, and this film provides some protection against rusting. However, when parts are stored for longer than a few days under conditions of high humidity, rust prevention should be given special consideration. Some oils, particularly paraffinic types, provide better protection and can be enhanced by using high molecular weight emulsifiers that promote a continuous oil film. A rust-retarding formula usually consists of the following: (*a*) 15 to 23% high molecular weight sulfonate; (*b*) a blending agent containing 4 to 6% of a suitable high molecular weight alcohol, ester, or ether ester; and (*c*) 71 to 80% petroleum oil with a viscosity at 38 °C (100 °F), 200 SUS or higher. Rust-retarding film must always be removed from parts before electroplating, painting, or phosphating.

Limitations. Emulsion cleaning should not be used for parts such as sintered powder metallurgy parts and parts with blind holes or deep recesses, which are difficult to rinse thoroughly if the parts are to be plated. Emulsified oil can contaminate a plating bath sufficiently to cause excessive rejections. Use of alkaline soak cleaners, however, can help remove emulsified oil and prevent contamination of plating solutions from occurring. The solvent discharge from emulsion cleaners is becoming a serious problem, and handling is governed by both local and Environmental Protection Agency (EPA) regulations.

Safety

The hazards of toxicity and irritation to the skin resulting from contact with emulsion cleaners are negligible.

The fire hazard is slight if the temperature of the cleaner is kept 8 to 11 °C (15 to 20 °F) below the flash point, and adequate ventilation is provided. The flash temperature of the cleaner should be known before the cleaner is used. Open-fire burners must be kept away from emulsion cleaning installations. Wiring used in the vicinity of emulsion cleaning operations should be explosion-resistant.

Waste Disposal

Disposing of spent emulsions is a problem. Local regulations must be followed if spent emulsions are allowed to enter sewers or streams. In most areas, the maximum concentration of oil permitted for discharge into sewers or streams is from 100 to 500 ppm.

Methods that have been successfully used to break emulsions before separation are heating, distillation, aeration, chemical coagulation, and chlorination. Separation can be effected by sedimentation, flotation, centrifuging, or filtration, with sedimentation being the least expensive of these techniques.

Solvent Cleaning

By the ASM Committee
on Solvent Cleaning*

SOLVENT CLEANING is a surface preparation process, which is especially adept at removing organic compounds such as grease or oil from the surface of a metal. Organic compounds are easily solubilized by solvent and then removed from the workpieces. In some cases, solvent cleaning before other surface preparations can extend the life of cleaning operations and reduce costs. In other cases, solvent cleaning prepares workpieces for the next operation, such as assembly, painting, inspection, further machining, or packaging. Before plating, solvent cleaning is usually followed by an alkaline wash or other similar process which provides an oil-free surface. Solvent cleaning can also be used to remove water from electroplated parts, a common procedure in the jewelry industry.

Solvent cleaning can be accomplished in room temperature baths or by using vapor degreasing techniques. Room temperature solvent cleaning is referred to as cold cleaning. Vapor degreasing is the process of solvent cleaning parts by condensing solvent vapors of a nonflammable solvent on workpieces.

Parts may also be degreased by exposure to the solvent vapor, as well as by immersion in the hot solvent. Drying is accomplished by evaporating the solvent from the parts as they are withdrawn from the hot solvent vapor. In cold cleaning, parts are dried at room temperature or by the use of external heat, centrifuging, air blowing, or an absorptive medium.

Cold Cleaning

Cold cleaning is a process for removing oil, grease, loose metal chips, and other contaminants from the surfaces of metal parts. Common organic solvents such as aliphatic petroleums, chlorinated hydrocarbons, chlorofluorocarbons, or blends of these classes of solvents are used. Cleaning is usually performed at, or slightly above, room temperature. Parts are cleaned by being immersed and soaked in the solvent, with or without agitation. Parts that are too large to be immersed are sprayed or wiped with the solvent. Ultrasonics are sometimes used in conjunction with solvent cleaning to loosen and remove soils, such as abrasive compounds, from deep recesses or other difficult-to-reach areas. This reduces the time required for solvent cleaning complex shapes.

Solvents

Table 1 lists aliphatic petroleums, chlorinated hydrocarbons, chlorofluorocarbons, alcohols, and other solvents commonly used in cold cleaning. Stoddard solvent, mineral spirits, and VM&P naphtha are widely used, because of their low cost and relatively high flash points. The chlorinated hydrocarbons and chlorofluorocarbons exhibit a wide range of solvency and are nonflammable, but most are far more expensive than the aliphatic petroleums. For some uses, mineral spirits, or a solvent of this type, is mixed

Table 1 Properties of cold cleaning solvents

Solvent	Flash point(a) °C	Flash point(a) °F	OSHA TWA, ppm(b)
Aliphatic petroleums			
Kerosine	63	145	...
Naphtha, hi-flash	43	110	...
Mineral spirits	14	57	500
Naphtha, VM & P	9	48	500
Stoddard solvent	41	105	100
Chlorinated hydrocarbons(c)			
Methylene chloride	None	None	500
Perchlorethylene	None	None	100
Trichloroethane (1,1,1)	None	None	350
Trichlorethylene	None	None	100
Trichlorotrifluoroethane	None	None	1000
Alcohols			
Ethanol, SD	14	57	1000
Isopropanol	10	50	400
Methanol	12	54	200
Other solvents			
Acetone	−18	0	750
Benzol	−11	12	10
Cellosolve(d)	40	104	50
Toluol	4	40	100

(a) Tag closed cup. (b) OSHA exposure values expressed as parts of vapor or gas per million parts of air by volume at 25 °C (77 °F) and 760 mm Hg pressure. These values should not be regarded as precise boundaries between safe and dangerous concentrations. They represent conditions under which it is believed that nearly all workers may be repeatedly exposed, day after day, without adverse effect. The values refer to time-weighted average concentrations for a normal workday. (c) Also used for vapor degreasing. (d) 2-ethoxyethanol

*Richard W. Clement, Sales Manager, Industrial Equipment Division, Detrex Chemical Industries, Inc.; John H. Rains, Applications Chemist, Ethyl Corp.; Robert Simmons, Research Supervisor, Chlorinated Solvents Section, Dow Chemical Co.; Kenneth Surprenant, Research Leader, Dow Chemical Co.

with 15 to 35% of a chlorinated hydrocarbon to reduce fire hazard by elevating flash points.

The alcohols are used alone, or in conjunction with chlorocarbons or chlorofluorocarbons, for special cold cleaning applications, such as removing activated soldering fluxes. Acetone and other solvents of low flash temperature are used for special purposes only, such as cleaning the components of precision instruments. Materials such as acetone and toluol constitute a serious fire hazard, and their storage and use require strict observance of all safety precautions. Solvents such as methylene chloride and 1, 1, 1-trichloroethane are frequently used in spray cleaning delicate components such as relay contactors before they are sealed by covers. Adequate ventilation must be supplied to all cold cleaning operations.

Control of Process Variables

Selection of the most appropriate solvent, operating temperature of the solvent, use of agitation, and the nature of the contaminant to be removed are the principal variables affecting the process. For example, in one application, miniature ball bearings were soaked in mineral spirits to remove light oil and miscellaneous foreign particles. Because the rejection rate was high, mineral spirits were replaced with trichlorotrifluoroethane or methylene chloride. The cycle also was changed from immersion only to a solvent spray, immersion with ultrasonic agitation, and a spray with clean solvent.

Operating Temperature. Higher operating temperatures invariably increase the cleaning efficiency of a solvent. However, unless strict precautions are observed, solvents should not be heated above their flash points (shown in Table 1); solvents with flash points below 21 °C (70 °F) are usually used at room temperature. The chlorinated hydrocarbons and chlorofluoronated hydrocarbons can be heated, but the fumes evolved, as with other solvents, require the use of properly designed equipment and adequate ventilation for safety and health of personnel.

Agitation. Except for parts with extremely heavy soil, hard-to-reach areas, or both, moderate agitation usually eliminates the need for prolonged soaking. Agitation allows more solvent dissolution of the oils. Mild mechanical action, such as brushing, is also very effective in enhancing the cleaning.

High boiling point solvents are sometimes agitated by air sparging. The evaporation rate of low boiling point solvents (<150 °C or <300 °F) makes this mode of mixing impractical. Recirculating pumps, movement of the parts, impellers, and ultrasonic cavitation provide other means of mixing.

Cleanness of Solvent. As contamination of the solvent increases, cleaning efficiency and the cleanness of processed parts decrease correspondingly. Cleanness requirements prescribe the time at which the solvent must be replaced.

Solvent Reclamation. All solvents can be reclaimed by either a factory-operated still or licensed reclamation service. In general, the reclamation process is one of simple distillation. Figures given in the following tabulation show cost of recovery as percentages of the cost of new solvent, for different percentages of reclamation.

Quantity reclaimed	Percentage of cost of new solvent(a) for various percentages of reclamation(b): High yield, % 90	High yield, % 50	Low yield, % 35	Low yield, % 20
Under 4 drums	25	25	35	40
Over 4 drums	18	20	25	30

(a) Percentage of the cost of replacing with an equivalent amount of unused solvent. (b) Reclamation percentage refers to the percentage (by volume) of pure solvent reclaimed from contaminated solvent

The above figures do not include shipping cost, which increases as yield decreases. Often, solvents are so dirty or contain such troublesome contaminants that the low yield makes reclamation by a reclamation service uneconomical. An in-house still is more economical for reclaiming large quan- of extremely dirty solvent.

Selection of factory distillation equipment must be made on the basis of volume of solvent used, flammable nature of the solvent, boiling point of the solvent, nature of the contaminants, and degree of purity required. Such stills are frequently incorporated into the large sizes of dip or soak equipment on a semiautomated basis.

Standards on recovered solvent usually relate to color, clarity, moisture content, and neutrality, although tests for specific contaminants may be included. Chlorinated hydrocarbons should contain stabilizers, added during manufacture; many times, distillation necessitates supplemental inhibition. Inadequate amounts of stabilizers or improper stabilizers may lead to dangerous situations, including the possibility of explosion.

The change out point of a solvent is determined by the degree of redeposition. Each part placed in a dip solvent comes out of the solvent with a thin film of soil redeposited on its surface. The permissible degree of redeposition determines the practical limit of usefulness of a solvent and the rate at which fresh solvent must be introduced. Alternatively, immersion in sequentially cleaner solvent baths can prolong the useful life of the solvent. In spray wipe applications in which the solvent is aided by strong mechanical action, there is a nearly continuous use of fresh solvent, which is seldom reused.

Tests of cleanness made directly on parts generally are more practical for determining the reclamation point than are measurements of soil buildup in the solvent. Although checking the cleaned item for satisfactory performance in subsequent operations is a practical method for determining if a required degree of cleanness has been obtained, various other methods of testing for cleanness are also available. These test methods, in order of increasing degree of cleaning requirements, are as follows:

1 Visual observation of parts and solvent condition
2 Wiping parts with a clean dry white cloth and then examining the cloth for adhering soil
3 Applying tape to the cleaned surface, removing it, and examining it for adhering soil (Scotch tape test)
4 Tests for the adhesion of paints ranging from special low-adhesion test paints to conventional paint
5 Microscopic examination of parts
6 Resoaking parts in fresh solvent and weighing the nonvolatile residue
7 Chemical analysis for specific soils
8 Electrical test (on combinations of conductors and nonconductors only)
9 Use of radioactive tracers

Methods from the above listing generally are used for specific purposes according to the following tabulation:

Method No.	Purpose of cleaning
1, 2, 3	Preclean only
4	Prepare for paint or adhesive
5, 6, 7, 8	Precision-instrument parts
6, 7, 8, 9	Initial studies on precision parts

Drying the Work. The cleaning solvent is sometimes permitted to remain on steel parts to serve as a mild rust-

Fig. 1 Spray unit for solvent cleaning of small precision parts before assembly

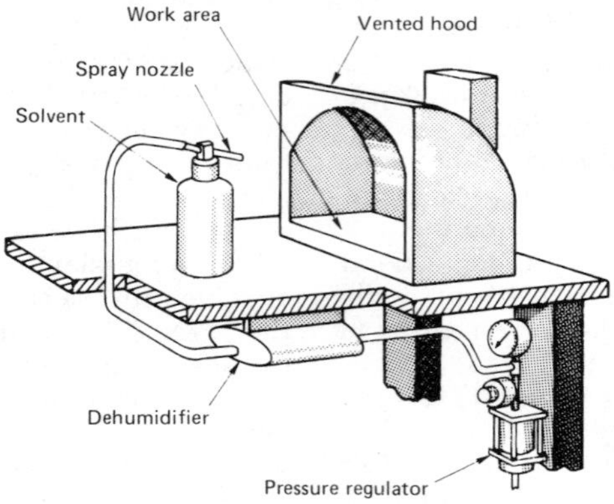

Solvent container should be in hood or located where no vapors can escape into the atmosphere

Table 2 Equipment for removing light soil by dip cleaning in high-flash naphtha

Operating conditions	Small pieces	Medium pieces	Large pieces
Production requirements			
Size of work, mm (in.)	25 by 25 by 25 max (1 by 1 by 1 max)	300 by 300 by 300 max (12 by 12 by 12 max)	300 by 300 by 300 min (12 by 12 by 12 min)
Number of pieces per hour	20 000	1000	1000
Area cleaned per hour, m^2 (ft^2)	25 (250)	95 (1000)	185 (2000)
Weight of each piece, kg (lb)	Up to 0.05 (Up to 0.1)	0.05 to 2 (0.1 to 5)	455 (Up to 1000)
Equipment requirements			
Size of dip tank, cm (ft)	30 by 30 diam (1 by 1 diam)	180 by 90 by 60 (6 by 3 by 2)	360 by 180 by 150 (12 by 6 by 5)
Number of tanks	2(a)	1	1
Fresh solvent consumption(b)	60 000	2000	200
Operating temperature	Room temperature	Room temperature	Room temperature

(a) Alternating tanks are used because of necessity for frequent movement of workpieces. (b) Number of pieces cleaned per 4 L (1 gal) of fresh solvent consumed; based on normal ventilation and an evaporation rate ten times slower than that of the standard carbon tetrachloride reference. High-flash naphtha used had Kauri-butanol value of about 40

preventive coating, particularly when solvents such as kerosine are used. However, a drying process usually is necessary to evaporate or remove the solvent from parts cleaned by solvent cleaning. The following procedures are frequently used for drying:

- Lighter or more volatile solvents can be removed by allowing them to evaporate at room temperature, using a blast of air from a fan or exhaust system
- Centrifuging of small parts is used to accelerate solvent removal. To remove heavier solvents, heat is introduced (usually from a steam coil) into the centrifuge dryer
- Drying of parts is accomplished also by bringing the parts into contact with absorptive materials, such as sawdust, crushed corncobs, or other commercially available drying materials

In all drying operations, solvent fumes must be exhausted to prevent the possibility of fire, explosion, or health hazards.

Equipment

Pails, tanks, and spray equipment are used in solvent cleaning. Pails with covers are the simplest containers and are often used to contain kerosine, mineral spirits, or chlorinated hydrocarbons for hand brush cleaning or wiping. Soaking tanks of various designs and sizes are used, depending on the nature of the work. Such tanks may be heated by steam coils, but more often they are used at room temperature. Agitation is sometimes provided by mixer impellers or forced air. For inprocess cleaning of small parts, such as those encountered on subassembly lines, a variety of specially made safety tanks is available. Some are designed to permit quick opening and closing by means of a foot pedal, minimizing evaporation and fire hazard. Some are equipped to supply fresh solvent quickly to the work zone and dispense contaminated solvent to another reservoir for subsequent discarding or reclamation.

Spraying is most often performed with conventional airless paint spraying equipment with the exception of air-aspirated spraying. However, air aspiration greatly increases evaporation and fire hazard. Halogenated solvent manufacturers do not recommend equipment of aluminum or aluminum alloy construction. In many installations, small bench sprayers, similar to the unit shown in Fig. 1, are used on assembly lines for cleaning delicate components. Such equipment can be made a part of an automated operation. However, a suction exhaust must be provided to carry away fumes that may be toxic, flammable, or both.

Washing machines also are available for cleaning small precision parts. Some of these machines are similar to home laundry machines in design. Parts are placed on trays, and the agitated solvent provides a constant washing action. In many applications in which the removal of oil and grease is not the main purpose, the equipment is used to remove the residue of polishing or lapping compounds. A filtering system on the machine continuously removes solid particles from the solvent as they are washed from the workpieces.

Equipment requirements for solvent cleaning vary with the size, shape, and quantity of workpieces, as well as the amount of soil to be removed. No matter what equipment is selected, proper covers to minimize solvent loss should be used. Examples are given in Tables 2 and 3.

Specific Applications

Solvent cleaning has traditionally been regarded as a method for precleaning or as one reserved for special applications. However, with the rise in the manufacture of electronic components and other assemblies that comprise many small parts, the use of solvents as a final cleaner has increased. At present, most solvent cleaning applications fall within one of the following categories:

- Inexpensive precleaning of parts that later are to be further cleaned by another method
- Hand cleaning of parts too large for immersion or spray machine cleaning
- Cleaning of heat-sensitive, water sensitive, or chemical-sensitive parts
- Removal of organic materials such as plating stop-offs, marking crayons, or soldering flux
- Cleaning of precision items in a succession of steps in which the work is first cleaned in nonpolar solvent to remove oil and then in water (polar solvent) to remove inorganic contaminants
- Temporary general cleaning where the cost of vapor degreasing equipment is not justified
- Cleaning electrical or electronic assemblies in which the presence of inorganic salt deposits may cause current leakage

Process Limitations

Virtually all common industrial metals can be cleaned in the commonly used cleaning grade solvents without harm to the metal, unless the solvent has become contaminated with acids or alkalis. Cleaning cycles should be adjusted to minimize the immersion time. Certain plastic materials can be affected by cleaning solvents, and tests must be conducted to determine compatibility.

Solvent degreasing is ineffective in removing such insoluble contaminants as metallic salts and oxides; sand; forging, heat treat or welding scale; carbonaceous deposits; and many of the inorganic soldering, brazing, and welding fluxes. Likewise, fingerprints can resist solvent removal.

Table 3 Equipment for removing heavy soil by soak cleaning in high-flash naphtha

Operating conditions	Small pieces	Medium pieces	Large pieces
Production requirements			
Size of work, mm (in.)	25 by 25 by 25 max (1 by 1 by 1 max)	300 by 300 by 300 max (12 by 12 by 12 max)	300 by 300 by 300 min (12 by 12 by 12 min)
Number of pieces per hour	3000	200	100
Area cleaned per hour, m^2 (ft^2)	5 (40)	20 (200)	75 (800)
Weight of each piece, kg (lb)	Up to 0.05 (Up to 0.1)	0.05 to 2 (0.1 to 5)	Up to 454 (Up to 1000)
Equipment requirements			
Size of soak tank, cm (ft)	30 by 30 diam (1 by 1 diam)	180 by 90 by 60 (6 by 3 by 2)	360 by 180 by 150 (12 by 6 by 5)
Number of tanks:			
Preclean	2(a)	1	1
Prepaint preparation	6(a)	3(b)	2(b)(c)
High degree of cleaning	6(a)(d)	(c)	(c)
Fresh solvent consumption(e):			
Preclean	15 000	600	100
Prepaint preparation	8500	320	...
High degree of cleaning	3800	...	...
Operating temperature	Room temperature	Room temperature	Room temperature

(a) Alternating tanks used because of necessity for frequent transfer of workpieces. (b) Tank contains an auxiliary spray attachment. (c) Work rarely cleaned by this procedure. (d) A spray operation with clean solvent is used between each of the two soak cleaning operations. (e) Number of pieces cleaned per 4 L (1 gal) of fresh solvent consumed; based on periodic downgrading of solvents to the next tank, normal ventilation, and an evaporation rate ten times slower than that of the carbon tetrachloride reference. High-flash naphtha used had Kauri-butanol value of about 40

Size and Shape. Size is never a limitation in solvent cleaning. Parts as small as sewing needles and as large as diesel engines are regularly cleaned with solvents. The shape of the workpiece is seldom a limitation. The most intricate parts are solvent cleaned by devising techniques of handling that allow the solvent to reach and drain from all areas.

Quantity of Work. Although there are many high-production applications that regularly use cold cleaning, the method is more likely to be used for maintenance and intermittent cleaning of small quantities. Because cold cleaning is usually done at or near room temperature, the problem of heating up, or otherwise preparing, equipment for a small quantity of work is eliminated. Unless there is some special requirement, other methods of cleaning such as vapor, alkaline, emulsion, or acid are usually cheaper and more satisfactory for cleaning large quantities in continuous production.

Lack of uniformity is often a severe limitation of cold cleaning. Because the process is basically one of dissolving a contaminant in a solvent, in immersion cleaning, as the solvent is used, the resoiling increases because the work does not receive a final rinse in pure solvent as it does in vapor degreasing. The work is seldom, if ever, perfectly clean. Therefore, except in special applications where spray techniques are used, solvent cleaning is more likely to be used as a preliminary, rather than as a final, cleaning method. The amount of soil that remains on the part depends on how much was there initially and how often the solvent was reclaimed. In some applications, the use of two or more consecutive solvent baths serves to provide more uniform cleaning results.

Applicability to Soils. Virtually all types of soil can be removed by solvent cleaning (at least to some degree of cleanness, as discussed above) by combinations of technique and solvent. The

Table 4 Vapor degreasing solvents

Solvent	Flash point	TLV, ppm(a)	Solvency	Photochemical reactivity	Vapor density (air = 1.0)	Volume of condensate L	gal	Stabilization	Boiling point °C	°F	Molecular weight
Trichlorethylene	None	100	Strong	Yes	4.5	3.8	1.00	Yes	88	190	131
1,1,1-trichloroethane	None	350	Moderate	No	4.6	3.3	0.86	Yes	74	165	133
Perchlorethylene	None	100	Moderate	Yes	5.7	5.9	1.60	Yes	120	250	166
Trichlorotrifluoroethane fluorocarbon 113	None	1000	Mild	No	6.5	2.0	0.54	No	49	120	187
Methylene chloride	None	100	Strong	No	2.9	0.72	0.19	Yes	41	105	85

(a) Adopted by the American Conference of Governmental Industrial Hygienists, 1981

range of soils on which solvents are highly effective is greater than for vapor degreasing because (*a*) lower temperatures permit a wider choice of solvents, and (*b*) lower drying temperatures usually used in solvent cleaning do not bake on insolubles, such as polishing or buffing compounds. Mechanical agitation, ultrasonics, and sometimes hand scrubbing are used in solvent cleaning to help loosen and float away insolubles.

Safety and Health Hazards

Fire and excessive exposure are the greatest hazards entailed in the use of solvents for cleaning. The flash points and permissible vapor concentrations (ppm) of the solvents adopted for specific operations must be known (Table 1). All flammable solvents should be stored in metal containers, such as groundable safety cans and must be used in metal containers.

Adequate ventilation should be provided to prevent accumulation of vapors or fumes. No solvents should be used close to an open flame or heaters with open coils.

Operators should be cautioned against repeated exposure of the skin to solvents. Protective gloves or protective hand coatings should be used to prevent extraction of natural oils from the skin, which can cause cracking of the skin and dermatitis.

Common solvents vary in relative toxicity, and the vapors of these solvents are capable of exerting a potentially lethal anesthetic action when excesses are inhaled. Common solvents have relatively slight toxic effect, but maintenance men have lost their lives after working inside tanks containing very high concentrations of vapor, as a result of their strongly narcotic effect. When working in an enclosed space, such as tanks or pits, workers should follow confined space entry procedures. Drain and vent thoroughly. Check air for adequate oxygen and the absence of flammable or toxic vapor concentrations. Always use an air supplying respirator and life belt. Any person working with a solvent should be familiar with its material safety data sheet that can be obtained from the supplier.

Vapor Degreasing

Vapor degreasing is a generic term applied to a cleaning process that uses the hot vapors of a chlorinated or fluorinated solvent to remove soils, particularly oils, greases, and waxes. A vapor degreasing unit consists of an open steel tank with a heated solvent reservoir, or sump, at the bottom and a cooling zone near the top. Sufficient heat is introduced into the sump to boil the solvent and generate hot solvent vapor. Because the hot vapor is heavier than air, it displaces the air and fills the tank up to the cooling zone. The hot vapor is condensed when it reaches the cooling zone, thus maintaining a fixed vapor level and creating a thermal balance. The temperature differential between the hot vapor and the cool workpiece causes the vapor to condense on the workpiece and dissolve the soil.

The soils removed from the workpieces usually boil at much higher temperatures than the solvent, which results in essentially pure solvent vapors being formed, even though the boiling solvent may be quite contaminated with soil from previous work parts. Vapor degreasing is an improvement over cold solvent cleaning, because the parts are always washed with pure solvent. By contrast, in cold cleaning, the solvent bath becomes more and more contaminated as repeated work loads are processed and redeposition of soil increases. In vapor degreasing, the parts are heated by condensation of the solvent vapors to the boiling temperature of the degreasing solvent, and they dry instantly as they are withdrawn from the vapor zone. Cold cleaned parts dry more slowly.

To supplement vapor cleaning, some degreasing units are equipped with facilities for immersing work in warm or boiling solvent and for spraying workpiece surfaces with clean solvent. The efficiency of the liquid phase of the cleaning cycle can be augmented by the application of ultrasonic energy.

Solvents

Only halogenated solvents are used in vapor degreasing and have all or most of the following characteristics:

- High solvency for oil, grease, and other contaminants to be removed
- Low heat of vaporization and low specific heat to maximize the amount of solvent that condenses on a given weight of metal and to minimize heat requirements
- Boiling point high enough so that sufficient solvent vapor is condensed on the work to ensure adequate final rinsing in clean vapor
- Boiling point low enough to permit the solvent to be separated easily from oil, grease, or other contaminants by simple distillation
- High vapor density, in comparison with air, and low rate of diffusion into air, to minimize loss of solvent to the atmosphere
- Chemical stability in the process
- Noncorrosiveness to metals used in workpieces and in construction of equipment for the process
- Nonflammability, nonexplosiveness, and controllability with respect to health hazards under operating conditions

Table 4 lists pertinent properties of halogenated solvents used for vapor degreasing. Table 5 is a comparative evaluation of these solvents for vapor degreasing applications.

Trichlorethylene (C_2HCl_3) historically has been the major solvent used in

Table 5 Comparative evaluation for vapor degreasing applications

Property	Trichlor-ethylene	Perchlor-ethylene	1,1,1-trichloro-ethane	Methylene chloride
General stability	Good	Excellent	Selective	Good
Solvency	Aggressive	Selective	Selective	Aggressive
Recoverability (steam stripping and carbon adsorption)	Good	Good	Unsuitable	Limited
Parts handling (based on temperature after vapor rinse)	Little delay	Delay	Little delay	Immediate
Removal of high melting waxes	Good	Excellent	Good	Fair
Removal of water (spot free dryer)	Fair	Excellent	Poor	Poor
Cooling water availability and cost	Good	Good	Good	Poor
Cost to vaporize (heat of vaporization)	Moderate	Good	Moderate	High
Cleaning of light-gage parts	Good	Excellent	Good	Poor
Use with water-soluble oils	Good	Excellent	Poor	Poor
Stability towards white metals	Good	Good	Fair	Good
Stability towards caustics	Hazardous	Good	Hazardous	Good
Nonflammability	Good	Excellent	Good	Good
Steam pressures needed	Moderate	High	Fair	Low
Temperature effect on work area	Good	Fair	Good	Excellent
Use history	Very extensive	Extensive	Very extensive	Very limited
Air pollution classification	Nonexempt some areas	Nonexempt	Exempt	Exempt
Cost per pound	Medium	Lower	Higher	Higher

industrial vapor degreasing and cleaning applications. However, air pollution control regulations have led to its replacement by 1, 1, 1-trichloroethane and other solvents. Trichlorethylene is still frequently an excellent solvent choice because of its very aggressive solvent action on oils, greases, waxes, tars, gums, and rosins and on certain resins and polymers. Its fast, efficient action leaves no residue or film to interfere with subsequent metal treatment such as welding, heat treating, electroplating, or painting.

Trichlorethylene can be safely used with iron, steel, aluminum, magnesium, copper, brass, and various plating metals, without harm to the parts or to the degreasing equipment. The listed vapor degreasing solvents should be used with some caution with titanium and its alloys. Residual solvent or chlorides could cause hot salt stress-corrosion cracking if the workpieces are subsequently welded or experience service temperatures of 280 °C (550 °F) or higher. Care must be taken to remove any residuals. Dipping in nitric or nitric-hydrofluoric acid is recommended.

Always avoid the use of strong caustic (sodium hydroxide) around the degreasing operation since trichlorethylene can react vigorously with this chemical to produce spontaneously flammable dichloroacetylene. Because of the moderate boiling temperature of trichlorethylene, the degreased parts can be handled soon after the vapor rinse is complete. Normal operation uses steam at 69 to 105 kPa (10 to 15 psig).

Perchlorethylene (C_2Cl_4) has been used for many years as an important specialized solvent for difficult industrial cleaning applications. For vapor degreasing, it effectively resists chemical decomposition under heavy work loads and adverse operating conditions. Steam at 345 to 415 kPa (50 to 60 psig) is required for heating. Because of its high boiling point, it has found particular use for removal of high melting waxes since these are melted for easy solubilization. Perchlorethylene has also been of particular value for spot-free drying of metal parts having a bright finish or an intricate design. Frequently, in such cases, water that is brought into the degreaser is trapped in recessed parts and blind holes even under normal operating conditions. Since the boiling solvent is at a higher temperature than the boiling point of solvent and water, water quickly forms an azeotrope and is swept away. The rather high operating temperature of perchlorethylene also aids in the degreasing of light-gage metals by permitting a longer and more thorough rinsing action with minimum staining. It can be used effectively with iron, steel, aluminum, magnesium, copper, brass, zinc, and various plating metals, without harm to the metal parts or to the degreasing equipment.

Because of the high boiling point of perchlorethylene, vapor degreasing produces work which is too hot for immediate hand processing. This can be dealt with if the work cycle is adjusted to allow for a cooling period after degreasing. Another related problem is that the degreaser itself, operating at the boiling point of perchlorethylene, is a source of extra heat in the work area. This may cause considerable discomfort (and even danger of burns) to the operating personnel. Often the best solution is to insulate the degreaser. At other times, a little extra local ventilation, coupled with the installation of a guard rail, is all that is needed.

1, 1, 1-trichloroethane ($C_2H_3Cl_3$) has found increasing usage in a variety of vapor degreasing applications and has become the most widely used degreasing solvent. This solvent has properties similar to trichlorethylene, but it is less regulated because it is a negligible source of ozone (smog). It is an excellent solvent for many oils, greases, waxes, and tars, while at the same time it has a unique specificity toward individual plastics, polymers, and resins. An economical way to use 1, 1, 1-trichloroethane in plants using both cold cleaning and vapor degreasing is to carry out a cold cleaning step with the virgin solvent and then charge the used solvent to a vapor degreasing unit where the distilling vapors readily produce clean solvent for this second cleaning operation. Steam pressure usually ranges from 20 to 40 kPa (3 to 6 psig) due to its lower boiling point compared with trichlorethylene.

As a general purpose vapor degreasing solvent, 1, 1, 1-trichloroethane hydrolyzes when boiled with free water to produce acidic by-products. Thus, in a vapor degreasing application, water being introduced on the workpieces should be limited by an efficiently operating water separator. Such a separator, with provisions for cooling the condensate as it leaves the trough or by a coil within the water separator, is recommended for all degreasers. As with all vapor-degreasing solvents, unless it is properly stabilized, it can react with reactive white metals such as aluminum, magnesium, and zinc; 1, 1, 1-trichloroethane suitably stabilized for vapor degreasing has been widely used with all types of metal parts and is the most frequently used degreasing solvent. Avoid using 1, 1, 1-trichloroethane in an operation where it may come into contact with harsh caustic chemicals (sodium hydroxide).

Methylene chloride (CH_2Cl_2) is a versatile solvent, aggressive towards many oils, fats, greases, waxes, tars, plastics, resins, polymers, lacquers, and both synthetic and natural rubber. Use of methylene chloride should be considered particularly where the work parts might be damaged by the higher boiling temperatures of the other chlorinated degreasing solvents or where its aggressive solvency powers are specifically required. In this latter connection, some plastics and elastomers normally used in chlorinated solvents service for hose, gaskets, and containers undergo degradation when continuously in contact with methylene chloride.

For general utility, methylene chloride has the inherent limitations deriving from its low boiling point. For economy of use, special considerations apply to condensing the solvent in the machine. In some applications, this solvent tends to puddle on the work, later leading to spotting on smooth or polished surfaces. Spotting can frequently be overcome by tumbling the work in baskets or perforated drums as it is undergoing final vapor rinse. Care should also be exercised that the parts are allowed to dry fully before leaving the freeboard area of the vapor degreaser. Recently, the use of methylene chloride has been boosted by the need for additional alternate solvents to replace trichlorethylene that do not contribute to ozone in the atmosphere.

Trichlorotrifluoroethane ($C_2Cl_3F_3$) is a highly stable solvent requiring little or no additives to maintain its stability in use. It is often referred to as fluorocarbon 113 (FC 113). Fluorocarbon 113 boils only slightly above methylene chloride and like methylene chloride, refrigeration is normally required for vapor condensation and control. While methylene chloride is the strongest solvent, fluorocarbon 113 is the gentlest. This property permits its use in cleaning some assemblies containing sensitive plastic components; however, the gentle solvency is not sufficient for some soils. To compensate and to provide special solvent properties, fluorocarbon 113 is available in azeotropic composition with methylene chloride and acetone.

Stabilization of the azeotropes is needed for vapor degreasing, particularly for zinc. Fluorocarbon 113 and its blends are more costly, so they are chosen for special applications where other solvents are not suitable. Fluorocarbon 113 is among the select group of solvents identified as not causing ozone.

Solvent stability is usually controlled by the addition of stabilizers when the solvent is manufactured. Trichlorethylene, methylene chloride, 1, 1, 1-trichloroethane, and perchlorethylene all require stabilizers to perform successfully in vapor degreasing. Acid acceptance levels of the chlorocarbon solvents range from 0.1 to 0.3 wt % as sodium hydroxide. The useful life of a vapor degreasing solvent can be determined by plotting fall-off of the acid acceptance percentage over operating time.

Severe degradation problems may result from permitting cross contamination solvents during transportation, storage, or use. For example, trichlorethylene has been found to develop an acidic condition when contaminated with 200 ppm or more of 1, 1, 1-trichloroethane. Methylene chloride may develop acid-based corrosivity when mixed with 1% or more of 1, 1, 1-trichloroethane. With increased emphasis in recent years on chlorocarbon solvent recovery and recycling, the inadvertent mixing of these solvents has become a more frequent problem for the user.

Degreasing Systems and Procedures

Procedures used for cleaning various classes of work and soils by degreasing systems are indicated schematically in Fig. 2. Regardless of the system used, the distinctive features of vapor degreasing are the final rinse in pure vapors, and a dry final product.

Vapor Phase Only. The simplest form of degreasing system uses the vapor phase only (Fig. 2a). The work to be cleaned is lowered into the vapor zone, where the relative coolness of the work causes the vapor to condense on its surface. The condensate dissolves the soil and removes it from the surface of the work by dripping back into the boiling solvent. When the work reaches the temperature of the hot vapor, condensation and cleaning action cease. Workpieces are dry when removed from the tank.

Vapor-Spray-Vapor. If the workpiece contains blind holes or recesses that are not accessible to the vapor, or if the soil cannot be removed by the vapor, a spray stage may be added. The system then consists of vapor, spray, vapor (Fig. 2b).

Usually, the work to be cleaned is lowered into the vapor zone, where the condensing solvent does the preliminary cleaning; when condensation ceases, the work remains in the vapor zone and is sprayed with warm solvent. The pressure of the spray forces the liquid solvent into blind holes and effects the removal of stubborn soils that cannot be removed by vapor alone. The warm spray also lowers the temperature of the work; after spraying, the work is cool enough to cause further condensation of vapor for a final rinse.

The hot vapor may bake on some soils, such as buffing compounds, and make them difficult to remove. For complete removal of these soils, the work must be sprayed immediately upon entering the vapor and before the heat of the vapor can affect the compounds. The spray nozzle must be below the vapor line, and all spraying takes place within the vapor zone. Normal spray pressures for standard degreasers is 40 kPa (6 psi) and should not exceed 55 kPa (8 psi). Excessive spray pressure disturbs the vapor zone, resulting in a high rate of emission.

Warm Liquid-Vapor. Small parts with thin sections may attain temperature equalization before the work is clean. For these parts, and for other small parts that are packed in baskets, the warm liquid-vapor system (Fig. 2c) is recommended. In the degreasing unit shown in Fig. 2(c), work may be held in the vapor zone until condensation ceases, and then be lowered into the warm liquid, or the work may be lowered directly into the warm liquid. Agitation of the work in the warm liquid mechanically removes some additional soil. From the warm liquid, the work is transferred to the vapor zone for a final rinse.

Boiling Liquid-Warm Liquid-Vapor. For cleaning parts with par-

Fig. 2 Principal systems of vapor degreasing

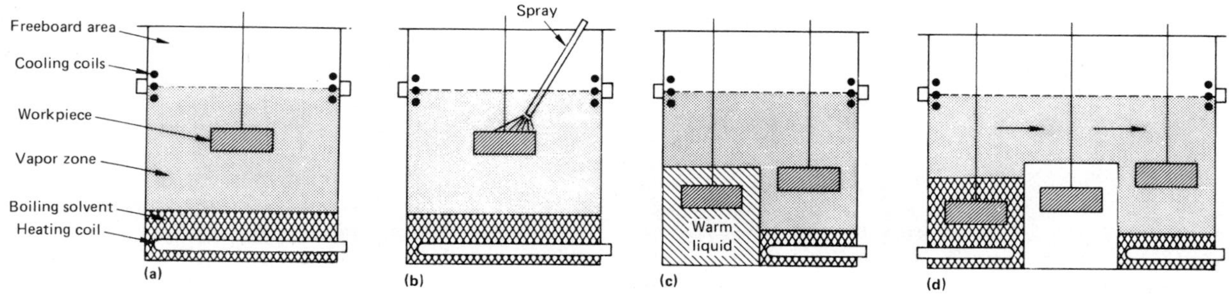

(a) Vapor phase only. (b) Vapor-spray-vapor. (c) Warm liquid-vapor. (d) Boiling liquid-warm liquid-vapor

Table 6 Physical properties of mineral oil-in-solvent mixtures

Solvent	Boiling point for vol % oil loading: 0 °C	0 °F	10 °C	10 °F	20 °C	20 °F	30 °C	30 °F	Specific gravity at 25/25 °C for vol % oil loading: 0	10	20	30
Perchlorethylene	121	250	122	252	124	255	126	259	1.619	1.542	1.464	1.395
Trichlorethylene	87	189	88	190	89	192	90	194	1.457	1.406	1.345	1.288
1,1,1-trichloroethane	74	165	76	169	77	171	79	174	1.320	1.272	1.227	1.180
Methylene chloride	40.0	104	40.6	105	41.1	106	41.5	107	1.320	1.274	1.228	1.182

ticularly heavy or adherent soil or small workpieces that are nested or packed closely together in baskets, the boiling liquid-warm liquid-vapor system (Fig. 2d) is recommended. In the unit shown in Fig. 2(d), the work may be held in the vapor zone until condensation ceases and then lowered into the boiling liquid, or the work may be lowered directly into the boiling liquid. In the boiling liquid, the violent boiling action scrubs off most of the heavy deposit, as well as metal chips and insolubles. Next, the work is transferred to the warm liquid, which removes any remaining dirty solvent and lowers the work temperature. Finally, the work is transferred to the vapor zone, where condensation provides a final rinse.

Ultrasonic Degreasing. Ultrasonic transducers, which convert electrical energy into ultrasonic vibrations, can be used in conjunction with the vapor degreasing process. The transducer materials used are of two basic types, electrostrictive (barium titanate) and magnetostrictive. The latter is capable of handling larger power inputs. Barium titanate transducers generally are operated over a range of 30 to 40 kHz; magnetostrictive transducers usually operate at about 20 kHz, but may operate at frequencies up to about 50 kHz.

Cleaning efficiency in the liquid phase of a vapor degreasing cycle can be considerably augmented by the application of ultrasonic energy. However, ultrasonic cleaning is expensive and is seldom used in a degreasing cycle unless other modifications have failed to attain the desired degree of cleanness. It is often applied to parts that are too small or too intricate to receive maximum benefit from conventional degreasing cycles.

The inside walls of hypodermic needles can be thoroughly cleaned by ultrasonic degreasing. Other examples of parts cleaned by ultrasonics because they failed to respond to conventional degreasing methods are (*a*) small ball bearing and shaft assemblies, (*b*) printed circuit boards for removal of soldering flux, (*c*) intricate telephone relays, (*d*) plug valve inserts (contaminated with lapping compounds), and (*e*) strands of cable for removal of oil and other manufacturing contaminants trapped between the strands.

Rustproofing. When a ferrous metal is vapor degreased, organic films are usually removed, and the metal is highly susceptible to atmospheric corrosion. If the surrounding atmosphere is humid or contains products of combustion or other corrosive contaminants, immediate steps must be taken to provide exposed metal surfaces with a protective film. When precision steel parts with a high surface finish (antifriction bearings, for example) are being degreased and complete rust prevention is desired, rustproofing by flushing or immersion should be included as an integral part of the degreasing system.

Control of Solvent Contamination

The cleanness and chemical stability of the degreasing solvent are important influences on the efficiency of vapor degreasing. For example, an excess of contaminant oil raises the boiling point of the solvent and detracts from its effectiveness in cleaning.

Oils. The chlorinated solvents used in degreasers are stabilized or inhibited to resist the harmful effects of many contaminants. However, certain cutting oils with a high content of free fatty acid can overcome the effects of stabilization and may contribute to a sour, acidic condition. Oils with high contents of sulfur or chlorine as additives have the same effect. These oils and greases accumulate in the boiling or vapor chamber and cause foaming and a reduction in solvent evaporation. Baked sludge accumulates on the steam coils and other heated areas, thus reducing the efficiency of the degreaser.

When the oil content of the solvent reaches 25 vol %, the solvent should be

Fig. 3 Vapor degreasing unit designed specifically for a vapor-spray-vapor system

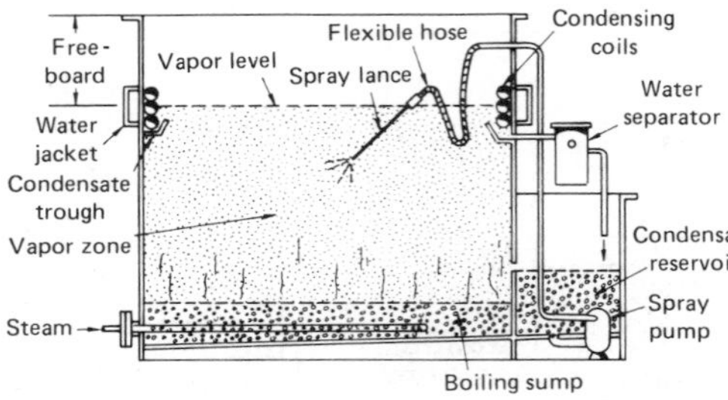

replaced and the oily solvent reclaimed. The percentage of mineral oil in trichlorethylene, perchlorethylene, 1,1,1-trichloroethane, and methylene chloride can be determined from the boiling temperatures given in Table 6.

Paint Pigments. Pigments from painted surfaces that are washed into the degreaser should be filtered or removed by other mechanical means. The oils in pigment or paint dissolve in the degreasing solvent, but the remaining material is insoluble. This material usually floats on the surface of the degreaser solution and adheres to the work. In addition to reducing cleaning efficiency, these pigments may bake out on the heating coils and the work.

Chips washed from parts into the degreaser should be removed periodically, because they contaminate other parts entering the degreaser. Such contamination is possible even in ultrasonic degreasers when the solution is not filtered continuously. An excessive amount of chips in the vapor or boiling tank reduces heat transfer and evaporation rates. An accumulation of fine aluminum particles may also result in solvent breakdown.

Water can be present in degreasers as a result of the presence of water on parts being degreased or of the accumulation of condensate on the cooling coil or jacket of the degreaser. Most chlorinated degreasing solvents are inhibited against the effects of hydrochloric acid formation in the presence of water; nevertheless, to avoid stains, spotting, and rusting of parts, all water must be removed from the degreaser. To accomplish this, degreasers equipped with one or more water separators, which continuously remove free water from the circulating recondensed solvent should be used (see Fig. 3).

Other contaminants, such as silicones, should not be allowed to enter the degreaser, because they cause foaming at the surface of the liquid solvent. All acids, oxidizing agents, cyanides, or strong alkalis must be prevented from entering the degreasing solvent.

Conservation of Solvent

The maintenance of an adequate volume of solvent in the degreasing tank also is important to the efficiency of the degreasing process. Loss of solvent can be minimized by observing the following precautions:

- The degreaser should not be located in an area subject to drafts from doors, windows, or fans
- Dragout loss should be minimized by proper drainage. Specially designed racks or rotating baskets made from wire mesh or round stock are effective
- Where the work is small and tightly packed into a basket, the basket should be allowed to drain in the vapor area before being removed from the degreaser
- Spraying, when required, should be held to a minimum and performed well below the vapor level
- Work should remain in the vapor until all condensation has ceased
- Work should not be rapidly introduced into or withdrawn from the degreaser. Vertical speed of mechanical handling equipment should not exceed about 3.4 m/min (11 ft/min)
- The degreaser should be covered when not in use. Well-designed manually operated degreasers are provided with suitable covers; conveyorized degreasers are provided with hoods
- Plumbing, valves, and pumps should be checked periodically for solvent leakage
- Introduction of moisture into the degreaser should be avoided. Work that has been wetted in a previous process should not be brought into the degreaser until it is completely dry
- Work loads should not occupy more than 50% of the open cross-sectional area of the degreasing tank. When work is lowered into the vapors, it absorbs the heat in the vapors, causing the vapor level to drop. Work load should be sized to minimize this fluctuation of the vapor level
- Porous or absorbent materials should not be degreased

Recovery of Solvent

Used solvent may be transferred to a still and recovered by distillation with or without steam. Also, the solvent may be recovered by using the degreaser as its own still and drawing off the distillate to storage.

Distillation in the degreasing unit may be accomplished by operating the degreaser with the solvent return line closed. After being passed through the water separator, the distilled solvent may be collected in a clean drum or tank, leaving the sludge behind in the boiling compartment. Some degreasers have built-in tanks for this purpose. As the concentration of high boiling oils in the sludge increases, the amount of solvent recovered decreases sharply until it is no longer profitable to continue distillation.

At no time during distillation should the heating element be exposed. Such exposure may be detected by the copious white fumes generated. The high surface temperature developed by an exposed heater destroys the heater, deteriorates the solvent, and, in extreme cases, may cause a flash fire.

Solvent Still. The use of a special still for solvent recovery is usually justified (*a*) when large amounts of soil must be removed from the solvent daily, (*b*) when it is required that work at all times be immersed in a solvent with very little contamination, or (*c*) when the degreasing unit is conveyorized and downtime for maintenance must be held to an absolute minimum.

When a still is available, the residue-containing solvent may be drained from the degreasing unit after the unit has been cooled and stored for future refining. After draining, the boiling compartment of the degreaser should be scraped clean. Where there have been acid contaminants, the cleaning procedure should be augmented by charging the compartment with water containing 30 g/L (4 oz/gal) sodium carbonate (soda ash), to a depth of about 300 mm (12 in.). The solution should be boiled for about 15 min, and the compartment rinsed and thoroughly dried. The degreasing unit is then ready for recharging with clean solvent. If acid conditions persist, contact the solvent supplier or degreaser manufacturer for detailed procedures to cope with the condition and to prevent its recurrence.

Vapor Degreasing Equipment

As in the case of cold cleaning, vapor degreasing equipment varies in complexity. All of the vapor degreaser designs provide for an inventory of solvent, a heating system to boil the solvent, and a condenser system to prevent loss of solvent vapors and to control the upper level of the vapor zone within the equipment. Heating the degreaser is usually accomplished by steam. However, electrical resistance ($\leq$3.8 W/cm^2 or $\leq$25 W/in.^{2}) heaters, gas combustion tubes, and hot water can be used. Gas combustion heaters with open flames located below the vapor degreaser are not recommended and are prohibited by Occupational Safety and Health Administration (OSHA) regulations. Specialized degreasers are designed to use a heat pump principle for both heating and vapor condensation. In this instance, the compressed gases from the heat pump are used for heating the vapor degreasing solvent and the expanded refrigeration gases are used for vapor condensation. Such a degreaser offers mobility that permits movement without connecting to water, steam, or gas for operation.

Normal vapor control is achieved with plant water circulation through the condensing coils. Refrigeration-cooled water or direct expansion of the refrigeration gases in the condenser coils are effective means of vapor control. Where a sufficient cool water supply is not available, or where plant water is excessively warm, a low boiling vapor degreasing solvent, such as methylene chloride or fluorocarbon 113, is chosen.

For safety, economy, and in some cases, to comply with regulations, degreasers are usually equipped with a number of auxiliary devices. These devices include:

- *Water separator*: a chamber designed to separate and remove water contamination from the degreaser. Solvent and water condensate collected by the condenser coils is carried by the condensate collection trough and exterior plumbing to the water separator. The water separator is designed to contain enough solvent and water condensate to last 5 to 6 min. This provides for nonturbulent flow and flotation of the insoluble water. This water is discharged from the equipment while the solvent condensate is returned to the degreasing equipment
- *Vapor safety thermostat*: located just above the condensing coils, detects the solvent vapors if they rise above the designed level in the equipment. This could occur with inadequately cool condensing water or condenser water flow interruption. When solvent vapors are detected, the heat input to the degreaser is turned off automatically. Manual resetting is preferred and used, because this demands attention and alerts the operator to a malfunction
- *Boiling sump thermostat*: in the cleaning operation, high boiling oils and greases are removed and collect in the boiling chamber. These contaminants elevate the boiling temperature of the solvent and could cause solvent decomposition if left to accumulate without control. The boiling sump thermostat is located in the boiling chamber solvent and, like the vapor safety thermostat, turns off the heat to the degreaser if it senses temperatures higher than those appropriate for the solvent being used
- *Condenser water thermostats and/or flow switches*: the water flow switch will not allow heat to be turned on unless condensing water is flowing into degreaser coils and will turn off heat source if flow stops during operation. The condensor water thermostat shuts off the heat source if condensing water leaving the degreaser is too warm. Turn off the degreaser heat if the water flow through the condenser system is inadequate or the water temperature is insufficiently cool to control the solvent vapors in the degreaser
- *Solvent spray thermostat*: a temperature sensing device, located just below the vapor air interface in the degreaser and designed to prevent manual or automatic spraying if the vapor zone is not at or above the thermostat level. This device has been required by some regulations, although no available devices are known to be effective. Spraying above the vapor zone can exaggerate solvent losses by causing air and solvent vapor mixing
- *Liquid level control*: this control shuts off heat source if liquid level in boiling chamber drops to within 2 in. of heaters. This control protects the heaters and reduces possibility of thermal breakdown of solvent

Modifications in this basic vapor degreaser are designed to permit various cleaning cycles, including spraying of the workpieces or immersing them in boiling or cool solvent. Further vapor degreaser designs are available to provide various conveyor and transport means through the cleaning cycles. Common conveyor systems include (*a*) the monorail vapor degreaser, (*b*) the crossrod vapor degreaser, (*c*) the vibratory conveyorized degreaser, and (*d*) the elevator degreaser. Open top degreasers constitute over 80% of the vapor degreasers used in industry. Their size ranges from benchtop models with perhaps 0.2 m^2 (2 ft^2) of open top area to tanks over 30 m (100 ft) long. The most common sizes range between 1.2 to 2.4 m (4 to 8 ft) long and 0.6 to 1.2 m (2 to 4 ft) wide. The most frequently used cleaning cycle is vapor-solvent spray-vapor. Among the conveyorized vapor degreasers, the monorail is the most prevalent. Generally, open top degreasers are much lower in cost, permit greater flexibility in cleaning different workloads, occupy much less floor space, and are adaptable to both maintenance and production cleaning. Because of their relative low cost and minimum space requirements, they are preferred for intermittent operations and for decentralized cleaning where transport of parts to be cleaned to a centralized location adds substantially to the cleaning cost.

Consideration must be given to capacity requirements, cost of equipment, and cost of operation. Tables 7 through 10 present comparisons of capacities and operating requirements for small, medium and large units for use in the four principal systems of vapor degreasing.

Installation of degreasing equipment should be supervised by a qualified individual. Some important considerations relating to installation are:

- A degreaser should never be installed in a location that is subjected

Table 7 Equipment-selection data for degreasing by vapor-only system

Operating conditions	Small	Size of unit Medium	Large
Maximum size of work or work basket, mm (in.)(a)	460 by 380 by 200 (18 by 15 by 8)	1220 by 610 by 610 (48 by 24 by 24)	2740 by 1070 by 1070 (108 by 42 by 42)
Production capacity, kg/h (lb/h)	450 (1000)	1360 (3000)	2720 (6000)
Size of vapor zone, mm (in.)(a)	610 by 510 by 460 (24 by 20 by 18)	1520 by 900 by 900 (60 by 36 by 36)	3050 by 1220 by 1220 (120 by 48 by 48)
Solvent in boiling sump, L (gal)	85 (22)	305 (80)	760 (200)
Heat required, kJ (Btu/h)	37 000 (35 000)	95 000 (90 000)	174 000 (165 000)
Water required, L/h (gal/h)	250 (66)	610 (160)	1060 (280)

(a) Dimensions refer, in order given, to length, width, and depth

Table 8 Equipment-selection data for degreasing by vapor- spray-vapor system

Operating condition	Small	Size of unit Medium	Large
Maximum size of work or work basket, mm (in.)(a)	460 by 380 by 200 (18 by 15 by 8)	1220 by 610 by 610 (48 by 24 by 24)	2740 by 1070 by 1070 (108 by 42 by 42)
Production capacity, kg/h (lb/h)	450 (1000)	1360 (3000)	2720 (6000)
Size of vapor zone, mm (in.)(a)	610 by 510 by 460 (24 by 20 by 18)	1520 by 900 by 900 (60 by 36 by 36)	3050 by 1220 by 1220 (120 by 48 by 48)
Solvent in boiling sump, L (gal)	70 (18)	230 (60)	720 (190)
Heat required, kJ/h (Btu/h)	73 000 (69 000)	116 000 (110 000)	232 000 (220 000)
Water required, L/h (gal/h)	460 (120)	800 (210)	1520 (400)

(a) Dimensions refer, in order given, to length, width, and depth

to drafts from ventilators, unit heaters, fans, doors, or windows. When units cannot be ideally located, such drafts should be reduced by the installation of baffles

- No degreaser should be installed near open flames unless the combustion products of these flames are exhausted outside the building. Location near welding or other operations using high temperatures must be avoided, because exposure of solvent vapors to high temperatures and high-intensity light results in decomposition to toxic and corrosive substances such as phosgene and hydrogen chloride
- The flue from the combustion chamber of a gas-fired unit should conform with local laws or ordinances. All exhausts should be discharged outside the building at an adequate distance from air intakes
- Water outlets from condenser jackets or coils should not be connected directly to sewer lines, but instead should drain freely into a funnel or other open-to-view collecting device that is connected to sewer lines. This prevents back pressure and assures maximum efficiency of the condensing coils. As water and sewage treatment costs continue to escalate, recirculating condenser water systems such as water chillers and cooling towers are being used. Many degreasers using low-temperature boiling solvents incorporate direct refrigeration. Several manufacturers offer heat recovery or heat recycling systems for use with low boiling temperature solvents
- All degreasers should have a legible, highly durable sign attached to them which bears solvent label information (see ASTM standard D 3698) and operating procedures as required by most state environmental protection agencies

Baskets and racks should be constructed of open-mesh, nonporous material. When baskets are completely filled with closely packed small items, basket size should not exceed more than 50% of the work area of the degreaser. For baskets handling large parts with generous open spaces, however, the 50% maximum may be exceeded slightly. Baskets that are too large may act as a piston as they enter the tank and displace the vapor level, thus forcing the vapor from the unit into the atmosphere.

The placement of work in the basket is critical, particularly when the parts have blind holes, which may entrap solvent. Precautions must be taken to ensure that entrapped air does not prevent liquid solvent or vapor from reaching all surfaces. After cleaning, the solvent must be completely drained from the parts to reduce dragout. To satisfy these requirements, specially designed racks or rotating baskets may be necessary.

Operating and Maintaining the Degreaser

One of the safest and most efficient degreasing operations is one that employs an effective operator training program and a routine maintenance program. By using proper education and maintenance practices, the solvent working life can be extended greatly with assurance of smooth production. Following the checklist provided below should aid in beginning an efficient degreasing operation.

Start-up

- Check the general cleanliness of the machine and work area. Remove all excess materials used in installing the machine such as tools, spare parts, and scraps of insulation

Table 9 Equipment-selection data for degreasing by warm liquid-vapor system

Operating condition	Size of unit Small	Medium	Large
Maximum size of work or work basket, mm (in.)(a)	200 by 200 by 150 (8 by 8 by 6)	360 by 360 by 200 (14 by 14 by 8)	560 by 460 by 300 (22 by 18 by 12)
Production capacity, kg/h (lb/h)	55 (125)	365 (800)	545 (1200)
Size of vapor zone and of warm liquid sump, mm (in.)(a)	300 by 300 by 300 (12 by 12 by 12)	510 by 510 by 380 (20 by 20 by 15)	760 by 610 by 610 (30 by 24 by 24)
Solvent in warm liquid sump, L (gal)	45 (12)	150 (40)	380 (100)
Solvent in vapor sump, L (gal)	23 (6)	55 (15)	135 (35)
Heat required, kJ/h (Btu/h)	19 000 (18 000)	47 000 (45 000)	100 000 (95 000)
Water required, L/h (gal/h)	95 (25)	265 (70)	585 (155)

(a) Dimensions refer, in order given, to length, width, and depth

Table 10 Equipment-selection data for degreasing by boiling liquid-warm liquid-vapor system

Operating condition	Size of unit Small	Medium	Large
Maximum size of work or work basket, mm (in.)(a)	200 by 200 by 150 (8 by 8 by 6)	360 by 360 by 250 (14 by 14 by 10)	460 by 460 by 300 (18 by 18 by 12)
Production capacity, kg/h (lb/h...	365 (800)	545 (1200)	680 (1500)
Size of vapor zone, boiling liquid sump and warm liquid sump, mm (in.)(a)	300 by 300 by 300 (12 by 12 by 12)	510 by 510 by 510 (20 by 20 by 20)	610 by 610 by 610 (24 by 24 by 24)
Solvent in boiling liquid sump, L (gal)	55 (15)	230 (60)	360 (95)
Solvent in warm liquid sump, L (gal)	45 (12)	150 (40)	265 (70)
Solvent in vapor sump, L (gal)	23 (6)	55 (15)	75 (20)
Heat required, kJ/h (Btu/h)	63 000 (60 000)	84 000 (80 000)	116 000 (110 000)
Water required, L/h (gal/h)	305 (80)	380 (100)	455 (120)

(a) Dimensions refer, in order given, to length, width, and depth

- Check room lighting, ventilation, and other utilities to see that all are in a condition of readiness for start-up
- Be sure the degreaser operator is equipped with the appropriate safety equipment and clothing. For emergency situations, such as power failures, condenser coolant stoppages, and ventilation interruptions, have organic vapor respirators or air-line masks available for immediate use. Also, be sure the operator knows how to use his personal protective equipment, understands first aid procedures, and is familiar with the hazards of the operation
- Turn on the condensing water. Observe the rate of flow and check for leaks. If the cooling water supply of the degreaser is equipped with an outlet water temperature control or a flow control safety shut-off, check these for proper operation
- Adjust the high temperature cut-off control and the vapor safety thermostat control to the temperatures recommended for the particular degreasing solvent to be used. The high temperature cut-off control setting should be about 11 °C (20 °F) higher than the boiling point of a 30% mineral oil-in-solvent mixture (see Table 6). The vapor safety control setting should be about 6 °C (10 °F) lower than the boiling point of the solvent-water azeotrope. This azeotrope might be present in a situation where a cooling water leak into the degreaser occurred at about the same time as the cooling water supply failure. Turn on the heat for a short period of time to the dry machine and observe for proper operation of the high-temperature cut-off thermostat. Do not turn on gas or electrically heated degreaser unless the heaters are covered by solvent. If the machine is steam heated, check steam pressure gages and traps for proper settings and functioning
- Add some solvent to the degreaser and check the operation of the low liquid level cut-out control, if the machine is so equipped. Finish filling the degreaser by adding enough solvent to cover the heating elements by 75 to 150 mm (3 to 6 in.), or up to the bottom of the work rest if the machine is so provided
- Turn on the heat and, as the temperature rises, assure proper operation of the various heat controls which may be in use
- As condensation begins, observe the flow of condensate from the coil and jacket, through the trough and water separator, and the returning stream to the degreaser. Interrupt the flow of condensing water and observe for proper operation of the vapor safety control
- Charge water to the water separator in accordance with the machine manufacturer's instructions. Observe for the appearance of water at the water overflow outlet
- Check the functioning of the degreaser's auxiliary equipment such as the sprayer, conveyor, still feed pump, and the still. Look at the solvent levels in each degreaser compartment and adjust as necessary

- Begin supplying work to the unit
- Check the first parts through for satisfactory cleanliness and for any signs of machine malfunction. Adjust the condenser discharge water temperature to about 8 to 11 °C (15 to 20 °F) above the dew point of the surrounding atmosphere, that is, about 32 to 46 °C (90 to 115 °F), for all the chlorinated solvents except methylene chloride. For methylene chloride or fluorocarbon 113, do not allow the discharge water temperature to go above about 29 °C (85 °F)

Fig. 4 Positioning of cup-shaped parts to drain solvent

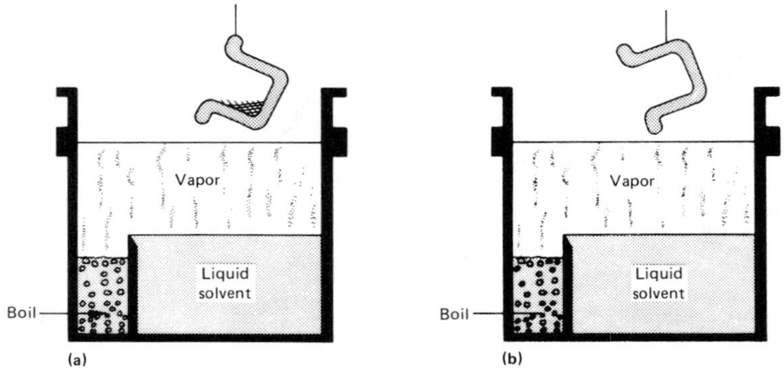

(a) Incorrect positioning. (b) Correct positioning

Operation

- While the degreaser is operating, maintain a routine surveillance to see that the work is being cleaned properly and the various systems continue to function satisfactorily
- Give some detailed attention to the arrangement of the work parts being cleaned. It may be necessary to reposition some of the parts to get proper cleaning and free-draining. Cup-shaped parts, for example, should be positioned as shown in Fig. 4
- Observe the spraying operation. Be sure that the vapor-air interface is not being unnecessarily disturbed
- Check to see that the amount of work being fed at one time is not too great, causing vapor shock. The vapor level should not recede a distance greater than the freeboard height when the work load is introduced. Be sure the rate of introduction of the work does not exceed 3.4 vertical m/min (11 vertical ft/min) because too fast a rate of entry increases the vapor shock
- Observe the vapor level as the work is being removed. The vapor level should not rise above the cooling coil or jacket. If the vapor level is rising too much, check the cross section of the work. This generally should not exceed 50% of the open area of the degreaser if the parts are traveling at a rate of about 3.4 vertical m/min (11 vertical ft/min). If the parts are larger than this, the rate of vertical movement should be reduced accordingly
- Check to see that the parts are within the vapor zone long enough for all condensation to cease before the parts are brought up into the freeboard area. Also, see that the parts are remaining in the freeboard area long enough for the solvent to evaporate completely
- After the degreasing operation has continued for several hours, observe the water separator to see that water is coming in from the degreaser as it accumulates and is being withdrawn efficiently by the separator. A cloudy ghost vapor in the vapor zone of the degreaser is a warning sign that water is not being properly removed. If water is allowed to accumulate in the degreaser, the boiling point of the solvent drops due to the formation of the solvent-water azeotrope. The direct results are poor cleaning, greater solvent losses and more odor complaints
- As the solvent level in the degreaser drops due to evaporation and leakage losses, fresh make-up solvent should be added to maintain a solvent level of about 150 mm (6 in.) above the heating elements. Particular care should be exercised that the solvent level in the boil chamber never drops lower than 25 mm (1 in.) or so above the heating elements. Make-up solvent should be added to the degreaser before start-up, that is, while cold. If it is necessary to add make-up solvent to an operating degreaser, it must be done very slowly and cautiously through water separation
- On a periodic basis, perhaps every few days during initial operation, the inhibitor level of the solvent should be checked for acid acceptance to see that a safe operation is being maintained. Should the inhibitor level show an unexpected drop, the trouble should be traced and eliminated. The problem might be excessive water in the degreaser, introduction of acid soils, soil buildup on the heating surfaces, or some other condition which proves in operation to be destructive of solvent or of one or more of the inhibitor components
- Based on the total soil load and type, and taking into account work scheduling, regular periodic degreaser cleanouts should be carried out. The frequency of cleanout can sometimes be extended by removal of particulate soils from the degreasing solvent by use of an external filtration system. Nevertheless, at intervals varying from a few days to a few weeks, it is necessary to shut down the degreaser and clean it out. The oily soil level of the degreaser should not be allowed to go higher than 30 vol %. If the actual soil is a mineral oil, the level of oil in the solvent can be conveniently checked by observing the temperature of the boiling solvent in the degreaser or by sampling the boil chamber solvent and running a specific gravity determination. The value obtained can be checked against the data given in Table 6 to get an approximate reading of the percentage of oil present

Shut-down

- For a scheduled shut-down, plan so that work is not inconveniently backlogged. The degreaser, of course, should be shut-down only after the last parts in process have cleared the machine
- Turn off the heat supply to the degreaser and wait for solvent condensation on the cooling surfaces to cease
- Turn off the cooling water and throw the switch on any unneeded pump or ventilating fan
- Any time work is not being processed in the degreaser, the cover should be

Table 11a Applications of vapor degreasing by vapor-spray-vapor systems

Parts	Metal	Production kg/h	Production lb/h	Soil removed	Subsequent operation	Notes on processing
Spark plugs	Steel	270	600	Machining oil	...	Special fixture and conveyor
Kitchen utensils	Aluminum	450	1 000	Buffing compound	Inspection	Special fixture and conveyor
Valves (automotive)	Steel	540	1 200	Machining oil	Nitriding	Automatic conveyor
Valves (aircraft)	Steel	590	1 300	Machining oil	Aluminum coating	Automatic conveyor
Small-bore tubing	Aluminum	680	1 500	Wax extrusion lubricant	Annealing	Hoist-operated unit
Builders' hardware	Brass	2 270	5 000	Buffing compound; rouge	Lacquer spray	Racked work on continuous monorail
Acoustic ceiling tile	Steel	2 720	6 000	Light oil (stamping lubricant)	Painting	Monorail conveyor
Gas meters	Terneplate	4 540	10 000	Light oil	Painting	Monorail conveyor
Continuous strip, 0.25-4.1 mm (0.010-0.160 in.)	Cold rolled and stainless steels; titanium	13 600	30 000	Oil emulsion (steels); palm oil (titanium)	Annealing	Continuous processing at up to 0.6 m/s (120 ft/min)
Automatic transmission components	Steel	18 100	40 000	Machining oil; light chips; shop dirt	Assembly	Double monorail conveyor

Note: degreasing by vapor only is applicable to the cleaning of flat parts with light soils and little contamination. Anything that can be cleaned by vapor degreasing usually can be cleaned better by liquid-vapor systems

closed. Degreaser manufacturers supply covers for their degreasers. The cover should be relatively tight fitting, but should allow the degreaser to breathe.

- If the degreaser is being boiled-down for a cleanout, before the heat supply is turned off, stop the condensate return to the boil chamber by diverting it to storage. Let the solvent level in the boil chamber then drop to within about 25 mm (1 in.) of the top of the heating element before cutting off the heat supply

Maintenance

- Outfit maintenance people with appropriate safety equipment and instruct them in its use. Also, be sure they are familiar with the hazards of the operation and that they have been instructed in first aid procedures. Their equipment should include, as a minimum, long-cuffed neoprene or PVA gloves, tight-fitting chemical goggles, an approved gas mask, and an ankle length neoprene or PVA apron. No one should enter a degreaser until solvent and residues have been mopped up and the various chambers have been thoroughly ventilated. OSHA regulations require that appropriate tests be made to ensure that the solvent vapors in the degreaser are below the regulated upper limits and that the oxygen concentration is 19.5% or greater. Then, the person entering the degreaser should wear an air-line supplied mask or an approved self-contained breathing mask, heavy neoprene gloves and boots, and should be equipped with a safety harness and life line. Someone outside the degreaser should be assigned to man the life line and keep the person in the degreaser under constant observation
- For a routine clean-out, allow the machine to cool off completely and then drain out the soil-laden solvent. An additional consideration is that some residues present a greater fire hazard when hot because they are more likely to be within their flash point range. If there is doubt on this last point, it should be checked before proceeding
- Remove any auxiliary equipment from the degreaser which may interfere with the cleaning or might be damaged in the process
- Clean out the trough, water separator, spray pump sump, and associated piping
- Scrape and brush out the metal fines and other particulate soils. Pay particular attention to corners and recesses where residues tend to collect. Take care not to damage the organic coating on machines so fabricated
- Clean off excess rust and corrosion, paying particular attention to the heating elements. If an organic coating on a degreaser shows small areas of peeled coating, these spots become sites for pitting and corrosion and should be repaired
- Inspect and repair any defective auxiliary equipment. Lubricate pumps and conveyor drives
- Install a new cleanout door gasket using as a sealant either plain or litharge-thickened, glycerol or ethylene glycol. Reinstall all auxiliary equipment items removed during cleanout

Process Applications

The wide range of applications in which vapor degreasing is used is indicated in Table 11, which lists parts and metals cleaned by the degreasing systems, as well as soils removed, production rates, and subsequent operations. The data in this table represent the experience of numerous manufacturing plants.

Removal of Magnetic Particles. Vapor degreasing is used for cleaning workpieces before and after magnetic particle inspection. When the wet method of magnetic particle testing is used, grease and dirt must be removed, or it may be difficult or impossible to produce indications of defects or discontinuities. Depending on the shape of the part, either the vapor-spray-vapor or the boiling liquid-warm liquid-vapor systems satisfactorily remove most shop soils.

Unless the workpieces are fully demagnetized after testing, no cleaning

Table 11b Applications of vapor degreasing by vapor-spray-vapor systems

Parts	Metal	Production kg/h	Production lb/h	Soil removed	Subsequent operation	Notes on processing
Degreasing by warm liquid-vapor system						
Aircraft castings	Magnesium	230	500	Polyester resin (from impregnating)	Curing	Solvent: methylene chloride
Speedometer shafts and gears	Steel; brass	340	750	Machining oil; chips	Inspection; assembly	Rotating baskets (drainage and chip-removal)
Screws	Steel; brass	680	1 500	Machining oil; chips	Painting; finishing	Flat and rotating baskets; conveyorized
Automotive die castings	Zinc-base	910	2 000	Light oils, grease; tapping lubricants; chips	Assembly	Flat and rotating baskets; conveyorized
Electron-tube components	Steel	910	2 000	Light oils	Dry hydrogen fire	Conveyorized unit
Tractor gears and shafts	Steel	910	2 000	Machining oil; chips; quenching oil	Nitriding	Elevator-type conveyor handling of work in heat treating trays
Flexible hose connectors	Steel; brass	1250	2 750	Machining oil; chips	Assembly	Conveyorized unit
Wire, 0.8-3.2-mm (0.030-0.125-in.) diam.	Aluminum	1810	4 000	Drawing lubricants; light oil	Shipment	Processed at 3 m/s (500 ft/min)
Hand power-tool components	Cast iron; aluminum	2270	5 000	Machining oil; chips; polishing; buffing compounds	Painting or plating; assembly	Rotating and flat baskets on conveyorized machine
Tubing, 6-76 mm (¼-3 in.) diam; 762-1270 mm (30-50 in.) long	Aluminum	5670	12 500	Drawing lubricants	Annealing	Hoist-operated 1134-kg (2500-lb) loads
Degreasing by boiling liquid-warm liquid-vapor system						
Transistors	Gold and tin plated	25	50	Silicone oil; light oil	Painting; branding	Manual; mesh basket
Electron-tube components	Stainless steel	90	200	Light oil	Dry hydrogen oil	Manual; mesh basket
Calculating-machine components	Steel	450	1 000	Stamping oil	Painting	Manual operation
Valves (automotive, aircraft)	Steel	450	1 000	Machining oil	Welding	Manual operation
Knife blades	Steel	820	1 800	Oil; emery	Buffing	Manual operation
Carbide-tip tool holders	Steel	910	2 000	Lubricant; chips	Recess milling	Conveyorized unit
Tubing, 60 cm (2 ft) long	Aluminum	910	2 000	Drawing lubricants; quench oil	Satin finishing	Conveyorized; tube handled vertically
Calculating-machine components	Steel	1360	3 000	Stamping oil	Plating	Conveyorized unit
Hand-tool housings, die-cast	Zinc-base	1360	3 000	Tapping oil; chips	Assembly	Automatic conveyor; racks
Screw machine products	Steel; brass	1360	3 000	Cutting lubricants; chips	Assembly	Flat and rotating basket; conveyor
Cable fittings	Steel	1810	4 000	Light oils	Inspection	Conveyorized
Stampings (miscellaneous)	Steel	2270	5 000	Light oil; chips	Furnace brazing	Small stampings nested in baskets
Wafers	Silicon	...	...	Sealing wax; paraffin	Acid etch; diffusing	Manual, in beakers; fixtured

Note: degreasing by vapor only is applicable to the cleaning of flat parts with light soils and little contamination. Anything that can be cleaned by vapor degreasing usually can be cleaned better by liquid-vapor systems

method can be completely effective in removing the magnetic particles. Passing the work through an alternating-current coil does not guarantee complete demagnetization. Because it is difficult to obtain complete demagnetization, some portions of parts following vapor degreasing must be wiped by hand.

Removal of Radioactive Soil. In most applications, radioactive soil is inorganic and is insoluble in detergent solutions or solvents. However, this soil often adheres to a film of oil or grease that is soluble, and by subjecting parts to degreasing, solids can be completely removed along with the oily film. The vapor-spray-vapor and boiling liquid-warm liquid-vapor methods are most effective. In the latter method, soil removal can be accelerated by the use of ultrasonic agitation in the solvent immersion bath.

Process Limitations

The principal limitations of the vapor degreasing process are related to (*a*) the materials it can clean without damaging effects, and (*b*) the soils it can remove effectively. Size and shape of workpieces, quantity of work, and degree of cleanness obtainable may also limit the applicability of vapor degreasing, but to a lesser extent; in general, these variables merely determine the degreasing system selected.

Materials. All metals can safely be degreased with a minimum of difficulty, provided the chlorinated solvent is properly stabilized for vapor degreasing, and the degreaser is properly operated. Ferrous metal parts are highly susceptible to rusting after degreasing, especially in humid atmospheres, and should be given a protective coating immediately after being degreased.

Some chlorinated solvents attack rubber, plastics, and organic dyes; this must be considered when degreasing assemblies with both metallic and nonmetallic components. When degreasing aluminum, the buildup of fines on heating coils should be avoided. These contaminants cause localized heating and accelerate decomposition of the solvent. If the solvent is allowed to become acidic, aluminum is attacked and forms aluminum chloride that serves as a catalyst to solvent decomposition. Use of an inhibited solvent, reasonable adherence to a planned cleaning schedule, and frequent examination of solvent prevents difficulty.

Aluminum and magnesium should be kept out of degreasers used for extremely critical parts, such as components for electron tubes. The chlorides formed by aluminum and magnesium contaminate electron-tube components, including those made of stainless steel and seriously hamper their electrical operating characteristics.

Size of Workpiece. In vapor degreasing, the size and weight of a workpiece usually are more significant than its shape. Although extremely small parts are commonly cleaned by vapor degreasing, the vapor phase alone is relatively ineffective if section thicknesses are less than about 3 mm (1/8 in.). The weight of the part is of less importance when section thickness exceeds 3 mm (1/8 in.). When small parts are vapor degreased, it is usually necessary to use the vapor-spray-vapor system or the boiling liquid-warm liquid-vapor system. In the latter, ultrasonic agitation may be added to the immersion cycle to increase the effectiveness of cleaning extremely small parts.

The maximum size of parts that can be efficiently cleaned by degreasing is limited only by the size of available equipment. Large machine tool and missile components, bars, and tubes are examples of large parts that are cleaned by vapor degreasing. Large parts receive maximum cleaning action in the vapor phase. Parts larger than the degreasing tanks have been processed by being exposed one half at a time to the vapor zone. Special techniques have been devised to clean the internal surfaces of large tanks.

Workpiece shape seldom limits the applicability of vapor degreasing, because the vapor or liquid reaches virtually all areas of highly complex shapes. However, effective cleaning of complex parts depends largely on positioning of workpieces and selection of the most appropriate degreasing system.

Although parts of simple shape soiled with readily soluble materials can be thoroughly cleaned by use of the vapor phase only, the vapor-spray-vapor system is adaptable to a greater variety of shapes and soils. This system is especially recommended for parts with blind recesses. Preferably, the parts are set in trays in a manner that allows drainage of the recesses, lowered into the vapor zone for initial condensation, and then sprayed vigorously, concentrating on the recesses. The spray washes out the recesses and cools the work; recondensation occurs when the work is returned to the vapor zone. This system is recommended also for cleaning larger articles, such as metal doors, kitchen equipment, and appliance hardware, after forming and before phosphatizing and painting, and for cleaning polished articles, such as lipstick cases, compacts, reflectors, and bezels, that can be held tightly on fixtures before lacquering.

Warm liquid-vapor is a simple and economical system for degreasing small parts during stages of manufacturing and before inspection, assembly, or packing. Small screws, nuts, bolts, washers, studs, springs, and related parts can be rotated in horizontal meshed cylinders, provided they are rugged enough to withstand the tumbling action. Delicate parts are individually racked. Deep drawn or cupped work of a size that cannot be readily rotated in cylinders can be racked to favor maximum drainage. In automatic equipment, the racked work can be periodically rotated or tilted to facilitate drainage of entrapped solvent.

The boiling liquid-warm liquid-vapor system, with the addition of a spray after immersion in boiling liquid and before immersion in the warm liquid, is the most versatile method. If the boiling phase is controlled to permit not more than 25% of contaminant in the liquid, a steady supply of vapor is ensured, and the unit operates on a minimum amount of heat input. This method has proved effective in removing oils and fats from parts with small recesses before plating or chemical finishing.

Parts, set in trays to favor entry of solution, are lowered into the vapor until condensation ceases and are then transferred into the boiling liquid for complete removal of soil. Parts are raised and sprayed vigorously until all recesses have been cleaned. They are then flushed in warm liquid and returned to the vapor zone for a final rinse and drying.

In one case, machining oil was removed from an electrical chassis 150 by 200 by 25 mm (6 by 8 by 1 in.) deep and containing 12 blind holes, by the following sequence of cycles:

- Boil for 15 min in trichlorethylene
- Flush holes with spray
- Soak in warm trichlorethylene for ½ h
- Hang in vapor until condensation ceases

This procedure produced a degree of cleanness that was satisfactory for plating with copper, silver, and gold.

Agitation during the immersion phase is helpful for effective cleaning of intricately shaped parts. Hardware items, such as U-shaped return bends, are examples of parts that require agitation for thorough cleaning. Such parts are commonly processed in rotating baskets that are attached to the crossrods of a conveyorized degreaser.

Quantity of work to be processed is not a significant factor when considering the use of vapor degreasing. Available units range from those that are suitable for occasional cleaning of a few parts to completely automated installations geared to high-production operations.

Degree of Cleanliness Obtainable. Under normal operating conditions, vapor degreasing provides a degree of cleanness that is suitable for subsequent polishing, passivating, assembly, phosphating, or painting. However, when parts are to be electroplated or subjected to other electrochemical treatments, vapor degreasing is seldom adequate and must be followed by another cleaning operation, such as electrolytic alkaline cleaning. Vapor degreasing is used immediately preceding the final cleaners to remove most of the soil, thus prolonging life of the final cleaners.

Radioactive and water-break testing techniques have indicated that a degree of cleanness between 0.1 and 1.0 monomolecular layers of soil is attainable in vapor degreasing. Under normal operating conditions, the degree of cleanness is usually near the upper level. Surface condition and section thickness may affect the degree of cleanness obtainable by vapor degreasing. For example, a polished surface is easier to clean than a grit blasted surface. Thin sections receive less cleaning action than heavy sections because the former equalize in temperature with the vapor zone in less time.

Removal of Difficult Soils

Virtually all ordinary oils and greases are soluble in chlorinated hydrocarbons and can be completely removed by one or more of the methods illustrated in Fig. 2. Other types of soils vary in responsiveness to vapor degreasing, from mild to almost total resistance to its cleaning action.

Frequently, vapor degreasing is used to remove soils that can be more effectively and more economically removed by other cleaning methods. This does not conform to recommended practice, but it may be justified when a small quantity of parts is involved and more suitable cleaning equipment is not available. Among these difficult soils are pigmented drawing compounds, water-based cutting fluids, chips, polishing and buffing compounds, and soldering fluxes.

Pigmented drawing compounds usually contain soaps, glycerides, fatty acids, and fillers such as talc and fine clay; these compounds are extremely difficult to remove. They have practically no solubility in chlorinated solvents, and at best are only partially removed by the vapor-spray-vapor system of degreasing. Furthermore, contact with the solvent vapor results in the formation of dry, inert residues, which are even more difficult to remove. Consequently, vapor degreasing is not recommended for removal of soap-based pigmented compounds.

If vapor degreasing is the only cleaning method available, consideration should be given to the use of drawing lubricants that can be removed by vapor-spray-vapor degreasing. Such lubricants include mineral oil plus extreme pressure (EP) additives, solvent-soluble wax-based materials, and phosphate-coated surfaces impregnated with mineral oil plus EP additives.

Cutting Fluids and Chips. Vapor degreasing has only limited usefulness for removing water-based cutting fluids and chips, because:

- Many cutting fluids contain substantial amounts of water, which contaminates trichlorethylene and 1,1,1-trichloroethane. Most stabilizer components are soluble in water. Water causes degradation of the solvent if not removed by an efficient water separator
- Cutting fluids often contain soaps, which are insoluble in degreasing solvents
- Sulfonated, fatty cutting oils accumulate in the solvent and stain some metals, such as brass and zinc
- Normal degreasing methods lack the mechanical action needed for removal of metal chips

When vapor degreasing is used to remove aqueous cutting fluids, perchlorethylene is the preferred solvent, because its boiling temperature is higher than that of water.

The vapor phase alone seldom removes soap-containing cutting fluids; under some conditions, it removes compounds based on mineral or vegetable oils. The vapor-spray-vapor system may be used to remove cutting fluids and chips if the workpieces are racked or fixtured in a manner that allows direct impingement of the spray on machined areas. Parts should be positioned to permit the chips to fall by gravity when the oil is removed. To flush away the chips, spray jets should be adjusted for direct impingement on specific areas.

Screw machine parts and small cast or fabricated bodies may require the use of rotating fixtures or baskets to provide adequate cleaning, drainage, and chip removal. Small castings, such as carburetor bodies, mixing valve bodies, and other types of regulator bodies, should be placed in partitioned racks or fixtures that separate the individual pieces and hold them in the same relative position throughout the cycle. This prevents damage while they are being rotated.

Polishing and Buffing Compounds. Vapor degreasing usually is not recommended for removal of polishing and buffing compounds, for two main reasons: (*a*) these compounds contain insoluble materials that require mechanical force for removal; and (*b*) frequently, workpieces that are polished or buffed are thin stampings that become heated too rapidly in the vapor phase, thus inhibiting its effectiveness.

However, if the limitations are considered, vapor degreasing may be effectively used in some applications for removing polishing and buffing compounds. The vapor-spray-vapor method is usually the most effective, because of the mechanical action by the spray. This method has successfully removed polishing and buffing compounds from parts such as cooking utensils and builders' hardware. The boiling liquid-warm liquid-vapor method also has been successfully used; workpieces were racked and agitated in the boiling solvent.

In the removal of polishing and buffing compounds, particularly when rouge is one of the soils being removed, the possibility of redeposition must be eliminated. Continuous filtration of the solvent through diatomaceous earth filters has proved to be an effective means of eliminating redeposition. These filters should be connected to the spray sumps when the vapor-spray-vapor system is used, or to the first immersion sump when the boiling liquid-warm liquid-vapor system is used.

To avoid the baking on of abrasives, the polished or buffed work should be subjected to minimum exposure to solvent vapors before entering the spray zone. Exposures of 10 to 15 s or more in the vapor zone are usually excessive and often result in incomplete degreasing and failure to remove the lubricating vehicle.

One degreaser manufacturer provides a conveyorized liquid-seal type of degreaser incorporating high-pressure sprays, 0.17 to 0.69 MPa (25 to 100 psi). This provides excellent polishing, buffing, and lapping compound removal. Ultrasonic methods are also used for effective removal of these compounds.

Soldering Fluxes. Mild soldering fluxes, such as rosin, tallow, and stearin, can be satisfactorily removed by vapor degreasing. The shape of the part determines whether the vapor-spray-vapor or the boiling liquid-warm liquid-vapor method should be used. The mechanical force either of a spray or of agitation in the boiling solution is required for rapid removal of fluxes.

Vapor degreasing should not be used for the removal of inorganic soldering fluxes, particularly those containing zinc chloride, ammonium chloride, or other activators. These materials have a harmful effect on both solvent and equipment. Water or alcohol, or mixtures of the two, are preferred for removing fluxes of this type.

Safety and Health Hazards

The chlorinated hydrocarbons used in vapor degreasing are modestly toxic when inhaled; gross overexposures result in anesthetic effects. Prolonged or repeated exposure of the skin to these solvents should be avoided because they extract oils from the skin, causing cracking and dermatitis.

Personnel operating degreasers or using chlorinated solvents should be warned of attendant potential hazards and observe proper operating instructions. They should be familiarized with the symptoms of excessive inhalation. These are headaches, fatigue, loss of appetite, nausea, coughing, and loss of the sense of balance. Maintenance men have lost their lives climbing inside tanks containing extremely high concentrations of solvent vapors. Death was attributed to the strong anesthetic power or asphyxiation.

Every effort should be made to clean or maintain a degreaser without entering the tank. However, if tank entry is necessary, strict procedures should be followed including wearing of a safety harness with another person stationed outside the tank and the use of approved air make-up systems. See ASTM STP360 Handbook on Vapor Degreasing under "Tank Entry".

The Occupational Safety and Health Administration (OSHA) has the primary responsibility for protecting worker health. Numerous general regulations exist which apply to open tanks or heated equipment. Examples of these include providing (*a*) a cover; (*b*) guardrails for platforms or walkways; (*c*) an open-top edge or guardrail 1050 mm (42 in.) high; and (*d*) enclosed combustion heaters with corrosion-resistant exhaust ducts. Where flammable solvents are used, special requirements such as explosion-resistant equipment and fusible link cover supports are required. Solvent spraying in general must be conducted in an enclosure to prevent spray discharge into the working area. Spraying in a vapor degreaser should be done only below the solvent vapor zone to prevent forcing air into the vapor zone. Welding and chlorinated solvent cleaning operations must be located separately so that the solvent vapors are not drawn into welding areas. Exposure of the chlorinated solvent vapors to the high-intensity ultraviolet light radiated by welding can cause solvent decomposition to corrosive and toxic products.

The primary health hazard associated with solvent cleaning is the inhalation of excessive vapor concentrations. Acceptable time-weighted average (TWA) vapor exposure standards have been adopted by OSHA for 8 h per day. OSHA requires that worker exposures be maintained at or below these concentration limits. Mechanical ventilation may be required to control exposures below these concentrations. The measurement of actual exposures to vapor concentrations can be accomplished by industrial hygiene surveys using activated carbon collection tubes and calibrated air pumps, continuous reading vapor detectors, and detector tubes. Additional information can be found in the 29 May 1971 Federal Register, p 10466 and in the 27 June 1974 ed., p 23540.

Solvent Fume Emission

The Environmental Protection Agency air regulations were developed to limit the formation of ozone or smog in the ambient atmosphere. Nearly all hydrocarbon solvents react with the oxides of nitrogen in the presence of sunlight to form ozone. Three particular solvents have been found to cause essentially no ozone formation. They are (*a*) methylene chloride; (*b*) 1, 1, 1-trichloroethane; and (*c*) fluorocarbon 113. For this reason, these three solvents, along with methane and ethane, have been excluded from air regulations in most states. The EPA requires a state regulation to control solvent emissions to the atmosphere if one or more localities within a state exceeds the national ambient air quality standard for ozone (0.12 ppm). Most industrialized states have at least one area that exceeds this standard. These regulations vary from state to state. However, some understanding of these regulations in general can be obtained by reviewing the regulation guidelines to the states provided by the EPA.

Disposal of Solvent Wastes

The Resource Conservation and Recovery Act, also known as the Solid Waste Disposal Act, promotes the protection of health and the environment and the conservation of valuable material and energy resources. Virtually all chemical wastes have the potential to be defined as hazardous, because the EPA defined solid waste as any solid, liquid, semisolid, or contained gaseous material resulting from industrial, commercial, mining, or agricultural operations, or from community activities. There are exceptions and a good background document is the May 1980 *Federal Register*, Vol 45, (No. 98), Identification and Listing of Hazardous Waste.

Most electroplating wastes, including solvent residues, require disposal according to these regulations. Quantity exemptions, such as less than 1000 kg (2200 lb) per month, exist in some states for some wastes providing relief from paperwork; however, proper waste disposal is still required. Solvent distillation can reduce the quantity of waste to a minimum, particularly with the nonflammable vapor degreasing solvents. Under some circumstances, still bottoms (residues) can be used as a fuel in industrial boilers. Nonhazardous waste such as paper should be segregated from hazardous wastes to minimize disposal costs. Incineration is the best known ultimate disposal method for wastes from solvent cleaning operations. However, wastes containing reasonable quantities of solvent may be saleable to local reclaimers.

Acid Cleaning of Iron and Steel

ACID CLEANING is a process in which a solution of a mineral acid, organic acid, or acid salt, in combination with a wetting agent and detergent, is used to remove oxide, shop soil, oil, grease, and other contaminants from metal surfaces, with or without the application of heat. The distinction between acid cleaning and acid pickling is a matter of degree, and some overlapping in the use of these terms occurs. Acid pickling is a more severe treatment for the removal of scale from semifinished mill products, forgings, or castings, whereas acid cleaning generally refers to the use of acid solutions for final or near-final preparation of metal surfaces before plating, painting, or storage. Acid pickling is discussed in the article on pickling of iron and steel in this Volume.

Mineral Acid Cleaning*

Cleaner Composition

A variety of mineral acids and solutions of acid salts can be used, either with or without surfactants (wetting agents), inhibitors, and solvents. The large number of compositions that are used may be classified as:

- Inorganic (mineral) acid solutions
- Acid-solvent mixtures
- Solutions of acid salts

Many acid cleaners are available as proprietary compounds, either as a liquid concentrate or a powder to be mixed with water. Compositions of several solutions used for cleaning ferrous metals are given in Table 1. Formulas suitable for use with nonferrous metals can be found in the section of this Volume on cleaning and finishing of nonferrous metals. Table 2 contains some possible operating conditions when cleaning ferrous metals.

Organic acids such as citric, tartaric, acetic, oxalic, and gluconic, and acid salts such as sodium phosphates, ammonium persulfate, sodium acid sulfate, and bifluoride salts, are used in various combinations. Solvents such as ethylene glycol monobutyl ether and other glycol ethers, wetting agents and detergents such as alkyl aryl, polyether alcohols, antifoam agents, and inhibitors may be included to enhance the removal of soil, oil, and grease.

Strength of the acid solutions varies from as weak as 5.5 pH for acid-salt mixtures to the equivalent of the strong acids used for pickling.

The phosphoric acid and ethylene glycol monobutyl ether mixtures (Table 1) are used for removing grease, oil, drawing compounds, and light rust from iron and steel. In various concentrations, these mixtures are adaptable to immersion, spray, or wiping methods and leave a light phosphate coating (110 to 320 mg/m^2 or 10 to 30 mg/ft^2) that provides a paint base or temporary resistance to rusting if the parts are to be sorted.

Chromic acid solutions are used occasionally to clean cast iron and stainless steel. A chromic acid formula used for cleaning stainless steel is 60 g/L (8 oz/gal) chromium trioxide, 60 g/L (8 oz/gal) sulfuric acid, and 60 g/L (8 oz/gal) hydrofluoric acid in water, and used at room temperature in an immersion system. Another solution, used frequently for cleaning stainless steel, is a solution of nitric acid (10 to 50 vol %) and hydrofluoric acid (1 to 3 vol %) in water. The steel is immersed in the solution at room temperature for 3 to 30 min.

Chromic acid solutions and mixtures containing chromic acid are often used as final rinses in acid cleaning-phosphating systems. The acid enhances the corrosion resistance of the coated surface and enables the rinse to clean excess and unattached phosphate from

*Revised by John Pilznienski, Senior Research Chemist, Kolene Corp.

Table 1 Typical composition of acid cleaners for cleaning ferrous metals

Composition of each constituent is given in percent by weight

Constituent	Immersion		Spray		Barrel	Wipe	Electrolytic
Phosphoric acid	70	···	70	···	···	15-25	···
Sodium acid pyrophosphate	···	16.5	···	16.5	16.5	···	···
Sodium bisulfate	···	80	···	80	80	···	···
Sulfuric acid	···	···	···	···	···	···	55-70
Nonionic wetting agent (a)	5	···	5	···	···	7-20	···
Anionic wetting agent	···	3	···	3	3	···	···
Other additives	(b)	(b)	(b)(c)	(b)(c)	(b)(c)	(b)(d)	(b)
Water	25(e)	···	25(e)	···	···	rem	rem

(a) Ethylene glycol monobutyl ether is used. (b) Inhibitors up to 1% concentration may be used to minimize attack on metal. (c) An antifoaming agent is usually required when the cleaner is used in a spray or barrel system. (d) Small additions of sodium nitrate are often used as an accelerator in cleaning rolled steel; nickel nitrate is used in cleaning galvanized steel. (e) Before dilution

Table 2 Operating conditions for acid cleaners for ferrous metals

Type of acid cleaner	Concentration g/L	Concentration oz/gal	Temperature °C	Temperature °F
Immersion	120	16	71	160
	60-120	8-16	60	140
Spray	60	8	60	140
	15-30	2-4	60	140
Barrel	15-60	2-8	Room	
Wipe	···	···	Room	
Electrolytic(a)	···	···	21	70

(a) Current density, 10 A/dm^2 (100 A/ft^2)

the surface of the treated metal without leaving hard water salts deposited on the surface. Paint applied following such a treatment gives greater protection against corrosion by salt water. Chromic acid is used in solutions of low pH when a strong oxidant is required. Nitric acid is also a strong oxidant, and a 10 to 20% nitric acid solution is used to brighten stainless steel.

Acid solutions of 40 to 60 vol % hydrochloric or 6 to 8 vol % sulfuric (often containing up to 1% inhibitor) are used at room temperature for removing soil and light rust. Stronger solutions of these acids are used in electrolytic baths for final cleaning of ferrous metals before electroplating (Table 1). Because of excessive evolution of gas, phosphoric acid is unsuitable for use in an electrolytic bath.

Various soils, including light rust, are removed by combining acid cleaning with mechanical action. Acid salts such as sodium acid pyrophosphate, sodium bisulfate, and mixtures of the two are widely used to clean ferrous metal parts in rotating barrels. (A formula is given in Table 1.) A solution with this formula may also be used for parts that are immersed or sprayed.

Additives such as oxalic acid occasionally are used with the acid salts when ferrous metal parts are being cleaned in rotating barrels. Oxalic acid attacks steel, but seldom to an objectionable degree. Thiourea is a good inhibitor, if inhibited oxalic acid solutions are required. The addition of fluoride salts to acid salts, such as 8 to 15 g/L (1 to 2 oz/gal) sodium fluoride or ammonium bifluoride, improves efficiency in the removal of silica sand from castings when parts are cleaned in a barrel or tank.

A formula used for wipe cleaning is also given in Table 1. Other cleaners used for wiping are (*a*) 6 to 8 vol % sulfuric acid in water; (*b*) 70% phosphoric acid, 5% wetting agent, and 25% water; and (*c*) a paste made of 85 to 95% ammonium dihydrogen phosphate and the remainder wetting agent, used on a wet cloth or sponge.

Inhibitors are often included in cleaners used on ferrous metals to minimize attack on metal and lower acid consumption. Composition of inhibitors varies widely. Numerous by-products such as sludge acid from oil refineries, waste animal materials, waste sulfite cellulose liquor, offgrade wheat flour, and sulfonation products of such materials as wood tar, coal tar, and asphaltum have been successfully used. These materials cost less than synthetic inhibitors, but vary widely in uniformity and effectiveness. For this reason, synthetic inhibitors have been increasingly used.

Synthetic inhibitors are usually complex organic compounds. Most often, a given compound or class of compounds will function most effectively with only one type of acid, so choosing the proper inhibitor should not be a haphazard process. Many proprietary compositions of these chemicals are available for use in various acid systems.

The amount of inhibitor used depends on the workpiece composition, acid cleaner formulation, temperature of operation, and nature of soil being removed. From ½ to 1% inhibitor before dilution with water is used. Higher percentages of inhibitor are used for higher acid concentrations and operating temperatures. Overusage is wasteful because it has little effect in reducing attack on metal. Increasing solution temperature increases the rate of attack in both inhibited and uninhibited acid cleaning solutions.

Antifoaming agents may be required in acid spray cleaners to prevent excessive foaming. Sometimes foaming can be reduced using naturally hard water or by adding small amounts of calcium chloride, up to 30 grains hardness. Addition of a plasticizer such as triethylhexylphosphate or one of the high molecular weight polyols (organic alcohols) reduces foaming. Because of variation in water and other conditions in a specific installation, several additives may need to be tried before foaming is brought under control. Silicones are usually effective as antifoaming agents, but they should not be used if parts are to be painted or plated, because of residual contamination. Paint or plating does not adhere to the contaminated areas resulting in a fisheye appearance at the contaminated spots.

Methods of Application

Wipe on/wipe off, spray, immersion, and rotating barrel methods are all used extensively for acid cleaning. Although heating greatly increases efficiency, cleaning is frequently done at room temperature for superior process control and economy of operation. When heat is used, the temperature range of the cleaner is usually 60 to 82 °C (140 to 180 °F) with temperatures up to 93 °C (200 °F) used occasionally. Time cycles for acid cleaning are short compared with acid pickling, especially when stronger acids are being used. Selection of method depends on the nature of soil being removed, the size and shape of the workpiece, quantity of similar pieces to be cleaned, and type of acid cleaner used.

Wipe on/wipe off is the simplest method of acid cleaning; virtually no equipment is required. Using a formula such as shown in Table 1, an operator suitably protected by rubber gloves and apron wipes the soiled workpieces with an acid-impregnated cloth or sponge. After the cleaner is allowed to react (2 or 3 min is usually sufficient), work is rinsed with water.

The wiping method is practical only for cleaning a few parts at a time or for large, bulky parts that cannot be immersed conveniently in a cleaning bath. Labor cost becomes excessive if many parts are cleaned. Cleaner concentrations are stronger than in dip and spray solutions, and cleaner is not usually recovered for further use.

Spray cleaning is more practical than wiping when larger quantities of bulky parts are acid cleaned. Multistage spray washers have been designed to accommodate a variety of work that can be racked or suspended from hooks. Large components, such as truck cabs and furniture, are usually cleaned by this method. Cost of labor is lower than for hand wiping. Also, consumption of cleaner ingredients is considerably less because concentrations are lower, and cleaner is recirculated for reuse. The capital investment for spray cleaning equipment is high, and large production quantities are usually needed to justify the expense. Steady or high production quantities are not always necessary to warrant the installation of spray equipment. It is sometimes feasible to accumulate parts for about 2 days and then operate the washer for part of a day.

In one automotive plant, a spray system replaced a hand wiping system with the following results. A wipe on/wipe off system using phosphoric acid-ethylene glycol monobutyl ether was used to prepare large steel stampings for painting. A total of 46 supervisory and production employees was required. Installation of an automatic spray system decreased cleaner consumption and provided the same productivity with only six employees. In addition, a heavier phosphate coat was obtained, 540 to 650 mg/m^2 (50 to 60 mg/ft^2) by spraying, compared to 110 to 220 mg/m^2 (10 to 20 mg/ft^2) by wiping.

Immersion is the most versatile of the acid cleaning methods. The operation may vary from hand dipping a single part or agitating a basket containing several parts in an earthenware crock at room temperature to a highly automated installation operating at elevated temperature and using controlled agitation. The type of cleaner that can be used in an immersion system is not restricted. Higher concentrations of the same type of cleaner are recommended for immersion cleaning than for spray cleaning (Table 1). Efficient cleaning by immersion depends on placing workpieces in baskets or on racks to avoid entrapment of air or nesting of parts.

Barrel cleaning is often used for large quantities of small parts. Perforated barrels containing 225 to 450 kg (500 to 1000 lb) of parts are immersed and rotated in tanks of cleaning solution. Solutions of acid salts (Table 1) are used for this method, although other cleaning solutions may be applicable. In most instances, a medium such as stones is added to the charge, frequently comprising up to two thirds of the total load. The medium aids in cleaning by providing an abrading action. It also prevents workpieces from damaging each other. Acid cleaning in barrels is usually performed at room temperature. Heated solutions can be used if required by the nature of the soil being removed.

Barrel methods can be used for cleaning in continuous high production. Several barrels can be arranged so that some can be loaded while others are in the cleaning tank. The chief limitation of the barrel method is the size and shape of workpieces. Parts such as bolts are ideal for barrel cleaning, while delicate stampings are not.

Electrolytic cleaning is effective because of the mechanical scrubbing that results from evolution of gas and the chemical reduction of surface oxide films when used anodically. Sulfuric acid baths are most commonly electrolyzed (Table 1) and are usually used as a final cleaner before plating. All grease and oil should be removed before electrolytic cleaning to reduce contaminating of the electrolytic bath. If alkaline cleaners are used as precleaners, the rinse must be thorough or the acid bath can be neutralized by the alkali. Time cycles in electrolyzed acid solutions must be short, usually less than 2 min, or excessive etching can occur. Current distribution must be uniform, or localized etching may damage the workpiece.

Selection Factors

In any acid cleaning operation, etching usually occurs. In many instances, this light etching is advantageous for final finishing operations. However, if etching is not permissible, some other cleaning process should be used.

Limitations of acid cleaning include:

- Inability to remove heavy deposits of oil or grease without large additions of expensive material such as ethylene glycol monobutyl ether
- Attack on the metal to some degree, even when inhibitors are used
- Requirement of acid-resistant equipment

If parts are soiled with heavy deposits of oil or grease, as well as rust, preliminary alkaline cleaning preceding acid cleaning may prove economical. Multiple rinses or a chromic acid rinse (1 to 2% in water) may be used to prevent carryover of alkali.

Selection of Process

Reasons for selecting acid cleaning and specific acids are illustrated in the following examples. Parts deep drawn from low-carbon sheet steel as received from the supplier were covered with pigmented drawing compound and other shop soil and frequently became rusty during transit. Alkaline cleaning even with hand scrubbing did not consistently remove the drawing compound and allowed most of the rust to remain. Acid cleaning in a multistage spray washer completely removed all soil and rust without hand scrubbing. A phosphoric acid and ethylene glycol monobutyl ether mixture (Table 1) was spray applied using a concentration of about 60 g/L (8 oz/gal) at 66 °C (150 °F). In addition to thorough cleaning, the process deposited the light phosphate coating that was desired as a base for subsequent painting.

Finish-machined surfaces on large castings showed a light blushing rust after a weekend in high humidity. Abrasive cleaning could not be used because of possible damage to finished surfaces. The rust was removed without etching by hand wiping with a paste-like compound of about 90% ammonium dihydrogen phosphate and 10% wetting agent, followed by wipe rinsing.

Combinations of alkaline and acid cleaning methods are often used advantageously. Machined parts having heavy deposits of oil, grease, and light blushing rust were being acid cleaned using phosphoric acid and ethylene glycol monobutyl ether in an immersion system at 60 °C (140 °F). Results were satisfactory, but the cleaner became contaminated from the oil and grease so rapidly that the replacement cost of cleaner became excessive. Adding a preliminary alkaline cleaning operation removed most of the soil. Parts were then rinsed, first in unheated water, then in an unheated neutralizing rinse containing 2% chromic acid. Immersion in the phosphoric acid and ethylene glycol monobutyl ether mixture removed the rust and provided a surface ready for painting. This practice prolonged the life of the acid cleaner by a factor of five or more.

In other instances, combining alkaline and acid cleaning does not prove economically feasible. In one plant, small steel stampings were being prepared for painting by removing light oil and some rust in a five-stage spray washer. The first stage was alkaline, followed by water rinsing, then two stages of phosphoric acid cleaning, followed by water rinsing and a rinse in chromic acid solution. Alkaline contamination of the first acid stage was excessive, necessitating weekly dumping of the acid cleaner. A change to three successive stages of acid cleaning followed by one plain water rinse and one rinse with chromic acid in water proved to be the more economical and satisfactory method. The practice was then to dump the cleaner periodically from the first stage and decant the second stage cleaner to the first stage, recharging the second stage while maintaining the third stage.

For small parts that are not easily bent or otherwise damaged, barrel methods often are the most satisfactory. Small miscellaneous parts having no deep recesses required removal of light oil and minor rust. They were placed in a horizontal barrel and rotated in a solution of acid salt cleaner (similar to the composition shown in Table 1) at room temperature using a concentration of 45 to 60 g/L (6 to 8 oz/gal). After tumbling for 10 to 20 min, the barrel was removed from the cleaner tank, drained, rinsed, drained, and tumbled for 30 to 60 min at room temperature in a tank containing 45 to 60 g/L (6 to 8 oz/gal) of alkaline cleaner. The charge was then rinsed in water, unloaded and dried. Tumbling in the alkaline solution neutralized residual acid and produced a shine on the workpieces.

If optimum equipment is not readily available, requirements may sometimes be met with available equipment. Box-shaped cast iron parts, 200 by 150 by 100 mm (8 by 6 by 4 in.) deep, open on one end and having several drilled holes, were covered with light mineral oil. Parts needed to be cleaned and provided with a phosphate coating suitable for painting. Available equipment was a two-stage alkaline spray washer. Parts were washed in this equipment and then dipped in a phosphating tank. Because the workpieces were heavy and bulky, this procedure was inadequate to meet the production demand of 2500 to 3000 parts per 8-h day. The problem was solved by changing the alkaline solution in the spray washer to an acid phosphate cleaner that contained low-foaming surfactants (wetting agents). Parts were sprayed for 1 min with a solution containing 110 g (4 oz) of acid phosphate cleaner per 4 L (1 gal) of solution, operated at 71 °C (160 °F). They were then sprayed with unheated water for 30 s, dipped in water base inhibitor, air dried, and painted.

For parts that are to be electroplated, electrolytic acid cleaning is often used. After precleaning small parts to remove most of the oil, the following cycle was established for small carbon steel parts before electroplating:

1. Water rinse at 82 °C (180 °F)
2. Immerse for ¾ to 2 min in 55 to 70% sulfuric acid at 21 °C (70 °F), using a current density of 10 A/dm^2 (100 A/ft^2)
3. Flowing water rinse for 15 to 30 s at room temperature
4. Repeat step 3 in a second tank
5. Dip in 20% hydrochloric acid for 15 s at room temperature
6. Flowing water rinse for 15 to 30 s at room temperature

Electrolytic cleaning was successfully used in this application. Auto bumpers were cold formed from phosphated and lubricated sheet steel. Alkaline cleaning was used to remove mill dirt and soap-type lubricant. Electrolytic acid cleaning followed the alkaline treatment to ensure removal of the phosphate coating and residual lubricant. Because of scrubbing action by the gas evolved at the work surface, the electrolytic bath assisted in removing adherent solid particles that were the residue of a polishing compound. Slight metal removal occurred which removed metal slivers and produced a microetch suitable for plating. The ability of this bath to remove tenacious oxide coatings permitted the electroplating of nickel with good adhesion. While this cleaning could have been done by other means, the electrolytic acid system proved to be the most satisfactory method for this application.

Equipment

Wipe on/wipe off cleaning requires only the simplest equipment. Acid-resistant pails and protective clothing, and common mops, brushes, and wiping cloths are all that is needed.

Immersion systems require equipment varying from earthen crocks for hand dipping at room temperature to fully automated systems using heat and ultrasonic or electrolytic assistance. The construction for an acid tank is shown in Fig. 1. Tanks for sulfuric acid may be lined with natural rubber and acid-resistant red shale or carbon brick joined with silica-filled hot poured sulfur cement. Liners or free-standing fabricated tanks of polypropylene are also used. Tanks intended to contain nitric or hydrofluoric acids may be lined with polyvinyl chloride and carbon brick joined with carbon-filled hot poured sulfur cement. Carbon brick liners are not needed for nitric acid, but they are usually used to contain hydrofluoric acid.

If the cleaning operation uses only acid solutions, an immersion installation would consist of (*a*) an immersion tank for the acid solution, capable of being heated to 82 °C (180 °F), (*b*) two rinse tanks for flowing cold water, and (*c*) drying facilities, either convection or infrared.

Various modifications can be made for specific conditions. If parts are precleaned in alkaline solutions, two water rinse tanks should precede the acid cleaning tank. One of these two rinses may be a still tank containing dilute chromic acid. The final may be a heated still tank containing dilute chromic acid or a hot water tank (up to 82 °C or 180 °F). One advantage in using heat in the final rinse is that subsequent drying is accelerated.

Various degrees of automation are feasible with immersion systems. Automated cleaning of racked parts can be applied to immersion systems by using an overhead monorail that raises and lowers racks according to a predetermined cycle.

Electrolytic acid cleaning tanks must be constructed to resist acids. Venting is recommended and usually required; otherwise, these tanks are no

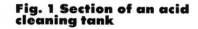

Fig. 1 Section of an acid cleaning tank

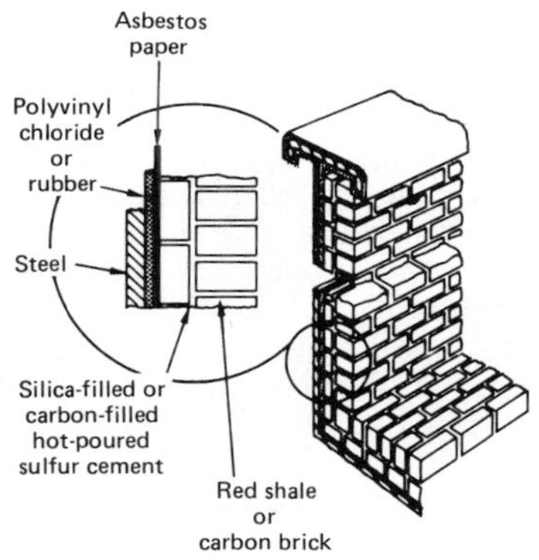

Inner lining of brick acts only as a thermal shield and as a protection against mechanical damage to the corrosion-resistant polyvinyl chloride or rubber membrane

different from tanks used for electrolytic alkaline cleaning. A typical electrolytic cleaning tank is shown in the article on alkaline cleaning. Various types of auxiliary equipment may be used for removing fumes from an electrolytic tank. One effective design is shown in the article on hard chromium plating in this Volume. Electrodes are preferably made of lead.

Rinse tanks should be as small as is compatible with easy handling of the largest load to be rinsed. For a given flow rate, smaller tanks allow better mixing and faster rinsing of impurities. If a series of rinse tanks is used, all should be uniform in size for simple flow rate control.

Polyvinyl chloride is a proven material for rinse tanks. Polypropylene, which can withstand higher temperatures than polyvinyl chloride, has also been used, as well as polyester, rubber, brick, lead, and plain carbon steel coated with protective paint. Stainless steel can be used in rinse tanks where chloride solutions are not used. Chlorides cause pitting of stainless steel, especially if tanks are used intermittently.

Rinse tanks can be equipped with automatic controls that flush tanks when impurities reach an established level, as monitored by continuous measurement of the electrical conductivity.

Spray systems are designed with special features for high-production acid cleaning. The number of stations varies, but a five-stage system is usually used for cleaning and phosphating parts such as large stampings. The first stage is acid cleaning (usually phosphoric and ethylene glycol monobutyl ether), and is followed by spray phosphating. The process is completed by using either three successive stages of unheated water rinsing or two stages of unheated water and one of unheated or heated mild chromic acid solution. Parts are conveyed from stage to stage singly on a belt or by using an overhead monorail system with parts hanging singly or on racks.

Heating Equipment. Acid cleaners are rarely heated above 82 °C (180 °F). Improved detergent systems in recent years have permitted a much wider range of work to be acid cleaned at room temperature with consequent energy savings, but removal of rust or stubborn soils such as buffing compounds usually benefits from the application of heat. The temperature range most frequently used when acid cleaners are heated is 60 to 71 °C (140 to 160 °F).

Drying is usually done by heated forced air. However, temperatures higher than about 76 °C (170 °F) are not used, for economic reasons. Infrared dryers may be used if controlled to proper operating temperature.

Acid Attack and Sludge Formation. In phosphoric acid cleaning and coating systems, acid attack on work is minor, although some metal is dissolved. Solutions used in phosphate coating are kept at saturation, and as the temperature drops, precipitation occurs. The amount of phosphate compounds in the sludge, as well as the severity of acid attack on the work, depends on the temperature.

Acid attack on the major items of equipment is almost negligible. For example, tanks and pipes used in one high-production installation have not been replaced during the first 16 years of operation and are still in serviceable condition. The tanks and pipes for this installation were made of low-carbon steel; pumps and nozzles were made of stainless steel. Most equipment deterioration is caused by erosion on parts such as pump impellers, riser pipe elbows, tees, and nipples. Some attack occurs initially, but once the steel surface has become coated with phosphate, attack ceases unless flakes of coating are dislodged by abrasion, exposing bare steel. Also, deposits of sludge serve as inhibitors of acid attack and further protect the metal from the acid. The major cause for replacing parts such as risers and nozzles is clogging by sludge.

The chemical reactions of phosphating continue in the equipment even when no work is going through the system, because aeration of the solution in the spray system causes some oxidation of soluble phosphate to insoluble phosphate. Thus, catalyst and chemicals are consumed, and sludge is deposited during standby periods.

In a spray system, sludge is usually removed by filters. In immersion systems, the sludge accumulated at the bottom of the tank is usually shoveled out after most of the still usable solution has been removed (decanted). A sludge pan is often helpful. Such a pan covers the entire bottom of the tank except for small areas at the edges. This permits easy removal. The pan is usually 75 to 125 mm (3 to 5 in.) deep. Rods with hooks extending above the solution level allow the pan to be lifted to remove sludge. Thus, the solution need not be decanted, downtime is minimized, and labor is saved.

Handling and Conveying. Parts such as nuts and bolts are most commonly cleaned in rotating barrels. However, if barrel equipment is not available, such parts can be cleaned in baskets. Conveyance may be by hand, by lift systems, or by belt when a spray is used, or by a combination of these systems. Small parts that cannot be tumbled in barrels may be placed in wire baskets, racked for immersing or spraying, or placed singly on belts in a spray system.

Racks, hooks, and baskets are usually made of a metal that will resist acids. Types 316L and 347 stainless steel are successful for these components. Where racks or hooks travel through a series of cleaning, phosphating, and painting systems, the racks are continually recoated making low-carbon steel an acceptable rack material. A rack used for cleaning and phosphating of small stampings, such as doors for automobile glove compartments, is illustrated in Fig. 2. Large components are usually hung singly on hooks and transported by an overhead monorail. Figure 3 illustrates an arrangement for carrying truck cabs through a five-stage spray cleaning installation.

Fig. 2 Rack used for cleaning and phosphate coating small stampings

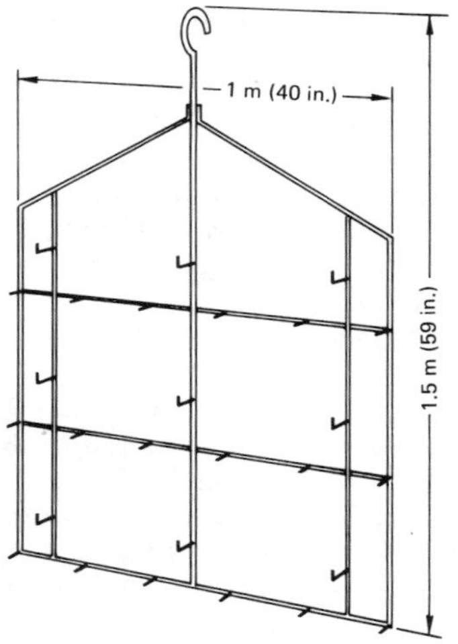

Fig. 3 Arrangement for conveying truck cabs through a five-stage spray cleaning installation

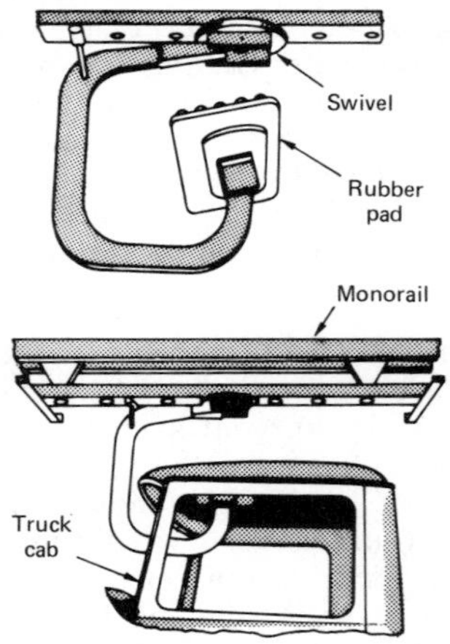

Control of Process Variables

Agitation, operating temperature, acid concentration, solution contamination, and rinsing are the principal variables that affect efficiency and results in acid cleaning.

Agitation, either of the solution or the workpieces, is usually necessary in all systems. In wipe on/wipe off methods, agitation is under direct control of an operator. In spray systems, agitation is provided by the impingement of the solution on the workpieces, and the impingement is basically controlled by the pressure. Pressures used in spray systems are commonly 100 to 170 kPa (15 to 25 psi), measured at the pump. Pressures up to 280 kPa (40 psi) are sometimes used for removing tenacious soils. For cleaning complex parts, some experimentation is usually required in adjusting the nozzles to achieve a spray pattern that reaches cavities and crevices.

Immersion systems use a variety of methods for agitation. In smaller production quantities, parts contained in baskets are hand agitated by raising, lowering, and turning. Underwater air jets or mechanical propellers are also effective for agitation in cleaning tanks, and they can decrease the soaking period. In automated immersion systems, the forward motion of parts often provides sufficient agitation. However, this can be enhanced if necessary by simultaneously agitating the solution. In barrel cleaning, agitation of both work and solution is provided by the rotation of the barrel.

Ultrasonic cleaning methods can be applied to acid cleaners in the same manner as is done with other cleaning methods. Because initial cost and maintenance of ultrasonic equipment is high, this form of energy is used only when simpler methods fail to achieve satisfactory cleaning either because the soil is extremely difficult to remove or because the shape of the workpiece is complex.

Electrolytic cleaning provides agitation from gas evolution, which produces a scrubbing action.

Operating Temperature. Although the efficiency of soil removal increases as temperature increases, a significant amount of acid cleaning is done in unheated solutions, because heated solutions may present the following disadvantages: (*a*) attack on workpieces increases with temperature; (*b*) cleaners deteriorate or are used up more rapidly, in part because of dissolved metal; (*c*) surfaces emerging from hot acid solutions are likely to dry and become streaked before they are rinsed; and (*d*) the life of the tanks and other equipment decreases as operating temperature is increased.

As mentioned previously, when acid solutions are heated, temperatures ranging from 60 to 71 °C (140 to 160 °F) are most frequently used. Higher temperatures (up to 82 °C or 180 °F) are sometimes required to remove soils such as drawing compounds that contain high-melting waxes or greases. In barrel cleaning with solutions of acid salt, temperatures up to 93 °C (200 °F) are sometimes used, but these cleaners are relatively mild so that problems of attack on workpieces and equipment are not great. Maintenance of temperature within ±3 °C (±5 °F) usually provides adequate reproducibility.

Control of cleaner composition is necessary for consistently satisfactory results. Depletion of cleaner by its reaction with workpieces or equipment, dragout, drag-in of alkali or other contaminants, and decomposition of the cleaner constituents are factors that affect cleaner life.

Chemical analysis using simple titrations for acid and metal content permit control of solution composition. Visual inspection of processed workpieces also indicates condition of the cleaner. In a new installation, when a new solution is being used, or when a different soil is being removed, the solution should be checked every hour until the required frequency of testing is established.

Control of rinsing is necessary for consistently good results. Cold water is adequate for most purposes except when high-melting waxes and greases are being removed. Residues of such soils may set from cold water rinsing. An initial rinse with demineralized water at 71 to 82 °C (160 to 180 °F) is often used when removing these soils. Rinsing qualities of water can be greatly improved by adding a wetting agent, usually 0.02 wt %. Agitation during rinsing is important and is achieved by the same means used with cleaning solutions.

Rinsing is expensive, but cost can be minimized by (*a*) using tanks as small as possible, (*b*) using tanks of uniform size if in a series, (*c*) using automatic flush control of contamination limit, and (*d*) using counterflow rinse tanks.

Sludge buildup is proportional to the amount and type of soils entering the system. Even though sludge buildup does not directly impair the efficiency of an immersion system, a large amount of sludge should not be allowed to accumulate because it may foul heating or control equipment. In spray systems, good filtration and screening are required to prevent fouling of nozzles and related equipment.

Maintenance

For obtaining consistently good results, a regular schedule of maintenance is recommended for any immersion or spray cleaning installation. The required frequency of maintenance varies considerably with the specific operation. Experience with a particular installation soon indicates the items that need close attention to prevent costly shutdowns or inadequate cleaning. The following list suggests a program for maintaining immersion and spray systems.

Daily

- Check temperature
- Check solution concentration
- Check and adjust spray nozzles
- Clean screens in spray systems

Weekly

- Decant or dump solutions and recharge
- Remove sludge from tanks, heating coils, and temperature regulators
- Flush risers in spray systems
- Remove and clean spray nozzles

Monthly

- Inspect exhaust hoods
- Clean tank exteriors
- Check temperature control systems
- Inspect pumps in spray systems
- Inspect spray nozzles, and replace if necessary

Semiannually

- Clean heating coils and exhaust hoods
- Clean and paint exterior components
- Clean riser scale
- Dismantle and repair pumps

Waste Disposal

Disposal of waste acid cleaners is a problem, regardless of whether the location is urban or rural. Several federal, state, and local groups regulate waste disposal. Laws and regulations, such as the Federal Resource Conservation and Recovery Act of 1976, as amended, are subject to change. Therefore, local authorities should be consulted about proposed and current operations.

Safety Precautions

Acids, even in dilute form, can cause serious injuries to the eyes and other portions of the body. Acids are destructive to clothing as well. Therefore operators should be protected with face shields and rubber boots and aprons. Eye fountains and showers adjacent to acid cleaning operations should be provided for use in case of accidents. Nonslip floor coverings in the vicinity of tanks or spray operations are also advised.

Precautions must be taken against cyanides entering the acid cleaning system to avoid formation of deadly gas.

Electrolytic cleaning systems are potentially dangerous because of splashing; therefore, rubber shoes and gloves are necessary to protect operators working near these installations. Electric power at 5 to 15 V is not hazardous to operators.

Mist from spray systems or from gassing can be a health hazard. Mist formation increases with the amount of work in process, the temperature, the acidity of the solution, and the current density in electrolytic cleaning. This mist contains all the ingredients of the acid solution. Adequate ventilation is important. Additional information concerning hazards in the use and disposal of acids is given in the article on pickling of iron and steel in this Volume.

Health and safety regulations are made and enforced by several groups within the federal, state, and local governments. Since the regulations vary and are subject to change, the several sets of regulations should be considered when planning an installation or major changes in operations.

Organic Acid Cleaning*

Organic acids are presently used in a variety of metal cleaning applications. Primary organic acids used in metal cleaning include acetic acid, citric acid, EDTA, formic acid, gluconic acid, and hydroxyacetic acid. Depending on the application, acids may be used alone, but often are formulated with bases and other additives. Organic acids often replace mineral acids, such as hydrochloric and sulfuric acid, in many metal cleaning applications. Advantages in using organic acids include:

- Efficiency in removing certain metal oxides
- Low corrosivity to base metal
- Safety and ease of handling
- Ease of disposal

Disadvantages of organic acids include longer cleaning times, higher temperature requirements, and higher costs compared to other cleaning operations.

Advantages of Organic Acids

Although organic acids are relatively weak, they remove metal oxides through the following mechanisms. As the organic acid reacts with the metal to produce citrates, acetates and other by-products, hydrogen gas is released. The hydrogen builds up under the scale and can often lift the remaining oxides off the metal. In addition, organic acids act as sequesterants by tieing up the dissolved metal ions and carrying them away from the surface being cleaned. With the use of heated solutions and proper circulation of cleaning solution, organic acids efficiently remove metal oxides.

Low corrosivity to the cleaned metal surface is another important reason for choosing an organic acid over a mineral acid. Mineral acids have high corrosion rates, and repeated cleanings with these solvents can significantly corrode fabricated metal parts. The low corrosion rates of organic acids can be reduced further with the use of corrosion inhibitors. In addition, the sequestering ability of the organic acids allows cleaning at a higher pH, reducing corrosion rates even further. The weak acidic nature of most organic acids and the use of a higher pH than mineral acid-based processes provide for safe, easy-to-handle compositions. The cleaning solutions can be used with hand-held steam and high-pressure spray equipment with little hazard to the operator. Most of the organic acids are nonvolatile; therefore, harmful vapors are not released during the cleaning operation.

Spent organic acid cleaning solutions can be disposed of with relative ease. A variety of methods, such as biodegradation, chemical treatment, and incineration, are being used for disposal of organic acid-based cleaning solutions. Spent solutions can be regenerated with techniques such as ion exchange, electrodialysis, and reduction of metal ions with reducing agents.

*By James D. Winchester, Technical Service Representative, Industrial Products Technical Service, Pfizer, Inc., and H.L. Gewanter, Manager, Industrial Products Technical Service, Pfizer, Inc.

Applications
Boiler Cleaning

A patented process (Ref 1) removes boiler deposits containing iron oxides, copper oxides, and copper metal with a single filling solution. For iron and copper oxide removal, a 3 to 5% citric acid solution is treated with sufficient ammonia to achieve a pH of 3.5. The boiler to be cleaned is filled with this solution, heated to 93 °C (200 °F), and the solution is circulated until iron oxide removal is complete. The progress of the iron removal operation is monitored analytically until the iron removal rate levels off. Any copper oxides present are rapidly dissolved in the low-pH citric acid solution; however, dissolved copper ions tend to plate out on the cleansed steel.

This plated copper is removed during the second stage of the cleaning process, which also results in a passive metal surface. The second stage cleaning solution is prepared by ammoniating the same filling solution to pH 9.5 and allowing the temperature to drop to 49 °C (120 °F). An oxidant, such as sodium nitrite at a level of 0.25 to 0.5% of the solution weight, is added to oxidize ferrous ions to ferric ions, which are responsible for dissolving the plated copper according to the following equation:

$$2\,Fe^{+3} + Cu^{\circ} \rightarrow 2\,Fe^{+2} + Cu^{+2}$$

The dissolved copper is stabilized as the copper-ammonium complex, $Cu\,(NH_3)_4^{+2}$. The high-pH solution is also responsible for producing a film of hydrated iron oxide, which results in a passivated surface that remains rust free while the citrate solution is removed and the boiler is rinsed with water. The unit is then ready to be placed back into service.

Stainless Steel Cleaning

Some of the uses for organic acids in the cleaning and finishing of stainless steels are presented below.

Acid Cleaning. Organic acid solutions are used to remove rust and mill scale from newly fabricated stainless steel stock. By removing embedded iron and scale from the stainless steel surface, appearance and corrosion resistance of the alloy is restored. A typical formulation for this application consists of 5% dibasic ammonium citrate containing 0.1% wetting agent at a temperature of 82 °C (180 °F). This solution finds particular use in cleaning equipment for storage and manufacture of foods, beverages, fine chemicals, and pharmaceuticals.

Steam Cleaning. A particularly useful technique for cleaning these types of fabricated stainless steel tanks, as well as stainless steel machinery, trucks, and railroad cars, involves steam cleaning. A concentrated organic acid solution is injected into a high-pressure jet of steam at a rate that yields 1 to 5% concentration by weight in the superheated solutions. A low-foaming nonionic wetting agent added to the acid solution removes oil and grease from the steel surfaces.

Alkaline Cleaning. Caustic gluconate solutions, prepared by dissolving gluconic acid or sodium gluconate in caustic soda, are useful for removing both organic soils and metal oxides with one solution. Also, because the solution is on the alkaline side, the cleaned metal surface has little tendency to rerust (Ref 2).

Nuclear Power Plant Decontamination. Oxidation products of alloys used in nuclear power plant construction must be dissolved and flushed out of the unit. Because these oxidation products often contain radioactive materials, solvent and rinse waters require care in disposal. The following considerations are important for proper disposal of waste material. Solvent volumes should be as low as possible; the solvent should be compatible with different waste disposal methods, and quantitative stabilization of the radioactive materials in solutions should be maintained throughout solvent transfer and sampling. Among the cleaning methods employed, most involve an oxidizing pretreatment with alkaline permanganate (AP) followed by a chelant removal of the deposit. Among the chelant treatments are the following:

- Alkaline permanganate-ammoniated citric acid (APAC)—citric acid, 5 to 10%, ammoniated to pH 5 to 7
- Alkaline permanganate-ammoniated citric acid-EDTA (APACE)—citric acid, 2%; dibasic ammonium citrate, 5%; disodium EDTA, 0.5%

Additional benefits of using organic acids in stainless steel cleaning solutions include: (*a*) they are chloride free, which eliminates the problem of chloride-stress cracking, and (*b*) their weakly acidic nature reduces the chances of hydrogen embrittlement (Ref 3). Additional applications include cleaning lube oil systems, heat exchanger surfaces, pendant superheaters and reheaters, and startup and operational cleaning of once-through boilers (Ref 4). Two new applications for organic acids have been developed.

Removal of Iron- and Copper-Bearing Deposits

A citric acid-based cleaning method is used to derust the steel shells of heat exchangers containing a high ratio of copper to iron, such as is found in marine air conditioning units. When the fluorocarbon refrigerant becomes contaminated with small amounts of water, corrosive hydrochloric and hydrofluoric acids are formed, causing significant corrosion of the steel shells. These corrosion products must be removed to restore the unit to its proper functioning.

Standard organic acid cleaning techniques are inadequate in this application due to the large amounts of copper oxides present, which are more easily dissolved than the iron oxide and tend to consume the organic acid before the iron oxides can be removed. To overcome this problem, a citrate-based cleaning formulation is modified to contain a reducing agent, which reduces the dissolved copper oxides to precipitated copper metal, which is filtered from the solution. This precipitated copper removal restores the citric acid content of the solution, making it available to dissolve the iron oxides. Any copper metal residue remaining in the system from the first solution is removed in a second step, which is also a citric acid-based solution. The second step also passivates the steel surfaces. The specific formulas for step one and step two are:

- **Step 1: iron and copper oxide removal**
  3% citric acid, 3% erythorbic acid, pH adjusted to 3.5 with triethanolamine (replaces ammonia which is corrosive to copper)
- **Step 2: copper metal removal and passivation**
  A second solution is prepared as follows: 3% trisodium citrate, 1.2% triethanolamine, 1% sodium nitrite (Ref 5). It is important that the order of additions to be followed precisely to avoid toxic nitrogen oxide gas generation

Another new application is the use of an EDTA-based solution to dissolve iron- and copper-bearing deposits from pressurized water reactor nuclear power plants. In pressurized water reactor nuclear power plant steam generators, the accumulation of secondary side corrosion deposits and impurities forms sludges that are composed primarily of metal oxides and metallic copper deposits. To remove these deposits, the following solution has been found effective:

Iron solvent

- 10% EDTA
- 1% hydrazine
- Ammonium hydroxide to pH 7
- 0.5% inhibitor CCI-80/1 applied at 90 to 120 °C (195 to 250 °F)

Copper solvent

- 5% EDTA
- Ammonium hydroxide to pH 7
- EDA (ethylenediamine) to pH 9.5 to 10.0
- 2 to 3% hydrogen peroxide applied at 32 to 43 °C (90 to 110 °F) (Ref 6)

REFERENCES

1 U.S. Patent No. 3,072,502, Process for removing copper-containing iron oxide scale from metal surfaces, Salvatore Alfano, Houston, TX, Chas. Pfizer and Co., Inc., Brooklyn, NY

2 W.J. Blume, Role of Organic Acids in Cleaning Stainless Steels, American Society for Testing and Materials, Special Technical Publication No. 538, 1973, p 43-53

3 Data Sheet No. 672, Chemical cleaning with citric acid solutions, Pfizer, Inc., 1981, p 13-14

4 A.H. Roebuck, Safe Chemical Cleaning—the Organic Way, Chemical Engineering, 31 July, 1978, p 107-110

5 D.R. Uhr, Jr., Citric Acid-Based Cleaning of Mixed Metal Systems, Paper No. 217, *Corrosion 80,* 3-7 March, 1980

6 D.J. Stiteler, *et al,* A Chemical Cleaning Process to Remove Deposits from Nuclear Steam Generators, Paper No. 32, *Corrosion 82,* 22-62 March, 1982

Pickling of Iron and Steel

By the ASM Committee on
Pickling of Iron and Steel*

PICKLING is the chemical removal of surface oxides (scale) and other contaminants such as dirt from metal by immersion in an aqueous acid solution. Wide variations are possible in the type, strength, and temperature of the acid solutions used. Because of its economy and adaptability to continuous operations, pickling is the most efficient method of scale removal for large-tonnage products such as merchant bar, blooms, billets, sheet, strip, wire, and tubing. Pickling is also applicable to many types of forgings and castings. A continuous pickling operation is shown in Fig. 1.

Pickling Solutions

Sulfuric acid is the most common pickling acid. Hydrochloric acid is used for many steels, particularly for special purposes, such as etching before galvanizing or tinning. Nitric-hydrofluoric acid is used in pickling stainless steel. Hydrofluoric acid is sometimes used to accelerate pickling baths and is used occasionally in pickling castings to remove sand.

Pickling is generally associated with the removal of scale with mineral acids. The general mechanism of removal is the penetration of acid through cracks in the scale and the reaction of the acid with the metal, which generates hydrogen gas. As the hydrogen gas pressure increases, a point is reached where the scale is blown off the metal surface. The pertinent reactions most commonly used for pickling with sulfuric (H_2SO_4) and hydrochloric (HCl) acids are:

$$H_2SO_4 + Fe \rightleftharpoons FeSO_4 + H_2\ (g)$$

$$2HCl + Fe \rightleftharpoons FeCl_2 + H_2\ (g)$$

Sulfuric acid, despite certain drawbacks, is still the most common pickling liquor. It produces satisfactory results when used (*a*) for batch descaling of carbon steel rod and wire (to 0.60% carbon), and (*b*) for continuous cleaning, provided the iron concentration in the bath is less than 8 wt %. Table 1 lists the types of carbon and alloy steel products that are pickled in sulfuric acid and also lists the ranges of acid concentrations and temperatures used.

In comparison to hydrochloric acid, sulfuric acid offers the advantages of lower cost, less fumes, and less volume of acid to handle. In tank-car quantities, sulfuric acid and hydrochloric acid cost about the same. Commercial sulfuric acid is usually supplied at a concentration of 96%, and hydrochloric acid is supplied at a concentration of 31%. Disadvantages of sulfuric acid are:

- Darker surfaces and smut produced, particularly on high-carbon steel
- Greater inhibiting effect on sulfuric acid of iron salts in the bath
- Higher heating costs, because of the higher operating temperatures of sulfuric acid baths

Hydrochloric acid is preferred for batch pickling of hot rolled or heat treated high-carbon steel rod and wire. This acid produces a uniform light gray surface and decreases the possibility of overpickling, which produces a black smut on the surface. Continuous pickling operations also use hydrochloric acid for producing the very uniform surface characteristics required for both low- and high-carbon steel. The surface requirements are obtained by pickling the heat treated steel for short times (1 to 20 s) in hot solutions of concentrated acid (6 to 14 wt %). The possibility of overpickling is eliminated in these short-time operations, which often precede the deposition of a metallic coating. The acid also dissolves lead oxides adhering to steel previously heat treated in molten lead baths.

Operating conditions for batch and continuous pickling in hydrochloric acid solutions are as follows:

*Robert M. Hudson, Research Consultant, Research Laboratory, U.S. Steel Corp.; Ronald J. Joniec, General Manager, Harry Miller Corp.; Stephen R. Shatynski, Professor, Rensselaer Polytechnic Institute

Fig. 1 Schematic diagram of a modern continuous strip pickling line

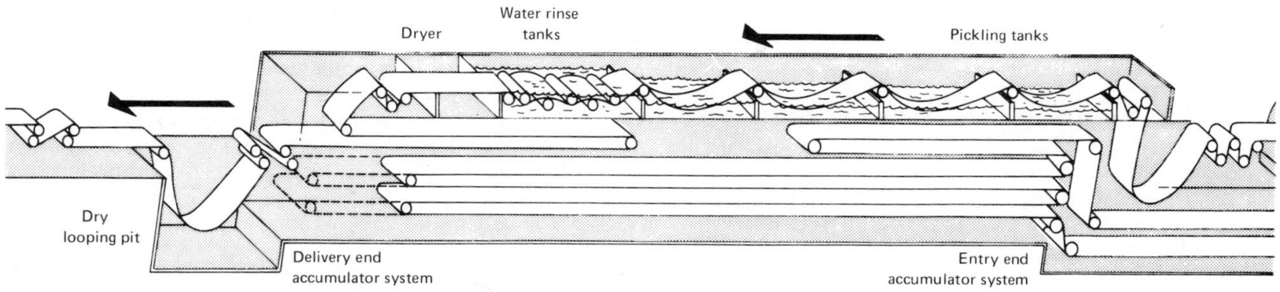

Operating variable	Batch	Continuous
Hydrochloric acid, wt %	8-12	6-14
Temperature, °C (°F)	38-40 (100-105)	77-93 (170-200)
Immersion time	5-15 min	1-20 s
Iron concentration(a), wt %	13	13

(a) Maximum allowable in bath

Hydrochloric acid offers the following advantages over sulfuric and other acids:

- It consistently produces a uniform light gray surface on high-carbon steel
- The possibility of overpickling is less than for other acids
- Effective pickling can be obtained with iron concentrations up to 13%
- Rinsing is easy, because of high solubility of chlorides
- Subsequent electroplated coatings are uniform and adherent
- It is safer to handle than sulfuric acid, although protective equipment is required
- Cost of heating the bath for batch-type operations is less because of lower operating temperatures

The chief disadvantage of hydrochloric acid is the necessity for a fume-control system.

Other Acid Mixtures. The addition of a suitable wetting agent to acid solutions is beneficial for cleaning steel that has been oiled. Continued use of this solution for removing oil and for pickling causes contamination of the bath and results in staining of the steel and nonuniform descaling. If a sufficient quantity of oiled steel is to be pickled, the material should be degreased before pickling. When pickling degreased steel, the use of a wetting agent in the acid solution increases the effectiveness and efficiency of the bath, thereby reducing immersion time.

Annealing smut and heavy metal ions can be removed from the surface of steel to be cold drawn, porcelain enameled, or tin plated by adding sodium ferrocyanide to the acid pickling solution. A solution of sulfuric and hydrofluoric acids can be used to pickle castings that have burned sand embedded in the surface. The relative concentration of each acid is determined by whether the primary objective is the removal of sand or the removal of scale.

Determining Concentrations of Acid and Iron. Analytical equipment should be installed near the pickling operation for monitoring acid concentration and to test the solution using chemical analysis at regular intervals. Testing after each acid addition and once or twice during an 8-h shift is usually sufficient.

Another procedure for determining the concentration of the acid consists of adding 75 to 100 mL of distilled water and 2 or 3 drops of methyl orange indicator to a 5-mL sample of the solution and titrating from a burette containing a 1.02*N* sodium carbonate solution until the red color changes to yellow. The sample solution must be agitated during titration. Each millilitre of sodium carbonate solution is equivalent to 1 g of acid per 100 mL of pickling solution. This relation may be used as a control test. The concentration of sulfuric acid is determined by the formula:

$$\text{wt \% } H_2SO_4 = \frac{\text{mL } Na_2CO_3 \times N \times 0.049 \times 100}{\text{mL sample} \times \text{specific gravity}}$$

where N is the normality of the sodium carbonate solution and specific gravity is that of the pickling solution being tested. The above procedure is applica-

Table 1 Solution concentrations and operating temperatures used for pickling carbon and alloy steel products

Product	Sulfuric acid concentration, wt% min	max	Bath temperature min °C	°F	max °C	°F
Bar, low-carbon	7	15	68	155	85	185
Bar, alloy	9	12	66	150	77	170
Billet, low-carbon	7	12	74	165	82	180
Billet, alloy	9	12	82	180	93	200
Pipe for galvanizing	7	15	71	160	88	190
Sheet for galvanizing	4	12	66	150	77	170
Sheet, tin plate (white pickle)	9	12	66	150	85	185
Strip, soft	6	12	77	170	88	190
Strip, alloy and high-carbon	7	12	66	150	77	170
Strip, continuous pickling	23	38	77	170	100	212
Tubing, low-carbon seamless	7	12	77	170	88	190
Tubing, high-carbon and alloy structural	9	12	71	160	93	200
Tubing (over 0.40% carbon)	9	12	60	140	71	160
Wire, soft	4	11	77	170	88	190
Wire, alloy and high-carbon	3	7	63	145	74	165
Fabricated parts (for tinning):						
Initial pickle	5	10	66	150	88	190
Final dip	(a)	(a)	...	...	38	100

(a) Concentrated hydrochloric acid, 1.14 to 1.16 sp gr

ble also to hydrochloric acid solutions; the formula is multiplied by 0.7449 to obtain the percentage of hydrochloric acid by weight.

To determine the concentration of iron in the pickling solution, a 1-mL sample is diluted to about 25 mL with distilled water, then 5 mL of a 50% solution of sulfuric acid and 5 drops of diphenylamine indicator are added. This mixture is titrated with standard potassium dichromate solution to a permanent deep blue end point. The titrating solution contains 8.9 g of potassium dichromate per litre and is standardized with pure iron so that 1 mL equals 0.01 g of iron; thus, 1 mL of titrating solution equals 1% iron for a 1-mL sample of pickling solution. A simpler, but only approximate, determination of iron content consists of hydrometer measurement of the solution (1% iron = 3.5° Baumé).

Plants frequently report acid and iron-salt concentrations in grams per 100 mL (g/100 mL). Although the units are sometimes referred to as percent, concentrations in g/100mL must be divided by the density of the solution in grams per millilitre (g/mL) to convert to true weight percent. For this purpose, approximate equations for calculating densities have been developed from published data on sulfuric acid-iron sulfate and on hydrochloric acid-iron chloride solutions:

$$D = 0.9971 + (6.33 \times 10^{-3})\, C_{H_2SO_4} + (9.90 \times 10^{-3})\, C_{FeSO_4}$$

$$D = 0.9971 + (4.46 \times 10^{-3})\, C_{HCl} + (8.15 \times 10^{-3})\, C_{FeCl_2}$$

where D is expressed in g/mL at 25 °C (77 °F) and $C_{H_2SO_4}$, C_{FeSO_4}, C_{HCl}, and C_{FeCl_2} are the concentrations expressed in g/100 mL.

Inhibitors

Inhibitors are added to acid pickling solutions primarily to protect the steel being cleaned by retarding or stopping the chemical action of the acid on the base metal. Inhibitors (*a*) minimize the loss of iron, (*b*) reduce the extent of hydrogen embrittlement, (*c*) protect the metal against pitting (caused by overpickling) and poor surface quality, (*d*) reduce acid fumes resulting from excessive reaction between the acid and base metal, and (*e*) reduce acid consumption. In production operations, inhibitors do not appreciably affect the rate of scale or rust removal.

Both natural and synthetic organic compounds are used as inhibitors. Natural compounds include low-grade bran flour, gelatin, glue, sludge from petroleum, animal wastes, sulfonated coal-tar products, asphaltum, and wood tars. The synthetic materials are nitrogen-based materials and their derivatives, pyridines, quinidines, aldehydes and thicaldehydes, and other sulfur-containing compounds.

In sulfuric acid pickling, the accumulation of iron sulfates in the acid solution also inhibits the activity of the acid and therefore reduces the effectiveness of the solution for cleaning and brightening the steel. Acid inhibitors should be stable at all possible operating temperatures and conditions and must not decompose in solution, stain, or contaminate the steel. An adequate inhibitor keeps acid and maintenance costs down and does not emit offensive odors. Most pickling operations involving steels that are reactive with acid require inhibited acid solutions. These solutions are also used in oil well drilling operations to clean the internal surfaces of pipe. They are also often used in continuous strip lines.

For plain carbon steels containing less than 0.40% carbon, a strong inhibitor should be used in pickling baths containing 10 to 14 wt % sulfuric acid (1.82 sp gr) and operating at 71 °C (160 °F) or higher. When the concentration of iron sulfate reaches 30%, the solution should be discarded, because the iron sulfate inhibits pickling and often causes smut to form on the surface of the product. Plain carbon steels containing 0.40% carbon or more are pickled in solutions of the same sulfuric acid and inhibitor content as for lower carbon steels. The operating temperature should be 60 to 66 °C (140 to 150 °F), and the concentration of iron sulfate should be less than 20%. More or less inhibitor can be used in any particular operation where time and speed of handling dictate pickling cycles.

Uninhibited acid solutions are more desirable for pickling high-alloy steels, because more chemical action is required to remove the oxide. Alloy or plain carbon steels for forming flat and shaped sections must be etched by uninhibited acid solutions to produce a surface that retains the die lubricant during cold working. These solutions are used also for conditioning steel with slivers and sharp corners and steel that is ground before coating for additional cold working.

If an inhibitor is required when pickling alloy steels, a concentration of 0.15 vol %, based on the volume of concentrated sulfuric acid (1.83 sp gr), is recommended. This solution should also contain a sodium chloride pickling compound (120 g/L or 16 oz/gal of 1.83 sp gr sulfuric acid). The sodium chloride addition increases the ionic concentration. Inhibitors are added to the bath at the time of acid additions and in the same proportion as in the original prepared solution. After the iron compounds build up to a level where they become effective as inhibitors, the addition of acid inhibitors is no longer required.

The efficiency of inhibitors is determined by comparing the loss of weight test on steel specimens or the hydrogen evolution test against uninhibited acid of equal strength. The loss of weight test is usually used, and by using scaled metal specimens, an indication of descaling rates can be determined. Subsequent loss of weight figures on base metal can be converted to acid consumption per unit area to provide an indication of process efficiency and pickling costs.

Precleaning

Alkaline precleaning before acid pickling is beneficial for removing soils that do not readily react with acid. These soils include grease, oil, soaps, lubricants, and carrier coatings. If such materials are carbonized by exposure to heat, they become more difficult to remove and usually contribute to the formation of smut on the surface of the metal during the pickling operation.

Precleaning is especially advantageous in applications where the acid pickling time is limited to 1 to 20 s, as in the continuous pickling of wire or controlled high-speed batch pickling processes. It is also used for cleaning materials that are to be electroplated, hot metal dipped, or painted following acid pickling. Precleaning is not required when the soil consists only of rust or scale and pickling time is not critical.

An alkaline cleaner can be composed of 20% sodium hydroxide, 30% organic chelating agents, 45% complex phosphates, and 5% surface-activating agents. The concentration of the cleaner in the precleaning solution is 30 to 45 g/L (4 to 6 oz/gal), and the operating temperature of the solution ranges from 82 °C (180 °F) to boiling. Immersion time varies from 1 to 20 s, after which the work is rinsed in water at room temperature. Generally, 3.8 L (1 gal) of alkaline solution with the above concentration cleans about 93 m^2 (1000 ft^2) of surface.

Chelating Agents. Frequently, the drawing lubricants used in cold working operations contain compounds of calcium, zinc, magnesium, iron, or some other metal. These lubricants are difficult to remove, but can be cleaned from the work with alkaline cleaners containing the proper chelating agent. Chelates improve and accelerate cleaning by reacting with the metal ion in the lubricant to form a metal chelate. For example, a calcium stearate lubricant is only slightly soluble in an alkaline solution, but when the proper chelate is present, the calcium ions that go into solution combine with the chelate to form metal chelate molecules. As the calcium ions are sequestered by the chelate, more calcium stearate dissolves and the cleaning process progresses rapidly to completion.

Chelates used in alkaline cleaning solutions include citric acid, ethylenediamine tetraacetic acid (EDTA), gluconic acid, and nitrilotriacetic acid. Sodium salts are compounded with each of these chelates to produce a dry, concentrated cleaning powder. The cleaning solution is produced by dissolving the powder in water.

The following conditions control the use of chelating agents:

- The pH of the cleaning solution should remain basic if possible
- Although all chelating agents react in a similar manner, certain chelates have greater affinity for specific metal ions, for example, the stearates with calcium
- The solution must contain an adequate amount of chelate and be of proper pH for effective cleaning
- Chelates must be stable with respect to the particular environment (oxidizing or reducing, acid or alkaline)

A procedure for determining the chelating power of the alkaline cleaning formula consists of mixing 10 mL of filtered cleaning solution with 10 mL of distilled water and 10 drops of saturated 5% ammonium oxalate solution. This mixture is then titrated carefully with a 2% solution of calcium chloride until one drop of the latter produces a faint, permanent turbidity. The number of millilitres of 2% calcium chloride required is equivalent to the chelating power. This procedure is being replaced with more modern analytical techniques. Infrared spectroscopy is now commonly used to monitor the alkaline cleaning formulation.

Carbon Steel

By far, the greatest tonnage of steel pickled is carbon steel. Because of the limited alloy additions, the oxides are simply iron oxides, which are easier to clean than the more complex oxides resulting from alloying additions to the steel. Carbon steel may be pickled by either batch or continuous methods. Batch processing generally involves lower acid cost than continuous processing, but the latter provides more uniformly clean surfaces.

Batch Pickling. Steel rod ranging from 5-mm (0.200-in.) to over 17-mm (0.675-in.) diam and in coils weighing 136 kg (300 lb) to more than 454 kg (1000 lb) is batch pickled, where from one to ten coils can be handled at a time. Steel wire is pickled in a similar manner, but as the diameter of the wire decreases, particularly below 3 mm (0.100 in.), the packing effect of the wire makes it difficult for the acid solution, usually sulfuric, to penetrate and pass through the coils. In this case, the penetration of the acid is particularly difficult because overpickling can easily occur in the outer coil. This problem is usually resolved by distributing a lesser weight of wire per coil, thus providing easier access of the acid to the inner coil. For example, if satisfactory pickling of 3-mm (0.1-in.) diam wire is obtained when the weight of the coil is 227 kg (500 lb), the batch weight of 1.3-mm (0.050-in.) diam wire should be about 113 kg (250 lb) for satisfactory results, and for 0.64-mm (0.025-in.) diam wire, the batch weight may be only 57 kg (125 lb).

To induce better cleaning, coils may be dropped (bounced) from several positions after rinsing, then pickled again. Other techniques include separating the layers by hand, straightening, and rewrapping. Wetting agents are also useful, but the coils still are usually washed, rolled, and bounced at least once during the pickling operation. Bar flats must be held apart with separators to improve solution contact. Patented rod (0.40 to 0.80% carbon) is usually covered with heavy scale that can be removed by pickling in sulfuric acid. The following example describes the cleaning operation in one plant.

One operation for pickling patented rod consists of loading 1362 to 1816 kg (3000 to 4000 lb) of material into a tank containing 19 680 L (5200 gal) of sulfuric acid solution, pickling for 12 to 15 min, rinsing in a tank overflowing with cold water, spray rinsing, coating with borax or lime, and drying in an oven. Halfway through the pickling cycle, the material is inspected and is turned over to permit thorough contact with the solution.

The initial makeup solution contains 8 vol % sulfuric acid (1.83 sp gr), and the bath is operated at 60 °C (140 °F) until the concentration of the acid drops to about 4 vol %. Acid is then added to increase the concentration to about 7%, and the bath is operated until the concentration decreases to about 3½ vol %. As the concentration decreases from 7 to 3½ %, the temperature is increased gradually to about 68 °C (155 °F). An inhibitor is used and is maintained at a concentration of ½ wt % of the sulfuric acid present. When the iron content of the solution attains a value of 6 %, the solution is discarded.

Continuous pickling of steel rod and wire consists of passing the products through an acid tank, which usually follows the heat treating furnace in a continuous heat treating and cleaning line. This method eliminates the separate cleaninghouse operation and improves the degree of cleanness of the product.

The cost of acid for continuous pickling is greater than that for batch pickling, because to accomplish complete descaling in the limited time available, the maximum permissible iron concentration in sulfuric acid (4 to 8%) is lower in the continuous operation. The increased acid cost, however, is more than offset by the advantage of uniformly clean surfaces. Continuous pickling is used in conjunction with continuous galvanizing of steel wire, as in the following example. In one plant, the sequence of operations for cleaning and galvanizing 0.40 to 0.80% carbon steel wire is as follows:

- Pickling in a 10 to 15 vol % solution of sulfuric acid (1.83 sp gr) at 71 to 82 °C (160 to 180 °F)
- Spray rinsing with cold water
- Pickling in a 40 to 50 vol % solution of hydrochloric acid (1.16 sp gr) at 54 to 60 °C (130 to 140 °F)
- Spray rinsing with cold water
- Fluxing and drying
- Dipping molten zinc

Immersion time in each of the pickling solutions ranges from 16 to 32 s; neither solution contains an inhibitor. The wire is usually bright drawn stock that has a light scale of lead oxide obtained from the molten lead bath following drawing. The pickling procedure can be used also for slightly rusted wire and wire with patenting scale.

Cast Iron and Cast Steel

Most iron and steel castings are cleaned by mechanical methods, such as shot blasting, sand blasting, and tumbling. When pickling is used, the castings are cleaned in solutions containing sulfuric and hydrofluoric acids. The concentration of each acid depends on whether the primary purpose is to remove sand or scale. Increased hydrofluoric acid is needed to remove embedded sand from the casting surface, while sulfuric or hydrochloric acids are sufficient for simple scale removal. Table 2 gives the operating conditions for pickling iron and steel castings. Before being pickled, castings must be free of oil, grease, and other contamination.

After being removed from the pickling solution, castings are rinsed thoroughly in hot water, dipped in a 10% sodium hydroxide solution at 93 °C (200 °F) to neutralize any residual acid, and rinsed again in hot water. Residual heat permits self-drying, but drying may be accelerated by using a fan. If the shape of the castings hinders drying, the process can be completed in a baking oven.

Stainless Steel

Stainless steel products should be free of oil, grease, or any other soil that rapidly contaminates the pickling solution. Common precleaners include alkaline cleaners and degreasers. Heating the steel to some temperature under 540 °C (1005 °F) to burn off contaminants such as light oils also may be used as a cleaning method before pickling. However, some austenitic stainless steels are sensitized when heated at 400 to 900 °C (750 to 1650 °F), and these steels, when sensitized, may undergo intergranular corrosion. If alkaline precleaning is used, adequate rinsing is required before pickling.

Surface contaminants such as oil, grease, and lubricants should be removed before an annealing operation; otherwise, the contaminants may stain or adversely affect the surface of the stainless steel. Precleaning is not required if oxide or scale is the only soil on the surface. Forgings and castings usually are not pickled, but forgings can be pickled as an inspection procedure to determine the presence of surface defects. If pickling is necessary, castings and forgings are pickled by the procedures used for rolled forms of stainless steel.

In many cases, special pickling procedures may be needed to adequately clean the 300 and 400 series stainless steels. A molten salt bath may be used to pretreat the part before pickling in acid. Other processes rely on the use of electrolytic sodium sulfate solutions before pickling.

Pickling cycles for 300 series austenitic stainless steels are given in Table 3. The lower immersion time values are for pickling the lower alloy steels, whereas the upper values are for the more highly alloyed steels, such as types 309, 310, 316, 317, and 318. When pickling a large volume of products, the immersion time in the acid solutions can be reduced substantially by prior treatment of the material in a salt descaling bath.

Pickling cycles for the low-carbon 400 series stainless steels are given in Table 4. As with 300 series steels, immersion time can be reduced substantially by prior treatment of steel in a salt descaling bath. The higher carbon grades of 400 series stainless steels are pickled as indicated in Table 4, except that the immersion time in the sulfuric and nitric acid solutions is reduced to 5 to 20 min and 5 to 10 min, respectively.

Precipitation-hardenable stainless steels, such as 17-7 PH, AM-350, and AM-355, may be pickled in the annealed condition in a manner similar to that for the 300 series grades. In the precipitation-hardened condition, the high hardness and the nature of the structure make the steel susceptible to strain cracking during pickling. Therefore, the immersion time in the acid solutions should be as short as possible.

For precipitation-hardenable steels in the fully hardened condition, grit blasting to remove scale, followed by passivation in a 30 to 50% nitric acid solution is recommended. Stainless steels may also be cleaned electrolytically.

Control of Process Variables. Overpickling, underpickling, and pitting usually are the direct results of lack of control over process variables in pickling stainless steel. This can cause permanent damage to the surface of the product, necessitating reprocessing or scrapping. A control program is justified on the basis of the value of the product involved and the inherent hazards of pickling.

Table 2 Operating conditions and solution compositions(a) for pickling iron and steel castings

Operating variable	Sand removal	Scale removal
Sulfuric acid, %	5	7
Hydrofluoric acid, %	5	3
Water, %	90	90
Temperature(b), °C (°F)	66-85 (150-185)	49 to over 85 (120 to over 185)
Average immersion time, h	4	4

(a) Percentages by volume. (b) 49 °C (120 °F) is for slow pickling, 66 to 85 °C (150 to 185 °F) for average pickling speed, and over 85 °C (185 °F) is for fast pickling

The immersion time required to pickle a particular product can best be determined by trial. After immersion time has been determined, it should be maintained for product uniformity. The influence of temperature on pickling time and iron buildup in the solution is pronounced. In 15% sulfuric acid, for example, each increase in temperature of 8 to 11 °C (15 to 20 °F) within the range of 21 to 93 °C (70 to 200 °F) doubles the pickling rate, and the rate of solution of iron at 82 °C (180 °F) is about five times that at 60 °C (140 °F) and about 100 times greater at boiling than at room temperature. Automatic control is required for maintaining temperature within specified limits.

The control of concentration of sulfuric acid pickling baths is important because the rate of pickling increases in direct proportion to the concentration of the acid from 0 to 25 wt %. The frequency of analysis of solutions is usually determined by the size of the pickling installation and the volume of material being processed. For pickling tanks of 11 400- to 15 100-L (3000- to 4000-gal) capacity operating 16 h per day, analysis of the solutions at the beginning of each day is usually adequate for normal handling. However, samples taken from the acid bath throughout the day should be titrated periodically to maintain a closer control over acid concentration.

Table 3 Sequence of procedures for pickling 300 series stainless steels

Cycle	Solution composition(a), vol %	Operating temperature °C	°F	Immersion time(b), min
Sulfuric acid dip	15-25 H_2SO_4(c)	71-82	160-180	30-60
Water rinse(d)	...	Ambient	Ambient	...
Nitric-hydrofluoric acid dip	5-12 HNO_3, 2-4 HF	49 max	120 max	2-20
Water rinse(d)	...	Ambient	Ambient	...
Caustic-permanganate dip(e)	18-20 NaOH, 4-6 $KMnO_4$(f)	71-93	160-200	15-60
Water rinse(d)	...	Ambient	Ambient	...
Sulfuric acid dip	15-25 H_2SO_4(c)	71-82	160-180	2-5
Water rinse(d)	...	Ambient	Ambient	...
Nitric acid dip	10-30 HNO_3	60-82	140-180	5-15
Water rinse (dip)	...	Ambient(g)	Ambient(g)	...

(a) Acid solutions are not inhibited. (b) Shorter times are for lower-alloy steels; longer times are for more highly alloyed types, such as 309, 310, 316, 317, and 318. (c) Sodium chloride (up to 5 wt %) may be added. (d) Dip or pressure spray. (e) Sometimes used to loosen scale. (f) Percent by weight. (g) Boiling water may be used to facilitate drying

Table 4 Sequence for pickling low-carbon 400 series stainless steels

Cycle	Solution composition(a), vol %	Operating temperature °C	°F	Immersion time
Sulfuric acid dip	15-25 H_2SO_4(b)	71-82	160-180	5-30 min
Water rinse(c)	...	Ambient	Ambient	...
Caustic permanganate dip(d)	18-20 NaOH, 4-6 $KMnO_4$(e)	71-93	160-200	20 min to 8 h(f)
Water rinse(c)	...	Ambient	Ambient	...
Sulfuric acid dip	15-25 H_2SO_4(b)	71-82	160-180	2-3 min
Nitric acid dip	30 HNO_3	Ambient	Ambient	10-30 min
Water rinse (dip)	...	Ambient(g)	Ambient(g)	...

(a) Acid solutions are not inhibited. (b) Sodium chloride (up to 5 wt %) may be added. (c) Dip, pressure hose, or spray. High-pressure spray or jets are more effective for removing scale and smut. (d) Sometimes used to loosen scale. (e) Wt %. (f) Immersion time may exceed this range. (g) Boiling water may be used to facilitate drying

The useful life of a pickling solution is determined by the amount of contamination, and a solution is discarded or diverted after specific limits are reached Solutions containing sulfuric acid and salt may either be diverted to the pickling of other material or be discarded when the iron content reaches 5 to 7 wt %. The nitric-hydrofluoric acid solution is discarded when the iron content reaches 5 wt%; the nitric acid solution, 2 wt%. The caustic permanganate solution (Tables 3 and 4) is desludged every three to five months, and makeup material is added to it. The analytical procedures and sampling of these solutions are the same as those described in most standard methods of chemical analysis. Simplified methods using standard solutions are available so that tests can be made by personnel with very little training.

In most installations, rinse tanks are of the overflowing variety. These require little or no attention, because there is a constant flow of fresh water into the tanks. However, when the rinse tanks are not of the overflowing type, frequent monitoring of the rinse water and pH is required to indicate if an appreciable buildup of acid is occurring. This determines how frequently the rinse water should be discarded.

Effects of Process Variables on Scale Removal in Sulfuric Acid

The degree to which pickling efficiency is affected by (*a*) sulfuric acid concentration, temperature, and agitation; (*b*) workpiece temperature; and (*c*) concentrations of ferrous sulfate and inhibitor was determined experimentally in one plant. Tests were made using 50-by-100 mm (2-by-4 in.) samples cut from the center and tail end of a low-carbon drawing quality sheet, 2 mm (0.080 in.) thick after scale breaking. The samples were cleaned, weighed, pickled for 2 min in sulfuric acid solutions which were discarded after pickling two samples, rinsed, dried, and reweighed. The pickling solutions were agitated by a mechanical stirrer at 100 rev/min, except when the effect of agitation was being tested.

Scale thickness was determined microscopically, and measured 0.00475 mm (0.000187 in.) on a center sample and 0.00953 mm (0.000375 in.) on a tail sample. The results of these tests are illustrated by the curves plotted in Fig. 2 through 7 and are discussed below.

Acid Concentration and Temperature. Increasing acid concentration (Fig. 2a) hastened the removal of scale, and the same effect was observed with increasing temperature (Fig. 2b). The most rapid removal occurred in 25% sulfuric acid at 100 °C (212 °F). Figure 3 shows the effects of acid concentration and temperature on weight loss after pickling for 2 min. Any weight loss in excess of the average weight of scale signifies overpickling. In production operations, excessive weight losses of base metal can be limited by the use of an optimum concentration of inhibitor.

Agitation. Increasing the speed of the mechanical stirrer from 100 to 350 rev/min caused an average decrease of 5.7 s (11%) in the pickling time for solutions of 10 to 25% sulfuric acid at 100 °C (212 °F), as shown in Fig. 4.

Preheating the test specimens at 90 °C (194 °F) caused an average decrease of 9.7 s (12%) in the pickling time for solutions of 10 to 25% sulfuric acid at 100 °C (212 °F), as Fig. 4 shows.

Ferrous Sulfate. Pickling efficiency decreases with time, because the reaction of scale and base metal with the solution lowers the acid concentration while proportionally increasing the ferrous sulfate concentration. The effect of ferrous sulfate on pickling time is shown in Fig. 5(a). Based on time required for scale removal, increased amounts of ferrous sulfate in the 5 and 10% sulfuric acid solutions seemed to inhibit pickling action more than in 15, 20, and 25% solutions. The effect of increased amounts of iron sulfate on the percentage weight loss for the 2-min pickling tests is shown in Fig. 5(b). These data indicate that effective scale removal can be accomplished at higher concentrations of ferrous sulfate without seriously affecting pickling efficiency.

Fig. 2 Effect of acid concentration and temperature of acid solution on pickling time required to remove scale from sheet steel

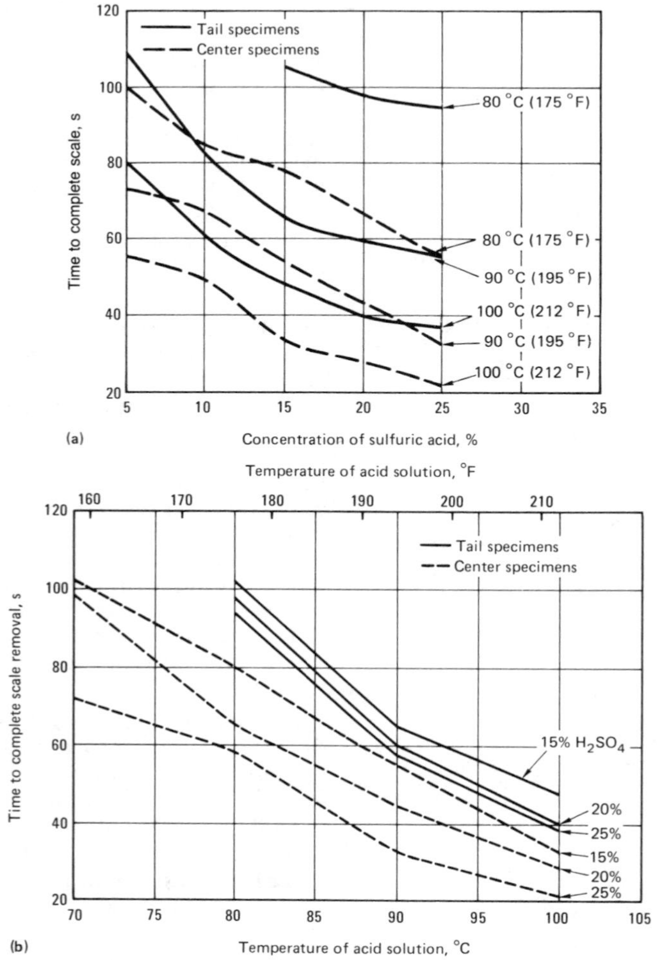

(a) Effect of acid concentration and (b) effect of temperature of acid solution, on the pickling time required to remove scale completely from low-carbon drawing quality sheet, 2 mm (0.080 in.) thick. Specimens taken from center and tail end of sheet and pickled for 2 min

Inhibitor. As shown in Fig. 6(a), increasing concentrations of inhibitor in acid solutions increase the time required for complete scale removal. In the tests, the time increase was greater for the 5 and 10% sulfuric acid solutions than for the 15, 20, and 25% solutions. The effect of inhibitor on weight loss during the 2-min pickling tests is shown in Fig. 6(b). In all the acid solutions tested, as the inhibitor concentration was increased from 0.0031 to 0.125 vol %, weight loss significantly decreased. The optimum concentration of inhibitor in the tests was between 0.02 and 0.064 vol %.

Ferrous Sulfate Plus Inhibitor. The effect of 0.0314 vol % inhibitor with increasing concentrations of ferrous sulfate on the time for complete scale removal in 15 and 20% sulfuric acid solutions is shown in Fig. 7(a). These data indicate that each compound affects the pickling time independently. The combined effect of 0.0314 vol % inhibitor with increasing concentrations of ferrous sulfate on weight loss in 15 and 20% sulfuric acid solutions is shown in Fig. 7(b). Ferrous sulfate had little or no effect on the power of the inhibitor in restricting weight loss to the scale weight of the samples.

The influence of temper mill scalebreaking (breaking up of the scale by imposing moderate room-temperature deformation to the workpiece) on descaling time of hot rolled strip in sulfuric acid solutions is pronounced. Descaling time is frequently about half that required in a given solution without temper mill scalebreaking, as illustrated in Fig. 8. The results of bench-scale experiments with a commercial hot rolled low-carbon steel with a scale weight of 3.4 mg/cm^2 (22 mg/in.^{2}) are also shown in Fig. 8. For nontemper-rolled material, descaling times were decreased as the temperature increased from 82 to 105 °C (180 to 220 °F). The pickling times achieved by increasing the temperature from 93 to 105 °C (200 to 220 °F) were about the same as those that resulted from maintaining the temperature at 93 °C (200 °F) and using temper mill scalebreaking (3%) before pickling.

Effect of Process Variables on Scale Removal in Hydrochloric Acid

The effect of hydrochloric acid and iron chloride concentrations, solution temperature, and temper mill scalebreaking on pickling rates of hot rolled low-carbon steel has been studied in the laboratory. The effect of strip speed on the pickling process was determined by use of an apparatus constructed to simulate the motion of strip through a continuous-pickling line. Steel specimens were mounted on a cylindrical holder which could be rotated through a pickling solution; the solution was contained in a holder provided with baffles to minimize bulk movement of the solution. Most tests were made with specimens that had not been subjected to temper mill scalebreaking, but this variable was studied under selected conditions.

Descaling times for hot rolled low-carbon steel 8005-809 (scale weight 3.6 mg/cm^2 or 23 mg/in.^{2}) were determined for strip velocities from 0 to 4 m/s (0 to 800 ft/min) at solution temperatures from 66 to 93 °C (150 to 200 °F). The pickling solution for this series of tests contained 4 g HCl/100 mL and 22.7 g $FeCl_2$/100 mL. Figure 9 shows that descaling time is decreased with an increase in strip velocity from 0 to about 1.3 m/s (0 to

Fig. 3 Effects of acid concentration and temperature on weight loss of specimens from low-carbon drawing quality sheet steel after pickling for 2 min

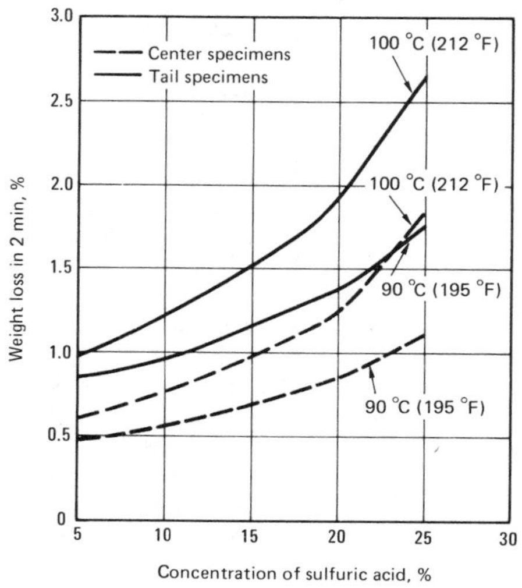

Fig. 4 Effects of increased agitation and of preheating steel to be pickled on the time required for complete removal of scale from low-carbon drawing quality sheet

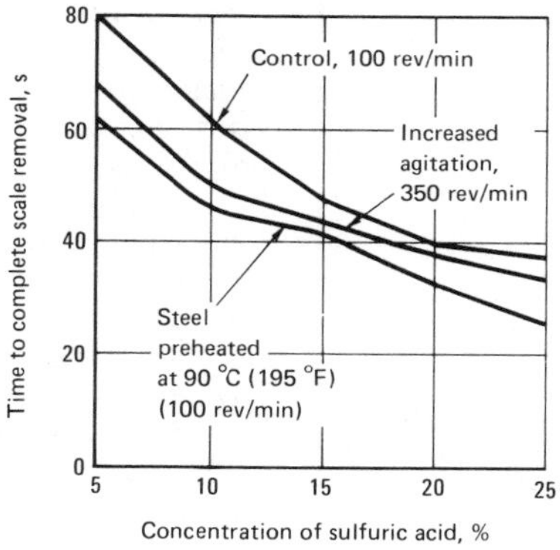

Sheet pickled for 2 min in sulfuric acid at 100 °C (212 °F)

250 ft/min). As strip speeds were increased from 1.3 to 4 m/s (250 to 800 ft/min), there was no further decrease in descaling time. The decrease in pickling time for speeds above 1.3 m/s (250 ft/min) was less by factors of 1.7, 2.2, and 2.7 for tests at 66, 80, and 93 °C (150, 175, and 200 °F), respectively, when compared with pickling time in still baths. If the observed velocity effects for hydrochloric acid are mainly related to the depletion of acid that occurs near the steel surface during pickling in a still bath, it is expected that the velocity effect is not as great for pickling in sulfuric acid because of the higher acid concentrations that are used; this has been verified by experimentation. Because present studies indicate that descaling time in hydrochloric acid does not change for strip velocities above 1.3 m/s (250 ft/min), experiments in which solution composition and other conditions were studied were carried out at 2 m/s (400 ft/min). These results are pertinent to commercial continuous strip-pickling operations in which line speeds can range from 1.5 to 6 m/s (300 to 1200 ft/min).

For both still baths and at a strip velocity of 2 m/s (400 ft/min) the functional relationship of descaling time with hydrochloric acid concentration and temperature was found to be similar. The logarithm of descaling time is a linear function of the logarithm of hydrochloric acid concentration and the reciprocal of absolute temperature of the solution:

$$\log t = A + B \log C_{HCl} + \frac{D}{T_F + 459}$$

where t is descaling time in seconds, C_{HCl} is hydrochloric acid concentration in grams per 100 mL, and T_F is the temperature in degrees Fahrenheit. This relationship is illustrated for steel 8005-809 in a still bath (Fig. 10) and at a strip velocity of 2 m/s (400 ft/min) (Fig. 11). Coefficients A and B are negative and D is positive. Similar plots, shown in Fig. 12, were obtained for other lots of hot rolled low-carbon steel, with differences in response to pickling largely reflected by differences in the coefficient A. Steels 8005-0044 and 8105-843 had scale weights of 3.1 and 4.8 mg/cm^2 (20 and 31 mg/in.^{2}), respectively. Equations of this type are useful for predicting the effect of changes in hydrochloric acid concentration and temperature on pickling rate. Separate studies showed that the influence of iron buildup in hydrochloric acid solutions is negligible until levels of about 34 g $FeCl_2$ per 100 mL are exceeded, a contrast with pickling behavior in sulfuric acid for which increasing iron-salt concentrations to levels much lower than this tends to decrease pickling rates.

Pickling rates in hydrochloric acid solutions at temperatures approaching 93 °C (200 °F) are not appreciably affected by temper mill scalebreaking or the use of inhibitors at moderate concentrations of up to 0.5 vol % based on concentrated hydrochloric acid, about twice the level normally recommended by inhibitor suppliers. At temperatures of 80 °C (175 °F) and below, an increase of not more than 5 s in pickling time may be attributed to inhibitor use. For most steels investigated, pickling times in hydrochloric acid were lowered by not more than about 5 s by use of temper mill scalebreaking (3% extension), an effect of much lower magnitude than that observed for sulfuric acid pickling.

Fig. 5 Inhibiting action of ferrous sulfate on low-carbon drawing quality sheet

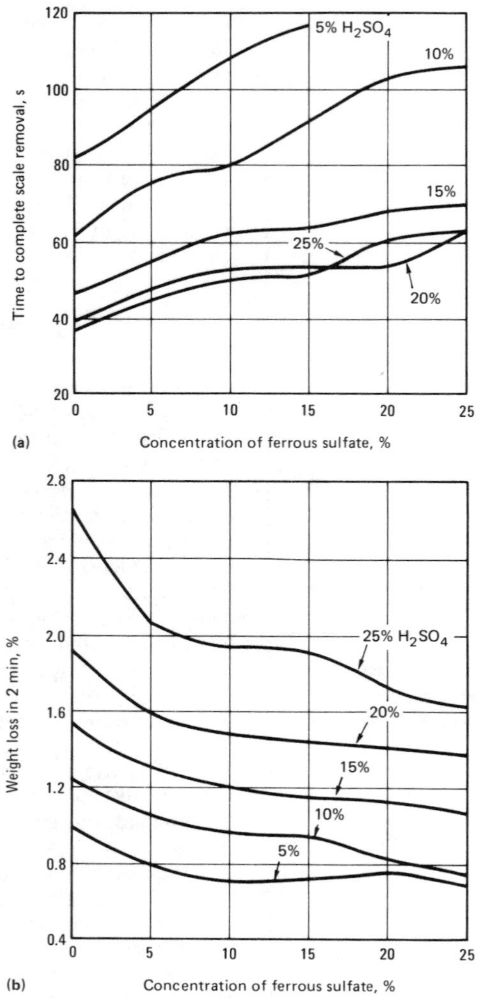

Specimens, taken from the tail end of a low-carbon drawing quality sheet, were pickled for 2 min in sulfuric acid solutions of concentrations indicated. (a) Pickling time for complete scale removal. (b) Weight loss

Although precise measurements of hydrogen chloride vapor pressure above iron-free aqueous solutions of hydrochloric acid are available, more accurate data for solutions that contain iron chloride are needed. One study reports an increase in vapor pressure with increases in temperature and increases in both hydrochloric acid and iron chloride concentration in solution. From the prediction equations just described relating pickling time in hydrochloric acid solutions to acid concentration and temperature, descaling within a given time can be accomplished by using a high hydrochloric acid concentration at a low temperature or a low hydrochloric acid concentration at a high temperature. The latter set of conditions tends to minimize hydrochloric acid fuming because the calculated vapor concentrations of hydrochloric acid above more dilute solutions is lower.

Electrolytic Pickling

Electrolytic pickling of iron and steel is used to overcome some of the difficulties encountered in still pickling. The removal of rust, $Fe(OH)_3$, and iron oxide, Fe_2O_3, is comparatively easy with still pickling methods, but the removal of the black magnetic oxide of iron, Fe_3O_4, which is only slowly soluble in sulfuric acid, is difficult to accomplish without the use of electrolytic pickling methods. In addition, stainless steel scale may also have chromium sesquioxide, Cr_2O_3, and iron chromite, $FeCr_2O_4$, oxides present. These oxides are difficult to remove during still pickling operations.

The black oxidation product of iron, commonly called fire scale or hammer scale, which forms during the heat treatment of iron or steel, is composed of an inner layer of ferrous oxide (FeO) accompanied by a thinner layer of the magnetic iron oxide (Fe_3O_4), and a top, very thin layer of ferric oxide (Fe_2O_3). The relative thicknesses of the various layers in fire scale on hot rolled stock and heat treated parts may vary considerably; if the scale is slowly cooled, the inner layer will be iron and iron oxide (Fe_3O_4) since ferrous oxide (FeO) is no longer stable under 570 °C (1060 °F). The composition of the scale always varies from an iron-rich inner layer at the metal interface to magnetic oxide with a thin layer of oxygen-rich ferric oxide at the surface. The magnetic oxide is attacked only slightly by sulfuric acid, but hydrogen, formed either electrolytically or by the attack of acid on the iron, leads to rapid removal of the scale. The removal of fire scale probably involves these successive reactions: (*a*) an undermining of the residual scale, with some helpful mechanical lifting action being provided by the force of the discharged hydrogen; (*b*) possible partial reduction of the magnetic oxide and the dissolution of the reduction products; (*c*) reduction resulting from the electrolytically formed hydrogen, or, in still pickling, by the action of the acid on the iron; and (*d*) minimal acid attack on the ferrous and ferric oxides.

Equipment

Construction materials for pickling tanks include wood, concrete, brick, plastic, and steel. Acid-resistant linings provide protection for the outer shell of the tank. These are commonly of natural, pure gum or synthetic rubber. Acid-resistant brick linings line the sides and floor of the tank. Figure 13 shows the materials used in construction of a tank of 3.9- to 4.9-t (4- to 5-ton) capacity used for pickling coiled steel.

Construction of Pickling Tanks. Details of tank construction for pick-

Fig. 6 Effect of inhibitor on low-carbon drawing quality sheet

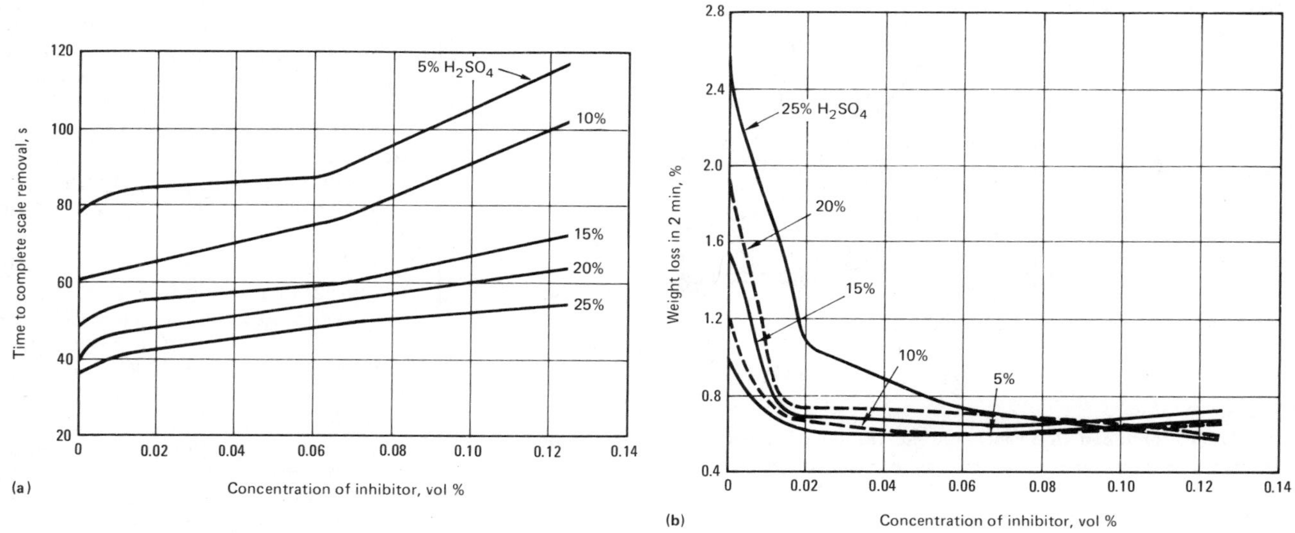

Specimens taken from the tail end of a low-carbon drawing quality sheet were pickled in sulfuric acid solutions at 100 °C (212 °F). (a) Pickling for complete scale removal. (b) Weight loss

Fig. 7 Effect of both inhibitor and ferrous sulfate on low-carbon drawing quality sheet

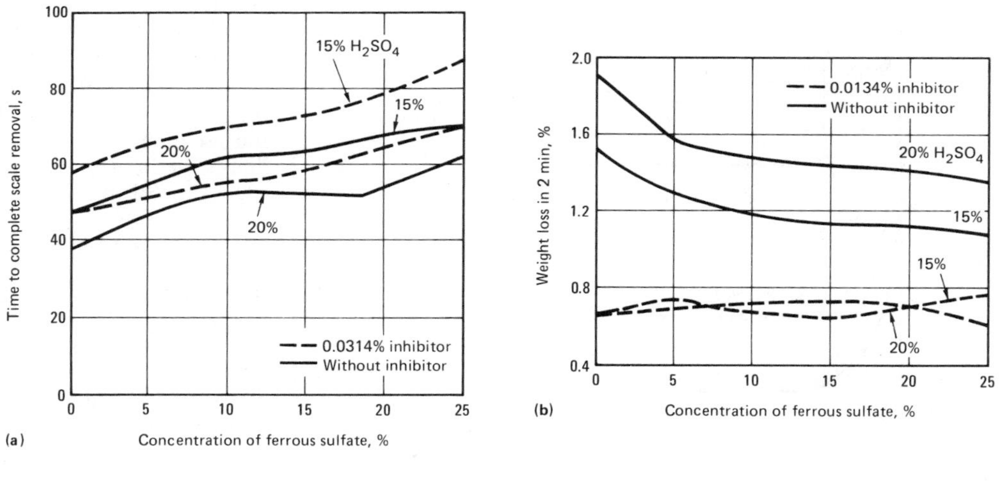

Specimens taken from a low-carbon drawing quality sheet and pickled in solutions of 15 and 20% sulfuric acid. (a) Pickling time for complete scale removal. (b) Weight loss

ling specific products are described in the following examples.

Example 1. Tanks for pickling bars are constructed of either 100-by-100 mm or 200-by-200 mm (4-by-4 in. or 8-by-8 in.) timbers of long-leaf yellow pine that are held together with 25-mm (1-in.) diam tie rods, nuts, and washers made of Monel. The tanks are set over previously prepared reinforced-concrete drain pits that are lined with acid-resistant brick 100 mm (4 in.) thick. The tanks sit on brick tiers or heavy cast iron sleepers.

After the tanks are placed in drain pits, the inside walls of the tanks are lined with a heavy grade of asphalt roofing paper and sealed with an asphaltic-

Fig. 8 Effect of solution temperature on pickling time for hot rolled low-carbon steel; comparison with temper mill scalebreaking

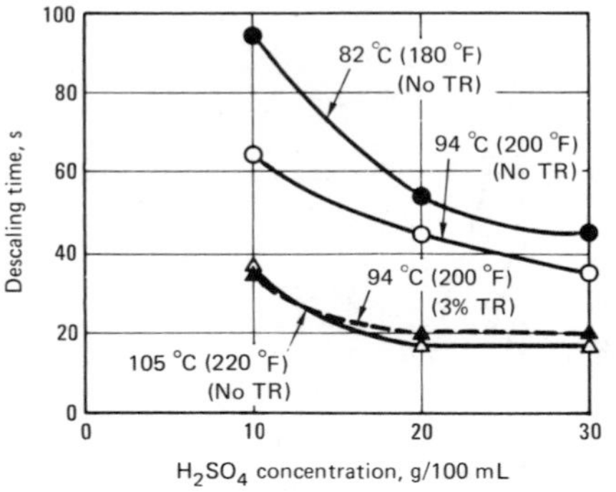

All solutions contained 15 g $FeSO_4$/100 mL. TR: temper rolled

Fig. 9 Effect of strip velocity on descaling time of hot rolled low-carbon steel in hydrochloric acid solutions

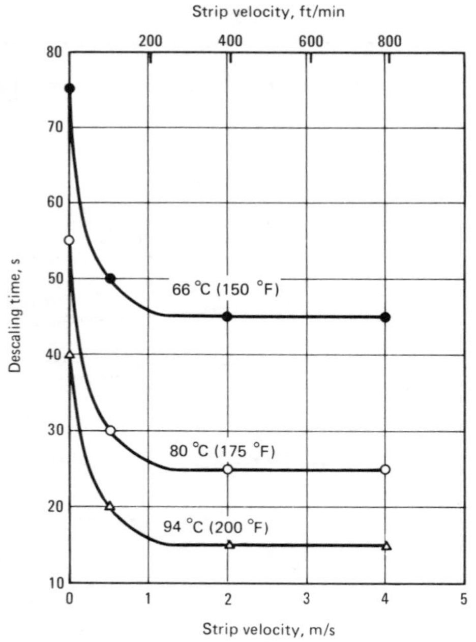

4 g HCl/100 mL. 22.7 g $FeCl_2$/100 mL. Strip not temper rolled. Sample of 8005-809

based trowel-on mastic. Liners of 50-by-150 mm (2-by-6 in.) tongue-and-groove long-leaf yellow pine lumber are nailed with copper nails over the roofing paper, and each joint is coated with mastic. The floors of the tanks are lined with a single course of acid-resistant brick, 100 mm (4 in.) thick, and mortared with poured sulfur cement or troweled-on acid-resistant resinous cement.

Example 2. Another tank for pickling bars consists of a welded and reinforced steel shell fabricated from hot rolled sheet steel 6.5 or 9.5 mm (¼ or ⅜ in.) thick. The shell, which is strong enough to carry the static load of the lining, brickwork, and solution, and to withstand some shock loading inherent to the operation, is prepared by trimming and grinding all sharp corners and welds. The interior and exterior surfaces are then sandblasted. The interior surfaces are then precoated with an acid-resistant sealer and adhesive over which a sheet of polyvinyl chloride, 3.2 to 4.8 mm (⅛ to 3⁄16 in.) thick, is applied.

The interior is then covered with a single layer (100 mm or 4 in. thick) of acid-resistant brick, for operating temperatures at 93 °C (200 °F) or less, or with a double layer (200 mm or 8 in. thick), for operating above 93 °C (200 °F). The single layer of brick lining has a buttered-on joint of resinous cement. Sponge-rubber pads, 10 mm (½ in.) thick, are placed at each end of the tank between the brick and the vinyl sheets, to permit expansion and contraction of the brickwork. Also, two unvulcanized rubber expansion joints, 10 mm (½ in.) thick, are installed across the width and evenly spaced along the length of the tank. The outer surface of the steel shell is painted with three coats (prime, intermediate, and finish) of vinyl for protection against spillage or fumes. Any commercially available acid-resistant or spray-on coating may be used for this purpose.

Example 3. This bar-pickling tank has a shell of steel-reinforced concrete, either monolithic or with keyed walls. The tank floor is pitched from a depth of 990 mm (39 in.) at one end to a depth of 1145 mm (45 in.) at the other, where a 100-mm (4-in.) deep, cleanout sump is located. The interior and exterior surfaces of the shell are sealed with an acid-resistant prime coat and coated with a specially compounded asphaltic membrane 6.5 to 9.5 mm (¼ to ⅜ in.) thick. This mastic coating can be reinforced with layers of glass or synthetic cloth.

The interior surfaces are lined with a double layer of 100-mm (4-in.) thick acid-resistant brick. The course next to the wall has joints of poured acid-resistant sulfur cement. The second course has troweled joints of acid-resistant, all-purpose, resinous cement.

Sponge rubber pad expansion joints, 10 mm (½ in.) thick, are placed between the brick and mastic at each end of the tank. Also, two unvulcanized rubber expansion joints, 10 mm (½ in.) thick, are installed across the width

Fig. 10 Effect of acid concentration and temperature on descaling time for hot rolled low-carbon steel in hydrochloric acid solutions (still bath)

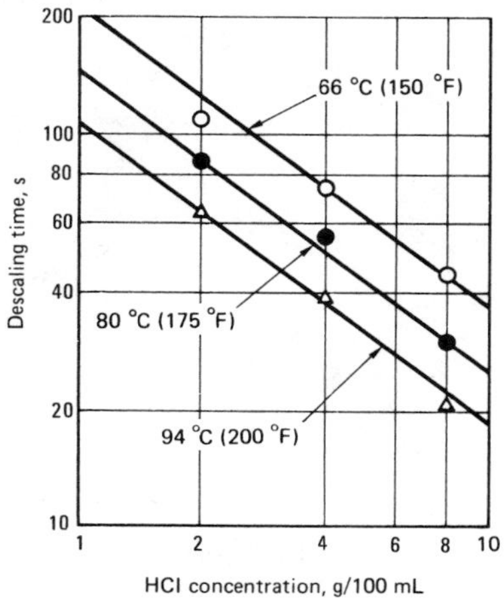

Solutions contained 22.7 g $FeCl_2$/100 mL. Still bath. Strip not temper rolled. Sample of 8005-809, $\log t = -1.438 - 0.749 \log C_{HCl} + 2282/T_F + 459$

Fig. 11 Effect of temperature and acid concentration on descaling time of hot rolled low-carbon steel in hydrochloric acid solutions as influenced by strip velocity

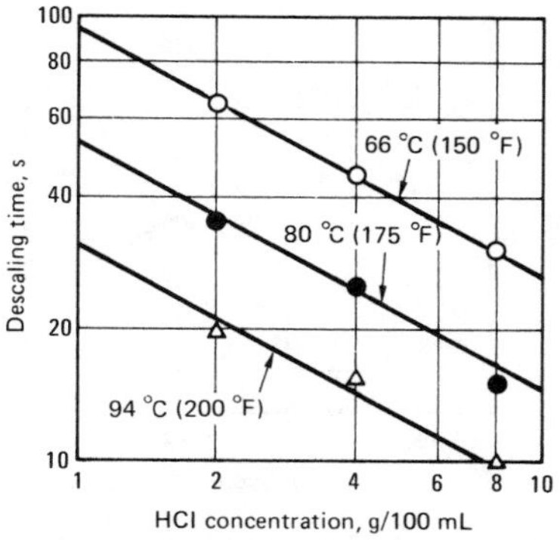

Solution contained 22.7 g $FeCl_2$/100 mL. Strip velocity 2 m/s (400 ft/min). Strip not temper rolled. Sample of 8005-809, $\log t = -4.455 - 0.556 \log C_{HCl} + 3916/T_F + 459$

and evenly spaced along the length of the tank.

A coping and veneer, both of 100 mm (4 in.) thick acid-resistant brick, and with poured sulfur cement or buttered-on resinous cement joints, are applied to the top and outside walls of the tank for protecting the concrete shell. If an agitator system is used, the agitator stands should be bricked in to the height of the tank walls with acid-resistant brick, using poured or buttered-on acid-resistant cement joint.

The tanks described in the preceding examples are representative of relatively large batch pickling tanks for pickling bars in sulfuric or hydrochloric acid solutions at 66 to 71 °C (150 to 160 °F). Other acids or higher operating temperatures may involve changes in construction.

Rinsing. Thorough rinsing is essential for obtaining the clean, stain-free and smut-free surface necessary for subsequent tinning or other metal-coating operation. For example, optimum rinsing conditions for strip steel require maximum maintenance of the high-pressure cold water sprays for thorough flushing of all acid from the strip before entry into the hot water rinse bath. Thus, acid contamination of the hot rinse bath is eliminated or minimized; the maximum acidity of the hot water permitted is a pH of 5 to 6. When pH determinations indicate higher acidity, the hot water input is increased to increase the overflow until the proper pH is attained. Spray nozzles, wringer rolls, and the wringer-roll adjustment are checked periodically to ensure minimum drag-in of acid to the hot water bath. Acid and pH determinations are made every 4 h, or more frequently if high acidity is indicated.

Heating Methods and Temperature Control. Most systems are now using strict computer control which carefully monitors the heating and temperature of baths. In addition, careful analysis of iron buildup can also be monitored continuously, allowing extremely accurate control. The most widely used method of heating pickling solutions is by direct injection of live steam through steam jets. Steam also provides some turbulence and agitation of the pickling solution. The steam condensate may accumulate to the point of overflow, necessitating some drain-off and loss of acid. Adequate trapping near the point of steam injection minimizes such drain-off loss.

If drain-off or dilution of the solution is unacceptable, other heating methods may be used, usually at increased cost. These include coil steam heat, submerged combustion of gases, heat interchangers, and, for small installations, electric immersion heaters. Heating equipment must be made of acid-resistant materials, such as carbon, lead alloy, or stainless steel or zirconium for H_2SO_4 and carbon and teflon covered heat exchangers for HCl. Acid-resistant indicating and regulating temperature-control instruments are available for pickling solutions. For agitated solutions, one temperature regulator placed at one end of a batch-type tank is usually sufficient.

Temperature regulators can control to within a few degrees of the desired operating temperature. During pickling in inhibited acid solutions at 66 to 71 °C (150 to 160 °F), fluctuations of 3 °C (5 °F) are acceptable, and a difference of 6 °C (10 °F) between opposite ends of a long tank for pickling bars is also acceptable. At higher operating temperatures, and particularly in non-inhibited solutions, heating and temperature control are more critical. Heaters can be centrally located, or in continuous pickling can be placed at strategic intervals. For long continuous pickling tanks, temperature controls should be spaced at proper intervals along the tank.

Storage Tanks for Acid. Hydrochloric acid can be stored in rubber-lined steel tanks, or reinforced plastic or teflon tanks. Hydrofluoric acid is generally stored in plastic-lined tanks, although passivated low-carbon steel is

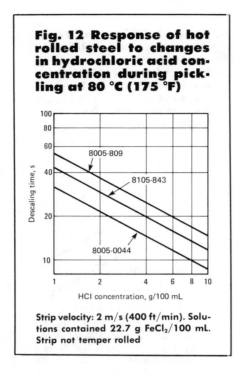

Fig. 12 Response of hot rolled steel to changes in hydrochloric acid concentration during pickling at 80 °C (175 °F)

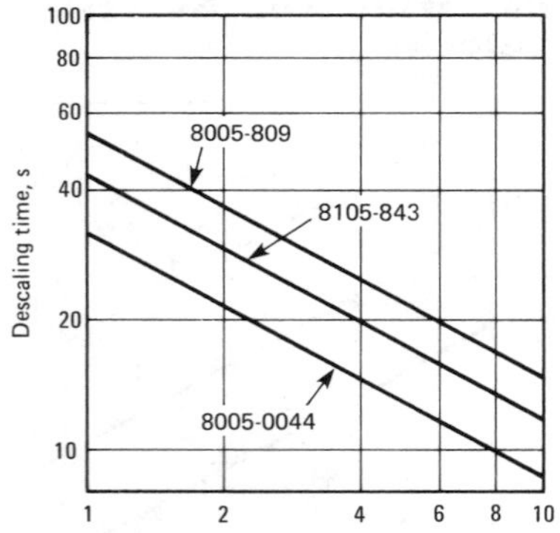

Strip velocity: 2 m/s (400 ft/min). Solutions contained 22.7 g $FeCl_2$/100 mL. Strip not temper rolled

Fig. 13 Materials used in construction of 3.9- to 4.9-t (4- to 5-ton) capacity tank for pickling coils of steel

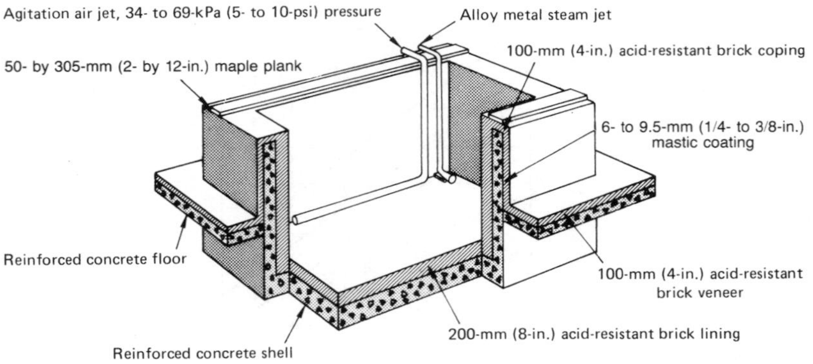

satisfactory for storing and handling 70% HF at temperatures up to 38 °C (100 °F).

Acid lines carrying sulfuric acid from the storage tank to the pickling tank may be of iron pipe with iron fittings, under the following conditions:

Concentration of sulfuric acid	Safe temperature for iron pipe
1.83 sp gr	Any temperature
1.71 sp gr	Under 38 °C (100 °F)
1.67 sp gr or less	None

For safety at any temperature, acid-resistant pipe and fittings are recommended for sulfuric acid of 1.71 sp gr or less. Care should be taken to calk all the joints with an acid-resisting material. Sections of pipe that are immersed in the pickling bath should be made of lead, copper, or bronze. Heavy rubber hose is sometimes used. Stainless steel lines are satisfactory for use in strong nitric acid.

When copper or bronze is used in connection with equipment for pickling, the possibility of deposition of copper on the pickled surface exists. Ordinarily this is of no consequence, but if the stock is to be carburized without machining after pickling, the copper film prevents carburization. Also, copper stains on full-finished sheets are objectionable. Copper or bronze should not be used in the pickling tank in such instances.

Drainage lines should be made of vitrified tile carefully calked with an acid-resistant cement to avoid leakage and should be large enough to provide adequate drainage at all times. A wooden lattice, preferably of cypress, should cover the floors around the pickling tanks.

Pickling Defects

Pickling is frequently blamed for certain defects that appear during the pickling operation but that have their inception elsewhere. The tendency is to blame all failures on overpickling or to condemn the steel as defective. Certain defects, however, come from neither of these causes, but are the result of earlier operations, such as rolling, heat treating, or forging. Care must be exercised in the pickling operation because many possibile defects arise from poor pickling practice. With the advent of computer controlled pickling, defects are minimized by continuous and accurate monitoring of the baths.

Overpickling causes porosity of the transverse surfaces and a roughening of the whole surface, accompanied by discoloration and a decrease in size and weight. Overpickling can be avoided by removing the material from the bath promptly when complete removal of scale has been accomplished. Inhibitors aid in preventing overpickling, but are not a complete guarantee that it cannot occur.

Pitting. Electrolytic pitting is by far the most prevalent and troublesome of the various types of pitting, particularly on heat treated alloy steels and forgings. It is characterized by a patchwork of irregularly shaped pitted areas; the depth is indirectly proportional to the area. Caused by an electrical potential between the scaled areas and the clean steel, electrolytic pitting occurs only where the scale has been removed from small areas before pickling, or at an early stage in the process. The pitted areas are often longitudinally aligned and the boundaries are usually rather sharply defined or channeled, probably because maximum potential exists at such locations. When inhibited acid is used, the pitted area is generally uniform in depth, with a channeled boundary; uninhibited acid makes the pitted area irregular in depth.

Although severe pitting is rare, it may be caused by overpickling, particularly when inhibitors are used, but in less-than-adequate amounts. This serious defect indicates carelessness on the part of supervisory personnel, because such pitting appears only when material is allowed to remain in the bath much longer than is necessary for complete scale removal. Pits caused by rolled-in scale or by refractories in the rolling process are intensified during pickling.

Nonmetallic inclusions, segregated carbides, or surface strains are frequently held responsible for pickle pitting. These factors may have some influence, but their importance has been overstressed. The more frequently occurring causes discussed in the preceding paragraphs should be investigated first.

Blistering is a troublesome defect on sheet and strip steel. The most reasonable explanation for blistering is that gaseous inclusions cause flaws in the steel by forming gas pockets just beneath the surface during rolling. Hydrogen generated in the pickling operation penetrates these pockets and lifts the surface, causing a blister. Properly selected inhibitors may minimize blistering, but it is doubtful that they can prevent it entirely.

Hydrogen embrittlement occurs when cold working operations follow too soon after pickling. Hydrogen embrittlement results from the penetration of the steel by nascent hydrogen. This type of embrittlement is not permanent and may be eliminated by aging. Baking cleaned parts at 200 to 240 °C (390 to 465 °F) for 3 to 4 h may be necessary for high-strength steels (>1380 MPa or >200 ksi). In such cases, cracking may occur within 4 h after pickling. Inhibitors are valuable in minimizing this effect.

Some inhibitors, especially those containing certain organic sulfur compounds, are highly effective in limiting base metal loss, but actually promote hydrogen pickup.

Safety

General safety practices to be followed during alkaline precleaning and acid pickling operations are as follows:

- Employees handling chemicals should wash their hands and faces before eating and before leaving at the end of a shift
- Only authorized employees familiar with safety rules regarding the handling of chemicals should be permitted to make additions of chemicals to pickling tanks
- Face shields, rubber gloves, and rubber aprons should be worn by employees cleaning or repairing tanks or making additions of chemicals
- If splashed or spilled chemicals come in contact with the body, employees should wash immediately and thoroughly with cool water and report promptly to the emergency hospital for treatment
- Any change in working procedures or any unusual occurrences relating to the use of chemicals should be brought to the attention of supervisory personnel
- An emergency shower should be available as well as facilities for eye care

Safe practices in the use of alkaline and acid solutions are as follows:

Alkaline solutions

- Alkalis should be added to water slowly, using a hopper or a shovel, to obtain an even distribution of the chemicals in the solution
- Adequate agitation should be provided after an alkali has been added, to ensure that the chemicals dissolve
- To prevent eruption of the solution caused by rapid solution of the alkali, temperature of the solution should not exceed 66 °C (150 °F) when alkaline chemicals are added

Acid solutions

- Always add acid to water; never add water to concentrated acid
- When preparing a new solution, add acid to cold water, and do not heat the solution until the acid has been added
- Additions of acid to hot pickling solutions should be made with extreme caution; preferably, the solution should be cooled before the acid is added
- Materials should be carefully immersed in and withdrawn from acid solutions to avoid splashing
- After being pickled, all material should be rinsed in water to remove all traces of residual acid
- Spent pickle liquor should be neutralized before being discarded or be pumped to receptacles to be processed for disposal

Disposal of Spent Pickle Liquor

Pickle liquor must be neutralized before being disposed of, to comply with environmental regulations and to avoid contamination of the water table and streams. Some plants have spent pickle liquor hauled away for treatment and disposal at approved dump sites.

Spent sulfuric acid solution contains 2 to 5 wt % sulfuric acid and either 21 to 32 wt % iron sulfate or 40 to 60 wt % copperas ($FeSO_4 \cdot 7H_2O$). As a minimum treatment, this and other acid solutions are neutralized with a suitable material such as lime; this neutralized, high-iron liquor may be piped to ponds or hauled to abandoned pits or quarries. For neutralizing small amounts of solution and to minimize the formation of solids, sodium hydroxide, sodium carbonate, or ammonia are sometimes used.

Neutralizing Sulfate Liquor. The complete treatment of sulfate pickle liquor consists of treating with hydrated lime to neutralize the acid and to precipitate iron hydroxide. Dolomite also may be used and has the advantage of producing more soluble precipitate salts. Lime is added as a slurry containing 25 to 35 wt % calcium hydroxide. Usually, the slurry is stored in a tank equipped with a means for agitation to prevent settling, but the lime may be slurried through dispensers and mixing tanks. The lime slurry is metered into a reaction tank containing a known quantity of spent liquor. When large volumes of rinse water are to be treated, the water should be combined with the pickle liquor in a large holding tank, so that uniformity of the raw product to be treated can be maintained.

The lime slurry is fed into the reaction tank until the pH of the solution reaches 5 to 7. The reaction mixture is agitated and aerated to produce a coarse, fast-settling precipitate. For batches of 5700 L (1500 gal), the lime metering and reaction time required is less than 1 h. Aeration of the reaction mixture is desirable because it provides some oxidation to a ferro-ferric iron hydroxide mixture that makes the combined gypsum-hydroxide slime settle more readily or filter more easily. Coagulating agents, such as polymers, are also helpful.

The reaction mixture may then be pumped to lagoons for settling, to mechanical clarifying tanks, or to mechanical filters. The supernatant liquid is clear enough for reuse, providing it does not contain high amounts of dissolved solids. The mixture still may contain substantial amounts of dissolved nickel and chromium and cannot be disposed of in sewers. The settled solids, or filter cake, are discarded, unless the solid is classified as hazardous waste.

Recovery of Sulfuric Acid and Iron Sulfate. Because ferrous sulfate becomes less soluble as the temperature of the solution is decreased or as the acid concentration is increased, it can be recovered as crystals by cooling or evaporating the spent liquor. Refrigeration may be used for cooling, and evaporation may be by heating or air flow, with or without the aid of vacuum systems. If recovery is from a cold solution (0 to 10 °C or 32 to 50 °F), the product is copperas ($FeSO_4 \cdot 7H_2O$); if from a hot solution, monohydrate ($FeSO_4 \cdot H_2O$) is formed, but this method is seldom used.

During concentration of the solution, the action of decreasing solubility of iron sulfate with increasing acid concentration can be augmented by the addition of concentrated sulfuric acid, especially when recovery is by high-temperature evaporation. After removing the iron sulfate crystals by centrifuging, the solution containing the sulfuric acid and a reduced iron content is returned to the pickling tanks, thus effecting a partial recovery of the unreacted acid. The iron sulfate crystals may be used in fertilizers or be treated to produce sulfuric acid and iron powder.

The concentration and iron content of the recovered acid depend on the degree of evaporation and supercooling for the heptahydrate (copperas) method and on the degree of evaporation and amount of concentrated acid added in the monohydrate method. Recovered acid with an iron content as low as 2% may be obtained with the heptahydrate method.

The acid in the solution from the crystallizers can be adjusted for batch pickling operations by the addition of water and fresh acids or returned to continuous pickling lines at a rate which maintains a fairly constant iron content. In the latter instance the amount of pickle liquor withdrawn for treatment must be adjusted to the volume of the pickling solution, the concentration of acid in the solution, and the operational efficiency of the recovery system. The life of a batch pickling solution is decreased when recovery acid is used for make-up of a fresh solution, because of the iron content in the acid.

Regeneration of Hydrochloric Acid. One of the advantages of pickling with hydrochloric acid is the complete regeneration possible from spent acid. In the spray roaster process, waste pickle liquors that typically contain from 1 to 4% hydrochloric acid and 18 to 27% ferrous chloride are first concentrated by evaporation of water. Then, the concentrated spent pickle liquor enters the roaster, which operates at a temperature of about 540 °C (1005 °F). The free hydrogen chloride is absorbed in water along with additional hydrogen chloride formed by oxidation of ferrous chloride. The red oxide of iron, Fe_2O_3, is a by-product of the operation. Regenerated acid typically contains from 16 to 20% HCl.

Abrasive Blast Cleaning

By the ASM Committee on Abrasive Blast Cleaning*

ABRASIVE blast cleaning entails the forceful direction of abrasive particles, either dry or suspended in a liquid, against the surfaces of parts or products. Abrasive blast cleaning removes contaminants and conditions the surfaces for subsequent finishing. Typical uses include:

- Removing rust, scale, dry soils, mold sand, or paint
- Roughening surfaces in preparation for bonding, painting, or other coating
- Removing burrs
- Developing a matte surface finish
- Removing flash from molding operations
- Carving in glass or porcelain

Types of workpieces that can be blast cleaned include: ferrous and nonferrous castings, forgings, and weldments; thermoplastic and thermosetting plastic parts, precision molded rubber parts; and miscellaneous parts of glass, wood, leather, and other materials. The abrasives used in blast cleaning can be propelled in several different ways, and a great variety of abrasive media may be used. The success of blast cleaning operations depends primarily on judicious selection of method and abrasive medium.

Abrasives for Dry Blast Cleaning

The materials used in dry abrasive blast cleaning may be categorized as metallic grit, metallic shot, sand, glass, and miscellaneous. Hardness, density, size, and shape are important considerations in choosing an abrasive for a specific application.

The selection of the type and size of the blast cleaning material will depend on the size and shape of the parts to be cleaned, the finish desired, and the treatment or operation that may follow blast cleaning.

The surface, especially ferrous surfaces, tend to be very active following abrasive cleaning, and any subsequent operation such as plating or painting should be performed as soon as possible after abrasive cleaning.

Metallics

Grit consists of angular metallic particles with high cutting power. Grit is usually made of crushed, hardened cast steel shot, which may be tempered, or of chilled white cast iron shot, which may be malleablized. Size specifications for cast grit are shown in Table 1.

Generally, three hardnesses are offered in steel grit: (*a*) 45 HRC, (*b*) 56 HRC, and (*c*) 65 HRC. The screen distribution and the velocity of the grit impacting on the part surfaces control the finish. Usually, grit blast produces a brighter finish than shot blast. Applications for grit include removal of heavy forging and heat treat scale, removal of rust, and controlled profiling of workpieces before bonding or coating. Hard grit is also used to provide a gripping surface on steel mill rolls.

Shot, normally made of the same materials as grit, is usually in the form of spherical particles. Shot removes scale, sand, and other surface contaminants by impact. Size specifications for cast shot are indicated in Table 2. Steel shot is the most widely used metallic abrasive medium and is least destructive to the components of the abrasive blast system. The matte finish produced by steel shot on metal surfaces can be controlled by the screen distribution of the operating mix and the velocity of shot impacting on part surfaces.

Shot also may be made of aluminum or cut steel wire. Cut steel wire deforms into rounded particles during usage; it is used frequently in the same manner as cast shot. Table 3 shows the specifi-

*R. M. Leliaert, Technical Manager, Wheelabrator-Frye, Inc.; Neil Weightman, Sales Manager, Comco Inc.; and Michael M. Woelfel, Technical Service Manager, Impact Products, Potters Industries, Inc.

Table 1 Size specifications for cast grit (SAE J444)

Size No.	Screen tolerances(a)	Screen opening mm	in.
G10	All pass No. 7	2.82	0.1110
	80% min on No. 10	2.00	0.0787
	90% min on No. 12	1.68	0.0661
G12	All pass No. 8	2.38	0.0937
	80% min on No. 12	1.68	0.0661
	90% min on No. 14	1.41	0.0555
G14	All pass No. 10	2.00	0.0787
	80% min on No. 14	1.41	0.0555
	90% min on No. 16	1.19	0.0469
G16	All pass No. 12	1.68	0.0661
	75% min on No. 16	1.19	0.0469
	85% min on No. 18	1.00	0.0394
G18	All pass No. 14	1.41	0.0555
	75% min on No. 18	1.00	0.0394
	85% min on No. 25	0.711	0.0280
G25	All pass No. 16	1.19	0.0469
	70% min on No. 25	0.711	0.0280
	80% min on No. 40	0.419	0.0165
G40	All pass No. 18	1.00	0.0394
	70% min on No. 40	0.419	0.0165
	80% min on No. 50	0.297	0.0117
G50	All pass No. 25	0.711	0.0280
	65% min on No. 50	0.297	0.0117
	75% min on No. 80	0.18	0.0070
G80	All pass No. 40	0.419	0.0165
	65% min on No. 80	0.18	0.0070
	75% min on No. 120	0.12	0.0049
G120	All pass No. 50	0.297	0.0117
	60% min on No. 120	0.12	0.0049
	70% min on No. 200	0.074	0.0029
G200	All pass No. 80	0.18	0.0070
	55% min on No. 200	0.074	0.0029
	65% min on No. 325	0.043	0.0017
G325	All pass No. 120	0.12	0.0049
	20% min on No. 325	0.043	0.0017

(a) Minimum cumulative percentages (by weight) allowed on screens of numbers and opening sizes as indicated

Table 2 Cast shot size specifications for shot peening or blast cleaning (SAE)

Screen No.	Screen size mm	in.	Screen opening(a)	Passing (a), %
7	2.82	0.111	780	All pass
8	2.38	0.0937	660	All pass
10	2.00	0.0787	780	85 min
			550	All pass
			460	All pass
12	1.67	0.0661	780	97 min
			660	85 min
			460	5 max
			390	All pass
14	1.41	0.0555	660	97 min
			550	85 min
			390	5 max
			330	All pass
16	1.19	0.0469	550	97 min
			460	85 min
			330	5 max
			280	All pass
18	1.00	0.0394	460	96 min
			390	85 min
			280	5 max
			230	All pass
20	0.841	0.0331	390	96 min
			330	85 min
			230	10 min
			170	All pass
25	0.711	0.0280	330	96 min
			280	85 min
			170	10 max
30	0.590	0.0232	280	96 min
			230	85 min
			110	All pass
35	0.500	0.0197	230	97 min
			110	10 max
40	0.419	0.0165	170	85 min
			70	All pass
45	0.351	0.0138	170	97 min
			70	10 max
50	0.297	0.0117	110	80 min
80	0.18	0.007	110	90 min
			70	80 min
120	0.124	0.0049	70	90 min

(a) Screen opening sizes and screen numbers with maximum and minimum cumulative percentages allowed on corresponding screens

Table 3 Specifications for cut steel wire shot (SAE J441)

Size No.	Diameter of wire mm	in.	Minimum hardness, HRC
CW-62	1.59 ± 0.05	0.0625 ± 0.002	36
CW-54	1.4 ± 0.05	0.054 ± 0.002	39
CW-47	1.2 ± 0.05	0.047 ± 0.002	41
CW-41	1.0 ± 0.05	0.041 ± 0.002	42
CW-35	0.89 ± 0.03	0.035 ± 0.001	44
CW-32	0.81 ± 0.03	0.032 ± 0.001	45
CW-28	0.71 ± 0.03	0.028 ± 0.001	46
CW-23	0.58 ± 0.03	0.023 ± 0.001	48
CW-20	0.51 ± 0.03	0.020 ± 0.001	48

cations relating standard size numbers for cut steel wire shot to diameter and minimum hardness.

Nonmetallics

Table 4 lists physical properties and comparative characteristics of a variety of nonmetallic abrasives.

Sand. This term is applied to diverse nonmetallic abrasives, in addition to ordinary silica sand. These materials are used when it is necessary to protect the surface of the workpiece from metallic contamination. They may be either natural materials, such as garnet, novaculite, dolomite, pumice, and flint quartz, or manufactured materials such as aluminum oxide, silicon carbide, and slag. The natural materials are lowest in initial cost; the manufactured materials, although somewhat more expensive than natural sands, cost less than metallic abrasives.

Glass is available as angular particles (ground glass) or spherical particles (glass beads), ranging in size from approximately 1 mm (0.039 in.) down to less than 40 μm (0.0016 in.). The particles are usually made from soda-lime-silica glass with a hardness of about 500 HK (100 g load) equivalent to 46 to 50 HRC. Ground glass is effective for deburring and relatively aggressive finishing. Glass beads produce a fine matte appearance and clean without removing base metal. Size and roundness specifications for glass beads are given in Table 5.

Miscellaneous Materials. Mild abrasive action is provided by the use of such agricultural products as crushed walnut or pecan shells, rice hulls, rye husks, corncobs, and sawdust. Smooth plastic beads serve a similar purpose.

Selection of a suitable abrasive for a specific application is influenced by the type of surface contamination to be removed, size and shape of the work piece, surface finish specified, type and efficiency of cleaning equipment, and required production rate.

Table 4 Physical properties and comparative characteristics of nonmetallic abrasives

Description	Glass beads(a)	Coarse mineral abrasives(b)	Fine angular mineral abrasives(c)	Organic soft grit abrasives(d)	Plastic abrasives(e)
Physical properties					
Shape	Spherical	Granular	Angular	Irregular	Cylindrical (diameter/length = 1)
Color	Clear	Tan	Brown/white	Brown/tan	Nylon: white, polycarbonate: orange
Specific gravity	2.45-2.50	2.4-2.7	2.4-4.0	1.3-1.4	Nylon: 1.15-1.17, polycarbonate: 1.2-1.65
Free silica content	None	100%	<1%	None	None
Free iron content	<1%	<1%	<1%	None	None
Hardness (MOH)	5.5	7.5	9.0	1.0	R-110 to R-120
Media comparisons					
Toxicity	None	High	Low	Low/none	None
Metal removal	Low/none	High	High	None	Deburring only
Cleaning speed	Medium/high	High	High	Low	Low
Peening ability	High	None	None	None	None
Finish achieved	Range (various matte)	Rough anchor	Various matte	Smooth	Smooth
Surface contamination	None	Medium	Medium	Medium/high	Low to none
Suitability for wet blasting	High	Low	Low	Low	Low
Suitability for dry blasting	High	High	High	High	High
Standard size ranges	20-325 U.S. mesh	8-200 U.S. mesh	80-325 U.S. mesh	60-325 U.S. mesh	0.76 by 0.76 mm (0.030 by 0.030 in.) 1.1 by 1.1 mm (0.045 by 0.045 in.) 1.5 by 1.5 mm (0.060 by 0.060 in.)
Consumption rate	Low	High	Medium	High	Very low
Cost comparison	Medium	Low	High/medium	High/medium	High/medium

(a) Glass beads are used for cleaning, finishing, light-to-medium peening, and deburring. (b) Coarse mineral abrasives such as sand are used where metal removal and surface contamination are not considered. (c) Fine angular mineral abrasives such as aluminum oxide are used in cleaning when smooth finish and surface contamination are not important. (d) Organic soft grit abrasives, for example, walnut shells, are used in light deburring and cleaning of fragile items. (e) Plastic abrasives such as nylon and polycarbonate are used to deflash themoset plastic parts and deburr finished machine parts

Type, size, and hardness of metallic abrasives recommended for some typical applications are given in Table 6. In addition to the recommendations in this table, the following general observations relating to the performance of abrasive particles may be helpful:

- The smaller the abrasive particle, the finer the surface finish and the greater the surface coverage
- The larger the abrasive particle, the greater the impact
- In general, the harder the abrasive particle, the faster is its cleaning action

Replacement of Abrasive. Production of a uniformly abraded surface depends on maintenance of a uniform working mix of abrasive in the machine at all times. Metal surfaces treated with hard grit are more sensitive to a change in working mix than those treated with a soft grit or with shot. A program that includes testing and periodic replacement of the abrasive is recommended. The working mix contains more abrasive fines than new abrasive. Examples of screen analyses of new and used abrasives are given in Table 7.

A practical method of maintaining a reasonable degree of consistency of the working mix is to keep a uniform level of abrasive in the supply tank, bin, or hopper. This is accomplished by adding new abrasive periodically, usually at least once every 8 h. If surface finish requirements are more critical, hourly additions of new abrasive may be needed. The total tonnage of abrasive required to fill the machine to operating capacity also may affect the frequency of additions. When surface finish requirements are critical, the use of an automatic abrasive-replenisher is indicated. This device maintains the abrasive level in the blasting machine by automatically feeding abrasive from a supply hopper.

For control purposes, a periodic analysis should be made of a representative sample of the abrasive used in the machine. The frequency of these tests depends largely on surface finish and production requirements. Dust fines should be removed from the mix by an air-separating system; proper performance of a separating device depends on a uniform flow of air through the separator and on maintenance of uniform abrasive sizes and quantities.

Control of Contaminants. After the abrasive medium makes contact with the workpiece, it is returned to a storage hopper for reuse. Coarse and fine contaminants picked up in the process are removed as the medium is returned to the hopper. Coarse contaminants include tramp metal, fins, core wire, core nails, slag, sand lumps, large flakes of rust and mill scale, and flash. These are usually removed by screening the abrasive mix through wire mesh, perforated plate, or expanded metal.

Fine contaminants include sand, fine mill scale and rust particles, metallic dusts, and disintegrated abrasive particles. Buildup of fine contaminants reduces blast efficiency, and a high sand content results in excessive wear of centrifugal blast wheel parts. Fine contaminants are removed from the mix by a current of air. In an expansion chamber, the heavier fine particles resist an upward turn of the air current and drop into the settling area to be discharged as refuse. Lighter particles remain suspended in the air current and are carried out through the ventilating lines to the dust collector or exhaust.

In many blast cleaning operations, small metallic particles resulting from the wearing or breaking down of abrasives assist in scouring small crevices or valleys. If such fines are to be retained, separators must be adjusted accordingly.

Table 5 Standard size and roundness specifications for glass beads (MIL-G-9954A)

U.S. standard screen	Bead size	Roundness, minimum, %	Passing, %
10	1	60	100
12	1	60	95-100
	2	60	100
14	1	60	0-15
	2	60	95-100
	3	65	100
20	1	60	0-5
	2	60	0-15
	3	65	95-100
	4	70	100
30	2	60	0-5
	3	65	0-15
	4	70	95-100
	5	70	100
40	3	65	0-5
	4	70	0-15
	5	70	95-100
	6	80	100
50	4	70	0-5
	5	70	0-15
	6	80	95-100
	7	80	100
60	5	70	0-5
	7	80	95-100
	8	80	100
70	6	80	0-15
	8	80	95-100
	9	80	100
80	6	80	0-5
	7	80	0-15
	9	80	95-100
	10	90	100
100	7	80	0-5
	8	80	0-15
	10	90	95-100
	11	90	100
120	8	80	0-5
	9	80	0-15
	11	90	95-100
	12	90	100
140	9	80	0-5
	12	90	95-100
	13	95	100
170	10	90	0-15
	13	95	95-100
200	10	90	0-5
	11	90	0-15
230	11	90	0-5
	12	90	0-15
325	12	90	0-5
	13	95	0-15
400	13	95	0-5

Table 6 Selection of abrasive

Application	Recommended abrasive(a) Type	Nominal diameter mm	Nominal diameter in.	Hardness, HRC
Blasting of ferrous metals				
Removal of light scale(b)	Shot or grit	0.2-0.71	0.007-0.028	30-66(c)
Removal of heavy scale(b)	Shot or grit	0.71-2.0	0.028-0.078	45-66
Cleaning of castings	Shot or grit	0.43-2.0	0.017-0.078	30-66(c)
Blasting of nonferrous metals				
Frosted appearance only	Grit	0.1-0.43	0.005-0.017	50-66
Preparation for other surface finishes	Shot or grit	0.2-0.71	0.007-0.028	30-66(c)
Nondirectional matte finish	Glass beads	0.05-0.2	0.002-0.007	46-50

(a) Cast iron or cast steel. (b) For phosphating or painting. (c) Cast steel abrasive usually is not available in a hardness of less than 40 HRC

Table 7 Screen analyses of three metal abrasives

Screen No.	Screen size mm	Screen size in.	Abrasive remaining on screen, % New abrasive	Abrasive remaining on screen, % Working mix
G25 cast steel grit				
16	1.19	0.0469	Trace	None
20	0.841	0.0331	82	23
30	0.589	0.0232	16	33
40	0.419	0.0165	1	19
50	0.297	0.0117	Trace	17
70	0.21	0.0083	Trace	6
Pan	...	...	Trace	2
G40 cast steel grit				
16	1.19	0.0469	None	None
20	0.841	0.0331	19	None
30	0.589	0.0232	76	18
40	0.419	0.0165	4	51
50	0.297	0.0117	Trace	26
70	0.21	0.0083	Trace	5
Pan	...	...	Trace	Trace
5280 cast steel shot				
16	1.19	0.0469	None	None
20	0.841	0.0331	40	27
30	0.589	0.0232	58	56
40	0.419	0.0165	1	11
50	0.297	0.0117	1	5
70	0.21	0.0083	None	1
Pan	...	...	None	None

Propelling Abrasive Media

Three basic methods are used to propel the abrasive media against the surfaces of the workpieces: (*a*) airless abrasive blast blade or vane-type wheels, (*b*) pressure blast nozzle systems, and (*c*) suction (induction) blast nozzle systems.

Abrasive blast cleaning began commercially with air or steam directed through a conduit of pipe or hose with a final nozzle to direct the impacting abrasive stream. Both pressure blast and suction blast nozzle systems require high power to generate pressurized air or steam used to accelerate and propel the abrasive. This requirement is due to aerodynamic inefficiencies in accelerating the spherical and angular abrasive particles, especially the higher density ferrous abrasives.

Airless abrasive wheels using vanes require about 10% of the horsepower required by air blast systems to throw equal volumes of abrasive at the same velocities. The power losses in an airless system are friction between the abrasive and vanes, and the impeller-control cage and the wheel-drive system.

Wheels. Airless abrasive blast wheels are generally of the slider blade type as shown in Fig. 1. These wheels may have one or two side plates, one of which is attached to a hub, shaft bearings, and belt drive, or the side plate may be attached directly to the shaft of a suitable motor. The side plate holds four to eight throwing blades, depending on the size of the wheel. Blade tip diameters range from 205 to 610 mm (8 to 24 in.) and blade widths from 40 to 125 mm (1.5 to 5 in.). Rotational speeds range from 500 to 4000 rev/min or more. Usable abrasive velocities range from 15 m/s (50 ft/s) up to 122 m/s (400 ft/s), with 75 m/s (245 ft/s) as the most widely used velocity. Abrasive flow rates with steel shot range from 23 kg/min(50 lb/min) up to 1040 kg/min (2300 lb/min) with a 100-hp motor drive.

Operation of an airless abrasive blast wheel of the slider type is shown in Fig. 1. A controlled flow of abrasive (through a valve not shown) is fed by gravity into an abrasive feed spout from which it flows into a rotating vaned impeller. The impeller rotates at the same speed as the bladed wheel and has an equal number of vanes to wheel blades. The impeller rotates in a stationary cylinder that is equipped with an opening that may be rotated and locked in a preferred position. As the impeller forces the abrasive out of the control cage opening, each of the blades picks up a metered amount of abrasive at the inner end of the blade and accelerates the abrasive to produce a tent blast pattern as shown. Airless abrasive blast machines are very simple and can be manually controlled with one low horsepower wheel. A simple barrel machine is usually equipped with a two-bladed wheel, a gravity abrasive hopper, a belt and bucket elevator, an airwash abrasive separator, and a fabric filter dust collector. Applications for blast machines include descaling, deburring, and deflashing.

Centrifugal blast wheel units are enclosed in housings to prevent the discharge of stray abrasive. The principal wearing parts of the blast wheel assembly are the impeller, control cage, wheel blades, and housing liners. These parts are most economically made of high-alloy cast iron, and each can be individually replaced. Unalloyed cast iron parts, although less expensive, have a very short life under normal operating conditions.

The life of these parts is influenced primarily by the type and condition of the abrasive medium and contaminants picked up in the cleaning process. Clean steel shot provides the longest useful life of wheel and guard housing liners. Much greater wear results from the use of nonmetallic abrasives such as sand, aluminum oxide, and silicon carbide. Table 8 shows effects of abrasives in various conditions on the life of components of a centrifugal blast wheel unit.

Glass beads, nonferrous shot, and the agricultural abrasives, frequently used in deburring and special finishing applications, cause relatively little wear on wheel parts and housing liners.

Pressure blast nozzle systems generally rely on a 685 kPa (100 psig) air supply to propel the abrasive through a special nozzle. A typical intermittent pressure tank has dimensions of 610 by 610 mm (24 by 24 in.) and an abrasive discharge capacity of 0.12 m^3 (4.2 ft^3). This capacity is adequate to operate one 6-mm (¼-in.) diam blast nozzle for 30 to 60 min. This type of tank is refilled through the filling valve by gravity when the air supply is shut off. Without air pressure in the tank, the filling valve is pushed down and open by the weight of the abrasive. When the air pressure is turned on again, the valve rises and stops the flow of abrasive into the tank. The abrasive in the now pressurized tank moves into a mixing chamber. Mixing chambers usually are equipped with an adjustable control to regulate the flow rate of abrasive into the mixing chamber and on through the hose and nozzle assembly.

Table 8 Effect of abrasives on life of components of a centrifugal blast wheel unit

	Life of components(a), h			
Abrasive	Blades	Impeller	Controller cage	Alloy housing liners
100% steel shot (few fines)	250	250	400	3000
Steel shot, 1% sand	80-100	100	100	2000
Steel shot, 3% sand	15-25	50	50	1500
100% steel grit(b)	125-150	150	150	1000-1500
100% sand	4-6	25	25	500

(a) Life based on running time of centrifugal blast wheel 495-mm (19½-in.) diam and 65 mm (2½ in.) wide, 30-hp drive and flow rate of 375 kg/min (830 lb/min). (b) G25 grit; 55 to 60 HRC

Fig. 1 Slider blade airless abrasive blast wheel

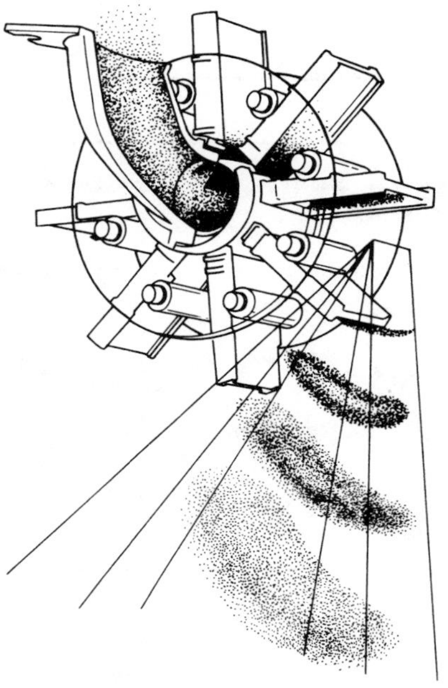

Airblast nozzles are used in a variety of shapes, some as simple as a piece of pipe. Most systems use replaceable nozzles of metal alloys or nozzles with abrasive-resistant ceramic inserts. The latter nozzles may be of straight bore or venturi cross section. All types of abrasive may be handled with the pressure blast system in a variety of environments. In exceptional cases, air pressure blasting is performed in an open field with sand as the abrasive. Protective clothing and a helmet with air supply are the only health precautions taken. Quite often the sand is not recovered after use.

Suction blast systems are generally considered the simplest form of abrasive blast equipment. The suction blast cabinets may be used manually or may have fixed or oscillating nozzles. Figure 2 illustrates a 1220 by 915 by 840 mm (48 by 36 by 33 in.) cabinet. The pressure tank and filling valve may be vertically doubled with a timer and proper valving to provide a continuous automatic pressure tank.

Figure 3 illustrates a suction blast nozzle assembly. The nozzle in the suction cabinet is an induction nozzle which creates a blasting mixture by the siphon effect of the air discharged through the nozzle body. This effect pulls abrasive through the abrasive hose from the cabinet hopper and the blast mixture is formed within the nozzle body. Since only compressed air flows through the air nozzle, the air consumption remains constant. The air nozzle is cast of a wear-resistant alloy. The nozzle can be used until considerably enlarged without affecting the efficiency of the blast. This cannot be done in a direct pressure blast nozzle without seriously affecting air consumption. The amount of abrasive or the mixture of air and abrasive can be controlled in the suction cabinet by changing the relative position of the end of the abrasive hose to the abrasive flowing from the cabinet hopper.

Equipment for Dry Blast Cleaning

Dry blast cleaning is probably the most efficient and environmentally ef-

Fig. 2 Suction blast cabinet

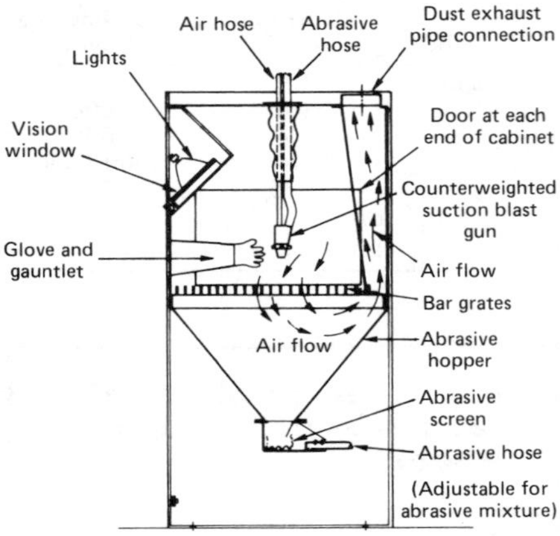

Fig. 3 Suction blast nozzle assembly

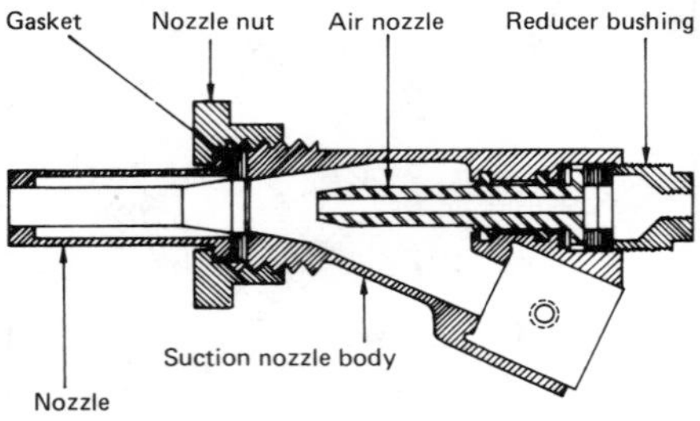

fective method for abrasion cleaning and finishing. Proper ventilation helps maintain a clean work area. No settling ponds or chemical treatment are required. Dust collectors provide dust disposal that is clean and simple, using sealed containers. Dry-blast systems need only be kept dry and can be started and stopped with minimum start-up or shut-down operations. Several types of equipment are available for dry-blast cleaning. Decision as to type of equipment is primarily based on the type of parts to be blasted and the relative throughput required.

Cabinet Machines. A high percentage of dry blast cleaning is performed using cabinet machines. A cabinet houses the abrasive-propelling mechanism, such as a centrifugal wheel or compressed air, holds the work in position, and confines flying abrasive particles and dust. Cabinets are available in a wide range of sizes, shapes, and types to meet various cleaning, production, and materials-handling requirements. Cabinet machines may be designed for manual, semiautomatic, or completely automated operation, to provide single-piece, batch, or continuous-flow blast cleaning.

The table-type machine (Fig. 4) is a form of cabinet machine that contains a power-driven rotating worktable; within the cabinet, the blast stream is confined to approximately half the table area. The unit shown in Fig. 4 is self-contained and mounted on the floor. The work is positioned on the slowly rotating table. The abrasive particles are propelled by an overhead centrifugal wheel. When the doors are closed, blast cleaning continues for a predetermined time cycle. Some table-type machines are designed with one or more openings in the cabinet. These openings are shielded by curtains and permit manual adjustment of the parts during the blast cycle as well as continuous loading and unloading.

With environmental regulations becoming more important and with the necessity of recovering expensive abrasives, environmentally effective room or cabinet enclosures are required. These systems have abrasive recovery systems which remove all foreign contaminants from the used abrasive to permit effective reuse. Removal of the contaminants and fines is performed with an airwash separator as shown in Fig. 5. Spent abrasive and contaminants are fed by a belt and bucket elevator to the helicoid conveyor. The abrasive is screened in the rotary screen, falls in a vertical curtain, and passes under a swinging baffle. The abrasive is then subjected to a controlled cross flow of air which cleans it and removes foreign contaminants and fines. The abrasive then gravitates to a storage hopper and is ready for reuse.

Continuous-flow machines equipped with proper supporting and conveying devices are used for continuous blast cleaning of steel strip, coil, and wire. These machines are also used to clean castings and forgings at a high production rate, making use of skew rolls, monorails, and other continuous work-handling mechanisms.

A continuous centrifugal blast cleaning machine, equipped with a monorail, is shown in Fig. 6. In operation, the work is loaded outside the blast cabinet and is conveyed into it through a curtained vestibule, which is designed with 90° turns to prevent the escape of flying abrasive particles. The conveyor indexes the work to the center of each blast station and rotates it for complete blast coverage. If the workpiece contains intricate pockets, it may be indexed to an off-center position and be slowly conveyed past the blast in a manner that most effectively exposes the pockets to the abrasive stream. To minimize cycle time, the work is moved at an accelerated rate between blast stations. As it is conveyed and rotated on a return passageway that follows along the back of the cabinet, the work is exposed to additional cleaning and acts as a barrier to protect the cabinet walls from wear. Continuous-flow machines incorporate abrasive-recycling facilities and an exhaust system for removing dust and fines.

Fig. 4 Table-type blast cleaning machine

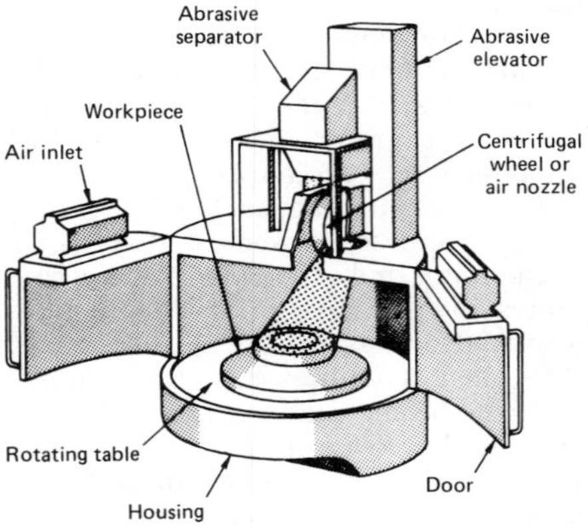

The centrifugal wheel propels the abrasive particles

Blasting-tumbling machines (Fig. 7) consist of an enclosed, endless conveyor, a blast-propelling device, and an abrasive-recycling system. These machines simultaneously tumble and blast the work and are made in various sizes, to accommodate work loads from 0.03 to 2.8 m^3 (1 to 100 ft^3). The work usually is loaded into the conveyor by means of a skip-bucket loader. As the conveyor moves, it gently tumbles the work and exposes all workpiece surfaces to the abrasive blast. At the end of the cleaning cycle, the conveyor is reversed and the work is automatically discharged from the machine.

Blasting-tumbling machines are used for cleaning unmachined castings, forgings, and weldments whose size, shape, and material permit them to be tum-

Fig. 5 Airwash separator

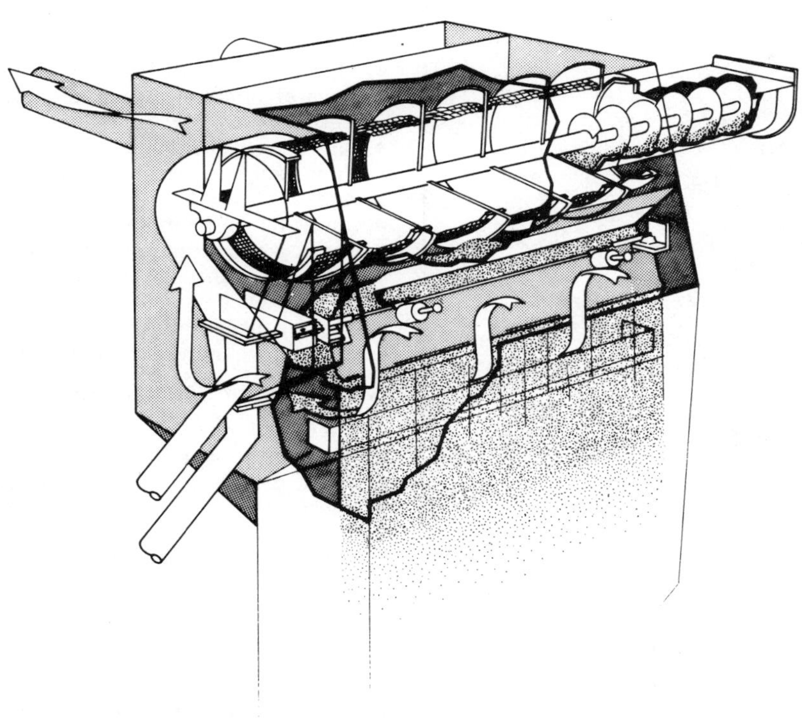

Fig. 6 Continuous centrifugal blast cleaning machine

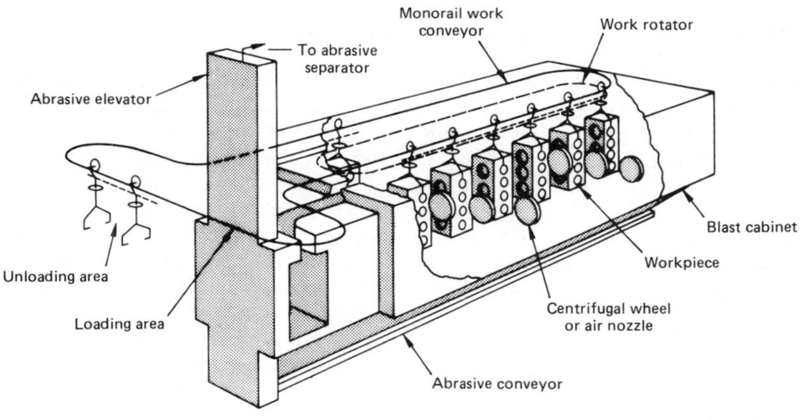

Fig. 7 Blasting-tumbling machine

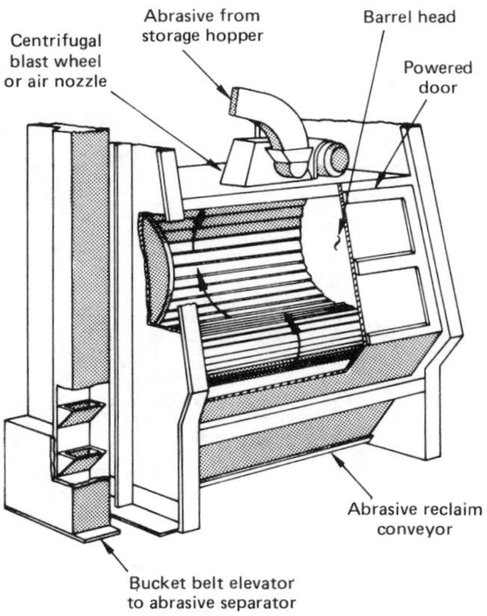

bled without damage. This equipment is not used for cleaning parts after machining, because tumbling damages machined surfaces. Blasting-tumbling machines remove dry contaminants such as sand, rust, scale, and welding flux and provide surface preparation for enameling, rubber bonding, electroplating, or etching before tinning. Blasting-tumbling machines can be integrated into automatic systems for high production rates. An example is the presort and tote box loader work-handling system shown in Fig. 8.

Portable Equipment. When parts to be cleaned are too large to be placed in blasting machines, portable equipment, such as air-blast equipment, can be brought to the workpiece. A low-cost sand usually is used, because it is difficult to reclaim or recirculate the abrasive with portable equipment. Also, it is necessary to prevent random scatter of flying particles. A new development in air pressure-blasting is portable recirculating equipment. This equipment uses a pressurized, internal media hose contained within a larger, evacuated hose. After impact, the media are returned through the outer hose to the central unit for reclaiming and recycling. A brush baffle prevents escape of media at the part surface. With this equipment, large external jobs may be done with specialized media without environmental problems.

Microabrasive blasting is another portable air blasting method. Both the abrasive particle size and nozzle opening are very small. Particle sizes are normally 10 to 100 μm (0.4 to 4 mils) and nozzle openings are 0.4 to 1.2 mm (0.015 to 0.045 in.) in diameter. The tungsten carbide nozzle tips are usually screwed into a pencil-shaped handpiece. Microabrasive blasting is normally a hand-held operation for precision deburring, cleaning, or surface preparation. The design of the handpiece and size of abrasive particle allow a large degree of control in pointing the blast at the work surface. This is advantageous in deburring or cleaning blind orifices, in-

Fig. 8 Presort and tote box loader work-handling system

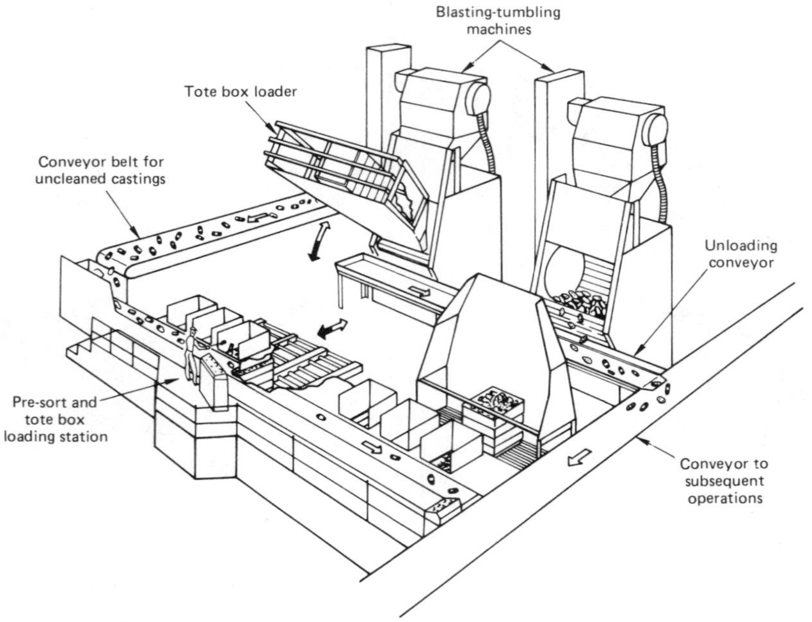

Sorter operates tote box shuttle to and from blasting-tumbling machine

tersecting slots, or internal bores with irregular surfaces. The process is not effective for gross material removal or to cover large areas. In microabrasive blasting, dryness and uniformity of particle classification are very critical and abrasives cannot be reused. Because of the small nozzle size and the types of applications, abrasive usage is not excessive and nonreclamation is reasonable. In continuous duty operation, 0.2 to 0.5 kg (½ to 1 lb) of abrasive is consumed per hour.

Environmental Control. Abrasive blasting operations can be kept environmentally effective by proper ventilation and maintenance as well as training of operating personnel. To ensure proper ventilation of the abrasive blast cabinets, a fabric filter dust collector is generally used with properly designed duct work. The fabric filters are generally equipped with exhaust fans on the clean air side of the dust collector. This location is preferred because it eliminates erosion of the exhaust fan parts.

A second type of fabric filter is the pulse jet. This system uses tubular filter bags made of natural or synthetic felt with an internal support cage of heavy wire and a venturi. Dust and foreign material accumulate on the outer surfaces of the bag and are removed by a short timed, high pressure pulse of compressed air into the top opening of the venturi. Both types of fabric filters can be designed for light or heavy dust loadings and have throughput capacities as low as 2.83 m^3/s (100 ft^3/min) up through several thousand m^3/s (or several million ft^3/min).

Maintenance. Because abrasive blasting machines are essentially self-destructive, every effort must be made to protect components from the violent action of the abrasive. Protect machine interiors with wear-resistant cast alloy metal liners or with heavy rubber mats or sheets, to prevent erosion of metallic surfaces. High-velocity particles bounce from the rubber, usually without damage to either the rubber or the abrasive. If the rubber receives the full impact of the blast, it may require periodic replacement.

Typical maintenance schedules that have proven satisfactory for the principal types of abrasive blasting machines are as follows:

Centrifugal wheel machines: weekly

- Grease and oil all machinery
- Check blades and wheel for wear. An unbalanced wheel can cause bearing wear and shaft bending. Install new blades if needed, check wheel balance, and test for cleaning pattern
- Check for loose buckets on elevator belt; loose buckets may catch on side of elevator shaft
- Check sprocket at top and bottom of elevator shaft for wear and broken teeth
- Check for wear on top plates of machine, rubber table tops, and table rings
- Check for leaks in suction lines
- Check entire machine for possible wear holes through which abrasive might escape
- Check rubber flaps at opening of machine for wear and escaping abrasives

Automatic air blast machines: daily

- Check all nozzles and air jets for wear and proper flow
- Check media and air hoses for leaks
- Check table plates and rubber table tops for wear
- Check suction lines for leaks
- Check belts and chain for wear or slippage
- Check shear pin; replace if necessary

Hand air blast machines: daily

- Check nozzles and air jets for wear and proper flow
- Check the following for leaks: media and air hoses; door gaskets; roof bellows and gauntlets; and suction lines
- Check gun bodies for uneven wear

Cycle Times for Dry Blasting

The amount of abrasive blasting required for a specific application depends on the workpiece material, surface finish requirements, and the performance characteristics of the blast equipment. No dependable formula exists for establishing minimum blasting cycles; the amount of blasting time required to produce a given result in a given machine is established by trial. Table 9 lists abrasives, equipment, and cycles that have been used for dry blasting a number of materials or products for specific purposes.

Applications and Limitations of Dry Blast Methods

Although virtually all metals can be abrasive blast cleaned by at least one of the available processes, the abrasive medium must be carefully selected for soft, ductile metals and their alloys,

Table 9 Abrasives, equipment, and cycles used for dry blasting

Material or product	Reason for blasting	Abrasive Type	Abrasive Size No.	Equipment Type	Horsepower	Nozzle diameter mm	Nozzle diameter in.	Blasting cycle
Ferrous metals								
Cast iron	Prepare for zinc impregnation	Iron grit	G80	Air, table(a)	...	6	1/4	1 h
	Remove molding sand	Steel shot	S230	Wheel, barrel	15	...	...	10 min
Cold rolled steel	Remove graphite for painting	Iron grit	G80	Wheel, barrel	15	...	...	10 min
				Air, table(a)	...	6	1/4	40 min
Gray iron exhaust manifolds, bearing caps	Clean for machining	Malleable iron shot	S460	Wheel, tumble(b)	80	...	...	1500 pieces/h
Gray iron motor blocks and heads	Remove sand and scale after heat treatment	Malleable iron shot	S550	Wheel, blast cabinet(c)	150	...	...	0.19-0.27 min/piece
Hardened steel screws	Remove heat treat scale	Iron grit	G80	Wheel, barrel	10	...	...	5 min
Hot rolled steel	Prepare for painting	Iron grit	G80	Air, table(a)	...	6	1/4	1 h
Malleable iron castings	Prepare for galvanizing	Steel grit	G50	Wheel, barrel	40	...	...	15 min
Pole-line hardware	Prepare for galvanizing	Steel grit	G50	Wheel, barrel	40	...	...	15-20 min
Round steel bar	Etch for adhesive coating	Iron grit	G80	Air, blast room	...	6	1/4	2 min
Soil pipe fittings	Remove molding sand	Steel shot	S330	Wheel, barrel	30	...	...	181 kg (400 lb) in 5 min
Steel drums	Prepare for painting	Iron grit	G80	Air, blast room	...	6	1/4	4 min
Steel rod	Clean for wiredrawing	Steel grit	G40	Wheel, continuous(d)	80	...	...	0.2-1.5 m/s (40-300 ft/min)
Steel screws	Prepare for plating	Iron grit	G80	Air, barrel(a)	...	8	5/16	20 min
Structural steel	Prepare for painting	Steel grit	G40	Wheel, continuous(d)	80	...	...	0.02 m/s (30 ft/min)
Weldments (steel)	Remove scale, welding flux, and splatter for painting	Steel grit	G25	Wheel, barrel	30	...	...	136-272 kg (300-600 lb) in 7 min
Engine parts for rebuilding	Remove paint, scale, and carbon deposits	Glass beads	60-100 mesh	Air	...	6	1/4	5-20 min
Nonferrous metals								
Aluminum	Produce frosted surface	Sand	50	Air, barrel	...	6	1/4	20 min
	Prepare for painting	Iron grit	G80	Wheel, barrel	15	...	...	5 min
Bronze	Produce frosted surface	Sand	50	Air, barrel	...	6	1/4	20 min
Aluminum and bronze	Prepare and condition surface	Glass beads	20-400	Air	...	6	1/4	5-20 min
Nonmetallic materials								
Clear plastic parts	Produce frosted surface	Sand	50	Air, barrel	...	6	1/4	15 min
Hard rubber	Improve appearance	Sand	50	Air, barrel	...	6	1/4	20 min
Molded plastic parts	Remove flash	Walnut shells	...	Wheel, barrel	10	...	...	8 min
Phenolic fiber	Produce frosted surface	Sand	50	Air, barrel	...	6	1/4	30 min
	Prepare for painting	Sand	50	Air, barrel	...	6	1/4	20 min

(a) Four air nozzles. (b) Two wheels, 40 hp each. (c) Six wheels, 25 hp each. (d) Four wheels, 20 hp each

such as aluminum, magnesium, copper, zinc, and beryllium. Otherwise, abrasive blasting may result in severe surface damage.

In some instances, abrasive blast cleaning induces residual compressive stresses in the surface of the workpiece. This is especially true with steel shot or glass beads. Although these stresses are highly desirable in terms of fatigue strength, they are detrimental to electrical components such as motor laminations, because they alter electrical and magnetic characteristics. Blasting at high pressures with a large particle size may produce warping in thin sections of steel and other metals as a result of induced stresses. The blasting of extremely hard and brittle materials may result in chipping and excessive media consumption. The corrosion resistance of stainless steels may be adversely affected by the adherence of dissimilar metals on the matte surface produced by abrasive blasting with metallic media. If this is a concern, grit blasting should be followed by chemical cleaning.

Because abrasive blasting usually roughens highly finished surfaces, particularly those of low hardness, the process is unsuitable for cleaning parts for which dimensional or surface finish requirements are critical. The peening effect of abrasive particles may distort flat parts, particularly those with a high ratio of surface area to volume, such as clutch disks, long thin shafts, and control bars.

Even when the application of abrasive blast cleaning is known to be advantageous for a specific part, the particular abrasives and process selected should be entirely compatible with part requirements. For example, because small fragile parts may break in a tumbling operation, they should be processed in a stationary position on a rotating table or in conveyor equipment. Shields or caps, made of abrasion-resistant rubber compounds, sheet metal, or plastics, are used to protect threaded sections from the abrasive blast. The tooth profiles of gear teeth may be protected from excessive blasting by positioning designed to control their exposure to the blast. Baffles and reflectors may be used to direct abrasive particles to certain areas, such as undercuts, which should not be exposed to the severity of direct impingement. Because it is usually difficult to adjust velocities of mechanical cleaners, a finer shot or grit size may be selected to modify cleaning characteristics.

Types of Soils. Mechanical dry blasting does not readily lend itself to the removal of viscous or resilient soils such as grease, oil, or tar. These materials not only resist the blast action but also cling to, or coat, the abrasive material and components of the abrasive-recycling system. In time, such soils disrupt proper recycling, reclamation, and airwash separation of the reusable abrasive. Therefore, parts coated with oil or other viscous soils must be thoroughly degreased, or scrubbed and dried, before the mechanical dry blast operation.

Dry surface soils, such as sand, scale, rust, paint, weld spatter, and carbon, are readily removed by the dry blast action. These friable contaminants are compatible with airwash separation for reclamation of usable abrasive. Dry contaminants can be present on a surface in any quantity. Sand cores and molding sand are removed by the centrifugal blast method during coreknockout operations. Large castings are processed with portable equipment, and small castings are processed in batch-type machines.

On a limited or intermittent production basis, using air or wet blast methods to remove soils that are not removable by wheel blasting is possible. For example, an air blast nozzle may be used with soft agricultural abrasives, which absorb viscous soils, to clean oily or greasy surfaces. Because the initial cost of the abrasive is relatively low, the material can be discarded when it becomes contaminated or saturated. This method is often used by maintenance personnel for cleaning motors and gear reducers.

Workpiece Shape. Parts of virtually any shape can be cleaned by some method of abrasive blasting, although complex parts with deep recesses or shielded areas present special problems. For example, it is often difficult for the abrasive to make contact with all surfaces of deep blind pockets with a velocity sufficient to loosen the soil to be removed. When direct impingement is impossible, deflection of the abrasive particles by means of baffles sometimes solves the problem. For effective cleaning of the inside surface of pipe, special air blast nozzles and lance air blast equipment must be used to deliver the abrasive with adequate velocity. Even these techniques have practical limitations, depending on the diameter and length of the pipe.

A second problem encountered in the cleaning of pockets or recessed areas is the buildup of abrasive in these areas. An accumulation of abrasive shields the surface from further blast action and interferes with cleaning. This problem is usually solved by positioning the work in a manner that permits the abrasive particles to drain by means of gravity. This positioning change may necessitate a corresponding change in the positioning of blast equipment. Cylinder blocks and valve bodies are typical examples of parts with recesses that catch and retain accumulations of abrasive.

Workpiece Size. The size of parts that can be cleaned by the centrifugal blast wheel method is limited principally by the size of the enclosure and the number of wheel units that can be applied economically. Wheel units are maneuverable to only a limited extent. Therefore, as part size increases, it is necessary to rotate or convey the part in a manner that properly exposes it to the available blast units. Castings and weldments 6 m (20 ft) in diameter and 5 m (16 ft) high, and weighing up to 136 t (150 tons), have been cleaned in mechanical blast rooms. These rooms are equipped with a rotary table or a car-mounted rotary table and with several centrifugal wheel units operating simultaneously. During cleaning, such extremely large parts frequently require repositioning to expose all surfaces to the blast. Intricately shaped large parts may also require auxiliary air blast touch-up cleaning.

Various types of continuous blast machines are used for the cleaning of repetitive work. These machines vary in size and design in accordance with the application and type of work-handling equipment required. Rolled steel products, such as sheet, strip, wire, rod, and structural shapes, lend themselves to continuous mechanical blasting at high production rates. For example, hot rolled strip up to 1830 mm (72 in.) wide is mechanically blasted on a continuous basis, to reduce the time required for acid pickling.

Structural shapes, including the largest sections rolled commercially, may be cleaned on continuous roll conveyor machines equipped with multiple wheel units for coverage of all surfaces. Such equipment is used for the removal of mill scale and rust before welding or painting. Hot rolled rod and bar stock are cleaned on single or multiple nozzle or wheel machines to remove surface scale and prepare the surface for cold drawing or cold heading.

Air blast equipment, by virtue of the flexibility provided by blast hose and nozzles, is widely used for cleaning extremely large parts and assemblies. Railroad cars, for example, can be reconditioned inside and outside by this method. Large storage tanks and vessels also are cleaned with air blast equipment, using inexpensive abrasives such as sand, which need not be reclaimed, or with reclaiming equipment previously described in this article.

In contrast, parts as small as 10 to 13 mm (3/8 to 1/2 in.) in diameter can be satisfactorily cleaned by abrasive blasting. Usually, these small parts are most efficiently handled in mechanical or air blasting equipment, either barrel machines or combination tumbling-and-blasting units. Auxiliary devices, such as wire cages or baskets, may be used to prevent very small parts from being lost in the abrasive.

Mixed Work Loads. In blasting with either fixed nozzles or centrifugal wheels, it is always more economical to process loads made up of parts of about the same size. Mixing large and small parts in the same load is basically inefficient, because it wastes abrasive and power and frequently results in overblasting some parts and underblasting others, although parts can be mixed within reasonable limits. In job-shop operations especially, a varied product mix can be cleaned in a single tumbling and blasting operation. However, parts with thin sections that may bend or seriously distort should not be processed with parts that are relatively compact.

Quantity and Flow of Work. Continuous airless blast cleaning equipment is generally used for medium-to-high production cleaning applications; however, there are no actual quantity limitations. For the most economical use of continuous blast equipment, not only must the work being cleaned be repetitive and similar in size and shape, but the quantity of work flowing through the blast cleaning machine must be uniform and constant.

Monorail conveyor equipment should be operated with all work hangers fully loaded and few gaps in the production flow. This type of equipment usually is designed so that conveyor speeds can be regulated or index times varied to match work flow requirements, and so that the feeding of abrasive into the blast wheels can be regulated to suit work flow conditions. Monorail conveyor equipment for cleaning gray iron motor blocks and similar parts is capable of cleaning from 400 to 500 workpieces per hour.

Continuous blasting-tumbling barrel machines also require a steady flow of work of relatively uniform size and shape. A constant level of work in the blast chamber makes the operation more economical and promotes uniform cleaning. In cleaning medium-size gray iron castings, these barrels have a capacity of over 23 t (25 tons) per hour. If a steady flow of work cannot be maintained, it is economical to stockpile work until a sufficient accumulation is available. Barrel blasting machines, some table machines, and spinner-hanger machines are suited to this type of operation.

Miscellaneous Applications. Dry abrasive blasting has proven useful in applications in which cleaning is of only secondary importance. One automotive manufacturer blasts induction hardened transmission pins with chilled iron grit to permit rapid visual inspection and segregation of improperly hardened pins. After blasting, hardened surfaces have a markedly shiny appearance and unhardened surfaces appear dull. This same inspection technique is used by a manufacturer of rolling-mill rolls to determine uniformity of heat treatment of the roll surface. A manufacturer of carburized gears uses the technique to detect areas of decarburization and case leakage.

In some applications, dry abrasive blasting serves to supplement other inspection techniques. Aircraft quality investment and sand castings are blasted before magnetic-particle inspection to reduce or eliminate glare caused by polishing or machining. Defects are more readily detected on the dull blasted surface.

Wet Blasting

Wet blasting differs from dry blasting in that the abrasive particles used are usually much finer and are suspended in chemically treated water to form a slurry. The slurry, pumped and continually agitated to prevent settling, is forced by compressed air through one or more nozzles, which are directed at the work.

Applications. In further contrast to dry blasting, wet blasting is not intended for the gross removal of heavy scale, coarse burrs, or soil, but is intended to produce only relatively slight effects on the workpiece surface. Wet blasting is most commonly used for:

- Removing minute burrs on precision parts
- Producing satin or matte finishes
- Inspecting finish ground, hardened parts
- Removing fine tool marks from hard parts
- Removing light mill scale or machining marks in preparation for plating
- Removing surface oxide in preparation for soldering of electronic components and printed circuits
- Removing welding scale

Many small parts, including hypodermic needles and electronic components, are deburred by wet blasting. The application of wet blasting to large parts is limited to the cleaning and finishing of forging dies from which a minimum of metal removal is desired; dies weighing up to 90 t (100 tons) have been wet blasted. Many cutting tools are wet blasted after final grinding.

Precleaning. In most instances, precleaning must precede wet blasting to prevent contamination of the recirculating slurry. Grease, protective coatings, and heavy oils may be removed by conventional degreasing methods. Heavy rust and dry soils may be removed by dry blasting. A very light layer of rust may be removed from machined parts by wet blasting without precleaning; in general, however, the finer the abrasive used in the wet blast slurry, the greater is the need for

Table 10 Characteristics and typical applications of abrasives used in wet blasting

Abrasive	Mesh size	Characteristics and applications
Silica	40-80	Fast-cutting. Used for deburring steel and cast iron, removing oxides from steel. Close tolerances cannot be held
Silica	80	Fast-cutting. Used for deburring steel and cast iron, roughening surfaces for plastic bonding or rough plating. Has peening action. Tolerances cannot be held
Quartz (ground)	80	Very fast-cutting. Used for removing heavy burrs, light or medium scale, excessive rust. Can be used on nickel alloy steels. Tolerances cannot be held
Novaculite	100	Fast-cutting. Used for cleaning carbon from piston and valve heads; deburring brass, bronze, and copper. Can be used on crankshafts. Tolerances cannot be held
Quartz (ground)	100, 140	Fast-cutting. Used for blending-in preliminary grind lines on steel, brass and die castings; removing medium-hard carbon deposits; blasting radii of 0.1 to 0.3 mm (0.005 to 0.010 in.)
Silica	140	Used for removing small burrs from steel, copper, aluminum, and die castings; rough cleaning of dies and tools, removing metal. Tolerances cannot be held
Novaculite	325	Slow-cutting. Used in first stage for cleaning master rods and glass, and in second stage for cleaning aluminum pistons, crankshafts, impellers, valves. Holds tolerances to 0.06 mm (0.0025 in.)
Aluminum oxide	400	Fast-cutting. Used on stainless steel and on zinc and aluminum die castings. Excellent for oil-contaminated surfaces
Novaculite	1250	Used in second stage for cleaning crankshafts, impellers, rods, pistons, valves, gears and bearings. Also for polishing metals, tools, dies and die castings. Tolerances can be held
Novaculite	5000	Used for obtaining extra-fine surfaces on parts
Glass beads	20-400	Used for removing scale or discoloration after heat treating, removing oxide from jet-engine and electronic components. Produces peening effect

precleaning to prevent contamination of the slurry.

Abrasives for Wet Blasting

Many different kinds and sizes of abrasives can be used in wet blasting. Sizes range from 20-mesh (very coarse) to 5000-mesh (which is much finer than face powder). Among the types of abrasives used are: organic or agricultural materials such as walnut shells and peach pits; novaculite, silica, quartz, garnet, and aluminum oxide; other refractory abrasives; and glass beads.

The organic or agricultural materials are used for mild blasting only. Novaculite, a soft type of silicon dioxide (99.46% silica; Mohs hardness, 6 to 6.5), is used to remove very light burrs and produces a fine matte finish. The silicas are slightly more aggressive than novaculite and are used to remove larger burrs and scale. Quartz is still more aggressive than silica and lasts longer; it is used for the rapid removal of tenacious burrs and scale.

Garnet abrasives are highly aggressive and have long life; for a comparable mesh size, garnet will produce a rougher surface finish than will silicon dioxide. Aluminum oxide, silicon carbide, and other artificial abrasives are the most aggressive and erosive. Round glass shot (beads) composed of 72% silicon dioxide, 15% sodium monoxide, 9% calcium oxide, and 1% alumina produce a brighter finish and provide the most peening action. The high ricocheting property of glass beads is of value in blasting areas that are hard to reach. Table 10 describes various abrasives used in wet blasting and lists typical applications.

Liquid Carriers. The liquid most commonly used to carry the abrasive particles is water containing additives such as rust inhibitors, wetting agents, and anticlogging and antisettling compounds. In a few applications, such as in the manufacture of spark plugs, petroleum distillates have been used as abrasive-carriers, for the removal of oil residues and fine chips and burrs. In these instances, the use of water would create difficulties. Petroleum distillates, however, can be used only with specially designed wet blasting units because of the fire hazard.

Fig. 9 Effect of variables on amount of metal removed during wet blasting of 1010 steel

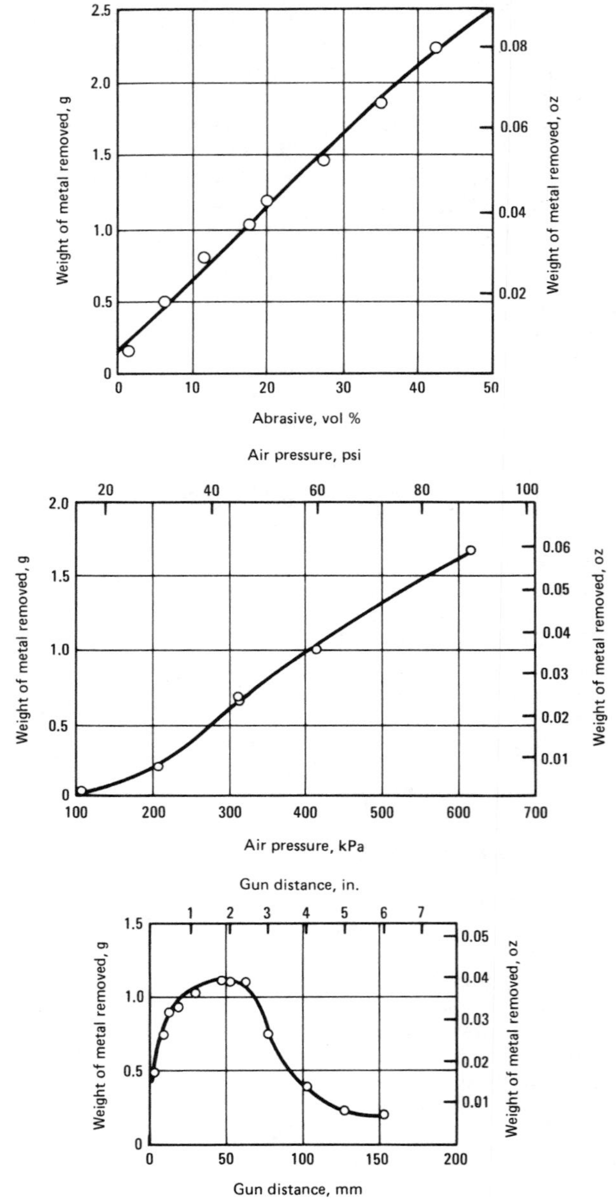

The proportion of abrasive to liquid in the wet abrasive slurry can be varied over a wide range, although certain limitations apply. The use of a very small percentage of abrasive results in slight cleaning action, while too large a percentage of abrasive might result in the formation of a paste that could not be properly circulated. Proportions should be fixed at a predetermined level for each application, to ensure uniform cleaning action and production of a uniform finish. A range of 20 to 35 vol % abrasive is satisfactory for most applications but may be modified because of particle size, surface tension, specific gravity, agitation, and the desired effect on the workpiece. Figure 9 illustrates the effects of three variables on metal removal during wet blasting of 1010 steel. Reclamation of wet abrasive is usually impractical. The mixture gen-

Fig. 10 Wet blasting machines

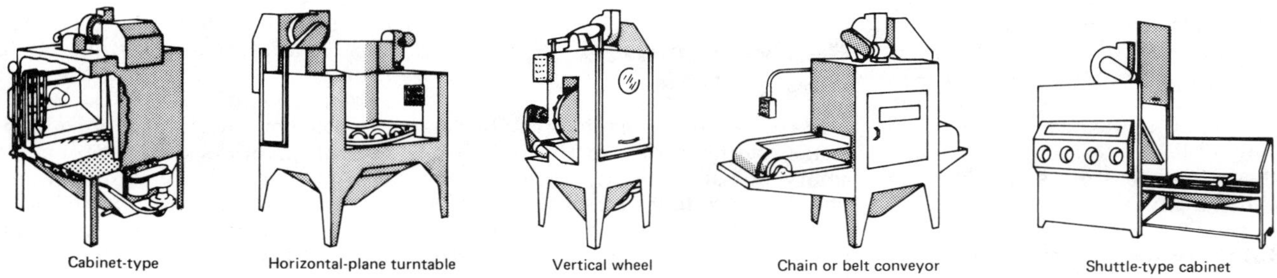

erally is used until its cleaning action becomes unsatisfactory and it is then discarded.

Equipment for Wet Blasting

Although equipment for wet blasting is often of special design for a particular application, several basic types of machines (Fig. 10) have been developed for general use. Among these are:

- Cabinet-type machines
- Horizontal-plane turntable machines with tables of various diameters
- Vertical wheel-type machines
- Chain or belt conveyor machines
- Shuttle-type cabinets with cars and rail extensions
- Car-mounted, self-contained turning mechanisms for shafts or tubular parts

These basic types may be equipped with strippers, takeoff conveyors, and wash-rinse-dry stations. The blast guns may be mounted in a stationary position or be made to oscillate, depending on the application.

Fig. 11 Amount of air required for various sizes of air-jet nozzles at different operating pressures

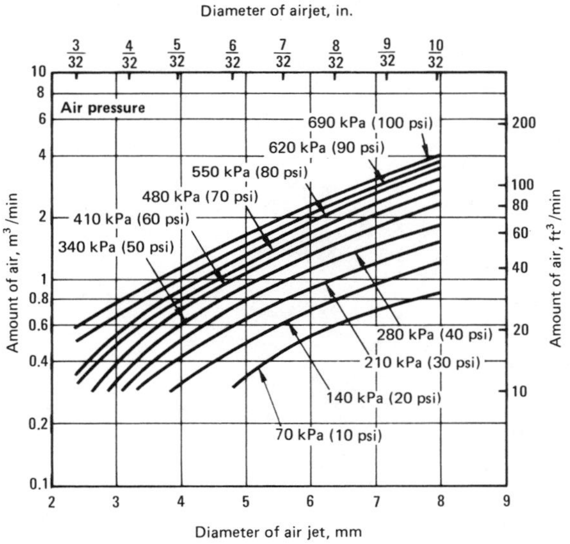

Combination tumbling-and-blasting cabinet-type machines also are available. These machines are made with an extension stand provided with rails so that a car-mounted tumbling barrel may be moved into the cabinet. The barrel, a self-contained unit, is provided with a driving mechanism for turning the barrel and oscillating the blast gun or guns inside the barrel. Barrels range in diameter from 305 to 660 mm (12 to 26 in.) and are perforated to allow the slurry to escape and recirculate.

Nozzles for wet blasting vary in form, shape, length, and diameter according to the type of abrasive used and the size of the parts to be blasted. The most common form is a cylindrical nozzle about 25 mm (1 in.) long, 13 mm (½ in.) ID and 19 mm (¾ in.) OD. It is made of low-carbon steel and is used in hand-operated cabinets for general cleaning purposes. Using silicates or quartz in mesh sizes ranging from 100 to 300 and at blasting pressures of 550 to 620 kPa (80 to 90 psi), this nozzle has an average life of about 40 h.

Special nozzles, such as fan-shaped types, usually are made of an alloy cast iron, chilled to a high hardness, although some special nozzles are made of rubber. Carbide nozzles are used in mechanized units that provide surface preparation for such processes as anodizing and phosphating. The higher cost of cylindrical carbide nozzles is justifiable in terms of long life and low maintenance. For special shapes, however, the cost of carbide is likely to be prohibitive. When used with fine abrasives, a cylindrical carbide nozzle has a life of several thousand hours. In one application, a carbide nozzle exhibited no measurable wear after 1000 h of service with 140-mesh quartz at a pressure of 620 kPa (90 psi). A nozzle for wet blasting is considered worn out when (*a*) its wall thickness has been so reduced as to be potentially damaging to the gun, or (*b*) the blast pattern provides inadequate coverage of the workpiece.

The amount of air required for wet blasting depends on the diameter of the air-jet nozzle and the operating pressure. Figure 11 illustrates the cubic feet of air needed to operate air-jet nozzles of various diameters 2.4 to 8 mm (3⁄32 to 5⁄16 in.), at pressures ranging from 70 to 690 kPa (10 to 100 psi). The data are based on operating a wet blasting gun with a slurry containing 40 vol % abrasive of 140 mesh.

Maintenance. Typical maintenance schedules for hand-operated wet blasting cabinets are as follows:

- *Weekly*: wash filter and pump-intake strainer, making sure all foreign particles are removed; grease the pump follower
- *Monthly*: check pump packing gland for leaks, and tighten if necessary
- *Semiannually*: remove wear plate from pump and inspect impeller and housing for wear; grease fan pillow-block bearings

Health Hazards and Safety

Blast cleaning operations can be performed without risk to the health or safety of personnel if precautionary measures are followed rigorously. The health hazard that accompanies blast cleaning is silicosis, a disease of the lungs that results from the prolonged breathing of very fine particles of silica sand. In past years, improperly done sandblasting has been the cause of many cases of silicosis, the effects of which can be permanently disabling and even fatal. No matter how good the equipment may be, the blaster should be x-rayed before he is employed and at least once a year thereafter. The films should be read by a physician specializing in this particular field.

Health Precautions. Sand blasting of sand-free steel or iron castings and shot blasting of sandy castings produces enough fine silica to make the air dangerous to breathe. The blaster must be protected by a helmet supplied with air, special gauntlets, an apron, and often special spats. The blasting equipment and the blasting room must be brought to a high degree of efficiency and include ventilation with cleaning and recirculation of the air that is withdrawn. The air supplied the worker (about 0.17 m^3/s or 6 ft^3/min) should be clean, odorless, dry, and free from gaseous contaminants. If taken directly from the compressed-air receiver, it is likely to be wet and malodorous, due to oil since most compressed air is oil pumped.

Salt Bath Descaling

By William G. Wood
Vice President
Technology/Research and Development
Kolene Corp.
and
Thomas F. McCardle
Supervisor-Sales
Kolene Corp.

SALT BATH DESCALING processes are classified as oxidizing, electrolytic, and reducing. Complete removal of scale requires the use of an acid dip or acid pickling after salt bath descaling. The oxidizing process is more important industrially because of wider applications and simplicity of operation. The electrolytic process usually provides more thorough scale removal than the reducing process, and it also has the capability of functioning both in an oxidizing and a reducing mode. Because the reducing process employs a bath that is operated at a lower temperature, it is sometimes advantageous for descaling metals that undergo a change in properties at higher temperatures.

The chemical reactions promoting salt bath conditioning are as follows:

$$2MO + NaNO_3 = M_2O_3 + NaNO_2 \quad \text{(Eq 1)}$$

$$NaNO_2 + \tfrac{1}{2}O_2 = NaNO_3 \quad \text{(Eq 2)}$$

$$MO + 2Na = M + Na_2O \quad \text{(Eq 3)}$$

$$MO + NaH = M + NaOH \quad \text{(Eq 4)}$$

In Equation 1, a lower valence metal oxide is converted to a more soluble higher valence oxide by oxidation in the molten salt. Equation 2 describes the process by which the oxidizing potential of the bath is maintained at a constant level by reoxidation in contact with the atmosphere. Equation 3 is a typical electrolytic bath reaction where the metal oxide is reduced by the active sodium metal, and Equation 4 indicates a type of reaction taking place in a sodium hydride reducing bath.

Fused Oxidizing Salt Process

Metal finishing operations involving stainless steels, superalloys, and titanium metals usually require an oxidizing salt conditioning treatment in the processing of bar, rod, wire, and strip on a production basis. The operating temperatures of these salts vary between 205 and 480 °C (400 and 900 °F), and in this range, the high chemical activity required for the removal of the complex oxides and scales developed in hot forming operations is ensured.

At the present time, molten salt baths are not complete cleaning systems. They must be used in conjunction with acid pickling solutions. The required concentration and temperature of pickling acids can be considerably reduced by the conditioning action of the fused salt. Table 1 represents the actual

Table 1 Acid concentrations and operating temperatures used for pickling

Tank No.	Solution	Acid	Amount, %	Operating temperature °C	Operating temperature °F
Before salt bath installation					
1	Electrolytic	HNO_3	12	...	...
2	...	HNO_3 + HF	18-25	66-88	150-190
3	...	HNO_3	6-8	21	70
After salt bath installation					
1	Electrolytic	H_2SO_4	6-8	...	...
2	...	HNO_3 + HF	8-10	52-66	125-150
3	...	HNO_3	1-2	21	70

Fig. 1 Sludge collection system without shutdown

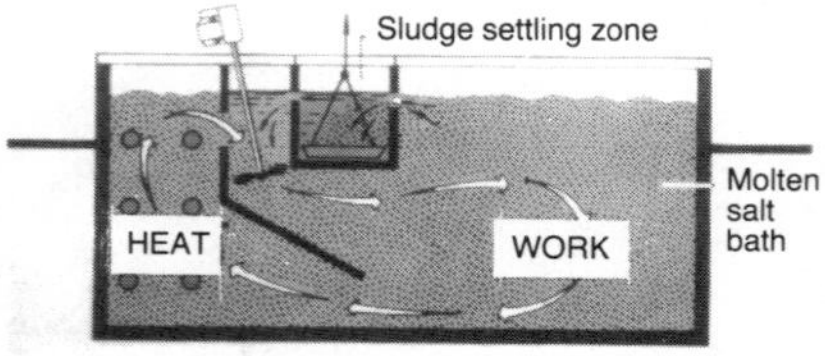

Fig. 2 Salt bath cleaning system for diesel locomotive cylinder lines

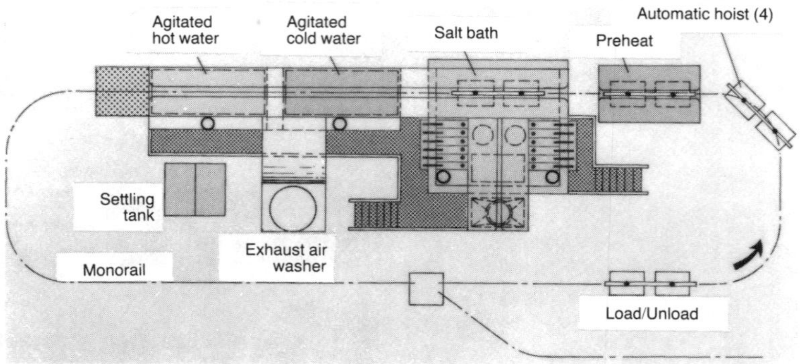

Table 2 Simultaneous analysis of salt and sludge

	Na_2CO_3, %	TiO_2, %	Fe, %
Salt	9.44	0.36	0.059
Sludge	7.56	16.42	0.98
Salt	10.67	0.057	0.435
Sludge	11.40	1.07	0.37
Salt	10.79	0.335	0.047
Sludge	9.15	3.39	8.72
Salt	6.99	0.317	0.039
Sludge	10.27	4.34	0.85

Note the high content of metallic impurities which are present in the sludge as compared to that in the working body of the salt

Fig. 3 Sand particles embedded in iron casting

improvement effected in a production facility by the installation of an oxidizing salt in a continuous anneal and pickle facility.

Systems employing oxidizing-type fused salts are used extensively in cleaning all grades of stainless steel in continuous anneal and pickle lines because of their very rapid reaction with the anneal scales, which is necessary for line speeds approaching 45 to 60 m/min (150 to 200 ft/min). These salts are also widely applied in the cleaning of forgings, castings, and stainless steel wire. Oxidizing salts are also preferred for the cleaning of titanium mill products and titanium fabrications because their reactions tend to minimize or eliminate the problems associated with hydrogen pickup in reducing processes.

Sludge Removal. During the bath operation, oxidizing-type fused salts develop sludge products consisting of carbonates and metal alkali combinations which are insoluble in the salt. These insoluble residues, however, are easily removed in a sludge collection system as shown in Fig. 1. Controlled directional agitation sweeps the molten salt through the heat zone and back into the work zone, thereby effecting constant heat transfer and uniform temperatures in the bath. The agitation produces a turbulent movement throughout the work zone and heat zone preventing any suspended particles from settling. This aids in rapid and efficient cleaning of the work.

Concurrent with the work cycles, some of the bath passes through a small opening from the work area into the settling zone. A gate controls the rate of flow to suit requirements. Movement of the bath in this area is slowed sufficiently for the sludge particles to settle into a pan which rests at the bottom of the compartment. The pan is periodically removed, emptied, and replaced without interruption of the cycle. The effectiveness of this sludge collection system is shown in Table 2. This method of purification not only maintains the fused salt in peak operating condition, but also in conjunction with fresh salt makeup and the regenerative oxidation reaction shown in Equation 2,

Fig. 4 Void in casting after removal of sand particles in electrolytic salt

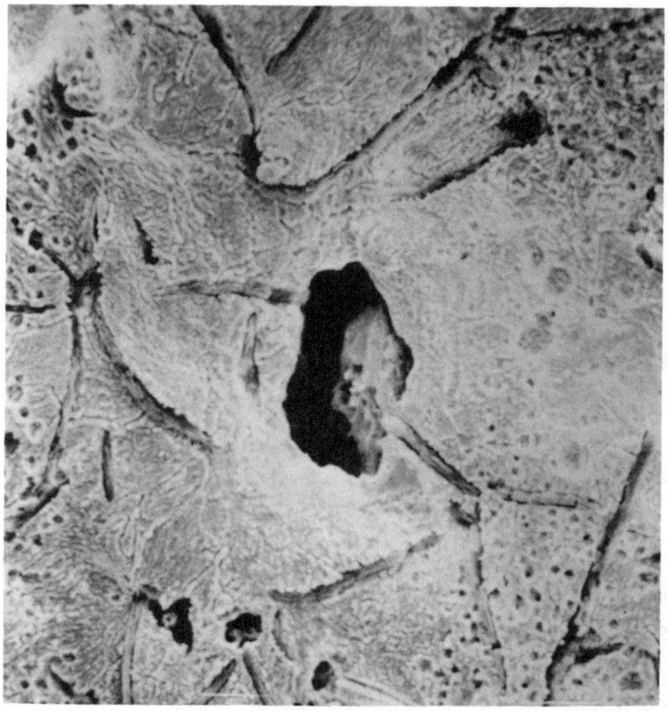

permits the continuous operation of oxidizing salt baths for an almost unlimited length of time.

Electrolytic Process

Fused salts that are neither chemically oxidizing nor chemically reducing can be activated to produce either of these conditions by the input of electrical energy. The electrical system involved is rather simple. It employs a direct current source, a reversing switch, a positive and a negative pole that can be either the furnace wall or the work load, and a conducting media that is the molten salt bath.

Electrolytic fused salt systems are used primarily for the removal of sand from iron castings, especially where designs have limited access to internal surfaces and dislodged particles of retained sand can cause substantial damage under operating conditions. For this purpose, current reversal is not required, and the electrolytic salt bath is operated with the work at a negative or reducing polarity. Cycles are usually developed through experience and may vary from 15 to 30 min, followed by a water rinse and drying. If the surface is being prepared for silver brazing, a 15-min reduction cycle is usually followed by a 20-min oxidation cycle and finally by a 10-min short reduction cycle. Cycle changes are easily effected by the simple reversing switch. Electrolytic salt bath cleaning is usually performed in batch processing equipment, although newer trends have been toward completely automated lines as shown in Fig. 2.

Sand Removal. The chemical composition of the soils that are normally removed in an electrolytic salt system is, either by nature or design, so resistant to attack by reactive solvents that even very strong acid or alkalis are an impractical approach. The silica or sand particle shown in Fig. 3 embedded in the iron casting is relatively harmless, providing it is on the exterior surface and can be removed by tumbling or mechanical abrasion. When it is allowed to remain in a crevice, a shielded area, or an internal surface, it may be dislodged in operations producing either premature wear or catastrophic failure in an automotive, farm implement, or aircraft structural part. The void shown in Fig. 4 remains after removal of the sand by a brief electrolytic molten salt treatment which is preferable to dissolution in hydrofluoric acid.

Electrolytic salt processing also completely removes oxide and scale from hydraulic parts. This is usually accomplished in conjunction with sand removal, but fused salt removal of oxides could be a requirement in complex castings. Acid pickling, although less expensive than the fused salt treatment, has a decided disadvantage of promoting excess metal loss if the cycle is sufficiently long to ensure that all internal passages are completely cleaned. It may also be assumed that, as acid disposal costs increase, the economics of acid pickling may not be quite as attractive.

The removal of sand and oxide in an electrolytic salt is accomplished by the following sequence of operations. The workpiece in the fused mixture of sodium and potassium hydroxides, fluorides, chlorides, and aluminates is negatively charged through the direct current reversing switch. The container or furnace wall is the opposite pole carrying the positive charges indicated. Under these conditions, the electrochemical system produces a concentration of ions in the work area that react selectively with sand particles and metallic oxides as shown in the following equations:

$$2NaOH + SiO_2 = Na_2SiO_3 + H_2O \quad \text{(Eq 5)}$$

$$2Na^+ + 2H_2O + 2e = 2NaOH + H_2 \quad \text{(Eq 6)}$$

$$6Na^+ + Fe_2O_3 + 3H_2O + 6e = 6NaOH + 2Fe \quad \text{(Eq 7)}$$

Current density requirements are normally not calculated as exact figures.

Carbon Removal. Another contaminant on cast iron surfaces, which is most effectively removed by fused salts, is graphitic carbon or graphite smears developed in machining. All traces of surface graphite should be removed prior to silver brazing of cast iron if good wetting and a fluid tight joint is to be obtained in fittings. Because carbon is chemically inert, it is resistant to attack by most chemicals. The electrolytic fused salt process removes carbon completely from the surface as shown in the photomicrograph of Fig. 5, permitting bonding of bearing metal to cast iron shells, joining of iron fittings to steel tubing, and overall casting sim-

Fig. 5 Gray cast iron prior to and after electrolytic salt bath cleaning

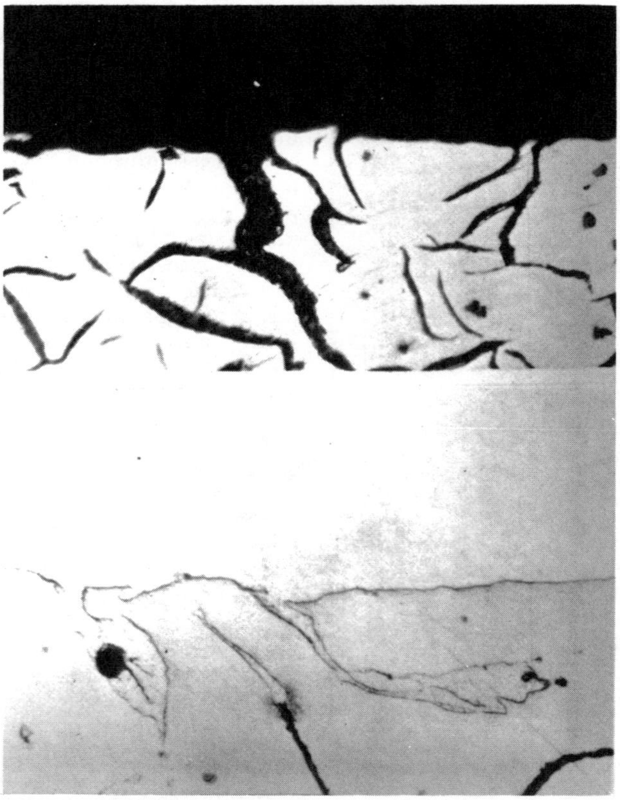

plification. The top section of the photomicrograph shows a finished surface of gray cast iron prepared for silver brazing. Note the predominance of graphite flakes at the surface, which in this condition does not permit good wetting by the silver brazing alloy. The bottom section of the photomicrograph shows that all of the surface carbon can be completely removed by electrolytic fused salt cleaning. In this condition, the cast iron is prepared to accept a uniform film of silver brazing alloy and thereby effect a completely uniform bond.

Carbon removal in molten salts is accomplished by charging the work load with a positive charge through the direct current reversing switch. At this polarity, an oxidizing condition is developed at the surface removes carbon according to the following reactions:

$$C + 4(OH)^- = 2H_2O + CO_2 + 4e$$

$$2Fe + 1\tfrac{1}{2}O_2 = Fe_2O_3$$

Under these conditions, the metallic surface iron is also oxidized, but this situation can be corrected by a short period of reduction with a simple reversal of the electrical switch.

The Reducing Process

The reducing or sodium hydride descaling process is performed in a fused caustic bath in which sodium hydride (NaH) is generated *in situ* from metallic sodium and dissociated ammonia. The reaction takes place in generators immersed in the molten sodium hydroxide, and the recommended concentration of 1.5 to 2% sodium hydride is adjusted and maintained by controlling the feed of sodium and gaseous hydrogen. This control of the sodium hydride concentration of the bath is necessary for efficient descaling and is achieved by regular analyses. The uniformity of sodium hydride throughout the bath is dependent on convection currents and the periodic immersion of products to be descaled.

Heat process scale can be removed from all metallic alloys that do not react with the required caustic base and that are not metallurgically affected by the temperature of operation of between 370 and 400 °C (700 and 750 °F). The scale present on the surface is reduced according to the following reaction:

$$4NaH + M_3O_4 = 4NaOH + 3M \quad \text{(Eq 8)}$$

This reaction is ideal and applies to simple oxides such as iron oxide. Where complex mixtures including iron, nickel, and chrome are present, the reducing action does not proceed to completion, but leaves a residual form of the oxide which must be removed by acid pickling.

A sodium hydride bath in constant contact with air can cause excessive consumption of sodium because of the reaction of sodium hydride with the oxygen in the air according to:

$$NaH + \tfrac{1}{2}O_2 = NaOH \quad \text{(Eq 9)}$$

This reaction tends to reduce the concentration of the active agent and therefore when the bath is idle or during long periods of work immersion, a graphite cover is maintained on the surface in order to prevent contact of the air with the molten sodium hydroxide. Another impurity generated in the bath is carbonate which forms according to the following reaction:

$$2NaOH + CO_2 = Na_2CO_3 + H_2O \quad \text{(Eq 10)}$$

The carbonate is an impurity and tends to increase the viscosity of the bath and thereby increases the dragout. In order to prevent excessive consumption of the salt, a periodic desludging operation is necessary.

The sodium hydride descaling process has limited application in modern fused salt descaling. It does not lend itself well to continuous lines because of the necessity of keeping the surface protected from contact with the atmosphere. It does not descale most high alloy austenite stainless or nickel-based grades as rapidly or effectively as the oxidizing molten salt processes. It is quite effective in removing heavy scales from 400 series stainless and some carbon steels. Also, the process is much more effective than oxidizing salts at descaling titanium. It should, however, be used with great caution because of the adverse effects of hydrogen pickup. Hydrogen becomes an increasingly serious problem with decreasing gage; thus, sodium hydride may be usable for plate, but not for sheet.

Fig. 6 Fused salt cleaning system

Fig. 7 Schematic of salt bath furnace

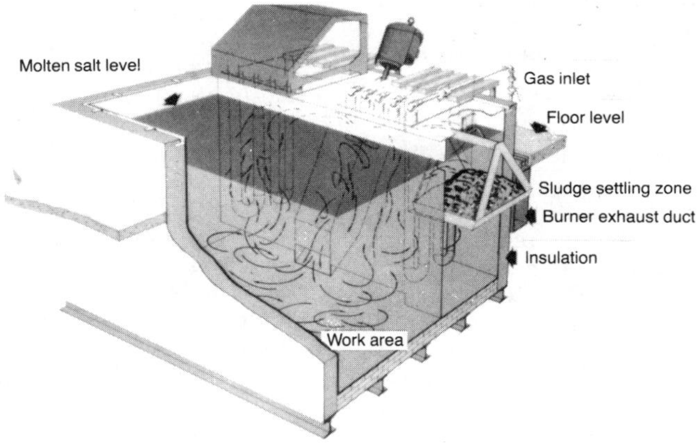

Agitated molten salt bath with sludge settling zone

Fused Salt Equipment

Fused salt equipment is becoming increasingly more complicated as the need for automation, safety, and environmental protection continues to increase. The fused salt cleaning system shown in Fig. 6 is completely hooded, including the wash and rinse tanks, in order to contain any gaseous products in an area where they can be transported to appropriate equipment for removal of solids or harmful materials prior to discharge into the atmosphere. The completely hooded system also protects the operator and any other workers in the area from any discomfort created by exposure to the salt bath or the hot water rinse tank. Equipment of this type can be heated by immersion gas burners or electrical resistance heaters immersed in the fused salt. The choice between electrical and gas frequently depends on availability and cost in the particular area where the installation is made. Salt bath equipment has no known size limitation providing the economic restrictions can be met. Large plates, bars, wire coils, continuous strip, and extremely long lengths of tubing have been effectively cleaned in fused salts.

Equipment of this type usually contains provisions for additions of makeup salt arranged in such a way that the operator is completely protected during additions. In the same way, allowances are made for the removal of the sludge products generated, and again, the equipment is such that the operator is completely protected from accidental splashing or spilling. The internal structure of a salt bath unit which could have any conceivable dimensions is shown in Fig. 7. Basically, it is imperative that the equipment be properly designed and engineered to fit the exact requirements if the maximum efficiency is to be obtained from a fused salt cleaning system.

Applications for Fused Salt Cleaning Systems

The need for fused salt cleaning and production systems has risen in direct proportion to the increase in the development of materials that are resistant to common acidic or alkaline solvents. Some general classes of the materials that require molten salts for removal are:

- *Oxides*: MO, M_2O_3 (hot work and heat treat scales)
- *Organics*: C (free graphite, lubricants, and binders)
- *Ceramics*: SiO_2 (glass-type lubricants, mold and core material)
- *Combination coatings*: MO + C (lubricant and protective coatings, core wash)

In addition to these, salt baths are in widespread demand for the removal of complex coring from investment castings, flame sprayed and plasma sprayed coatings, synthetic fibers from spinnerettes, and all types of inorganic and organic coatings.

More specifically, the heat processing of stainless steel and superalloys develops very complex surface oxides. When these oxides are present on a high-speed continuous strip, the cleaning portion of the anneal and pickle line becomes the governing factor in line speed. Without the use of molten descaling salt baths, required strip travel could never be achieved.

Cleaning of Stainless Steel Strip. In this process, strip is immersed in the molten oxidizing salt for periods ranging from 3 to 20 s, immediately following the annealing stage. Rolls immersed in the salt maintain the strip below the salt level for the required time. Although the fused oxidizing salts have a low viscosity at operating temperature, they do have excellent wetting properties and completely coat the metal surface. At strip travel speeds of 15 to 30 m/min (50 to 100 ft/min), a continuous strip would tend to drag out large volumes of salt. Originally, wiper rolls were used at the exit end of strip lines to force some of the salt back into the furnace. Because their effectiveness was limited by the temperature of the operation requiring the roll construction to be of cast iron or steel, the rolls lacked the flexibility to produce good wiping action. In addition, any entrapped waste products in the salt bath could create surface problems when dragged between the roll and the moving strip. Another approach to minimize salt dragout is the use of a high-pressure air blast directed at the top and bottom surfaces of the strip at the exit end of the salt bath. This system has been tried numerous times, and although quite effective in reducing salt consumption, it has the adverse effect of changing the chemistry of the salt bath according to the following reaction:

$$2NaOH + CO_2 = Na_2CO_3 + H_2O \quad \text{(Eq 11)}$$

Use of a high-pressure stream that is inert to the salt bath and can be generated at a temperature just above the melting temperature of the salt has been found to be the most effective process. Steam is relatively inexpensive and can be brought to the required temperature at a minimum cost. It has also been observed and recorded that contact of steam with a converted salt bath oxide produces a synergistic effect which aids oxides solubility and subsequent acid pickling.

Rinsing is usually accomplished by immersion in room temperature water or by low-pressure spray nozzles, or combinations of both. Rinsing accomplishes both the removal of residual salt and the reduction of the strip temperature prior to immersion in the subsequent acid pickle. Typically, a mild sulfuric acid used primarily to neutralize minute amounts of caustic adhering to the strip is usually followed by nitric hydrofluoric pickling to achieve the brightness required.

Cleaning of Wire Products. In the field of wire processing, molten salts are not a novelty. They have been employed for many years in the areas of metal conditioning, fluxing, heat treatment, and cleaning. The number of molten salt wire applications has increased impressively during the past few years, especially in the metal cleaning and conditioning operations. This increase is closely correlated to the rapid rise in use of higher alloyed metals that present difficult cleaning problems.

The importance of salt conditioning in the manufacture of stainless steel wire products is the result of numerous operations inherent in the product process. Several typical problems associated with wire cleaning are:

- Frequent annealing is required because of the severe cold work involved
- Pressures involved in wire drawing require high film strength lubricants which are difficult to remove
- Maximum cleanliness is mandatory for good die life and surface finish
- Production requirements and handling methods create dense coils which are difficult to penetrate in precleaning operations, resulting in nonuniform anneal scale
- Density of scaled coils also inhibits pickling action and can result in selective gauge loss

The severity of cold work involved in finishing the hot mill wire product in the size range of 6.4 to 9.0 mm (0.25 to 0.35 in.) in diameter requires a minimum of one and a maximum of six furnace anneals. The oxidized surface resulting from this process must be removed prior to further working or final shipment. This is accomplished by several separate pickling operations. The high film strength lubricants necessary to prevent metal to metal contact during wire drawing naturally have good adherence properties. Removal of these compounds from the surface area of dense coils is a major cleaning problem. The condition is further aggravated when conversion coatings are used or when the lubricant reacts chemically with the wire under the pressures and temperatures involved. In the latter case, the soils involved are an integral part of the surface and must be removed by chemical cleaning or mechanical abrasion.

Deformation of stainless and alloy wire products under the best conditions requires very high unit pressures. Even very minute particles of scale can effect the precision dies involved. In many cases, the net result of residual scale is not only an expensive die repair operation, but it can also generate an appreciable amount of scrap wire because of the processing speeds involved.

Wire products, unlike continuous strip, are annealed for longer periods of time because they are processed in coils and in batch operations. By the same token, cleaning facilities are usually constructed to allow a dwell time comparable to the annealing time, and it is not unusual for conditioning operations in wire production to take 15 to 20 min dwell time in the fused salt. Following the salt treatment, the wire coils are pressure rinsed and pickled either in sulfuric, nitric, or nitric hydrofluoric acid, depending on the grade and the preference of the pickling house personnel. There is, as indicated in the description of continuous strip processing, considerable freedom in choosing the proper acid pickling solution following salt bath conditioning with the one notable exception, nitric hydrofluoric solutions are not used on 400 series stainless steels.

Stainless steel in superalloy forgings, castings, and extrusions are all processed in cleaning salts in a manner similar to that which is used for wire coils. All of these products are batch process and therefore have longer dwell times in both the salt and the acid. In addition, many of these products have heavily contaminated surfaces such as scale lubricant combinations and glass scale combinations which are most difficult to remove without salt bath conditioning. In general, salt dwell time is in the range of 15 to 20 min, with recycling occasionally required, especially where glass-type lubricants are extremely heavy on the surface. Acid pickling solutions are similar to and have the same restrictions placed on them as those recommended for processing wire products.

Other Metal Cleaning Salt Applications. Titanium and titanium alloy mill products and fabrications also require salt bath conditioning for efficient and effective cleaning after heat processing. Fused salt treatment of titanium products is described in the article on cleaning and finishing of reactive and refractory metals and alloys in this Volume. Specialized cleaning involving investment material, ceramic cores, and hard-faced coating removal always requires longer dwell times and some multiple cycling in the fused salt; however, the decided advantages in time cycles and costs that are inherent in salt bath processing usually dictate the selection of this form of cleaning.

Organic Coating Removal. The specialized area of cleaning that has traditionally involved fused salt systems is the removal of organic coatings from paint hooks and racks. Cleaning of these fixtures in a molten salt evolved over a period of years as the organic coatings employed in industry were progressively developed to be more solvent resistant. Quality control requirements for clean fixtures increased, and finally environmental concerns limited the application of the few water solvents capable of removing modern organic finishes. Molten salts are a natural solvent for even the most durable organic coatings because the reactivity achieved at the operating temperature of 480 °C (900 °F) reacts with organics according to the following:

$$C + 2NO_3^- = CO_2 + 2NO_2^- \qquad \text{(Eq 12)}$$

Complete destruction of the organic binder is achieved in this way, and any pigmented material is dispersed in the body of the salt bath to be eventually removed as a waste product with the sludge. The regenerative steps of reoxidation, replenishment, and sludge removal described under stainless steel descaling also apply, and again, most of these fused salt baths operate for an almost indefinite period of time without changing. There are some gaseous and dust waste products generated in the stripping operation, but these are easily controlled through the use of baghouses or water scrubbers. Fused salts are today the most efficient method of removing complex organics from metal surfaces, although some less efficient compromises are available where cleaning requirements are not as stringent.

Mechanical Finishing

Polishing and Buffing

By the ASM Committee on Polishing and Buffing*

POLISHING AND BUFFING improve edge and surface conditions of a product for decorative or functional purposes. These techniques are abrading operations although some plastic working of surfaces may occur, particularly in buffing. Buffing is a process using abrasives or abrasive compounds that adhere loosely to a flexible backing, such as a wheel. Polishing is the use of abrasives firmly attached to a flexible backing, such as a wheel or belt. Grinding is the use of abrasives firmly adhering to a rigid backing.

Polishing operations usually follow grinding and precede buffing. In general, polishing permits more aggressive abrading action than buffing. It has greater capability to modify the shape of a component and has greater definition of scratches. Buffing achieves finer finishes, has greater flexibility, and follows contours of components.

Applications

Polishing and buffing processes are used on most metals and many nonmetals for refining edges and surfaces of castings, forgings, machined and stamped components, and molded and fabricated parts. Traditionally, these processes have been considered as means of developing attractive, decorative surface appearance and for generating suitable preplate and prepaint surfaces.

Products with improved edge and surface finishes are generally safer, more durable, and more efficient. Many products must have sharp edges removed to avoid cutting the user. Improvement of edge and corner radiusing and of surface finish remove stress risers from which fatigue cracks are most likely to propagate. Most mechanisms have reduced friction if edge and surface finishes are improved, resulting in greater efficiency and requiring less dynamic energy to function at design limits. Radiused edges also make assembly of components easier and more efficient.

Polishing and buffing equipment and materials have been improved enormously, and these techniques are now used for flexible machining applications. Polishing processes can be utilized for substantial metal removing purposes, maintaining close tolerances while improving surface finish. Buffing operations can be utilized for edge and corner radiusing, can generate extremely fine surfaces to match those associated with honing and lapping, and can maintain equally very close tolerances associated with modern machining techniques.

Equipment for buffing is similar to that used for polishing, for manual, semiautomatic, or fully automatic applications. Polishing equipment may require greater precision and rigidity than buffing machines because the polishing process is less flexible and is more frequently used to modify the shape of parts and to machine them to specific tolerances. Consequently, polishing equipment requires more precise component fixturing and setup of the system than does buffing equipment. Semiautomatic and automatic machines consist of one or more buffing or polishing heads, a conveyor to enable components to pass across the wheels or belts, and fixturing to locate the components in correct position on the conveyor. For large parts, the wheels or belts might traverse across the component.

Material. Abrading materials for both buffing and polishing are the consumable materials. The abrasives and the backers for polishing with these materials are either abrasive belts or abrasive wheels. The abrasives are firmly attached to the backing materials. Occasionally, water-soluble lubricants are sprayed on the wheels or belts to lubricate, to cool the workpiece and polishing materials, and to prevent the belts from being overloaded with the material being finished.

The two materials required for buffing are buffs and buffing compounds. Buffs are usually cloth wheels which act as a carrier for the compound. Compounds may be liquid or in bar form. In either case, they contain bonding and lubricating agents. Liquid compounds are sprayed on the buff, and bar com-

*J. Bernard Hignett, *Chairman*, Vice President, The Harper Co.; Thomas C. Andrew, Executive Vice President, The Harper Co.; William Downing, Process Engineer, Lea Manufacturing Co.; Ernest J. Duwell, Corporate Scientist, Industrial Abrasives Division, 3M Co.; James Belanger, President, Belanger Inc.; Edward H. Tulinski, Regional Manager, The Harper Co.

pounds are applied by hand or with an automatic applicator.

Labor is a major variable. Manual polishing and buffing require highly skilled labor and are very dirty operations. For these reasons and because mechanical finishing has become a more important part of the manufacturing operation, new materials and equipment have been developed rapidly. Initially, automated equipment was designed and built to mimic the manual finishing operations, but today mush buffing, precision buffing, and precision polishing yield results unobtainable by hand operations.

Polishing

Traditionally, polishing is used for improving surface condition, and to remove unwanted protrusions such as casting and forging flash, or parting lines. Modern polishing processes are also used as flexible machining centers, capable of substantial metal removal and precise control of the shape and size of components being polished. Polishing processes are used for improving the shape and precision of a component, as well as for improving its edge and surface conditions.

Polishing is an abrading operation used to remove or smooth grinding lines, scratches, pits, mold marks, parting lines, tool marks, stretcher strains, and surface defects that adversely affect the appearance or function of a part. Polishing is performed with either a wheel or belt to which an abrasive is bonded. The process causes some plastic working of the surface as metal is removed. The resulting polished surface usually has a finish of 0.4 μm (16 μin.) or less.

In operational sequence, polishing usually follows grinding and precedes buffing; however, in some applications, polishing is used as the final surface-finishing operation. Grinding is generally considered a machining operation that produces a surface rougher than 0.4 μm (16 μin.). Polishing removes less metal than grinding, but more than buffing or lapping.

Polishing Abrasives

Emery and corundum are natural abrasives commonly used on polishing wheels. Emery is a naturally occurring mixture containing between 57 and 75% aluminum oxide. The remainder is iron oxide and impurities. Corundum is a naturally occurring fused aluminum oxide (Al_2O_3). It is tougher and harder than emery. Natural abrasives usually cost less than synthetic abrasives and are useful for polishing some softer metals. When applied to the harder metals, the natural abrasives wear smooth at a rapid rate. Consequently, these natural materials have been replaced largely by synthetic abrasives.

Aluminum oxide and silicon carbide are the most widely used synthetic abrasives. They are harder, more uniform, longer lasting, and easier to control than natural abrasives. Because aluminum oxide grains are very angular and have excellent bonding properties, they are particularly useful in polishing tougher metals, such as alloy steels, high-speed steels, and malleable and wrought iron. Silicon carbide is harder than aluminum oxide; it fractures easily, providing new cutting surfaces which extend the useful life of the abrasive. Silicon carbide is usually used in polishing low-strength metals, such as aluminum and copper. It is applied in polishing hard, brittle materials, such as carbide tools, high-strength steels, and chilled and gray irons. Standard grit sizes for aluminum oxide and silicon carbide abrasives are as follows:

Type	Grit size, mesh
Screened sizes	24,30,36,40(a),46(b), 50(a), 54(b), 60, 70(b), 80, 90(b), 100, 120, 150, 180, 220, 240
Classified flours	280, 320, 400, 500 600

(a) Standard for abrasive belts only. (b) Standard for loose, abrasive grains only

Polishing Wheels

Polishing or setup wheels were the original means of abrading a product for finishing purposes. Still used in many organizations, they have been substantially replaced by abrasive belts which offer better performance, greater consistency, and reduced labor. Polishing wheels are disks of materials, such as cloth and leather, sewed and/or glued together. The periphery is coated with alternating layers of adhesive and abrasive grain to the desired thickness. The type of material and construction determines the hardness and flexibility of the wheel. Muslin, canvas, various weaves of cotton, leather, sheepskin, and felt are among the materials used. The canvas is usually glued and pressed together to make it the hardest. The solid felt wheel is also hard, but its hardness is dependent on the density of the felt. Leather and sheepskin can be glued and pressed, machine sewn, or hand sewn to give the flexibility desired. Differences in types of leathers also determine the final effect. A leather strip can also be glued to the periphery of a cloth or wooden wheel to combine the hardness of the underlying wheel with the flexibility of the leather.

The wheel periphery is trued or shaped by applying moderate pressure to the rotating wheel with an abrasive stone or sheet. The wheel is balanced statically and dynamically by adding weights. A sizing coat is then applied with the same adhesive that bonds the abrasive to the wheel. After this sizing has almost dried or is tacky, another coating is applied, and the wheel is rolled in a tray holding the loose abrasive. The wheel is then rolled on a hard surface to drive the abrasive into the adhesive. The adhesive is formulated to anchor the abrasive approximately half way into the final dried adhesive. It is allowed to become almost dry. Another coating of adhesive is then applied; the wheel is rolled in the abrasive, and the abrasive is driven into the adhesive. This is repeated until the desired number of layers are applied. Finally, the polishing wheel is allowed to dry completely. The polishing wheel is rechecked for static balance, then the face is struck at a 45° angle with a steel bar from both sides of the wheel completely around the wheel. This creates small diamond-shaped cutting tools by taking advantage of the flexibility of the wheel. The polishing wheel is then ready for use.

Two main types of adhesives are hide glue and silicate cements. The hide glue is the more resilient. The silicate cements are formulated to give a range of effects. The hide glue melts under heavy pressure, must be soaked and made up daily, requires controlled heating between 60 and 65 °C (140 and 150 °F), and must be formulated to the size of the abrasive. Despite these disadvantages, hide glue is still popular because it gives the softness, flexibility, or cushion required for certain applications. Final drying for glue polishing wheels usually requires 24 h for each layer at 50% relative humidity and 27 °C (80 °F).

Silicate cement does not usually melt under heavy pressures when high heats are developed. It can be used as received from the manufacturer. No heating of these cements is required but their containers, like the water and hide glue mixture, should be closed tightly to prevent loss of solvent when not in use. The silicate cement should be selected and formulated for the particular range of abrasive sizes to be used, otherwise the cement will not hold the larger grits or may produce deeper scratches than the smaller grits. Drying can be accomplished by allowing the wheels to hang in a well-ventilated area or by hanging them in an oven at 65 °C (150 °F). Final drying time is 8 to 16 h.

The technique and care required to make the polishing or setup wheel can be eliminated by using polishing belts, but in certain applications, polishing wheels cannot be substituted. An obscure or confined area, such as an automobile wheel that is to be plated, is readily polished with a polishing wheel. The cutlery industry and the jewelry industry use polishing wheels and bobs for special effects, where some areas are difficult to reach and where polished finish must be confined to certain areas.

During polishing with setup wheels, efficiency, quality of results obtained, and cost are most significantly affected by the following operating variables:

- Type and density of the polishing wheel
- Preparation of the wheel before application of adhesive
- Wheel-curing (drying) procedures
- Wheel balancing
- Wheel abuse by inexperienced operators
- Wheel speeds
- Type and grit size of abrasive

Wheel speeds for buffing and polishing are shown in Table 1. Table 2 gives grit sizes used on polishing wheels in specific applications.

Table 1 Setup wheel speeds for polishing and buffing

Material	Wheel speed: Polishing m/s	Polishing sfm	Buffing m/s	Buffing sfm
Aluminum	31-38	6000-7500	38-43	7500- 8 500
Carbon steel	36-46	7000-9000	31-51	6000-10 000
Chromium plate	26-38	5000-7500	36-46	7000- 9 000
Brass and other copper alloys	23-38	4500-7500	36-46	5000- 9 000
Nickel	31-38	6000-7500	31-46	6000- 9 000
Stainless steel and Monel	36-46	7000-9000	31-51	6000-10 000
Zinc	26-36	5000-7000	15-38	3000- 7 500
Plastics	...	...	15-26	3000- 5 000

Table 2 Aluminum oxide grit sizes used on polishing wheels in specific applications

Products	First stage(a)	Second stage	Third stage	Fourth stage
Aluminum, die cast	150(b),(c)	...	...	...
Aluminum, sand cast:				
Internal surfaces	34-46	...	...	...
External surfaces	60-80	120-180(c)	...	...
Aluminum, sheet	120(b)	180(b),(c)	...	...
Automotive bumpers	60-90	120	150-180(b)	220(b)
Automotive headlight trim	180-220(b),(c)	...	...	...
Axes	46-60	70-90	120	150-180(b)
Band saw, steel	60-80	120-150	...	...
Brass, sand cast	60-80(b)	150-180(b)	220(d)	...
Brass, sheet	180-220(b),(c)	...	...	...
Electric irons	(e)	120(b)	150(b)	180-240(b)
Gray iron, not pickled	70	120-150	...	...
Gray iron, pickled	80	120-150	...	...
Hammer heads	46-60	100-120(b)	...	...
Locomotive side rods	36	60-70	120	...
Machetes, steel, edges	46-60	...	...	...
Machetes, steel, faces	80	120(b)	...	...
Monel, cast	80	120	150	150(b),(c)
Monel, deep drawn	120	150	180(b),(c)	...
Monel, full-finish sheet	180	180(b)	220(b),(c)	...
Plow disks	30-46	70-90	...	...
Plow moldboards	24-36	80	120-150(b)	180-220(b)
Plowshares	36-46	...	...	...
Shears, tinsmith	46	60	120-150	180
Shovel blades	(f)	120	...	...
Shovel straps	(g)	120(h)	...	...
Stainless steel sheet:				
Mirror finish	60-80	100-120(b)	150	220(b),(c)
Commercial finish	80	120-150(b)	220(c)	150(b)
Table knife blades	80-90	120-150(b)	220(c)	...
Table knife handles, steel	46-60	...	...	...
Wrenches	(j)	(k)	120(b)	...
Automotive fenders	90	120-150	...	...
Automotive hardware	36-54	90	120	220(b)
Cutlery	80	120	220(b)	...
Hayforks	60-70	100-120	...	...
Hoes, first quality(m)	36-46	70	100-120	...
Spading forks	24	...	...	...

(a) First stage may or may not be necessary where multistage polishing is indicated, depending on surface roughness. (b) Grease or oil wheel used. (c) Buffing follows this stage. (d) Only aluminum oxide used for third stage. (e) Natural abrasive, 80 mesh. Aluminum oxide, 60 to 80 mesh. (f) Natural abrasive, 30 to 46 mesh. Aluminum oxide, 36 to 46 mesh. (g) Natural abrasive, 36 to 70 mesh. Aluminum oxide, 36 to 60 mesh. (h) Abrasive belt used for second-stage polishing. (j) Natural abrasive, 30 to 60 mesh. Aluminum oxide, 30 to 46 mesh. (k) Natural abrasive, 80 to 90 mesh. Aluminum oxide, 80 mesh. (m) Second-quality hoes are not polished beyond first stage

Abrasive Belt Polishing

Endless cloth belts, precoated with abrasive, are widely used in polishing operations. In comparison to setup wheels, they offer the following advantages:

- Closer control of surface finish
- Smaller inventory of polishing accessories
- Less heat generation in polishing because of the larger surface area of belts
- Elimination of the need for costly, temperature-controlled wheel setup rooms, and for the care, preparation, and upkeep of setup wheels, adhesive, and abrasives
- More uniform finish than can be achieved by setup wheels
- Less labor, because no labor is required to recoat contact wheels because belts offer more consistent performance
- Greater flexibility than setup wheels because of various densities of contact roll or wheels which can make belts hard or soft. In some applications, the slack of the belt can be used effectively

Construction. Coated abrasives are a composite of three components. These are backing, adhesive, and mineral. In general, the backing for a belt is either paper or cloth. Different weights of paper and types of cloth are used depending on requirements for strength, flexibility, and water resistance. The adhesive is usually resin or glue and is used to bond the mineral to the backing. It is applied in two coats, with the first layer or make coat anchoring the mineral in place on the backing. The second adhesive layer or size coat is applied over the mineral to strengthen further its bond to the backing. The most common minerals used on coated abrasive belts are aluminum oxide and silicon carbide. The mineral is applied to the backing by electrostatic deposition. The mineral is passed through an electric field where an electrostatic force propels it into the resin with a controlled orientation of each mineral grain on the backing. This results in a sharp coated abrasive product containing many exposed, well-oriented cutting points.

Selection Factors. The selection of coated abrasive grades for a given operation depends on the amount of stock to be removed, the desired final finish, and the type of material being finished. An efficient operation uses the coarsest grade that still produces an acceptable finish. Depending on the objective of an operation, more than one abrasive grade may be needed. When both stock removal and finish are required, a sequence of abrasive grades is used. The coarsest grade is determined by the amount of stock removed and by the type of material. It should be the finest possible grade that still removes stock at an acceptable rate so that no excessively coarse scratches are produced. The finishing grade is determined by final finish requirements. The number of grit sizes of intermediate grades are important for quality and for cost. If too many intermediate grades are skipped, some deep scratches remain which decrease buffability. In general, more grades can be skipped in the coarser grades than in finer ones and when finishing a soft material rather than a hard one. Parts with small contact areas allow more grades to be skipped and require fewer finishing grades because of increased buffability.

Abrasive grades used for a given operation and the approximate number of grades that can be skipped are shown in Table 3.

Table 3 Abrasion grades for various operations

Operation	Grade range	Condition
Stripping	12-20	Removal of old finishes, rust, and other materials which tend to load belt
Heavy stock removal	24-30-36-40	Used for rapid stock removal and large depths of cut. Can skip 3 grades
Medium stock removal	50-60-80	Average stock removal and progression from rougher finishes. Skip 2 grades
Light stock removal	100-120-150	Used in operations requiring minor dimensional changes. Skip 1 or 2 grades
Finishing	180-220-240	Negligible stock removal. Produces a desired appearance or surface for plating. Skip 0 or 1 grade
Polishing	(280-320)-(360-400)-(500-600)	Preparation for mirror or near mirror finishes

Fig. 1 Surface finishes obtained with varying grades of belts

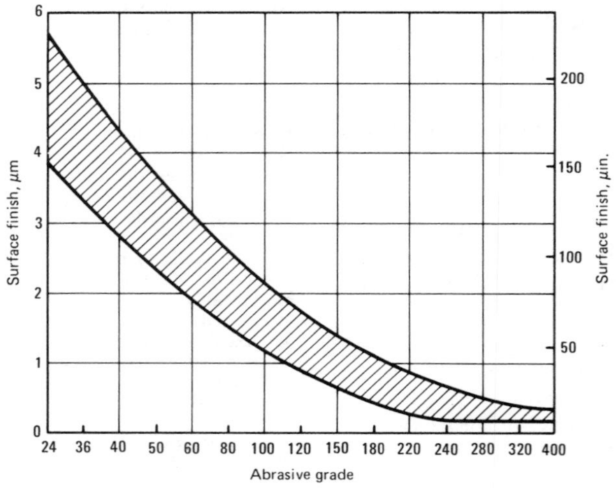

In many operations, the surface finish specifications are determined by scratch depth. Scratch depth is affected by factors such as abrasive grade and type, lubricant, material to be finished, and finishing conditions, such as contact wheel and abrasive speed. Figure 1 shows a range of surface finishes that can be obtained with a given abrasive grade. Specific applications give finishes outside of this range; therefore, the graph should be used only as a general guideline.

Contact wheels over which the abrasive belt rides provide pressure of the belt against the workpiece. Depending on its hardness, the contact wheel can provide either high unit pressure, hard wheel, or low unit pressure, soft wheel. Selection of contact wheel directly affects the rate of stock removal, the ability to blend in polishing, the surface finish obtained, and the cost of the polishing operations.

Manufacturers of contact wheels are responsible for speed testing and balancing the wheels, as well as dressing the wheel faces for the operating speed at which they are run. Although it is possible to operate a contact wheel satisfactorily at speeds of from 10 to 50 m/s (2000 to 10 000 sfm), normal operating speeds usually range from 18 to 38 m/s (3500 to 7500 sfm). Table 4 illustrates and describes principal types of contact wheels and indicates their applicability to various grinding or polishing operations.

Applicability of Belts. Originally, the use of coated abrasive belts was limited to the polishing of flat surfaces, using straight-face contact wheels. However, after resin bonding of abrasives became practical, it was possible

Table 4 Characteristics and uses of abrasive belt contact wheels

Wheel	Type	Material	Hardness	Purpose	Characteristics
1	Knurled or spiral grooved(a)	Steel	Rockwell C 52 to 55	Heavy grinding	Provides most aggressive action
2	Cog tooth(b)	Rubber	70 to 90 durometer	Grinding(c)	Fast-cutting, allows long belt life
3	Standard serrated(d)	Rubber	30 to 50 durometer(e)	Grinding(f)	Leaves rough-to-medium surface, excellent life
4	X-shaped serrations(g)	Rubber	30 to 60 durometer	Grinding, polishing(h)	Flexibility allows entry to contours
5	Plain face	Rubber	20 to 40 durometer(j)	Grinding, polishing(k)	Allows controlled penetration of abrasive grain
6	Flexible	Compressed canvas	(m)	Grinding, polishing(n)	Tough and durable
7	Flexible	Rubber-coated canvas	Medium	Contour polishing	Contours well, yet gives substantial stock removal
8	Flexible	Solid section canvas	Soft, medium, hard	Polishing(p)	A low-cost wheel with uniform face density
9	Flexible	Buff section canvas(q)	Soft	Contour polishing	For fine polishing and finishing, low-cost
10	Pneumatic drum	Inflated rubber	(r)	Grinding, polishing	Gives uniform finishes, adjusts to contours
11	Plastic foam	Polyurethane	Extremely soft	Fine polishing	Most flexible, for extreme contours

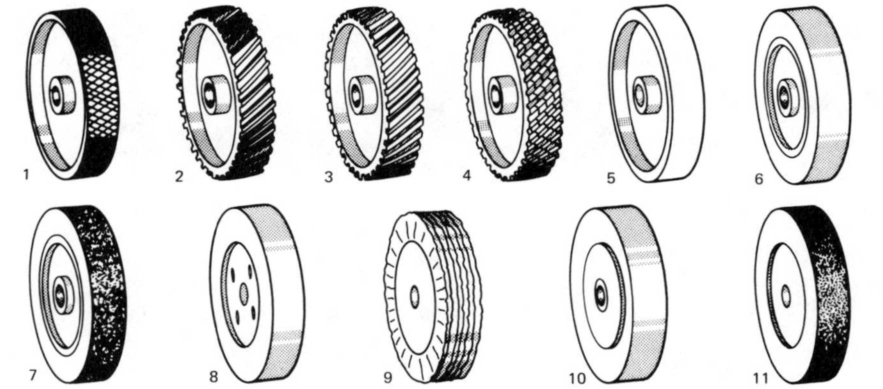

(a) No. 14 standard face; 4-pitch 2 by 2 mm (1/16 by 1/16 in.). (b) Land, 5 mm (3/16 in.); groove, 14 mm (9/16 in.); depth, 1.5 mm (1/16 in.); cushion, 19 mm (3/4 in.). (c) For cutting down projections, such as weld beads, gates, risers, and sprues. (d) Land, 10 mm (3/8 in.); groove, 10 mm (3/8 in.); depth, 10 mm (3/8 in.); cushion, 22 mm (7/8 in.). (e) Wheel also may be of dual density, with hard rubber, 60 durometer, at hub, softer rubber, 20 to 40 durometer, at working surface. (f) For smoothing or blending cut-down projections or surface defects. (g) Land, 5 mm (3/16 in.); groove, 14 mm (9/16 in.); depth, 8 mm (5/16 in.); slit, 13-mm (1/2-in.) spaced; cushion, 22 mm (7/8 in.). (h) For light stock removal and medium polishing; preferred to standard serrated wheel for softer nonferrous materials. (j) Softer wheels give better finishes. (k) For flat surfaces. (m) Nine densities (very hard to very soft). Hard wheels can remove metal, but slower than wheel 2; softer wheels can polish to fine smoothness. (n) Good for medium-range grinding and polishing. See footnote m. (p) Handles all types of polishing, giving uniform results without leaving abrasive pattern on work; adjusts to contours or can be pre-formed for contours. (q) Can be widened or narrowed by addition or removal of sections. (r) Hardness controlled by air pressure

to produce belts with greater flexibility and improved joints and to introduce improvements in polishing machinery that permitted the use of coated abrasive belts for contour polishing operations. The selection of tough, flexible polyester backings also contributes to the ability of coated abrasives to contour polish.

In its simplest form, manual belt polishing with contact wheels is performed on two spindle polishing jacks, with backstand idlers mounted to the rear of each jack. The backstands provide tension on the belts by means of springs or air cylinders, and they track the belts accurately over the contact wheel. Jacks usually are powered by motors of 1 to 10 hp, capable of giving belt speeds of 18 to 38 m/s (3500 to 7500 sfm).

Another means of belt polishing utilizes a machine in which the contact wheel is an idler mounted on a yoke in front of the machine. Uniform belt tension is maintained by a screw-actuated or air-tensioned idler wheel mounted over the drive pulley. Because they use lower belt speeds and have more uniform belt tension, these machines can handle a variety of deep contouring operations and can accommodate wheels of a more complex shape. Consequently, these machines are used to polish workpieces with complicated shapes, such as surgical instruments, scissors, jet-engine blades, plumbing fixtures, cutlery, and propellers. Even narrow belts, from 13 to 75 mm (1/2 to 3 in.) in width, can be readily used.

In formed contact wheel abrasive belt polishing, bias buff wheels, compressed canvas wheels, and muslin polishing wheels made from spiral-sewed buff sections are most widely used. Formed rubber and felt wheels are usually reserved for special jobs.

Belt speeds and abrasive grit sizes used for a variety of applications are given in Table 5. Too low a belt speed may result in rapid fracture of abrasive grains, lowering cutting efficiency. Too fast a belt speed results in glazing or dwelling of the abrasive.

Abrasive Wheels

Nylon wheels are flexible, resilient, and made of abrasive-impregnated, nonwoven nylon or other synthetic fibers. Abrasive is dispersed evenly throughout these materials, so that wheel performance remains substantially constant despite wheel wear. These wheels, in many cases, can be shaped to conform to curved, irregular surfaces. They are water resistant, range from 25 to 2030 mm (1 to 80 in.) in width and from 25 to 405 mm (1 to 16 in.) in diameter. A wide variety of densities and grit sizes are available. Nylon wheels may be used wet or dry or with grinding lubricants, such as waxes, oils, or greases.

These wheels are especially useful for imposing uniform scratch patterns or satin finishes on nonferrous metals, plastics, and wood. They are recommended for various types of blending, as well as for the removal of light surface stains, scale, rust, or old coatings. They are not suitable for the removal of large amounts of stock, deep defects, or surface irregularities; however, these wheels are also effective in removing burrs and radiusing sharp edges. They operate with various speeds ranging

Table 5 Grit sizes and belt speeds recommended for abrasive belt roughing and polishing applications

Abrasive belts are coated with aluminum oxide unless otherwise noted

Product	Obstruction or roughness removed	Roughing: Grit size, mesh	Roughing: Belt speed m/s	Roughing: Belt speed sfm	Rough polishing: Grit size, mesh	Rough polishing: Belt speed m/s	Rough polishing: Belt speed sfm	Polishing: Grit size, mesh(a)	Polishing: Belt speed m/s	Polishing: Belt speed sfm	Polishing aid(b)
Aluminum											
Die castings	Flash	200	28	5500	···	···	···	320	28	5500	G
Extrusions	Draw marks	180	28	5500	(c)	···	···	···	···	···	G
Pots and pans	···	···	···	···	80	28	5500	180, 240	···	5500	O
Sand castings	Gates	50	28	5500	80	28	5500	320	28	5500	G
Sheets	···	···	···	···	220	27	5200	(c)	···	···	G
Tubes	Die marks	···	···	···	···	···	···	220, 320	26	5000	G
Armatures, rotors, laminated	···	···	···	···	···	···	···	50	18	3500	WSO
Automotive bumpers, sheets	Inclusions	100	18	3500	120	18	3500	150, 180(d)	18	3500	G
Axes, edging	Scale	36	27.5	5400	···	···	···	80	28	5500	G
Axes, sharpening	···	···	···	···	50	···	···	···	···	···	G
Axles, cylindrical	···	···	···	···	50	28	5500	···	···	···	O
Band saws, steel	Pits	100	18	3500	···	···	···	150	18	3500	O
Bearings, inner or outer race	···	···	···	···	···	···	···	240	(e)	(e)	O
Beryllium, sheets	Pits	60(d)	11	2100	100(d)	···	2100	150(d)	11	2100	WSO
Billets, alloy or stainless steel	···	···	···	···	50	18	3500	···	···	···	G
Bits, steel, auger or flute	···	···	···	···	50	18	3500	150	27	5200	G
Brake lining(f)	···	···	···	···	36(d)	18	3500	···	···	···	···
Brass											
Die castings	Flash	220	28	5500	···	···	···	320	28	5500	G
Sand castings	Gates	60	28	5500	···	···	···	180	28	5500	G
Sheet or strip	···	···	···	···	···	···	···	220, 320	23	4500	WSO
Bronze											
Sand castings	Scale	60	18	3500	···	···	···	180	28	5500	G
Sheet or strip	···	···	···	···	···	···	···	220	23	4500	G
Tubes or bars	···	···	···	···	···	···	···	220, 320	26	5000	G
Cams, lobes(g)	···	···	···	···	80	26	5000	···	···	···	O
Cast iron, small castings	Gates	50	18	3500	···	···	···	80	18	3500	···
Chilled iron rolls	···	···	···	···	50	18	3500	80, 120, 150, 220(d), 320(d)	18	3500	WSO
Chisels, woodworking, steel	···	···	···	···	80	26	5000	150	26	5000	G
Copper:											
Plumbing fixtures	Gates	60	26	5000	80	27	5200	220	27	5200	G
Rolls	···	···	···	···	···	···	···	320(d), 400(d)	28	5500	WSO

(continued)

(a) For successive stages. (b) G indicates grease. O indicates oil. WSO indicates water-soluble oil. (c) Operation completed with abrasive-impregnated nylon wheel. (d) Silicon carbide. (e) Fixtured abrasive. (f) Molded or metal mix. (g) Hardened steel cams for diesel engines. (h) Drawing and sinking dies. (j) Diamond abrasive. (k) Centerless. (m) Fixtured abrasive, reciprocating. (n) Grinding of crown. (p) Grind end. (q) Solution of potassium tribasic or nitrite amine. (r) And excess weld metal

from 3 to 33 m/s (500 to 6500 sfm) depending on the specific wheel and application.

When nylon wheels are used to finish flat sheet stock, wheel oscillation is recommended to attain a uniform finish free of streaks. An oscillation of about 200 cycles/min with 10-mm (3/8 in.) amplitude gives good results in a variety of applications. Greater amplitudes, up to 50 mm (2 in.) can be used on steel or other hard metals, but on softer metals, amplitudes greater than 13 mm (1/2 in.) are likely to create snake marks.

Flap Wheels. There are basically two types of metal hub flap wheels—reloadable and throwaway. Either style can be used with coated abrasive materials for polishing and various types of materials for buffing. Basic construction of flap wheel incorporates a metal hub. A reloadable hub, depending on its diameter, has located around its outer perimeter a specific number of keyhole-shaped slots. Flap wheel packs are inserted into the slots utilizing a bulbous-shaped root member that matches the profile contour of the keyhole-shaped opening in the reloadable flap wheel hub. Materials used for this bulbous-shaped root member vary from new alloys of plastic to high-strength steel. Newer developments of the root member allow the inserted pack to swing or flex, enabling the packs to pivot independently so that they may find their own balance points. This pivoting ability neutralizes the stresses induced when the pack is subjected to abnormal impacts.

Throwaway mechanical flap wheels have the same advantages that reload-

Table 5 (continued)

Abrasive belts are coated with aluminum oxide unless otherwise noted

Product	Obstruction or roughness removed	Roughing: Grit size, mesh	Roughing: Belt speed m/s	Roughing: Belt speed sfm	Rough polishing: Grit size, mesh	Rough polishing: Belt speed m/s	Rough polishing: Belt speed sfm	Polishing: Grit size, mesh(a)	Polishing: Belt speed m/s	Polishing: Belt speed sfm	Polishing aid(b)
Sand castings	Scale	60	28	5500	80	28	5500	180	28	5500	G
Sheet or strip	...	...	...	...	...	...	...	320, 400	26	5000	O
Tubes or bars	...	...	...	...	50	18	3500	180	26	5000	WSO
Crankshaft pins and journals	...	...	...	...	...	...	...	320	(e)	(e)	O
Cutlery, stainless steel:											
Blades, tapering	...	50	28	5500	80	28	5500	180	28	5500	G
Forks or knives	...	...	...	...	80	27	5200	150	27	5200	G
Spoons	...	...	...	...	...	...	...	150, 220	27	5200	G
Dies, steel(h)	...	...	...	...	180	26	5000	320(d)	26	5000	WSO
Dies, tungsten carbide(h)	...	...	...	...	200(j)	26	5000	52 μm (2 μin.)(j), 20 μm (0.7 μin.)(j)	26	5000	WSO
Drive shafts, forged(k)	...	...	...	...	50	18	3500	150	18	3500	O
Electric irons, aluminum	...	...	...	...	150(d)	18	3500	220(d), 280(d), 320(d)	18	3500	O
Files, hand, hardened steel	...	...	...	...	...	...	...	50	28	5500	...
Gears, steel	Burrs	...	...	...	50	26	5000	...	...	...	...
Golf clubs, irons, forgings	Scale	120	30	5800							
Golf clubs, shafts	Weld excess	150	23	4500	...	...	...	220, 320	28	5500	WSO
Hammers, forged	...	...	...	...	50	26	5000	80	27	5200	G
Hypodermic needles, stainless	...	...	...	...	...	...	...	320(d), 400(d)	23	4600	WSO
Jet blades, Inconel	...	...	...	...	50	28	5500	120	28	5500	O
Jet blades, stainless steel:											
Airfoil	...	...	...	...	50	28	5500	120	28	5500	O
Longitudinal	...	...	...	...	...	...	...	80, 100, 120, 150, 180	(m)	(m)	O
Metallographic specimens	...	...	...	...	120	26	5000	220, 320, 400, 500	26	5000	WSO
Plow disks	Scale	50	18	3500	...	...	...	80	21	4200	G
Plowshares	Scale	50	18	3500	80	18	3500	...	...	...	G
Railroad track, butt welded	...	...	...	...	50	18	3500	...	...	...	...
Razor blade strip(n)	Pits	100(d)	11	2100	120(d)	11	2100	...	...	...	O
Refrigerator doors, steel	Weld excess	80	26	5000	...	...	...	...	...	...	G
Relays, nickel	...	...	...	...	...	...	...	320	18	3500	...

(continued)

(a) For successive stages. (b) G indicates grease. O indicates oil. WSO indicates water-soluble oil. (c) Operation completed with abrasive-impregnated nylon wheel. (d) Silicon carbide. (e) Fixtured abrasive. (f) Molded or metal mix. (g) Hardened steel cams for diesel engines. (h) Drawing and sinking dies. (j) Diamond abrasive. (k) Centerless. (m) Fixtured abrasive, reciprocating. (n) Grinding of crown. (p) Grind end. (q) Solution of potassium tribasic or nitrite amine. (r) And excess weld metal

able flap wheels have except that the pack becomes a permanent subassembly of the hub. A different holder is used for the polishing and buffing materials. The holder is designed so that a metal pin can be fitted through the holder and welded to a set of metal plates or a series of metal plates depending on the width of the wheel. The holder pivots on the metal pin allowing the packs to find their own balance point as the wheel is revolving.

In polishing and buffing operations, the metal hub containing coated abrasive packs or buffing packs conveys the material to the workpiece presenting a longer lasting, stronger, and perfectly balanced tool that imparts a uniform finish from the time the wheel is put on until the packs are completely worn. This is accomplished by continually allowing new grain or buffing fabric to be exposed to the workpiece as the flap wheel packs wear. This constant revelation of new grain and buffing material renders a consistency of finish.

Some features of modern flap wheels include contouring the face, slashing the packs, reversing the lead sheet, preoiling the packs, and combining the coated abrasive sheets with materials that cushion the cut. Although flap wheels are ideally used in automatic operations, they are quite effective in manual or off-hand polishing operations.

Table 5 (continued)

Abrasive belts are coated with aluminum oxide unless otherwise noted

Product	Obstruction or roughness removed	Roughing: Grit size, mesh	Roughing: Belt speed m/s	Roughing: Belt speed sfm	Rough polishing: Grit size, mesh	Rough polishing: Belt speed m/s	Rough polishing: Belt speed sfm	Polishing: Grit size, mesh(a)	Polishing: Belt speed m/s	Polishing: Belt speed sfm	Polishing aid(b)
Rifle barrels(k)	...	...	...	...	50	18	3500	80, 120, 180, 220, 320	23	4500	O
Rifle levers	...	...	...	...	80	27	5200	120	27	5200	G
Rifle receivers	...	...	...	...	50	23	4500	80, 120, 180, 240	23	4500	O
Saws, circular	Oxide	150	18	3500	...	...	...	...	...	...	WSO
Saws, strip	Scale	100	18	3500	...	...	...	...	...	...	O
Shears (bows, necks, sides)	...	...	...	...	80	26	5000	150, 240	26	5000	G
Shears, rings	Flash	60	15	3000	...	...	...	150	20	4000	G
Shovel blades	...	...	...	...	50	18	3500	...	...	...	...
Silicon steel, rod or sheet	...	...	...	...	50(d)	15	3000	150(d)	15	3000	WSO
Skate blades, forged	Pits	80	18	3500	150	18	3500	240, 320	20	4000	G
Springs, coil(p)	...	...	...	...	...	...	...	120	18	3500	G
Springs, leaf	...	...	...	...	50	18	3500	...	...	...	O
Stainless steel:											
Coil, series 300	Pits	60	17	3400	80	17	3400	120, 150	17	3400	O
Pots and pans	Wrinkles	80	26	5000	...	...	...	220, 320(d)	26	5000	G
Press plates	Scratches	80	20	4000	100	20	4000	120, 150, 180, 240, 320	20	4000	G
Sheet, No. 3 finish	...	...	...	...	80	20	4000	100	20	4000	G
Sheet, No. 4 finish	Inclusions	100	20	4000	120	20	4000	150(d)	20	4000	G
Sheet, No. 7 finish	Inclusions	100	20	4000	150	20	4000	180(d), 240(d), 280(d)	20	4000	G
Tubes	...	...	...	...	150	18	3500	220, 280, 320	23	4500	O
Turbine nozzles and buckets	...	...	...	...	80	26	5000	120	23	4500	G
Steel, low-carbon:											
Centerless and cylindrical	...	...	...	...	50	18	3500	150, 220, 320	18	3500	O
Stampings	...	...	...	...	...	...	...	120	23	4500	G
Titanium, sheets	...	...	...	...	80(d)	14	2800	100(d), 120(d), 150(d)	14	2800	(q)
Tools, hand, forged	Flash	50	18	3500	80	23	4500	100, 150, 240, 320	23	4500	G
Wrought iron	Burrs(r)	...	...	...	50	18	3500	100	20	4000	...
Zinc-based die castings	Flash	80	26	5000	...	...	...	220	26	5000	G
Zinc sheet	Pits	100	26	5000	150(d)	26	5000	220(d), 240(d), 280(d), 400(d)	26	5000	O

(a) For successive stages. (b) G indicates grease. O indicates oil. WSO indicates water-soluble oil. (c) Operation completed with abrasive-impregnated nylon wheel. (d) Silicon carbide. (e) Fixtured abrasive. (f) Molded or metal mix. (g) Hardened steel cams for diesel engines. (h) Drawing and sinking dies. (j) Diamond abrasive. (k) Centerless. (m) Fixtured abrasive, reciprocating. (n) Grinding of crown. (p) Grind end. (q) Solution of potassium tribasic or nitrite amine. (r) And excess weld metal

Contouring the face of flap wheel to fit the configuration of a particular workpiece eliminates costly break-in time. Sheets or packs are die cut to the configuration of a cross section of the workpiece. Slashing or prescoring of the packs softens the wheel and allows the wheel to conform more readily to irregular surfaces. Staggered slashing means that the slashing in every other pack of coated abrasive is alternately offset from the slashing of those packs that immediately precede or follow. Staggered slashing can also be accomplished by offsetting the slash of the alternate coated abrasive sheets within the pack. Reversing the lead sheets of packs, especially in the coarser grits, is a definite enhancement to the longevity of flap wheels. Reversing the lead sheet in the pack eliminates the coated abrasive from cutting through the backing of the sheets in the preceding pack. Preoiling wheels can increase wheel life by as much as 30%. Lubrication of the wheel helps prevent the natural oils inherent in the cotton backing from drying out. Lubrication also tends to prevent linting. Much finer finishes can be obtained from oiled wheels than from dry wheels, and preoiling tends to cause the wheel to run cooler.

Combination wheels mix or interleaf a sheet or stack of sheets of coated abrasive with three-dimensional abrasive or other materials. Three dimensional abrasive is a nonwoven synthetic product that is impregnated with grain. It is advantageous to use the combination wheel when involved in an off-hand operation. The combining of coated abrasive sheets with a three-dimensional product creates a wheel with a great amount of resiliency. Due to the resiliency created, it greatly reduces operator fatigue. Because this wheel is easier to use, less operator training is required to produce excellent finishes. The cushioning action of this wheel enables operators to finish obscure areas without edge cutting and tearing the abrasive sheets.

Polishing Lubricants

Lubricants are applied to wheels and belts to extend their life and to improve surface finish. Choice of lubricant is influenced largely by the type of metal being polished and the degree of metal cutting entailed.

Wheel Lubricants. Grease stick lubricants for use on polishing wheels are usually made of a mixture of animal tallow, fatty acids, and waxes. Because setup wheels tend to run hotter than coated abrasive belts, the melting points of grease used on wheels are considerably higher than those of greases used on belts. Some wheel lubricants use a mild abrasive to add luster or color to the polished surface. In addition to their polishing action, grease stick lubricants prevent chips loading on the wheel and minimize wheel glazing. They are used also to modify the harsh cut of a newly set-up wheel.

Belt lubricants are available in grease stick tube form for offhand grinding and polishing operations and in liquid form for automatic applications. Their principal function is to eliminate excessive heat, which causes discoloration, distortion, and warping. In polishing low melting point metals, selection of the proper lubricant prevents surface fusion and greatly simplifies any subsequent buffing operations.

Within limits, abrasive belt lubricants help to control surface finish. For example, a thick lubricant loads between the abrasive grains and serves as a cushion; consequently, only the top of the abrasive grit is permitted to penetrate the work surface, reducing the depth of cut and refining the surface finish. In contrast, a light-bodied lubricant keeps the belt open, assisting in stock removal. The belt surface stays wet and oily, and grinding debris is prevented from redepositing and clogging the coated abrasive surface.

Polishing time is sometimes reduced and surface finish is improved by using a coarse-grit belt with a lubricant, rather than a fine-grit belt without a lubricant. By eliminating dry patches and providing a more uniform cutting action over the whole surface of the belt, the lubricant permits greater stock removal without burning.

Application. Used on wheels or belts, lubricants should be applied in small amounts and at frequent intervals as the polishing operation progresses. A majority of lubricants for belts and wheels are water soluble, biodegradable, and maintain the same lubricating characteristics of kerosine mixtures. Automatic lubrication systems have been developed to effectively apply lubricants to belts on semiautomatic and automatic polishing machines. Care should be taken to ensure that low flash point lubricants are not used, to promote operator and equipment safety.

Limitations of Polishing

The limitations of polishing generally are associated either with part size and shape or with the ability of the process to attain a specific surface finish. A part with hooked edges or sharp projections can cut or snag the polishing belt or wheel. Limitations in terms of surface finishes obtainable are related primarily to the abrasive grit sizes available.

It is relatively simple to polish a part that is small enough to manipulate manually against a stationary wheel, but it is exceedingly difficult to achieve a uniform finish over a large surface with a traveling wheel. Extremely small parts present a different problem. As parts become too small to be manipulated manually, they must be held in jigs and fixtures. At this point, part configuration also becomes a controlling factor, because the complexity of the shape of a part frequently dictates the complexity of jigs and fixtures required. A simple geometric shape, such as a ball, can be readily polished to a high finish in sizes down to 0.25 mm (0.010 in.) diameter by locating it in a circular groove of appropriate size and allowing it to move in contact with a firm wheel. Very small cylindrical parts can be polished in a similar manner. However, as part configuration becomes more complex, the polishing of small parts becomes increasingly difficult unless a mass finishing method can be used.

The limitations of automatic or semiautomatic polishing are different from those associated with manual operations. Large surfaces can be uniformly finished by affixing the part to a moving conveyor and having a polishing wheel exert even pressure against this component as it is drawn past the polishing head. Pneumatic controls allow automatic polishing equipment to be extremely exact in repeating finish generation or work to be accomplished. The limitation falls on the tolerance of the repeated component coming to the machine. Complex parts, large or small, can be accommodated on automatic equipment, but the compromise is the initial cost of the equipment as well as the productivity gained in pieces finished per hour. Automation in polishing has greater repeatability, can accommodate complex parts, but lacks the flexibility associated with salvage operations that are accomplished by hand.

Buffing

Buffing produces smooth reflective surfaces. It is the rearrangement and refinement of component scratch and has the ability to level surfaces where the scratch is only visible under high magnification. Buffing is accomplished by bringing a workpiece into direct contact with a revolving cloth or sisal buffing wheel which is charged with a suitable compound. Depending on the type of wheel and compound, substantial stock can be removed, radii generated consistently, and smooth, bright, and lustrous surfaces can be produced with precision.

Hard buffing is used to cut down or smooth the surface of a material that may or may not have been previously polished. Frequently, the finish produced by hard buffing is adequate for a particular application. Buffs used for hard buffing are usually made of cloth with a high thread count (86/80). The density of the thread count varies with the requirements of the job. Because hard buffing entails both a cutting action and a smoothing action, an aggressive compound is generally used.

Color buffing refines a surface to a very low micro-inch finish and produces a lustrous, scratch-free condition. Any marks or scratches left on a surface after hard buffing can be removed by color buffing. Color buffs are always softer than buffs used for hard buffing and are made of cloths with a lower thread count (64/68). Color buffing compounds must produce a high finish and permit the buffing wheel to shed or to throw off the small particles of metal and compound that build up during buffing. If a compound does not provide this shedding action, the buildup of particles forms a solid cutting head, requiring the buff to be raked frequently.

Contact Buffing. Most conventional modern automatic buffing equipment is set up for contact buffing. For contact buffing, a machine imitates the movement of a manual buffing operation, and densely packed wheels are used, rotating at shaft speeds usually higher than 1200 rev/min. Because the wheel is very hard when rotating at high speeds, the area covered is relatively small.

To increase versatility of finishing equipment and to maintain more precise control, hydraulic drives for conveyors and for the buffing and polishing heads are readily available. Hydraulic drives facilitate precise control of speed, which is readily changed when there is a change in the product to be finished. Other developments include high-speed and straight-line indexing conveyors, programmed controls and three-dimensional fixture manipulation.

Mush buffing is a technique in which a broad wheel is used to cover much larger areas than possible in contact buffing. Shafts can be up to 3.6 m (12 ft) long, with buffs mounted along the shafts, usually separated with 13 to 75 mm (½ to 3 in.) of space between each of the buff sections.

In mush buffing, the wheel itself flexes and envelops the contours of components being finished. Parts travel across the face of the wheel without precise manipulation of the fixtures. In general, simple rotation or oscillation is all that is required to ensure a uniform finish on component faces. Shaft speeds for mush buffing are almost always much slower than for contact buffing. The slower speed of rotation keeps the working temperature of the buff sections lower, resulting in longer buff life. Productivity is generally considerably higher than with conventional methods. Uniformity and quality of finish are enhanced by the increased flexibility of the slower speed buffs.

Refinement of the mush buffing process led to the development of modular construction of buffing equipment. Modular conveyors may be assembled from standard carriages mounted on a rigid framework. Modular design of equipment increases versatility in both short and long term. The suspended mush buffing heads can be placed inside or outside the conveyor and can be relocated for quick application changes. Hence, a single machine can finish a range of components simultaneously and then can be set up to handle a completely different range of parts, just by changing positions of heads and component fixtures. When requirements change or production increases, modification of the shape or size of the conveyor can be made, and more heads of the same or a different type can be added.

Fig. 2 Mechanism for applying bar compound to buffing wheel of automatic buffing machine

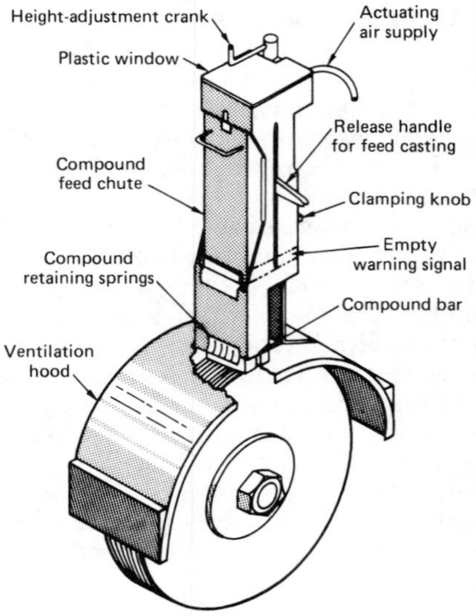

Compound Systems

The prime purpose for the buffing wheel is to carry the compound, which is usually sprayed on the wheel with the use of an air-operated spray gun. The following finishing compound systems find wide usage in the finishing industry.

Bar Compound Application System. The compound is compacted into bar form and is applied to the wheel by using a compound applicator that pushes the bar into the surface of the wheel at timed intervals. A bar compound application is shown in Fig. 2. The disadvantage of this system, of course, is the nubbin that is left over, due to requirement to hold the end of the bar. Bars are cast in standard forms to suit the several types of solid applicators or bar applicators that are used.

Low-pressure liquid compound systems utilize pressure tanks or pumps. Pressure tanks up to 190 L (50 gal) can be purchased and pressurized to approximately 345 kPa (50 psi), depending on the viscosity of the compound. Compound made for this type of system is fairly fluid so that the compound constantly seeks its own level and does not develop cavitation which introduces air to the compound system. The other low-pressure system utilizes the 3:1 ratio pump that is placed in a 210-L (55-gal) drum. Compound is pumped directly to the system, which includes a compound line or manifold, air manifold, electric solenoids for firing the guns, and timers. In this low-pressure system, depending on viscosity of the compound, pressures of up to 825 kPa (120 psi) are used. Low-pressure paint application type guns with chrome tip or carbide tip needles are used. These protective tips are added to resist wear. The gun is used with an open and shut valve, energized by an air solenoid which opens the valve and releases the needle from the seat at the nose of the gun. Guns should be adjusted and locked out so that they are open fully or closed fully at any time. Any attempt to control the volume of compound by adjusting the needle closer to the seat at the nose of the gun creates a straining effect between the liquids and solids in the compound and causes plugging in the guns.

Low-Pressure/High-Pressure Systems. Low-pressure/high-pressure guns are used in conjunction with the low-pressure systems. These guns have a 27 to 1 ratio between the air cylinder and the firing mechanism of the gun. An advantage of this type of system is that measured amounts of compound are fired into the wheel at pressures up to 8300 kPa (1200 psi). Tripoli and calcined alumina compounds are the only compounds that can be used successfully in this type of gun because the rate of wear using aluminum oxide compounds is too great.

High-Pressure Systems. High-pressure systems utilizing a 30 to 1 pump, high pressure lines, and high pressure manifolds are also used. Guns are fired in the same manner as the low-pressure systems, utilizing timers and solenoid valves. High-pressure and the low-pressure/high-pressure process find the widest application for nonferrous buffing.

Compounds

Grains generally used in conjunction with all finishing compounds include the following:

- Tripoli, and/or silica, for most nonferrous materials and plastics
- Fused aluminum oxide for finishing ferrous products
- Calcined aluminas for ferrous and nonferrous materials
- Red rouge for high coloring or finishing of brass products
- Green chrome oxide flours for high coloring or finishing of nonferrous materials and stainless steel, and finishing of plastic products

All of these materials are used with fatty acids and wetting agents in liquid form and are made to suit all of the respective cleaning processes that are necessary prior to any plating or anodizing operation.

Most buffing compounds consist of an abrasive that is immersed in a binder carrier. The abrasive serves as the principal cutting medium; the binder provides lubrication, prevents overheating of the work, and firmly cements the abrasive to the wheel face. Binders must not chemically etch, corrode, or mar the metal surface.

The buffing action of any individual compound can be altered by increasing or decreasing the particle size of the abrasive, or by varying the amount and types of grease used in the binder. If a more intense cutting action is required of a given compound, the particle size of the same abrasive or the quantity of binder may be increased. Modern buffing compounds use water-soluble binders. The grease used in the binder is generally a mineral, animal, or vegetable base.

Selection of a buffing compound is influenced mainly by the type of metal to be buffed, the initial surface condition, type of buffing equipment, and the surface finish desired. The size and design of the part to be buffed may also influence compound selection, particularly if these characteristics dictate the lineal speed of the work as it passes the buffing wheel (dwell time).

Types of Compounds. Many buffing compounds derive their names from the abrasives used in their formulas. Others are designated to reflect the type of metal on which they are most commonly used or the function they perform. Several of the more widely used buffing compounds include the following:

- *Tripoli compound* contains a cryptocrystalline silica, of which there is approximately 75% free or crystalline silica. It has the ability to cut sharply at first and then break down in size, resulting in a finer cut and a higher coloring action. It is used extensively for buffing nonferrous metals, particularly copper, zinc, aluminum, and brass
- *Bobbing compounds* usually contain some form of coarse silica such as flint or quartz. These abrasives are considerably harder and sharper than tripoli compounds, and are used for heavy cutting down and the removing of pits from extruded or cast aluminum and silver alloys
- *Cut or cutdown compounds* usually contain tripoli for buffing nonferrous metals, and fused aluminum oxide abrasives for buffing carbon or stainless steel
- *Cut and color compounds* combine fast cutdown and coloring operations, sacrificing maximum cutting properties of cutdown buffing, and the extra brilliance of separate color buffing. Abrasive mixtures are selected to suit the particular metal being buffed, and the desired levels of cut and color
- *Color or coloring compounds*, through the use of a very fine soft abrasive, produce maximum brilliance and freedom from scratch on the buffed piece. Choice of buff hardness, buffing compound, peripheral speed, rate of speed, contact pressure, and buffing time will vary with different types of pieces being colored, and the degree of luster required by the customer. Very fine soft white silica powders, very fine alumina powders with highly lubricated binders, or red rouge are usually used in compositions for coloring precious metals such as silver or gold. Fine alumina with lubricated binders and a minimum of buffing pressure is usually the choice for aluminum, pewter, thermoplastic and thermosetting plastics, magnesium and their alloys. Fine calcined aluminum oxide (alumina), chromic oxide abrasives, and lime substitutes, combined with clean working binders, color nickel, chrome and zinc. Fine fused or calcined aluminum oxide, or combinations of the two, are the abrasives generally used to color carbon and stainless steel.
- *Stainless steel buffing compounds* generally contain fused or unfused aluminum oxide powder or a mixture of both. These abrasives are available in a wide range of particle sizes, hardness, and oil-absorbing qualities. The cutting and coloring action obtained depends on the type of aluminum oxide used and the amount of grease binder in the formulation
- *Steel buffing compounds* generally contain a mixture of fused and unfused aluminum oxides. They provide a sharp cutting action, as well as a lustrous finish
- *Chromium buffing compounds* consist of fine unfused alumina. These abrasives are also used as secondary coloring compounds on stainless steel. They were originally developed to color buff chromium plated parts that had been stained or frosted in the electroplating process
- *Rouge compounds* are prepared from almost pure red iron oxide powder and are intended for finishing the noble metals. Another form is made with green chromium oxide powder for buffing hard ferrous metals to produce an extremely high finish without stock removal
- *Emery paste* refers to a grease or tallow stick impregnated with emery. Although it can be used on setup polishing wheels during the breaking-in period, it is used extensively on tampico brush wheels for producing a light satin or brush finish
- *Greaseless compounds* are a special variety of compounds that are entirely free of grease binders. The abrasive is blended with water and gelatine-glue. It is packaged in airtight containers, and should not be exposed to the air for a period of time as it becomes dry and too hard for efficient

wheel transfer. When these compounds are applied to cloth buffing wheels, a flexible, dry cutting abrasive head is formed. By varying the type of abrasive, a variety of finishes, ranging from an almost bright luster to an almost totally nonreflective finish can be obtained. Applied to loose buffs, greaseless compounds are effective in producing satin finishes or performing light polishing operations. Their cutting action is increased when they are applied to sewed cloth buffs or felt wheels. Because greaseless compounds are completely free of tallows, oils, or waxes, workpieces are left clean and dry and require no subsequent cleaning. Storage temperature of the product should be between 4 and 21 °C (40 and 70 °F) because it is perishable

- *Liquid buffing compounds* contain abrasives that are suspended in a liquid binder carrier. The abrasives are identical to those used in solid bar compounds. Liquid compounds are applied to the buff or the work surface by means of gravity feed, spraying, brushing, or dipping. Liquid compounds are preferred for large volume production on semiautomatic or automatic buffing equipment. They offer the following advantages:

  Reduce direct labor costs, because less time is involved in applying or replacing compounds

  Reduce waste of buffing compound, because they can be applied in small quantities with manual or automatic controls, and they leave no unusable waste

  Provide a wider range of cutting and coloring characteristics because they can be formulated from a wider range of grease binders

  Extend buff life by wetting and penetrating the buff to a greater depth, by maintaining an optimum amount of buffing compound on the surface of the buff at all times, and by permitting a cooler buff operating temperature

  Saponify and emulsify faster than bar compounds and therefore are less likely to back-transfer on the workpiece or to setup solidly in crevices or recesses; thus, they are easier to remove in subsequent cleaning operations

  Formulated precisely to meet the requirements of a specific application. Unlike bar compounds, they are not limited by forming a bar that is physically strong enough to survive application shock

Fig. 3 Construction of buffing wheels

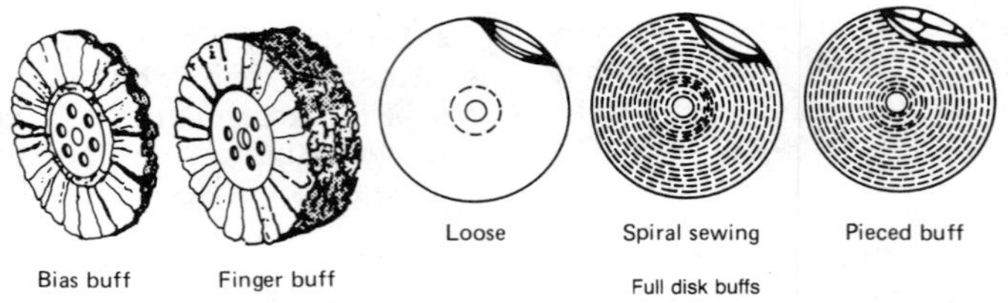

Buffing Wheels

A buffing wheel carries the abrasive buffing compound across the work surface and produces a cutting and coloring action. Selection of an improper buffing wheel may not result in the production of an unacceptable surface finish. However, improper selection invariably results in increased costs because of (*a*) reduced buff life, (*b*) excessive use of buffing compounds, (*c*) increased direct labor, and (*d*) a higher rejection rate. Buffing wheels in general use are classified under the designations of (*a*) bias buffs, (*b*) finger buffs, (*c*) full-disk buffs, and (*d*) pieced buffs. The construction of these buffs is illustrated in Fig. 3. Each type has certain characteristic advantages.

Bias Buffs. In automatic buffing, the use of bias buffs is popular. Buff sections are usually composed of sections approximately 13 mm (½ in.) wide and made concentric to the spindle by the use of center plates. The method of construction and assembly is such that air circulates through from the center to cool the buffs and to prevent buildup of heat at the surface of the buff or wheel in contact with the part. Resilience of the buff section is controlled by (*a*) varying the number of plies of cloth in each section, (*b*) the diameter of the center retaining plate, and (*c*) the degree of pleating. Heavy buffing operations require buffs with a greater number of plies and a greater number of pleats in their construction. The gathering of the pleats at the center of the buff maintains the aggressiveness of the buff as the wheel wears out and the surface speed decrease. Use of cloth treatments can increase the effectiveness of the buff to retain compound, increase the stiffness and aggressiveness of the wheel, or lengthen the time before the wheel wears down to an unusable condition. Cutting wheels are composed of cloth with 86/80 threads per square inch, and lighter duty wheels are made up of cloth of 64/68 threads per square inch. All of the materials are cut on the bias so that the threads are continuous from the center out and do not fray during use. Bias buffs used on automatic equipment with a properly designed buff shaft, draw air down the shaft and expel it through the periphery of the buff itself. This action cools the part being buffed and the wheel, which prolongs the service life of the wheel.

Treated bias buffing wheels impregnated with abrasive usually reduce the finishing cost per piece. Because the cloth is treated with abrasive and binder, a more efficient action at the point of contact between the buffing wheel and the work are realized. This type of buff generally works cleaner, lasts longer, increases metal removal, uses less buffing compound, cuts faster, produces a brighter finish; all at reduced buffing pressures and/or buffing time.

Finger buffs are composed of individual segments of bias-cut cloth assembled on a buff hub. When buffing pressure is applied, these segments yield and flex as they follow the work contour. The freedom of movement afforded each segment provides the flexibility and pliancy required for buffing highly contoured surfaces. The density of finger buffs is varied by (*a*) altering the number of layers stitched into each segment unit, (*b*) increasing or decreasing the number of rows of sewing on each segment, (*c*) changing the number of fingers mounted around the center ring, or (*d*) changing the diameter of the buff hub. Finger buffs are used when other types of buffs are unable to make adequate contact with the work surface.

Full-disk buffs are constructed so that each ply in the buff is a full one-piece circular layer of cloth. A 20-ply section is considered standard and will average about 6 mm or ¼ in. in thickness. However, any reasonable number of plies can be furnished when a thicker or thinner buff is desired. Loose buffs, in which the layers of plies are assembled and held together with a single row of stitching around the arbor hole, are generally used for light cutting down or coloring operations and other applications in which a great deal of looseness and flexibility is desirable. Various patterns of stitching can be used to control the hardness, flexiblity, resiliency, and compound-holding ability of full-disk buffs. Spiral stitching is most widely used and is readily obtainable in three widths, which determine buff hardness. These include soft density (10 mm or ⅜ in.), medium (6 mm or ¼ in.), and hard (3 mm or ⅛ in.). Even harder buffs can be produced by using the block pattern of a square-sewn buff with sewings tightly together.

Pieced buffs are constructed from new, but irregularly shaped remnants and strips of cloth that are by-product cuttings from other textile articles. Pieces are laminated between an outside layer of full disks of cloth. Various designs are stitched in to tie the plies together into a completed buff section. A variety of grades and qualities of pieced buffs are available. These consist of buffs manufactured from all colored pieced material or of mixtures of unbleached, bleached, and mixed weights and cloth counts of fabrics. These buffs are generally used in heavy cutting down operations or in operations that severely attack the buff face with a raking effect.

Flannel and Sisal Buffs. Several major types of buffs are available also in Canton flannel or sisal. Because of its softness, flannel lends itself to light color buffing of silver, gold, and other precious metals. It is used for light-duty buffing of celluloid, plastic, wood, and lacquered surfaces.

Sisal is a slender, hard, cellulous strand of vegetable fiber which is tough and resilient. Because of its natural abrading characteristics, it is ideal for cutting down operations. Sisal buffs are used extensively in the finishing of steels, including stainless steels. In these applications, they can sometimes eliminate the fine grit size polishing operations that would usually be performed when surface imperfections are too deep to be removed effectively or economically by cotton buffing wheels. Sisal buffs can often remove or blend stretcher strains, orange peel, directional scratch patterns imparted by polishing wheels or abrasive belts, and light die marks.

Sisal buffs are available in almost the same variety of types and sizes as the more familiar cloth buffs. Occasionally, cloth is used in combination with sisal. Because of its open weave and coarseness, a good head of buffing compound is difficult to build up and to retain on a sisal buff. Consequently, compounds for sisal buffs have lower melting points and a higher percentage of grease bond. Sisal buffs operate effectively in the speed range of 38 to 50 m/s (7500 to 10 000 sfm). Their effectiveness decreases sharply when speeds drop below 30 m/s (6000 sfm).

Wheel speeds are influenced to some degree by the contour of the workpiece. A workpiece with excessively deep recesses or a complex shape cannot be adequately buffed unless centrifugal force, buff stiffness, or speed of the wheel is reduced to the point where the buff twists, or mushes to conform with the contours. Speeds must also be lowered to avoid glazing. A glazed wheel burnishes (slips on the work), rather than polishing it.

Machines for Polishing and Buffing

A variety of standard finishing machines are available for simple manual polishing and buffing, and specially designed machines are available for highly automated finishing lines. Equipment and production rates for specific polishing and buffing applications are given in Table 6. Table 7 supplies equipment and operating conditions for various buffing applications. The type of machine used for a specific finishing operation depends on the (*a*) shape and size of the part, (*b*) surfaces to be finished, (*c*) material of the part, (*d*) finish specifications, and (*e*) production rates. Investment in the finishing equipment is also a major factor and must be justified by cost savings, improved quality of the product, product consistency, and production requirements. Major elements of semiautomatic and automatic finishing machines are heads, conveyors, and work-holding mechanisms.

Heads

There are several types of heads, including those that are fixed in a set position for semiautomatic or hand finishing operations, and those that are adjustable to permit angling of the finishing wheel spindle to suit a specific relationship between the wheel and the part. Heads may be spring, pneumatically, or hydraulically balanced so that the wheel follows the contour of the component being finished. These are broad wheels for mush buffing.

The fixed head, called a conventional or backstand head, may use a single spindle on which a finishing wheel or belt can be mounted, or it may incorporate a double-end spindle so that two separate finishing operations may be performed, one on each end of the drive spindle. The latter is generally called a conventional double-end head or jack. The adjustable type of finishing head is more versatile and can be used for both semiautomatic and automatic operations. As with the conventional head, the spindle diameter, length, and horsepower of the adjustable head are determined by the finishing requirement of the part. Wheel speeds also are important to each finishing application, especially for jobbing work. Variable speed drive arrangements are helpful and substantial time savers for resetting adjustable head speeds to obtain the most efficient finishing. The cost of variable speed controls should be weighed against the estimated savings from improved efficiency of wheels and compounds and reduced job changeover time.

For many operations, adjustable finishing heads with stroking wheel spindles or stroking heads operate more efficiently and economically. Stroking heads improve wheel and belt life as well as producing a more uniform quality of finish. Usually, lift mechanisms for head retraction are required on adjustable finishing heads for use during part transfer operations in automatic and semiautomatic finishing machines. Retracting heads eliminate leave-off blemishes, dubbing, and excessive pressures during transfer of the parts from station to station.

Mechanical, pneumatic, and hydraulic floating head devices permit the finishing wheel or belt to ride over-contoured areas. Floating head attachments can control finishing pressures on contoured and out of round parts and eliminate burning on the part and excessive wear on the wheel in contact buffing examples.

Buff heads can be supplied with low speeds suitable for mush buffing operations. In mush buffing, wide, up to 3.6 m (12 ft) long buffing heads are used. Large diameter bias buffs are operated at spindle speeds from 300 to approximately 1200 rev/min. Individual buff sections are separated by spacers to provide flexibility so that parts can be forced into the buff surface, enabling

Table 6 Equipment used and production rates obtained in specific polishing and buffing applications

Products	Finish desired	Type of machine	Finishing medium	Production, pieces/h
Aluminum products				
Architectural extrusions	Bright or satin	Reciprocating straight-line	Sisal and cloth buffs	(a)
Automobile wheels	Bright	Single-spindle semiautomatic(b)	Cloth buffs	40-60
Electric-tool parts, die cast(c)	Bright and colored	Universal straight-line	Cloth buff	300-600
	Bright and colored	Semiautomatic	Cloth buff	60-120
	Bright and colored	Continuous rotary index	Cloth buff	300-600
Flashlight cases	Bright	Four-spindle semi-automatic	Cloth buff	200-600
Hi-fi speakers	Bright	Single-spindle semi-automatic	Cloth buff	100-120
Stampings, contoured	Bright	Rotary index	Cloth buff	300
Stove burner heads	Bright and colored	Rotary index	Sisal and cloth buffs(d)	300-600
Brass products				
Coffeepot bodies	Bright	Semiautomatic; special fixtures	Cloth buffs	60-100
Doorknobs	Bright	Universal straight-line	Cloth buffs	4000
Faucet aerators	Bright	Rotary index	Bias cloth buffs	500-1000
Fire extinguishers	Bright	Rotary index	Cloth buffs	40-60
Fireplace fixture trim	Bright	Vertical double-roll buffer	Cloth buffs(e)	200-750
Plumbing fittings(f)	Bright	Universal straight-line	Cloth buffs	Up-1000

(continued)

(a) 90 to 300 m/h (300 to 1000 ft/h) for 45- by 100-mm (1¾- by 4-in.) extrusions buffed on four sides. (b) Special machine. (c) Motor housings, gear housings, handles. (d) Bias and full-disk cloth buffs. (e) 915 mm (36 in.) wide. (f) Valve bodies, spouts, J-bends. (g) 0.05 to 0.2 m/s (10 to 30 ft/min). (h) 0.1 to 0.3 m/s (20 to 60 ft/min). (j) Types 302 and 430 stainless steel. (k) 1018 and 4140 steel forgings. (m) 0.25 to 0.4 μm (10 to 16 μin.). (n) 0.25 to 0.3 μm (10 to 12 μin.). (p) 0.2 m/s (30 ft/min)

contoured areas to be buffed in a single stage. Mush buffing is particularly applicable to rectangular machines on which extremely wide buff heads can be used. Mounting of heads inside and outside of conveyor doubles the effective use of conveyor length. The majority of indexing, straight-line, and rotary table machines use follower buff heads that can be withdrawn after the finishing operation. The buffing head is usually mounted on a pivoted arm and can therefore follow the contour of the surface being finished as the article passes the buff in a continuous motion.

Wear-compensating units for automatic finishing wheels can be provided on adjustable heads to eliminate manual adjustments. The cost of compensators can be justified on the higher production automatic finishing machines where multiple head operations are performed by the increase in effective machine utilization and even in fairly low production applications by improvement in product quality. Cartridge finishing wheel assemblies can be used on wide wheels to reduce the time required for replacing worn-out wheels. Reduction of the change-over time results in greater operating efficiency.

Work-holding mechanisms accomplish rotation, cam positioning, or fixing of parts as they contact the polishing or buffing wheel. In addition, these mechanisms are used to convey parts through successive operations and into the unloading station. Work-holding mechanisms are usually designed to hold fixtures by means of chuck spindles, conveyor platens, fixed table supports, or conveyor belts. Figure 4 illustrates four types of work-holding mechanisms used in buffing on semiautomatic machines.

Semiautomatic Machines

Semiautomatic equipment is used for low-cost polishing and buffing operations where quantities of production may be limited or a large variety of parts are polished or buffed. Compared with manual methods, semiautomatic machines may provide an economical method of producing these parts. Most semiautomatic machines are single or double spindled; however, eight work spindles are utilized for some specialized applications. These units may be used in conjunction with single or double wheels, with conventional buffing lathes, or adjustable heads. They may be designed for hand or automatic indexing. Tooling costs of semiautomatic equipment are usually considerably less than for full automatics. Rapid changes in setups are possible because of the use of one single buffing head and the relatively small number of fixtures to be changed.

Semiautomatic machines are also suitable for oval or rectangular pieces, such as hardware items, plumbing goods, electrical parts, and small utensils. A cam-operated chuck follows the contour of the piece, which makes it possible to finish these various shapes. Production rates of up to 600 pieces per hour can be attained.

Automatic Buffing and Polishing

For larger outputs, where many polishing stations are necessary, fully automatic equipment is available. The type of equipment required depends on the size and shape of the components, the output requirements, the number of polishing stages, and finishing techniques required. An automatic buffing machine consists of one or more buffing heads with a transfer system on a carrier or track on which the work is mounted on jigs or fixtures. The work carriers or jigs are usually provided with rotation or controlled articulation to ensure effective buffing on all relevant edges and surfaces of the product. In some cases, the whole part may be buffed in a single operation. In other operations, a series of buffs is used to finish different areas.

Rotary automatic machines consist basically of a series of spindles mounted on a rotary table. Polishing

Table 6 (continued)

Products	Finish desired	Type of machine	Finishing medium	Production, pieces/h
Stainless steel products				
Automobile exterior trim	Bright	Horizontal-return straight-line and cutaway roller-feed	Sisal and cloth buffs	(g)
Coils	Polished	Sheet-polishing mills	Abrasive belt	(h)
Fire extinguishers	Bright	(b)	Cloth buffs	120-180
Kitchen sinks, inside walls (j)	Polished	Special sink-polishing	Bob-type abrasive wheel	40
Kitchen sink tops	Bright or satin	Reciprocating straight-line	Buffing wheels	40-60
Rocker-arm valve buttons	Mirror	Continuous rotary	Sisal buffs	1600
Carbon and alloy steel products				
Automobile bumpers	Bright	Open-center straight-line	Setup wheels; sisal and cloth buffs	80-300
Automobile grilles	Bright	Straight-line	Sisal and cloth buffs	(g)
Piston rod, piston, cylinder tube(k)	Polished(m)	Special lathe-type polishing	Abrasive belt; cloth buffs	Up to 180
Sheets, flat	Polished(n)	Sheet-polishing mills	Abrasive belt	(p)
Zinc-based die castings				
Automobile door handles	Bright	Rotary automatic(b)	Cloth buffs	1000
Automobile hubcaps	Bright	Universal straight-line	Cloth buffs	800
	Bright	Semiautomatic	Cloth buffs	80
Automobile instrument housing	Deburred; polished	Roller-feed table	Abrasive belt	40-60

(a) 90 to 300 m/h (300 to 1000 ft/h) for 45- by 100-mm (1¾- by 4-in.) extrusions buffed on four sides. (b) Special machine. (c) Motor housings, gear housings, handles. (d) Bias and full-disk cloth buffs. (e) 915 mm (36 in.) wide. (f) Valve bodies, spouts, J-bends. (g) 0.05 to 0.2 m/s (10 to 30 ft/min). (h) 0.1 to 0.3 m/s (20 to 60 ft/min). (j) Types 302 and 430 stainless steel. (k) 1018 and 4140 steel forgings. (m) 0.25 to 0.4 μm (10 to 16 μin.). (n) 0.25 to 0.3 μm (10 to 12 μin.). (p) 0.2 m/s (30 ft/min)

and buffing operations are performed by adjustable heads placed around the periphery of the rotary table. These machines are adaptable to many different sizes and shapes of parts and provide an economical and versatile means for high production operations. To meet particular part configuration and production finishing requirements, rotary automatic machines are available with indexing table rotation, continuous table rotation, combination indexing and continuous table rotation with individual chuck spindle motor drives, and special table arrangements in a wide range of sizes.

Indexing rotary automatic machines, as shown in Fig. 5a, index parts from station to station through various finishing operations. Dwell time is variable and can be preset to suit finishing requirements. The work or chuck spindles mounted on the rotary table are driven. Parts are rotated at each work station. Indexing machines are used to (*a*) permit sufficient polishing and buffing time on the part at each finishing station for producing the quality of finish desired, (*b*) reduce the amount of tooling required for fixturing parts on an automatic finishing machine, (*c*) permit finishing heads to retract and eliminate dubbing and finishing blemishes that might occur, (*d*) permit the use of special attachments required for finishing noncircular parts, and (*e*) permit finishing heads to retract from recessed and hollow areas of parts, thus eliminating excessive finishing pressures during transfer of parts and increasing wheel life. All indexing equipment uses contact buffing concepts.

Continuous rotary automatic machines, shown in Fig. 5b, provide continuous flow of parts through each finishing station. The driven work spindles mounted on the table rotate each part as it passes through each finishing operation. This type of machine is utilized to (*a*) meet high production requirements, (*b*) finish parts that do not require a great amount of polishing or buffing time, and (*c*) permit high production finishing of part shapes that lend themselves to continuous travel under the finishing heads without causing excessive wheel pressures and undesirable wheel marks. Continuous rotary equipment may have a tendency to shape buff wheels. It may use contact or mush buffing techniques.

Combination rotary automatic machines combine the features of both the indexing and the continuous types into one machine. This permits high production flexibility; one machine can be used for both long and short production runs. These machines can be easily and quickly converted from indexing to continuous rotary motion, or from continuous to indexing for production finishing a variety of parts.

Individual chuck-spindle-drive rotary automatic machines are similar in operation to indexing rotary automatics. The spindles can be driven either clockwise or counterclockwise at each finishing station for greater coverage on recessed or embossed parts. Reversing the rotation of the work spindle at successive stations permits the finishing wheels to enter recessed areas of parts in opposite directions and assures uniform coverage. The individual spindle drives also permit heavier loads on the work spindles.

Straight-Line Machines

Straight line machines are the most versatile of all buffing and polishing equipment. They polish and buff practically any kind of part. As indicated by their name, all parts pass through finishing heads in a straight line, thus maintaining a straight face on buff wheels. They can be equipped with heads for polishing, sisal buffing, cutting, and for final coloring. They can be made almost any size and do a great number of operations. Point contact heads are possible, as well as wide face heads. Indexing can be added to many types of straight-line machines to obtain a great advantage over standard indexing rotary equipment. The number of point contact heads can be increased as desired.

Reciprocating straight-line machines, as shown in Fig. 6, have a table on which work moves back and

Fig. 4 Work-holding mechanisms for buffing on semi-automatic machines

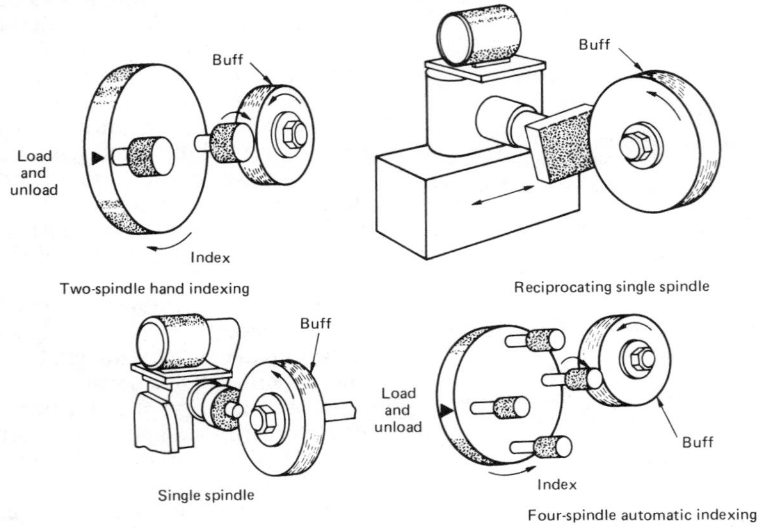

Fig. 5 Rotary automatic finishing machines

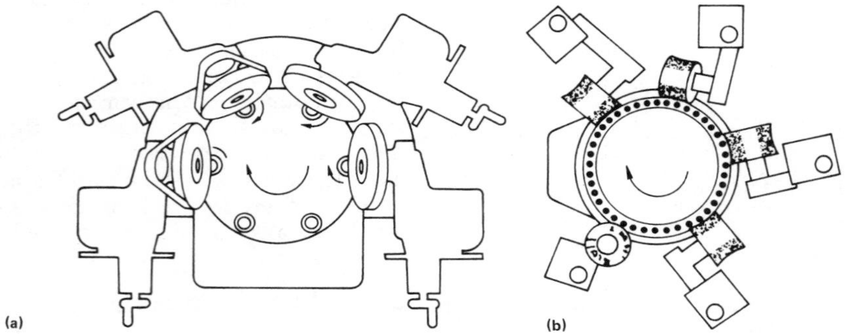

(a) 760-mm (30-in.) indexing machine for finishing clutch plates. (b) 1270-mm (50-in.) continuous machine for finishing tumblers

Fig. 6 Reciprocating straight-line machine using one finishing wheel on adjustable lathe

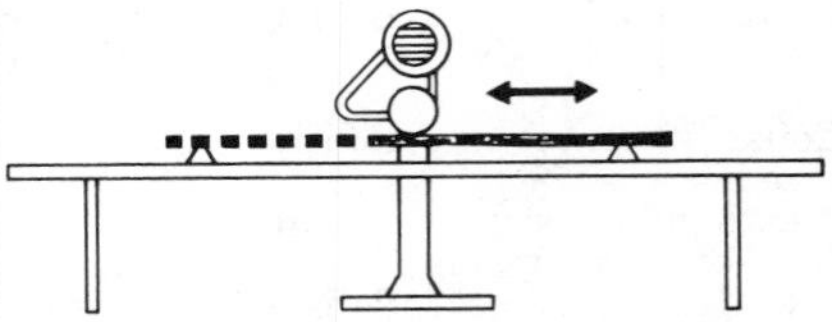

forth under one or more buffing heads. The parts are placed on a fixture at a load station, passed through the required finishing heads, and returned to the operator station for unloading. Parts such as towel bars, bathroom basin legs, moldings, extrusions, stove tops, various door ornaments, and castings are finished on this equipment.

Horizontal return straight-line machines, shown in Fig. 7, are high production equipment that can finish automotive trim, appliance trim, and other manufactured products requiring a finish. Parts are loaded and unloaded at the operator station at the ends of the conveyor. The fixtures can be turned at 180°, or the machine can be designed with a flop-over fixture for buffing both sides of an item. Horizontal return machines also lend themselves to camming fixtures for irregularly shaped parts.

Over-and-under type machines usually are used for smaller parts such as various die cast automotive parts, bumper ends, electric appliances, and cigarette lighters. This equipment conserves floor space by permitting heads to be mounted on each side of the conveyor so that parts buffing can be done on both sides simultaneously. Loading and unloading can be accomplished automatically or by a single operator at one end of the machine.

Universal or modular rectangular type straight-line equipment is popular for finishing a variety of parts from small die castings to large automotive parts. Examples are die cast automotive ornaments, hub caps, door handles, plumbing fixtures, telephone parts, builders' hardware, hollow ware, electrical appliances, and car bumpers. This machine can be designed for indexing, or it may be cammed with fixtures that turn the part in any direction or any motion.

Automation. Modern automated buffing and polishing equipment frequently utilizes pick-and-place mechanisms for loading and unloading of components. They reposition them on the machine to expose other surfaces to be finished or to convey parts to and from feeders or transfer lines. Robots are used to reload some automatic machinery. They are also used occasionally to manipulate components against buffing and polishing wheels to simulate motions of a manual polishing operator.

For flexible automated polishing and buffing, there are several styles of rectangular conveyor machinery where all heads may be quickly repositioned by a programmable controller. With this equipment, an operator selects the program for a given part number and the machine automatically configures itself to the operating condition necessary to produce finished parts. Fixtures may have to be changed by the operator or by a pick-and-place mechanism. Occasionally, fixturing may have sufficient versatility to handle more than one style of component.

Flat Part Polishing Machines

Abrasive belt machines perform grinding, polishing, and deburring operations on bar, strip, coils, blanks, stampings,

Table 7 Equipment and operating conditions for buffing applications

Part and material	Previous operation	Equipment	Compound	Wheel Diameter mm	Wheel Diameter in.	Wheel Speed, rev/min	Production, pieces/h	Subsequent operation
Andiron column, brass	Polish	Double-end lathe(a)	Grease stick	355	14	2200	19	Color buff(b)
Automotive parts								
Headlight bezel, aluminum	Press form	Rectangular straight line(c)	Liquid tripoli	405	16	1000-1400	1150-1200	Anodize
Spotlight body, brass	Press form	Indexing rotary(d)	Liquid tripoli	405	16	1400-1600	520-560	Plate(e)
Taillight housing, zinc die casting	Hand deburr	Retangular straight line(c)	Liquid tripoli	405	16	800-1200	700-740	Plate(e)
Frying pan, aluminum(f)	Rotary polish	Indexing rotary(g)	Liquid tripoli	430	17	1750	450	...
Lipstick shell, brass	Press form	Rotary automatic(h)	(j)	355	14	1750-2200	6000	...
Screw, steel(k)	Hand lathe polish	Hand lathe(m)	Stick	150	6	...	300	...
Shaft, steel(n)	Fine grind	Hand lathe(p)	Stick	100	4	...	500	...
Toaster shell, copper-plated steel	Copper plate	Indexing rotary(q)	Liquid tripoli	430	17	(r)	150	...

(a) Variable speed. Buff used was 16-ply, 86/93, ventilated airflow mop 38 mm (1½ in.) thick, loosely assembled from three 13 mm (½ in.) sections. (b) Color buffing performed on same double-end, variable-speed lathe, using 255-mm (10-in.) diam buffing mop 50 mm (2 in.) thick, assembled from six 8-mm (⅓-in.) sections, which was cross-stitched to 150 mm (6 in.) from center and only had radial stitching on outside 100 mm (4 in.). (c) Stitched 6-mm (¼-in.) concentrically sewed buffs and No. 4, 86/93 bias buffs. (d) No. 4 and No. 6, 86/93 bias buffs. (e) Nickel-chromium decorative plating. (f) Square. (g) With four heads, buffs were No. 4, 14-ply, 16-spoke sewed, 86/93. (h) For buffing, machine had three heads. Buffs were No. 2, 16-ply, ventilated, 86/93, with face 305 mm (12 in.) by 380 mm (15 in.). Color buffing performed with single head, using 355-mm (14-in.) diam, closed face, 14-ply, 64/68 buff of 380 mm (15 in.) face. (j) Liquid tripoli used during buffing, liquid rouge for color buffing. (k) Round head instrument screw, 6-mm (¼-in.) diam, of 1% carbon steel, 52 HRC. (m) Soft felt wheel. (n) Instrument shaft of 1.2% carbon steel. (p) Pivot end of shaft buffed to mirror finish with medium-hard felt wheel. (q) With five heads, buffs on four heads were No. 8, 12-ply. For color buffing, single buff used on one head was full disk buff of dommet flannel. (r) 1750 rev/min for buffing, 1500 rev/min for color buffing

Fig. 7 Horizontal-return straight-line finishing machine

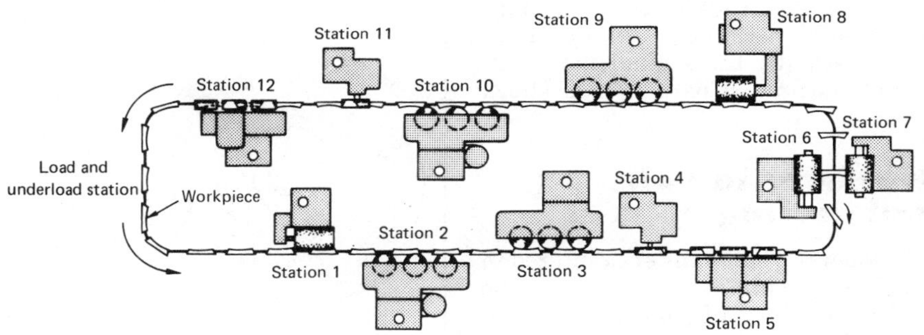

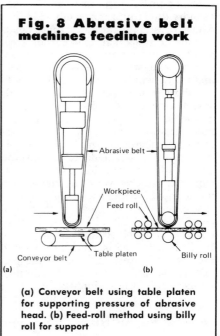

Fig. 8 Abrasive belt machines feeding work

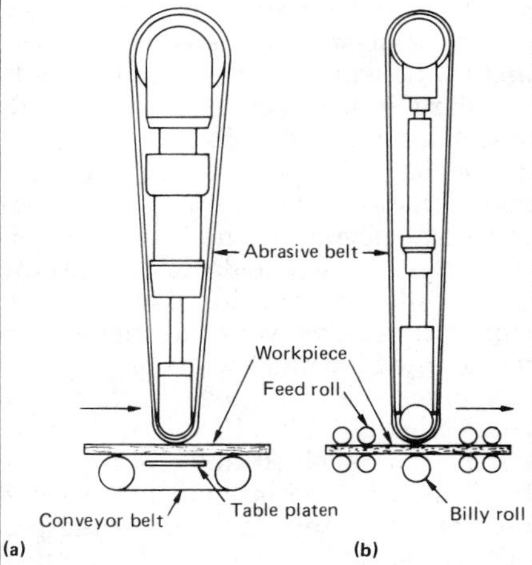

(a) Conveyor belt using table platen for supporting pressure of abrasive head. (b) Feed-roll method using billy roll for support

forgings, die castings, and sand castings. Parts made of metal, plastic, ceramic, wood, or rubber can be handled on this equipment. Any flat surface that requires finishing, sizing, deburring, and descaling can be processed. Parts can be fed against abrasive belt polishing heads by conveyor belts or feed rolls, shown in Fig. 8. Conveyor belts can be made from oil-resistant rubber, sponge rubber, or abrasive-coated cloth. Cleats or fixtures sometimes must be mounted on the belt to prevent slippage of parts beneath the polishing heads. Hold-down fingers and cleats are used to hold nonferrous materials for grinding or polishing operations. Feed rolls are made of steel or rubber-covered steel. As shown in Fig. 8, table platens on conveyor belt machines or retractable or rigidly-mounted billy rolls on feed-roll machines are mounted beneath the polishing heads to support the grinding pressure applied on the part.

Abrasive belts can be tensioned by mechanical or pneumatic means, and manual or automatic tracking devices are available. Automatic tracking devices are generally used for belts 305 mm (12 in.) or more in width. Manual or automatic adjustments for grinding pressure can be used to suit specific applications. Tandem arrangements of any of the machines mentioned can be made using as many polishing units as are necessary to produce the required successive grit finish in one pass through the machine. Conveyor belt machines can be furnished with an automatic turnover arrangement to polish or deburr both sides of parts. Abrasive belt heads or flap wheels can be mounted on the bottom side of the pass line to permit two-sided simultaneous polishing operations. Flat surface polishing machines can be equipped for wet or dry finishing. A means for the application of oil spray, oil mist, or other wet polishing media may be provided.

Wide-Sheet Polishing Machines. A few of the applications for wide-sheet polishing mills include:

- Extrusion sizing, grinding and deburring
- Prepolishing of stock before forming
- Polishing sheets to a high quality satin finish
- Obtaining engineered finishes on stainless steel
- Upgrading finish of commercial cold rolled stock
- Sizing operations on all types of sheet and coil stock

Sheet-polishing heads can be used as individual units or assembled in tandem arrays. Applications of these heads include (a) prepolishing of carbon steel sheet for plating applications, (*b*) conditioning of stainless steel coil stock, and (*c*) polishing of sheet and plate to standard finishes. A broad, 2 m (7 ft) wheel, angled to the sheet to be buffed, can achieve high degrees of microfinish and reflectivity by allowing the sheet to be transferred by the head or heads in a continuous mode or by having sheets placed individually on a reciprocating conveyor. The heads can be mounted top and bottom of the sheet in a pinch roll configuration, and with some degree of oscillation, buffing streaks are avoided and uniform fine finish is achieved on both ferrous and nonferrous materials for finishing prior to the sheet being blanked, stamped, or used full dimensionally for other purposes.

Tooling for Buffing

The success of semiautomatic and automatic buffing operations depends largely on how effectively the area to be buffed is presented to the wheel. Good presentation entails a combination of motion and effective tooling. Circular parts that comprise only two areas to be buffed are sometimes buffed on two semiautomatic machines, used with the standard buffing head. One machine is positioned at the right side of the head, with the spindle at right angles to the face of the buff. This machine is used to buff the outer periphery of the part. A second machine is positioned at the left side of the head, with the spindle positioned to buff the top or face of the part. The operator transfers the part from the right-hand machine to the left-hand machine to complete one cycle. This setup is referred to as a double-chuck operation.

Straight-line buffing machines can be designed to handle many heads to buff parts of any shape. With the part fixtured in a vertical position, special camming arrangements, such as those shown in Fig. 9, can be used to rotate the part horizontally by any fraction of 180° as the part contacts the wheel. Movement of the part may be controlled with close precision.

Noncircular parts can be buffed efficiently on semiautomatic machines equipped with cam-following arrangements. Cams are designed to the same profile as the part. The machine is designed to permit the head to float as the cam maintains contact with the cam follower, as shown in Fig. 10. This method allows the part to remain tangent to the buff, regardless of the shape of the outer periphery. If more exactness is desired in the semiautomatic buffing of noncircular parts, a gear chuck is used.

Fig. 9 Straight-line camming-action mechanism for rotating part up to 180° during buffing

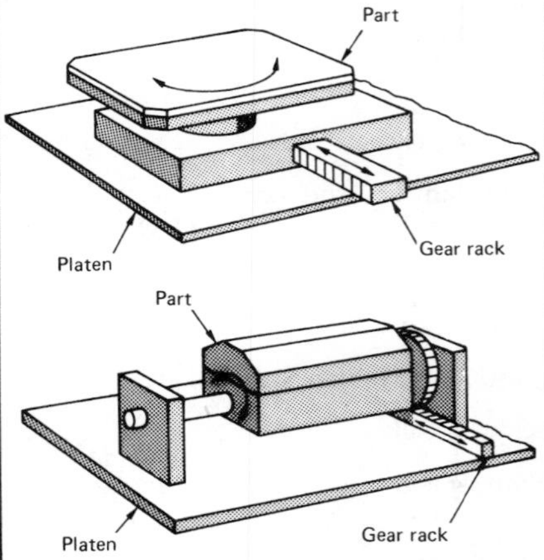

Fig. 10 Setup for buffing noncircular parts on semiautomatic machines

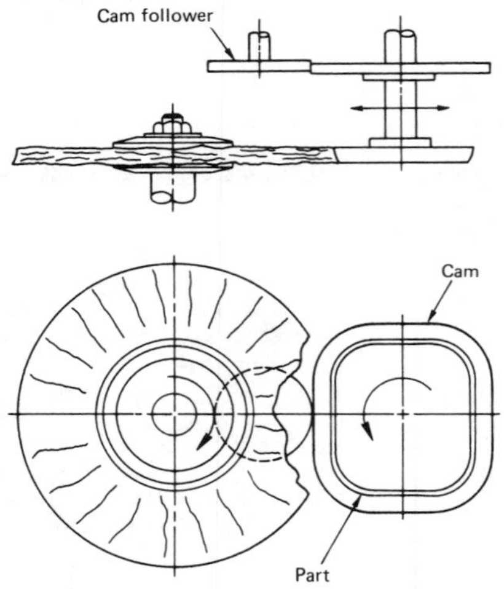

Problems in Polishing and Buffing

Frequently, problems encountered in polishing and buffing are caused by the surface condition of parts prior to finishing. To avoid these problems, rigid control must be maintained over incoming materials. Tools or dies that are in poor condition, abusive handling, and coarse or nonuniform grain size are factors that may cause difficulties in polishing or buffing to a desired finish. For example, decorative aluminum parts are difficult to buff if grain size is not controlled. The appearance of heavy orange peel on the surface of such parts is the result of large and erratic grain structure and requires aggressive buffing.

The grain size of brass is important when this metal is used for decorative parts. Brass is frequently work hardened during drawing or forming, and this may contribute to nonuniformity of finish. Polishing prior to buffing is desirable if the surface finish is over 0.8 μm (30 μin.) or contains die marks or other imperfections. Polishing reduces the surface to the lowest possible micro-inch finish. The cut of the abrasives must be sharp and clean to avoid dragging or tearing the metal. Tiny ruptures produced by dragging or tearing are not always visible after polishing, but they show up quickly when the part is subsequently buffed.

Surface imperfections caused by tools and handling are most effectively controlled by dies, fixtures, and conveyors. Fixtures that produce scratches can be cushioned by floating. Conveyor carriers must not permit contact between parts. When transported, parts must never be stacked or placed loosely in large containers. Many problems in buffing operations result from misapplication of buffing wheels and compounds. If the workpiece surface is to specification as received and the desired results are not produced, a systematic check of these factors indicates the cause of difficulty.

Equipment should be designed to process the most poorly conditioned part before die or mold change. If equipment is designed to handle marginal quality components, high quality parts entering the automatic finishing systems are finished to the same degree.

Buffing Wheels. A buff that is too hard may not buff the entire surface uniformly or may cut too aggressively, resulting in a work surface with a streaked or cloudy appearance. A buff that is too hard may also cut through high spots on plated work or create a wavy surface, especially on the softer metals. Conversely, a buff that is too soft may not permit the cutting particles in the buffing compound to come in direct contact with the work surface. The buff mushes, turning the ply over at the face so that a skidding or sliding action takes place. The density or flexibility of a given buff can be changed by increasing or decreasing wheel speed. A soft buff, such as No. 2 bias, can increase in density at higher wheel speeds. Most metal center buffs have the same cloth length at the wheel periphery for all diameters. For example, a 405 mm (16 in.) diameter, No. 4 bias buff may have the density of a No. 8 bias buff at its worn down 305 mm (12 in.) diameter.

Excessive buffing wheel pressure against the work may be used, assuming that increased pressure aids cutting action. If the surface is buffed too aggressively, severe distortion, discoloration, or pitting and dragout of the work surface may result. Excessively heavy wheel pressures and high speeds can trap some of the buffing debris beneath the work surface. The entrapped debris is overlaid with smears of the metal being buffed. Such conditions interfere with subsequent cleaning or electroplating operations.

Buffing Compounds. Differences in surface characteristics of the workpiece can be affected by varying the binder of the compound. The same compound with less binder is drier, with less cutting action, and leaves a higher luster. Increasing the binder ratio results in a greasier compound that, in most instances, cuts more sharply, provides more lubrication, and reduces the frequency of application. Too much grease causes smeared surfaces that are difficult to inspect and clean. Small but frequent compound applications are best, whether the compound is in bar form or liquid. If the buff receives insufficient compound, poor cutting and finishing result. Conversely, too much compound may create slippage and glazing between the buff and the work surface and results in poor surface finishing, compound waste, and increased costs.

Contact time required to achieve the desired results is very important. The time a given component must be in or under a particular wheel to meet the finish specification is generally measured in seconds. It can be determined by the following formula: contact time in seconds equals the total buff face in feet times 60 divided by the conveyor speed in feet per minute. Conveyor speeds can be calculated as conveyor speed in inches divided by 720. For example, if 15 s contact time is required to finish a component, and the production rate should be 1200 pieces/h, with parts based every 16 in. apart on a straight-line conveyor, then 7 ft of buff face is necessary at a conveyor speed of 28 ft/min. These formulas are commonly used when determining the size and number of buffing heads for both inline and continuous rotary equipment.

For indexing equipment, contact time is simply calculated. Most equipment of this type is arranged so that the head is raised during times of table rotation; therefore, contact time is the time in which the conveyor or table is stationary. The speed of index and the dwell time of the component under the wheel control the output and impose the limits. Indexing time is not productive and should be minimized.

Glossary of Terms Relating to Polishing and Buffing

backstand. A free-running adjustable idler pulley mounted at the rear of a polishing head and used to support the abrasive belt

bias buff. A buffing wheel consisting of a rectangular strip of continuous bias-cut cloth that is fastened to a rigid center of metal, plastic, or fiber so that the layers or plies of cloth in the finished product are on the same plane as the rigid center and so that the pleats or folds of cloth form puckered pockets. Also called puckered buff, bias buffs are designated by a type or pack number, ranging from 0 to 10 on the basis of the length of the cloth strip used in the manufacture of each buff section. The amount of cloth packed into a buff section determines its type, number, and life expectancy

binder. A mixture of tallows, fatty acids or hydrogenated fatty acids, hydrogenated fish oil, hydrocarbons, waxes, or other organic compounds, used as a carrier for the abrasive in bar or liquid buffing compounds

bob. A polishing or buffing wheel with a short stub and a rounded end, made of felt, leather, or cloth, such as muslin, used for finishing the interior surfaces of hollow ware or similar products

bond. Same as binder

buff designations. An example of the standard designations used in specifying buffs is 14 by 5, number 2, 16 ply, 86/80, 2.50 weight, and 1.25 arbor hole. This means a buff 14 in. in outside diameter on a 5 in. diameter center plate, of a type or degree of pleating number 2; 16 plies of cloth in each buff section, the cloth constructed of 86 warp (lengthwise) threads per inch and 60 woof or weft (crosswise) threads per inch, with a total weight of one pound of cloth equaling 2.50 running yards 40 in. wide, and a 1.25 in. diameter arbor hole in the center plate

buffing head. A piece of machinery, usually part of an automatic buffing machine, with a motor-driven spindle that is capable of carrying a number of buff sections and is arranged to be held at any given angle

buffing lathe. Old term for a machine having a driving motor, at constant or variable speeds, one or two spindles that are arranged in a horizontal position with suitable threads, flanges, and nuts so that wheels are attached or removed easily

buff sewings. Designs of stitching used to hold the plies of a full-disk, pieced, or folded buff together and to provide required density. Stitching can be concentric, radial, spiral, or square cross stitching, with spacing between rows as desired. More than one design may be used on a single buff

butler finishing. See satin finishing

calcined. A term indicating that an abrasive (usually aluminum oxide) has been heated to a temperature below its melting point. This type is much softer in abrasive properties than a fused variety (heated to melting) of aluminum oxide

Canton flannel buff. A soft face buffing wheel with each ply a full disk of cotton flannel having one surface of heavily combed cotton

closed-face buff. A bias buff in which the strip fastened to the rigid center plate is 2-ply; this is done by folding the strip and fastening the fold to the center

color buffing. (1) A final buffing operation performed before plating, in which a milder compound and a softer wheel are used to remove wheel marks from hard buffing. (2) A luster-producing buffing operation performed on high quality stock or on plated surfaces, in which the work travels with the direction of wheel rotation

contact wheel. A wheel made of steel, canvas, rubber, polyurethane foam, or other materials, and has one of a variety of densities and surface configurations, that is mounted on the head spindle to back up the working area of an abrasive belt against the pressure of the workpiece

cracking. An operation performed on a newly headed or reheaded polishing wheel that consists of striking the wheel face at a 45° angle with a piece of pipe. The wheel is struck at four different locations approximately 90° apart. The operation produces a diamond-shaped pattern on the working surface of the wheel, shortens break-in time, and provides a cushioned surface

crocus. A purple, coarse rouge occasionally used for buffing cutlery and some nonferrous metals

cross buffing. Buffing at an angle, usually 45° or more, to the direction of the polishing lines

cut-and-color compound. A blend of amorphous silica and tripoli mixed with a grease binder, used for cutting down and color buffing in a single operation. Known also as double-duty compound

cutting down. A preliminary buffing operation performed with a sharp or fast-cutting compound to smooth the surface and to bring up as bright a finish as possible. The work travels against the direction of wheel rotation

double-duty compound. Same as cut-and-color compound

dragging. Producing grooves by overbuffing areas around holes and other depressions

dry fining. A second polishing operation using a grit size that is from 50 to 100 mesh sizes finer than the roughing grit used, or from 50 to 100 mesh sizes coarser than the subsequent greasing or oil-out abrasive to be used

dubbing. An operation performed on a polishing wheel. It removes stock from the wheel edges adjacent to the wheel face to form a radius

face. The outside edge or periphery of a polishing or buffing wheel

fanning. Buffing workpieces very lightly to bring up luster. Similar to wiping, work travels with the direction of wheel rotation

finger buff. A buffing wheel constructed of many fingers of folded cloth, which extend radially from the metal center plate to which they are fastened

folded buff. A buffing wheel assembled from single plies of cloth that have been folded over two or three times to form quarters or eighths of a circle. The wedge-shaped pieces are dispersed evenly between two full disks of cloth and are often sewn spirally from the arbor hole to the periphery, although sewing may be far in from the periphery and just close enough to the arbor hole to hold the fold firmly

full-disk buff. A buffing wheel in which each of the plies of muslin that comprise a section is a full circular disk. In loose buffs, all disks in section are held together by a single circle of stitching near and concentric with the arbor hole. In sewn buffs, the disks are sewn to the periphery by one of the methods described under buff sewings

glaze. (1) The mixture of abraded metal, worn abrasive grains, and other debris which lodges between the unused abrasive grains on a belt or setup wheel in polishing. (2) The mixture of abraded metal, used buffing compound, used buff and other debris that accumulates on the face of a buffing wheel during operation

greaseless compound. A compound containing an abrasive or a mixture of abrasives in a glue-water gel binder. When this type of compound is applied to a revolving wheel, sufficient heat is generated by friction to cause it to melt and be transferred to the wheel

grease wheel. A third polishing operation, following dry fining, that produces a fine finish and uses a grit size ranging from 180 to 320, depending on the result desired. Sufficient polishing grease stick is used so that a light film of grease remains on each part polished

head. The compound and abrasive attached to the face of a polishing buffing wheel

highlighting. Buffing oxidized or sulfided parts to remove the oxide or sulfide and to impart a lustrous finish to the highlights that contrasts with colored or unfinished areas

leave-off. An abraded streak that appears on a buffed surface where the wheel leaves the work. This streak is accentuated by the presence of grease from the binder in the buffing compound

nubbins. Ends of bar compound that are too short to be of further use unless remelted

oiling. Lubricating a polishing wheel or belt with materials such as kerosine, oil, grease stick, or tallow, usually to aid in the removal of coarse lines produced during dry polishing

oiling wheel. See grease wheel

open-face buff. A bias buff in which the strips fastened to the rigid center plate are single ply

pieced buff. A buffing wheel made from clippings of odd-shaped pieces of various grades and types of fabrics, which are held between full-disk cover sheets of muslin by various buff sewings

ply. One of the layers of cloth that compose the thickness of a buff section

puckered buff. Same as bias buff

puckers. The pleats, folds or corrugations formed by pulling the cloth toward the center of the wheel, as in the manufacture of a bias buff

raking. Removing the glaze from a wheel, or shaping a wheel, by applying a sharp instrument, lump pumice, or stone against the face of the wheel as it revolves

satin finishing. A surface finish that behaves as a diffused reflector and is lustrous, but not mirror-like. The following finishes are various types of satin finishing:

- Scratch brush finish shows a combination of coarse lines with a slight underlying luster. Prior to the development of greaseless compound, the scratch brush finish was produced by a brass wire or nickel silver wire wheel revolving over a tray. A mixture of water and bran meal was applied liberally over the work being finished and over the wheel. The bran meal provided a slight lubrication. The old scratch brush finish was used only on nonferrous metals. It has been superseded in practically all cases by greaseless compound

- Butler finish is composed of soft, fine parallel lines produced by a lubricated greaseless compound. This finish is generally found in the silver industry and is classified as a dull butler, butler, and bright butler, according to the degree of brilliance achieved
- Satin finish is the equivalent of the formerly used scratch brush finish. It is a more economical finish to produce and has considerably more eye appeal and underlying luster. It is now widely accepted as an excellent finish for consumer products
- Colonial finish is achieved by relieving an oxidized surface, emphasizing the lack of luster adjacent to the relieved area. The finished highlights are satin finished and create an attractive decorative effect against the oxidized background. Articles with raised or embossed designs are often given a colonial finish
- Matte finish is produced by the use of a sand blast or acid dip, creating on the metal a frosted and nonreflective surface completely free of parallel lines. On metals such as aluminum, the greaseless compound method can produce a comparable finish by eliminating lubricating and by operating at high speeds
- Sanded finish is applicable to wood and plastics and is produced in the same way as a satin finish would be produced on metal. In many cases, this method is used in place of belt sanders for finishing ornamental wood earrings, wood heels and other irregularly shaped articles

section. A number of plies of cloth attached together. A buff section may have designations such as 12-ply, 14-ply, and 16-ply

strapping. A polishing operation performed with an abrasive belt. The workpiece is placed in contact with an area of the belt that is not backed up by a comtact wheel or platen

vienna lime compounds. This is currently almost completely replaced by alumina buffing compounds due to availability. Lime compounds, white finish, or nickel buffing compounds were particularly effective for producing a high color on articles plated with nickel or copper, for secondary coloring operations on aluminum and brass and, to a lesser extent, for coloring cadmium, tin, and zinc plates. Vienna lime abrasive was a form of limestone made up primarily of calcium and magnesium carbonates. By means of crushing, grading, and calcining, the abrasive is converted to virtually pure calcium and magnesium oxides. In this state, because of its raw lime content, it can slake and disintegrate if exposed to air. Thus, it is usually packaged in airtight containers

wiping. Buffing with a dry cloth wheel or one lightly dressed with compound, to improve the color of the buffed surface. The work travels with the direction of wheel rotation

Mass Finishing

By the ASM Committee
on Mass Finishing*

MASS FINISHING normally involves the loading of components to be finished into a container together with some abrasive media, water, and compound. Action is applied to the container to cause the media to rub against the surfaces, edges, and corners of the components, or for components to rub against each other, or both. This action may deburr, generate edge and corner radii, clean the parts by removing rust and scale, and modify the surface stress. The basic mass finishing processes include:

- Barrel finishing
- Vibratory finishing
- Centrifugal disc finishing
- Centrifugal barrel finishing
- Spindle finishing

Mass finishing is a simple and low-cost means of deburring and surface conditioning components. Consistent results from part to part and batch to batch are generally assured. All metals and many nonmetals in a variety of sizes and shapes can be handled. Processes range from heavy radiusing and grinding operations to very fine finishing.

A basic advantage of mass finishing is that the action is effective on all the surface edges and corners of the part. Normally, preferential treatment to one area is impossible. Action is greater on corners than other similarly exposed surfaces. Action in holes and recesses is less than on exposed areas.

The mass finishing processes are used:

- To clean, descale, and degrease
- To deburr
- To radius edges and corners
- To change surface condition
- To remove surface roughness
- To brighten
- To inhibit corrosion
- To dry

Barrel Finishing

The rotary barrel, or tumbling barrel, utilizes the sliding movement of an upper layer of workload in the tumbling barrel, as shown in Fig. 1. The barrel is normally loaded about 60% full with a mixture of parts, media, compound, and water. As the barrel rotates, the load moves upward to a turnover point; then the force of gravity overcomes the tendency of the mass to stick together, and the top layer slides toward the lower area of the barrel.

Although abrading action may occur as the work load rises, about 90% of the rubbing action occurs during the slide. In the case of a horizontal barrel which is just over half full, the most effective action occurs to produce the longest slide. As may be judged from Fig. 1, the faster the rotation of the barrel, the steeper the angle of the slide. With all but the shallowest of angles of slide, there is invariably more tendency for tumbling of the load, as well as sliding. The faster the rotation of the barrel, the faster the action will take place. However, the faster the action, the poorer the surface and edge condition will be, and the greater the likelihood of parts being damaged.

Fig. 1 Action of media and parts within a rotating barrel

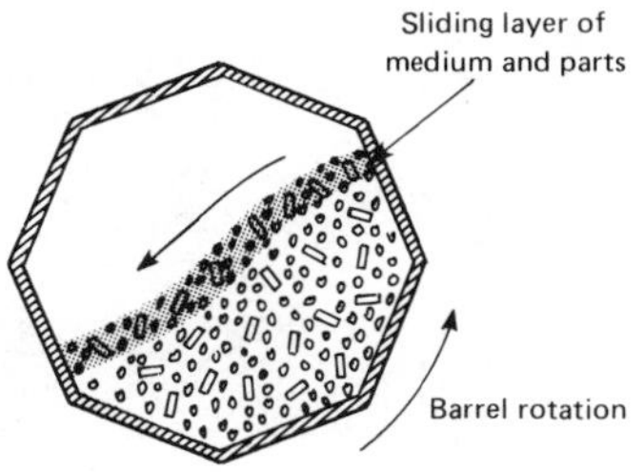

Typical equipment for barrel tumbling is shown in Fig. 2 and includes:

- Open-ended, tilted-type barrel (concrete mixer)
- An adaptation known as the bottle-neck barrel
- Horizontal barrel
- Submerged barrel
- Fixtured barrel

The open-ended, tilting barrel is used for light finishing and also for drying. The bottle-neck shape provides essentially the same aciton, but more effectively, because parts and media remain better mixed. The horizontal barrel, usually of octagonal cross section, is the standard or most used tumbling barrel, capable of achieving a variety of results and maintaining real consistency. Rotational speeds of these barrels range from about 50 to 200 sfm; the higher

*J. Bernard Hignett, *Chairman*, Vice President, The Harper Company; John Coffield, Presidnet, Osro Company; Robert W. Johannesen, Consultant; John B. Kittredge, Consultant; Gary A. Snell, Wisconsin Porcelain Co.

Fig. 2 Barrels available in a variety of sizes

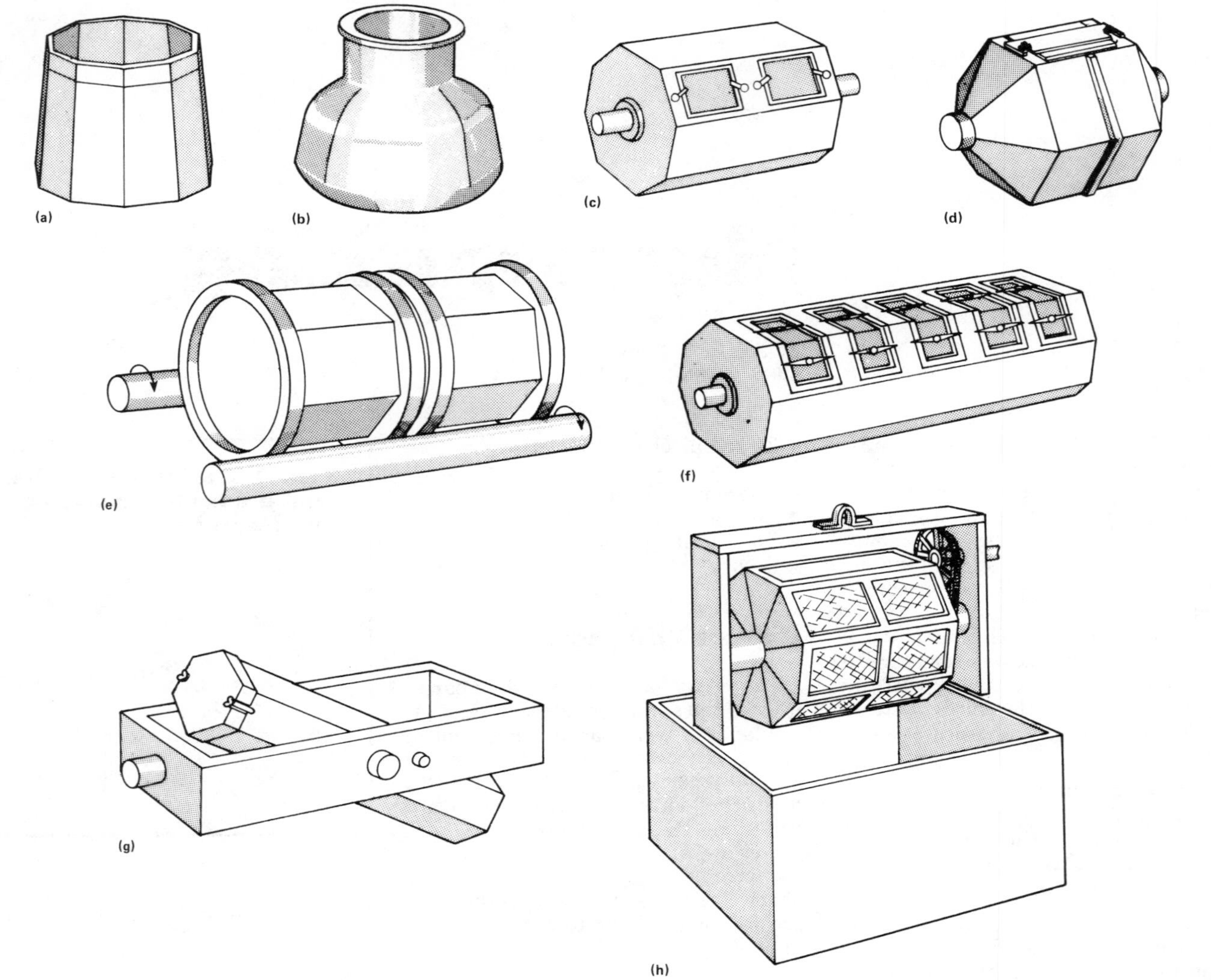

(a) Standard open-end, tilting. (b) Bottlenecked. (c) Horizontal octagonal. (d) Triple-action, polygonal. (e) Multiple drums. (f) Multi-compartment. (g) End loading. (h) Submerged

speeds for cutting, the slower speeds for burnishing and fine finishing. Ratio of media to parts ranges from about 3-to-1 to 15-to-1 by volume. Rough work can be loaded only with parts, that is where parts self-tumble against each other. Some of the factors to be considered in determining the media to parts ratio are:

- Size and complexity of the workpiece
- Possibility of media lodging in the parts
- Possibility of the parts nesting
- Quality of final finish

Automation of barrel tumbling equipment is possible. The barrel can rotate in a clockwise direction to deburr and finish the parts. At the end of the process, the barrel rotation is reversed and parts feed out through a scroll, then through a screener into a material handling unit.

The capital cost of barrel tumbling equipment is low and generally maintenance costs are also low. Because barrel tumbling is a very well-established process, the basic conditions are well understood and guidance on best process techniques is readily available. Barrel tumbling is, however, a slow process, almost invariably involving several hours and on occasion, several days. The process is space consuming with a high level of work in progress. More modern mass finishing processes offer greater versatility and convenience, with better use of labor and consistency of quality in production. For these reasons, and because higher operator skills are needed, barrel tumbling has been replaced in most modern production facilities.

Fig. 3 Tub vibrators

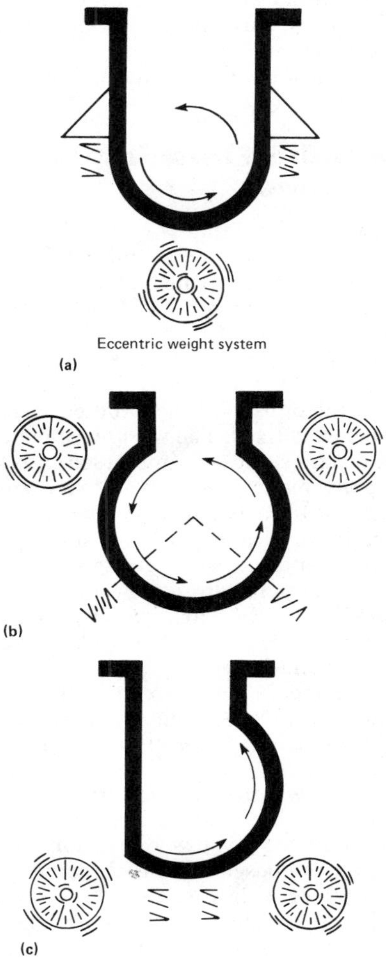

Vibratory Finishing

A vibratory finishing machine is an open-topped tub or bowl mounted on springs, usually lined with polyurethane. Parts and media are loaded in a fashion similar to that of a tumbling barrel. With a vibratory machine, the container can be almost completely filled. Vibratory action is created either by a vibratory motor attached to the bottom of the container, by a shaft or shafts with eccentric loads driven by a standard motor, or by a system of electromagnets operating at 50 or 60 Hz. The action of media against components takes place throughout the load, so that process cycles are substantially shorter than conventional tumbling in barrels. Components can be inspected and checked during the process cycle. This process is faster, more convenient, and more versatile than tumbling barrels. The vibratory machine is able to handle larger parts, and is more readily automated. The process achieves more abrasive action in the recessed areas of components, is easier to operate, and is cleaner.

Two important variables for operation of vibratory equipment are frequency and amplitude of vibration. Frequency may range from 900 to 3000 cycles/min. Amplitude can range from 2 to 10 mm (1/16 to 3/8 in.). Most equipment operates in the range of 1200 to 1800 cycles/min and 3 to 6 mm (1/8 to 1/4 in.) amplitude.

The tub vibrator consists of an open container where the cross section is either U-shaped, a round inverted keyhole shape, or a modification of these. Figure 3 shows three types of tub vibrators. The modifications from the U-shaped cross section are constructed to develop more uniform flow of the mass, because the U-shaped cross section has a tendency for some back flow against one of the walls. The means of creating the vibratory action is with a single shaft mounted directly underneath the tub. At each end of the shaft, or along its length, are eccentric weights and a facility to add or reduce some of the weights. The greater the weight, the more amplitude. The faster the action, the rougher the surface finish and the greater the media wear. The shaft may be driven from a motor coupled to the shaft through a flexible coupling or by belts. Variable speed of rotation of the shaft is fairly easily obtained and gives variable frequency of vibration. The higher the frequency, the faster the cut, but the rougher the surface. Modifications of the drive system include:

- Use of double shaft, as shown in both Fig. 3(b) and 3(c), to maintain greater uniformity of the parts and media mix and offer somewhat faster action
- Use of eccentric shaft, rather than eccentric weights at the ends of a shaft, to give greater consistency of action throughout the load
- Location of eccentric weights or shaft on one side, at top of tub wall, or at both sides

Tub vibratory equipment can be sized to suit the largest components. Units are in production to handle parts as long as 12 m (40 ft), and other equipment has been built to handle parts of cross section as much as 2 m (6 ft) by 2 m (6 ft).

Separtion of parts and media is sometimes accomplished by dumping the total contents of the vibratory tub into a material handling system. Most tub vibrators have unload doors at one end, which may be opened manually or automatically and through which the total load is fed. Some units have a system for emptying by inverting the complete tub.

Tub vibratory equipment is readily automated for continuous production. A long tub vibrator may have parts and media fed in at one end of the container. These proceed along the tub until finished parts and media exit at the other end. The discharge end is lower than the parts' entry end, causing the media and parts to move to the discharge end. Parts are then screened out from the media. The media is returned with a simple belt conveyor directly or through a storage unit. Such systems are suitable for processes where satisfactory results are achieved with process cycles of up to 30 min.

Other automated tub vibratory systems include storage for different media which may be automatically selected for different processing tasks. Tub vibratory machines can be highly versatile and permit process cycles normally 1/5 to 1/10 of those achieved with tumbling barrels. They are capable of handling all sizes of workpieces, including complex shapes. Material handling is usually accomplished by units added onto the vibrator itself, designed and built to meet specific applications.

Bowl vibrators are round bowl- or doughnut-shaped, and like the tub vibrator, have a container mounted on springs. The principle of this type of equipment is shown in Fig. 4. Vibratory action is imparted to the bowl by eccentric weights mounted on a vertical shaft at the center of the bowl. Eccentric weights at either end of the shaft are adjustable in their relationship to one another. This is referred to as the lead angle. When set within the normal range established by the manufacturer, a spiral motion is imparted to the mass of parts and media. Changing the relationship of the weights alters the spiral path, the speed at which the load moves around the bowl, and the amplitude of vibration. The bowl vibrator is capable of more gentle action than tub equipment because parts hold their relationship to one another as they proceed around the bowl; consequently there is less chance of part-on-part impingement. Another advantage of bowl vibrators is

Fig. 4 Bowl vibrators

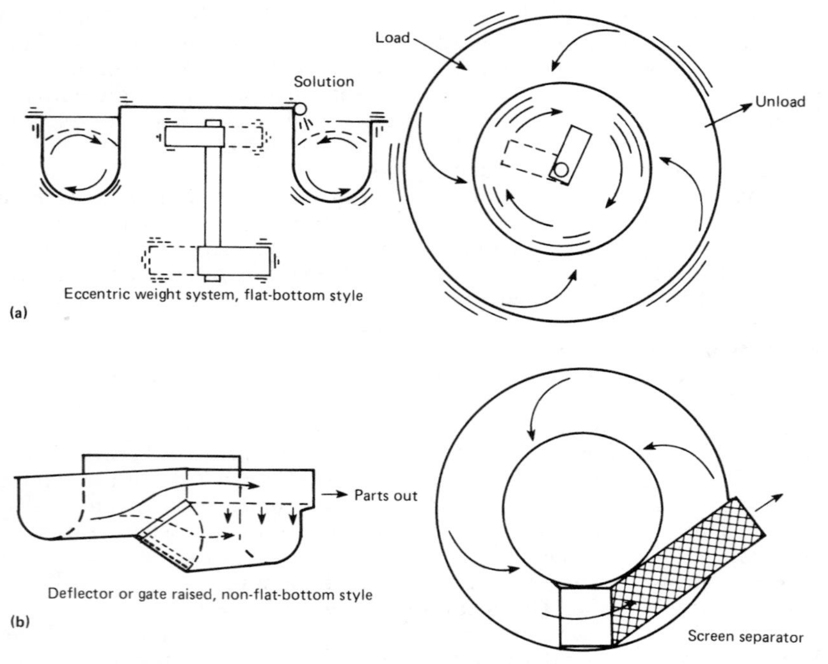

that they require less floor space than most tub units of similar capacity.

The major advantage of bowl vibrators is that an integral separation system can be incorporated, such as that shown in Fig. 4(b). To separate parts from media, a dam is placed in the channel so that parts and media are forced up and over. At the top of the dam is a screen over which parts and media pass. Media falls through the screen back into the vibrator bowl. Parts are deflected off the screen into a collection hopper or conveyor. Bowl vibrators with integral separation may have a non-flat-bottom base, so the dam has a shorter distance to travel. Parts and media do not have to lift as high to reach the screen, excellent separation is achieved, and there is less likelihood of parts piling up at the dam with possible danger of impingement. Action of the dam may be automatic; some equipment has automatic reversal of the action and increased vibration to promote faster and more complete separation. Therefore, the round vibrators can be fully automated. For short process cycles up to about 30 min, flow-through automation can be achieved with the dam permanently in place.

Bowl vibratory equipment is preferred if integral separation meets all requirements, and if there is no need for frequent media change. Bowl vibrators are somewhat slower than the tub units. They cannot handle very large parts, but are gentle and can handle parts in individual compartments.

Causes of Problems. Some causes of sluggish, ineffective vibratory action include:

- Water flow rate is too high or drains are clogged, resulting in excessive dampening
- Use of wrong type of compound, little or too much compound
- Frequency of vibratory action may not be best for the parts
- Amplitude is too great
- Equipment is underpowered and therefore does not develop enough vibratory action for effective performance

Unacceptable finishes may be the result of:

- Insufficient cutdown of surface
- Improper media for amount of cut needed on workpieces
- Dirt, oil, or metal fragments in the solution may work back into surface of parts
- Insufficient flushing and cleaning of media from processing tub
- Use of wrong type of finishing compound and/or wrong concentration
- Vibratory action too harsh
- Ratio of parts to media may be incorrect
- Careless handling in auxiliary operations

Spindle Finishing Machines

Spindle finishing is categorized as a mass finishing process, although parts to be deburred or finished are mounted on fixtures. The process uses fine abrasive media for finishing. The spindle machine is a circular rotating tub which holds the abrasive media, and a rotating or oscillating spindle to which the part is fixed. The workpiece mounted on the spindle is immersed into the rapidly moving abrasive slurry, causing the abrasive to flow swiftly over rough edges and over the surfaces of components. In some designs, the media container is stationary, and the fixtures move the parts rapidly through the media. Figure 5 shows a layout of a spindle machine.

The media used are usually small aluminum oxide nuggets, although all forms of small finishing media are applicable. Most operations are carried out in a water solution, although very fine finishing of some components can be achieved by dry operation using media composed of fine abrasives and corncob or walnut shells.

Process cycles in spindle equipment rarely exceed 20 min and are frequently less than 30 s. The equipment is well suited for parts such as gears, sprockets, and bearing cages where parts cannot be allowed to contact one another. Equipment can deburr, edge radius, and produce very fine surface finishes. Because parts are fixtured, part-on-part impingement does not occur during the process or at reload time. The limitations result primarily from the need to fixture the workpieces. Where parts can be handled entirely satisfactorily in bulk in vibratory equipment, centrifugal barrel machines, or centrifugal disc equipment, the operation is more economical, convenient, and versatile.

Centrifugal Disc Finishing

The centrifugal disc process is a high energy mass finishing process. The basic design is a vertical cylinder with

Fig. 5 Two spindle finishing machine

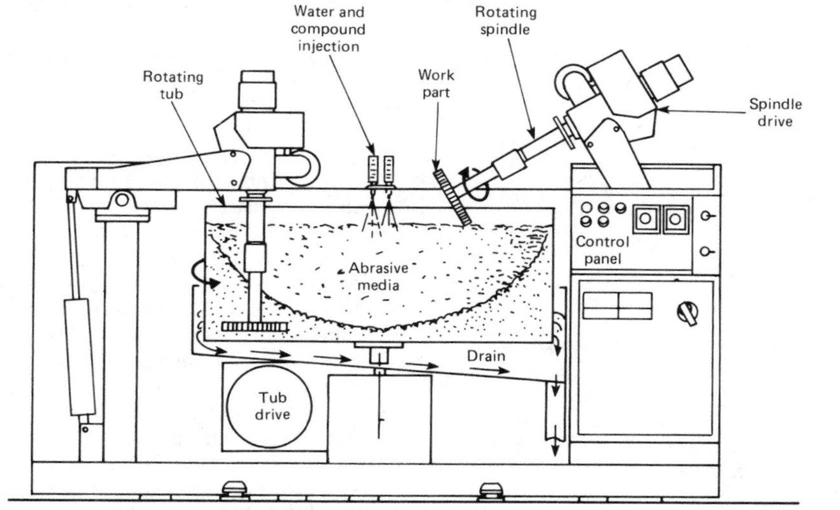

Fig. 6 Centrifugal disc machine

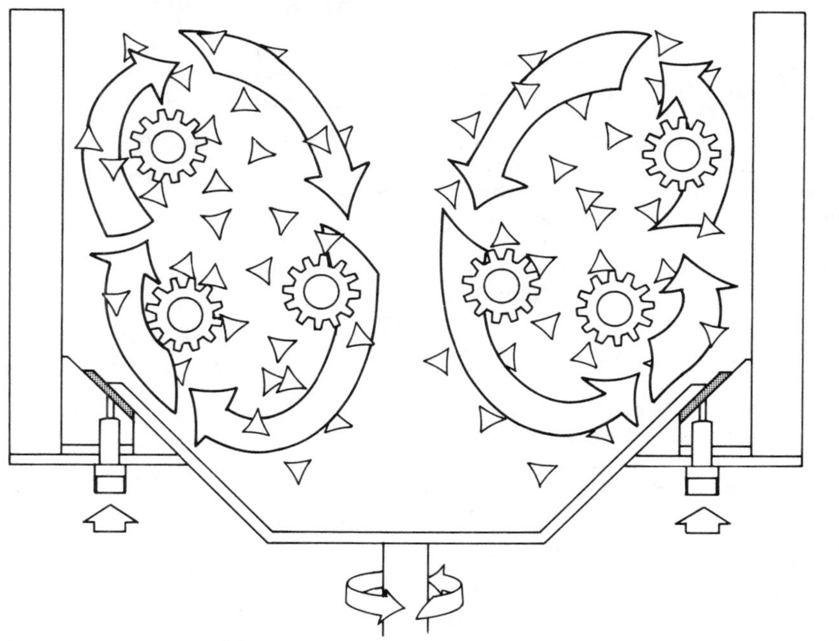

side walls which are stationary. The top of the cylinder is open. The bottom of the cylinder is formed by a disc which is driven to rotate at a high speed. Media, compound, and parts are contained in the cylinder. As the disc rotates with peripheral speeds of up to 10.2 m/s (2000 ft/min), the mass within the container is accelerated outward and then upward against the stationary side walls of the container which act as a brake. The media and parts rise to the top of the load and then flow in towards the center and back down to the disc.

The action achieved in the centrifugal disc machine, shown in Fig. 6, is substantially faster than in vibratory equipment, because of the centrifugal forces of as much as ten times gravity pressing abrasive media against components. Process cycles are up to 1/20 those of vibratory processing. The short process cycles result in reduced floor space requirements in the finishing department, increased versatility, and less work in progress. As with vibratory equipment, parts can be readily inspected during the process cycle and variable speed can occasionally combine deburring with a final, more gentle surface refinement operation. The faster process speeds create a much greater rate of media wear and increase demands on the compound solution and its flow rate.

Centrifugal disc equipment is available with capacities ranging from 0.04 to 0.6 m^3 (1.5 to 20 ft^3). This equipment is easily automated. The load is emptied through a door in the side of the container through which parts and media can be fed, or by tilting the whole bowl through 180° to dump the load. Subsequent separation, classification, washing systems, and the return of media to the container for the next operation are similar to those techniques used with tub vibratory machines.

Centrifugal Barrel Finishing

Centrifugal barrel equipment is comprised of containers mounted on the periphery of a turret. The turret rotates at a high speed in one direction while the drums rotate at a slower speed in the opposite direction. The drums are loaded in a manner similar to normal tumbling or vibratory operations with parts, media, water, and some form of compound. Turret rotation creates a high centrifugal force, up to 100 times gravity. This force compacts the load within the drums into a tight mass. Rotation of the drums causes the media to slide against the work load, to remove burrs and to refine surfaces. The action of a centrifugal barrel machine is shown in Fig. 7.

The abrading action, under high centrifugal force, results in short process cycles, generally less than 1/50 of the time taken in vibratory equipment. Because of the counter rotation of drums to turrets, a completely smooth sliding action of media against components is generated, with little possibility of one part falling or impinging against another. This completely smooth action achieves consistent and reproducible results. Very high tolerances are maintained even with fragile parts, and very high surface finishes are achieved.

Fig. 7 Centrifugal barrel machines

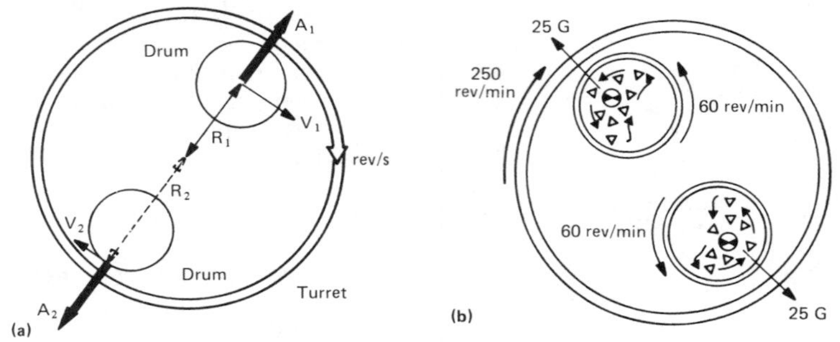

(a) Principle of centrifugal barrel process. (b) Action within a centrifugal barrel machine

Process variables are similar to those of other mass finishing processes, but an advantage of centrifugal barrel finishing is the ability to control the force with which media are pressed against components. This gives greater latitude in choice of media. For example, hard and low abrasion media may be used to deburr by running at high speed. Then the same media refines surfaces when the machine is switched to a low speed. Therefore, two operations can be combined into a single cycle. Particle size of media is selected to meet requirements of uniformity and to ease separation without increasing the time of the process cycle.

One important application of centrifugal barrel equipment is to impart a high compressive stress to the surface layer of components and to increase resistance to fatigue failure. The capability of imparting improved fatigue strength is used in bearings, aircraft engine parts, springs, and compressor and pump components. The improved fatigue strength is generally greater than that of the other finishing processes combined with shot peening. It is almost always significantly lower in cost.

Economic considerations frequently dictate the choice between centrifugal barrel equipment and the other mass finishing processes. If satisfactory results are achieved with a process cycle of less than 1 h, a vibratory process is the more economical method. If the process cycle is much longer, if there is a wide variety of components to be handled, or if there are special finishing requirements or parts of very high precision, centrifugal barrel machines are usually better suited.

Centrifugal barrel equipment is available in sizes ranging from less than 0.01-m^3 (¼-ft^3) capacity to 1.4-m^3 (50-ft^3) capacity. The process is fully batch automated, but not subject to continuous flow-through of workpieces.

Other Types of Mass Finishing Equipment

The mass finishing processes discussed so far are the well established mechanical finishing techniques throughout industry. Equipment comes in a variety of shapes and sizes to meet special requirements. Equipment has been developed to suit changing requirements and is frequently built to meet special purposes. Other mass finishing machines offered with slightly different actions from those that have been discussed include the following:

- *Vibratory rotary barrel machines*: conventional tumbling barrels which are vibrated as they rotate. They were originally developed to avoid some of the problems of parts migrating out of the media in tub vibratory equipment. These machines do not appear to have any substantial merit for modern general-purpose deburring and finishing
- *Reciprocal finishing:* as in spindle finishing systems, parts are attached to a holding device and placed into a tub of media, water, and compound. The component is moved through the mass of media in a reciprocating motion. This process permits the handling of parts too large for spindle finishing machines. Action is concentrated on just one side of the component for specialized finishing. Process cycles are generally much longer than with spin finishing machines and equipment is special and of limited purpose
- *Chemically accelerated centrifugal barrel finishing:* a combination of chemical polishing with centrifugal barrel finishing, has found many worthwhile applications, particularly with high precision, complex-shaped ferrous components. The process is substantially more costly than normal centrifugal barrel finishing, and is only suited to very special purposes
- *Electrochemically accelerated mass finishing equipment:* for special purposes, a combination of electropolish and vibratory or tumbling barrel action achieves very fast stock removal. Expensive and inconvenient, not commercially used in the United States
- *Orboresonant cleaning and finishing:* parts clamped on fixtures are oscillated at very high frequency in a bed of fluidized media, with some capability for removing internal burrs and cleaning internal surfaces

Selecting Mass Finishing Equipment

Factors to be considered when selecting the most suitable mass finishing process include:

Production requirement

- Size and configuration of parts
- Batch size
- Part material
- Variety of parts
- Hourly production
- Annual production

Quality requirements

- Consistency of quality entering department
- Consistency of quality leaving department
- Surface finish
- Edge condition required
- Cleanliness of parts
- Uniformity over edges and surfaces
- Uniformity part to part

Process variables

- Relationship to other manufacturing processes
- Automation requirements
- Process time
- Total investment
- Operation and maintenance costs
- Consumable materials
- Energy
- Water and effluent removal and treatment
- Preventive maintenance and repairs
- Available floor space
- Inventory requirements

Table 1 Advantages and disadvantages of mass finishing processes

Process	Advantage	Disadvantage
Industry standard		
Tumble barrel	Low initial cost Low operating cost for supplies Very low equipment maintenance cost	Slow process Skillful operator essential Automation impractical Wet working area No in-process inspection
Vibratory tub	Fast operation Handles all part sizes Open for in-process inspection Practical full batch automation Practical in-line automation	Slower than high energy process External material handling required
Vibratory bowl	Open for in-process inspection Practical full batch and continuous operations Requires no auxiliary equipment to automate Internal separation Space saving Lowest cost for general purpose work Simple selection and operation with little operator skill required Can produce cleanest parts and excellent surface finish	Somewhat slower than vibratory tub
High energy		
Spindle	Fast processing Possible automation, robot load and unload No part-on-part impingement	Limited part geometry Parts must be fixtured High labor cost
Centrifugal disc	Fast processing Open for in-process inspection Practical automated batch process Compact operation	Part size limitation High initial investment
Centrifugal barrel	Fast processing Fragile part handling High precision part handling Potential automated batch processing Automatic change from grind to super finish Produces finest finish Improves fatigue strength	No in-process inspection High initial investment

- EPA and OSHA considerations
- Labor, direct, supervisory, and quality control
- Future and current needs

Traditionally, mechanical finishing has been a centralized service operation for all manufacturing operations within an organization. Although the centralized finishing department is still suitable in many organizations, deburring and mechanical finishing is frequently incorporated in the production line. In reviewing mechanical finishing requirements, as in buying any metal-forming equipment, consider that equipment as part of a system with all controls between the system and all ancillary and handling equipment.

Some of the advantages and disadvantages of mass finishing processes that should be considered are included in Table 1.

Mass Finishing Consumable Materials

For the vast majority of mass finishing processes, equipment is loaded with components to be finished, media, compound, and water. Media are solid stones or chips. Compounds are the materials that dissolve in water to form solutions to facilitate or modify the action of media against components.

There are very few occasions when media may be used with no water and, therefore, no compound. There are some occasions when the parts themselves act as the media; this is a self-tumbling process.

Clearly, the selection of media and compounds for the mass finishing operation is as important as the selection of the correct tooling for any forming operation, and an understanding of the materials available is essential for effective use of equipment.

Mass Finishing Compounds

The correct use of compound-water solutions is vital to good and consistent mass finishing processing in any equipment. The compound solution has the following functions:

- To develop and maintain cleanliness of parts and media during the process
- To control pH, foam, and water hardness
- To wet surfaces
- To emulsify oily soils
- To remove tarnish or scale
- To control part color
- To suspend soils
- To control lubricity
- To prevent corrosion
- To provide cooling
- To ensure effluent, meeting EPA, OSHA, and plant standards

Compounds may be liquid or powder. Addition of compound to the mass finishing equipment may be made in three ways:

- *Batch*: this is the simplest technique, used in closed machines, barrels, small centrifugal barrels, and vibrators with no drains. The machine is charged with compound and water, which is flushed away at the end of the process cycle
- *Recirculation*: the solution is mixed in a tank and pumped into the process, allowing it to drain back into the tank for reuse. This process is simple and has the basic benefit that there is no continuous drainage, and the effluent can be treated on a batch basis. The major disadvantage of this system is that the solution deteriorates during its life and therefore, results vary. For this reason, recirculation shouldbe avoided if at all possible
- *Flow-through*: this system pumps fresh solution into the machine, allows it to act, and then drains it out. Modern compound solutions can be very dilute and economical. For example, a 0.3 m^3 (10 ft^3) machine normally uses less than 4 L (1 gal) of compound per shift. When using flow-through systems with liquid compounds, automatic addition through a compound metering pump is possible, further reducing waste and improving consistency of the process

Because the use of a flow-through compound solution system permits close control and is readily automated, liquid compounds are normally preferred for consistent results. When mass finishing in closed-batch systems, powdered compounds can be economical. Powdered compounds can include loose abrasives which enhance cutting capabilities of media. Abrasives are seldom recommended in vibratory systems.

Finishing Media

The media in the mass finishing operation are equivalent to the tooling used in any machining operation. The function of media is to abrade or burnish edges and surfaces of components to be finished and to keep parts separate from one another to avoid or limit any part-on-part impingement. Media may be selected from any of the following materials:

- *Natural media*: stones that have been quarried, crushed, and graded were the original media for mass finishing operations. Natural media have been largely replaced by synthetic materials which are harder with longer life, greater consistency of cut, wear, and dimension, and greater variety of capabilities
- *Agricultural materials*: sawdust, corncob, and walnut shells are frequently used in mass finishing machines for drying. Mixed with fine abrasives, these materials are suited for some fine polishing operations, particularly in the jewelry industry. Wood pegs may be coated with fine abrasives or waxes for edge radiusing and finishing of some wooden and plastic components
- *Synthetic random media*: fused and sintered aluminum-oxide media, crushed and graded are available in a number of grades, both heavy cutting and fine finishing. Generally they are much tougher than natural media and more consistent
- *Preformed ceramic media*: porcelain or other vitreous material is mixed with abrasives and formed into shapes, then fired to vitrify. These media are available in a large range of shapes and sizes. The selection of a proper grade of abrasive, the proportion of abrasive to binder, and the type of binder enables selection of material of the degree of abrasion and surface finishing capability to suit virtually every application. The consistency of quality can be ensured. This type of material is the present standard for the mass finishing industry
- *Preformed resin-bonded media*: these are abrasives bonded into polyester or urea-formaldehyde resins. Like the ceramic materials, these are available in a broad range of shapes and sizes with different types, grades, and quantities of abrasive to meet a range of applications. Plastic media are somewhat softer than ceramic and for a given degree of abrasion, usually have shorter life but also somewhat lower cost and lower density. These softer materials achieve better preplate finish than other media and are better suited for handling soft metals
- *Steel*: hardened steel preformed shapes are available in a variety of shapes and sizes and are well suited for burnishing. Steel applications also include cleaning and light deburring. The basic benefit is that these media wear very little. Although there is a high initial investment, they are not consumable and do not need reclassification. Steel pins and tacks used with abrasive compounds can be useful means of removing somewhat inaccessible burrs

Final selection of the best media must be made on a trial and error basis, but these factors affecting selection should be understood and considered before any form of testing procedure is started:

Shape and size of media

- To remove burrs
- To achieve uniform edge and surface finishes
- To avoid jamming in holes and recesses
- To achieve ease of separation
- To achieve shortest cycle time

Availability and cost of media

- From consistent supplier
- For economical cost per pound or unit volume
- For consistent quality

Ability and versatility of media

- For minimum wear and reclassification
- For handling a range of products within a given machine
- For minimum break in requirements
- For cushioning action between parts

Optimum media to part ratio is another consideration for choosing the best media. The following table shows media to part ratios typical for most vibratory and tumbling equipment:

Media to part ratio, by volume	Commercial application
0:1	No media, part-on-part, used for beating off burrs, no media for cutting; sometimes suitable for burnishing
1:1	Equal volumes of media and parts, forgings, and castings, crude, very rough surfaces
2:1	More gentle, more separation, still severe part-on-part damage is possible
3:1	About minimum for nonferrous parts, considerable part-on-part contact, fair to good for ferrous metals
4:1	Probably average conditions for nonferrous parts, fair to good surfaces, good for ferrous metals
5:1	Good for nonferrous metals, minimal part-to-part contact
6:1	Very good for nonferrous parts, common for preplate work on zinc with plastic media
8:1	For higher quality preplate finishes
10:1 to 15:1 or more	Better, used for very irregularly shaped parts or parts that tangle or bend
No contact	Absolutely no part-on-part contact, one part per machine or compartment, part fixturing

Mass Finishing Process Considerations

The mechanical finishing department in virtually every metal-working plant is used as a general rectification shop. This is inevitable because metal forming machinery does not produce consistent burrs, nor truly consistent surface and edge finish conditions. Equipment in the mechanical finishing department should be sufficiently versatile to meet the changing quality of parts.

Cleanliness. Most mass finishing processes tolerate some oil, scale, and dirt on components. In fact, mass finishing is frequently used as a highly effective means of cleaning components, having both mechanical and chemical action; however, some operations, particularly super finishing on very high precision parts, demand cleanliness or at least consistency of any surface contamination. The finishing department should be notified of any changes in prior operations, such as change of machining lubricants, or heat treatment methods.

Ancillary Equipment. Most mass finishing processes are wet operations. After parts are unloaded from the mass finishing equipment, parts are rinsed, corrosion protection is applied, and the parts are dried. A decision must be made whether these processes are separate from the mass finishing machine or if combination compounds could incorporate these steps in the finishing machine.

There is a requirement to separate parts from media by screening or magnetic separation. Media reclassification is also frequently incorporated into a screening separator.

Automation. Opportunities to automate mass finishing should be investigated. Mass finishing operation requires (*a*) measured quantities of parts and media to be placed into the equipment, (*b*) controlled addition of compound solution with change of solution at some point during the process cycle, (*c*) unloading of the equipment, (*d*) separation of parts from media, (*e*) washing and drying and conveying parts, and (*f*) classification, conveying, and storage of media. Such a sequence may represent a substantial labor involvement and many opportunities for errors that are difficult to control. The cost of automation is repaid by improved work flow, improved quality, and lower labor costs.

Process Instructions and Control. Process instructions should cover all variables. The variables include:

- Machine cycle, including process times, barrel speeds, or vibratory frequency and amplitude, or G force and speed, depending on type of equipment
- Load levels, total load, and proportion of parts to media
- Compound and water flow rates
- Media and compound variables
- Pretreatment and post-treatment variables

Masking and Fixturing. Holes in components with rubber or plastic are plugged to keep media from being jammed into those holes during processing. Corners and edges that must be left sharp may also be masked. Some components are processed on fixtures in the mass finishing equipment. This is essential for spindle finishing. In some mass finishing equipment, large components are better processed in individual compartments within the machine. Occasionally, threads are masked that might get damaged or abraded too much during processing.

Causes of Difficulties Commonly Encountered in Mass Finishing

The following is a list of difficulties encountered during mass finishing processes:

Excessive impingement on part surface

- Insufficient amount of media in chamber
- Parts in work load are too large for the machine
- Excessive equipment speed or amplitude
- Low solution level or flow rate
- Insufficient or wrong compounds
- Undesirable handling methods
- Media too large for part size
- Wrong type of media

Roll over of edges, corners, and burrs

- Abrasive action too slow
- Wrong type of media
- Media particles too large for part size
- Excessive equipment speed or amplitude
- Excessive work load for volume of media
- Incorrect water level
- Processing time too long

Persistent lodging of media in holes, slots, or recessed areas

- Wrong size of media
- Wrong shape of media
- Excessive wear or depreciation of media
- Media fracturing during operation
- Failure to classify media as required

Poor surface finish

- Insufficient cut-down in process
- Inadequate flushing and cleaning of equipment
- Incorrect compound
- In-process corrosion
- Water flow not correct
- Media too aggressive or too hard
- Part-on-part impingement

Safety Precautions

In addition to the normal hazards with rotating machinery, other hazards in finishing are (*a*) generation of gas, usually flammable, in a closed barrel, and (*b*) handling of acid or alkaline compounds. Accidents caused by a buildup of gas pressure within a closed barrel have ranged from the blowing off of a cover during its removal to the fracture and fragmentation of portions of a barrel during operation. All closed barrels should have a cover-locking device that permits loosening the cover, but restrains it so that the cover cannot be blown off. Ventilation of the barrel, preferably automatic, is also highly recommended.

Hazards connected with the use of acids and alkalis are generally recognized. The use of rubber gloves and face masks is recommended.

Waste Disposal

Some compounds used in barrel finishing are harmful and their disposal may pose problems. In all instances, local regulations should be checked before establishing any procedure for waste disposal.

Most compounds are alkaline when mixed with water, although some are acid. Compounds containing cyanides or chromates are toxic and are not recommended.

Methods of waste treatment include:

- Before disposal, acid or alkaline waste should be neutralized
- Oils can be removed by running the solution under a skimming bar
- Solid wastes should be settled out in a basin, preferably after neutralization

Shot Peening

By the ASM Committee on
Shot Peening*

SHOT PEENING is a method of cold working in which compressive stresses are induced in the exposed surface layers of metallic parts by the impingement of a stream of shot, directed at the metal surface at high velocity under controlled conditions. It differs from blast cleaning in primary purpose and in the extent to which it is controlled to yield accurate and reproducible results. Although shot peening cleans the surface being peened, this function is incidental. The major purpose of shot peening is to increase fatigue strength. The process has other useful applications, such as relieving tensile stresses that contribute to stress-corrosion cracking, forming and straightening of metal parts, and testing the adhesion of silver plate on steel.

Peening Action

When individual particles of shot in a high-velocity stream contact a metal surface, they produce slight, rounded depressions in the surface, stretching it radially and causing plastic flow of surface metal at the instant of contact. The effect usually extends to about 0.13 to 0.25 mm (0.005 to 0.010 in.), but may extend as much as 0.50 mm (0.02 in.) below the surface. The metal beneath this layer is not plastically deformed. In the stress distribution that results, the surface metal has induced or residual compressive stress parallel to the surface, while metal beneath has reaction-induced tensile stress. The surface compressive stress may be several times greater than the subsurface tensile stress. This compressive stress offsets any service-imposed tensile stress, such as that encountered in bending, and improves fatigue life of parts in service markedly.

Peening action improves the distribution of stresses in surfaces that have been disturbed by grinding, machining, or heat treating. It is particularly effective on ground or machined surfaces, because it changes the undesirable residual tensile stress condition that they usually impose in a metal surface to a beneficial compressive stress condition. Shot peening is especially effective in reducing the harmful stress concentration effects of notches, fillets, forging pits, surface defects, and the low-strength effects of decarburization, and the heat affected zones of weldments.

Strain Peening. The magnitude of residual stress that can be induced by shot peening is limited. In hard metals, it is slightly more than half the yield strength. A higher residual stress, approaching the full yield strength, can be obtained by strain peening, which consists of peening the surface as it is being strained in tension. The effectiveness of strain peening is limited to parts, such as springs, gears, and shafts, that are subjected to unidirectional service loads.

Surface Coverage and Peening Intensity

The workpiece surface being peened is affected by the amount of the target surface peened and the effectiveness of the peening action on that target surface.

Surface coverage is a measure of how completely an area has been hit by the myriad of impinging shot particles. Without 100% coverage or saturation, the improvement in fatigue characteristics conventionally produced by shot peening is not obtained.

As stated in SAE Recommended Practice J443, Procedure for Using Shot Peening Test Strip, a definite and quantitative relationship between coverage and exposure time exists, which may be expressed as follows:

$$C_n = 1 - (1 - C_1)^n$$

Where C_1 is the percentage of coverage (decimal) after 1 cycle, C_n is the percentage of coverage (decimal) after n cycles, and n is the number of cycles.

This relationship indicates that coverage approaches 100% as a limit. Accurate measurements above 98% coverage are difficult to obtain, but a measurement at a lower degree of coverage serves as a means of determining the exposure time or equivalent time required to obtain any desired coverage. Because accurate measurement can be made up to 98% coverage, this value is arbitrarily chosen to represent full coverage or saturation. Peening at less than saturation is ineffective because of the amount of unpeened surface. Beyond this value, the coverage is expressed as a multiple of the exposure time required to produce saturation. For example, 1.5 coverage represents a

*Clifford S. Mehelich, Senior Vice President, Metal Improvement Co., Inc.; L. Van Kuiken, President, Progressive Blasting Systems; Michael M. Woelfel, Technical Services Manager, Impact Products, Potter Industries

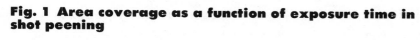

Fig. 1 Area coverage as a function of exposure time in shot peening

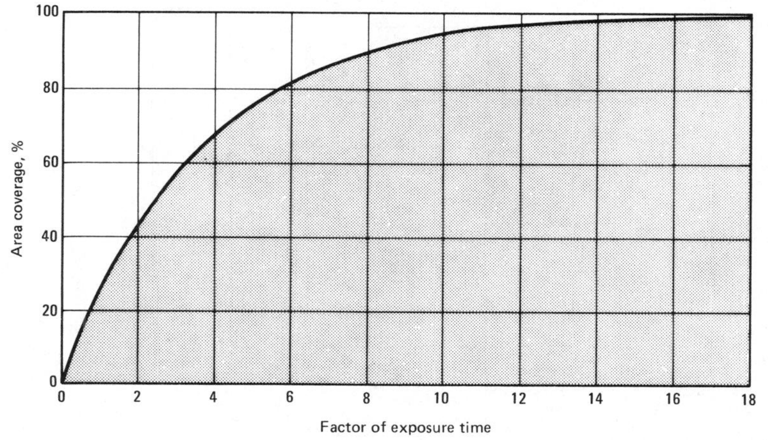

Fig. 2 Relation of measuring coverage to peening time

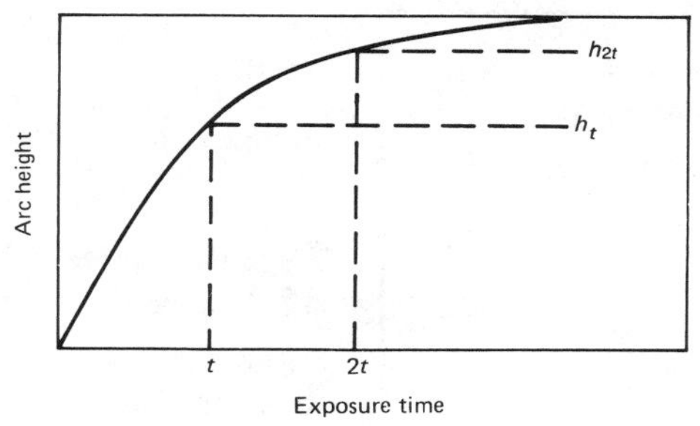

Coverage is considered full at time *t*, if doubling exposure to time 2*t* results in change in arc height less than 10%

condition in which the specimen or workpiece has been exposed to the blast 1.5 times the exposure required to obtain saturation. Figure 1 shows the relationship between exposure time and coverage and indicates that after a measurement of a low percentage of coverage has been established, the correct exposure time for any percentage of coverage can be readily determined.

Measurement of Coverage. Direct methods for measuring coverage include visual methods and the Straub method. One of the indirect methods is the Valentine method, which involves layer removal.

Visual methods, although not quantitative, are almost universally used. The simplest of these consists of visual inspection, with or without the aid of optical (10×) magnification of the surface of the peened part. This method may be supplemented by a series of reference photographs illustrating various percentages of coverage.

Another visual method consists of preparing a transparent plastic replica of the peened surface and comparing it, by means of photographic projection, with reference replicas having various percentages of coverage.

The Straub method consists of (*a*) exposing a polished surface to the shot stream, (*b*) projecting the surface at a magnification of 50 diameters on the ground glass of a metallographic camera, (*c*) tracing the images of the indented areas on translucent paper, and (*d*) measuring the total area and the indented area with a planimeter. Percentage of coverage is expressed as the ratio of indented area to total area multiplied by 100. About 15 min is required to make one measurement.

The Peenscan method is offered in lieu of visual inspection in MIL-S-13165 and consists of (*a*) painting a part before peening with a dye sensitive to ultraviolet light, (*b*) shot peening the part, (*c*) inspecting the part under the ultraviolet light for any missed areas, (*d*) shot peening the part, and (*e*) reinspecting the part under ultraviolet light. Complete removal of the dye indicates 100% coverage of the part.

The Valentine method consists of (*a*) making a duplicate of the part from low-carbon steel, (*b*) peening the part, (*c*) annealing it for several hours to promote recrystallization and grain growth, and (*d*) relating peening coverage to the amount and continuity of grain growth, by metallographic examination of cross-sectional areas. For additional information, see the article on recrystallization annealing in *Metals Handbook*, 9th ed., Vol 4.

Because of the difficulty in quantitatively measuring coverage by these methods, percent coverage is usually estimated from the curve of Almen arc height against the duration of shot exposure. The Almen test is described in the Appendix to this article.The graph in Fig. 2 shows the relation between shot coverage and doubling exposure time. A change in arc height of 10% or less indicates saturation peening.

Peening intensity is governed by the velocity, hardness, size, and weight of the shot pellets, and by the angle at which the stream of shot impinges against the surface of the workpiece. Intensity is expressed as the arc height of an Almen test strip at or more than saturation coverage. Arc height is the measure of the curvature of a test strip that has been peened on one side only. At or above saturation of the Almen strip, arc height is a measure of the effectiveness of the peening operation on a specific part. The Almen test is the primary standard of quality control and should be used at regular intervals, often on a day-to-day basis, and in the same location in the peening setup. Used correctly, a lower arc height indicates a reduction in peening intensity caused by (*a*) a reduction in wheel speed or a drop in air pressure, (*b*) ex-

Fig. 3 Relation of peening intensity to cross-sectional thickness of parts peened

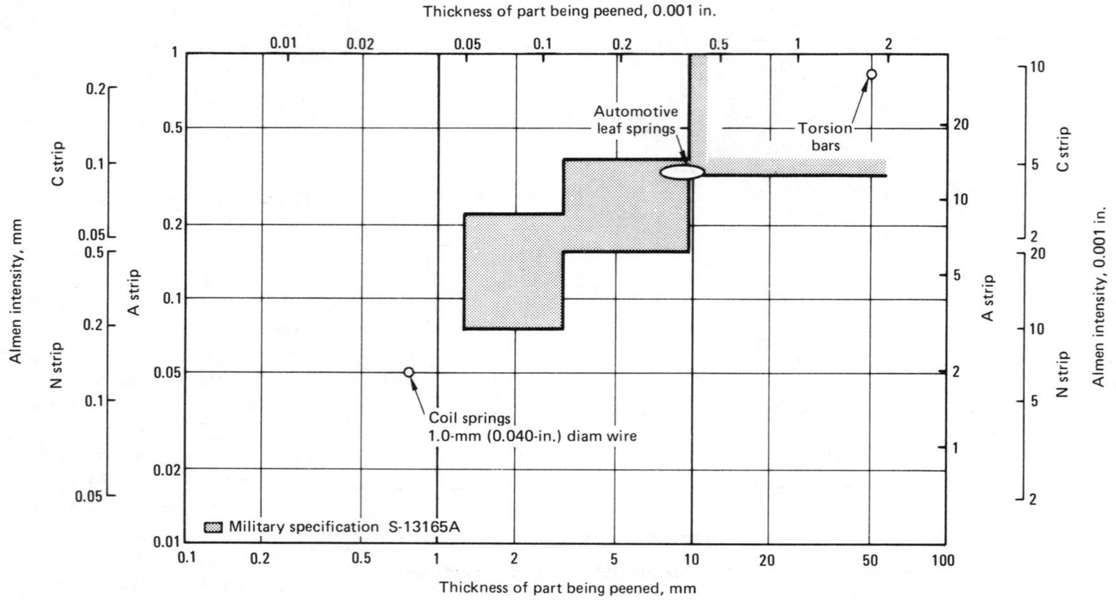

Fig. 4 Relation of depth of compressed layer to peening intensity for steel of two different hardnesses

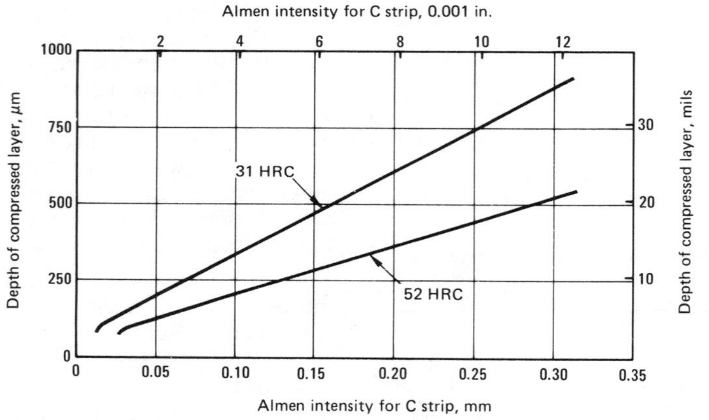

cessive breakdown of shot, or (*c*) other operational faults, such as undersized shot in the machine or clogged feed valves.

Selection of Intensity. The lowest peening intensity capable of producing the desired compressive stress is the most efficient and least costly, because the peening process can be achieved with the minimum shot size in the minimum exposure time. Conversely, an intensity may be considered excessive if, as with very thin parts, a condition is produced in which the tensile stresses of the core material outweigh the beneficial compressive stresses induced at the surface. Figure 3 presents data that indicate the relation of peening intensity to cross-sectional thickness.

The depth of compressed layer to be produced by peening is a factor in selecting peening intensity. For instance, a heavy steel component with a partially decarburized skin requires a peening intensity high enough to induce a compressive stress beneath the decarburized layer. The relation between peening intensity and depth of compressed layer, for steel hardened to 31 and 52 HRC, is shown in Fig. 4.

Types and Sizes of Shot

Shot used for peening is made of iron, steel, or glass. Metallic shot is designated by numbers according to size. Shot numbers, as standardized by MIL-S-13165, range from S70 to S930. The shot number is approximately the same as the nominal diameter of the individual pellets in ten thousandths of an inch. Standard size specifications for cast iron and steel shot are given in Table 1.

Glass shot, used primarily for peening nonferrous material, is available in a wider range of basic diameters. Hardness of glass shot is equivalent to 46 to 50 HRC. For further information on glass beads, see the article on abrasive blast cleaning in this Volume. Table 2 shows the effect of shot size and peening intensity on fatigue life.

Cast steel shot is made by blasting a stream of molten steel with water and forming globules that rapidly solidify into nearly spherical pellets. This process is also called atomizing. The pellets are screened for sizing, reheated for hardening, quenched, and tempered

Table 1 Cast shot numbers and screening tolerances

Numbers in parentheses measured in inches

Peening shot size No.	All pass U.S. screen size	Maximum 2% on U.S screen	Maximum 50% on U.S. screen	Cumulative minimum 90% on U.S. screen	Maximum 8% on U.S. screen	Maximum number of deformed shot acceptable
930	5 (0.157)	6 (0.1320)	7 (0.1110)	8 (0.0937)	10 (0.0787)	5(a)
780	6 (0.132)	7 (0.1110)	8 (0.0937)	10 (0.0787)	12 (0.0661)	5(a)
660	7 (0.111)	8 (0.0937)	10 (0.0787)	12 (0.0661)	14 (0.0555)	12(a)
550	8 (0.0937)	10 (0.0787)	12 (0.0661)	14 (0.0555)	16 (0.0469)	12(a)
460	10 (0.0787)	12 (0.0661)	14 (0.0555)	16 (0.0469)	18 (0.0394)	15(a)
390	12 (0.0661)	14 (0.0555)	16 (0.0469)	18 (0.0394)	20 (0.0331)	20(a)
330	14 (0.0555)	16 (0.0469)	18 (0.0394)	20 (0.0331)	25 (0.0280)	20(b)
280	16 (0.0469)	18 (0.0394)	20 (0.0331)	25 (0.0280)	30 (0.0232)	20(b)
230	18 (0.0394)	20 (0.0331)	25 (0.0280)	30 (0.0232)	35 (0.0197)	20(b)
190	20 (0.0331)	25 (0.0280)	30 (0.0232)	35 (0.0197)	40 (0.0165)	20(b)
170	25 (0.0280)	30 (0.0232)	35 (0.0197)	40 (0.0165)	45 (0.0138)	20(b)
130	30 (0.0232)	35 (0.0197)	40 (0.0165)	45 (0.0138)	50 (0.0117)	30(c)
110	35 (0.0197)	40 (0.0165)	45 (0.0138)	50 (0.0117)	80 (0.0070)	40(c)
70	40 (0.0165)	45 (0.0138)	50 (0.0117)	80 (0.0070)	120 (0.0049)	40(c)

(a) Per area, 1-in. square. (b) Per area, ½-in. square. (c) Per area, ¼-in. square

to the desired hardness. According to the SAE Recommended Practice J827, Cast Steel Shot, 90% of the hardness measurements made on a representative sample should fall within the range equivalent to 40 to 50 HRC. To maximize the peened effect, shot should always be at least as hard as the workpiece. For hard metals, special hard-cast steel shot, 57 to 62 HRC, should be used.

Cast steel shot is the most widely used peening medium.With suitable heat treatment, cast steel shot has a useful life many times that of cast iron shot. Its improved impact and fatigue properties markedly lower the rate of shot breakage, increase peening quality, and extend the life of components of peening machines.

Cast iron shot or chilled iron is brittle, with an as-cast hardness of 58 to 65 HRC. It breaks down rapidly; however, its inherently high hardness yields higher peening intensities for a given shot size, in comparison to softer materials. A high rate of shot breakage complicates the control of peening quality and increases the cost of equipment maintenance and cost of shot, because broken shot must be eliminated for best results.

Glass beads are used for peening stainless steel, titanium, aluminum, magnesium, and other metals that might be contaminated by iron or steel shot. They are also used for peening thin sections. Relatively low Almen A shot peening intensities, seldom exceeding 0.15 to 0.25 mm (0.006 to 0.010 in.), are used. Glass beads can be used in either wet or dry peening processes.

Control of Process Variables

Major variables in the shot peening process are shot size and hardness, shot velocity, surface coverage, angle of impingement, the resulting peening intensity, and shot breakdown. The quality and effectiveness of peening depend on the control of each of these interdependent variables.

Size of Shot. When other factors, such as shot velocity and exposure time, are constant, an increase in shot size results in an increase in peening intensity and depth of the compressed layer, plus a decrease in coverage. Selecting the minimum shot size capable of producing the required intensity is preferable to take advantage of the more rapid rate of coverage obtained with smaller shot.

The selection of a particular shot size may be dictated by the shape of the part to be peened. In shot peening the firtree serrations of steel compressor blades, complete coverage can be obtained only if the radius of the shot does not exceed the radius of the serrations. The same principle applies to the selection of shot size for peening the root radius of threads. When peening fillets, the diameter of the shot used should not exceed one half the radius of the fillet.

Hardness of Shot. Variations in the hardness of shot do not affect peening intensity, provided the shot is harder than the workpiece. If the shot is softer than the workpiece, a decrease in intensity occurs.

Velocity of Shot. Peening intensity increases with shot velocity; however, when velocity is increased, shot must be inspected for breakdown more frequently for purging the system of broken shot.

Angle of Impingement. By definition, the angle of impingement is the angle, 90° or less, between the surface of the workpiece and the direction of the blast. As this angle is decreased from 90°, peening intensity is reduced. Peening intensity varies directly as the sine of the angle of impingement. When a low impingement angle is unavoidable, increases in shot size and velocity may be required to attain a desired intensity.

Breakdown of Shot. To maintain the required intensity and to provide consistent peening results, a production peening unit must be equipped with a separator that continuously removes broken or undersized shot from the system. Rate of removal should approximate the rate of wear and breakdown. The percentage of full-sized and rounded shot in the system should never fall below 85%. Higher percentages are preferred. Sharp-edged broken media can scratch a part, generating a stress raiser; therefore, rounded shot, like miniscule peening hammers, is mandatory. Integral shot conditioners on peening machines consist of screens or air wash systems, neither of which can fully discriminate whether shot is broken or not because these devices are designed to handle only the specific weight of the shot. The only practical method of maintaining 85% good shot in the machine is to remove the entire shot load and to reclassify it in a separate machine that distinguishes both size and shape.

Equipment

The equipment used in shot peening is essentially the same as that used in abrasive blast cleaning, except for certain auxiliary equipment made necessary by the more stringent controls imposed in the shot peening process. For a description of basic equipment, such as cabinets, wheels, nozzles, and conveyors, see the article on abrasive blast cleaning in this Volume.

The principal components of shot peening equipment are a shot-propelling device, shot-cycling arrangements, and a work-handling conveyor. All portions of equipment that are exposed to

Table 2 Effect of shot peening on fatigue strength of aluminum alloys and carbon and low-alloy steels

Metal tested	Type of specimen	Stress cycle	Surface condition as received
Aluminum alloys			
2014-T6	Plain, 38-mm (1.5-in.) diam	Reversed bending	Smooth turned(a)
2024-T4	Plain, 38-mm (1.5-in.) diam	Reversed bending	Turned(a)
7079-T6	Plain, 38-mm (1.5-in.) diam	Reversed bending	Turned(a)
7075-T6		Reversed bending	Turned(a)
Carbon and low-alloy steels			
5160 spring steel(c)	Flat leaf, 38 mm (1.5 in.) wide, 4.88 mm (0.192 in.) thick	Unidirectional bending	Machined before heat treatment(d)
1045 steel (165 HB)	Plain (R. R. Moore)	Rotating bending	Machined
1045 steel (285 HB)	Plain (R. R. Moore)	Rotating bending	Machined
9260 steel (526 HB)	Plain (R. R. Moore)	Rotating bending	Machined
Ingot iron (121 HB)	Plain (R. R. Moore)	Rotating bending	Machined
4340 steel (277 HB)	Plain (R. R. Moore)	Rotating bending	Machined
4118 steel (60 HRC)	Single gear tooth	Unidirec. bending	Machined
8620 steel (58 HRC)	Single gear tooth	Unidirec. bending	Machined
S-11 steel(r)	Grooved, 7-mm (0.3-in.) D(s)	Rotating bending	Machined
0.54% C steel(u)	Plain, 8-mm (0.315-in.) diam	Rotating bending	Decarburized
	Plain, 5.99-mm (0.236-in.) diam	Reversed torsion	Decarburized
	10.0-mm (0.394-in.) diam bars:		
	Smooth	Rotating bending	Polished
	Round-notched(v)	Rotating bending	Machined
	V-notched(w)	Rotating bending	Machined(x)
Music wire(y)	Coil spring	Not reversed	...
4340 steel(aa)	14-mm (0.560-in.) diam(bb)	Reversed torsion	Smooth turned
4340 steel(dd)	6.4-mm (0.250-in.) diam(bb)	Rotating bending	Highly polished
4340 steel(gg)	6.4-mm (0.250-in.) diam(bb)	Rotating bending	Highly polished

(a) 0.50 μm (20 μin.). (b) Values at 1 000 000 cycles. (c) Oil quenched and tempered at 370 °C (700 °F), hardness is 46 to 50 HRC. (d) 0.2 to 0.3 μm (7 to 12 μin.). (e) Shot peened only on side subjected to tension in fatigue test. (f) Peened under a strain +0.60, 1240 MPa (180 ksi). (g) Fatigue limit based on 5 000 000 cycles. (j) Peened under zero strain. (k) Equal parts of 170 and 280. (m) Depth of cold work, peening, 150 μm (6 mils). (n) Fatigue limit based on 10 000 000 cycles. (p) Stress relieved at 205 °C (400 °F) for 20 min after peening. (q) Fatigue limit in kilograms (pounds) load based on 5 000 000 cycles. (r) 3% nickel-chromium steel, oil quenched from 830 °C (1525 °F), tempered at 600 °C (1110 °F), tensile strength 930 MPa (135 ksi). (s) 0.79-mm (0.031-in.) semicircular groove. (t) Air pressure, 345 kPa (50 psi). (u) Tensile strength, 1410 MPa (205 ksi). (v) Notch depth 1.0 mm (0.040 in.), radius 0.051 mm (0.002 in.). (w) Notch depth 1.0 mm (0.040 in.). (x) Root of notch not hit by shot. (y) 0.99-mm (0.039-in.) diam. (z) For 400 000-cycle life. (aa) Tensile strength 1860 MPa (270 ksi). (bb) Chromium plated. (cc) Same value obtained for peened and chromium plated, not peened and plated is less than 275 MPa (40 ksi). (dd) Tensile strength 1520 MPa (220 ksi). (ee) Fatigue limit for chromium plated and baked. (ff) Fatigue limit for peened, chromium plated, and baked.

the stream of shot are enclosed to confine the shot and permit it to be recycled.

Propulsion of Shot. Two methods of propelling the shot are used widely in shot peening. One uses a motor-driven bladed wheel, rotating at high speed. The other uses a continuous stream of compressed air.

In the wheel method, shot is propelled by a bladed wheel that uses a combination of radial and tangential forces to impart the necessary peening velocity to the shot. The position on the wheel from which the shot is projected is controlled to concentrate the peening blast in the desired direction. Among the advantages of the wheel method of propulsion are easy control of shot velocity when equipped with a variable speed drive, high production capacity, lower power consumption, and freedom from the moisture problem encountered with compressed air.

The air blast method introduces the shot, either by gravity or by direct pressure, into a stream of compressed air directed through a nozzle onto the work to be peened. Aside from being more economical for limited production quantities, the air blast method can develop higher intensities with small shot sizes, permits the peening of deep holes and cavities by using a long nozzle, consumes less shot in peening small areas on intricate parts, and has lower initial cost, especially when a source of compressed air is already available.

In the late 1970's, another peening method was developed which uses gravitational force to propel the shot. Kinetic energy of the peening media is closely controlled by requiring the media to pass through a labyrinth before falling on the substrate from specified heights. By using 1 to 2 mm (0.04 to 0.08 in.) hardened and polished steel balls as a peening media, surface finishes of less than 0.38 μm (15 μin.) are obtained while peening in the range of 0.23 to 0.38 mm (0.009 to 0.010 in.) Almen N shot peening intensities. Gravity peening has the drawback of requiring much more time for saturation, due to the much lower impact velocities and the greater shot diameter used.

Cycling of Shot. Equipment for shot recycling consists of devices for the separation and removal of dust and undersize shot from the used shot mix. An air current or air wash, adjusted to the size of shot being processed, passes through a thin curtain of used shot as it falls through the separator. The air blows the dust and undersize shot from the mix, allowing the reusable shot to fall into a hopper for storage until needed at the peening machine. The major differences between the low-volume and the high-volume separators, illus-

Table 2 (continued)

Metal tested	Shot: Type	Peening conditions: Size No.	Intensity: 0.025 mm	Intensity: 0.001 in.	Fatigue strength: As received: 7 MPa	As received: 1 ksi	Polished: 7 MPa	Polished: 1 ksi	Peened: 7 MPa	Peened: 1 ksi	Strength gain by peening, %: Over as recd	Over polished	Ref
Aluminum alloys													
2014-T6	Cast steel	70	0.15	6 A	215	31(b)	···		260	38(b)	23	···	···
		230	0.76	30 A	215	31(b)	···		260	38(b)	23	···	···
		550	0.33	13 A	215	31(b)	···		260	38(b)	23	···	···
2024-T4	Cast steel	230	0.25	10 A	180	26(b)	···		240	35(b)	34	···	···
7079-T6	Cast steel	230	0.25	10 A	195	28(b)	···		250	36.5(b)	30	···	···
7075-T6		MIL-5	0.15	6 A	220	32(b)	···		275	40(b)	25	···	(1)
Carbon and low-alloy steels													
5160 spring steel(c)	Chilled iron	230(e)	0.15	6 C(f)	880	128(g)	···		1340	194(g)	51	···	(2)
		230(e)	0.15	6 C(h)	880	128(g)	···		1215	176(g)	37	···	···
		230(e)	0.15	6 C(j)	880	128(g)	···		970	141(g)	10	···	···
1045 steel (165 HB)	Chilled iron	(k)	(m)		···		275	40(n)	305	43.8(n)	··	10	(3)
1045 steel (285 HB)	Chilled iron	(k)	(m)		···		560	81(n)	515	75(n)	··	−7	(3)
9260 steel (526 HB)	Chilled iron	(k)	(m)(p)		···		750	109(n)	730	106(n)	··	−2	(3)
Ingot iron (121 HB)	Chilled iron	(k)	(m)(p)		···		185	27(n)	185	27(n)	··	0.7	(3)
4340 steel (277 HB)	Chilled iron	(k)	(m)		···		455	66(n)	540	78(n)	··	18	(3)
4118 steel (60 HRC)	Cast steel	110	0.20-0.25	8-10A	16 200	2 350(q)	···		20 900	3 025(q)	29	···	(4)
8620 steel (58 HRC)	Cast steel	230	0.41	16 A	86 185	12 500(q)	···		105 150	5 250(q)	22	···	(4)
S-11 steel (r)	···	280	(t)		260	38	···		420	61	62	···	(5)
0.54% C steel(u)	Chilled iron	460	0.48	19 A	310	45	···		475	69	54	···	(6)
	Chilled iron	460	0.48	19 A	225	33	···		325	47	43	···	···
	Chilled iron	460	0.48	19 A	585	85	···		600	87	3	···	(6)
	Chilled iron	460	0.48	19 A	285	43	···		395	57	33	···	···
	Chilled iron	460	0.48	19 A	185	27	···		325	47	73	···	···
Music wire(y)	···	110	···		825	120(z)	···		1310	190(z)	58	···	(7)
4340 steel(aa)	Cast steel	170	0.20	8 A	275	40(g)	···		515	75(g)(cc)	87	···	(8)
4340 steel(dd)	···	···	0.25	10 A	570	83	270	39(ee)	675	98(ff)	··	150	(9)
4340 steel(gg)	···	···	0.25	10 A	725	105	380	55(ee)	710	103(ff)	··	87	(9)

(gg) Tensile strength 1990 MPa (288 ksi). **References:** (1) Fatigue strength of 7075-T6 Aluminum Alloy when Peened with Steel Shot or Glass Beads, Potters Industries PII-I-74, 1974. (2) R.L. Mattson and J.G. Roberts, The Effect of Residual Stresses Induced by Strain-Peening Upon Fatigue Strength, Internal Stresses and Fatigue in Metals; Elsevier, Amsterdam, 1958. (3) J.M. Lessells and W.M. Murray, *Proc ASTM,* 41, 659 (1941). (4) J.A. Halgren and D.J. Wulpi, *Trans SAE,* 65, 452 (1957). (5) W.J. Harris, *Metallic Fatigue,* Pergamon, 1961. (6) S. Takeuchi and M. Honma, Effect of Shot Peening on Fatigue Strength of Metals, Reports of the Research Institute for Iron, Steel and Other Metals, Tohoku University, Sendai, Japan, 1959. (7) H.C. Burnett, *Proc ASTM,* 58, 515 (1958). (8) Effect of Chromium Plate on Torsion Fatigue Life of Shot Peened 4340 Steel, Douglas Aircraft Co. Report No. MP 20.005 (Sept. 13, 1960); available through SAE. (9) B. Cohen, Effect of Shot Peening Prior to Chrome Plate on the Fatigue Strength of High Strength Steel, WADC Technical Note 57-178, U.S. Air Force, June 1957

trated in Fig. 5 and 6, are the use of a trash removal screen in the low-volume separator (Fig. 5) prior to the air wash and the direct dumping of used shot from the elevator buckets into the separator (not shown) in contrast to the hopper and screw conveyor feed of used shot into the high-volume separator (Fig. 6). The effectiveness of the separator depends on careful control of the velocity of the air.

Shot-adding devices automatically replenish to maintain an adequate quantity of shot in the machine at all times. They are equipped with a capacitance switch or similar device to control the level of shot in the storage hopper and to add shot, as required, from a supply hopper.

Work Handling. The effectiveness of shot peening depends largely on peening intensity and adequate exposure of the workpiece to the blast stream.

Fig. 5 Shot separator for use with a low-volume shot peening machine

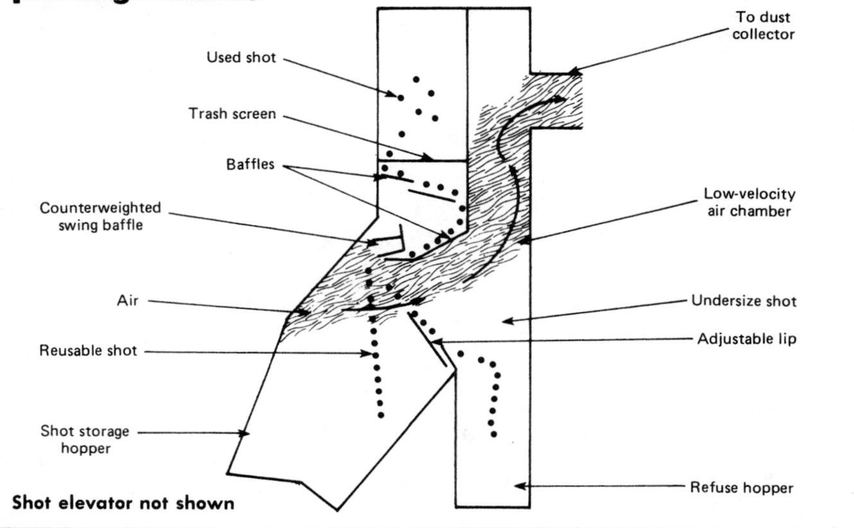

Fig. 6 Shot separator for use with a high-volume shot peening machine

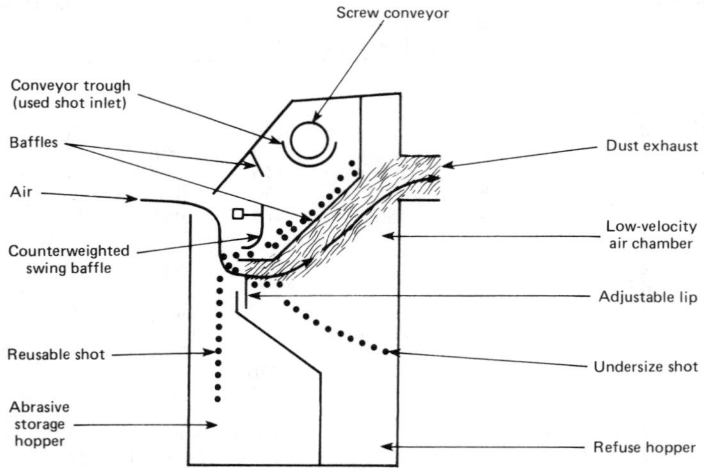

Shot elevator and overflow not shown

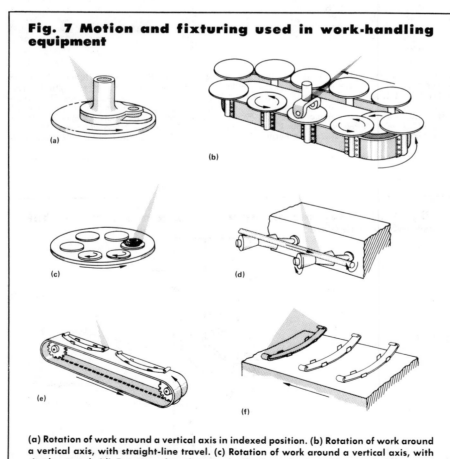

Fig. 7 Motion and fixturing used in work-handling equipment

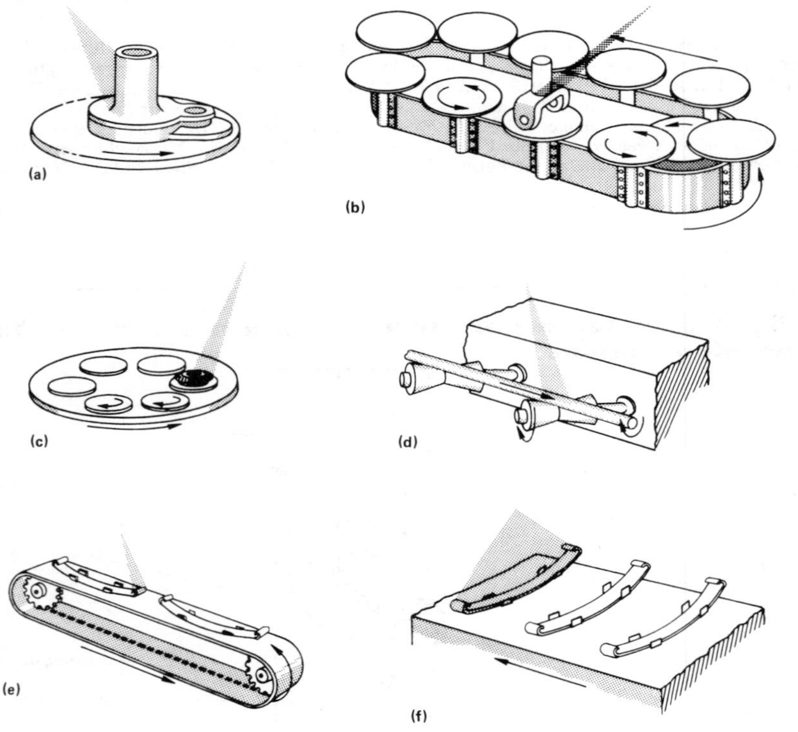

(a) Rotation of work around a vertical axis in indexed position. (b) Rotation of work around a vertical axis, with straight-line travel. (c) Rotation of work around a vertical axis, with circular travel. (d) Rotation of work around a horizontal axis, with straight-line travel parallel to the axis of rotation. (e) Straight-line travel of work with no rotation. (f) Straight-line travel of work in a transverse direction

Proper exposure is facilitated by using efficient work-handling fixtures, conveyors, and mechanisms. Figure 7 shows six types of work-handling mechanisms, which incorporate several basic motions for effective exposure of parts with a variety of shapes.

Stop-Offs. Various methods and materials have been developed for masking parts that require shot peening on localized areas. Masking with tape is economical when low production quantities are involved, but its cost is prohibitive on a large-scale basis.

When fabrication of special masks is warranted, masks usually are designed to serve as holding fixtures as well as stop-offs. Ordinarily, masks of this type are made of molded urethane or rubber as shown in Fig. 8. Large parts sometimes are protected with masks made of steel, carbide, polypropylene, or urethane.

Testing. Control of the shot peening process depends on systematic, periodic testing to determine intensity, coverage, and other important control factors. Standardized equipment for measuring peening intensity is illustrated and described in the SAE Standard J442, Holder and Gage for Shot Peening. A digest for measuring intensity is given in the Appendix to this article.

Dry Peening with Glass Beads. The methods used for dry peening with glass beads are comparable to the methods that use dry metallic shot. Separation of broken or damaged beads, as well as dust and other contamination, is accomplished by a centrifugal cyclone air separator, with final classification performed by a vibrating screen.

Wet Peening with Glass Beads. Wet glass peening is performed with glass beads, usually mixed in water and contained in a suitable hopper. In the automatic machine shown in Fig. 9, a mixer pump maintains a slurry of beads in water, and a feed pump forces the flow of slurry to the nozzle. The movement of slurry through the nozzle is accelerated by compressed air. The nozzles are attached to an oscillating bar that directs the flow of slurry at the workpiece. After making contact with the workpiece, the slurry is fed back to the hopper and recycled.

The principal controls in wet peening with glass are similar to those used in conventional shot peening. The peening pattern of the slurry is controlled

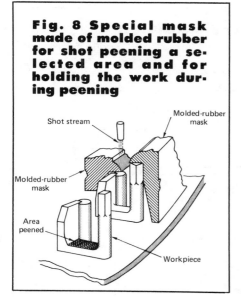

Fig. 8 Special mask made of molded rubber for shot peening a selected area and for holding the work during peening

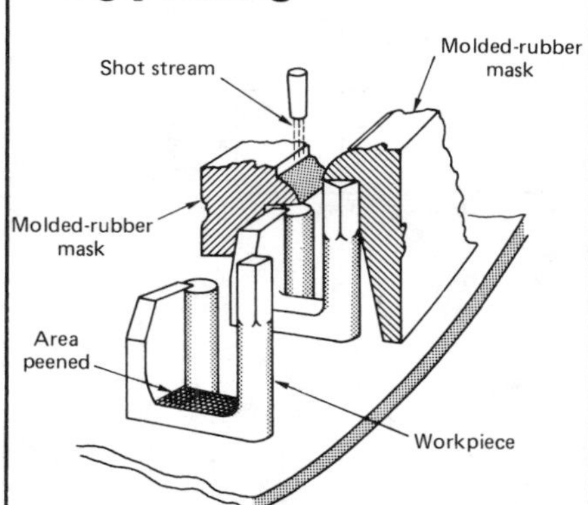

by the oscillating nozzles. Air pressure is controlled at each nozzle by separate regulators. Exposure time for the peening cycle is controlled by automatic timing devices. Intensity of peening must be carefully matched to bead size, both to sustain maximum fatigue life of the peened part and to minimize bead breakage.

Applications

Although the major application of shot peening is related to improvement of fatigue characteristics, other useful applications have been developed, such as metal forming, straightening, improving resistance to stress corrosion, and testing the adhesion of plated deposits of silver on steel.

Improving Fatigue Properties. The improvement in fatigue strength obtained on several aluminum alloys and carbon and low-alloy steels is indicated in Table 2, which lists the type and size of shot, the peening intensity used on most of the materials, and data on the fatigue test specimens, the type of fatigue test, and the surface condition of specimens.

Forming. Shot peening is well suited for certain operations in the forming of thin sections. It has been used to form, as well as to strengthen, structural components of aircraft. An example is integrally stiffened aircraft wing panels. These were machined from slabs of aluminum alloy 2024-T6, 7075-T6, or others, which had to be curved for aerodynamic reasons. The large size of these panels, 10 m (32 ft) by 1.2 m (46 in.), precluded the use of hot forming. Cold forming produced surface tensile stresses of 140 MPa (20 ksi) or more, which were alleviated by shot peening the panels on the tension side. Proper curvature of the panels could be obtained by shot peening alone, with careful control of intensity. The need for conventional cold forming methods was avoided, and the high compressive stresses induced by peening reduced the probability of early fatigue failure.

Fig. 9 Automatic machine for wet peening with glass beads

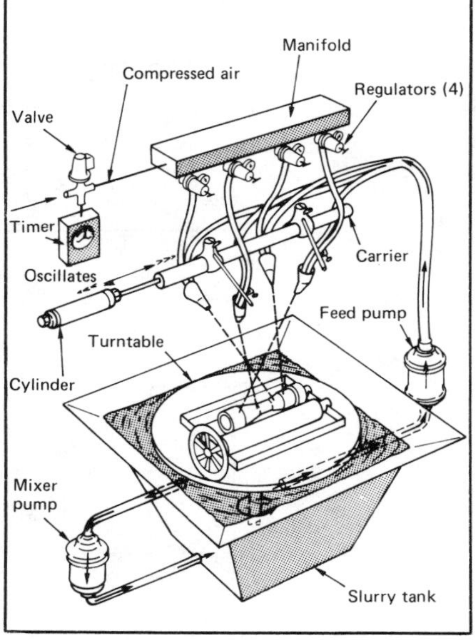

Other parts that have been successfully formed by peening techniques include precision collets and large aluminum tubes that were preformed in halves in a press brake and peened to the desired diameter.

Straightening and correcting of distortion by peening have been used in salvaging parts. For example, large ring gears, 915-mm (36-in.) OD by 19 mm (¾ in.) thick, developed 3-mm (⅛-in.) out-of-roundness as a result of heat treating. Shot peening restored the gears to within 0.13 mm (0.005 in.) of perfect roundness. In another instance, shafts 50-mm (2-in.) OD by 2 m (80 in.) long developed a 19-mm (¾-in.) bow, which was straightened to within 0.8 mm (1/32 in.) by shot peening.

Improving Resistance to Stress Corrosion. Stress corrosion is a complex interaction of sustained tensile stress at a surface and corrosive attack that can result in brittle failure of a ductile material. Cracking due to stress corrosion has been associated with several metals, including brass, steel, stainless steel, aluminum, zinc, titanium, and magnesium. The surface tensile stresses that cause stress corrosion can be effectively overcome by the compressive stresses induced by shot peening, with either steel shot or glass beads. An example is a large 7079-T6 aluminum alloy die forging that had a deep, moisture-retaining pocket machined at the flash line. The machining exposed grain flow perpendicular to the surface, at which a residual tensile stress of about 50% of the yield strength was present.

In another case, test bars, 11.1 mm (0.437 in.) in diameter, were cut in the short transverse direction from a 7075-T6 aluminum alloy hand forging and stressed to 75% of the yield strength. During alternate immersion tests in 3½% sodium chloride solution, unpeened specimens failed in 1, 5, 17, and 28 days, respectively. Specimens peened in the unstressed condition with S230 cast steel shot lasted 365 and 730 days, when failure occurred in the unpeened grip outside the test area. During exposure to an industrial atmosphere, similar unpeened test bars failed in 20, 37, 120, and 161 days, respectively, whereas a peened specimen under the same conditions as above was uncracked when it was removed from testing after an exposure of 8½ years.

Salt-fog tests on axial tension-test specimens of martensitic stainless steel showed that failure could be expected in a few days at stresses between 275 MPa (40 ksi) and 965 MPa (140 ksi). Shot peened specimens stressed at 690 MPa (100 ksi) lasted 14 to 21 days, as compared to 2 to 4 days for unpeened specimens. At a stress of 415 MPa (60 ksi), no failure of a peened specimen had occurred in 75 days, at which point the test was discontinued. Peening was beneficial, but it could not prevent stress corrosion at high stress levels. Table 3 presents stress-corrosion data indicating the life of peened and unpeened specimens of magnesium alloys, brass, and stainless steel in various corrosive mediums. All of these materials showed a high degree of improvement in resistance to stress corrosion as a result of peening.

Testing Adhesion of Silver Plate. The successful use of silver as a heavy-duty bearing material depends on a uniform high-strength bond between the silver plate and the steel substrate. Evaluating the integrity of the bond by peening has been accomplished with a high degree of reliability. Use of this technique on other electrodeposits is unknown. In the poorly bonded areas, the silver deforms plastically under the peening action of the shot and forms wrinkles or blisters.

Shot peening intensities required for revealing defectively bonded areas may be determined experimentally, using the data in Fig. 10 as a guide. Figure 10(a) shows the minimum shot peening intensity required to blister poorly bonded silver plate in relation to the thickness of the plate. Silver is plated at least 60% thicker than the finished dimensions require. The plate is then machined to a uniform extra thickness for peen testing. The intensity is adjusted to +0.004, −0 of that indicated in Fig. 10(a). Uniform coverage and exposure time should be maintained. Masking is applied to the unplated areas. After peening, the surface is machined again to final dimensions. Data in Fig. 10(b) show the relation between the minimum thickness of silver for peen testing and the maximum finished thickness of silver.

Fig. 10 Peen testing of silver plate on steel

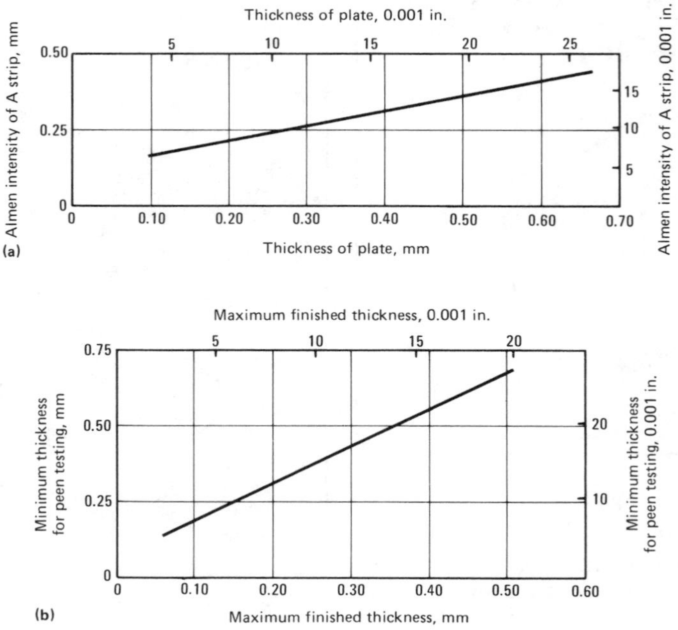

(a) Minimum shot peening intensity required to blister poorly bonded silver plate, shown as related to plate thickness. (b) Relation between minimum thickness of silver plate for peen testing and maximum finished thickness of plate

Table 3 Effects of shot peening on stress-corrosion life of alloys

Material	Solution to which exposed	Time to failure Unpeened	Time to failure Peened
Magnesium, AZ31B-H	Potassium chromate and sodium chloride	110 s	>10 days
Magnesium, AZ61A-H	Potassium chromate and sodium chloride	9¼ min	430 h
Brass cups, cold drawn	Ammonia	2½ h	19 and 47 h
Stainless steel, type 309	Hydrated magnesium chloride	270 h	>3000 h

Limitations

Shot peening has few practical limitations in terms of the materials or of the size, shape, quantity, surface condition, and surface hardness of parts that can be peened. Major limitations are not related to the mechanical aspects of the peening process, but to subsequent processing, such as the effects of post-peening elevated temperature and of machining, that can nullify the beneficial results of shot peening.

Size and Shape of Workpiece. The size of the peening cabinet is usually the only limitation on the size of workpiece that can be peened. To some extent, even this limitation can be overcome by the use of portable mechanized peening equipment. Provided the surface to be peened is accessible to the blast, shape of a workpiece is seldom a limitation. The peening of small radii in fillets and thread roots is limited by the smallest available media size, currently 0.0200-mm (0.001-in.) diameter glass beads. Sharp edges that must retain their sharpness should not be peened.

Surface condition, provided the workpiece surface is free of gross contaminants, is seldom a limitation in shot peening. Water, oil, and grease seriously contaminate the shot and interfere with peening quality and effectiveness. An as-forged surface usually shows greater improvement in fatigue strength than a polished surface as a result of peening. Cast surfaces respond as well to peening as wrought surfaces. Peened aluminum parts may be bright-dipped peened before being anodized.

Temperature Limitations. Low tempering temperatures, such as those conventionally used for carburized parts, have no adverse effect on peening stresses. Low-alloy steels can be heated to about 175 to 230 °C (350 to 450 °F) for about a half hour before a significant decrease in the compressive stresses occurs. Steels intended for elevated-temperature application usually withstand temperatures of 260 to 290 °C (500 to 550 °F) without under-

Fig. 11 Production parts that presented problems in shot peening

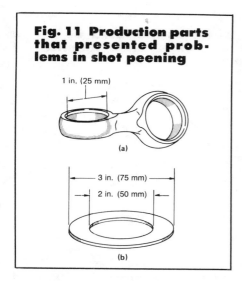

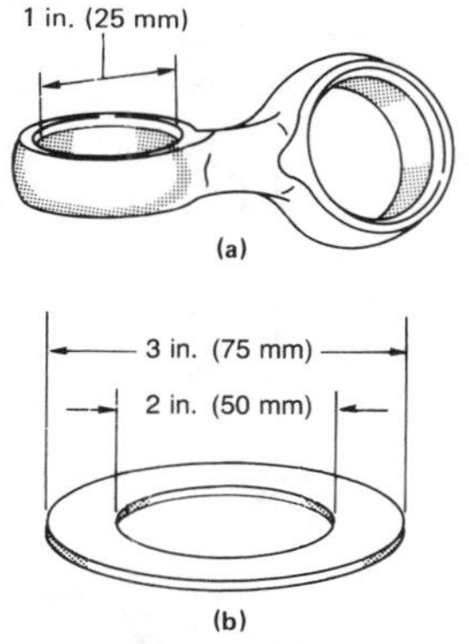

Fig. 12 Relation of nozzle angle, angle of load face, and resulting angle of impingement in peening root serrations of compressor blades

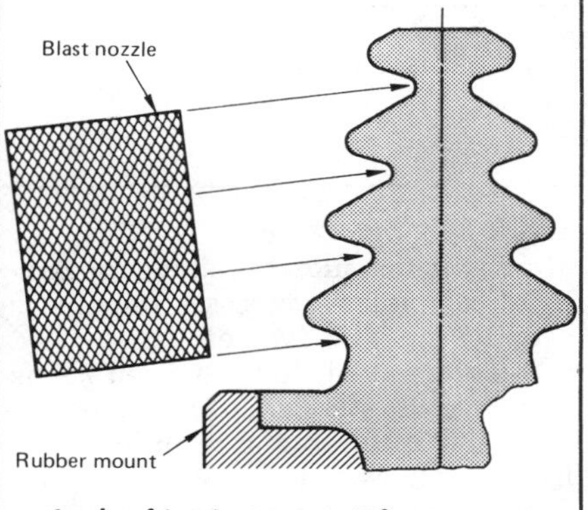

Angle of impingement, 83°

Fig. 13 Peening intensity as a function of angle of impingement

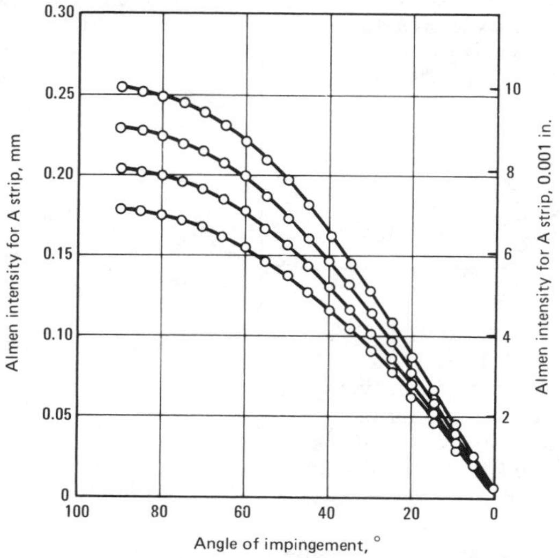

Table 4 Peening intensity as affected by nozzle and aspirator sizes and distance from nozzle to work

Test No.(a)	Nozzle size, mm	Nozzle size, in.	Aspirator size, mm	Aspirator size, in.	Almen intensity, A strip, 0.025 mm (0.001 in.) — Distance from nozzle to work, mm (in.): 200 (8)	180 (7)	150 (6)	120 (5)	100 (4)
1	9.5	3/8	4	5/32	5.5	4	5	5.5	5
2	9.5	3/8	5.5	7/32	4	5.5	5	5.5	5
3	13	1/2	4	5/32	3	2.5	3	3	3.5
4	13	1/2	5.5	7/32	5	6	7	7	7
5	16	5/8	4	5/32	7	6	5.5	7	7-8
6	16	5/8	7	9/32	6	6.5	7	7	7.5

(a) Tests made using suction shot peening test cabinet, 70 steel shot; pressure 635 kPa (92 psi), flow of shot to nozzle 3.3 kg/min (7¼ lb/min)

going a significant stress-relieving effect; however, exposure at 540 °C (1000 °F) or above relieves induced stresses in all high-temperature alloys. Exposure to temperatures above 175 °C (350 °F) can eliminate the induced compressive stresses in some alloys of aluminum.

Problems in Production Peening

Problems in the shot peening of production parts have been corrected by a variety of solutions. A small forging, as shown in Fig. 11(a), contained two holes with thin wall sections. The inner bearing surfaces had to be held to a tolerance of 0.013 mm (0.0005 in.). In shot peening a hole with a heavy walled section, the size of the hole is reduced by the peening action. With the thin wall section here, the size of the hole was not reduced, but the hole became oval shaped. The solution to this problem required the establishment of new dimensions before shot peening that permitted a light honing operation after shot peening to bring the hole size to within dimensional requirements.

In another instance, a flat ring, as shown in Fig. 11(b), failed in fatigue either by cracking from the inside diameter to the outside diameter or by flaking. Shot peening eliminated the fatigue failures, but caused the parts to dish as well as to warp. These parts were required to retain flatness to within a tolerance of 0.013 mm (0.0005 in.). After various intensities were tried, the distortion problem was solved by peening one side of the ring at a higher intensity than that used on the other side, depending on warpage direction.

In shot peening the fir-tree serrations of type 410 stainless steel compressor blades used in a jet aircraft engine, as shown in Fig. 12, it was determined that a maximum improvement of 26% in the fatigue life of serrated blade roots could be obtained only if a closely controlled peening procedure was followed. Certain variations from this procedure actually proved harmful to blade life. An alteration in peening intensity had a marked effect on fatigue characteristics. Variations in blast intensity that could be obtained by changing nozzle size, aspirator size, or the distance from the nozzle to the work are given in Table 4.

The procedure adopted consisted of peening the root serrations with S70 steel shot at an intensity of 0.007 to 0.008 A at an impingement angle of 83°. The intensity indicated was measured

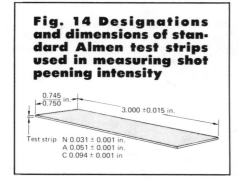

Fig. 14 Designations and dimensions of standard Almen test strips used in measuring shot peening intensity

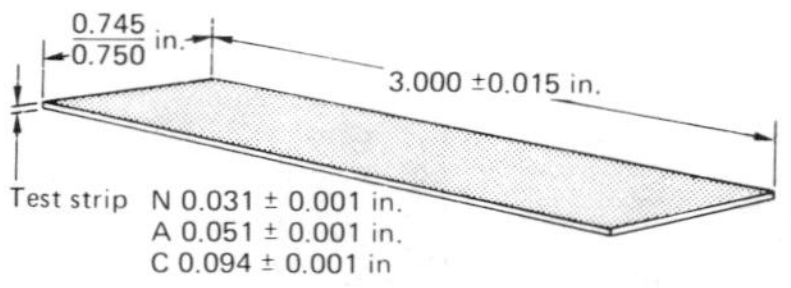

at an impingement angle of 90°. The true intensity on various portions of the root serrations was considerably less, depending on impingement angle; however, maximum intensity was obtained at the critical root radius, using the 83° impingement angle calculated to have this effect. The relation between impingement angle and intensity is shown in Fig. 13.

Peening was performed in a gravity feed, continuous conveyor, production cabinet, using a 16-mm (5/8-in.) bore diameter nozzle, a 4-mm (5/32-in.) aspirator, 635-kPa (92-psi) air line pressure, and a shot flow of 3.4 kg/min (7.5 lb/min). The distance from the nozzle to the work was set at 100 mm (4 in.). The cabinet conveyor moved at a fixed speed that exposed the work to the blast for a period of 5 s. The peening operation required two passes under the nozzle, one for each side of the blade root. The airfoil sections of the blade were protected from the blast by a sheet rubber covering.

Fig. 15 Correlation of intensities as indicated by arc heights of A, C, and N strips peened under identical blast and exposure conditions

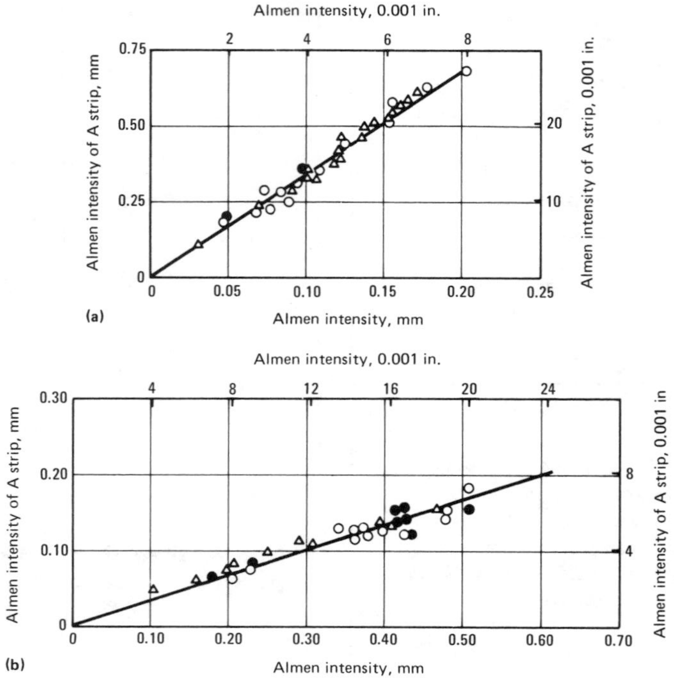

(a) C strip. (b) N strip (SAE J442)

Costs

The cost of shot peening on a production basis depends on several factors, including the size, shape, and hardness of parts, the total area to be peened, and the required intensity and coverage. Shot type, size, and velocity also influence costs, because they affect peening intensity, rate of shot breakdown, and the rate at which the desired coverage is obtained.

With an increase in shot velocity, the rate of shot breakdown increases far more rapidly than the intensity of the peening blast; however, the weight of a shot pellet varies directly, while the number of pellets per kilogram (pound) of shot varies inversely, with the cube of the diameter of the pellet. Consequently, for a given Almen intensity, coverage is obtained much more rapidly with smaller shot and higher velocity. Because of these opposing factors, the number of kilograms (pounds) of shot used per part is virtually the same for various shot sizes, assuming similar intensity, coverage, and control of uniform shot size. The use of small shot at high velocity increases production rate and reduces labor costs.

Processing After Peening

Shot peening itself is a finishing treatment, and usually no further processing of peened work is required, except for the application of a rust preventive on low-alloy steels. The as-peened surfaces of these steels are clean and chemically active and are highly susceptible to corrosion from fingerprints and other contaminants. Such surfaces are also highly receptive to oils for rust prevention and lubrication, and provide an excellent base for organic or inorganic coatings that do not require thermal treatment other than low-temperature baking. Temperatures high enough to relieve the beneficial compressive stresses imposed by peening must be avoided.

Stainless steel that has been peened with iron or steel shot should be passivated to counteract contamination by iron particles which causes rusting. Passivation is not required on superalloys intended for use at elevated temperature. Secondary peening with glass beads, after peening with steel shot, removes contaminating ferrous residue and increases the fatigue life of the peened part.

Because the compressive layer induced by peening is relatively thin, subsequent grinding or machining of peened surfaces should be avoided, except for aluminum or magnesium alloys that have been peened to a greater depth. As much as 0.13 mm (0.005 in.) may be removed from the surface of these alloys without harmful effect to the peened layer, and the improved surface finish may prove beneficial to fatigue properties; however, a knowledge of stress gradients must be available before stock removal.

Fig. 16 Assembled Almen test strip and holder

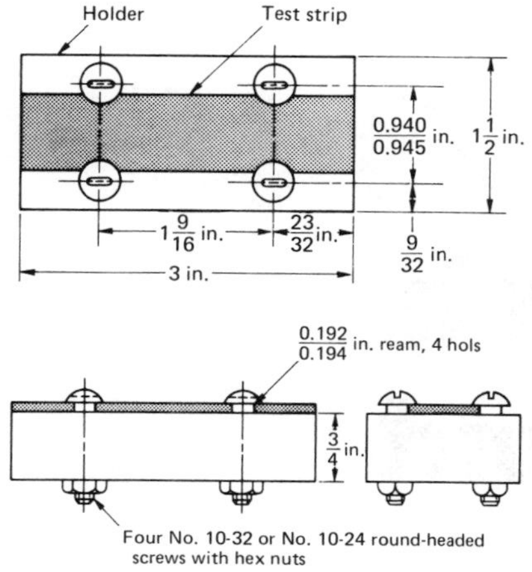

Fig. 17 Almen gage for measuring arc height for test strip

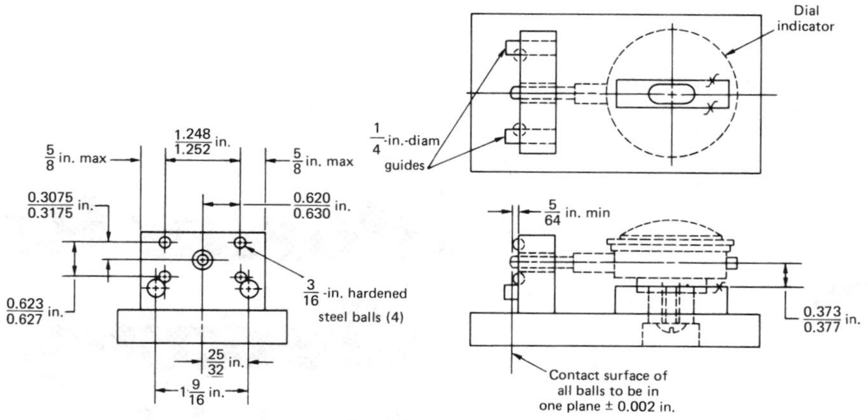

Steels may be lightly honed or lapped after peening. There is limited evidence that these operations for fine-particle abrasive blasting have a beneficial effect where maximum fatigue resistance is desired. After peening, straightening or cold forming by conventional methods should be avoided. These operations may result in a complete reversal of the stress pattern. Peen straightening and peen forming, however, are permissible, because these processes do not introduce harmful residual tensile stresses.

Appendix

Peening Test Strips, Holder, and Gage

The SAE standard J442 describes the test strips, strip holder, and gage used in measuring shot peening intensity.

If a thin flat piece of steel is clamped to a solid block and exposed to a blast of shot, it will be curved after removal from the block. The curvature is convex on the peened side. The extent of this curvature on a standard sample serves as a means of measurement of the intensity of the peening.

Standard test strips of three thicknesses used are shown in Fig. 14. Made of 1070 cold rolled spring steel, these strips have a specified hardness of 44 to 50 HRC. Strip A is recommended for testing intensities that produce curvatures having arc heights ranging from 0.15 to 0.60 mm (0.006 to 0.024 in.). For lesser intensities, the N strip is recommended. For greater intensities, the C strip is used. The relationship between strips A, C, and N is shown in Fig. 15. The data represent readings for conditions of identical blast and exposure.

During peening, the test strip is mounted on a holder as shown in Fig. 16. After being peened, the strip is removed and placed in an Almen gage shown in Fig. 17, located so that the dial indicator stem bears against the unpeened surface. The curvature of the strip is determined by a measurement of the height of the combined longitudinal and transverse arcs across standard chords. This arc height is obtained by measuring the displacement of a central point on the unpeened surface from the plane of four balls forming the corners of a rectangle.

The standard designation of intensity includes the gage reading or arc height and the test strip used. For example, 0.013 A, signifies that the arc height of the peened test strip is 0.33 mm (0.013 in.), and that the test strip used was of the A thickness. Uniform procedures for using the standard test strips described in the above digest of SAE J442 are provided in SAE Recommended Practice J443.

Power Brush Cleaning and Finishing

By Alfred F. Scheider
Vice President
Research & Development
The Osborn Manufacturing Corporation

POWER-DRIVEN wire, abrasive filament, and synthetic and natural fiber-filled rotary brushes are widely used for cleaning, deburring, edge blending, polishing, and surface finishing of metals. Selecting a power brush for a specific operation depends both on compliance with proper safety standards and the proper combination of variables in brush construction and operating procedure. This article considers some of the more common applications for power brushing, selection factors involved in choosing a suitable brush and brushing speed, and the advantages and limitations of the process.

The performance characteristics of power brushes that determine their suitability for specific applications are governed by three principal variables of brush construction:

1 *Type of fill material:* straight wire, twisted wire, crimped wire, abrasive filament, and synthetic or natural fiber
2 *Trim:* the length of fill material extending from the retaining member to the brush face, designated as short, medium, or long
3 *Density of fill:* heavy, medium, or light

The adjustments needed for optimal conditions for power brush performance are described in Table 1.

The speed at which the brush is rotated is another important factor in brush performance. The recommended speeds for a number of different power brushing operations are presented in the sections of this article that follow.

The position of the brush on the workpiece should also be considered. The full face of the brush should be in contact with the work to avoid grooving the brush. When this requirement cannot be met for a particular operation, provisions for dressing the brush face must be made. The tips of the brush can become dull during use, and working clearance can be lost. The necessary reconditioning and resharpening can be done by alternately reversing the direction of rotation during use.

In conventional power brushing operations, the material being removed, such as burrs, scale, dirt, weld slag, and broken brush filaments, can fly off the brush with considerable force. Therefore, a risk of serious injury exists for the brush operator and others in the immediate work area. Power brush safety requirements have been established by the American Brush Manufacturers Association, and they include the following:

- Keep all machine guards in place
- All operators and other employees working in the area of power brush operations must wear safety goggles, safety glasses with side shields, or face shields. Comply with the requirements of ANSI Z87.1-1979, Occupational Eye and Face Protection
- Observe all speed restrictions indicated on brushes, containers, or labels or printed in pertinent literature
- Use appropriate clothing and equipment where there is a possibility of injury that can be prevented by such clothing or equipment
- Comply with the Safety Standards of the Industrial Division of the American Brush Manufacturers Association and the American National Standards Institute Standard B165.1-1979, Safety Requirements for the Design, Care and Use of Power Driven Brushing Tools

Dry Brush Cleaning

For the removal of surface contaminants and imperfections, such as excess weld metal, flash, oxide films, scale,

Table 1 Adjustments for eliminating undesirable conditions in power-brush finishing

Undesirable condition	Possible adjustments for eliminating condition
Brush works too slowly	Decrease trim length and increase fill density Increase filament diameter Increase surface speed by increasing rev/min or outside diameter
Brush works too fast	Reduce filament diameter Reduce surface speed by reducing rev/min or outside diameter Reduce fill density Increase trim length
Action of brush peens burr to adjacent surface	Decrease trim length and increase fill density If wire brush tests indicate metal too ductile (burr is peened rather than removed), change to nonmetallic brush such as treated tampico brush used with burring compound or abrasive filament brush
Finer or smoother finish required	Decrease trim length and increase fill density Decrease filament diameter Try treated tampico or cord brushes with suitable compounds at recommended speeds Use auxiliary buffing compound with brush
Finish too smooth and lustrous	Increase trim length Reduce brush fill density Reduce surface speed Increase filament diameter
Brushing action not sufficiently uniform	Devise hand-held or mechanical fixture or machine which will avoid irregular off-hand manipulation Increase trim length and decrease fill density

Source: Erik Oberg, Franklin D. Jones, Holbrook L. Horton, *Machinery's Handbook,* 20th ed., New York: Industrial Press, 1975, p 2040

rust, and old paint, wire brushes are used without coolant in a variety of applications. The brushes most commonly used for dry surface cleaning are the knot-type twisted wire radial and crimped wire radial (both with medium or long trim), cup-type, wide-face, and strip-type wide face. Some of the uses, characteristics, and operating conditions of these brushes are described in the following paragraphs.

Knot-Type Twisted Wire Radial Brushes. Surface speed is highly important to the performance characteristics of knot-type twisted wire radial brushes, shown in Fig. 1(a). At low surface speeds, these brushes are very flexible, while at high surface speeds they are extremely hard and fast cutting.

Medium-trim brushes 75 to 200 mm (3 to 8 in.) in diameter are used for weld-cleaning operations at speeds of 6000 to 9500 rev/min. For most weld cleaning, brushes filled with straight twisted wire are used. For example, straight twisted wire brushes are used on portable tools in automobile body assembly plants for cleaning welds prior to leading various areas on body assemblies.

Similar brushes are used in pipeline construction because the narrow face of the brush, developed at speeds of 4500 to 6000 rev/min, is well suited for cleaning restricted welds.

In the rubber industry, 380-mm or 15-in. diam medium-trim knot-type brushes are operated at 60 m/s (12 000 sfm) for removing rubber flash. At this high surface speed, the brush face is narrowed to an almost knifelike edge and thus develops the considerable cutting action required to strip flash from molded products. Operated at 60 m/s (1200 sfm), a 380-mm (15-in.) diam brush can be used for cleaning wire mesh conveyor belts that have become loaded with various resins and other materials, including rock wool or glass fiber insulation. At this low surface speed, the brush is very flexible and will dislodge material from a wire mesh belt quite readily.

Knot-type brushes 250 and 300 mm (10 and 12 in.) in diameter are used primarily for cleaning and finishing operations. These brushes, when operated at speeds of 23 to 28 m/s (4500 to 5500 sfm), produce a deep satin or orange peel finish and are used for removing paint, varnish, and loose scale.

Crimped wire radial brushes, shown in Fig. 1(b), have a softness or cushion desirable for brushing parts held by operators. This cushion also allows the brush to conform to irregular surfaces and difficult to reach areas.

Crimped wire radial brushes 100 to 200 mm (4 to 8 in.) in diameter, of light to medium density and with medium to long trim, are used primarily as utility tools on bench grinders. These brushes can remove feather grinding or machining burrs, clean, or produce a satin finish. These brushes are suitable for jobs in which the operator, rather than a machine, holds the part for brushing. The cushion inherent in this type of construction allows the operator to hold and control the part easily. These smaller diameter brushes are not normally recommended for high-production machine operations.

Fig. 1 Radial wire brushes used for dry brush cleaning

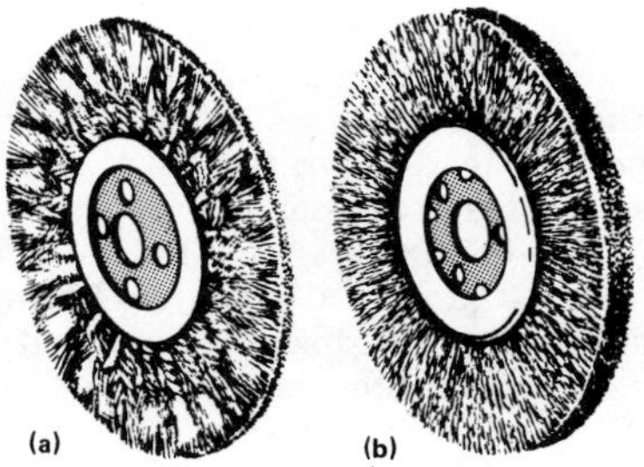

(a) Knot-type twisted wire brush. (b) Crimped wire brush

The larger brushes, from 250 to 380 mm (10 to 15 in.) in diameter, are used for machine operations and are especially suited for removing feathery grinding burrs, producing uniform finishes, and removing oxide where finish is important. These brushes usually produce a satin or matte finish, and they are operated at 23 to 30 m/s (4500 to 6000 sfm) for most applications.

Fig. 2 Cup-type brush with twisted wire fill

Fig. 3 Wide face brushes

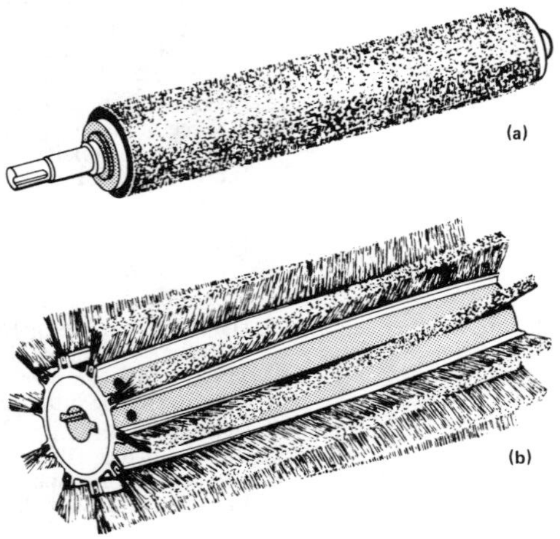

(a) Solid face brush roll. (b) Strip-type wide face brush used for cleaning conveyor belts

Wire sizes of 0.20 to 0.50 mm (0.008 to 0.020 in.) are used for the majority of work; however, 0.125-mm (0.005-in.) wire is used for producing fine finishes and is sometimes used with deburring compound to improve the micro-inch finish on a part.

Deburring compound can be a stick, liquid, or slurry, usually a grease base. Sufficient compound is applied to the brush face to give additional cutting action to the brush. The compound prevents rolling a grinding burr or rolling an edge of a soft, ductile metal.

Cup-Type Brushes. Wire-filled cup-type brushes, shown in Fig. 2, are extremely fast cutting wheels that can be used on portable tools for removing scale, rust, and oxides formed by welding.The wire size used in a majority of these applications is 0.50 mm (0.020 in.), and the wheels are operated at speeds of 40 to 50 m/s (8000 to 10 000 sfm).

Shipyards and the structural steel building industry are the most frequent users of cup-type brushes. The pipeline construction industry also uses cup-type brushes to clean the outside surface of pipe prior to coating. For this application, machines designed to use a number of cup brushes on a fixture that revolves around the outside of the pipe are used.

Wide face brushes, similar to the brush illustrated in Fig. 3(a), are made to specific customer requirements and are used for a number of operations in the aluminum, steel, copper, and brass industries. The brushes are dynamically balanced at the speed at which they will operate and often have a ground face to ensure uniform contact across the face of the brush. For most operations, these brushes are used at surface speeds ranging from 10 to 20 m/s (2000 to 4000 sfm).

In the aluminum industry, wide face wire or abrasive grit filament brushes are used to control the buildup of aluminum oxide on the work rolls in the hot mill. The brushes are operated at speeds of 600 to 1200 rev/min, depending on the line speed of the mill. For this operation, a soluble oil solution is used with the brushes. Brushing pressure of approximately 1100 N per lineal metre (6 lb per lineal inch) of face provide satisfactory cleaning and maximum brush life.

In the steel industry, wide face brushes filled with wire are used in combination with water to remove lime or magnesia coatings from certain types of steel. The operation is similar to scrubbing steel with nonmetallic-filled or abrasive filament brushes. In addition, wide face brushes can be used for removing carbon smut from steel following abrasive blast cleaning and pickling operations.

With wide face brush rolls, individual brush drives should be used for each brush roll. Brush pressure can be estimated from the reading on an ammeter installed in the drive-motor circuit. When brushing strip products, use a steel backup roll opposite the brush roll to provide adequate support for the surface to be brushed. Do not use brush rolls opposite each other for wide face setups.

Strip-type wide face brushes have an interrupted brush face, shown in Fig. 3(b), and are used primarily in cleaning operations that would cause a solid-face brush to become loaded and unusable. The most important application of strip-type wide face brushes is in the cleaning of conveyor belts. See Fig. 4 for a typical installation.

When certain materials are conveyed on rubber and fabric belts, some of the material can adhere to the belt after it has dropped its load. The adhering carry-back material loads the snubber pulley and return idlers, necessitating an expensive maintenance cleaning operation. Brushes can be used to remove the carry-back to eliminate the problem. For this procedure, the brushes are normally operated at a surface speed two to three times the surface speed of the belt in a direction opposite to the travel of the belt. Synthetic materials that can withstand both hot and cold environments are used as fill material for these units.

Wet Brush Cleaning

The brushes normally used in wet brush cleaning are 300-mm (12-in.) diam wide face brushes filled with natural fibers, synthetic fibers, or a mixture of both. Synthetic fibers with abrasive grit interspersed throughout each filament are used for applications that require surface improvement. These brushes are most commonly used in conjunction with hot alkaline cleaning solutions and warm water rinses in

Fig. 4 Use of strip-type wide face brush for cleaning of conveyor belt

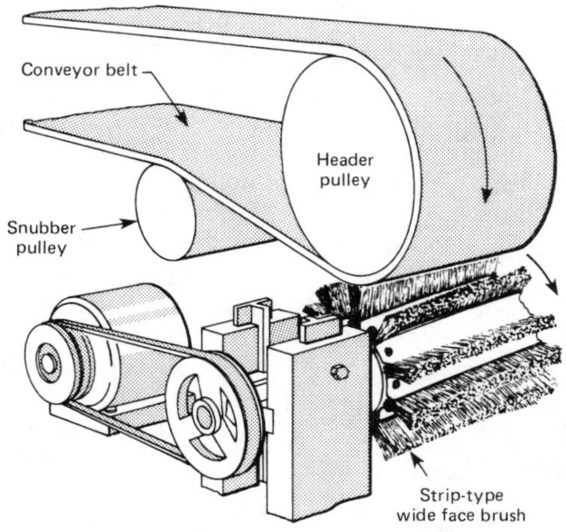

such steel industry applications as electrolytic tin plating lines, continuous annealing lines, continuous paint lines, silicon steel cleaning lines, as well as galvanizing lines.

For most applications employing nonmetallic-filled brushes, the recommended speed is in the range of 10 to 18 m/s (2000 to 3600 sfm). Higher speeds will not improve the operation appreciably, but will accelerate wear of the brushes. Because the tips of the fibers do the work, excessive pressure should be avoided. Brush pressure can be estimated from the reading on an ammeter installed in the drive-motor circuit.

To obtain the best cleaning action and maintain brushes at a maximum level of efficiency, brushes are rotated in the opposite direction from the travel of the steel strip. A backup roll is used opposite the brush, and the line is designed so that the strip is flexed over the backup roll where possible. A spray of rinse water is directed to the area where the brush and strip make contact.

A brush mounting arbor with a rotating coupling that permits rinse water to be circulated through the mounting arbor and out through the face of the brush should be used. A water pressure of 350 kPa (50 psi) is adequate to permit the water to flow through the mounting and out through the face of the brush. The water, which washes the brush filaments and prevents heating at the base of the brush, is especially important when synthetic fill materials are used. Synthetic materials do not wear as rapidly as natural fibers, but pick up more dirt. The circulation of water through the face of the brush minimizes loading.

Nonmetallic wide face brushes are used with slurries, such as a mixture of pumice and water, for producing a dull or matte finish on stainless steel sheets. The brushes are rotated at approximately 600 rev/min. The slurry is pumped onto the surface of the sheets before they enter the brushing stations. The brushes dull rub the surface and produce a uniform matte finish.

Nonmetallic wide face brushes are also used in the material preparation departments of stamping plants in the automotive and appliance industries. The brushes are frequently used in roller levelers to remove dirt and oil from the surface of steel prior to leveling.

Deburring

Various types of wire and abrasive filament-filled radial brushes, cup-type brushes, wheel-type and side-action brushes, tube brushes, and end brushes are used for deburring and for developing a radius on sharp edges. These brushes and the procedures for their use are described in the following paragraphs.

Short-trim dense wire-filled radial brushes, shown in Fig. 5, have very little impact action but instead have vigorous cutting action because of their short trim and dense construction. These brushes are extremely fast cutting at speeds of 33 m/s (6500 sfm). The most widely used fill-wire diameter is 0.3 mm (0.012 in.).

Short-trim radial brushes are used in the gear cutting industry to remove burrs and to produce radii on the edges of gear teeth. The following is a typical sequence of operations:

- Gears are hobbed, and internal or external splines are broached
- After hobbing and broaching, the parts are finished with a radial brush
- Gear teeth are shaved, and the face sides of broached areas are ground
- Parts are heat treated

Brushing the parts after hobbing or broaching removes burrs and produces a large radius. On helical gears, a large radius is required on the acute-angled tooth edge locations because of the extremely sharp edge produced by shaving. The shaving process can be performed after brushing without the danger of compacting the metal at the edges in a manner that will cause blistering during heat treating. The radii produced by brushing are large enough to provide a blended edge after shaving.

Several advantages are claimed for the use of a short-trim radial brush in deburring: (*a*) uniformly finished parts are produced throughout the life of the brush when it is used with suitable finishing equipment; (*b*) no fragmented metal or sharp edges are left on the part; and (*c*) metal is removed from the edge without changing the dimensions of machined surfaces because the brush is selective to edges.

In deburring, part and brush contact is extremely important. The brush should be set to brush across the edge as shown in Fig. 6(a). Line contact brushing or brushing parallel to the edge should be avoided, if possible. The brush face will not wear uniformly, and the flexible wire points will flare to the side, minimizing the effectiveness of the brushing operation. Line contact brushing is shown in Fig. 6(b).

When brushing gears, space the brushes to contact the tooth profile at the centerline of the gear as shown in Fig. 7(a). This setup allows the brushes to be reversed periodically to maintain the wire points at their maximum cutting efficiency. The setup for brushing spline bores differs in that the brushes are located off center, shown in Fig. 7(b). When helical gears are brushed, it may be necessary to favor the acute-angled portion of the gear tooth to develop a large radius prior to shaving, shown in Fig. 7(c).

Fig. 5 Short-trim dense wire-filled radial brushes

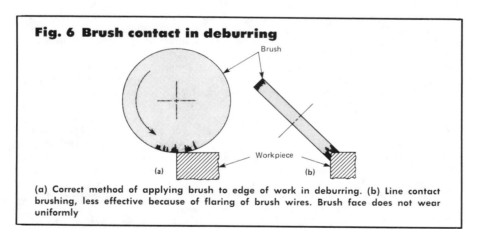

Fig. 6 Brush contact in deburring

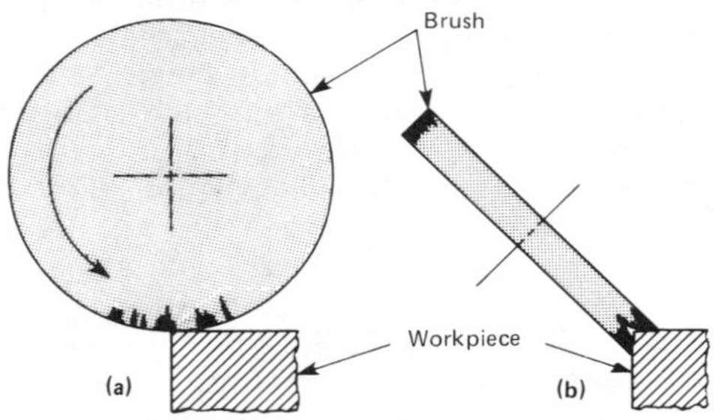

(a) Correct method of applying brush to edge of work in deburring. (b) Line contact brushing, less effective because of flaring of brush wires. Brush face does not wear uniformly

Cup-type brushes, shown in Fig. 2, are also useful for deburring and producing a radius around the periphery of holes, such as those on tube sheets used in heat exchangers. The action of a cup-type brush on the edge of a hole permits the formation of a uniform radius without elongating the hole. When radial brushes are used, elongation may occur.

Cup-type brushes are also used for removing burrs and producing a radius on the edge of internal gear teeth. A cup-type brush is selected for this work because a radial brush cannot contact the area as effectively. In addition, the radial brush, owing to its angle of approach, is restricted to an extremely small diameter. In a high-production operation, use as large a diameter brush as possible to minimize downtime for replacing brushes.

The surface speeds of cup brushes for deburring range from 23 to 33 m/s (4500 to 6500 sfm). Brush rotation should be alternated both clockwise and counterclockwise to obtain uniformity of the finished edges.

Elastomer-bonded wire-filled radial brushes, shown in Fig. 8, have characteristics similar to those of dense wire-filled brushes. Their filaments can move and still retain a certain cushion in the brush. These brushes can be used effectively for the following:

- Removing insulation from copper leads in the electrical industry. Speeds in the range of 20 to 30 m/s (4000 to 6000 sfm) are used for this application. Periodic dressing of the brush face with abrasive will increase brush life. Brush fill-wire may vary in diameter from 0.063 to 0.200 mm (0.0025 to 0.008 in.)
- Removing burrs from keyslots and other restricted areas where it is important for the brush to maintain a uniform face throughout its life
- Removing burrs from tubing and stampings. Speeds of 25 to 35 m/s (5000 to 7000 sfm) are used for this application. Brushes filled with 0.30-mm (0.0118-in.) diam wire are normally used for this operation

Small power-driven brushes, some with a working diameter of only 4 mm (5/32 in.), are made in numerous shapes to provide a variety of brushing capabilities. These brushes are capable of deburring intersecting holes, cleaning internal threads, removing heat treating and welding scale, cleaning molds, and removing insulation from electrical wire. In aircraft assembly applications, these brushes can be used for spot facing. Small radial wheel-type brushes are most effective in diameters of 38 to 75 mm (1½ to 3 in.). The fill materials can be either wire abrasive filament or nonmetallics such as tampico or synthetics. Wire diameters range from 0.063 to 0.036 mm (0.0025 to 0.014 in.), with most applications requiring wire sizes of 0.125 to 0.30 mm (0.005 to 0.0118 in.). Surface speeds of 13 to 25 m/s (2500 to 5000 sfm) are commonly used with these brushes.

Fig. 7 Optimum relations of the centerline of the brush to the centerline of the work in deburring

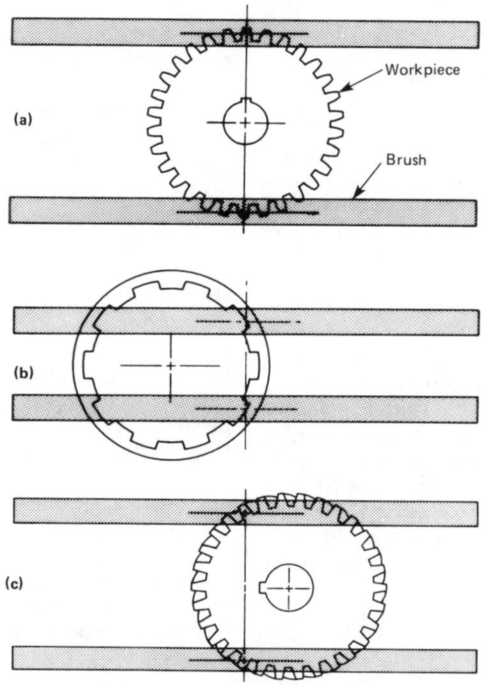

(a) Spur gears. (b) Spline bores. (c) Helical gears

Fig. 8 Elastomer-bonded wire-filled radial brushes

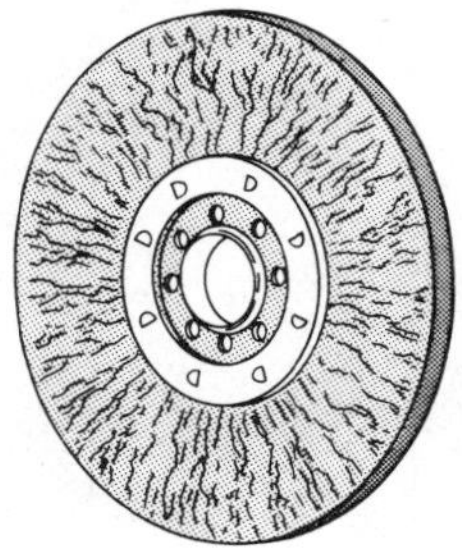

Fig. 9 Radial brush filled with treated tampico

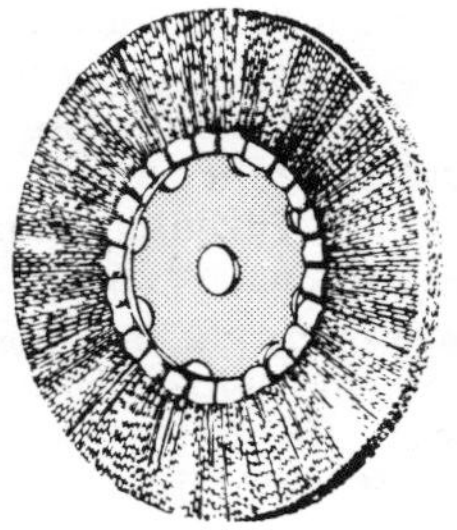

End brushes, circular end brushes, and side-and-bottom brushes are used for cleaning the walls of cavities without marking or scoring the bottom of the cavity. These brushes can be used for removing carbon from cylinders, cleaning valve seats, cleaning recesses in combustion chambers, and polishing dies and molds.

Side-action brushes are used to remove burrs from internal threads. These brushes can be made with a working diameter as small as 4 mm (5/32 in.) and can be mounted in special fixtures for cleaning the outside surface of copper and aluminum tubing to be soldered. Side-action brushes are operated at speeds of 1750 to 3400 rev/min. Reusable holders are available to hold these brushes securely and prevent the stems from being bent. With the holders, brushes can be reversed without damaging either the stem or fill material.

All small-diameter brushes are designed for use in areas where large brushes cannot contact the area to be brushed. For maximum efficiency, use the largest brush practical for a given job. Using rotational speeds over 6000 rev/min for tools 75 mm (3 in.) or less in diameter does not appreciably increase the cutting action.

Edge Blending

Radial brushes filled with treated tampico or cord (see Fig. 9) are used with a grease-base deburring or buffing compound to remove sharp edges, to blend tool and grinding marks, to eliminate areas of stress concentration, and to increase production finishing rates. This type of brush finishing is capable of improving a finish that originally measured 0.6 to 0.9 μm (24 to 35 μin.) to 0.1 to 0.2 μm (4 to 7 μin.).

Manufacturers of jet engines have reported that certain failures experienced in well-designed jet engine parts are the result of stress concentrations producing progressive fractures that occur at scratches, sharp edges, or burrs. With the repeated stresses imposed by engine operation, microscopic cracks are propagated, and the entire member will rupture. Data have been presented to show that a sharp corner or edge may reduce the endurance limit of a part by as much as 50%. A distinct V-notch, such as might be made with a sharp tool, can reduce the endurance limit of a part by as much as 60%. This knowledge has resulted in high standards for surface finishing jet engine parts. Power brush finishing methods, using treated tampico brushes with deburring compound, are widely used to produce radii and improve the microfinish. These brushes are operated at 28 to 33 m/s (5500 to 6500 sfm).

The effectiveness of power brush finishing jet engine parts can be seen in this example. Cams for aircraft magnetos were being polished by hand to a surface finish of 0.025 μm (1 μin.), using crocus cloth and oil. These parts had an operating life of 300 h. Subsequently, parts were brush finished with treated tampico and deburring compound as they came off a contour grinder with a 0.25-μm (10-μin.) finish. After brushing for 5 s, they had a 0.100-μm (4-μin.) finish. Tests showed no wear after 1000 h.

Fig. 10 Centerless wire-filled radial brush

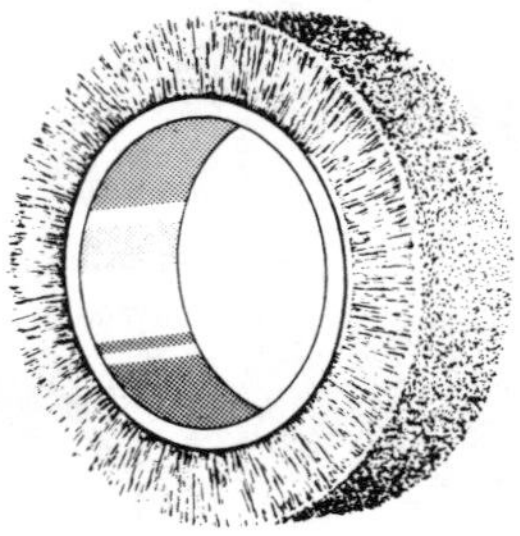

Finishing Flat Surfaces

In preparing a surface for plating, sharp peaks left on the surface by fixed abrasives must be removed to allow the surface to accept a uniform electrodeposit. Polishing marks, drawing marks, and scratches can be blended successfully with treated tampico and cord radial brushes as shown in Fig. 9.

For this application, brushes are usually operated at 38 m/s (7500 sfm). The brushes are used on the first heads of in-line or rotary finishing machines, and sisal or cloth buffs are used on the succeeding heads to produce a high-luster finish.

Treated cord brushes perform extremely well on stainless steel parts having a relatively flat surface. Cord-filled brushes are fast cutting, but less flexible than those filled with treated tampico, producing a better finish and higher luster. When used on stainless steel, this wheel also should be placed on the first station of an in-line or rotary machine, followed by sisal or cloth buffs.

Finishing Round Surfaces

The development of brushes for use on conventional centerless grinders has increased the number of applications for power-driven brushes. Fine wire-filled brushes used with conventional grinding coolants are being used for removing feathery grinding burrs and improving surface finish on such parts as rocker-arm shafts, piston pins, and control valves for automatic transmissions. In addition to removing feathery grinding burrs, these brushes also improve the finish from 0.18 to 0.10 μm (7 to 4 μin.), without altering dimensions. Centerless brushes will not remove metal from a cylindrical surface. The part must be ground to size prior to brushing.

Centerless brushes are large radial brushes filled with wire abrasive filaments 0.13 to 0.25 mm (0.005 to 0.010 in.) in diameter or with treated tampico, or treated cord. (See the example in Fig. 10.) Wire-filled centerless brushes are used for removing grinding burrs and for producing a minimum edge blend of 0.13 mm (0.005 in.). The brushes can be used dry, but should preferably be used with grinding coolant. A grease-base polishing compound is applied periodically to the face of brushes filled with tampico or cord to permit the brushes to produce finishes in the range of 0.05 to 0.18 μm (2 to 7 μin.).

Selection of Equipment for Driving Brushes

The use of power-driven brushes in high-production operations requires the selection of suitable finishing machines that can be integrated with other automated equipment in a line. These machines must have ample horsepower, proper-sized shafts, and good bearings. Methods for determining brushing pressure and flexibility to accommodate changes in operating techniques or in part design are also necessary considerations.

Standard machines have been designed and built to meet these requirements. Ammeters are installed in the drive-motor circuit to permit brushing pressure to be estimated. Reversing switches are used to control brush rotation clockwise or counterclockwise to maintain the maximum cutting efficiency of the wire points. Workpiece dwell time is changed to suit individual job requirements. Automatic and motorized brush-wear feed is an option that may be available.

Finishing machines for driving brushes should be selected on the basis of the following considerations:

- Compliance with all current safety standards and requirements
- Operation to be completed (deburring, radiusing, improving surface finish)
- Production rate
- Number of surfaces to contact
- Required brush type
- Required horsepower
- Integration of power brushing equipment with other equipment

Plating and Electropolishing

Copper Plating

By L. M. Weisenberger
Product Development Manager
Allied-Kelite Division
WITCO Chemical Corp.

COPPER electrodeposits are used widely as underplates in multiple-plate systems, as stop-offs for heat transfer, in electroforming, and in plating printed circuit boards. Although copper is relatively corrosion-resistant, it tarnishes and stains rapidly when exposed to the atmosphere. Electrodeposited copper is rarely used alone in applications where a durable and attractive surface is required. Bright electrodeposited copper is used as a protective underplate in multiplate systems or is protected against tarnishing and staining by an overcoating of clear lacquer when used as a decorative finish.

Copper can be electrodeposited from numerous electrolytes. Cyanide and pyrophosphate alkalines, plus sulfate and fluoborate acid baths are the primary electrolytes used in copper plating.

Alkaline Plating Baths

Pyrophosphate alkaline baths are used primarily to produce thick deposits, because they exhibit good plating rates. However, they must be carefully controlled. The cyanide alkaline baths can be easily controlled to produce thin deposits of relatively uniform thickness on all surfaces. They have the best macrothrowing power and are the most widely used baths.

Dilute cyanide and Rochelle cyanide baths are used to deposit a strike coating of 1.0 to 3.0 μm (0.05 to 0.1 mil) of copper before copper plating or electrodepositing other metals. The high-concentration Rochelle cyanide bath can be used efficiently for plating up to about 8 μm (0.3 mil) thickness. With a modification in composition, the Rochelle electrolyte may be used for barrel plating. The Rochelle cyanide can be used for still-tank plating, with mechanical agitation, or more efficiently with air agitation. These baths can also be used with current interruption or periodic-reverse plating.

Cyanide copper plating baths, characterized by low copper metal and high free-cyanide contents, clean the surface of parts during the plating operation. Although plating baths should not be used intentionally for cleaning purposes, the cleaning action of these cyanide baths can be an advantage, because difficult-to-clean parts can be given a copper strike in one of these baths with a high degree of success. Plating in other baths could cause poor adhesion and incomplete coverage.

High-Efficiency Sodium and Potassium Cyanide Baths. With proprietary additives, the high-concentration baths are used to produce deposits of various degrees of brightness and leveling, in thicknesses ranging from 8 to 50 μm (0.3 to 2.0 mils). Thick deposits that are ductile and bright can be produced in routine operations. Under most plating conditions, the high throwing power of the electrolyte produces adequate coverage of sufficient thickness in recessed areas. Antipitting additives are generally used in these baths to promote pore-free (nonpitted) deposits.

Before being plated in the high-efficiency baths, parts must first receive a strike coating of copper, about 1.3 μm (0.05 mil) thick from a dilute cyanide copper electrolyte.

The high-efficiency baths are characterized by relatively high operating temperature, high copper content, and rapid operation. Deposition rates are three to five times faster than the rates for the dilute cyanide and Rochelle cyanide baths. Parts to be plated in the high-efficiency electrolytes must be cleaned thoroughly, or the plate will be of inferior quality and the bath will require frequent purification for the removal of organic contaminants.

The potassium complexes formed by the combination of potassium cyanide and copper cyanide are more soluble than those formed when sodium cyanide is used; therefore, a higher metal content and higher rates of deposition are possible than with the sodium cyanide high-concentration bath. The potassium bath has more operating flexibility than the sodium bath and is favored because it raises the resistance to deposit burning and accordingly permits the use of higher current densities (faster plating rates).

Current interruption is used frequently for operating high-efficiency electrolytes to produce greater leveling and uniform distribution of copper on complex shapes and to reduce plating time and the amount of metal required for plating complex shapes to a specified minimum thickness. Periodic reversal may be used to provide even higher leveling and better metal distribution than can be obtained with current interruption. Periodic reversal

Table 1 Compositions and operating conditions of cyanide copper plating baths

Constituent or condition	Dilute cyanide	Standard barrel	Rochelle cyanide: Low concentration(a)	Rochelle cyanide: High concentration(a)	High-efficiency: Sodium cyanide(b)	High-efficiency: Potassium cyanide(b)
Bath composition, g/L (oz/gal)						
Copper cyanide	22 (3)	45 (6)	26 (4)	60 (8)	80 (11)	80 (11)
Sodium cyanide	33 (4)	68 (9)	35 (5)	80 (11)	105 (14)	105 (14)
Sodium carbonate	15 (2)	...	30 (4)	30 (4)	...	...
Sodium hydroxide	To pH	...	To pH	To pH	30 (4)	...
Rochelle salt	...	45-75 (6-10)	45 (6)	90 (12)	...	...
Potassium hydroxide	...	8-15 (1-2)	...	...	...	35 (5)
Bath analysis, g/L (oz/gal)						
Copper	16 (2)	32 (4)	18 (2)	43 (6)	56 (7)	56 (7)
Free cyanide	9 (1)	18 (2)	7 (0.8)	15 (2)	18 (2)	18 (2)
Operating conditions						
Temperature, °C (°F)	30-50 (86-120)	55-70 (130-160)	55-70 (130-160)(c)	60-75 (140-170)	60-75 (140-170)	60-75 (140-170)
Current density, A/dm^2 (A/ft^2)	1.0-1.5 (10-15)	...	1.0-4.0 (10-40)	2.0-5.0 (20-50)	2.0-6.0 (20-60)	2.0-6.0 (20-60)
Cathode efficiency, %	30-50	...	40-60	60-90	70-100	70-100
Voltage, V	6	6(d)	6	6	6	6
pH	12.0-12.6	...	12.0-12.6(c)	13	>13	>13
Anodes	Copper, steel	Copper	Copper	Copper	Copper	Copper

(a) Low concentration typical for strike; high concentration typical for plating. (b) Used with addition agents, as proprietary or patented processes. (c) For zinc-based die castings, maintain temperature at 60 to 71 °C (140 to 160 °F) and a pH between 11.6 and 12.3. (d) At 6 V, the bath draws approximately 0.3 A/L (2 A/gal) through the solution. At 12 V, the bath draws 0.4 A/L (3 A/gal)

Table 2 Concentration limits and operating conditions of copper pyrophosphate plating baths

Concentration limits, g/L (oz/gal)	
Copper	22-38 (3-5)
Pyrophosphate	150-250 (20-33)
Ammonia	1-3 (0.1-0.3)
Nitrate	6-12 (0.8-1.6)
Weight ratio, pyrophosphate to copper	7.0-8.5 (0.9-1.2)
Operating conditions	
Temperature	40-60 °C (105-140 °F)
Current density	1.0-7.0 A/dm^2 (10-70 A/ft^2)
Cathode efficiency	95-100%
Voltage at tank	2-5 V
pH, electrometric	8.0-8.8(a)
Anodes	Copper

(a) May be maintained with pyrophosphoric acid and potassium hydroxide

also improves the pore-filling characteristics of the high-efficiency electrolytes. Compositions and operating conditions of cyanide copper plating baths are given in Table 1.

The operation of high-efficiency electrolytes can be improved by the use of proprietary additives, which improve anodic and cathodic bath efficiency and anode corrosion. These additives produce matte to full-bright, fine-grained deposits. Proprietary additives are also used to control the effects of organic and inorganic contaminants.

Copper pyrophosphate baths are used for decorative multiplate applications, through-hole plating of printed circuit boards, and as a stop-off in selective case hardening of steels. Concentration limits and operating conditions for the pyrophosphate bath are given in Table 2.

Copper pyrophosphate bath characteristics are intermediate between cyanide and acid baths and are very similar to the high-efficiency cyanide bath. Electrode efficiencies are 100%, and throwing power and plating rates are good while the bath operates at almost neutral pH. Deposits from pyrophosphate baths are fine-grained and semibright. For pyrophosphate plating on steel, zinc die castings, magnesium, or aluminum, a preliminary strike should be used. For striking, a dilute cyanide or pyrophosphate copper, nickel, or other solution may be used.

Acid Plating Baths

Electrodeposition of copper from acid baths is used extensively for electroforming, electrorefining, manufacturing of copper powder, and decorative electroplating. Acid copper plating baths contain copper in the bivalent form and are more tolerant of ionic impurities than alkaline baths, but have less macrothrowing power and poorer metal distribution. Acid baths have excellent microthrowing power, which can be effective in sealing porous die castings.

A cyanide copper or nickel strike must be applied to steel or zinc-alloy die castings before they are plated in acid copper. A nickel strike prevents deposition of copper by immersion and precludes peeling plate. Concentration limits and operating conditions of acid copper plating baths are given in Table 3.

Copper Sulfate Bath. The copper sulfate bath is the most frequently used of the acid copper electrolytes and has its primary use in electroforming. It is also used extensively for the application of copper as an undercoating for bright nickel-chromium, as on automotive bumper bars, and for plating the cylinders used in rotogravure printing. By altering the composition of the copper sulfate bath, it can be used in through-hole plating of printed circuit boards where a deposit ratio of 1 to 1 in the hole to board surface is desired. With additives, the bath produces a bright deposit with good leveling characteristics or a semibright deposit that is easily buffed. Where copper is used as an undercoating, deposit thicknesses will generally range up to about 50 μm (2.0 mils). Similar thicknesses are utilized on printed circuit boards. In electroforming and on rotogravure printing rolls, thicknesses of 500 μm (20 mils) or more are not uncommon.

Copper Fluoborate Bath. The copper fluoborate bath produces high-speed plating and dense deposits to any required thickness, usually up to 500 μm (20 mils). This bath is simple to prepare, stable, and easy to control. Operating efficiency approaches 100%. Deposits are smooth and attractive. De-

Table 3 Compositions and operating conditions of acid copper plating baths

Constituent or condition	Copper sulfate bath: General	Copper sulfate bath: Printed circuit through-hole	Copper fluoborate bath: Low copper	Copper fluoborate bath: High copper
Bath composition, g/L (oz/gal)				
Copper sulfate, $CuSO_4 \cdot 5H_2O$	200-240 (27-32)	60-110 (8-15)	...	...
Sulfuric acid, H_2SO_4	45-75 (6-10)	180-260 (24-35)	...	...
Copper fluoborate, $Cu(BF_4)_2$	...	...	225 (30)	450 (60)
Fluoboric acid, HBF_4	...	...	To pH	40 (5)
Bath analysis, g/L (oz/gal)				
Copper	50-60 (7-8)	15-28 (2-4)	8 (1)	16 (2)
Sulfuric acid	45-75 (6-10)	180-260 (24-35)	...	...
Specific gravity at 25 °C (77 °F)	...	...	1.17-1.18	1.35-1.37
Operating conditions				
Temperature, °C (°F)	20-50 (68-120)	20-40 (68-105)	20-70 (68-160)	20-70 (68-160)
Current density, A/dm^2 (A/ft^2)	2.0-10.0 (20-100)	0.1-6.0 (1-6)	7.0-13.0 (70-130)	12-35 (120-350)
Cathode efficiency, %	95-100	95-100	95-100	95-100
Voltage, V	6	6	6	6-12
pH	...	...	0.8-1.7	<0.6
Anodes	Copper(a)	Copper(a)	Copper(b)	Copper(b)

(a) Phosphorized copper is recommended. (b) High-purity, oxygen-free, nonphosphorized copper is recommended

posits from the low copper bath operated at 49 °C (120 °F) are soft and are easily buffed to a high luster. The addition of molasses to either the high copper or the low copper bath operated at 49 °C (120 °F) results in deposits that are harder and stronger.

Good smoothness of coatings up to 500 μm (20 mils) thick can be obtained without addition agents. For thicknesses greater than 500 μm (20 mils), addition agents must be used to avoid excessive porosity.

Surface Preparation

Careful cleaning and preparation of the surface of the basis metal are important for the effective electrodeposition of copper. Among the cleaning methods used to prepare substrate surfaces for copper plating are soak or electrolytic alkaline cleaning, vapor degreasing, and solvent cleaning. More information about the techniques used in these processes is found in the articles on metal cleaning in this Volume. Specifications and practices for copper electroplating are given in Table 4.

Cyanide Baths. Although the dilute cyanide and the Rochelle cyanide baths exert a significant cleaning action on the surface of the parts during the plating operation, thorough cleaning of parts to be plated in these baths is still necessary.

The high-efficiency sodium cyanide and potassium cyanide electrolytes have virtually no surface-cleaning ability during plating because of the absence of hydrogen evolution. Parts to be plated in these electrolytes must be thoroughly cleaned. Parts also must receive first a dilute cyanide copper strike about 1.3 μm (0.05 mil) thick.

Pyrophosphate Bath. If pyrophosphate electrolytes are to be used, conventional cleaning cycles are generally satisfactory. A preliminary strike should be applied to steel, zinc-based die castings, magnesium, and aluminum. The strike solution may be a dilute cyanide copper, dilute pyrophosphate copper, or nickel. If a cyanide copper strike is used, adequate rinsing or, preferably, a mild acid dip following the strike is recommended before final pyrophosphate copper plating.

Acid Baths. When sulfate or fluoborate copper is to be deposited, steel or zinc must first receive a cyanide copper or nickel strike. With complete coverage, the strike may be as thin as 2 μm (0.08 mil). After the strike, the parts should be dipped in a dilute solution of sulfuric acid to neutralize solution retained from the alkaline strike bath. The parts should then be rinsed before acid copper plating. Nickel or nickel alloy parts, when surface activated by reverse-current etching in sulfuric acid, can be plated directly, provided contact is made to the work before immersion.

Table 4 Specifications and recommended uses for copper electroplating

Specification	Uses
Copper plating	
AMS 2418	Copper plating
MIL-C-14550 (Ord)	Copper plating
Copper plating in multiplate systems	
ASTM B456	Electrodeposited coatings of copper plus nickel plus chromium and nickel plus chromium
ASTM B200	Electrodeposited coatings of lead on steel
AMS 2412	Plating silver, copper strike, low bake
AMS 2413	Silver and rhodium plating
AMS 2420	Plating, aluminum for solderability, zincate process
AMS 2421	Plating, magnesium for solderability, zincate process
QQ-N-290	Nickel plating (electrodeposited)
Surface preparation	
ASTM A380	Descaling and cleaning stainless steel surfaces
ASTM B183	Preparation of low-carbon steel for electroplating
ASTM B242	Preparation of high-carbon steel for electroplating
ASTM B252	Preparation of zinc-based die castings for electroplating
ASTM B253	Preparation of and electroplating on aluminum alloys
ASTM B254	Preparation of and electroplating on stainless steel
ASTM B281	Preparation of copper and copper-based alloys for electroplating
ASTM B319	Preparation of lead and lead alloys for electroplating
ASTM B322	Cleaning metals before plating
ASTM B480	Preparation of magnesium and magnesium alloys for electroplating
ASTM B481	Preparation of titanium and titanium alloys for electroplating
MIL-HDBK-132 (Ord)	Military handbook, protective finishes

Table 5 Estimated time required for plating copper (valence 1) to a given thickness at 100% cathode efficiency

Cyanide baths contain copper with a valence of 1. For baths containing copper with a valence of 2, such as sulfate, pyrophosphate, and fluoborate baths, double the time values given in this table. Values must be corrected for losses in cathode efficiency by adding the difference between the actual cathode efficiency and 100%; for example, for 70% cathode efficiency, add 30% to values in table to determine estimated time

Thickness of plate		Plating time, min(a) at current density, A/dm^2 (A/ft^2)							
μm	mils	1.0 (10)	1.5 (15)	2.0 (20)	2.5 (25)	3.0 (30)	3.5 (35)	4.0 (40)	4.5 (45)
2	0.08	4	3	2	2	2	1	1	1
5	0.2	11	8	6	5	4	3	3	2
10	0.4	23	15	11	9	8	6	6	5
20	0.8	45	30	23	18	15	13	11	9
30	1.2	68	45	34	27	23	19	17	14
40	1.6	90	60	45	36	30	26	23	18
50	2.0	113	75	57	45	38	32	28	23
60	2.4	136	90	68	54	45	39	34	27
70	2.8	158	106	79	63	53	45	40	32
80	3.1	181	120	90	72	60	52	45	36

(a) To nearest whole value

Bath Composition and Operating Variables

The compositions and analyses given in Tables 1, 2, and 3 for cyanide, pyrophosphate, and acid copper plating baths may be varied within the control limits to satisfy requirements for specific applications.

Current density can be altered to effect more efficient control and to increase the deposition rate of copper. The data in Table 5 can be used as a guide to the selection of current density.

Impurities. The degree of control required to protect copper plating baths from impurities varies with the type of bath and the method of processing used. Known causes of roughness in copper deposits are (*a*) dragover from cleaners which results in the formation of insoluble silicates in the electrolyte, (*b*) poor anode corrosion, (*c*) insoluble metallic sulfides because of sulfide impurities, (*d*) organic matter in the water used for composition especially in rinse tanks, and (*e*) insoluble carbonates because of calcium and magnesium in hard water.

Purity of Water Used in Composition. The purity of the water used in the composition of the baths is important for all plating operations. Iron in the water causes roughness in the deposit, if the pH of the electrolyte is above 3.5 where iron can be precipitated. Chlorides, in concentrations greater than about 0.44 g/L (0.05 oz/gal), promote the formation of nodular deposits. Calcium, magnesium, and iron precipitate in the bath. Organic matter may cause pitting of deposits.

When plating in sodium or potassium high-efficiency electrolytes and distilled, deionized, softened, or good quality tap water may be used for solution composition and for replenishment. Tap water with high contents of calcium and/or iron should not be used, because it may cause roughness of the deposit. Softened water should be used with care, especially in plating baths where chloride contents are critical, such as bright copper sulfate baths.

Agitation during plating permits the use of higher current densities, which create rapid deposition of copper. The amount of increase permissible in current density varies for the different baths. Preferred methods of agitation for the types of baths are:

Cyanide baths... Cathode or air, or both
Pyrophosphate baths .... Air
Acid baths..... Cathode or air, or both

When air agitation is used, all air-line pipes should be made of inert material or coated with an inert material to prevent attack by the electrolytes. The air used for agitation must be clean to avoid bath contamination. Filtered air from a low-pressure blower is required.

Ultrasonic vibration also has been used for the agitation of copper plating baths. This method does not largely improve the properties or appearance of electroplates, but it can improve plating speed by permitting an increase in the current density without the hazard of burning the parts. Increased plating speed does not necessarily justify the increased cost and complexity of ultrasonic operation, because the high-speed baths can usually be operated with a fairly high current density at nearly 100% efficiency.

Plating in the Dilute Cyanide Bath

In the dilute cyanide bath, corrosion of the anodes increases with increasing concentration of free cyanide. Low free cyanide may cause rough deposits; however, excessive free cyanide lowers cathode efficiency, resulting in thinner deposits per unit time.

The dilute copper cyanide bath can be operated at room temperature, but the general practice is to operate the bath between 32 and 49 °C (90 and 120 °F) to increase the rate of deposition and to improve anode dissolution. This electrolyte is usually operated with a cathode current density of 1 to 1.5 A/dm^2 (10 to 15 A/ft^2). The tank voltage is normally between 4 and 6 V.

Agitation of the bath produces more uniform composition throughout the electrolyte, more uniform anode corrosion, and an increase in current densities where the brightest deposits are obtained. Current densities in excess of 5 A/dm^2 (50 A/ft^2) have been applied successfully by using air agitation of the solution and agitating the work.

Continuous filtration is preferred for dilute cyanide baths. Organic contamination or suspended matter in the strike is frequently responsible for roughness of copper plate subsequently deposited in the cyanide copper plating bath. Hexavalent chromium in the strike causes blistering of the deposit. Proprietary additives can be used to improve the bath operation, as well as aid in the control of organic and inorganic contaminants.

Plating in Rochelle Cyanide Baths

Rochelle electrolytes with lower metal concentrations can be used for striking with higher metal concentrations for plating. Rochelle salt aids anode corrosion and reduces carbonate formation. The Rochelle electrolyte can also be used for periodic-reverse plating with good results. Barrel plating with a Rochelle bath requires a different solution.

Rochelle baths usually are operated at a current density between 2 and 5 A/dm^2 (20 and 50 A/ft^2). Substituting potassium salts for sodium salts in the baths with higher metal concentration, up to 38 g/L (5 oz/gal) copper, can increase the allowable current density to 6 A/dm^2 (60 A/ft^2), but at the penalty of lowering cathode efficiency.

Fig. 1 Buffer curve for adjusting the pH of Rochelle electrolytes

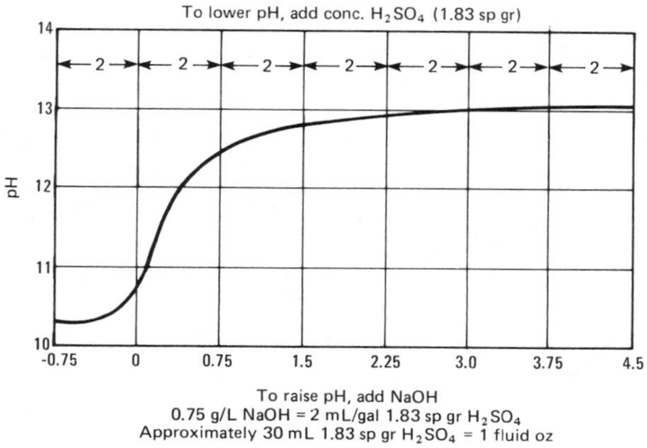

Source: *Modern Electroplating*, New York: Wiley, 1974, p 173

Fig. 2 Stress in thin copper plate deposited on stainless steel spirals

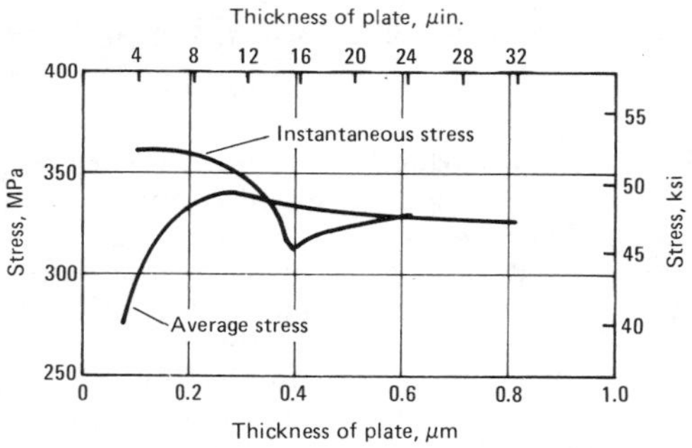

Stainless steel spirals are 0.127 mm (0.005 in.) thick. Based on *Journal of the Electrochemistry Society*, Nov 1961

The Rochelle baths are usually operated between 54 and 71 °C (130 and 160 °F) for best efficiency. The rate of deposition is higher at the higher temperatures. A high-efficiency electrolyte having a higher metal concentration can be operated up to 77 °C (170 °F). For copper plating of zinc-based die castings, the electrolyte is best operated at 60 to 71 °C (140 to 160 °F), provided the pH of the bath is maintained between 11.6 and 12.3. An increase in the operating temperature of Rochelle cyanide baths increases the efficiency of the anode and cathode; however, free cyanide decomposes more rapidly and carbonates form. Agitation and higher operating temperature cause an increase in anode efficiency, but increase carbonate formation.

Rochelle copper baths should be maintained at a pH between 12.2 and 13.0. Anode efficiency may be prohibitively low if the pH is too high. Raising the pH also decreases the voltage drop across the anode film. Figure 1 shows a buffer curve for adjusting the pH of Rochelle electrolytes.

Conductivity of the bath is improved by raising the free alkali cyanide and the concentration of the copper complexes. When depositing copper directly on steel, brass, or copper, conductivity can be improved by the addition of 2 to 15 g/L (¼ to 2 oz/gal) of sodium hydroxide. Sodium hydroxide must not be added if the electrolyte is used to deposit copper onto zinc-based die castings, aluminum, or magnesium.

Rochelle baths can become contaminated during plating of zinc-based die castings. Zinc contamination can be removed by electrolysis of the bath at room temperature, at the current density that produces the most brassy or off-color deposit, usually 0.2 to 0.3 A/dm^2 (2 to 3 A/ft^2). Iron cannot be removed readily from the bath and causes a reduction in current efficiency. Drag-in of chloride ion from acid dips must be kept very low to prevent iron buildup. Bipolarity of steel tanks or heat exchangers should be avoided.

The Rochelle bath is susceptible to organic contamination, which can be controlled by the use of wetting agents. Organic contaminants should be removed by periodic batch treatment of the electrolyte with activated carbon, followed by filtration. Organic contamination is especially high in barrel plating. A low-foaming, free-rinsing surfactant or a dispersing agent must be used in barrel plating baths to prevent organic contamination from adversely affecting the quality of the plated deposit. Continuous filtration of cyanide electrolytes is recommended to eliminate particulate matter or salts, which can result in rough deposits.

Increase in the current density or the presence of lead in the Rochelle cyanide bath causes an increase in the stresses of copper plate. These stresses can be reduced by increasing the concentration of copper in Rochelle baths. The addition of 15 g/L (2 oz/gal) of potassium thiocyanate, produces an expansion stress, instead of the usual contraction stress. Figure 2 shows stress in thin copper electrodeposits plated from a cyanide solution onto stainless steel.

Plating in High-Efficiency Sodium and Potassium Cyanide Baths

High-efficiency sodium and potassium cyanide baths operate at nearly 100% efficiency. The composition of the baths may vary by ±10% without introducing adverse plating characteristics. The sodium or potassium constituent improves the conductivity of the bath. Copper is present as cuprous ions.

Operation of the sodium cyanide and potassium cyanide electrolytes at 66 to 74 °C (150 to 165 °F) produces quality deposits. Temperatures in excess of 74 °C (165 °F) allow the use of higher current densities but breakdown of the cyanide becomes excessive at elevated temperatures. The anode current densities are limited by polarization, resulting in poor anode efficiency and higher voltage requirements. The cathode current densities are limited by burning of the deposit, resulting in reduced efficiency, loss of brightness, and roughness. These limits are higher in the potassium cyanide electrolyte.

Agitation of sodium cyanide and potassium cyanide high-efficiency baths is important for achieving maximum plating speed. Agitation can be accomplished by (*a*) solution movement, (*b*) cathode-rod movement, or (*c*) use of air. Each type of agitation improves the maximum allowable current densities, with air agitation providing the greatest improvement. All three types of agitation may be used within a single bath. Solution movement can be accomplished by mixing or by the flow of solution through filtration equipment. Cathode-rod movement of about 1 to 2 m/min (3 to 7 ft/min) allows increased plating rates. Gentle air agitation should be supplied by the use of a low-pressure blower which has a clean, filtered air source. Care must be taken to use clean, oil-free air for agitation to avoid contamination of the plating solution.

Filtration is also essential when operating high-efficiency cyanide copper electrolytes, especially for plating deposits thicker than 13 μm (0.5 mil). Filtration equipment should have the capability of one to two complete turnovers of the solution each hour while removing particulate matter from the electrolyte. Roughness of the copper deposits from particulate matter is often caused by faulty cleaning or by the formation of metallic copper or cuprous oxide particles at the anodes. Suspended dirt or solid matter in the cyanide copper electrolyte also causes surface roughness. Anode bags of proper size, material, weight, and weave are beneficial in retaining particulate matter formed at the anode. Other foreign particles introduced into the cyanide copper electrolyte are removed by the filtration equipment.

Carbonate buildup in high-efficiency copper cyanide baths can adversely affect the bath operation. High concentrations of carbonate reduce plating efficiency and speed. Excessive carbonates also affect the smoothness of the deposits. Carbonate contents of 120 to 150 g/L (16 to 20 oz/gal) or more may result in lower plating efficiency and plating speed. Excessive carbonates can also lower and reduce the acceptable plating range. These effects are more pronounced in a sodium cyanide bath than in a potassium cyanide bath.

Fig. 3 Cycle efficiency during periodic current reversal copper plating

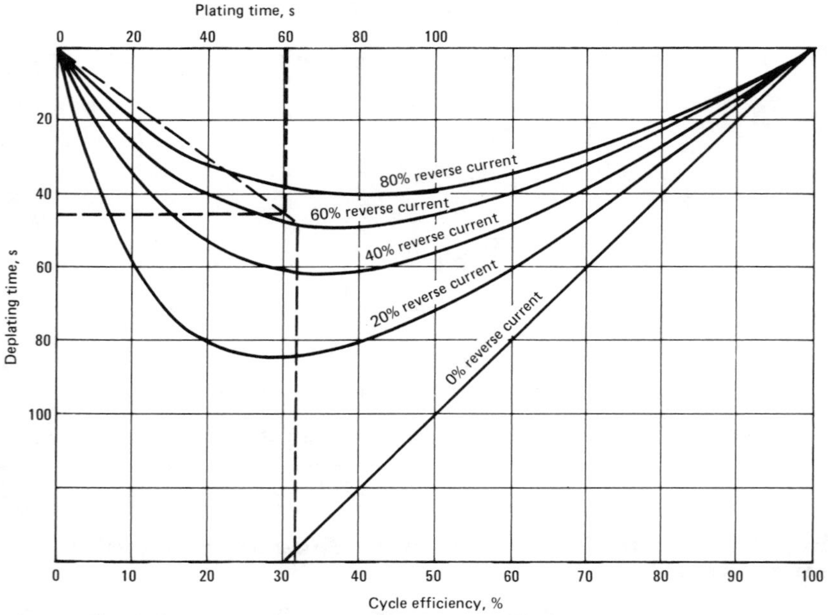

Source: *Electroplating Engineering Handbook*, New York: Reinhold, 1971, p 748

The primary source of carbonate formation is the breakdown of cyanide as a result of poor anode efficiency. Operating cyanide electrolytes at temperatures above the recommended levels can also result in carbonate formation. Operating temperatures above about 74 °C (165 °F) cause decomposition of the cyanide ion. Air containing high levels of carbon dioxide (CO_2) should not be used in air-agitated systems, because the carbon dioxide is dissolved by the alkaline plating solution, also forming carbonate. The air source for air-agitated systems should be placed where it provides a clean, fresh supply.

Excessive carbonates can be removed by freezing or precipitation with lime or proprietary additives. Sodium cyanide baths can be treated either by precipitation or freezing. Potassium cyanide baths can only be treated by precipitation. Freezing is not effective for potassium cyanide baths because of the high solubility of the carbonate salts.

Current interruption cycles frequently improve the operating range of high-efficiency sodium or potassium copper cyanide plating solutions. Current interruption cycles generally allow the use of higher current densities while maintaining bath efficiency. Current interruption cycles also improve the brightness of the copper deposits, and in some cases give excellent deposit brightness from bright plating baths which are so contaminated that acceptable deposits cannot be produced when using continuous direct current.

Current interruption cycles in the range of 8 to 15 s plating time followed by interrupting the current for 1 to 3 s are generally used. Plating times of less than 8 s and current interruptions of more than 3 s lower the net plating rate. Plating times of more than 15 s and current interruption of less than 1 s reduce the benefits obtained by using a current interruption cycle.

The use of periodic current reversal can also be used to great advantage in high-efficiency copper cyanide plating solutions. This technique involves plating parts in the conventional manner for a selected time and then deplating for a shorter period by reversing the current. Shorter periodic reversal cycles, such as 2 to 40 s of plating followed

Fig. 4 Thickness of copper deposit as a function of cycle efficiency and current density during periodic current reversal plating

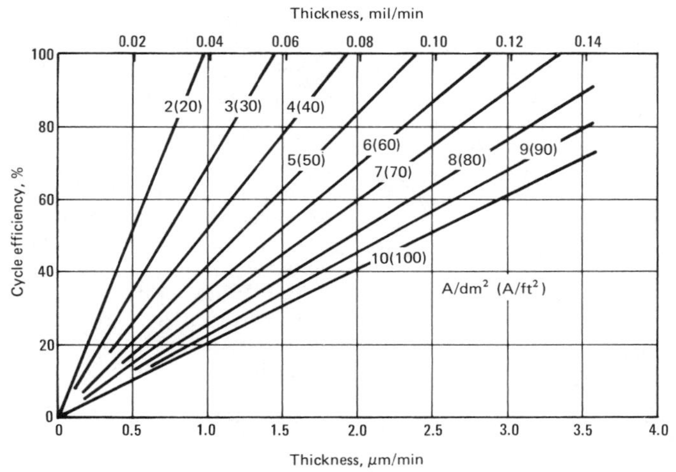

***Electroplating Engineering Handbook*, New York: Reinhold, 1971, p 750**

by 1 to 10 s of deplating (reversal) results in improved deposit brightness similar to that obtained with current interruption. A major advantage in using periodic reversal is the degree of leveling that can be achieved, particularly when relatively long reversal cycles are used. These longer cycles, in excess of 45 s direct with reverse cycles in excess of 10 s, can provide leveling characteristics in excess of 50%. The use of periodic reversal permits the use of higher plating and deplating current densities.

The leveling characteristics of the deposit are improved by increasing the reversal current, whereas cycle efficiency is increased by lowering the reversal current. Figure 3 shows cycle efficiency for periodic-reverse plating. Figure 4 indicates thickness of deposit as a function of cycle efficiency.

Other current interrupting schemes being used for this and other copper plating systems include pulse plating. This normally involves a pulse power source (rectifier) that produces square wave current pulses. Square wave conventionally suggests a pulse with a rise-and-fall time of approximately 10 to 85 μs and a standard frequency of 150 and 10 000 cycles. The periodic interruption of the current with proper time sequences allows much faster plating without surface burning, produces finer grain deposits, and increases throwing power and distribution.

Proprietary additives are used in high-efficiency copper cyanide baths to improve anode corrosion, to increase both anode and cathode efficiencies, and to control contamination. Organic and metallic additives are also used to improve deposit characteristics and brightness. These additives produce deposits ranging from matte to full-bright.

Plating in Pyrophosphate Baths

Copper pyrophosphate plating baths offer a number of desirable features. Copper pyrophosphate forms a highly soluble and conductive complex when dissolved in potassium pyrophosphate solution. Potassium salts are preferred because of their higher solubilities. Copper pyrophosphate plating baths operate at nearly 100% cathode efficiency and provide good throwing power. They are noncorrosive because the operating pH is near neutral. Concentration limits and operating conditions for copper pyrophosphate baths are given in Table 2.

Pyrophosphate forms a highly soluble complex with copper. Excess pyrophosphate is necessary to increase the conductivity of the bath and to effect proper corrosion of the anodes. Ammonia assists anode corrosion, helps enhance the luster of the deposit, and aids pH control. Nitrate allows the use of higher operating current densities by inhibiting the reduction of hydrogen at the upper end of the current density range. The pH of the pyrophosphate bath is maintained between 8.0 and 8.8. A high pH reduces anode efficiency, and a low pH reduces the throwing power of the solution and the stability of the complex compound in solution with the formation of orthophosphate. The pH of the bath can be lowered with pyrophosphoric acid and raised with potassium hydroxide. Good agitation is also essential for consistent operation. Air agitation provides for good performance of the anodes and cathodes and is preferred to cathode agitation.

Pyrophosphate electrolytes can be operated at current densities up to 7.0 A/dm^2 (70 A/ft^2) or higher. The operating current density can be increased by agitating the solution or by increasing the temperature of the bath. The anode current density should be maintained between 2 and 4 A/dm^2 (20 and 40 A/ft^2).

High bath temperatures should be avoided, because excessive formation of orthophosphate occurs. Orthophosphate formed by the hydrolysis of pyrophosphate is beneficial up to about 90 g/L (12 oz/gal), because it promotes anode corrosion and acts as a buffer. Above this concentration, conductivity and bright plating range are decreased and banded deposits are obtained. Orthophosphate cannot be removed chemically from the solution. The concentration can be reduced only by discarding the bath or diluting and rebuilding the pyrophosphate plating solution.

Copper pyrophosphate plating baths are susceptible to organic contamination including oil and excess or decomposed addition agents. These are removed by treatment with activated carbon and filtration. Cyanide and lead also contaminate the bath. Cyanide is removed by treatment with hydrogen peroxide and lead by electrolysis. Precautionary methods, such as proper cleaning, adequate rinsing, and good solution control and maintenance, prevent these contaminants from entering or building up in the bath, avoiding the need for frequent purification. Copper pyrophosphate solutions are tolerant of other metallic contamination.

Proprietary brighteners are available which refine the grain structure, impart leveling characteristics, and act as brighteners. However, decomposition products from an excessive additive concentration cause stress and brittle deposits. Thus, for quality deposits, additives should be added to the bath on an as-consumed basis.

Plating in Acid Sulfate Baths

Copper sulfate and sulfuric acid are the primary constituents of the copper sulfate electrolyte. The metal ions are furnished by the copper sulfate. Sulfuric acid increases solution conductivity and anode corrosion and helps avoid the formation of basic cuprous or cupric salts. With an efficiency of 95 to 100%, the copper sulfate bath is easy to operate and control.

If solution agitation or work movement is minimal, current densities should not exceed about 4.5 A/dm^2 (45 A/ft^2) because excessive anode polarization may occur and the deposits can be spongy. Where higher current densities are desired, such as for electrotypes or wire plating, air agitation is used. Air agitation is necessary to accelerate ionic diffusion and produce high-quality, fine-grained deposits where current densities are in excess of 10 A/dm^2 (100 A/ft^2).

The effect of temperature changes on grain structure and surface smoothness of deposits plated from the copper sulfate bath is less significant than the effect of changes in cathode current densities. An increase in temperature results in higher conductivity and reduced anode and cathode polarization. Increased temperature also reduces the tensile strength of deposits and increases grain size. Excessive temperatures should be avoided in copper sulfate baths where proprietary brightener formulations are used, because reduced plating ranges, excessive additive use, and solution contamination from additive breakdown result.

Care must be taken to avoid accelerated buildup of copper metal as in cases where dragout rates are low or improper anode to cathode ratios are maintained. An increase in the concentration of the copper sulfate increases the solution resistivity and slightly reduces the anode and cathode polarization. Copper sulfate concentrations in excess of 248 g/L (33 oz/gal) may result in salt crystallization in the plating solution. Normal bath composition is restored by discarding a portion of the bath and adding water and sulfuric acid.

Many copper sulfate plating solutions require the use of additives to produce smooth, fine-grained, bright, leveled, and ductile deposits. Most of the addition agents used in copper sulfate plating solutions are proprietary formulas. These proprietary additives are capable of producing the desired characteristics in the copper deposit, and deposit hardness can be increased where necessary.

In copper sulfate systems that produce bright deposits, a catalyst must be added to avoid streaky deposits, in addition to the primary constituents. This catalyst is chloride and is maintained between 0.02 to 0.1 g/L (0.003 to 0.01 oz/gal), or 20 to 100 ppm. The chloride is usually added as hydrochloric acid.

To improve the throwing power of some bright copper sulfate baths used for plating printed circuit boards, a low copper sulfate and high sulfuric acid electrolyte is used. The use of this electrolyte allows a near equal deposit distribution when plating the through-holes of the printed circuit board.

In sulfate electrolytes, impurities such as silver, gold, arsenic, and antimony can codeposit with copper. Arsenic and antimony cause copper deposits to be brittle and rough, and silver may cause roughness. Nickel and iron impurities reduce the conductivity of the plating bath. Lead impurities do not codeposit with copper; however, they precipitate in the electrolyte. Soluble silicates may precipitate onto the work. Organic contamination from decomposition products of addition agents, tank linings, and anode bags can cause brittle or discolored deposits. These organics can be removed from the electrolyte by treating it with activated carbon.

Plating in Fluoborate Baths

Copper fluoborate and fluoboric acid are the primary constituents of the copper fluoborate electrolyte. The metal ions are furnished by the copper fluoborate, which is more soluble than copper sulfate used in the sulfate bath, and the anode current density is not critical. Therefore, the metal-ion concentration in the fluoborate bath can be more than twice that in the copper sulfate solution, and this permits higher cathode current densities. The cupric salts in the fluoborate bath are highly ionized, except for small amounts of less ionized complex salts formed with certain addition agents.

In the copper fluoborate bath, the anode current density can be as high as 40 A/dm^2 (400 A/ft^2) without excessive anode polarization. The effect of temperature changes on grain structure and surface smoothness of deposits plated from the copper fluoborate bath is less significant than the effect of changes in cathode current density.

Agitation is preferred for the fluoborate bath, although acceptable deposits 25 μm (1 mil) thick have been produced in a high-concentration bath without agitation and with current density maintained at 35 A/dm^2 (350 A/ft^2). When agitation is used, a low-concentration bath operated at a current density of 4 to 5 A/dm^2 (40 to 50 A/ft^2) is preferred.

Although fluoborate baths containing no additives can produce dense and smooth deposits up to 500 μm (20 mils) thick, additives may be used to aid in the deposition of brighter or more uniform coatings or to assist in control of plating conditions. Although deposits from fluoborate baths are easily buffed to a high luster, brighteners of acetyl thiourea can be added to the electrolyte to produce bright coatings. The addition of free acid to the bath increases solution conductivity, reduces anode and cathode polarization, and prevents the precipitation of basic salts. Hard deposits and minimum edge effects result when molasses (1 mL/L or 0.1 fluid oz/gal) is added to the electrolyte. If the pH of these baths exceeds 1.7, deposits become dull, dark, and brittle.

The resistivity of fluoborate electrolytes is reduced if the concentration of fluoboric acid exceeds 15 g/L (2 oz/gal), or if the concentration of copper fluoborate exceeds 220 g/L (29 oz/gal). In the fluoborate bath, the metal-ion concentration can be more than double that in a copper sulfate solution containing 50 to 75 g/L (6.7 to 10 oz/gal) of sulfuric acid.

In the fluoborate electrolytes, silver, gold, arsenic, and antimony may codeposit with copper, but the effects of such impurities in this electrolyte have not been reported. Lead is the only metallic impurity known to interfere with the deposition of ductile copper deposits. Additions of sulfuric acid precipitate the lead. As with the sulfate electrolytes, organic impurities sometimes cause deposits to be brittle or discolored. They can be removed by treating the bath with activated carbon.

Table 6 Materials of construction for equipment basic to copper plating

Tank linings are of rubber or plastic(a)

Plating bath	Heating coils	Filters	Filter aids
Dilute cyanide	Low-carbon steel Teflon(b)	Low-carbon or stainless steel; cast iron	Diatomite Cellulose
Rochelle cyanide	Low-carbon steel Teflon(b)	Low-carbon or stainless steel; cast iron	Diatomite Cellulose
High-efficiency cyanide	Low-carbon steel Teflon(b)	Low-carbon or stainless steel; cast iron	Diatomite Cellulose
Pyrophosphate	Stainless steel Teflon(b)	Stainless steel Rubber- or vinyl-lined steel	Diatomite Cellulose
Acid copper sulfate	Titanium(c) Teflon(b)	Rubber- or vinyl-lined steel	Diatomite Cellulose
Fluoborate	Carbon(c) Teflon(b)	Rubber- or vinyl-lined steel	Diatomite Cellulose

(a) Of approved compositions; in the absence of data on bath contamination and effects on deposits, compatibility tests are required. (b) Dupont trademark. (c) Also for cooling coils, if bath is used below 32 °C (90 °F)

Table 7 Materials for anodes and racks for use in copper plating

Racks are made of copper(a)

Plating bath	Anodes
Dilute cyanide	Copper; steel
Rochelle cyanide	Copper(b)(c)(d)
High-efficiency cyanide	Copper(b)(c)(d)
Pyrophosphate	Copper(b)(c)(d)
Acid copper sulfate	Copper(e)
Fluoborate	Copper(d)

(a) Racks are generally coated with an inert plastic coating to prevent plating. (b) Cast copper, high purity. (c) Rolled copper, high purity. (d) Oxygen-free high purity copper. (e) Phosphorized copper

Wastewater Control and Treating

Increasing regulations governing discharge water have led to improved techniques for reducing the quantities of wastes which must be treated. These techniques have not only reduced the quantity of wastewater to be treated, but have also reduced the quantity of chemicals used and have lowered water consumption. These methods can be applied to any plating operation.

The use of counterflow rinses have reduced water consumption and wastewaters while maintaining adequate rinsing between plating operations. Reduced dragout of plating electrolytes can be accomplished by allowing processed parts leaving the plating solution to drain into the plating solution. Drip pans also reduce the amount of electrolyte dragout.

Closed-loop systems have dramatically reduced wastewater, lowered water consumption, and diminished chemical usage. Closed-loop systems allow recovery of rinse waters and chemicals by evaporative, reverse osmosis, or ion exchange recovery methods. Care must be exercised when using closed-loop systems, especially with copper plating, to minimize the impurities and contaminants from preplate operations entering the copper plating bath and being trapped by the closed-loop operation.

In any plating operation, there are wastewaters which must be treated to reduce the hazardous materials to meet regulations. The general procedures used for waste treating copper plating electrolytes and rinse waters resulting from copper plating systems are as follows:

- Cyanide-bearing solutions require oxidation of the cyanide with an oxidizing agent such as chlorine or hypochlorite, followed by precipitation of the heavy metals
- Pyrophosphate wastes require low pH hydrolysis to orthophosphate, followed by precipitation of the heavy metals
- Acid sulfate and fluoborate wastes are pH adjusted to precipitate the copper

Plating Equipment

Construction materials for equipment are indicated in Table 6. Construction materials for racks and anodes are given in Table 7.

Tanks. For cyanide copper solutions, low-carbon steel tanks are suitable. Polypropylene tanks with adequate reinforcing may also be used, provided the operating temperature is not excessive. Low-carbon steel tanks should be lined with rubber, polyvinylchloride, or another synthetic material which is not susceptible to attack by the cyanide plating solution to prevent bipolar effects, which may rob current from significant areas of the work. Tanks for copper pyrophosphate, acid copper sulfate, and copper fluoborate solutions should be of similar construction. Low-carbon steel tanks used for these solutions must be lined with the above materials to prevent the solutions from attacking the low-carbon steel, resulting in short tank life and immersion deposits. New tanks, as well as all other equipment coming in contact with the plating solution, should be leached before use to remove any materials which may leach into the plating solution and cause poor quality deposits. Leaching solutions should be similar to the plating solution to be used, such as a 15 to 30 g/L (2 to 4 oz/gal) caustic solution for copper cyanide equipment or a 5 to 10% sulfuric acid solution for acid copper sulfate.

Barrels. High-speed copper plating solutions for barrel plating are being used in product operations. Polypropylene barrels have been used successfully for prolonged periods.

Anodes. The types of copper anodes used in each of the copper plating solutions are indicated in Table 7. High-purity copper anodes are recommended. Anodes with a lesser purity may form heavy sludges during electrolysis and contribute appreciably to roughness of the deposit. Copper anodes are available in many forms, such as bars, balls, or chips. Bars are suspended from the anode bar. Balls or chips are placed in titanium baskets. The anode area in a copper plating solution should be controlled and maintained. If the anode area is not maintained, it decreases as the copper is dissolved and the anode current density rises, resulting in increased polarization and formation of undesirable films. These films can restrict current flow or sluff from the anode and cause roughness in the plating solution.

Anode Bags. Bags made of cotton, Dynel, or polypropylene are used in copper plating solutions. Cotton bags are preferred for cyanide copper solutions and Dynel or polypropylene are used in the acid copper solutions. Bags are used to prevent fine particles formed at the anode from migrating to the cathode and resulting in roughness.

The weave and weight of the anode bag are most important. The bag material must be capable of retaining the particles formed at the anode and at the same time allow the plating solution to flow freely around the anode. Anode bags are not generally used in pyrophosphate baths, because they interfere with dissolution of the anode by decreasing the circulation of the solution around the anode.

Fig. 5 Copper required for covering an area with a specific thickness

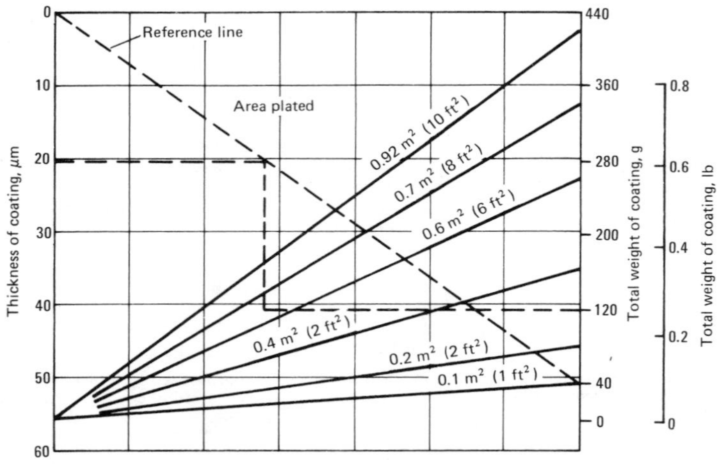

Plate Characteristics

Variations in processing during surface preparation or during plating have significant effects on the quality of the copper electrodeposit. Certain variations can adversely affect the adhesion of copper to the substrate metal. Variations also can affect brightness, porosity, blistering, roughness, hardness, solderability, and leveling.

Brightness. Bright copper coatings are generally obtained by the addition of brighteners to the electrolyte, although buffing of the electrodeposited coating provides a high luster. Plating from high-concentration cyanide baths with current interruption or periodic reversal of current also improves the luster of the copper coating.

Buffing or electropolishing the work before plating it in an electrolyte not containing a brightener results in the deposition of a smooth and sometimes semibright coating. If an electrolyte containing a brightener is used, the luster of the coating is enhanced. The high cost of labor is a primary concern when buffing is considered as the method of brightening the coatings. Plating from high-efficiency cyanide baths with current interruption or periodic reversal of current also improves the luster of the deposits. Improved casting techniques and mechanical finishing before plating can improve the quality of the copper deposit.

Adhesion. The type of substrate surface and proper preparation of the surface before plating are important for good adhesion. In general, cast and other porous surfaces are less receptive to good-quality electrodeposited coatings than wrought surfaces.

The kind of material to be electroplated with copper is another important consideration. For magnesium-based or aluminum-based die castings, the zincate layer between the substrate and the copper deposit is a critical control factor. For a properly activated stainless steel surface, a controlling factor for ensured adhesion of copper is the speed with which the workpiece is immersed in the bath. Some brighteners, especially organic brighteners, may adversely affect adhesion of subsequent electrodeposited coatings. Adhesion of copper electrodeposits from acid baths can be ensured only if a strike from a cyanide copper bath precedes copper plating.

Porosity. The degree of porosity in a copper coating can be controlled by the kind of copper plating bath selected, the composition and control of the electrolyte, the basis material to be plated, and the condition of the surface to be plated. The degree of porosity on the surface of the metal to be plated also dictates the techniques needed to minimize porosity in the coating. A porous surface has high surface area and requires high current density for efficient plating.

Blistering of copper plate, particularly when the plated work is subjected to heat, occurs mostly on zinc-based die castings. Blistering can also occur on parts made of magnesium or aluminum in any form as a result of poor quality of castings, poor surface preparation, or both. Blistering of copper plate on zinc-based die castings plated in a cyanide strike electrolyte and then subjected to heat can be reduced by lowering the pH of the cyanide strike bath from the range of 12.0 to 12.6 to about 10. Caution must be used because operation at a pH value this low may result in the release of poisonous hydrogen cyanide gas. It is imperative that the plating bath be thoroughly vented.

Blistering of copper-plated magnesium and aluminum, especially during subsequent soldering or heating in service, is caused by poor adhesion at the zincate-copper interface. Unfortunately, blistering often does not become evident until subsequent electrodeposits have been applied and the coating has been subjected to heat. Exposing all copper-plated magnesium and aluminum parts to controlled heat representative of that to be subsequently encountered is good practice. This causes blistering before deposition of subsequent metal coatings if there is poor adhesion at the interface.

Roughness in copper deposits is often caused by foreign particles present in the bath as the result of faulty cleaning, or by the migration to the cathode of metallic copper or cuprous oxide particles that form at the anode. Such roughness is especially likely to occur with the sodium cyanide high-concentration electrolytes and can be prevented by using anode bags.

Solderability of the coating is good when (*a*) the copper surface is free of oxide, (*b*) the coating is thick enough, and (*c*) the adhesion of the copper plate is superior. Direct soldering of electrodeposited copper is not unusual for parts that are subsequently contained in hermetically sealed units.

Soldering is a routine operation for aluminum and magnesium electronic parts used in aerospace applications. A copper strike and copper plate frequently comprise the initial metal coating over the zincated surfaces of these parts, after which electrodeposits of other metals are applied before soldering. A top coat of tin, or of cadmium plate that has been chromate conversion coated, is a particularly effective means of producing a good combination of solderability and corrosion resistance for parts exposed to the atmosphere.

Hardness. Without the use of addition agents, cyanide electrolytes produce harder coatings than acid baths. With the use of addition agents, the hardness of copper deposits from any electrolyte can be increased. Hardness of the electrodeposit is generally associated with fine grain, but hardness can be increased by introducing preferred crystal orientation in the absence of grain refinement. Changes in the copper sulfate or sulfuric acid concentration of acid baths have little effect on the hardness of copper plate.

Leveling has a significant effect on the appearance of the copper coating, as well as on the appearance of the final product when other metals are subsequently plated over the copper. Often, the substrate metal does not have the degree of smoothness that is desired of the plated surface. Metal substrate surfaces can be mechanically or chemically worked to reduce surface roughness before electroplating; however, some copper electrolytes can produce substantial leveling in the deposited coating, thus reducing cost related to elaborate prepolishing or other means of smoothing the surface. The high-concentration potassium cyanide electrolytes produce excellent leveling when certain addition agents are added and interrupted current or periodic reversal is used during plating. Although somewhat less effective, high-concentration sodium cyanide baths, mixed sodium and potassium electrolytes, and Rochelle cyanide electrolytes also have good leveling characteristics with interrupted or periodically reversed current.

Copper in Multiple-Plate Systems

Electrodeposited copper is widely used as a basis for subsequent plated coatings in multiple-plate systems. The use of copper plate in copper-nickel-chromium systems is discussed in the article on decorative chromium plating in this Volume.

Cost

The cost of copper plating is influenced largely by the type of installation. In a modern, automated, multiple-phase shop, brighteners and wetting agents probably are the greatest cost factor. In a still-tank operation, the cost of labor is of major importance. An increase in current density reduces cost, because of the reduction in time required to deposit a given thickness of coating.

For a routine plating operation, the cost of the copper deposited can be estimated with the aid of Fig. 5. For example, Fig. 5 shows that 120 g (4 oz) of copper is required for plating an area of 0.7 m^2 (7 ft^2) with a coating 20 μm (0.8 mil) thick. The cost of the copper coating is obtained by multiplying the weight of copper required by the cost of copper anodes. For areas larger than 2.4 m^2 (10 ft^2), multiply by the proper factor.

Hard Chromium Plating

By Hyman Chessin
Senior Research Associate
Plating Division
M & T Chemicals Inc.
and
Everett H. Fernald, Jr.
President
Induplate, Inc.

HARD CHROMIUM PLATING is produced by electrodeposition from a solution containing chromic acid (CrO_3) and a catalytic anion in proper proportion. The metal so produced is extremely hard and corrosion resistant.

The process is used for rebuilding mismachined or worn parts, for automotive valve stems, piston rings, shock rods, MacPherson struts, the bores of diesel and aircraft cylinders, and for hydraulic shafts.

Hard chromium plating is also known as industrial, functional, or engineering chromium plating and differs from decorative chromium plating in the following ways:

- Hard chromium deposits are intended primarily to increase service life of functional parts by increasing their resistance to (*a*) wear, (*b*) abrasion, (*c*) heat, or (*d*) corrosion. Deposits are also applied to restore dimensions of undersized parts
- Hard chromium normally is deposited to thicknesses ranging from 2.5 to 500 μm (0.1 to 20 mils) and for certain applications to considerably greater thicknesses, whereas decorative coatings seldom exceed 1.3 μm (0.05 mil)
- With certain exceptions, hard chromium is applied directly to the basis metal; decorative chromium is applied over undercoats of nickel or of copper and nickel, and is either buffed or used in the as-plated condition

Principal Uses

The major uses of hard chromium plating are (*a*) wear-resistance applications, (*b*) improvement of tool performance and tool life, and (*c*) part salvage. Table 1 lists parts to which hard chromium plate is applied and presents data regarding plate thickness and plating times. Plating times can be reduced by using high speed or mixed catalyst baths.

Wear Resistance. Extensive performance data indicate the effectiveness of chromium plate in reducing the wear of piston rings because of scuffing and abrasion. The average life of a chromium-plated ring is approximately five times that of an unplated ring made of the same basis metal. Piston rings for most engines have a chromium plate thickness of 100 to 175 μm (4 to 7 mils) on the wearing face, although thicknesses up to 250 μm (10 mils) are specified for some heavy-duty engines.

In the automotive industry, hard chromium is also applied to shock absorber rods and struts to increase their resistance to wear and corrosion. Valve stems are plated with a flash coating (about 0.25 μm or 0.01 mil) to reduce wear. Hydraulic shafts for all kinds of equipment are plated with 2 to 3 μm (0.08 to 0.12 mil) of hard chromium to increase service life.

Tooling Applications. Various types of tools are plated with chromium for one or more of the following reasons: (*a*) to minimize wear, (*b*) to prevent seizing and galling, (*c*) to reduce friction, and (*d*) to prevent or minimize corrosion. Steel or beryllium copper dies for molding of plastics are usually plated with chromium, especially when vinyl or other corrosive plastic mate-

Table 1 Applications of hard chromium plating

Part	Base metal	Thickness of plate μm	mils	Plating time
Computer printer type	Carbon steel	25	1	60 min
Face seals	Steel or copper	75-180	3-7	10 h
Aircraft engine parts	Nickel-based alloys, high strength steel	75-180	3-7	10 h
Plastic molds	Tool steel	5-13	0.2-0.5	30 min
Textile guides	Steel	5-100	0.2-4	20-240 min
Piston rings	Steel or cast iron	150-255	6-10	8 h
Balls for ball valves	Brass or steel	7.5-13	0.3-0.5	20 min
Micrometers	Steel	7.5-13	0.3-0.5	20 min
Golf ball molds	Brass or steel	7.5-25	0.3-1	20-60 min
Lock cases	Brass	5-7.5	0.2-0.3	20 min
Cylinder	Cast iron	255	10	300 min
Bushing	1018 carburized, 56 HRC	25	1	45 min
Crankshafts	Steel	255-3800	10-150	...
Cutting tools	Tool steel	1.3	0.05	5 min
Forming and drawing dies	Steel	25	1	60 min
Gage	Steel	125	5	150 min
Gun barrels, 30 caliber(a)	Steel	25	1	40 min
Hydraulic cylinder	1045 steel	13	0.5	40 min
Pin	Steel	13	0.5	30 min
Pin	1045 steel, 60 HRC	125	5	40 min
Plug gage	1040 steel, 55 HRC	125	5	150 min
Relief-valve plunger	1113 steel, soft	100	4	60 min
Ring gage	Steel	205	8	240 min
Rolls	Steel	13-255	0.5-10	20-300 min

(a) M-16 rifle, barrel and chamber

rials are to be molded. Plating thicknesses of 2.5 to 125 μm (0.1 to 5 mils) usually are recommended for preventing wear in parts sticking in molds and for reducing frequency of polishing when plastics that attack steel or beryllium copper are being molded. Chromium-plated dies should not be used when plastics containing fire-retardant chlorides are molded.

Service life of plug gages and other types of gages may be prolonged by hard chromium plating. Most gage manufacturers provide chromium-plated gages.

Records in one plant indicated that plug gages made from hardened O1 tool steel wore 0.0025 mm (0.0001 in.) after gaging 5000 cast iron parts. Hard chromium plating of these gages allowed the gaging of 40 000 parts per 0.0025 mm (0.0001 in.) of wear.

Worn gages can be salvaged by being built up with hard chromium plate. Also, chromium plate provides steel gages with good protection against rusting in normal exposure and handling. Chromium plating is not recommended, however, for gages that are subjected to impact at exposed edges during operation.

Deep drawing tools often are plated with chromium, in thicknesses up to 100 μm (4 mils), for improvement of tool performance or building up of worn areas, or for both reasons. The life of draw rings and punches may be prolonged by plating. In addition, plating reduces frictional force on punches and facilitates removal of workpieces from punches in instances where sticking is encountered with plain steel surfaces. If deep drawing tools are chromium plated, the basis metal should be harder than 50 HRC. Steel dies used for drawing bars and tubes are often plated with relatively heavy thicknesses (up to 250 μm or 10 mils) of chromium to (*a*) minimize die wear, (*b*) reduce friction, and (*c*) prevent seizing and galling.

The service life of cutting tools is often extended by chromium plate, in thicknesses ranging from less than 2.5 to 13 μm (0.1 to 0.5 mil). Taps and reamers are examples of tools on which chromium plate has proved advantageous. A flash plate on ½-in. −20 taps used to thread cold worked 1010 steel improved tap life from 250 (for unplated taps) to 6000 parts per tap. The poor tool life of the unplated taps has been caused by buildup of metal on the cutting edges. Hard chromium plating is not recommended for cold extrusion tools for severe applications where extreme heat and pressure are generated, because the plate is likely to crack and spall and may be incompatible with phosphate-soap lubricants.

Part Salvage. Hard chromium plating is sometimes used for restoring mismachined or worn surfaces. Since 1970, the use of this process for part salvage has been frequently replaced by thermal spraying and plasma coatings, which can be applied more quickly. The fact that a chromium deposit can significantly reduce fatigue strength must be considered in determining whether or not chromium plating can be safely used.

Hard chromium plating is used to restore to original dimensions the worn surfaces of large crankshafts for diesel and gas engines and for compressors. In these applications, in which coating thickness usually ranges from 125 to 1250 μm (5 to 50 mils), the excellent wearing qualities and low coefficient of friction of chromium are highly advantageous. The plate is prevented from depositing in fillet areas as a precaution against fatigue failure.

The extremely close dimensional tolerances specified for components of compressors for jet aircraft engines are not always correctly met in machining. Hard chromium plating is sometimes used to salvage mismachined parts. Most frequently mismachined are the diameters of rotor disks and spacers. Maximum thickness of plate on these components, which are made of 4130 and 4340 steels, generally does not exceed 380 μm (15 mils).

Other Applications. Hard chromium plate is applied to printing plates and stereotypes, especially to those intended for long runs, because it (*a*) wipes cleaner, (*b*) provides sharper reproduction than other coatings, and (*c*) increases length of press run. It is used on press rams because of its excellent resistance to corrosion, seizing, galling, and other forms of wear.

Selection Factors

The decision to use hard chromium plating on a specific part should be based on the following considerations:

- Inherent hardness and wear resistance of electrodeposited chromium
- Thickness of chromium required
- Shape, size and construction of part to be plated, and kind of metal of which it is made
- Masking requirements for parts that are to be selectively plated
- Dimensional requirements (that is, whether or not mechanical finishing is required and can be accomplished in accordance with desired tolerances)

Hardness of chromium electrodeposits is a function of plating conditions. In general, chromium plated in the

Table 2 Sulfate baths for hard chromium plating

Type of bath	Chromic acid(a) g/L	Chromic acid(a) oz/gal	Sulfate(a) g/L	Sulfate(a) oz/gal	Current density A/dm^2	Current density $A/in.^2$	Bath temperature °C	Bath temperature °F
Low concentration	250	33	2.5	0.33	31-62	2-4	52-63	125-145
High concentration....	400	53	4.0	0.53	16-54	1-3.5	43-63	110-145

(a) Concentration usually can deviate ±10% without creating problems. It is recommended that adjustments be such that the concentrations listed above lie in the middle of the range permitted. For example, chromic acid can fluctuate by ±23 g/L (±3 oz/gal); therefore, the concentration range should be 225-270 g/L (30 to 36 oz/gal), rather than 205-250 g/L (27 to 33 oz/gal)

bright range is optimally hard. Bright chromium deposits from conventional plating baths have Vickers diamond pyramid hardnesses of 900 to 1000; those from mixed catalyst baths have hardnesses of 1000 to 1100 or higher.

Size. Frequently, a very large part can be plated in sections or can be rotated so that only a portion of the part is immersed in the plating bath at any given time. The latter method has been used to plate large cylinders up to 4 m (12 ft) in diameter and up to 18 m (60 ft) long. When this technique is used, all of the surface to be plated that is exposed to the atmosphere must remain wet with plating solution.

Journal surfaces of the largest diesel crankshafts and bores of large naval guns and M-16 rifles have been hard chromium plated on a production basis. Internal surfaces of very large cylinders have been plated by closing off the ends of each cylinder and retaining the plating solution within it.

Basis Metal. Most hard chromium deposits are applied to parts made of ferrous alloys; however, numerous aerospace applications require the chromium plating of aluminum and nickel base alloys. From the standpoint of processing, hard chromium plate may be applied to steels regardless of their surface hardness or chemical composition, provided the basis metal is hard enough to support the chromium layer in service. Similarly, cast irons can be plated provided the surface is capable of conducting the required current and is reasonably free of (*a*) voids, (*b*) pits, (*c*) gross silicate inclusions, (*d*) massive segregation, (*e*) slivers, and (*f*) feather edges.

Plating Baths

Chromic acid is the source of metal in hard chromium-plating baths. However, a chromic acid solution does not deposit chromium unless a definite amount of catalyst is present. If there is either too much or too little catalyst, no chromium metal is deposited. Catalysts that have proved successful are acid anions, the first of which to be used was sulfate. Substitution of fluoride ions present in complex acid radicals for a portion of the sulfate marginally improves the chromium-plating operation. Reflecting this difference in catalyst, the principal types of baths are designated as conventional sulfate and mixed catalyst.

The fatigue limit of high-strength steel (1240 MPa or 180 ksi ultimate tensile strength and above) may be reduced by about 50% when chromium is plated in either a conventional sulfate or a mixed catalyst bath, because of the inherent crack structure of hard chromium. However, special techniques using shot peening and postplate heat treatments may be used to retain most of the original fatigue strength.

Conventional Sulfate Baths

Composition of conventional chromic acid baths catalyzed by sulfate can vary widely, provided the ratio by weight of chromic acid to sulfate radical is within a range between 75 to 1 and 120 to 1. Throwing power, or distribution of plate, is optimum at ratios between 90 to 1 and 110 to 1; however, in the range from 75 to 1 to 90 to 1, brighter deposits are obtained, less burning occurs, and a higher current density can be used.

Solutions containing chromic acid in a concentration as low as 50 g/L (7 oz/gal) have been reported but are not practical for production for several reasons: (*a*) plating range is too limited, (*b*) such solutions are more sensitive to contamination, (*c*) they have a higher electrical resistance, and (*d*) they require a higher voltage for operation. Compositions and operating conditions for two chromic acid-sulfate baths (low and high concentrations) for hard chromium plating are given in Table 2.

The low-concentration bath is widely used for hard chromium plating because (*a*) it is capable of plating faster than the high-concentration bath and (*b*) it minimizes dragout losses. The high-concentration bath has the advantage of being less sensitive to concentration changes; it is also easier to control and has better throwing power. Because the high-concentration bath is more conductive, it can be operated at lower voltages and heats up less in operation.

If available power is limited and maximum available voltage does not exceed 6 V, it may be preferable to operate a solution with higher conductivity in the range of 300 to 405 g/L (40 to 54 oz/gal) of chromic acid, even though current efficiency is lower than with the less concentrated solutions. The lower current efficiency is offset by increased current density obtained because of better conductivity. Usually, the best range of operation is from 195 to 300 g/L (26 to 40 oz/gal) of chromic acid. Even within this range, the solution containing 195 g/L (26 oz/gal) has the highest current efficiency, while the solution containing 300 g/L (40 oz/gal) has the best conductivity.

Mixed Catalyst Baths

The mixed catalyst solutions are similar to conventional sulfate baths in conductivity but produce harder deposits and have higher intrinsic current efficiency than conventional baths under identical conditions, as well as a higher current efficiency at higher concentration. Mixed catalyst baths can increase production rate 40 to 60% over that obtainable with conventional baths, because of their greater current efficiency and ability to operate at higher current densities without creating harmful effects on the deposit. In addition, these baths can plate thick deposits without building up heavy nodules and trees and give harder deposits. One limitation of mixed catalyst baths is that they may cause etching of unplated surfaces at areas of low current density. This etching can be prevented on significant surfaces by masking areas not to be plated.

Solution Control

Chromium-plating baths of all types must be subjected to periodic chemical analyses for control of bath composition. Solution control can be simplified if a record is kept of the way a particular chromium bath changes in composition during its use. Change in composition depends on (*a*) number

Table 3 Determination of chromic acid in chromium baths with a Baumé hydrometer

Direct conversion of degrees Baumé to ounces of chromic acid per gallon, at 25 °C (77 °F)

°Bé	Chromic acid g/L	Chromic acid oz/gal
10.5	113	15.0
11.0	119	15.8
11.5	124	16.5
12.0	130	17.3
12.5	137	18.2
13.0	144	19.1
13.5	149	19.8
14.0	153	20.4
14.5	159	21.2
15.0	165	22.0
15.5	172	22.9
16.0	178	23.7
16.5	184	24.5
17.0	191	25.4
17.5	198	26.3
18.0	204	27.2
18.5	211	28.1
19.0	218	29.0
19.5	224	29.8
20.0	230	30.6
20.5	237	31.5
21.0	244	32.4
21.5	250	33.3
22.0	257	34.2
22.5	264	35.1
23.0	271	36.0
23.5	279	37.1
24.0	287	38.2
24.5	294	39.1
25.0	301	40.0
25.5	308	40.9
26.0	315	41.9
26.5	323	42.9
27.0	331	44.0
27.5	338	45.0
28.0	346	46.0
28.5	354	47.1
29.0	362	48.2
29.5	370	49.2
30.0	378	50.2
30.5	387	51.5
31.0	399	53.0
31.5	406	54.0
32.0	415	55.2

Note: Impurities will cause the Baumé hydrometer reading to be higher than is warranted by the actual content of chromic acid

of ampere-hours of current passed through the bath, (*b*) dragout of solution, and (*c*) spray losses. Also, evaporation losses usually cause the concentration to change by about 2 to 5% during a day's operation. Change in bath composition is sometimes proportional to the change in chromic acid content. A complete solution analysis should be made at periodic intervals (weekly or monthly, depending on production rate) and the bath should be brought into proper balance.

Chromic acid content can be determined simply by placing a hydrometer in the plating bath provided other contaminants, such as iron or copper, are not present in significant quantities. Hydrometers calibrated in ounces of chromic acid per gallon at the operating temperature of the bath are available. Baumé hydrometers can be used in samples cooled to the calibration temperature marked on the hydrometer; the reading can be converted to concentration by using Table 3. Impurities cause the Baumé reading to be higher than is warranted by the actual chromic acid content. Periodically, the chromic acid content should be determined by chemical analysis and a notation made of the difference between the contents as shown by chemical analysis and by hydrometer reading. Hydrometer readings should then be corrected by that amount. When a hydrometer reading shows 30 to 38 g/L (4 to 5 oz/gal) more chromic acid than is actually present, the bath should be discarded. Adjustments in chromic acid concentration are simplified by the use of Table 4.

Chromic acid content can be determined also by various analytical procedures. Some of these procedures use standard solutions that can be used for other determinations; this may be the deciding factor as to which procedure to use. The following procedure is simple and quite rapid:

Reagents

- Acid mixture: mix 1 part sulfuric acid, 1 part phosphoric acid, and 1 part water
- Ferrous ammonium sulfate (FAS) 0.1N: dissolve 40 g/L (5.3 oz/gal) $FAS \cdot 6H_2O$ and add 25 mL/L concentrated sulfuric acid. When in use, keep about 6500 mm^2 (10 $in.^2$) of aluminum metal in the container to maintain constant normality
- Orthophenanthroline ferrous sulfate complex indicator
- Potassium dichromate, 0.1000N: dissolve 4.900 g (0.173 oz) of $K_2Cr_2O_7$ (CP grade) and dilute to a litre

Standardization of FAS

- Pipette 25 mL of potassium dichromate standard into a 250-mL Erlenmeyer flask and dilute to 85 mL
- Add 15 mL of acid mixture and 3 drops of indicator
- Titrate with FAS to clear orange endpoint:

$$\text{Normality of FAS} = \frac{2.5}{\text{mL of FAS}}$$

Procedure

1a Pipette 10 mL of chromium bath into a 250-mL volumetric flask and dilute to volume. Pipette a 10-mL aliquot into a 250-mL Erlenmeyer flask

1b Pipette 0.4 mL of chromium bath sample into a 250-mL Erlenmeyer flask, using a 1-mL pipette graduated in hundredths of a millilitre

2 Add 50 mL of water, 15 mL of acid mixture and 3 drops of indicator

3 Titrate with FAS to clear orange end point:

$$\text{g/L chromic acid} = (\text{mL of FAS}) \times (N \text{ of FAS}) \times (83.3)$$

$$\text{oz/gal chromic acid} = (\text{mL of FAS}) \times (N \text{ of FAS}) \times (11.12)$$

Use either step 1a or 1b, followed by steps 2 and 3. Step 1b is slightly less accurate than 1a, but is much faster.

Sulfate in a chromium bath can be determined to a high degree of accuracy by a gravimetric method, or with reasonable accuracy using the much faster centrifuge method. Unless there is some reason for great precision, the centrifuge method is entirely satisfactory. Most graduated centrifuge tubes can be calibrated for sulfate; however, the technique should be checked regularly by analysis of a standard solution, or by the gravimetric technique.

With the centrifuge method, the amount of insoluble material in the sample must be determined before sulfate is precipitated. This can be done by running a blank sample or centrifuging the sample before precipitating the sulfate. Because sulfate determinations are made in duplicate, it is advisable to determine how closely the two tubes match in calibration. This can be done by pouring a small amount of mercury into first one tube and then the other. These tubes should be marked to be used together and a notation made if necessary as to the amount of correction needed in the reading. Adjustments of sulfate concentration in chromium baths can be simplified by using Table 5.

Contamination. In the operation of any chromium bath, the bath should be kept free of excessive amounts of contamination. Introduction of (*a*) copper, (*b*) iron, or (*c*) trivalent chromium seriously decreases conductivity of the bath and requires the use of a higher operating voltage to produce a given current density.

Table 4 Conversion equivalents for chromic acid concentration in chromium baths

Values in table are based on the following formula, which may be used to compute values for which conditions are not given: Chromic acid required, kg(lb) = g/L(oz/gal) CrO_3 to be added × bath volume, in litres (gallons)/16

Chromic acid to be added g/L	oz/gal	Chromic acid required, kg (lb); Volume of bath, L (gal): 380 (100)	760 (200)	1140 (300)	1510 (400)	1890 (500)	2270 (600)	2650 (700)	3030 (800)	3410 (900)	3790 (1000)
4.0	0.5	1.4(3.1)	2.9(6.3)	4.3(9.4)	5.7(12.5)	7.1(15.6)	8.5(18.8)	9.9(21.9)	11.3(25.0)	12.7(28.1)	14.2(31.2)
7.5	1.0	2.8(6.2)	5.7(12.5)	8.5(18.8)	11.3(25.0)	14.2(31.2)	17.0(37.5)	19.9(43.8)	22.7(50.0)	25.5(56.3)	28.3(62.5)
11.0	1.5	4.3(9.4)	8.5(18.8)	12.8(28.2)	17.0(37.5)	21.3(46.9)	25.5(56.3)	29.8(65.6)	34.0(75.0)	38.3(84.4)	42.5(93.8)
15.0	2.0	5.7(12.5)	11.3(25.0)	17.0(37.5)	22.7(50.0)	28.3(62.5)	34.0(75.0)	39.7(87.5)	45.3(100.0)	51.3(113.0)	56.7(125.0)
19.0	2.5	7.1(15.6)	14.2(31.2)	21.3(46.9)	28.3(62.5)	35.4(78.1)	42.5(93.8)	49.4(109.0)	56.7(125.0)	64.0(141.0)	70.8(156.0)
22.0	3.0	8.5(18.8)	17.0(37.5)	25.5(56.3)	34.0(75.0)	42.6(93.8)	51.3(113.0)	59.4(131.0)	68.0(150.0)	76.7(169.0)	85.3(188.0)
26.0	3.5	9.9(21.9)	19.8(43.7)	29.8(65.6)	39.7(87.5)	49.4(109.0)	59.4(131.0)	69.4(153.0)	79.4(175.0)	89.4(197.0)	99.3(219.0)
30.0	4.0	11.3(25.0)	22.7(50.0)	34.0(75.0)	45.3(100.0)	56.7(125.0)	68.0(150.0)	79.4(175.0)	90.7(200.0)	102.0(225.0)	113.0(250.0)
34.0	4.5	12.7(28.1)	25.5(56.2)	38.3(84.4)	51.3(113.0)	64.0(141.0)	76.7(169.0)	89.4(197.0)	102.0(225.0)	115.0(253.0)	127.0(281.0)
37.0	5.0	14.2(31.2)	28.3(62.5)	42.5(93.8)	56.7(125.0)	70.8(156.0)	85.3(188.0)	99.3(219.0)	113.0(250.0)	128.0(281.0)	142.0(312.0)
41.0	5.5	15.6(34.4)	31.2(68.7)	46.7(103.0)	62.6(138.0)	78.0(172.0)	93.4(206.0)	109.0(241.0)	125.0(275.0)	140.0(309.0)	156.0(344.0)
45.0	6.0	17.0(37.5)	34.0(75.0)	51.3(113.0)	68.0(150.0)	85.3(188.0)	102.0(225.0)	119.0(262.0)	136.0(300.0)	153.0(338.0)	170.0(375.0)
49.0	6.5	18.4(40.6)	36.8(81.2)	55.3(122.0)	73.9(163.0)	92.1(203.0)	111.0(244.0)	129.0(284.0)	147.0(325.0)	162.0(365.0)	184.0(406.0)
52.0	7.0	19.8(43.7)	39.7(87.5)	59.4(131.0)	79.4(175.0)	99.3(219.0)	119.0(262.0)	139.0(306.0)	159.0(350.0)	179.0(394.0)	198.0(437.0)
56.0	7.5	21.3(46.9)	42.5(93.8)	64.0(141.0)	85.3(188.0)	106.0(234.0)	128.0(281.0)	149.0(328.0)	170.0(375.0)	191.0(422.0)	213.0(469.0)
60.0	8.0	22.7(50.0)	45.4(100.0)	68.0(150.0)	90.7(200.0)	113.0(250.0)	136.0(300.0)	159.0(350.0)	181.0(400.0)	204.0(450.0)	227.0(500.0)
64.0	8.5	24.1(53.1)	48.1(106.0)	72.1(159.0)	96.6(213.0)	121.0(266.0)	145.0(319.0)	169.0(372.0)	193.0(425.0)	216.0(477.0)	241.0(531.0)
67.0	9.0	25.5(56.2)	51.3(113.0)	76.7(169.0)	102.0(225.0)	127.0(281.0)	153.0(338.0)	179.0(394.0)	204.0(450.0)	230.0(506.0)	255.0(562.0)
71.0	9.5	26.9(59.4)	54.0(119.0)	80.7(178.0)	108.0(238.0)	135.0(297.0)	161.0(356.0)	189.0(416.0)	215.0(475.0)	243.0(535.0)	269.0(594.0)
75.0	10.0	28.3(62.5)	56.7(125.0)	85.3(188.0)	113.0(250.0)	142.0(312.0)	170.0(375.0)	199.0(438.0)	227.0(500.0)	255.0(563.0)	284.0(625.0)

Table 5 Conversion equivalents for adjusting sulfate concentration in chromium baths

Values in table are based on the following formula, which may be used to compute values for which conditions are not given: Sulfuric acid (66 °Bé) required, fluid oz = 0.522 × oz/gal H_2SO_4 to be added × bath volume, gal

Sulfuric acid to be added g/L	oz/gal	Sulfuric acid (66 °Bé) required, mL (fluid oz); Volume of bath, L (gal): 380 (100)	760 (200)	1140 (300)	1515 (400)	1890 (500)	2270 (600)	2650 (700)	3030 (800)	3410 (900)	3790 (1000)
0.08	0.01	15(0.5)	30(1.0)	47(1.6)	62(2.1)	77(2.6)	92(3.1)	109(3.7)	124(4.2)	139(4.7)	154(5.2)
0.15	0.02	30(1.0)	62(2.1)	92(3.1)	124(4.2)	154(5.2)	186(6.3)	216(7.3)	249(8.4)	278(9.4)	308(10.4)
0.22	0.03	47(1.6)	92(3.1)	139(4.7)	186(6.3)	231(7.8)	284(9.6)	323(10.9)	370(12.5)	417(14.1)	465(15.7)
0.30	0.04	62(2.1)	124(4.2)	186(6.3)	249(8.4)	308(10.4)	370(12.5)	432(14.6)	494(16.7)	556(18.8)	619(20.9)
0.37	0.05	77(2.6)	154(5.2)	231(7.8)	308(10.4)	385(13.0)	462(15.6)	539(18.2)	619(20.9)	696(23.5)	773(26.1)
0.45	0.06	92(3.1)	186(6.3)	284(9.6)	370(12.5)	462(15.6)	556(18.8)	648(21.9)	740(25.0)	835(28.2)	926(31.3)
0.53	0.07	110(3.7)	216(7.3)	323(10.9)	432(14.6)	539(18.2)	648(21.9)	758(25.6)	864(29.2)	974(32.9)	1080(36.5)
0.60	0.08	124(4.2)	249(8.4)	370(12.5)	494(16.7)	619(20.9)	740(25.0)	864(29.2)	989(33.4)	1113(37.6)	1237(41.8)
0.67	0.09	139(4.7)	278(9.4)	417(14.1)	556(18.8)	696(23.5)	835(28.2)	974(32.9)	1143(38.6)	1252(42.3)	1391(47.0)
0.75	0.10	154(5.2)	308(10.4)	465(15.7)	619(20.9)	773(26.1)	926(31.3)	1080(36.5)	1237(41.8)	1391(47.0)	1545(52.2)

Note: To neutralize excess sulfuric acid (thereby lowering the sulfate content) in a chromium bath, add approximately 1.5 g/L (0.2 oz/gal) of barium carbonate for each 0.8 g/L (0.1 oz/gal) of excess sulfuric acid

Trivalent chromium can be formed by the decomposition of organics, but more often it results from too low a ratio of anode-to-cathode area, a condition always encountered in plating interior surfaces of cylinders. Trivalent chromium can be reoxidized to hexavalent chromium by electrolyzing the solution at 60 to 66 °C (140 to 150 °F), with an anode-to-cathode area ratio of about 30 to 1, and using a cathode current density of about 60 A/dm^2 (4 A/in.^{2}). About 50 A·h/L (200 A·h/gal) is required for reoxidizing 15 g/L (2 oz/gal) of trivalent chromium; ordinarily, this operation is performed during the weekend shutdown period.

Removal of copper, iron, and other contaminants is usually accomplished by disposing of part or all of the bath, or where permissible, by ion exchange. In addition to reducing the conductivity of the bath, impurities, particularly iron and trivalent chromium, also reduce current efficiency. Normally, the presence of 10 g/L (1.5 oz/gal) of iron in solution reduces cathode current efficiency by about 30%. Iron and trivalent chromium can produce rougher, more treed deposits. Other contaminants affect the throwing power and coverage.

Mandrel Test. A test procedure that may be useful when difficulties are experienced with the hard chromium production bath is known as the mandrel test, because it is usually performed by plating a low-carbon steel rod or mandrel (9.5 mm in diameter by 75 mm long or 3/8 in. in diameter by 3 in. long) with a conforming circular anode to give uniform current distribution. A convenient procedure is to use 540 mL of the bath solution in a 600-mL beaker made of heat-resistant glass. The steel mandrel is buffed to a bright finish, and the top 25 mm (1 in.) that projects out of the solution and into a holder is stopped-off to indicate the exact size of the plating area.

The mandrel is immersed in the test solution, which is heated to 55 °C

(130 °F) in a water bath, treated anodically at 15 A/dm^2 (1 A/in.^{2}) for 15 s, and then plated at 30 A/dm^2 (2 A/in.^{2}) for 1 h. The temperature of the water bath should be reduced to about 52 °C (125 °F) during plating to compensate for the heat generated by the current. The mandrel is then (*a*) removed, (*b*) rinsed, (*c*) dried, and (*d*) inspected.

Better temperature control through the plating process is obtained by using 1 L of solution in a 1.5-L beaker. Heat the solution with a quartz heater (125 W) controlled by a thermoregulator and relay. Operate at 55 °C (130 °F).

A hard chromium bath in proper adjustment gives a bright plate under these conditions. If the deposit is dull, the solution is contaminated or out of balance. Burning on the bottom edge of the mandrel also indicates that the solution is out of adjustment. A high content of trivalent chromium causes dullness and growth of metal whiskers from the bottom edge of the mandrel.

Current efficiency and plating speed can be checked with the same setup, but this is more conveniently done with a flat steel panel and flat anodes. A polished steel panel, 25 by 75 mm (1 by 3 in.), can be used, with the bottom 50 mm (2 in.) marked off as the plating area. Plating conditions would be 30 A/dm^2 (2 A/in.^{2}) for 15 min at 55 °C (130 °F), but other conditions can be tested as desired. The steel panel is accurately weighed before and after plating (without current reversal). Current efficiency, thickness of chromium, and plating speed are calculated as follows:

$$\text{Current efficiency, \%} = \frac{\text{weight (in grams) of chromium} \times 100}{\text{amperes} \times \text{hours} \times 0.323}$$

$$\text{Thickness of deposit, mils} = \frac{\text{weight (in grams) of chromium}}{4 \times 0.116}$$

$$\text{Plating speed, mils/h} = \frac{\text{thickness of deposit} \times 60}{\text{plating time, min}}$$

Process Control

In addition to bath composition, the principal variables that must be controlled for satisfactory hard chromium plating are (*a*) anodes, (*b*) current density, and (*c*) bath temperature.

Anodes. In contrast to other plating baths, which use soluble anodes to supply the bath with a large part of the metal ion being plated, chromium-plating baths are operated with insoluble lead alloy anodes. Therefore, additions of chromic acid must be made as required to keep the chromium-plating bath supplied with chromium metal ions.

A coating of lead peroxide forms on the lead alloy anodes during electrolysis. This coating is usually dark charcoal brown on anodes that are functioning correctly; the presence of an orange-to-yellow lead chromate coating indicates the anodes are not passing current properly. Periodic cleaning of the anodes and their hooks is mandatory for efficient operation.

Insoluble antimonial lead (93% Pb-bal Sb plus Sn) and lead-tin alloy (93% Pb-7% Sn) are the most widely used anode materials in chromium-plating solutions. These alloys minimize corrosion. Each anode must have sufficient cross section to prevent overheating. The bottom of each anode should be at least 150 mm (6 in.) above the bottom of the plating tank. For mixed catalyst baths, anodes must be heavier than for conventional baths to carry the increased current without overheating.

Conductivity of the chromium bath is based on chromic acid concentration. Concentrations higher than 250 g/L (33 oz/gal) require lower operating voltage; however, current efficiency decreases with increasing chromic acid concentration. Lower concentrations are usually impractical unless 9- or 12-V power is available.

Current Density and Efficiency. Cathode current efficiency varies with current density and temperature of the plating bath. Efficiency increases with increasing current density and decreasing temperature. These two variables have a definite effect on appearance and hardness of the deposit. A high bath temperature results in a milky, dull, and softer deposit at lower current efficiencies, unless the current density is increased substantially. Raising current density causes the deposit to change successively at specific temperatures (Table 6). Because tank time is an important economic factor, the highest rates of deposition that are produced by the highest available currents may determine which plating bath temperature is most useful.

Deposition Rates. Times required to plate hard chromium deposits of various thicknesses are shown as a function of current density in Table 7 (for low-concentration baths) and Table 8 (for high-concentration baths).

Bath temperature affects both the conductivity and the current required. If limited power is available, satisfactory hard chromium plating can be obtained at lower temperatures (43 to 49 °C or 110 to 120 °F); but if power supply is adequate, it is advantageous to work at higher temperatures up to 66 °C (150 °F), because of the faster deposition rate and the improved durability of the deposit. At 43 to 49 °C (110 to 120 °F), current densities of 8 to 30 A/dm^2 (½ to 2 A/in.^{2}) are satisfactory; at 60 to 66 °C (140 to 150 °F), 45 to 60 A/dm^2 (3 to 4 A/in.^{2}) may be required, and as high as 80 to 90 A/dm^2 (5 to 6 A/in.^{2}) can sometimes be used. At all temperatures, increased agitation will allow higher current densities.

Control of bath temperature to within a narrow range is necessary because of the marked influence of temperature on deposition rate. An increase of ±2 °C (±5 °F) in mean bath temperature, for example, can cause a reduction of 5% or more in mean chromium thickness, thus necessitating a sizable adjust-

Table 6 Effect of bath temperature and current density on appearance and hardness of chromium deposits

Bath contained 406 g/L CrO_3; ratio of CrO_3 to SO_4 ranged from 90-to-1 to 100-to-1

Current density, A/dm^2	Current density, A/in.^{2}	Appearance of deposit	Hardness(a), DPH
Plating bath at 43 °C (110 °F)			
Below 8.53	Below 0.55	Dull matte	(b)
8.53	0.55	Semibright	695
17.1	1.10	Bright	900
25.6	1.65	Bright, pebbly	Over 940
34.1	2.20	Dull, nodular(c)	Over 940
Plating bath at 49 °C (120 °F)			
Below 10.9	Below 0.70	Dull matte	510-595
10.9	0.70	Semibright	695
21.7	1.40(d)	Bright	900
32.6	2.10	Bright, pebbly	Over 940
43.4	2.80	Dull, nodular(c)	Over 940
Plating bath at 54 °C (130 °F)			
Below 14.0	Below 0.90	Dull matte	510-595
14.0	0.90	Semibright	695
27.9	1.80(d)	Bright	900
41.9	2.70	Bright, pebbly	Over 940
55.8	3.60	Dull, nodular(c)	Over 940

(a) Hardness of mounted and unmounted specimens was determined with a tester employing a Vickers diamond, a load of 200 g, and a magnification of 200X. Mounted specimens were mounted flat (not cross section). (b) Current density too low to plate a sufficient amount of chromium for hardness test. (c) Specimens polished lightly to smooth out nodules. (d) Optimum current density.

ment in either plating time or current density. The bath temperature should be maintained within ±1 °C (±2 °F). Bath temperature is controlled automatically. Manual control is impractical for a production operation.

For automatic control, it is important that the thermostat be placed in the plating bath in a location where it can readily sense any significant change in bath temperature. Obviously, the thermostat must not be in close proximity to a heating or cooling pipe or to an electric heating element. Location of the thermostat is greatly simplified when an external heat exchanger is used and the solution is pumped from the heat exchanger to the bath. Rapid movement of the solution helps to promote temperature uniformity.

Table 7 Rates of deposition of hard chromium from low-concentration baths

Thickness of plate µm	mils	Plating time, h:min, at current density of: A/dm² 31 (A/in.² 2.0)	47 (3.0)	62 (4.0)
Conventional sulfate bath(a)				
25	1	1:05	0:40	0:25
50	2	2:05	1:20	0:55
125	5	5:20	3:20	2:20
Mixed catalyst bath(b)				
25	1	0:50	0:30	0:20
50	2	1:40	1:00	0:40
125	5	4:05	2:25	1:45

(a) Bath containing 250 g/L (33 oz/gal) of chromic acid and with 100-to-1 ratio of chromic acid to sulfate, operated at 54 °C (130 °F). (b) Proprietary bath containing 250 g/L (33 oz/gal) of chromic acid, operated at 54 °C (130 °F)

Table 8 Rates of deposition of hard chromium from high-concentration baths

Thickness of plate µm	mils	Plating time, h:min, at current density of: 23 A/dm² (1.5 A/in.²)	31 A/dm² (2.0 A/in.²)	39 A/dm² (2.5 A/in.²)	47 A/dm² (3.0 A/in.²)	54 A/dm² (3.5 A/in.²)
Conventional sulfate bath(a)						
25	1	2:20	1:35	1:15	0:55	0:45
50	2	4:35	3:10	2:30	1:55	1:30
125	5	11:30	8:00	6:15	4:40	3:50
255	10	23:00	16:00	12:30	9:25	7:35
380	15	34:30	24:00	18:45	14:05	11:25
510	20	46:00	32:00	25:00	18:50	15:10
Mixed catalyst bath(b)						
25	1	1:25	0:55	0:45	0:35	0:25
50	2	2:50	1:50	1:25	1:05	0:50
125	5	7:00	4:40	3:35	2:45	2:10
255	10	14:00	9:20	7:10	5:25	4:20
380	15	21:00	14:00	10:45	8:10	6:25
510	20	8:00	18:40	14:20	10:55	8:35

(a) Bath containing 400 g/L (53 oz/gal) of chromic acid with 100-to-1 ratio of chromic acid to sulfate, operated at 54 °C (130 °F). (b) Bath containing 400 g/L (53 oz/gal) of chromic acid, 1.5 g/L (0.20 oz/gal) of sulfate, and sufficient fluoride catalyst to give 100-to-1 ratio results; operating temperature, 54 °C (130 °F)

Problems and Corrective Procedures

Faulty operation of a chromium-plating bath can result in slow plating speed or in deposits with undesirable characteristics. The problems encountered in hard chromium plating, their possible causes, and suggested corrective procedures include:

Poor coverage

- Low chromic acid content
- Low ratio of chromic acid content to total catalyst content. Correct by adding chromic acid or by precipitating sulfate, if too high, with barium carbonate
- Temperature too high
- Current density too low
- Passive or scaled anodes. Correct by cleaning and reactivating anodes; use high current density until uniform gassing is obtained; check for good anode contact
- Rack contacts too heavily built up with metal causing rack to rob plate
- Thieves too large or too close
- Open holes preventing uniform plate in adjacent areas. Correct by use of nonconducting plugs in holes
- Gas entrapment preventing plating solution from reaching some areas. Correct by positioning parts in bath so all gases can escape or by agitating parts while they are plating

Burnt deposits

- Ratio of chromic acid to total catalysts too high. Correct by adding necessary catalysts or lowering chromic acid content
- Current density too high
- Temperature too low
- Large parts were colder than bath temperature when plating began
- Some parts, in a load of different parts, receive too much current. Correct by adjusting anoding and contacts to make sure that each part receives correct current density
- Anode spacing too close (current density too high)
- Excessive amount of anode within a given area for the part being plated. Correct by eliminating all excess anodes in the tank; design the shape of conforming anodes to minimize current density at high-density areas and, if necessary, use nonconducting shields at these areas

Slow plating speed

- Chromic acid content too high
- Ratio of chromic acid to total catalyst too high
- Temperature too high
- Current density too low
- Scaled anodes
- Insufficient or inadequate sizes of conductors in anode or cathode circuits
- Thief obtains too much current. Correct by redesigning thief or by removing nodules from thief
- Current leakage
- Single phasing of rectifier caused by partial burnout. Correct by repairing rectifier
- High contact resistance on busbars, racks, or jigs. Correct by cleaning contacts to lower contact resistance. Do not always rely on voltage for control, because it does not indicate the conditions present on the part; instead, control by amperage
- Large variety of parts in same tank prohibiting proper current density for each
- Tank overcrowded with parts
- High content of metallic impurities. Correct by discarding a portion of bath; readjust after dilution

Nodular deposits

- Insufficient etching before plating in relation to thickness of deposit. Correct by increasing etching time

Table 9 Process and equipment requirements for hard chromium plating

Item	Area of part mm²	Area of part in.²	Area of load mm²	Area of load in.²	No. of pieces/ 8 h	Thickness of plate µm	Thickness of plate mil	Current density A/dm²	Current density A/in.²	Plating time, min	Temperature of bath °C	Temperature of bath °F	No. of work rods	Tank dimensions mm	Tank dimensions in.
Small cutting tools	4800	7.5	967000	1500	10000	1.3	0.05	30	2	5	50	120	1	1500 by 760 by 910	60 by 30 by 36
Shafts.......	20000	30	600000	930	200	25	1	30	2	63	50	120	2	1800 by 910 by 910	72 by 36 by 36
Gun barrels(a)..	15000	23	534000	828	180	25	1	45	3	40	54	130	2	2400 by 910 by 610	96 by 36 by 24

(a) Plating of inside diameter of 30-caliber gun barrels

- Rough surface before plating
- Chromic acid content too high. Correct by removing portion of bath; dilute the remainder, and adjust
- Low temperature
- Low sulfate content
- Current density too high

Pitted deposits

- Pitted basis metal
- Marking dye not completely removed
- Material suspended in solution. Correct by filtering out suspended material
- Surface-activating agents used cause deep pits when plating thick deposits. Correct by discarding as much of bath as necessary to eliminate pits and replace with fresh solution. Prevented by discontinuing use of mist suppressors; plastic parts, such as floats, that decompose can form decomposition products that also create pits
- Gas bubbles adhering to part. Correct by improving surface finish before plating; agitate part occasionally during plating
- Part is magnetized. Correct by demagnetizing
- Magnetic particles in the bath. Correct by removing particles with magnet
- Insufficient cleaning prior to plating
- Particles falling on work from anodes or thieves. Correct by improving design of anodes and thieves; clean both regularly, to remove loose particles
- Carbon smut on surface. Correct by scrubbing before plating
- Excessively etched surface during reverse-etch or stripping operation

Poor adhesion

- Insufficient or no etching before plating
- Contaminants not completely removed from surfaces during cleaning
- Excessive grinding rate at edges or sharp projections, where basis metal fractures and may appear as though plated material did not adhere
- Single phasing of rectifier caused by partial burnout. Correct by repairing rectifier
- Current interruption during plating
- Cold solution

Macrocracks

- Highly stressed basis metal; cracks are visible during grinding or when heat is applied. Correct by relieving stresses in basis metal
- Grinding at too fast a rate produces heat checks

Equipment

The discussion of equipment that follows is confined largely to considerations that are specific to chromic acid plating processes. Mixed catalyst baths have essentially the same equipment requirements as conventional sulfate baths, except all parts of the electrical system may need to be heavier, to accommodate the increased current used. Equipment requirements for plating three specific parts are given in Table 9.

Tanks and Linings. Figure 1 illustrates a hard chromium plating tank arrangement. Most tanks for chromium plating are made of steel and lined with an acid-resisting material. Because of their excellent resistance to corrosion by chromic acid, lead alloys containing antimony or tin may be used as tank linings.

Acid-resistant brick has been used as a lining material. Because of its electrical insulating characteristics, acid-resistant brick lining has the advantage over metal linings of reducing possible current losses or stray currents. Some installations combine a lead lining or plastic sheet lining with an acid-resistant brick facing. With fluoride-containing solutions, a brick lining is suitable only for temporary use.

Almost invariably, plasticized polyvinyl chloride is used for both sulfate and mixed catalyst baths, provided the bath temperature does not exceed 66 °C (150 °F). Sheets of this plastic are cemented to tank walls and welded at joints and corners. Other plastic materials are equally resistant to chemical attack, but are more likely to fail at the welds when exposed to an oxidizing acid. Fiberglass utilizing either polyester or epoxy is unsatisfactory, because exposed fiberglass will be attacked by the secondary catalyst.

Design specifications for low-carbon steel tanks for chromium plating are given in Table 10. Lining materials for low-carbon steel tanks are given in Table 11. Steel tanks should be supported at least 100 mm (4 in.) from the floor; steel I-beams are used to provide this support and are mandatory when side bracing is required. To provide insulation, reinforced strips of resin-bonded glass fiber can be placed between the floor and the I-beams. Glass brick can be used as insulation between electrodes and the plating tank.

Heating and Cooling. Steam heating coils and cooling coils can be made of antimonial lead or silver-bearing lead. Titanium coils are preferred for conventional plating solutions. Tantalum- or niobium-clad coils are suitable for mixed catalyst baths as well. These coils are mounted on tank walls behind the tank anodes. Steel pipes carrying steam and cooling water to the tank must have a nonconducting section in each leg, so that the coils cannot become an electrical ground back through the power plant system.

Fig. 1 Tank and accessory equipment used for hard chromium plating

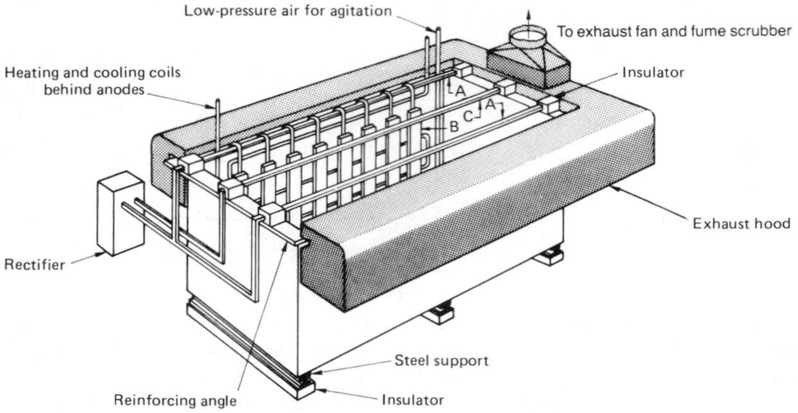

A: anode rods; B: lead or lead-tin anodes; C: cathode rod

Electric immersion heaters sheathed in fused quartz are suitable for heating chromic acid baths. The quartz is fragile and must be handled with care. Similar immersion heaters are sheathed in either tantalum, titanium, or lead alloy. It is sometimes feasible to heat and cool a chromic acid bath by piping the liquid to a tube bundle, concentric, or tube heat exchanger located outside the plating tank. Preferably, heat exchanger tubes should be made of tantalum or titanium. This method has the disadvantage of requiring pumping of the solution.

Temperature-control planning should begin with selection of the volume of solution required in the plating bath. An ideal volume consists of 1 L or more of solution for each 13 W of plating power (1 gal or more of solution for each 50 W of plating power). About 60% of this plating power (30 W) produces heat and maintains the solution at temperature in an uninsulated tank of standard design. Power applications in excess of 13 W/L (50 W/gal) require cooling of the plating bath, and cause relatively rapid changes in solution composition.

Agitation. A chromium-plating bath should be agitated periodically and particularly when the bath is being started to prevent temperature stratification. For manual agitation, a hoe or a paddle is suitable. For mechanical agitation, an electrical stirrer with a plastisol-coated shaft and blade is recommended. Air agitation is effective, but oil from an air pump must not be permitted to leak into the air system. Preferably, the air should come from an oil-free low pressure blower. A perforated pipe, of rigid polyvinyl chloride, may be used to distribute air in the solution.

Rods and Insulators. Anode and cathode rods are usually made of round or rectangular copper bar stock. These rods should be adequately supported to prevent them from sagging under the weight of anodes and work. Generally, selection of rod size is determined by allowing 1 cm^2 of cross-sectional area for each 150 A (1 in.^{2} of cross-sectional area for each 1000 A), although mechanical strength for load support is also a factor in determining rod size. Anode and cathode rods are supported above the tank rim by insulators, which may be made of (*a*) brick, (*b*) porcelain, or (*c*) plastic. Even metallic supports can be used, if a strip of electrical insulating material is placed between the plating tank and the busbar.

Power Sources. Although dynamos or motor-generator sets were the usual sources of power for low-voltage direct current for plating, rectifiers are now regularly used. In general, use of motor-generator sets is now restricted to larger and more permanent installations. Originally, plating rectifiers were of copper oxide or magnesium-copper sulfide types, but these have been largely replaced by silicon rectifiers. Silicon is favored for plating rectifiers because of its high resistance to thermal overload and small space requirement. Hard chromium platers often start plating on a piece by sweeping up applied voltage and current from very low values to the high values used for platings. Because silicon-controlled rectifiers have high ripple at low outputs, the output should be filtered. Tap-switch controls, however, produce relatively low ripple over the entire output range.

A 6-V power source can be used for chromium plating, but it is generally desirable or necessary to operate with 9 to 12 V available. Chromium plating requires full-wave rectification with a three-phase input and full control, giving a ripple less than 5% and no current interruptions. If a rectifier becomes partially burned out, it may single phase to some degree, and this can cause dull plate, or a laminated, peeling deposit.

Fume Exhaust. A chromium-plating process produces a chromic acid mist (chromates in the form of CrO_3), which is highly toxic. The maximum allowable concentration for 8-h continuous exposure is 0.1 mg of chromic acid mist per cubic metre of air. This concentration value is in accordance with recommendations by the American Conference of Governmental Industrial Hygienists. Because of the extreme toxicity of this mist, it is mandatory to provide adequate facilities for removing it. The minimum ventilation rate should be 60 m^3/min per square metre (200 ft^3/min per square foot) of solution surface area.

Generally, fumes are exhausted from a chromium-plating tank by means of lateral exhaust hoods along both long sides of the tank. For narrow tanks, up to 600 mm (24 in.) wide, a lateral exhaust on one side of the tank should be adequate unless strong cross drafts exist. Velocity of the air at the lateral exhaust hood slots should be 600 m/min (2000 ft/min) or more.

In the design of ductwork, condensate duct traps should be included to capture chromic acid solution. Drains from these traps should be directed to a special container and not to the sewer. In this way, chromic acid solutions can be returned to the tank or recovery system, or safely destroyed. A fume scrubber or a demister should also be included in the system to remove most of the chromic acid fumes before exhausted air is emitted to the atmosphere. Many communities have air pollution regulations requiring fume scrubbers. Fume exhaust ductwork may be made of carbon steel and coated with acid-resistant paint. Modern construction uses chlorinated polyvinyl chloride (CPVC).

Rinse Facilities. Rinsing the work after chromium plating prevents it from becoming stained or discolored. Insufficient rinsing can result in contamination of cleaning solutions during subsequent cycling of racks. Multiple rinsing facilities are recommended. After being plated, parts should be rinsed in a nonrunning reclaim tank, which can be used to recover part of the chromium-bath dragout. After they are rinsed in the reclaim tank, plated parts should be rinsed in counterflowing cold water and hot water tanks. Water should cascade from the hot water to the cold water tank. A multiple counterflowing arrangement requires much less water than is required for two separate rinsing tanks.

If rinse water is being returned to a chromic acid waste disposal unit, the flow of water into the hot water tank should be controlled automatically by a conductivity-sensing element in the cold water tank. At a predetermined concentration of chromic acid in the cold water, the water inlet to the hot water tank should flow, causing an overflow of cold water to the waste disposal unit. This arrangement decreases amount of water consumed and minimizes required capacity of the waste disposal unit.

Cold water rinse tanks may be coated, sprayed, or otherwise lined with plasticized polyvinyl chloride. Hot water rinse tanks may be constructed of types 347, 304, or 316 stainless steel, or they may be made of carbon steel and lined with lead. Reinforced polyester glass fiber also may be used for either hot water or cold water rinse tanks.

Spray rinsing also serves effectively to remove residual chromic acid. Because spraying does not always reach recessed areas, sprays should be positioned above a dip rinse. As parts are removed from the dip rinse, they may be sprayed with clean water, which, in turn, is returned to the dip tank.

Maintenance. Following is a maintenance schedule for a still tank installation for hard chromium plating. This schedule is intended only as a guide; local conditions determine exact requirements. The rate of variation of bath constituents depends on (*a*) volume of solution, (*b*) method of operation for bath, and (*c*) type of work. The maintenance schedule for a still tank installation for hard chromium plating includes the following:

- *Daily*: check temperature. Check concentration of bath by density measurements. Clean busbars and electrical connections. Remove any parts which fall from racks
- *Weekly*: analyze for chromic acid and sulfate contents
- *Monthly*: remove all sludge and parts from tank using a hoe and dragging the bottom. If tank is used for plating inside diameters, analyze for trivalent chrome
- *Semiannually*: check tanks for leaks and condition of lining. Clean and inspect rectifiers or motor-generating units. Check ammeter calibration
- *As necessary*: analyze for trivalent chromium, iron, nickel, copper, and zinc. Check condition of anodes

Table 10 Design specifications for low-carbon steel tanks for hard chromium plating

Size of tank: Length, m	Length, ft	Depth, m	Depth, ft	Thickness of low-carbon steel, mm	in.	Width of rim, mm	in.	Tank reinforcing
Up to 1	Up to 4	Under 0.9	Under 3	5	3/16	50	2	No
Up to 1	Up to 4	Over 0.9	Over 3	5	3/16	50	2	Yes
1-4	4-12	All		6	1/4	75	3	Yes
Over 4	Over 12	All		10	3/8	75	3	Yes

Table 11 Lining materials for low-carbon steel tanks for hard chromium plating

Tank length, m	ft	Lining material: Lead alloy(a), kg/m²	Lead alloy(a), lb/ft²	PVC(b), mm	PVC(b), in.	Brick(c), mm	Brick(c), in.
Up to 2	Up to 6	40	8	5(d)	3/16(d)	100	4
2-4	6-12	50	10	5(d)	3/16(d)	100	4
Over 4	Over 12	60	12	5(d)	3/16(d)	100	4

(a) Antimonial lead, or lead-tin alloy. (b) Plasticized polyvinyl chloride. (c) Acid-resistant brick. For further protection, brick may be backed up with 39 kg/m² (8 lb/ft²) of antimonial lead or lead-tin alloy, or with plasticized polyvinyl chloride sheet. (d) Lining should be 10 mm (3/8 in.) thick at top to 0.3 m (1 ft) below top of tank

Racks and Fixtures

The following recommendations are offered regarding the design and use of plating racks:

- Racks should be designed to hold workpieces in a favorable position for plating uniformly on significant surfaces and to facilitate racking and unracking
- Workpieces with protruding sections should be racked so that parts shield each other, or if this is not possible, a current thief or stealer should be used to reduce current density at the protruding points
- Electrical contact with the part should be made on a nonsignificant surface
- Contact or rack tip should be rigid enough to hold workpieces securely and maintain positive contact. When the work is heavy enough to ensure positive contact, a hook often suffices
- To minimize solution losses due to dragout, the work should be hung as nearly vertical as possible, with the lower edge of the work tilted from the horizontal to permit runoff at a corner rather than a whole edge. When recessed areas cannot be racked to allow proper runoff, provision should be made for drain holes or perhaps tilting of the rack when it is being withdrawn from the solution

Although the design of racks and the methods of racking vary greatly, two basic types of racks are generally used. The first type consists of a single high-conductivity bar on which suitable supports have been mounted for holding the work to be plated; this rack is the cathode side of the plating circuit. The second type consists of two elements, the cathode and the anode; the work is held by the cathode and the cathode is attached to, but insulated from, the anode. Both types of racks are illustrated in Fig. 2. To prevent deposition of chromium or attack by the plating solution on parts of the rack that are immersed in the bath, these parts are covered with nonconducting material such as (*a*) water-resistant tape, (*b*) special insulating lacquer, or (*c*) plastisol coatings.

Fig. 2 Racks used in hard chromium plating

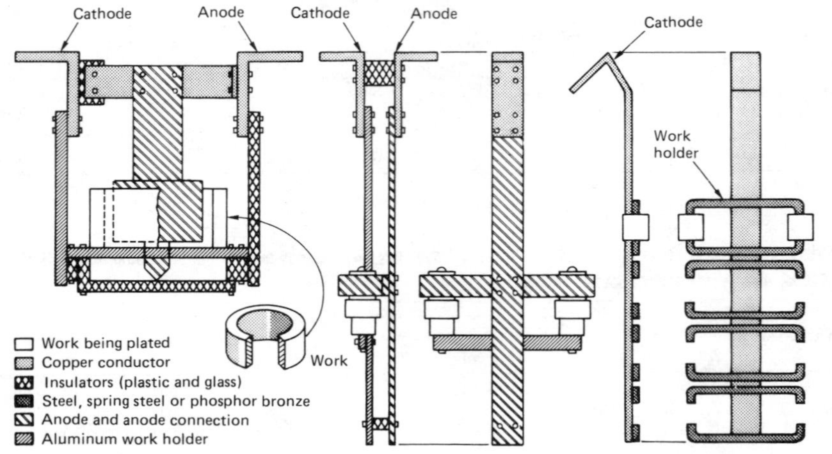

Barrel Plating

Parts to be barrel chromium plated should tumble freely and should not be permitted to nest or lock together. They should be heavy enough to make good electrical contact with the plating barrel; sheet metal parts less than about 0.75 mm (0.030 in.) thick are too light to make adequate contact. Screw machine parts are usually ideally suited to barrel plating. Watch crowns and business machine parts are barrel plated with hard chromium for resistance to wear and corrosion; an outstanding application is dental burrs. Other parts that have been successfully barrel plated include (*a*) electric razor cutters, (*b*) sewing machine parts, (*c*) carburetor shafts, (*d*) bearings for electric meters, and (*e*) ball and roller bearings.

Surface Preparation

All soils and passive films must be removed from surfaces of ferrous and nonferrous metals before they are hard chromium plated. In addition to cleaning, certain surface-activating processes are often important in preparing the basis metal for hard chromium plating. The processes include (*a*) etching of steel, (*b*) preplate machining, and (*c*) nonferrous metals preparation.

Etching of steel before plating is needed to ensure adherence of the chromium deposit. Anodic etching is preferred for this purpose. Slight etching by acid immersion may be used for highly finished surfaces, but with possible sacrifice of maximum adherence.

Steel can be etched anodically in the chromium-plating bath at its operating temperature for plating. A reversing switch is used to make the steel to be plated the anode for 10 s to 1 min (usually 30 s to 1 min) at a current density of about 15 to 45 A/dm^2 (1 to 3 A/in.^{2}). Tank voltage should ordinarily be 4 to 6 V. Because mixed catalyst baths etch more rapidly, a shorter etching time is required. This process has the disadvantage of causing the bath to become contaminated with iron from the work and with copper from the conductors.

As an alternative, steel may be anodically etched in a separate chromic acid bath without sulfate additions and containing 120 to 450 g/L (16 to 60 oz/gal) of chromic acid. Bath temperature may range from room temperature to that of the chromic acid plating bath, or be even higher, provided current density and time of treatment are adjusted to suit the type of work being processed.

A sulfuric acid solution (sp gr 1.53 to 1.71) may be used for anodic etching, provided the bath temperature is held below 30 °C or 86 °F (preferably below 25 °C or 77 °F). The time of treatment may vary from 30 to 60 s and the current density from about 15 to 45 A/dm^2 (1 to 3 A/in.^{2}) at tank voltages ordinarily between 4 and 6 V. A lead-lined tank with lead cathodes should be used. With the use of a sulfuric acid solution, however, two difficulties may be encountered: (*a*) if the rinsing following etching is incomplete, the drag-in of sulfuric acid throws the chromium-plating bath out of balance with respect to the ratio of chromic acid to sulfate; and (*b*) in handling parts that are difficult to manipulate, there is danger of rusting of surfaces exposed to air more than a very short time and of over etching finely finished surfaces.

For high-carbon steel, a sulfuric acid solution of 250 to 1000 g/L (33 to 133 oz/gal), used at a temperature of not more than 30 °C (86 °F) and preferably below 25 °C (77 °F), is effective for anodic etching. The addition of 125 g/L (16.6 oz/gal) of sodium sulfate, based on the anhydrous salt, is of benefit for many grades of steel. Anodic treatment in this solution for a time usually not exceeding 1 min at a current density of about 15 A/dm^2 (1 A/in.^{2}) (range of 15 to 45 A/dm^2 or 1 to 3 A/in.^{2}) is sufficient. High acid content, high current density, and low temperature (within the ranges specified) minimizes the attack on the basis metal and produces a smoother surface. This sulfuric acid solution is stable and not appreciably affected by iron buildup.

Preplate Machining. Metal debris on the surface should be removed before etching (an activation procedure). The use of abrasive-coated papers is common as is the use of successively finer grit stones in honing and grinding. To prepare a sound surface in superfinishing, 600-grit stones may be used. Electropolishing is sometimes used to remove highly stressed metal and metal debris from the surface of cold worked steel. This process improves bond strength and corrosion resistance of electroplated coatings. It accomplishes this function without formation of smut, which may result from anodic etching. This treatment is not recommended for parts that are subjected to critical fatigue stresses and that are expensive to manufacture.

Nonferrous Metals. Aluminum, in common with certain other metals, quickly develops a natural, passive oxide film after exposure to preplating cleaning cycles. This film must be removed before aluminum is plated. The most widely used method of preparing aluminum for plating involves a zincating treatment, which may be followed by a thin 5 μm (0.2 mil) copper electrodeposit. However, it is possible to plate chromium directly over the zincate.

Aluminum parts used in hydraulic systems require a nickel undercoat before being plated to provide corrosion protection to all plated surfaces that are not completely and constantly immersed in hydraulic fluid or similarly

Fig. 3 Variations in plate thickness

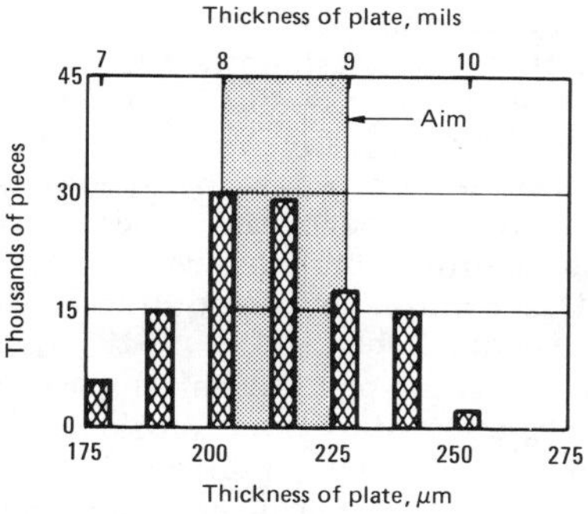

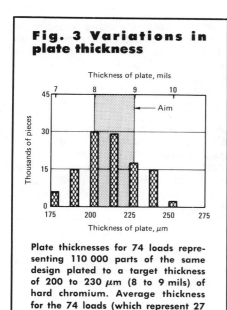

Plate thicknesses for 74 loads representing 110 000 parts of the same design plated to a target thickness of 200 to 230 μm (8 to 9 mils) of hard chromium. Average thickness for the 74 loads (which represent 27 days of operation) was 215 μm (8.4 mils)

Fig. 4 Variation in thickness of chromium plate on feedworm, as a function of distance of anode from part

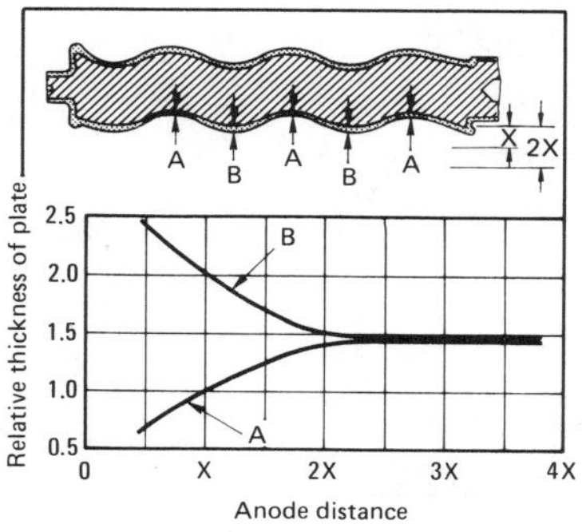

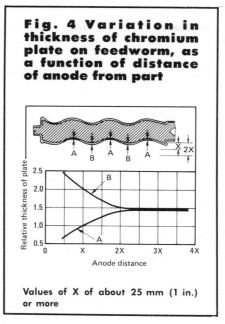

Values of X of about 25 mm (1 in.) or more

protective fluids. A minimum thickness of 10 to 15 μm (0.4 to 0.6 mil) of nickel is usually specified. This undercoat may also be required for steel parts in similar applications.

Titanium and titanium alloys, as well as magnesium, also form a tight, stable oxide coating and are therefore difficult to plate. These metals can be pretreated with an electroless nickel plate or a coating deposited from a high-chloride nickel strike bath.

Variations in Plate Thickness

Variations in the thickness of hard chromium plate depends primarily on potential field distribution. Potential field is controlled by the placement of anodes, shields, thieves, and other parts, as well as the relative position of the sides and surface of the tank. Variations in plate thickness also depend on (*a*) surface preparation, (*b*) control of bath conditions, and (*c*) uniformity of the power source.

Methods of Measuring Plate Thickness. Several methods and types of instruments are available for determining the thickness of plate. These include (*a*) electrolytic stripping, (*b*) microscopic measurements of cross sections, (*c*) torsion-dynamometer measurements made with magnets of various strengths, (*d*) measurement by eddy-current instruments, and (*e*) accurate measurement of dimensions of the part before and after plating to determine thickness by difference.

Electrolytic stripping and microscopic measurements of cross sections are destructive methods that are most frequently used for purposes of (*a*) verification, (*b*) calibration, and (*c*) sampling of production runs. When calibrating instruments with prototype plated parts, using microscopic measurements of cross sections as umpire checks, several calibration reference curves may be required, depending on the parts being plated.

Measurements by properly calibrated eddy current or torsion-dynamometer instruments are affected by (*a*) surface finish of the deposit, (*b*) width and thickness of the piece, (*c*) surface contour, and (*d*) composition of the basis material. With a properly calibrated instrument, thickness measurements are usually within 10% of the actual thickness. Individual thickness measurements should not be used as a basis for acceptance or rejection; however, an average of several determinations from a well-calibrated instrument provide an acceptable measure of the mean thickness from a controlled process.

Normal variation in plate thickness that can be expected when plating the outside diameter of (*a*) cylinders, (*b*) rods, or (*c*) round parts racked as cylinders is ±0.2 μm/μm (±0.2 mil/mil) of plate intended. This has been determined over a period of several years by average quality level thickness measurements on piston rings racked as cylinders.

This normal variation of ±20% was confirmed in an actual production situation. In plating identical parts to a consistent plate-depth requirement, sample checks from 74 loads (110 000 parts) representing 27 days of operation were made to determine the plating tolerances that could be expected. The plating cycle was set to provide a plate thickness of 200 to 230 μm (8 to 9 mils) to meet a final requirement for minimum plate thickness of 150 μm (6 mils) after light stock removal during the subsequent finishing operation. Results of this analysis are shown in Fig. 3.

Throwing power of chromium-plating solutions is related to the ratio of chromic acid concentration to the catalyst concentration. Higher ratios give better throwing power at a given temperature and current density. This is evidenced by the fact that when a very low current density is present on certain areas of irregularly shaped parts, the cathode efficiency at that low current density is less for a solution high in sulfate than for a lower sulfate content. Therefore, less metal is deposited on the areas of low current density from a solution of high catalyst content.

The current density at which no metal deposits is greater for high catalyst solutions than for lower catalyst solutions. Also, metal deposits from a solution of low catalyst concentration at a current density that would be too low for depositing from a solution with high catalyst concentration. Thus, the following factors must be considered to ensure successful plating of complex shapes: (*a*) chemical balance, (*b*) operating variables, (*c*) type of anode, and (*d*) design of fixtures or racks.

Chromium plating requires far more attention to those variables which affect current distribution than (*a*) cadmium, (*b*) zinc, (*c*) copper, or (*d*) nickel plating. It is theoretically impossible to obtain the same current density at an inside corner as on the flat adjacent to it. An outside corner without shielding or thieving always has the highest current density and hence greatest plate thickness. Conforming anodes, shields, and thieves may be used to minimize thickness variation, but except on the simplest shapes, do not eliminate it.

Although some metal is deposited at low current densities in most other plating solutions, in chromic acid solutions there is a minimum current den-

Fig. 5 Parts difficult to plate uniformly even with use of specially contoured anodes

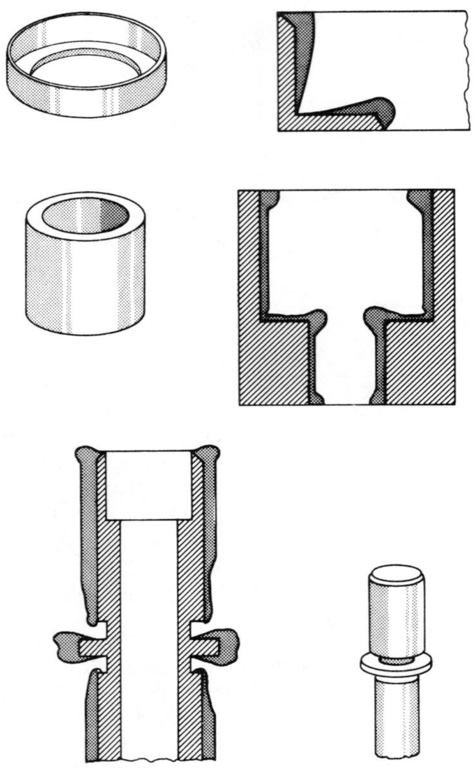

Variations in plate thickness shown are approximately to scale

sity for a given solution at a given temperature below which no metal is deposited. If an area of an internal or irregular shape receives less than this minimum current density, no deposition of metal occurs in this area. This explains why it is so difficult to chromium plate recesses and internal shapes without special anodes. Special hardware, in the form of thieves or shields, is required for lowering the current density on areas such as edges to prevent excessive buildup of deposit.

In most electroplating baths, the primary current distribution on an irregular object can be improved by increasing the tank anode-to-cathode distance. However, beyond a maximum distance, which depends on shape of the part, no further improvement can be attained. This maximum distance is proportional to both the size of the part and of the tank.

Because of the low throwing power of hard chromium-plating baths, an increase in the anode-to-cathode distance does not result in adequate thicknesses of deposit in sharp reentrant surfaces such as those formed by internal angles. For plating parts containing shapes of this type, conforming anodes or current shields, or both, must be used.

Figure 4 illustrates the relation between thickness of deposit and distance of the anode from the part being plated. In this instance, an alternative to an increase in the anode distance is the use of an anode contoured to the curvature of the part.

Special Anodes. When the part contains sharp, narrow recesses, such as grooves, a reduction of the anode distance may help to increase the thickness of the deposit at the bottom of the grooves. However, some parts with sharp-cornered grooves, bosses, and undercuts cannot be uniformly covered even when contoured anodes are used. Examples of parts in this category and the areas of heavy deposits are illustrated in Fig. 5.

Anodes used for plating recesses can be directly connected to the power supply or can be bipolar in nature. The bipolar anode has no direct electrical connection and takes advantage of the fact that current follows the path of least resistance. Bipolar anodes are an interesting curiosity which may have application in rare instances; however, direct connection of the anode to the positive direct current through a rheostat and ammeter, if required, is far more controllable.

The deposit on internal shapes can also be affected by the evolution of gas that occurs during plating. Gas can cause streaked deposits or produce a taper in a long bore. To minimize this effect, the parts should be positioned in a manner that permits the gas to move rapidly away from the part.

Because of fabrication problems encountered with lead alloys, complex-shaped anodes are made of steel, then coated with lead to produce the effect of solid lead anodes. These composite anodes are more economical and lighter in weight. However, the basis metal can be destroyed if there are pores or through holes in the lead alloy coatings. Brass or copper should never be used on the anode side, as it dissolves rapidly and seriously contaminates the solution. Low-carbon steel may be used alone for short runs, and lead-coated steel used for longer service.

Crack Patterns and Other Characteristics of Hard Chromium Plate

Quality of hard chromium plate is evaluated chiefly from the standpoint of (*a*) thickness and thickness distribution, (*b*) appearance, (*c*) crack pattern, (*d*) crack size, (*e*) porosity, (*f*) roughness, and (*g*) adhesion of the plate to the basis metal.

Surface Cracks. Normal deposits of chromium exhibit cracks in the surface. The pattern usually consists of crack-free areas, plateaus, completely surrounded by crack boundaries. The plateaus from an average conventional sulfate bath are 2 to 3 times larger than those from a mixed catalyst bath; that is, there are more cracks per inch in a deposit from the average mixed catalyst bath. When the ratio of chromic acid to catalyst in either type of bath is raised, size of the plateaus is increased. This increase in size can be brought about also by increasing temperature of the plating bath. When both ratio and temperature are high enough, a surface almost completely free of cracks is produced and is retained as long as current density remains below a certain critical level.

Chromium plate virtually free of cracks is smooth and silvery rather than bright. The speed of plating is about 25 μm/h (1 mil/h), using a current density of 45 A/dm^2 (3 A/in.^{2}), a bath temperature of 66 °C (150 °F), and a solution containing 282 g/L (37.5 oz/gal) of chromic acid. This deposit is softer than a bright chromium plate. It has a diamond pyramid hardness number of about 600, compared to about 900 or more for bright plate. The deposit has good lubricity and resistance to shock. Applications include (*a*) broaches, (*b*) cams, (*c*) dies for metal forming, (*d*) metalworking rolls, and (*e*) stamping dies for embossing silverware. Complicated shapes create a large range of current density and are difficult to plate with a crack-free surface. Corners, edges, or other high-current-density areas are most likely to crack during plating.

Porous Chromium. The cracks or porosity that characterize chromium deposits are undesirable for resistance to corrosion. Furthermore, they lower the fatigue resistance of the plated part. However, a porous structure can be advantageous in wear applications in which lubrication is required, because it promotes wetting action and provides oil retention after initial lubrication. Engine cylinders are the outstanding application.

Most chromium-plated cylinder surfaces consist of some form of interrupted surface, generally porous chromium. An interrupted surface may be obtained by electrolytic or chemical etching of chromium after it is plated on a smoothly honed bore, as with porous chromium, or by preroughening the bore by (*a*) shot blasting, (*b*) knurling, or (*c*) tooling, and then reproducing this roughness in the final chromium plate.

Two distinct types of porous chromium are produced. One has pinpoint porosity with many microscopic depressions in a honed chromium surface. This has been used in all types of engine cylinders except aircraft. Channel porosity is used for aircraft. This surface is also finish honed but is broken by random connected channels leaving isolated bearing plateaus.

For both types, the percentage porosity is generally controlled between 20 and 50% of the total area. Average plateau size is further controlled between 0.25- to 0.75-mm (0.010- to 0.030-in.) diam with the channel type of porosity. Porosity as low as 5% approaches dense chromium and is susceptible to scoring because of sparse oil distribution. High porosity, such as 75%, may cause high initial ring wear and high oil consumption. In normal engine service, cylinders coated with chromium of optimum porosity give wear rates of ⅓ to ⅒ better than those of uncoated cast iron or steel, hardened or unhardened.

Several methods, electrochemical, mechanical, or combinations of both, have been developed to provide controlled porosity in heavy chromium deposits. Mechanical methods entail either severe grit blasting of the surface to be plated or roughening of it with a fine knurling tool. The roughened surface is reproduced by the deposit. Using a patterned mask, the surface can also be roughened by chemical or electrochemical means before plating. The most widely used techniques, however, involve chemical or electrochemical etching of the chromium deposit after plating. Note that the pattern or crack density and the size of the plateaus are largely determined by the composition (ratio) of the bath, and the plating temperature.

Etching is performed on plated thicknesses ranging from 120 to 180 μm (5 to 7 mils). Porosity is developed after plating by electrochemically etching anodically in chromic acid solution. The etched surface is finished by (*a*) honing, (*b*) polishing, or (*c*) lapping. Metal removal that exceeds the depth of porosity must be avoided. To avoid accelerated wear in service, finished surfaces must be thoroughly cleaned of abrasive and chromium particles.

Quality-Control Tests. Usually, visual examination is sufficient for determining appearance and roughness of the surface of hard chromium plate. Magnetic particle inspection can be used to examine chromium plate up to 100 μm (4 mils) thick for cracks after grinding. The plate should be as smooth as the basis metal before plating and should be free of pits and nodules. The deposit should not exhibit excessive thickness variation. Particularly, deposits which have dendritic growths (trees) should be rejected. Adequate plating control requires that such dendritic deposits occur on thieves rather than in functional areas.

For process development and quality verification, destructive testing may be used to determine crack pattern and bond between the plate and basis metal. The crack pattern can be developed by etchants such as a hot 50 vol % hydrochloric acid aqueous solution, or by short-time deplating in a chromium-plating solution. The quality of the bond can be determined by (*a*) a punch test, (*b*) bend testing, (*c*) examining the bond line metallographically, or (*d*) judgment of ground or hammered samples. Well-bonded chromium, because of its low ductility, does not fail by pulling away from the bond line; however, it fails by cracking and spalling if it is subjected to excessive stress or distortion in 45° diagonal tension.

Excessive porosity of thin chromium plate, less than 25 μm (1 mil) thick, on steel can be determined by applying an acidified copper sulfate solution to the plated areas. The pores permit the solution to copper coat steel by displacement, and the degree of copper coating thus indicates the degree of porosity. Porosity can also be determined by the ferroxyl test described in *Metal Finishing Guidebook,* 1982.

The mandrel test can also be used in quality control. If a portion of the chromium plate is made anodic for 3 min at 15 A/dm^2 (1 A/in.^{2}) in a solution containing 250 g/L (33 oz/gal) chromic acid at 60 °C (140 °F), the crack pattern is developed. Counting the crack density under the microscope is an excellent procedure for noting the constancy of the composition (mainly ratio) and temperature of the solution.

Hardness of Plate

Because valid hardness measurements are difficult to make, and values are dependent on test conditions, hardness values should not ordinarily be used as quality control specifications and routine criteria. Rather, the as-plated brightness can serve as an indication of hardness.

The hardness of chromium plate cannot be accurately determined by the common hardness testers such as Brinell and Rockwell because hard chrome is generally too thin for these tests. The indentation produced in these tests distorts the base metal and is influenced by it.

The most reliable and most widely accepted hardness values are those obtained with the Vickers 136° diamond pyramid indenter or the Knoop indenter. With these the hardness test must be made on a carefully prepared and polished surface, preferably on a cross section of the plate to eliminate any possible influence of the basis metal on the hardness values obtained.

Cracks in the chromium plate influence the hardness values, depending on the type of indenter used and the load applied. In general, lighter loads are more sensitive to hardness variations and result in higher hardness values. Cracks influence values obtained with heavy loads more than values obtained with light loads. Also, because of the smaller area covered by the Vickers 136° diamond pyramid indenter, hardness values determined with this indenter are influenced less by underlying cracks than are values obtained with the Knoop indenter.

When conducting microhardness tests, it is important to make sharp and accurate impressions, particularly when using light loads. Also, when hardness values are reported, the following should be stated: (*a*) load, (*b*) type of indenter, and (*c*) optical system used. The importance of stating the load and type of indenter is indicated by the following data obtained on chromium plate from one plating cycle (each range or average represents 25 tests):

Load, g	Hardness value: 136° diamond pyramid	Hardness value: Knopp indenter
100	950-1110 (1040 avg)	940-1090 (1025 avg)
500	780-905 (850 avg)	685-890 (830 avg)

In an investigation of the scratch hardness of chromium deposits of all types, it was observed that bright or semibright deposits had the best combination of hardness and wear resistance regardless of plating conditions. In this investigation, the wear resistance was measured by means of a specially constructed abrasion hardness machine. The machine contained a small grinding wheel that revolved at 18 rev/min, and the number of 2-s cuts to grind through a 25 μm (1 mil) deposit on steel was an indication of the abrasion hardness. Results are summarized in Table 12, which shows the relation between appearance of deposits and their hardness and resistance to abrasive wear.

The effect of temperature on the hardness of electrodeposited chromium is often a significant factor in applications involving wear resistance. The electrodeposited metal begins to decrease in hardness when it is exposed to temperatures above about 205 °C (400 °F). Hardness decreases progressively with an increase in temperature, as shown in Fig. 6. As the hardness of chromium plate decreases, its resistance to wear may be affected adversely. Chromium plate should not be used for wear resistance in applications where service temperature exceeds 420 °C (790 °F).

Cost

Although the cost of hard chromium plating is high, this cost is justified when hard chromium plating is the only treatment that permits a part to meet service requirements. Several factors contribute to the high cost of hard chromium plating. These include (*a*) shape of the part, (*b*) design and arrangement of anodes, and (*c*) special finishing techniques. Labor charges are higher than for other plating processes largely because most chromium plating operations are performed manually; automatic or semiautomatic equipment is rarely used. Also, more labor is required to prepare some parts for plating or for finishing after plating.

Cost of plating increases with the complexity of the design of parts. Parts containing sharp notches, section changes, and unplated areas of cross holes necessitate more handling during preparation and finishing than do parts of simple shape. Also, anodes must more precisely conform to the shape of such parts for accurate control of plating thickness. Defects associated with chrome plate, such as dendritic growths and excessive buildups on external corners, can be minimized with tooling and special techniques. Depending on the nature of the part, these may increase cost. Special finishing techniques, primarily grinding and lapping, are required. Often, a heat treatment of plated parts is necessary before they are ground or lapped.

For surfaces that are difficult to finish mechanically, even greater care must be taken in design and construction of anodes to approximate desired results closely. The cost of anode design and construction must be amortized over the number of parts plated. If sufficient production is required, refinement of masking and anodizing generally pays with decreased afterplate finishing.

Costs are affected also by the quality of the plate. Hard, high-quality plating

Fig. 6 Effect of annealing temperature on the hardness of chromium plate

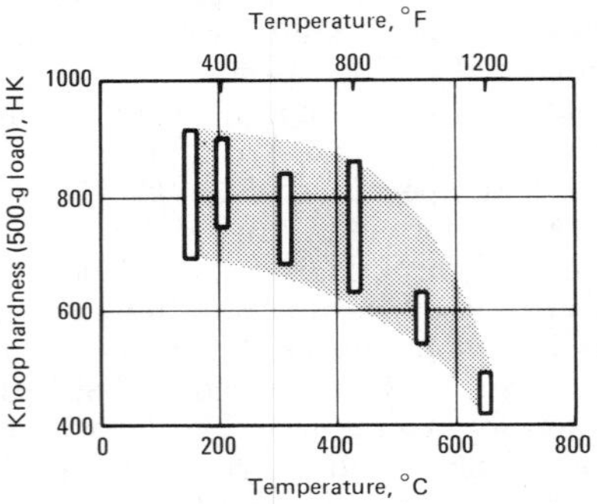

Chromium plate deposited during a single cycle in a mixed catalyst bath. All data represent 25 measurements of each condition; every readable impression was accepted as valid. 1-h heating cycles were used

Table 12 Relation between appearance and hardness of hard chromium plate

Average appearance	Average scratch hardness(a), HB	Average relative abrasive hardness
Matte (cold bath)	640	25
Milky	830	100
Slightly milky	990	290
Bright	1000	300
Slightly frosty	1005	300
Frosty (smooth)	1020	235
Frosty (rough)	1060	125
Burnt	1165	110

(a) Converted to Brinell scale from values obtained with a Bierbaum microcharacter using a 9 g (0.3 oz) load

finished to a bright luster is very expensive because of rigid controls required in its production. As with other plating processes, costs vary because of local utility charges and ordinances pertaining to disposal of wastes. Availability of power and water, and quality of the water available, can significantly affect cost. Safety requirements and waste disposal methods can be responsible for at least a 20% difference in plating costs among different plants.

Removal of Chromium Plate

Most customers require salvage of misplated parts. Further, there are significant numbers of parts in the aircraft industry, business machine industry, and plastic mold industry, where parts are run for the life of the plating and then overhauled by stripping worn plating and replating.

Methods of Stripping. Chemical, electrochemical, or mechanical methods are used to remove hard chromium deposits. When the basis material is steel, brass, copper, or nickel, hydrochloric acid at any concentration over 10 vol % and at room temperature or above removes chromium. In some operations, inhibitors are added to the acid solution to minimize attack on the steel substrate.

Chromium is removed electrochemically from steel or nickel by the use of any convenient heavy-duty alkaline cleaner at room temperature or above, at 5 to 6 V with reverse current. This method is unsatisfactory for nickel-based alloys which should be stripped chemically in hydrochloric acid. Chromium may be stripped from aluminum by making the part the anode in a cold chromium plating bath or in conventional chromic acid or sulfuric acid anodizing baths. Because aluminum alloys with a high alloy content and alloys subjected to various heat treatments all react differently in stripping solutions, precautions must be taken to prevent attack on the basis metal. Anodic stripping operations result in formation of oxide films on the basis metal. These films should be removed by one of the conventional deoxidizing processes prior to replating.

Stripping of chromium deposits from high-strength steel must be performed electrochemically in an alkaline solution. The parts are then stress relieved at 190 °C ± 14 °C (375 °F ± 25 °F) for a minimum of 3 h. The following solutions and operating conditions are recommended for removing chromium deposits from the materials indicated:

Removal from steel or nickel plated steel

- Sodium hydroxide, 45 to 230 g/L (6 to 30 oz/gal); anodic treatment at 3 to 8 A/dm^2 (0.2 to 0.5 A/in.^{2}); bath temperature 21 to 71 °C (70 to 160 °F)
- Anhydrous sodium carbonate, 45 to 60 g/L (6 to 8 oz/gal); anodic treatment at 2.5 to 5.5 A/dm^2 (0.15 to 0.35 A/in.^{2}); bath temperature, 21 to 66 °C (70 to 150 °F). Use 2.3 A/dm^2 (0.15 A/in.^{2}) with bath temperature of 66 °C (150 °F), to reduce possibility of pitting alloy steel
- Sodium hydroxide, 52 g/L (7 oz/gal); sodium carbonate, 30 g/L (4 oz/gal); anodic treatment at 8 A/dm^2 (0.5 A/in.^{2})
- Concentrated hydrochloric acid at room temperature
- Hydrochloric acid, 50 vol % at room temperature

Removal from aluminum and aluminum alloys

- Sulfuric acid, 67 vol %; glycerine, 5 vol %; anodic treatment at 1 to 3 A/dm^2 (0.1 to 0.2 A/in.^{2}); bath temperature, 21 to 27 °C (70 to 80 °F)

Removal from magnesium and magnesium alloys

- Anhydrous sodium carbonate, 50 g/L (6.5 oz/gal); anodic treatment at 2 to 5 A/dm^2 (0.15 to 0.30 A/in.^{2}); bath temperature, 66 °C (150 °F)

Removal from zinc and zinc alloys

- Sodium sulfide, 30 g/L (4 oz/gal); sodium hydroxide, 20 g/L (2.7 oz/gal); anodic treatment at 1.9 to 2.3 A/dm^2 (0.12 to 0.15 A/in.^{2}); bath temperature, 21 to 27 °C (70 to 80 °F)

Grinding is used occasionally to remove heavy chromium deposits. Because most defective chromium deposits are observed during subsequent grinding for finishing, it is sometimes expedient to continue grinding to remove all of the plate and then replate. In the grinding of heavy deposits for the removal of several thousandths of an inch of chromium to attain required dimensions or surface finish, the most important requisites for successful results are:

- A soft grinding wheel
- A sufficient amount of coolant
- A light cut
- Correct peripheral speed
- Freedom from vibration
- Frequent wheel dressing

Because chromium is hard and brittle, a soft grinding wheel is essential. A hard wheel forms a glazed surface, which results in a temperature rise that causes the chromium to crack. A soft wheel breaks down rapidly enough to prevent formation of a glaze; however, too soft a wheel is not economical because of rapid wheel wear. Good performance can be obtained with an aluminum oxide resin-bonded wheel made with a grit of about size 60 and of H-grade (hardness).

To prevent or minimize glazing, the contact area should be flooded with a coolant. Usually, the coolant is water with a small amount of soluble oil. Because of its hardness, excess chromium cannot be removed as rapidly as when grinding most other materials. The maximum thickness of metal removed should not exceed 5 μm (0.2 mil) per pass, and this amount should be reduced if there is any evidence of cracking. The optimum grinding speed is about 20.4 m/s (4000 sfm).

Effective grinding requires a rigid machine. Any appreciable vibration can cause cracking of chromium because of uneven contact pressure, and also results in a wavy surface. Several factors are essential for a rigid machine: (*a*) well-fitting spindle bearing, (*b*) balanced wheel, (*c*) heavy bed, and (*c*) well-supported workpiece. Whenever there is the least indication of glazing or nonuniform wheel surface, the wheel should be dressed with a diamond point. Adherence to the preceding recommendations will result in a good surface with a finish of 0.35 to 0.5 μm (14 to 20 μin.). Subsequent lapping (240 grit) will produce a finish of 0.1 to 0.3 μm (5 to 10 μin.).

Special care should be taken when grinding chromium-plated high strength steel parts (>1240 MPa or >180 ksi ultimate tensile strength) used in stressed applications. Numerous failures have occurred due to formation of untempered martensite caused by the heat of the grinding operation. For information and guidelines on grinding chromium-plated high strength steel parts, see military specification MIL-STD-866B.

Hydrogen Embrittlement

Factors that increase susceptibility to hydrogen embrittlement are (*a*) hardening of the steel, (*b*) grinding, (*c*) surface defects, (*d*) pickling, (*e*) cathodic cleaning, and (*f*) depth of plate relative to thickness and hardness of the part being plated. Unless the hydrogen absorbed during pickling and cathodic cleaning is removed, subsequent plating further embrittles the part to such an extent that breakage can occur during plating. This effect becomes more prevalent with increasing hardness of steel and on parts of thin cross section.

The thickness of plate on thin sections is of importance from the standpoint of notch effect. This was illustrated in an actual production setting. Thin sections of steel, 2.5 by 0.5 mm (0.10 by 0.020 in.) and 25 to 38 mm (1 to 1½ in.) long, were plated with chromium to a thickness of 8 to 13 μm (0.3 to 0.5 mil). The hardness of the steel was 57 to 59 HRC. These parts were aligned by being bent until permanently set. When the thickness of the chromium plate was increased to range from 15 to 23 μm (0.6 to 0.9 mil), the parts would break before taking a set. Baking them at 205 °C (400 °F) for 4 days did not relieve this condition. It was necessary to decrease the hardness of the steel to 53 to 55 HRC to prevent breakage of parts with heavier plate.

Stress Relieving Before Plating. Surfaces to be chromium plated must be free from stresses induced during (*a*) machining, (*b*) grinding or, (*c*) hardening. Stresses from the hardening operation may be further increased during grinding and result in microcracks. If the hardness of the steel is less than 40 HRC, it is unlikely that any damaging effect will occur as a result of residual stress. Steel with a hardness exceeding 40 HRC should be stress relieved before it is plated by heating at 150 to 230 °C (300 to 450 °F).

Baking After Plating. Steel parts with a hardness above 40 HRC should be baked at least at 190 °C (375 °F) for 4 h, after plating to ameliorate the effects of hydrogen embrittlement. This treatment should be started as soon as possible, preferably within 15 min after plating. The fatigue strength of parts subjected to alternating stresses is reduced by the baking treatment; such parts should be shot peened before plating.

The applications of shot peening and baking, as related to hardness of steel to be chromium plated, are described in Federal Specification QQ-C-320B (Amendment 1) as follows:

- Plated parts below 40 HRC and subject to static loads or designed for limited life under dynamic loads, or combinations thereof, shall not require shot peening prior to plating or baking after plating
- Plated parts below 40 HRC which are designed for unlimited life under dynamic loads shall be shot peened in accordance with military specification MIL-S-13165 before plating. Unless otherwise specified, the shot peening shall be accomplished on all surfaces for which the coating is required and on all immediately adjacent surfaces when they contain notched fillets, or other abrupt changes of section size where stresses will be concentrated
- Plated parts which have a hardness of 40 HRC, or above, and are subject to static loads or designed for limited life under dynamic loads, or combinations thereof, shall be baked at 190 ± 14 °C (375 ± 25 °F) for not less than 3 h after plating
- Plated parts which have a hardness of 40 HRC, or above, and are designed for unlimited life under dynamic loads, shall be shot peened in accordance with military specification MIL-S-13165 before plating. Unless otherwise specified, the shot peening shall be accomplished on all surfaces for which the coating is required and on all immediately adjacent surfaces when they contain notched fillets, or other abrupt changes of section size where stresses will be concentrated. After plating, parts shall be baked at 190 ± 14 °C (375 ± 25 °F) for not less than 3 h

Safety Precautions

Potential hazards that attend the chromium-plating process are associated with various chemicals used in (*a*) acid and alkaline cleaners, (*b*) vapor degreasers, and (*c*) chromium plating bath. The mixing of certain of these chemicals can result in a violent reaction. Consequently, all tanks and containers should be plainly marked to indicate their contents and to prevent the inadvertent addition of an improper chemical to a solution. Chromic acid flakes should not be stored near alcohols or similar solvents, because the oxidizing action of the flakes may ignite the solvents. Empty chromic acid cans should be flushed before they are discarded, to avoid fire hazards.

Personnel should wear rubber gloves, rubber aprons, and face shields when making additions to any plating or cleaning bath. Tank operators should wear proper protective clothing at all times. Face shields are usually mandatory; protective hand creams are recommended. The operator should spray liquid vaseline into his nose at the beginning and middle of each shift, to reduce irritation from acid fumes.

Recovery and Disposal of Wastes

Chromic acid wastes may be either recovered or destroyed, and the decision regarding which process to select should be based on a comparison of (*a*) initial costs, (*b*) labor for operation and maintenance, (*c*) chemical costs, (*d*) space requirements, and (*e*) utility costs. The volume of wastes and the value of the chromic acid and rinse water saved can greatly influence the choice.

Local, state, and federal authorities are constantly increasing their attention to antipollution programs. Strict regulations are being enforced regarding the allowable limits for chromic acid wastes that are dumped into rivers, lakes, and sewage disposal systems. The prevailing limits for chromic acid contamination of waste water range from about 0.05 to 5 ppm. These limits vary for each locality, depending on (*a*) uses of the receiving body of water, (*b*) supplementary water flows that affect dilution, or (*c*) ability of sewage plant to handle wastes.

Preventive Measures. The problem of waste disposal can be greatly minimized if suitable measures are taken to reduce to a minimum the amount of wastes produced. The following practices contribute to minimizing wastes:

- Extend drainage periods to permit more solution to return to the tank. In hand operations, this is made possible by providing a drainage bar over the tank to hold racks
- Provide drip boards to return solution lost when going from tank to tank
- If possible, rack parts in such a way as to eliminate cupping action
- Use reclaim rinse tanks. The rinse solution can be used to maintain the level of liquid in the processing tank. Sometimes concentration methods may be profitable to facilitate use of rinse waters
- Control drag-in of water to permit use of reclaim rinse tanks
- Complete recovery should be used only in conjunction with removal of metallic impurities (cationic resin). For regeneration of cationic resins, use only sulfuric acid, never hydrochloric acid

Disposal of chromic acid wastes is based on reduction of hexavalent chromium to the trivalent form and, in either a batch or a continuous operation, precipitating the metal by means of an alkali. The actual chemicals used vary from locality to locality, depending on cost and availability. Chromic acid is first acidified to a suitable pH and is then reduced with (*a*) one of the sulfite compounds (sodium sulfite, sodium metabisulfite), (*b*) sulfur dioxide, (*c*) ferrous sulfate, (*d*) iron, (*e*) copper, or (*f*) brass. After completion of reduction, trivalent chromium is precipitated as hydroxide with alkali. The amount of chemicals required to complete reduction can be governed by laboratory analysis, or, because the reaction is solely one of oxidation-reduction, it may be controlled automatically by use of electrodes.

The most regularly used reducing agent is sulfur dioxide gas. It can be obtained in liquid form in cylinders of various sizes, is comparatively inexpensive, and can be fed directly into the treatment tank. The rate of addition is easily controlled and gas is delivered from the cylinder under its own pressure. A lower initial acidity is required, because the gas forms sulfurous acid when dissolved in water. The operating pH is 2 to 3, and the ratio of sulfur dioxide to chromic acid used commercially is slightly under 3 to 1. The sulfur dioxide method lends itself readily to an automatic system because the gas feed can be controlled by a flowmeter, and the reaction can be controlled by oxidation-reduction potentials.

Ferrous sulfate also is a widely used reducing chemical, especially in localities where large quantities are available from pickling plants. The quantity required can be easily determined by titration. The ratio of ferrous sulfate to chromic acid varies from 5 to 1 to 16 to 1. Reduction of chromium is followed by neutralization with lime or caustic. Above a pH of 7, the metals precipitate as hydroxides, together with calcium sulfate. The main disadvantage of the ferrous sulfate method is the large volumes of sludge that have to be handled. The sulfite-containing compounds generally are slightly more expensive than sulfur dioxide or ferrous sulfate. In addition, several difficulties are involved in sulfite treatment, such as (*a*) solubility, (*b*) loss of hydrogen sulfide through hydrolysis, (*c*) slightly lower pH, and, occasionally, (*d*) need for additional treatment to complete the process.

Regardless of the chemical treatment selected, all chromic acid disposal systems require collection, treatment, and settling tanks. Operating procedure consists of (*a*) chemical additions, (*b*) mixing, (*c*) separation of precipitated metal, (*d*) clarification, and (*e*) sludge disposal. Variations in equipment design affect economy, time and labor requirements, and equipment costs.

Appendix

Stop-Off Media for Selective Plating

During plating, part surfaces that are not to be plated are protected from the solution by stop-off media, such as (*a*) lacquers, (*b*) foils, (*c*) tapes, (*d*) waxes, and (*e*) machined reusable fixtures. Stop-off media must (*a*) adhere well to the metal surface; (*b*) not become soft at temperature of the solution or be brittle at room temperature; (*c*) be resistant to solutions used for cleaning, etching, and plating; and (*d*) easy to remove after plating.

Lacquers used to prevent surfaces from being plated can be easily applied by brushing, spraying, or dipping. After plating, the lacquer can be stripped off or dissolved in an appropriate solvent.

Lead sheet, foil, and wire not only provide a positive stop-off, but also act as stealers to aid in leveling current distribution. Lead can be pounded into holes, keyways, or slots and trimmed with a sharp knife.

Tapes of several kinds are used as stop-off media. They vary from adhesive tapes backed with lead foil to tapes made of vinyl and other plastics. Lead foil tapes combine a specially compounded lead foil with a highly pressure-sensitive adhesive to provide a quick and convenient stop-off for short runs. The lead backing is useful as a stealer in areas of high current density or it can be lacquered when used in areas of low current density for equalizing current distribution. It is soft enough to conform to various configurations. Vinyl and other plastic tapes are soft and pliable, and have extruded edges for providing a leak-free seal on almost any contour.

Sheet Materials. For large production runs it is convenient to make stop-off forms that can be reused many times. Plastic sheet, generally 0.1 to 0.15 mm (0.004 to 0.006 in.) thick, is excellent for masking simple plates, cylinders, or other configurations. Steel sheet is sometimes substituted for plastic if it is desirable to equalize current distribution. To prevent plating or corrosion of the basis metal, the plastic or steel stop-off must adhere firmly to the area being masked. Snug-fitting cylinders can be made to fit inside or outside diameters. To mask areas that are flat or of irregular shape, lacquer may be used to glue the stop-off material to the part; the lacquer may be removed with a thinner after plating.

Waxes. Several waxes designed for use as stop-offs are obtainable commercially. The use of a dip tank thermostatically controlled to maintain temperature −1 to 4 °C (30 to 40 °F) above the melting point of the wax makes use of these materials comparatively fast and simple. The portion of the part to be plated can be covered with masking tape to prevent wax from adhering; or, if desired, the whole part can be coated and the wax stripped with a knife from areas to be plated. Choose a wax mixture with a melting point low enough to allow removal of the bulk with boiling water. Because they evolve poisonous fumes when heated, waxes containing chlorinated naphthalene must be provided with exhaust equipment. High-melting mineral and vegetable waxes are not dangerous to use.

SELECTED REFERENCES

- Joseph P. Branciaroli, Control of Chromium Plating Baths, *Product Finishing* (Cincinnati), Vol 25, December 1960, p 50-56
- J.K. Dennis and T.E. Such, *Nickel and Chromium Plating,* New York: John Wiley & Sons, 1972
- George Dubpernell, *Modern Electroplating,* 2nd ed., New York: John Wiley & Sons, 1963, p 80-140
- George Dubpernell, The Development of Chromium Plating, *Plating,* Vol 47, January 1960, p 35-53
- R.J. Girard and Edward F. Koetsch, Jr., Chromium Plating of Rifle Barrels, *Tech Proc Am Electroplaters' Soc,* Vol 47, 1960, p 199-206
- J.P. Hoare, *Journal of the Electrochemical Society,* Vol 126, 1979, p 190
- J.P. Hoare, A.H. Holden and M.A. LaBoda, *Plating and Surface Finishing,* Vol 67, March 1980, p 42
- M.A. LaBoda, A.H. Holden and J.P. Hoare, *Journal of the Electrochemical Society,* Vol 127, 1980, p 1709
- I. Laird Newell, Waste Disposal for Metal Finishing Industries, *Plating,* Vol 48, April 1961, p 373-378
- J.G. Poor, H. Chessin and C.L. Alderuccio, Adherence of Thick Chromium Deposits as Affected by Surface Preparation, *Plating,* Vol 47, July 1960, p 811-813
- Ing. Robert Weiner and Adrian Walmsley, *Chromium Plating,* England: Finishing Publications Ltd., 1980

Decorative Chromium Plating*

DECORATIVE CHROMIUM PLATING is differentiated from hard chromium plating in thickness and the type of undercoating used. Decorative chromium coatings are very thin, usually not exceeding an average thickness of 1.25 μm (50 μin.). Decorative chromium is applied over undercoatings, such as nickel or copper and nickel, which impart a bright, semibright, or satin cosmetic appearance to the chromium. The choice of undercoatings, as well as the type of chromium applied, can also provide corrosion protection. Currently, most decorative chromium coatings are applied from hexavalent chromium processes, based on chromic anhydride (CrO_3). Since 1975, however, trivalent chromium processes have become available commercially.

Hexavalent Chromium Plating

Deposits from decorative chromium plating baths generally range from 0.13 to 1.3 μm (0.005 to 0.05 mil) in thickness. Because of the thinness, deposits from properly operated baths generally reproduce the finish of the substrate. To obtain optimum luster of the final chromium deposit:

- Plate the substrate coating to a uniformly bright condition
- If the substrate is nonuniform, grainy, hazy, or dull, polish and buff it to a uniformly high luster before plating chromium
- If the final chromium coating over a uniformly bright substrate is hazy on certain areas, strip and replate the chromium, or buff the hazy areas on a wheel. Buffing of chromium is not allowed where corrosive service conditions are to be encountered

In addition to being lustrous, the final chromium deposit should cover all significant areas. If these areas are not adequately covered because of an improperly operated chromium bath, the chromium can be stripped, the substrate reactivated if necessary, and the part chromium plated again.

Excessively high current densities, improper temperatures, and passivated substrates can result in hazy, nonuniform chromium deposits. Proper operating conditions, properly operated nickel baths, and other similar precautions ensure uniformly lustrous chromium deposits.

Adhesion of chromium to an active or properly prepared substrate is not usually a problem. However, if substrate coatings are improperly applied and their adhesion is questionable, the chromium plating operation may cause the intrinsically poor adhesion of the substrate coating to show up as blistering or exfoliation, either immediately after chromium plating, in storage or in service. Chromium plating, can be a good way of accentuating faulty adhesion of basis coatings and indicating inadequate adhesion.

Porosity and Cracking. Up to a thickness of about 0.13 μm (0.005 mil), chromium deposits are highly porous. When these coatings are applied to a basis coating such as nickel, a relatively large area of nickel is exposed through the pores. Porosity decreases with increasing thickness, and at about 0.5 μm (0.02 mil), deposits are almost completely pore-free (Fig. 1).

When chromium is deposited from solutions operated below 50 °C (120 °F), the deposit begins to craze when it exceeds 0.5 μm (0.02 mil) in thickness. A macrocrack pattern visible to the unaided eye appears. This pattern generally has 12.5 to 25 cracks per centimetre (5 to 10 cracks per inch).

When a bath is prepared and operated to provide a crack-free deposit, porosity decreases with increasing thickness, so that at 1.3 μm (0.05 mil) or more a substantially pore-free deposit is attained. High bath temperatures of 50 to 55 °C (120 to 130 °F) promote crack-free deposits.

With a composition that creates a microcracked structure, the number of cracks per unit of lineal distance increases with increasing thickness. A microcracked deposit generally has more than 275 cracks per centimetre (700 cracks per inch), usually ranging

*Revised by Louis Gianelos, Manager, Technical Service, Industrial Finishing, Harshaw Chemical Co.

Fig. 1 Porosity in chromium plate as a function of plate thickness

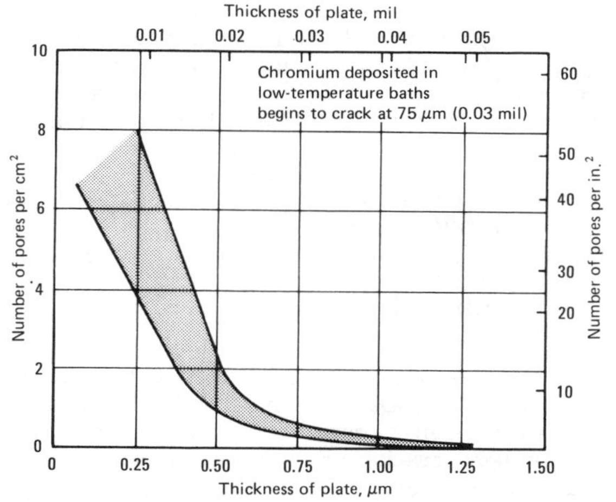

from 395 to 985 cracks per centimetre (1000 to 2500 cracks per inch).

Microcracked chromium plate is is chromium that is electrodeposited over conventional copper-nickel or all-nickel substrates in such a manner that the chromium contains a multiplicity of fine cracks that expose the underlying nickel.

In an outdoor corrosive environment, as well as in accelerated corrosion tests, corrosion has been observed to proceed by galvanic cell action between nickel and the chromium, with nickel acting anodically. Because the rate of penetration of corrosion through the nickel layer is a function of the anodic current density of the corrosion cell, the reduction of current density obtained by the increase in exposed nickel area prolongs the time required to penetrate a given thickness of nickel. Microcracked chromium deposits can be obtained from systems using single specially formulated chromium solutions, dual specially formulated chromium solutions such as duplex chromium or by the use of special stressed nickel strike deposits preceding conventional chromium solutions.

The advantage of microcracked chromium lies in its ability to provide long-term corrosion protection without developing the fine surface pitting associated with long exposure of duplex nickel. The finish as applied has a very slight haze when viewed critically, but not to an objectionable degree.

Crack Pattern. Microcracked chromium should exhibit a uniform pattern of cracks and contain a minimum of 275 cracks per centimetre (700 cracks per inch), though 475 cracks per centimetre (1200 cracks per inch) is preferable. Crack density can be as high as 790 cracks per centimetre (2000 cracks per inch). Gross cracks extending through the bright nickel cannot be tolerated. Factors favoring a desirable crack pattern include (*a*) relatively high bath temperature, (*b*) low chromic acid concentration, (*c*) relatively high fluoride content, and (*d*) thickness. Although an adequate microcrack pattern can be obtained with a thickness of 0.5 to 0.64 μm (0.02 to 0.025 mil) of chromium, 1.27 μm (0.05 mil) might be required for most bath compositions. Unfortunately, conditions that favor formation of the microcrack pattern usually have an adverse effect on the covering power of the chromium solution. This has been resolved through the use of two successive chromium baths; the first is used to obtain coverage and thickness and the second to create the microcrack pattern.

Satisfactory coatings are not too difficult to obtain on steel parts with relatively simple shapes, but zinc-based die castings can present a serious problem because of the difficulty of obtaining adequate thickness of chromium in low current density areas. A matter of major concern is the possibility of long plating times, 8 min or more, that may be necessary to obtain the required thickness for the development of the crack pattern. Long plating times are not required, nor desirable, where the stressed nickel strike system is used. All plating bath parameters must be closely controlled for optimum performance when plating complicated shapes. The fluoride content of the bath must be closely controlled, within 1.5 to 2.25 g/L (0.20 to 0.30 oz/gal) for steel parts and 1.9 to 2.63 g/L (0.25 to 0.35 oz/gal) for die castings, to form the optimum crack pattern in such areas. It may also be necessary to extend the plating time and to use surge plating, which permits the use of a higher average current density.

Microdiscontinuous chromium deposits, either microcracked or microporous, may also be produced by altering the nature of the preceding nickel deposit. Briefly, microcracking can be produced by using a thin layer, about 1.25 μm (0.05 mil), of a highly stressed nickel deposit preceding chromium plating. The advantage of this procedure is that microcracking can be obtained with the usual decorative thicknesses of chromium, for example, 0.25 to 0.5 μm (0.01 to 0.02 mil). Crack density can be varied from approximately 300 cracks/cm (750 cracks/in.) to 800 cracks per centimetre (2000 cracks per inch) by varying the conditions in the nickel bath.

Microporous chromium can be obtained by, again, using a thin layer of nickel deposited from a solution containing large quantities of very fine inert particles followed by chromium plating. Pore densities averaging 16 000 pores per square centimetre (100 000 pores per square inch) can be obtained easily, and variation is obtained by varying the quantity of particles or type of additives used in the nickel bath. A disadvantage of these processes is the addition of an extra plating tank and rinse tanks between nickel and chromium. More information can be found in the article on nickel plating in this Volume.

Chromium Bath Composition

Sometimes referred to as the ordinary or conventional bath, the oldest chromium plating bath used for decorative plating consists of an aqueous solution of chromic anhydride (CrO_3) that also contains a small amount of soluble sulfate ($SO_4^=$), referred to as a catalyst, added as sulfuric acid or as soluble sulfate salt such as sodium sulfate. When dissolved in water, the chromic anhydride forms chromic acid, which is believed to exist in the following equilibrium:

Table 1 Compositions and operating conditions for two chromium plating baths

Constituent or condition	General decorative	Bright crack-free
Chromic acid	250 g/L (33 oz/gal)	260-300 g/L (35-40 oz/gal)
Ratio of chromic acid to sulfate	100:1 to 125:1	150:1
Operating temperature	38-49 °C (100-120 °F)	52-54 °C (125-130 °F)
Cathode current density	7.5-17.5 A/dm² (75-175 A/ft²)	25-30 A/dm² (250-300 A/ft²)

$$H_2Cr_2O_7 + H_2O \rightleftarrows 2H_2CrO_4$$

The ratio of chromic acid to sulfate, generally given as the weight ratio of chromic anhydride to sulfate, governs the current efficiency for chromium metal deposition.The cathode current efficiency is also affected by solution variables such as (*a*) concentration of chromic acid, (*b*) temperature, and (*c*) content of metallic impurities. The content of metallic impurities is an important consideration for commercial operation, because an excessively high content of impurities such as copper, iron, zinc, and nickel seriously affect bath conductivity, cathode current efficiency, throwing power, and covering power, even if the ratio of chromic anhydride to sulfate is within optimum limits for the application.

Most chromium is deposited within the following operating limits:

Chromic anhydride.... 200 to 400 g/L (27 to 54 oz/gal)
Chromic anhydride-to-sulfate ratio........ 80:1 to 125:1
Cathode current density...........7.5 to 17.5 A/dm² (75 to 175 A/ft²)

These wide limits encompass a broad variety of decorative applications. The current trend is toward using chromic anhydride concentrations of 250 to 300 g/L (33 to 40 oz/gal) and avoiding the more highly concentrated baths. A few favor a low limit of 150 g/L (20 oz/gal); others the high concentration of 400 g/L (54 oz/gal). In any event, the ratio of chromic acid to sulfate ion is usually maintained close to 100 to 1.

With the development of duplex, microcracked, and crack-free applications, specialized bath compositions and operating conditions have come into use. Many of these, however, are either proprietary or are not subjects of general agreement. Table 1 provides information for a general decorative chromium-plating bath and a bright crack-free bath.

To meet specific requirements of plating speed, activation of nickel, and crack pattern, the chromic anhydride and sulfate concentrations should be properly correlated with temperature and cathode current density limits. In preparing a bath and establishing operating conditions, the following relationships should be considered:

- An increase in the temperature of the bath, except for mixed catalyst baths, (*a*) decreases the cathode efficiency, (*b*) decreases the number of cracks per unit length, (*c*) decreases coverage at low current density, (*d*) increases the limiting current density at which burning occurs, and (*e*) increases passivating action on nickel
- An increase in the ratio of chromic anhydride to sulfate decreases the number of cracks per unit length, and increases the nickel passivity, which can cause whitewash
- At constant chromic anhydride concentration, temperature, and cathode current density, an increase in sulfate increases the cathode current efficiency to a maximum; beyond this point, any further increase in sulfate concentration can cause a decrease in cathode current efficiency

In dilute chromic acid solutions, any small carryover of soluble sulfates from earlier solutions can quickly upset the balance of the solution. However, dilute solutions have a higher cathode efficiency and a slightly wider bright range, although these solutions require higher tank voltages to maintain desired current density.

With other plating conditions (such as temperature and current density) held constant, the plating operation can be seriously disrupted by any change in the ratio of chromic acid to sulfate. Raising the chromic acid concentration to 400 g/L (54 oz/gal) greatly reduces the cathode efficiency and the bright range of chromium plating solutions. However, these high concentration solutions have greater tolerance to metallic impurities and higher conductivity which requires lower voltages to maintain current density. Also, without proper and sufficient reclaim rinses, the cost of processing supplies is higher, because of greater dragout on the plated article. However, reclaiming dragout also reclaims contaminants.

Because there are advantages and disadvantages for using either high or low chromic acid content, some compromise is necessary. The size and shape of the article to be plated and the equipment and power available often determine exactly what solution is used. In decorative chromium plating, constant attention must be given to keeping all variables in the proper relationship; frequent bath analysis and prompt adjustments are essential to maintain balanced conditions.

Mixed Catalyst Baths. Since the mid 1950's, a number of mixed catalyst proprietary chromium plating baths have been developed. These baths offer the following advantages: (*a*) increased cathode current efficiencies, (*b*) increased activating action on nickel and stainless steel, (*c*) improved coverage at low current density, (*d*) broader bright plating ranges, and (*e*) capability of improved decorative chromium applications including dual, microcracked, and bright or dull crack-free.

Mixed catalyst compositions contain chromic acid, sulfate and fluoro compounds (frequently fluosilicate ions) as active ingredients. Proprietary baths, formulated to regulate the concentration of the catalyst ions, contain strontium or calcium salts to control the solubility of sulfate ions, or potassium salts to control the solubility of fluosilicate ions. For details concerning compositions of mixed catalyst compositions see U.S. patent No. 2,640,022 (1953), 2,841,540 (1958), 2,841,541 (1958), and British patent No. 813,445 (1959). Most control requirements applicable for standard baths also apply for proprietary baths.

With the exception of ratio control, the problems for low, medium, or high chromic acid concentrations in mixed catalyst baths are the same as those in conventional chromium baths. After optimum conditions are found, the same close control must be maintained to prevent mischromes (absence of plate) in areas of low current density, and blue, matte, or burnt deposits at high current density areas. Because supplies for mixed catalyst solutions are more expensive than chromic acid alone, using less concentrated solutions can be an advantage. However, if proper reclaim rinses or recovery methods are used, this condition may be unimportant.

Table 2 Bath compositions and conditions for plating microcracked chromium

Part	Chromic acid g/L	Chromic acid oz/gal	Fluoride g/L	Fluoride oz/gal	Chromic anhydride to sulfate ratio	Temperature °C	Temperature °F	Current density A/dm^2	Current density A/ft^2
First plating bath									
Steel	338-375	45-50	...	...	100:1	46-52	115-125	10-15	100-150
Zinc	375-413	50-55	...	...	140:1	46-52	115-125	12-16	120-160
Second plating bath									
Steel	165-195	22-26	1.5-2.25	0.20-0.30	180:1	43-54	120-130	9-12.5	90-125
Zinc	210-233	28-31	1.9-6.23	0.25-0.35	180:1	52-57	125-135	9-15	90-150

Baths for Microcracked Chromium. Usually, two chromium solutions are used successively to produce microcracked chromium plate. The first chromium solution may be either conventional or proprietary and is op- erated in a normal manner. A plating time of 8 min is preferred when recessed areas are involved, although plating times of 5 to 6 min are often used. Surging of the current may be used to increase plating speed. Composition ranges and operating conditions for nonproprietary first plating solutions used to plate steel and zinc parts are given in Table 2.

The second chromium solution, similar to the first solution, may be proprietary or nonproprietary, but the chromic acid concentration is lower, and fluosilicate ions must be present in the bath. Plating time is approximately the same as in the first solution, 5 to 8 min. Again, surging can be used if desired. Composition ranges and operating conditions for the second chromium bath are indicated in Table 2. The plating conditions used are governed by the nature of the parts being plated. Solutions for parts having deep recesses have higher chromic acid and fluoride content and lower sulfate. Thickness, however, must be weighed against other influences on microcrack formation, and for this reason, operating conditions can be established on a firm basis only by actual operation with the parts to be processed. On simple shapes, the second plating bath formulation may be used alone.

The use of a rinse or rinses between the two chromium solutions is not essential to the process, but it may help avoid control problems because of dragout from the first chromium solution to the low concentration second solution. When used, these rinses may be operated as reclaim tanks to minimize dragout losses.

Solution Control. Regardless of the chromium bath used, periodic analyses are required. For information on control procedures, see the article on hard chromium plating in this Volume.

Temperature of Chromium Baths

All chromium plating solutions require control of temperature, current density, and solution composition. The exact temperature at which bright, milky, frosty, or burnt deposits occur depends on solution composition and current density.

Chromium plating is performed within the range of 38 to 60 °C (100 to 140 °F); 46 to 52 °C (115 to 125 °F) is the most common operating range. At room temperature, the bright plating range is impractically narrow.

In a process set up to plate at 50 °C (120 °F) with all variables properly controlled, the temperature need vary only 1.5 or 2 °C (3 or 4 °F) up or down to move the electrodeposit out of the clear, bright range. Consequently, an accurate temperature controller and facilities for rapid cooling and heating of the bath are essential. Variation outside the ±1.5 °C (±3 °F) range may cause an unacceptably high rejection rate or may necessitate costly stripping and replating operations.

Current Density in Chromium Plating

The standard sulfate bath is usually operated in the range of 10 to 16 A/dm^2 (100 to 160 A/ft^2). A current density of about 10 A/dm^2 (100 A/ft^2) is used for solutions maintained at 38 °C (100 °F). A higher current density, sometimes as high as 30 A/dm^2 (300 A/ft^2), is required for solutions at 55 °C (130 °F). The choice for a specific use depends on such variables as the complexity of the article being plated, and the equipment available. After the current density has been established, close control must be maintained.

Changes in the ratio of chromic acid to sulfate require compensating adjustments in current density. If the sulfate content increases (lower ratio), current density must be increased to maintain full coverage in areas of low current density. If sulfate is decreased (higher ratio), current density must be decreased to prevent burning on areas of high current density.

An increase in temperature may require an increase in current density to ensure full coverage in areas of low current density. A decrease in temperature may require a decrease in current density to prevent gray (burnt) deposits in areas of high current density.

As chromic acid content increases, higher current densities can be used. Cathode efficiencies of most conventional chromium solutions are about 13% over a wide range of concentration, making it possible to plate for shorter times when using the most concentrated solution.

Anodes for Chromium Plating

In chromium plating, insoluble lead or lead-alloy anodes are almost always used. Chromium metal is supplied by chromic acid in the electrolyte.

Pure lead anodes are often attacked excessively by idle baths, which causes the formation of a heavy sludge of lead chromate on the bottom of the tank, making pure lead anodes impractical for all but continuous operations. During plating, a coating of lead peroxide forms on the anode. This coating favors oxidation of trivalent chromium at the anode. However, when the bath is idle, the coating dissolves to some extent in the solution, making attack on the anode possible.

To reduce the attack of the chromic acid bath on the anode, several lead alloys are being used. For conventional sulfate baths, 6 to 8% antimonial lead is preferred. For solutions containing fluoride, lead alloys containing 4 to 7% tin are used.

For an anode to provide optimum throwing power and coverage, the anode must be positioned properly in relation to the work, and must have a continuous and uniform film of lead peroxide on the entire surface. Anodes with crusty surfaces have low conductivity and should be cleaned periodically by wire brushing or alkaline cleaning to ensure proper current distribution. The function of the anode is

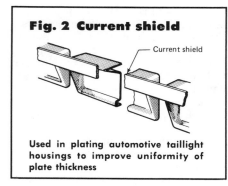

Fig. 2 Current shield

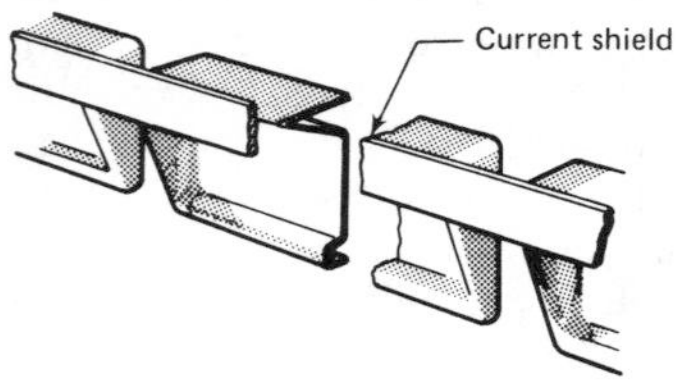

Used in plating automotive taillight housings to improve uniformity of plate thickness

Fig. 3 Use of auxiliary anode for a part difficult to plate to uniform thickness because of concave surface

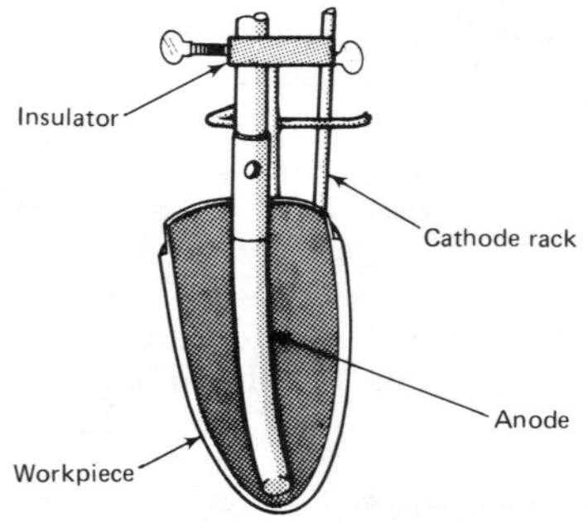

Fig. 4 Rack assembly for decorative chromium plating

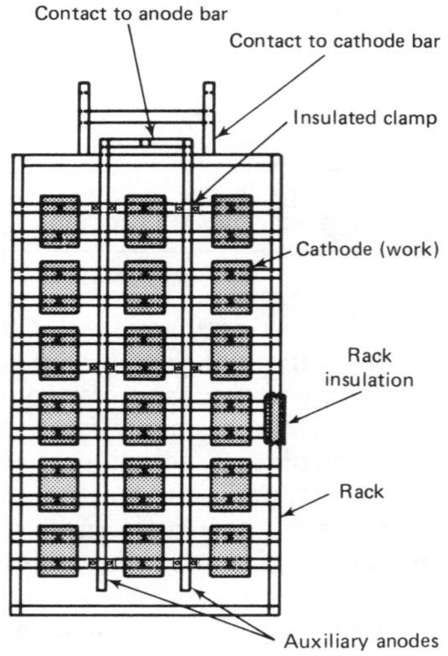

Shows position of cathodes and auxiliary anodes

not only to conduct the plating current, but also to oxidize trivalent chromium formed at the cathode back to hexavalent chromium. To accomplish this, the anode area should be adjusted to provide the optimum anode current density for the oxidation necessary to keep the trivalent chromium at the desired level, usually 0.25 to 1.0 g/L (0.033 to 0.13 oz/gal). In decorative chromium plating, an anode-to-cathode area ratio of 2-to-1 is common for proper reoxidation and balance. If trivalent chromium continues to increase above the desired level, the anode area should be increased to the point where the trivalent chromium concentration remains stable. Overheating of the bath may occur if the anode area is so small that resistive heating becomes a factor.

Anodes with round cross sections are most commonly used. When maximum anode area is desired, corrugated, ribbed, ridged, and multiedged anodes are used. The round anode is preferred because its surface is active on its entire circumference, enabling it to carry higher amperage at lower voltage. The absence of inactive areas on this anode minimizes the formation of lead chromate film, reducing maintenance requirements. If the weight of the anode presents a problem, hollow round anodes can be used. Such anodes provide a 25 to 40% reduction in weight; however, their current-carrying capacity is less than that of solid anodes.

Anodes are manufactured by extrusion. The contact with the busbar may be provided by a copper hook homogeneously burned to the extruded anode. Pure nickel, nickel-plated copper, and lead-coated copper are also used for hooks. Several hook styles are used; the knife-edge hook is preferred. The hook and the top of the anode are covered with plastisol for protection against corrosion by fumes and dragout drip.

Control of Current Distribution in Chromium Plating

Chromium plating baths have poorer throwing and covering power than most other plating baths (such as nickel plating baths). To obtain thickness and coverage in low current density areas, special auxiliary anodes are sometimes used.

Any workpiece of intricate shape constitutes a problem of proper current distribution when nonconforming anodes are used. The current density, and the thickness, on a workpiece varies from highest on corners, edges, and areas closest to the anode to lowest on recesses and areas distant from the anode. Variations in current density result in differences of cathode efficiency, which accentuates the problems of uneven plate, burning, or complete absence of plate. These problems can be overcome to some extent by special racking and shielding techniques such as:

- Wide spacing of concave parts on rack
- Increasing the distance between workpiece and anode
- Intentional shielding of a projection on one workpiece with a depression on an adjacent one
- Orienting of areas of low current density toward the periphery of the fixture
- Moving the parts in the center of the rack closer to the anodes than those on the periphery of the rack

Improved coverage on areas of low current density can be achieved with striking, that is plating at high current density for a short period of time. The striking time duration is kept to a minimum, usually 5 to 20 s, to avoid burning. Plating is continued at normal current density after the strike.

Current Shields. A nonconducting plate or panel, termed a current shield, may be mounted on the plating rack to direct current away from areas of high current density or to direct additional current into areas of low current density. Figure 2 illustrates the use of a device to shunt away some of the current that would otherwise cause excessive current density and possible burning at work areas closest to the anodes. Positioning and size of current shields are extremely important for their effective use and can be established best by trial and error. The use of shields is always accompanied by some increase in dragout.

Thieves or robbers made of metal conductors are positioned near edges and points to divert current from these areas. Rods with a diameter of 9.5 to 16 mm (3/8 to 5/8 in.) sometimes are suspended vertically on both ends of a plating rack to prevent burning or rough plate on the edges of the cathodes. Maintenance of robbers is of the utmost importance, because robbers can be the source of large dragout losses if allowed to build up excessively.

Special racks and auxiliary anodes are used only when conventional techniques fail to produce satisfactory coatings. Parts with deep recesses, such as coffeepots and small appliance hous-

Fig. 5 Racking arrangement to prevent plating of chromium in the hole of a shift lever

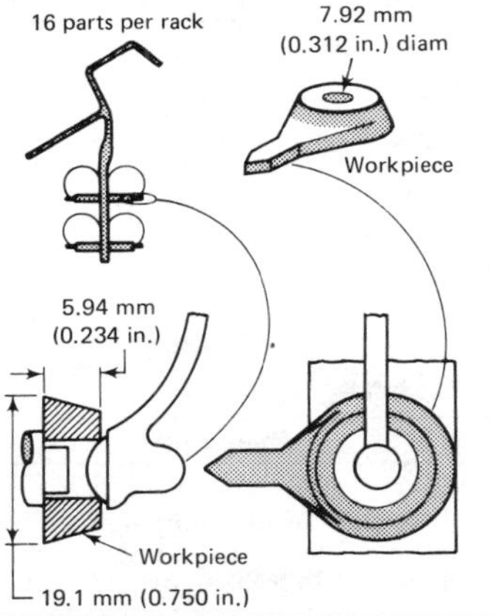

Table 3 Typical system cycles

Usage for specific systems — System	Cycles	Total plate, μm	Total plate, mils	Total time, min(a)
Cu + Ni + Cr	A, B, D, F	50	2.01	48
Cu + Cr	A, B, F	20	0.81	14
Ni + Cr	D, F	30	1.21	36
Ni + Cr + Cr	D, G	32	1.25	41.5
Ni + Ni + Cr	C, E, F	30	1.21	36
Ni + Ni + Cr + Cr	C, E, G	32	1.25	41.5

Operating parameters

A—copper strike	
Current density	3 A/dm² (30 A/ft²)
Plating time	2 min
Heat(b)	49-65 °C (120-150 °F)
Filtration	Yes
Agitation	Optional
B—acid copper plate, high speed, bright (20 μm or 0.8 mil)	
Current density	4 A/dm² (40 A/ft²)
Plating time (100% efficiency)	10 min
Heat(b)	21-27 °C (70-80 °F)
Filtration and agitation	Yes
C—nickel plate, semibright (23 μm or 0.9 mil)	
Current density	4 A/dm² (40 A/ft²)
Plating time (100% efficiency)	26 min
Heat(b)	55-65 °C (130-150 °F)
Filtration	Yes
Agitation	Usually
D—nickel plate, bright (30 μm or 1.2 mils)	
Current density	4 A/dm² (40 A/ft²)
Plating time (100% efficiency)	34 min
Heat(b)	55-65 °C (130-150 °F)
Filtration	Yes
Agitation	Usually
E—nickel plate, bright (8 μm or 0.3 mil)	
Current density	4 A/dm² (40 A/ft²)
Plating time (100% efficiency)	8 min
Heat(b)	55-65 °C (130-150 °F)
Filtration	Yes
Agitation	Usually
F—chromium plate (0.3 μm or 0.01 mil)	
Current density	14.4 A/dm² (144 A/ft²)
Plating time:	
Conventional (10% efficiency)	2 min
High speed (25% efficiency)	54 s
Heat(b)(c)	46-65 °C (115-150 °F)
Filtration and agitation	No
Ventilation(d)	Yes
G—chromium plate, microcracked (0.64 μm or 0.025 mil)	
Current density	21.6 A/dm² (216 A/ft²)
Plating time (25% efficiency)	2.5 min
Heat(b)(c)	46-65 °C (115-150 °F)
Filtration and agitation	No
Agitation	No
Ventilation(d)	Yes

(a) Power requirements and plating times given are theoretical values for perfect coverage. In practice, these values would be approximately doubled to ensure adequate thickness of plate in all areas. Table 4 has data for practical conditions. (b) For operating temperature indicated. (c) Cooling as well as heating may be required. (d) Chemical suppressants (mist or spray) may be used in addition to ventilation

ings, require auxiliary anodes. Auxiliary anodes are also used for parts with concave surfaces that are difficult to plate uniformly (Fig. 3). Auxiliary anodes offer the additional possibility of cost reduction by directing the plate into areas of minimum plate thickness without the penalty of overplating areas of high current density. The use of such devices should be considered even for some parts that do not present serious problems in meeting specifications for plate thickness. The shapes of many die castings make the use of auxiliary anodes particularly applicable. The current supply for auxiliary anodes may be the same as the major plating circuit with separate current control such as a rheostat. Greater flexibility is obtained if a separate current source is used for the auxiliary anodes.

Auxiliary anodes are mounted on the plating rack, insulated from the cathode current-carrying members and provided with means of direct connection to the anodic side of the electrical circuit (Fig. 4). In still tanks, the connection may simply be a flexible cable equipped with battery clamps. In fully automatic machines, cables are permanently mounted on the carriers, and contact brushes riding on an anode rail are provided to pick up the current. Connections must be positive; interruption or drastic reduction of current could cause the auxiliary anode to function as a robber or shield, resulting in local interruption of plate with consequent darkening or loss of adhesion.

The auxiliary anode need not follow the contour of the part closely, and anode-to-work spacing of 13 mm (½ in.) or slightly more is usually effective. The auxiliary anode mounting must be designed carefully to prevent the anode from interfering with efficient racking and unracking of parts. The anode may be designed to be removed while parts are being loaded on the rack, but good contact must be preserved. The auxiliary anode should be held rigidly to prevent it from short circuiting against the cathode.

Some users connect auxiliary anodes electrically only during chromium plating, a practice that is usually satisfactory for still tank operation, where an anode can be physically mounted immediately before chromium plating. In an automatic plating machine, however, the auxiliary anode should be connected in the acid-copper bath, the nickel bath (at least in the last half of the tank), and the chromium bath to avoid low thicknesses and low current density effects that could detract from appearance and cause difficulty in chromium plating. Because of the excellent throwing power of cyanide copper, auxiliary anodes need not be connected in the tank, although electrical connection to produce some current flow is sometimes desirable.

Auxiliary Anode Material. Unless made of insoluble material, auxiliary anodes are used up in plating, and the design should permit easy replacement. Plastisol-coated steel bushings with locking screw heads protected by stop-off lacquer are satisfactory. As anodes become thin, they must be carefully inspected for replacement to avoid shorting out. The diameter of rod used should be as large as is compatible with the size of the part and with construction requirements to minimize the need for frequent replacement. A diameter of 13 mm (½ in.) has been found suitable for a variety of parts ranging from small brackets to instrument panels and moldings. On larger parts, diameters as large as 25 mm (1 in.), or specially cast sections, may be useful.

Bags should not be used on auxiliary anodes because of the resulting contamination of solutions from drag-in. Avoiding roughness from bare anodes is a matter for serious consideration. Roughness is not a problem when the anode is to be immersed only in the chromium solution. Lead-alloy or steel anode material has been used satisfactorily for this purpose. Graphite rods have also been used to a limited extent. Auxiliary anodes are most frequently made of platinized titanium.

Bipolar anodes are a special variant of auxiliary anodes in which current is not supplied by external connection. In use, collector plates are mounted at the cathodic end (the end closest to the tank anodes) of the bipolar anode to draw current from a larger section of bath. Bipolar anodes may be used on conveyorized systems where a special busbar is unavailable.

Although adequate for some purposes, bipolar anodes are usually less effective than other auxiliary anodes and must be carefully maintained to avoid the problems of roughness from loosely adherent deposits of nickel and chromium on the collector plates. The metal deposited on the collector plates is often not reusable.

Stop-offs are not widely used in decorative chromium plating. However, when selective plating is required, a number of materials have the necessary qualifications, including (*a*) ease of application and removal, (*b*) resistance to hot cleaners and plating solutions, and (*c*) excellent adherence and electrical insulation characteristics during use.

Special racks are sometimes used to prevent plating solution from entering tapped holes and areas where plate is not wanted. Figure 5 shows a plated lever with a 7.92-mm (0.312-in.) diam hole that had to be free of plate. If conventional racking had been used, the hole would have had to be reamed to remove the plate.

Table 4 Design basis of equipment for continuous plating of zinc die castings

Metal deposited	Designed current density, A/dm²	Designed current density, A/ft²	Minimum thickness of plate, μm	Minimum thickness of plate, mil(a)	Nominal thickness of plate, μm	Nominal thickness of plate, mil	Plating time, min
Copper cyanide strike	10.0	100	...	...	0.25	0.01	3-4
Bright copper	3.0	30	15	0.6	20	0.8	25-30
Semibright nickel	7.5	75	15	0.6	20	0.8	30
Bright nickel	10.0	100	5	0.2	7.5	0.3	17
First chromium	20.0	200	3	0.012	0.5	0.020	6.5
Second chromium	15.0	150	2.5	0.010	0.5	0.020	6.5

(a) On parts of moderate complexity

Table 5 Plating problems and corrections

Defect	Possible cause	Possible remedy
Poor covering power or low deposit thickness	1 Temperature too high	1 Adjust temperature to standard range
	2 Current density too low	2 Increase current density
	3 Low chromic acid	3 Adjust chromic acid to standard range
	4 Fluoride catalyst too high	4 Reduce concentration (by dilution)
	5 Low chromic acid to sulfate ratio	5 Adjust ratio
	6 Poor electrical contact	6 Correct electrical contact
Burning in high current density areas	1 Temperature too low	1 Adjust temperature to standard range
	2 Current density too high	2 Reduce current density
	3 Chromic acid low	3 Raise chromic acid
	4 High chromic acid/sulfate ratio	4 Adjust ratio
	5 Fluoride catalyst too low	5 Adjust concentration of fluoride catalyst
Deposit color nonuniform	1 Underlying surface not clean or active	1 Remove any interfering films and provide active surface
	2 Bipolarity during entrance to chromium	2 Enter bath with precontact

Equipment for Chromium Plating

Tanks for chromium plating may be constructed of steel and lined with one of the following materials:

- Flexible plastic-type materials, such as fiber glass or polyvinyl chloride, either in sheet form or sprayed
- Lead alloy (6% antimony)
- Acid-resistant, high-temperature baked brick or tile, set in a silica cement

Lead-alloy linings are approximately 3.2 mm (⅛ in.) thick; plastics, from 2.4 to 4.8 mm (3/32 to 3/16 in.); and brick or tile, from 65 to 115 mm (2½ to 4½ in.). Brick or tile linings have all but disappeared from commercial use. At present, plastic linings are preferred, particularly for proprietary baths with fluoride-containing anions, which may have a greater rate of attack on lead and brick linings with siliceous cements. Rubber mats or plastisol-coated metal ribs are often used to protect the sides or bottoms of lead-lined tanks from shorts caused by accidental contact or from being punctured by dropped anodes or workpieces. Lead linings can cause serious bipolarity problems due to their electrical conductivity.

Heating. Chromium plating tanks may be heated internally or externally. Internal heating, by steam coils or electric immersion heaters, is usually used for small tanks; external heating, by heat exchangers, is used for large tanks. Coils for internal heating may be made of lead or a lead alloy (such as 4% tin or 6% antimony) or of tantalum; titanium may be used for baths that do not contain fluoride ions. Consultation with vendors of these processes is suggested. Immersion heaters are quartz covered. Heat exchangers may be made of tantalum, lead alloy (4% tin or 6% antimony), high-silicon cast iron or heat-resistant glass. Tantalum is preferred for heating coils or heat exchangers for proprietary solutions containing fluoride ions, because titanium is attacked by the fluoride. Consultation with vendors of these processes is suggested.

Typical system cycles for the application of six decorative chromium plating systems to identical work loads are outlined in Table 3. Each system is identified by (*a*) specific combination of metals successively deposited, (*b*) total thickness of plate, and (*c*) total plating time. Plating times and power requirements listed in Table 3 are theoretical values for perfect coverage; in practice, these values would be considerably greater to ensure adequate thickness of plate in all areas. In Table 4, the requirements for the design of several installed machines for continuous plating of automotive zinc die castings of average complexity is shown. The higher than normal designed current density is related to possible future needs in excess of present requirements.

Maintenance. The importance of proper maintenance of solutions and electrical, mechanical, and other equipment used in plating processes cannot be overemphasized. An outline of daily, weekly, monthly, and annual inspect and correct operations follows, which should be of assistance in setting up an adequate maintenance program for chromium plating:

Daily

- Fill plating tank with solution from save-rinse or boil-down tanks
- Stir solution thoroughly, using low-pressure air agitation
- Check solution for chromic acid, sulfate, and antispray additives; make corrective additions
- Check temperature controls for satisfactory operation; adjust temperature to proper range
- Inspect plating racks; repair as necessary
- Check ground lights to see that plating circuits are clear; do not start plating until grounds are cleared
- Put dummy cathodes in tanks and electrolyze solution at maximum voltage for 15 to 30 min at start of each day
- Hull cell check

Weekly

- Boil down save-rinse solution
- Check auxiliary catalyst; make additions as necessary

Monthly

- Check solution for metallic impurity content (iron, zinc, copper, nickel)
- Clean and straighten anodes
- Check solution for trivalent chromium content

Annually

- Check all ammeters and ampere-hour meters
- Inspect and adjust all temperature controllers
- Clean and repair outside of all tanks; clean and repair all ventilation hoods and ducts
- Pump out solution, remove sludge. Clean and inspect tank and heating coil; repair as needed. Disconnect all busbar connections; clean, draw-file and reconnect, including all anode and cathode joints. Inspect anodes; clean, straighten, or replace, as required

Plating problems can develop even with proper maintenance. Some typical plating problems and solutions are presented in Table 5. Actual solutions to plating problems can be drawn from production experience. Examples of the problems and corrections used with a variety of chromium-plated parts are discussed in the following paragraphs.

Although bath composition and temperature were carefully controlled, burnt chromium occurred on die castings plated at moderate amperage. This resulted from nonuniform distribution of current caused by the corrosion of mounting blocks attaching the busbar to the anode bar. The situation was corrected by welding the busbar to the anode rail, eliminating mounting blocks.

Mischromes (absence of chromium in certain areas) occurred on die cast window frames that occupied the lower portions of racks during plating in a full return type automatic plating machine. These defects were caused by the presence of short anodes at the exit end of the nickel plating tank. Replacement with anodes of the correct length solved the problem.

Fig. 6 Plating cost-time relationship

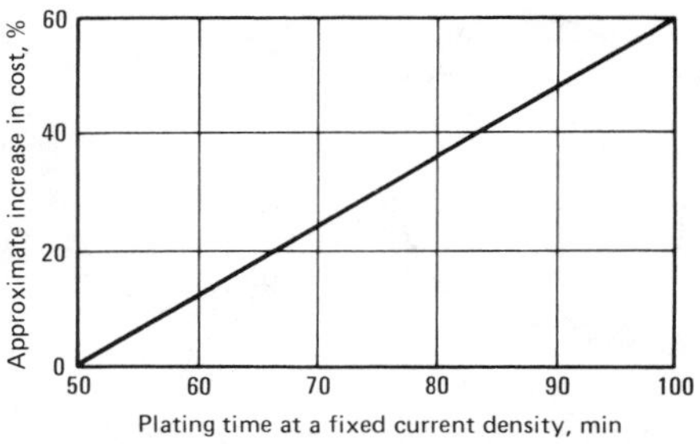

Nickel-plated business machine parts were stored submerged in cold water to await barrel chromium plating. Although these parts were acid activated before chromium plating, chromium coverage was poor to nil on parts that had been stored for only a few hours. To remedy the problem, parts were stored submerged in a 10% solution of potassium bitartrate (cream of tartar). After several days of storage, parts could be electrolytically reactivated and barrel chromium plated satisfactorily.

Inadequate rinsing after chromium plating, which failed to remove small amounts of bath impurities, resulted in nonuniform appearance of the plate on automotive zinc-based die castings for exterior service. The parts were given a hot rinse at 93 °C (200 °F) before customary room temperature rinses to remove bath impurities.

L-shaped die cast frames approximately 0.09 m^2 (1 ft^2) in area, although plated in identical racks in an automatic plating machine, exhibited nonuniformity of plate and, in some racks, burnt deposits. This was found to be caused by variations in current from rack to rack in the plating machine. To correct this, mechanical joints were eliminated from the electrical circuit. The mechanical joints were replaced by welding cables from the carrier contact brushes to the rack mounting bar.

Influence of Design on Quality and Cost

The cost of electroplating is often greatly influenced by the complexity of the workpiece. Simple shapes can be processed through all cleaning and plating sequences, with a minimum of about 33 μm (1.3 mils) of copper plus nickel, and 0.25 μm (0.01 mil) of chro-

Fig. 7 Die cast frames for headlights, representing degrees of difficulty in plating

(a) Easy. (b) Intermediate. (c) Difficult

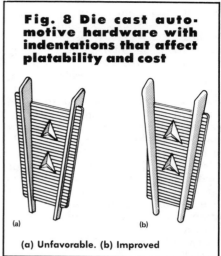

Fig. 8 Die cast automotive hardware with indentations that affect platability and cost

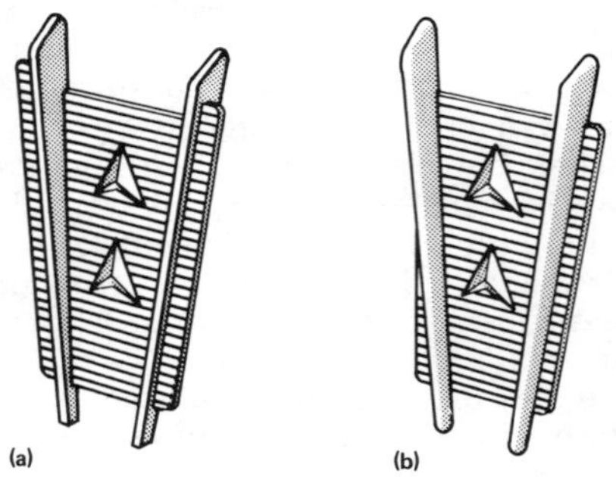

(a) Unfavorable. (b) Improved

mium, in approximately 50 min. To provide these minimum thicknesses on complex shapes, longer plating periods, special fixturing, special anodes, and current shields are required. Plating costs are increased in each case, although buffing and cleaning costs may be unchanged. Equipment and overhead costs per work unit increase in direct proportion to the plating time when longer than normal plating times are necessary to compensate for complexity of shape. In addition, the cost of materials is increased, because more metal is plated unnecessarily on projections and other areas of high current density. Figure 6 indicates how plating costs increase with plating time at a fixed current density.

Extreme complexity may preclude the application of truly corrosion-protective electroplate by conventional plating procedures. The sketches and comments in Table 6 demonstrate the influence of some design features on platability and plating cost. The distribution of electroplate on the shapes sketched in Table 6 is indicated in an exaggerated fashion by the solid black outlines. Crosshatched areas indicate the part before plating. Figures 7, 8 and 9 illustrate these general principles as they relate to specific items of die cast automotive hardware. Figure 10 shows a complex steel assembly that was extremely difficult to plate and a simple formed part that was easy to plate.

Trivalent Chromium Plating

The commercial plating of decorative chromium from trivalent chromium solutions began in England in 1975 and in the United States in 1976. Several unsuccessful attempts at commercially acceptable trivalent chromium processes were tried in both countries before these dates.

One of the main difficulties with the development of trivalent baths was the formation of hexavalent chromium at the insoluble anode, resulting in the cessation of chromium plating. All of the trivalent chromium electrolytes currently used incorporate some form of patented and/or proprietary means of eliminating or reducing the formation of hexavalent chromium.

Solution Composition. The nominal composition for the trivalent chromium bath is shown below:

Constituent	Quantity
Chromium (total)	18-25 g/L (2.4-3.3 oz/gal)
Conductivity salts	250-300 g/L (33-40 oz/gal)
Total salts	350-400 g/L (46.7-54 oz/gal)
Surfactant	0.5-2.0 mL/L (0.07-0.26 fluid oz/gal) (0.05-0.2 vol %)
Complexant	60-80 mL/L (7.8-10 fluid oz/gal) (6-8 vol %)
Specific gravity	1.21

The chromium is introduced as a water-soluble salt and forms a stable complex with the mixture of salts (such as potassium chloride and ammonium chloride) known as the conductivity salts and the complexant. The chromium content of the bath is very low compared to hexavalent baths and the total of the chromium salt and conductivity salts is approximately 375 g/L (50 oz/gal). In spite of the high salt content, the trivalent bath has less conductivity than most hexavalent baths because of the low pH of the latter. Somewhat higher voltages are usually required for trivalent baths.

The complexant is usually an organic acid selected to complex the trivalent chromium suitably for plating purposes but not so tightly as to preclude deposition of the chromium from the complex. A characteristic of trivalent chromium baths is that upward facing surfaces frequently plate thicker than downward facing surfaces. The surfactant, although usually added to provide mist suppression, is present to improve on this type of deposit distribution.

Operating Conditions. The optimum operating conditions for trivalent chromium are summarized below:

pH	3.2
Temperature	23 °C (72 °F)
Current density:	
Cathode	10 A/dm^2 (100 A/ft^2)
Anode	5 A/dm^2 (50 A/ft^2) max
Anodes	Graphite
Agitation	Mild air
Specific gravity	1.21

The pH must be controlled closely because both covering power and plating speed are affected, with lower pH contributing to reduced covering power and increased plating speed. The buffering ability of the solution is strong enough that large pH fluctuations do not occur. Temperature should also be controlled closely, because high temperature reduces covering power. In some cases, some degree of cooling is required to offset the wattage used for deposition and maintain the temperature at the preferred point. Cathode current densities are frequently lower than those used with hexavalent chromium baths, but there is no upper limit for using higher current densities because burning does not occur. Cathode efficiency decreases with increasing current density, and plating speed is not

Fig. 9 Die cast automotive ornaments of various levels of platability

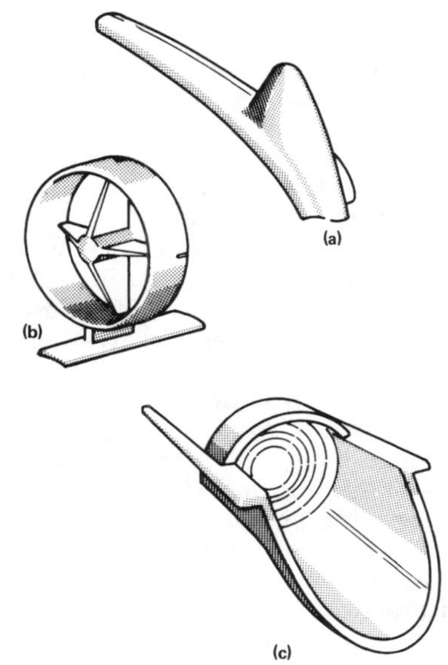

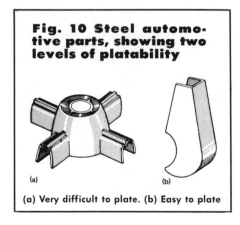

(a) Easy. (b) Intermediate, a three-part assembly with each part plated separately. (c) Difficult

Fig. 10 Steel automotive parts, showing two levels of platability

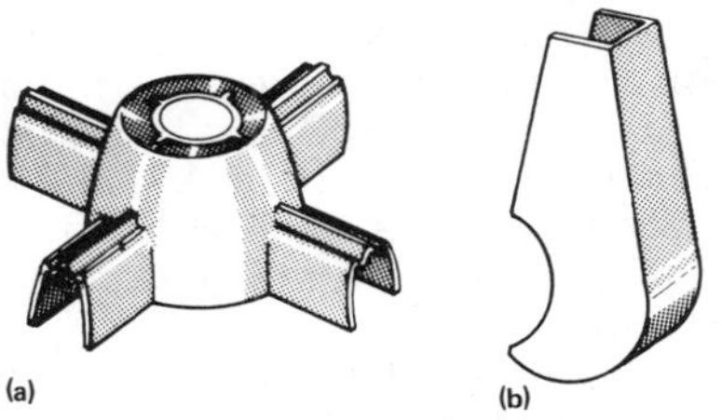

(a) Very difficult to plate. (b) Easy to plate

Table 6 Influence of design on platability of zinc-based die castings

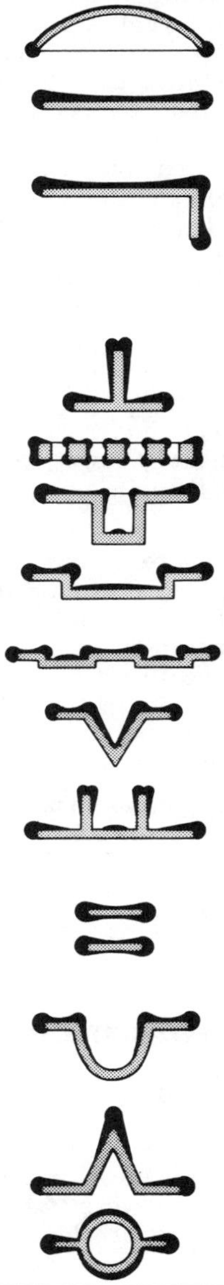
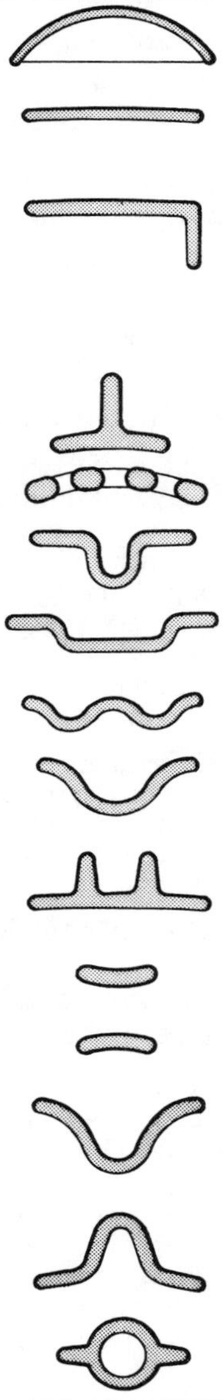

Design feature	Influence on platability	Better design
	Convex surface. Ideal shape. Easy to plate to uniform thickness, especially where edges are rounded	
	Flat surface. Not as desirable as crowned surface. Use 0.015-mm/mm (0.015-in./in.) crown to hide undulations caused by uneven buffing	
	Sharply angled edge. Undesirable. Reduced thickness of plate at center areas; requires increased plating time for depositing minimum thickness of durable plate. All edges should be rounded. Edges that contact painted surfaces should have a 0.8-mm (1/32-in.) min radius	
	Flange. Large flange with sharp inside angles should be avoided to minimize plating costs. Use generous radius on inside angles, and taper abutment	
	Slots. Narrow, closely spaced slots and holes cannot be plated properly with nickel and chromium unless corners are rounded	
	Blind hole. Must usually be exempted from minimum thickness requirements	
	Sharply angled indentation. Increases plating time and cost for attaining a specified minimum thickness and reduces the durability of the plated part	
	Flat-bottom groove. Inside and outside angles should be rounded generously to minimize plating costs	
	V-shaped groove. Deep grooves cannot be plated satisfactorily; should be avoided. Shallow, rounded grooves are better	
	Fins. Increase plating time and costs for attaining a specified minimum thickness and reduce the durability of the plated part	
	Ribs. Narrow ribs with sharp angles usually reduce platability; wide ribs with rounded edges impose no problem. Taper each rib from its center to both sides and round off edges. Increase spacing, if possible	
	Deep scoop. Increases time and cost for plating specified minimum thickness	
	Spearlike jut. Buildup on jut robs corners of electroplate. Crown base and round all corners	
	Ring. Platability depends on dimensions. Round corners; crown from center line, sloping toward both sides	

improved by raising current density. As a result, the deposition rate is really related to time and averages 3 to 5 μm/min (0.12 to 0.2 μin./min). Anode current densities should be maintained below 5 A/dm^2 (50 A/ft^2) to promote anode life. Properly operated anodes should last indefinitely. Some electrolytes may produce halogen gases at the anode during operation of the bath and require ventilation. Other patented electrolytes produce no halogen, because of the solution chemistry. Mild and uniform air agitation is used to assist the surfactant in promoting good deposit distribution and to prevent increasing limiting current density.

Physical Properties. With the exception of color and porosity, the physical properties of chromium plated from current trivalent baths are similar to those plated from decorative hexavalent chromium baths. Trivalent chromium deposits have a deeper, slightly darker color than the blue and white of hexavalent chromium. In most cases, the color difference is objectionable only when the part is going to be near a hexavalent chromium-plated part. Microporosity of the trivalent chromium deposit is present at all thicknesses unlike hexavalent chromium which requires special pretreatments to induce microporosity or microcracking in heavy 1.4-μm (35-μin.) deposits from special baths to form microcracks. The microdiscontinuities, micropores or microcracks are important for providing corrosion protection when used with appropriate nickel-plated undercoatings.

Plating Problems and Corrections. Some of the effects that can be obtained from trivalent chromium baths and their usual causes are listed below:

Poor coverage

- pH too low
- Temperature too high
- Current density too low
- Lead contamination
- Zinc contamination

Dark smudges on work

- Metallic contamination
- Low surfactant concentration
- Low complexor concentration

White patches on work

- High surfactant
- Lead contamination
- Drying on nickel solution because of excessively hot nickel bath or prolonged drying
- Incorrect precleaning before nickel plating
- High nonpitter concentration in nickel solution

In the case of metallic contamination, patented purification procedures or low current density electrolysis, 3 A/dm^2 (30 A/ft^2), may be used for removal.

Appendix

Stripping of Defective Electrodeposits

Stripping of defective decorative electrodeposits should be avoided if at all possible. The operations are time consuming and expensive and often result in excessive etching of basis metals. This etching necessitates expensive polishing and buffing to restore a finish suitable for replating. Defective chromium, however, can be stripped because its deposits are thin and easily removed.

Stripping of Chromium. Chromium electrodeposits can be stripped electrolytically in a solution of sodium hydroxide, 45 g/L (6 oz/gal), operated at room temperature with reverse current at 6 V. Chromium can also be stripped rapidly in a room temperature solution consisting of equal parts, by volume, of concentrated hydrochloric acid and water. During the stripping of chromium, the plated nickel undercoating may be superficially etched, particularly in a hydrochloric acid solution and may require buffing and cleaning before being replated with chromium.

Alkaline electrolytic stripping solutions often have a pronounced passivating action on the nickel undercoating. To reactivate the nickel before chromium plating, one of the following procedures may be used:

- Cathodically clean parts in mild alkaline solution; water rinse; dip in 2 to 5 vol % solution of concentrated sulfuric acid; water rinse
- Treat parts cathodically in 5 to 10 vol % solution of concentrated sulfuric acid at room temperature and 6 V, using lead anodes; water rinse

To improve activation, potassium iodide or other additives, in amounts equivalent to 0.5 kg/400 L (1 lb/100 gal), are sometimes added to the sulfuric acid dips.

Methods of Evaluating Copper-Nickel-Chromium Combinations

Durability of deposits on articles designed for relatively severe corrosion service is evaluated by surveys of service performance and by accelerated corrosion tests. The two most widely used accelerated corrosion tests are the CASS (copper—accelerated acetic acid—salt spray) fog-chamber test (ASTM B368) and the Corrodkote test (ASTM B380).

Two other fog tests, less widely used, use the same type of fog chamber and many of the same conditions as the CASS test; these are the salt spray test (ASTM B117) and the acetic acid—salt spray test (ASTM B287). A sulfur dioxide test, called the Kesternich test, (described in 46th Annual Technical Proceedings AES, 1959; p 154) is used in Europe, but has been largely limited to research use in the United States.

Deposit Thickness. Thicknesses of copper and nickel deposits, as well as chromium deposits of more than 1 μm (0.05 mil), may be determined by metallographic procedures, such as cross sectioning, polishing, and microscopic examination, described in ASTM A219. In practice, however, magnetic ASTM B499 and B530, coulometric ASTM B504 and high-frequency eddy current instruments are used for determining the thickness of copper-nickel. Thickness of chromium may also be determined by the hydrochloric acid spot test (ASTM B556) or the coulometric method.

Appearance is judged by visual observation of the significant surfaces. A desirable appearance is one that is uniform; cracks, pits, blisters, or variations in luster are undesirable.

Ductility can be judged in a qualitative manner by any of several nonstandard bend tests. Quantitative tests include bend tests such as ASTM B489 for plated coatings on sheet or strip or ASTM B490 for foils produced from the electroplating bath.

Leveling is determined by measuring the decrease in surface roughness after plating.

Nickel Plating*

NICKEL PLATE, with or without an underlying copper strike, is one of the oldest protective-decorative electrodeposited metallic coatings for steel, brass, and other basis metals. More recently, it has been applied to plastics of various kinds. The first applications of nickel plate were for stove and bicycle components. Unless polished occasionally, nickel plate tarnishes, taking on a yellow color during long exposure to mildly corrosive atmospheres, or turning green on severe exposure. The introduction of chromium plate in the late 1920's overcame the tarnishing problem and led to a great increase in the use of nickel as a component of protective-decorative coatings, in various combinations with copper and chromium.

Electrodeposits of nickel possess a wide variety of properties, depending on composition of the plating bath and operating conditions. They may be classified according to application or appearance as general purpose, special purpose, black, and bright.

General-purpose nickel deposits, produced by Watts, sulfamate, and fluoborate baths, are essentially sulfur-free. They are used primarily to protect alloys based on iron, copper, or zinc against corrosive attack in rural, marine, and industrial atmospheres (see Table 1 for recommended thicknesses). To a lesser degree, they are used also for decorative purposes. Heavy deposits from these baths are used to build up worn or undersized parts and to provide protection against corrosive chemical environments.

Special-purpose nickel deposits are selected either because of an unusual property of the deposit, such as extreme hardness, or because the plating bath is particularly well suited for a special application, such as barrel plating or electroforming.

Black nickel deposits, derived from baths containing zinc sulfate or zinc chloride, have little protective value and are used primarily to obtain a dark, nonreflective, decorative finish.

Bright nickel deposits are used primarily to provide decorative finishes on metals that corrode or tarnish without suitable protection. These deposits contain various amounts (0.02 to 0.13%) of sulfur, which reduces their corrosion resistance. They are most frequently applied as undercoatings for chromium or one of the precious metals, or as overcoatings for a sulfur-free nickel deposit (a combination known as duplex nickel).

In addition to the major applications, many other uses have been found for electrodeposits of nickel; for example, nickel deposits have been used to protect molybdenum and uranium against oxidation at elevated temperature. Also, in many applications, the selection of nickel plating is based on more than one functional requirement.

General-Purpose Plating Baths

Table 2 lists ranges of compositions and operating conditions for three general-purpose nickel plating baths (Watts, sulfamate, and fluoborate) and indicates mechanical properties of deposits produced by these solutions. These baths are sometimes referred to as gray nickel baths.

The Watts bath and its higher nickel-chloride modification (Table 2) are the standard general-purpose nickel plating solutions. The major portion of the nickel ion content is contributed by the relatively inexpensive nickel sulfate; a high nickel sulfate concentration is used when high current densities are required.

Nickel chloride serves primarily to improve anode corrosion. Higher chloride concentrations increase the limiting current density, decrease pitting and nodular growths, and increase conductivity, cathode efficiency, and throwing power. However, baths with higher chloride concentrations also increase stress and are more corrosive. The hardness and tensile strength of the deposit are at a minimum, and elongation at a maximum, with a nickel chloride content of 55 to 70 g/L (7 to 9 oz/gal).

Boric acid is used in a nickel plating solution for buffering purposes. Normally, no effect is noticed in high concentration up to saturation (around 45 g/L, or 6 oz/gal, depending on other variables). Below 30 g/L (4 oz/gal), however, noticeable conditions appear in areas of high current density; first, the deposit becomes frosty, and then as the boric acid approaches 15 to 23 g/L (2 to 3 oz/gal), the deposit may be burnt and cracked.

The effects of pH and temperature on physical and mechanical properties are

*Revised by Louis Gianelos, Manager, Technical Service, Industrial Finishing, Harshaw Chemical Co. and Warren H. McMullen, Commercial Development Manager, Plating Division, M&T Chemicals Inc.

Table 1 Recommended thicknesses of nickel plate for resistance to atmospheric corrosion

Basis metal	Recommended thickness of nickel plate — Indoor exposure, μm	Indoor exposure, mils	Rural exposure, μm	Rural exposure, mils	Industrial exposure, μm	Industrial exposure, mils	Marine exposure, μm	Marine exposure, mils
Steel	5-18	0.2-0.7	20-30	0.8-1.2	38-50	1.5-2.0	38-50	1.5-2.0
Copper and copper-based alloys(a)	5-18	0.2-0.7	8-18	0.3-0.7	13-18	0.5-0.7	13-25	0.5-1.0
Zinc and zinc-based alloys(b)	5-18	0.2-0.7	20-30	0.8-1.2	38-50	1.5-2.0	38-50	1.5-2.0

(a) Copper-based alloys containing more than 40% zinc must be plated with 8 μm (0.3 mil) of copper before being nickel plated. (b) An undercoating of 5 to 10 μm (0.2 to 0.4 mil) of copper is required

Table 2 General-purpose nickel plating baths

Compositions, operating conditions, and typical mechanical properties of deposits

Constituent, condition, or property	Watts bath(a)	Sulfamate bath	Fluoborate bath
Composition			
Nickel sulfate, $NiSO_4 \cdot 6H_2O$	225-410 g/L (30-55 oz/gal)	...	...
Nickel chloride, $NiCl_2 \cdot 6H_2O$	30-60 g/L (4-8 oz/gal)(a)	0-30 g/L (0-4 oz/gal)	0-15 g/L (0-2 oz/gal)
Nickel sulfamate, $Ni(SO_3NH_2)_2$	...	263-450 g/L (35-60 oz/gal)	...
Nickel fluoborate, $Ni(BF_4)_2$	...	...	225-300 g/L (30-40 oz/gal)
Total nickel as metal	55-105 g/L (7.7-14.2 oz/gal)	62-113 g/L (8.2-15 oz/gal)	55-80 g/L (7.6-10.5 oz/gal)
Boric acid, H_3BO_3	30-45 g/L (4-6 oz/gal)	30-45 g/L (4-6 oz/gal)	15-30 g/L (2-4 oz/gal)
Antipitting additives,	(b)	(b)	(b)
Operating conditions			
pH	1.5-5.2	3-5	2.5-4
Temperature	46-71 °C (115-160 °F)	38-60 °C (100-140 °F)	38-71 °C (100-160 °F)
Current density	1-10 A/dm^2 (10-100 A/ft^2)	2.5-30 A/dm^2 (25-300 A/ft^2)	2.5-30 A/dm^2 (25-300 A/ft^2)
Mechanical properties of deposits(c)			
Tensile strength	345-690 MPa (50-100 ksi)	380-1070 MPa (55-155 ksi)	380-830 MPa (55-120 ksi)
Vickers hardness	100-250 HV	130-600 HV	125-300 HV
Elongation in 50 mm (2 in.)	10-35%	3-30%	5-30%
Stress	105-205 MPa (15-30 ksi)	3-110 MPa (0.5-16 ksi)	90-205 MPa (13-30 ksi)
References			
See list following footnotes	6, 7	1, 2, 3, 6	4, 5, 6, 7

(a) A high-chloride modification of the Watts bath may be made by increasing content of nickel chloride to 113 g/L (15 oz/gal). (b) In many instances, antipitting additives are not required if solution is pure. When required, either of the following may be added: 0.03 to 0.075 mL/L (0.1 to 0.25 mL/gal) of solution (not to be added to solutions containing wetting agents or organic stress-reducers), or a surfactant and wetting agent of approved type to give 50 to 32 dynes/cm. (c) Wide ranges of values reflect variations in composition and degree of purity of plating baths

References: **(1)** R. Barrett, *Plating,* Vol 41, 1954, p 1027. **(2)** M. Diggin, *Metal Progress,* Vol 66, Oct 1954, p 132. **(3)** D. Fanner and R. Hammond, *Trans Inst Met Fin,* Vol 36, 1958-1959, p 32. **(4)** C. Struyk and A. Carlson, *Plating,* Vol 37, 1950, p 1242. **(5)** E. Roehl and W. Wesley, *Plating,* Vol 37, 1950, p 142. **(6)** A. K. Graham, *Electroplating Engineering Handbook,* New York: Reinhold Publishing Corp., 1962. **(7)** AES Research Report No. 20

not straight-line functions. High-pH Watts solutions, operated at relatively low current density, are generally used where maximum throwing power is required. However, for many industrial applications, high-pH solutions are not desirable in terms of stress and ductility. Therefore, a compromise pH range of 3 to 4 is usually selected.

In some applications, it is advantageous to use a pH as low as 1.5. The low pH makes possible the plating of large steel parts requiring 1 to 5 min of transfer time between rinsing and plating. During the long transfer period, an oxide film forms on steel parts before they can be immersed in the nickel solution. A low-pH bath removes this light oxide, and good adhesion of the deposit is obtained. A low-pH, high-chloride solution can be used as a nickel strike before plating in a normal pH (3 to 4 pH) solution.

The sulfamate bath (Table 2) is a general-purpose bath that yields deposits of low stress, has a wide operating range, and is easy to control. Because of the very high solubility of nickel sulfamate, maintaining a higher nickel metal concentration in this solution than in other nickel baths is possible, permitting the use of lower operating temperatures and higher plating rates. A small amount of nickel chloride is usually added to the bath to minimize anode passivity, especially when higher plating rates are used.

If nickel chloride is not used, and as a safeguard where it is used, only rolled oval depolarized nickel anodes with anode bags or sulfur-containing nickel anode forms (approximately 0.02% sulfur in squares or rounds) in titanium anode baskets should be used. The use of any other nickel anode material may result in the anodic oxidation of the sulfamate ion, resulting in the production of sulfur-containing compounds that act as stress reducers and cannot easily be removed from the solution.

When wetting agents are added to the bath to prevent pitting, surface tension may be checked qualitatively with a wire loop approximately 100 mm (4 in.) in diameter. If the loop can be withdrawn slowly from the solution without breaking the film of solution across the loop, the surface tension should be low enough to prevent hydrogen pitting. Special low-foaming wetting agents are used in air-agitated solutions, precluding the use of this procedure.

Because nickel sulfamate is so soluble that it cannot be readily recrystallized from solution, it must be purchased as a concentrated solution or prepared in place by the reaction of nickel carbonate and sulfamic acid. For this reason, as well as the higher cost of sulfamic acid, sulfamate baths have a higher initial cost and a higher replenishment cost than other commonly used nickel electroplating baths using easily available commercially chemical salts. In addition, prolonged use at temperatures above 60 °C (140 °F) or at a pH of less than 3.0 can hydrolyze the nickel sulfamate to the less soluble form of nickel ammonium sulfate. The ammonium and sulfate ions produced from the hydrolysis also act to increase the tensile stress of the deposit.

Nickel electrodeposited from a well-purified sulfamate bath containing no stress-reducing agent and operated at 46 °C (115 °F), a pH of 4.0, and a current density of 2.0 A/dm^2 (20 A/ft^2) has a residual tensile stress varying from 15 to 40 MPa (2 to 6 ksi). The stress in a deposit produced from a similarly operated Watts bath would be about 170 MPa (25 ksi); and in one from an all-chloride bath (see Table 3), about 275 MPa (40 ksi).

Sulfamate nickel plating baths are especially useful for applications requiring low residual stress in the electrodeposited nickel, such as in electroforming, and for coating objects that are susceptible to fatigue cracking. Steel crankshafts that are nickel plated for resistance to corrosion and wear should be coated with a low-stress nickel deposit, such as sulfamate nickel, to minimize loss of fatigue strength. The fatigue limit of nickel plated steel is reduced almost proportionally to the amount of residual tensile stress in the nickel plate.

The fluoborate bath (Table 2) can be operated over a wide range of nickel concentrations, temperature, and current density. However, because the bath is well buffered, the maximum pH is limited, even for high-pH treatments. The reactivity of the fluoborate ion with some materials of construction requires some consideration. Silica filter aids cannot be used on a continuous basis, although cellulose filters are satisfactory. Lead, titanium, and high-silicon cast iron are readily attacked. Stainless steels containing 20% chromium, 25 to 30% nickel, and 2 to 3% molybdenum are suitably resistant. Vinyon, polypropylene, or Orlon anode bags may be used; nylon bags are unsuitable. Only sleeve-type glass electrodes for pH measurement should be used because of the formation of relatively insoluble potassium fluoborate with permanent junction electrodes; however, standardized pH paper is usually satisfactory.

The mechanical and physical properties of deposits produced by the fluoborate bath are similar to those produced by Watts baths. Therefore, the fluoborate bath may be used where properties similar to those of Watts nickel and/or higher plating speeds are required. It is used primarily for heavy deposits, for electroforming, and for plating wire, rod, and strip that are subsequently drawn to vacuum tube filament size.

Special-Purpose Plating Baths

Many nickel plating applications require baths of special composition. The electrotyping bath, for example, is intended for plating over wax and lead molds. The hard nickel bath permits control of the hardness, strength, and ductility of the deposit. The compositions and operating conditions of nine special-purpose nickel plating baths are given in Table 3. Each of these baths is discussed in the following section.

The chloride-sulfate bath is an extension of the high-chloride Watts solution; its composition is midway between the Watts and all-chloride solutions. The properties of its deposits also are about midway between those of deposits from these two solutions. Where high conductivity and high permissible current densities are desired, the chloride-sulfate bath has definite advantages. By the same token, it produces deposits of higher stress than the Watts bath when stress-reducing agents are not used. Because of the high chloride concentration, lead may not be used for equipment in contact with the solution.

The all-sulfate bath (Table 3) is used primarily in applications requiring the use of insoluble conforming anodes, such as in plating small steel pipes and fittings. When the use of nickel anodes is impossible or impractical because of the shape or size of a part, the use of a nonconsumable conforming anode is extremely advantageous.

The all-sulfate solution does not contain nickel chloride; otherwise, its composition is similar to that of the Watts solution. To prevent pitting, hydrogen peroxide may be added to all-sulfate solutions, provided they contain no wetting agents or organic stress-reducers.

Although nickel deposits at the cathode, oxygen is evolved at the insoluble anode (chlorine would be evolved if the chloride ion was present). The nickel concentration and pH decrease during plating. The addition of nickel carbonate or nickel hydrate is the usual method of replenishing the nickel concentration and maintaining the pH.

Occasional additions of nickel sulfate are made to compensate for dragout losses. Continuous addition of nickel carbonate is made most effectively to the first compartment of a replenishment tank equipped with baffles, over and under which the plating solution passes. This allows carbon dioxide to evolve from the solution before it passes through the filter on its return to the plating tank. The size of the replenishment tank, relative to that of the plating tank, depends on the current consumption, solution temperature, and operating pH range of the plating tank. The lower the pH, the faster the reaction with the nickel carbonate occurs. Another procedure that has been used in low-pH solutions replenishes the nickel electrolytically, using a reversing current cycle and nickel anodes.

The insoluble anodes used in the all-sulfate bath may be of lead, carbon, graphite or platinum. If a small anode area is required, solid platinum (in the form of wire) may be used; for large anode areas, platinum-plated or platinum-clad titanium is recommended. In some forms, carbon and graphite have the disadvantage of being fragile; lead has the disadvantage of forming loose oxide layers, especially if it is immersed in other solutions in the course of a plating cycle. In a relatively chloride-free solution, pure nickel is almost insoluble and may be used as the anode if properly bagged to collect loose particles.

One use for the all-sulfate solution is in plating the inside of steel pipe and fittings. The plating cycle is begun with pH at 4.5, and thereafter pH is maintained between 2.5 and 4.5. Lead-covered copper anodes are used. Bath temperature is maintained at 54 to 65 °C (130 to 150 °F), using a cathode current density of 4 to 6 A/dm^2 (40 to 60 A/ft^2). The anode current density may be 10 to 60 A/dm^2 (100 to 600 A/ft^2), depending on the relative size of pipe and anode.

All-Chloride Bath. The principal advantage of the all-chloride bath (Ta-

Table 3 Special-purpose nickel plating baths

Compositions, operating conditions, and typical mechanical properties of deposits

Constituent, condition, or property	Chloride sulfate	All sulfate	All chloride	High sulfate	Cold	Barrel plating	Electro-typing	Hard nickel	Nickel phosphorus(a)
Composition									
Nickel sulfate, $NiSO_4 \cdot 6H_2O$	150-225 g/L (20-30 oz/gal)	225-410 g/L (30-55 oz/gal)	...	75-110 g/L (10-15 oz/gal)	120 g/L (16 oz/gal)	150 g/L (20 oz/gal)	35-65 g/L (5-9 oz/gal)	180 g/L (24 oz/gal)	170 or 330 g/L (23 or 44 oz/gal)
Nickel chloride, $NiCl_2 \cdot 6H_2O$	150-225 g/L (20-30 oz/gal)	...	225-300 g/L (30-40 oz/gal)	...	...	...	...	...	35-55 g/L (5-7 oz/gal)
Total nickel as metal	70-105 g/L (9.4-14 oz/gal)	50-105 g/L (7-14 oz/gal)	16-25 g/L (2.2-3.4 oz/gal)	16-25 g/L (2.2-3.4 oz/gal)	25 g/L (3.6 oz/gal)	34 g/L (4.5 oz/gal)	8-15 g/L (1.0-2.0 oz/gal)	40 g/L (5.4 oz/gal)	50-85 g/L (6.4-11.5 oz/gal)
Boric acid, H_3BO_3	30-45 g/L (4-6 oz/gal)	30-45 g/L (4-6 oz/gal)	30-35 g/L (4-5 oz/gal)	15 g/L (2 oz/gal)	15 g/L (2 oz/gal)	30 g/L (4 oz/gal)	...	30 g/L (4 oz/gal)	0 or 4 g/L (0 or 30 oz/gal)
Ammonium chloride, NH_4Cl	...	...	...	15-35 g/L (2-5 oz/gal)	15 g/L (2 oz/gal)	30 g/L (4 oz/gal)	5-15 g/L (0.7-2 oz/gal)	25 g/L (3.3 oz/gal)	...
Sodium sulfate, Na_2SO_4	...	...	...	75-110 g/L (10-15 oz/gal)	...	...	...	...	...
Phosphoric acid, H_3PO_4	...	...	...	...	...	...	...	...	50 or 0 g/L (6.7 or 0 oz/gal)
Phosphorous acid, H_3PO_3	...	...	...	...	...	...	...	...	2-40 g/L (0.3-5.3 oz/gal)
Antipitting additives(b)	(c)	(d)	(c)	(d)	(d)	(d)	(d)	(d)	(e)
Operating conditions									
pH	1.5-2.5	1.5-4	1-4	5.3-5.8	5-5.5	5-5.5	5.6-6.2	5.6-5.9	0.5-3
Temperature	43-52 °C (110-125 °F)	38-71 °C (100-160 °F)	38-60 °C (100-140 °F)	21-32 °C (70-90 °F)	Room temperature	21-32 °C (70-90 °F)	35 °C (95 °F)	43-60 °C (110-140 °F)	60-95 °C (140-205 °F)
Current density	2.5-15 A/dm^2 (25-150 A/ft^2)	1-10 A/dm^2 (10-100 A/ft^2)	2.5-30 A/dm^2 (25-300 A/ft^2)	0.5-2.5 A/dm^2 (5-25 A/ft^2)	0.5-1 A/dm^2 (5-10 A/ft^2)	(f)	1-2 A/dm^2 (10-20 A/ft^2)	2.5-5 A/dm^2 (25-50 A/ft^2)	2-5 A/dm^2 (20-50 A/ft^2)
Mechanical properties of deposits(g)									
Tensile strength	480-720 MPa (70-105 ksi)	410-480 MPa (60-70 ksi)	620-930 MPa (90-135 ksi)	...	...	...	...	990-1100 MPa (115-160 ksi)	...
Vickers hardness	150-280 HV	180-275 HV	200-390 HV	...	...	...	...	350-500 HV	300-900 HV
Elongation in 50 mm (2 in.)	5-25%	20%	4-20%	...	...	...	...	4-9%	...
Stress	210-280 MPa (30-40 ksi)	120 MPa (17 ksi)	210-310 MPa (30-45 ksi)	...	...	...	...	...	...
References									
See list following footnotes	1, 10, 11	11, 12	2, 10, 11	3, 4	...	...	5, 6	7, 10	8, 9, 10

(a) Brenner and Knapp solutions. Where a choice is indicated for constituent quantity, the first value listed is for the Brenner solution, the second for the Knapp; where a range is indicated, this applies to both solutions. (b) In many instances, antipitting additives are not required if solution is pure. (c) Either of the following may be added: 0.03 to 0.075 mL/L (0.1 to 0.25 mL/gal) of 30% hydrogen peroxide of solution (not to be added to solutions containing wetting agents or organic stress-reducers), or a surfactant and wetting agent of approved type to give 50 to 32 dynes/cm. (d) 0.03 to 0.075 mL/L (0.1 to 0.25 mL/gal) of 30% hydrogen peroxide. See restriction in (c). (e) Surfactant and wetting agent of approved type to give 50 to 32 dynes/cm. (f) Varies widely, depending on barrel size and design and on tumbling action of workpieces. (g) Wide ranges of values reflect variations in composition

References: **(1)** W. Pinner and R. Kinnaman, *Monthly Rev Am Electroplaters' Soc,* Vol 32, 1945, p 327. **(2)** W. Wesley, *Monthly Rev Am Electroplaters' Soc,* Vol 33, 1946, p 504. **(3)** E. Roehl, *Metal Finishing,* Vol 45, 1947, p 63. **(4)** M. Thompson, *Trans Am Electrochem Soc,* Vol 47, 1925, p 163. **(5)** E. Peters, *Proc Am Electroplaters' Soc,* 1950, p 74. **(6)** Electrotyping in the Government Printing Office, GPO-PIA Joint Research Bulletin PL-Z. **(7)** W. Wesley and E. Roehl, *Trans Electrochem Soc,* Vol 25, 1942, p 37. **(8)** A. Brenner and D. Couch, U.S. Patent 2,643,221 (1953). **(9)** B. Knapp and D. Karr, U.S. Patent 2,594,933 (1952). **(10)** A. K. Graham, *Electroplating Engineering Handbook,* New York: Reinhold Publishing Corp., 1962. **(11)** AES Research Report No. 20. **(12)** W. Wesley, et al, *Plating,* Vol 38, 1951, p 1243

ble 3) is its ability to operate effectively at high cathode current densities. Other advantages include (*a*) high conductivity, (*b*) slightly better throwing power, and (*c*) less tendency to form nodular growths on edges, to pit, or to form nonadherent deposits if the current is interrupted. Deposits from this solution are harder and stronger than those from Watts solutions. Compared to the Watts solution, the all-chloride bath has two major disadvantages, namely, the lower ductility and higher stress of its deposits. Lead, because of the partial solubility of lead chloride, cannot be used in contact with the all-chloride solution, and mists from this solution are corrosive to the superstructure, vents, and other plant equipment, if not well protected.

Mainly because of the relatively high stress of its deposits, the all-chloride bath has not been widely used for industrial plating purposes. Stress has been compensated for by mold dimensioning and by annealing the nickel deposit in a few applications. The solution has been used to some extent for salvaging undersize or worn shafts and gears.

The following cycle has been used to deposit 25 μm (1 mil) of nickel on undersize threads at one end of a mild steel shaft:

- Anodic alkaline clean
- Rinse in cold water
- Dip in aqueous hydrochloric acid solution (20 vol % HCl) at room temperature for 15 to 30 s
- Rinse
- Plate in all-chloride nickel solution at 49 °C (120 °F) for 2 min at 5 A/dm^2 (50 A/ft^2); then increase to 10 A/dm^2 (100 A/ft^2) for 14 to 15 min. The actual area of the threads is roughly twice the apparent area
- Rinse and dry
- Rechase threads

In general, the all-chloride solution may be used where better throwing power, little edge buildup, and high plating speeds are required, and where the stress in the deposit is not an important consideration.

The high-sulfate bath (Table 3), although seldom used in the United States, was designed especially for plating nickel directly on zinc-base die castings. It may be used also to plate nickel on aluminum that has been given a zincate or comparable surface bonding treatment. The high sulfate and low nickel contents, together with the high pH, provide good throwing power with little attack of the zinc. The deposits are less ductile and more highly stressed than Watts nickel; for this reason, high-sulfate nickel is sometimes used as a thin undercoating for a more ductile nickel. Bath variations such as higher nickel content, less sodium sulfate, lower pH, and higher bath temperature yield a more ductile, lower stressed deposit, but usually also reduce the throwing power of the bath and increase the attack on zinc. Frequently, the solution must be adjusted to give optimum plating results on given shapes.

The cold bath (Table 3), sometimes called also the double salt bath, probably because the ammonium ion originally was added as double nickel salt, has few uses other than as the initial strike solution for wax recordings and electrotypes. It is normally operated at room temperature, low current density, and high pH—conditions that provide relatively good throwing power. The cold bath may be used where warm solutions would be detrimental to the substrate being plated. However, this is not a unique advantage, because other solutions also can be used at room temperature.

Barrel Plating Bath. The cathode current density for the barrel plating bath is not indicated in Table 3 because it can vary over a wide range, depending on the design and size of the barrel and the tumbling action of the workpieces within the barrel. A good electrical connection must be made with the work, and the barrel should be provided with openings that are large enough to permit adequate circulation of the plating solution.

Because of the restriction to current flow by the barrel, and the relatively high contact resistance at the cathode pieces, the voltage drop is relatively high in barrel plating. Therefore, for normal power sources, and to avoid overheating of the solution, the average cathode current density is usually lower than when plating in tanks and ranges from 0.3 to 0.5 A/dm^2 (3 to 5 A/ft^2).

Although the composition given in Table 3 is commonly called the barrel plating solution, other compositions may be used with the same or better results. The Watts and sulfamate solutions probably are used more frequently in barrel plating than is the solution designated for barrel plating in Table 3. By composition, the barrel plating solution given in Table 3 is not a bright plating solution, although bright solutions have been developed specifically for barrel plating.

Electrotyping Bath. This bath (Table 3) is characterized chiefly by the absence of boric acid, which if present under the conditions used causes cracking and exfoliation of the deposit.

Although other solutions, particularly the sulfamate, are gradually supplanting it for the nickel plating of electrotype molds, the electrotyping bath does have advantages over some common solutions such as the Watts bath, because it produces hard, low-stress deposits and has good covering power. One disadvantage is its limiting current density; but for this solution, only 6.3 to 25 μm (0.25 to 1.0 mil) of nickel is deposited, which in turn is followed by a heavy copper deposit.

The following is a typical cycle for nickel plating wax or plastic molds in an electrotyping bath:

- With a soft brush, gently scrub the mold with a solution of stannous chloride in isopropyl alcohol to clean and sensitize the surface
- Spray rinse
- Spray silver (an operation whereby a thin, specular layer of silver is deposited by simultaneously spraying the mold with an ammoniacal silver nitrate solution from one nozzle and with a reducer, such as a solution of formaldehyde or hydrazine, from another nozzle)
- Rinse
- Nickel plate in electrotyping bath at 1 A/dm^2 (10 A/ft^2) for 5 min; then increase to 1.5 to 2 A/dm^2 (15 to 20 A/ft^2) for ½ to 2 h to obtain the thickness desired
- Rinse
- Copper plate

Hard Nickel Bath. Developed especially for engineering applications, this solution (Table 3) is used where controllable hardness, tensile strength, and ductility are required. Hardness and tensile strength can be increased, with a corresponding decrease in ductility, by increasing pH or decreasing bath temperature. The disadvantages of the hard nickel bath are its low operating current density, its tendency to form nodules on edges, and the low annealing temperature of 230 °C (450 °F) of its deposits.

Table 4 Compositions and operating conditions of two black nickel plating baths

Constituent or condition	Sulfate bath	Chloride bath
Composition		
Nickel sulfate, $NiSO_4 \cdot 6H_2O$	75 g/L (10 oz/gal)	...
Nickel chloride, $NiCl_2 \cdot 6H_2O$	...	75 g/L (10 oz/gal)
Zinc sulfate, $ZnSO_4 \cdot 7H_2O$	30 g/L (4 oz/gal)	...
Zinc chloride, $ZnCl_2$	...	30 g/L (4 oz/gal)
Ammonium sulfate, $(NH_4)_2SO_4$	35 g/L (5 oz/gal)	...
Ammonium chloride, NH_4Cl	...	30 g/L (4 oz/gal)
Sodium thiocyanate, NaSCN	15 g/L (2 oz/gal)	15g/L (2 oz/gal)
Operating conditions		
pH	5.6	5.0
Temperature	21-24 °C (70-75 °F)	21-24 °C (70-75 °F)
Current density	0.15 A/dm^2 (1.5 A/ft^2)	0.15-0.6 A/dm^2 (1.5-6 A/ft^2)

Hard nickel deposits are used primarily for buildup or salvage purposes; their stress is slightly higher than that of Watts deposits. In many applications, a part salvaged by plating with hard nickel has a greater life than the original. Although the physical and mechanical properties of its deposits can be equaled or surpassed, the hard nickel solution represents a good compromise when strength, hardness, ductility, and stress are all important. For optimum results, the ammonium ion concentration should be maintained at 8 g/L (1.1 oz/gal).

The following is an outline of procedures used to repair and hard nickel plate a worn steel crankshaft:

- Grind worn surface smooth and about 180 μm (7 mils) undersize
- Rack part and mask all areas not to be plated, using stop-off paint and tape
- Clean anodically in alkaline solution at 7 A/dm^2 (70 A/ft^2) and 82 °C (180 °F) for 1 min
- Rinse in hot water
- Etch anodically in aqueous sulfuric acid solution (60 vol % H_2SO_4) at 10 A/dm^2 (100 A/ft^2) at room temperature for 2 min
- Rinse
- Plate in hard nickel solution at 49 °C (120 °F) for about 10 h at 2.5 A/dm^2 (25 A/ft^2) to obtain a minimum plate thickness of about 250 μm (10 mils)
- Rinse and remove masking and rack
- Rinse and dry
- Grind to size

Nickel-Phosphorus Alloy Baths. These solutions (Table 3) have found little commercial use, probably because deposits of maximum hardness are brittle and highly stressed. Unlike the hard nickel deposit, the hardness can be increased by heat treatment, with the maximum hardness occurring at 400 °C (750 °F). The Brenner solution uses phosphate as the buffer, with low or high phosphite to supply the cathode phosphorus. The Knapp solution uses boric acid as the buffer. For both solutions, the phosphide content of the deposits is best controlled by frequent additions of phosphite or phosphorous acid.

Black Nickel Plating Baths

Black nickel deposits are used primarily for decorative effect and to provide nonreflecting surfaces. Uses include (*a*) typewriter and camera parts, (*b*) military instruments, (*c*) clothes fasteners, and (*d*) costume jewelry. Deposits are brittle and readily chip or flake on bending or impact; for this reason, deposits usually are not permitted to exceed 1.0 to 1.5 μm (0.04 to 0.06 mil) in thickness.

The exact composition of these deposits is not known, but because they contain zinc and sulfur, sulfides of nickel and zinc may be involved. Oxides of these metals may also be present, as well as occlusions of other compounds. A typical black nickel electrodeposit contains 50% nickel, 7% zinc, and 15% sulfur.

On ferrous materials, a substantial undercoating of zinc, cadmium, copper, brass, or nickel must be applied before black nickel plating to provide corrosion protection. A nickel undercoating is preferred when greater abrasion resistance of the black nickel finish is required. On brass or copper, an undercoat is usually not required unless the service environment is severe, in which case a nickel deposit may be applied before black nickel. On stainless steels, a nickel flash from a low-pH nickel strike bath is used to enhance adhesion.

There are several successful compositions for producing black nickel deposits; all incorporate zinc (Zn^{++}) and thiocyanate (CNS^-) ions. Table 4 gives the compositions and operating conditions for a sulfate and a chloride black nickel plating bath. The sulfate bath is more commonly used, but the chloride formula is said to permit higher current densities and a wider pH range, enhancing adhesion by permitting the use of an acid-nickel chloride strike pretreatment without rinsing. Table 5 summarizes a sequence for chloride-bath black nickel on steel.

Bright Plating Baths

Bright nickel plating baths are modifications of the Watts nickel solution and contain organic or a combination of organic and inorganic brightening agents; these additions serve to produce a high degree of brightness, leveling reflectivity, and hardness. A great variety of additives is used, usually in highly specific combinations. Their function is to produce as brilliant and ductile a deposit as possible over a wide range of current densities and operating conditions. Some of these additives are consumed very slowly during electrolysis, others more rapidly.

Although the nomenclature for bright plating additives is not standardized, with various names and descriptions being used for the same type of substance, in general they may be classified as carriers, brighteners, and auxiliary brighteners.

Carriers. Known also as brighteners of the first class, secondary brighteners, control agents, and ductilizers, these organic materials are usually aromatic compounds but can also (infrequently) be unsaturated aliphatics. Their primary function is to refine grain structure and provide deposits with increased luster in comparison with the usual nonuniform dullness of deposits from baths containing no additives. Some of these additives can be used in Watts or modified high-chloride Watts compositions. This class of brighteners widens the bright range when used in conjunction with brighteners. Some examples of carriers are:

- Saccharin (o-sulfobenzoic imide)
- Paratoluene sulfonamide
- Benzene sulfonamide
- Benzene monosulfonate (sodium salt)

- Ortho sulfobenzaldehyde (sodium salt)
- Naphthalene 1, 3, 6-trisulfonate (sodium salt)

Carriers are used in concentrations of about 1 to 25 g/L (0.1 to 3 oz/gal), either singly or in combination. They are not consumed rapidly by electrolysis, and consumption is primarily by drag-out and losses by carbon treatment.

Carriers containing sulfur in combination with oxygen and carbon reduce the tensile stress of the bright nickel deposit but do not, by themselves, produce brilliant deposits. Consumed relatively slowly during the plating operation, their concentration in the nickel-plating solution is many times that of the brightener. When added to a solution containing a brightener-leveler, they widen the range of current density within which bright deposits are obtained.

The stress-reducing property of carriers is increased if they contain amido or imido nitrogen. For example, saccharin is a most effective stress reducer and often helps to decrease or eliminate hazes. It is generally used as sodium saccharin at a concentration of 0.5 to 4.0 g/L (0.07 to 0.5 oz/gal).

Saccharin cannot be used in the presence of zinc contamination greater than about 0.035 g/L (0.004 oz/gal), because it promotes the formation of a dark, streaky nickel deposit in areas of low current density. The effect may be decreased by lowering the pH of the solution, but leveling is decreased simultaneously. Proprietary additives are available that can eliminate this effect.

Auxiliary brighteners may be either organic or inorganic. Their function is to augment the luster attainable with the carriers and brighteners and to increase the rate of brightening and leveling. Some examples are:

- Sodium allyl sulfonate
- Zinc, cobalt, cadmium (for rack and barrel plating)
- 1, 4-butyne 2-diol

The concentration of these additives may vary from about 0.1 to 4 g/L (0.01 to 0.5 oz/gal). The rate of consumption depends on the type of compound and may vary widely. These compounds may be of aromatic or aliphatic types and usually are heterocyclic or unsaturated.

Brighteners. Known also as brighteners of the second class, primary brighteners, and levelers, the brighteners in this category produce, in conjunction with the other classes of additives, bright-to-brilliant deposits having good ductility and a high rate of brightening and leveling over a wide range of current densities. A wide variety of materials have been used as brighteners, some of which are:

- Reduced fuchsin
- Phenosafranin
- Thiourea
- *N*-allylquinolinium bromide
- 1, 4-butyne 2-diol
- 5-aminobenzimidazolethiol-2

Materials of this type generally are used in concentrations of 0.005 to 0.2 g/L (0.0006 to 0.02 oz/gal); an excess of brighteners may cause serious embrittlement. The rates of consumption of these materials may vary within wide limits.

Commercially successful brightener-levelers have the following capabilities and operating characteristics:

- Produce deposits that are bright over a wide range of current density
- Produce significant leveling or scratch-filling
- Produce deposits with fair ductility and low stress
- Produce bright deposits in areas of low current density
- Permit use of high average current densities and bath temperatures
- Are not sensitive to metallic contaminants, particularly zinc
- Permit continuous purification of the plating solution by use of activated carbon on filters
- Produce breakdown products that are not harmful or that can be removed by activated carbon
- Are not sensitive to anode effects
- Are not highly toxic or malodorous

A recent development in decorative nickel plating has been the use of a bath containing suitable organic additives and a suspended finely divided inorganic material. Deposits from this bath under optimum operating conditions are satiny and can be used as a basis for chromium, brass, gold or other metals, or can be lacquered.

Table 5 Sequence of cycles for black nickel plating on steel

Cycle	Time, min	Temperature °C	Temperature °F	Constituent	Concentration	Current density A/dm²	Current density A/ft²
Electroclean	½-1	82-105	180-220	...	60-75 g/L (8-10 oz/gal)	...	...
Anodic acid etch(a)	1-3	21	70	H_2SO_4	50 vol%	10-40	100-400
Acid dip	¼-1	21-27	70-80	HCl	10-20 vol%	...	...
Nickel strike	3-6	21-24	70-75	(b)	(b)	3	30
Nickel plate (Watts)	5-30	43-54	110-130	(c)	(c)	1-5	10-50
Black nickel plate	20-30	21-24	70-75	(d)	(d)	0.1-0.15	1-1.5

(a) Optional. (b) 8 to 10% hydrochloric acid in a nickel chloride solution of 180 to 240 g/L (24 to 32 oz/gal). (c) See "Watts bath" in Table 2. (d) See "Chloride bath" in Table 4.

Barrel Nickel Plating

Barrel nickel plating may be performed in a variety of Watts or high-chloride electrolytes. Magnesium sulfate is sometimes added to the Watts bath to increase bath conductivity or to promote whiteness in the deposit.

A variety of additives, some proprietary, are used to increase the luster of barrel nickel deposits. Examples of some of these are:

- Cobalt sulfate plus cadmium sulfate
- Naphthalene 1, 3, 6-trisodium sulfonate
- Saccharin

Brighteners are at times used with the second and third compounds listed above. These brighteners should be insensitive to current interruption; otherwise, trouble can be encountered in double-plate peeling, that is, peeling of one layer of nickel from another.

If barrel nickel-plated parts are to be barrel chromium plated, it is imperative that the nickel deposits be as active and receptive to chromium as possible. Precautions include the following:

- Parts should be chromium plated as soon as possible after nickel plating. A dip of the nickel-plated parts in a 10 to 25 vol % solution of concentrated hydrochloric acid and water, followed by a water rinse, usually provides sufficient activation before chromium plating
- The concentration of organic and metallic impurities in the nickel-plating

baths must be maintained as low as possible. Circulation of the nickel solution through a carbon-packed filter using 0.12 to 0.24 g/L (0.1 to 0.2 lb per 100 gal) activated carbon is helpful in maintaining low levels of organic impurities. Low current-density electrolysis and occasional high-pH activated carbon treatments are also helpful

Antipitting Agents. During electrolysis, hydrogen bubbles are formed at the cathode surface due to the approximate 93 to 95% cathode efficiency. To prevent the bright centered pits caused by the adherence of tiny hydrogen bubbles to the plated surface, wetting agents, such as the salts of sulfated alcohols, are customarily added to Watts nickel plating solutions. Wetting agents eliminate pitting by reducing surface tension to the point that hydrogen bubbles do not adhere to the surfaces being plated. This is normally accomplished at about 30 to 40 dynes-cm; above this range, hydrogen pitting may still occur. Using too much wetting agent adds to cost, increases foam, and complicates batch purification. Wetting agents are used also in other nickel plating solutions. Low-foaming wetting agents are used in air-agitated baths.

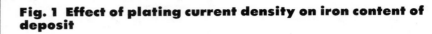

Fig. 1 Effect of plating current density on iron content of deposit

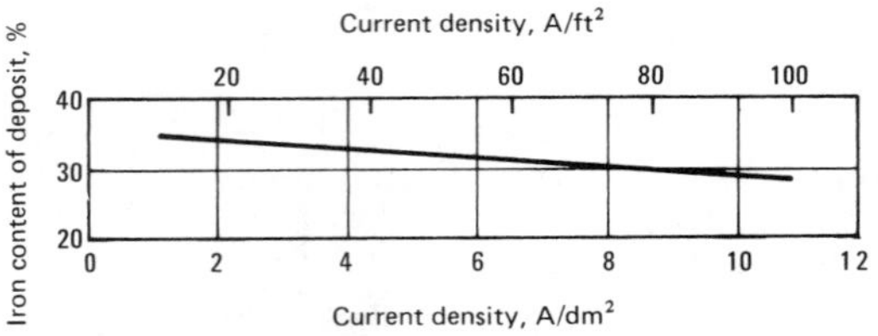

Fig. 2 Ratio of nickel ions (Ni^{++}) to ferrous ions (Fe^{++}) in solution versus percent iron in deposit

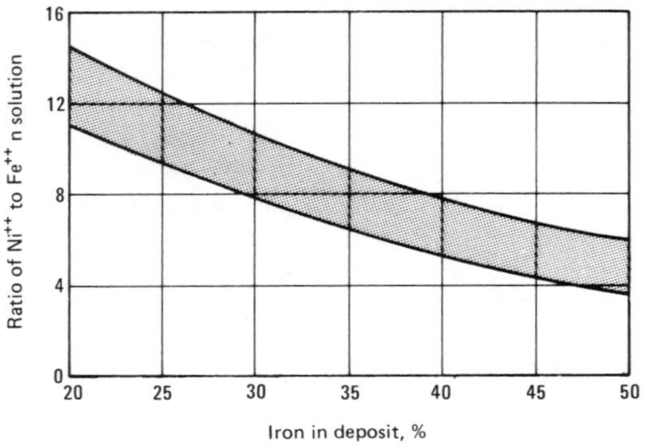

Nickel-Iron Plating

Nickel-iron alloy plating has been introduced as a means of reducing production costs. This primarily decorative nickel-iron alloy provides a deposit having full brightness, high leveling, excellent ductility, and good receptivity for chromium. Nickel-iron can be plated on steel, brass, aluminum, zinc die castings, or plastic substrates in either barrel or rack equipment. The operation of the process is similar to nickel plating, and the composition of most of the baths in production falls within the following concentration ranges:

Nickel sulfate	35-225 g/L (5-30 oz/gal)
Nickel chloride	75-225 g/L (10-30 oz/gal)
Boric acid	45-50 g/L (6-7 oz/gal)
Ferrous sulfate	10-75 g/L (1-10 oz/gal)
Nickel (Ni^{++})	30-75 g/L (4-10 oz/gal)
Iron (Fe^{++})	2-15 g/L (0.2-2 oz/gal)
Proprietary additives	As required
pH	3.0-4.0
Temperature	43-71 °C (110-160 °F)
Cathode current density	0.5-10 A/dm² (5-100 A/ft²)
Agitation	Cathode rod or air

The proprietary additive systems are similar to those for bright nickel plating in that they consist of carrier, secondary auxiliary, and brightener additives. In addition, the bath requires additives to stabilize the ferrous and ferric ions so that hydroxide compounds do not form and precipitate.

The pH should be maintained within the recommended limits. High values can produce highly stressed deposits and nonuniform brightness. Low values reduce leveling and promote chemical dissolution of iron anodes and reduced cathode efficiency. Higher temperatures can yield hazy and stressed deposits. Lower temperatures can cause high current density burning. Air agitation is used in most installations, although the process works well with mechanical agitation. Sulfur depolarized nickel anodes must be used to minimize anode polarization. Iron anodes should be of high purity.

As plated in industry, deposits usually contain from 20 to 50% iron. The effect of current density on alloy composition is shown in Fig. 1; it is fairly constant across the normal operating range. The composition of the alloy is determined, to a great degree, by the ratio of nickel ions to ferrous ions in the bath, as shown in Fig. 2.

The procedures for converting a nickel bath to a nickel-iron bath are extremely simple. It is possible to slide-convert a nickel bath. Some nickel anodes are removed, replaced with iron anodes, and ferrous sulfate is allowed to build up in the bath through electrolysis to an equilibrium value. Another method is to dilute the bath, exchange a percentage of nickel anodes with iron anodes, and add ferrous sulfate.

The hardness of bright nickel-iron, containing 20 to 50% iron, is 550 to 700 DPH. The stress of the deposit, as measured by the contractometer, is normally on the tensile side, and the ductility is generally 80 to 100% as measured by the foil test (ASTM B490).

The pH, additive concentration, temperature, and current density have an effect on the mechanical properties of the deposit.

Deposits applied on steel or on copper and subsequently chromium plated have excellent life in a humidity test, showing no stain or failure to the substrate after 1000 h. Consequently, they have had good acceptance for use in interior applications as a substitute for bright nickel coatings. A neutral salt spray test of chromium plated over nickel-iron is comparable to bright nickel-chromium coatings, both showing staining and failure to red rust at the same time. In severe accelerated testing (CASS and Corrodkote), nickel-iron shows brown staining on parts for exterior automotive applications such as bumpers. Mobile tests and outdoor tests in marine and industrial environments, however, show that microdiscontinuous chromium utilizing a pure nickel undercoat can alleviate the staining effect.

Duplex Nickel Plating

Duplex nickel plating involves the deposition of two layers of nickel, one semibright and one bright, before the application of chromium. This practice has become widely accepted in recent years.

Semibright Layer. The first layer (semibright) is deposited from a Watts-type formulation containing one or more sulfur-free organic additives. Ideally, this semibright deposit is sulfur free, has a predominantly columnar structure, has good ductility and leveling, and is uniformly fine-grained over a wide range of current density. Composition and operating conditions for a semibright nickel plating bath are given below:

Nickel sulfate	300 g/L (40 oz/gal)
Nickel chloride	35 g/L (4.7 oz/gal)
Boric acid	45 g/L (6 oz/gal)
Organic additive	0.05-0.15 g/L (0.007-0.02 oz/gal)
Antipitting agent, vol %	0.5
Temperature	54 °C (130 °F)
pH (electrometric)	3.5-4.5
Agitation	Mechanical or air

Stress will increase with increasing nickel chloride content; deposits also tend to be nonuniform in color and leveling. Organic additives are of the sulfur free type.

Semibright deposits may range in thickness from 13 to 30 μm (0.5 to 1.2 mil). They should be uniformly fine-grained, because a coarse-grained deposit in any portion of the range of current density makes it difficult to obtain full luster in the second (bright) nickel deposit.

Semibright plating additives produce semilustrous, smooth, uniformly fine-grained, ductile deposits over a wide range of current densities. They are sulfur free and can be organic compounds of aromatic or aliphatic nature. Concentrations are usually fairly low, from 0.05 to 0.5 g/L (0.006 to 0.06 oz/gal). Examples of these additives are:

- 1, 4-butyne 2-diol (or other aliphatic compounds with olefinic or acetylenic unsaturation)
- Formaldehyde

The baths may also contain anionic surfactants as antipitting agents, singly or in combination.

Bright Layer. The second layer, which may be one of a variety of bright nickels, may range in thickness from 5 to 8 μm (0.2 to 0.3 mil), or about 20 to 35% of the total nickel thickness. Ideally, it should be plated from a bath that is compatible with the semibright additive, or additives, because in most duplex systems the semibright additive functions as either a brightener or an auxiliary brightener in the bright nickel bath. This compatibility is especially important when no intermediate rinsing is used between the semibright and bright plating tanks. Rinsing between semibright and bright baths is infrequently used for steel parts and often used for die cast parts. The rinse is usually acidified, particularly for die cast parts.

Advantages of duplex nickel plating are primarily:

- Corrosion protection of the plated article is enhanced by the composite coatings
- The semibright coating, and therefore the composite, usually has superior ductility

Disadvantages of duplex nickel are:

- Some of the semibright additives yield decomposition products that, unless otherwise controlled, adversely affect deposit ductility and residual stress; this may necessitate regular purification treatments
- Adhesion between the semibright and bright layers may be unsatisfactory unless transfer time is rapid or, in automatic plating when rapid transfer is impractical, an intermediate rinse is added. Also, precautions must be taken to avoid bipolar effects. This is accomplished by applying a lower than normal current density at low voltage while the workpieces are removed from the semibright bath and immersed in the bright bath
- Some semibright baths are sensitive to metallic contaminants such as zinc and require electrolytic purification to minimize the concentration of such impurities

Corrosion Protection of Duplex Nickel. Due to the surface content, bright nickel is anodic to the semibright nickel; therefore, the bright nickel corrodes sacrificially. Because of this, during corrosion, penetration is retarded at the bright-semibright interface, and subsequent corrosion proceeds mostly laterally in the bright layer. Corrosion protection is further improved when a microdiscontinuous chromium overlay is used because corrosion then occurs at many sites in a minor amount rather than at a few sites in a major amount.

Corrosion protection of parts plated with duplex nickel is governed by the total thickness of the nickel deposits (see ASTM B456) and by the relative thickness of the two layers. A satisfactory ratio is 65 to 75% semibright to 25 to 35% bright nickel. As parts become increasingly complex, greater total thickness, particularly in recesses, is required. Therefore, the minimum thickness for total nickel should be raised, from 13 to 20 μm (0.5 to 0.8 mil), for example. Compositions and operating conditions of semibright and bright nickel baths used in duplex plating are shown in Table 6.

Nickel Baths for Producing Microdiscontinuous Chromium (MD) Deposits. Microdiscontinuous chromium, (MD) overlays serve to improve the corrosion protection of nickel and copper nickel systems. Maximum protection is obtained when MD chromium is applied over duplex nickel deposits. The improvement in corrosion protection is brought about by the distribution of the corrosion current, in the nickel-chromium couple, over a larger

Table 6 Compositions and operating conditions of semibright and bright nickel baths used in duplex plating

Composition or operating condition	Semibright	Bright
Nickel sulfate	300-335 g/L (40-45 oz/gal)	300-375 g/L (40-50 oz/gal)
Nickel chloride	38-41 g/L (5-5.5 oz/gal)	60-90 g/L (8-12 oz/gal)
Boric acid	38-41 g/L (5-5.5 oz/gal)	41-45 g/L (5.5-6 oz/gal)
pH	4.0-4.2	4.0-4.2
Temperature	57-63 °C (135-145 °F)	66-71 °C (150-160 °F)
Current density	5-7 A/dm^2 (50-70 A/ft^2)	6-8 A/dm^2 (60-80 A/ft^2)
Agitation	Air	Air

area. The resulting reduction in corrosion current density at each corrosion site reduces corrosion penetration.

Special nickel solutions are used after normal bright nickel plating, for example, the top layer of duplex nickel, and prior to chromium plating. The thin layer of special nickel, usually 1 to 2 μm (0.04 to 0.08 mil) produces either microcracked (MC) or microporous (MP) chromium when subsequently chromium plated.

Microcracked chromium is produced by applying the thin layer of nickel from a bath designed to produce high tensile stresses in the nickel deposit. When chromium plated, the thin nickel and chromium will crack. Varying the conditions under which the nickel layer is deposited can provide variations in the crack density over a range of from 300 cracks per centimetre (750 cracks per inch) to 800 cracks per centimetre (2000 cracks per inch). The nickel bath usually consists of a basic nickel chloride electrolyte with additives that provide additional stress such as the ammonium ion. Boric acid is not used, but other buffers such as the acetate ion are frequently used. Proprietary organic additives are also used to enhance the brightness and the ability of the deposit to crack, especially in the low current density areas. Temperature and pH are controlled to provide the variation in crack density; low temperature, 23 °C (73 °F) and high pH (4.5) favor higher crack densities; high temperature, 36 °C (97 °F) with low pH (3.5) favors lower crack densities. Cracking of the chromium deposit should occur subsequent to chromium plating. Aging or the use of a hot water dip may be necessary to promote the formation of all microcracks.

Microporous chromium is produced from Watts-type nickel baths using air agitation and containing very fine inert particles, usually inorganic, containing also the normal additives used for bright nickel plating. Chromium, plated over the resulting nickel particle matrix, deposits around the particles creating pores. The nickel baths are operated much like bright nickel solutions with the exception that filtration, obviously, cannot be performed. In some instances, auxiliary additives are used to allow for reduction of the particle concentration in the plating bath and still provide high pore densities. Pore densities can vary according to the concentration of particles, agitation rates and additives. Generally, a pore density of 10 000 pores per square centimetre (64 000 pores per square inch) is desirable.

In either case, chromium thicknesses should not be allowed to exceed about 0.5 μm (0.02 mil) or the cracks/pores will start to heal and defeat the purpose of the coating. Microcracked systems produced directly by dual or single layer chromium baths require about 1.0 μm (0.04 mil) or more to produce microcracking and the resulting deposit is milky or hazy in appearance. Properly applied MC or MP chromium produced by the aforementioned nickel baths should possess only a very slight haze.

In service, environmental corrosion will in time produce a slight dulling of all MD chromium surfaces, however, the benefits to be gained by the additional protection provided by the MD chromium should far outweigh the slight cosmetic loss in appearance.

Contaminants

Commercial rolled or cast nickel anodes may contain enough iron and silicon to cause a buildup of these elements in the solution. Contamination by zinc, aluminum, and copper is most often caused by the dissolution of zinc-based die castings that have fallen from racks into the plating tank and been permitted to remain there. Inadequate rinsing before nickel plating increases the drag-in of metallics. The presence of cadmium and lead may be attributed to a number of sources, including lead-lined equipment and tanks, impure salts, and drag-in of other plating solutions on poorly rinsed racks. Chromium is almost always carried into the nickel solution on rack tips that have not been chromium stripped or on poorly maintained racks that have been used in the chromium tank and have trapped chromium plating solution in holes, pockets, and tears in the rack coating. The metallic contaminants affect bright nickel deposition in the following ways.

Aluminum and silicon produce hazes, generally in areas of medium-to-high current density. Aluminum and silicon may also cause a fine roughness called salt and pepper or stardust.

Iron produces various degrees of roughness, particularly when the pH of the solution is above about 3.8.

Calcium contributes to needle-like roughness as a result of the precipitation of calcium sulfate when calcium sulfate exceeds the saturation point of about 0.5 g/L (0.06 oz/gal) at 60 °C (140 °F) as calcium.

Chromium. A few parts per million of chromium as chromate causes dark streaks, high current density gassing, and may cause peeling. After reduction to the trivalent form by reaction with organic materials in the solution or at the cathode, chromium may produce hazing and roughness effects similar to those produced by iron, silicon, and aluminum.

Copper, zinc, cadmium, and lead affect areas of low current density, producing hazes and dark-to-black deposits.

Organic contaminants may produce hazes or cloudiness on a bright deposit or result in a degradation of physical properties. Haze defects may appear at any current density area, or they may be confined to narrow current density ranges. Zones of haze may be produced also by agitation patterns, and combination effects are not unusual. Mechanical defects producing hairline cracks may be encountered if the coating is sufficiently stressed as a result of solution contamination. These cracks usually appear in areas of heavier plating thickness (higher current density), but are not necessarily confined to these zones.

Bath Purification. A widely used method for removing organic contaminants from bright nickel plating solutions consists of filtering the solutions through activated carbon. Where difficulty is encountered, an oxidizing agent, such as hydrogen peroxide, may be used to assist in purification. Batch treatment in a separate treatment tank, using 4.8 to 7.2 g/L (4 to 6 lb/100 gal) activated carbon, is used in cases of more severe organic contamination. These treatments are usually performed at 60 °C (140 °F) for a period of 8 to 16 h. Filtration is used to remove the activated carbon, and any addition agents removed by the carbon are replaced.

Metallic contaminants are removed by high-pH precipitation or low current density electrolysis. High-pH precipitation, a method for removing or substantially reducing iron, aluminum, trivalent chromium, and silicon, consists of raising the pH of the plating solution to 5.0 to 5.2 (electrometric) with nickel carbonate or nickel hydrate in a separate treatment tank. The bath is mixed for several hours, during which it is held at about 60 °C (140 °F). During the high-pH treatment, activated carbon often is added, at the rate of 6 g/L (5 lb/100 gal) of plating solution. After purification, the solution is filtered to remove suspended matter; the pH is lowered to the normal operating range, and plating is resumed.

To remove copper, zinc, cadmium, or lead, the solution is electrolyzed at a low current density of 0.2 to 0.5 A/dm^2 (2 to 5 A/ft^2), using nickel plated corrugated steel cathodes. Agitation of the solution and lower pH (about 3 to 3.5) increase the rate of impurity removal, shortens the time required for purification. At the beginning of this treatment, the cathodes may assume a dark appearance, but brighten when the metallic contaminants have plated out. Organic contaminants that cause difficulties at low current density may also be removed by low current density electrolysis, sometimes referred to as dummying. If the solution is saturated with calcium sulfate at the plating temperature, heating the solution to about 71 to 77 °C (160 to 170 °F) decreases solubility allowing removal by settling or filtration while the solution is held at the high temperature.

Control of Bath Composition

Control of the composition of the plating bath is probably the most important single factor contributing to the quality of the nickel deposit. Initially, the bath must be carefully prepared to the desired composition and pH, and it must be purified before use. Thereafter, the composition and pH of the solution must be controlled within prescribed limits in terms of both replenishment and contamination by foreign substances.

Initial Purification. Before any freshly made up nickel plating bath is used, contaminants such as iron, copper, zinc, and organics present in trace quantities in commercial salts must be removed to obtain the best results. The following methods are used for purifying the unused plating solution:

- *High-pH treatment* consists of adding nickel carbonate to the hot solution until a pH of 5.0 to 5.5 is obtained. This precipitates the hydroxides of metals such as iron, aluminum, and silicon, which in turn frequently absorb other impurities
- *Peroxide treatment* oxidizes iron to the ferric state, making it more easily precipitated at high pH. This treatment frequently destroys organics
- *Electrolytic purification* removes most of the harmful metallic and organic impurities

A purification procedure for a Watts solution, using all three of the above methods, would comprise the following operations:

- Use a storage tank — not the plating tank — to dissolve the nickel sulfate and nickel chloride in hot water at 38 to 49 °C (100 to 120 °F) to about 80% of desired volume
- Add 1 to 2 mL (0.8 to 1.6 pints/100 gal) of 30% hydrogen peroxide; agitate briefly and allow to stand 1 h
- Add 1.2 to 2.4 g/L (1 to 2 lb/100 gal) activated carbon and agitate thoroughly
- Heat to 66 °C (150 °F) and add nickel carbonate, 1.2 to 2.4 g/L (1 to 2 lb/100 gal) of solution, with agitation, to bring the pH to 5.2 to 5.5. Allow to settle 8 to 16 h
- Filter to the plating tank
- Add and dissolve boric acid and add water to bring bath up to 100 vol %
- Electrolytically purify by using a large area of nickel plated corrugated steel sheets as cathodes. The average cathode current density should be 0.5 A/dm^2 (5 A/ft^2) and the time sufficient to pass 0.5 to 1.3 A·h/L (2 to 5 A·h/gal) of solution. The solution should be agitated and the temperature held at 49 to 60 °C (120 to 140 °F)
- Remove the dummy cathodes
- Adjust pH

Continuing Control. To ensure satisfactory performance, the four following basic constituents of nickel plating baths must be regularly controlled: (*a*) the nickel salts; (*b*) the boric acid; (*c*) the pH; and (*d*) any addition agents used.

Nickel ions are furnished by the nickel sulfate, chloride, sulfamate, or fluoborate, or by combinations of these nickel-containing salts, in the plating bath. For most commercial applications, the nickel metal concentration should be maintained between 60 and 80 g/L (8.0 and 10.5 oz/gal) through the addition of suitable quantities of nickel salts. However, if the plating current density is low, the bath temperature high, and the solution agitated over the surface of a part, it is possible to nickel plate with less than 60 g/L (8 oz/gal) of nickel metal in the solution. It is desirable to have a minimum of 23 g/L (3 oz/gal) of nickel chloride in the solution to promote nickel anode corrosion. Water used for preparing nickel plating solutions and for replacing evaporation losses should be as pure as possible. Demineralized water should be used if local tap water has a high (>200 ppm as calcium) hardness. For additional assurance against rough nickel deposits, make-up water and additions should be filtered into the plating tank to eliminate particles from pipelines, storage tanks or other sources of sediment.

Boric Acid. Boric acid is the most commonly used buffering agent for nickel plating baths. Boric acid is effective in stabilizing the pH in the cathode film within the ranges normally required for best plating performance. It is available in a purified form and is inexpensive. Ranges of boric acid concentration for various nickel plating baths are shown in Tables 2, 3, and 6.

pH. The pH of the nickel plating solution will rise during normal operation of the bath necessitating regular additions of acid to maintain the pH within the prescribed limits.

Addition Agents. Addition agents must be replenished due to losses from dragout, electrolytic consumption and the effects of carbon filtration (or batch treatment). Quantities replenished will vary with the individual addition agent but should be based on analysis and/or experience or, if proprietary, on the vendors recommendations.

Contaminants, either inorganic or organic, may be present in nickel plating solutions in the form of liquids, solids, or gases. Their presence even in small quantities may result in plating defects such as brittleness, cracking,

Table 7 Limits of metallic contaminants in nickel plating baths

Contaminant	Normal g/L	Normal oz/gal	Maximum g/L	Maximum oz/gal
Aluminum	0.005	0.0007	0.060	0.008
Chromium	0.002	0.0003	0.020	0.003
Copper	0.001	0.0001	0.010	0.001
Iron	0.005	0.0007	0.050	0.007
Lead	0.0008	0.0001	0.002	0.0003
Zinc	0.001	0.0001	0.010	0.001

peeling, pitting, burning, roughness, or poor covering power.

A general procedure for removing many contaminants consists of oxidation with hydrogen peroxide at high pH and adsorption with activated carbon at high temperature in a spare tank. The solution is then filtered back into the plating tank, where it is electrolyzed at low current density at the normal operating temperature and pH.

Inorganic contaminants arise from numerous sources, including nickel salts of technical grade, hard water, carry-over from acid dip tanks, airborne dust, and bipolar attack of metallic immersion heaters. Table 7 lists optimum limits of metallic contaminants in nickel plating baths and also indicates approximate maximum limits consistent with the production of high-quality deposits.

The degree of contamination by many inorganic materials may be controlled by continuous filtration and dummying, that is, electrolysis of the plating solution at 0.2 to 0.5 A/dm^2 (2 to 5 A/ft^2). This may be accomplished on a batch basis or continuously by installing a dummy compartment and overflow dam at one end of the plating tank. Solution from the filter is pumped into the bottom of the dummy compartment, up past the corrugated cathode sheets, over the dam, into the plating section of the tank, out through a bottom outlet at the far end of the tank and back to the filter. Solid particles and soluble metallic impurities (for example, copper, zinc, lead) are removed simultaneously by this procedure.

Organic contaminants may arise from many sources and may be introduced by the following examples:

- Buffing compounds
- Lubricating oil dropped from overhead equipment
- Sizing in anode bags
- Weaving lubricants on plastic anode bags
- Uncured rack coatings or stop-off lacquers
- Adhesives on certain types of masking tapes
- Decomposition products from wetting agents
- Organic stabilizers in hydrogen peroxide
- Paint spray
- New or patched rubber tank linings

In general, everything coming into contact with the nickel plating solution should be clean, and any new material to be used in the solution, such as masking materials and rack coatings, should be pretested for compatibility with the bath.

Many organic contaminants may be effectively removed from nickel plating solutions by adsorption on activated carbon on either a batch or a continuous basis. On a batch basis, the solution is transferred to a spare tank, heated to 60 to 71 °C (140 to 160 °F), stirred for several hours with a slurry of 6 g/L (5 lb/100 gal) of activated carbon, permitted to settle, and then filtered back into the plating tank. For solutions in which organic contamination is a recurring problem, continuous circulation of the solution through a filter, coated at frequent intervals with small amounts of fresh activated carbon, is recommended. When continuous carbon filtration is used, the wetting agent in the solution must be replenished and controlled more carefully, to prevent pitting of the nickel deposits.

Gaseous contamination of nickel plating solutions usually consists of dissolved air or carbon dioxide. Dissolved air in small amounts may lead to a type of pitting characterized by a teardrop pattern. Dissolved air in the plating solution usually can be traced to entrainment of air in the pumping system when the solution is circulated. The sudden appearance of foam on the surface of a nickel bath is one indication of possible air entrainment in the solution coming from the circulating pump. If the pump is being starved because inlet size is insufficient, or because a valve on the inlet line is partly closed, air is sucked through the packing gland along the impeller shaft and becomes entrained in the plating solution. Accordingly, flow valves should be placed only on the outlet side of centrifugal pumps, and the pump packing should be tightened occasionally. As a safety precaution, a water seal can be placed on the pump shaft so that water, instead of air, is drawn into the solution in the event of pump cavitation. The solution should be purged of dissolved air by being heated to a temperature at least 6 °C (10 °F) higher than the normal operating temperature for several hours. The solution is cooled to the operating temperature before plating is resumed.

Dissolved carbon dioxide in a nickel plating solution is usually found after nickel carbonate has been added to raise the pH. This is usually a temporary condition, because the carbon dioxide is liberated from warm nickel plating solutions after several hours. However, when solutions containing carbon dioxide are scheduled for immediate use, they should be purged by a combination of heating and air agitation for approximately 1 h at 6 °C (10 °F) or more above the normal plating temperature.

Tests for Determining Bath Composition. Two general methods of determining the composition of nickel plating solutions are used: (*a*) chemical analyses and (*b*) electroplating tests. The quantitative analysis of a plating solution is normally required only weekly or monthly, whereas plating tests ordinarily are performed daily; however, the frequency of control testing depends on the nature of the work, the daily volume of work processed per litre (gallon) of solution, and the plating solution used.

Because the properties of nickel deposits vary with the acidity of the plating solution, the pH of the solution should be maintained within ±0.15 pH units, to obtain consistent results. Control checks of pH should be made daily or more frequently, depending on the nature of the plating process. Nickel deposits are more likely to be burned at high plating rates when the pH reaches or exceeds 5.0. Figure 3 shows the effect of pH on the hardness of nickel electrodeposits from several nickel plating solutions operated at a temperature of 54 °C (130 °F), and current density of 5 A/dm^2 (50 A/ft^2).

Electroplating tests may be used to

Fig. 3 Effect of pH on hardness of deposits produced by six nickel baths

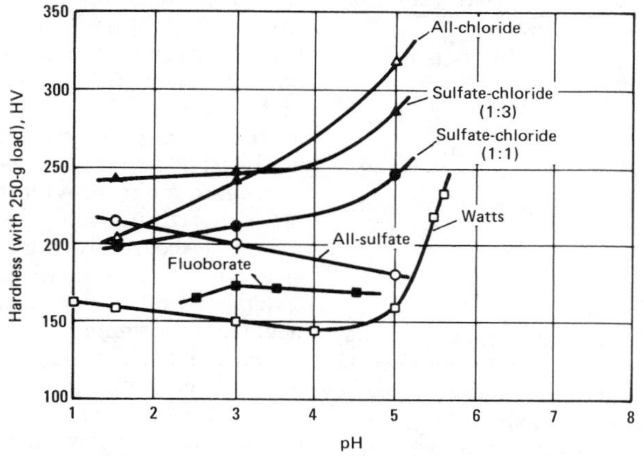

Bath temperature is 54 °C (130 °F), with a current density of 5 A/dm² (50 A/ft²)

Fig. 4 Effect of bath temperature on hardness of deposits produced by six nickel plating baths

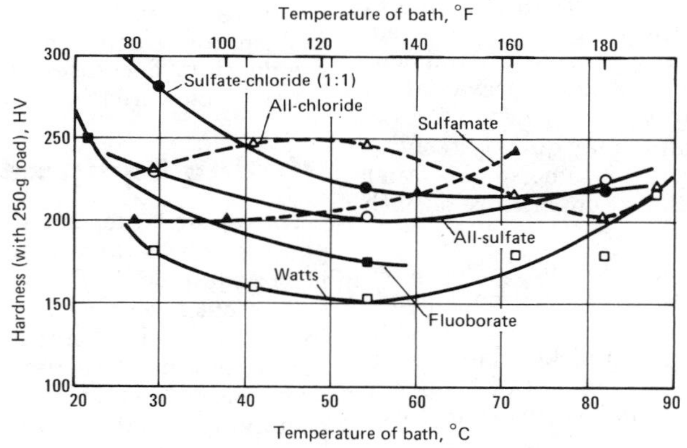

Current density is 5 A/dm² (50 A/ft²), and pH is 3.0

estimate the degree of contamination, and the tendency toward pitting of the plating bath, and the degree of stress and other physical properties of plated deposits. A Hull cell, containing a nickel anode, a sample of the nickel plating solution, and a cathode panel at a specific angle to the anode, reveal the nature of nickel deposits over a wide range of current densities. Dark deposits in the low current density region of the panel usually indicate metallic contamination of the solution. At the high current density region of the panel, peeling indicates high stress, bright streaks indicate organic contamination, and pitting indicates insufficient wetting agent in the solution.

To determine the residual stress in electrodeposited nickel under a given set of operating conditions, a spiral contractometer may be used. The outside of a thin stainless steel helical coil is plated at the current density under investigation, and the degree of deflection is shown on a dial. The amount of residual stress in the nickel may be calculated from the magnitude of the deflection and from the direction of the deflection. The stress may be characterized as either tensile (contractile) or compressive (expansive).

Anodes

A chill cast anode containing 99% nickel came into acceptance with the widespread use of the Watts bath; subsequently, a rolled, depolarized anode was developed that contained small amounts of nickel oxide but retained a 99% nickel content. Rolled depolarized anodes are used in high pH baths, although they may be used throughout the entire pH range of nickel plating solutions. To avoid the formation of rough deposits, these anodes usually are covered with cotton or synthetic fiber anode bags while in use.

Cast or rolled carbon anodes (99% nickel anodes containing about 0.2% carbon) may be used in baths with a pH of 4.5 or less. Although they are capable of forming an adherent carbon-silica film that retains loose anode particles, they are covered with anode bags during use to prevent the formation of rough deposits.

Virtually all nickel anodes made today are of high purity and contain a minimum of 99% nickel, the remainder consisting of various elements added to prevent anode passivity and to ensure uniform solution into the plating bath. Table 8 presents analyses of anode materials currently available.

Insoluble Anodes. The selection of a suitable material for insoluble anodes in nickel plating depends primarily on the composition and operating characteristics of the plating bath. Lead, in wire, rod, or sheet form, or wrought or electrolytic nickel, is a suitable anode material provided the solution does not contain chlorides. When chlorides are present in the bath, the most suitable anode materials are carbon or platinum; platinum may be in solid form or may be clad over titanium or some other material. Because the use of insoluble anodes depletes the nickel content of the bath and may release chlorine gas, adequate provision must be made for replenishment of nickel and adequate ventilation.

Bath Temperature

Variation in the operating temperature of a nickel plating bath can have a marked effect on the properties of electrodeposited nickel. To obtain consistent results, the temperature of a nickel plating bath should be maintained within ±2 °C (±4 °F) of the recommended temperature for a given application. In general, most industrial nickel electro-

Table 8 Analyses of nickel anodes

These analyses should be considered typical and do not constitute specifications

Significant constituent	Rolled depolarized(a), %	Rolled carbon(a), %	Electrolytic(b), %	Sulfur-containing electrolytic (b), %
Nickel and cobalt	99.75	99.40	>99.90	99.95
Carbon	...	0.20	0.01	<0.01
Silicon	0.01	0.25	...	...
Sulfur	0.01	0.005	0.001	0.02
Manganese	...	0.002	...	...
Oxygen	0.15	...	...	...
Magnesium	0.08	...	...	...

(a) Obtainable as bars. (b) Obtainable in various forms and usually used in pierced or expanded titanium metal anode baskets

plating baths are operated in the range of 38 to 60 °C (100 to 140 °F).

Electroplating baths that contain highly soluble nickel salts, such as the sulfamate and fluoborate baths, may be operated at temperatures as low as 38 °C (100 °F) without adverse effects. The higher nickel concentration obtainable with these salts makes it possible to produce sound, ductile, low-stress deposits at lower operating temperatures than those required in plating baths composed of nickel chloride and nickel sulfate. Baths composed of the less soluble nickel salts must ordinarily be operated at higher temperature, usually from 49 to 60 °C (120 to 140 °F) to yield sound deposits at rapid plating rates.

Figure 4 shows the effect of plating-bath temperature on the hardness of electrodeposited nickel for six different plating baths, which were operated under identical conditions of pH and current density. As indicated, the hardness of nickel electrodeposited by these baths (except for the all chloride) approaches a minimum at about 54 °C (130 °F).

Current Density

Without agitation, Watts nickel plating solutions generally are used at average current densities in the range of 0.5 to 3.5 A/dm^2 (5 to 35 A/ft^2). With mechanical agitation this range can be increased significantly depending on the rate of agitation. When these baths are operated at room temperature, plating must be done at very low current densities usually less than 1.0 A/dm^2 (10 A/ft^2). The pH can have a significant influence on the current densities that can be used. From pH 4.8 to 5.4, mid-range current densities must be used. Low-pH solutions may be used at current densities from the middle to high end of the range.

In high-chloride gray nickel baths, where the nickel chloride content is 90 g/L (12 oz/gal) and above, current densities may be greatly increased—up to 18 A/dm^2 (175 A/ft^2) and higher, depending on part shape and other factors.

For maximum production capacity, the highest current densities consistent with the shape of the part and other plating practices are used. Grain refinement usually results from increased current density, and as the maximum amperage in the range is approached, the deposit may be harder and less ductile. Treeing is more prevalent in areas of highest current density, however, and burning may quickly result if control is lost over some of the other variables, such as temperature or bath composition.

The weight of nickel electrodeposited during a given period of time depends on the number of amperes flowing in the circuit. For most nickel plating solutions, the cathode current efficiency is about 93 to 95% of the theoretical value. At 100% cathode efficiency, the calculated weight of nickel deposited by 1 A in 1 h is 1.095 g (0.039 oz). From the specific gravity of nickel (8.90), it can be calculated that 21 g (0.74 oz) of nickel is required to form a deposit 25 μm (1 mil) thick over an area of 0.09 m^2 (1 ft^2). To deposit this weight of nickel requires 21/1.095 (0.74/0.039), or 19 A·h (1140 A·min). Thus, at 1.9 A/dm^2 (19 A/ft^2) cathode current density, a period of 1 h is required to deposit nickel to a thickness of 25 μm (1 mil). At twice the current density (3.8 A/dm^2 or 38 A/ft^2), 25 μm (1 mil) is deposited in ½ h.

Because 100% cathode efficiency is not attained in commercial practice, the above theoretical values must be adjusted for practicality; generally, 1200 A·min (20 A·h) should be allowed for the deposition of 25 μm (1 mil) of nickel per 0.09 m^2 (1 ft^2) of surface.

At high current densities, deposits at the edges and protruding areas of the work being plated can be of greater than average thickness. This effect can be minimized by proper rack design and by the use of nonconducting baffles or conducting thieves. However, each application generally has its own best arrangement to provide the most uniform distribution of electrodeposited nickel. Excessively high current density can produce burning, resulting in a deposit that is nonuniform, cracked, peeled, and sometimes powdery. Low current densities, on the other hand, are time consuming.

The hardness of nickel deposits increases at very low current densities, usually below about 2 A/dm^2 (20 A/ft^2). The effect of increasing current density on the hardness of nickel deposits obtained in various plating baths is shown in Figure 5. Additives are used in many nickel plating solutions to produce special effects such as brightening and leveling (scratch filling). Quite often, the hardness and stress of the nickel deposit from these solutions increase, and the ductility decreases, as the current density is lowered.

Plating Equipment

Nickel can be deposited in still tanks, plating barrels, and a variety of automatic equipment. The design of such equipment need not vary significantly from that of the conventional plating equipment used in depositing other widely used metals, including copper, cadmium, and zinc. Significant differences, where they occur, are likely to be found in the materials used and the means provided for filtering, agitating, and heating the plating solutions.

Tank Linings. Most tanks used for nickel plating are made of steel and are lined with hard rubber or polyvinyl chloride. The lining protects the steel from attack by the plating bath, minimizing contamination of the solution, and prevents some bipolar effects. Plastisol-coated steel grids or heavy rubber mats are sometimes placed in the bottom of the tanks to prevent puncture of the linings by workpieces, racks, anodes, or other objects inadvertently dropped into the tank. Brick bottoms, with the bricks placed loosely, are also used.

Small tanks can be dipped in polyvinyl chloride compound and baked to

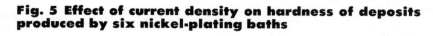

Fig. 5 Effect of current density on hardness of deposits produced by six nickel-plating baths

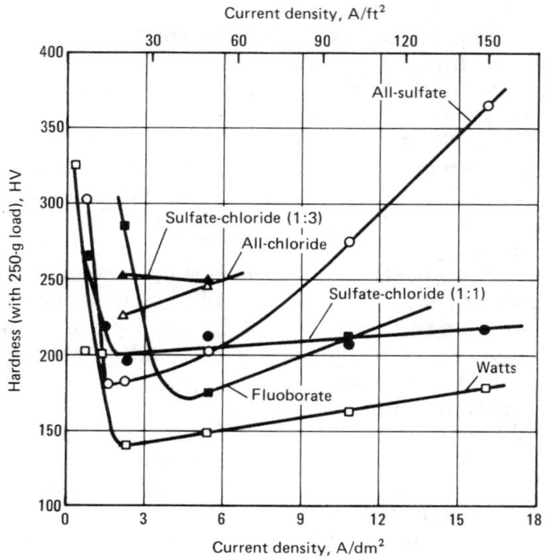

Bath temperature is 54 °C (130 °F), with a pH of 3.0

remove the solvents. Large tanks are usually lined by cementing sheets of hard rubber or polyvinyl chloride to the inside surfaces. Seams are tested with a high-voltage discharge unit (spark testing) to ensure that no pinpoint pores remain to allow attack on the steel tanks.

A lining material should be evaluated for compatibility with each solution because some solutions or brighteners may leach out the plasticizer in the liner, which can result in embrittled or rough deposits. When proprietary bright nickel solutions are used, the supplier of the solution should be consulted regarding suitable lining materials.

Testing of Linings. In choosing a lining for a nickel plating tank, regardless of the type of nickel solution it is to contain, the effect of the lining material on the nickel solution to be used should be tested. Such a test involves comparing the physical properties, for example, color, ductility, and stress of a sample deposit taken from a sample of the nickel solution before and after several days of leaching the proposed tank lining.

Plating Barrels. Many material systems have been used for the construction of nickel plating barrels, including rubber-covered steel. Recent developments in plastic materials have made available several compositions that have excellent mechanical properties over a wide temperature range, and resistance to many corrosive chemicals.

Of the more common barrel materials, hard rubber is acceptable for resistance to heat and chemical corrosion, but it is likely to chip in service. Methacrylates are frequently used, and polypropylene is becoming a common material for barrel construction.

Power-Generating Equipment. Although early plating rectifiers were relatively inefficient and could not tolerate overloading, recent developments have overcome these deficiencies by providing overload protection and by improving the rectifying elements to increase efficiency and stability. The initial cost of a rectifier is generally less than that of a comparable motor generator. Motor generators are suitable for supplying the power required in nickel plating. Motor generators have proved efficient and reliable even when operated under overload conditions for extended periods of time.

Heating and Cooling Systems. For small nickel plating tanks containing 100 to 1900 L (25 to 500 gal) of solution, in-tank heating and cooling units are most often used. These units are of four types:

- Electric immersion heaters
- Hairpin tubes of tantalum, titanium, lead, or glass, using steam for heating (lead must not be used with high-chloride solutions)
- Plate coils
- Bonded graphite immersion heat exchangers, using steam for heating

For large tanks, external heat exchangers are customary. These are preferably in separate circuits, or they may precede or follow the filter in the filter circuit. When the heat exchanger is in the filter circuit, boiling should be avoided, because the resultant bumping in the filter can release particles that negate the filtering process. External heat exchangers are used with steam for heating and with water for cooling, and may be made of bonded graphite, tantalum, high-silicon cast iron, or titanium high-temperature glass.

Pumping Equipment. Because they do not require pressure controls and bypasses for constant pumping conditions, suitably lined centrifugal pumps are recommended for nickel plating filtering systems. Pumps and motors should be properly insulated to prevent current losses and electrolytic corrosion.

Filtration is used to remove solid particles, which, for example, may originate from anodes, be dragged in as particles of unremoved buffing and polishing compounds, or drift into the bath from the atmosphere. If not removed, these solid particles cause roughness in the plate. Organic impurities may be dragged-in oil, grease or buffing compound, or the products of brightener breakdown. Filtering is normally accomplished by pumping the solution through filter cloth or paper coated with filter aids such as diatomite or alpha cellulose.

Activated carbon is used to remove soluble organic impurities. The carbon particles are held on the filter in the same fashion as other solids. In batch treatment, 1.2 to 6 g/L (1 to 5 lb per 100 gal) of carbon is frequently used. The need for and nature of treatment are best predicated on Hull cell examinations. Temperature and the time interval after carbon is added can be factors affecting removal.

The bath is usually pumped to a storage tank, given the prescribed treatment with carbon, and then returned to the plating tank through filters. Agitation is provided in the storage tank to avoid settling of the carbon. The filtration is accomplished through a suitable filter aid, so that no carbon particles pass back to the plating tank.

Although batch filtration is completely satisfactory for some operations, many large installations depend on continuous filtration. In continuous filtering, a complete turnover of solution, preferably every hour, is often used as a rule-of-thumb. The filter aid is frequently supplemented with a coating of activated carbon, in an amount up to 0.12 g/L (1 lb/1000 gal) of solution.

The filter intake should be positioned at or near the bottom of the solution to scavenge the entire solution, rather than only a wedge or segment. Large tanks require multiple intakes. A weir-type arrangement may be used to skim oil or grease from the top of the solution. The weir is often incorporated in a tank through which the solution flows in a maze-like path for electrolytic purification at a low cathode current density to remove metallic impurities such as iron, zinc, and copper. There is less agreement on the location of the filter discharge. Some filters discharge at the solution level; others below the solution level; some discharge into a header; others use a weir-cascading arrangement.

Commercial filters are of three types: (*a*) leaf (vertical or horizontal), (*b*) horizontal plate, and (*c*) cartridge. The cartridge type may be used for baths of moderate volume, up to about 38 000 L (10 000 gal). The materials of construction are steel, stainless steel, or rubber-lined steel. Plastic filter plates are in use. Cloth-backed filter paper can be used in the horizontal types. Cloth bags are used in conjunction with screens, and should be leached before being used. Paper is often used without a cloth backing, and should be checked for compatibility with the solution.

Agitation in Nickel Plating

Agitation is required in nickel plating to replenish the spent cathode film and to help dislodge gas bubbles from the work. Agitation provided by movement of the work is often supplemented by blowing air through perforated plastic tubes that are placed in the bottom of the plating tank and anchored in a position such that the air bubbles pass over the surfaces being plated. The air is obtained from a low-pressure blower that does not introduce oil into the bath. Work must be racked so that air is not trapped in recessed areas; otherwise, thin deposits result there.

Mechanical agitation can be accomplished by oscillating movement of horizontal or vertical work rods, mechanical stirring, pumping, or (as in automatic plating machines) continuous unidirectional movement of the work. For work-rod agitation, a 75-mm (3-in.) stroke and 20 to 30 strokes per minute in either direction are normally adequate.

The type of agitation used is governed by the size and shape of the part and the placement of the parts on the racks. More complex parts usually require more attention to agitation. The relations among current density, plating time, and plate thickness are important considerations. Tank size and shape also influence the selection of type of agitation. Air agitation is generally used, in preference to mechanical agitation, as long as the workpiece makes good positive contact with the plating rack.

Maintenance Schedules for Solutions and Equipment

Still Tank Plating. Maintenance routines for a 38 000-L (10 000-gal) high-chloride (low-pH) Watts nickel plating bath kept warm at all times to prevent precipitation of salts, and for auxiliary equipment comprise the following:

Daily

- Check pH of nickel solution and correct, if necessary, by adding sulfuric acid

Semiweekly

- Measure sulfuric acid reverse etching bath (when used) with a hydrometer, and adjust the bath if necessary

Weekly

- Perform chemical analyses of nickel bath for nickel sulfate, nickel chloride, and boric acid contents. Adjust, if necessary
- Check hydrochloric acid pickling solution

Semimonthly

- Check soak cleaner and adjust or replace, if necessary
- Check electrocleaner and adjust or replace, if necessary

Semiannually

- Completely check power supply and filtering equipment
- Replace filter bags

Barrel Plating. The following schedule of maintenance is applied to barrel nickel plating equipment:

Daily

- Check nickel bath temperature
- Check pH and adjust, if necessary

Weekly

- Perform chemical analysis of nickel solution; make necessary adjustments
- Dummy plate at low voltage over weekend, to plate out impurities

Monthly

- Filter solution; or if solution is under continuous filtration, clean filtering bags and recoat
- Examine anodes, anode holders, and anode bags; replace, if necessary

Semiannually

- Filter solution into storage tank. Oxidize solution with hydrogen peroxide and purify with activated carbon
- Examine tank and coils for cracks and leaks
- Clean all electrical connectors
- Filter solution back to tank and adjust to operating pH range

Automatic Plating. The following schedule of maintenance is applied to an automatic plating machine:

Daily

- Perform chemical analysis of nickel plating solution
- Adjust height of solution to working level
- Visually check meters and temperature controls
- Remove faulty racks from service, and set aside for repair

Weekly

- Change cleaner and acid solutions
- Replace anodes, as necessary
- Clean anodes in cleaner tank
- Filter plating solution
- Calibrate ammeters and voltmeters
- Check and clean electrical contacts on carrier arms, pivots, and racks

Monthly

- Check rectifier stacks and control knobs for proper operation; clean rectifiers, if necessary
- Calibrate temperature controls

- Repair accumulation of faulty racks
- Check stock of plating supplies

Semiannually

- Check ventilating equipment and air washer for proper operation and for corrosion; repair as required
- Check baking oven for accuracy of temperature control and for efficiency of elements
- Check filtering and solution-transfer pumps, motors, seals, hoses, and gaskets; repair or replace, as necessary
- Check automatic-stop controls, guards, and other safety equipment for proper operation; replace fire-extinguisher liquid
- Clean all drains
- Check system for corrosion; apply paint and oil for protection
- Replace worn machinery parts

Selection of Method

Both technical and cost considerations are involved in the selection of plating method. Some of the technical considerations are discussed below.

Still Tank Plating. The principal technical factors that influence the selection of still tank plating are part size and shape and required plate thickness and distribution. Still tanks are particularly well adapted to the plating of very large parts and unusual shapes that also may require the use of complicated conforming anodes, thieves, and other special accessories. They are also ideal for plating thicknesses of nickel in excess of about 50 μm (2 mils), particularly if an evenly distributed coating is required, for plating small lots, and when special anode placement and control are desired.

Barrel plating is especially suitable for plating large quantities of small parts that are difficult and costly to wire or rack and that, by virtue of their size and shape, can be readily tumbled without denting, excessive shielding, or nesting. Barrel plating is largely restricted to the depositing of thin coatings, usually less than 13 μm (0.5 mil) thick, in applications where uniform distribution of plate is not critical. As part size increases, distribution of plate in a barrel becomes increasingly less uniform.

Automatic plating, in which parts are racked, is used most efficiently in plating large quantities of parts that are too large, or are otherwise unsuitable for barrel plating, and that require a plating thickness of less than 50 μm (2 mils). Ideally, the parts should be small enough and light enough to be easily loaded and unloaded. If conforming anodes are required, they should be of a simple design that can be fastened permanently to the rack. Complex anodes that require considerable adjustment increase loading and unloading time excessively. Production capacity of equipment should be closely related to actual requirements, thus avoiding excessive idle time.

Fig. 6 Still tank plating of 150-mm (6-in.) diam disklike steel production parts

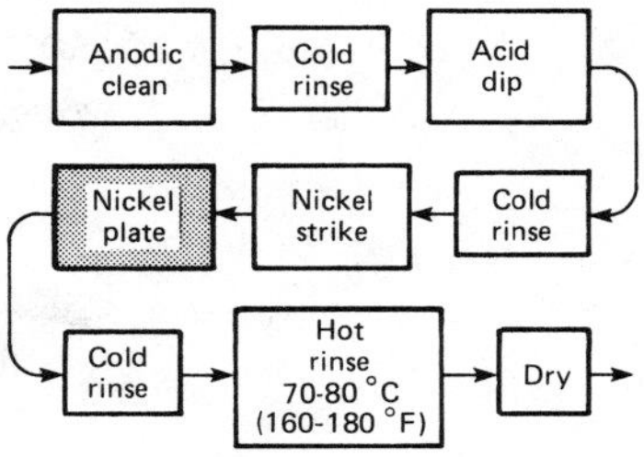

Production requirements

Load area .... 0.05 m^2 (70 $in.^2$)
Load weight ...... 0.52 kg (1.10 lb)
No. of pieces per hour 180
Minimum thickness of plate .... 25 μm (1 mil)

Plating conditions

Bath temperature ... 66 °C (150 °F)
Current density (9 V) ... 4.5 A/dm^2 (45 A/ft^2)(a)
Plating time, ... 40 min
Ratio, minimum to average thickness 2-3

Equipment required

16 plating racks ...... 12 pieces per rack(b)
1 tank .... 2 m long, 1 m wide, 1 m deep(6 ft long, 3.5 ft wide, 4 ft deep)(c)
6 tanks ...... 2 m long, 0.9 m wide, 1 m deep (3.5 ft long, 3 ft wide, 4 ft deep) for hot and cold rinses, acid dip, and nickel strike
1 tank .... 3 m long, 1 m wide, 1.2 m deep (10 ft long, 3.5 m wide, 4 ft deep) for nickel plating

(a) 270 A per rack, 3780 A per tank of 14 racks. (b) Double-lane operation; racks processed two at a time, in sequence, to use full plating time. (c) For electrocleaner

Fig. 7 Barrel plating of assorted nuts, washers, and bolts

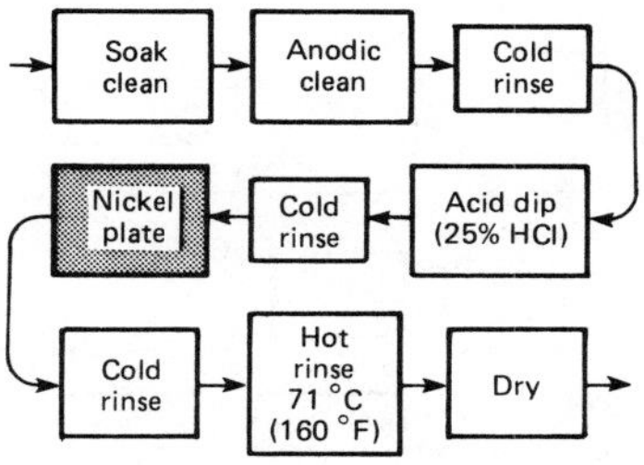

Production requirements

Load area .... 5 m^2 (50 ft^2)
Load weight ... Determined by area
Area plated per hour .... 45 m^2 (500 ft^2)
Minimum thickness of plate .... 25 μm (1 mil)

Plating conditions

Bath temperature ... 55 °C (130 °F)
Current density .... 0.5 A/dm^2 (5 A/ft^2)
Plating time .... 60 min

Equipment required

Barrel size ... 360-mm (14-in.) diam by 760 mm (30 in.)
Plating tank .... 10 station, 9500 L (2500 gal) capacity

The figures that follow illustrate the use of still tank, barrel, and automatic methods for nickel plating on a production basis. Figure 6 shows the flow of processing steps and gives details of operating conditions for still tank plating of 150-mm (6-in.) diam disklike parts, which were plated to a minimum thickness of 25 μm (1 mil) at the rate of 180 pieces per hour. Figure 7 describes the use of barrel plating for depositing a minimum of 5 μm (0.2 mil) of nickel on assorted fasteners, at the rate of 45 m^2 (500 ft^2) per hour. Figure 8 deals with the use of a return automatic-plating machine for depositing 30 μm (1.2 mils) of nickel on 1-m^2 (10-ft^2), 14-kg (30-lb) workpieces at the rate of 50 pieces per hour.

Process Limitations

The size and shape of parts are seldom limiting factors in the application of nickel plating, provided the required plating equipment is available and plating cost is not prohibitive. Similarly, the hardness of the basis metal does not interfere with the successful deposi-

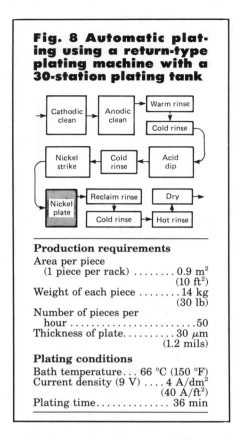

Fig. 8 Automatic plating using a return-type plating machine with a 30-station plating tank

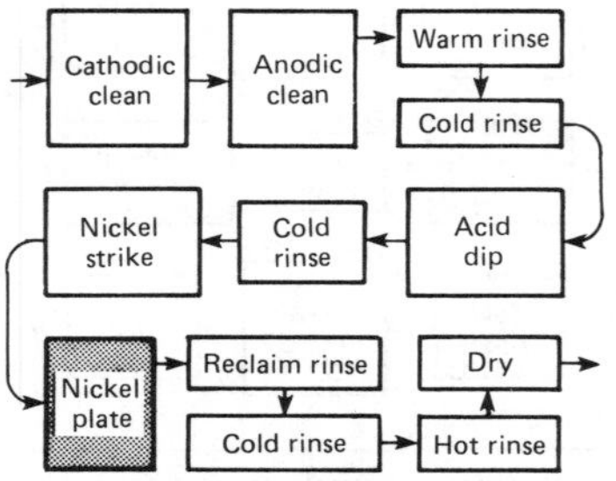

Production requirements

Area per piece (1 piece per rack)	0.9 m^2 (10 ft^2)
Weight of each piece	14 kg (30 lb)
Number of pieces per hour	50
Thickness of plate	30 μm (1.2 mils)

Plating conditions

Bath temperature	66 °C (150 °F)
Current density (9 V)	4 A/dm^2 (40 A/ft^2)
Plating time	36 min

Fig. 9 Cross section of box plated in sulfamate nickel bath, showing variations in plate thickness

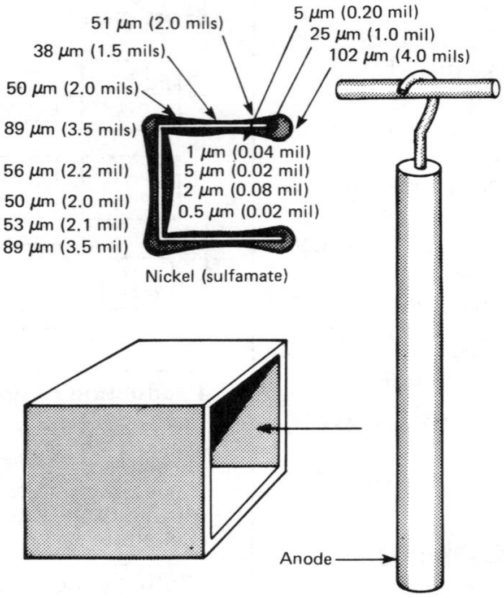

Plating bath	Thickness ratio(a) Side	Bottom
Silver (cyanide)	1:2.5	1:5
Copper (cyanide)	1:3.0	1:6
Cadmium (cyanide)	1:4.3	1:12
Nickel (sulfamate)	1:10.0	1:33

(a) Ratio of average plate thickness on inside to average plate thickness on outside

100-mm (4-in.) cube having open end pointed toward anode during plating. Numbers indicate thickness at each location. For ratios of average thickness, see table

Fig. 10 Effect of radius in recessed area on uniformity of thickness of nickel plate

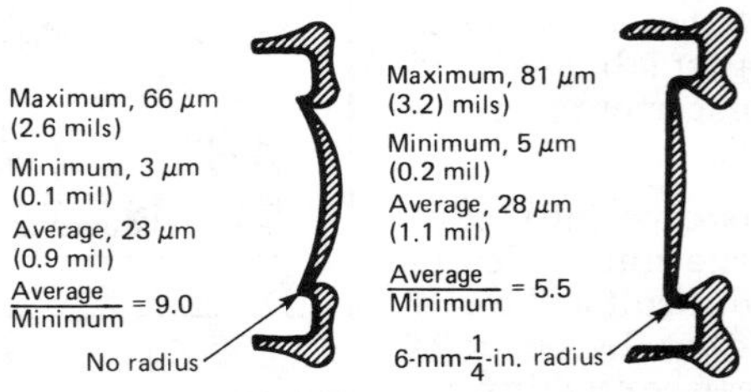

tion of nickel, provided the surface is properly prepared for nickel plating. Nickel however, cannot be deposited directly on all basis metals.

Limitations of Basis Metal. Although nickel can be deposited directly on plain carbon steels and on most low-alloy steels, copper, and brass, it cannot be plated directly on lead, zinc, zinc-based alloys, and copper alloys containing more than 40% zinc. According to Federal Specification QQ-N-290, zinc and zinc-based alloys must have an undercoating of copper 5 to 10 μm (0.2 to 0.4 mil) thick, and copper alloys containing more than 40% zinc must have an undercoating of copper 8 μm (0.3 mil) thick, before a nickel plate is applied. If required to promote good adhesion, a copper strike is permitted on low-alloy steel before nickel plating. Other applicable specifications relating to the nickel plating of these metals include ASTM B456.

Nickel can be successfully plated over stainless steels and heat-resisting chromium-nickel alloys if the basis metal is first activated and given a light coating of nickel in a nickel chloride-hydrochloric acid bath. Several baths for simultaneous activation plating are described in ASTM B254; composition and operating conditions for one of these baths (U. S. Patent 2,437,409) are:

Nickel chloride	240 g/L (32 oz/gal)
Hydrochloric acid (20° Bé)	120 g/L (16 oz/gal)
Temperature	Room temperature to 32 °C (Room temperature to 90 °F)
Current density	5-20 A/dm^2 (50-200 A/ft^2)
Time	2-4 min
Anodes	Nickel

For specific instructions regarding the plating of nickel on aluminum and magnesium, see the articles on finishing of those metals in this Volume and refer also to ASTM B253.

Fig. 11 Annealing of nickel deposits produced in Watts and all-chloride nickel solutions

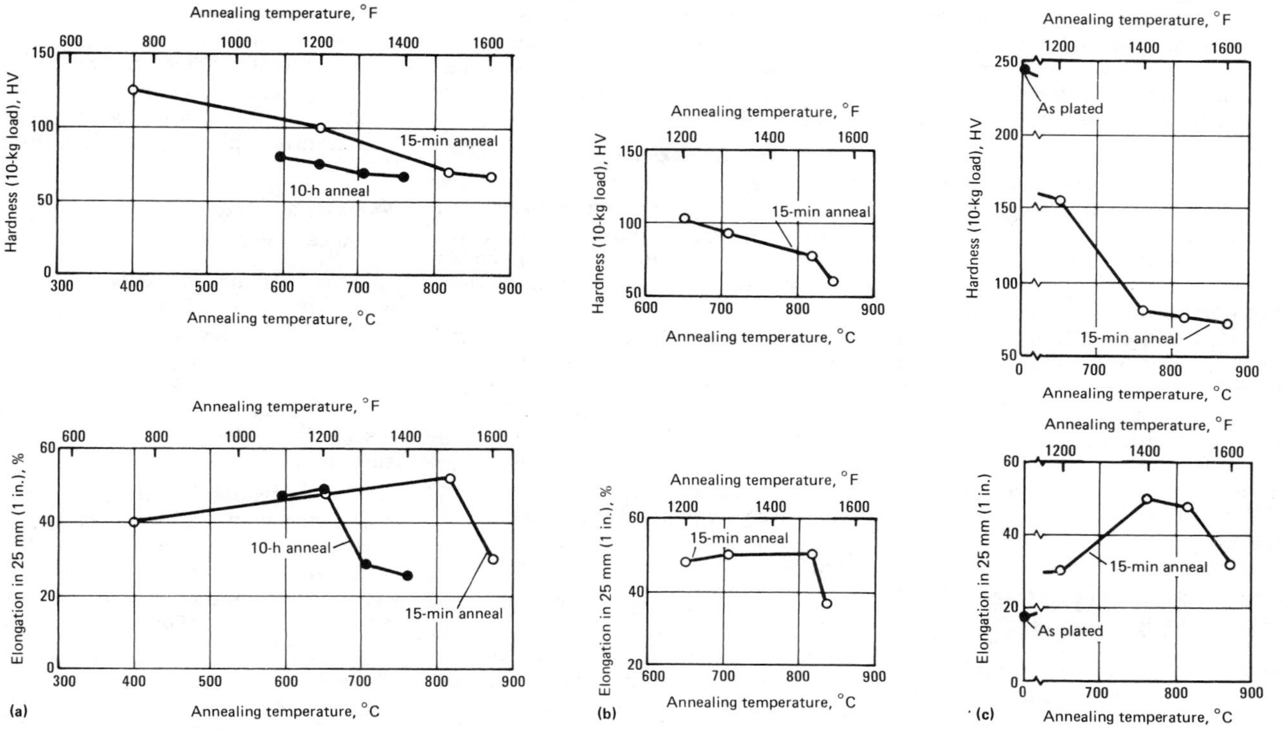

(a) Watts bath at pH of 2.0. Current density, 5 A/dm² (50 A/ft²) at 55 °C (130 °F). (b) Watts bath at pH of 5.0. Current density, 4 A/dm² (40 A/ft²) at 55 °C (130 °F). (c) Chloride bath at pH of 2.0. Current density, 5.5 A/dm² (55 A/ft²) at 55 °C (130 °F)

Variations in Plate Thickness

The relatively poor throwing power of nickel plating baths affects thickness (Fig. 9). Uniformity of thickness of nickel electrodeposits is influenced by several factors, including shape of parts, plating method, average current density, proximity to tank anodes, and solution composition, pH, and temperature. In barrel plating, the speed of rotation of the barrel and the area of the load being plated also affect uniformity of plate thickness.

The influence of shape is exemplified by an open-ended box as shown in Fig. 9 and the table accompanying it. Arranged in the order of least effect of shape on distribution of electrodeposit (gold plating) to greatest effect of shape (chromium), the various widely used solutions compare as follows: (*a*) gold, (*b*) cyanide silver, (*c*) alkaline tin, (*d*) cyanide copper, (*e*) cyanide cadmium, (*f*) cyanide zinc, (*g*) acid tin, (*h*) high-pH nickel, (*i*) acid cadmium, (*j*) low-pH (ordinary) nickel, (*k*) acid copper or zinc, (*l*) iron, and (*m*) chromium. This order can be altered only slightly by changes in bath composition and plating process.

As the design of a part becomes more complex, uniformity of plate thickness is more difficult to achieve without the use of special conforming anodes and similar accessories. Figure 10 shows the effect of radius in a recessed area on the ratio of average to minimum plate thickness.

Conforming Anodes. The use of conforming anodes in nickel plating baths is similar to their use in other common plating solutions. For a discussion of the applications of conforming anodes, the reader may refer to other plating articles in this Volume, especially those on hard chromium plating.

Normal variations in plate thickness are inherent in all plating processes, and unless they exceed the specified range of allowable variation, they are not usually considered objectionable.

Hydrogen Embrittlement

Nickel plating, as well as acid cleaning and cathodic electrocleaning, is accompanied by the deposition of some atomic hydrogen. Atomic hydrogen is absorbed by steel and may cause hydrogen embrittlement, particularly in steels with a tensile strength value above about 1100 MPa (160 ksi). To minimize possible hydrogen pickup, parts to be plated may be descaled in molten salt or by mechanical blasting, and acid pickling should be avoided or minimized. Electrocleaning may be performed anodically. If acid dips are used to activate the surface before plating, immersion times should be held to a minimum. Nickel plating in baths at the high end of the pH range also helps to reduce hydrogen pickup.

Baking Treatments. Because nickel, unlike cadmium, is quite permeable to hydrogen, embrittlement can readily be avoided by baking nickel plated steel at moderate temperatures for relatively short periods of time. Baking should be done immediately after plating, although this is not essential unless the plated component is subjected to stress in the interim between plating and baking. The following are among the accepted practices for baking electroplated nickel on carbon and alloy steels:

1 For parts cold worked after hardening and tempering, regardless of hardness, heat to 190 ± 5 °C (375 ± 10 °F) and hold for 3 h, except as noted in item 3, below
2 For parts tempered to a hardness of 33 HRC and over, and for springs, heat to 190 ± 5 °C (375 ± 10 °F) and hold for not less than 3 h, except as noted in item 3
3 For parts affected adversely by heating to 190 °C (375 °F), including carburized parts, heat to 135 ± 5 °C (275 ± 10 °F) and hold for 5 h or more

Chemical Stripping of Nickel Deposits

Procedures for the chemical stripping of nickel deposits vary with substrate material, as indicated below.

Iron-Based Alloys. Concentrated nitric acid strips nickel coatings from steel and other iron-based alloys. The acid must not be diluted with water, and adequate provision should be made to remove the toxic fumes that evolve. Stripping can be accelerated by adding a small amount of hydrochloric acid to the nitric acid bath; this procedure is covered by U.S. Patent 2,200,486.

Heavy nickel deposits can be stripped electrolytically in an aqueous solution containing 50 vol % sulfuric acid. With the workpiece made the anode, this procedure requires lead cathodes and controlled voltage not exceeding 2 V. Solution temperature must not be permitted to rise above 21 °C (70 °F).

Another electrolytic method has the advantage of operating at a common plating voltage level, namely, 6 V. The proportions of the aqueous bath are 4 L (1 gal) concentrated sulfuric acid, 0.5 L (1 pint) of water, and 28 g (1 oz) of glycerin or 112 g (4 oz) of copper sulfate. This procedure also uses lead cathodes, and the bath is operated at room temperature.

Copper-based alloys, including brass, may be electrolytically stripped of nickel in a room temperature solution containing 15 g/L (2 oz/gal) hydrochloric acid. The work is stripped anodically, using carbon cathodes, at 6 to 12 V.

An alternative anodic stripping procedure uses an aqueous solution containing 60 to 80 vol % sulfuric acid and 0.5 to 1.5% glycerin. Operated at 6 V, this method requires either lead or nickel cathodes. Both solutions corrosively attack the basis metal unless the workpiece is removed from the bath immediately after the nickel has been stripped.

Aluminum Alloys. Nickel coatings may be chemically stripped from aluminum and aluminum alloys by immersion in concentrated nitric acid. Adequate ventilation is required to dispose of fumes.

Zinc-Based Alloys. Because of the difficulty of chemically stripping nickel from copper-nickel-chromium plated zinc without repolishing, or without damaging the basis metal, this procedure is seldom attempted. However, it is possible to anodically strip both chromium and nickel from zinc in an aqueous solution containing 50 to 93 wt % sulfuric acid. Solutions containing the higher percentages of acid do the least damage to the copper undercoating, which can subsequently be polished and replated.

Many decorative nickel deposits are covered with a thin layer of chromium, and this layer must be removed before the nickel is stripped. Chromium can be stripped electrolytically in an aqueous solution containing 45 g/L (6 oz/gal) of sodium hydroxide. The workpiece is made the anode, and current is supplied at 6 V. Chromium can be stripped also by the use of hydrochloric acid, either in concentrated form at room temperature, or a 12.5 vol % solution and heated to 52 °C (125 °F). See also the articles on chromium plating in this Volume.

Annealing of Nickel Plate

Under certain conditions, it may be desirable to anneal nickel deposits to improve ductility and decrease hardness. The effect of elevated temperature on nickel deposits is of interest in annealing and in evaluating the usefulness of these deposits in high temperature applications.

The degree of structural change encountered in nickel deposits varies directly with temperature. Although structural changes below about 595 °C (1100 °F) are minor, a marked change occurs above this temperature.

Some softening of hard nickel deposits is experienced at 260 °C (500 °F) and above; below 260 °C (500 °F), a range of 300 to 450 HV is retained. The annealability of the deposit is partly dependent on freedom from contaminants. Heating a deposit that contains lead in excess of 0.01% or sulfur in excess of 0.002% results in embrittlement of the coating. These contaminants may be introduced from lead-lined equipment or from sulfur-contributing addition agents.

Response to annealing varies also with the composition and operating characteristics of the nickel plating bath. These variations can be observed in the data given in Fig. 11, which compare the annealing characteristics of deposits obtained in two Watts baths operating at a pH of 2.0 and 5.0, respectively, and in an all-chloride bath at a pH of 2.0.

Electroless Nickel Plating

By William D. Fields
Vice President
ELNIC, Inc.
Ronald N. Duncan
Director
Research & Engineering
ELNIC, Inc.
Joseph R. Zickgraf
Technical Services Director
ELNIC, Inc.
and
The ASM Committee on Electroless
Nickel Plating*

ELECTROLESS NICKEL PLATING is used to deposit nickel without the use of an electric current. The coating is deposited by an autocatalytic chemical reduction of nickel ions by hypophosphite, aminoborane, or borohydride compounds. Two other methods have been used commercially for plating nickel without electric current, including (*a*) immersion plating on steel from solutions of nickel chloride and boric acid at 70 °C (160 °F) and (*b*) decomposition of nickel carbonyl vapor at 180 °C (360 °F). Immersion deposits, however, are poorly adherent and nonprotective, while the decomposition of nickel carbonyl is expensive and hazardous. Accordingly, only electroless nickel plating has gained wide acceptance.

Since gaining commercial use in the 1950's, electroless nickel plating has grown rapidly and now is an established industrial process. Currently, hot acid hypophosphite reduced baths are most frequently used to plate steel and other metals, whereas warm alkaline hypophosphite baths are used for plating plastics and nonmetals. Borohydride reduced baths are also used to plate iron and copper alloys, especially in Europe.

Electroless nickel is an engineering coating, normally used because of excellent corrosion and wear resistance. Electroless nickel coatings are also frequently applied on aluminum to provide a solderable surface and are used with molds and dies to improve lubricity and part release. Because of these properties, electroless nickel coatings have found many applications, including those in petroleum, chemicals, plastics, optics, printing, mining, aerospace, nuclear, automotive, electronics, computers, textiles, paper, and food machinery (Ref 1). Some advantages and limitations of electroless nickel coatings include:

Advantages

- Good resistance to corrosion and wear
- Excellent uniformity
- Solderability and brazability
- Low labor costs

Limitations

- High chemical cost
- Brittleness
- Poor welding characteristics of nickel phosphorus deposits
- Need to copper strike plate alloys containing significant amounts of lead, tin, cadmium, and zinc before electroless nickel can be applied
- Slower plating rate, as compared to electrolytic methods

Bath Composition and Characteristics

Electroless nickel coatings are produced by the controlled chemical reduc-

*Donald W. Baudrand, Vice President, New Market Development, Allied-Kelite, Division of the Richardson Co.; Russell A. Henry, Jr., President, Wear-Cote International, Inc.

tion of nickel ions onto a catalytic surface. The deposit itself is catalytic to reduction, and the reaction continues as long as the surface remains in contact with the electroless nickel solution. Because the deposit is applied without an electric current, its thickness is uniform on all areas of an article in contact with fresh solution.

Electroless nickel solutions are blends of different chemicals, each performing an important function. Electroless nickel solutions contain:

- A source of nickel, usually nickel sulfate
- A reducing agent to supply electrons for the reduction of nickel
- Energy (heat)
- Complexing agents (chelators) to control the free nickel available to the reaction
- Buffering agents to resist the pH changes caused by the hydrogen released during deposition
- Accelerators (exultants) to help increase the speed of the reaction
- Inhibitors (stabilizers) to help control reduction
- Reaction by-products

The characteristics of an electroless nickel bath and its deposit are determined by the composition of these components.

Table 1 Hypophosphite-reduced electroless nickel plating solutions

Constituent or condition	Alkaline Bath 1	Alkaline Bath 2	Alkaline Bath 3	Acid Bath 4	Acid Bath 5	Acid Bath 6
Composition						
Nickel chloride, g/L (oz/gal)	45 (6)	30 (4)	30 (4)	...	...	...
Nickel sulfate, g/L (oz/gal)	...	...	...	21 (2.8)	34 (4.5)	45 (6)
Sodium hypophosphite, g/L (oz/gal)	11 (1.5)	10 (1.3)	10 (1.3)	24 (3.2)	35 (4.7)	10 (1.3)
Ammonium chloride, g/L (oz/gal)	50 (6.7)	50 (6.7)	...	...	...	...
Sodium citrate, g/L (oz/gal)	100 (13.3)	...	...	...	...	...
Ammonium citrate, g/L (oz/gal)	...	65 (8.6)	...	...	...	...
Ammonium hydroxide,	To pH	To pH	...	...	...	...
Lactic acid, g/L (oz/gal)	...	...	...	28 (3.7)	...	...
Malic acid, g/L (oz/gal)	...	...	...	...	35 (4.7)	...
Amino-acetic acid, g/L (oz/gal)	...	...	...	...	...	40 (5.3)
Sodium hydroxyacetate, g/L (oz/gal)	...	...	10 (1.3)	...	...	...
Propionic acid, g/L (oz/gal)	...	...	...	2.2 (0.3)	...	...
Acetic acid, g/L (oz/gal)	...	...	...	...	...	10 (1.3)
Succinic acid, g/L (oz/gal)	...	...	...	...	10 (1.3)	...
Lead, ppm	...	...	...	1	...	...
Thiourea, ppm	...	...	...	...	1	...
Operating conditions						
pH	8.5-10	8-10	4-6	4.3-4.6	4.5-5.5	4.5-5.5
Temperature, °C (°F)	90-95 (195-205)	90-95 (195-205)	88-95 (190-205)	88-95 (190-205)	88-95 (190-205)	88-95 (190-205)
Plating rate, μm/h (mil/h)	10 (0.4)	8 (0.3)	10 (0.4)	25 (1)	25 (1)	25 (1)

Reducing Agents

A number of different reducing agents have been used in preparing electroless nickel baths, including (*a*) sodium hypophosphite, (*b*) amino-boranes, (*c*) sodium borohydride, and (*d*) hydrazine.

Sodium Hypophosphite Baths. The majority of electroless nickel used commercially is deposited from solutions reduced with sodium hypophosphite. The principal advantages of these solutions over those reduced with boron compounds or hydrazine include (*a*) lower cost, (*b*) greater ease of control, and (*c*) better corrosion resistance of the deposit.

Several mechanisms have been proposed for the chemical reactions which occur in hypophosphite reduced electroless nickel plating solutions. The most widely accepted of these mechanisms are illustrated by the following equations:

$$(H_2PO_2)^- + H_2O \xrightarrow[\text{Heat}]{\text{Catalyst}} H^+ + (HPO_3)^= + 2H_{abs} \quad \text{(Eq 1)}$$

$$Ni^{++} + 2H_{abs} \rightarrow Ni + 2H^+ \quad \text{(Eq 2)}$$

$$(H_2PO_2)^- + H_{abs} \rightarrow H_2O + OH^- + P \quad \text{(Eq 3)}$$

$$(H_2PO_2)^- + H_2O \rightarrow H^+ + (HPO_3)^= + H_2 \quad \text{(Eq 4)}$$

In the presence of a catalytic surface and sufficient energy, hypophosphite ions are oxidized to orthophosphite. A portion of the hydrogen given off is absorbed onto the catalytic surface (Eq 1). Nickel at the surface of the catalyst is then reduced by the absorbed active hydrogen (Eq 2). Simultaneously, some of the absorbed hydrogen reduces a small amount of the hypophosphite at the catalytic surface to water, hydroxyl ion, and phosphorus (Eq 3). Most of the hypophosphite present is catalytically oxidized to orthophosphite and gaseous hydrogen (Eq 4) independently of the deposition of nickel and phosphorus, causing the low efficiency of electroless nickel solutions. Usually 5 kg (10 lb) of sodium hypophosphite is required to reduce 1 kg (2 lb) of nickel, for an average efficiency of 37% (Ref 2, 3).

Early electroless nickel formulations were ammoniacal and operated at high pH. Later acid solutions were found to have several advantages over alkaline solutions. Among these are (*a*) higher plating rate, (*b*) better stability, (*c*) greater ease of control, and (*d*) improved deposit corrosion resistance. Acordingly, most hypophosphite reduced electroless nickel solutions are operated between 4 and 5.5 pH. Compositions for alkaline and acid plating solutions are listed in Table 1 (Ref 2-5).

Aminoborane Baths. The use of aminoboranes in commercial electroless nickel plating solutions has generally been limited to two compounds: (*a*) N-dimethylamine borane (DMAB) —$(CH_3)_2\ NHBH_3$, and (*b*) N-diethylamine borane (DEAB)—$(C_2H_5)_2\ NHBH_3$.

DEAB is used primarily in European facilities, whereas DMAB is used principally in the United States. DMAB is readily soluble in aqueous systems. DEAB must be mixed with a short chain aliphatic alcohol, such as ethanol, before it can be dissolved in the plating solution.

Aminoborane reduced electroless nickel solutions have been formulated over wide pH ranges, although they are usually operated between 6 and 9 pH. Operating temperatures for these baths range from 50 to 80 °C (120 to 180 °F), but they can be used at temperatures as low as 30 °C (90 °F). Accordingly, aminoborane baths are very useful for plating plastics and nonmetals, which is their primary application. The rate of deposition varies with pH and temperature, but is usually 7 to 12 μm/h (0.3 to 0.5 mil/h). The boron content of the deposit from these baths varies between 0.4 and 5%. Compositions and operating conditions for aminoborane baths are listed in Table 2 (Ref 2, 5, 6).

Table 2 Aminoborane and borohydride reduced electroless nickel plating solutions

Constituent or condition	Aminoborane Bath 7	Aminoborane Bath 8	Borohydride Bath 9	Borohydride Bath 10
Composition				
Nickel chloride, g/L (oz/gal)	30 (4)	24-48 (3.2-6.4)	...	20 (2.7)
Nickel sulfate, g/L (oz/gal)	...	...	50 (6.7)	...
DMAB, g/L (oz/gal)	...	3-4.8 (0.4-0.64)	3 (0.4)	...
DEAB, g/L (oz/gal)	3 (0.4)	...	...	...
Isopropanol, mL (fluid oz)	50 (1.7)	...	...	...
Sodium citrate, g/L (oz/gal)	10 (1.3)	...	...	...
Sodium succinate, g/L (oz/gal)	20 (2.7)	...	...	...
Potassium acetate, g/L (oz/gal)	...	18-37 (2.4-4.9)	...	...
Sodium pyrophosphate, g/L (oz/gal)	...	...	100 (13.3)	...
Sodium borohydride, g/L (oz/gal)	...	...	...	0.4 (0.05)
Sodium hydroxide, g/L (oz/gal)	...	...	...	90 (12)
Ethylene diamine, 98%, g/L (oz/gal)	...	...	...	90 (12)
Thallium sulfate, g/L (oz/gal)	...	...	...	0.4 (0.05)
Operating conditions				
pH	5-7	5.5	10	14
Temperature, °C (°F)	65 (150)	70 (160)	25 (77)	95 (205)
Plating rate, μm/h (mil/h)	7-12 (0.5)	7-12 (0.5)	...	15-20 (0.6-0.8)

Sodium Borohydride Baths. The borohydride ion is the most powerful reducing agent available for electroless nickel plating. Any water-soluble borohydride may be used, although sodium borohydride is preferred.

In acid or neutral solutions, hydrolysis of borohydride ions is very rapid. In the presence of nickel ions, nickel boride may form spontaneously. If the pH of the plating solution is maintained between 12 and 14, however, nickel boride formation is suppressed, and the reaction product is principally elemental nickel. One mol of sodium borohydride can reduce approximately one mol of nickel, so that the reduction of 1 kg (2 lb) of nickel requires 0.6 kg (1 lb) of sodium borohydride. Deposits from borohydride reduced electroless nickel solutions contain 3 to 8% boron.

To prevent precipitation of nickel hydroxide, complexing agents, such as ethylene diamine, that are effective between 12 to 14 pH must be used. Such strong complexing agents, however, decrease the rate of deposition. At an operating temperature of 90 to 95 °C (195 to 205 °F), the plating rate of commercial baths is 25 to 30 μm/h (1 to 1.2 mil/h). Compositions of a borohydride reduced electroless nickel bath are also shown in Table 2 (Ref 6).

During the course of reduction, the solution pH decreases, requiring constant additions of an alkali hydroxide. Spontaneous solution decomposition may occur if the bath pH is allowed to fall below 12. Because of the high operating pH, borohydride plating baths cannot be used for aluminum substrates (Ref 2, 5, 7).

Hydrazine Baths. Hydrazine has also been used to produce electroless nickel deposits. These baths operate at 90 to 95 °C (195 to 205 °F) and 10 to 11 pH. Their plating rate is approximately 12 μm/h (0.5 mil/h). Because of the instability of hydrazine at high temperatures, however, these baths tend to be very unstable and difficult to control.

Whereas the deposit from hydrazine reduced solutions is 97 to 99% nickel, it does not have a metallic appearance. The deposit is brittle and highly stressed with poor corrosion resistance. Unlike hypophosphite and boron reduced nickels, hardness from a hydrazine reduced solution is not increased by heat treatment. At present, hydrazine reduced electroless nickel has very little commercial use (Ref 2).

Energy

The amount of energy or heat present in an electroless nickel solution is one of the most important variables affecting coating deposition. In a plating bath, temperature is a measure of its energy content.

Temperature has a strong effect upon the deposition rate of acid hypophosphite reduced solutions. The rate of deposition is usually very low at temperatures below 65 °C (150 °F), but increases rapidly with increased temperature (Ref 5). This is illustrated in Fig. 1, which gives the results of tests conducted using bath 3 in Table 1 (Ref 7). The effect of temperature on deposition in boron reduced solutions is similar. At temperatures above 100 °C (212 °F), electroless nickel solutions may decompose. Accordingly, the preferred operating range for most solutions is 85 to 95 °C (185 to 205 °F).

Complexing Agents

To avoid spontaneous decomposition of electroless nickel solutions and to control the reaction so that it occurs only on the catalytic surface, complexing agents are added. Complexing agents are organic acids or their salts, added to control the amount of free nickel available for reaction. They act to stabilize the solution and to retard the precipitation of nickel phosphite.

Complexing agents also buffer the plating solution and prevent its pH from decreasing too rapidly as hydrogen ions are produced by the reduction

Fig. 1 Effect of solution temperature on the rate of deposition

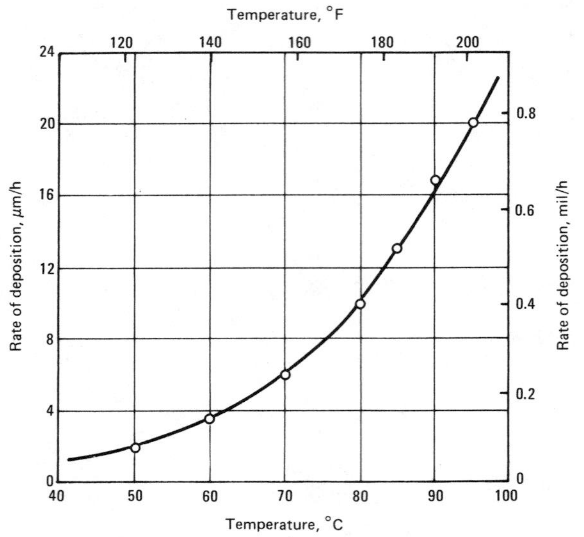

Tests conducted on bath 3 at 5 pH

reaction. Ammonia, hydroxides, or carbonates, however, may also have to be added periodically to neutralize hydrogen.

Original electroless nickel solutions were complexed with the salts of glycolic, citric, or acetic acids. Later baths were prepared using other polydentate acids, including succinic, glutaric, lactic, propionic, and aminoacetic. The complexing ability of an individual acid or group of acids varies, but may be quantified by the amount of orthophosphite which can be held in solution without precipitation (Ref 2, 8). This is illustrated in Fig. 2, which shows the maximum solubility of orthophosphite in solutions complexed with citric and glycolic acids as a function of pH (Ref 9). The complexing agent used in the plating solution can also have a pronounced effect on the quality of the deposit, especially on its phosphorus content, internal stress, and porosity (Ref 8).

Accelerators

Complexing agents reduce the speed of deposition and can cause the plating rate to become uneconomically slow. To overcome this, organic additives, called accelerators or exultants, are often added to the plating solution in small amounts. Accelerators are thought to function by loosening the bond between hydrogen and phosphorus atoms in the hypophosphite molecule, allowing it to be more easily removed and absorbed onto the catalytic surface. Accelerators activate the hypophosphite ion and speed the reaction shown in Eq 1 (Ref 2, 3). In hypophosphite reduced solutions, succinic acid is the accelerator most frequently used. Other carbonic acids, soluble fluorides, and some solvents, however, have also been used (Ref 2). The effect of succinate additions upon deposition rate is illustrated in Fig. 3 (Ref 3).

Inhibitors

The reduction reaction in an electroless nickel plating bath must be controlled so that deposition occurs at a predictable rate and only on the substrate to be plated. To accomplish this, inhibitors, also known as stabilizers, are added. Electroless nickel plating solutions can operate for hours or days without inhibitors, only to decompose unexpectedly. Decomposition is usually initiated by the presence of colloidal, solid nuclei in the solution. These particles may be the result of the presence of foreign matter (such as dust or blasting media), or may be generated in the bath as the concentration of orthophosphite exceeds its solubility limit. Whatever the source, the large surface area of the particles catalyzes reduction, leading to a self-accelerating chain reaction and decomposition. This is usually preceded by increased hydrogen evolution and the appearance of a finely divided black precipitate throughout the solution. This precipitate consists of nickel and either nickel phosphide or nickel boride.

Spontaneous decomposition can be controlled by adding trace amounts of catalytic inhibitors to the solution. These inhibitors are absorbed on any colloidal particles present in the solution and prevent the reduction of nickel on their surface. Traditionally, inhibitors used with hypophosphite reduced electroless nickel have been of three types: (*a*) sulfur compounds, such as thiourea; (*b*) oxy anions, such as molybdates or iodates; and (*c*) heavy metals, such as lead, bismuth, tin, or cadmium. More recently, organic compounds, including oleates and some unsaturated acids, have been used for some functional solutions. Organic sulfide, thio compounds, and metals, such as selenium and thallium, are used to inhibit aminoborane and borohydride reduced electroless nickel solutions.

The addition of inhibitors can have harmful as well as beneficial effects on the plating bath and its deposit. In small amounts, some inhibitors increase the rate of deposition and/or the brightness of the deposit; others, especially metals or sulfur compounds, increase internal stress and porosity and reduce ductility, thus reducing the ability of the coating to resist corrosion and wear (Ref 2, 3, 5).

The amount of inhibitor used is critical. The presence of only about 1 mg/L (4 mg/gal) of HS^- ion completely stops deposition, whereas at a concentration of 0.01 mg/L (0.04 mg/gal), this ion is an effective inhibitor. The effect of lead additions on a hypophosphite reduced succinate bath at pH 4.6 and 95 °C (205 °F) is shown in Fig. 4 (Ref 3). The tests illustrated in Fig. 4 also showed that baths containing less than 0.1 mg/L (0.4 mg/gal) Pb^{++} decomposed rapidly, whereas baths containing higher concentrations were stable. Excess inhibitor absorbs preferentially at sharp edges and corners, resulting in incomplete coverage (edge pull back) and porosity.

Reaction By-Products

During electroless nickel deposition, the by-products of the reduction, orthophosphite or borate and hydrogen ions, as well as dissolved metals from the

Fig. 2 Limits of solubility for orthophosphite in electroless nickel solutions

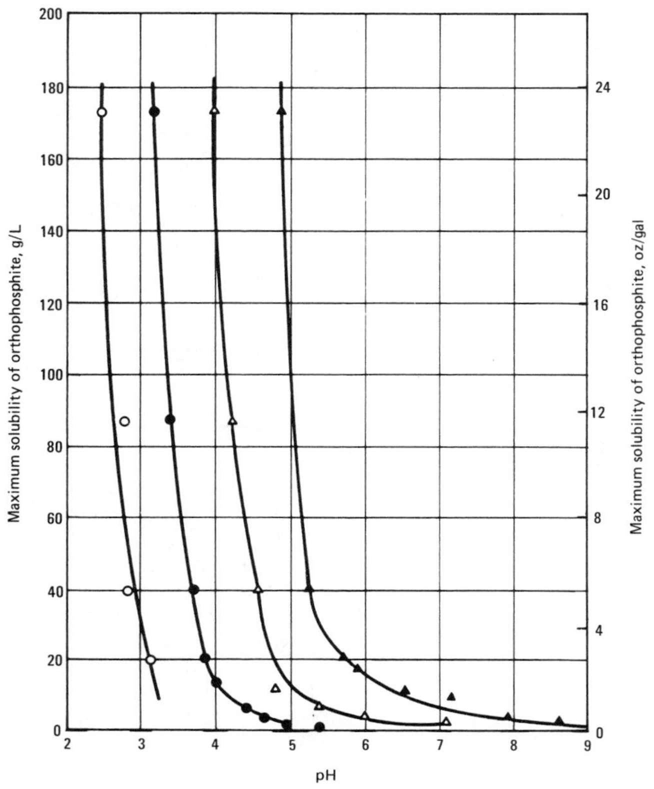

Solutions contain 30 g/L (4 oz/gal) nickel chloride ($NiCl_2$) and 10 g/L (1.3 oz/gal) sodium hypophosphite (NaH_2PO_2). ○: without a complexing agent; ●: with 15 g/L (2 oz/gal) citric acid; △: with 39 g/L (5.2 oz/gal) glycolic acid; ▲: with 78 g/L (10 oz/gal) glycolic acid

substrate accumulate in the solution. These can affect the performance of the plating bath.

Orthophosphite. As nickel is reduced, orthophosphite ion ($HPO_3^{=}$) accumulates in the solution and at some point interferes with the reaction. As the concentration of orthophosphite increases, there is usually a small decrease in the deposition rate and a small increase in the phosphorus content of the deposit. Ultimately the accumulation of orthophosphite in the plating solution results in the precipitation of nickel phosphite, causing rough deposits and spontaneous decomposition. Orthophosphite ion also codeposits with nickel and phosphorus, creating a highly stressed, porous deposit.

The solubility of phosphite in the solution is increased when complexing agents, such as citric or glycolic acids, are added. This effect is shown in Fig. 2. However, the use of strong complexors, in other than limited quantities, tends to reduce the deposition rate and increase the porosity and brittleness of the deposit (Ref 8).

Borates. The accumulation of metaborate ion (BO_2^-) from the reduction of borohydride or of boric acid (H_3BO_3) from the reduction of aminoboranes has little effect on electroless nickel plating baths. Both borohydride and aminoborane baths have been operated through numerous regenerations with only a slight decrease in plating rate and without decomposing. With aminoborane reduced solutions, the solubility of boric acid is probably increased by the presence of amine through the formation of a complex aminoborate (Ref 10).

Hydrogen ions (H^+), produced by the reduction reaction, cause the pH of the bath to decrease. The amount of hydrogen produced, however, depends on the reducing agent being used. Because they are less efficient, hypophosphite reduced solutions tend to generate more hydrogen ions than those reduced with boron compounds.

The pH of the bath has a strong effect on both solution operation and the composition of the deposit. This is illustrated in Fig. 5, which shows the plating rate and deposit phosphorus content resulting from varying solution pH values in a bath containing 33 g/L (4.4 oz/gal) of nickel sulfate and 20 g/L (2.7 oz/gal) of sodium hypophosphite at 82 °C (180 °F) (Ref 11).

To retard pH changes and to help keep operating conditions and deposit properties constant, buffers are included in electroless nickel solutions. Some of the most frequently used buffers include acetate, propionate, and succinate salts. Additions of alkaline materials, such as hydroxide, carbonate solutions, or ammonia, are also required periodically to neutralize the acid formed during plating.

Properties of Electroless Nickel-Phosphorus Coatings

Hypophosphite reduced electroless nickel is an unusual engineering material, because of both its method of application and its unique properties. As applied, nickel-phosphorus coatings are uniform, hard, relatively brittle, lubrious, easily solderable, and highly corrosion resistant. They can be precipitation hardened to very high levels through the use of low temperature treatments, producing wear resistance equal to that of commercial hard chromium coatings. This combination of properties makes the coating well suited for many severe applications and often allows it to be used in place of more expensive or less readily available alloys.

Structure. Hypophosphite reduced electroless nickel is one of the very few metallic glasses used as an engineering material. Depending on the formulation of the plating solution, commercial coatings may contain 6 to 12% phosphorus dissolved in nickel, and as much as 0.25% of other elements. As applied, most of these coatings are amorphous; they have no crystal or phase structure. Their continuity, however, depends upon their composition. Coatings containing more than 10% phosphorus and less than 0.05% impurities are typically continuous. A cross section of one of these coatings is shown in Fig. 6.

Fig. 3 Effect of succinate additions on the plating rate of an electroless nickel solution

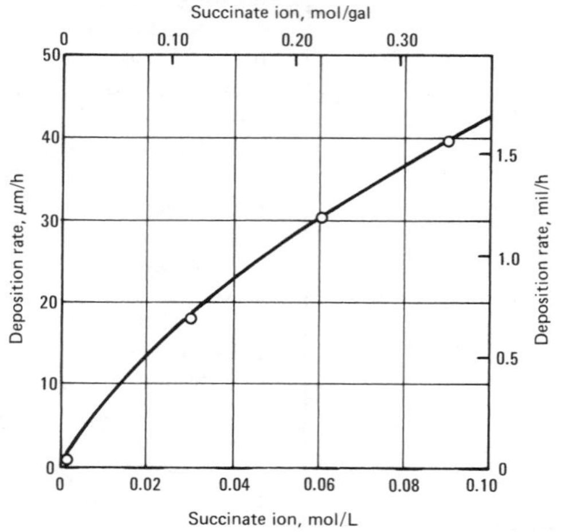

Solution contains 16 g/L (2.1 oz/gal) nickel chloride, 24 g/L (3.2 oz/gal) sodium hypophosphite (NaH_2PO_2), 5 g/L (0.7 oz/gal) ammonium hydroxide (NH_4OH) and 1 mg/L (4 mg/gal) lead at 5 pH and 95 °C (205 °F)

Fig. 4 Effect of lead additions on plating rate in a hypophosphite reduced succinate based bath

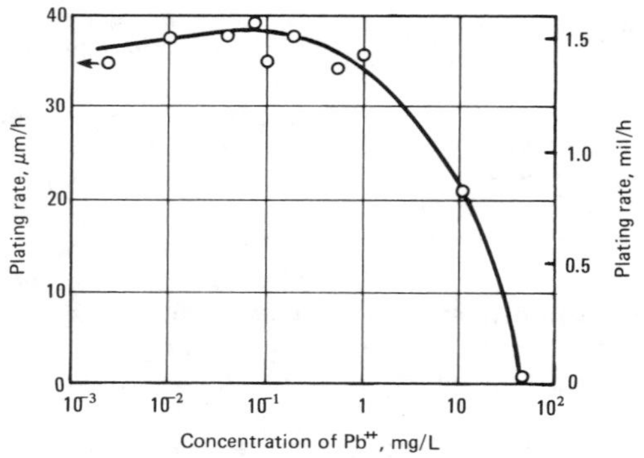

Bath at 4.6 pH and 95 °C (205 °F). Solutions containing less than 0.1 mg/L (0.4 mg/gal) lead^{++} were unstable

Fig. 5 Effect of solution pH on deposition rate and deposit phosphorus content

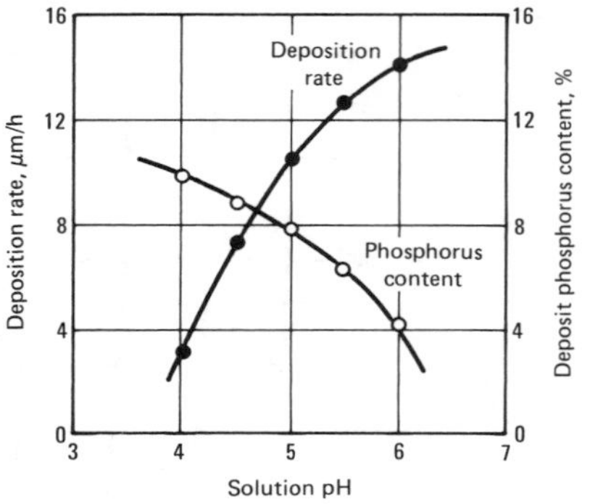

Coatings with lower phosphorus content, especially those applied from baths stabilized with heavy metals or sulfur compounds, are often porous. These deposits consist of columns of amorphous material separated by cracks and holes. The presence of such discontinuities has a severe effect on the properties of the deposit, especially on ductility and corrosion resistance.

As electroless nickel-phosphorus is heated to temperatures above 220 to 260 °C (430 to 500 °F), structural changes begin to occur. First, coherent and then distinct particles of nickel phosphite (Ni_3P) form within the alloy. Then, at temperatures above 320 °C (610 °F), the deposit begins to crystallize and lose its amorphous character. With continued heating, nickel phosphite particles conglomerate and a two-phase alloy forms. With coatings containing more than 8% phosphorus, a matrix of nickel phosphite forms, whereas almost pure nickel is the predominant phase in deposits with lower phosphorus content. These changes cause a rapid increase in the hardness and wear resistance of the coating, but cause its corrosion resistance and ductility to be reduced (Ref 2, 12-14).

Internal stress in electroless nickel coatings is primarily a function of coating composition. As illustrated in Fig. 7, stress in coatings used on steel containing more than 10% phosphorus is neutral or compressive (Ref 15). With lower phosphorus deposits, however, tensile stresses of 15 to 45 MPa (2.2 to 6.5 ksi) develop because of the difference in thermal expansion between the deposits and the substrate. The high level of stress in these coatings promotes cracking and porosity (Ref 12).

The structural changes during heat treatment at temperatures above 220 °C (430 °F) cause a volumetric shrinkage of electroless nickel deposits of up to 4 to 6% (Ref 16). This increases tensile stress and reduces compressive stress in the coating.

Deposit stress can also be increased by the codeposition of orthophosphites or heavy metals, as well as by the presence of excess complexing agents in the plating solution. Even small quantities of some metals can produce a severe increase in stress. The addition of only 5 mg/L (20 mg/gal) of bismuth and antimony to most baths can cause the deposit tensile stress to increase to as much as 350 MPa (50 ksi). High levels of internal stress also reduce the ductility of the coating and increase cracking (Ref 2, 16).

Uniformity. One especially beneficial property of electroless nickel is uniform coating thickness. With electro-

Fig. 6 Cross section of a 75-μm (3-mils) thick electroless nickel deposit

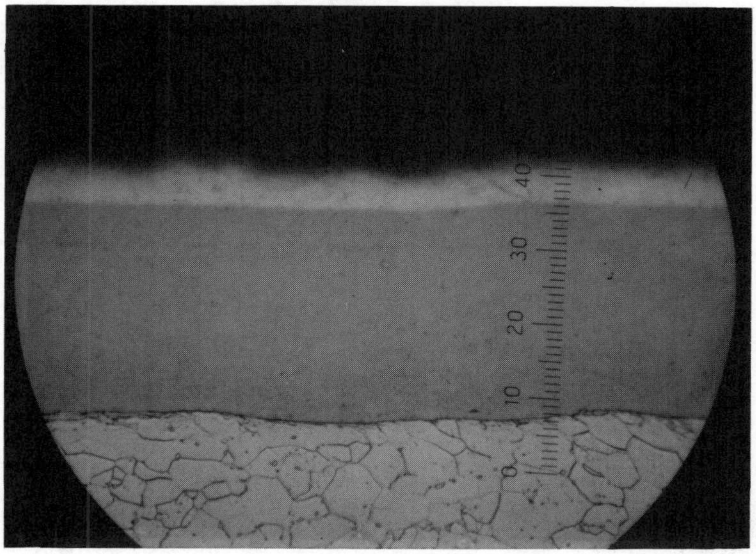

400X. Contains approximately 10½% phosphorus and less than 0.05% other elements

Fig. 7 Effect of phosphorus content on the internal stress of electroless nickel deposits on steel

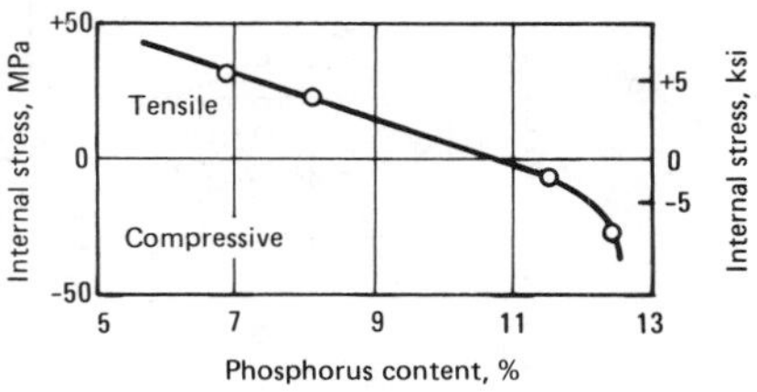

Fig. 8 Effect of deposit phosphorus content on coating density

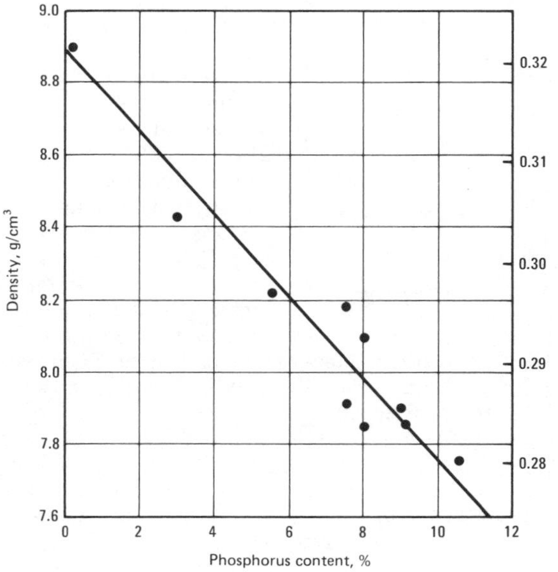

plated coatings, thickness can vary significantly depending on the shape of the part and the proximity of the part to the anodes. These variations can affect the ultimate performance of the coating, and additional finishing may be required after plating. With electroless nickel, the plating rate and coating thickness are the same on any section of the part exposed to fresh plating solution. Grooves and blind holes have the same amount of coating as the outside of a part.

With electroless nickel, coating thickness can be controlled to suit the application. Coatings as thin as 2.5 μm (0.1 mil) are applied for electronic components, whereas those as thick as 75 to 125 μm (3 to 5 mils) are normally used in corrosive environments. Coatings thicker than 250 μm (10 mils) are used for salvage or repair of worn or mismachined parts (Ref 12).

Adhesion of electroless nickel coatings to most metals is excellent. The initial replacement reaction, which occurs with catalytic metals, together with the associated ability of the baths to remove submicroscopic soils, allows the deposit to establish metallic as well as mechanical bonds with the substrate. The bond strength of the coating to properly cleaned steel or aluminum substrates has been found to be at least 300 to 400 MPa (40 to 60 ksi).

With noncatalytic or passive metals, such as stainless steel, an initial replacement reaction does not occur, and adhesion is reduced. With proper pretreatment and activation, however, the bond strength of the coating usually exceeds 140 MPa or 20 ksi (Ref 2, 12, 13). With metals such as aluminum, parts are baked after plating for 1½ h at 190 to 210 °C (375 to 410 °F) to increase the adhesion of the coating. These treatments relieve hydrogen from the part and the deposit and provide a very minor amount of codiffusion between coating and substrate. Baking parts is most useful where pretreatment has been less than adequate and adhesion is marginal. With properly applied coatings, baking has only a minimal effect upon bond strength (Ref 2, 12, 14).

Physical Properties. The density of electroless nickel coatings is inverse-

Fig. 9 Effect of deposit phosphorus content on coefficient on thermal expansion

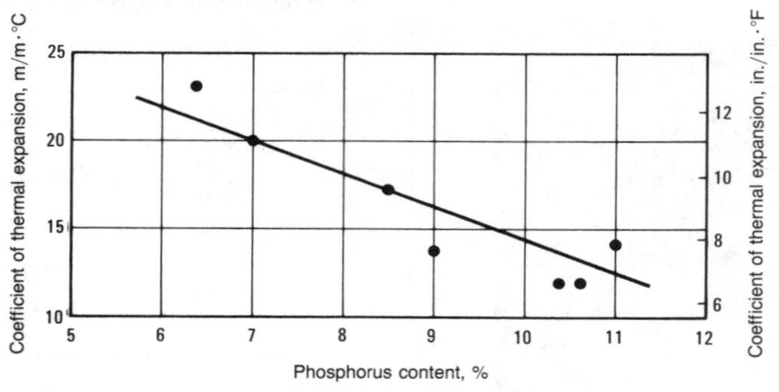

Fig. 11 Effect of heat treatment on the ductility of a 6% phosphorus electroless nickel

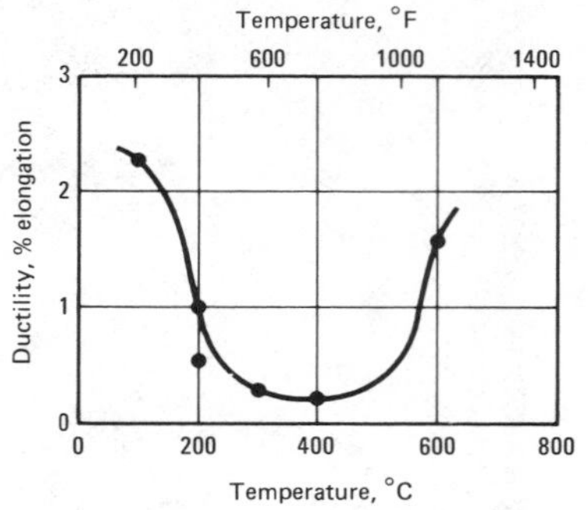

Fig. 10 Effect of phosphorus content on strength and strain at fracture

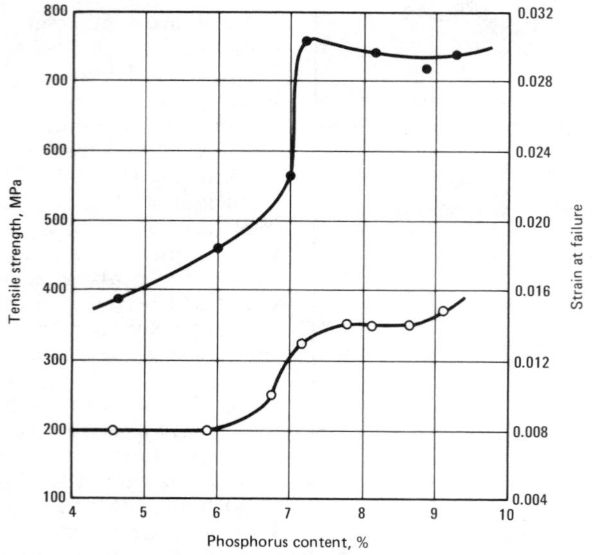

ly proportional to their phosphorus content. As shown in Fig. 8, density varies from about 8.5 gm/cm^3 for low phosphorous deposits to 7.75 gm/cm^3 for coatings containing 10 to 11% phosphorus (Ref 2, 13, 17 to 19).

The thermal and electrical properties of these coatings also vary with composition. For commercial coatings, however, electrical resistivity and thermal conductivity are generally about 50 to 90 $\mu\Omega \cdot$cm and 0.010 to 0.013 cal/cm·s·°C (2.4 to 3.1 Btu/ft·h·°F), respectively. Accordingly, these coatings are significantly less conductive than conventional conductors such as copper or silver.

Heat treatments precipitate phosphorus from the alloy and can increase its conductivity by three to four times (Ref 2, 13). The formulation of the plating solution can also affect conductivity. Tests with baths complexed with sodium acetate and succinic acid, slowed electrical resistivities of 61 and 84 $\mu\Omega \cdot$cm, respectively (Ref 2).

Phosphorus content also has a strong effect on the thermal expansion of electroless nickel. This is shown in Fig. 9, which is based on deposit stress measurements on different substrates (Ref 15). The coefficient of thermal expansion of high phosphorus coatings is approximately equal to that of steel. As deposited, coatings containing more than 10% phosphorus are completely nonmagnetic. Lower phosphorus coatings, however, have some magnetic susceptibility. The coercivity of 3 to 6% phosphorus coatings is about 20 to 80 Oe (1592 to 6366 A/m), while that of deposits containing 7 to 9% phosphorus is typically 1 to 2 Oe (80 to 160 A/m). Heat treatments at temperatures above 300 °C (570 °F) improve the magnetic response of electroless nickel and can provide coercivities of about 100 to 300 Oe (7958 to 23 873 A/m) (Ref 11, 20).

Mechanical Properties. The mechanical properties of electroless nickel deposits are similar to those of other glasses. They have high strength, limited ductility, and a high modulus of elasticity. The ultimate tensile strength of commercial coatings exceeds 700 MPa (102 ksi) and allows the coating to withstand a considerable amount of abuse without damage. The effect of phosphorus content upon the strength and strain at fracture of electroless nickel deposits is shown in Fig. 10 (Ref 21).

The ductility of electroless nickel coatings also varies with composition. High phosphorus, high purity coatings have a ductility of about 1 to 1½% (as elongation). Although this is less ductile than most engineering materials, it is adequate for most coating applications. Thin films of deposit can be bent completely around themselves without fracture. With lower phosphorus deposits, or with deposits containing metallic or sulfur impurities, ductility is greatly reduced and may approach zero (Ref 12, 14).

Fig. 12 Effect of heat treatment at different temperatures on the hardness of 10½% phosphorus electroless nickel coating

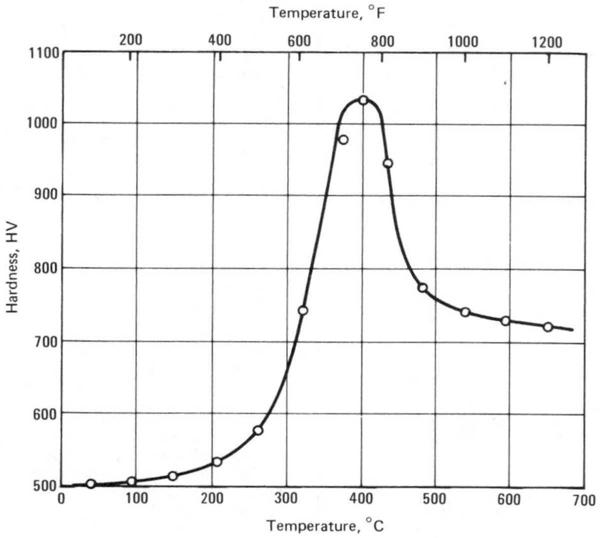

Fig. 13 Effect of different heat treatment periods on hardness of a high phosphorus electroless nickel

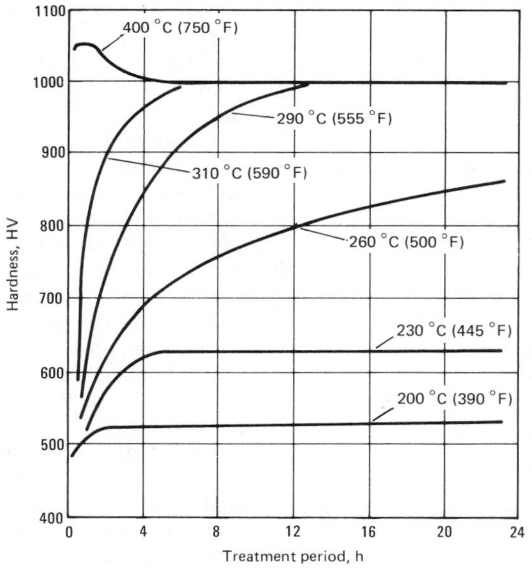

Hardening type heat treatments reduce both the strength and ductility of electroless nickel deposits. Exposure to temperatures above 220 °C (428 °F) causes an 80 to 90% reduction in strength and can destroy ductility. This is illustrated by Fig. 11, which shows the effect of different 1-h heat treatments on the elongation at fracture of brass panels coated with a 6% phosphorus electroless nickel (Ref 11). The modulus of elasticity of electroless nickel coatings containing 7 to 11% phosphorus is about 200 GPa (29 000 ksi) and is very similar to that of steel.

Hardness and wear resistance are extremely important properties for many applications. As deposited, the microhardness of electroless nickel coatings is about 500 to 600 HV_{100}, which is approximately equal to 48 to 52 HRC and equivalent to many hardened alloy steels. Heat treatment causes these alloys to age harden and can produce hardness values as high as 1100 HV_{100}, equal to most commercial hard chromium coatings (Ref 2, 12). Figure 12 shows the effect of different 1-h heat treatments on the hardness of electroless nickel containing 10½% phosphorus (Ref 2).

For some applications, high temperature treatments cannot be tolerated because parts may warp, or the strength of the substrate may be reduced. For these applications, longer times and lower temperatures are sometimes used to obtain the desired hardness. This is illustrated in Fig. 13, which shows the effect of different treatment periods on the hardness of a coating containing 10½% phosphorus (Ref 12).

Electroless nickel coatings also have excellent hot hardness. To about 400 °C (750 °F), the hardness of heat treated electroless nickel is equal to or better than that of hard chromium coatings. As-deposited coatings also retain their hardness to this temperature, although at a lower level. The effect of elevated temperature on a 10% phosphorus coating is shown in Fig. 14 (Ref 22, 23).

Because of their high hardness, electroless nickel coatings have excellent resistance to wear and abrasion, both in the as-deposited and hardened conditions. Taber Abraser Index values for electroless nickel and for electrodeposited nickel and chromium are summarized in Table 3 (Ref 24, 25, 29).

Tests with electroless nickel-coated vee-blocks in a Falex Wear Tester have shown a similar relationship between heat treatment and wear and confirmed the coating to be equal to hard chrome under lubricated wear conditions (Ref 14, 24). The effect of phosphorus content upon the wear experienced by electroless nickel coatings under lubricated conditions is summarized in Fig. 15. These rotating ball tests showed that after heat treatment, high phosphorus deposits provide the best resistance to adhesive wear (Ref 6, 26).

Frictional properties of electroless nickel coatings are excellent and simi-

Fig. 14 Effect of temperature on the elevated temperature hardness of a 10% phosphorus electroless nickel

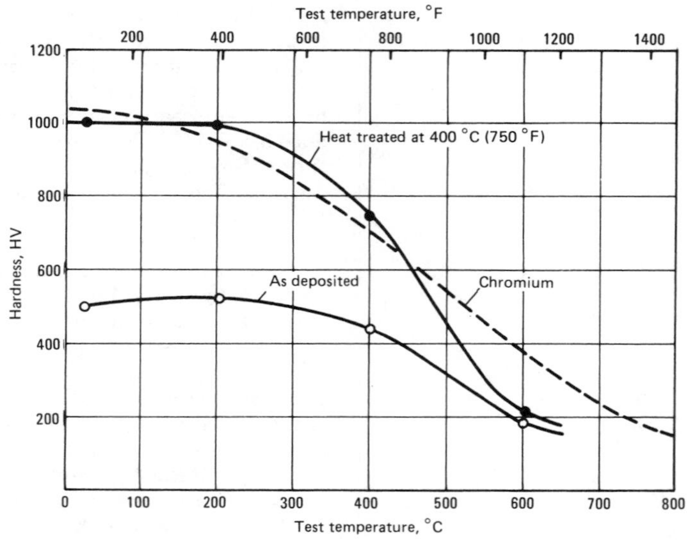

Fig. 15 Effect of phosphorus content on the wear of electroless nickel coatings in rotating ball tests

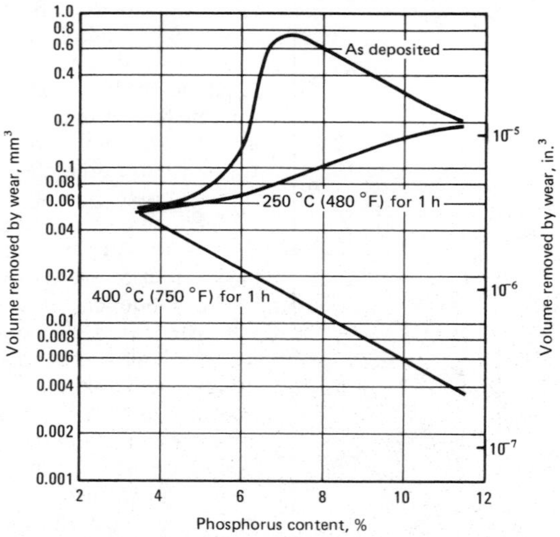

lar to those of chromium. Their phosphorus content provides a natural lubricity, which can be very useful for applications such as plastic molding. The coefficient of friction for electroless nickel versus steel is about 0.13 for lubricated conditions and 0.4 for unlubricated conditions. The frictional properties of these coatings vary little with either phosphorus content or with heat treatment (Ref 2, 24, 26).

Solderability. Electroless nickel coatings can be easily soldered and are used in electronic applications to facilitate soldering such light metals as aluminum. For most components, rosin mildly activated (RMA) flux is specified along with conventional tin-lead solder. Preheating the component to 100 to 110 °C (212 to 230 °F) improves the ease and speed of joining. With moderately oxidized surfaces, such as those resulting from steam aging, activated rosin (RA) flux is usually required to obtain wetting of the coating (Ref 2, 27).

Table 3 Comparison of the Taber abraser resistance of different engineering coatings

Coating	Heat treatment for 1 h °C	°F	Taber wear index, mg/1000 cycles(a)
Watts nickel	None	None	25
Electroless Ni-P(b)	None	None	17
Electroless Ni-P(b)	300	570	10
Electroless Ni-P(b)	500	930	6
Electroless Ni-P(b)	650	1200	4
Electroless Ni-B(c)	None	None	9
Electroless Ni-B(c)	400	750	3
Hard chromium	None	None	2

(a) CS-10 abraser wheels, 1000-g load, determined as average weight loss per 1000 cycles for total test of 6000 cycles. (b) Hypophosphite reduced electroless nickel containing approximately 9% phosphorus. (c) Borohydride reduced electroless nickel containing approximately 5% boron

Corrosion Resistance. Electroless nickel is a barrier coating, protecting the substrate by sealing it off from the environment, rather than using sacrificial action. Because of its amorphous nature and passivity, the corrosion resistance of the coating is excellent and, in many environments, superior to that of pure nickel or chromium alloys. Amorphous alloys have better resistance to attack than equivalent polycrystalline materials, because of their freedom from grain or phase boundaries, and because of the glassy films which form on and passivate their surfaces. Some examples of the corrosion experienced in different environments are shown in Table 4 (Ref 2, 16, 28, 29).

Effect of Composition. The corrosion resistance of an electroless nickel coating is a function of its composition. Most deposits are naturally passive and very resistant to attack in most environments. Their degree of passivity and corrosion resistance, however, is greatly affected by their phosphorus content. Alloys containing more than 10% phosphorus are more resistant to attack than those with lower phosphorus contents (Ref 16, 17).

Often the tramp constituents present in an electroless nickel are even more important to its corrosion resistance than its phosphorus content. Most coat-

Table 4 Corrosion of electroless nickel coatings in various environments

Environment	Temperature °C	Temperature °F	Corrosion rate: Electroless nickel-phosphorus(a) μm/yr	mil/yr	Electroless nickel-boron(b) μm/yr	mil/yr
Acetic acid, glacial	20	68	0.8	0.03	84	3.3
Acetone	20	68	0.08	0.003	Nil	Nil
Aluminum sulfate, 27%	20	68	5	0.2	...	...
Ammonia, 25%	20	68	16	0.6	40	1.6
Ammonia nitrate, 20%	20	68	15	0.6	(c)	(c)
Ammonium sulfate, saturated	20	68	3	0.1	3.5	0.14
Benzene	20	68	Nil	Nil	Nil	Nil
Brine, 3½% salt, CO_2 saturated	95	205	5	0.2	...	...
Brine, 3½% salt, H_2S saturated	95	205	Nil	Nil	...	...
Calcium chloride, 42%	20	68	0.2	0.008	...	...
Carbon tetrachloride	20	68	Nil	Nil	Nil	Nil
Citric acid, saturated	20	68	7	0.3	42	1.7
Cupric chloride, 5%	20	68	25	1	...	...
Ethylene glycol	20	68	0.6	0.02	0.2	0.008
Ferric chloride, 1%	20	68	200	8	...	...
Formic acid, 88%	20	68	13	0.5	90	3.5
Hydrochloric acid, 5%	20	68	24	0.9	...	...
Hydrochloric acid, 2%	20	68	27	1.1	...	...
Lactic acid, 85%	20	68	1	0.04	...	...
Lead acetate, 36%	20	68	0.2	0.008	...	...
Nitric acid, 1%	20	68	25	2	...	...
Oxalic acid, 10%	20	68	3	0.1	...	...
Phenol, 90%	20	68	0.2	0.008	Nil	Nil
Phosphoric acid, 85%	20	68	3	0.1	(c)	(c)
Potassium hydroxide, 50%	20	68	Nil	Nil	Nil	Nil
Sodium carbonate, saturated	20	68	1	0.04	Nil	Nil
Sodium hydroxide, 45%	20	68	Nil	Nil	Nil	Nil
Sodium hydroxide, 50%	95	205	0.2	0.008	...	...
Sodium sulfate, 10%	20	68	0.8	0.03	11	0.4
Sulfuric acid, 65%	20	68	9	0.4	...	...
Water, acid mine, 3.3 pH	20	68	7	0.3	...	...
Water, distilled, N_2 deaerated	100	212	Nil	Nil	Nil	Nil
Water, distilled, O_2 saturated	95	205	Nil	Nil	Nil	Nil
Water, sea (3½% salt)	95	205	Nil	Nil	...	...

(a) Hypophosphite reduced electroless nickel containing approximately 10½% phosphorus. (b) Borohydride reduced electroless nickel containing approximately 5% boron. (c) Very rapid. Specimen dissolved during test

Table 5 The effect of heat treatment on the corrosion of a 10½% phosphorus electroless nickel in 10% hydrochloric acid

Heat treatment	Deposit hardness, HV_{100}	Corrosion rate μm/yr	mil/yr
None	480	15	0.6
190 °C (375 °F) for 1½ h	500	20	0.8
290 °C (550 °F) for 6 h	900	1900	75
290 °C (550 °F) for 10 h	970	1400	55
340 °C (650 °F) for 4 h	970	900	35
400 °C (750 °F) for 1 h	1050	1200	47

ings are applied from baths inhibited with lead, tin, cadmium, or sulfur. Codeposition of these elements in more than trace amounts causes the corrosion resistance of the coating to be increased by 5 to 40 times (Ref 16).

Effect of Heat Treatment. One of the most important variables affecting the corrosion of electroless nickel is its heat treatment. As nickel-phosphorus deposits are heated to temperatures above 220 °C (430 °F), nickel phosphide particles begin to form, reducing the phosphorus content of the remaining material. This reduces the corrosion resistance of the coating. The particles also create small active/passive corrosion cells, further contributing to the destruction of the deposit. The deposit also shrinks as it hardens, which can crack the coating and expose the substrate to attack. The effect of these changes is illustrated in Table 5, which shows the results of tests with a 10½% phosphorus deposit heat treated to represent different commercial treatments and then exposed to 10% hydrochloric acid at ambient temperature (Ref 16). Baking at 190 °C (375 °F), similar to the treatment used for hydrogen embrittlement relief, caused no signifi-

Table 6 Physical and mechanical properties of electroless nickel-boron and nickel-phosphorus deposits(a)

Property	Electroless nickel-boron(b)	Electroless nickel-phosphorus(c)
Density, g/cm³ (lb/in.³)	8.25 (2.98)	7.75 (2.8)
Melting point, °C (°F)	1080 (1980)	890 (1630)
Electrical resistivity, μΩ·cm	89	90
Thermal conductivity, W/m·K (cal/cm·s·°C)	...	4 (0.01)
Coefficient of thermal expansion (22-100 °C, or 72-212 °F), μm/m·°C (μin./in.·°F)	12.6 (7.1)	12 (6.7)
Magnetic properties	Very weakly ferromagnetic	Nonmagnetic
Internal stress, MPa (ksi)	110 (16)	Nil
Tensile strength	110 (16)	700 (100)
Ductility, % elongation	0.2	1.0
Modulus of elasticity, GPa (10^6 psi)	120 (17)	200 (29)
As-deposited hardness, HV_{100}	700	500
Heat treated hardness, 400 °C (750 °F) for 1 h, HV_{100}	1200	1100
Coefficient of friction vs steel, lubricated	0.12	0.13
Wear resistance, as deposited, Taber mg/1000 cycles	9	18
Wear resistance, heat treated 400 °C (750 °F) for 1 h, Taber mg/1000 cycles	3	9

(*a*) Properties are for coatings in the as-deposited condition, unless noted. (*b*) Borohydride reduced electroless nickel containing approximately 5% boron. (*c*) Hypophosphite reduced electroless nickel containing approximately 10½% phosphorus

Fig. 16 Effect of heat treatments at 400 °C (752 °F) on the strain at fracture of electroless Ni-5% B and Ni-9% P coatings

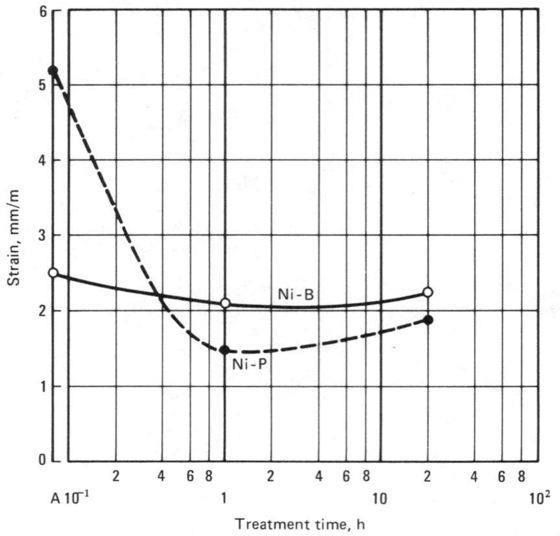

Fig. 17 Effect of different 1-h heat treatments on the hardness and wear resistance of borohydride reduced electroless nickel

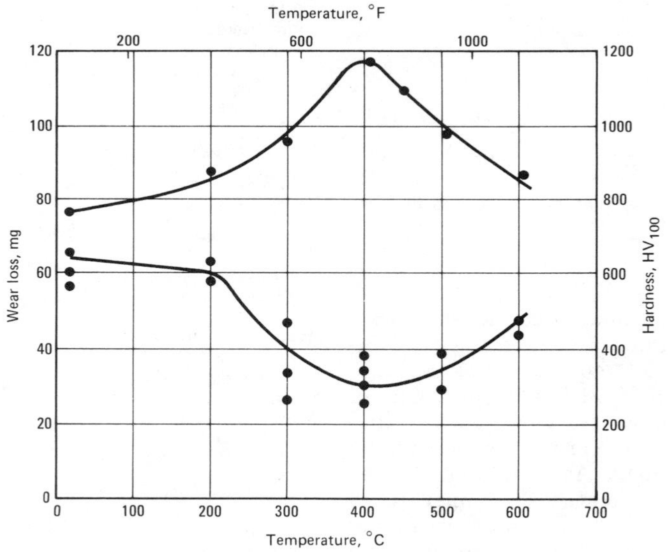

Table 7 Effect of boron content and complexing agent on internal stress in DMAB reduced deposits

Complexing agent	Boron content, %	Internal stress MPa(a)	Internal stress ksi
Malonate	4.3	120	17.4
Malein-glycine	1.2	310	44.9
Pyrophosphate	0.4	480	69.6

(a) Based on tests with 12½ μm thick coatings on a Brenner-Senderoff Spiral Contractometer. Stresses are tensile

cant increase in corrosion. Hardening, however, caused the corrosion rate of the deposit to increase from 15 μm/yr (0.6 mil/yr) to more than 900 μm/yr (35 mils/yr). Tests in other environments showed a similar reduction in resistance after hardening. Where corrosion resistance is required, hardened coatings should not be used (Ref 16).

Properties of Electroless Nickel-Boron Coatings

The properties of deposits from borohydride or aminoborane reduced baths are similar to those of electroless nickel-phosphorus alloys with a few exceptions. The hardness of nickel-boron alloys is very high, and these alloys can be heat treated to levels greater than that of hard chromium. Nickel-boron coatings have outstanding resistance to wear and abrasion. These coatings, however, are not completely amorphous and have reduced resistance to corrosive environments, as well as being much more costly than nickel-phosphorus coatings. The physical and mechanical properties of borohydride reduced electroless nickel are summarized in Table 6 (Ref 2, 6, 30). For comparison, the properties of a hypophosphite reduced coating containing 10½% phosphorus are also listed (Ref 12).

Structure and Internal Stress. The boron content of electroless nickel reduced with DMAB or DEAB can vary from 0.2 to 4% depending on bath formulation and operation. Commercial borohydride reduced coatings typically contain 5% boron. Unlike nickel-phosphorus coatings in the as-deposited condition, electroless nickel-boron contains crystalline nickel mixed with nickel-boron (typically Ni_2B) glass. These coatings also are not totally homogeneous and consist of phases of different composition (Ref 2, 7, 31).

During heating, electroless nickel-boron age hardens in the same manner as nickel-phosphorus alloys. At temperatures over 250 °C (480 °F), particles of nickel boride (Ni_3B) form, and at 370 to 380 °C (700 to 715 °F), the coating crystallizes. The final structure of hardened nickel-boron coatings consists of nickel-boron intermetallic compounds (principally Ni_3B and Ni_2B) and about 10% nickel (Ref 2, 7, 30).

The internal stress level of nickel-boron deposits is generally high. The effect of boron content and complexing agent on the stress in DMAB reduced electroless nickel coatings is shown in Table 7 (Ref 10). The internal stress of borohydride reduced coatings is typically 110 to 200 MPa (16 to 29 ksi) tensile (Ref 30).

Physical and mechanical properties of borohydride reduced electroless nickel are summarized in Table 6 (Ref 2, 7, 31). For comparison, the properties of a hypophosphite reduced coating containing 10½% phosphorus is also listed (Ref 12). The density of electroless nickel-boron is very similar to that of nickel-phosphorus coatings of equal alloy content. The density of borohydride reduced coatings containing 5% boron is 8.25 g/cm^3 in both the as-deposited and heat treated condition (Ref 2, 30).

The melting point of nickel-boron coatings is relatively high and can approach that of metallic nickel. Sodium borohydride reduced coatings melt at 1080 °C (1975 °F), while the melting point of DMAB reduced coatings varies from about 1350 to 1390 °C (2460 to 2535 °F) (Ref 2, 30).

The electrical resistivity of 5% boron coatings is similar to that of nickel-phosphorus alloys, ranging from 89 $\mu\Omega \cdot$ cm in the as-deposited condition to 43 $\mu\Omega \cdot$cm after heat treatment at 1100 °C (2010 °F). In the as-deposited condition, nickel-boron coatings are very weakly ferromagnetic, with coercivities about 10% of that of metallic nickel. Their magnetic susceptibility, however, can be increased by heat treatments at temperatures above 370 °C (700 °F) (Ref 2, 7, 30).

The strength and ductility of nickel-boron coatings is only about one fifth of that of high phosphorus deposits. Guided bent tests of panels coated with 5% nickel-boron showed its strain at fracture to be 2.5 mm/m (2.5 mils/in.). In the same test, the breaking strain of a hypophosphite reduced electroless nickel containing approximately 9% phosphorus was 5.3 mm/m (5.3 mils/in.). Unlike nickel-phosphorus coatings, however, heat treatment has little effect on the ductility of nickel-boron. As illustrated by Fig. 16, even after 12 h at 400 °C (750 °F), strain at fracture declines by only 15% (Ref 7). The modulus of elasticity of borohydride reduced coatings ranges from 120 GPa (17 000 ksi) in the as-deposited condition to 180 GPa (26 000 ksi) for coatings heat treated at 400 °C (750 °F) for 1 h (Ref 30).

Hardness and Wear Resistance. The principle advantage of electroless nickel-boron is its high hardness and superior wear resistance. In the as-deposited condition, microhardness values of 650 to 750 HV_{100} are typical for borohydride reduced coatings. After 1-h heat treatments at 350 to 400 °C (660 to 750 °F) hardness values of 1200 HV_{100} can be produced. This is illustrated by Fig. 17 which shows the effect of heat treatment temperature on hardness (Ref 2, 7, 30).

Long-term treatments (30 to 40 weeks) at temperatures between 200 and 300 °C (390 to 570 °F) can produce hardness values of 1700 to 2000 HV_{100}. These low temperature treatments result in a finer dispersion of nickel boride than do higher temperatures and in the formation of iron borides (such as Fe_2B and $Fe_3C_{0.2}B_{0.8}$ within the coating (Ref 2, 30).

The wear resistance of electroless nickel-boron is exceptional and after heat treatment equals or exceeds that of hard chromium coatings. Typical Taber wear test results for a 5% boron coating is shown in Tables 3 and 6. The effect of heat treatment and hardness on the wear experienced in rotating ring and block tests (similar to the Alpha LFW-1 test described in ASTM D2714) (Ref 31) under nonlubricated conditions is also shown in Fig. 17.

Electroless nickel-boron coatings are naturally lubrious. Their coefficient of friction versus steel is typically 0.12 to 0.13 in the lubricating conditions, and 0.43 to 0.44 for dry wear (Ref 2, 30).

Corrosion Resistance. In general, the corrosion resistance of electroless nickel-boron coatings is less than that of high phosphorus alloys. This is illustrated by Table 4 which compares the attack experienced by hypophosphite and borohydride reduced coatings in different media. In environments that cause little corrosion of nickel-phosphorus, such as alkalis and solvents, electroless nickel-boron is also very resistant. In environments, however, that cause moderate attack of nickel-phosphorus, such as acids and ammonia solutions, nickel-boron coatings can be severely corroded. In strongly oxidizing media, of course, neither coating is satisfactory.

Effect of Electroless Nickel Coatings on the Fatigue Strength of Steel

Because of their tendency to crack under cyclic loads, electroless nickel coatings can cause a significant reduction in the fatigue strength of steel substrates. The magnitude of the reduction, however, depends on the composition, heat treatment, and thickness of the coating, as well as original fatigue strength of the steel. Several investigations have shown that the use of electroless nickel coatings cause a 10 to 50% reduction in the fatigue strength and endurance limit of steel substrates (Ref 6, 7, 32-34). In these tests, fatigue strength of notched specimens was reduced by at least 15%, whereas unnotched samples showed relatively small reductions.

Table 8 Effect of heat treatment of an electroless nickel-5% boron coating on the fatigue strength of steel

Condition	Fatigue strength MPa	ksi
Not coated	350	51
As-deposited	270	39
250 °C (480 °F) for 1 h	260	38
350 °C (660 °F) for 1 h	245	36
400 °C (750 °F) for 1 h	270	39

The loss of fatigue strength has principally been a problem with hypophosphite reduced coatings, containing less than 10% phosphorus, and with nickel-boron alloys. These deposits contain high levels of internal tensile stress and under cyclic stress conditions tend to crack and initiate fatigue failures. Recent tests have implied that this is not a significant problem with high phosphorus deposits (Ref 35, 36). Coatings containing 10½% or more phosphorus are compressively stressed on steel and tend to resist cracking.

Heat treatment of electroless nickel coatings tends to increase the effect of these coatings on fatigue strength. Heat treated coatings tend to be more highly stressed than as-deposited coatings and have a greater tendency to crack. Heat treating a high phosphorus, compressively stressed coating can cause it to become tensilely stressed (Ref 14). Coatings heat treated at temperatures above 340 °C (650 °F) also tend to be cracked because of the shrinkage of the alloy. These cracks act as stress risers and further reduce fatigue resistance. Table 8 shows the effect of different 1-h heat treatments on the fatigue strength of a 0.42% carbon steel (Werkstoff C45) coated with 30 μm (1.2 mils) of borohydride reduced electroless nickel (Ref 6). Heat treatments at very high temperatures, 650 to 800 °C (1200 to 1470 °F), produce a thick diffusion zone between the coating and the substrate, which may eliminate or at least greatly reduce the effect of the coating on fatigue strength.

Fig. 18 Effect of coating thickness on the fatigue strength of a carbon-manganese steel

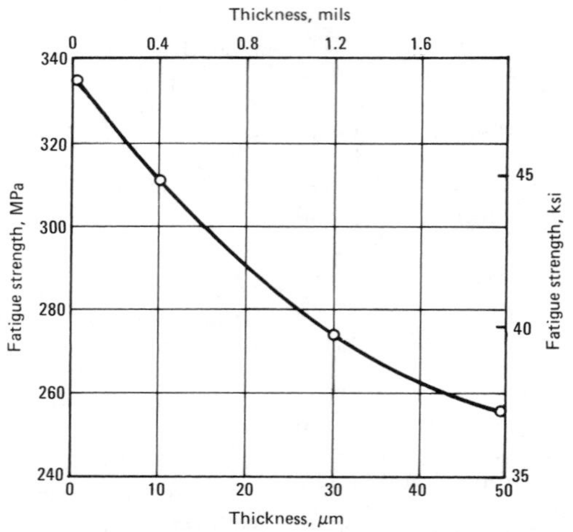

The reduction in fatigue strength produced by electroless nickel deposits is also affected by the thickness of the coating. Thicker deposits have the greatest effect on fatigue strength. This is illustrated in Fig. 18, which shows the reduction in strength of a carbon-manganese steel (Werkstoff St52) produced by different thicknesses of a 5% boron-nickel (Ref 6).

Pretreatment for Electroless Nickel Coatings

Proper pretreatment can be as important to the successful application of an electroless nickel coating as the actual deposit. Inadequate cleaning can result in lack of adhesion, roughness, coating porosity, and early failure. The methods used to clean and prepare a metal surface for electroless nickel plating are similar to those used for conventional electroplating, although more care and control is required. One penetrant that is unique to electroless nickel plating is the application of a strike copper plate to alloys containing significant amounts of lead, tin, cadmium, or zinc. This ensures adequate coverage and prevents contamination of the electroless solution.

Pretreatment for Ferrous Alloys

To properly prepare ferrous alloys for electroless nickel plating, the combination of solvent and alkaline degreasing, acid activation, and electrocleaning are required, with intermediate water rinses. These steps are discussed in other articles of this Volume. Recommended pretreatment procedures for different ferrous alloys are summarized below:

Carbon and low-alloy steel

1. Degrease
2. Soak clean for 10 to 30 min
3. Rinse
4. Electroclean at 5 V for 60 to 120 s
5. Rinse
6. Dip in 30% hydrochloric acid for 30 to 60 s
7. Rinse
8. Electroclean at 5 V for 30 to 60 s
9. Rinse
10. Plate to thickness

Alloy steel (Cr or Ni >1½%)

1. Degrease
2. Soak clean for 10 to 30 min
3. Rinse
4. Electroclean at 5 V for 60 to 120 s
5. Rinse
6. Dip in hydrochloric acid for 30 to 60 s
7. Rinse
8. Electroclean at 5 V for 30 to 60 s
9. Rinse
10. Dip in 30% HCl for 30 to 60 s
11. Rinse
12. Nickel strike at 2 A/dm^2 (20 A/ft^2) for 60 s
13. Rinse
14. Plate to thickness

300 or 400 series stainless steel

1. Degrease
2. Soak clean for 10 to 30 min
3. Rinse
4. Electroclean at 5 V for 60 to 120 s
5. Rinse
6. Dip in 30% hydrochloric acid for 60 s
7. Rinse
8. Nickel strike at 2 A/dm^2 (20 A/ft^2) for 60 s
9. Rinse
10. Plate to thickness

300 series stainless steel (complex shapes)

1. Degrease
2. Soak clean for 10 to 30 min
3. Rinse
4. Electroclean at 5 V for 60 to 120 s
5. Rinse
6. Dip in 30% hydrochloric acid for 60 s
7. Rinse
8. 10% H_2SO_4 at 60 °C (140 °F) for 30 s
9. Plate to thickness

400 series stainless steel (complex shapes)

1. Degrease
2. Soak clean for 10 to 30 min
3. Rinse
4. Electroclean at 5 V for 60 to 120 s
5. Rinse
6. Dip in 30% hydrochloric acid for 60 s
7. Rinse
8. Dip in 20% hydrochloric acid at 50 °C (120 °F) for 30 s
9. Rinse with deionized water
10. Plate to thickness

In Step 1, all alkaline cleaners should be operated at their supplier's maximum recommended temperature, typically 60 to 80 °C (140 to 175 °F). Unless otherwise indicated, all other processes are at ambient temperature. In Step 2, electrocleaning is with at least three reversals of current (part, cathodic/anodic, three times) at 3 to 5 A/dm^2 (30 to 50 A/ft^2). Except for 300 series stainless steel, the final current cycle should be with the part anodic; with 300 series stainless steels the final current cycle should be with the part cathodic to minimize the formation of an oxide film on its surface.

Activation for Alloy Steels. Before electroless plating, stainless and alloy steels parts must be chemically activated to obtain satisfactory adhesion. For this, a low pH nickel strike is normally used. Two common strike baths are listed below:

Nickel sulfamate strike

Nickel sulfamate	165-325 g/L (22-43 oz/gal)
Nickel (as metal)	35-75 g/L (5-10 oz/gal)
Sulfamic acid (approx 20 g/L or 2.7 oz/gal)	to pH 1-1.5
Boric acid	30-34 g/L (4-4.5 oz/gal)
Hydrochloric acid (20° Bé)	12 mL/L (1.5 fluid oz/gal)
Temperature	Room temperature
Cathode current density	1-10 A/dm^2 (10-100 A/ft^2)
Time	30-60 s
Anodes (bagged)	Sulfur depolarized nickel

Woods nickel strike

Nickel chloride	240 g/L (32 oz/gal)
Hydrochloric acid	250 mL/L (32 fluid oz/gal)
Temperature	Room temperature
Cathode current density	2-10 A/dm^2 (20-100 A/ft^2)
Time	30-120 s
Anodes	Sulfur depolarized nickel

Nickel strikes should not be used to cover up improper pretreatment of plain or low-alloy steel. Nickel strike activation should be considered, however, when processing steel with chromium or nickel contents of over 1.5%, carburized or nitrided steels, and stainless steels. Nickel strike processing should follow acid activation to avoid drag-in of alkaline materials into the strike (Ref 2, 37-42).

Pretreatment for Aluminum Alloys

Like steel, aluminum is catalytic to electroless nickel deposition, and could be plated after only a simple cleaning. Aluminum is very reactive, however, and oxides form very rapidly on its surface during rinsing or exposure to air. The oxide films that develop prevent metallic bonds from forming between the coating and the substrate and can result in adhesion failure. To avoid this problem, special processing procedures are required, including deoxidizing and zincating. Processing procedures for aluminum alloys are discussed in the article on cleaning and finishing of aluminum and aluminum alloys in this Volume.

Pretreatment for Copper Alloys

Copper-based alloys are prepared for electroless nickel plating using procedures similar to those for steel, alkaline cleaning and acid deoxidizing. Two important differences exist, however:

1 Copper is not catalytic to the chemical reduction of electroless nickel, and its alloys must be activated chemically or electrolytically before they can be plated
2 Lead in amounts of ½ to 10% is often added to copper alloys to make them easier to machine. Unless the free lead present on the surface of the part is removed, adhesion failures and coating porosity result

Processing procedures for copper alloys are given in the article on cleaning and finishing of copper and copper alloys in this Volume.

Activation. Once a copper alloy surface is clean and oxide-free, it must be activated before electroless nickel can deposit. The preferred method for initiating deposition is an electrolytic strike in the electroless nickel bath. Using a nickel anode, the parts are made cathodic at 5 V for 30 to 60 s. This applies a thin, electrolytic nickel-phosphorus coating and provides a catalytic surface. After the current is removed, the electroless deposition can continue.

Another method for initiating electroless deposition on copper alloy surfaces is to preplate surfaces with electrolytic nickel. One disadvantage of this method is that blind holes, internal surfaces, or low current density areas may not be coated by the strike, resulting in incomplete coverage or unplated areas. The use of nickel chloride strikes also may result in chloride contamination of the electroless nickel bath through drag-in.

A third method of activating copper alloys in electroless nickel solutions is to touch them with a piece of steel or with another part already coated with electroless nickel after they have been immersed in the bath. This creates a galvanic cell, producing an electric current to initiate the electroless reaction. Deposition spreads until the whole part is covered with electroless nickel. However, two problems can occur with galvanic activation:

1 Galvanic currents do not travel well around sharp curves, such as those on threads or corners, and can leave bare spots or areas of reduced thickness
2 Passivation of the copper can occur before the deposit spreads across the entire surface leading to poor adhesion

Leaded Alloys. Unlike other elements added to brass or bronze, lead does not combine with copper to form an alloy. Instead, it remains in the metal as globules. The lead exposed during cutting or machining acts as a lubricant by flowing or smearing across the surface. Electroless nickel does not deposit on lead. Unless lead smears are removed, the applied coating is porous with poor adhesion. Lead remaining on the surface of parts can also contaminate electroless nickel solutions, causing a rapid decline in plating rate and deposit quality.

Surface lead is best removed by immersing parts for 30 s to 2 min in a 10 to 30% solution of fluoboric acid at room temperature. Sulfamic acid and dilute nitric acid have also been reported to be effective solutions for removing lead. The removal of lead must occur before deoxidizing or bright dipping in the pretreatment cycle, and it is not a substitute for these steps (Ref 2, 37, 43).

Equipment for Electroless Nickel Plating

Because electroless nickel is applied by a chemical reaction rather than by electrolytic deposition, unique tanks and auxiliary equipment are required to ensure trouble-free operation and quality coatings.

Plating Tanks

Cylindrical or bell-shaped tanks have been used for electroless nickel plating, although rectangular tanks have been found to be the most convenient to build and operate. Rectangular tanks have been constructed from various materials in many different sizes. A common electroless nickel plating system is shown in Fig. 19.

Physical Dimensions. The following factors should be considered when selecting the size of an electroless nickel plating tank:

- Size of the part to be plated
- Number of parts to be plated each day
- Plating thickness required
- Plating rate of the solution (most conventional electroless nickel solu-

Fig. 19 Twin tank system for electroless nickel plating

Tanks are used alternately. While one tank is being used to plate, the second is being passivated. Cylindrical tank is used to store 30% nitric acid for passivation

tions deposit between 12 and 25 μm/h or 0.5 and 1 mil/h)

- Type of rack, barrel, or basket used to support parts
- Number of production hours available each day to process parts
- Maximum recommended work load of 1.2 dm^2/L (0.5 ft^2/gal) of working solution

The size of the part or the size of the supporting rack, barrel, or basket usually defines the minimum size tank that can be used. The minimum dimension of the tank should be at least 15 cm (6 in.) greater than the maximum dimension of the part or its support to allow proper agitation and the flow of fresh solution to all surfaces. The size of the tank may have to be increased, however, to accommodate the volume of parts required or to provide a more suitable work area to solution volume ratio.

Construction Materials. The following factors should be considered when selecting construction materials for a plating tank:

- Operating temperature of the electroless nickel plating solution, usually 85 to 95 °C (185 to 205 °F)
- Tendency of tank material to become sensitized to the deposition of electroless nickel
- Cost of tank material, including both initial construction cost and its life in a production environment

With continued exposure to heated electroless nickel solutions, almost any surface eventually becomes sensitized or receptive to deposition of the coating. The more inert or passive the material selected, the less likely that plate out can occur. All material in contact with the plating solution must be repassivated periodically with 30 vol % nitric acid to minimize deposition on its surface.

The most widely used materials for tank construction have been polypropylene, stainless steel, and steel or aluminum with a 635 μm (25 mil) thick polyvinyl chloride bag liner. Although all of these materials have been used successfully, a 6 to 12 mm (0.25 to 0.5 in.) thick polypropylene liner installed in a steel or fiberglass support tank, has proven to be the most trouble-free material and has gained the widest acceptance. Polypropylene is relatively inexpensive and is very resistant to plate out. The smooth surface of polypropylene also reduces the possibility of deposit nucleation.

When constructing a polypropylene tank, only stress relieved, unfilled virgin material should be used. Welds should be made under an inert gas shield, such as nitrogen, to prevent oxidation of the polypropylene and incomplete fusion. All welds should be spark tested at 20 000 V before use to ensure integrity.

Heating the Solution

Steam and electricity are the two most common sources of power for heating plating solutions. Although the capital expenditures for steam or pressurized hot water are somewhat higher than that for electricity, the operating costs for steam are considerably less.

Steam. Heating with steam is accomplished using immersion coils or external heat exchangers. The most common immersion coils are those made of Teflon or stainless steel.

Teflon heat exchanger coils are made of many small diameter Teflon tubes looped into the tank between manifolds. Because of the poor conductivity of the plastic, a much larger coil surface area must be used than would be needed with a metal heater. Teflon tubes are delicate, and the tubes must be protected from mechanical damage.

Stainless steel panel coils are constructed of plates joined together with internal passages for the flow of heating medium. These coils are very efficient and economical. Their primary disadvantage is that they are easily galvanically activated and are prone to plate out. To prevent this, coils are often coated with Teflon. This, however, reduces their heat transfer and their efficiency.

Anodic passivation is also sometimes used to prevent stainless steel coils from plating. With this technique, a slight positive charge is applied to the

Fig. 20 Electric immersion heater

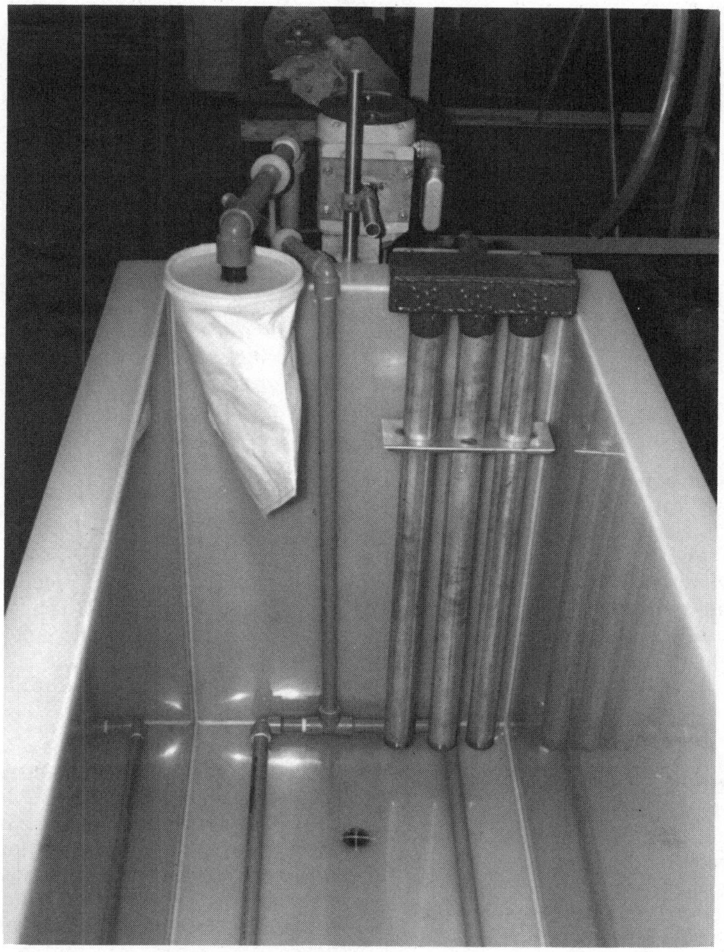

coil, preventing the deposition of electroless nickel. If the work is suspended too close to an anodically passivated coil, however, stray currents from the coil may affect the quality of the plating. Static electricity discharges from steam coils to the work can also cause nonuniform or pitted coatings. To avoid this, coils should be isolated from the steam piping with dielectric couplings.

Steam can also be used to heat the plating solution through a heat exchanger, which is mounted outside the tank. The heat exchangers are usually of shell and tube or plate coil design and are constructed of stainless steel. The solution is pumped through exchangers and returned to the tank, often through a filter. To prevent the inside of the exchanger from plating, the solution velocity must be maintained above 2½ m/s (8 ft/s).

Electric. Heating with electricity is usually accomplished with tube immersion heaters. The resistance heating elements are sheathed in quartz, titanium, or stainless steel. Stainless steel is the most economical material and is usually preferred. Either type 304 or 316 stainless steel is acceptable. Occasionally electropolished stainless steel or Teflon-coated heaters are also used. The cost of these additions, however, cannot be justified for most applications. An electric immersion heater is shown in Fig. 20.

Pumps

Pumps are used in electroless nickel plating systems for solution transfer and filtration. The following factors should be considered when selecting pumps for electroless nickel plating systems:

- Operating temperature of the plating solution, usually 85 to 95 °C (185 to 205 °F)
- Chemicals being handled in both the electroless nickel plating solution and the 30% nitric acid solution used for passivation
- Volume flow rate (litres per minute) required to allow the total tank volume to be filtered approximately ten times each hour

Two materials, CPVC plastic and type 304 stainless steel, have been proven to be satisfactory for electroless nickel pumps. CPVC plastic is more resistant to plate out than stainless steel and is less expensive. However, no plastic pump with a power rating over 1 hp is presently built which is adequate for this service. Larger plastic pumps lack the capacity and mechanical strength needed to provide proper filtration in electroless nickel systems. Accordingly, plastic pumps are used for flow rates less than 300 L/min (80 gal/min), whereas stainless steel is used for higher flow applications.

Vertical Pumps. Vertical centrifugal pumps are now the most commonly used pumps for electroless nickel systems. These pumps can be mounted so only the impeller is below the solution level and shaft seals are not required. Consequently, maintenance of this pump is minimized. Some vertical pumps can also be mounted outside the tank, providing the maximum area for plating.

With CPVC plastic pumps, the impeller should be machined or molded; glued impellers should not be used. All gaskets and O-rings for electroless nickel systems should be fluorocarbon rubber.

The velocity of the solution through the pump should be at least 2½ m/s (8 ft/min) to prevent the solution from plating out on the pump housing, especially when stainless steel is used. To accomplish this, a pump speed of 1750 rev/min is required.

Piping and Valves

Piping and valves available for electroless nickel systems are of four principal types: (*a*) stainless steel; (*b*) polyvinylidene fluoride; (*c*) CPVC plastic; and (*d*) polypropylene. The advantages and disadvantages of each of these materials are summarized in Table 9.

Piping components in electroless nickel plating systems are used for air agitation spiders, tank outlet, pump inlet, and discharge pipes, solution man-

Table 9 Comparison of piping and valve materials for electroless nickel plating systems

Material	Resistance to plating temperatures	Resistance to plate out	Relative cost	Availability
Piping				
Stainless steel	High	Low	High	Good
Kynar	High	High	Moderate	Poor
CPVC	Moderate	Moderate	Low	Good
Polypropylene	Low	High	Low	Limited
Valves				
Stainless steel	High	Low	Moderate	Good
CPVC	Moderate	Moderate	Moderate	Good
Polypropylene	Moderate	High	Moderate	Good

ifolds, and deionized water fill lines. These pipes must be sized to minimize restrictions and provide proper agitation and filtration. The diameter of the tank outlet piping should be at least as large as the pump inlet connection to avoid cavitation and increased pump wear. CPVC plastic is normally used for pipe exposed to the plating solution.

Although CPVC or other plastic pipe may be joined by solvent welding, threaded joints are preferred. Threaded connections are easier to make and more trouble-free, allowing repairs or modifications to be accomplished quickly. When threading plastic pipe, a plug should be inserted inside the pipe end to support the pipe and prevent collapse or thread breakage. Threads should be wrapped with Teflon tape before joining to prevent potential leakage from the galling of the plastic.

Valves. Almost all of the valves used for electroless nickel systems are a ball and seat design. Because of prolonged exposure to stagnant plating solutions, inertness or resistance to deposit plate out is of primary importance with these valves. Accordingly, polypropylene is used most often. The reduced strength of polypropylene at plating temperatures is not a problem with valves, because of their compactness and greater thickness.

CPVC plastic valves are also used occasionally for electroless nickel systems, although their reduced resistance to deposit plate out makes them more prone to seizure and failure due to deposit buildup than polypropylene. Because of their somewhat higher cost and tendency to activation and deposition, stainless steel valves are not normally used. For valves in agitation air supply lines, plain PVC plastic valves may be used if they are mounted at least 200 mm (8 in.) away from hot plating solution. Valves and piping for steam services should be steel or stainless steel.

Agitation

Agitation of parts and solution is necessary during electroless nickel plating to provide a fresh supply of solution to the part and to remove the hydrogen produced during deposition. Without consistent renewal of plating solution, localized depleted areas can occur, resulting in nonuniform coating thickness. Hydrogen bubbles, if allowed to remain on the surface of the part, tend to mask plating and can cause pitting or fisheyes in the coating.

Agitation is accomplished by moving the part mechanically through the solution or by bubbling air through the bath to move the solution past the part. Both methods provide satisfactory results. Air agitation, however, usually requires a lower investment and is easier to install and maintain. Accordingly, it is used more often. A typical air agitation spider is shown in Fig. 20. For air agitation, a clean low pressure air source, such as is provided by centrifugal blowers, is preferred. High pressure air from compressors can introduce oil or other contaminants into the bath and affect deposit quality.

Filtration

Two types of filtration are used for electroless nickel systems, cartridge filters and filter bags. Both require the use of an external circulation pump, and both should be capable of removing particles larger than 5 μm (0.2 mil) in size. Wound cartridge filters are supported in CPVC or polypropylene chambers located outside of the tank. The installation cost of these filters is high, however, and replacement of the cartridges is a large maintenance cost. Also the added back pressure of the filter can significantly reduce the flow of the pump and often its life.

Woven polypropylene bags are now being used to filter electroless nickel solutions. These bags are mounted above the plating tank itself, allowing the solution to flow through the bag by gravity. Filter bags are relatively inexpensive and result in only a minimum restriction on the discharge of the pump. When bags become soiled or begin to plate out, the change is obvious to the operator, and the bags can be quickly and easily replaced. Filter bags with stainless steel support rings rather than plated steel rings should be used. Plated rings can introduce cadmium or zinc into the bath and slow or stop deposition. A filter assembly is shown in Fig. 20.

Racking for Electroless Nickel Plating

Because electroless nickel is applied by chemical reduction, anode to cathode area relationships and current density considerations, usually of concern in electrolytical applications, are usually not important. This simplifies rack design.

Construction Materials. Racks for plating ferrous and copper alloys should be capable of carrying 3 to 6 A/dm^2 (30 to 60 A/ft^2) of part surface during electrocleaning and striking without overheating or excessive voltage loss. Suitable materials for racks include steel, stainless steel, copper, and titanium. Of these, steel or plastic-coated steel is most often used. Stainless steel and titanium can be cleaned easily in the nitric acid, but are rarely used because of high cost and limited current carrying capability. The cost of copper racks is reasonable and current capacity is excellent. With copper, however, all submersed surface, except the contact points, should be coated to avoid copper contamination of the cleaning and plating solutions and to minimize stripping of the coating from the frame.

Because electrolytic steps are not required when processing aluminum alloys, plastics as well as metals can be used to support parts. The materials used for racks for aluminum alloys include polypropylene, CPVC, aluminum, and stainless steel. Polypropylene and CPVC are especially useful, because they are easily constructed, inexpensive, and highly resistant to plating. Iron, nickel, or copper alloys are not suitable, because they are rapidly attacked by the oxidizing and desmutting solutions used for aluminum alloys.

Coatings for racks and fixtures used in electroless nickel plating have only limited life. The high temperatures and harsh chemicals used during pretreatment and stripping can cause rapid degradation of vinyls, epoxies, and phenolics. Coatings, however, do reduce current requirements during cleaning and striking operations and can reduce unwanted deposition on the racks.

Fixturing. When fixturing and positioning a part, the following factors should be considered:

- *Hydrogen evolution*: during the deposition of electroless nickel, hydrogen gas is evolved at the surface of the part. As the hydrogen bubble grows and rises, it should be able to free itself from the part. If hydrogen becomes trapped in any area of the part, such as an inverted hole, it masks the surface and can reduce or prevent plating
- *Electrical contact*: good contact is needed between the support and the part to ensure adequate and uniform current for electrocleaning and striking. Proximity to anodes is not usually very important with these operations, although in extreme cases, such as deep holes, internal anodes may be required
- *Rinsing*: easy rinsing is necessary to minimize dragout of the pretreatment cleaners and to prevent dragin of contaminants to the electroless nickel bath

A rack should be designed to allow blind holes to drain easily or to allow holes to be rinsed thoroughly with a hose. Some racks are designed to be tipped or turned upside down to ensure rinsing and to control dragout. During plating, these holes must be positioned vertically to allow hydrogen gas to escape.

Bulk and Barrel Plating

The uniform plating thickness of electroless nickel coatings allows many parts that would have to be racked if they were finished electrolytically to be bulk plated. Because of the resulting labor savings, coatings such as chromium can sometimes be replaced with electroless nickel at a lower over-all finished cost, although the chemical cost is higher. Four principal types of bulk plating are used:

1 *Soldier style racking*: parts are placed so close together that complete coverage would be difficult, if not impossible, with an electrolytic process
2 *Baskets*: many bulk plating jobs can be run efficiently in baskets made of polypropylene or stainless steel, especially in smaller electroless nickel tanks. Baskets occupy much less space than barrels and allow more loads to be run. When compared to using barrels, baskets have the disadvantage of not mechanically agitating parts during plating. Accordingly, baskets should be shaken and moved periodically to allow fresh plating solution to circulate around parts
3 *Trays*: many jobs, such as small shafts and bars, can be run most easily using egg crate or test tube rack trays. In addition, many parts, because of their finish or design, must be separated during processing to keep them from touching or nesting. Separated trays accomplish this successfully and allow good solution transfer, minimizing the labor required for fixturing. Trays are most often constructed of polypropylene, steel, or stainless steel
4 *Barrels*: where very large volumes of parts are to be plated or continuous mechanical agitation is necessary, barrels usually provide the most efficient and economical method of processing

Barrels for electroless nickel plating should be made from natural polypropylene. If added strength is required, glass-filled polypropylene construction is preferred. Polypropylene gears, rather than a belt drive, should be used to turn the barrel. Plastisol-coated steel barrels are not successful for electroless nickel plating, because they are prone to coating failures, plate out, and occasional drive failures. For electroless nickel plating, the barrel speed should be 1 to 2 rev/min. Higher speed barrels may be required, however, where the solution must be pumped through internal passages or holes in a part. The drive mechanism should allow the barrel to rotate, both in the processing tanks and in transfer stages, to ensure free rinsing and minimize dragout. To allow adequate solution transfer in and out of the barrel, the hole size should be as large as possible and should be just capable of containing parts.

Table 10 Heat treatment of steels to relieve hydrogen embrittlement

Maximum specified tensile strength MPa	ksi	Heat treatment at 190 to 210 °C (375 to 410 °F), h
Up to 1050	Up to 152	Not required
1051-1450	152-210	2
1451-1800	210-260	18
Over 1800	Over 260	24

All racks, baskets, trays, and barrels used for electroless nickel plating should be used exclusively for this operation. The use of equipment from other plating systems can result in contamination of the electroless nickel plating solution, in decomposition, or in reduced deposit quality.

Solution Control

To ensure a quality deposit and consistent plating rate, the composition of the plating solution must be kept relatively constant. This requires periodic analyses for the determination of pH, nickel content, and hypophosphite and orthophosphite concentrations, as well as careful temperature control. With modern premixed solutions, only checks of nickel content and pH are required. The frequency with which these analyses should be made depends on the quantity of work being plated and the volume and type of solution being used.

Hydrogen Embrittlement Relief

Hydrogen embrittlement is the failure that results from the absorption of hydrogen into metals. Hydrogen embrittlement usually occurs in combination with residual or applied stresses in a part, happening most frequently in high-strength steels and occasionally in other high-strength alloys.

Hydrogen can be introduced into a metal by processes such as pickling, electrocleaning, acid activation, electroplating, or electroless deposition. Although the hydrogen produced by electroless nickel plating is much less than that produced by an electrolytic process, such as cadmium or hard chrome plating, it can be enough to cause cracking of high-strength steels. To prevent this, components are baked at 200 °C ± 10 °C (390 °F ± 18 °F) to diffuse the absorbed hydrogen out of the steel. This usually restores the me-

Table 11 Applications of electroless nickel plating

Application	Basis metal	Coating thickness(a) µm	Coating thickness(a) mils	Reason for use
Automotive				
Heat sinks	Aluminum	10	0.4	Corrosion resistance, solderability, uniformity
Carburetor components	Steel	15	0.6	Corrosion resistance
Fuel injectors	Steel	25	1.0	Corrosion and wear resistance
Ball studs	Steel	25(b)	1.0(b)	Wear resistance
Differential pinion ball shafts	Steel	25(b)	1.0(b)	Wear resistance
Disc brake pistons and pad holders	Steel	25(b)	1.0(b)	Wear resistance
Transmission thrust washers	Steel	25(b)	1.0(b)	Wear resistance
Syncromesh gears	Brass	30	1.2	Wear resistance
Knuckle pins	Steel	38(b)	1.0(b)	Wear resistance
Exhaust manifolds and pipes and mufflers	Steel	25	1.0	Corrosion resistance
Shock absorbers	Steel	10	0.4	Corrosion resistance and lubricity
Lock components	Steel	10	0.4	Wear and corrosion resistance and lubricity
Hose couplings	Steel	5	0.2	Wear and corrosion resistance
Gears and gear assemblies	Carburized steel	25(c)	1.0(c)	Buildup of worn surfaces and wear resistance
Aircraft				
Bearing journals	Aluminum	38(d)	1.5(d)	Wear resistance and uniformity
Servo valves	Steel	18	0.7	Corrosion resistance, uniformity and lubricity
Compressor blades	Alloy steel	25(e)	1.0(e)	
Hot zone hardware	Alloy steel	25	1.0	Corrosion and wear resistance
Piston heads	Aluminum	25	1.0	Wear resistance
Engine main shafts and propellers	Steel	>38	>1.5	Buildup of worn surfaces and wear resistance
Hydraulic actuator splines	Steel	25(b)	1.0(b)	Wear resistance
Seal snaps and spaces	Steel	20(e)	0.8(e)	Wear and corrosion resistance
Landing gear components	Aluminum	>125	>5.0	Buildup of mis-machined surfaces
Struts	Stainless steel	>25	>1.0	Buildup of mis-machined or worn surfaces
Pitot tubes	Brass/stainless steel	12	0.5	Corrosion and wear resistance
Gyro parts	Steel	12	0.5	Wear resistance and lubricity
Engine mounts	4140 Steel	25	1.0	Wear and corrosion resistance
Oil nozzle components	Steel	25	1.0	Corrosion resistance and uniformity
Printing				
Printing rolls	Steel/cast iron	38	1.5	Corrosion and wear resistance
Press bed	Steel/cast iron	38	1.5	Corrosion and wear resistance
Textiles				
Feeds and guides	Steel	50(b)	2.0(b)	Wear resistance
Fabric knives	Steel	12(b)	0.5(b)	Wear resistance
Spinnerettes	Stainless steel	25	1.0	Corrosion and wear resistance
Loom ratchets	Aluminum	25	1.0	Wear resistance
Knitting needles	Steel	12	0.5	Wear resistance
Molds and dies				
Zinc die cast dies	Alloy steel	25	1.0	Wear resistance and part release
Glass molds	Steel	50	2.0	Wear resistance and part release
Plastic injection molds	Alloy steel	15	0.6	Corrosion and wear resistance and part release
Plastic extrusion die	Alloy steel	25	1.0	Corrosion and wear resistance and part release
Military				
Fuse assemblies	Steel	12	0.5	Corrosion resistance
Mortar detonators	Steel	10	0.4	Corrosion resistance
Tank turret bearings	Alloy steel	30	1.2	Wear and corrosion resistance
Radar wave guides	Aluminum	25	1.0	Corrosion resistance and uniformity
Mirrows	Aluminum/beryllium	>75	>3.0	Uniformity and reflectivity
Firearms				
Commercial and military	Steel	8	0.3	Corrosion and wear resistance and lubricity
Marine				
Marine hardware	Brass	25	1.0	Corrosion resistance
Pumps and equipment	Steel/cast iron	50	2.0	Corrosion and wear resistance
Electronics				
Heat sinks	Aluminum	10	0.4	Corrosion resistance and solderability
Computer drive mechanisms	Aluminum	18	0.7	Corrosion and wear resistance
Memory drums and discs	Aluminum	25	1.0	Corrosion and wear resistance and uniformity
Terminals and lead wires	Alloy steel	2	0.1	Solderability
Chassis	Aluminum/Steel	12	0.5	Corrosion resistance and solderability
Connectors	Steel/Aluminum	25	1.0	Corrosion and wear resistance and solderability
Diode and transistor cans	Steel	5	0.2	Corrosion resistance and solderability
Interlocks	Steel/brass	12	0.5	Corrosion and wear resistance
Junction fittings	Aluminum/plastic	10	0.4	Corrosion and wear resistance, solderability and conductivity
Printed circuit boards	Plastic	5	0.2	Solderability and weldability
Railroad				
Tank cars	Steel	90(f)	3.5(f)	Corrosion resistance

(continued)

Table 11 (continued)

Application	Basis metal	Coating thickness(a) µm	Coating thickness(a) mils	Reason for use
Diesel engine shafts	Steel	>25	>1.0	Wear and fretting resistance and buildup of worn surfaces
Car hardware	Powder iron	20	0.8	Corrosion and wear resistance
Electrical				
Motor shafts	Steel	12	0.5	Wear and corrosion resistance
Rotor blades	Steel/aluminum	25(b)	1.0(b)	Wear and corrosion resistance
Stator rings	Steel/aluminum	25	1.0	Wear and corrosion resistance
Chemical and petroleum				
Pressure vessels	Steel	50	2.0	Corrosion resistance
Reactors	Steel	100(f)	4.0(f)	Corrosion resistance and product purity
Mixer shafts	Steel	38	1.5	Corrosion resistance
Pumps and impellers	Cast iron/steel	75	3.0	Corrosion and erosion resistance
Heat exchangers	Steel	75	3.0	Corrosion resistance
Filters and components	Steel	25	1.0	Corrosion and erosion resistance
Turbine blades and rotor assemblies	Steel	75	3.0	Corrosion and erosion resistance
Compressor blades and impellers	Steel/aluminum	125(d)	5.0(d)	Corrosion and erosion resistance
Spray nozzles	Brass/steel	12	0.5	Corrosion and wear resistance
Ball, gate, plug, check and butterfly valves	Steel	75	3.0	Corrosion resistance and lubricity
Valves	Stainless steel	25(b)	1.0(b)	Wear and galling resistance and protection against stress corrosion cracking
Chokes and control valves	Steel/stainless steel	75	3.0	Corrosion and wear resistance and protection against stress corrosion cracking
Oil field tools	Steel	75	3.0	Corrosion and wear resistance
Oil well packers and equipment	Alloy steel	75	3.0	Corrosion and erosion resistance
Oil well tubing and pumps	Steel	50	2.0	Corrosion and wear resistance
Drilling mud pumps	Alloy steel	75	3.0	Corrosion resistance and protection against stress corrosion cracking
Hydraulic systems and actuators	Steel	75	3.0	Corrosion and wear resistance and lubricity
Blowout preventers	Alloy steel	75	3.0	Corrosion and wear resistance
Medical and pharmaceutical(g)				
Disposable surgical instruments and equipment	Steel/aluminum	12	0.5	Corrosion resistance and ease of operation
Sizing screens	Steel	20	0.8	Corrosion resistance and cleanliness
Pill sorters	Steel	20	0.8	Corrosion resistance and cleanliness
Feed screws and extruders	Steel	25	1.0	Corrosion and wear resistance and cleanliness
Food(g)				
Pneumatic canning machinery	Steel	25	1.0	Corrosion and wear resistance and cleanliness
Baking pans	Steel	25	1.0	High temperature resistance, cleanliness, and ease of release
Molds	Steel	12	0.5	Cleanliness, corrosion resistance and ease of release
Grills and fryers	Steel	12	0.5	Cleanliness, corrosion resistance and ease of release
Mixing bowls	Steel	25	1.0	Cleanliness and corrosion and wear resistance
Bun warmers	Steel	12	0.5	Cleanliness and ease of release
Feed screws and extruders	Steel	25	1.0	Cleanliness and corrosion and wear resistance
Material handling				
Hydraulic cylinders and shafts	Steel	25	1.0	Corrosion and wear resistance and lubricity
Extruders	Alloy steel	75(b)	3.0(b)	Wear and corrosion resistance
Link drive belts	Steel	12	0.5	Wear and corrosion resistance and lubricity
Gears and clutches	Steel	>25	>1.0	Wear resistance and buildup of worn surfaces
Mining				
Hydraulic systems	Steel	60	2.4	Corrosion and abrasion resistance
Jetting pump heads	Steel	60	2.4	Corrosion and erosion resistance
Mine engine components	Steel/cast iron	30	1.2	Corrosion and wear resistance
Piping connections	Steel	60	2.4	Corrosion resistance
Framing hardware	Steel	30	1.2	Corrosion resistance
Wood and paper				
Knife holder corer plates	Steel	30	1.2	Corrosion and abrasion resistance
Abrading plates	Steel	30	1.2	Corrosion and abrasion resistance
Chopping machine parts	Steel	30	1.2	Corrosion and abrasion resistance
Miscellaneous				
Chain saw engines	Aluminum	25	1.0	Wear and corrosion resistance
Drill and taps	Alloy steel	12(b)	0.5(b)	Wear resistance and ease of use
Precision tools	Alloy steel	12	0.5	Wear resistance and cleanliness
Shaver blades and heads	Steel	8	0.3	Wear resistance and smoothness
Pen tips	Brass	5	0.2	Corrosion resistance

(a) Many components are heat treated at 190 to 210 °C (375 to 410 °F) for 1 to 3 h to improve adhesion or to relieve hydrogen embrittlement. (b) Heat treated for 1 h at 400 °C (750 °F) for maximum hardness. (c) Heat treated for 6 h at 135 °C (275 °F) for hydrogen embrittlement relief. (d) Heat treated for 10 h at 290 °C (550 °F) for maximum hardness. (e) Cadmium plated after electroless nickel and then heat treated for 2 h at 340 °C (640 °F) to diffuse cadmium into the nickel. (f) Heat treated for 1 h at 620 °C (1150 °F) to diffuse coating into basis metal. (g) For medical, pharmaceutical, and food applications, coatings must be free of toxic heavy metals like lead, cadmium, mercury, or thallium

chanical properties of the steel almost completely, helping to ensure against failure.

The time required to remove hydrogen from a steel and avoid embrittlement depends on the strength of the steel. Longer relief treatment periods or higher temperatures are needed as the strength of a steel increases. Recommendations for embrittlement relief of steels on different strength levels are summarized in Table 10. Hydrogen embrittlement relief treatment should begin within 4 h of the completion of electroless nickel plating (Ref 2, 44, 45).

Applications

Electroless nickel is applied for five different applications: (*a*) corrosion resistance, (*b*) wear resistance, (*c*) lubricity, (*d*) solderability, or (*e*) buildup of worn or over-machined surfaces. To varying degrees, these properties are used by all segments of industry, either separately or in combination. Applications of these coatings are given in Table 11.

Specifications

The three published specifications for electroless nickel-phosphorus currently available in the United States include:

1 AMS 2404B, Electroless Nickel Plating (Ref 46)
2 ASTM B656, Autocatalytic Nickel Deposition on Metals for Engineering Use (Ref 39)
3 MIL-C-26074B, Military Specification Requirements for Electroless Nickel Coatings (Ref 47)

In addition, an international standard is now being drafted by the International Standards Organization (Ref 45). Published standards for electroless nickel-boron coatings for engineering purposes are not available.

Although these standards are good guidelines for testing and quality control, none include any real requirements for structural quality, corrosion resistance, or wear resistance. The standards consist primarily of a visual examination and simple tests for thickness and adhesion. Often this forces industrial users to develop their own internal specifications for coating quality. These inhouse specifications can be relatively simple with requirements for only a few desired properties, or very detailed with requirements for substrate pretreatment, bath operation, equipment design, deposit chemistry, and properties.

Electroless Nickel Composite Coatings

Composites are one of the most recently developed types of electroless nickel coatings. These cermet deposits consist of small particles of intermetallic compounds, fluorocarbons, or diamonds dispersed in an electroless nickel-phosphorus matrix. These coatings have a high apparent hardness and superior wear and abrasion resistance.

Chemistry. Most composite coatings are applied from proprietary baths. Typically, they consist of 20 to 30 vol% of particles entrapped in an electroless nickel containing 7 to 11% phosphorus. Most commonly silicon carbide and diamond particles are used, although PTFE Teflon and calcium fluoride are also occasionally codeposited. The particles are carefully sized and are normally 1 to 3 μm in diameter (Ref 48-50). A micrograph of a typical silicon carbide composite coating is shown in Fig. 21 (Ref 51). The baths used for composite plating are conventional sodium hypophosphite reduced electroless nickel solutions, with the desired particles suspended in them. These baths, however, are heavily stabilized to overcome or inhibit the very high surface area produced by the particles. The baths otherwise are operated normally and the nickel-phosphorus matrix is produced by the traditional hypophosphite reduction of nickel. The particles are merely caught or trapped in the coating as it forms. Their bond to the coating is purely mechanical.

Hardness and Wear. The primary use for electroless nickel composite coatings is for applications requiring maximum resistance to wear and abrasion. The hardness of diamond and silicon carbide is 10 000 and 4500 HV, respectively. In addition, the coatings are normally heat treated to provide maximum hardness (1000 to 1100 HV_{100}) of the electroless nickel matrix. The resulting apparent surface hardness of the composite is 1300 HV_{100} or more (Ref 48, 51).

The wear surface of a composite coating consists of very hard mounds separated by lower areas of hard electroless nickel. During wear, the mating surface usually rides on the particles and slides over the matrix. Thus, the wear characteristics of these coatings approach that of the particle material (Ref 48). Typical wear test results for a silicon carbide composite coating are shown in Table 12 (Ref 51).

Frictional properties of composite coatings are similar to those of other electroless nickels. Typically, the coefficient of friction of these materials is about 0.13 in the lubricated condition and 0.3 to 0.4 in the unlubricated condition (Ref 48, 49).

Corrosion Resistance. In general, the corrosion resistance of composite coatings is significantly less than that of other electroless nickel coatings. The electroless nickel matrix contains large amounts of codeposited inhibitor, which reduces the alloy's passivity and corrosion resistance. Also, heat treated coatings are less protective than are as-applied coatings, both because of the conversion of the amorphous deposit to crystalline nickel and Ni_3P and because of cracking of the coating (Ref 48, 51). With composites, this problem is amplified because of the presence of the diamond or intermetallic particles. The mixture of phosphides, nickel, and particles creates a very strong galvanic couple accelerating attack. For applications requiring good corrosion resistance, electroless nickel composite coatings are not normally used.

Plating on Plastics

Except for ferrous alloys, plastics are probably the substrate most commonly electroless nickel plated. The coating is typically applied to nonmetallics as a conductive base for subsequent electroplating of both decorative and functional deposits. Occasionally, electroless nickel is used by itself for applications requiring resistance to abrasion or environmental attack (Ref 2). Because plastics are nonconductive and are not catalytic to the chemical reduction of nickel, special processing steps are required to ensure adequate adhesion and to initiate deposition. With synthetics, metallic bonds cannot form between the coating and the substrate. Thus, adhesion results only from mechanical bonding of the coating to the substrate surface. To improve adhesion plastics are typically etched in acidic solutions or organic solvents to roughen their surface and to provide more bonding sites.

In order to initiate electroless nickel plating on plastics (or other nonmetals) their etched surface must first be catalyzed with stannous chloride and pal-

Fig. 21 Cross-sectional view of a typical silicon carbide composite coating. 1000X

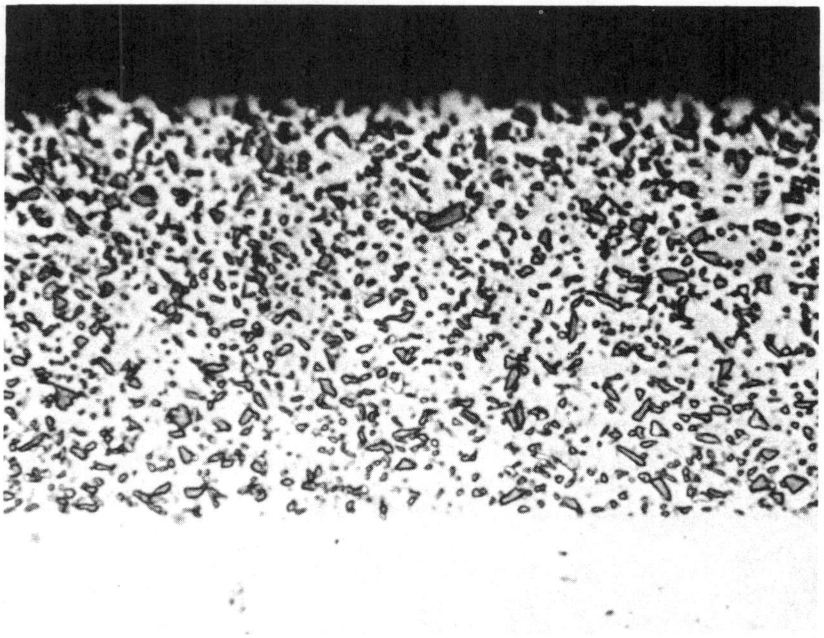

Heater mounted in a 200-L (50-gal) electroless nickel plating tank. A bag filter is mounted on the filtration pump discharge

ladium chloride and then accelerated in an acid. This produces palladium nucleation sites on the surface for deposition. A typical pretreatment sequence for plastics is:

- Degreasing
- Etching
- Neutralization
- Catalyzation
- Acceleration
- Electroless nickel deposition

Thorough rinsing after each processing step is essential. After the electroless nickel layer has been completed, the part may be plated conventionally with any desired electrolytic coating (Ref 2, 52).

Degreasing. When necessary, light soils or fingerprints can be removed from plastic parts by immersion in a mildly alkaline soak cleaner for 2 to 5 min. A typical degreasing solution contains 25 g/L each of sodium carbonate and trisodium phosphate and is operated at 50 to 70 °C (120 to 160 °F). Alkaline cleaning is not always required, provided the plastic is carefully handled after molding and is not allowed to become excessively soiled. Fingerprints and loose dust or dirt are normally removed by the etching solution.

Etching solutions for plastics are typically strongly oxidizing acids which cause a microscopic roughening of the part's surface. These solutions also alter the chemical character of the surface and cause it to become hydrophylic. Etching not only improves mechanical bonding and adhesion of the coating to the plastic substrate, but also improves access of subsequent processing solutions to the surface. Most commercially used etching solutions are formulated with either chromic acid or mixtures of sulfuric acid and chromic acid or dichromate salt. These solutions are typically operated at 50 to 70 °C (120 to 160 °F) with immersion times of 3 to 10 min. Chromic acid based solutions are particularly effective with ABS plastics, but are also used for polyethylene, polypropylene, PVC, polyesters, and other common polymers.

Table 12 Comparison of the Taber abraser resistance of silicon carbide composite coatings with other engineering materials

Material	Hardness	Taber wear index, Mg 11 000 cycles
400-C stainless steel	57 HRC	5.6
A-2 tool steel	60-62 HRC	5.0
Electroless nickel (hardened)	900-1000 HV	3.7
Hard chromium	1000-1100 HV	3.0
Tungsten carbide	1300 HV	2.0
Electroless nickel Silicon carbide composite	1300 HV	0.18-0.22

Note: Taber wear index determined for an average of three 5000-cycle runs with 1000-g load and CS17 abrasive test wheels

Neutralizing. After the plastic has been properly etched and rinsed, it should be neutralized to remove residual chromium ions, which may interfere with subsequent catalyzation. Neutralizers are rinsing aids and are typically dilute acid or alkaline solutions, often containing complexing and reducing agents. Ionic surfactants are sometimes also added to increase the absorption of the catalyst on the surface. Neutralizing solutions are normally operated at 40 °C (105 °F) with immersion times of 1 to 2 min.

Catalyzing. In order to initiate deposition of the electroless nickel coating on plastics their surface must be catalyzed. This is normally accomplished by chemically depositing small amounts of palladium. The original commercial catalyzing procedures required two processing steps. In the first step, stannous chloride was absorbed onto the surface from a solution of $SnCl_2$ and HCl. After rinsing, the part was immersed in a solution of $PdCl_2$ and HCl, and palladium chloride was absorbed onto the surface. The stannous ions then reduced the palladious ions leaving discrete sites of metallic palladium. Currently, a one-step catalyzing procedure is normally used. For this, a solution of stannous chloride and palladium chloride in hydrochloric acid is used. The solution consists of tin/palladium complexes and colloids stabilized by excess stannous chloride. The chloride content of the solution is critical and must be carefully controlled. During immersion, globules of tin/palladium colloid absorb onto the plastic surface. After rinsing, nuclei of metallic palladium, surrounded by hydrolyzed stannous hydroxide, are left attached to the surface.

Acceleration. With one-step catalyzation, a further step is required to remove excess stannous hydroxide from the surface and to expose the palladium nuclei. This step is called acceleration and is accomplished by immersing the part in a dilute solution of hydrochloric acid or an acid salt. The acid reacts with the insoluble stannous hydroxide forming soluble stannous and stannic chloride. After rinsing, the surface is free of tin and active catalytic sites are present. Acceleration solutions are typically operated at a temperature of 50 °C (120 °F) and are agitated with air. The parts are normally immersed for 30 to 60 s.

Electroless Nickel Deposition. Most electroless nickel solutions operate at too high a temperature for plastics. High temperatures may cause plastics to warp. In addition, the large difference in coefficient of thermal expansion between plastics and electroless nickel may cause adhesion failures during cooling from bath temperatures. Electroless nickel solutions for plating on plastics, thus, are formulated to operate at low temperatures—typically 20 to 50 °C (70 to 120 °F). These solutions are normally alkaline and reduced with sodium hypophosphite, although some DMAB solutions are also used. Ammonia based plating baths are preferred because of their ability to complex excess palladium dragged in with the part and to avoid spontaneous decomposition. While most of these solutions are proprietary, some typical formulations (Ref 2) are:

	Bath 1 g/L	Bath 1 oz/gal	Bath 2 g/L	Bath 2 oz/gal
Composition				
Nickel chloride	119	15	...	...
Nickel sulfate	...	...	50	6.5
Sodium hypophosphite	106	14	50	6.5
Sodium pyrophosphate	...	...	100	13
Ammoniun citrate	65	8	...	...
Ammonia, mL/L (fluid oz/gal)	...	...	45	5.8
Sodium hydroxide	to pH		...	...

	Bath 1	Bath 2
Operating conditions		
pH	10	10
Temperature, °C (°F)	30-50 (85-120)	25
Typical plating rate, μm/hr (mils/h)	3-11 (0.12-0.44)	3 (0.12)

Plastic parts are normally immersed in the electroless nickel solution for 5 to 10 min to provide a uniform metal film about 0.25 to 0.50 μm thick. This coating is sufficient to cover the surface of the plastic and to make them conductive for subsequent electroplating. These deposits typically contain 2 to 6% phosphorus. After proper pretreatment the peel strength of 25 mm (1 in.) width strips of these coatings on plastics like ABS and polypropylene is on the order of 50 to 100 N (Ref 2, 52).

REFERENCES

1. K. Parker, Recent Advances in Electroless Nickel Deposits, 8th Interfinish Conference, Basel, 1972
2. G.G. Gawrilov, *Chemical (Electroless) Nickel Plating,* Redhill, England: Portcullis Press, 1979
3. G. Gutzeit, An Outline of the Chemistry Involved in the Process of Catalytic Nickel Deposition from Aqueous Solution, *Plating and Surface Finishing,* Vol 46 (No. 10), 1959, p 1158
4. A. Brenner and G. Riddell, Deposition of Nickel and Cobalt by Chemical Reduction, *Journal of Research of the National Bureau of Standards,* Vol 39 (No. 11), 1947, p 385
5. G.O. Mallory, The Electroless Nickel Plating Bath, Electroless Nickel Conference, Cincinnati, Nov 1979
6. K. Stallman and H. Speckhardt, Deposition and Properties of Nickel-Boron Coatings, *Metalloberflaeche Angewandte Elektrochemie,* Vol 35 (No. 10), 1981, p 979
7. K.M. Gorbunova and A.A. Nikiforova, *Physicochemical Principles of Nickel Plating,* Moscow: Izdatel'stvo Akademii Nauk SSSR, 1960
8. G.O. Mallory, Influence of the Electroless Plating Bath on the Corrosion Resistance of the Deposits, *Plating,* Vol 61 (No. 11), 1974, p 1005
9. C.E. deMinjer and A. Brenner, Studies on Electroless Nickel Plating, *Plating,* Vol 44 (No. 12), 1957, p 1297
10. G.O. Mallory, The Electroless Nickel-Boron Plating Bath; Effects of Variables on Deposit Properties, *Plating,* Vol 58 (No. 4), 1971, p 319
11. C. Baldwin and T.E. Such, The Plating Rates and Physical Properties of Electroless Nickel/Phosphorus Alloy Deposits, *Transactions of the Institute of Metal Finishing,* Vol 46 (No. 2), 1968, p 73
12. R.N. Duncan, Properties and Applications of Electroless Nickel Deposits, *Finishers' Management,* Vol 26 (No. 3), 1981, p 5
13. W.H. Metzger, Characteristics of Deposits, *Symposium on Electroless Nickel Plating,* STP 265, American Society for Testing and Materials, Philadelphia, 1959
14. K. Parker, Effects of Heat Treatment on the Properties of Electroless Nickel Deposits, *Plating and Surface Finishing,* Vol 68 (No. 12), 1981, p 71
15. K. Parker and H. Shah, Residual Stresses in Electroless Nickel Plating, *Plating,* Vol 58 (No. 3), 1971, p 230
16. R.N. Duncan, Performance of Electroless Nickel Coatings in Oil Field Environments, CORROSION/82 Conference, National Association of Corrosion Engineers, Houston, March 1982
17. G.D.R. Jarrett, Electroless Nickel Plating, *Industrial Finishing (London),* Vol 18 (No. 218), 1966, p 41
18. L.G. Fitzgerald, Chemical Nickel Plating, Products *Finishing* (London), Vol 13 (No. 5), 1960, p68
19. W.H. Roberts, Coating Beryllium with Electroless Nickel, U.S. Atomic Energy Commission, Report RFP478, 1964
20. L.F. Spencer, Electroless Nickel Plating—A Review, *Metal Finishing*, Vol 72 (No. 12), 1974, p 58
21. A.H. Graham, R.W. Lindsay, and H.J. Read, The Structure and Mechanical Properties of Electroless Nickel, *Journal of The Electrochemical Society*, Vol 112 (No. 4), 1965, p 401
22. K. Nemoto, et al, The Study on Hardness of Non-Electrolytically Plated Ni-P Deposits at High Tem-

peratures and Effects Given by Heat Treatments, *Journal of the Metal Finishing Society of Japan,* Vol 16 (No. 3), p 106
23. L. Domnikov, Chromium and Electroless Nickel Deposits, Hardness at High Temperatures, *Metal Finishing,* Vol 60 (No. 1), 1962, p 67
24. K. Parker, Hardness and Wear Resistance Tests of Electroless Nickel Deposits, *Plating,* Vol 61 (No. 9), 1974, p 834
25. Industrial Nickel Plating and Coating, International Nickel Company, New York, 1976
26. J.P. Randin and H.E. Hintermann, Electroless Nickel Deposited at Controlled pH; Mechanical Properties as a Function of Phosphorus Content, *Plating,* Vol 54 (No. 5), 1967, p 523
27. D.W. Bandrand, Use of Electroless Nickel to Reduce Gold Requirements, *Plating and Surface Finishing,* Vol 68 (No. 12), 1981, p 57
28. R.N. Duncan, Corrosion Control with Electroless Nickel Coatings, Electroless Plating Symposium, American Electroplaters' Society, St. Louis, March 1982
29. H.G. Klein, et al, *Metalloberflaeche. Angewandte Elektrochemie,* Vol 25 (No. 9), 1971, Vol 26 (No. 1), 1972
30. Technical Information About Electroless of Chemical Nickel Plating by the Nibodur Method, Paul Anke KG, Essen, West Germany
31. ASTM Standard D2714, Calibration and Operation of The Alpha Model LFW-1 Friction and Wear Testing Machine, Part 24, American Society for Testing and Materials, 1978
32. H. Spahn, The Effect of Internal Stress on the Fatigue and Corrosion Fatigue Properties of Electroplated and Chemically Plated Nickel Deposits, Proceedings of the 6th International Metal Finishing Conference, *Transactions of the Institute of Metal Finishing,* Vol 42, 1964, p 364
33. E.F. Jungslager, Electroless Deposition of Nickel, *Tijdschr, Oppervlakte Tech. Metalen,* Vol 9 (No. 1), 1965, p 2
34. E.W. Turns and J.W. Browning, Properties of Electroless Nickel Coatings on High Strength Steels, *Plating,* Vol 60 (No. 5), 1973
35. Gus Reinhardt, Potential Applications of Electroless Nickel in Airline Maintenance Operations, Electroless Nickel Conference, Cincinnati, 6-7 Nov 1979
36. Sunada Izumi and Kondo, The Fatigue Strength of Electroless Nickel Plated Steel, The 18th Japan Conference on Materials Research, March 1975
37. S. Spring, *Industrial Cleaning,* Melbourne: Prism Press, 1974
38. John Kuczma, How to Operate Electroless Nickel More Efficiently, *Products Finishing (Directory),* Vol 44 (No. 12A), 1980, p 158
39. ASTM Standard B656, Autocatalytic Nickel Deposition on Metals for Engineering Use, American Society for Testing and Materials, Part 9, 1981
40. ASTM Standard B183, Preparation of Low-Carbon Steel for Electroplating, American Society for Testing and Materials, Part 9, 1981
41. ASTM Standard B242, Preparation of High-Carbon Steel for Electroplating, American Society for Testing and Materials, Part 9, 1981
42. ASTM Standard B254, Preparation of and Electroplating on Stainless Steel, American Society for Testing and Materials, Part 9, 1981
43. ASTM Standard B281, Preparation of Copper and Copper-Base Alloys for Electroplating, American Society for Testing and Materials, Part 9, 1981
44. *Metals Handbook,* 9th ed., Vol 1, American Society for Metals, 1978
45. International Standard (Draft) ISO/TC107/SC3/WG1-N3, Autocatalytic Nickel-Phosphorus Coatings, International Standards Organization
46. AMS 2404B, Electroless Nickel Plating, Society of Automotive Engineers, 1977
47. MIL-C-26074B, Military Specification — Coatings, Electroless Nickel, Requirements for, U.S. Government Printing Office, 1959 and 1971
48. J.M. Scale, Wear Resistance of Silicon Carbide Composite Coatings, *Metal Progress*, Vol 115 (No. 4), 1979, p 44
49. D.J. Kenton *et al*, Development of Dual Particle Multifunction Electroless Nickel Composite Coatings, Electroless Plating Symposium, American Electroplaters' Society, St. Louis, March 1982
50. N. Feldstein *et al*, The State of the Art in Electroless Composite Plating, Electroless Plating Symposium, American Electroplaters' Society, St. Louis, March 1982
51. W.B. Martin *et al*, Electroless Nickel Composites — The Second Generation of Chemical Plating, Electroless Nickel Conference, Cincinnati, Nov 1979
52. J.K. Dennis and T.E. Such, *Nickel and Chromium Plating*, London: Newres-Butterworths, 1972, p 287

Zinc Plating

By H. H. Geduld
Board Chairman
Columbia Chemical Corp.
and
The ASM Committee on Zinc Plating*

ZINC is anodic to iron and steel and therefore offers more protection when applied in thin films of 7 to 15 μm (0.3 to 0.5 mil) than similar thicknesses of nickel and other cathodic coatings, except in marine environments where it is surpassed by cadmium. Because it is relatively inexpensive and readily applied in barrel, tank, or continuous plating facilities, zinc is often preferred for coating iron and steel parts when protection from either atmospheric or indoor corrosion is the primary objective. Normal electroplated zinc without subsequent treatment becomes dull gray in appearance after exposure to air. Bright zinc that has been subsequently given a bleached chromate conversion coating or a coating of clear lacquer (or both) is sometimes used as a decorative finish. Such a finish, although less durable than heavy nickel chromium, in many instances offers better corrosion protection than thin coatings of nickel chromium, and at much lower cost.

Plating Baths

Commercial zinc plating is accomplished by a number of distinctively different systems: cyanide baths, alkaline noncyanide baths, and acid chloride baths. In the 1970's, most commercial zinc plating was done in conventional cyanide baths, but the passage of stringent pollution control laws throughout the world has led to the continuing development and widespread use of other processes. Today, bright acid zinc plating (acid chloride bath) is possibly the fastest growing system in the field. Approximately half of the existing baths in developed nations use this technology and most new installations specify it.

The preplate cleaning and postplate chromate treatments are quite similar for all zinc processes; however, the baths themselves are radically different. Each separate system is reviewed in detail in this article, giving its composition and the advantages and disadvantages.

Cyanide Zinc Baths

Bright cyanide zinc baths may be divided into four broad classifications based on their cyanide content: (*a*) regular cyanide zinc baths, (*b*) midcyanide or half-strength cyanide baths, (*c*) low-cyanide baths, and (*d*) microcyanide zinc baths. Table 1 gives the general composition and operating conditions for these various systems.

Most cyanide baths are prepared from zinc cyanide, sodium cyanide, and sodium hydroxide, or from proprietary concentrates. Sodium polysulfide or tetrasulfide, commonly marketed as zinc purifier, is normally required in the standard, midcyanide, and occasionally in the low-cyanide baths, to precipitate heavy metals such as lead and cadmium which enter the baths as an anode impurity.

Standard cyanide zinc baths have a number of advantages. Standard cyanide baths have been the mainstay of the bright zinc plating industry since the early 1940's. A vast amount of information regarding standard cyanide bath technology is available, including information on technology of operation, bath treatments, and troubleshooting.

The standard cyanide bath provides excellent throwing and covering power. The ability of the standard cyanide zinc bath to cover at very low current densities is greater than that of any other zinc plating system. This capability depends on the bath composition, temperature, base metal, and proprietary additives used, but it is generally superior to the acid chloride systems. This advantage may be critical in plating complex shapes.

Despite its complex chemistry, the cyanide zinc bath can be easily controlled by a simple 5-min analysis for three basic constituents: zinc metal, total sodium cyanide, and sodium hydroxide.

Cyanide zinc formulas are highly flexible and a wide variety of bath compositions can be prepared to meet diverse plating requirements. Zinc cyanide systems are highly alkaline and pose no corrosive problems to equipment. Steel tanks and anode baskets can be used for the bath, substantially reducing initial plant investment.

*Juan Hajdu, Vice President, Technology, Enthone Inc.; Vincent Paneccasio, Supervisor, Organic Research, Enthone Inc.; and Roger Marce, Manager, Technical Services, Zinc Institute Inc.

Table 1 Composition and operating conditions of cyanide zinc baths

Cathode current density: limiting 0.002 to 25 A/dm² (0.02 to 250 A/ft²); average barrel 0.6 A/dm² (6 A/ft²); average rack 2.0 to 5 A/dm² (20 to 50 ft²). Bath voltage: 3 to 6 V, rack; 12 to 25 V, barrel

Constituent	Standard cyanide bath(a) Optimum g/L	Optimum oz/gal	Range g/L	Range oz/gal	Mild or half-strength cyanide bath(b) Optimum g/L	Optimum oz/gal	Range g/L	Range oz/gal
Preparation								
Zinc cyanide	61	8.1	54-86	7.2-11.5	30	4.0	27-34	3.6-4.5
Sodium cyanide	42	5.6	30-41	4.0-5.5	20	2.7	15-28	2.0-3.7
Sodium hydroxide	79	10.5	68-105	9.0-14.0	75	10.0	60-90	8.0-12.0
Sodium carbonate	15	2.0	15-60	2.0-8.0	15	2.0	15-60	2.0-8.0
Sodium polysulfide	.2	0.3	2-3	0.3-0.4	2	0.3	2-3	0.3-0.4
Brightener	(g)	(g)	1-4	0.1-0.5	(g)	(g)	1-4	0.1-0.5
Analysis								
Zinc metal	34	4.5	30-48	4.0-6.4	17	2.3	15-19	2.0-2.5
Total sodium cyanide	93	12.4	75-113	10.0-15.1	45	6.0	38-57	5.0-7.6
Sodium hydroxide	79	10.5	68-105	9.0-14.0	75	10.0	60-90	8.0-12.0
Ratio: NaCN to Zn	2.75	0.37	2.0-3.0	0.3-0.4	2.6	0.3	2.0-3.0	0.2-0.4

Constituent	Low-cyanide bath(c) Optimum g/L	Optimum oz/gal	Range g/L	Range oz/gal	Microcyanide bath(d) Optimum g/L	Optimum oz/gal	Range g/L	Range oz/gal
Preparation								
Zinc cyanide	9.4(b)	1.3(e)	7.5-14(b)	1.0-1.9	(f)	(f)	(f)	(f)
Sodium cyanide	7.5	1.0	6.0-15.0	0.8-2.0	1.0	0.1	0.75-1.0	0.4-0.13
Sodium hydroxide	65	8.7	52-75	6.9-10.0	75	10.0	60-75	8-10
Sodium carbonate	15	2.0	15-60	2.0-8.0	...	...	...	...
Sodium polysulfide	...	...	...	...	...	...	...	...
Brightener	(g)	(g)	1-4	0.1-0.5	(g)	(g)	1-5	0.1-0.7
Analysis								
Zinc metal	7.5	1.0	...	0.8-1.5	7.5	1.0	6.0-11.3	0.8-1.5
Total sodium cyanide	7.5	1.0	6.0-15.0	0.8-2.0	1.0	0.1	0.75-1.0	0.1-0.13
Sodium hydroxide	75	10	60-75	8.0-10.0	75	10.0	60-75	8-10
Ratio: NaCN to Zn	1.0	0.1	1.0	0.1	...	...	...	...

(a) Operating temperature: 29 °C (84 °F) optimum; range of 21 to 40 °C (69 to 105 °F). (b) Operating temperature: 29 °C (84 °F) optimum; range of 21 to 40 °C (69 to 105 °F). (c) Operating temperature: 27 °C (79 °F) optimum; range of 21 to 35 °C (69 to 94 °F). (d) Operating temperature: 27 °C (79 °F) optimum; range of 21 to 35 °C (69 to 94 °F). (e) Zinc oxide. (f) Dissolve zinc anodes in solution until desired concentration of zinc metal is obtained. (g) As specified

The cyanide system also has a number of disadvantages, including toxicity. The standard cyanide zinc bath containing 90 g/L (12 oz/gal) of total sodium cyanide, with the possible exception of cadmium cyanide baths, is the most potentially toxic bath used in the plating industry. This high cyanide content and the cost for treating cyanide wastes has been the primary reason for the development of the lower cyanide baths and the switch to alkaline noncyanide and acid baths. Although the technology for waste treatment of cyanide baths is well developed, the cost for the initial treatment plant may be as much as or more than for the installation.

Another disadvantage is the relatively poor bath conductivity. The conductivity of the cyanide bath is substantially inferior to the acid bath, thus substantial power savings may be had by the use of the latter.

The plating efficiency of the cyanide system varies greatly depending on such factors as bath temperature, cyanide content, and current density. In barrel installations at current densities up to 2.5 A/dm² (25 A/ft²), the efficiency can range within 75 to 90%. In rack installations, the efficiency rapidly drops below 50% at current densities above 6 A/dm² (60 A/ft²).

While the depth of brilliance obtained from the cyanide zinc bath has increased steadily since 1950, none of the additives shows any degree of the intrinsic leveling found in the acid baths. The ultimate in depth of color and level deposits reached in the newer acid baths cannot be duplicated in the cyanide bath.

Midcyanide Zinc Bath. In an effort to reduce cyanide waste as well as treatment and operating costs, most cyanide zinc baths are currently at the so-called midcyanide, half-strength, or dilute cyanide bath concentration, indicated in Table 1. Plating characteristics of midcyanide baths and regular cyanide baths are practically identical. The only drawback of the midcyanide bath when compared with the standard bath is a somewhat lower tolerance to impurities and poor preplate cleaning. This drawback is seldom encountered in practice in the well-run plant. The greater ease of rinsing, substantially less dragout, and savings in bath preparation, maintenance, and effluent disposal costs for midcyanide baths make this type of bath the most prominent among cyanide baths.

Low-cyanide zinc baths, which are generally defined as those baths operating at approximately 6 to 12 g/L (0.68 to 1.36 oz/gal) sodium cyanide and zinc metal, are substantially different in plating characteristics from the midcyanide or standard cyanide baths. The plating additives normally used in regular and midstrength cyanide baths do not function well with low metal and cyanide contents. Special low-cyanide brighteners have been developed for these baths.

Low-cyanide zinc baths are more sensitive to extremes of operating temperatures than either the regular or midbath. The efficiency of the bath may be similar to a regular cyanide bath initially, but tends to drop off more rapidly (especially at higher current densities) as the bath ages. Bright throwing power and covering power are slightly inferior to that of a standard or midcyanide bath. However, the majority of work which can be plated in the higher cyanide electrolytes, can usually be plated in the low-cyanide bath. Despite the fact that low-cyanide baths have significantly lower metal and cyanide contents, they are less sensitive to impurity content than the standard or mid-bath. Heavy metal impurities are much less soluble at lower cyanide contents. The deposit from a low-cyanide bath is usually brighter than that from a regular or midcyanide system, especially at higher current densities. These baths are used extensively for rack plating of wire goods. Unlike the other cyanide systems, low-cyanide baths are quite sensitive to sulfide treatments to reduce impurities. Regular sulfide additions may be deleterious to the plating brightness produced.

Microcyanide zinc baths are essentially a retrogression from the alkaline noncyanide zinc process discussed in the following section. Since the alkaline bath is often difficult to operate within its somewhat limited parameters, many platers use an absolute minimal amount of cyanide in these baths, 1.0 g/L (0.13 oz/gal), for example. This acts essentially as an additive, increasing the overall bright range of the baths and simplifying operation. It negates the purpose of the alkaline noncyanide bath, which is to totally eliminate cyanide, and should be viewed as an expedient crutch rather than a normally recommended zinc plating system.

Preparation of Cyanide Zinc Baths

Cyanide zinc liquid concentrates which are diluted with water, and to which sodium hydroxide is normally added, or baths may be prepared as follows:

- Fill the makeup and/or plating tank approximately two-thirds full of tap water
- Slowly stir in the required amount of sodium hydroxide
- Add the required amount of sodium cyanide and mix until dissolved
- Prepare a slurry of the required amount of zinc oxide or zinc cyanide and slowly add to the bath. Mix until completely dissolved. Instead of zinc salts, the bath may be charged with containers of zinc anode balls which are allowed to dissolve into the solution until the desired metal content is reached
- Add an initial addition of 15 g/L (2.0 oz/gal) sodium carbonate for rack plating baths
- Add approximately 1 to 2 g/L (0.1 to 0.2 oz/gal) of sodium polysulfide or zinc purifier for regular and mid-cyanide baths
- Run plating test panels and add the necessary amount of brightener to the bath. If a satisfactory deposit is obtained, anodes are placed in the bath and it is ready for production

Zinc baths prepared from impure zinc salts may require treatment with zinc dust and/or low current density dummying. Zinc dust should be added at the rate of 2 g/L (0.26 oz/gal) and the bath agitated for about 1 h. After settling, the bath should be filtered into the plating tank. Dummying, the process of plating out bath impurities, is preferably done on steel cathode sheets at low current densities of 0.2 to 0.3 A/dm^2 (2 to 3 A/ft^2) for a 12 to 24 h period.

Cyanide Zinc Plating Brighteners

Zinc plating bath brighteners are almost exclusively proprietary mixtures of organic additives, usually combinations of polyepoxyamine reaction products, polyvinyl alcohols, aromatic aldehydes, and quaternary nicotinates. These materials are formulated for producing brightness at both low and high density areas and for stability at elevated temperatures. Metallic brighteners based on nickel and molybdenum are no longer commercially used in zinc systems, because their concentration in the deposit is highly critical. Proprietary additives should be used following the manufacturer's recommendations of bath operation. Some incompatibility between various proprietary additives may be encountered, and Hull Cell plating tests should always be used to test a given bath and evaluate new brighteners.

Alkaline Noncyanide Baths

Alkaline noncyanide baths are a logical development in the effort to produce a relatively nontoxic, cyanide-free zinc electrolyte. Approximately 10 to 15% of zinc plated at present is deposited from these baths. Bath composition and operating parameters of this electrolyte are given in Table 2. The operating characteristics of an alkaline noncyanide system depend to a great extent on the proprietary additives and brightening agents used in the bath, because the zinc deposit may actually contain 0.3 to 0.5 wt % carbon which originates from these additives. This is ten times as much carbon as is found in deposits from the cyanide system.

Although alkaline noncyanide baths are inexpensive to prepare and maintain and produce bright deposits and cyanide-free effluents, they have a rather narrow optimum operating range of zinc metal content and for this reason have not found extensive use in industry. An alkaline noncyanide zinc bath with a zinc metal content of 7.5 to 12 g/L (1.0 to 1.6 oz/gal) used at 3 A/dm^2 (30 A/ft^2) produces an acceptable bright deposit at efficiencies of approximately 80%, as shown in Fig. 1. However, if the metal content is allowed to drop 2 g/L (0.26 oz/gal), efficiency drops to below 60% at this current density. Raising the metal content much above 17 g/L (2.3 oz/gal) produces dull gray deposits from this electrolyte.

Increasing sodium hydroxide concentration increases efficiency, as shown in Fig. 2. However, excessively high concentrations will cause metal buildup on sharp cornered edges.

Alkaline noncyanide zinc is a practical plating bath having hundreds of thousands of gallons in use now. Because of the closer analytical control required of this bath, most installations are found in large, chemist-supervised, captive plating installations.

Operating Parameters of Standard Cyanide and Midcyanide Zinc Solutions

Anodes. Almost every physical form of zinc anode material has been used in cyanide zinc plating, the type and prevalence varying from country to country. In the United States, cast zinc balls of approximately 50-mm (2-in.) diam contained in spiral steel wire cages are by far the most common anode material. A practical variation of this is the so-called flat top anode, with a flat surface to distinguish it from cadmium ball anodes. The use of ball anodes provides maximum anode area, ease of maintenance, and practically complete dissolution of the zinc anodes with no scrap anode formation.

One of the most economical forms of anode material is the large cast zinc slabs which form the prime material for

Table 2 Composition and operating characteristics of alkaline noncyanide zinc baths

Constituent	Optimum(a) g/L	Optimum(a) oz/gal	Range(b) g/L	Range(b) oz/gal
Preparation				
Zinc oxide	9.4	1.3	7.5-21	1-2.8
Sodium hydroxide	65	8.6	65-90	8.6-12
Proprietary additive	(c)	(c)	3-5	0.4-0.7
Analysis				
Zinc metal	7.5	1.0	6.0-17.0	0.8-2.3
Sodium hydroxide	75.0	10.0	75-112	10.0-14.9

(a) Operating conditions: temperature, 27 °C (81 °F) optimum; cathode current density, 0.6 A/dm² (6 A/ft²); bath voltages, 3 to 6 rack. (b) Operating conditions: temperature, 21 to 35 °C (69 to 94 °F) range; cathode current density, 2.0 to 4.0 A/dm² (20 to 40 A/ft²); bath voltages, 12 to 18 barrel. (c) As specified

Fig. 1 Cathode current efficiency of alkaline noncyanide zinc baths as related to zinc metal contents

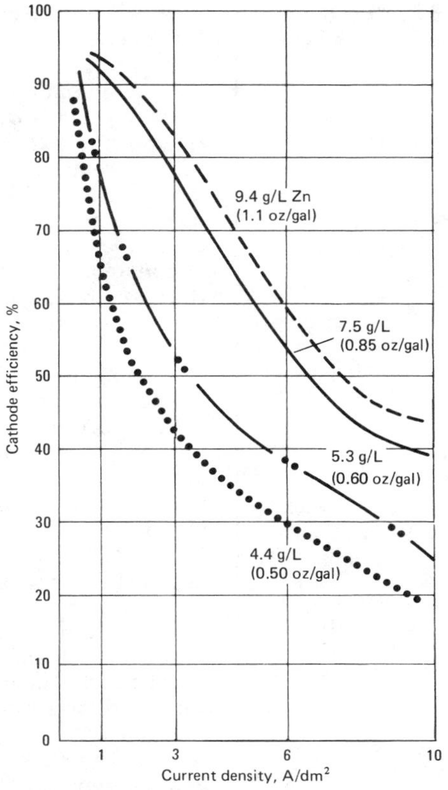

NaOH, 80 g/L (11 oz/gal); Na_2CO_3, 15 g/L (2 oz/gal)

subsequent ball or elliptical anode casting. Although these have the disadvantage of bulky handling and the construction of specially fabricated anode baskets, their lower initial cost makes their use an important economic factor in the larger zinc plating shop.

Three grades of zinc for anodes are conventionally used for cyanide zinc plating: prime western, intermediate, and special high-grade zinc. The zinc content of these are approximately 98.5%, 99.5%, and 99.99%, respectively. The usual impurities in zinc anodes are all heavy metals, which, unless continuously treated, cause deposition problems; thus, nearly trouble-free results can consistently be obtained through the use of special high-grade zinc.

A typical composition of special high-grade zinc is:

Constituent	Amount, %
Zinc	99.9930
Lead	0.0031
Cadmium	0.0017
Iron	0.0010
Copper	Trace

Control of Zinc Metal Content. Zinc anodes dissolve chemically as well as electrochemically in cyanide baths, and therefore, effective anode efficiency will be above 100%, causing a buildup in zinc metal content, because cathode efficiencies are usually substantially less than 100%. There are a number of procedures which have been developed to control this tendency.

In a conventional new zinc cyanide installation, approximately ten spiral anode ball containers should be used for every metre of anode rod. These should be filled initially and after 1 or 2 weeks of operation adjusted to compensate for anode corrosion and dragout losses so that the metal content remains as constant as possible. During shutdown periods in excess of 48 h, most cyanide zinc platers remove anodes from the bath. In large automatic installations, this may be done by using a submerged steel anode bar sitting in yokes which can be easily lifted by hoist mechanisms.

One of the prime causes of zinc metal buildup is the very active galvanic cell between the zinc anodes and the steel anode containers. This is evidenced by intense gassing in the area of anodes in a tank not in operation. Zinc buildup from this source can be eliminated by plating the anode containers with zinc before shutdown. This eliminates the galvanic couple.

Temperature. Probably no operating variable is as important and as often overlooked in the operating of cyanide zinc baths as operating temperature. In actual practice, cyanide zinc solutions have been reported operating between the rather wide limits of 12 to 55 °C (54 to 130 °F) with the vast majority of baths operating between 23 to 32 °C (73 to 90 °F). The exact operating temperature for a given installation depends on the type of work processed, finish desired, and engineering characteristics of the plating system. Bath temperature has an effect on a great many variables in the cyanide zinc systems so that the optimum temperature is generally a compromise. Increasing the bath temperature:

- Increases cathode efficiency
- Increases bath conductivity
- Increases anode corrosion
- Produces duller deposits over a broad range of current densities
- Reduces covering power

Fig. 2 Effect of zinc and sodium hydroxide concentration on the cathode efficiency of noncyanide zinc solutions

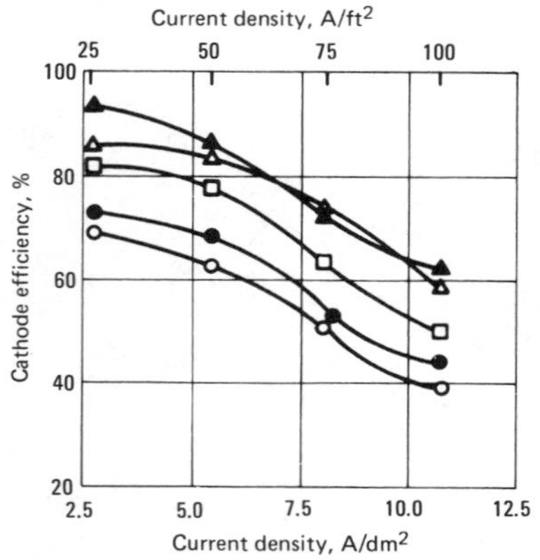

Temperature: 26 °C (77 °F). ○: 7.5 g/L (1 oz/gal) Zn, 75 g/L (10 oz/gal) NaOH; ●: 7.5 g/L (1.0 oz/gal) Zn, 150 g/L (20 oz/gal) NaOH; △: 11 g/L (1.5 oz/gal) Zn, 110 g/L (15 oz/gal) NaOH; ▲: 15 g/L (2.0 oz/gal) Zn, 150 g/L (20 oz/gal) NaOH; □: 11 g/L (1.5 oz/gal) Zn, 150 g/L (20 oz/gal)

- Reduces throwing power
- Increases breakdown of cyanide and addition agents

Lowering the bath temperature has the opposite effects. Thus, if a plater is primarily concerned with plating of pipe or conduit where deposit brilliance is not of great importance and covering and throwing power are not critical, operating the bath at the highest practical temperature to give optimum conductivity and plating efficiency would be preferred. For general bright plating of fabricated stampings, a lower bath temperature, permitting the required excellent covering and throwing power and bright deposits, should be used.

The effects of higher bath temperature can be compensated to a substantial extent by increasing the total cyanide to zinc ratio of the solution. The exact ratio varies slightly for a given proprietary system, as shown in Table 3.

Cathode Current Densities. Bright cyanide zinc solutions operate at wide ranging cathode current densities varying from extremely low, less than 0.002 A/dm² (0.02 A/ft²), to above 25 A/dm² (250 A/ft²) without burning. Usable current density limits depend on bath composition, temperature, cathode film movement, and addition agents used.

Average current densities used vary and are approximately 0.6 A/dm² (6 A/ft²) in barrel plating and 2 to 5 A/dm² (20 to 50 A/ft²) in still or rack plating. Barrel zinc plating is a complex phenomenon in which a large mass of parts is constantly tumbled in the plating cylinder at varying distances from the cathode contact surfaces. At any given time, a part may have an infinitesimally low current density or it may even be deplating, and in another instant, appearing near the outer surface of the tumbling mass, may approach as high as 20.0 A/dm² (200 A/ft²) in current density. In general, the bulk of deposition takes place in the lower current density range of 0.2 to 1 A/dm² (2 to 10 A/ft²).

Average cathode current densities are generally easier to maintain in rack and still line operations and range from approximately 2 to 5 A/dm² (20 to 50 A/ft²). The actual current density of any particular area of a given part will, however, vary greatly depending upon part configuration, anode to cathode distance, bath shape, and other factors affecting the primary and secondary current distribution characteristics. In most cases, with proper attention to racking and work shape, current density variations can be kept within practical limits on fabricated parts so that if a minimum average thickness of 4 μm (0.15 mil) is required on a specific part, variations from approximately 2.5 to 8 μm (0.09 to 0.3 mil) occur at various areas on the part.

Cathode current efficiencies in barrel cyanide zinc plating can vary between 75 and 93% depending on temperature, formulation, and barrel current densities.

In rack or still plating, however, there is quite a wide variation in current efficiencies when higher current densities, especially above 3 A/dm² (30 A/ft²), are used. The effects of zinc metal content, sodium hydroxide content, and the cyanide-to-zinc ratio on cathode current efficiency are shown in Fig. 3. As can be seen from the graphs, the current efficiency in the most commonly used baths drops dramatically from approximately 90% at 2.5 A/dm² (25 A/ft²) to 50% at 5 A/dm² (50 A/ft²). An improvement in current efficiency can be obtained by using a high strength bath; however, this is offset by the relatively poor throwing power of the solution, higher brightener consumption, operating costs, and maintenance difficulties. The lower standard bath concentration or the midbath, which gives practically identical results, is used for practically all plating installations except a selected few rack tanks which plate conduit or large flat surfaces with no critical recessed areas.

Table 3 Effect of bath temperature on total cyanide to zinc ratio

Temperature °C	°F	Total NaCN to Zn Ratio (standard cyanide bath)	Total NaCN to Zn ratio (midcyanide bath)
22	72	2.6	2.2
26	79	2.7	2.3
30	86	2.8	2.4
34	93	2.9	2.5
38	100	3.0	2.6
42	108	3.2	2.7
46	115	3.3	3.0

Sodium carbonate is present in every cyanide and alkaline zinc solution. It enters the bath initially in one of two ways: (*a*) as an impurity from the makeup salts; sodium hydroxide and sodium cyanide may contain anywhere from 0.5 to 2% of sodium carbonate, or (*b*) deliberately added in 15 to 30 g/L (2.0 to 4 oz/gal) additions to initial baths as previously noted.

The harmful effects of sodium carbonate in cyanide zinc plating are not as critical as in cyanide cadmium plating. Sodium carbonate does not begin to affect normal bath operation until it builds up to above 75 to 105 g/L (10 to 14 oz/gal). Depending on over-all bath composition and the type of work being done, a carbonate content in this range results in a slight decrease in current efficiency, especially at higher current densities, decreased bath conductivity, grainier deposits, and roughness, which becomes visible when the carbonate crystallizes out of cold solutions.

The carbonate content of zinc baths builds up by decomposition of sodium cyanide and absorption of carbon dioxide from the air reacting with the sodium hydroxide in the bath. Carbonates are best removed by one of the common cooling or refrigeration methods rather than chemical methods, which are simple in theory but extremely cumbersome in practice. When an operating zinc bath has reached the point where excessive carbonates presents a problem, it undoubtedly is contaminated with a great many other dragged in impurities, and dilution is often a much quicker and wiser over-all method of treatment.

Fig. 3 Effects of bath composition variables and cathode current density on cathode efficiency in cyanide zinc plating

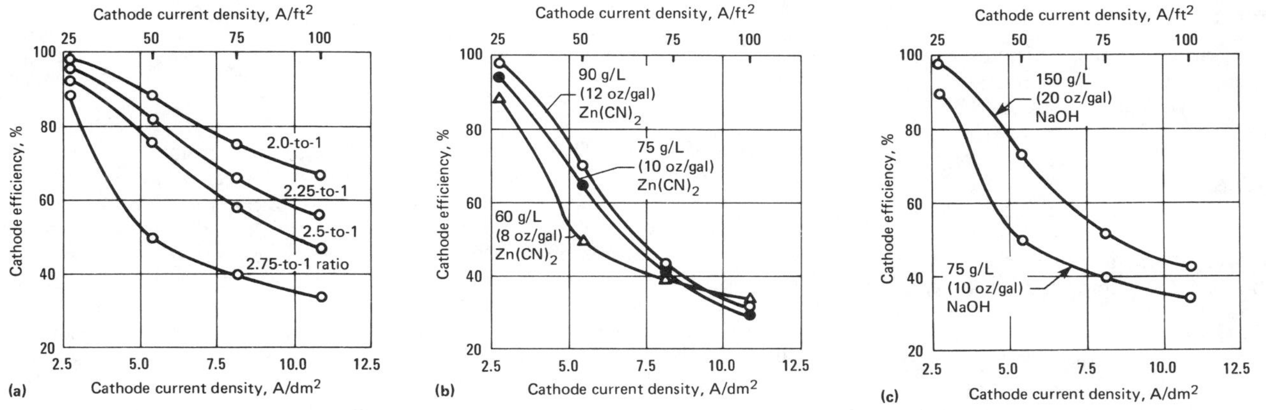

(a) Effect of NaCN/Zn ratio. 60 g/L (8 oz/gal) Zn (CN); 17.5 to 43.7 g/L (2.33 to 5.82 oz/gal) NaCN; 75.2 g/L (10 oz/gal) NaOH; 2.0-to-1 to 2.75-to-1 ratios of NaCN to Zn. Temperature: 30 °C (86 °F). (b) Effect of Zn metal content. 60.1, 75.2, and 90.2 g/L (8, 10, and 12 oz/gal) Zn (CN); 43.7, 54.6, and 65.5 g/L (5.82, 7.27, and 8.72 oz/gal) NaCN; 75.2 g/L (10 oz/gal) NaOH; 2.75-to-1 ratio of NaCN to Zn. Temperature: 30 °C (86 °F). (c) Effect of NaOH content 60.1 g/L (8 oz/gal) Zn(CN); 43.6 g/L (5.8 oz/gal) NaCN; 150.4 and 75.2 g/L (20 and 10 oz/gal) NaOH; 2.75-to-1 ratio of NaCN to Zn. Temperature: 30 °C (86 °F)

Operating Parameters of Low-Cyanide Zinc Systems

Temperature control is as critical, if not more critical, in the low-cyanide bath as in the regular or midcyanide bath. The optimum operating temperature for most proprietary baths is 29 °C (84 °F) with permissible range more restricted than the standard cyanide bath. Adequate cooling facilities are, therefore, mandatory and more critical with the low-cyanide than with the standard system.

Cathode Current Density. The average cathode current densities used in most low-cyanide processes are the same as in the standard cyanide bath; however, some proprietary baths do not have the extreme high current density capabilities of the standard cyanide bath and burning on extremely high current density areas may be more of a problem with the low-cyanide bath than with the conventional baths.

Agitation. Unlike the standard cyanide bath where agitation is usually nonexistent, air or mechanical agitation of the low-cyanide bath is common and is often quite useful in obtaining the optimum high current density plating range of the bath.

Filtration. Most low-cyanide baths appear to be much cleaner operating when compared to the standard or mid-cyanide bath. The bath is a poor cleaner, and soils which may be removed and crystallized out of high cyanide baths are not as readily affected by the low-cyanide bath.

Efficiency. The efficiency of the low-cyanide bath on aging is, to a much greater extent than the standard cyanide bath, dependent on the particular addition agent used, there being a substantial difference in various proprietary systems. In a new low-cyanide bath, current efficiency is slightly higher than that of a standard or mid-cyanide system. However, as the bath ages, current efficiency tends to drop, possibly because of formation of additive breakdown products, and the efficiency of a bath after 2 or 3 months of operation may be as much as 30% below that of a cyanide system, especially at higher current densities. As in the standard cyanide bath, increasing the sodium hydroxide and zinc metal content as well as operating temperature increases the efficiency of the low-cyanide bath. However, in low-cyanide baths, increasing the sodium hydroxide and zinc metal content and increasing operating temperature have marked, harmful effects on the bright operating range and usually override the benefit of increased efficiency. The effects of bath constituents and temperature on the plating characteristics of the bright low-cyanide zinc systems are given in Table 4. Figure 4 shows the effect of sodium cyanide concentration on cathode efficiency.

Bright Throwing Power and Covering Power. The bright covering power of a low-cyanide bath operated at low current density is intrinsically not as good as that of a standard or mid-cyanide bath. In most operations, the degree of difference, however, is negligible and only on extremely deep recessed parts is this a factor of any significance. The vast majority of parts which can be adequately covered in a standard cyanide bath can be similarly plated in a low-cyanide bath without any production problems, such as excessively dull recessed areas or stripping by subsequent bright dipping.

Increasing the brightener and cyanide content within limits improves the bright low current density deposition to a visible degree. Where extreme low current density bright throwing power is a problem, this is often solved in practice by raising the cyanide content to approximately 15 g/L (2 oz/gal) which in effect, returns the system to the lower range of the midcyanide bath.

Operating Parameters of Alkaline Noncyanide Zinc Baths

Temperature control is more critical in noncyanide zinc than in the regular and low-cyanide baths. The optimum temperature for most baths is approximately 29 °C (84 °F) with permissible operating ranges of 21 to 35 °C (70 to 95 °F). Low operating temperatures will result in no plating, or at most, very thin, milky white deposits. High operating temperatures rapidly narrow the bright plating current range, cause dullness at low current density, and result in very high brightener consumption. Because these temperature limitations for noncyanide zinc are, however, within those commonly used in regular cyanide zinc, no additional refrigeration or cooling equipment is required for conversion to the process.

Operating Voltages. Normal voltages used in standard cyanide zinc plating are adequate for the noncyanide zinc bath both in rack and barrel range. Normal voltage will be approximately 3 V with a range of 2 to 20 V, depending on part shape, anode to cathode relationship, temperature, barrel hole size, and similar variants which are unique for each individual operation.

Fig. 4 Effect of sodium cyanide concentration on the cathode efficiency of low-cyanide zinc solutions

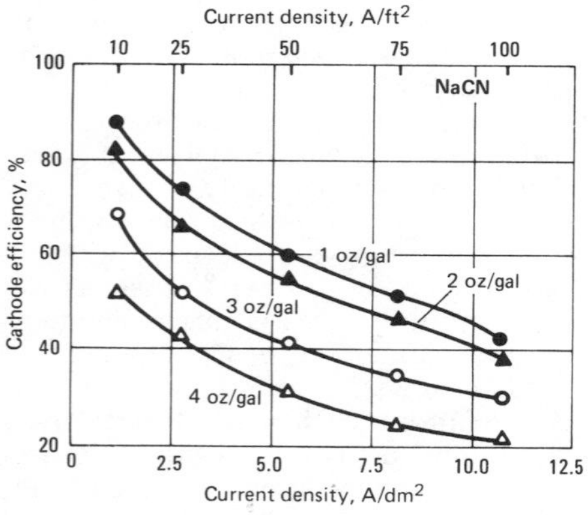

NaCN; ○: 20 g/L (2.5 oz/gal); ●: 8 g/L (1 oz/gal); △: 30 g/L (4 oz/gal); ▲: 15 /L (2 oz/gal)

Table 4 Effect of bath constituents and temperature on plating characteristics of bright, low-cyanide zinc plating

Variable	Cathode efficiency	Bright plating range	Bright low current density throwing power
Increasing sodium hydroxide	Increases	Slightly decreases	Negligible
Increasing zinc metal	Increases	Decreases	Decreases
Increasing sodium cyanide	Decreases	Increases	Increases
Increasing brightener	Increases	Increases	Increases
Increasing temperature	Increases	Decreases	Decreases

Cathode Current Densities. The maximum allowable cathode current densities of the noncomplexing noncyanide bath closely approximate those of a standard cyanide bath. Current density ranges of 0.1 to more than 20 A/dm^2 (1 to 200 A/ft^2) as a maximum limiting value can be obtained. This extremely wide plating range permits operation at an average current density of 2 to 4 A/dm^2 (20 to 40 A/ft^2) in rack plating, which makes a noncyanide system practical for high production work.

Anodes. Standard zinc ball or slab anodes in steel containers are used in the noncyanide electrolyte. During the first 2 or 3 weeks of installation of noncyanide zinc baths, anode area should be watched carefully to determine the appropriate anode area to maintain a stable analysis of zinc in the system. A large steel to zinc anode ratio should be avoided, and the zinc anode area should be as high as possible. Zinc anodes, whenever possible, should be removed during weekend shutdown periods to avoid excessive metal buildup.

Filtration of noncyanide baths is not an absolute necessity. However, the occurrence of roughness in these baths presents a greater potential problem than in regular cyanide baths. This is because of both the nature of the deposit, which, if the brightener is not maintained at an optimum level, may become amorphous at very high current densities, and anode polarization problems, which result in sloughing off of anode slimes, a more common occurrence in these baths.

Filtration is also the preferred method for removing zinc dust used to treat any metallic impurities in the system.

The bright plating range of the alkaline, noncyanide zinc bath is totally dependent on the particular additive used. Without any additive, the deposit from an alkaline, noncyanide bath is totally useless for commercial finishing, with a powdery, black amorphous deposit over the entire normal plating range.

Proper maintenance of the addition agent at the recommended level is extremely important in noncyanide alkaline zinc. A plater does not have the liberty of maintaining low levels of brightener in the bath and still obtaining passable bright deposits as is the case in cyanide systems. Low brightener content rapidly leads into high and medium current density burning because in the noncyanide, as in the low-cyanide bath, burning and brightness are interdependent.

Cathode Current Efficiency. The efficiency of a noncyanide bath is, as indicated in Fig. 1, a very critical function of the metal content. At lower metal concentrations of approximately 4 g/L (0.5 oz/gal), efficiency is less than that of a standard cyanide bath while at a metal content of approximately 9 g/L (1.2 oz/gal), efficiency is somewhat higher than both regular and low-cyanide baths. Thus, if a plater can maintain metal content close to the 9 g/L (1.2 oz/gal) value, there will be no problem in producing deposition rates similar to cyanide baths.

Acid Baths

The continuing development of acid zinc plating baths based on zinc chloride has radically altered the technology of zinc plating since the early 1970's. Acid zinc plating baths now constitute 40 to 50% of all zinc baths in most developed nations and are the fastest growing baths throughout the world. Acid zinc formulas and operating limits are given in Table 5. Bright acid zinc baths have a number of intrinsic advantages over the other zinc baths:

- Waste disposal is minimized, consisting only of neutralization, at pH 8.5 to 9, and precipitation of zinc metal, when required
- Acid zinc baths operate at current efficiencies of 95 to 98%, normally much higher than cyanide or alkaline processes, especially at higher current densities as shown in Fig. 5

Table 5 Composition and operating characteristics of acid chloride zinc plating baths

Constituent	Ammoniated bath Barrel Optimum	Ammoniated bath Barrel Range	Ammoniated bath Rack Optimum	Ammoniated bath Rack Range
Preparation				
Zinc chloride	18 g/L (2.4 oz/gal)	15-25 g/L (2.0-3.8 oz/gal)	30 g/L (4.0 oz/gal)	19-56 g/L (2.5-7.5 oz/gal)
Ammonium chloride	120 g/L (16.0 oz/gal)	100-150 g/L (13.4-20.0 oz/gal)	180 g/L (24.0 oz/gal)	120-200 g/L (16.0-26.7 oz/gal)
Potassium chloride	...	...	...	...
Sodium chloride	...	...	...	...
Boric acid	...	...	...	...
Carrier brightener(a)	4 vol%	3-5%	3.5%	3-4%
Primary brightener(a)	0.25%	0.1-0.3%	0.25%	0.1-0.3%
pH	5.6	5.5-5.8	5.8	5.2-6.2
Analysis				
Zinc metal	9 g/L (1.2 oz/gal)	7.5-25 g/L (1.0-3.8 oz/gal)	14.5 g/L (1.9 oz/gal)	9-27 g/L (1.2-3.6 oz/gal)
Chloride ion	90 g/L (1.2 oz/gal)	75-112 g/L (10.0-14.9 oz/gal)	135 g/L (18.0 oz/gal)	90-161 g/L (12.0-21.5 oz/gal)
Boric acid	...	...	...	...
Operating conditions				
Temperature	24 °C (75 °F)	21-27 °C (69-79 °F)	24 °C (75 °F)	21-27 °C (69-79 °F)
Cathode current density	...	0.3-1.0 A/dm^2 (3-10 A/ft^2)	...	2.0-5 A/dm^2 (20-50 A/ft^2)
Voltage	...	4-12 V	...	1-5 V

(a) Carrier and primary brighteners for acid chloride are proprietary, and exact recommendations of manufacturer should be followed. Values given are representative

Constituent	Potassium bath Optimum	Potassium bath Range	Mixed sodium ammonium Barrel bath Optimum	Mixed sodium ammonium Barrel bath Range
Preparation				
Zinc chloride	71 g/L (9.5 oz/gal)	62-85 g/L (8.3-11.4 oz/gal)	34 g/L (4.5 oz/gal)	31-40 g/L (4.1-5.3 oz/gal)
Ammonium chloride	...	...	30 g/L (4.0 oz/gal)	25-35 g/L (3.3-4.7 oz/gal)
Potassium chloride	207 g/L (27.6 oz/gal)	186-255 g/L (24.8-34.0 oz/gal)	...	...
Sodium chloride	...	...	120 g/L (16.0 oz/gal)	100-140 g/L (13.3-18.7 oz/gal)
Boric acid	34 g/L (4.5 oz/gal)	30-38 g/L (4.0-5.1 oz/gal)	...	...
Carrier brightener(a)	4%	4-5%	4%	3-5%
Primary brightener(a)	0.25%	0.1-0.3%	0.2%	0.1-0.3%
pH	5.2	4.8-5.8	5.0	4.8-5.3
Analysis				
Zinc metal	34 g/L (4.5 oz/gal)	30-41 g/L (4.0-5.5 oz/gal)	16.5 g/L (2.2 oz/gal)	15-19 g/L (2.0-2.5 oz/gal)
Chloride ion	135 g/L (18.0 oz/gal)	120-165 g/L (16.0-22.0 oz/gal)	110 g/L (14.7 oz/gal)	93-130 g/L (12.4-17.4 oz/gal)
Boric acid	34 g/L (4.5 oz/gal)	30-38 g/L (4.0-5.1 oz/gal)	...	...
Operating conditions				
Temperature	27 °C (79 °F)	21-35 °C (69-94 °F)	27 °C (79 °F)	25-35 °C (76-94 °F)
Cathode current density	...	2.0-4 A/dm^2 (20-40 A/ft^2)	...	0.3-1 A/dm^2 (3-10 A/ft^2)
Voltage	...	1-5 V	...	4-12 V

- Acid zinc baths are the only zinc baths possessing any leveling ability. This, combined with their superb out-of-bath brightness, produces the most brilliant zinc deposits available
- Cast iron, malleable iron, and carbonitrided parts, which are difficult or impossible to plate from cyanide or alkaline baths, can be readily plated in the acid zinc baths
- Acid zinc baths have much higher conductivity in comparison to cyanide and alkaline baths, producing substantial energy savings
- Because of their high cathode efficiency, acid chloride baths produce minimal hydrogen embrittlement when compared to other zinc baths

The negative aspects of the acid chloride bath are:

- The acid chloride electrolyte is corrosive. All equipment in contact with the bath such as tanks and superstructures must be coated with corrosion-resistant materials
- Bleedout of entrapped plating solution occurs to some extent with every plating process. It can become a serious and limiting factor, prohibiting the use of acid chloride baths on some fabricated, stamped, or spot welded parts which entrap solution. Bleedout may occur months after plating and the corrosive electrolyte will ruin the part. This potential problem should be carefully noted when plating complex assemblies in acid chloride electrolytes

Acid chloride zinc baths currently in use are principally of two types: those based on ammonium chloride and those based on potassium chloride. The ammonium-based baths were the first to be developed. They can be operated at higher current densities than potassium baths. Both systems depend upon a rather high concentration, 4 to 6 vol %, of wetting agents to solubilize the primary brighteners. This is more readily accomplished in the ammonia systems, which makes bath control somewhat easier. Ammonium ions, however, act as a complexing agent in waste streams containing nickel and copper effluents and in many localities must be disposed of by expensive chlorination. This was the essential reason for the development of the potassium chloride bath.

All bright acid chloride processes are proprietary and some degree of incompatibility may be encountered between them. Conversion from an existing process should be done only after a Hull Cell plating test evaluation.

The latest acid chloride zinc baths to become available to the industry are those based on salt (sodium chloride) rather than the more expensive potassium chloride. In many of these baths, salt is substituted for a portion of either ammonium or potassium chloride, producing a mixed bath. Sodium acid chloride baths at present are generally restricted to barrel operation, because burning occurs much more readily in these baths at higher current densities. With the continuing development of additive technology, they may challenge the widely used nonammoniated potassium bath in the near future.

A number of zinc baths based on zinc sulfate and zinc fluoborate have been developed, but these have very limited applications. They are used principally for high-speed, continuous plating of wire and strip and are not commercially used for plating fabricated parts. Table 6 shows compositions and operating conditions for some typical fluoborate and sulfate baths.

Fig. 5 Comparison of cathode current efficiencies of bright zinc plating electrolytes

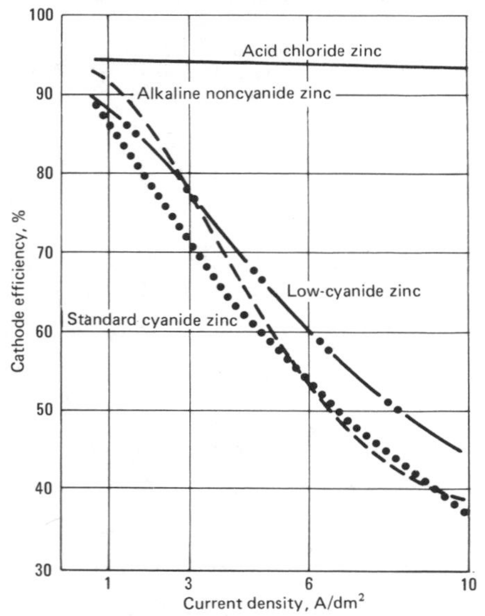

Operating Parameters of Acid Chloride Zinc Baths

Anodes for acid chloride zinc should be special high grade, 99.99% zinc. Most installations use zinc ball or flat top anodes in titanium anode baskets. Baskets should not be used if the applied voltage on an installation exceeds 8 V, as there may be some attack on the baskets. Baskets should be kept filled to the solution level with zinc balls.

Slab zinc anodes, drilled and tapped for titanium hooks, may be used. Any areas of hooks or splines exposed to the solution should be protective coated.

Anode bags are optional but recommended for most processes, especially for rack plating where they are useful to minimize roughness. Bags may be made of polypropylene, dynel, or nylon. They should be leached for 24 h in a 5% hydrochloric acid solution containing 0.1% of the carrier or wetting agent used in the particular plating bath before being used.

Chemical Composition. Zinc, total chloride, pH, and boric acid, when used, are to be controlled and maintained in the recommended ranges (see Table 5) by periodic replenishment using chemically pure materials. Too high zinc content causes poor low current density deposits, while too low concentrations cause high current density burning. High chloride may cause separation of brightener, while low chloride concentrations reduce conductivity of solutions. High pH values cause the formation of precipitates and anode polarization, while too low pH values cause poor plating. Insufficient boric acid reduces the plating range. Brighteners have to be replenished by periodic additions. Since the nature of these brighteners is proprietary, the suppliers specify concentrations and control procedures.

Agitation is mandatory in acid chloride baths to achieve practical operating current densities. Solution circulation is recommended in barrel baths to supplement barrel rotation. In rack baths, solution circulation is usually accomplished by locating the intake and discharge of the filter at opposite ends of the plating tank.

Cathode rod agitation is suitable for many hand-operated rack lines. Air agitation is the preferred method for most installations. A low pressure air blower should be used as a supply source.

Temperature control is more critical in acid zinc baths than in cyanide zinc baths, and auxiliary refrigeration should be provided to maintain the bath at its maximum recommended operating temperature, usually 35 °C (95 °F). Cooling coils in the bath itself should be of Teflon or Teflon-coated tubing. Titanium coils may be used if they are isolated from the direct current source.

Operating an acid chloride bath above its maximum recommended temperature causes low over-all brightness, usually beginning at low current densities and rapidly progressing over the entire part. High temperatures may also bring the bath above the cloud point of the brightener system. As the acid bath gets hot, it reaches a point where additives start coming out of solution, giving the bath a milky or cloudy appearance. This causes total bath imbalance. Conversely, low temperatures, usually below 21 °C (70 °F), cause many baths to crystallize out and the organic additives to separate out of solution causing roughness and, in extreme cases, a sticky globular deposit on the bath and work, which clogs filters and completely curtails operations.

Cathode Current Efficiency. The high cathode current efficiencies exhibited by acid chloride zinc baths are one of the most important properties of these baths. As shown in Fig. 5, the average cathode current efficiency for these baths is approximately 95 to 98% over the entire range of operable current densities. No other zinc plating system approaches this extremely high efficiency at higher current densities. In practice, this high efficiency can lead to productivity increases of 15 to 50% over cyanide baths. In barrel plating, barrel loads may often be doubled in comparison with cyanide baths and equivalent plating thickness achieved in half the time.

pH control of acid zinc baths is usually done on a daily basis. Electro-

Table 6 Fluoborate and sulfate electroplating bath compositions

Constituent	Fluoborate(a) g/L	Fluoborate(a) oz/gal	Sulfate(b) g/L	Sulfate(b) oz/gal
Zinc	65-105	9-14	135	18
Zinc fluoborate	225-375	30-50	...	...
Zinc sulfate	...	...	375	50
Ammonium fluoborate	30-45	4-6	...	...
Ammonium chloride	...	...	7.5-22.5	1-3
Addition agent	(c)	(c)	(c)	(c)
pH	3.5-4		3-4	

(a) At room temperature; 3.5 to 4 pH; at 20 to 60 A/dm² (200 to 600 A/ft²). (b) At 30 to 52 °C (85 to 125 °F); 3 to 4 pH; at 10 to 60 A/dm² (100 to 600 A/ft²). (c) As needed

metric methods are preferred over papers. The pH of a bath is lowered with a hydrochloric acid addition and when required, the pH may be raised with a potassium or ammonium hydroxide addition.

Iron contamination is a common problem in all acid chloride zinc baths. Iron is introduced into the bath from parts falling into the tank during operation, from attack by the solution on parts at current densities below the normal range, such as the inside of steel tubular parts, and from contaminated rinse waters used before plating. Iron contamination usually appears as dark deposits at high current densities and in barrel plating as stained dark spots reproducing the perforations of the plating barrel. A high iron content turns the plating solution brown and murky.

Iron can be readily removed from acid chloride baths by oxidizing soluble ferrous iron to insoluble ferric hydroxide. This is accomplished by adding concentrated hydrogen peroxide to the bath, usually on a daily basis. Approximately 10 mL (0.34 fluid oz) of 30% hydrogen peroxide per every 100 L (26.4 gal) of bath should be used. The peroxide should be diluted with 4 to 5 parts water and dispersed over the bath surface. The precipitated iron hydroxide should then be filtered from the bath using a 15 μm (0.6 mil) or smaller filter coated with diatomaceous earth or a similar filter aid.

Control of Plate Thickness

The thicknesses of zinc specified for service in various indoor and outdoor atmospheres are discussed in the section of this article on applications. Many combinations of variables must be considered in attempting to plate to a given thickness. For conventional operations, each variable may change considerably and result in thickness changes. To hold each variable at a steady value is virtually impossible under production conditions. Thus, as one variable changes spontaneously, others must be adjusted to maintain uniformity of plate thickness. In automatic plating this is impractical, hence the process is set up to give a certain minimum thickness under a great variety of conditions. This accounts for much of the thickness variation normally encountered in automatic plating of a run of identical pieces.

The shape and size of parts that may be plated all over with or without the use of conforming anodes to attain uniformity of plate thickness are essentially the same in zinc plating as in cadmium plating. (See the article on cadmium plating in this Volume.)

Normal Variations. Preferred thicknesses in automatic zinc plating are usually minimum specified thicknesses, and there is little concern regarding the maximum thicknesses obtained. Thickness variations encountered should therefore be over the established minimum thickness.

For example, as shown in Fig. 6, tests were made on 75 samples, selected over a one-week period, of 100 mm (4 in.) long, 39 g (1.375 oz) parts that were automatically plated to a minimum specified thickness of 3.8 μm (0.15 mil). Although actual thickness of plate was found to range from 2.5 to 7.5 μm (0.1 to 0.3 mil), over 80% of the parts examined exceeded the minimum aimed at.

Thickness variations obtained in barrel plating are markedly affected by the tumbling characteristics of the part and by the density of the load in the plating barrel. Parts that can be tumbled readily are more likely to develop a uniform coating.

As shown in Fig. 7, a minimum plate thickness of 12.5 μm (0.5 mil) was the aim in barrel plating a 0.12 kg (0.26 lb) S-shaped part made from 3-mm (0.125-in.) flat stock. Of 75 parts examined, all were found to be plated to thicknesses that exceeded the aimed-at minimum—a few in excess of 23 μm (0.9 mil).

Similarities Between Cadmium and Zinc Plating

Except for differences in plating baths and in such operational details as current density and rates of deposition, alkaline cadmium and zinc plating are essentially similar processes. Reference should be made to the article on cadmium plating for a detailed discussion of plating methods, equipment, and processing.

Exceptions with respect to equipment and processing are described below.

Plating Equipment. The equipment requirements for zinc plating are the same as those noted for cadmium plating, except for the following:

- In barrel plating, zinc solutions require higher voltage and current density and therefore must be provided with greater cooling capacity to prevent overheating. Also, because the cyanide zinc bath generates much larger amounts of hydrogen, barrel design should incorporate safety features to prevent explosions
- Fume hoods should be used on cyanide, low-cyanide, and especially, alkaline noncyanide baths to exhaust caustic spray and toxic fumes
- Barrels, tanks, and all superstructures coming into contact with acid chloride zinc plating baths should be coated with a material able to resist acid corrosion. Polypropylene, polyethylene, polyvinyl chloride, and fiberglass are commonly used materials. Lead-lined tanks should never be used in these systems. Heating and cooling coils should be built of titanium electrically isolated from the tank, or high temperature Teflon may be used

Hydrogen embrittlement of steels is a major problem in all types of cyanide zinc plating. These formulas should not be used for spring tempered or other parts susceptible to this type of embrittlement. Spring tempered parts and other parts susceptible to hydrogen embrittlement should be plated in the acid chloride electrolyte. When no embrittlement whatsoever can be tolerated, the use of mechanically deposited zinc, is the preferable alternative.

Processing Steps. Time requirements for various operations involved in still tank, barrel, and automatic methods of plating zinc to a thickness of less than 12.5 μm (0.5 mil) are:

Processing cycle	Time for each operation
Hand- or hoist-operated still tank	
Electrolytic cleaning	1-3 min
Cold water rinse	10-20 s
Acid pickle	30 s-2 min
Cold water rinse	10-20 s
Cold water rinse	10-20 s
Zinc plate	6-8 min
Cold water rinse	10-20 s
Cold water rinse	10-20 s
Chromate conversion coat	15-30 s
Cold water rinse	10-20 s
Hot water rinse	20-30 s
Air dry	1 min
Hand- or hoist-operated barrel line	
Soak clean	4 min
Electroclean	4 min
Cold water rinse	1-2 min
Acid pickle	2-3 min
Zinc plate	20-30 min
Cold water rinse	1-2 min
Cold water rinse	1-2 min
Chromate conversion coat	30 s-1 min
Cold water rinse	1-2 min
Hot water rinse	2-3 min
Centrifugal dry	3-5 min
Automatic barrel line	
Soak clean	6 min
Electroclean	3 min
Cold water rinse	2 min
Cold water rinse	2 min
Acid pickle	1 min
Neutralize dip	3 min
Cold water rinse	2 min
Zinc plate	30-40 min
Dragout rinse	2 min
Neutralize rinse	2 min
Cold water rinse	2 min
Nitric acid dip	30 s
Cold water rinse	2 min
Chromate dip	30 s
Cold water rinse	2 min
Hot water rinse	2 min
Centrifugal dry	3 min

Applications

In the presence of moisture, zinc becomes a sacrificial protecting agent when in contact with iron and other metals below zinc in the galvanic series. Attack is most severe when the electrolyte has high electrical conductivity (as in marine atmospheres) and when the area ratio of zinc to the other metal is small.

Plate Thickness. The life of a zinc coating in the atmosphere is nearly proportional to the coating thickness. Its rate of corrosion is highest in industrial areas, intermediate in marine environments, and lowest in rural locations. Corrosion is greatly increased by frequent dew and fog, particularly if the exposure is such that evaporation is slow.

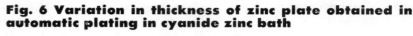

Fig. 6 Variation in thickness of zinc plate obtained in automatic plating in cyanide zinc bath

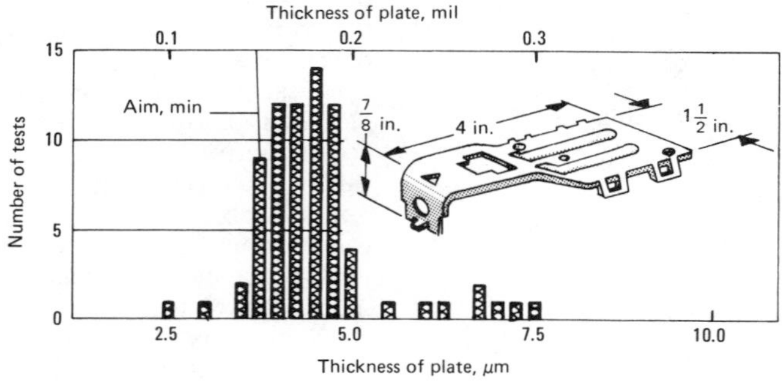

Automatic plating, 75 tests

Fig. 7 Variation in thickness of zinc plate obtained in barrel plating a 3.2 mm (1/8 in.) thick part in a cyanide zinc bath

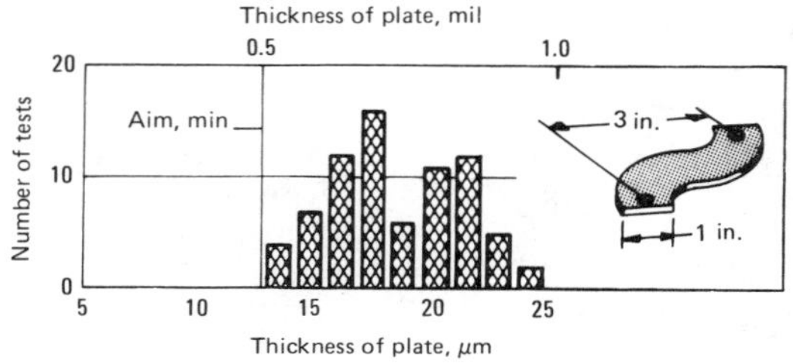

Barrel plating, 75 tests

Table 7 gives estimated life of different thicknesses of unprotected zinc coatings on steel in different outdoor atmospheres. The majority of zinc plated parts are coated with a thickness of 7.5 to 12.5 μm (0.3 to 0.5 mil). Typical applications employing thicknesses less than or greater than the usual 7.5 to 12.5 μm (0.3 to 0.5 mil) are given in Table 8.

Supplementary Coatings. Because corrosion is rapid in industrial or marine locations, zinc plated parts that must endure for many years are usually protected by supplementary coatings (Table 8). Steel with 5 μm (0.2 mil) of electroplated zinc is often painted to obtain a coating system for general outdoor service; a phosphate or chromate postplating treatment ensures suitable adherence of paint to zinc.

In uncontaminated indoor atmospheres, zinc corrodes very little. A 5 μm (0.2 mil) coating has been known to protect steel framework on indoor cabinets for more than 20 years. Atmospheric contaminants accelerate corrosion of zinc if condensation occurs on cooler parts of structural members inside buildings; 12.5 μm (0.5 mil) of zinc may be dissipated in 10 years or less. Zinc-plated steel in such locations is usually given a protective coating of paint.

A satisfactory coating for parts such as those on the inside of an office machine must afford protection in storage, assembly, and service. The cost is also important. Gears, cams, and other parts of the working mechanism can be plated with 3.8 to 6.3 μm (0.15 to 0.25 mil) of zinc to meet these requirements.

Chromate conversion coatings, colored or clear, are almost universally applied to zinc-plated parts for both indoor and outdoor use to retard corrosion from intermittent condensation, such as may occur in unheated warehouses. Chromate films minimize staining from fingerprints and provide a more permanent surface appearance than bare zinc.

Limitations. Zinc-plated steel is not used for equipment that is continually immersed in aqueous solutions and

Table 7 Estimated average service life of unprotected zinc coatings on steel in outdoor service

Condition	Coating thickness μm	Coating thickness mil	Service, yr
Rural	5	0.2	3
	13	0.5	7
	25	1.0	14
	38	1.5	20
	50	2.0	30
Temperate marine	5	0.2	1
	13	0.5	3
	25	1.0	7
	38	1.5	10
	50	2.0	13
Industrial marine	5	0.2	1
	13	0.5	2
	25	1.0	4
	38	1.5	7
	50	2.0	9
Severe industrial	5	0.2	0.5
	13	0.5	1
	25	1.0	3
	38	1.5	4
	50	2.0	6

must not be used in contact with foods and beverages because of dangerous health effects.

Although zinc may be used in contact with gases such as carbon dioxide and sulfur dioxide at normal temperatures if moisture is absent, it has poor resistance to most common liquid chemicals and to chemicals of the petroleum and pharmaceutical industries.

Fasteners. Steel fasteners, such as screws, nuts, bolts, and washers, are often electroplated for corrosion resistance and appearance. If protection against atmospheric corrosion is the sole objective, zinc is the most economical coating metal. Coatings of 5 to 7.5 μm (0.2 to 0.3 mil) give protection for 20 years or more for indoor applications in the absence of frequent condensation of moisture. Chromate coatings are used to retard corrosion from condensates, to provide a more permanent surface appearance, and to prevent staining from fingerprints. For indoor use in industrial areas and in locations where condensation is prevalent, as in unheated buildings, corrosion may be rapid, and the zinc surface should be phosphated and then painted to extend its service beyond the few years that would be obtained by the unpainted coating. Unprotected zinc plated screws should not be used to fasten bare parts if the service is to include marine exposure.

The dimensional tolerance of most threaded articles, such as nuts, bolts, screws, and similar fasteners with complementary threads, does not permit the application of coatings much thicker than 7.5 μm (0.3 mil). The limitation of coating thickness on threaded fasteners imposed by dimensional tolerances, including class or fit, should be considered whenever practicable, to prevent the application of thicker coatings than are generally permissible. If heavier coatings are required for satisfactory corrosion resistance, allowance must be made in the manufacture of the threaded fasteners for the tolerance necessary for plate buildup. If this is not practicable, phosphating before assembly, and painting after assembly, will increase service life. Approximate durability of 5 μm (0.2 mil) untreated coatings is given in Table 7.

Appearance. The appearance of electrodeposited zinc can be varied over a wide range, depending on bath composition, current density, the use of brighteners, and postplating treatments. The appearance of electroplated zinc is bright and silvery, and the deposit from the acid chloride baths is often initially indistinguishable from bright nickel chrome when plated.

Table 8 Applications of zinc plating above 7 to 13 μm (0.3 to 0.5 mil)

Application	Plate thickness μm	Plate thickness mil
Less than 7 μm (0.3 mil) of zinc		
Automobile ashtrays(a)	5-7	0.2-0.3
Birdcages(b)	5	0.2
Electrical outlet boxes(c)	4-13	0.15-0.5
Tacks	5	0.2
Tubular rivets(d)	5	0.2
More than 13 μm (0.5 mil) of zinc		
Conduit tubing(e)	30	1.2
Pipe fittings and couplings	25	1.0

(a) Chromated after plating. (b) Chromated after plating; some parts dyed and lacquered. (c) Bright chromated after plating. (d) Chromated, clear or colored, after plating. (e) Dipped in 0.5% HNO_3 or chromated after plating

Currently, nearly all plated zinc is followed by some type of chromate dip. These preserve the appearance of the part and vastly increase the bright shelf life of the surface. The cost of chromating is so minimal that its use has become practically universal. Presently, bright zinc deposits are used for a wide variety of low-cost consumer goods such as children's toys, bird cages, bicycles, and tools. Refrigerator shelves are a common item bright zinc plated, chromated, and lacquered. Without lacquer protection, even chromated bright zinc will tarnish and discolor quite rapidly when handled, and unlacquered bright zinc plate is not a good substitute for nickel chrome when a long-lasting bright finish is desired. The vast majority of zinc plate is deposited primarily to impart corrosion resistance. Brightness is not the primary factor for these applications.

Cadmium Plating

By Roger E. Marce
Manager, Technical Services
Zinc Institute Inc.

ELECTRODEPOSITS OF CADMIUM are used extensively to protect steel and cast iron against corrosion. Because cadmium is anodic to iron, the underlying ferrous metal is protected at the expense of the cadmium plate even if the cadmium becomes scratched or nicked, exposing the substrate.

Cadmium is usually applied as a thin coating (less than 25 μm or 1 mil thick) intended to withstand atmospheric corrosion. It is seldom used as an undercoating for other metals, and its resistance to corrosion by most chemicals is low.

Besides having excellent corrosion protective properties, cadmium has many useful engineering properties, including natural lubricity. When corrosion products are formed on cadmium electroplated parts, they are not voluminous, and there is no dimensional change. These two properties are responsible for the wide use of cadmium on moving parts or threaded assemblies.

Cadmium has excellent electrical conductivity and low contact resistance. Noncorrosive fluxes can be used to produce top quality soldered sections. Steel that is coated with cadmium can be formed and shaped because of the ductility of the cadmium. Malleable iron, cast iron, powdered metals, and other hard-to-plate surfaces can be coated with cadmium, and materials used for adhesives bond very well to cadmium-coated surfaces.

Plating Baths

Most cadmium plating is done in cyanide baths, which generally are made by dissolving cadmium oxide in a sodium cyanide solution. Sodium cyanide provides conductivity and makes the corrosion of the cadmium anodes possible.

Cyanide Baths. Compositions and operating conditions of four cyanide baths are given in Table 1. Note that for each of these baths a ratio of total sodium cyanide to cadmium metal is indicated; maintenance of the recommended ratio is important to the operating characteristics of the bath.

For still tank or automatic plating of steel, selection of a bath on the basis of cyanide-to-metal ratio depends on the type of work being plated and the results desired:

- For parts with no recesses and when protection of the basis metal is the sole requirement, Solution 1 in Table 1 (ratio, 4 to 1) is recommended
- For plating parts with deep recesses and when a bright, uniform finish is required, Solution 2 in Table 1 (ratio, 7 to 1) is recommended
- For all-purpose bright plating of various shapes, Solution 3 in Table 1 (ratio, 5 to 1) is recommended
- For high-speed, high-efficiency plating, Solution 4 in Table 1 (ratio, 4.5 to 1) is recommended

Although the use of brighteners produces maximum improvement in uniformity and throwing power in Solution 3 in Table 1, brighteners also improve these properties in Solutions 1 and 2.

Normally, the sodium hydroxide content of cyanide baths is not critical. Usual limits are 22 g/L (3 oz/gal); the preferred concentration for best results is 15 ± 4 g/L (2 ± 0.5 oz/gal). Sodium hydroxide contributes to conductivity and, in excess, affects the current-density range for obtaining bright plate. Analytical procedures useful in the maintenance of cyanide baths are outlined in Appendix 1 of this article.

In recent years, the need for pollution control of cyanide solutions has led to the development of noncyanide cadmium electroplating baths, shown in Table 2. Noncyanide baths generate little hydrogen embrittlement and are used to electroplate hardened, high-strength steels. Only the fluoborate bath has been used for some time as a substitute for cyanide baths and working data are available. The fluoborate bath is characterized by high cathode efficiency, good stability, and relatively little production of hydrogen embrittlement (Appendix 2). The major disadvantage of the fluoborate bath is its poor throwing power. It is widely used in barrel plating operations. If this bath is used for still plating at high current density, air agitation is desirable. Wire and strip can readily be plated in a fluoborate bath. Practically all of the oth-

Table 1(a) Compositions of cadmium plating cyanide solutions

Solution No.	Ratio of total sodium cyanide to cadmium metal	Cadmium oxide g/L	Cadmium oxide oz/gal	Cadmium metal g/L	Cadmium metal oz/gal	Sodium cyanide g/L	Sodium cyanide oz/gal	Sodium hydroxide(b) g/L	Sodium hydroxide(b) oz/gal	Sodium carbonate(c) g/L	Sodium carbonate(c) oz/gal
		Composition(a)									
1	4:1	23	3	19.6	2.62	78	10.4	14.2	1.90	30-75	4-10
2	7:1	23	3	19.6	2.62	138	18.4	14.2	1.90	30-45	4-6
3	5:1	26	3.5	22.9	3.06	115	15.3	16.4	2.19	30-60	4-8
4	4.5:1	40	5.5	36.1	4.82	163	21.7	25.8	3.44	30-45	4-6

(a) Metal-organic agents are added to cyanide solutions to produce fine-grained deposits. The addition of excessive quantities of these agents should be avoided, because this will cause deposits to be of inferior quality and to have poor resistance to corrosion. The addition of these agents to solutions used for plating cast iron is not recommended. (b) Sodium hydroxide produced by the cadmium oxide used. In barrel plating, 7.5 g/L (1 oz/gal) is added for conductivity. (c) Sodium carbonate produced by decomposition of sodium cyanide and absorption of carbon dioxide, and by poor anode efficiency. Excess sodium carbonate causes anode polarization, rough coatings, and lower efficiency. Excess sodium carbonate may be reduced by freezing, or by treatment with calcium sulfate

Table 1(b) Operating conditions of cadmium plating cyanide solutions

Solution No.	Current density(a) Range A/dm^2	Range A/ft^2	Average A/dm^2	Average A/ft^2	Operating temperature °C	°F	Remarks
1	0.5-6	5-60	2.5	25	27-32	80-90	For use in still tanks. Good efficiency, fair throwing power Also used in bright barrel plating
2	1-8	10-80	2.5	25	27-32	80-90	For use in still tanks and automatic plating. High throwing power, uniform deposits, fair efficiency. Not for use in barrel plating
3	0.5-9	5-90	3.5	35	24-29	75-85	Primarily for use in still tanks, but can be used in automatic plating and barrel plating. High efficiency and good throwing power
4	0.5-15	5-150	5.0	50	27-32	80-90	Used for plating cast iron. High speed and high efficiency(b)

(a) For uniform deposits from cyanide solutions, the use of a current density of at least 2 A/dm^2 (20 A/ft^2) is recommended. Agitation and cooling of solution are required at high current densities. (b) Agitation and cooling are required when current density is high (above 2 A/dm^2, or 20 A/ft^2)

er acid-type baths shown in Table 2 are supplied to electroplaters as proprietary baths. Because each proprietary bath has its own peculiarities, it is advisable to obtain all proper operating information from the supplier to obtain the desired results.

Brighteners. The most used, and probably the safest, brightening agents for cyanide baths are organics such as:

- Aldehydes
- Ketones
- Alcohols
- Furfural
- Dextrin
- Gelatin
- Milk sugar
- Molasses
- Piperonal
- Some sulfonic acids

These materials form complexes with the electrolyte in cyanide baths and influence the orientation and growth of electrodeposited crystals, resulting in the formation of fine longitudinal crystals, and hence a bright deposit. Care should be taken not to add the brighteners in too large an amount. Too much brightener can result in dullness, pitting, blistering, and general poor quality and appearance. It is difficult to remove the excess brightener. Many organic brighteners are available as proprietary materials. When these are used, manufacturers' recommendations regarding amounts and other conditions of use should be followed.

Another method of brightening consists of the use of trace quantities of metallic nickel, cobalt, molybdenum, and selenium. The concentration of these elements in the bath is much more critical than the concentration of the organic brighteners. Poor bright dipping qualities or poor ductility and corrosion resistance of the coating may result from an excess of these metals. Certain proprietary brighteners contain both metallic and organic compounds. Brighteners for the noncyanide baths are also proprietary products.

Table 2 Noncyanide cadmium plating baths

Bath	Proprietary(a) g/L	Proprietary(a) oz/gal	Fluoborate(b) g/L	Fluoborate(b) oz/gal	Acid sulfate(c)
Ammonium chloride	11-23	1.5-3.0	...	...	...
Ammonium fluoborate	...	...	60	8	...
Ammonium sulfate	75-115	10-15	...	...	...
Boric acid	...	...	25	3.6	...
Cadmium	4-11	0.5-1.5	95	12.6	...
Cadmium fluoborate	...	...	242	32.2	...
Cadmium oxide	...	...	...	...	8-11 g/L (1.0-1.5 oz/gal)
Sulfuric acid	...	...	...	...	4.5-5.0%

(a) Proprietary requires a current density of 0.2 to 1.5 A/dm^2 (2 to 15 A/ft^2), and an operating temperature of 16 to 38 °C (61 to 100 °F). (b) Fluoborate requires a current density of 3 to 6 A/dm^2 (30 to 60 A/ft^2), and an operating temperature of 21 to 38 °C (70 to 100 °F). (c) Acid sulfate requires a current density of 1.0 to 6.0 A/dm^2 (10 to 61 A/ft^2), and an operating temperature of 16 to 32 °C (61 to 90 °F)

Rough or pitted deposits should not be encountered in a well-balanced, carefully operated bath. However, if the concentration of metal is too low or the ratio of metal to cyanide varies from recommended values, roughness may result. Other factors that may contribute are contamination by dust, dirt, oil, metallic particles, or soap. Excessive concentrations of sodium carbonate and too high a temperature or current density also promote surface roughness.

Pitted deposits usually are the result of metallic impurities or an excessive amount of decomposed organic addition agents. The interfering metals are: antimony, lead, silver, arsenic, tin, and thallium. Pitting may result also from the presence of nitrates.

Correction of roughness or pitting may require a complete solution cleanup, including removal of excess sodium carbonate, purification with zinc dust, treatment with activated carbon, and filtration.

Formation and Elimination of Carbonate. Sodium carbonate forms in the cyanide bath as a result of the decomposition of sodium cyanide and the reaction of sodium cyanide with carbon dioxide from the air. When air agitation is used, the buildup of sodium carbonate is accelerated. The buildup also results from failure to keep ball-anode racks full or from the use of a large area of insoluble steel anodes.

Maximum concentrations of sodium carbonate that can be present in the bath without adverse effect on operating efficiency and deposit characteristics depend on the metal content of the bath. For example, carbonate can be present in concentrations up to 60 g/L (8 oz/gal) if the metal content is 19 g/L (2.5 oz/gal), and up to 30 g/L (4 oz/gal) if metal content is 30 g/L (4 oz/gal), without deleterious effect. Exceeding these concentrations results in anode polarization, in depletion of the metal content of the bath, and in poor, irregular, and dull deposits.

To remove carbonate, either calcium sulfate or calcium cyanide may be added to the bath. Calcium sulfate forms a bulky precipitate, whereas the product formed by calcium cyanide is beneficial.

Purification and Filtration. Whenever it is convenient, continuous filtration is advisable. If a solution is contaminated by impurities such as copper, tin, lead, or other metals, the following treatment is recommended.

Transfer the solution to an auxiliary tank of the same size as the plating tank; stir in 0.7 to 1 kg (1.5 to 2 lb) of purified zinc dust per 400 L (100 gal). Continue to stir for about 1 h, then allow to settle for no more than 6 h. Filter through a well-packed filter. If the solution contains excess organic impurities, such as decomposed brighteners, it should be treated with activated carbon and filtered. Pumps and filter parts should be made of iron or steel for alkaline cyanide baths. The solution attacks brass or bronze, and heavy copper contamination results.

Fig. 1 Still plating tank with spiral steel holders for cadmium ball anodes

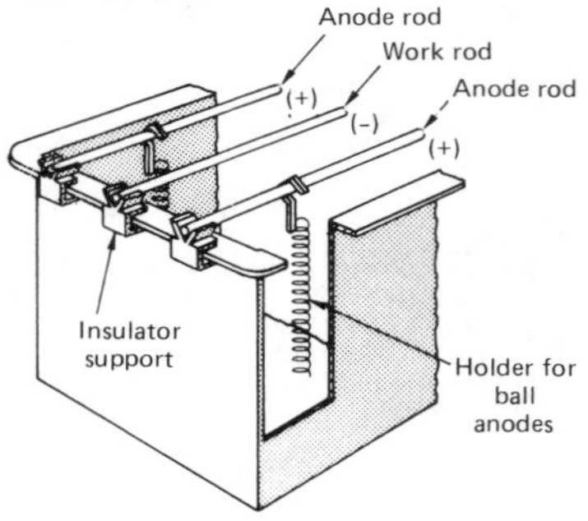

Anodes

The anode system for cadmium plating from a cyanide solution consists of ball-shaped cadmium anodes in a spiral cage of bare steel (Fig. 1). The spherical shape provides a large surface area in relation to weight, without a large investment in cadmium. Ball anodes also make it possible to maintain an approximately constant anode area, and little or no anode scrap is produced. Cadmium balls are usually 50 mm (2 in.) in diameter and weigh 0.57 kg (1¼ lb) per ball.

If a cadmium cyanide solution is to be left idle for an extended period of time (a week or more), the steel anode cages should be removed from the solution, because the galvanic cell set up between the steel and the cadmium anodes will accelerate chemical dissolution of the anodes when the current is off.

When cadmium is plated from an acid solution, such as the fluoborate bath, ball anodes in uncoated steel cages cannot be used, because the steel would dissolve. Rather, bar anodes of elliptical or oval cross section, 460 to 2440 mm (18 to 96 in.) long are used.

The use of bar anodes in a cyanide solution results in a high percentage of waste, because they must be removed and replaced when the cross-sectional area decreases, or they will dissolve preferentially at the solution level and drop to the bottom of the plating tank.

Purity of the anode is of great importance, especially if a bright deposit is to be produced. Typical composition range for cadmium anodes is as follows:

Element	Composition, %
Cadmium	99.95-99.97
Lead	0.008-0.03
Iron	0.005-0.008
Copper	0.002-0.01
Arsenic	0-0.001
Zinc	0-0.001

Anode composition complying with Federal Specification QQ-A-671 is:

Cadmium	99.9%, min
Silver, lead, tin	0.05%, total max
Arsenic, antimony, thallium	0.005%, max

Insoluble anodes, which are made of low-carbon steel strip or wire, offer no particular advantage except where inside anodes are necessary or for special applications in which they are required because of a need to reduce metal concentration in the plating bath. When insoluble anodes are used, their total area should be 10 to 15% of the total anode area. Insoluble anodes accelerate the formation of carbonate.

Current Density

Cyanide cadmium baths may be operated over a wide range of cathode current densities, as indicated in Table 1. In a properly formulated bath operated within its intended current-density range, the cathode efficiency is 90% ± about 5%. Thus, to apply a 25 μm (1 mil) deposit of cadmium requires 1.1 $A \cdot h/dm^2$ (11 $A \cdot h/ft^2$).

The ranges of current density given in Table 1 are suggested limiting values. Choice of current density is governed mainly by the type of work being plated; for example, low current densities are suitable for small lightweight parts, current densities up to 4 A/dm^2 (40 A/ft^2) for medium-weight parts of fairly uniform shape and high current densities for uniform heavy parts like cylinders and shafts.

Baths containing 22 g/L (2.5 oz/gal) of cadmium are suitable for general use at current densities up to 2.5 A/dm^2

(25 A/ft²); higher concentrations of cadmium, up to 44 g/L (5 oz/gal), permit operation at higher current density.

A bath containing 19 g/L (2.5 oz/gal) of cadmium is suitable for barrel plating, where average current density may be about 0.5 A/dm² (5 A/ft²). Such a bath is suitable also for many still tank or automatic plating applications in which current densities do not exceed 2.5 A/dm² (25 A/ft²). At higher current densities, burning may result, with attendant dull, rough deposits that lack decorative and protective qualities. Where higher current densities are required, baths of higher metal content should be used.

Too low a current density (less than 0.5 A/dm² or 5 A/ft²), particularly in still tank or automatic plating, can result in excessively long plating times and inferior appearance of deposits.

The recommended range of current densities for plating with a fluoborate bath is 3 to 6 A/dm² (30 to 60 A/ft²). Even near 6 A/dm² (60 A/ft²), however, the bath has poor throwing power.

Deposition Rates

Among plating baths used commercially to deposit common metals (other than precious metals), cadmium cyanide baths are high in both throwing and covering power; only alkaline tin and cyanide copper have greater throwing power; based on Haring-Blum cell measurements, the throwing power of cadmium cyanide baths is rated between 40 and 45% with a distance ratio of 5. Therefore, the distance between the anode and the work is not critical, although as the distance is increased, current density and efficiency decrease, and current distribution is altered.

Table 3 lists the times required to plate cadmium deposits from 3 to 18 μm (0.1 to 0.7 mil) thick. These times are predicated on 90% cathode efficiency.

Bath Temperature

Typical operating temperature ranges for cyanide baths are given in Table 1. Data for noncyanide baths are shown in Table 2. In general, satisfactory plating results are obtained by controlling bath temperature within ±3 °C (±5 °F) during plating. When greater precision is required, temperature should be controlled within ±1 °C (±2 °F).

Table 3 Time for plating cadmium to a given thickness at various current densities
Based on 90% cathode efficiency, in a cyanide bath

Thickness of plate		Plating time in minutes at current density of:					
μm	mil	0.5 A/dm² (5 A/ft²)	1.0 A/dm² (10 A/ft²)	1.5 A/dm² (15 A/ft²)	2.0 A/dm² (20 A/ft²)	2.5 A/dm² (25 A/ft²)	3.0 A/dm² (30 A/ft²)
3	0.1	13.0	6.5	4.3	3.2	2.6	2.2
5	0.2	26.0	13.0	8.6	6.4	5.2	4.4
8	0.3	39.0	19.5	13.9	9.6	7.8	6.6
10	0.4	52.0	26.0	18.2	12.8	10.4	8.8
13	0.5	65.0	32.5	22.5	16.0	13.0	11.0
15	0.6	78.0	39.0	25.8	19.2	15.6	13.2
18	0.7	91.0	45.5	30.4	22.4	18.2	15.4

Plating Equipment

Considerations specific to the operation of cadmium cyanide baths in conventional plating equipment are discussed here, with attention to the materials of construction used.

Still Tanks. Usually, unlined steel tanks are used for alkaline cadmium plating; however, steel tanks with rubber linings are useful in preventing stray tank currents. Rubber and plastics used for tank linings should be tested for compatibility with the plating bath, to prevent contamination from constituents of the lining. Vinyl plastisols are compatible, commercially available, and require no further testing.

Filters and cooling coils also may be made of steel. Equipment for fume control should be used; such equipment in some cases is required by local ordinances. Typical tank arrangement is shown in Fig. 1. Equipment for baths other than the cyanide must be made acid-resistant.

Barrels may be made of hard rubber, polypropylene, acrylic resins, phenol-formaldehyde or melamine-formaldehyde laminates, or of expanded or perforated sheet steel coated with vinyl plastisol. The plastisol coating is about 3.2 mm (⅛ in.) thick and is resistant to the standard barrel plating solutions and temperatures. Usually, doors and wall ends are of the same material.

Perforated cylinders for oblique barrels also have perforated bottoms and are made of the same materials used for perforated cylinders of horizontal barrels.

Anodes used for barrel plating may be bar- or ball-shaped. For maximum current density, the anodes are curved to shorten the path of the current. Curved solid anodes are placed on insulated supports, whereas anode balls are placed in curved holders tied together at the lower ends.

Figure 2 illustrates schematically the use of barrel equipment for cadmium plating. Although not shown in the illustration, barrel installations are equipped with plate coils to remove the excess heat caused by the high current used in the plating bath.

Automatic plating machines may be of either the straight line or the return type. In straight line plating machines, the work is loaded at one end, carried through the various phases of the cleaning and finishing cycles, and unloaded at the opposite end. Such a machine is considered a heavy-duty unit, because it can be designed for large racks and heavy loads.

Loading and unloading of the return machine is performed in the same area; the work follows an elliptical path, as indicated by the schematic layout of Fig. 3. This unit can be designed for either light or heavy loads.

Both types of automatic machines may be continuous, with the work load in constant motion, or intermittent, in which case the motion of the carriers stops for a predetermined time after the work is immersed in each solution.

Power for cadmium plating is provided by rectifiers, which are low in initial cost, flexible, and have a favorable power factor and efficient load characteristics. Three-phase, full-wave rectifiers should be used, because bright cadmium plating is sensitive to a high ripple current, which produces a dull matte finish on coatings plated in the medium and high ranges of current density.

The rectifier elements may be copper sulfide, copper oxide, silicon, or selenium. At present, most of the rectifiers have selenium elements. Selenium rectifiers are more compact than the copper types and are designed with higher inverse voltage limits, resulting in greater service life under normal plating conditions.

Fig. 2 Installation for cadmium plating by the barrel method

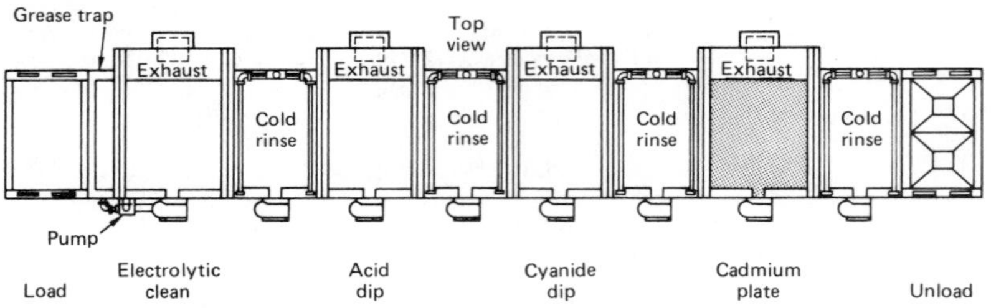

Fig. 3 Installation for automatic cadmium plating

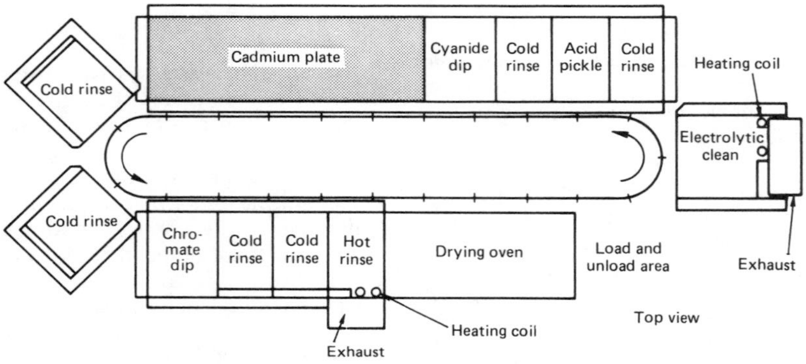

Rinse Tanks. Although longer tank life will be obtained if rinse tanks are lined or coated with polyvinyl chloride or rubber, all rinsing, with the exception of the rinse following hydrochloric acid pickling, may be done in unlined steel tanks. The use of unlined steel tanks for rinsing following pickling or acid plating is not recommended.

Racking of parts for cadmium plating is subject to the same considerations as in the electrodeposition of other metals. Information on design and use of plating racks is given in the article on hard chromium plating in this Volume.

Maintenance. The following is a typical schedule of maintenance for plating and auxiliary equipment:

Daily

- Check anodes; replenish when necessary
- Check all contacts, anode and cathode
- Check solution levels
- Check bath temperatures and controls
- Check bath composition, if possible, using chemical analysis (see Appendix 1) and plating cell test
- Probe tank bottom for lost parts
- Check motors for signs of overheating, arcing or failure
- Check amperage and voltage to work
- Check lubrication on automatic equipment

Weekly

- Probe tank bottom for lost parts, if not checked daily
- Check rubber tank linings for damage
- Filter plating bath, unless constant filtration is used
- Check bath analysis, chemically and with plating cell, and make additions and corrections, if these functions are not performed more frequently
- Oil equipment
- Clean all contacts
- Check for preventive-maintenance items that cannot be repaired during the week
- Dump and replenish cleaning lines where necessary

Monthly

- Pump plating solution to purification tank; treat for impurities, if necessary
- Inspect tank linings while plating tanks are empty; repair if necessary
- Inspect and clean heat exchangers or plate coils if accumulation or buildup exists
- Blow out and check rectifier stacks for condition and power delivery
- Check for arcing or scored armatures on generators. Blow out coils
- Perform general preventive maintenance examination of all equipment

Semiannually

- Clean out exhaust systems
- Repair exhaust fans
- Check all motors
- Repaint where necessary
- Inspect and clean out all floor drains
- Check for leaks and cross connections between cyanide and acid drains
- Check all items usually covered on annual or semiannual overhaul, such as solenoid valves, limit switches, relays, and automatic electrical equipment

Selection of Method

Selection of plating method involves both technical and economic factors. Still plating, with parts racked, is the oldest and most universally used plating method. Barrell plating, limited to smaller parts that can be tumbled in the plating bath, is popular because many parts can be plated at one time. Automatic hoist units offer a means of mechanizing the still-tank rack or the barrel.

Still tanks are suitable for all types of work. They are used for small production quantities, in general, and for all quantities of parts that cannot be plated in barrel or automatic systems, because of a need for auxiliary anodes or special handling or because plating dimensions are critical.

Example 1. Valve bodies and baffle plates are typical of many parts that are plated in still tanks (see Table 4). Details of the equipment needed to cadmium plate the production quantities of these parts given in Table 4 are given below:

Plating tank (1440 L or 380 gal) .....270 by 75 by 75 cm (9 by 2½ by 2½ ft)
Other tanks (420 L or 110 gal) ......90 by 75 by 75 cm (3 by 2½ by 2½ ft)
Power rectifier (600 A)..................1.5 to 6 V
Dimensions of rectifier........75 by 90 by 210 cm (2½ by 3 by 7 ft)

Total floor space of equipment and access area .... 195 by 450 cm (6½ by 15 ft)
Number of racks 15

Other tanks include a cleaning tank, an acid pickle tank, a hot water rinse tank, and three cold water rinse tanks.

Barrel plating may be used for parts up to 100 mm (4 in.) long and 50 mm (2 in.) thick. Parts such as machine bolts, nuts, and washers are ideal for barrel plating. Conversely, intricate shapes such as ornaments and complex castings of brittle metals with small sections that fracture easily should not be barrel plated; the tumbling action may damage these parts, and variation in plating thickness and appearance may result. Intricate designs incorporating recessed or shielded areas may present problems in plating coverage, luster, and appearance. Barrel plating is not applicable for parts requiring heavy plate. Usually, 8 to 13 μm (0.3 to 0.5 mil) is the maximum thickness of plate applied.

Example 2. Small coil springs and brush holders are illustrative of parts suitable for barrel plating. Production requirements for plating these parts in horizontal barrels are given in Table 5. Equipment specifications are as follows:

Plating tank (1330 L or 350 gal) .... 180 by 120 by 75 cm (6 by 4 by 2½ ft)
Other tanks (610 L or 160 gal) ..... 90 by 120 by 75 cm (3 by 4 by 2½ ft)
Power rectifier (2000 A) .... 9 to 15 V
Dimensions of rectifier .... 90 by 120 by 240 cm (3 by 4 by 8 ft)
Centrifugal dryer .. 60 by 60 by 75 cm (2 by 2 by 2½ ft)
Baking oven ..... 120 by 90 by 240 cm (4 by 3 by 8 ft)
Equipment floor space .... 12 m^2 (125 ft^2)
Access area behind line .... 6 m^2 (68 ft^2)
Access area in front .... 9 m^2 (100 ft^2)

Other tanks in the list above refer to cleaning tanks, acid pickle tanks, hot water tanks, and three cold water rinse tanks.

Automatic Plating. The primary selection factor for automatic plating is cost. Volume of work must be sufficient to warrant installation of the equipment.

Table 4 Equipment requirements for cadmium plating valve bodies and baffle plates in still tanks

Production requirements	Valve body	Baffle plate
Weight per piece	1.1 kg (2½ lb)	0.2 kg (0.5 lb)
Pieces plated per hour	210	175
Area plated per hour	6.5 m^2 (70 ft^2)	11.1 m^2 (120 ft^2)
Minimum thickness	8 μm (0.3 mil)	4 μm (0.15 mil)

Example 3. Voltage-regulator bases were cadmium plated, to a minimum thickness of 3.8 μm (0.15 mil), in automatic equipment at the rate of 2640 pieces per hour. Equipment requirements are given below:

- **Production requirements**

Weight per piece .... 170 g (6 oz)
Pieces plated per hour 2640
Area to be plated per hour .... 53 m^2 (570 ft^2)
Minimum plate thickness .... 4 μm (0.15 mil)

- **Equipment requirements**

Dimensions of full automatic plating unit ...... 2100 by 330 by 270 cm (70 by 11 by 9 ft)
Width of access space on sides of unit .... 90 cm (3 ft)
Width of access space on load end of unit .... 300 cm (10 ft)
Motor-generator set ... 15 V, 7500 A
Dimensions of motor-generator set ...... 300 by 300 by 240 cm (10 by 10 by 8 ft)

Example 4. A quantity of 12 000 to 14 000 electrical-outlet receptacles per 8-h day was required in order to justify the use of a small automatic plating system of 3800-L (1000-gal) solution capacity with a single lane of rods and workpieces and plating 4 to 5 μm (0.15 to 0.20 mil) of cadmium. When the size and shape of the parts are such that either automatic or still tank plating processes may be used, the requirements for racking often are the most important factor in determining the relative economy of still tank and automatic plating. Two kinds of automated plating equipment are available: (*a*) the regular return machine; and (*b*) the programmed hoist unit, which is an automated straight line unit. The latter equipment is much less expensive to purchase.

Table 5 Production requirements for cadmium plating coil springs and brush holders in a horizontal barrel

Production requirements	Coil spring	Brush holder
Weight per piece	14 g (½ oz)	9 g (5⁄16 oz)
Pieces plated per hour	7200	3800
Area plated per hour	22 m^2 (240 ft^2)	17 m^2 (180 ft^2)
Minimum thickness	4 μm (0.15 mil)	8 μm (0.3 mil)

Cleaning and rinsing are essential operations in any plating sequence. Figures 2 and 3 show the number of tanks or stations required for such operations in typical barrel and automatic processes. In Fig. 4, where cleaning, rinsing, and postplating operations are indicated for various initial conditions of the work surface, the plating step itself is a rather inconspicuous item in the flow chart of the total finishing process. Table 6 shows variations in processing techniques for still tank, barrel, or automatic plating to a thickness of less than 13 μm (0.5 mil).

Variations in Plate Thickness

For adequate protection of steel, the thicknesses of cadmium in Table 7 are recommended. The shape of a part can markedly influence uniformity of the electrodeposit. Parts of simple design, such as socket wrenches and bathroom hardware, can be plated with a high degree of uniformity of plate thickness. On such parts, about 90% uniformity would be anticipated.

Threaded fasteners present a special problem, because of variations in contour and because of tolerance requirements. These items ordinarily are barrel plated, and thicknesses of 3 to 4 μm (0.10 to 0.15 mil) are usually specified.

Throwing Power. The effect of shape on uniformity of deposit thickness is exemplified by the open-ended box (100-mm, or 4-in. cube) of Fig. 5. The open end of the box is pointed toward one of the anodes to produce the most desirable condition for this shape with-

Fig. 4 Cleaning, cadmium plating, and post-treatments of steel and cast iron parts

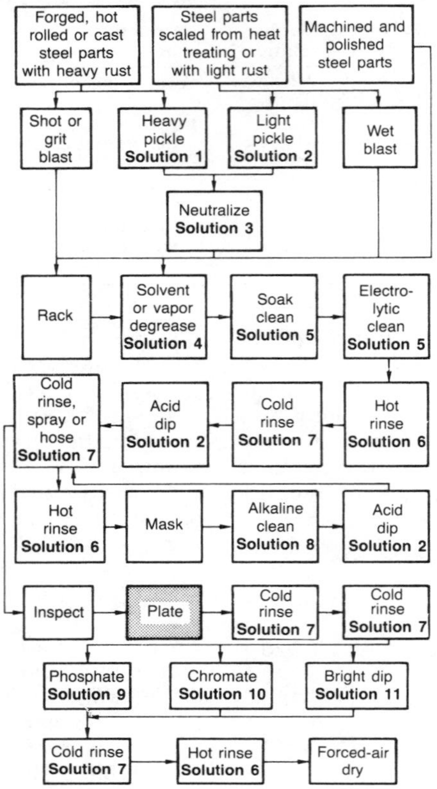

Solution No.	Composition	Amount	Temperature °C	Temperature °F	Immersion time
1	H_2SO_4	8-12 vol %	71-93	160-200	10-120 min
2	HCl	20-50 vol %	Room temperature		1/6-3 min
3	Na_2CO_3	75-90 g/L (10-12 oz/gal)	Room temperature		15-60 s(a)
4	Petroleum solvent	...	Room temperature		1/2-3 min
5	Alkali(b)	60-75 g/L(b) (8-10 oz/gal)	82-93(b)	180-200(b)	1/2-3 min
6	Water	...	82-93(c)	180-200(c)	5-15 s
7	Water(d)	...	Room temperature		5-15 s
8	Alkali	60-75 g/L (8-10 oz/gal)	66 max	150 max	1/2-1 min
9	(e)	(e)	(e)	(e)	1/2-1 min
10	(e)	(e)	(e)	(e)	30 s
11	NaCN	45-60 g/L (6-8 oz/gal)	Room temperature		5-15 s

Note: For cast iron, solutions, conditions, and procedure are the same as for steel, except that cast iron parts, after being thoroughly washed in cold water following the acid dip, are dipped for 5 s in a room-temperature cyanide solution (NaCN, 45 to 60 g/L or 6 to 8 oz/gal) and then again rinsed in cold water, before proceeding to inspection, plating, and post-treatments

(a) When solution is sprayed, time is 5 to 15 s. (b) Heavy-duty cleaner. For electrolytic cleaning, concentration of alkali is 45 to 60 g/L (6 to 8 oz/gal), temperature is 82 °C (180 °F), and time is 1 to 3 min. (c) When a spray rinse is used, water temperature is 71 to 82 °C (160 to 180 °F). (d) Immersion or spray rinsing. (e) Proprietary compounds

out auxiliary thief rings, shields, bipolar anodes, insoluble anodes, or other devices. Results of plating such boxes with cadmium, silver, and copper, all deposited from cyanide baths, are shown in Fig. 5. These diagrams illustrate two facts: (*a*) thickness of plate varies significantly from place to place on the simplest shape; and (*b*) various plating baths have different throwing power or ability to plate uniformly over the surface, regardless of shape.

The data on cyanide baths tabulated in Fig. 5 show that cadmium has appreciably less throwing power than silver or copper. However, cyanide cadmium has greater throwing power than nickel, chromium, iron, cyanide zinc, acid tin, acid cadmium, acid copper, or acid zinc. Normally, metals plated from cyanide or alkaline baths are more uniformly distributed than metals from acid baths. As design becomes more complex, uniform thickness of plate is more difficult to achieve without the use of special conforming anodes.

Example 5. A cylindrical, cup-shaped production part that was plated without the use of conforming anodes is shown in Fig. 6. Thickness of plate varied from a minimum of 6 μm (0.25 mil) to a maximum of 25 μm (1 mil).

Conforming Anodes. Parts of complex shape with stringent dimensional requirements, such as those shown in Fig. 7 and 8, require the use of special techniques, conforming anodes, and shields, in order to obtain the required uniformity of plate thickness.

Example 6. A shim, 300 mm (12 in.) long by 40 mm (1½ in.) wide by 2.4 mm (0.095 in.) thick, is shown in Fig. 7. Parallelism of all sides, as well as plate thickness, was extremely critical. When this part was plated in a simple rack, plate thickness varied from 13 μm (0.5 mil) at the center to 50 to 75 μm (2 to 3 mils) at the edges and ends.

By using shields that approximated the outline of the shim, it was possible to cadmium plate all over to a depth of 13 ± 5 μm (0.5 ± 0.2 mil). The part was gently agitated in a still bath.

Example 7. A coupling that required 8 to 13 μm (0.3 to 0.5 mil) of cadmium all over, except for the last 6 mm (¼ in.) of the outside diameter of the small end is shown in Fig. 8. The internal splines on both large and small bores were checked with plug gages and a single-tooth gage to assure uniformity of plate thickness. To obtain the required uniformity, a 6-mm (¼-in.) diam anode was centered in the bore during plating. Although the outer surface of the large end of the coupling accumulated a heavier coating than other areas, general plate-thickness uniformity met requirements.

Example 8. A coupling that, except for the external teeth, was cadmium plated all over to a specified depth of 8

Table 6 Conditions for plating cadmium to a thickness of less than 13 μm (0.5 mil)

Process variable	Still tank	Barrel	Automatic
Soak cleaning			
Alkali	53 g/L (6 oz/gal)	106 g/L (12 oz/gal)	70 g/L (8 oz/gal)
Temperature	82 °C (180 °F)	82 °C (180 °F)	82 °C (180 °F)
Time, min	2-3	5	3-5
Rinsing			
Temperature	Ambient	Ambient	Ambient
Time, min	¼	3	½
Electrolytic cleaning			
Alkali	70 g/L (8 oz/gal)	...	70 g/L (8 oz/gal)
Temperature	82 °C (180 °F)	...	82 °C (180 °F)
Time, min	½-1	...	1-3
Rinsing			
Temperature	Ambient	Ambient	Ambient
Time, min	¼	3	1
Acid dipping			
HCl, vol %	10-50	10-50	10-50
Temperature	Ambient	Ambient	Ambient
Time, min	⅙-1	3	½ to >1
Rinsing			
Temperature	Ambient	Ambient	Ambient
Time, min	¼	3	1
Cyanide dipping			
NaCN	30-45 g/L (4-6 oz/gal)	30-45 g/L (4-6 oz/gal)	30-45 g/L (4-6 oz/gal)
Temperature	Ambient	Ambient	Ambient
Time, min	¼	3	1
Plating			
Temperature	29 °C (85 °F)	29 °C (85 °F)	29 °C (85 °F)
Amperage	2.5 A/dm^2 (25 A/ft^2)	9-15 V	2.5 A/dm^2 (25 A/ft^2)
Time, min	10	30	10
Rinsing			
Temperature	Ambient	Ambient	Ambient
Time, min	¼	3	½
Rinsing			
Temperature	Ambient	Ambient	Ambient
Time, min	¼	2	½
Bright dipping			
HNO_3, vol %	¼-½	¼-½	¼-½
Temperature	82 °C (180 °F)	Ambient	Ambient
Time, min	⅙	2	½
Rinsing			
Temperature	... (160-180 °F)	71-82 °C	82 °C (180 °F)
Time, min	...	2	½
Drying			
Temperature	82-105 °C (180-220 °F)	82-105 °C (180-220 °F)	82-105 °C (180-220 °F)
Time, min	1-3	5	1-3

to 13 μm (0.3 to 0.5 mil) is also shown in Fig. 8. Spline and internal bore dimensions were critical and had to be held to a tolerance of ±5 μm (±0.2 mil) after plating. Again, uniformity of plate thickness was achieved by centering a 6-mm (¼-in.) diam anode in the bore during plating.

Simple cylindrical, cuboid, and channel shapes, such as those shown in Fig. 9, usually require conforming anodes in order to achieve complete coverage of plate and reasonable plating uniformity. Dimensional limits that definitely require the use of an internal anode are indicated for each geometric shape.

Normal Variations. Even under preferred production conditions, some variation in plate thickness must be anticipated. Usually, this normal scatter is acceptable and falls within the specified range of allowable variation.

In general, barrel plating produces greater variations in thickness than still plating. In barrel plating, factors such as the weight, size, and shape of the part usually exert a greater influence on uniformity of plate thickness than they do in still or automatic plating.

Screws, nuts, and other small parts of fairly regular shape will usually coat uniformly in barrel plating. Parts that are likely to nest because they have large flat areas or cup-shaped recesses exhibit wide variations in coating thickness. Variations decrease somewhat as the thickness of plate increases.

Variations in plate thickness obtained on production parts are detailed in the example that follows:

Example 9. The small cylindrical part shown in Fig. 10 was plated in a horizontal barrel. The load contained about 5000 pieces. Thickness of plate was measured with a magnetic gage on 90 parts from each load. Plating thickness ranged from 5 to 14 μm (0.2 to 0.6 mil).

Other Application Factors

Aside from considerations of cost, there are no size limitations on parts that can be cadmium plated, provided a tank of adequate size and other essential equipment are available. When a very large part is to be plated, jet plating methods may sometimes be used, rather than constructing a very large plating tank. In the jet technique, a steady stream of solution is impinged against the part to be plated until the required thickness of plate is obtained. Because of the rapid movement of the solution, very high current densities can be used. The quality of the plate is comparable to that obtained by conventional methods.

Hardness. The hardness of the basis metal has little or no effect on the successful deposition of cadmium. However, the harder steels are likely to be more highly alloyed and may produce difficult-to-remove smuts from excessive pickling or chemical cleaning. Pickling is also a source of hydrogen embrittlement, which may be particularly harmful to hardened and stressed parts.

Springs often are electroplated with cadmium for protection against corrosion and abrasion. The following example deals with failure of a cadmium

Table 7 Recommended thicknesses of cadmium

Environmental exposure	Description	Thickness μm	Thickness mil	Uses
Mild	Exposure to indoor atmospheres with rare condensation. Minimum wear and abrasion	5	0.2	Springs, lock washers, fasteners
Moderate	Exposure mostly to dry indoor atmospheres. Subject to occasional condensation, wear or abrasion	8	0.3	Television and radio chassis, threaded parts, screws, bolts, radio parts instruments
Severe	Exposure to condensation, infrequent wetting by rain, cleaners	13	0.5	Washing machine parts, military hardware, electronic parts for tropical service
Very severe	Subject to frequent exposure to moisture, saline solutions, and cleaners	25	1	...

plated compression spring that was not properly treated to release hydrogen.

Example 10. A spring used in a high-temperature relief valve under intermittent loading had dimensions and specifications as follows: wire size, 9 mm (0.345 in.); outside diameter of spring, 50 mm (2 in.); length, 75 mm (3 in.); six coils; 6150 alloy steel at 43 HRC; stress relieved immediately after coiling. The plating sequence was: (*a*) alkaline clean, (*b*) rinse in cold water, (*c*) electroplate with cadmium 8 μm (0.3 mil) thick, (*d*) rinse in hot water, and (*e*) relieve hydrogen embrittlement in boiling water ½ h.

The spring broke with a shatter fracture typical of that caused by hydrogen embrittlement. The corrective action was to bake the spring at 190 °C (375 °F) for 5 h.

For additional information on this subject, refer to the section on hydrogen embrittlement in Appendix 2 of this article.

Service Temperature. Cadmium-plated high-strength steel parts that are subjected to heavy loading should never be used at temperatures above 230 °C (450 °F). Cadmium melts at 320 °C (610 °F); at temperatures approaching 260 °C (500 °F), damage occurs that adversely affects mechanical properties.

Diffused Coatings. During the past few years, the aviation industry has developed an application for cadmium. The substrate is first plated with 10 μm (0.4 mil) of nickel and then 5 μm (0.2 mil) of cadmium. The alloy is diffused at 340 °C (645 °F) for about 1 h. Coverage with nickel must be complete, because cadmium can detrimentally affect the steel substrate when heated above the melting point of cadmium. In this way, an alloy with a very high melting point can be formed. Low-alloy steel parts that operate in jet engines at a temperature of 540 °C (1005 °F) were coated with this diffused alloy. After operating for 1 h at 540 °C (1005 °F), the parts withstood 100 h of salt spray without rusting. Cadmium can also be plated on copper and zinc, as well as on nickel.

Solderability. Although cadmium usually solders well with solders of the 60% tin, 40% lead type, using an inactive rosin flux, its performance may sometimes be unaccountably erratic. Solderability can be improved and made more consistent by predepositing a thin (3 to 4 μm or 0.10 to 0.15 mil) layer of copper. If the final cadmium deposit is at least 4 μm (0.15 mil) thick, the copper coating will not adversely affect corrosion resistance in mild indoor atmospheres.

Cadmium on Stainless. Cadmium can be successfully plated over stainless steels and heat-resisting chromium-nickel alloys if the basis metal is first activated and given a light coating of nickel in a nickel chloride-hydrochloric acid bath (U. S. Patent 2,437,409). Composition and operating conditions for this bath are as follows:

Nickel chloride	240 g/L (32 oz/gal)
Hydrochloric acid (1.16 sp gr)	120 g/L (16 oz/gal)
Temperature	Room temperature
Current density	5 to 20 A/dm^2 (50 to 200 A/ft^2)
Time	2 to 4 min
Anodes	Nickel

Plating of Cast Iron

Cast iron is difficult to plate because of the graphite flakes or nodules in the microstructure. The larger the graphite inclusions, the more difficult the plating operation. Cast iron parts with unmachined surfaces should be cleaned by mechanical methods such as shot blasting or tumbling before plating. Heavy pickling should be avoided if possible, because it produces smut that is difficult to remove. However, light pickling is required after abrasive cleaning, to activate the surface for plating.

Pickling should be followed by a thorough water rinse and a cyanide dip (see note in the tabulation below Fig. 4). Any carryover of acid to the cyanide dip must be avoided, because the combination of these chemicals generates a highly poisonous hydrocyanic gas. The fluoborate solution described in Table 1 is excellent for plating cast iron parts without deep recesses. The cyanide solutions in Table 1 also may be used, provided no metal-organic grain-refining agents have been added. Current density on the high side of the indicated ranges is recommended, to establish a continuous film of cadmium on the iron as soon as possible.

Cadmium and Zinc Compared

In rural areas, cadmium and zinc offer equal protection. However, zinc is superior to cadmium in industrial environments (Table 8). In uncontaminated marine atmospheres, zinc and cadmium give approximately equal protection. When the comparison is made at a distance of 24 m (80 ft) from the ocean, cadmium gives significantly greater protection than zinc. Although it is used to a limited extent in the paper and textile industries, cadmium plate has poor resistance to the common chemicals and also to the chemicals of the petroleum and pharmaceutical industries.

One reason for preferring cadmium to zinc is that cadmium plate forms a smaller amount of corrosion products than zinc, particularly in marine atmospheres. Cadmium also retains its initial appearance for a longer time. This is an important consideration in applications where a buildup of corrosion products would have a detrimental effect, such as preventing the flow of current in electrical components or the movement of closely fitting parts such as hinges. For such applications, cad-

Fig. 5 Deposit thickness from cyanide baths

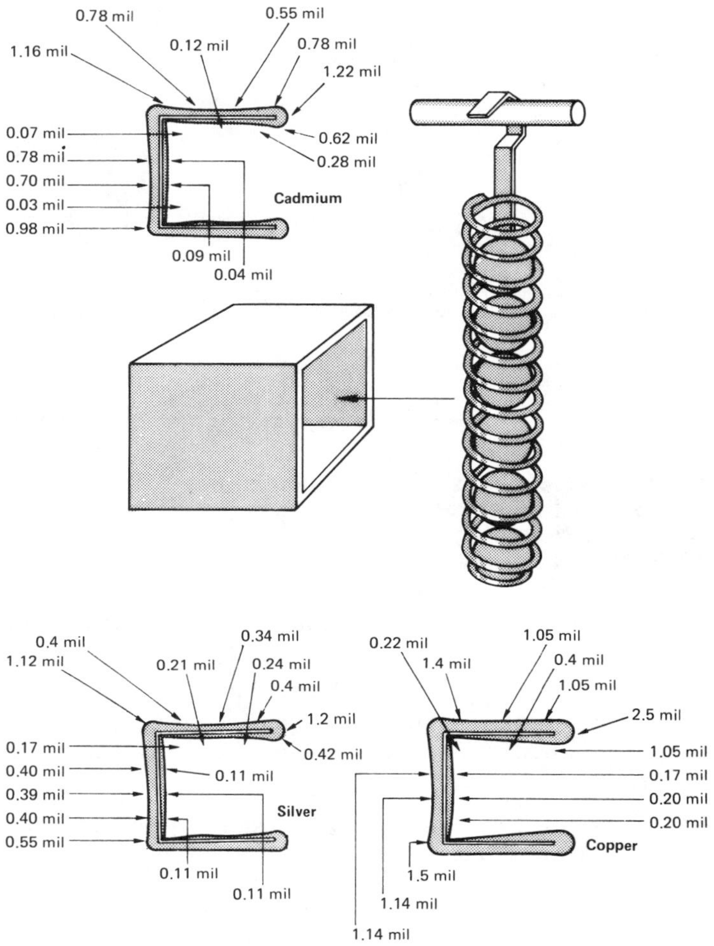

Plating bath	Thickness ratio(a) Side	Bottom
Cadmium	1:4.25	1:12
Copper	1:3.0	1:6
Silver	1:2.5	1:5

Note: Cross sections of boxes (100-mm, or 4-in. cubes) plated in cadmium, copper, and silver cyanide baths, with open ends pointed toward ball anodes during plating

(a) Ratio of average plate thickness on inside to average plate thickness on outside

Fig. 6 Plate-thickness variations obtained on a production part plated without the use of conforming anodes

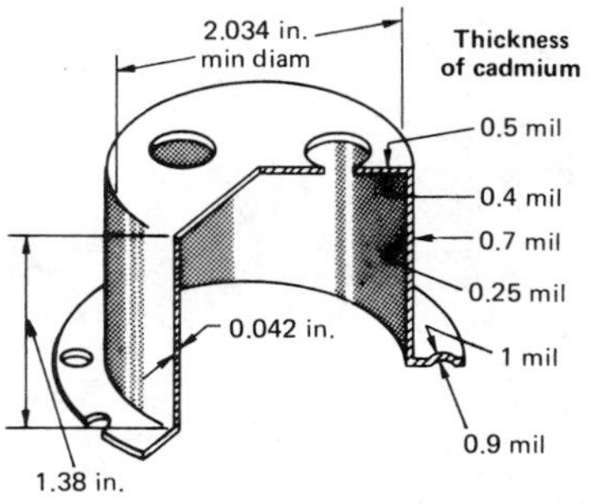

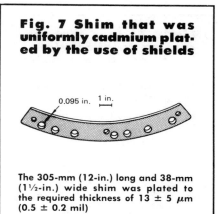

Fig. 7 Shim that was uniformly cadmium plated by the use of shields

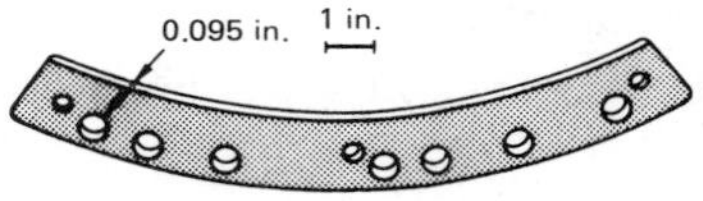

The 305-mm (12-in.) long and 38-mm (1½-in.) wide shim was plated to the required thickness of 13 ± 5 μm (0.5 ± 0.2 mil)

mium should be chosen in preference to zinc. Cadmium is preferable to zinc for plating cast iron.

Appendix 1

Chemical Analysis of Cyanide Cadmium Plating Baths

The following analytical procedures may be applied to cyanide cadmium plating baths to determine their contents of cadmium metal, sodium cyanide, sodium hydroxide, and sodium carbonate.

Determination of Cadmium Metal

Reagents

- Hydrochloric or sulfuric acid (concentrated)
- Ammonium hydroxide (concentrated)
- Eriochrome black "T" indicator (0.5% solution in alcohol)
- Formaldehyde (8% solution in water)
- Disodium dihydrogen ethylenediaminetetraacetate dihydrate (EDTA), 0.575 molar solution (21.4 g/L or 2.85 oz/gal)

Procedure

- Pipette exactly 2 mL of plating bath into a 250-mL Erlenmeyer flask, and dilute to about 100 mL with distilled water
- Neutralize this dilution to a faint white precipitate with hydrochloric or sulfuric acid. This can be conveniently done from the burette of standard sulfuric acid (0.94*N*) used for the caustic titration, or by the addition of a 50% solution of hydrochloric acid from an eyedropper. If no precipitate appears, as may happen with a new bath, thymolphthalein can be used as an indicator and will change from blue to colorless on neutralization

Fig. 8 Couplings that were uniformly cadmium plated

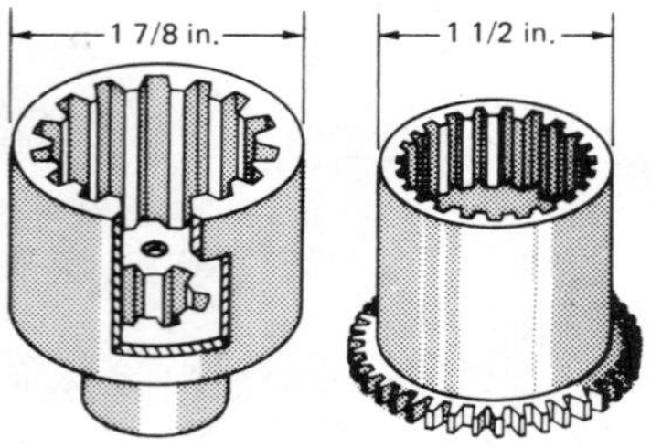

Plating thickness of 8 to 13 μm (0.3 to 0.5 mil) was achieved by centering a 6.4-mm (¼-in.) diam anode in the bore during plating

Fig. 10 Distribution of thickness of cadmium plate among 90 test parts

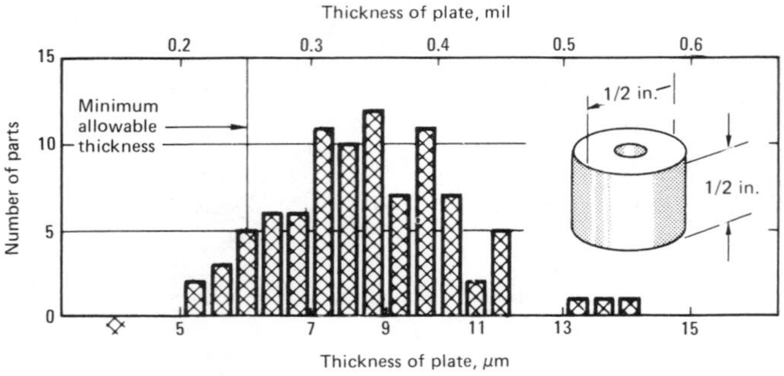

Fig. 9 Shapes that, according to dimensional limits indicated, require the use of conforming anodes to ensure uniform plate

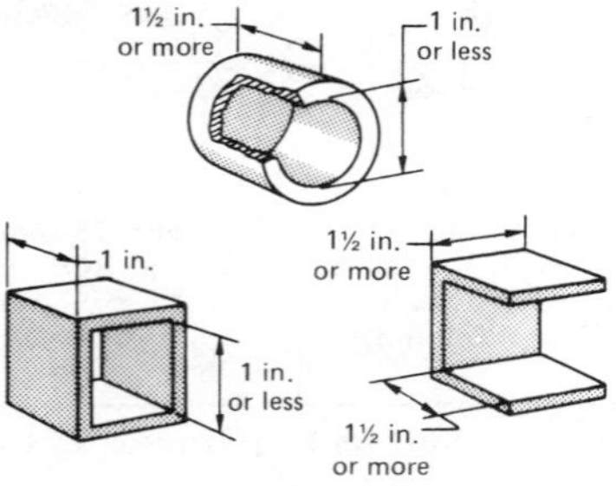

- Add 10 mL of concentrated ammonium hydroxide and about ¾ mL of Eriochrome indicator

- Zero the burette

- Add 8 mL of 8% formaldehyde solution

- Titrate immediately with EDTA solution. The color change is from red to blue and is sharpest when the solution is titrated as soon as possible after the formaldehyde has been added. A rapid titration will also give a sharper end point. Occasionally, the presence of impurities in the bath will prevent the attainment of a clear blue end point, but the color change is still sharp, from a red to a purplish blue

Table 8 Protection against rusting imparted to steel in various atmospheres by 25 μm (1 mil) of cadmium plate or zinc plate (ASTM)

Test location	Atmosphere	Time required for 5 to 10% rusting, yr: Cadmium	Zinc
New York, NY	Industrial	2	4
Pittsburgh, PA	Industrial	3	4
Sandy Hook, NJ	Marine, industrial	6	5
State College, PA	Rural	>11	>11
Key West, FL	Marine	>7	>9

Calculation

- Millilitres of EDTA solution used × 0.432 = ounces per gallon, cadmium metal

Determination of Total Sodium Cyanide

Reagents

- Ammonium hydroxide (concentrated)
- Potassium iodide (10% solution in water)
- Silver nitrate (13 g/L or 1.7 oz/gal solution in water)

Procedure

- Pipette a 2-mL sample of plating bath into a 250-mL flask
- Add to the sample about 50 mL distilled water, 5 to 7 mL of ammonium hydroxide, and 2 to 3 mL of potassium iodide solution
- Titrate with silver nitrate solution to the first stable faint yellowish turbidity

Calculation

- Millilitres of silver nitrate used × 0.5 = ounces per gallon, total sodium cyanide

Determination of Sodium Hydroxide

Reagents

- LaMotte sulfo-orange indicator
- Sulfuric acid, standard (0.94*N*)

Procedure

- Pipette 10 mL of the plating bath into a 250-mL flask
- Add to the sample about ½ mL of indicator solution
- Titrate with the sulfuric acid to the color change from orange to yellow

Calculation

- Millilitres of sulfuric acid used × 0.5 = ounces per gallon, sodium hydroxide

Determination of Sodium Carbonate

Reagents

- Barium chloride (10% solution in water)
- Methyl orange-xylene cyanole indicator solution
- Hydrochloric acid, standard (0.7*N*)

Procedure

- Pipette 10 mL of plating bath into a 250-mL beaker, add to it about 100 mL of water, and heat to boiling
- Stir into boiling bath dilution about 20 mL of barium chloride solution; cover mixture; allow to stand warm for about ¼ to ½ h

- Filter, using No. 41 Whatman filter paper, and wash precipitate and flask at least 2 or 3 times with hot distilled water
- Place paper and precipitate in the original beaker, add about 10 mL of hot distilled water and 3 or 4 drops of indicator
- Titrate with the hydrochloric acid (while stirring) to the first permanent color change from green to purple

Calculation

- Millilitres of hydrochloric acid used × 0.5 = ounces per gallon, sodium carbonate

Methods for Measuring Thickness of Cadmium Plate

Three widely used methods for measuring the thickness of cadmium deposits on iron and steel are microscopic examination, the chemical drop test, and magnetic testing.

Microscopic examination of a section cut at right angles to the plated surface is a method used on small and irregularly shaped parts that do not lend themselves to measurement by the drop test or magnetic methods. Preparation of the specimen follows standard metallographic procedures, but particular care must be taken not to scrape or smear the soft cadmium deposit. If the basis metal is iron or steel, dilute nitric acid is a suitable etchant.

The chemical drop test is performed on parts with flat surfaces or large diameters. It entails the use of a solution consisting of the following:

Ammonium nitrate	10 g/L (1.3 oz/gal) water
Hydrochloric acid(a)	10 m/L (0.08 pint/gal) water
Potassium ferricyanide	300 g/L (40 oz/gal) solution(b)

(a) Chemically pure, 1.19 sp gr. (b) Small amount

This solution is dropped on the plated surface at the rate of 95 to 105 drops per minute and eventually strips off the cadmium deposit. The dissolution of the plate is indicated immediately by the appearance of a deep blue color on the test piece. This is the result of a reaction of the solution with the iron or steel basis metal. Coating thickness is calculated from the length of time required for the solution to strip the plate.

The magnetic method measures the force required to pull a permanent magnet away from the plated ferrous metal and calibrates the result in terms of plate thickness. Because cadmium is nonmagnetic, force decreases as plate thickness increases.

Solutions for Stripping Cadmium Plate

Electrodeposited cadmium can be stripped chemically from the basis metal by immersion in one of the following solutions: ammonium nitrate, inhibited hydrochloric acid, chromic acid with a sulfuric acid addition, and ammonium persulfate with an ammonium hydroxide addition. Electrolytic stripping is performed in a solution of sodium cyanide with an addition of sodium hydroxide. Compositions of these stripping solutions and immersion times to be used with them are given in Table 9.

Toxicity of Cadmium

Two hazardous consequences attend the use of cadmium in contact with food products: (*a*) acute poisoning resulting from the ingestion of cadmium dissolved from containers or from food-handling equipment, and (*b*) poisoning from the inhalation of fumes of cadmium oxide, if cadmium plated vessels or food-handling equipment is heated.

Acute poisoning has resulted from the ingestion of cadmium salts derived from cadmium plated vessels in which any acid foods have been stored for even short periods of time; therefore, cadmium should not be used on food containers of any kind. Fatal poisoning is more apt to result from the inhalation of dust or fumes of cadmium salts and cadmium oxide. These are the kinds of exposure encountered in industrial operations when cadmium plated parts are heated. Exposure to dust or fumes of cadmium should be avoided.

Deposits of cadmium on the sides or bottom of a tank previously used for cadmium plating should not be burned off, because the fumes from this operation are highly toxic. These deposits should be removed mechanically or deplated. For high-efficiency deplating, the solution used contains 45 to 60 g/L (6 to 8 oz/gal) of sodium cyanide and 23 to 30 g/L (3 to 4 oz/gal) of sodium hydroxide in water; the tank is the anode, and steel sheets or scrap steel parts are the cathodes.

Appendix 2*

Selective Plating

When plating must be applied to only certain areas of parts, the areas not to be plated must be stopped off or masked, which means they must be covered with materials that will not conduct current, such as waxes, lacquers or rubber tape.

Waxes. Ordinarily, a petroleum-derived wax is used for stopping off. The wax must not contain any oil or other organic materials that will dissolve in the plating solution and contaminate it. It must also be capable of adhering tightly to the part, to prevent the plating solution from coming in contact with the stopped-off area.

Before being applied, the wax is heated in a pot to about 28 °C (80 °F) above its melting point, so that it does not solidify too rapidly and will adhere more readily. Still better adhesion is obtained if parts are warmed on a hot plate before the wax is applied.

Parts must be positioned so that only the area to be coated is placed in the molten wax. This means that, normally, only end areas or protrusions can be stopped off with wax. The wax can be applied with camel's hair brushes, but this is time consuming if many parts are to be treated. For a large number of similar parts, a fixture can be used that will dip each part to the proper depth.

A sharp, uniform demarcation between plated and nonplated areas can be obtained by the use of pressure-sensitive tape and wax, following either of two procedures:

- Apply the tape to the part so that the trailing edge of the tape follows the demarcation line; dip that portion of the part to be left unplated in molten wax so as to overlap the trailing edge of the tape slightly; and then remove the wax when it has solidified
- Apply the tape to the part so that the leading edge follows the demarcation; dip that portion of the part to be left unplated in molten wax so as partly to overlap the trailing edge of the tape; and then, when the wax has solidified, plate the part without removing the tape

Waxing must be done carefully, so that areas that are to be plated have no wax on them. If wax does get on areas

*Applicable to the plating of cadmium and other metals

Table 9 Solutions for stripping electrodeposited cadmium

Solution(a)	Composition	Amount		Immersion time, min(b)
1	Ammonium nitrate	105-135 g/L	14-18 oz/gal	10-20
2(c)	Hydrochloric acid (1.18 sp gr), undiluted	...	...	10-20
	Antimony trioxide	15 g/L	2 oz/gal	...
3	Chromic acid	200 g/L	26.8 oz/gal	5-10
	Sulfuric acid (95%)	50 mL/L	6.4 fluid oz/gal	
4	Ammonium persulfate	50 g/L	6.7 oz/gal	5-10
	Ammonium hydroxide	100 mL/L	12.8 fluid oz/gal	...
5(d)	Sodium cyanide	60-90 g/L	8-12 oz/gal	10-20
	Sodium hydroxide	15-30 g/L	2-4 oz/gal	...

(a) Solutions are listed in order of preference; all solutions are used at room temperature. (b) Immersion times are for deposits 8 to 13 μm (0.3 to 0.5 mil) thick. (c) Solution should not be used on stressed or hardened parts. (d) Solution for stripping electrolytically; 5 to 10 A/dm^2 (50 to 100 A/ft^2) and 6 to 8 V; part to e stripped is made the anode

to be plated, it must be thoroughly removed with steel wool. After plating and postplating treatments, the wax is removed from parts by placing them in hot water.

Lacquers may be used instead of wax as stop-off coatings, but their use is generally limited to instances in which the plating bath is operated at a temperature at which the wax would melt. Lacquer is applied by dipping or painting the areas to be stopped off. Normally, two to four coats of lacquer must be applied. One disadvantage of lacquer is that it is difficult and time consuming to get all of it off. Heavier coatings prevent leakage and make stripping easier.

Rubber Tape. For stopping off irregular areas of heavy parts that cannot be dipped or that are too large to be painted, for example, splines, large shafts, or bearing shoulders, a rubber tape is used. The tape is wound tightly and stretched over the irregular areas. To prevent leakage, each turn should overlap the preceding one at least halfway. At the edge of the stop-off area, a pressure-sensitive tape is used to form a sharp line and to prevent the leakage of plating solution under the rubber tape.

Rubber tape is expensive to use. When many similar parts are to be selectively plated, rubber sheet, held in place by pressure-sensitive tape, may be used for stopping off areas not to be plated. Rubber stoppers or corks, sealed with wax, are used for stopping off internal areas of cylindrical parts. Rubber tubing can be used to stop off areas of small cylindrical parts.

Rinsing and Drying

Although one of the simplest, least expensive, and apparently least critical operations in plating, rinsing is often the most difficult to accomplish. The primary requirements are that the rinsing be effective in removing the solutions used in the preceding tank and that no contaminants be introduced into the subsequent tank. Rinse baths, whether hot or cold, usually are provided with some means for constant changing of the water, good agitation, and skimming of the surface. Agitation of both the water and part is usually necessary. The surface skimmer may consist of jets of water shooting across the surface to rinse surface films into an overflow trough at the far side of the rinse tank. Water should enter at the bottom of one side of a rinse tank and escape over a weir outlet along the top at the opposite side of the tank. Constant chemical checks of the rinse bath should be made to assure that the rinse water is clean. The amount of contamination in rinse tanks can be regulated by controlling the flow of fresh water into the rinse through a valve actuated by a conductivity cell.

The temperature of the postplating rinse bath depends to some extent on the mass of the work being rinsed, because the workpiece must supply all the heat of evaporation for drying. Thin-gage materials require rinse temperatures of 93 to 100 °C (200 to 212 °F); otherwise, the workpiece cools before evaporation is complete. Parts made from thicker materials may be rinsed in water at 82 to 88 °C (180 to 190 °F).

Rapid and thorough drying of the plated work is important, to prevent water marks and stains and to eliminate the moisture from residual salt that may not have been entirely removed from crevices or recesses by rinsing. Residual salt and moisture can be a source of corrosion.

Drying practice is influenced also by the shape and orientation of the workpiece as it leaves the final rinse. In many applications, hot water rinsing is followed with oven drying, wherein hot air is blown directly against the work. In automatic installations, oven temperatures are maintained at 105 °C (220 °F) or higher and the work passes through in 3 to 5 min. Centrifuges with a hot air blast are used for barrel plated work.

Hydrogen Embrittlement

If an electrodeposited coating is to be applied to a highly stressed part or a high-strength (over 1240 MPa or 180 ksi) heat treated steel part, it is important that the processing should not decrease the static or fatigue strength of the part. Hydrogen embrittlement does not affect fatigue life. Coatings having high residual stresses, such as chromium, will affect fatigue life; however, this is not the case with cadmium.

Cadmium deposited from a cyanide solution is more likely to produce hydrogen embrittlement than any other commonly plated metal. Heat treated steels, particularly those plated and used at 35 HRC and above, are susceptible to hydrogen embrittlement. Most susceptible is spring steel that has not been adequately stress relieved after forming.

Although the thickness of the plated deposit appears to have no direct bearing on hydrogen embrittlement, it is always more difficult to release the hydrogen (by baking) from heavy deposits.

By adhering to the following procedures, hydrogen embrittlement can be minimized or made inconsequential:

- Use mechanical cleaning methods, such as brushing, blasting, and tumbling
- Wherever possible, avoid the use of strong acid pickling solutions and extended exposure to acid pickling
- If pickling is essential to the preparation of medium-strength and high-strength steel parts, bake the parts at 175 to 205 °C (350 to 400 °F) for 3 h after pickling and before plating
- In plating, use the higher current densities to produce a more porous deposit; 7 A/dm^2 (70 A/ft^2) in a cyanide bath without brighteners has been satisfactory for steel at 46 HRC
- After plating, bake parts at 175 to 205 °C (350 to 400 °F) for 3 to 24 h.

The shorter baking periods are generally adequate for parts with a tensile strength below about 1520 MPa (220 ksi); longer baking periods are recommended for steel of tensile strength above about 1520 MPa (220 ksi) or lower strength parts if sharp notches or threads exist. Parts greater than 25-mm (1-in.) thickness should also be baked for 24 h. The elapsed time between plating and baking must never exceed 8 h and should be carried out as soon as possible, preferably within 4 h.

- Plate parts to a thickness of about 5 μm (0.2 mil), bake for 3 h at 195 °C (385 °F), activate in cyanide, and then complete the plating to required final thickness

The applications of shot peening and baking, as related to the hardness of the steel to be plated, are described in Federal Specification QQ-C-320 (Amendment 1), and are summarized in the article on hard chromium plating in this Volume.

Tests for Adhesion of Plated Coatings

The tests used for evaluating adhesion of plated coatings are largely qualitative. A bend test, described in Federal Specification QQ-P-416, involves observation of the degree of flaking that occurs as a specimen is bent. In another test, a pressure-sensitive tape, such as surgical adhesive or masking tape, is attached to the plated surface. The tape is quickly stripped from the specimen by pulling it at right angles to the surface. If adhesion is poor, loose plate or blisters will appear as flecks on the surface of the adhesive.

Another good test for adhesion, on parts that have been baked after being plated, is a visual inspection for blisters in the plate. If a good bond has not been established, the plate will always pull away from the basis metal and form blisters.

Chromate Conversion Coatings

The corrosion of cadmium plate can be retarded by applying a supplemental chemical conversion coating of the chromate type. The chromate films, produced by immersing the plated article in a solution containing chromic acid or other chromates and catalytic agents, provide protection against initial corrosion through the inhibitive properties of the water-soluble chromium compounds present in the films. However, the chromate finish must not be applied before stress relieving or baking, because its beneficial effect will be destroyed by the elevated temperature.

Chromate conversion coatings are used in some instances to improve the bond between paint and cadmium-plated surfaces and to provide the plate with resistance to corrosion if gaps should occur in the paint film. However, wash primers will not adhere to chromate finishes, and baking painted chromate finishes will produce poor bonding.

Plate Discoloration. Cadmium tarnishes easily from handling and, at a lesser rate, from normal oxidation. Both types of tarnish may be prevented by the use of chromate conversion coatings. For maximum prevention of tarnish, an unmodified chromate film should be applied, if the iridescence or the light yellow coloration it imparts is not objectionable. Such a surface film also provides resistance against salt spray and humidity, and its application for this purpose is frequently standard practice. The clear film obtained by bleaching a chromate coating affords much poorer protection, but it is superior to an as-plated cadmium surface with respect to resistance to tarnishing, humidity, and salt spray.

With a plate thickness of 13 to 18 μm (0.5 to 0.7 mil) and a chromate conversion coating, cadmium will provide adequate service in marine and humid tropical atmospheres. When long-term exposure is anticipated, a paint coating is desirable.

If a chromate treatment is used, only two cold water rinse tanks are necessary after plating. The first may be for reclaiming the cadmium solution or for the treatment of waste. The second rinse should be provided with sufficient flow and agitation to prevent carryover of cyanide into the chromate solution. After chromate dipping, three rinse tanks are required. Again, the first tank may be for reclaiming or waste treatment.

Yellow chromate finish is obtained by dipping in acidified sodium or potassium dichromate. Excellent corrosion protection and a superior base for organic finishing are obtained.

Clear chromate finish bath consists of 117 g (13.3 oz) of chromic acid and 1.2 g (0.14 oz) of sulfuric acid per litre (gallon) of water and provides good passivation and attractive appearance. Although the protective film is very thin, it prevents the formation of a white, powdery corrosion product on cadmium-plated parts in indoor or internal-component use.

Olive green coating is obtained in an acidified dichromate solution and is easily colored by any of the acid dyes.

Other Postplating Processes

Bright Dipping. The solution for bright dipping consists of ¼ to 1% of commercial grade nitric acid (1.41 sp gr) and is used at room temperature. The acid neutralizes any alkaline salts on the surface and provides some passivation. It is used extensively because it does not interfere with solderability. Immersion times vary from 2 to 30 s.

A solution of acidified hydrogen peroxide is also used for bright dipping. It consists of 6 to 7% commercial-grade (35%) hydrogen peroxide acidified with about 0.25% sulfuric acid. It produces a bright luster and uniform finish but adversely affects resistance to atmospheric corrosion, ultimately resulting in the formation of a white powder. The solution is rather expensive and has a short life.

Phosphate treatment produces a supplementary conversion coating. The solution consists of 3 to 4% equivalent phosphoric acid at a pH of 3.5 to 4.2. The solution is maintained at a temperature of 71 to 88 °C (160 to 190 °F); immersion time ranges from 3 to 5 min. Following the acid dip, parts are water rinsed and then passivated for 2 to 3 min in a solution of sodium dichromate (1 to 2 g/L or 0.1 to 0.2 oz/gal) or chromic acid (pH, 3.5 to 4.0) at a temperature of 66 to 77 °C (150 to 170 °F). The coating provides a good basis for organic finishes.

Molybdenum coating is performed in a proprietary bath containing molybdenum salts dissolved in a highly concentrated solution of ammonium chloride at 54 to 66 °C (130 to 150 °F). An attractive, adherent black finish is obtained.

Tin Plating

By Paul E. Davis
Manager
Tin Research Institute, Inc.

TIN is a versatile, low-melting point, nontoxic metal possessing valuable physical properties. It alloys readily with most other metals, and because it is amphoteric in nature (for example, capable of reacting chemically as an acid or a base), tin forms many useful inorganic and organic chemical compounds. It has the largest melting point to boiling point range (from 230 to 2370 °C or 450 to 4300 °F) of any metal. Evaporation from a pot of liquid tin does not occur in conventional metallurgical applications. Tin is used in a multitude of products, although the percentage of tin present is relatively small as a percentage of the total. Most manufacturers use some tin, and it is an essential material to industries such as communications, transportation, agriculture, food processing, and construction. Pewter containing at least 90% tin is the only consumer product where tin is the major constituent.

Electrodeposits

Electrodeposits are very useful, because a thin coating can provide the desirable properties of tin, such as excellent solderability, ductility, softness, and corrosion or tarnish resistance. Thus, stronger materials, required for their engineering properties, may exhibit the desirable properties of tin on their surface. A tin deposit provides sacrificial protection to copper, nickel, and many nonferrous metals and alloys. It does not provide the same properties to steel in a normal atmosphere. Thick, nonporous coatings provide long-term protection in almost any application. The required coating thickness is established by the application. Recommended thicknesses for tin coatings on steel- and copper-based materials are shown in Table 1.

Applications. The largest use of tin electrodeposits is the production of tinplate at steel mills, primarily for use as food preservation containers. A thin tin coating protects the steel inside a tin can as long as an oxygen-free environment is produced. The second largest use of tin electrodeposits is in the electronics industry where many surfaces requiring good solderability and corrosion or tarnish resistance are coated. Electrodeposited tin is also used on food handling equipment as bearing surfaces and in many other applications where the unique properties protect other materials or render them usable in applications for which they are otherwise unsuited.

Types of Electrolytes

Tin may be deposited from alkaline or acid solutions. The electrolyte compositions and process operating details are readily available in published references (Ref 1, 2, and 3). Basic details of electrolyte composition and operating conditions are shown in Table 2 for alkaline solutions and in Tables 3 and 4 for acid solutions. Tin ions in the alkaline electrolytes have a valence of +4. Those in the acid electrolytes have a valence of +2. Consequently, the alkaline systems require the passage of twice as much current to deposit one gram molecule of tin at the cathode.

Alkaline electrolytes usually contain only an alkaline stannate and the applicable hydroxide to obtain satisfactory coatings. Unlined steel equipment is acceptable, but the electrolyte must be heated. Factors such as operating temperature, solution constituent concentration, and operating current density affect the efficiency and plating rate of the system and must be properly balanced and controlled.

Unusual operating features of the alkaline electrolytes involve (*a*) tin anode control and electrochemical solution mode, (*b*) cathodic deposition occurring from Sn^{+4}, and (*c*) solubility of the alkaline stannate in water. Ninety percent of the problems encountered in alkaline tin plating are the result of improper anode control. Conversely, operation of the alkaline electrolytes is simple if the anode behavior is understood, because there are no electrolyte constituents except the applicable stannate and hydroxide.

Tin anodes must be properly filmed, or polarized, in alkaline solutions to dissolve as the required cation (Sn^{+4}). Anode filming is accomplished by providing a surge of electrical current when plating is started. Once established, the anode film continues to provide the Sn^{+4} cations. The film may be lost because of a low operating current density with the tin then dissolving as Sn^{+2}; or the anode can become passive as a result of excessive current density, in which case no tin dissolves. Film formation is confirmed by (*a*) a sudden increase in the electrolyte cell voltage, (*b*) a drop in the amperage passing through the cell, and (*c*) the observance of a film on the anodes. The film is yellow-green in color for pure tin anodes and turns somewhat darker when high-speed anodes are used. Because the anodes do not function at 100% efficiency when filmed, moderate gassing also occurs as the result of visible O_2 gas being present near the anodes.

Unfilmed tin anodes do not produce gassing, but dissolve electrochemically in an undesirable form, as the Sn^{+2} cation. The presence of the Sn^{+2} ion, stannite, in the electrolyte produces unsatisfactory plating conditions, and the deposit becomes bulky, rough, and porous. The addition of hydrogen peroxide to the electrolyte oxidizes the Sn^{+2} ions to the Sn^{+4} form to return it to a usable condition, but if this remedy is required frequently, it indicates other problems that must be remedied. The anode film becomes passive when the anode current density is higher than the maximum value of the conventional operating range. When passive, the film becomes black, and excessive anode gassing occurs. In this condition, no tin dissolves electrochemically and the electrolyte is losing tin ions. The anode current density should be reduced until the normal film color returns. The addition of a small amount of aluminum to the anodes greatly increases the current density range over which they function satisfactorily. These are known as high-speed anodes (Ref 1 and 2).

Two factors tend to restrict the usable current density range and limit the deposition rate with the alkaline systems. One is the solubility of the stannates in hydroxide solutions. With the sodium formula, the normal increase is not possible, because sodium stannate is one of the unusual salts which has a reverse temperature coefficient of solubility. Examples of this process are given in Table 2. Less sodium stannate dissolves as the electrolyte temperature is increased, and the usable current density and the plating rate are reduced. Potassium stannate is more soluble with increasing temperature, but as the stannate is increased, the potassium hydroxide must also be increased, and the stannate solubility decreases as the hydroxide content increases.

The second factor for these electrolytes is that cathode efficiency decreases as the current density is increased. Eventually, a point is reached where these factors offset, and a further increase in current density does not increase the deposition rate. This limits the rate at which tin can be deposited.

In specialized applications, such as plating the inside of oil well pipe, it is not possible to have sufficient anode surface to avoid passivity. A higher current density can be used if insoluble anodes are utilized, but tin deposited on the cathode must then be replaced by the addition of chemicals. The addition of stannate to provide the tin cations also adds sodium or potassium to the electrolyte. Although the resulting additional alkalinity can be neutralized by the addition of a calculated amount of an alkaline acetate, the sodium or potassium ion concentration continues to increase and the alkaline stannate solubility equation is pushed quickly to the reactant side. This reduces the available Sn^{+4} ion to a low enough concentration that the plating rate decreases rapidly, and the electrolyte must be discarded. A potassium-based composition has been developed where the necessary Sn^{+4} ions are added to the electrolyte as a soluble, colloidal, hydrated tin oxide (Ref 2). Because the potassium ion concentration builds up more slowly in this composition, the electrolyte life is nearly indefinite.

Acid Electrolytes. Several acid tin plating electrolytes are available. Two of these can be considered as general systems adaptable to almost any application, and electrolytes such as Halogen (chloride-fluoride based) or Ferrostan (a special sulfate-based system) have been developed for highly specialized use. The acid electrolytes differ from the alkaline in many respects. The solution of a stannous salt in a water solution of the applicable acid does not produce a smooth, adherent deposit on a cathode, and a grain refining addition agent (gelatin or peptone) must be used. Usually such materials are not directly soluble in water solution and a wetting agent type material (such as β-naphthol) is also necessary.

Phenolsulfonic or cresolsulfonic acid is added to a sulfate-based system, and hydroquinone is added to a fluoboric acid-based system to retard the oxidation of the stannous tin ions to the stannic form. Although the acid electrolytes can contain large amounts of stannic ions without affecting the operation of the system, only the stannous ions are

Table 1 Recommended thicknesses of tin deposits on steel and copper substrates (ASTM designation B545-72)

Service condition(a)	Thickness on steel: Local minimum, µm	µin.	Thickness on steel: Average minimum, µm	µin.	Thickness on copper and copper alloy: Local minimum, µm	µin.	Thickness on copper and copper alloy: Average minimum, µm	µin.
Mild	5	200	8	300	5	200	8	300
Moderate	10	400	15	600	8	300	12	500
Severe	20	800	30	1200	15	600	20	800
Very severe	30	1200	45	1800	30	1200	45	1800

(a) Mild conditions include use to provide solderability, surface preparation for painting, or antigalling; moderate conditions include service in interior atmospheres; severe conditions include exposure to dampness and use in mild industrial atmospheres; very severe conditions require coatings which are pore-free for use in corrosive atmospheres or under abrasive conditions

Table 2 Composition and operating conditions for stannate plating electrolytes

Values of composition are for electrolyte start-up; operating limits for the electrolyte composition are approximately −10 to +10% of start-up values

Baths	Composition: Potassium stannate, g/L	oz/gal	Sodium stannate, g/L	oz/gal	Potassium hydroxide, g/L	oz/gal	Sodium hydroxide, g/L	oz/gal	Tin metal(a), g/L	oz/gal	Operating conditions: Temperature, °C	°F	Cathode current density, A/dm²	A/ft²
A	105	14	...	...	15(b)	2(b)	...	...	40	5.3	66-88	150-190	3-10	30-100
B	210	28	...	...	22	3	...	...	80	10.6	77-88	170-190	to 16	to 160
C	420	56	...	...	22	3	...	...	160	21.2	77-88	170-190	to 40	to 400
D	...	...	105(c)	14	...	...	10(b)	1.3(b)	42	5.6	60-82	140-180	0.5-3	6-30

(a) As stannate. (b) Free alkali may need to be higher for barrel plating. (c) $Na_2SnO_3 \cdot 3H_2O$; solubility in water is 61.3 g/L (8.2 oz/gal) at 16 °C (60 °F) and 50 g/L (6.6 oz/gal) at 100 °C (212 °F)

deposited at the cathode. As a result, oxidation depletes the stannous ions available, and these must be replaced by the addition of the corresponding stannous salt to the bath. To limit oxidation of stannous ions, sufficient anode area must be maintained; the operating temperature must be kept as low as possible, and the introduction of oxygen into the solution, because of a filter leak or the use of air agitation, must be avoided. An antioxidant is usually added to the solution.

The basic differences of operating characteristics are related to the type of tin ion present in the alkaline and acid electrolytes. Acid systems require that the stannous ions are not oxidized to stannic and are operated at lower temperatures. The acid electrolytes require only one half as much current passage to deposit a gram molecule of tin. The tin dissolves directly from the metallic anodes, and the control of an anode film is not involved. Acid electrolytes are 100% efficient, both anodically and cathodically. This avoids the necessity to add tin regularly as chemicals, and the problems of oxygen gas evolution at the anode surface and hydrogen gas at the cathode surface. Some particulate matter is produced as sludge from (*a*) anode slime products, (*b*) the precipitation of addition agents and their breakdown products, and (*c*) basic tin compounds formed by oxidation. These must be removed during operation. In a still tank, the precipitates gradually settle, but agitated solutions require continuous filtration.

Acidic electrolytes require lined or plastic tanks. The throwing power of acid tin electrolytes is inferior to that of the alkaline tin solutions, but they are sufficiently superior to other acid processes so that special anode systems are not usually required.

The stannous sulfate electrolyte is much used because of its general ease of operation. The rate of deposition is somewhat limited by optimum metal concentration in the electrolyte. A still bath is operated at a cathode current density of 1 to 2 A/dm^2 (10 to 20 A/ft^2). Currents densities of up to 10 A/dm^2 (100 A/ft^2) are possible with suitable electrolyte agitation. Anode surface area must be increased when higher current densities are used. Addition agent control is not quantitative in nature, but deficiencies are easily recognized by the experienced plater. An electrolyte may be prepared from readily available chemicals, or a proprietary system may be purchased from suppliers. All commercial bright acid tin processes and the more recent matte acid tin systems are based on the stannous sulfate solution. They also suffer greatly from any factor which produces oxidation in the solution, and the factors involved must be controlled. Precise information on operation and control can only be obtained directly from the specific suppliers.

The stannous fluoborate electrolyte is a good general-purpose electrolyte. It is capable of operation at higher current densities, because of the conductivity provided by the fluoboric acid. Cathode current densities of 20 A/dm^2 (200 A/ft^2) and higher are possible with suitable solution agitation. The need to increase anode surface area at high current densities and the control of the addition agents is parallel to that necessary for the use of stannous sulfate. The three solution compositions are listed in Table 4, because each meets a specific need. The solution conductivity lost because of the lower metal content in the high throwing power bath is compensated for by the higher concentration of fluoboric acid. The lower total metal in the solution reduces the variance in deposit thickness usually associated with varying areas of cathode current density. Boric acid is listed as a constituent of the fluoborate solutions because of its presence in the stannous fluoborate and fluoboric acid used to prepare them. It is not a necessary ingredient in the electrolyte.

Table 3 Composition and operating conditions for sulfate electrolyte

Temperature for sulfate electrolytes are 21 to 38 °C (70 to 100 °F); they do not require heating; cooling may be considered if temperature rises to reduce adverse effects of temperature on the electrolyte constituents; cathode current density is 1 to 10 A/dm^2 (10 to 100 A/ft^2)

Constituent	Amount g/L	Amount oz/gal	Operating limits g/L	Operating limits oz/gal
Stannous sulfate	80	10.6	60-100	8-13
Tin metal, as sulfate	40	5.3	30-50	4-6.5
Free sulfuric acid	50	6.7	40-70	5.3-9.3
Phenolsulfuric acid(a)	40	5.3	30-60	4-8
β-naphthol	1	0.125	1	0.125
Gelatine	2	0.25	2	0.25

(a) Phenolsulfuric acid is most often used. Cresolsulfonic acid performs equally well and is a constituent of some proprietary solutions

REFERENCES

1. Frederick A. Lowenheim, Ed., *Modern Electroplating*, 3rd ed., New York: Wiley-Interscience, Electrochemical Society Series, 1974
2. *Metal Finishing Guidebook and Directory*, annual issue, Hackensack, NJ: Metals and Plastics Publications, Inc., 1982
3. Publications of the International Tin Research Institute, Greenford, England with offices in Palo Alto, CA, and Columbus, OH

Table 4 Composition and operating conditions for fluoborate electrolyte

Three electrolyte compositions are given; the standard is generally used for rack or still plating; the high speed for applications like wire plating; the high throwing power for barrel plating or applications where a great variance exists in cathode current density as a result of cathode configuration

Electrolyte	Stannous fluoborate g/L	Stannous fluoborate oz/gal	Tin metal(a) g/L	Tin metal(a) oz/gal	Free fluoboric acid g/L	Free fluoboric acid oz/gal	Free boric acid g/L	Free boric acid oz/gal	Paptone(b) g/L	Paptone(b) oz/gal	β-naphthol g/L	β-naphthol oz/gal	Hydroquinone g/L	Hydroquinone oz/gal	Temperature °C	Temperature °F	Cathode current density A/dm^2	Cathode current density A/ft^2
Standard	200	26.8	80	10.8	100	13.4	25	3.35	5	0.67	1	0.13	1	0.13	16-38(c)	60-100(c)	2-20	20-200
High speed	300	39.7	120	16.1	200	26.8	25	3.35	5	0.67	1	0.13	1	0.13	16-38	60-100	2-20	20-200
High throwing power	75	9.9	30	4.0	300	40.2	25	3.35	5	0.67	1	0.13	1	0.13	16-38	60-100	2-20	20-200

(a) As fluoborate. (b) Dry basis. (c) Electrolytes do not require heating. Cooling may be considered if temperature rises to reduce adverse effects of temperature on the electrolyte constituents

Lead Plating

By Nicholas J. Spiliotis
Supervisor—Technical Service
Allied Chemical Company

LEAD has been deposited from a variety of electrolytes among which are fluoborates, fluosilicates, and sulfamates. Fluoborate baths are the most widely used because of the availability of lead fluoborate and the simplicity of bath preparation, operation, and stability. Fluoborate baths provide finer grained, denser lead deposits. Fluosilicate baths, although less costly to use for large operations, are difficult to prepare for small-scale plating. They are not suitable for plating directly on steel and are subject to decomposition, which produces silica and lead fluoride. Use of sulfamate baths is almost nonexistent in the United States, because neither lead silicofluoride nor lead sulfamate is available commercially. These salts must be prepared by the plater using litharge (PbO) and the corresponding fluosilicic or sulfamic acids. Sulfamate baths are subject to decomposition, which produces lead sulfate.

Preparing Basis Metal

Low-Carbon Steel. Lead may be plated directly on steel from the fluoborate bath using the following cycle:

- Degrease with solvent (optional)
- Alkali clean (anodic)
- Water rinse
- Dip in 10% fluoboric acid*
- Water rinse
- Lead plate
- Rinse

Lead may be plated on steel from the fluosilicate and sulfamate baths using the following cycle:

- Degrease with solvent (optional)
- Alkali clean (anodic)
- Rinse
- Dip in 5 to 25% hydrochloric acid
- Rinse thoroughly
- Dip in 30 to 75 g/L (4 to 10 oz/gal) sodium cyanide
- Rinse
- Copper cyanide strike
- Rinse thoroughly
- Dip in 10% fluoboric acid*
- Rinse
- Lead plate
- Rinse

Copper. Lead may be plated directly on copper from fluoborate, fluosilicate, or sulfamate baths using the following cycle:

- Alkali clean (anodic or cathodic/anodic)
- Rinse
- Dip in 10% fluoric acid*
- Rinse
- Lead plate
- Rinse

Fluoborate Baths

Lead fluoborate baths are prepared by adding the required amount of lead fluoborate concentrate and fluoboric acid to water followed by peptone as the preferred addition agent.

Many different types of glue and gelatin additives are available, but no one type is manufactured specifically for lead plating. Depending on the method of manufacture, each may exhibit different levels of solubility and impurities that may be of concern to the plater.

Glue and gelatin addition agents must be swelled and dissolved in water by the plater just prior to addition to the bath. The resultant colloidal solution has a limited shelf-life and is prone to bacterial degradation on standing. Glue and hydroquinone are relatively expensive. Often, it is a by-product of an industrial process and can contain organic and inorganic impurities detrimental to the lead plating process. No grade is manufactured and sold specifically for lead plating.

Concentrates of lead fluoborate and fluoboric acid contain free boric acid to ensure bath stability. An anode bag filled with boric acid in each corner of the plating tank is recommended to maintain a stable level of boric acid in the bath solution. The concentration of boric acid in the bath is not critical and can vary from 1 g/L (0.13 oz/gal) to saturation. The water used in the bath preparation must be low in sulfate and chloride, as these lead salts are insoluble.

Table 1 provides the compositions and operating conditions of high-speed and high throwing power fluoborate plating baths. The high-speed bath is useful for plating of wire and strip where high current densities are used. The high throwing power formulation is used in applications such as barrel plating of small parts or where thickness distribution on intricate or irregularly shaped parts is important. The high throwing power bath should be operated at a lower current density because of the lower lead content of the bath.

*Hydrochloric or sulfuric acid should not be used in order to avoid precipitating insoluble lead sulfate or chloride on the work in the event of poor rinsing

Table 1 Compositions and operating conditions of lead fluoborate baths

Bath	Lead g/L	Lead oz/gal	Fluoboric acid (min) g/L	Fluoboric acid (min) oz/gal	Peptone solution, vol %	Free boric acid g/L	Free boric acid oz/gal	Temperature °C	Temperature °F	Cathode current density(a) A/dm²	Cathode current density(a) A/ft²	Anode composition	Anode/cathode ratio
High-speed	225	30	100	13.4	1.7	1 to saturation	0.13 to saturation	20-41	68-105	5	50	Pure lead	2:1
High throwing power	15	2	400	54	1.7	...	...	24-71	75-105	1	10	Pure lead	2:1

(a) Values given are minimums. Current density should be increased as high as possible without burning the deposit; this is influenced by the degree of agitation

Table 2 Compositions and operating conditions of lead fluosilicate baths

Bath	Lead g/L (oz/gal)	Animal glue g/L (oz/gal)	Peptone equivalent g/L (oz/gal)	Total fluosilicate g/L (oz/gal)	Temperature °C	Temperature °F	Cathode current density A/dm²	Cathode current density A/ft²	Anode current density A/dm²	Anode current density A/ft²	Anode composition
1	10 (1.3)	0.19 (0.025)	5 (0.67)	150 (20)	35-41	95-105	0.5-8	5-80	0.5-3	5-30	Pure lead
2	180 (24)	5.6 (0.75)	150 (20.1)	140 (18.75)	35-41	95-105	0.5-8	5-80	0.5-3	5-30	Pure lead

Fluoborate baths rank among the most highly conductive plating electrolytes; this results in low voltage requirements for the amperage used.

Maintenance and Control. The very high solubility of lead fluoborate in solution with fluoboric acid and water accounts for its almost universal use for lead plating. In the high-speed bath formulation of Table 1, neither the lead nor acid content is critical, and the bath can be operated over a wide range of lead and acid concentrations.

The high throwing power bath formulation of Table 1 must be operated fairly close to the guidelines given. Lowering the lead concentration improves the throwing power characteristics; however, a reduction in lead concentration must be followed by a corresponding decrease in the cathode current density. On the other hand, an increase in lead content above the optimum permits the use of higher current densities, with a corresponding decrease in throwing power.

Sludge may form in the fluoborate bath, resulting from use of impure lead anodes that contain bismuth or antimony or from the drag-in of sulfates. Fluoborate baths should be constantly filtered through dynel or polypropylene filter media to remove any sludge which may be formed. Anodes must be bagged in dynel or polypropylene cloth. Absence of gas bubbles at the cathode or anode while plating indicates all electric energy is theoretically being used to transfer lead from the anode to the workpiece, thus there is 100% anode and cathode efficiency. The plating bath concentration therefore remains unchanged except for concentration from evaporation and dilution from placing wet parts in the bath in combination with dragout when the parts are removed from the bath. Methods are available for the analysis of lead and fluoboric acid. Additive concentration can be adequately evaluated through the use of the Hull cell. Low concentration of additive results in loss of throwing power, coarse grained deposits and treeing, the formation of irregular projections during electrodeposition, on a cathode especially at edges and other high current density areas.

Fluosilicate Baths

Fluosilicic acid is formed by the action of hydrofluoric acid on silicon dioxide. The lead fluosilicate ($PbSiF_6$) electrolyte is formed when fluosilicic acid is treated with litharge. No great excess of silicic acid can be held in solution; therefore, the fluosilicate solution is less stable than the fluoborate solution. Table 2 lists compositions and operating conditions for two lead fluosilicate baths.

Although at low current densities it is possible to secure smooth deposits of lead from the fluosilicate bath without additive agents, higher current densities are likely to produce treeing, especially in heavy deposits. Therefore, an additive agent, such as peptone glue or other colloidal materials or reducing agents, is always used. The use of excess glue in lead plating baths, however, may result in dark deposits. Maintenance and control procedures for the fluosilicate baths are similar to those described for the fluoborate baths.

Sulfamate Baths

Sulfamate baths consist essentially of lead sulfamate with sufficient sulfamic acid to obtain a pH of about 1.5. Sulfamic acid is stable, non-hygroscopic, and is considered a strong acid. Compositions and operating conditions of two typical sulfamate baths are given in Table 3.

Because the acid and the salt used in the solutions in Table 3 are highly soluble in water, sulfamate baths may be prepared either by adding constituents singly or as formulated salts to water. Solutions are usually formulated to concentrations that allow bath operation over a wide range of current densities. Lead concentration may vary from 112 to 165 g/L (15 to 22 oz/gal), while the pH is held at about 1.5. As in other lead plating solutions, additive agents (peptone gelatin or other colloids, alkyl or alkyl aryl polyethylene glycols) are required to produce smooth, fine grained deposits.

Spongy deposits are obtained if (*a*) the lead concentration is too low, (*b*) the current density is too high, or (*c*) the concentration of additive agent is too low. At low pH or high temperature, sulfamate ions hydrolyze to ammonium bisulfate to subsequently form insoluble lead sulfate. Ordinarily, this hydrolysis presents no problem, provided the bath is correctly operated.

Maintenance and Control. Sulfamate baths do not require much attention other than to maintain the correct proportion of additive agents to produce the desired deposit quality. Additive agent content is evaluated by the use of the Hull cell. The pH is easily adjust-

Table 3 Compositions and operating conditions of lead sulfamate bath

Bath	Composition: Lead g/L (oz/gal)	Composition: Animal glue g/L (oz/gal)	Composition: Peptone equivalent g/L (oz/gal)	Free sulfamic acid	Temperature °C	Temperature °F	Cathode current density A/dm^2	Cathode current density A/ft^2	Anode current density A/dm^2	Anode current density A/ft^2	Anode/cathode ratio	Anode composition
1	140 (18.75)	5.6 (0.75)	150 (20.1)	pH 1.5	24-49	75-120	0.5-4	5-40	0.5-4	5-40	1:1	Pure lead
2	54 (7.25)	5.6 (0.75)		50 g/L (6.75 oz/gal)	24-49	75-120	0.5-4	5-40	0.5-4	5-40	1:1	Pure lead

Table 4 Solutions and operating conditions for stripping lead from steel

Method A	
Sodium hydroxide	100 g/L (13.4 oz/gal)
Sodium metasilicate	75 g/L (10 oz/gal)
Rochelle salt	50 g/L (6.7 oz/gal)
Temperature	82 °C (180 °F)
Work is made anodic at	1.9-3.7 A/dm^2 (18.5-37 A/ft^2)
Method B	
Sodium nitrite	500 g/L (67 oz/gal)
pH	6-10
Temperature	20-82 °C (68-180 °F)
Work is made anodic at	1.9-18.5 A/dm^2 (18.5-185 A/ft^2)
Method C(a)	
Acetic acid (glacial)	10-85 vol %
Hydrogen peroxide (30%)	5 vol %
Method D(a)(b)	
Fluoboric acid (48-50%)	4 parts
Hydrogen peroxide (30%)	1 part
Water	2 parts
Temperature	20-25 °C (68-77 °F)

(a) Formulations should be made up fresh daily. (b) Alternate method for stripping lead or lead-tin deposits. Work must be removed as soon as the lead is stripped; otherwise, the base metal will be attacked

ed with sulfamic acid or ammonia and can be measured with a glass electrode. Lead concentration can be determined with sufficient accuracy by hydrometer readings or an occasional gravimetric analysis.

Anodes

Lead of satisfactory purity for anodes may be obtained either as corroding lead or chemical lead. Chemical lead anodes are preferred because of their higher purity. Impurities in the anodes such as antimony, bismuth, copper, and silver cause the formation of anode slime or sludge and can cause rough deposits if they enter the plating solution. These impurities can also cause anode polarization if present in the anode, especially at higher anode current densities. Small amounts of tin and zinc are not harmful. Anode efficiency in acid baths is virtually 100%.

Equipment

Anodes should be bagged in dynel or polypropylene cloth to prevent sludge from entering the plating bath. Bags should be leached in hot water to remove any sizing agents used in their manufacture before use in the plating bath. Nylon and cotton materials deteriorate rapidly and should not be used in any of the baths. Fluoborate and fluosilicate baths attack equipment made of titanium, neoprene, glass, or other silicated material and should not be used in these solutions. Anode hooks should be made of Monel metal.

Tanks or tank linings should be made of rubber, polypropylene, or other plastic materials inert to the solution. Pumps and filters of type 316 stainless steel or Hastalloy C are satisfactory for intermittent use; for continuous use, however, equipment should be made from or lined with graphite, rubber, polypropylene, or other inert plastic. Filter aids used for the fluoborate solution should be cellulose type rather than asbestos or diatomaceous earth.

Applications

The appearance and properties of lead limit its commercial use in electroplating largely to corrosion-protection and bearing applications — two fields in which the physical and chemical properties of lead render it unique among the commercially plated metals. Lead has not been extensively electroplated because its low melting point of 325 °C (620 °F) facilitates application by hot dipping. Electrodeposited lead has been used for (*a*) the protection of metals from corrosive liquids such as dilute sulfuric acid; (*b*) the lining of brine refrigerating tanks, chemical apparatus, and metal gas shells; and (*c*) for barrel plating of nuts and bolts, storage battery parts, and equipment used in the viscose industry.

Electroplated lead has been used for corrosion protection of electrical fuse boxes installed in industrial plants or where sulfur-bearing atmospheres are present. Presently, lead is also codeposited with tin for wire plating, automotive crankshaft bearings, and printed circuits.

Stripping of Lead

Table 4 identifies solutions and operating conditions for stripping lead from steel. Method C, at about 16 °C (60 °F), strips 25 μm (1 mil) of lead in 6 or 7 min with very slight etching of the steel. With Method B, voltage increases suddenly when the lead coating has been removed; at room temperature and 9.3 A/dm^2 (93 A/ft^2), the voltage may be about 2.7 V during stripping, but increases to 4.6 V when stripping is complete.

With the solutions used in Method A or B, a stain occasionally remains on the steel after stripping. The stain can be removed by immersion for 30 s in the solution used in Method C, leaving the steel completely clean and unetched (unless the nitrate solution of Method B was used at less than about 2 V).

Tin-Lead Plating

By Nicholas J. Spiliotis
Allied Chemical Co.

ELECTRODEPOSITION (plating) of tin-lead alloys is used to protect steel against corrosion, serve as an etch resistant, and facilitate soldering. Tin-lead plating is a relatively simple process because the standard electrode potentials of tin and lead differ by only 10 mV. Tin-lead alloys have been deposited by electrolytes such as sulfamates, fluosilicates, pyrophosphates, chlorides, fluoborates and infrequently, phenosulfonates of benzenesulfonates. Of these, fluoborate is available commercially and is generally used to plate tin-lead alloys.

A fluoborate bath plates tin from the stannous valence state. Stannous valence state refers to the valence of tin in solution. In the case of fluoborate, the tin is in the +2 valence state as Sn^{+2}. Tin will plate only from the +2 state in acid solution. Alkaline stannate baths plate tin from the +4 valence state. In fluoborate baths, the stannous tin requires only two electrons to reduce it to metal as shown below:

$$Sn^{+2} + 2e \rightarrow Sn^0 \text{ (metal)}$$

Stannous fluoborate, with lead fluoborate, fluoboric acid, and an addition agent, comprises the plating bath. These components, as well as various addition agents, are available in commercial quantities. The bath operates at 100% cathode and anode efficiency.

Uses of Tin-Lead

Electrodeposition of tin-lead alloys was first patented in 1920 and 1921, when these alloys were used to protect the interior of air flasks of torpedos against corrosion. When air was pumped into a flask under pressure, moisture in the air condensed and corroded the flask, weakening it. Lead coatings had been previously used for protecting the interior against corrosion, but tin-lead alloy was found to be superior in corrosion resistance. Tin-lead deposits, which usually contain 4 to 15% tin, are used as corrosion-resistant, protective coatings for steel. The composition of the alloy varies with the application. Automotive crankshaft bearings are plated with tin-lead or tin-lead-copper alloys containing 7 to 10% tin, whereas an alloy containing 55 to 65% tin is plated onto printed circuit boards. Tin-lead plating on circuit boards acts as an etch resistant and facilitates soldering of board components after they have been inserted into the board. Copper wire is often plated with the same alloy composition when used in the manufacture of electronic components undergoing soldering operations.

Plating Bath Parameters

Any desired tin-lead alloy composition can be plated from a fluoborate bath. Composition of deposits depends on (*a*) amount of stannous tin and lead in the solution, (*b*) type and amount of addition agent, (*c*) current density, and (*d*) tin-lead content of anodes. Bath temperature and degree of agitation affect composition to a lesser degree.

Bath Components. Because concentrated solutions of stannous and lead fluoborates and fluoboric acid are available commercially, alloy plating baths are made by mixing and diluting concentrates. Compositions of concentrates are given in Table 1. The fluoborates of tin and lead contain free or excess fluoboric and boric acids for stability, and fluoboric acid contains free boric acid for the same reason.

The reason excess boric and fluoboric acids provide stability in the fluoborate concentrates can best be shown by the reactions described below with lead fluoborate used as an example, although the same is true for all other fluoborate concentrates. In the absence of boric acid, the metal fluoride will form. To stabilize the lead fluoborate, the following reaction takes place:

$$4PbF_2 + 2H_3BO_3 \rightarrow Pb(BF_4)_2 + 3Pb(OH)_2$$

The reaction is incomplete unless fluoboric acid is added to produce the result:

$$3Pb(OH)_2 + 6HBF_4 \rightarrow 3Pb(BF_4)_2 + 3H_2O$$

The overall reaction is then:

$$4PbF_2 + 2H_3BO_3 + 6HBF_4 \rightarrow 4Pb(BF_4)_2 + 6H_2O$$

Commercially, fluoboric acid is made by reacting hydrofluoric acid with

Table 1 Composition of alloy plating bath concentrates

Constituent	wt %	Amount g/L	Amount oz/gal
Lead fluoborate			
Lead fluoborate, $Pb(BF_4)_2$	51.0	893	119
Lead, Pb(a)	27.7	485	65
Fluoboric acid, free HBF_4	0.6	10.5	1.4
Boric acid, free H_3BO_3	1.0	18	2.4
Stannous (tin) fluoborate			
Stannous fluoborate, $Sn(BF_4)_2$	51.0	816	109.0
Tin, Sn^{+2}(a)	20.7	331	44.3
Fluoboric acid, free HBF_4	1.8	29	3.9
Boric acid, free H_3BO_3	1.0	16	2.1
Fluoboric acid			
Fluoboric acid, HBF_4	49	671	89.9
Boric acid, free H_3BO_3	0.6	8.3	1.1
Hydrofluoric acid, free HF	None	...	...

(a) Equivalent

Table 2 Composition of anode and bath for deposits up to 50% tin

Plated at 3.2 A/dm² or 30 A/ft²; composition of all baths contains a minimum of 100 g/L (13.3 oz/gal) of free HBF_4, 25 g/L (3.3 oz/gal) of free H_3BO_3, and 5.0 g/L (0.7 oz/gal) of peptone

Composition of deposit and anode, % Tin	Lead	Stannous tin g/L	Stannous tin oz/gal	Lead g/L	Lead oz/gal
5	95	4	0.5	85	11.3
7	93	6	0.8	88	11.8
10	90	8.5	1.1	90	12.0
15	85	13	1.7	80	10.7
25	75	22	2.9	65	8.7
40	60	35	4.8	44	5.8
50	50	45	6.0	35	4.7

boric acid as shown in the following equation:

$$4HF + H_3BO_3 \rightarrow HBF_4 + 3H_2O$$

When excess boric acid is added beyond the amount required to react stoichiometrically with the hydrofluoric acid present, the reaction is driven far to the right, thus stabilizing the fluoboric acid and preventing the formation of fluorides.

A tin-lead plating bath deficient in free boric acid can precipitate insoluble lead fluoride. To guard against this possibility, anode bags filled with boric acid should be hung in the corners of the plating tank and immersed in solution. Bags should be refilled when the boric acid has dissolved.

Addition agents are important for the production of dense, fine-grained deposits and the improvement of throwing power in a tin-lead bath operation. Many organic addition agents have been used in tin-lead baths, among which are bone glue, gelatin, peptone, aldehyde condensation products, glycols, sulfonated organic acids, beta-naphthol, hydroquinone, and resorcinol. Peptone is the addition agent most frequently used because of its commercial availability as a stabilized solution specifically prepared for tin-lead plating baths.

Bath Compositions and Operating Conditions. The following are tin-lead bath compositions used most frequently:

7% tin, 93% lead

- Uses: bearings, corrosion protection of steel
- Bath composition:
  - Stannous tin, 6.0 g/L (0.80 oz/gal)
  - Lead, 88.0 g/L (11.8 oz/gal)
  - Fluoboric acid, 100 g/L (13.4 oz/gal) min
  - Boric acid, 25 g/L (3.4 oz/gal)
  - Peptone, dry basis, 5 g/L (0.67 oz/gal)
- Operating conditions:
  - Temperature, 18 to 38 °C (65 to 100 °F)
  - Current density, 3.2 A/dm² (30 A/ft²)
  - Agitation, mild, mechanical
  - Anodes, 7% tin, 93% lead

60% tin, 40% lead

- Uses: printed circuit boards, barrel plating of small parts and applications requiring high throwing power
- Bath composition:
  - Stannous tin, 15 g/L (2 oz/gal)
  - Lead, 10 g/L (1.3 oz/gal)
  - Fluoboric acid, 400 g/L (53.4 oz/gal)
  - Boric acid, 25 g/L (3.4 oz/gal)
  - Peptone, dry basis, 5 g/L (0.7 oz/gal)
- Operating conditions:
  - Temperature, 18 to 38 °C (65 to 100 °F)
  - Current density, 2.1 A/dm² (20 A/ft²)
  - Agitation, mild, mechanical
  - Anodes, 60% tin, 40% lead

60% tin, 40% lead

- Uses: high-speed wire and strip plating or for general plating where throwing power is not of prime importance
- Bath composition:
  - Stannous tin, 52 g/L (7.0 oz/gal)
  - Lead, 30 g/L (4.0 oz/gal)
  - Fluoboric acid, 100 g/L (13.4 oz/gal) min
  - Boric acid, 25 g/L (3.4 oz/gal)
  - Peptone, dry basis, 5 g/L (0.7 oz/gal)
- Operating conditions:
  - Temperature, 18 to 38 °C (65 to 100 °F)
  - Current density, 3.2 A/dm² (30 A/ft²)
  - Agitation, mild, mechanical
  - Anodes, 60% tin, 40% lead

The above bath formula can be used to deposit 60% tin, 40% lead on wire or strip. In this case, current densities in excess of 32 A/dm² (300 A/ft²) can be used if the wire or strip is continuously moved through the plating bath at a relatively high rate of speed.

Composition of anodes and baths for deposits up to 50% tin are listed in Table 2. Composition of the anode should be the same as that desired in the deposit. If deposits do not have the desired composition, anode composition should be maintained as indicated and adjustments made to the bath formula.

Table 2 is based on an operating density of 3.2 A/dm² (30 A/ft²). Higher or lower current densities may result in deposition of alloys of compositions differing from those given in the table. It is then necessary to make compensating corrections in bath composition. Deposition rates of tin-lead coatings can be controlled by current density. Table 3 shows that as current density of a fluoborate bath is increased, the rate of 60% tin to 40% lead deposition also increases.

Temperature. Tin-lead fluoborate baths operate efficiently in a temperature range of 18 to 38 °C (65 to 100 °F). Upper temperatures slightly increase tin in deposits, but lower temperatures can decrease tin.

Current densities below the specified amount for a particular bath formula can decrease tin content of deposits. Higher current densities can increase tin content.

Agitation is an important factor in tin-lead plating. Optimum conditions exist when mild agitation is used. Use of a still bath results in nonuniform deposits because of local exhaustion of the bath at the cathode surface. Vigorous agitation may cause increases in stannic tin content of a bath resulting in a decrease of tin in deposits. Cathode rod agitation or circulation through an outside pump provides suitable agitation

Table 3 Rate of 60% tin to 40% lead deposition from the fluoborate bath
100% cathode efficiency

Current density A/dm²	A/ft²	Time in bath, min, at thickness of: 2.5 μm (0.0001 in.)	7.5 μm (0.0003 in.)	12.5 μm (0.0005 in.)	25 μm (0.001 in.)
1.0	10	4.5	13.5	22.5	45
1.5	15	3.0	9.0	15.0	30
2.0	20	2.3	6.8	11.3	22.5
2.5	25	1.8	5.4	9.0	18
3.0	30	1.5	4.5	7.5	15

for a tin-lead plating bath. Air agitation should not be used because it can oxidize stannous tin.

Boric acid is added to maintain bath stability. Approximately 25 g/L (3.4 oz/gal) of boric acid has been found desirable, but its concentration is not critical. An anode bag filled with boric acid may be hung in a corner of the tank to maintain the required concentration. Based on the following formula:

$$4HF + H_3BO3 \rightleftharpoons HBF_4 + 3H_2O$$

Hydrofluoric acid + Boric acid ⇌ Fluoboric acid

The reaction is reversible if the stoichiometric amount of boric acid is used to react with the HF present. As the amount of boric acid in the above reaction is increased, the reaction is driven far to the right so that the reaction becomes irreversible and no free hydrofluoric acid is regenerated. This is important since if free hydrofluoric acid was present, their insoluble fluorides, especially lead fluoride, would precipitate. Thus, all fluoborate concentrates and plating baths contain free boric acid. Although 25 g/L of boric acid is optimum (close to its solubility) any amount of free boric acid is desirable to prevent the formation of fluorides.

Free fluoboric acid is maintained in the bath to provide requisite acidity and to raise conductivity. In conjunction with peptone, it can prevent treeing and leave a fine-grained deposit. Free fluoboric acid can be added in amounts ranging between 100 to 500 g/L (13.4 to 67 oz/gal), depending on the bath formula used.

Peptone is added to a bath to promote formation of fine-grained adherent deposits and prevent treeing. Peptone solution is available commercially, and proper amounts can be measured and poured directly into baths. A bath can then be used immediately, after gently stirring to ensure complete mixing of peptone.

During the operation of a bath, a loss of peptone can result because of dragout, chemical breakdown, and codeposition with the metal. As peptone is depleted, it must be replenished. Replenishment amounts should be determined by experience with baths. As a guide, 1 L (2.1 pint) of peptone solution per 380 L (100 gal) of bath, per week, can be used. A Hull cell operated at 1 A for 10 min can be used to control peptone content of a plating bath.

Carbon Treatment. Tin-lead fluoborate baths containing peptone should be carbon treated at least four times per year to ensure removal of organic breakdown products and to avoid buildup of peptone from indiscriminate additions. A bath should be treated with about 4.5 kg (10 lb) of activated carbon per 380 L (100 gal) of solution until, after filtration, the solution is water white.

Fresh peptone is added after carbon treatment. Because there is no simple analytical method for determining peptone concentration in this bath, carbon treatment and replenishment of peptone every 3 or 4 months ensures proper amounts of peptone in a bath.

Metallic impurities are removed by low current density electrolysis, but in a tin-lead bath, low current density favors deposition of lead, which may unbalance the bath. Metallic impurities can be removed by dummying a bath at a current density of 0.2 A/dm² (2 A/ft²) for at least 8 h. A bath should then be analyze and brought up to specification with stannous or lead fluoborate. Iron, nickel, and other metals above hydrogen in the electromotive series are not removed by dummying, although copper is easily removed.

Filtration. A tin-lead fluoborate bath should be filtrated constantly, because it can keep the bath clear. If constant filtration is not used, a bath can turn cloudy because of sulfates entering the solution and precipitating as lead sulfate. Stannic salts can also precipitate out of the solution. Anode sludge or breakdown products from peptone solutions can contribute to a cloudy appearance as well.

Polypropylene filter spools or cartridges can be used as filters, but they must first be leached in hot water (65 °C or 150 °F) to remove organic agents used in their manufacture. The end of the return hose from the filter must be submerged in the bath to prevent aeration of the bath.

Anodes. Tin-lead alloy anodes of at least 99.9% purity must be used. The most objectionable anode impurities are arsenic, silver, bismuth, antimony, copper, iron, sulfur, nickel, and zinc. Extruded anodes are preferred over cast anodes, because cast anodes have a larger grain size and suffer from intergranular corrosion, which causes large pits or depressions to form on the anode surface. The finer grain size of extruded anodes provides uniform and efficient corrosion during plating. Tin-lead anodes should be left in an idle tin-lead fluoborate bath because they exercise a reducing effect on tin in solution, thus maintaining the bath in a stannous valence state. Tin-lead anodes should be bagged with dynel fiber or polypropylene cloth to contain any anode sludge which may form. Anode sludge suspended in solution can cause rough deposits.

Materials of construction for tin-lead plating baths include the following:

- Tanks/pumps: steel lined with rubber or polypropylene or made entirely of polypropylene
- Anode hooks: Monel metal
- Anode bags: polypropylene
- Filter spools: polypropylene
- Filter aid: pure paper pulp (alpha cellulose)

The following materials should not be used in contact with a fluoborate bath: (*a*) glass, (*b*) quartz or other silicated materials, (*c*) nylon, (*d*) neoprene, or (*e*) titanium. Equipment in contact with fluoborates should have the recommendation of the manufacturer for use in a fluoborate bath.

Silver Plating

By A. Korbelak
Consultant
Smith Precious Metals Company

STAINLESS STEEL and pewter designs have gained in popularity as flatware and tableware, causing a significant increase in the volume of silver that is being electrodeposited onto base metal alloys. This decorative application constitutes the largest usage of this white metal. However, with rapid escalation of the price of gold, the electronics industry has taken to increased usage of silver in areas where its tendency to migrate, tarnish or destabilize electrically poses no performance problems.

The unique corrosion-resistant characteristics of silver make it applicable as a coating for storage containers for specific chemicals, such as tank cars for bulk transport of specific chemicals. In medical instrument applications, silver plating has been used frequently on electrocardiogram probes. More recently, electrodeposition of alloys, identified as low-karat golds, have been applied to jewelry, watch cases and similar items. Although these alloys are referred to as gold alloys, the silver content is substantially greater than the gold content.

Solution Formulations

For well over 100 years, silver has been plated out of alkaline cyanide electrolytes produced by dissolving silver cyanide in a solution of sodium cyanide. Currently, a double cyanide salt (potassium silver cyanide containing 54.2% metallic silver) is used for bath makeup and replenishment. The product dissolves readily, does not present the dusting hazard of silver cyanide, and simplifies bath replenishment of metal and cyanide (2 avoir oz of potassium silver cyanide yield about 1 troy oz of silver). Filtration after additions of potassium silver cyanide is practically eliminated because essentially no residue of insoluble material occurs through its use. The added step of filtration is necessary when silver cyanide is dissolved for metal replenishment. Use of a double cyanide salt eliminates the need of recalculating extra cyanide additions, which would be necessary if silver cyanide were used, to maintain a proper free cyanide content.

Table 1 summarizes the bath ingredients and operating conditions of silver strike solutions for plating ferrous and nonferrous materials. Strikes, low concentration baths operated at high cathode current densities, are necessary because of the tendency of silver to form poorly adherent immersion coatings. An overlong striking time should be avoided to prevent the formation of a powdery layer of silver. Typical composition and operating conditions for conventional and high-speed electroplating solutions are listed in Table 2.

Two separate cost-cutting objectives have resulted in the development of proprietary silver plating processes. One process restricts deposition of silver to selected areas only on substrate materials, where it is functionally necessary. Required thicknesses that range from 2½ to 10 μm (0.1 to 0.4 mils) are applied in a few seconds at current densities to 100 A/dm^2 (1000 A/ft^2). Silver content in the electrolytes is between 50 and 80 g/L (6 and 10 oz/gal). Bath temperatures are higher than those listed in Tables 1 and 2 and range between 60 and 70 °C (140 and 160 °F).

Another type of electrolyte reduces costs associated with cyanide waste treatment. These alkaline, noncyanide baths are designed to plate silver over copper, copper alloys, and copper-plated metals. Cyanides must be avoided in cleaning and other preparatory steps; consequently, pyrophosphate copper is used to plate over steel and aluminum. Good results are obtained by use of these baths on copper without the necessity of striking with silver. Bath makeup chemistry is similar to that of cyanide systems. The pH of solutions must be maintained between 8.0 and 9.0, and succinimide complex must be between 15 and 30 g/L (2 and 4 oz/gal). Potassium nitrite is added to maintain a concentration between 15 and 30 g/L (2 and 4 oz/gal) for conductivity purposes and to prevent oxidation at the silver anodes.

U.S. Patents 4,126,524 and 4,246,077 have been granted on both process and manufacture of the succinimide complex.

Functions of Bath Ingredients

In cyanide formulations, free cyanide ensures a stable metal ion concentration by providing good anode corrosion. Electrolyte conductivity is also increased by cyanide, carbonate, and hydroxide. The latter reduces rate of cyanide decomposition.

Brightening of the white matte silver deposits is raised to image reflecting

Table 1 Suggested silver strike solutions for silver plating

Material	Potassium silver cyanide, $KAg(CN)_2$ g/L	oz/gal	Copper cyanide, CuCN g/L	oz/gal	Potassium cyanide, KCN g/L	oz/gal	Temperature °C	°F	Current density A/dm^2	A/ft^2	Time, s
Ferrous metals	3	0.4	10-12	1.14-1.36	75-90	10-12	25±2	75±4	1.6-2.7	16-27	10-25
Nonferrous metals	5-10	0.7-1.3	...	...	75-90	10-12	25±2	75±4	1.6-2.7	16-27	10-25

Table 2 Suggested plating solutions for silver plating

Solution type	Potassium silver cyanide, $KAg(CN)_2$ g/L	oz/gal	Potassium cyanide, KCN g/L	oz/gal	Potassium carbonate, K_2CO_3 g/L	oz/gal	Temperature °C	°F	Current density A/dm^2	A/ft^2
Conventional	45-60	6-8	30-45	4-6	30-90	4-12	20-25	68-77	0.55-1.6	5.5-16
High speed	150-225	20-30	110-150	15-20	15-75	2-10	20-30	68-86	5.4-10.8	54-108

levels by additions of such agents as ammonium thiosulfate, selenium, and/or antimony. Succinimide complex performs this function in noncyanide systems. Small amounts of antimony increase the hardness of silver coatings to values approaching that of a 90/10 wrought alloy (coin silver).

Impurities that occur in silver solutions may cause varying degrees of deleterious results, such as dull or off-color deposits, rough plate, poor adhesion, and lower cathode efficiencies. Metallic impurities may be removed by low current density electrolysis or by chelation. Most troublesome are organic materials that are dragged in from cleaning solutions, from buffing compound residues that remain on workpieces, from rack coatings, or from decomposition of brightening agents. Organic contamination tends to (*a*) reduce throwing power of a bath, (*b*) limit its permissible current density range, and (*c*) cause coarser grained and more porous deposits. Removal of such interfering materials is a must. Activated carbon may be effectively employed to remove contaminants, although some may be more persistent and impossible to remove at a reasonable cost. This situation may require salvage through a refinery of precious metals or an in-house separation of silver solutions through standard procedures.

Specifications

In addition to numerous in-house silver plating specifications, three federal documents detail requirements of silver coatings on industry products. Federal Specification QQ-S-365C (November 2, 1972) gives general requirements for silver plating, electrodeposited. Another government document (RR-T-51D) covers requirements for silver plating on tableware and flatware. A third government document covers electroplating of silver on jewelry products. These Guides for the Jewelry Industry were formerly known as Trade Practice Rules for the Jewelry Industry, promulgated by the Federal Trade Commission June 28, 1957 and amended November 17, 1959. These were incorporated into Guides on February 27, 1979. They contain sections on the entire industry (such as gemstones and imports) of which the one on electroplating is of special interest to practitioners of this particular discipline. The former two documents list specific minimum thicknesses of the white metals for particular end-point requirements. The latter calls for a substantial thickness of silver on all significant surfaces, without defining the terms substantial and significant.

A Standard Specification for Electrodeposited Coatings of Silver for Engineering Use, issued by the American Society for Testing and Materials, carries the ASTM designation B 700-81. The standard lists requirements for purity, thickness, appearance, and adhesion of silver coatings. Also included is a definition of significant surfaces and references to testing procedures on thickness, adhesion, and purity. Information on sampling procedures and data on surface preparation of the substrate are provided.

SELECTED REFERENCES

- W. Carriero, Silver Plating, *Metal Finishing Guidebook*, 48th ed., Metals & Plastics Publications, Hackensack, p 304, 1980
- D. G. Foulke and A. Korbelak, Silver Plating, Illustrated slide lecture, American Electroplaters' Society, Winter Park, FL, 1972
- A. K. Graham, *Electroplating Engineering Handbook*, 3rd ed., New York: Van Nostrand-Reinhold Publishing, 1971, p 250
- L. Greenspan, U.S. Patent No. 2,735,808 and 2,735,809 (1956)
- O. Kardos, U.S. Patent No. 2,666,738 (1954)
- F. Lowenheim, *Electroplating Fundamentals of Finishing,* Silver Plating, 1st ed., New York: McGraw-Hill, 1978, p 257-260
- C. A. Hampel, Ed., *The Encyclopedia of Electrochemistry,* Silver Plating, New York: Reinhold Publishing, 1964, p 1046
- R. Fabian, Ed., *The Encyclopedia of Engineering Materials,* Silver Coatings, 4th ed., New York: Reinhold Publishing, 1963, p 627
- Method of Making Silver Complex, U.S. Patent No. 4, 126, 524 (1978)
- Noncyanide Bright Silver Electroplating Bath and Method of Making Silver Compounds, U.S. Patent No. 4, 246, 077 (1981)

Gold Plating

By Fred I. Nobel
Executive Vice President and
Technical Director
LeaRonal, Inc.

GOLD PLATING research has received great impetus in the period since the early 1950's, encouraged by the needs of the electronics-related industries. These industries required thick, low-porosity coatings with greater hardness, wear resistance, and corrosion resistance. The need for gold deposits with these improved properties has led to coating processes that also offer high purity, silicon eutectic-forming ability, low electrical contact resistance, good solderability and weldability, and high infrared emissivity. In the decorative field as well, processes have been developed which can plate thick, ductile, and wear-resistant deposits in a variety of colors and karats.

Pure gold deposits are those that are classified as 24 karat. The deposits may be hard or soft, of high purity or containing traces of codeposited metals or polymers. The latest military specification, MIL G 45204 B and Amendment 2 (1971) has outlined the following standards for three types of purity and four ranges for hardness:

Purity

- Type I: 99.7% gold minimum
- Type II: 99.0% gold minimum
- Type III: 99.9% gold minimum

Hardness

- Grade A: 90 HK maximum
- Grade B: 91 to 129 HK inclusive
- Grade C: 130 to 200 HK inclusive
- Grade D: 201 HK and above

Purity-hardness relationship

- Type I: permissible grades A, B, or C
- Type II: permissible grades B, C, or D
- Type III: permissible grade A only

For type I, grade A and type III, the maximum metallic impurity level for chromium, copper, tin, lead, silver, cadmium, or zinc is 0.1%. For iron, nickel, and cobalt, the combined maximum is 0.05%, and the individual maximum is 0.03%. For type I, grades B and C and type II, the maximum individual metallic impurity level is 0.1%. Metallic hardening agents purposely added are not considered to be impurities.

Plating Solutions

Hot Cyanide Baths. Typical formulations of hot gold cyanide baths can be found in Table 1. Gold metal is added as potassium gold cyanide, and the solutions are made more conductive by adding potassium cyanide, potassium carbonate, or dipotassium phosphate, either singly or in combination. These baths are used to plate 24-karat pure soft gold deposits.

In general, high gold contents are used for heavy gold thickness, because this permits higher current densities and higher cathode efficiencies. Higher temperatures and increased agitation also favor higher plating speeds. Insoluble stainless steel anodes are generally used, and this necessitates replenishing the gold with potassium gold cyanide to maintain the gold content.

Either potassium or sodium salts can be used in preparing gold plating electrolytes, but for concentrated solutions, potassium salts are preferred because of their higher solubility. The addition of carbonates and phosphates buffer the solution and increase both the conductivity and throwing power.

Potassium carbonate is not replenished, because it tends to build up in the baths during use. Control of the carbonate content is important for high-quality deposits. When the potassium carbonate content of the standard hot cyanide 24-karat gold solutions reaches 60 to 90 g/L (6.7 to 10 oz/gal), the free cyanide content is permitted to decrease to 12 to 24 g/L (1.3 to 2.7 oz/gal). When potassium carbonate rises to 115 to 150 g/L (15 to 20 oz/gal), it should be reduced to 60 g/L (8 oz/gal) by precipitation with calcium nitrate, barium cyanide, or calcium acid phosphate. Compounds that introduce hydroxyl ions should not be used. The precipitation should be both slow and from a hot solution to promote the growth of large crystals and facilitate subsequent filtration with a minimum loss of gold.

Potassium cyanide is added frequently in formulas 1 and 2 of Table 1, although it is not required in formula 3. Dipotassium phosphate, however, helps to brighten the deposit and is replenished periodically.

The pH of gold cyanide baths is regularly maintained at less than 11.8. As the pH exceeds 12, the bright plating range of any gold bath is reduced, and the bath tends to plate red or smutty deposits are eventually produced. At a pH below 10.5, solutions containing more than 6 g/L (0.7 oz/gal)

Table 1 Formulas for hot cyanide baths

Formula No.	Gold(a) g/L	Gold(a) troy oz/gal	Potassium cyanide g/L	Potassium cyanide oz/gal	Potassium carbonate g/L	Potassium carbonate oz/gal	Dipotassium phosphate g/L	Dipotassium phosphate oz/gal	Temperature °C	Temperature °F	Current density A/m²	Current density A/ft²
1	1-8	0.1-1.0	30	4.0	...	...	...	...	48-66	120-150	11-54	1-5
2	4-12	0.5-1.5	30	4.0	30	4.0	30	4.0	48-66	120-150	11-54	1-5
3	10	1.2	...	...	...	...	120	16	48-71	120-160	11-54	1-5

(a) As potassium gold cyanide

free cyanide darken on electrolysis and eventually form a black precipitate consisting of cyanogen derivatives, while solutions containing 5 g/L (0.6 oz/gal) or less free cyanide slowly liberate hydrogen cyanide without changing color.

Agitation is preferred, and mechanical agitation is used rather than air agitation.

For barrel plating, a gold concentration of 1 to 4 g/L (0.1 to 0.5 troy oz/gal) and a current density of 11 to 32 A/m² (1 to 3 A/ft²) are preferred. Current efficiency during barrel plating is greatly affected by gold concentration, current density, size of the load, size and shape of the work, ease of tumbling, rotation rate of the barrel, and number and size of the perforations in the barrel. Under ideal conditions—such as low current density, small loads (short distance to center of load), and adequate solution around each piece—a current efficiency of 90 to 95% can be attained, and the deposit will be more uniformly distributed throughout the load.

The principal advantage of hot cyanide solutions is economy, because gold is purchased in its most economical form. Hot cyanide solutions have their disadvantages as well. The baths have poor stability; the solutions cannot be used with printed circuit boards; the deposits are subject to staining; and heavy deposits tend to be dull. The solutions also tend to become contaminated, because of the attack on copper alloy substrates.

Neutral Baths. In order to overcome the disadvantages of the hot cyanide solutions, researchers have developed solutions with lower pH values. A neutral bath proposed by Volk in 1957 and its composition is:

Potassium gold cyanide ....6 g/L (0.8 oz/gal)
Monosodium phosphate ....15 g/L (2.0 oz/gal)
Dipotassium phosphate ....20 g/L (2.7 oz/gal)
Nickel (as potassium nickel cyanide) ....0.5 g/L (0.05 oz/gal)
Voltage ....2 to 3
Temperature ....65 to 75 °C (150 to 165 °F)
pH ....6.5 to 7.5
Current density ....53.8 A/m² (5 A/ft²)

Acid Baths. An acid gold bath was used successfully for gold flashing strainless steel pen points and was patented in 1936. These deposits were bright and showed good adhesion, but could not be used for heavy deposits. The composition for this acid bath is:

Concentrated hydrochloric acid ....125 mL/L (1 pint/gal)
Sodium cyanide ....60 g/L (8 oz/gal)
Sodium gold cyanide ....1 g/L (⅛ oz/gal)
Temperature .... ambient
Voltage 6V

The quantity of acid is not critical as long as the bath is acid to litmus paper. If the deposit became dull during use, sodium cyanide was added and if the bath went above neutral, acid was added.

A more recent and commercially successful acid gold bath was more stable than the bath patented in 1936 and could be used for heavy deposits as well as flash coatings. Deposits were smooth, lustrous and fine grained. The composition for this acid gold bath is outlined below:

Gold (as potassium gold cyanide) .... 4 to 12 g/L (0.5 to 1.5 oz/gal)
Potassium citrate and citric acid ....90 g/L (12 oz/gal)
pH (electrometric) ....3 to 6
Current density ...... up to 107 A/m² (10 A/ft²)
Anodes .... carbon or platinum

This bath has been shown to be stable and useful in plating electronic parts such as printed circuit boards, and in overcoming many disadvantages of hot cyanide baths. Another acid bath was developed based on phosphoric acid. Deposits were said to be satisfactory from the following bath:

Potassium gold cyanide ....10 g/L (1.3 oz/gal)
Phosphoric acid (85%) .... 20 ml/L (0.16 pint/gal)
Potassium hydroxide .... to pH 1.8
Temperature .... 10 °C (50 °F)

Generally, acid baths have reduced cathode efficiencies, especially if the baths become contaminated with base metals and organic impurities. Their advantages include brightness of deposits, ease of plating, absence of staining, and low porosity.

Alloying metals are added to industrial gold plating baths in order to obtain alloy gold deposits with specific characteristics. In most cases, alloyed golds are harder, brighter, and less ductile than pure gold deposits.

Nickel and cobalt can be added to gold baths to increase the hardness and wearability required for electrical contact surfaces. The amount of codeposited nickel or cobalt varies from 0.05 to 0.25%, depending on the hardness requirement, which can vary from 120 to 250 HK. In most cases, the purity of the deposit is high to ensure good solderability and low contact resistance.

When plating in alkaline cyanide baths, nickel and cobalt are added as soluble cyanide complex salts. In acid baths, these metals are added as salts or chelates.

Silver is added to gold baths to increase hardness and brightness. The silver is added as the soluble cyanide complex to alkaline gold baths and can increase hardness up to 150 HK. Silver-gold alloys with 8% silver have been described by Haar, who reduced the brittleness of the deposit by plating from hot solutions.

Copper: Copper-gold alloys that are hard and exhibit good toughness characteristics have been described by Fluhmann. He showed that 18-karat gold alloys with copper could have wearability, hardness, and ductility suitable for industrial use. Solutions are of the alkaline cyanide type.

Other metals, such as antimony, indium, and gallium have been alloyed with gold for industrial use, mainly for specialized applications in the transistor industry. The use of these metals is not as widespread as the alloying metals described previously.

Processing Considerations

Preparation is very important before any type of gold plating is done. This is especially true in industrial gold plating, because the part is often subjected to severe test conditions, and rejects are costly. Faulty preparation has caused parts to fail even though the gold deposit is sound.

When plating semiconductor parts, preparation is particularly important, because of the large variety of basis metals and combinations of basis metals encountered. A single part may contain kovar, steel, copper, silver brazing alloy, and nickel, which creates a complicated cleaning and preparation problem. Parts containing multiple basis metals challenge the plater to develop a preparation procedure that will ensure good adhesion of the subsequent gold deposit to all of the surfaces.

In general, all parts should receive an initial gold strike from an acid gold bath prior to gold plating. Nickel alloys require strong hydrochloric acid pickles and, sometimes, cyanide activation with direct current prior to gold plating. Copper alloys often require acid pickling followed by conventional bright dips and/or cyanide dips prior to gold plating. Many techniques are available, but for best results, the preparation cycle should be as simple as possible to remove all traces of soil and oxide from the part and to activate the metal surface so that it will readily receive gold with good adhesion.

Metallic Impurities

Where high-purity deposits are required or when specific physical and metallurgical properties are sought, metallic impurities must be closely controlled. The effect of base metal contamination varies with the nature of the bath. For example, in an alkaline cyanide solution, iron is complexed to such a degree that it does not codeposit with the gold, whereas in an acid type citrate or phosphate bath a few parts per million can codeposit and adversely affect metallurgical characteristics of the deposit, such as heat resistance. Contamination should be prevented by the judicious application of good basic plating practice, such as the use of gold strikes prior to gold plating, thorough rinsing before and after striking, and prompt removal of parts from the plating baths which may fall from barrels or racks.

Lead is one of the worst metallic contaminants, because only a few parts per million may have severe deleterious effects, particularly in low-current density areas. Many solutions contain specific chelating agents to render lead inactive. A badly contaminated bath is usually best discarded and replaced, although for some proprietary processes the supplier may recommend a purification procedure based on precipitation and filtration.

Copper contamination can cause dullness in low-current density areas and in some solutions, will codeposit with the gold, resulting in reduced corrosion resistance and increased contact resistance. Purification procedures are available from suppliers for removal of copper from solution.

Iron, nickel, and cobalt have a greater tendency to codeposit in acid baths than in alkaline cyanide solutions and have deleterious effects on deposit purity and color. Coatings may also show poor solderability and brittleness. Codeposition may be reduced by the use of proprietary chelating agents, and techniques of purification are available from suppliers.

Silver is the most common contaminant in gold salts, and it tends to plate out very rapidly with gold. When plating high purity, soft gold deposits, it is important to prepare and replenish the bath with high purity-gold salts because traces of silver in the deposit tend to harden the gold and change its metallurgical properties.

Metal Distribution. Because of the high cost of gold, it is of utmost importance that the gold deposit be uniform over the surface of the part being plated. In addition, the thickness of deposit should be as uniform as possible from one part to the next, whether plating on racks or in barrels. It is both wasteful and costly to overplate any parts on the rack or in the barrel in order to make certain that all or most of the parts meet minimum thickness requirements. The heavier the deposit thickness, the more important it is to have uniform metal distribution.

In every gold plating operation where heavy deposits are required, the plater should study the effect that all of the important variables in the process have on metal distribution in order to establish optimum conditions for obtaining uniform metal distribution. The best, and often the only, way to accomplish this is by measuring the thickness of a representative sample and applying simplified statistical methods to determine how to adjust process variables to best advantage.

The following are several important variables that affect good metal distribution:

- Plating at current densities that are as low as practical
- Proper anode area and anode spacing (especially important for rack plating)
- Increased speed of rotation in barrels (10 to 12 rev/min)
- Proper barrel design to accentuate good load mixing, solution transfer, and cathode contact
- Proper rack design
- Proper selection of gold solution for the particular parts to be plated

Gold can be stripped anodically from nickel substrates in a warm cyanide solution. A nonelectric method consists of immersing the plated nickel part in a proprietary stripping solution containing cyanide and a strong organic oxidizing agent.

Commercial gold-stripping solutions also can be used for determining average plate thickness. For measuring local thicknesses of 0.25 to 2.5 μm (0.01 to 1.0 mil), backscatter radiation instruments can be employed. Considerable use has also been made of x-ray diffraction and fluorescence methods for determining gold thickness.

Selective Plating. Because of the high price of gold, considerable savings can be realized by confining the deposit only to the surface of a plated part that actually requires gold. This can be achieved most economically in continuous selective plating equipment through the use of an insulating mask of some type. Partial immersion in the plating bath can also be used. Selective plating operations are generally done on highly specialized machines that are specifically engineered for each application. The machines are designed to perform all the required cleaning, plating, and finishing operations on the metal as it unwinds from a reel at one end of the machine and rewinds onto a reel at the other end after all plating processes are completed.

In order to maintain a high production rate, the steps in the plating sequence must be as short as possible. Exceptionally high plating speeds are accomplished by using a plating bath specifically designed for the particular machine. A high degree of solution agitation can be achieved by pumping the plating bath solution as vigorously as possible through an orifice. This forms a jet stream that impinges on the metal being plated and is confined by masks to plate gold only onto the desired area of the part. Current densities can vary from 5382 to 26 910 A/m^2 (500 to 2500 A/ft^2), with cathode efficiencies of 50 to 100%.

Gold savings as high as 95% can readily be achieved through selective plating, depending on the degree of selectivity. The baths are mostly proprietary, and applications can vary from soft pure 24-karat gold used for semiconductor work to hard gold used for electrical contacts and connectors. Selective gold plating has become a significant gold-plating process, widely accepted commercially.

Recovery. Because gold is expensive, one or more dragout tanks are used to recover the solution dragged out with the work. Recovered dragout solutions from hot baths can be used to replace evaporation losses. Dragout from concentrated solutions of room temperature baths can also be used to replenish the strike bath, provided the two baths are compatible.

When recovery by these methods is not possible, gold from cyanide solutions may be recovered by (*a*) electrolysis, (*b*) precipitation, (*c*) the use of ion-exchange resins, or (*d*) processing at a refining plant.

Decorative Gold Plating

In the plating of jewelry and other decorative items, flash or thin deposits of gold (up to 0.175 μm or 7 μin. thick) are applied to bright nickel-plated brass or to other basis metal parts in order to achieve a desired color. Gold flashing is also done on jewelry parts made from rolled gold plate or gold-filled strip or sheet. Gold-filled strip or sheet consists of a thin layer of 10- or 12-karat gold over a nickel-copper-zinc alloy base. Gold is flashed over these parts to cover edges that do not contain gold, to prevent the low-karat gold from tarnishing, and to obtain a desired color.

The list of gold baths for producing colored gold finishes is very long, indicating that this is an art requiring much experience. Practically all flashing baths are based on alkaline cyanide solutions containing gold, free cyanide, and one or more of a variety of metals. Some of the solutions also contain buffers such as carbonates, phosphates, sulfites, ferrocyanides, and tartrates. Many proprietary formulations are available to produce the variety of colors needed, and suppliers have stock solutions for maintaining the baths.

In general, the following variables enhance gold deposition in any alloy bath used for flashing, making the deposit richer in color:

- Increased agitation
- Increased temperature
- Increased free cyanide
- Decreased current density (except for silver alloys)
- Adding copper makes the deposit pinker; deposits that are too pink (less than 18 karats) should be avoided because they tend to tarnish
- Adding nickel makes the deposit paler or whiter; nickel is a desirable additive because it hardens the deposit and improves wear resistance
- Adding silver makes the deposit greener
- Adding cadmium makes the deposit greener

The effectiveness of the base metal increases with its higher concentration, or with lower levels of free potassium cyanide. Time is kept to a minimum in flashing, allowing just enough time for the part to be well covered with deposit. This usually occurs in about 5 to 15 s.

Brass Plating

By Henry Strow
Oxyphen Products Co.

BRASS PLATING is one of the most common alloy plating processes. All alloys from pure zinc to pure copper can be plated. Alloys with compositions of 80% zinc and 20% copper are used commercially under the name of white brass; this article covers only the colored alloys of copper and zinc that contain 60% or more copper.

The color of brass alloys varies. A 70% copper and 30% zinc alloy possesses a yellow coloration, while a rich yellow color is demonstrated by an 80% copper and 20% zinc alloy. An 85% copper and 15% zinc alloy is light bronze in color, and a 90% copper and 10% zinc alloy is a dark bronze. A virtually copper-colored alloy is achieved by alloying 95% copper and 5% zinc. All of these alloys can be plated consistently. They are true alpha brasses and are ductile. The plated alloys match the wrought alloys in color, composition, and properties. The 60% copper and 40% zinc alloy (Muntz metal) has a brown cast and is in the beta phase. The standard phase diagram for copper and zinc alloys applies to plated brass.

Applications

Brass plating is widely used in many applications. For decorative use, especially wire goods, it is used as a very thin flash plate over bright nickel or other suitable bright plates. The nickel plate supplies a bright leveling plate, and the brass, which is usually plated from 20 s to 1 min, provides a bright brass surface. While yellow brass is most often used, gold colored brasses (85% copper and 15% zinc to 90% copper and 10% zinc) are used on items such as picture frames, cosmetic items, and builders hardware. This technique is also used for bulk (barrel) plating.

Brass is also plated to heavier thickness on items that are burnished, brushed, buffed, or bright dipped after plating to secure a suitable finish. Brass plating is often followed by various processes to produce antique or other dark finishes. These treatments are usually followed by brushing or other mechanical treatment to achieve the desired effect.

Brass plating is used for many engineering applications. Brass plate acts as a good drawing lubricant on steel sheet and provides the best adhesion of rubber to steel. Brass plating of steel tire cord wire is essential to secure maximum adhesion of the tire tread rubber to the steel wire cord. Careful control of the alloy is necessary to achieve the best adhesion. Copper content is generally in the 65 to 70% range. Brass plating is also used to produce rubber hose with wire braid reinforcement. Brass plate may be applied in any thickness as dictated by the application.

Plating Baths

Commercial brass plating solutions are cyanide based. Noncyanide solutions have enjoyed limited utilization because they suffer from lack of stability, difficulties with alloy control, or wide ranges of current density.

The basic ingredients of a cyanide brass plating solution are sodium (or infrequently potassium) cyanide, copper cyanide, and zinc cyanide. Also present in the bath make-up are ammonia (added intentionally or by decomposition of cyanide) and carbonate. Sodium carbonate is necessary in a new plating solution to provide buffering action without which the solution will not produce a consistent color. A mixture of sodium carbonate and bicarbonate is used to produce buffering action in the proper range of approximately 10 pH. Table 1 provides compositions and operating conditions of various brass plating solutions.

The copper content of the solution is generally between 15 g/L (2 oz/gal) and 30 g/L (4 oz/gal). Copper is present as a complex cyanide as $Na_2Cu(CN)_3$. The zinc content of the plating solution is generally between 4 g/L (0.5 oz/gal) and 10 g/L (1.3 oz/gal).

The zinc is present in an equilibrium between the complex cyanide $Na_2Zn(CN)_4$ and the zincate Na_2ZnO_2, alkali complex. The cyanide content should be carefully controlled. The term "free cyanide" is used in brass plating as well as in cyanide copper plating; however, it needs to be redefined in reference to brass plating, because of the cyanide-alkali equilibrium. In brass plating, free cyanide refers to the cyanide in excess of the copper complex. This can be determined accurately with simple analytical methods.

The critical element in the control of color and alloy composition is the ratio of cyanide to zinc. The lower ratios that are sufficient to form the cyanide complex are used for the yellow brass alloys. A slight excess of cyanide is necessary to maintain some anode solubility and to prevent the formation of insoluble zinc and copper salt on the anodes which will stop the flow of current. A

Table 1 Composition and operating conditions of typical brass plating solutions

Type	Sodium cyanide, g/L (oz/gal)	Potassium cyanide, g/L (oz/gal)	Copper cyanide, g/L (oz/gal)	Copper (a), g/L (oz/gal)	Zinc cyanide, g/L (oz/gal)	Zinc (b), g/L (oz/gal)	Sodium carbonate, g/L (oz/gal)	Sodium hydroxide, g/L (oz/gal)	Potassium hydroxide, g/L (oz/gal)	Ammonia (c), %	Addition agents, %	Temperature, °C (°F)	pH	Current density A/dm^2 (A/ft^2)
Conventional brass for flash or rack or still plating	36 (4.8)	...	26 (3.5)	18 (2.4)	11 (1.5)	6 (0.8)	10 (1.3)	...	...	0.5	...	32 (90)	10.0	0-3 (0-30)
Conventional brass for bulk (barrel) plating	60 (8)	...	36 (4.9)	26 (3.5)	14 (1.7)	8 (1.1)	10 (1.3)	...	...	0.5	...	32 (90)	10.0	0-2 (0-20)
Conventional brass for high-speed plating	60 (8)	...	36 (4.9)	25 (3.3)	14 (1.7)	8 (1.1)	10 (1.3)	...	...	...	1	70 (158)	10.0	0-7 (0-70)
High-speed, high alkali brass for strip plating	120 (16)	...	90 (12)	63 (8.3)	7 (0.9)	4 (0.5)	...	65 (8.6)	...	...	...	82 (180)	...	3-16 (30-160)
High-speed, high alkali brass for general purpose	125 (16.6)	...	75 (10)	53 (7)	5 (0.7)	3 (0.4)	...	45 (6)	...	...	6	70 (158)	...	1-8 (10-80)
High alkali brass to produce gold colored (high copper) plating	...	110 (14.6)	42 (5.6)	30 (4)	5 (0.7)	2.5 (0.3)	...	...	22 (2.9)	...	11	45 (113)	...	0.5-5 (5-50)
Conventional brass for gold colored (high copper) plating	70 (9.3)	...	42 (5.6)	30 (4)	7 (0.9)	4 (0.5)	10 (1.3)	...	...	0.1	...	40 (104)	10.0	0-2 (0-20)

(a) As Cu. (b) As Zn. (c) Aqua

larger excess of cyanide will produce the higher copper alloys and reduce plating efficiency even to a complete lack of plate. The 85% copper and 15% zinc and 90% copper and 10% zinc alloys require careful balancing of the cyanide to zinc ratios, especially because small changes in alloy composition change the color of the plate.

Efficiency of brass plating is controlled by the copper content of the solution, with higher copper contents providing higher efficiency. Efficiency is also controlled by the ratio of cyanide to zinc, with higher ratios giving lower efficiencies. Temperature is also an important variable. Plating at 24 °C (75 °F) is only about half as efficient as plating at 35 °C (95 °F) in the lower concentration solutions.

Ammonia is an essential ingredient in brass plating solutions. Ammonia increases the amount of zinc in the deposit. Large amounts produce bands of white zinc plate. A new solution requires the addition of 1/4 to 1/2% of aqua ammonia to the solution. Larger amounts are required if the solution is operated at higher temperatures. Operation below 35 °C (95 °F) normally requires only irregular additions after down periods.

Carbonate is necessary in small amounts up to 30 g/L (4 oz/gal), but above this is an impurity. Normally, an equilibrium is formed where the drag-out balances the formation. Trouble is encountered when the solubility limit is exceeded usually over 150 g/L (20 oz/gal). The best method of removal is by cooling and removing the crystals that form.

pH control is also important in maintaining plating solutions. Normal pH is in a range of 9.8 to 10.3, although higher pH values are frequently usable. Higher pH levels may produce redness in the recesses. While the pH is buffered and changes only slowly, additions should be made to maintain the desired range. Additions of sodium hydroxide are used to raise the pH, while additions of sodium bicarbonate are used to lower it.

Proprietary additions for conventional brass plating solutions are available, but close control of the solution composition provides good deposits that are fully bright over bright base metals with a deposit thickness of up to 0.0025 mm (0.0001 in.).

Current densities of over 2 A/dm^2 (20 A/ft^2) are easily possible with the low current densities remaining with the same color below 0.1 A/dm^2 (1 A/ft^2). Poor performance at higher current densities is an early indication that solution stability requires checking.

Impurities are not a troublesome worry in plating bath solutions. Lead and tin cause red recesses even at very low concentrations. Their presence is usually due to impure anodes. Hexavalent chromium produces dark deposits at low current densities. Extreme amounts may cause complete lack of plate. Drag-in of chromium on racks or parts is usually responsible. Organic impurities such as soaps, buffing dirt, many wetting agents, and other impurities may cause a brown smut on the plate. These impurities may be removed by treatment with activated carbon either through batch treatment or filtration through the activated carbon.

Anodes for brass plating should be of the approximate alloy being plated. For yellow brass, 70% copper and 30% zinc are recommended. Anodes may be ball, slug, or other shapes used in steel or titanium baskets. Bar anodes are also used. Purity is essential with lead content below 0.01% and all foreign metals below 0.02%.

Equipment is of standard design with steel being suitable for tanks, coils, and filters. Preference should be given to rubber- or plastic-lined tanks with stainless or titanium coils, because iron forms ferrocyanides that precipitate as zinc ferrocyanide which causes a grayish sludge. Filters also may be of plastic construction.

High-Speed Plating

In about 1938, brass solutions of high alkalinity were developed and commercially introduced. These solutions used higher temperatures and higher concentrations of components than conventional solutions in use at that time and provided high efficiencies and higher speeds of plating. Originally, solutions used potassium cyanide, but were later modified to use sodium salts. Without addition agents, the solutions produced redness at low current densities, and the color of the alloy could not be controlled as well as with conventional brass plating solutions. Most data pertaining to conventional brass plating solutions are applicable to high-speed plating solutions, except that at high alkalinity levels and high temperatures, ammonia is dissipated as rapidly as it is formed. Because of this rapid dissipation, ammonia is not a factor in the operation of the solution.

Conventional brass plating solutions may be used at high temperatures to obtain high-speed plating, but large amounts of ammonia added frequently or the use of additives which are stable are necessary to obtain good results with uniform color.

SELECTED REFERENCES

- A. Brenner, *Electrodeposition of Alloys,* Vol I, New York: Academic Press, 1963
- W. Safranek, *Properties of Electrodeposited Metals,* New York: American Elsevier Publishing Co., 1974
- Technical Bulletin on Brass Plating, C. P. Chemicals, Sewaren, NJ
- Bulletin 59021 Lea Ronal Golden Brass, Lea Ronal Co., Freeport, NY
- Bulletin 50016 Lea Ronal Bright High Speed Brass Process, Lea Ronal Co., Freeport, NY

Bronze Plating

By Henry Strow
Oxyphen Products Co.

BRONZE PLATING is sometimes applied to high copper brasses, but true bronzes are alloys of copper and tin. These plated alloys are usually ductile when the tin content is less than about 20%, but higher alloys such as speculum (which contains approximately 40% tin) are brittle, and heavy deposits may crack or flake off. Intermetallic compounds that are not found in wrought alloys are formed within copper-tin alloy plates. Thus, the normal phase diagrams are not applicable. The intermetallic compounds cause no change in color or brittleness, but limit temperatures to which the bronze plate may be heated.

Ternary alloys of copper and tin that are alloyed with other metals can be plated, but control of the plating process is so difficult that they have found very limited use. For additional information on plating ternary alloys, see the selected references at the end of this article.

Commercial uses of bronze plating are varied. Alloys containing from 10 to 15% tin are attractive and are used for decorative wares. These alloys have a gold color that is browner than true gold; equivalent copper-zinc alloys are pinker in color. Ternary alloys of copper-tin-zinc can closely match gold but are difficult to control.

Bronze plating is used on builders hardware, locks, and hinges to provide an attractive appearance and excellent corrosion resistance. Bronze-plated steel or cast iron bushings replace solid bronze bushings for many uses. Bronze plating is used where improved lubricity and wear resistance against steel are desired. Good corrosion resistance makes it desirable as an undercoat on steel for bright nickel and chromium plate. Speculum alloys are almost entirely used for decorative uses where they are similar in appearance to silver.

Plating Considerations

Bronze plating solutions have copper as the cyanide complex and tin as the stannate complex, with excess cyanide and hydroxide. Both sodium and potassium salt formulations give higher efficiencies and enable lower concentrations and temperatures.

While solutions of materials other than cyanide and stannate complexes have been explored, none have been commercially successful. Table 1 gives the composition for several typical bronze plating solutions. Considerable variation from the listed values may be necessary to meet individual needs.

Alloy composition is controlled by four factors: (*a*) the ratio of copper to tin in the solution, (*b*) the excess of free cyanide, (*c*) the hydroxide content, and (*d*) the temperature. The amount of copper in the deposit is increased when free cyanide is lowered. Tin content of the solution is increased by a lowering of the hydroxide content. The copper content of the deposit is increased by lowering the temperature. With other variables easily held constant, normal control of the alloy is achieved by varying the cyanide and hydroxide contents which are easily analyzed.

Tin exists in two valence forms. The stannate has a valence of 4 and is the form from which plating is done in alkaline solutions. The other valence of 2 forms an alkali complex known as stannite which can be formed in the solution by reduction of the stannate.

Temperature of the solution is an important plating variable. Temperatures below 41 °C (105 °F) produce poor deposits almost always higher in copper. Higher temperatures create higher efficiencies and greater permissible current densities. Normal temperatures are from 60 to 80 °C (140 to 175 °F). Bulk (barrel) plating solutions usually use lower temperatures. Addition of Rochelle salt or similar products improves plate appearance and assists anode corrosion.

Anodes are usually made of pure copper. Bronze anodes dissolve poorly; the use of separate tin and copper anodes has been tried, but this arrangement is almost impossible to control. Thus, the normal practice is to use copper anodes and add the necessary tin as sodium or potassium stannate.

Equipment used for bronze plating may be steel, but rubber- or plastic-lined tanks are preferable. Heating coils may be steel or stainless steel. Barrels for bulk plating should be of any standard design suited for the elevated temperature.

SELECTED REFERENCES

- A. Brenner, *Electrodeposition of Alloys*, Vol I, New York: Academic Press, 1963
- W. Safranek, *Properties of Electrodeposited Metals*, New York: American Elsevier Publishing Co., 1974
- Bulletin High-speed Bronze Plating Process, M & T Chemicals Inc., Rahway, NJ

Table 1 Compositions and operating conditions of typical bronze plating baths

Solution	Sodium cyanide, g/L (oz/gal)	Potassium cyanide, g/L (oz/gal)	Copper cyanide, g/L (oz/gal)	Copper (a), g/L (oz/gal)	Sodium stannate, g/L (oz/gal)	Potassium stannate, g/L (oz/gal)	Tin(a), g/L (oz/gal)	Sodium hydroxide, g/L (oz/gal)	Potassium hydroxide, g/L (oz/gal)	Rochelle salt(c), %	Addition agent, %	Temperature, °C (°F)	Current density, A/dm^2 (A/ft^2)
Standard bronze plating solution, 15% tin	70 (9.3)	...	40 (5.3)	28 (3.7)	20 (2.7)	...	9 (1.2)	19 (2.5)	...	...	...	60 (140)	0.1-8 (1-80)
Bright bronze plating solution, 12% tin	...	90 (12)	40 (5.3)	28 (3.7)	...	60 (8)	23 (3.1)	...	7.5 (1.0)	5	0.25	60-71 (140-160)	to 11 (to 110)
Speculum plating solution, 40% tin	37 (4.9)	...	20 (2.7)	14 (1.9)	103 (13.7)	...	46 (6.1)	10 (1.3)	...	...	...	66 (150)	3 (30)

(a) As metal. (b) Or equix

Rhodium Plating

By A. Korbelak
Consultant
Smith Precious Metals Co.

RHODIUM PLATING was not used to any appreciable extent until 1930, when the superior properties of rhodium as a noble, white finish for jewelry were first recognized. Since that time, rhodium plate has been used for decorative and, especially since 1945, engineering applications.

Decorative Applications. Use of rhodium continues in a wide range of applications. Rhodium is used to provide a whiter finish on platinum goods and to impart a nontarnishing coating over sterling silver products. Decorative rhodium solutions are detailed in Table 1.

Engineering Applications. Two properties of rhodium make it suitable as a coating in many electronic applications: (*a*) low electrical resistivity and (*b*) high hardness. Thicker deposits from conventional jewelry electrolytes tend to stress crack, causing coatings on silver to tarnish the substrate at exposed areas. This is controlled by a thin layer of nickel (1.25 μm or 0.05 mil) or a thinner layer of palladium (0.1 μm or 0.004 mil) on the silver and under the rhodium. For electronic applications where undercoatings of nickel are undesirable, low-stress compositions have been developed. One electrolyte contains selenic acid and another contains magnesium sulfamate. Deposit thicknesses obtained from these solutions range from 25 μm (1 mil) to 200 μm (8 mils), respectively. Rhodium solutions for engineering applications are summarized in Table 2. The low-stress sulfamate process is used to electroplate rhodium on small electronic parts through use of plating barrels. Table 3 details data on deposit thicknesses that are obtained in barrel plating under varying conditions.

Specifications. In addition to numerous in-house rhodium plating specifications that are mandatory on work jobbed to outside sources, federal specifications also supply requirements for coatings of rhodium on metal products. Military specification MIL-R-46085A (23 March 1972) lists five thickness classes: 0.05, 0.25, 0.5, 2.5, and 6.25 μm (0.002, 0.01, 0.02, 0.1, and 0.247 mil).

Table 1 Decorative rhodium solutions

Solution type	Rhodium g/L	Rhodium oz/gal	Phosphoric acid (concentrated) fluid mL/L	Phosphoric acid (concentrated) fluid oz/gal	Sulfuric acid (concentrated) fluid mL/L	Sulfuric acid (concentrated) fluid oz/gal	Current density A/dm^2	Current density A/ft^2	Voltage, V	Temperature °C	Temperature °F	Anodes
Phosphate	2(a)	0.3(a)	40-80	5-10	...	...	2-16	20-160	4-8	40-50	105-120	Platinum or platinum-coated(b)
Phosphate-sulfate	2(c)	0.3(c)	...	...	40-80	5-10	2-11	20-110	3-6	40-50	105-120	Platinum or platinum-coated(b)
Sulfate	1.3-2(c)	0.17-0.3(c)	...	...	40-80	5-10	2-11	20-110	3-6	40-50	105-120	Platinum or platinum-coated(b)

(a) Rhodium, as metal, from phosphate complex syrup. (b) Platinum-coated products are also known as platinized titanium. (c) Rhodium, as metal, from sulfate complex syrup

Table 2 Low-stress engineering rhodium solutions

Solution	Selenic acid process	Magnesium sulfamate process
Rhodium (sulfate complex)	10 g/L (1.3 oz/gal)	2-10 g/L (0.3-1.3 oz/gal)
Sulfuric acid (concentrated)	15-200 mL/L (2-26 fluid oz/gal)	5-50 mL/L (0.7-7 fluid oz/gal)
Selenic acid	0.1-1.0 g/L (0.01-0.1 oz/gal)	...
Magnesium sulfamate	...	10-100 g/L (1.3-13 oz/gal)
Magnesium sulfate	...	0-50 g/L) (0-7 oz/gal)
Current density	1-2 A/dm^2 (10-20 A/ft^2)	0.4-2 A/dm^2 (4-22 A/ft^2)
Temperature	50-75 °C (120-165 °F)	20-50 °C (68-120 °F)

The specification outlines thickness, adhesion, and reflectivity test requirements for these coatings. It also includes the requirements of Missile Purchase Description MPD-592B (31 Jan 1961) and cross references the above five thickness classes with seven thickness classes of MPD-592B.

Revisions recommended for Guides for the Jewelry Industry (16CFR Part 23, 27 Feb 1979) before the Federal Trade Commission refer to the anti-tarnishing effect imparted to silver products by electroplated layers of rhodium. Though used on sterling silver, no provision in the recommendations has been made to identify that the surface coating of rhodium is present. The recommendations include a note which provides for an undercoating of nickel under the rhodium without requiring identification of the base metal sandwich on goods to be marked as sterling silver. ASTM B 634-78 is a standard specification for electrodeposited coatings of rhodium for engineering use. Six thickness classes are listed: 0.2, 0.5, 1, 2, 4, and 6.25 μm (0.008, 0.02, 0.04, 0.08, 0.157, and 0.247 mil). Sampling procedures are referenced as are thickness and adhesion test methods.

SELECTED REFERENCES

- R. J. Fabian, Rhodium Coatings, *Encyclopedia of Engineering Materials & Processes,* New York: Reinhold Publishing Corp., p 578
- A. Graham, Ed., Rhodium Baths, *Electroplating Engineering Handbook,* 3rd ed., 1971, New York: Reinhold Publishing Corp., p 248
- C. A. Hampel, Ed., Precious Metal Electroplating, *Encyclopedia of Electrochemistry,* New York: Reinhold Publishing Corp., and London: Chapman and Hall Ltd., p 987
- A. Korbelak, Precious Metal Platings, 6th International Metal Finishing Conference, London, May 1964
- F. Lowenheim, *Electroplating (Fundamentals of Surface Finishing),* New York: McGraw-Hill Book Company, p 294
- E. Parker, Rhodium Plating, *Metal Finishing Guidebook,* 48th ed., Hackensack, NJ: Metals & Plastics Publications, 1980, p 298
- F. Reid, *Trans Inst Met Finishing,* Vol 36, 1959, p 74
- K. Schumpelt, U.S. patent 2,895,890, 1959

Table 3 Rhodium barrel plating data

Solution	Required thickness, μm	Required thickness, mil	Thickness of plate, μm	Thickness of plate, mil	Apparent current density(a), A/dm^2	Apparent current density(a), A/ft^2	Calculated current density(a), A/dm^2	Calculated current density(a), A/ft^2	Plating time
Low-stress sulfamate	1	0.04	0.5-1.5	0.02-0.06	0.55	5.5	1.6-2.2	16-22	35 min
	2.5	0.1	1.75-3.25	0.07-0.127	0.55	5.5	1.6-2.2	16-22	1¼ h
	12.5	0.5	11.25-13.75	0.44-0.54	0.35	3.5	1.1-1.3	11-13	7 h

(a) Calculated current density is an estimate of the amount of current being used by those parts that are making electrical contact and are not being shielded by other parts in the rotating load in the barrel. Calculated current density is considered to be about three times the apparent current density, that is, the actual current used for the load divided by the surface of that load

Selective Plating

By Douglas W. Maitland
Applications Engineer
Vanguard Pacific, Inc.
and
Marshall J. Deitsch
Vice President
Vanguard Pacific, Inc.

SELECTIVE PLATING, also referred to as electrochemical metallizing, is a method of depositing metal from a concentrated electrolyte solution without using immersion tanks. The solution is held in an absorbent material covering a portable anode or stylus which is connected by a lead to a specialized direct current power pack. The cathode lead on the power pack is connected to the workpiece, completing the plating circuit. Metal is deposited by contact of the solution-saturated anode with the work area. Constant motion between anode and work and appropriate amperage are required to produce high-quality deposits. The system may be portable, semiportable, or permanent, and plating procedures may be manual, semiautomatic, or automatic.

Brush Plating

Selective plating in its current form has been in existence for over 25 years This process has often been confused with a much older technique, known as brush or swab plating, which was used to repair a part, or portion of a part that had not plated properly. These defects on one or two parts were not worth the expense of reprocessing parts through the entire plating procedure, so a technique was developed to solve the problem in the most economical way possible.

In brush or swab plating, the positive lead of the plating rectifier was connected to a metal rod, file, or similar metal object and became the anode. The lead was then wrapped in an absorbent material, such as cloth, and dipped into tank plating solution. The negative lead of the plating rectifier was then clipped to the part with the missing plating, becoming the cathode. Swabbing the area with missing plating with the wrapped and soaked anode resulted in a cosmetically acceptable deposit of metal in the precise area that it was needed. This touch-up took place outside the tank and required a minimum amount of effort. These brush-plated repairs are normally quite thin with marginal adhesion.

Selective plating uses a uniquely different chemistry and far higher current densities than those used in tank plating. The process involves hand-held anodes and absorbent wrapping and also became known as brush or swab plating. However, selective plating is capable of building up significant thicknesses of metals at relatively high speeds, with excellent adhesion and sound metallurgical properties (Fig. 1).

Selective plating has a number of advantages over other, more familiar metal restoration processes, such as tank plating, welding, and plasma or flame spraying. A comparative analysis of selective plating in relation to these other metal buildup processes is presented in Table 1.

Selective Plating Process

Selective plating is a process similar to a combination of arc welding and electroplating. A flexible cable, or lead, connects the positive output terminal of the power pack, a highly specialized rectifier, with a hand-held, or fixtured, insulated handle called the stylus. The end of the stylus is connected to a shaped graphite block, called the anode. Another flexible cable, or lead, connects the negative output terminal of the power pack with a clamp that is attached to the metal part on which the buildup is to occur. The metal part, or workpiece, becomes the cathode.

The anode is wrapped with an absorbent material, such as cotton batting or polyester sleeving, and dipped into a liquid solution of the metal to be deposited. When this saturated anode wrap is brought into contact with the cathodic workpiece, electrical current permits the flow of metal ions from the solution in the wrap onto the surface of

Fig. 1 Photomicrograph of a selective plating deposit

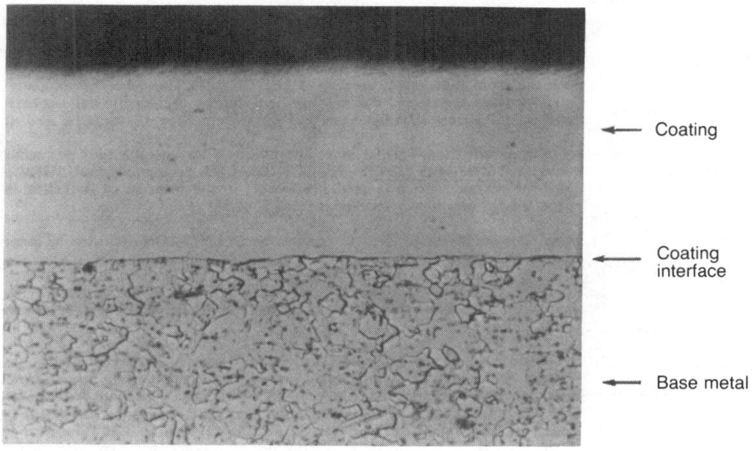

Magnification, 400X. Courtesy of Selectrons Ltd., Waterbury, CT

Table 1 Selective plating in comparison with other processes of metal deposition

Characteristic	Selective plating	Welding	Flame spray or plasma metallizing	Electroplating
Precision buildup capability	Excellent	Poor	Poor	Fair to good
Quality of bond	Excellent	Excellent	Fair to good	Good
Heat distortion or stresses	None	Frequently	Sometimes	None
Heat cracking	None	Frequently	Sometimes	None
Speed of deposit	Fast	Very fast	Very fast	Slow
Density of deposit (porosity)	Very dense(a)	Very dense, but with blowholes	70-90% of theoretical density	Moderately dense
Portability	Yes	Yes	Sometimes, but over-spray precludes its use	No
Requirement for post-machining	Not required on thicknesses up to 0.254 mm (0.010 in.) on smooth surface	Always required	Almost always required	Usually required
Hydrogen embrittlement	No(b)	No	No	Yes

(a) Generally 25% less porous than electroplating and 70% less porous than flame spray or plasma metallizing. (b) Specific cadmium, nickel, and nickel-tungsten deposits have been tested on high-strength steel and were found to be nonembrittling. Other deposits may or may not cause embrittlement
Data provided courtesy of Selectrons Ltd., Waterbury, CT

the workpiece, creating a buildup of metal on the workpiece surface. A schematic diagram of this operation is shown in Fig. 2. This technique allows the metal buildup to be limited only to the area of contact between wrapped anode and workpiece. Little or no masking of workpieces is required, and the thickness of buildups can be controlled very accurately.

Equipment

The equipment required for selective plating, in addition to its very specialized solutions, consists of the following: (*a*) a specially designed power pack, (*b*) an assortment of working tools, called styli, which include shaped graphite anodes, (*c*) accessories and auxiliary equipment which provide flow of solution and/or motion of the anode or part, and (*d*) instruction manuals and operator training.

Power packs used in selective plating are specially designed rectifiers that include features that are unique to the process. These features are required to maintain the current control necessary to produce the desired results.

Voltmeters and ammeters are common to most rectifiers. However, the voltage range of power packs must extend to 20 to 30 V, much higher than most tank plating rectifiers, with infinitely variable voltage control. When the selective plating process is operated manually, control of voltage is the method of regulating the current used most frequently. When mechanized or flow-plating methods are used, the current is controlled by the amperage being drawn. With motion and solution flow being held constant, amperage remains fairly constant, permitting more accurate control of the process by monitoring the amperage. Most higher quality power packs have a dual range amp meter (high and low scale) for control of both large and small parts, which draw correspondingly high and low current.

Accuracy of buildup thickness is controlled by the total amount of current consumed as a factor of time. This measurement, expressed in ampere hours (A · h), is directly related to the thickness of the metal deposited. For this reason, power packs contain a precision digital readout ampere-hour meter, to eliminate the need for separate monitoring of amperage and elapsed time, a method that has proven to be very inaccurate. An additional feature of the ampere-hour meter is a predetermining ampere-hour counter which can be preset to activate a warning light and sound an alarm when the desired thickness of deposit is achieved.

Power packs should also contain a polarity reversing switch to simplify the use of preparatory or stripping solutions that require the current to be used in a reverse direction. This eliminates the need for an operator to change connections to the workpiece and anode in the middle of the procedure.

Another necessary feature found in power packs is an instantaneous safety cutoff switch. This switch quickly shuts off the machine within ½ cycle, or 1/120 s, if a short between the anode and the workpiece occurs. Such a short, which can occur if the absorbent wrap over the anode wears through, would

Fig. 2 Manual operation of selective plating process

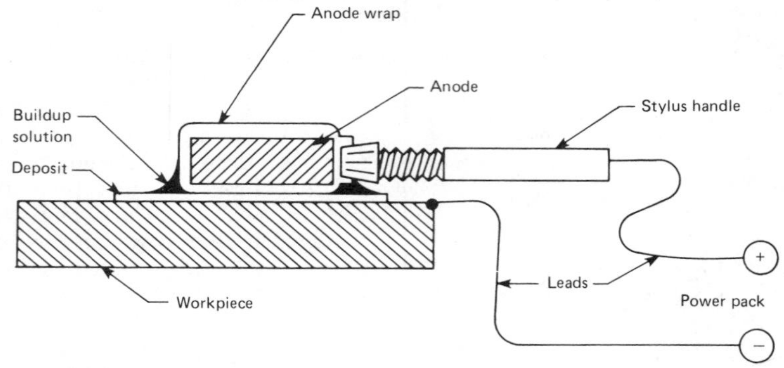

Courtesy of Selectrons Ltd., Waterbury, CT

Fig. 3 Power pack

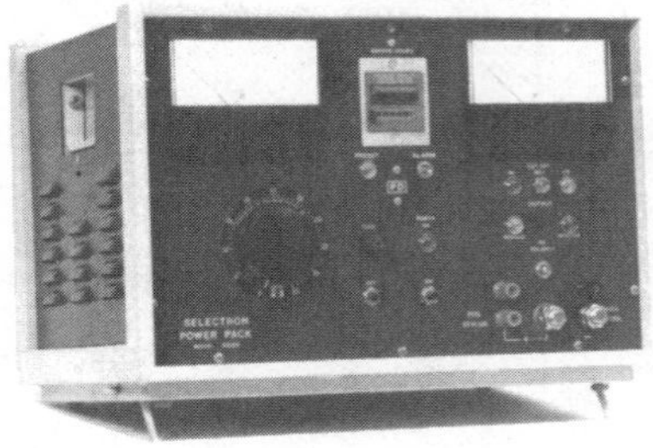

Courtesy of Selectrons Ltd., Waterbury, CT

otherwise cause an arc, possibly damaging the workpiece, a potentially expensive component. The safety cutoff feature also protects an operator with wet hands if the rectification within the power pack itself should break down. The features of a power pack are illustrated in Fig. 3.

Styli and Anodes. Styli are the insulated handles with cooling fins that are attached to the specially shaped graphite anodes required for selective plating. Handles are sometimes made of aluminum, but most often are made of stainless steel, and are insulated to keep the operator out of the circuit. The base of each handle is equipped with a receptacle where a banana plug or twist-lock terminal from the end of the anodic power pack lead can be attached. The interior of the handle contains a current-carrying rod that joins the receptacle at one end, with a threaded fitting at the other end, which can be adapted to fit into anodes of various sizes and shapes.

Anodes used for selective plating, unlike most tank plating anodes, are inert. They play no active part in the deposition of metal other than supplying current. The material from which these anodes are constructed is usually a special grade of prepurified, high-density graphite. This graphite should contain none of the bonding, or hardening agents or metallic impurities, such as iron, copper, or zinc, that are present in most commercial grades of graphite. The presence of any of the metallic impurities acts as a contaminant to solutions and codeposit with the desired metal on the workpiece. If a codeposit of this nature should occur, the risk of galvanic corrosion of the metal deposit would greatly increase. For the same reasons, a graphite anode, because of its inherent porous nature, must be used with only one solution, or contamination can occur.

Graphite anodes are very easy to sand, file, or machine into a desired shape. This offers a significant advantage, because selective plating solutions perform best when the gap between anode and workpiece is approximately 3 to 5 mm ($\frac{1}{8}$ to $\frac{3}{16}$ in.). This gap is created by controlling the thickness of the absorbent wrap over the graphite surface. Concave or convex parts (inside or outside diameters) should have anodes shaped to match.

For smaller sizes, however, platinum or a 90% platinum, 10% irridium alloy should be used, because graphite anodes in thin sections are very brittle and tend to break easily. Although far more expensive than graphite, platinum anodes do not break. They also do not deteriorate with extended use as graphite does, tending to break down into a granular form as it is eroded by the current, which itself may become a contaminant. Platinum anodes are used most often for production applications and small areas, whereas graphite is used for all other applications.

Unlike tank plating, anodes should never be made of the metal being plated. At the high voltages and current densities used in selective plating, anodes tend to polarize very rapidly, causing the current to become limited and greatly slowing down the rate of deposition. Attempts to use stainless steel as an anode material have not been successful because of the tendency of stainless steel to be attacked by many of the solutions used, rapidly eroding the anode and contaminating the solution.

Anodes must be wrapped with an absorbent material, such as cotton batting, or polyester felt. The purpose of the wrap is twofold: (*a*) holding the solution in place, which is especially important for applications where solution is not being pumped, and (*b*) establishing the prescribed gap between anode and workpiece and preventing shorting and arcing of current between them. When cotton batting is used as a wrap, it is generally covered with sleeving (surgical tubing) made of either all polyester or a polyester-cotton blend. Sleeving provides the cotton with better wear resistance and prevents snagging.

For production buildups, especially where thicker deposits are desired and solution is supplied to the anode by a small pump, the use of an abrasive cloth such as Scotch-Brite instead of the cotton batting or polyester felt is recommended. Because it is slightly abrasive, this wrap not only polishes the surface, but also reduces stresses and produces a much finer grained deposit.

Accessories and Auxiliary Equipment. In selective plating, the rate of motion between the anode and the cathode (workpiece), known as anode-cathode motion, must be controlled. When anode-cathode motion is too slow for the amount of current being applied, the resulting deposit is dark, rough, porous, and heavily oxidized (burnt). When the anode-cathode motion is too fast for the amount of current being applied, the rate of deposition is decreased, and the resulting deposit becomes very high

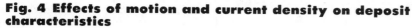

Fig. 4 Effects of motion and current density on deposit characteristics

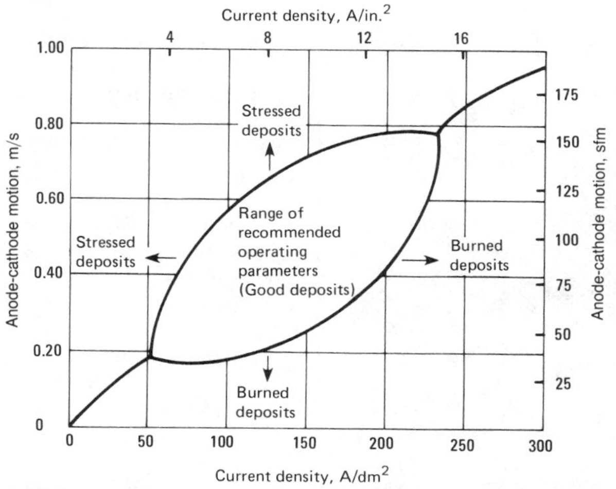

Current density measured from anode contact area. Illustration not intended to represent a specific solution

Table 2 Anode-cathode motion and current density for selective plating solutions

Selective plating solution	Anode-cathode motion m/s	ft/min	Current density A/dm²	A/ft²
Cadmium (acid)	0.26-0.561	50-110	86.4	864
Cadmium LHE (low hydrogen embrittlement formula)	0.20-0.41	40-80	86.4	864
Chromium	0.02-0.03	4-6	86.4	864
Cobalt (machinable)	0.13-0.26	25-50	115.2	1152
Copper (high speed, acid)	0.20-0.51	40-100	144.0	1440
Gold	0.15-0.31	30-60	28.8	288
Lead	0.15-0.26	30-50	86.4	864
Nickel (acid)	0.10-0.26	20-50	86.4	864
Nickel (high speed)	0.20-0.41	40-80	144.0	1440
Nickel-tungsten alloy	0.10-0.15	20-30	72.0	720
Rhodium	0.03-0.05	5-10	43.2	432
Silver (heavy build)	0.10-0.31	20-60	72.0	720
Tin (alkaline)	0.10-0.41	20-80	86.4	864
Zinc (alkaline)	0.15-0.612	30-120	115.2	1152

Data provided courtesy of Selectrons Ltd., Waterbury, CT

in internal stresses, tending to become shiny in appearance and subject to internal stress cracks. Proper anode-cathode motion is needed to achieve metallurgically sound deposits. For each solution, a range of anode-cathode motion, relative to current density, exists, as can be seen in Fig. 4. Specific anode-cathode motion ranges and current densities for a given solution are included in the instruction manual provided by the manufacturer of the solution. Ranges for some of the more frequently used solutions are listed in Table 2.

Mechanical devices available as auxiliary equipment can automatically provide a constant control of anode-cathode motion for circular parts. One such device, called a turning head, consists of a variable speed motor with a forward/reverse switch and an electrically isolated chuck for holding various sized parts (Fig. 5). Such a device must be capable of holding parts to be worked and must provide sufficient torque to rotate at speeds from about 5 to 600 rev/min. Smooth rotation and stepless speed control throughout this range are required. Turning heads are useful for buildups on both outside diameters, such as shafts, and inside diameters, such as bearing fits, where the surface to be built up can be rotated.

The heavy-duty tilt turntable was recently developed to rotate large parts. An approximately 1-m (3-ft) diam face plate with a variety of hole locations to which heavy parts can be mounted replaces the chuck present on the turning head. The face plate is driven by a heavy-duty, SCR-controlled motor and can be rotated in either a forward or reverse direction at controllable speeds between 2 and 100 rev/min.

A rotostylus is another piece of auxiliary equipment for mechanically providing rotary motion. This device, shown in Fig. 6, is useful when the surface to be built up is an inside diameter which does not lend itself to rotation, such as a machined bore in a stationary component. A rotostylus is also useful for reducing operator fatigue when heavy buildups are required and high rates of anode-cathode motion are needed. A rotostylus consists of a variable speed motor with reversing switch, a flexible cable shaft, and a special rotating stylus handle that rotates at speeds from 5 to 600 rev/min and provides current to the anode by means of a rotary electrical connection. Whereas a turning head rotates the workpiece, a rotostylus rotates the anode to provide the required anode-cathode motion. Whenever possible, rotating the workpiece is the preferred method.

A solution pump is an accessory that is useful for building deposits greater than 0.050 to 0.075 mm (0.002 to 0.003 in.) in depth. By providing a fresh supply of solution to the anode-cathode interface by pumping rather than dipping the anode in a tray of solution, the deposition rate is significantly increased. The pump used for this purpose must have no metal parts that could come into contact with the deposition solution, or harmful contamination of solution could result. Plastic impeller or peristaltic pumps are used for this purpose. Although solutions may be flowed over the top surface of the anode, drilling and fitting the anode with a rigid plastic insert is preferred to connect tubing from the pump directly to the anode. This ensures an adequate supply of fresh solution at the surface where it is needed.

Flow plating is a mechanized version of selective plating that uses a pump to flow the solution through the anode (Fig. 7) in combination with a mechanical rotation device, such as a turning head. Because of the high speed of deposition and the generally heavier buildups obtained by flow-plating, a solution cooler (to keep the temperature of the solution below 60 °C or 140 °F at all times) and an in-line solution filter (to remove graphite particles resulting from electrical erosion

Fig. 5 Turning head

Fig. 6 Rotostylus

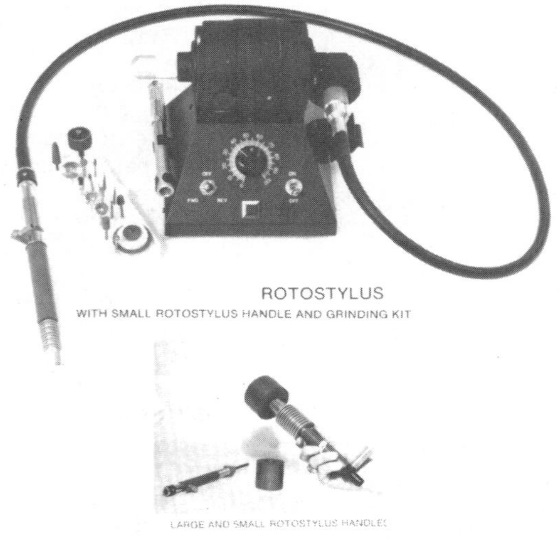

of the anodes) are additional accessories that have proven useful. The use of red Scotch-Brite can be used as the anode wrap for flow-plating to obtain a smoother, more stress-free deposit.

ROTOSTYLUS
WITH SMALL ROTOSTYLUS HANDLE AND GRINDING KIT

Solutions

The formulas used for selective plating are proprietary. Some variations in chemistry and rates of deposition exist between otherwise similar solutions produced by different manufacturers. Some similar characteristics do apply to most high-quality solutions available. Most of these solutions contain high concentrations of metal salts, generally referred to as organo-metallic chelates, dissolved in aqueous media. Characteristics of these solutions can be divided into three classes: (*a*) preparatory solutions, (*b*) bonding solutions, and (*c*) buildup solutions.

Preparatory solutions are used to remove surface contaminants, such as fingerprints, light oils, and oxides, from workpieces or to allow a stronger bond with the metal being deposited. These aqueous solutions generally contain none of the metal salts present in bonding or buildup solutions, but instead use acidic or alkaline additions in addition to any wetting agents or additives necessary.

Because preparatory solutions are proprietary, differences in chemistry between solutions of different manufacturers result. For best results from a standpoint of adhesion and consistency, preparatory solutions from one manufacturer should never be intermixed or used in combination with the preparatory, bonding, or buildup solutions of another manufacturer.

Preparatory solutions are basically light-duty cleaners and should always be preceded with a precleaning operation to remove any heavy amounts of grease, oil, rust, or general tarnish and corrosion as necessary. Solvents should be used to remove grease and oil, and mechanical methods, such as abrasive blasting or silicon-carbide abrasive paper, should be used to remove tarnish or corrosion.

Bonding solutions are used to deposit a preplate or strike of metal onto the surface of a workpiece, before depositing the desired metal. The purpose of this preplate, or bonding layer, is to increase the molecular adhesion for greater bond strength (adhesion) of the buildup metal to the workpiece. The preplate is not always used for adhesion purposes. It can be used to prevent intermetallic diffusion between deposit and workpiece, or to prevent a relatively corrosive solution from reacting with the surface of the workpiece.

A bonding layer is not always necessary, but the following examples do need a bonding layer:

- Surfaces consisting of nickel, chromium, or their alloys, stainless steel, or alloy steel require a preplate of a special highly acidic nickel to maximize bond strength
- Steel surfaces, other than stainless steel, that are to be built up with an acidic copper solution require a preplate of alkaline copper or nickel to prevent nonadhering immersion deposits
- The deposition of silver on most surfaces requires a preplate of gold, palladium, or silver strike to prevent contamination of the silver solution

Fig. 7 Flow-through deposition

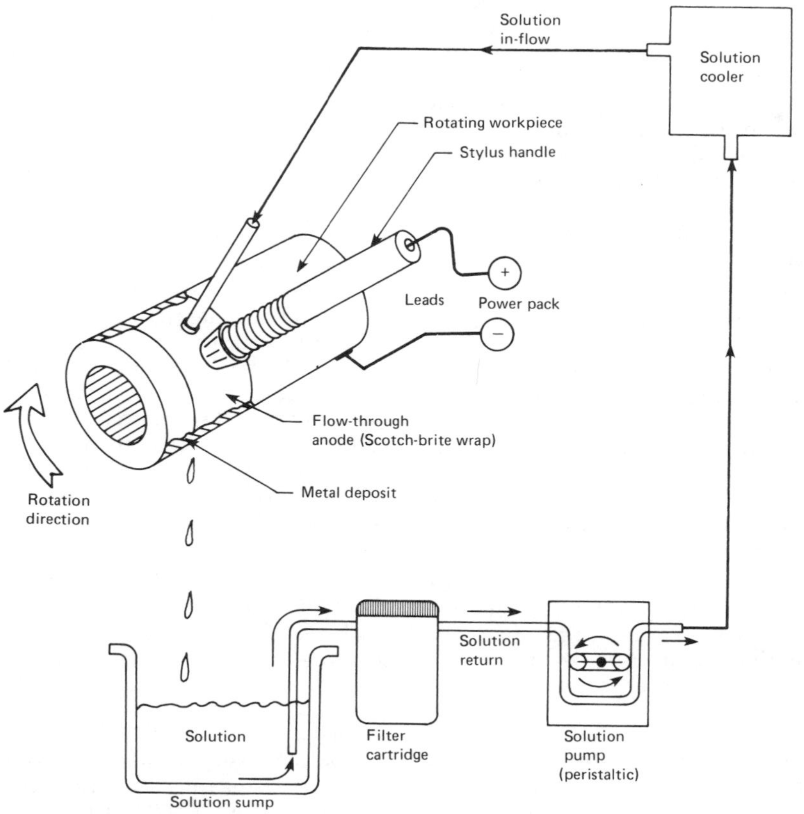

Courtesy of Vanguard Pacific, Inc., Santa Monica, CA

and to prevent immersion deposits

- The deposition of gold onto copper or copper alloy surfaces generally requires the presence of a preplate of nickel to prevent intermetallic diffusion between the copper and gold
- White metals, such as tin, zinc, or cadmium, require a preplate with a neutral or alkaline solution before a buildup deposit with an acidic solution can be achieved. Acid solutions tend to attack these metals

Buildup solutions are generally aqueous solutions containing high concentrations of metal salts, known as organo-metallic chelates. Although the precise formulas are proprietary, common characteristics distinguish these solutions from those used in tank electroplating.

Selective plating solutions have very high concentrations of metal. For example, an alkaline copper solution for tank electroplating contains approximately 20 g/L (3 oz/gal) of metal salt, and selective plating solutions contain 83 g/L (11 oz/gal). For gold plating solutions, approximately 8 g/L (1 oz/gal) is used in tank electroplating, compared with more than 90 g/L (12 oz/gal) for selective plating. Examination of other metals, such as cadmium, tin, and nickel would yield similar comparisons.

High metal concentrations result in high speeds of deposit. Buildups that require an hour or longer using tank electroplating can often be achieved in 60 to 120 s with selective plating solutions. Higher metal concentration permits higher levels of current to be used without excessive oxidation (burning) of the surface of the deposit, and these higher current levels create faster deposition rates. Higher metal concentration and higher current levels result in increased bond strength, because of the higher level of molecular adhesion, and a decrease in porosity of the deposit, approximately 25% less porous than equivalent thicknesses of tank electroplated deposits.

Other characteristics of selective plating solutions include: (*a*) ability to operate well at or slightly above room temperature; (*b*) efficient operation over a wide range of current densities; (*c*) lack of toxic materials, such as cyanides; and (*d*) excellent shelf life. Brighteners are not usually added to these solutions, but nickel and cobalt solutions often have additive agents to reduce internal stresses. Most deposits, however, are extremely fine grained and relatively ductile, with a matte surface appearance.

Selective plating solutions remain balanced during the deposition process. As the metal from the solution is depleted, the liquid portion evaporates at a rate that is comparable to the deposition rate. Additions or corrections to solutions during use are not needed. These solutions do not require mixing or blending. They are poured directly from original containers into the tray or sump from which they will be used. When deposition is complete, any remaining solution is filtered and is saved for reuse. It is not added back to the original container, however, because contamination could result.

Control of Deposit Thickness

Deposit thickness in selective plating is controlled by the ampere-hour meter that is built into the power pack. Each solution has a thickness factor printed on its label and in the instruction manual that tells how many ampere-hours (or fractions thereof) are required to build up a given thickness of deposit, usually 0.025 mm (0.001 in.) on a given area of the workpiece, 6.45 cm^2 (1 $in.^2$). This is expressed mathematically, by the formula: $A \cdot h = F \times A \times T$, where $A \cdot h$ is the desired ampere-hour reading on the power pack, F is the factor printed on the label of the solution bottle, A is the area to be built up with the deposit expressed in square centimetres (square inches), and T is the desired thickness of deposit, expressed in microns (inches). This calculation is demonstrated in the following example. A copper (alkaline, heavy build) deposit of a thickness of 406.4 μm (0.016 in.) is needed in an area of a part that measures 25.8 cm^2 (4 $in.^2$).

$$\begin{aligned} A \cdot h &= F \times A \times T \\ &= 0.00079 \times 25.8 \times 406.4 \\ &= (130 \times 4 \times 0.016) \\ &= 8.32 \text{ A} \cdot \text{h} \end{aligned}$$

Safety Considerations

In general, the selective plating process is relatively safe. It meets the safety requirements set by the Occupational Safety and Health Administration (OSHA) for in-plant use. Electrical protection of the power pack, workpiece, and operator are provided by a direct current circuit breaker within the power pack. If the anode should create a direct short to the workpiece, or if the maximum capacity of the power pack is exceeded, the circuit breaker automatically shuts off the current in ½ cycle, or 1/120 s. This causes little or no arcing to the workpiece. If a rectification failure occurs within the power pack, a circuit breaker prevents the full input alternating current line voltage of the power pack from reaching the operator.

Manufacturers provide Material Safety Data Sheets with solutions, which conform to OSHA requirements. The solutions are, however, still industrial chemicals and conventional safety guidelines should be observed. The work area should be adequately ventilated, and when working in a small enclosed room, a supplemental fan should be used to remove fumes from the area of the operator. Safety glasses and plastic or rubber gloves should be worn when working with the solutions. After using solutions, operators should wash their hands thoroughly before smoking or eating.

None of the solutions are flammable, and flash points and fire hazards are not a concern. However, because gold solutions and some silver solutions contain small percentages of free cyanide, gloves, eye protection, and adequate ventilation are necessary when using these materials. Some metal salts contained in the solutions would be harmful if taken internally and should never be ingested.

The acidity and alkalinity of the various solutions have been carefully controlled to keep them as safe as possible for use when the process is applied by hand. Many of the solutions are no more acidic than soft drinks, nor more alkaline than conventional household cleaners.

Applications and Limitations

Selective plating can be used to add additional plating to portions of already plated parts that have either thin or missing plating in localized areas, as well as to repair areas of a plated part where damage or intentional remachining has occurred. Missing plating can be added and blended into the surrounding areas without separating the part from its surrounding structure. Production plating of parts that require plating only in a localized area and would require extensive masking before tank plating is a cost effective use of selective plating, because the need for masking is either eliminated or greatly minimized.

Selective plating can be successfully used for cadmium plating of high-strength steels. Extensive tests conducted by the U.S. government and by major manufacturers of aircraft have shown that the unique nature of specialized cadmium solutions allows them to be deposited onto high-strength steels without the danger of hydrogen embrittlement, eliminating expensive baking operations.

Selective plating has also proven useful for adding plating to localized areas of parts that are too large to fit into conventional plating tanks or that contain components that should not be immersed in a plating tank. Parts that could be treated in this way include electronic assemblies, electrical motors, or riveted metal assemblies where solution entrapment or attack would create problems.

Plating is often required in localized areas of large, relatively immobile objects such as church domes or oil rigs, or in localized areas of equipment components in the field, such as aircraft landing gear or valves installed on submarines. For these applications, portable selective plating equipment makes it possible to work directly in the field, thus eliminating costly disassembly and downtime.

When considering substituting selective plating for conventional tank plating, the higher costs of proprietary selective plating solutions make over-all plating of parts done in plating tanks more economical. Selective plating should be used only for specialized applications when a large part or localized area needs plating.

Electrochemical Metallizing. With the development of higher speed, heavier buildup solutions for selective plating, additional applications have evolved. Because of the precise thickness control that is possible and the ability to bond to difficult-to-plate metal surfaces such as aluminum, chromium, and passive alloys, selective plating is no longer considered strictly a plating process. Electrochemical metallizing is the most recent use of the process, and metal parts are being rebuilt with this process for purposes not related to a need for the surface of the part to be electroplated.

Electrochemical metallizing (selective plating) is used when an expensive metal part is either worn, damaged, or overmachined on an area where precise dimensional control is required. These applications include rebuilding of worn shafts and bearing fits, repair of scored hydraulic seal surfaces, and restoring worn parting lines on metal molds. In these situations, metal surfaces can be rebuilt with metal whose metallurgical properties are equivalent or, in many cases, superior to those of the original. Because of the precise thickness control possible with electrochemical metallizing, the desired buildup of metal can often be achieved without remachining the deposit.

Electrochemical metallizing has thickness limitations. Buildups of the hard, wear-resistant deposits are under 0.25 mm (0.010 in.) in thickness. Buildups of 0.25 to 1.0 mm (0.010 to 0.040 in.) can be achieved by more experienced operators, using the proper flow-plating techniques. For the softer, fill-in metals, such as copper or tin, buildups of as much as 2.5 mm (0.10 in.) or greater have been achieved. A summary of applications using selective plating and electrochemical metallizing applications includes:

- Touch-up of damaged or missing plating
- Production plating in localized areas
- Adding metal to repair worn or overmachined parts
- Filling scores, scratches, or pits, and adding wear resistance to prevent them
- Adding corrosion protection to localized areas
- Plating to aluminum for solderability
- Localized plating to various base metals to facilitate brazing and welding
- Plating in the field, on large or assembled parts
- Repair of dies, molds, and fixtures
- Deburring small parts
- Repair of printed circuit boards and electrical components
- Localized plating on electrical contacts and busbars (copper and aluminum)

Automation. Electrochemical metallizing has potential for high volume production applications, and automated tooling has been designed for a number of applications. A specially constructed, computer-controlled, and robot-assisted metallizing lathe, designed in Sweden, has been introduced into the United States. This unit can be programmed to deposit a predesignated amount of metal on a specified area of virtually any workpiece of conventional shape, and to do so repeatedly, without needing human labor, to an accuracy that eliminates the need for postmachining in all but the most critical cases.

Appendix 1

Approval Specifications and Standards

The selective plating (electrochemical metallizing) process has been extensively tested and approved for various industrial and military applications. The following is a partial listing of some of these approvals:

- MIL-STD-865A (U.S. Air Force)
- NAVSHIPS 0900-038-6010 (U.S. Navy)
- BAC 5849 and BAC 5854 (Boeing)
- DPS 9.89 (Douglas Aircraft)
- MPS 1118A (Lockheed Aircraft)
- SS 8413 (Sikorsky Aircraft)
- BPSFW4312 (Bell Helicopter)
- FPS 1046 (General Dynamics)
- HA 0109-018 (North American Rockwell)
- GSS PO60B (Grumman)
- Standard Practice Manual 70-45-03 and T.M. No. 72-191 (General Electric)
- PS 137—Issue 1 (Dowty Rotol Ltd., England)
- ITF 40-839-01 (Messier-Hispano, France)
- TCMK-5 (Saab-Scania, Sweden)
- Report No. NAEC-AML 1617 (Naval Air Engineering Center, Philadelphia)
- M-967-80 (Association of American Railroads)
- AC No. 43.13-1A, Chg. 1 (F.A.A.)

In addition to the above, many U.S. Navy and Air Force activities have generated their own subordinate process instruction for selective plating of specific hardware or components at their specific facility.

In the electronics field, the repair of printed circuit boards for military applications is authorized by MIL-STD-865A. For commercial applications, the Institute for Interconnecting and Packaging Electronic Circuits (IPC) approves the use of selective plating for repair of both bare and assembled printed circuit boards. It describes recommended procedures in its IPC-R-700B Manual, *Modification and Repair for Printed Boards and Assemblies*. In addition, many electronics companies have issued their own specifications, for inhouse repair of circuit boards and electronic components and assemblies, by use of selective plating procedures.

Mechanical Coating

By Art O'Cone
Marketing Manager
Minnesota Mining &
Manufacturing Co.

MECHANICAL COATING is a method which utilizes kinetic energy to deposit metallic coatings onto parts. It is also known as mechanical plating or peen plating when the coating applied is less than 25 μm (1 mil) and as mechanical galvanizing for thicker coatings. In general, parts, glass beads, water, chemicals, and metal powder are tumbled together in rotary barrels similar to those used for mass finishing to obtain the coating desired. See Fig. 1 for complete operation sequence. Coatings from 5 μm (0.2 mil) to 75 μm (3 mils) can be deposited. Applicable specifications include: ASTM B695, B696; MIL 81562; QQ-325B, QQ-P416C.

The process is used primarily to provide ferrous-based parts with coatings of zinc, cadmium, tin, and alloys of these metals in various combinations. Layered, or sandwich, coatings can also be applied using this method. Other soft, ductile metals such as copper, lead, and indium can also be plated out, although applications for these are limited. Parts treated by mechanical coating are most often small parts, which are handled in bulk. The appearance of the finish is matte as opposed to the lustrous finishes obtained from electroplating.

Mechanical Plating

Mechanical plating, plating of less than 25 μm (1 mil), is widely used in the automotive industry on hardened (greater than 32 HRC) steel parts as a replacement for electroplating because of hydrogen embrittlement problems. Spring steel parts found in electrical components and appliances are examples of other applications of the process. Mechanical coatings respond to chromating chemicals in the same way as electroplated coatings. Salt spray corrosion resistance is similar to chromated electrodeposits (Table 1).

Mechanical Galvanizing

Where coatings of 25 μm (1 mil) or greater are required for extended outdoor corrosion protection, mechanical galvanizing has gained wide acceptance as an alternative to hot dip galvanizing of fasteners and other small components. The primary benefits are the uniformity of the coating and the elimination of stickers, parts that are welded together. When small parts are coated in bulk, uniform thickness is difficult to control using hot dip methods. Coatings of up to 75 μm (3 mils) of zinc can be mechanically deposited. The process is allowed as an alternate method to hot dip galvanizing in ASTM specifications for fasteners. Corrosion resistance is equivalent to hot dip galvanized coatings and quite often improved because of better uniformity (Table 2).

Fig. 1 Mechanical coating operation sequence

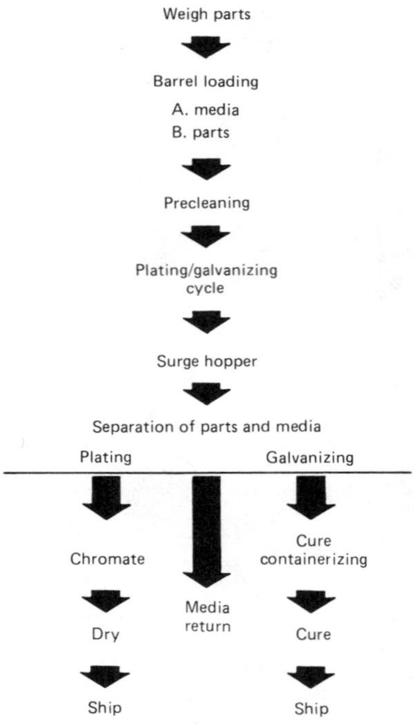

Applicable Parts

The suitability of mechanical coating process for a part to be coated is determined by part size and shape. Size and

Table 1 Corrosion resistance of mechanically plated coatings

Salt spray test results, ASTM B117-62

Type of coating	Time to red rust for thickness in microns (mils), h					
	5.1 (0.2)	6.4 (0.25)	7.6 (0.3)	10 (0.4)	13 (0.5)	25 (1.0)
Plain zinc	36	...	48	...	96	192
Zinc and clear chromate	36	...	48	...	96	192
Zinc and colored chromate	108-336	...	120-384	...	168-480	264-650
Plain cadmium	50	...	120	200	320	...
Cadmium and chromate	200-500	200-500	500-2000	500-2000	2000-20 000	2000-20 000
Cadmium/tin (50/50)	50	...	120	200	320	...
Cadmium/tin (50/50) and chromate	200-500	200-500	500-2000	500-2000	2000-20 000	2000-20 000

Table 2 Estimated average service of unprotected zinc coatings on steel outdoors

Coating thickness μm	Coating thickness mils	Rural	Temperate marine	Industrial marine	Severe industrial
38	1.5	20	10	7	4
51	2.0	30	13	9	6

shape must be carefully considered as they greatly affect resultant coating performance and economics. Generally, parts that will fit into an imaginary cube with 150-mm (6-in.) sides and weighing less than 0.5 kg (1 lb) are suitable for the process. Parts of greater dimensions can also be processed, but these require special handling techniques. Other important considerations are blind holes or recesses. To obtain a satisfactory coating, solution and media must flow freely across the surfaces of the part. If recesses are too deep, the kinetic energy of the impact media may not be sufficient to deform and peen the individual metal particles onto the base metal. Impact media must also be chosen carefully to avoid lodging. Media of the wrong size could become lodged in holes and crevices. Parts suitable for mechanical coating include:

Mechanical plating

- Screws
- Bolts
- Nuts
- Washers
- J-nuts
- U-clips
- Pall nuts
- Self-tapping screws
- Brake pins
- Nails
- Chain
- Seat belt hardware

Mechanical galvanizing

- Structural bolts
- Washers
- Nuts
- Tank bolts
- U-bolts
- Anchor bolts
- Stud bolts
- Nails
- Pole line hardware
- Chain hardware
- Conduit hardware
- Threaded rod

Combination and Alloy Coatings

Combinations of zinc, cadmium, and tin are widely used such as:

- 50% cadmium, 50% tin
- 75% cadmium, 25% zinc
- 50% cadmium, 50% zinc
- 75% zinc, 25% tin
- 50% zinc, 50% tin

Combination coatings are obtained by adding the individual powders of the metals to the coating barrel. The rotating action of the barrel generates sufficient dispersion of the powders to obtain a uniform mixture in the coating. These are often referred to improperly as alloy coatings, because a true alloy coating can be obtained only by alloying the metals at the time of manufacture of the powder, and few of these powders are available commercially. Corrosion resistance may vary between a combination coating and an alloy coating of the same composition.

Combination coatings provide exceptionally good corrosion resistance on steel parts which are galvanically coupled with aluminum. Test results of U-clips plated with various combinations of metals are given in Table 3.

Layered or Sandwich Coatings

Because of the nature of the mechanical coating process, layered coatings can be obtained. This is done by adding the individual metal powders separately during the coating step. For example, if a layered coating of tin and zinc is desired with the zinc as the bottom layer, the zinc powder is added during the first part of the coating step. When essentially all of the zinc has plated out, the tin powder is added. The coating step is continued until essentially all of the tin is deposited.

Processing Powder Metallurgy (P/M) Parts

Powder metallurgy parts having a minimum density of 83 to 87% can be mechanically coated successfully without impregnating before coating. When electroplating is used, impregnation with resins or waxes is required to prevent absorption of the electroplating solution.

Impregnation is normally done under vacuum. Air is evacuated from pores or voids and is replaced by resins or waxes, which are cured by heat or air. This process prevents interconnecting pores that may trap the plating solution used in electroplating. Trapped solution eventually bleeds out, which means the electrolyte will seep out over time causing staining and premature corrosion of the coating. This phenomenon does not occur with mechanical coatings provided the minimum density of the finished part, after sintering, is observed. Density is normally specified in grams per cubic centimetre. For example, 5.8 g/cm^3 is the minimum density that an iron part would be produced by powder metallurgy. Below that it is difficult to get a part to hold together.

With borderline densities, a simple test will reveal if bleedout has occurred. Place a freshly plated and dried sample in a nonporous plastic bag and seal the opening. If the part traps any liquid, condensation becomes evident inside the bag. Time limits vary depending on part porosity and other relating factors, but if the P/M has entrapped liquid it usually appears within an hour. Bor-

Table 3 Test results of U-clips plated with various metals

All U-clip fasteners tested received a chromate dip treatment after plating; the fasteners were firmly attached to the aluminum panels

Coating thickness mm	in.	Type of coating	Type of aluminum	Fastener, hours to red rust	First noted exfoliation or pitting damage on aluminum panels, h
0.0063	0.00025	25% cadmium, 75% zinc	Plain 6061	2160	1512
0.0127	0.0005	25% cadmium, 75% zinc	Plain 6061	No red rust at 6720	1512
0.0063	0.00025	50% cadmium, 50% zinc	Plain 6061	5904	1512
0.0127	0.0005	50% cadmium, 50% zinc	Plain 6061	No red rust at 6720	1512
0.0063	0.00025	25% cadmium, 75% zinc	7029 buffed	No red rust at 2064	No visible damage at 2064
0.0127	0.0005	50% cadmium, 50% tin	Plain 6061	1360	1080
0.0127	0.0005	25% zinc, 75% solder	Plain 6061	1656	1176

derline parts of this nature must be spread out after drying and allowed to stand, preferably without contact, for 24 h.

Elimination of Hydrogen Embrittlement

Steel hardened to 32 HRC or greater is susceptible to hydrogen embrittlement. Exhaustive testing has shown that mechanical coatings applied to hardened steel do not induce permanent hydrogen embrittlement. This is quite often a deciding factor in the selection of the process as opposed to electroplating or hot dip galvanizing, both of which can cause failures on fasteners, for example, at levels below the designed load. Fasteners and other small components, which require a sacrificially protective coating and are used in critical safety items on automobiles, such as seat belt assemblies and brake mechanisms, are often mechanically plated for this reason.

Process Cycle

Parts which have been degreased, descaled, and copper flashed are tumbled in rubber-lined barrels with water, glass bead impact media, promoter chemicals, and a finely divided powder of the metal to be plated. The promoter chemical cleans the metal powder and controls the size of the metal powder agglomerates that form. It also acts as a catalyst. The mechanical energy generated from the rotation of the barrel is transmitted through the glass impact media and causes the clean metal powder to be cold welded to the clean metal parts, that is, joined below their melting points, providing an adherent metallic coating. Chemicals used in this process are available commercially.

Degreased parts, parts free from a water break, are loaded together with water and glass impact media into the coating module. An acidic surface conditioner is added to the rotating coating module. This surface conditioner removes any oxide present on the parts. A coppering solution is then added to provide an adherent copper flash on the parts. The copper flash forms an excellent base for adhesion of subsequent coatings and also presents the same metal surface for the coating step regardless of the metallurgy of the substrate material. The promoter chemical is then added. The addition of a small amount of metal powder provides a flash coating on the parts. Metal powder, the amount of which depends on the surface area of the parts and the coating thickness and weight desired, is then added to the barrel.

Equipment

A mechanical coating system is comprised of a series of operations which include (*a*) weighing, (*b*) loading, (*c*) cleaning, (*d*) galvanizing or plating, (*e*) rinsing, (*f*) surging, (*g*) separation, (*h*) media handling, (*i*) post-treatments, (*j*) rinsing, and (*k*) curing or drying. The components in the system are matched and sized to ensure compatibility for optimum material handling and productivity. As with any system, capacity is limited by the component with the lowest productivity. The labor required depends on the sophistication of the material handling equipment.

A system layout can be designed in any number of shapes to accommodate available space. The following are considerations that affect operational efficiency:

- *Capacity*: production requirements, cleaning requirements, and part size
- *Space available*: minimum and maximum requirements and ceiling height
- *Manpower*: labor intensive or automated
- *Services*: air, water, steam, and electrical
- *Waste treatment*: local, state, and federal Environmental Protection Agency (EPA) requirements
- *Expansion*: future requirements

Mechanical plating or galvanizing can be done in barrels ranging from 0.03 m^3 (1 ft^3) nominal capacity to 0.8 m^3 (27 ft^3).

The system can be as simple as a modified cement mixer or as sophisticated as a fully automated finishing line. The size or sophistication of the system dictates the over-all operating cost per pound processed, the capital required, and ultimately the return on investment.

Waste Treatment

The effluent from a mechanical coating process is acidic and contains metal ions and anions. It contains no cyanide. The effluent can be treated with lime or sodium hydroxide to precipitate metals. Clarifying and final neutralization are required to meet local and EPA regulations. If chromating is performed, the hexavalent chrome must be converted to the trivalent state before precipitation.

Electro-polishing*

ELECTROPOLISHING is the process of smoothing a metal surface anodically in a concentrated acid or alkaline solution. During the process, products of anodic metal dissolution react with the electrolyte to form a film at the metal surface. Two types of films have been observed: (*a*) a viscous liquid that is nearly saturated, or is supersaturated, with the dissolution products; and (*b*) anodically discharged gas, usually oxygen. Both types of films exist simultaneously in most commercial electropolishing solutions. The gas appears to be a blanket on the outside of the viscous film. Which type of film predominates depends on (*a*) the kind of metal and (*b*) the nature of the electrolyte. For example, when brass is electropolished in orthophosphoric acid, a gaseous film forms, whereas in chromic acid, a viscous liquid film forms.

Neither film conforms closely to the microroughness of a metal surface. Both essentially conform to the macrocontour. Thus, the film is effectively thinner over microprojections and thicker at microdepressions. Resistance to the flow of electric current is less at microprojections; hence, more current can flow there than in microdepressions. This situation is augmented by relatively shorter diffusion paths for acceptor (solubilizing) anions to the projections and by-products of dissolution from the projections. The result is more rapid dissolution of projections to cause microleveling of the surface. The ultimate result is a metal surface with so little scattering of incident light that polishing is accomplished, and a glossy appearance is attained.

Nonuniform disturbance of anode film results in local variations in the metal dissolution rate, which in turn results in a nonuniform polished surface appearance. This effect can be minimized or prevented by moving the work during electropolishing.

Applications

Metals are electropolished for one or more of the following purposes:

- Improve appearance and reflectivity
- Improve resistance to corrosion
- Prepare metals for plating, anodizing or conversion coating
- Remove edge burrs produced by mechanical cutting tools
- Remove the stressed and disturbed layer of surface metal caused by the cutting, smearing, and tearing action of mechanical stock removal or of abrasive finishing
- Inspect for surface imperfections in cast, forged, or wrought metal
- Remove excess material as desired for milling metal parts
- Improve surfaces for use of sterilization (drug industry)
- Removal of radioactive surface contamination

Many commercial products made from various materials are electropolished. Industrial installations for electropolishing handle daily quantities ranging from less than 100 to more than 30 000 pieces per facility. Size of products electropolished in these facilities ranges from less than 650 mm^2 (1 $in.^2$) for screws and electronics parts, to several hundred square feet. Some of these products and the materials involved include:

- *Aluminum alloys* (including 3003, 5457, 5557, and 6463): nameplates, wave guides, furniture, reflectors, and watch cases
- *Brass* (alloys include 95-5, 70-30, 60-40, and 85-15): pins, tie clasps, buckles, lighting fixtures, surgical instruments, and tubing
- *Beryllium copper*: pressure transducers, stators, rotors, and connectors
- *Nickel silver*: watch cases, watch bands, and surgical instruments
- *High-temperature alloys* (including molybdenum, Nimonic alloys, Waspaloy, and tungsten): turbine blades, impellers, turbine wheels, missile and rocket engine parts
- *Stainless steels* (types 202, 302, 304, 316, 317, 347, 410, 420, 430, and 440, plus 600 and 700 series; also AMS 5366 and 5648): automotive piston rings, wheel covers, insignia, hood ornaments, lock covers, trim strips, appliance nameplates, wire baskets, percolators, restaurant accessories, plumbing fixtures, dairy and food equipment, surgical instruments, cutlery, missile and rocket parts, cast pumps and valves
- *Low-alloy steels* (4130, 4140, and similar steels): automotive piston rings, crankpins, cotton-picker spin-

*Revised by Louis S. Winter, Vice President, Hydrite Chemical Co.

dles, hand tools, gears, television chassis, and paper knives

- *Zinc die casting alloys*: automotive door handles, window lifts, and panel knobs

Nuclear Applications. Electro-polishing removes radionuclides from the surface of metals and holds contaminants in the solution. In the nuclear power industry, piping, valves, and reactor parts can be decontaminated and returned to use or for safe disposal. Further, used and obsolete laboratory equipment, such as glove boxes, can be cleaned to safe levels. The primary purposes for electropolishing in this nuclear field are reduction of exposure and waste volume.

Electropolishing Solutions and Operating Conditions

The most widely used electropolishing solutions contain one or more of the concentrated inorganic acids–sulfuric, phosphoric, and chromic acids. Sometimes, an inorganic acid, such as hydrofluoric or hydrochloric acid, or an organic acid, such as acetic, citric, tartaric, or glycolic acid, may be used with one or more of the concentrated inorganic acids mentioned above.

The organic acids are said to (*a*) assist throwing power, (*b*) permit electropolishing at lower current densities and temperature, and (*c*) extend the useful life of the solution.

Electropolishing solution compositions started out as proprietary, but many of the patents have expired. Those still active relate to organic constituents or to special equipment. The toxicity of any solution constituent should be pointed out by the supplier, who also should provide information on protection of operators and on analytical control.

Conditions of industrial electropolishing operations using acid electrolytes are indicated in Table 1. Figure 1 shows a typical processing sequence used in electropolishing.

Process Control

During electropolishing, the bath is consumed by metal dissolving from the workpieces. The acids (or alkalis) are converted to metal salts. At some stage, the concentration of dissolved salt becomes high enough to (*a*) prevent polishing, (*b*) precipitate insoluble salts, or (*c*) plate metals onto cathodes. Thus,

Fig. 1 Process cycles typically employed in electropolishing

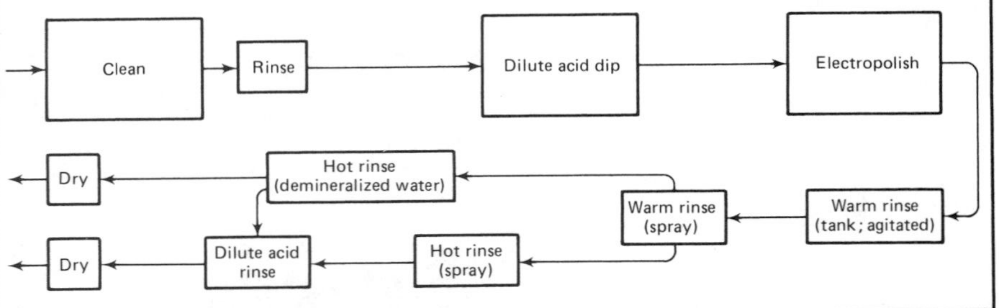

chemical control of the bath composition is required.

Control during establishment of operating routines is accomplished by chemical analysis. Afterwards, experience and use of an ampere-hour meter may adequately indicate the quantity of additions needed to hold the electrolyte at a proper working level. If metal from the work remains dissolved in the electrolyte, some solution can be decanted at regular intervals and replaced with fresh electrolyte.

Equipment and Processing Procedures

Equipment and layout for electropolishing generally resemble an electroplating installation. A work bar or other means is provided for electrical connection to and for supporting the work, as well as for centering the work between two rows of cathodes or two work bars are provided between three cathode rows. The work bar is connected to the positive terminal of a source of direct current; cathode bars are connected to the negative terminal. Meters and current-control devices are of conventional design, and similar to those used in electroplating.

Racks. Items to be electropolished are placed with firm electrical contact on a rack to be positioned on the work bar. The parts are placed so that current distribution will be uniform when they are immersed between cathodes in the electropolishing solution. Usually, racks are made of copper or titanium and have spring clips or fingers of bronze, steel, or titanium to hold the workpieces. The racks are insulated by a plastisol coating. In phosphoric-sulfuric acid electropolishing solutions, bare copper racks can be used for holding nickel and ferrous metal parts. In solutions for brass and aluminum, bare titanium racks can be used.

Tanks are made of stainless steel, low-carbon steel, plastic, or fiberglass. Tanks may require linings of rubber, polyvinyl chloride, polypropylene, or lead, depending on the electrolyte used. The use of plastic-lined tanks is limited to electropolishing in which operating temperature is below that at which the lining can soften.

Cathodes are made of copper, lead, stainless steel, or carbon. The material to be used depends on the temperature and the kind of electropolishing bath. Cathodes are individual rods, flat strips, or expanded metal in strips, which can be suspended from the cathode bar in a pattern for good current distribution. The tank should not be a cathode.

Heating and Cooling. Temperature is maintained within ±1 to ±3 °C (±2 to ±5 °F) of specified value by means of coils for conveying water for cooling or steam for heating. Coils are thermostatically controlled. Also, electric immersion heaters with quartz or carbon sheaths can be used for heating. Immersion-heating coils, plates, or heaters should be positioned so that they do not (*a*) receive stray currents, (*b*) become bipolar in the electrolytic circuit, or (*c*) become covered with sludge. Sometimes chemical attack can occur; it is avoided by making heating coils cathodic at about 0.5 A/dm^2 (5 A/ft^2). Current is taken from the polishing circuit.

Direct current is supplied from rectifiers or motor-generator sets at 12 to 50 V, depending on size of installation and electrolyte. The resistance of the electropolishing solution is the major contributor to the voltage required. The voltage drop through the solution is in direct proportion to the distance between the anode and the cathode. Therefore, the plane of the workpieces

Table 1 Conditions for electropolishing in acid electrolytes

Type of metal (and product)	Purpose of treatment	Bath volume		Installed power		Current density		Polishing cycle, min	Daily production			Oper-ators
		L	gal	A	V	A/dm^2	A/ft^2		No. of parts	Area m^2	Area ft^2	
Sulfuric-phosphoric acid electrolytes												
Monel (fishline guides)	Smooth	750	200	500	12	...	...	5-8	2000	1.5-2.5	15-25	2
302 and 430 stainless (job-shop work)	Bright finish	1150	300	1500	15	30	300	3-8	3500	25	250	2
302 and 202 stainless (plumbingware)	Bright finish	1150	300	1500	12	30	300	3-4	1000-2000	...	...	1
303 stainless (food-processing equipment)	Bright finish	2650	700	2500	18	20	200	4-10	400-500	190-370	2000-4000	2
Series 300 and 400 stainless (job-shop work)	Various	2250	600	3000	18	...	...	10(avg)	25-400	5-45	50-500	2
304 stainless	Brighten; deburr	1300	350	2000	12	30	300	4	3000	55-75	600-800	1
Stainless steel (aircraft components)	...	2650	700	3000	18	25-30	250-300	5	200	30	300	1
430 stainless	Bright finish	1500	400	1500	14	...	...	1-2	7000	75	800	1
430 stainless (trim items)	Brighten; deburr	1500	400	750	18	...	...	3	12 000	230	2500	1
430 stainless (automotive trim)	Bright finish	3800	1000	3000	18	30	300	4	250/h	...	...	1
430 stainless (automotive rain shields)	Bright finish	3800	1000	3000	18	25	250	5	450/h	...	...	1
Stainless and carbon steels (job-shop work)	Brighten; deburr	1500	400	1500	12	25-40	250-400	Varies	Varies	...	...	1
4140 steel	Prepare for chromium plate	3200	850	4000	12	15	150	10	7000-10 000	45-75	500-800	3
Phosphoric-chromic acid electrolytes												
Brass (lighting fixtures)	Final finish(a)	1500	400	1500	18	30-40	300-400	8-15	100-1500	9 or 10	100	1
Brass and copper (electronics contactors)(b)	Smooth	50	10	100	12	...	...	5	100	0.2	2	1(c)
Low-carbon steel, Nitralloy, 440 (paper knives)(d)	Smooth; sharpen	5700	1500	4000	18	...	...	8	75-80	...	...	1
Steel (aircraft instrument and control parts)	...	1700	450	2000	18	...	...	2-4	50	0.5-0.7	5-8	2
Sulfuric-phosphoric-chromic acid electrolytes												
Aluminum	Prepare for anodizing	1900	500	300	18	...	...	3-4	5000-6000	35	400	1
Aluminum (nameplates)	Prepare for anodizing	2250	600	1500	18	10	100	6-12	4000	9 or 10	100	2
Aluminum (eyeglass frames)	Prepare for anodizing	1150	300	1000	24	10	100	6-12	9000	12	130	2
302 stainless (surgical instruments)	Smooth; polish	3600	950	3000	18	30	300	5	5000	...	...	1
Carbon steel	Smooth; deburr	1500	400	...	...	30	300	2	5000	90	1000	1
4130 steel (tools)	Bright finish	950	250	1500	9	17.5	175	4	2000-5000	...	...	2

(a) Some hand coloring after electropolishing. (b) Also electropolished in phosphoric acid bath. (c) Part time. (d) Knives made from all three metals

on the rack should be as near to the cathode as permitted by handling rack loads in and out of the tank.

Installations in the 380- to 1500-L (100- to 400-gal) size should have an 18 V rectifier, while 24 V is recommended for 1900- to 19 000-L (500- to 5000-gal) systems. Those applications above 19 000 L (5000 gal) may require 30 to 50 V. When electropolishing, amperage should be sufficient to handle the work load at 15 to 50 A/dm^2 (150 to 500 A/ft^2). A rule of thumb is to provide 4 L (1 gal) of bath for each 2 to 4 A of current to be passed in electrolytes at 65 to 95 °C (150 to 200 °F).

Each electropolishing tank should have a separate source of direct current placed at some distance from the site of operation. A means for adjusting current by remote control, as well as an ammeter and a voltmeter, should be provided at the tank.

Agitation. Back-and-forth movement of the work bar is usually provided for most uniform electropolishing. Oscillation of the work bar should range from about 15 cycles per min for a 150-mm (6-in.) travel to about 30 cycles per min for a 50-mm (2-in.) travel. This agitation of the work prevents gas streaks at holes and protrusions in the parts.

Cleaning. Facilities for vapor degreasing and alkaline cleaning are required for preparing the work for electropolishing. Oil, grease, solid particles, and, especially, die lubricants must be removed prior to polishing; otherwise, an unsatisfactory appearance will result. Dipping in concentrated hydrochloric acid or nitric acid, or in both acids used separately, may be required for the removal of residual surface contaminants.

Rinsing is an important phase of the process. After electropolishing, equipment is needed for agitated tank rinsing in warm water, followed by warm spray rinsing, and, finally, either hot spray rinsing or rinsing in hot demineralized water, to prevent water spots during drying. The foggy appearance of electropolished and rinsed stainless steel can be removed by dipping in a fresh unused electropolishing electrolyte or in dilute nitric acid solution (10 to 20% HNO_3) and rinsing.

Ventilation is recommended and is provided by operating small electropolishing baths under a fume hood or by installing slot-opening draft boxes along one side of a tank 460 mm (18 in.) or less in width, or along both sides of wider tanks. An exhaust capacity of 60 m^3/min per square metre (200 ft^3/min per square foot) of bath surface is adequate unless strong drafts occur across the tank, in which case an exhaust of 90 m^3/min per square metre (300 or more ft^3/min per square foot) is required.

Electropolishing baths containing only inorganic acids, or alkalis, discharge only hydrogen at the cathode and some oxygen at the work (anode). No poisonous gases are evolved. The bubbling action of hydrogen and oxygen may form a froth on the surface of the electrolyte. A spark resulting from breaking contact between workload rack and anode bar may cause reuniting of hydrogen and oxygen in the froth to form water. This is accompanied by a loud but mild report in the open atmosphere. This action in the exhaust ducts should be avoided by providing enough air velocity to keep the ratio of hydrogen to air at less than about 1 to 30.

Special Equipment. Certain requirements have prompted new equipment which allows processing without immersion. Special brush techniques are used to polish spots and nonsubmersible areas. Electrified liquid streams are used for selected areas. Moving cells are now used for very large areas such as metal walls.

Preparation for Electroplating or Anodizing

Electropolishing prepares a metal surface to accept a uniform, less porous electroplate having maximum bond strength with the basis metal. Structure of the initial layers of electroplate is influenced by crystal structure and stress in the surface of the basis metal. This is important to the protective quality of electroplates, particularly chromium and nickel, applied for (*a*) wear resistance, (*b*) building up of worn or mismachined parts, and (*c*) protection against erosion or against corrosion fatigue.

Coatings formed by anodizing are more uniform chemically and in appearance on an electropolished surface than on a buffed surface. The oxide film is clear and brilliant for reflectors, and it can be dyed to a uniform, clear color of a deep and rich tone. This is attributable partly to the absence of embedded buffing compound and partly to the metallurgically clean surface. Benefits of electropolishing are most notable with high-purity (99.9%) aluminum.

Effect on Appearance

Metals electropolished to a glossy appearance usually have greater total reflectance of light or heat than do metals that have been mechanically polished and buffed. Mechanically buffed surfaces, however, have better specular reflectance.

A superior appearance is produced when surfaces are subjected to both color buffing and electropolishing. This procedure is practiced commercially in the finishing of aluminum and stainless steel reflectors. Silverware, however, is electropolished and then color buffed. The as-electropolished surface is too brilliant in comparison with the softer, buffed color tone that traditionally represents quality. When finished only by buffing, the loss of silver is greater.

Table 2 Typical depth of surface disturbance by mechanical surface treatments of 18-8 stainless steel(a)

Treatment	Depth of disturbed layer, mm	Depth of disturbed layer, in.
Dry lathe turning (0.025-mm or 0.001-in. cut)	0.045	0.0018
Machine grinding (60-grit wheel; 0.013-mm or 0.0005-in. cut, with compound)	0.035	0.0014
Hand grinding (60-grit wheel, dry)	0.045	0.0017
Coarse filing (dry)	0.045	0.0018
Fine filing	0.035	0.0014
100-mesh belt polishing	0.015	0.0005
150-mesh aluminum oxide polishing	0.015	0.0005
Abrasion with 220-grade silicon carbide paper	0.006	0.00024
Abrasion with 400-grade silicon carbide paper	0.003	0.0001
Abrasion with 600-grade silicon carbide paper	0.002	0.00009
Electropolishing	None	None

Effect on Corrosion Resistance

An electropolished surface is essentially equipotentialized. Thus, only a minimum number of local corrosion cells, or none at all, can be set up, because local galvanic differences caused by stress in the metal surface have been eliminated. Initial corrosion rates and oxidation rates of electropolished surfaces differ from those of surfaces of the same metal finished mechanically, and the rates are reproducible. Corrosion and oxidation rates for mechanically finished surfaces are variable.

Some electropolishing processes impart a thin, transparent and notably protective oxide (or salt) film on the surface of some metals, such as stainless steel and aluminum. The film on aluminum, however, is less protective than that formed by anodizing. On other metals, the protective film considerably improves service life of blades for steam and jet engine turbines, pump elements, and electrical-resistance wire.

Electropolishing passivates stainless steel to a greater extent than does any other passivation treatment. This feature is an advantage of electrodeburring of stainless steel. Storage life of carbon steel parts, from the standpoint of corrosion, is aided by electropolish-

Fig. 2 Influence of surface smoothness on coefficient of friction and amount of wear

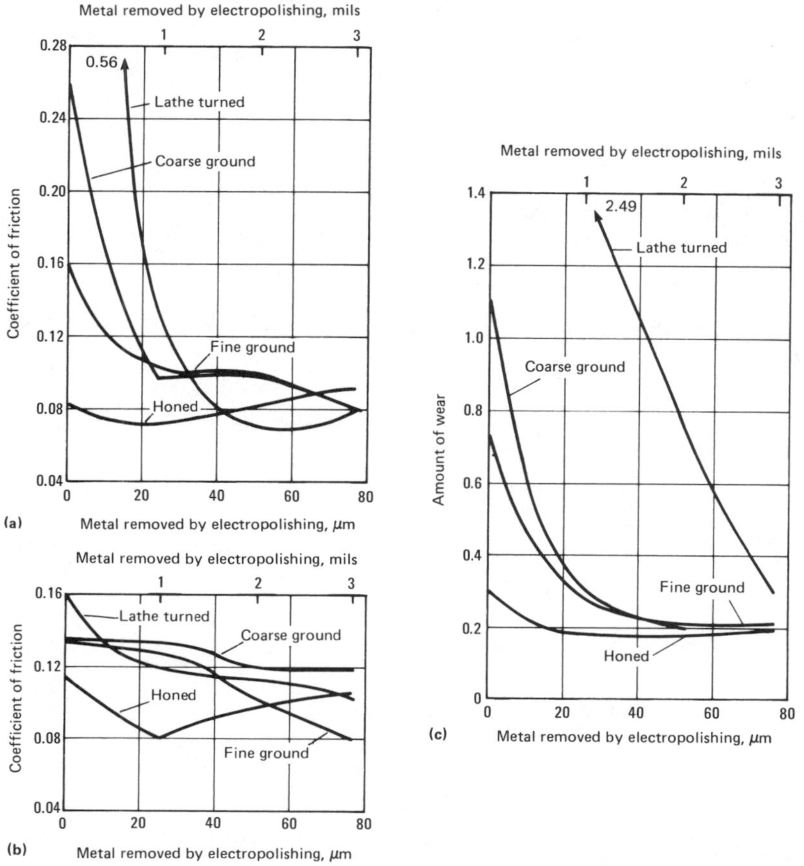

Influence of surface smoothness as related to amount of metal removed in electropolishing surfaces in various previous conditions. (a) Start of test. (b) End of test. (c) Roller wear.

Fig. 3 Influence of surface compressive stress by mechanical polishing and by electropolishing

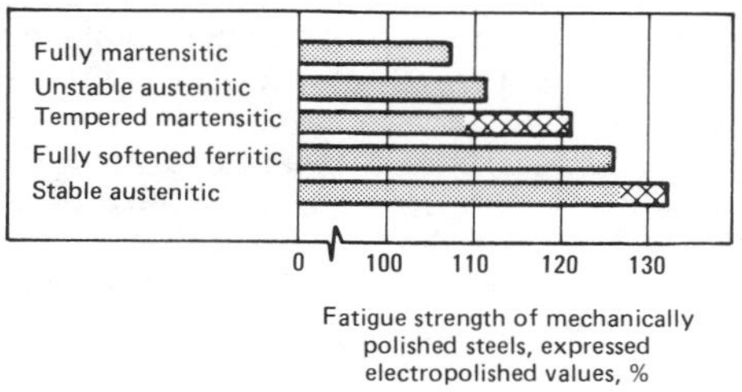

Cross-hatched areas indicate ranges of maximum values reported by different sources for different steels of the same type. Based on data assembled from several sources and presented to SAE Division 20, 6 May 1958, by W. S. Hyler

ing. For example, electropolished carbon steel parts have been stored at 60 to 70% relative humidity for over 6 months without visible rust.

Brass electropolished in phosphoric-chromic acid solution tarnishes only slightly, if at all, whereas buffed brass turns dark in the same atmospheric environment. Although electropolishing does not make brass or steel tarnish-resistant or rust-resistant, the improvement in resistance to corrosion and oxidation can be a practical advantage during manufacturing.

Certain high-temperature alloys, especially, nickel-based alloys, are subject to surface loss by oxidation, which increases initially because of stresses induced during mechanical finishing. Also, recrystallization of the stressed surface takes place at a lower temperature. Both effects are detrimental to best performance in service and are avoided by electropolishing.

Electropolishing has been found to enhance oxidation resistance of Udimet 700 nickel-based alloy. Cold working results in intergranular attack, eventually leading to spalling of mechanically finished material subjected to high temperature. Electropolished material was not attacked intergranularly, and only a thin scale was observed after heating at 980 °C (1800 °F) in still air for 100 h. Under the same test environment, a surface-ground specimen lost 0.013 mm (0.0005 in.) by oxidation.

Oxide formation as a result of electropolishing can be detrimental. Electropolishing of iron-chromium (over 11% chromium) alloys forms an oxide film that leads to a markedly higher initial reaction rate in high temperature oxidation than that induced by other methods of surface preparation.

Effect on Physical and Mechanical Properties

Because it uses no mechanical action, electropolishing provides the means for studying the true effect of mechanical treatments of the metal surface. Smoothing without mechanical action permits correct and reproducible microhardness measurements on the surface. The depth to which mechanical surface cutting, grinding, and abrasive finishing cause structural disturbance can be determined by (*a*) electropolishing followed by measuring microhardness, (*b*) measuring distortion of a specimen because of removing stressed surface metal, or (*c*) metallographic examination of successively exposed levels.

Table 2 shows depths of disturbance by various means of mechanically processing a metal surface. Much less disturbance can result by successive stages of finishing to eliminate the prior worked surface of each preceding stage. As little as 1% of compressive stress can be attained. Surfaces of about 1.3 μm (50 μin.) roughness or less are smoothed to about half that value by electropolishing off 25 μm (1 mil) of metal. The limit of smoothing depends on the grain size and microstructure of the metal and is usually about 0.05 to 0.13 μm (2 to 5 μin.). Carbon steel with a 0.36-μm (14-μin.) finish is electropolished in 8 to 10 min to a finish of 0.08 μm (3 μin.). Extremely fine-grained metals have been electropolished to 0.003 μm (0.1 μin.).

The microroughness of a surface is a major factor in friction and abrasion. The data plotted in Fig. 2 show lower coefficients of friction for surfaces of a finer finish. Figure 2 also shows the appreciable decrease in coefficient of friction and in wear that can be effected as a result of electropolishing.

Seizing of bronze nuts is prevented by electropolishing 13% chromium stainless steel jackscrews. Electropolishing can impart a smoothness to 1045 steel about equal to hand finishing with 320-grit paper and oil. For bearing service, the electropolished surface can be superior to that finished mechanically. Electropolished knives pass more smoothly through other materials, such as wood for veneer and paper, because there is no drag by scratches and burrs that are introduced by grinding wheels and abrasive finishing papers.

Table 3 Typical surface compression stress and fatigue strength of various carbon steels finished mechanically or by electropolishing

Finishing method	Fatigue strength, % of value for mechanical polishing	Depth of cold work, mm	Depth of cold work, in.	Surface compressive stress, MPa	Surface compressive stress, ksi
Mechanical polishing	100	<0.050	<0.002	620	90
Electropolishing	70-90	None	None	None	None
Lathe turning	65-90	0.50	0.02	...	...
Milling	...	0.18	0.007	...	...
Grinding	80-140	Up to 0.25	Up to 0.01	760	110
Surface rolling	115-190	1.00	0.04	900	130
Shot peening	85-155	0.50	0.02	1030	150

Table 4 Electrodeburring bath compositions and operating conditions

Metal to be deburred	Electrolyte	Operating temperature, °C	Operating temperature, °F	Current density, A/dm²	Current density, A/ft²
Ferrous metals	A: 50 vol %, 1.83 sp gr sulfuric acid + 50 vol %, 85% phosphoric acid(a)	55-95	130-200	15-30	150-300
	B: 85% orthophosphoric acid + 5 wt % to saturation chromic acid and 5 wt %sulfuric acid(b)	55-120	130-250	15-30	150-300
Copper(c) and brass(d)	85% orthophosphoric acid + 5 wt % to saturation chromic acid	55-65	130-150	15-30	150-300
Aluminum	85% orthophosphoric acid + 5 wt % saturation chromic acid and 5 wt % sulfuric acid(b)	60-80	140-180	10	100
Nickel and alloys	Phosphoric-sulfuric acid mixture(e)	45-75	110-170	5-20	50-200

(a) U.S. Patent 2,334,699 (1943). (b) U.S. Patent 2,338,321 (1944). (c) U.S. Patent 2,336,714 (1943). (d) U.S. Patent 2,407,543 (1946). (e) U.S. Patent 2,440,715 (1948)

Effect on Fatigue Limit

Removal of not more than 25 μm (1 mil) on the diameter of steel fatigue specimens by electropolishing can lower the endurance limit from 10 million cycles without failure to failure at 100 000 to 120 000 cycles at 520 MPa (75 ksi). The decrease in diameter is not responsible for the loss. Grinding and hand finishing to the same undersize has no adverse effect.

Static stress-strain values on electropolished specimens showed little or no scatter, and when the electropolished surface was rubbed with used 000 emery paper, the original value for fatigue limit was obtained. This mild treatment indicates that any detrimental effect of an electropolished surface results from removal of a compressively stressed skin. It has been observed that wet blasting raised fatigue strength of electropolished specimens to the level attained by polishing mechanically.

The lower fatigue strength of electropolished specimens appears to be because of removal of, or inability to produce, compressive stress in metal surfaces. Thus, electropolishing is comparable in effect to a stress-relieving anneal. For example, mechanical polishing of a chromium-vanadium steel produced compressive stress of 350 to 520 MPa (50 to 75 ksi) at a depth of 0.02 to 0.05 mm (0.0008 to 0.002 in.) below the surface. The stress was relieved by heating at 500 °C (930 °F) for 2 h, which resulted in lowering the fatigue limit from approximately 580 to 560 MPa (85 to 80 ksi). Similar treatment of a low-carbon steel lowered fatigue limit from approximately 730 to 630 MPa (105 to 90 ksi). Electropolishing lowered fatigue limit comparably.

Fatigue strength is not always lower after electropolishing. In an alternating torsion test, a nickel-chromium-molybdenum steel heat treated to 1450 MPa (210 ksi) had 34% higher fatigue strength after electropolishing a ground surface. The same steel heat treated to a lower tensile strength showed lower fatigue strength after electropolishing.

Because fatigue data for electropolished specimens show considerably less scatter than for mechanically polished specimens, electropolishing tends to show true fatigue value characteristic of a particular metal and metallurgical condition. Thus, any irregularly stressed surface can be removed by electropolishing, and a uniform compressive stress can then be applied by controlled working, such as shot peening, wet blasting, or mild abrasive polishing.

Removal by electropolishing of 100 μm (4 mils) of metal from the surface of mechanically polished low-alloy steel (0.44 carbon, 0.61 manganese, 2.48 nickel, 0.82 chromium, and 0.48 molybdenum; heat treated to a tensile strength of 1100 MPa or 160 ksi) changed the residual surface compressive stress of approximately 170 to 200 MPa (25 to

30 ksi) to a tensile stress of 0 to 41 MPa (0 to 5 ksi). The fatigue limit was lowered about 28 MPa (4 ksi). The stress gradient was approximately 140 to 200 MPa (20 to 30 ksi) in 0.100 mm (0.004 in.) in the mechanically polished surfaces. The residual surface stress in mechanically polished specimens differed by approximately 200 to 240 MPa (30 to 35 ksi) from that of electropolished specimens with comparable surface roughness; fatigue limits differed by approximately 580 to 550 MPa (85 to 80 ksi).

Figure 3 shows fatigue strength of mechanically polished steels as a percentage of the values for the same steels as electropolished. Variability of fatigue strength after certain mechanical operations and after electropolishing is shown in Table 3.

Electrodeburring

Electrodeburring is a special application of electropolishing. Burrs are removed from cut edges because electrolytic current flow is greater at edges and protrusions. The ideal situation would be one wherein all the current flowed only to the burrs, and no anodic action occurred elsewhere on the part. This ideal is not attained by any known process. The nearest approach is attained with electropolishing solutions, which are effective because the major surface is not unduly etched or discolored as the burr is being removed.

Electrodeburring can be a practical, low-cost operation. Size of the part and size and location of the burrs determine whether electrodeburring can be accomplished technically and practically. Curled chips left attached, for example, as from a milling operation, are not removed by a practical amount of electrodeburring. When electrodeburring is economically feasible, the problem of change in dimensions must be considered.

Processing Conditions. Electrodeburring solutions usually are modifications of an electropolishing solution for the type of metal involved. Table 4 lists compositions and operating conditions of several patented solutions.

Installations for electrodeburring range from 100 to 5700 L (25 to 1500 gal) of solution, capable of treating from 20 to 30 000 parts per day. Direct current equipment for these installations ranges from 150 to 4000 A at 12 to 18 V for operating at current densities of 10 to 30 A/dm^2 (100 to 300 A/ft^2). Processing time depends on size and location of the burr. The process may not be practical if deburring cannot be successfully completed within 5 to 15 min.

Applications. Electrodeburring is used successfully in the production of hydraulic pump cylinders and some office machine parts. Burrs are removed from the inside bored surface of sideports drilled into hydraulic pump cylinders. These burrs are not easily accessible for removal by mechanical means; only ten parts per day can be deburred by hand operations. Electrodeburring, however, processes 10 parts simultaneously in 10 min and offers the added advantage of producing slightly rounded corners and edges.

Gears and other office-machine parts blanked from flat steel plate are electrodeburred during manufacture. Burrs from the blanking operation are present on one side of (*a*) the periphery of teeth, (*b*) all locating holes, and (*c*) the shaft hole of gears. These burrs cause gears to nest and prevent independent movement. Forty-five man-minutes are required for assembling a set of gears deburred by hand, and the set must be run in for as long as 60 h before being assembled in a machine. Electrodeburred gears can be assembled in the same set in 15 man-minutes and require no run-in before final assembly.

Electrodeburring is also used in the manufacture of various other products, including agricultural equipment, aircraft and missile components, paper knives, textile shuttles, stainless steel hypodermic needles, gyro housings, drive gears, and drive shafts.

Treatment of Plating Wastes

By Craig J. Brown
Vice President and General Manager
Eco-Tec Limited

TREATMENT OF PLATING WASTES involves one of two possible approaches — destruction or recovery. Destructive-type techniques convert toxic or environmentally harmful wastes into legally acceptable forms for subsequent disposal. Recovery techniques recycle plating wastes for further use. Either approach must be used to comply with government regulations, which determine the extent and quality of the waste treatment operation.

Regulations governing the discharge to publically owned treatment works (POTW) of pollutants leaving electroplating facilities have been revised several times since the passing of the Pollution Control Act Amendments in 1972. Although limits have yet to be finalized, they will likely resemble those shown in Table 1. Note that these are proposed minimum requirements. Local authorities may enforce more stringent regulations if desired.

Slightly more lenient limits may be allowed as a result of the presumed ability of the POTW to further treat wastewater before final discharge to the environment. Proposed electroplating limits after application of the POTW Removal Credits are shown in Tables 1 and 2. These limits are substantially higher than those originally promulgated in 1972. For the most part, they are readily achievable through a conscientiously applied program of good housekeeping practices and conventional waste treatment technology.

Since passing of the Resource Conservation and Recovery Act, disposal of sludges resulting from waste treatment has become a major concern. Some electroplating sludges are classified as hazardous and as such necessitate considerable recordkeeping and must be disposed of in a hazardous waste disposal facility. Various sludges have been exempted from the hazardous waste category (de-listed). These include sludges from sulfuric anodizing of aluminum, tin plating of steel, zinc plating (segregated basis) on carbon steel, aluminum or zinc-aluminum plating on carbon steel, cleaning and stripping associated with these operations, and chemical etching.

If an electroplating sludge contains no cadmium, chromium, nickel, or cyanide or it can be proven that these contaminants, although present in the waste, do not leach from the waste at significant levels, it may be possible to have the sludge de-listed. However, this is an involved and time-consuming process and the chances of success are not necessarily good.

Regardless of the confusion surrounding pollution control regulations, at the present time it is safe to say that all electroplating facilities should, as a minimum requirement, implement an effective pollution control program using the conventional treatment technology currently available. If necessary, this can be upgraded at a later time to meet more stringent requirements. At the same time, everything should be done to minimize the amount of sludge produced, thereby reducing the cost and amount of paperwork associated with its disposal.

Minimizing Treatment Load

Primary sources of liquid waste in the plating shop include rinsewaters; spent process solutions; and accidental spills, leaks, drips, and splash. Good housekeeping and well-designed finishing equipment can do much to minimize the latter. Several steps can be taken to reduce the amount of rinsewater-borne waste because reduction of dragout and rinsewater flows can substantially reduce treatment load.

Reduction of Dragout. The amount of solution dragged out of the process bath can be significantly reduced by:

Table 1 Proposed electroplating pretreatment requirements(a)

Pollutant	Daily max mg/L	Daily max mg/gal
Cadmium	1.2	4.6
Chromium	7.0	26.6
Copper	4.5	17.1
Nickel	4.1	15.6
Lead	0.6	2.3
Zinc	4.2	16.0

(a) Subject to revision; operators of electroplating facilities should ascertain regulatory status

Table 2 Electroplating pretreatment requirements after application of removal credits(a)

Pollutant	Daily max mg/L	Daily max mg/gal
Cadmium	1.9	7.2
Chromium	20.0	76.0
Copper	12.7	48.3
Nickel	5.1	19.4
Lead	1.2	4.6
Zinc	12.0	45.6

(a) Subject to revision

Fig. 1 Electroplating treatment flowsheet

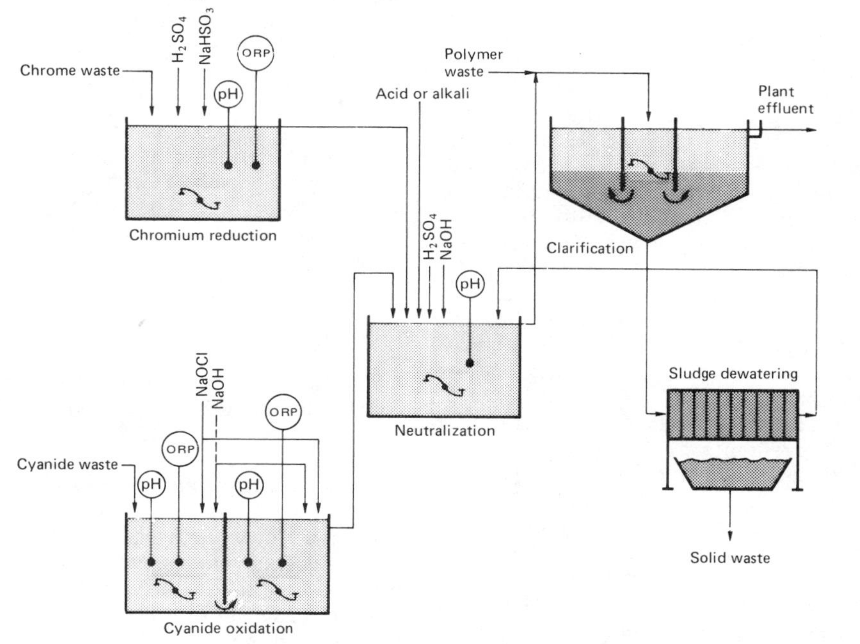

- Minimizing concentration of chemicals in the bath
- Operating bath as hot as possible
- Withdrawing parts slowly from the bath and allowing for adequate drainage over the bath—with a given, fixed time period allowed for withdrawal and drainage, use the largest time interval for withdrawal
- Racking parts to avoid cupping, minimizing drainage paths, and tilting all parts with horizontal surfaces

Minimizing Rinsewater Flows. Because the cost and effectiveness of most treatment techniques depends to a large degree on hydraulic loading and because the majority of water consumed in a plating shop is for rinsing, it is well worth the effort to do everything possible to minimize rinsewater flow rates. Following are some ways this can be done:

- Multistage countercurrent flow rinses
- Conductivity type rinsewater controllers
- Air agitation
- Maximize withdrawal and drainage times (see above)
- Spray or fog rinses
- Warm water

Many pollution problems can be avoided or significantly reduced through use of alternate treatment processes or chemicals. The best example of this is the use of noncyanide plating baths where possible.

Conventional Wastewater Treatment

The wastewater treatment flowsheet shown in Fig. 1 is suitable for the majority of plating shops. It consists of the following unit operations:

- Chromium reduction of segregated chromium waste streams to chemically reduce the chromium from its hexavalent state to the trivalent state
- Cyanide oxidation of segregated cyanide-bearing waste streams to oxidize the toxic cyanides to harmless carbon and nitrogen compounds
- Neutralization of combined wastewaters to adjust the pH within acceptable discharge limits and to precipitate dissolved metals as hydroxides
- Clarification of neutralized wastewater streams to separate decontaminated wastewater from the metal hydroxide sludge
- Dewatering of sludge to reduce volume of solid waste that must be ultimately disposed

Chromium Reduction

Chromium can exist in either of two aqueous valence states: hexavalent or trivalent. In chromic acid and chromate salts, the most common chromium containing chemicals used, chromium is in the hexavalent form. All hexavalent chromium-bearing wastewaters are normally segregated and collected in a chrome sump. They are fed from the chrome sump to the chrome reduction tank.

Hexavalent chromium, as chromate, is normally very soluble under all pH conditions. Trivalent chrome is quite insoluble in the pH range of 6 to 10. Effective removal of chromium therefore necessitates reduction to the trivalent form. Reduction is normally accomplished with either sulfur dioxide or sodium meta-bisulfite, but some other sulfur bearing compounds (including sulfur dioxide) can also be used. Ferrous sulfate, which has been used extensively in the past, is not recommended because of the additional quantities of iron hydroxide sludge which are ultimately generated. The reduction reaction with sulfur dioxide is:

$$2CrO_3 + 3SO_2 \rightarrow Cr_2(SO_4)_3 \qquad \text{(Eq 1)}$$

With sodium bisulfite, the reaction is:

$$4CrO_3 + 6\ NaHSO_3 + 3H_2SO_4 \rightarrow 2Cr_2(SO_4)_3 + 3Na_2SO_4 + 6H_2O \quad \text{(Eq 2)}$$

These reactions are predictable, repeatable, and rapid if carried out under the correct pH conditions. The rate is inversely proportional to the pH; therefore, it is fastest at a low pH level. Optimum pH for this reduction is 2.5 to 4.0. Sulfuric acid is required to complete the reaction with bisulfite. While the reaction is faster at a lower pH level, considerable odors are generated from the release of SO_3 to the atmosphere. At a pH level much greater than 4.5, the rate of reaction is too slow. Above pH of 5.5, little reduction of Cr^{6+} (hexavalent chromium) occurs.

Oxidation-reduction potential (ORP) measurements are normally used to control the additions of reducing agent. In practice, about 0.9 kg (2.0 lb) of sulfur dioxide or 1.4 kg (3.0 lb) of sodium bisulfite plus 0.7 kg (1.5 lb) of sulfuric acid are required to reduce each pound of hexavalent chromium. At a given pH, an exact ORP level can be determined to indicate completion of reaction.

The ORP controller continuously monitors the reaction and automatically controls the addition of reducing agents to maintain the ORP potential less positive than about +300 mV. The pH controller controls the addition of sulfuric acid to maintain pH 2.5 to 4. The absence of any traces of yellow color is a good indication that the reduction is being effectively accomplished. Simple colorimetric tests are also available to check completeness of the reaction. The chrome reduction tank should be well agitated and sized to provide a retention time of 10 to 20 min. Tank size should be based on maximum load conditions. A procedure for the batch treatment of hexavalent chromium bearing water is:

- Reduce pH to 2.5 with dilute acid
- Add bisulfite or sulfur dioxide until ORP becomes less positive than about +300 mV, continuously adding dilute acid to maintain pH below 4
- Check for presence of hexavalent chrome, adding more reducing agent if necessary
- Raise pH to 8.5

Cyanide Oxidation

Although there are various techniques now available for destruction of cyanide, alkaline chlorination using either chlorine gas or sodium hypochlorite solution (industrial bleach) is the most common. Three chemical reactions are used to oxidize the cyanide ultimately to carbon dioxide and nitrogen.

The cyanide is oxidized by chlorine to cyanogen chloride:

$$NaCN + Cl_2 \rightarrow CNCl + NaCl \quad \text{(Eq 3)}$$

Cyanogen chloride is a highly toxic gas, but by maintaining the pH above 11, the cyanogen chloride is quickly hydrolyzed to cyanate. The cyanate ion is much less toxic than the cyanide ion, as shown by:

$$CNCl + NaCl + 2NaOH \rightarrow NaCNO + 2NaCl + H_2O \quad \text{(Eq 4)}$$

The final reaction is the oxidation of cyanate to nitrogen and carbon dioxide. This reaction is also pH dependent, its rate being accelerated by decreasing pH. As this reaction occurs slowly at the optimum pH condition required for oxidation of cyanide to cyanate, it is normally necessary to lower the pH to 7 to 8 with dilute acid. At pH levels below 7, cyanate converts to nitrate which may be objectionable:

$$2NaCNO + 4NaOH + 3Cl_2 \rightarrow 6NaCl + N_2 + 2CO_2 + 2H_2O \quad \text{(Eq 5)}$$

It may be necessary to perform only single-stage treatment of cyanide-cyanate oxidation for discharge to some publicly owned treatment works because the cyanate should pose no problems to its treatment process. This should be confirmed with local authorities.

Cyanides of sodium, potassium, cadmium, copper, and zinc are oxidized to cyanate rapidly and efficiently. However, cyanides of nickel, silver, and gold require a long time for treatment, and as soluble free cyanide is destroyed in the chlorination, there is danger that metal cyanides will precipitate as the insoluble salts. Sludges containing slowly soluble metal cyanides would result, leading to problems in ultimate disposal. Special care must be exercised when treating wastes containing these compounds. Iron cyanides do not respond to the standard alkaline chlorination procedure.

Approximately 1.2 kg (2.7 lb) of chlorine are required to oxidize 0.45 kg (1 lb) of cyanide to cyanate. If chlorine gas is used, approximately 1.4 kg (3.1 lb) of caustic soda are required to maintain the alkalinity at the proper level. If sodium hypochlorite is used, approximately 8.7 L (2.3 gal) of 15% NaOCl is required for each pound of CN. Because hypochlorite contains considerable alkalinity, very little caustic is needed, although provision should be made for its use. Oxidation of cyanate to carbon dioxide and nitrogen requires an additional 1.9 kg (4.1 lb) of chlorine and 1.4 kg (3.1 lb) of caustic or 10 L (2.65 gal) of hypochlorite. The above chemical requirements are for oxidation of the cyanide only. There may be many other materials present which would react with chlorine and significantly increase consumption.

In actual practice, the process is carried out on a continuous basis in a single tank with an internal baffle across the length of the tank dividing it into two compartments. The volume would allow 1 h total retention: 20 min in the first stage, and 40 min in the second stage. In the first stage, the first two reactions occur virtually simultaneously to oxidize cyanide to cyanate. An oxidation reduction potential controller automatically regulates chlorine additions to maintain the potential more negative than about +300 mV. The ORP controller can be equipped with an adjustable alarm contact and a recorder. A pH controller maintains the pH at 10.5. The pH controller should be equipped with an alarm and possibly a recorder. In the second stage, the pH is held at a level of 8 with another pH controller to allow oxidation of cyanate to carbon dioxide and nitrogen. An ORP controller controls the addition of chlorine to maintain a potential of typically +700 mV.

A typical procedure for the batch treatment of cyanide bearing wastes is:

- Raise pH to 10.5 with dilute caustic
- Add hypochlorite or chlorine until ORP becomes more positive than about +300 mV while maintaining pH of 10.5 with caustic
- Agitate for 30 min to ensure ORP remains stable
- Reduce pH to 8 with dilute acid
- Add hypochlorite or chlorine until ORP reaches about +700 mV

Neutralization

Acidic and alkaline wastewater from the various cleaning and plating processes that do not contain hexavalent chromium or cyanide are combined in the neutralizer tank with treated wastewater from the chromium reduction and cyanide oxidation tanks. Alternatively, acids may be collected in the chrome sump and processed along with hexavalent chromium wastes (alkaline elec-

Table 3 Sludge production
Basis: 1.0 kg (2.2 lb) metal treated

Metal	Theoretical alkali requirements(a) NaOH	$Ca(OH)_2$(b)	Sludge produced 1% L	gal	35% kg	lb	Dry weight kg	lb
Cadmium	0.712	0.658	130	34.0	3.71	8.18	1.30	2.87
Chromium	2.31	2.14	198	52.3	5.67	12.5	1.98	4.36
Copper	1.25	1.16	153	40.4	4.37	9.63	1.53	3.37
Nickel	1.36	1.26	158	41.7	4.51	9.94	1.58	3.48
Lead	0.386	0.357	116	30.6	3.32	7.33	1.16	2.56
Zinc	1.22	1.13	152	40.1	4.34	9.57	1.52	3.35

(a) Alkali requirements are for metal precipitation only and do not include that requirement for free acid neutralization. (b) Normally a 10 to 25% excess of lime is required

trocleaner rinses which contain hexavalent chromium are also collected in the chrome sump), and alkalines may be collected in the cyanide sump. In any event, the combined wastewaters normally pass through the neutralizer tank.

If the combined effluent is alkaline, dilute sulfuric or hydrochloric acid may be used to lower the pH. If the effluent is acidic, which is more often the case, pH may be adjusted upward with caustic soda (NaOH) or lime, $Ca(OH)_2$. Although more expensive, liquid caustic soda is more convenient to use. Lime is used where large amounts of acid must be neutralized, and the saving can justify additional equipment and inconvenience. Lime should be considered where consumption would exceed about 0.45 t (0.5 ton) per day. Because of its low solubility, lime must be prepared and fed as a slurry. Quicklime (CaO), which is still less expensive, requires an additional hydrating or slaking step to convert it to the usable $Ca(OH)_2$ form and is only justified where consumption exceeds about 2.7 t (3 ton) per day.

Lime neutralization also has a tendency to produce more sludge than caustic. This is because normally a 10 to 25% excess must be used and because it reacts with sulfuric acid, when the feed pH is less than about pH of 2, to produce the slightly soluble gypsum ($CaSO_4$):

$$H_2SO_4 + Ca(OH)_2 \rightarrow CaSO_4 + 2H_2O \quad \text{(Eq 6)}$$

Additional sludge produced by lime neutralization is fortunately somewhat easier to settle out and dewater. As discussed below, lime neutralization also has some additional advantages in the treatment of complex wastewaters. Metals react with caustic or lime to produce relatively insoluble metal hydroxides according to the following equations, where M^{++} represents a heavy metal cation:

$$M^{++} + 2NaOH \rightarrow M(OH)_2\downarrow + 2Na^+ \quad \text{(Eq 7)}$$

$$M^{++} + Ca(OH)_2 \rightarrow M(OH)_2\downarrow + Ca^{++} \quad \text{(Eq 8)}$$

Metals precipitate out at various pH levels. Figure 2 shows theoretical solubility curves for several metals at various pH levels. It is apparent that when two or more heavy metals are found in the same waste stream, the optimum pH for precipitation may be different for each metal. It must be determined if one pH can be found that will produce satisfactory, though not optimum insolubility, for each of the metal ions present in the wastewater. In some instances, it may be necessary to precipitate one or more of the metals separately at one pH and treat the remaining stream at another pH. This would necessitate duplication of the neutralization system and subsequent clarification tanks.

Equipment required for neutralization normally consists of a simple tank equipped with an agitator and a pH controller. A low set point (7.5) on the pH controller allows addition of dilute caustic or lime slurry. A high set point (8.5) controls the addition of dilute sulfuric or hydrochloric acid. If wastewater is subject to rapid change in flow rate or pH, neutralization may be accomplished with more precision by utilization of two tanks. The pH is raised to about pH of 6 in the first tank, and final adjustments are made in the second tank. Alternatively, a more sophisticated pH controller with an analog output can vary the feed rate of acid or caustic in proportion to the deviance from the set point.

Neutralization tanks designed for use with caustic should have a retention of 10 to 30 min. Lime neutralization normally requires longer retentions. Good agitation is important and, as a rule of thumb, should be less than 20% of retention time (for example, for 10 min retention, turnover should be less than 2 min).

Fig. 2 Precipitation of metal salts versus pH

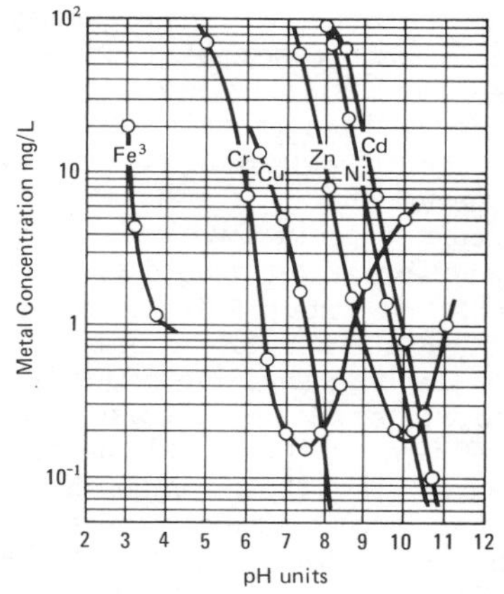

A typical procedure for batch neutralization and precipitation is:

- Slowly add dilute acid, caustic, or lime to adjust pH to correct value for metal precipitation (Fig. 2). Be careful temperature does not rise beyond limitation of tank material; make sure pH is stable for at least 10 min
- Add polyelectrolyte to experimentally determined dosage
- Continue agitation for about 1 min
- Shut off agitator and allow sludge to settle 2 to 4 h
- Slowly decant supernatant liquid from top, being careful not to disturb sludge layer
- Withdraw sludge from bottom of tank

All values of ORP cited should be considered as rough guides only. Exact values must be experimentally determined for each individual system. Tanks should be well agitated for all treatment steps, except sludge settling.

Clarification

After neutralization, precipitated metal hydroxide solids must be separated from decontaminated wastewater. This is usually accomplished by gravity sedimentation or clarification. The suspended metal hydroxide solids leaving the neutralizer are very fine particles with a density very close to that of water. As such, they do not settle out very efficiently. To enhance their settling characteristics, flocculating agents such as polymers, alum, or ferrous sulfate are added. The particles then tend to coagulate. Dosages of 1 to 5 ppm of anionic polymer have been effectively used in some cases. This process is fostered in a flocculation zone in the clarifier or a separate flocculator tank where gentle agitation, over a retention period of 5 to 10 min, allows the particles to contact with each other, but avoids any shearing action which may break them up. After flocculation, wastewater must not be pumped because the high shearing action of the pump will break up flocculated solids or floc. Gravity flow should be used.

Floc settles to the bottom of the clarifier and clarified wastewater overflows from the top. A clarifier overflow will usually contain 10 to 50 ppm of suspended solids. The floc accumulates in the bottom of the clarifier, agglomerates further, and forms a sludge blanket across the bottom. Sludge must be periodically withdrawn from the bottom of the clarifier to maintain the level of the sludge blanket about 760 mm (30 in.) below the level of the overflow weir. Presence of a sludge blanket in the clarifier aids in the efficiency of the solids separation so that a sludge blanket several feet in depth should always be maintained. The bottom of the clarifier is often shaped to a 60° cone so that sludge can be withdrawn by gravity. Larger units may be equipped with a mechanical rake to facilitate sludge collection. A clarifier underflow sludge usually has a concentration of 1 to 2% dry solids and has the consistency of creamed soup. The volume of 1% sludge produced for each kilogram of metal treated is shown in Table 3.

Retention time in the clarifier is not as important as the cross-sectional area. Retention times of about 4 h with liquid rise rates in the order of 20 L/min per square metre (0.5 gal/min per square foot) of tank area are typical. The clarifier should also be designed so that liquid velocity over the final weir does not exceed 90 L/min per lineal metre (7 gal/min per lineal foot), and the inlet should be baffled to dissipate the inlet velocity to less than 0.6 m/s (2 ft/s).

Various clarifier design modifications and operational techniques are used to reduce size and improve the separation efficiency. In the solids contact clarifier, the feed is contacted with the sludge blanket in a gently agitated zone located in the lower central portion of the clarifier. The sludge blanket acts somewhat like a filter, entrapping the solids. Tube settlers and lamella-type clarifiers use slanted tubes or plates to increase the effective settling area and reduce turbulence. Rise rates can then be increased to 65 L/min per square metre (1.5 gal/min per square foot) or more and residence times decreased to 1 h or even less. Overflow suspended solids levels can be reduced to less than 20 ppm.

In many instances, the clarifier reduces metal concentrations below required limits. In other applications, additional purification is required. This can be accomplished by filtering the remaining suspended solids carried over. Any one of a number of designs, including sand filters and precoated pressure leaf filters, can be used for this purpose. Relatively high flow rates can be used so that these filters are usually fairly small. The clarifier underflow, bearing the metal hydroxide sludge, is directed to a sludge dewatering system to reduce the volume of waste requiring disposal. The settling characteristics of neutralized wastewaters vary widely from plant to plant. Flocculent selection and dosage, as well as clarifier rise rates, should be based upon laboratory tests for each installation.

Sludge Dewatering

Precipitated metal hydroxides leave the bottom of the clarifier as a slurry or sludge at a concentration of 1 to 2 wt % dry solids. As shown in Table 3, 0.455 kg (1 lb) of precipitated nickel generates about 72 L (19 gal) of sludge. This sludge must usually be dewatered to reduce its volume and produce a solid or semisolid consistency suitable for landfill. Because of difficulties encountered with hazardous waste disposal, the trend is to produce as dry a final waste as possible.

The first step in dewatering sludge may be to thicken it by gravity sedimentation in a separate thickener tank. These tanks are usually 2.4 to 3.0 m (8 to 10 ft) in height and are fitted with 60° conical bottoms to facilitate sludge withdrawal and a device for manually decanting clarified liquid from the top at periodic intervals. A thickener normally increases sludge concentration to the 4 to 8% range. There are several devices available to further dewater the sludge.

Centrifuges, once very popular, are falling into disfavor because the sludge produced contains only 15 to 25% solids, having the consistency of thick mud. Centrifuges have the advantage of small space requirements, relative low cost, and ease of operation. Because the solid removal efficiency of the centrifuge is only about 85 to 90%, the centrate must be returned to the clarifier.

Rotary drum vacuum filters normally increase sludge concentration to 15 to 25% dry solids. Usually, feed to these units must be thickened to about 4 to 8%. As with the centrifuge, filtrate from the vacuum filter contains significant levels of solids and must be recycled to the clarifier. Its major advantage is that it operates continuously. Its major disadvantages are limited dewatering capabilities and rather high capital cost.

Pressure Filters. Various types of pressure filters are available, the most common being the traditional filter press. Use of these units has become increasingly popular because of their ability to produce a filter cake containing 35 to 50% dry solids. Other advantages include simplicity, moderate cost, and the ability to produce a relatively pure filtrate. Primary disadvantages of the filter press are downtime for cleaning and its high labor requirements, although it is possible to automate it to varying degrees. Thickening of the clarifier underflow is not usually required with these units.

Treatment of Complex Wastewaters

The presence of complexing ions such as phosphates, tartrates, EDTA, and ammonia that are commonly found in cleaning and plating formulations may have an adverse effect on metal removal efficiencies when hydroxide precipitation is used. A modification of the hydroxide precipitation process has been effective for treating electroless plating wastewaters which are heavily chelated. The basic method of treatment is to:

- Lower the pH to 2.7 to 5 with dilute sulfuric acid
- Add ferrous sulfate
- Agitate for about 5 min
- Raise the pH to a level greater than 9 with lime. An anionic polymer aids settling of the resulting precipitate

Precipitation techniques, other than hydroxide, that use reagents such as insoluble starch xanthate, sulfide, and dithio carbamate, although expensive to operate, have been shown effective

Fig. 3 Direct recovery system

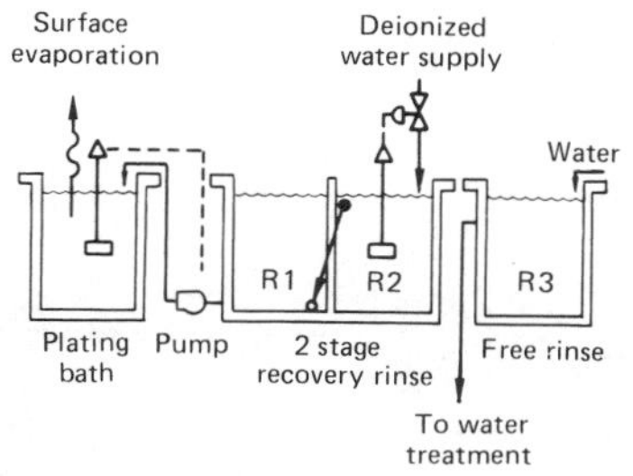

△: level controller

Fig. 4 Surface evaporation plating baths with no aeration

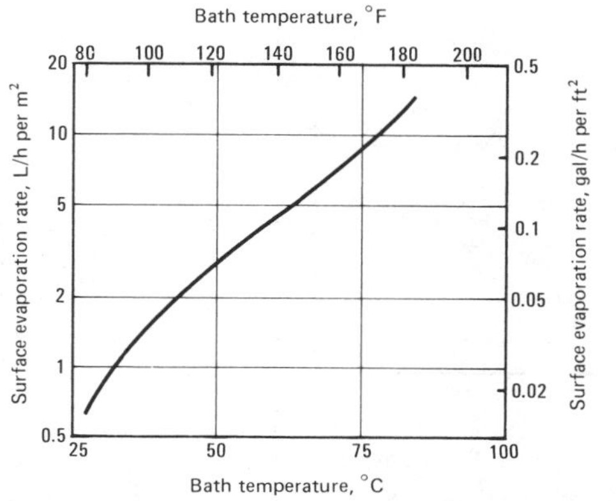

Ambient conditions are 24 °C (75 °F), 75% relative humidity. Plating solution is 95% mole fraction H_2O

when used as polishers to remove residual metals left after hydroxide precipitation. Ion exchange resins capable of breaking metal complexes and removing metals can also be used.

Recovery Systems

Recovery systems are attractive because they can abate pollution more economically than conventional destruction-type treatment techniques. Savings in purchases of recovered chemicals and water, as well as substantial reductions in solid waste disposal costs, make recovery worth considering wherever technically feasible. Installation of a recovery system can often reduce hydraulic and chemical loading on an existing waste treatment system and consequently improve its operation. In some instances, installation of a properly designed, efficient recovery system may obviate the need of installing a conventional waste treatment plant. The number of recovery systems available has expanded dramatically during the late 1970's and early 1980's. The more common and established systems include direct recovery, evaporation, and ion exchange procedures.

Direct (Natural) Recovery

The simplest and least expensive recovery technology is direct dragout recovery. Dragout tanks or still rinse tanks have been used for many years, but through use of air agitation, spray or fog rinsing, multiple countercurrent flow tanks, and automatic solution transfer, the efficiency of these tanks can be dramatically improved. A suitable system design is shown in Fig. 3. Level controls in the plating tank control the transfer of solution from the first rinse tank (R1) to the plating bath

Fig. 5 Surface evaporation rate from aerated plating baths

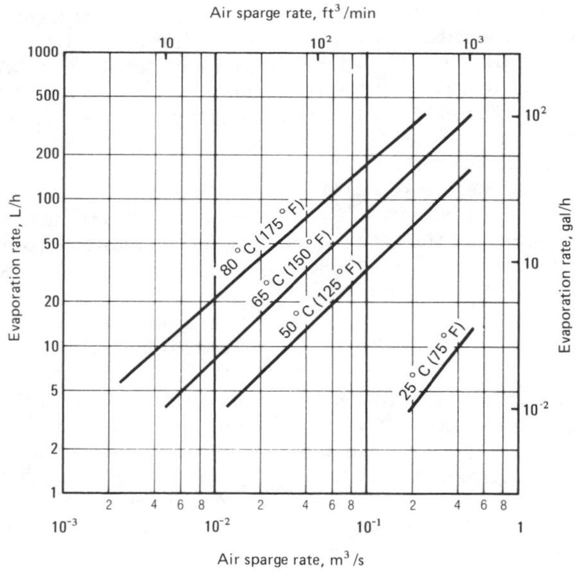

Supply air at 24 °C (75 °F), 75% relative humidity. Plating solution is 95% mole fraction H_2O

to replace evaporation losses. Another level control mounted in R1 controls addition of deionized water to R2 which counterflows to R1. The use of deionized water is advisable to prevent a buildup of tap water-borne contaminants in the plating bath. Chrome plating baths are particularly sensitive to chloride, and nickel plating baths are sensitive to calcium. Both contaminants are major constituents of most city water supplies.

Most plating baths tend to generate their own contaminants, either through drag-in of chemicals from previous processes, corrosion of the substrate metal, or breakdown of bath constituents. Dragout provides a natural purge for these contaminants. Recycling via a direct recovery system (or any other tech-

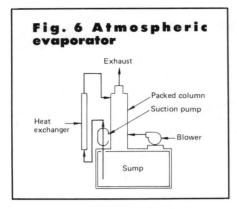

Fig. 6 Atmospheric evaporator

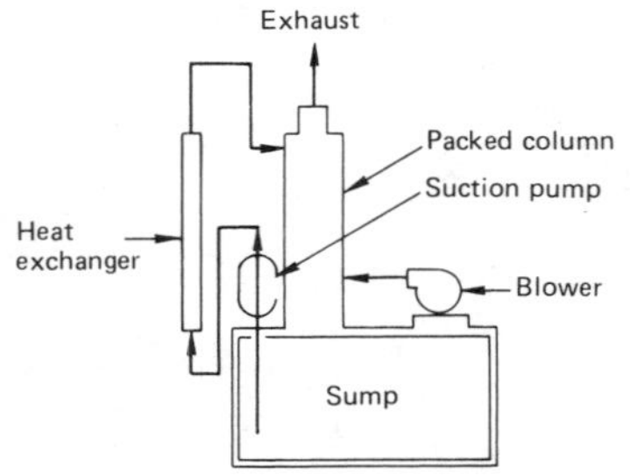

Fig. 7 Rising film evaporator

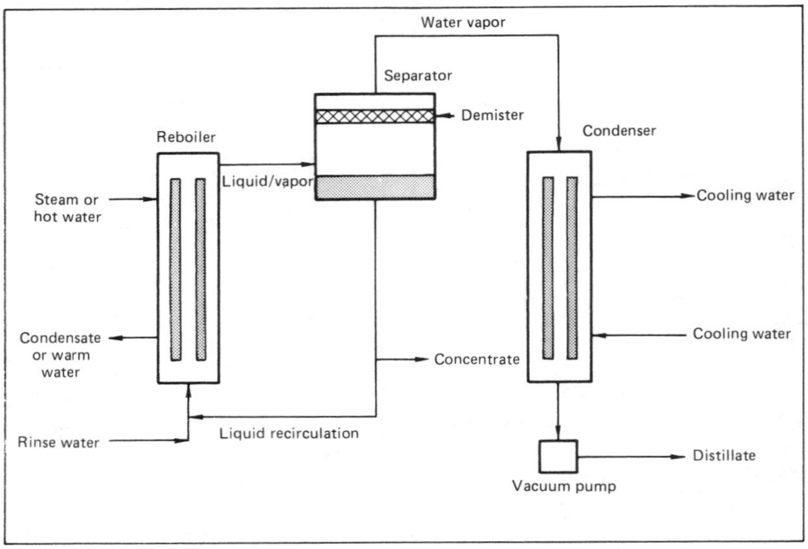

nique that does not allow for contaminant removal) may ultimately foul the plating bath. Such contamination can be minimized through good housekeeping practices so that the remaining loss may allow for sufficient contaminant purging.

An obvious limitation to the direct recovery method is the amount of evaporation from the plating bath. Plating baths should be operated as hot as possible to maximize the rate of evaporation. Ventilation and air agitation also tend to increase bath evaporation rates. Additional energy expended in operating in this manner will be far outweighed by chemical savings. Evaporation rates can be estimated from Fig. 4 and 5. A free-running rinse (R3) is recommended following the recovery rinses to provide adequate final rinse quality and to avoid downstream contamination.

Evaporation

A logical extension of the direct recycle method is the evaporative type recovery system. An evaporator can be used to artificially increase the evaporation rate of the plating bath or to concentrate the solution collected from the first rinse. There are two basic evaporator designs commonly used in the recovery of electroplating wastes: atmospheric evaporator and vacuum evaporator.

The principle of the atmospheric evaporator is similar to that of a heated open tank except that the warm liquid from the tank is sprayed over plastic packing in a tower similar to a wet fume scrubber to increase the exposed liquid surface area. An induced or forced draft of air passing over wetted packing removes the moisture. This air is exhausted from the plant. A schematic is shown in Fig. 6.

Atmospheric evaporators are particularly suitable for chrome plating applications because they can also serve as plating bath fume scrubbers. Another advantage of the atmospheric evaporator is its relatively low capital cost. Its major disadvantage is that evaporated water is not recovered. A separate source of purified water should be provided to avoid contamination of the plating bath with tap water-borne contaminants.

There are several different vacuum evaporator designs used including rising film or natural thermosyphon, forced circulation, submerged tube, and flash. A schematic of the rising film evaporator is shown in Fig. 7. This design uses a steam or hot water-heated vertical tube reboiler, a liquid-vapor separator which also holds accumulated bath concentrate, a vertical shell, and tube water-cooled condenser and vacuum pump.

Basic operation of the rising film evaporator begins when rinse water is drawn by vacuum into the tubes of the reboiler where some water evaporates. A high velocity mixture of water vapor and concentrate enters the separator which allows vapor to pass through a mesh de-entrainment pack to the condenser. Condensed water vapor or distillate is withdrawn from the condenser by the vacuum pump which also maintains a vacuum in the system. Concentrate which accumulates in the separator is allowed to recycle back through the reboiler to mix with incoming feed to be further concentrated. When the concentration reaches a predetermined value, the system vents to atmosphere and discharges the concentrate to storage.

Approximately 75 to 190 L (20 to 50 gal) of cooling water are required in the condenser of a vacuum evaporator per gallon of water evaporated. This cooling water can normally be reused elsewhere or recycled by a cooling tower.

Steam is the most common source of heat energy used in industrial evaporators. Evaporation, regardless of which type of evaporator is used, requires approximately 2300 kJ/kg (1000 Btu/lb) of water evaporated. This can be cut by almost 50% by reusing the energy in a second evaporator effect, or by employing a process called vapor recompression. However, the additional capital cost and complexity can usually only be justified for larger installations.

Because of the large amounts of energy required to evaporate water, it is essential that a well-designed multitank countercurrent flow rinse system be used to minimize the required flow rate. An open-loop or partial recovery layout, similar to that used for the direct recovery system shown in Fig. 3, using one or two recovery rinses flowing to the evaporator followed by a free-flowing rinse directed to the plant waste treatment system, frequently provides a chemical recovery efficiency of 80 to 90% at reasonable operating and capital expense.

Because evaporation does not remove contamination, it is essential that a means of handling the inevitable buildup of metallic and organic contaminants in the plating bath be incorporated into the system. Other recovery techniques such as ion exchange can be effective in this regard. Evaporation is probably the most versatile recovery technique available and, although not necessarily the optimum choice in each application, can be applied to a wide variety of electroplating solutions. Some of these include the following:

- Chromic acid-based plastic preplate etch
- Chromium plating solutions
- Nickel plating solutions
- Brass cyanide
- Acid copper
- Copper cyanide
- Cadmium cyanide
- Zinc cyanide
- Zinc chloride
- Tin-lead fluoborate
- Silver cyanide
- Gold cyanide

Ion Exchange

The ion exchange technique has two major advantages over most other recovery technologies. It has the ability to purify as well as concentrate chemicals, and it can treat large flows of dilute solutions at low operating cost. The major disadvantage of ion exchange is that there are certain limitations to the concentration of solutions that can be treated and produced by the process. Another limitation is that different designs and operating conditions must be established for each application.

The heart of any ion exchange system is the ion exchange resin. Although there are basically only two classes of resins available, cation and anion, there have been a wide variety of different resins developed, with different ionization strengths and selectivities for specific ions. Ion exchange occurs in a very reproducible manner under a given set of conditions, and the exchange can be written like a chemical reaction, as shown below:

$$2RH + Ni^{++} \rightarrow R_2Ni + 2H^+ \qquad \text{(Eq 9)}$$

RH represents a cation resin in the freshly regenerated hydrogen or acid form. The resin can be regenerated by reversing the above reaction through use of an excess of a moderately concentrated regenerant solution. Anion resins are usually regenerated with sodium hydroxide, cation resins with sulfuric or hydrochloric acid. Regeneration can thus yield a more concentrated solution of the exchanged ion. After regeneration, residual regenerant chemicals must be rinsed from the bed with water.

Resins contain finite numbers of exchange sites, and their capacity is most easily expressed in equivalents per litre of resin or kilograins per cubic foot as $CaCO_3$. For example, a strong acid cation resin may have a total exchange capacity of 2 eq/L, a strong base anion resin 1.4 eq/L. Unfortunately, in practice, it is difficult to use all of this available capacity. Usually, at reasonable regeneration dosages, only 50 to 80% of the total capacity can be effectively used. In fact, efficiencies in uptake and regeneration can significantly improve if only the most accessible exchange sites, representing 10 to 35% of the total, are used.

Resins show a preference for ions to varying degrees, depending on valence (charge in solution) and a number of other factors. In general, resins prefer ions of higher valence. An exception occurs with weak acid cation and weak base anion resins which greatly prefer H^+ and OH^- respectively over all others. As a result, operating pH of these resins is limited, although they regenerate very easily.

Chelating resins behave in a manner analogous to plating shop chelators such as EDTA. They hold certain transition metal ions (for example, Cu^{++} and Fe^{+++}) very tightly except under low pH conditions. Lighter metals such as sodium, magnesium, and calcium are held very weakly. As a result, it is possible to recover low concentrations of valuable or toxic metals from solutions containing high background salt levels or complexing agents.

In addition to exchanging ions, resins can adsorb and/or absorb chemicals present in the surrounding solution to varying degrees. The chemicals can often be desorbed with water. Examples of reversible sorption processes are ion exclusion (cation resin), ion retardation (special resin), and acid retardation (anion resin).

Acid retardation has recently seen extensive application in the metal finishing industry for acid recovery. The resins sorb strong acids such as H_2SO_4, HCl, and HNO_3, but do not sorb salts such as $Al_2(SO_4)_3$, $FeCl_2$, and $Cu(NO_3)_2$. Purified acid can be recovered by desorption with water.

There are several types of ion exchange systems available ranging from

Fig. 8 Reciprocating flow ion exchanger operating cycle for chromic acid rinse recovery

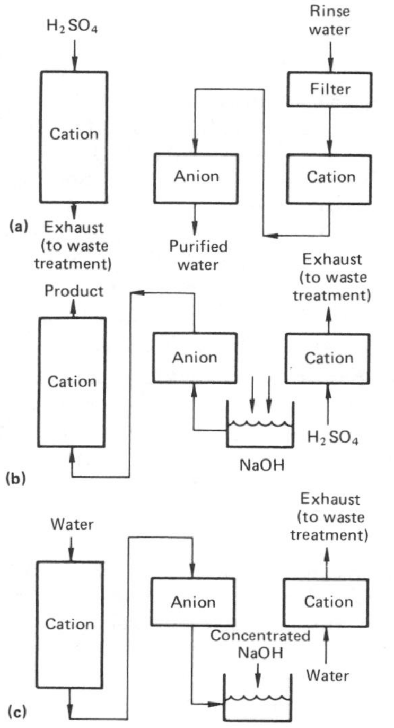

(a) On stream (loading): Rinse water is pumped from the chromium plating rinse tanks through the prefilter to remove any solids, then through the first cation bed where cationic contaminants (Fe^{+3}, Cr^{+3}) are removed by the resin. The rinse water then passes through the anion bed where the chromate ions are removed. The purified rinse is returned to the rinse system. While the unit is on stream, the second cation bed is regenerated. (b) Regeneration: After a preset period of time, the unit goes off stream. The first cation bed is regenerated with sulfuric acid and washed with water. The anion bed is regenerated with sodium hydroxide, and the effluent is passed through the second cation bed; the concentrated chromic acid solution resulting is returned to the plating tank. (c) Washing: The three beds are then washed with water; because the product from the anion bed contains the excess caustic used in the regeneration step, it is mixed with concentrated NaOH and used in the next regeneration cycle. The unit then goes back on stream

inexpensive water softener units to large continuous countercurrent units. A process called reciprocating flow ion exchange, using short (75 mm or 3 in.) resin beds and countercurrent regeneration, has been extensively applied in the electroplating industry with excellent success. Ion exchange has been used for many years prior to this in the water treatment field.

Three major areas of application for ion exchange in the treatment of electroplating wastewater have been demonstrated. These include chemical recovery, water recycle, and effluent polishing. Chromic acid purification, chrome rinsewater recovery, and acid bath purification have also been used as ion exchange techniques.

Chromic acid purification involves using cation exchange to remove cationic metal contamination, such as iron trivalent chrome, directly from hard chrome plating baths or from rinsewaters, in conjunction with an evaporative recovery system.

Chrome rinsewater recovery, a combination of anion exchange and cation exchange, is used to recover a purified chromic acid concentrate from dilute rinsewaters (see Fig. 8).

Nickel, copper, zinc, or tin rinsewater recovery uses cation exchange to recover various metal salts from rinsewaters. In the case of nickel sulfate recovery, acid retardation is used to remove residual acid from the recovered product. This process is only applicable to acid plating processes (not cyanide).

Acid Bath Purification. The acid retardation process (acid bath purification) has been successfully used to remove metal contamination from sulfuric, hydrochloric, or nitric acid process baths used in pickling, anodizing, and rack stripping.

Water Recycle. Although many rinsewaters contain no valuable chemicals, the water can be easily deionized by cation/anion exchange and reused (water recycle). Indeed, because deionized water is required for several rinses and makeup of many baths, the same unit can serve both purposes. Recovery of rinsewaters in this manner simplifies waste treatment because it is much easier and cheaper to treat a concentrated, low-volume stream than a dilute, high-volume stream. Reaction tanks, reagent pumps, and clarifiers are all reduced in size. In some instances, water recycle may allow installation of a small batch treatment system in lieu of a large continuous system.

Fig. 9 Simplified reverse osmosis schematic

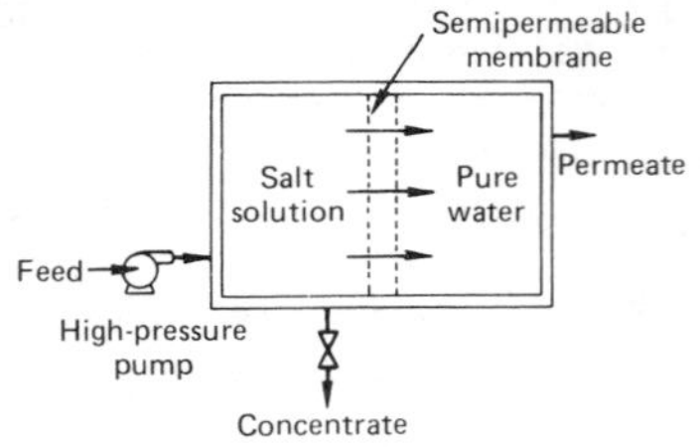

Effluent Polishing. Most plating shops can achieve sufficient metal removal to comply with discharge regulations by using conventional hydroxide precipitation processes described earlier. In applications where complexing agents interfere with hydroxide precipitation or where effluent restrictions are unusually strict, selective ion exchangers can be used to polish the effluent by removing residual metals from the final effluent. Various weak acid cation and chelating resins take up heavy metals very efficiently but allow high background levels of sodium, calcium, and magnesium to pass through. Regeneration of the exhausted resins yields a simple metal salt concentrate that can be treated in the conventional manner. However, wastewaters bearing EDTA cannot normally be treated with this technique.

Reverse Osmosis

Although not technically correct, reverse osmosis (RO) can be considered a molecular scale filter and indeed is sometimes referred to as hyperfiltration. Water and nonionic chemicals of molecular weight less than about 100 are forced through semipermeable membranes at pressures of 2800 kPa (400 psi). Ionic materials, such as metal salts, are rejected by the membranes and their concentration can therefore be increased. Usually, the higher the ionic charge, the greater the rejection efficiency. Nickel for example has a rejection efficiency in the order of 98 to 99% (see Fig. 9).

Because the flux or flow per unit of membrane area is very low, it is necessary to construct membrane modules with large membrane areas per unit volume. The two most common configurations are the spiral wound and the hollow fibre configuration. The hollow fibre units offer the greatest surface area to volume ratio but are very susceptible to fouling by suspended solids and are very fragile. All RO membranes have limited life spans on the order of 1 to 5 years and are very temperature sensitive.

Fig. 10 Electrodialysis flow schematic

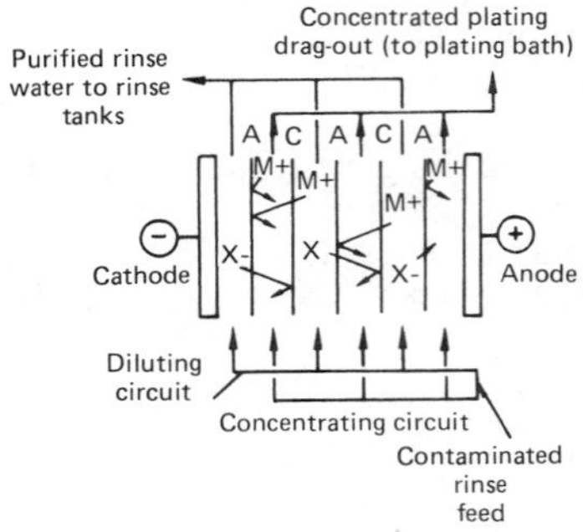

C, cation-selective membrane; A, anion-selective membrane; M^+, cations; X, anions

The only widespread application of RO to date in the electroplating industry is nickel salt recovery. This is because the rinsewater is of acceptable pH (pH 4 to 5), and the relatively high evaporation losses from the plating bath allow recycle of the fairly dilute (Ni = 1.3 oz/gal or 10 g/L) recovered product. A running rinse must normally be used following one or two recovery rinses, as permeate water of adequate purity for a final rinse before chrome plating cannot be produced by RO.

Electrodialysis

Electrodialysis (ED), a relatively new process, has only been applied to the recovery of electroplating wastes during the late 1970's and early 1980's. Like reverse osmosis, it is a membrane process, but uses ion exchange membranes and an electrical driving force. Cation exchange membranes allow cations such as copper or nickel to pass, while anion exchange membranes pass anions such as sulfate, chloride, or cyanide.

To achieve sufficient membrane area, the membranes are normally stacked parallel between spacers, alternating cation and anion to provide alternating concentrating and diluting compartments. Electrodes are placed at each end. Passage of a direct current from a rectifier through the stack causes ions in the solution to move in the direction of the oppositely charged electrode; therefore cations move toward the cathode and anions move toward the anode (see Fig. 10).

The ED unit is normally operated on the first rinse only. Solution from this tank is continuously circulated through the stack which removes and concentrates the ionic species from the plating solution and returns ion-depleted solution back to the tank. Because flow through the stack must be turbulent, flow rates are high and only 20 to 30% of metal in solution is removed with each pass. Partially ionized species such as boric acid are only partially recovered and nonionics such as some organic brighteners are recovered only to a very limited extent. Buildup of these materials in the rinse tank could cause problems in some applications.

Theoretically, there are few limitations on the solutions that may be recovered with ED. Acid, alkaline, and cyanide plating solution can be recovered. Solutions reportedly being recovered include: chrome, nickel, copper, zinc, silver, gold, platinum, palladium, and cadmium.

Electrowinning

Electrowinning, or the recovery of elemental metals from solution via the electroplating process, can be used for the recovery of a wide variety of metals including copper, silver, gold, tin, and cadmium. In the basic electrowinning cell design, cathodes are placed alternately about 50 to 150 mm (2 to 6 in.) apart in an open tank. Cathodes are either thin starter sheets of the metal being plated or stainless steel blanks from which the recovered metal can be stripped. Anodes used include lead alloys, graphite, or platinized titanium. Air agitation is very beneficial in most applications, improving plate quality and current efficiency.

The major problem with the electrowinning process for recovery applications is the decreasing current efficiency and quality of plate at lower metal concentrations in the solution. This is alleviated through a variety of proprietary techniques that provide highly efficient agitation and/or a high surface area cathode.

REFERENCES

1. AES Environmental Compliance and Control Course, American Electroplater's Society, Winter Park, FL, 1980
2. C. Calmon and H. Gold, Ed., *Ion Exchange for Pollution Control, Volume I,* Boca Raton, FL: CRC Press, Inc., 1979
3. P.S. Cartwright, Reverse Osmosis and Ultrafiltration in the Plating Shop, *Plating and Surface Finishing,* p 40, April 1981
4. P. Crampton, Reverse Osmosis in the Metal Finishing Industry, *Plating and Surface Finishing,* p 21, March 1982
5. "Economics of Wastewater Treatment Alternatives for the Electroplating Industry", EPA-625/5-79-016, U.S. Environmental Protection Agency, 1979
6. H.S. Hartley, The Evolution of Evaporative Recovery, *Plating and Surface Finishing,* p 40, January 1982
7. J.B. Kushner, *Water and Waste Control for the Plating Shop,* 2nd ed., Cincinnati: Products Finishing
8. Proceedings of the First Annual Conference on Advanced Pollution Control for the Metal Finishing Industry, EPA-600/8-78-010, U.S. Environmental Protection Agency, January 17-19, 1978
9. Proceedings of the Second Annual Conference on Advanced Pollution Control for the Metal Finishing Industry, EPA-600/8-79-014, U.S. Environmental Protection Agency, February 5-7, 1979
10. Proceedings of the Third Annual Conference on Advanced Pollution Control for the Metal Finishing Industry, EPA-600/2-81-028, U.S. Environmental Protection Agency, April 14-16, 1980
11. "Summary Report — Control Technology for the Metal Finishing Industry — Evaporators", EPA-625/8-79-002, U.S. Environmental Protection Agency, 1979
12. "Summary Report — Control Technology for the Metal Finishing Industry — Ion Exchange", EPA-625/8-81-007, U.S. Environmental Protection Agency, 1981
13. "Summary Report — Control Technology for the Metal Finishing Industry — Sulfide Precipitation", EPA-625/8-80-003, U.S. Environmental Protection Agency, 1980

Metallic Coating Processes Other Than Plating

Hot Dip Galvanized Coatings

By the ASM Committee on Hot Dip Galvanized Coatings*

HOT DIP GALVANIZING is a process in which an adherent, protective coating of zinc and zinc compounds is developed on the surfaces of iron and steel products by immersing them in a bath of molten zinc. The protective coating usually consists of several layers (see Fig. 1). Those closest to the basis metal are composed of iron-zinc compounds; these, in turn, are covered by an outer layer consisting almost entirely of zinc.

The complex structure of layers that comprise a galvanized coating varies greatly in chemical composition and physical and mechanical properties, being affected by chemical activity, diffusion, and subsequent cooling. Small differences in coating composition, bath temperature, time of immersion, and rate of cooling or subsequent reheating can result in significant changes in the appearance and properties of the coating.

Hot dip galvanized coatings are produced on a variety of steel mill products, using fully mechanized, mass production methods. This article, however, is concerned primarily with the hot dip galvanizing of fabricated articles in manual or semiautomatic batch operations.

Applications

Galvanized coatings are applied to iron and steel primarily to provide protection against corrosion of the basis metal. Some major applications of hot dip galvanized coatings include (*a*) structural steel for power generating plants, petrochemical facilities, heat exchangers, cooling coils, and electrical transmission towers and poles; (*b*) bridge structural members, culverts, corrugated steel pipe, and arches; (*c*) reinforcing steel for cooling towers, architectural precast concrete, and bridge decks exposed to chlorides; (*d*) pole line hardware and railroad electrification structures; (*e*) highway guard rail, high-rise lighting standards, and sign bridge structures; (*f*) marine pilings and rails; and (*g*) grates, ladders, and safety cages. Hot dip galvanized tower and high-strength bolts are produced and used in large quantities for service conditions where long-term integrity of bolted joints is required. In short, wherever steel is exposed to atmospheric, soil, or water corrosion, hot dip galvanized zinc coatings are a standard, effective, and economical method of protection.

The usefulness of hot dip galvanized coatings depends on the following: (*a*) the relatively slow rate of corrosion of zinc as compared with that of iron, (*b*) the electrolytic protection provided to the basis steel when the coating is damaged, (*c*) the durability and wearing properties of the zinc coating and the intermetallic iron-zinc alloy layers, and (*d*) relative ease and low cost of painting the zinc coating when it is necessary to further extend the life of the structure, which is usually done after 15 to 25 years of maintenance-free service in rural and light industrial atmospheres.

Hot dip galvanized zinc coatings have their longest life expectancy in rural areas where sulfur dioxide and other industrial pollutant concentrations are low. These coatings also give satisfactory service in most marine environments. Although the life expectancy of hot dip galvanized coatings in more severe industrial environments is not as long as for less aggressive environments, the coatings are still used extensively in those exposures, because

*Daryl E. Tonini, *Chairman,* Manager, Technical Services, American Hot Dip Galvanizers Association, Inc.; Serge Belisle, Research Engineer, Noranda Research Centre; Hart F. Graff, Principal Research Engineer, Armco Inc.; David C. Pearce, Senior Research Metallurgist, ASARCO, Inc.

Fig. 1 Photomicrograph of a typical hot dip galvanized coating X250

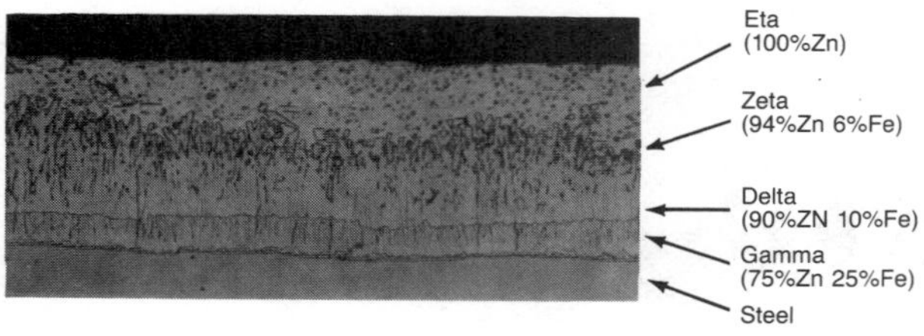

The molten zinc is interlocked into the steel by the alloy reaction which forms zinc iron layers and creates a metallurgical bond

in general, no more effective and economical method of protection is available. In cases involving particularly severe exposure conditions, coatings slightly heavier than the standard 610 g/m^2 (2.0 oz/ft^2) minimum in ASTM standard specifications A123 or paint over galvanized coatings are often selected as the preferred protective system.

When hot dip galvanized after fabrication coatings are painted for protective or decorative reasons, a primer coat is usually needed to etch the surface of the galvanized coating in preparation for the top coating system. A wide range of proprietary top coat systems are available for use with materials hot dip galvanized after fabrication material.

Metallurgical Characteristics of Coatings

Iron and Steel Substrates. The chemical composition of irons and steels, and even the form in which certain elements such as carbon and silicon are present, determines the suitability of ferrous metals for hot dip galvanizing and may markedly influence the appearance and properties of the coating. Steels that contain less than 0.25% carbon, less than 0.05% phosphorus, less than 1.35% manganese, and less than 0.05% silicon, individually or in combination, are generally suitable for galvanizing using conventional techniques.

To avoid brittleness of the iron-zinc alloy layer in cast iron materials, substrate iron must be low in phosphorus and silicon; a preferred composition may contain about 0.01% phosphorus and about 0.12% silicon.

Quality of Zinc. Any grade of zinc in ASTM B6 can be used in galvanizing; Prime Western zinc contains the highest allowable percentage of total impurities, for example, 1.65% for lead, iron, and cadmium combined. In hot dip galvanizing, high-purity high grade and special high grade zinc causes sufficient dissolution from the surface of steel work and the tank walls so that the equilibrium iron content of the bath nearly equals that of Prime Western. These coatings have the same metallurgical properties as those obtained with Prime Western zinc in the bath. Thus, high-purity zinc has little metallurgical advantage for use on fabricated items.

Bath Alloying Elements. Cadmium and iron are usually present in zinc baths as contaminants, but are not intentionally added to the bath as alloying elements. An aluminum concentration as low as 0.02% will improve drainage and increase the brightness of the galvanized coating. Small amounts of lead may be added to promote proper spangle and better drainage and to aid with drossing the bath.

An aluminum concentration less than 0.01% aluminum is generally maintained in the zinc bath when a preflux and/or a bath flux are used. The high chloride content of the fluxes reacts with the aluminum in the bath, producing a surface film of dross, oxide, and chloride on the bath surface.

Coating Thickness. In addition to base metal chemistry and surface profile, the thickness of coatings applied by hot dipping is primarily a function of (*a*) the duration of immersion, which controls the thickness of alloy layer; (*b*) the speed of withdrawal from the bath, which controls the amount of unalloyed zinc adhering, and (*c*) the temperature of the bath, which affects both the alloy and free zinc layers. Coating weight can be further affected by the amount of zinc removed by wiping, shaking, or centrifuging after the dipping process.

The protection against corrosion provided by zinc coatings is essentially determined by the thickness of the coating. Many comprehensive studies have shown that all other factors, such as method of applying the zinc coating, purity of the zinc, and the extent to which it is alloyed with the iron, are minor in determining life, as compared with the thickness of the coating.

Zinc coatings are measured in ounces of coating per square foot (grams per square metre) of surface. However, the weight of galvanized coatings on sheet is stated in ounces per square foot (grams per square metre) of sheet. Since the sheet is coated on both sides, the coating weight per square foot (metre) of surface on each side is approximately one half the average weight of coating per square foot (grams per square metre) of sheet. Occasionally, zinc coatings on steel are specified in terms of thickness in thousandths of an inch (micrometres), or in terms of the weight of the coating expressed as a percentage of the weight of the steel base. ASTM specifications A123 and A153 give coating weight requirements as a function of thickness and type of material to be coated.

Effect of Galvanizing Process on Substrate Materials

Tensile Strength, Impact Toughness, and Formability. The tensile strength, yield strength, elongation at rupture, and reduction of area of hot rolled steels remain virtually unchanged after hot dip galvanizing. In welded structures, the weld stresses may be reduced by 50 to 60% as a result of hot dip galvanizing. Increased strength levels induced by cold working

Fig. 2 Photomicrograph of a galvanized coating on a steel containing 0.40% silicon X250

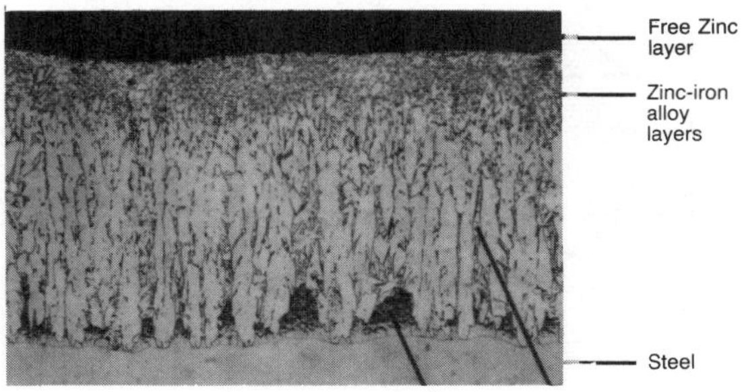

or heat treatment are generally reduced by hot dip galvanizing. The degree of strength reduction depends on such factors as the amount of working, the nature of heat treatment, and base steel chemistry. Impact toughness is slightly reduced, but not so much that the applicability of the steel is affected.

The formability of the steel is not affected. However, if the steel is sharply bent, the zinc coating may craze or crack on the tension side of the bend, depending on thickness of coating and bend radius.

Fatigue strength of various types of steels is affected differently as a result of the hot dip galvanizing process. Rimmed and aluminum-killed steels exhibit relatively little reduction of fatigue strength, whereas the fatigue strength of silicon-killed steels can be reduced considerably by hot dip galvanizing.

The reason for this difference in fatigue strength for silicon-killed steels is attributable to the different structure of the coating (see Fig. 2). Under the influence of fatigue stresses, cracks may form in the iron-zinc layer and act as crack initiators in the steel surface.

Fatigue strength is typically determined by laboratory tests where untreated new steels with mill scale are compared with hot dip galvanized material. If a steel structure is exposed outdoors in the untreated state, it is immediately attacked by rust. The corrosion pits that form result in a loss in fatigue strength. Thus, under inservice exposure conditions, the fatigue strength of ungalvanized steel declines rapidly as a result of the rust attack that occurs at the point of damage. For hot dip galvanized steel, however, the fatigue strength of the steel does not appreciably change during the exposure period as long as the zinc coating remains on the steel surface.

Hydrogen embrittlement does not result from the hot dip galvanizing of ordinary unalloyed and low-carbon mild steels. Any hydrogen absorbed during pickling is effectively eliminated on immersion in the zinc bath because of the relatively high temperature of about 460 °C (860 °F). Hardened steels can become brittle because of hydrogen diffusion into the steel. Such materials should always be tested for embrittlement after pickling before large lots are hot dip galvanized.

Intercrystalline cracking because of the penetration of zinc into the intergranular boundaries in steel can sometimes occur in connection with hot dip galvanizing, but only in cases where large stresses have been induced in the steel by welding or hardening. The risk of intercrystalline cracks and embrittlement failure because of zinc penetration is negligible in connection with hot dip galvanizing of ordinary low-carbon structural steels. However, hardened materials may be sensitive and should be tested for susceptibility to zinc penetration and cracking before hot dip galvanizing large quantities is undertaken.

Cleaning Before Galvanizing

Iron and steel pieces to be hot dip galvanized after fabrication must be free of oil, grease, drawing lubricants, mill scale, and other surface contaminants before fluxing and immersion in molten zinc. Inadequate or improper surface preparation is the most frequent cause of defects and bare spots in galvanized coatings.

In batch hot dip galvanizing, the material to be galvanized is first degreased and then pickled in sulfuric or hydrochloric acid. Since any iron salts or particles left on the surface of the material form dross in the galvanizing kettle, each of the degreasing and pickling steps is followed by a water rinse.

Degreasing. Organic contaminants can be removed from the work by several methods. The most common of these in the after fabrication hot dip galvanizing process is the use of heated alkaline cleaning baths. The alkaline cleaning process performs five basic functions: (*a*) dispersion by washing soil from the work, (*b*) emulsification by breaking up the soil and suspending it in solution, (*c*) film shrinkage by forming beads of oil to remove oil films, (*d*) saponification by converting animal and vegetable oils to water-soluble soaps, and (*e*) aggregation by collecting soil particles away from the work where they can be more easily removed from the solution.

The alkaline cleaning solutions should be heated to between 65 to 82 °C (150 to 180 °F). Live steam heating may be used for the added benefit of providing agitation energy for the solution, which is an important factor in accelerating the organic contaminant removal process.

Control of the strength of the heated alkaline solution is essential to an effective degreasing operation. These solutions lose strength because of chemical cleaning action effects and are diluted by the use of live steam for heating and by make-up water added to replace dragout losses. Although experience can be a good indicator of cleaning solution activity, a better method is to test the alkaline solution periodically for strength and to make periodic additions to maintain the solution at the desired concentration. Alkaline cleaning processes are described in detail in separate articles in this Volume.

Acid Pickling. Aqueous solutions of sulfuric acid or hydrochloric acid are generally used to remove mill scale and rust from steel parts before galvanizing. These pickling solutions may either be sulfuric acid, 3 to 10 wt %, or hydrochloric acid, 5 to 15 wt %. To increase effectiveness, sulfuric acid solutions are always used hot at 60 to 79 °C (140 to 175 °F); hydrochloric acid solutions are usually used at about room temperature, 24 to 38 °C (75 to 100 °F), to avoid excessive fuming. To avoid overpickling, inhibitors are often used with both sulfuric and hydrochloric acid solutions.

Nitric acid pickling, when used, is employed primarily for removing silicates from malleable and gray iron castings. Silicates are insoluble in hydrochloric or sulfuric acid.

Abrasive Cleaning. Some assemblies are made up of both cast and wrought materials; these assemblies require additional surface preparation prior to fluxing and galvanizing. All assemblies of cast iron, cast steel, and malleable iron with wrought steel should be abrasively cleaned after assembly and before pickling. Many other parts may also be abrasively cleaned to minimize or eliminate pickling. Procedures and equipment for abrasive cleaning are described in the article on abrasive blast cleaning in this Volume.

In general, cast materials require less pickling than hot rolled steel products, unless scale is removed from steel by blast cleaning. Therefore, care must be exercised to avoid burning or damaging the castings in an assembly during the time required to pickle the hot rolled components.

Fig. 3 Phase diagram for zinc chloride and ammonium chloride

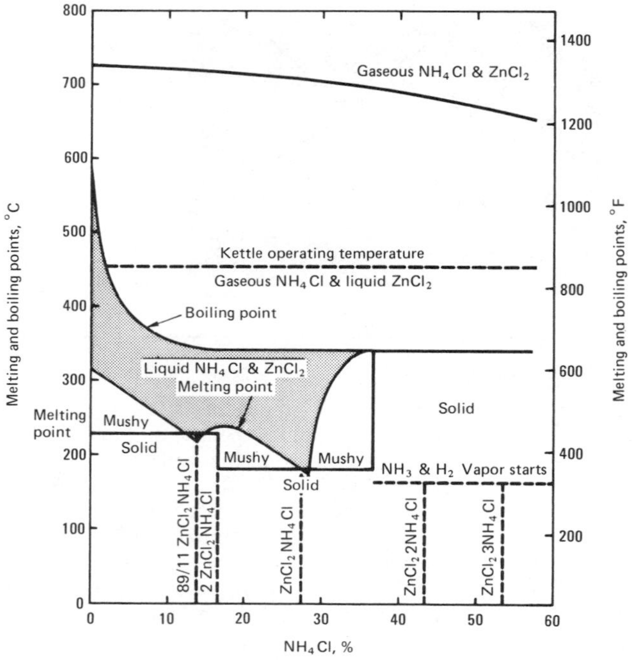

General Batch Galvanizing Procedures

Wet and Dry Galvanizing. Two types of conventional batch galvanizing practices currently in use are the wet and the dry process. Dry galvanizing was developed and refined in Europe. Wet galvanizing is more commonly used in North America. The dry process is generally considered to be less energy intensive than the wet process, but it is more sensitive to surface preparation deficiencies.

Wet galvanizing usually involves pickling with heated sulfuric acid in conjunction with a kettle top flux blanket; dry galvanizing is commonly associated with hydrochloric acid pickling, preflux washes, and no flux blanket on the kettle. However, sulfuric acid pickling has been used in conjunction with dry galvanizing. The choice of the wet or dry process can be largely attributed to differences in the availability and cost of acid and energy.

In the dry galvanizing process, after the material is degreased and pickled, the workpieces are immersed in an aqueous flux solution, dried, and then immersed in the molten zinc bath. In the wet galvanizing process, the work is not usually prefluxed after cleaning and pickling but is placed in the molten zinc bath through a top flux blanket on the kettle. However, an aqueous preflux may be used in conjunction with a top flux on the zinc bath.

Workpieces are handled either mechanically, using overhead hoists, or with hand tools. Small items such as washers, fasteners, and nails are handled in baskets. The baskets are usually shaken, jarred, or centrifuged as the work leaves the molten zinc bath to remove excess zinc and to distribute the coating evenly. In all cases, workpieces must be handled properly until the coating has completely solidified, as a result of either quenching or air cooling.

Surface Conditioning Requirements. Although degreasing, pickling, water rinsing, and other cleaning procedures remove most of the surface contamination and scale from iron and steel, small amounts of impurities in the form of oxides, chlorides, sulfates, and sulfides are retained. Unless removed, these impurities can interfere with the iron-zinc reaction when the iron or steel part is immersed in molten zinc.

In the wet galvanizing process, a flux blanket on the surface of the molten zinc bath is used to remove these impurities and to keep that portion of the surface of the zinc bath through which the steel is immersed free from oxides. The flux blanket floats on the surface, and when workpieces are immersed in the bath, their surfaces are wetted by the molten flux. The flux must have sufficient chemical stability to maintain a chemically active foam at the galvanizing temperature and to perform its cleaning function at a high rate of speed.

Zinc ammonium chloride is generally used to provide a flux blanket on the molten zinc bath. There are several procedures for preparing the flux blanket. One generic method consists of mixing ammonium chloride (sal ammoniac) and zinc oxide to form the monoamine of zinc chloride. In the ensuing reaction, hydrogen and nitrogen are released, and the flux takes the form of a foam.

To be fully effective and to attain minimal fume galvanizing conditions, the flux should not contain an excess of ammonium chloride. To prevent abnormally rapid chemical breakdown of the

Fig. 4 Effect of kettle immersion time on silicon-coating weight for killed and unkilled steels

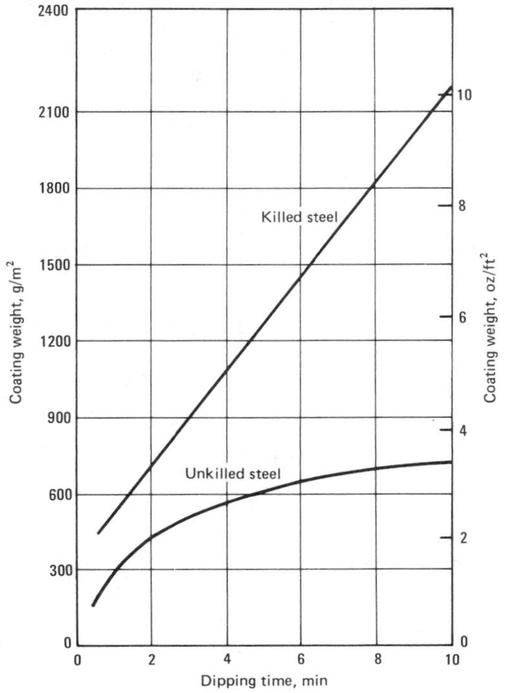

Coating weights on silicon-killed and unkilled steels at a galvanizing temperature of 455 °C (850 °F). Killed steel: 0.35% carbon, 0.26% silicon, 0.46% manganese. Unkilled steel: 0.13% carbon, trace silicon, 0.40% manganese

flux, glycerin and other organic substances are added to the flux in small amounts of 1 to 2 vol %. These substances increase the foaming action, markedly reduce the loss of ammonia, and serve as insulators. Because the boiling point of a mixture of zinc chloride and ammonium chloride decreases as the ammonium chloride content increases (Fig. 3), no more than 2 to 3% ammonium chloride will remain in the molten salt solution at a kettle temperature of 455 °C (850 °F). However, to function effectively, a zinc chloride-ammonium chloride top flux must normally contain more than 5 to 10% dissolved ammonium chloride. To minimize fuming, it is necessary to reduce the top flux surface temperature, maintain an optimal level of dissolved ammonium chloride, and to disturb this equilibrium as little as possible.

For dry galvanizing, it was once common practice to take the work directly from the hydrochloric acid pickle, dry it, and then put it in the molten zinc. However, the practice of going from the pickle tank to drying and then to the molten zinc produces more dross than the procedures of rinsing, prefluxing, and drying.

A more preferred method, widely used today, is to pickle, rinse, flux in an aqueous zinc ammonium chloride, dry, and dip in the molten zinc bath. By using a preflux step, better control of the fluxing is possible resulting in a more consistent finish. Also, the material may be held for 1 or 2 h before dipping which gives the galvanizer some flexibility in work flow on the galvanizing line.

Galvanizing Bath. The molten zinc bath is operated at temperatures usually in the range of 445 to 465 °C (830 to 870 °F). At 480 °C (900 °F) and above, the dissolution rate of iron and steel in zinc is extremely rapid, and the effects of these temperatures on both workpiece and galvanizing tank are generally harmful. Within the conventional galvanizing temperature range, an increase in temperature (*a*) increases the fluidity of molten zinc; (*b*) accelerates the formation of oxides on the bath surface; (*c*) heats the part to a higher temperature, thus increasing the time required for the zinc to solidify when the part is withdrawn; and (*d*) reduces immersion time, thereby increasing the kettle utilization factor. Each of these considerations has a distinct effect and may be used to control the galvanizing process.

An increase in the fluidity of the bath improves drainage and is desirable provided the bath temperature does not exceed the normal operating range. An increase in bath temperature produces a much sharper temperature gradient from the surface to the center of the part, which depending on shape may result in an increase in distortion.

Unless the bath contains aluminum or unless its surface is well protected by a foam blanket of flux, an increase in bath temperature will accelerate the formation of an oxide film (or ash) on the surface of the bath. Some of this oxide film may cling to the workpiece when it is withdrawn from the bath, interfering with drainage and contributing to the formation of a coating with less desirable aesthetic properties. The effects of these oxides are most apparent on parts of thin cross section and large surface area.

Depending on the chemical composition of the iron or steel, the bath temperature may have significant metallurgical effects on the galvanized coating. The temperature at which the iron-zinc alloy layers are formed affects the relative amounts of each iron-zinc phase formed and the depth or total thickness of alloy layer.

In the hot dip galvanizing of fabricated articles, the thickness of the coating is controlled by immersion time (Fig. 4). Although timing is to some extent dependent on ease of handling and must be established by trial for each design of part being coated, the duration of immersion is usually in the range of 1 to 5 min. The speed of immersion influences the uniformity of the coating, particularly with long articles for which the difference in immersion time between the first and last areas to enter the bath may be considerable.

The reaction between clean low silicon steel and molten zinc proceeds rapidly for the first 1 or 2 min after the work has been immersed, producing an alloy layer that continues to grow at decreased rate the longer the article is left in the bath. However, for steels con-

Fig. 5 Effect of withdrawal rate on coating weight

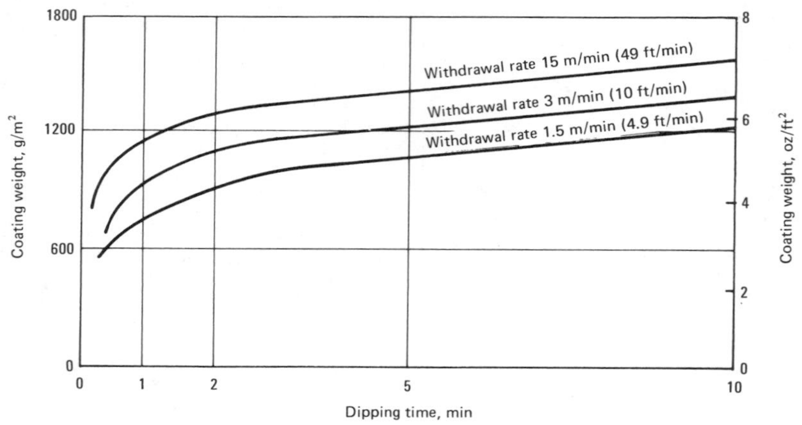

Temperature was 435 °C (815 °F)

Fig. 6 Comparison of coating weight as a function of silicon content for conventional and Polygalva galvanizing processes

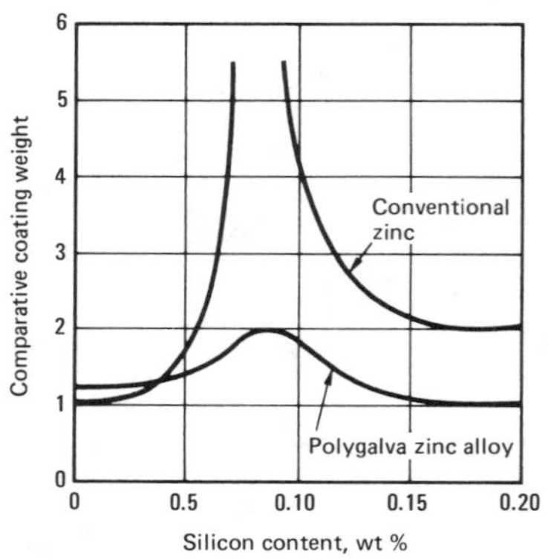

taining silicon in excess of 0.05%, the coating weight increases linearly with respect to the time of immersion producing, in general, heavier coatings. Therefore, it is important to minimize immersion time for silicon-bearing steels to prevent excessive alloy growth and coating weight.

To provide a uniform coating of minimum thickness, work that is not subsequently to be centrifuged is withdrawn from the bath slowly and at a controlled rate, thus permitting maximum drainage. Two-speed hoists are usually employed, permitting the work to be immersed rapidly and withdrawn slowly. The rate of withdrawal, which determines the thickness of the unalloyed zinc layer left on the work (Fig. 5), varies according to the type of process being operated. The optimum withdrawal rate for most articles is about 1.5 m/min (5 ft/min).

With long articles, for which the withdrawal occupies a large part of the total cycle time, higher speeds may be necessary to maintain a reasonable rate of production. If possible, however, it is better to overcome this difficulty by using special jigs and carriers for dipping and withdrawing the work in batches. Provided the work is not withdrawn faster than the rate at which the zinc drains freely from the surface, the unalloyed zinc layer of the coating is uniformly distributed. With faster rates of withdrawal, the surplus zinc carried clear of the bath runs down the surface until it solidifies, and the resultant coating may be lumpy and uneven.

When withdrawal from the bath at a slow controlled rate of speed is not feasible or economical, as it generally is not for small parts, the withdrawal rate may be greatly increased, and drainage is accomplished by spinning the parts in a basket protected by a closed container. The excess zinc is drained from the parts by centrifugal force and retained within the container or cover. To promote fluidity and reduce oxidation, the parts may be sprinkled with a small amount of flux before spinning. Mechanically wiping the excess zinc from the parts with tools designed for this purpose is an alternative method of forced drainage.

Cooling. Because of retained heat, the iron-zinc reactions can continue to occur even after the surface layer of zinc has frozen. This type of post-immersion reaction may occur if cooling is hindered by the stacking of parts in close proximity and by the heat capacity of the part. Some or all of the pure zinc layer may be converted to iron-zinc alloy, thus discoloring the surface and altering its properties.

To avoid delayed cooling, parts should be spaced adequately after immersion to permit the free circulation of air. Parts with large cross-sectional areas or parts fabricated of non-silicon-bearing steel may require forced cooling with air or water.

Galvanizing of Silicon-Killed Steels. The difficulties encountered in galvanizing silicon-bearing steels have been referred to under the preceding discussions on conventional galvanizing practice. Many techniques have been studied in an effort to find a better way to control the iron-zinc reaction kinetics in the presence of silicon, none of which have proven to be fully acceptable over the silicon ranges being encountered. At temperatures not exceeding 460 °C (860 °F), aluminum additions of 0.02 to 0.04% to the bath may be of advantage for steels containing up to 0.05% silicon. Shot blasting the steel surface is beneficial for steels containing less than 0.12% silicon. The most common method, however, has been to use conventional techniques at a temperature not exceeding 440 °C (825 °F) coupled with a short immersion time. For light structural shapes, this is possible without any additional measures. Heavy structural shapes do require preheating to reduce the immersion time to an acceptable level.

However, there are two technologies, developed during the 1970's, which permit better control to be maintained in the galvanizing at reactive silicon-bearing steels. One is a patented process using the zinc alloy Polygalva for galvanizing. The second is by galvanizing at high temperatures, 550 °C (1020 °F), instead of the temperatures used for the conventional process, 450 °C (840 °F). Neither of these processes are commercially available at this time in North America.

Polygalva Process. Polygalva is essentially a zinc alloy containing controlled amounts of aluminum, magnesium, tin, and lead. The aluminum

Fig. 7 Micrograph of silicon bearing steel (0.08% Si) galvanized in a conventional bath

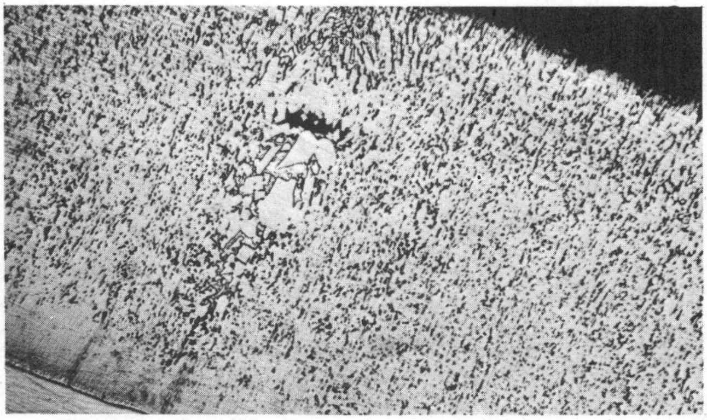

Fig. 8 Micrograph of silicon bearing steel galvanized in Polygalva

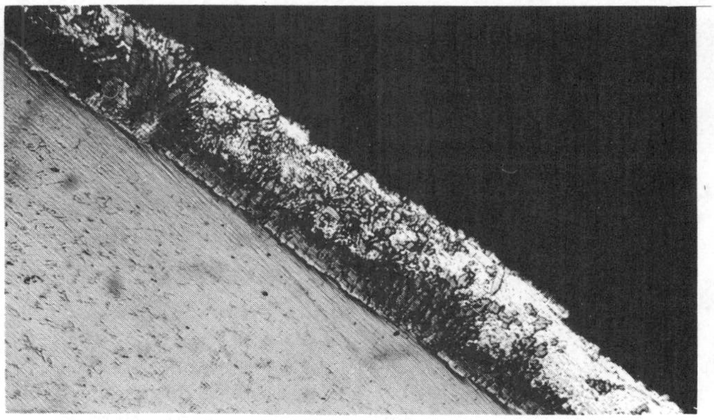

Fig. 9 Effect of time and temperature in a high temperature galvanizing bath containing 0.22% iron

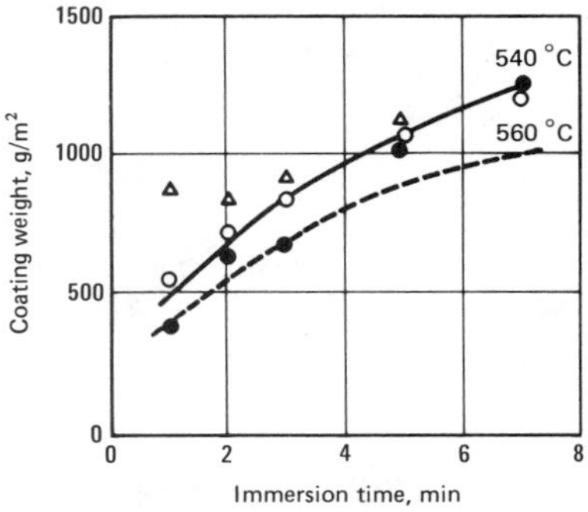

○: Steel D, 0.02% silicon; ●: steel N, 0.22% silicon; △: steel S, 0.42% silicon

is used to retard the formation of the intermetallic layer, and the other elements help to insure continuity of the galvanized coating. As with the conventional process, thorough surface preparation prior to galvanizing is essential for good results and must include the following steps unless otherwise indicated:

- Degreasing, alkaline bath heated to 80 to 90 °C (176 to 194 °F)
- Rinse, running water
- Pickle, 50% hydrochloric acid with inhibitor
- Rinse, running water
- Pickle, 70% hydrochloric acid without inhibitor (in most situations this step is optional
- Routine maintenance of pretreatment facilities is important

Also, a weekly zinc bath chemical analysis is required to ensure that the alloy composition is being maintained within the working range. During galvanizing operations, two master alloys are added to the bath to compensate for losses of aluminum and magnesium. By doing this, proper alloy balance is reportedly readily maintained.

As shown in Fig. 6, Polygalva is most effective in galvanizing steels with silicon in the range of 0.05 to 0.20%. Comparative micrographs of a steel with 0.08% silicon galvanized using conventional and Polygalva techniques are shown in Fig. 7 and 8. Above 0.20% silicon, the Polygalva alloy reportedly loses at least part of its effectiveness.

High-Temperature Galvanizing. It has been found that, when galvanizing is performed at a temperature of approximately 550 °C (1020 °F), a coating weight to immersion time relationship is obtained which is much less sensitive than for silicon-bearing steels at conventional galvanizing temperatures. The effect of immersion time at elevated galvanizing temperatures is shown in Fig. 9; the coating weight increases at a rate less than linear with time. Doubling the time of immersion from 4 to 8 min increases the coating weight by about 30%.

Because of the high reactivity between molten zinc and steel at these temperatures, a ceramic vessel must be used instead of a steel kettle. While the available ceramic kettle technology is adequate for this application, the state of the art of ceramic kettle heating currently lacks the efficiency of flat flame burners used to heat steel kettles. The development of immersion heaters to heat the ceramic baths is proceeding and offers good promise of improved efficiencies. Pending development of an operationally suitable immersion heater system, ceramic galvanizing kettles are top heated. This technique is relatively inefficient and interferes with the handling of materials into and out of the kettle.

In the bath, the solubility of iron increases from 0.03% at 450 °C (840 °F) to 0.3% at 550 °C (1020 °F). By controlling the iron content, the coating weight can be controlled. As shown in Fig. 10, increasing the iron content from zero to

Fig. 10 Coating weight as a function of iron content

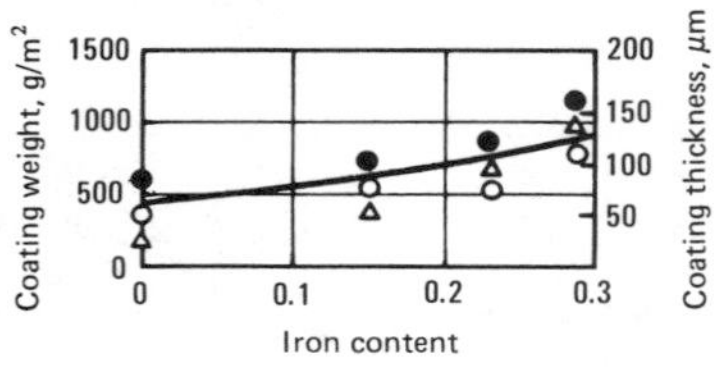

○: Steel D, 0.02% silicon; ●: steel N, 0.22% silicon; △: steel S, 0.42% silicon; galvanized 3 min at 550 °C (1020 °F)

Fig. 11 Electron-scanning image of high-temperature galvanized coating, 375X

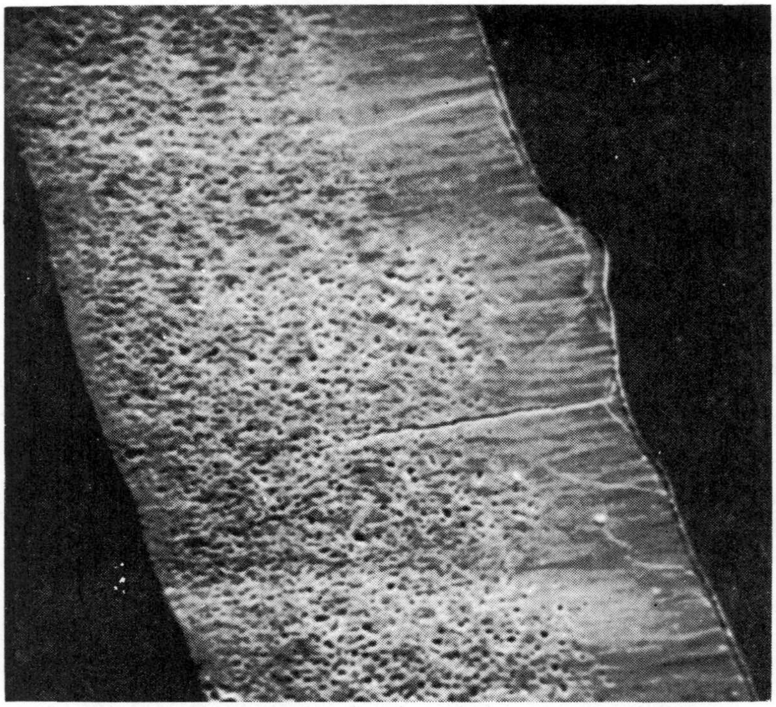

Innermost, compact, and outermost, duplex, layer

0.3%, the solubility limit at 550 °C (1020 °F) increases the coating thickness by a factor of 2. Controlling the iron content within the range 0.1 to 0.2% produces coating weights which meet specifications and are not excessively heavy.

In a high temperature galvanizing bath, aluminum up to 0.5% does not appear to have any systematic effect on coating weight. However, above 0.3%, it produces a floating dross which can mar the coating appearance. An aluminum addition of 0.03% is sufficient to brighten the coating if this is desired.

There have been reports of occasional adherence deficiencies of high temperature galvanized coatings. At this time, it is believed that this is the result of a lead deficient bath. As a result, the bath lead level should be maintained at 1%. Therefore, the following bath conditions are considered ideal:

- Temperature at 560 °C (1040 °F)
- Iron content between 0.1 and 0.2%
- Lead content about 1%
- Aluminum content of 0.05%

The coating has a light gray, uniform appearance which does not vary with silicon content of the basis steel. Brighter coatings may be obtained by the aluminum addition to the bath described above and by quenching instead of air cooling. Coating adhesion and ductility are equivalent to coatings galvanized at conventional temperatures. The metallographic structures of all high-temperature coatings are similar, the only variation being in the constituent proportions of some of the layers (Fig. 11).

Mechanical property tests of the basis steel subjected to high-temperature galvanizing reveals no significant differences from the results obtained using conventional galvanizing techniques. These tests include tensile properties on plain, punched, and welded specimens, reverse bend tests for strain age embrittlement, ductility bend tests, and pulsating tension fatigue tests.

Because of their recent development, the performance of high-temperature galvanized coatings is not as well documented as are conventional coatings. Based on a limited body of longer term exposure data and accelerated weathering tests in marine, urban, and industrial environments, it appears that the performance of these coatings is at least equal to that of coatings produced by conventional techniques.

Batch Galvanizing Equipment

Because the galvanizing kettle is the most important piece of equipment used in galvanizing, its selection should be based on the careful evaluation of several major variables, such as size, shape, wall thickness, tank material, source of heat, and auxiliary equipment requirements.

Size and Shape. Although the size and shape of the galvanizing kettle are governed primarily by the parts to be galvanized in it, other factors must also be considered. The kettle must be large enough to be an adequate heat container; that is it must possess sufficient heat capacity in the molten zinc to compensate for the loss of heat encountered when cold workpieces are immersed in the tank. The minimum and maximum operating temperatures that must be maintained depend on production requirements; usually, the weight of zinc in the tank should be equivalent to 15 to 20 times the weight of parts that are to be galvanized in 1 h. In many production installations, the weight ratio of zinc to workpieces is more likely to approach 40 to 1.

Although the shape of the kettle must accommodate the workpieces that are to be immersed, it should also be designed to expose a minimum of bath surface. If the size of a kettle is to be increased to accommodate a particular part, the depth of the kettle rather than its length or width should be increased to minimize the exposed surface area of the bath. A minimum surface area conserves heat and produces less surface oxide than a larger area. Avoid kettles of complicated shape, because they are susceptible to damage by severe thermal stresses. Simple, rectangular kettles are most widely used.

Wall Thickness. Theoretically, the selection of wall thickness of a galvanizing kettle should be governed by these factors: (*a*) the rate of corrosive attack by liquid zinc, (*b*) the hydrostatic load imposed against the kettle walls by the volume of the zinc bath, (*c*) the strength of the kettle wall material at the operating temperature of the bath, and (*d*) the support afforded the kettle walls by the surrounding brickwork or by other reinforcing elements. Because the variables are so numerous and complex, accurate calculation of a required wall thickness is literally impossible, and selection is based entirely on empirical data. Depending on the size of the kettle and its reinforcing elements, wall thickness usually varies from 20 to 50 mm (¾ to 2 in.).

Kettle Material. Aside from strength, the principal requirement of a galvanizing tank material is the ability to resist the corrosive attack of molten zinc. The most widely used material is boiler plate of flange and firebox quality with low silicon. The chemical composition of this steel ensures a minimum rate of attack by molten zinc; also, the good welding and bending characteristics of this material are essential features in kettle fabrication. The chemical composition of the welding rods used in kettle fabrication should also be of low carbon and low silicon.

If a flux layer is to be maintained on the bath surface, a collar of firebrick or other suitable ceramic material should surround and abut the top 150 or 180 mm (6 or 7 in.) of the tank to retard heat transfer in this area and thus reduce attack by the flux on the steel kettle wall.

Source of Heat. Galvanizing kettles can be suitably heated by combustion of oil or gas, by electrical resistors, or by electromagnetic induction. The source of heat is of minor importance provided the heating installation satisfies the following requirements:

- High efficiency factor
- Good adjustability and control to maintain an even temperature
- Ability to maintain the minimum temperature required on the outside walls of the kettle
- Uniform heating along the outer walls, without hot or cold spots

Failure to satisfy all these requirements severely curtails the life of the kettle and may result in unexpected kettle failure.

Temperature Controls. When a new galvanizing tank is installed, a complete temperature survey should be made of the molten zinc bath. Based on this survey, control thermocouples may be located in the bath to maintain temperature uniformity and control.

Fig. 12 Galvanized material stacked on an incline with spacers to prevent wet storage stain

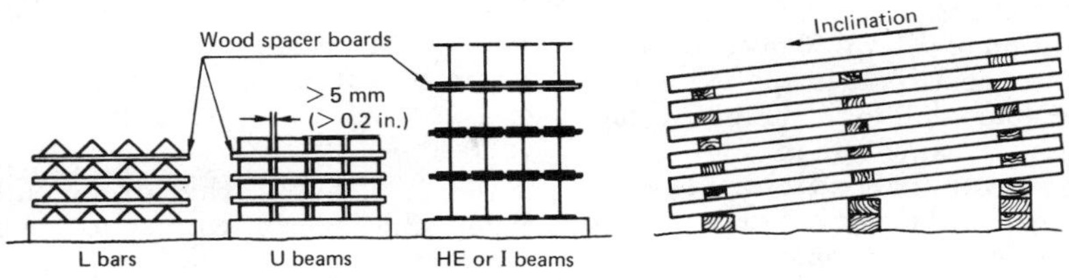

Post Treatments

Wet Storage Stain Inhibitors. A white stain, commonly called white rust or wet storage stain, may appear on zinc surfaces during storage or shipment. The stain is found on material with newly galvanized, bright surfaces and especially in such areas as crevices between closely packed sheets and angle bars if the surfaces come into contact with condensate or rainwater and the moisture does not dry quickly. Zinc surfaces that have developed a normal protective layer of corrosion products are seldom attacked.

When zinc coatings corrode openly in air, zinc oxide and zinc hydroxide are normally formed. In the presence of atmospheric carbon dioxide, these compounds are transformed to basic zinc carbonate. If the supply of air to the surface of the zinc coating is restricted, as in a narrow crevice, then sufficient carbon dioxide is not supplied for the formation of the normal layer of zinc carbonate.

The layer of zinc oxide and zinc hydroxide is voluminous and porous and adheres loosely to the zinc surface. Consequently, it does not protect the zinc surface against oxygen in the water. Corrosion can therefore proceed as long as there is moisture left on the surfaces. When wet storage staining occurs, arrange the objects so their surfaces dry rapidly. The attack ceases, and with a free supply of air to the surfaces, the normal protective layer of corrosion products forms. The white corrosion products gradually wash off, and the surface of the coating takes on the normal appearance of a hot dip galvanized, exposed object.

Because the corrosion products are very voluminous (about 500 that of the zinc that has been consumed), any attack may appear serious. Usually, however, such an attack of wet storage stain is of little or no importance to the durability of the corrosion protection.

Wet storage stain is best avoided by preventing newly galvanized surfaces from coming into contact with rain or condensate water during storage and transport. Arrange materials stored outdoors so that water can easily run off the surfaces and so that all surfaces are well ventilated (Fig. 12).

Temporary protection against wet storage staining is obtained by chromating or phosphating. Painting after galvanizing also provides effective protection.

Where the surface staining is light and smooth without growth of the zinc oxide layer as judged by lightly rubbing fingertips across the surface, the staining will gradually disappear and blend in with the surrounding zinc surface as a result of normal weathering in service. When the affected area will not be fully exposed in service or when it will be subject to a humid environment, wet storage staining must be removed, even if it is superficial, to allow formation of the basic zinc carbonate film which normally contributes to the corrosion resistance of galvanized coatings.

Medium to heavy buildup of white corrosion product must be removed, otherwise the essential protective film of basic zinc carbonates cannot form in affected areas. Light deposits can be removed by brushing with a 5% solution of sodium or potassium dichromate with the addition of 0.1 vol % of concentrated sulfuric acid. This is applied with a stiff brush and left for about 30 s before thoroughly rinsing and drying.

Paint Over Galvanizing. Hot dip galvanized steel may need to be painted for the following reasons:

- Additional corrosion protection for exposure to aggressive environments, especially if future maintenance will be difficult or if the zinc coating is thin, such as on sheet metal
- Another color of coating is desired for aesthetic reasons, for warning purposes, or for camouflage
- Protection against galvanic corrosion if hot dip galvanized steel is to be in contact with another metal such as copper

Hot dip galvanizing combined with painting offers good corrosion protection, even in very aggressive environments. The durability of such a duplex system is approximately 1.2 to 1.5 times that of the durability of either the painted bare steel or the zinc coating alone.

The zinc coating can be painted immediately after hot dip galvanizing or after some time of exposure. In most cases, painting immediately after hot dip galvanizing is preferable, since the surfaces are least contaminated.

Regardless of whether the paint is applied to a fresh, bright coating surface or to an exposed surface with corrosion products, the surfaces must be cleaned carefully prior to painting. The paint on zinc surfaces is more sensitive in this respect than many other materials, because even small quantities of impurities on the surfaces can affect the adhesion of the paint film.

However, the surfaces of zinc coatings are often much easier to clean than steel surfaces. It is important that an appropriate cleaning procedure be used for the particular impurities present on the surfaces.

Exposed Matte Surfaces. When zinc coatings are exposed, the surface corrodes and is covered with corrosion products. The basic zinc carbonate that forms in clean air can be painted over. This is the reason for the traditional recommendation to wait from 6 months to 1 year before painting hot dip galvanized objects.

Today, however, the air is seldom clean. The layer of corrosion products contains such substances as sulfides, sulfites, sulfates, and chlorides. Many of these compounds are water soluble and some are even hygroscopic. To achieve good results when painting, all water-soluble impurities must be removed.

Cleaning and Surface Preparation. Wash heavily contaminated surfaces, both fresh and exposed, with a suitable organic solvent such as white mineral spirits, and then bristle brush to remove solid particles and corrosion products. Follow this washing with a thorough rinsing with water at high pressure, if possible.

Moderately contaminated surfaces, for example, fresh newly galvanized surfaces and surfaces that have been exposed for a longer period of time but have not been contaminated with oil and grease, can be washed with water to which 5 to 10% ammonia, caustic soda (NaOH), or acetic acid has been added. Afterwards, buff the surface with a soft brush. This treatment must be followed by very thorough rinsing with water at high pressure, if possible.

Chromated surfaces, on continuously hot dip galvanized sheet, for example, can also be washed with ammonia, caustic soda, or acetic acid in water and buffed, followed by thorough rinsing. The alkaline or acid solution dissolves the chromate layer. In general, when galvanized after fabrication material is to be painted as a post-treatment, it should not be chromate treated.

Surfaces that have been exposed, moderately contaminated surfaces, or newly galvanized surfaces can also be brush blasted, that is, blasted with low pressure and a rapid motion of the nozzle, for example 0.3 MPa (0.04 ksi) at 6 mm (0.2 in.) nozzle diam and 250 to 300 mm (9.8 to 11.9 in.) nozzle distance. Abrasives consisting of silicates and slags of 0.2 to 0.5 mm (0.008 to 0.02 in.) are recommended. Glass beads and fine-grained aluminum oxide can also be used.

Sweep blasting effectively removes any corrosion products and provides an advantageous roughening of the surface of newly applied bright zinc coatings. However, brush blasting must be carried out carefully so that the zinc coating is not destroyed and large stresses are not built into the coating. These stresses may subsequently cause flaking of the paint coat.

Choice of Paint. Paints suitable for direct application to properly cleaned hot dip galvanized steel are discussed below. As with most other paints, first apply a suitable primer to the zinc surface.

Paints consist of 10 to 20 different components and each different manufacturer has its own formula for a certain type of paint. Paints of the same type but from different manufacturers can have different properties. Discuss detailed recommendations with the manufacturer.

In moderately corrosive atmospheres paints based on acrylate and PVAc-latex are suitable. However, it takes about 10 to 14 days for these paints to achieve maximum hardness and adhesion. If the objects are to be handled or transported within this time, special care must be observed to avoid damage.

Under severe chemical conditions, such as in industry, and in aggressive atmospheres, paints with better chemical resistance than latex paints are required. Such paints are based on PVC, vinyl copolymers, chlorinated rubber, polyurethane, and epoxy.

In water and soil, tar/bitumen paints are recommended, preferably in combination with epoxy and polyurethane. Certain aluminum-pigmented asphalt solutions can also be used for structures in water, but they have relatively poor mechanical strength.

Aluminum Coating of Steel

By the ASM Committee on Aluminum Coating of Steel*

ALUMINUM-COATED STEEL PRODUCTS may be classified with respect to their intended service, according to their properties, their behavior, and the economy of their non-rusting surfaces. Properties of aluminum-coated steel are beneficial for products which require the advantages of good corrosion resistance, bright metallic appearance, receptiveness to finishes, high reflectivity, and good electrical conductivity. For some products, the behavior of the aluminum-iron interfacial compound is relied upon for resistance to oxidation, scale formation, abrasion and high hardness. For low-cost nonrusting products, aluminum-coated steel provides the mechanical properties, formability and weldability of steel in parts where finishes peculiar to aluminum surfaces can be applied.

Methods of applying aluminum coatings include batch or continuous hot dipping, pack diffusion, slurry process, thermal spraying, cladding, vacuum or chemical vapor deposition, ion vapor deposition, electroplating, and electrophoresis. Choice of method is determined by coating performance required, intended use of the coated product, size and shape of the product, production volume, and cost.

Two important factors in successful coating are (*a*) proper preparation of the steel surface, and (*b*) in the case of high temperature processes, control of the formation and growth of the intermetallic compound of aluminum and iron that forms at the interface of the aluminum coating and the steel substrate. Although there are many possible methods of preparing the surface, the method selected must remove the iron oxide scale, either mechanically or chemically, and also remove adsorbed moisture and gas from the surface. When an intermetallic compound is desired, growth of the iron-aluminum interfacial layer can be controlled by silicon, beryllium, or chromium, present in the steel or in the aluminum coating. In some applications, a very thin phosphate film permits bonding and hinders growth of the interfacial layer.

Applications

Aluminum-coated steel products are used successfully in corrosive and oxidizing environments in which the temperature ranges from that of outdoor exposure to 1150 °C (2100 °F).

Atmospheric Exposure. Some of the corrodents encountered in outdoor exposure are:

- High sulfur content industrial atmospheres
- Nitrate-phosphate chemicals from fertilizers and manures in rural atmospheres
- Salt compounds used for ice removal on streets and roads
- Organic acids in food wastes
- Marine environments

Exposed to severe industrial environments, aluminum-coated steel products, such as pole line hardware, corrugated roofing and siding and prefabricated steel buildings, have been found in excellent condition after 15 to 20 years. Galvanized products in the same applications have red rusted in less than four years.

Examples of outdoor applications of aluminum-coated low-carbon steel are given in Table 1. In general, these applications relate to designs requiring the high modulus and strength of steel,

*Robert Baboian, *Chairman*, Head, Electrochemical and Corrosion Laboratory, Materials and Electrical Products Group, Texas Instruments, Inc.; Hart Graff, Principal Research Engineer, Armco Inc.; Farrell Kilbane, Senior Staff Physicist, Armco Inc.; Carl F. Mietzner, Manager, Corrosion and Coatings Research Section, Bethlehem Steel Corp.; Richard A. Nickola, Supervising Research Engineer, Inland Steel Co.; D. J. Schardein, Director, Finishing Technology Section, Reynolds Metals Co.

but with the corrosion resistance of aluminum.

For parts made of fabricated sheet, the design must allow for the bare sheared edges. Rusting of sheared edges, uncoated spots, or areas of mechanical damage to the coating is usually nonprogressive. The initial rust film at coating discontinuities of aluminum-coated steel form a protective, adherent, corrosion-product scale, which acts as a barrier film to stifle further corrosion and prevent base metal attack or undercutting of the aluminum coating.

Aluminum-coated steel fasteners can be used to maintain the appearance and structural integrity of an assembly. Only heavy aluminum coatings, 230 to 380 μm (9 to 15 mils) thick, are recommended for use in environments of continuous condensation or for immersion in liquids. Thinner coatings are subject to pitting attack, which perforates the coating and corrodes the basis metal. An exception is aluminum coatings applied by ion vapor deposition, which is being used in the aerospace industry. Here 25 μm (1 mil) thick coatings are applied to high-steel components such as landing gears.

Elevated-Temperature Exposure. Successful application of aluminum-coated steel for resistance to oxidation and corrosion at elevated temperatures depends on the physical and mechanical properties of the alloy chemical bond between the aluminum and the steel. It is important that the hot strength of the steel be suitable for the stress and temperatures encountered. Recently, low-carbon steels alloyed with titanium or columbium have been introduced to offer improved high-temperature creep resistance when used as substrates for aluminum coatings.

Aluminum containing from 5 to 11% silicon (ASTM A463) applied to sheet steel minimizes the thickness of the iron-aluminum alloy bond and improves formability. Undiffused, this type of coating retains excellent heat reflectivity at temperatures to 480 °C (900 °F).

Above 480 °C (900 °F), further alloying occurs between the aluminum coating and steel base. Because the rate of alloying is dependent on time and temperature, all coating converts to aluminum-iron-silicon alloy with sufficient time at temperature. The refractory alloy formed is extremely heat resistant and not prone to spalling up to 680 °C (1250 °F). This spalling tendency at service temperatures above 680 °C (1250 °F) is overcome in a new heat-resistant aluminized steel which contains sufficient titanium to stabilize carbon and nitrogen as well as maintain excess titanium in solution (Ref 1).

Table 2 lists applications for steels that have been prepared by batch hot dipping in aluminum, and then heat treated to diffuse the aluminum into the steel. This treatment eliminates spalling and provides an impervious protective coating during high-temperature service.

The use of aluminum-coated plain carbon steel for complicated heat treating fixtures subjected to temperatures lower than 870 °C (1600 °F) may decrease the overall cost of fixtures as compared to that for fixtures made of the highly alloyed austenitic steels normally used for this application. Figure 1 shows the effects of coating thickness

Table 1 Outdoor applications for aluminum-coated low-carbon steels

Applications	Method of coating
Fabricated bar, strip, or fasteners	
Anchor bolts for aluminum railing, lighting posts, fasteners for aluminum fabricated items, pole line hardware for electrical transmission, high-temperature studs for chemical plants and oil refineries	Batch hot dip
Sheet or strip	
Air conditioner housings, awnings, building panels, corrugated roofing and siding, ductwork, outdoor signs, roof decking, tractor muffler outer sheets, truck body rocker panels, weather shields, welding tubing for fence posts, automotive exhaust components	Continuous hot dip(a)
Welded assemblies	
Agricultural implements, clothes driers, frames for air filters, furnace heater casings	Batch hot dip or spray
Wire products	
Chain link fencing, cores for aluminum electrical transmission lines, barbed wire, guy strands	Continuous hot dip
Nails	Batch hot dip

(a) Vapor deposition is used when steel is thinner than 0.378 mm (0.0149 in.)

Table 2 Applications of diffused aluminum coatings for resistance to oxidation and corrosion at 455 to 980 °C (850 to 1800 °F)

Product and basis metal	Type of service
Heat treating equipment	
Burner pipes, 5Cr-0.5Mo	Oxidation, 870 °C (1600 °F)
Fixtures, low-carbon and medium-alloy steels	Carburizing, carbonitriding
Flue stacks, low-carbon steel	Oxidation, sulfur corrosion
Furnace insulation supports, low-carbon and medium-alloy steels	Oxidation, 540-650 °C (1000-1200 °F)
Pyrometer protection tubes, 310 and 316 stainless steel, low-carbon steel	Oxidation, 980 °C (1800 °F)
Heat exchanger components	
Boiler soot blowers, 1Cr-0.5Mo	Oxidation, sulfur attack
Boiler tubing, 2Cr-0.5Mo	Oxidation, 540-595 °C (1000-1100 °F)
Cylinder barrel, air-cooled engine, Nitralloy	Oxidation to 480 °C (900 °F)
Preheater tubing, 1Cr-0.5Mo	Oxidation, 650 °C (1200 °F)
Tubing, low-carbon steel, 1.5Cr-0.5Mo	Hydrogen sulfide gases
Fasteners	
Steel fasteners for chemical piping and boilers	Oxidation to 480 °C (900 °F)
High-temperature fasteners	Oxidation to 760 °C (1400 °F)
Studs, 4140, for chemical and oil refineries	Oxidation and ease of removal after service at 480 °C (900 °F)
Miscellaneous equipment	
Chemical reactor tubing, low-carbon steel	Carbonization, iron contamination
Chimney caps, low-carbon steel	Oxidation and corrosion
Recuperator tubing, 2.5Cr-0.5Mo	Oxidation and sulfidation
Refinery tubing, 304 stainless steel, 2.25Cr-1Mo	Oxidation and sulfidation
Sulfuric acid converters, 5Cr-0.5Mo	Sulfur dioxide corrosion, 705 °C (1300 °F)

Fig. 1 Effects of coating thickness and exposure temperature on oxidation of coated and uncoated steel

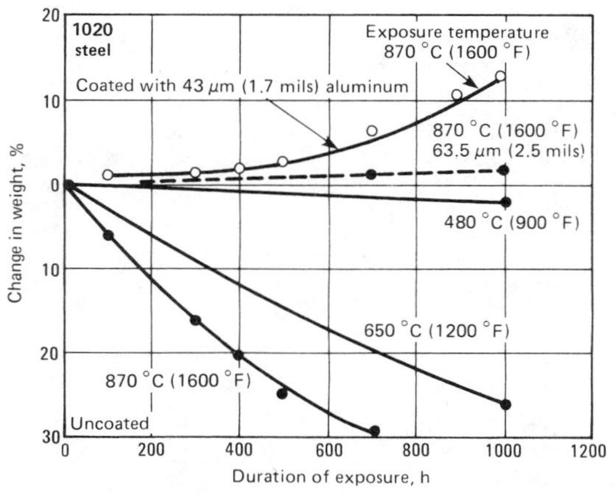

Oxidation at 480 to 870 °C (900 to 1600 °F). Steel 6.4 mm (¼ in.) thick was completely oxidized at end of 700 h at 870 °C (1600 °F)

Table 3 Aluminum-coated parts for high-stress high-temperature applications

Part and basis metal	Type of service	Coating method
Exhaust poppet valves, Silicrome XB	Oxidation and corrosion by leaded fuel combustion products at 760 °C (1400 °F)	Spray and diffuse
Intake poppet valves, 8640	Oxidation resistance	Spray and diffuse
Turbine blades, Iconel 713, GMR-235	Oxidation and thermal shock at 815-925 °C (1500-1700 °F)	Hot dip and diffuse, spray and diffuse
Turbine vanes, HS-31	Oxidation and thermal shock at 900-1150 °C (1650-2100 °F)	Spray and diffuse, pack diffuse

and operating temperature on oxidation resistance for coated and uncoated heat treating fixtures made of 1020 steel.

Aluminum-coated carbon steel offers greater resistance to attack by hydrogen sulfide than does solid stainless steel. One set of test data indicated that aluminized carbon steels are more than 100 times as resistant to pure hydrogen sulfide as 18-8 stainless steel at 595 °C (1110 °F). Another stated that they were 25 times as resistant as straight chromium steel.

The greatest increase in the use of diffused or impregnated aluminum coatings has occurred in ultrahigh-temperature applications for automotive and aircraft components, some of which are listed in Table 3. Coated automotive exhaust poppet valves have improved performance and durability because they are capable of resisting high-temperature erosion from high-velocity exhaust gases containing sulfur dioxide and combustion products from tetraethyl lead. Aluminum coatings are useful also in reducing scaling oxidation of medium-alloy steels used for intake valves that are subjected to the temperatures generated in high-compression engines. In a similar corrosive environment, but with temperatures in excess of 815 °C (1500 °F), blades and nozzle vanes for gas turbine engines are coated with aluminum alloy for effective resistance to thermal shock and erosion.

Problems may occur with welded aluminum-coated steel parts. The joining process is not usually the problem, but the alloying of aluminum and iron can create a loss of ductility and lowering of corrosion resistance in the weld and in the heat-affected zone.

For example, welded shrouds of hot dip aluminum-coated 18-8 stainless steel, 3.2 mm (⅛ in.) thick sheet, were used in a 2.4-m (8-ft) ID petroleum reformer reactor. During service, the shell became very hot, indicating that the shroud was not keeping hot hydrogen gases away from the internal refractory lining of the shell.

During welding of the shrouds, aluminum had been absorbed into the 18-8 stainless steel weld and had formed a nonductile, hardenable alloy, in effect a precipitation-hardenable stainless steel. The welds cracked in service, allowing hot gases to overheat the shell.

This problem was corrected by grinding the prepared weld groove and depositing pure nickel in the root pass, the most highly stressed part of the weld. The remainder of the weld was filled with 18-8 stainless steel electrode.

Continuous Hot Dip Coating of Mill Products

Aluminum hot dip coating of steel strip and wire is performed by a number of patented processes. The Sendzimir method, which is the most widely used for sheet, consists of oxidizing the surface of the steel, reducing the oxidized surface in a reducing atmosphere, and immersing the steel in molten aluminum. This procedure expedites wetting, or formation of the alloy between the aluminum and the steel.

The use of a nonoxidizing, direct-fired furnace in place of an oxidizing furnace offers substantial benefits in increased production, reduction of hearth roll pickup, and ability to maintain a positive pressure throughout the furnace line, preventing air leaks into the furnace atmosphere. Combustion products of a direct-fired furnace are maintained with a slight excess of combustibles assuring fast removal of oil and smut and reduction of surface oxides on the incoming strip.

Other processes are also based on the use of a reducing atmosphere, but without preliminary oxidation of the strip or wire. In a broad sense, the reducing atmosphere may be considered as a gaseous flux.

The Lundin process uses aqueous fluxes that are applied to the wire before immersion in molten aluminum. This eliminates the need for a reducing atmosphere. In this process, adequate ducts are necessary to remove corrosive salt fumes from the coating surface. Fumes are carried through wash systems to the outside surface.

Procedures and Control. Most commercial hot dipped aluminum-coated steel strip is produced on continuous, anneal-in-line equipment similar to that used for galvanizing. The process consists essentially of three operations: surface preparation, heat treatment of the steel base, and aluminum coating.

Surface preparation is a two-phase operation. First, all soil is removed from the surface by oxidizing at elevated temperature or by chemical cleaning. Then, the surface oxides are reduced in a suitable atmosphere to prepare the strip for coating.

Because the reaction between aluminum and steel is extremely rapid, the immersion time, the temperature of the molten aluminum, and the temperature of the strip before and after coating must all be controlled to prevent the formation of an excess of iron-aluminum interfacial alloy (Table 4). Unless a void layer separates the alloyed coating from the base metal, the amount of iron in the alloyed coating increases with time as the aluminum continues to diffuse into the base metal.

The amount of brittle interfacial alloy layer can be altered also by the addition of silicon to the coating bath (Table 4). This increases the apparent ductility of the coating, enabling more severe fabrication of the sheet without peeling of the coating. As shown in Fig. 2, there is a rapid decrease in the thickness of the interfacial layer as the silicon content increases to about 2.5%. A smaller decrease occurs as the silicon content is further increased.

Equipment for a continuous line of hot dip aluminum coating consists of a feeding section, a furnace section, and a delivery section. In the feeding section, equipment uncoils incoming strip and feeds it into the coating line at a designated constant speed under specified tension. The furnace section contains the preheating or oxidizing furnace, the annealing furnace, the cooling furnace, and the coating pot. If chemical cleaning is used, alkaline cleaner and water rinse tanks are substituted for the preheating furnace. The cooling furnace is connected directly with the annealing furnace and extends to the coating bath with its end sealed by means of a snout extending into the molten aluminum bath. A dry reducing atmosphere of hydrogen and nitrogen is maintained within the annealing and cooling furnaces. The delivery section is equipped to provide rapid cooling and sufficient time for setting the coating before the strip contacts the support roll over the coating bath. Drive rolls and equipment for looping, roller leveling, coiling and shearing, and stretch leveling and

Table 4 Effect of silicon and diffusion treatment on thickness of coating

Condition	Thickness: Intermetallic layer, μm	Intermetallic layer, mils	Total coating, μm	Total coating, mils
Pure aluminum coating(a)				
As coated	23	0.9	51	2.0
Diffused for:				
1100 h at 480 °C (900 °F)	33	1.3	51	2.0
1100 h at 540 °C (1000 °F)	36	1.4	51	2.0
1000 h at 595 °C (1100 °F)	43	1.7	51	2.0
456 h at 675 °C (1250 °F)	66	2.6	66	2.6
120 h at 760 °C (1400 °F)	66	2.6	66	2.6
360 h at 845 °C (1550 °F)	71	2.8	71	2.8
3 min at 1090 °C (2000 °F)	71	2.8	71	2.8
Aluminum-silicon alloy coating(b)				
As coated	8	0.3	28	1.1
Diffused for:				
1100 h at 480 °C (900 °F)	28	1.1	30	1.2
1100 h at 540 °C (1000 °F)	33	1.3	33	1.3
1000 h at 595 °C (1100 °F)	38	1.5	38	1.5
456 h at 675 °C (1250 °F)	56	2.2	56	2.2
24 h at 790 °C (1450 °F)	69	2.7	69	2.7
5 min at 1120 °C (2050 °F)	76	3.0	76	3.0
3 min at 1260 °C (2300 °F)	114	4.5	114	4.5

(a) Coating weight, 0.3 kg/m^2 (1 oz/ft^2). (b) Coating weight, 0.15 kg/m^2 (0.5 oz/ft^2), alloy contained 9% silicon

Table 5 Effect of hot dip aluminum coating on tensile strength of steel wire

Steel type and condition(a)	Tensile strength: Before coating, MPa	Before coating, ksi	After coating, MPa	After coating, ksi
0.10% carbon, cold drawn	896	130	551	80
0.45% carbon, cold drawn	1241	180	896	130
0.75% carbon, cold drawn	1896	275	1379	200
0.45% carbon, air patented	745	108	724	105
0.75% carbon, air patented	1131	164	1117	162

(a) 1.62-mm (0.064-in.) diam specimens

Fig. 2 Effect of silicon on the formation of iron-aluminum interfacial layer

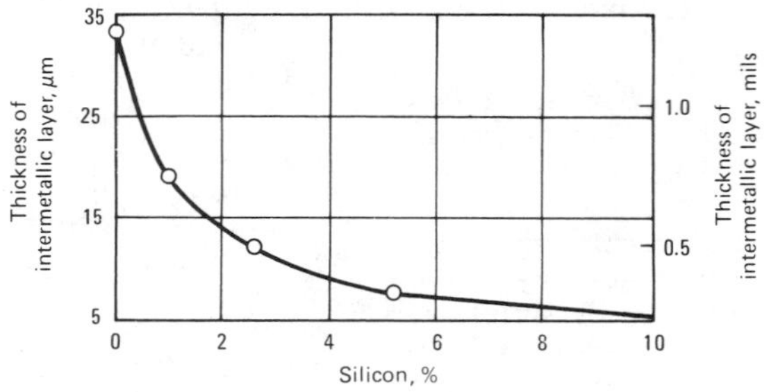

Immersed 15 s at 700 °C (1290 °F)

surface conditioning are all contained in the delivery section.

Effect of Coating on Strength and Fabricability. For different reasons, aluminum-coated steel strip and wire decreases measurably in strength. The strength of strip decreases because it is normally annealed prior to coating in order to improve its fabricability. Wire strength decreases as a result of the high temperature, above 650 °C (1200 °F), of the hot dip coating bath. An example is aluminum conductor, steel reinforced ACSR wire, which is either aluminum coated or galvanized. Its tensile strengths range from 1140 to 1450 MPa (165 to 210 ksi) depending on the type and class of coating specified. The data in Table 5 illustrate this effect for cold drawn or air patented steel wire of various carbon contents.

Steel sheet coated with aluminum-silicon alloy withstands moderate forming, drawing and spinning operations without flaking or peeling of the coating. Steel sheet coated with commercially pure aluminum withstands moderate brake and roll forming operations and can be spun or embossed, but is not suitable for drawing. Sheet with either type of coating can be given a 180° bend around a diameter equal to twice the thickness of the material; however, in any forming operation, it is advisable to allow liberal radii to prevent crazing of the coating.

Because sheared edges are susceptible to corrosion, the use of aluminum-coated sheet for fabricated assemblies may be limited for appearance considerations. Corrosion protection is not impaired, because there is no undercutting of the coating.

Coated sheet should be fabricated before the coating is diffused. Diffusion converts the coating to an iron-aluminum compound, which is very brittle.

Typical Applications. A wide variety of industrial, farm, and consumer products are fabricated from steel sheet aluminum coated at the mill. The following products require resistance to oxidation and corrosion at temperatures from 95 to 680 °C (200 to 1250 °F): combustion chamber and outer casings, agricultural crop dryers, automotive mufflers, space heaters, furnace flues, oven interiors, barbecue grills, and wrappers for water heater elements. Fabrication of these parts requires moderate drawing, forming, punching and spot welding.

Fig. 3 Automatic line for high-production aluminum coating of small parts by batch hot dip method

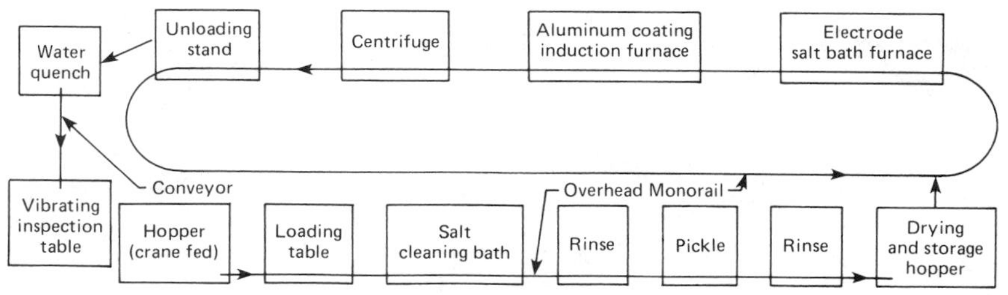

Batch Hot Dip Coating of Parts

Soil adhering to the surface of parts is removed by hot alkaline cleaning and water rinsing. Steel parts are then descaled by abrasive blasting or acid pickling, rinsed and dried. Gray or malleable iron parts are given an additional cleaning in molten salt to remove carbon smut.

For coating parts by the fused-salt fluxing method, an electrically heated fluxing furnace is used in conjunction with a coating furnace. The fluxing furnace is lined with a porous refractory brick, such as mullite, and insulated with porous silica brick surrounded by a steel shell. The brick lining must be compatible with both aluminum and halide salts of sodium and potassium, at temperatures up to 790 °C (1450 °F). The fluxing furnace contains aluminum for maintenance of the salt bath.

The coating furnace is preferably a low-frequency induction furnace with a monocast or rammed lining. To prevent oxidation, a layer of molten salt 25 to 75 mm (1 to 3 in.) thick floats above the molten aluminum. A cast iron pot can be used for the coating furnace, but the pot must be coated with a wash of iron oxide and titanium dioxide in a silicate binder to prevent attack by the molten aluminum.

Procedures and Control. An example of an automatic conveyorized line for high-production batch hot dip coating of small parts such as fasteners is shown in Fig. 3. The procedure consists of cleaning, preheating, fluxing, and coating.

In this cleaning process, a 113 kg (250 lb) load of parts is put into a basket and immersed in reducing salt at a temperature of 540 °C (1000 °F) for 20 to 25 min. The parts are dip rinsed in cold water, then pickled for 15 to 20 min in 8 to 10% H_2SO_4 at 70 °C (160 °F). The parts are dipped and sprayed with cold water, and dried in circulating hot air.

For the preheating and fluxing processes, a basket is loaded with 9 to 18 kg (20 to 40 lb) of cleaned parts. The parts are dipped in salt (40% NaCl, 40% KCl, 10% AlF_3, 10% Na_3AlF_6) for 8 to 15 min at 705 °C (1300 °F). The basket is in constant motion during the preheating, fluxing and coating processes to remove any trapped air and to flux all surfaces of the part thoroughly.

The basket of parts proceeds to the coating process where the parts move through the molten aluminum alloy at 700 °C (1290 °F) for 1 to 2½ min. Excess aluminum is removed from parts by centrifuging, air blasting or shaking. After the basket is unloaded, the parts are quenched in water at 70 to 80 °C (160 to 180 °F). Then they are air dried and inspected.

In another coating process, the pickling and fluxing operations are eliminated. Parts are abrasive grit blasted and then are immersed directly into molten aluminum. The bath surface is skimmed before parts are immersed and before they are removed. Unless it is carefully controlled, this procedure has the disadvantages of (*a*) heating parts to coating temperatures in the molten aluminum, producing a heavier alloy layer; and (*b*) dissolving additional iron into the bath.

An iron content in excess of 2% may produce rough coatings. The iron content of the coating bath can be controlled to some extent by reducing bath temperature. At lower temperatures, aluminum high in iron becomes mushy and concentrates along the sides and corners of the pot, from which it can

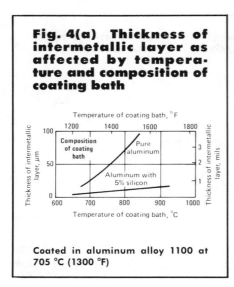

Fig. 4(a) Thickness of intermetallic layer as affected by temperature and composition of coating bath

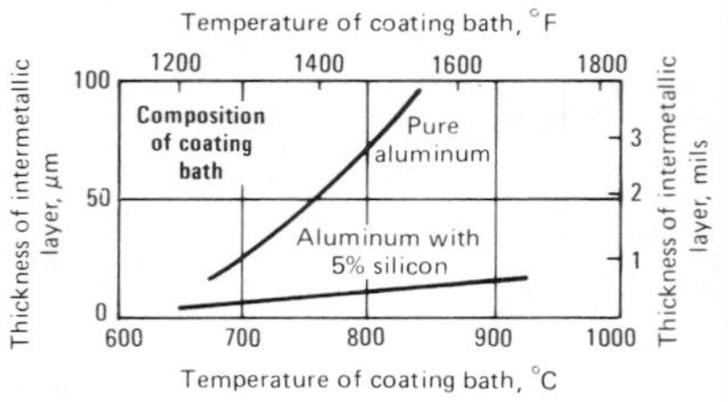

Coated in aluminum alloy 1100 at 705 °C (1300 °F)

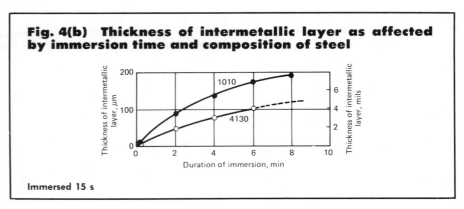

Fig. 4(b) Thickness of intermetallic layer as affected by immersion time and composition of steel

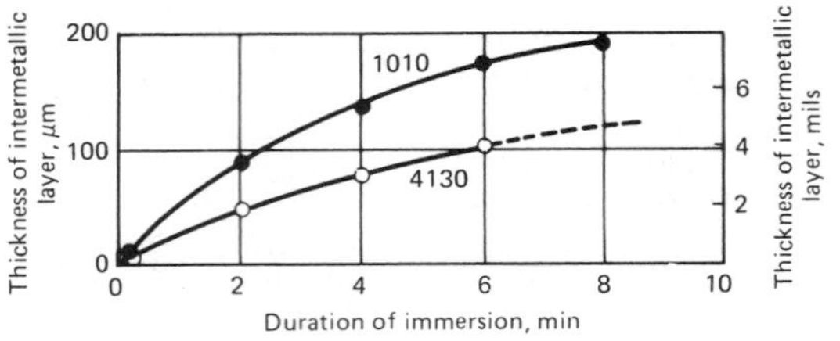

Immersed 15 s

Fig. 5 Effects of elements added to aluminum coating bath on the thickness of the interfacial layer of coatings on plain carbon steel

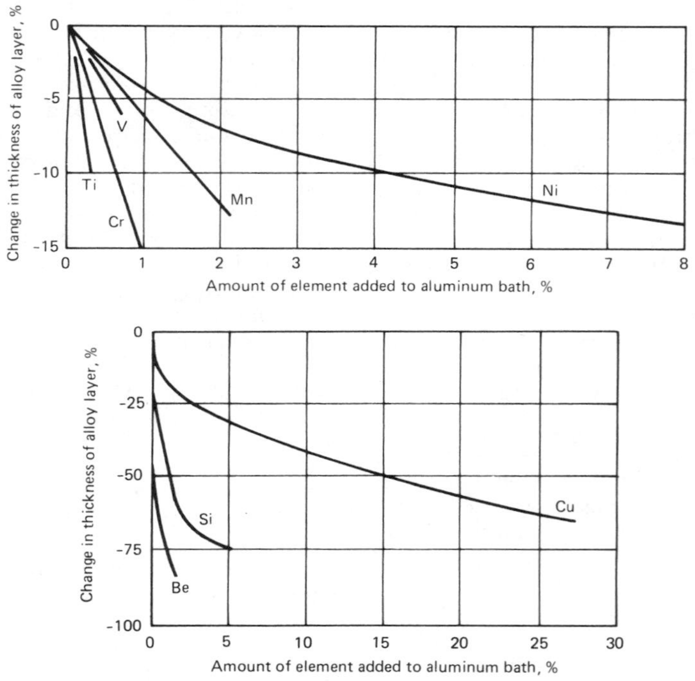

be removed rapidly. Another method of reducing the iron content is to remove part of the bath and add iron-free aluminum.

Temperature control of the molten aluminum is important. As indicated in Fig. 4(a), higher temperatures increase the thickness of the intermetallic layer and decrease the thickness of the pure aluminum overlay. A wider range of temperatures can be tolerated when the aluminum contains 5% silicon.

Immersion time must be controlled closely. The thickness of the intermetallic layer increases with immersion time (Fig. 4b). The pure aluminum overlay is not affected, because its thickness depends on the viscosity of the aluminum and on the speed with which the part is withdrawn from the bath.

Two other variables that affect coating thickness are the composition of the steel being coated and the composition of the aluminum coating bath. As indicated in Fig. 4(b), steels of higher carbon and alloy content produce thinner coatings. Figure 5 shows the effects of various elements added to the aluminum coating bath on the thickness of the intermetallic layer of coatings on plain carbon steel. These data, determined experimentally, indicate that silicon and beryllium have the greatest effect in preventing the buildup of this layer.

Limitations. Small parts with fine threads are considered impractical to coat by dip methods. Washers and cup-shaped parts that may nest offer some difficulty in batch processing. Continuous, uniform coatings are difficult to produce on items with complicated configurations involving blind holes and re-entrant angles, in which air can be entrapped.

Long slender parts of thin cross section may warp at coating temperatures 660 to 720 °C (1225 to 1325 °F). Such parts must be well supported or immersed in the vertical position to minimize warping; however, some deformation still may occur, as a result of the stress-relieving effect of the bath temperature. The use of the lower temperature aluminum-silicon alloy may help to minimize distortion.

The strength of coated parts, especially those made of cold worked material, may be reduced by the coating operation. In certain instances, this condition can be alleviated by the use of higher strength material. One producer of aluminum-coated bolts made of 4140 steel states that a tensile strength of 760 to 790 MPa (110 to 115 ksi) is the maximum practical strength that can be obtained after coating without resorting to high-alloy steels. The effect of hot dip coating with aluminum alloy 1100 at 700 to 720 °C (1300 to 1330 °F) on the hardness of several basis metals is indicated in Table 6.

The composition of the steel does not limit the batch hot dip coating process. The choice of steel depends on the strength and service requirements.

Aluminum coatings applied by hot dipping are more costly than hot dip coatings of other metals. This is because the basis metal requires more thorough cleaning, more heat is needed for coating at about 700 °C (1300 °F), and electricity must be used for heating the ceramic pot. Fuel costs may be decreased by the use of iron pots fired by gas or oil, but ceramic pots are preferred for this process.

Table 6 Hardness values of basis metals before and after hot dip coating

Coatings were aluminum alloy 1100; average thickness was 152 μm (6 mils), 50% of which was intermetallic compound

Basis metal	Coating time, min	Coating temperature °C	Coating temperature °F	Hardness, HRC Before coating	Hardness, HRC After coating
4140	5	720	1330	32-36	29-32
4340	5	720	1330	32-36	29-32
410 stainless	4	715	1320	40-42	28-30
17-22AS	6	705	1300	46-48	40-41
17-7PH (a)	5	705	1300	24-25	28-37(b)
17-7PH (a)	5	705	1300	42-44	31-34(c)
17-7PH (a)	5	705	1300	38-40	27-29(c)
Air hardening tool steels	5	715	1320	50-55	40-45
Air hardening tool steels	4	705	1300	55-60	45-50
Cast iron	5	720	1330	32	32

(a) Hardness of steel depends on prior and final heat treatments. (b) Increase in hardness due to aging at the hot dip temperature and final furnace aging at 510 °C (950 °F) for 1½ h. (c) Decrease in hardness caused by stress relieving effect of the hot dip coating operation on the precipitation-hardened steel

Coating of High-Production Parts

Automotive poppet valves, blades and nozzle vanes for gas turbine engines, and fasteners used in connection with pole line hardware and aluminum assemblies are examples of high-production parts that are coated with aluminum.

Corner Castings. In contrast to the complexity of the high-production line for fasteners and complicated small parts (Fig. 3) is the procedure for coating large corner castings used on aluminum fabricated cargo containers. These castings, each weighing about 14 kg (30 lb), are made of medium-carbon steel. The procedure for aluminum coating a production run of 12 000 of these castings is as follows:

- *Inspection*: incoming castings are inspected visually for flaws. If cavities are present, a fixture is used to check their size. Because the finished product must mate with another steel fixture, the size of part and buildup of excess aluminum are critical
- *Cleaning*: castings are oven-baked in batches at 200 °C (400 °F) for 5 h to remove gases and grease contamination, then blasted with clean steel grit, 25 to 40 mesh
- *Racking*: castings are handled with clean cotton gloves to prevent surface contamination during racking. Castings must be racked so that air pockets do not develop during immersion in molten aluminum. An air pocket causes oxidation of the ferrous metal surface and interferes with coating
- *Immersion*: each rack containing 30 pieces is dipped into molten aluminum, 99.5%, at 720 °C (1330 °F) for 12 min. The iron and silicon contents of the bath are determined weekly. When the sum of the two exceeds 3%, uncontaminated aluminum is added to reduce the concentration, or the bath is scrapped and recharged
- *Cleaning*: immediately on removal from the molten aluminum, the rack is shaken and the castings are air blasted to remove excess aluminum
- *Inspection*: each piece is inspected to ensure fit

Turbine Blades and Vanes. These parts, usually made of superalloys for use at 700 to 1100 °C (1300 to 2000 °F), can be coated satisfactorily or impregnated with aluminum by the slurry method, by pack diffusion, or by hot dipping and diffusion treatments. The earliest work used a hot dip process that recognized the importance of a thin alloy surface layer for optimum resistance to thermal shock and corrosion.

The following are basic steps in the processing of turbine blades or turbine vane segments by the hot-dip-and-diffuse method.

Surface preparation

- Immerse in molten caustic, oxidizing at 500 °C (925 °F) for 15 min
- Rinse in water at 80 °C (180 °F) for 2 to 3 min
- Wet blast with 240-mesh grit at gage pressure of 690 kPa (100 psi)

Aluminum coating

- Immerse in molten salt flux: 35 to 37% KCl, 35 to 45% NaCl, 0.5 to 12% AlF_3, 8 to 20% Na_3AlF_6, at 720 °C (1325 °F) for 3 to 5 min. This salt flux should be maintained above the molten aluminum for the purpose of activation
- Immerse in aluminum bath at 700 °C (1300 °F) for 3 to 10 s
- Rinse in molten flux, then air blast or centrifuge and wash in water
- Dip in 25% HNO_3 solution at 21 °C (70 °F) to brighten and clean the surface
- Leach in 10 to 12% HCl solution at 65 to 75 °C (150 to 170 °F) when necessary to remove aluminum
- Wash in water at 100 °C (212 °F) and dry

Heat treating for diffusion

- Load in furnace at temperature
- Diffuse at 1090 to 1150 °C (2000 to 2100 °F) for 2 h at temperature in air, argon, helium or endothermic generator gas
- Cool slowly in furnace to below 760 °C (1400 °F)

Final cleaning of the diffusion heat treated blades is necessary if an accurate dye penetrant or fluorescent oil penetrant inspection is required. Wet blasting is suitable for removing the oxide prior to such inspection.

Automotive Poppet Valves. The most advanced method of coating poppet valves consists of these operations:

- Clean and degrease after finish machining and grinding
- Induction preheat surfaces to be coated
- Spray aluminum onto preheated surfaces. Spray gun should be at a fixed position from the preheating station.

Valves are rotated to obtain uniform mechanically bonded coating

- The rotating valve progresses to an induction heating station, where the coating is heated to bond it metallurgically to the substrate metal by diffusion to a depth of about 25 μm (1 mil)
- Cool in air cooling chamber
- Inspect visually and metallurgically

This method results in minor discoloration of the valve head and a slight roughening of the aluminum coating. Subsequent surface conditioning is unnecessary, and even undesirable, because the full thickness of heat-resistant surface alloy is an advantage during service. The slightly roughened surface, consisting of undiffused aluminum, becomes smooth in the first few seconds of engine operation. Most aluminum-coated valves are made by this method, which is the most economical and the most effective for developing soundness, uniformity and durability.

Operating conditions for the primary process stations of an aluminum coating line for poppet valves are given in Table 7. The arrrangement of the coil around the valve during diffusion is shown in Fig. 6.

Valves may be coated also by the hot-dip-and-diffuse method or by spraying and subsequently diffusing in a salt bath, but the cost of either of these procedures is greater than for the method described above.

Pack Diffusion Processes

Pack diffusion processes are analogous to pack carburizing and can be referred to as cementation or impregnation processes. Alloys of iron, nickel, cobalt and copper are commonly coated by these methods. Because of the high temperature of the substrate, the aluminum being deposited alloys immediately with the basis metal; thus, a pure aluminum overlay never forms. If required, a separate diffusion treatment may be used after the cementation process has been completed.

Reaction Agents and Product. Cementation aluminum coating processes are performed in a pack consisting of (*a*) aluminum, in the form of powder or a ferroalloy, (*b*) a ceramic phase, to prevent agglomeration of the metallic components, and (*c*) a volatile halide, to act as a chemical transfer medium for the aluminum. Precleaned parts are placed in a metal retort together with the pack material, the composition of which varies widely depending on the process used and the parts being coated. The aluminum content of the pack may vary from 5% to over 60%. Two reactions that employ halide salts in the pack are as follows:

- *Displacement reaction:* the pack material consists of aluminum, aluminum oxide and aluminum chloride. The following reaction sequence occurs:

$$NH_4Cl \rightarrow NH_3 + HCl$$
$$6HCl\ (vapor) + 2Al \rightarrow 2AlCl_3 + 3H_2$$
$$AlCl_3 + 2Al \rightarrow 3AlCl$$
$$2AlCl\ (vapor) + 3Fe\ (substrate) \rightarrow 2AlFe\ (alloy) + FeCl_2\ (vapor)$$
$$3FeCl_2\ (vapor) + 5Al\ (pack) \rightarrow 3AlFe\ (alloy) + 2AlCl_3\ (vapor)$$

- *Disproportionation reaction:* the pack material consists of aluminum, aluminum oxide, and aluminum iodide. The reaction sequence is:

$$AlI_3 + 2Al \rightarrow 3AlI$$
$$3AlI + 2Fe\ (substrate) \rightarrow AlI_3 + 2AlFe\ (alloy)$$

The rate of the reactions in the pack is controlled by the difference in concentration of aluminum at the surface of the source material and at the surface of the alloy being coated, the substrate. The rate is also controlled by the mobility of the packs, and the mobility of the vapor species.

Calorizing, by definition, is a term relating to all pack diffusion processes for coating metal with aluminum. It is also a trade name relating to a particular two-stage pack diffusion process, in which (*a*) the parts are coated with a high concentration of aluminum, which penetrates the material to a depth of up to 150 μm (6 mils), and (*b*) the aluminum coating is diffused into the material to a depth of 1000 μm (40 mils) to form an alloy with the basis metal.

In current practice, the process consists of packing the parts with halide salts and aluminum powder in a closed container. The container is held at 820 to 980 °C (1500 to 1800 °F) for 6 to 24 h, after which the parts are removed from the container and heated in air to diffuse the aluminum. The concentration of aluminum in the surface alloy layer is about 50 to 60% after coating and about 25% after the diffusion cycle.

This method is capable of coating tubular products up to 12 m (40 ft) long and ranging in outside diameter from 13 to 910 mm (½ to 36 in.). Parts can be packed statically in containers or loaded into rotating retorts. Because of the tumbling action and more uniform heating, parts processed in rotating retorts have a more uniform thickness of diffused alloy.

Procedures and Equipment. The following operational sequence applies to all pack diffusion processes, particularly those related to the coating of nickel-base and cobalt-base alloy parts:

- *Cleaning:* parts must be chemically clean prior to coating. Clean as follows: degrease, blast with aluminum oxide (30 to 120 mesh) at 552 to 620 kPa (80 to 90 psi), and remove dust; handle with tongs or lint-free cotton gloves
- *Packing:* place parts in retorts and pack with the aluminum powder blend so that parts do not touch each other or the retort walls. Pack carefully to prevent bridging of compound. Vibrate to complete packing, close and seal retort
- *Coating:* charge retorts into diffusion furnace employing a reducing atmosphere. Heat at 870 to 1200 °C (1600 to 2190 °F) to obtain the required coating thickness in a practical period of time. Then cool in the reducing atmosphere. Reducing atmosphere is employed as a precautionary measure in the event that the sealing of retorts is not completely effective. When retorts are effectively sealed, air atmosphere may be used. An air atmosphere may result in some decrease in the life of furnace parts and retorts
- *Finishing:* open retorts, unpack, and wet blast lightly if necessary to clean parts

Typical equipment needed for pack diffusion coating includes equipment for vapor degreasing or acid pickling and abrasive blasting, to prepare the articles for coating; powder preparation equipment, consisting of blenders, screens, and measuring devices; furnace and retort equipment usually made of Hastelloy X; and a wet blasting apparatus to aid in cleaning the coated articles. Proper ventilating and dust collecting equipment must be used to prevent dust explosions of aluminum powders in the air. In addition, laboratory equipment is required for controlling the composition of the pack material and coating quality.

Ferrous articles are coated in retorts as large as 12 by 3 by 1.2 m (40 by 10

Fig. 6 Induction work coil for fusing sprayed aluminum coating on automotive poppet valves

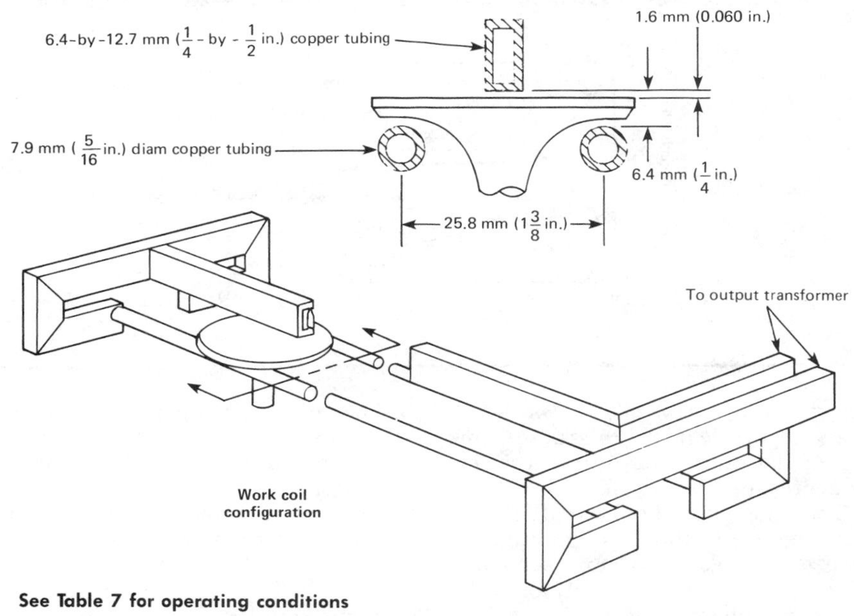

See Table 7 for operating conditions

Table 7 Operating conditions for coating automotive poppet valves

For a production rate of 3000 to 4000 valves per hour

Degreasing unit	
Conveyor speed	0.06 m/s (11 ft/min)
Temperature of liquid solvent	82 °C (180 °F)
Boiling temperature of solvent	85 °C (185 °F)
Control temperature of vapor	66 °C (150 °F)
Distilled liquid temperature	85 °C (185 °F)
Steam pressure	55 kPa (8 psi)
Spray coating unit	
Motor generator, 10 kc	150 kW, 800 V
Preheat spindle speed	540 rev/min
Coating spindle speed	650 rev/min
Diffusion spindle speed	540 rev/min
Conveyor speed	0.052 m/s (10.2 ft/min)
Length of preheat coil	216 mm (8½ in.)
Length of diffusion coil	978 mm (38½ in.)
Distilled water tank	1211 L (320 gal)
Acetylene pressure at manifold	100 kPa (14.5 psi)
Oxygen line pressure	138-145 kP (20-21 psi)
Control panel readings	
Acetylene pressure	96 kPa (14 psi)
Oxygen pressure	138 kPa (20 psi)
Flowmeter readings	
Oxygen	16-18
Acetylene	16-18
Air pressure for coating gun	448 kPa (65 psi)(a)
Air pressure for wire-feed drive	241 kPa (35 psi)(a)
Size of 1100 aluminum wire	15 gage
Wire-feed rate	0.05 m/s (10 ft/min)
Nozzle-to-work distance	50 mm (2 in.)

(a) Air pressure during operation

by 4 ft); nickel-base and cobalt-base alloys may not be coated in retorts larger than 0.3 m^2 (3 ft^2), if close control of the coating thickness is desired. Gas- or oil- fired rotating retorts are sometimes used to obtain uniform coatings on tubes.

Control of Coating. With pack diffusion methods, it is possible to obtain a maximum of about 60% aluminum at the surface, but the processes are controlled to limit the aluminum concentration to less than 25% on iron-base alloys and to less than 12% on nickel-base and cobalt-base alloys. Sometimes, a diffusion treatment subsequent to the cementation process is used to decrease the surface concentration of aluminum and to increase the depth of diffusion.

Coatings from 25 to 1000 μm (1 to 40 mils) thick can be obtained by pack diffusion. The thicker coatings are commonly applied to steel and copper. The thinner coatings are usually applied to nickel-base and cobalt-base alloys.

Slurry Processes

Slurry or powder-paint methods of coating metal parts and assemblies with aluminum are widely used for high-temperature processing equipment used in the chemical and petroleum industries, and for aircraft parts operating in the environment of combustion gases of reciprocating and gas turbine engines.

Configuration and size of the workpiece do not limit the application of these processes, provided the workpiece can be cleaned by pickling or sand blasting and the furnace is large enough to contain it. Materials with cross sections of less than 0.25 mm (0.010 in.), however, are embrittled and distorted because of the hardness of the coating and the difference in expansion between the coating and substrate.

Types of Processes. The three basic types of slurry processes for aluminum coating differ with respect to the mechanism by which aluminum is transferred from the applied and dried slurry or bisque to the substrate.

In the Type I process, aluminum in the bisque reacts with and diffuses into the substrate. The substrate may be any series 300 stainless steel not less than 0.25 mm (0.010 in.) thick; typical thickness of the diffusion layer is 50 to 100 μm (2 to 4 mils).

In the Type II process, the aluminum in the bisque melts and flows over the substrate, producing complete coverage during a low-temperature firing cycle. After the removal of the required flux, a second high-temperature firing cycle may be used for complete diffusion of the coating. Some applications do not require fully diffused coatings. The substrate may be alloys based on iron, nickel, or cobalt. Typical thickness of the diffusion layer is 25 to 75 μm (1 to 3 mils) in the cobalt-base and nickel-base alloys.

In the Type III process, vapor-phase reaction is combined with solid-state diffusion to produce fully diffused, thin layers of low aluminum content. A non-melting alloy of aluminum and a gaseous halide carrier are fired at a high temperature above 1040 °C (1900 °F). The aluminum alloy acts as a spot diffusion source, as in the Type I process, and also as the source of aluminum to sustain the following gas-phase reactions:

AlXn (halide) + Fe, Ni, or Co (substrate) → FeAl alloy (coating) + FeXm (secondary halide)

FeXm (secondary halide) + Aluminum alloy (Al source) → AlXn (Al halide) + Fe, dissolved in aluminum alloy (by-product)

The substrate in the Type III process may be nickel-base or cobalt-base alloys. Thickness of the diffusion layer can be controlled to 6.3 to 19 μm (0.25 to 0.75 mil).

Processes of Types II and III are used for coating high-temperature austenitic stainless steels and superalloys for which thin precision coatings are required to minimize the reduction in mechanical properties of the substrate.

Figure 7 compares the room-temperature tensile strengths of coated and uncoated type 321 stainless steel, L-605 cobalt-base alloy and Inconel 600 after exposure at high temperatures for extended periods of time. In another high-temperature test, coated and uncoated coupons of Inconel 713 showed no significant difference in creep.

Fig. 7 Effect of extended high-temperature exposure on the room-temperature tensile strength of coated and uncoated L-605 Inconel 600 and type 321 stainless steel

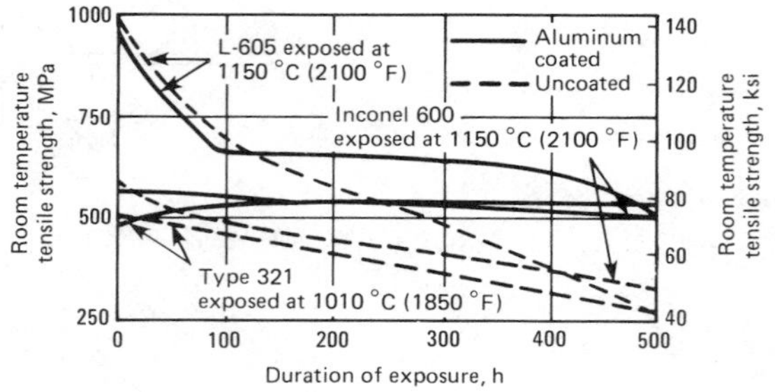

Composition of slurry for the Type I process may be either of two types as shown in Table 8. A corrosion inhibitor is added to the class A slurry to prevent reaction of water with the finely divided aluminum. The inert material is optional, but is usually used to prevent agglomeration of aluminum and pitting of substrate.

Slurries used in the Type II process differ primarily with respect to the composition of the aluminum used. If the product is to be placed in service after low-temperature fusion, an aluminum alloy containing 5 to 12% silicon is used to inhibit diffusion into the substrate. Pure aluminum powder is used when fully diffused coatings are required. A typical composition of a slurry for the Type II process uses an organic solvent for the vehicle. The aluminum source is atomized pure aluminum or 5 to 12% silicon alloy. The flux is a type of halide used in welding or brazing aluminum. A flow-control additive and a binder are required, and a suspending agent is optional.

The materials are blended together in a ball mill to a spray or dip consistency. Flow-control additives, usually metallic, are required to prevent flowing of the aluminum during the low-temperature firing cycle.

The slurry for the Type III process consists of an alloy of aluminum and iron or nickel, an organic solvent, a binder, a suspending agent, and a halide vapor-phase carrier.

Procedures and Equipment. The application of slurry coatings consists of the following operations:

Type I process

- Degrease by vapor or alkaline methods, or by heating to 540 °C (1000 °F)
- Blast with sand or garnet (grit size, 35 to 60), or acid pickle in a solution containing 1 to 3% HF and 10 to 20% HNO_3
- Spray or dip to obtain a bisque 150 to 300 μm (6 to 12 mils) thick
- Dry at 120 °C (250 °F) max on a moving disk
- Fire at 1040 °C (1900 °F) for 1 to 3 h in an electric or gas-fired furnace; atmosphere composition is not critical
- Cool parts
- Remove loose bisque by lightly blasting with 35- to 60-grit garnet

Type II process

- Degrease
- Acid pickle. Composition of solution depends on alloy. Sand blast with 35- to 60-grit abrasive
- Spray or dip to obtain a bisque 50 to 300 μm (2 to 12 mils) thick. Thickness depends on alloy
- Dry in recirculating-air oven at 120 °C (250 °F) max
- Place parts in retort and purge with nitrogen or argon
- Fire at 700 to 760 °C (1300 to 1400 °F) for 30 min
- Cool and remove parts from retort
- Remove flux with hot water. Scour with a stiff-bristle brush as required, and dry
- Place parts in an electric or gas-fired furnace (air atmosphere) and heat at 930 to 1200 °C (1700 to 2200 °F) for 1 to 3 h. Time-temperature cycle depends on alloy
- Cool parts
- Wet blast with corundum (200 to 325 grit), or wire brush, to remove loose oxide

Type III process

- Degrease
- Blast with 35- to 60-grit sand or garnet, or acid pickle. Composition of solution depends on alloy
- Spray to obtain a bisque thickness of 100 to 150 μm (4 to 6 mils)
- Dry in recirculating-air oven at 120 °C (250 °F)
- Place parts in retort and purge with argon or other inert gas
- Fire at 1040 to 1200 °C (1900 to 2200 °F) for 1 to 3 h, depending on alloy
- Cool parts
- Remove loose bisque by wet blasting with 200- to 325-grit corundum, or by wire brushing

The minimum equipment required for preparing the basis metal for application of slurry, for coating, and for heat treating the coated product is the following:

- Ball mill to prepare the slip
- Sand blasting, pickling, paint spraying and wet blasting equipment
- Dip and storage tanks
- Electric or gas-fired furnace, preferably one capable of operating at 650 °C (2200 °F) and controllable to ±14 °C (±25 °F)

Table 8 Slurry compositions for type I process

Slurry	Vehicle	Aluminum source	Corrosion inhibitor	Binder	Suspending agent	Inert material
Class A	Water	Pure aluminum(a)	Required	Clay or gum	Clay or gum	Alumina or other ceramic oxide
Class B	Organic solvent	Pure aluminum(a)	Not required	Soluble resin or thickener	Organic thickener	Alumina or other ceramic oxide

(a) Atomized or flaked

- Special retorts and gas metering equipment for processes of Types II and III. Retorts should be made of Inconel

Effect of Variables on Coating Depth. For slurry-type coatings, temperature and substrate composition significantly affect the thickness of the diffusion layer, as indicated in Fig. 8. For chromium stainless steels, the thickness of the diffusion layer increases with temperature (Fig. 8a). Nickel and cobalt cause a decrease in the diffusion rate, thereby limiting the thickness of the diffusion layer (Fig. 8b). The stability of the intermetallic compounds NiAl, Ni_3Al, CoAl and Co_3Al, and the low solid solubility of aluminum in cobalt and nickel, are responsible for the low diffusion rate.

Time at temperature has less effect on the thickness of the diffusion layer, as indicated in Fig. 8(c). Diffusion time has a more pronounced effect on depth of diffusion in alloys of high iron content than in stainless steels.

Fig. 8 Effect of diffusion temperature, diffusion time, and composition of basis metal on the thickness of the diffusion layer of slurry-type aluminum coatings

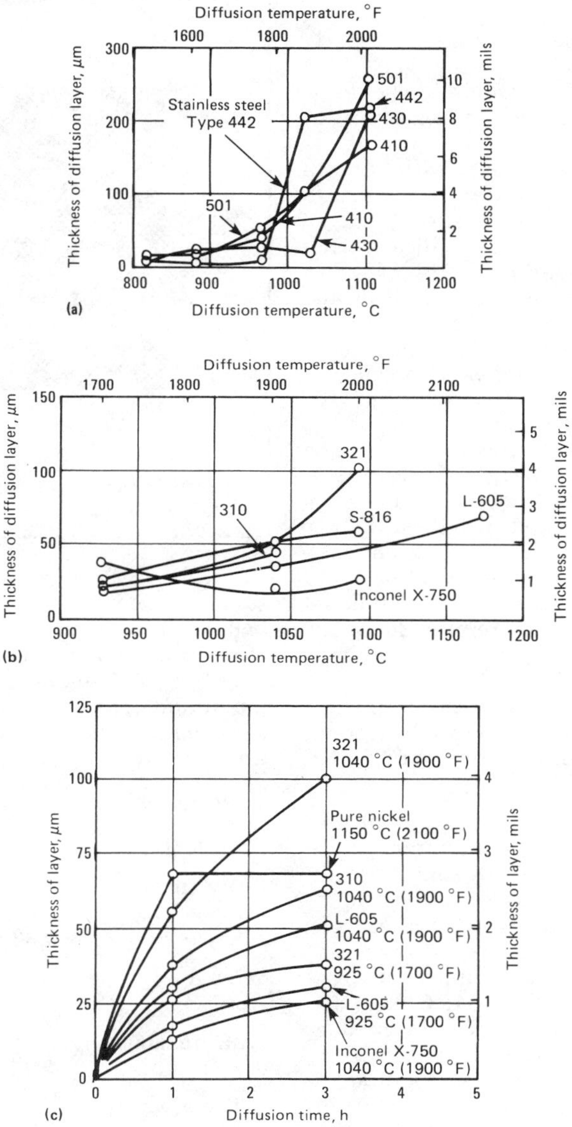

Coated with 50 μm (2 mils) aluminum and diffused for 1 h in air at temperature. (a) Diffusion rate of aluminum in chromium stainless steels is greater than in (b) materials containing nickel or cobalt. (c) Diffusion time has less effect than basis metal composition on thickness of the interfacial layer

Spray Coating

Sprayed aluminum coatings are obtained by melting aluminum wire with an oxyacetylene or oxypropane flame, or by electrical resistance, and then atomizing the droplets and propelling them, by use of an air blast, against the surface to be coated. Upon impact with the surface, the droplets deform to flattened or flake-like particles. The coating has a theoretical density of 85 to 90% and an oxide content of 0.5 to 3.0%. Spraying to a coating thickness of 230 μm (9 mils) eliminates continuous pores.

Heating to above 480 °C (900 °F) metallurgically bonds the coating to the steel. If all of the aluminum is not completely diffused during the heat treatment, the coating will consist of an outer layer of aluminum superimposed on the alloy layer. Figure 9 compares the effects of time and temperature of two different diffusion treatments on the structure, thickness and hardness of aluminum coatings sprayed on plain carbon steel.

Cleaning. Good bond strength between the aluminum and the substrate can be obtained only when the surface to be sprayed is chemically clean and moisture free. For some articles, surface preparation may consist simply of degreasing a precision-machined surface. Usually, after degreasing the surface is ground, machined, knurled or, most frequently, blasted with sand, grit or shot. When blasting is used, the following conditions and precautions are recommended:

- Grit must be maintained properly and free of oil contamination
- Steel grit, 25 to 45 mesh, should be controlled closely and free of rust or dust
- Sand, 20 to 40 mesh, with a minimum of 40% retained on a 30-mesh screen, should be washed, hard, of angular shape, and of silica, garnet or similar material, free of softer minerals that will break down and possibly contaminate the surface
- Blasted surfaces should not be touched by hand, should be kept clean and dry until sprayed, and should be sprayed within 4 h after blasting
- Compressed air used for blasting and spraying must be free of moisture or oil. An adequate number of moisture and oil traps must be in the air line to ensure this condition, and one trap must be close to the operation
- The source of compressed air must maintain a pressure of at least 550 kPa (80 psi) when the blasting nozzle is wide open
- Higher air pressure and larger grit are needed for stainless or heat treated steels with a hardness of 250 HB or more

Preheating should be accomplished as rapidly as possible to minimize distortion and oxidation. A slightly reducing flame may be used, but heating by induction is preferred when the size of the part permits its use.

Typical preheating temperatures are from 150 to 320 °C (300 to 600 °F). Occluded moisture can be prevented by preheating the surface to 35 °C (200 °F). Preheating at 260 to 370 °C (500 to 700 °F) results in a more favorable stress distribution in the finished coating, and improves the bond strength by promoting greater particle deformation and increased oxide cementation.

Spraying. The gun mechanism generally employs oxyacetylene gas as the source of heat to melt the aluminum and relies on a closely controlled source of compressed air to atomize and provide kinetic energy for impact of the aluminum droplets. Typical spray gun conditions (with air turbine wire drive) are as follows:

Fig. 9 Microstructure, thickness and hardness of sprayed aluminum coatings on plain carbon steel

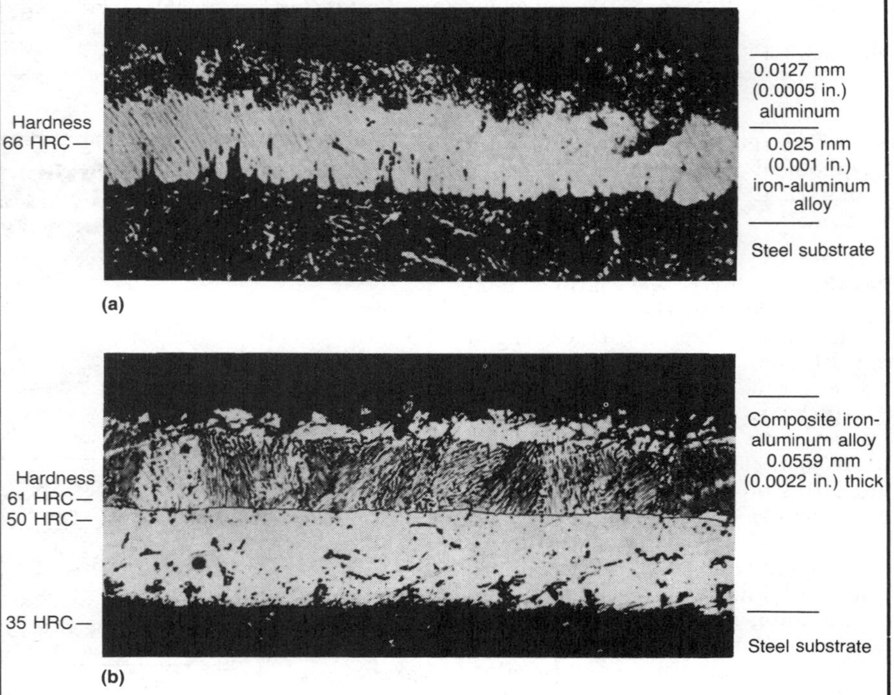

Etched in nital. 100X. (a) Diffused at 780 °C (1450 °F) for 15 s. (b) Diffused at 1090 °C (2000 °F) for 45 s

Tank pressure
Oxygen .... 275 kPa (40 psi)
Acetylene .... 140 kPa (20 psi)

Flow rate
Oxygen .... 0.62 m^3/h (22 ft^3/h)
Acetylene .... 0.62 m^3/h (22 ft^3/h)

Wire feed
Air .... 170 kPa (25 psi)

Gun nozzle
Air .... 450 kPa (65 psi)

The spray of metal droplets is cooled by the air blast during transfer to the receiving surface. The aluminum strikes the steel surface as a spray of semisolid, partially oxidized particles that flatten on impact. The coating is composed of numerous overlapping leaves that are mechanically keyed to each other and to the microscopically irregular steel surface. Little or no interfacial metallic bond is formed, because of the rapid cooling of the atomized aluminum particles. Within the coating, there is oxide cementation between particles.

The thickness of sprayed aluminum coatings ranges from 75 to 380 μm (3 to 15 mils). For resistance to atmospheric corrosion in rural and industrial locations, the thickness should be from 130 to 300 μm (5 to 12 mils). For coastal locations or immersion in salt or fresh waters, a thickness of 230 to 380 μm (9 to 15 mils) is recommended. All surfaces submerged in water must be completely covered with aluminum; if basis metal is exposed, rapid galvanic corrosion occurs and the aluminum will soon disappear.

Sealing. Spray coatings, regardless of thickness, are porous. Unless the pores are sealed, damaging oxidation and corrosion will occur. Oxidation and

Table 9 Effects of diffusion treatment and aluminum composition on sprayed coatings

Composition of aluminum(a)	Diffusion time, s	Diffusion temperature(b) °C	°F	Intermetallic layer Total thickness µm	mils	Inner thickness µm	mils	Hardness, HRC Outer	Inner	Color of coating
No silicon	15	790	1450	28	1.1	...	...	66	...	Dull gray
5% silicon	15	790	1450	18	0.7	...	...	67	...	Dull gray
12% silicon	15	790	1450	13	0.5	...	...	60	...	Dull gray
No silicon	30	980	1800	38	1.5	2.5	0.1	69	44	Dark gray
5% silicon	30	980	1800	30	1.2	3.8	0.15	66	44	Dark blue-gray
12% silicon	30	980	1800	20	0.8	2.5	0.1	63	43	Dark blue-purple
No silicon	45	1090	2000	46	1.8	20	0.8	65	41	Dark gray
5% silicon	45	1090	2000	41	1.6	18	0.7	64	39	Dark blue-gray
12% silicon	45	1090	2000	38	1.5	18	0.7	62	42	Dark blue-purple

(a) Composition of wire used for spray coating. (b) Induction heating used for diffusion

corrosion occur in storage at room temperature, in the diffusion furnace before the conversion temperature has been reached, or in service if the coating has not been previously diffused. The pores are sealed immediately after spraying by means of burnishing, rolling, swaging, wire brushing, shot peening, or by effective sealers, such as waxes, vinyl copolymers, vinyl alkyds, bitumastic sealers, and silicon-base materials.

Depending on their composition, sealing will effectively protect as-sprayed coatings for extended periods at temperatures up to about 650 °C (1200 °F). The bitumastic sealers are the least expensive, but they burn off at temperatures well below 480 °C (900 °F), leaving a slight film of flake aluminum on the sprayed aluminum surface. These sealers provide useful protection up to about 430 °C (800 °F). Silicone-base sealers withstand temperatures up to about 650 °C (1200 °F). In applications where the service temperature does not exceed 650 °C (1200 °F), it is economical to use these sealers rather than to diffuse the sprayed coating at 730 °C (1350 °F) or above.

For elevated-temperature applications in which oxidation resistance is the primary requirement, coatings 130 to 200 µm (5 to 8 mils) thick, sealed with aluminum silicone sealer, are recommended to 480 °C (900 °F). For operating temperatures up to 820 °C (1500 °F), coatings should be 100 to 200 µm (4 to 8 mils) thick, diffused at temperatures up to 820 °C (1500 °F), but preferably not below the service temperature, and sealed with bitumastic aluminum sealer.

Diffusion. Sprayed materials may be subsequently heated at 730 to 1090 °C (1350 to 2000 °F) to provide a dense, uniform coating metallurgically bonded to the basis metal. At these temperatures, the coating melts, coalesces, homogenizes, and begins to diffuse into the basis metal to form an interfacial alloy layer. As indicated by the data in Table 9, the thickness and hardness of this layer vary with time and temperature of diffusion and with the composition of the sprayed coating.

Table 10 Diffusion treatments for valve alloys

Alloy	Diffusion temperature °C	°F	Time to reach temperature, s	Time held at temperature, s
Silcrome 1, 8440, 8645, and 1041	790	1450	5-10	5-15
Silcrome XCR and XB	845	1550	15-20	10-20
21-4N, Silcrome 142, and Silcrome 10	870	1600	15-20	30-40

Recommended procedures for diffusing aluminum sprayed coatings on several valve alloys are shown in Table 10. The thickness of the sprayed coating and of the alloy layer, after diffusion of valves made of two of the alloys, is given in Table 11. The coatings on the 21-4N valves were diffused by induction heating at 900 °C (1650 °F). Those on the XCR valves were diffused by heating under a salt flux at 730 °C (1350 °F) for 3½ min. For additional information on sprayed aluminum coatings see the article on thermal spraying in this Volume.

Cladding by Rolling

Steel sheet can be clad with aluminum by either hot or cold rolling. German producers developed clad sheet by using 0.06% carbon steel on which aluminum sheet containing about 0.7% silicon was hot rolled at about 200 °C (400 °F) to a 40% reduction. The aluminum and steel surfaces were roughened by scratch brushing before rolling. No intermediate anneals were used, and the final anneal was controlled at 530 to 550 °C (995 to 1020 °F).

In recent years, stainless steel clad aluminum has become widely used for automotive trim, bumpers and cookware. Aluminum strip is bonded to stainless steel strip by cold rolling in a single pass. The strips are preconditioned by abrading and 50% of the reduction is exhibited in the aluminum leaving the stainless steel free from work hardening. The bonded strip is subsequently heated or sintered in a protective atmosphere to strengthen the bond. The stainless steel surface is then buffed to a high luster for automotive trim and bumper applications.

One commercial mill product used for anodes in vacuum tubes contains a copper core in a sandwich of aluminum-clad steel. Aluminum strip or foil is bonded to steel strip (0.08% carbon max) by a cold rolling reduction of 60% in a single pass followed by heating or sintering in a protective atmosphere to improve the bond. The aluminum alloy contains 1.0 to 1.5% silicon. After sintering, the bonded strip is cleaned by scrubbing the aluminum side and abrading the steel side. The steel side is then bonded to copper by a 60% reduction in the same manner as the aluminum was bonded to the steel. After being rolled to the final gage, the coils of the composite material are soak annealed at 540 °C (1000 °F).

A silicon content of 1.2% in the aluminum alloy prevents the formation of a brittle iron-aluminum alloy interfacial layer during annealing at up to 570 °C (1050 °F) for 16 h. A silicon content of 0.7% is not effective in preventing the formation of the brittle layer.

When clean surfaces of aluminum and steel sheet are pressed together at a pressure of 70 MPa (10 ksi) and at a temperature near 620 °C (1150 °F) for several minutes, bonding occurs by diffusion and with the formation of an interfacial layer. This type of bonding can be accomplished without the formation of an interfacial layer by rolling at a temperature as low as 230 °C (450 °F), if the aluminum and steel are previously silver plated. Such a composite has been used in the manufacture of steel-backed aluminum bearings.

Wire is clad commercially, without formation of the undesirable aluminum-iron compound at the interface, by the application of compacted aluminum powder to high-strength steel rod. The rod is cleaned, aluminum powder is applied, and the composite rod is subjected to heat and pressure to form a solid aluminum coating on the steel core. The bimetallic rod is cold drawn to finished wire sizes. A typical aluminum thickness is 25% of the cross-sectional area of composite wire. Aluminum powder is used also for cladding steel hardware products by a peening action in a ball mill.

Table 11 Thickness of coating and intermetallic layer after diffusion

Thickness range	Total coating thickness µm	Total coating thickness mils	Intermetallic layer thickness µm	Intermetallic layer thickness mils
Silcrome XCR valve alloy(a)				
High	20	0.79	9.1	0.36
Low	12	0.49	4.1	0.16
Average	19	0.63	6.1	0.24
21-4N valve alloy(b)				
High	36	1.40	16	0.64
Low	8.1	0.32	5.8	0.23
Average	25	1.00	10	0.41

(a) Sprayed coating diffused under salt flux at 730 °C (1350 °F) for 3½ min. (b) Sprayed coating diffused by induction heating at 900 °C (1650 °F)

Vacuum Deposition

Vacuum deposition or vacuum metallizing is widely used for depositing very thin coatings of aluminum on substrates for decorative or optical applications. Because the coatings are very thin, less than 2.6 µm (0.1 mils), this process is not often used to apply aluminum coatings on steel for functional reasons. For similar reasons, chemical vapor deposition, sometimes referred to as gas plating, also has limited applications on steel. For a discussion of these processes, see the articles on vacuum coating and vapor deposition coatings in this Volume.

Ion Vapor Deposition

Ion vapor deposition of aluminum coatings on steel components for corrosion protection is a recent development. The process is being used by the aircraft industry for the protection of high-strength steel components, such as landing gear, and as a fastener coating to reduce the galvanic effects of steel fasteners inserted in an aluminum structure.

Ion vapor deposition has several advantages over the other aluminum coating processes. The aluminum deposit is very adherent and passes the bend-to-break test. The deposit does not affect the mechanical properties of the substrate and can be applied with precise thickness control on a wide range of shapes. Both coating and coating process are nontoxic and do not contribute to the pollution of our environment. Because of these advantages, ion vapor deposition aluminum can be used in a wide range of applications and is particularly effective as a replacement for cadmium coatings.

Ion Vapor Deposition Process. The ion vapor deposition (IVD) of aluminum is similar to the conventional vacuum metallizing process used for applying aluminum to details for decorative purposes in that aluminum is vaporized and allowed to condense on the surface of the parts being coated. However, IVD aluminum coatings are thicker, denser, and more adherent, basic requirements for a finish for corrosion protection.

The denser, more adherent coating from ion vapor deposition is obtained by applying a high negative potential between the part being coated and the source of evaporation. An inert gas is introduced into the vacuum system and becomes ionized. The positively charged ions are attracted to the negatively charged part surface, and their bombardment of the surface performs final cleaning. The clean surfaces result in better adhesion.

Following glow discharge cleaning, aluminum is evaporated. As it passes through a glow discharge region, a portion of it becomes ionized and is accelerated toward parts. This results in denser coatings and also contributes to better adhesion. Ionization also provides better throwing power and allows complex shapes to be uniformly plated.

Table 12 Minimum corrosion resistance requirements for MIL-C-83488

Class	Minimum thickness µm	Minimum thickness mils	Corrosion resistance, h
1	25	1	672
2	13	0.5	504
3	8	0.3	336

The thicker coatings are obtained by feeding aluminum wire continuously into the evaporation source, rather than the processes used in decorative vacuum metallizing, which flash evaporate aluminum staples that are placed on resistance-heated filaments.

Equipment used in ion vapor deposition consists of (*a*) a vacuum chamber, (*b*) pumping system, (*c*) evaporation source, and (*d*) a high voltage power supply. Two types of equipment are used in this process. One, used for detailed parts, is designed for plating individually racked parts. The second type of equipment, called a barrel coater, is for small parts, such as fasteners, which are normally barrel electroplated.

Coating Performance. For aircraft applications, three classes and two types of coatings of IVD aluminum are used. Type I refers to as coated, and Type II, as coated with a supplementary chromate treatment. Type II treatment provides additional protection against corrosion and forms a good base for paint adhesion.

Classes reflect coating thickness. Class I coatings are used for high temperature and exterior applications where severe corrosion environments are encountered. Class 2 is recommended for interior parts where less severe environments are encountered. Class 3 is used only when close tolerances are required, such as fine threaded parts. Table 12 illustrates the relative corrosion resistance of the three coating classes.

Electroplating

Aluminum is electroplated on steel by the use of anhydrous electrolytes composed of fused mixtures of aluminum chloride and alkali chlorides. Pure chemicals and high-purity aluminum anodes are required, because all metallic impurities below aluminum in the electromotive series interfere with the production of a smooth, bright, adherent coating. This type of coating is free from any interfacial alloy layer.

A typical fused-salt electrolyte contains 80% aluminum chloride and 20% sodium chloride, and is operated at 180 °C (350 °F) and at a current density of 1.5 A/dm^2 (15 A/ft^2). Higher current densities may be used if the bath is agitated. Containers for the fused salt are constructed of aluminum to avoid contamination of the bath. Before aluminum deposition begins, dissolved moisture and metallic impurities must be removed from the bath by electrolysis or by treatment with scrap aluminum. Because fuming is severe, ventilation is required. Aluminum chloride must be added frequently to maintain proper concentration. Aluminum can be electroplated also from several anhydrous organic electrolytes. Another plating bath consists of aluminum chloride and lithium hydride (or lithium aluminum hydride) in an ethyl ether solvent. Preparation, use, and storage of such a bath must be within a gas tight enclosure containing a dry inert atmosphere. Work to be plated proceeds through an air lock chamber, which can be purged with dry nitrogen gas. After plating, work is removed in a similar fashion. A deposition rate of 25 μm (1 mil) to 50 μm (2 mil) per hour is used.

Electrophoresis

One process for coating aluminum is based on the electrophoretic deposition of aluminum powder from a bath of spheroidal aluminum particles in alcohol. This powder is consolidated into a solid layer by rolling (at least 7% reduction) and is then sintered and bonded to the steel by heating slowly at 500 °C (930 °F).

REFERENCE

1. Yong-Wu Kim and Richard A. Nickola, A Heat Resistant Aluminized Steel for High Temperature Applications, SAE Technical Paper, Series 800316, SAE Congress, 1980

55% Aluminum-Zinc Alloy Coated Steel Sheet and Wire

By H. E. Townsend
Supervisor
Corrosion and Surface Research
Bethlehem Steel Corp.
and
T. W. Fisher
Supervisor
Metallic Coatings Development
Bethlehem Steel Corp.

ALUMINUM-ZINC alloy coatings of steel sheet and wire are applied on continuous hot dip coating lines with in-line gas cleaning and heat treating of the steel substrate. With a nominal composition of 55% aluminum, 43.4% zinc, and 1.6% silicon, the coating provides the durability and high temperature resistance of aluminum coatings with the sacrificial protection characteristics of zinc coatings. Silicon is added to the coating bath to control growth of an intermetallic layer. Steel sheet and wire products coated with 55% aluminum-zinc alloy are especially useful in applications requiring superior atmospheric corrosion resistance, cut edge protection, and/or heat oxidation resistance.

Microstructure

The 55% aluminum-zinc coating has a cored dendritic microstructure (see Fig. 1), composed of about 80 vol % aluminum-rich dendrite arms and about 20 vol % zinc-rich interdendritic material. A few silicon particles are also present in the overlay. The coating is bonded to the steel substrate by a thin layer of an intermetallic phase comprised of aluminum, iron, zinc, and silicon, which gives diffraction lines similar to that of $Al_{13}Fe_4$.

This multiphase dendritic microstructure leads to a mechanism of corrosion in which the zinc-rich interdendritic regions corrode preferentially, providing sacrificial protection to the steel substrate and the other components of the coating. The fine aluminum-rich dendrites form a labyrinth that acts to mechanically trap and retain zinc corrosion products. Because of this unique corrosion mechanism, the 55% aluminum-zinc coating provides at least two to four times the durability of an equal thickness zinc coating and retains the edge protection characteristics of zinc coatings in industrial and rural environments.

Coating Process

A 55% aluminum-zinc coating line for either sheet or wire is very similar to continuous annealed hot dip galvanizing lines. (See the article on hot

Fig. 1 Aluminum-zinc coated sheet

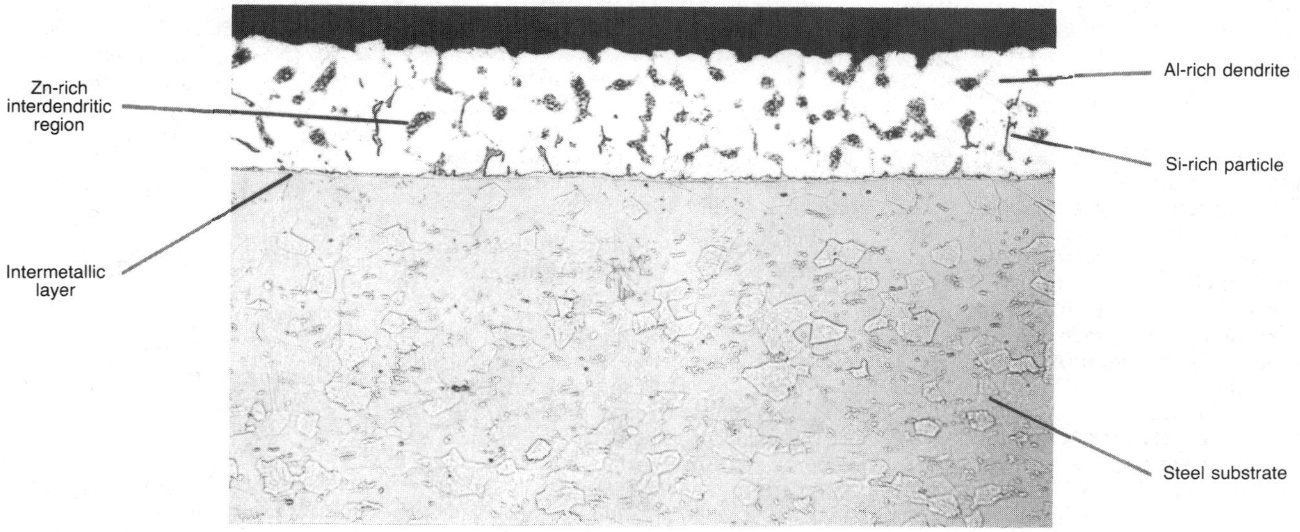

Amy nital etch, 500X

dip galvanizing in this Volume.) The coating process involves the following functions:

- Cleaning
- Heat treating
- Coating
- Accelerated cooling

Cleaning the steel surface before coating is important to avoid surface imperfections on the coated product. Because residual surface contaminants differ significantly for sheet and wire-feed stock, removal techniques vary. Residual rolling oils are the principal surface contaminants on full-hard, cold rolled sheet. Some commercial lines use a hot, electrolytic alkaline cleaning treatment to remove the surface contaminants, whereas in other direct-fire type lines, the residual lubricants are volatilized during the initial heating pass.

Residual rod coating and drawing lubricants are the principal surface contaminants on full-hard, cold drawn wire. These compounds are usually metal phosphates and stearates. The cleaning system designed to remove surface residuals combines sequential alkaline and acid electrolytes. For more information on cleaning steel surfaces, see the article on selection of cleaning process in this Volume.

Heat Treating. A critical phase of the 55% aluminum-zinc coating process is the dual purpose heat treating of the steel before coating. The elevated temperature serves to (*a*) promote gas-metal reactions between the protective atmosphere and the steel surface, and (*b*) achieve metallurgical conditioning of the steel in the form of recrystallization or stress relieving.

Coating. After heat treating, the hot steel enters the coating bath which is nominally 55% aluminum, 1.6% silicon, and the remainder zinc. The bath temperature is maintained at approximately 600 °C (1110 °F). An induction-heated, ceramic-lined pot is used to contain the spelter. A similar but smaller pot is used as a premelt container to ensure that the melt has been homogenized before being added to the coating pot. Coating weight is controlled to the following limits by wiping the excess coating with gas or jet wipers:

Application	Average coating weight, min g/m^2	$mg/in.^2$
Conventional sheet products	150	97
Culvert sheet	215	140
Class 30 wire	98	63
Class 45 wire	144	93

Cooling. To obtain maximum corrosion resistance, the coated steel is forced-air cooled from the coating temperature to about 370 °C (700 °F), at a minimum rate of 11 °C (20 °F) per second, after exiting the coating bath.

Aluminum-Zinc Coated Sheet

Coated sheet can be roll formed, lock seamed, bent, stretched, welded, and painted to produce a variety of products. Applications include metal building roofing and siding, patio roofs, awnings, rainware, ductwork, metal studs, automotive exhaust components, heat shields, school bus flooring, appliance parts, factory-built fireplaces, water collection tanks, and highway culverts.

Aluminum-zinc coated sheet is produced with physical properties similar to continuous-annealed, hot dipped galvanized ASTM specification A446, grades A, B, C, D, and E. Low-carbon, low-residual steel is the base for A446, grades A, B, and C. In some cases, aluminum-killed and/or titanium-stabilized steels are used for these grades. During processing, the cold rolled sheet is recrystallized and the resulting aluminum-zinc coated product meets the equivalent mechanical property requirements of 230 MPa (33 ksi) minimum yield strength for grade A; 255 MPa (37 ksi) minimum yield strength for grade B; and 275 MPa (40 ksi) minimum yield strength for grade C.

Classification A446 grade D is produced by making appropriate carbon and/or phosphorus additions to steel. During processing, the steel is recrystallized and the resulting aluminum-zinc coated product meets the equivalent grade D mechanical property requirements of 345 MPa (50 ksi) minimum yield strength.

Low-carbon, low-residual steel also is the steel base for A446 grade E. During processing, the heat treating cycle is controlled to effect a stress-relieved, essentially unrecrystallized aluminum-zinc coated product meeting the equivalent grade E mechanical property requirements of 550 MPa (80 ksi) minimum yield strength.

Special Applications. For applications requiring better formability than that provided by as-coated, commercial quality 55% aluminum-zinc coated sheet, the ductility of substrate and coating can be improved by low-temperature heat treatment. Heating the coated sheet to temperatures of 345 to 385 °C (655 to 725 °F) for 16 h, followed by cooling at less than 0.6 °C (1 °F) per minute, provides modest improvements in formability without impairing the corrosion resistance of coated sheet.

In uses involving prolonged exposure to temperatures above 315 °C (600 °F), embrittlement of steel substrate because of intergranular diffusion of zinc from the coating is prevented by use of steel containing greater than 0.04 wt % phosphorus. Phosphorus in the base steel segregates preemptively to the ferrite grain boundary, preventing intergranular diffusion of zinc.

Aluminum-Zinc Coated Wire

Coated wire can be woven or stranded to produce a broad range of products. Aluminum-zinc coated wire can be used in high-tensile, fabric, and chain link fencing, wire cloth, conveyor belts, well screens, heating element supports, guy and messenger strand, static strand, aluminum conductor steel reinforced core wire and strand, and highway guard cable.

Aluminum-zinc coated wire is produced with mechanical properties similar to equivalent zinc-coated wire. The steel base being coated varies from the low-carbon grades, AISI 1008, up through the high-carbon grades, AISI 1080, depending on application. Aluminum-zinc coated wire is produced to the following ASTM specifications:

- ASTM A783: chain link fence fabric
- ASTM A784: highway guard rail
- ASTM A785: guy and messenger strand

Low-carbon steel serves as the base for grade A783. During processing, the steel is recrystallized with the resulting product meeting mechanical property requirements of 520 MPa (75 ksi) minimum yield strength. A broad range of strength levels is delineated by specifications A784 and A785. Depending on the specific application, a medium- to high-carbon steel is the base for these products. During processing, the heat treatment is controlled to effect a stress relieved, essentially unrecrystallized coated product meeting product mechanical property requirements. See the appropriate ASTM specification for mechanical property requirements.

SELECTED REFERENCES

- L. Allegra and J.C. Zoccola, Aluminum-Zinc Coated Sheet Steel for Automotive Exhaust Applications, *Materials Performance,* Vol 28 (No. 5), 1979, p 18-21
- L. Allegra and H.E. Townsend, Undervehicle Corrosion Resistance of 55% Al-Zn Coated Steel Sheet, *Metal Progress,* Vol 119 (No. 5), 1981, p 33-35
- L. Allegra, J.B. Horton, and H.E. Townsend, Zinc-Alloy Coated Ferrous Product Resistant to Embrittlement, U.S. Patent 4,264,684, 28 April 1981
- L. Allegra, H.E. Townsend, and A.R. Borzillo, Method of Producing an Aluminum-Zinc Alloy Coated Ferrous Product to Improve Corrosion Resistance, U.S. Patent 4,287,009, 1 Sept 1981
- D.J. Blickwede, 55% Al-Zn-Alloy-Coated Sheet Steel, *Tetsu-To-Hagane,* Vol 66 (No. 7), 1980, p 821-834
- A.R. Borzillo and J.B. Horton, Ferrous Metal Article Coated with an Aluminum Zinc Alloy, U.S. Patent 3,343,930, 26 Sept 1967
- A.R. Borzillo and J.B. Horton, Method of Forming Improved Zinc-Aluminum Coating on Ferrous Surfaces, U.S. Patent 3,393,089, 16 July 1968
- L.B. Caldwell and L.P. Pellatiro, Method of Treating Ferrous Strand for Coating with Aluminum-Zinc Alloys, U.S. Patent 4,053,663, 11 Oct 1977
- H.J. Cleary and J.B. Horton, Corrosion Resistant Aluminum-Zinc Coating and Method of Making, U.S. Patent 3,782,909, 1 Jan 1974
- G.J. Harvey, Structure and Corrosion Resistance of Zincalume Coatings, *BHP Technical Bulletin,* Vol 25 (No. 2), 1981, p 63-67
- T.E. Torok, P.W. Shin, and A.R. Borzillo, Method of Improving the Ductility of the Coating of an Aluminum-Zinc Alloy Coated Ferrous Product, U.S. Patent 4,287,008, 1 Sept 1981
- H.E. Townsend and J.C. Zoccola, Chromate Passivation Protection of Zn and Al-Zn-Coated Steel Sheet Against Wet-Storage Stain, *Journal of the Electrochemical Society,* Vol 125 (No. 8), 1978, p 1290-1292
- H.E. Townsend and J.C. Zoccola, Atmospheric Corrosion Resistance of 55% Al-Zn Coated Sheet Steel: 13-Year Test Results, *Materials Performance,* Vol 18 (No. 10), 1979, p 13-20
- J.C. Zoccola, H.E. Townsend, A.R. Borzillo, and J.B. Horton, Atmospheric Corrosion Behavior of Aluminum-Zinc Alloy-Coated Steel, *Atmospheric Factors Affecting the Corrosion of Engineering Materials, ASTM STP 646,* S.K. Coburn, Ed., American Society for Testing and Materials, 1978, p 165-184

Hot Dip Tin Coating of Steel and Cast Iron*

By Daniel J. Maykuth
Manager
Tin Research Institute, Inc.

HOT DIP tinning is accomplished by applying a thin coating of molten tin to a metallic object. Such coatings are applied to iron and steel to (*a*) provide a nontoxic, protective, or decorative coating for food-handling, packaging, or dairy equipment; (*b*) facilitate the soldering of a variety of components used in electronic and electrical equipment; and (*c*) assist in bonding another metal to the basis metal, as in the tinning of cast iron bearing shells prior to lining with lead-base or tin-base alloy. The usual thickness range of hot dip tin coatings is 3.8 to 18 μm (0.15 to 0.7 mils).

Steels for Hot Dip Tinning

Low-carbon steels, containing less than 0.2% carbon, are intrinsically well suited for hot dip tinning. Medium- to high-carbon steels (0.3 to 1.0% carbon) may require greater care in pickling, but ordinarily there is little difficulty in processing them. The higher alloy steels, especially those with high chromium content (such as 18-8 stainless), are difficult to tin satisfactorily, and special procedures may have to be used in the relatively few instances where tinning is required.

Within the ranges normally encountered, the alloying elements and metalloids commonly present in low-carbon steels, such as phosphorus, sulfur, silicon, manganese, and copper, do not affect tinning quality significantly. The small amounts of nickel and chromium, which may be present because of the use of alloy scrap in steel melting, have no significant effect either, but increased pickling may be necessary. This may also be true of medium-carbon steels.

Occasionally, a batch of low-carbon steel sheet or strip is more difficult to tin than others. The tin coatings on it de-wet despite careful cleaning and pickling by recommended methods. This condition is characteristic of steels that have been exposed to lubricants during cold working and have been subsequently annealed without removal of surface lubricants. The lubricants polymerize or oxidize to form a nonreactive skin that resists removal by normal cleaning methods.

To ensure good tinning characteristics, steel mill products should be purchased to a specification that includes suitability for hot dip tin coating. If optimum material characteristics are not available, tests should be performed on small sample pieces to determine tinning quality.

Cast Irons for Hot Dip Tinning

Cast irons with chemical compositions within the following ranges are generally suitable for hot dip tinning:

Element	Composition, %
Total carbon	3.2-3.5
Silicon	1.7-2.7
Manganese	0.5-0.8
Sulfur	0.05-0.12
Phosphorus	Up to 1.3

Annealed irons containing subsurface oxides of silicon may be less suited to tinning than as-cast irons.

The presence of graphite in the structure of cast irons necessitates the use

*The principal source of information for this article was *Practical Hot-Tinning* by C.J. Thwaites, published by the International Tin Research Institute, 1981

of surface preparation methods that differ from those used for steel. Iron castings may also have a hard casting skin that is high in silica and must be removed prior to tin coating. Avoid the difficulty of tinning over a graphite-contaminated surface by electroplating a coating of iron, nickel, or copper on the surface. Common methods for overcoming the surface effects that make tinning of cast iron difficult are discussed in the sections on cleaning and fluxing later in this article. After cleaning and fluxing, cast iron may be tinned in the same manner as steel.

Cleaning Before Hot Dip Tinning

Iron and steel parts must be free of surface contaminants such as oil, grease, drawing lubricants, and mill scale before fluxing and immersion in molten tin. Inadequate or improper surface preparation is a frequent cause of defects such as poor adhesion in hot dip tin coatings.

Degreasing. Oil, grease, soap, and other lubricants used in machining, drawing, and forming can be removed by one or more of several methods, including vapor degreasing, solvent cleaning, alkaline cleaning, and emulsion cleaning. Some details of various processes are given below.

The organic solvents generally used are trichlorethylene or, occasionally, perchlorethylene. Degreasing is effected by placing the articles in the hot liquid, or in the vapor, or in both in turn. In the liquid process, the solvent is continuously circulated and purified by distillation. In the vapor process, the cold articles are cleaned by the condensed vapor of the boiling solvent condensing on them. Commercial equipment for solvent degreasing is available.

Solvent procedures are ideal for removing mineral oils, greases, and many types of vegetable oils, but are less effective with certain types of drawing compounds and spinning soaps which leave behind solid constituents. In a liquor-vapor plant, a short cooling period should follow immersion in the boiling solvent to ensure that adequate condensation of solvent occurs on the articles which should be so racked or supported that the condensed solvent runs off them completely and does not collect in recesses.

Wet articles must not be loaded into a solvent degreaser, because corrosion of the units may cause decomposition of the solvent. Smoking and naked flames must be prohibited close to trichlorethylene degreasing plants to avoid the risk of phosgene poisoning.

Alkaline detergents act by penetrating the contaminant layer and removing it by emulsification, saponification, or flocculation. Appropriate commercial salts are available in powder or crystal form, which are dissolved in water at a concentration of from 1 to 10%, according to the instructions. Alkaline cleaning solutions may be made more effective by employing electrolysis at the same time, but this is not commonly practiced.

Alkaline cleaners usually contain sodium hydroxide with other constituents added to render the grease soluble. Proprietary cleaners are recommended because they are formulated to deal with specific types of contaminants and basis metal. The temperature of alkaline solutions should not be below 85 °C (185 °F).

A simple 5% sodium hydroxide solution at 80 to 90 °C (185 to 195 °F) often is adequate for the anodic electrolytic cleaning of steel, but a specifically formulated proprietary solution is preferred. The articles are suspended to form one electrode, and a plain steel tank containing the alkaline solution is the other electrode. A 6 to 12 V direct current is applied between busbar and tank to obtain a current density on the work of between 2 and 5 A/dm^2 (20 to 50 A/ft^2) of surface. Generally, the use of electrolysis allows more latitude in the temperature of the alkaline bath, but the hotter the solution, the more efficient the cleaning. Tenacious oil films causing slow pickling and tinning difficulty are sometimes best removed by an electrolytic cleaning treatment.

Articles should be rinsed immediately as they are removed from alkali to avoid deposition of salts or de-emulsification of the grease on the surface. The best procedure consists of a hot water rinse, followed by a final rinse in a cold water tank provided with running intake and overflow. The rinse water should not contain acids or salts likely to bring about the breakdown of emulsions adhering to articles removed for rinsing. For this reason, avoid using the same rinse tank for degreasing and pickling.

Ultrasonic cleaning involves the use of high frequency mechanical vibrations from a transducer device in a cleaning solution to achieve a higher degree of cleanliness and at a much faster rate than is possible by conventional methods. Frequencies of 20 to 40 kHz are frequently used. The scouring effect penetrates into crevices, holes, and complex contours that are inaccessible to mechanical action such as brushing.

The temperature of the cleaning liquid is important as it affects density and volatility. Aqueous solutions are used at about 50 °C (120 °F) instead of the more conventional 80 to 95 °C (175 to 205 °F), and emulsifying agents may be present which are effective at lower temperatures. Organic-solvent, ultrasonic, degreasing plants usually have a special compartment containing cool solvent in which the transducers are fitted. Articles are initially degreased in boiling solvent before passing to the ultrasonic treatment chamber. A final degreasing in vapor alone may be used.

Pickling of steel, usually done in aqueous solutions of hydrochloric or sulfuric acid, can be used to remove mill scale and rust before hot tinning. Hydrochloric acid pickles efficiently at room temperature, and in most applications, no provision is made for heating it. Dilutions range from one part acid in two parts water to three parts acid in one part water. Immersion times range from 10 to 60 min. When pickling for hot dip tinning, prolong immersion in the pickling bath for a few minutes beyond that required for the total removal of visible scale and rust. This gives the steel a light etch, which will promote wetting of the basis metal during the tinning process.

Depending on the condition of the surface being treated, the composition of the aqueous sulfuric acid pickling solution varies from about 4 to 12% sulfuric acid. The recommended operating temperature range for these solutions is 80 to 85 °C (175 to 185 °F). Removing light scale or rust normally requires an immersion time of 1/2 to 2 min; even heavy scale should not require an immersion time of more than 15 min. In sulfuric acid pickling baths, inhibitors are commonly added to concentrate the attack on the scale and reduce acid consumption, metal loss, spray, and risk of hydrogen absorption by the steel.

Difficult steels, such as those having surface layers formed by decomposing lubricants, often require oxidizing conditions and a hydrochloric or sulfuric acid immersion sufficient to remove surface oxides followed by 1 to 3 min in 10 to 25 vol % nitric acid to achieve a tinnable surface.

Table 1 Composition of flux solutions used in hot dip tinning

Solution	Zinc chloride kg	Zinc chloride lb	Ammonium chloride kg	Ammonium chloride lb	Sodium chloride kg	Sodium chloride lb	Hydrochloric acid(a) cm^3	Hydrochloric acid(a) oz	Water L	Water gal
A	11	25	0.7	1.5	...	...	296-591	10-20	38	10
B	11	24	1.4	3.0	3	6	296-591	10-20	45	12

(a) Commercial grade, 28%

Table 2 Composition of flux covers for hot dip tinning baths

Mixture	Melting point °C	Melting point °F	Constituent	Composition, wt %
A	260	500	Zinc chloride	78
			Sodium chloride	22
B	260	500	Zinc chloride	73
			Sodium chloride	18
			Ammonium chloride	9

Cast iron should not be acid pickled for tinning, because a heavy carbon smut derived from graphite forms over the whole surface and prevents tinning.

Details of operating procedures and equipment required for pickling in hydrochloric and sulfuric acid, as well as the use of inhibitors to minimize acid attack, are given in the article on pickling of iron and steel in this Volume. Removal of scale from iron and steel in molten salt is discussed in the article on salt bath descaling in this Volume.

Abrasive blast cleaning must be done on castings and all assemblies of cast iron, cast steel, and malleable iron with wrought steel prior to hot dip tinning. Iron castings to be tinned by the direct chloride method or by wiping should be blasted with fine (70-mesh) angular chilled iron grit. Blasting should be thorough with all surfaces to be tinned treated for 30 to 60 s. For a description of the equipment and techniques used in abrasive blasting, see the article on abrasive blast cleaning in this Volume.

Fluxing

Fluxing facilitates and speeds the reaction of molten tin with iron or steel, promoting the formation of a continuous thin layer of tin-iron or other intermetallic phases on which the liquid tin coating can spread in an even, smooth, continuous film. In hot dip tinning, fluxes may be used in three different ways: (*a*) as aqueous solutions in which the work is briefly dipped before it is immersed in the molten tin; (*b*) as a molten fused layer or cover on the top of the molten tin bath; and (*c*) as a solution, paste, or admixture to tin powder that is applied to the surface of the work prior to wipe tinning. The material compositions of two aqueous flux solutions are given in Table 1.

Flux Covers. A cover of molten flux should be maintained on the surface of the first tin dipping bath. The flux cover, which must be molten at the operating temperature of the bath, is normally regenerated by absorbing aqueous flux solutions on the surface of the incoming work. The addition of water as a fine spray to the surface of the flux may still be necessary at intervals to rejuvenate it. Compositions of two effective flux covers are given in Table 2. The salt components of the various flux solutions given in Table 1 form suitable flux covers.

Fluxing Procedure. After being pickled, rinsed, and dried of excess rinse water, the steel (or in the case of iron castings, the dry, shot-blasted workpiece) is immersed in an aqueous flux solution. Workpieces may require some movement in the flux bath to remove all air locks and to ensure that all surfaces are fully wetted by the flux. Work should be immersed in the flux only long enough to ensure complete coverage; nothing is gained by prolonged immersion. When the workpiece is withdrawn from the bath, permit the excess flux to drain briefly. The workpiece is then ready for immediate immersion in the tin bath.

Single-Pot Tinning

Single-pot tinning is used to provide a preliminary coating for bonding or soldering, or to coat workpieces that do not require the highest quality finish. The process involves a single immersion of fluxed workpieces in a molten tin bath heated to 280 to 325 °C (535 to 615 °F). The average operating temperature of the bath is maintained at about 300 °C (575 °F). When the workpiece is withdrawn from the bath, the surface of the work may have spots of flux which must be removed by suitable washing.

Place enough flux on the surface of the molten tin bath to provide a molten flux layer that covers about two thirds of the surface area. The use of partial rather than total flux coverage helps eliminate excessive pickup of flux when the work is withdrawn.

After being coated with aqueous flux, the workpieces are picked up by pretinned tongs, hooks, or perforated baskets and lowered gently into the molten tin, passing through the portion of the bath that is covered with flux. The optimum immersion rate normally ranges from 13 to 51 mm/s (½ to 2 in./s). Heavy sections are immersed at a slower rate than light sections. The work need only be immersed long enough to reach the temperature of the bath.

In order to minimize flux pickup and to shed any particles of iron-tin compound that have accumulated in the bath, workpieces are withdrawn from the bath with a clean, rapid movement through an area not covered with flux. Flux may be moved to one side with a paddle.

The operations that follow withdrawal from the tin bath vary considerably. The work may be shaken or swung by hand to remove excess tin. To remove tears or droplets of tin that collect at lower or outer edges of the workpiece, allow the teared edge barely to touch the surface of the clean metal in the pot. This slight contact pulls the droplets away from the work by surface tension. Centrifuging or spinning of jigged work is also used to remove excess molten tin.

After the excess tin has been removed, allow the work to cool in air, then remove flux residues by rinsing in cold water acidified with about 1 vol % hydrochloric acid or 5 to 10 vol % citric acid. Follow with a water rinse. Quenching the work in water or kerosine is also possible.

Two-Pot Tinning

In two-pot tinning, the work is dipped first in a tin bath with a flux cover and then in a tin bath covered with oil or molten grease. The process develops high-quality thick coatings and offers the following advantages over single-pot tinning:

- Because no reactive tinning involving iron takes place during the second dip, metal in the second tinning pot can be kept lower in iron content. This results in final coatings that are low in contaminants
- Flux residues entrained on the work from the first dip are absorbed into the oil cover of the second dip
- The finished work retains a thin film of oil that protects the coating during shipment and storage

In two-pot tinning, follow the procedures described for single-pot tinning up to the withdrawal of the work from the first tinning pot. At this point, immerse the work immediately in the second tinning pot. The bath surface should be about two thirds covered with palm oil, tallow, or a synthetic mineral oil-base tinning oil. Hold the work in the second tinning pot only long enough for it to cool to the bath temperature; then withdraw the workpiece using a rapid movement similar to that employed in the first tinning operation.

The working temperature of the second tinning bath should be high enough to provide sufficient time for the work to be manipulated after it is withdrawn from the pot, but not so high as to cause the oil cover to deteriorate rapidly and undue yellow staining of the coatings from oxidation to occur. The temperature of the second tinning pot varies according to the type of work being handled, ranging from 235 to 270 °C (455 to 520 °F). A temperature of 250 °C (480 °F) is commonly employed.

After being withdrawn from the tin bath, work tinned by the two-pot process can be cooled in air (after listing off, if required), quenched in oil, or centrifuged. Listing off is a process in which heavily tinned articles are suspended over the bath so that the molten tin drains to the lowest corner. The droplets which form are then touched to the molten bath surface and removed by surface tension forces.

Wipe Tinning

Wipe tinning usually is employed only when the workpiece is too large to be dipped, or when a specific area, such as the inside of a container, needs to be coated. The method may be applied to wrought steel products or iron castings, but it generally produces a less uniformly thick and therefore less satisfactory coating. It is also more difficult to perform than dip-tinning procedures.

After the usual surface preparation, coat the surfaces to be tinned with a concentrated flux solution. Mount the work so that it may be heated by gas burners or other means to a temperature of at least 275 °C (525 °F). Place a small pellet of tin at a convenient location on the surface to serve as an indicator. As soon as the pellet melts, apply additional tin to the surface by pouring it from a ladle or wiping the surface with a stick of tin. Additional flux solution can be added, if required. Work the tin and spread it over the surface using a wire brush or scraper until the entire surface is coated continuously. Finally, wipe off the excess metal with a soft, absorbent wad; allow the article to cool, and remove the flux residue by washing. Solid ammonium chloride powder may be used as the flux in wipe tinning, and tin may be applied as a coating of tinning paint.

Ingot Tin Quality

The three recognized grades of ingot tin used in hot dip tin coatings and their tin contents are (*a*) standard tin containing over 99% tin; (*b*) refined tin with over 99.8% tin; and (*c*) highest purity tin with over 99.9% tin. The impurity most frequently present in common tin is lead. As little as 0.2 to 0.5% lead, in addition to small amounts of other impurities, may cause the coating to be spangled. The effect is similar to that observed in galvanizing and results in a decrease in luster. For types of work that do not require the brightest finish, this condition is acceptable. However, in the majority of coating applications, the use of grade A tin (99.8% tin) is virtually mandatory. In those applications where tin is applied exclusively to facilitate soldering, impurities contained in standard tin (99% tin) are not considered harmful.

Equipment and Materials for Fluxing and Tinning

A hot dip tinning installation requires equipment for cleaning, pickling, fluxing, and tin dipping. An efficient ventilating system is also necessary, because fumes are produced at nearly every stage of the tinning process. The equipment requirements for cleaning and pickling operations, such as vapor degreasing, solvent cleaning, alkaline cleaning, emulsion cleaning, abrasive blast cleaning, and acid pickling, are dealt with in the articles on pickling of iron and steel and abrasive blast cleaning in this Volume.

Fluxes. Aqueous fluxes used in hot dip tin coatings are usually slightly acid solutions of zinc chloride to which other halides, such as the chlorides of ammonium, sodium, and tin may be added. These solutions do not require heating and may be conveniently contained in nonmetallic or rubber-lined steel tanks. Acid-resistant vitreous enamel is equally suitable, but is more likely to be damaged by heavy workpieces. Do not use unlined steel tanks for holding flux, because the steel is attacked by the chlorides present in the flux. Iron chlorides form, which when transferred to the tin bath result in contaminating iron-tin compounds.

Tinning tanks or pots used to contain molten tin may be made of low-carbon steel or cast iron. Most of the tinning tanks in current use are constructed from welded low-carbon steel. Some corrosion does occur at the flux line, but it is not severe enough to affect tank life seriously or to cause appreciable contamination of the bath.

The size and shape of the tank are determined by production requirements and the size and shape of the articles to be processed. However, it is essential that tank size have a load clearance of at least 152 mm (6 in.) from the bottom and sides of the tank. This clearance provides a reasonable space for manipulating work loads.

Tinning tanks may be heated by gas, electricity, or oil; the choice depends primarily on economic considerations. A temperature controller and indicator are essential. Iron contamination in iron or steel tanks can be minimized by uniform heating, best provided by arranging the burners or heaters in such a way that the maximum amount of heat is directed against the sides of the tin pot, rather than against the bottom.

Selective Tin Coating

Several materials, including ordinary whitewash, may be used as stop-offs to permit selective tin coating. A very effective stop-off is a slurry containing 0.5 kg (1 lb) magnesium oxide,

1 L (1 qt) sodium silicate (water glass), and 0.5 L (1 pint) water. This slurry may be thinned with more water to aid in application.

Procedure. Work to be selectively tin coated is first cleaned, pickled, and rinsed in the normal manner, and then dried quickly. The surfaces to be tin coated are next painted with flux solution, and surfaces to be protected are painted with stop-off. The work is then dried in an oven. A fresh coat of flux is applied to the surfaces to be tinned, and the work is hot dip tinned by any of the several processes. Following tinning, the stop-off coating is removed by wire brushing.

Precautions

Certain hazardous chemicals are used in the tinning processes discussed in this article. Exercise great care when handling these materials, especially the acids. Acids should be diluted by adding the acid slowly to water; water should not be added to acid.

Wet, flux-coated articles should not be lowered directly into a bath of molten tin, or dangerous spitting will occur. Instead, immerse articles carefully through a flux cover. The rate of immersion should be slow enough to prevent dangerous spluttering or explosions caused by trapped water or moisture being added to the molten tin.

Fumes are produced at nearly every stage of the tinning process, and an efficient ventilating system is therefore essential.

Babbitting

By Edward A. Barsditis
Senior Manufacturing Engineer,
Metals Joining
Westinghouse Electric Corp.

BABBITTING is a process by which softer metals (basically a tin/lead combination) are bonded chemically or mechanically to a shell or stiffener, which supports the weight and torsion of a rotating, oscillating, or sliding shaft. The babbitt, being softer and having excellent antifrictional qualities, prevents galling and/or scoring of the shaft over longer period of usage. The application of babbitt to steam turbine generator bearings, automotive connecting rods and journal bearings, ship drive shaft bearings, steam shovel bearings, and myriad other uses, is common. Additionally, babbitt has been used for jewelry, shot, filler metals, and various other applications.

Babbitting of bearing shells can be accomplished by several methods: (*a*) static babbitting or hand casting, (*b*) centrifugal babbitting, or (*c*) metal spray babbitting. Whichever method is used, bond quality is an important factor, particularly when heat transfer through the babbitt into the shell is expected to contribute to the extended life of the bearing. Bonding the babbitt to the shell can be done either mechanically or chemically.

Mechanical Bonding

Mechanical bonding of a babbitt to a shell is a simple fastening process. Mechanical methods used in babbitting shops include anchor groove dovetails and drilled and tapped holes into which molten babbitt flows, locking the babbitt in place. Copper or brass screws inserted into threaded holes also help hold the babbitt to the shell. In some instances, brass or copper bars are tinned and recessed into the bore and bolted or screwed into the bore of the casting. In most cases, these methods are used on cast iron when caustic treatment is not available. The heads of the screws must be recessed into the babbitt to prevent the screw head from becoming exposed when the babbitt is machined to a final bore size. Figure 1 shows one example of babbitting practice.

Fig. 1 Screw recessed in babbitt

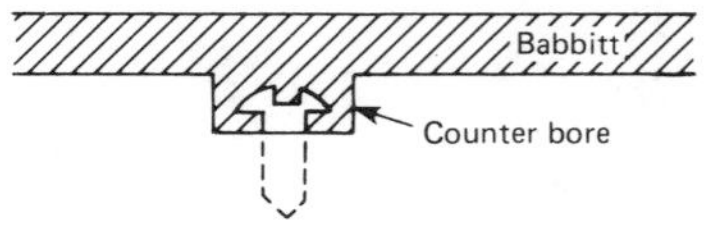

A method used exclusively on cast iron when caustic treatment is not available

Cleaning and Degreasing. Before the babbitt has been bonded to the shell, clean the shell by degreasing. Use a steam jenny with an industrial strength detergent for the cleaning. After cleaning, rinse in clean hot water or clean steam to remove all detergent. Vapor degreasing is also an excellent way to clean the shell. Although shot, sand, or grit blasting do not remove oil or grease deposits, these processes remove light scale or oxides. For further details on tinning various materials, see the article on hot dip tin coating of steel and cast iron in this Volume.

Static Babbitting

Static babbitting requires a babbitting mandrel to form the babbitt in the bearing shell. Mandrels can be designed for use in a vertical or horizontal position. Plates are placed at each end when the mandrel is designed for use in horizontal position. Heat the mandrel to 300 °C ± 10 °C (570 °F ± 18 °F). If a riser is used at the top of the vertical mandrel to hold extra babbitt, the riser must be heated to the same temperature. Heat the tinned bearing shells to 325 °C ± 25 °C (615 °F ± 45 °F), preferably by submersion into a large tinning pot. If submersion is not practical, heat the shell with natural gas flame applied to the back of the shell. Do not apply heat directly to the tinned surface of the bore and disperse the flame to prevent hot spots.

When the bearing shell and mandrel are in position and at the proper temperature, the shell is clamped to the mandrel. Skim the dross from the babbitt pot, which is maintained at 475 °C ± 15 °C (885 °F ± 27 °F). Stir the babbitt to prevent segregation of the alloys. Ladle sufficient babbitt into the assembly, carefully moving the ladle around the periphery of the mandrel/shell opening to prevent hot spots from forming. If a leak occurs, quickly apply any commercially available mud to solidify the leak area. Fill the opening between shell and mandrel, including the riser, if used, to the top.

Solidification of the babbitt must begin at the base of the mandrel, continuing to the top of the bearing. Solidification must also occur at the shell, proceeding toward the mandrel. Keep the back of the mandrel hot with flame, to promote this directional cooling, until definite evidence indicates the solidification direction is being maintained.

Table 1 Rotating speeds for centrifugal babbitting bearings

Rough bore diameter, mm	Rough bore diameter, in.	Speed, rev/min	Rough bore diameter, mm	Rough bore diameter, in.	Speed, rev/min
73.2	2.88	925	197	7.75	564
76.2	3.00	907	200	7.88	559
79.2	3.12	890	203	8.00	555
82.6	3.25	871	210	8.25	547
85.9	3.38	855	216	8.50	539
88.9	3.50	840	222	8.75	531
91.9	3.62	825	229	9.00	523
95.3	3.75	810	235	9.25	516
98.6	3.88	798	241	9.50	510
102	4.00	785	248	9.75	503
105	4.12	773	254	10.00	496
108	4.25	762	260.4	10.25	490
111	4.38	752	266.7	10.50	484
114	4.50	740	273.1	10.75	478
117	4.62	732	279.4	11.00	472
121	4.75	720	285.8	11.25	467
124	4.88	710	292.1	11.50	462
127	5.00	703	298.4	11.75	457
130	5.12	694	304.8	12.00	452
133	5.25	685	317.5	12.50	444
137	5.38	677	330.2	13.00	435
140	5.50	670	342.9	13.50	428
143	5.62	662	355.6	14.00	419
146	5.75	655	368.3	14.50	412
149	5.88	647	381	15.00	405
152	6.00	640	393.7	15.50	398
155	6.12	635	406.4	16.00	392
159	6.25	628	419.1	16.50	386
162	6.38	622	431.8	17.00	380
165	6.50	616	444.5	17.50	375
168	6.62	610	457.2	18.00	370
171	6.75	605	469.9	18.50	365
175	6.88	599	482.6	19.00	360
178	7.00	594	495.3	19.50	355
181	7.12	589	508	20.00	350
184	7.25	584	520.7	20.50	346
187	7.38	579	533.4	21.00	342
191	7.50	573	546.1	21.50	338
194	7.62	569	558.8	22.00	334

If the babbitt solidifies at the mandrel before solidifying at the shell, two things happen. First, the babbitt pulls away from the shell, and then the molten babbitt draws from the shell toward the mandrel, creating excessive porosity and voids in the babbitt face.

As the babbitt cools, use a 5 mm (0.196 in.) diam steel rod, long enough to reach the bottom of the bearing to break up voids in the semiliquid babbitt. This allows the molten babbitt to fill the voids. Add more babbitt to the riser area as the babbitt shrinks.

Immediately after the babbitt is poured, apply an air or water mist to the bearing shell to hasten cooling and to help control the direction of solidification. Continue this cooling action until the bearing shell cools to 150 °C (300 °F), then remove shell from mandrel.

Babbitting of bronze bearing shells is done at slightly lower temperatures, maintaining mandrel and shell temperatures at 275 °C (525 °F). Procedures for cooling of the shell and maintaining mandrel temperatures also apply to bronze bearing shells.

Centrifugal Babbitting

Centrifugal babbitting requires a machine expressly designed and built or modified for this purpose. No mandrel is used in horizontal applications. Submersion of the bearing shell in tin alloy, 40% tin and 60% lead, is recommended for preheating because this provides closer control of temperature. Use fit plates to center the bearing in the machine.

Steel shim plates, oxidized by heating with a torch to a blue color condition, are placed between the partings to allow separation of the bearing halves after centrifugal babbitting is completed. Drill holes approximately 5 mm (0.196 in.) diam through the shim plates at the bearing bore line. These holes allow the babbitt to circulate between halves and equalize the babbitt level.

The following is an example of a centrifugal babbitting procedure:

- Select bearings and fit plates
- Assemble fit plates to machine
- Place bearing halves into tin pots and heat to 300 °C (570 °F)
- Remove bearing top and bottom half from tin pot when tinning alloy becomes molten and flows freely on surface of castings. Match and bolt together halves using required shims between partings
- Return assembled bearing to tin pot and heat
- Remove assembled bearing from tin pot, place between fit plates, and clamp into position
- Close hood and start machine
- Observe tachometer and select proper revolutions per minute as shown in Table 1
- Ladle babbitt into bearing bore
- Apply water spray as shown below:

Unbabbitted shell, kg	Unbabbitted shell, lb	Water mist, min
0.4-14	1-30	5
14-18	31-40	6
19-27	41-60	7
28-36	61-80	8
37-45.4	81-100	9
45.8-54.4	101-120	10
54.9-63.6	121-140	11
64.0-72.6	141-160	12

For best results, assembly and disassembly of bearing shell into machine must be done quickly. Start rotation and pour babbitt immediately. When cycle stops, remove babbitted bearing from machine and separate bearing halves.

Metal-Spray Babbitting

Babbitting by the metal-spray method requires special equipment consisting of an acetylene/oxygen flame spray gun using a high tin-based babbitt in wire form. The molten alloy is sprayed on a bond coating, which has been previously sprayed on the bearing shell. The buildup of the babbitt is relatively slow, and bond strengths are somewhat lower than with other methods; however, voids are eliminated and a high quality product results. Spray babbitt thicknesses to 26 mm (1.02 in.) are possible, and spray control is sufficient to require only a small excess of deposit for subsequent machining. Overspray losses are minimal. Mechanical aids to hold the babbitt to the bearing shell are not necessary and are not recommended for this process, and skill levels for operators are not as high as those required for static babbitting.

Hot Dip Lead Alloy Coating of Steel*

HOT DIPPING with lead alloys containing 2 to 25% tin requires close control of the surface preparation of the steel base, the composition of the chloride flux, the composition and temperature of the molten lead-tin alloy coating bath, the time of immersion in the bath, and regulation of the amount of coating left on the steel.

Because lead alone does not alloy with iron, tin or other elements such as antimony must be added to form on the surface of the steel an iron-tin alloy to which the lead will adhere. The iron-tin metallurgical bond provides the coating with excellent adhesion, and the lead acts as a lubricant that facilitates forming and drawing. The lead-tin alloy, which is usually called a terne coating, also provides excellent solderability and good corrosion resistance in various environments and corrosive media.

Terne Coatings

The terms "long" and "short" ternes are often used. Short ternes are very light gage sheet, similar in thickness to tinplate (below about 0.30 mm or 0.01 in.). They are rarely produced today. Long ternes, whether produced as sheet or as continuous strip, are normally produced in steel base thicknesses between 0.25 and 2 mm (0.01 to 0.08 in.). A range of coating thicknesses from 60 to over 250 g/m^2 (0.28 to 1.17 oz/ft^2) may be specified as shown in Table 1. Of the coatings listed, the most commonly used are the four lighter coating grades. Coating weight values refer to the total coating on both surfaces. Normally, it can be expected that not less than 40% of the single spot check limit will be found on either surface, when tested according to ASTM Specification A 308.

Terne-coated sheet made with commercial, drawing, or physical qualities of carbon steel is available in both coil and cut-length sheet forms. Terne-coated stainless steel sheet is also available.

Formed Steel. In hot dipping formed and fabricated articles, the steel is first cleaned of oils and greases using solvents or detergents. Then, the steel is pickled in a 6 to 12 vol % sulfuric acid solution at 71 °C (160 °F), or an inhibited 5 to 10% hydrochloric acid solution at 49 to 65 °C (120 to 150 °F). The iron content of these solutions should not exceed 5%. Next, the steel is immersed for 5 to 20 s in either a zinc-chloride or zinc-ammonium-chloride flux, used either as a molten salt or as a water-base solution prior to dipping in the molten lead alloy bath.

The lead bath is maintained in the range 325 to 390 °C (620 to 735 °F). The bath temperature and duration of immersion of the work is increased as the mass of the article being coated increases. When withdrawn from the bath, the coated item is centrifuged or shaken to remove excess coating metal and flux. The workpiece may be quenched in water to solidify the coating before contacting other articles. The thickness of the terne coating on formed articles normally varies from 5 to 15 μm (0.2 to 0.6 mil) depending on the application. The thickness may be controlled by variations in (*a*) the temperature of the bath, (*b*) the duration of immersion, and (*c*) the amount of shaking or centrifuging. Many fabricated articles lend themselves to hand dipping, either individually or in batches, because of their shape or use. Bolts, washers, nuts, plates, brackets, and other fixtures used in industry can be effectively coated.

Sheet and Strip. Substantial tonnages of flat-rolled sheet and strip are terne coated by steel producers and subsequently formed, stamped, or drawn by customers in several industries.

*Revised by D. O. Gittings, Chief Research Engineer — Coated Sheets, U. S. Steel Research

Fig. 1 Continuous long terne coating line

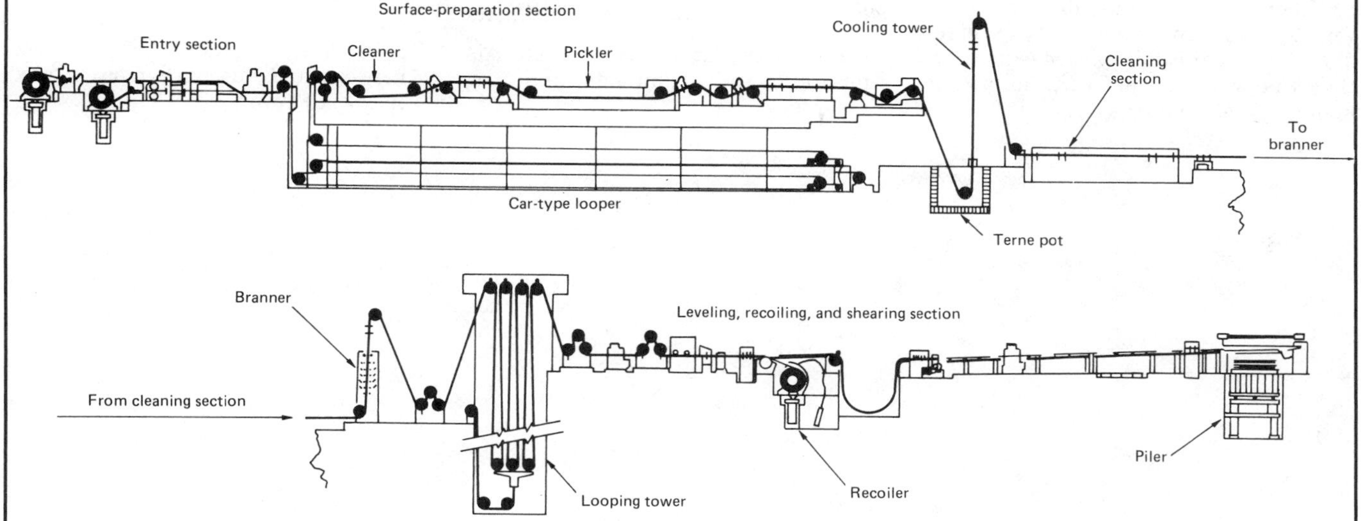

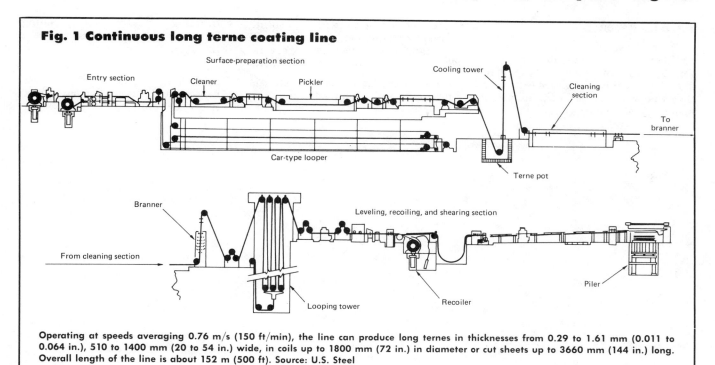

Operating at speeds averaging 0.76 m/s (150 ft/min), the line can produce long ternes in thicknesses from 0.29 to 1.61 mm (0.011 to 0.064 in.), 510 to 1400 mm (20 to 54 in.) wide, in coils up to 1800 mm (72 in.) in diameter or cut sheets up to 3660 mm (144 in.) long. Overall length of the line is about 152 m (500 ft). Source: U.S. Steel

Table 1 Coating designation and minimum coating test limits

Coating designation	Corresponds to obsolete class		Minimum coating check limits(a)			
			Triple spot test		Single spot test	
	g/m²	oz/ft²	g/m²	oz/ft²	g/m²	oz/ft²
LT01	Commercial		No check		No check	
LT25	75	0.35	76	0.25	61	0.20
LT35	105	0.45	107	0.35	76	0.25
LT40	120	0.55	122	0.40	92	0.30
LT55	170	0.75	168	0.55	122	0.40
LT85	260	1.10	259	0.85	214	0.70
LT110	340	1.45	336	1.10	275	0.90

(a) Total of both sides

Hot dip coating lines for the production of terne strip may be of the continuous or semicontinuous type (Fig. 1). In the semicontinuous production line, each incoming coil is processed as a separate entity; whereas in fully continuous lines, the leading edge of each new coil is welded to the trailing edge of the preceding one. This entails more complex and expensive handling equipment at both entry and exit ends of the line, but leads to higher productivity. After coating, the strip is either recoiled or cut into sheets. The continuous-coil lines use a process that includes alkaline cleaning, acid pickling, fluxing with a water-base zinc-ammonium-chloride solution and dipping in a molten terne-metal bath consisting of lead and 5 to 15% tin.

The sheet is conveyed through the coating metal via a submerged roller and emerges vertically from the molten coating bath into either: (*a*) an oil bath, squeegee-roll system or (*b*) an air-knife system which regulates the amount of terne metal applied to the sheet or strip.

Hot dip coating lines for the production of sheet and strip products may be of the semicontinuous or continuous type. In the semicontinuous production line, each incoming coil is processed as a separate entity. In fully continuous lines, Fig. 1, the leading edge of each new coil is welded to the trailing edge of the preceding coil, which leads to higher productivity.

When the oil bath, squeegee-roll system is used, the viscosity of the liquid oil (palm, fish, or mineral oil), and the pressure of the squeegee rolls result in metering of the terne metal on the sheet to the intended weight.

Because the liquid oil is a major fire hazard and also must be removed after cooling, which requires expensive cleaning and branning machines, the oil bath system is being replaced by the air-knife system.

In the air-knife system, the sheet emerges from the molten terne metal and passes between two slots on opposing sides of the sheet. Air or other gas, such as steam or nitrogen, is blown from the slots against the molten coating on the sheet. The amount of terne left on the sheet is controlled by regulating the pressure of the air emerging from the slots. Coating thicknesses on sheet products range from about 2 to 5 μm (0.08 to 0.2 mil).

On emerging from the coating-control system, the sheet travels vertically up and down in a high tower section in which air blasts are used to solidify and cool the terne coating. Subsequently, the coated steel passes through: (*a*) cleaners and branners (if oil must be removed), (*b*) roller levelers to flatten the steel, (*c*) recoilers (when the product is shipped in coil form), or (*d*) shears and pilers (when the product is shipped in cut-length pieces).

Applications

Terne-coated sheet products possess good corrosion-resistant properties, as well as excellent weldability, solderability, and paintability. Consequently, they find wide usage in the following applications:

- *Transportation industry:* gasoline tanks for automobiles, trucks, and tractors; radiator parts, valve covers, oil-filter cans, and air filter containers
- *Electrical appliance industry:* chassis for radios, television sets, and tape recorders
- *Building industry:* roofing, siding, flashing, gutters, downspouts, and fire doors
- *Miscellaneous:* fuel tanks for lawn mowers and outboard motors; electrical hardware

Thermal Spray Coatings

By James H. Clare
Principal Research Scientist
Battelle Columbus Laboratories
and
Daryl E. Crawmer
Research Scientist
Battelle Columbus Laboratories

THERMAL SPRAY is a generic term for a group of commonly used processes for depositing metallic and nonmetallic coatings. These processes, sometimes known as metallizing, are flame spray, plasma-arc spray, and electric-arc spray. Coatings can be sprayed from rod or wire stock, or from powdered material. In the form of wire or rod, material is fed into a flame axially from the rear, where it is melted. The molten stock is then stripped from the end of the wire or rod and atomized by a high velocity stream of compressed air or other gas which propels the material onto a prepared substrate or workpiece. In powder form, the material is metered by a powder feeder or hopper, into a compressed air or gas stream which suspends and delivers the material to the flame where it is heated to a molten or semi-molten state and propelled to the workpiece, where a bond is produced upon impact.

As the molten or semi-molten plastic-like particles impinge upon the substrate, one or more of three possible bonding mechanisms cause a coating to build up:

- Mechanical bonding occurs when the particles splatter on the substrate, as shown in Fig. 1. The particles interlock with the roughened surface and/or other deposited particles, forming a coating similar to the example in Fig. 2
- With some combinations of substrates and coating materials, localized diffusion or alloying can occur
- Some bonding may occur by means of Van der Waals forces. This is similar to the mutual attraction and cohesion which occurs between any two clean surfaces in contact, for example, two optical flats or two gage blocks

Depending on the process, coating material, and substrate composition, any or all of these bonding mechanisms can occur. Two noted exceptions to this are the transferred-arc and flame spray and fuse processes which produce metallurgical bonding throughout the workpiece. Substrate cleanliness is of very great importance to all coating applications to ensure bonding.

Applications. Thermal spray metal and ceramic coatings are used to produce:

- *Catalytic surfaces:* coatings used as catalyst generally have high porosity which presents a greater surface area to the medium being processed. Catalytic coatings are commonly applied by powder flame spray although plasma-arc can be used
- *Corrosion resistance:* wire flame spray coatings of zinc or aluminum are commonly used to protect structures from oxidation or salt water corrosion. Other materials can be applied for unique applications
- *Electrical conductivity:* plasma-arc and flame spray have been used to apply metals for electrical contacts or contact areas for applications such as electrical resistance heaters, capacitors and grounding connections
- *Electrical resistance:* plasma-arc and flame spray are widely used to apply oxide coatings as insulating layers in applications such as induction heating coils, high-temperature strain gages, and for dielectric layers
- *Electromagnetic interference shielding:* electromagnetic or radio frequency interference (EMI, RFI) which can damage electronic components or in-

Fig. 1 Deformation of molten or semi-molten particles resulting from spray impacting on a substrate

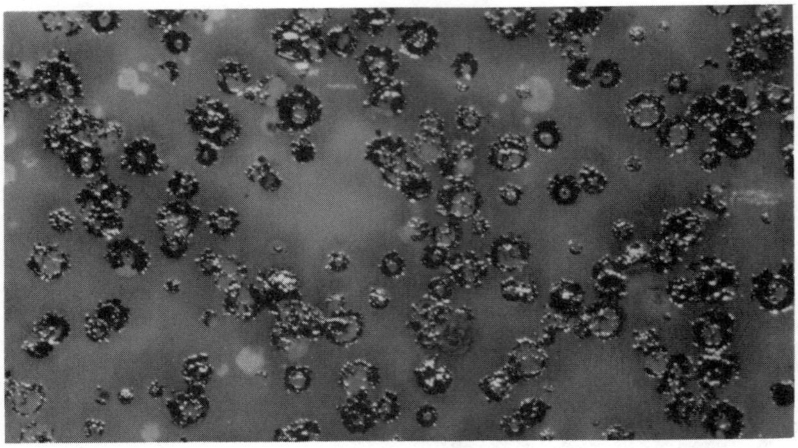

Fig. 2 High density thermal spray coating of tantalum

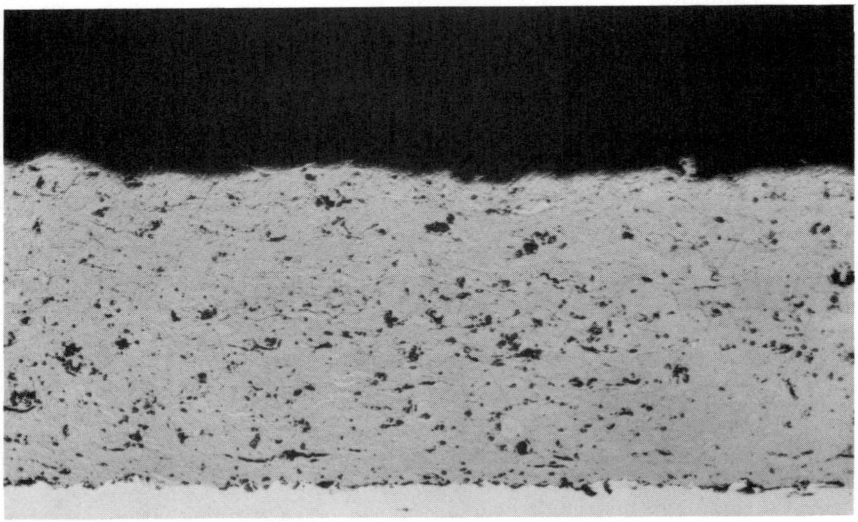

terfere with low level signals can be reduced by shielding. Commonly done with screening metal flake paints, shielding can be accomplished more effectively and economically using electric arc spray. Thin, sprayed layers of zinc, tin, and other low melting materials are applied to the inside surfaces of plastic or fiberglass instrumentation cabinets for shielding

- *Free-standing shapes:* parts can be fabricated from hard to machine materials by building up a coating on a removable form. In this way, parts can be made to near final shape and dimension. Rocket nozzles and ion engine components are two examples of applications for this technique. Plasma-arc spray is generally used to fabricate free-standing shapes
- *Molding making:* using electric-arc spray, short run or prototype molds can be constructed at a substantial reduction in labor and cost. A prototype part is coated with a mold release, then sprayed with tin-zinc or a similar suitable metal to form the mold face. Backup is provided by a filled epoxy cast onto the mold face
- *Nuclear moderators:* many thermal spray coatings have been approved for use in nuclear reactors. Examples are a plasma-arc spray oxide wear surface for control-rod drive seals and graded Z coatings for radiation hardening
- *Oxidation protection:* nickel and cobalt base, and nickel, chrome and cobalt alloys are frequently used to provide oxidation protection. Areas of application include heat treating fixtures, and exhaust components. These coatings are applied using plasma-arc and flame spray
- *Parting films for hot isostatic pressing (HIP):* a thin oxide coating applied using plasma-arc spray is used to provide a parting layer between the HIP mandrel and component being hipped
- *Dimensional buildup for salvage of worn parts:* the most common use of thermal spray is salvaging worn parts. Worn parts are restored to their original dimensions with a material which will equal or exceed design requirements. All thermal spray processes are used in salvage work
- *Thermal barriers:* zirconia, magnesia, and alumina are used singly or together to provide thermal protection or to improve the efficiency of thermal systems. Rocket engines and adiabatic engines use thermal barrier coatings to improve efficiency and reduce metal temperatures or cooling requirements. Heat treating and brazing fixtures sometimes utilize these coatings to confine heat to the part being treated
- *Thermal conductivity:* thermal conductivity can be improved by applying coatings which will improve heat transfer by increasing surface area or by applying coatings whose physical properties have better thermal characteristics. Copper and aluminum are used frequently, beryllium oxide can be used where electrical conductivity can be harmful
- *Wear:* the wear properties of low-alloy steel and nonferrous metal surfaces can be improved by using the appropriate combinations of thermal spray processes and coating materials. Hard metal coatings such as the chrome-nickel-boron alloys and carbide bearing coatings are two mainstays in thermal spray coatings

Fig. 3 Typical plasma-arc spray gun

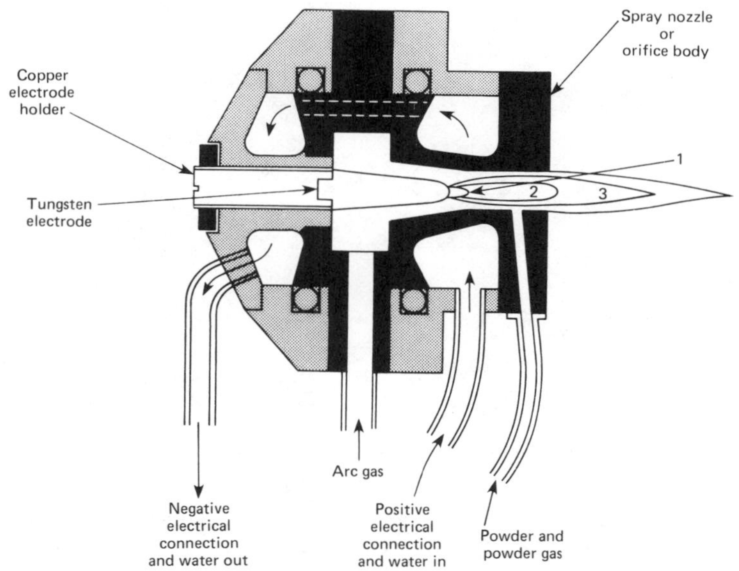

Approximate plasma temperatures: (1) 7800 °C (14 000 °F); (2) 10 000 to 13 400 °C (18 000 to 24 000 °F); (3) 7800 to 10 000 °C (14 000 to 18 000 °F)

Processes

Plasma-Arc Spray. The use of plasma-arc spray coatings has grown significantly in recent years primarily because of higher original equipment manufacturer cost, and also as a result of more stringent coating requirements. The plasma-arc process produces higher temperatures and higher powder particle velocities than most of the flame spray processes. Plasma-arc spray coatings exhibit higher densities and higher bond strengths. The oxide content of metal coatings is inherently lower due to the use of inert arc gases.

A plasma is an excited gas, often considered to be a fourth state of matter, consisting of an equal ratio of free electrons and positive ions. This forms an electrically neutral flame. A plasma-arc gun is a water-cooled device which has an open-ended chamber in which the plasma is formed, shown in Fig. 3. The primary arc gas, usually argon or nitrogen, is introduced into the chamber and is ionized by the electrical discharge from a high frequency arc starter. Once discharge is initiated, the plasma can conduct currents as high as 2000 A direct current, with voltage potentials ranging from approximately 30 to 80 V direct current. Standard plasma guns are rated at up to 40 kW. The latest high-energy guns are rated at up to 80 kW, producing spray velocities in excess of Mach 2.

A plasma is heated by resistance. In monatomic gases, higher temperatures are generated by simply passing more current through the plasma. To achieve even higher temperatures, secondary gases such as nitrogen, helium, and hydrogen are added to the plasma. These gases raise the ionization potential of the arc gas mixture, raising the plasma heat content to produce higher temperatures at lower power levels.

Figure 4 shows the modules constituting a basic plasma-arc spraying system and the variables that must be controlled. The power level, the pressure and flow of the arc gases, and the rate of flow of powder and carrier gas are controlled at the console of the system. The spray gun position and gun-to-work distance are usually preset. The movement of the workpiece is controlled by using automated or semi-automated tooling. Substrate temperatures can be controlled by preheating and by limiting the temperature increase during processing by interrupted spraying.

In a manually operated system, care should be taken when introducing secondary gases. The increase in ionization potential causes a corresponding decrease in current. To avoid a flame out, the idling current level should be increased to 300 to 400 A before adding secondary gases. As more secondary gas is added, the current will have to be adjusted to achieve the desired operating conditions. When powering down the plasma gun, current and secondary arc gas flow should be decreased simultaneously. Decreasing the gas flow more rapidly than the current can cause the current to exceed the ratings of some equipment.

Transferred Plasma-Arc. The transferred arc process adds to plasma-arc spray the capability of substrate surface heating and melting. Figure 5 is a schematic representation of the transferred plasma-arc process. In this process, a secondary current is established through the plasma and substrate. Surface melting and depth of penetration are controlled by the secondary arc current. Several advantages result from this direct heating: metallurgical bonding; high density coatings; high deposition rates; and high thicknesses per pass. Coating thicknesses of 0.50 to 6.35 mm (0.020 to 0.250 in.) and widths up to 32 mm (1.25 in.) can be made in a single pass at powder feed rates of 9 kg/h (20 lb/h). In addition to this, less electrical power is required than with nontransferred arc processes. For example, an 88% tungsten-carbide, 12% cobalt material, plasma spray deposited 0.30 mm (0.012 in.) thick and 9.50 mm (0.375 in.) in width, might require 24 passes at 40 to 60 kW to achieve maximum coating properties. This same material can be applied, using the transferred arc process, in one pass at approximately 2.5 kW.

The method of heating and heat transfer in the transferred arc process eliminates many of the problems related to using powders with wide particle size distributions or large particle sizes. Larger particle size powders, for example in the 50-mesh range, tend to be less expensive than closely classified, −44 μm + 10 μm powders.

Some limitations of the process should be considered for any potential application. Because substrate heating is a part of the process, some alteration of its microstructure is inevitable. Applications are also limited to substrates which are electrically conductive and can withstand some deformation. The

Fig. 4 A complete plasma-arc spray system

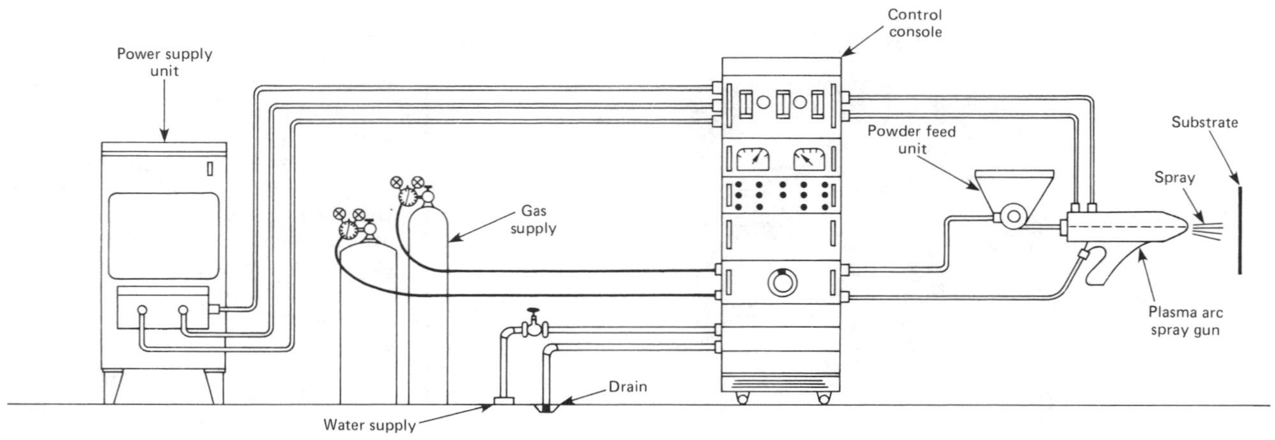

Fig. 5 Transferred arc plasma spraying

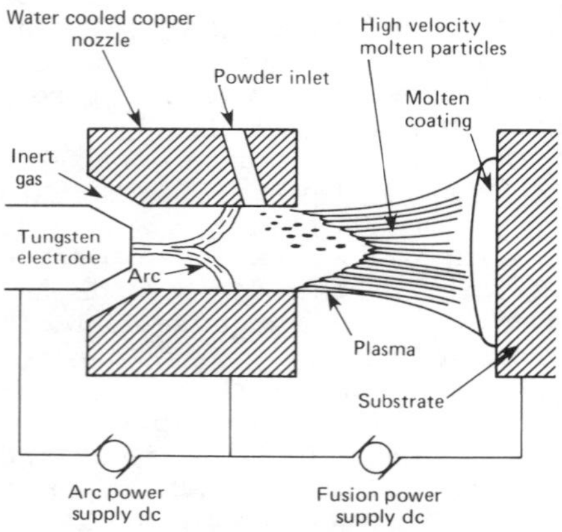

transferred plasma-arc process is used in hardfacing applications for valve seats, farm equipment such as plowshares, oil field components, and mining machinery.

Inert Atmosphere and Low-Pressure Chamber Spray. Thermal spray done in an inert atmosphere and/or low-pressure chamber has become a widely accepted practice throughout the coatings industry, particularly the aircraft engine industry. High output aircraft engines that operate at high temperatures require a high temperature coating system. Complex coating alloys for high temperature applications have been developed to protect and extend the life of these costly components. Inert atmosphere and low-pressure plasma spray systems have proven to be an effective means for applying complex coatings of nickel-cobalt-chromium-aluminum-yttrium (Ni, Cr, Al, Y) and other MCrAlY-type alloys to high temperature aircraft engine components. Although the coatings produced by each of the processes bear some similarities, the physical properties can vary extensively.

Inert atmosphere chamber spraying is used for two purposes, to confine hazardous materials and to restrict formation of oxides that would occur in open air spraying. Hazardous materials are grouped into two categories, toxic and pyrophoric. Toxic materials include beryllium and its alloys. Pyrophoric materials include magnesium, titanium, lithium, sodium, and zirconium, which tend to burn readily when in a finely divided form or when purified by the plasma process. An inert chamber spray system may include a jacketed, water-cooled box, an air lock, a plasma system, workpiece handling equipment, glove parts, a vacuum pumping system, and an inert gas backfill manifold. Usually, the chamber is pumped down to a pressure of 10^{-4} to 10^{-5} torr then backfilled with high purity, dry argon. In any good inert gas chamber, oxygen levels can be easily maintained below 30 ppm. Metal powders tend to cleanup when sprayed in an inert gas chamber by the reduction of surface oxides. By the same mechanism, some oxide powders tend to reduce when sprayed in an inert gas chamber.

Inert atmosphere spraying in a low-pressure chamber offers several unique advantages over conventional plasma spraying in an inert atmosphere at atmospheric pressure. To deposit a coating with optimum physical properties, the spray material must maintain its original composition. These conditions are rarely achieved when depositing coatings in atmospheric conditions. In low-pressure plasma spraying, the bond strength is increased because higher substrate temperatures allow the coating to diffuse into the substrate. Deposition efficiency can be increased because of increased particle dwell time in the longer heating zone of the plasma, and minimal changes in chemistry of the coating result because of the inert atmosphere. The closed system also minimizes environmental problems such as dust and noise.

Normal operating procedures require the spray chamber be pumped down to approximately 0.4 torr Hg (55 Pa) and then backfilled with inert gas to 300 torr. Once the system has been sufficiently purged to achieve an acceptable inert atmosphere, the plasma spray operation is activated and the chamber pressure adjusted to the desired level for spraying. The entire spray operation is accomplished in a soft vacuum of approximately 50 torr. The optimum spray condition exists when the plasma temperature at the substrate approximates the melting point of the powder particles; however, the optimum spraying conditions will vary with the chemistry and particle size of each spray material. These variables are similar to conventional plasma spraying. Because of the complexity of low-pressure spraying, the entire process is best controlled

Fig. 6 Inert atmosphere and/or low pressure plasma chamber

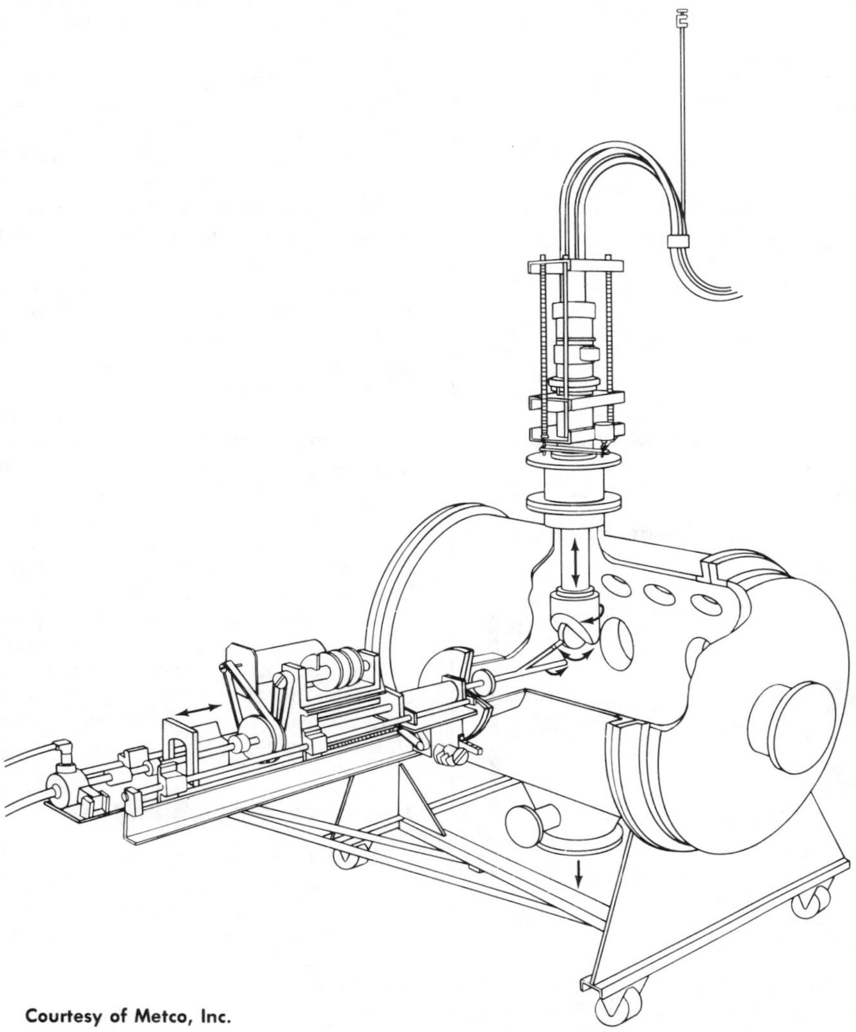

Courtesy of Metco, Inc.

by a computer, to ensure complete reproducibility and uniformity throughout the coating. Figure 6 shows a typical inert atmosphere and/or low-pressure plasma chamber.

Electric-Arc Spray. The electric-arc spray process utilizes metal in wire form. This process differs from the other thermal spray processes in that there is no external heat source such as gas flame or electrically induced plasma. Heating and melting occur when two electrically opposed charged wires, comprising the spray material, are fed together in such a manner that a controlled arc occurs at the intersection. The molten metal is atomized and propelled onto a prepared substrate by a stream of compressed air or gas (Fig. 7).

Electric-arc spray offers several advantages over other thermal spray processes. In general, this process exhibits higher bond strengths, in excess of 69 MPa (10 000 psi) for some materials. Deposition rates of up to 55 kg/h (120 lb/h) have been achieved for some nickel-base alloys. Substrate heating is lower than in other processes due primarily to the absence of flame touching on the substrate. The electric-arc process is in most instances less expensive to operate than the other processes. Electrical power requirements are low. With few exceptions no expensive gas, such as argon, is necessary. By using dissimilar wires, it is possible to deposit pseudo-alloys. A less expensive wear surface can be deposited using this technique. One wire, or 50% of the coating matrix, can be an inexpensive filler material.

Metal-face molds can be made using a fine spray attachment available from some manufacturers. Molds made in this way can duplicate extremely fine detail, such as the relief lettering on a printed page.

The electric-arc process is limited to relatively ductile, electrically conductive wire about 1.5 mm (0.60 in.) in diameter. Electric-arc spray coatings of carbides, nitrides, and oxides are therefore not currently practical.

Flame spray utilizes combustible gas as a heat source to melt the coating material. Flame spray guns are available to spray materials in either rod, wire, or powder forms. Most flame spray guns can be adapted to use several combinations of gases to balance operating cost and coating properties. Acetylene, propane, Mapp gas, and oxygen-hydrogen are commonly used flame spray gases. In general, changing the nozzle and/or air cap is all that is required to adapt the gun. Figures 8 and 9 depict wire flame and powder flame guns. For all practical purposes, the rod and wire guns are similar.

Flame temperatures and characteristics depend on the oxygen to fuel gas ratios as illustrated in Table 1. The flame spray process is characterized by low capital investment, high deposition rates and efficiencies, and relative ease and cost of maintenance. In general, flame sprayed coatings exhibit lower bond strengths, higher porosity, a narrower working temperature range, and higher heat transmittal to the substrate than plasma-arc and electric-arc spray. Notwithstanding, the flame spray process is widely used by industry for the reclamation of worn or out-of-tolerance parts.

Flame spray and fuse is a modification of the powder-flame spray method previously mentioned. The materials used for coating are self fluxing, fusible materials which require post-spray heat treatment. In general, these are nickel- or cobalt-base alloys which use boron, phosphorus, or silicon, either singly or in combination, as melting point depressants and fluxing agents. In practice, parts are prepared and coated as in other thermal spray processes and then fused. Fusing is accomplished us-

Fig. 7 Typical electric-arc spray device

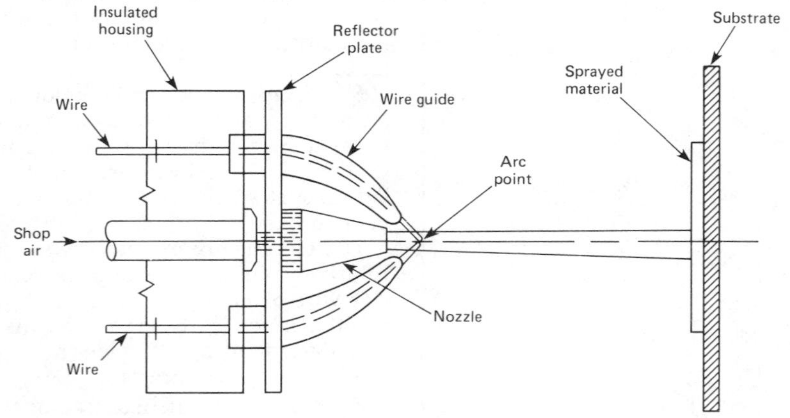

Fig. 8 Cross section of typical wire or rod flame spray gun

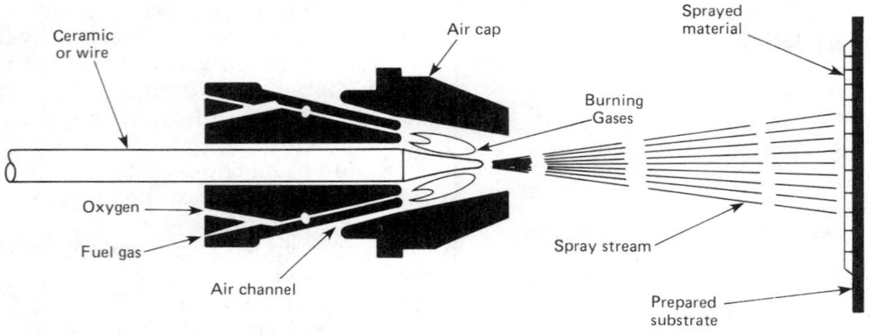

ing one of several techniques such as flame or torch, induction, or in vacuum, inert, or hydrogen furnaces. These alloys generally fuse between 1010 to 1175 °C (1850 to 2150 °F) depending upon composition. Reducing atmosphere flames should be used to ensure a clean, well-bonded coating. In vacuum and hydrogen furnaces, the coating may have a tendency to wick or run onto adjacent areas. Several brushable stop-off materials are commercially available to confine the coating. It is recommended that test parts be fused whenever the shape, coating alloy, or lot of material is changed, to establish the minimum and maximum fusing temperatures. Fusing temperature is known to vary slightly between lots of spray material. On vertical surfaces, coating material may sag or run off if the fusing temperature is exceeded by more than a few degrees. Excessive porosity and nonuniform bonding are usually indicative of insufficient heating. Spray and fuse coatings are widely used in applications where excessive wear is a problem. These alloys generally exhibit good resistance to wear and have been successfully used in the oil industry for sucker rod and in agriculture for plowshares. In most applications, fusible alloys make possible the use of less expensive substrate materials. Coating hardnesses can be as high as 65 HRC. Some powder manufacturers offer these alloys blended with tungsten carbide or chromium carbide particles to increase resistance to wear from abrasion, fretting, and erosion. As mentioned previously, these coatings are fully dense and exhibit metallurgical bonds. Grinding is usually necessary for machining a fused coating because of the high hardness. Use of spray and fuse coatings is limited to substrate materials which can tolerate the 1010 to 1175 °C (1850 to 2150 °F) fusing temperatures. Fusing temperatures may alter the heat treated properties of some alloys. However, the coating will usually withstand additional heat treatment of the substrate.

Process and Equipment Comparison. Thermal spray coatings vary over a wide range of characteristics, depending on the operating conditions used. Therefore, the information listed in Table 2 is a generalization and should be used only for comparison. For example, the rise in substrate temperature during processing varies considerably with the mass, coating thickness, deposition rate, materials involved, and especially, the precautions taken to control temperatures.

Thermal spray devices can be hand held or machine mounted. Guns are commonly mounted on a lathe compound to spray cylindrical parts. Large flat parts are usually sprayed with guns mounted to two axis positioners, such as those used by the welding industry. Complex parts requiring three or more axes of freedom can now be coated using commercially available, multiple-axis, automated computer-controlled systems. Using these techniques, parts ranging from simple cylinders to complex airfoils can be coated.

Surface Preparation

Thermal spray coatings rely mostly upon mechanical bonding to the substrate surface. It is critical that a substrate be properly prepared to ensure coating quality. Surface cleanliness is very important in all coating systems, especially for flame spray and fuse coatings. Surface roughness is the next most critical factor. It is especially necessary that the surface remain uncontaminated by lubricants from handling equipment or body oils from hands and arms. It also is recommended that the prepared surface be coated as soon as possible after preparation to prevent the possibility of contamination or surface oxidation.

Cleaning and Degreasing. Heavy deposits of corrosion, rust, paint, or grease must be removed before any spraying is started. Deposits can be removed by scraping, wire brushing, machining, grit blasting, or by chemical action.

Degreasing is usually the most economical and safest way to remove lubricants and body oils. Large parts, and parts with attached hardware which

Fig. 9 Cross section of typical powder flame gun

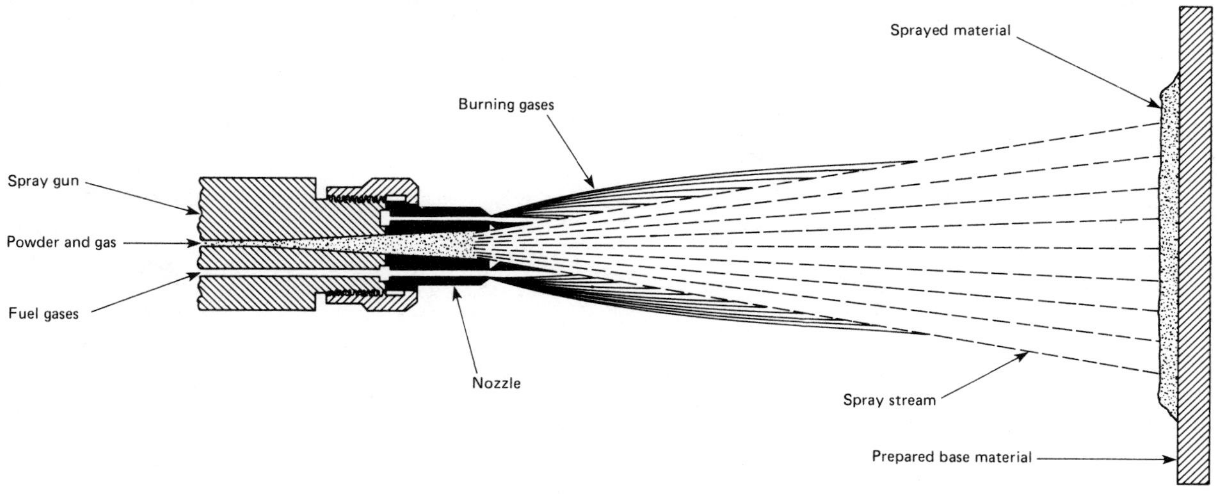

Table 1 Oxygen to fuel gas ratios

Ratio (oxygen: fuel gas)	Temperature °C	°F	Flame condition	Result
1:1	3000	5400	Carburizing	Insufficient heat
1:1	3000	5400	Reducing	Good for some metal
1.1:1	3050	5500	Neutral	Recommended for general use
1.1:1	3350	6000	Oxidizing	Good for some ceramics

may be damaged by vapor degreasing, should be degreased manually. Freon TF is recommended for these applications. Freon TF is more expensive than other widely used degreasers, but it presents fewer health hazards. All solvents should be used only in well-ventilated areas, by properly protected personnel trained in their use. Consult plant safety personnel for local regulations for the use, care, and handling of solvents.

Surface Roughening. After degreasing has been accomplished, surface roughening can follow. Three methods of surface roughening for thermal spray are widely used. They are, in order of their ability to improve bond strength, rough threading, grit blasting, and a combination of rough threading, then grit blasting. Two methods have fallen into disuse because of relative ineffectiveness or substrate overworking. These are electric bonding, a method of disrupting the surface by electrical discharge, and knurling. A method is used where mechanical deformation of the surface is not possible or practical. This method uses materials which are exothermic and somewhat self bonding. Nickel-aluminum alloys are widely used for this purpose. Molybdenum is also used as a bond coat material. In these applications, coating bond strength is lower than with grit blasting and/or threading techniques. Bond coating will almost always enhance the bond strength of coatings applied to grit blasted and/or rough threaded surfaces. This holds true especially for metal oxide coatings.

Rough threading is generally used for cylindrical surfaces. The part to be prepared is securely mounted in a lathe and a single thread cut is taken. Tools for this purpose have a 60 to 70° point with a slight negative back rake. Screw feeds are approximately 0.80 to 1.25 mm (0.032 to 0.050 in.) or 0.78 to 1.26 threads/mm (20 to 32 threads/in.). This technique is limited to thicker substrate sections. It is not recommended for thin coatings. Higher bond strengths are obtained when threading is followed by grit blasting. When cutting fluid is necessary for threading, the part must be degreased before grit blasting or coating.

Grit Blasting. Contaminated substrate surfaces cause coating failures. For this reason, grit blasting equipment used for thermal spray should not be used for other purposes since dirt, paint, and lubricants contaminating the grit can be redeposited on blasted parts. The air supply to the grit blast equipment must be clean and dry.

Aluminum oxide and chilled iron are the most widely used abrasive grits for thermal spray surface preparation. Practical grit size ranges are −10 +30 mesh, −14 +40 mesh, and −30 +80 mesh. Because surface roughness is primarily the result of grit particle size, the selection of the grit size is determined by the required coating thickness. Table 3 gives general recommendations for grit size selection. Surface roughness can also be varied slightly by air pressure. This factor should be considered on an individual basis for each combination of grit size, type, and substrate material.

Sand, crushed steel, and silicon carbide are also used as abrasive grit. Sand is commonly used on large exterior structures such as bridges, towers, and piping where recovery of the grit is impractical. Crushed steel grit, obtained commercially in hardnesses to 65 HRC, is used on hardened steels. Silicon carbide is used by some but has several drawbacks. It is relatively ex-

Table 2 Comparison of thermal spray equipment and processes

Equipment Type	Applications	Feed material	Surface preparation	Process temperature °C	Process temperature °F	Particle velocity m/s	Particle velocity ft/s	Ceramic kg/h	Ceramic lb/h	Metal kg/h	Metal lb/h
Direct current plasma-arc, 40-kW unit	Metallic, ceramic, plastics, and compounds	Powder	Grit blasting or rough threading	95-120	200-250	490	1600	5.4-6.8	12-15	3.2-9.1	7-20
Powder flame spray	Metallic, ceramic, and fusible coatings	Powder	Grit blasting or rough threading	105-160	225-325	24-36	80-120	1.4-2.3	3-5	3.2-9.1	7-20
Wire flame spray	Metallic coatings	Wire	Grit blasting or rough threading	95-135	200-275	240	800	...	...	2.3-29.5	5-65
Ceramic rod spray	Ceramic and cermet coatings	Rod	Grit blasting	95-135	200-275	240	800	0.45-0.9	1-2	...	...
Two-wire electric-arc	Metallic coatings	Wire	Grit blasting or rough threading	50-120	125-250	240	800	...	...	4.5-54.4	10-120
Transferred-arc	Metallic, fusible coatings	Powder	Light grit blasting or chemical cleaning	Fuses base metal	Fuses base metal	490	1600	...	...	3.2-9.1	7-20

pensive, breaks down quickly and tends to embed in the substrate.

Consideration should be given to the substrate material in the selection of grit type. Traces of residual grit may adversely affect some coatings. Chemical compatibility in the finished coating system must be considered. Alumina, sand, and especially silicon carbide may embed in softer metals such as aluminum, copper, and their alloys. For these metals, lower air pressures are recommended to minimize embedding. Chilled iron or crushed steel should be used in preparing surfaces to be flame sprayed and fused. Alumina, silica, or silicon carbide may act as a stop off to the fusible coatings.

Grit blasting air pressure varies from 210 to 620 kPa (30 to 90 psi), with stand off or working distances of 50 to 150 mm (2 to 6 in.). Grit blast nozzle openings are generally 6 to 10 mm (0.25 to 0.375 in.) in diameter. The blasting angle to the substrate should be about 90°.

The substrate should be cleaned following grit blasting to remove residual dust. It is recommended that the surface be rinsed with Freon TF alcohol, or a similar solvent. Allow the solvent and dust to run off onto an area which will not be coated. Air drying may be done using clean, dry compressed air.

Again, it is very important that the surface remain uncontaminated by lubricants from handling equipment or body oils from hands and arms. It is recommended that the prepared surface be coated as soon as possible after rinsing to prevent surface oxidation or contamination.

Table 3 Grit size vs coating thickness recommendation

Roughness	Grit size, mesh		Sieve openings mm	Sieve openings in.	Applications
Course	−10	+30	2.007/0.610	0.079/0.024	For coatings exceeding 0.254 mm (0.010 in.) and best adherence
Medium	−14	+40	1.422/0.432	0.056/0.017	For fair adherence and smoother finishes of coatings less than 0.25 mm (0.010 in.) thick
Fine	−30	+80	0.610/0.175	0.024/0.007	For smoothest finishes on coatings les than 0.25 mm (0.010 in.) thick to be used in as-sprayed condition

Finishing Treatment

Thermal spray coatings usually exhibit two common features in the as-deposited condition, a sandpaperlike surface finish and a structure with inherent porosity. A typical surface roughness of an as-sprayed coating ranges from 5.0 to 13 μm or 200 to 500 μin. arithmetic average (AA). The porosity usually ranges from 2 to 17 vol %, depending on the process by which the coating is deposited and the material sprayed. In many cases, thermal spray coatings are applied to machine elements where the coated component must conform to close dimensional tolerances, or have a high surface finish. In many applications, coatings are exposed to corrosive solutions or hydraulic fluids, which can infiltrate the pores, resulting in fluid leakage or corrosion of the base material. These conditions can contribute to the premature failure of the coating. Many applications require the coating to be sealed before finishing by machining or grinding, to protect the thermal spray coating.

Sealing is the process by which the pores of a coating are filled to eliminate the possibility of infiltration by fluids or corrosive media that can contribute to premature failure. In thermal spray coatings, the porosity may range as high as 17 vol % and is usually interconnected, making the coating permeable to gases and liquids. The porosity can leave the substrate available to corrosive attack and may result in undesirable leakage of the operating medium. Surface porosity can cause considerable difficulty in obtaining a smooth finish when machining or grinding. Sealing a coating also helps to reduce particle pullout from the surface during finishing.

Sealing a coating will prevent the corrosive attack of the base material and leakage of fluids. To ensure complete sealing of the coating, it is necessary to apply the seal material prior to surface finishing. Seal materials such as waxes, phenolics, and inorganics are readily available and easily applied. The wax sealers are useful in preventing infiltration of liquids at low service temperatures. Resin-base sealers are effective at temperatures ranging from 90 to 260 °C (200 to 500 °F). Some silicone-base sealers have proven to be effective protection in salt-spray tests conducted in accordance with military standards up to 480 °C (900 °F). Epoxy and phenolic sealers are usually more effective on coatings with higher porosity and are generally effective within the temperature limits of stability.

One of the most effective methods of sealing coating porosity is vacuum impregnation. This method will usually fill all interconnected pores. To vacuum impregnate, immerse the coated part in a container of epoxy resin and place the container in a vacuum chamber. A vacuum is drawn to pull air from the pores. When the vacuum is released, atmospheric pressure forces sealer into the evacuated pores. Generally, most applications do not require such thorough penetration. Another effective means for sealing coating porosity is the use of low-viscosity sealers. These sealers set by an anaerobic reaction or by heat curing. They can be brushed or sprayed onto a coating surface at room temperature and are drawn into pores by capillary action. The depth of sealer penetration may be as much as 1.8 mm (0.070 in.) in some instances. Regardless of the method used to seal the coating, it is important that the sealer penetrate the coating sufficiently to prevent removal during the finishing operation. Table 4 is a brief summary of commercially available sealant materials.

Table 4 Description of sealing materials

Description	Method of application	Form	Curing method
Petroleum-base wax	Rubbing	Stick	Air dry
Air drying, oil modified phenolic varnish resin	Brush	Liquid	Air dry
Baking type, clear synthetic phenolic	Brush or pressure impregnate	Liquid	Bake at 175 °C (350 °F)
Dimethacrylate, polyester resin (anaerobic curing)	Brush	Liquid	Air dry or bake at 120 °C (250 °F)
Phenolic base	Brush	Liquid	Air dry
Aluminum silicate	Brush	Liquid	Air dry
Clear vinyl	Brush	Liquid	Air dry
Aluminum flake, suspended in a vinyl lacquer	Brush	Liquid	Air dry

Surface Finishing. Thermal spray coatings can be finished by using standard metal finishing techniques such as machining, grinding, lapping, or polishing. Because of the unique structure of thermal spray coatings, considerable care must be taken when grinding or machining to avoid damage to the coating. A sprayed coating is composed of an aggregation of individual particles, with some metallic oxides, or metal coatings. Improper techniques during the finishing operation may result in excessive particle pullout, singly or in clusters, producing a severely pitted surface. It is essential that the sprayed particles be cleanly sheared and not pulled from the surface. Although a coating may be properly finished, the surface may not be shiny, as wrought material, but have a matte finish because of surface porosity.

Because the thermal spray coating is primarily mechanically bonded to the substrate, special attention should be given to the selection of the finishing method. The particular selection of the finishing method depends on the type of coating, hardness, and thickness.

Machining. High speed steel tools can be used to cut the softer sprayed materials, such as low-carbon steel, copper-base alloys, and aluminum. Harder materials usually require the use of carbide tools or grinding. If surface finish is critical, it is advisable to use carbide tools for all materials other than those requiring grinding wheels.

There are no standard rules for determining the proper cutting feeds, speeds, and tool angles. It is generally accepted practice to machine coatings at a much lower feed rate than that used for wrought materials. Usually, rough cutting the coating is limited to less than 0.50 mm (0.020 in.) per pass. This practice will limit the stress on the coating-to-substrate material bond.

Table 5 lists typical ranges of feeds and speeds used in machining sprayed metal coatings. These data may be used as a guide; the first cut will indicate the need for adjustments to obtain the desired results.

The angles for high speed steel and carbide tools are important for satisfactory tool life and surface finish. For metals such as Monel, nickel, and bronze, a properly ground and mounted tool produces a burnished finish that greatly improves the surface condition. Recommended shapes for carbide and high speed steel tools are given in Fig. 10.

The roughing cut serves as a guide in determining what cutting changes are necessary. Hand honing of the carbide cutting edge will aid in maintaining surface finish as the tool wears.

Grinding of thermal spray coatings can be done either wet or dry; wet grinding is preferred, when possible. In either method, loading of the grinding wheel can result in surface checking and smearing. Selection of proper grinding wheels and techniques is important for producing satisfactory surface finishes. In general, a wheel of low bond strength and relatively coarse structure is required. Usually, wheels for grinding the substrate do not produce a satisfactory finish on sprayed material.

Extremely light infeed is used for the finishing pass. Dwelled cuts are often used during wet grinding. The surface of the finish-ground, sprayed metal coating is dull and finely porous, not bright or shiny. Lack of visible fine porosity may indicate surface smearing, caused by an improper wheel or inadequate dressing.

Table 5 Typical ranges of speeds and feeds used in machining sprayed metal coatings

	High speed steel tool				Carbide tool(a)			
	Speed		Feed		Speed		Feed	
Coating Metal	m/s	sfm	mm/rev	in./rev	m/s	sfm	mm/rev	in./rev
Steels								
Low-carbon, medium-carbon, low-alloy	0.25-0.50	50-100	0.075-0.125	0.003-0.005	0.25-0.50	50-100	0.075-0.125	0.003-0.005
High-carbon, stainless	...	...	...	...	0.15-0.200	30-40	0.075-0.100	0.003-0.004
Nonferrous metals								
Brass, bronze, nickel,copper, Monel	0.50-0.75	100-150	0.075-0.125	0.003-0.005	1.25-1.80	250-350	0.050-0.150	0.002-0.006
Lead, tin, zinc, aluminum, babbitt	0.75-1.00	150-200	0.075-0.175	0.003-0.007	1.25-1.80(b)	250-350(b)	0.050-0.100	0.002-0.004(b)

(a) Composition: 6% Co, 94% WC. (b) Aluminum only

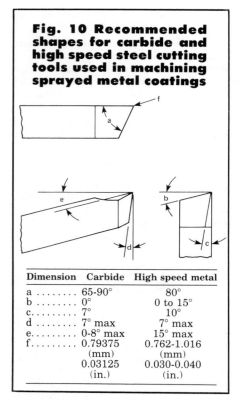

Fig. 10 Recommended shapes for carbide and high speed steel cutting tools used in machining sprayed metal coatings

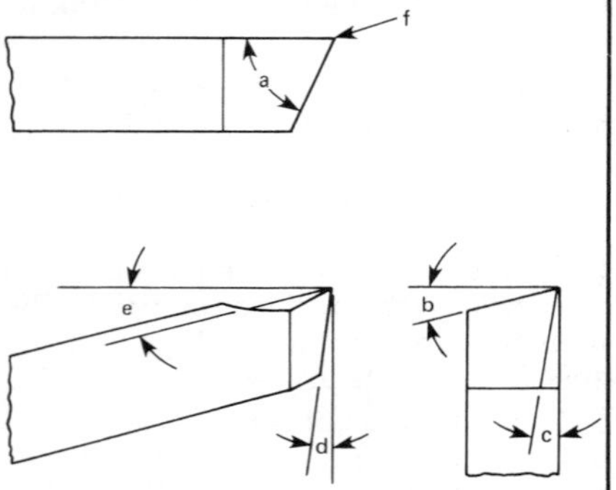

Dimension	Carbide	High speed metal
a	65-90°	80°
b	0°	0 to 15°
c	7°	10°
d	7° max	7° max
e	0-8° max	15° max
f	0.79375 (mm) 0.03125 (in.)	0.762-1.016 (mm) 0.030-0.040 (in.)

For many applications, it is desirable to treat the sprayed coating with a sealer prior to grinding. Sealing prevents the fine particles of grinding sludge from entering the pores of the coating. The particles could be expelled during service, causing damage to mating components. Sealing before grinding also permits the use of higher grinding rates with less wheel wear, and results in a better surface finish. If sealers are not used, the ground surfaces, especially those of bearings, must be thoroughly cleaned to remove all grinding particles.

Burnishing and Polishing. Burnishing is used especially on softer materials, such as tin, zinc, and babbitt, to produce a dense, smooth finished surface. When subsequent finishing operations such as painting and plating are to be used, burnishing is recommended for eliminating surface pitting and porosity.

All sprayed metal coatings can be polished, but the results obtained depend on the hardness of the sprayed metal. The softer materials are easily polished and buffed. Copper, brass, bronze, and steel are more difficult. Steel coatings should not be used when bright, shiny surfaces are required. Blistering of the coating may occur during burnishing, polishing, or buffing, when high, localized temperatures are developed.

Coating Repair. Repair of thermal spray coatings is not generally recommended. No reliable nondestructive testing techniques are currently available to examine the bond line and intracoating quality of thermal spray coatings. Routine surface preparation methods, when used on existing coatings, tend to degrade the overall coating quality with embedded grit, contamination, and stresses induced by grit blasting. Coatings which have been in service and have become damaged should be replaced. When new coatings are damaged in production, it is usually less expensive to replace the entire coating than to risk coating failures in the field.

Quality Assurance

Quantitative information is almost always required for the evaluation of thermal spray coatings. It is desirable to determine percentages of porosity, oxides, and other phases present in the sprayed coating. The grid area point-count method and the use of television scanning equipment are common methods used to make these determinations. Microhardness data can provide considerable information about the nature of thermal spray coatings. Differences in hardness can be a valuable method for distinguishing between and identifying phases of a sprayed coating. The type of hardness testing, whether superficial Rockwell, Vickers, or Knoop, should be chosen according to the application. Microprobe and/or x-ray diffraction may be used to identify phases in thermal spray coatings during the research and development stage of a coating, but are seldom needed for production applications.

Hardness. Macrohardness and microhardness testing are sometimes used to qualify thermal spray coatings. Hardness readings cannot be reliably converted to other mechanical properties such as yield strength, wear resistance, or porosity. However, hardness testing may be used to qualify coatings relative to production standards and as a general indication of the other mechanical properties.

Macrohardness measurements are normally made using the Rockwell B and C scales. As a rule of thumb, the coating thickness should be ten times greater than the depth of penetration of the penetrator. This precaution should eliminate anomalous readings from the substrate material.

Conversion values of microhardness measurements tend toward higher values than macrohardness measurements made on the same coating. This is because smaller sample areas minimize the effects of porosity. Another factor in high microhardness values is the possible presence of oxides. Macrohardness measurements may be made on relatively rough surfaces, but microhardness measurements require a metallographic quality surface. The testing surface must be clean and free of imperfections in either case. Measurements should be made allowing a distance of at least three impression diameters from edges, imperfections, and other penetrator impressions. Refer to ASTM E18-74 and ANSI Z115.6 for further explanation of this hardness testing.

Metallographic examination is a primary tool used to evaluate the quality of thermal spray coatings. Because approval of a coating for a given application depends on the results of metallographic study, it is absolutely necessary that the specimen studied be representative of the coating and that the specimen be prepared properly.

The metallographic procedures for the preparation of thermal spray coatings are similar to those used for most wrought materials. However, specimens of thermal spray coatings require some special techniques. The structure of thermal spray coatings is usually nonhomogeneous, consisting of particles mechanically and metallurgically bonded to each other. Coatings generally contain oxides and metallic phases not normally found in wrought materials. Care must be taken in sectioning and polishing to prevent pullout, thus avoiding false evaluations. Ultrasonic cleaning should be avoided, and vibratory polishing used rarely, to avoid the possibility of loosening surface particles.

Sectioning and Mounting. After careful selection, samples can be sectioned on any type of metallurgical cutoff machine. A continuous-rim diamond cutoff wheel, 1.0 mm (0.040 in.) in thickness can be used for sectioning test pieces with ceramic coatings. Coolant, usually tap water or water and cutting oil in a recirculating system, is used in all cutting operations. To avoid chipping the unsupported coating, it is sometimes advisable to mount the samples in epoxy resin before sectioning.

Samples can be mounted in a liquid casting plastic, using vacuum impregnation procedures to fill the pores. This material is easy to handle and has good adherence, hardness, and dimensional stability. Thermosetting powders can be used, but mount size is limited, and pores are not always completely filled. Two specimens can be mounted together, positioning the coatings face to face. The substrate material serves as a backup to protect the coating.

Grinding. Samples can be ground on a 250-mm (10-in.) disk grinder using 120-grit silicon-carbon grinding disks to level the mount, and using water as a coolant. Wet grinding may continue on a 200-mm (8-in.) disk grinder, using successively finer silicon-carbon disks of 180, 240, 400, and 600 grit. With each change of grit, the mount is turned 90°, and the sample is ground until all scratches left by the preceding grit have been removed. After each step, wash the mount with running water. Ultrasonic cleaning is never used for cleaning specimens of thermal spray coatings. Diamond laps of 240, 400, and 600 grit are used instead of silicon-carbon disks to grind very hard materials such as ceramics.

Polishing. Preliminary polishing of soft metallic samples is done by hand on a low-speed (100 to 300 rev/min) wheel, using a medium-nap cloth charged with 3-μm diamond paste. Kerosine or lapping oil is used as a lubricant, and polishing proceeds until all grinding scratches have been removed. The sample can then be hand polished using 1-μm diamond on a medium-nap cloth. If better flatness and edge retention are desired, vibratory polishing may be used with 1½-μm diamond on a silk cloth. Tungsten carbide and metal oxide coatings should not be polished on a vibratory polisher.

Diamond polishing causes excessive porosity in ceramics, cermets, and carbides. Overpolishing of tungsten carbide and hard metal coatings with diamond create relief of the harder constituents and pullout of any loosely bonded constituents. Therefore, an alternative preliminary polishing technique is used for polishing some semihard to very hard materials. Grinding scratches are removed on a fast wheel (1750 rev/min) covered with a low-nap cloth (nylon, silk, or lintless) and a slurry of 200 cm^3 (7 fluid oz) water, 4 g (0.15 oz) chromium trioxide and 25 g (0.90 oz) of polishing abrasive. If chromium trioxide preferentially attacks some phases, it can be reduced or eliminated from the slurry.

Final polish is obtained on a low speed (100 to 300 rev/min) wheel covered with a fine-nap cloth and using a slurry of 200 cm^3 (7 fluid oz) water, 4 g (0.15 oz) chromium trioxide, and 20 g (0.7 oz) ferric oxide. As in the preliminary polishing, chromium trioxide can be reduced or eliminated as necessary from the slurry to obtain the most satisfactory results. For some materials, 0.05 μm gamma alumina on a fine-nap cloth is used. Care should be taken not to polish too long, thus preventing excessive relief of carbide and solid solution phases. When there are problems with galvanic attack due to the difference of potentials between brass or copper wheels and the substrate and microconstituents in the coating, it can be eliminated by placing plastic wrap between the wheel and the cloth.

Metallographic Evaluation. In many preliminary metallographic investigations, qualitative information is all that is necessary. On a production basis, a good method for determining whether a sprayed coating is acceptable is to compare its appearance with that of a standard material known to perform adequately in service.

Interpretation of microstructure and correlation of that information with the overall properties of the sprayed coating is a problem. Metallographic techniques are widely used to judge the porosity of thermal spray coatings. The quantitative approaches to metallographic evaluation are based on the relationships between measurements on the two-dimensional plane of polish and the magnitudes of the microstructural features in three-dimensional materials. The volume fraction occupied by a microstructural feature, such as a void or pore, is equal to the areal, linear, or point ratio of the selected feature as seen on random sections through the microstructure.

Television scanning equipment is gaining popularity for applications where the number of specimens to be examined justifies automation. Linear ratios are commonly measured in a semiautomated fashion utilizing a manually operated intercept counter. The point-counting method is accurate and requires very little special equipment.

With point-counting, the test points to be counted are those falling within the images of the microstructural field. The number of points counted, divided by the total number of available test points, gives the volume fraction of the specimen occupied by voids. Ordinarily, the array of available test points is provided by the intersections of a grid inserted in the eyepiece of a microscope. Alternatively, a clear plastic grid can be placed over micrographs and used for point counting. In either case, the grid spacing should be close to the spacing of the microstructural feature of interest, for example, pores. Grid points falling on the boundary of a pore should be counted as one half. Care must be taken in counting phases or voids not to overlook any or to count some twice.

Because point counting is tedious, and special instruments for quantitative metallography are expensive, some organizations judge porosity by comparing the microstructures with standard photomicrographs. This method is simpler, quicker, and suitable for control and acceptance purposes. Direct measurement of weight to volume ratios can be made using wet density techniques. The principal sources of errors in estimating pore volumes are inadequately prepared metallographic specimens and taking too few measurements.

Other aspects of coating structures can be examined by selective etching of the metallographic mount. Etching reveals the coating components by preferential attack or by staining the various phases. Chemical etching is done by swabbing or by immersing the mount. Some coatings require electrolytic etching to reveal the structure. Structures of certain coatings are revealed by the proper use of the chromium trioxide etch polish. Another method is to heat tint the polished surface. The color of the oxides formed on the polished surface relates to the composition of the carbide phases present.

Bond Testing. Several bond test schemes have been developed by different organizations to evaluate coating quality. One method, ASTM 633-69 Standard Test Method for Adhesive or Cohesive Strength of Flame-Sprayed Coatings, measures tensile properties. Specimens, or bond caps, used for this test are nominally 25-mm (1-in.) diameter cylinders. One end is drilled and threaded for attachment to the loading fixture, the other is ground or machined perpendicular to the axis of the cylinder. The finished end is prepared for coatings using the method intended for the process being tested. The bond cap is then coated to a thickness of 0.40 to 0.45 mm (0.015 to 0.018 in.) with the selected material. The coated cap is then cemented to the machined or ground end of an uncoated bond cap using any structural adhesive having 69 MPa (10 000 psi) or greater bond strengths.

The cemented bond caps are pulled in a tensile testing machine at a controlled crosshead speed of 1.0 mm/min (0.04 in./min) and the ultimate strength recorded. Generally, sets of three to seven identical bond caps are tested to obtain an average bond strength. Other coating strength tests have been developed to qualify a particular application; however, none is universally accepted.

Health and Safety

Some hazards to health and safety are present with thermal spray processes, as with most industrial processes. In general, the hazards associated with thermal spray are similar to those encountered with welding processes. Plant safety and local, state, and federal Occupational Safety and Health Administration (OSHA) directives should be followed.

Hazards related to equipment are best avoided by following the particular manufacturer's recommended procedures for inspection and maintenance. Proper care of gas, water, and electrical supply lines and fittings will reduce many hazards.

Dust and Fumes. All thermal spray processes produce fumes. Filtered exhaust hoods are used to collect dust or overspray and to vent the process fumes. A 60 m/min (200 ft/min) air velocity at the workplace is usually adequate air movement to protect personnel. When materials being sprayed are toxic, or when personnel must be in the spray facility during operation, respirators are recommended.

The type of respirator needed is dictated by the particular hazardous material. Again, the plant safety office should be consulted for recommendations and compliance with government regulations.

Noise is generated in each of the thermal spray processes. The range may be 80 dB for some flame spray processes to +140 dB for the supersonic, plasma-arc guns. These values would be measured at the distance from the thermal spray device to the operator's ear. Threshold limit values for unprotected exposure to noise are as follows:

Sound level, dB(a)	Duration per day, h
80	16
85	8
90	4
95	2
100	1
105	1/2
110	1/4
115	1/8

(a) No exposure to continuous or intermittent sound in excess of 115 dB

Ear plugs and/or ear muffs are required according to regulation. Several equipment manufacturers market welding face shields with ear muffs attached. It is recommended that workers who may be exposed to sound levels exceeding threshold limit values be given hearing tests upon assignment, followed by periodic check-ups.

Light Radiation. The spectrum of light emitted by thermal spray devices ranges from far infrared to extreme ultraviolet. Adequate eye and skin protection must be taken when spraying. Shade 5 lenses will usually provide adequate eye protection from flame spray devices. Shade 12 is required for spraying with plasma-arc guns. Fire retardant, closely woven fabrics should be worn to protect the operator from burns. Burns can be caused by heated particles which bounce from the substrate, hot gases, and light. The worst burns are from ultraviolet light. Ultraviolet radiation will burn exposed skin, and will penetrate loosely knit fabrics to cause skin burns similar to sunburn, in minutes.

Terms Related to the Thermal Spray Industry

air cooler. A workpiece cooler, see *workpiece cooler*

anode. The electrode maintained at a positive electrical potential. In most plasma-torch designs, this is the front electrode, constructed as a hollow nozzle and fabricated from copper

apparent density or density ratio. The ratio of the measured density of an object to the absolute density of a perfectly solid material of the same composition, usually expressed as a percentage

arc. A luminous discharge of electrical current crossing the gap between two electrodes

arc chamber. The confined space enclosing the anode and cathode in which the arc is struck

arc gas. The gas introduced into the arc chamber and ionized by the arc to form a plasma

base material. A substrate, see *substrate*

base metal. A substrate, see *substrate*

blasting. A method of cleaning and/or surface roughening by a forcibly projected stream of sharp angular abrasive

bond. (1) To join securely. (2) A uniting force. (3) In thermal spraying, the junction between the material deposited and the substrate, or its strength

bond coating or bonding coat. A thin, intermediate plasma-sprayed layer of a material, for example, molybdenum, applied on the substrate to enhance the adherence of a subsequently sprayed coating

bond strength. The force required to pull a coating free of a substrate

carrier gas. In thermal spraying, the gas used to carry the powdered materials from the powder feeder or hopper to the gun

cathode. The electrode maintained at a negative electrical potential. Usually the rear electrode, conically shaped and fabricated from tungsten or thoriated tungsten

coating density. The ratio of the determined density of the coating to the theoretical density of the material used in the coating process. Usually expressed as percent of theoretical density

coating strength. A measure of the cohesive bond within a coating, as opposed to coating-to-substrate bond; the tensile strength of a coating

coating stress. The stresses in a coating resulting from rapid cooling of molten or semi-molten particles as they impact the substrate

composite coating. A coating consisting of two or more layers of different spray materials

control console. The instrumented unit from which the plasma torch is operated and operating conditions are monitored. Functions controlled and monitored are power level, stabilizing gas pressure and flow, powder-feed gas pressure and flow, and cooling water flow

controlled atmosphere chamber. An inert gas-filled enclosure or cabinet in which plasma spraying or welding can be performed to minimize (or prevent) oxidation of the coating or substrate. The enclosure is usually fitted with viewing ports, glove ports to permit manipulations, and a small separate airlock for introducing or removing components without loss of atmosphere

density. The mass per unit volume of a material, usually expressed as grams/cubic centimeter or pounds/cubic inch

density ratio. An apparent density, see *apparent density*

deposit. A spray deposit, see *spray deposit*

deposition rate. The speed with which material is deposited on a substrate, usually expressed in grams/minute or pounds/hour

dwell time. The length of time the particles spend in the plasma stream

edge effect. Loosening of the bond between the sprayed material and the base material at the edges, due to stresses set up in cooling

electrode. An electrical conductor for leading current into or out of a medium. In arc and plasma spraying, the current carrying components which support the arc

enclosure. For metal spraying, a chamber used for minimizing contamination. See *controlled atmosphere chamber*

exhaust booth. A mechanically ventilated, semi-enclosed area in which an air flow across the work area is used to remove fumes, gases, and overspray material during thermal spraying operations

feed rate or spray rate. The quantity of material passed through the gun in a unit of time

fines. Those particles at the lower end of the specified mesh size

flame spraying. A process in which materials are melted or softened in a heating zone and propelled in a molten or heat-softened (plastic) condition onto a target to form a coating. The term flame spraying is usually used when referring to a combustion-spraying process, as differentiated from plasma spraying or plasma-flame spraying

flow meter. Device for indicating the rate of gas flow in a system

fusion spray. The process in which the coating is completely fused to the base metal, resulting in a metallurgically bonded, essentially void free coating

gradated or gradient coating. A deposit which changes continuously but almost imperceptibly in composition from one surface to another, for example, from 100% material A at substrate to 100% material B at top of the coating

graded coating. A coating consisting of several successive layers of different materials; for example, starting with 100% metal, followed by one or more layers of metal-ceramic mixtures, and finishing with 100% ceramic

grit blasting. The same as abrasive blasting

gun. A term used to identify a thermal spraying device, especially the types used for depositing coatings by the plasma-arc

inert gas. A gas, such as helium, argon, or neon, which is stable and does not form reaction products with other materials

interface. A surface forming a common boundary between adjacent layers, usually the surface between the spray deposit and the substrate

mask. A device for protecting a surface from the effects of blasting and/or coating. Masks are generally either reusable or disposable

mechanical bond. Adherence of a coating to a base material, accomplished mainly through mechanical interlocking with roughened surfaces

metallizing. Forming a metallic coating by spraying with molten metal or by vacuum deposition

metallurgical bond. Adherence of a coating to the base material characterized by diffusion, alloying, or intermolecular or intergranular attraction at the interface between the sprayed particles and the base material; usually stronger than a mechanical bond

open circuit voltage. The potential difference applied between the anode and cathode prior to initiating the arc

overspray. The excess spray material that is not deposited on the part being sprayed

particle-size range. Classification of spray powders defined by an upper and lower size limit; for example, −200 +325 mesh: a quantity of powder, the largest particles of which will pass through a 200-mesh sieve and the smallest of which will not pass through a 325-mesh sieve

pass. A single progression of the thermal spray device across the surface of the substrate

plasma. An electrically neutral, highly ionized gas composed of ions, electrons, and neutral particles

plasma flame. The zone of intense heat and light emanating from the orifice of the arc chamber resulting from energy liberated as the charged gas particles (ions) recombine with electrons

plasma-forming gas. The gas, in the plasma gun, which is heated to the high temperature plasma state by the electric arc

plasma gases. A plasma forming gas, see *plasma forming gas*

plasma gun. See *plasma torch*

plasma spraying. Producing a coating by passing a material in particulate or powder form through a plasma flame and depositing the subsequently heat-softened particles onto a base material or substrate

plasma torch or plasma gun. A device for producing a plasma flame, consisting of an arc chamber, an anode and a cathode, and equipped with cooling water, stabilizing gas, and powder-feed inlets and external power leads

powder feeder. A mechanical device designed to introduce a controlled flow of powder into the plasma-spray torch

powder-feed rate. The quantity of powder introduced into the arc per unit time; expressed in pounds/hour or grams/minute

ppm. Parts per million; 1 ppm is equal to 0.0001%

preheat. Heat applied to the substrate, prior to starting the spray operation, to eliminate condensation on the substrate surface as the first particles are deposited or to minimize residual stresses

primary gas. In thermal spraying, the gas constituting the major constituent of the arc gas fed to the gun to produce the plasma

rotary roughening. A method of surface roughening wherein a revolving roughening tool is pressed against the surface being prepared, while either the work, or the tool, or both, move

rough threading. A method of surface roughening which consists of cutting threads with the sides and tops of the threads jagged and torn

seal coat. Material applied to infiltrate the pores of a thermal spray deposit

secondary gas. In thermal spraying, the gas constituting the minor constituent of the arc gas fed to the gun to produce the plasma

self-bonding materials. Those materials that exhibit the characteristics of forming a metallurgical bond with the substrate in the as-sprayed condition

self-fluxing alloys. Certain materials that wet the substrate and coalesce when heated to their melting points, without the addition of a fluxing agent

sfm. Surface feet per minute; linear pass velocity in surface speed per minute

shadow mask. Method of partially shielding an area during the spraying operation, thus permitting some overspray to produce a feathering at the coating edge

shielding gas. A stream of inert gas directed at the substrate during spraying so as to envelop the plasma flame and substrate; intended to provide a barrier to the atmosphere in order to minimize oxidation

spalling. Flaking or separation from the substrate of a sprayed coating

spray. A moving mass of dispersed liquid droplets or heat softened particles

spray angle. The angle of particle approach, measured from the surface of the substrate to the axis of the spray nozzle

spray deposit. A coating applied by any of the thermal spray methods

spray distance. The distance maintained between the gun nozzle and the substrate surface during spraying

spray rate. The same as feed rate, see *feed rate*

spraying sequence. The order in which different passes of similar or different materials are applied in a planned relationship, such as overlapping, superimposed, or at certain angles

stabilizing gas. The arc gas, which is ionized to form the plasma. Introduced into the arc chamber tangentially, the relatively cold gas chills the outer surface of the arc stream, tending to constrict the arc, raise its temperature, and force it out of the front anode nozzle in a steady, relatively unfluctuating stream

substrate. The material, workpiece, or substance on which the coating is deposited

substrate preparation. The set of operations, including cleaning, degreasing, and roughening, applied to the base material prior to applying a coating; intended to ensure an adequate bond to the coating

substrate temperature. The temperature attained by the base material as the coating is applied. Proper control of the substrate temperature by intermittent spraying or by the application of external cooling will minimize stresses caused by substrate and coating thermal expansion differences

surface preparation. The operations necessary to prepare a surface for thermal spraying

surface roughening. The process of producing irregularities on the surface to be thermal sprayed. See *blasting, rotary roughening, rough threading*, and *threading and knurling*

thermal spray. Any coating process in which particles are heated to a molten or plastic state and propelled onto a substrate to form a coating; includes flame and plasma spraying using wire or powder processes

threading and knurling. A method of surface roughening wherein spiral threads are made and the tops of the threads are spread with a knurling tool

torch. Usually, a gas burner with feed lines for fuel and oxygen used to braze, cut, weld, or to heat material to be sprayed

transferred arc. A plasma-spray process in which the workpiece is the anode for the plasma arc

traverse speed. The lineal velocity at which the torch is passed across the substrate during the spraying operation

undercoat. A deposited coat of material which acts as a substrate for a subsequent thermal sprayed deposit. See *bond coat*

water wash. The forcing of exhaust air and fumes from a spray booth through water so that the vented air is free of thermal sprayed particles or fumes

wire flame spraying. Flame spraying metallic material fed to the torch or gun in wire or rod form

wire flame spray gun. A flame spraying device utilizing an oxyacetylene flame to provide the heat, and the metallic material to be sprayed in wire or rod form

wire speed. The length of wire sprayed in a unit of time

wire straightener. A fixture for taking the curve out of coiled wire to enable it to be fed into the gun without binding

workpiece cooler. A device used to direct an air blast onto a part being sprayed to prevent overheating of the part

REFERENCES

1. *A Plasma Flame Spray Handbook*, Naval Station, Louisville, KY, 1977
2. Thermal Spraying, *Welding Handbook*, Chapter 29, American Welding Society, Miami, 1973
3. Ingham and A. P. Shepard, *Flame Spray Handbook*, Vol I-II-III, Metco, Inc., Westbury, NY

Oxidation Protective Coatings for Superalloys and Refractory Metals

By Edwin S. Bartlett
Senior Researcher
Battelle Columbus Laboratories
and
Harry A. Beale
Associate Section Manager
Battelle Columbus Laboratories

OXIDATION protective coating development for superalloys and refractory metals has been spurred by advances in propulsion technology for aircraft and space vehicles since about 1960. These advances have placed increasing temperature and structural demands on materials for service at temperatures of 1010 °C (1850 °F) and above. This, coupled with weight-associated penalties for flight systems, has driven materials and design technology to maximize the hot strength of structural components.

To achieve the mechanical properties demanded by modern gas turbine technology, superalloys (complex nickel- and cobalt-based alloys) have had to sacrifice the oxidation and corrosion resistance that was inherent in previously used heat-resistant alloys with higher chromium contents. Consequently, coatings are relied on to protect superalloy components such as turbine blades and vanes from environmental attack. This advanced technology was initially required for advanced military aircraft gas turbines. The performance and fuel economy advantages associated with advanced turbine technology proved sufficiently attractive to cause advanced materials technology application, including the use of coatings, to be used in commercial aircraft, marine, and stationary turbines.

Refractory metals and their alloys are not oxidation resistant. For service of any duration in air or other oxidizing media at high temperatures, they must be protected by coatings, although uncoated tungsten bodies are used as oneway rocket nozzle inserts. The refractory metals, primarily niobium, molybdenum, and tungsten, are used to some extent for space propulsion and vectoring systems involving hot, oxidizing gases. A coated niobium alloy is now in production in at least one military aircraft engine, although it is not being used in the turbine section.

For both superalloy and refractory metal substrates, the exclusive purpose of the coating is to prevent corrosive attack of the substrate for the maximum possible time with the maximum degree of reliability. Coatings that are used for protection against environmental attack are not in equilibrium with the substrate. At the high temperatures involved, interdiffusion between coating and substrate occurs, although at a relatively low rate in well-balanced coating/substrate systems. Mechanically, physically, and chemically, coatings and substrates differ. In the design, development, and application of coated systems, the objective is not only to apply a coating material that provides corrosion protection, but also to result in a coated hardware system that is adequate in all respects for the intended service.

Coatings for Superalloys

Advanced, high-strength superalloys do contain chromium and/or aluminum additions that confer to them some resistance to corrosion at high temperatures by the formation of protective chromium or aluminum-rich oxides. For either chromia or alumina formers to exhibit satisfactory lives under cyclic conditions, the chromium content in the original alloy should be about 20 wt%. This level of chromium is not compatible with the high-temperature strength and microstructural stability demanded for the most rigorous service conditions. Alloys with 5 to 15 wt % chromium, typical of advanced superalloys, require protective coatings.

Requirements. Gas turbine applications are the major impetus of coating technology for superalloys. The two basic types of turbine section environmental attack that must be combated are oxidation and hot corrosion. Hot corrosion, or sulfidation, occurs in two forms: (*a*) a high-temperature (900 to 1050 °C, or 1650 to 1920 °F); and (*b*) a low-temperature (680 to 750 °C, or 1255 to 1380 °F) manifestation. Hot corrosion is triggered by the presence of sulfur in fuel and impurities, such as sea salt, in the ingested air. Both hot corrosion and oxidation involve rapid attack and consumption of hardware, and coatings are designed to resist the specific attack expected for particular turbine applications. Marine propulsion turbines, for example, are susceptible to both forms of hot corrosion, whereas aircraft gas turbines are more subject to oxidation, and sometimes high-temperature hot corrosion, if the aircraft operates consistently in a coastal or marine environment.

In addition to corrosion resistance, coatings for superalloys are required to resist both thermal cycling, often at rapid rates, and mechanical forces acting on hardware, without cracking. The protective oxide film (alumina), on which coatings for superalloys rely for protection, does not readily span even small gaps. Cracking of protective coatings is soon followed by a breakdown in protection, and localized substrate attack follows shortly. The thermal expansion match between coating and substrate is important, and coatings with a modicum of ductility are desirable from a corrosion standpoint. The coatings also must withstand combustion char and solid particulates ingested during turbine operations that result in erosive action.

Coatings must be reasonably compatible with the substrates to which they are applied. Components of the coating and the method of coating application should be selected to avoid undesirable reaction phases between coating and substrate and rapid penetration of the substrate by coating elements. The presence of such phases and/or interdiffusion leads to void formation, or cracking at the interface, and coating spall. These imperfections compromise the mechanical performance capability of the system and the protection it provides.

Types of Coatings

As a result of many years of development and use in government and industry, two basic types of coatings for superalloys have emerged. These are diffusion coatings and overlay coatings. Both depend on the sacrificial oxidation of aluminum in the coating to provide a spall-resistant, protective alumina coating to prevent or strongly inhibit further attack.

Diffusion Coatings. In the diffusion coating application process, aluminum is made to react at the surface of the substrate, forming a layer of monoaluminide (known generically as MAl). For coatings applied over nickel-base superalloys, nickel aluminide (NiAl) is the resulting species, and over cobalt, it is cobalt aluminide (CoAl). This type of coating is modified to some extent by the elements contained in the substrate as the diffusion reaction proceeds, and usually is further modified by other metallic elements intentionally added during the coating process to improve coating and/or system performance for specific expected service conditions.

Overlay coatings do not rely on reaction with the substrate for their formation, although some moderate interdiffusion usually occurs during service. Rather, the material applied over the substrate during the coating process and the specific method of processing determine the composition and microstructure of the coating (see Fig. 1). Coatings of this type in current use are generically called MCrAlY coatings, and essentially comprise a monoaluminide (MAl) component contained in a more ductile matrix of solid solution (γ) in the case of cobalt-chromium-aluminum-yttrium (CoCrAlY) coatings, or a mixture of γ and $\gamma'(Ni_3Al)$ phase in the case of nickel-chromium-aluminum-yttrium (NiCrAlY) coatings. Nickel-cobalt-chromium-aluminum-yttrium (NiCoCrAlY) and iron-chromium-aluminum-yttrium (FeCrAlY) modifications are also available. The matrix is nickel- or cobalt-based, and contains a rather large amount of chromium and an intermediate amount of aluminum. The supply of aluminum for formation of protective alumina scales comes largely from the dispersed MAl phase during the useful life of such coatings. The contained chromium is important in combating hot corrosion. The solid solution matrix provides ductility in this coating class that is generally not possible with diffusion coatings (see Fig. 2), and imparts much improved resistance to thermal fatigue cracking (see Fig. 3). A small amount of yttrium is usually included in overlay coatings to improve the adherence of the oxidation product. CoCrAlY coatings are recognized as being superior in hot corrosion resistance, whereas NiCrAlY coatings possess the better oxidation resistance. The range of NiCoCrAlY coatings may be tailored for a desired compromise between oxidation and hot corrosion resistance. The composition of overlay coatings, as well as that of the substrate, also affects the extent of coating-substrate interdiffusion during service. Coating composition may also be adjusted to achieve a good thermal expansion fit with a given superalloy substrate. MCrAlY coatings may contain 15 to 25% chromium, 10 to 15% aluminum, and 0.2 to 0.5% yttrium.

Service Characteristics. Coatings for superalloys, whether diffusion or overlay types, are applied with thicknesses of 50 to 150 μm (2 to 6 mils). Protective life increases in proportion to thickness. Thicker coatings in a given system are more prone to cracking in response to thermal cyclic-induced stresses. Thus, a hot military aircraft engine that is required to provide rapid acceleration and consequent rapid and extreme temperature changes in the turbine section may preclude the use of thicker coatings. Frequent inspection and short time between overhauls (TBO) of 100 to 500 h may be needed.

Fig. 1 Microstructure of two-phase cobalt-chromium-aluminum-yttrium (CoCrAlY) overlay coating

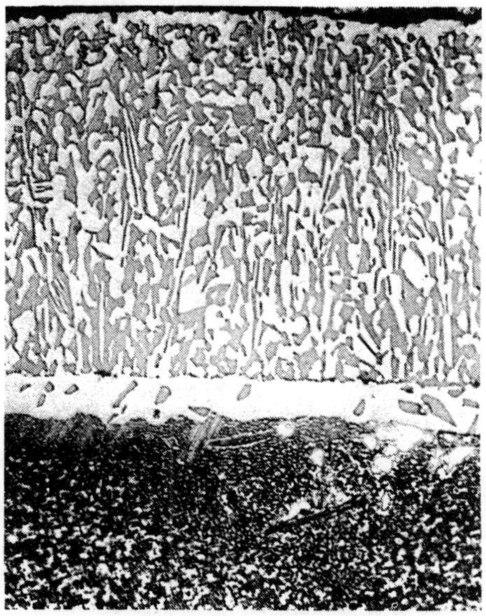

Source: "Protective Coatings for High Temperature Alloys State of the Technology", written by G. William Goward, *Proceedings of the Symposium on Properties of High Temperature Alloys*, 1977, p 806. This figure was originally presented at the 1976 Fall Meeting of The Electrochemical Society, Inc. held in Las Vegas, Nevada

Fig. 2 Ductility of cobalt-chromium-aluminum-yttrium (CoCrAlY) overlay and diffusion aluminide coatings

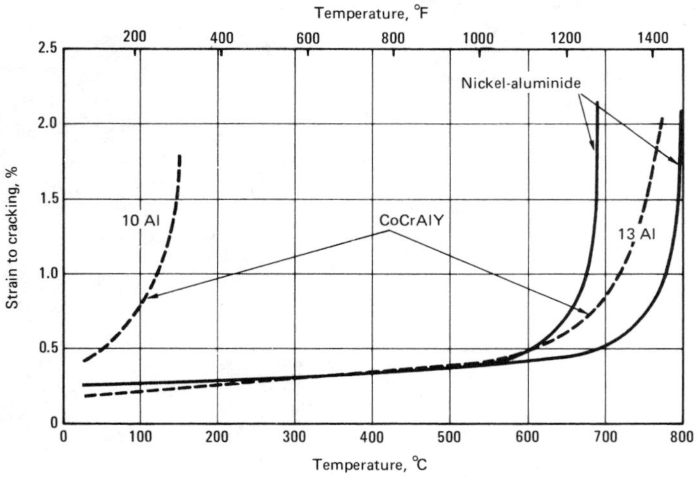

CoCrAlY overlay coatings show significant ductility (1 to 3%) at lower temperature ranges, ambient to 650 °C (1200 °F), with ductility continuously increasing as the volume fraction of CoAl is decreased

Engines for commercial aircraft, some military aircraft, and power generation service may be designed for more moderate temperature and loading conditions, and protective coatings can approach the higher end of the thickness range with less risk of damage through cracking and spalling. Longer lifetimes are achieved, and the time between overhauls can run into thousands of hours. During overhaul, turbine blades and vanes that have not exceeded creep limits and are not otherwise severely eroded or damaged are refurbished for reuse. Coatings are stripped. The parts are reworked and cleaned as necessary, recoated, and returned to service. Some coatings need not be completely stripped before recoating if weld repair is not required.

Methods of Applying Diffusion Coatings

Many organizations have the capacity and facilities to apply diffusion coatings to superalloys. These include the manufacturers of gas turbine engines, suppliers of turbine hardware, specialty coating houses, and users such as military aircraft maintenance facilities and commercial airlines. Two basic methods for the application of diffusion coatings are used in these operations, but many variations of the methods exist. One basic operation is the application and diffusion of the coating in essentially one operation, although a postapplication heat treatment is sometimes used to complete the diffusion process. The other comprises first applying a green coating, which may be applied by mechanical means before diffusing and which contains unreacted coating species, and then heat treating to accomplish the final formation of the coating by fusion/diffusion processes. For some coatings, both methods are combined.

Some aspects of most processes for coating superalloy components are proprietary to the coating organization. Specific processes, and sometimes operating personnel, must be properly certified for processing gas turbine components.

Pack cementation is the most widely used process for applying diffusion coatings to superalloys. See the article on aluminum coating of steel in this Volume. Briefly, parts to be coated are cleaned by pickling and/or dry grit blasting. Blade roots or other portions of hardware not requiring coating are masked. Parts are then embedded in a blended powder mixture of the active coating species (aluminum or a high-melting, aluminum-rich compound, such as Co_2Al_5), inert filler material (alumina powder), and a halide activator or carrier (for example, ammonium chloride or ammonium fluoride) contained in a retort (Inconel or Hastelloy). The aluminum source usually constitutes between 5 and 25% of the powder weight, and the activator between 1 and 5%.

A lid is placed on the retort, which may or may not be sealed, evacuated, or purged with inert gas. With ammonium halides, packs are self-purging. Some variations include an expurgator, such as urea, in the pack mix. The retort is then placed in a furnace and heated for a few hours at temperatures between about 705 and 1095 °C (1300 to 2000 °F). When the cycle is completed, retorts are removed from the furnace, cooled, opened, and coated parts are recovered and cleaned by brushing or light blasting. Recesses, for example, passages in cooled vanes or blades, may require reaming with a wire to remove impacted and partly sintered pack mix. The pack mix sometimes is reused with rejuvenating additions.

Other methods for providing aluminum in the vicinity of the part to be coated include brushing, dipping, or spraying to apply a green cocoon containing coating elements to the part. The cocoon may also contain elemental

Fig. 3 Thermal mechanical fatigue behavior of brittle (high aluminum) and ductile (low aluminum) overlay coatings, compared with aluminide diffusion coating

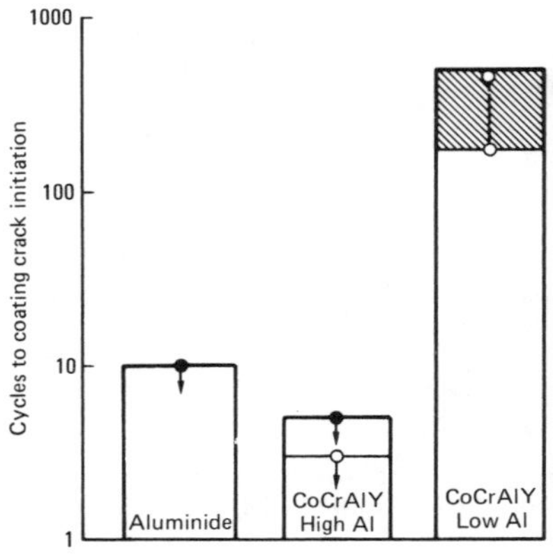

When thermal strain reaches its maximum value at relatively low temperatures, substantial improvement in thermal fatigue behavior can be achieved with a ductile MCrAlY coating. Cycle: ○: 425 °C (800 °F), 925 °C (1700 °F) ±3% strain ●: 425 °C (800 °F), 925 °C (1700 °F) ±0.25% strain

Fig. 4 Typical diffusion aluminide coatings on nickel-based superalloys

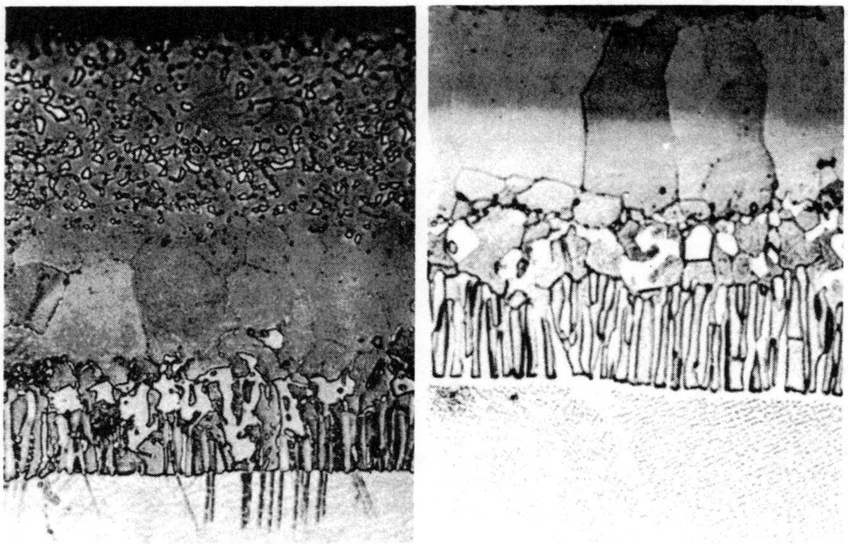

Magnification: 400×. (a) Inward, (b) Outward. (Source: "Protective Coatings for High Temperature Alloys State of the Technology", written by G. William Goward, *Proceedings of the Symposium* on Properties of High Temperature Alloys, 1977, p 806. These figures were originally presented at the 1976 Fall Meeting of The Electrochemical Society, Inc. held in Las Vegas, Nevada)

aluminum or an aluminum-rich compound. Activators may or may not be required, and if required, may either be included in the cocoon, or slip cast mix, or delivered in a heat treating retort. Electrophoretic deposition of coating elements to provide a very uniform green coating has recently come into use. All such methods require heating the green-coated parts to carry out the diffusion reaction and develop the final desired coating.

To a limited extent, pack cementation processes can incorporate modifiers such as silicon, manganese, and chromium into the coating for control of properties. Preapplication methods followed by diffusion are somewhat more versatile in modification capabilities. Some processes have been developed that first apply a modifying layer, for example, electrodeposition of nickel-chromium or platinum, then treat via pack cementation to aluminize the surface-modified substrate, altering the coating composition and performance.

Diffusion coatings for superalloys may be classed as either inward or outward diffusion coatings (see Fig. 4). For inward diffusing coatings, aluminum is the net diffusing specie, diffusing through the coating as it forms, and reacting at the interface of coating and substrate. This class is usually formed by conducting the diffusion reaction process at a relatively low temperature (about 700 to 800 °C or 1290 to 1475 °F) and results in an as-applied coating approximating the δ Ni_2Al_3 phase. Such coatings must be heat treated at a higher temperature, such as 1040 to 1095 °C (1900 to 2000 °F), to convert the low-melting δ phase to the more refractory monoaluminide phase for service. Because the coating is initially formed by the inward diffusion of aluminum, substrate modification of the coating is maximized as substrate elements are locked in place. The outer layer of final-processed inward diffusion coatings is multiphased and, depending on the final heat treatment, can be quite rich (hyperstoichiometric) in aluminum. In practice, inward diffusion coatings are applicable only to nickel-based superalloys. Cobalt-based superalloys, which do not contain aluminum can form brittle, refractory phases at the substrate coating interface during service. This compromises coating integrity.

Outward diffusion coatings are formed by applying and diffusing coatings at relatively high temperatures from pack mixes or preapplied green coatings with lower aluminum activity. Under these conditions, coatings are formed by the selective diffusion of nickel or cobalt outward through the monoaluminide layer. These coatings are modified to a lesser degree by the slower diffusing refractory elements in the substrate than are inward diffusion coatings, and the overall aluminum contents are hypostoichiometric, providing somewhat improved ductility in the coating when compared to inward diffusion coatings. Outward diffusion coatings can be modified by elements supplied during the coating process, such as silicon, manganese, or chromium added to the cocoon or pack mix, whereas inward diffusion coatings usually cannot. Organizations that apply diffusion coatings to superalloys have several possible options in selecting methods of application, compositional modifications, and devices for structural control. This allows a coating to be tailored to a particular substrate to meet particular service targets in the design of turbine hardware.

Methods of Applying Overlay Coatings

Three methods have been developed for applying overlay coatings to superalloys. Of these methods, two are used routinely. Physical-vapor-deposition (PVD) by melting and evaporating a coating material source bar with a focused electron beam (EB) in an evacuated chamber is one commercial method. Plasma-spraying prealloyed powders is the other (see the article on thermal spray coatings in this Volume). Sputtered overlay coatings of good quality can be achieved, but this method of application is not yet used commercially. Other processes, such as slurry sinter or slurry fusion, followed by aluminizing, have been studied in considerable detail. Although they possess potential economic and operational advantages, at present it does not seem possible to use them to apply the compositional, structural, and performance characteristics desired in overlay coatings of the established MCrAlY class.

All current methods for applying overlay coatings are limited to line-of-sight application. Only surfaces directly exposed to the flow of material emanating from the plasma torch or vapor flux can be coated with acceptable uniformity. Vapor-phase diffusion coatings, by contrast, allow internal passages and recessed details to be quite effectively coated.

Physical-Vapor Deposition. In applying EB-PVD overlay coatings, the cleaned and masked component to be coated is heated and rotated on a spindle within a vacuum chamber above the streaming vapor generated by electron beam heating of the molten source pool contained in a cooled cylinder. The rate of deposition is determined by the power input to generate superheat in the molten pool. Coating time to deposit a predetermined coating thickness depends on the deposition rate. The substrate temperature is important in establishing coating quality and adherence. As the coating material evaporates from the source pool, the solid source feed rod is continuously fed upward to maintain a constant location for the surface of the molten pool. In depositing MCrAlY coatings, the streaming vapors and composition of the coating closely approximate the composition of the source material, although some control over composition can be exercised by controlling the temperature of the source pool (EB power level). EB-PVD-coated hardware is usually heat treated after the coating is applied for final development of coating microstructure and substrate properties, as well as to stabilize the interface between coating and substrate.

Plasma-sprayed overlay coatings are applied by entraining prealloyed powders in the plasma jet and transferring them to an unheated substrate where the molten or semimolten droplets impact and solidify. This can be done in an ambient environment, but superior quality coatings applied in a closed and pumped chamber with reduced pressure of inert gas have been developed. Such controlled atmosphere spraying minimizes oxidation of the coating during the manipulation required for spraying. As-deposited coatings are rough and exhibit a laminated texture. Postdeposition operations include glass bead peening to smooth the surface and heat treatment to homogenize the coatings and develop the desired microstructure. In the usual manual deposition method, considerable operator skill is required to obtain coatings of uniform thickness. Research to develop automated systems for plasma spraying overlay coatings on airfoil shapes is in progress.

Sputtering is a chamber process wherein thermally emitted electrons collide with inert gas (argon) atoms, which accelerate toward and impact a negatively charged electrode that is a target of the coating material. The impacting ions dislodge atoms of the target material, which are in turn projected to and deposited on the substrate to form the coating. As with EB-PVD, coating adherence, structure, and quality depend on the degree to which the substrate is preheated. The development of economic, high-rate sputtering processes for the deposition of high-quality overlay coatings of the MCrAlY type has not yet been achieved, although the potential for such a process is attractive.

Economics. The per-part costs of coating application may be generically ranked according to capital equipment requirements and the complexity of processes. As a class, diffusion coatings for superalloys are not as expensive as overlay coatings. Principal equipment requirements, in addition to cleaning and inspection facilities needed for all types of coatings, are facilities for blending pack mixes (cementation coatings) or for mixing and dipping, brushing, or spraying coating materials, retorts, and furnaces.

Cost for units for the application of EB-PVD coatings is quite high, and as a consequence, EB-PVD coatings are probably the most expensive. Plasma-sprayed overlay coatings are intermediate in cost between diffusion coatings and EB-PVD coatings, principally because these are applied manually, with each piece handled individually. The expected development of automated plasma spray facilities may increase the equipment complexity and cost, but should decrease the processing cost per piece because of reduced labor requirements and reworking.

Coatings for Refractory Metals

Refractory metals and their alloys have long been of interest for service in oxidizing environments at temperatures greater than 1095 to 1205 °C (2000 to 2200 °F). At these high temperatures, consumption of these materials by oxidation, and degradation of strength and toughness by the absorption of oxygen, occur at very high rates of some one to three orders of magnitude more rapidly than uncoated superalloys at their maximum service temperatures. Alloying, to the extent that metallic behavior is retained, has little influence on this natural behavior. Coatings are thus essential for service in oxidizing environments.

Requirements. For coated superalloys, a crack or other breakdown of the protective system is not apt to be immediately catastrophic, because the substrate itself possesses some resistance to corrosion. This condition is not present with refractory metal alloys. For critical high-temperature refractory metal components, the protection system must exhibit extremely high reliability.

The smallest defect in a coating that has no self-healing capability cannot be tolerated. For this reason, coatings that react to some extent with oxygen to form products of oxidation that plug small defects and prevent access of the oxidizing environment to the refractory metal substrate are necessary. Most applications involving high temperatures are intermittent and require repeated heating and cooling. The refractory metals as a class possess low thermal expansion properties, which poses a severe limitation on the selection of coating species for cyclic service.

Other characteristics that a coating may be required to possess depending upon the service spectrum include:

- Ability to withstand rapid corrosive attack at all temperatures up to the maximum service temperature
- Ability to accommodate deformation without loss of protective ability
- Stability under various environmental pressure and velocity regimes, and under anticipated thermal cycle spectra
- Good chemical and mechanical compatibility with the substrate

Substrate Characteristics. Niobium, molybdenum, tantalum, tungsten, and their alloys are the refractory metals. Niobium and tantalum are ductile, low-modulus refractory metals, and alloying with tungsten, hafnium, zirconium, or rhenium is required for structural usefulness at high temperatures. Such alloys are reactive; they readily dissolve oxygen and are embrittled by this phenomenon. Molybdenum and tungsten are inherently strong, but they are not as tough as niobium or tantalum alloys at low temperatures and are not reactive; however, they readily oxidize. Niobium and molybdenum alloys possess relatively low densities and melting temperatures of roughly 2400 to 2595 °C (4350 to 4700 °F). The densities of tantalum and tungsten alloys are roughly twice those of the lower melting refractory metals, but their melting temperatures are higher (roughly 2980 to 3400 °C or 5400 to 6150 °F). Because of these considerations, niobium and molybdenum alloys are considered to be structurally useful up to perhaps 1400 °C (2550 °F), and only tantalum and tungsten alloys are considered useful at higher temperatures. On a strength/density basis, tantalum and tungsten alloys are not particularly competitive with alloys of niobium or molybdenum at temperatures less than about 1400 °C (2550 °F). Thus, the probable maximum use temperatures for structural application are between about 1100 °C (2010 °F), which is near the upper limit for superalloys, and 1400 °C (2550 °F) for alloys of niobium or molybdenum, and from about 1400 °C (2550 °F) upwards for tantalum and tungsten alloys.

Coating Types and Service Characteristics. Intensive development of refractory metal and protective coating technology was conducted during the 1950's, 1960's, and early 1970's. Most of the coating development activity was concentrated on silicide and aluminide diffusion coatings applied by various modifications of pack cementation, fluidized-bed deposition, and spray- or dip-and-sinter processes. Both aluminide and silicide coatings have been used for space vehicle applications, the most publicized being a fused slurry aluminide coating on a niobium alloy for the large Apollo service propulsion system nozzle extension skirt. More than 40 coatings for niobium alloys, and about half that number of molybdenum alloys, had been brought to at least near-commercial status by the early 1970's by numerous organizations. For several of these, coatings lasting 100 h or more in cyclic service at temperatures between 1100 and 1400 °C (2010 and 2550 °F) were demonstrated to have reliabilities in excess of 90%. Because of the difficulties of protecting tantalum alloys and tungsten for reasonable times at the very high temperatures where these alloys would be used, no truly long-lived, reliable protection systems were developed, although some systems with capabilities for several to a few tens of hours were demonstrated.

For cyclic service, aluminide coatings for refractory metals are inferior to silicide coatings because of (*a*) the thermal expansion match is much less favorable; and (*b*) silicide coatings can be more easily modified to impart some self-healing at cracks via viscous flow of a protective silica glass. Although silicide coatings do crack during cyclic service, the presence of cracks does not necessarily result in the early loss of protective capability.

Coated refractory metal heat shields, hot structures, and re-entry and hypersonic aircraft thermal protection systems have been demonstrated in several government funded efforts. Although these programs demonstrated the capability for coatings in protecting structural substrates, sufficient reliability for mission-critical components for man-rated systems was generally not demonstrated. Laboratory studies on the best coating/substrate systems have indicated reliabilities between 99 and 99.9% or 1 to 10 premature failures per 1000 parts. At present, only a few specialty coating organizations offer protective coatings for niobium and molybdenum alloys on a commercial or semicommercial basis. No commercial coatings or processes for coating tantalum or tungsten alloys are available in 1982.

Current Coating Systems. For niobium alloys, aluminide and silicide coatings are commercially available. The aluminide coating is a modified $NbAl_3$ coating applied by slurry dipping and fusion sintering which diffuses the melt with the substrate to form the coating. Two companies commercially apply silicide coatings; both are slurry fusion coatings. One is modified with iron and chromium, and the other with hafnium and tantalum. Both, in the final-diffused coating, contain $NbSi_2$ phase and other modified silicides. The modified silicides are important to performance, suppressing pesting, or a form of rapid attack that occurs at low-to-intermediate temperatures, and modifying the silicon dioxide protective layer for viscous flow. Both aluminide and silicide coatings may be applied by dipping or spraying, and special techniques have been developed to provide uniform coating coverage at the edges of components. The green coatings are cured and then fired at high temperatures to fuse the coatings and to allow the required interdiffusion with the substrate to optimize performance.

Coatings for molybdenum and its alloys that are currently available are applied by pack cementation processes from powder mixes containing silicon, a halide carrier or activator, inert filler, and modifiers, for example, chromium or boron, which are codeposited with silicon. Coatings are applied at high temperatures, and a coating of predominantly molybdenum silicide ($MoSi_2$) results from displacement reaction and diffusion during the coating process.

Numerous organizations possess capabilities for applying a range of coatings to refractory metal alloys on a custom or experimental basis. In many cases, past experience may be useful in defining coating compositions and processes that might be expected to meet service requirements. However, these operations remain largely experimental in nature.

Recommended Reading. Technical literature and government reports describe many aspects of coating processes and performance evaluations on a selective basis. For additional, more detailed background, a Materials Advisory Board report: High-Temperature Oxidation-Resistant Coatings, ISBN 0-309-01769-6, L.C. catalogue 78-6062-78 (1970), is recommended.

Chemical Vapor Deposition

By John M. Blocher, Jr.
CVD Consultant

CHEMICAL VAPOR DEPOSITION (CVD) is a process which in some ways is like the gas carburizing and the carbonitriding processes. In the process, a reactant atmosphere gas is fed into the processing chamber where it decomposes at the surface of the workpiece, liberating one material for either absorption by or accumulation on the workpiece. A second material is liberated in gas form and is removed from the processing chamber, along with excess atmosphere gas, as a mixture referred to as off-gas.

Metals

- Cu, Be, Al, Ti, Zr, Hf, Th, Ge, Sn, Pb, V, Nb, Ta, As, Sb, Bi, Cr, Mo, W, U, Re, Fe, Co, Ni, Ru, Rh, Os, Ir, Pt

Graphite and carbides

- C, B_4C, SiC, TiC, ZrC, HfC, ThC, ThC_2, VC, NbC, Nb_2C, TaC, Ta_2C, CrC, Cr_4C, Cr_7C_3, Cr_3C_2, MoC, Mo_2C, WC, W_2C, V_2C_3, VC_2

Nitrides

- BN, TiN, ZrN, VN, NbN, TaN, Si_3N_4

Boron and borides

- B, AlB_2, TiB_2, ZrB_2, ThB_4, ThB, NbB, TaB, MoB, Mo_3B_2, WB, Fe_2B, FeB, NiB, Ni_3B_2, Ni_2B

Silicon and silicides

- Si and various silicides of Ti, Zr, Nb, Mo, W, Mn, Fe, Ni, Co

Oxides

- Al_2O_3, BeO, SiO_2, ZrO_2, Cr_2O_3, SnO_2

Reactant atmospheres used in CVD include chlorides, fluorides, bromides, and iodides; carbonyls, organometallic compounds, hydrides, and hydrocarbons. Hydrogen is often included as a reducing agent. The reactant atmosphere must be reasonably stable until it reaches the substrate, where reaction must occur with reasonably efficient conversion of the reactant. It is sometimes necessary to heat the reactant to produce the gaseous atmosphere.

A few reactions for deposition occur at substrate temperatures below 200 °C (390 °F). Some organometallic compounds deposit at temperatures below 600 °C (1110 °F). Most reactions and reaction products require temperatures above 800 °C (1470 °F).

Attention must be given to the compatibility of the coating material, the substrate material, and the coating atmosphere in terms of chemical, mechanical, and physical properties, particularly as related to temperature. Because of such limitations, plasma-assisted chemical vapor deposition (PACVD), in which an electrical discharge is maintained in the gas phase adjacent to the substrate, is being used increasingly. The presence of activated and ionic components in the discharge permits deposition at temperatures lower than are otherwise required. However, because of the low deposition rate of PACVD, measured in angstroms per minute rather than micrometres per minute for conventional CVD, the use of PACVD is generally limited to films of less than micrometre thickness, such as those widely used in microelectronic integrated circuits.

Advantages and Limitations

Because of the wide range of properties of the reactants used, a single piece of equipment can only be used for a limited variety of coatings and substrate shapes.

Advantages and limitations of chemical vapor deposition are:

Advantages

- Deposition of refractory materials at temperatures far below their melting points or sintering temperatures
- Near-theoretical or controlled density of deposit

- Preferred grain orientation of deposit possible
- Controlled grain size of deposit
- Epitaxial grain growth possible
- Best method for depositing in holes and recessed areas
- Processing at atmospheric pressure possible
- Generally good bonding

Limitations

- High cost of some reactants
- Corrosive, toxic, or moisture sensitive characteristics of most reactants necessitating a closed system
- Scarcity of reactions available for low-temperature processing (<300 °C or <570 °F)
- Low utilization of material

Table 1 Typical properties of CVD nickel

Source of carbonyl(a)	Yield strength 10^{-3} kPa	Yield strength kpsi	Tensile strength 10^{-3} kPa	Tensile strength kpsi	Elongation in 50 mm (2 in.), %	Hardness, HRB
A	545	80	700	100	17	95
B	485	70	640	92.5	19	100
C	605	85	820	120	16	91

(a) A and B are two batches from one supplier; C is one batch from a second supplier, not otherwise identified
Source: W. C. Jenkin, A New Engineering Material, Nickel Deposited by Chemical Deposition, Pyrolytic Co., Barberton, OH

Common CVD Coatings

Deposits produced by the CVD process are most frequently applied for purposes of corrosion resistance, wear resistance, or the production of shapes, especially of difficult to machine materials. Of the potentially useful CVD metal coatings, nickel, tungsten, and chromium are the most important commercially. Of the nonmetal coatings on metal substrates, titanium carbide is the most important. Titanium carbide is used on alloy steel substrates as a wear-resistant coating for punching and embossing tools, and on a cemented carbide cutting tool to impart wear resistance. About 60% of the disposable inserts in use in 1981 were CVD coated, using either titanium carbide, titanium nitride, aluminum oxide, or combinations of these.

Nickel. CVD nickel is generally separated from a nickel carbonyl, $Ni(Co)_4$, atmosphere. The properties of the deposited nickel (Table 1) are equivalent to those of sulfamate nickel deposited electrolytically. Because of the greater cost of CVD nickel, coating or forming with CVD nickel is limited to applications such as coatings on the inside of long narrow tubes or as a mold material for plastics. In the latter case, the vapor-formed molds are more uniform in thickness and structure at inside corners as compared to electroforms, thereby avoiding weakness encountered in those areas. Although the CVD nickel mold may be roughly twice as expensive as its electroformed counterpart, its longer life more than offsets the cost differential.

Because of the low deposition temperature of 150 to 180 °C (300 to 355 °F), the retort may be heated resistively or by heat lamps. Deposition may be carried out at atmospheric pressure with a gas mixture consisting of 80 mol% nickel carbonyl and 20 mol% carbon monoxide. The purpose of the carbon monoxide is to suppress precipitation of nickel from the processing atmosphere. Under these conditions, a deposition rate of 2 μm (0.08 mil) per minute is typical. Because of the low temperature used in carbonyl deposition, diffusion of nickel atoms into the substrate is suppressed, and the deposit remains fine-grained in the order of 5 μm (0.2 mil) across.

The waste atmosphere must be stripped of unreacted nickel carbonyl and preferably of carbon monoxide before being vented outside. This is done by passing the waste atmosphere through a tube packed with copper or steel wool held at 250 to 500 °C (480 to 930 °F), which decomposes most of the unreacted carbonyl. The gas is then passed through a burner to convert the carbon monoxide to carbon dioxide and to destroy the last traces of nickel carbonyl, which is more toxic than the carbon monoxide it contains.

Tungsten may be deposited by thermal decomposition of tungsten carbonyl at 300 to 600 °C (570 to 1110 °F), or it may be deposited by hydrogen reduction of tungsten hexachloride at 700 to 900 °C (1290 to 1650 °F). The most convenient and most widely used reaction is the hydrogen reduction of tungsten hexafluoride.

Each tungsten deposition method has advantages and disadvantages. An advantage of using the carbonyl process is limiting the stress resulting from differential contraction of the coating and substrate in cooling from the deposition temperature which may be as low as 300 °C (570 °F). The advantage of lower corrosivity of the reactant atmosphere is particularly important in the case of reactive substrates such as titanium or its alloys. The unavoidable codeposition of up to 1 at.% carbon may be unacceptable.

Tungsten deposition from the hexachloride at higher temperatures develops a preferred, surface-parallel, (110) grain orientation, which gives maximum emission in thermionic converters. Deposition from the hexafluoride produces the less favorable (100) grain orientation, which is unstable in heating to 2000 °C (3630 °F). Deposition from the hexafluoride is perhaps more convenient, because the hexafluoride is an easily vaporized liquid.

In contrast to the noncorroding atmosphere of the carbonyl decomposition processes where almost any material capable of withstanding the process temperature can be used for the processing equipment, the hexafluoride atmosphere is aggressive. For example, glass and silica are attacked, so nickel or nickel-based alloys are used to contain the process. Axially symmetric parts are rotated during deposition to promote uniformity. An atmosphere can consist of 5 mol% tungsten hexafluoride with hydrogen used at atmospheric pressure or at a fraction of an atmosphere. A cold wall reactor in which the substrate is heated preferentially to temperatures in the range of 500 to 600 °C (930 to 1110 °F), with the retort heated only enough to prevent condensation of the reactants and reaction by-products is used.

Because of limitations on the permissible discharge of fluorides into the environment, the waste reactant atmosphere from hydrogen reduction of tungsten hexafluoride is converted to sodium fluoride by reaction with soda lime and is stored as a solid waste.

When applying CVD composite coatings, deposition reaction and deposition sequence should be considered. In vapor-forming a tungsten-silicon carbide composite coating, it is better to deposit tungsten from the hexafluoride at 550 °C (1020 °F) on the silicon carbide than to deposit silicon carbide from silicon tetrachloride and methane at 1200 °C (2190 °F) on tungsten. At that temperature, tungsten silicide forms at the interface with a decrease in volume that destroys the bond.

Thin tungsten coatings tend to be fine-grained initially. However, because grain growth has a preferential direction, favorably oriented grains tend to dominate the others. When a deposit approaches a millimetre thickness, it may have developed a columnar structure with strong texture; for example, pyramidal (111) facets on the surface. Such a texture may be acceptable or even desirable for some applications, but intergranular strength perpendicular to growth direction is generally lower than in wrought tungsten.

Although CVD tungsten coatings have various applications in industry, such as on the copper targets of x-ray tubes, most CVD tungsten is in the form of vapor-formed tubing and crucibles. It is much cheaper to vapor-form a tungsten shape on a removable mandrel than to machine it from a powder metallurgy billet. A nickel interlayer is used to improve the adhesion of tungsten to iron and ferrous alloys.

Chromium. Coating of steel and other alloys with chromium may be done by pack cementation, a process similar to pack carburizing, or by a dynamic, flow through CVD process (open-tube CVD). Pack cementation may be considered CVD, because coating occurs by thermal decomposition of a gaseous chromium carrier. The packing mixture consists of elemental chromium, an inert filler, and an activator such as ammonium iodide. The pack is heated to 1000 to 1100 °C (1830 to 2010 °F) causing chromium to interdiffuse at appreciable rates; chromous iodide is formed in the vapor phase as the chromium carrier. Chromium is transported from the mixture at unit chemical activity to the alloy case at lower chemical activity. The alloy case forms in increasing thickness by diffusion of the chromium into the substrate. Alloy cases containing from 15 to 25 at.% chromium may be obtained.

Some manufacturers chromize sheet steel by circulating a mixture of chromous chloride vapor and hydrogen through a loose or expanded coil of steel in a retort. The average chromium content of the coating ranges from 15 to 50 at.% or more depending on the reduction and displacement reactions. Chromium coatings may be obtained by thermal decomposition of organometallic compounds such as dicumene chromium $(C_9H_{12})_2Cr$ at 300 to 350 °C (570 to 660 °F). Decomposition may occur either in the vapor alone at a pressure of 0.001 to 0.01 atm or with hydrogen or an inert gas bringing the total pressure to 1 atm. Normally, chromium carbide, predominantly Cr_7C_3, is deposited from dicumene chromium. The addition of 0.01 mol% hydrogen chloride or hydrogen iodide gas to the reactant suppresses the deposition of carbon and yields a deposit of predominantly elemental chromium.

Although chromium coatings from the dicumene have been used for specialized applications, no established commercial use appears to exist. High-purity chromium in crystalline form, CVD-refined via chromous iodide, is available commercially for use in alloying and as sputtering targets for preparation of masks for microelectronic circuitry manufacture. High-purity titanium, hafnium, and zirconium from iodide processing are available commercially.

Titanium Carbide. Wear-resistant coatings of titanium carbide are formed by the hydrogen reduction of titanium tetrachloride in the presence of methane or some other hydrocarbon. The substrate temperature ranges from 900 to 1010 °C (1650 to 1850 °F), depending on the substrate. Although the process can be carried out at atmospheric pressure, reduced pressure (0.1 atm) is used to improve uniformity of deposition over the parts being coated. In coating cemented carbide tool inserts, thread guides, and other small parts, an externally heated retort of Inconel or similar alloy is used. The parts to be coated are suspended in the retort or laid on the spaced, coarse wire-mesh screens of a fixture placed in the retort. The direction of gas flow through the retort may be reversed periodically to promote coating uniformity by offsetting the effect of downstream reactant depletion. Partially automated equipment is available commercially for depositing titanium carbide, titanium nitride, aluminum oxide, and chromium carbide coatings.

Titanium nitride deposition is accomplished by hydrogen reduction of titanium tetrachloride in the presence of ammonia gas at 900 to 1100 °C (1650 to 2010 °F). Aluminum oxide is deposited via hydrolysis of aluminum chloride vapor at 1050 to 1100 °C (1920 to 2010 °F). The water vapor for hydrolysis is provided indirectly by reaction of hydrogen and carbon dioxide, a reaction that is surface catalyzed at the substrate, to avoid unwanted precipitation of alumina snow that occurs with the direct addition of water vapor. Chromium carbide is applied at 900 to 1100 °C (1650 to 2010 °F) with a hydrogen reduction of chromous chloride generated *in situ* by reaction of chromium with hydrogen chloride. The carbon is supplied by diffusion from the substrate.

Equipment for applying these hard coatings consists of multiple retorts that sequence through all stages of the operation, including loading, processing, cooling, and unloading. Figure 1 depicts a processing sequence. The reaction by-products are scrubbed with a caustic solution, diluted, and flushed down the drain as harmless sodium chloride and hydrated oxides.

Although titanium carbide is very hard (about 4000 kg/mm^2), it is not the hardest material that can be deposited by CVD. Its effectiveness as a tool coating lies in the decrease of coefficient of friction (decreased chemical affinity) between the tool and the work. An uncoated cobalt-cemented tungsten-carbide tool exhibits cratering on the upper surface behind the cutting edge, caused by galling between the cobalt binder and the hot chip from the workpiece. Titanium carbide of 7 μm (0.3 mil) in thickness is an effective coating for low-speed machining.

A problem frequently encountered in titanium carbide coating of cobalt bonded substrate is the formation of a weak eta phase ($W_xCo_{(12-x)}C$) at the substrate-coating interface. If insufficient carbon is supplied from the gas atmosphere, the substrate can become decarburized by diffusion of carbon to the interface to form titanium carbide. However, there is a limit to which the methane content of the vapor phase can be raised to offset this tendency, because of the possibility of elemental carbon in the deposit. A careful balance of concentration must be maintained, depending primarily on the processing temperature. Titanium nitride coatings are recommended for intermediate duty, depending on the material being machined and the conditions.

Although aluminum oxide may be deposited directly on cemented carbide tool inserts, it tends to form preferentially on the tungsten carbide and to leave areas of exposed cobalt. Much denser overall formation with improved coating adherence is obtained by coating first with 5 μm (0.2 mil) titanium carbide, followed by the functional 1 to 2 μm (0.04 to 0.08 mil) aluminum oxide coating. Aluminum oxide is effective for high-speed machining. The greater chemical inertness of the aluminum oxide and its lower thermal conductivity are believed to contribute to its effectiveness.

Process Conditions

The choice of coating reaction and process conditions depends on the substrate material and its shape. Heat treatment of the substrate imposed by the CVD process must be considered. Optimum conditions exist which vary widely between applications. Generalizations can be made regarding CVD processing.

Surface preparation for CVD coating by degreasing or grit blasting is similar to methods used for other dry coating processes such as vacuum evaporation and sputtering. In addition, a CVD precoating treatment may be given. For substrates whose superficial oxides may be reduced by hydrogen, a soak in a hydrogen atmosphere at an appropriate temperature is used. In some cases, such as silicon wafer processing, a gaseous etch with hydrogen chloride or hydrogen fluoride is used. In other cases, an intermediate coating may be applied to improve adhesion or to protect the substrate from an aggressive coating atmosphere, such as a nickel layer on ferrous materials to be tungsten coated or a titanium carbide layer on ferrous materials before tantalum coating.

Reactant Concentration. Concentration, temperature, and pressure conditions for CVD reactions vary widely. Because the gas phase is a dilute process stream at best, production is maximized by maintaining as high a reactant concentration as possible, other factors being equal. However, the vapor pressure of the reactant or reaction byproducts may be limiting when it may not be practical or convenient to heat the connecting lines to the temperature required to prevent condensation from high partial pressure of reactants. Also, if an excess of reducing gas, such as hydrogen, is needed for efficient reaction, maximum utilization of reactant is obtained at some limited concentration of the reactant being reduced.

In some cases, it may be necessary to limit reactant concentration to avoid unwanted precipitation from the reactant gas before the substrate surface is reached. The point at which such precipitation occurs is dependent on reactant concentration, temperature, and pressure. Temperatures at which a concentration of silane (SiH_4) in hydrogen precipitates finely divided silicon from the gas phase at atmospheric pressure are given in the following table:

Temperature °C	°F	Silane concentration, vol %
780	1440	0.248
840	1540	0.080
890	1630	0.042
940	1720	0.024
990	1810	0.013
1040	1900	0.007
1090	1990	0.003
1140	2080	0.002

Source: F.C. Eversteijn, *Philips Res. Repts.*, Vol 26, 1971, p 134–144

Deposition Temperature. With few exceptions, the rate of deposition from CVD reactions increases with temperature in a manner specific to each reaction. Deposition at the highest possible rate is preferable; however, there are limitations which require a processing compromise. As temperature is increased, the deposition goes through two and sometimes three stages. At low temperatures, the rate of deposition depends strongly on temperature and is limited by the rate of chemical reaction at the surface. Concentration gradients in the reactant atmosphere are small at sufficiently high flow rates, and the deposition rate is uniform from point to point on substrates of uniform temperature.

As the temperature and deposition rate are increased, transport of material through the gas phase by diffusion adjacent to the substrate becomes a limiting factor, and the deposition rate becomes less dependent on temperature. The attendant concentrations of the reactant atmosphere become very dependent on gas flow dynamics. Raising the temperature further may lead to another problem such as the decomposition of nickel carbonyl or silane. The gas atmosphere temperature adjacent to the substrate may become high enough to result in precipitation of a finely divided reaction product. Incorporation of the precipitate into the coating disrupts growth of the coating and in most cases must be avoided.

In the lowest temperature range, where the deposition rate is limited by the rate of chemical reaction at the surface, concentration gradients in the reactant gas are minimized. Processing

Fig. 1 Time-temperature sequence for CVD titanium carbide coating on D2 tool steel

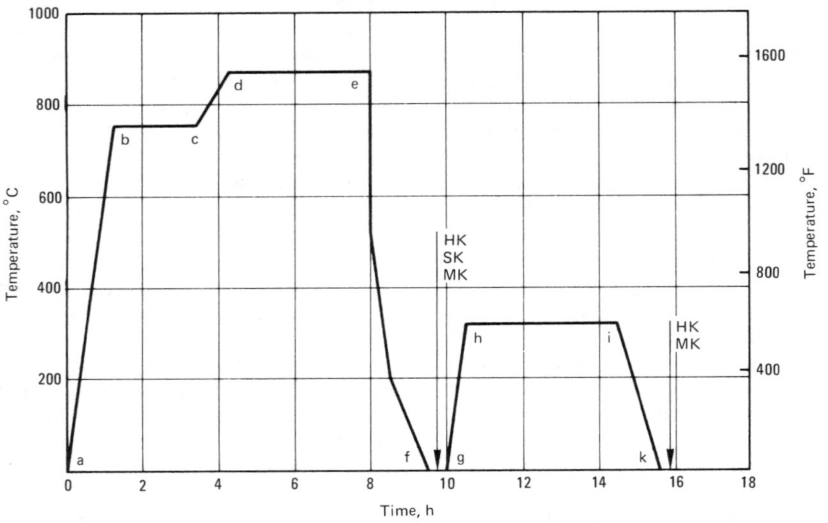

Line segment a-b may incorporate a step-up anneal consistent with the substrate alloy and prior condition. Line segment b-c is generally as indicated. Line segment d-e generally ranges from 900 to 1010 °C (1650 to 1850 °F), depending on the substrate alloy. a-d heating and solution heat treatment according to DIN 1.7014; d-e reaction time; e-f hardening; g-k tempering; HK hardness check; SK check of layer thickness; MK dimension check. (Courtesy of the Bernex Division of Sylvester and Co., Beachwood, OH)

in the low temperature range has the effect of:

- Minimizing thermal changes in the substrate
- Promoting uniformity of the coating thickness
- Avoiding accelerated growth of protrusions, where a surface irregularity penetrates the concentration gradient and grows faster at the tip exposed to higher reactant gas concentration
- Avoiding reactant gas precipitation

Accordingly, raising the temperature to increase the rate of deposition (and thus decrease the cost) implies a processing compromise.

Deposition Pressure. Atmosphere pressure in the processing chamber is of interest in CVD because of (*a*) its effect on the rate and efficiency of the deposition reaction, and (*b*) its effect on diffusivity in the gas phase. For most CVD reactions, increasing the partial pressure of reactants tends to increase the amount of material transported per unit of atmosphere gas volume and thus the rate of deposition. Because diffusivity in the reactant gas decreases with increased pressure, maximizing deposition rate in this way also leads to decreased coating uniformity as described in the preceding section. Despite the convenience of operating at atmospheric pressure, use of a gas that is not one of the reactants should be avoided, for the sake of deposit uniformity at a given partial pressure of reactants.

Elimination of the nonreactant gas and operation at the corresponding reduced pressure actually results in an increase in deposition rate because of increased diffusivity. However, when the partial pressure of reactants is reduced, the compromise between increased uniformity and decreased deposition rate must be faced. Because of decreased concentration gradients in the reactant atmosphere that accompanies increased diffusivity, operation at reduced pressure also tends to suppress the growth of surface protrusions and the incidence of gas phase precipitation.

Coating Characteristics

The characteristics of CVD coatings range widely, depending on the materials involved and the coating conditions. Dense, fine-grained coatings have properties consistent with the wrought forms of that coating material. Porous or preferentially oriented coatings have characteristics of that material fabricated or produced by other methods.

Coating Structure. The structure of CVD coatings depends primarily on deposition temperature and to a lesser extent, on reactant concentration as reflected in deposition rate. Low-temperature coatings tend to be either amorphous or fine grained. At high temperatures where diffusion is significant, equiaxed grains can result. Adhesion, a function of interface structure, depends on the affinity of the substrate for the coating and the presence or absence of intentional or unintentional interlayers. To promote dense nucleation, the substrate should be exposed initially to a full concentration of coating reactants rather than to a gradually increasing concentration, as from a reactant vaporizer being brought to operating temperature. Gradual increase in reactant gas in the retort is eliminated by bypassing the reactant flow until full concentration is being produced before routing it to the coating retort. The growth of protrusions from the substrate is suppressed by avoiding diffusion-limited deposition as noted above.

Control of grain structure in CVD coatings has been given considerable attention, particularly the suppression of the growth of columnar grains perpendicular to the substrate. Because of defects and impurities at the grain boundaries, the strength of such coatings parallel to the substrate is low relative to that of conventional wrought material. This factor is particularly important in vapor-forming tungsten and tungsten-based deposits. Interruption of grain growth and renucleation, leading to a finer grained, more equiaxed structure has been accomplished by contacting the workpiece surface during deposition using a wire brush, or by codeposition of a second phase. Surface contact during deposition is accomplished automatically in the fluidized-bed coating of objects where a fine-grain coating is obtained. The potential for codeposition of a second phase is inherent in some systems, such as tungsten from its carbonyl in association with hydrogen plus water vapor. Temperature/concentration conditions may be found for obtaining the desired structure based on that principle. Another route to grain refinement appears to be important in some systems, such as the deposition of silicon carbide from silicon tetrachloride and methane in hydrogen. Droplets of liquid-polymer intermediates may form in the reactant gas, settle on the substrate, and undergo final decomposition. This competing growth mechanism leads to grain refinement and may explain the barely discernible banded structure of such deposits.

Stresses. Although intrinsic stresses, usually tensile, may be found in material formed by CVD, differential thermal expansion is the greatest source of residual stress. To avoid coating failure from this source, the differential should be minimized by the choice of materials and conditions leaving, where possible, the coating in compression.

The ability of CVD to form preferentially oriented material can lead to anisotropic stresses in thick deposits. For example, in pyrolytic graphite where the coefficient of expansion perpendicular to the substrates is many times that of the coefficient of expansion parallel, delamination on cooling from the deposition temperature can occur for deposits whose ratio of thickness to radius of curvature exceeds about 0.07. Pyrolytic boron nitride is another example of highly anisotropic CVD material. For materials having a lower degree of anisotropy, such problems are usually masked by differential expansion of the coating relative to the substrate.

SELECTED REFERENCES

- D. T. Hawkins, Ed., *Chemical Vapor Deposition, 1960-1980, A Bibliography*, New York: IFI/Plenum, 1981
- *Proceedings of the International Conferences on Chemical Vapor Deposition*: **CVD I**, A. C. Schaffhauser, Ed., American Nuclear Society, Hinsdale, IL, 1967; **CVD II**, J. M. Blocher, Jr. and J. C. Withers, Ed., Electrochemical Society, Pennington, NJ, 1970; **CVD III**, F. Glaski, Ed., American Nuclear Society, 1972; **CVD IV**, G. F. Wakefield and J. M. Blocher, Jr., Ed., Electrochemical Society, 1973; **CVD V**, J. M. Blocher, Jr., H. E. Hinterman, and L. H. Hall, Ed., Electrochemical Society, 1975; **CVD VI**, L. F. Donaghey, P. Rai-Choudhury, and R. N. Tauber, Ed., Electrochemical Society, 1977; **CVD VII**, T. Sedgwick and H. Lydtin, Ed., Electrochemical Society, 1979; **CVD VIII**, J. M. Blocher, Jr., G. E. Vuillard, and G. Wahl, Ed., Electrochemical Society, 1981
- H. E. Hinterman, Ed., *Proceedings of the Third European Conference on Chemical Vapor Deposition*, Lab.

Suisse de Recherches Horlogères, Neuchâtel, Switzerland, 1980

- H. Schäfer, *Chemical Transport Reactions,* New York: Academic Press, 1964
- C.F. Powell, J.H. Oxley, and J.M. Blocher, Jr., Ed., *Vapor Deposition,* New York: John Wiley and Sons, 1966
- H.E. Hinterman and H. Gass, *Schweizer Archiv.,* Vol 33, No. 6, p 157, 1967
- R.A. Holzl, Vapor Deposition Techniques, *Tech Metals Rev,* Vol 1, No. 3, p 1377-1405, 1968
- Anonymous, *Design Engineering,* Nov 1971, p 76
- H.E. Hinterman and H. Gass, *Oberfläche Surface,* Vol 12, No. 10, p 177, 1971
- J.M. Blocher, Jr., *Soc Auto Eng.,* May 1973, p 1780; reprint No. 730543; *Machine Design,* Vol 43, No. 16, p 58, 1971
- P.N. Baker, Preparation and Properties of Tantalum Thin Films, *Thin Solid Films,* Vol 14, No. 1, p 3-25, 1972
- R.W. Haskell and J.G. Byrne, *Treatise on Materials Science and Technology,* H. Herman, Ed., New York: Academic Press, Vol 1, p 239, 1972
- W.A. Bryant, The Fundamentals of Chemical Vapor Deposition, *J Mater Sci,* Vol 12, p 1285-1306, 1977
- C. Béguin, *Metall,* Vol 1, p 21, 1974
- W. Ruppert, Influence of Deposition Temperature on the Hardness and Dimensions of Steels with Hard Coatings Produced by Chemical Vapor Deposition, *Thin Solid Films,* Vol 40, p 27-40, 1977
- C. Hayman, Introduction to Chemical Vapor Deposition, Inst Met Course, Series 3, Surface Treatment Prot, 130-8, 1978
- K.K. Yee, Protective Coatings for Metals by Chemical Vapor Deposition, *International Metals Reviews,* No. 226, Vol 1, 19-42, 1978

Vacuum Coating

By David V. Rigney
Manager, Advanced Programs
General Electric Co.

VACUUM COATING is the process of depositing metals and metal compounds from a source in a high-vacuum environment onto a substrate. Three principal techniques used to accomplish the deposition process are evaporation, ion plating, and sputtering. In each technique, the transport of vapor is carried out in an evacuated, controlled environment chamber of 1 to 10^{-5} Pa (10^{-2} to 10^{-7} torr) residual air pressure.

In the evaporation process, vapor is generated by heating the source material to a temperature such that the vapor pressure significantly exceeds the ambient chamber pressure and produces sufficient vapor for practical deposition. The process is carried out at pressures of less than 10^{-1} Pa (10^{-3} torr), and usually in vacuum levels of 10^{-2} to 10^{-3} Pa (10^{-4} to 10^{-5} torr). Direct resistance heating is used to produce deposits of high vapor pressure materials, or to produce thin coatings. Both resistance and induction heating methods are used to deposit thicker coatings and deposits of intermediate vapor pressure materials. More refractory substances, or high-rate evaporation to deposit thick films, require high-power density heating methods such as electron beam heating and radiation heating sources. Laser beam heating methods have been used on occasion. Evaporation rates as high as 250 000 Å/min (1000 μin./min) are used; rates of up to 750 000 Å/min (3000 μin./min) have been achieved. The evaporant is in the electrically neutral state and is expelled from the surface of the source at thermal energies of 0.1 to 0.3 eV.

The substrates are mounted over the source in a position that exposes the surfaces to be coated to the vapor source. For most applications, the substrate must be moved in a complex motion to deposit a uniform coating on the substrate, because the vapor flux from the source is localized and directional.

In the ion plating process, coating source material is vaporized in a manner similar to the evaporation process. The vapor is passed through a region of ionized gas (glow discharge region) between the source and the substrate, ionizing a fraction of the vapor ($\sim$5%) that is then accelerated to the substrate. The substrate is electrically biased at a high negative potential (3000 to 4000 V), to produce the glow discharge from a gas (usually argon), introduced into the chamber at 10 to 10^{-1} Pa (10^{-1} to 10^{-3} torr). The ionized atoms (which have an energy of 10 to 40 eV) strike the substrate and sputter off some of the material on the surface. This continuous sputtering of the surface results in a cleaner, more adherent deposit, but reduces the net deposition rate. In addition, the substrate may be heated to an undesirably high temperature as a result of the intensity of the ionic bombardment. In such cases, the glow discharge is generated by providing a separate negatively biased tungsten filament. The high gas pressure results in considerable scattering of the deposit vapor, and results in a much more uniform deposit on the substrate than observed in normal evaporative processes. Substrates to be coated are arranged in a manner similar to that used for evaporation coatings; however, translation requirements to achieve deposit uniformity are substantially reduced. For more information, see the article on ion plating in this Volume.

In the sputtering process, ionized gas (usually argon), is produced in a glow discharge at 10 to 10^{-1} Pa (10^{-1} to 10^{-3} torr). The coating source material is negatively biased at a high potential. The ions produced bombard the target with energies of up to 1000 eV. Momentum interchanges between the impinging ions and the source result in dislodging of atoms from the surface into the vapor state. These atoms are then transported to the substrate with energies of 10 to 40 eV. Sputtering is a controllable, but inefficient way to produce vapor. Typically, only one or two atoms may be dislodged from the surface as a result of a single ion impingement. Energy costs range from three to ten times that of evaporation. Extraction of heat developed during ionic bombardment limits the sputtering rate; with exceptions of materials with high conductivity, sputtering rates of 500 to 1000 Å/min (2 to 4 μin./min) are achieved. For more information, see the article on sputtering in this Volume.

Evaporation

Evaporation is a surface phenomenon, and does not necessarily constitute boiling. Surface evaporation occurs from conduction, not from formation of subsurface vapor bubbles at a depth below the surface, if the liquid has sufficient thermal conductivity such as that possessed by metals. The rate of evaporation, assuming that none of the vaporized material returns to the surface, is given by the Langmuir equation:

$$w = 0.0585\, P \left(\frac{M}{T}\right)^{1/2} \text{ g/cm}^2\text{ s}$$

where P is the vapor pressure of the evaporant in torr, M is the molecular weight of the evaporating material, and T is absolute temperature, K, and w is the amount of vapor material evaporated per unit time.

The evaporating metal atom (molecule) leaves the surface in a straight line. Collision of the evaporating atoms with residual gas molecules randomly moving about the vacuum chamber lowers the energy of the evaporant and also changes its direction. A sufficient number of collisions results in too great a loss in energy and too great a randomization of the flow from the source to produce an adherent deposit. Therefore, proper positioning of the substrate with respect to the source and the quality of the vacuum in the chamber must be considered for each application. Highest quality coatings (electronic and optical applications) are deposited when the source-to-substrate distances are less than the mean path distance between collisions of a gas molecule and the equipment. Deposits acceptable for decorative purposes can be deposited at distances of several mean free paths. At a 10^{-1} Pa (10^{-3} torr) chamber pressure, the source-to-substrate distance should be less than 500 mm (20 in.). At 10^{-2} Pa (10^{-4} torr), the source-to-substrate distance can be increased to over 4000 mm (160 in.).

To coat the entire surface of a substrate, it must be rotated and translated over the vapor source. Deposits made at substrates positioned at low angles to the vapor source result in fibrous, poorly bonded structures. Deposits resulting from excessive gas scattering are poorly adherent, amorphous, and generally dark in color. Thus, if only one surface is to be coated, other surfaces should be masked to prevent coating deposition. The highest quality deposits are made on surfaces nearly normal to the vapor flux. Such deposits faithfully reproduce the substrate surface texture. Highly polished substrates produce lustrous deposits, and the bulk properties of the deposits are maximized for the given deposition conditions.

For practicable deposition rates, source materials should be heated to a temperature so that its vapor pressure is at least 1 Pa (10^{-2} torr) or higher. Temperatures at which the vapor pressure is 1 Pa (10^{-2} torr) for several common elements are shown in Table 1. The vapor pressure of many of the elements have been compiled by several authors (Ref 1-5). Deposition rates for evaporating bulk vacuum coatings can be very high. Commercial coating equipment can deposit up to 500 000 Å/min (2000 μin./min) using large ingot material sources and high-powered electron beam heating techniques.

As indicated previously, the directionality of evaporating atoms from a vapor source generally requires the substrate to be articulated (positioned in a specific fashion) within the vapor cloud. To obtain a specific film distribution on a substrate, the shape of the object, the arrangement of the vapor source relative to the component surfaces, and the nature of the evaporation source must be accounted for. The distribution of the evaporant leaving the surface of the source is given by the cosine law, as illustrated in Fig. 1 for a small surface source of area (dA), emitting a vapor flux (m). The material passing through a solid angle ψ at a direction of angle ϕ with the normal to the surface dA, per unit time is given by:

$$dm = \frac{m}{\pi} \cos\phi \; d\psi \qquad \text{(Eq 1)}$$

Material arriving at a small substrate area (dA_2), inclined at an angle θ to the vapor stream direction, at a distance (r) from the source can be expressed in the following manner. The projection of solid angle ($d\psi$) onto area (dA_2) is:

$$d\psi = \frac{\cos\theta}{r^2}\, dA_2 \qquad \text{(Eq 2)}$$

Combining Eq 1 and 2 gives:

$$dm = \frac{m}{\pi r^2} \cos\phi \, \cos\theta \; dA_2 \qquad \text{(Eq 3)}$$

These relationships have held true at relatively low rates of evaporation.

At high rates of evaporation, such as those experienced in the practical production of thick films using high-power electron beam sources, the mass distribution is much steeper than predicted by Eq 3. This has been attributed to: (*a*) evaporant interactions above the source, (*b*) formation of a depression in the surface by the impact of energetic electrons from the heating source, and (*c*) formation of waves on the surface of the molten source.

Table 1 Temperature for vapor pressure of 1 Pa (10^{-2} torr)

Element	Temperature °C	°F
Aluminum	1150	2100
Beryllium	1245	2270
Bismuth	680	1260
Cadmium	265	510
Carbon	2460	4460
Chromium	1400	2550
Cobalt	1520	2770
Copper	1260	2290
Germanium	1400	2550
Gold	1400	2550
Indium	945	1730
Lead	715	1320
Magnesium	440	825
Manganese	940	1720
Molybdenum	2350	4260
Nickel	1530	2780
Osmium	2910	5270
Platinum	2090	3790
Rhenium	3065	5550
Silicon	1470	2680
Silver	1630	2970
Tantalum	3060	5540
Tin	1250	2280
Titanium	1740	3160
Tungsten	3230	5840
Uranium	1930	3510
Vanadium	1850	3360
Yttrium	1630	2970
Zinc	345	650
Zirconium	2400	4350

Evaporation Sources

Most elemental metals, semiconductors, compounds, and many alloys can be directly evaporated in vacuum. The principal difficulties associated with evaporating material systems are:

- Dissociation of oxides, halides, and alloys, under the conditions required for evaporation
- Reactions with containment materials
- Toxicity of the evaporant
- Degassing and soundness of the source material
- Adapting process techniques to the physical properties of the evaporant

Support materials used to contain and assist in evaporation of materials are critical to successful deposition. The source must be able to heat the evaporant to a sufficient temperature

Fig. 1 Surface element (dA_2) receiving deposit from a small-area source (dA_1)

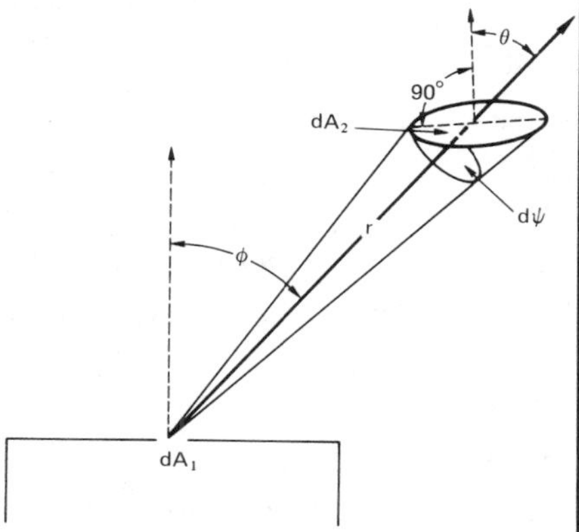

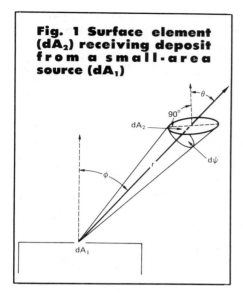

Fig. 2 Wire and metal-foil sources

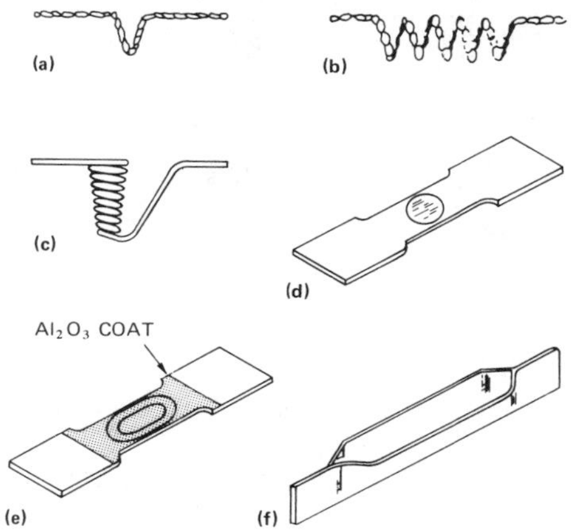

(a) Hairpin source. (b) Wire helix. (c) Wire basket. (d) Dimpled foil. (e) Dimpled foil with alumina coating. (f) Canoe-type

to ensure evaporation. The container must have negligible evaporation rates and dissociation pressures to avoid contamination of the deposited film. Alloying of the container and the evaporant can drastically lower the evaporation rates and contaminate the deposit. Chemical reactions between compounds and the container tend to produce volatile contaminants in the film deposit. Additional factors that affect the choice of containment materials are availability in the form desired and the capacity required. Table 2 lists some common materials along with containment materials commonly used for evaporation sources (Ref 6).

Resistance Sources. The simplest sources are resistance wires and metal foils of various types (Fig. 2). They are generally constructed of refractory metals such as tungsten, molybdenum, and tantalum. The filaments serve the dual function of heating and holding the material for evaporation. Filament heaters are made from multistranded wire, usually from 0.5 to 1.5 mm (0.020 to 0.060 in.) diam individual filaments. These sources depend on the evaporant wetting the wire to increase the evaporation area. The materials used for the filament are determined by the wetting characteristics of the evaporant and the evaporation temperature. If the evaporant does not readily wet the filament, and a greater flux density is required, a basket coil can be used. The evaporant can be electroplated onto the wire itself to increase the charge weight. Basket shapes also are used to evaporate pellets or chips of dielectric. Wetting the surface shorts out the filament. The maximum charge that filaments a few centimetres long can hold is about 1 g. Sheet metal foils 0.1 to 0.04 mm (0.005 to 0.0015 in.) thick, as shown in Fig. 3, have capacities of a few grams and are the most widely used sources for small-scale evaporation. Wetting the surface of the boat-shaped heaters can lower the resistivity of the circuit, and result in a consequential drop in temperature. Plasma-sprayed oxides (such as alumina) may be used to coat the foil surfaces to prevent wetting. Under these circumstances, formation of volatile oxides from the deposit must be considered. The power requirements for these sources are modest, usually between 1 and 3 kW. Nearly all the elements except for the refractories can be evaporated in small quantities using these sources. The refractory metals require either electron beam evaporation from a water-cooled hearth or vacuum-arc evaporation. Carbon is deposited using a vacuum arc evaporation source or electron bombardment.

Sublimation Sources. Several elements, such as chromium, palladium, molybdenum, vanadium, iron, and silicon, can be evaporated directly from the solid phase. A practical chromium sublimation source developed by Roberts and Via (Ref 7) is shown in Fig. 3. The nonuniformity and variability of temperature experienced with basket-type sources are largely eliminated by this design. The potential difficulties resulting from expulsion of small granules of the evaporant from the surface can be avoided by providing a heated reflector made from refractory metals to deflect the vapor into a secondary source. This type of source has been successfully used to deposit SiO, and sulfides, selenides, tellurides of zinc, as well as MgF_2.

Crucible sources comprise the greatest applications in high-volume production for evaporating refractory metals and compounds. The crucible materials are usually refractory metals, oxides and nitrides, and carbon. Crucibles offer the capability to evaporate intermediate quantities of materials.

Heating can be accomplished by radiation from a secondary refractory heating element (Fig. 4), by a combination of radiation and conduction (Fig. 5), and by radio frequency induction heating. As with direct resistance heating, refractory metal crucibles are used for those metals that interact minimally with the evaporant. Refractory oxides (ThO_2, Al_2O_3, BeO, and MgO) can be used to deposit arsenic, antimony, bismuth, tellurium, calcium, manganese, and other metals with sufficient vapor pressures below 1000 °C (1830 °F), and cobalt, iron, palladium, platinum, and rhodium, which require temperatures between 1500 and 2100 °C (2730 and 3810 °F). The refractory metals, tungsten, molybdenum, and tantalum, cannot be evaporated from oxide crucibles. Pyrolytic boron nitride and a sintered 50% BN-50% TiB_2 have been

Table 2 Containment materials for evaporants(a)

Element or compound	Dissociation produced or vapor species	Evaporation temperature °C	Evaporation temperature °F	Containment materials Wire-foil	Containment materials Crucibles(b)	Remarks
Aluminum	Al	1220	2230	W	C, BN, BN-TiC, (Cu)	Wets material readily. Alloys with tungsten nitrides preferred
Antimony	Sb_2, Sb_4	630	1170	Mo, Ta, Ni	Oxides, BN, C	Toxic material
Bismuth	Bi, Bi_2	610	1130	W, Mo, Ta, Ni, Fe	Oxides, metal (Cu)	Toxic in vapor form
Cadmium	Cd	265	510	W, Mo, Ta, Ni, Fe	Oxides, metal	Sublimates. Deposits contaminate vacuum chamber
Carbon	C_3, C_2, C	2600	4710	C	...	Carbon arc or electron beam
Chromium	Cr	1400	2550	W, Ta	...	Sublimates. Electrodeposits release H_2
Cobalt	Co	1520	2770	W	Al_2O_3, BeO, (Cu)	Alloy reaction with tungsten. Electron beam melting preferred
Copper	Cu	1260	2300	W, Ta, Mo	Mo, C, Al_2O_3	Practically no reaction with refractories
Gallium	Ga	1130	2070	...	BeO, Al_2O_3	Oxides attached above 1100 °C (2010 °F)
Germanium	Ge	1400	2550	W, Mo, Ta	W, C, Al_2O_3	Wets refractory metals, but little reaction. Electron beam preferred
Gold	Au	1400	2550	W, Mo	Mo, C	Molybdenum preferred
Iron	Fe	1450	2640	W	BeO, Al_2O_3, ZrO_2, (Cu)	Alloys with refractories. Electron beam
Lead	Pb	715	1320	W, Mo, Ni, Fe	Metals	Does not wet refractories. Toxic
Manganese	Mn	940	1720	W, Mo, Ta	Al_2O_3	Wets refractories
Molybdenum	Mo	2530	4590	...	(Cu)	Electron beam evaporation
Nickel	Ni	1530	2790	W foil, W Al_2O_3	Refractory oxides (Cu)	Alloys with refractories. Electron beam preferred
Platinum	Pt	2100	3810	W	ZrO_2, ThO_2 (Cu)	Alloys with refractory metals. Electron beam preferred
Silicon	Si	1350	2460	...	BeO, ZrO_2	Oxides attached by silicon. Electron beam preferred
Silver	Ag	1030	1890	Mo, Ta	Mo, C	Molybdenum crucibles preferred
Tantalum	Ta	3060	5540	...	(Cu)	Electron beam heating
Titanium	Ti	1750	3180	W, Ta	(Cu), C, ThO_2	Electron beam heating required
Zinc	Zn	345	650	W, Ta, Ni	Fe, Al_2O_3, C, Mo	Deposits contaminate vacuum system
Zirconium	Zr	2400	4350	W	(Cu)	Electron beam heating preferred
Al_2O_3	Al, O, AlO, Al_2, O_2, $(AlO)_2$	1850-2200	3360-3990	...	W, Mo, (Cu)	Tungsten containment results in small O_2 dissociation. Electron beam heating

(a) Ref 6. (b) (Cu) indicates water-cooled copper crucible

well established as crucible materials, especially when heated using a radio-frequency heating source, as illustrated in Fig. 6. Boron nitride crucibles are used extensively for the deposition of aluminum.

Electron beam heating provides a flexible heating method that can concentrate heat on the evaporant. Portions of the evaporant next to the container can be kept at lower temperatures, thus minimizing interaction. Two principal electron guns in use are the linear focus, which uses magnetic and electrostatic focusing methods (Fig. 7a), and the bent-beam magnetically focused gun (Fig. 7b). The bent-beam gun has the advantage of compactness; a large filament area allows operation below 10 kV without sacrificing power, as experienced with the linear system. The disadvantages of electron beam heating methods are (*a*) interference with evaporation rate monitors as a result of charge accumulation, and (*b*) electrical charging of substrates by stray electrons; interference of electronic rate monitors can be minimized by placing a grounded grid in front of the monitor. Substrates may be allowed to float or be maintained at the filament voltage to resist charging. The guns must be operated in a vacuum environment of 10^{-2} Pa (10^{-4} torr) or less to preserve filament life and to prevent gas scattering of the emitted electrons.

The water-cooled copper hearth is the general crucible containment system, although many other materials, including refractory metals, have been used successfully. This system allows a skin or skull of solid evaporant to be maintained next to the containment surface. Film deposits from electron-beam heated sources are the purest. Nearly any material that does not undergo dissociation can be directly electron-beam evaporated. However, electron beam heating often is not used when small quantities of materials are to be evaporated, and more easily controlled resistance heating methods are available.

Evaporation of compounds, mixtures, and alloys requires experimental determination of the conditions under which the deposit with the desired composition and structure are obtained. As indicated in Table 2, only a

Table 2 Containment materials for evaporants(a)

Element or compound	Dissociation produced or vapor species	Evaporation temperature °C	Evaporation temperature °F	Containment materials Wire-foil	Containment materials Crucibles(b)	Remarks
BaO	Ba, BaO, O_2, Ba_2O $(BaO)_2$, Ba_2O_3	1200-1500	2190-2730	...	Al_2O_3, Pt	...
BeO	Be, O, $(BeO)_n$ (n = 1-6)	2070-2200	3760-3990	W	(Cu)	Electron beam heating at 2400-2700 °C (4350-4890 °F)
MgO	Mg, MgO, O, O_2	1900	3450	...	(Cu)(Mo)	Electron beam heating
MoO_3	$(MoO_3)_3$, O_2, $(MoO_3)_n$, MoO_2	500-700	930-1290	...	Mo	500-700 °C (930-1290 °F) gives MoO_3, 1000 °C (1830 °F) gives MoO_2 + O_2
NiO	Ni, O_2, O, NiO	1300-1450	2370-2640	...	Al_2O_3	Heavy decomposition
SiO	SiO	1150-1250	2100-2280	...	Ta, Mo	Dissociates above 1250 °C (2280 °F)
SiO_2	SiO, O_2	1500-1600	2730-2910	...	(Cu), Ta, Mo	Electron beam heating. Tantalum, molybdenum react to form volatile oxides
TiO_2	TiO, Ti, O_2, TiO_2	...	...	...	(Cu)	TiO_2 decomposes; pulsed electron beam heating can give TiO_2 films
ZrO_2	ZrO, O_2	...	...	...	W(Cu)	ZrO_2 loses O_2 when heated by electron beam, or by tungsten resistance
ZnS	...	1000	1830	Mo	Mo	Small deviations from stoichiometry
ZnSe	...	820	1510	...	...	...
CdS	S_2, Cd, S, S_3, S_4	600-740	1110-1360	Ta, Mo, Al_2O_3	Cu, SiO_2	Platinum oven at 740 °C (1360 °F). Others, 600-700 °C (1110-1290 °F)
CdSe	Se_2, Cd	660	1220	...	Al_2O_3	...
CdTe	...	750-850	1380-1560	Ta	...	Film stoichiometry depends on condensation temperature
PbS	PbS, Pb, S_2, $(PbS)_2$	625-925	1160-1700	...	SiO_2	Quartz crucibles or oven
Sb_2Se_3	Sb, $(SbSe)_2$, SbSe	500-725	930-1340	...	C, Ta	Graphite at 725 °C (1340 °F), tantalum oven 500-600 °C (930-1110 °F). Fractionation. Variable stoichiometry
NaCl	NaCl, $(NaCl)_2$, $(NaCl)_3$	550-800	1020-1470	...	Ta, Cr, Mo	Oven sources
KCl	KCl, $(KCl)_2$	500-740	930-1360	...	Ni, Cu	Oven source
AgCl	AgCl, $(AgCl)_2$	710-770	1310-1420	Mo	Mo	...
MgF_2	MgF_2, $(MgF_2)_2$, $(MgF_2)_3$	950-1230	1740-2250	...	Pt	Oven source
CaF_2	CaF_2, CaF	980-1400	1800-2550	Mo	Ta	Tantalum oven

(a) Ref 6. (b) (Cu) indicates water-cooled copper crucible

few compounds can be directly evaporated without disproportionation or dissociation. Reactive evaporation, flash evaporation, and multiple and continuous feed have been developed to provide the desired compositions.

Reactive evaporation, or evaporation in which a chemical reaction takes place, is most often used to produce fully oxidized coatings. It is used when the oxide cannot be evaporated because it disproportionates or decomposes. A controlled leak of oxygen or nitrogen is used to form either the oxides or nitrides. The amount of the gas used is sufficient to form the stoichiometric oxide desired. Because the residual gas pressure ranges between 10^{-3} and 1 Pa (10^{-5} and 10^{-2} torr), vapor phase reactions are minimal, and the principal reactions occur at the deposit surface. The rate and completeness of reactions are dependent upon the temperature of the substrate and the amount of reactive gas introduced. Film microstructure is dependent on both temperature and reactive gas content. Stoichiometric Al_2O_3, Cr_2O_3, Fe_2O_3, SiO_2, Ta_2O_3, TiO_2, ZrO_2, and Y_2O_3- ZrO_2 alloys have been formed by reactive oxidation evaporation. TiN and ZrN have been produced using a nitrogen reactive gas.

Multiple-Source Evaporation. The use of two or more sources to deposit multilayer and alloy films has enabled manufacture of both multilayer and complex alloy films that are not easily deposited from a single source. Heating sources utilized are the same as those used for single component evaporation, such as resistance-heated boats and electron beam heating. To prevent contamination of the sources, each is isolated from the other, and each source is independently controlled to produce the desired vapor flux (Fig. 8). The vapor flux can be controlled independently by ionization rate monitors, or other evaporation rate monitors. With feedback control and proper manipulation of the substrates within the vapor clouds, deposit control of ±1 to 2% have been achieved. Examples of binary alloys and compounds produced by dual source evaporation are CdS, CdSe, PbSe, PbTe, Bi_2Te_3, AlSi, GeAs, InAs, InSe, Cr-SiO, Au-SiO, Cr-MgF_2, Nb_3Sn, and V_3Si.

Flash evaporation is used to deposit small quantities of alloys with widely variant vapor pressures. Unlike the multiple-source method, vapor flux need not be monitored. The objective of the method is to completely evaporate small quantities of materials at one

Fig. 3 Chromium sublimation source (Ref 7), with electric current flowing through the tantalum cylinder

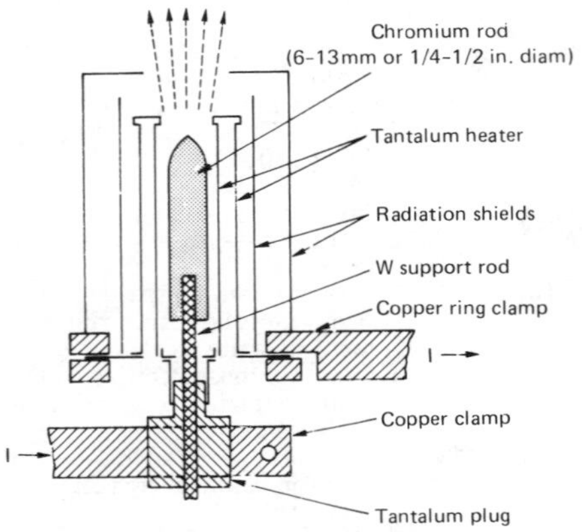

Fig. 4 Molybdenum crucible source with tantalum sheet filament

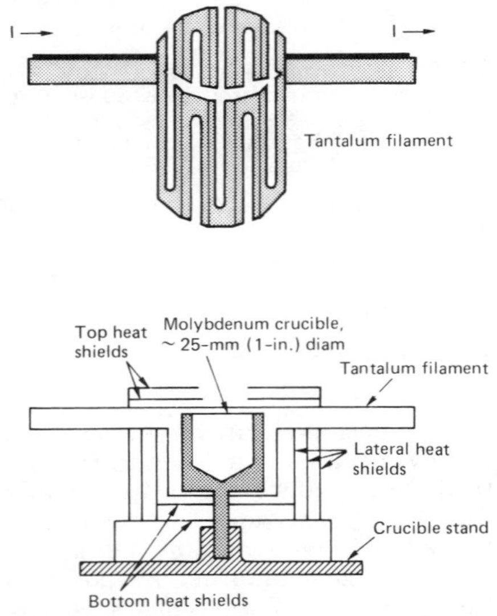

Fig. 5 Oxide crucible with wire-coil heater

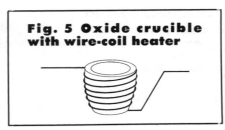

Fig. 6 Radio frequency heated aluminum source with boron nitride-titanium diboride crucible (Ref 8)

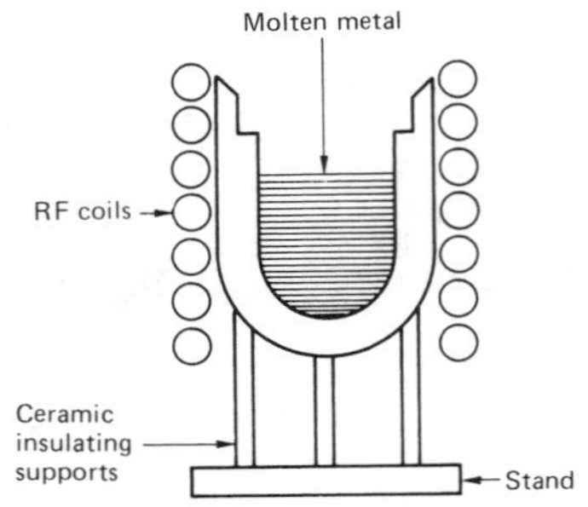

time. Although fractionation occurs, the deposit exhibits compositional variations only within a few atom layers. By providing a slow, constant feed of the evaporant, the inhomogeneity can be further minimized. To obtain maximum homogeneity, the evaporation temperature should be set to attain the vaporization of the least volatile constituent. Both powder and wire feed mechanisms have been used to deliver the alloy to the evaporator. Powder feeders are most often used and consist of a hopper to hold the powder reservoir, and a conveyer that transports the powder from the hopper to a trough that delivers the powder to the heater at a constant rate. A typical feeder is shown in Fig. 9.

Continuous Feed. High-rate evaporation of alloys to form film thicknesses of 100 to 150 μm (4 to 6 mil) require electron beam heating sources and large quantities of evaporation source material. Electron beams of 45 kW or higher are used to melt evaporants in water-cooled copper hearths up to 150 by 450 mm (6 by 18 in.) cross section. More typically, the hearth diameter is about 100 mm (4 in.), and rod diameters about 50 mm (2 in.). Continuous rod feeding from beneath the hearth replenishes the evaporated material. At a given temperature, once a steady state equilibrium has been attained, the vapor composition is equivalent to the rod composition. The molten pool in the hearth has a composition roughly inversely proportional to the vapor pressure of each constituent. The more volatile the constituent, the less is present in the molten pool. For alloys that do not obey Raoult's Law, and for multiconstituent alloys, experimental determination of the equilibrium pool composition is required. To maintain equilibrium, the volume, height, and temperature of the melt must be closely controlled. The method has been used to deposit Ni-20Cr, Ti-6Al-4V, Ag10Cu, Ag5Cu, Ni20Cr-10Al-0.5Y, and a number of other four- and five-component alloys used for oxidation and hot corrosion protection coatings for gas turbine superalloys.

Unlike the flash evaporation method, where the constituents are fully evaporated, there is a substantial thermal gradient in the pool of the hearth. Thus, the composition of the vapor above the molten pool varies from location to location. This variation is minimized by rastering the path of the electron beam around the pool to minimize the temperature differential within the melt. As with other vapor sources, the substrates have to be articulated throughout the deposition cycle to deposit coatings of the same thickness. This movement also tends to result in compositional variations in the most volatile constituents of the alloy. Usable NiCrAlY alloy coating deposits allow variations of ±3 wt % chromium, ±1.0 wt % aluminum, and ±0.2 wt % yttrium. Deposition rates achievable for the continuous feed process can be substantially higher than for the previous sources discussed. Rates of 5 μm/min (0.2 mil/min) are normally achieved in production coating of thick film deposits.

Applications

Coatings applied by vacuum deposition may be broadly classified as decorative and functional. Decorative coatings are widely used in the automotive, home appliance, hardware, costume jewelry, and novelty fields. Functional coatings have numerous applications such as reflection, antireflection, filter, and beam-splitter coatings on optical instruments; current-carrying, dielectric, and semiconductor coatings on electronic components; hot corrosion-resistant coatings on aircraft and missile parts; and oxidation-resistant coatings for gas turbine blades and vanes. Table 3 gives functions of some of the metals, alloys, and compounds that are evaporated.

Aluminum deposits comprise about 90% of the volume of evaporation-deposited components. The element is widely used for decorative deposits, optically useful coatings, electrical and electronic coatings, and corrosion-resistant coatings. Its abrasion resis-

Fig. 7 Two principal guns

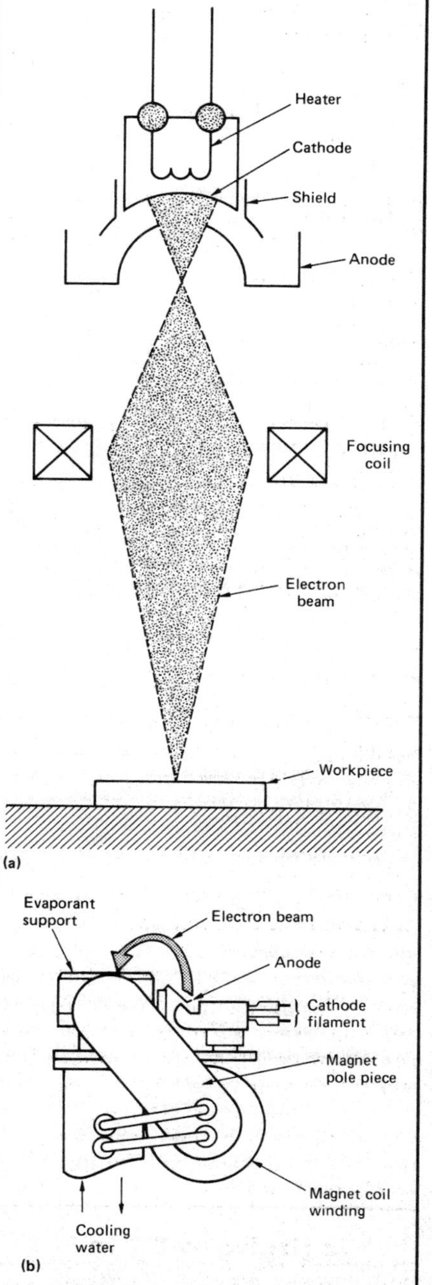

(a) Linear focusing gun. (b) Bent-beam electron gun with water-cooled evaporant support

Fig. 8 Two-source evaporation arrangement yielding variable film composition

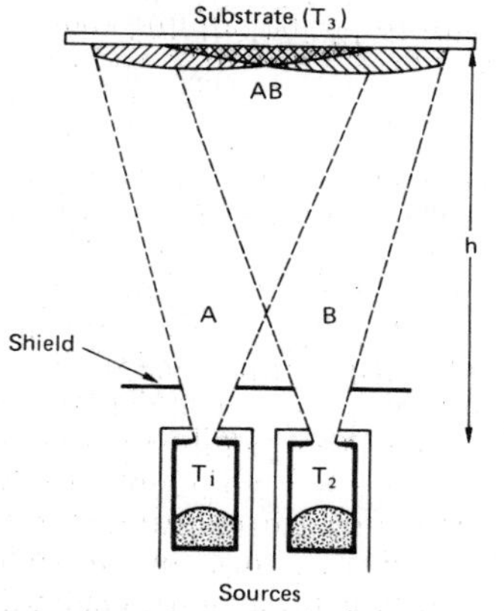

Fig. 9 Powder feeder for dispersing alloy powder for flash evaporation

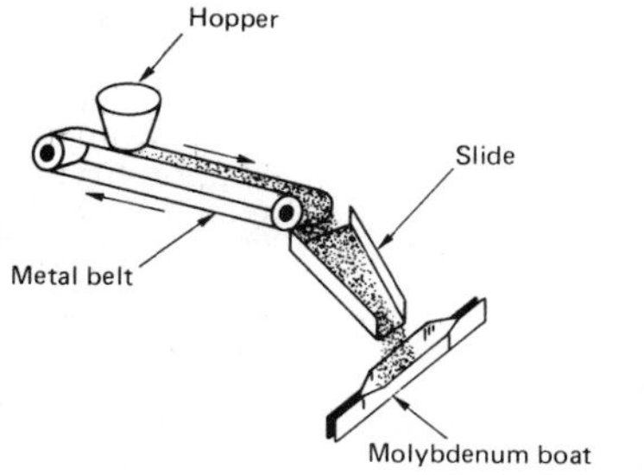

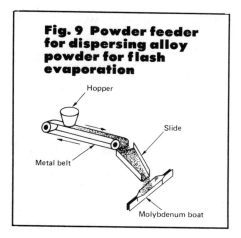

tance is low, and, therefore, aluminum deposits are often overcoated to improve erosion and abrasion resistance, condition the surface for optimum reflectance in a limited spectral region, and prevent slow environmental oxidation of the deposit that results in a decrease in the ultraviolet reflectance.

Aluminum films are usually protected with an overcoat of evaporated silicon monoxide. A transparent film has no effect on the reflectance of a metal at the particular wave length at which the optical thickness (refractive index × physical thickness) of the film is exactly one-half wave length. Consequently, the evaporated silicon monoxide overcoating for an aluminum mirror is made one-half wave length thick optically for a mean wave length of the visible spectrum.

In practice, the silicon monoxide coating is applied immediately after the deposition of the aluminum coating, without interruption of the vacuum. Actually, a silicon monoxide film is often first deposited directly on the work surface, and is then followed by the aluminum deposit and the half-wave protective silicon monoxide overcoating. Because the aluminum film is opaque, the first silicon monoxide film serves no optical purpose, but it does condition the silicon monoxide in the vapor source and may promote better adherence of the reflecting film of aluminum.

Freshly deposited aluminum films lose their vacuum ultraviolet reflectance value rapidly (a decrease of 20% at a wave length of 0.12 μm after 1 h in air); therefore, an overcoating of magnesium fluoride is used to prevent oxidation and maintain high reflectance at these very short wave lengths. Actually, a magnesium fluoride film 0.025 μm (0.6 mil) thick enhances the reflectance of aluminum from a value of 62% for freshly deposited aluminum to 80% at a wave length of 0.12 μm after overcoating. There is no measurable decrease in this high value after exposure to air for six months.

Decorative Coatings. Aluminum is the most widely used material for vacuum-deposited decorative coatings. Because these coatings, in the thickness to which they are usually applied (0.08 to 0.1 μm or 0.003 to 0.005 mil), offer negligible protection against wear and corrosion, the finished part must depend for this protection entirely on a protective coating applied over the metallic film.

Although the durability of decorative coatings on plastics has been improved by new formulations of the plastic basis materials and the higher curing temperatures permitted, the wear resistance and corrosion resistance of these coatings do not compare with those of conventional electroplated finishes. For this reason, the application of vacuum deposits as first-surface coatings is confined to interior parts subject to limited wear and mild atmospheric conditions.

Second-surface vacuum coatings on transparent materials, however, have given satisfactory service for many years on exterior as well as interior parts. In these applications, one surface of the metallic film has the protection of the transparent base material of the part and the other side of the film has the protection of a heavy coat of corrosion-resisting paint.

Automotive rearview mirrors provide an example of a specific advantage of vacuum coatings over those deposited by other processes. Formerly, all mirrors were produced by chemically

Table 3 Applications of evaporative coatings

Coating material	Function	Coating thickness μm	mil	Substrate material	Pretreatment	Post-treatment
Al	Decorative	0.03-0.1	0.001-0.004	Metal(a)	Lacquered	Lacquered
	Reflective	0.03-0.1	0.001-0.004	Glass	None	(b)
	Protective	13	0.5	Steel	None	Anodized(c)
	Decorative	0.03	0.001	Plastic sheet	None	Laminated
	Decorative	13	0.5	Aluminum(d)	None	Anodized
	Electrodes(e)	0.01-0.03	0.0005-0.001	Plastic shot	Baked	None
Cd	Protective	13	0.5	Steel	None	Painted(c)
	Electrical resistance	0.003-0.03	0.0001-0.001	Glass or plastic	None	Multiple layers
Cr	Electrical resistance	0.003-0.03	0.0001-0.001	Glass or plastic	Cleaned	Air baked
Au	Electrodes for piezo-electric crystals	0.03-0.1	0.001-0.004	Organic or inorganic crystals	Cleaned	None
MgF_2	Nonreflective	0.1	0.004	Glass	None	None
SiO	Abrasion resistance	0.1	0.004	Glass	(b)	Oxidized
TiO_2	Decorative; optical	0.1	0.004	Glass	None	Oxidized
Zn	Electrical resistance	0.03-0.1	0.001-0.004	Paper	Lacquer	None
CoCrAlY	Oxidation/corrosion resistance(f)	51-152	2.0-6.0	Superalloys	Cleaned	Peened and heat treated

(a) Automotive trim. (b) Aluminum coating for reflectance, followed by silicon monoxide coating for abrasion resistance, for mirrors. (c) If required. (d) Die casting. (e) For capacitors. (f) For turbine blades

Table 4 Electronic applications of vacuum-deposited metals and metal compounds

Coating material	Application	Coating thickness μm	mil
Metals			
Al	Conductor(a)	0.01-0.2	0.0004-0.008
Bi	Conductor(b)	0.05-0.5	0.002-0.02
Cd	Conductor	0.05-1	0.002-0.04
Cr(c)	Resistor	0.002-0.1	0.00008-0.004
Nb(d)	Superconductor	0.05-0.1	0.002-0.004
Cu	Conductor	0.01-0.2	0.0004-0.008
Ge(d)	Semiconductor	0.5-10	0.02-0.4
Au	Conductor	0.01-0.2	0.0004-0.008
In	Conductor	0.05-0.2	0.002-0.008
Pb	Conductor	0.05-0.2	0.002-0.008
Mo	Conductor(e)	0.05-0.2	0.002-0.008
Ni	Conductor	0.05-0.2	0.002-0.008
Pt(f)	Conductor	0.01-0.2	0.0004-0.008
Se	Semiconductor	0.5-100	0.02-4
Si(d)	Semiconductor	0.5-10	0.02-0.4
Ag	Conductor	0.01-0.2	0.0004-0.008
Ta(f)	Resistor	0.01-0.2	0.0004-0.008
Sn	Superconductor	0.05-0.2	0.002-0.008
Metal compounds			
Al_2O_3(f)	Capacitor	0.1-2	0.004-0.08
Cadmium sulfide	Semiconductor; photoconductor	0.1-2	0.004-0.08
Cerium oxide	Capacitor	0.1-2	0.004-0.08
Silicon oxide	Capacitor; insulator	0.1-2	0.004-0.08
Tantalum oxide(g)	Capacitor	0.01-0.2	0.0004-0.008
Titanium oxide(h)	Capacitor	0.03-0.2	0.001-0.008
ZnO	Semiconductor	0.1-2	0.004-0.08

(a) Good counterelectrode for capacitors. (b) Good counterelectrode for rectifiers. (c) Adheres exceptionally well to glass. (d) Difficult to obtain purity desired. (e) Good conductor for high-temperature applications. (f) Difficult to vaporize except by electron-bombardment heating. (g) Produced by anodizing tantalum films. (h) From thermal oxidation of titanium metal

depositing either silver or lead sulfide on the second surface of polished glass blanks. A high-quality plate glass was necessary to produce mirrors without detectable double-image effect. Many of the automotive rearview mirrors now produced consist of a first-surface vacuum coating of chromium or chromium alloy on glass. First-surface coating completely eliminates the double-image effect and provides the additional advantage that only one side of the glass blank must meet the specifications regarding surface finish before coating.

The processing of mirrors differs somewhat from the conventional procedure for decorative coatings in that no base coat or top coat is used; the vaporized metal is deposited directly on the surface of the glass. The cleanness of the glass is important, because a contaminated surface results in poor film properties. Films deposited on properly cleaned glass have good resistance to abrasion, wear and corrosion.

Vacuum-deposited metal films may be used in conjunction with electroplating. For example, a paint mask for a thermoplastic part may be made by first vacuum depositing a conductive film of copper or silver on the part. The part is transferred to an electroplating tank, where a heavy layer of metal, usually nickel, is deposited over the vacuum-deposited film. The plastic is removed from the plated shell by heating or dissolution in a solvent. This provides a conforming metal negative replica of the part, in which openings can be machined or otherwise made.

When electroplating on a vacuum-deposited film, the part is made the cathode, and the current is applied before the part is immersed in the electroplating bath. Application of the current after the part is immersed in the electroplating bath frequently etches the thin vacuum coating and destroys its continuity before electrical contact is established.

Electrical Coatings. Table 4 lists some metals and metal compounds that are vacuum deposited in various thicknesses for use as conductors, resistors and capacitors, and for other electronic-circuit applications.

For conductors, selection of coating material usually is determined by the desired electrical properties. A high-chromium alloy, or chromium plus silicon dioxide, is deposited when high resistance is required; aluminum or silver is selected when high electrical conductivity is needed. Aluminum is superior to other metals for maximum electrical conductivity either per unit weight deposited or per unit of heat of condensation.

Gold is used as a transparent conductive coating in one specific application in which the gold is deposited over a surface that has been coated with an oxide or sulfide material. This treatment of the substrate permits the deposition of the gold as a continuous film, rather than as the discontinuous layer characteristic of gold deposited directly onto a plastic. Gold coatings with a light transmission in the visible range of 60 to 70% can be obtained by this procedure.

Silicon monoxide and magnesium fluoride are used for dielectric layers. These materials are used as deposited, and do not require subsequent oxidation. Silicon monoxide is used particularly in miniaturized electronic components in which one segment of a circuit is deposited through a mask, after which an insulating layer is deposited, also through a mask.

All the common noble metals, as well as certain other metals more difficult to evaporate (such as tin and indium), have been used for circuit terminations. Gold, platinum, copper, aluminum, and silver are the most frequently used coatings for this application, in about the order listed. Gold, platinum, and silver are used also as coatings on contact electrodes.

Various nickel-chromium-based alloys, as well as platinum-gold and other alloys of platinum, are used for resistive elements. Certain oxides and metallic compounds, such as tin oxide, are also in use, but these require more elaborate processing other than that concerned with vacuum techniques alone. The resistance of the film is inversely proportional to its thickness.

Iron-nickel alloy films are vacuum deposited for use as thin-film magnetic memories of electronic circuits. The composition of the vacuum-deposited alloy has to be closely controlled; consequently, evaporation is accomplished by simultaneous deposition of iron and nickel under controlled rates. This procedure requires great skill for consistent results. In contrast, the simultaneous electroplating of iron and nickel for this application can be performed with less trouble, and some reports indicate that more consistent results can be obtained than with vacuum deposition.

Vacuum coating is widely used for thin-film integrated electronic circuits because of the good control that is possible and because the conditions under which the deposits are made permit a great deal of flexibility in the number and kinds of materials that can be deposited one after another without an intervening cleaning of the surface of the work.

Many applications, especially large-scale integrated circuits (LSI), require use of multilevel circuits to decrease the access time (speed). Construction of these devices requires the use of photo-resist masking and multiple-step processing to produce the circuit. High density magnetic bubble memory devices (256 K memory) use seven-level fabrication methods. A number of metallization and dielectric film depositions are required.

Optical Coatings. Because of their reflective properties, vacuum-deposited metal films are useful for first-surface mirror applications ranging from automotive rearview mirrors and reflectors for sealed-beam lamps to scientific instruments, such as microscopes, monochromators, and astronomical telescopes. The optical characteristics of semitransparent metal films also are useful in certain applications. The vacuum deposition method permits film thickness to be controlled with sufficient accuracy that the ratio of reflectance to transmittance may be assigned any selected value in applications using semitransparent films as beamsplitters.

In other applications the reflectance is immaterial, and the semitransparent metal films serve as light-attenuators or neutral-density filters to reduce intensity levels. Some examples are coatings for sunglasses, neutral-density filters to control microscope illumination without loss of aperture, neutral-density wedges to obtain a continuously variable density, and specially graded filters to make the illumination of a wide-angle lens more uniform. Evaporated films of certain alloys, such as Inconel, can be produced with a relatively constant transmittance over a wide spectral range.

The thermal radiative power of surfaces is of interest in the design of space devices. The infrared reflectance of metal films is important here, because the reflectance of an opaque metal coating controls the emittance of the surface on which it is deposited. For radiation of long wave lengths, the reflectance of all metals approaches 100% and consequently the emittance approaches zero.

Coatings for Resistance to Corrosion and Wear. Cadmium and aluminum are usually applied for corrosion resistance. Coatings vary in thickness from 6 to 25 μm (0.25 to 1 mil). Unlike the electroplating process, vacuum coating of high-tensile steel does not subject the steel to hydrogen embrittlement.

Silicon monoxide is the material most commonly used for abrasion resistance. The coating is heated in air for conversion to silicon dioxide. Coatings approximately 0.1 μm (0.004 mil) thick used as top coatings over aluminum improve abrasion resistance by a factor of 10^4.

Reflectors may be coated with silver, particularly when the coating is not subject to atmospheric corrosion, but are usually coated with aluminum, which has reflectivity approaching that of silver. For reflectors that are subject to severe atmospheric exposure, rhodium, chromium, or chromium-nickel coatings are used because of their greater abrasion resistance, although their reflectivity is slightly lower than that of aluminum.

Thick Films for High-Temperature Protection. Four- and five-component corrosion and oxidation resistant M-CrAlY (where M indicates nickel, cobalt, iron or a combination, such as nickel-cobalt) alloys are electron beam evaporation-deposited onto the high-pressure turbine components of gas turbine engines. The deposits range in thickness from 64 to 305 μm (2.5 to 12 mils), depending on the particular application. The deposited alloy provides a several-fold increase in oxidation and hot corrosion resistance compared to the aluminide coating it replaces. By varying the composition of the alloy deposit, hot corrosion resistance can be balanced against the required oxidation resistance and the mechanical property needs of the components. The coating method enables incorporation of reactive elements that assist in maintaining the protective oxide on the surface of the deposited alloy. Turbine engine components coated with these alloys exhibit useful lives in terms of coating degradation of 10 000 operation hours or more, with exposure temperatures that can locally exceed

1150 °C (2100 °F) during high power portions of the engine operation (takeoff, climb and thrust reverse during landing).

Requirements of the Substrate Material

The primary requirement of the material to be coated is that it be stable in vacuum. It must not evolve gas or vapor when exposed to the metal vapor. Gas evolution may result from:

- Release of gas adsorbed on the surface
- Release of gas trapped in the pores of a porous substrate
- Evolution of a material such as plasticizers used in plastics
- Actual vaporization of an ingredient in the substrate material

If the gas is evolved from an adsorbed layer, the substrate may be degassed by being warmed in vacuum before being coated. However, if the substrate is porous and gas is being released from the pores, the surface may be lacquered to seal the pores and prevent the liberation of gas during the vacuum coating operation.

Plasticizers cannot be completely removed without increasing the brittleness of the substrate. Lacquering before coating may be satisfactory, although heavily plasticized materials usually are not subjected to vacuum coating.

If the substrate material is volatile, or if one of its constituents is volatile, the surface must be chilled before vacuum coating, to suppress the vaporization of the substrate. As a rough approximation, the vapor pressure or decomposition pressure at the surface of a substrate should be less than 10^{-1} Pa (10^{-3} torr) at approximately 35 °C (95 °F) for effective vacuum coating.

Gas evolution increases the pressure within the vacuum chamber and causes an excessive number of collisions between the metal vapor and the gas evolved from the surface. This results in condensation of the vapor before it makes contact with the substrate surface, and causes a darkened deposit with poor adhesion. This deposit consists of small discrete powder particles rather than the continuous film characteristically developed by vacuum deposition. Also, difficulty occurs if the evolution is a result of a reaction between the substrate and the metal vapor. This frequently occurs with water or hydrocarbons. These gas-evolving materials react with the vapor to liberate hydrogen, which causes an excessive load on the pumping system, and the entire vacuum system deteriorates as the result of the buildup of noncondensable gases.

Although many of the problems occurring with a gas-evolving substrate may be overcome by coating the substrate with an impervious layer of lacquer, the use of lacquers is limited to the deposition of films for decorative purposes only. Lacquers do not have sufficient thermal stability to withstand the heat of condensation of heavier deposits.

Thermal Stability. In general, the more thermally stable the substrate, the more readily it may be vacuum coated. This indicates that a metal substrate is preferable to a plastic substrate. Most metals are satisfactory for coating with decorative films; however, metals containing a volatile constituent are not satisfactory for coating with protective deposits. Copper can be coated readily with heavy deposits, whereas brass cannot, because of the loss of zinc. Zinc-based die castings do not permit thick deposits.

Before the deposition of thick films, the substrate must be preheated to ensure adhesion of the deposited layer. Preheating, however, may cause difficulties. If the substrate forms a stable oxide that is not adherent, this oxide must be removed before deposition of the metal. If steel is to be coated with aluminum, excessive preheating causes diffusion of the aluminum into the steel, resulting in the formation of an iron-aluminum compound at the interface. This compound is extremely brittle, and bending of the coated substrate results in failure of the coating at the interface. Therefore, any combination of substrate metal and coating must be considered in terms of the minimum preheating temperature required to produce adhesion and the maximum preheating temperature permissible to prevent diffusion, particularly if the part is to be flexed or shaped after coating.

Substrates for decorative coatings are usually glass, metal, or plastic. The surface to be decorated must be free of deep scratches, pits, or other defects that cannot be completely covered by the resinous precoat used to provide an adherent, smooth, high-gloss surface for the vaporized metal.

Most metals can be decorated by vacuum vapor deposition provided they meet the surface requirements. Some castings and slush moldings present a problem, however, if excessive porosity is present. The expulsion of entrapped gases may cause blistering of the precoat during the heat curing cycle.

Articles made of glass may require a precoat. If the surface is free of imperfections, however, the evaporated metal may be deposited directly on the surface, but the surface must be absolutely clean to ensure adhesion and good coating quality.

Substrates for Optical Coatings. Metallic films can be vacuum deposited on all glass, crystal, ceramic, plastic, and lacquered surfaces of interest in optics. Glass surfaces may be fire polished or optically polished, or may even be in the ground state. However, the degree of specular reflectance, as well as the brilliance of the metal coating, is adversely affected by lack of smoothness of the substrate.

Semiconductors (for example, silicon and germanium, which are used for infrared windows) and crystalline materials, unless they are water soluble, can be satisfactorily coated. Water-soluble materials, such as rock-salt plates, require special precautions to maintain a dry surface during deposition, including rapid transfer to the coating chamber after cleaning, rapid pumpdown (but avoidance of a Wilson cloud), prompt firing, and the use of small trays of phosphorus pentoxide in the coating chamber. The subsequent protection against water vapor offered the surface by coatings of water-soluble materials may be of doubtful value.

Negligible gas evolution is a requisite for the satisfactory coating of plastic or lacquered surfaces. Because there is some increase in the temperature of the work during vacuum deposition, thermally stable plastics must be used and the duration of exposure to the vapor source must be held to a minimum. Occasionally, arrangements for cooling the base are also necessary.

Although evaporated metal coatings are readily formed on all these materials, adequate and satisfactory adherence may be difficult to obtain. For example, chromium films bond very tenaciously to glass, and aluminum films are moderately adherent, but silver films have only fair adherence.

Substrates for Electrical Coatings. In the selection of substrate materials for electronic-circuit films, consideration must be given to gas evolution in the vacuum process, heating effects, film adhesion, film crystallization, optimum circuit-panel area versus uniformity of deposition, and physical requirements of the functional package. Metal substrates have the advantage of better heat dissipation and are used wherever substrate capacitance either may be purposely incorporated in the circuit or has negligible functional effects.

Although glass has the disadvantage of being more fragile, it is favored for the following reasons:

- Ease of cleaning
- Can be obtained with surface roughness of less than 50 Å (0.2 μin.)
- Favorable coefficient of expansion
- Inherent insulating properties
- Minimum capacitive reaction with electronic circuitry
- Unaffected by oxidation

Typical specifications call for borosilicate glass with low alkali content; no elements of lime, magnesium, or zinc; no defects, cracks, or pit marks; and no scratches visible with a 25-power microscope.

Typically, vendors supply glass substrates with ±0.4 mm (±0.015 in.) planarity and ±0.01 mm (±0.0005 in.) thickness. Other substrates for electronic components are fused silica, glass-ceramics, and various grades of alumina, beryllia, sapphire, and barium titanate. The relative costs of these substrates range from 1 for soda-lime glass to 3 for borosilicate glass, and 400 for polished sapphire. Substrates with desirable bulk properties may be upgraded by suitable premanufacturing treatments, such as polishing and glazing of ceramics, and application of thick films such as SiO to glass to allow removal of scratches, pits, and other minor surface defects.

Substrates for High-Temperature Coatings. Substrates for high temperature corrosion- and oxidation-resistant coatings are selected for their high-strength and other thermomechanical properties for specific application in the turbine engine, such as turbine blades and vanes. They consist of the nickel- and cobalt-based superalloys such as René 80, René 77, B-1900, DS Mar-M200 + Hf, and X-40 and Mar-M509. Coating alloys are specifically designed for compatibility with the substrate, for the environmental use intended, and the thermal-mechanical regime experienced by the component.

Material Compatibility. If incompatible materials are placed in contact by means of vacuum deposition, one or both of the materials may be destroyed by processes related to electrochemical corrosion or to the formation of alloys whose properties are undesirable. In general, the rules that govern the use of dissimilar metals in other applications can be used to determine the compatibility of vacuum-deposited coatings. However, even less tolerance to unfavorable combinations can be expected in vacuum-deposited coatings, because the contact between layers is more intimate than that achieved between the surfaces of ordinary solid bodies.

Also, certain combinations of metal films can form alloys when subjected to high temperatures as part of a bonding or stabilizing process. Such alloying increases resistivity at metal interfaces and can result in embrittlement.

Processing Techniques and Limitations

Vacuum deposition is used to accomplish (*a*) bulk coating of powders or small parts (encapsulation), (*b*) batch coating of individual parts, or (*c*) continuous or semicontinuous coating of rolled materials.

Encapsulation. Tiny parts or finely divided powders are transferred through a coating zone while held in a monolayer on a supporting oscillating tray. The material to be coated must be free-flowing and not excessively agglomerated. The material to be coated also should preferably be spherical, although irregular crystals of powder can be coated, provided reentrant angles are not present at the edges of the crystal. Sharp edges are undesirable but are not impossible to coat. The minimum particle size of powder that can be encapsulated is approximately 10 μm (0.4 mil) in diameter. Particles of smaller size resist rotation during the coating operation, and therefore only one surface is exposed.

Batch Coating. Irregularly shaped finished parts are coated for either decorative or functional applications by batch coating. These parts are individually fixtured and are rotated during the coating operation to expose all surfaces. The minimum size of parts that can be successfully processed is about 3 mm (1/8 in.) in diameter by 6 mm (1/4 in.) long; smaller parts are more difficult to grip individually. Maximum size is limited only by the size of the evaporator. At present, commercial evaporators are available that are approximately 1.7 m (5½ ft) in diameter by 1.8 to 2.4 m (6 to 8 ft) long. For large parts, the distance between the surface to be coated and the metal source should not vary by more than a factor of 2; otherwise, coating uniformity is affected. Large concave surfaces do not lend themselves to uniform coating.

Figure 10 shows various shapes of parts whose exterior surfaces (except for the center recess of Fig. 10a), can be readily coated by proper fixturing and mechanical rotation. Surfaces with blind holes or deep holes become progressively less well coated. Holes should not exceed two diameters in depth for reasonable coating uniformity. Although it is possible to deposit to a depth of five diameters, the coating is thin.

Multiple sources of metal evaporation are required for uniform coating of some parts. If the projections on the underside of Fig. 10b are comparable in length to the closest spacing to the metal source, multiple sources with elevations similar to the contour of the part are necessary for uniform coating. Multiple sources of metal vapor permit the uniform coating of complex shapes and of larger parts with simple shapes.

Continuous and Semicontinuous Coating. Rolls of material that vary from monofilaments a few μm (mils) in diameter to sheets 1.8 m (6 ft) in width may be continuously coated. For coating these products, materials of evaporation that are at least as volatile as gold should be used. With metals of this type, the beam of metal vapor may be directed downward if necessary, so that complex surfaces can be more readily coated from multiple sources. With the more refractory metals, evaporation is usually done only in an upward direction. This requirement limits the complexity of surface that can be coated and reduces the productivity of equipment.

For semicontinuous coating, production is limited by the size of the roll of material that can be installed in the evaporator at one time. Coating for decorative applications may proceed at a rate in excess of 5 m/s (1000 ft/min) for the full width of a roll. For thick-film functional applications, the speed may be reduced to as low as a few centimetres (inches) per minute. The productivity of the semicontinuous coater is associated also with the time for

Fig. 10 Parts whose exterior surfaces, except for deep recess of (a), can be readily coated when properly fixtured and rotated. Coating of interior surfaces depends on the conditions given below

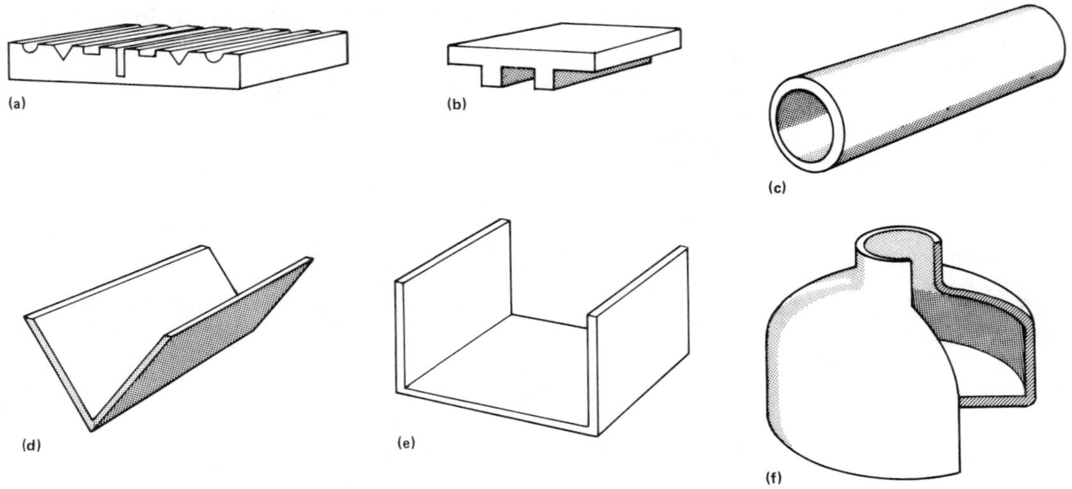

(a) Deep recess is difficult to coat. (b) Coating on irregular side of transparent part appears as a dark area over tab locations when viewed from the uncoated flat side. (c), (d), and (e) Coating of inside surfaces depends on the size of part, racking, and rotation. (f) Coating of inside surface requires use of an internal filament

pumpdown, although coating times may be for periods up to 2 h.

Decorative Coatings. The chambers of industrial vacuum coaters for applying decorative coatings vary in size from the 460-mm (18-in.) diam vertical bell-jar type to the more common horizontal type that is 1 to 2 m (3 to 6 ft) in diameter and 1.5 to 2 m (5 to 6 ft) in length. The heaters for evaporating the metal usually are located in the center and extend from the front to the back of the chamber. The minimum heater-to-work distance is critical, especially for parts that have been lacquered; excessive heating of the lacquered surface during metal deposition can cause off-color coatings. Small parts present no particular problem provided they can be racked to permit effective lacquering and deposition of the metal.

The relative ease of decorating a part by this process is closely related to its shape. Complex shapes can be coated if precoat and topcoat materials can be applied by dipping or flow coating. However, if only certain areas are to be decorated, precoat and topcoat materials must be applied by spraying. Deep, narrow recessed areas are difficult to spray and also present a problem in the deposition of the metal vapor. Because the vapor emanating from the metal source travels in straight lines, the comparative amount of metal deposited on a surface is a function of the angle that the surface makes with the vapor stream; the optimum angle is 90°.

The material being coated can affect the production rate by increasing the pumping period as a result of gas evolution. This occurs with parts made of acetates and butyrates, especially when these materials contain plasticizers for softer flow. Also, atmospheric conditions may affect the time necessary to evacuate the coating chamber and work load. During warm, humid weather, the inner surfaces of the chamber, the coating racks, and other associated equipment absorb moisture that must be removed by the pumping system. The water vapor, unless trapped by refrigerated coils or other means, contaminates the pump oils, thereby reducing their efficiency. The accumulation of metals on the chamber walls and the accumulation of lacquers and metal on the work-holding racks prolongs the pumpdown period, because the surface area exposed to moisture is increased. Therefore, periodic cleaning of racks and chamber is an important factor in the overall efficiency of the process.

Electrical Coatings. The limitations on large parts are set by the amount of film-thickness tolerance that is acceptable. In some applications, a 2% variation in thickness over a coated substrate is acceptable. The variations of film thickness from a metal source are governed by laws of geometry. In general, for a uniform thickness over a large area, these laws require that the substrate be set at greater distances from the source or that multiple sources be used. Thus, the size of the chamber that is feasible becomes a matter of economic calculation.

Limitations imposed by the smallness of the workpiece depend primarily on the need for holding the workpiece during deposition. For extremely small workpieces, other methods, such as fluid techniques, are feasible.

In electronics applications, most workpieces are rectangular, flat, or cylindrical (solid or hollow). When the workpiece is flat, the size limitations depend on the amount of area that can be covered uniformly. When coating the exterior of cylindrical workpieces, the limitations on size of the part are essentially the same as those applying to the long dimension of a flat piece.

Coating the inside of a hollow cylinder, especially when the inside diameter is small, is subject to severe limitations. The only materials that can be deposited are those that can be evaporated from a straight wire-filament heater that is threaded through the tube. There is also a limitation on the material of the tube, because it must withstand the proximity of a hot source.

In microelectronic applications, crucible, wire, and electron beam sources are used to deposit films for ohmic contacts, conductors, and insulators, often requiring masks with gaps for conductive deposits as narrow as 2.5 μm (0.1 mil) wide. Limitations of the substrate temperature and compositional consistency must be considered for each application. These processing methods require precise control of the evaporation rate, thickness of deposit, and composition and microstructure of the deposit. Sputtering of oxide systems has tended to replace vacuum evaporation because of the greater uniformity of the deposit.

Optical Coatings. Ordinarily, optical coating is a batch process, and the size of the chamber is governed by batch size for economical processing. In common practice, a 460-mm (18-in.) diam chamber is economical for batches with work diameters of 50 mm (2 in.) or less, a 915-mm (36-in.) diam chamber for work diameters of 200 mm (8 in.), and a 1520-mm (60-in.) diam chamber for work diameters of 510 mm (20 in.).

The contours of the surfaces to be coated must be simple and have no abrupt changes in profile, so that exposure to the vapor beam is relatively uniform. This precludes the presence of deep holes, cavities, and other surface intricacies.

Opaque first-surface reflector coatings must not vary in thickness by more than about 50% if the original optical value of the surface is to be preserved; and for the highest-quality optics, the tolerance may be much less. If the coating is a multilayer, the metallic and nonmetallic films comprising it must be uniformly thick, with permissible variation of as little as 1%, depending on the coating structure.

Any small area on the receiving surface that is to be coated with this high degree of uniformity must be in direct line to the vapor source, and the film thickness is proportional to the cosine of the angle of incidence on the small area, as well as to the vapor density.

If possible, angles of incidence greater than about 45° should not be permitted. When larger angles are used, the films generally are soft and do not attain their maximum brilliance or reflectance. At very large angles, films can become slightly bireflectant with a preferred direction in the plane of incidence. Nonmetals also are likely to be soft and porous, with lower values of refractive index than at normal incidence. Basically, these effects are ascribed to the texture or structure of films deposited from vapor beams at high angles of incidence.

Thick Films for High-Temperature Protection. The chambers for the application of thick (64 to 305 μm or 2.5 to 12 mil) alloy films for the high-temperature protection of superalloy components are several metres (feet) across, with extensions containing load locks and preheaters to load the parts onto a shaft and heat them to deposition temperature. Parts are then cycled into the vapor chamber and are rotated and oscillated in the vapor stream. The complexity of the design and the size of the component determines how many components are coated at the same time. The alloy vapor, heated by an electron beam, is condensed onto the component. The temperature of deposition (870 to 980 °C or 1600 to 1800 °F) is sufficiently high that auxiliary heating methods are used to augment the heat of condensation to maintain the temperature of deposition. Excessively low substrate temperatures can result in spalling of the deposit and exaggerated film growth defects. Limitations on the quality of the deposit are dictated by (*a*) the angle of orientation of the surface to the vapor flux, (*b*) the surface finish quality of the substrate, and (*c*) the spurious deposition of solid contaminants. Angles of incidence of less than 45° show excessive growth defects. Components as small as 13 by 25 mm (0.5 by 1 in.) long, and as large as 250 mm (10 in.) long by 150 mm (6 in.) wide weighing 11 kg (25 lb) have been coated. Substantial variations in the composition of the vapor resulting from thermal variations in the source limit the precision of alloy composition. Aircraft components are coated on a batch basis with about 100 components per batch. Several batches may be coated during continuous operation of the coater extending as long as 100 h.

Electrical coatings often require application of an alloy. As indicated in Table 6 wire can be used for flash evaporation sources as well as powder, provided that the vapor pressures of the constituents are not widely variant.

Typical Vapor Sources

Decorative and Electrical Coatings. The form of the evaporant is important from the standpoint of the thickness of the deposit and the mode by which the coating is to be made. Aluminum, for example, is extensively used for decorative, electrical, and corrosion-resistant coatings. For batch processing of decorative items or for applications of thin coatings for ohmic contacts, wire-wrapped tungsten wire may be used to deposit the thin coatings required. In the case of the decorative deposit, there may be little need to gage the actual thickness of the deposit during deposition; however, the rate of evaporation and the resistance of the deposit will be monitored for electrical deposits.

Once the desired resistance has been achieved, the power to the heater is then turned off. For continuous processing of parts, such as in the application of aluminum to paper, plastic, or steel sheet, or continuous barrel coating of small parts, a larger source, such as a crucible (BN or BN-TiC) which is indirectly heated by induction, may be used. Replenishment of the aluminum source material is achieved by feeding wire into the crucible at a rate consistent with evaporation losses. Typical operating conditions for decorative coating depositions are shown in Table 5.

Table 5 Operating conditions for producing decorative coatings(a)

Evaporant metal	Form of evaporant	Evaporation temperature(b) °C	°F	Time to reach evaporation temperature, s	Exposure time, min
Al	Ingot/crucible(c)	1000	1830	...	1/6
	Wire staples	1000	1830	3-6	1/6-1/3
Cr	Powder; pellets	1400	2550	5-30	1-5
Cu	Staples(d); pellets	1270	2320	20-60	1-3
Au	Wire; small chunks	1400	2550	20-60	1-3
Ag	Wire; chunks; pellets	1020	1870	20-60	1-3

(a) Coating thickness ranges from 0.08 to 0.1 μm (0.003 to 0.004 mil). (b) Temperature at which vapor pressure is 10^{-1} Pa (10^{-3} torr). (c) Continuous coating of sheet. (d) Of small-diameter wire

Optical Coatings. Some of the metals, metal compounds, and semiconductors most commonly used to form optical coatings on the surfaces of glass, plastic, and other materials, are listed and described in Tables 7, 8, and 9.

Most of these coating materials must be melted before their vapor pressure is sufficient for a reasonable rate of deposition in vacuum. Some of the materials sublime. The metals are opaque at a thickness of 0.08 or 0.1 μm (0.003 or 0.004 mil). This thickness can be obtained in a few seconds of deposition time. For example, the deposition time for opaque aluminum films should not exceed 10 s if the highest ultraviolet reflectance is needed. Metal films more than about 3 μm (0.1 mil) thick are usually gray or have a rough surface.

Nonmetals or transparent materials, and semiconductors that are transparent in the infrared, are evaporated to produce coatings ranging in thickness from a fraction of a microinch to 1 to 13 μm (0.05 to 0.5 mil), depending on the intended optical application. Each material has a limiting coating thickness above which it becomes unsuitable for optical purposes because the film cracks as the result of high internal stresses or fails from other causes. The limiting thickness of the coating can often be increased by heating the receiving surface; therefore, the typical maximum coating thicknesses in Table 8 should be regarded only as guides.

Thick Films for High-Temperature Protection. Vapor sources used to produce thick film deposits on high temperature components are generally vacuum-induction melted and cast into ingots 150 to 915 mm (6 to 36 in.) long. Powder consolidation by hot isostatic pressing (HIP) has often been successful in producing ingots of controlled composition. Because the composition of the deposit is dependent on the composition of the ingot, the composition of the ingot is determined at intervals along the ingot length. The ingots are x-rayed to determine porosity levels, and then are precision ground to dimension to fit through seals at the base of the hearth. The solid ingot forms the base of the hearth. Hearth material corresponding to the equilibrium composition in the hearth at the desired evaporating temperature is similarly cast in chunks. This material is added to the hearth before deposition to shorten the time to stabilization of the composition of the vapor. Special care must be exercised to minimize the gas and porosity content of the ingot to prevent spitting of the evaporant from the molten pool.

Table 6 Operating conditions for producing electrical coatings

Evaporant metal	Form of evaporant	Time to reach evaporation temperature, s	Exposure time, min	Coating thickness μm	Coating thickness mil
Pd-Au	Wire	20-30	½-2	0.01-0.03	0.0004-0.001
Cd	Clips	60-120	1-5	0.015-0.05	0.0006-0.002
	Plated on heater	30-60	1-3	0.02-0.06	0.0008-0.002
	Pellets	Up to 60	2-5	0.01-0.03	0.0004-0.001
	Pellets	60-120	3-6	0.01-0.03	0.0004-0.001
Ni-Fe-Cr alloy	Wire	20-30	1-4	0.01-0.08	0.0004-0.003
	Clips	60	2-3	0.015-0.08	0.0006-0.003
Cu-Ni	Wire	30-90	1½-4	0.02-0.045	0.0008-0.002
Ta	Pellet(a)	30-60	...	0.004-0.006	0.0002-0.0002

(a) Electron beam heated

Table 7 Metals for optical coatings

Evaporant	Form of evaporant	Function of coating	Remarks
Al	Wire	High reflection of ultraviolet, visible, and infrared light	Coated with SiO for mechanical protection
Cr	Chunks	Semireflecting	Extremely durable
Cu	Wire; shot	Infrared reflection(a)	Soft coating
Au	Wire	Infrared reflection(a)	Soft coating
Inconel	Wire	Neutral-density filters	Very durable; semitransparent
Pt	Wire; plate	Vacuum ultraviolet reflectors	Difficult to evaporate(b)
Pt-Pd alloy	Wire	Shadow-casting	Replicas for electron microscopy
Rh	Wire; sheet	Specially durable first-surface reflectors	Sublimes from direct-heated sheet
Ag	Wire; shot	High reflection of visible and infrared light	Soft coating
Ti	Wire	Beam splitters (after conversion to oxide)	Easily oxidized to TiO_2 when heated in air

(a) Also for decoration. (b) Induction or electron beam heating is required

Pretreatment of Substrate Surface

The primary prerequisite for any surface to be vacuum coated is that it be extremely clean. Any loosely adhering soil prevents adhesion of the deposited coating. Cleaned parts should be handled with cotton gloves to prevent fingermarks on the surface.

Surface finish affects the appearance of the finished coat, but does not affect the physical properties or adhesion of the coating, provided the surface has been properly cleaned. Coatings on a rough surface, such as a sand blasted surface, have a matte finish.

Oil causes a darkening of the coating at the precise location of the oil and in the area adjacent to it. Darkening is caused by the evolution of gases from the oil during deposition, which causes a condensation of the vapor before it contacts the surface.

Nonvolatile oils prevent adhesion in areas where they are present, in addition to causing discoloration of the coating. This behavior of nonvolatile oils provides a means of masking, whereby a pattern can be printed using an oil, the entire surface vacuum coated, and the coating readily removed from those areas on which the oil is intentionally placed.

The surfaces of substrates that evolve gas may be sealed with a lacquer before vapor coating. Lacquering is also a rapid, inexpensive process for improving surface quality, and is used in the manufacture of inexpensive costume jewelry made of zinc-based die castings.

A glow discharge can be used as the final cleaning process during the pump-down of the vacuum chamber at a residual air pressure of about 10 Pa (10^{-1} torr). The discharge is produced by an aluminum electrode that is mounted in a space in the vacuum

Table 8 Metal compounds for optical coatings

Evaporant(a)	Film refractive index	Typical maximum thickness µm	mil	Useful range of wavelength µm	mil	Remarks
Ceric oxide (CeO_2)	2.1	2.0	0.08	0.4-3	0.02-0.1	Hard film
Cerous fluoride (CeF_2)	1.6-1.8	2.0	0.08	0.3-8	0.01-0.3	Evaporant oxidizes
Cryolite ($3NaF \cdot AlF_3$)	1.35	2.5	0.1	0.15-10	0.006-0.4	Evaporant sublimes
Lead fluoride (PbF_2)	1.8	2.0	0.08	0.25-12	0.01-0.5	Platinum heater
Lithium fluoride (LiF)	1.4	2.5	0.1	0.1-9	0.004-0.4	Soft film
Magnesium fluoride (MgF_2)	1.38	1.2	0.05	0.12-4	0.005-0.2	Hard film
Silicon monoxide (SiO)(b)	1.5-1.9	8.0	0.3	0.2-8	0.008-0.3	Hard film
Thorium fluoride (ThF_4)	1.5	2.0	0.08	0.2-9	0.008-0.4	Heat darkens film
Titanium dioxide (TiO_2)(c)	2.3	1.5	0.06	0.4-3	0.02-0.1	Hard film
Zinc sulfide (ZnS)	2.3	12.5	0.5	0.4-15	0.02-0.6	Evaporant sublimes
Zirconium oxide (ZrO_3)	2.1	2.0	0.08	0.4-3	0.02-0.1	Hard film

(a) Materials evaporated from powder form, unless footnoted to indicate otherwise. (b) In chunk form. (c) Evaporated from TiO or titanium in oxygen

Table 9 Semiconductors for optical coatings

Evaporant	Form of evaporant	Film refractive index	Minimum useful wavelength µm	mil	Function
Cadmium sulfide	Powder	2.4	0.5	0.02	Photoconductivity
Carbon(a)	Rods	...	0.001	0.00004	Replicas for electron microscopy
Germanium(b)	Chunks(c)	3.8	1.5	0.06	High-index multilayer film
Lead telluride	Chunks(c)	5.1	3.5	0.1	High-index multilayer film
Selenium	Powder	2.4	0.6	0.02	Photoconductivity
Silicon(d)	Chunks(c)	3.4	1.0	0.04	High-index multilayer film
Stibnite (Sb_2S_3)	Black powder	2.7	0.8	0.03	High-index multilayer film

(a) Evaporated from vacuum arc. (b) Evaporated from carbon crucible. (c) Polycrystalline material of semiconductor grade. (d) Evaporated from boron nitride crucible

chamber near the surface to be cleaned. When a potential of approximately 5000 V is applied between the wire and the chamber (at ground potential), a glow is established within the system and the surface is bombarded by ions that mechanically remove any residual or loosely adhering soil. Glass surfaces are cleaned rapidly without difficulty. The glow discharge cleans loosely adhering material from plastic surfaces and successfully removes plasticizers from the surface.

Decorative Surfaces. Many plastic parts require pretreatment to cover mold flaws, die marks, scratches, and other surface blemishes. Pretreatment also provides a clean surface, thereby improving the adhesion of the vacuum coating. Parts made of transparent materials molded in highly polished dies and to be coated on the second surface usually are not pretreated unless a metallic color tint is desired, in which instance pretreatment consists of applying a tinted precoat material on the second surface before the deposition of the metal.

Pretreatment of plastic parts consists of applying a resinous material on the cleaned surface. This material should have good buildup and flow characteristics, and should impart a mirrorlike gloss. The resinous film must also have good adhesion, because it is the foundation of a three-coat system — base coat, metallic coat, and topcoat.

Some plastic materials require a surface treatment before the application of the resinous precoat, to effect adhesion of the vacuum coating. For these materials, of which the polyethylenes and acetals are typical examples, surfaces are roughened by chemical etching or flame treatment.

Metal parts are generally cleaned in solvent vapors to remove oil and any other soluble foreign materials. Plastic parts usually are not cleaned before application of the precoat, except for a blow-off with a jet of ionized air to remove dust and lint. However, plastic parts should be processed through the finishing department as soon as possible after being molded, and should be kept clean until finishing is completed.

The resinous precoat materials are usually nonpigmented and are applied by spraying (air or electrostatic), dipping, or flow coating. They are heat cured at temperatures ranging from 60 to 180 °C (140 to 350 °F). The curing temperatures for precoats on metals can be much higher than those for precoats on most plastics; precoats applied on plastics must be cured at a temperature below that which may cause distortion of the molded part. Parts molded of acrylics, acetates, butyrates, and polystyrene usually can be heated to 60 to 71 °C (140 to 160 °F) without distortion; the heat-resisting modified acrylics and polystyrenes, to 82 °C (180 °F); and parts molded of acetal or polycarbonate resins or of nylon, to 150 °C (300 °F).

The curing of the resinous material is a function of time and temperature. A 2-h curing period may be required for those plastics that may distort if heated above 71 °C (160 °F), whereas 15 to 20 min may be sufficient for curing the precoats on metal and the high-temperature plastics. Overcuring of the precoat may adversely affect the adhesion of a subsequently applied protective coating, as a slight solvent bite is desirable for maximum adhesion of the laminate.

Thermoplastic parts with intricate shapes of varying wall thicknesses may be susceptible to crazing or cracking due to residual stress. The solvents in the precoat material may attack highly stressed areas to the extent that a dull surface, rather than a high-gloss surface, results. Therefore, before being precoated, parts that are highly stressed must be heated below the heat-distortion temperature of the particular plastic for a sufficient time to relieve the stresses.

An additional pretreatment that is desirable, particularly for decorative applications, is the chilling of the surfaces of plastic parts, to reduce the vapor pressure of any evolvable materials the parts may contain and thus prevent their evolution during vacuum coating.

Optical Surfaces. The simplest test for a clean surface is the breath pattern. The breath pattern on a contaminated glass surface appears as a gray condensate, because of the scattering of light on the globules into which the moisture collects on such a surface. A clean surface reveals a black breath pattern, because the moisture condenses as a uniform film. This moisture film of low refractive index has antireflection properties for glass surfaces. As it vanishes, the colors of antireflection films can be distinguished in reverse sequence.

More definitive tests for cleanliness are defined in ASTM F21-62J, which defines the atomizer test and the water-break test. In the atomizer test, the dry substrate is exposed to a fine water spray. On a contaminated surface, the water coalesces into larger droplets, whereas, on a clean surface, the mist deposit is stable. This test is able to detect contaminants as thin as 1⁄10 monolayer. The water-break test consists of immersing a substrate in water and withdrawing it. A clean surface leaves a continuous film, whereas breaks in the water film are observed on a contaminated surface. The method is sensitive to one monolayer.

Before coating, the glass is cleaned by conventional methods. Vapor degreasing may be used as a final conventional operation in mass-production cleaning of glass, provided the workpiece can withstand the thermal shock and has a shape that permits draining of the solvent. Isopropyl alcohol is used as the solvent for the initial dip and in the degreaser. Chlorinated compounds usually used as degreasing solvents leave residues that adversely affect the vacuum coating operation.

Vigorous scrubbing or polishing with white rouge and water is a common procedure for cleaning the surface of glass, if the workpiece is large enough to be grasped firmly. This method cannot be used for cleaning optically figured glass surfaces, because two wipes with white rouge on wet cotton produce a detectable furrow in a fine optical surface. After cleaning by conventional methods, the glass workpieces are mounted in the vacuum chamber, where final cleaning to remove traces of surface contamination is accomplished by glow discharge during evacuation.

Thick Film for High-Temperature Protection. As with other physical vapor deposition substrates, components to be evaporation coated with thick films require clean, grease- and oil-free surfaces. Components such as turbine blades and vanes may have an oxidized surface resulting from prior manufacturing operations. Such surface scale is generally removed by blasting with an aerated high pressure water slurry or by dry blasting with fine aluminum oxide. The parts are then degreased in either freon or chlorinated solvent. As with other vacuum-coated components, parts are stored and handled in a dust- and oil-free environment before coating.

Post-Treatment of Vacuum Coatings

The need for post-treatment of vacuum coatings, and the type of treatment if needed, is determined by whether the coatings are applied for decorative, protective, electrical, or optical purposes, or are applied as thick film deposits for high-temperature applications.

Decorative Coatings. Except for chromium and certain chromium alloys, the metals most often used as decorative coatings are relatively soft and offer little resistance to corrosion and wear. Protective coatings usually are applied over the vacuum-deposited films, to provide this resistance.

First-surface coatings are top coated with a clear or a color-tinted resinous material. The coating applied over the metallic film on second-surface applications is usually an opaque pigmented material called back coating.

The protective coating must meet requirements regarding adhesion, heat, humidity, weathering, and wear. It must also have good clarity and, when a chromium or silver appearance is required, must be water white and not discolor on aging.

Protective coatings are either air-drying or heat-curing materials. In either instance, the protective coating must not attack the precoat to the extent of causing film distortion or blush, but it should provide some bite into the precoat for adhesion.

When heat-curing materials are used, temperature limitations are similar to those for curing precoat materials. As a general rule, however, the curing period for protective coatings is somewhat shorter than that required for curing the precoats. Protective coatings used over vacuum-deposited copper should either be air-drying or be capable of curing below 93 °C (200 °F), to prevent darkening of the copper film by oxidation.

Coatings applied over the vacuum-deposited film on the first surface of decorative parts may serve a dual purpose; with appropriate masks, certain areas of a part may be coated with a clear material and others with a transparent color-tinted material to simulate gold or other colors.

Protective Coatings. When aluminum is used as a thick coating or as a corrosion-resistant coating, its abrasion resistance can be increased by anodizing. The thickness of the aluminum coating must be at least 13 μm (0.5 mil) to permit anodizing.

Vacuum-deposited cadmium for aircraft applications is frequently treated with a chromate paint to increase the corrosion resistance. Also, the chromate primes the cadmium surface for subsequent painting.

Post-treatment also includes the use of electroplating to increase the thickness of vacuum-deposited coatings. This procedure is used with vacuum coated nonconductive base materials, such as used for printed circuits.

Electrical Coatings. Electronic circuitry usually requires the application of insulating layers after final deposits of metallic or conductive materials. This is accomplished by vacuum coating silicon monoxide or an equivalent material, or by coating with a plastic.

Certain resistor compositions, such as chromium-silicon monoxide, require heat treatment after vacuum coating, to stabilize the predetermined ohmic value. Typical resistors of this material are treated by holding the panel at about 430 °C (800 °F) for 2 to 5 h in an inert atmosphere.

Other resistors, consisting of thin films of metals such as chromium, chromium-aluminum, chromium-copper, nickel, nickel-iron-chromium, and palladium-gold, are artificially aged at various temperatures and times in air. Because the resistance tolerance specified for these resistors can be as close as $\pm 0.01\%$, the object of the heat treatment is to oxidize the surface of the film, thereby stabilizing and controlling the resistance, which can otherwise change asymptotically under natural aging conditions. The resistance can be adjusted also by mechanically cutting a longer resistance path into the film.

A typical processing cycle for producing a thin-film resistor by vacuum coating a nickel-chromium alloy on heat-resistant glass consists of these steps:

1. Vacuum coat the substrate
2. Heat treat at 125 °C (255 °F) for 16 h, then at 180 °C (355 °F) for 80 h
3. Adjust resistance to 2% below the lower tolerance limit by mechanically cutting to effect a change in resistance value. The amount of change is determined empirically. This adjustment is required to prevent postadjustment drift in resistance beyond the upper tolerance limit
4. Heat treat at 125 °C (255 °F) for 24 h. The resistors should be within tolerance after this step
5. Final adjustment—used to produce high-precision resistors (tolerance of ±0.1% or less); the aging and adjustment in steps 2 and 3 are set to produce a resistance value just below tolerance limits (1 or 1.5%). A final cut is then made to bring the resistance value to about 0.5% below the lower limit. The resistor is then aged at 125 °C (255 °F) for 48 h, after which it is within tolerance

Artificial aging of thin-film resistors requires these operating conditions:

- Accurate furnace temperatures
- Admission of atmospheric air through vents to ensure a sufficient supply of oxygen for the oxidizing process
- Scrupulously clean furnaces used only for thin-film resistors. Furnaces should be in a separate room

Optical Coatings. Reflectors are removed from the vacuum chamber as finished products and do not require a post-treatment for optical purposes. However, to prevent marring during handling, mass-produced metal films are frequently rinsed with water containing soap or detergent after being coated, to create a water-repellent surface by the adsorption of an organic monolayer.

Semitransparent films of bismuth, cerium, copper, iron, lead, manganese, nickel, samarium, and titanium may be converted to optically acceptable oxide films by being baked in air. These oxides offer a useful series of dielectric films of various, usually high, refractive indexes. Oxide films cannot always be obtained directly by evaporation, because the oxide decomposes to the lower oxide state under heat in vacuum. However, the resulting lower oxide films may be completely oxidized by heating in air. In practice, the lower oxide itself is often preferably used as the evaporant.

Baking cycles are established by trial. When optical-glass substrates are involved, furnace temperatures should not exceed 400 °C (750 °F). The baking time can be as short as a few minutes (as required for cerium), or it may be more than an hour (as required for oxidizing titanium).

The high refractive index (2.7) of titanium oxide films obtained by baking metallic titanium films deposited on glass makes them desirable as optical beam splitters. At a quarter-wave optical thickness, the value of the reflectance-to-transmittance ratio (R/T) is 45/55. This thickness is obtained from titanium films originally monitored to a transmittance of 5 to 20%, depending on the form (wire, powder, or sponge) of the titanium used in the source and on vacuum conditions. Often 1 or 2% absorption, which results from incomplete conversion to the oxide, is tolerated in the interest of substantially shortening the baking cycle.

Thick Films for High-Temperature Protection. Post-deposition treatment of thick film, high-temperature coatings such as CoCrAlY, NiCrAlY, and NiCoCrAlY may include glass or shot peening to compress the rapidly-grown grains, and to induce sufficient surface stresses to cause recrystallization when subsequently heat treated. Specific intensity and coverage depends on the design of the component, and the shot composition, and size. Generally, Almen intensities of 10N to 6A may be successfully used. The procedure also provides an *ad hoc* test of the bond quality between the coating and the substrate. Heat treatments at 1050 °C (1950 °F) for 2 to 4 h in vacuum generally are sufficient to induce recrystallization and to promote better diffusion bonding with the substrate.

Equipment

An arrangement of a vacuum system is illustrated in Fig. 11. At the beginning of a pumpdown cycle, the diffusion pump is isolated from the atmosphere by closing the main vacuum valve and the foreline valve. The mechanical pump rough-pumps the system through the roughing line until the pressure is reduced to a level at which the diffusion pump can be used. At this point, the roughing valve is closed and the main vacuum valve and the foreline valve are opened.

Vacuum gauges can be mounted within the chamber, away from the inlet, in the pump inlet, and in the foreline to measure the pressures accurately while the system operates. A coating chamber connected to a pumping system capable of achieving pressures of 10^{-1} to 10^{-3} Pa (10^{-3} to 10^{-5} torr) or less is usually required for evaporation deposition. After evacuation, tooling that holds the parts is rotated to expose the substrates to the vapor. The tooling is rotated through a vacuum-tight rotary seal by a motor mounted outside the vacuum chamber wall. If, as in the most common process, the metal to be deposited is an aluminum decorative coating, wire clips are hung on tungsten filaments connected to copper busbars mounted in the axis of rotation. In a large system, power at 6 to 20 V and 200 to 1000 A is applied through vacuum-sealed insulated feed throughs by a step-down transformer. Evaporant temperature control can be achieved using a pyrometer mounted on a sight port; evaporation rates may be determined by quartz oscillator methods depending on the precision required.

Schematic illustrations of semicontinuous equipment for coating substrates in roll form are shown in Fig. 12. The setup shown in Fig. 12(a) and 12(b) permits the coating of paper, textile, and plastic substrates in 1.5-m (5-ft) widths at speeds ranging from 0.5 to 10 m/s (100 to 2000 ft/min). The substrate can be exposed to two sources of metal vapor, to coat both sides in the same operation. Evaporation of metal must be maintained at a uniform rate for a period of several hours. The drums at each coating station must be water-colled, to remove the heat of condensation and radiation from the vapor sources.

When thin metal in roll form is coated, it must be cleaned and preheated before exposure to the metal vapors. Temperature must be carefully controlled to promote adhesion of the coating without causing it to diffuse into the substrate.

The system shown in Fig. 12(c) is useful when substrates contain large amounts of volatile materials. These materials can be partially removed while the roll is unwinding in the upper chamber. The substrate progresses through vacuum seals; and because only a small area of the substrate is exposed for a short time, a low coating pressure can be maintained in the lower chamber.

Small parts and powders are coated by being discharged from a hopper at a controlled rate onto a vibrating tray, where they are exposed to metal vapors. When a heavy coating is required, the parts may roll about on an oscillating tray suitably mounted to hold the parts in the coating zone for a sufficient time, or the parts may be recycled for a second exposure. The tray may be heated or cooled, depending on the type of substrate. The vapor source can be designed so that metal vapors are directed only to the coating area.

Pumping Equipment. Vacuum pumping systems usually include at least one roughing pump and one diffusion pump. The roughing equipment may consist of a positive-displacement mechanical pump, a rotating water-sealed pump, and water and steam ejectors. The pressure within the chamber can be reduced to 10^3 to 10 Pa (10 to 10^{-1} torr) with this equipment. The use of a rotary blower, an oil diffusion pump, and an oil diffusion booster pump further reduces the pressure from 10^{-1} to 10^{-8} Pa (10^{-3} to 10^{-10} torr).

Oil diffusion pumps must be operated in series with a mechanical pump, to prevent oxidation of oil in the diffusion pump. Usually, pump forepressures should be maintained well below 10^2 Pa (1 torr). Rotary blowers are operated in series with a mechanical pump to permit operation of the blowers with a minimum power requirement. The blower is equipped with an electrical interlock that prevents its operation until the mechanical pump lowers the pressure of the system to a safe operating pressure of 10^3 Pa (10 torr).

Figure 13 is a schematic representation of a continuous coating system. The substrate in roll form is unwound in air and passes through a series of vacuum seals into a high-vacuum chamber, where it is exposed to metal vapors, after which it passes through a second series of vacuum seals and is rewound in air.

Figure 14 is a schematic representation, pictured in Fig. 15, showing a large electron beam coater designed to deposit thick oxidation and hot corrosion-resistant alloys on turbine components. The components are loaded into load locks that are then evacuated. The parts are heated to a predetermined temperature, and then coated in the central chamber. This particular coater has three locks, so that parts can be continuously cycled over the vapor, fully utilizing the evaporant.

Traps and baffles provide a balance between increased pumping rate and the prevention of backstreaming of vacuum pump oils. The function of the trap is to condense low-vapor pressure gases. The efficiency of the trap depends on its operating temperature, designed surface area, and degree of surface contamination. The most effective coolant for traps is liquid nitrogen; however, freon or organic solvents, such as acetone cooled with dry ice, are also used. Traps should have a large surface area with cooled walls to prevent surface diffusion of condensables into the vacuum chamber. Periodic cleaning and baking of the trap with the vacuum chamber isolated is necessary to remove accumulated condensates that generally consist of water and diffusion pump oil.

Baffles are designed to further reduce backstreaming of pump oils while providing minimal obstruction to the evaluation of the vacuum chamber. The ultimate pressures obtained with oil diffusion pumps are not limited by the pump's compression ratio but by the desorption of gases in the vacuum system itself. A properly designed vacuum system that has been degassed and in which suitable adsorbant traps and cooled baffles are used is capable of achieving pressures of 10^{-9} Pa (10^{-11} torr).

Vacuum Seals. Permanent seals for metal-to-metal connections may be made by welding, silver soldering, or soft soldering. The requirement for sealing demountable connections in vacuum systems is directly related to the use of the systems. For most vacuum systems, in which ultimate pressures of 10^{-5} Pa (10^{-7} torr) or higher are contemplated, elastomeric seals made from teflon, viton, or rubber are used for flat gaskets and O-rings to seal bolted flanges. Flat rings and O-rings are used for inspection ports and flanges that are opened after each coating cycle. Depending on the formulation, elastomers can be heated at up to 93 °C (200 °F) when the vacuum system is baked out. All mating surfaces should be clean and scratch-free. Greases, oils, and gasket materials must have a low vapor pressure. Small seals for temporary use in emergencies can be made with Glyptal or Teflon tape on the high-pressure side of the vacuum system, and with Teflon tape on the low-

Fig. 11 Vacuum system for depositing vapor coatings

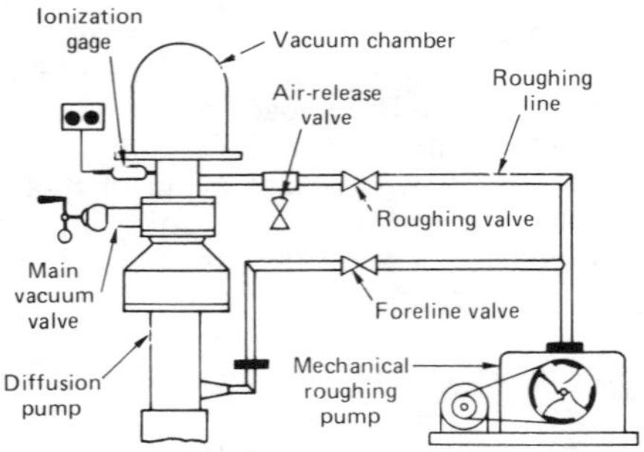

Fig. 12 Equipment for semicontinuous coating

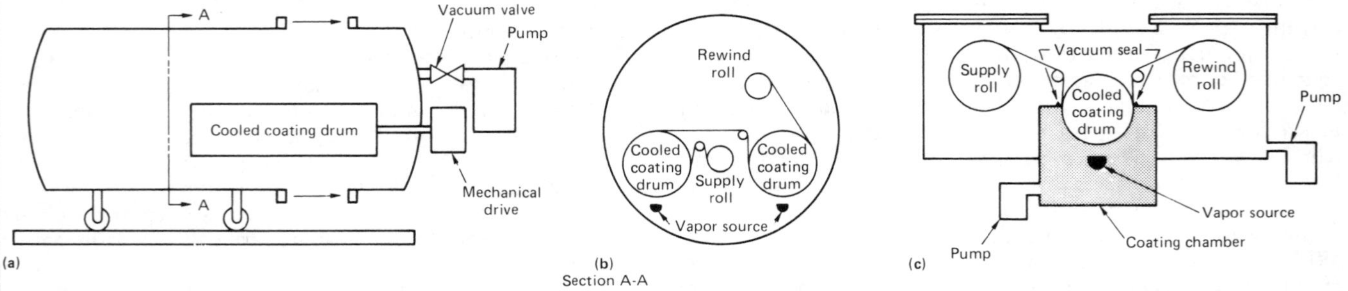

temperature side. Seals for vacuum systems in which ultimate pressures of less than 10^{-6} Pa (10^{-8} torr) are contemplated are made from oxygen-free high-conductivity copper that is deformed between grooved mating flange sections. These seals are expensive, and are not readily reuseable. Rotating shafts may be sealed as follows:

- *A direct shaft* is usually sealed through a chamber of low vapor-pressure oil or grease contained by a suitable synthetic rubber grease seal. The seals should rotate as slowly as possible (less than 200 to 300 rev/min). Torque limitations are seldom encountered. Sliding shafts also may be sealed by this method
- *A direct shaft that rotates between two O-rings* mounted in a body can minimize leaks by evacuating the space between the O-rings. As in the example above, a low vapor-pressure grease assists in sealing
- *A direct shaft that rotates only part of a turn* can be sealed by coupling it to a concentric tubular piece welded to the chamber by a short length of flexible tubing
- *A simple indirect shaft coupling* can be built with two permanent magnets that are mounted on shafts inside and outside the vacuum chamber. The magnets are separated by a thin, nonmagnetic metal plate. Rotary motion is transmitted by magnetic flux. The amount of torque that can be produced with this arrangement is limited by the practical size of the components

Fig. 13 A continuous coating system

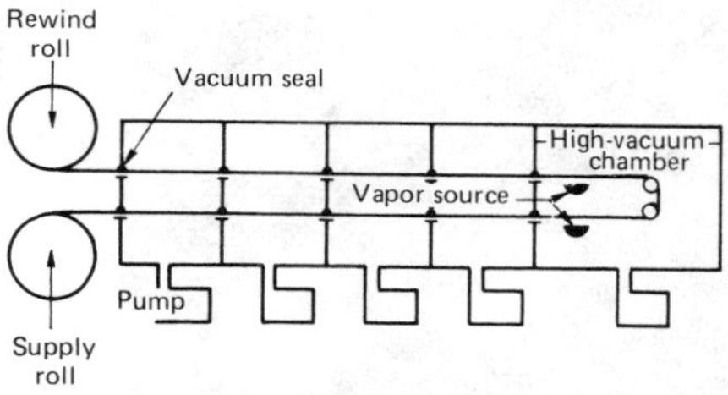

Work-holding racks are one of the most important items with respect to the overall cost of vacuum coating. Because of the variety of part shapes, the design of racks cannot be standardized, but the following points are basic to their proper design:

- Maximum number of pieces that can be accommodated on the rack and still be adequately covered by the vapor deposit specified
- Ease with which parts can be mounted on and removed from the racks
- Positioning of the mounted parts to provide best runoff of lacquer if dipping or flow coating is used
- Positioning of the parts to provide maximum efficiency for applying the precoat and top coat material if spraying (manual, automatic, or electrostatic) is used

Fig. 14 200-kW electron beam coater capable of gas scattering deposition

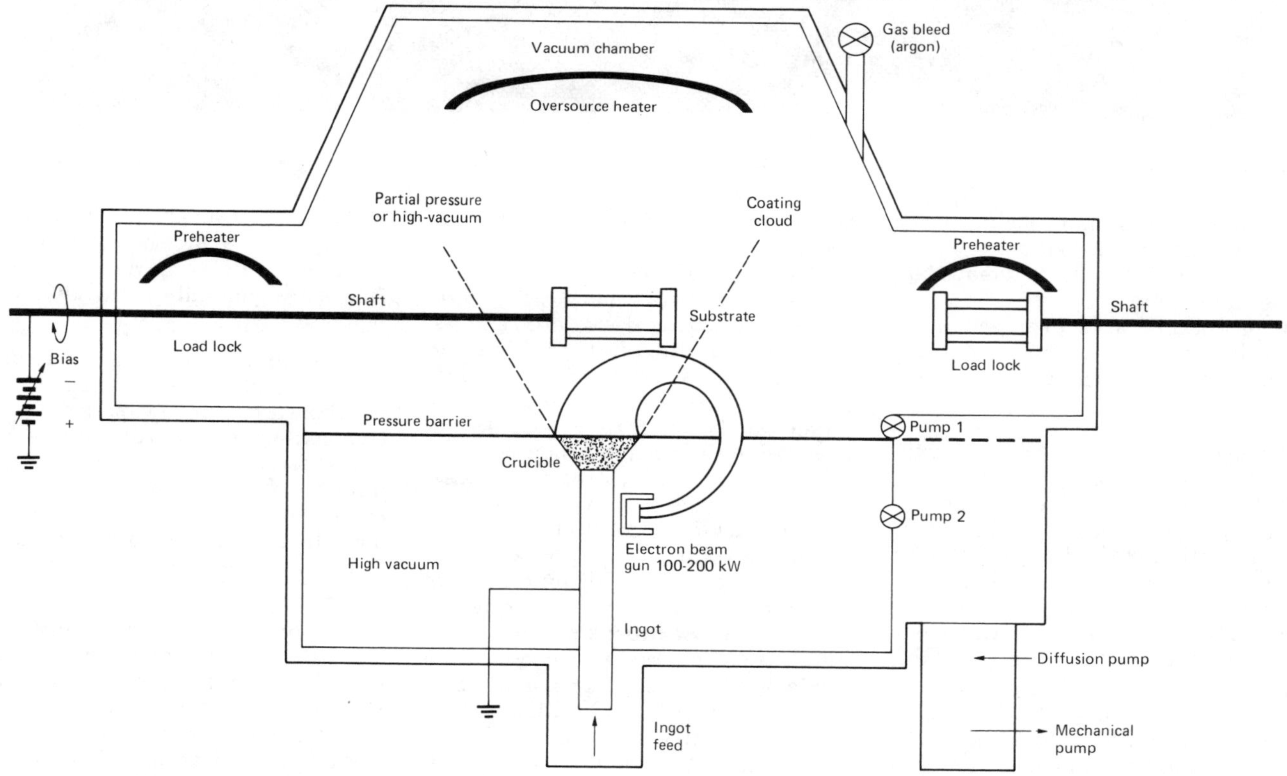

Parts are loaded in load lock, preheated to coating temperature and then put into chamber. Parts can be sputter-cleaned before coating, coated under high vacuum, or with gas scattering. High substrate temperature aids in maintaining good metallurgical structure during coating

Fig. 15 Electron beam coater designed to apply M-CrAlY coatings to turbine blades and vanes

(Courtesy of Airco-Temescal, Inc.)

- Provision for minimum shadowing or rack marks on finished parts
- Ease of cleaning racks to remove lacquer and metal
- Ruggedness of racks, to prevent damage during cleaning and handling
- Simplicity, to reduce cost and weight

The material used for the construction of work-holding fixtures, racks, and clips must: (*a*) not evolve gas at reduced pressures; (*b*) have the necessary strength and rigidity to carry the weight of the parts; (*c*) not react with the solvent of the precoat and top-coat materials; (*d*) withstand the temperatures attained during curing of the lacquers and during the cleaning process to remove accumulated lacquer and metal, and (*e*) not be attacked by the acids and solvents used for this cleaning. The selection of rack material is governed also by the metal being evaporated, especially when the parts being processed on the racks are not precoated.

The shape of the workpiece determines whether stationary racks or rotating, mechanized racks are required for coating specified surfaces. Figure 16 shows parts that are coated on stationary and rotating racks. Generally, a larger number of parts can be accommodated in the vacuum chamber if the parts are fastened to rotating racks. Stationary racks usually are less costly to construct than mechanized racks, but if the volume of a specific part is large enough, the additional cost of mechanized racks may be warranted.

Carriage Mechanism. The loaded rotors (rotating racks) are held by a carriage mechanism that usually consists of a movable frame with appropriately spaced fixtures to accommodate the rotors. The carriage may be mechanized to permit the rotation of the entire work load about the central axis of the chamber as each individual rotor revolves about its axis, or it can be stationary, permitting the rotation of the rotors only. Rotating mechanized carriages are generally referred to as the planetary type.

The planetary carriage is required for maximum utilization of the chamber capacity during evaporation of materials from boat or crucible-type heating sources, because vapors emerge from these sources at an angle of less than 180°. Stationary carriages are used in conjunction with filaments, from which the vapors of the evaporant are emitted in all directions.

Carriages usually are constructed of steel, but several factors must be considered in their design and construction. The carriage should cause a minimum of interference with the maximum utilization of the chamber space for the work load. The rotation of the carriage as well as of the rotors must be smooth; excessive vibration may cause breakage of the heaters and loss of the evaporant.

Because of the uniformity required in optical coating, the workpiece often is rotated, especially if multilayer film deposition is involved. If rotation is not feasible because of size or weight of the workpiece, an array of vapor sources can be used for deposition of simple coatings. The rotating mechanism may consist of the planetary carriage described above or a rotating table on which the workpieces are mounted. The justification for rotation of the work during coating is twofold:

- Irregularities of the vapor-emission time, as well as irregularities and asymmetries in the emission with respect to azimuth, are averaged out. (It is assumed that the work makes many revolutions during the application of the coating)
- Rotation of the work enables the use of masks to obtain either a uniformly thick film or some preselected distribution of thickness

Fig. 16 Shapes of parts that can be coated on (a) stationary racks, and (b) rotating racks

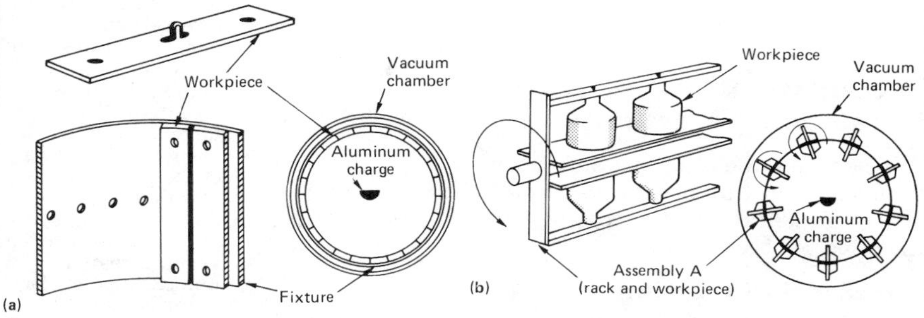

A filament-type heater is used with stationary racks; boat-type or filament heaters are used with rotating racks

Masking for Uniform Coating. The distribution of the vapor density around a source is influenced by its size, shape, and position. At any point on a surface, the thickness of film deposited is proportional to the density of the vapor stream, to the cosine of the angle of incidence, and to the evaporation time.

The point source emits uniformly in all directions, the vapor density falling off inversely as the square of the throw (the distance between the source and the surface being coated). However, in practice, there are no point sources and emission is not uniform in all directions; even filaments cast shadows.

To coat the inner surface of a cylinder uniformly, the cylinder is oscillated relative to the short axial line of the source. The momentary deposition area is restricted by shields that limit the angle of incidence to a few degrees. The difficulty of source miniaturization limits the diameter of small-bore tubing that can be coated internally.

The external surface of a rotating cylinder can be coated through a slot mask to limit the angle of deposition. Wire or other small-diameter work is difficult to coat by the usual directed vapor stream of the vacuum deposition process.

For coating the inside surface of approximately spherical enclosures, the source of metal vapor is placed at the center of the enclosure if the source behaves like a point or line source, as is the condition with tungsten filaments loaded with the evaporant. Rotation is used to eliminate the effects of source asymmetry. If the source behaves more like a small-area surface emitter, it is located eccentrically.

The convex surfaces of spheres require the use of a mask. Generally, only a restricted annular zone of the spherical surface can be coated uniformly using simple rotation of the work around an axis perpendicular to the plane of the zone. The mask limits the maximum angle of deposition. The shape of the mask is determined empirically, and the source is located on a radius of the sphere passing through the mean latitude of the zone.

Masking must often be used when large areas are to be coated uniformly from a centrally located vapor source. Because the film thickness is proportional to the exposure time and to the vapor stream, the angular openings of a mask interposed between the rotating work and the vapor source controls the relative film thicknesses along a radius.

Distribution masking is used when the coating must be given a special profile (Fig. 17). For the two-step mask in Fig. 17(a), with a relative rotation between the work and mask, the film thicknesses t_1 and t_2 deposited from a distant vapor source is in the ratio of the angles $2\alpha_1$ and $2\alpha_2$ of the two openings. The mask in Fig. 17(b) is cut to give several film steps of equal thickness.

Because masking involves blocking out some of the evaporating film material, the capacity of the vapor source must be sufficient to produce the desired coating thickness. Masks are usually positioned at a small distance of about 3 mm (about 1/8 in.) from the work or the worktable, to feather out small machining irregularities. The azimuth of the vapor source can sometimes be positioned to advantage in this regard, because the emission is seldom symmetrical. Masks with small openings may cause difficulty by curtailing the deposition rate below the limit at which a given metal deposits.

Masks are usually made of steel, so they are not warped by heat radiating from vapor sources. Defining edges of the openings should be beveled.

Maintenance. The following schedule is recommended for vacuum equipment:

Daily

- Check oil level on mechanical pump

Weekly

- Clean vacuum chamber
- Check blank-off pressure of mechanical pumps; change oil if pressure is above 5 Pa (5×10^{-2} torr)
- Check blank-off pressure of diffusion pumps; change oil if pressure is above 10^{-3} Pa (10^{-5} torr)
- Check leak rate of coating chamber
- Lubricate rotary seals and moving parts in vacuum chamber

Semiannually

- Grease motors on pumps and generators

When electrical power fails, air and mechanical-pump oil can leak into the high-vacuum pumps before they can cool. All lines and pumps must then be cleaned and new oil installed. The diffusion pump should be interlocked electrically so that it shuts off when the mechanical pump stops.

Examples of Processing Cycles

Cycle times and other operating conditions for applying decorative, protective, or electrical vacuum coatings are described in these examples.

Decorative Coating. A single 1.7-m (5.5-ft) diam evaporator with removable racks is adequate for batch coating plastic parts on a production basis. About 15 min is required for the evaporation of an aluminum coating; 8 to 9 min to evacuate the system to a pressure of less than 10^{-1} Pa (10^{-3} torr); about 1½ min to melt and evaporate the aluminum; and about 5 min to vent the chamber, remove the rack of coated parts, and install a second rack previously loaded. Personnel requirements consist of an operator and a helper for the vacuum coating operation. Additional personnel are required for precoating and postcoating with lacquer.

Process cycles for decorative aluminum coating of plastic parts are:

Operation	Description of operation
Stress relieve	Hold parts at 74 °C (165 °F), 2 h
Rack	Load parts on work-holding fixtures
Clean	Blow off parts with clean ionized air
Precoat	Spray, dip, or flow coat with precoat
Bake precoat	At 71 °C (160 °F), 2 h
Vacuum coat	Evacuate loaded vacuum chamber to pressure of 5×10^{-2} Pa or less; fire aluminum charge
Topcoat	Apply topcoat
Bake topcoat	Bake at 71 °C (160 °F) for 1 h
Unload	Unload parts
Inspect and package	Inspect for surface defects and quality of coating; pack

Injection-molded parts of styrene-type thermoplastic resin are 125 by 50 by 25 mm (5 by 2 by 1 in.). Surfaces of these parts are coated to a thickness of from 0.08 to 0.1 μm (0.003 to 0.005 mil). The process sequence for decorative second-surface aluminum coating of transparent parts is:

Operation	Description of operation
Stress relieve	Hold at 71 °C (160 °F), 4 h
Multicolor	Mask clean, dust-free part; spray-coat with air-drying lacquer
Dry	Air dry to remove solvents
Rack	Load parts on work-holding fixtures
Vacuum coat	Evacuate loaded vacuum chamber to pressure of 10^{-1} Pa or less; vaporize aluminum charge
Protective coat	Spray air-drying synthetic enamel over the vacuum-coated multicolored surface
Dry	Air dry synthetic-enamel coating
Unload	Unload parts from work-holding fixtures
Clean first surface	Remove oversprayed lacquer and protective enamel, and stray vapor coating from first surface by buffing
Inspect and package	Inspect parts for surface defects and quality of coating; pack

Parts are injection-molded of methyl methacrylate. Coating thickness ranges from 0.08 to 0.1 μm (0.003 to 0.005 mil).

Protective Coating. The application of an aluminum coating, 13 to 25 μm (0.5 to 1 mil) thick, on high-strength bolts requires about 55 min. Cycles consist of about 10 min for evacuation of the system, 30 min for preheating the parts, 10 min for aluminum coating, and 5 min for cooling under vacuum.

Electrical Coating. The process sequence followed for the deposition of thin-film electronic circuits on 25-by-25-mm (1-by-1-in.) glass substrates is as follows:

Substrate cleaning

- Wash substrate in hot detergent solution, with ultrasonic agitation, for 5 to 10 min
- Rinse in hot tap water
- Rinse in hot deionized water
- Rinse in hot deionized water, with ultrasonic agitation, for 5 to 10 min
- Degrease for 30 min in refluxing vapor of isopropyl alcohol
- Place in vacuum chamber while hot

Deposition of conductor films

- Place appropriate mask over substrate
- Prepare aluminum evaporation source (source consists of a strand of braided tungsten wire around which aluminum wire is wound for evaporation)
- Evacuate chamber to a pressure of less than 5×10^{-3} Pa (5×10^{-5} torr)
- Heat the substrate to 99 °C (210 °F)
- Deposit the aluminum film; monitor deposition by measuring the resistance between the terminals of a monitor substrate

Deposition of resistor films

- Place appropriate mask over substrate
- Prepare nickel-chromium-based alloy evaporation source (source con-

Fig. 17 Two types of distribution masks, and cross sections of the film thicknesses obtained

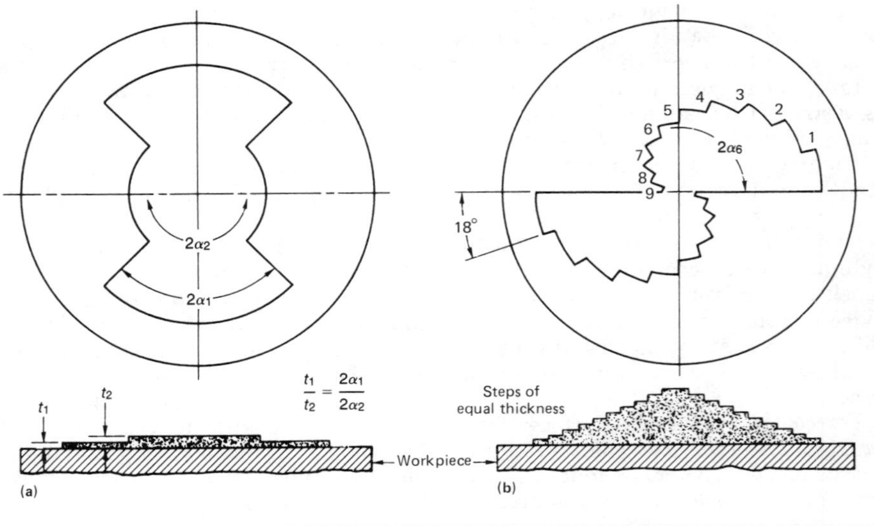

sists of a helix of tungsten wire into which a bundle of alloy wires is placed)

- Evacuate chamber to a pressure of less than 5×10^{-3} Pa (5×10^{-5} torr)
- Heat substrate to 300 °C (570 °F)
- Deposit the nickel-chromium-based alloy film, using a resistance monitoring method

Deposition of dielectric films

- Mask for dielectric films
- Prepare source for evaporation of silicon monoxide; use indirectly heated crucible
- Evacuate chamber to a pressure of less than 5×10^{-3} Pa (5×10^{-5} torr)
- Heat substrate to 300 °C (570 °F)
- Deposit the silicon monoxide film, monitoring by optical-interference maximums and minimums detected by a photomultiplier and recorded on a strip chart

Post-coating heat treatment

- After the coated substrate is removed from the vacuum chamber, heat it in air to 350 °C (660 °F), and hold at temperature for 8 h

When a second deposition of conductor films is required, the procedure is identical to that described here for the first deposition of conductor films. A second deposition of silicon monoxide is used to apply a protective layer over all parts of the deposited circuit except those conductor areas where soldering or other means of bonding must be subsequently performed. Procedures for this second deposition are identical to those described here for first deposition of dielectric films.

Protective Thick Film Coatings. The process sequence followed for the deposition of a thick (0.1 mm or 0.004 in.) M-CrAlY alloy deposit on a gas turbine blade is described below:

Substrate cleaning

- Slurry hone surfaces using 1200-mesh Novocite slurry
- Flush with water (5 min)
- Ultrasonic degrease in Freon or vapor degrease in trichlorethylene
- Dry
- Store under plastic in dust-free area

Deposition

- Load ingot sources in magazine
- Load crucible with compensating alloy
- Evacuate chamber to 10^{-3} Pa (10^{-5} torr) or less
- Initiate melting of ingot, stabilize evaporation rate, liquid level, ingot feed rate; check for spitting, traverse rate of electron beam
- Place weighed test coupons in holders on shafts in transfer chamber
- Evacuate load lock, preheat test specimens to coating temperature 1000 °C (1830 °F)
- Cycle test coupon into chamber; coat
- Remove coated test coupon from chamber; weigh to determine weight deposit, inspect surface, determine composition by x-ray fluorescence analysis
- Weigh blades and load into holders
- Place holders on shafts in level lock; evacuate load lock
- Preheat blades to 1000 °C (1830 °F)
- Cycle blades into coating chamber and coat (8 to 10 min)
- Cycle blades into load lock; cool
- Remove blades, holders

Post-deposition

- Remove blades and reweigh to determine weight of deposit
- On a sampling basis, determine deposit composition by x-ray fluorescence analysis, thickness by metallographic examination
- Inspect parts
- Peen parts, if required
- Heat treat in vacuum or hydrogen at 1038 to 1100 °C (1900 to 2000 °F), depending on substrate, for 2 to 4 h
- Visually inspect blades for surface defects

Process and Quality Control

Process control requirements for evaporation processes are directly related to the needs of the individual application. For example, in the deposition of aluminum for decorative purposes, control of the plasticizer content of the substrate and the precoating lacquering process is essential for defect-free coating. Surface smoothness has a substantially greater effect on deposit quality than, for example, absolute deposit thickness control, which may be ±25% of a nominal value of 0.1 μm (0.004 mil). Elaborate environmental control systems have been developed for the electronic and optical coating industries, in which lock-controlled clean rooms provide a humidity-controlled, nearly dust-free environment for preparation and processing of substrate materials.

Deposition Rate Monitors. The objective in most precision vacuum-evaporated film deposition is to produce films of a known thickness. If the requirements for the deposit are such that they may be affected by the rate of deposition, such as density, stress, resistivity, or grain size, the rate of deposition, and in many cases, the temperature of the substrate, have a substantial influence. Sensing monitors measure a variety of physical properties, including mass, ionization characteristics, and optical properties, and allow precise film deposits to be obtained.

Vapor stream monitors are divided into two classes: ionization monitors and vapor-impact pressure monitors. Both types measure the instantaneous vapor flux, and the thickness of the deposit must be integrated from instantaneous values. Both systems must be experimentally calibrated for the individual application by measuring actual film thicknesses against the output of the monitor. Ionization monitors are essentially highly modified ionization gauges, in which the tungsten filament ionizes the evaporant, producing an ion current. The ion current is made up of both the vapor-induced current and the current resulting from the residual atmosphere in the chamber. In electron-beam heated systems, a grid must be placed in front of the monitor to collect stray electrons from the gun. With proper calibration, precisions of ±2% can be achieved for these systems.

Impact pressure-type monitors use an aluminum cylinder, in which one-half of the cylinder is shielded from the vapor longitudinally along its axis. One type is mounted on a torsion wire within a damping magnetic field; the other is mounted on a low-friction pivot within a magnetic field. In the first case, the angular displacement of the cylinder from a rest position determines the vapor flux momentum. In the second case, the angular velocity determines the rate, and the total amount evaporated is proportional to the total number of turns accumulated. These devices and others using similar principles are sensitive to less than 0.008 μin./s (2 Å/s) and are independent of residual gas effects as well as the effect of previous deposits.

Deposit mass monitors can be used for all evaporants. The deposit mass can be detected by a torsion microbalance, or by the change in oscillating frequency of a quartz crystal upon which the vapor condenses. The microbalance is affected by system vibration, and in some cases, by brownian movements in the chamber. Quartz oscillators, on the other hand, are unaffected by mechanical vibrations, and are simple mechanical constructions with about the same sensitivity as the microbalance. The frequency change as a function of thickness is linear until the deposit thickness is an appreciable fraction of the crystal thickness. Even though there is a low temperature coefficient, the crystal is protected from the heat of condensation or radiation from the evaporation source by water cooling and radiation shields. Changes in weight are usually measured by comparing the frequency against a comparable reference crystal so that the changes in frequency can be measured with an audio analyzer. The thickness and rate of deposition can be continuously monitored to a limit of ±2%.

Monitoring Specific Properties. Optical methods that use transmittance, light absorption reflectance, and interference are used to measure the deposit thicknesses of coatings used for optical systems applications. The equipment consists of a photocell and a light source, and the mode of measurement depends on the film to be deposited, its thickness, and the substrate. With such methods, film thickness can be controlled to within ±2%. Resistance monitors are used to automatically control the construction of resistors. Using a wheatstone bridge, resistors of 0.01% accuracy or greater are produced.

Electron Beam Heated Sources. Evaporation rates from electron-beam sources may be controlled by the emission current of the electron beam gun. In the evaporation of complex alloys such as the M-CrAlY systems, the volume of the pool is monitored by measurement of the height of the liquid in the hearth, and the rate of movement of the ingot feed stock through the base of the hearth.

Gas Analysis. Continuing measures of the residual gas content of chambers provides a monitor of the amount of various residual gases in the environment. In reactive deposition, continuous analysis of the reactive partial gas pressure is necessary to provide the proper deposit stoichiometry.

Microprocessor-Controlled Systems. Integration of sensor technologies with real time computer control technology provides the capability to control the overall process cycle by automation, and thus eliminate operator judgment in control of the process. This technology enables the performance of the pumpdown cycles and deposition cycles by combining pressure and temperature measuring devices, evaporation rate monitors, and servodrives with programmable logic sequences and feedback control loops, so that each step of the process is initiated only when predetermined conditions of pressure, temperature, and evaporation rate are established and maintained. The result is a substantial improvement in reliability and consistency of the deposit.

Quality and performance testing methods vary substantially according to the needs of industry and the product being produced. Test procedures used include measurement of film thickness, composition, entrapped gas content, microstructure, adhesion, stress, strength, and hardness.

Thickness. Many of the films deposited are in the thickness ranges of light, and may be measured by interferrometric techniques. Therefore, film thickness in the range of about 0.1 to 5 μm (0.004 to 0.2 mil) may be measured by depositing the film on a glass substrate without completely covering the substrate. A fairly sharp edge must exist where the film ends and the bare substrate begins. The film is prepared for measurement by coating both the film and the substrate in the vicinity of the edge with a thin film of aluminum or silver to render the surfaces reflective.

An optical flat is then placed in contact with this metal film in the region where it crosses the edge referred to above. The interface between the optical flat and the coated layer is then illuminated at normal incidence with monochromatic light. A series of approximately parallel interference fringes is formed as a result of the wedge of air between the optical flat and the metal film. The shift in this fringe pattern on the film to be measured relative to the pattern on the substrate gives a measure of the film thickness. For example, in blue-green light with a wavelength of 5080 Å, a shift of one fringe corresponds to a film thickness of 2540 Å (10.2 μin.), or 0.3 μm (0.01 mil). This method has a precision of about ±200 Å (0.8 μin.).

Adhesion test procedures for vacuum-deposited coatings have not been standardized. In general, the test for adhesion consists of the tape test, in which a 19-mm (3/4-in.) wide strip of adhesive cellophane tape is applied to the coating and quickly removed. The quick removal of the tape should leave the finish intact. Some specifications require that the tape test be applied after first scoring the finished surface in two directions with a sharp instrument. Another specification requires that the surface be scored with a sharp instrument and that a jet of air under a specified pressure be directed at a definite angle to this area. None of the finish should be removed by these tests.

Scratch methods, in which a probe is impressed onto the deposit surface and drawn across the surface at varying loads until lifting is achieved, has been shown to correlate with the degree of adhesion. Residual stress may be recovered by coating a thin, stress-free substrate and observing the deflection resulting from the stress of the deposit by x-ray diffraction or electron diffraction methods in crystalline deposits.

Process Economics

The shape of the part to be coated usually has the greatest influence on the cost of vacuum deposition. Other important influences on cost are total number of parts processed, design and construction of racks, loading and unloading of parts, precoating and post-treatments, cleaning, heat-source material, evaporant material, and maintenance. The cost of precoating and post-treatment can vary over a considerable range, depending on the materials used and the method of application. For example, lacquer costs depend on the solids, their content, and the solvent blend. When parts are sprayed individually (often required when certain areas are masked), labor cost is greater than when parts are loaded on racks and the entire unit is sprayed, dipped, or flow coated.

In general, material costs per unit area covered are higher when precoated and top-coat materials are applied by manual or automatic air-spray methods, because a considerable amount of material is lost as overspray and exhausted through the spray-booth stack. Electrostatic spraying, when applicable, reduces materials loss considerably. The dip method is probably the most economical, provided the material has good tank life. Flow coating is an effective means of applying these materials on parts whose shapes make them difficult to spray or dip; however, material cost per unit area covered may be somewhat higher than for the dip method, because of greater solvent loss and shorter tank life.

The percentage of parts that may be rejected by the inspection department is difficult to estimate, but must be considered in computing the overall cost of operation. Even if rejected parts are not a total loss, considerable expense for material and labor may be involved in salvaging them. When considering the costs of materials for vacuum deposition of aluminum, actual cost for aluminum is almost negligible. However, the cost of filaments is significant. Filaments are generally stranded tungsten and have a useful life of approximately 12 evaporations.

Capital investment is essentially the same regardless of the volume of production. However, as the following examples indicate, volume of parts produced greatly influences cost per piece.

Automotive dome light reflectors are produced at the rate of 40 000 per day. Each dome light reflector intercepts 4516 mm^2 (7 in.^{2}) of deposit. On the other hand, reflectors for photographic enlargers use the same process cycle and arc of similar design, but are 65 times larger (291 612 mm^2 or 452 in.^{2}), and are produced at the rate of only a few thousand per day. The photographic reflector enlarger coating costs 100 times the cost of the automotive dome light reflector. The 50% increase in cost per area coated is directly proportional to the quantity produced.

Turbine blades are coated at the rate of 30 000 per year in large groupings of 1000 to 3000 parts per run. By increasing the number to 120 000 parts per year, using the same coating, the per blade cost of the coating can be reduced by more than 40%. This cost reduction is achieved by amortizing fixed costs of the chamber, support equipment, and overhead (analytical traceability control over a much larger number of parts).

Photographic enlarger reflector. The equipment cited above did enable a substantial reduction in cost compared to electroplated reflectors the previous process used. Electroplating the reflectors (0.6 m or 2 ft across), cost some four times the aluminization costs. Moreover, the evaporation coating allowed a change from spun copper sheet to spun aluminum sheet, resulting in a savings of 62% in the manufacturing of the preform. The overall cost savings realized for the reflector was more than 65%.

REFERENCES

1. A. N. Nesmayenov, *Vapor Pressure of the Chemical Elements,* New York: Elsevier Publishing Co., 1963
2. O. Kubaschewski and E. L. Evans, *Metallurgical Thermochemistry,* New York: Pergamon Press, 1965
3. R. Hultgren, *et al, Selected Values of Thermodynamic Properties of Metals and Alloys,* New York: John Wiley & Sons, 1963
4. D. R. Stull, "JANAF Thermochemical Data", The Dow Chemical Co., U. S. Clearinghouse, Springfield, VA, 1965/1966
5. R. E. Honig, *RCA Rev, 7* Vol 23 (No. 567), 1962
6. R. Glang, Vacuum Evaporation, *Handbook of Thin Film Technology,* New York: McGraw-Hill, 1970, p 1-30
7. G. C. Roberts and G. G. Via, U. S. Patent 3,313,914, 1967
8. I. Ames, L. H. Kaplan, and P. A. Roland, *Review Science Institute,* Vol 37, p 1737, 1966

SELECTED REFERENCES

- R. W. Berry, P. M. Hall, and M. F. Harris, *Thin Film Technology,* New York: D. Van Nostrand, 1968
- L. Holland, *Vacuum Deposition of Thin Films,* London: Chapman and Hall, 1970
- L. I. Maissel and R. Glang, *Handbook of Thin Film Technology,* New York: McGraw-Hill, 1970
- *New Trends in Materials Processing,* American Society for Metals, 1976
- *Thin Films,* American Society for Metals, 1964

Sputtering

By John A. Thornton
Vice President
Research and Development
Telic Company
and
Wolf-Dieter Münz
Manager, Research and Development
Department for Coating
Leybold-Heraeus GmbH

SPUTTERING is a process wherein material is ejected from the surface of a solid or liquid because of the momentum exchange associated with bombardment by energetic particles. The bombarding species are generally ions of a heavy inert gas. Argon is most commonly used. The source of ions may be an ion beam or a plasma discharge into which the material to be bombarded is immersed.

Sputtering may be used as a method of surface etching or coating. This article describes sputter coating by the plasma discharge method. This is the most widely used application of sputtering. Sputter etching is described in Ref 1. Ion beam sputtering is described in Ref 2.

In the plasma discharge sputter coating process, a source of coating material called a target is placed into a vacuum chamber which is evacuated and then backfilled with a working gas, such as argon, to a pressure adequate to sustain a plasma discharge. A negative bias is then applied to the target so that it is bombarded by positive ions from the plasma. The most direct method for providing the plasma and the target ion bombardment is to make the target the cathode of the electric discharge. Such discharges operate at voltages in the range of 0.5 to 5 kV. Ions accelerated by this voltage impact on the target and cause ejection sputtering of the target material. The substrates are positioned so as to intercept the flux of sputtered atoms, which may collide repeatedly with the working gas atoms before reaching the substrate where they condense to form a coating of the target material.

The most striking characteristic of the sputtering process is its universality. Because the coating material is passed into the vapor phase by a mechanical process (momentum exchange) rather than a chemical or thermal process, virtually any material is a candidate for coating. Direct current discharges are generally used for sputtering metals. A radiofrequency (RF) potential must be applied to the target to sputter a nonconducting material.

After a short equilibration period the flux of sputtered material leaving a target will be identical in composition to the target, provided that (*a*) the target is kept cool enough to avoid interdiffusion of the constituents, and (*b*) the target does not decompose. If the sputter emission directions, the gas phase transport, and the substrate sticking coefficients are the same for all the constituents, then the coating composition will be identical to that of the target. Coatings of alloys such as stainless steel, compounds such as aluminum oxide and even bone (Ref 3), and a polymer polytetrafluoroethylene (Ref 4), have been successfully deposited by sputtering.

Process Controls

The sputtered species are primarily neutral atoms. They are ejected from polycrystalline-type targets in a near-cosine distribution with relatively high energies (average of 10 to 40 eV). The sputtering process is quantified in terms of the sputtering yield, defined as the number of atoms ejected per incident ion. The sputtering yields of most materials are about unity and within an order of magnitude of one another. The sputtering rates for all materials at a given target power input are therefore similar. By contrast, in evaporation, the rates for different materials at a given power input to the source can differ by several orders of magnitude.

Sputter coating chambers are typically evacuated to pressures in the 10^{-3} to 10^{-5} Pa range before backfilling with argon to pressures in the 0.1 to 10 Pa range. The intensity of the plasma discharge, and thus the ion flux and sputtering rate that can be achieved, depends on the shape of the cathode electrode, and on the effective use of a magnetic field to confine the plasma electrons.

Deposition Rates. The deposition rate depends on the target sputtering rate and the apparatus geometry; for example, the target shape and the position of the substrates relative to the target or targets. It also depends on the working gas pressure, since high pressures limit the passage of sputtered flux to the substrates.

For many years, most sputtering was done using simple planar electrode systems of the type shown in Fig. 1. Even when operated under the most favorable conditions (pressures in the 1 to 10 Pa range), these devices yield relatively weak discharges with power densities of about 5 W/cm^2 (33 W/in.^{2}) and deposition rates that are typically less than 2 nm/s (0.08 μin./s). Consequently, sputtering developed a reputation for very low deposition rates.

The development of a class of sputtering sources with magnetic plasma confinement, called magnetrons, has greatly enhanced the capabilities of the sputtering process (Ref 5). These devices can provide order-of-magnitude increases in sputtering rates compared to the planar diodes.

Magnetron sputtering sources can be defined as diode devices in which magnetic fields are used in concert with the cathode surface to form electron traps which are so configured that the $\vec{E}\times\vec{B}$ electron drift currents can close on themselves. Ions are trapped electrostatically by the confined electrons. In the case of cylindrical magnetrons, an axial magnetic field, generated by magnetic field coils that are external to the vacuum system, is used to confine an intense plasma sheet over the outer surface of a cylindrical-post target or on the inner surface of a cylindrical hollow target. In the case of the very important planar magnetrons, a circular ring-shaped plasma is confined over a planar target by the magnetic field which is generated by a system of permanent magnets or field coils which are located in a water-cooled cavity behind the target.

In the magnetron sources, the electrons that maintain the discharge ionization are effectively confined in the vicinity of the target (cathode) surface until they diffuse away by engaging in energy exchange processes of the type that produce ionization. Consequently, ion production in magnetrons is very efficient, and high discharge currents, and therefore high sputtering rates, can be achieved at relatively low discharge voltages in the 200- to 800-V range.

Magnetrons can be operated at such low pressures (0.1 to 0.5 Pa) that the sputtered atoms pass line of sight to the substrates. They can be scaled to large sizes that provide uniform deposition over very large areas.

Fig. 1 Planar electrode system used for sputtering

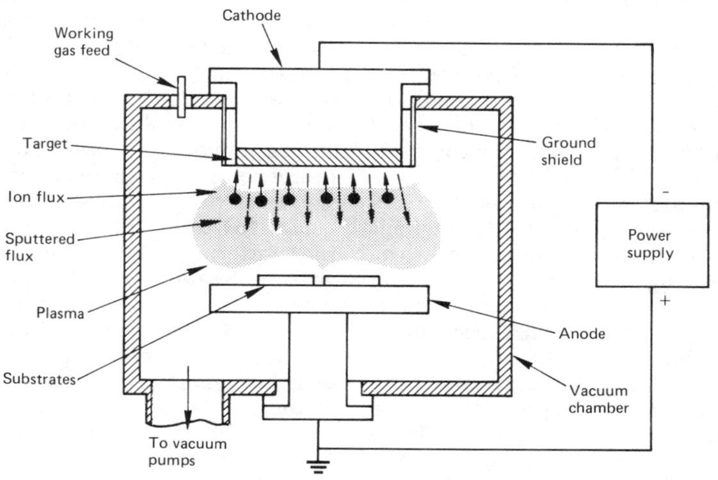

Substrate temperature depends on the apparatus configuration and the application. Substrate heating depends on the deposition rate and on whether the substrates are in contact with the plasma. In the case of planar diode apparatuses of the type shown in Fig. 1, the substrates in contact with the plasma can reach temperatures in the 300 to 500 °C (570 to 930 °F) range, even at moderate deposition rates (~1 nm/s or ~0.04 μin./s). The magnetic plasma confinement permits the substrates in magnetron sources to be removed from the plasma. Thus, magnetrons can yield the low substrate temperatures of 49 to 250 °C (120 to 480 °F) that are necessary to deposit coatings on heat-sensitive substrates like plastics or porous casting materials. Substrate heating to provide higher temperatures is used for many applications, such as applying hard titanium nitride coatings on tools.

Process Variations. The basic sputtering process has many variations. In the process of sputter cleaning, the substrates are biased as electrodes before coating, so that contamination is removed by sputtering, and coating nucleation sites are generated on the surface. In the process of bias sputtering, the substrates are biased to cause ion bombardment during deposition for the purpose of removing loosely bonded contamination or modifying the structure of the resulting coating. In the process of reactive sputtering, a gas is used to introduce one or more of the coating constituents into the chamber.

Reactive sputtering can be performed with virtually any metal/reactive-gas combination. A working gas, consisting of argon plus the reactive component, is generally used. Thus, compound semiconductor coatings of CdS have been formed by sputtering from a Cd target in an Ar + H_2S atmosphere, and hard TiN wear coatings for tools have been formed by sputtering from a Ti target in an Ar + N_2 atmosphere.

During reactive sputtering, reactions can occur on the cathode surface, following which the reacted material is sputtered, as well as at the substrate. The stoichiometry of the coating is determined by the partial pressure of reactive gas that is maintained in the target-substrate region. It is possible to control the coating compositon by controlling the sputtering rate and the injection rate or partial pressure of the reactive gas.

Coating Contoured Substrates. When the substrate is contoured, special problems arise. These are especially important if reactive processes are involved, or if an automated in-line process is being used where there is no provision for rotating the parts to be coated as they pass through the coating area.

Conservation of mass requires that the sputtering deposition rate decreases as the distance from the target increases. This means that the coating thickness cannot be equal at all points on a contoured substrate. The coating thickness will also be less in recessed regions where shadowing limits the coating flux.

If a reactive process is involved, different deposition rates can result in different compositions of the deposited material over the surface of a complex-shaped part. This condition occurs because the partial pressure of the reactive gas is almost constant throughout the entire discharge area, while the metal coating flux is not. Considering the particular case of reactive sputtering titanium nitride, these effects cause the hardness of the surface layers on the side of the substrate facing the target to be different from that of the layers at the rear of the substrate. If the substrate is a watch case, the gold tones on the front and the back of the case may not be identical.

An additional complication results when ion bombardment is used for sputter cleaning or bias sputtering, since the ion current density will, in general, be nonuniform on a substrate of complex shape. The current density will also depend on the position of the substrate, relative to the sputtering plasma.

At least a partial solution to the problem of coating substrates of complex shape exists in the selection of a sputtering arrangement that is symmetrical to the contoured substrate. For example, a spherical substrate can be coated by using a cylindrical-hollow magnetron source or by using two high-rate planar magnetron sources to coat the sphere from two positions at the same time (Ref 6). Referring to the latter case, for each individual source, the composition depends on the power density, distance, and the magnetic field strength. However, the cathodes can be arranged in such a manner that each cathode will supply coating material to spots on the substrate not reached by material sputtered from the other, thus facilitating a uniform coating of the substrate.

By use of paired cathode arrangements of the type described above, the amount of substrate ion bombardment, the temperature rise, and the self-cleaning effect can be effectively increased. In one experiment, a substrate having an area one third that of the cathode was biased at a voltage of 250 V. The current flow was about 2 A, thereby delivering a power to the substrate (500 W) that was about 10% of that delivered to the cathode (4.6 kW). In such a system, substrate surface loads near 3.6 W/cm^2 (23 W/in.^{2}) that are comparable to those in an ion plating system are obtained.

In summary, the width of the zone of constant condensation and discharge parameters exhibits interdependence on the individual process parameters. Process optimization and apparatus design ensure the most useful zone width.

Limitations. Often, application of the sputtering process is limited by target considerations. Metal targets are, in general, relatively easy to fabricate and to cool. However, magnetron sources can accommodate only relatively thin targets of magnetic materials because of the influence of the target in distorting the magnetic field used for plasma confinement. Targets of compounds such as oxides, carbides, and nitrides can be formed by hot-pressing powders. Poor heat transfer, coupled with vulnerability to target cracking or the loss of volatile constituents, generally limits the power levels that can be delivered to nonmetallic targets.

Metal deposition rates with direct current-driven magnetron sources are typically in the range from 5 to 30 nm/s (0.2 to 1.2 μin./s) (Ref 7). Target cracking and decomposition effects, along with the basic character of radio frequency discharges, typically limit the deposition rates for directly sputtered compounds to the range 1 to 10 nm/s (0.04 to 0.4 μin./s), even when magnetrons are used (Ref 7). Reactive sputtering with metal targets greatly simplifies the target fabrication problem. However, deposition rates are still generally limited to the 1 to 10 nm/s (0.04 to 0.4 μin./s) range because of surface layers of modified composition which tend to form on the target and effectively reduce the sputtering yields.

Apparatus Shape

Vacuum chambers and pumping systems are required for sputtering, as well as the sputtering sources and power supplies, and a feed and control system for the working gas. Chambers with vacuum interlocks which permit the substrates to be introduced without exposing the chamber walls and sputtering sources to the atmosphere are being increasingly used. The rate of sputtering from a given target is proportional to the power delivered to the target, as described above, and can be controlled relatively accurately. Deposit thicknesses are often established simply by the time of deposition rather than by the use of a rate monitor.

Electrode shapes of the type shown schematically in Fig. 1 are known as planar diodes. Although limited to moderate deposition rates, planar diodes have been widely used in the electronics industry to deposit a range of coating materials onto flat wafer substrates using direct current or radiofrequency power. Radiofrequency sputtering is generally done at the industrially allowed frequency of 13.56 MHz. Power levels are typically in the range of 500 to 5000 W. Voltages (direct current or radio frequency peak-to-peak) are typically 2 to 4 kV. In the case of the magnetron sources (Ref 2, 7), magnetic fields are used to confine intense plasma rings or sheets, containing circulating currents, over planar or cylindrical targets, as described previously. The planar magnetrons are often configured with racetrack-shaped plasma rings confined on rectangular targets. Sputtered atoms are emitted from beneath the plasma ring and deposited onto substrates. The cylindrical targets may be hollow cathodes, where the sputtered atoms pass inward to axially mounted substrates. They may also be in the form of central posts from which the sputtered flux passes radially outward to circumferentially mounted substrates. Cylindrical-post magnetrons are particularly effective for batch process applications, because large deposition areas having a near uniform coating flux are available surrounding the central source. Magnetrons are usually driven by constant current power supplies at current levels in the 5- to 50-A range with voltages of 0.5 to 1 kV.

Sputtering is an inefficient process from the viewpoint of energy. Most of the power input appears as target heating. Accordingly, sputtering targets are generally water cooled. Substrates may be heated or cooled depending on the application. Sputtering, by its basic nature, is a nonpolluting process.

Coating Properties

A schematic representation showing the dependence of the microstructure of sputtered coatings on the argon pressure and substrate temperature is given in Fig. 2 (see Ref 8). The same basic diagram (at zero argon pressure) also applies to evaporated coatings. The structures of sputtered coatings may differ from those of evaporated coatings in three important ways.

First, the sputtered atoms can become scattered by the working gas so that they approach the substrates at very oblique angles. Thus, elevated working pressures at low T/T_m, where T is the substrate temperature and T_m is the melting point of coating material, tend to promote the Zone 1 type structure, which is characterized by columnar voids. The voids result from self-shadowing, which occurs because high points on the growing coating surface receive more coating flux than valleys do.

Second, significant numbers of ions can be neutralized and reflected at the target surface and then pass back in the direction of the substrates with energies as high as several hundred eV. At low working pressures, these atoms reach the substrates and subject them to a bombardment which tends to promote the dense, fine-grained Zone T structure. Third, the substrates may be in contact with the plasma. Plasma ion bombardment tends to suppress the Zone 1 structure, even at elevated argon pressures, and to promote a dense, fine-grained structure of the Zone T type. Ion bombardment on uncooled substrates also promotes higher temperature structures. These bombardment effects have given sputtering a reputation for producing dense, high-quality deposits.

Referring again to Fig. 2, at $T/T_m > 0.5$, dense columnar grains are formed, because at these substrate temperatures, the adatom mobility during coating growth is sufficient to overcome the self-shadowing effects which give rise to the underdense Zone 1 structure. At very high T/T_m, recrystallization can result in large columnar or equiaxed grains. This is Zone 3.

Coatings that are subjected to energetic reflected atom or ion bombardment during growth and that have the Zone T structure often exhibit high compression type internal stresses. Coatings deposited under conditions that yield the Zone 1 structure are often in tension, apparently because they are underdense with wider than equilibrium interatomic spacings.

Fig. 2 Dependence of the microstructure of sputtered coatings on argon pressure and substrate temperature

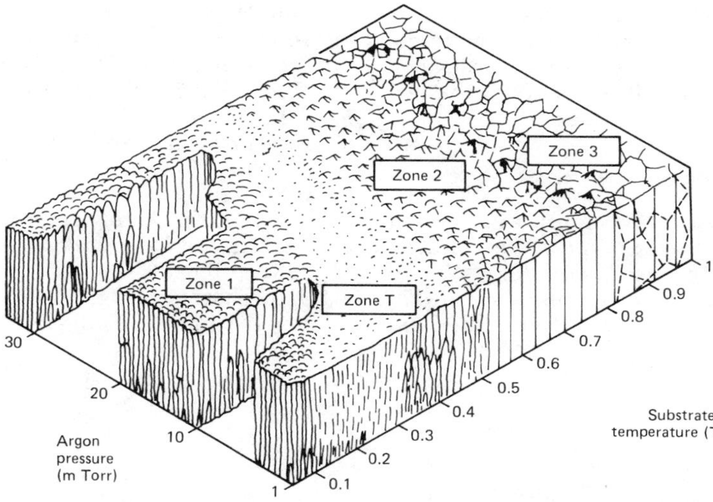

Applications

The ability to control coating composition makes sputtering useful in the electronics industry for applications such as:

- Aluminum alloy microcircuit metallization layers
- Refractory metal microcircuit metallization layers
- Oxide microcircuit insulation layers
- Transparent conducting electrodes
- Amorphous optical films for integrated optics devices
- Piezoelectric transducers
- Photoconductors for display devices
- Luminescent films for display devices
- Optically addressed memory devices
- Amorphous bubble memory devices
- Thin film resistors
- Thin film capacitors
- Video discs
- Solid electrolytes
- Thin film lasers
- Microcircuit photolithographic mask blanks

Other Applications. The combination of hardness (wear resistance) color, corrosion resistance for sputtered films, and the possibility to coat in a broad range of temperatures (from 75 to 650 °C or 165 to 1200 °F) offers the use of sputtered films in the field of decorative applications. One example is the golden colored 0.5 μm (19.7 μin.) thick TiN film which can be substituted for a 20 μm (787 μin.) gold layer.

Another application involves the deposition of hard, protective coatings on high-speed steel cutting tools. This increases cutting speeds (productivity) and cutting length (tool life time). A 3 μm TiN film of a hardness of more than 2000 K/mm^2 on hobs shows a typical decrease in flank wear by a factor of 10, compared to the untreated tool. This results in a lifetime factor of 4 for coated hobs.

The use of solid lubricants is also an effective method of reducing wear in many applications. Films of MoS_2 sputtered in crystalline structure exhibit good adhesion and reduce the coefficient of friction down to 0.04. Even under hardest space conditions, these films offer more than 10^4 the lifetime of conventional brushed-in MoS_2 films.

Another new field of cathode sputtering is the production of multilayer stacks of dielectric materials, such as TiO_2, Ta_2O_5, Al_2O_3, SiO_2, and others. Compared with evaporated layer-systems, sputtered stacks are harder and more dense. This results also in higher refractive indies. The multilayer films therefore show negligible aging effects, which means the optical spectral characteristics are very stable. This technique is successfully introduced for the production of laser mirrors, narrow band-filters, line-filters, and edge-filters.

Magnetron sources have opened up more new applications because of their large area capability and reduced substrate heating. Large in-line systems with vacuum interlocks and roll coating systems are used to coat 2- by 3.5-m (7- by 11-ft) architectural glass plates or thin plastic films, the deposited films thus acting as solar control or heat reflecting films. Magnetrons are also used on a production basis to deposit chromium decorative coatings on plastic automobile grilles and other exterior trim.

REFERENCES

1. R. E. Lee, Microfabrication by Ion Beam Etching, *Journal of Vacuum Science and Technology,* Vol 16 (No. 2), March/April 1979, p 16-170
2. J. L. Vossen and W. Kern, Ed., *Thin Film Processes,* New York: Academic Press, 1978, p 175-206
3. B. J. Shaw and R. P. Miller, Sputtering of Bone on Prostheses, U. S. Patent 3 918 000 (November 11, 1975)
4. D. T. Morrison and T. Robertson, RF Sputtering of Plastics, *Thin Solid Films,* Vol 15 (No. 1), 1973, p 87
5. J. L. Vossen and W. Kern, Ed., *Thin Film Processes,* New York: Academic Press, 1978, p 76-110, p 115-128, p 131-170
6. W. D. Münz, G. Hessberger, *Industrial Research & Development,* Sept 1981, p 130-135
7. J. A. Thornton, High Rate Sputtering Techniques, *Thin Solid Films,* Vol 80 (No. 1-3), June 1981, p 1
8. J. A. Thornton, High Rate Thick Film Growth, *Annual Reviews of Material Science,* Vol 7, 1977, p 239-60

SELECTED REFERENCES

- G. Kienel, B. Meyer, W. D. Münz, Vakuumtechnik, 30. Jahrg, Heft 8, 1981, p 236-246
- W. D. Münz, S. R. Reineck, SPIE's Los Angeles Technical Symposium and Instrument Exhibit, 25-29 Jan 1982, Proceedings to be published
- G. Kienel, *Thin Solid Films* 77, 1981, p 213-224

Ion Plating

By Robert F. Hochman
Associate Director for Metallurgy
Georgia Institute of Technology
and
D. M. Mattox
Supervisor
Surface Metallurgy Division
Sandia National Laboratories

ION PLATING is a generic term applied to atomistic film deposition processes in which the substrate surface and/or the depositing film is subjected to a flux of high-energy particles (usually gas ions) sufficient to cause changes in the interfacial region or film properties. Such changes may be in film adhesion to the substrate, film morphology, film density, film stress, or surface coverage by the depositing film material. The above definition of ion plating refers only to processes that affect the film and/or substrate and does not define either the source of the depositing material or the origin of the bombarding species. The ion-plating technique (U. S. Patent No. 3,329,601, 1967) was first reported in the technical literature in 1964 (Ref 1).

Process Basics

Ion plating is typically done in an inert-gas discharge system similar to that used in sputter deposition (see the article on sputtering in this Volume), except that the substrate is the sputtering cathode and the bombarded surface often has a complex geometry.

A schematic of a system, typical for many ion-plating operations, is shown in Fig. 1. Basically, the ion-plating apparatus is comprised of a vacuum chamber and a pumping system, which is typical of any conventional vacuum deposition unit. There is also a film atom vapor source and an inert-gas inlet. For a conductive sample, the workpiece is the high-voltage electrode, which is insulated from the surrounding system. In the more generalized situation, a workpiece holder is the high-voltage electrode and either conductive or nonconductive materials for plating are attached to it. Once the specimen to be plated is attached to the high-voltage electrode or holder and the boat or filament vaporization source is loaded with the coating material, the system is closed and the chamber pumped down to a pressure in the range of 10^{-3} to 10^{-4} Pa (10^{-5} to 10^{-6} torr). When a desirable vacuum has been achieved, the chamber is backfilled with argon to a pressure of approximately 1 to 0.1 Pa (10^{-2} to 10^{-3} torr). The argon influx is controlled by a variable leak valve, and it can be further controlled by a selected closure or a baffle valve between the work chamber and the vacuum system. A potential of -3 to -5 kV is then introduced across the high-voltage electrode (specimen or specimen-holder) and the ground for the system. A glow discharge occurs between the electrodes which results in the specimen being bombarded by the high-energy argon ions produced in the discharge, which is equivalent to direct current sputtering. The argon ion bombardment effectively cleans the specimen surfaces by removing any adsorbed layers or surface contamination. The cleaning step usually can be completed in a few minutes and provides a clean work surface for receipt of the plating atoms and ions. The coating source is then energized and the coating material is vaporized into the glow discharge. The above process provides the workpiece with a uniform ion bombardment, and gas scattering along with ion deflection gives rise to the effectively high throwing power of this technique and results in uniform coating of even rather intricate variations in surface contour. For a film to form, the deposition rate must exceed the sputtering rate, and ion bombardment may or may not be continued after the interfacial region has been formed.

After the interfacial region has been formed, the coating can continue to be developed at a low rate if bombardment effects on film properties are desired. Film properties may be varied by controlling the substrate potentials and atom/ion bombardment ratio during deposition. The coating may be deposited at higher rates or in a vacuum if ion bombardment effects on film properties are not desired.

For nonconductive workpieces, the bias potential must be produced between anode and the workpiece holder or a conductive screen or similar device held to or around the workpiece. This results in a uniform system for developing the argon plasma, which is the principal source of bombarding ions. A wide variation in the geometries of holders

Fig. 1 Simple ion-plating apparatus

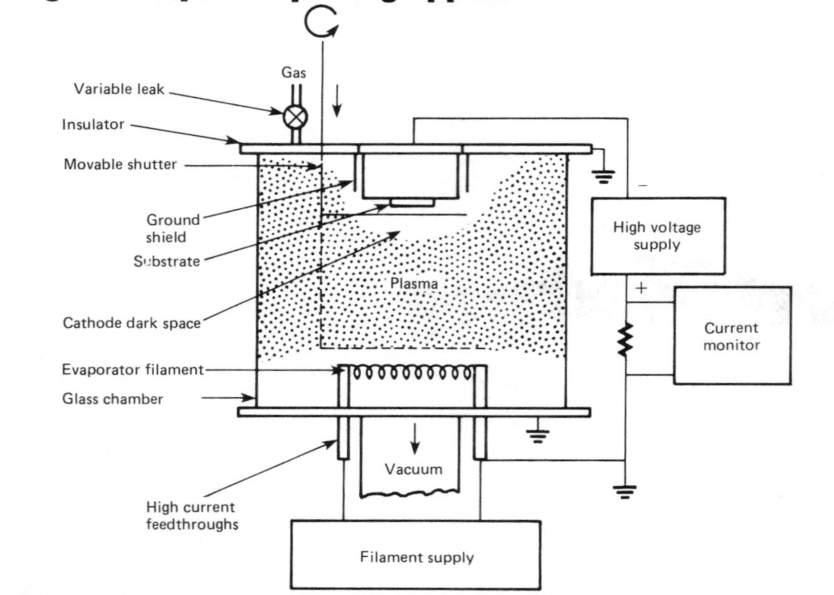

Utilizing a dc gas discharge and an evaporator filament

and even entire systems has been developed for conductive as well as nonconductive materials, and a system is designed to be generally unique to a specific series of coating workpieces.

Ionization and Evaporation Sources

Depending on the type of ion plating to be performed and the desired properties of the ion-plated part, a broad range of evaporation or ionization techniques is available. The major evaporation techniques are summarized below.

Resistance Heating Vaporization. Either a refractory metal filament holding the metal to be plated or a refractory metal boat containing the material to be plated is used for resistance heating evaporation. This technique is limited to plating materials with a melting temperature of less than 1500 °C (2730 °F). The technique also has limitations when evaporating alloys or compounds, although some difficulties have been overcome by the development of "flash" evaporation.

In flash evaporation, the material to be coated is in powdered form and is fed continuously into a preheated boat. The temperature of the boat is set as high as necessary so as to instantaneously evaporate the least volatile component of the material being used. The constituents of the powder, when coming in contact with the boat, are vaporized instantly so as to prevent any fractional or partial decomposition of the material. The powdered material is fed into the boat continuously with a continuous flash evaporation reaction. No material should accumulate in the boat and the vapor which is produced from the uniform powder feed should generally provide a surface coating with the same composition as the original material. Parameters to be controlled in flash evaporation are particle size, rate of powder delivery, and preheat temperature of the reaction boat.

Electron Beam Vaporization. High melting point materials, with melting temperatures up to 3500 °C (6330 °F), have been evaporated in ion-plating systems by using conventional electron beam guns. The electron beam gun is a principal evaporation source in many experimental and commercial ion-plating systems. However, because of the necessity of a lower vacuum for operation of the electron beam gun, a system utilizing an electron beam must have a conductance baffle to separate the plasma area from the higher vacuum of the electron beam area.

Sputtering Sources. Another source for depositing material in an ion-plating system is the use of sputtering targets. The production of the vapor is very similar to the normal sputtering process, with the exception of the higher voltages used. Some variations in sputtering targets can be achieved so that a broad range of potential ionization sources can be realized.

Reactive Ion Plating. In this technique, a reactant gas is introduced into the ion-plating system in controlled amounts. The reactant gas, when introduced into the plasma, undergoes dissociation and ionization of the atoms of the reactant gas molecules. In instances where alloy or compound coatings are desired, the coating composition can be maintained by careful control of the process parameters, particularly the concentration of the reactant gas. This system has been used for deposition of selected nitrides, carbides, silicides, and oxides.

Radio Frequency (RF) Induction Ionization. It is possible to introduce a bare induction coil directly into the glow discharge without producing an arc, if the operating frequency has been reduced to about 75 kHz. This can be accomplished by using an induction generator reduced to this level from the more normal metallurgical operating frequency of about 450 kHz. The latter frequency cannot be used in ion plating because severe arcing between the coils through the plasma would occur.

Other lesser used ion sources and ionization techniques include: (*a*) the promotion of evaporation by high frequency induction heating; (*b*) the use of electron-emitting filaments to form a triode, ion evaporation system; and (*c*) the use of an ion beam source similar to that of ion implantation, with the ion accelerated through a suitable acceleration system (vacuum ion plating).

Process Control

Because there are generally a large number of interactive variables in the ion-plating process, it is often necessary to have more extensive process controls than are used in other vacuum processes. This is particularly true when using reactive gases to form compound coatings. Process variables in ion plating include:

- Precleaning/handling processes
- In-situ predeposition processes
    - Residual contaminants
    - Sputter cleaning parameters
    - Substrate temperature
- Deposition process
    - Substrate temperature
    - Vapor source, atoms/unit time

Bombardment process (ions and neutrals)
    Species
        Energy distribution
        Dose rate
- Reactive gas (ions and neutrals)
    Species
    Energy distribution
    Dose rate

Generally, predeposition and deposition variables are not controlled directly, but rather are reproduced and held constant by controlling process parameters such as cathode current density, cathode voltage, chamber pressure, substrate temperature, fixture geometry, and residual gas contaminants. Recently, the use of vacuum ion plating, where the ion-bombardment source is an ion gun and the substrate is in a moderately good vacuum, has allowed the disassociation of the ion-bombardment parameters from the deposition parameters. This has led to a more controllable process, where the interface and coating properties are very process sensitive. Figure 2 shows such an ion-gun system, using a Kaufman ion source.

Fig. 2 Ion-gun ion-plating apparatus

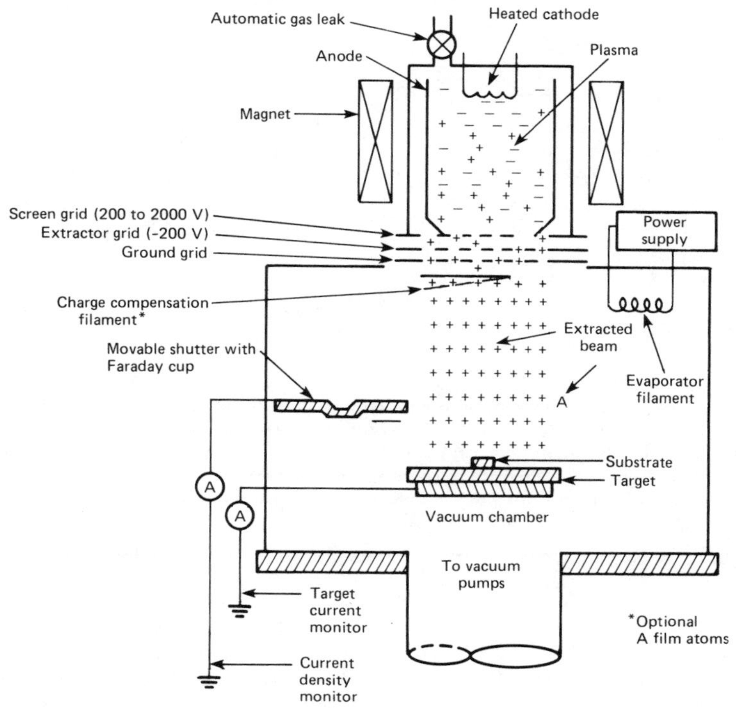

Ion-beam ion-plating system, using a Kaufman ion source and an evaporator filament source

Plasma density near the substrate surface is important, and in the direct current glow discharge technique, the plasma density may be highly variable due to the substrate geometry. Figure 3 shows some enhanced plasma sources that increase the plasma density by techniques other than ionization from secondary electrons arising from the ion bombardment, as in the direct current glow discharge technique. In Fig. 3(a), (b), and (d), the ionizing electrons are generated by a hot electron-emitting filament. In Fig. 3(d), the electron path length is increased by the use of a small magnetic field, and the shape is similar to that of triode sputtering. In Fig. 3(c), the secondary electrons emitted by the electron beam evaporation process are extracted toward the positively biased electrode and create ionization in the region above the evaporation hearth.

The most important region in the ion plating plasma is so-called dark space near the cathode. It is in this region that there is the largest potential drop in the plasma. Hence the significance of the dark space in relation to ion plating is that in this region the ions within the plasma will achieve the highest acceleration towards the cathodic target or workpiece. The size of this dark space can be expanded or contracted by increasing or decreasing the pressure of the gas used for forming the plasma.

In Fig. 1, an electrically conductive substrate was assumed to exist. If the substrate is an electrical insulator, the negative bias may be applied by using a radio frequency (RF) potential, similar to RF sputtering, or a high transmittance grid may be used in front of the substrate surface. An ion-gun system with a neutralizer filament also may be used to electrically bombard insulating surfaces.

Substrate Temperature. In a typical direct current diode ion-plating operation, the power input may be several watts per square centimetre. This may lead to a high substrate temperature if the power dissipation ability of the part is poor. Part temperatures may be lowered by improving power dissipation, using pulsed ion bombardment, precleaning to reduce the need for sputter cleaning, or changing the ion-energy spectrum to improve sputtering efficiency. High substrate temperature also can be alleviated by using the supported discharge ion-plating process, where the substrate is at a lower negative potential and the electrons necessary for supporting the discharge come from an auxiliary heated tungsten filament (Fig. 3d).

Beneficial Effects

Ion plating has been used most often to provide good adhesion between a film and a surface. The principal benefits obtained from the ion-plating process are its ability to:

- Modify the substrate surface in a manner conducive to good adhesion and maintain this condition until the film begins to form
- Provide a high-energy flux to the substrate surface, giving a high effective surface temperature, thus enhancing diffusion and chemical reactions without necessitating bulk heating
- Alter the surface and interfacial structure by introducing high defect concentrations, physically mixing the film and the substrate material, and influencing the nucleation and growth of the depositing film

In addition to modifying the substrate surface and influencing the film-sub-

Fig. 3 Enhanced ionization sources used for ion plating

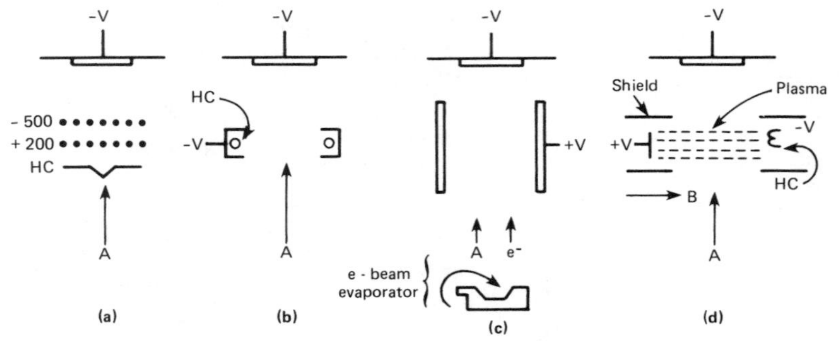

(a) Hot electron emitting filament with accelerating grid. (b) Electron emitting filament. (c) Positive cylindrical electrode to attract secondary electrons from evaporation hearth; may be used with a magnetic field. (d) Hot cathode magnetically confined plasma. In each configuration, the electrons enhance the plasma density near the surface to be coated. The substrate potential extracts ions from the plasma to bombard the surface. A, atoms from source; B, magnetic field; HC, hot cathode

strate interface, ion bombardment of the growing film may cause modification of the morphology of the deposited material, changes in the internal stress of the deposited film, or modification of other physical and electrical properties.

In some situations, another advantage of ion plating is the high throwing power, or ability to cover a surface, as compared with that obtained by vacuum evaporation. This high throwing power results from gas scattering, entrainment, and redeposition from the sputtered film surface. This allows coatings to be formed in recesses and areas remote from the source substrate line of sight, thus giving more complete surface coverage than is attained with vacuum processes.

Ion-Plating Applications

Ion plating in production applications is limited, but because of its potential, many applications are becoming apparent with its development. Each independent process requires its own set of processing variables, and therefore, growth has been in new and difficult coating applications rather than in replacing a standard process. A number of typical production applications now in use, as well as a number of potential applications, are given below. It is too early in the development of this process to provide a systematic, overall set of data that can be applied in general to the technique. Thus, the individual user must evaluate variations in plasma voltage, ionization source, and vacuum pressure within the system before and during the plasma phase and adjust these to his needs. Deposition rates can produce coatings as rapid as 25 μm/min (1000 μin./min), but generally about 3 to 5 μm/min (120 to 200 μin./min) is more common. The applications discussed give the potential user an idea of the broad range of applicability of the technique and/or where and how to apply the process.

Processes that are being used commercially include coating of aircraft and spacecraft steel and titanium fasteners. Ion plating was much preferred to other coating processes because of excellent uniformity of thickness, aluminum adhesion, and good throwing power. Fasteners, which were previously coated with cadmium, are now coated with the more compatible aluminum through the ion-plating process. The aluminum coating has excellent uniformity in thickness, strength, and mechanical properties. This aids in the improvement of stress corrosion resistance, precludes embrittlement of cadmium-coated titanium fasteners, and reduces the potential for general corrosion of this type of part. A broad range of steel and titanium fasteners, as well as ion-plated thin-walled stainless steel tubes for interconnect sections of small heat exchangers, can be coated by ion-plated aluminum. When coated

Table 1 Applications of ion plating

Coating material	Substrate material	Use
Optical		
Cr	Plastics	...
In(Sn)O	Glass	...
	Plastics	...
TiN	Metals	...
ZrC	Metals	...
Electrical		
Pt, Al, Au, Ag	Si	Contact
Al	GaAs	Contact
In(Ga)	CdS	Contact
Cu	Al_2O_3	Contact
Si_3N_4	Si	Passivation
Mechanical		
Au, Ag, MoS_2	Metals	Lubrication
Ti(C, N), Zr (C, N), Cr (C, N)	Metals, carbides	Wear/erosion
Ag	Be	Bonding
Cu	Ta, W, Nb, oxides	Bonding
Au-Cr	Mo	Bonding
Pt	Ti	Creep
Corrosion		
Al	U, Ti, steel	Gas/liquid
Cr	Steel	Gas/liquid
Al_2O_3	Steel	Gas/liquid
Si_3N_4	Mo	Gas/liquid
Cd	Steel	Gas/liquid
Ti	Steel	Gas/liquid
C	Metals	Biological
Ti	Metals	Biological
Ta	Steel	Chemical

with aluminum and subjected to 12-month endurance tests, no indication of aluminum flaking, contamination of the system, or indications of corrosion were found.

A series of more unique applications of ion plating include spacecraft telepoint rotors, missile launchers, high-strength steel aircraft components, and aircraft actuator door extension springs. A telepoint rotor normally is coated with cadmium when used in the earth's atmosphere but, because cadmium has an extremely high vapor pressure, it cannot be used when exposed to the hard vacuum of space. Ion-plated aluminum was tested along with nickel and gold coatings, which were applied by vapor deposition and electro-deposition. It was shown that the ion-deposited aluminum coatings were superior to all other coatings for this application. Ion-plated aluminum protection for steel springs for aircraft actuator extension springs was another application successfully tested.

Aluminum coating of uranium reactor parts was one of the first applications of ion plating. Investigation of this process was based on the knowledge that aluminum provides a low neutron cross section, has high thermal conductivity, and has electronegativity close to that of uranium so that galvanic corrosion is minimized. The reactor elements were first cleaned by sputtering, after which the aluminum was ion plated. Deposits of 0.013 mm (0.0005 in.) could be produced in 2 to 5 min. It was further shown that subsequent heating of this deposit layer resulted in UAl_3 providing a thermodynamic stable barrier to any reactor corrosion.

Examples of four different titanium alloys having platinum ion coatings showed improved high-temperature fatigue resistance, oxidation resistance, and cyclic fatigue properties.

Electroplating. Application of ion plating is used as a precursor to electroplating or to provide a surface on a metal or alloy not normally compatible with electroplating. Because of the exceptional bonding that occurs, and because the ion plating process ensures a method whereby a prior ion bombardment can produce a clean, generally oxide-free coating, the resultant electro overplate is very effective. For example, the use of ion-plated copper on titanium can serve as the precursor for electrodeposited copper. The prospects of successfully electrodepositing copper directly to titanium is extremely poor, but electroplated coatings to titanium that had a prior ion-plated copper surface showed excellent integrity and overall bonding. This method can be used to produce electroplated coatings on metals that are not normally compatible, such as ion plating of gold onto beryllium oxide gyro rotors used in guidance systems. These individual units were found to be most effectively gold coated using ion plating (Ref 2).

Production of ion-plated titanium carbide and titanium nitride coatings on hardened steel surfaces resulted in a series of patents that have dealt principally with the formation of titanium carbide on the surface of a hardened steel (Ref 3). Commercially available meat cutting band saws show a marked increase in life (minimum of three to five times that of normal band saw blades) as well as improved corrosion resistance and cutting characteristics.

The technique involved in ion plating on high-strength plain carbon steels also has the potential for improving many alloy steels used in cutting operations. The voltage used was typically in the neighborhood of 3 to 4 kV for titanium plating. This was followed by carburization at temperatures resulting in the formation of titanium carbide through diffusion. The surfaces had hardness in excess of 1400 HK, with the potential for several hundred points higher with increased carburization.

Improved corrosion-resistant stainless steels for biomedical applications also were reported (Ref 4). The ion plating of titanium and molybdenum on the surface of cold worked stainless steels showed direct improvement of fatigue life and corrosion resistance. The corrosion resistance of cold worked 316 stainless steel was improved most effectively through ion plating of molybdenum. This was the result of the ability of the molybdenum to preclude the initiation of pitting and/or crevice corrosion due to the precipitation of an insoluble molybdenate, as the pH levels occurring in localized corrosion. In addition, marked improvement in fatigue life was found, particularly for the molybdenum ion-plated materials. This included both air and corrosion fatigue.

The technique of ion plating opens up a broad range of new types of surface coatings which, when plated normally, have poor adhesion. Coating of complex shapes, which have been a major problem in both electrochemical and chemical vapor deposition, has been solved because of the excellent throwing power of the ion-plating technique. The potential for ion plating spans a much broader range than the applications presented, such as corrosion-resistant films, fatigue-resistant films, wear-resistant coatings, low-friction coatings, optical applications, and catalyst surfaces. Table 1 lists some of these applications in more detail. Reference 5 contains a bibliography of ion plating and its applications.

REFERENCES

1. D. M. Mattox, *Electrochemical Technology*, Vol 2, 1964, p 295
2. Stupp, B. C., *Plating Journal,* 1974, p 1090
3. Engel, N. E., U. S. Patent No. 3,915,757 (1975) and 3,925,116 (1975)
4. Hochman, R. F. and Marek, M., Proceedings of World Biomaterial Congress, Society of Biomaterials, Vienna, 1980
5. R. F. Bunshah, Ed., Ion Plating Technology, in *Deposition Technology*, Noyes Publications, 1982, p 244

SELECTED REFERENCES

- D. M. Mattox, *Journal of Vacuum Science and Technology*, Vol 10, 1973, p 47
- D. M. Mattox, Thin Film Adhesion and Adhesive Failure—A Perspective, ASTM Special Publication 640, American Society for Testing and Materials, Philadelphia, 1978, p 54
- D. M. Mattox, *Thin Solid Films*, Vol 53, 1978, p 81
- D. J. Sharp, J. K. G. Panitz, and D. M. Mattox, *Journal of Vacuum Science and Technology*, Vol 16, 1979, p 1879
- R. F. Bunshah, Hard Coatings for Wear Resistance by Physical Deposition Processes, *Specialized Cleaning, Finishing, and Coating Processes*, American Society for Metals, 1981

Ion Implantation

By Robert F. Hochman
Associate Director for Metallurgy
Georgia Institute of Technology

ION IMPLANTATION is the process of modifying the physical or chemical properties of the near surface of a solid by embedding appropriate atoms into it from a beam of ionized particles. The properties to be modified may be electrical, optical, or mechanical, and they may relate to the semiconducting behavior of the material or its corrosion behavior. The solid may be crystalline, polycrystalline, or amorphous and need not be homogeneous (Ref 1).

Ion implantation allows the controlled introduction of one or more species into the surface of a substrate by using high energy ion beams of the species to be introduced. When alternatives such as diffusion coatings or coevaporation are impractical, ion implantation offers a straightforward and reproducible method for attaining a desired result. Because this is a nonequilibrium process, solubility limits may be exceeded with or without subsequent precipitation. This makes it possible to incorporate any kind of ion without developing a set of diffusion conditions or considering the control of chemical constraints. Ion implantation was initially concerned with the fabrication of semiconductor components. For this application, ions were introduced at a relatively low concentration into the substrate. However, it has been discovered that by introducing large quantities of an implanted species, a broad range of chemical and mechanical properties of substrate material may be modified (Ref 1, 2, 3).

Capabilities

The use of accelerated electron beams can produce penetrations into a substrate surface on the order of 0.1 to 0.2 μm (1000 to 2000 Å) at 100 kV and with higher accelerating voltages, the potential for depth of penetration is increased. Surfaces may be treated by ion implantation and produce an effective alloyed surface layer where the composition varies as a function of depth. An ion beam implanter, which can provide a range of ion energies, can produce remarkable variations in surface characteristics and for limited depths below the surface. In addition, unique alloys not possible through normal alloying can be produced through this technique. Theoretically, two or more metals, completely insoluble in each other, can be alloyed in this manner. Reactive substances can also be added to the ion beam, which can produce surface and subsurface compounds. Post implant heat treatment can be used to anneal out defective structures or produce surface products with unique characteristics. The equipment necessary for treatments limits the application to high-cost parts where a specific surface is extremely important, or to applications in fail-safe systems where high reliability is necessary. Such reliability can be achieved because of the improved corrosion and fatigue resistance, as well as the much reduced friction and or wear which are characteristic of this process. An ion beam accelerator is illustrated schematically in Fig. 1.

Although relatively thin, an implanted ion layer can alter mechanical and chemical properties significantly. For example, the optimum surface layer thickness for wear resistance under high load conditions can be quite thin, only 2 μm (0.08 mil) for a molybdenum disulfide compound applied in a conventional manner (Ref 4). A conventional lubricating film applied with a binder over an oxide film is not in close contact with the underlying substrate; however, an ion implanted surface is in excellent atomic contact with the substrate.

With the restricted distance ions travel in a substrate, ion implantation cannot be expected to modify bulk properties of materials. Ion implantation is restricted to modification of the outer surface layer and can alter only surface properties significantly.

Ion Penetration

As an ion beam enters a solid target, it undergoes collisions with free electrons in the target and atom nuclei on lattice sites. Each collision causes a loss of energy, referred to as electronic and nuclear stopping. The relevant parameter is the stopping power, dE/dx, representing rate of energy loss with distance of penetration. When the energy of an ion falls to about 20 eV, it ceases to move through the solid, becoming trapped by the cohesive force of

Fig. 1 Ion accelerator for ion implantation applications

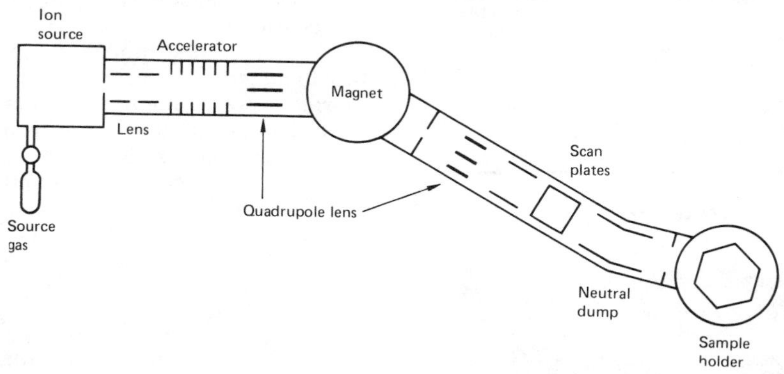

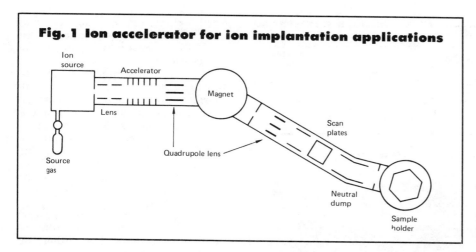

the material. The distance travelled from the surface to the trapping point is called the total range. In many cases, subsequent motion because of electrostatic forces (ion drift) or thermal diffusion may occur, but these are exceedingly slow mechanisms when compared with the 10^{-12} to 10^{-15} s involved in stopping an incident ion from a conventional ion beam source.

The behavior of a high energy ion, losing energy principally by electronic collisions, is not appreciably influenced by a few atomic layers of contaminant, and conversely the ion does not appreciably alter the surface composition. A slow moving ion, however, losing energy primarily by nuclear collisions, collides with surface atoms. As a result, any contaminant or oxide layers influence the progress of a slow ion, and the surface of the target erodes through sputtering and loss of ions from surface backscattering. Contaminant atoms may also be driven into the substrate, causing an additional, unwanted, implanted species. By ensuring that the substrate targets are atomically cleaned by sputtering, many surface difficulties can be avoided.

Range. High energy ions undergo many collisions as they penetrate the substrate. Each collision results in a change in scattering angle, as well as fluctuations in energy loss. A number of statistically related variables are used to describe the probability distribution of penetration depth. A most probable range is defined as R, a median range R_m, a mean range $\overline{R}$, and whenever the distribution shows a definable cutoff, a maximum range R_{max}.

A projected ion changes directions frequently. The path length of a projected ion is the total range, R_{tot}. Because the concern is depth of penetration, not total length of path, a more useful parameter for ion implantation is the projection of total range parallel to the incident ion direction, R_p (projected range). The total range of a particle entering a target at energy E_o is simply related to the rate of energy loss by the equation:

$$R_{tot} = \int_0^{E_o} \frac{dE}{-(dE/dx)}$$

Because ion penetration is an inherently statistical process, parameters such as the mean square fluctuation for each of the range parameters given above must also be defined.

Depth of Distribution. An implanted species forms a broad layer some depth from the surface. For incoming energies of a few hundred keV, this depth may be in the order of microns, and essentially none of the implanted species appear at the surface. At sufficiently low energies (a few keV), the projectiles come to rest close to the surface, and the scattering of ranges ensures that many of the implanted species lie at the surface itself. By implanting a single substrate with ions of both low and high energies, a uniform depth distribution extending to a few microns can be developed.

Evaluation of the order of magnitude of the quantities involved in ion implantation is important. Table 1 shows the various energies, the projected range (R_p), and standard deviation (ΔR_p) of Ti^+ ions implanted into iron at a beam current of 10 μm/cm^2. Also shown is a rough estimate of percentage concentration after 10 s irradiation calculated on the assumption that all particles come to rest within $\pm\Delta R_p$ of the projected range and neglecting diffusion after the ion has come to rest. The ranges of implanted particles are clearly rather small. Despite this, the implanted region cannot be regarded as a surface layer in the atomic sense because it occupies many atomic layers and, for the higher energy implants, few atoms may be at the surface. The ion implanted surface is defined as a surface in a mechanical sense of hundreds of atomic layers.

Lattice Damage. At high ion projectile energies, the principal energy loss mechanism is collision with electrons; at low energies, collisions with substrate nuclei predominate. An energetic ion entering a target workpiece first loses energy by electronic stopping and nuclear collisions predominate only as the ion reaches the end of its range. Lattice damage occurs by displacement of substrate atoms at the end of the incident ion range. Such damage may be removed by subsequent annealing of the implanted material. The properties of ion implanted specimens change because of radiation damage and changes as a result of the implanted species. In ion implanted materials, the location of the implanted species must be determined to understand the effects observed. Experimental work (Ref 5) has provided general formulas that are useful for estimating ion penetration in noncrystalline materials, and this has been refined to include electron shell structure (Ref 6). However, most existing evaluations of energy loss and range are formulated for a single ion entering an undamaged substrate. Although these evaluations are adequate where the implanted species represent only a small concentration in the target, as in the fabrication of semiconductor devices, these analyses have not been modified to treat implantation conditions with large concentrations of implant ions entering a lattice or structure already damaged by impinging ions.

Channeling. When ions are directed into a single crystal along a crystallographic path, implants may enter channels between rows of atoms and lose energy only by electronic stopping, penetrating considerably further than in an amorphous sample. The ions cause little radiation damage. However, an ion that does not enter close to the center of a channel undergoes a large angle deviation through collisions with occupied lattice sites and subsequently moves into the crystal in

Table 1 Energies, projected ranges, and standard deviations involved in ion implantation

Energy, keV	R_p, μm	R_p, Å	ΔR_p, μm	ΔR_p, Å	Concentration in 10 s, %
10	0.0052	52	0.0029	29	3.7
100	0.0342	342	0.0166	166	0.7
1000	0.3750	3750	0.0825	825	0.13

Fig. 2 Wear track of nitrogen implanted and unimplanted titanium disks

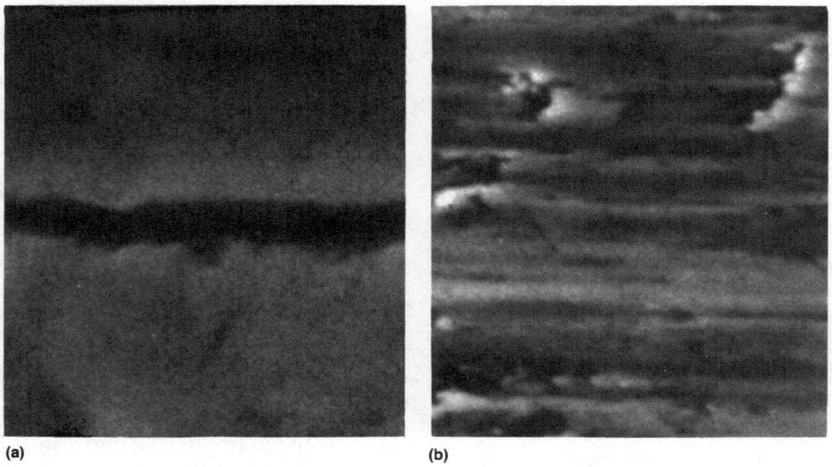

(a) Nitrogen implanted and (b) unimplanted titanium disks after 12 000 revolutions of sliding. Titanium implanted with nitrogen showed a greater improvement in wear resistance than nitrogen implanted iron disks. Plowing grooves are visible only with the aid of a scanning electron microscope (SEM)

a random direction, rapidly coming to rest. A potential for very deep implants exists from the channeling of ions in an oriented structure associated with very little radiation damage. However, channeling usually occurs in only single crystals or high oriented polycrystalline samples and may have very limited technological application.

Diffusion. In implanted materials, an important consideration is radiation-enhanced diffusion, first suggested nearly 20 years ago (Ref 7, 8). The distribution of implanted ions is modified by the diffusion of atoms after they have been reduced to thermal velocities. This modification may occur even for species where diffusion does not usually occur at room temperature. The additional vacancies and interstitials created in implantation may far exceed those created thermally, thus excess vacancies may enhance the rate of diffusion at relatively low temperatures. Radiation-enhanced diffusion has also been observed in metals (Ref 9), but the phenomenon is complex and not fully understood. Radiation damage may induce diffusion of the implanted species and result in a final profile that extends several factors deeper than that expected from a simple ion range calculation.

Applications of Ion Implantation

The potential of ion implantation was well documented in a recent edited version of a National Material Advisory Board (NMAB) report in the *SAMPE Journal* (Ref 10). The NMAB Committee on Ion Implantation and New Surface Technologies report also succinctly covers the advantages of the process, which are included in the following list:

- A variety of ion species can be implanted with the same basic apparatus. Almost all elements of the periodic table have been implanted
- Novel nonequilibrium structures and metallurgical phases with properties that cannot be duplicated in bulk material can be produced at the surface. In certain cases, amorphous or glassy phases can be formed
- If the implanted atoms are mobile, inclusions and precipitates can be formed. For example, implanted argon and helium atoms are insoluble in metals and may form bubbles
- Cooperative effects of two implanted species can occur; for example, implantations of both molybdenum and sulfur into steels seem to have an effect similar to lubrication with molybdenum silicide
- Ion implantation is a low-temperature process. It can often be added to the end of a production line without affecting existing operations
- The surface of finished products can be treated without introducing significant dimensional changes and without changing bulk properties
- The absence of a discontinuous interface between the implanted surface layer and the bulk leads to excellent adhesion of the implanted layer
- The process is easily controlled through the electrical signals applied to the accelerator
- Ion implantation creates no problems of disposal of waste products, as does electroplating

Disadvantages. These advantages must also be weighed against the known disadvantages of ion implantation. These are:

- Shallow surface layer
- Process strictly line-of-sight or beam-line
- Skilled labor and expensive equipment are required for this high technology process
- Necessity for target or substrate to be manipulated *in vacuo* to implant surfaces of varying design
- Study and development of ion sources needed to produce a high beam concentration that is relatively clean
- High maintenance costs of the vacuum system and limitations on the vapor pressure materials to be used

Major industrial applications of ion implantation have been in the electronics industry. Ion implantation has served as a useful method for doping semiconductors to produce controllable, homogeneous concentrations of doping agents with excellent reproducibility compared to conventional diffusion processing. The use of ion implantation in the semiconductor industry has led to development of more reliable and more

Fig. 3 Change in friction coefficient with sliding distance of unimplanted iron, Al+ implanted, and N+ implanted iron systems

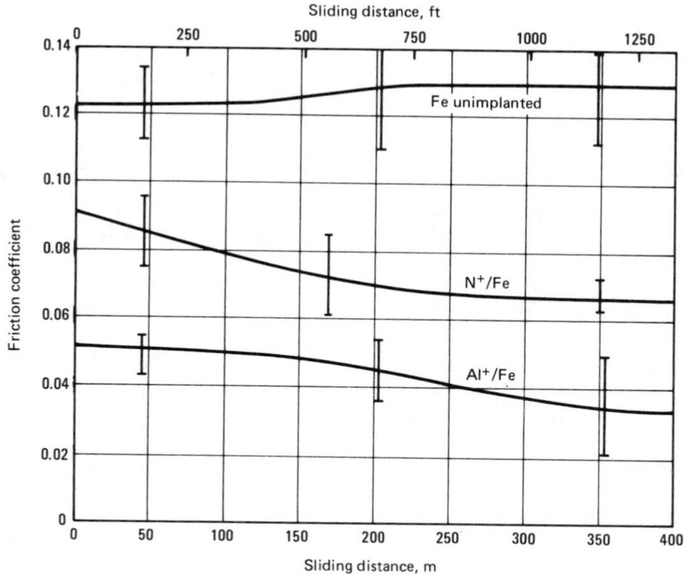

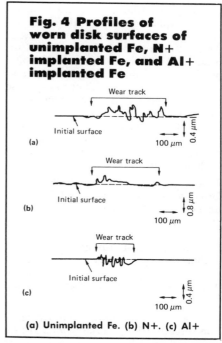

Fig. 4 Profiles of worn disk surfaces of unimplanted Fe, N+ implanted Fe, and Al+ implanted Fe

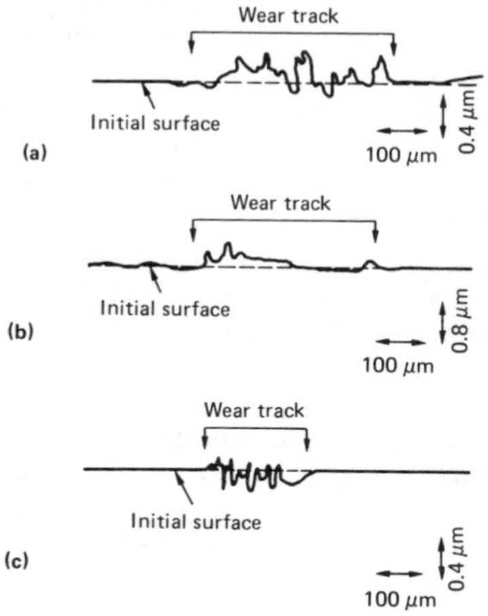

(a) Unimplanted Fe. (b) N+. (c) Al+

easily operated ion implantation equipment. Ion accelerators have moved out of the laboratory and into industry, and equipment to do a broad range of industrial implants is available or can be easily adapted to a range of new applications without extensive modification.

Potentially fruitful nonelectronic applications involve improved properties and life of metal tools, cemented carbide tools and more recently, ceramic tools; improvements are because of increased surface hardness, and reduced wear and friction. Also, improved aqueous corrosion resistance and improved high temperature corrosion and oxidation resistance can be achieved. Other possible benefits include improved optical properties, unique catalyst surfaces, bonding and adhesion between highly dissimilar materials, and any other process where a unique metallurgical surface or surface alloy is warranted.

Wear Resistance. In pin-on disk friction and wear tests of nitrogen and aluminum-implanted iron and titanium, significant reduction in friction and wear of the iron and titanium disks can be attributed to a hard layer formed during the ion implantation process (Ref 11). This hard layer minimizes plowing (see Fig. 2) and subsurface deformation, and reduces the delamination wear process, which consists of crack nucleation, crack propagation, and the formation of delamination wear sheets. The formation of an alloyed surface via ion implantation produces a substantial change in hardness, which minimizes the friction coefficient. Figure 3 compares change of friction coefficient with sliding distance for unimplanted iron, nitrogen-implanted, and aluminum-implanted iron disks. Figure 4 illustrates the resulting wear profiles of these three disk surfaces.

Several developing industrial applications involving nitrogen-implanted iron have led to improved properties of steel surfaces for tools used in steel forming in the automotive industry (Ref 12). Examples include a steel press tool, which showed an improved performance after nitrogen ion implantation of nearly ten times that of its unimplanted counterpart. A treated forming die resulted in the production of nearly 5000 automobile parts compared to the normal 2000 part life from a similar tool hard faced with electrodeposited chromium. Another example is a ring cutter for tin plate treated by nitrogen ion implantation of a tool steel, resulting in extending the life of the tool by a factor of three. Improved hot rolling dies for nonferrous rod and steel gear cutters were also produced by ion implanting nitrogen. All of these have shown increased life from 200 to 1000% following nitrogen ion implantation with an ion flux in the neighborhood of 5×10^{17} nitrogen ions/cm^2 (3×10^{18} ions/in.^{2}) with implantation voltages in the neighborhood of 100 keV.

Improved life of nitrogen ion implanted tools is also achieved by injection molds used in plastic production. Injection molding screws and dies have shown marked life increases as the result of ion implantation of nitrogen. Cobalt-cemented tungsten carbide, used for many dies and wear-resistant tools, can be effectively improved by ion implantation of nitrogen and carbon. Wire drawing dies having three to five times the usual lifespan as a result of nitrogen ion implantation treatment have been documented (Ref 12).

Work at the U.S. Naval Research Laboratories (Ref 2) has resulted in improved surfaces for beryllium hardened by boron implantation at energies between 90 and 250 keV. The total dosage of boron ions was approximately 6×10^{17} ions/cm^2 (4×10^{18} ions/in.^{2}).

Corrosion Resistance. Improved corrosion and wear resistance of burner tips through which fuel oil and air are injected into oil-fired furnaces in power generating plants has also been achieved (Ref 13). Although nitrogen ion implantation was ineffective, cerium implantation significantly reduced the degree of sulfidization attack because of the reaction with the sulfur-rich combustion products. Additional work in the United Kingdom has also shown that stainless steel pumps and valves used in the food processing industry have poor wear resistance characteristics, but these materials have shown a marked reduction in adhesive wear because of nitrogen ion implantation.

Another area that has been considered by several authors and discussed in a paper on thermal oxidation (Ref 13) has been the improvement of turbine blades for jet aircraft engines. The potential for corrosion-resistant coatings, either through reactive ion implantation or through the production of refractory surfaces, is being explored and developed.

In a new and unique application of ion implantation, mercury ions are introduced into aluminum to activate the surface so the material may be used as a sacrificial anode (Ref 14). Once mercury is introduced at the aluminum surface, an electrode in the presence of oxygen forms a mercury-aluminum (HgAl) amalgam, and then reoxidizes to alumina and mercury with the alumina being nonprotective. This product is being tested as a potentially inexpensive sacrificial anode compared to zinc or magnesium.

REFERENCES

1. G. Dearnaley, J. H. Freeman, R. S. Nelson, and J. Stephen, *Ion Implantation*, NY: North Holland Publishing Co., 1973, p 6
2. J.K. Hirvonen, Ed., Treatise on Materials Science and Technology, *Ion Implantation*, Vol 18, New York: Academic Press, 1980
3. C.M. Preece and J.K. Hirvonen, Ed., *Ion Implantation Metallurgy*, AIME Conf. Proceedings, 1980
4. V. Hopkins and M. Campbell, *Lubrication Engineering 23*, 1969, p 288
5. J. Lindhard, M. Scharft, and H. E. Schitt, Kgl. Danske Vid. Selsk., *Matt.-Fys. Medd., 36*, No. 10, 1963
6. I. G. Cheshire, G. Dearnaley, and J. M. Poate, *Proc. Roy. Soc.*, A311, 47, 1969
7. J. C. Pfister and P. Baruch, *Journal of Phys. Soc. Japan, 18*, Suppl. 251, 1962
8. H. Strock, *Journal Applied Phys.*, 34, 3405, 1963
9. S.M. Myers and R.A. Langley, *Applications of Ion Beams to Metals*, New York: Plenum Press, 19
10. "Prospects for Ion Implantation as a New Surface Treatment Technology," NMAB Committee Report, *SAMPE Journal*, March/April 1980, p 12-18
11. SR. Shepard and N.P. Suh, The Effects of Ion Implantation on Friction and Wear of Metals, *Journal of Lubrication Technology,* Transactions of the ASME, Vol 104, Jan 1982, p 29-38
12. G. Dearnaley, *Ion Implantation Metallurgy*, 1-20, AIME Conf. Proceedings, 1980
13. G. Dearnaley, Treatise on Materials Science and Technology, *Ion Implantation*, Vol 18, New York: Academic Press, 1980
14. R. Procter, Private Communication

Nonmetallic Coating Processes

Corrosion Theory

By Dean M. Berger
Corrosion Specialist
Gilbert/Commonwealth Companies

CORROSION of metal is a chemical or electrochemical process in which surface atoms of a solid metal react with a substance in contact with the exposed surface. The corroding medium is usually a liquid, but can be a gas or a solid.

All structural metals corrode to some extent in natural environments. Bronze, brass, most stainless steels, zinc, and pure aluminum corrode so slowly in service conditions that long service life is expected without protective coatings. Corrosion of structural grades of iron and steel, the 400 series stainless steels, and some aluminum alloys, however, proceeds rapidly unless the metal is protected against corrosion. Corrosion of iron and steel is of particular concern because annual losses attributed to corrosion of steel have been estimated at nearly 10 billion dollars. Corrosion of metal is commonly identified by appearance, the method by which it takes place, or the method by which it is accelerated, as shown in the following (Ref 1):

- *Uniform*: uniform surface effect as opposed to localized pitting
- *Electrochemical*: corrosion occurring by chemical dissolution
- *Galvanic*: corrosion accelerated by a difference in potential between metals
- *Concentration cell*: corrosion accelerated by a difference in concentration of an ion or other dissolved substance
- *Erosion corrosion*: corrosion accelerated by flow of liquid or gas
- *Embrittlement*: corrosion which causes a ductile material to fail in a manner such that no significant localized shear or yield occurs
- *Stress corrosion*: corrosion accelerated or activated by stress
- *Filaform*: a form of corrosion producing porosity resembling worm holes
- *Corrosion fatigue*: fatigue accelerated by the presence of a corrosive environment
- *Intergranular*: corrosion which proceeds along grain boundaries
- *Fretting*: metal to metal wear which is accelerated by corrosion
- *Impingement*: corrosion accelerated by fluid impingement
- *Dezincification*: corrosion of brass in which zinc is preferentially leached out of a copper matrix
- *Chemical reaction*: corrosion which takes place by dissolution of or reaction of the metal and the corrosive medium

A detailed discussion of these different forms of corrosion can be found in the articles on corrosion failure, stress corrosion cracking, hydrogen damage failure, and corrosion-fatigue failure in *Metals Handbook*, 8th ed., Vol 10.

Electrochemical Corrosion Basics

Electrochemical corrosion in metals in a natural environment, whether atmosphere, in water, or underground, is caused by a flow of electricity from one metal to another, or from one part of a metal surface to another part of the same surface where conditions permit the flow of electricity. For the flow of energy to take place, either a moist conductor or an electrolyte must be present. An electrolyte is an electricity conducting solution containing ions which are atomic particles or radicals bearing an electrical charge. Charged ions are present in solutions of acids, alkalies, and salts. The presence of an electrolyte is necessary for corrosion to occur. Water, especially salt water, is an excellent electrolyte.

Electricity passes from a negative area to a positive area through the electrolyte. For corrosion to occur in metals, one must have (*a*) an electrolyte, (*b*) an area or region on a metallic surface with a negative charge, (*c*) a second area with a positive charge, and (*d*) an electrically conductive path between (*b*) and (*c*) (Ref 2). These components are arranged to form a closed electrical circuit. In the simplest case, the anode would be one metal, such as iron, the cathode another, perhaps copper, and the electrolyte might or might not have the same composition at both anode and cathode. The anode and cathode could be of the same metal under conditions described later in this article.

The cell shown in Fig. 1 illustrates the corrosion process in its simplest form. This cell includes the following essential components: (*a*) a metal an-

Fig. 1 Simple cell showing components necessary for corrosion (Ref 3)

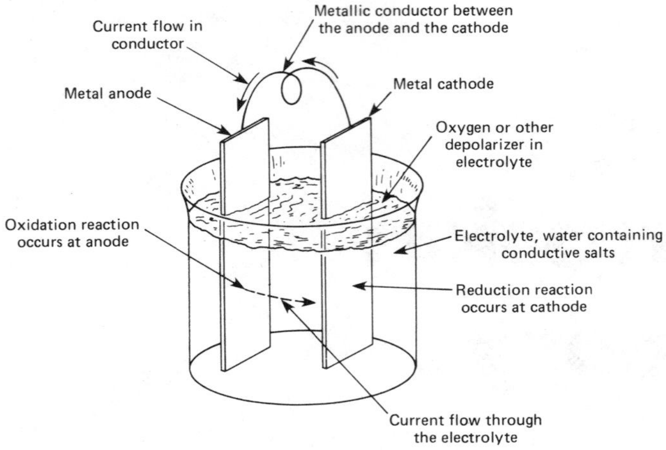

ode, (*b*) a metal cathode, (*c*) a metallic conductor between the anode and the cathode, and (*d*) an electrolyte in contact with the anode and the cathode. If the cell were constructed and allowed to function, an electrical current would flow through the metallic conductor and the electrolyte, and if the conductor were replaced by a voltmeter, a potential difference between the anode and the cathode could be measured. The anode would corrode. Chemically, this is an oxidation reaction. The formation of hydrated red iron rust by electrochemical reactions may be expressed as follows:

$$4Fe \rightarrow 4Fe^{++} + 8\ e^{-}$$

$$4Fe + 3O_2 + H_2O \rightarrow 2\ Fe_2O_3 \cdot H_2O \quad \text{(Eq 1)}$$

$$4Fe + 2O_2 + 4H_2O \rightarrow 4Fe\ (OH)_2$$

$$4Fe\ (OH)_2 + O_2 \rightarrow 2Fe_2O_3 \cdot H_2O + 2H_2O \quad \text{(Eq 2)}$$

During metallic corrosion, the rate of oxidation equals the rate of reduction. Thus, a nondestructive chemical reaction, reduction, would proceed simultaneously at the cathode. In most cases, hydrogen gas is produced on the cathode. When the gas layer insulates the cathode from the electrolyte, current flow stops, and the cell is polarized. However, oxygen or some other depolarizing agent is usually present to react with the hydrogen, which reduces this effect and allows the cell to continue to function.

Contact between dissimilar metallic conductors or differences in the concentration of the solution cause the difference in potential that results in electrical current. Any lack of homogeneity on the metal surface or its environment may initiate attack by causing a difference in potential, and this results in localized corrosion. The metal undergoing electrochemical corrosion need not be immersed in a liquid, but may be in contact with moist soil, or may have moist areas on the metal surface.

Corrosive Conditions

If oxygen and water are both present, corrosion will normally occur on iron and steel. Rapid corrosion may take place in water, the rate of corrosion being accelerated by several factors such as: (*a*) the velocity or the acidity of the water, (*b*) the motion of the metal, (*c*) an increase in temperature or aeration, and (*d*) the presence of certain bacteria. Corrosion can be retarded by protective layers or films consisting of corrosion products or adsorbed oxygen. High alkalinity of the water also retards the rate of corrosion on steel surfaces. Water and oxygen remain the essential factors, however, and the amount of corrosion is generally controlled by one or the other. For example, corrosion of steel does not occur in dry air and is negligible when the relative humidity of the air is below 30% at normal or lower temperatures. This is the basis for prevention of corrosion by dehumidification (Ref 4).

Water can readily dissolve a small amount of oxygen from the atmosphere, thus becoming highly corrosive. When the free oxygen dissolved in water is removed, the water becomes practically noncorrosive unless it becomes acidic, or anaerobic bacteria incite corrosion. If oxygen-free water is maintained at a neutral pH or at slight alkalinity, it is practically noncorrosive to structural steel. Steam boilers and water supply systems are effectively protected by deaerating the water. Additional information can be obtained in the articles on corrosion in fresh water and corrosion in seawater in *Metals Handbook*, 9th ed., Vol 1.

Soils. Dispersed metallic particles or bacteria pockets can provide a natural electrical pathway for buried metal. If an electrolyte is present, and the soil has a negative charge in relation to the metal, an electrical path from the metal to the soil will occur, resulting in corrosion. Differences in soil conditions, such as moisture content and resistivity, are commonly responsible for creating anodic and cathodic areas (Fig. 2). Where a difference exists in the concentration of oxygen in the water or in moist soils in contact with metal at different areas, cathodes develop at points of relatively high-oxygen concentrations and anodes at points of low concentration. Further information is available in the article on soil corrosion in *Metals Handbook*, 9th ed., Vol 1.

Chemicals. In an acid environment, even without the presence of oxygen, the metal at the anode is attacked at a rapid rate. At the cathode, atomic hydrogen is released continuously, to become hydrogen gas. Corrosion by an acid can result in the formation of a salt, which slows the reaction because the salt formation on the surface is then attacked.

Corrosion by direct chemical attack is the single most destructive force against steel surfaces. Substances having chlorine or other halogens in their composition are particularly aggressive. Galvanized roofing has been known to corrode completely within six months of construction, the building being downwind of an aluminum ingot plant where fluorides were always present in the atmosphere. Consequently, galvanized steel should not have been specified. Selection of materials and evaluation

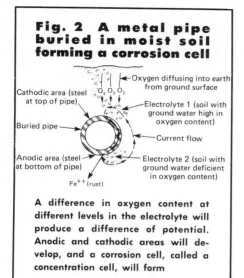

Fig. 2 A metal pipe buried in moist soil forming a corrosion cell

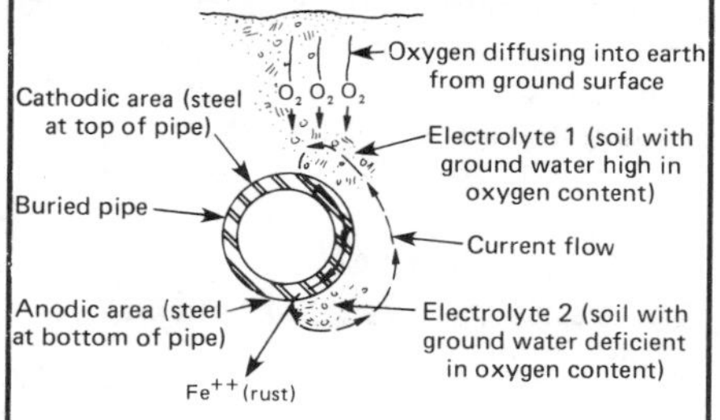

A difference in oxygen content at different levels in the electrolyte will produce a difference of potential. Anodic and cathodic areas will develop, and a corrosion cell, called a concentration cell, will form

of service conditions are extremely important in combating corrosion.

Atmospheric corrosion differs from the corrosive action that occurs in water or underground because sufficient oxygen is always present. In atmospheric corrosion, the formation of insoluble films and the presence of moisture and deposits from the atmosphere control the rate of corrosion. Contaminants such as sulfur compounds and salt particles can accelerate the corrosion rate. Nevertheless, atmospheric corrosion occurs primarily through electrochemical means and is not directly caused by chemical attack. The anodic and cathodic areas are usually quite small and close together so that corrosion appears uniform, rather than in the form of severe pitting which can occur in water or soil. A more detailed discussion can be found in the article on atmospheric corrosion in *Metals Handbook*, 9th ed., Vol 1.

Galvanic Corrosion

The potential available to promote the electrochemical corrosion reaction between dissimilar metals is suggested by the galvanic series which lists a number of common metals and alloys arranged according to their tendency to corrode when in galvanic contact (Table 1). Metals close to one another on the table generally do not have a strong effect on each other, but the farther apart any two metals are separated, the stronger the corroding effect on the one higher in the list. It is possible for certain metals to reverse their positions in some environments, but the order given in Table 1 is maintained in natural waters and the atmosphere. The galvanic series should not be confused with the similar electromotive force series which shows exact potentials based on highly standardized conditions which rarely exist in nature (Ref 3).

Table 1 Galvanic series

These metals are arranged by their tendency to corrode galvanically; consult article for further details

Corroded end (anodic)
Magnesium
Magnesium alloys
Zinc
Aluminum 1100
Cadmium
Aluminum 2017
Steel or iron
Cast iron
Chromium-iron (active)
Ni-Resist
18-8 chromium-nickel-iron (active)
18-8-3 chromium-nickel-molybdenum-iron (active)
Lead-tin solders
Lead
Tin
Nickel (active)
Inconel (active)
Hastelloy C (active)
Brass
Copper
Bronzes
Copper-nickel alloys
Monel
Silver solder
Nickel (passive)
Inconel (passive)
Chromium-iron (passive)
18-8 chromium-nickel-iron (passive)
18-8-3 chromium-nickel-molybdenum-iron (passive)
Hastelloy C (passive)
Silver
Graphite
Gold
Platinum
Protected end (cathodic)

The three-layer iron oxide scale formed on steel during rolling varies with the operation performed and the rolling temperature. The dissimilarity of the metal and the scale can cause corrosion to occur, with the steel acting as the anode in this instance. Unfortunately, mill scale is cathodic to steel, and an electric current can easily be produced between the steel and the mill scale. This electrochemical action will corrode the steel without affecting the mill scale (Fig. 3).

A galvanic couple may be the cause of premature failure in metal components of water-related structures or may be advantageously exploited. Galvanizing iron sheet is an example of useful application of galvanic action or cathodic protection. Iron is the cathode and is protected against corrosion at the expense of the sacrificial zinc anode. Alternatively, a zinc or magnesium anode may be located in the electrolyte close to the structure and may be connected electrically to the iron or steel. This method is referred to as cathodic protection of the structure. Iron or steel can become the anode when in contact with copper, brass, or bronze; however, they corrode rapidly while protecting the latter metals. Also, weld metal may be anodic to the basis metal, creating a corrosion cell when immersed (Fig. 4).

Fig. 3 Mill scale forming a corrosion cell on steel

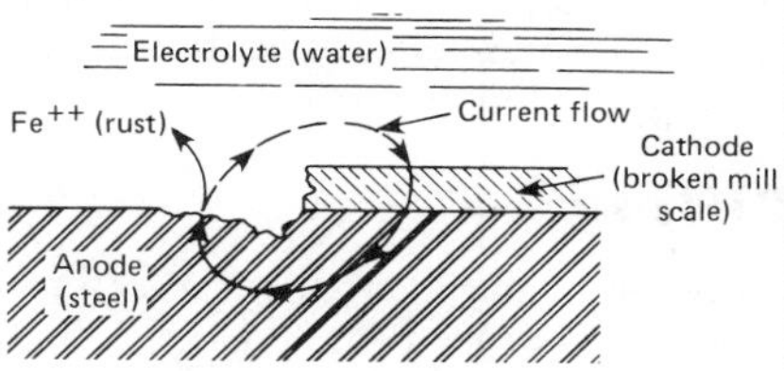

While the galvanic series (Table 1) represents the potential available to promote a corrosive reaction, the actual corrosion is difficult to predict. Electrolytes may be poor conductors, or long distances may introduce large resistance into the corrosion cell circuit. More frequently, scale formation forms a partially insulating layer over the anode. A cathode having a layer of adsorbed gas bubbles, as a consequence of the corrosion cell reaction, is polarized. The effect of such conditions is to reduce the theoretical consumption of metal by corrosion. The area relationship between the anode and cathode may also strongly affect the corrosion rate; a high ratio of cathode area to anode area produces more rapid corrosion. In the reverse case, the cathode polarizes, and the corrosion rate soon drops to a negligible level.

The passivity of stainless steels is attributed to either the presence of a corrosion-resistant oxide film or an oxygen-caused polarizing effect, durable only as long as there is sufficient oxygen to maintain the effect, over the surfaces. In most natural environments, stainless steels will remain in a passive state and thus tend to be cathodic to ordinary iron and steel. Change to an active state usually occurs only where chloride concentrations are high, as in seawater or reducing solutions. Oxygen starvation also produces a change to an active state. This occurs where the oxygen supply is limited, as in crevices and beneath contamination on partially fouled surfaces.

Pitting

Pitting is a type of localized cell corrosion. It is predominantly responsible for the functional failure of iron and steel water-related installations. Pitting may result in the perforation of water pipe, rendering it unserviceable, even though less than 5% of the total metal has been lost through rusting. Where confinement of water is not a factor, pitting causes structural failure from localized weakening while considerable sound metal still remains.

Pitting develops when the anodic or corroding area is small in relation to the cathodic or protected area. For example, pitting can occur where large areas of the surface are covered by mill scale, applied coatings, or deposits of various kinds, and breaks exist in the continuity of the protective coating. Pitting may also develop on bare, clean metal surfaces because of irregularities in the physical or chemical structure of the metal. Localized, dissimilar soil conditions at the surface of steel can also create conditions that promote pitting (Ref 3).

Electrical contact between dissimilar materials or concentration cells (areas of the same metal where oxygen or conductive salt concentrations in water differ) accelerates the rate of pitting. In closed-vessel structures, these couples cause a difference of potential which results in an electric current flowing through the water or across the moist steel from the metallic anode to a nearby cathode. The cathode may be copper, brass, mill scale, or any portion of a metal surface that is cathodic to the more active metal areas. In practice, mill scale is cathodic to steel and is found to be a common cause of pitting. The difference of potential generated between steel and mill scale often amounts to 0.2 to 0.3 V. This couple is nearly as powerful a generator of corrosion currents as is the copper-steel couple. However, when the anodic area is relatively large compared with the cathodic area, the damage is spread out and usually negligible, but when the anode is relatively small, the metal loss is concentrated and may be very serious.

Fig. 4 Weld metal forming a corrosion cell on steel

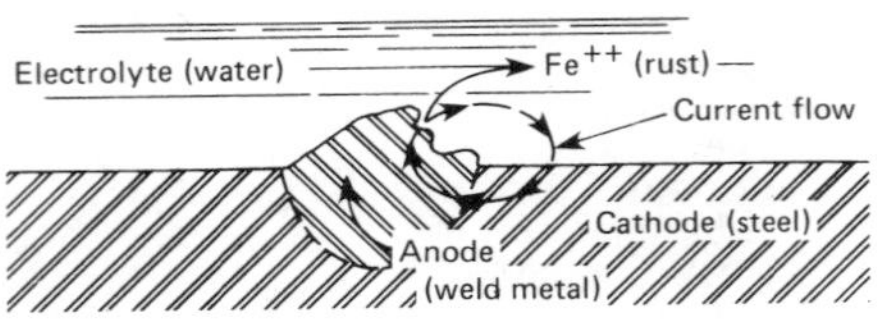

Weld metal may be anodic to steel, creating a corrosion cell when immersed

On surfaces having some mill scale, the total metal loss is nearly constant as the anode is decreased, but the degree of penetration increases. Figure 3 shows how a pit forms where a break occurs in mill scale. When contact between dissimilar materials is unavoidable and the surface is painted, it is preferred to paint both materials. If only one surface is painted, it should be the cathode. If only the anode is coated, any weak points such as pinholes or holidays in the coating will probably result in intense pitting (Ref 5).

Severe pitting is often caused by concentration differences in the electrolyte, especially dissolved oxygen. When part of the metal is in contact with water relatively low in dissolved oxygen, it is anodic to adjoining areas in contact with water higher in dissolved oxygen. The lack of oxygen may be caused by exhaustion of dissolved oxygen as in a crevice (Fig. 5). This figure illustrates another type of concentration cell. This cell, at the mouth of a crevice, is caused by a difference in concentration of the metal in solution. These two effects sometimes blend together as in a re-entrant angle in a riveted seam.

As a pit, perhaps at a break in mill scale, becomes deeper, an oxygen concentration cell is started by depletion of oxygen in the pit. The rate of penetration by such pits is accelerated proportionately as the bottom of the pit becomes more anodic. Fabrication operations may crack mill scale and result in accelerated corrosion.

Stray Currents. Accelerated corrosion of steel and iron can be produced by stray currents. Direct currents in the soil or water associated with nearby cathodic protection systems, industrial activities, or direct current electric railways can be intercepted and carried for considerable distances by buried steel structures. Corrosion takes place where the stray currents are discharged from the steel to the environment, and damage to the structure can occur very rapidly (Ref 3).

Coatings and Corrosion Prevention

The problems of corrosion should be approached in the design stage, and the selection of a protective coating is important. Paint systems and lining materials exist which slow the corrosion rate of carbon steel surfaces. High-performance organic coatings such as epoxy, polyesters, polyurethanes, vinyl, or chlorinated rubber help to satisfy the need for corrosion prevention. Special primers are used to provide passivation, galvanic protection, corrosion inhibition, or mechanical or electrical barriers to corrosive action (Ref 6).

Corrosion Inhibitors. A water-soluble corrosion inhibitor reduces galvanic action by making the metal passive or by providing an insulating film on the anode, the cathode, or both. A very small amount of chromate, polyphosphate, or silicate added to water creates a water-soluble inhibitor. A slightly soluble inhibitor incorporated into the prime coat of paint may also have a considerable protective influence. Inhibitive pigments in paint primers are successful inhibitors except when they dissolve sufficiently to leave holes in the paint film. Most paint primers contain a partially soluble inhibitive pigment such as zinc chromate which reacts with the steel substrate to form the iron salt. The presence of these salts slows corrosion of steel. Chromates, phosphates, molybdates, borates, silicates, and plumbates are commonly used for this purpose. Some pigments add alkalinity, slowing chemical attack on steel. Alkaline pigments, such as metaborates, cement, lime, or red lead, are effective provided that the environment is not too aggressive. In addition, many new pigments have been introduced to the paint industry such as zinc phosphosilicate and zinc flake.

Sacrificial Coatings. Zinc-rich primers are applied at 75 μm (3.0 mils) dry film thickness to provide galvanic protection. These primers are very effective, even in chemical environ-

Fig. 5 Corrosion caused at crevices by concentration cells

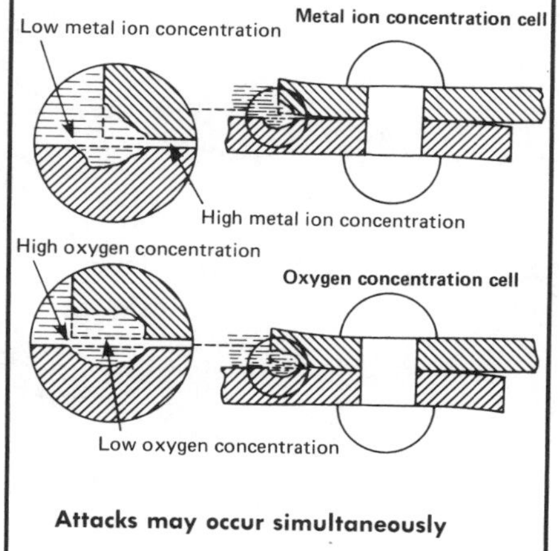

ments, because the zinc is affected before the steel is attacked. Adequate high-performance topcoats are recommended to prolong the life of this coating system.

Barrier Coatings. Protective coatings are the most widely used and recognized forms of barrier materials in engineered and remedial construction. Barrier coatings may vary in thickness from thin paint films of only a few mils to heavy mastic coatings applied from 6 to 13 mm (0.25 to 0.50 in.) thicknesses to acid-proof brick linings where they may be several inches thick. Barrier coatings are effective because they keep moisture, oxygen, and corrosive chemicals away from the structure. Protective barrier coatings vary considerably in composition, performance, and applied cost. Various rubberlike materials, plastics, tars, and waxes are used.

Cathodic protection involves the reversal of electric current flow within the corrosion cell. Cathodic protection can reduce or eliminate corrosion by connecting a more active metal to a metal that must be protected. The use of cathodic protection to reduce or eliminate corrosion is a successful technique of long-standing use in marine structures, pipelines, bridge decks, sheet piling, and equipment and tankage of all types, particularly below water or underground. Typically, zinc or magnesium anodes are used to protect steel in marine environments, and the anodes are replaced after they are consumed.

Cathodic protection uses an impressed direct current supplied by any low output voltage source and a relatively inert anode. As is the case in all forms of cathodic activity, an electrolyte is needed for current flow. Cathodic protection and the use of protective coatings are most often employed jointly, especially in marine applications and on board ships where impressed current inputs do not usually exceed 1 V (Ref 7). Beyond 1 V, many coating systems tend to disbond. Current source for cathodic protection in soils is usually 1½ to 2 V.

Choice of anodes for buried steel pipe depends on soil conditions. Magnesium is most commonly used for galvanic anodes; however, zinc can also be used. Galvanic anodes are seldom used when the resistivity of the soil is over 30 Ω·m (3000 Ω·cm); impressed current is normally used for these conditions. Graphite, high-silicon cast iron, scrap iron, aluminum, and platinum are used as anodes with impressed current. The availability of low-cost power is often the deciding factor in choosing between galvanic or impressed current cathodic protection.

Protective coatings are normally used in conjunction with cathodic protection and should not be disregarded where cathodic protection is contemplated in new construction. Because the cathodic protection current must protect only the bare or poorly insulated areas of the surface, coatings that are highly insulating, very durable, and free of discontinuities lower the current requirements and system costs. A good coating also enables a single-impressed current installation to protect many miles of piping. Coal-tar enamel, epoxy powder coatings, and vinyl resin are examples of coatings which are most suitable for use with cathodic protection. Certain other coatings may be incompatible, such as phenolic coatings which may deteriorate rapidly in the alkaline environment created by the cathodic protection currents. Although cement mortar initially conducts the electrical current freely, polarization, the formation of an insulating film on the surface as a result of the protective current, is believed to reduce the current requirement moderately.

Cathodic protection is used increasingly to protect buried or submerged metal structures in the oil, gas, and waterworks industries and can be used in specialized applications, such as for the interiors of water storage tanks. Pipelines are routinely designed to ensure the electrical continuity necessary for effective functioning of the cathodic protection system. Thus, electrical connections or bonds are required between pipe sections in lines using mechanically coupled joints, and insulating couplings may be employed at intervals to isolate some parts of the line electrically from other parts. Leads may be attached during construction to facilitate the cathodic protection installation when needed.

REFERENCES

1. D.M. Berger, Corrosion Principles Can Never Be Forgotten in Organic Finishing, *Metal Finishing*, Nov 1974
2. "Introduction to Corrosion", Carboline Co., 1968
3. "Paint Manual", Bureau of Reclamation, Denver, CO, 1976
4. Theory of Corrosion, Vol I, Steel Structures Painting Council, 1968
5. F.N. Speller, *Corrosion Causes and Prevention*, New York: McGraw-Hill, 1951
6. F.L. Whitney, Designing to Prevent Corrosion in the Process Industry, 59-SA-58, American Society of Mechanical Engineers, May 1959
7. W.A. Anderton, The Aluminum Vinyl for Ships Bottoms, JOCCA, Nov 1970

Phosphate Coating

By the ASM Committee on Phosphate Coating*

PHOSPHATE COATING is the treatment of iron, steel, galvanized steel, or aluminum with a dilute solution of phosphoric acid and other chemicals in which the surface of the metal, reacting chemically with the phosphoric acid media, is converted to an integral, mildly protective layer of insoluble crystalline phosphate. The weight and crystalline structure of the coating and the extent of penetration of the coating into the basis metal can be controlled by (*a*) method of cleaning before treatment, (*b*) use of activating rinses containing titanium and other metals or compounds, (*c*) method of applying the solution, (*d*) temperature, concentration, and duration of treatment, and (*e*) modification of the chemical composition of phosphating solution.

The method of applying phosphate coatings is usually determined by the size and shape of the article to be coated. Small items, such as nuts, bolts, screws, and stampings, are coated in tumbling barrels immersed in phosphating solution. Large fabricated articles, such as refrigerator cabinets, are spray coated with solution while on conveyors. Automobile bodies are sprayed with or immersed in phosphating solution. Steel sheet and strip can be passed continuously through the phosphating solution or can be sprayed.

Phosphate coatings range in thickness from less than 3 to 50 μm (0.1 to 2 mils). Coating weight (grams per square metre of coated area), rather than coating thickness, has been adopted as the basis for expressing the amount of coating deposited.

Phosphate Coatings

Three principal types of phosphate coatings are in general use: (*a*) zinc, (*b*) iron, and (*c*) manganese. A fourth type lead phosphate, more recently introduced, is operated at ambient temperatures.

Zinc phosphate coatings encompass a wide range of weights and crystal characteristics, ranging from heavy films with coarse crystals to ultrathin microcrystalline deposits. Zinc phosphate coatings vary from light to dark gray in color. Coatings are darker as the carbon content of the underlying steel increases, as the ferrous content of the coating increases, as heavy metal ions are incorporated into the phosphating solution, or as the substrate metal is acid pickled prior to phosphating. Zinc phosphating solutions containing active oxidizers usually produce lighter colored coatings than solutions using milder accelerators.

Zinc phosphate coatings can be applied by spray, immersion, or by a combination of the two. Coatings can be used for any of the following applications of phosphating: (*a*) base for paint or oil; *(b)* aid to cold forming; tube drawing, and wire drawing; *(c)* increasing wear resistance; or *(d)* rustproofing. Spray coatings on steel surfaces range in weight from 1.08 to 10.8 g/m^2 (100 to 1000 mg/ft^2), immersion coatings, from 1.61 to 43.0 g/m^2 (150 to 4000 mg/ft^2).

Iron phosphate coatings were the first to be used commercially. Early iron phosphating solutions consisted of ferrous phosphate/phosphoric acid used at temperatures near boiling and produced dark gray coatings with coarse crystals. The term iron phosphate coatings refers to coatings resulting from alkali-metal phosphate solutions operated at pH in the range of 4.0 to 5.0, which produce exceedingly fine crystals. The solutions produce an amorphous coating consisting primarily of iron oxides and having an interference color range of iridescent blue to red-dish-blue color.

Although iron phosphate coatings are applied to steel to provide a receptive surface for the bonding of fabric, wood, and other materials, their chief application is as a base for subsequent films of paint. Processes that produce iron phosphate coatings are also available for treatment of galvanized and aluminum surfaces. Iron phosphate coatings have excellent adherence and provide good resistance to flaking from impact or flexing when painted. Corrosion resistance, either through film or

*James W. Davis, Group Leader, Amchem Products, Inc.; James I. Maurer, Director, Technology, Parker Division, Occidental Chemical Corp.; Lester E. Steinbrecher, Director, Research and Development, Amchem Products, Inc.; Carl M. Varga, General Sales Manager, Man-Gill Chemical Co.; James K. White, Chemical Process Engineer, Caterpillar Tractor Co.

scribe under-cut is usually less than that attained with zinc phosphate. However, a good iron phosphate coating often outperforms a poor zinc phosphate coating.

Spray application of iron phosphate coatings is most frequently used, although immersion application also is practical. The accepted range of coating weights is 0.21 to 0.86 g/m^2 (20 to 80 mg/ft^2). Little benefit is derived from exceeding this range, and coatings of less than 0.21 g/m^2 (20 mg/ft^2) are likely to be nonuniform or discontinuous. Quality iron phosphate coatings are routinely deposited at temperatures from 25 to 65 °C (80 to 150 °F) by either spray or immersion methods.

Manganese phosphate coatings are applied to ferrous parts (bearings, gears, and internal combustion engine parts, for example) for break-in and to prevent galling. These coatings are usually dark gray. However, because almost all manganese phosphate coatings are used as an oil base and the oil intensifies the coloring, manganese phosphate coatings are usually black in appearance. In some instances, a calcium modified zinc phosphate coating can be substituted for manganese phosphate to impart break-in and antigalling properties.

Manganese phosphate coatings are applied only by immersion, requiring times ranging from 5 to 30 min. Coating weights normally vary from 5.4 to 32.3 g/m^2 (500 to 3000 mg/ft^2), but can be greater if required. The manganese phosphate coating usually preferred is tight and fine-grained, rather than loose and coarse-grained. However, desired crystal size varies with service requirements. In many instances, the crystal is refined as the result of some pretreatment (certain types of cleaners and/or conditioning agents based on manganese phosphate) of the metal surface.

Manganese-iron phosphate coatings are usually formed from high-temperature baths from 85 to 95 °C (190 to 200 °F).

Composition of Phosphate Coating

All phosphate coatings are produced by the same type of chemical reaction; the acid bath, containing the coating chemicals, reacts with the metal to be coated, and at the interface, a thin film of solution is neutralized because of its attack on the metal. In the neutralized solution, solubility of the metal phosphates is reduced, and they precipitate from the solution as crystals. Crystals are then attracted to the surface of the metal by the normal electrostatic potential within the metal, and deposit on the cathodic sites.

When an acid phosphate reacts with steel, two types of iron phosphate are produced: (*a*) a primary phosphate, which enters the coating; and (*b*) a secondary phosphate, which enters the solution as a soluble iron compound. If this secondary ferrous phosphate were oxidized to a ferric phosphate, it would no longer be soluble and would precipitate from the bath. Oxidizing agents are incorporated to remove the soluble secondary ferrous phosphate because the ferrous phosphate inhibits coating formation.

Although all phosphating baths are acid in nature and to some extent attack the metal being coated, hydrogen embrittlement seldom occurs as a result of phosphating. This is primarily because all phosphating baths contain depolarizers or oxidizers that react with the hydrogen as it is formed and render it harmless to the metal. In some instances, however, zinc-phosphate processes, intended for use with rust-inhibiting oils for corrosion resistance or manganese-phosphate treatments, can cause hydrogen embrittlement because they may contain a minimum amount of depolarizers and oxidizers. A dwell time before use or mild heating may be needed to relieve embrittlement.

The acidity of phosphating baths varies, depending on the type of phosphating compound and its method of application. Immersion zinc phosphating baths operate in a pH range of 1.4 to 2.4, whereas spray zinc phosphating solutions can operate at a pH as high as 3.4, depending on the bath temperature. Iron phosphating baths usually operate at a pH of 3.8 to 5. Manganese phosphating baths operate in a pH range comparable to that of the immersion zinc phosphating solutions. Lead-phosphate solutions are usually more acidic than any of the others.

Zinc, iron, and manganese phosphating baths usually contain an accelerator, which can range from the mild oxidants, such as nitrate, to the more vigorous nitrite, chlorate, peroxide, or organic sulfonic acids. The purpose of these accelerators is to speed up the rate of coating, to oxidize ferrous iron, and to reduce crystal size. This is accomplished because of the ability of the accelerators to oxidize the hydrogen from the surface of the metal being coated. Phosphating solution can then contact the metal continuously, permitting completeness of reaction and uniformity of coverage. Accelerators have an oxidizing effect on the dissolved iron in the bath, thus extending the useful life of the solution. Some zinc and iron phosphating processes rely on oxygen from the air as the accelerator. Zinc phosphating baths for aluminum usually contain complex or free fluorides to accelerate coating formation and to block the coating inhibiting effect of soluble aluminum.

Applications

On the basis of pounds of chemicals consumed or tons of steel treated, the greatest use of phosphate coatings is as a base for paint. Phosphate coatings are also used to provide: (*a*) a base for oil or other rust-preventive material; (*b*) lubricity and resistance to wear, galling or scoring of parts moving in contact, with or without oil; (*c*) a surface that facilitates cold forming; (*d*) temporary or short-time resistance to mild corrosion; and (*e*) a base for adhesives in plastic-metal laminations or rubber-to-metal applications.

Phosphate Coatings as a Base for Paint

The useful life of any painted metal article depends mainly on (*a*) the durability of the organic coating itself, and (*b*) the adherence of the film to the surface on which it is applied. The primary function of any protective coating of paint is to prevent corrosion of the basis metal in the environment in which it is used. To accomplish this purpose, the method of preparing the metal should reduce the activity of the metal surface, so that underfilm corrosion is prevented at the interface between paint and metal.

When used as a base for paint films, phosphate coatings promote good paint adhesion, increase the resistance of the films to humidity and water soaking, and substantially retard the spread of any corrosion that may occur. A phosphate coating retards the amount of corrosion creep, because the coating is a dielectric film that insulates the active anode and cathode centers existing over the entire surface of the basis metal. By insulating these areas, corrosion of the surface is arrested or at least substantially retarded.

Table 1 Low-carbon sheet steel parts spray phosphate coated for paint finishing

Part	Area m²	Area ft²	Production, pieces/h
Zinc phosphate(a)			
Automobile body	74	800	50
Dryer shell	4.0	42.5	400
Cabinet back panel	1.2	12.7	700
Cabinet top	0.7	7.9	1400
Compressor housing	0.4	4.5	600
Motor access panel	0.2	1.6	4500
80-mm mortar shell	0.1	1.0	1000
Iron phosphate(b)			
Washing machine shell	4.9	52.9	330
Dryer top	1.8	12.7	660
Range side panel	0.9	9.7	660
Dishwasher door	0.8	8.4	660
Wiring channel	0.5	4.9	4950
Control housing	0.3	3.8	1980
Condenser cover	0.2	1.6	3330
Range gusset plate	0.1	0.84	4950
Conduit cover plate	0.03	0.31	8900

(a) 1.6 to 2.1 g/m² (150 to 200 mg/ft²). (b) 0.4 to 0.9 g/m² (40 to 80 mg/ft²)

Zinc phosphate coatings of light to medium weight (1.6 to 2.1 g/m² or 150 to 200 mg/ft²) and lightweight iron phosphate coatings (0.3 to 0.9 g/m² or 30 to 80 mg/ft²) are generally used for paint bases. Examples of products so treated are steel, galvanized and aluminum stampings for automobiles, household appliances, metal cabinets, and metal furniture (Table 1). Enamel that is baked at a temperature of 205 °C (400 °F) or higher can be successfully applied to phosphate-coated steel.

Phosphated surfaces to be painted should not be touched by bare hands or other parts of the body to ensure good adherence of the paint film. Body salts can contaminate phosphate coatings. Contaminated areas can be reflected as surface imperfections of the paint film and can decrease corrosion and humidity resistance.

Difference Between Metals Cleaned in Phosphoric Acid and Those Coated with Phosphate. Phosphoric acid metal cleaners usually consist of phosphoric acid and a water-soluble solvent, with or without a wetting agent. In the preparation of metal with such solutions, the purpose is to complete the following steps in a single operation: (*a*) remove oil, grease, and rust, and (*b*) provide a slight etch of the metal to promote the adhesion of paint. The cleaning solution must contain enough acid (15 to 20% phosphoric acid) and solvent to remove rust, oil, and grease. This concentration of phosphoric acid prevents the formation of any substantial phosphate coating.

When metal surfaces are to be phosphate coated, articles are first freed of rust and grease by suitable cleaning methods. Articles are then treated with a balanced dilute acid phosphate salt with a slight excess of acid (0.6 to 1.0% phosphoric acid), so that reaction of the acid with the metal results in the conversion of the surface to a refined crystalline phosphate coating. Tests conducted on steel cleaned with phosphoric acid have revealed that the phosphate film remaining from the cleaning operation averages only 0.05 to 0.10 g/m² (5 to 10 mg/ft²) of surface. In contrast, when steel is phosphated for painting by using standard zinc phosphating solutions, coating weights usually range from 1.08 to 4.3 g/m² (100 to 400 mg/ft²), depending on the solution and method.

Corrosion Protection

Conversion of a metal surface to an insoluble phosphate coating provides a metal with a physical barrier against moisture. The degree of corrosion protection that phosphate coatings impart to surfaces of ferrous metals depends on (*a*) uniformity of coating coverage, (*b*) on coating thickness, density, and crystal size and (*c*) the type of final seal employed. Coatings can be produced with a wide range of thickness, depending on (*a*) method of cleaning before treatment, (*b*) composition of the phosphating solution, (*c*) temperature, and (*d*) duration of treatment. In phosphating, no electric current is used, and formation of the coating depends primarily on contact between the phosphating solution and the metal surface and on the temperature of the solution. Consequently, uniform coatings are produced on irregularly shaped articles, in recessed areas, and on threaded and flat surfaces, because of the chemical nature of the coating process.

The affinity of heavy phosphate coatings for oil or wax is used to increase the corrosion resistance of these coatings. Frequently, phosphate-coated articles are finished by a dip in nondrying or drying oils that contain corrosion inhibitors. The articles are then drained or centrifuged to remove the excess oil.

Medium to heavy zinc phosphate coatings, and occasionally, heavy manganese phosphate coatings are used for corrosion resistance when supplemented by an oil or wax coating. Zinc phosphate plus oil or wax is usually used to treat cast, forged, and hot rolled steel nuts, bolts, screws, cartridge clips, and many similar items. Manganese phosphate plus oil or wax is also used on cast iron and steel parts.

Phosphate Coating as an Aid in Forming Steel

The power used in deep drawing operations sets up a great amount of friction between the steel surface and the die. The phosphate coating of steel as a metal-forming lubricant, before it is drawn (*a*) reduces friction, (*b*) increases speed of the drawing operations, (*c*) reduces consumption of power, and (*d*) increases the life of tools and dies. When phosphate-coated steel is used in drawing seamless steel tubing, the resulting decrease in friction is so pronounced that greater reduction of tube size per pass is possible. This reduction may be as great as one half.

Reduction in the number of draws and anneals in deep forming results in economy of operation. Conversion of a steel surface to a nonmetallic phosphate coating permits the distribution and retention of a uniform film of lubricant over the entire surface of the steel. This combination of lubricant and nonmetallic coating prevents welding and scratching of steel in the drawing operations, and greatly decreases rejections.

Zinc phosphate coatings of light to medium weight are applied to steel to aid in drawing and forming operations. The phosphated surface is coated with a lubricant (such as soap, oil, drawing compound, or an emulsion of oil and fatty acid) before the forming operation. The zinc phosphate surface, which prevents metal-to-metal contact, makes it practical to cold form and extrude more difficult shapes than is possible without the coating. Table 2 lists and describes some products that are zinc phosphate dip or spray coated before being cold formed.

Wear Resistance

Phosphating is a widely used method of reducing wear on machine elements. The ability of phosphate coating to reduce wear depends on (*a*) uniformity of the phosphate coating, (*b*) penetration of the coating into metal, and (*c*) affinity of the coating for oil. A phosphate coating permits new parts to be broken

Table 2 Applications of zinc phosphate dip or spray coating to facilitate cold forming

Bath used for all applications listed is zinc phosphate bath accelerated with nitrous oxide

Part	Steel	Area m^2	Area ft^2	Coating weight g/m^2	Coating weight mg/ft^2	Production, pieces/h	Sequence of operations
Spark plug body	1110	0.0002	0.002	4.3-6.5	400-600	500	(a)
Universal-joint bearing cup	1010	0.005	0.05	4.3-6.5	400-600	2000	(a)
Truck wheel nut	1008	0.007	0.08	4.3-6.5	400-600	1000	(a)
Piston pin	5015	0.009-0.01	0.1-0.15	5.4-7.0	500-650	2600	(a)
Standard-transmission output shaft	4028	0.009-0.01	0.1-0.15	4.3-6.5	400-600	300	(a)
Rocket-nozzle plate (69.85 mm or 2.75 in.)	4130, 4140	0.05	0.5	4.3-6.5	400-600	150	(b)
Mortar shell (80-mm or 3.2-in.)	1010	0.07-0.1	0.8-1.2	4.3-6.5	400-600	4000	(c)
Cartridge cases (75-mm or 3-in.)	1030	0.08-0.3	0.9-3.0	4.3-6.5	400-600	1000	(b)

(a) Alkaline wash, rinse, activating rinse, phosphate, rinse, neutralizing rinse, oil dip. (b) Sulfuric acid pickle, rinse, rinse, phosphate, rinse, rinse, oil dip. (c) Alkaline wash, rinse, sulfuric acid pickle, rinse, rinse, phosphate, rinse, oil dip

in rapidly by permitting retention of an adequate film of oil on surfaces at that critical time. In addition, the phosphate coating itself functions as a lubricant during the high stress of break-in.

Heavy manganese phosphate coatings (10.8 to 43.0 g/m^2 or 1000 to 4000 mg/ft^2), supplemented with proper lubrication, are used for wear-resistance applications. Parts that are manganese phosphate coated for wear resistance are listed in Table 3.

When two parts, manganese phosphated to reduce friction by providing lubricity, are put into service in contact with each other, the manganese coating is smeared between the parts. The coating acts as a buffer to prevent galling or, on heavily loaded gears, to prevent welding. The phosphate coating need not stand up for an extended length of time, because it is in initial movements that parts can be damaged and require lubricity. For example, scoring of the mating surfaces of gears usually takes place in the first few revolutions. During this time, the phosphate coating prevents close contact of the faces. As the coating is broken down in operation, some of it is packed into pits or small cavities formed in gear surfaces by the etching action of the acid during phosphating.

Long after break-in, the material packed into the pits or coating that was originally formed in the pits prevents direct contact of mating surfaces of gear teeth. In addition, it acts as a minute reservoir for oil, providing continuing lubrication. As work hardening of the gear surfaces takes place, the coating and the etched area may disappear completely, but by this time scoring is unlikely to occur.

Table 3 Parts immersion coated with manganese phosphate for wear resistance

Coating weights range from 10.8 to 43.0 g/m^2 (1000 to 4000 mg/ft^2)

Part	Material	Coating time, min	Supplementary coatings
Components for small arms, threaded fasteners(a)	Cast iron or steel; forged steel	15-30	Oils, waxes
Bearing races	High-alloy steel forgings or bar stock	7-15	Oils, colloidal graphite
Valve tappets, camshafts	Low-alloy steel forgings or bar stock	7-15	Oils, colloidal graphite
Piston rings	Forged steel, cast iron	15-30	Oils
Gears(b)	Forged steel, cast iron	15-30	Oils

(a) Coating may be applied by barrel tumbling. (b) Coating weights from 5.4 to 43.0 g/m^2 (500 to 4000 mg/ft^2)

Phosphate-Coated Ferrous Alloys

In general, the stainless steels and certain alloy steels cannot be successfully phosphate coated. All other steels accept a coating, with difficulties experienced in the coating process varying with alloy content. Most cast irons are readily coated, and alloy content has little effect on their coatability.

Steels

Most phosphate-coated steel is low-carbon flat rolled material used for applications such as sheet metal parts for automobiles and household appliances, and phosphating processes have been designed for coating such material. Steels with carbon contents in the range specified for 1025 to 1060 inclusive are suitable for phosphating if the silicon content is held to normal limits. Steels with higher carbon contents, in the range from 1064 to 1095, may require the following modifications of phosphating processes to produce satisfactory results: (*a*) increasing time, (*b*) increasing temperature, or (*c*) increasing solution strength. Copper content up to 0.3% in low-carbon steel, the normal limit for copper-bearing steel, is not a deterrent to phosphate coating. The addition of copper, by itself, at about 0.5% causes surface checking of steel during hot rolling. This acts as a restriction on the amount of copper that may be present to serve as a deterrent to phosphating.

Low-alloy high-strength steel, provided nickel or chromium does not exceed 1%, can be successfully phosphated. Generally, with some modification, chromium content of up to 9% can be tolerated while still depositing a phosphate coating. Nickel-chromium and chromium stainless steels are not recommended for phosphate coating. For some applications, however, oxalate coating processes are used.

Because electrical steels used in motor laminations and electrical transformers have a silicon content in the range of 1.2 to 4.5%, they are not recommended for phosphate coating by normal phosphating processes. These require processes accelerated by the use of fluoride compounds or special dried in-place salts of phosphates.

Low-carbon steels annealed in a properly controlled atmosphere to provide a clean, oxide-free surface, are readily phosphated. Temper rolled annealed low-carbon steels are the most readily phosphated of all steels. Cold reduced or cold rolled, full-hard, low-carbon steels readily accept phosphate coatings.

Low-carbon hot rolled steel, and normalized and pickled steel, if thoroughly rinsed after pickling, phosphates well. Excessive amounts of residual pickling salts (sulfates) can interfere with normal phosphating. Pickling residue on cold reduced or cold rolled steels seldom presents problems, because of the extensive processing that follows the pickling operation. Cold reduction of 30 to 70% spreads the residue over large areas. Cleaning and scrubbing of a cold reduced strip, followed by annealing and temper rolling, remove or dilute surface contaminants. Phosphating processes that provide for relatively long-time and high-temperature treatments are the least sensitive to small variations in alloy composition and surface conditions.

Galvanized Steel. Many parts produced from galvanized sheet steel, such as certain automotive stampings, and some appliances, require a phosphate coating as a base for a subsequent paint film. Phosphating imparts superior resistance to corrosion and greater ability to retain paint to galvanized sheet and strip steel by converting the surface to an insoluble phosphate coating. Galvanized steel can be readily phosphated provided the surface of the plate has not been passivated by a chromate-based solution. The passivated surface of the chromate-treated material resists the action of a phosphating solution. Treatment of such passivated surfaces requires the use of an alkali-permanganate solution or, depending on the age and degree of passivation, removal with strong alkaline cleaners.

Cast Irons

Gray, ductile or malleable iron castings are readily phosphated. The ability of a cast iron to accept a phosphate coating is not affected by alloy content, but hinges primarily on two requirements: (*a*) a clean surface, and (*b*) a metal temperature approximately equal to that of the phosphating bath. Machined surfaces need no further cleaning; however, cast surfaces can be prepared by removing scale and sand by blasting or other cleaning.

Phosphating bath temperatures are not critical for cast iron. Acceptable coatings can be obtained in baths ranging from 71 to 96 °C (160 to 205 °F). Often, lower temperatures are viable. A problem usually exists in raising the temperature of a casting, particularly one with heavy sections, to approximate the temperature of the bath. Preheating heavy castings to the temperature of the bath minimizes or eliminates excessive pickling action in areas that require a long time to reach the temperature of the phosphating solution.

Manganese phosphate coatings, applied only by immersion, are easily deposited on cast iron surfaces. The normally coarse crystal breaks down readily to provide temporary lubrication during break-in. If this is not sufficient, castings may be given a supplemental oil dip. Interstices between the coarse crystals hold sufficient oil to provide adequate short-time lubrication.

Because manganese phosphate crystals on cast iron build up rapidly to thicknesses of as much as 25 μm (1 mil), machined dimensions carrying close tolerances may be altered significantly by coating. If this is not acceptable, the dimension can be reduced by removing some of the coating thickness. If a fine crystalline structure is necessary, the presence of an appropriate oil on the surface before phosphating, such as film remaining after emulsion cleaning, refines the normally coarse phosphate crystal. Preferably, special manganese phosphate activating chemicals are used in a water rinse preceding the manganese phosphate process.

Aluminum

Zinc phosphate coatings, applied via spray, are easily deposited upon aluminum surfaces provided fluoride ion is present in the bath. Sodium or potassium salts are also present to prevent the buildup of soluble aluminum in the bath, which inhibits coating formation. In the processing of a metal mix of aluminum, steel, and galvanized steel, separate fluoride additions to the bath may be required if the metal mix consists of greater than 10% aluminum. Coating weights range from 0.27 to 2.2 g/m^2 (25 to 200 mg/ft^2).

Process Details

The application of a phosphate coating for paint-based application normally comprises five successive operations: (*a*) cleaning, (*b*) rinsing, (*c*) phosphating, (*d*) rinsing, and (*e*) chromic acid rinsing. Some of these operations may be omitted or combined, such as cleaning and coating in one operation. Additional operations may be required, depending on the surface condition of parts to be phosphated or on the function of the phosphate coating. Parts exemplifying these exceptions are:

- Heavily scaled parts, which may require pickling before cleaning
- Parts with extremely heavy coatings of oil or drawing compounds, which may require rough cleaning before the normal cleaning operation
- Parts that are tempered in a controlled atmosphere before being phosphated, may not require cleaning and rinsing before phosphating
- Parts that are phosphated and later oiled for antifriction purposes, which may have the chromic acid rinse omitted, because corrosion resistance is not required. Some rust-preventive oils negate the need for a chromic rinse while still providing excellent corrosion resistance
- Automotive parts when electrodeposition of a primer is involved. A deionized water rinse is required after the chromic acid rinse

Hexavalent chromic acid for a passivating rinse is no longer used in some plants because of strict effluent controls imposed by the Environmental Protection Agency (EPA). Other, less restrictive materials, such as phosphoric acid and various proprietary compounds are being used.

Cleaning

Because the chemical reaction that results in the deposition of a phosphate coating depends entirely on the phosphating solution making contact with the surface of the metal being treated, parts should always be sufficiently clean to permit the phosphating solution to wet the surface uniformly. Cleaning may involve chemical action or mechanical action, or both. Precautions must be taken to avoid carryover of cleaning materials into phosphating tanks. This is particularly true for alkaline cleaners, which can neutralize the acid phosphating solutions, rendering them useless. Other cleaning compounds can contaminate the bath, causing poor quality coatings, such as complex phosphates.

Soil that is not removed can act as a mechanical barrier to the phosphating solution, retarding the rate of coating, interfering with the bonding of the crystals to the metal, or completely preventing solution contact. Some soils can be coated with phosphate crystals, but adherence of the coating will be poor and affect the ability of a subsequent paint film to remain continuous.

Ordinary mineral oil is usually easy to remove and presents no problem to the phosphating processes. However, with the use of more complex materials in forming metal, in rustproofing metal, in stripping paint, and in removing scale, cleaning has become a major consideration in any phosphating operation. Materials, such as cutting oils, drawing compounds, coolants, and rust inhibitors can react with the basis metal and form a deleterious film.

Several solutions to phosphating problems that arose because of improper or inadequate surface preparation, have been used in actual production situations. In one plant, irregularities occurred in the thickness and crystal size of phosphate coatings on deep drawn parts. These irregularities varied in severity, but were sometimes acute enough to cause roughness in the subsequently applied paint film. When fresh cleaning solutions were used, the problem was somewhat alleviated, but the irregularities recurred after cleaners had been in use for only a short time. Investigation revealed a variety of drawing compounds were being used on the parts, and that each contributed in some degree to the contamination of the cleaner and the inability of the cleaner to remove soil. Some drawing compounds react with the steel surface, forming oxides not removable by regular cleaners. Other drawing compounds cause excessive, undesirable foaming when in contact with cleaners. Still other drawing compounds form into small globules in the cleaner and are redeposited on the metal and not completely removed in the subsequent rinse. The degree of cleanness of the parts was reflected in the degree of variation of the phosphate coating. The problem was solved by using a different cleaner with a different detergent system plus an increase in caustic content, and by selecting drawing compounds that, while still performing as required, would be effectively removed by this cleaner without contaminating it.

Another solution was used during an extended production run of 80-mm (3.2-in.) mortar shell cases. In this situation, dip phosphoric acid pickling was replaced by spray sulfuric acid pickling as the final cleaning operation before phosphating for a paint base. Pickling was a three-stage operation: (*a*) pickling, and (*b*) two rinses. Between pickling and phosphating, a 100% hydrostatic test was required for assurance that the pickling solution had not opened any pinholes through the soldered tail plugs of the shell cases. With sulfuric acid pickling, chromic acid was added to the second rinse as a rust inhibitor to protect parts during test and transfer. The phosphate coatings obtained after this changeover were of poor quality because little or no coating was deposited on the outsides of the shell cases, and inside surfaces rusted badly. All materials and processes were checked and found to be in order. Sample panels that were processed, but not pickled, accepted a satisfactory coating. The problem was traced to a passive film on the shell surfaces, left there by the chromic acid in the final rinse after pickling. Replacing the chromic acid rinse with an alkaline sodium nitrite rinse solved the problem.

In another plant, manganese phosphate coatings had been successfully applied to a variety of machined steel parts. One of these parts, produced in high volume, was a valve tappet of low-alloy steel, for which a surface finish of 0.1 to 0.3 μm (4 to 10 μin.) was required. For no apparent reason, difficulty was suddenly encountered in the form of mottled, noncrystalline coatings, and occasional bare spots on the tappets. Other parts were satisfactorily coated. It was discovered that a change in polishing compound had been made to facilitate obtaining the required finish on the tappets. Carrier wax in the new compound contained a larger amount of unsaturated material than was in the wax in the previous polishing compound. These unsaturates are more readily oxidized to insoluble compounds by the heat generated in polishing than are fully saturated material. Reverting to the original polishing compound corrected the difficulty.

In another instance, an alkaline cleaner at a concentration of 7.5 g/L (1 oz/gal) was used in the cleaning stage in a zinc phosphating line for processing sheet steel stampings. Stampings were coated with mill oil and drawing oil, and cleaning was satisfactorily accomplished. On certain new parts, however, because of a difficult drawing operation, a pigmented drawing compound was required that consisted of emulsified palm oil and powdered French talc. The cleaner would not remove this drawing compound sufficiently to permit acceptable coatings to be deposited. It was found that the cleaner was removing the oil but leaving the talc on the parts. Increasing the concentration of the alkali in the cleaner to 15 g/L (2 oz/gal) resulted in no improvement. However, when the temperature of the cleaner was lowered to 71 °C (160 °F), both oil and talc were removed, permitting satisfactory phosphate coatings. This is counter to the concept that cleaning efficiency increases with the temperature of the cleaner.

Parts that have been tempered in air or a controlled atmosphere, as the last operation before phosphating, usually require no cleaning before being phosphated. The blue oxide film imparted by the normal tempering operation is not detrimental to phosphating. However, if tempering produces a scaly or sooty surface, or if scale or soot is produced in the heat treating furnace and is not removed before tempering, the parts must be descaled by acid pickling, tumbling, or blasting. Incorporation of crystal refiners (titanation) into the alkaline cleaner promotes the deposition of a dense, finely crystalline zinc phosphate coating. Overheating (greater than 66 °C or 150 °F) of the activated cleaner stage inhibits the crystal refinement effect.

Rinsing After Cleaning

In the past, water at 71 to 82 °C (160 to 180 °F) ordinarily was used for rinsing parts after cleaning and before phosphating. Hot water is in effect an additional cleaner, serving to remove cleaning compounds that adhere to part surfaces. Ambient temperature rinses are now often used. Parts may be rinsed by immersion or spray. A single rinse tank or spray stage is usually adequate for rinsing simple parts, and a minimum rinse time of 30 s is normally required. An additional spray rinse stage should be added for parts with blind holes or deep recesses.

Immersion Rinsing. Rinse tanks should be equipped to provide adequate agitation of rinse water to increase rinsing efficiency. Agitation may be accomplished using compressed air at low pressure, distributed through evenly spaced holes in pipes laid along the bottom of the tank. However, where compressed air is not available, pumping rinse water through similar pipes can provide suitable agitation. Fresh water may be continuously added to the tank through such pipes to provide agitation, but a siphon-breaker must be installed in the supply line to prevent siphoning contaminated rinse water into the water-supply system. The supply of make-up water should be planned to provide adequate rinsing without wasting water. Relatively pure waters, containing less than 150 ppm total solids, require less replacement water than harder or impure waters containing 400 to 600 ppm total solids.

Use of solenoid valves, controlled by conductivity meters in water supply lines to rinse tanks, can maintain adequate rinse water purity at minimum waste. Another effective method of improving rinse water quality is to supply make-up water through spray nozzles. This process causes the fresh water to be the last water to hit the parts as they are being withdrawn from a dip rinse or carried from a spray rinse station. Rinse water make-up added to a water rinse tank other than by the methods discussed above should be supplied to the end of the tank opposite the overflow trough. Water containing appreciable quantities of chlorides, fluorides, or sulfides may not provide good rinsing. The length of time parts are allowed to remain in the rinse tank depends on their complexity and on material to be rinsed away.

Spray Rinsing. Vertical and horizontal spacing and size of nozzles in a spray rinse tunnel are determined by size and nature of parts being processed and speed of conveyor carrying parts through the tunnel. A pressure of 70 to 140 kPa (10 to 20 psi) at the nozzle is normally adequate. Minimum pump volume capacity is determined by multiplying the volume capacity of the nozzle at the desired pressure by the number of nozzles required and adding an allowance for losses because of piping length and restrictions. A spray rinse tank should hold a minimum volume of 2½ times the volume of solution piped through the nozzles per minute.

Rinse solutions should be piped from the pump or pumps through large main headers to vertical drop lines containing the nozzles. To assist in scale removal, drop lines should be fitted on the bottom with removable caps. Pressure that is too high may be relieved by drilling suitable holes in the center of bottom caps. This also enables excess rinse solution to flush out scale and sludge continuously. A hot water spray rinse station is shown schematically in Fig. 1. The rinse before phosphating should be maintained slightly alkaline (7.5 to 9.0 pH) to prevent rust-blushing of parts. A rinse containing crystal refiners is recommended just before the phosphate stage, especially when a nontitanated cleaner is used.

Fig. 1 Hot water rinse station in a spray phosphating line

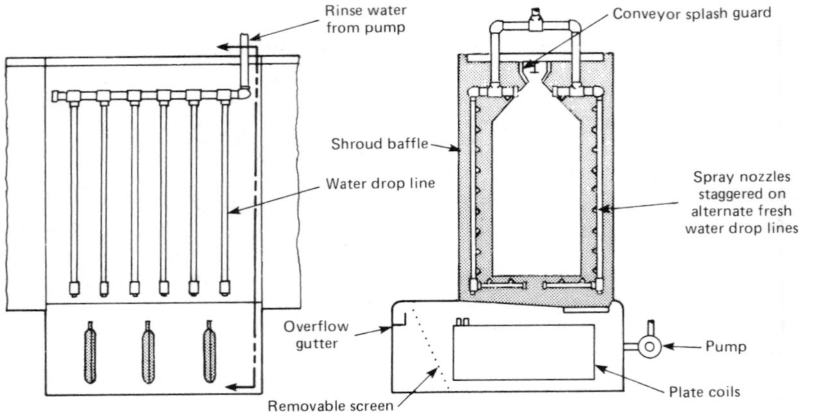

Phosphating Methods

Phosphate coatings may be applied to a surface by either immersion or spray, or a combination of immersion and spray. There has been a modern trend worldwide to the immersion treatment of automobile bodies using zinc-phosphate coating processes. This is usually a combination of spray and immersion. The work is sprayed as it enters the phosphate tank as well as when it exits. Occasionally, a surface may be coated by brushing or wiping, but these methods are seldom used.

Immersion. All three types of phosphate coatings: (*a*) zinc, (*b*) iron, and (*c*) manganese, can be deposited by immersion. Immersion is applicable to racked parts, barrel coating of small parts, and continuous coating of strip. In general, smaller parts are more economically coated by immersion than by spraying. Small parts, such as springs, clips, washers, and screws that are produced in large volume, can be coated efficiently only in an immersion system. Such parts are loaded into drums that are rotated at approximately 4 rev/min after they are immersed in the phosphating solution. Small parts may be placed in a basket and immersed, without rotation, in the bath for coating. This method generally is not completely satisfactory, because no phosphate is deposited where parts contact each other or the basket. It is used as a stopgap method or when volume is too low to justify the use of rotating drums.

Low-volume larger parts are immersed manually in a tank. Certain large parts produced in large volume, but whose shape does not encourage complete coverage by spray phosphating, may be coated by immersion. Intricately shaped parts, such as hydraulic valves or pump bodies that have areas inaccessible to spray, are immersion coated. Either the immersion or the spray method may be used to deposit heavy zinc phosphate coatings used as aids in cold extruding or drawing. However, the immersion system usually provides a heavier coating. Shell casings formed by cold extrusion are first coated by the immersion system to produce the heavy coating required for the cold extrusion of the metal. After finish machining, the shell casing can be either spray or immersion coated for a paint base. The immersion system usually is preferred, because of the necessity of coating internal areas.

Although a manually operated immersion system requires very little floor space, a conveyorized immersion system requires more floor space than a conveyorized spray system for comparable production quantities. When parts are of comparable size, the immersion system cannot equal the production output of a spray system. An advantage of an immersion system is that heat required is much less than for a spray system because of the heat lost from the sprays. The use of an immersion system for automotive bodies provides phosphate coverage in areas not accessible by spraying, namely, box sections. In the immersion process, agitation of the solution accelerates coating formation.

Spray. Zinc and iron phosphate coatings are applied by the spray method, although manganese phosphate is not. The spray method is used to apply a phosphate coating to racked parts, such as panels for household appliances, or to a continuous strip. Occasionally, baskets of parts are passed through a spray system, but this is not a preferred method. Spray phosphating, because of the equipment required, is usually most applicable to high-volume coating of parts.

It is easier to control the coating solution for iron phosphating in a spray system, and the resulting coating is generally of better quality than the coating obtained from an immersion system. Zinc phosphate coatings produced by a spray system are usually lighter in weight than those produced by immersion. In addition, different zinc phosphate crystalline structures may result from spraying compared with immersion.

Phosphating Time. In general, the spray method produces a given coating weight at a faster rate than the immersion method. In spray zinc phosphating, a coating of 1.6 to 2.1 g/m^2 (150 to 200 mg/ft^2) normally can be obtained in 1 min or less, whereas obtaining a coating of this weight by the immersion method may require as much as 2 to 5 min. For galvanized steel treated in coil lines, zinc phosphate coatings are produced in times as short as 3 to 5 s and iron phosphate coatings on cold rolled steel in from 5 to 10 s. A 1-min spray application of one iron phosphating solution would result in a coating of approximately 0.3 to 0.4 g/m^2 (27.9 to 37.2 mg/ft^2). It is estimated that it would require 2 to 3 min to produce the same coating weight by immersion. Bath parameters are all interrelated, however. In some operations, coatings can be effectively deposited in 3 to 5 s.

Table 4 Operating temperature ranges for phosphating solutions in phosphating applications

Phosphate coating/method	Metal treated	Reason for treatment	Operating temperature °C	Operating temperature °F
Medium iron, immersion	Steel	Paint bonding	60-82	140-180
Heavy zinc, immersion	Steel	Corrosion resistance	88-96	190-205
Medium zinc, immersion	Steel	Paint bonding	32-82	90-180
Medium zinc, immersion	Steel plated with zinc or cadmium	Paint bonding	60-82	140-180
Medium zinc, spray	Sheet steel	Paint bonding	38-60	100-140
Medium zinc, spray	Steel	Cold drawing	60-74	140-165
Medium zinc, spray	Galvanized steel	Paint bonding	49-60	120-140
Manganese, immersion	Steel	Wear resistance	93-99	200-210

The weight of manganese phosphate coatings on steel surfaces is a function of immersion time, as indicated by the curve in Fig. 2. The slope of this curve can vary. The time required to obtain a specific coating weight in a range of 5.4 to 32.3 g/m^2 (500 to 3000 mg/ft^2) can vary from 2 to 40 min, depending on such factors as the type and hardness of the steel being coated, and methods of precleaning and pretreatment. Exposure to the phosphating solution for a shorter time than recommended usually results in a coating that is incomplete or too thin, or both.

Operating Temperature. Although operating temperatures of different phosphating solutions may range from 32 to 99 °C or 90 to 210 °F (Table 4), individual solutions are compounded to operate at maximum efficiency within specific temperature limits. The trend in recent years has been to lower operating temperatures. A phosphating solution should be held within the specified operating temperature range. If the solution is permitted to operate below the minimum recommended temperature, the phosphate coating is thin or nonexistent. If the temperature of the solution exceeds the recommended maximum, the coating builds up excessively and has a nonadherent, powdery surface, and the bath solution may become unbalanced, with excessive sludge and scale the result. Special low-temperature solutions are available for applying iron or zinc phosphate (see discussion later in this article on low-temperature coatings).

Solution temperature influenced the results obtained in manganese phosphating small cast iron parts. Coatings deposited at the beginning of each day were thin and red-tinged. However, after the tank had been in operation for about 1½ h, conventional coatings (heavy and dark gray) were obtained, and no further trouble was experienced for the rest of the day. Bath analysis revealed the phosphating solution had an unusually high concentration of free acid at the start of each day's operation, but the concentration was normal at shutdown time. A review of past records indicated that this condition had not previously existed. Investigation revealed the condition resulted from improper and excessive preheating of the bath before parts were processed. Only the steam bypass was being turned on, thus circumventing automatic temperature controls, which caused the solution to boil, upsetting composition and thereby affecting coating characteristics. A return to correct preheating procedures ended the problem.

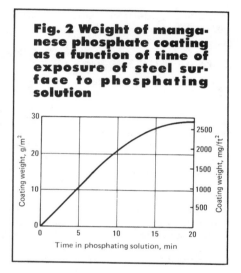

Fig. 2 Weight of manganese phosphate coating as a function of time of exposure of steel surface to phosphating solution

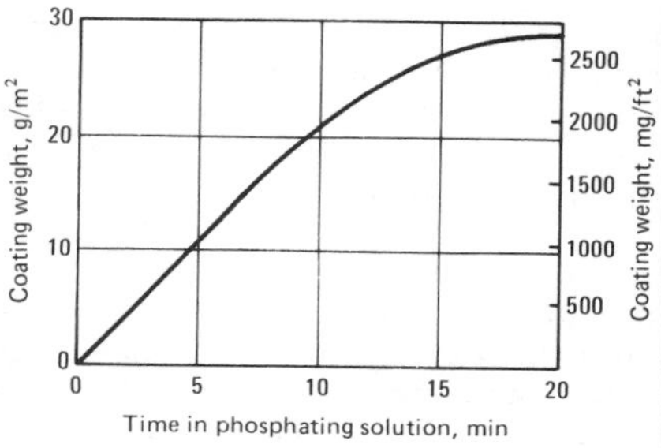

Rinsing After Phosphating

Parts must be rinsed after being phosphated to remove active chemicals from the phosphating solution that remain on the surface of coated parts. Any chemicals not removed may cause corrosion of parts, or blistering of a subsequent paint film. Any phosphating chemicals carried over into the chromic acid rinse may contaminate the solution used as a rinse. Rinsing after phosphating never should be hot. The temperature should range from 21 to 49 °C (70 to 120 °F), preferably maintained on the low side. A rinse that is too warm may set the residual chemicals and cause them to adhere to phosphate crystals, resulting in (*a*) a rough coating, (*b*) whitish appearance of coating, and (*c*) lower corrosion resistance. Usually, only one rinse is required. If the water supply is so high in mineral content that a residue remains on the parts after rinsing, a rinse in deionized water may be required.

Chromic Acid Rinsing

Most phosphated parts that are used as a base for paint are given a post-treatment following the post-phosphating rinse. These post-treatments vary from simple chromic-acid solutions to complex proprietary formulas that may be free of chromium entirely. Because of difficulty experienced when these post-treatments are allowed to dry on a phosphated part because of concentrations of the post-treatment at the lower edges and around openings such as holes or slots, the excess post-treatment should be removed with a deionized water rinse. Better proprietary post-treatments allow removal of the excess with deionized water without substantially decreasing corrosion resistance of the painted system while, at the same time, retaining good humidity and physical test results associated with conventional post-treatment. Environmental and health concerns have resulted in increasing interest in and the development of improved chromium-free post-treatments for paint-based applications. In the case of heavy zinc-phosphate coating used with oil for corrosion resistance, chromic acid post-treatment may or may not be used, depending on the quality and nature of the rust-preventative oil applied thereafter.

Zinc or manganese phosphate coatings applied to reduce friction usually do not receive a chromic acid rinse, because they are not applied for corrosion resistance. Rather, oil films are normally applied after phosphating to increase antifriction properties of coatings. On parts phosphated to assist in cold extrusion or drawing, application of drawing lubricants usually supplants the chromic acid rinse.

Chemical Control of Phosphating Processes

An efficiently operated phosphating line includes close chemical control of all materials used. Even the mineral content of plain water rinses may need to be controlled to avoid leaving a residue on parts. To obtain satisfactory phosphate coatings on steel surfaces, phosphating solutions must be chemically controlled within limits. These limits vary depending on the specific phosphating concentrate used. Solutions should be tested on a regular schedule. Frequency of tests is determined by the work load of the phosphating line.

Zinc Phosphating Solutions. When zinc phosphating solutions become unbalanced, the results are poor coatings, excessive sludge buildup, and insufficient coating weights. Several chemical tests are usually made on a zinc phosphating solution, used for paint-based application, to determine its suitability for coating. These are tests of (*a*) total acid value; (*b*) accelerator content; (*c*) free acid; (*d*) in the case of heavy zinc phosphate coatings for use with oil, the iron concentration (ferrous iron); and (*e*) zinc concentration.

Total Acid Value. Zinc phosphate solutions have a total acid value established that should be maintained for good performance. One regularly used solution is controlled at 25 to 27 points. To determine the total acid value, a 10-mL sample of the solution is titrated with 0.1*N* sodium hydroxide (1 mL equals 1 point), 1 using phenolphthalein as an indicator. The end point is reached when the solution changes from colorless to pink.

Free-Acid Value. Zinc phosphate solutions have a free-acid value established that should be maintained for satisfactory performance. To determine the free-acid value, a 10-mL sample of solution is titrated with 0.1 normal sodium hydroxide, using bromphenol blue or methyl orange indicator. The end point is reached when the solution changes from yellow to greenish-blue, for the former.

Accelerator Test. Sodium nitrite is used as an accelerator in some zinc phosphate solutions. It is usually controlled at 3.0 points. Before the test for sodium nitrite is made, phosphate solution should be tested for absence of iron. This is done by dipping a strip of iron-test paper in phosphate solution. If paper does not change color, no iron is present in the solution. If the paper changes to pink, however, iron is present, and small additions of sodium nitrite are then made until an iron-test paper shows no change. The sodium nitrite test is made using a 25-mL sample of the phosphate solution. From 10 to 20 drops of 50% sulfuric acid are added carefully to the solution and it is then titrated with 0.042*N* potassium permanganate. The end point is reached when the solution turns from colorless to pink (1 mL equals 1 point). The sodium nitrite test may also be performed using a gas evolution apparatus. After filling the apparatus with the bath solution, sulfamic acid 4 g (0.14 oz) is added to the solution. Evolution of gas into the calibrated section of the apparatus provides the direct reading (in millilitres) of sodium nitrite in the bath. The millilitre of gas is equivalent to the millilitre reading obtained with the potassium permanganate ($KMnO_4$) titration procedure.

Iron Concentration. Because iron is constantly being dissolved from parts being zinc phosphated, the concentration of iron may build up until the efficiency of the solution is impaired. Some zinc phosphating solutions operate best when the iron concentration is maintained between 3 and 4 points. Production experience with a particular solution will indicate whether the iron content can be expanded without affecting the quality of the coating. To determine the iron content, a 10-mL sample of solution is first acidified with a sufficient amount of a 50% mixture of sulfuric and phosphoric acid to assure a low pH while titrating (2 or 3 drops may be sufficient). The solution is then titrated with 0.2*N* potassium permanganate until a permanent pink color is obtained (1 mL equals 1 point). This titration is used for immersion zinc phosphating solution. Spray zinc phosphating usually does not involve a buildup of iron in the solution because of the oxidizers that are present. Immersion zinc phosphate solutions generate, in situ, sufficient nitrite to prevent iron buildup. Phosphate coatings formed with iron in the bath usually do not prevent galling during cold heading processing as well as phosphate baths operated without iron in the bath (Toner side). Immersion zinc phosphating baths are usually controlled by a total and free-acid titration and acid ratio, total acid divided by free-acid values.

Iron Phosphating Solutions. If recommended chemical limits are not maintained in iron phosphating solutions, results are (*a*) low coating weights, (*b*) powdery coatings, or (*c*) incomplete coatings. To maintain required balance in iron phosphating solutions, titration checks are made to determine total acid value and acid consumed value.

Total acid value is determined by titration of a 10-mL sample of phosphating solution with 0.1*N* sodium hydroxide, using phenolphthalein as an indicator. The end point is reached when the solution changes from colorless to pink. The number of millilitres of the 0.1*N* sodium hydroxide is the total acid value, in points, of the phosphating solution. A normal concentration would be 10.0 points.

Acid consumed value is determined by titration of a 10-mL sample of phosphating solution with 0.1*N* hydrochloric acid, using bromocresol green indicator. The end point is reached when the solution changes from blue to green. A normal range for the acid consumed value for a solution with a 10-point total acid value would be from 0.0 to 0.9 mL (0.03 fluid oz) of 0.1*N* hydrochloric acid.

Manganese phosphating solutions used to produce wear-resistant and corrosion-protective coatings are maintained in balance by control of (*a*) total acid value, (*b*) free-acid value, (*c*) acid ratio, and (*d*) iron concentration. Because the phosphate solutions are acid, these values are determined by titration methods using a standard basic solution. Frequency of control checks on manganese phosphating solutions depends on the amount of work being processed through the tank and on the volume of the solution. However, one to two checks per shift would be sufficient.

Total acid value is determined by titration of a 2-mL sample of phosphating solution with 0.1*N* sodium hydroxide, using phenolphthalein as an indicator. The end point is reached when the solution changes from colorless to pink.

Free-acid value is determined by titration of a 2-mL sample of phosphating solution with 0.1*N* sodium hydroxide, using bromophenol blue indicator. The end point is reached when the solution color changes from yellow to blue/violet.

Acid Ratio. To obtain satisfactory coatings, the ratio of total acid to free acid contents of manganese phosphating solutions should be maintained within certain limits. For a solution with a 60-to-70 point total acid value, this ratio should be between 5.5 to 1 and 6.5 to 1. Low-ratio solutions produce (*a*) incomplete coatings, (*b*) poorly adherent coatings, or (*c*) coatings with a reddish cast. High ratio solutions also result in poor coatings.

Iron Concentration. Because iron is continually dissolved from parts going into the phosphating bath, the concentration of ferrous iron in the bath gradually builds up. Some manganese phosphate coating problems that can be traced to high iron concentrations are (*a*) light gray instead of dark gray to black coatings, (*b*) powdery coatings, and (*c*) incomplete coatings in a conventional time cycle.

Concentration limits of iron depend on the type, hardness, and surface condition of the steel being treated. A manganese phosphating bath operates satisfactorily with an iron concentration ranging from 0.2 to 0.4%. Production experience indicates whether iron concentration limits can be expanded without affecting the quality of the coating. To determine iron concentration, a 10-mL sample of phosphating solution is used. To this sample 1-mL of 50% sulfuric acid is added. Solution is then titrated with 0.18*N* potassium permanganate. The end point is reached when solution changes from colorless to pink. One millilitre of the 0.18*N* potassium permanganate is equivalent to 0.1% iron.

Iron Removal. If iron removal becomes necessary, the ferrous iron in the solution is oxidized with hydrogen peroxide, which causes iron to precipitate as ferric phosphate and also liberates free acid in the bath. Because free acid in the bath increases and lowers the acid ratio, the liberated free acid should be neutralized by adding manganese carbonate. The approximate amount of hydrogen peroxide needed to lower the concentration of iron by 0.1% is 125 mL/100 L of solution. In this instance, about 450 g (1 lb) of manganese carbonate is needed to neutralize the liberated free acid. Iron removal may be unnecessary if the square footage of steel being processed and the volume of the phosphate bath limit the amount of iron buildup.

Chromic Acid or Other Post-Treatment Solutions. Control of chromic acid or other post-treatment solutions vary considerably because of the wide variety of chemicals and uses. The following procedures are often used with conventional nonreactive chromic-acid post-treatments.

Total acid value is determined by titrating a 25-mL sample of chromic acid solution with 0.1*N* sodium hydroxide, using phenolphthalein indicator. The end point is reached when the color changes from amber to a reddish shade that lasts at least 15 s. Each millilitre of 0.1*N* sodium hydroxide required equals 1 point total acid.

Free-acid value is determined by titrating a 25-mL sample of chromic acid solution with 0.1*N* sodium hydroxide, using bromocresol green indicator. The end point is reached when the color changes from yellow to green. Each millilitre of 0.1*N* sodium hydroxide required equals 1 point free acid. The concentration of free acid in chromic acid solutions is usually maintained between 0.2 to 0.8 mL.

Chromate concentration may be determined by placing a 25-mL sample of solution into a 250-mL beaker, adding 25-mL of a 50% sulfuric acid solution, 2 drops of orthophenan-throline ferrous complex indicator, and titrating with a 0.1*N* ferrous sulfate solution. Each millilitre of 0.1*N* ferrous sulfate solution of the amount required to change the solution from blue to a reddish-brown color is 1 point of chromate concentration. Chromate concentration ranges from 200 to 400 ppm (as chromium) and may also be determined with a test kit containing diphenylcarbazide.

In reactive chromate post-treatments, those that can be post-rinsed with deionized water, pH is often used as a means of determining whether the post-treatment solution is in proper balance.

Each supplier has his own particular means of checking the concentration in chromium-free post-treatments, and in some instances, the acidity of the solution.

Solution Maintenance Schedules

Frequency and extent of solution maintenance is dictated by the materials used and the work load of the line. In one plant, a schedule was established for (*a*) solution testing, (*b*) solution maintenance, and (*c*) tank cleaning. The company revised solution control procedures to correct problems experienced with zinc phosphate coatings produced in a large automatic line. Coatings periodically were coarse and nonuniform, making it necessary to interrupt production to change phosphating solution. A comprehensive program of testing and checking was inaugurated. Routines for solution maintenance and tank cleaning were established that eliminated defective coatings and downtime for changing phosphating solutions. It was determined that the phosphating bath maintained good coating ability for 9 to 10 weeks. However, tanks required cleaning every 4 to 5 weeks because of sludge buildup. The schedule established includes tank cleaning every 4 weeks and solution change every 8 weeks, coinciding with tank cleaning. Work is done on weekends so production is not interrupted. The solution is allowed to cool and sit idle for 8 h.

For tank cleaning only, the entire contents of the tank are pumped to a reserve tank, the sludge removed, and the solution returned to the phosphating tank. When the solution requires changing, the procedure is as follows: (*a*) top half of solution is pumped to a reserve tank, (*b*) bottom half is discarded, and (*c*) sludge is cleaned from the tank. Salvaged solution is then pumped back into the tank, and sufficient water and phosphating concentrate are added to bring the bath to correct concentration and operating level. Iron, in the form of steel wool, and soda ash are added to the solution to adjust it to proper operating condition, so satisfactory coatings can be produced when production is resumed.

Schedules for operating control and tank and solution maintenance were established by another company for a multistage phosphating process in which automobile bodies were phosphated. Surface area of the bodies approximated 70 to 80 m^2 (750 to 860 ft^2). Production rate could range as high as 75 car bodies an hour. The spray chamber of the phosphating line, including drain area, was 12 m (39 ft) in length. The phosphating solution tank, at operating level, holds 40 000 L (11 000 gal). A continuous desludging system is incorporated in the system (hydromation unit). Phosphating and accelerator solutions are replenished continuously and automatically via pH and redox measurements through variable-feed metering pumps. Corrective additions of caustic soda are made manually. Coating weights are maintained at 2.7 to 3.2 g/m^2 (250 to 300 mg/ft^2).

Operating Control

Hourly

- Phosphating solution titrations:
  Total acid
  Free acid
  Visual coating appearance
  Chemical-supply pumps
- All stages:
  Temperature
  Pressure
  Solution level
- Neutralizing rinse:
  Add 30-mL activator, as indicated by visual check of coating continuity
- Neutralizing rinse: Check pH
- Check film thickness of soap lubricant

Twice per shift

- Alkaline cleaner stage titrations
- Alkalinity of water rinses before phosphating
- Acidity of water rinse after phosphating
- Activating rinse concentration

Once per shift

- Conductivity of recirculated deionized rinse, 100 μmho maximum
- Conductivity of fresh deionized harness rinse

Tank and Solution Maintenance

Daily

- Empty and clean stages corresponding to solutions 4 and 6 (See Fig. 9)

Weekly

- Empty and clean stages corresponding to solutions 7 and 8 (See Fig. 9)

Twice monthly

- Empty and clean stages corresponding to solutions 1 and 2 (See Fig. 9)

Monthly

- Empty and clean stage corresponding to solution 3 (See Fig. 9)

Quarterly

- Cleaner, water rinse, and phosphate stages should receive heated acidic cleanout on a quarterly basis. Blocked nozzles should be removed and cleaned or replaced. Heated acidic cleanout may involve inhibited hydrochloric acid

Phosphating Tank Maintenance

Phosphate tank should be desludged on a continuous, automatic basis. Depending on work appearance, nozzles and spray pressure at the nozzle may require checking on a monthly basis, rather than quarterly. Phosphate heat exchangers require a heated acidic cleanout to maintain heating efficiency. Acidic cleanout usually involves the following procedure:

- Pump out solution to holding tank
- Flush tank and spray piping with water
- Fill to pumping level with water, add hydrochloric acid (1*N* or 10% volume acid/volume water, v/v); add inhibitor
- Heat to 50 °C (120 °F), circulate spray system for 1 h
- Empty tank and flush with water
- Fill to pump level, add sodium hydroxide to pH of 10 to 12, circulate 5 to 10 min
- Empty tank, flush with water, and restore phosphate solution

Break-In of Phosphating Solutions

Some zinc and manganese phosphating solutions, although mixed to recommended concentrations, must be broken in by the addition of ferrous salt, such as ferrous sulphate, before they can operate properly. Iron phosphating solutions require no break-in. After being mixed to proper concentration, iron phosphating solutions need only be raised to operating temperature to be ready for use. Most zinc phosphate processes used for paint base or for metal forming operate free of ferrous iron, and the break-in of these phosphating solutions is not a factor.

Zinc. One method of breaking in a zinc phosphating solution is to tolerate a poor phosphate coating until some iron has gone into solution from the chemical reaction between the bath and the parts being coated. Some iron also may be present in sludge that has settled to the bottom of the tank or crusted on the sides from a previous bath. The coating on first parts or batch are of poorest quality, getting progressively better as more iron goes into solution. A simple method is to suspend clean steel wool or scrap in the bath, or to introduce a small quantity of clean iron powder. Another method is to add 170 g (6 oz) of salt, such as ferrous sulphate, to each 380 L (100 gal) of solution. This is applicable to bath spray and immersion baths.

Manganese. Careful attention should be given to breaking in of a manganese phosphating bath because of its higher acid concentration in comparison to that of a zinc bath. For the best quality of manganese phosphate coatings, 0.2 to 0.4% of ferrous iron in solution is the proper range. Usually, breaking in of a new bath is begun by the addition of 70 g (6 oz) of a ferrous salt, such as ferrous sulphate, powder to each 380 L (100 gal) of bath. This is followed by treatments using clean steel wool, powdered iron, or scrap iron to build up ferrous iron content. Manganese phosphating baths operate to best advantage when they have a steady, heavy work load. This permits considerable dissolution of iron, which usually maintains the ferrous iron content at a suitable level.

Equipment for Immersion Systems

An immersion phosphating system for all types of coatings (zinc, manganese, and iron) should include (*a*) required number of tanks, (*b*) temperature and solution-level controls, (*c*) overflow and drainage systems, (*d*) vapor-exhaust systems, and (*e*) materials-handling devices. When drums are used to contain the parts, devices are required at each tank to rotate the drums at approximately 4 rev/min while the drums and the parts within are submerged.

Phosphating tanks are usually made from low-carbon steel plate about 6 mm (1/4 in.) thick. A tank and drum for immersion phosphating small parts are shown in Fig. 3. The normal life of a low-carbon steel tank for zinc phosphating solution under average operating conditions is about 1 year. However, some companies report 2- to 3-years service. One company fabricates zinc phosphating tanks from 9.5 mm (3/8 in.) low-carbon steel plate. This tank lasts 4 to 5 years. Stainless steel may give longer life, but its greater cost generally is not justified for a zinc or iron phosphating line, unless acidic solutions contain high levels of chloride. Because of greater acid concentration in manganese phosphating solutions, 6 to 10% v/v as compared with 1 to 3% v/v in zinc phosphating solutions, and higher operating temperatures used, low-carbon steel tanks for manganese phosphating solutions have a shorter life than when used with zinc or iron phosphating solutions. For this reason, stainless steel tanks may be economically practical for manganese phosphating solutions. Stainless steel should also be considered for the heating coils. Plastic, fiberglass, and rubber-lined tanks have also been used successfully.

Table 5, representing data from the experience of one company, shows a comparison of expected tank life, in years, for the three types of phosphating solutions using both low-carbon steel and type 316 stainless steel. These figures are approximately correct for all solutions. Some solutions permit extended tank life, and others shorten the tank life. One tank made of 6-mm (1/4-in.) mild steel has been in continuous operation for 15 years in an iron phosphating line.

Many phosphating tanks made of low-carbon steel are lined with glass fiber impregnated with polyester resins. Phosphating compounds have no effect on this material, and it will last indefinitely in normal service. It is, however, susceptible to damage from impact, and careless handling of equipment while loading or unloading the tanks may cause fractures or cracks. Care must be exercised in the placement of heating coils when using polyester-impregnated glass fiber liners. The maximum temperature this material can withstand is about 105 °C (225 °F). Many polyester resins have little resistance to alkaline materials and should not be used where more than casual contact with strong alkaline cleaners is possible.

Tank accessories, including (*a*) steam coils or other heating mediums, (*b*) piping, (*c*) screens, (*d*) drum trunnions, and (*e*) drum-rotating mechanisms, may be made of low-carbon steel or stainless steel. Electropolished stainless steel steam coils permit less sludge buildup on the coils.

Tank Design. Tanks should have sufficient capacity to stabilize solution temperature and solution concentration, and to prevent rapid buildup of solution contamination. Tanks for the phosphating stage should have a sloping bottom, with at least 0.46 m (1.5 ft) of space below the lowest work level to accommodate sludge buildup.

Rinse Tanks. Water rinse tanks and associated equipment, including (*a*) steam coils or other heating mediums, (*b*) piping, and (*c*) screens, may be constructed of low-carbon steel. Rinse tanks for certain parts sometimes require drum-rotating devices. Immersion rinse tanks should include a method for solution agitation to assist rinsing action. This can be accomplished by use of low-pressure air distributed through evenly spaced holes in pipes laid along the bottom of the tanks. Another method is to recirculate rinse water through a similar piping arrangement. For a clear water rinse, the pump housing, bearings, impeller, and any other part in contact with the water, may be of normal material. Acidulated rinses containing chromium preclude the use of brass or bronze in any part of the pump or valving that is in contact with the solution.

Drying equipment for immersion phosphating systems can be of several types. For small parts, such as washers, a centrifuge may be used to spin off moisture. If parts are hot enough, no additional heated air is required. However, if parts are cold, heated air may be introduced into the centrifuge. When parts are centrifuged, the phosphate coating may be damaged on some parts, rendering them unacceptable. Such parts may be dried in a basket or on a rack, in the same manner that larger parts are dried. This is done in a final tank or enclosure in which the parts

Fig. 3 Immersion phosphating tank for batch coating small parts

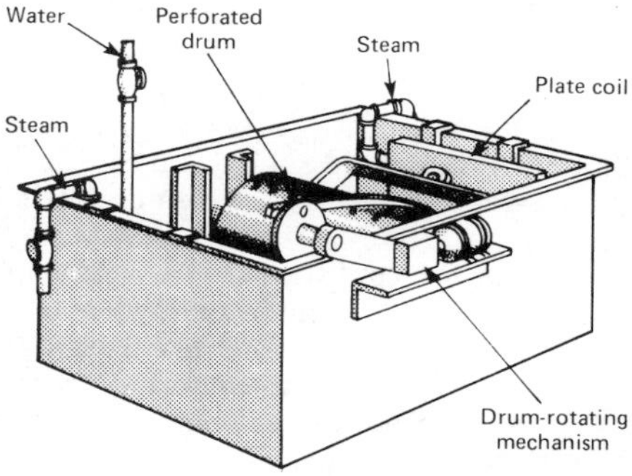

Drum into which parts are loaded is shown in immersion position

Table 5 Expected service life of low-carbon steel and type 316 stainless steel phosphating tanks

Process	Service life, yr: Low-carbon steel	Service life, yr: Type 316 stainless steel(a)
Iron phosphating	10(b)	20
Zinc phosphating	4-5(c)	10-20
Manganese phosphating	Not used	10

(a) Any thickness that provides mechanical strength required. May be used as liner for low-carbon steel tank. (b) 6 mm (1/4 in.) thick. (c) 9.5 mm (3/8 in.) thick

are held while heated air (at 120 to 175 °C or 250 to 350 °F) is blown on them. Heat sources may be (*a*) steam coils, (*b*) gas burners, or (*c*) electric heaters. Drying time usually ranges from 2 min, for simple parts, to 5 min for complex parts. If rinse solutions are retained in pockets or seams, drying requires additional time or temperature, or a mechanical aid such as an air blast directed at the pocket or seam, or tilting the part.

Drums for containing and rotating parts are usually made of low-carbon steel. To obtain longer life, stainless steels may be used; however, one company reports a life expectancy of approximately 10 years from similar drums made from low-carbon steel. This long life is attributed to a hard coating of phosphate that develops on surfaces of the drum. Drums should have a loading-and-unloading door with a positive latch to prevent accidental opening, and loss of load, during a processing cycle. A drum for containing small parts during batch phosphating is shown in Fig. 4.

Baskets for handling parts too large for drums or too small or too heavy for racks can be made of either low-carbon or stainless steel. The choice is dictated by cost-life relationships.

Conveying equipment for the immersion process may be of any type that can transport work from the loading to the unloading stage. It must be capable of lowering work into and raising it out of various tanks in the proper sequence and at the proper time, either automatically or manually. Various types of conveying equipment are:

- Overhead monorail conveyors with manual or electric hoists. For very small production, the work can be moved manually from tank to tank
- Chain-driven conveyors that lower and raise work into and out of each tank while it is continuously moving
- Automatic equipment, similar to automatic plating equipment but without equipment necessary for supplying electric current

Conveying equipment can be of varied design, but it must allow sufficient time for solution to drain from the work as it is raised from the tank. This solution should drain back to the original tank so it will not contaminate the next tank. Drainage and transfer time should not exceed 30 s, or the work may become discolored because of partial or complete drying between stages. The conveyor need not be made of acid-resistant material.

Work-supporting equipment, such as racks, hooks, and baskets, is similar in design and function to that used in electroplating, with the exception that it need not be electrically insulated. For phosphating, however, racks, hooks, and baskets should be resistant to alkaline cleaners, acid phosphating solutions, and other materials used in a phosphating line. Low-carbon steel is usually satisfactory. Stainless steel may be used where its additional life justifies the greater cost. Work-supporting equipment does not need to have tight contact with the work to be phosphated. Light contact with work-supporting equipment is more desirable, particularly on significant surfaces of the work, because coating may be thin or nonexistent at the point of contact, depending on degree of insulation of the surface by hook, rack, or basket.

Equipment for Spray Systems

Spray systems usually are completely enclosed in a continuous, chambered tunnel or cabinet for better control of the process and cleanliness of the operation. Parts or panels to be processed are hung on racks or hooks, or placed in baskets, and automatically carried through the various stages of the spray phosphating line (Fig. 5). Temperature and pressure gages and controls are required at all stations, as are pumps of adequate capacity. Time and space intervals between stages, and between final rinse and the drying oven must provide sufficient drain time to minimize carryover of solutions to succeeding stages. However, the time must also be as short as possible, because dried-on solutions cause blotchy coatings and can reduce final corrosion resistance or adhesion.

Spray cabinets are usually made from low-carbon steel, as are the reservoirs from which the cleaners, phosphating solutions and various rinses are pumped. Steam coils or other heating mediums, piping, screens, and valves also may be made of low-carbon steel. Spray nozzles may be made of low-carbon stainless steel, or polypropylene. Pumps may be of all-iron construction with stainless steel impellers. Valves may be all iron. As in immersion systems, because acidulated rinses usually contain chromium, no brass or bronze should be used in contact with these rinses. Also, no brass should be used in contact with alkaline cleaners or phosphate solutions. Potential suppliers of the chemicals should

Fig. 4 Drum used in batch phosphate coating of small parts

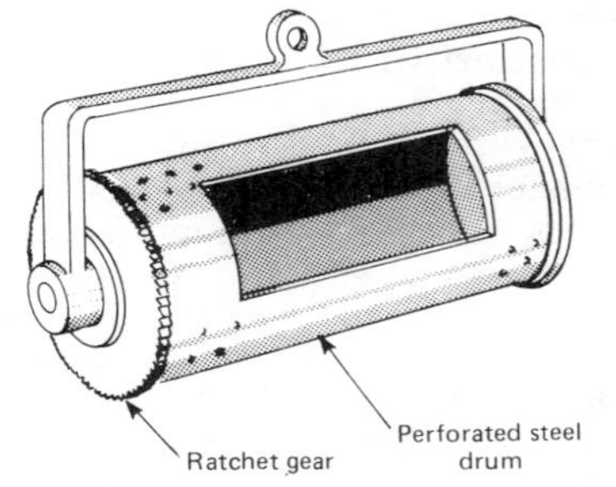

Fig. 5 Continuous conveyorized spray line for phosphating

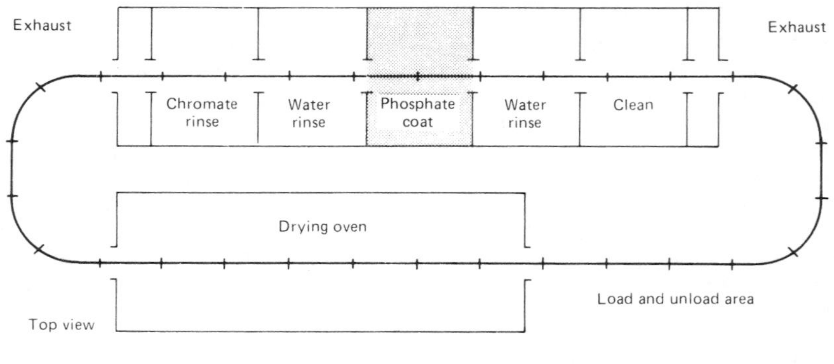

be consulted to assure compatibility with the specific process to be used.

One large manufacturer doing extensive phosphate coating recommends all parts and accessories in the phosphating stage of a spray zinc phosphating line should be made of stainless steel except the storage tank. These tanks usually give satisfactory service when made from 9.5 mm (3/8 in.) thick low-carbon steel. Heating coils should be made from electropolished stainless steel to discourage a buildup of zinc phosphate sludge on the coils.

In a continuous spray phosphating line, baffle ridges on the floor and baffle doors or curtains are essential. Baffling between stages eliminates much of the mixing and contamination from carryover of the solutions. The storage tanks from which various solutions are pumped should have a minimum capacity of 2 to 2½ times the volume pumped out per minute. Some process lines provide cleaner tanks with a minimum capacity of 2 to 2½ times the volume pumped out per minute and phosphate solution tanks with a minimum capacity of 2½ to 3½ times the volume pumped out per minute, to provide room for sludge to collect.

Drying equipment for spray lines usually consists either of indirect-heated convection ovens, fired by gas or oil, or electric-fired or gas-fired infrared ovens capable of raising the temperature of parts to 150 to 205 °C (300 to 400 °F). If parts tend to hold rinse solutions in pockets or seams, it may be necessary to direct a blast of heated air at the pocket, or to tilt the part automatically to drain off the retained solution. Time required for drying varies from 2 min for simple, thin-gage parts, to 6 min for complex parts. For phosphate coatings used ahead of electropaints, the dry-off oven is often omitted, and parts go into the electropaint tank either air-dried or wet.

Work-Supporting Equipment. The size and design of hooks, racks, and other work-supporting equipment used in spray phosphating depend on the size and contour of parts being processed. These supports should be designed so that significant surfaces of parts receive the full impingement of phosphating solution as parts are conveyed through the unit. As in immersion phosphating, light or point contact is desirable so as not to mask off any surfaces of parts from the phosphating and rinsing solutions. Intricately contoured parts should be suspended in such a manner as to eliminate or minimize the entrapment of solution, so that as little as possible is carried from one tank to another. Parts that, because of their shape, are impossible to suspend with mechanical racks or hooks can be suspended with magnetic hooks.

Work-supporting equipment usually is fabricated from low-carbon steel, although stainless steel can be used. Selection generally is predicated on economics, weighing the greater cost of stainless steel equipment against its longer life. Special care must be exercised in handling finish machined parts and other easily damaged parts in phosphating. Although such parts may be processed in special baskets, this often is unsatisfactory because coatings may be thin or completely absent at the numerous areas of contact. Racks usually are preferable.

Racks must be designed to hold parts in such a way that all significant surfaces are satisfactorily coated and parts are separated to prevent them from bumping and damaging each other during processing. On closely conforming racks, an accumulation of scale can hamper proper hanging or holding of parts or cause the rack to mask more than normal areas, causing imperfect coatings. These racks must be descaled frequently. Descaling can be done either by pickling in an inhibited solution of muriatic acid or by cracking the scale from the rack. Conveyor equipment carrying these racks must have a gentle motion to avoid knocking parts against each other. At the same time, conveyor speed must be rapid enough to prevent solutions from drying on parts as they are being moved between stations.

Conveying equipment used in spray phosphating may be of any type that can transport work through the various processing and draining stages. Continuously moving chain-driven conveyors, either overhead or floor-level, are usually used. Conveyor chains and part-carrying accessories can create problems by dragging out solutions and carrying them from one tank to another or from tank to part. The use of conveyor shielding can minimize some of these problems. One such problem occurred in a spray zinc phosphating line. Parts were hung on hooks suspended from a conveyor chain and remained on these hooks while being carried through the subsequent painting cycle. In the drying stage, phosphating solution (a proprietary solution containing sodium bifluoride) dripped from the conveyor chain and hooks onto the phosphated parts. Blisters in the paint film occurred in areas on which the solution had dripped. To correct this problem, a thorough system of rinsing, drying, and cleaning of the conveyor chain and hooks was initiated:

- A conveyor-chain washer was added; this consisted of several nozzles to spray fresh water on conveyor chain between the phosphating stage and the subsequent rinse
- Additional spray drop lines were added in the rinse stage, including a final drop line that sprayed fresh, rather than recirculated water
- Compressed-air nozzles for blowing off the conveyor chain and hooks were added immediately following the acidulated rinse stage
- A final short rinse with unrecirculated demineralized water was added to remove any remaining contaminants from conveyor and parts
- Frequency of removing accumulated paint from part hooks was increased to avoid entrapment of contaminants in built-up paint

Use of Equipment. Equipment required for phosphate coating can vary from the simple to the elaborate. Some of the factors that influence equipment requirements include (*a*) work load, (*b*) size of products to be phosphated, (*c*) material tc be phosphated, and (*d*) processing method. An example of equipment requirements and process cycles can be found in a company that produces threaded fasteners, zinc phosphates, and oil dips at an average of 3600 kg (8000 lb) of these parts each 8-h shift. One man operates the entire immersion phosphating line. All parts are cleaned and pickled before being phosphated. Production requirements for the manual immersion zinc phosphating of fasteners are listed below:

Weight of each piece......... 0.013 kg (0.029 lb)
Weight of each load.... 193 kg (425 lb)
Average weight processed per hour........... 454 kg (1000 lb)
Average number of pieces per hour.................... 34 483

The equipment required for manual immersion zinc phosphating includes:

Work handling

- Two workbaskets (for cleaning); stainless steel; 230-kg (500-lb) capacity each

- Three perforated drums (for phosphating); motor-rotated; 230-kg (500-lb) capacity each
- One drum-loading stand, 230-kg (500-lb) capacity
- One loading chute, 227-kg (500-lb) capacity
- Two hoists, 910-kg (1-ton) capacity each

Cleaning

- One alkali soak tank, 760-L (200-gal) capacity
- One acid tank, sulfuric acid; 760-L (200-gal) capacity
- Three rinse tanks, 760-L (200-gal) capacity each

Phosphating

- Three phosphating tanks, each of 760-L (200-gal) capacity and heated by stainless steel steam plate coil; automatic temperature control
- One rinse tank, 760-L (200-gal) capacity
- Chromic acid rinse tank, 760-L (200-gal) capacity heated by steam plate coil; temperature automatically controlled
- Centrifuge for drying, equipped with hot-air blower driven by 3-hp motor
- Oil dip tank, corrosion-resistant; 380-L (100-gal) capacity

The equipment requirements for zinc phosphate coating of cast iron cylinder heads include the use of an automatic indexing immersion phosphating machine. These parts, which weigh 121 kg (267 lb) each, are processed in baskets, three to a basket, and are loaded standing on their sides to facilitate drainage of solutions from inner passages. A coating weight of 3.8 g/m^2 (350 mg/ft^2) is obtained. The machine includes a phosphating tank that accommodates three workbaskets, thus allowing processing time equal to three times that of any other tank plus the time required to index the machine twice. Details of the equipment comprised by this automatic machine together with production requirements and operating conditions for phosphating the cast iron cylinder heads are listed below:

Production requirements

Weight of coating . . . . . . 3.8 g/m^2 (350 mg/ft^2)
Size of each piece(a) . . . . . . 1143 by 406 by 152 mm (45 by 16 by 6 in.)
Weight of each piece . . . . . . 121 kg (267 lb)
Pieces per load 3
Load weight . . . . . . 363 kg (800 lb)
Production per hour . . . . . . 20 loads
Immersion time, min:
Cleaning (each tank) 2
Cold water rinse 2
Hot water rinse (each tank) 2
Phosphating 8
Cold water rinse 2
Chromic acid rinse 2
Oil dip 2

Equipment requirements

Size of work-basket(a) . . . . . . 1220 by 610 by 660 mm (48 by 24 by 26 in.)
Cleaning tanks (two tanks):
Size(a), each . . . . . . 1525 by 890 by 1525 mm (60 by 35 by 60 in.)
Material . . . . . . Low-carbon steel
Method of heating . . . . . . Steam plate coils
Temperature(b) . . . . . . 93 to 96 °C (200 to 205 °F)
Liquid level control . . . . . . Automatic
Cold water rinse tank:
Size(a) . . . . . . 1525 by 1070 by 1525 mm (60 by 42 by 60 in.)
Material . . . . . . Low-carbon steel
Water level control . . . . . . Overflow
Hot water rinse tanks (two tanks):
Size(a), each . . . . . . 1525 by 890 by 1525 mm (60 by 35 by 60 in.)
Material . . . . . . Low-carbon steel
Method of heating . . . . . . Steam plate coils
Temperature(b) . . . . . . 27 to 93 °C (80 to 200 °F)
Water level control . . . . . . Overflow
Phosphating tank:
Size(a) . . . . . . 1525 by 2720 by 1525 mm (60 by 107 by 60 in.)
Material . . . . . . Stainless steel
Method of heating . . . . . . Steam pipe coil
Temperature(b) . . . . . . 93 to 96 °C (200 to 205 °F)
Liquid level control . . . . . . Automatic
Cold water rinse tank:
Size(a) . . . . . . 1525 by 890 by 1525 mm (60 by 35 by 60 in.)
Material . . . . . . Low-carbon steel
Water level control . . . . . . Overflow
Chromic acid rinse tank:
Size(a) . . . . . . 1525 by 1070 by 1525 mm (60 by 42 by 60 in.)
Material . . . . . . Low-carbon steel
Method of heating . . . . . . Steam plate coils
Temperature(b) . . . . . . As recommended by manufacturer of solution
Liquid level control . . . . . . Automatic
Oil dip tank:
Size(a) . . . . . . 1525 by 890 by 1525 mm (60 by 35 by 60 in.)
Material . . . . . . Low-carbon steel
Oil control . . . . . . Drain out carryover water and add oil as needed
Size of drip pan (attached to oil dip tank)(a) . . . . . . 1525 by 890 by 1525 mm (60 by 35 by 60 in.)
Phosphate sludge settling tank:
Size(a) . . . . . . 1525 by 2720 by 1525 mm (60 by 107 by 60 in.), or equivalent volume
Material . . . . . . Stainless steel
Method of transfer . . . . . . Centrifugal pump
Settling time . . . . . . 24 h

(a) Length, width, and depth, in the order listed. (b) Automatically maintained

Automobile or truck fenders and hoods are zinc phosphate coated in an automatic, conveyorized spray phosphating line. Table 6 lists the sequence of processing stages and indicates operating conditions for each station. Although this example is based on a specific application, the data are applicable to the processing of similar parts.

The sequence of operations involved in spray iron phosphating panels, brackets, and miscellaneous parts for household appliances is indicated in Table 7. Solutions, operating temperatures, and cycle times are also shown. The entire process is completed in 8½ min. Coating weight ranges from 0.4 to 0.6 g/m^2 (40 to 60 mg/ft^2).

Manganese phosphate coatings are applied to military equipment to provide increased resistance to scuffing, galling, and corrosion. Table 8 lists the progressive stages of a phosphating indexing line, which is completely automatic, and indicates the operating conditions.

Figure 6 shows a schematic layout of an automatic, conveyorized line for immersion zinc phosphating and lubricating of blanks from which casings for 80-mm (3.2-in.) mortar shells are cold formed. These blanks, made of 1010 steel, are coated at the rate of 4000 pieces per hour. Each blank has an area of approximately 0.1 m^2 (1 ft^2). Conveyor speed is 2.0 m (6.5 ft) per minute. Details of operating conditions and solutions used are presented in the table accompanying Fig. 6.

Figure 7 compares processing stages involved in manganese phosphating to two different ranges of coating thickness, 2.5 to 7.6 μm (0.1 to 0.3 mil) and 7.6 to 15 μm (0.3 to 0.6 mil) for heavy, or conventional coatings. The same phosphating compound is used in each line. Difference in coating weights depends on cleaner used and time in the phosphating solution. For conventional, heavy manganese phosphate coatings, parts are cleaned in an alkaline cleaning solution, providing a surface that permits good contact between metal and the phosphating bath. The resulting coating is heavy and coarse-grained, and can readily absorb oil. For light coatings, a kerosine-based, or similar solvent, emulsion cleaner is used. A thin residue of oil left on the metal after two rinses acts as a buffering agent or grain-refiner, to produce a thinner, finer-grained coating. Usually, less lubricating oil is desired in conjunction with a fine-grained coating. Consequently, an additional step is involved for removing excess oil. The additional step is not usually necessary with a coarse-grained coating.

A phosphating line for spray zinc phosphating of automobile bodies is shown in Fig. 8. The bodies average 80 m^2 (861 ft^2) in area and the line speed ranges from 7.3 to 8.5 m/min (24 to 28 ft/min). Coating weights range from 1.6 to 2.4 g/m^2 (148.7 to 223 mg/ft^2). Production rate may reach 75 bodies per hour. The table accompanying Fig. 8 gives details of solutions used in this line and lists cycle times for various stages.

Figure 9 shows an automatic spray line with operations involved in zinc phosphate coating 80-mm (3.2-in.) mortar shell casings as a base for paint. Total area coated on each piece is 0.1 m^2 (1 ft^2). Coating weight ranges from 1.7 to 2.1 g/m^2 (160 to 200 mg/ft^2). Conveyor speed is 0.03 m/s (5 ft/min) and production rate is 1000 pieces per hour. Details of solutions used, together with cycle times, are given in the table with Fig. 9.

Control of Coating Weight

Tables 1, 2, and 3 present phosphate coating applications and weights. Table 1 deals only with spray application, but covers both iron and zinc phosphate coatings as bases for paint films. By comparing the area of the parts and the production per hour controlled to obtain the uniform coating weights shown, the interrelation of size, production time, and coating weight is apparent. In all applications, material being coated was low-carbon steel sheet. Table 3 lists applications for manganese phosphate coatings for wear resistance.

As indicated by the curve in Fig. 10, based on experience of one processor of small threaded parts, the consumption of phosphating solution concentrate is directly proportional to the area and thickness of the coating applied. These parts were immersion zinc phosphated, processed in batches in a rotating drum, to a coating weight of approximately 10.8 g/m^2 (1000 mg/ft^2). Figure 10 shows the direct proportionality of area coated to concentrate consumed does not begin until an initial coat is deposited. At the time when parts are immersed, there is an immediate reaction in which an irregular coating is quickly deposited. Because the maximum area of bare steel is exposed to the bath at that time, maximum efficiency takes place. The remaining time in the bath serves to refine the coat by depositing crystals to fill gaps between existing crystals and to increase coating weight to uniform thickness by depositing crystals over previously deposited crystals.

Control of Crystal Size

The crystalline structure of the chemically bonded phosphate coating (Fig. 11) provides a suitable base for

Table 6 Sequence of operations in automatic spray application of zinc phosphate to automobile or truck small parts

Operation	Solution	Concentration	Temperature °C	Temperature °F	Time, s	Pressure kPa	Pressure psi
1 Clean	Alkaline titanated cleaner	4-6 mL(a)	60-65	140-150	60	100-140	15-20
2 Rinse	Water	1.0 mL max(b)	57-60	135-140	30	100-140	15-20
3 Clean	Alkaline titanated cleaner	4-6 mL(a)	60-65	140-150	60	100-140	15-20
4 Rinse	Water	1.0 mL max(b)	57-60	135-140	30	100-140	15-20
5 Phosphate	Accelerated zinc phosphate	20-25 mL(c)	52-55	125-130	60	55-83	8-12
6 Rinse	Water	1.0 mL max(c)	35-40	95-105	30	69-100	10-15
7 Acidulated rinse	Partially reduced chromic and/or phosphoric acids	150-250 ppm Cr^{6+} pH 4.0-5.0	35-40	95-105	30	69-100	10-15
8 Demineralized rinse	Distilled water, deionized 100 μmho max						
9 Dry	Air(d)						

(a) Number of millilitres required to titrate a 10-mL sample to the phenolphthalein end point using 0.1*N* hydrochloric acid. (b) Number of millilitres required to titrate a 10-mL sample to the bromocresol green end point using 0.1*N* hydrochloric acid. (c) Number of millilitres required to titrate a 10-mL sample to the phenolphthalein end point using 0.1*N* sodium hydroxide. (d) Or dry at 170 to 180 °C (340 to 355 °F) for 4 min

Table 7 Sequence of spraying operations in iron phosphating of panels, brackets, and other parts for household appliances

Coating weight ranges from 0.4 to 0.6 g/m^2 (40 to 60 mg/ft^2)

Operation	Solution	Concentration	Temperature °C	Temperature °F	Time, s
1 Clean	Alkaline cleaner	5.6-9.40 g/L (0.75-1.25 oz/gal)	66-74	150-165	60
2 Rinse	Water	...	66	150	30
3 Rinse	Water	...	66	150	30
4 Phosphate	Iron phosphate	19 kg/380 L (42 lb/100 gal)	68-74	155-165	60
5 Rinse	Water	...	Ambient		30
6 Acidulated rince	Chromic acid	110 g/380 L (4 oz/100 gal)	54-86	130-150	30
7 Rinse (optional)	Deionized water	25 ppm (max) impurities	Ambient		30
8 Dry (optional)	...	...	150-230	300-450	240

Fig. 6 Immersion zinc phosphating

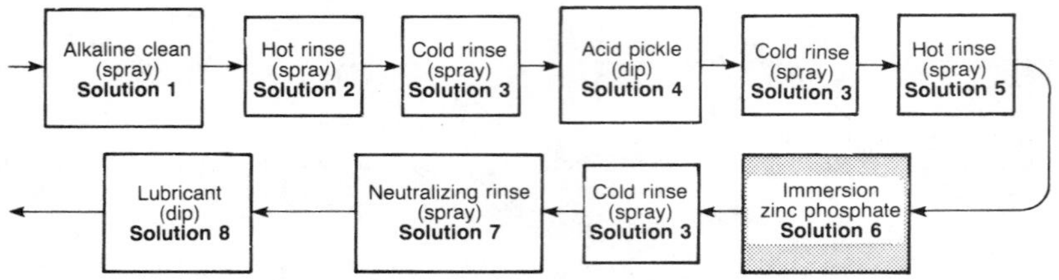

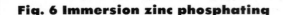

Solution No.	Solution	Composition	Operating temperature °C	°F	Cycle time, min
1	Alkaline cleaner	Alkali, 3.8 g/L (0.5 oz/gal)	82	180	1
2	Hot rinse	Water(a)	77	170	0.75
3	Cold rinse	Water(b)	Room		1
4	Acid pickle	H_2SO_4, 15-18 wt %(c)	66	150	10
5	Hot rinse	Water	71	160	0.75
6	Zinc phosphate	Chlorate-accelerated(d)	82	180	6
7	Neutralizing rinse	$NaNO_2$, 0.4 kg/380 L(e) (1 lb/100 gal)	Room		1
8	Lubricant	Soap, 10 wt %	66	150	6

(a) When lime is present in water, sequestering agent is added in concentration of 1.9 g/L (¼ oz/gal). (b) Purity maintained by overflow. (c) Solution is discarded when iron content reaches 5%. (d) Contains 36 points total acid, 7.5 points free acid, based on titration of 10-mL sample; concentration of accelerator, 3.5 g/L (0.47 oz/gal). (e) NaOH added to establish pH in range of 10 to 11

Automatic, conveyorized cleaning, immersion zinc phosphating, and lubricating of 80-mm (3.2-in.) mortar shell blanks (1010 steel) before cold forming. Average area of shell blanks was 0.1 m^2 (1 ft^2); coating weight, 16 g/m^2 (1500 mg/ft^2). Conveyor speed was 0.03 m/s (6.5 ft/min), and the production rate was 4000 pieces per hour

subsequent paint or oil films. Crystals permit the paint to penetrate, providing the paint with exceptional adherence. When oil is the rust preventive, the interstices of the crystalline structure function efficiently as an oil-retaining reservoir. The adhesion of phosphate coating to the basis metal, as determined by flexing of the metal, varies with the type and thickness of the coating. Generally, heavier coatings are composed of large crystals, which do not bond to each other or to the surface of the metal as well as do fine-grained, thinner coatings. Consequently, where adhesion and flexibility may be a problem because of nature of the application, phosphating material is selected that produces a thin, fine-grained coating. However, this may not result in maximum corrosion resistance. Organic additives, special accelerators and/or calcium added to a zinc phosphate process provides a microcrystalline structure that exhibits optimum paint adhesion and corrosion resistance.

Zinc phosphate coatings (Fig. 11a and b) are widely used as bases for paint or oil. A fine uniform crystal is necessary when gloss is desired for the paint film. Coarse crystals promote dullness and often require higher paint thickness to gain uniform and acceptable coverage. However, when coating is applied to provide lubricity, a coarse crystal may be preferable. With few exceptions, zinc phosphating concentrates are proprietary materials designed to produce, within a specified time and with available equipment, coatings within a specific range of weight and with a desired crystal size and texture.

Usually, strongly acid baths build coating at a slow rate but deposit large crystals. This is due to the longer time needed to develop neutralization at the coating interface. A bath that has been activated by an accelerator deposits coating more quickly and with a smaller crystal. Up to a point, as long as the part stays in the bath, solution continues depositing crystals by building up thinner sections of coating and filling interstices with more crystals. Because depositing coating gradually insulates metal from fresh solution, no crystals are deposited beyond a certain point. These characteristics are inherent in proprietary solutions, and little control can be exerted other than to maintain the bath as prescribed.

The surface condition of the parts being coated is a factor that influences coating characteristics which can be controlled. Certain oils, residues from solvent cleaning or vapor degreasing, or solvent emulsions, when retained on the surface in very thin films, function as crystal refiners. An example is the film left on parts that are cleaned using a kerosine-based emulsion cleaner. Although this cleaning operation is usually followed by a hot water rinse, enough oil is retained on the surface to have a beneficial influence, if a fine crystal is desired. Conversely, if a strong alkaline cleaner is used and completely rinsed away, or if blast cleaning is used, a coarser crystal is obtained. Another method of refining or decreasing crystal size is the use of a proprietary titanium phosphating conditioner. These titanium phosphate salts can be used either in the water rinse that precedes phosphating or with certain alkaline-based cleaners.

Iron phosphate coatings (Fig. 11c) are generally of a very fine structure, and amorphous in appearance. Because these coatings are used primarily as bases for paint or to assist in bonding of metal to a nonmetallic surface, fine structure is desirable. With iron phosphate coatings, the problem is one of adherence and powdery coatings rather than crystal size or coating weight. Attention must be directed to (*a*) surface cleanliness, (*b*) maintenance of the bath within the prescribed limits, and (*c*) proper processing.

Manganese phosphate coatings (Fig. 11d) are usually heavy and coarse. Because these coatings are generally used for their lubricating qualities and often incorporate a supplementary oil film, a continuous coating may not be mandatory. The length of time in bath may be varied, within limits, to vary the film thickness to meet the functional requirements of the coating. Crystal size and coating thickness are controlled by condition of the surface to be coated. Oils, and residues from alkaline cleaning, solvent cleaning, vapor degreasing, and emulsion cleaners, serve as crystal refiners and reduce coating thickness. Proprietary crystal refiners

Fig. 7 Operations for light and heavy manganese phosphate coatings

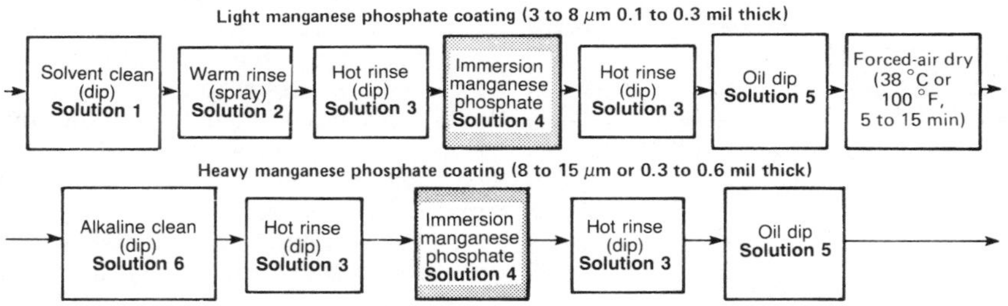

Solution No.	Type	Composition	Operating temperature °C	Operating temperature °F	Cycle time, min
1	Solvent cleaner	...	...	...	3-10
2	Warm rinse	Water	38	100	1-3
3	Hot rinse	Water	82	180	1-3
4	Manganese phosphate	(a)	93	200	(b)
5	Oil	Soluble oil, 5%	60	140	1-3
6	Alkaline cleaner	...	93	200	3-10

(a) Contains 12 points total acid, as measured by titration of a 2-mL sample. (b) For light coating, 8 to 12 min; for heavy coating, 10 to 20 min

Coating weight is determined by cleaner used and immersion time in phosphating solution

are available and usually contain a heavy metal phosphate. They are used in the water rinse just before phosphating.

To meet severe requirements, manganese phosphate coatings may be produced to extremely heavy weights. It is difficult, however, to obtain uniformity in such coatings. Pretreatment, including (*a*) cleaning, (*b*) rinsing, and (*c*) etching, as well as the equilibrium of the phosphating solution, is critical. The cycles and operating conditions used in one plant for applying extremely thick and heavy manganese phosphate coatings to carburized and hardened differential gears are listed below. These gears had been failing through localized surface seizure caused by extremely high unit loading. Extreme-pressure gear lubricants, zinc and manganese phosphate coatings of conventional weight, and various other surface treatments proved ineffective in preventing metal-to-metal contact. The manganese phosphate coatings which were 25 to 75 µm (1 to 3 mils) thick and weighed 110 to 325 g/m^2 (10 000 to 30 000 mg/ft^2), prevented seizure under the most adverse conditions. The cycles for application of heavy manganese phosphate coating for lubrication of carburized and hardened differential gears are as follows:

Alkaline immersion clean	
Time	15 min
Temperature	99 °C (210 °F)
Concentration of proprietary cleaner	30-45 g/L (4-6 oz/gal)
Cold water rinse, sulfuric acid etch	
Time	5 min
Temperature	Ambient
Concentration of sulfuric acid	10-20%
Cold water rinse, manganese phosphate	
Time	15-60 min
Temperature (min)	91 °C (195 °F)
Concentration:	
Total acid	55-85 points(a)
Free acid	8-17 points(a)
Ferrous iron	0.05-0.40 g/L (0.007-0.05 oz/gal)

(a) Total acid refers to total ions in solution. Free acid refers to hydrogen ions in solution. Points are the minimum millilitres of the titrating solution required to cause a reaction with a definite quantity of the solution being tested. Reaction is indicated by a color change of the solution

Inspection Methods

The majority of phosphate coating quality control methods are based on visual inspections. For zinc and manganese phosphate, the coating must be continuous, adhere well to the surface, and be of uniform crystalline texture suitable for the intended use. Color should be from gray to black. Causes for rejection include (*a*) loose smut or white powder (because of inclusion of ferric phosphate by-product into the phosphate coating or dried phosphate solution), (*b*) blotchiness, (*c*) excessive coarseness, and (*d*) poor adhesion. Crystal size may be observed by using micrographs at magnifications of 10X to 500X, depending on the coating. Iron phosphate coatings have no apparent crystalline texture. Instead they appear to be amorphous. Their color varies from iridescent yellow to blue to brown. Loose or patchy coatings are cause for rejection.

Determination of coating weight on ferrous surfaces can be made by a stripping procedure, such as follows:

1 Phosphate a part of known surface area
2 Thoroughly clean part to remove all oil
3 Weigh part to nearest tenth of a milligram
4 Strip phosphate coated part in a 2½% chromic acid solution at 71 °C (160 °F), immersing for 10 min (zinc phosphate coating), 15 min (manganese phosphate coating), or 5 min (iron phosphate coating). Time and concentration of the chromic acid solution may require adjustment for specific coatings
5 Rinse in clean water
6 Dry
7 Reweigh stripped part to nearest tenth of a milligram; difference in weight from that in third step equals total coating weight
8 Calculate weight of coating per unit of area; standard units are grams per square metre

If the size or shape of items being coated preclude the performance of the above procedure, test specimens of identical material, heat treatment, and surface finish may be substituted. An accurate measurement of coating weight cannot be obtained by weighing the part, applying the phosphate coating, and then reweighing the part. Because the phosphating solution attacks steel, a measurable, but not always predictable, amount of steel is removed. This condition can vary with the acidity of the bath as well as with the type of metal being coated.

Coating voids or spots not covered may be checked by using a clean, dry phosphated specimen. Next, soak a piece of filter paper, 40 to 50 cm^2 (6.2 to 7.7 in.^{2}) in area, in a solution containing 7.5 g/L (1.0 oz/gal) potassium ferricyanide and 20 g/L (3 oz/gal) sodium chloride. Allow excess solution to drain

Fig. 8 Spray zinc phosphating

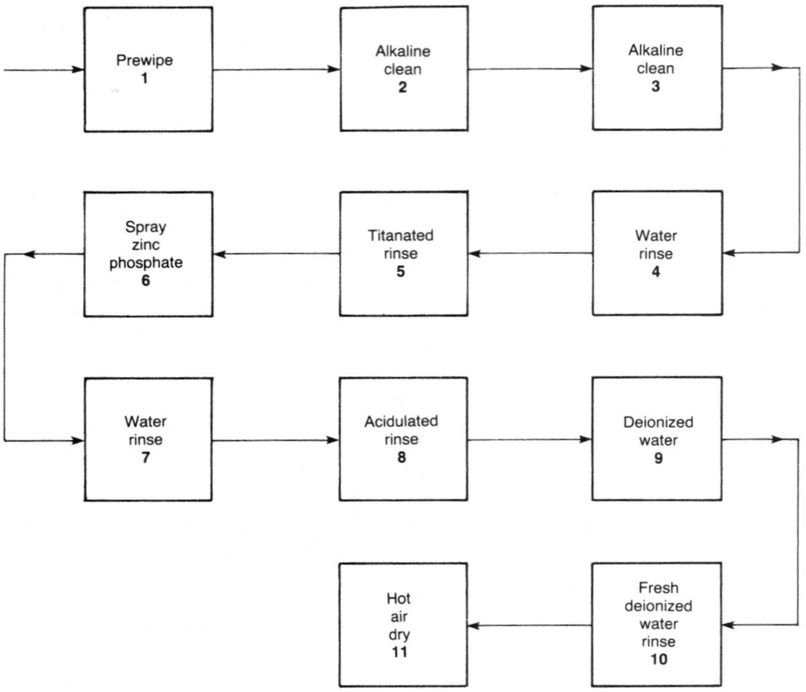

Stage	Type	Composition	Temperature °C	°F	Time, s
1	Organic solvent	Mineral spirits	30	86	60
2	Alkaline cleaner	Titanated, alkali 6.0 g/L (0.8 oz/gal)	60-65	140-150	70
3	Alkaline cleaner	Titanated, alkali 6.0 g/L (0.8 oz/gal)	60-65	140-150	70
4	Hot rinse	Water	55-60	130-150	150
5	Activated water rinse	Titanated, 7.5-8.5 pH 0.5 g/L (0.07 oz/gal)	40-45	104-115	35
6	Zinc phosphate	ClO_3 accelerated(a)	50-55	122-130	70
7	Rinse	Water	35	95	15
8	Acidulated rinse	Partially reduced chromic acid (150 to 200 ppm Cr^{6+})	35	95	35
9	Rinse	Deionized water (100 μmho max	35	95	70
10	Rinse	Deionized water (10 μmho)	35	95	15
11	Dryer	Hot air	...	...	...

(a) Total acid 20 mL, free acid 0.9 mL, nitrite accelerator 1.5 mL; acid checked with 10-mL sample, accelerator checked with gas evolution apparatus

off. Apply wet filter paper to phosphated sample 5 min. Remove and observe blue spots, which indicate noncoated areas. The method of rating may vary with different processes and requirements. One general method is as follows:

- *Excellent*: none to three fine spots up to 1 mm (0.04 in.)
- *Good*: not more than 10 fine spots
- *Satisfactory*: not more than 20 fine spots or up to 3 large spots

Repair of Phosphate Coatings

Small parts that did not accept a satisfactory phosphate coating can easily be stripped, cleaned, and rephosphated. Large parts with a faulty coating or with a coating that was damaged in processing are less easily handled, and repair of phosphated surface may be preferable to stripping and rephosphating. The simplest method is to sand the phosphate film until all defective coating is removed and clean, bare metal is exposed. A proprietary phosphating solution compounded for this application is brushed or wiped on the area to be rephosphated and is allowed to remain for a prescribed length of time (usually measured in seconds). Surplus solution is then removed by thoroughly water rinsing and wiping dry with clean rags. These solutions range from simple systems—phosphoric acid, butyl cellosolve, and a suitable wetting agent, plus 50 to 70% water—to accelerated systems that produce a crystalline zinc phosphate coating. If volume of repairs is considerable, a portable steam spray unit can be used. This will spray hot phosphating solution, water, and chromic acid rinsing solution through a hose and nozzle.

Limitations of Phosphating

Limitations Imposed by Shape. It is seldom impossible to phosphate a part because of its shape. However, shapes can restrict production or limit the choice of process. Parts with complex passages must be immersion coated, because spray phosphating cannot reach all areas of the passages. Cup-shaped parts, phosphate coated by either method, present problems of liquid retention and require special handling to achieve complete drainage. Blind holes or cavities may entrap air, preventing phosphating solution in the immersion process from contacting all areas to be coated.

At one company, hydraulic pump components such as gears, vanes, and valves, and hydraulic valve bodies were manganese phosphated to provide break-in lubrication. Many of these parts had blind tapped holes on several surfaces. Although all critical wearing surfaces of these components were adequately coated, investigation showed that many of the blind holes were only partially phosphated because of air entrapment. Similar parts, with holes or cavities that did require phosphate coating, would require special handling, to ensure coating of these areas.

In another instance, large cylinder heads weighing approximately 120 kg (265 lb) as cast were coated with a microcrystalline zinc phosphate coating to prevent rusting during storage. These parts were placed in baskets in such a way that no air was entrapped. All internal surfaces were permitted to come in contact with the phosphating solution.

Tanks that require a phosphate coating on the inside after fabrication, and that have few drain holes or openings of any size, must be phosphated by immersion. However, if these tanks require phosphate coating on the outside only, spray coating is more appropriate.

Limitations Imposed by Size. Size of parts that can be phosphated is limited only by size and type of equipment available. However, part size does generally determine the method of application. Very small parts such as springs, clips, nuts, bolts, and washers, are almost always coated by immersion. Spray phosphating of these parts on a volume basis would be impractical. Conversely, for extremely large parts, such as transformer housings, which may be as much as 6 m (20 ft) high, spray phosphating is the only practical method for volume production. On extremely large parts produced in low volume, however, coatings are usually applied by brushing or wiping. Parts in between these extremes of sizes are coated by either spray or immersion, depending on (*a*) equipment available, (*b*) quantity to be coated, and (*c*) complexity of parts. Examples of parts satisfactorily coated by both methods include automobile bodies, castings, panels, and machined parts.

Table 8 Sequence and details of operations for an automatic manganese phosphating indexing line

Operation	Temperature °C	Temperature °F	Time, min	Function	Concentration
1 Alkaline cleaner	95	125	5	Degrease, remove soils	6.0-8.0 mL(a)
2 Alkaline cleaner	70	160	5	Degrease, remove soils	4.0-6.0 mL(a)
3 Water rinse	70	160	0.5	Remove alkali from parts	...
4 Water rinse	90	195	0.5	Remove alkali from parts	...
5 Water rinse	90	195	0.5	Remove alkali from parts	...
6 Water rinse	35	95	0.5	Remove alkali from parts	...
7 Pickle	40	105	5	Remove scale or rust	Phosphoric acid, 20%
8 Water rinse	90	195	0.5	Remove acid from part	...
9 Activating rinse	35	95	3	Refine phosphate crystal	0.2-4% wt/vol
10 Phosphate	98	210	20	Manganese phosphate coating	FA 1.5-2.0 mL(b) TA 9.5-12.0 mL(b) Iron, 1.5-2.0 mL
11 Water rinse	35	95	0.5	Remove acidic phosphate solution	...
12 Acidulated rinse	35	95	0.75	Remove water salts, provide rust-proofing	...
13 Dryer	...	...	...	...	...
14 Oil dipping	...	...	0.75	Provide corrosion resistance	...

(a) Number of millilitres required to titrate a 10-mL sample to the phenolphthalein end point using 0.1*N* hydrochloric acid. (b) 2-mL sample size for FA and TA titrations, 10-mL sample size for iron titration

Supplemental Oil Coatings

Unless they are to be painted, parts usually receive a supplemental coating of oil after being phosphated. This coating is applied to increase corrosion resistance, and also neutralizes any residual acid that might remain on parts from the phosphating bath. The type of oil used depends on (*a*) degree of corrosion protection desired, (*b*) subsequent operations to be performed on phosphated parts, and the handling involved in these operations, (*c*) appearance requirements, and (*d*) compatibility of the oil with other lubricants in assemblies. Materials commonly used are (*a*) water-soluble oils, (*b*) nondrying oils, (*c*) non- hard-drying greaselike materials, and (*d*) oils that are dry to the touch.

Water-soluble oils provide both short-term and long-term protection against corrosion, depending on their composition. Water-soluble oils offer the advantage of allowing parts to go into water-soluble oil in a wet state. An additional advantage of the water-soluble oil is that it eliminates a fire hazard from the operation.

Figure 12 shows an immersion tank used in the coating of lightweight parts with soluble oil for applications in which subsequent handling or assembly requirements require a virtually dry part.

Flash points of water-soluble oils are sometimes lower than those of petroleum-based oils or synthetic organic oils. However, after water-soluble oils are mixed with water to the 5 to 25% concentration range, little fire hazard attends their use.

Nondrying oils vary in type and viscosity and are selected on the basis of requirements of in-process handling or ultimate service. An advantage of this type of material is its ability to self-heal any scratches that may occur in bulk handling. Corrosion protection may be increased by adding a commercially available rust inhibitor that is compatible with the oil. Petroleum-based oils can be reduced with petroleum solvents to form a thinner film if desired. If parts are not completely dry before the application of oil, water-displacing additives may be used.

Nonhard-drying materials are greaselike substances that have melting points above room temperature. These materials may be applied by (*a*) dipping, (*b*) spraying, or (*c*) brushing. When necessary, these materials are readily removed by petroleum solvents. Figure 13 shows a tank for dip application of greaselike materials. The tank is provided with facilities for heating and cooling to maintain temperature control of the coating material.

Safety Precautions

Safety precautions on a phosphating line must begin with the basic design of the equipment involved.

Immersion Phosphating. Proper ventilation of immersion tanks is necessary to eliminate concentration of vapors from the tanks in buildings or work areas. Local regulations in some areas, however, may prohibit exhausting directly to the outside, and special filtering equipment may be required. Tanks containing acid must be resistant to the acid they hold to eliminate the possibility of the acid corrod-

Fig. 9 Spray zinc phosphating

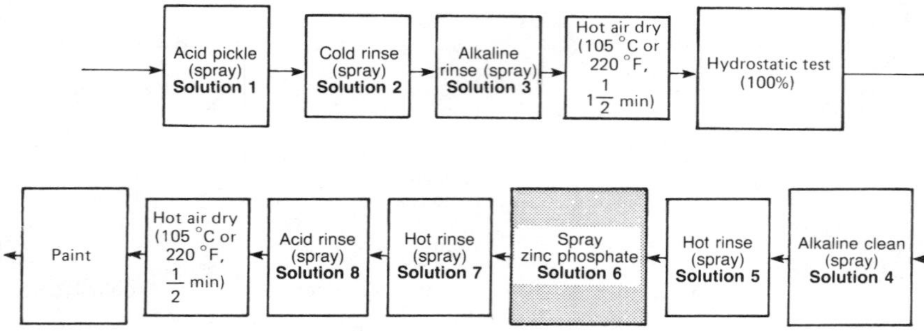

Solution No.	Type	Composition	Operating temperature °C	°F	Cycle time, min
1	Acid pickle	H_2SO_4, 25 wt %	71	160	2
2	Cold rinse	Water(a)	Room		1¼
3	Alkaline rinse	$NaNO_2$, 0.9 kg/380 L (2 lb/100 gal)(b)	66	150	1¼
4	Alkaline cleaner	Alkali, 0.7 g/L (0.1 oz/gal)	71	160	½
5	Hot rinse	Water	66	150	½
6	Zinc phosphate	NO_2, accelerated(c)	60	140	1½
7	Hot rinse	Water(a)	60	140	½
8	Acid rinse	Chromic and phosphoric acids(d)	71	160	½

(a) Purity maintained by overflow. (b) Sodium hydroxide added to establish pH of 11. (c) Total acid, 10 points; free acid, 0.7 to 1.1 points; acid checked using 10-mL sample. NO_2 accelerator, 1.5 to 2.0 points, determined using 25-mL sample. (d) Free acid, 0.4 to 0.6 points; total acid, less than 5 points; checked using 25-mL sample

Spray zinc phosphating 80-mm (3.2-in.) mortar shell casings before painting. Total area, inside and outside, of each shell was 0.1 m² (1 ft²); coating weight ranged from 1.7 to 2.1 g/m² (160 to 200 mg/ft²). Conveyor speed was 0.03 m/s (5 ft/min); production rate was 1000 shells per hour

Fig. 10 Consumption of zinc phosphating concentrate as a function of area

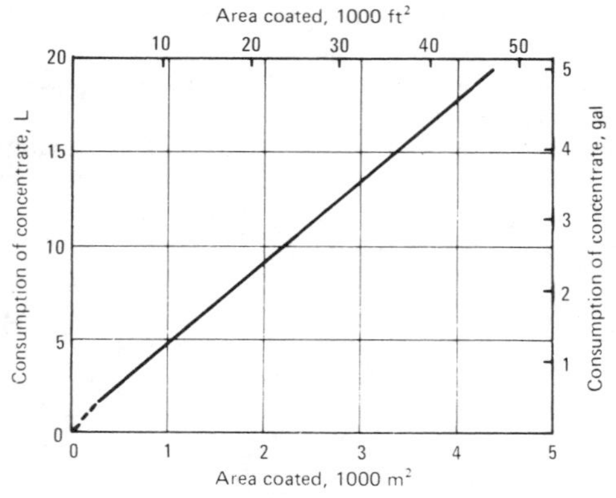

Small threaded parts coated to 10.8 g/m² (1000 mg/ft²) in barrel phosphating

ing through the tanks and spilling on the floor. Curbing should surround tanks to retain spilled or leaked solutions.

Spray Phosphating. Equipment used in spray phosphating must be properly vented for removing vapors. A heavy grating should surmount each of the various tanks for protecting the personnel cleaning and repairing tanks, risers, and spray nozzles. These gratings also prevent workpieces from falling into tanks from conveyors. Access doors in drain areas, which are used to check carryover and condition of work, and for access during breakdown, should be easily opened from the inside.

Handling of Alkalis and Acids. All alkaline cleaners should be handled with care. Rubber gloves and face or eye shields should be worn when these materials are added to cleaning tanks. Should these materials contact the skin, it should be flushed with water as soon as possible. Repeated or prolonged contact can cause skin irritation.

These precautions apply also to handling acids used in phosphating, phosphoric acid, and chromic acid. Although chromic acid is oxidizing, it does not burn the skin immediately, as do common mineral acids, but severe irritation results from prolonged exposure to the skin. Goggles or face shields should be worn at all times during handling of chromic acid, because contact with the eyes can cause serious damage. All contaminated clothing should be removed and washed before reuse.

To transfer liquid from a carboy, a carboy-tilter, commercial siphon, or bulb siphon should be used. Liquid should never be drawn from a carboy by using air pressure to force it out, even when using the so-called air pressure reducers. Danger is always present that the carboy will break, or even explode, and spray or splash acid on the operator. This also holds true for drums. All drums should be specified to be plugged with one-way breather plugs. If this is not possible, the solid plug must be removed carefully to avoid acid spray. The plug should be opened only enough to permit the compressed gases to escape slowly. Once the inside and outside pressures are equalized, the plug may be removed.

Disposal of Solutions

Nearly all municipalities have maximum limitations on the (*a*) acidity,

Fig. 11 Three principal phosphate coatings

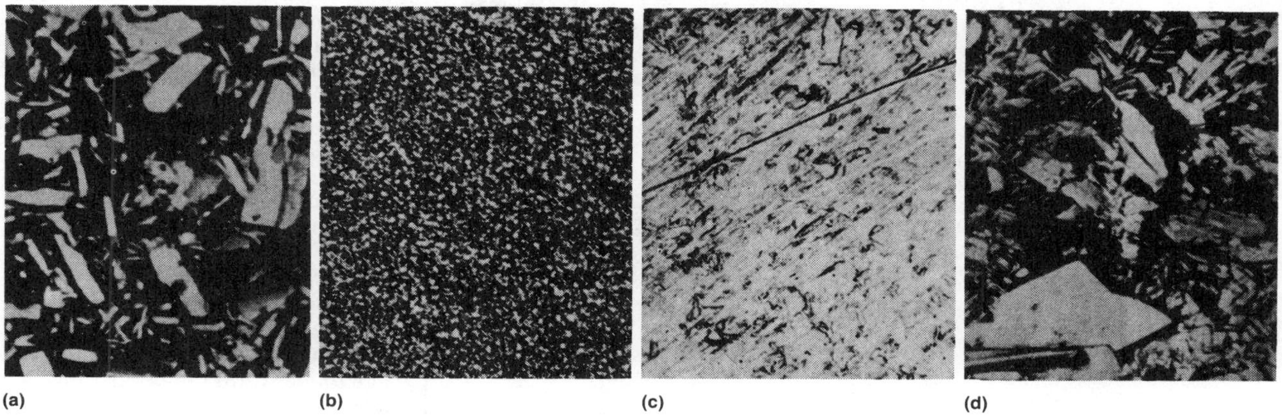

Magnification, 125X. (a) Heavy zinc phosphate. (b) Microcrystalline zinc phosphate. (c) Iron phosphate, principally iron oxide. (d) Manganese oxide

(*b*) solids, (*c*) chromium, (*d*) zinc, (*e*) nickel, (*f*) iron, and (*g*) oil that may be drained to a stream or sewer system. Because these limits vary widely, local governmental authorities should be consulted for specific restrictions on each toxic or nontoxic material in a phosphating system so that proper waste treatment plans can be made. The trend is toward stricter controls, and waste treatment programs should include a margin of safety to accommodate future changes. Limits and suggested treatments for the more serious contaminants from a phosphating system are indicated in Table 9. The waste treatment methods described are widely used, but other methods also are available.

Two sets of limits need to be considered: (*a*) direct discharge to streams, lakes, and rivers, and (*b*) discharge to publicly owned treatment works. Table 9 lists the limits of contaminants.

Soda ash is not the best neutralizing agent. In order of preference, lime, caustic and soda ash are suggested. Recommended neutralization pH values are:

Chemical	pH
Zinc	7.0-9.0
Nickel	8.5-9.5
Chromium	8.0-9.0

Limits and Treatments of Contaminants

acidity. Influenced by cleaning stage (alkaline), phosphating stage (acid), and final rinse (acid)

chromium. Most of the chromium in final rinse stage is in the more toxic hexavalent form but can be readily reduced to the less toxic trivalent form with sodium bisulfite or sodium metabisulfite in a pH range of 3.5 to 5.0. To remove the trivalent chromium, add an alkaline material such as soda ash, mixing thoroughly, to a pH of 8.5 to 10.0, allow to settle for 1 h, and draw off the liquid, leaving the chromium in the sludge. The drawn-off liquid is essentially water and may be discharged into a sewer or stream in the normal manner, in compliance with local regulations

zinc. Usually, it is economically desirable to avoid any loss of zinc phosphating solution, and therefore a problem exists only in disposal of the rinse solution (contaminated water) following the phosphating stage and the sludge periodically removed from the phosphating tank. The sludge should be hauled to a suitable dump, rather than be discharged into a sewer or stream. If it is necessary to dispose of a zinc phosphating solution, or if the rinse water, following the phosphating stage, contains enough zinc phosphating solution from carryover to create a disposal problem, first precipitate the zinc by raising the pH to 8.5 to 10.0 with any available alkaline material, such as soda ash. Then, after thorough mixing, allow to settle for 1 h and draw off the liquid, leaving the zinc to be handled as a sludge. This drawn-off liquid also is essentially water, and may be discharged into a sewer or stream in the normal manner in compliance with local regulations

Fig. 12 Tank for oil coating

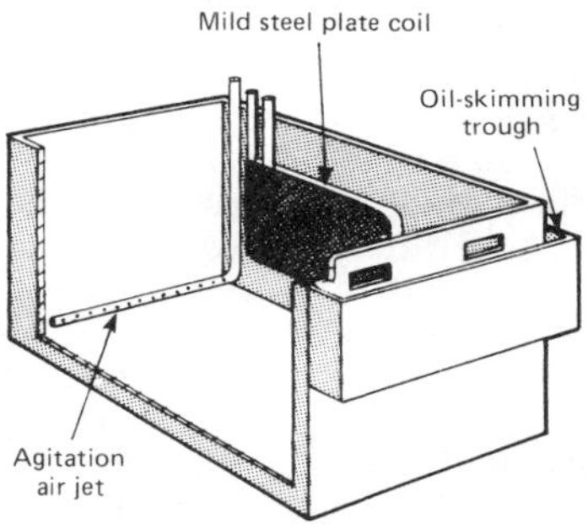

Used in immersion application of a thin coating of soluble oil to lightweight phosphated parts. Skimming trough removes floating globules of oil that might cling to parts

nickel. Some of the newer zinc phosphating compounds that coat both bare and galvanized steel contain appreciable quantities of nickel. This presents no in-plant problems of disposal because nickel is precipitated along with zinc when disposal of contaminated rinse water or phosphating solution is required. However, because nickel may be more harmful in a sewage disposal plant, some municipalities that have accepted small quantities of zinc in the drain water demand almost complete free-

Table 9 Limits and treatments of contaminants from a phosphating system

Contaminant	Direct discharge, ppm	Publicly owned treatment works, ppm(a)
Chromium:		
Hexavalent	0.05	0.05-5.0
Total	0.10-0.5	1.0-10.0
Zinc	0.1-0.5	1.0-5.0
Nickel	0.1-0.5	1.0-5.0
Oil	1.0-10.0	10.0-100

(a) Environmental protection laws are subject to change. Suppliers of the metal pretreatment should be contacted to determine needs in terms of treatments of these chemicals

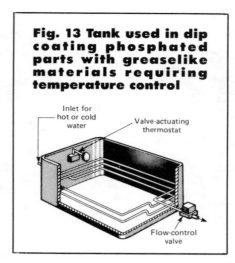

dom from nickel. More complete processing of solutions is required in these situations

oil. Most phosphating systems do not present a serious problem in treatment of oil waste, but some manganese of zinc phosphating systems contain a final oil or soluble-oil dip. A few systems have sufficient oil removed from parts in the cleaning stage to require treatment. Most soluble-oil and water solutions can be broken with calcium chloride, often assisted by small additions of aluminum sulfate or flocculating agents. The separated oil layer is floated off for removal in a tank truck or in drums, and the water, if sufficiently clear, is metered to a sewer system. Oils must be physically removed for such nonharmful uses as dust-laying treatments of dirt or gravel roads or parking lots

Low-Temperature Coatings

Both iron and zinc phosphate coatings can be applied at much lower temperatures than have been traditional, thus reducing heat energy costs significantly. Manganese phosphate coatings still require solution temperatures around 93 °C (200 °F). Solutions are available to apply iron phosphate by either dip or spray at 24 °C (75 °F), producing coating weights from 0.44 to 0.66 g/m^2 (40 to 60 mg/ft^2). Zinc phosphate baths can be compounded to produce 3.3 to 5.5 g/m^2 (300 to 500 mg/ft^2) coatings at 38 °C (100 °F).

Fig. 13 Tank used in dip coating phosphated parts with greaselike materials requiring temperature control

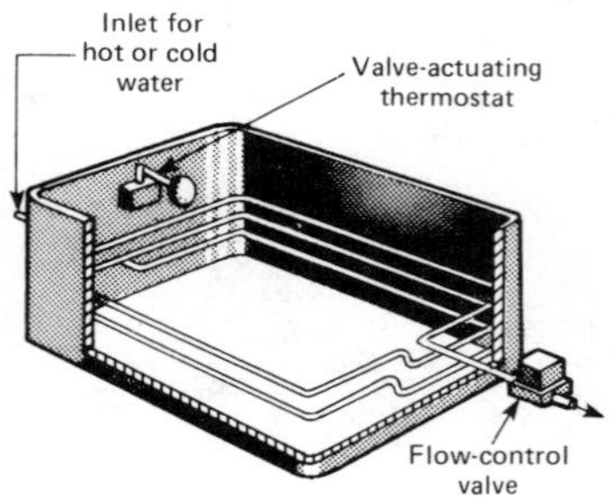

Energy demands of hot spray systems are determined primarily by their temperature and recirculation rate. A paint preparation line using iron phosphate sprayed at 71 °C (160 °F) was found to be using 8.5 million kg·cal/day (33.9 million Btu/day). The iron phosphating stage is part of the pretreatment of an electrocoat prime paint line for castings and forgings. The energy requirements for the spray iron phosphate are:

- Tank capacity: 8000 L (2100 gal)
- Pump rate: 2300 L/min (600 gal/min)
- Temperature: 70 °C (160 °F)
- Operating time: 16 h/day
- Heat up energy: 0.55 million kg·cal (2.22 million Btu)
- Temperature maintenance energy: 0.45 million kg·cal/h (1.98 million Btu/h)
- Total energy requirements per day: 8.5 kg·cal/day (33.9 million Btu/day)

Changing to a product operating at 24 °C (75 °C) allowed the system to operate with no heat required beyond the pump energy, which maintained the required 24 °C (75 °F). With no requirement for heating coils or heat exchangers, new installations save those capital expenditures. Both systems produced coating weights of 0.44 to 0.55 g/m^2 (40 to 50 mg/ft^2).

Although immersion baths have much lower heat losses than spray applications, worthwhile heat savings still can be realized. In general, heat losses rapidly increase at about 60 °C (140 °F).

A system producing fine grain immersion zinc phosphate coatings on miscellaneous formed and machined parts was investigated, and actual steam usage to maintain temperature was measured. Energy demands did not include heating the parts but can be calculated from their mass. The following is a description of the system:

- Tank capacity: 4200 L (1100 gal)
- Tank surface area: 2.8 m^2 (30 ft^2)
- Operating time: 16 h/day

At 88 °C (190 °F) operating temperature

- Heat up energy: 0.30 million kg·cal (1.21 million Btu)
- Temperature maintenance energy: 0.04 million kg·cal/h (0.160 million Btu/h)
- Total energy required per day: 0.95 million kg·cal (3.77 million Btu)

At 55 °C (130 °F) operating temperature

- Heat up energy 0.022 million kg·cal (0.088 million Btu)
- Temperature maintenance: 0.0027 million kg·cal (0.011 million Btu/h)
- Total energy required per day: 0.065 million kg·cal (0.26 million Btu/day)
- Energy Savings 55 ° versus 88 °C (130 ° versus 190 °F): 0.88 million kg·cal or 212 million kg·cal (3.51 million Btu/day or 842 million Btu/yr)

Please note that less energy was required to maintain the tank at temperature than to shut down for 8 h and reheat.

Chromate Conversion Coating

By the ASM Committee on Chromate Conversion Coating*

CHROMATE CONVERSION COATINGS are formed on metal surfaces as a result of the chemical attack that occurs when a metal is immersed in or sprayed with an aqueous solution of chromic acid, chromium salts such as sodium or potassium chromate or dichromate, hydrofluoric acid or hydrofluoric acid salts, phosphoric acid, or other mineral acids. The chemical attack causes the dissolution of some surface metal and the formation of a protective film containing complex chromium compounds.

A variety of metals and electrodeposited metal coatings, including zinc, cadmium, magnesium, and aluminum, may be chromate conversion coated. Several articles in this Volume contain details of the procedures used to apply chromate conversion coatings to specific metals and metal coatings. The conversion coating of cadmium electrodeposits is discussed in the article on cadmium plating; the articles on cleaning and finishing of aluminum and magnesium alloys contain information relating to the application of chromate conversion coatings to these metals; and procedures for applying conversion coatings to electrodeposited zinc are described in the article on zinc plating. Chromate conversion coating of zinc is also treated in ASTM B201. Chromate conversion coatings on aluminum have been approved for military specification MIL-C-5541B.

Process Mechanism. Chromate treatments are of two types: (*a*) those that are complete in themselves and deposit substantial chromate films on the substrate metal, and (*b*) those that are used to seal or supplement oxide, phosphate, or other types of nonmetallic protective coatings. Two chromate treatments for aluminum are complete in themselves, chromium chromate and chromium phosphate. To form the chromium phosphate coating, phosphoric acid or phosphoric acid salts are required in addition to chromic and hydrofluoric acids or their corresponding salts.

For a chromate film to be deposited, the passivity that develops on a metal in a solution of strictly chromate anions must be broken down in solution in a controlled way. This is achieved by adding other anions, such as sulfate, nitrate, chloride, or fluoride, as activators to attack the metal. When attack occurs, some metal is dissolved, the resulting hydrogen reduces some of the chromate ion, and a slightly soluble hydrated chromium chromate, is formed:

$$[Cr(OH)_2] + [HCrO_4] - \cdot Al(OH)_3 \cdot 2H_2O$$

This compound is deposited on the metal surface unless the solution is sufficiently acid to dissolve it as soon as it is formed. The two types of chromium chromate coatings are accelerated and nonaccelerated (shown above). The accelerated process usually involves the addition of ferricyanide ion or molybdate to the bath.

The stability of the natural oxide film reinforced by the chromate ion determines the conditions of pH, ratio of activating anion to chromate, and temperature at which the oxide is broken down and a chromate film deposited. Thus, magnesium alloys can be chromated in nearly neutral solutions, whereas aluminum alloys can be treated only in solutions of appreciable acidity or alkalinity.

Properties. Most conversion coatings dissolve very slowly in water and provide limited protection against corrosion in this medium. Chromate coatings are beneficial under marine atmospheric conditions or in high

*James W. Davis, Group Leader, Amchem Products, Inc.; James I. Maurer, Director, Technology, Parker Division, Occidental Chemical Corp.; Lester E. Steinbrecher, Director, Research and Development, Amchem Products, Inc.

humidity. These coatings can retard the formation of white corrosion products on zinc or cadmium when these metals are subjected to prolonged outdoor exposure, particularly in rural and seacoast environments. The protective value of chromate conversion coatings in water or air increases directly with the thickness of the coating.

Conversion coated aluminum parts are tested for corrosion resistance according to Federal Test Method Standard 151, Method 811.1. Chromate conversion coatings are both decorative and functional. They can be produced in a variety of colors, ranging from the very bright coatings obtained on zinc and cadmium that simulate the appearance of bright nickel and chromium, to the olive drab frequently applied to military equipment. Chromate coatings provide an excellent nonporous bond for all paints that have good molecular adhesion. Clear conversion coatings on cadmium, copper, and silver do not interfere with soldering when rosin fluxes are used. However, the soldering of zinc that has been conversion coated requires prior removal of the chromate coating or use of an acid flux.

Aluminum. Chromium chromate coatings on aluminum range in weight from 0.088 to 3.52 g/m^2 (5 to 200 mg/ft^2) and may be stable up to temperatures of 300 to 320 °C (570 to 610 °F), depending on exposure time. The corrosion resistance of the coatings on aluminum decreases after heat treatment. Chromium phosphate coatings on aluminum range in weight from less than 0.0176 g/m^2 (1 mg/ft^2) to over 8.8 g/m^2 (500 mg/ft^2) and range in color from clear to emerald green.

To pretreat aluminum for painting, especially coil and preformed can processing, an alternative to chromate conversion coatings are processes based on zirconium and titanium, with additions of modifying fluoride and phosphate compounds. Such processes are more ecologically acceptable than chromium-containing coatings. End products include: aluminum siding, gutters, and decorative panels, and beer and beverage cans.

Rust-Preventive Compounds*

RUST-PREVENTIVE COMPOUNDS are removable coatings used to protect iron and steel surfaces against corrosive environments during fabrication, storage, or use. These compounds are blends of various basis materials, additives, and inhibitors. Rust-preventive compounds vary in physical characteristics, coating properties, and degree of rust protection afforded, depending on composition.

These compounds occupy potential corrosion sites and prevent electrolytes such as water and oxygen or oxygen donor compounds from coming into contact with the metal surface and causing corrosion. Compounds usually are applied as a relatively thick barrier film, a relatively thin film of a polar material with a greater affinity for the metal than the potential electrolyte, or a combination of the two types.

Although the rust preventives discussed in this article are divided into seven types, two or more of these types are often combined into a single material when reduced handling and processing time or enhanced rust protection are required. Petrolatums, which provide barrier films, are often fortified with polar materials to increase the degree of rust protection provided. Fingerprint-neutralizing and water-displacing materials are frequently combined in one product to provide an in-process rust preventive and washing fluid.

Types of Compounds

Rust-preventive compounds may be divided into seven general categories:

1 *Petrolatum compounds* have a greaselike consistency. These compounds provide protection by the inclusion of effective corrosion inhibitors and by the continuous physical barrier presented to corrodents by the film. The film is self-healing and stable over a wide range of ambient storage temperatures. These materials can be applied at room temperature by brushing or wiping, but usually they are heated to a fluid state and applied by dipping or spraying
2 *Oil compounds* are similar to lubricating oils but contain active rust-inhibitors and corrosion-inhibitors, in amounts to afford effective protection under various conditions of exposure
3 *Hard dry-film compounds* establish a film either by the evaporation of a solvent diluent or by chemical reaction after application. The coatings generally are thin but fairly hard, being similar in appearance to varnish. These films offer excellent protection, but some are extremely difficult to remove; usually, the films cannot be removed by wiping with petroleum solvents, such as kerosine, Stoddard solvent, and naphtha. They withstand abrasion and handling, and several types are suitable for use as an undercoat for paint
4 *Solvent-cutback petroleum-based compounds* are deposited as residual coatings through evaporation of the solvent. This class encompasses materials ranging from those that leave a thin, transparent film to those forming a relatively heavy asphaltic or bitumastic film that hardens sufficiently to form an impervious barrier. The wax films are included in this class of materials. Solvent-cutback coatings may be used on precision surfaces for storage and between processing stages. The heavier coatings are applied to articles with relatively rough surfaces, such as bull gears, pipe and structural components, for protection against severe outdoor exposure during storage
5 *Emulsion compounds* rely on the polarity and preferential attraction of emulsifiers or inhibitors to metal surfaces for protection of the surfaces after the water phase has evaporated. Without water dilution, emulsifying oils or soluble oils are effective short-term protective coatings and are generally used for in-process protection. They are economical and easy to apply and remove
6 *Water-displacing polar compounds* contain preferential wetting agents that have an affinity for metal sur-

* Reviewed by Howard H. Hovey, Chief, Standardization Branch, Packaging Division, Department of the Army

faces. They displace moisture and establish a protective coating directly on the metal

7 *Fingerprint removers and neutralizers* are low-viscosity compounds containing suitable solvents to dissolve, suppress, and neutralize acids, salts, and residues from handling and from other sources of surface contamination. They are used for in-process or other short-term protection, or as a pretreatment for longer term protection. They must be removed with a solvent or mild cleaning solution before a long-term rust preventive is applied

Table 1 lists the characteristics, applications, and properties of rust-preventive compounds covered by military specifications.

Many rust-preventive materials are formulated to function also as lubricants or drawing compounds, or as a base for paint. Table 1 contains many examples of corrosion-preventive oils and greases that also provide lubrication. MIL-C-6529, an aircraft-engine corrosion preventive, can also be used as an engine lubricant for flight service, when used in the appropriate range of viscosity. MIL-C-8188 describes another preservative that provides adequate lubrication for jet and turboprop engines. Conversely, many lubricating oils are capable also of preventing corrosion. A low-temperature lubricating oil meeting specification VV-L-800 has water-displacing and general-purpose preservative qualities. A military lubricating oil for instruments, MIL-L-6085, simultaneously protects instruments from corrosion. A hydraulic fluid, MIL-H-6083, also functions as a corrosion preventive.

In industrial operations, corrosion preventives serve dual purposes in metal forming operations. For example, in steel mills, a slushing oil (a light preservative oil) is used to coat hot rolled steel strip after pickling. This oil displaces the residual moisture from the pickling operation and protects the steel from corrosion. Subsequent cold rolling operations are assisted by the lubricating properties of the slushing oil on the surface of the sheet or strip. In metal forming operations, residues of properly selected oil coolants can function as rust preventives.

Selection of Material

The occurrence of rust on the surfaces of steel or other iron-base alloys depends on the contact of the surfaces with moisture and oxygen. The extent of rusting is a function of the duration of this contact and of temperature. Temperature also affects the rate of rusting, which may be accelerated by the presence of salt chlorides, oxides of sulfur, and other electrolytes, which enter into or modify the chemical reactions involved.

Composition and metallurgical characteristics of the alloy also influence the extent and rate of rusting, and frequently determine whether the rusting reaction affects the surface in a general and uniform way or by the development of localized pitting. Two important factors in determining the extent to which a surface must be protected from rusting are (*a*) the degree of finish, for example, whether the surface is ground, rough machined, or as-forged; and (*b*) the extent to which rust impairs function. The selection of rust-preventive compounds is influenced by:

- Environment, which includes climatic conditions, geographic location, and type of storage facilities
- Anticipated time in storage
- Material or object to be protected
- Necessity for removing compound

The severity of climatic, geographic, and storage environments varies considerably. Mild conditions are encountered in a rural inland area with a daily temperature that ranges from −1 to 16 °C (30 to 60 °F) and a range of relative humidity of 25 to 70%. Severe conditions are those of a southwestern seacoast area where the temperature differential may exceed 33 °C (60 °F) and where 100% relative humidity is reached once or twice each day. When such conditions result in the dew point being reached, the result is moisture condensation. Therefore, the requirements for a successful rust preventive for any specific surface stored for six months inside a completely enclosed building would range from a very light oil for the rural area, to a heavy-film petrolatum for the seacoast location.

Outside storage in either location would require the use of a material designed to cope with the erosive effect of rainfall. In an industrial inland area, the corrosion-preventive requirement is increased to provide resistance to the increased reactivity of the fumes present in the industrial atmosphere; for example, a very light oil might need to be replaced by a soft petrolatum. Careful determination of the environmental factors against which the rust preventive must perform is of major importance in selecting a proper rust-preventive material. Duration of exposure to the environment is another important consideration.

The requirements of a rust preventive can be affected significantly by the degree of auxiliary protection afforded by the type of enclosure used for storing coated articles. Maximum auxiliary protection is provided by storage in a fully enclosed, humidity-controlled area in a permanent building, or in a container sealed against moisture and holding desiccants to absorb any moisture originally present. There are four general categories of parts or assemblies to which rust-preventive materials are applied when they are stored:

1. Assembled machinery or equipment in standby storage
2. Finished parts in stock, or spare parts for replacement
3. Tools, such as drills, taps, dies, and gages
4. Mill products such as sheet, strip, rod, and bar

Machinery or equipment in standby storage requires two distinctly different types of protection:

- Exterior surfaces require protection from the atmospheric conditions to which they are exposed, and the rust-preventive material used must be able to withstand normal handling
- Interior surfaces of assembled machinery require the use of preservative oils that must act as temporary lubricants when the equipment is put into service, and that must be completely compatible with the lubricating oils, hydraulic fluids, or greases that are to be used in service

In an adequately heated, well-ventilated building, oil-film protection is usually sufficient for external and internal surfaces of exposed machinery parts. In loosely constructed, inadequately heated buildings, where machinery may be subjected to extreme changes in humidity, petrolatum or solvent-cutback, rust preventives are essential for all exposed surfaces. Interior surfaces must be protected with a good corrosion-preventive oil. For outdoor storage or storage in open-side or unheated sheds in dry, clean atmospheres, appropriate petrolatum outer coatings, and corrosion-preventive oils for interior surfaces should provide adequate protection. In marine atmospheres or where there are corrosive industrial fumes, hard-film corrosion preventives or heavy-duty petrolatum materials must be used on all exposed surfaces; specially designed oil compounds are available for internal or physically shielded surfaces.

Machinery or equipment exposed to rough handling or mechanical abrasion that would damage or remove an oil or petrolatum film should be protected with a hard-film rust-preventive compound of a type that may be removed when required.

One large manufacturer of machine tools whose main plant is situated in an inland industrial location with a daily temperature differential of 11 °C (20 °F) and with frequent 100% humidity during spring and fall months, uses three basic types of rust-preventive products for in-plant preshipment protection of parts and assemblies. For in-process periods of two to three weeks, a very light fingerprint-neutralizer suffices. For longer periods of in-plant storage, an oil rust preventive with a viscosity of about 200 SUS at 43 cSt at 38 °C (100 °F) is necessary. For very long or indeterminate periods of storage of inventory items, or for covered outside storage, a solvent-cutback rust preventive is used. For shipment of assembled machine tools, products similar to those described in MIL-C-16173 (Table 1) are used for the protection of exterior surfaces. Interior protection of lubricating systems and reservoirs is provided by preservative oils similar to grades 1 and 2 of MIL-L-21260 (Table 1).

Finished machined parts and spare parts, such as gears, mill rolls, and mandrels, should be cleaned of coolants, machine oils, cutting oils, and other foreign matter before being coated with rust preventives. Highly sulfurized fatty or chlorinated cutting oils in prolonged contact with steel can cause staining. Covering these cutting oils and coolants with a rust-preventive compound does not prevent this corrosive action.

For short-term indoor protection, fingerprint-neutralizing compounds usually provide adequate protection for up to two weeks. Although test results for some products indicate protection for up to six weeks, the short-term recommendation should be followed. For long-term indoor storage or storage in corrosive atmospheres, fingerprint neutralizers are first applied to prevent corrosion or staining from body acids or salts deposited on the finished surfaces from finger contact. These neutralizers are then removed, and an appropriate long-term rust preventive is applied.

In an atmosphere that is not particularly corrosive, parts stored in a weatherproof building with no humidity control are protected with a light coating of an oil-based compound. In marine or industrial areas with corrosive atmospheres, appropriate petrolatum or solvent-cutback materials usually provide protection for 6 months. Machined parts stored outdoors should be protected by a heavy coating of a petrolatum material.

Tools. Before drills, reamers, taps, cutter heads, tool holders, and other metal-cutting tools and accessories are placed in storage after use, corrosion-protective measures may be required. Water-emulsion or soluble-oil cutting fluids should be completely removed before tools are coated with a rust preventive to prevent entrapment of moisture. One tool manufacturer suggests cleaning tools with 1,1,1-trichloroethane before applying the protective film. Machine oils, used at the machine, may provide adequate protection for short-term storage of tools and accessories. Surplus oil should be wiped off to minimize runoff in storage.

Although the oily, nondrying rust preventives provide adequate protection for tools stored in the normal crib, the runoff or ruboff of oil may be undesirable. Dry films present no problem in handling or oil runoff, and because the films are dry, normal shop dirt does not adhere to them. If the dry film is not thick, it may be unnecessary to remove it when the tools or accessories are again placed in service. This protective material, if it is not to be removed, should be compatible with fluids used at the machines.

Tools such as drills, reamers, and milling cutters sometimes are protected from rust and abrasion by being coated with a strippable cellulose acetate compound. The coating is applied by dipping the tool in the compound at a temperature of about 120 °C (250 °F).

Hard-film coatings are not recommended for use on adjustable chucks and other tool accessories from which the incomplete removal of a hard film could cause malfunction. For such equipment, thinner, oily films are usually recommended; if shop dirt is a problem, a protective wrapping may be applied. In this case, it is necessary for the wrapping and the preservative to be compatible, so that accelerated corrosion of the metal is not initiated.

Before being placed in storage, molding, stamping, or forming dies may require only a light coating of rust-preventive oil, if the corrosive environment is not severe. They may require a petrolatum compound if the atmosphere is corrosive. It is often unnecessary to provide any additional rust-preventive coating once these dies have been used in production; the oily film of die lubricant remaining from the production run may provide sufficient protection. If residual die lubricants do not provide a complete film, a thin oil usually is sprayed on the die; in extreme cases, a greaselike material is brushed on.

Dies for molding or extruding plastic parts are highly polished. These dies are run dry; consequently, a rust-preventive material, usually a thin oily compound, is sprayed on the die or mold before storage. This material is then removed with a solvent before the die or mold is placed in production.

Mechanical gages require special consideration in the selection of rust preventives. Complete protection is required, yet the film must be such that the gage can be used without the necessity for removing the film. Thin oil preservatives are applied in a light film. Periodic renewal of the film may be necessary.

During temporary storage, used gages often can be protected also by being brushed with a soft petrolatum, which can be removed easily by wiping. Specially treated papers for wrapping precision tools and gages for maximum protection are required during overseas shipment.

Mill products as received from the mill usually are not protected against corrosion. They are available with light oil coatings that protect for short periods, and it is often necessary to provide additional protection. For an extended period of outdoor storage, heavy oils or petrolatums may be required; for short-term outdoor storage, light oils usually provide adequate protection, although exposure to corrosive atmospheres may necessitate more protection.

The use of corrosion-preventive materials on mill products must be considered from the standpoint of the manufacturing operations that the products are to undergo. In the protection of rod and bar stock, rust preventives selected should be compatible with cutting fluids that are to be used, so that the compounds do not require removal. Sheet and strip that are to go through a punch press may be coated with a material that does not require removal and that can function also as the lubricant to facilitate manufacturing.

Finished tubing usually is protected with one of the following products, depending on the severity of the expected

Table 1a Characteristics and applications of rust-preventive materials covered by military specifications

Specification No.	Title	Description	Intended use
MIL-C-450 (Types I, II, and III)	Coating compound, bituminous, solvent type, black	Black asphalt, dry coating; solvent cutback	Protect battery racks; coat interior of projectiles
MIL-C-4339	Corrosion-preventive soluble oil, for water-injection systems	Mineral oil with emulsifying agents and inhibitors	Protection against corrosion by water and alcohol
MIL-C-5545	Compound, corrosion preventive, aircraft engine, heavy oil type	Thixotropic preservative oil; leaves soft, greasy film that becomes fluid when engine is started	Corrosion-protection of internal parts and surfaces of engines and equipment
MIL-C-6529	Corrosion preventive, aircraft engine	Concentrate intended for 1-to-3 dilution with Navy symbol oil	Preservation of turbojet and reciprocating engines
MIL-C-8188	Corrosion-preventive oil, gas turbine engine, aircraft synthetic base	Synthetic oil plus corrosion inhibitors	Preservation of turbojet and turboprop engines
MIL-C-10382	Corrosion preventive, petrolatum, spraying application for food-handling machinery and equipment	Thin film, easily removable when dry; solvent cutback	Corrosion-protection of food-handling equipment
MIL-C-11796 Classes 1 and 1A(a)	Corrosion-preventive compound, petrolatum, hot application	Thick, nondrying, dark, firm, greaselike film; leaves oil slick on salt water(a)	Unshielded outdoor storage of gun tubes; long-term protection of highly finished parts of simple design
Class 2	Corrosion-preventive compound, petrolatum, hot application	Thick, dark, medium-firm, greasy film	Unshielded outdoor storage in moderate climates, at below flow point of compound (64 °C or 145 °F); general packaging of automotive parts
Class 3	Corrosion-preventive compound, petrolatum, hot application	Fairly thin, soft, greaselike material	Protection of highly finished, cleanable parts of complex design; preservation of antifriction bearings
MIL-C-15074	Corrosion preventive, fingerprint remover	A mixture of organic solvents, petroleum solvents, and inhibitors	Removal of fresh fingerprint residues; temporary corrosion prevention
MIL-C-16173 Grades 1 and 1A(b)	Corrosion-preventive compound, solvent cutback, cold application	Hard, dark film about 50 to 100 μm (2 to 4 mils) thick; solvent cutback; dries to touch in 4 h(b)	On metals under outdoor conditions; general-purpose preservation, indoor or outdoor, with or without cover; domestic or overseas shipment where a dry-to-touch film is required; not for intricate assemblies
Grade 2	Corrosion-preventive compound, solvent cutback, cold application	Amber-colored soft film about 25 μm (1 mil) thick; solvent-cutback; drying time, 4 h; 200 μm (8 mils) maximum thickness	Extended undercover protection of interior or exterior surfaces of machinery, instruments, bearings, or materials with or without overwrap
Grade 3	Corrosion-preventive compound, solvent cutback, cold application	Nondrying film about 8 to 20 μm (0.3 to 0.8 mil) thick; solvent cutback	Where water or saline solution must be displaced; protection of materials under cover for limited periods; protection of critical bare or phosphated steel surfaces for extended periods when overwrap is used
Grade 4	Corrosion-preventive compound, solvent cutback, cold application	Transparent, nontacky film; 25 μm (1 mil) maximum thickness	Shed or indoor storage where dry transparent coating needed. Permits stacking
MIL-C-22235	Corrosion-preventive oil, nonstaining	Lubricating oil with added inhibitors; pour point, −12 °C (10 °F)	Rust preventive for hot and cold rolled steel in storage, in stacks
MIL-C-40084	Corrosion-preventive compound, water emulsifiable, oil type	Mineral oil with corrosion inhibitors in an emulsion; soft film; fire resistant	Low-cost corrosion preventive, fire resistant, thin film
MIL-G-10924	Grease, automotive and artillery	Preservative grease	Lubricant for automotive and artillery equipment, at −54 to 51 °C (−65 to 125 °F)

(continued)

(a) Class 1A is similar in all respects to Class 1, except that it will not leave an oil slick on salt water. (b) Grade 1A is similar in all respects to Grade 1, except that drying time is 72 h

Table 1a (continued)

Specification No.	Title	Description	Intended use
MIL-G-18458	Grease, wire rope, exposed gear	Petroleum oil and soaps plus inhibitors	Lubrication and corrosion-protection for running ropes and exposed gear
MIL-H-6083	Hydraulic fluid, petroleum base, preservative	Light petroleum hydraulic fluid with inhibitors	As a preservative oil in aircraft hydraulic systems and shock-absorber struts
VV-L-800	Lubricating oil, general purpose, preservative (water-displacing, low temperature)	Lubricating oil with added inhibitors; pour point, −57 °C (−70 °F) max	Lubrication and corrosion-protection of small arms; general application wherever a multipurpose, low-temperature oil is required
MIL-L-3150	Lubricating-oil preservative, medium	Lubricating oil with added inhibitors; similar to SAE 30; pour point, −7 °C (20 °F)	Corrosion-protection of highly finished internal and external surfaces; not for internal-combustion engines
MIL-L-6085	Lubricating oil, instrument, aircraft, low volatility	Light, low-volatility, synthetic lubricant containing rust preventive; pour point, −57 °C (−70 °F)	Protection of bearings in instruments, electronic equipment or wherever a low-evaporation oil is required for both high and low temperatures
MIL-L-11734	Lubricating oil, synthetic (for mechanical time fuses)	Synthetic oil with inhibitors	Lubricate and protect mechanical time fuses at normal and below-freezing temperatures
MIL-L-14107	Lubricating oil, for aircraft weapons	Synthetic lubricating oil with inhibitors; pour point, −59 °C (−75 °F)	Lubrication of aircraft automatic weapons; for use below −18 °C (0 °F)
MIL-L-17331	Lubricating oil, steam turbine (noncorrosive)	Petroleum oil plus inhibitors	Lubrication of main turbine gears, auxiliary turbines, air compressors, hydraulic equipment
MIL-L-19224 (Grades A, B, and C)	Lubricating oil, mineral, preservative; pour point −34 °C (−30 °F)	Lubricating oil in three grades with inhibitors	Lubricant and preservative where uninhibited oils do not afford sufficient protection
MIL-L-21006	Rust-retarding compound, flotation type, ballast tank protection	Petroleum-base fluid compound with inhibitors; pour point, −7 °C (20 °F)	Rust-retarder in tanks used for salt-water ballast
MIL-L-21260			
Grade 1	Lubricating oil, internal-combustion engine, preservative	Engine-lubricating oil (SAE 10W) with corrosion inhibitors, including hydrobromic acid neutralizers; pour point, −29 °C (−20 °F)	Lubricant in spark-ignition and compression-ignition reciprocating internal-combustion engines
Grade 2	Lubricating oil, internal-combustion engine, preservative	Same makeup as Grade 1, but SAE 30W and pour point of −18 °C (0 °F)	Same as Grade 1
Grade 3	Lubricating oil, internal-combustion engine, preservative	Same makeup as Grade 1, but SAE 50 and pour point of −9 °C (15 °F)	Same as Grade 1
MIL-L-46000	Lubricating oil, semifluid, automatic weapons	Synthetic gel-like oil plus inhibitors	Operation of M61, M39, and related types of automatic weapons
MIL-L-46002			
Grade 1	Lubricating oil, contact and volatile corrosion inhibited	Lubricating oil with added contact and volatile corrosion inhibitors, SAE 5	Preservation of enclosed systems
Grade 2	Lubricating oil, contact and volatile corrosion inhibited	Same as Grade 1, but SAE 30	Preservation of enclosed systems
MIL-P-3420	Packaging materials volatile corrosion inhibitor, treated, opaque	Barrier material, impregnated with volatile corrosion inhibitors	Protection of ferrous metal parts
MIL-W-3688	Wax emulsion (rust inhibiting)	Dry, hard, nontacky, flexible surface emulsion of wax in water	Dry lubricant and rust-inhibiting coating for general weatherproofing

(a) Class 1A is similar in all respects to Class 1, except that it will not leave an oil slick on salt water. (b) Grade 1A is similar in all respects to Grade 1, except that drying time is 72 h

Table 1b Properties of rust-preventive materials covered by military specifications

Specification No.	Minimum melting point (MP) or flash point (FP), °C	Method(a)	Application tempera-ture, °C	Coverage(b) m^2/L	Coverage(b) ft^2/gal	Method of removal	Viscosity	Penetration
MIL-C-450 (Types I, II, and III)	...	B, S	Ambient	15	750	Petroleum solvent	I: 15-28 s(d) II: 120-190 s(d)	III: 150-250 at 25 °C
MIL-C-4339	...	D, S	Ambient	...	...	Petroleum solvent	100-400 SUS at 38 °C	...
MIL-C-5545	FP: 260	F(e)	96	10	500	Not required(f)	115-150 SUS at 99 °C	...
MIL-C-6529	FP: 205	B, D, S, or fill	Ambient	30	1500	Not required(f)	...	...
MIL-C-8188	FP: 180	B, D, S	Ambient	40	2000	Petroleum solvent	3.0 cSt at 99 °C 11.0 cSt at 38 °C 18 000 cSt at −54 °C	... ...
MIL-C-10382	FP: 38 MP: 66	B, D, S	Ambient	20	1000	Petroleum solvent	...	...
MIL-C-11796								
Classes 1 and 1A(g)	MP: 68 FP: 180	D, S	79-93(h) 93-105(j)		15/lb	Petroleum solvent	...	30-80 at 25 °C
Class 2	FP: 180 MP: 66	D, S	71-88(h) 85-90(j)		15/lb	Petroleum solvent	...	90-150 at 25 °C
Class 3	FP: 180 MP: 57	B, D, S	16-49(k) 66-82(h) 78-85(j)		15/lb	Petroleum solvent	...	200-325 at 25 °C
MIL-C-15074	FP: 38	B, D	Ambient	...	...	Petroleum solvent	30 cSt at 38 °C	
MIL-C-16173								
Grades 1 and 1A(m)	FP: 38 MP: 66-71	B, D, S	7-35	9	450	Petroleum solvent	...	25(n) at 25 °C
Grade 2	FP: 38	B, D, S	4-35	20	1000	Petroleum solvent	...	200 at 25 °C
Grade 3	FP: 38	B, D, S	4-32	20	1000	Petroleum solvent	...	...
Grade 4	FP: 38	B, D, S	4-32	20	1000	Petroleum solvent	...	...
MIL-C-22235	FP: 140	B, D, S	Ambient	30	1500	Petroleum solvent	100-140 SUS at 38 °C	...
MIL-C-40084	FP: 96	B, D, S	25-82	30	1500	Petroleum solvent	...	...
MIL-G-10924	160(p)	Gun	Ambient	...	...	(f)(q)	...	265-295 at 25 °C
MIL-G-18458	66(p)	Gun	Ambient	...	...	(f)(q)	...	200-300 at 25 °C

(continued)

(a) B, brush; D, dip; F, fog; S, spray. (b) Average anticipated coverage; subject to wide variation. (c) Viscosity conversions: 100 SUS (20.55 cSt); 115 SUS (23.85 cSt); 150 SUS (31.70 cSt); 400 SUS (86.2 cSt). According to ASTM D217 penetration is recorded in tenths of a millimetre. Penetration values for greases are frequently reported without the units. (d) No. 4 Ford cup reading. (e) For cylinders. For crankcase, dilute 1-to-1 with grade 1100 Navy symbol oil, fill, run engine until hot, then drain. (f) Removal, if desired, may be effected with petroleum solvent. (g) Class 1A is similar in all respects to Class 1, except that it will not leave an oil slick on salt water. (h) For dip application. (j) For spray application. (k) For brush application. (m) Grade 1A is similar in all respects to Grade 1, except that drying time is 72 h. (n) On solids; needle penetration. (p) Dropping point, minimum. (q) Flush with fresh grease. (r) Into cylinders or crank case. (s) Remove powderlike residues with methyl alcohol rinse

Table 1b (continued)

Specification No.	Minimum melting point (MP) or flash point (FP), °C	Method(a)	Application temperature, °C	Coverage(b) m²/L	Coverage(b) ft²/gal	Method of removal	Viscosity	Penetration
MIL-H-6083	FP: 93	Fill	16-49	...	...	Not required(f)	10 cSt at 54 °C 800 cSt at −40 °C	...
VV-L-800	FP: 135	B, D, S	16-49	50	2500	Not required(f)	12 cSt at 38 °C 60 000 cSt at −54 °C	...
MIL-L-3150	FP: 150	B, D, S	16-49	30	1500	Not required(f)	185-255 SUS at 54 °C	...
MIL-L-6085	FP: 185	D, S	Ambient	50	2500	Not required(f)	8 cSt at 54 °C 2 000 cSt at −40 °C 12 000 cSt at −54 °C	...
MIL-L-11734	...	B, D, S	Ambient	40	2000	Not required(f)	12.5 cSt at −38 °C 15 000 cSt at −57 °C	...
MIL-L-14107	FP: 160	B, D, S	Ambient	50	2500	Not required(f)	5.8 cSt at 38 °C 900 cSt at −59 °C	...
MIL-L-17331	FP: 180	B, D, S	Ambient	...	...	Not required(f)	8.2 cSt at 99 °C 82-110 cSt at 38 °C	...
MIL-L-19224 (Grades A, B, and C)	(A)FP: 160 (B)FP: 200 (C)FP: 210	B, D	−30 to 74	30	1500	Not required(f)	Grade A: 90-120 SUS at 54 °C; Grade B: 45-55 SUS at 99 °C; Grade C: 60-70 SUS at 99 °C	...
MIL-L-21006	FP: 160	...	Ambient	24	1200	Not required(f)	...	...
MIL-L-21260								
Grade 1	FP: 180	F, S(r)	Ambient	30	1500	Not required(f)	44-50 SUS at 99 °C 12 000 SUS at −18 °C	...
Grade 2	FP: 200	F, S(r)	Ambient	30	1500	Not required(f)	58-70 SUS at 99 °C 200 000 SUS at −18 °C	...
Grade 3	FP: 205	F, S(r)	Ambient	30	1500	Not required(f)	85-110 SUS at 99 °C	...
MIL-L-46000	...	B, D	Ambient	15	750	Not required(f)	...	350-385 at 25 °C
MIL-L-46002								
Grade 1	FP: 115	B, D, S	Ambient	40	2000	Not required(f)	12 cSt at 38 °C 10 000 cSt at −40 °C	...
Grade 2	FP: 120	B, D, S	Ambient	30	1500	Not required(f)	9.65-12.98 cSt at 99 °C 95-125 cSt at 38 °C	...
MIL-P-3420	...	...	Ambient	...	...	(s)	...	...
MIL-W-3688	74(p)	D, S, or wipe	Room temperature	...	...	Petroleum solvent	...	...

(a) B, brush; D, dip; F, fog; S, spray. (b) Average anticipated coverage; subject to wide variation. (c) Viscosity conversions: 100 SUS (20.55 cSt); 115 SUS (23.85 cSt); 150 SUS (31.70 cSt); 400 SUS (86.2 cSt). According to ASTM D217 penetration is recorded in tenths of a millimetre. Penetration values for greases are frequently reported without the units. (d) No. 4 Ford cup reading. (e) For cylinders. For crankcase, dilute 1-to-1 with grade 1100 Navy symbol oil, fill, run engine until hot, then drain. (f) Removal, if desired, may be effected with petroleum solvent. (g) Class 1A is similar in all respects to Class 1, except that it will not leave an oil slick on salt water. (h) For dip application. (j) For spray application. (k) For brush application. (m) Grade 1A is similar in all respects to Grade 1, except that drying time is 72 h. (n) On solids; needle penetration. (p) Dropping point, minimum. (q) Flush with fresh grease. (r) Into cylinders or crank case. (s) Remove powderlike residues with methyl alcohol rinse

storage conditions or on the user's requirement:

- Well-refined, neutral mineral oil (No. 1 or No. 2 color) with a viscosity of 100 SUS at 20.55 cST at 38 °C (100 °F) and containing a suitable rust-preventive additive
- Rust-inhibited petrolatum, of soft, greaselike consistency, dip applied at 82 °C (180 °F). The compound should be amber to green in color and have a melting point ranging from 51 to 57 °C (125 to 135 °F) and a viscosity of about 90 SUS at 17.98 cST at 100 °C (210 °F)
- Transparent varnishlike rust preventive, a solvent-cutback hard-film product that is spray applied at room temperature. The coating sets in 25 min and is dry to the touch in 2 to 3 h

Physical Suitability. The physical form of a rust preventive, that is, whether it is an oil-based or petrolatum material or is an aqueous emulsion, may dictate or prevent its use for a specific application. For example, a petrolatum compound is unsuitable for the corrosion-protection of internal components of high-speed business machines, because the heavy make-up of a compound of this type could cause excessive sticking and eventual malfunction of the machine, even though the components were well protected from corrosion. The use of a solvent cutback rust preventive also is undesirable, because minute traces of the solvent might vaporize after assembly of the components, and the resulting hard film could interfere with proper machine function. Solvent-diluted rust preventives should not be used in a confined system unless ventilation is provided to disperse the combustible solvent vapors and prevent internal degreasing. For this application, a rust preventive based on a mineral oil of suitable viscosity would be a proper selection.

The physical characteristics of the film left by a rust preventive should be considered if the surface to which it is applied is to receive additional coatings. The use of an oily material of any type is less desirable than the use of a material that forms a hard, dry film, if the surface is to be painted without an intermediate cleaning operation. Extremely viscous or semisolid rust preventives should not be used on intricate assemblies that may present a cleaning problem, or on highly polished surfaces. For example, the polished surfaces of plastic molding dies must remain mirror bright. Rust preventives applied for storage or shipping purposes must be readily removed without any abrading of the surface before the die can be used.

The protection of antifriction bearings with viscous or semisolid rust preventives is impractical, because removal of such materials is difficult. Antifriction bearings should be protected by a rust preventive that is thin enough to permit satisfactory operation after addition of the proper lubricant, and that does not require removal. In one plant, antifriction bearings are coated with a rust-inhibited petrolatum that has a melting point of 40 to 43 °C (105 to 110 °F), and an application temperature of 115 °C (235 °F). The petrolatum coating produces a thin film compatible with any lubricant; afterwards, the bearings are wrapped with treated paper.

Chemical Suitability. Chemical reactions that may take place between a rust-preventive compound and the surface to which it is applied must be considered in selection. Some aqueous-based rust preventives passivate the surface instead of forming a protective film. If these materials are based on alkaline solutions of organic chemicals in water, they should not be used in contact with magnesium, zinc, or other metals susceptible to alkaline attack. An example of this material is the rust preventive or corrosion inhibitor frequently used to regenerate automotive antifreeze solutions.

Mineral oil-based rust preventives used on components of hydraulic systems may be difficult to remove in preparing the system for operation, because many of the newer hydraulic fluids are incompatible with mineral oils. The U.S. Navy uses hydraulic fluids based on phosphate ester systems in various types of operating equipment. At the time of manufacture, such equipment should be protected with a rust preventive based on a phosphate ester similar to that to be used in operation. Many government and industrial installations use water-glycol fluids in hydraulic equipment; any rust preventives used on such equipment should be compatible with the water-glycol fluid.

Nonmetallic Materials. Some assemblies incorporate nonmetallic materials that may deteriorate on contact with hydrocarbon-based materials or preservatives containing acids and bases. For example, polymeric materials may swell, harden, crack, soften, or otherwise deteriorate so as to be unfit for service. Product storage and shipping requirements should be anticipated in the beginning stages of design, so that a proper preservative is selected to protect the metallic units of the assembly, without attacking the nonmetallic components.

Methods of Application

Rust preventives are applied by spraying or fogging, dipping, flowing or slushing, and brushing or wiping. The method used is determined by the type of preventive and by the quantity, size, complexity, and surface finish of the articles to be coated. The equipment and methods of application used are similar to those used in painting.

Petrolatum compounds may be applied either hot or cold. Cold application generally is restricted to parts too large and bulky for practical tank immersion, or to parts on which only localized protection is desired, such as the ways for a lathe bed. When these compounds are applied cold, their consistency requires that brushing or wiping be used. When brushing is used, the bristles of the brush should be stiff enough to permit brush-out of the material, but not so stiff as to leave deep brush marks in the preservative. It is often advantageous to build up coatings of these materials to the desired final thickness by the successive application of thin layers.

When petrolatum rust-preventive materials are applied hot, dipping is the most practical method, although parts that are too large to be dipped may be coated by brushing, wiping, or spraying. Small parts to be dipped are placed in baskets; larger parts are individually dipped. Dipping may be manual, if volume is low, or conveyorized, for parts produced in large quantities.

Most dipping tanks for petrolatum compounds are heated by steam coils, a hot water jacket, electrical immersion heaters, or plate coils using either hot water or steam. Although precise temperature control is not usually required, overheating must not occur, because overheating can cause decomposition of the preservative, and the resulting products may act as corrosion agents.

Oil compounds can be applied by dipping, spraying, flowing, brushing, or wiping. The thickness of the coating depends on the viscosity, fluid characteristics (whether the fluid is Newtonian or non-Newtonian), and surface

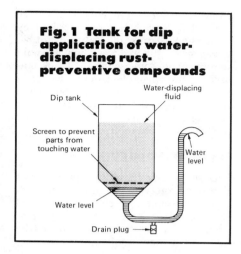

Fig. 1 Tank for dip application of water-displacing rust-preventive compounds

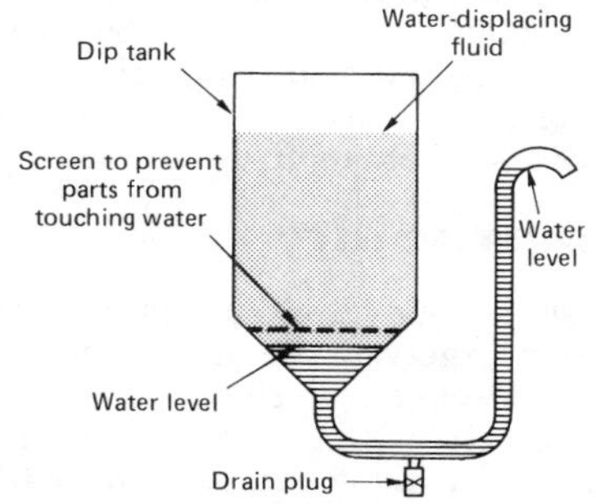

tension of the oil compound. Oils of high viscosity usually are heated before application. Parts or assemblies coated by immersion in oil preservatives should be turned or agitated if necessary to permit all trapped air to escape. The parts require immersion only long enough to ensure complete coverage, if the material is applied at room temperature.

For spray application, oil rust preventives usually require no dilution; they are applied as received. Moderate air pressure is used to avoid misting and overspraying of the material. A wetting spray is usually sufficient. It is more difficult to control coating weight and uniformity by spray application than by dipping. To ensure an adequate coat, the material should be applied until it just begins to flow.

Emulsion compounds, most widely used for small items, are oil-in-water emulsions containing 8 to 12% solids. The compounds are available as concentrates, and are diluted with water at ratios of 1 part concentrate to from 4 to 10 parts water, as indicated by specification MIL-C-40084.

The compounds are applied by dipping or spraying. The effectiveness of the application may be increased by heating either the parts or the compound. For example, the unheated compound may be applied in the third stage of a power washer in which the parts retain residual heat produced during the first two stages; or if dipping is used, the compound may be heated to a gentle boil before immersion of unheated parts. Oil-in-water emulsions are fire resistant during application; however, after the water has been removed by drying, the residual film has flammability characteristics comparable to petroleum oils or waxes.

Solvent-cutback compounds essentially are waxes modified with high-melting-point polar additives, such as soaps. Compounds that use organic solvents as the diluent are applied at room temperature. Small parts usually are batch coated in wire baskets, which are immersed, withdrawn, and centrifuged to remove the excess material. Commercial equipment is available for this method of application. The parts are then removed from the basket and spread out to dry, or conveyed through an air-drying stage, which may be mildly heated.

Because some solvent-cutback compounds develop dry, hard films, parts may become bonded together by the film. Separation of these parts often results in tears in the coating. Large parts may be spray or dip coated, but all solutions are not necessarily applied equally well by both methods. Parts too large for dipping are coated by spraying or brushing.

Water-displacing compounds are most effective when dip-applied, although spray application also may be used. These materials actually remove films or droplets of water from the metal surfaces by preferential wetting; that is, the attraction of the rust-preventive compound to the metal surface is greater than that of water, and thus the preservative displaces the water. Complete immersion of the part makes the coverage more positive.

A schematic representation of a suitable dip tank for the automatic removal of water from a water-displacing preservative system is shown in Fig. 1. Because the specific gravity of the preservative is less than that of water, the height of the column of water is somewhat less than the column of preservative. Parts too large to coat by immersion and too impractical to spray coat may be coated by wiping. Wiping is used also when only certain areas of parts are to be coated.

Fingerprint removers and neutralizers may be applied by spraying, dipping, or brushing; brushing usually is used when only a section of a part is to be coated. These materials are applied at room temperature. Sufficient fluid should be present in the system to prevent gross contamination or preferential depletion of the active ingredients that dissolve or neutralize the corrosive materials.

Control of Film Thickness

The film thickness of rust-preventive compounds must be controlled in order to maintain uniformity of corrosion prevention and to maintain a prescribed level of efficiency of application. Films that vary in thickness permit no accurate forecast of storage life or material cost. The film thickness of oil compounds is controlled by viscosity and surface tension; that of solvent-cutbacks, by solids content; and that of petrolatum compounds, by application technique and temperature.

Fluid types applied by any method drain off to a thin film. Low-viscosity oils drain quickly; oils of higher viscosity provide thicker films originally, but eventually drain to thin films. Solvent-cutback rust preventives are compounded to provide a specific film thickness when the solvent evaporates. This film thickness is controlled by the percentage of solids contained in the compound.

Petrolatum or grease compounds may be applied by brushing to almost any thickness; coating thickness is limited only by the skill and intent of the operator. However, if the material is almost colorless, determining the uniformity of the coating is difficult. One company has found that adding coloring matter to the preservative improves uniformity of coating.

When compounds are applied by hot dipping or hot spraying, the temperature of the coating and the temperature of the part both influence film thickness. Parts at room temperature, if dipped and withdrawn quickly so that no appreciable increase in part temperature is accomplished, congeal thick coatings. However, if a part is immersed until its temperature approximates that of the heated rust preventive, a thinner coat results. In general, the greater the difference between the temperature of the rust preventive and the temperature of the part at the time of with drawal from the material, the heavier the coating. Figure 2 shows the influence of the temperature of a petrolatum rust preventive and the temperature of dip-coated panels on the film thickness obtained when the panels are dipped and withdrawn rapidly (1 s for immersion, 1 s for withdrawal). The curve shown is for panels at 27 °C (80 °F); comparisons of film thicknesses are presented for panels at 21 and 32 °C (70

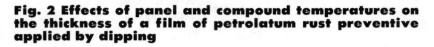

Fig. 2 Effects of panel and compound temperatures on the thickness of a film of petrolatum rust preventive applied by dipping

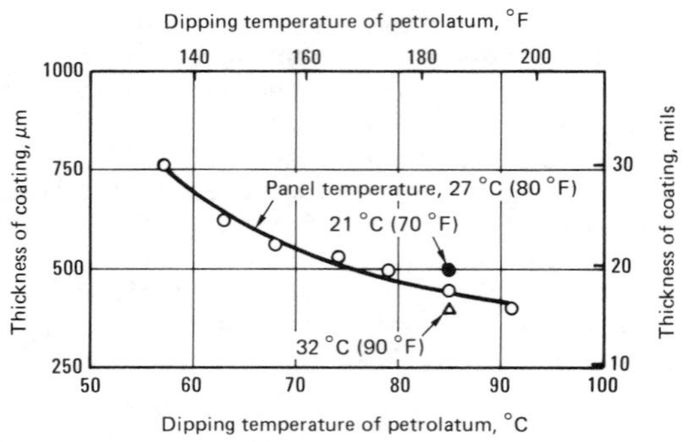

and 90 °F) dipped in the rust preventive at 85 °C (185 °F).

Specific gravity of the compound also influences coating thickness; for a specific weight of compound, the film thickness increases as the specific gravity of the compound decreases. The results of a test to determine the influence of specific gravity on film thickness are shown in Fig. 3. The panels used in this test were 75 mm (3 in.) long, 50 mm (2 in.) wide, and 3.2 mm (⅛ in.) thick. They were at an ambient temperature of about 24 °C (75 °F) when they were rapidly dipped into and withdrawn from (1 s for immersion, 1 s for withdrawal) the petrolatum compound, which was heated to various temperatures to produce different film thicknesses. Five petrolatum preservatives, each with a different specific gravity, as indicated, were used.

Tests were conducted to determine the influence of four variables on coating thickness obtained under laboratory conditions using a petrolatum rust-preventive compound. Variables were:

- Temperature of preservative
- Mass of metal specimen
- Duration of immersion
- Rate of withdrawal

The conclusions drawn from data obtained in these tests were as follows:

- As the temperature of the material was increased, its fluidity increased. Thus, progressively thinner coatings are obtained with petrolatum rust preventives as the temperature is increased
- The thickness of the panel (the mass of the metal specimen) influenced the thickness of the coating. With a greater metal mass, a temperature lag existed between the specimen panel and the heated preservative. The coating thickness was greatest with the greatest temperature differential. Thinner coatings developed as the temperature of the panel approached that of the compound
- Duration of immersion has great influence on the coating thickness. As immersion time was increased, the coating first reached a maximum value, then melted away uniformly, as the panel temperature approached that of the preventive
- A slow rate of withdrawal left the surface of the coating smooth and regular, indicating uniform film thickness. Rapid withdrawal resulted in distorted, irregular surfaces with numerous small, shallow areas

The curves in Fig. 4, obtained from these tests, show the influence of mass and withdrawal rate of the panel on film thickness for one petrolatum rust preventive. For both curves, immersion time was 10 s.

Viscous rust preventives usually are applied by immersion at elevated temperature. These materials react in approximately the same manner as the petrolatums. However, any tendency of viscous materials to flow when cold results in wedge-shaped films. If viscous materials are applied too thick, the coatings may sag. Film thickness of rust-preventive coatings is controlled in the same manner when compounds are applied by spray as when applied by immersion. Because of loss of solvent between spray gun and work, which would reduce runoff, some solvent-cutback materials may provide thicker coatings when applied by spraying than when they are applied by dipping. Hard-film coatings, however, are likely to crack and check when applied too thick.

Removability

Ease of removal of rust preventives depends largely on the thickness, hardness, and chemical characteristics of the protective film. Nondrying oil or grease compounds can be removed by simple petroleum solvents, preferably those with a flash point above 40 °C (105 °F). Vapor degreasing, hot power-spray washing with alkalis or strong detergents, and steam-alkali or steam-detergent blasting are also used to remove nondrying oil or grease compounds. If the desired degree of removal permits, wiping can be used. The solvent-cutback asphaltic or dry-film compounds usually require high-solvency organic solvents, vapor degreasing, or vigorous extended treatment with steam or hot-spray cleaners.

The removal of hard-film preservatives that are completely or partially impervious to solvents may require the use of scraping, blasting, or wire brushing. This can result in damage to machined or polished surfaces. Hard-film, viscous-oil, and heavy-grease compounds used to protect parts in outdoor storage or highly corrosive atmospheres present problems if not thoroughly removed, because their residues usually are incompatible with lubricating oils and greases. Components of hydraulically actuated systems, and precision machine tool spindles are examples of parts with which such problems can arise. Also, these rust preventives can clog filters and screens in circulating oil systems. Additionally, hard wax rust preventives can cause serious problems in sliding bearing mechanisms if incompletely removed.

Costs

Rust-preventive compounds rarely are selected on the basis of cost alone. Table 2 lists typical duration of protection provided in three storage conditions by four general types of rust-preventive compounds. In general, cost per litre (or gallon) increases as the protection increases.

Fig. 3 Influence of specific gravity of a petrolatum rust preventive on film thickness for a specific weight of rust preventive applied to test panels

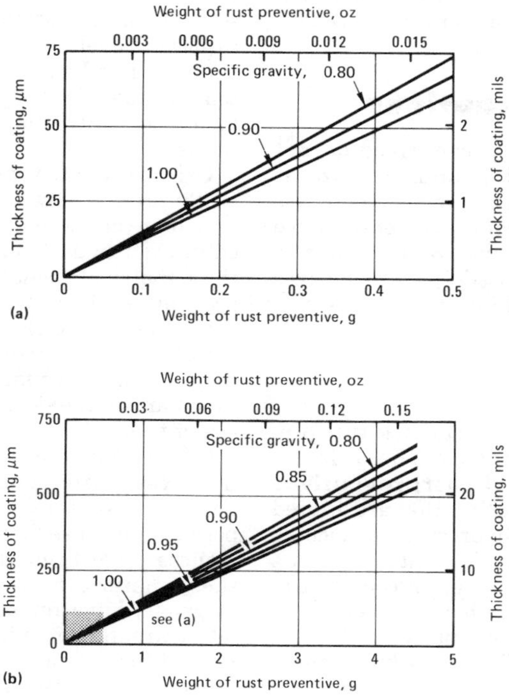

Fig. 4 Comparison of coating thicknesses obtained using the same withdrawal rate of panels from a petrolatum rust-preventive compound

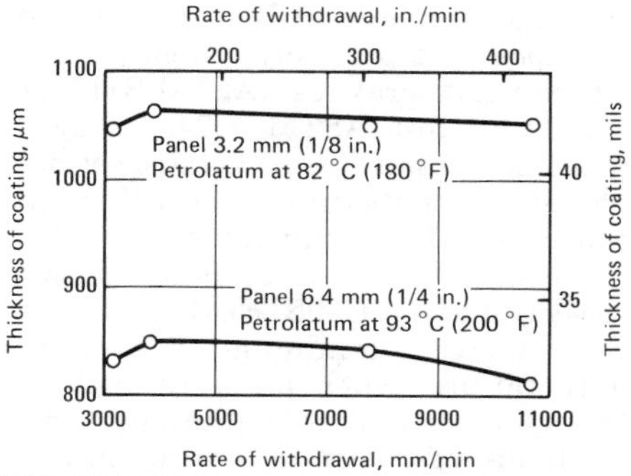

Table 2 Duration of protection afforded by rust-preventive compounds

Type of compound	Duration of protection, months: Indoor storage	Shed storage	Outdoor storage
Petrolatum	Not used	Over 24	6-24
Oil	12-36	3-24	3-12
Solvent-cutback	6-18	3-12	1-6
Emulsion	3-12	Under 3	Not suited

Quality Control

Quality control of rust-preventive compounds is necessary for maintenance of established levels of efficiency and performance.

The extent and frequency of the quality control program depends on the rust-preventive compound being used and the degree of control required. It is often advisable to start with frequent checks to develop a history of a system and then to adjust to safe intervals.

The checks should include concentration level of the active ingredients, contaminant levels for dirt and water, and degree of oxidation by some means of pH and neutralization number. The nature of the processing operations prior to the application of the rust-preventive compound may make it necessary to run other special quality control checks, which may include viscosity, flash point, copper strip corrosion, and infrared spectra.

The significance of changes in the quality control checks is best interpreted by performing the same tests on the rust preventive as received from the supplier. Quality control limits for the rust-preventive compounds may be established in cooperation with the supplier.

Viscosity and Specific Gravity. Oil and solvent-cutback rust preventives are checked as received for viscosity or specific gravity, or for both, at a specific temperature. Standard values, against which the results of checks are compared, either are obtained from the supplier or are established from material that has performed satisfactorily in controlled salt spray, humidity, or field tests. The viscosity is checked using effluent cups, or instruments that measure the resistance of the material to a moving circular spindle or to a falling ball. Methods for testing viscosity are described in detail in the article on painting in this Volume.

Specific gravity, sometimes used to indicate solids content, is determined by the correct hydrometer. Usually, the specific gravity is taken at a specific temperature of the preservative. The viscosity of the preventive should be checked when processing begins and rechecked at definite intervals so that changes resulting from the conditions of use can be corrected, for example, so that solvent lost by evaporation can be replaced. Some materials, however, have approximately the same viscosity as the material used to dilute it; for these, periodic checking of the specific gravity or other characteristics of the preservative is necessary for maintaining the proper balance between solids and solvents.

Flow characteristics of petrolatum rust preventives may be determined empirically by methods described in ASTM D937. The results obtained by this method are of maximum value when correlated with the flow characteristics of similar materials. To determine flow characteristics of materials that are solid at ambient temperatures, a specific amount of the material is placed into a cup at the high end of a precisely inclined plane, heated to a prescribed temperature above its melting point, and then permitted to flow down a graduated groove. The time it takes for the material to flow from the cup to the bottom of the inclined plane is indicative of the flow characteristics of the material. Flow characteristics results must also be correlated with known materials.

Film Characteristics. The efficiency of the film ordinarily is determined by exposing a precisely coated part to either a salt spray test (ASTM B117) or a humidity test (ASTM D1748). Projection of test results to an expected service life can be done accurately only through experience. One company makes this statement for a material: in a 20% salt spray test at 35 °C (95 °F), the material will survive a minimum of 10 days until failure; when properly applied, this same preventive will protect for a minimum of 1 year in severe outdoor exposure. For another material, the minimum protection afforded is 24 h in 20% salt spray, 30 days in a humidity test at 38 °C (100 °F), and no corrosion of steel, brass, copper, or aluminum at the end of 120 days at room temperature. Measurement of film thickness is necessary to ensure that minimum corrosion protection and maximum efficiency of coverage will be obtained.

Surface Preparation. Preparation or condition of surfaces to be protected has an important bearing on the effectiveness of a rust preventive. Prior surface treatment ensures freedom from incompatible processing oils, waxes, hygroscopic or chemically reactive salts, or random dirt. These materials could modify the adhesion or thickness of the film or counteract the inhibitors in the film.

Precleaning, during routine processing of parts, should be kept under effective process control to minimize variance from acceptable limits of surface cleanness. When a temporary setup is used, extra precautions against contamination are essential. Contamination of the rust preventive itself from careless handling or storage procedures should be prevented.

Safety Precautions

Although rust preventives do not present abnormal or unusual hazards to personnel or equipment, their use and storage do require the observance of safety precautions. With respect to the hazards presented, rust preventives may be separated into three classes:

- Solvent-cutback materials
- Oil-based materials
- Water-based materials

Of these, the solvent-cutback rust preventives present the most hazards.

Solvent-cutback rust preventives include those materials that are diluted with low-boiling hydrocarbons such as kerosine and gas oil, volatile solvents such as the ketones and alcohols, aromatic solvents such as toluene and xylene, and halogenated materials such as chloro-fluoro hydrocarbons. Inhalation of these solvents over a period of time can result in respiratory difficulties or produce the appearance of intoxication. The aromatic solvents probably are the most hazardous physiologically. In selection of a solvent, regulations of the Occupational Safety and Health Administration should be consulted.

Solvent-cutback materials also may present fire and explosion hazards. Flash points of the solvents may range from −12 to 40 °C (10 to 105 °F). Those with flash points below 38 °C (100 °F) are considered hazardous, and some spraying shops do not use them. Whenever solvent cutbacks are used, adequate ventilation is essential to reduce solvent concentration below the lower explosion level, thus eliminating explosion and fire hazards. Ventilation systems should be designed to maintain vapor concentration below explosive limits.

Oil-based rust preventives generally present few health hazards and may have no more ill effect physiologically than household soaps. However, these materials usually owe their rust-preventive characteristics to surface-active agents, which may contain heavy metals such as barium. These may be toxic if excessive quantities are allowed to remain on the skin for extended periods of time.

Oil-based rust preventives usually do not present fire or explosion hazards; nevertheless, they should not be exposed to intense heat. Numerous instances have been reported in which these materials, through spillage or drippage, have impregnated the insulation around heat pipes and have been ignited through a combination of spontaneous combustion and heat from the enclosed pipe, even with pipes operating as low as 150 °C (300 °F). Mineral seal oil has a flash point of 121 °C (250 °F)

Water-based rust preventives are rarely hazardous in terms of ignition or explosion. The additives they contain, however, can cause dermatitis, depending on the degree of sensitivity of the individual exposed to them. These materials should be kept away from the face, particularly from the eyes.

Painting

By Melvin H. Sandler
Consulting Services for Protective Coatings
and
The ASM Committee on Painting*

PAINTING is a generic term for the application of a thin organic coating to the surface of a material for decorative, protective, or functional purposes. Painting offers the following advantages over other processes used for the protection or decoration of metal parts and assemblies:

- The equipment required for applying paint is usually less expensive to buy and install, is simpler to operate, and requires less control
- Material and labor costs per unit area of surface coated often are much lower
- Organic coatings are available in a wide range of pigments and vehicles and can meet practically any coating requirement for color, gloss, or surface texture
- Paints have been developed that can withstand most corrosive conditions, and unlike many metallic protective coatings, organic films can simultaneously resist more than one corrosive condition, such as combinations of marine atmosphere and acid fumes
- Conventional paint films have good dielectric properties, which enable them to inhibit galvanic action between dissimilar metals. Conversely, paints are available that contain special pigments to provide conductivity suitable for grounding induced or static electricity

Types of Paint

The general terms "paint" and "organic coating" are essentially interchangeable and are used to designate certain coatings having an organic base. Most organic coatings are based on a film former or binder which is dissolved or dispersed in a solvent or water. This film forming liquid constitutes the vehicle in which pigments are dispersed to give color, opacity, and other properties to the dried film. Many other ingredients may be added to the vehicle to achieve specific film properties. These would include such things as driers to aid curing, plasticizers to impart flexibility and other properties, stabilizers to lessen the deleterious effects of heat or sunlight. A wide variety of film-forming materials is available and includes oils, varnishes, synthetic resins and polymers such as cellulose, vinyl, epoxy, and polyester. In general, major performance characteristics depend on the binder used.

Enamels are topcoats characterized by their ability to form a smooth surface that is typically of high gloss, but may also include lower degrees of gloss such as flat enamels. Enamels may air dry or bake. Air-dry enamels are cured essentially by a combination of solvent evaporation and oxidation. Baking enamels incorporate catalysts and cross-linking agents which require heat for polymerization.

Lacquers are compositions based on synthetic thermoplastic film-forming materials dissolved in organic solvent. These dry primarily by solvent evaporation. Lacquers are generally characterized by fast drying properties. Typical lacquers include those based on nitrocellulose, other cellulose derivatives, vinyl resins, and acrylic resins.

Water-borne paints are dilutable with water. There are three principal types: solutions, colloidal dispersions, and emulsions. Solution coatings are based on water-soluble binders. Many conventional binders (alkyds, acrylics, and epoxies) can be made water soluble by chemically attaching polar groups such as carboxyl, hydroxyl, and amide, which are strongly hydrophilic. Some hydrocarbon solvents are usually necessary, up to 20% of the total, to improve solubility.

Colloidal dispersions are very small particles of binder, less than 0.1 μm (2.5 mils) diameter, dispersed in water. Normally, these dispersions contain water-soluble polar groups to partially solubilize a portion of the resin. Emulsions, or latexes, are water dispersions that differ from colloidal dispersion by having much larger particle size on the order of 0.1 μm (2.5 mils) or larger. They are made by precipitation in water and therefore do not need to be dispersed.

* Albert O. Hungerford, Chemical and Coatings Specialist, Butler Manufacturing Co.; Roy C. Kissler, Program Manager, Environment & Energy, General Electric Co.; Floyd I. Young, Principal Project Engineer, Motor Wheel Corp.

Pigments must be compatible with water. Metallic particles are usually coated before being mixed into the paint to prevent chemical reaction with water, which would cause the mixture to generate gas. Water-reducible paints have a low volatile organic content (VOC) and comply with most environmental regulations. The advantages of water-borne paints include:

- Low flammability
- Reduced toxicity and odor
- Easy cleanup with water
- Good film continuity, with continuous film similar to conventional solvent systems
- Good mechanical stability, can be pumped in all types of equipment similar to conventional solvent paints
- Application by air spraying, dipping, flow coating, electrodeposition, and roller coating

The disadvantages of water-borne paints include:

- Application by electrostatic spraying requires complete electrical isolation because of the water conductivity
- Coatings require a longer flash tunnel before curing
- Temperature must be raised more slowly to evaporate water at a slow enough rate to prevent the coating from blistering
- Coatings are more susceptible to dirt pickup
- Proper temperature and humidity control are vital. If the humidity is too high or the temperature too low, coating can sag or run off the workpiece

Electrophoretic paints are special water-reducible paints. Resin and pigment materials are shipped and stored as concentrates to be added to the production tank as needed. Electrophoretic films are always deposited from a dip tank. The operating bath consists of resin concentrate and pigment concentrate mixed with deionized water and small amounts of solubilizers and defoamers. The concentration of nonvolatile solids in the bath varies from about 10 to 20%, depending on type and composition. Paint films are deposited on the work by electrophoretic action. Immediately after the film has been deposited, the work is removed from the bath and rinsed with water to remove the excess paint bath, leaving a uniform, tightly adhering film of paint on the workpiece. The workpiece is then baked. Paints can be prepared to deposit films on either the anode or cathode. The resins used most frequently are epoxies and acrylics, including numerous modifications and hybrids. For more information on this process, see the section on electrocoating in this article.

Autophoretic paints are water-reducible paints deposited on metal surfaces by the catalytic action of the metal on the paint materials in the bath. Currently, only ferrous surfaces activate the autophoretic paints available commercially. Tubular automotive frames are coated with this method, because the entire length of the tubing can be coated inside and outside with equal ease.

High-solids paints contain 50% or more solids by volume. One method of obtaining high-solids paints is to use lower molecular weight polymers, which require less solvent to attain the desired application viscosity. Another method of reducing viscosity of high-solids paints is by heating the paint material to a temperature of about 32 to 52 °C (90 to 125 °F). Many two-component systems use a catalyst to increase the rate of the curing reaction. Fast-reacting two-component systems are usually applied with special spray guns that mix the two components at the spray nozzle. Single-component resins in high-solids paints include (*a*) epoxy, (*b*) acrylic, (*c*) polyester, and (*d*) alkyd; whereas two-component resins may be (*a*) urethanes, (*b*) acrylic-urethane, or (*c*) epoxy-amine. The advantages and disadvantages of using high-solids paints are listed below:

Advantages

- Color control and color matching is no more difficult than with conventional solvent paints
- These paints can be applied by higher speed (6000 to 30 000 rev/min) electrostatic bells and disks requiring minimum facility conversion from existing bell or disk systems
- Performance properties are equivalent to those of conventional solvent paints
- Applied cost per square foot is lower than that of conventional solvent paints
- In many cases, these paints require less energy for curing than conventional solvent paints

Disadvantages

- High-solids paints require specialized pumping and transport equipment
- Cleanup of overspray is much more difficult than with conventional solvent paints
- Toxicity of the isocyanates used with urethanes and amines used with epoxies can be a problem

Powder paint consists of plastic resins, color pigments, and additives. In a mixing and grinding unit, the ingredients are combined in a homogeneous mixture that is heated to the melting point. The molten material is extruded into a thin sheet, which is cooled and crushed. The chips are pulverized to a fine powder of carefully controlled particle size, ensuring optimum fluidity and efficient flow through the finishing system. For more information on powder painting, see the section on methods of paint application in this article.

Selection of a Paint System

To select a paint system consistent with the production and economic requirements of the product being coated, a general knowledge of coatings is necessary, including favorable and unfavorable characteristics, available forms, relative costs, and application methods. Table 1 lists some of the major resins used and their general properties. Many of the resins listed are compatible with others, and when blended, undergo changes of properties which provide performance not available with the individual resins. From a performance standpoint, equally good choices may be available. Many resins are supplied as both solvent and water-borne types. These resins can be formulated for a wide variety of application methods, including (*a*) conventional air atomized, airless, and electrostatic spray; (*b*) roller coating; (*c*) dip coating; and (*d*) flow coating. Other resins have properties that make them suitable for special application techniques such as powder or electrophoretic coating. Although a wide variety of coatings is available, the ideal coating system, one with all desired performance properties, simple application, and low cost, is difficult to find. Factors such as (*a*) regulations, (*b*) service environment, (*c*) substrate and service condition, (*d*) basic function, (*e*) application limitations, and (*f*) cost usually must be compromised.

Service Environment. When selecting a paint system, many factors con-

Table 1 Coating resins and properties

Resin	Forms available	Drying method	Favorable characteristics	Unfavorable characteristics	Cost	Uses
Acrylic	Solvent, waterborne, powder	Air dry, bake	Water white, outdoor durability, chemical, heat resistance	Poor-fair adhesion, tendency to be brittle	Moderate, high	Automotive topcoats, appliances, coil coatings, aluminum siding, general industrial use
Alkyd	Solvent, waterborne	Air dry, bake	High gloss, flexibility, good durability, versatility	Poor alkali resistance, generally not hard, tendency to yellow, depending on resin	Low, moderate	Trade sales enamels, trim paints, exterior enamels, general metal finishing
Chlorinated rubber	Solvent	Air dry	Water, alkali, acid resistance	Abrasion resistance, hardness, gloss, sensitivity to solvents	Moderate	Maintenance coatings, ship bottom paints, swimming pool paints, chemical process equipment
Epoxy	Solvent, waterborne, powder	Air dry, bake	Excellent adhesion, chemical resistance, flexibility, abrasion resistance, hardness	Rapid chalking on exterior exposure, poor resistance to oxidizing acids, yellows in clears	Moderate, high	Maintenance paints, automotive primers, appliances, metal products
Fluorocarbon	Solvent, powder	Bake	Highest exterior durability, chemical resistance	Adhesion, recoatability, high baking temperatures	High	Coil coatings, siding
Nitrocellulose(a)	Solvent	Air dry, bake	Extremely fast drying, good hardness, abrasion resistance	Low solids content, fair to good exterior durability, low flash point solvents	Low, moderate	Furniture finishes, touch-up lacquers, general-purpose product finishes, aerosol lacquers
Phenolic	Solvent, waterborne	Air dry, bake	Hardness, adhesion, resistance to chemicals, corrosion	Darkens, can only be used in dark-colored coatings	High	Can linings, tank linings, maintenance paint on metals
Polyester	Solvent, waterborne, powder	Air dry, bake	High gloss, hardness, chemical resistance, high film build	Fair adhesion, may hydrolize under certain conditions	High	Wood finishes, coil coatings, specialty bake coats
Polyurethane	Solvent, waterborne, powder	Air dry, bake	Chemical resistance, abrasion resistance, hardness, exterior durability	Some types yellow and chalk readily on exterior exposure	Moderate, high	Aircraft finishes, metal and plastic coatings
Silicone	Solvent, waterborne	Air dry, bake	High heat resistance, exterior durability, gloss and color retention	Tendency toward brittleness. Unmodified types require high baking temperatures	High	Any finish for high heat resistance, exterior metal coatings
Vinyl	Solvent, powder	Air dry, bake	Chemical resistance, flexibility, fast air dry, formability, resistance to acid, alkali, abrasion	Generally low solids, low flash points	Moderate	Can, tank linings, maintenance paints, metal decorating paints

(a) Must be modified with other resins

cerning service environment must be considered. Needs are different for interior and exterior coatings. The specific properties needed, such as resistance to heat, cold, sunlight, and weathering must be determined. If the coating needs to be resistant to chemicals, the specific chemical, such as acid, alkali, solvent, or water immersion should be specified. Mechanical properties required, including hardness, flexibility, impact resistance, and abrasion resistance, should also be decided.

Substrate and Surface Condition. The type of substrate must be considered. A coating capable of giving excellent performance on one metal may fail badly on another. Smoothness, porosity, dimensional stability, and corrodibility affect the choice of a proper finishing system. Smooth, clean metal surfaces that cannot be phosphate coated lack sufficient tooth for good adhesion of some air-dried coatings.

No special primer coats are required on surfaces containing small quantities of tightly adhering rust, if the finished parts are intended for indoor service in a mildly corrosive atmosphere. Attractive and durable finishes can be obtained over such surfaces with a single coat of a special paint that produces a textured finish, such as wrinkle finishes and pebble finishes. These paints are relatively inexpensive and hide surface irregularities with their own irregular appearance.

On parts intended for outdoor use where tightly adhering rust is present and not economically removable, a rust-inhibiting primer with good penetrating qualities must be used to prevent, or at least substantially retard, further rusting in service. Parts with heavy rust or mill scale should not be painted unless the loose rust and mill scale are removed.

The diverse requirements of substrate and environment often necessitate a dual or multiple coating system. In these systems, a primer with one composition is used to satisfy substrate adhesion and corrosion resistance, and a coating with a different composition is used as a topcoat to withstand environmental conditions, for example, a vinyl wash primer followed by an epoxy primer with an acrylic topcoat.

Basic Function. Coatings may be applied for appearance, to meet functional requirements, or to meet combined function and appearance needs. If the basic purpose is appearance, the gloss, color, and retention of these properties in service are emphasized. In some applications, functional requirements are of equal importance to appearance. On office furniture, for example, paint films must provide attractive appearance and resist marring and abrasion. On automobiles, paint films must be attractive in appearance, easily applied, and readily repaired, but be resistant to abrasion, marring, and impact as well as capable of protecting the underlying metal from corrosion. In other applications, such as corrosion protection of tanks or chemical equipment, the functional requirements of the paint film are of prime concern. Corrosion resistance is the most important of functional requirements.

Corrosion of steel and cast iron occurs in all environments. The rate and extent of corrosion vary from mild attack in dry, clean environments, to highly accelerated attack in marine or industrial areas where corrosive fumes are present in the air. Table 2 lists paints selected for service in a wide range of corrosive conditions.

Table 2 Organic coatings selected for corrosion resistance in various environments

Coatings	Applications
Outdoor exposure	
Oil paints	Buildings, vehicles, bridges; maintenance
Alkyds	Trim paints, metal finishes, product finishes
Amino resin-modified alkyds	Automotive, metal awnings, aluminum siding
Nitrocellulose lacquers	Product finishes, aerosol lacquers
Acrylics	Automotive finishes
Marine atmosphere	
Alkyds, chlorinated rubber, phenolics, vinyls, vinyl-alkyds	Superstructures and shore installations
Urethanes	Clear marine varnishes
Water immersion	
Phenolics	Ship bottoms
Vinyls	Ship bottoms, locks
Chlorinated rubber	Ship bottoms, swimming pools
Urethanes	Clear marine varnishes
Chemical fumes	
Epoxies, chlorinated rubber, vinyls, urethanes	Chemical-processing equipment
Extreme sunlight	
Vinyls	Metal awnings
Acrylics	Automotive finishes
Silicone alkyds	Petroleum-industry processing equipment
High humidity	
Amino resin-modified alkyds	Refrigerators, washing machines
Epoxies	Air conditioners
Catalyzed epoxies, chlorinated rubber, phenolics	Maintenance; chemical and paper plants
High temperature	
Epoxies	Motors, piping, 120 °C (250 °F) max
Modified silicones	Stove parts, roasters, 205 °C (400 °F) max
Silicones	Stove parts, roasters, 290 °C (550 °F) max; aluminum-pigmented paints 650 °C (1200 °F) max

Table 3 lists the paints and methods of application used for prime and finish coats on three steel parts required to withstand 3 years of outdoor exposure in a marine environment and on a steel part required to withstand 10 years of indoor exposure. All four parts, sketched in Table 3, would be phosphate coated before being painted.

In service, paint films are frequently required to resist exposure to highly deleterious materials. For example, decorative finishes, such as those on home laundry equipment, must resist detergents, and paint films on equipment powered by gasoline engines must withstand attack from gasoline.

Paint films also may be required to resist acids and alkalis, solvents, staining, heat, impact, marring, and abrasion. Some coatings must be able to withstand flexing without cracking or flaking. Table 4 lists paints that have proven successful in withstanding mechanical and chemical action.

Application Limitations. In selecting a coating for a given purpose, application properties in relation to available facilities and conditions in which the coating is to be applied must be considered. The most suitable application, such as spraying, dipping, roller coating, or brushing, as well as drying speed, storage stability, flammability, and toxicity should all be determined. For example, toxic and flammable materials should not be used in areas without adequate ventilation and safety equipment. Heat-convertible coatings cannot be used unless adequate baking facilities are available. Two-component coatings, such as epoxies and polyurethanes, have limited pot life after mixing and must be used within a determined period or discarded, unless two-component spray equipment is used.

Cost must be weighed against the performance required of the coating system. A low-cost coating that fails to perform its function is a wasted expense. With the relatively high cost of labor involved in applying protective coatings, a short service life of a low-cost, inferior material makes ultimate coating and maintenance costs far higher than if a more expensive durable coating had been used originally. Other factors must also be considered in arriving at an over-all cost estimate, including: spreading rate, the area adequately covered by a unit of volume of paint; probable application time and resulting labor costs; equipment required for application; and expected service life.

Examples of Selection. Paints are seldom selected to meet only one requirement. Most organic coatings must perform several functions. The considerations that governed the selection of paint in three different applications are described in the following production examples.

One plant annually produces over a million ballast cases for fluorescent lighting fixtures. Paint applied to these cases, which are rectangular steel boxes, must provide acceptable appearance, resist mildly corrosive exposure, and withstand a certain amount of handling and abuse during assembly. Pro-

Table 3 Coating recommendations for environmental protection of sheet steel parts

Part No.	Priming	Finishing	Application
For 3-yr exposure to marine atmosphere			
1	Epoxy-ester	Modified acrylic	Electrostatic spray
2	Epoxy-ester	Epoxy-ester	Dip or spray
3	Zinc chromate	Phenolic or vinyl	Dip, spray, flow coat
For 10-yr (minimum) indoor exposure			
4	Alkyd or modified epoxy	Urea-alkyd or modified acrylic	Spray or flow coat

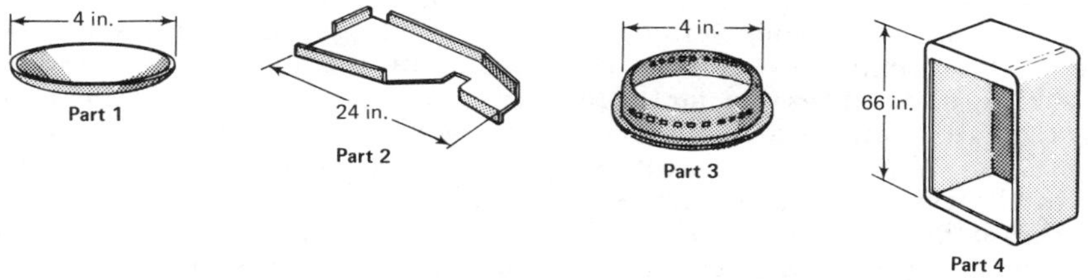

All parts should receive an appropriate phosphate coating before being painted

Table 4 Paints selected for resistance to mechanical or chemical action

Action	Paint
Abrasion	Vinyls; plastisols; polyurethanes
Impact	Epoxies; vinyls; polyurethanes
Marring	Thermosetting acrylics; vinyls
Flexing	Epoxies; vinyls
Acids	Chlorinated rubber; vinyls; epoxies
Solvents	Epoxies; phenolics
Detergents	Thermosetting acrylics; epoxies
Staining	Thermosetting acrylics
Gasoline	Alkyds; lacquers
Alkalis	Phenolics
Heat	Alkyd-amines; silicone resins

cessing and material costs must also be low. These requirements are met by a water-emulsion paint, applied by conveyorized dipping and subsequently baked. This paint is low in initial cost and because the solvent is water eliminates the usual cost of solvent replacement. It has good dipping qualities, provides adequate corrosion resistance, and stands up well in assembly. The paint has good tank stability and, unlike volatile-solvent paints, is nonflammable.

Electric motor shells, 75-mm (3-in.) long by 90 mm (3½ in.) in diameter, required a thin paint film for corrosion protection. Dip painting was desired, because both the inside and the outside of each shell were to be coated. To avoid the necessity of removing paint from the portion of the shell that mated with the motor end bells, the thickness of the paint film had to be closely controlled. The maximum permissible thickness was 50 μm (2 mils), to avoid misalignment of the bearings in the end bells and maintain a uniform air gap between stator and rotor in final assembly. No runs or beads could be tolerated.

A vinyl butyrate lacquer was selected for its good flow characteristics and hiding power. It provided the required corrosion protection and was compatible with cost limitations. By standing the parts vertically in coarse-mesh baskets and carefully controlling withdrawal from the dip tank, 36 parts at a time were successfully painted to coating thicknesses that met tolerances.

One company that had been painting cold rolled steel products with one coat of baking primer and one topcoat of baking enamel needed to expand production facilities. One method would have been to set up a duplicate finishing line. It was found, however, that an acceptable finish could be obtained by priming with an air drying, flash primer and finish painting with a topcoat of baking enamel. Rather than adding a complete painting line, only two prime booths for applying the primer and a flash-off leg on the conveyor, to permit the primer to air dry, had to be added, saving a capital expenditure of several thousand dollars.

Surface Preparation

The importance of proper surface preparation to the durability of any coating system cannot be overemphasized. Without proper surface preparation, the finest paint, applied with the greatest of skill, will fall short of its maximum performance or may even fail miserably. A coating can perform its function only so long as it remains intact and firmly bonded to the substrate.

An adequately prepared surface not only provides a good anchor for the coating but also ensures a surface free of corrosion products and contaminants which might shorten the life of the film by spreading along the coating/substrate interface and destroying adhesion or by actually breaking through the coating.

Cleaning. Before being painted, metals usually are exposed to one or more fabricating processes, such as rolling, stamping, forming, forging, machining and heat treating. In these processes, the metal surfaces pick up various contaminants that can either interfere with the adhesion of the paint film or allow corrosion to progress beneath the paint film and cause it to fall prematurely.

The principle surface contaminants that adversely affect the performance of paint films include oils, greases, dirt, rust, mill scale, water, and salts such as chlorides and sulfides. These contaminants must be removed from the surface before paint is applied.

Selection of cleaning process is governed by the soil or contaminant to be removed, the degree of cleanness required, the type of paint to be applied, and the size, shape, material and end use of the part.

Methods of cleaning ferrous metal surfaces may be classified as mechanical and chemical. To meet rigid requirements for surface cleanness, mechanical and chemical cleaning methods may be used in combination. For example, before structural steel intended for an application involving exposure to corrosive chemical environments is painted, oil, grease, rust, mill scale and any other surface contaminants must be completely removed. Chemical cleaning alone would be inadequate, and mechanical cleaning, if heavy coatings of oil or grease are present, may cause contaminants to adhere even more stubbornly to the surface of the steel.

Mechanical cleaning methods include power brushing, grinding, abrasive blasting.

Power brushing is an abrasive cleaning operation utilizing a power-driven rotary brush. Different types of brushes, and various lengths and gages of wire or fiber provide a wide range of abrasive action. For heavy abrasion, steel wire brushes are used. Mild abrasion is obtained with fiber, horsehair or other bristle-type brushes. Power brushing may be used to remove surface rust, dirt and loose mill scale; it is unsuitable for removing embedded oxides or tight mill scale. (For a discussion of this process, see the article on power brush cleaning and finishing in this Volume.

Grinding, using abrasive wheels of various shapes and grit sizes, is used for the removal of coarse irregularities, such as burrs or flash, and coarse mill scale and heavy rust. Selection of the proper grade of abrasive wheel is important: Too coarse a wheel may produce deep abrasions that are difficult to hide by paint; a wheel with grit that is too fine will clog easily and make the process inefficient.

Abrasive blast cleaning is accomplished by bombarding a surface with abrasive particles propelled at high velocity by air, water, or centrifugal force. The effects of blasting are influenced by the type, hardness, particle size, velocity, and angle of impact of the abrasive. Blasting is a rapid method of removing rust and mill scale. Virtually any degree of cleanness of the blasted surface is obtainable. (For a more detailed discussion of this process, see the article on abrasive blast cleaning in this Volume.

Chemical cleaning includes emulsion cleaning, solvent cleaning, vapor degreasing, alkaline cleaning, acid cleaning, pickling and steam cleaning.

Emulsion cleaning, solvent cleaning and vapor degreasing employ common organic solvents for the removal of oil, grease, loose metal chips, and other contaminants from metal surfaces. For a description of processing methods, see the articles on emulsion cleaning and solvent cleaning in this Volume.

Alkaline cleaning is an effective method for removing oils and greases, water-soluble residues, heat treating salts, acid deposits, and other inorganic dirt. Alkaline cleaners work by detergent action and saponification, and are usually used in soak tank, pressure spray, or electrolytic cleaning. Thorough rinsing or neutralizing of the cleaned surface is necessary after alkaline cleaning prior to painting. For a more detailed discussion of processes, see the article on alkaline cleaning in this Volume.

Acid cleaning is used for removal of light soil or rust. It is unsuitable for removing heavy coats of oil, grease, dirt, and mill scale. Acid cleaners used prior to painting usually are water solutions of phosphoric acid, organic solvents, acid-stable detergents and wetting agents. These solutions are used either hot or cold, in soak tanks or spray cleaning systems. Cleaning is accomplished by emulsifying the oils that are on the surface and by dissolving or undercutting oxide films. For additional information on processing, see the article on acid cleaning of iron and steel in this Volume.

Pickling also utilizes acid, for the removal of rust, mill scale, and some types of soil. Wide variations are possible in the type, strength, and temperature of the acid solutions. The acids most commonly used for pickling ferrous metals are sulfuric and hydrochloric. Inhibitors usually are added to pickling solutions to retard acid attack on the metal. Some parts are precleaned to remove films of oil, grease or other contaminants that would prevent the pickling solution from contacting the metal surface. After pickling, thorough rinsing and neutralizing of the metal surface are necessary. For a more detailed discussion see the article on pickling of iron and steel in this Volume.

Steam cleaning may be used on parts too large or too heavy to be cleaned by conventional methods. In this process, a jet of live steam is directed against the metal surface. The cleaning power of the steam is enhanced by the addition of detergents or alkaline cleaners to the water being vaporized. When such additives are used, cleaning should be followed by thorough rinsing or neutralizing, and drying. Paint adhesion may be poor on steam-cleaned surfaces. If no further surface treatment is feasible, a paint such as one of the asphalt types is recommended. These paints adhere well to a wide variety of surfaces and usually have good adhesion on steam-cleaned metals.

The effectiveness of abrasive blast cleaning compared with wire brushing in removing mill scale is illustrated in the following example. A laboratory study was conducted to compare the effectiveness of abrasive blast cleaning and wire brushing in removing mill scale from hot rolled carbon steel panels. Panel surfaces were cleaned by each method. Surfaces were then individually coated, under controlled conditions, with alkyd, vinyl, epoxy ester, or catalyzed epoxy-based paints. The panels were then exposed to an industrial atmosphere. Within 12 months, all coatings on the panels that had been wire brushed indicated failure in the form of either blistering of the paint or rusting of the steel surface. After 24 months, however, no signs of paint failure were evident on the panels that had been blast cleaned. The poor performance of the paint on the wire brushed surfaces was the result of incomplete removal of mill scale by wire brushing.

A high degree of cleanness can be obtained by combining mechanical and chemical cleaning methods. Before being painted, steel surfaces of tank cars used to carry corrosive materials were alkaline or solvent cleaned and then abrasive blasted. This removed all contaminants. A corrosion-inhibiting primer was applied, followed by a chemical-resistant topcoat. This paint film, controlled to a minimum thickness of 75 μm (3 mils), has a life expectancy of 4 to 6 years.

The influences of part size, shape, and required paint film durability, on the method of surface preparation are shown in Table 5.

Surface Smoothness. If a smooth, high-quality paint finish is desired, imperfections such as scratches, die marks, indentations, roughness, and localized porosity must be repaired before finish painting.

If the imperfections are deep (3.2 mm or ⅛ in. or more), they may be filled with wiping solders, which are lead-tin alloys prepared for this application. The metal surface is first degreased with a nonflammable solvent such as 1,1,1-trichloroethane. When dry, the surface is coated with a flux and heated using a torch. The solder is flowed into the indentation or depression and blended with the contour of the part. All traces of flux are removed with solvents or neutralizers. After the surface is dry, it is primed and sanded smooth and then given the prime and finish coats of paint.

Significant surface imperfections can also be primed and filled with a suitable putty. The putty used must be chemically compatible with both the prime coat and the finish coat and should adhere well to both. Putties are usually highly pigmented pastes, containing such materials as vinyls, epoxies, alkyds, or drying oils as binding and drying agents. Air drying or baking putties can be used, although air drying putty is most frequently used. Baking of thick putty films, either for curing the putty or for curing subse-

Table 5 Surface preparation and painting procedures for providing three steel parts with two degrees of durability in industrial atmospheres

Part No.	Size	Procedures for 1-yr durability: Surface preparation	Procedures for 1-yr durability: Prime painting	Procedures for 1-yr durability: Finish painting	Procedures for 3-yr durability: Surface preparation	Procedures for 3-yr durability: Prime painting	Procedures for 3-yr durability: Finish painting
1	Under 0.76 m (2.5 ft) long	(a)(b)(c)(d)	None	(h)(j)	(f)(b)(g)(b)(c)(d)	(h)(j)	(n)(m)
	Over 0.76 m (2.5 ft) long	(a)(b)(c)(d)	None	(k)(m)	(f)(b)(g)(b)(c)(d)	(k)(m)	(h)(j)
2	100 mm (4 in.) long	(a)(b)(d)	None	(n)(m)	(a)(b)(c)(d)	(n)(m)	(r)(p)
	0.6 m (2 ft) long	(a)(b)(d)	None	(h)(p)	(a)(b)(c)(d)	(h)(j)	(h)(p)
	3.0 m (10 ft) long	(e)(a)(b)(d)	None	(k)(m)	(e)(g)(b)(c)(d)	(k)(m)	(k)(m)
3	100-mm (4-in.) diam	(e)(a)(b)(d)	None	(n)(m)	(e)(a)(b)(c)(d)	(n)(m)	(n)(m)
	0.3-m (1-ft) diam	(e)(a)(b)(d)	None	(h)(q)	(e)(a)(b)(c)(d)	(h)(p)	(h)(q)
	1-m (4-ft) diam	(e)(a)(b)(d)	None	(h)(q)	(e)(a)(b)(c)(d)	(h)(p)	(h)(q)

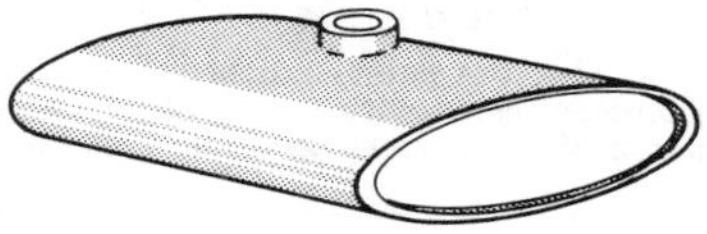
Part 1

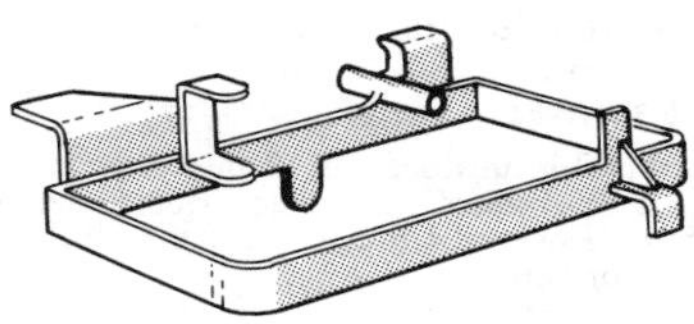
Part 2

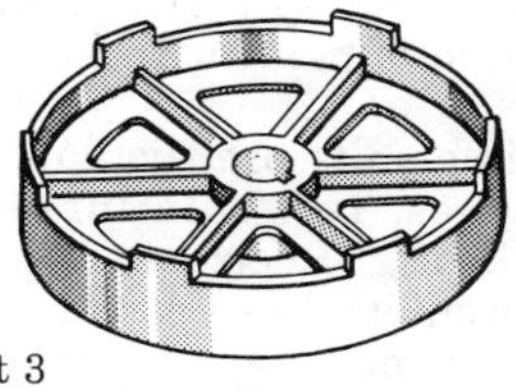
Part 3

(a) Phosphoric acid clean. (b) Water rinse. (c) Chromic acid rinse. (d) Forced-air dry. (e) Abrasive blast clean. (f) Alkaline clean. (g) Zinc phosphate coat. (h) Flow coat with baking alkyd. (j) Bake at 150 °C (300 °F) for 15 min. (k) Hand spray with air drying alkyd. (m) Air dry. (n) Dip in air-drying alkyd. (p) Bake at 150 °C (300 °F) for 20 min. (q) Bake at 150 °C (300 °F) for 25 min. (r) Hand spray with baking alkyd

quent paint films, usually results in blisters or pinholes because of the release of entrapped solvents during baking. Catalytically cured putties should be used when subsequent baking is required, because these putties cure with an exothermic reaction, driving the solvent from the film of putty before it is set and minimizing the formation of pits or blisters during subsequent baking of paint films.

Prepaint Treatments. In addition to cleaning and smoothing, the metal surface is often given a prepaint treatment to improve paint adhesion and reduce corrosion. A prepaint treatment may be a phosphate coating or an organic pretreatment known as wash primer or etch primer.

Phosphate coatings are formed by chemical reaction during immersion of the metal in the phosphate solution that deposits a nonmetallic and nonconducting coating. Both coating weight and crystal size can be varied to provide almost any surface condition required for the appearance or function of the paint. For a detailed discussion of coatings, equipment, and processing methods, see the article on phosphate coatings in this Volume.

The advantages of a phosphate coating as a base for paint are indicated in Table 6, which outlines the results of comprehensive tests conducted on panels for a large manufacturer of home appliances. In these tests, three methods of surface preparation, including cleaning only, cleaning plus iron phosphating, and cleaning plus zinc phosphating, were evaluated by their effect on properties of subsequent paint films. Alkyd-melamine paints from various suppliers were used in the tests and prepared specifically for use following each of the prepainting surface-preparation methods. All panels were spray painted to the same film thickness and were baked and cured in the same manner to achieve polymerization.

Organic pretreatments or wash primers are materials which provide the properties of an inhibitive wash coat or metal conditioner in an organic film. The essential components of wash primers include (*a*) polyvinylbutyral resin, (*b*) chromate pigment, and (*c*) phosphoric acid. Wash primers are described in Military Specifications DOD-P-15328 and MIL-P-8514. Organic pretreatments are formulated for spray application and should be used in the following situations: (*a*) where phosphating equipment is not available, (*b*) where size and shape of parts preclude use of the phosphating process, or (*c*) where parts containing mixed metal components are assembled before painting.

Spraying

Spraying is adaptable to either large-volume or low-volume production. Applications may be limited because of solvent emissions, possible fire hazards, or potential damage from overspray. Spraying methods include (*a*) conventional air spraying, in which the paint is atomized and propelled against the work by means of compressed air, (*b*) hot spraying, (*c*) hydraulic airless spraying, and (*d*) air and airless electrostatic spraying.

Spraying is used for applications in which good appearance and uniformity of coating are desired, such as automobile bodies. Figure 1 shows large and medium-size metal housings that are examples of parts painted by conventional (air atomized) spraying to obtain a good appearance and uniform coating. In the application of metallic or polychromatic enamels, the spray process achieves the necessary dispersion of the particles. Also, very large objects such as bridges are usually spray painted.

Spray painting generally consumes more paint than other painting processes, because of overspray losses. In air atomized spraying, only a small amount of the air at the nozzle is used for atomizing. The remainder of the air pushes the paint and controls the pattern and droplet size. The remainder of the air also causes overspray, as the atomized paint bounces off, or is driven past, the part being painted. The efficiency of the spray operation depends on (*a*) part shape and size; (*b*) operator ability; (*c*) setup of automatic guns; and (*d*) type of equipment whether air atomized, electrostatic, or hydraulic airless spray systems are used.

Hot spraying is a method in which air atomized spraying equipment is used in conjunction with a heat exchanger to heat the paint to a predetermined temperature. The temperature range for hot spraying is usually from 60 to 82 °C (140 to 180 °F). In air atomized spraying, viscosities suitable for application are obtained with the use of solvents. The hot spraying method uses heat to lower the viscosity to the optimum range for spraying, allowing the application of paint with higher solids content.

Heating units are available in two basic types, recirculating and nonrecirculating. Recirculating heating units circulate heated paint between the heater and spray gun, maintaining a constant temperature. Nonrecirculating units heat the paint only once. After the paint passes through the heater, it is subject to various degrees of cooling, depending on hose lengths and conditions of application.

When properly prepared paint materials are used, the hot spray method may have three distinct advantages over conventional spraying:

1 Little or no solvent is used for thinning, reducing labor and solvent costs.
2 Thicker films can be applied using fewer gun passes or fewer coats.
3 Paint can be stored in unheated areas or at very low temperatures

The principal disadvantages of hot spraying are:

- The purchase and operation of the heating unit add to the over-all cost of painting
- Spraying with paint containing a greater solids content can result in higher overspray losses if improper spraying techniques are used

Airless spraying, with either heated or unheated paint, uses hydraulic (pumping) pressure to propel the paint through the hoses and atomizing nozzle. The main advantages of airless spray painting stem from the elimination of air as the force for atomizing and propelling the paint. Airless spraying requires the release of considerably less energy at the nozzle, and overspray is minimized. Heavier film thickness without runs or sags can be obtained when more viscous material is sprayed.

Table 6 Effect of prepainting surface treatments on performance of paint films in tests

Alkyd-melamine paints were from various suppliers and specially prepared for each of the prepainting treatments; after being applied, paints were baked for 30 min at 150 °C (300 °F) and cured for 72 h at 49 °C (120 °F); dry film thickness was 38 to 46 μm (1.5 to 1.8 mils); gloss as measured with 60° glossmeter, was 82°

	Results, for prepainting treatments		
		Cleaning plus	
Test	Cleaning only	Iron phosphating	Zinc phosphating
Pencil hardness	HB(a)	HB-F	HB-F
Adhesion	Poor	Excellent	Excellent
Impact resistance (face and back)(b)	7.2 N·m (64 in.·lb)	7.2 N·m (64 in.·lb)	7.2 N·m (64 in.·lb)
Bend(c)	Fair to good	Excellent	Excellent
Crimp	100% failure	Excellent	Excellent
100% relative humidity at 43 °C (110 °F), h(d)	300 (max)	1000	1000
5% salt spray, h(d)	72 (max)	500 (max)	750

(a) Pencil of this hardness cut through and removed paint cleanly; when cut, paint scratched cleanly with fingernail. (b) Impacting surface 1.59 mm (0.0625 in.) in diameter tested with 0.9-kg (2-lb) weight. (c) 180° bend over 6.4-mm (¼-in.) mandrel. (d) Hours to 65-mm (2.6-in.) creepage from scribe or to blistering, or both

Fig. 1 Parts where conventional (air atomized) spray painting is used to meet requirements of good appearance and uniform coating

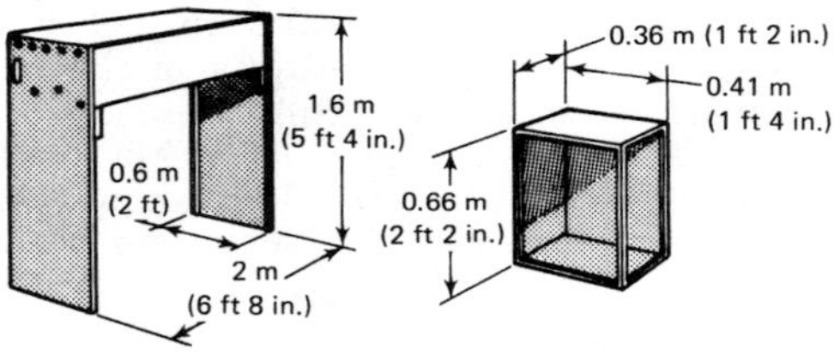

As in air atomized spraying, however, airless spraying requires an increase in pressure as the paint increases in viscosity.

Elimination of air also (*a*) permits the use of comparatively simple and inexpensive spray booth exhaust systems, (*b*) permits full-coverage spraying into corners and recesses with a minimum of bounce-back, and (*c*) reduces masking requirements. Airless systems use simpler, lighter weight guns, and fewer hoses than are required in air systems.

The chief disadvantage of airless spraying is that, unlike air spray guns, airless spray guns cannot be throttled. Because a full flow of paint comes constantly from the airless gun, greater operator skill is required for controlling coating application in difficult-to-reach areas.

Electrostatic Spraying. In electrostatic spray painting, electrically charged atomized particles of paint are attracted to the grounded part. Paint for electrostatic application can be atomized by conventional air, airless, or rotational techniques.

One of the chief advantages of electrostatic spraying is the small loss of paint from overspray. Whereas in conventional spray painting as much as 70% of the paint sprayed may be lost because of overspray, as little as 10% may be lost in electrostatic spraying. Another advantage is the ability of this method to produce a consistent paint film over a long production run.

The main disadvantage of electrostatic spraying is that the electrostatic attraction of the part for the paint draws the paint to the nearest edge or surface of the part, and it is difficult or impossible to get paint into deep recesses, corners, and shielded areas. Another disadvantage is that the necessity for grounding the surfaces to be coated

Table 7 Equipment requirements for conventional spray painting of two production parts

Procedure or equipment	Shell for fire extinguisher(a)	Enclosure panel(b)
Production requirements		
Surface preparation	Clean and phosphate(c)	Clean and phosphate(c)
Prime coating	Zinc chromate primer	Red oxide primer
Finish coating	Alkyd baking enamel	High-luster lacquer
Production per hour	500	40
Equipment requirements		
Overhead conveyor	305-mm (12-in.) hook spacing	610-mm (24-in.) hook spacing
Cleaning and phosphating(c):		
Capacity and solution of rinse tanks	379 L (100 gal)	379 L (100 gal)
Solution and rinse temperature	66 °C (150 °F)	66 °C (150 °F)
Prime and finish coating:		
Spray booth(d)	2.4 by 2.1 by 1.5 m (8 by 7 by 5 ft)	3.7 by 2.4 by 1.8 m (12 by 8 by 6 ft)
Spray gun	Hand operated	Hand operated
Paint supply tank (with agitation)	95-L (25-gal), pressurized	95-L (25-gal), pressurized
Drying	Infrared oven(e)	Air dry(f)

(a) Shell, 75 mm (3 in.) in diameter and 355 mm (14 in.) long, of 1.0-mm (0.040-in.) steel. (b) Panel 915 by 760 mm (36 by 30 in.), of 12-gage (2.657 mm or 0.1046 in.) sheet steel. (c) Proprietary phosphate cleaner coater used; 2-min immersion in solution and rinse tanks, followed by drying. (d) Open-ended water-wash. (e) Drying cycle, 10 min at 160 °C (325 °F). (f) Drying of prime coat may be forced by brief infrared heating

makes it impossible to paint assemblies in which parts are insulated from other parts by a dielectric component.

Equipment for Air Atomized Spray Painting

Basic equipment for air atomized spray painting consists of (*a*) a spray gun, (*b*) a container for the paint, (*c*) an air compressor, (*d*) an air regulator or transformer, (*e*) connecting pipes and hoses, and (*f*) a spray booth, and (*g*) an air filter and moisture trap.

Spray guns are available commercially to fit virtually any requirement. An air cap at the front of the gun atomizes the paint and forms the desired spray pattern. Air caps may be interchanged to meet the requirements of different applications.

A fluid tip or nozzle, located directly behind the air cap, directs and meters the paint into the air stream. These tips vary in material and size according to the type and viscosity of the paint being applied and with the speed and volume requirements of the application. Larger orificed nozzles are required for heavy, coarse, or fibrous paints. Smaller orificed nozzles are used for thin paints. Because atomization of the paint improves as the nozzle size decreases, the smallest orifice that permits proper passage of the paint should be selected. Small nozzle sizes also permit more effective control of fluid pressures, a necessity when applying thin paints that have a tendency to sag. Abrasive or corrosive paints require tips made of materials with resistance to wear or corrosion.

Air compressors provide the force for air atomized spray painting. Air compressors may supply compressed air throughout the shop or only to the paint line. A fluctuating air pressure at the gun, usually caused by inadequate compressor capacity or an inadequate distribution system, can result in improper atomization of the paint and defective paint films.

Air regulators or transformers regulate air pressure as required. Air filters and the moisture trap remove oil, dirt, and moisture from the compressed air. Air regulators, filters, and the moisture trap are installed in the air-supply lines between the compressor and the paint container and between the compressor and the spray gun. The regulators and filters should be as close to the tanks and guns as possible.

Pipes and hoses that distribute the compressed air or paint must be of adequate size to handle peak loads without starving any station and must be able to withstand any abrasive or chemical effects of the paint.

Spray booths are fire-resistant enclosures that confine overspray and fumes and use ventilating systems to draw in fresh air and exhaust the contaminated air after filtering out the solids. A booth face air velocity of 0.559 to 0.635 m/s (110 to 125 ft/min) is required. Dry or water-wash filters may be used. In most installations, spray booth and exhaust stack are protected with automatic sprinklers.

In dry spray booths, fumes and overspray are forced through special filters that remove solids before exhausting the air and fumes. Dry spray booths with replaceable filters are usually less costly to install than water-wash booths, because no water or drain lines are required.

In water-wash spray booths, contaminated air is drawn through a series of water curtains and baffles to remove solids from overspray before air and fumes are exhausted. These booths also reduce fire hazards by collecting overspray in water. However, water-wash spray booths are usually more costly to install, operate, and maintain than dry booths. Some localities prohibit discharging spray booth water into sewers, and adequate recirculation systems and controls must be installed.

Spray booths exhaust large volumes of air, and to ensure a sufficient quantity of replacement air without creating drafts or heating problems in other areas of a plant, air replacement units must be installed. These units, which are placed next to the spray booths, draw in air from the outside, filter it, heat it, and deliver it to the spray booth. This permits proper functioning of spray booth exhaust systems and maintains the desired booth temperature for suitable spraying conditions. Few manufacturing operations are dust free, and air replacement units greatly reduce the possibility that airborne contaminants will be drawn into the paint area and deposited on wet paint surfaces.

Table 7 lists and describes the equipment required for air atomized spray painting of two production parts: a fire extinguisher shell 355 mm (14 in.) long by 75 mm (3 in.) in diameter, constructed from 1.0 mm (0.040 in.) thick steel; and an enclosure panel 915 mm (36 in.) long by 760 mm (30 in.) wide, produced from 0.268 mm (0.105 in.) thick sheet.

Equipment for Electrostatic Spray Painting

Basic equipment for electrostatic spray painting consists of (*a*) a paint supply system, (*b*) an atomizer (or gun), (*c*) a source of electrical power, (*d*) a spray booth with a ventilating system, (*e*) a conveyor, and (*f*) properly designed racks.

Paint supply systems deliver paint under pressure to one or more atomizers. In such systems, paint is prepared and stored in a central tank, from which it is distributed to the atomizers by air or variable-speed positive-displacement pumps.

Atomizers (guns) can be either air or airless. An air gun uses compressed air to atomize the paint delivered to it. In one air gun, the nozzle is insulated from the metal atomizer body by a plastic sleeve. The fluid tip and needle are charged to a high negative electrical potential that is imparted to the atomized paint particles as they leave the nozzle.

Another air electrostatic spray painting system uses ordinary air spray painting guns for atomization. Parts on the conveyor are surrounded by a grid of fine wires charged to a high negative potential. The paint is sprayed between the grid wires and the parts, where it picks up the negative charge and is repelled by the grid wires and attracted by the grounded parts on the conveyor.

Guns that are operated automatically may be mounted in a fixed, but adjustable position, or they may be attached to a mechanism that causes them to reciprocate vertically or horizontally as required by the size and shape of the part being painted.

Airless atomizers use a rotating component, either disk-shaped or bell-shaped, charged to a high negative electrical potential. Paint, forced up through the center of the rotating member, is charged with this same high negative potential. The centrifugal force of the spinning component moves the paint to the outer edge, where the combination of centrifugal and electrostatic forces atomizes the paint and disperses it. The paint is attracted electrostatically to the grounded parts on the conveyor.

Bell-shaped atomizers generally have the plane of the opening of the bell parallel to the surface being coated, and the work is carried past on a straight-line conveyor. The bell may be fixed or may reciprocate and can be used in any position. Hand-operated portable guns with bell-shaped atomizers are available.

Disk atomizers are operated as automatic or manual units. The disks operate in an essentially horizontal plane; however, in nonreciprocating installations, the disk may be positioned at up to a 35° angle to produce a greater vertical spray area. Because the paint is dispersed around the entire periphery of the disk, conveyors usually carry the parts in a circular path around the atomizer. For long parts, the disk may be reciprocated up and down to facilitate coating the entire length of the parts as they are conveyed around the loop.

Fig. 2 Parts that can be efficiently coated by dip painting

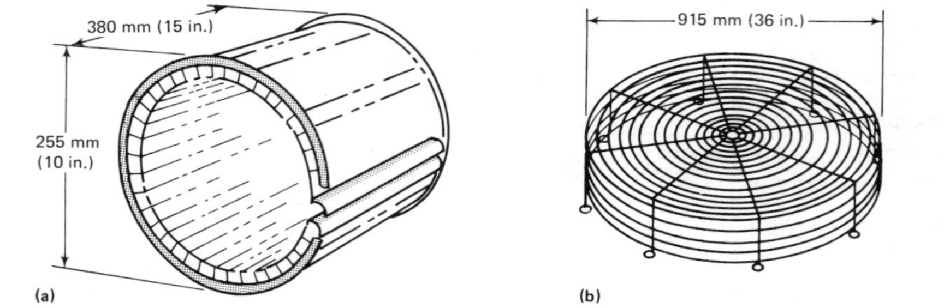

If considerably larger, parts like these could be painted more efficiently by the flow coating process. (a) Blower wheel. (b) Wire fan guard

Power supply units must be capable of providing the high static potential required for each atomizer. These units can operate on 110, 220, or 440 V at 60 cycles. The power output is usually 90 000 to 120 000 V and 5 mA. One power supply unit can handle up to ten atomizers. Units produce a very small current to eliminate any serious hazard to the operator. Usually, an automatic cutoff control is incorporated for added safety.

Spray booths and ventilating systems are used to confine and collect the overspray and to exhaust the fumes from the vaporizing solvents. Minimum velocity of air through the booth is desired to avoid interference with deposition of paint on the part. However, velocity must be adequate to maintain the solvent concentration below its lower explosive level and toxic concentrations. Air velocities of 0.25 to 0.5 m/s (50 to 100 ft/min) are usually adequate for electrostatic spraying applications.

Racks and conveyors used in electrostatic spray painting, in addition to holding and moving the parts, must provide the electrical ground necessary to develop the electrostatic attraction. Parts should be mounted as close to one another as possible on the conveyor to minimize overspraying. Some parts require rotation as they pass the atomizer to ensure coating of all surfaces.

Dip Painting

Dip painting consists of submerging a part in paint contained in a tank, withdrawing the part, and permitting the part to drain. Parts with complex surfaces may be coated efficiently by dipping. Larger parts, produced in quantity, are racked or hung on conveyors, which carry the parts to the paint tank, automatically immerse and withdraw the parts, and carry them over drip troughs into which the excess paint runs off.

Dip painting is seldom used where uniformity of paint thickness is required. This procedure may be unsatisfactory for painting parts having machined holes or surfaces where masking is impractical. Usually, paint films applied by dipping are heavier at the bottom than at the top, because of paint runoff, and are thin at sharp edges. In addition, bubbles and bumps may be found at the bottom edges of painted pieces. Parts may be dip painted to obtain complete coverage and then spray painted to obtain a required surface appearance.

The blower wheel shown in Fig. 2(a) is an example of a part that requires coating of all surfaces, but for which a variation in the coating thickness is acceptable. Dip painting, followed by spinning to remove excess paint, is the preferred method for coating this part. If a part of this type were so large that dip painting would require a prohibitive volume of paint, flow coating could be substituted, although rotation of the part as it passes the nozzles might be required. If the size were increased still further, spray painting might be used.

The wire fan guard shown in Fig. 2(b) is an example of a part that is difficult to cover efficiently by conventional spraying without considerable loss of paint by overspray, but can be efficiently coated by dipping. The part in Fig. 2(b) also could be painted by flow coating, electrostatic spraying, or electrodeposition. If the part were larger, flow coating might be preferred to dipping because of the smaller volume of paint required.

Equipment for Dip Painting

Basic equipment for dip painting consists of (*a*) a tank to contain the paint, (*b*) a device to agitate the paint and prevent settling or separation of ingredients, (*c*) a device for lowering the parts into the paint and raising them out of it, (*d*) a drip trough, and (*e*) ventilators.

Dip tanks should be as small as possible, consistent with the maximum size of parts to be dipped. Smaller tanks require less paint to fill them, and the higher rate of turnover helps to maintain paint stability. Dip tanks should be designed to expose the smallest possible area of paint to prevent evaporation of paint thinner.

Dip tanks for conveyorized dipping should be designed with ends that slant according to the angle parts travel as they are conveyed into and out of the paint. This angle should be not less than 30°, measured from the horizontal, and should be 45° if possible. The sharper the angle of travel, the shorter the dip tank, and there is consequently less exposure of paint to air.

Introduction of fine air bubbles into the tank should be avoided, because these bubbles can cause pinholing in the final film. Pumps are used to dislodge trapped air bubbles from blind spots in parts. Mechanical devices, such as tipping bars which change the position of parts during immersion, are also used to dislodge bubbles.

Dip tanks should be designed with provision for paint agitation. The bottom of the tank should be just wide enough to accommodate the agitator.

Agitators in dip tanks prevent the paint from separating. Horizontal screw or horizontal paddle agitators may be used. Usually, agitators extend the full length of the tank and are no greater than 0.3 m (1 ft) in diameter. Tanks with a capacity of 7570 L (2000 gal) or more may require agitators of larger diameter.

Agitator speed is usually between 60 and 120 rev/min. Correct speed is usually determined by visual inspection of the tank surface. The entire surface of the tank should show agitation with no clear vehicle forming. Agitation should be of sufficient force to keep the paint well mixed and prevent the pigment from settling. Protective screening should be installed over the agitator to protect the blades from being damaged by parts that may fall into the tank.

A pump circulator added to an agitator system permits the use of strainers to remove lumps that may form or foreign material that may be introduced into the tank.

In tanks with a capacity of about 1135 L (300 gal) or less, complete recirculation of the paint every 20 min usually provides sufficient agitation for preventing settling or separation of ingredients. Paint should be withdrawn from the bottom of the tank and pumped back into the tank just under the minimum working level.

Ventilation. Dip tanks should have a means of ventilation to remove fumes from the tank. Hoods should not interfere with the conveyor system or the device for immersing parts.

Drip troughs catch the excess paint that drips from the parts after they have been removed from the tank. A conveyorized dip painting system usually includes a sloped drip trough to allow excess paint to flow back into the tank. Working areas around drip troughs and dip tanks must be equipped with adequate fire control equipment.

Table 8 lists equipment requirements for dip painting two production parts. One part is an angle bracket about 25 mm (1 in.) wide, produced from 3.2 mm (1/8 in.) thick stock and with each leg 75 mm (3 in.) long. The other, a spring hanger produced from 4.8-mm (3/16-in.) stock, is 100 mm (4 in.) wide with 150-mm (6-in.) and 100-mm (4-in.) legs.

Table 8 Equipment requirements for dip painting two steel production parts

Procedure or equipment	Angle bracket(a)	Spring hanger(b)
Production requirements		
Surface preparation	Clean and phosphate(c)	Clean and phosphate(c)
Prime coating	None	Red oxide primer
Finish coating	Lacquer; air dry	Lacquer; bake
Production per hour	1200	650
Equipment requirements		
Work handling	Manual; baskets(d)	Automatic; conveyor(e)
Cleaning and phosphating:		
Capacity of solution and rinse tanks	190 L (50 gal)	190 L (50 gal)
Solution and rinse temperature	66 °C (150 °F)	66 °C (150 °F)
Prime coating:		
Capacity of dip tank	...	190 L (50 gal)
Drying	...	12-m (40-ft) infrared oven(f)
Finish coating:		
Capacity of dip tank	190 L (50 gal)	190 L (50 gal)
Drying	Air dry	12-m (40-ft) infrared oven(f)

(a) 75 by 75-mm (3 by 3-in.) L-shaped bracket, 3.2 mm (1/8 in.) thick, 25 mm (1 in.) wide. (b) L-shaped bracket, of 4.8-mm (3/16-in.) stock, 100 mm (4 in.) wide and with one 100-mm (4-in.) leg, one 150-mm (6-in.) leg. (c) Proprietary phosphate cleaner coater used; 1-min immersion in solution and rinse tanks, followed by drying. (d) 90 to 100 pieces per load. (e) Hooks 305 mm (12 in.) apart. (f) Parts conveyed through oven; drying cycle, 10 to 12 min at 160 °C (325 °F)

Flow Coating

In flow coating, paint is pumped from a storage tank through properly positioned nozzles, onto all surfaces of parts, as they are conveyed. Excess paint drains back to the storage tank for recirculation. Paint films applied by flow coating are wedge shaped, thinner at the top and thicker at the bottom of painted parts. In flow coating, utilization of paint approaches 95% as opposed to about 50% for atomized air spraying and 70 to 80% for dipping. Properly designed flow coat machines with vapor chamber flow-out reduce solvent losses, and eliminate tears, sags, and curtains.

Flow coating is used extensively to paint panels for home appliances. Flow

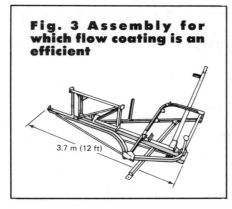

Fig. 3 Assembly for which flow coating is an efficient

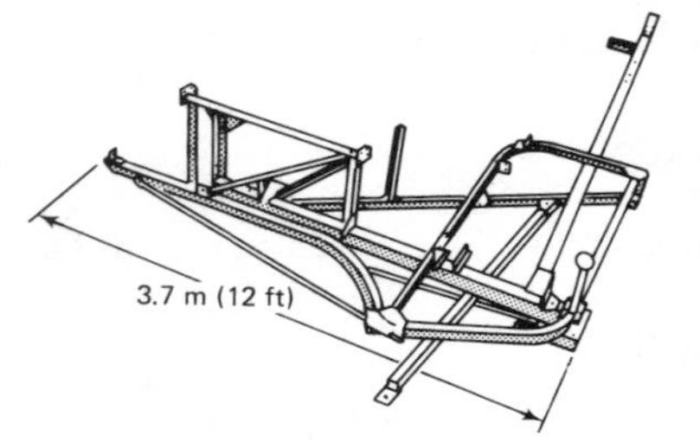

coating is also used (*a*) to coat parts with recesses inaccessible to spraying, (*b*) to coat parts (such as bedsprings) for which good appearance is desirable but is secondary to complete coverage and economical application, and (*c*) to coat intricate parts that are too open in design to permit efficient spray painting and too large to be practical for dip painting.

Figure 3 shows a large assembly which, because of its size and construction, is an example of a part for which flow coating is the most efficient method of painting. Spraying would be inefficient because of overspray, and dip coating would require a large quantity of paint to fill the dip tank. However, flow coating would be impractical for a similar assembly twice as large, and spray painting would be preferred.

Equipment for Flow Coating

Basic equipment for flow coating consists of (*a*) a chamber, (*b*) a paint-storage tank, (*c*) a pump, (*d*) a drain-off section, and (*e*) continuous conveyors. Because flow coating ordinarily is used only for high-volume production parts, equipment usually is set up as part of a continuous process, which may include cleaning, phosphating, drying, prime flow coating, finish flow coating, and baking. One conveyor may carry the parts through all operations. Process variables that require close control are nozzle pressure, viscosity and temperature of the paint, and hanger design and spacing of the parts.

Pressure control of paint in the circulating system is critical, particularly at the nozzle. Excessive pressure causes the paint to flow off the parts at too high a rate, resulting in high solvent losses and bubbles in the paint. Too low a pressure leaves areas on the work uncoated or improperly coated. Pressures vary from 1 to 14 kg (3 to 30 lb). The higher pressures are used to coat difficult-to-reach or recessed areas. The lowest pressure that can be used in a particular operation is the most economical when solvent loss is considered.

Paint viscosity must be closely controlled. Viscosity that is too high causes poor flow-off of paint from the work. This results in sags, beads, blistering, and other defects associated with excessive paint thickness. Viscosity that is too low results in excessive solvent loss and inadequate film thickness. Viscosities are held within the range of 18 and 32 s (No. 2 Zahn cup), although paints with a No. 2 Zahn cup viscosity as high as 100 s have been used. Viscosity must be adjusted for each paint and each differently shaped part. Once the optimum viscosity has been determined, it should be maintained.

Temperature control of paint is essential. Too high a temperature results in excessive solvent loss and may cause instability of the paint. Too low a temperature requires increased use of solvents to maintain proper viscosity and can result in inadequate film thickness. Although temperatures as low as 16 °C (60 °F) and as high as 38 °C (100 °F) have been used, 21 to 32 °C (70 to 90 °F) is the recommended temperature range.

Drain-off chambers, enclosed tunnels immediately adjacent to the coating chamber, are incorporated in many installations. In drain-off chambers, a high concentration of solvent vapor is maintained to retard drying. This eliminates beads, bubbling, and other surface defects that result when flow-off of paint is incomplete. Some installations include an electrostatic detearing device to remove any beads or drops of paint still clinging to the edge of the part.

The equipment and production requirements for using flow coating to paint two steel production parts are discussed below. One part is a step hanger (an angle bracket) produced from 4.8-mm (3⁄16-in.) stock 50 mm (2 in.) wide, with two legs 355 and 125 mm (14 and 5 in.) long, respectively. The other, a bracket for holding a fire extinguisher, is produced from 1.5-mm (0.060-in.) stock and has legs 305 and 75 mm (12 and 3 in.) long; a metal clamp is attached to the 305-mm (12-in.) leg to retain the extinguisher. The production requirements are as follows:

Part A

- Step hanger, 4.8 mm (3⁄16 in.) thick; maximum dimensions 125 by 355 mm (5 by 14 in.)
- Production rate, 540 pieces per hour

Part B

- Bracket for holding fire extinguisher, 1.5 mm (0.060 in.) thick; maximum outside dimensions, 75 by 305 mm (3 by 12 in.)
- Production rate, 720 pieces per hour

Both parts

- Clean and phosphate, using proprietary phosphate cleaner coater with 2-min immersion in solution and rinse tanks before drying; color coat with alkyd baking enamel (no prime coat)

Equipment requirements for both parts are as follows:

Work handling	Overhead conveyor(a)
Cleaning and phosphating(b)	
Solution and rinse tanks	190 L (50 gal)
Solution and rinse temperature	66 °C (150 °F)
Flow coating	
Flow coater	Conventional; 2 nozzles
Nozzle pressure	275 kPa (40 psi)
Drying	Infrared; 7 to 8 min; 160 °C (325 °F)

(a) 305-mm (12-in.) hook spacing. (b) Proprietary phosphate cleaner coater used; 2-min immersion in solution and rinse tanks, then drying

Roller Coating

Roller coating is a high-speed machine painting process used for continuous coating of sheet and strip stock. The process consists of transferring an organic coating from a revolving applicator roller to the surface of sheet or strip as it is passed through the machine. The top, the bottom, or both surfaces may be coated in one pass. Paint can be rolled on with excellent control of film thickness.

Roller coating is one of the most economical painting processes. Steel strip can be roller coated, baked, and coiled for later fabrication into parts, such as the slats used in the manufacture of Venetian blinds.

Because this process is similar to rotary printing, designs can be reproduced and repeated in any pattern necessary. Thus, bottle caps, food cans, toys, and similar painted parts can be produced economically. The designs are roller coated on the metal strip before it is stamped or formed to shape.

Fig. 4 Examples of curtain coating equipment

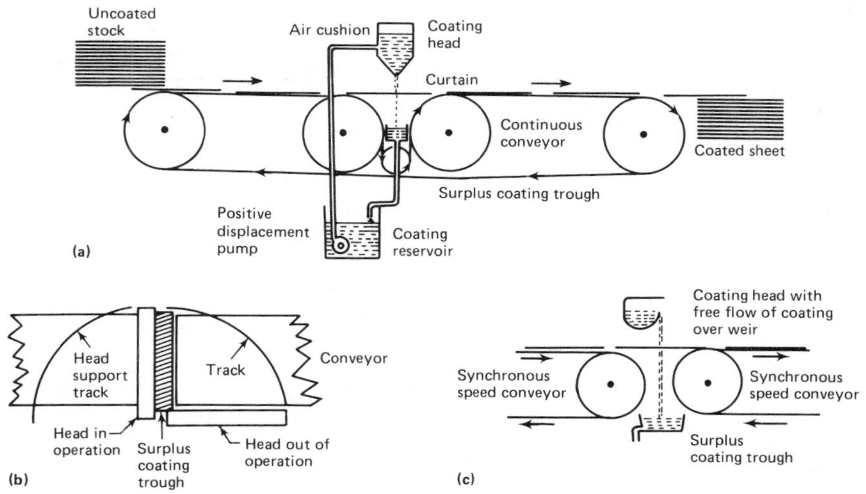

(a) Pressure head curtain coater. (b) Double head machine gives fast color changes, applies two component coatings. (c) Gravity flow coater with synchronized conveyors

Equipment for Roller Coating

Equipment for roller coating, similar in function to a rotary printing press, may vary from a relatively simple machine to an elaborate complex installation several hundred feet long. The equipment transfers an organic coating from the revolving applicator roll to sheet or strip as it travels through the machine. The thickness of the film is regulated by a metering roll that controls the amount of paint transferred to the applicator roll.

Applicator rolls must be made of material that does not swell, soften, or dissolve from contact with paint. Usually, applicator rolls are made of, or faced with, a resilient material, such as neoprene or polyurethane. Resiliency permits rolls to conform to irregularities found in commercial stock.

Almost every organic coating material can be applied by roller coating. Usually, the paints use slow-evaporating solvents and are applied at higher viscosities than in other painting methods, which permits close control of paint flow and film thickness.

Curtain Coating

Curtain coating is a method of applying finishes at high speeds with little paint loss. In curtain coating, flat or shaped strip and sheet are moved by conveyor belt beneath a pump-fed reservoir with an adjustable slot opening at its lower edge. The slot opening provides a controlled continuous wet curtain of coating flowing onto the work. Some curtain coaters have heads that function by use of gravity overflow of material over one side of the reservoir.

The pressure-fed curtain coater applies a uniform wet film to flat or shaped surfaces of practically any material. Coatings can be applied between 13 and 2540 μm (1/2 and 100 mils) thick, depending on coating characteristics, conveyor speed, and parts being coated. Deposited films are smooth and uniform with no ridges, washboarding, or similar effects that can be found with roller coating. No sagging or uneven thickness occurs as with flow coating and dipping.

Equipment for Curtain Coating

Curtain coating equipment consists of a conveyor belt that moves the items to be coated under a special coating head. This coating head may be either pump fed or gravity overflow. In the former, the reservoir has an adjustable slot opening at its lower edge that provides a controlled continuous wet curtain of coating on the work as it moves past. In the latter, the material flows over one side of the reservoir. Viscosity must be closely controlled and flow properties adjusted to form a good curtain. A collection trough is located under the head and any of the coating that does not fall on the work is returned to the pump for reuse. Figure 4 illustrates a curtain coater.

Tumble Coating

Tumble or barrel coating is one of the most economical means of coating large numbers of small pieces. It is particularly useful for such articles as nails, screws, buttons, and other metal objects weighing less than 0.5 kg (1 lb). Tumble coating is also a good method for pretreating objects to be coated by other systems. In barrel coating, a predetermined weight of coating, adequate to produce a uniform finish over each item without causing drops to form, is poured over the items. The barrel is rotated to distribute the coating evenly over the parts. Drying air is then forced through the center axle and circulates through the continuously moving charge, finally escaping through the other end of the axle and removing solvent vapors. After sufficient air has passed through, parts are dry and ready for packing. In some cases, drying is not done in the barrel. Pigmented coatings are a good example, because the rubbing process during drying damages the appearance of the coatings. In this case, the parts are discharged wet onto a screen and allowed to dry by forced air or natural air circulation. The slight blemishes that occur at the points of contact are virtually invisible.

Electrocoating

Electrocoating is a process in which the object to be coated is dipped into a tank of water-borne paint, and a current is passed through to charge the paint particles electrically. The charged paint particles migrate to the object to be coated, which has an opposite charge. When the object is reached, the paint materials come out of solution and coat the surface. The article is then withdrawn from the tank, rinsed with water to remove any undeposited paint, and baked. This process is also known as electrodeposition, electrophoretic deposition, or electropainting.

Coating Thickness. Paint particles migrate to the part to be coated when a sufficient potential difference, usually between 80 and 180 V, is applied between the part and the tank or separate electrodes. Coating thickness depends on complexity of the part, time, temperature, voltage, solids content, throwing power of the paint, and its ability to penetrate recessed areas. Coating thicknesses up to 38 μm (1.5 mils) are obtainable. The applied coating has a solids content of about 95% and is essentially water insoluble. The relatively high electrical resistance of the coating drastically reduces the rate of deposition as thickness increases. The thickness on easily reached portions of a part builds up rapidly to a near-maximum value and then levels off, and the thinner coating at recessed portions continues to increase in thickness until a very uniform paint film is deposited on even the most complex parts.

Resins used for electrocoating include:

- Maleinized drying oils
- Styrenated and vinyl toluenated maleinized oil
- Phenolic modified maleinized oils
- Alkyds
- Epoxy esters
- Styrene allyl alcohol copolymer esters
- Acrylic copolymers
- Polyesters

Fig. 5 Electrocoating finishing system

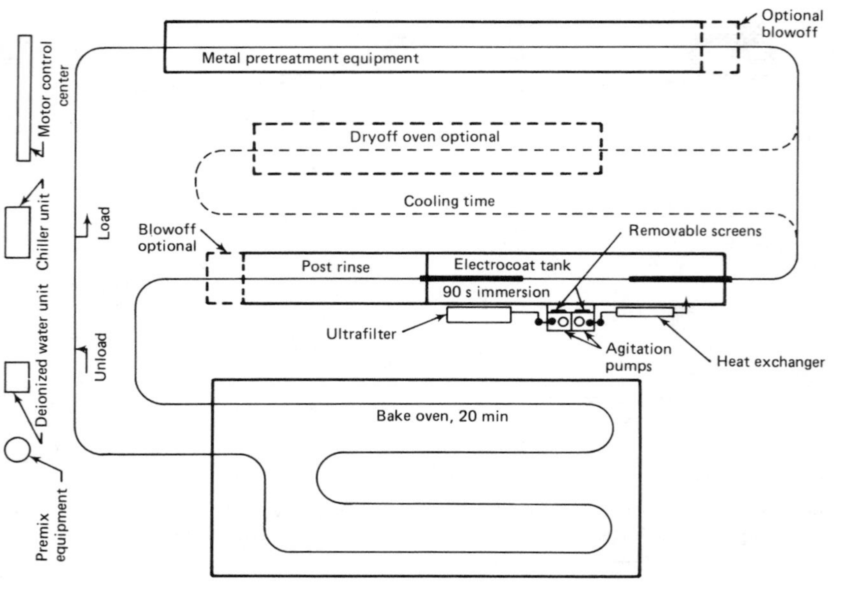

Advantages and Limitations. Electrocoating offers several advantages over more conventional coating processes. These include:

- The process lends itself to total automation, reducing labor costs
- Intermixed parts with different shapes and sizes can be coated
- A more uniform coating thickness is obtained
- Good edge and recess coverage without heavy buildup produces better corrosion resistance
- Absence of runs and sags minimizes rework
- Significant paint savings of as much as 30% are obtainable
- Absence of fire hazard

The disadvantages of the process should also be considered:

- Cost of equipment is high
- Temperature, pH, and alkali content of the coating material must be controlled closely
- Pretreatment and rinsing requirements are more stringent than with conventional coatings
- Surface defects in the substrate are visible through the coatings
- Only a single film can be applied
- Changing colors is difficult and expensive

Cathodic Systems. All early electrocoating systems were anodic. However, many have recently been converted to cathodic. The advantages of cathodic over anodic systems are as follows:

- In cathodic systems, unlike anodic, metal dissolution does not occur at the cathode. The absence of electrodissolved metal in the film results in better film properties, especially in the case of white electrodeposits over steel
- Cathodic deposition tends to deposit over contaminants in the metal surface, and they do not appear in the film
- Salt spray and humidity resistance is improved
- Cathodic coatings have better color consistency over welded areas

Equipment for Electrocoating

Electrocoating systems usually include a dip tank, power supply, heat exchanger, filters, plane replenishment tanks, and a baking oven. Dip tanks are equipped with stirrers and circulating pumps to keep the paint homogeneous. Tanks can range in size from several hundred to several thousand gallons. In an anodic system, parts may be made the anode or the cathode, and the tank the other electrode, or separate metal electrodes can be placed in an insulated tank. For cathodic systems, tanks must be insulated and kept at ground potential. A relatively large power supply is needed, with voltage ranging from 50 to 400 V and amperage from 50 to 4000 A. Heat exchangers are needed to keep the bath at optimum plating temperature, because the electrodeposition process generates heat. Ultrafilters are used for reclaiming dragout paint. Figure 5 illustrates an electrocoating system layout. Figure 6 depicts a basic layout of a cathodic electrodeposition system.

Powder Coating

Powder coatings are paint films which are applied to parts as a dry powder. The powders are basically the same type of polymers and resins (see Table 1) as used in liquid coatings except no solvent is used. Instead, the coating composition is ground to a fine powder. After application the film is formed by fusing the powder particles at temperatures above the melting point of the powder. Powder coatings are used on a wide range of products including metal furniture, wire goods such as baskets and racks, appliance housings, chemical and laboratory equipment, and aircraft and automotive components. The advantages and disadvantages of powder coating systems are given below:

Fig. 6 Cathodic electrodeposition system

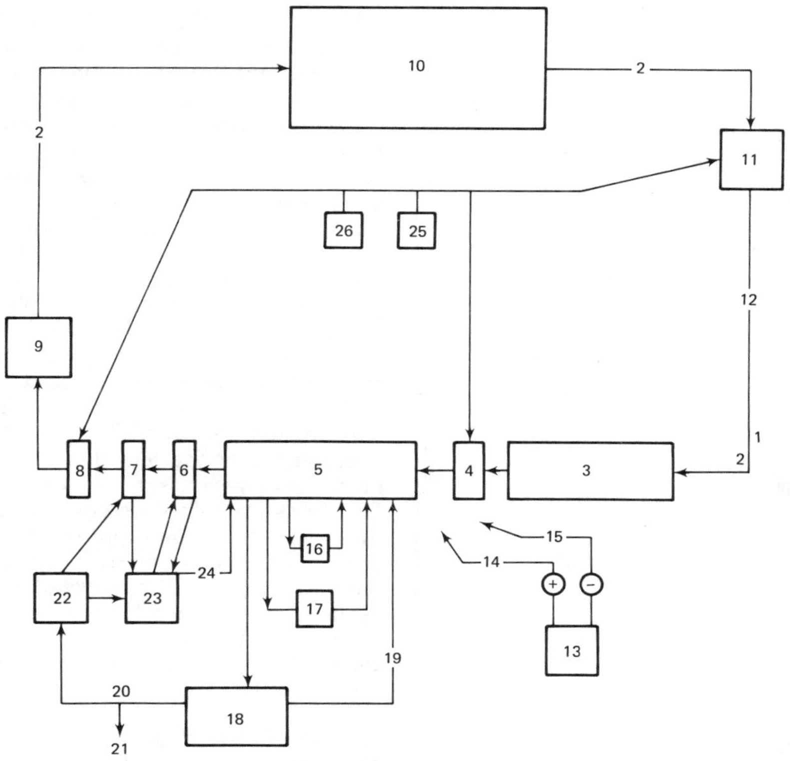

(1) Load area. (2) Conveyor. (3) Pretreatment. (4) Deionized water rinse. (5) Electrodeposition tank. (6) Recirculated permeate rinse. (7) Fresh permeate rinse. (8) Deionized water rinse. (9) Dryoff. (10) Curing oven. (11) Deionized quench for cooling. (12) Offload area. (13) Source for direct current. (14) To anodes in paint bath. (15) To work on conveyor. (16) Particle filter. (17) Cooler. (18) Ultrafilter. (19) Paint solids return. (20) Permeate. (21) Controlled flow to waste. (22) Permeate storage. (23) Recirculated permeate. (24) Overflow permeate. (25) Source of deionized water. (26) Deionized water storage

Fig. 7 Electrostatic spray system

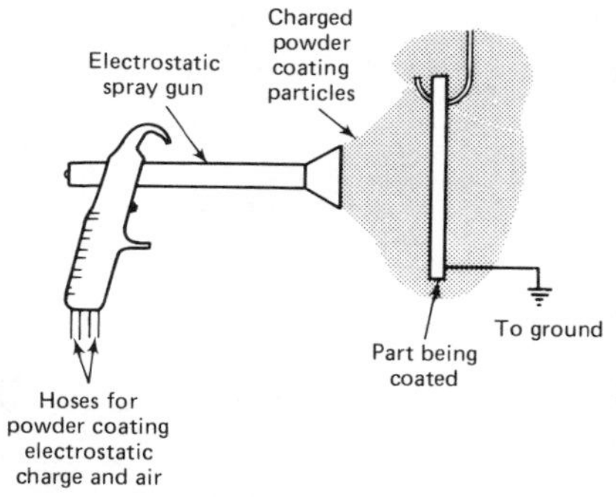

Advantages

- Meets all current EPA requirements for reduction in VOC emissions
- Material use can approach 100% if powder can be collected and reused
- Maintenance is less, because powder can be vacuumed from any unbaked surface
- Exhaust air volume is greatly reduced from that used for solvent-borne systems

Disadvantages

- Color change is difficult, because a separate booth is usually required for each color
- Color matching is more difficult with powder coatings than with solvent coatings
- Powder coating materials are discrete particles, each of which must be the same color. No tinting or blending by the user is possible
- Applying films of 25 μm (1 mil) or less is extremely difficult, and at times, impossible

Equipment for Powder Coating

The three primary methods of applying powder paints include: electrostatic spray, fluidized bed, and electrostatic fluidized bed. In addition, several variations can be applied to the basic electrostatic spray process. The best process for a particular application depends on such factors as: end-use of coating, coating thickness, size and shape of parts, rate of production, and material handling techniques.

Electrostatic spray is the most versatile and flexible application process. The electrostatic disk, cloud chamber, and Gourdine tunnel are variations of the basic electrostatic spray process.

The basic electrostatic spray process uses spray guns to apply powder to the parts (see Fig. 7). These guns may be manual or automatic. Powder is delivered to the guns through flexible tubing from a supply hopper. At the guns, the powder is charged electrostatically. Coating parts are then carried to the oven, where the powder melts, flows, and fuses to the surface. Air exhausted from the booth serves three functions: maintains the powder-air concentration below the minimum explosive concentration (MEC), keeps powder from drifting outside the booth, and begins the powder recovery process. The dust collection equipment used may be a cyclone, a bag filter, or a cartridge filter. Air exhausted from the dust collector is returned to the plant. The moving filter belt is a modification of the basic powder recovery technique. An endless belt of porous fabric forms the floor of the spray booth. Booth exhaust air passes through the belt into a plenum beneath, and the powder is left on the belt. The belt carries the powder to the end of the booth beneath a vacuum pickup head.

The electrostatic disk is a variation of the electrostatic disk for liquid paint. The disk propels the charged particles outward by air force as it reciprocates. The workpiece passes around the disk in an omega booth.

In a cloud chamber, automatic electrostatic spray guns create a cloud of electrostatically charged powder in an almost totally enclosed booth. Powder not adhering to parts conveyed through the booth falls to a fluidized bed in the bottom of the booth. The powder is withdrawn from the fluidized bed and recycled to the guns. Only enough air is drawn into the booth to keep powder from drifting from the workpiece openings. Because there is the constant po-

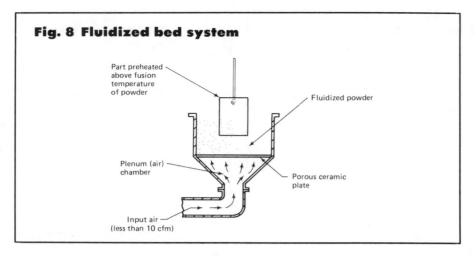

Fig. 8 Fluidized bed system

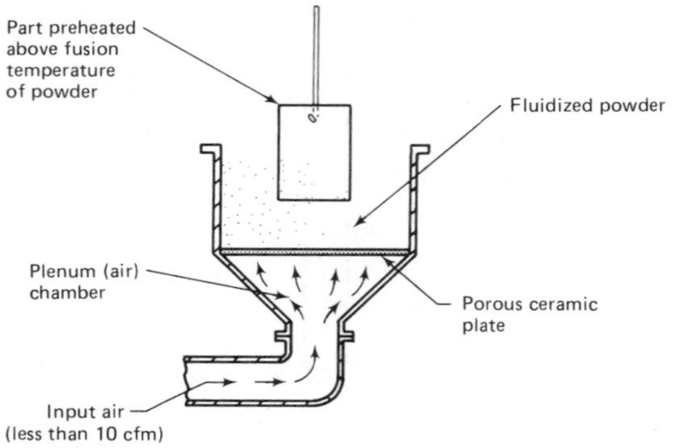

Fig. 9 Electrostatic fluidized bed

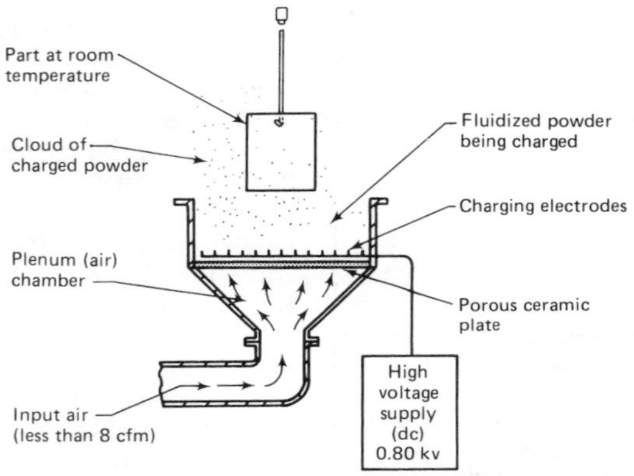

tential for exceeding the MEC, cloud chambers must be equipped with an explosion suppression system.

In the Gourdine tunnel, powder is sprayed in a nonconductive plastic tunnel. The tunnel is not grounded, and charged particles are not attracted to its surface. The parts themselves are grounded, and they move through the tunnel, first passing through the charging section where powder is sprayed from electrostatic guns on both sides of the workpiece. Parts then pass into a precipitation chamber, with interior walls lined with conductive plates charged at the same polarity as the powder particles. Air introduced into the tunnel carries particles not adhering to the part into the precipitation chamber. The particles are repelled by the charged plates and directed back toward the workpiece. The system is said to achieve 90% deposition efficiency, needing no recovery system. Like the cloud chamber, the Gourdine tunnel also needs explosion protection.

Conventional Fluidized Bed. In the fluidized bed process, dry air is forced through a porous plate or membrane into an open top tank which is about half filled with powder (see Fig. 8). The air suspends the powder, increasing its volume, and makes the powder act like a fluid. The item to be coated is preheated above the fusion point of the powder and is dipped in the powder for a few seconds. The powder which comes in contact with the item fuses and forms a coating. The item is then removed, shaken or blown to remove any loose powder, and either recooled or reheated to obtain a more uniform film. Reheating is used if a thermoplastic powder is used or if a thermoset powder is used, and the film is to be cured. With the fluidized bed method, film thicknesses of 150 to 1525 μm (6 to 60 mils) can be obtained. Films of less than 50 μm (6 mils) are difficult to obtain with this method. This procedure does not coat recesses well and cannot coat one side of an object without the use of masking.

Electrostatic Fluidized Bed. In this system, an electrode is incorporated in the powder chamber (see Fig. 9). The powder is fluidized as in the conventional procedure. However, high voltage is applied, and the particles receive a charge and rise in a fine cloud above the fluidized bed. When a grounded part is passed into or through the cloud, the charged particles are attracted to and adhere to the part. With this system, preheating is not necessary. A smaller amount of powder is necessary, and the less dense powder suspension permits better control of film thickness. This system also eliminates the need for dipping, because powder rises to the part. Because the part is not preheated, powder can be removed from areas which are to be left bare, eliminating the need for mechanical masking and the problems associated with it.

Radiation Cure Coatings

Radiation cure coatings are organic monomer or polymer resin binders of low viscosity which polymerize to a cured film when subjected to radiation. Two main types of radiation curing are used: (*a*) electron beam (EB) and (*b*) ultraviolet (UV). In both processes, the materials used are solvent free and 100% reactive, giving off little vapor, creating no pollution problems. Curing time ranges from a fraction of a second to minutes depending on the source of radiation. Curing is achieved at or slightly above room temperature, which allows heat sensitive materials, such as plastics, wood, and electronic components to be coated without harm. Because no baking ovens are needed, less floor space is required for finishing. The differences between EB and UV curing systems must be considered:

- Electron beam radiation is much stronger than UV. Ultraviolet coatings usually require activators to initiate curing, potentially shortening storage life
- Electron beam coatings cure almost instantaneously, whereas UV coatings may require several seconds to several minutes

Fig. 10 Direct fired and indirect fired convection ovens

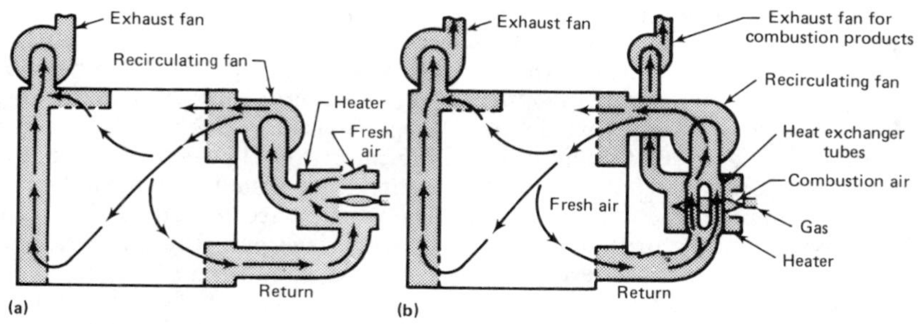

(a) Direct fired. (b) Indirect fired

Fig. 11 Indirect fired continuous convection oven

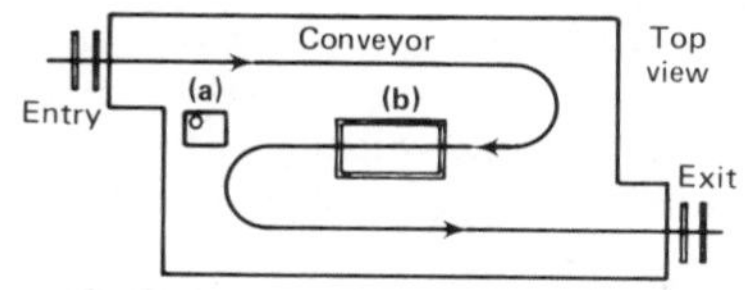

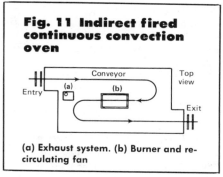

(a) Exhaust system. (b) Burner and recirculating fan

- Ultraviolet coatings are generally limited to thin clear films (up to 75 μm or 3 mils); EB can be clear or pigmented (up to 255 μm or 10 mils)
- Ultraviolet lamps generate heat from 38 to 49 °C (100 to 120 °F), EB does not. Thermally sensitive substrates may cause problems if UV cured
- Electron beam curing has higher equipment cost than UV
- Ultraviolet coatings are more readily available than EB coatings

Curing of Paint Films

Curing is the process of converting an applied coating to a dry film. Baking or thermosetting coatings require heat to cure. Those coatings that dry by evaporating the solvent are air drying or thermoplastic. Ovens are used to supply the energy needed to cure the film and/or evaporate the solvent.

Heat is transferred by conduction, convection, or radiation. Convection and radiation are used in ovens for baking paint. Conduction takes place in shielded areas only incidentally. Convection baking ovens usually are heated by gas, although oil, electricity, or steam may also be used. Radiation ovens may be heated by gas or electricity, although electricity is used more frequently.

Batch ovens are used for baking paint films on parts produced intermittently, in limited quantities, or on parts not easily conveyorized. These ovens, which transfer heat by convection, consist of one or more insulated compartments equipped for heating, recirculating, and exhausting air. These compartments also regulate and control temperature.

Batch ovens range in size from small single-compartment units that accommodate only a few small pieces, to large units with single or multiple compartments for baking several bulky parts simultaneously. Multiple-compartment batch ovens may be used for simultaneous production of separate runs of parts that require different baking cycles. Batch ovens may be direct fired or indirect fired.

Direct fired batch ovens (Fig. 10) are used more often than indirect fired ovens. In direct fired ovens, fuel is burned directly in the circulated air. Direct fired ovens are more economical to operate, because the heat does not have to be conducted through the walls of a heat exchanger. The disadvantage of direct fired ovens is that the paint films are exposed to the products of combustion, which may harm some coatings.

The solvent vapors extracted from the paint are burned when they are passed through the flame with the recirculated air. Proper circulation and introduction of a sufficient supply of fresh air are essential—particularly during the early minutes of the cycle, when the rate of solvent evaporation is greatest. The required dilution rate for safe operation depends on the total quantity of solvent present and on its lower explosive limit. Insurance and safety organizations have established minimum exhaust rates for safe operation of direct fired baking ovens for most of the solvents used in paint films.

Most regulations require maintaining 25% or less of the lower explosive limit (LEL) of solvent in the oven. This is done by calculating solvent emission at full oven load and exhausting sufficient air to keep solvent level at or below 25% of LEL. New solvent level monitoring equipment allows automatic damper control of oven exhaust to maintain safe solvent levels, depending on actual load. This reduces energy requirements by reducing exhaust rates and provides richer solvent exhaust for more efficient use of incineration for VOC emission control.

Indirect fired batch ovens (Fig. 10) burn the fuel outside the oven walls, and the oven is heated by means of heat-exchange surfaces. Indirect fired ovens are less efficient and consume more fuel than direct fired ovens. The advantages of indirect fired ovens are that the paint film is protected from the products of combustion, and solvent vapors are not exposed directly to the flame.

Continuous ovens are used in finishing parts produced in large quantities. Parts are conveyed through the ovens as a cycle in a continuous finishing process that may include (*a*) cleaning, (*b*) phosphating, (*c*) drying, (*d*) painting, and (*e*) baking. Continuous ovens may use either convection or radiation.

Convection continuous ovens, direct fired or indirect fired, may use gas, oil, electricity, or steam as the source of heat, although steam usually limits oven temperature to 150 °C (300 °F). Figure 11 illustrates an indirect fired continuous convection oven. Two 180° turns have been incorporated in the conveyor path to shorten the distance the heat must travel from the burner. Ductwork may be installed to direct and control the distribution of the heated air. A large volume of hot air can be directed to the area where the cold work enters and the volume of air can be limited to the area where the work has attained the desired temperature.

Heat retention is a problem in these ovens, because parts being conveyed must be permitted to enter and leave the oven in a continuous process without restriction. One method for heat retention is to slope entrance and exit chambers downward from the oven section, trapping rising heated air. For ovens into which the conveyor enters and exits at the same level as the hot zone, heat retention can be accomplished by high-velocity air seals (air curtains) across the entrance and exit of the oven.

Radiant continuous ovens are widely used. Using infrared radiation, radiant ovens require less insulation or heat sealing, as well as less complicated ventilation and exhaust systems. Many infrared continuous baking ovens have an enclosed tunnel construction to take advantage of the air heated by contact with the parts. A simple baking tunnel may consist only of racks of lamps on each side of the conveyor line. Heat is produced as required as parts are conveyed through the system, permitting a more flexible use of the baking facilities.

Infrared generators may be either electric or gas fired. Electric infrared generators are glass-lamp, ceramic-coated resistance wire, or quartz. Gas-fired infrared generators consist of ceramic grids heated by gas until the grids are irradiant.

In infrared baking, electrical energy or heat energy from the burning gas is converted into radiant energy, which, when directed to the wet paint surface, is rapidly absorbed by both the paint film and the metal beneath it. The surface beneath the paint heats up as rapidly as the paint. This accomplishes solvent evaporation simultaneously throughout the thickness of the paint film, or, in some cases, from the inside to the outside. Pits and blisters are thus minimized, and a uniform polymerization of the paint film is accomplished. This is a distinct advantage over convection baking, in which the outside paint surface is first to heat. Infrared baking decreases the tendency to form a skin or shell that traps the solvents under the surface, creating pressure and leading to the formation of pinholes or blisters when the vapors escape. Because it penetrates, infrared also bakes faster than convection heat, if the entire painted surface is exposed to the radiant heat.

High velocity ovens expose parts to very high temperatures for short times. 205 to 315 °C (400 to 600 °F) for 1 to 3 min is typical. If parts can tolerate high surface temperatures for a short time, high velocity ovens may be an economical alternative.

Heat Recovery. The combination of decreased availability of gasoline, gas, and electrical energy with the resulting increased energy costs and stricter air pollution regulations in recent years makes the recovery of oven exhaust gases increasingly important. As much as 80% of the fuel energy may be exhausted from the oven. Considerable progress is being made in heat exchangers, afterburners, fume incinerators, and catalytic converters to recover some of the oven exhaust energy. This energy may be used to provide added heat source, to preheat incoming prepared air for the oven, or to supplement the oven fuel supply. Oven manufacturers should be consulted for most recent developments in energy recovery.

Convection Baking Time and Temperature. Time and temperature of the baking cycle in convection ovens are dictated by the polymerization characteristics of the paint. Although baking temperatures often are expressed as oven temperatures, these may not be sufficient to ensure complete polymerization. Until the metal reaches baking temperature, it is conducting heat away from the paint at the interface. This may reduce the time at temperature of this part of the film enough to prevent complete polymerization. Thus, baking cycles are specified in terms of time, which makes allowance for the metal to attain the proper temperature.

The graph in Fig. 12 illustrates the importance of metal temperature in baking. Metal temperatures have been plotted against time for two panels of equal area but of different mass. Assuming that a baking temperature of 150 °C (300 °F) is desired, the 24-gage (0.607 mm or 0.0239 in.) cold rolled steel panel can reach 150 °C (300 °F) approximately 10 min before the 13-mm (½-in.) thick casting.

Adjusting the cycles so that the areas under the curves are approximately equal usually gives comparable baking results. This is not true, however, if the paint being baked has a critical curing temperature. Some urea-alkyds, for example, cannot polymerize at temperatures below 120 °C (250 °F).

Prebake Solvent Evaporation. If paints are exposed to the heat of the oven too soon after being applied, particularly when infrared heat is used, the solvent materials vaporize too fast and disturb the continuity of the paint film. Vapors may also become entrapped because of incomplete ventilation in the oven. This results in a partial breakdown of the paint film and a chalky appearance. If vapor concentration is high enough, the paint film may break down completely.

To avoid these problems, sufficient time and ventilation should be allowed to permit some solvent evaporation and dispersion before baking. For batch baking, parts are allowed to stand in a well-ventilated area for a period of time before they are placed in the oven. In conveyorized painting and baking, the conveyor should pass through a well-ventilated area before carrying the parts into the oven. In this situation, forced circulation of clean air hastens the evaporation and dispersion of the solvent vapors, permitting a shorter interval of time and distance for conveyor travel.

Defects Attributable to Improper Baking. Table 9 lists some of the defects in paint films that may result from improper baking, the causes, and corrective measures that should be taken. Equipment and operating requirements for baking two specific parts are discussed in the following paragraphs.

The upper half of a 64-mm (2½-in.) long shouldered wing bolt was spray painted to a wet film thickness of 30 μm (1.2 mils) with glossy bronze baking enamel, using toluene as the solvent. Before being painted, the parts were placed in aluminum racks incorporating 156 holes that accepted the wing bolts up to the shoulder on the body of the bolt, serving to mask the lower part from the paint. Parts were painted, air dried for 5 to 6 min, then baked at a production rate of 1500 pieces per hour. The operating conditions were the following:

Weight of each piece	0.032 kg (0.0702 lb)
Number of pieces per rack	156
Weight of rack plus load	5.089 kg (11.22 lb)
Pieces processed per hour	1500
Type of baking enamel	Glossy bronze
Solvent used	Toluene
Baking cycle	20 min at 150 °C (300 °F)

The requirements for the baking equipment included the following:

- *Racks*: aluminum sheet, 710 by 660 mm (28 by 26 in.) and 1.0 mm (0.040 in.) thick, with 13 rows of 12 equally spaced 9.5-mm (⅜-in.) diam holes (156 holes); weight 0.12 kg (0.27 lb)
- *Oven*: electrically heated single-compartment batch oven, 750 mm (30 in.) square and 622 mm (24.5 in.) high (inside dimensions), with shelves accommodating five racks of parts, which are placed in oven every 30 min

- *Heat input*: 13.3 kW/h, assuming that oven is at heat and that 15 min is required for load to reach oven temperature
- *Exhaust air*: 0.46 m^3/s (98 ft^3/min) at room temperature; 0.670 m^3/s (142 ft^3/min) at 150 °C (300 °F)
- *Air recirculation rate inside oven*: 3.8 m^3/s (800 ft^3/min) at 150 °C (300 °F)
- *Space for oven and controls*: width, 1625 mm (64 in.); depth, 915 mm (36 in.); height, 2135 mm (84 in.)

A welded frame assembly, angular in shape, 2 m (8 ft) long with each leg approximately 0.9 m (3 ft) long, was produced from 6.4-mm (¼-in.) steel plate, and weighed approximately 410 kg (900 lb). The parts were spray painted to a wet film thickness of 30 μm (1.2 mils), using a glossy yellow baking enamel with xylene as the thinner. Before baking, the parts were air dried 10 to 20 min. The parts were supported during baking by special dollies designed to have minimum contact with the parts. Three parts were processed each hour. Operating conditions used in this process included:

Weight of each assembly	408 kg (900 lb)
Assemblies painted	3 per hour
Baking enamel	Glossy yellow
Solvent used	Xylene
Baking cycle	45 min at 135 °C (275 °F)

The requirements for the baking equipment included the following:

- *Dollies*: of steel construction, designed to support frame assemblies with minimum contact during finishing; 3 m (9 ft) long, 1 m (4 ft) wide, 380 mm (15 in.) high; weight, 90 kg (200 lb). Nine required, three each in spray room, oven, and cooling area
- *Oven*: electrically heated walk-in batch oven with three compartments, each 1.5 m (5 ft) wide, 3.0 m (10 ft) deep, 2 m (6 ft) high
- *Heat input*: 95.6 kW/h, assuming that one dolly and part are replaced every 20 min and that 15 min is required for one dolly and part to reach oven temperature
- *Exhaust air*: 1.26 m^3/s (267 ft^3/min) at room temperature; 1.75 m^3/s (370 ft^3/min) at 135 °C (275 °F)
- *Air recirculation rate inside oven*: 30.68 m^3/s (6500 ft^3/min) at 135 °C (275 °F)
- *Space for oven and controls*: width, 6.1 m (20 ft); depth, 3.4 m (11 ft); height, 3.0 m (10 ft)

Fig. 12 Times required for interfaces of two steel parts to reach baking temperature of 150 °C (300 °F)

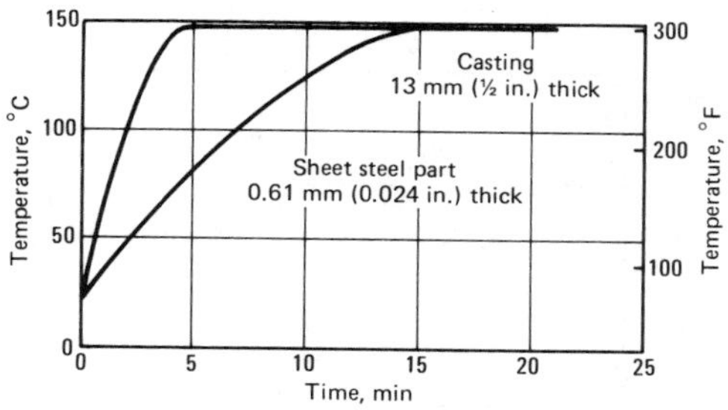

Two steel parts of equal area but different mass

Table 9 Causes and corrections of paint film defects associated with baking

Cause	Correction
Soft film	
1 Baking cycle inadequate	1 Increase time or temperature of baking
2 Metal not reaching curing temperature	2 Increase time or temperature of baking(a)
Pinholing or blistering	
1 Entrapment of solvent vapors because of quick drying at baking temperature	1 Extend air-drying time before baking to increase evaporation of solvent
Wrinkling	
1 Unequal curing, because of extreme temperature difference of film and metal	1 Extend air-drying time before baking, or change from convection to radiation
Hazing or poor gloss	
1 Temperature too high	1 Reduce temperature; increase time
2 Combustion fumes in oven attacking paint	2 Increase input of fresh air and air circulation in oven
3 Venting of solvent vapors from oven not fast enough	3 Increase input of fresh air and air circulation in oven
Discoloration	
1 Temperature too high	1 Reduce oven temperature
2 Hot spots in oven	2 Control circulation of air in oven
Discontinuity of coating	
1 Solvent vapors collecting in oven to high concentration	1 Increase air input and circulation in oven; allow longer air-drying time
2 Same as 1 under pinholing or blistering	2 Extend air-drying time before baking
Brittleness	
1 Overbaking (excessive time or temperature)	1 Careful control of time and temperature

(a) In convection ovens, adjust louvers to deliver greatest concentration of heated air to area in which parts have not attained baking temperature, or change to radiation heating

Quality Control

Once a coating has been chosen for production, it is necessary to ensure that future supplies of the coating are consistent from batch to batch, maintaining satisfactory application properties, appearance, stability, and performance characteristics. Only by so doing can production quality be maintained. Testing can be costly; consequently, quality control programs should be designed using the simplest test methods and the least number of tests necessary to ensure essential quality levels. Numerous tests and equipment for evaluating and monitoring of coatings are available, most of which are described in the following publications:

- American Society for Testing and Materials (ASTM) Standards
  - Part 27, Paint—Tests for Formulated Products and Applied Coatings
  - Part 28, Paint—Pigments, Resins, Polymers
  - Part 29, Paint—Fatty Oils and Acids, Solvents, Miscellaneous; Aromatic Hydrocarbons; Naval Stores

Fig. 13 Viscosity of one specific point as a function of temperature

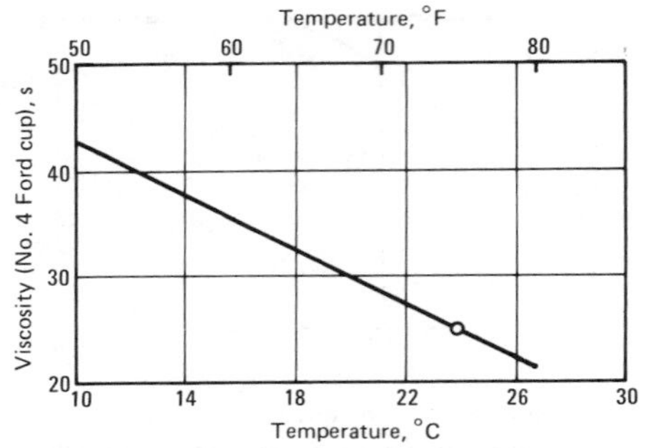

Fig. 14 Spectrophotometric curves

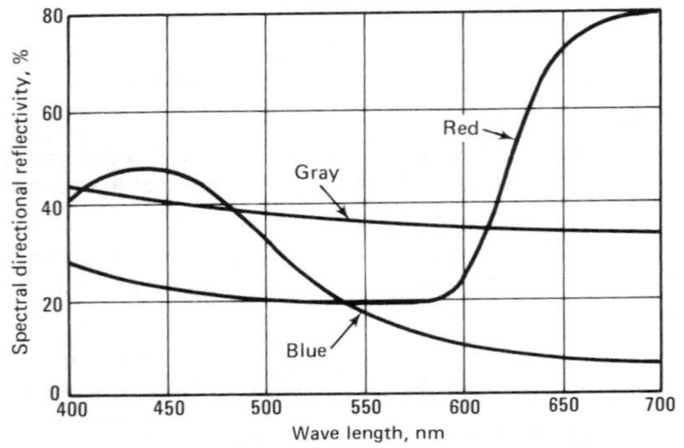

- Federal Test Method Standard No. 141a Paint, Varnish, Lacquer and Related Materials; Methods of Inspection, Sampling and Testing
- Paint Testing Manual, Physical and Chemical Examination of Paint, Varnishes, Lacquers, and Color STP500, by American Society for Testing and Materials

Table 10 lists a number of tests for paint, which present a sample of tests available.

Viscosity. Controlling paint viscosity is necessary to maintain the desired properties of the coating and to ensure that the process operates at the maximum possible efficiency. Pigmented materials require close control of solids content to eliminate that influence on viscosity readings. Close control of the temperature is also necessary, because viscosity varies inversely with temperature. In one manufacturing company, for example, paint was mixed in one area, then transferred in batch lots to other areas for application. Temperatures varied widely between locations, particularly in the winter. The application required a viscosity of 25 s at 24 °C (75 °F), using a No. 4 Ford cup. To coordinate temperatures of mixing and application, tests were conducted, and an isoviscosity line was plotted as shown in Fig. 13. Viscosity may be checked by:

- The efflux cup method
- The torsional method
- The bubble viscometer

Commercial equipment is available to assist in performing any of these tests for viscosity.

The efflux cup method uses containers closely controlled in size having a precise orifice in the bottom. After the cup is filled with paint to be checked, paint is permitted to drain through the orifice. The length of time, in seconds, to the first break in the flow stream of the paint, is the viscosity. Several commercial cups are available, including the Zahn and Ford cups, designed to permit rapid testing of viscosity on the production line.

Table 10 Test methods

Test	ASTM	Federal method 141
Wet or liquid tests		
Nonvolatile (solids content)	D2369	4041
Vehicle solids	D2698	4052
Pigment content	D2698	4022
Weight per gallon	D1475	4184
Flash point	D56, D93, D92	4291, 4293, 4294
Viscosity	...	...
Ford cup	D1200	4282
Brookfield	D2196	4287
Stormer	D562	4281
Zahn	D3794	...
Gardner Holt tubes	D1545	4271
Fineness of grind	D1210	4411
Reducibility and dilution stability	...	4203
Drying time	...	4061
Dry film performance tests		
Hiding power	D344, D2805	...
Color (pigmented coatings)	D1729	4250
Specular gloss	D523	6101
Abrasion resistance	...	...
Falling sand	D968	6191
Taber abraser	D1044	6192
Adhesion	D2197	6303
Dry film thickness	...	...
Magnetic gage	D1186	6181
Mechanical gage	D1005	6183
Hardness	...	...
Pencil	D3363	...
Sward Roeker	D2143	...
Indentation	D1474	6212
Humidity resistance	D2247	6201
Salt spray resistance	B117	6061
Immersion resistance	D1303 Sec D	6011

The torsional method measures the resistance of paint to rotation of a disk immersed in the paint. Several instruments for measuring the rotational resistance require that the paint be placed in a special container, and other instruments permit the viscosity to be measured in the shipping container.

With the bubble viscometer, the viscosity of the liquid is measured by the speed with which a bubble of air rises in the liquid. The material is confined in a glass tube, which is completely filled, except for a small bubble, and stoppered. The viscosity determination may be made in two ways:

1. Comparing the rate of rise with that of a material of known viscosity contained in a tube of the same size
2. Measuring the time required for the bubble to travel between two marks on the tube. In this method the tube must be calibrated with one or more liquids of known viscosity

Color of paint is most often controlled by visual comparison against a reference standard. One deficiency of this type of control, however, is the difficulty of retaining permanent color standards. Because of this difficulty, stabilized dry drift control panels have come into use. These panels can be either paper chips coated with paints designed to have a minimum of color change on aging, or porcelain or ceramic panels properly coated with the appropriate paint. Each control has its advantages; however, both are subject to soiling. All comparisons should be made under a standardized source of light to eliminate extraneous influences of various sources of light.

Visual comparisons do not permit the assignment of numerical values to differences and are subject to wide variations of opinion. It is possible, however, to analyze color on various instruments, assign numerical tolerances, and eliminate subjective judgment to some degree.

Colorimeters measure the three attributes of color: (*a*) hue, (*b*) saturation, and (*c*) brightness. Colorimeters require the use of some reference standard, although not necessarily of exactly the same color as the paint being tested. Because the numerical values established with the colorimeter must still be related to a reference standard, they cannot be considered as absolute.

Spectrophotometers, which measure and analyze color throughout the entire visual spectrum of 400 to 700 mμ, come nearest to being absolute measuring devices. Curves obtained on the spectrophotometer are used as a permanent, reproducible reference. Figure 14 shows spectrophotometric recordings of red, gray, and blue flat paints prepared by extending cadmium red, ivory black, and Prussian blue, respectively, with zinc white. These curves indicate the variation in the light reflectance of the three colors across the entire visible spectrum.

Gloss. The procedure used for measuring gloss of a painted surface is by visual comparison with a known standard. By comparing the sharpness of an image reflected by a sample surface with the image reflected by a standard surface, even relatively small differences can be detected. However, because this procedure is based on human judgment and does not lend itself to the assignment of numerical values, instruments are often used for the measurement of gloss.

Photoelectric glossmeters measure gloss from various fixed angles. The viewing angle used most often is 60° from the vertical, but a viewing angle of 85° is more sensitive for low-gloss paints. The scale of glossmeters is calibrated from 0 to 100 with the higher numbers indicating higher gloss. Readings of 0 to 15 are generally considered flat, 15 to 80 are semigloss, and 80 to 100, high gloss. Instruments accurate to less than one unit are available. However, it is difficult to apply paint films to this degree of accuracy with any consistency, and a five-unit variation in gloss is acceptable.

Abrasion resistance of organic films may be determined by test methods using either falling sand or an abraser.

The falling-sand method uses a funnel-shaped hopper, which feeds sand to a vertical tube of 19-mm (¾-in.) ID and 915 mm (36 in.) long. Sand is permitted to flow down the tube and impinge on the test panel, which is placed at a 45° angle beneath the tube. The test is complete when the sand abrades through the paint film, exposing a spot of bare metal 4 mm (5/32 in.) in diameter. The abrasion coefficient in litres per mil is found by dividing the volume (in litres) of sand used by thickness of the paint film (in mils).

The Taber Abraser method uses abrasive wheels of various grits, a method of applying loads of 250, 500, or 1000 g on the wheels, and a turntable to which the test panel can be clamped. This test can be used to obtain either the wear index (rate of wear) or the wear cycles (amount of wear) of the paint film.

Elongation properties of an applied organic film may be measured by bending a test panel over a tapered cone and measuring the length of the first continuous crack. The apparatus and test methods used are described in detail in ASTM D522.

Blistering. A water immersion test may be conducted to determine the resistance of organic films to failure when immersed in water in an accelerated manner. Distilled water is used to eliminate the influence of any chemicals contained in tap water. The test procedure is set forth in ASTM D870. The method for evaluating degree of blistering is given in ASTM D714.

Environmental tests may be required to evaluate paint films in a particular service environment. For example, a detergent immersion test is used to determine the suitability of a particular paint for use on a home laundry machine. Tests for resistance to acids, alkalis, industrial fumes, and other corrosive mediums may be established, with the criterion for failure being predetermined by agreement.

Exterior-exposure tests may be conducted in accordance with ASTM D1014 to determine the resistance of a paint film to exposure. These tests are usually conducted in specified areas to obtain information on the influence of various atmospheres, such as industrial fumes, arid but intensely sunny climates, or salt air.

Artificial weathering tests using apparatus for exposing specimens to water and carbon-arc light are detailed in ASTM E42. Fluorescent UV-condensation type tests are described in ASTM G53. These test methods simulate conditions of atmospheric exposure that act in a highly accelerated manner on the test panels. ASTM D822 is concerned with the variations in test conditions and the evaluation of test results. This test predicts the results of more time-consuming exterior-exposure tests with some degree of reliability.

Dry film thickness bears a direct relation to product cost and product performance. Film must be measured accurately for control of cost and performance. Several types of equipment and methods may be used for measuring the thickness of dry paint films on ferrous and nonferrous metals. Among these are dial comparators, eddy current and magnetic thickness gages, and penetration and microscopic methods.

Dial comparators (dial micrometers) are accurate depth-measuring devices with a calibrated dial and a pressure foot with a maximum diameter of 3.2 mm (⅛ in.). The test panel is clamped firmly to a base. The pressure foot of the dial comparator is brought into contact with the paint film, and a reading is taken. Without disturbing the panel, the paint film is carefully removed from the panel where the reading was taken. The pressure foot is then brought into contact with the panel at the point of the previous reading. The difference in gage readings is the thickness of the paint film. This procedure, set forth in ASTM D1005, is not adaptable to paint films thinner than 13 μm (0.5 mil) or unusually soft films.

Eddy current thickness gages work on the principle of induced current changes in a high-frequency alternating current coil in a probe held in close proximity to a metal surface. Thickness of nonconductive films may be measured on any metal substrate.

Magnetic thickness gages are available in several types, the most useful being the portable gage using permanent magnets. These instruments measure the reduction of magnetic forces by a nonmagnetic coat of paint between a permanent magnet and the magnetic base to which the paint is applied. This reduction in magnetic force is calibrated in terms of paint-film thickness. This method is suitable for films less than 13 μm (0.5 mil) thick, but not for unusually soft films.

The penetration method is practical where the paint film is applied to a surface that conducts electricity. This method measures the depth of travel of a small drill. The drill and the painted metal are connected electrically to a signal light that lights when the drill tip touches the metal panel. The drill is set with the tip just touching the surface of the paint film. The drill is then rotated and advanced slowly into the paint film until the signal light indicates contact. The measured distance of travel of the drill is a direct measurement of the paint film thickness. The accuracy of this method is approximately ±10% of the film thickness. A portable scratch thickness gage also is available for determining the thickness of a dry paint film by the penetration method.

The microscopic method of measuring dry film thickness is the most accurate of the methods described. A cross section of the painted panel or part is mounted and polished. Using a calibrated eyepiece or screen, the magnified image of the paint film is measured. This method has the disadvantages of being a destructive test as well as requiring more specimen preparation than other methods. The accuracy of the microscopic method is limited only by the optical equipment used and the care exercised in preparing the specimen.

Wet Film Thickness. Inspection gages may be used to measure the thickness of wet paint films. With these gages, it is possible to determine whether the wet film is of adequate thickness to develop the desired dry film thickness.

Salt spray (fog) tests are arbitrary performance tests useful in establishing and maintaining certain standards of quality for the organic finish, particularly when correlated with field tests. For example, if spring clips, phosphate coated and painted with two coats of phenolic-based zinc chromate primer, can withstand 100 h in salt spray before failure, they may be expected to last 5 years or more in applications such as license plate brackets, molding retainers, and wire retainers on automobiles.

The test consists of placing parts or panels to be tested in a chamber in which a 5 wt % solution of sodium chloride is atomized. The exposure zone of the salt spray chamber must be maintained at 33 to 36 °C (92 to 97 °F).

Hardness of a paint film may be approximately determined by scratching it with pencils of different hardness sharpened in a mechanical sharpener. This test does not reveal a specific hardness, but enables one paint film to be compared either to another paint film or to an acceptable standard film. The test is run to determine the softest lead that can penetrate the surface of the paint film. If only one manufacturer's pencils are used, more consistent comparisons are obtained. The disadvantage of this test is the possible variability of the force exerted by the operator. It is a useful test, however, when one skilled operator is making empirical comparisons of two panels side by side.

Adhesion of the paint film may be determined by a simple test. Using a sharp razor blade, parallel cuts 1.6 mm (1/16 in.) apart are made in two directions perpendicular to each other. The cuts should penetrate the paint film to the base metal, and should be made to produce a grid section with 1.6-mm (1/16-in.) squares. The over-all size of the incised area should be approximately 9.5 mm (3/8 in.) square. A piece of cellophane tape is pressed firmly over the incised area, then quickly stripped off. Adhesion is determined and compared by the number of small squares of paint that remain on the test area.

Adhesion testing can be simplified and standardized by using commercially available instruments, which have a solid cutting head that makes several parallel lines with one pass. Spacing is always identical, providing a uniform test. Cutters with different spacings are available. Some specifications require three sets of parallel lines, two sets at right angles to form squares and a third set making diagonals to form triangles. The spacing of scribed lines is usually correlated to total film thickness with thicker films requiring spacing of cuts to be wider.

Fig. 15 Gravelometer

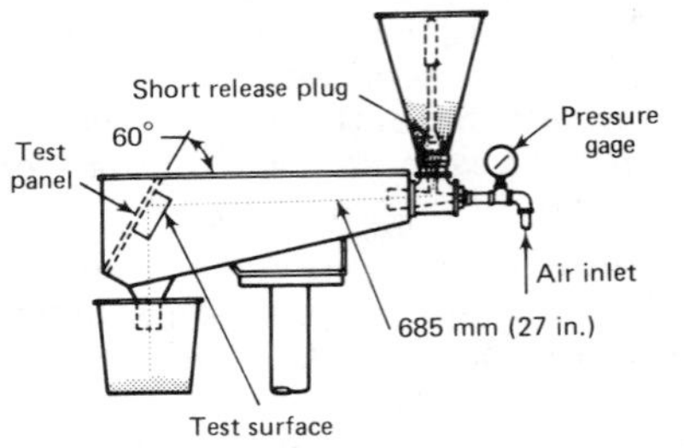

Used to measure impact resistance of paint films

Impact resistance may be determined by the use of an instrument that consists of a 25-mm (1-in.) diam impact rod that tapers to an impact nose with a 6.4-mm (1/4-in.) spherical radius, a tube 26-mm (1 1/32-in.) ID that serves to guide the impact rod in its downward fall, a base plate with a 13-mm (1/2-in.) diam hole through it, and a bracket to support the tube and position the base plate. The tube is graduated in inch-pounds and is slotted, so that a pin, protruding from the impact rod, can be used to raise the rod to a specific inch-pound location. The base is positioned so that its 13-mm (1/2-in.) diam hole can engage the nose of the rod at the bottom of its fall.

To test the impact characteristics of a paint film, a test panel, or an actual part, if made of sheet metal, is placed over the base, and the impact rod is permitted to fall from a height that generates the desired force. Direct impact is obtained by facing the paint film toward the falling rod; reverse impact is obtained by facing the paint film away from the rod. The impact makes a 6.4-mm (1/4-in.) spherically radiused impression in the test panel. Results are measured by the force the paint film can withstand without cracking, chipping, or flaking.

The Gravelometer (Fig. 15) is another device which is used to compare impact resistance of organic films against an accepted standard. Using air at 70 kPa (10 psi), this device propels 14 kg (30 lb) of steel shot a distance of 685 mm (27 in.) against a painted area 125 by 150 mm (5 by 6 in.). The amount of paint that is retained on the panel is indicative of the impact resistance and the adhesion of the paint film. This is a more severe test than the rod impact tester, but the Gravelometer is more revealing, because of the greater area involved. The SAE J400 test uses graded gravel instead of steel shot and is gaining wide acceptance as a standard test.

Causes of Paint Film Defects

Defective paint films are usually the result of either (*a*) improper preparation of the paint or substrate surface, or (*b*) lack of control of processing variables. Paint film defects attributable to improper baking procedures are listed in Table 9, and paint film defects and frequent causes are listed below:

Poor adhesion (peeling, flaking, poor bonding)

- Paint applied over grease, oil, water, rust, alkali residues, or other foreign materials on surface (dip, flow, roller, spray)
- Finish coat applied over incompletely dried undercoat (dip, flow, roller, spray)
- Paint applied to surface that is too hot (above 71 °C or 160 °F) or too cold (below 16 °C or 60 °F), (dip, flow, spray)
- Film too thick (dip, flow, roller, spray)

Beading

- Solvent vaporizes too rapidly (dip, flow)
- Parts drawn from paint too rapidly (dip)

Blistering (pimpling, bubbling, pinholing, pitting)

- Moisture entrapped on surface of substrate (dip, flow, roller, spray)
- Air entrapment
- Improper solvents (dip, flow, roller, spray)
- Wide temperature differential between paint and work (dip, flow, roller, spray)
- Insufficient drying time between coats; solvent trapped in undercoat escapes through partially dried finish coat (dip, flow, roller, spray)
- Excessive air pressure and dry spraying of undercoat, causing porosity and air pockets under finish coat (spray)
- Water in air line (spray)

Brittleness

- Improper composition of paint (dip, flow, roller, spray)

Checking (alligatoring, crazing, crowfooting, hairlining)

- Application of finish coat over incompletely dried undercoat (dip, flow, roller, spray)
- Insufficient mixing of material to blend all pigment (dip, flow, roller, spray)

Color separation

- Incomplete mixing of paint before application (dip, flow, roller, spray)
- Poor agitation during application (dip, roller, spray)

Cracking (shrinking, splitting)

- Paint not thoroughly mixed before application (dip, flow, roller, spray)
- Surface not completely clean (dip, flow, roller, spray)
- Insufficient thinner (dip, flow, roller, spray)
- Undercoat too thick (dip, flow, roller, spray)
- Surface too hot or too cold (dip, flow, spray)
- Oil or water in air line (spray)

Cratering

- Surface air bubbles during withdrawal (dip)
- Silicone contamination (dip, flow, spray)

Slow drying

- Application over grease, oil, or fingerprints (dip, flow, roller, spray)
- Application of too much paint (dip, flow, roller, spray)
- Poor ventilation, air drying only (dip, flow, roller, spray)
- Drying atmosphere too cold or too humid, air drying only (dip, flow, roller, spray)
- Oil in air line (spray)
- Solvent vaporizes too slowly (dip, flow, roller, spray)

Dusty or gritty appearance (dry spray)

- Insufficient solvent (spray)
- Solvent vaporizes too rapidly (spray)
- Excessive air pressure (spray)
- Spray gun too far (over 305 mm or 12 in.) from work (spray)
- Excess air movement in spray booth (spray)
- Spray pattern too wide

Edge pull-away

- Solvent vaporizes too slowly (dip, flow)
- Too much solvent (dip, flow)

Poor flow-out

- Viscosity of paint too high (dip, flow, roller, spray)
- Temperature of paint too low (dip, flow, roller)
- Solvent vaporizes too rapidly (dip, flow, roller, spray)
- Temperature of surface too low (spray)

Poor gloss

- Paint film too thin (dip, flow, roller, spray)
- Application over incompletely dried undercoat (dip, flow, roller, spray)
- Drying atmosphere too cold or too humid, air drying only (dip, flow, roller, spray)
- Application over alkali residues (dip, flow, roller, spray)
- Too much solvent vapor in drying oven (dip, flow, roller, spray)
- Improper solvent

Poor hiding

- Pigment settling because of poor agitation (dip, flow, roller, spray)
- Solvent vaporizes too slowly (dip, flow)
- Too much solvent (dip, flow)
- Withdrawal from paint too slow (dip)
- Vapor concentration too high (dip, flow)

Nonuniform film thickness (or uncoated areas)

- Solvent vaporizes too slowly (dip, flow)
- Nonuniform roll pressure; roll out-of-round (roller)
- Improper hanging of panels (dip, flow)
- Paints not lapping properly when spraying

Orange peel

- Excessive film thickness (dip, flow, roller, spray)
- Surface temperature too hot or too cold (flow, dip, spray)
- Improper adjustment of spray gun pattern (spray)
- Air pressure too high (spray)
- Solvent vaporizes too rapidly (dip, flow, roller, spray)

Undesirable pattern

- Swelling of roll (roller)
- Roll improperly ground (roller)
- Foreign material on roll (roller)
- Poor spray gun technique (spray)

Runs (curtains, sags)

- Too much solvent (dip, flow, spray)
- Paint surface or drying atmosphere too hot or too cold (dip, flow, spray)

Table 11 Theoretical spreading rate for coatings

Volume solids, %	13 μm (0.5 mil) m²/L	13 μm (0.5 mil) ft²/gal	25 μm (1.0 mil) m²/L	25 μm (1.0 mil) ft²/gal	38 μm (1.5 mils) m²/L	38 μm (1.5 mils) ft²/gal	51 μm (2.0 mils) m²/L	51 μm (2.0 mils) ft²/gal	64 μm (2.5 mils) m²/L	64 μm (2.5 mils) ft²/gal	76 μm (3.0 mils) m²/L	76 μm (3.0 mils) ft²/gal
10	6.4	320	3.2	160	2.14	107	1.6	80	1.3	64	1.1	54
20	12.8	640	6.4	320	4.26	213	3.2	160	2.56	128	2.14	107
30	19.2	962	9.62	481	6.4	320	4.8	240	3.94	197	3.2	160
40	25.68	1284	12.8	641	8.54	427	6.42	321	5.14	257	4.28	214
50	32.1	1605	16.0	802	10.7	535	8.02	401	6.42	321	5.34	267
60	38.52	1926	19.2	962	12.8	641	9.62	481	7.7	385	6.42	321
70	44.8	2240	22.4	1120	14.9	747	11.2	560	8.96	448	7.46	373
80	51.2	2560	25.6	1280	17.1	853	12.8	640	10.2	512	8.54	427
90	57.6	2880	28.8	1440	19.2	960	14.4	720	11.5	576	9.6	480
100	64.16	3208	32.08	1604	21.38	1069	16.0	802	12.8	642	10.7	535

Table 12 Average application efficiency

Method	Efficiency, %
Conventional spray	
Air atomized	50
Airless atomized	65
Electrostatic spray	
Air atomized	70
Airless atomized	80
Centrifugally atomized	90
Dip, flow, curtain coat	80
Coil and roller coat	90
Electrodeposition	90
Powder coat	90

- Solvent vaporizes too slowly (dip, flow, spray)
- Excessive paint applied (dip, flow, spray)
- Poor spray gun technique (spray)
- Distorted spray gun pattern (spray)
- Air pressure too low (spray)
- Withdrawal from paint too rapid (dip)
- Automatic equipment is jerky (dip, flow)
- Drafts (dip)
- Improper racking (dip, flow, spray)

Streaking

- Metal too hot or too cold (dip, flow, spray)
- Poor spray technique; insufficient overlap between passes, should be 50% (spray)
- Distorted spray pattern (spray)

Water spotting

- Rain or dew settling on finish (dip, flow, roller, spray)
- Washing before completely dry (dip, flow, roller, spray)

Wrinkling

- Excessive film thickness (dip, flow, roller, spray)
- Abnormally hot or humid drying environment, air drying only (dip, flow, roller, spray)

Calculating Coating Coverage and Costs

The cost of the applied coating is directly related to (*a*) cost per gallon of coating material, (*b*) application efficiency, (*c*) required thickness of film, and (*d*) spreading rate. To calculate the gallons needed and cost for coating an object the following information is needed:

- Volume solids of coating
- Spreading rate of coating at application viscosity
- Dry film thickness required
- Area to be coated
- Application efficiency
- Cost per gallon

Volume solids content of the coating is the volume percentage of vehicle solids and pigments in a gallon of paint. The remaining volume percentage is the solvent, which evaporates in the drying or curing process. The volume solids should be given by the supplier and should not be confused with weight solids.

Spreading Rate. One U.S. gallon of any liquid has a measurement of 3785 cm^3 (231 $in.^3$). Thus, 100% volume solids liquid can cover 149 m^2 (1604 ft^2) at a thickness of 25 μm (1.0 mil). This figure is derived from the following formula:

$$\text{Spreading rate} = \frac{\text{in.}^3/\text{gal} \times \text{vol solids}}{\text{in.}^2/\text{ft}^2 \times \text{in.}}$$

Using this formula, the spreading rate for any liquid coating can be determined. For example, if the volume solids of a product is 40% at application viscosity and a 38-μm (1.5-mil) dry film is required, the area covered by 1 gal of the paint is approximately 39.7 m^2 (427 ft^2) determined as follows:

$$\frac{231 \times 0.40}{144 \times 0.0015} = \begin{array}{c}427\ \text{ft}^2/\text{gal}\\ \text{(spreading rate)}\end{array}$$

Table 11 is an approximation of spreading rates at various volume solids and dry film thicknesses. Because cost per gallon is determined at application viscosity, any solvent, thinner, or reducer added to the paint must be included in the final cost as well as the volume solids determined at application viscosity. The cost per gallon is equal to the cost per gallon of paint plus the cost per gallon of thinner divided by the total number of gallons.

Application Efficiency. In addition to cost per gallon at application viscosity, the percent of the coating deposited on the substrate by the application method must also be considered. Table 12 lists the average efficiency of various types of application methods. The cost of paint for a part can be calculated using the following formula and the data in Tables 11 and 12.

$$\text{Cost per square foot} = \frac{\text{cost per gal at application viscosity}}{\text{spreading rate} \times \text{application efficiency}}$$

Safety and Environmental Precautions

The Clean Air Act administered by the EPA and the Occupational Safety and Health Act administered by OSHA has had a major impact on the finishing industry. These acts together with regulations issued by EPA, OSHA, and local authorities establish limits on (*a*) solvent emissions to the atmosphere, (*b*) levels of certain contaminants in liquid effluent, (*c*) restraints on use of certain solvents and chemicals, (*d*) standards for minimizing hazards that might produce physical injuries to workers, and (*e*) permissible levels of noise. Although all aspects of the OSHA and EPA requirements and the subsequent changes cannot be covered within this article, the following standards published by OSHA in the *Federal Register,* Vol 36, No. 105 dated 29 May 1971 should be studied:

- Subpart H 1910.106, Flammable and Combustible Materials

- Subpart H 1910.107, Spray Finishing Using Flammable and Combustible Materials
- Subpart H 1910.108, Dip Tanks Containing Flammable and Combustible Liquids

From the Guideline Series, Control of Volatile Organic Emissions from Existing Stationary Sources, the following should be obtained from EPA, Office of Air Quality Planning and Standards, Research Triangle Park, NC 27711:

- Volume II, Surface Coating of Cans, Coils, Paper, Fabrics, Automobiles, and Light-Duty Trucks, EPA 450/2-27-008 (OAQPS No. 1.2-073), May 1977
- Volume VI, Surface Coating of Miscellaneous Metal Parts and Products, EPA, 450/2-78-015 (OAQPS No. 1.2-101), June 1978

Maintain contact with OSHA, EPA, and local authorities for current information on requirements.

Definitions of Terms Relating to Painting

acid number. The number of milligrams of potassium hydroxide required to neutralize the free fatty acid in 1 g of oil, resin, varnish, or other substance

acrylic resins. Synthetic resins polymerized from acrylic or methacrylic acid. The more important acrylic resins are polymers of the methyl and ethyl esters of these acids or copolymers of mixtures of these monomers. Acrylic resins are thermoplastic and are characterized by their water-white color, resistance to discoloration, and transparency. Most of them are resistant to water, alcohol, acids, alkalis, and mineral oils. Acrylic resins are soluble in aromatic hydrocarbons, chlorinated hydrocarbons, esters, and ketones. They are compatible with nitrocellulose and with plasticizers such as dibutyl phthalate and tricresyl phosphate, but are incompatible with drying oils, oleoresinous varnishes, many types of alkyd resins, and most hard resins

aldehyde resins. Resins produced by interaction of an aldehyde with another substance. The most important of this group are phenolic resins. The extreme reactivity of aldehydes gives rise to the amine-aldehyde resins, also of great commercial value, as well as to a nearly unlimited number of condensation products that can result from the reaction of aldehydes with a wide variety of organic substances

alkyd. A synthetic resin resulting from condensation involving (*a*) a polybasic acid, such as phthalic, maleic, or succinic, and (*b*) a polyhydric alcohol, such as glycerin and the glycols—usually with the addition of a modifying agent. Alkyds are used in paints, varnishes, and lacquers

antioxidant. A material that, when added to a varnish or an oil, prevents or retards oxidation and drying

blistering. A paint-film failure usually caused by the application of paint on a surface containing an excessive amount of water or other volatile material

bloom. A bluish fluorescent cast that forms on the surface of some paint films, caused by deposition of a thin film of materials such as smoke, dust, or oil

blushing. Partial opacity, cloudiness or translucency upon application or drying; usually associated with lacquers. Blushing has two main causes: (*a*) solid ingredients become partially or wholly precipitated as a result of condensed moisture, caused by the lowering of the temperature by evaporation (blushing caused in this way can usually be eliminated by the addition of solvents of higher boiling range and slower evaporation rate); (*b*) improper solvent balance, with solvents evaporating faster than diluents, leaving solid ingredients (nonvolatile content) in a precipitated state, which imparts a cloudy appearance to the finish

chlorinated rubber. A synthetic resin, made by chlorinating rubber under specified conditions, which is characterized by (*a*) a neutral acidity, (*b*) resistance to acids and alkalis, (*c*) solubility in aromatic and aliphatic hydrocarbons and turpentine, and (*d*) insolubility in lacquer solvents and alcohol

coal tar resins. Synthetic resins usually produced from crude coal tar naphthas in the approximate boiling range of 150 to 205 °C (300 to 400 °F). Coumarone-indene coal tar resins are thermoplastic, range from a very faint yellow to darker shades, and are sold in flake, chip, or liquid form. Coal tar resins possess good resistance to alkalis and acids, and are used in varnish formulation

crackle finish. A novelty textured finish, usually produced by a lacquer resulting from applying a topcoat designed to shrink, crack, and expose a more flexible undercoat, usually of a different color

crazing. Fine lines of minute surface cracks occurring on painted or enameled surfaces as the result of unequal contraction in drying or cooling

drier. Any catalytic material that when added to a vegetable or marine animal drying or semidrying oil accelerates the drying or hardening of the film. Driers are usually in the form of organic salts of lead, cobalt, manganese, and zinc, including naphthenates, resinates, and linoleates

drying oils. Oils, usually obtained from plant sources, that harden when exposed to air

earth pigments. Pigments mined directly from the earth. Frequently called natural or mineral pigments or colors. Examples include red and yellow iron oxides, yellow ochre, raw and burnt siennas and umbers. These pigments are quite stable, unaffected by alkalis, heat, light, moisture, and most vehicles

enamel. A broad classification of free-flowing clear or pigmented varnishes, treated oils, or other forms of organic coatings, which usually dry to a hard glossy or semiglossy finish

epoxy. Synthetic resins formed by the condensation of epichlorohydrin and bisphenol-A

extender. A low-cost pigment, usually inert, used to extend or increase the bulk of a paint, reducing its unit cost. Extenders are also used to adjust the consistency of a paint and reduce the intensity of colored pigments of great tinting strength. The use of an extender in a paint does not necessarily mean that the resulting product is inferior; extenders sometimes improve certain characteristics of a film

flash point. The temperature at which the vapor of a thinner or solvent ignites when exposed to a small flame under controlled test conditions

fluorocarbon resins. This family of resins includes polytetrafluoroethylene, vinylidene fluoride, hexafluoropropylene, and similar compounds

Ford cup. An efflux viscometer, used extensively in paint laboratories, to measure viscosity by the time required for the sample being tested to flow through an orifice of a specific diameter

glossmeter (glossimeter). Any instrument that measures the regular reflecting power (objective gloss) of a surface under test

hi-flash solvent naphtha. An important solvent in industrial enamels that consists of aromatic hydrocarbons related to xylene. This substance is slower in evaporation rate and therefore of higher flash point and safer

homogeneous. The same in structure or quality; identical

kauri-butanol value. The solvency power of hydrocarbon solvents used in paint, defined as the number of millilitres of the solvent that can just cause turbidity in a standard solution of hard kauri in normal butyl alcohol

lacquer. Coating composition which is based on synthetic film forming material dissolved in organic solvents and which dries primarily by solvent evaporation. Lacquers include those based on nitrocellulose, other cellulose derivatives, vinyl resins, and acrylic resins

lacquer diluent. A volatile liquid, which is not a solvent for cellulose derivatives, used primarily to reduce the cost of lacquers and sometimes to minimize the lifting tendency of undercoats. Aliphatic and aromatic hydrocarbons, and their derivatives, are used as diluents and are used in conjunction with active solvents

lacquer thinner. A combination of organic solvents and diluents used to reduce the viscosity of cellulose lacquers

lifting. The softening and penetration of a film by the solvents of a subsequent film, which results in raising and wrinkling

livering. An increase in the consistency of a paint or enamel, characterized by rubberiness or coagulation. Is caused as the result of a chemical reaction between the vehicle and a solid dispersed material. May also result from the polymerization of the vehicle

melamine resins. Synthetic resins that are condensation products of formaldehyde and melamine. Their preparation is similar to that of urea-formaldehyde resins. They are characterized by high heat resistance, high speed of conversion, and stability of color. Because they cure quickly at low temperatures, their greatest use is in baking enamels

mineral spirits. See preferred term, petroleum spirits

oil length. Ratio of oil to resin in a varnish or other coating, expressed as number of gallons of oil cooked with 100 lb of resin. Varnishes are usually classified with respect to oil length as short oil, medium oil, or long oil varnishes

oleoresinous. Made by the combination of an oil and a resin, such as a varnish

orange peel. A pebbled film surface caused by too rapid drying after spraying, or by failure of the coating material to exhibit the desired leveling effects

petroleum spirits. Straight-run or blended petroleum naphthas with a boiling range usually between 150 and 205 °C (300 and 400 °F), used as thinners for paints, enamels, and varnishes. As originally manufactured, petroleum spirits were fractionated to match the evaporation rate of gum turpentine. Most petroleum spirits have a Tag closed-cup flash point of slightly over 38 °C (100 °F)

phenolic resins. A class of synthetic oil-soluble resins, either fusible or heat hardening, produced by condensation of phenol or substituted phenols with formaldehyde or some similar aldehyde

pine oil. A naval store's product consisting of terpene alcohols, ketones, and ethers, with very small quantities of high-boiling terpene hydrocarbons, all oil distilling between 195 and 200 °C (380 and 390 °F)

pinholing. Film defect characterized by small porelike flaws in a coating, which extend entirely through the applied film and have a general appearance of pinpricks when viewed by reflected light. Applied to holes caused by solvent bubbling, moisture, other volatile products, or the presence of extraneous particles in the applied film

plasticizers. Materials that are miscible with resins and that when compounded with resins produce films that are less brittle than the resin alone

polyamide resins. Polymers made by condensation of diamines with dibasic acids

polyester resins. A group of synthetic resins produced by the reaction of dibasic acids with dihydric alcohols

polystyrene resins. Synthetic resins formed by the polymerization of styrene, which occurs when heated with or without catalysts. Polystyrene resins, which vary in color from water-white to light yellow, are soluble in aromatic hydrocarbons, chlorinated hydrocarbons, and esters. Their physical properties can be made to vary widely by the conditions under which polymerization takes place. These resins are thermoplastic and have a high dielectric value. They are resistant to water, alcohol, and most acids and bases

polyurethane resins. Synthetic resins that may be either thermoplastic or thermosetting, usually made by action of toluene diisocyanate or another diamine with polyols, polyethers, polyesters, or other materials containing hydroxyl groups

polyvinyl acetate resins. Polymers of vinyl acetate characterized by colorlessness, tastelessness, odorlessness, permanent thermoplasticity, light color, flexibility, stability toward light, transparency to ultraviolet rays, high dielectric strength, toughness, and hardness. The higher the degree of polymerization, the higher the softening temperature of these resins

reducer. See preferred term, thinner

resin. A solid or semisolid organic or amorphous material, either of vegetable origin or synthesized. Range in color from yellow or amber to dark brown, and are usually transparent or translucent. Resins are nonconductors of electricity and are soluble in many organic solvents but insoluble in water. See also synthetic resins

rosin. A resinous material obtained from various pine trees and containing principally abietic acid. Wood rosin is obtained from stumps or other dead wood by steam distillation. Gum rosin is obtained from the sap that exudes from the living tree. When used as a raw material for varnish, rosin is generally, but not always, hardened by the addition of lime or resinates of zinc, zinc oxide, calcium oxide, and similar substances, or by prolonged heating

synthetic resins. Complex, substantially amorphous organic semisolid or solid materials (usually a mixture of substances) built up by chemical reaction of comparatively simple compounds. Synthetic resins resemble natural resins in luster, fracture, comparative brittleness, insolubility in water, fusibility or plasticity when heated or exposed to heat and pressure, and at a certain more or less narrow temperature range before fusion, a degree of rubberlike extensibility. They deviate widely from natural resins in chemical construction and behavior with reagents

tack. Slight stickiness of the surface of a paint, varnish, or lacquer, apparent when the film is pressed with the finger

tall oil. A dark brown, viscous oily liquid obtained as an alkaline waste-liquor in the sulfate process of making wood pulp; often called liquid rosin

thermoplastic. Synthetic resins that may be softened by heat, but regain original properties when cooled

thermosetting. Synthetic resins that solidify or set on heating and cannot be remelted. This property is usually associated with a cross-linking reaction of constituents to form a three-dimensional network of polymer molecules

thinner. A somewhat volatile compound that dries by evaporation and penetration and is used to bring coatings to the proper tack and consistency

thixotropy. The property possessed by certain gels or dispersions of becoming liquid on being shaken and coagulating again when left in an undisturbed condition

toner. An organic pigment, usually a heavy metal salt of a water-soluble dye, that does not contain inorganic pigment or inorganic carrying base

urea-formaldehyde resins. Synthetic resins obtained by the chemical reaction between urea and formaldehyde in the presence of acid or alkaline catalysts. Characterized by light color and rapid development of hardness by baking

varnish. Any homogeneous transparent or translucent liquid that when applied as a thin film dries on exposure to air to a continuous film, giving a decorative and protective coating to the surface to which it is applied. A spirit varnish is usually one that dries by evaporation alone and consists of a solution of a resin in a solvent. An oil varnish is one that dries by evaporation in combination with oxidation or polymerization, or with both, and that usually consists of a combination of resins, vegetable oils, driers, and solvent or thinner

vinyl resin. Any of a class of synthetic resins resulting from he polymerization of vinyl compounds, such as polyvinyl acetate, polyvinyl chloride, and polyvinyl acetals. Vinyl resins are thermoplastic, odorless, tasteless, colorless, and either nonflammable or slow-burning

viscometer (viscosimeter). Any instrument that measures viscosity, or the internal friction of fluidity of a liquid

VM&P naphtha (varnish maker's and painter's naphtha). A straight-run petroleum spirit with a boiling point usually between 93 and 160 °C (200 and 320 °F) and with an evaporation rate similar to, or somewhat faster than, 10 °C xylene

weatherometer. A device for evaluating weather resistance of decorative and protective coatings and finishes, in which a generator of synthetic sunlight correlates the thermal shock of a cooling spray with the actinic action of the June sun at noon. In its combined effects, accelerates weathering of films on rotating test panels

whiting. An inert white crystalline pigment or extender composed principally of calcium carbonate

wrinkle finish. A novelty film of enamel or varnish characterized by fine wrinkles or irregular ridges

SELECTED REFERENCES

- George E.F. Brewer, Electrocoating—Theory and Reduction to Practice in *Pigment Handbook,* Vol 3, Edited by Temple C. Pattan, New York: John Wiley & Sons, 1973
- Gordon E. Cole, Jr., Powder Coatings, *Modern Paint and Coatings,* March 1982
- Henry R. Friedberg, Regulating VOC's, *Products Finishing,* Feb 1982
- C.O. Hutchinson, How Can I Paint My Product, *Products Finishing,* 1981 Directory
- Timothy Keating, Automation in Finishing, *Industrial Finishing,* March 1981
- Sidney B. Levinson, Electrocoat, Powder Coat, Radiate, *Journal of Paint Technology,* Part I, II, III, June, July, Aug 1972
- *Metal Finishing Guidebook and Directory,* Hackensack, NJ: Metals and Plastics Publications, Inc., published annually
- Raymond R. Meyers and J.R. Long, *Treatise on Coatings,* NY: Marcel Dekker
  Vol 1, Part I: Film Forming Compositions, 1967
  Vol 1, Part II: Film Forming Compositions, 1968
  Vol 1, Part III: Film Forming Compositions, 1972
  Vol 2, Part I: Characterization of Coatings, Physical Techniques, 1969
  Vol 2, Part II: Characterization of Coatings, Physical Techniques, 1976
  Vol 3, Part I: Pigments, 1975
  Vol 5, Part I: Formulations, 1975
- Modern Finishing Methods, M5 W1A7, Canadian Paint and Finishing, June 1972
- *Paint and Coatings Dictionary,* Philadelphia: Federation of Societies for Coating Technology, 1978
- Steve Suslik, Coating Equipment for the '80's, *Industrial Finishing,* Jan 1981
- Steve Suslik, Coatings of the Future Which Will They Be, *Industrial Finishing,* July 1981
- E.P. Tripp III and J. Weisman, EB and UV Curable Coatings, *Modern Paint and Coatings,* March 1982

Painting of Structural Steel

By Howard G. Lasser
Materials Engineer
Materials Research Consultants

Painting of structural steel is done to protect the environmental area affected by the structure being coated, to preserve manufacturing or plant functionality, and to provide aesthetic appeal. Painted structures may have to withstand high-operating temperatures, adverse weather conditions, and marine, industrial (urban) and rural exposure as well as chemical exposure. All these conditions dictate the frequency of coating application, selection of the method of surface preparation, and coating system selected.

The introduction of restrictive regulations has had an impact on the use of painting coating systems. For example, the federal Environmental Protection Agency (EPA) has limited the amount of volatile organic compounds that can be emitted from painting facilities. For the protection of the applicator, the use of active pigments in primers, such as lead compounds and chromates, has been limited by the Occupational Safety and Health Administration (OSHA). Local jurisdictions of the (EPA) should be consulted for regulations concerning the environment. Guidance from these authorities changes rapidly and should be updated often.

Table 13 Characteristics of the resins for coating structural steel

Resin	Curing method	Solvents	Chemical and weather resistance: Acid	Alkali	Solvent	Water	Weather	Remarks
Raw and boiled linseed oil	Air drying Oxidative polymerization	Aliphatic hydrocarbons	Fair	Bad	Poor	Fair	Fair	Vehicle for corrosion inhibitive primers for wire-brushed steel, slow drying
Oleoresinous varnishes	Air drying Condensation and/or oxidative polymerization	Aliphatic hydrocarbons and/or aromatic hydrocarbons	Fair	Bad	Poor	Good	Good	Pale-colored finishes that yellow on exposure
Alkyds	Air drying Oxidation Polymerization	Aliphatic hydrocarbons	Fair	Bad	Poor	Fair	Very good	Long oil alkyds are generally used, although these alkyds may be blended with with medium oil alkyds
Modified alkyds	Air drying Oxidative polymerization	Dependent on modification A wide variety of solvents	Fair	Fair	Fair	Good	Very good	...
Epoxy, aliphatic amine or polyamide blends	Air drying Addition polymerization	Blends rich in high ketones	Fair	Very good	Very good	Poor	Good	Two-component compositions
Epoxy, fatty acid esters	Air drying Oxidative polymerization	Aliphatic and/or aromatic hydrocarbons	Fair	Fair	Poor	Fair to good	Poor to fair	...
Polyester urethane	Addition polymerization	Blend rich in ketones and esters	Fair to good	Good	Very good	Fair to good	Very good	Two-component compositions
Vinyl resins	Air drying solvent evaporation	Blends usually rich in ketones	Very good	Very good	Poor	Very good	Good	Fire hazard, unless high solids compositions are used
Chlorinated rubber	Air drying solvent evaporation	Aromatic hydrocarbons	Good	Good	Poor	Good	Very good	Very poor heat resistance
Acrylic resins (water emulsion)	Water evaporation and coalescing	Water dispersant	Fair	Good	Very good	Very good	Good	Used as a maintenance coating system. Porosity of film results in poor chemical and weather resistance

Composition and Characteristics of Organic Coatings

Coatings have three components: the volatile vehicle (solvent or dispersant), the nonvolatile vehicle (resin), and the pigment. The volatile vehicle is the portion of the coating that allows the coating to be spread or applied. This component can be ketone, ester alcohol, petroleum solvent, water, or a combination of these materials. Water, for example, acts as the dispersant in water reducible coating systems.

The pigment portion provides opacity and color, as well as viscosity control and reduced water permeability. Other pigments also provide corrosion resistance. Materials that are considered pigments include: (*a*) driers, (*b*) plasticizers, (*c*) ultraviolet light absorbers, (*d*) emulsifiers, and (*e*) dispersing agents. These materials are added to modify coating properties as required.

A number of resins are used in the preparation of coatings. These include plant derived oleoresins (linseed oil, safflower oil, and tung oil); fish oils (menhaden, sardine, and pilchard); lacquers (nitrocellulose, cellulose acetate, ethyl cellulose, and acrylic polymers); and synthetic resins (alkyds, aminoplast-alkyd blends, phenolic, epoxy, acrylic, vinyl, urethane, silicone, and chlorinated rubber). Table 13 provides information on some frequently used resins.

Linseed oil is the most important drying oil used in oil varnishes and in the preparation of alkyd resins because of its ready availability and relatively low cost. The main advantages of linseed oil are good surface wetting properties, easy preparation of surface, good application properties, and ease in coating preparation. Raw linseed has disadvantages as well, including (*a*) very slow drying, (*b*) lack of gloss when formulated into coatings, and (*c*) poor leveling qualities.

Linseed oil is composed of a mixture of linoleic and linolenic triglycerides. Both of these acids contain a chain of 17 hydrocarbon members with an acid radical and 2 or 3 double-bonded positions. On drying (oxidizing) the double bonds cross link with other molecules to form a high molecular weight film.

Heat treated oils fall into three categories: boiled oils, stand oils, and blown oils. Boiled oils are prepared by heating linseed oils in the presence of metal driers, such as lead naphthenate and cobalt naphthenate. Boiled linseed oils have a higher viscosity and better drying properties because of higher molecular weight and more complex molecular structure. Boiled linseed oil is used in oil-based primers and in conjunction with oil varnishes and undercoats.

Stand oils have an even higher viscosity than boiled linseed oil. These oils are prepared by heat-polymerizing linseed oil alone or in a mixture with tung oil. Stand oils are mostly used in combination with oil varnishes and alkyd resins to improve application properties and increase total oil.

Blown oils are partially oxidized in addition to being polymerized. As the name implies, air is blown through the heated oil, which results in a poorly drying product. Generally, the blown oil is used as a plasticizer.

Oleoresinous varnishes are varnishes and paint media prepared from drying oils and natural or synthetic resins, such as rosins, gum congo, rosin-modified phenolics, and 100% oil-soluble phenolics. Oleoresinous varnishes are used to improve drying and film-forming properties.

The phenolic varnishes require surface preparations to provide at least a commercial blast-cleaned surface (SSPC-SP-6). (For more information on Steel Structures Painting Council specifications, see Table 14 in the section on surface preparation in this article.) Tung-oil phenolic resin has very good water resistance and is mildly acid- and alkali-resistant.

Alkyd resins are produced by reacting a polyhydric alcohol with a monobasic and polybasic fatty acid to yield an ester. The more frequently used alcohols include ethylene glycol, glycerol, and pentaerythritol. Phthalic anhydride is the acid used most often because of the plentiful supply of the petrochemical orthoxylene from which it is made. Isophthalic acid is used in primers or where a harder, more weather-resistant coating is desired. Unsaturated dibasic acids, such as maleic anhydride, are used to provide higher molecular weight polymers.

For coating structural steel, air-drying alkyd resins are most frequently used. The films are formed by the oxidation of the drying oils these alkyds contain. The fatty acids are obtained from linseed, soya, sunflower, cottonseed, safflower, tung oil, and fish oils, in the form of a glycerol triester. These triesters, or the oils chemically separated, are used to prepare primers, intermediate coatings, and finishing coatings. The shorter (more reactive) the acid oil, the more brittle and less forgiving the primer. Short oil resins require more meticulous surface preparation, and they dry more quickly. Soya oil is used with other faster drying oils to produce non-yellowing white and pastel color compositions.

When additional hardness, chemical resistance, and durability may be required, alkyd resins are modified with vinyl, acrylic, silicone, and urethane and other adducts to the acid oil constituent. The resins resulting from these modifications are higher in molecu-

Table 14 SSPC-Vis 2 for pictorial representation of rust classification

Paint system condition	Cleaning and painting recommended	Rust grades	Area of example, %
Nondeteriorated, 0 to 0.1% rust			
Paint almost intact; some primer may show; rust covers less than 0.1% of the surface	Solvent clean (SSPC-SP1) entire repaint area, and spot prime, if necessary. If required to maintain film thickness or continuity, spot apply finish coat, then apply 38-51 μm (1.5-2.0 mils) of finish coat over entire repaint area	10-8	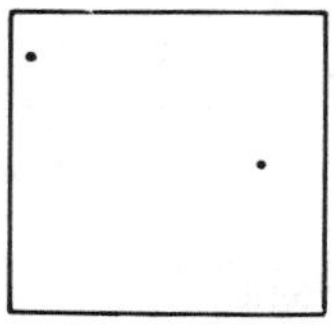
Slightly to moderately deteriorated, 0.1 to 1% rust			
Finish coat somewhat weathered; primer may show slight staining or blistering; after stains are wiped off, less than 1% of area shows rust, blistering, loose mill scale, or loose paint film	Spot clean (minimum SSPC-SP2) entire repaint area, and spot prime. If required to maintain minimum film thickness or continuity, spot apply finish coat, then apply 38-51 μm (1.5-2.0 mils) of finish coat over entire repaint area	8-6	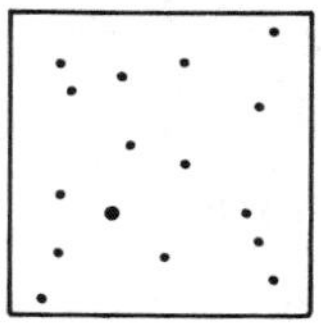
Deteriorated, 1 to 10% rust			
Paint thoroughly weathered, blistered, or stained; up to 10% of surface is covered with rust, rust blisters, hard scale or loose paint film, very little pitting visible	Spot clean (minimum SSPC-SP2) entire repaint area, feather edges, and spot prime. If required to maintain film thickness or continuity, spot apply finish coat, then apply 38-51 μm (1.5-2.0 mils) of finish coat over entire repaint area	6-4	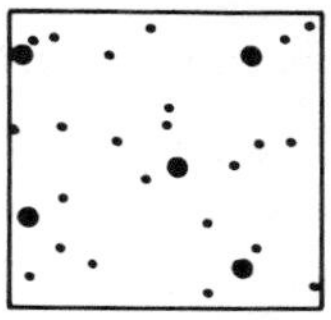
Severely deteriorated, 10 to 50% rust			
Large portion of surface is covered with rust, pits, rust nodules, and non-adherent paint. Pitting is visible	Spot clean (minimum SSPC-SP6) entire repaint area, feather edges, and spot prime. If required to maintain film thickness or continuity, spot apply finish coat, then apply 38-51 μm (1.5-2.0 mils) of finish coat over entire repaint area	4-1	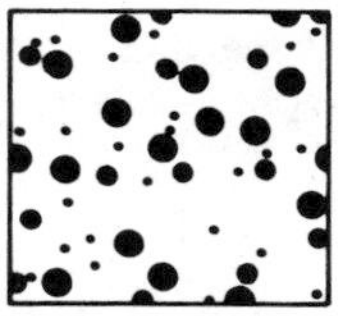
Totally deteriorated, 50 to 100% rust			
	Clean (minimum SSPC-SP6) entire repaint area and apply primer, intermediate, and finish coats over entire repaint area	1-0	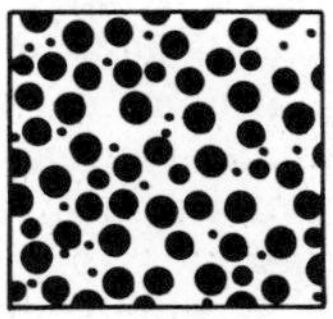

Table 15 Paint compatibility

	Topcoat												
	Solvent-thinned				Lacquer			Water thinned (latex)		Chemically reactive			
Primer or weathered paint	Oleo-resinous	Alkyd	Silicone alkyd	Phenolic oleo-resinous	Vinyl	Chlori-nated rubber	Styrene-butadiene Styrene-acrylate	Acrylic	Poly-vinyl acetate	Epoxy	Coal tar epoxy	Poly-ester	Urethane
Solvent thinned													
Oleoresinous	C	C	C	C	NR	NR	NR	CT	CT	NR	NR	NR	NR
Alkyd	C	C	C	C	NR	NR	NR	CT	CT	NR	NR	NR	NR
Silicone alkyd	C	C	C	C	NR	NR	NR	CT	CT	NR	NR	NR	NR
Phenolic oleoresinous	C	C	C	C	NR	NR	NR	CT	CT	NR	NR	NR	NR
Lacquer													
Vinyl	C	C	NR	NR	CT	CT	CT	CT	CT	CT	CT	NR	NR
Chlorinated rubber	C	C	C	C	CT	CT	NR	CT	CT	NR	NR	NR	NR
Styrene-butadiene Styrene-acrylate	C	C	C	C	CT	NR	CT	CT	CT	NR	NR	NR	NR
Bituminous	NR	NR	NR	NR	CT	CT	CT	CT	NR	NR	NR	NR	NR
Water thinned (latex)													
Acrylic	C	C	C	NR	CT	CT	CT	CT	CT	NR	NR	NR	NR
Polyvinyl acetate	C	C	C	NR	CT	CT	CT	CT	CT	NR	NR	NR	NR
Chemically reactive													
Catalyzed epoxy	NR	NR	NR	NR	CT	NR	NR	NR	NR	CT	CT	CT	CT
Coal tar epoxy	NR	NR	NR	NR	NR	NR	NR	NR	NR	CT	CT	NR	NR
Zinc-rich epoxy	NR	NR	NR	NR	NR	CT	CT	NR	NR	CT	CT	NR	NR
Polyester	NR	NR	NR	NR	NR	NR	NR	NR	NR	CT	NR	CT	CT
Inorganic zinc	NR	NR	NR	NR	CT	CT	CT	CT	NR	CT	NR	NR	NR
Cementitious	NR	NR	NR	NR	CT	CT	CT	CT	CT	C	C	C	C
Urethane	NR	NR	NR	NR	NR	NR	NR	CT	NR	NR	NR	C	CT

Note: C: normally compatible. CT: compatible with special surface preparation and/or application. NR: not recommended because of known or suspected problems. Certain combinations marked NR may be used provided a suitable tie coat is applied between the two coatings. Specifications and/or manufacturer's literature should be consulted for guidance

lar weight than the original alkyd and may require oxygenated solvents such as ketones.

Epoxy resins are long-chain polyhydric alcohols with epoxy end groups. The alcohol and epoxy groupings are available for reaction with the aliphatic polyamines, amine adducts, and polyamides to provide film formers with excellent chemical resistance at room temperature. The epoxy resins may be separated from their curing adducts such as polyamines and polyamides, as two component coating systems. These components are mixed in the required proportions immediately before use. Epoxy coatings are used where chemical resistance is required. A surface preparation of SSPC-SP-10 or better is required.

Polyurethane resins are characterized by having isocyanate groupings at the end of the molecular chain, acting as the reactive groups that combine with moisture or a reactive polyol such as glycerol, glycol, phenols, alkyds, and many others. The latter resins are called two-component polyurethane coatings, and the former are classified as single-package or moisture-curing coatings. Two-pack urethane resins provide a chemical-resistant film of excellent quality and high gloss and are used to coat structural steel as a lining coating for petroleum storage tanks. Moisture-cured polyurethane coatings are aliphatic urethanes that are less chemically resistant than the two-component systems.

Vinyl coatings, for most purposes, are prepared from polyvinyl formal, polyvinyl acetal, and polyvinyl butyral.

Vinyl resin paints are inert and are used for coatings applied to tanks, pipelines, petroleum equipment, offshore drilling rigs, railroad hopper cars, dairy and brewery equipment, and tanks that require acid and alkali resistance. Vinyl films are abrasion resistant, with low water permeability, high dielectric resistance, and high speed drying capabilities. Vinyl paints require steel to be cleaned to an SSPC-SP10 or better, a near-white abrasive blast cleaning. Vinyl paints are ideal for immersion in fresh water and brackish water, but do not impart the same degree of protection in salt water, where coal tar or epoxy coatings would be preferred.

Chlorinated rubber resins are prepared by the chlorination of isoprene in solution to approximately 65 wt % chlorine. Natural rubber is used as the raw material, and when combined with chlorine, it results in a trichloro- and a tetrachloro-polymeric mixture. The resulting films provide excellent corrosion resistance for ferrous metals in marine environments, strong acids, and weak alkalis, but films chalk when exposed to sunlight. For a paint to be designated as a chlorinated rubber paint, it should contain not less than 60 wt % chlorinated rubber in the vehicle solids.

Table 16 Classification of coatings according to methods of cure

Method of curing	Generic type	Comments
Air oxidation of drying oils (solvent thinned)	Oleoresinous	Good wetting, slow curing, soft film recommended in normal environments only
	Alkyd	Good wetting and appearance, poor in alkaline or solvent environments
	Silicone alkyd	Improved durability, gloss, and chemical resistance compared to alkyds, but still poor in alkaline or solvent environments
	Phenolic oleoresinous	Good resistance to abrasion and mild chemical environments; however, dark color of binder precludes use in white or light tints
Solvent evaporation (lacquers)	Vinyl (polyvinyl chloride-acetate)	Good water resistance, limited solvent resistance, poor adhesion unless surface has been properly prepared with abrasive blast cleaning
	Chlorinated rubber	Good water resistance, limited solvent resistance
	Styrene-butadiene, styrene-acrylate	Good water resistance, limited solvent resistance
	Coal tar	Soft, black only; of limited use, mostly on mechanically cleaned surfaces
	Polyvinyl-butyral	Exclusively used in pretreatment (wash) primers
Evaporation of water (latex, emulsion, water-thinned)	Acrylic	Recommended in normal environments only
Chemical reaction	Epoxy	Good water, chemical, abrasion and solvent resistance, chalks freely on exterior exposure, difficult to topcoat
	Coal tar epoxy	Improved water resistance and lower raw material costs compared to epoxies, black only. Difficult to topcoat
	Polyester	Frequently used with glass fibers to give abrasion and water-resistant coating. Only fair alkali resistance
	Zinc inorganic	Requires adequate surface preparation (SSPC No. 10, Near White Blast Cleaning), adequate curing time required, excellent corrosion protection, good abrasion, solvent, and high temperature resistance, must be topcoated in aggressive environments, reacts with alkali-sensitive topcoats
	Cementitious	Inexpensive, requires adequate curing for best performance, and tends to chalk with aging, poor corrosion resistance
	Urethane	Good water, chemical, abrasion, and solvent resistance. Difficult to topcoat

Acrylic resins are formed by the esterification of acrylic or methacrylic acids with alcohols. The major homopolymers are methyl, ethyl, butyl and isobutyl methacrylate, and methyl, ethyl and butyl acrylate. Lacquers and enamels, based on acrylic resins, have been developed for both ferrous and nonferrous metals. Latex emulsions are used in home painting.

Types of Paint

Paints are generally described by their resin (binder) designation and sometimes their pigment composition, such as red lead alkyd or epoxy polyamide. Paints may be applied in single or multiple layers. When multiple layers are used, each layer has a special purpose.

The primer is the first layer to be applied. The main purpose of the primer is to wet the surface of the substrate, and provide adhesion and corrosion protection. Pigments such as red lead, zinc chromate, or zinc molybdate are active corrosion inhibitors. Titanium dioxide, chrome oxide, and ferric oxide pigments are less active.

Intermediate coatings are sometimes applied to improve adhesion and impact strength between primer and topcoat and to provide a barrier layer between coats, as well as surfacers and sealers. Intermediate film also inhibits light penetration, reducing actinic degradation.

The final coating is described as the top or finish coat enamel, which provides environmental and chemical resistance. Also, the finish coat provides aesthetics of color and gloss, as well as film characteristics such as hardness and abrasion resistance. Table 15 shows the compatibility of paints applied as multiple layers.

Enamels may be cured by air drying or oven baking. Air dry enamels are cured essentially by a combination of solvent evaporation and oxidation. Baking enamels incorporate catalysts and cross-linking agents that require heat for polymerization. Coatings may be classified according to curing method, as is shown in Table 16. Each generic type is discussed separately in the following sections.

Air-Oxidizing Coatings. Oleoresinous (oil-drying) coatings were among the earliest to be used to provide protection from environmental deterioration. These coatings contained natural vegetable or fish oils which cure to a solid by reacting with oxygen from the air. This usually slow reaction is accelerated by using driers in the coating formulation. Paints using these oils have excellent wetting properties and are used on poorly prepared surfaces such as SSPC-SP-1 or 3.

Alkyd coatings are developed by reacting drying oils with phthalic anhydride or other polybasic acids that increase durability and hardness. Alkyd coatings are currently the most widely used coatings, because of their good wetting, flexibility, curing, and application properties. Alkyd coatings are among the most suitable for general atmospheric exposure.

Silicone alkyd coatings have a silicone resin reacted with the alkyd resin to form the resin binder. When compared to alkyd coatings, silicone alkyd coatings have superior chemical resistance, color retention, and gloss.

Phenolic coatings are prepared using resin binders that incorporate phenol-formaldehyde in dry oils. The use of phenol-formaldehyde improves water resistance, but lowers exterior durability.

Epoxy resins combined with drying oils improves the corrosion resistance of epoxy ester coatings. Epoxy ester coatings are single-component coatings and should not be confused with two-component chemically reacting epoxies.

Lacquers are coatings that dry by solvent evaporation. The major resins used are acrylics, cellulosics, and vinyl. Lacquer coatings contain dissolved solid resins that form a continuous hard film after the solvent evaporates from the coating. Because coatings are not chemically active during curing or weathering, they can be redissolved in the same solvent. This allows lacquers to be readily overcoated, resulting in excellent intercoat adhesion. Lacquer coatings have poor solvent resistance, but have excellent chemical and water resistance.

Acrylic resins are frequently used in lacquers, because of their fast drying properties and chemical and water resistance. Chlorinated rubber coatings are fast drying and are being used increasingly in this country. Chlorinated rubber coatings have excellent moisture resistance. Coal tar and asphaltic coatings are used on below-grade structural steel because of their moisture resistance and abrasion resistance. Polyvinyl-butyral coatings, such as those conforming to military specification DOD-P-15328D, are used as pretreatments for alkyd and vinyl coatings applied to steel and galvanized structures.

Two-Component Coatings. In general, coatings that cure by chemical reaction have the best combination of durability and water, solvent, and chemical resistance. Chemically cured coatings are packaged in two separate containers, and the chemical reaction is initiated after the two components are combined. Epoxy coatings have either an amine or polyamide curing agent although polyester and polyurethane constituents have been used. Amino-cured epoxies tend to have better chemical and solvent resistance. Polyamide-cured epoxies have better flexibility and water resistance. Because epoxy coatings cure to a hard, smooth, solvent-resistant finish, they are difficult to overcoat. To ensure good bonding, topcoats are applied to incompletely cured undercoats, allowing the topcoats to chemically react with the undercoats. If the solvent has not evaporated by the time the topcoat is applied the solvent can be entrapped, resulting in the blistering of the topcoat. If the undercoat is completely cured before the topcoat is applied, a fog or mist coat (thinned topcoat) is first applied to improve intercoat adhesion. Epoxy coatings chalk during weathering, although chalking occurs to a smaller extent with newer materials. When chalking occurs, it can be removed by sanding or brush-off abrasive blasting. Epoxies are exothermic when they cure; they cure slowly when applied below 8 °C (50 °F) and extremely rapidly above 32 °C (90 °F). Coal tar epoxies have coal tar pitch added to the epoxy resin. This combination increases the water resistance and makes the coating more tolerant to poor surface preparation. A coal tar epoxy cured with a low molecular weight amine is especially resistant to an alkaline environment, such as occurs on a cathodically protected structure. Some coal tar epoxy systems become brittle when exposed to the sun and must be protected. Coal tar epoxies are more difficult to overcoat than epoxies and come in colors ranging from tan to black.

Urethane coatings provide the tough, durable, smooth finish that is typical of chemically cured coatings. Aliphatic urethanes provide bright, chalk-resistant finishes with exceptional physical properties. Urethane coatings may be air dried or heat cured.

Polyester coatings are used most frequently with glass fibers or flakes for reinforcement. Polyester provides a thick coating that is tough and durable, with good resistance to abrasion.

Inorganic zinc coatings are available as primers or complete coating systems. Solvent-borne inorganic zinc coatings are based upon an ethyl silicate vehicle supplied as a two-component coating with finely divided zinc dust added to the binder just before application. An inorganic zinc coating applied to a well prepared ferrous substrate can provide good protection without being topcoated at film thicknesses from 0.75 to 1.25 mm (0.03 to 0.05 in.). All topcoats applied to inorganic zinc must be alkaline resistant, or the topcoat can saponify and become water soluble. Zinc coatings should not be allowed to come in contact with gasoline, because the zinc is slightly soluble and will be leached from the film.

Organic zinc-rich coatings may be formulated with a number of resins but the epoxy resins are most widely used. Zinc-rich coatings are not abrasion resistant and should be overcoated when required with a coating system containing a compatible binder similar to that used in the original coating. Zinc-rich organic coatings are more tolerant to poor surface preparation and are easier to topcoat.

Paint Application

Basic application procedures must be followed to obtain optimum performance from a coating system regardless of the equipment selected for applying the coating. Cleaned, pretreated surfaces must first be coated within specific time limits established to prevent corrosion products, dirt, and moisture from accumulating and interfering with the coating process. Surface and ambient temperatures must generally be between 8 and 32 °C (50 and 90 °F) for water-borne coatings and 7 and 35 °C (45 and 95 °F) for solvent coatings. Some coatings that are catalyzed, such as epoxy and polyurethane coatings, dry within 4 h at 21 °C (70 °F), within 2 h at 27 °C (80 °F), and within 1 h at 32 °C (90 °F). These coatings in some circumstances may gel in the container or in the spray hoses unless two-component spray equipment is used. Paint should not be applied when temperature is expected to drop below freezing or when the relative humidity is higher than 80% and a temperature drop of more than 3 °C (5 °F) is expected. When successive coats of the same paint are used, each coat should be tinted differently to aid in determining proper application and to ensure complete coverage. Sufficient time must be allowed for each coat to dry thoroughly before overcoating. Allow the final coat to dry for as long as is practical before service is resumed.

Brush application procedures require brushes of first quality, maintained in perfect working condition. Brushes are identified by the type of bristle used, natural, synthetic, or mixed. Chinese hog bristles are the finest natural bristles, because of their length, durability, and resiliency. Hog bristles are unique; the bristle end forks out, resembling a tree branch. This flagging permits more paint to be carried on the brush and leaves finer brush marks on the applied coating which flow together more readily with the over-all result of a smoother finish. Horsehair bristles are used in cheap brushes and are a very unsatisfactory substitute. Horsehairs do not flag; the bristles quickly become limp; they hold far less paint; and they do not spread the paint well. Badger hair brushes make good varnish brushes, and squirrel and sable bristle brushes are used for fine work in lining and lettering. Nylon brushes are used for water-thinned coatings, and these brushes are superior to horsehair. Nylon brushes cannot be used for lacquer materials, because the solvents may dissolve or soften the bristles.

Roller application involves the use of a roller, which consists of a cylindrical sleeve or cover that slips over a rotatable cage with attached handle. The sleeve, or cover, is generally 38 to 64 mm (1 1/2 to 2 1/4 in.) in inside diameter and 75, 100, 175, or 230 mm (3, 4, 7, or 9 in.) in length. Special rollers are available in unusual shapes for corners and in lengths of 38 to 455 mm (1 1/2 to 18 in.) for painting pipes, fences, and other hard-to-reach places. Pressure rolling equipment is available. The paint is fed to the roller under pressure, and the paint flow is controlled by a valve.

Fabrics used in covering the rollers include the following:

- *Lamb's wool* (*pelt*): this material is the most solvent resistant and is available in nap lengths of up to 31.8 mm (1 1/4 in.). Lamb's wool is recommended for applying synthetic finishes on semismooth and rough surfaces. Lamb's wool mats badly in water and should not be used in water-thinned paints
- *Mohair* (*angora*): mohair is solvent resistant and may be used with water-thinned paints. Mohair is supplied in 4.8 and 6 mm (3/16 and 1/4 in.) nap lengths
- *Dynel* (*modified acrylic fiber*): Dynel has excellent water resistance and is somewhat tolerant to most solvents, except such strong solvents as ketones. Dynel is best for conventional water-thinned water paints and solvent systems, except ketone-containing lacquers. Dynel rollers are available in nap lengths from 6 to 31.8 mm (1/4 to 1 1/4 in.)
- *Dacron* (*polyester*): Dacron is a synthetic fiber softer than Dynel, suitable for exterior oil or water-thinned paints. It is available in nap lengths from 7.9 to 13 mm (5/16 to 1/2 in.)
- *Rayon*: Rayon fabric should not be used because poor results have frequently occurred. Rayon also mats badly in water

Table 17 is a guide that can assist in the selection of a suitable roller.

Table 17 Roller selection guide

Product standards do not exist in the paint roller industry, and quality varies greatly among manufacturers; table is based on experience with first-line, high quality products

Paint	Smooth metal	Surface Blasted metal	Pitted and weathered metal
Aluminum	C	A	A
Enamel or semigloss alkyd	A or B	A	...
Enamel undercoat	A or B	A	...
Epoxy coatings	B or D	D	D
Urethane coatings	B or D	D	D
Latex (water-thinned) paint	A	A	A
Metal primers	A	A or D	...
Varnishes, all types	A or B	...	...

Roller cover key	Material	Nap length mm	in.	mm	in.	mm	in.
A	Dynel (modified acrylic)	6.4-9.5	1/4-3/8	9.5-19	3/8-3/4	25-32	1-1 1/4
B	Mohair	4.8-6.4	3/16-1/4	...	...	...	...
C	Dacron (polyester)	6.4-9.5	1/4-3/8	13	1/2	...	...
D	Lamb's wool pelt	6.4-9.5	1/4-3/8	13-19	1/2-3/4	25-32	1-1 1/4

Table 18 Resin designation and surface preparation

NACE designation	SSPC designation	Definition	Remarks
1	5	Zinc-rich primer with topcoats on near-white metals	Topcoating inorganic zinc requires the use of nonoleoresinous paints and the exclusion of linseed oil and alkyd resin-based coatings; this coating system is initially the most costly, but provides both active and passive protection to the steel substrate; used in the most aggressive exposures, such as marine industrial
2	10	Vinyls, polyurethanes, epoxy, and other synthetic polymers on near-white metal	Provides the best passive coating systems without the active protection of the zinc-rich primer and is the next most costly category; used in marine exposure
3	6	Lead-containing or zinc chromate pigmented oleo-resinous paints on a commercial blast surface	Most economical when only original cost is considered and should be used in rural exposures only
4	2 or 3	Lead-containing or zinc chromate pigmented paints on a hand- or machine-cleaned surface	Restricted to hard-to-paint surfaces because of shape of structure or inaccessibility because of height

Coating System Selection

Selection of resin-type designations of coatings should be based on such factors as solvent limitations and chemical and weather resistance properties as shown in Table 13, the category of the coating system based on surface preparation (see Tables 18 and 20), and the estimated life of the paint (see Table 19). Special treatment must be provided for galvanized structural steel which normally has a nominal thickness of 0.13 mm (5 mils). Data indicate that in rural exposures rusting will not occur for 74 years, and in a marine environment, 33 years of rust-free service can be expected. For a marine-industiral exposure, 16 years of rust-free use can be expected. Normally, the galvanized structure must be degreased and acid phosphate etched using a vinyl butyral pretreatment similar to Mil-Spec DOD-P-15328D. A primer, intermediate coat and topcoat is then applied to the prepared surface. The primer coat must be selected to be compatible with the wash primer; the coatings manufacturer should be consulted at this point in this selection process. Careful control as to coating thickness of wash primer must be followed to provide a cohesive film. The wash primer must be between 0.006 and 0.013 mm (0.3 and 0.5 mil) thick followed by the appropriate thickness of the subsequent coating system.

When painting steel structures, many economic factors need to be considered in preparing a job estimate, including:

- The area to be painted
- The surfaces to be coated, such as galvanized steel or ungalvanized steel or the combination of the two

Table 19 Estimated life of paint systems in years

Paint system	Cleaning SSPC designation	Average dry film thickness, µm	Average dry film thickness, mils	Climatic conditions: Mild	Climatic conditions: Moderate	Climatic conditions: Severe	Immersion service: Fresh water	Immersion service: Salt water	Immersion service: Petroleum products	Splashes and spills: Acid	Splashes and spills: Alkaline	Splashes and spills: Halogens
Alkyd:												
3 coat	SP3	114	4.5	4	2	1.5	...	...	...	...	...	...
3 coat	SP6	114	4.5	6	4	2	...	...	...	...	...	...
Latex (acrylic):												
3 coat	SP3	127	5.0	6	3	1.5	...	...	...	...	...	...
3 coat	SP6	127	5.0	10	5	3	...	...	...	...	...	...
Epoxy polyamide:												
2 coat	SP6	152	6.0	7	6	5	...	...	...	...	...	...
3 coat	SP6	254	10.0	10	8	5	5	...	...	5	6	5
Inorganic zinc:												
+3 coat	SP10	254	10.0	12	10	6	6	5	...	6	7	6
3 coat	SP5	305	12.0	14	10	7	6	...	12	5	6	5
Urethane-epoxy:												
Inorganic zinc + 2 coat epoxy + urethane	SP10	305	12.0	15	12	10	...	...	...	...	...	4
2 coat epoxy + urethane	SP5	254	10.0	15	10	8	...	...	20	7	7	5
Vinyl:												
Inorganic zinc + 3 coat	SP10	305	12.0	15	10	8	8	6	...	...	...	...
3 coat	SP10	254	10.0	12	8	6	6	4	...	3	5	4
Chlorinated rubber	SP6	305	10.0	10	9	8	5	5	...	9	6	9
Coal tar epoxy	SP6	406	16.0	8	7	6	8	6	...	4	...	4

- The complexity of the structure that is bolted, riveted, or welded
- The type of structural member to be encountered
- The accessibility of surfaces with the tools required
- The downtime for the facility, including the time to erect scaffolds, place drop cloths, rope off the area, provide alternate parking and alternate storage if required
- The cost of scaffolding and drop cloths, which should include horticultural care such as plantings and grass
- Surface preparation and removal of debris
- Painting

Actual costs vary depending on the season and the location of the structure. Preparing some surfaces for specified paints may add to the cost. Steel that is pitted, gouged, or has discontinuous welds, requires filling with epoxy putty or similar filler material and sometimes even rewelding and grinding are required for high value structures.

The cost of deferred (maintenance) painting should be taken into account, because these costs can be higher than original painting at the time of plant start-up. The tax write-off can be more than the cost of refinishing the structure because of the deterioration of the primer and subsequent coats required for plant start-up.

Surface Preparation

After the selection of a coating system, the most important factor to be considered is surface preparation. Surface preparation must be compatible with the primer and topcoating. Surface preparation is often the most costly phase of the corrosion prevention process for steel, averaging over half the cost of paint application exclusive of scaffolding.

Surface preparation should remove mill scale, rust, oil, grease, atmospheric materials, weld spatter, and old coatings. Surface preparation also provides an anchor pattern to allow the primer and following coats of paint to key into the surface for a good bond. Table 20 shows minimum surface preparation requirements for steel with commonly used coatings. Table 21 summarizes various methods of surface preparation used before painting and provides the SSPC designation for each method.

Inspection of Surface. Before proper surface preparation and painting method can be applied, the condition of the surface must be determined. For new or previously uncoated steel surfaces, the visual standard (SSPC-VIS1/SIS 05 5900) defines the four rust grades of structural steel and contains colored photographs presenting surface preparation standards. Shop primed coated steel would be expected to have no rusting except for small areas abraded during handling.

However, most of the steel structures encountered have been coated previously. The coating layers may be concealing as much rust as is showing on the surface. The surface condition of a previously coated steel substrate may be classified in accordance with ASTM D 610/SSPC-VIS2 for the degree of rusting as shown in Table 14 (SSPC-VIS2 for Pictorial Representation of Rust Classification). The chalking, blistering, flaking, erosion, checking, and cracking of the coating film may be classified by the corresponding ASTM visual standards. Instrumentation is available for determining the presence of pinholes, the adhesion of the coating film to the substrate, and the film thickness.

Surface profile allows the coating system to key into the metal substrate. Surface profile is determined by the abrasive material, including hardness, mass, and firmability, and the force with which the abrasive material impinges upon the surface. The selection of appropriate surface profile provides the bases for good primer adhesion. The coating formulator takes into account these factors: (*a*) the viscosity of the primer which allows the coating to fill the contact surface, (*b*) the number of polar groups to come in contact with this surface, and (*c*) the mechanical anchor or tooth which facilitates adhesion of the primer to the contact surface. Surface profile cannot expose the peaks of the metal once the primer is applied.

Therefore, the abrasive blaster must use the correct angle of attack and distance from the work during the blasting operation. An abrasive-blasted surface profile of from 0.038 to 0.089 mm (1.5 to 3.5 mils), measured from the top of the highest peak to the bottom of the lowest valley, is used for most coating systems.

Surface inspection includes the determination of the condition of the surface before surface preparation and the results of the surface preparation. SSPC VIS-1 provides a standard for surfaces before preparation and prepared surfaces. Surface comparison may also be made by using a surface-profile comparator and the NACE Standard TM-01-70, prepared by the National Association of Corrosion Engineers.

Structural design may limit access to the sections being prepared for painting. Access for abrasive blasting requires approximately 455 mm (18 in.) of clearance for the blast nozzle, although 150 mm (6 in.) can be sufficient if superficial cleaning can be tolerated. Sharp edges resulting from corrosion pits, deep gouges, or cut edges should be properly prepared. Pits and gouges over 3.18-mm (⅛-in.) deep should be filled with weld metal and ground flush to the surface or prepared for painting with an epoxy grout. Cut edges should be chamfered to a 3.18-mm (⅛-in.) radius by grinding. Bolts and rivets should be tight against the steel plates, not allowing crevices; areas surrounding bolts and rivets should be hand brushed with primer after the areas have been abrasive blasted. Discontinuous weldments or tack welds should be properly prepared. Continuous welds should be required, or an epoxy group should be used to eliminate water accumulation. Where water may accumulate, in areas such as cross members joining channels and L sections that are directly exposed to the sky, weep holes are needed that are sufficiently large to allow water to drain and debris to be flushed through.

Quality Control

Paint materials should be purchased to meet the needs of the job at hand. Storage should be minimal to avoid material deterioration because of temperature variations. Coating materials, pretreatment primer, primer, intermediate coating, and topcoating, should be obtained from the same supplier and prepared by the same manufacturer to avoid incompatibility of materials and abrogation of the manufacturer's warranty. When using new and unfamiliar materials, the manufacturer's representative should be consulted and should supervise critical portions of the surface preparation and coating application.

Paint in freshly opened containers should not require straining. However, if skins, lumps, color flecks, or foreign materials are present, paints should be strained after mixing. First, remove any skins from the paint surface, thoroughly mix the paint, thin to application viscosity necessary, and strain through a fine sieve. Use straining as a standard procedure when paint is to be applied by spraying to avoid clogging the spray gun.

Paints should be ready for application by brush or roller when received. Unnecessary thinning or excessive thinning results in an inadequate film thickness and drastically reduces the longevity and protective qualities of the applied coating. In all instances, measure the viscosity of the material to determine that it is correct for the method of application established by the manufacturer. When thinning is necessary, it must be done by competent personnel using the compatible thinning agents recommended in label or specification instructions. Do not thin to improve brush or rolling of paint materials which are cold. Paint materials should be preconditioned to bring them to between 15 and 29 °C (65 and 85 °F) for application.

Sampling and testing of the coating material may be required before it is applied. This is done to confirm that the materials that have been supplied meet specifications. Tests should be performed by the supplier and if confirmation is required, additional testing should be performed by a qualified independent testing laboratory retained by the purchaser of the material.

Sending paints out to an independent testing laboratory requires accurate selection and labeling. Select samples from each lot of coating material supplied by the painting contractor, if more than 380 L (100 gal) of material of each kind is to be used on the job. A representative of the contractor should take these samples. The samples should be two full gallons if supplied in gallon containers or two 1-qt containers properly labeled. Inspect the containers to determine that full measure has been received. Record the following on each sample container: (*a*) manufacturer's name and address, (*b*) trade name and manufacturer's designation of the material, (*c*) contractor's name and address and contract number when applicable, (*d*) date and weather conditions when the sample was taken if put in a 1-qt container, (*e*) batch or lot number, (*f*) date of manufacture, and (*g*) number of gallons represented by the sample. Forward the samples to the laboratory with a written request for the tests required, either full compliance or specific test desired. Include the above information in the request form.

Field and paint shop testing should be done if there is any doubt that materials meet specification requirements. When preparing the paint for use and during painting operations by contractors, limited testing should be done to determine if paints have been adulterated. However, limited field testing should not be considered as a substitute for standard laboratory techniques. Field testing is used to discover major flaws or adulteration in a coating material. Sampling on the job is done by the contractor in the presence of the inspector, unless other arrangements have been made.

Table 20 Minimum surface preparation requirements for steel with commonly used coatings

Listed coatings should not be used unless minimum surface preparation requirements can be met

Coating	Minimum surface preparation
Drying oil	Hand or power tool cleaning (SSPC-SP2 or 3)(a)
Alkyd	Commercial blast (SSPC-SP6)
Oleoresinous phenolic	Commercial blast (SSPC-SP6)
Coal tar	Commercial blast (SSPC-SP6)
Asphaltic	Near white or commercial blast (SSPC-SP10 or 6)
Vinyl	Near white or commercial blast (SSPC-SP10 or 6)
Chlorinated rubber	Near white or commercial blast (SSPC-SP10 or 6)
Epoxy	Near white or commercial blast (SSPC-SP10 or 6)(b)
Coal tar epoxy	Near white or commercial (SSPC-SP10 or 6)
Urethane	Near white or commercial (SSPC-SP10 or 6)
Organic zinc	Near white or commercial (SSPC-SP10 or 6)
Inorganic zinc	White or near white (SSPC-SP5 or 10)

Note: No established criteria are available for the latex paints finding increasing use on steel
(a) SSPC-SP: Steel Structures Painting Council Surface Preparation. (b) Polyamide-cured epoxies require only a commercial blast

Table 21 Surface preparation methods

SSPC designation	Method of surface preparation	NACE designation(a)	Equipment and materials	Remarks
SP1	Solvent cleaning	...	Mineral spirits, chlorinated solvents, coal tar solvents, using tack rags or dip tanks	For the removal of grease, oil, or other soluble materials before removing mill scale, rust, and coatings by other methods. Alkaline cleaners saponify oils and greases, but these cleaners must be neutralized with 0.1 wt % chromic acid, sodium dichromate, or potassium dichromate
SP2	Hand tool cleaning	...	Hand scrapers	Hand tool cleaning should be limited to removing loose material for materials for maintenance and normal atmospheric exposure; coatings with good wetting properties are brush applied
SP3	Power tool cleaning	...	Power wire brushes, grinders, sanders, impact tools, needle guns	For the removal of loose rust, loose mill scale, and loose paint by power tool chipping, descaling, sanding, wire brushing, and grinding without excessive roughing that causes ridges, burrs, or burnishing. Used when primer is to be brush applied
SP4	Flame cleaning	(b)	...	Removal of contaminants by high-velocity oxyacetylene flame burners. Usually followed by wire brushing
SP5	White metal blast	1	Abrasive blasting	Removal of 100% of oil, grease, dirt, rust, mill scale, and paint. Cleaning rate 9.3 m^2/h (100 ft^2/h), using 7.94-mm (5/16-in.) nozzle with 690 kPa (100 psig) at nozzle. Because of atmospheric contamination, maintaining this degree of cleanliness before primer application is difficult
SP6	Commercial blast	3	Abrasive blasting	Removal of 67% of oil, grease, dirt, rust, mill scale, and paint. Cleaning rate of 34 m^2/h (370 ft^2/h), using 7.94-mm (5/16-in.) nozzle with 690 kPa (100 psig) at nozzle. Used for general purpose blast cleaning to remove all detrimental matter from the surface, but leaves staining from rust or mill scale
SP7	Brush-off blast cleaning	4	Abrasive blasting	All loose mill scale and rust are removed, with tight mill scale, paint and minor amounts of rust and other foreign matter remaining. The remaining rust is an integral part of the surface. This level of surface preparation is used for mild exposure and is suitable where a temperature change of less than 11 °C/h (20 °F/h) can be anticipated. Cleaning rate of 81 m^2/h (870 ft^2/h) using 7.94-mm (5/16-in.) nozzle
SP8	Pickling	...	Hydrochloric acid, sulfuric acid with inhibitors, or phosphoric acid with a final phosphate treatment	A shop method of surface preparation for removal of rust and mill scale from structural shapes, beams, and plates where there are few pockets or crevices to trap acid. Excess acid must be rinsed off with water, and painting is required as soon as possible to prevent recontamination of the surface
SP9	Weathering	(c)	...	Although mill scale is weathered away, this process is detrimental because surface contamination is more difficult to remove when weathered
SP10	Near-white blast	2	Abrasive blasting	Removal of 95% of oil, grease, dirt, rust, mill scale, and paint. A cost savings of 25% can be realized on average where this level of cleanliness can be tolerated. Shadows, streaks, or discolorations are distributed over the surface, but not concentrated in any area or particular spot. Cleaning rate 16 m^2/h (175 ft^2/h) using a 7.94-mm (5/16-in.) nozzle and 690 kPa (100 psig) at nozzle
	Water blasting	...	Inhibited water at pressures of 6900 to 69 000 kPa (1000 to 10 000 psig) used	Removal is slow and the degree of cleaning must be specified. High pressures may cause damage to substrate or structures

(a) NACE: National Association of Corrosion Engineers. (b) Discontinued as of Jan 1982. (c) Discontinued in 1971

Material storage should be at 24 °C (75 °F) plus or minus 6 °C (10 °F), because paint is a temperature sensitive product. Low temperatures cause paints to increase in viscosity and may require conditioning for 24 h before use. Freezing temperatures may ruin water-borne paints and cause containers to bulge or burst. High temperatures result in lower viscosities, causing pigment to settle and poor flow characteristics. Coating materials may be extremely sensitive to heat. At temperatures over 38 °C (100 °F), gelation may occur, resulting in unusable material. At these high temperatures, pressure can build up within the containers enough to cause lids to blow off, creating a serious fire hazard. Application is seriously affected when coating materials are used after being stored at very high or low temperatures. Additional conditioning time and effort are required in these cases to ensure proper application and optimum surface protection.

Other factors to be considered are high humidity, which causes containers to corrode and labels to deteriorate, and poor ventilation, which allows the collection of excessive concentrations of solvent vapors that are both toxic and combustible. Pumps for drawing liquids from steel drums must be approved by fire underwriters. Gravity spigots, other than self-closing types, should not be used because of the possibility of accidental spillage. Stock should be stored so that all labels can be easily read and containers can be rotated to use older material first. Materials should be issued for each work shift in amounts that are consumed during that time without loss or spoilage.

Maintenance Program

The most economical approach to maintaining a coating system is the establishment of a periodic maintenance program. If the appropriate structural steel alloy is selected, uniform corrosion should be the only corrosion encountered without a protective coating system. Proper surface preparation and selection of a coating system suitable to the ambient conditions can provide a 10-year life cycle for the coating. An annual touch-up and the application of a full topcoat after 5-year of exposure can provide this 10-year life cycle.

A spotty appearance can result from periodic touch-ups, especially if the topcoat has chalked. In addition, catalyzed coating systems are difficult to topcoat, because only cohesive forces hold the freshly applied topcoat to the aged coating, and the aged coating must be roughened by brushoff blasting before applying a fresh coating. A maintenance coating must be compatible with existing coatings.

When programmed maintenance is instituted, all structural exteriors should be inspected annually to establish the following:

- Update the condition of the structure to determine the priority for painting

Table 22 Safety related publications for painting structural steel

Publication	Edition	Source
Accident Prevention Manual for Industry International Book No. 0.87912-024-X	7th Edition 1974	National Safety Council 425 North Michigan Avenue Chicago, IL 60611
29CFR1910, OSHA Safety and Health Standards for General Industry	Current	Superintendent of Documents U.S. Government Printing Office Washington, DC 20402
ANSI Z87.1 (Industrial Eye Protection)	1968	American National Standards Institute 1430 Broadway New York, NY 10018
ANSI Z88.2 (Respiratory Protection)	1969	See above
ANSI Z89.1 (Industrial Head Protection)	1968	See above
ANSI Z9.2 (Design and Operation of Local Exhaust Systems)	1960	See above
ANSI Z9.3 (Design, Construction, and Ventilation of Spray Finishing Operations)	1964	See above
B.S. 5493 (Code of Practice for Protective Coating of Iron and Steel Structures Against Corrosion)	1977	British Standards Institution 2, Park Street London, WI, Great Britain
Hazardous Substance Guide for Construction	1977	Associated General Contractors of America 1957 E Street, NW Washington, DC
Guide on Hazardous Materials	1981	National Fire Protection Assoc. 470 Atlantic Avenue Boston, MA 02210
Safety Guide SG10	1979	Manufacturing Chemists Assoc. 1825 Connecticut Avenue, NW Washington, DC 20009
Industrial Ventilation–A Manual of Recommended Practice	1980	American Conference of Governmental Industrial Hygienists P.O. Box 16153 Lansing, MI 48901
Play It Safe	1976	International Brotherhood of Painters and Allied Trades AFL-CIO United Union Building 1750 New York Avenue, NW Washington, DC 20009
Threshold Limit Values of Chemical Substances in Work Room Air	1981 Annually Revised	American Conference of Governmental Industrial Hygienists P.O. Box 1397 Cincinnati, OH 45201

(continued)

Table 22 (continued)

Publication	Edition	Source
OSHA Reference Manual	1982	Painting and Decorating Contractors of America 7223 Lee Highway Falls Church, VA 22046
ANSI A14.1 (Safety Code for Wood Ladders)	1975	See above
ANSI A14.2 (Safety Code for Metal Ladders)	1973	See above
Civil Works Construction Guide Specification CW-09940 Painting Hydraulic Structures and Appurtenant Works	August 1981	OCE Publications Depot 890 South Pickett Street Alexandria, VA 22304
Paint and Protective Coatings (Army TM 5-618, NAVFAC MO 110 and Air Force AFM 85-3	June 1981	See above
ANSI A10.8 (Scaffolding)	1974	See above
EM 385-1-1 (Safety Requirements)	1980	See above

- Determine if washing is the only process needed to improve or maintain appearance
- Determine what portions of the structure require touch-up painting
- Determine if the structure requires a totally new coating system

Specifications and Industrial Guidance

Specifications and industrial guidance have been developed by the Steel Structures Painting Council committees. These specifications are used for industrial applications. A second source of standardization data and guidance is the Department of Defense publications. The standardization documents include military handbooks which reference federal and military specifications. All these federal and military documents are listed in the Department of Defense Index of Specifications and Standards (DODISS). The American Society for Testing and Materials issues consensus standards on paint constituents and the testing of paints; however, these standards currently do not cover paints as supplied by industrial producers or vendors. Other specifications and guidance are available from the National Association of Corrosion Engineers and the American Society of Naval Architects. Most of the documents are available in various libraries throughout the United States.

Safety

Safety hazards are of two types—those resulting from the location of and access to the structural steel and hazards occurring as the result of exposure to paint as a possible toxic material. Hazards in the work place are regulated by Department of Labor in the Occupational Safety and Health Administration (OSHA), which is responsible for establishing safety rules for workers and the materials they use. Hazards that are emitted to the environment are regulated by the Environmental Protection Agency.

Federal laws govern all works, and these laws established by congressional action are defined by OSHA rulings. These rulings are established by priority which is at the state and local level; therefore, the rulings can vary depending where the work takes place and on the inspector who enforces the ruling. State safety requirements exert control in relevant areas whenever federal law or rulings do not cover projects. Municipal or township ordinances should be followed whenever specific rulings impose restrictions beyond the federal and state controls. It is at this level that safety requirement priorities are established and are moved up to the state, federal district, and national levels. Table 22 provides a listing of safety related publications.

SELECTED REFERENCES

- Gordon H. Brevoort and A.H. Roebuck, Simplified Method for Calculating Cost of an Applied Paint System, *Materials Performance,* Vol 19 (No. 6), June 1980, p 24-32
- P.G. Campbell, et al, Army TM5-618, NAVFAC MO-110, Air Force AFM85-3, *Paints and Protective Coatings,* Philadelphia: U.S. Naval Publications and Forms Center
- *Department of Defense Index of Specifications and Standards,* Philadelphia: U.S. Naval Publications and Forms Center
- C.P. Dillon, The Economic Advantage of Deferred Topcoat Zinc-Base Painting Systems, *Materials Performance,* Vol 14 (No. 5), May 1975, p 29-31
- M. Hess, et al, *Hess's Paint Film Defects,* 3rd ed., London: Chapman and Hall, 1979
- W. Seiter, Hot Dip Galvanizing Plus Paint—An Outstanding Corrosion Protection for Steel Structures in Electrical Supply Undertakings, Proceedings, 9th International Galvanizing Conference, 1970, p 9-14
- *Steel Structures Painting Manual,* Vol 2, Systems and Specifications, 1982

Porcelain Enameling

By the ASM Committee on Porcelain Enameling*

PORCELAIN ENAMELS are glass coatings applied primarily to products made of sheet steel, cast iron, or aluminum to improve appearance and protect the metal surface. Porcelain enamels are distinguished from other ceramic coatings by their predominantly vitreous nature and the types of applications for which they are used, and from paint by their inorganic composition and the fusion of the coating matrix to the substrate metal. Porcelain enamels of all compositions are matured at 425 °C (800 °F) or above.

The most common applications of porcelain enamels are for major appliances, water heater tanks, sanitary ware and cookware; in addition, porcelain enamels are used in a wide variety of applications ranging from chemical processing vessels, agricultural storage tanks, piping and pump components and barbeque grills to architectural panels, signing, specially executed murals and microcircuitry components. Normally, porcelain enamels are selected for products or components where there is a need for one or more special service requirements that porcelain enamel can provide — chemical resistance, corrosion protection, weather resistance, specific mechanical or electrical properties, appearance or color needs, cleanability or thermal shock capability.

Types of Porcelain Enamels

Porcelain enamels for sheet steel and cast iron are classified as either ground-coat or cover-coat enamels. Ground-coat enamels contain oxides that promote adherence of the enamel to the metal substrate. Cover-coat enamels are applied over ground coats to improve the appearance and properties of the coating. Cover coats may also be applied directly to properly prepared decarburized steel substrates. The color of ground coats is limited to various shades of blue, black, brown, and gray. Cover coats, which may be clear, semiopaque, or opaque, may be pigmented to take on a great variety of colors. Colors may also be smelted into the basic coating material. Opaque cover coats are usually white.

For aluminum, neither ground coats nor adherence-promoting oxides are required. Single-coat systems are used for most applications. When two coats are desired, the first coat can be of any color. Porcelain enamels for aluminum are usually transparent and can be pigmented and opacified inorganically to produce the desired appearance.

The basic material of the porcelain enamel coating is called frit; it is a special glass of small friable particles produced by quenching a molten glassy mixture. Because porcelain enamels are usually designed for specific applications, the compositions of the frits from which they are made vary widely. A number of compositions of frits for enamels for sheet steel, cast iron, and aluminum are discussed below; however, many variations of these compositions are used commercially.

Enamel Frits for Sheet Steel. All the frits for which compositions are given in Table 1 are classified as alkali borosilicates for use as ground coats on sheet steel. Their compositions differ depending on the application environment of the enameled product. For example, acid resistance is obtained by the addition of titanium dioxide and a large increase in silicon dioxide with a corresponding decrease in boron

*L. N. Smith, *Co-Chairman,* Technical Director, Porcelain Metals Corp.; Larry L. Steele, *Co-Chairman,* Research Metallurgist, Armco Inc.; Clifton G. Bergeron, Professor and Head, Department of Ceramic Engineering, University of Illinois at Urbana; G. Thomas Cavanaugh, Manager of Finishing Engineering, Jenn-Air Corp.; John E. Cox, Special Projects Manager, O. Hommel Co.; James W. Elliott, Vice President, Manufacturing, Porcelain Industries, Inc.; Wayne L. Gasper, Chief Process Engineer, Maytag Co.; M. B. Gibbs, Assistant Superintendent, Technical Service Department, Inland Steel Co.; Albert L. Gugeler, Manager, Frit Quality Control, Ferro Corp.; Robert K. Laird, Ceramic Engineer, American Standard, Inc.; Daniel H. Luehrs, Ceramic Engineer, Clyde Division, Whirlpool Corp.; Dennis E. McCloskey, Production Manager, White Consolidated Industries-Mansfield; John C. Oliver, Executive Vice President, Porcelain Enamel Institute, Inc.; James F. Quigley, Manager, Porcelain Enamel Coatings and Frit Operations, Ferro Corp.; Donald R. Sauder, Division Finishing Manager, Tappan Co.; Albert J. Schmidt, Director of Research, American Porcelain Enamel Co.; Howard F. Smalley, Technical Service Coordinator, Pemco Products, Mobay Chemical Corp.; Thomas L. Stalter, Manager, Porcelain Enamel Development, Pemco Products, Mobay Chemical Corp.; Carl G. Strobach, Director, Advanced Engineering and Technical Services, Rheem Water Heating Division, City Investing Co.; James D. Sullivan, Manager, Ceramic Research Laboratory, A. O. Smith Corp.; Donald A. Toland, Associate Research Consultant, U. S. Steel Corp.; Donald B. Tolly, Senior Application Engineer – Ceramics, General Electric Co.; Daniel R. Yearick, Senior Manufacturing Engineer, Caloric Corp.

Table 1 Melted-oxide compositions of frits for ground-coat enamels for sheet steel

Constituent	Composition, wt %: Regular blue-black enamel	Alkali-resistant enamel	Acid-resistant enamel	Water resistant enamel
SiO_2	33.74	36.34	56.44	48.00
B_2O_3	20.16	19.41	14.90	12.82
Na_2O	16.74	14.99	16.59	18.48
K_2O	0.90	1.47	0.51	...
Li_2O	...	0.89	0.72	1.14
CaO	8.48	4.08	3.06	2.90
BaO	9.24	8.59	...	...
ZnO	...	2.29	...	...
Al_2O_3	4.11	3.69	0.27	...
ZrO_2	...	2.29	...	8.52
TiO_2	...	...	3.10	3.46
CuO	...	...	0.39	...
MnO_2	1.43	1.49	1.12	0.52
NiO	1.25	1.14	0.03	1.21
Co_3O_4	0.59	1.00	1.24	0.81
P_2O_5	1.04	...	...	0.20
F_2	2.32	2.33	1.63	1.94

Table 2 Melted-oxide compositions of frits for cover-coat enamels for sheet steel

Constituent	Composition, wt %, Titania white enamel: Fused at 815 °C (1500 °F)	Alkali resistant	Weather-resistant blue enamel
SiO_2	41.55	43.10	43.97
B_2O_3	12.85	13.81	6.51
Na_2O	7.18	5.99	13.83
K_2O	7.96	10.12	0.21
Li_2O	0.59	0.57	2.37
CaO	...	...	2.68
PbO	...	...	14.96
ZnO	1.13	...	...
Al_2O_3	...	...	0.43
ZrO_2	...	2.05	...
TiO_2	21.30	19.39	5.86
P_2O_5	3.03	0.54	...
Co_3O_4	...	...	3.72
F_2	4.41	4.43	5.46

trioxide. Resistance of the enamel to alkalis or to water can be improved by adding zirconium oxide, usually as zircon, to the frit and maintaining a high content of silicon dioxide.

Weather resistance is usually a function of acid resistance. Porcelain enamels for use outdoors are made from various types of frits that produce the resistance and color desired. Resistance to thermal shock and to high temperature is obtained by controlling the expansion of the glass coating which is accomplished by adjusting the frit composition and by the addition of refractory materials, such as silicon dioxide, aluminum oxide, and zircon to the mill formula.

Table 3 Melted-oxide compositions of frits for enamels for cast iron

Constituent	Composition, wt %: Ground coat(a)	Cover coats: Zirconium-opacified enamel(a)	Antimony-opacified enamel(b)	Acid-resistant enamel (a)(b)
SiO_2	77.7	28.0	22.9	37.0
B_2O_3	6.8	8.8	11.2	4.9
Na_2O	4.3	10.0	12.3	16.8
K_2O	...	4.1	6.0	1.7
PbO	4.0	17.8	9.8	8.8
CaO	...	8.7	8.0	2.0
ZnO	...	6.1	7.5	5.9
Al_2O_3	7.2	4.5	6.4	1.9
Sb_2O_3	...	...	13.9	13.1
ZrO_2	...	6.1	...	...
TiO_2	...	...	...	7.9
F_2	...	5.9	2.0	...

(a) For dry process. (b) For wet process

Cover coats for sheet steel are applied over ground coats or directly to properly prepared decarburized steel. Compositions of frits for cover-coat enamels are given in Table 2. Electrostatic dry powder cover coats may be applied over an electrostatic dry powder ground coat and the entire two-coat/one-fire system matured in a single firing.

Cover-coat enamels made from titania-opacified frits are generally quite acid resistant; even in amounts too small to impart any opacity, titania imparts acid resistance. For alkali resistance, zirconium oxide is a desirable constituent. Clear frits containing 8 to 11% titanium dioxide are used for strong to medium strength colors. Semiopaque frits containing 12 to 15% titanium dioxide are used for medium-strength colors, and opaque frits containing 17 to 20% titanium dioxide are used for pastel colors. The blue frit listed in Table 2 produces an enamel that has excellent weather resistance.

Enamel Frits for Cast Iron. Compositions of frits for enamels for cast iron vary depending on whether the frit is applied by the dry process or the wet process (Table 3). Dry process enamels are commonly used for large cast iron fixtures because of their brilliance and ability to cover small surface irregularities. Acid resistance is imparted to these enamels by reducing the alumina content, increasing silica, and adding up to about 8% titanium dioxide. Dry process enamels are seldom used in applications requiring resistance to severe thermal shock.

Ground coats are usually necessary to fill surface voids in castings. Ground coats for wet process enamels often are mixtures of frit, enamel reclaim, and refractory raw material used at very low application weight. Ground coats for dry process enamels are applied by the wet process and fused to thin, viscous coatings that protect the casting surface from excessive oxidations while it is heated to enameling temperature.

Enamel frits for aluminum are usually based on lead silicate and on cadmium silicate, but may be based on phosphate or barium. Table 4 gives the compositions of some frits for aluminum.

The high-lead enamels for aluminum have a high gloss, good acid and weather resistance, and good mechanical properties. The phosphate enamels generally are not alkali resistant or water resistant, but may have good acid resistance. They melt at relatively low temperatures and are useful in many applications. The barium enamels are not as low melting as the lead or phosphate glasses, but they do have good chemical durability.

Preparation of Frits

Frits, the major constituents of porcelain enamels, are smelted complex glass or ceramic systems. Frits generally are compounded of 5 to 20 or more components, which are thoroughly mixed together and melted into a glassy system. The molten glass is then quenched to a friable (easily broken up) condition, by being either poured into water or rolled into a thin sheet between water-cooled rolls. If quenched in water, the frit is dried before use. If quenched in sheet form by water-cooled rolls, the sheet ordinarily is shattered into small flakes by mechanical means before shipment or use.

Grinding and Blending. Porcelain enamel is usually applied as a suspension of finely milled frit in water; however, it may also be applied as a dry powder by electrostatically spraying on sheet steel or by dredging on cast iron. The wet process frit is reduced to a fine powder in a ball mill. For milling, the ball charge should occupy 50 to 55% of the mill volume. After loading the frit charge and mill additions such as clay, bentonite, electrolytes, and coloring oxides, the water is added. Frits for dry electrostatic application are ground without water by the frit supplier and furnished to the porcelain enameler in a ready-to-use form. Small amounts of proprietary additives are included during grinding to aid in the electric charge retention during application and handling. Particle size is fitted to the appropriate level for each particular application.

The fineness of various types of porcelain enamels for sheet steel, cast iron, and aluminum is shown below:

Type of enamel	Milled fineness, % on 200-mesh screen(a)
Sheet steel	
Ground coat	4-9
Cover-coat:	
Non-acid resistant	4-8
Titania, acid resistant	0.5-3.0
Colored, non-acid resistant	0.5-5.0
Colored, acid resistant	0.5-3.0
Wet electrostatic	1-6(b)
Dry powder, electrostatic	0-6
Cast iron	
Ground coat	0.5-4.0
Wet process cover coat	3-6
Dry process	60-80(c)
Aluminum	
Ground and cover coats	0.2-1.5(b)

(a) Percentage of washed and dried wet-milled enamel sample remaining on 200-mesh screen. (b) Percentage on 325-mesh screen, which replaces the 200-mesh screen in standard fineness test, see PEI Bulletin P-305. (c) Through a 200-mesh screen, dry test

Applications using these enamels and service criteria are shown below:

Product	Service criteria
Wet process and dry powder, electrostatic	
Architectural panels	Acid and weather resistant
Chemicalware	Very high acid resistant
Cookware	Acid and thermal shock resistant
Heaters and heat exchangers	Thermal shock resistant
Laundry units:	
Exteriors	Alkali resistant
Interiors	Alkali resistant
Oven liners:	
Conventional	Acid resistant
Pyrolytic	Acid and thermal shock resistant
Ranges:	
Exteriors	Acid resistant
Top	Acid and thermal shock resistant
Grates and burners	Acid and thermal shock resistant
Reflectors, electric light	High reflectance, white
Refrigerators	Acid resistant
Sanitary ware, sheet steel	Acid resistant
Signs	Acid and weather resistant
Water heater tanks	Water resistant
Dry process	
Cast iron sanitary ware	Acid resistant
Chemicalware	Very high acid resistant

The enamel slip is unloaded either by gravity flow or by applying air pressure. Centrifugal or vibratory screening and magnetic separation of the slip as it is transferred from the mill to the storage tank is a recommended practice. Use separate mills for grinding frits for ground coats, white cover coats, and colors.

Frit ground to the correct size for some single-coat dark colors can be purchased from the manufacturer. This frit is combined with the usual mill additions and water. The mixture is blunged, that is, amalgamated and blended by the enameler into a suitable slip in a high-speed blunger operating at 900 to 2000 rev/min. Frit applied by the dry process is milled in a ball mill without water. More complete information on ball milling is given in the Porcelain Enamel Institute Bulletin P-305.

Table 4 Melted-oxide compositions of frits for enamels for aluminum

Constituent	Composition, wt % Lead-base enamel	Barium enamel	Phosphate enamel
PbO	14-45	...	...
SiO_2	30-40	25	...
Na_2O	14-20	20	20
K_2O	7-12	25	...
Li_2O	2-4	...	4
B_2O_3	1-2	15	8
Al_2O_3	...	3	23
BaO	2-6	12	...
P_2O_5	2-4	...	40
F_2	...	...	5
TiO_2	15-20	(a)	(a)

(a) TiO_2, 7 to 9 wt %, added to frit during mill preparation of the enamel slip

Mill Additions. Clays and electrolytes are used with frit to make enamels applied by the wet process to control the properties of the slip. For frits for aluminum, clays or electrolytes are not used. Refractory materials and pigments may be added to impart desired properties to the fired enamel. To increase the refractories or to change the color of the enamel, color pigments, opacifiers, or silica are mixed with frits for the dry process. Table 5 lists some of the common mill additions used with wet process frits; also shown is the amount commonly used, along with the effect of each mill addition.

The soluble electrolytes that are highly alkaline, such as potassium carbonate and sodium aluminate, are responsible for deflocculation of the clays. Deflocculation is necessary to permit proper suspending action by the clays. Minor adjustments in the flow properties of the slip are made by varying the content of the clay and the electrolytes.

Table 5 Mill additions for wet-process enamel frits for sheet steel and cast iron

Addition material	Amount added, %	Effect of addition
Clay	2-8	Suspends glass, increases set, hardens bisque
Bentonite	0-0.5	Suspends glass, increases set, hardens bisque
Borax	0-0.75	Stabilizes suspension
Gum tragacanth	0-0.06	Hardens bisque
Alginates	0-0.06	Hardens bisque
Urea(a)	0-1	Reduces tearing
Sodium nitrite(a)	0-0.75	Increases set; reduces tearing
Magnesium carbonate	0-0.25	Increases set; moderately softens bisque
Potassium carbonate(a)	0-0.5	Hardens bisque; retards poppers; reduces tearing
Sodium aluminate	0-0.5	Increases set; stabilizes suspension
Potassium chloride(a)(b)	0-0.5	Increases set
Potassium nitrite(a)	0-0.75	Strongly increases set; reduces tearing
Potassium nitrate	0-0.06	Slightly increases set
Tetrasodium pyrophosphate(a)	0-0.33	Strongly decreases set
Lithium titanate	0-2	Fluxes enamel for lower firing temperature
Zinc oxide	0-2	Fluxes enamel for lower firing temperature
Formaldehyde(a)	0-0.1	Prevents bacterial growths
Titanium dioxide	0-3	Increases opacity
Silica	0-15	Lowers gloss, increases resistance to chemicals and heat; increases refractoriness
Pigments	0-4	Produces desired color in fired enamels
Opacifiers	0-4	Produces opacity in fired enamels

(a) May be added to the slip after milling. (b) Never used in direct-to-steel enamel slips

Mill additions for wet process enamel frits for aluminum consist of boric acid, potassium silicate, sodium silicate, and other additives. These materials are used to control the wet suspension of the frits and to contribute to the characteristics of the fired enamel. Also, titanium dioxide and ceramic pigments are added to produce opacity and the desired color, respectively.

Steels for Porcelain Enameling

Typical compositions of the various grades of low-carbon sheet steel that are commercially available for porcelain enameling are listed in Table 6. A description of the area of application of each grade is given below.

Cold Rolled Rimmed Steel. This nonpremium grade is used for such nonappearance applications as interior parts of appliances. These components are normally porcelain enameled with ground coat only. Cold rolled rimmed steel is also used for two-coat enameling where quality standards are lower than those for such components as exterior parts of appliances. Because of the high tendency of cold rolled rimmed steel to warp and to carbon boil during porcelain enameling, the use of this material is generally restricted to (*a*) parts for which appreciable distortion during firing can be tolerated, or (*b*) applications in which a low enamel firing temperature or sufficient metal thickness to prevent warpage, is used.

Enameling iron is a low-metalloid rimmed steel that is widely used for porcelain enameling exterior parts that must meet high standards of visual inspection and that require greater resistance to warpage than is afforded by cold rolled rimmed steel. Enameling iron is a premium grade of steel that requires a porcelain enamel ground coat before cover coating.

Titanium-stabilized steel is a premium-grade steel developed primarily for direct cover coating. However, because of its excellent resistance to warpage, the greatest use of this steel is for large flat panels when maximum flatness after porcelain enameling is essential.

Special low-carbon steel is also a premium grade steel developed for direct cover coating. However, it can be ground coated or porcelain enameled with ground and cover coats with results that are superior to those obtained with cold rolled rimmed steel or enameling iron. Special low-carbon steel is a cold rolled rimmed or fully killed steel that has been thoroughly decarburized by open-coil annealing. The resultant low carbon content of 0.005% or less eliminates primary boiling and consequent defects such as black specks, pull-through, and dimples caused by the evolution of the carbon monoxide and dioxide through the porcelain enamel during firing. This grade has excellent resistance to warpage and exhibits good resistance to defects such as ground-coat reboiling during firing of the cover coat, and fishscale both of which are caused by the evolution of hydrogen gas.

Although special low-carbon steel has superior enamelability compared to the other steels described, during porcelain enamel firing it is subject to grain coarsening in those areas that have been strained between 4 and 16% during forming. Porcelain enamel over those areas is susceptible to chipping during assembly because of the lower yield strength of 70 to 105 MPa (10 to 15 ksi) after firing of the grain-coarsened steel. The use of lower firing temperatures and redesign of the parts helps alleviate this problem.

Interstitial-Free Enameling Steel. Interstitial-free enameling steel is an aluminum deoxidized, vacuum decarburized steel. A combination of niobium and titanium alloy additions are utilized to fully stabilize the steel. This steel has superior formability to conventional enameling iron and decarburized steels. Interstitial-free enameling steel exhibits excellent strength retention after strain and firing, as well as base metal sag properties superior to enameling iron and decarburized enameling steels. This type of steel does not exhibit stretcher strain and is nonaging and nonfluting. It is suitable for either ground coat or direct-on cover coat porcelain enamel systems.

Table 6 Composition of low-carbon sheet iron and steel for porcelain enameling

Enameled metal	Element, wt % Carbon	Manganese	Phosphorus	Sulfur	Aluminum	Titanium	Niobium
Enameling iron	0.03	0.05(a)	0.01	0.02	...(b)	...	...
Decarburized enameling steel	0.005	0.20-0.30	0.01	0.02	...(b)	...	...
Titanium-stabilized enameling steel	0.05	0.30	0.01	0.02	0.05	0.30	...
Interstitial-free enameling steel	0.005	0.20	0.01	0.02	...	0.04	0.09
Cold rolled steel	0.06	0.35	0.01	0.02	...(b)	...	...

(a) Some enameling iron may have manganese contents of 0.20 wt %. (b) Some steels may be supplied as aluminum-killed products

Drawing-Quality Special-Killed Steel. Aluminum-killed steel may be made by either ladle-killing or special ingot-killing practices when intended for use in porcelain enameling applications. Thus, for drawn parts that require sheet steel with better than average formability, drawing-quality special-killed (DQSK) grades in cold rolled enameling iron and special low-carbon sheet are available.

Steel plate, tubes, pipes, and rolled sections also may be porcelain enameled for special applications. Some examples are low-alloy steel tubing that is fabricated and glass-lined for chemical pressure vessels and low-carbon hot rolled steel sections that are enameled for use as window sash and other architectural parts. Some of the steels used for these special applications are identified in Table 7. The sheet should be thick enough so that sagging is reduced to an economically tolerable level. Effect of sheet thickness on sag resistance at 870 °C (1600 °F) is shown in Fig. 1(a).

Factors in Selecting Steel

The most important factors in the selection of steel for porcelain enameling are enamelability, freedom from surface defects, formability, sag characteristics, strength, and weldability. The relative importance of each depends on the requirements of the finished product.

Enamelability. During enamel firing, carbon reacts with oxides in the glass-metal system and with the furnace atmosphere to form carbon monoxide and carbon dioxide which evolve through the molten enamel. A limited amount of such gas evolution can be tolerated in two-coat porcelain enameling because the ground-coat enamel has high fluidity and tends to heal over

Fig. 1 Sag characteristics of enameled sheet steels

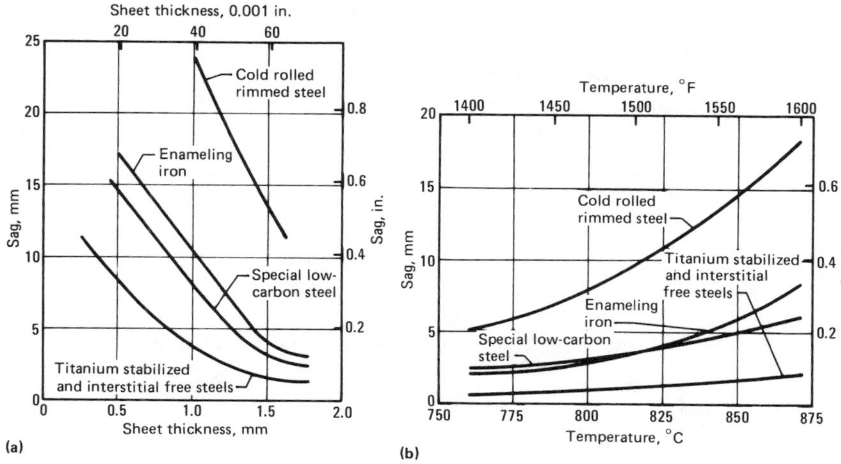

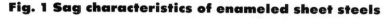

Each of the test specimens measured 305 by 50 mm (12 by 2 in.). They were mounted on supports spaced 255 mm (10 in.) apart. (a) Sag characteristics as a function of sheet thickness. Tested at 870 °C (1600 °F) (b) Sag characteristics as a function of temperature, 19-gage, 1.0617-mm (0.0418-in.) steel used.

Fig. 2 Yield strength after strain and firing

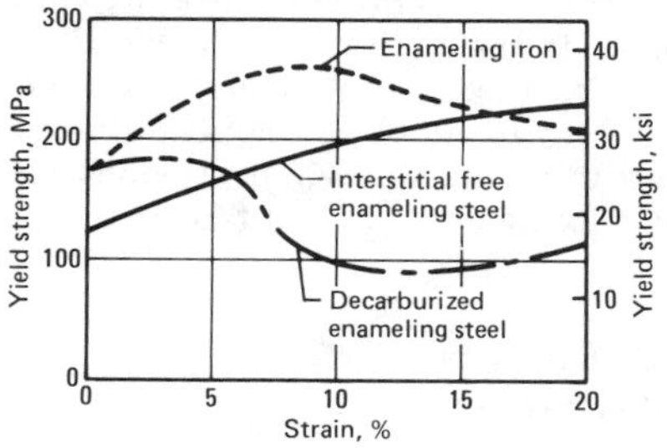

Table 7 Composition of low-alloy and stainless steels for porcelain enameling

Steel	Carbon	Manganese	Silicon	Nickel	Chromium	Molybdenum
	Principal constituents, %					
Low-alloy steels						
ASTM A203, grade A	0.17	0.80	0.13-0.32	2.13-2.67	...	...
ASTM A203, grade D	0.17	0.80	0.13-0.32	3.18-3.82	0.06	0.02
ASTM A225, grade A	0.18	1.45	0.13-0.32	...	...	...
ASTM A285, grade A	0.15	0.45	...	...	...	...
ASTM A387, grade B	0.17	0.65	0.13-0.32	...	0.95	0.50
Stainless steels						
Austenitic	0.08	2.00	...	8.0-12.0	16.0	...
Ferritic	0.12	0.50	0.35	0.30	16.0	...

defects. Because primary boiling does not recur on subsequent firing, a defect-free cover coat can be obtained. In direct-cover-coat enameling, however, very little gas evolution can be tolerated. Therefore, select a steel that is relatively free of carbon or one that has the carbon stabilized by the addition of titanium or niobium (Table 6). When the higher carbon steels (Table 7) are porcelain enameled, surfaces must be free of large angular cementite (iron carbide) particles.

Although it is available as enameling iron or low-carbon rimmed steel, avoid hot rolled sheet if possible because of its high susceptibility to fish-scaling. When the use of hot rolled sheet is necessitated by thickness requirements for a given application, such as for glass-lined tanks for water heaters, limit porcelain enameling to only one side of the sheet. This permits hydrogen gas, which produces the fish-scaling, to escape from the unenameled side of the sheet.

Surface Defects. It is imperative that the surface of the steel be free of defects such as deep scratches, pits, slivers, scale, and open laminations. These defects are particularly troublesome when present in steel that is direct-cover-coat enameled.

Jagged metal burrs along scratches protrude through the ground-coat enamel. Oxidation of these burrs during firing causes copperheads in the ground coat and blistering in the cover coat. Folded-over burrs, deep scratches, and surface laminations entrap dirt, drawing compounds, or cleaning and pickling solutions, all of which promote the formation of blisters in the porcelain enamel during firing.

Formability. Sheet steels for porcelain enameling are available in conventional commercial quality, drawing-quality, and drawing-quality special-killed grades for varying severity of draw. The steels usually are supplied unoiled to avoid cleaning difficulties.

Sag Characteristics. Sag is defined as the permanent bending or creep of a material during enamel firing. The sag resistance of steel is related to its strength at elevated temperature and to the temperature at which the steel starts to transform from ferrite to austenite on heating. In general, the higher the strength and the higher the transformation temperature, the better the sag resistance of the steel.

A comparison of the sag characteristics of the types of sheet steel commonly porcelain enameled is shown in Fig. 1. The data in Fig. 1(b) show that in the enamel firing temperature range, cold rolled rimmed steel has inferior sag resistance, enameling iron and special low-carbon steel are intermediate, and titanium stabilized steel and interstitial-free steels have the best sag resistance.

Strength. Because the fired porcelain enamel is invariably fractured when the base metal is strained beyond the elastic limit of more than 0.002 mm/mm (0.002 in./in.), consideration should be given to the yield strength of the steel as enameled. In addition, the contour and thickness of the metal are major factors for consideration. Figure 2 shows relative yield strength levels of common base metals after straining and firing.

Weldability. Good welding characteristics are required for fabricated shapes in the appliance industry. Welds must be free of inclusions, blowholes, and laminations. Porcelain Enamel Institute Bulletin P-306 provides useful information on steel fabrication.

Cast Iron and Aluminum

Cast iron for enameling usually has a composition within the limits stated below:

Constituent	Amount, %
Total carbon	3.20-3.60
Silicon	2.30-3.00
Manganese	0.30-0.60
Sulfur	0.05-0.12
Phosphorus	0.40-0.80

Total carbon and silicon should vary in opposite directions within the ranges shown. If both are low, the iron tends to be brittle and blisters during porcelain enameling. If total carbon and silicon are high, the iron is soft and warps easily when reheated for porcelain enameling.

Manganese and sulfur should range in the same direction so that all of the sulfur is converted to manganese sulfide. Within the normal range, phosphorus has a negligible effect on the strength of the iron at porcelain enameling firing temperatures.

Aluminum. The common porcelain enameling alloys for the various forms of aluminum are:

- Sheet: 1100, 3003, and 6061
- Extrusion: 6061
- Casting alloys: 43 and 356

Of the wrought alloys, only 6061 alloy is heat treatable. Because of its higher strength, 6061 alloy has better handling characteristics before and during porcelain enameling. It is stronger after porcelain enameling. The non-heat treatable alloys are easier to form before porcelain enameling and are used for small parts for which the amount of distortion and low strength encountered after firing are acceptable; however, non-heat treatable alloys are unsuitable for more than one coat of porcelain because of crazing after a second firing.

Fig. 3 Process for preparing steel surfaces for ground-coat porcelain enameling

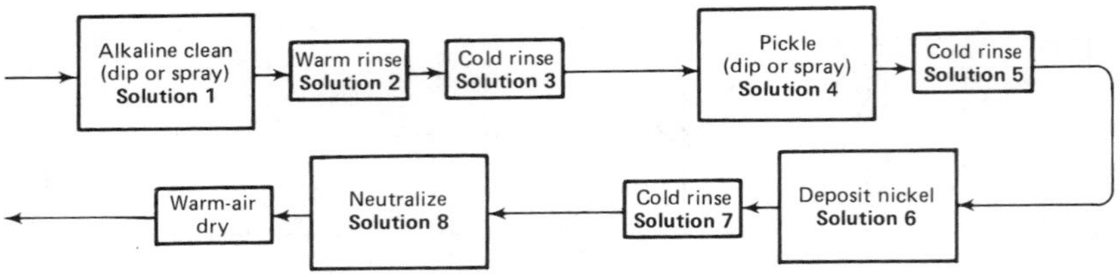

For dip or spray application

No.	Solution	Composition	Temperature °C	Temperature °F	Cycle time, min Dip	Cycle time, min Spray
1	Alkaline cleaner(a)	Cleaner, 15-60 g/L (2-8 oz/gal)(b)	Ambient to 100(c)	Ambient to 212(c)	6-12	1-3
2	Warm rinse	Water	49-60	120-140	½-4	½-1
3	Cold rinse	Water	Ambient		2-4	½-1
4	Pickle(d)	H_2SO_4, 6-8%	66-71	150-160	5-10	3-5
5	Cold rinse	Water, H_2SO_4(e)	Ambient		½-4	½-1
6	Nickel-deposition(f)	$NiSO_4 6H_2O$, 5.6-7.5 g/L (0.75-1.0 oz/gal)(e)	60-82	140-180	5-10	4-6
7	Cold rinse	Water, H_2SO_4(e)	Ambient		½-4	½-1
8	Neutralize	2/3 Na_2CO_3 and 1/3 borax, 0.60-2.10 g/L (0.008-0.28 oz/gal) as Na_2O	49-71	120-160	1-6	1-2

(a) For spray cleaning, use a two-stage process. (b) For spray cleaning, use 3.8 to 15 g/L (0.5 to 2.0 oz/gal). (c) 60 to 82 °C (140 to 180 °F) for spray cleaner. (d) Weight loss of metal is 3 to 5 g/m^2 (0.3 to 0.5 g/ft^2). (e) Solution pH, 3 to 3.5, to prevent formation of ferric iron. (f) Nickel deposit should be 0.2 to 0.6 g/m^2 (0.02 to 0.06 g/ft^2). Continuous filtration is commonly used to remove $Fe(OH)_3$

Metal Preparation

The bond and appearance of porcelain enamel depend on closely controlled cleaning and roughening of the metal surface. Complete removal of oil, sand, drawing compounds, weld oxide, and other surface contaminants is required. Steel may be prepared by chemical or mechanical procedures.

Preparation of Steel for Porcelain Enameling

Chemical Treatment. When chemical treatments are used, mechanized equipment is usually used in production operations. The parts are placed on corrosion-resistant racks and dipped in or sprayed with a series of solutions. The sequence of processing steps and the solutions used in conventional production operations are indicated in Fig. 3. After drying at 93 to 150 °C (200 to 300 °F), the parts have a light straw color.

When special low-carbon decarburized steel is direct-cover-coat enameled, it must be etched to remove a minimum of 22 g/m^2 (2 g/ft^2) of metal surface, and a nickel deposit of 0.9 to 1.3 g/m^2 (0.08 to 0.12 g/ft^2) of surface is required. Table 8 indicates modifications of the solutions and operating conditions for the acid pickling and nickel deposition cycles shown in Fig. 3 that have been used to provide the increased metal removal and to increase the amount of nickel deposited. The use of ferric sulfate etching solution to remove 22 g/m^2 (2 g/ft^2) of metal surface results in a better bond than that obtained after etching with sulfuric acid.

A modification of the ferric sulfate system called oxy-acid is also used. Oxy-acid etchant solution is a mixture of sulfuric acid, ferric and ferrous sulfates. Reactions involved are the same as with the ferric sulfate system; that is, the sulfuric acid and the ferric sulfate etch the metal, producing ferrous sulfate. The ferrous sulfate is then oxidized to ferric sulfate in the presence of sulfuric acid. The advantage of this system is that all of the reactions take place in one etching tank where two etching tanks (sulfuric acid and ferric sulfate) are required for the ferric sulfate method.

Table 8 Pickling and nickel-deposition solutions for preparing special low-carbon steel for direct cover coating(a)

Solution	Composition of solution	Operating temperature °C	°F	Cycle time, min Dip	Spray
Pickling solutions(b)					
1	H_2SO_4, 6-8 wt %	71	160	15-30(c)	8-15(c)
2(d)	H_2PO_4, 17-20 wt %	60	140	4-8	2-5
3(e) First stage	Ferric sulfate(f), 5 wt %	65	150	2-4	1½-3
Second stage	H_2SO_4, 6-8 wt %	65	150	2-4	1½-3
4 Oxy-Acid	H_2SO_4, 6-9 wt %; $Fe_2(SO_4)_3$, 3-5 wt %; $FeSO_4$, 3-20 wt %	74	165	1½-4	1½-3
Nickel deposition solution(g)					
5	$NiSO_4$, $6H_2O$, 7.5-9.4 g/L (1-1¼ oz/gal)(h)	60-71	140-160	3-8	3-8

(a) Except for the use of these solutions, preparation entails processing as indicated in Fig. 3. (b) Any of these solutions may be used in place of solution 4 in Fig. 3. Minimum metal removal required is 22 g/m² (2 g/ft²) of metal. (c) Cycle time may be reduced to that indicated in Fig. 3 for ground coats by oxidizing the metal 16 g/m² (1½ g of iron/ft²) at 680 °C (1250 °F) in an air atmosphere, or by blasting with sand or steel grit to remove metal prior to pickling. (d) Equipment containing lead or Monel cannot be used. (e) Equipment containing lead cannot be used. (f) Convert ferrous to ferric by adding hydrogen peroxide, sulfuric acid, and water. (g) Nickel deposit should be 0.6 to 1 g/m² (0.06 to 0.10 g/ft²). (h) pH of solution, 3.2 to 3.5. Sulfuric acid or sodium hydroxide is used to adjust pH. The addition of 0.3 to 0.8 g/L (1 to 3 g/gal) of sodium hypophosphite to solution will increase the rate of nickel deposition and permit the use of the lower end of the temperature range without excessive cycle time

Fig. 4 Process for preparing steel for "no nickel/no pickle" enameling

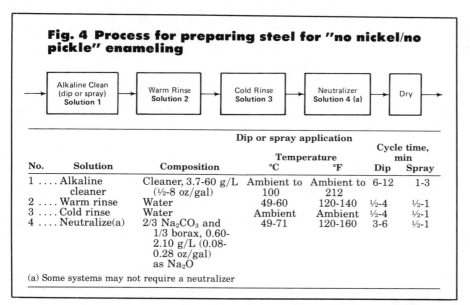

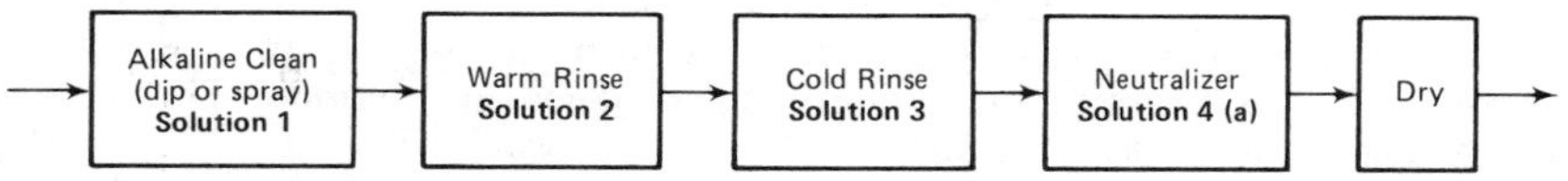

Dip or spray application

No.	Solution	Composition	Temperature °C	°F	Cycle time, min Dip	Spray
1	Alkaline cleaner	Cleaner, 3.7-60 g/L (½-8 oz/gal)	Ambient to 100	Ambient to 212	6-12	1-3
2	Warm rinse	Water	49-60	120-140	½-4	½-1
3	Cold rinse	Water	Ambient	Ambient	½-4	½-1
4	Neutralize(a)	2/3 Na_2CO_3 and 1/3 borax, 0.60-2.10 g/L (0.08-0.28 oz/gal) as Na_2O	49-71	120-160	3-6	½-1

(a) Some systems may not require a neutralizer

Metal preparation for "no nickel/no pickle" enameling requires at least the same amount of cleaning as conventional metal preparation does for conventional enamel. There is, however, no acid etching or nickel deposition required with the "no nickel/no pickle" system. An advantage of utilizing this system lies in reduction of wastewater treatment problems. Figure 4 illustrates the metal preparation cycle for the system.

Mechanical preparation consists of abrasive blasting using steel shot or steel grit. Grit or shot blasting is used on parts designed without pockets or crevices when configuration and thickness permit blasting without distortion. The flat areas of parts made with 16-gage, 1.52 mm (0.0598 in.) sheet steel or thinner distort excessively when cleaned by this method.

Abrasive blasting is used particularly for preparing hot rolled steel and parts that are to be enameled on one side only. The process is also used for preparing large parts and when enamels with poor bonding characteristics are used. Before blasting, oil and drawing compounds are removed by alkaline cleaning or by heating at 425 to 455 °C (800 to 850 °F) to burn off the organic contaminants.

Preparation of Cast Iron and Aluminum for Porcelain Enameling

Cast iron is prepared by blasting to remove adhering mold sand and the thin surface layer of chilled iron. Because the surface contains more combined carbon than is present in the remainder of the casting, it must be removed to prevent excessive evolution of gas during firing of the enamel.

Quartz sand of nearly spherical grains, propelled by compressed air, is commonly used for abrasive cleaning of cast iron; however, steel shot, steel grit, and chilled cast iron grit, propelled centrifugally from rotating wheels, are generally used for cleaning sanitary ware. Zircon sand and fused alumina grit are used for special purposes.

After blasting, inspect the casting for cracks, sand holes, slag holes, blowholes, fins, and washes. Fill cracks and larger holes by welding. After this, spot grind to blend the area with the surrounding surface. Remove fins and washes by grinding. The repaired casting is blasted a second time prior to enameling. Small holes need not be repaired by welding; usually, they are filled with a ceramic paste after final blasting.

Enamel cast iron within a few days after cleaning, especially during periods of high humidity, because even a very thin layer of rust reduces the adherence of the ground-coat enamel. Parts that have rusted excessively can be reconditioned by being heated to a red heat, cooled to room temperature, and abrasively cleaned.

Aluminum. The preparation of parts made of heat treatable aluminum alloys for porcelain enameling involves the removal of soil and surface oxide and the application of a chromate coating. Figure 5 shows the sequence of these surface preparation treatments and gives operating conditions. Final drying removes all surface moisture; drying must be accomplished without contaminating the cleaned surface of the aluminum. Parts made of non-heat treatable aluminum alloys require only the removal of soil, which can be done by alkaline cleaning or vapor degreasing.

The Porcelain Enameling Process

Several basic methods are used to apply the porcelain enamel to the base metal. Included are dipping, low-coating, electrodeposition, manual spray, electrostatic spray, and dry powder spray. The best method of application for a particular part is determined by quantity and quality requirements, the type of material being applied, units produced per hour, capital investment, labor cost, and ultimately, part cost.

Application techniques can be manual or mechanized. Mechanization is used for high-volume part requirements of the same or similar shape. Hand application is necessary if a variety of parts must go through the same process system.

Regardless of the method of application, good porcelain enameling techniques must be used to ensure uniform coverage in the areas requiring the porcelain enamel protection. Excessive thickness, beads, or pooling of the porcelain enamel reduce product quality, and the product is more prone to chip. Areas where the coating is too thin do not receive the full protection and decorative capabilities of the porcelain enamel.

Fig. 5 Process for preparing heat treatable aluminum alloys for porcelain enameling

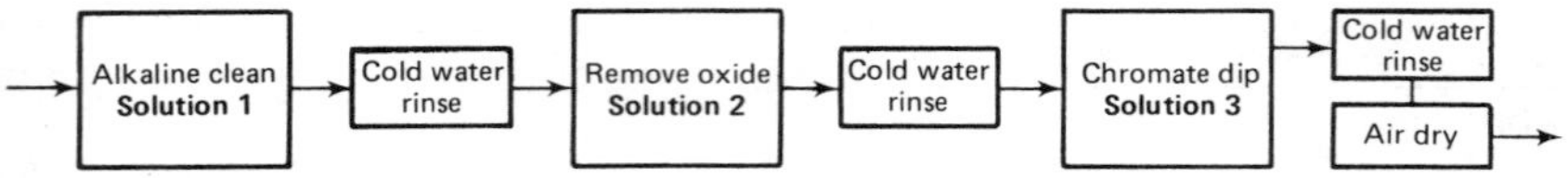

No.	Type	Composition of solution: Constituent	wt %	Operating temperature: °C	°F	Cycle time, min
1	Alkaline cleaner(a)	(b)	(b)	60-82	140-180	2-5
2	Oxide removal	Chromic acid	3.5	82	180	3-10
		Sulfuric acid	18.0			
3	Chromate dip	Chromic sulfate	0.2	Ambient		1-6
		Potassium dichromate	14.4	...		...
		Sodium hydroxide	7.75	...		...

(a) Vapor degreasing may be used instead of alkaline cleaning. (b) Either inhibited or mildly etching (uninhibited) cleaners can be used

Porcelain Enameling of Sheet Steel

The porcelain enamel slip is applied to sheet steel in a variety of ways. These include dipping, flow coating, spraying of wet slip, electrostatic powder spraying, and electrodeposition.

Dipping is widely used as a method to apply the porcelain enamel, particularly when both sides of the parts require coverage. Dipping can be used for both ground-coat application and cover-coat application. It is performed by immersing the part in the prepared porcelain enamel slip, then withdrawing it and allowing the excess material to drain from the part. Sometimes it is necessary to rotate, tilt, spin, or shake complex shapes to ensure uniform coverage. In areas where excessive porcelain enamel slip is retained on the part after draining, remove the excess with a siphon or a wiping device before the part is dried. Dip porcelain enamel films are normally applied at a thickness of 50 to 100 μm (2 to 4 mils) to provide adequate coverage.

Figure 6 shows a typical hand dipping, conveyorized draining setup for applying porcelain enamel slip to lids for automatic washing machines. Automatic dipping equipment is available for a dipping application when part configuration allows and production volume warrants the investment. The rheology of the porcelain enamel slip can be better controlled if the dip tank is constructed with a double wall (as shown in Fig. 6) to allow the insertion of the heat exchanger to maintain the porcelain enamel slip at a uniform temperature. The walls should slope toward a central sump to facilitate recirculation and drainage from the tank. Equipment to provide recirculation of the porcelain enamel slip consists of a diaphragm pump that continuously circulates the porcelain enamel slip through a magnetic separator and screens to remove iron particles and other foreign materials. The recirculation system helps maintain the consistency and uniformity of the porcelain enamel slip. The AISI 300 series stainless steels are the preferred materials for the tank, piping, separators, and screens.

Flow Coating. In flow coating, the porcelain enamel slip is flowed onto the surface of the part. The process is applicable to high-volume continuous operations for parts requiring the same porcelain enamel. In automatic flow coating, the parts are placed on hangers at the correct angle for draining and carried by conveyor through the flow coating chamber. The porcelain enamel slip is pumped at a high volume, 570 L/min (150 gal/min), and low pressure, 70 to 105 kPa (10 to 15 psi) through a series of nozzles that are directed at various areas of the part to ensure complete coverage. A schematic representation of a typical automatic flow coating chamber is shown in Fig. 7.

When flow coating a surface which has high appearance requirements such as the top of a kitchen range, it is necessary to incorporate a curtain of porcelain enamel slip near the exit of the flow coating chamber. This curtain, which is provided by flood plates (Fig. 7), flows out the porcelain enamel slip into an even coating free of drain lines and splatter marks. On emerging from the flow coating chamber, flow-coated parts drain in a manner similar to that of the dip method. Various devices can be added along the length of the conveyor to cause repositioning of the parts and provide a more even coating.

Another version of automatic flow coating involves the use of a constant-head tank to supply slip at a constant velocity to headers and nozzles which flood parts with slip as they are conveyed through the flow-coat chamber. The advantage of this system is that the flow to the nozzles is constant and not subject to variations present in pumped systems (see Fig. 8).

Automatic flow coating is favored over hand dipping because it offers (*a*) higher rates of production, (*b*) improved coating quality, and (*c*) reduced cost of the applied film. Control of the porcelain enamel slip and proper operation of the machine are important functions of flow coating. It is common practice to check the specific gravity and pickup of the porcelain enamel slip three to four times each hour. All parts of the flow coating machine which come in contact with the porcelain enamel

Fig. 6 Setup for applying ground coat to lids for automatic washing machines by hand dipping and conveyorized draining

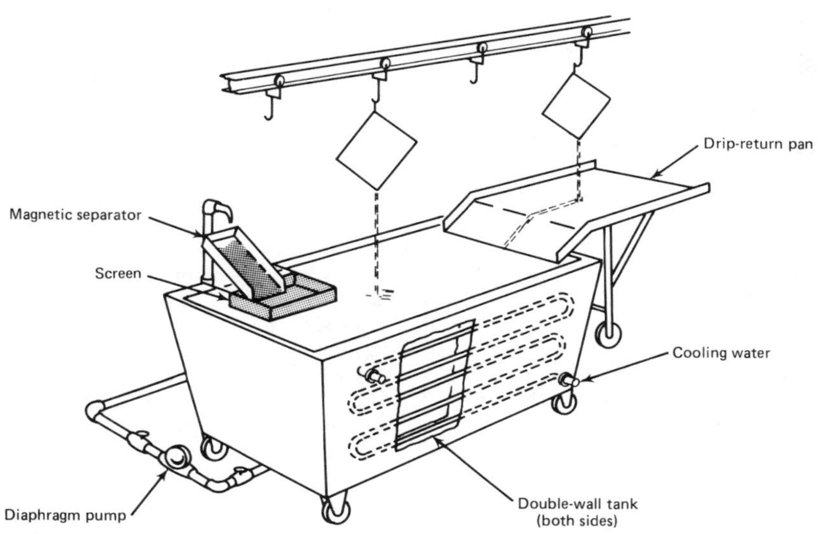

Fig. 7 Automatic equipment for applying enamel slip by flow coating

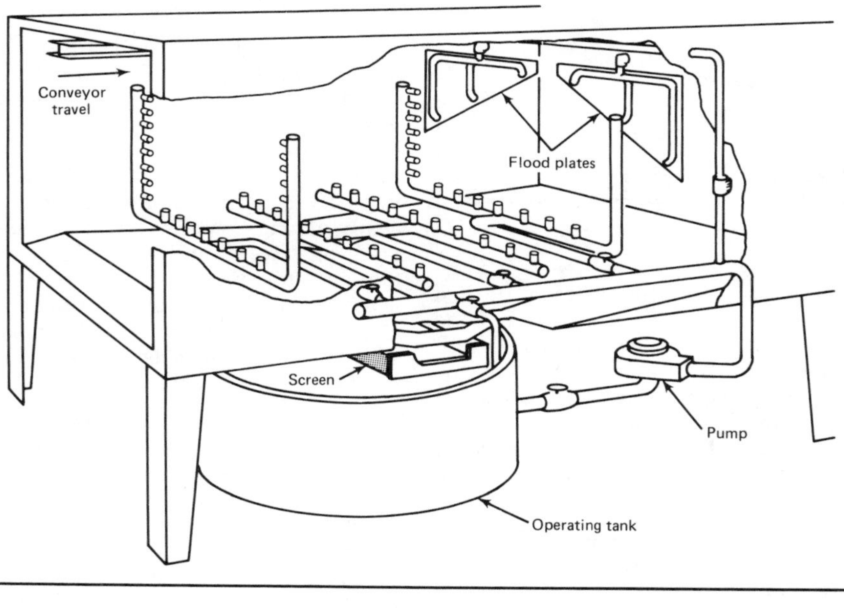

slip should be constructed of 300 series stainless steel. Additional details are available from the Porcelain Enamel Institute Bulletin, P-302.

Spraying. Spraying of the porcelain enamel slip is done primarily for one-side coverage. It is also used for reinforcing enamel bisque and for making repairs on enameled surfaces. Spraying is ideal for parts that are too large for hand or mechanical manipulation, particularly where service and appearance requirements permit no drain lines beading, or buildup of the porcelain enamel.

Figure 9 schematically illustrates equipment for the manual spray application of porcelain enamel slip. The prepared porcelain enamel slip is placed in the pressure feed tank and constantly agitated to keep the material homogeneous. Regulated air pressure is applied to the pressure feed tank at 55 to 125 kPa (8 to 18 psi) to force the porcelain enamel slip through the fluid hose to the tip of the spray gun nozzle, where the slip is atomized by clean compressed air, regulated at 170 to 415 kPa (25 to 60 psi), and then directed to the part. The amount of air pressure required depends on the specific porcelain enamel slip and the shape of the part being sprayed.

Wet electrostatic spraying of porcelain enamel is used to reduce losses in material by charging the porcelain enamel slip during atomization to a potential of 100 000 to 120 000 V. The electrostatically charged droplets are attracted to the grounded parts being sprayed. A well-operated electrostatic unit can deposit up to 85% of the sprayed material on the part as compared to 30 to 50% in conventional spraying operations. Additional details are available from the Porcelain Enamel Institute Bulletin P-301.

Electrostatic Powder Spray. Electrostatic powder spray is another method when a large volume of parts is being produced which require the application of the same porcelain enamel. The parts must be of the configuration that can be properly and evenly coated by this process. When these conditions are met, this is a very efficient method of applying porcelain enamel. Up to 99% of the material is utilized with little or no direct labor required for the application operation. Smooth running conveyors are required with this method of application to prevent loss of powder prior to firing the parts.

Powder is delivered to the spray guns from a feeder unit where it is diffused by clean compressed air into a fluid-like state. The fluidized powder is then siphoned by the movement of high-velocity air flowing through a venturi and is propelled through powder feed tubes to the spray gun. The powder feeder provides a steady, controlled flow of powder to the guns. Independent control of powder and air volume ensures the proper ratios to provide desired thickness coverage on the product.

The powder leaves the spray gun in the form of a diffused cloud, being

Fig. 8 Constant-head flow coat system

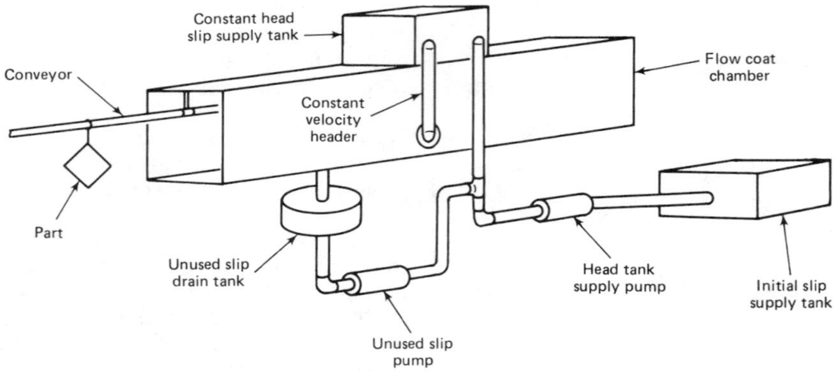

Head tank supply pump: maintains level of slip in constant-head tank. Constant velocity header: flows from constant-head tank to nozzles inside both sides of flow coat chamber. Initial slip supply tank: primary source of slip supply for constant-head tank

Fig. 9 Equipment for spraying cover-coat enamel slip on lids for automatic washing machines

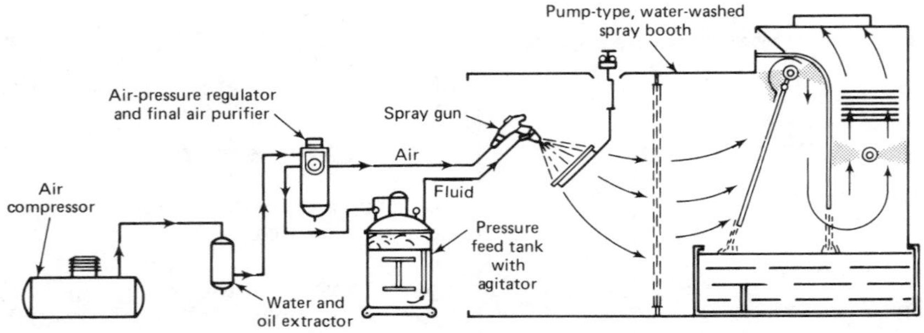

propelled toward the workpiece. A high-voltage, low-amperage power unit supplies current to the charging electrode, causing powder to seek out and attach itself to the grounded workpiece. Variable voltage allows the operator to overcome faraday cages and adjust for the type of powder being sprayed (by lowering voltage) to provide wraparound (by increasing voltage).

The recovery equipment booth serves to collect and return the powder that is not held on the workpiece; it moves through a closed loop system with the use of filters and final filters, so none of the airborne powder escapes into the environment.

Electrodeposition is another process which can be used to apply enamel to steel. The process uses a series of tanks in which the parts are submerged, and enamel is deposited electrophoretically. This process is basically limited to direct-on enameling, but can be considered for two-coat/one-fire applications. The main advantage of this system is very uniform appearance and exceptionally thin enamel layers.

Drying of Porcelain Enamel Slip. Parts coated with porcelain enamel slip are dried before firing to:

- Permit the application of additional porcelain enamel slip when required without disturbing the previously applied coating
- Permit brushing of the coated parts, if required
- Allow parts to be handled more easily for transfer to the holding fixture used during firing
- Reduce the amount of water vapor introduced into the firing furnace; high level of moisture cannot be tolerated in every instance

Coated parts may be air dried or placed in dryers using radiant heating or heated circulating air. Air drying is an inefficient procedure for production operations. Drying by radiant heating is at least 20% faster than by convection.

For convection dryers, the drying cycle consists of gradually increasing the air temperature to 120 °C (250 °F). A cycle time of 2 to 5 min is required to dry the coating completely in continuous dryers. Batch and intermittent dryers are used also; cycle time for these dryers ranges from 10 to 20 min, depending on the size of the load and the type of dryer used. There should be sufficient circulation of the air for uniform drying of the porcelain enamel slip. Moisture-laden air from both convection heating and radiant heating dryers must be exhausted from the dryers.

If drying is too rapid, a hard film forms on the coating and traps moisture under the film. This condition results in tearing in the finished enamel. The rate of drying can be varied by controlling the temperature or the humidity (dew point). Humidity generally is controlled by regulating the amount of outside air entering the dryer.

Drying can directly affect the bisque strength of the porcelain enamel film, which determines the relative ease with which the dried piece can be brushed or handled. In general, low dryer temperatures produce a film that is easily brushed, but which may be easily damaged when the piece is handled; on the other hand, high dryer temperatures produce a hard film that is easily handled but difficult to brush. Mill additions also affect the bisque strength and the brushability of the bisque.

The coating is absorbent during drying and collects gases that are present in the dryer. Most porcelain enamels do not tolerate the absorption of sulfur gases. Sulfur, from any source, causes a scum on the surface of the parts or pitting in the enamel surface. Therefore, dryers usually are indirect-fired to minimize the absorption of products of combustion of fuel gases.

During the initial stages of drying, the coating is still wet and is subject to contamination from dirt and other foreign particles that cannot be removed before firing. Therefore, drying must be accomplished in an atmosphere free of dirt, scale, and dust.

Auxiliary Coating Procedures. After drying and before firing, the coating may require reinforcing or brushing.

Reinforcing. The application of more slip to areas where the coating is of insufficient thickness usually is performed on parts after the application of ground coat by dipping or flow coating. The coating thickness must be built up to prevent burn-off. Reinforcing is accomplished by manual spraying, usually with a spray gun having a nozzle with a smaller opening than is used for full coverage spraying.

Because the additional porcelain enamel slip is applied immediately after drying while the part is still hot, exercise care to prevent an excessive amount of porcelain enamel slip from being deposited. The water in the porcelain enamel slip flashes off during spraying of the hot part and the reinforcing coat appears thin. Add black oxide to the reinforcing slip to hide streaks in the ground coat where the cover coat is to be brushed off.

Brushing usually is performed for one of three reasons: (*a*) to reduce the over-all thickness of the porcelain enamel and thus reduce the possibility of chipping, (*b*) to produce a texture or pattern, or (*c*) to produce an area of bare metal for electrical contact. Brushing consists of removing the dried porcelain enamel slip, called bisque, and can be performed on both ground-coat and cover-coat bisques.

Brushes are available in many shapes and sizes for various purposes. Edging brushes up to a 50-mm (2-in.) brushing margin are standard, as are certain sizes of bolt-hole brushes, tube brushes, and circular brushes.

Some of the problems encountered in brushing are (*a*) removal of the coating dust after brushing, (*b*) films being too hard to brush as a result of an excessively high drying temperature, and (*c*) brittle or soft films resulting in flaking off along a brushed edge.

When brushing of the cover coat is required, the ground coat must be smooth and free of pits; otherwise, the brushings from the cover coat fill the pits and cause the fired porcelain enamel to have a speckled appearance.

Porcelain Enameling of Cast Iron. Cast iron is porcelain enameled by the dry process or by the same wet process as used for enameling sheet steel.

In the dry process, a very thin coat of ground-coat enamel slip is applied to the cold casting, generally by spraying, but sometimes by dipping or other methods. After the ground coat is dry, the casting is put in a furnace and heated to a bright red heat. It is then withdrawn from the furnace and, while the casting is still hot, the cover coat, in the form of dry powder, is sprinkled by means of a vibrating sieve over the surfaces to be covered. The enamel melts as it falls on the hot surface. The application of powdered enamel continues until the temperature of the casting drops to the point at which the enamel will not melt. Then the piece is returned to the furnace and heated until the enamel is properly fused. For some types of products such as lavatories, one application of powdered enamel is sufficient, but other products may require several applications; bathtubs and combination sinks require two or more.

In the wet process, the enamel is applied to the part when it is cold. Handling of the part is easier, and it is easier to apply uniform coats of enamel than in the dry process. However, because of the composition of cast iron and because of casting irregularities, it is very difficult to enamel large articles such as sanitary ware satisfactorily by the wet process.

During the heating of an article made of gray iron, gas is evolved from surface reactions of carbon in the iron after the temperature reaches approximately 675 °C (1250 °F). The rate of gas evolution increases as the temperature increases until the completion of the reaction that produces the gas. Gases produce bubbles or blisters in any enamel whose maturing temperature is higher than the temperature at which the gas evolution begins.This varies with the composition and processing history of the iron. Flaws in the casting, such as sand holes and blowholes, act as focal points for the evolution of the gas and produce larger blisters at these areas than at areas where no flaws are present.

The dry process can be used for castings of almost any size or shape. The thin ground coat allows the gas to escape from the iron during the initial firing without forming large blisters. The ground coat is so thin that the bubbles break while they are still very small. Many castings that can be enameled by the dry process without difficulty would be severely blistered if coated by the wet process. The dry process also results in savings of fuel, labor for handling, floor space, and equipment for cooling the coated part.

Thicker coats of enamel can be applied by the dry process so that the normal roughness of iron castings can be hidden with one or two applications of cover-coat enamel. Thicker coats have an additional advantage in that they can be fired out to a finish that has more gloss and less waviness than is common when a wet process slip is applied by spraying.

Cast iron enameled by the wet process represents only a small portion of the total tonnage of enameled cast iron products. Some of these products are enameled with one coat, applied by the wet process, especially those that can be covered with a dark colored enamel whose maturing temperature is below the temperature of gas evolution.

Porcelain Enameling of Aluminum. Porcelain enamel slips for aluminum usually are applied by spraying, using either manual or automatic equipment with agitated pressure tanks. Slips for aluminum are not self-leveling and, therefore, must be deposited smoothly in an even thickness and without runs or ripples.

Many aluminum parts are coated satisfactorily by the one-coat/one-fire method. Although the heat treatable alloys can be recoated one or more times, the opacity and color of the coating change with the thickness of the porcelain and with repeated firing.

The desirable minimum fired enamel thickness is 65 μm (2.5 mils), and the desirable maximum 90 μm (3.5 mils).

Enameling Furnaces

Firing is accomplished in continuous, intermittent, or batch furnaces heated by oil, natural gas, propane gas, or electricity. With oil heating, a muffle furnace is used to prevent the products of combustion from contaminating the enamel coating. Gas-fired furnaces are either muffle, radiant-tube, or luminous wall types.

Continuous furnaces are of either straight-through or U-type design; furnaces of both designs use air curtains to prevent heat losses through the end openings. A spray-coating and firing installation using a U-type continuous furnace is illustrated schematically in Fig. 10. This setup is used for the application of white and colored porcelain enamel cover coats to sheet steel parts.

A laydown wire-mesh-belt conveyor is used for products such as small signs, dials, microcircuitry, and other flat pieces. Most continuous furnaces, how-

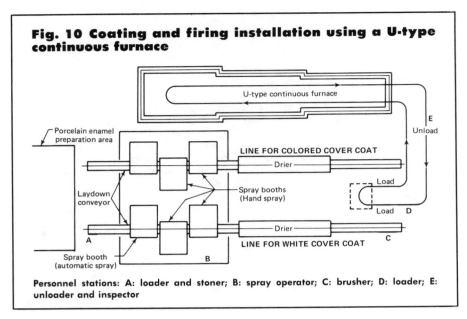

Fig. 10 Coating and firing installation using a U-type continuous furnace

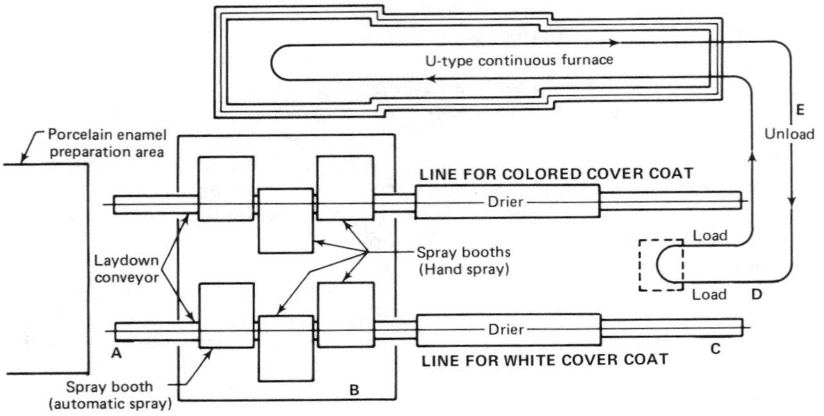

Personnel stations: A: loader and stoner; B: spray operator; C: brusher; D: loader; E: unloader and inspector

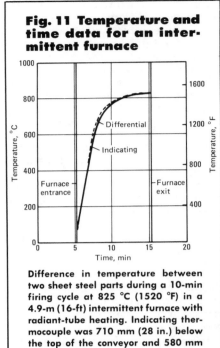

Fig. 11 Temperature and time data for an intermittent furnace

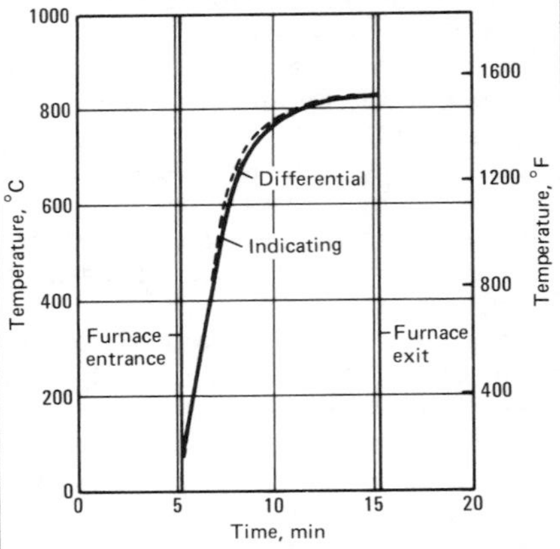

Difference in temperature between two sheet steel parts during a 10-min firing cycle at 825 °C (1520 °F) in a 4.9-m (16-ft) intermittent furnace with radiant-tube heating. Indicating thermocouple was 710 mm (28 in.) below the top of the conveyor and 580 mm (22.75 in.) above the differential thermocouple. Both thermocouples were attached to the parts

Table 9 Cycles for firing ground-coated and cover-coated sheet steel parts in a continuous furnace

Type of part	Gage of steel	Operating temperature(a) °C	°F	Firing time, min(b)
Architectural panels	16-22	805	1480	4-5
Home laundry equipment	18-22	805	1480	4-5
Water heater tanks	7-16	870	1600	8-12
Range ware	18-24	805	1480	3
Refrigerator liners	20-22	805	1480	2½-3
Sanitary ware	14-18	815	1500	5-8
Signs	16-22	805	1480	3-5

(a) Temperature varies with composition of frit. (b) Time in hot zone of furnace

ever, are equipped with overhead monorail conveyors which are located above the furnace roof. Alloy hook or drop rods extend down from the conveyor trolleys through a narrow slot in the furnace roof to transport the parts. Sealing of this slot is accomplished by the use of articulated steel or alloy seal plates carried by the conveyor. When over the furnace these plates ride on case slides on the furnace roof. Cycles for continuous-furnace firing of ground coats and cover coats on a number of different types of steel parts are given in Table 9.

Intermittent furnaces are equipped at both ends with split side-opening doors, air-cylinder operated, and have an overhead monorail conveyor similar to that of the continuous furnaces. With this type of furnace, however, the conveyor moves in increments so that when one load is discharged, a new load enters the furnace where it remains until the firing cycle is completed. Doors and conveyor are electronically interlocked so that the firing cycle can be timer controlled.

Figure 11 illustrates the small difference in temperature between the center and bottom of an intermittent 5-m (16-ft) radiant tube furnace set at 825 °C (1520 °F) for firing a load of sheet steel parts.

Two types of batch furnaces are used. One type has a vertical-lift single door through which the load is charged and withdrawn by means of a charging fork. The second type has a slot in the roof and a manually operated overhead conveyor. The load for this furnace design is supported by alloy rods from overhead trolleys and is manually pushed into the furnace for firing. Table 10 shows cycles for firing ground and cover coats of acid-resistant enamels on cast iron parts in a batch furnace.

Current technology for either retrofitting existing furnaces or erecting new porcelain enameling furnaces, both continuous and batch units, uses a thin wall lining, typically 150 to 205 mm (6 to 8 in.) of ceramic fiber in conjunction with radiant tubes in both the preheat and hot zones. This permits very short heat-up and cool-down periods with no resulting refractory damage. Consequently, these furnaces can be turned off during nonproduction hours which substantially contributes toward achieving dramatic fuel savings.

Forced convection is the preferred method of heating furnaces for firing porcelain enamel on aluminum. The heat is provided by electric package heaters, quartz-tube electric heaters, or metal-sheath heaters, all specially designed for operation at high ambient air temperature. Quartz-tube and metal-sheath heaters are adapted to the furnace so that radiant heat is available in the firing zone along with forced circulation. Package heaters are placed remote from the firing zone; this is the most effective method of eliminating direct radiation and hot spots. Heat imparted to the work from the package heater is derived completely from adequate air circulation to maintain a temperature uniformity throughout the furnace of ±1% of the nominal operating temperature.

Table 10 Cycles for firing cast iron parts in a batch furnace

Part	Type of enamel	Firing of ground coat: Time, min(a)	Temperature °C	Temperature °F	Sieving time, min	Application of cover coat, Melting: Time, min	Temperature °C	Temperature °F
Bathtub	Acid resistant	9-15	950-970	1740-1780	2¼-2½	2½-3	855-890	1575-1640
Sink (40 lb)	Acid resistant	5-15	930-940	1700-1730	½-¼	1¼-1½	855-890	1575-1640
Sink (80 lb)	Acid resistant	7-15	940-955	1730-1750	1½-2	2-2½	855-890	1575-1640

(a) Time, in minutes, required to fire the ground coat and for the cast iron part to attain proper temperature for subsequent application of the cover coat

Table 11 Cycles for firing porcelain enamel on aluminum

Type of part	Section thickness, 0.025 mm (0.001 in.)	Firing time, min	Firing temperature °C	°F
Any configuration	26-40	5-6½	540	1000
Any configuration	51-64	7-8	540	1000
Extrusions	125	10	550	1020
Extrusions	187 up	12-15	550	1020

Forced-convection heating is also accomplished with gas-fired radiant tubes as the heat source. The tubes are baffled from the work or firing zone so that air circulation provides the same advantages as in electric-package forced-convection heating.

Furnace construction for aluminum enameling generally requires the use of stainless steel inner liner sheets, low density wall insulation, and plain carbon steel exterior shell. This type of fabrication eliminates long heat-up and cool-down periods.

Firing of enamel on aluminum is accomplished between 525 and 550 °C (980 and 1020 °F); cycles are shown in Table 11. To control the color and gloss of the enamel within acceptable limits, the temperature throughout the work must be held to ±1½ °C (±2½ °F).

Process Variables

Thickness of the applied layer of porcelain enamel, firing time, and firing temperature markedly affect the properties of the coating. Increasing the thickness of the coating increases the resistance to burn-off and produces truer colors; however, thin coatings have the greatest flexibility.

Coating Thickness. The optimum thickness of porcelain enamel depends on the substrate metal and the service requirements of the part. On aluminum, porcelain enamel is applied to produce a fired enamel thickness ranging from about 65 to 125 μm (2.5 to 5 mils). A tolerance of ±13 μm (±0.5 mil) is required for a white enamel coating 115 μm (4.5 mils) thick, to maintain uniform opacity.

On sheet steel, a ground coat about 50 to 100 μm (2 to 4 mils) thick is used to promote adhesion. To cover the ground coat, a very opaque white or pastel cover coat about 100 to 150 μm (4 to 6 mils) thickness is required. Thus, a two-coat system on these products has a thickness ranging from 150 to 255 μm (6 to 10 mils). Brightly colored porcelain enamels are produced by applying less opaque coats with more saturated colors over a white intermediate coating. For these the total thickness of the coating system usually ranges from 255 to 380 μm (10 to 15 mils) and is sometimes as much as 635 μm (25 mils). Some decorative finishes are textured; the thickest parts of these coatings are 635 μm (25 mils) thick. A thickness of more than 635 μm (25 mils) is usually undesirable. Coating thickness is about 125 to 150 μm (5 to 6 mils) for cover coat porcelain enamels applied directly to decarburized or specially stabilized steels.

Coatings for cast iron products are much thicker than those for sheet steel or aluminum. Dry process coatings on cast iron products such as sanitary ware range from 1020 to 1780 μm (40 to 70 mils) in thickness. Coatings applied by the wet process are thinner than dry process coatings; wet process coatings range from 255 to 635 μm (10 to 25 mils) in thickness.

Coating thickness for hot water tanks normally ranges from 150 to 230 μm (6 to 9 mils) with 150 μm (6 mils) a generally accepted minimum thickness. Heat exchanger surfaces, depending on end use, are sometimes double coated for added durability.

Uniform thickness of the enamel coating depends on the method of application and on the configuration of the part. Figure 12 shows the distribution of the thickness of enamel on steel washing machine tubs, steel sinks, and cast iron lavatories. The ground coat on the washing machine tubs is applied by dipping or spraying, the cover coat by spraying. Spraying may be manual or semiautomated. Hand spraying is used to apply both the ground coat and the cover coats to the sinks. The cast iron lavatories are enameled with a spray-applied ground coat about 115 μm (4.6 mils) in thickness, after which the cover coat is applied to the hot castings by being dusted through a sieve.

The thickness of the porcelain enamel on a large part of simple configuration can be closely controlled when application is by a mechanical spraying system that is adapted to the part. For example, mechanical-applied porcelain enamel on curved silo panels 2 by 3 m (5 by 9 ft), can be maintained within ±13 μm (±0.5 mil); however, when application is by hand spraying, the variation in enamel thickness is ±50 μm (±2 mils).

An enamel thickness of 65 to 180 μm (2.5 to 7 mils) is desirable for aluminum architectural panels. When white or light colored enamel is used, however, the enamel thickness ranges above 75 μm (3 mils), to produce acceptable opacity. Two coats with a total thickness of about 125 μm (5 mils) result in more uniform opacity than one coat 125 μm (5 mils) thick. Additional details are available from the Porcelain Enamel Institute Bulletin P-302.

Firing Time and Temperature. Firing of porcelain enamel involves the flow and consolidation of a viscous liquid and the escape of gases through the coating during its formation. Within limits, time and temperature are varied in a compensating manner. For example, similar properties and appearance develop when firing liners for household refrigerators at 805 °C (1480 °F) for 2½ min or at 790 °C (1450 °F) for 4 min. In all instances, there is a minimum practical temperature for the attainment of complete fusion, acceptable adherence, and desired appearance. Most ground-coat enamels for high production steel parts exhibit acceptable properties over a firing range of 55 °C (100 °F) at an optimum firing time. However, control within 11 °C (20 °F) is ordinarily maintained to produce uniform appearance and allow interchangeability of parts. As

Fig. 12 Distribution of thickness of porcelain enamel on steel washing machine tubs, steel sinks, and cast iron lavatories

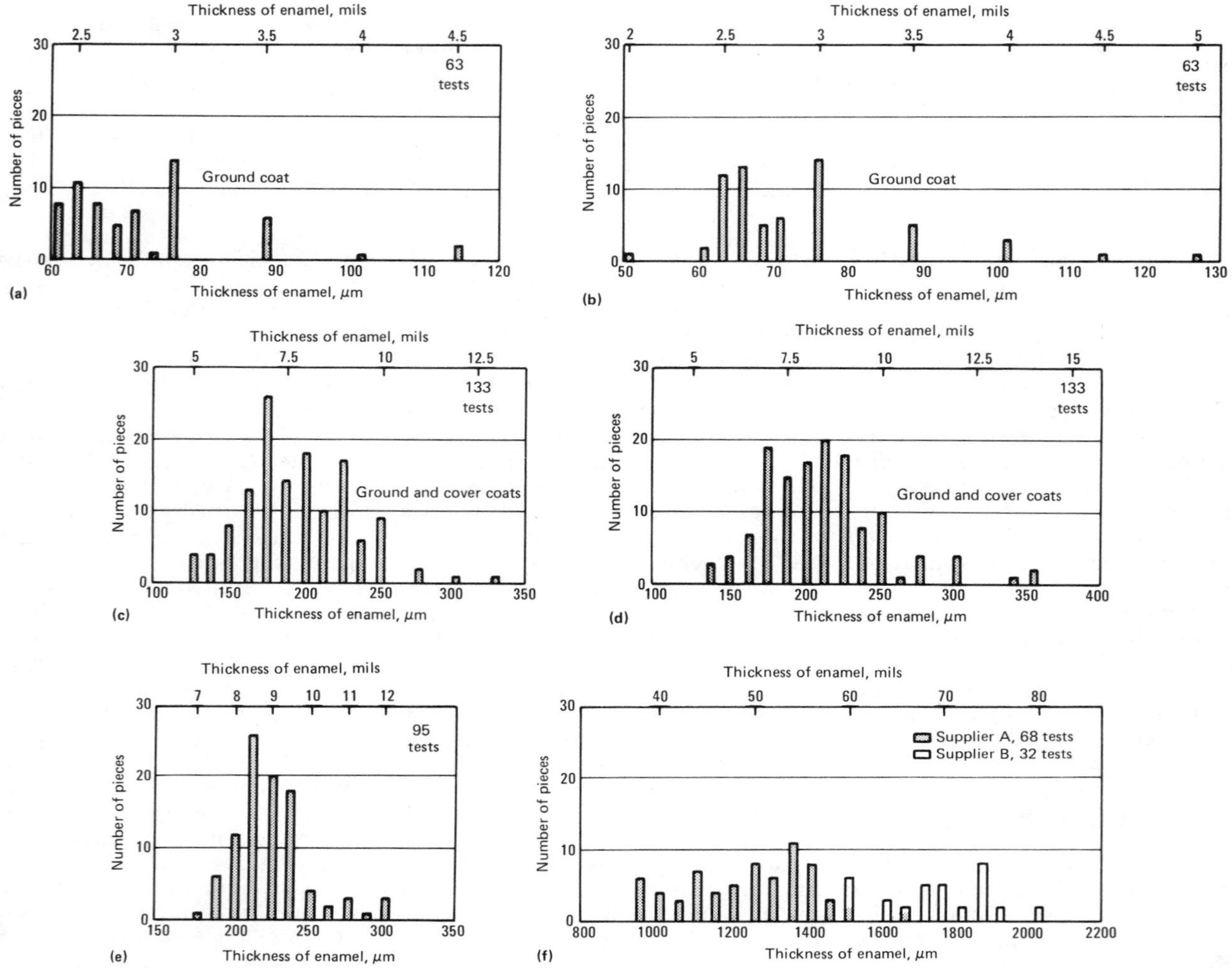

(a) Washing machine tubs, inside surface. Ground coat, 63 tests. (b) Washing machine tubs, outside surface. Ground coat, 63 tests. (c) Washing machine tubs, inside surface. Ground and cover coats, 133 tests. (d) Washing machine tubs, outside surface. Ground and cover coats, 133 tests. (e) Sinks, ground and cover coats, 95 tests. (f) Lavatories, ground and cover coats

the combined effects of firing time and temperature increase, resulting in more thorough firing, up to a maximum, the following conditions occur:

- Colors shift dramatically, particularly reds and yellows. In general, white and colors shift toward yellow. Usually, furnace temperature is changed to achieve minor adjustments in color matching
- Gloss of the enamel coating increases
- Chemical resistance of the enamel coating increases
- Gas bubbles are eliminated
- Enamel coating becomes more dense and brittle and less resistant to chipping
- Maximum adherence is attained in the optimum portion of the firing range

Color Matching and Color Control. In color matching, primary coloring oxides are used; although preblended oxides for a specific color are available, they are more difficult to adjust. In most instances, two or three oxides are sufficient to match any specific color. Use a minimum number of oxides; for example, a stable green oxide is preferred to a blend of blue and yellow oxides. Usually, the proper color intensity is obtained first, then adjustment is made for the desired color shade. Cadmium-sulfoselenide pigments, for red and yellow, are generally used with cadmium-stabilized clear frits.

Color stability can be adversely affected by improper mill additions. However, a color with only fair stability may be improved by the proper mill additions, and minor color adjustments, particularly white, are possible.

Sometimes gum must be used to control bisque strength; sodium nitrite and urea are used to control tearing. Because these additions have a marked effect on some colors, they should be used in initial color matching.

Finer grinding reduces the intensity of the color. It is imperative that the fineness of the milled color be controlled within specified limits. Milling is usually stopped before completion to permit sample firing of the enamel and comparison with a color standard. Adjustments to the mill may then be made. Color may be controlled to some extent by variations in fineness of grinding. The thickness of the fired enamel coating affects many colors. In general, thick coatings produce lighter colors, thin coatings result in darker colors.

The set and specific gravity of a colored enamel slip are important to the finished results. Mottling or color separation is possible if the colored enamel is applied too wet or too dry. Color corrections of electrostatic dry powder cannot be made by the enameler.

Flatness and Distortion. Sag and distortion of sheet steel parts result from low metal strength at the firing temperature, thermal stresses due to nonuniform heating and cooling, and transformation to austenite. Changes in design of the parts and firing practice alleviate the first two causes, and the use of extra low carbon content or of special stabilized steels minimizes transformation to austenite.

Ground-coat enamels have a limited effect on distortion of parts because their coefficients of thermal expansion approach that of steel. When ground coat is applied to both sides of the metal, there is a counter balancing of expansion and contraction stresses.

The effect of cover-coat enamels on the configuration and flatness of porcelain enameled parts can be pronounced as a result of low coefficients of expansion and one-side application. The likelihood of distortion is greatly increased when multiple or thicker coats of cover-coat enamels are necessary on one surface. Sometimes cover coats must be applied to the back side of parts to equalize the stresses.

Adjustments in the firing cycle can sometimes help to minimize distortion. A cycle with relatively slow heating and cooling rates is preferable to rapid heating and cooling.

Variations in the method of supporting the work during firing can often change the sagging characteristics to an appreciable degree. Furnace supports and fixtures can be designed to distribute the load and equalize heating and cooling rates. Porcelain Enamel Institute Bulletin P-306 discusses design and fabrication of sheet steel parts for porcelain enameling.

Process Control

Proper workability during application of wet process porcelain enamels depends on (*a*) control of the porcelain enamel slip, particularly with respect to stability of suspension; (*b*) weight of enamel slip deposited and retained per unit area; (*c*) specific gravity, consistency, and particle size of the enamel slip, and (*d*) stability on aging at ambient temperature.

Stability of suspension, or the ability of the various mill additions to keep the milled frit in suspension, is determined by both slip measurements and visual observation of any separation that occurs; the enameler should note the accumulation of clear liquid on top of the enamel or of a heavier sludge on the bottom.

Stability of the suspension is a function of many factors, but is usually controlled by the quantity of colloid, in the form of clay or bentonite, and electrolytes used to deflocculate the clay. Enamel slips for aluminum have a shorter shelf life than those for sheet steel.

Pickup weight of enamel and retained per unit area is measured by draining the enamel on a flat or cylindrical shape of known weight and area and actually weighing the pickup of enamel in wet or dry form. This is a most useful test, particularly for dipping enamels, and one that closely simulates actual production operations. During the test, the operator can observe any tendency toward sliding, excessively long or short drain time, and variations in setting time. The pickup of an enamel is a function of specific gravity, colloid content, total salts content and type, and consistency. These are controlled by varying the water content, addition of salts, and fineness of grind.

Specific gravity of enamel slips is measured either by weighing a known volume in comparison with the weight of an equivalent volume of water or by the use of a hydrometer. Control of specific gravity is almost entirely a function of the ratio of water to solids. To ensure uniformity, testing for specific gravity is required for the preparation of all porcelain enamel slips.

Consistency of a porcelain enamel slip for spraying is commonly determined by the slump test. In this test, a fixed volume of the porcelain enamel slip is allowed to flow out suddenly in a circular pattern on a calibrated plate, and the diameter of the resulting pool is measured immediately. This is a simple and useful test for porcelain enamels to be sprayed because it indicates uniformity of slip conditions between various millings. Other tests for consistency involve the use of viscosimeters of various types, including those that use the rotational, flow, and falling piston methods. However, enamel slips do not behave like ordinary liquids and do not follow the laws of viscous flow. With enamel slips, a certain amount of force is required to start the flow. This force is the yield value, and the rate of flow is referred to as the mobility. These can be determined with use of a consistometer; however, in plant practice careful control of specific gravity and slump should be adequate. A study of the flow properties of porcelain enamel slips indicates that they are non-Newtonian liquids and consequently show variable rates of shear with varying stresses.

For porcelain enamel slips applied by dipping, a measure of drain time is a useful test. Drain time is the total elapsed interval between the time a standard size sample plate is removed from a container of well-stirred porcelain enamel slip and the time at which the draining motion of the slurry on the sample has stopped.

Particle size of the frit for porcelain enamel slips is commonly determined by standard screen analysis. Reproducible measurements are easily obtained when a standardized shaking device is used. The particle size of frit is important to the suspension characteristics of the porcelain enamel slip, and slight solubility of the frits shows a major change with variation in the size of the particles.

Stability toward aging of porcelain enamel slips is measured by exposing a tested sample of the enamel to whatever temperatures are expected in normal service. Exposure is for many hours and days, and retests of the critical properties are made at intervals during testing. Aging usually has an effect on stability of the suspension pickup, setting time, and the consistency of the porcelain enamel slip.

Aging causes bubbly glass and poor surface quality of the fired enamel. Leaching of soluble elements such as sodium or boron from the frit is a cause of aging. This problem is encountered more with less water-resistant frits. The effect is greater at higher temperatures.

In-Process Repairs

Complexity of processing limits the probability of producing 100% acceptable parts in one processing cycle. Reprocessing of 2 to 25% of parts is common, depending on severity of the specifications on appearance and uniformity of coverage.

Repair of Ground Coat. Parts rejected at the ground-coat stage are repaired and refired, if necessary, before the cover coat is applied. It is good practice to confine repair techniques to the defective area and not to grind or stone adjacent surfaces indiscriminately.

Remove dirt particles, scale, and similar contaminants with a sharp-pointed instrument. Stone disturbed area around the site lightly with an alundum rubbing stone. The dust generated in this operation should be blown off. Ground-coat enamel is spotted-in at the repair area, and the entire piece is lightly dust coated. The piece is then dried and fired.

Lumps, handling defects, chips, and similar flaws are usually stoned if not too severe and then cover coated. Otherwise, they are treated as described above.

Defects that cover more extensive areas or are located at or near the steel-enamel interface, such as burn-off, blisters, copperheads, embedded grit particles, and dents, require different repair techniques because it is often necessary to grind down to bare steel. Grinding may be accomplished with a power disk grinder, using first a 24-grit silicon carbide disk and then an 80-grit silicon carbide disk. A damp sponge placed under the work area prevents overheating which adversely affects the quality of repair.

After firing, the repaired ground-coat area is hand stoned to blend it into the surrounding area. Do not stone the repaired area if blisters are present in the ground coat.

Repair of Cover Coat. Defects that are missed during inspection of the ground coat usually become visible in the cover coat. Specks, blisters, copperheads, and dents are typical of such transmitted defects, and they are removed by grinding into the ground coat. The ground coat is then repaired as described above, before another cover coat is applied. Cover-coat defects such as lumps, handling defects, chips, and thin coating areas are repaired by stoning, respraying lightly, and refiring the piece.

Direct cover-coat applications are repaired in much the same manner as ground coats. If the defect area is small, a light grinding or stoning is all that is required. This is followed by a localized application of cover coat and then a full dust coat, drying, and firing. If the repair requires grinding into the steel, then a ground coat patch in that area is required for adherence. A half coat of cover-coat enamel over the fired ground-coat patch is required, followed by a dust coat over the entire piece.

Design for Porcelain Enameling

The glasslike nature of porcelain enamel and the high firing temperatures used in its application impose limitations on the design of articles to be enameled to ensure that finished work is within dimensional tolerances and durable enough for intended service.

In general, the size of products that can be porcelain enameled is limited only by the ability of facilities to accommodate them. Table 12 indicates the maximum dimensions of workpieces that can be enameled in several types of conventional facilities.

Steel Parts. The size relations for steel sheet given in Table 13 are recommended as a guide to provide adequate flatness, rigidity, and sag resistance.

Distortion during firing is minimized by uniformity of stress and temperature during the cycle. Nonuniform thicknesses heat at different rates and cause distortion and variations in enamel maturity. Specific aspects of design that should be considered are:

- Flatness requirements in a porcelain enameled part must be maintained at all processing stages. Wavy or buckled areas in the original sheet and those introduced in forming and handling persist in the fired part
- Bend and corner radii should be at least 4.7 mm (3⁄16 in.). In drawn parts, the use of symmetrical embossed ridges and panels increases resistance to distortion caused by uneven residual forming stresses
- Flanges increase strength and flatness, but they can cause irregular stresses in firing. Flanges on one side only may require welded braces. Flanges meeting at a corner must be welded there and should not vary in depth by a factor of more than three
- Cutouts should have round corners; and when cutouts are located on flange edges, at least 6.3 mm (1⁄4 in.) of flange should remain
- Welded lugs and ears for attachment and assembly result in double metal thickness. They should be of the same or lighter gage as the main part, and as small as possible
- Spot and seam welds also result in double metal thickness. Spot welds are difficult to enamel because of movement during firing and entrapment of solutions used in preparing the metal for enameling. Flatten seam welds to prevent burrs, rough projections, and protruding edges, all of which enamel poorly
- Fusion welds must be free of crevices and oxide seams. Weld spatter must be removed to avoid coating defects
- Sharp edges are subject to burn-off of enamel in which the iron oxide formed during firing exceeds the amount of iron oxide that is soluble in the coating. The tight, matte-finish layer that results is protective, but may not meet appearance or severe corrosion requirements
- Holes are designed up to 1.5 mm (1⁄16 in.) oversize to allow for enamel buildup. To avoid sharp areas on the

Table 12 Maximum dimensions accommodated by enameling facilities

Type of enameling facility	Maximum dimensions of workpiece: Length, m	Length, ft	Width, m	Width, ft	Height, m	Height, ft
Dial, small sign, or art	0.3	1	0.3	1	0.08	1⁄4
Range or laundry equipment	1.2	4	0.8	2½	1.2	4
Refrigerator	1.2	4	0.8	2½	1.8	6
Architectural; job shop	3.7	12	0.6	2	1.5	5
Special; chemical	18	60	4.6	15	4.6	15
Cast iron, dry process	1.8	6	0.9	3	0.8	2½
Aluminum	7.3	24	1.5	5	0.2	½

Table 13 Suggested sizes of steel sheet for adequate flatness, rigidity, and sag resistance

Sheet gage and width	Maximum total area m²	ft²
24-gage sheet		
150 mm (6 in.) wide	0.05	½
305 mm (12 in.) wide	0.3	3(a)
455 mm (18 in.) wide	0.5	5(a)
22-gage sheet		
150 mm (6 in.) wide	0.09	1
305 mm (12 in.) wide	0.3	3½
455 mm (18 in.) wide	0.5	6(a)
610 mm (24 in.) wide	0.7	8(a)
20-gage sheet		
150 mm (6 in.) wide	0.1	1½
305 mm (12 in.) wide	0.5	5
455 mm (18 in.) wide	0.7	8(a)(b)
610 mm (24 in.) wide	0.9-1.4	10-15 (a)(b)

(a) Should be embossed, flanged or otherwise reinforced. (b) All parts exceeding this area should be made from 18-gage 1.21 mm (0.0478 in.) or heavier thicknesses of steel sheet

appearance side of punched holes, the holes may be extruded inward, allowing maximum coverage on the face side, but intensifying the burn-off on the reverse side
- Fasteners should be provided with flexible washers or gaskets to distribute stresses and prevent crazing of the enamel
- Processing requires some means of attachment for handling, usually by holes which may be required also for draining of metal preparation solutions and excess enamel slip

For further reference, see Porcelain Enamel Institute Bulletin P-306.

Cast Iron Parts. Uniformity of section, simplicity of design, and the minimizing of lugs and braces are desirable characteristics for cast iron parts to be enameled. Radii of curved sections should be as generous as design limitations permit. The minimum radius may be 6.3 mm (¼ in.) for decorative beading on a flat or slightly curved surface, or as large as 38 mm (1½ in.) for one of the components of a compound curve on a large casting.

Aluminum Parts. Because enamel ordinarily is applied to aluminum to only about half the thickness to which it is applied to steel, freedom from surface scratches, burrs, and irregularities is doubly important for aluminum. Most shaping of aluminum is done before enameling, but the thin coating permits some bending, shearing, punching, and sawing of the enameled part.

Surfaces to be enameled should have generous inside radii of not less than 4.8 mm (3⁄16 in.). Surfaces should have outside radii of not less than 1.6 mm (1⁄16 in.); 3.2 mm (⅛ in.) is preferred for dark colors, and 3.2 mm (⅛ in.) is preferred for light colors. Weld attachments to the unenameled back side of enameled heavy-gage aluminum sheet or extrusions. The visible metal surfaces must not be overheated; overheating causes the aluminum to blister and alters the color and gloss of the enamel. Welding can be done before enameling, provided the weld area is cleaned properly before coating.

Properties of Porcelain Enamels

Enamels are prepared to ensure satisfactory properties for specific environments, consistent with ease of processing and minimum cost.

Chemical Resistance. Porcelain enamel is extensively used because of its resistance to household chemicals and foods. Mild alkaline or acid environments are generally involved in household applications. Table 14 presents examples of corrosive environments for which porcelain enamels are widely used for long periods of service. Special enamels or glass compositions are available to resist most acids, except for hydrofluoric or concentrated phosphorics, to temperatures of 230 °C (450 °F). These glasses also resist alkali concentration to pH 12 at 93 °C (200 °F).

Weather Resistance. The important factors that determine the weather resistance of porcelain enamels are chemical durability, color stability, cleanability, and continuity of coating. Gloss and enamel texture do not necessarily affect weather resistance.

Appearance for Indoor Exposure. Where corrosive attack is unlikely to limit the life of a given part, and attractive appearance is the principal requirement, enamel selection and processing are directed toward providing reproducible color matching and optimum gloss and smoothness. Frequently, different enamels are used on different parts of the same product to assure the best balance of properties and cost, particularly if high volume is involved. For example, range tops and sidepanels or refrigerator food-compartment liners and crisper pans utilize somewhat different enamel compositions because of differing property and appearance requirements, even though processed in the same plant. Appearance standards, in particular, are established according to the component and its use and location as the end product. Small surface defects may be tolerated in areas not heavily used or readily seen in the finished part, provided they do not affect basic serviceability.

Parts exhibiting defects may be re-enameled one or more times as required, but this increases the enamel thickness and consequently decreases resistance to chipping; therefore, every effort is made to achieve realistic standards and high quality on the first enameling cycle.

Service Temperatures. Softening of the glassy matrix limits the temperature to which porcelain enamels can be exposed. Softening releases gases remaining from reactions between the enamel and ferrous metal, producing random defects known as reboil. Service-temperature limits for porcelain enamels are shown in Table 15.

Thermal shock intensifies the effect of elevated temperature, as does operation under severe temperature gradients. Enamels are formulated so that expansion characteristics place the enamel in compression under service conditions. If combinations of mechanical stress and elevated temperature place the enamel in tension, crazing forms a pattern of fine cracks perpendicular to the tensile stress. The relation between enamel and metal expansion patterns on heating and cooling is shown in Fig. 13.

Mechanical Properties. The hardness of porcelain enamels ranges from 3.5 to 6.0 on the Mohs scale. Porcelain enamels show a high degree of abrasion resistance. Under abrasive test conditions where plate glass retains 50% specular gloss, porcelain enamel compositions retain from 35 to 85% specular gloss. Subsurface abrasion resistance varies with processing variables that affect the bubble structure of the enamel, that is, gas bubbles frozen in during cooling of the enamel. A decrease in abrasion resistance occurs with an increase in the number or size of gas bubbles. Enamel compositions are available containing crystalline particles in mill additions or by devitrification heat treatment that increase abrasion resistance as much as 50%.

The porcelain enamel coating contributes to the strength of sheet metal parts. Table 16 indicates the increased resistance to torsion provided by porcelain enamel on metal angles made

Table 14 Applications in which porcelain enamels are used for resistance to corrosive environments

Application	Corrosive environment: Temperature °C	°F	pH	Corrosive medium
Bathtubs	To 49	To 120	5-9	Water; cleansers
Chemical ware	To 100	To 212	12	Alkaline solutions
	To 100	To 212	1-2	All acids except hydrofluoric
	175-230	350-450	1-2	Concentrated sulfuric acid, nitric acid, and hydrochloric acid
Home laundry equipment	To 71	To 160	11	Water; detergents; bleach
Range exteriors	21-66	70-150	2-10	Food acids; cleaners
Range oven liners, conventional	66-315	70-600	2-10	Food acids; cleaners
Range burner grates	66-590	70-1100	2-10	Food acids; cleaners
Refrigerators	18-66	0-70	2-10	Food acids; cleaners
Kitchen sinks	To 71	To 160	2-10	Food acids; water; cleansers
Water heaters	To 71	To 160	5-8	Water

Table 15 Service-temperature limits for porcelain enamels

Service temperature °C	°F	Limiting conditions
425	800	Usual limit for enamels maturing at about 815 °C (1500 °F)
540	1000	Maximum for enamels maturing at about 815 °C (1500 °F), without reboil
760	1400	Operating limit for special high-temperature enamels
1095	2000	Refractory enamels useful for short periods for protection of stainless steels and special alloys

Source: Porcelain Enamel Institute Bulletin E-6

of three different materials. Metal failure occurs at the point of maximum stress, which is followed by buckling of the angle.

Electrical Properties. Porcelain enamels are electrical insulators. The electrical resistance per unit area is a function of thickness and enamel composition. In addition to resistance, usually expressed as resistivity, other electrical properties of interest are dielectric constant, dissipation factor, and dielectric strength.

As with many glassy materials, these properties vary with temperature. In general, as the temperature increases the resistivity and dielectric strength decrease, while the dielectric constant and dissipation factor increase. The dielectric constant and the dissipation factor also vary with frequency.

When porcelain enamel is used for its electrical properties, the selection of the enamel composition and the enameling process requires careful attention. For electronic applications, such as porcelain enameled substrates for hybrid circuits, special electronic grade enamels are used. These electronic grade compositions have considerably higher resistivities and dielectric strengths and are less sensitive to temperature changes than conventional porcelain enamels. Such specialty porcelain coatings are currently being used increasingly in sophisticated electronic circuitry.

Evaluation of Porcelain Enameled Surfaces

Specifications and quality control for porcelain enamel coatings require the evaluation of a range of properties depending on the intended service of the porcelain enameled product. Although material and process variables can be brought into approximate control using small test panels, process control is maintained by the evaluation of finished parts, even though the mechanical and chemical tests entailed are destructive.

Standard test procedures are available for most porcelain enamels. A listing of specific test methods for various properties is given in Table 17. Some of these properties and tests are discussed in the following paragraphs.

Adherence refers to the degree of attachment of enamel to the metal substrate. While none of the adherence tests in common use gives the force per unit area required to detach the enamel by tensile force normal to the interface, various tests aimed to evaluate adherence are regularly used in the industry.

The standard adherence tests for porcelain enamel on steel are ASTM C313, "Adherence of Porcelain Enamel and Ceramic Coatings to Sheet Metal", and PEI Bulletin T-29, "Test for Adherence of Porcelain Enamel Cover Coats Direct to Steel". ASTM C313 is applicable only to steel substrates between 0.4 mm (0.016 in.) and 2 mm (0.087 in.) in thickness. PEI T-29 applies to direct-on cover coats on substrates with a thickness range of 0.7 mm (0.0284 in.) to 1.3 mm (0.0508 in.).

Both adherence tests for porcelain enamel on steel include deforming the metal and measuring the area from which the porcelain enamel is removed. The indicator of adherence is the adherence index, which is the ratio of the porcelain enamel remaining in the deformed area to that in the same measured area prior to deformation.

Fig. 13 Enamel/metal expansion and stress patterns

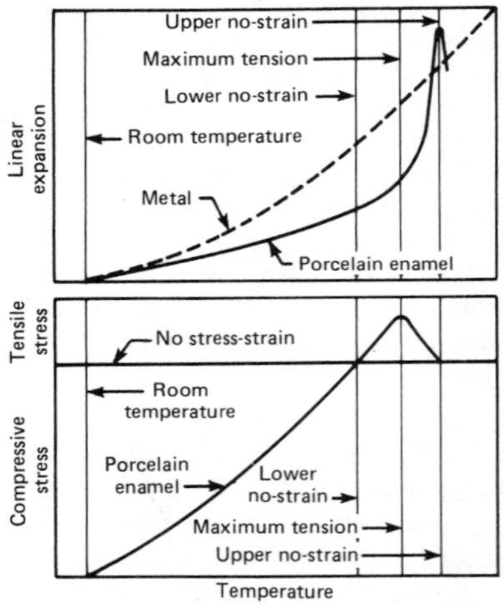

Linear expansion and stress in a composite of acid-resistant cover-coat porcelain enamel and low-carbon steel

Enamels for cast iron pose a special problem because of the relatively greater thickness and rigidity of the metal substrate and the brittleness of the iron. Here, simple unstandardized impact tests are used.

Resistance to spalling, a defect characterized by separation of porcelain enamel from the base metal without apparent external cause, is the indicator used to measure adherence of porcelain enamel on aluminum. Spalling

Table 16 Effect of porcelain enamels on torsion resistance of metal angles

Material	Increase of stress, %(a) Ground coat	One coat	Two coats
Cold rolled steel	27	18	33
Electric steel	22	13	32
Enameling iron	22	16	42

(a) Increase of stress for metal failure, as provided by porcelain enamel, over stress for failure of annealed unenameled material

Table 17 Test methods, specifications, and standards for porcelain enamels

Designation(a)	Title of test, specification, or standard
Adherence	
ASTM C313	Adherence of Porcelain Enamel and Ceramic Coatings to Sheet Metal
ASTM C703	Spalling Resistance of Porcelain Enameled Aluminum
PEI T-29	Adherence of Porcelain Enamel Copper Coats Direct To Steel
Thickness	
ASTM C664	Thickness of Diffusion Coating
ASTM D1186	Dry Film Thickness of Non-Magnetic Organic Coatings Applied on a Magnetic Base, Measurement of
ASTM E376	Coating Thickness by Magnetic Field or Eddy-Current (Electromagnetic) Test Methods, Rec. Practice for Measuring
Color and gloss	
ASTM C346	Gloss of Ceramic Materials, 45 deg Specular
ASTM C540	Image Gloss of Porcelain Enamel Surfaces
ASTM E97	Reflectance Factor of Opaque Specimens by Broad-Band Filter Reflectometry, 45 deg, 0 deg Directional
ASTM C347	Reflectivity and Coefficient of Scatter of White Porcelain Enamels
ASTM D2244	Color Differences of Opaque Materials, Instrumental Evaluation of
ASTM D1535	Color by the Munsell System, Specifying
ASTM C538	Color Retention of Red, Orange and Yellow Porcelain Enamels
Chemical resistance and weather characteristics	
ASTM C282	Acid Resistance of Porcelain Enamels (Citric Acid Spot Test)
ASTM C614	Alkali Resistance of Porcelain Enamels
ASTM C756	Cleanability of Surface Finishes
ASTM C283	Boiling Acid, Resistance of Porcelain Enameled Utensils to
ASTM D1567	Detergent Cleaners for Evaluation of Corrosive Effects on Certain Porcelain Enamels, Testing
ASTM C872	Lead and Cadmium Releases from Porcelain Enamel Surfaces
Chipping resistance	
ASTM C409	Torsion Resistance of Laboratory Specimens of Porcelain Enameled Iron and Steel
Abrasion resistance	
ASTM C448	Abrasion Resistance of Porcelain Enamels

(continued)

(a) ASTM, American Society for Testing and Materials; PEI, Porcelain Enamel Institute; ANSI, American National Standards Institute

can result from the use of improper alloys or enamel formulations, incorrect pretreatment of the base metal, or faulty application and firing procedures.

ASTM C703, "Spalling Resistance of Porcelain Enameled Aluminum", outlines two methods for determining resistance to spalling. Method A, using a 5% solution of ammonium chloride, requires a 96-h immersion of the test specimen. Method B uses a 1% solution of antimony trichloride and requires a 20-h immersion of the test specimen.

The spall test is a pass/fail test, with failure determined by (*a*) the existence of spall areas of specified dimensions at specimen edges, or (*b*) the existence on the specimen interior of spots exceeding specified dimensions or a spot level exceeding a specified density, usually spots/m^2 (spots/ft^2).

Thickness. The specifications for architectural porcelain enamel (PEI S-100 for steel and PEI ALS-105 for aluminum), the specification for porcelain enameled aluminum cookware (PEI ALS-106), and the ANSI standards for porcelain enameled formed steel plumbing fixtures (ANSI A112.19.4) and porcelain enameled cast iron plumbing fixtures (ANSI A112.19.1M) all require a specific thickness for the porcelain enamel coating.

The procedure used for measurement of the porcelain enamel thickness depends on the type of base metal used. For porcelain-on-steel products, enamel thickness is measured through procedures outlined in ASTM D1186, "Standard Method for Measurement of Dry Film Thickness of Non-magnetic Organic Coatings Applied on a Magnetic Base". For porcelain-on-aluminum products, coating thickness is measured according to test procedures specified in ASTM E376, "Recommended Practice for Measurement of Coating Thickness by Magnetic Field or Eddy Current (Electromagnetic) Test Methods". In some cases, primarily laboratory investigations, measurement of porcelain enamel thickness is accomplished by using the procedures specified in ASTM C664, "Thickness of Diffusion Coatings".

Color and Gloss. Porcelain enamel finishes are produced in literally hundreds of colors and many textures. This capability provides the manufacturers of appliances, cookware, outdoor grills, architectural panels, signs, decorative products, and many other applications with unusual design versatility as well as desirable performance properties.

The common method of specifying color is based on the capacity of the observed article to reflect light of different wave lengths and hence of different colors. A physical standard such as a plaque of porcelain enameled steel is provided as the color to be matched within stated limits. The difference in color between the control standard and the test specimen can be measured by a color difference meter, using the procedures specified in ASTM D2244, "Color Difference of Opaque Materials, Instrumental Evaluation".

Gloss, which is defined in ASTM C286, as "the shine or luster of porcelain enamel", is particularly desirable in some products such as appliances. However, high-gloss enamels capable of reflecting distinct images are not recommended for architectural porcelain enamel for exterior use.

Gloss of porcelain enamel may be measured by following procedures from two primary ASTM standards: ASTM C346, "Standard Test Method for 45%

Table 17 (continued)

Designation(a)	Title of test, specification, or standard
Thermal shock	
ASTM C385	Thermal Shock Resistance of Porcelain Enameled Utensils
Tests related to preparation of coatings and substrates	
ASTM C374	Fusion Flow of Porcelain Enamel Frits (Flow-Button Methods)
ASTM C539	Linear Thermal Expansion of Porcelain Enamel and Glaze Frits and Ceramic Whiteware Materials by the Interferometric Method
ASTM C285	Sieve Analysis of Wet-Milled and Dry-Milled Porcelain Enamel
ASTM C839	Compressive Stress of Porcelain Enamels by Loaded-Beam Method
ASTM C715	Nickel on Steel for Porcelain Enameling by Photometric Analysis
ASTM C810	Nickel on Steel for Porcelain Enameling by X-Ray Emission Spectrometry
ASTM C632	Reboiling Tendency of Sheet Steel for Porcelain Enameling
ASTM C694	Weight Loss of Sheet Steel During Immersion in Sulfuric Acid Solution
ASTM C774	Yield Strength of Enameling Steels After Straining and Firing
ASTM C660	Production and Preparation of Gray Iron Coatings for Porcelain Enameling
Tests related to continuity of coating	
ASTM C536	Continuity of Coatings in Glassed Steel Equipment by Electrical Testing
ASTM C743	Continuity of Porcelain Enamel Coatings
Specifications	
PEI S-100	Specification for Architectural Porcelain Enamel on Steel for Exterior Use
PEI ALS-105	Recommended Specifications for Architectural Porcelain Enamel on Aluminum for Exterior Use
PEI ALS-106	Recommended Specifications for Porcelain Enamel Finishes on Aluminum Cookware
WH-196-J	Federal Specification — heater, water, electric and gas fired residential. (This covers resistance of porcelain enamels to hot water under "Solubility of Glass Lining.")
Plumbing fixtures standards	
ANSI A 112.19.4	Porcelain Enameled Formed Steel Plumbing Fixtures
ANSI A 112.19.1M	Enameled Cast Iron Plumbing Fixtures

(a) ASTM, American Society for Testing and Materials; PEI, Porcelain Enamel Institute; ANSI, American National Standards Institute

Specular Gloss of Ceramic Materials", and ASTM C540, "Image Gloss of Porcelain Enamel Surfaces". The ASTM test chosen for evaluating gloss depends on the purpose for which the porcelain enamel is used. For example, appliances are evaluated according to ASTM C346 because they are required to have a relatively high gloss; architectural porcelain enamel is evaluated primarily according to ASTM C540 because it is not expected to reflect distant images.

Acid Resistance. Porcelain enamels can be formulated to exhibit high resistance to all acids except hydrofluoric or concentrated phosphoric. This resistance to food acids and certain chemical cleaners is of particular importance to manufacturers of ranges, refrigerators, and other kitchen and household appliances, as well as plumbing fixtures and various industrial and chemical processing products.

The acid resistance of enamels under boiling conditions is important in cookware. Similarly, acid resistance is a major consideration for architectural porcelain enamel components, because extensive testing under controlled exposure conditions has shown a distinct correlation between acid resistance and weatherability.

There are two standard tests for determining acid resistance: ASTM C282, "Acid Resistance of Porcelain Enamels", commonly referred to as the citric acid spot test, and ASTM C283, "Resistance of Porcelain Enameled Utensils to Boiling Acid", generally known as the boiling acid test. The citric acid spot test is used as a testing and grading system for such porcelain enamel applications as appliances, plumbing fixtures, and architectural products; evaluation is based upon visual examination following wet or dry rubbing and blurring-highlight tests. The boiling acid test is designed primarily for cookware applications, and the results are expressed in weight loss per unit area.

Alkali Resistance. Home laundry equipment, dishwashers, and other porcelain enamel applications where the surface is normally exposed to an alkaline environment at elevated temperatures require an alkali-resistant coating.

The standard test for alkali resistance is ASTM C614, "Alkali Resistance of Porcelain Enamels", which covers the measurement of the resistance of a porcelain enamel to a hot solution of tetrasodium pyrophosphate. Alkali resistance is expressed as mg/cm^2 ($mg/in.^2$) in terms of weight loss for the area exposed to the test solution.

Weather Resistance. The long-term weatherability of porcelain enamel is of key interest to specifiers and manufacturers of architectural porcelain enamel products for exterior use. Actual weathering performance of the material has been conducted and documented in a lengthy series of on-site-exposure tests conducted by the National Bureau of Standards in cooperation with the Porcelain Enamel Institute.

Weathering of porcelain enamel is evaluated in terms of the changes in gloss and color which occur during outside exposure. Changes in gloss are measured using the procedures specified in ASTM C346, "Gloss of Ceramic Materials, 45° Specular", while weather-induced color changes may be measured through procedures described in ASTM D2244, "Instrumental Evaluation of Color Differences of Opaque Materials". Weathering tests of up to 30 years show porcelain enamels have considerable inherent gloss and color stability, with rates of change influenced primarily by (*a*) enamel compositions; (*b*) choice of colors, with reds and yellows being most susceptible to color change; and (*c*) severity of the exposure site (seacoast and corrosive industrial environments have been found to have the most aggressive effect).

Spalling of enamels during exposure to weather occurs if improper materials and processing procedures have been used. Spalling resistance of weathered porcelain enamel on aluminum, for instance, is best ascertained through use of ASTM C703, "Spalling Resistance of Porcelain Enameled Aluminum".

In exposure tests, good correlation between acid resistance as determined by ASTM C282, the acid spot test, and the color retention of steel enamels has been observed. However, use of the acid spot test has shown an even stronger reliability in predicting color change in porcelain-on-steel. The correlation between acid resistance and color retention in porcelain-on-aluminum is somewhat less definitive but still observable. For prediction of the weatherability of red, orange, and yellow porcelain enamels, ASTM C538, "Color Retention of Red, Orange and Yellow Porcelain Enamels", is the best indicator.

Because architectural porcelain enamel for exterior use is subjected to weathering for long periods of time, both the specifications for porcelain enamel on aluminum (PEI ALS-105) and porcelain enamel on steel (PEI S-100) require compliance with specific levels of acid resistance, which are considered indicators of good weatherability.

Resistance to Organic Solvents. Porcelain enamels are inert towards all common organic solvents. There is no standard procedure for determining the resistance of porcelain enamel to common organic solvents.

Resistance to Chipping. Relatively thick layers of porcelain enamel cannot be subjected to severe bending or other substrate deformation without fracture. However, thin coatings of 125 μm (5 mils) or less that are well bonded to a relatively thin metal substrate, for example 26 gage 0.4546 mm (0.0179 in.) or less, can withstand the bending of the substrate to radii of curvature within its elastic limit and to return to the original shape with little or no apparent damage. Chipping of typical porcelain enamel on sheet iron occurs at about the strain required for permanent deformation of the base metal.

Torsion Resistance. In transit and during service, porcelain enameled ware is subjected to distortion through bending, twisting, or a combination of both. This can result in coating failures resulting from fractures originating at the outer surface and normal to the tensional stress. ASTM C409, "Torsion Resistance of Laboratory Specimens of Porcelain Enameled Iron and Steel", covers the evaluation of the relative resistance to torsion-induced failure of laboratory test specimens, which may include steel substrate thicknesses ranging from 24 to 12 gage.

Laboratory test data developed from ASTM C409 have been found to correlate with service and transit failure data. Since enamel thickness has a significant effect on torsion resistance, evaluation is primarily expressed in terms of the degree of twist that causes failure for a given enamel of given thickness.

Abrasion Resistance. The test for determining the resistance of porcelain enamel to various types of abrasion is ASTM C448, "Abrasion Resistance of Porcelain Enamels". The test consists of three parts; the first determines the resistance to surface abrasion of porcelain enamels for which the unabraded 45° specular gloss is more than 30 gloss units. Here the specular gloss of the specimens is measured before and after a specified abrasive treatment of the surface; the percentage of the original specular gloss that is retained is the surface abrasion index.

The second part of this test determines the resistance to surface abrasion of porcelain enamels for which the unabraded 45° specular gloss is 30 gloss units or less. The weight loss by a specific abrasive treatment—modified by an adjustment factor for each abrasive tester, lot of abrasive, and lot of calibrated plate glass standards used—results in an adjusted weight loss value. This is recorded as the index of resistance to surface abrasion.

The third and final portion of the test measures the resistance of porcelain enamels to subsurface abrasion. In this case, the scope of the linear portion of the abrasion time-weight loss curve is determined and then multiplied by an adjustment factor associated with each abrasion tester, lot of abrasive, and lot of calibrated plate glass standards used. The adjusted scope is taken as an index of the resistance to subsurface abrasion.

Thermal Shock Resistance. Thermal shock resistance of a porcelain enamel varies inversely with the thickness of the enamel. It is also affected by the compressive stress present in the enamel at room temperature. Most porcelain enamels can be quenched in ice water from a temperature of 205 °C (400 °F) without thermal shock failure. Some porcelain enamels specifically designed for resistance to thermal shock will not fail when quenched in ice water from 650 °C (1200 °F).

There is a standard for evaluating the durability of porcelain enameled utensils when subjected to thermal shock, ASTM C385, "Thermal Shock Resistance of Porcelain Enameled Utensils". In this test, cooking utensils are subjected to a series of dry heating and quenching cycles until the utensil fails by removal of the enamel from the utensil.

Continuity of Coatings. Ensuring continuity of coating after manufacture is important in porcelain enamel or so-called glassed steel applications where a prime purpose of the coating is to protect the substrate against corrosion.

There are two principal test methods used to determine either discontinuity of coverage or potential discontinuity through too-thin coverage. For glassed steel equipment, the prescribed test procedure is ASTM C536, "Continuity of Coatings in Glassed Steel Equipment by Electrical Testing". For conventional porcelain enameled products such as appliances, plumbing fixtures, and architectural panels, the test most commonly used is ASTM C743, "Continuity of Porcelain Enamel Coatings". Both tests essentially involve the use of electrical probes of relatively high voltage to discern either discontinuities in the coating or insufficient coverage for coating integrity in service use.

Resistance to Hot Water. Federal specification W-H-196 J, "Heater Water, Electric and Gas Fired, Residential", specifies a solubility test to determine the resistance of porcelain enamels to hot water. In this test, 89 by 89 mm (3½ by 3½ in.) sections cut from the outer wall of a water heater are tested at a rolling boil for 8 cycles of 18 h each in a special apparatus containing 400 mg of reagent-grade sodium bicarbonate dissolved in one litre of water. The resistance of the porcelain enamel is specified in terms of weight loss in milligrams per square inch exposed area. The weight loss allowed by W-H-196 J is 2.3 mg/cm^2 (15 mg/in.^{2}).

Definitions of Terms Relating to Porcelain Enameling (ASTM C286)

aging. The storing of porcelain enamel slips or powders before use. The change occurring in slips or powders with the lapse of time

ball mill. In porcelain enamels, a dense ceramic-lined rotating cylinder in which ceramic materials are wet or dry ground, generally using pebbles or porcelain ball as grinding media

bisque. A coating of wet-process porcelain enamel that has been dried, but not fired

blister. A defect consisting of a bubble that forms during coating fusion and remains when the porcelain enamel solidifies

boiling. A defect visible in the fired porcelain enamel that may take the form of numerous blisters, pinholes, black specks, dimples, or spongy surface

bubble structure. Size and spatial distribution of voids within the fired porcelain enamel

chipping. Fracturing and breaking away of fragments of a porcelain enameled surface

color oxide. A material used to impart color to a porcelain enamel

continuity of coating. The degree to which a porcelain enamel or ceramic coating is free of defects such as bare spots, boiling, blisters, or copperheads, that could reduce its protective properties

copperhead. A defect occurring in sheet metal ground coat that appears as a small freckle or pimple-like spot, reddish brown in color

cover coat. A porcelain enamel finish applied and fused over a ground coat or directly to the metal substrate

deflocculating. Thinning the consistency of a slip by adding a suitable electrolyte

delayed fishscaling. A fishscaling defect that occurs after the final porcelain enamel processing

dipping. The process of coating a metal shape by immersion in slip, removal, and draining. In dry process enameling, the method of coating by immersing the heated metal shape for a short time in powdered frit

direct fire. A method of maturing porcelain enamel so that the products of combustion come in contact with the ware

draining. The part of the dipping or flow coating process in which the excess slip flows from suitably positioned ware

drain time. Time required for porcelain enamel slip applied by dipping, slushing, or flow coating to complete movement across the surfaces of the coated part

dredge, dredging. In dry process enameling, (1) the application of dry, powdered frit to hot ware by sifting. (2) The sieve used to apply powdered porcelain enamel frit to the ware

dry process enameling. A porcelain enameling process in which the metal article is heated to a temperature above the maturing temperature of the coating, usually 870 to 955 °C (1600 to 1750 °F); the coating materials are applied to the hot metal as a dry powder and fired

dry weight. The weight per unit area of the bisque

film strength. The relative resistance of the bisque to mechanical damage

fineness of enamel. A measurement of the degree to which a frit has been milled in wet or dry form, usually expressed in grams residue retained on a certain mesh screen from a 50-cm^3 or a 100-g sample

firing. The controlled heat treatment of ceramic ware in a kiln or furnace to develop the desired final properties

firing time. The period during which the ware remains in the firing zone of the furnace to mature the coating

fishscaling. A defect appearing as small half moon-shaped fractures somewhat resembling the scales of a fish

flocculating. The thickening of the consistency of a slip by adding a suitable electrolyte

flow coating. The process of coating a metal shape by causing the slip to flow over its surface and allowing it to drain

flux. A substance that promotes fusion in a given ceramic mixture

frit. The small friable particles produced by quenching a molten glassy material

glass. A term sometimes used for porcelain enamel or frit

glass-coated steel, glass-lined steel, glassed steel. Designations generally applied to a class of porcelain enamels that has high resistance to chemical attack at elevated temperatures and pressures

gloss. The shine or luster of a porcelain enamel

ground coat. (1) A porcelain enamel applied directly to the base metal to function as an intermediate layer between the metal and the cover coat. (2) On sheet steel, a porcelain enamel coating containing adherence-promoting agents which may be used either as an intermediate layer between the metal and the cover coat or as a single coat over the base metal

mill addition. Any of the materials added to the ball mill charge of a frit

one-coat ware, one-coat work. (1) Articles finished in a single coat of porcelain enamel. (2) Sometimes a contraction of one-cover coat ware, in which the finish consists of a single cover coat applied over ground coat

opacifier. A material that imparts or increases the diffuse reflectance of porcelain enamel

opacity. The property of reflecting light diffusely and nonselectively; properly defined in ASTM Method C 347, "Test for Reflectivity and Coefficient of Scatter of White Porcelain Enamels", under the term **contrast ratio**

orange peel. A surface condition characterized by an irregular waviness of the porcelain enamel resembling an orange skin in texture; sometimes considered a defect

pickup. The amount of slip retained per unit area on dipped ware

pinhole, pinholing. (1) In sheet steel enameling, a porcelain enamel surface defect characterized by a depression which looks like it is made by a pin. (2) In cast iron enameling, a defect characterized by a small conical hole that apparently extends through the porcelain enamel to the base metal

pit. A defect similar to a dimple but slightly smaller

porcelain enamel. A substantially vitreous or glassy, inorganic coating bonded to metal by fusion at a temperature above 425 °C (800 °F)

primary boiling. The evolution of gas during the initial firing of porcelain enamel; sometimes a defect

reboiling. Gas evolution occurring and recurring during repeated firing of the ground coat; sometimes a defect

reclaim. Overspray that is removed from the spray booth and reconditioned for use

sagging. (1) A defect characterized by a wavy line or lines appearing on those surfaces of porcelain enamel fired in a vertical position. (2) A defect characterized by irreversible downward bending in an article insufficiently supported during the firing cycle

sanitary ware. Porcelain enameled ware such as sinks, lavatories, and bathtubs

screen test. A standard test for fineness of porcelain enamel slip or powder

scumming. A defect characterized by areas of poor gloss on the surface of porcelain enamel

set. A flow property of porcelain enamel slip affecting the rate of draining, residual thickness, and uniformity of coating

setting-up agent (set-up agent). An electrolyte used to increase the measured pickup of a slip

slip, slurry. A suspension of finely divided ceramic material in liquid

slump test. A test to determine consistency of slip whereby measurement is made of the spreading of a specified volume of slip over a flat plate

smelt. A specific batch or lot of frit

smelter. A furnace in which the raw materials of the frit batch are melted

spall, spalling, or spontaneous spalling. A defect characterized by chipping that occurs without apparent external causes

stippled finish. A pebbly textured porcelain enamel, often multicolored

tearing. A defect in the surface of porcelain enamel, characterized by short breaks or cracks which have been healed

wet milling. The grinding of porcelain enamel materials with sufficient liquid to form a slurry

wet process enameling. A method of porcelain enameling in which slip is applied to a metal article at ambient temperature, dried, and fired

SELECTED REFERENCES

- Alloy, Design and Fabrication Considerations For Porcelain Enameling Aluminum, Bulletin P-402, Porcelain Enamel Institute, Inc., Washington, D. C.
- A. I. Andrews, *The Preparation, Application and Properties of Enamels*, 2nd ed., Champaign: Garrard Press
- Ball Mill Wet Grinding of Slips for Porcelain Enameling, Bulletin P-305, Porcelain Enamel Institute, Inc., Washington, D. C.
- Design and Fabrication of Sheet Steel Parts For Porcelain Enameling, Bulletin P-306, Porcelain Enamel Institute, Inc., Washington, D. C.
- Enamel Preparation, Application and Firing for Porcelain Enameling Aluminum, Bulletin P-404, Porcelain Enamel Institute, Inc., Washington, D. C.
- Manual of Dipping and Flow Coating for Porcelain Enameling, Bulletin P-302, Porcelain Enamel Institute, Inc., Washington, D. C.
- Manual of Spraying for Porcelain Enameling, Bulletin P-301, Porcelain Enamel Institute, Inc., Washington, D. C.
- 1939 Exposure Test of Porcelain Enamels on Steel, 30-Year Inspection, Building Science Series 38, National Bureau of Standards
- 1982 Annual Book of ASTM Standards, Part 17, Refractories, Glass, Ceramic Materials; Carbon and Graphite Products
- Preparation of Sheet Steel for Porcelain Enameling, Bulletin P-307, Porcelain Enamel Institute, Inc., Washington, D. C.
- Pretreatment of Alloys for Porcelain Enameling Aluminum, Bulletin P-403, Porcelain Enamel Institute, Inc., Washington, D. C.
- Quality Control Procedures for Porcelain Enameling Aluminum, Bulletin P-405, Porcelain Enamel Institute, Inc., Washington, D. C.
- Weathering of Porcelain Enamels on Aluminum, 12-Year Inspection, Exposure period 1964-1976, Department of Ceramic Engineering, University of Illinois at Urbana-Champaign, 1977
- Weather Resistance of Porcelain Enamels, 15-Year Inspection of the 1956 Exposure Test, Building Science Series 50, National Bureau of Standards

Ceramic Coating*

CERAMIC COATINGS include the super porcelains, which are based on silicates and oxides. Ceramic coatings also refer to high-temperature coatings based on oxides, carbides, silicides, borides, nitrides, cermets and the super porcelains, and other inorganic materials. Ceramic coatings are applied to metals to protect them against oxidation and corrosion at room and at elevated temperatures. Special coatings have been developed for specific uses, including wear resistance, chemical resistance, high reflectivity, electrical resistance, and prevention of hydrogen diffusion. Ceramic-coated metals are used for furnace components, heat treating equipment, chemical processing equipment, heat exchangers, rocket motor nozzles, exhaust manifolds, jet engine parts, and nuclear power plant components.

Selection Factors

Several factors must be considered when selecting a ceramic coating. These include:

- Service environment to be encountered by the coated metal
- Mechanisms by which the coatings provide protection at elevated temperature
- Compatibility of the coating with the substrate metal
- Method of applying the coating
- Quality control of the coating
- Ability of coating to be repaired

Service environment may involve a wide range of conditions. The intended operating life of a coated part may range from a few seconds to several hundred hours. Conditions may involve exposure to atmospheric gases at various mass flows with velocities up to, or even beyond, Mach 10. Components made of the refractory alloys may be subjected to very high stresses, or they may be used as heat shields or furnace windings, for which the only load is the mass of the component. Heating and cooling rates may be gradual or rapid, and one or more thermal cycles may be involved. For any specific service environment, the coating selected must protect the metal from oxidation and the effects of hydrogen pickup by preventing or minimizing the diffusion of oxygen, nitrogen, and hydrogen from the atmosphere through the coating into the substrate.

Mechanisms of Protection. Ceramic coatings have two mechanisms to protect metals at elevated temperature. One type of coating is applied as a layer of stable oxide on the surface of the metal which prevents or delays contact between metal and atmosphere. The other type of ceramic coating is an intermetallic compound that forms a thin oxide film on its surface. The composition of the intermetallic is such that it provides the optimum combination of metallic elements for forming a stable and adherent protective oxide film on its surface and for healing the oxide film in the event the film is broken. Thus, this type of coating depends on the formation and preservation of the oxide film for protection of the substrate material.

Chemical and Mechanical Compatibility. Chemical compatibility of the ceramic coating with the substrate metal is important, especially when the coating is applied to refractory metals and nickel-based alloys for high-temperature service. The so-called stable oxide coatings, such as alumina, are not stable in the presence of some of the refractory metals, such as niobium (columbium) and tantalum, at temperatures above 1370 °C (2500 °F). Also, alumina reacts with metals such as titanium and zirconium, and the protective characteristics of the coating are soon exhausted.

The coating must also be mechanically compatible with the underlying metal, so that undesirable mechanical stresses are not induced in either material. Because most stable coatings are brittle at low temperature, the coefficients of thermal expansion of the coating and substrate should not be greatly different; however, the coefficient of expansion of the coating should be somewhat less than that of the substrate, so that the coating will be in compression at low (room) temperature. The mismatch in expansion should be greater for parts subjected to thermal cycling.

The effect of the coating on fatigue life and on the brittle transition temperature of the composite material should also be considered. In general, the coating is more brittle than the substrate metal, and cracks that form in the coating during service act as stress raisers on the substrate, thus reducing the low-temperature ductility and fatigue life.

Application Method. The method of applying the coating is restricted by the type of coating, the type of metal to be coated, and the size and configuration of the work. Many of the coating processes include heat treatment to promote bonding and sealing. The at-

*Revised by Thomas A. Taylor, Consultant, Materials Development, Engineering Products Division, Union Carbide Corp.; Clifton G. Bergeron, Professor and Head, Department of Ceramic Engineering, University of Illinois at Urbana; Richard A. Eppler, Scientist, Research and Development, Inorganic Chemicals Division, Pemco Products, Mobay Chemical Corp.

mospheres used for spraying and heat treating must be closely controlled to prevent any deterioration in properties of the substrate metal.

Control of Coating Quality. It is important to ensure that the coating is capable of protecting the substrate. Thickness measurements and visual observations are two methods of determining coating quality. However, these methods are not satisfactory for coatings on complex shapes and internal passages that are difficult to see or reach. A preliminary oxidation test of a few minutes or hours in an oxidizing atmosphere at the operating temperature is also an acceptable method for determing quality of the coating.

Ability of coating to be repaired is an important consideration in coating selection. The ideal coating should be repairable if coverage is insufficient in the initial application or if the coating is damaged during handling or service. Repair procedures and their effectiveness differ for the various coatings, methods of application, substrate metals, and size and shape of work.

Table 1 Compositions of unmelted frit batches for high-temperature service silicate-based coatings

Constituent	Parts by weight for specific frits(a)						
	UI-32	UI-285	UI-346	UI-418	NBS-11	NBS-331	NBS-332
Quartz	29.3	21.2	18.3	31.2	18.0	38.0	37.5
Feldspar	42.0	30.2	47.4	...	31.0	...	...
Hydrated borax	28.9	21.0	17.9	...	37.1	...	...
Sodium carbonate	7.7	5.3	6.1	...	5.9	...	...
Sodium nitrate	5.0	4.0	4.4	...	3.8	...	...
Fluorspar	4.5	3.2	2.8	...	3.0	...	...
Tricobalt tetroxide	0.6	...	0.4	...	0.5	...	...
Nickel oxide	0.6	...	0.4	...	0.6	...	...
Manganese dioxide	1.8	...	1.1	...	1.1	...	...
Barium carbonate	...	...	...	26.3	...	56.6	56.6
Zinc oxide	...	...	...	4.2	...	5.0	5.0
Whiting	...	...	...	7.5	...	7.1	6.3
Vanadium pentoxide	...	...	1.3	...	...	...	...
Aluminum hydrate	...	15.1	...	...	...	...	1.5
Boric acid	...	...	...	12.0	...	11.5	11.5
Cerium oxide	...	...	...	4.2	...	...	...
Titania	...	...	...	4.2	...	...	...
Bismuth nitrate	...	...	...	4.2	...	...	...
Bismuth oxide	...	...	...	6.2	...	...	...
Beryllia	...	...	...	...	...	2.5	...
Zirconia	...	...	...	...	...	...	2.5

(a) UI numbers designate frit compositions developed at the University of Illinois; NBS numbers, frits developed at the National Bureau of Standards

Coating Materials

The nonmetallic, inorganic materials used as ceramic coatings have several characteristics in common. Among these are relatively good chemical stability at elevated temperature, hardness, brittle behavior under load, and mechanical continuity in thin cross section.

Silicates. Coatings prepared from silicate powders (frits), with or without mill-added refractories, have the greatest industrial usage of all ceramic coatings. Silicate-frit coatings are used for such long-duration elevated-temperature applications as aircraft combustion chambers, turbines and exhaust manifolds, and heat exchangers. Variations in composition of the silicate frits are virtually unlimited. Frits range from alkali-alumina borosilicate glasses, which are relatively soft, low melting, and highly fluxed, to barium crown glasses.

Crystallized glass coatings have been developed from silicate frits. In these coatings, crystallization of the glass is controlled by formulation and heat treatment and by the presence of nucleating agents added to the glass during melting.

Several different refractory materials may be combined with silicate frits to produce satisfactory coatings for elevated-temperature service. The addition of a refractory material depends on (*a*) service requirements, and (*b*) compatibility of the refractory material with the frit, other mill-added materials, and the substrate metal. Compositions of frits before and after melting, and of slips (mixtures consisting of frit and water which have a creamy consistency) for some of the coatings developed by these organizations, are indicated in Tables 1, 2, and 3, respectively.

Silicate coatings can be applied by spraying or air brushing (for which the material is atomized and carried by compressed air), dipping and draining (which may be followed by spraying), slushing and draining, filling and draining, and flow coating. Under certain conditions, electrostatic spraying also can be used.

Spraying is the most commonly used method, except when the configuration of the part prevents complete coverage or when production requirements are great enough so that a saving in material costs would be realized by the use of dipping or slushing. Dipping and draining is the most economical procedure for coating small parts with simple shapes. For larger parts with restricted areas, filling and draining may be the best coating method. Rotation or shaking of parts is often necessary when dipping or filling is used to distribute the coating and obtain uniform draining.

Silicate coatings are brittle, but when applied at the usual thickness of 25 to 50 μm (1 to 2 mils), they will withstand considerable abuse even at edges and unavoidable sharp corners. Mechanical roughening of the metal surface before applying the coating helps promote adherence. After application, coatings are dried at slightly elevated temperature and subsequently fired at higher temperatures to provide them with the desired performance and appearance characteristics.

Another type of silicate coating consists of water-soluble silicates of sodium, potassium, or lithium. These coatings are applied by spraying, brushing, or dipping, are cured at room temperature or at temperatures ranging from 95 to 315 °C (200 to 600 °F), and are chemically bonded to the substrate. Coating thickness ranges from 25 to 125 μm (1 to 5 mils). These water-soluble alkali-silicate coatings are tough, withstand limited deformation of the substrate, are resistant to shock and fatigue, and provide protection against oxidation in air at temperatures up to 1370 °C (2500 °F). These coatings are also used for control of thermal radiation and permeability.

Metals coated with soluble alkali silicates are aluminum, magnesium, copper, steel, stainless steel, and superalloys. Because of their chemically combined water content, coatings must be dried carefully to prevent separation from the substrate by blistering, cracking, or peeling.

Oxides. Coatings based on oxide materials provide underlying metals, except refractory metals, with protection against oxidation at elevated temperature and with a high degree of thermal insulation. Flame-sprayed oxide coatings do not provide refractory metals with the necessary protection against oxygen because of their inherent porosity. Oxide coatings can be readily applied in thicknesses up to 6.4 mm (0.25 in.), but their resistance to thermal shock decreases with increasing thickness.

Alumina (Al_2O_3) and zirconia (ZrO_2) are the oxides most commonly used as coatings. Alumina coatings are hard and have excellent resistance to abrasion and good resistance to corrosion. Zirconia is widely used as a thermal barrier, because of its low thermal conductivity.

Table 4 lists the principal oxides used for coatings and gives their melting points. Basic oxide is the major constituent of an oxide coating, usually being present in excess of 95 wt %. Other materials, such as calcium oxide (CaO), chromium oxide (Cr_2O_3), and magnesium oxide (MgO), are added in small percentages for stabilization, increase of as-sprayed density, modification of surface-emittance characteristics, and improvement of resistance to thermal shock. Physical properties for alumina and zirconia coatings flame sprayed from rod are given in Table 5.

Oxide coatings are usually applied by the flame or plasma-arc spraying methods. Before spraying by either method, the substrate surface should be clean and rough; abrasive blasting provides a satisfactory surface condition. Sprayed coatings usually range in thickness from 25 to 2500 μm (1 to 100 mils).

Flame spraying, using either oxyhydrogen or oxyacetylene systems, can deposit any refractory oxide whose melting point is below 2760 °C (5000 °F).

Table 2 Compositions of melted silicate frits for high-temperature service ceramic coatings

Constituent	Percentage for specific frits(a) UI-32	UI-285	UI-346	UI-418	NBS-11	NBS-331	NBS-332
Silicon dioxide (SiO_2)	56.5	51.1	57.0	37.0	42.0	38.5	37.5
Aluminum oxide (Al_2O_3)	8.4	19.8	10.7	...	5.6	...	1.0
Boron oxide (B_2O_3)	10.6	9.5	8.0	8.0	25.8	6.5	6.5
Sodium monoxide (Na_2O)	12.0	10.9	10.8	...	17.9	...	...
Potassium monoxide (K_2O)	5.1	4.6	6.7	...	3.4	...	...
Calcium fluoride (CaF_2)	4.5	4.1	3.4	...	3.0	...	...
Cobalt oxide (CoO)	0.5	...	0.45	...	0.5	...	...
Nickel monoxide (NiO)	0.6	...	0.45	...	0.6	...	...
Manganese dioxide (MnO_2)	1.8	...	1.3	...	1.2	...	...
Vanadium pentoxide (V_2O_5)	...	...	1.2	...	...	...	...
Bismuth dioxide (BiO_2)	...	...	...	10.0	...	...	...
Calcium oxide (CaO)	...	...	...	5.0	...	3.5	3.5
Barium oxide (BaO)	...	...	...	25.0	...	44.0	44.0
Zinc oxide (ZnO)	...	...	...	5.0	...	5.0	5.0
Cerium dioxide (CeO_2)	...	...	...	5.0	...	...	...
Titanium dioxide (TiO_2)	...	...	...	5.0	...	...	...
Beryllium oxide (BeO)	...	...	...	...	...	2.5	...
Zirconium oxide (ZrO_2)	...	...	...	...	...	...	2.5

(a) UI numbers designate frit compositions developed at the University of Illinois; NBS numbers, frits developed at the National Bureau of Standards

Table 3 Compositions of slips for high-temperature service silicate-based ceramic coatings

Constituent	Parts by weight for specific coatings(a) UI-32-22	UI-32-53	UI-285-1	UI-346-2	UI-346-4	UI-418-1	UI-418-4	A-19H	A-418	A-417	A-520
Frit(b)	88(c)	100(c)	100(d)	100(e)	100(e)	100(f)	100(f)	100(g)	70(h)	70(j)	90(j)
Diaspore	12	...	...	15	...	...	...	...	...	...	...
Clay	7	10	7	7	10	7	6	10	5	5	6
Hydrated borax	0.75	0.5	0.75	0.75	0.5	0.75	...	...	...	...	...
Water	50	50	50	55	50	50	50	50	48	48	45
Sodium pyrophosphate	...	0.05	...	...	0.05	...	...	...	...	...	...
Calcined aluminum oxide	...	25	...	...	25	...	...	25	...	...	...
Tricobalt tetroxide	...	...	...	...	...	...	...	1	...	...	...
Citric acid	...	...	...	...	...	...	...	0.05	...	...	...
Chromic oxide	...	...	...	...	...	...	25	...	30	30	...
Copper oxide	...	...	...	...	...	...	...	...	...	...	10
Sodium nitrite	...	...	...	...	...	...	...	...	...	...	0.025

(a) UI numbers designate coatings developed at the University of Illinois; A numbers, coatings developed at the National Bureau of Standards. (b) See Tables 1 and 2 for batch and melted compositions of frits. (c) UI-32 frit. (d) UI-285 frit. (e) UI-346 frit. (f) UI-418 frit. (g) NBS-11 frit. (h) NBS-332 frit. (j) NBS-331 frit

Table 4 Melting points of principal oxides used in ceramic coatings

Oxide	Melting point °C	°F
Aluminum oxide (Al_2O_3)	2070 ± 28	3760 ± 50
Aluminum titanate ($Al_2O_3TiO_2$)	1860	3380
Beryllium oxide (BeO)	2570 ± 84	4660 ± 150
Calcium zircanate ($CaZrO_3$)	2345	4250
Cerium oxide (CeO_2)	Over 2600	Over 4710
Chromium oxide (Cr_2O_3)	2265 ± 110	4110 ± 200
Cobalt oxide (CoO)	1805 ± 56	3280 ± 100
Forsterite ($2MgO \cdot SiO_2$)	1885	3425
Hafnium oxide (HfO_2)	2900 ± 110	5250 ± 200
Magnesium oxide (MgO)	2850 ± 28	5165 ± 50
Mullite ($3Al_2O_3 \cdot SiO_2$)	1810	3290
Nickel oxide (NiO)	1980 ± 110	3600 ± 200
Silicon dioxide (SiO_2)	1720 ± 14	3130 ± 25
Spinel ($MgO \cdot Al_2O_3$)	2135	3875
Thorium oxide (ThO_2)	3220 ± 56	5830 ± 100
Titanium oxide (TiO_2)	1870 ± 28	3400 ± 50
Uranium oxide (UO_2)	2875 ± 56	5210 ± 100
Yttrium oxide (Y_2O_3)	2455 ± 56	4455 ± 100
Zircon ($ZrO_2 \cdot SiO_2$)	1775 ± 11	3225 ± 20(a)
Zirconium oxide (ZrO_2)	2710 ± 110	4910 ± 200

(a) Decomposes

However, certain refractory oxides, particularly silicon dioxide, do not spray well even though their melting points are considerably below 2760 °C (5000 °F).

An oxidation-resistant nickel chromium alloy often is applied to the substrate before an oxide coating is deposited by flame spraying. Without such a base coat, the adhesion of the oxide may be inadequate. Coating rates during flame spraying are slow, usually in the range of 16 to 410 cm^3/h (1 to 25 $in.^3/h$).

All oxides that can be flame sprayed and those with higher melting points can be applied by plasma spraying. In general, plasma spraying produces coatings of greater density (porosity of sprayed oxide coatings ranges from 5 to 15%, depending on method of application), greater hardness, and smoother finish than those obtained by flame spraying. Also, the temperature of the substrate remains lower, because deposition is faster. Because of the inert gases used during plasma spraying, oxidation of the substrate is minimized.

In addition to spraying, any of the oxides may be applied by troweling. Troweled coatings usually are thicker than sprayed coatings and are designed to provide maximum thermal protection to the underlying metal. A bonding medium, such as sodium silicate, calcium aluminate, phosphoric acid, or glass, is used for coatings applied by troweling. In addition, the use of expanded-metal reinforcements greatly improves troweled coatings.

Carbides as ceramic coatings are principally used for wear and seal applications, in which the high hardness of carbides is an advantage. These applications include jet engine seals, rubber-skiving knives, paper machine knives, and plug gages. Carbide coatings for wear resistance are applied by flame spraying or detonation-gun techniques. Table 6 gives the melting points of ten carbides.

Silicides are the most important coating materials for protecting refractory metals against oxidation. Silicide-based coatings protect by means of a thin coating of silica that forms on the coating surface when heated in an oxygen-containing atmosphere. To improve the self-healing, emittance, chemical stability, or adherence of this thin silica coating, other elements, such as chromium, niobium, boron, or aluminum, are added to the coating formula.

Table 7 lists and describes several silicide coatings. These materials are usually applied to a substrate by some variation of the vapor-deposition process. Deposition, diffusion, and reaction of silicon (and any other elements added in small quantities) with the substrate metal at a high temperature produce the silicide-based coating.

Vapor-deposited and diffused silicide coatings are characterized by their superior adhesion to the substrate. Fairly precise control of coating thickness is obtained through this process. Uniform silicide coatings with a thickness of a few tenths of a mil to several mils are produced on both simple and complex shapes by either the pack-cementation of the fluidized-bed technique.

The slurry fusion process is the most commonly used method for the deposition of silicide coatings on refractory metals. A slurry of fine silicon powder with desired additives (iron, chromium, hafnium, or titanium) in an organic liquid is applied to the part by dipping, spraying, or brushing. The coated part is heated in a vacuum or inert atmosphere at 1300 to 1400 °C (2370 to 2550 °F) for 30 to 60 min. An excellent coating to substrate bond is developed.

Because they are more brittle than the substrate metals, silicide coatings are highly susceptible to crack formation, which can act as a stress raiser on the substrate. In general, silicide coatings have an adverse effect on all room-temperature mechanical properties of the substrate; the thicker the coating, the greater the effect. Silicide coatings generally embrittle the metals to which they are applied, but do not necessarily impair the usefulness of the coated metals for structural applications.

Phosphate-Bonded Coatings. Phosphates for metal protective coating systems are formed by the chemical reaction of phosphoric acid and a metal oxide such as aluminum oxide, chromium oxide, hafnium oxide, zinc oxide, and zirconium oxide. The phosphate-bonded materials are used to protect metals against heat and to act as a binder in thin ceramic paint films. Thicker composites are troweled, rammed, or sprayed to the desired thickness. Phosphate-bonded coatings have low density, low thermal conductivity, and high refractoriness after curing in place at temperatures ranging from 21 to 425 °C (70 to 800 °F), and they can be applied in greater thicknesses than other ceramic coatings. Thus, a thick refractory composite can be used to protect lower temperature-resistant metal systems. Phosphate-bonded composites, depending on composition, withstand temperatures up to 2425 °C (4400 °F) and have been applied in thicknesses up to 50 mm (2 in.).

Reinforcements, bonded or welded to the metal substrate, usually are used within phosphate-bonded coatings to facilitate bonding to the substrate and to provide resistance to vibration and impact. Reinforcements are corrugated metal screen, expanded metal, open metal strips, and metal and nonmetallic honeycomb.

When phosphate-bonded composites are prepared, one of the strongest bonds between the metal oxide particles is obtained with 85% orthophosphoric acid (H_3PO_4). However, composites bonded with orthophosphoric acid have lower

Table 5 Physical properties of alumina and zirconia flame sprayed from rod

Coating	Bulk density g/cm³	Bulk density lb/in.³	Porosity, %	Color	Typical compressive strength MPa	Typical compressive strength ksi	Thermal expansion(a) μm/m·K	Thermal expansion(a) μin./in.·°F	Thermal conductivity(b) W/m·K	Thermal conductivity(b) Btu·in./ft²·h·°F
Alumina	3.3	0.12	8-12	White	255	37	7.4	4.1	33	19
Zirconia	5.2	0.19	8-12	Light tan	145	21	9.7	5.4	14	8

(a) 20 to 1230 °C (70 to 2250 °F). (b) 540 to 1095 °C (1000 to 2000 °F)

maximum service temperatures than composites formed by the reaction of metal oxide and fluorophosphoric acid (H_2PO_3F). The use of fluorophosphoric acid also permits the use of lower curing temperatures.

After preparation, the composites are aged for 24 h or more to permit reaction between the acid and the metal oxide. The aged composite is troweled either directly onto the substrate or over another protective coating. The coating is then cured with close control of time and temperature. Oxides bonded with orthophosphoric acid are cured for 1 h at each of the following temperatures successively: 93, 120, 150, 215, 315, and 425 °C (200, 250, 300, 420, 600, and 800 °F). Oxides bonded with fluorophosphoric acid are cured for 3 h at room temperature and then for 1 h at 120, 150, and 205 °C (250, 300, and 400 °F). Table 8 identifies several common phosphate-bonded coatings and gives their densities and maximum service temperatures.

A reaction between the acidic coating and the substrate may cause bloating or blistering upon deposition or after initial curing as the result of the release of hydrogen from the acid. The volatilization of phosphorus pentoxide (P_2O_5), a decomposition product of the acid, also can cause blistering.

Various compounds, such as chromic oxide, ammonia compounds, or ferric phosphate, are added to the coating materials to prevent phosphorus pentoxide from corroding the substrate. These additives increase the pH of the coating without affecting the bonding action. Chromic acid may also be added to improve heat emission of the coating. Coatings are usually thixotropic and appear to have a greater viscosity than is actual because slight agitation causes the material to flow.

Coatings are formulated to possess optimum physical and thermal properties. Particle size and filler-to-binder ratio have a great influence on the final properties, including shrinkage, resistance to thermal shock, bond strength, porosity, and thermal conductivity. The common range of particle size for phosphate-bonded coatings is −14 to −325 mesh.

Phosphate-bonded coatings are used primarily to prevent deterioration of the substrate metal during high temperature service. Applications include combustion-chamber linings, re-entry leading edges, hot gas ducts, and high-temperature insulation repairs.

Cermets. Table 9 lists the constituents of electrodeposited coatings based on cermets and indicates thicknesses, service life at elevated temperature, and suitable substrates for these materials. Electrodeposited cermet coatings currently have only a few commercial applications.

Cermet coatings, consisting of a mixture of metal and ceramic oxides, protect metallic substrates against oxidation and erosion. The electrodeposition process used for applying these coatings is a combination of electroplating (for metals) and electrophoresis (for ceramics). The amount of ceramic that can be deposited depends on particle size, density, and composition. Ceramic particles ranging in size from less than 1 μm to 44 μm (40 to 1730 μin.) can be plated. These particles are suspended in any common electroplating bath by agitation. With ordinary procedures, coatings containing about 20 wt % ceramic can be obtained in a deposit; with special procedures, this can be increased to 50 to 60 wt %. Because most cermet coatings are for erosion-resistance applications such as rocket nozzles, coatings are relatively thick (>75 μm or >3 mils). Thinner coatings can be obtained, however, and thickness can be controlled to 25 μm (1 mil).

Cermets applied by plasma spraying or detonation gun processes are the basis for increasing the wear resistance of metals and superalloys. The most important cermets are metal bonded carbides and borides, especially tungsten carbide with 8 to 15% cobalt. At the lower cobalt content, high hardness and wear resistance are produced. Increasing the cobalt content increases the toughness necessary for wear plus impact service. Tungsten carbides wear well to about 590 °C (1000 °F) in air. At higher temperatures, chromium carbide and nichrome are used because of self-lubricating qualities. Coatings based on aluminum oxide, refractory carbides, and an oxidation-resistant metallic binder are in use at temperatures above 870 °C (1600 °F).

Table 6 Melting points of carbides

Carbide	Melting point °C	Melting point °F
Boron carbide (B_4C)	2470	4480
Chromium carbide (Cr_3C_2)	1900	3440
Niobium carbide (NbC)	3480	6295
Hafnium carbide (HfC)	3890	7030
Molybdenum carbide (Mo_2C)	2410	4375
Silicon carbide (SiC)	2540	4605
Tantalum carbide (TaC)	3980	7200
Titanium carbide (TiC)	2940	5325
Tungsten carbide (WC)	2790	5050
Zirconium carbide (ZrC)	3400	6150

Coating Methods

Ceramic coatings may be applied by brushing, spraying, dipping, flow coating, combustion flame spraying, plasma-arc flame spraying, detonation gun spraying, pack cementation, fluidized-bed deposition, vapor streaming, troweling, and electrophoresis. Most of these methods have been used for coating production parts.

Selection of coating method depends on the following factors:

- Substrate metal
- Coating material (some materials are restricted as to method of application)
- Size and shape of the part to be coated
- Cost
- Service conditions (coating method

Table 7 Silicide coatings for protection of refractory metals against oxidation

Constituents of as-applied coating: Silicide	Additives	Suitable substrate metal	Oxidation protection(a): Temperature °C	°F	Life, h	Application: Method	Thickness μm	mil
Molybdenum silicide ($MoSi_2$)	None	Mo-0.5 Ti	1480	2700	10	Fluidized bed	25-50	1-2
	Nb	Mo-0.5 Ti	1540	2800	12	Pack cementation	75	3
	Cr, Al	Mo-0.5 Ti	1540	2800	8	Pack cementation(b)	60	2
	Cr	Mo-0.5 Ti	1480	2700	36(c)	Pack cementation	60	2
	Cr, Al, B, Nb, Mn	Mo	1540	2800	19-45	Pack cementation	60-100	2-4
Niobium silicide ($NbSi_2$)	None	Nb-33 Ta-0.8 Zr	1480	2700	3	Fluidized bed	25-50	1-2
		Nb-10 Ti-10-10 Mo	1425	2600	15-25	Pack cementation	50	2
	Cr, Ti	Nb-10 Ti-10 Mo	1370	2500	Over 100	Vacuum pack(b)(d)	100	4
	Cr, B	Nb-10 Ti-10 Mo	1370	2500(e)	Over 15(e)	Pack cementation(b)	50	2
Niobium silicide ($NbSi_2$), ($NbAl_3$)(f)	...	Nb	1370	2500	396	Pack cementation	Over 150	Over 6
Tantalum silicide ($TaSi_2$), plus others(f)	...	Ta	1370	2500	275	Pack cementation	Over 150	Over 6

(a) Representative data only; can vary depending on test conditions. (b) Multiple-cycle processing. (c) 95% confidence. (d) Variation of pack-cementation process; pack is elevated to remove residual air before heating. (e) Life of coating system is at least 10 h at 1425 °C (2600 °F). (f) Proprietary

can modify the properties of the coating)

Spraying and Dipping

Spraying and dipping are two methods of applying ceramic coatings in a slip or slurry form. Spraying and dipping methods are used to apply silicates and other coatings onto engine exhaust ducts, space heaters, radiators, and other high production parts.

Spraying can be used when the shape of the work permits direct access to all surface areas to be coated. This method is usually used for applying a closely controlled thickness of coating to exterior surfaces only.

Dipping can be used for almost all parts. This includes riveted or spot welded assemblies, except those in which faying surfaces would be inadequately covered by the slurry. For a uniform coating thickness, a handling cycle must be established for each part to produce drainage of each surface at the proper angle.

Surface Preparation. Parts must be thoroughly cleaned before spraying or dipping. Oily spots prevent adherence of the coating and cause blistering or spalling during firing. When sand blasting is used, the abrasive must be free of contaminants. Sharp workpiece edges should be rounded because they are difficult to coat. If sharp edges are coated without being rounded off, the coating will often spall after firing.

The principal cleaning processes used are chemical (acid or alkaline) and abrasive. Chemical cleaning methods for metals of low-alloy content are similar to those used before porcelain enameling. (Additional information can be found in the article on porcelain enameling in this Volume.) For high-alloy materials, such as stainless steels and heat-resisting alloys, heat scaling or trichlorethylene-vapor degreasing, followed by grit blasting, is the preferred cleaning method, except for parts made of thin-gage material or with inaccessible areas. For these parts, chemical cleaning is required. Table 10 shows the sequence of chemical cleaning solutions and immersion times used for stainless steels and heat-resisting alloys. The use of chemical cleaning with acid or alkaline solutions is limited to parts that permit good drainage. All cleaning solutions must be removed from the work by water rinsing before the coating is applied.

Abrasive blast cleaning should be used on parts that will not be distorted by the blasting action and whose surfaces are accessible to the blasting medium. Abrasive cleaning is particularly applicable when an extremely strong mechanical bond between the coating and substrate is required. Silica sand, the most commonly used blasting medium, has low initial cost. However, a high breakdown rate and its highly detrimental effect on blasting equipment results in high equipment maintenance costs. Use of materials with a higher initial cost, such as steel grit or shot, aluminum oxide, garnet, and glass shot, often results in lower overall cost. For additional information about materials, equipment, and tech-

Table 8 Characteristics of phosphate-bonded ceramic coatings

Type of phosphate	Constituents	Density kg/m³	lb/ft³	Maximum service temperature °C	°F
Aluminum	85% $H_3PO_4 + Al_2O_3$	3040-3600	190-225	1925	3500
	$H_2PO_3F + Al_2O_3$	3040-3600	190-225	1980	3600
Hafnium	85% $H_3PO_4 + HfO_2$	4490-4810	280-300	1925	3500
	$H_2PO_3F + HfO_2$	4490-4810	280-300	2205	4000
Zinc	85% $H_3PO_4 + ZnO$	...	...	1650	3000
Zirconium	85% $H_3PO_4 + ZrO_2$	3200-4650	200-290	1925	3500
	$H_2PO_3F + ZrO_2$	3200-4650	200-290	2315	4200

Table 9 Cermet electrodeposited coatings for high-temperature oxidation protection

Constituents of coating as applied	Suitable substrate metal	Service temperature °C	°F	Service life, min	Thickness μm	mil
$Cr + ZrB_2$	Mo-0.5 Ti	2130	3865	20	75-150	3-6
	Tantalum	2130	3865	20	75-150	3-6
	Tungsten	2205	4000	10	75	3
$Pt\text{-}Rh + ZrB_2$	Tungsten	2870	5200	1	510-760	20-30
$Cr + HfO_2$	W or Ta	2620	4750	5	125	5

Table 10 Descaling of stainless steels and heat-resisting alloys before ceramic coating

Alloy	Immersion time, min Solution 1 Sodium hydroxide(a)	Solution 2 Sodium hydride(b)	Solution 3 Nitric-hydrofluoric acid(c)	Solution 4 Nitric acid(d)
410, 430	1-2	10-30	None	5-15
321, 347, 316; 19-9 DL	1-2	10-30	10-30	5-15
Inconel; Nimonic 75, 80A	1-2	10-30	5 max	5-15
S-816; N-155	1-2	10-30	2 max	3 max

(a) Molten sodium hydroxide at 400 to 425 °C (750 to 800 °F). (b) Molten sodium hydroxide containing 0.1 to 2 wt % sodium hydride; bath at 370 to 400 °C (700 to 750 °F). (c) Aqueous solution containing 1 to 4 vol % 70% hydrofluoric acid and 15 to 25% nitric acid (1.41 sp gr); temperature 60 to 82 °C (140 to 180 °F). Solution may be used at ambient temperature by increasing immersion time. (d) Aqueous solution containing 12 to 20 vol % nitric acid (1.41 sp gr); temperature, 60 to 82 °C (140 to 180 °F)

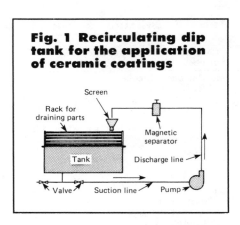

Fig. 1 Recirculating dip tank for the application of ceramic coatings

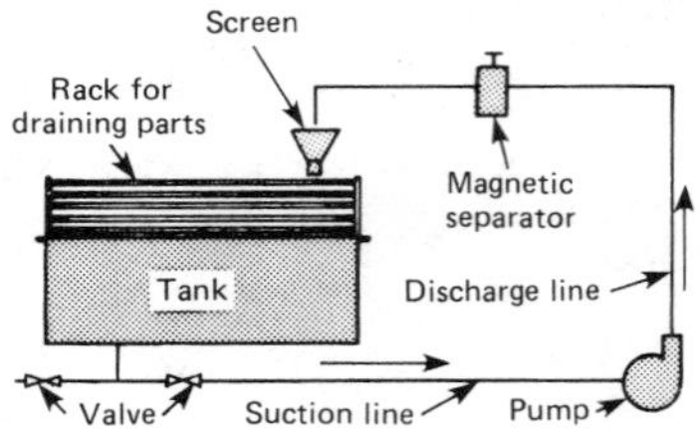

niques used in abrasive blast cleaning, see the article on abrasive blast cleaning in this Volume.

Processing. Ceramic coating materials may be purchased as slips, in which form only the specific gravity requires adjusting by either adding or pouring off water before application. Coatings may also be obtained as frits. Frits are milled in porcelain-lined mills with water to which refractory oxides or other inert materials, clay, and set-up agents are added to produce the required analysis.

Changes from the recommended composition and specific gravity of a slip may produce undesirable results. A low specific gravity causes limited coverage or running of the applied coating. A high specific gravity results in thick coatings that spall on firing. To obtain satisfactory results, test pieces, 25 by 75 mm (1 by 3 in.) and of the same composition and gage as the work material, should be used to check the dry film weight of the slip and the characteristics of the coating as fired. The slip should be checked at the beginning of each working period and whenever an adjustment is made or when a new batch is prepared.

Application of the slip to the work is also critical because of the coating thickness. For stainless steels and heat-resisting alloys, coating thickness ranges from 13 to 75 μm (0.5 to 3 mils). No specific tolerances for coating thickness exist. If the coating is too thin, it oxidizes during firing and loses its protective value; if too thick, it spalls.

Dipping is the preferred method of coating for production operations, although spraying is also extensively used. In some instances, manual debeading is necessary to remove excess coating from points of buildup to prevent spalling. Complex shapes must be sprayed, because dipping builds up excessive beads or fillets in inaccessible areas.

After being applied, the slip is dried in forced circulating air at 60 to 120 °C (140 to 250 °F) for 10 to 15 min. If the drying temperature is too low, waterlines appear on the coating; if too high, the coating tears. All water must be removed, or the coating will blister during firing.

Firing is accomplished in a gas or electrically heated furnace. Firing temperature and time depend on the thickness of both the coating and the substrate. The temperature and furnace atmosphere are controlled to produce the required as-fired appearance of the coating with maximum adherence. Overfiring causes excessive oxidation of the substrate, resulting in poor adhesion or a decrease in coating properties. Underfired coatings have poor adhesion and strength and do not develop maximum coating protection.

Equipment for spraying ceramic coatings is available commercially. The spray gun should have a nozzle with an orifice diameter of 1.30 to 2.80 mm (0.050 to 0.110 in.). Efficient nozzles have a spraying capacity of 260 000 to 330 000 mm^3 (16 to 20 in.^{3}) of coating material per minute using an air pressure of 345 kPa (50 psi) to propel the coating material to the work surface. The compressed air supply should be filtered to remove dirt, rust, oil, and moisture. A reliable air pressure regulator should be used to permit accurate adjustment of pressures, particularly in the range of 205 to 550 kPa (30 to 80 psi).

For most dipping applications, the equipment consists of a tank such as that shown in Fig. 1. The tank should be large enough to permit complete submersion of the part into the slip. An easel or rack is required for draining. Dipping equipment can be elaborated to include temperature-controlled dip tanks and recirculating systems with screens and separators in the line for removing contaminants. Automatic equipment incorporating positioners for proper drainage is often used.

Flow coating is modified dipping and draining in which slip is flowed onto conveyorized parts. Slip flows from nozzles designed to flush all surfaces of the work, after which it drains into a catch basin and is recirculated.

Flame Spraying

Most ceramic coating materials used currently can be applied by flame spraying. Silicates, silicides, oxides, carbides, borides, and nitrides are among the principal materials deposited by this process. Three methods of heating and propelling the particles in the plastic condition to the substrate surface include: (*a*) combustion flame spraying, (*b*) plasma-arc flame spraying, and (*c*) detonation gun spraying. The first two methods use coating materials in powder or rod form. Detonation gun spraying uses only powder materials.

Applicability. Flame-sprayed ceramic coatings can be applied to workpieces in a wide range of sizes and shapes. Practically all metals that can be adequately cleaned, textured by standard abrasive blasting equipment, and safely heated to 150 to 205 °C (300 to 400 °F) can be coated.

Spray equipment can be fitted with extensions having deflecting heads that can turn the spray direction up to 45°. Thus, any shape can be coated if the spray head can be placed within a few inches of the substrate and at an angle of ±45° to the surface.

From a practical standpoint, the maximum size limits for coating the outside and inside surfaces of workpieces depend only on the preparation and handling equipment. In general, the minimum size of the internal diameter is limited to 50 mm (2 in.), and the length should not exceed 3.7 m (12 ft) unless the diameter is large enough to accommodate the entire gun and the supply lines. The coating of curved passages is limited to sizes and shapes that permit approach of the gun at the angles and distances already prescribed. For example, satisfactory coatings have been applied to wires as small as 0.10 mm (0.004 in.) in diameter, to 6.4-mm (0.25-in.) diam-throat rocket nozzles 13 mm (0.5 in.) in length, and to large ducting 2 m (6 ft) in diameter by 8.2 m (27 ft) long.

Combustion Flame Spraying

Processing variables of flame spraying that directly affect the serviceability of a coating are principally: (*a*) surface preparation, (*b*) gun operation, (*c*) spraying distance, (*d*) temperature of the workpiece, and (*e*) type of coating.

The serviceability of the coating depends on the surface preparation. If a surface is not absolutely clean or is not roughened sufficiently to offer maximum mechanical anchorage, bond strength may be reduced 50% or more.

Optimum adherence of the spray particles to each other (cohesion) depends on the following: (*a*) fineness of the spray, (*b*) uniformity of the spray pattern, (*c*) correct adjustment of gas ratios and pressures, and (*d*) proper material feed rate. These can be accomplished only by proper adjustment of the spray system.

The temperature and size of the spray particles must be closely controlled. If a rod gun periodically produces large spray particles, they are not sufficiently heated and are consequently less plastic, resulting in poor bonding to adjacent particles or to the substrate and creating a weak point or area in the coating.

Spray guns should be maintained at the prescribed distance from the substrate for the type of coating desired. If the gun is too close, the coating becomes crazed and has low thermal shock resistance. An excessive gun-to-work distance can result in soft, spongy deposits with low physical properties and decreased deposit efficiency.

Surface Preparation. Because flame-sprayed particles adhere to the substrate surface primarily by mechanical bonding, suitable methods of surface roughening are essential. These consist of undercutting, grooving, threading, knurling, abrasive blasting, and applying sprayed metal undercoats. Abrasive blasting and metal undercoats provide optimum surface conditions for ceramic coatings. When blasting is used, abrasives must be clean and sharp. Roughening should be uniform and should produce as many re-entrant angles and sharp peaks as possible.

Steel grit is one of the most satisfactory blasting abrasives. It disintegrates slowly and offers maximum life. The grit should be screened periodically to remove dirt and fines. Angular steel grit is available in many sizes. A G25 grit propelled by 275-kPa (40-psi) air pressure is used in many applications.

Fused-alumina grit may be used when surface contamination by a steel abrasive is objectionable. Optimum surface preparation is obtained with a No. 24 grit propelled at a pressure of 275 to 345 kPa (40 to 50 psi). This abrasive cuts faster than steel, but some breakdown of the grit occurs, and the fines should be removed before reuse. For light-gage materials, finer grit (No. 46) and lower blasting pressures are recommended to prevent distortion of the work. Silicon carbide also produces a satisfactory surface for ceramic coatings.

Sprayed metal coatings provide an anchoring base for flame-sprayed ceramics equal to that obtained by abrasive blasting. Sprayed molybdenum undercoatings are used as a bonding coat for subsequent application of ceramic coatings to metal substrates that are too hard or too thin to receive adequate surface roughening through abrasive blasting.

Nickel-chromium or nickel-chromium-aluminum alloy sprayed undercoatings are used as an adherent base for flame-sprayed ceramic coatings that are repeatedly subjected to high temperatures. An undercoat in thicknesses of 50 to 330 μm (2 to 13 mils) develops a surface roughness that provides an optimum physical bond for the ceramic coating. When the metal alloy is used, the substrate surface is first roughened by abrasive blasting. Areas that do not require coating can be protected with masking tape, rubber, or sheet metal, depending on the severity of the surface roughening operation.

Processing. After surface preparation, the spray gun is loaded with ceramic coating material of proper size, and the gun is ignited according to the procedure recommended by the manufacturer. Techniques used in flame spraying ceramics are similar to those used in spraying paint. Successful application depends primarily on the skill of the operator.

Spraying distance and rate of gun traverse across the work should be held as nearly constant as possible. The distance and rate of traverse depend on: (*a*) spraying equipment, (*b*) composition of the coating material, (*c*) substrate metal, and (*d*) desired physical characteristics of the coating. Powder guns have a relatively long, bushy flame to heat the ceramic powder during its passage through the extensive heat zone. Consequently, powder guns may need to be placed 150 to 200 mm (6 to 8 in.) from the workpiece and traversed quite rapidly to minimize overheating. Rod guns using the same type of heating operate with a very short flame and heat zone, because heating of the ceramic always takes place at a fixed location at the end of the rod. For rod guns, the optimum spraying distance is about 75 mm (3 in.).

The gun should be moved continuously across a surface in such a manner that each pass slightly overlaps the preceding one. When the surface is completely coated, succeeding passes to increase thickness should be at right angles to those used for initial coverage.

When possible, spray at a 90° angle to the workpiece surface to produce the smoothest coating at the fastest rate. Spraying angles up to 45° from the preferred gun position can be tolerated if the slight reduction in physical characteristics of the coating is acceptable.

To obtain optimum coating properties, the workpiece temperature should be controlled. Adherence of the coating is greatly reduced if the substrate is heated over 260 °C (500 °F). Substrate temperatures can be measured on the reverse sides of panel specimens by applying temperature-sensitive paint or crayon that melts when a specific temperature is exceeded.

When a rod gun is used for coating flat surfaces, use the following practices to avoid overheating the substrate:

- Move the gun across the face of the work in a smooth motion and at about 0.3 m (1 ft) every 5 s
- Maintain the proper distance between gun and work (about 75 mm or 3 in.) during spraying passes

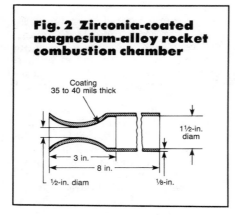

Fig. 2 Zirconia-coated magnesium-alloy rocket combustion chamber

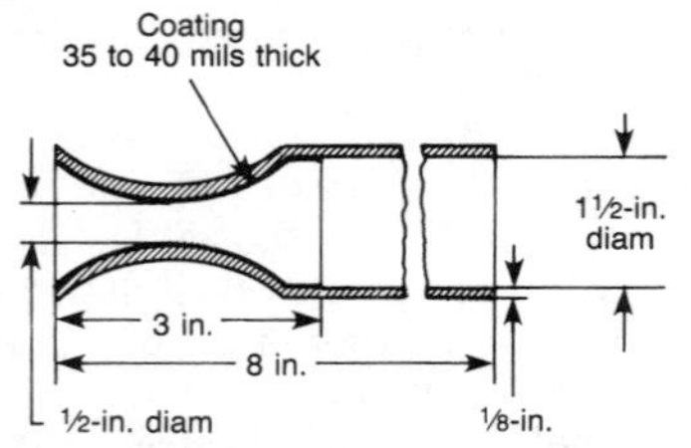

Fig. 3 Metal nozzle coated with alumina and zirconia

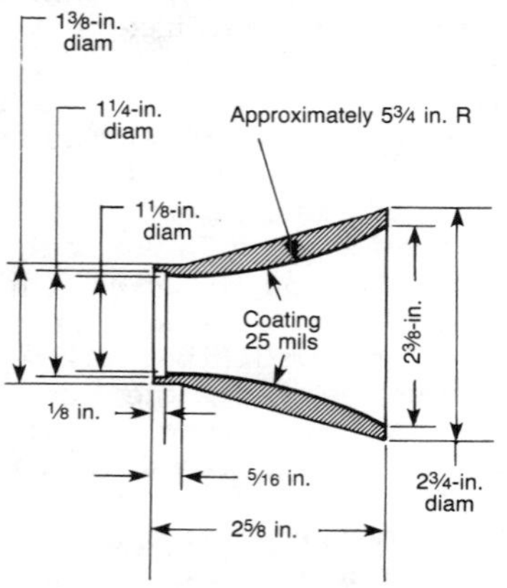

- If a workpiece is small, pause to the side of the work after a coating pass to permit the workpiece to cool slightly

Overheating substrate was overcome in one plant by fixturing the work and spraying for a limited time. The conditions of this operation are illustrated in Fig. 2. Combustion flame spray coating of the inside surface of the rocket combustion chamber shown in Fig. 2 caused melting or burning of the magnesium-alloy substrate when the zirconia coating was applied in a continuous operation. Destruction occurred before the required coating thickness of 890 to 1020 μm (35 to 40 mils) could be applied.

This problem was solved by using a fixture comprised of three friction-loaded thin steel fingers that extended from a standard rotatable chuck. The fingers gripped the exterior of each combustion chamber with just enough force to hold the workpiece during rotation and to permit rapid interchange of the workpieces.

The coating operation consisted of rotating the workpiece at about 30 to 50 rev/min, spraying for not longer than 25 s, then removing the workpiece to permit cooling to room temperature, during which time uncoated or partly coated workpieces would be processed in the same manner. Each combustion chamber required eight or more cycles for producing a coating of the specified thickness.

Figure 3 illustrates a metal nozzle to which a coating of alumina and zirconia was applied 635 μm (25 mils) thick. The operating conditions were as follows:

Conditions	Coating material Alumina	Zirconia
Size of ceramic rod	4.8 by 610 mm (3⁄16 by 24 in.)	4.8 by 455 mm (3⁄16 by 18 in.)
Rods per nozzle coated	1	1½
Average feed rate	180 mm/min (7 in./min)	100 mm/min (4 in./min)

The total area coated on each nozzle was 7100 mm^2 (11 in.^{2}). The time required for preparation, sand blasting, coating, and handling is broken down as follows:

Process	Cycle time, min Alumina	Zirconia
Fixturing	5	5
Masking	5	5
Sand blasting	1	1
Coating	3	6
Unpacking, repacking, transportation, paper work	6	6
Inspection, individual packaging	1	1
Total time	21	24

When large areas require coating, it may be more economical to use more than one spray gun. With the proper mechanical setup, one operator can operate four spray guns efficiently.

Equipment. Most parts require fixturing. For simple shapes that are hand coated, only a simple clamping device is needed for rigidly supporting the part within an exhaust hood during coating. Sheet metal is used for masking areas that do not require coating. A lathe is a suitable fixture for coating parts such as cylinders and nozzles. The chuck rotates the part, and the tool post carriage mechanically moves the spray gun. This setup requires a movable exhaust system for removal of the combustion products and excess spray material.

Gravity-feed or pressure-feed spray guns for powder or electric-feed or air-motor-feed rod guns are used in combustion flame spraying.

A typical gravity-feed powder spray installation consists of (*a*) a fuel gas-control unit, including regulators, to provide a supply of oxygen and acetylene or hydrogen fuel gas; (*b*) a meter for accurate measurement of aspirating gas flow; and (*c*) a spray gun with a nozzle and a canister for containing powder. The principle of operation for this gun is illustrated in Fig. 4. Powder falls through a metering valve in the bottom of the canister into a stream of aspirating gas, which propels it to a stream of fuel gas that has been diverted through a valving system in the gun. The flow rate of the powder is controlled by the size of the metering valve and the amount of aspirating gas metered through the nozzle. This gun usually has a vibrator to maintain uniform powder flow.

In the pressure-feed system, the powder container is separated from the gun and connected by means of a hose through which powder and carrier gas flow. The carrier gas may be compressed air, fuel gas, or inert gas. Hydrogen is commonly used as both carrier and fuel gas.

Control of particle size is important in both gravity-feed and pressure-feed systems. However, the pressure-feed system requires less control of distribution of particle size because of the higher velocity of the carrier gas. Compared to rod spraying, powder spraying has lower initial equipment costs and greater flexibility of coating properties, as well as being adaptable to a wider variety of coating materials.

A typical rod spray installation is illustrated in Fig. 5. In addition to the auxiliary equipment required for the powder spray process, rod spraying requires the following: (*a*) a supply of compressed air, (*b*) an air-control unit that includes a filter and a regulator, and (*c*) an air flowmeter. A good grade of acetylene should be used, and at least two tanks should be manifolded so that withdrawal rates can be kept below a maximum of one-seventh of the volume of the tank capacity per hour to prevent acetone withdrawal. This is

Fig. 4 Operational principle of a gravity-feed powder spray gun

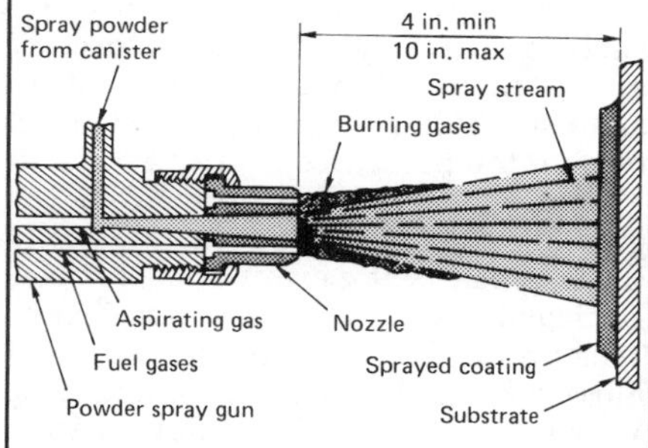

Fig. 5 Rod spray installation

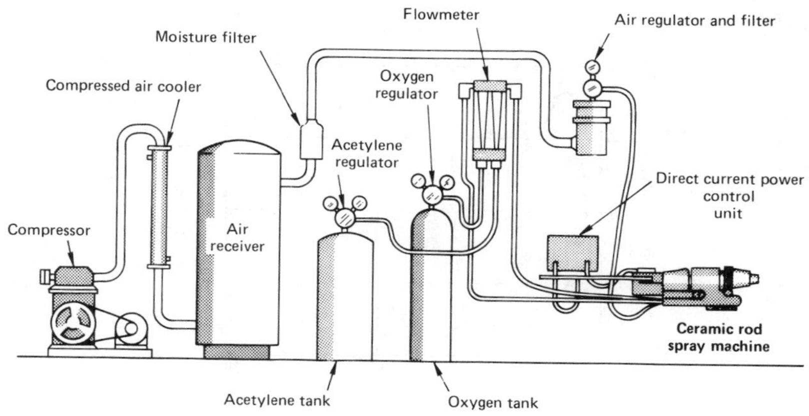

recommended because of the cooling effect that acetone vapor has on flame temperature.

The operation of a ceramic rod spray gun is shown in Fig. 6. The ceramic rod is fed through the center of the nozzle and atomized by the surrounding oxyacetylene flame and compressed air. Compressed air is used to cool the nozzle, increase the velocity of the sprayed material, and control the spray pattern. Control of the diameter and straightness of the rod is required to eliminate problems such as rod sticking and blowback. Control of rod speed is important for control of the density and surface characteristics of the coating. The rod and powder guns can be equipped with extensions and 45° angle air caps for coating inside diameters. Velocity of the sprayed particles from a rod gun as compared to those from a powder gun is shown in Fig. 7.

Rod spraying causes less heating of a workpiece than powder spraying, and it produces a coating with higher density and better bond between the coating and substrate.

Control of coating thickness is related directly to the method used for handling the workpiece and the spray gun. Hand-applied coatings can easily be held within a tolerance of ±50 μm (±2 mils). Mechanical systems for handling both workpiece and gun decrease this tolerance by 50% or more. The variation in coating thickness obtained by hand spraying alumina and zirconia on one side of steel test coupons (25 by 25 by 3.2 mm, or 1 by 1 by ⅛ in.) with a rod gun is shown in Fig. 8.

Flame-sprayed coatings are applied relatively slowly; therefore, after a uniform surface coverage system has been set up, control of the coating thickness depends on timing the duration of coating application with sufficient accuracy to achieve the desired tolerances. If closer tolerances or finer surface finishes are required, most flame-sprayed ceramic coatings can be ground by conventional grinding techniques.

Plasma-Arc Flame Spraying

In the plasma process, a gas or a mixture of gases, such as argon, hydrogen, or nitrogen, is fed into the arc chamber of the plasma generator and heated by an electric arc struck between an electrode and the nozzle. The gas is heated to temperatures as high as 8300 °C (15 000 °F) to form a plasma, or ionized gas, that is accelerated through the nozzle. The ceramic powder, carried by a gas stream, is injected into the plasma, where it is heated, melted, and propelled toward the workpiece.

The higher temperature heat source of the plasma arc imparts an energy content to the ceramic particles that is different from that in the combustion flame process. This necessitates some modification of the gun position. When the plasma process is used, higher melting ceramic materials, such as the refractory metal carbides, can be deposited with a greater deposition rate.

The processing operations for plasma-arc spraying are similar to those discussed in the section on combustion flame spraying. For a more detailed discussion of plasma-arc spraying, see the article on thermal spraying in this Volume.

Thermal barrier coating is one current application using plasma-arc spraying. Applied to certain high-temperature components, such as the inside of combustion chambers or the first-stage vane or blade of a gas turbine engine, thermal barrier coatings act to insulate the metal substrate thermally. Coatings are designed to provide as much as a 110 °C (200 °F) drop in temperature at 980 to 1095 °C (1800 to 2000 °F), but should be used in a temperature gradient, such as is provided by air cooling the substrate or metal side. A thermal barrier could be a 150 to 200 μm (6 to 8 mils) undercoat of a high-temperature nickel-cobalt-chromium-aluminum-yttrium alloy, followed by 255 to 305 μm (10 to 12 mils) of yttria-stabilized zirconia or magnesium zirconate ($MgO \cdot ZrO_2$). If greater thickness for greater insulation is desired, thermal stresses resulting from appli-

Fig. 6 Operational principle of rod gun

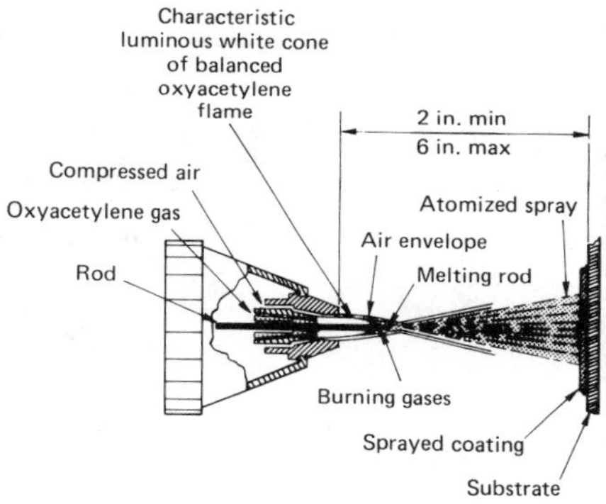

Fig. 7 Comparison of spray particle velocity from rod and powder guns

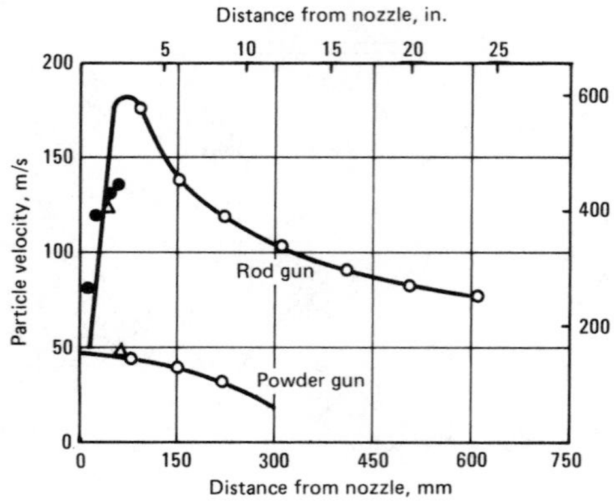

○: Velocimeter; ●: streak camera, fast particle; △: high-speed motion pictures

Fig. 8 Variation in thickness of hand-sprayed alumina and zirconia coatings on steel test coupons

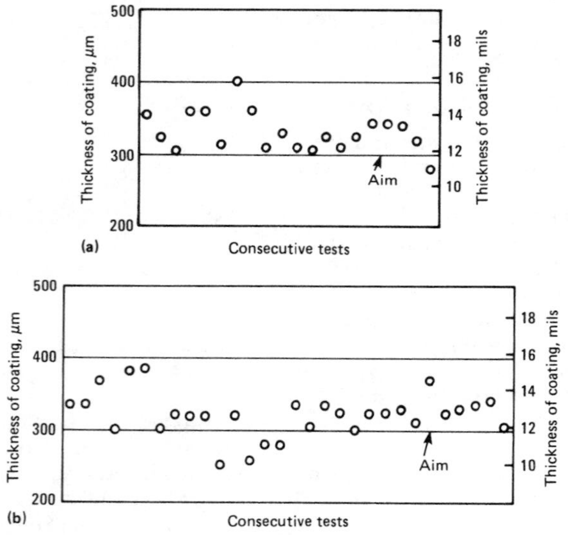

Coatings flame sprayed from rod. (a) Alumina on steel, 20 tests. (b) Zirconia on steel, 30 tests

cation should be carefully considered. In laboratory applications, a thermal cycling test is used followed by a bench engine evaluation to qualify the coating and estimate service life.

Detonation Gun Flame Spraying

Detonation gun flame spraying is markedly different from other flame spraying processes and was developed specifically for the deposition of hard, wear-resistant materials, such as tungsten carbide. Detonation gun spraying uses controlled detonations of acetylene and oxygen to melt and propel the particles onto the substrate.

Powder materials sprayed by this process are carbides containing a small amount of metal binder, and oxides or oxide mixtures. Coatings are usually less than 255 μm (10 mils) thick and are used primarily in applications requiring wear resistance under extreme service conditions. Applications include aircraft jet engine seals (for protection against high-temperature dry rubbing wear) and aircraft compressor and turbine blades (for protection against fretting corrosion at medium to high temperatures).

Coating particles emerge from the gun at supersonic speeds, and only those areas that permit the particles sufficient access are plated uniformly. This limitation prevents the coating of narrow holes, blind cavities, and deep V-grooves. Internal diameters over 9.7 mm (0.38 in.) and open at both ends can be coated to a depth of 1½ times the diameter.

Cementation Processes

Pack cementation, the fluidized-bed process, and vapor streaming are three types of cementation processes used in ceramic coating. These processes are used to produce impervious, oxidation-protective coatings for refractory metals and nickel-based, cobalt-based, and vanadium-based alloys. The principal types of coating applied by the cementation processes are silicides, carbides, and borides, usually of the basis metal although frequently of codeposited or alternately deposited other metals such as chromium, niobium, molybdenum, and titanium.

Pack Cementation

Preparation of the substrate surface for application of a ceramic coating by pack cementation consists of (*a*) removing burrs, (*b*) rounding edges (0.125 mm or 0.005 in. minimum radius to half the edge thickness, for foil), and (*c*) rounding corners (preferably to a minimum radius of 3.2 or 0.125 in.). Edges and corners must be rounded to prevent cracking of the coating (Fig. 9). This can be accomplished by manual sanding with fine-mesh cloth or with a small motor-driven fine-mesh conical grinding wheel. Mass (barrel) finishing can be used for removing burrs and rounding edges and corners of small articles such as rivets.

The next operation consists of cleaning the work by vapor degreasing followed by mechanical or chemical cleaning. Mechanical cleaning is usually the more desirable cleaning method and may consist of wet blasting, abrasive blasting with 200-mesh alumina, or

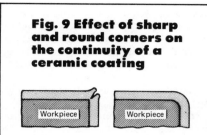

Fig. 9 Effect of sharp and round corners on the continuity of a ceramic coating

Table 11 Cycles for application of silicide and other oxidation-resistant ceramic coatings by pack cementation

Processing cycles suitable for depositing coatings of silicon, chromium, boron, aluminum, titanium, zirconium, vanadium, hafnium, and iron

	Processing cycle		
	Temperature(a)		
Substrate metal	°C	°F	Time, h(b)
Niobium alloys	1040-1260	1900-2300	4-16
Molybdenum alloys	1040-1150	1900-2100	4-16
Tantalum alloys	1040-1150	1900-2100	4-12
Tungsten alloys	1040-1370	1900-2500	3-16

(a) Tolerances: ±6 °C (±10 °F) at 1040 °C (1900 °F); ±14 °C (±25 °F) at 1260 °C (2300 °F). (b) Tolerance, ±10 min

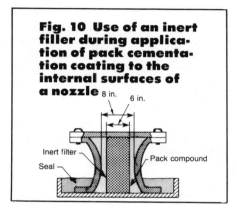

Fig. 10 Use of an inert filler during application of pack cementation coating to the internal surfaces of a nozzle

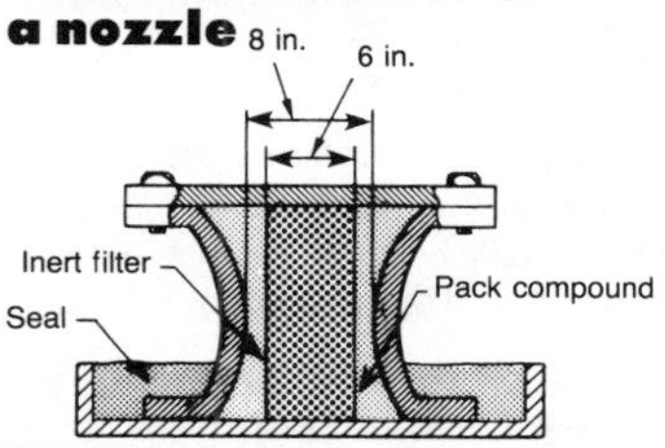

buffing. Parts that are buffed should be washed in acetone, and precautions should be taken to prevent adherence of the buffing compound. Chemical cleaning is used when the shape of the part is not suited to blasting or buffing. Parts must be rinsed and dried thoroughly after removing them from chemical solutions, and precautions must be taken to avoid contamination of cleaned parts during subsequent handling.

Processing. After cleaning, parts are packed in a retort with the desired coating material. Parts should be placed about 25 mm (1 in.) from the retort walls; spacing may be from 3.2 to 13 mm (1/8 to 1/2 in.) between parts, and from 6.4 to 25 mm (1/4 to 1 in.) between layers. Packing material must fill all cavities or areas that may entrap air. Sufficient packing material must be placed between the bottom of the retort and the first layer of parts, and over the top layer. The packed retort should not be handled roughly or be vibrated before or during the thermal process cycle.

An inert filler (aluminum oxide) is used to obtain the most efficient use of packing material when large assemblies or components are being coated. The filler should be no closer than 13 mm (1/2 in.) from the substrate surface. Figure 10 shows the use of a filler for filling space within the throat of a nozzle, the internal surfaces of which were being coated by the pack cementation process.

The packing material usually consists of coating materials (in elemental or combined form), a suitable activator or carrier-gas-producing compound, and inert filler material. A standard siliconizing packing material contains silicon powder (100 to 325 mesh), a halide salt (ammonium chloride, sodium fluoride, or potassium bromide), and an inert filler (aluminum oxide, 100 to 325 mesh). Occasionally, urea is incorporated in the pack material to purge entrapped air before the cementation reaction begins.

The processing temperatures used for pack cementation coating of refractory metals depend on the substrate metal and the desired coating characteristics. In general, temperature controls the rate of deposition, and time is varied to control the thickness of the coating. A low processing temperature results in a coarse, columnar structure and an uneven deposit. High processing temperatures result in deposits of uniform thickness and dense structure. The recrystallization temperature of molybdenum-based and tungsten-based substrates should not be exceeded because of resulting embrittlement. Table 11 gives time-temperature cycles adequate for applying oxidation-resistant coatings.

After thermal treatment is completed, the retort may be cooled in the furnace or in air. The coated parts can be removed from the retort when they are cool enough to handle. Loose packing material is removed by washing the parts in warm water, bristle brushing, and spray rinsing. Water under pressure may be used to remove packing material from difficult-to-clean areas. If a second pack cementation operation is required for the addition of other coating elements, parts should be handled with clean gloves or plastic-tipped tongs. If contaminated, parts must be vapor degreased just before packing for the next coating cycle.

When a second coating cycle is not required, the coated parts may be subjected to a high temperature (about 1095 °C or 2000 °F) to form a protective oxide surface. Normally, 15 to 30 min at this temperature is sufficient to form a protective film on refractory alloys.

Components of assemblies are coated individually, then assembled and packed for the second cycle to protect the joint areas. If assembling causes discontinuities or cracks in the coating, areas are wet blasted and dried or are lightly sand blasted before packing.

The optimum thickness of coating on refractory metals is from 25 to 100 μm (1 to 4 mils). In general, oxidation resistance increases with coating thickness; however, the sharp radii of foils do not permit a coating thickness of much over 25 μm (1 mil). The usual thickness of pack cementation coatings is 38 ± 13 μm (1.5 ± 0.5 mils) for machined components, formed parts, and sheet materials; for foils of 0.250 mm (0.010 in.) or less, the coating thickness is usually 25 ± 8 μm (1.0 ± 0.3 mils).

Equipment for pack cementation consists of a retort and a furnace of suitable size to accommodate the retort. Furnace atmosphere is not critical and may be air, endothermic, exothermic, or inert gas. When a specific atmosphere around the retort is essential, an atmosphere housing may be incorporated.

Retorts are either top-loaded or inverted, and may be designed for shallow or deep sealing (Fig. 11). The type of material from which retorts are made depends on the operating temperature and furnace atmosphere. For operating temperatures between 980 to 1260 °C (1800 to 2300 °F), Inconel and types 310, 321, and 347 stainless steel

Fig. 11 Designs of retorts used in the pack cementation process

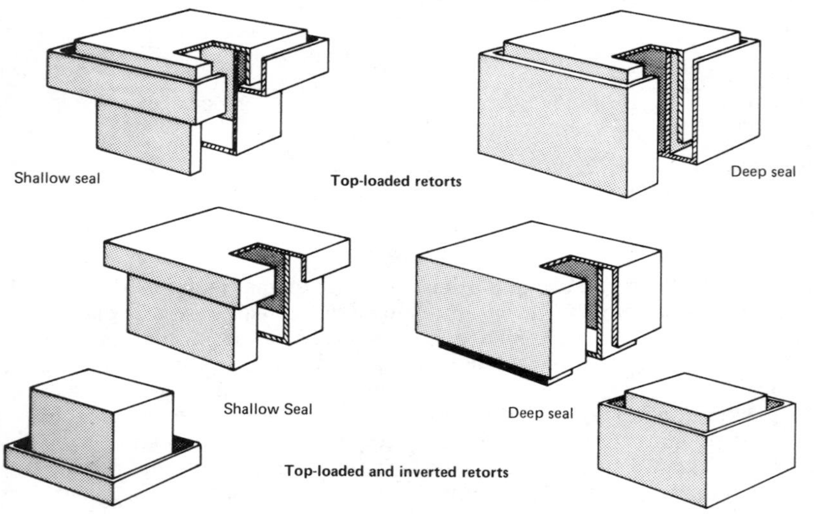

Fig. 12 Fluidized-bed cementation process

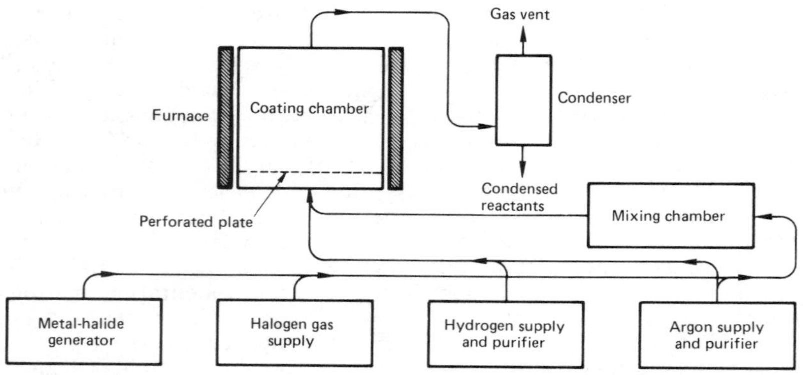

provide satisfactory service. When the furnace atmosphere is oxidizing or carburizing, a stop-off slip ceramic coating on exposed areas of the retort prolongs its service life. Materials for sealing the retort may be sand, alumina, or garnet, with or without oxide scavengers such as silicon or titanium, or low melting point materials like sodium orthosilicate.

Fluidized-Bed Cementation Process

The fluidized-bed process for applying ceramic coatings involves:

- Thermal decomposition and displacement reactions of metal halides
- Presence of hydrogen to reduce the halides
- Diffusion of deposited materials into the substrate metal to produce an intermetallic compound, such as molybdenum disilicide

In this process, a bed of metal powder reactant and inert material is fluidized or floated at elevated temperature by an inert or reactive gas. The finely divided particles of reactant and inert material are constantly agitated by the fluidizing gas. Thus, the transfer of heat between the object to be coated, the coating material, and the gas, is greatly increased by the diffusion of vapor and gas and by the relatively high flow rates. Vapors of coating material can be prepared within the fluidizing chamber by the reaction of particles in the bed with the gases, or they can be prepared and evaporated in a separate vessel. A schematic flow diagram of the fluidized-bed process is shown in Fig. 12.

Processing. Preparation of the surface of the work consists of rounding the edges, buffing the surfaces and edges, and etching. The following etching procedure is used for molybdenum-base substrates:

- Dip in 80% nitric acid solution at room temperature for several seconds
- Rinse in cold water (three rinses)
- Dip in 50% hydrochloric acid at room temperature for several seconds
- Rinse in cold water (three rinses)
- Wash in acetone

After etching, parts are placed into the fluidizing chamber and processed at 1065 °C (1950 °F) for 1 h. Coated parts are removed from the furnace when cool.

Effect of Process Variables on Coating Characteristics. The control of time, temperature, and carrier-compound concentration is important in the fluidized-bed process, because these variables control the thickness and uniformity of the coating, as well as the rates of deposition and diffusion. Temperature should be controlled to within ±14 °C (±25 °F).

Coating thickness as a result of time and temperature is shown in Fig. 13 for a silicide coating applied to Mo-0.5Ti alloy. The coating thickness represented by these data was calculated from the change in weight of the coated part, using the average density of molybdenum silicide ($MoSi_2$). Although data for operating temperatures below 925 °C (1700 °F) are included, coatings applied to refractory metals at these low temperatures have poor oxidation resistance.

Applicability. Complex shapes can be coated by the fluidized-bed process. With special techniques, inside surfaces of long small-diameter closed-end tubes can be coated. However, coatings will form only on edges with a radius of 0.125 mm (0.005 in.) or more.

Cracks around rivet heads and joints between sheets cannot be bridged by ceramic coating during elevated-temperature service. Therefore, double processing cycles are required, one before joining and one after assembly of the component parts.

Fig. 13 Effects of time and temperature on the thickness of a silicide coating applied by the fluidized-bed process to Mo-0.5Ti alloy

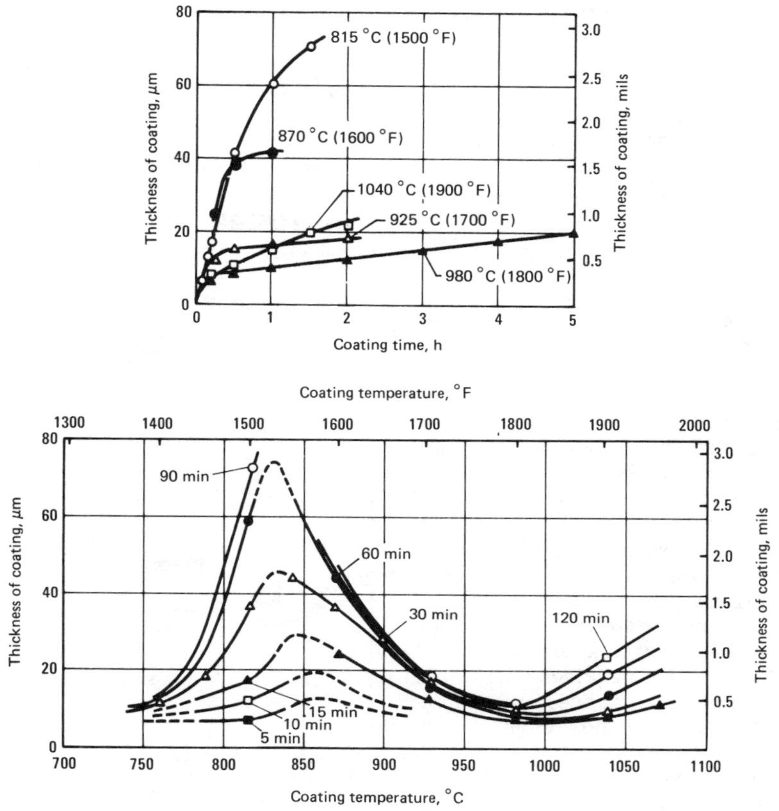

Service life of a coating 50 to 75 μm (2 to 3 mils) thick on flat surfaces is about 1 to 2 h at 1650 °C (3000 °F). The effects of edges and corners combine to reduce this life, because coating thickness for satisfactory coverage is less at these locations.

Vapor-Streaming Cementation

Vapor streaming is a cementation process in which a vapor of the coating material is decomposed on the surface of a heated part. For example, silicide coatings are produced by passing silicon halide vapor in a hydrogen atmosphere over heated substrate. The silicon halide is reduced, and silicon deposits on the substrate and diffuses to form an intermetallic compound. Commercial application of this process has been insignificant.

Trowel Coating

Coatings applied by troweling are (*a*) acid-bonded systems (phosphates and sulfates), (*b*) hydraulic-setting cements (calcium aluminate and portland cement), (*c*) soluble silicate-bonded systems (sodium, potassium, and lithium), and (*d*) colloidal metal oxide-bonded ceramic oxides and carbides. Troweled coatings are used for furnace linings, hot gas ducts, and certain repair patches on other coatings for relatively short service exposure. The resistance to heat transfer of these coatings depends on the porosity, density, and thermal conductivity of the solid phase, and on the thermal shorts caused by any reinforcement metal present. Coatings applied by troweling consist of filler, binder, carrier, and additives. Coating constituents are blended in a muller or other suitable mixer to a uniform consistency. Some materials, such as the acid-bonded coatings, require aging before application to permit reaction between constituents.

Surface Preparation. Surfaces to be coated must be free of contaminants such as oil and grease that may interfere with the wetting and bonding of these water-based coatings. Most coatings applied by troweling are chemically bonded or hydraulic-setting materials. Because these materials do not form a strong metallurgical bond with substrate metals, and because of the differences in coefficients of expansion, substrate surfaces must be roughened for maximum mechanical bonding and to minimize the effects of expansion, vibration, and impact during service. Surface roughening is accomplished by grit blasting or chemical cleaning, or by attaching mechanical reinforcements such as wire mesh, corrugated metal, angular clips, or honeycomb structures. Reinforcement is usually required for surfaces having a finish of less than 6.35 μm (250 μin.).

Processing. Application of coating material, in thicknesses ranging from 3 to 25 mm (0.1 in. to over 1 in.), is accomplished by standard troweling techniques. The material is worked under, around, and through the reinforcements. The smoothed thickness can be measured by a depth gage or with prefixed height gages. Vibration of the coating followed by retroweling produces a denser coating. Hydraulic-setting coatings must be applied immediately after mixing with water, because bonding occurs during dehydration.

After application, hydraulic-setting coatings may be cured at ambient temperature or by being heated at less than 100 °C (212 °F). If heat curing is used, the coated work should be raised to temperature at a slow rate to prevent the coating from blistering. The acid-bonded composites are cured at temperatures ranging from 20 to 425 °C (68 to 800 °F), depending on composition and thickness of the coating. Well-ventilated facilities must be used when working with acid-bonded coatings.

Soluble silicate materials are cured at temperatures from 21 to 425 °C (70 to 800 °F), depending on the system and special additives that produce air-drying properties. To remove entrapped moisture, chemical-setting materials are dried in air. Colloidal metal oxides require only the removal of excess water either by air drying or by heating to 100 °C (212 °F).

Electrophoresis

Electrophoresis is the migration of electrically charged particles suspended in a colloidal solution under the influence of an applied electric field.

Table 12 Hardness of three ceramic coatings deposited by three processes

Coating material	Hardness of coating, HV Flame sprayed	Plasma sprayed	Detonation gun sprayed
Tungsten carbide + 8% cobalt	...	600-700	1200-1450
Tungsten carbide + 12% cobalt	...	600-700	1050-1200
Chromium oxide	900-1100	1200-1350	900-1150

Deposition occurs at one of the electrodes where the charge on the particle is neutralized. The particles acquire a static charge during milling, or they can be charged artificially by absorption of certain additives or electrolytes. This coating process, now being used commercially with increasing frequency, is applicable to practically all substrates, including tool steels, stainless steels, superalloys, refractory metals, oxides, and graphite.

Coatings as applied are soft, and densification is sometimes required. Densification, if needed, may be accomplished by isostatic pressing, hot pressing, or a combination of these methods, depending on the substrate metal. The coating is sintered, usually in a controlled atmosphere.

During sintering, metal coatings are bonded to the substrate by diffusion; oxide coatings, by mechanical and electrochemical bonding. Coating thickness rarely exceeds 75 μm (3 mils), and the thermal expansivity of coating and substrate should be closely matched to prevent spalling. The electrophoresis coating process is simple and easily automated while providing better control of coating thickness and composition than is possible with the slurry and pack cementation processes.

Quality Control

No single nondestructive method is adequate for evaluating the quality of a ceramic coating. Although visual inspection or comparison is only of limited usefulness, many plants prepare samples of coating with surface defects that are known to be harmful to the protective value and service life of the coating and use these samples as visual comparators.

High-Temperature Test. The most reliable test procedure for determining coating continuity and oxidation resistance on complex structures made of refractory metals is to subject the structure to a high-temperature test environment under carefully controlled conditions. First, exposed surfaces are inspected visually under low-power magnification. The work is then heated to 1095 to 1205 °C (2000 to 2200 °F) in air. After a 15-min heating period, surfaces are examined while hot for evidence of evolution of an oxide gas (molybdenum trioxide when molybdenum is the substrate) or for the discoloration that accompanies oxidation of a niobium, tantalum, or tungsten substrate. If no evidence of oxidation is observed, the work is removed from the furnace, cooled, and examined under a magnification of 15 diam; areas that may indicate oxidation of the substrate surface are examined at a magnification of 100 diameters. If no defects are observed, the work is reheated for an additional 45 min, cooled, and reexamined. Accessible defects observed after the 15-min heating period are repaired. Inaccessible defects, such as those on faying surfaces, may necessitate disassembling of the structure for reprocessing the defective area.

Fluorescent-penetrant inspection is useful for detecting cracks, pits, and similar discontinuities in coating surfaces. The work is immersed in a penetrant, the excess penetrant is removed from the surfaces, and the surfaces are coated with a colloidal suspension known as a developer. Penetrant that has been entrapped by a defect seeps through the developer and reveals the outline of the defect when the surface is exposed to ultraviolet light. This is a sensitive test, and it frequently reveals very tight surface defects. When flexible-handle magnifying mirrors are used, this test method can be extended to the inspection of complex shapes and tubes.

Destructive tests can be performed on a workpiece or on specimens prepared and coated simultaneously with the workpieces. Standard test methods, such as for tensile strength, modulus of rupture, transverse bending, density, hardness, metallographic and chemical analysis, can be used on specially prepared sections obtained from a thick section of the coating. An example is illustrated in Fig. 14(a.) The tensile specimen of the coating is prepared as follows:

- Grind slots in coating with a cut off wheel
- Remove 13-mm (½-in.) wide sections of coating from the substrate and grind them flat to form 13-mm (½-in.) wide beam samples 0.125 mm (0.050 in.) in thickness. Beam samples may be tested in transverse rupture by a standard beam test. The coating can be removed by force when applied to a graphite substrate, because graphite has very low strength. A more widely used procedure is to use a substrate that can be chemically dis-

Fig. 14 Sectioning and testing of ceramic coatings

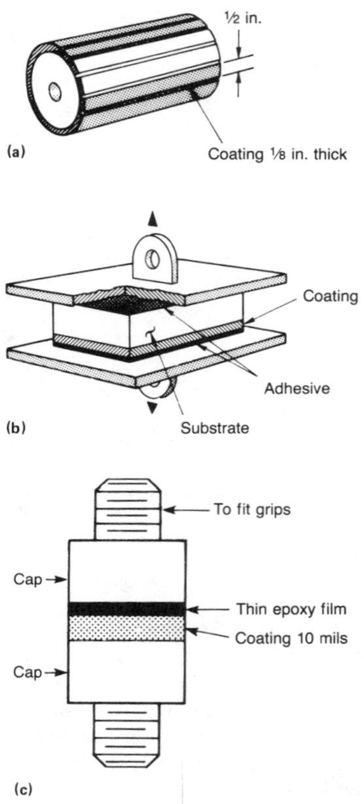

(a) Sectioning of 3 mm (⅛ in.) thick coating on a cylinder for preparation of specimens for determination of tensile strength, transverse bending, and other properties by standard test methods. (b) Testing the bond strength of coatings applied by plasma-arc or combustion flame spraying. (c) Testing bond strength of coatings applied by detonation gun process

solved by a solvent that will not attack the coating

- Grind samples to form a tensile specimen

Bond Strength. A simple test for bond of a coating to a substrate is diagrammed in Fig. 14(b). This test, which makes use of an epoxy adhesive, is applicable to most coatings applied by plasma-arc or combustion flame spraying. A bond cap arrangement is illustrated in Fig. 14(c).

Impact Strength. Although conventional impact values can be obtained for a coating by an Izod or Charpy test on a specimen of the coating, a more useful impact test consists of projecting a pellet from an air gun with sufficient velocity to cause a measurable deformation of the substrate metal. The coating is then visually inspected for chipping and cracking. This test is best suited to coatings less than about 125 μm (5 mils) thick.

Wear Properties. In the dry-rubbing test, two specimens are mated and rubbed together with a load and relative surface speed selected on the basis of service severity. A similar set-up can be used for determining wear properties at elevated temperature, compatibility of the coating with lubricants or corrosives, and the effects of abrasives.

Structure and Hardness. The microscope is a useful tool for observing bond, binder, and metallic or oxide inclusions in a coating. Hardness testing provides a direct measurement of interparticle bond strength. For example, the true hardness of aluminum oxide usually ranges from 1800 to 2200 HV. Accepted Vickers hardness values of aluminum oxide deposited by various methods are 600 to 800 HV for flame-sprayed coatings, 700 to 1000 HV for plasma-sprayed coatings, and 1000 to 1200 HV for detonation gun-sprayed coatings. The maximum values represent the highest hardness obtained by these processes and thus, the highest degree of interparticle bond. Accepted Vickers hardness values for three other ceramic coatings are indicated in Table 12. Hardness readings obtained with a Knoop indenter can be converted to Vickers for comparison.

SELECTED REFERENCES

- M. H. Ault and W. M. Wheildon, Modern Flame-Sprayed Ceramic Coatings, in *Modern Materials,* Edited by H. H. Hausner, New York: Academic Press, 1960
- J. M. Blocher, *et al,* Chemical Vapor Deposition, DMIC Report 170, Battelle Memorial Institute, Columbus, OH, 1962
- A. G. Eubanks and D. G. Moore, Development of Erosion-Resistant Coatings for Aircraft Engine Components, WADC Technical Report 56-109, Dayton, OH, Jan 1956
- J. J. Gangler, Current Developments in Coated Refractory Metals and Their Alloys, *Metals Eng Quart,* Aug 1963, p 39-44
- W. N. Harrison, D.G. Moore, and J. C. Richmond, Ceramic Coatings for High-Temperature Protection of Steel, *J Res Nat Bur Std,* Report 1773, March 1947, p 293-307
- John Huminik, Jr., *High-Temperature Inorganic Coatings,* New York: Reinhold Publishing Corp., 1963
- V. A. Lamb and W.E. Reid, Electrophoretic Deposition of Metals, Metalloids and Refractory Oxides, *Plating,* March 1960, p 291-296
- S. J. Paprocki, *et al,* Gas Pressure Bonding, DMIC Report 159, Battelle Memorial Institute, Columbus, OH, 1962
- Report of the Subpanel on Coatings, Report No. MAB-181-M, Refractory Metals Sheet Rolling Panel, Materials Advisory Board

Cleaning and Finishing of Stainless Steel and Heat-Resisting Alloys

Cleaning and Finishing of Stainless Steel

By the ASM Committee on Cleaning and Finishing of Stainless Steel*

STAINLESS STEEL is finished to provide a variety of smooth finishes, ranging from dull to highly reflective. The techniques used to produce standard mill finishes on fabricated stainless steel parts and assemblies have been summarized in the article below.

Mill Finishes

Stainless steel mill products are supplied in a variety of standard finishes, ranging from a dull, hot rolled, and descaled finish to a highly reflective finish from which all grit lines have been eliminated.

Sheet. Distinctions are made in the finishes of stainless steel sheet by a system of numbers:

No. 1 is a very dull finish produced by hot rolling the steel on hand sheet mills to specified thicknesses, annealing, and descaling. It is used in industrial applications for resistance to heat or corrosion, where smooth finish is not of particular importance.

No. 2D is a dull finish produced on either hand sheet mills or continuous mills by cold rolling to the specified thickness, annealing, and descaling. The dull finish may result from the descaling operation or may be developed by a final light cold roll pass on dull rolls. This finish is favorable to the surface retention of lubricants in deep drawing operations. It generally is used in forming deep drawn articles that may be polished after fabrication.

No. 2B is a bright, cold rolled finish commonly produced in the same way as No. 2D, except that the annealed and descaled sheet receives a final light cold roll pass on polished rolls. It is a general-purpose finish used for all but exceptionally difficult deep drawing applications. This finish is more readily polished than No. 1 or No. 2D.

No. 3 is an intermediate polished finish for use where a semifinished polished surface is required and a further finishing operation follows fabrication. For sheets or articles that will not be subject to additional finishing or polishing, No. 4 finish is recommended.

No. 4 is a general-purpose polished finish widely used for architectural panels and trim, and for dairy, restaurant, and kitchen equipment. Following initial grinding with coarser abrasives, sheets are finally finished with lubricated 120- to 150-mesh abrasive belts.

No. 6 is a dull satin finish having lower reflectivity than No. 4. It is produced by tampico brushing No. 4 finished sheets in a medium of abrasive and oil and is used for architectural applications and ornamentation where a high luster is undesirable. It also is used to contrast with brighter finishes.

No. 7 is a finish with a high degree of reflectivity, produced by buffing a finely ground surface without removing the grit lines. It is used chiefly for architectural and ornamental purposes.

No. 8 is the most reflective finish and is obtained by polishing with successively finer abrasives and buffing extensively with very fine buffing rouges. The surface is essentially free of grit lines from preliminary grinding operations. This finish is most widely used for press plates, as well as for small mirrors and reflectors. Sheets can be produced with one or both sides polished. When polished on one side only, the other side may be rough ground in order to obtain the necessary flatness. The relationship between abrasive grit numbers and surface roughness in terms of microinches is sometimes a basis for specification. The values are approximately as follows:

*Joseph A. Douthett, Senior Staff Engineer, Armco Inc.; Robert R. Gaugh, Senior Staff Engineer, Armco Inc.; Bob Srinivasan, Research Manager, Diversey Wyandotte Corp.

Abrasive grit No.	Surface roughness μm	μin.
500	0.10 to 0.25	4 to 10
320	0.15 to 0.38	6 to 15
240	0.20 to 0.51	8 to 20
180	0.64 max	25 max
120	1.14 max	45 max
60	3.56 max	140 max

Strip finishes are designated only as No. 1 or No. 2, although considerable latitude in appearance is allowed. For the No. 1 finish, strip is cold rolled to the specified thickness, then annealed and pickled. The appearance varies from a dull matte to a fairly reflective finish, depending on the composition of the steel. This finish is used principally for formed parts, or in applications where the brighter No. 2 finish is not required.

To produce the No. 2 finish, a final and light cold roll pass is applied to the No. 1 finish, generally using highly polished rolls. The No. 2 finish is a general-purpose finish and also minimizes stretcher strains that may appear in ferritic grades.

Bright annealed stainless strip is available in commercial quantities in both No. 1 and No. 2 finishes. The product generally is superior to conventional annealed strip in reflectivity, buffability, and corrosion resistance.

Most stainless strip is produced with a No. 2 finish. Applications include automotive moldings, cookware, appliances, and architectural trim. The finish is suitable for moderate drawing and forming and is easy to buff. A No. 2 finish should be used for those grades (such as type 430 or type 302 clad with low-carbon steel) that are susceptible to stretcher strains and are to be used for light drawing or forming. Additional special cold rolled finishes may be imparted to the strip by conditioning the roll surface by grinding, blasting, etching, or machining.

Grade Limitations. Not all of standard compositions of stainless sheet and strip are available in each of the standard mill finishes. Surface finishes of these products depend on end use rather than any restriction imposed by the supplier.

Table 1 lists the finishes most often applied to sheet and strip of a number of standard grades of austenitic, martensitic, and ferritic steels.

Bar and Tube. Surface finishes applicable to stainless steel bars include:

Hot worked only

- Scale not removed (excluding spot conditioning)
- Rough turned
- Pickled, or blast cleaned and pickled

Annealed or otherwise heat treated

- Scale not removed (excluding spot conditioning)
- Rough turned
- Pickled, or blast cleaned and pickled
- Cold drawn or cold rolled
- Centerless ground
- Polished

Annealed and cold worked to high tensile strength

- Cold drawn or cold rolled
- Centerless ground
- Polished

Rough turned, centerless ground, and polished surface finishes are applicable to round bars only. Bars of the 400 series stainless steels that are highly hardenable, such as types 414, 420, 431, 440A, 440B, and 440C, are annealed before rough turning. Other hardenable grades, such as types 403, 410, 416, 416(S), and 416(Se), also may require annealing, depending on their composition and size. Bars that are annealed and cold worked to high tensile strength are produced in types 302, 303(S), 303(Se), 304, and 316.

The technology of tubing is not completely standardized, but the entire range of finishes is generally found, including (*a*) dull and smooth, (*b*) white pickled, (*c*) ground finishes from abrasives up to a fineness of 500 mesh, (*d*) rough polished, and (*e*) tumbled.

Preservation of Mill Finishes. In making a finished part, the original mill finish may be retained with little or no modification depending on fabrication requirements. If this finish is satisfactory, no additional finishing operations are necessary. Therefore, every effort should be made to preserve the mill finish while the steel is in storage or being processed. The following preventive measures will serve to minimize additional cleaning and polishing:

- Steel should be kept in original containers or wrappers until fabrication actually begins
- Steel should be stored indoors, on clean racks, shelves, or platforms, and should be covered wherever possible
- Storage areas should be kept free of shop dirt, pickling or plating fumes, particles of scale from steel fabrication, and other contaminants
- Storage areas should not be located beneath line shafting from which lubricating oils and grease may fall
- Fabricating equipment should be cleaned of all residues before being used for a new operation on stainless steel
- Steel should be handled with clean gloves or cloths, to avoid finger marks
- Only marking materials that leave no permanent blemishes on work should be used
- Whenever feasible, paper or other protective covering should be placed on all surfaces between processes
- Chips should not be removed by compressed air, which may contain oil or other contaminants

Table 1 Mill finishes available on stainless steel sheet and strip

See text for explanation of numerical designations of finishes

Type	Sheet Unpolished	Sheet Polished	Strip
Austenitic steels(a)			
201	2D, 2B	(b)	1, 2
202	2D, 2B	3, 4	1, 2
301	2D, 2B	(b)	1, 2
302	2D, 2B	3, 4, 6, 7	1, 2
302B	2D	(b)	...
304	2D, 2B	3, 4, 6, 7	1, 2
304L	2D, 2B	4	1, 2
305	2D, 2B	(b)	1, 2
309, 309S	2D	(b)	1, 2
310	2D	(b)	1, 2
316	2D, 2B	4	1, 2
316L	2D, 2B	(b)	1, 2
321	2D, 2B	(b)	1, 2
347	2D, 2B	(b)	1, 2
348	2D, 2B	(b)	1, 2
Martensitic steels			
403	2D, 2B	(b)	1, 2
410	2D, 2B	(b)	1, 2
420	(c)	(c)	1, 2
440A, B, and C	(c)	(c)	(d)
Ferritic steels(a)			
430	2D, 2B	3, 4	1, 2
446	2D, 2B	(b)	1, 2

(a) All grades listed, in both sheet and strip form, are regularly available in the smooth rolled and bright annealed condition. (b) Usually not polished. (c) Not available in sheet form. (d) Material available in strip form on special order only; finish negotiated with supplier

Abrasive Blast Cleaning

Aspects of abrasive blast cleaning that apply particularly or exclusively to stainless steel include sand blasting, shot blasting, and wet blasting. For more complete information on abrasive blast cleaning methods and equipment, see the article on abrasive blast cleaning in this Volume.

Sand blasting is effective for rapidly removing heavy or tightly adhering scale before acid pickling. Applied to stainless steel, it is not a complete cleaning procedure. Types of work for which it is frequently used include heavily scaled plate sections, forgings and castings, and parts made of straight chromium steel that have developed a tightly adhering scale during annealing. It is fast and economical when used in conjunction with a final pickling treatment.

Only clean silica sand should be used. If the sand is iron-bearing or becomes contaminated with scale, minute particles of these contaminants can become embedded in the metal. The only sure way to remove such contaminants and to produce a thoroughly clean, rust-free surface is to follow the blasting with an acid pickling treatment.

Sand blasting should not be used on materials that are too light to stand the blast pressure, because distortion results from the local stretching caused by impingement of the sand. The blast should be kept moving to avoid excessive cutting at localized points.

Shot Blasting. The use of carbon steel shot, steel wire, or iron grit as blasting media are not recommended as they may cause particles of iron to become embedded and seriously detract from the corrosion resistance of stainless steel surfaces. Unless these contaminants are completely removed by acid pickling, they can rust and begin pitting.

The use of stainless steel shot or grit reduces the danger of rusting, but it cannot eliminate the possibility of residual oxide scale. Final pickling is mandatory for maximum corrosion resistance of surfaces so treated.

Stainless steels, and particularly those of the 300 series, work harden when they are subjected to cold working. Therefore, if work hardening of the surface is undesirable for a particular application, shot blasting should not be used.

Wet blasting is adaptable for use with stainless steel. Various abrasives conveyed in liquid carriers are discharged at the work by compressed air. A variety of finishes can be obtained through selection of abrasives and adjustment of pressures. Finishes that are much smoother than those resulting from blasting with dry sand and that are similar in appearance to a No. 6 (tampico brushed) finish may be obtained by wet blasting. As in sand blasting, a final pickling treatment and water washing are required.

Acid Pickling

Nitric acid solutions of 15 to 20% concentration, used at 38 to 50 °C (100 to 120 °F), remove iron particles but not residual oxide scale. For pickling following abrasive blast cleaning, an aqueous solution containing 10 vol % nitric acid and 1 to 2% hydrofluoric acid is suggested. This solution should be used at a maximum temperature of 60 °C (140 °F) and a maximum immersion time of 3 min. The operation should be used with care on the martensitic alloys in the fully hardened condition because of the danger of hydrogen embrittlement cracking.

Immersion should be followed with (*a*) thorough scrubbing and water washing, (*b*) treatment with 20% nitric acid solution at 50 to 60 °C (120 to 140 °F), and (*c*) a final rinsing operation with clean hot water. In certain applications, (b) is omitted.

If this final pickling operation must be done by swabbing, a nitric-hydrochloric acid solution of the following approximate composition by volume should be applied: 20% nitric acid, 2% hydrochloric acid, and 78% water. This mixed acid solution should be discarded before its ferric chloride content increases enough to cause pitting of the chromium-nickel grades. In cleaning of low-carbon steels, 8% $FeCl_3$ can cause pitting. The concentration of $FeCl_3$ sufficient to cause pitting depends on acid concentration, temperature, and type of steel being pickled, however.

Before a stainless steel is subjected to acid pickling, the following information should be obtained: (*a*) the composition of the alloy, (*b*) its metallurgical and physical condition, and (*c*) the nature of the scale that is to be removed by pickling. The acids most commonly used for scale removal are nitric, hydrofluoric, sulfuric, and hydrochloric. However, there is no single acid solution or process that is effective for removing oxide scale from all grades of stainless steel. A more complete discussion of the acid pickling of stainless steel appears in the article on pickling of iron and steel in this Volume.

Salt Bath Descaling

The removal of oxide scale may be accelerated by using baths of molten sodium hydroxide to which certain reagents are added. These baths can be used with virtually all grades of stainless steel. Salt bath descaling has several advantages:

- It acts only on the scale and does not result in metal loss or etching
- It does not preferentially attack areas in which intergranular carbides are present
- It is particularly useful in descaling the straight chromium grades without the preliminary sand blasting that is frequently required prior to acid descaling

Use of molten salts is not recommended for those stainless steels that precipitation harden at the operating temperature of the bath. Procedures and equipment for descaling in molten salt are described in the article on salt bath descaling in this Volume.

Sequence of Cleaning Methods

Usually, more than one method of cleaning is used to remove scale from stainless steel. For instance, in mill processing sheet, salt bath descaling and acid pickling often are used for a cold rolled product and abrasive blast cleaning and acid pickling for a hot rolled product.

In a continuous annealing and pickling line, multistep cleaning usually occurs twice in producing stainless steel sheet. The sequence of operations is: hot roll, anneal, descale, pickle, cold reduce 30 to 50%, anneal, descale, and pickle. As shown in Fig. 1, salt descaling is used for cold rolled sheet, whereas the salt is bypassed and abrasive blast cleaning is used when a hot rolled product is being cleaned. Salt also can be eliminated on low-chromium cold rolled sheet.

Cold rolled sheet is left in the salt bath 10 to 15 min. Oxides are not removed from the surface. The rack of hot sheet is transferred to a water bath, and the violent reaction that results on immersion blasts much of the scale from the steel. The surface is essentially free of scale, but quite dull and dirty. To remove the last remnants of oxide, the rack is transferred to a 10 to 15% sulfuric acid solution at 70 °C (160 °F) for 3 to 5 min. After removal

Fig. 1 Sequence of annealing and scale-removing operations in mill processing of stainless steel sheet

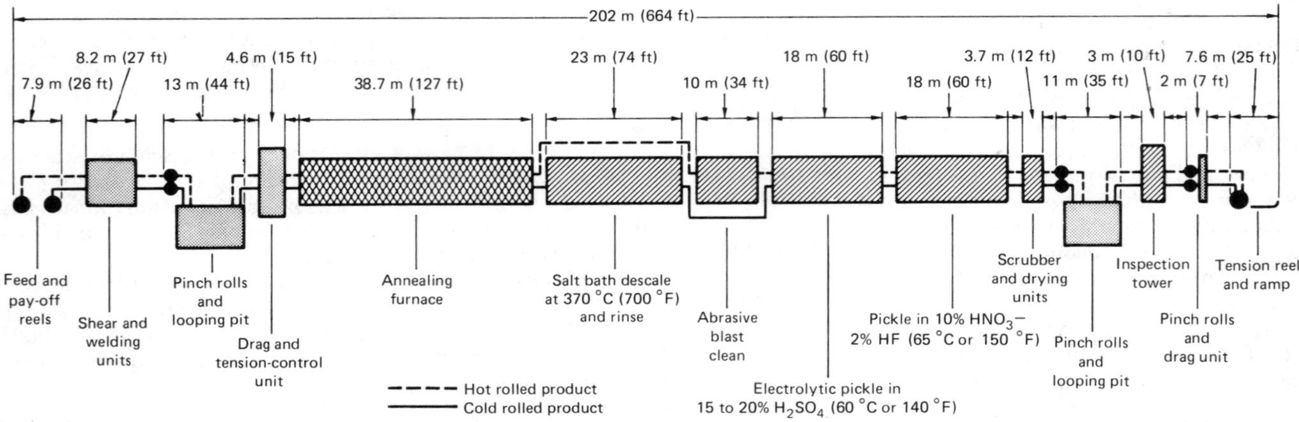

from the sulfuric acid tank, the rack of sheet steel is rinsed and then immersed for 1 to 2 min in nitric-hydrofluoric acid (8 to 10% HNO_3, 1 to 2% HF) at 60 °C (140 °F). This removes all smut and residue and produces a white finish. The sheet is finally rinsed, removed from the rack, scrubbed, and dried with an air blast.

Mass (Barrel) Finishing

Mass (barrel) finishing provides a combined tumbling and abrasive action that can be advantageously used for surface treating stainless steel parts. It is adaptable to removal of burrs, scale, and residual flux, and it also can be used for light surface treatment, such as cleaning, burnishing, or coloring. Methods and equipment for mass finishing are discussed in detail in the article on mass finishing in this Volume.

Post-Treatments. Once a mass finishing treatment is completed, stainless steel parts are rinsed thoroughly in water and dipped in a 20 vol % solution of nitric acid at 50 °C (120 °F) for 10 to 15 min. This is followed by another water rinse. If any oxide scale remains on the work after mass finishing, it must be removed by acid pickling.

Mass Finishing Processes

The following examples describe mass finishing procedures that are applied to stainless steel production parts.

Eggcups. Because of its shape, a stainless steel eggcup cannot be finished economically by polishing and buffing. Mass finishing is much less expensive and provides a commercially acceptable finish.

In one plant, eggcups are tumbled in a horizontal, two-compartment barrel, which has a steel shell and is completely lined with hardwood. All interior walls, partitions, and trapdoors are covered with hardwood to prevent contact between the stainless cups and the steel shell.

After preliminary washing of the eggcups, granite chips (sizes 4 or 5) are placed in the barrel with them; the ratio of cups to chips is maintained at 1 to 1. The barrel is then completely filled with warm water and a compound containing lye and a detergent. For a normal finish, the tumbling cycle is continued for 2 h. If at the end of this cycle a still brighter finish is desired, the following procedure is used:

- Drain water and compound from barrel
- Refill barrel with clear water
- Tumble load for several minutes
- Drain liquid from barrel
- Add water and compound as for initial tumbling cycle
- Tumble load for a time depending on the degree of brightness desired

Impellers. Used in milk-processing equipment, the small impeller shown in Fig. 2 is made of type 316 sheet and requires a 2B finish. This finish is obtained by fixture tumbling in two stages.

First, the parts and holding fixture are secured to the finishing barrel, and a mixture of fused alumina media of various sizes to accommodate large and small radii is added until the barrel is about 70% full. A slightly alkaline coarse abrasive compound, which serves to increase cutting action, is also added. Water level is about 25 mm (1 in.) above the tumbling mixture. The barrel is rotated clockwise for 2 h and then counterclockwise for 2 h to ensure positive action of the mixture on all surfaces. The 4-h cycle removes sharp edges, small nicks and scratches, and provides a dull satin finish.

Fig. 2 Impeller polished to 2B finish by fixtured tumbling

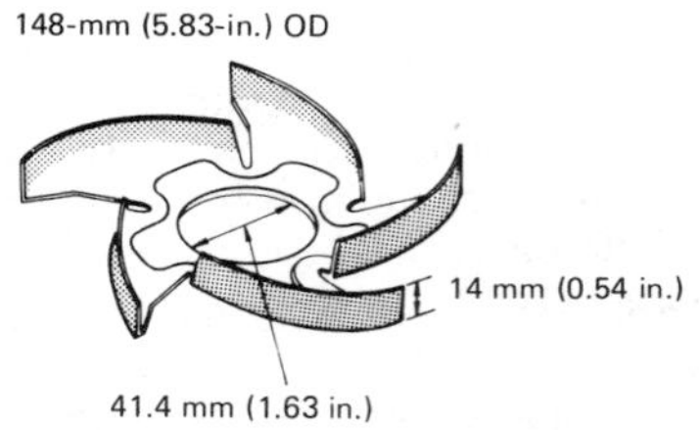

The impeller (type 316 sheet) is used in milk processing equipment

To obtain the required 2B finish, the barrel is thoroughly flushed with water until all abrasives and dirt have been removed. A completely soluble nonabrasive (burnishing) compound is then added to serve as a lubricant and to color the parts, and the water level is increased to 75 mm (3 in.) above the mixture to aid dirt suspension. The same 4-h, two-direction tumbling cycle is then repeated.

Heat-exchanger plates press formed of 1.27-mm (0.050-in.) thick type 316 stainless steel sheet are fixture tumbled to restore original mill finish and color. Twenty-six of these plates (230 mm or 9 in. wide, 800 mm or 31½ in. long) are held in the fixture, which is designed to retain them on edge and at right angles to the axis of the barrel. Fixture and parts are loaded into the barrel together with 225 kg (500 lb) of No. 6 bonded chips, which fill the barrel to 65% of its capacity. Then a low pH descaling compound is added to remove oxidation and discoloration, and, finally, water is added to 25 mm (1 in.) above the chip level.

The barrel rotates, at 15 rev/min, for ½ h clockwise and ½ h counterclockwise. Following this cycle, the barrel is flushed with water to remove all residue and is recharged with a nonabrasive burnishing compound that produces a high color and luster. The water level is increased to 75 mm (3 in.) above the chip level to allow for heavy suds and to improve dirt suspension. The same rotation cycle is then repeated, except that running time is increased to 2 h. After burnishing, the parts and fixture are removed from the barrel and flushed with hot water.

Grinding

Although stainless steel can be readily ground, polished, and buffed, certain modifications of normal processing conditions are required:

- Because stainless steels have comparatively high tensile strength, more power is required for metal removal
- Low rates of heat conductivity require precautionary measures to avoid excessive heating
- With the high work-hardening chromium-nickel grades, grinding speeds and feeds should be low enough to prevent introducing residual stresses, which may increase susceptibility to stress corrosion cracking

Grinding is used for preliminary surface conditioning before polishing, and as a means of removing excess metal from weld beads and flash from forgings and castings. It also is used for dressing gas-cut welding edges. These operations usually require coarse or rough grinding. However, as the refinement of the surface increases with the use of finer abrasives, grinding begins to approach polishing; the dividing line between fine grinding and polishing is seldom clear cut.

Solid wheels are used for coarse grinding and include the vitreous and rubber-bonded or plastic-bonded types. They should be free-cutting to avoid loading and glazing. Abrasives frequently used are aluminum oxide and silicon carbide, in grit sizes ranging from 20 to 36 (for initial or coarse cutting) to 60 (for subsequent finishing work).

Surface speeds for solid wheels usually range between 25.5 and 30.6 m/s (5000 and 6000 sfm). For safety, wheels should never be operated above their maximum permissible speed. To realize the maximum cutting efficiency of which they are capable, they should not be run at less than recommended speeds.

Wheel Operation. Grinding wheels should never be forced; forcing causes excessive wheel breakdown and localized high temperature in the workpiece. In addition, the wheel should not be allowed to ride on the workpiece with insufficient pressure, because this causes rapid glazing of the wheel.

The elimination of heat buildup in localized zones is of major importance with stainless steel. The low thermal conductivity of stainless steel contributes to increased thermal distortion. This applies particularly to the chromium-nickel grades, of which the coefficients of thermal expansion are relatively high. Holding the metal at low temperature avoids heat tinting, which becomes evident at 230 to 260 °C (445 to 500 °F) and above. Marked increases in metal temperature can (*a*) reduce the hardness of heat treated grades, and (*b*) precipitate carbides in the unstabilized chromium-nickel grades, which are susceptible to intergranular corrosion.

Rough surfaces on weldments, castings, and forgings are cut down to the general finish contour with a solid wheel of either aluminum oxide or silicon carbide of grit size from 20 to 40, and with a bond loose enough to allow the wheel to remain open and free-cutting without excessive wheel wear. Again, it is important to prevent local overheating, which can cause either mechanical or metallurgical damage, and to avoid excessive metal removal, allowing sufficient material for further finishing. Surface contamination is not important at this stage of finishing.

The next roughing operation, which in many applications may replace the solid-wheel operation, requires the use of a portable disk grinder, powered by air or electricity, using abrasive disks coated with aluminum oxide ranging in grit size from 80 to 120. Localized overheating may be prevented by applying a small stream of water or water-soaked rags to the side not being ground.

Weld Beads. Excess metal in weld beads ordinarily is removed by grinding, although an initial cut may be made with a cold chisel when the size of the bead warrants. Grinding procedures and precautions conform to those previously described, except that the width of weld beads precludes right-angle cutting with successive grit sizes. Usually, a raised bead will take a fairly coarse grit at the outset. Canting of wheels must be avoided; otherwise, grooves may be cut parallel to the bead and undesirable thinness will result. Limiting stops attached to portable grinders may be installed to prevent canting and excessive metal removal.

When grinding is to be followed by polishing, as for weld joints on polished sheet, the grinding operation must terminate sufficiently above the level of the base metal to allow enough metal for final polishing to finish flush, without a ridge or groove. Limiting stops on grinding machines are helpful.

For economic reasons, grinding of weld joints in cold rolled or polished sheet should be held to a minimum by using welding procedures that avoid high beads of excess metal. The grinding step can often be eliminated, allowing the workpiece to go directly into the polishing operations where finer abrasives are used.

Metal adjacent to beads that are being ground should be protected from flying bits of metal cuttings by shields of material such as paper. Wet rags may be laid on the workpiece to absorb heat and reduce thermal distortion, particularly on thin-gage material.

Progressive Grinding. To remedy an existing surface condition, such as removing scale patterns or indentations from hoisting clamps on an annealed plate, a series of wheels of decreasing grit size is often needed. The initial grit size is selected on the basis of which coarseness is needed to remove the major portion of the unwanted condition. After the workpiece has been partly dressed down, the operation is completed by using a graduated series of successively finer wheels until the desired final finish is attained. The use of a relatively soft plastic wheel impregnated with fine, sharp grit makes it possible to reduce the number of finishing operations, because of the combination of free cutting and wheel resilience for ease in blending.

The direction of wheel traverse across the work is changed by 90° with each change in grit size to remove residual grinding lines. As each change is made, workpiece surfaces should be brushed thoroughly to remove any particles of the preceding abrasive or of metal cuttings that mar the performance of the finer grit to follow. The progression from coarse to fine grit size may be made in steps of 20 to 40 mesh.

When using a flat disk grinder, with which cutting is performed against the face (instead of the rim) of the wheel, a rotary or circular traversing motion is most frequently used. This eliminates the need for reversing the direction of grinding with each change of grit size.

Belt Grinding. Belts carrying abrasives of various grit sizes are widely used for grinding and polishing stainless steel surfaces. Although many complex shapes can be belt ground, a simple projection may make belt grinding impossible; for example, the studs welded to a cookware pot or pan to which the handle is affixed. In this application, the finishing operations, from grinding to color buffing, must be completed before the studs are welded in place.

Mechanized Belt Grinding. In mechanized belt grinding, longer belts provide a longer belt life and dissipate heat more effectively. Thus, longer belts frequently are operated without a coolant, eliminating a post-grinding cleaning operation.

In belt grinding rectangular stainless steel sinks, for example, it was necessary to use short belts to reach all internal surfaces. Each sink was ground in a machine with four grinding heads indexing about the main column of the machine while the sink was held in a cradle that rotated in a horizontal plane. A narrow belt ground the radii between the bottom and the side walls of the sink. Wider belts were used to grind the side walls; these belts were comparatively short and required the use of a coolant. Belt abrasives varied in grit size from 80 to 220 mesh, depending on the desired finish.

On a typical grinding machine using a 2200-mm (86-in.) wide abrasive belt and powered by a 250-hp ac motor, a positive hydraulic reciprocating drive permits instantaneous variations in table speed; belt speed is fixed at 25.5 m/s (5000 sfm). The 2200-mm (86-in.) belt travels over a conventional vertical-head assembly consisting of a dynamically balanced upper idler roll of steel and a rubber-covered serrated lower contact roll. A pneumatic belt-centering device ensures positive tracking of the abrasive belt and is adjustable to compensate for belts of different widths. Incorporated into the entire worktable is a 2200-mm (86-in.) wide vacuum chuck 4.9 m (16 ft) long. To produce single or compound tapers, a worktable can be tilted to any angle, for either right-hand or left-hand tapered sheets.

Type 302 stainless steel sand cast heat treating fixtures, each 1500 mm (60 in.) long, 810 mm (32 in.) wide, and 200 mm (8 in.) thick, and weighing 270 kg (600 lb), were tested to determine if fixtures of this type and weight could be refinished by coated abrasive methods. Specified finish for the working surfaces of the fixtures was 2.5 μm (100 μin.). With belt speed at 25.5 m/s (5000 sfm) and table speed at 0.03 to 0.04 m/s (6 to 8 sfm), the fixtures were rough ground and finish ground, using waterproof cloth belts coated with aluminum oxide abrasive. Grit size was 36-mesh for rough grinding, 60-mesh for finish grinding. Coolant was water-soluble oil.

To attain the specified 2.5-μm (100-μin.) surface, approximately 1.3 mm (0.050 in.) of material was removed from each side of each fixture during rough grinding, and 0.25 mm (0.010 in.) in finish grinding. The rough grinding cycle required about 2½ h for each side and finish grinding about ½ h. Planing, previously used for refinishing these fixtures, had required 4 h for each side.

Type 347 stainless steel sheet was tested to determine if a desired scratch pattern and an 0.8-μm (30-μin.) finish could be obtained using this equipment. Each sheet was 1800 by 910 by 0.9 mm (72 by 36 by 0.035 in.). Because of thinness, using the vacuum-chuck worktable to hold the sheet down during grinding would have been impractical. A tension fixture was devised especially for this purpose.

Grinding was done at a belt speed of 25.5 m/s (5000 sfm) and at table speeds of 0.04 and 0.05 m/s (8 and 10 sfm). Belts used were waterproof cloth coated with aluminum oxide abrasive (120-mesh for roughing, 150-mesh for finishing). A water-soluble oil coolant was used.

Material removed, one side only, was approximately 0.05 mm (0.002 in.) in rough grinding and 0.025 mm (0.001 in.) in finish grinding. Rough grinding and finish grinding cycles each required 20 min. The desired scratch pattern and the 0.75-μm (30-μin.) surface finish were obtained on the sheet.

Type 310 stainless steel surface plates (nonmagnetic) were tested to determine if required finish of 0.5 μm (20 μin.) could be obtained and also to compare the time required for abrasive belt grinding with that for resurfacing these plates with a surface grinding operation. Approximately 25 plates, most ranging in size from 760 by 150 by 32 mm to 460 by 760 by 25 mm (30 by 6 by 1¼ in. to 18 by 30 by 1 in.), were ground; the largest plate processed was 3 m (10 ft) long and 460 mm (18 in.) wide.

Grinding was performed using waterproof cloth aluminum oxide abrasive belts of 36-mesh grit size for roughing, 60-mesh for finishing. Coolant was water-soluble oil. Belt speed was 25.5 m/s (5000 sfm); table speed was 0.025 to 0.03 m/s (5 to 6 sfm). In rough grinding, 0.3 to 0.4 mm (0.012 to 0.016 in.) of material was removed, which included removal of hard scale on the top surface of each plate. In finish grinding, 0.15 to 0.2 mm (0.006 to 0.008 in.) was removed.

The required surface finish of 0.1 μm (20 μin.) was attained. Total grinding time for the largest plate was 2½ h; this compared with 6 h required for resurfacing these plates on a large surface grinder.

Belt life is influenced primarily by (*a*) belt speed, (*b*) type of material being ground and its hardness, (*c*) pressure of the belt against the work, (*d*) type of contact roll, (*e*) type of lubricant (if a lubricant is required), and (*f*) uniformity of the finish desired. In belt grinding stainless steel, the recommended belt speed is approximately 20.4 m/s (4000 sfm).

With the exception of the precipitation-hardening stainless steels, the hardness of the steel has a greater effect on the life of a grinding belt than its composition. Hardness also affects surface finish, and a high-quality finish is easier to obtain on a harder stainless than on one that is softer. With precipitation-hardening alloys, best grinding results are obtained with a waterproof cloth abrasive belt and a water-soluble oil lubricant.

Pressure of the workpiece against the abrasive belt is probably the most important factor affecting belt life. Excessive pressure on a new belt causes glazing of the abrasive and greatly reduces cutting action. Therefore, new belts should be subjected to very light pressures during the break-in period and until the belt is capable of maintaining a uniform cutting action. A light belt pressure is preferred, and increased stock removal should be obtained by changing to a coarser belt.

The contact roll that backs up the abrasive belt is another important factor; a properly serrated contact roll may increase belt life by as much as 60%. The angle of the serration affects both belt life and finish. To obtain fine finishes, contact rolls serrated at an angle of 75° to the axis of the spindle should be used. Rough finishing requires a 30° angle. For general work, contact rolls may be made with a 45° serration, using a 9.5-mm (3/8-in.) groove and 9.5-mm (3/8-in.) land. Hardness of the contact roll should range between 50 to 65 on the Shore scleroscope A-scale.

When lubricants are used, they should (*a*) maintain free-cutting (nonloading) edges, (*b*) add color to the finished product, (*c*) maintain a cool cutting surface, and (*d*) be easy to apply. Lubricants may be in the form of grease sticks, waxes, or cutting oils. Cutting oils generally are more effective when they are diluted as much as 4 to 1 with kerosine.

Safety. Metal fines collected in a container near machines during the belt grinding of stainless steel should be removed regularly, because the fines, together with polishing compounds or oils that are collected with them, constitute a potential fire hazard. Fires in the duct system can be extremely serious because of the high air flow in the ducts. Fire extinguishing equipment should be close to any machines using abrasive belts.

When wide abrasive belts are used, equipment with automatic tracking to center the belt in relation to the work is advantageous. All abrasive belt machinery should be equipped with motors that are totally enclosed.

Safety training for operators should begin with thorough instruction in the proper use of equipment, because most severe injuries result from improper use of equipment. Most common injuries are burns, cuts, and eye injuries. Serious accidents may arise from the snagging of parts because of improper loading, improper use of lubricating devices, and careless placement of hands and arms while the machine is in operation.

Polishing

Polishing operations use abrasives that are mounted on prepared shaped wheels or on belts that provide a resilient backing. The stainless steel to be polished may be in either a smooth rolled or a previously ground condition. For the smooth rolled condition, the starting grit size should be selected in a range of 150 to 220. For the ground condition, the initial grit should be coarse enough to remove or smooth out any residual cutting lines or other surface imperfections left from grinding. In either instance, the treatment with the initial grit should be continued until a clean, uniform, blemish-free surface texture is obtained. The initial grit size to use on a preground surface may be set at about 20 numbers finer than the last grit used in grinding and may be changed, if necessary, after inspection. A tallow lubricant may be used to reduce the sharpness of cutting. With broad-belt grinding of stainless steel sheet, the use of a lubricant or coolant is mandatory.

After completion of the initial stage of polishing, wheels or belts are changed to provide finer grits. The step-up in fineness is usually by 30 to 40 numbers. Each succeeding treatment is continued until all residual marks of the preceding cut are removed. Grease in stick form is applied to wheels carrying abrasives of 150-mesh and finer. Aluminum oxide buffing compounds and powdered pumice are preferred for use with abrasives of 200-mesh and finer grit size.

Polishing speeds are generally somewhat higher than those used in grinding. A typical speed for a coated-wheel operation is 38.3 m/s (7500 sfm).

The same precautions that must be observed in the grinding of stainless steel are equally applicable to polishing:

- Avoidance of iron or other contamination
- Care of wheels and belts when not in use
- Restriction of the use of wheels and belts to stainless steel only
- Avoidance of excessive pressure while polishing
- Operation at proper speeds
- Avoidance of localized heat buildup because of dwelling at one spot
- Removal of loose cuttings and bits of abrasive from work surfaces before changing from one grit size to another

Buffing

Buffed finishes are produced on stainless steel surfaces by using equipment, materials, and techniques that are similar to those used on other materials (see the article on polishing and buffing in this Volume). However, the skill needed for producing the high lusters obtainable on stainless steel is gained only through actual experience. Buffed finishes are not recommended for the stabilized grades of stainless steel, such as types 321 and 347, because these materials contain fine, hard particles of titanium or columbium compound that show up as pits on bright finishes.

The first step in applying a buffed finish of desired luster and color is that of providing a smooth surface, free of scratches and any other defects. For this reason, buffing is generally performed in two stages: the first is known as hard buffing (or cutting down); the second, as color buffing.

Hard buffing follows polishing, which generally ends with the use of abrasives of 200- to 250-mesh grit size. The fine scratches left by polishing are cut down with a buff that carries no previously glued on abrasive. Instead, such abrasive as is needed is applied intermittently to the buffing wheel as it rotates, either by rubbing a cutting compound in bar or stick form against it or by spraying it with a liquid compound. These cutting compounds contain (*a*) very fine artificial abrasives, such as aluminum oxide, of about 300-mesh grit size; and (*b*) a stiff grease or other material that acts as a binder. They adhere to the wheel by impregnating the cloth disks. Hard buffing may be conducted at from 33.2 m/s (6500 sfm) up to a maximum of 51 m/s (10 000 sfm).

Color buffing is performed in the same manner as hard buffing, except that a coloring compound is substituted for the cutting compound and speeds are held below 35.7 m/s (7000 sfm). Various compounds (rouges and other extremely fine abrasives) for use on stainless steel are available commercially, both in bar form for hand application and in liquid form for automatic application. The use of any material that may result in loss of corrosion resistance by stainless steel surfaces should be avoided.

Direct color buffing, without previous polishing or hard buffing, may be satisfactory for certain applications, such as:

- A color-buffing wheel may be applied directly on type 430 that has been given a finishing pass on a polished mill roll after final pickling
- Small articles blanked from bright finished straight chromium steel strip and then tumbled for burr removal may have a satisfactory appearance if run under a color buff for brightening
- Smooth, defect-free surfaces that have been electrolytically polished provide a good base for color buffing. By masking before buffing, contrasting surface effects can be obtained, as a result of the difference in reflectivity obtainable by electrolytic polishing and by color buffing

Color buffing, however, does not remove scratches or other surface defects because cloth wheels without coarse abrasives, which are used for color buffing, do not remove surface imperfections. Therefore, the continued presence of such imperfections on finished products must be expected.

Effect of Polishing and Buffing on Corrosion Resistance. In addition to altering the appearance of stainless steels, polishing and buffing may have a considerable effect on the corrosion resistance of these materials. For example, a steel with a No. 2B finish as received from the mill has excellent corrosion resistance. This can be adversely affected by polishing with coarse abrasive, but it can be fully restored by polishing to a No. 4 finish or higher. Polishing to a No. 7 or 8 finish, by removing very fine pits and other surface defects, improves corrosion resistance over that afforded by the original No. 2B finish.

Tanks for storing raw milk provide a commercial example of the importance of a polished finish to sanitary and corrosion-resistance properties. According to the sanitary codes, these tanks must be made of series 300 stainless steel, and all surfaces that come in contact with the milk must be polished to a pit-free No. 4 finish or better. The high finish not only promotes sanitary properties, but also provides improved resistance to corrosion by the chlorine-bearing chemical used in scouring the tanks after each use.

Several polishing and buffing compounds contain iron and iron compounds, which can be highly deleterious to the corrosion resistance of stainless steel. The amount of iron in these compounds that can be tolerated is extremely small (for maximum protection, less than 0.01% iron). If more than one polishing operation is involved, slightly more iron can be tolerated in the early stages of polishing, but the final stage should be virtually iron-free. Magnetic oxides of iron are as damaging as iron powder; the oxides generally occur in Turkish emery as well as in several synthetic abrasives. Their presence is most accurately determined by chemical analysis.

Cleaning and Passivation After Buffing. Cleaning is always required after a final buffing operation in which a surface finish of No. 4 or finer is achieved. The workpiece is vapor degreased or is cleaned with whiting (precipitated calcium carbonate), powdered chalk, or dehydrated lime, which is applied with a soft flannel cloth. This picks up the grease or lubricant from a color-buffing operation. The workpiece must then be protected from damage in handling.

Passivation is not required after buffing or fine polishing if the surface obtained is chemically clean and free of oil, grease, or adhesives used in the polishing media. A clean surface passivates itself naturally when it is exposed to air. However, if foreign metal, such as iron, has been picked up in the buffing operation, it must be removed by pickling or passivation.

Procedures and Equipment for Grinding, Polishing, and Buffing

Mechanical finishing procedures, which include operational details of grinding, polishing, and buffing, are summarized in Table 2, which outlines in normal sequence the major steps entailed in matching the principal mill finishes or in obtaining a desired surface effect. Table 2 also serves as a general guide to wheel selection, polishing compounds, and wheel speeds. For additional information on equipment, procedures, and materials, see the article on polishing and buffing in this Volume.

Matching Mill Finishes

In the fabrication of No. 4 polished sheet, it frequently is necessary to refinish weld zones to blend them with the original finish. Although it is virtually impossible to match a machine polished surface except by duplicating the original polishing, a close blending may be obtained by skillful use of manual methods.

If the original machine polished lines are parallel with the line of the weld, the bead can be dressed down by grinding with a hard or soft wheel and then finished by polishing with, progressively, No. 80 and No. 120 (and possibly No. 150) grit on a setup wheel driven by a portable machine. The traversing of this wheel should be kept in line with the run of the bead so that its cut lines are kept parallel with those of the original machine polished surface.

To avoid residual ridges or grooves, the metal of the joint should be brought flush with that of the basis metal. For a given starting grit size, the depth of the scratches produced depends on the amount of use it has received; thus, samples should be run before starting on finish work.

If the machine polished lines are not parallel with the line of the weld, final manual polishing should be done in the direction of the machine polishing. If the original polish lines on the two sides of a joint are not parallel with each other (for example, if they are parallel with the bead on one side and perpendicular to the bead on the other), the best procedure is to run the polishing cut lines along (not across) the bead. The girth weld between a tank shell and head exemplifies this problem. The cut lines of the shell extend around the unit and lie parallel with the girth joint, whereas the cut lines on the head are perpendicular, parallel, and at an angle around the periphery. Swinging such an assembly on the faceplate of a large lathe would permit repolishing the head and dressing the weld joint on the same setup, thus rendering parallel all of the cut lines on the head, joint, and shell. For this application, abrasive paper (or pieces cut from a belt) may be backed up with a block of wood or some softer material and guided by hand along the line of the weld joint.

Table 2 Grinding, polishing, and buffing of stainless steel sheet and strip

Operation	Abrasive grit size	Type of wheel	Lubricant	Compound	Wheel speed m/s	Wheel speed sfm	Finish produced
1 Roughing(a)	80(b)	Neoprene serrated(c)	Water-soluble oil	...	20.4-25.5	4 000-5 000	Rough ground
2 Polishing(d)	100(b)	Neoprene serrated(c)	Water-soluble oil	...	20.4-25.5	4 000-5 000	No. 3
3 Polishing(e)	120 or 150(b)	Neoprene serrated(c)	Oil and kerosine	...	20.4-25.5	4 000-5 000	Base for No. 4
4 Polishing(f)	180(b)	Neoprene serrated(c)	Oil and kerosine	...	20.4-25.5	4 000-5 000	No. 4
5 Polishing(g)	220 or 240(b)	Neoprene serrated(c)	Oil and kerosine	...	20.4-25.5	4 000-5 000	Base for No. 7 or 8
6 Hard buffing(h)	...	Bias or pocket buff	...	Alumina, hard buffing	40.8-51.0	8 000-10 000	No. 7
7 Color buffing(j)	...	Loose-disk or Canton flannel buff	...	Alumina and/or chromium oxide, color buffing	15.3-30.6	3 000-6 000	No. 8
Special surface conditioning operations							
Tampico brushing(k)	...	Tampico brush	...	Powdered pumice and oil	5.1-7.7	1 000-1 500	No.6
Butler finishing(k)	...	Bias or pocket buff	...	Proprietary greaseless bar	25.5 max	5 000 max	Butler
Wire brushing(k)	...	Stainless steel wire brush	...	...	25.5 max	5 000 max	Wire brushed
Cleaning(m)	...	Canton flannel cloth (hand operation)	...	Whiting, precipitated chalk, or dehydrated lime	...	...	...

(a) Follows grinding with 36- to 60-mesh abrasive, in preparing a welded area or as the initial operation on heavy-gage sheets or plates. (b) Coated abrasive belt. (c) Contact wheel. (d) May be used as initial operation for light-gage sheets. (e) Used only long enough to develop a uniform finish. (f) Follows operation 3 for production of a No. 4 finish. (g) Should continue long enough to remove all grit lines from previous operations and to develop considerable reflectivity. (h) Should continue long enough to develop a mirror finish. (j) Relatively short operation, for development of maximum color and reflectivity. (k) Used after operation 4. (m) Required for removing any film of lubricant remaining after polishing or buffing. Work may be vapor degreased, alternatively to being hand-wiped with materials listed in table

Electropolishing

Electropolishing is used primarily to produce a very smooth, bright, easily cleaned surface with maximum corrosion resistance. It removes the surface layer of a metal by anodic treatment in an acid bath. The process is discussed in the article in this Volume on electropolishing, where examples of applications involving stainless steel are cited.

Electropolishing is applicable to all stainless steel grades, hot or cold finished, cast or wrought. The amount of metal removed is subject to close control, depending on the desired result. The resulting surfaces have a bright, passive finish. The process is most frequently applied to cold finished surfaces, because they yield a smoother finish than can conventionally be obtained on hot finished surfaces. As in electroplating, the results depend on the contour and shape of the part. The end-grain surfaces of the free-machining stainless grades, such as types 303 and 416, will appear frosty after electropolishing due to removal of the sulfide inclusions.

Electropolishing can be used as a preliminary brightening operation before final buffing, particularly on drawn parts with burrs, sharp radii, or recessed areas, and it serves to reduce the amount of buffing required. Electropolishing is applied to decorative automotive parts and accessories, conveyor systems for food handling equipment, animal cages, and pharmacy equipment. It provides an economical finish on many parts that are difficult or impossible to finish by conventional polishing, such as items made from wire.

In contrast to mechanical finishing methods, electropolishing may make inclusions in the material more visible. Some types of inclusions are dissolved out, whereas others remain in relief. Electropolishing has been used as a surface inspection technique to reveal residual foreign material, such as embedded scale and particles of iron, carbide precipitation, and weld defects. The surface obtained by electropolishing is directly related to the original surface quality — the process cannot be used to remove digs, gouges, scratches, and the like.

Chemical Polishing

Chemical polishing is another method for providing a smooth and bright surface on stainless steel. Unlike electropolishing, chemical polishing can be done without the use of electricity and without racking of individual parts. Thus, chemical polishing offers significant savings in capital investment and labor. In addition, chemical polishing offers a greater degree of freedom in polishing items with blind holes and other recessed areas. However, it does not produce the high specular reflectivity (brightness) obtained with electropolishing.

Proprietary products for chemical polishing are available in the market. Generally, they are based on combinations of phosphoric acid, nitric acid, sulfuric acid, hydrochloric acid, organic acids, and special surfactants and stabilizers to promote a high degree of brightness and long bath life. Unlike the nitric and hydrofluoric acid mixtures which are used in chemical cleaning, the proprietary chemical bright dips do not cause severe attack on the grain boundaries or intergranular corrosion.

Passivation

During handling and processing operations such as forming, machining, tumbling, and lapping, particles of iron or tool steel may be embedded in or smeared on the surfaces of stainless steel components. If allowed to remain, these particles may corrode and produce rust spots on the stainless steel. To prevent this condition, semifinished or finished parts are given a passivation treatment. This treatment, which consists of immersing stainless steel parts in a solution of nitric acid, or of nitric acid plus oxidizing salts, dissolves the embedded or smeared iron and restores the original corrosion-resistant surface. Passivation of machined parts made of free-machining austenitic or low-chromium grades must be done carefully. If the parts become etched when immersed in the standard solution, passivation can be accomplished by increasing the nitric acid content of the solution to 50% nitric acid (HNO_3) or by adding 4 to 6 wt % sodium dichromate ($Na_2Cr_2O_7$). The passivating treatment provides the incidental benefit of revealing components that were inadvertently made of the wrong material. For example, carbon steel and Monel can be readily detected because they are severely attacked during the operation, and vigorous etching occurs.

It generally is considered good practice to passivate parts after forming, machining, grinding, or lapping, although, in some applications, experience may prove it to be unnecessary. When the need for passivation is in doubt, the question may be resolved by testing. For example, passivated and nonpassivated parts may be subjected to 100% humidity at 38 °C (100 °F) and compared for the presence of rust spots at the end of exposure time. Intermittent wetting and drying over a 24-h period also may be used as an evaluation test. Another simple test consists of wetting the surface with a copper sulfate solution. Free iron, if present, plates out the copper from the solution, and the surface develops a copper cast or color. Several sensitive tests for the presence of iron on stainless steel surfaces are given in ASTM Standard A-380.

Passivating either should not be employed or may be omitted in some specific cases. For example, nitrided or carburized stainless steel should not be passivated, because the treatment severely corrodes the case. Machined or ground parts that are to be plated or electropolished need not be passivated, because any iron contamination will be removed by these operations. Parts that are to be soldered or brazed should be passivated before, but not after, these operations have been performed, because the passivating solution may attack the solder or bronze metal. Also, parts should not be passivated after assembly, because removal of the acid may be quite difficult.

Prior Cleaning. Flux or slag from welding or high-temperature brazing should be removed by chipping, brushing with a stainless steel wire brush, grinding, polishing with an iron-free abrasive, or sand blasting. Machining, forming, or grinding oils must be removed in order for passivation to be effective. Cleaning should begin with solvent cleaning, which should be followed by alkaline soak cleaning and thorough water rinsing. Best results are obtained in passivation when the parts to be treated are as clean as they would have to be for plating.

When large parts or bulky vessels are to be cleaned, it may be necessary to apply cleaning liquids by means of pressure spray; exterior surfaces may be cleaned by immersion or swabbing.

Solutions. Listed below are several typical solutions used to passivate the various types of stainless steel, together with their operating temperatures and exposure times.

- **Solution A**

  For series 200, 300, and 400 grades containing 17% chromium or more, except free-machining grades and polished surfaces:

  70% nitric acid (HNO_3) .... 20 to 40 vol %
  Water .... rem
  Operating temperature . . Up to 60 °C (140 °F)
  Contact time .... 30 to 60 min

- **Solution B**

  For free-machining grades, polished surfaces, and series 400 containing less than 17% chromium:

  70% nitric acid (HNO_3) . . . 20 vol %
  Sodium dichromate ($Na_2Cr_2O \cdot 2H_2O$ ..... 4 to 6 wt %
  Water .... rem
  Operating temperature........Room temperature or 50 °C (120 °F)
  Contact time .... 30 min

These solutions also may be used to remove lead, copper, cadmium, or zinc applied to stainless steel wire for cold heading, wiredrawing, or spring winding.

Control of passivating solutions consists mainly of replenishing the nitric acid, sodium dichromate (if used), and water that are lost as a result of dragout. Because the amount of iron dissolved is extremely small compared with the area of surface processed, the solution remains effective for a long time. However, its acid content should be measured periodically by means of a hydrometer or by simple volumetric determination. Frequency of these measurements would depend on the volume of work processed and the size of the tank. A solution is discarded only if it becomes seriously contaminated. If the solution is used to remove metallic lubricants, such as lead, copper, or cadmium, and if the volume of this work is high, the life of the passivating solution is shortened appreciably.

Precautions. The usual precautions for working near strong oxidizing agents must be observed. For their protection, operating personnel should wear rubber aprons, gloves, and eyeshields or goggles.

Thorough rinsing of recessed parts must precede air blow-off in order to avoid the dispersal of acid solutions.

Handling equipment for parts preferably should be constructed of stainless steel or plastic. Monel baskets or containers must not be used.

Etching

Stainless steels may be etched by dry or wet methods. The dry methods include sand blasting and other dry surface marking methods; the wet methods use acid solutions.

Dry Etching. Areas to be etched are marked off by cut stencils, metal templates, or adhesive materials. Edges of the masking material must be held in close contact with the metal to ensure sharp, well-defined border lines. Etching is performed by means of sand blasting, grinding wheels, tampico polishing wheels, or stainless steel wire brushes.

Wet Etching. Although etching solutions must be strongly acid to act on stainless steels, they must be varied in composition and strength to suit the corrosion resistance of a specific alloy. Stop-off materials must be sufficiently acid-resistant to prevent undercutting. To ensure uniformity of acid attack, metal surfaces must be uniformly clean before being etched.

Assuming that suitable solution strengths and treatment times are determined by experimentation, the following acid solutions may be used to etch the straight-chromium types (except hardenable grades in the hardened condition) and chromium-nickel types of stainless steel:

Solution 1
- 25 parts water by volume
- 75 parts hydrochloric acid (HCl) by volume
- 5 parts nitric acid (HNO_3) by volume
- 20 wt % ferric chloride ($FeCl_3$)

Solution 2
- 10 to 20 vol % nitric acid (HNO_3)
- 3 to 5 vol % hydrofluoric acid (HF)
- Remainder water

Solution 1 rapidly produces a uniform white appearance. As it becomes exhausted, its concentration of ferric chloride, a strong pitting agent, increases. This is evidenced by the generation of a heavy brown sludge, which is an indication that the bath must be discarded. Solution 2 should be operated below 60 °C (140 °F), to reduce fuming and consequent loss of the hydrofluoric content.

Electroplating

Stainless steels may be plated with copper, chromium, nickel, cadmium, and the precious metals for such purposes as: (*a*) color matching; (*b*) lubrication during cold heading, spring coiling, or wire drawing; (*c*) reduction of scaling at high temperature; (*d*) improvement of wettability or of conductance of heat or electricity; (*e*) prevention of galling; (*f*) decorative uses in such applications as jewelry; and (*g*) prevention of superficial rusting.

Details of plating baths and their operation, and of applicable cleaning procedures and other pretreatments, are given in separate articles in this Volume on nickel, copper, chromium, and cadmium plating.

Although a stainless steel surface may be clean and scale-free, an adherent electrodeposit cannot be obtained until the surface is activated for removal of its normally ever-present oxide film. Activation, which is performed immediately before plating, may be accomplished by cathodic treatments, immersion treatments, or simultaneous activation-plating treatments. These treatments, together with other procedures necessary for preparing stainless steel for electroplating, are fully described in ASTM B254.

Figure 3 summarizes some of the data from ASTM B254. The same cleaning procedures would be appropriate before electrodeposition of other metals. When preparing stainless steel for electroplating, the following should be considered:

Fig. 3 Electroplating copper on stainless steels

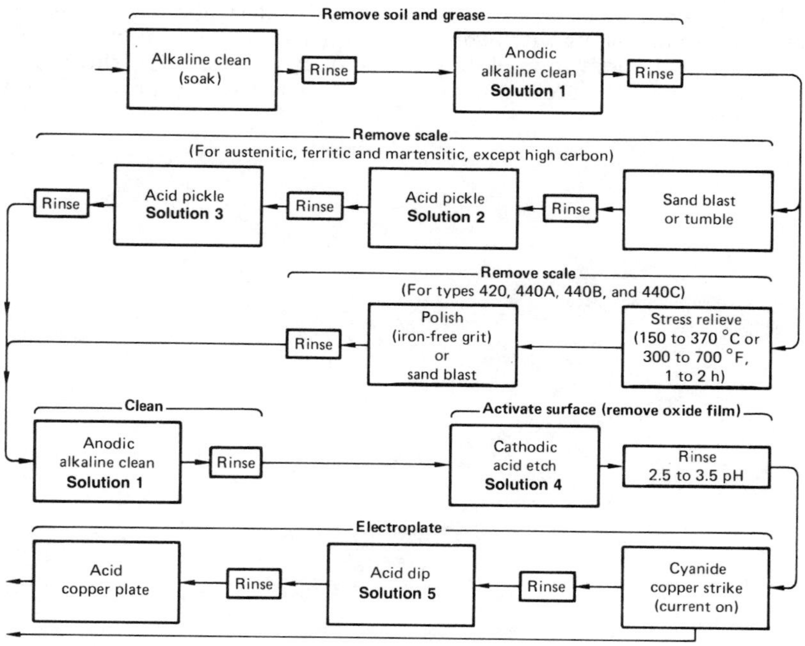

Solution No.	Type of solution	Composition of solution	Operating temperature °C	°F	Cycle time, min
1	Anodic alkaline cleaner(a)	Alkali, as low as possible	(b)	(b)	(b)
2	Acid pickle	H_2SO_4, 8-11 vol %	65-70	150-160	10-45
3	Acid pickle	HNO_3, 6-10 vol %; HF, 1.5 vol %	Room temperature	Room temperature	10-30
4	Cathodic acid etch(c)	H_2SO_4, 5-15 vol %	Room temperature	Room temperature	1-5(d)
5	Acid dip	H_2SO_4, 1 vol %	Room temperature	Room temperature	1/6-1/3

(a) Current density, as low as possible. (b) As low as possible. (c) Current density, 54 A/m^2 (5 A/ft^2). (d) After subsequent rinse, place parts into copper electrolyte while parts are still wet

- *Electrocleaning:* when bright plating is desired, the alkalinity, current density, and temperature of the cleaning bath should be kept as low as possible, especially for the high-chromium alloys. To minimize the severity of electrocleaning when a bright finish is desired, a short electropolishing treatment may be used in lieu of electrocleaning, provided the preliminary cleaning is adequate
- *Acid pickling:* freshly prepared pickling baths should be activated by dissolving some iron in them before using the bath; otherwise, the rate of pickling is slow until the bath has dissolved some iron from the steel being treated
- *Activating:* the activation step is the most important in surface preparation of stainless steel for electroplating. If the simultaneous activation-

plating treatment is used, where the anodes are nickel, the nickel content of the solution gradually increases, because of the low cathode efficiency. This can be compensated for by removing a portion of the solution and replenishing with hydrochloric acid. The activation-plating treatment is usually preferred for stainless steel parts that are to be copper plated

- *Plating:* after activating and rinsing, the work should be entered directly into the electrolyte. When possible, the rinse water should be slightly acid (2.5 to 3.5 pH); the required acidity is usually provided by the dragout of acid from the activation treatment. Stainless steel has much lower electrical conductivity than most other metals, and adequate attention, therefore, should be paid to racking when introducing work into copper electrolyte, to ensure an adequate number of contacts to obtain complete coverage in minimum time

Cleaning and Finishing of Heat-Resisting Alloys*

CLEANING is required to remove contaminants from the surfaces of parts made of heat-resisting alloys. Shop soils such as oil, grease, and cutting fluids can be removed by conventional solvents or soaps. Metallic contaminants, tarnish, and scale resulting from hot working or heat treating operations must also be removed.

Removal of Metallic Contaminants

Parts made of heat-resisting alloys may accumulate traces of other metals on their surfaces as a result of contact with cutting tools, forming dies, and heat treating fixtures. In some instances, these other metals are not harmful, but in others the presence of the contaminants can have a highly deleterious effect. For example, Inconel X-750 may be unaffected by traces of zinc from drawing dies, but even the smallest particle of aluminum will readily alloy with Inconel X-750 at elevated temperatures and degrade the affected areas. Also, copper may affect some nickel-base alloys, when the alloys are subsequently exposed to high temperatures. This article primarily focuses on the nickel, cobalt, and iron-base heat-resisting alloys. For cleaning and finishing information on refractory alloys, see the article on cleaning and finishing of reactive and refractory alloys in this Volume.

In operations such as cutting and forming, metal contamination can be avoided or sharply reduced by the use of lubricants. This is preferred practice, because lubricants can be removed easily and cheaply. If contamination cannot be avoided, testing is recommended to determine the seriousness of the contamination. After heat treating operations, metallographic tests to detect alloy diffusion or bend tests to detect embrittlement are commonly used. Lead may be detected by the yellow precipitate that forms when the following test solution at 21 to 60 °C (70 to 140 °F) is applied to the suspected surface:

Chromic acid	10 wt %
Sodium chlorate	1.5 wt %
Water	88.5 wt %

Although metal contamination is not always harmful, all surface contaminants should be removed before heat-resisting alloys are heat treated or subjected to service at elevated temperatures.

Mechanical Removal. Dry or wet abrasive blasting with metal-free abrasives is an effective means of removing metal contamination, as are polishing with ceramic materials and wet tumbling. The shape of a part, the surface finish required, and the allowable loss of gage will determine the suitability of these mechanical methods.

Chemical Removal. Chemical methods are used more often than mechanical methods for removing metal contamination. A typical procedure for the chemical removal of iron, zinc, and thin films of lead is as follows:

- Vapor degrease or alkaline clean
- Immerse in a 1-to-1 solution (by volume) of nitric acid (1.41 sp gr) and water for 15 to 30 min at about 35 °C (95 °F)
- Water rinse and dry

Another procedure, which has proved successful for removing brass, lead, zinc, bismuth, and tin from nickel-base and cobalt-base alloys, is given below:

- Vapor degrease or alkaline clean
- Soak at room temperature in solution of:

Nitric acid	54 g/L (7.22 oz/gal)
Acetic acid	150-375 g/L (20-50 oz/gal)

*Revised by Leonard Rozenberg, Research Project Engineer, Wyman-Gordon Company, and J. D. VanDevender, Chemist, Huntington Alloys, Inc.

Hydrogen peroxide	19-64 g/L (2.5-8.5 oz/gal)

Soaking time will vary from 20 min to 4 h, depending on the severity of the contamination, and is determined by visual observation of the reaction. After this treatment, parts must be rinsed thoroughly in water and dried. When possible, immerse a test specimen for the maximum time anticipated and examine it for chemical attack before processing the first load of parts.

Nickel-base alloys should be acid etched to prepare for subsequent nondestructive inspection. The etching process removes smeared metal which may be present as a thin surface layer following machining and/or shot blast cleaning. Parts to be examined may be etched by immersion in a bath of 80% hydrochloric acid, 13% hydrofluoric acid, and 7% nitric acid to remove the disturbed or smeared layer. This bath may leave smut that must be removed in a second bath containing 22% iron chloride, 75% hydrochloric acid, and 2% nitric acid and water. Following rinsing and drying, the parts are ready for optical macro inspection and penetrant inspection. The extent of etching depends on the depth of the smeared layer. All of the disturbed layer should be removed, but overetching causes excessive penetrant retention.

Removal of Tarnish

Tarnish, thin oxide film, may not always have a harmful effect on the end use of parts made of heat-resisting alloys. It may even prove useful. For example, tarnish may function as a bond for paint, as a barrier to prevent diffusion from another alloy, or as a retardant to further oxidation. However, considerations such as functional requirements may indicate that tarnish should be removed from parts. Tarnish always should be removed before welding or brazing.

Abrasive cleaning methods such as those used for removing metallic contaminants are used also for removing tarnish. The applicability of these methods is determined by the shape of parts, the surface finish required, and the allowable loss of gage.

Flash pickling is used more often than abrasive cleaning for tarnish removal since abrasive cleaning removes some metal and can degrade the surface finish. A typical formula for flash pickling is:

Nitric acid (1.41 sp gr)	23 vol %
Hydrofluoric acid (1.26 sp gr)	4 vol %
Water	73 vol %

Parts are immersed in the above solution at about 52 °C (125 °F) for 1 to 5 min. Warming the parts in hot water before flash pickling speeds tarnish removal. Water rinsing and drying must follow flash pickling.

Removal of Reduced Oxide and Scale

Acid pickling, abrasive cleaning by tumbling or blasting, and descaling in molten salt are the most widely used methods for removing reduced oxides or scale from heat-resisting alloys, listed above in order of decreasing preference based on economic considerations. Alkaline scale conditioning is helpful in modifying scale so that it may be more readily removed by these methods. When extremely heavy oxide layers must be removed, grinding is an appropriate preliminary operation. Combinations of two or more methods are often used.

Reduced oxide films are formed on parts heated in reducing atmospheres out of contact with air. Sometimes these oxides can be removed by immersing parts for 5 to 15 min in a tarnish-removing flash-pickling bath (see above formulation for flash pickling). However, most heat-resisting alloys, because of their high content of oxide-forming metals such as nickel, cobalt, and chromium, form a tenacious coating in the presence of carbon monoxide or water. The oxides formed vary widely with composition of the alloy and furnace atmosphere. Usually, pickling is required for their removal. Scale conditioners facilitate removal of these oxides (see typical procedure in Table 1).

Scale develops on hot forged, hot formed, or heat treated parts that are processed in air. Usually, scale is tenacious and occurs in all gradations including severe coatings that result from heating in an oxidizing furnace using high-sulfur fuels. Scale formed under such conditions has a dull, spongy appearance. Fine cracks may be present, and patches of scale may break from the surface. The underlying metal is rough and cannot be corrected by pickling. In these extreme conditions, grinding or abrasive blasting to sound metal, followed by flash pickling, is recommended.

Scale conditioning is used to soften, modify, or reduce scale for easier and more uniform acid pickling, but is seldom required for removal of discoloration or of interference coatings.

A scale-conditioning bath consists of a highly alkaline aqueous solution, sometimes containing complexing and chelating compounds. The main purpose of these agents is to solubilize the scale as much as possible. The performance of a particular chelating agent depends on the affinity of the compound for the metal ions present, the pH of the scale-conditioning solution, and the physical and chemical composition of the scale.

Little scale removal occurs during treatment in the alkaline scale-conditioning bath. Further treatment in highly alkaline solutions containing a strong oxidizing material, such as potassium permanganate, is often necessary. Scale on heat-resisting alloys sometimes contains carbon and incompletely burned and polymerized residues in addition to metallic oxides. These organic components react with the oxygen released by the alkaline oxidizing bath.

Acid Pickling. The conditioned scale finally is subjected to acid pickling, during which most of the high-temperature scale breaks away or becomes so loosely attached that pressure rinsing with water completes descaling. The acid pickle is usually a dilute nitric acid or a hydrofluoric acid—nitric acid solution. In addition to removing scale, pickling solutions containing nitric acid remove many surface contaminants through oxidation. However, because the acid solution attacks the basis metal, it is necessary to limit the time of pickling to prevent excessive metal loss and excessive roughening of the metal surface.

A special procedure is used for treating alloys high in aluminum and titanium, such as M-252 and René 41, before welding or brazing. When the alloy is in the solution-treated condition and descaling is required, it is treated in a scale-conditioning solution as previously described, after which it is immersed in a solution of 30 wt % nitric acid and 3 wt % hydrofluoric acid for 5 to 10 min. Alloys in the aged condition are descaled anodically in an acid solution. A solution for this procedure contains 75 wt % sulfuric acid and 3 wt % hydrofluoric acid and is operated with a current density of 2 to 4 A/dm^2 (20 to 40 A/ft^2) and graphite cathodes. The material is immersed in the electrolytic cleaning bath for 3 to 12 min; the opera-

Table 1 Procedure for removing scale from heat-resisting alloys

Operation	Time	Temperature °C	Temperature °F	Solution	Concentration
Precleaning cycle					
Vapor degreasing	5-10 min	87-88	185-190	Stabilized trichlorethylene	...
Emulsion cleaning	10-20 min	54-66	130-150	Emulsion cleaner	20 vol %
Scale-conditioning cycle					
Alkaline chelating	15-30 min	125-135	260-275	Caustic solution containing alkanol amines and aliphatic hydroxy acids(a)	...
Alkaline oxidizing	1-2 h	95-105	205-220	Potassium permanganate	5 wt %
				Sodium hydroxide	20 wt %
				Water	rem
Pickling cycle					
Acid pickling	5-30 min	49-60	120-140	Hydrofluoric acid	4 wt %
				Nitric acid	18 wt %
				Water	rem

(a) For use and composition refer to U. S. Patents 2,843,509, 2,992,946, 2,992,995, and 2,992,997

tion is complete when the amperage drops almost to zero. Sodium sulfite, 1.6 wt %, is used to reactivate the solution after a period of operation.

Alloys containing less than about 12% chromium, unstabilized austenitic stainless steels, and martensitic and ferritic stainless steels can undergo high metal loss or develop intergranular attack in pickling. When the susceptibility of a material to excessive metal loss or intergranular attack by acids is unknown, mechanical descaling is safer than acid pickling. In one specific instance, it was proved that acid descaling caused intergranular attack and subsequent loss of ductility in aged René 41.

Weld areas normally vary in composition and structure from the basis metal and do not react to the conditioning and pickling cycles in the same manner as the basis metal. Weld areas or the heat-affected zones are often susceptible to selective attack during pickling. Although inhibitors may eliminate or reduce selective attack, abrasive or other mechanical descaling methods are preferable to acid pickling for removing scale from welded parts, unless a safe pickling procedure has been found for a given application.

Hydrogen embrittlement does not occur in nickel-base, cobalt-base, or austenitic iron-base alloys as a result of aqueous descaling. Details of typical multicycle descaling operations are given in Tables 1 and 2.

Abrasive blasting with dry aluminum oxide can be used for removing oxide and scale from all types of wrought and cast heat-resisting alloys. Silicon carbide is more expensive than aluminum oxide and is seldom used. Silica (silicon dioxide or sand) has a limited application because of its lack of cutting ability; it is sometimes used for cleaning refractory metal forgings prior to pickling. Grit sizes as coarse as No. 30, of 0.59 mm (0.0232 in.) are recommended for cleaning forgings and castings. Finer grits like No. 90 and No. 100, of 0.17 and 0.15 mm (0.0065 and 0.0059 in.) are used for general blasting.

Metallic shot and grit should not be used for descaling heat-resisting alloys unless their use is followed by pickling to remove metal contamination. For parts to be welded or brazed, or where furnace atmospheres have produced highly tenacious scale, pickling after dry abrasive cleaning is recommended, regardless of the abrasive used.

Wet abrasive blasting is also used for cleaning heat-resisting alloys. This process uses silica abrasive particles (No. 200 to No. 1250; 0.074 to 0.010 mm or 0.0029 to 0.0004 in.) mixed with water to produce a slurry that removes loose scale, discoloration, and soils. Metal loss is not excessive when normal pressures and exposure times are used.

Stock loss is kept to an absolute minimum when spherical beads made of high-quality optical crown glass are used as the abrasive. Bead sizes of 0.038 to 0.074 mm (0.0015 to 0.0029 in.) are generally used, and blasting pressures are kept below 410 kPa (60 psi) to prevent bead fracturing.

Surfaces that have been wet blasted are usually suitable for welding, brazing, electroplating, and final inspection; further cleaning is seldom necessary. Exceptions are alloys with a high titanium and aluminum content, which require the special procedure previously discussed.

Most of the general advantages and limitations of abrasive cleaning of steel apply also to heat-resisting alloys. An exception is the risk of contamination from metallic abrasives or from cleaning parts with abrasives that have been used for other metals of widely different compositions. For example, heat-resisting alloys should not be blasted with abrasive material that has been used for cleaning low-alloy steel, aluminum, copper, or magnesium. However, abrasives that have been used to clean titanium and corrosion-resistant steels have been used for cleaning heat-resisting alloys without serious contamination. Flash pickling of heat-resisting alloys after abrasive cleaning provides additional assurance that no harmful surface contamination remains. Detailed information on equipment and procedures for dry and wet blasting is given in the article on abrasive blast cleaning in this Volume.

Wet tumbling by the barrel or vibratory method can be used for descaling heat-resisting alloys if the parts are of suitable shape and size; removal of burrs and sharp edges is accomplished in the same operation. Shop soils are also removed, thus eliminating the need for preliminary degreasing.

Parts are tumbled or vibrated in a mixture of acid descaling compound and metal-free abrasives, after which they undergo a neutralizing cycle. Precautions regarding metal contamination are similar to those noted above for abrasive blasting.

Pickling is required after tumbling and before resistance welding to remove residual smut, which can cause poor quality weldments. There is less need for pickling prior to fusion welding unless inspection of weldments re-

Table 2 Procedure for removing scale from Inconel alloys

Operation	Time	Temperature °C	Temperature °F	Solution(a)	Concentration(a)
Alkaline conditioning	1-2 h	96-105	205-220	Sodium hydroxide	20 wt %
				Potassium permanganate	5 wt %
Rinse	15-30 s	Not heated	Not heated	Water quench and water spray	...
Acid pickling	5-10 min	60-71	140-160	Sulfuric acid (1.83 sp gr)	7.5 vol %
				Hydrochloric acid (1.16 sp gr)	12 vol %
Rinse	15-30 s	Not heated	Not heated	Water	...
Acid pickling	10-20 min	60-71	140-160	Nitric acid (1.41 sp gr)	20 vol %
Rinse	15-30 s	Not heated	Not heated	Water	...
Acid pickling	5-60 min(b)	49-54	120-130	Hydrofluoric acid (1.26 sp gr)	3.7 vol %
				Nitric acid (1.41 sp gr)	22 vol %
Rinse	15-30 s	Not heated	Not heated	Water	...

(a) Remainder is water. (b) Type of oxide will determine immersion time required; until immersion is established, inspect frequently to avoid overpickling

veals porosity or inclusions that are a result of pickup from tumbling.

Wire brushing is sometimes used for removal of very light scale or surface discoloration. Brushes, when used for all heat-resisting alloys, must have stainless steel bristles.

Salt bath descaling is an effective first step in removing scale from heat-resisting alloys. The process is generally more costly than acid pickling, particularly if production is intermittent, because of the cost of maintaining the bath during idle time. The electrolytic salt bath used for descaling heat-resisting alloys does not contain sodium hydride, but uses fused caustic soda; the workpiece and tank are alternately negative and positive poles of a direct current.

A fused caustic soda bath containing oxidizing salts such as sodium nitrate has been used as the initial step in descaling heat-resisting alloys. This oxidizing bath, which is operated at 425 to 540 °C (800 to 1000 °F), is slightly more effective than the sodium hydride bath on high-chromium alloys such as type 310 stainless steel, and on the cobalt-chromium-nickel-base alloys such as L-605. Processing steps are similar. The work load is immersed in the oxidizing bath for 5 to 15 min, quenched in water, soaked in a solution of 5 to 10% sulfuric acid at 71 °C (160 °F) for 1 to 5 min, and then dipped in a solution of 15 to 20% nitric acid and 2 to 4% hydrofluoric acid at 54 to 60 °C (130 to 140 °F) for 2 to 15 min.

A typical sequence for hydride descaling and pickling is given in Table 3. Details of equipment and other information on salt bath descaling are given in the article on salt bath descaling in this Volume.

Applicable Finishing Processes

Many finishing operations commonly used for steel and other metals are not required for heat-resisting alloys because: (*a*) heat-resisting alloys are inherently resistant to corrosion in a wide range of environments; and (*b*) normally, the end use of parts made of heat-resisting alloys does not require them to have a polished finish.

The oxide coating obtained during processing is frequently of value to heat-resisting alloys that are subjected to elevated temperatures in service. Consequently, the dense, tenacious oxide that forms on formed or machined finished parts during final heat treatment is allowed to remain as protection against further oxidation.

Polishing of heat-resisting alloys is sometimes used to obtain a desired surface finish as well as to remove light scale or oxide from parts that are to be welded or brazed. Silicon carbide in various grit sizes is commonly used to prepare surfaces for brazing.

Surfaces are usually prepared for welding by polishing with a No. 90 aluminum oxide set up with sodium silicate on a cloth wheel. Discoloration can be removed by polishing with No. 120 aluminum oxide used with a greaseless compound and a cloth wheel. Buffing is seldom required for finishing heat-resisting alloys.

Electroplating. Chromium, copper, nickel, and silver are sometimes electroplated on heat-resisting alloys for one of the following reasons: (*a*) as preparation for brazing; (*b*) to deposit brazing metal; (*c*) to provide antigalling characteristics; and (*d*) to repair expensive parts or correct dimensional discrepancies. Nickel is most often used to assist in brazing. Conventional nickel plating processes are used, and deposits vary in thickness from 2.5 to 25 μm (0.1 to 1 mil). Alloys containing titanium or aluminum require the thicker deposits.

Silver and copper are the metals most often deposited as actual brazing materials. Some brazing alloys are deposited as separate layers of their various constituent metals on a weight-percentage basis. Thickness of plate depends on the amount of metal needed for brazing.

Electrodeposited silver or nickel is used in special applications for antigalling. Either nickel or chromium may be used to repair worn parts or to build up dimensions. Procedures and electrolytes for plating these metals are described in the articles on electroplating in this Volume.

Ceramic coatings are frequently applied to heat-resisting alloys for increasing the oxidation resistance of parts exposed to extremely high temperatures, such as gas turbine and missile components. Several types of ceramics are currently used for these purposes. For procedures, see the article on ceramic coatings in this Volume.

Diffusion coatings of aluminum, chromium, silicon, or combinations of these elements, also are used to protect heat-resisting alloy parts from high-temperature corrosion and to prolong their life. Aluminum coating of engine exhaust valves is a common example. Procedures for diffusion coating are described in the article on aluminum coating of steel in this Volume.

Shot peening is currently used for improving mechanical properties of compressor blades, turbine blade dovetails, and latter-stage turbine blade air-foils. Glass beads are favored over metallic shot for shot peening, because: (*a*) their use involves no risk of metal contamination; (*b*) almost no metal is removed; and (*c*) they are available in

Table 3 Procedure for sodium hydride descaling and acid pickling of heat-resisting alloys

Operation	Time, min	Temperature °C	Temperature °F	Solution(a)	Concentration(a), %
Sodium hydride descale	5-20	370-390	700-730	Sodium hydride	1.5-2.0
Quench	¼-½	Not heated	Not heated	Water	...
Neutralizing rinse	1-3	Room temperature to 60	Room temperature to 140	Sulfuric acid	2-10
Brightening pickle	5-15	54-60	130-140	Hydrofluoric acid	2-4
				Nitric acid	15-20
Rinse	¼-½	Not heated	Not heated	Water	...
High-pressure spray wash	(a)	Not heated	Not heated	Water(b)	...

(a) Governed by shape of part; 1 min on parts with accessible surfaces. (b) Water pressure of 690 kPa (100 psi)

smaller sizes than metallic shot and can therefore be used to peen areas that are difficult to reach with metallic shot. However, glass beads are not equivalent to metal shot for improvement of mechanical properties. Details on peening are given in the article on shot peening in this Volume.

Cleaning and Finishing Problems and Solutions

The complex oxides and scale that form on heat-resisting alloys often create production problems and require the use of special procedures to obtain desired surfaces. The examples that follow, all of which are drawn from actual production experience, present details of cleaning and finishing problems and the procedures used to solve them.

Example 1. After heat treatment, turbine combustion chambers made of Hastelloy X sheet exhibited irregularity of scale adherence, variations in surface finish, and loss of formability. Investigation disclosed that residual shop soils such as lubricants, marking inks, and handprints remained on the parts despite the solvent cleaning and vapor degreasing to which the parts were subjected before being heat treated and that these soils were decomposing during heat treatment and causing carbon diffusion.

The substitution of electrolytic alkaline cleaning for the methods previously used eliminated the difficulty. This procedure consisted of immersing parts for 5 min in a bath compounded to Federal Specification P-C-535 and operated at 82 to 93 °C (180 to 200 °F) using 6 to 8 V; parts were anodic in the electrical circuit. Heat treating scale was easily removed by subjecting the heat treated parts to a 5-min immersion in a room-temperature acid pickling bath composed of 70% nitric acid (20 to 30 vol %), 60% hydrofluoric acid (10 to 15 vol %), and water (55 to 70 vol %).

Example 2. Descaling of A-286 and S-816 alloys was inadequate when the following procedure was used:

- Descale in oxidizing salt bath at 480 to 495 °C (900 to 925 °F) for 25 min; quench in cold water
- Immerse in solution of 23 vol % sulfuric acid for 10 to 15 min; follow with cold water rinse
- Immerse in solution of 25% nitric acid and 3% hydrofluoric acid for 2 min; rinse in cold water
- Immerse in 30% nitric acid for 5 min, rinse in cold water; rinse in hot water; dry

A change to the following procedure proved successful:

- Descale in oxidizing salt bath at 480 to 495 °C (900 to 925 °F) for 10 min; quench in cold water
- Immerse in 23% sulfuric acid for 5 min; rinse in cold water
- Immerse for 1 min in solution of 25% nitric acid and 3% hydrofluoric acid; rinse in cold water; rinse in hot water; dry
- Reimmerse in oxidizing salt bath at 480 to 495 °C (900 to 925 °F) for 10 min; quench in cold water
- Immerse for 30 s in 23% sulfuric acid; rinse in cold water
- Immerse for 30 s in solution of 25% nitric acid and 3% hydrofluoric acid; rinse in cold water
- Immerse for 5 min in 30% nitric acid; rinse in cold water; rinse in hot water; dry

Example 3. Excessive zinc contamination of formed parts made of 19-9 DL alloy was traced to the contamination of drawing lubricants that were stored too close to where zinc dies were being finished by sanding. The zinc was removed from the parts by chemicals (see section on chemical removal of metal contaminants in this article); and a recurrence of this difficulty was prevented by improved housekeeping and by instructions to personnel.

Example 4. It was not feasible to broach turbine disks, made of D-979 alloy, in the solution treated and aged condition. The disks were broached in the solution treated condition; the surfaces were then protected by flash nickel (Wood's) plating and a subsequent overlay of copper, for aging above 730 °C (1350 °F) in a reducing gas atmosphere. After aging, the plated copper and nickel were removed from the disks by the procedures detailed below:

Stripping of copper plate

Composition of bath	
Chromic oxide	360 g/L (48 oz/gal)
Sulfuric acid (concentrated)	39-55 cm^3/L (5-7 fluid oz/gal)
Water	rem
Temperature	Room temperature
Time	Varies with thickness of plate

Stripping of nickel plate

Composition of bath	
Sodium cyanide	90 g/L (12 oz/gal)
Nitroaromatics	60 g/L (8 oz/gal)
Water	rem
Temperature	49-66 °C (120-150 °F)
Time	Varies with thickness of plate

Chemical stripping and a light vapor blast to remove smut resulted in clean, dimensionally unaltered parts and had no detrimental effect on mechanical properties. The procedure indicated in the above table was based on available facilities. A cheaper and faster method is stripping the copper and nickel in a nitric acid solution, 50 vol %, at room temperature.

Example 5. Conventional salt bath descaling and pickling failed to remove all annealing scale from stampings made of 19-9 DL, Hastelloy X, Inconel 600, and Inconel X-750. The sequence of operations performed on these stampings was forming, degreasing, removal of metal contamination, annealing, and descaling, or immersion in molten salt and pickling. The difficulty was traced to the open-hearth, gas-fired annealing furnaces. It was found that the atmosphere was reducing while the burners were on and that a thin, tight scale was produced.

A satisfactory remedy was to adjust the burners to bring the oxygen content to 3%. The resulting scale was loose and easily removed by the usual descaling and pickling procedures.

Example 6. Jet engine combustion liners made of Hastelloy X required three in-process anneals. Forming without removal of scale required expensive carbide dies, and the liners produced were unacceptable. The problem was solved by installing a salt bath descaling line. The following cycle proved satisfactory for descaling:

- Immerse in oxidizing salt bath at 480 °C (900 °F) for 10 to 20 min; rinse, first in cold water and then in water at 82 °C (180 °F)
- Immerse in solution of 24% hydrochloric acid for 2 to 10 min; rinse in cold running water
- Immerse in nitric-hydrofluoric acid for 2 to 10 min; rinse in cold running water; hose off and dry

Example 7. Small cracks appeared in welded 19-9 DL tubing (1.3-mm or 0.050-in. wall) after annealing. Processing sequence was as follows: form tubing from flat stock, degrease, automatic seam weld, anneal, descale. The tubing was formed on zinc-alloy dies and was not dezinced before being welded. Small amounts of zinc on the surface near the weld melted during welding. This initiated zinc diffusion, and the residual stresses around the weld were sufficient to crack the embrittled material. The problem was solved by pickling the tubing in 20% nitric acid to remove the zinc before welding.

Cleaning and Finishing of Nonferrous Metals

Cleaning and Finishing of Aluminum and Aluminum Alloys

By the ASM Committee on Finishing of Aluminum*

ALUMINUM OR ALUMINUM ALLOY products have various types of finishes applied to enhance appearance, improve functional properties of the surfaces, or both. This article discusses the methods employed in cleaning and finishing.

Abrasive Blast Cleaning

One of the simplest and most effective methods for cleaning aluminum surfaces is by blasting with dry nonmetallic or metallic abrasives. Although this method is normally associated with the cleaning of aluminum castings, it is also used to prepare surfaces of other product forms for subsequent finishes, such as organic coatings. In addition to cleaning, blasting is used to produce a matte texture for decorative purposes.

Abrasive blasting is most efficient in removing scale, sand, and mold residues from castings. It is readily adaptable to the cleaning of castings, because they are usually thick enough so that no distortion results from blasting.

Blast cleaning of parts with relatively thin sections is not recommended, because such parts are readily warped by the compressive stresses that are set up in the surface by blasting. The blasting of thin sections with coarse abrasive is not recommended because the coarse abrasive can wear through the aluminum. Typical conditions for dry blasting with silica abrasive are given in Table 1.

Washed silica sand and aluminum oxide are the abrasives most commonly used for blast cleaning of aluminum alloys, but steel grit is sometimes used. Because of the fragmenting characteristics of silica, steel grit often is preferred. It has longer life which lowers cleaning cost. However, when aluminum is blasted with grit, steel particles become embedded in the surface and, unless removed by a subsequent chemical treatment, will rust and stain the aluminum surface. It is good practice to remove particle contamination with a nitric acid pickle to prevent degradation of its corrosion resistance. A 20-min soak in 50% nitric acid solution at ambient temperature will dissolve embedded or smeared iron particles, but it will not remove silica or aluminum oxide. When aluminum is blasted with No. 40 or 50 steel grit, a 9.5-mm (3/8-in.) diam nozzle and air pressure at about 276 kPa (40 psi) are commonly used. Organic materials such as plastic pellets and crushed walnut shells also are used in blast cleaning aluminum, often for the removal of carbonaceous matter.

Steel shot is seldom used for cleaning aluminum surfaces, although stainless steel shot is sometimes used. Shot blasting is used as a preliminary operation for developing a surface with a hammered texture. An attractive finish is produced when this textured surface is bright dipped and anodized. In addition, the varying degrees of matte texture that can be produced by blasting offer many decorative possibilities. When maximum diffuseness of reflec-

*Howard G. Lasser, Chemical Engineer, Materials Research Consultants; D. J. Schardein, Director, Finishing Technology Section, Reynolds Metals Co.; Allan W. Morris, Technical Specialist, Material and Process Development, McDonnell Aircraft Co.; Bob Srinivasan, Manager, Research and Development, Industrial Products Group, Diversey Wyandotte Corp.; Walter G. Zelley, Technical Manager, Surface Technology Division, Aluminum Company of America

Table 1 Conditions for abrasive blast cleaning with silica

Grit size	Mesh	Nozzle diameter mm	Nozzle diameter in.	Nozzle to work(a) mm	Nozzle to work(a) in.	Air pressure kPa	Air pressure psi
20-60	Coarse	10-13	3/8-1/2	300-500	12-20	205-620	30-90
40-80	Medium	10-13	3/8-1/2	200-350	8-14	205-620	30-90
100-200	Fine	6-13	1/4-1/2	200-350	8-14	205-515	30-75
Over 200	Very fine	13	1/2	200-300	8-12	310	45

(a) Nozzle approximately 90° to work

tivity is required, blasting can produce it. For example, aluminum army canteens are blasted as a final finish to reduce glare. Glass bead blasting offers another approach to cleaning and producing diffuse surfaces.

Sand blasting with a fine abrasive produces a fine-grained matte finish on wrought or cast aluminum products. For plaques, spandrels, and related architectural decorative applications, sand blasting the background and polishing or buffing the raised portions of the surface produces an effect known as highlighting.

The matte finish produced by abrasive blasting is highly susceptible to scratching, and to staining from fingerprints. Therefore, matte-finish surfaces usually are protected by an anodic coating or clear lacquer. Anodizing is the more popular protective treatment, because it reproduces the original texture of a surface. Clear lacquers smooth out the roughened surface and produce various degrees of gloss, which may be undesirable. When the blasted surface is anodized, a gray color results because of embedded abrasive particles in the aluminum surface. This color frequently is nonuniform because of variations in blasting conditions such as nozzle-to-work distance, direction or movement of nozzle, and air pressure.

Close control of blasting conditions can be obtained by the use of specially designed equipment. Uniform movement of the work on conveyors, established nozzle movement, constant velocity of the abrasive, and controlled size of grit contribute to better color uniformity of subsequently anodized surfaces.

The nonuniform appearance that results from blasting can be corrected by bleaching prior to anodizing. Bleaching is done by deep etching in a solution of 5% sodium hydroxide at 38 to 66 °C (100 to 150 °F) to remove metal that contains embedded abrasive. Some trial and error may be necessary to determine etching time for specific conditions. If the surface is not etched enough, a mottled appearance may result. Embedded abrasive can also be removed with a solution of nitric acid and fluoride used at room temperature.

Care must also be used when selecting the aluminum or aluminum alloy to be sand blasted. For example, AA 1100 alloy, which contains 99% aluminum, provides a transparent anodic finish; whereas alloys rich in manganese, silicon, and copper are colored when anodized. In high-magnesium alloys, alloy segregation can occur, and pitting will appear unless special pretreatments are used. Table 2 lists some typical applications for abrasive blast cleaning of aluminum products, indicates type and size of abrasive used and typical production rates.

Wet blasting mixes a fine abrasive with water to form a slurry that is forced through nozzles and is directed at the part. Abrasive grits from 100 to 5000 mesh may be used. Wet blasting is generally used when a fine-grained matte finish is desired for decorative purposes.

An attractive two-tone finish on appliance trim can be obtained by contrasting a buffed finish with a wet blasted finish. Aluminum firearm components and ophthalmic parts such as frames and temples are wet blasted to produce fine matte finishes. In these applications, anodic coatings, either plain or colored, are used to protect without distorting the intended surface texture.

Typical wet blasting procedures are given in Table 3. Wet blasting is used also for preparing surfaces for organic or electroplated coatings. Ultrafine glass bead blasting can also be used in place of wet blasting.

Barrel Finishing

Many small aluminum stampings, castings, and machined parts are cleaned, deburred, and burnished by barrel finishing. In most instances, the main objective is deburring, burnishing, or both, with cleaning being an incidental benefit of the treatment. Barrel finishing is a low-cost method of smoothing sharp edges, imparting a matte finish, and preparing surfaces for anodizing, painting, or plating. Deburring sometimes is the final barrel operation, but more often it is followed by burnishing to obtain a smoother finish or one that is better suited to anodizing or plating. Parts that have been deburred only are often painted. Burnished parts are frequently anodized for protection.

Small aluminum parts are sometimes tumbled dry in media such as pumice and hardwood pegs, hardwood sawdust, or crushed walnut shells to remove burrs and improve the finish. However, this method is relatively inefficient, compared to the more widely used wet process.

All aluminum alloys can be safely finished by wet barrel methods. Limitations imposed by size and shape of workpieces are essentially the same as for steel and other metals. There are two general areas in which wet barrel finishing of aluminum parts is more critical than in processing similar parts made of steel. First, there is danger of surface contamination by ferrous metals, caused by the use of either a steel barrel or a steel medium. Second, pH of the compounds is more critical when processing aluminum, because the metal is susceptible to etching by both acids and alkalis, and because gas generated during chemical attack can build up pressure in the barrel and cause serious accidents. Barrels must be vented when processing aluminum. Compounds that are nearly neutral (pH of about 8) are recommended, although some alloys can be safely processed in compounds having a pH as high as 9.

Barrels used for aluminum are basically the same as those used for processing steel. However, barrels made of steel or cast iron should be lined with rubber or similar material to prevent contamination. A preferred practice is to use specific barrels for processing aluminum only.

Deburring is done by tumbling the work in a nonlubricating compound that contains abrasives. In most instances, media also are used to cushion the workpieces and increase the abrasive action. Synthetic detergents mixed with granite fines or limestone chips are usually preferred as the compound for deburring aluminum; aluminum oxide and silicon carbide are not desirable because they leave a smudge that is difficult to remove. High water levels, completely covering the mass, are used during deburring to assist in maintaining fluidity of the mass and to help prevent the medium from be-

Table 2 Applications for abrasive blast cleaning of aluminum products

Rotary automatic equipment with five nozzles used for blasting of all parts except cake pan, for which a hand-operated single nozzle setup was used

Product	Size, mm	Size, in.	Abrasive Type	Abrasive Mesh size	Pieces, h
Blasting to prepare for organic coating					
Cake pan	280 by 380 by 51	11 by 15 by 2	Alumina	100	60
Frying pan	250-mm diam	10-in. diam	Alumina	100	260
Griddle	6775 mm^2	10.5-in.^{2}	Alumina	100	225
Sauté pan	200-mm diam	8-in. diam	Alumina	100	250
Blasting for appearance produced					
Army canteen(a)	...	...	Steel	80	420
Cocktail-shaker body	100-mm diam by 180(b)	4-in. diam by 7(b)	Steel	80	375
Tray	300-mm diam(b)	12-in. diam(b)	Steel	80	180

(a) 1 qt army canteen blasted for reduction of light reflectivity. (b) Blasted for decorative effect

coming glazed and losing cutting action. Deburring can also be accomplished by using vibratory units with synthetic abrasives.

Barrel burnishing is used to produce a smooth, mirror-like texture on aluminum parts. Bright dipping immediately prior to burnishing will aid in producing a better finish. Other preliminary treatments also are helpful in specific instances, particularly for cast aluminum parts. One of these pretreatments entails etching the castings for 20 s in an alkaline solution at 82 °C (180 °F) and then dipping them for 2 to 3 s in a solution consisting of 3 parts by volume nitric acid (36° Bé) and 1 part hydrofluoric acid at 21 to 24 °C (70 to 75 °F).

The principle of barrel burnishing is to cause surface metal to flow, rather than to remove metal from the surface. Burnishing compounds must have lubricating qualities. Soaps made especially for burnishing are usually used. They are readily obtainable, and many of them have a pH of about 8, although more acidic materials can be used.

When burnishing aluminum, it is important to control the pH of the burnishing compound. This is accomplished by frequent titration of the burnishing compound followed by adding small amounts of borax or boric acid when titration indicates the need. Steel balls and shapes are the most commonly used burnishing media. Several examples of conditions used in barrel finishing applications are detailed in Table 4. It will be noted that deburring and burnishing are sometimes accomplished in a single operation.

Self-tumbling is an effective means of cleaning, deburring, or burnishing small aluminum parts. Procedures for self-tumbling are not basically different from those for other methods of barrel finishing, except that media are not used. Parts actually serve as the medium. Compounds for self-tumbling of aluminum should be of nearly neutral pH, and oxides should be removed from the aluminum parts before tumbling. Size and shape of the parts usually determine whether self-tumbling is suitable. Interior surfaces receive little or no action during self-tumbling.

Vibratory finishing is a newer method used for deburring and burnishing metal parts. In the use of this method for aluminum parts, compounds and media are subject to the same restrictions as discussed previously for conventional barrel finishing.

Polishing and Buffing

Because aluminum is more easily worked than many other metals, few aluminum parts require polishing prior to buffing for final finish. In some instances, polishing may be required for the removal of burrs, flash or surface imperfections. Usually, buffing with a sisal wheel prior to final buffing is sufficient. For general information on polishing and buffing, including machinery and equipment, abrasives, compounds, and binders, see the article on polishing and buffing in this Volume.

Polishing. Most polishing operations can be performed using either belts or setup wheels. Setup wheels may be superior to belts for rough cutting down when canvas wheels in a relatively crude setup can be used. For fine polishing work, a specially contoured wheel may be more satisfactory than a belt. Setup wheels have two main disadvantages in comparison with belts: (*a*) time, skill, and equipment are necessary for setting up wheels though the actual time required may be as short as 10 min, this time is spread over several hours, because of intermediate drying steps; and (*b*) wheels may be costly. Inventory thus becomes an important factor when wheels of several different types of abrasives or grit sizes are needed.Considerable operator skill is required for wheel polishing whereas unskilled labor may be used for belt polishing. Recently flap wheels have in many cases replaced setup wheels. The use of flap wheels tends to overcome the above-mentioned disadvantages. Typical conditions for polishing aluminum parts are discussed in the following examples.

Table 3 Conditions for wet blasting of aluminum-base materials

At nozzle-to-work distance of 75 to 100 mm (3 to 4 in.), operating pressure of 550 kPa (80 psi)

Operation	Abrasive Type	Abrasive Mesh size
Deburr and clean	Alumina	220
Blend and grind	Silica flour	325
Lap and hone	Glass	1000
	Diatomite	625-5000

The conditions for wheel polishing die cast aluminum soleplates for steam irons are as follows:

Type of polishing wheel	Felt
Setup time	10 min
Wheel speed	1800-2000 rev/min
Lubricant	Tallow grease stick

The medium-hard felt polishing wheel is 350 to 400 mm (14 to 16 in.) in diameter, with a 125 mm (5 in.) face. The surface of the wheel is double coated with 240-mesh alumina abrasive bonded with hide glue. Setup time, spread out over several hours of operation, totals 10 min. The polishing wheel can cover 34 to 43 m/s (6600 to 8400 sfm).

The soleplates are made of alloy 380.0 and the sides are polished to remove holes or other surface defects. Buffing follows to produce the required mirror finish. The polishing conditions given in the list above are based on a production rate of 115 pieces per hour per wheel. Each wheel has a service life of 5000 to 6000 pieces.

Table 5 gives the conditions and sequence of operations for belt polishing die cast steam-iron soleplates made of aluminum alloy 380.0. Ten polishing heads are used to produce a bright finish on sides and bottom of soleplates.

Buffing. Selection of procedure for buffing depends mainly on cost, because it is usually possible to obtain the desired results by any one of several different procedures. For example, in hand buffing, combinations of the various influences might call for the use of equipment ranging from simple, light-duty machines to heavy-duty, variable-speed, double-control units. These machines represent a wide range in capital investment.

Automatic buffing requires custom-made machinery or special fixtures on standard machinery. Size and complexity of the machinery are determined by the required production rates and by the size or shape of the workpieces. High production requires more stations, heavier equipment, and more power. The configuration of the part may be so simple that one buff covers the total area to be finished, or it may be so complex as to require the use of many buffs set at angles and advanced toward the workpiece by cam action.

For cut and color work, buffs are bias types with a thread count of 86/93. For severe cut-down, treated cloth is used with the same thread count. The final color work is accomplished with a buff with very little pucker and a low thread count of 64/64 (see Table 6). A number of procedures that have proved successful for high-luster buffing of specific aluminum parts are summarized in Table 6; others are described in the following examples.

Table 7 gives the conditions and sequence of operations for automatic buffing of wrought aluminum frying-pan covers. A specular finish was required. In another case, the sides of the die cast aluminum frying pans made from alloy 360 were buffed to a bright finish by an automatic machine with four buffing heads. The buffing wheel of each head consisted of a 14-ply, 16-spoke-sewed bias buff 430 mm (17 in.) outside diameter, 230 mm (9 in.) inside diameter, and with a 44-mm (1¾-in.) diam arbor hole. Wheel speed was 1745 rev/min, equal to 39 m/s (7700 sfm). Each buff was made up of four sections. A liquid buffing compound was applied by one gun per wheel at the rate of 3 g per shot (0.1 oz per shot) for the first wheel, 2.5 g per shot (0.09 oz per shot) for the second and third wheels, and 1 g per shot (0.04 oz per shot) for the fourth wheel. The gun operated for 0.1 s on time and 5 s off time. Service life of each buffing wheel was 1600 to 2100 pieces.

Die cast aluminum soleplates for steam irons (Table 8) were buffed to a bright finish on an automatic machine with eight buffing heads. The soleplates were made of alloy 380.0 and were prepolished with 320-mesh grit. A liquid buffing compound was applied by one gun per wheel for the first four heads and by two guns per wheel for the last four heads. The guns were on for 0.12 s and off for 13 s. Service life was 72 000 pieces for each buff of the first four heads, and 24 000 pieces for each buff of the last four heads.

Satin Finishing

Mechanical satin finishing is an established method for obtaining an

Table 4 Conditions for wet barrel finishing of aluminum products

Product	No. of pieces per kg	No. of pieces per lb	No. of pieces per load	Cycle time, min	Barrel speed, rev/min
Clean, deburr and brighten(a)					
Percolator spout	60	130	630	20	33
Measuring spoon	20	50	750	20	33
Flame guard	40	90	1500	15	33
Leg	35	80	2700	30	33
Toy spoon	80	180	5000	17	33
Handle	105	225	5800	25	33
Deburr and brighten(b)					
Die cast handles	11	25	600	120	15
Burnish to high gloss(c)					
Die cast housing	.5	1	60	45	14

(a) Rubber-lined steel, single-compartment drum, 560-mm (22-in.) diam, 760 mm (30 in.) long. Processing cycle: load drum with medium (20 kg or 50 lb of 3-mm or ⅛-in. steel balls per load), parts, and compound (154 g or 5.5 oz of burnishing soap per load); cover load with water (66 °C or 150 °F); rotate drum for specified time; unload, rinse, separate parts from medium; tumble-dry parts in sawdust for 4 min. (b) Rubber-lined steel, double-compartment drum; each component 740 mm (29 in.) long, 915 mm (36 in.) in diameter. Processing cycle: load deburring compartment with parts, compound (2 kg or 5 lb of burnishing soap), and medium (365 kg or 800 lb of No. 4 granite chips), using hoist; cover load with cold tap water; rotate drum for 1 h; unload, and rinse with cold water; separate parts and medium, and transfer parts to burnishing compartment; add burnishing compound (0.9 kg or 2 lb of burnishing soap) and medium (680 kg or 1500 lb of 3-mm or ⅛-in. steel balls); cover load with water (71 °C or 160 °F) and rotate drum for 1 h; separate parts and medium, and rinse parts in hot water and then in cold water; spin dry in a centrifugal hot-air drier. (c) Single-compartment drum, 1.5 m (5 ft) long, 1.2 m (4 ft) in diameter. Processing cycle: load drum with parts (parts are fixtured, to prevent scratching), medium (2040 to 2270 kg or 4500 to 5000 lb of steel balls 3, 6, and 8 mm or ⅛, ¼, and 5⁄16 in. in diameter, and 2 and 3 mm or 1⁄16 and ⅛ in. steel diagonals), and compound (5 kg or 12 lb of alkaline burnishing soap, pH 10); cover load with cold tap water; rotate drum for 22½ min in one direction, then 22½ min in reverse direction; rinse and unload; dip-rinse parts, and hand wipe

Table 5 Conditions of belt polishing for bright finishing die cast soleplates

Operation	Area polished	Polishing head, No.	Contact wheel: Type	Contact wheel: Size, mm	Contact wheel: Size, in.	Contact wheel: Hardness, durometer	Belt(a): Size, mm	Belt(a): Size, in.	Belt(a): Abrasive mesh size	Belt(a): Life, pieces
1	Side	1,2	Plain face	50 by 380	2 by 15	60	50 by 3050	2 by 120	280(b)	600
2	Side	3,4	Plain face	50 by 380	2 by 15	60	50 by 3050	2 by 120	320(b)	600
3	Bottom	5	Serrated(c)	150 by 380	6 by 15	45	150 by 3050	6 by 120	120(b)	1200
4	Bottom	6	Serrated(c)	150 by 380	6 by 15	45	150 by 3050	6 by 120	150(b)	2000
5	Bottom	7	Serrated(c)	150 by 380	6 by 15	45	150 by 3050	6 by 120	220(b)	2000
6	Bottom	8	Serrated(c)	150 by 380	6 by 15	45	150 by 3050	6 by 120	280(b)	2000
7	Bottom	9	Plain face	150 by 380	6 by 15	60	150 by 3050	6 by 120	320(b)	2000
8	Bottom	10	Plain face	150 by 380	6 by 15	60	150 by 3050	6 by 120	320(d)	600

(a) Belt speed for all operations was 35 m/s (6900 sfm). All belts were cloth; bond, resin over glue. (b) Aluminum oxide abrasive. (c) 45° serration, 13-mm (½-in.) land, 10-mm (⅜-in.) groove. (d) Silicon carbide abrasive

Table 6 Equipment and operating conditions for high-luster buffing of aluminum products

Product	Size, mm	Size, in.	Type of buffing machine	Buffing wheel: Type	Overall diameter, mm	Overall diameter, in.	Center diameter, mm	Center diameter, in.	Ply, mm	Ply, in.	Thread count	Wheel speed, m/s	Wheel speed, sfm	Type of compound	Production, pieces per hour
Biscuit pan	340 by 240	13¼ by 9	Semiautomatic	Bias	360	14	125	5	410	16	86/93	40	8250	Bar	205
Burner ring	75 diam by 20	3 diam by ¾	Continuous rotary(a)	Radial, vented	(b)	(b)	30	1⅛	510	20	64/64	(c)	(c)	Liquid	297
Cake-carrier base	270 diam by 20	11¼ diam by 1³⁄₁₆	Continuous rotary(a)	Bias	Two 360	Two 14	125	5	460	16	86/93	50	9550	Liquid	278
				Bias	Two 330	Two 13	75	3	50	2	64/68	45	8850	Liquid	...
Cake pan	350 by 241 by 65	14 by 9½ by 2½	Hand buffing (handles)	Bias	330	13	75	3	50	2	64/68	40	7650	Bar	438
			Semiautomatic (sides)	Bias	360	14	125	5	410	16	86/93	40	8250	Bar	200
Cake pan	200 by 203 by 50	8 by 8 by 2	Semiautomatic	Bias	360	14	125	5	410	16	86/93	40	8250	Bar	127
Cup	60 diam by 65	2⅜ diam by 2½	Semiautomatic	Bias	360	14	125	5	410	16	86/93	40	8250	Bar	450
Pan bottom	280 by 280	11⅛ square	Semiautomatic	Bias	360	14	125	5	410	16	86/93	40	8250	Bar	106
Pan cover	285 by 285	11¼ square	Semiautomatic (d)	Bias (sides)	360	14	125	5	410	16	86/93	40	8250	Bar	95
				Loose, vented (top)	(e)	(e)	50	2	510	20	64/64	(f)	(f)	Bar	95
Toy pitcher	65 diam by 90	2½ diam by 3½	Continuous rotary(g)	Bias	360	14	125	5	410	16	64/68	50	9550	Liquid	817
Toy tumbler	50 diam by 65	1⅞ diam by 2½	Continuous rotary(g)	Bias	360	14	125	5	410	16	64/68	50	9550	Liquid	864

(a) Five spindle machine; four buffing heads, one load-unload station. (b) Each of the four wheels used had one 330-mm (13-in.) and three 360-mm (14-in.) sections. (c) For 330-mm (13-in.) section, 45 m/s (8850 sfm); for 360-mm (14-in.), 49 m/s (9550 sfm). (d) Two machines, run by one operator. (e) Buff made up of 360-mm (14-in.), 381-mm (15-in.), and 410-mm (16-in.) sections. (f) 42 m/s (8250 sfm) for 360-mm (14-in.) sections, 45 m/s (8800 sfm) for 380-mm (15-in.) sections, and 48 m/s (9400 sfm) for 410-mm (16-in.) sections. (g) Eight-spindle machine

attractive surface texture on aluminum hardware items such as knobs, hinges, rosettes, and drawer pulls. Satin finishes are used also for architectural, appliance, and automotive trim. The satin finish results from small, nearly parallel scratches in the metal surface, which give the surface a soft, smooth sheen of lower reflectivity than that of polished or buffed surfaces.

Satin finishes can be applied by fine wire brushing. Other methods use a greaseless abrasive compound in conjunction with a conventional buffing head, tampico brush, cord brush, string buff, or brush-backed sander head. Abrasive-impregnated nylon disks mounted like buffs, and abrasive cloth sections mounted on a rotating hub are also used. All of these methods produce about the same type of finish, therefore the use of any particular one depends on the surface contour of the workpiece.

Surfaces of workpieces to be satin finished should be free of grease and oil, and low contact pressures should be used. Wire brushes must be kept free of oxide and accumulations of aluminum metal. This is accomplished by frequently bringing a pumice stone or soft brick in contact with the rotating brush. A combination of conditions commonly used in wire brushing consists of 0.4-mm (0.015-in.) diam wire filled 250-mm (10-in.) diam wheel having a surface speed at about 8.0 m/s (1600 sfm). Undue pressure on a rotating wire wheel will bend the wires and cause excessive tearing of the aluminum surface.

Stainless steel wires are recommended, because other metals such as brass or steel may become embedded in the aluminum surface to produce discoloration or corrosion. If brass or steel wire wheels are used, the particles embedded in the surface can be removed by immersing the work in a nitric acid solution (1 part water to 1 part acid by volume) at room temperature.

The satin finish processes in which a greaseless abrasive compound is used are essentially dry. Water is required for softening the binder in the abrasive compound so that it will adhere to the surface of the buff, and after the binder dries, the buff is ready for operation. At this stage a lubricant, such as a buffing compound or tallow, may be used. The use of lubricant produces a higher reflection.

Table 9 describes equipment and techniques employed in mechanical satin finishing processes. If the satin finished parts are to be anodized, etching or bright dipping should not precede anodizing, because the satin appearance will be lost. Cleaning treatments that do not etch or that only slightly etch the metal should be used before anodizing.

Table 7 Sequence and conditions of automatic buffing operations for obtaining specular finish on aluminum frying-pan covers

Operation	Area buffed	Buffing head No.	Buffing wheel: Type	Diameter: Overall mm	Overall in.	Center mm	Center in.	Arbor hole mm	Arbor hole in.	Ply	Thread count	Density	No. of sections	Speed m/s	Speed sfm	Life, pieces	Application of compound(a): No. of guns	Cycle, s On	Cycle, s Off	g per shot	oz per shot
1	Sides (4)	1,2,3,4	Bias, air cooled	430	17	180	7	45	1¾	...	86/93	2,4	20	40	7750	40 000	3	0.1	7	0.5	0.02
2	Corners (2)	5,6,7,8	Bias, 20-spoke sewed	430	17	180	7	45	1¾	16	86/93	4	4	40	7750	35 000	1	0.1	7	0.5	0.02
3	Sides (2)	9,10, 11,12	Bias, 20-spoke sewed	430	17	180	7	45	1¾	16	86/93	4	4	40	7750	50 000	1	0.1	7	0.5	0.02
4	Sides (2)	13,14, 15,16	Bias, 20-spoke sewed	430	17	180	7	45	1¾	16	86/93	4	4	40	7750	35 000	1	0.1	8	0.5	0.02
5	Top	17	Bias, 45° spoke sewed	430	17	180	7	45	1¾	16	86/93	4	15	40	7750	65 000	3	0.1	8	0.2	0.01
6	Top	18	Bias	430	17	180	7	45	1¾	12	86/93	8	18	40	7750	70 000	3	0.1	8	0.2	0.01
7	Top bias	19	Bias	430	17	180	7	45	1¾	12	86/93	8	18	40	7750	70 000	3	0.1	8	0.2	0.01
8	Top bias	20	Bias	410	16	125	5	45	1¾	14	64/68	2	19	37	7300	45 000	3	0.1	8	0.2	0.01
9	Corners (4)	21,22, 23,24	Bias	430	17	180	7	45	1¾	12	86/93	8	4	40	7750	65 000	1	0.1	10	0.5	0.02
10	Sides (4)	25,26, 27,28	Bias	430	17	180	7	45	1¾	12	86/93	8	4	40	7750	80 000	1	0.1	10	0.5	0.02
11	Top bias	29	Bias	410	16	125	5	45	1¾	14	64/68	2	15	23	4600	80 000	3	0.1	10	0.5	0.02
12	Sides (4)	30,31 (b)	Domet flannel	430	17	180	7	70	2¾	20	(d)	(d)	40	20	4000	80 000	6	0.1	11	0.2	0.01
13	Top	32	Domet flannel (c)	430	17	180	7	40	1⅝	32	(d)	(d)	24	25	4900	30 000	3	0.1	11	0.2	0.01
14	Top	33	Domet flannel (c)	430	17	180	7	40	1⅝	32	(d)	(d)	24	18	3550	30 000	3	0.1	11	0.2	0.01

(a) Liquid tripoli compound applied to buffing heads No. 1 through 29; stainless steel buffing compound applied to heads No. 30 through 33. (b) Each head buffs two sides. (c) Domet flannel sections interleaved with 180-mm (7-in.) diam disks of Kraft paper. (d) Inapplicable to flannel buff

Table 8 Automatic bright-finish buffing of aluminum soleplates

Operation	Area buffed	Buffing head No.	Buffing wheel: Type	Size	Diameter(a): Overall mm	Overall in.	Arbor hole mm	Arbor hole in.	No. of Sections	Speed m/s	Speed sfm	Life, pieces	Application of compound(b): No. of guns	g per shot	oz per shot
1	Side	1,2	Sisal	10-mm (⅜-in.) spiral sewed	410	16	40	1¼	2	37	7350	72 000	1	0.5	0.02
2	Side	3	Bias	16-ply, 20-spoke sewed	430	17	45	1¾	2	40	7800	72 000	1	0.5	0.02
3	Side	4	Bias	16-ply, 20-spoke sewed	430	17	45	1¾	2	40	7800	72 000	1	0.5	0.02
4	Top	5	Sisal	10-mm (⅜-in.) spiral sewed	410	16	45	1¾	15	37	7350	24 000	2	3.0	0.1
5	Top	6	Sisal	10-mm (⅜-in.) spiral sewed	410	16	45	1¾	15	37	7350	24 000	2	3.0	0.1
6	Top	7,8	Bias	16-ply, 20-spoke sewed	430	17	45	1¾	10	40	7800	24 000	2	3.0	0.1

(a) All wheels had 180-mm (7-in.) diam centers. (b) Proprietary liquid compound was used. Cycle time: 0.12 s on, 13.0 s off

Chemical Cleaning

The degree and nature of cleanliness required are governed by the subsequent finishing operations. For example, the cleaning requirements for plating or for the application of chromate or other mild-reaction conversion coatings are somewhat more stringent than for anodizing.

When establishing a cleaning cycle or when testing different cleaners or cleaning conditions, it is desirable to test the cleanness of the processed surface. Wetting an aluminum surface with water, known as the water break test, does not always provide an indication of cleanliness, if oxides are of concern, because oxide-coated surfaces free of oil or grease can be wetted uniformly. Also, a surface that has been processed with a detergent containing a wetting agent can be wetted even though not thoroughly clean, because of the film of wetting agent remaining on the unclean surface. Two other methods of testing aluminum for cleanliness are as follows:

- Spray or coat the work surface with, or dip a test panel into, an unheated aqueous solution containing 30 g/L (4 oz/gal) of cupric chloride and 29 mL/L (3.8 fluid oz/gal) of concentrated hydrochloric acid. Uniform gassing or a deposit of copper is an indication that the surface is chemically clean
- Spray or coat the work surface with, or dip a test panel into, an unheated chromate conversion coating bath of the acid type, until an orange-colored film is formed. A uniform orange film indicates a chemically clean surface

Additional methods of testing for cleanliness are discussed in the article on selection of cleaning processes in this Volume.

Table 9 Methods, equipment, and conditions for mechanical satin finishing of aluminum

Method	Suitable equipment: Buffing lathe	Portable power head	Power required	Speed: m/s	sfm	Lubricant
Wire brushing(a)	Yes	Yes	(b)	6-11	1200-2250	None
Sanding with brush-backed head(c)	Yes	No	(d)	900-1800 rev/min	900-1800 rev/min	Optional
Tampico or string brushing(e)	Yes	No	(b)	15-31	3000-6000	Pumice(f)
Finishing with abrasive-coated cloth(g)	Yes	Yes	(d)	31-36	6000-7000	Optional
Finishing with nylon disks(h)	Yes	Yes	(j)	23-33	4500-6500	Optional
Buffing with compounds(k)	Yes	Yes	(b)	15-26	3000-5000	(m)

(a) 305-mm (12-in.) diam brush of stainless steel wire 0.125 mm (0.005 in.) in diameter. (b) 1 hp per 25 mm (1 in.) of brush width. (c) Using 60- to 600-mesh abrasive cloth loadings. (d) 1 hp per head. (e) 300-mm. (12-in.) diam brush. (f) With oil or water; emery cake also may be used. (g) Cloth is mounted radially on rotating hubs; coated with 50- to 320-mesh emery abrasive. (h) Disks impregnated with silicon carbide abrasive, coarse to ultrafine. (j) 1/4 hp per 25 mm (1 in.) of disk width. (k) Greaseless satin-finishing compounds containing aluminum oxide abrasive (200 or 240 mesh) used with unstitched or loosely stitched buffs (360-mm or 14-in.) or with string brush. (m) Dry, or with buffing compound or grease stick

Solvent Cleaning. The primary function of solvent cleaners is the removal of oil and grease compounds. Organic solvents alone rarely provide sufficient cleaning to permit final finishing operations; solvents usually are used to remove large amounts of organic contaminants to minimize overloading of the subsequently used alkaline cleaners.

Greases and oils vary as to solubility in specific solvents. Fish oils are more difficult to remove than oils of other types. In the dried condition, some oxidizing oils, such as linseed oil, form a leathery film that is difficult to remove with any solvent.

Polishing and buffing compounds are readily removed by most solvents when cleaning is performed immediately after buffing. If the compounds are permitted to harden, they may be difficult to remove. Heated solutions, agitation, or mechanical action (ultrasonics or physical force) may be required for satisfactory cleaning. To remove compounds burned in the surface, it is necessary to soak the parts in a liquid, not vapor degrease, using organic degreaser such as trichlorethylene or methylene dichloride or in an inhibited alkaline cleaner.

If polishing and buffing compounds cannot be removed immediately after buffing, the application of a neutral mineral oil over the buffed surface will maintain the compounds in a more soluble condition for subsequent removal by a solvent. The sequence of operations usually required for buffed aluminum surfaces is as follows: solvent cleaning, nonetching alkaline detergent cleaning, rinsing, removing of surface oxides, rinsing, and finally the applying of the desired finish. Some of these steps may be omitted, depending on the type and quality of the buffing compound, the quality of workmanship in buffing, and the quality of solvents and cleaners used.

Emulsifiable solvents also are used to clean aluminum. These are organic solvents, such as kerosine, Stoddard solvent, and mineral spirits, to which small amounts of emulsifiers and surfactants are added. In use, this type of cleaner emulsifies the oil or grease on the surface. The soil and cleaner are removed with water, preferably applied by spraying.

This type of degreasing is satisfactory prior to anodizing, etching, removal of surface oxides, chemical conversion coating, plating and painting. In some instances, intermediate treatments are required, such as the removal of surface oxides before etching.

The emulsifiable solvent should have a pH of 8 or less; otherwise, it will stain or corrode the aluminum if permitted to remain on the surface for a period of time prior to rinsing or additional cleaning. However, emulsifiable solvents with higher pH are more efficient cleaners, and can be used if the surfaces are rinsed or are cleaned by additional methods within 2 or 3 min after degreasing.

A lower cost cleaning solution can be obtained by adding water to the emulsifiable solvent. This less efficient type of solution is limited to the removal of light oil and grease. A more detailed discussion on cleaning with solvents is given in the articles on solvent cleaning, emulsion cleaning, and vapor degreasing in this Volume. It is now common to use alkaline cleaners to remove oil and grease instead of solvents, the use of which is under increasing scrutiny by the Environmental Protection Agency (EPA).

Alkaline cleaning is the most widely used method for cleaning aluminum and aluminum alloys. This method is easy to apply in production operations, and equipment costs are low. Aluminum is readily attacked by alkaline solutions. Most solutions are maintained at a pH between 9 and 11, and they are often inhibited to some degree to minimize or prevent attack on the metal. The most frequently used cleaner is the mildly inhibited type.

Cleaners of either the etching or the nonetching type have some ability to emulsify vegetable and animal oils or greases, but not mineral oils or greases. Therefore, they can sometimes remove fresh buffing compounds and the lard oils used in spinning operations.

The nonetching types of cleaners may be classified as silicated and nonsilicated cleaners. The silicated cleaners are based on aqueous solutions of sodium carbonate, trisodium phosphate, or other alkalis, to which small amounts of sodium silicate are added to inhibit etching. The main disadvantage of the silicated types, aside from their inability to emulsify and remove mineral oils, is that the silicate may react with the aluminum to form an insoluble aluminum silicate, especially when the temperature of the bath exceeds 82 °C (180 °F). However, lower operating temperatures decrease the efficiency of the solution for the removal of certain soils. There are now available silicated alkaline cleaners used at 50 to 60 °C (120 to 140 °F) that reduce energy consumption.

The nonsilicated cleaners are often based on the use of relatively large concentrations of surfactants. High operating temperatures are required, but some cleaners used above 71 °C (160 °F) etch the aluminum surface. Cleaners containing a large quantity of surfactants, particularly those types of surfactants that resist complete rinsing, must not be carried into baths used for bright dipping, anodizing, or chemical conversion coating.

Neither the silicated nor the nonsilicated cleaners remove aluminum oxide uniformly. Because the removal of oxide is essential for the application of decorative or functional finishes, the best procedure is to clean, remove oxide with an acid solution, and then proceed with finishing.

Nonetching cleaners may be used after solvent cleaning to produce waterwetable surfaces, or they may be used alone when soils are light and easily removed. The surfaces should be treated to remove oxides afterwards. When sodium orthosilicate or sodium metasilicate are used, the concentration of carbonates must be kept at a minimum to minimize the formation of floc, which may redeposit on the work. Unlike sodium hydroxide, the alkali silicates have good wetting, emulsifying, and rinsing properties. The ratio of silicon dioxide to sodium oxide in the compound determines the effectiveness of the alkali silicates. Sodium orthosilicate has good detergency and is effective in the cleaner at a ratio of 1 to 2, whereas sodium metasilicate should be 1 to 1.

Agitation of the cleaner increases the cleaning action and is best created by pumps, propellers, or movement of the work. Air agitation, although easier to install and most convenient to operate, has the following disadvantages:

- Air can reduce the solution temperature
- The additional oxygen may cause staining and tarnishing on some alloys
- Air agitation introduces carbon dioxide, which may increase the carbonate content

Rinsing should be accomplished immediately after the work is removed from the alkaline bath to prevent dryon. Warm water is preferred; if lowtemperature cleaners are used, then rinsing with cold water is satisfactory.

Aluminum surfaces sometimes contain areas of localized corrosion, referred to as atmospheric etch, caused by contaminants in the air during storage. The corroded areas are more visible after alkaline cleaning or etching than before. When the corrosion spots are present, the work may be dipped in a sodium bisulfate solution 45 g/L (6 oz/gal), or in a cold 70% nitric acid solution, to minimize the effect of the subsequent alkaline cleaning.

During alkaline cleaning, especially if etching occurs, some alloys containing copper, iron, manganese or silicon develop a black smut on the surface. Compositions and operating conditions of common alkaline cleaners are given below:

- **Etching cleaners**

Sodium hydroxide ....22 to 75 g (3 to 10 oz)
Sodium phosphate.......... 0.8 to 4 g (0.1 to 0.5 oz)
Water, to make............ 4 L (1 gal)
Temperature of bath ......60 to 82 °C (140 to 180 °F)
Immersion time........30 s to 10 min

Sodium hydroxide............2 to 6 g (0.25 to 0.75 oz)
Sodium phosphate ....8 to 60 g (1 to 8 oz)
Sodium carbonate...........8 to 60 g (1 to 8 oz)
Water, to make .... 4 L (1 gal)
Temperature of bath ......60 to 82 °C (140 to 180 °F)
Immersion time........... 2 to 5 min

- **Nonetching cleaners**

Sodium pyrophosphate and sodium metasilicate........Total of 15 to 75 g (2 to 10 oz)
Water, to make............ 4 L (1 gal)
Temperature of bath ......60 to 71 °C (140 to 160 °F)
Immersion time........... 2 to 5 min

Trisodium phosphate and sodium metasilicate........Total of 15 to 75 g (2 to 10 oz)
Water, to make............ 4 L (1 gal)
Temperature of bath ......60 to 71 °C (140 to 160 °F)
Immersion time........... 2 to 5 min

Sodium carbonate............4 to 8 g (0.5 to 1 oz)
Sodium metasilicate..........4 to 8 g (0.5 to 1 oz)
Water, to make............ 4 L (1 gal)
Temperature of bath ......60 to 71 °C (140 to 160 °F)
Immersion time........... 2 to 5 min

Borax...........22 to 38 g (3 to 5 oz)
Sodium pyrophosphate ..... 4 to 8 g (0.5 to 1 oz)
Water, to make .... 4 L (1 gal)
Temperature of bath ......60 to 71 °C (140 to 160 °F)
Immersion time........... 2 to 5 min

For more information, see the article on alkaline cleaning in this Volume.

Electrocleaning is seldom used for cleaning aluminum and aluminum alloys, because it offers no advantage over an etching cleaner. However, a few processes are used in production operations. These use low-voltage, usually in the range of 6 to 12 V. Cathodic cleaning, in which the work is the cathode, is more common than anodic cleaning. Common practice is to reverse the current during the last 5 to 10 s of the cleaning operation.

After removal from the cleaner, the work is rinsed in warm or hot water, dipped in acid to neutralize any residual alkali, and finally rinsed in cold water. The work can then be finished as desired. The composition of two solutions that are recommended for electrocleaning are given below:

Constituents	Composition, %
Solution A	
Sodium orthosilicate	85
Sodium carbonate (anhydrous)	10
Sodium resinate	5
Solution B	
Sodium carbonate (anhydrous)	46
Trisodium phosphate	32
Sodium hydroxide	16
Rosin	6

Note: For typical operating conditions, see text

Acid Cleaning. Acid cleaners may be used alone or in conjunction with other acid, alkaline, or solvent cleaning systems. Vapor degreasing and alkaline cleaning may be required for the removal of heavy oils and grease from workpieces before they are immersed in an acid bath. One of the main functions of an acid cleaner is the removal of surface oxides prior to resistance welding, painting, conversion coating, bright dipping, etching or anodizing.

A mixture of chromic and sulfuric acids is commonly used to remove surface oxides, burnt-in oil, water stains, or other films, such as the iridescent or colored films formed during heat treating. This acid mixture cleans and imparts a slightly etched appearance to the surface, preparing it for painting, caustic etching, conversion coating, or anodizing. Nonpolluting, proprietary products free of chromic acid are available for acid cleaning and deoxidizing.

Oxide films must be thoroughly removed before spot welding. A mixture of phosphoric and chromic acids is another solution that may be used for this purpose. Because of the corrosive nature of the chlorides and fluorides in welding fluxes, the fluxes should be removed as soon as possible after welding. Mixtures of nitric and hydrofluoric acids are best for removing fluxes. Most fluxes can also be satisfactorily removed by a dilute (5 to 20 vol %) nitric acid solution.

$6 \times 3 \times 2 = 36 in^2$

$6 \times 1 = 6 in^2$

$42 \times 2 = 84 in^2$

$12 \times 12 = 144 in^2$ ft^2

$84/144 = .583$

$25 \times .583 = 14.6$ amps

50 √

Proprietary nonetching acid cleaners are available for cleaning aluminum and aluminum alloys. Operating temperatures of these solutions range from 54 to 82 °C (130 to 180 °F) and the pH usually ranges from 4.0 to 5.7.

Compositions and operating conditions for typical acid cleaning solutions are given as follows:

- **Solution 1**

Chromic acid ....45 to 90 g (6 to 12 oz)
Sulfuric acid (66° Bé).........150 to 190 mL (19 to 24 fluid oz)
Water, to make .... 4 L (1 gal)
Temperature of bath ......43 to 82 °C (110 to 180 °F)
Immersion time.........Up to 20 min

- **Solution 2**

Nitric acid (42° Bé).... 500 to 750 mL (64 to 96 fluid oz)
Hydrofluoric acid (48%)............25 to 190 mL (3 to 24 fluid oz)
Water, to make .... 4 L (1 gal)
Temperature of bath ....Room temperature
Immersion time........... 1 to 5 min

- **Solution 3**

Sulfuric acid (66° Bé)... 100 mL (13 fluid oz)
Hydrofluoric acid (48%)....... 25 mL (3 fluid oz)
Chromic acid .... 40 g (5 oz)
Water, to make .... 4 L (1 gal)
Temperature of bath ......65 to 70 °C (150 to 160 °F)
Immersion time........... 2 to 5 min

- **Solution 4**

Nitric acid (42° Bé)......38 to 125 mL (5 to 16 fluid oz)
Sodium sulfate (hydrate)....60 to 120 g (8 to 16 oz)
Water, to make .... 4 L (1 gal)
Temperature of bath ......76 to 79 °C (170 to 175 °F)
Immersion time........... 4 to 8 min

- **Solution 5**

Phosphoric acid.... 70 mL (9 fluid oz)
Chromic acid ....20 g (2.75 oz)
Water, to make .... 4 L (1 gal)
Temperature of bath ......43 to 65 °C (110 to 150 °F)
Immersion time........... 1 to 5 min

Aluminum parts should be insulated from ferrous metal baskets or supports when immersed in acid cleaning solutions, because contact of these two metals can produce a galvanic action that causes corrosion. Materials such as vinyl plastisols, epoxy, polyethylene, and polypropylene may be used for insulation. When practical, baskets or rods should be of the same or similar material as the workpieces.

Chemical Brightening (Polishing)

Chemical brightening, known also as bright dipping and chemical polishing, smoothens and brightens aluminum products by making use of the solution potential of the aluminum surface in the various baths employed and of the local differences in potential on the aluminum surface.

In general, chemical brightening baths can be concentrated or dilute acid solutions containing oxidizing agents. The acids commonly used are sulfuric, nitric, phosphoric, acetic and, to a lesser extent, chromic and hydrofluoric. Ammonium bifluoride is used when it is desirable to avoid the hazards that attend the use of hydrofluoric acid. Fluoboric and fluosilicic acids may also be used as alternates for hydrofluoric acid. An alkaline bath, such as Alupol, can also be used for chemical etching. This bath consists of 20 kg (44 lb) sodium nitrate, 15 kg (33 lb) sodium nitrite, 25 kg (55 lb) sodium hydroxide, and 20 kg (44 lb) water. An aluminum part is immersed for 1 to 5 min at a bath temperature of 90 to 140 °C (195 to 285 °F). Protrusions, valleys, and scratches are eliminated, and reflectance is increased.

Phosphoric-Nitric Acid Baths. Among the various types of concentrated baths, the phosphoric-nitric acid baths are the most widely used in the United States. Compositions and operating conditions for two commercial baths of this type are given below:

Constituent or condition	Range
Phosphoric-nitric(a)	
Phosphoric acid (85%)	45 to 98 wt %(b)
Nitric acid (60%)	0.5 to 50 wt %(b)
Water	2 to 35 wt %
Temperature	87-110 °C (190-230 °F)
Immersion time	½-5 min
Phosphoric-acetic-nitric(c)	
Phosphoric acid(85%)	80 vol %
Acetic acid (glacial, 99.5%)	15 vol %
Nitric acid (60%)	5 vol %
Temperature	87-110 °C (190-230 °F)
Immersion time	½-5 min

(a) U. S. Patent 2,729,551 (1956). (b) Recommended volumetric make-up consists of 93.5 parts of 85% phosphoric acid and 6.5 parts of 60% nitric acid. (c) U. S. Patent 2,650,157 (1953)

Additionally, certain proprietary chemical bright dips can be operated at 76 to 82 °C (170 to 180 °F), which is significantly lower than the normal 87 to 110 °C (190 to 230 °F) for conventional baths. The low temperature baths, however, are limited in allowable water drag-in from prior rinse operations. Excessive water drag-in results in poor brightening.

Alkali nitrates may be used as a substitute for nitric acid. Acetic acid, copper salts and other additives are used in some phosphoric-nitric acid baths. As content of the additives increases, solution control becomes more complex.

For economy, some phosphoric-nitric acid baths are operated with an aluminum phosphate content near the tolerable maximum of 10 to 12%, with a dissolved aluminum content of about 40 g/L (5 oz/gal). This is close to the saturation point, at which precipitation of this compound on the work produces etch patterns.

The addition of surfactants increases the amount of metal removed under a given set of operating conditions. Surfactants help to enhance the chemical polishing as well as suppressing the evolution of fumes. Acetic and sulfuric acids alter the physical property-composition relationship in the concentrated acid baths and also complicate control problems. Acetic acid volatilizes rapidly from the bath.

Small concentrations of heavy metals in the bath enhance the brightening effect, particularly on alloys with negligible copper content. Copper can be introduced into the bath by one of three methods: (*a*) the direct dissolution of metallic copper; (*b*) the addition of a small amount of a copper compound, such as 0.01 to 0.02% cupric nitrate; or (*c*) the use of racks made of aluminum-copper alloys. Copper is added to the bath when brightening aluminum alloys such as 2024 and 7075, which contain high percentages of copper. Excess copper can plate out of the bath. In some baths, however, excess copper causes etching, and sometimes nickel or zinc is used instead of copper.

The phosphoric-nitric acid baths are not recommended for brightening alloys containing silicon. Excessive dissolution causes dispersion of undissolved silicon, which deposits on the work surfaces and is difficult to remove by rinsing. When high silicon alloys are used, the addition of 1 to 2% hydrofluoric acid to the bath is recommended. The gradual buildup of other metals in the bath from the aluminum alloys processed usually causes no difficulty unless the amount of aluminum dissolved exceeds the solubility limit. When this occurs, excess aluminum precipitates and causes coprecipitation of trace elements, which may be difficult to remove from the work.

Contamination of the bath by more than trace amounts of buffing or polishing compounds and other soils should be avoided. These compounds may cause the bath to foam excessively and may interfere with its polishing action. Food-grade or NF (National Formulary) phosphoric acid should be used. Lower grades contain fluorides, arsenic and other impurities that are harmful to the process.

Close control of the nitric acid and water contents, necessary for optimum chemical brightening, is difficult because of the rapid volatilization of these liquids and because of the time required for chemical analysis of the bath. A control method based on an electronic device that monitors the nitric acid content and on the physical measurement of specific gravity and viscosity has been developed.

Dragout is a major factor in the cost of chemical brightening. The amount of solution and the weight of chemicals lost by dragout are related to the specific gravity and viscosity of the solution. Dragout may be minimized by operating the bath at higher temperatures, but this condition may increase the amount of transfer etch while moving to the rinse tank the rates of aluminum dissolution and evaporation of nitric acid and water. However, transfer etch may be avoided by rapid transfer into the rinse, and the rate of aluminum dissolution can be minimized by a shorter period of immersion. In general, an operating temperature in the range of 87 to 100 °C (190 to 212 °F) is satisfactory, provided an optimum bath composition, including additives, is maintained. Also, evaporation of nitric acid and water is not excessive at this temperature. Acetic acid also reduces transfer etch, but this acid volatilizes rapidly from the bath.

Surfactants are employed in some baths to suppress the evolution of fumes; however, they may cause foaming and an increase in the amount of dragout. Surfactants also increase the rate of workpiece dissolution. The generation of heat accompanying high dissolution rates must be considered when providing for the control of bath temperatures within the specified range.

Agitation is useful for maintaining a uniform temperature and composition throughout the bath, and for fast removal of reaction products and replenishment of reactants at the surfaces of the work. The most satisfactory method is mechanical agitation and movement of the work in an elliptical pattern. Air agitation is commonly used, but it must be properly controlled. Small air bubbles cause excessive loss of volatile acids by evaporation and an excess of nitrous oxide fumes. Large air bubbles sufficient to create uniform bath temperature provide satisfactory agitation. Excessive solution agitation can cause pitting and streaking on work, so the agitation should always be moderate.

The bath must be well vented to remove the noxious fumes; an exhaust of about 90 m^3/min per square metre (300 ft^3/min per square foot) of bath surface is recommended. Fumes evolved during transfer of the parts to the first rinse tank should likewise be vented, and it is good practice also to vent the first rinse tank, for which an exhaust of about 60 m^3/min per square metre (200 ft^3/min per square foot) of water surface is satisfactory. Water should be warm and air agitated.

The fumes may be exhausted by fan or steam jet. Fume-separators are required when the fumes cannot be exhausted into the atmosphere. Dilute caustic soda solutions are used to scrub the fumes and neutralize the acid.

Phosphoric and Phosphoric-Sulfuric Acid Baths. Concentrated solutions of phosphoric acid at operating temperatures above 79 °C (175 °F) were the first baths used for brightening aluminum. A more effective bath, which combines some smoothening or polishing with brightening action, is one containing 75 vol % phosphoric acid and 25 vol % sulfuric acid. This bath, which is operated at 90 to 110 °C (195 to 230 °F) for ½ to 2 min, produces a diffuse but bright finish.

Under some conditions of composition and bath temperature, a white film of phosphate salts remains on the metal after treatment in either of these baths. The film must be removed, and this can be accomplished with a hot (60 to 71 °C, or 140 to 160 °F) aqueous solution of chromic and sulfuric acids. The composition of this acid solution is not critical and may range from 2 to 4% CrO_3 and 10 to 15% H_2SO_4 by weight.

Electrolytic Brightening (Electropolishing)

Electrolytic brightening, or electropolishing, produces smooth and bright surfaces similar to those that result from chemical brightening. After pretreatment, which consists of buffing, cleaning in an inhibited alkaline soak cleaner, and thorough rinsing, the work is immersed in the electrobrightening bath, through which direct current is passed. The work is the anode.

Solution compositions and operating conditions for three commercial electrolytic brightening processes, as well as for suitable post-treatments, are given in Table 10. Operating conditions for electrolytes used in electrobrightening are selected to produce the desired selective dissolution, and may vary for optimum results on different aluminum alloys.

Fluoboric acid electrobrightening and suitable post-treatments are given below:

Constituent or condition	Range
Electrobrightening(a)	
Fluoboric acid	2.5 wt %
Temperature of bath	29 °C (85 °F)
Current density	1-2 A/dm^2 (10-20 A/ft^2)
Voltage	15-30 V
Immersion time	5-10 min
Agitation	None
Smut removal	
Phosphoric acid	1.0 wt %
Chromic acid	0.5 wt %
Temperature of bath	88-93 °C (190-200 °F)
Immersion time	30 s
Anodizing	
Sulfuric acid	7-15 wt %
Current density	1.3 A/dm^2 (12 A/ft^2)
Temperature of bath	21 °C (70 °F)
Immersion time	10 min
Sealing	
Distilled water	100%
Temperature of bath	93-100 °C (200-212 °F)
Immersion time	10 min

(a) U.S. Patent 2,108,603 (1938)

This process is applicable for specular and diffuse reflectors, and products made of super-purity aluminum (99.99%) in combination with up to 2% Mg, or of high-purity aluminum (99.7 to 99.9%).

Sodium carbonate electrobrightening is applicable for specular reflectors, automotive trim, decorative ware and jewelry. It can also be used for products made of super-purity aluminum (99.99%) in combination with up to 2% Mg, of high-purity aluminum (99.7 to 99.85%), or of the following commercial alloys (in approximate order of decreasing quality of finish: 5457, 5357, 6463, 6063, 5052, 1100, 5005, 3003, and 6061). Sodium carbonate electrobrightening and suitable post-treatments are as follows:

Constituent or condition	Range
Electrobrightening	
Sodium carbonate (anhydrous)	15 wt %
Trisodium phosphate	5 wt %
pH	10.5
Temperature of bath	79-82 °C (175-180 °F)
Current density	2-3 A/dm² (20-30 A/ft²)
Voltage	9-12 V
Initial immersion without current	20 s
Immersion time with current	5 min
Agitation	Work rod only
Smut removal(a)	
Sulfuric acid	10 vol %
Temperature of bath	21-27 °C (70-80 °F)
Immersion time	15-30 s
Anodizing(b)	
Sodium bisulfate	20 wt %
Temperature of bath	35 °C (95 °F)
Current density	0.5 A/dm² (5 A/ft²)
Voltage	10 V
Immersion time	15 min
Sealing	
Distilled water	100%
Temperature of bath	85 °C (185 °F)
Immersion time	20 min

(a) Smut may also be removed mechanically. (b) The anodizing treatment in the preceding list may be used as an alternative

Sulfuric-phosphoric-chromic acid electrobrightening and post-treatments are listed below:

Constituent or condition	Range
Electrobrightening(a)	
Sulfuric acid	4-45 wt %
Phosphoric acid	40-80 wt %
Chromic acid	0.2-9.0 wt %
Trivalent metals	6 wt % max
Temperature of bath	71-93 °C (160-200 °F)
Viscosity of bath at 82 °C (180 °F)	9-13 cP
Current density	2.5-95 A/dm² (25-950 A/ft²)
Voltage	7-15 V
Agitation	Mechanical
Smut removal	
Phosphoric acid	3.5 wt %
Chromic acid	2.0 wt %
Temperature of bath	88-93 °C (190-200 °F)
Anodizing	
Sulfuric acid	7-15 wt %
Current density	1.2 A/dm² (12 A/ft²)
Temperature of bath	21 °C (70 °F)
Immersion time	10-20 min
Sealing	
Distilled water	100%
Temperature of bath	93-100 °C (200-212 °F)
Immersion time	10 min

(a) U.S. Patent 2,550,544 (1951)

The process is used primarily for macrosmoothing to replace mechanical polishing wholly or in part. Other applications include architectural trim, decorative ware, jewelry, and products made of commercial alloys.

Selection of Chemical and Electrolytic Brightening Processes

Chemical and electrolytic brightening are essentially selective-dissolution processes, in which the high points of a rough surface are attacked more rapidly than the depressions. An important feature of these processes is their ability to remove a surface skin of metal that is contaminated with oxides and with traces of residual polishing and buffing compounds, or other inclusions, while at the same time brightening the surface.

Metallurgical Factors. The composition, orientation, and size of the individual grains within the workpiece have a direct effect on the uniformity of dissolution during brightening. Fine-grained material is the most desirable for chemical or electrolytic brightening. Best results are obtained with alloys that are of uniform chemical composition and that do not precipitate constituents of different potential from the matrix during any necessary heating or heat treatment. Also, the alloys should be such that forming operations cause only relatively minor detrimental effects.

Mill operations must be controlled to produce material that can be brightened satisfactorily. It is important that the material be fine-grained and that surfaces be free of all imperfections, such as segregation, oxide inclusions, laps, die marks, and stains.

Optical Factors. In general, the highest total and specular reflectance of the brightened surface is obtained on pure aluminum having a fine grain structure. Reflectance, both total and specular, decreases as alloy content increases, however at a given alloy content the decrease will vary with the process. Magnesium has a very small effect on reflectance. The effect of alloying elements varies greatly with different brightening processes.

In a few applications, chemically or electrolytically brightened surfaces are protected by a clear organic coating. However, most surfaces brightened by these methods are anodized to produce a clear, colorless, protective oxide coating. For many decorative uses, the anodic coating is subsequently dyed.

Applications of chemical and electrolytic brightening processes are functional and decorative, and include jewelry, razor parts, automotive trim, fountain pens, searchlight reflectors, natural-finish or brightly colored giftware, architectural trim, household appliances, and thermal reflectors for components of space vehicles.

Chemical and electrolytic brightening may be used to replace buffing, either completely or partly. Brightening may be used before or after buffing, or as an intermediate operation. In processes where brightening is used to replace buffing completely, aluminum is dissolved at relatively rapid rates, and 25 μm (1 mil) or more of metal is removed. In processes where brightening is used as the final operation of the finishing sequence, metal is dissolved more slowly, and total metal removal usually ranges from about 3 to 13 μm (0.1 to 0.5 mil). Such procedures are used primarily on super-purity aluminum with up to 2% Mg and on high-purity aluminum.

Chemical Versus Electrolytic Brightening. Because of recent improvements in chemical brightening processes, brightening results are equivalent to those obtained by the electrolytic processes, with the exception of reflector-type finishes on super-purity and high-purity aluminum.

Initial and operating costs for equipment are lower for chemical brightening than for electrolytic brightening, because electrical power and associated equipment are not required. Chemical brightening can be used on a variety of alloys.

Electrobrightening processes can have low chemical costs, because the chemicals used are less expensive, and because baths operate well at high levels of dissolved aluminum. Other advantages of some baths used in electrobrightening are chemical stability of the solution and the ability of the bath to operate continuously for long periods at optimum efficiency with relatively simple control.

Advantages Over Buffing. In performance and economy, chemical and electrolytic polishing processes offer the following advantages over buffing:

Performance

- Contaminants are not introduced into the metal surface. Chemical or electrolytic processes remove trace amounts of contaminants initially

Table 10 Electrolytic brightening and polishing

Bath	Percentage	Temperature °C	Temperature °F	Duration, min	Voltage, V	Current density A/dm²	Current density A/ft²	Film thickness, μm	Film thickness, mils	Appearance properties	Remarks
Seignette salt brightener (alkaline process)											
Sodium potassium tartrate	15 wt %	38-42	100-108	2-15	10	3-5	30-50	...	...	High luster, mirror-like reflectivity	For pure aluminum, Raffinal, Reflektal, and for jewelry and reflectors
Sodium hydroxide	1.2 wt%										
Aluminum powder	0.2 wt %										
Water	83.6 wt %										
Acid brightening											
Sulfuric acid	70 vol %	75-85	167-185	2-10	10-20	10-15	100-150	...	...	High luster, mirror-like reflectivity	For pure aluminum, and its alloys and for reflectors, architectural and structural shapes, and appliance parts
Phosphoric acid	15 vol %										
Nitric acid	1 vol %										
Water	14 vol %										
VAW brightener											
Sodium bisulfate	20 wt %	87-93	188-199	8-10	8-10	10-15	100-150	...	...	Colorless, transparent oxide film with effect on reflectivity	Used as a post-treatment after conventional anodizing is accomplished
Sodium sulfate	10 wt %										
Sodium hydroxide	1 wt %										
Water	69 wt %										
Smudge remover											
Sodium carbonate	2 wt %	92-97	198-207	...	...	...	...	...	...	Preserve reflectivity	For removing the thin film produced on the aluminum surface by electro-brightening, which otherwise would impair reflectivity
Sodium dichromate	1.5 wt %										
Water	96.5 wt %										
Anodic post-treatments											
Sulfuric acid	71 wt %	23-27	74-81	10	12	1	10	4	0.1	Colorless, transparent film	Without anodizing good results; best results with anodizing
Water	93 wt %										
Sulfuric acid	10-20 wt %	18-22	64-72	10	12	1	10	4	0.1	Colorless, transparent film	Without anodizing good results; best results with anodizing
Water	90-80 wt %										
Sodium bisulfate	20 wt %	33-37	91-98	10	10	0.5	5	2	0.08	Colorless, transparent film	Without anodizing good results; best results with anodizing
Water	80 wt %										

present in the surface skin or embedded in the metal during preceding operations. Surfaces brightened by these processes have better total and specular reflectance

- Anodized and dyed surfaces that have been chemically or electrolytically brightened have a brilliance, clarity, and depth not attainable with buffed surfaces. Anodizing reduces the reflectance values of chemically or electrolytically brightened surfaces less than it reduces the reflectance of buffed surfaces
- Chemical or electrolytic brightening of aluminum prior to electroplating provides better adhesion and continuity of the plated deposits. This improves corrosion resistance and serviceability

Economy

- Labor costs are lower than for buffing
- Processes are readily adaptable to high-production parts that, because of their shape, cannot be finished on automatic buffing machines, and to parts that require buffing of a large percentage of the total surface area. Modification of automatic buffing machines to accommodate parts of different shapes may be more expensive than changes in racking for chemical or electrolytic brightening of these parts
- Incorporation of processes into an automatic anodizing or electroplating line can result in economies in terms of space, equipment and operations, and may eliminate one or more cleaning or pickling operations in the pretreatment cycle. Deburring can sometimes be completely eliminated, because of the high rate of metal removal on edges and corners

Chemical Etching

Chemical etching, using either alkaline or acid solutions, produces a matte finish on aluminum products. These processes may be used for final finishing, but are more often used as intermediate treatments prior to lacquering, conversion coating, anodizing, or other finishing treatments. Chemical etching also is used extensively in conjunction with buffing or chemical brightening.

Etching with alkaline or acid solution prior to anodizing: (*a*) removes oxide films and embedded surface contaminants that would discolor the anodic coating if not removed; (*b*) roughens the surface slightly, to produce a less glossy anodized surface, and to minimize slight differences in the mill finish of different production lots; and (*c*) minimizes color-matching differences, which are more apparent with glossy or specular surfaces. Wrought and cast aluminum alloys on which matte finishes are readily produced by chemical etching are listed below:

Wrought alloys

- *Sheet and plate*: 1100, 2014, 2024, 3003, 5005, 5052, 5457, 6061, 7075
- *Extrusions*: 2014, 2024, 6061, 6063, 6463, 7075

Casting alloys

- 242, 295, 514, A514, B514, F514, 518, 520

Cleaning prior to etching is recommended for attainment of the highest quality finish. The need for prior cleaning, however, is determined by the amount and type of soil present on the surface of work being processed; in many instances, the etching solution serves both as a cleaner and as a finishing medium.

Post-Treatments. Subsequent treatments, such as anodizing or chromate conversion coating, are required for protection against corrosion, and to protect the soft, easily marred surface against mechanical damage. Clear lacquer may be applied to protect the matte finish produced by the etching process. Before being lacquered, the work must be cleaned of etching smut, thoroughly rinsed in clean cold water, and dried in warm air. Lacquering, or painting, should be done as soon as possible, in a clean atmosphere.

Alkaline Etching

Alkaline etching reduces or eliminates surface scratches, nicks, extrusion-die lines, and other imperfections. However, some surface contaminants, if not removed before the work enters the etching solution, may accentuate these imperfections during etching. Oxides, rolled-in dirt, and many other surface contaminants can sometimes be eliminated by deoxidizing the work with a 2 to 4 wt% chromic acid, 10 to 15 wt% sulfuric acid etchant at 60 to 71 °C (140 to 160 °F) prior to alkaline etching. This treatment removes stains resulting from heat treatment and other causes without removing much metal.

Solution Make-Up and Control. A hot (49 to 82 °C, or 120 to 180 °F) solution of sodium hydroxide, potassium hydroxide, trisodium phosphate or sodium carbonate is used for alkaline etching. The solution may contain more than one alkali. The use of uninhibited alkaline solutions (such as sodium hydroxide solutions) is not recommended for high strength 7XXX and 2XXX aluminum alloys in certain artificially aged tempers because of the danger of intergranular attack.

Sequestrants, such as gluconic acid, sodium gluconate, the glucamines, and sorbitol, are added to alkaline solutions to prevent the formation of hydrated alumina. If permitted to form, this compound coats tank walls and heating coils with a difficult-to-remove scale. Sequestrants increase the life of the bath by preventing the formation of scale, and by reducing the accumulation of sludge in the tank. They are added in concentrations of 1 to 5%.

Sodium hydroxide is the alkali most commonly used. Its reaction with aluminum is exothermic, produces hydrogen gas and sodium aluminate, and may cause a rise in the temperature of the bath, depending on the relationship between rate of metal removal and tank volume. Uniform finishes thus may be more difficult to obtain with large loads or rapid dissolution rates in small tanks, because the increase in temperature causes faster etching and more rapid depletion of the chemical constituents in the bath.

The concentration of sodium hydroxide in the etching solution usually ranges from 15 to 60 g/L (2 to 8 oz/gal). For most applications, a concentration of 30 to 45 g/L (4 to 6 oz/gal) is adequate. The choice of concentration is influenced by the finish desired, operating temperature of the bath, quality of water, transfer time between the etchant and rinse, and the amount of dragout.

Solution control is guided by regular titration of samples to determine free sodium hydroxide and sodium aluminate (aluminum). In a common method of operation, the concentration of free sodium hydroxide is not permitted to fall below 26 or 30 g/L (3.5 or 4 oz/gal) when a uniform, medium-deep etch is required. The normal working concentration of aluminum is about 30 g/L (4 oz/gal), or about 2.5 wt %. When the aluminum content of the solution approaches 55 to 75 g/L (7 to 10 oz/gal) and free sodium hydroxide about 40 g/L (5 oz/gal), the finish may become brighter and more reflective which indicates that the solution is nearly exhausted and should be partly or completely replaced.

Determination of specific gravity also is useful in solution control. A solution that has a specific gravity of 1.15 to 1.18 while maintaining a free sodium hydroxide content of 30 to 38 g/L (4 to 5 oz/gal) is considered to be approaching exhaustion. When this condition is reached, the finish being produced should be carefully observed for nonuniform etching and shiny appearance. As the aluminum content of the solution increases, the solution becomes more viscous, which may result in poor rinsing and may increase the amount of dragout. Special proprietary rinse additives are available, which help to reduce the dragout and streaking problems caused by high viscosity of the etchant.

Environmental regulations have led to the development of waste recovery technology for used caustic etching solutions. These baths can be operated without any chelating agents, such as sodium gluconate. Recovery processes depend on a controlled precipitation of dissolved aluminum with an accompanying regeneration of free sodium hydroxide. Closed loop recovery systems of this type also reduce chemical costs and provide more uniform etching. Because of high capital investment, these recovery processes are mainly applicable and economical for large installations.

Equipment and Operating Procedures. Tanks and heating coils for alkaline etching may be made of low-carbon steel. Ventilation is required for the etching tanks, because the mist-like fumes generated are a health hazard to personnel, and because alkali-contaminated air can corrode or etch unprotected aluminum in the work area, especially during periods of high humidity. Efficient venting should be provided to exhaust the fumes and spray evolved during the transfer of the parts to the first rinse tank.

Sometimes a blanket of foam on the solution is used to reduce the amount of mist. Foam is usually created by the addition of surface-active, or wetting, agents to the bath. A layer of 25 or 50 mm (1 or 2 in.) of foam on the surface of the bath is usually adequate.

Work to be processed may be placed on appropriate racks or loaded in baskets for immersion in the etching solution. Dipping is the method most often used for etching, although in some instances spray etching has been used. Workpieces to be bulk processed in baskets must be positioned to prevent the formation of air or gas pockets. For best results, it is desirable to agitate the solution by air or by movement of the work.

Racks and baskets are usually used when etching is followed by subsequent treatments, such as chemical brightening, or conversion coating. Stainless steel is a suitable material for bulk-etching baskets, because it withstands the corrosive conditions of the various solutions used in the cleaning and finishing processes. Baskets for bulk etching cannot be used for anodizing, because an electrical contact cannot be made. Bulk parts must be transferred to specially designed containers with a pressure contact prior to anodizing.

In general, bath temperatures range from 49 to 82 °C (120 to 180 °F). Specific operating temperature is determined by the final finish desired, the time cycle, available equipment, and the concentration of the bath constituents.

After etching, the work should be rinsed immediately. A high etching temperature and a long transfer time from the etching tank may cause dry-on of the etchant. This condition produces a nonuniform finish characterized by cloudy, pitted or stained areas.

An air-agitated rinse would be beneficial. Also, a double rinse in cold water flowing in a countercurrent is recommended. This type of rinsing uses smaller tanks and less water, and produces better rinsing, than when warm water or only one rinse tank is used. Warm water may cause staining as a result of post-etching, especially when only one rinse tank is used. The work should not remain too long in the first rinse tank following etching, because the first rinse tank usually contains sufficient residual sodium hydroxide to cause staining or a cloudy finish.

Spot welds, riveted areas, or folded edges may contain small cracks or crevices that entrap the alkaline solution. Rinsing may not remove the entrapped solution, with the result that the alkaline solution will bleed out and leave a residue of white powder after the finishing process is completed. Bleed-out can occur also after subsequent anodizing of the work. Bleed-out is unattractive and can cause failure of organic films, such as lacquers and paints, applied for added protection.

Dimensional Changes. Etching in alkaline solutions can remove a considerable amount of metal. Figure 1 shows the dimensional changes that occurred when sheet materials of various aluminum-base alloys were etched for 1, 2 or 3 min in an air-agitated sodium hydroxide solution (5 wt% NaOH) operated at 70 ± 5 °C (160 ± 5 °F). The etching cycle must be carefully controlled when clad materials are being treated, to prevent loss of the cladding.

Desmutting. During the cleaning and etching operation, smut (a gray-to-black residual film) is deposited on the surface of the work. This deposit usually consists of iron, silicon, copper or other alloying constituents (in an aluminum-base material) that are insoluble in sodium hydroxide. When etching is to be followed by anodizing, the smut can sometimes be removed by the anodizing solution (current flowing); this practice, however, generally cannot be controlled to produce a finish of uniform appearance. Copper and iron smuts dissolved in the anodizing electrolyte can accumulate until they make necessary premature disposal of the electrolyte. The recommended procedure is to remove the smut in a solution prepared specifically for this purpose. A nitric acid solution (10 to 15 vol % HNO_3 or more) will remove smut. A solution containing 0.5 to 1 wt % chromic acid plus 4 to 6 wt % sodium bisulfate is similarly effective. Solutions of proprietary compounds, which are nonchromated and hence nonpolluting, are also used. Fluorides are usually added to solutions to aid the removal of smut from high-silicon aluminum alloys and aluminum die castings. Good results have been obtained with a room-temperature solution of 3 parts nitric acid and 1 part hydrofluoric acid.

The following example describes the solution to a problem encountered in the desmutting of die castings of a high-silicon alloy that has been etched in a sodium hydroxide solution.

Because of the high silicon content of alloy 380.0 (7.5 to 9.5% silicon), desmutting to obtain an attractive, uniform finish on die castings is difficult. A chromate-type desmutting solution had been used after etching in sodium hydroxide, but had not been entirely effective. The addition of an acid fluoride etch provided the desired finish. The sequence of operations used is as follows:

- Soak in nonetching aluminum cleaner at 60 to 65 °C (140 to 150 °F) for 5 to 10 min
- Rinse in cold water
- Etch for 60 to 90 s in a sodium hydroxide etching solution at 60 to 65 °C (140 to 150 °F)
- Rinse in cold water
- Remove part of smut by immersing in an air-agitated chromate-type desmutting solution at room temperature
- Rinse in cold water
- Rinse in hot deionized water
- Immerse in a room-temperature acid etching solution containing fluoride and nitric acid for 30 to 60 s to remove remaining smut
- Rinse in cold water

After being prepared in this manner, the castings are chromate conversion coated and dipped in lacquer.

Proprietary chromic-sulfuric acid desmutting solutions generally require a tank made of type 302 or 304 stainless steel, although some solutions may require type 316 or 347. They are usually operated at room temperature and normally do not require ventilation. This is an advantage over nitric acid solutions. A disadvantage of some proprietary solutions is the need for treatment of the wastes to remove the harmful effects of chromium salts before the wastes are discharged from the plant. EPA regulations and local ordinances regulate the disposition of waste solutions.

Hexavalent chromium compounds and nitric acid are especially undesirable from an EPA environmental viewpoint. As a result, these traditionally effective oxidizing agents are frequently replaced with such compounds as ferric salts, persulfates, and peroxides.

Acid Etching

Acid solutions are commonly used for finishing castings, especially those made of high-silicon alloys. Acid etching can be done without heavy smut problems, particularly on aluminum die castings. Hydrochloric, hydro-

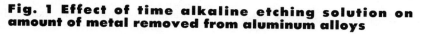

Fig. 1 Effect of time alkaline etching solution on amount of metal removed from aluminum alloys

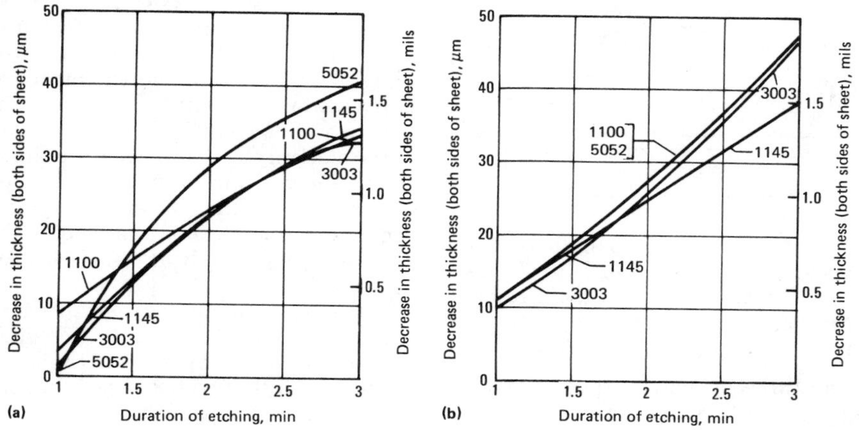

(a) By micrometer measurement. (b) Calculated from loss in weight. Both solutions contain 5 wt % NaOH at 71 ± 5 °C (160 ± 5 °F)

fluoric, nitric, phosphoric, chromic and sulfuric acids are used in acid etching.

Combinations of these acids and mixtures of acids and salts are often used for specific applications. Sulfuric-chromic acid solutions remove heat treating stains with little etching of the metal; dilute hydrofluoric-nitric acid solutions produce bright, slightly matte-textured surfaces; and hydrochloric acid containing sodium chloride and ferric chloride is used for deep etching of designs. Additions of cobalt and nickel salts to the hydrochloric acid solution accelerate etching, but do not affect the ability of the solution to produce a sufficiently smooth surface. Alloys containing silicon, such as sand castings, should be acid etched with a 2 to 5% aqueous solution of hydrofluoric acid prior to anodizing.

Compositions and operating conditions for three acid etching solutions are given below:

- **Solution 1**

Nitric acid............. 3 parts by vol
Hydrofluoric acid .... 1 part by vol
Temperature of bath .... 21 °C (70 °F) max
Immersion time........... ¼ to 1 min

- **Solution 2**

Chromic acid .... 80 g (10.5 oz)
Sulfuric acid... 675 mL (22.4 fluid oz)
Water, to make .... 4 L (1 gal)
Temperature of bath ...... 60 to 71 °C (140 to 160 °F)
Immersion time........... ½ to 2 min

- **Solution 3**

Chromic acid .... 175 g (23.5 oz)
Sulfuric acid..... 75 mL (2.5 fluid oz)
Water, to make .... 4 L (1 gal)
Temperature of bath ...... 60 to 71 °C (140 to 160 °F)
Immersion time........... ½ to 2 min

Figure 2 is a flow chart of the operations used in acid etching.

Fumes from most acid etching solutions are corrosive, and the mist or spray carried up by the gases evolved constitutes a health hazard. Ventilation is required, even for the solutions operated at room temperature. Tanks are made of stainless steel or plastic, or are plastic lined; plastic or plastic-lined tanks are used with solutions containing hydrochloric or hydrofluoric acid. Cooling coils may be required, because etching generates heat. Heating coils are required for solutions operated at elevated temperature.

Acid etching is often used alone, but it may be used in conjunction with alkaline etching, either preceding or following it. It is usually used before alkaline etching when oxides are to be removed, and after alkaline etching when smut removal is a problem. Acid etching solutions, especially those containing fluorides, are excellent smut and scale removers. After acid etching and thorough rinsing, the work is ready for further processing (Fig. 2).

An

Th
proc
min
whil
troly
outli

- *In*
nu
im
water attack. The anodic coating protects the underlying metal by serving as a barrier to corrodents. The amorphous aluminum oxide produced by anodizing is sealed by treating in acidified hot water or a hot dichromate solution. Sealing is discussed in a subsequent section
- *Increase paint adhesion*: the tightly adhering anodic coating offers a chemically active surface for most paint systems. Anodic films produced in sulfuric acid baths are colorless and offer a base for subsequent clear finishing systems. Aluminum-base materials that are painted for service in severe corrosive environments are anodized before being painted
- *Permit subsequent plating*: the inherent porosity of certain anodic films enhances electroplating. Usually, a phosphoric acid bath is used for anodizing prior to plating
- *Improve adhesive bonding*: a thin phosphoric acid coating improves bond strength and durability
- *Improve decorative appearance*: all anodic coatings are lustrous and have relatively good abrasion resistance. Therefore, these coatings are used as the final finishing treatment when the natural appearance of the aluminum is desired or when a mechanically induced pattern is to be preserved. The degree of luster of anodic coatings depends on the condition of the basis metal before anodizing. Dull etching decreases luster; bright etching, chemical or electrolytic brightening, and buffing increase luster, either diffuse or specular. Most of the aluminum used in architectural applications is anodized
- *Provide unique, decorative colors*: colored anodic coatings are produced by different methods. Organic dyes can be absorbed in the pores of the coatings to provide a whole spectrum of colored finishes. Certain mineral pigments can be precipitated within

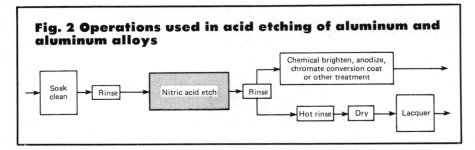

Fig. 2 Operations used in acid etching of aluminum and aluminum alloys

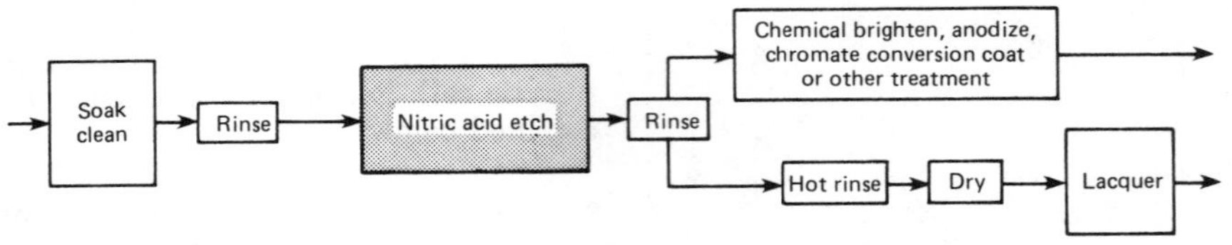

the pores to yield a limited range of stable colors. Integral color anodizing, depending on the alloy composition, is used to provide a range of stable earth-tone colors suitable for architectural applications. Electrolytic coloring is a two-step process involving conventional anodizing followed by electrodeposition of metallic pigments in the pores of the coating to achieve a range of stable colors useful in architecture. Coloring is discussed in a subsequent section

- *Provide electrical insulation*: aluminum oxide is a dielectric. The breakdown voltage of the anodic film varies from a few volts to several thousand volts, depending on the alloy and on the nature and thickness of the film
- *Permit application of photographic and lithographic emulsions*: the porosity of the anodic film offers a mechanical means of holding the emulsion
- *Increase emissivity*: anodic films more than 0.8 μm (0.032 mil) thick increase the emissivity of the aluminum. When dyed black, the film has excellent heat absorption up to 230 °C (450 °F)
- *Increase abrasion resistance*: the hard anodizing processes produce coatings from 25 μm (1 mil) to more than 100 μm (4 mil) thick. These coatings, with the inherent hardness of aluminum oxide, are thick enough for use in applications involving rotating parts where abrasion resistance is required. Although all anodic films are harder than the substrate material, the coatings produced by chromic acid and some sulfuric acid baths are too thin or too soft to meet the requirements for abrasion resistance
- *Detection of surface flaws*: a chromic acid anodizing solution can be used as an inspection medium for the detection of fine surface cracks. When a part containing a surface flaw is removed from the anodizing bath, then washed and dried quickly, chromic acid entrapped in the flaw seeps out and stains the anodized coating in the area adjacent to the flaw

The three principal types of anodizing processes are (*a*) chromic, in which the active agent is chromic acid; (*b*) sulfuric, in which the active agent is sulfuric acid; and (*c*) hard processes that use sulfuric acid alone or with additives. Other processes, used less frequently or for special purposes, use sulfuric-oxalic, phosphoric, oxalic, boric, sulfosalicylic, or sulfophthalic acid solutions. Except for those produced by hard anodizing processes, most anodic coatings range in thickness from 5 to 18 μm (0.2 to 0.7 mils). Table 11 lists some conventional anodizing processes. Table 12 describes a few applications in which anodizing is used as a step in final finishing. The succession of operations typically employed in anodizing is illustrated in Fig. 3.

Surface Preparation. A chemically clean surface free of oxides is a basic requirement for successful anodizing. The cleaning method is selected on the basis of the type of soils or contaminants that must be removed. Usually, the cleaning cycle consists of removing the major organic contaminants by vapor degreasing or solvent cleaning, and then making the surface chemically clean so that acid pickles or etch solutions can react uniformly over the entire surface. The various types of cleaners used are discussed in the section on chemical cleaning in this article.

After cleaning, the work is etched or pickled to remove oxides, or when specular surfaces are required, treated in a brightening solution. The procedures for etching and for brightening are discussed in the sections of this article on chemical etching, alkaline etching, acid etching, chemical brightening, and electrolytic brightening. After etching or brightening, desmutting usually is required, for the removal of heavy metal deposits resulting from the preceding operations. Desmutting is discussed in the section on alkaline etching.

Chromic Acid Process. The sequence of operations used in this process depends on the type of part, the alloy to be anodized, and the principal objective for anodizing. This process is used where it is difficult to remove all of the electrolyte. It will color the finish from yellow to dark olive depending on the thickness. Table 13 gives a typical sequence of operations that meets the requirements of military specification MIL-A-8625.

Chromic acid anodizing solutions contain from 3 to 10 wt % CrO_3. A solution is made up by filling the tank about half full of water, dissolving the acid in the water, and then adding water to adjust to the desired operating level.

A chromic acid anodizing solution should not be used unless:

- pH is between 0.5 and 1.0
- Concentration of chlorides (as sodium chloride) is less than 0.02%
- Concentration of sulfates (as H_2SO_4) is less than 0.05%
- Total chromic acid content, as determined by pH and Baumé readings, is less than 10%. When this percentage is exceeded, part of the bath is withdrawn and is replaced with fresh solution

Figure 4 shows the amount of chromic acid that is required for reducing the pH from the observed value to an operating value of 0.5.

When the anodizing process is started, the voltage is controlled so that it will increase from 0 to 40 V within 5 to 8 min. The voltage is regulated to produce a current density of not less than 0.1 A/dm^2 (1.0 A/ft^2) and anodizing is continued for the specified time period (usually, 30 to 40 min). At the end of the cycle the current is gradually reduced to zero, and the part is removed from the bath within 15 s, rinsed, and sealed. Weight of coating after sealing should be 19 mg/m^2 (200 mg/ft^2) minimum.

Sulfuric Acid Process. The basic operations for this process are the same as for the chromic acid process. Parts or assemblies that contain joints or recesses that could entrap the electrolyte should not be anodized in the sulfuric acid bath. The concentration of sulfuric acid (1.84 sp gr) in the anodizing solution is 12 to 20 wt %. A solution containing 36 L (9.5 gal) of H_2SO_4 per

Table 11 Conventional anodizing processes

Bath	Amount	Temperature °C	°F	Duration, min	Voltage, V	Current density A/dm²	A/ft²	Film thickness μm	mils	Appearance properties	Remarks
Sulfuric acid bath											
Sulfuric acid	10 wt %	18	65	15-30	14-18	1-2	10-20	5-17	0.2-0.7	Colorless, transparent films	Hard, unsuitable for coloring, tensile strength design 250-370 Kgf/mm (2450-3630 N/mm)
Water	90 wt %										
Alumilite											
Sulfuric acid	15 wt %	21	70	10-60	12-16	1.3	13	4-23	0.1-0.9	Colorless, transparent films	Good protection against corrosion
Water	85 wt %										
Oxydal											
Sulfuric acid	20 wt %	18	65	30	12-16	1-2	10-20	15-20	0.6-0.8	Colorless, transparent films	Good protection against corrosion, suitable for variegated and golden coloring
Water	80 wt %										
Anodal and anoxal											
Sulfuric acid	20 wt %	18	65	50	12-16	1-2	10-20	20-30	0.8-1.5	Colorless, transparent films	For coloring to dark tones, bronze and black
Water	80 wt %										
Bengough-Stuart (original process)											
Chromic acid	3 wt %	40	105	60	0-50	0.3	3	5	0.2	Colorless to dark brown	Good chemical resistance, poor abrasion resistance; suitable for parts with narrow cavities, as residual electrolyte is not detrimental
Water	97 wt %										
Commercial chromic acid process											
Chromic acid	5-10 wt %	40	105	30-60	0 to increasing limit controlled by amperage	0.5-1.0	5-10	4-7	0.2-0.3	Gray to iridescent	Good chemical resistance, poor abrasion resistance; suitable for parts with narrow cavities, as residual electrolyte is not detrimental
Water	95-90 wt %										
Eloxal GX											
Oxalic acid	2-10 wt %	20-80	68-175	30-80	20-80	0.5-30	5-300	5-60	0.2-2.4	Colorless to dark brown	Hard films, abrasion resistant, some self-coloring dependant on alloy, 450-480 Kgf/mm (4410-4710 N/mm) for tensile design
Water	98-90 wt %										
Oxal											
Oxalic acid	2-10 wt %	20-22	68-72	10-240	60	1.5	15	10-20 for 30 min	0.4-0.8 for 30 min	Colorless to dark brown	Hard films, abrasion resistant, some self-coloring dependent on alloy
Water	rem							30-40 for 50 min	1.5-1.6 for 30 min		
Ematal											
Oxalic acid	1.2 wt %	50-70	120-160	30-40	120	3	30	12-17	0.5-0.7	Not transparent gray opaque enamel-like	Hard and dense type film possessing extreme abrasion resistance
Titanium salt (TiOC₂O₄K₂·H₂O)	40										
Citric acid	1 g (28 oz)										
Boric acid	8 g (224 oz)										
Water	4 L (1 gal)

Table 12 Typical products for which anodizing is used in final finishing

Product	Size, mm	Size, in.	Alloy	Finishing before anodizing	Anodizing process	Post-treatment	Service requirements or environments
Auto head lamp	215-mm diam, 30	8½-in. diam, 1¼	5557-1125	Buff, chemical brighten	Sulfuric acid(a)	Seal	Atmospheric exposure
Canopy track	760-mm T-extrusion	30-in. T-extrusion	7075	Machine	Hard	None	Resist wear, sea air
Gelatin molds	150-205 mm overall	6-8 overall	1100-O	Chemical brighten as-drawn	Sulfuric acid	Dye, seal	Food
Landing gear	205-mm diam by 1.4 m	8-in. diam by 4½ ft	7079-T6	(b)	Chromic acid	Paint	Corrosion resistance
Mullion	3.7 m by 180 mm by 100 mm(c)	12 ft by 7 by 4	6063-T6	(d)	Sulfuric acid(e)	Seal, lacquer(f)	Urban atmosphere
Name plates	Various sizes	Various sizes	3003-1114	(g)	Sulfuric acid	Dye, seal	Atmospheric exposure
Percolator shell	125-mm diam by 150	5-in. diam by 6	...	Buff, chemical brighten	Sulfuric acid	Seal	Coffee
Seaplane-hull skin	2850 by 1020	112 by 40	Clad 2014-T6	(g)	Chromic acid	None	Erosion; corrosion(h)
Seat-stanchion tube	50-mm diam by 610	2-in. diam by 24	7075-T6	Machine	Hard	None	Wear resistance
Signal-cartridge container	190 by 140 by 165	7½ by 5½ by 6½	3003-O	As drawn	Chromic acid	Prime, paint	Marine atmosphere
Tray, household	430-mm diam	17-in. diam	...	Butler	Sulfuric acid	Seal, buff	Food
Utensil covers	Up to 0.20 m^2 total area	up to 2 ft^2 total area	1100	Buff, chemical brighten	Sulfuric acid(j)	Dye, seal	Steam, cooked foods(k)
Voice transmitter	50-mm diam	2-in. diam	5052-O	Burnish, alkaline etch	Sulfuric acid	Dye, seal(m)	Gas mask
Wheel pistons	Up to 5200-mm^2 area	Up to 8 $in.^2$ area	6151	Machine	Sulfuric acid(n)	Seal	Wear and corrosion(p)

(a) Anodic coating 8 μm (0.3 mil) thick. (b) Partially machine, clean with nonetching cleaner, and remove surface oxide. (c) 5 mm (0.2 in.) thick. (d) Lined finish (180-mesh grit) on 100-mm (4-in.) face; other surfaces alkaline etched. (e) Anodized for 80 min; minimum coating thickness, 30 μm (1.2 mils). (f) Sealed for 20 to 30 min. Methacrylate lacquer, 8 μm (0.3 mil) minimum. (g) Clean with nonetching cleaner; remove surface oxide. (h) Maximum resistance required. (j) Anodic coating 5 μm (0.2 mil) thick. (k) Must not discolor during service. (m) Sealed in dichromate solution. (n) Anodized in sulfuric acid solution (30% H_2SO_4) at 21 °C (70 °F) for 70 min at 2.5 A/dm^2 (25 A/ft^2). (p) In presence of hydraulic brake fluids

Fig. 3 Anodizing operations

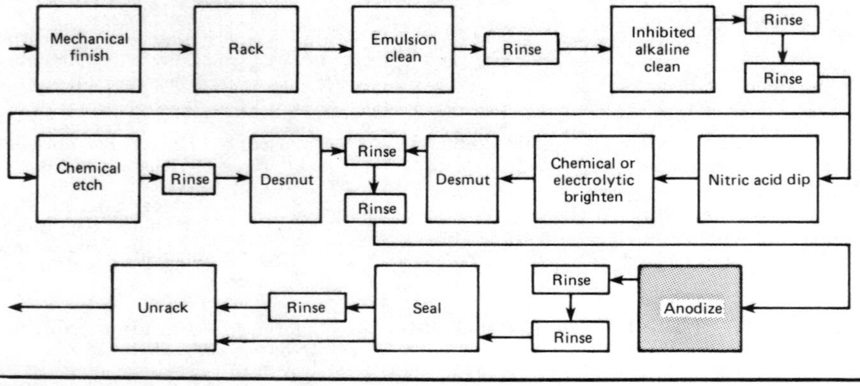

380 L (100 gal) of solution is capable of producing an anodic coating that, when sealed in a boiling dichromate solution, meets the requirements of MIL-A-8625.

A sulfuric acid anodizing solution should not be used unless:

- The concentration of chlorides (as sodium chloride) is less than 0.02%
- Aluminum concentration is less than 20 g/L (2.7 oz/gal)
- Sulfuric acid content is between 165 and 200 g/L (22 to 27 oz/gal)

At the start of the anodizing operation, the voltage is adjusted to produce a current density of 0.9 to 1.2 A/dm^2 (9 to 12 A/ft^2). Figure 5 shows the voltage required to anodize at two different temperatures with current density of 1.2 A/dm^2 (12 A/ft^2). The voltage will increase slightly as the aluminum content of the bath increases. The approximate voltage required for anodizing various wrought and cast aluminum alloys in a sulfuric acid bath at 1.2 A/dm^2 (12 A/ft^2) are as follows:

Alloy	Volts
Wrought alloys	
1100	15.0
2011	20.0
2014	21.0
2017	21.0
2024	21.0
2117	16.5
3003	16.0
3004	15.0
5005	15.0
5050	15.0
5052	14.5
5056	16.0
5357	15.0
6053	15.5
6061	15.0
6063	15.0
6151	15.0
7075	16.0

Alloy	Volts
Casting alloys	
413.0	26.0
443.0	18.0
242.0	13.0
295.0	21.0
514.0(a)	10.0
518.0(a)	10.0
319.0	23.0
355.0	17.0
356.0	19.0
380.0	23.0

(a) Current density, 0.9 A/dm^2 (9 A/ft^2)

When a current density of 1.2 A/dm^2 (12 A/ft^2) is attained, the anodizing process is continued until the specified weight of coating is produced, after which the flow of current is stopped and the parts are withdrawn immediately from the solution and rinsed. Figure 6 shows the effect of time on the weight of the coating developed on automotive trim anodized in 15% sulfuric acid solutions at 20 and 25 °C (68 and 77 °F), operated at a current density of 1.2 A/dm^2 (12 A/ft^2).

A flow chart and a table of operating conditions for operations typically used in anodizing architectural parts by the sulfuric acid process are presented in Fig. 7; similar information, for the anodizing of automotive bright trim, is given in Fig. 8.

Hard Anodizing. The primary differences between the sulfuric acid and hard anodizing processes are the operating temperature and the current density at which anodizing is accomplished. Hard anodizing produces a considerably heavier coating than conventional anodizing in a given length of time. Coating weights obtained as a function of time are compared for the two processes in Fig. 9.

The hard anodizing process uses a sulfuric acid bath containing 10 to 15 wt % acid, with or without additions. The operating temperature of the bath ranges from 0 to 10 °C (32 to 50 °F), and current density is between 2 and 3.6 A/dm^2 (20 and 36 A/ft^2). High temperatures cause the formation of soft and more porous outer layers of the anodic coating. This change in coating characteristics reduces wear resistance significantly and tends to limit coating thickness. Excessive operating temperatures result in dissolution of coating and can burn and damage the work.

Proprietary processes are commonly used. One of the more common of these processes uses a solution containing 120 to 160 g (16 to 21 oz) of sulfuric acid and 12 to 20 g (1.6 to 2.8 oz) of oxalic acid ($H_2C_2O_4$) per 3.8 L (1 gal) of water. This solution is operated at 10 ± 1 °C (50 ± 2 °F) and a current density of 2.5 to 3.6 A/dm^2 (25 to 36 A/ft^2) (voltage is increased gradually from zero to between 40 and 60 V); treatment time is 25 min/25 μm (1 mil) of coating thickness. Additional proprietary processes for hard anodizing are listed in Table 14.

A recent development in hard anodizing utilizes an intermittent pulse current which reduces tank time and makes it possible to use a 20 vol % sulfuric acid solution as the electrolyte.

Special Anodizing Processes. Table 15 gives the operating conditions for three anodizing baths that are

Table 13 Sequence of operations for chromic acid anodizing

Operation	Solution	Solution temperature °C	Solution temperature °F	Treatment time, min
Vapor degrease	Suitable solvent	...	...	...
Alkaline clean	Alkaline cleaner	(a)	(a)	(a)
Rinse(b)	Water	Ambient	Ambient	1
Desmut(c)	HNO_3, 10-25 vol %	Ambient	Ambient	As required
Rinse(b)	Water	Ambient	Ambient	1
Anodize	CrO_3, 46 g/L (5¼ oz/gal)(d)	32-35	90-95	30(e)
Rinse(b)	Water	Ambient	Ambient	1
Seal(f)	Water(g)	90-100	190-210	10-15
Air dry	...	105 max(h)	225 max(h)	As required

(a) According to individual specifications. (b) Running water or spray. (c) Generally used in conjunction with alkaline-etch type of cleaning. (d) pH 0.5. (e) Approximate; time may be increased to produce maximum coating weight desired. (f) Optional. (g) Water may be slightly acidulated with chromic acid, to a pH of 4 to 6. (h) Drying at elevated temperature is optional

Fig. 4 Control of pH of chromic acid anodizing solutions

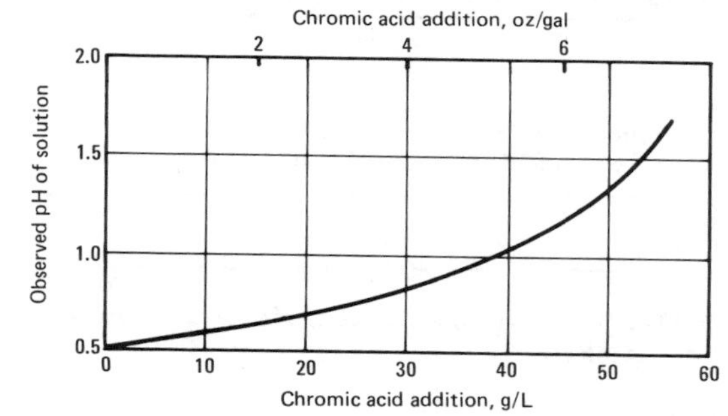

Amount of chromic acid required to reduce pH to 0.5 from observed pH

Fig. 5 Voltages required during sulfuric acid anodizing

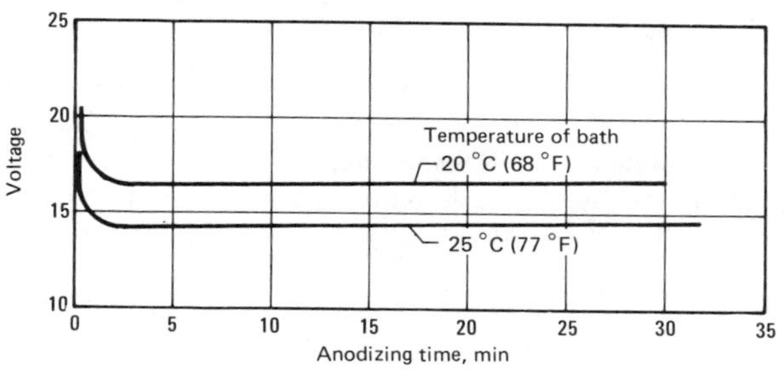

To maintain a current density of 1.2 A/dm^2 (12 A/ft^2), a bath temperature of 20 and 25 °C (68 and 77 °F) must be maintained

Fig. 6 Effect of anodizing time on weight of anodic coating

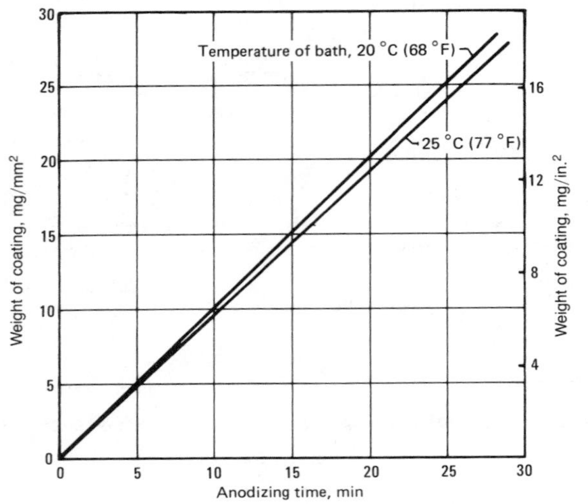

Data derived from aluminum-alloy automotive trim anodized in 15% sulfuric acid solutions at 20 and 25 °C (68 and 77 °F) and at 1.2 A/dm² (12 A/ft²)

Fig. 7 Operations sequence in sulfuric acid anodizing of architectural parts

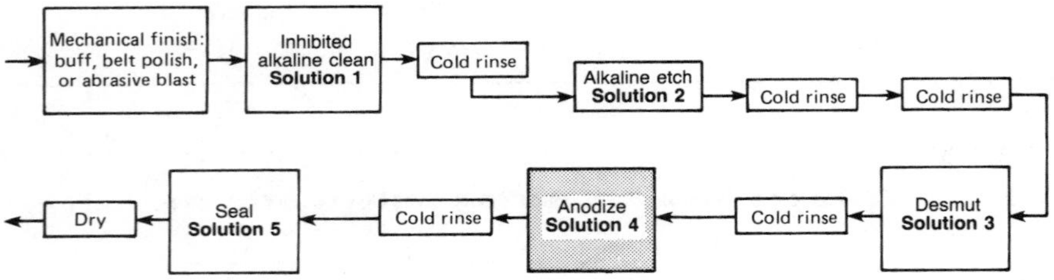

Solution No.	Type of solution	Composition	Operating temperature °C	Operating temperature °F	Cycle time, min
1	Alkaline cleaning	Alkali, inhibited	60-71	140-160	2-4
2	Alkaline etching	NaOH, 5 wt %	50-71	120-160	2-20
3	Desmutting	HNO_3, 25-35 vol %	Room	Room	2
4	Anodizing	H_2SO_4, 15 wt %	21-25	70-75	5-60
5	Sealing	Water, (pH 5.5-6.5)	100	212	5-20

used to produce an anodic coating with a hardness and porosity suitable for electroplating, or to produce anodic coatings of hardness or thickness intermediate to those obtainable from chromic acid, sulfuric acid and hard anodizing baths.

Process Limitations. Composition of the aluminum alloy, surface finish, prior processing, temper or heat treatment, and the use of inserts influence the quality of anodic coatings. The limitations imposed by each of these variables on the various anodizing processes are listed below.

Alloy Composition. The chromic acid process should not be used to anodize aluminum casting alloys containing more than 5% copper or more than 7.5% total alloying elements, because excessive pitting, commonly referred to as burning, may result. The sulfuric acid process can be used for any of the commercially available alloys, whereas the hard anodizing process is usually limited to alloys containing less than 5% copper and 7% silicon. Choice of alloys is important when maximum corrosion and/or abrasion resistance is required. Alloys such as 6061 are superior to the copper and copper-magnesium alloys in their ability to produce a hard, corrosion-resistant coating.

Two or more different alloys can be anodizing at the same time in the same bath if the anodizing voltage requirements are identical. This condition is more critical for the sulfuric acid process than for the chromic acid process.

Surface Finish. Anodic films accentuate any irregularities present in the original surface. However, surface irregularities are emphasized more by the chromic acid bath than by the sulfuric acid bath. Additionally, the sulfuric acid anodizing process should be used instead of the chromic acid process where optimum corrosion- and/or abrasion-resistant surfaces are required. Clad sheet should be handled with care to prevent mechanical abrasion or exposure of the core material. Anodizing magnifies scratches, and if the core material is exposed, it will anodize with a color different from that of the cladding.

Anodizing grade must be specified for extruded products so that mill operations are controlled to minimize longitudinal die marks and other surface blemishes. Surface irregularities must be removed from forgings, and the surfaces of the forgings must be cleaned by a process that removes trapped and burned-in die lubricants. Special attention is required when polishing the flash line if this area is to appear similar to other areas of the forging after anodizing.

Castings can be anodized provided their composition is within the process limits described under alloy composition. From the standpoint of uniform appearance, however, anodizing usually is undesirable for castings because of their nonuniform surface composition and their porosity. Improved results often can be obtained by soaking castings in boiling water after cleaning and before anodizing. This treatment, however, merely attempts to fill surface voids with water, so that voids do not entrap anodizing solution.

Usually, permanent mold castings have the best appearance after anodizing, then die castings, and finally

Fig. 8 Operations sequence in sulfuric acid anodizing of automotive bright trim

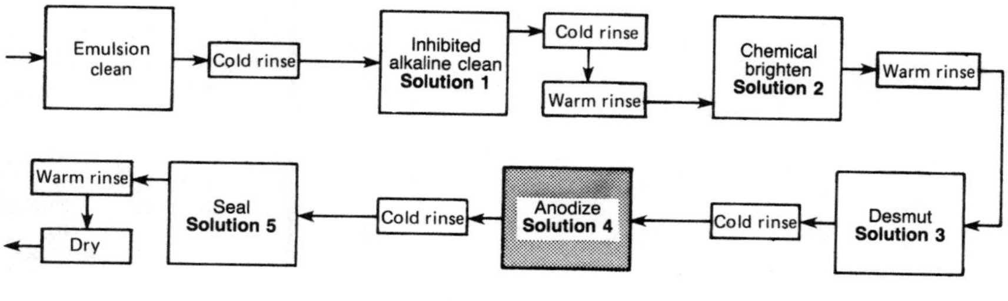

Solution No.	Type of solution	Composition	Operating temperature °C	Operating temperature °F	Cycle time, min
1	Alkaline cleaning	Alkali, inhibited	60-71	140-160	2-4
2	Chemical brightening	H_3PO_4 and HNO_3	88-110	190-230	½-5
3	Desmutting	HNO_3, 25-35 vol %	Room	Room	2
4	Anodizing	H_2SO_4, 15 wt %	21-25	70-75	5-60
5	Sealing	Water (pH 5.5-6.5)	100	212	5-20

Fig. 9 Effect of anodizing time on weight of hard and conventional anodic coatings

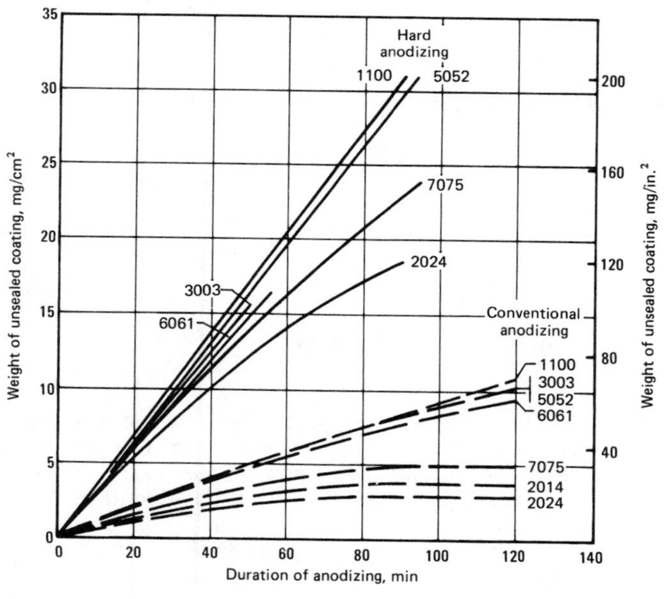

Hard anodizing solution contained (by weight) 12% H_2SO_4 and 1% $H_2C_2O_4$, was operated at 10 °C (50 °F) and 3.6 A/dm² (36 A/ft²). Conventional anodizing solution contained 1% (by weight) H_2SO_4, was operated at 20 °C (70 °F) and 1.2 A/dm² (12 A/ft²).

sand castings. Permanent mold castings should be specified if an anodic coating of uniform appearance is required. Anodizing usually reveals the metal flow lines inherent in the die-casting process, and this condition is objectionable if uniform appearance is desired. In general, solution heat treatment prior to anodizing is beneficial for producing the most uniform and bright anodized finish obtainable on castings.

Regardless of the product form, rough finishing should be avoided when maximum corrosion resistance or uniformity of appearance of the anodic coating is desired. Rough surfaces, such as those produced by sawing, sand blasting and shearing, are difficult to anodize and should be strongly etched prior to anodizing to ensure even minimal results. The machined areas of castings or forgings may have an appearance different from that of the unmachined surfaces.

Prior Processing. Because of their effect on surface finish, welding, brazing, and soldering affect the appearance of the anodic coating, for the reasons discussed above. In addition, the compositions of solders usually are not suited to anodizing. Spot, ultrasonic pressure, or other types of welding processes where there is no introduction of foreign metal, fluxes or other contaminants do not affect the appearance of the anodic coating. However, the sulfuric acid anodizing process should not be used for coating spot welded assemblies or other parts which cannot be rinsed to remove the electrolyte from lap joints.

Temper or Heat Treatment. Differences in temper of non-heat-treatable alloys have no marked effect on the uniform appearance of the anodic coating. The microstructural location of the alloying elements in heat treatable alloys affects the appearance of anodic coatings. Alloying elements in solution have little effect, but the effect is greater when the elements are precipitated from solid solution. The annealed condition should be avoided when maximum clarity of the anodic film is desired.

Inserts or attachments made of metals other than aluminum must be masked off, both electrically and chemically, to prevent burning and corrosion in surrounding areas. The masking must completely seal the faying surface between the insert and parent metal, to prevent adsorption of solution, which may result in corrosion and staining. Therefore, it is desirable to install inserts after anodizing.

Anodizing Equipment and Process Control

Chromic Acid Anodizing. Low-carbon steel tanks are satisfactory for chromic acid baths. It is common prac-

Table 14 Process and conditions for hard coating

Process	Bath	wt %	Temperature °C	Temperature °F	Duration, min	Voltage, V	Current density A/dm²	Current density A/ft²	Film thickness μm	Film thickness mils	Appearance properties	Remarks
Martin Hard Coat (MHC)	Sulfuric acid Water	15 85	−4 to 0	25-32	80	20-75	2.7	29	50	2	Light to dark gray or bronze	Very hard, wear resistant
Alumilite 225 and 226	Sulfuric acid Oxalic acid Water	12 1 ...	10	50	20, 40	10-75	28	301	25, 50	1, 2	Light to dark gray or bronze	Very hard, wear resistant, allows a higher operating temperature over MHC
Alcanodox	Oxalic acid in water	...	2-20	36-68	(a)	(a)	(a)	(a)	20-35	0.8-1.4	Golden to bronze	...
Hardas	Oxalic acid Water	6 94	4	39	(a)	60 dc plus ac override	2.0	22	...	...	Light yellow to brown	...
Sanford	Sulfuric acid with organic additive	...	0-15	32-58	(a)	15-150 dc	1.2-1.5	13-16	...	...	Light to dark gray or bronze	...
Kalcolor	Sulfo-salicylic acid Sulfuric acid Water	7-15 0.3-4 rem	18-24	64-75	...		1.5-4	16-43	15-35	0.6-1.4	Light yellow to brown to black	A self-coloring process, colors are dependent on alloy chosen, the colors produced are light fast
Lasser	Oxalic acid Water	0.75 99.25	1-7	35-44	to 20	From 50-500 rising ramp	Voltage controlled	Voltage controlled	700	28	Colorless	Hard, thick coatings produced with special cooling processes

(a) Proprietary information available to licensees only. Also, the entire Toro process is proprietary information available to licensees only

Table 15 Compositions and operating conditions of solutions for special anodizing processes

Type of solution	Composition	Current density A/dm²	Current density A/ft²	Solution temperature °C	Solution temperature °F	Treatment time, min	Use of solution
Sulfuric-oxalic	15-20 wt % H_2SO_4 and 5 wt % $H_2C_2O_4$	1.2	12	29-35	85-95	30	Thicker coating(a)
Phosphoric	20-60 vol % H_3PO_4	0.3-1.2(b)	3-12(b)	27-35	80-95	5-15	Prepare for plating
Oxalic	3 wt % $H_2C_2O_4$	1.2	12	22-30	72-86	15-60(c)	Harder coating(d)

(a) Coating is intermediate in thickness between coating produced by sulfuric acid anodizing and coating produced by hard anodizing. (b) Potential, 5 to 30 V. (c) Depends on coating thickness desired. (d) Hardness greater than by other processes except hard anodizing

tice to line up to half of the tank with an insulating material, such as glass, to limit the cathode area with respect to the expected anode area (a 1-to-1 ratio is normal). In nonconducting tanks, suitable cathode area is provided by the immersion of individual lead cathodes; however, these require the installation of additional busbars to the tanks for suspension of individual cathodes. Provision must be made for heating the anodizing solution to 32 to 35 °C (90 to 95 °F); electric or steam immersion heaters are satisfactory for this purpose. Electric heaters are preferred, because they are easy to operate and do not contaminate the bath.

The anodizing process generates heat; therefore, agitation is required to prevent overheating of the bath and especially of the electrolyte immediately adjacent to the aluminum parts being anodized. Exhaust facilities must be adequate to trap the effluent fumes of chromic acid and steam.

Sulfuric Acid Anodizing. Tanks for sulfuric acid anodizing may be made of low-carbon steel lined throughout with plasticized polyvinyl chloride and coated on the outside with corrosion-resistant synthetic-rubber paint. Other suitable materials for tank linings are lead, rubber and acid-proof brick. Tanks made of special sulfuric acid-resistant stainless steel containing copper and molybdenum, or made entirely of an organic material, may be used. The tank should have controls for maintaining temperature at between 20 to 30 °C (68 to 85 °F). Requirements for agitation and ventilation are the same as for chromic acid solutions. The surface of the floor under the tank should be acid resistant. The bottom of the tank should be about 150 mm (6 in.) above the floor on acid-resistant and moisture-repellent supports.

A separate heat exchanger and acid make-up tank should be provided for sulfuric acid anodizing installations. This tank should be made of lead-lined steel. Lead is preferred over plastic for the lining because lead withstands the heat generated when sulfuric acid is added. Polyvinyl chloride pipes are recommended for air agitation of the solution and for the acid-return lines between the two tanks. Cooling coils should be made of chemical lead or antimonial lead pipe.

Hard Anodizing. Most of the hard anodizing formulations are variations of the sulfuric acid bath. The requirements for hard anodizing tanks are substantially the same as those for sulfuric acid anodizing tanks, except that cooling, rather than heating, maintains the operating temperature at 0 to 10 °C (32 to 50 °F).

Temperature-control equipment for all anodizing processes must regulate the overall operating temperature of the bath and maintain the proper temperature of the interface of the work surface and electrolyte. The operating temperature for most anodizing baths is controlled within ±1 °C (±2 °F). This degree of control makes it necessary for the temperature-sensing mechanism and heat lag of the heating units to be balanced.

When electric immersion heaters are used, it is common practice to have high and low heat selection so that the bath can be heated rapidly to the operating temperature and then controlled more accurately on the low heat setting. Standard thermistor thermostats are used for sensing the temperature within the bath and activating the heating elements.

In steam heated systems, it is advantageous to have a throttling valve to prevent overheating. An intermediate heat exchanger is used in some installations to prevent contamination of the electrolyte and the steam system by a broken steam line within the anodizing bath.

Agitation may be accomplished by stirring with electrically driven impellers, by recirculation through externally located pumps, or by air. In some installations, the anode busbars are oscillated horizontally, thus imparting a stirring action to the work.

The two primary requirements of an agitation system are that it is adequate and that it does not introduce foreign materials into the solution. With air agitation, filters must be used in the line to keep oil and dirt out of the solution.

Power requirements for the principal anodizing processes are as follows:

Process	Voltage	Current density A/dm^2	A/ft^2
Chromic	42	0.1-0.3	1-3
Sulfuric	24	0.6-2.4	6-24
Hard	100	2.5-3.6	25-36

Direct current is required for all processes. Some hard anodizing procedures also require a superimposed alternating current or a pulsed current. At present, most power sources for anodizing use selenium or silicon rectifiers. Compared to motor generators, the selenium rectifiers have greater reliability, are lower in initial cost and maintenance cost, and have satisfactory service life.

Voltage drop between the rectifier and the work must be held to a minimum. This is accomplished by using adequate busbars or power-transmission cables. Automatic equipment to program the current during the entire cycle is preferred. Manual controls can be used, but necessitate frequent adjustments of voltage. The presence of a recording voltmeter in the circuit ensures that the time-voltage program specified for the particular installation is being adhered to by operating personnel. Current-recording devices also are advantageous.

Masking. When selective anodizing is required, masking is necessary for areas to be kept free of the anodic coating. Masking during anodizing may be required also for post-anodizing operations such as welding, for making an electrical connection to the basis metal, or for producing multicolor effects with dye coloring techniques.

Masking materials are usually pressure-sensitive tapes and stop-off lacquers. The tapes may have a metal, vinyl, or other plastic backing. Stop-off lacquers provide satisfactory masking, but some types are difficult to remove. For effective masking when either material is used, work surfaces must be very clean.

Racks for Anodizing

Anodizing racks or fixtures should be designed for efficiency in loading and unloading of workpieces. Important features that must be included in every properly designed rack are as follows:

- *Current-carrying capacity*: the rack must be large enough to carry the correct amount of current to each part attached to the rack. If the spline of the rack is too slender for the number of parts that are attached to the rack, the anodic coating will be of inadequate thickness, or it will be burned or soft as the result of overheating
- *Positioning of parts*: the rack should enable proper positioning of the parts to permit good drainage, minimum gassing effects and air entrapment, and good current distribution
- *Service life*: the rack must have adequate strength, and sufficient resistance to corrosion and heat, to withstand the environment of each phase of the anodizing cycle

The use of bolt and screw contacts, rather than spring or tension contacts, is a feature of racks designed for anodizing with the integral color processes. These processes require high current densities and accurate positioning of workpieces in the tank. Bolted contacts are used also on racks for conventional hard anodizing. However, bolting requires more loading and unloading time than tension contacts.

Materials for Racks. Aluminum and commercially pure titanium are the materials most commonly used for anodizing racks. Aluminum alloys used for racks should contain not more than 5% copper and 7% silicon. Alloys such as 3003, 2024 and 6061 are satisfactory. Contacts must be of aluminum or titanium. Racks made of aluminum have the disadvantage of being anodized with the parts. The anodic coating must be removed from the rack, or at least from contacts, before the rack can be reused. A 5% solution of sodium hydroxide at 38 to 65 °C (100 to 150 °F), or an aqueous solution of chromic and phosphoric acids (40 g or 5⅓ oz CrO_3 and 40 mL or 5⅓ fluid oz of H_3PO_4 per litre or per gallon of water) at 77 to 88 °C (170 to 190 °F), can be used to strip the film from the rack. The chromic-phosphoric acid solution does not continue to attack the aluminum rack after the anodic film is removed.

Caustic etching prior to anodizing attacks aluminum spring or tension contacts, causing a gradual decrease in their strength for holding the parts securely. This condition, coupled with vibration in the anodizing tank, especially from agitation, results in movement and burning of workpieces.

On many racks, aluminum is used for splines, crosspieces and other large members, and titanium for the contact tips. The tips may be replaceable or nonreplaceable. Although replaceable titanium tips offer versatility in racking, the aluminum portions of the rack must be protected with an insulating coating. However, if the anodizing electrolyte penetrates the coating, the aluminum portion of the rack may become anodized and thus become electrically insulated from the replaceable titanium contact. A more satisfactory rack design uses nonreplaceable titanium contacts on aluminum splines that are coated with a protective coating. Titanium contacts that are welded to replaceable titanium crossbars offer a solution to many racking problems created by the variety of parts to be anodized. These crossbar members can be rapidly connected to the spline. Titanium should not be used in solutions containing hydrofluoric acid.

Plastisol (unplasticized polyvinyl chloride) is used as a protective coating for anodizing racks. This material has good resistance to chemical attack by the solutions in the normal anodizing cycle; however, it should not be used continuously in a vapor degreasing operation or in chemical bright dip solutions. Furthermore, if the coating becomes loose and entraps processing solution, the solution may bleed out and drip on the workpieces, causing staining or spotting.

Bulk Processing. Small parts that are difficult to rack are bulk anodized in perforated cylindrical containers made of fiber, plastic or titanium. Each container has a stationary bottom, a threaded spindle centrally traversing its entire length, and a removable top that fits on the spindle to hold the parts in firm contact with each other.

Anodizing Problems

Causes and the means adopted for correction of several specific problems in anodizing aluminum are detailed in the following examples.

Example 1. Anodic coatings were dark and blotchy on 80 to 85% of a production run of construction workers' helmets made of alloy 2024. After drawing, these helmets had been heat treated in stacks, water quenched, artificially aged, alkaline etched with sodium hydroxide solution, anodized in sulfuric acid solution, sealed, and dried. The dark areas centered at the crowns of the helmets and radiated outward in an irregular pattern. Examination disclosed the presence of precipitated constituents and lower hardness in the dark areas. The condition proved to be the result of restricted circulation of the quench water when the helmets were stacked, which permitted precipitation of constituents, because of a slower cooling rate in the affected areas. The problem was solved by separating the helmets with at least 75 mm (3 in.) of space during heating and quenching.

Example 2. Pieces of interior trim made from alloy 5005 sheet varied in color from light to dark gray after anodizing. Rejection was excessive, because color matching was required. Investigation proved that the anodizing process itself was not at fault; the color variation occurred because the workpieces had been made of cutoffs from sheet stock obtained from two different sources of supply. To prevent further difficulty, two recommendations were made:

- All sheet metal of a given alloy should be purchased from one primary producer, or each job should be made of material from one source. In the latter instance, all cutoffs should be kept segregated
- More rigid specifications should be established for the desired quality of finish. Most producers can supply a clad material on certain alloys that gives better uniformity in finishing

Example 3. The problem was to improve the appearance of bright anodized automotive parts made of alloy 5357-H32. Deburring was the only treatment preceding anodizing. An acceptable finish was obtained by changing to an H25 temper. The H25 had a better grain structure for maintaining a mirror-bright finish during anodizing.

Example 4. After alkaline etching, web-shaped extrusions made of alloy 6063-T6 exhibited black spots that persisted through the anodizing cycle. These extrusions were 3 m (11 ft) long and had cross-sectional dimensions of 100 by 190 mm (4 by 7½ in.) and a web thickness of 5 mm (3⁄16 in.). Cleaning had consisted of treatment for 1 to 4 min in 15% sulfuric acid at 85 °C (185 °F) and etching for 8 min in a sodium hydroxide solution (40 g/L or 5 oz/gal) at 60 °C (140 °F). The spots occurred only on the outer faces of the web. Affected areas showed subnormal hardness and electrical conductivity. Metallographic examination revealed precipitation of magnesium silicide there.

The defects were found to have occurred in areas where cooling from the extrusion temperature was retarded by the presence of insulating air pockets created by poor joints between the carbon blocks that lined the runout table. The extrusion had only to remain stationary on the runout table (end of extrusion cycle, flipped on side for sawing) for as little as 1 min for $MgSi_2$ to precipitate at locations where cooling was retarded. This type of defect is not limited to a particular shape; it can result from a critical combination of size and shape of the extrusion, or from extrusion conditions and cooling rate.

The solution to the problem was to provide uniform cooling of the extrusion on the runout table; this was accomplished by modifying the table and employing forced-air cooling.

Sealing of Anodic Coatings

When properly done, sealing in boiling deionized water for 15 to 30 min partially converts the as-anodized alumina of an anodic coating to an aluminum monohydroxide known as Boehmite. It is also common practice to seal in hot aqueous solution containing nickel acetate. Precipitation of nickel hydroxide helps in plugging the pores.

The corrosion resistance of anodized aluminum depends largely on the effectiveness of the sealing operation. Sealing will be ineffective, however, unless the anodic coating is continuous, smooth, adherent, uniform in appearance, and free of surface blemishes and powdery areas. After sealing, the stain resistance of the anodic coating also is improved. For this reason, it is desirable to seal parts subject to staining during service.

Tanks made of stainless steel or lined low-carbon steel and incorporating adequate agitation and suitable temperature controls are used for sealing solutions.

Chromic acid anodized parts are sealed in slightly acidified hot water. One specific sealing solution contains 1 g (0.1 oz in 100 gal) of chromic acid in 100 L of solution. The sealing procedure consists of immersing the freshly anodized and rinsed part in the sealing solution at 79 ± 1 °C (175 ± 2 °F) for 5 min. The pH of this solution is maintained within a range of 4 to 6. The solution is discarded when there is a buildup of sediment in the tank or when contaminants float freely on the surface.

Sulfuric acid anodized parts may also be sealed in slightly acidified water (pH 5.5 to 6.5), at about 93 to 100 °C (200 to 212 °F). At temperatures below 88 °C (190 °F), the change in the crystalline form of the coating is not

Table 16 Sealing processes for anodic coatings

Process	Bath	Amount	Temperature °C	Temperature °F	Duration, min	Appearance, properties	Remarks
Nickel-cobalt	Nickel acetate Cobalt acetate Boric acid Water	0.5 kg (1.1 lb) 0.1 kg (0.2 lb) 0.8 kg (1.8 lb) 100 L (380 gal)	98-100	208-212	15-30	Colorless	Provides good corrosion resistance for a colorless seal after anodizing bath buffered to pH of 5.5 to 6.5 with small amounts of acetic acid sodium acetate
Dichromate	Sodium dichromate Water	5 wt % 95 wt %	98-100	208-212	30	Yellow color	Cannot be used for decorative and colored coatings where the yellow color is objectionable
Glauber salt	Sodium sulfate Water	20 wt % 80 wt %	98-100	208-212	30	Colorless	...
Lacquer seal	Lacquer and varnishes for interior and exterior exposure	...	...	...	...	Colorless to yellow or brown	Can provide good corrosion resistance provided that the correct formulation is selected. Formulations for exterior exposure use acrylic, epoxy, silicone-alkyds resins and for interior exposure the previously mentioned resins plus urethanes, vinyls and alkyds

satisfactorily accomplished within a reasonable time.

Dual sealing treatments are often used, particularly for clear anodized trim parts. A typical process involves a short-time immersion in hot nickel acetate 0.5 g/L (0.06 oz/gal) solution followed by rinsing and immersion in a hot, dilute dichromate solution. Advantages of dual sealing are less sealing smudge formed, greater tolerance for contaminants in the baths, and improved corrosion resistance of the sealed parts in accelerated tests, for example, the CASS test, ASTM B368.

One specific sealing solution contains 5 to 10 wt % potassium dichromate, and sufficient sodium hydroxide to maintain the pH at 5.0 to 6.0. This solution is prepared by adding potassium dichromate to the partly filled operating tank and stirring until the dichromate is completely dissolved. The tank is then filled with water to the operating level and heated to the operating temperature, after which the pH is adjusted by adding sodium hydroxide (which gives a yellow color to the bath).

For sealing, the freshly anodized and rinsed part is immersed in the solution at 100 ± 1 °C (210 ± 2 °F) for 10 to 15 min. After sealing, the part is air dried at a temperature no higher than 105 °C (225 °F). The dichromate seal imparts a yellow coloration to the anodic coating.

Control of this solution consists of maintaining the correct pH and operating temperature. The solution is discarded when excessive sediment builds up in the tank or when the surface is contaminated with foreign material. Sealing is not done on parts that have received any of the hard anodized coatings. Some other sealing processes are given in Table 16.

Water for sealing solutions can significantly affect the quality of the results obtained from the sealing treatment, as evidenced in the following example.

Strips for automotive exterior trim that were press formed from 5457-H25 sheet were found to have poor corrosion resistance after anodizing, even though appearance was acceptable. The strips had been finished in a continuous automatic anodizing line incorporating the usual steps of cleaning, chemical brightening, desmutting, and anodizing in a 15% sulfuric acid electrolyte to a coating thickness of 8 μm (0.3 mil). They had been sealed in deionized water at a pH of 6.0 and then warm air dried. Rinses after each step had been adequate, and all processing conditions had appeared normal.

Investigation eliminated metallurgical factors as a possible cause, but directed suspicion to the sealing operation, because test strips sealed in distilled water showed satisfactory corrosion resistance. Although the deionized water used in processing had better-than-average electrical resistance (1 000 000 Ω·cm or 10 000 Ω·m), analysis of the water showed that it contained a high concentration of oxidizable organic material. This was traced to residues resulting from the leaching of ion-exchange resins from the deionization column. The difficulty was remedied by the use of more stable resins in the deionization column.

Color Anodizing

Dyeing consists of impregnating the pores of the anodic coating, before sealing, with an organic coloring material. The depth of dye absorption depends on the thickness and porosity of the anodic coating. The dyed coating is transparent, and its appearance is affected by the basic reflectivity characteristics of the aluminum. For this reason, the colors of dyed aluminum articles should not be expected to match paints, enamel, printed fabrics or other opaque colors.

Shade matching of color anodized work is difficult to obtain. Single source colors usually are more uniform than colors made by mixing two or more dye materials together. Maximum uniformity of dyeing is obtained by reducing all variables of the anodizing process to a minimum and then maintaining stringent control of the dye bath. Fresh dye baths should be made daily when shade matching is required.

Mineral pigmentation involves precipitation of a pigment in the pores of the anodic coating before sealing. An example is precipitation of iron oxide from an aqueous solution of ferric ammonium oxalate to produce gold-colored coatings. Shade matching is somewhat more difficult to achieve than in dyeing.

Integral color anodizing is a single-step process in which the color is produced during anodizing. Pigmentation is caused by the occlusion of microparticles in the coating, resulting from the anodic reaction of the electrolyte with the microconstituents and matrix of the aluminum alloy. Thus, alloy composition and temper strongly affect the color produced. For example, aluminum alloys containing copper and chromium will color to a yellow or green when anodized in sulfuric or oxalic acid baths, whereas manganese and silicon alloys will have a gray to black appearance. Anodizing conditions such as electrolyte composition, voltage, and temperature are important and must be controlled to obtain shade matching. One electrolyte frequently used consists of 90 g/L (10 oz/gal) sulfophthalic acid plus 5 g/L (0.6 oz/gal) sulfuric acid.

Another method for coloring anodic coatings is the two-step, or electrolytic coloring process. After conventional anodizing in sulfuric acid electrolyte, the parts are rinsed and transferred to an acidic electrolyte containing a dissolved metal salt. Using alternating current, a metallic pigment is electrodeposited in the pores of the anodic coating. There are various proprietary electrolytic coloring processes. Usually tin, nickel, or cobalt is deposited, and the colors are bronzes and black. The stable colors produced are useful in architectural applications.

Evaluation of Anodic Coatings

Coating Thickness. The metallographic method measures coating thickness perpendicular to the surface of a perpendicular cross section of the anodized specimen, using a microscope with a calibrated eyepiece, and is the most accurate method for determining the thickness of coatings of at least 2.5 μm (0.1 mil). This method is used to calibrate standards for other methods, and is the reference method in cases of dispute. Because of variations in the coating thickness, multiple measurements must be made and the results averaged.

The micrometer method determines coating thickness of 2.5 μm (0.1 mil) or more by micrometrically measuring the thickness of a coated specimen, stripping the coating using the solution described in ASTM B137, micrometrically measuring the thickness of the stripped specimen, and subtracting the second measurement from the first.

Effectiveness of Sealing. The sulfur dioxide method comprises exposure of the anodic coating for 24 h to attack by moist air (95 to 100% relative humidity) containing 0.5 to 2 vol % sulfur dioxide, in a special test cabinet. The method is very discriminative. Coatings that are incompletely or poorly sealed develop a white bloom.

Abrasion Resistance. The Taber abrasion method determines abrasion resistance by an instrument that, by means of weighted abrasive wheels, abrades test specimens mounted on a revolving turntable. Abrasion resistance is measured in terms of either (*a*) weight loss of the test specimen for a definite number of cycles or (*b*) the number of cycles required for penetration of the coating. This procedure is covered by Method 6192 in Federal Test Method Standards No. 141.

Lightfastness. The fade-O-meter method is a modification of the artificial-weathering method, in that the cycle is conducted without the use of water. Staining and corrosion products thus cannot interfere with interpretation of results. A further modification entails the use of a high-intensity ultraviolet mercury-arc lamp and the reduction of exposure to a period of 24 to 48 h. Table 17 lists the various ASTM and ISO methods that can be used to evaluate the quality of anodic coatings.

Table 17 ASTM and ISO test methods for anodic coatings

Method	ASTM	ISO
Coating thickness		
Eddy current	B244	2360
Metallographic	B487	...
Light section microscope	B681	2128
Coating weight	B137	2106
Sealing		
Dye stain	B136	2143
Acid dissolution	B680	3210
Impedance/admittance	B457	2931
Voltage breakdown	B110	2376
Corrosion resistance		
Salt spray	B117	...
Cooper-accelerated, acetic acid salt-spray	B368	...

Effects of Anodic Coatings on Surface and Mechanical Properties

As the thickness of an anodic coating increases, light-reflectance, both total and specular, decreases. This decrease is only slight for pure aluminum surfaces, but it becomes more pronounced as the content of alloying elements other than magnesium, which has little effect, increases. The decrease in reflectance values is not strictly linear with increasing thickness of anodic coating; the decrease in total reflectance levels off when the thickness of the coating on super-purity and high-purity aluminum is greater than about 2.5 μm (0.1 mil).

Data comparing the reflectance values of chemically brightened and anodized aluminum materials with those of other decorative materials are given in "Anodic Oxidation of Aluminium and Its Alloys", Bulletin No. 14 of the Aluminium Development Association (now the Aluminium Federation), London, England, 1949.

The effect of anodized coatings 2 to 20 μm (0.08 to 0.8 mil) thick on the reflectance values of electrobrightened aluminum of three degrees of purity is shown in Table 18. This table also includes specular reflectance values for surfaces after removal of the anodic coating. These data show that the degree of roughening by the anodizing treatment increases as the purity of the aluminum decreases. The reflectance values of the anodized surfaces are influenced by the inclusion of foreign constituents or their oxides in the anodic coating.

Metallurgical factors have a significant influence on the effect of anodizing on reflectance. For minimum reduction in reflectance, the conversion of metal to oxide must be uniform in depth and composition. Particles of different composition do not react uniformly. They produce a nonuniform anodic coating and roughen the interface between the metal and the oxide coating.

Anodizing Conditions. The composition and operating conditions of the anodizing electrolyte also influence the light reflectance and other properties of the polished surface. Figure 10 shows the effect of sulfuric acid concentration, temperature of bath, and current density on the specular reflectance of chemically brightened aluminum alloy 5457. These data show that a particular level of specular reflectance can be produced by varying operating conditions.

Thermal Radiation. The reflectance of aluminum for infrared radiation also decreases with increasing thickness of the anodic coating, as shown in Fig. 11. These data indicate that the difference in purity of the aluminum is of minor significance. A comparison between anodized aluminum surfaces and polished aluminum sur-

Table 18 Effect of anodizing on reflectance values of electrobrightened aluminum

Thickness of anodic coating µm	mil	Specular reflectance, %: Electro-brightened	Electro-brightened and anodized	After removal of anodic coating(a)	Total reflectance after anodizing, %
Aluminum, 99.99%					
2	0.08	90	87	88	90
5	0.2	90	87	88	90
10	0.4	90	86	88	89
15	0.6	90	85	88	88
20	0.8	90	84	88	88
Aluminum, 99.8%					
2	0.08	88	68	83	89
5	0.2	88	63	85	88
10	0.4	88	58	85	87
15	0.6	88	53	85	86
20	0.8	88	57	85	84
Aluminum, 99.5%					
2	0.08	75	50	70	86
5	0.2	75	36	64	84
10	0.4	75	26	61	81
15	0.6	75	21	57	77
20	0.8	75	15	53	73

(a) Anodic coating removed in chormic-phosphoric acid
Source: Aluminum Development Council

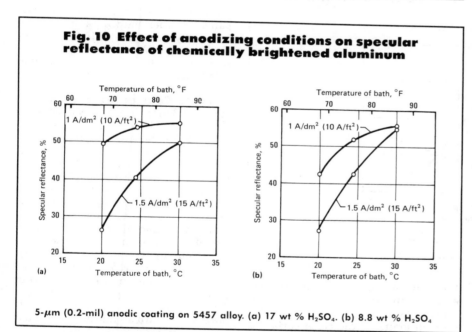

Fig. 10 Effect of anodizing conditions on specular reflectance of chemically brightened aluminum

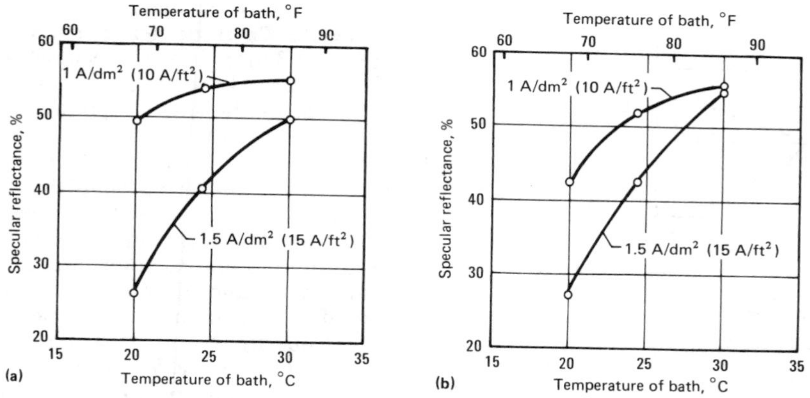

5-µm (0.2-mil) anodic coating on 5457 alloy. (a) 17 wt % H_2SO_4. (b) 8.8 wt % H_2SO_4

faces at 21 °C (70 °F) with respect to absorptance when exposed to black-body radiation from sources of different temperatures is presented in Fig. 12. Although anodized aluminum is a better absorber of low-temperature radiation, as-polished aluminum is a more effective absorber of blackbody radiation from sources at temperatures exceeding 3300 °R (1850 K).

Fatigue Strength. Anodic coatings are hard and brittle, and will crack readily under mechanical deformation. This is true for thin as well as thick coatings, even though cracks in thin coatings may be less easily visible. Cracks that develop in the coating act as stress raisers and are potential sources of fatigue failure of the substrate metal. Typical fatigue-strength values for aluminum alloys before and after application of a hard anodic coating 50 to 100 µm (2 to 4 mil) thick are given in Table 19.

Chemical Conversion Coating

Chemical conversion coatings are adherent surface layers of low-solubility oxide, phosphate, or chromate compounds produced by the reaction of suitable reagents with the metallic surface. These coatings affect the appearance, electrochemical potential, electrical resistivity, surface hardness, absorption and other surface properties of the material. They differ from anodic coatings in that conversion coatings are formed by a chemical oxidation-reduction reaction at the surface of the aluminum, whereas anodic coatings are formed by an electrochemical reaction. The reaction that takes place in chemical conversion coating involves the removal of 0.3 to 2.5 µm (0.01 to 0.1 mil) of the material being treated.

Conversion coatings are excellent for: (*a*) corrosion retardation under supplementary organic finishes or films of oil or wax; (*b*) improved adhesion for organic finishes; (*c*) mild wear resistance; (*d*) enhanced drawing or forming characteristics; (*e*) corrosion retardation without materially changing electrical resistivity; and (*f*) decorative purposes, when colored or dyed.

Conversion coatings are used interchangeably with anodic coatings in organic finishing schedules. One use of conversion coating is as a spot treatment for the repair of damaged areas in anodic coatings. Conversion coatings should not be used on surfaces to which adhesives will be applied because of the low strength of the coating. Anodic coatings are stronger than conversion coatings for adhesive bonding applications.

Table 19 Effect of anodizing on fatigue strength of aluminum alloys(a)

Alloy	Fatigue strength at 1 000 000 cycles: Not anodized MPa	ksi	Anodized MPa	ksi
Wrought alloys				
2024 (bare)	130	19	105	15
2024 (clad)	75	11	50	7.5
6061 (bare)	105	15	40	6
7075 (bare)	150	22	60	9
7075 (clad)	85	12	70	10
Casting alloys				
220	50	7.5	52	7.5
356	55	8	55	8

(a) Sulfuric acid hard coatings 50 to 100 µm (2 to 4 mils) thick applied using 15% sulfuric acid solution −4 to 0 °C (25 to 32 °F) and 10 to 75 V dc
Source: F.J. Gillig, WADC Technical Report 53-151, P.B. 111320, 1953

Fig. 11 Effect of anodic coating thickness on reflectance of infrared radiation

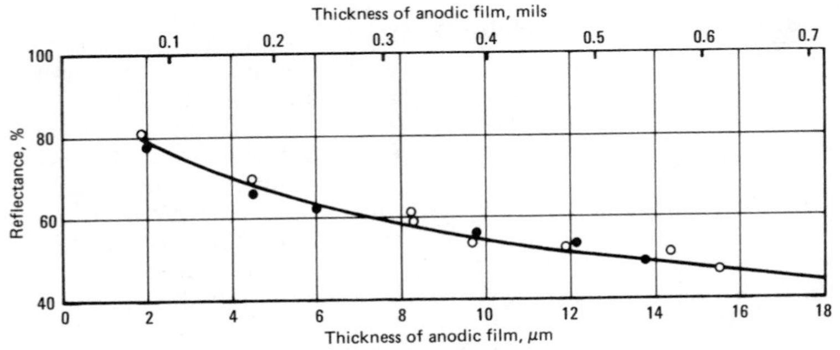

Temperature of infrared radiation source, 900 °C (1650 °F). ○:99.99% aluminum, ●:99.50% aluminum. Courtesy of Aluminum Development Council

Fig. 12 Comparison of absorptance of blackbody radiation by anodized aluminum and polished aluminum

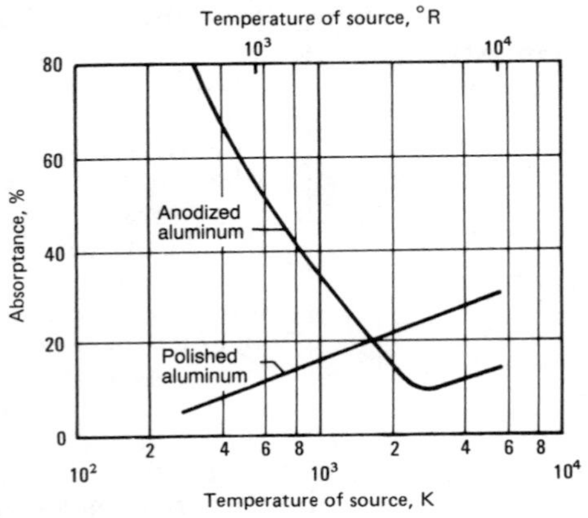

Temperature of aluminum surface, 530 °R (21 °C or 70 °F)

The simplicity of the basic process, together with the fact that solutions may be applied by immersion, spraying, brushing, wiping or any other wetting method, makes conversion coating convenient for production operations. Some applications using chemical conversion coatings on various aluminum alloys are given in Table 20. In most installations, conversion coating offers a cost advantage over electrolytic methods. Moreover, unlike some anodic coatings, chemical conversion coatings do not lower the fatigue resistance of the metal treated.

Procedure. The sequence of operations for applying a satisfactory conversion coating to aluminum-base materials is as follows:

- Removing of organic contaminants
- Removing of oxide or corrosion products
- Conditioning of the clean surface to make it susceptible to coating
- Conversion coating
- Rinsing
- Acidulated rinsing (recommended if supplementary coating is to be applied)
- Drying
- Applying of a supplementary coating, when required

Surface preparation entails the same procedures as are used in preparation for anodizing. However, the cleaning procedure for preparing aluminum for conversion coating is much more critical than for anodizing. After cleaning, removal of the natural oxide film is accomplished in any of the standard aqueous solutions, such as chromate-sulfate, chromate, or phosphate.

Pretreatment immediately prior to the coating operation is required for the development of extremely uniform conversion coatings. Solutions of either acid or alkaline type are used. Subsequent to the above operations, the work is subjected to the conversion coating solution. The addition of a wetting agent, such as sodium alkyl aryl sulfonate, to the solution helps to produce a uniform and continuous coating. After coating, the work is thoroughly rinsed and dried. The final rinse is usually hot (60 to 82 °C, or 140 to 180 °F) to aid drying. Drying is important because it prevents staining. Drying at temperatures higher than 65 °C (150 °F) usually dehydrates the coatings, and so increases hardness and abrasion resistance.

Supplementary coatings of oil, wax, paint, or other hard organic coatings frequently are applied. If the conversion coating is intended to improve subsequent forming or drawing, the final supplementary coating may be soap or a similar dry-film lubricant.

Oxide Coating Processes. The modified Bauer-Vogel (MBV), Erftwerk (EW), and Alrok processes are the principal methods for applying oxide-type conversion coatings. Nominal compositions of the solutions used and typical operating conditions are given in Table 21. The MBV process is used on pure aluminum, as well as on aluminum-magnesium, aluminum-manganese, and aluminum-silicon alloys. The coating produced varies from a lustrous light gray to a dark gray-black color. The EW process is used for alloys containing copper. The film produced is usually very light gray. The Alrok process is for general-purpose use with all alloys, and is often the final treatment for aluminum products. Coatings vary in color from gray to green, and are sealed in a hot dichromate solution.

Phosphate Coating Process. The range of operating conditions and a formula for a standard solution for phosphate coating are given below:

- **Specific formulation**

Ammonium dihydrogen phosphate ($NH_4H_2PO_4$) 61.7%
Ammonium bifluoride (NH_4HF_2) 22.9%
Potassium dichromate ($K_2Cr_2O_7$) 15.4%

Table 20 Applications using chemical conversion coatings

Application	Aluminum alloy	Subsequent coating
Oxide conversion coatings		
Baking pans(a)	1100, 3003, 3004, 5005, 5052	Silicone resin
Phosphate conversion coatings		
Screen cloth	5056	Clear varnish
Storm doors	6063	Acrylic paint (b)
Cans	3004	Sanitary lacquer
Fencing	6061	None applied
Chromate conversion coatings		
Aircraft fuselage skins	7075 clad with 7072	Zinc chromate primer
Electronic chassis	6061-T4	None applied
Cast missile bulkhead	356-T6	None applied
Screen	5056 clad with 6253	Clear varnish
Extruded doubler	6061-T6	Clear lacquer

(a) Baking pans of these alloys may alternatively be chromate conversion coated prior to the application of silicone resin. (b) Thermosetting

Operating temperature....43 to 49 °C (110 to 120 °F)
Treatment time........... 1 to 5 min

- **Desired operating range**

Phosphate ion .... 20 to 100 g/L (2.6 to 13.2 oz/gal)
Fluoride ion ....2 to 6 g/L (0.26 to 0.80 oz/gal)
Dichromate ion .... 6 to 20 g/L (0.80 to 2.6 oz/gal)
Operating temperature....18 to 49 °C (65 to 120 °F)
Treatment time......... 1½ to 5 min

Each litre (gallon) of solution contains 75 to 150 g (10 to 20 oz) of a mixture consisting of the above formulations (U.S. Patent 2,494,910 1950).

Phosphate coatings vary in color from light bluish-green to olive-green, depending on the composition of the aluminum-base material and operating conditions of the bath. The phosphate-chromate conversion coatings are used extensively on aluminum parts or assemblies giving galvanic protection from components of different kinds of materials, such as bushings or inserts made of steel.

Chromate Coating Process. Solution compositions and operating conditions for two chromate conversion coating processes are given in Table 22. Chromate coatings vary from clear and iridescent to light yellow or brown, depending on the composition of the aluminum-base material and on the thickness of the film. The chromate coatings are selected among the various conversion coatings when maximum resistance to corrosion is desired.

Chromate coatings exhibit low electrical resistivity. At a contact pressure of 1380 kPa (200 psi), in a direct-current circuit, the resistivity of a normal chromate film varies from 0.30 to 3.0 $\mu\Omega/mm^2$ (200 to 2000 $\mu\Omega/in.^2$). This resistivity is low enough so that a chromate-coated article can be used as an electrical ground. The conductivity of the films at radio frequencies is extremely high. This permits the use of a chromate film on electrical shields and wave guides. Thus, chromate conversion coating is widely used for treatment of aluminum articles for the electronics industry.

Processing equipment, tanks and racks for conversion coating solutions must be made from acid-resistant materials. Tanks may be made of type 316 stainless steel, or of low-carbon steel if lined with polyvinyl chloride or other suitably protective material. Tanks for solutions that do not contain fluorides may be made of acid-resistant brick or chemical stoneware. Racks may be made of low-carbon steel, but must be coated with acid-resistant compound. Heating coils or electrical immersion heaters should be made of stainless steel or stainless-clad material.

Some conversion coating solutions cause a sludge to form in the bottom of the tank. To prevent contact between the sludge and the workpieces, the tank may be equipped with a false bottom through which sludge can fall.

Adequate ventilation must be provided to remove vapors. The inhalation of fluoride vapors is dangerous. Solutions should not contact the skin, but if they do, the area affected should be washed immediately with running water and then treated by a physician. Respirators, goggles, and gloves should be worn when handling all chemicals used to make up solutions. When brushes are used for applying solutions, they should be made of natural bristles. Synthetic bristles are attacked by the solutions.

Control of Solution. Most users of conversion coating solutions purchase prepared formulations for make-up and solution adjustment. In general, the solutions require control of pH and of the concentration of the critical elements. Direct measurement of pH is made with a glass-cell electric pH meter. The percentage concentration of active ion is obtained by direct titration with a suitable base.

Solution control becomes more critical as the size of the bath decreases with respect to the amount of work treated. Experienced operators of a conversion coating process can detect changes in the composition of the solution by observing the color and appearance of the treated work. A skilled operator often can control the bath by this method alone.

During use, coating solutions are depleted by consumption of the basic chemicals, and by drag-in and dragout. In one plant, drag-in of alkaline cleaner into the conversion coating bath adversely affected the appearance of the conversion coating. Details of this problem and the method adopted for correcting it are given in the following example.

Aluminum screen cloth made from wires of alloy 5056 clad with alloy 6253 had a rejection rate as high as 3% because of the presence of sparklers on the product after chemical conversion coating. Sparklers, known also as shiners, are areas that have higher metallic reflectance than the rest of the conversion-coated surface; they are merely an appearance defect, and do not affect the adherence of organic coatings. The following processing cycle was being used:

1. Alkaline cleaning for 1 min in an inhibited solution at 71 °C (160 °F)
2. Rinsing for 30 s in overflowing cold water
3. Conversion coating for 2½ min in a phosphate-chromate solution at 38 to 46 °C (100 to 115 °F)
4. Rinsing for 30 s in overflowing cold water
5. Second rinsing for 30 s in overflowing cold water
6. Drying
7. Application of a clear varnish (baked at 135 °C or 275 °F for 1½ to 2 min) or of a gray pigmented paint. (For material to be painted, conversion coating required only 1½ min)

The coating defects were found to be caused by contamination (and neutralization) of the acid conversion coating solution by drag-in from the alkaline cleaner. Although the use of a rotating beater to shake droplets of cleaning solution out of the screen openings had

Table 21 Oxide conversion coating

Process	Bath composition	Amount g/L	Amount oz/gal	Temperature °C	Temperature °F	Duration, min	Uses
I MBV	Sodium chromate	15	1.7	96	205	5-10 or 20-30	Corrosion protection or foundation for varnishes or lacquers
	Sodium carbonate	50	5.7				
II MBV	Sodium chromate	15	1.7	65	150	15-30	In situ treatment of large objects with paint brush or spray. When 8 g/L (0.9 oz/gal) sodium hydroxide is added, MBV oxidation may be carried out at 35 °C (95 °F) for 30 min
	Sodium carbonate	50	5.7				
	Sodium hydroxide	4	0.5				
EW	Sodium carbonate	56	6.4	88-100	190-212	8-10	For copper-containing alloys
	Sodium chromate	19	2.2				
	Sodium silicate	0.75-4.5	0.09-0.5				
Alrok	Sodium carbonate	20	2.3	88-100	190-212	20	Final treatment for aluminum products
	Potassium dichromate	5	0.6				

reduced dragout from that bath, it did not eliminate it.

To prevent neutralization of the acid conversion coating solution by contamination with alkali from step 1, the slightly acid overflow from the rinse in step 4 was piped back into the rinse tank in step 2, thus keeping it slightly acid. Elimination of rejects resulted. This procedure also reduced the amount of overflow rinse water needed to operate the line.

Control of Coating Quality. A properly applied coating should be uniform in color and luster and should show no evidence of a loose or powdery surface. Poor luster or powdery surfaces are caused by low pH, improper cleaning and rinsing, excessive treatment temperature or treatment time, a contaminated bath, or insufficient agitation. Light and barely visible coatings are caused by high pH, low operating temperature, insufficient treatment time or high ion concentrations. Usually, the quality of a conversion coating is established on the basis of its appearance, corrosion resistance, hardness and adherence. These qualities may be determined by the ASTM test methods described in the standards listed below:

Corrosion resistance

- Salt spray: B117
- Copper-accelerated acetic acid salt spray (fog): B368
- Evaluation of painted or coated specimens subjected to corrosive exposure: D1654

Resistance to blistering

- Evaluation of blistering of paints: D714

Adherence

- Elongation of attached organic coatings with conical mandrel apparatus: D522

Hexavalent chromium compounds are especially effective components of solutions that form conversion coatings on aluminum. However, environmental regulations often make the handling of chromate-containing rinses a high-cost operation. Two types of technology address this problem. One is a dried-in-place chromate coating system which eliminates the need for subsequent rinsing. The second involves the use of chromium-free treatments which form oxide films containing selected metal ions. Where either of these processes can be used, the need for expensive chrome destruction is eliminated.

Electroplating

Aluminum-base materials are more difficult to electroplate than the common heavier metals, for the following reasons: (*a*) aluminum has a high affinity for oxygen, which results in a rapidly formed, impervious oxide film; and (*b*) most metals used in electroplating are cathodic to aluminum; therefore, voids in the coating lead to localized galvanic corrosion. The electrolytic potentials of several common metals are compared to pure aluminum in the table below:

Metal	Potential, mV(a)
Magnesium	−850
Zinc	−350
Cadmium	−20 to 0
Aluminum (pure)	0
Aluminum-magnesium alloys	+100
Aluminum-copper alloys	+150
Iron, low-carbon steel	+50 to 150
Tin	+300
Brass	+500
Nickel	+500
Copper	+550
Silver	+700
Stainless steel	+400 to 700
Gold	+950

Note: Metals above aluminum in this list will protect it; those below cause aluminum to corrode preferentially. Cathode and anode polarization, however, can cause a reversal of these relationships
(a) In a 6% sodium chloride solution. Source: *Metal Finishing,* Nov 1956

Electrodeposits of chromium, nickel, cadmium, copper, tin, zinc, gold or silver are used for various decorative and functional applications. For example, automotive aluminum bumpers get a zincate treatment, copper strike, and a plating of copper, nickel, and chromium. A copper strike coated with cadmium and chromate or by flowed tin enables the soft soldering of electrical terminals to an aluminum chassis. Brass enhances the bonding of rubber to aluminum. Silver, gold, and rhodium provide specific electrical and electronic surface properties. Examples of applications of plated aluminum with typical finishing sequences are given in Table 23.

Effect of Substrate Characteristics on Plating Results. Each aluminum alloy depending on the metallurgical structure behaves differently from others during electroplating. Alloying elements may be in solid solution in the aluminum, or they may be present as discrete particles or as intermetallic compounds. These microconstituents have different chemical or electrochemical reactivities, and their

Table 22 Chromate conversion coating

Process	Solution composition	Amount g/L	Amount oz/gal	pH	Temperature °C	Temperature °F	Treatment time
Process A(a)......	CrO_3	6(b)	0.80(b)	1.2-2.2	16-55	60-130	5 s - 8 min
	NH_4HF_2	3	0.40				
	$SnCl_4$	4	0.6				
Process B(c)......	$Na_2Cr_2O_7 \times 2H_2O$	7(b)	1(b)	1.2-2.2	16-55	60-130	5 s - 8 min
	NaF	1	0.1				
	$K_3Fe(CN)_6$	5	0.7				
	HNO_3(48° Bé)	(d)	(d)				
Process C(e)......	H_3PO_4	64	8.5	1.2-2.2	40-80	105-175	1-10 min
	NaF	5	0.6				
	CrO_3	10	1.3				

(a) U. S. Patents 2,507,956 (1950) and 2,851,385 (1958). (b) Desired range of hexavalent chromium ion, 1 to 7 g/L (0.13 to 0.90 oz/gal). (c) U. S. Patent 2,796,370 (1957). (d) 3 mL (0.1 fluid oz). (e) Process for Alodine, Alochrome, and Bonderite

surfaces do not respond uniformly to treatment. Also, variations in response occur between different lots or product forms of the same alloy.

Surface Preparation Methods. The three established methods for surface preparation prior to electroplating are surface roughening, anodizing, and immersion coating in zinc or tin solutions.

Surface roughening, which is accomplished either by mechanical abrasion or by chemical etching, assists in mechanically bonding the electrodeposits to the aluminum surface. Surface roughening is sometimes used in preparation for the application of hard chromium to aluminum engine parts such as pistons. A water blast of fine quartz flour may be used to remove surface oxides and to abrade the surface. The adherent wet film protects the aluminum surface from further oxidation before plating. The quartz film is dislodged by the evolution of hydrogen that occurs during plating. Chemical etching produces undercut pits that provide keying action for the electrodeposited metal. In general, mechanical bonding of electrodeposits is not reliable—particularly for applications involving temperature variations. Therefore, preparation by surface roughening is not recommended.

Anodizing is sometimes used as a method of surface preparation prior to electroplating. However, the adherence of the subsequent electrodeposit is limited; plated deposits over anodic films are highly sensitive to surface discontinuities, making the time, temperature, and current density of the anodizing process critical. Phosphoric acid anodizing has been used for the aluminum alloys listed in Table 24; the sequence of operations is as follows:

- Vapor degrease or solvent clean
- Mild alkaline clean
- Rinse
- Etch for 1 to 3 min in a solution containing sodium carbonate (23 g/L or 3 oz/gal) and sodium phosphate (23 g/L or 3 oz/gal), at 65 °C (150 °F)
- Rinse
- Dip in nitric acid solution (50% HNO_3 by volume) at room temperature
- Rinse
- Phosphoric acid anodize according to the conditions given in Table 24; the anodic coating should not be sealed
- Rinse
- Electroplate in a copper pyrophosphate or nickel sulfamate bath

Immersion coating in a zincate solution is a traditional method of preparing aluminum surfaces for electroplating. It is simple and low in cost, but it is also critical with respect to surface pretreatment, rinsing, and strike sequence used. The principle of zincating is one of chemical replacement, whereby aluminum ions replace zinc ions in an aqueous solution of zinc salts. Thus, a thin, adherent film of metallic zinc deposits on the aluminum surface. Adhesion of the zinc film depends almost entirely on the metallurgical bond between the zinc and the aluminum. The quality and adhesion of subsequent electrodeposits depend on obtaining a thin, adherent and continuous zinc film. The electrolytic Alstan strike is coming into general use as a more dependable method than the zincate process for obtaining good adhesion. It is followed by a bronze strike.

Another immersion process is based on the deposition of tin from a stannate solution. This offers the opportunity for improved corrosion resistance because of the more favorable electrolytic potential of tin versus zinc in chloride solutions.

Immersion Procedures. To obtain consistently good results with zinc or tin immersion procedures, it is essential that cleaning and conditioning treatments produce a surface of uniform activity for deposition. Vapor degreasing or solvent cleaning followed by alkaline cleaning is used for removing oil, grease, and other soils. The alkaline cleaner may be a mild etching solution containing 23 g sodium carbonate and 23 g sodium phosphate per litre (3 oz sodium carbonate and 3 oz sodium phosphate per gallon) of water. The solution temperature should range from 60 to 82 °C (140 to 180 °F), and the material should be immersed for 1 to 3 min, after which it should be thoroughly rinsed. After cleaning, the material is further treated to remove the original oxide film and to remove any microconstituents that may interfere with the formation of a continuous film or that may react with the subsequent plating solutions.

Castings present special problems, because their surfaces are more porous than those of wrought products. Solutions entrapped in pores are released during subsequent processing resulting in unplated areas, staining, or poor adhesion of the electrodeposit. Sometimes the trapped solutions become evident much later, during storage or further processing, such as heating for soldering. Furthermore, even if pores are free of solution, the deposit may not bridge them, thus creating a point of attack for corrosion of the basis metal. This is of particular significance in the electroplating of aluminum castings, because the electrodeposited metal is electrolytically dissimilar to aluminum and hence every opening in a casting surface will be a source of corrosion. To avoid these problems, it is essential when preparing cast aluminum surfaces for electroplating that all processing steps be carefully controlled and that electroplating of cast surfaces that have excessive porosity be avoided.

In zincating, the procedures used for removal of the original oxide film and for applying a film of zinc depend to a considerable degree on the aluminum alloy. Several methods are available for accomplishing this surface condi-

Table 23 Applications using electroplated coatings on aluminum products

Product	Form	Preplating treatment	Electroplating system	Thickness μm	Thickness mils	Reason for plating
Automotive applications						
Bearings	Sheet	None	Pb-Sn-Cu alloy	6 + 32	0.25-1.25	Prevent seizing
Bumper guards	Castings	Buff and zincate	Cu + Ni + Cr	2.5 + 51 + 0.8	0.1 + 2 + 0.03	Appearance; corrosion resistance
Lamp brackets; steering-column caps	Die castings	Buff and zincate	Cu + Ni + Cr	0.8 + 20 + 1.3	0.03 + 0.8 + 0.05	Appearance; corrosion resistance
Tire molds	Castings	None	Hard Cr	51	2	Appearance; corrosion resistance
Aircraft applications						
Hydraulic parts; landing gears; small engine pistons	Forgings	Machine and zincate	Cu flash + Cu + hard Cr	2.5 + 25 + 76	0.1 + 1-3	Wear resistance
Propellors	Forgings	Conductive rubber coating	Ni	203	8	Resistance to corrosion and erosion
Shell	Extrusion	Double zincate	Cu flash + Cd(a)	8 + 13(a)	0.3-0.5(a)	Dissimilar-metal protection
Electrical and electronics applications						
Busbars; switchgears	Extrusions	Zincate	Cu flash + Cu + Ag(b)	8 + 5(b)	0.3 + 0.2(b)	Nonoxidized surface; solderability; corrosion resistance
Intermediate-frequency housings	Die castings	Zincate	Cu flash + Cu + Ag + Au(c)	13 + 13 + 0.6(c)	0.5 + 0.5 + 0.025(c)	Surface conductivity; solderability; corrosion resistance
Microwave fittings	Die castings	Zincate	Cu flash + Cu + Ag + Rh	0.25 + 13 + 0.5	0.01 + 0.5 + 0.02	Smooth, nonoxidized interior; corrosion resistance of exterior
Terminal platters	Sheet	Zincate	Cu flash	...	...	Nonoxidized surface; solderability; corrosion resistance
General hardware						
Screws; nuts; bolts	Castings	Buff and zincate	Cd (on threads)	13; 0.5 on threads	0.5; 0.2 on threads	Corrosion resistance
Die cast spray guns and compressors	...	Buff and zincate	Hard Cr	51	2	Appearance
Die cast window and door hardware	...	Barrel burnish and zincate	Brass(d)	8(d)	0.3(d)	Appearance; low cost
Household appliances						
Coffee maker	Sheet	Buff and zincate	Cr	5	0.2	Appearance; cleanness; resistance to food contamination
Refrigerator handles; salad makers; cream dispensers	Die castings	Buff and zincate	Cu + Ni + Cr	2.5 + 13 + 0.8	0.1 + 0.5 + 0.03	Appearance; cleanness; resistance to food contamination
Personal products						
Compacts; fountain pens	Sheet	Buff and zincate	Cu flash + brass	5	0.2	Appearance; low cost
Hearing aids	Sheet	Zincate	Cu flash + Ni + Rh	19 + 0.25	0.75 + 0.01	Nonoxidizing surface; low cost
Jewelry	Sheet	Buff and zincate	Brass + Au	8 + 0.25	0.3 + 0.01	Appearance; low cost

(a) Chromate coating applied after cadmium plating. (b) Soldering operation follows silver plating. (c) Baked at 200 °C (400 °F) after copper plating and after silver plating. Soldering operation follows gold plating. (d) Brass plated in barrel or automatic equipment

Table 24 Conditions for anodizing prior to electroplating
Electrolyte solution of aqueous H_3PO_4

Alloy(a)	Specific gravity	Temperature °C	Temperature °F	Voltage	Time, min
1100	1.300	30	87	22	5
3003	1.300	29	85	22	5
5052	1.300	29	85	22	10
6061	1.300	29	85	22	7

(a) With special care, phosphoric acid anodizing may be used also for aluminum-copper or aluminum-silicon alloys (Wittlock, *Tech Proc AES*, Vol 48, 52, 1961).

Fig. 13 Preplating surface preparation of casting alloys, and alloys which contain high silicon or interfering microconstituents

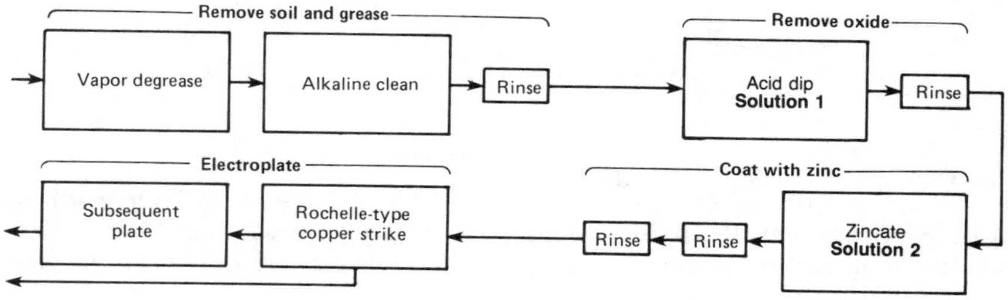

Solution No.	Type of solution	Composition	Amount	Operating temperature °C	Operating temperature °F	Cycle time, s
Alloys 1100 and 3003						
1	Acid dip	HNO_3	50 vol %	Room	Room	15
2	Zincating	NaOH	525 g/L (70 oz/gal)	16-27	60-80	30-60
		ZnO	98 g/L (13 oz/gal)			
Alloys 413, 319, 356 and 380						
1	Acid dip	HNO_3	75 vol %	Room	Room	3-5
		HF	25 vol %			
2	Zincating	NaOH	525 g/L (70 oz/gal)	16-27	60-80	30 max
		ZnO	98 g/L (13 oz/gal)			

tioning. Some alloys can be conditioned by more than one procedure. In these instances, the order of preference follows the order of discussion of these procedures in the following paragraph.

Wrought alloys without interfering microconstituents and casting alloys containing high silicon are prepared for electroplating according to the procedure shown by the flow chart in Fig. 13. The flow chart in Fig. 14 represents the procedure for alloys that contain interfering microconstituents; this procedure is suitable for all wrought alloys, most casting alloys, and especially aluminum-magnesium alloys. Figure 15 indicates the procedure for treating most casting alloys, wrought alloys that contain less than about 3% magnesium, and alloys of unknown composition.

Table 25 gives details of three zincating solutions that may be used alternatively to the solution indicated in the tables accompanying Fig. 13, 14, and 15. The modified solution in Table 25 is recommended when double-immersion zincating (Fig. 15) is required; it is not essential for alloys 2024 and 7075. This solution produces more uniform coverage than the unmodified solution, and also provides the treated work with greater resistance to corrosion. Dilute solution No. 1 in Table 25 is recommended when there are problems in rinsing and dragout. Dilute solution No. 2 provides a greater reserve of zinc for high-production operations, but at a slight sacrifice in effectiveness of rinsing. In some special operations, triple zincate treatment is used. This is essentially stripping of the second zinc film formed in double zincate and adding a zincate from a third solution. Triple zincate gives even more uniform and fine-grained zinc coating than double zincate.

With correct procedure, the resulting zinc deposit is uniform and firmly adherent to the aluminum surface. The appearance of the surface will vary with the alloy being coated, as well as with the rate at which the coating forms. The weight of zinc deposit should be from 1.5 to 5.0 mg/dm^2 (0.1 to 0.3 mg/in.^{2}). Generally, it is desirable to limit the deposit to 3 mg/dm^2 (0.2 mg/in.^{2}).

The thinner and more uniform zinc deposits are the most suitable for plating preparation and for the service performance of plated coatings. Heavy zinc deposits usually are spongy and less adherent, and undesirable from the standpoint of corrosion resistance.

Plating Procedures. Copper is one of the easiest metals to electrodeposit on zincated aluminum surfaces. For this reason, copper is used extensively as an initial strike over which other metals may be subsequently deposited. An advantage of the copper strike is that it protects the thin zinc film from attack by the plating solutions; penetration of the zinc film and attack of the underlying aluminum surface by the plating solutions result in a poorly bonded electrodeposit.

The copper strike bath should be a Rochelle-type copper cyanide solution. The composition and operating conditions recommended for this bath are as follows:

Copper cyanide .... 40 g/L (5.5 oz/gal)
Total sodium cyanide .... 50 g/L (6.5 oz/gal)
Free sodium cyanide .... 4 g/L (0.5 oz/gal) max
Sodium carbonate .... 30 g/L (4.0 oz/gal)
Rochelle salt .... 60 g/L (8.0 oz/gal)
Operating temperature .... 38-55 °C (100-130 °F)
pH .... Varies with alloy; see Table 26

A brass strike is sometimes used in place of copper; however, a bronze strike is frequently used on a tin immersion coating. Table 26 gives operating

Table 25 Zincating solutions

Solution type	Sodium hydroxide		Zinc oxide		Ferric chloride crystals		Rochelle salt		Sodium nitrate		Operating temperature		Procesing time, s
	g/L	oz/gal	g/L	oz/gal	g/L	oz/gal	g/L	oz/gal	g/L	oz/gal	°C	°F	
Modified(a)	525	70	100	13	1.00	0.13	9.8	1.3	...	...	15-27	60-80	30-60
Dilute 1(b)	50.3	6.7	5	0.7	2.03	0.30	50	6.7	0.98	0.13	21-24	70-75	30 max
Dilute 2(b)	120	16	20	2.7	2.03	0.30	50	6.7	0.98	0.13	21-24	70-75	30 max

(a) U.S. Patent 2,676,916 (1954). (b) U.S. Patents 2,676,916 (1954) and 2,650,886 (1953)

conditions for electrodepositing different metals on zincated aluminum surfaces. Environmental considerations sometimes necessitate substitution for cyanide-containing solutions. Nickel strikes, which are successful as a result of careful control of composition and operation conditions, permit this.

Immersion Plating

Immersion plating refers to processes in which another metal is deposited from solution on an aluminum surface under the influence of the potential that exists between the solution and the immersed aluminum material. An external potential is not required. Deposits produced by immersion plating are thin and of little protective value.

Zincating, the procedure used for coating aluminum surfaces with zinc prior to electroplating (see the preceding section), is an example of immersion plating. Brass deposits may be produced by adding copper compounds to the sodium zincate solutions used in zincating.

Fig. 14 Preplating surface preparation for all wrought alloys, most casting alloys, and magnesium-containing alloys that have interfering microconstituents

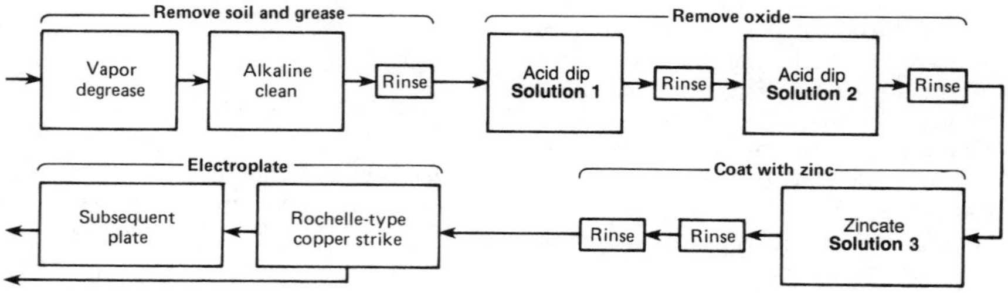

Examples are alloys 1100, 3003, 3004, 2011, 2017, 2024, 5052, 6061, 208, 295, 319, and 355

Solution No.	Type of solution	Composition	Amount	Operating temperature °C	°F	Cycle time, s
1.	Acid dip	H_2SO_4	15 vol %	85 min	185 min	120-300
2.	Acid dip	HNO_3	50 vol %	Room	Room	15
3.	Zincating	NaOH	525 g/L (70 oz/gal)	16-27	60-80	30-60
		ZnO	100 g/L (13 oz/gal)			

Tin may be deposited from solutions containing potassium stannate, stannous chloride, or stannous sulfate-flouride. The lubricating qualities of these tin deposits are desirable for piston and engine components made of aluminum alloys. Immersion tin coatings also are used to facilitate soft soldering and as a base coating for building up electrodeposits. The composition and operating conditions of a successful immersion tin bath are given below:

Potassium stannate ....100 g/L (13.40 oz/gal)
Zinc acetate ....2 g/L (0.27 oz/gal)
m-Cresol sulfonic acid.........35 g/L (4.40 oz/gal)
Temperature of solution .... 60 °C (140 °F)
Immersion time................2 min

Degreasing is the only pretreatment required. The thickness of the tin coating is about 1.3 μm (0.05 mil); solution life is about 0.75 m^2/L (30 ft^2/gal).

Chemical Plating

Chemical plating, often called electroless plating, refers to nonelectrolytic processes that involve chemical reduction, in which the metal is deposited in the presence of a reducing agent. Deposition may take place on almost any type of material, even the container of the solution. For a variety of applications in the aircraft industry, nickel is chemically plated on aluminum parts of shapes for which electroplating is not practical. However, chemical plating is too expensive to be used when conventional electroplating is feasible. The composition and operating conditions of a bath for successfully depositing nickel are given below:

Nickel chloride ....30 g/L (4 oz/gal)
Sodium hypophosphite .... 7.5 g/L (1 oz/gal)
Sodium citrate................72 g/L (9.60 oz/gal)
Ammonium chloride ....48 g/L (6.40 oz/gal)
Ammonium hydroxide (0.880 sp gr)................13 g/L (1.74 oz/gal)
pH..............................10
Temperature of solution ..... 82-88 °C (180-190 °F)
Immersion time..................1 h

Deposits produced contain about 6 wt % phosphorus and usually are not considered suitable as a base for chromium plate.

Table 26 Conditions for electroplating various metals on zincated aluminum surfaces

Electroplate	Minimum deposit µm	mils	Plating time, min	Bath temperature °C	°F	Current density A/dm²	A/ft²	Type of electrolyte
Copper:								
1 Copper strike(a)	7.5	0.3	2(b)	34-54	100-130	2.4(b)	24(b)	Rochelle cyanide(c)
2 Brass strike(a)	7.5	0.3	2-3	27-32	80-90	0.5	5	Cyanide
3 Copper plate(d)	12.5	0.5	40 s-2 min	76-83	170-180	3-6	30-60	High-speed NaCN or KCN
Brass	12.5	0.5	3-5	27-32	80-90	1	10	Cyanide
Cadmium:								
1 Copper strike(a)	12.5	0.5	2(b)	34-54	100-130	2.4(b)	24(b)	Rochelle cyanide(c)
2 Cadmium plate	12.5	0.5	8-20	21-35	70-95	1.4-4.5	14-45	Cyanide
Chromium, decorative:								
1 Copper strike(a)	7.5	0.3	2(b)	34-54	100-130	2.4	24(b)	Rochelle cyanide(c)
2 Brass strike(a)	7.5	0.3	2-3	27-32	80-90	0.5	5	Cyanide
3 Nickel undercoat	2.5-5	0.1-0.2	(e)	(e)	(e)	(e)	(e)	(e)
4 Chromium plate	.25-.50	0.01-0.02	10-12	43-46	110-115	0.07-0.15	0.7-1.5	Conventional
Chromium, decorative (direct on zincate)	0.75	0.03	5-10	18-21	65-70	0.07-0.15	0.7-1.5	Conventional
Chromium, hard:								
1 Copper strike	7.5	0.3	2(b)	34-54	100-130	2.4(b)	24(b)	Rochelle cyanide(c)
2 Chromium plate	1.25	0.05	5	54	130	0.07-0.15	0.7-1.5	Conventional
Chromium, hard (direct on zincate)	1.25	0.05	10-20 then 54(f)	18-21; then 130(f)	65-70;	0.07-0.15; then 0.3(f)	0.7-1.5 then 3 A/in.²	Conventional
Chromium, hard (for corrosion protection):								
1 Copper strike	7.5	0.3	2(b)	34-54	100-130	2.4(b)	24	Rochelle cyanide(c)
2 Brass strike	7.5	0.3	2-3	27-32	80-90	0.5	5	Cyanide
3 Nickel undercoat	25-50	1-2	(e)	(e)	(e)	(e)	(e)	(e)
4 Chromium plate	2.5-5	0.1-0.2	10-12	43-46	110-115	0.07-0.15	0.7-1.5	Conventional
Gold:								
1 Copper strike	7.5	0.3	2(b)	34-54	100-130	2.4(b)	24(b)	Rochelle cyanide(c)
2 Brass strike	7.5	0.3	2-3	27-32	80-90	0.5	5	Cyanide
3 Nickel undercoat	17.5	0.7	(e)	(e)	(e)	(e)	(e)	(e)
4 Gold plate	0.625	0.025	10 s-1 min	49-71	120-160	0.5-1.5	5-15	Potassium cyanide
Nickel (for minimum corrosion protection):								
1 Copper strike	7.5	0.3	2(b)	34-54	100-130	2.4(b)	24(b)	Rochelle cyanide(c)
2 Brass strike	7.5	0.3	2-3	27-32	80-90	0.5	5	Cyanide
3 Nickel plate	7.5-12.5	0.3-0.5	(e)	(e)	(e)	(e)	(e)	(e)
Nickel (for maximum corrosion protection):								
1 Copper strike	7.5	0.3	2(b)	34-54	100-130	2.4(b)	24(b)	Rochelle cyanide(c)
2 Brass strike	7.5	0.3	2-3	27-32	80-90	0.5	5	Cyanide
3 Nickel plate	25-50	0.3-0.5	(e)	(e)	(e)	(e)	(e)	(e)
Silver:								
1 Double silver strike	0.625	0.025	10 s(g)	30(g)	80(g)	1.5-2.5(g)	15-25(g)	Cyanide(h)
2 Silver plate	1.25-2.5	0.05-0.1	18-35	27-32	80	0.5	5	Cyanide
Silver (alternative method):								
1 Copper strike	7.5	0.3	2(b)	34-54	100-130	2.4(b)	24(b)	Rochelle cyanide(c)
2 Silver strike	0.50	0.02	10 s	27-32	80	1.5-2.5	15-25	Cyanide(j)
3 Silver plate	1.25-2.5	0.05-0.1	18-35	27-32	80	0.5	5	Cyanide
Tin:								
1 Copper strike	7.5	0.3	2(b)	34-54	100-130	2.4(b)	24(b)	Rochelle cyanide(c)
2 Tin plate (k)	17.5	0.7	15-30	93-99	200-210	4.5-6.5	45-65	Sodium stannate
Zinc:								
1 Copper strike	7.5	0.3	2(b)	34-54	100-130	2.4(b)	24(b)	Rochelle cyanide(c)
2 Zinc plate	12.5	0.5	18-45	24-30	75-86	1-3	10-30	Pyrophosphate
Zinc (direct on zincate)	12.5	0.5	10	24-35	75-95	0.5-5(m)	5-50(m)	Pyrophosphate

(a) An initial cyanide copper strike is generally used to achieve complete metal coverage of zincated aluminum parts prior to plating, because of the excellent throwing power of the copper electrolyte. A copper strike is not, however, recommended as the initial coating for alloys 5056, 214, 218, and others that contain substantial amounts of magnesium; these will achieve a better initial coverage in a brass strike. Neither copper strike nor brass strike should be used as a final finish; both should always have an electroplated top coat. (b) The copper strike is achieved during the first 2 min while the current density of the electrolyte is maintained at 2.4 A/dm² (24 A/ft²). Instead of being transferred from the strike bath to a high-speed sodium or potassium electrolyte for subsequent copper plating, the work may be allowed to remain (3 to 5 min) in the Rochelle-type electrolyte to be copper plated, provided the current density is lowered to 1.2 A/dm² (12 A/ft²). (c) Colorimetric pH of electrolyte is 12.0 for all treatable alloys except 5052, 6061, and 6063, for which pH is 10.2 to 10.5. (d) Work for which copper strike plating may be used may be left in the copper strike for copper plating, instead of being transferred to the high-speed sodium or potassium cyanide electrolyte (see footnote c). (e) As discussed in the article on nickel plating, various electrolytes are used, depending on the specific purpose of the plated deposit. If the nickel is to be deposited directly on the zincated surface, a bath must be selected that is suitable for application over zinc (examples of such baths are fluoborate and sulfamate nickel electrolytes). (f) The transition from low-temperature to high-temperature plating may be accomplished either by heating the electrolyte to 54 °C (130 °F) after deposition has started at 18 to 21 °C (65 to 70 °F) or by transferring the work (without rinsing) from an electrolyte at 18 to 21 °C (65 to 70 °F) to one at 54 °C (130 °F) and holding the work in the high-temperature electrolyte without current until the work reaches bath temperature. Current density is 0.07 to 0.15 A/dm² (0.7 to 1.5 A/in.²) in the electrolyte at 18 to 21 °C (65 to 70 °F), 1935 A/mm² (3 A/in.²) at 54 °C (130 °F). (g) Each bath. (h) First strike bath contains 1 g (0.11 oz) of AgCN and 90 g (10.2 oz) of NaCN per litre (gallon); second bath, 5.3 g (0.60 oz) of AgCN and 67.5 g (7.7 oz) of NaCN per litre (gallon). (j) Contains 5.3 g (0.60 oz) of AgCN and 67.5 g (7.7 oz) of NaCN per litre (gallon). (k) After the aluminum material has been copper strike plated, tin may be applied also by immersion for 45 min to 1 h in a sodium stannate solution at 49 to 74 °C (120 to 165 °F). Time and temperature depend on solution used. (m) Current is applied as work is being immersed in electrolyte

The immersion time given is for deposits 50 μm (2 mils) or more thick.

Silver may be chemically plated on anodized aluminum-base materials. The procedure consists of first degreasing the anodized surface, and then dipping in dilute hydrochloric acid, water rinsing, and then immersing the object in the silvering solution. A mixture of two solutions is required for silvering. The first consists of 3.33 mL (0.113 fluid oz) of a 10% solution of silver nitrate to which a 7.5 vol % solution of ammonium hydroxide is added until the precipitate first formed just redissolves, after which an excess of 40 mL (1.3 fluid oz) of ammonium hydroxide solution is added. The second solution is made up by adding 80 g (3 oz) of Rochelle salt or 40 g (1.4 oz) of potassium citrate to water to a total volume of 330 mL (11 fluid oz). Solutions are filtered and mixed immediately before they are to be used.

Fig. 15 Preplating surface preparation for most casting alloys, for wrought alloys containing less than about 3% magnesium and for alloys whose identities are not known

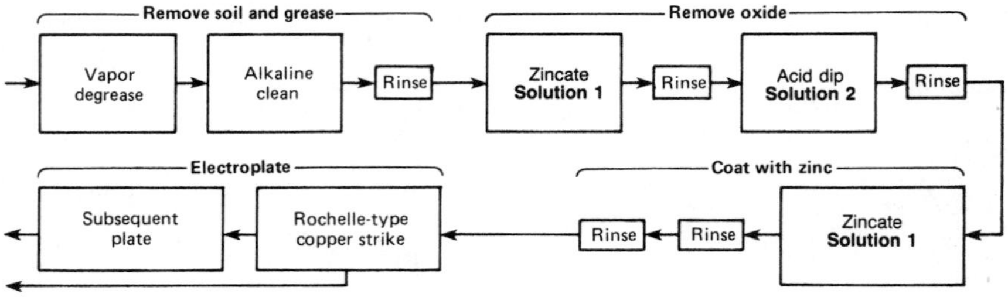

Solution No.	Type of solution	Composition	Amount	Operating temperature °C	Operating temperature °F	Cycle time, s
1.	Zincating	NaOH ZnO	525 g/L (70 oz/gal) 100 g/L (13 oz/gal)	16-27	60-80	30-60
2.	Acid dip	HNO_3	50 vol %	Room	Room	15

Examples of wrought alloys containing less than 3% magnesium are 1100, 3003, 3004, 2011, 2017, 2024, 5052, 6061, 413, 43, 208, 295, 319, 355, and 356

Painting

The difference between painting of aluminum and painting of iron and steel is primarily in the method of surface preparation. Therefore, the information on materials and methods of application in the article on painting of steel and cast iron in this Volume is, in general, applicable to aluminum.

Aluminum is an excellent substrate for organic coatings if the surface is properly cleaned and prepared. For many applications, such as indoor decorative parts, the coating may be applied directly to a clean surface. However, a suitable prime coat, such as a wash primer or a zinc chromate primer, usually improves the performance of the finish coat.

For applications involving outdoor exposure, or for indoor applications exposing the part to impact or abrasive forces, a surface treatment such as anodizing or chemical conversion coating is required prior to the application of a primer and a finish coat. These processes have been discussed in previous sections of this article.

Anodizing in sulfuric or chromic acid electrolytes provides an excellent surface for organic coatings. Usually, only thin anodic coatings are required as a prepaint treatment. Decorative parts for home appliances generally are anodized before painting to ensure good paint adhesion over an extended period. Sulfuric acid anodic coatings are used when painting of only part of the surface is required for decorative effects; the anodic coating protects the unpainted portions of the surface.

Conversion coatings usually are (*a*) less expensive than anodic coatings, (*b*) provide a good base for paint, and (*c*) improve the life of the paint by retarding corrosion of the aluminum substrate material. Adequate coverage of the entire surface by the conversion coating is important for good paint bonding.

Porcelain Enameling

Aluminum products may be finished by porcelain enameling to enhance appearance, chemical resistance, and weather resistance. The wrought aluminum alloys that can be enameled are 1100 and 3003 (sheet), and 6061 (sheet and extrusions). Special designations of these alloys that provide excellent response to porcelain enameling because of the compositional control used in fabrication. Castings made of alloys 43 and 356, which can be porcelain enameled, are commercially available.

Before the porcelain enamel frit is applied, aluminum surfaces are prepared by being (*a*) cleaned in an alkaline cleaner, to remove soil; (*b*) etched in a solution of chromic and sulfuric acids, to remove surface oxide; and (*c*) dipped in a chromate solution, to apply a chromate-oxide film. Details of processing procedures for the application of porcelain enamel coatings on aluminum may be found in the article on porcelain enameling in this Volume.

Aluminum Finish Standard

MIL-F-7179. Finishes and coatings; general specification for protection of aircraft and aircraft parts. Covers finishing procedures for corrosion protection of aircraft and component parts. This specification is not applicable to the purchase of mill products but is a good reference for recommended finishes for aluminum.

Aluminum Anodic Coating Standards

ASTM test methods for the evaluation of various properties and qualities of anodic coatings are listed in Table 17.

AMS 2468. Hard coating of aluminum alloys.

AMS 2470. Anodic films, corrosion-protective for aluminum alloys. Covers chromic acid anodizing of aluminum alloys.

AMS 2471. Anodic treatment for aluminum-base alloys. Covers anodizing of aluminum alloys by the sulfuric acid process.

AMS 7222. Rivets, 2117-T4. Covers 2117-T4 rivets that are to be anodized.

AMS 7223. Rivets, 2024-T4. Covers 2024-T4 rivets that are to be anodized.

MIL-A-8625. Anodic coatings for aluminum alloys. Covers the procedure and requirements for anodizing. Type I is a chromic acid coating, Type II a sulfuric acid coating.

MIL-B-6812. Aircraft bolts, 2024. Covers aluminum alloy 2024-T4 anodized bolts.

MIL-I-8474. Anodized process for inspection of aluminum parts. Covers the use of chromic acid anodizing for the surface inspection of cracks and other defects.

Aluminum Chemical Conversion Coatings

ASTM test methods that may be used for the evaluation of various properties and qualities of chemical conversion coatings are listed in the section on control of coating quality in this article.

AMS 2473. Chemical treatment for aluminum alloys; general purpose. Covers the requirements for chemical films applied for increased corrosion resistance or for paint adhesion.

AMS 2474. Chemical treatment for aluminum alloys; low electrical resistivity. Covers the requirements for chemical films applied for increased corrosion resistance or for paint adhesion where a conductive coating is required.

MIL-C-5541. Chemical films for aluminum and aluminum alloys. Covers the minimum requirements for qualification of chemical films on aluminum applied to aid paint adhesion or corrosion resistance. Specific acceptable treatments are listed in the table of qualified products.

Aluminum Painting Standards

ASTM test methods that may be used for the evaluation of various properties and qualities of paints are listed in the article on painting of steel and cast iron in this Volume.

DOD-P-15328. Wash primer pretreatment. Covers phosphoric acid wash primer to prepare aluminum surfaces for painting.

MIL-C-81706. Chemical conversion materials for coating aluminum and aluminum alloys.

MIL-C-83266. Coating, urethene, aliphatic, isocyanate, for aerospace applications.

MIL-P-23377. Prime coatings: epoxy-polyamide chemical and solvent resistant.

MIL-P-6808. Application of zinc chromate primer. Covers the procedures in applying zinc chromate primer to aircraft.

MIL-P-15930. Vinyl-zinc chromate primer. Covers zinc chromate primer suitable for use on aluminum surfaces that have been treated with a wash primer conforming to DOD-P-15328D.

MIL-S-5002. Surface treatments for metal parts in aircraft. Covers various treatments required for airframes prior to priming and painting.

MIL-T-704. Treatment and painting for construction and engineering equipment. Covers finishing schedules for various metals, including aluminum.

Electroplating Standards

ASTM B253. Recommended practice for preparation of and electroplating on aluminum alloys. Covers practices for electroplating of gold, silver, tin and zinc on aluminum alloys.

AMS 2420. Plating aluminum for solderability. Covers the zincate process for preparing aluminum parts for soldering.

Porcelain Enameling Standards

ASTM B244. Method of measuring thickness of anodic coatings on aluminum with eddy-current instruments.

Other ASTM test methods for determining properties of porcelain enamel coatings are listed in the article on porcelain enameling in this Volume.

PEI A1-2a (4th ed.). Recommended processing methods for porcelain enamel on aluminum alloys. (Porcelain Enamel Institute, Inc.)

PEI ALS-105 (57). Tentative specification for porcelain enamel on aluminum used for sign and architectural applications. (Porcelain Enamel Institute, Inc.)

SELECTED REFERENCES

General sources

- S. Wernick and R. Pinner, "The Surface Treatment and Finishing of Aluminum and Its Alloys", 4th ed., Robert Draper Ltd., Teddington, England, 1972
- G. H. Kissin, Ed., "Finishing of Aluminum", New York: Reinhold, 1963
- Kent R. Van Horn, Ed., "Aluminum, Vol. 3, Fabrication and Finishing", American Society for Metals, 1967

Chemical cleaning

- When Cleaning Aluminum, *Mod Metals,* Sept 1959
- W. B. Stoddard, Jr., U.S. Patent 3,041,259 (1962). (Cleaning of aluminum surfaces, with particular reference to a method of electrolytic alkaline cleaning.)

Chemical and electrolytic brightening

- W. J. Tegart, "The Electrolytic and Chemical Polishing of Metals in Research and Industry", New York: Pergamon Press, 1959
- N. P. Fedot'ev and S. Grilikhes, "Electropolishing, Anodizing, and Electrolytic Pickling of Metals", Moscow, 1957 (Translation by A. Behr, England: Robert Draper Ltd., 1959)
- R. M. Rall and H. Yoshimoto, Radiant Cooling or Shielding, *Prod Eng,* Nov 14, 1960
- F. A. Champion and E. E. Spillett, Super Purity Aluminum and Its Alloys, *Sheet Metal Ind,* Jan 1956
- W. H. Tingle and F. R. Potter, New Instrument Grades Polished Metal Surfaces, *Prod Eng,* March 27, 1961
- G. D. Nelson and C. J. Knapp, How to Control the Aluminum Bright Dipping Process, *Mod Metals,* Sept 1961
- A. W. Brace, "The Electrolytic and Chemical Polishing and Brightening of Aluminum and Its Alloys", Aluminum Development Assn. RP 54; reprinted from *Metal Finishing J* (London), June and July 1955
- W. R. Meyer and S. H. Brown, Cleaning, Etching, Chemical Polishing and Brightening of Aluminum, *Tech Proc Am Electroplater's Soc,* 36, 163 (1949)
- A. V. Davis, U.S. Patent 528,513 (1894). (Improving the surface of aluminum by treating with a mixture of hydrofluoric acid and nitric acid.)
- J. C. Lum *et al,* U.S. Patent 2,428,464 (1947). (An acid composition, of essentially phosphoric and nitric acids, for etching of metals. The specification and some claims refer to copper, copper alloys and ferrous metals. The specification states that a bright

surface free from pitting will be obtained.

- H. A. H. Pray *et al,* U.S. Patent 2,446,060 (1948). (A concentrated acid bath containing nitric, phosphoric and acetic acids and covering a broad range of concentrations, for chemically polishing metal surfaces. Metals given in some claims are nickel, silver, copper, brass, nickel and Monel.)
- F.H. Hesch, U.S. Patent 2,593,447 (1952). (A bath containing hydrogen, nitrate, fluoride, ammonium, and chromate ions.)
- F.H. Hesch, U.S. Patent 2,593,449 (1952). (A bath containing hydrogen, nitrate, fluoride, chromate, and noble-metal ions.)
- W.C. Cochran, U.S. Patent 2,613,141 (1952). (A bath for brightening aluminum in phosphoric acid and hydrogen peroxide.)
- H.J. Reindl and S. Prance, U.S. Patent 2,614,913 (1952). (A dilute bath containing nitric acid, soluble fluorides and various additives.)
- E. Shelton-Jones, U.S. Patent 2,650,156 (1953). (A process consisting of immersing aluminum in a bath of concentrated phosphoric and sulfuric acids.)
- W.C. Cochran, U.S. Patent 2,650,157 (1953). (Nitric and phosphoric acids with or without acetic or sulfuric and acetic acids.)
- J. T. Ferguson, U.S. Patent 2,671,717 (1954). (Method of brightening aluminum in a hot solution of alkali metal hydroxide and an alkali nitrate.)
- E. R. DuFresne *et al,* U.S. Patent 2,673,143 (1954). (Forming a reflective surface on aluminum using a bath of sodium hydroxide, sodium gluconate and an oxidizing agent.)
- R.C. Spooner, U.S. Patent 2,678,875 (1954). (Brightening of aluminum in hot concentrated phosphoric and nitric acids and silicic acid as an inhibitor.)
- J.F. Jumer, U.S. Patent 2,705,191 (1955). (A solution for polishing aluminum, containing phosphoric acid, alkali metal nitrate and alkali metal sulfate.)
- J.F. Murphy, U.S. Patent 2,719,079 (1955). (Bath containing complex fluoride ions, nitric acid and a noble metal.)
- F.H. Hesch, U.S. Patent 2,719,781 (1955). (A simple fluoride bath containing nitric acid, fluoride ions and noble-metal ions.)
- C.C. Cohn, U.S. Patent 2,729,551 (1956). (Method of treating aluminum with a broad range of phosphoric acid, nitric acid and water, or with phosphoric acid and water.)
- W. Helling *et al,* U.S. Patent 2,746,849 (1956). (An Erftwerk type of bath.)

Chemical etching

- R.H. Elliott, Jr., U.S. Patent 2,882,135 (1959). (Composition and operating conditions for an alkaline etching solution.)
- F.C. Prescott, J.K. Shaw, and Jay Lilken, Sequestering Agents in Aluminum Etching, *Metal Finishing,* 51, 1953
- R.J. Meyer, W.F. Johnson, and R.A. Wodehouse, U.S. Patent 4,136,026 (1979). (Recovery process for caustic etch solutions.)

Anodizing

- Arthur Brace, *Technology of Anodizing,* Teddington, England: Robert Draper, Ltd., 1968
- James H. Weaver, ASD-TDR-62-918, "Anodized Aluminum Coatings for Temperature Control of Space Vehicles", Aeronautical Systems Div., Wright-Patterson AFB, 1963
- F.J. Gillig, WADC Technical Report 53-151, 1953, P.B. 111320
- J. Herenguel and R. Segond, *Rev Met (Paris),* 46, 377 (1949), and 42, 258 (1945)
- J. Herenguel, Trans Inst Metal Finishing, 31 (1954), Advance Copy No. 1
- J.E. Bunch, Racks for Anodizing, Metal Finishing, July 1957, p 45
- A.K. Graham and H.L. Pinkerton, "Electroplating Engineering Handbook", New York: Reinhold, 1955
- "Anodic Oxidation of Aluminum and Its Alloys", Aluminum Development Assn. Bulletin No. 14
- J.M. Kape, Unusual Anodizing Processes and Their Practical Significance, *Electroplating Metal Finishing,* Nov 1961
- F. Howitt and I.H. Jenks, Improved Aluminum Alloys for Bright Anodizing, *Metal Progress,* Jan 1960
- W.E. Cooke, Factors Affecting Loss of Brightness and Image Clarity During Anodizing of Bright-Trim Aluminum Alloys in Sulfuric Acid Electrolyte, *Plating,* Nov 1962
- W.C. Cochran, "Integral Color and Hardcoat Anodizing," Aluminum Finishing Seminar, Paper A7, The Aluminum Association, Washington, DC, 1973
- P.G. Sheasby and W.E. Cooke, The Electrolytic Colouring of Anodized Aluminum, *Trans. Inst. Metal Finishing, 52,* Summer 1974
- N.T. McDevitt and J.S. Solomon, AD-A083202, "Thin Anodic Oxide Films On Aluminum Alloys and Their Role In The Durability of Adhesive Bonds", U.S. Air Force Materials Lab., Feb 1980
- W.C. Cochran, Standards for Anodized Aluminum, *Plating and Surface Finishing,* Jan 1981
- "Chemical Finishing of Aluminum", *Modern Metals, 15,* Sept 1959
- C.A. Vassey *et al,* U.S. Patent 3,706,603 (1972). (Dry in place, no rinse chromate conversion coating.)
- N. Das, U.S. Patent 3,964,936 (1976). (Chromate-free conversion coating solution containing zirconium.)
- Y. Matsushima *et al,* U.S. Patent 4,017,334 (1977). (Chromium-free conversion coating process with titanium and tannin.)

Electroplating

- F. Pearlstein, *AES Shop Guide,* American Electroplaters Society, Winter Park, FL 32789
- R.F. Hafer, Electroplating on Aluminum, *Metal Progress,* May 1955
- W. Wernick and R. Pinner, *Surface Treatment and Finishing of Aluminum and its Alloys,* 3rd ed.
- R. Spooner and D. Seraphim, Nickel-Chromium Plated Aluminum Sheet, *Metal Finishing,* Jan 1961
- "Metal Finishing Guidebook", 49th ed., (1981), Westwood, N.J.: Metals and Plastics Publications, Inc.
- W.G. Zelley, Plating on Aluminum, Chapter 24 in "Modern Electroplating", 2nd ed., New York: Wiley, 1963, p 556-564
- H. Uhlig, "The Corrosion Handbook", New York: Wiley, 1948
- "Recent Developments in the Plating of Aluminum", Aluminum Development Assn.
- F. Keller and W.G. Zelley, Conditioning Aluminum Alloys for Electroplating, *J Electrochem Soc,* April 1950
- F. Keller and W.G. Zelley, Plating on Aluminum Alloys, *Tech Proc Am Electroplaters' Soc,* 36, 1949, p 149-162
- H.V. Wittrock, Nickel-Chromium Plating Upon Anodized Aluminum, *Tech Proc Am Electroplaters' Soc,* 48, 52 (1961)

- A. E. Wyszynski, An Immersion Alloy Pretreatment for Electroplating on Aluminum, *Trans. Inst. Metal Finishing 45,* Part 4, 1967
- J. C. Jongkind, Bright Copper-Nickel-Chromium Plating on Aluminum Bumper Bar Stock, *Plating,* Dec 1975
- J. C. Jongkind, U. S. Patent 4,100,038 (1978). (Tin immersion process.)
- L. D. Brown, A New Method for Plating on Aluminum, *Trans. Inst. Metal Finishing 56,* Part 4, 1978
- Leo Missel, "Simple and Stable Strike for Plating on Aluminum", AES Symposium on Plating on Difficult To Plate Metals, Paper 15, New Orleans, 1980
- H. J. Wittrock, Adhesion of Sulfamate Nickel Electrodeposits on Aluminum-Zinc-Magnesium Alloys Anodized in Phosphoric Acid, *Plating and Surface Finishing,* Jan 1980
- D. S. Lashmore, Plating on Anodized Aluminum, *Plating and Surface Finishing,* April 1981

Painting

- MIL-C-81706 5 Chemical Conversion Materials for Coating Aluminum and Aluminum alloys
- R. I. Wray, Painting of Aluminum and Magnesium, *Metal Progress,* Dec 1954
- R. Burns and W. Bradley, "Protective Coatings for Metals", Reinhold, New York, 1955
- M. Hess, "Paint Film Defects", Reinhold, New York, 1952
- P. G. Campbell and H. Lasser, Army TM 5-618, NAVFAC MO-110, USAF 85-3, Paints and Protective Coatings, June 1981
- K. E. Martin and R. Rolles, Durability of Precoated Aluminum Products, *J. of Paint Technology,* July 1967
- MIL-P-23377D 2 Primer Coatings: Epoxy-Polyamide, Chemical and Solvent Resistant
- MIL-C-83266 Coating, Urethane, Aliphatic, Isocyanate for Aerospace Applications

Porcelain enameling

- W. A. Farrell, Glass on Aluminum, *Mod Metals,* Oct 1956
- N. H. Stradley, Porcelain Enamels for Aluminum and Magnesium, *Am Ceram Soc Bull,* Aug 1959
- F. L. Church, Porcelain on Aluminum, *Mod Metals,* June 1962
- D. Lamarche, Porcelain Enameling of Aluminum Alloys, *J Can Ceram Soc,* 31 (1962)
- A. L. Sopp and others, Chemical and Weather Resistance of Porcelain Enamels on Aluminum, *Ceram Age,* April 1960

Other finishing processes

- A. Cahne, U. S. Patent 2,944,917 (1960). (Method of Coating a Metal Base With Polytetrafluoroethylene.)
- C. A. MacNeill, Gas Plating—an Introduction, *Metal Finishing,* July 1963, p 40-44, 47

Designation System for Aluminum Finishes*

Finishes used on aluminum are categorized as mechanical or chemical finishes or as coatings. Types of coatings that can be applied on aluminum include anodic, resinous and other organic coatings, and vitreous coatings. In addition, electroplated and other metallic coatings, or laminated coatings can be used on aluminum. In this designation system, each of these categories is assigned a letter, and the various finishes in each category are designated by two-digit numerals. Specific finishes of the various types thus are designated by a letter followed by two numbers, as shown in Table 27. Two or more designations may be combined into a single designation to identify a sequence of operations covering all the important steps leading to a final complex finish.

When designations for chemical coatings are used alone, other processing steps normally used ahead of these finishes are at the option of the processor. Where a finish requires two or more treatments of the same class, the class letter is repeated, each time being followed by the appropriate two-digit numeral.

Designations for specific coatings have been developed for only the anodic coatings. Coatings of the four other classes may be tentatively designated by the letters respectively assigned for them; detailed designations for these four categories may be developed and added to the system later.

The examples that follow show how the designation system for aluminum finishes is used; each designation is preceded by the letters "AA" to identify it as an Aluminum Association designation.

*Developed by The Aluminum Association to assist users in specifying finishes. This presentation is an abridgment of the detailed explanation of the designation system that was published by the Association in Sept 1980

Smooth Specular Finish. A finish obtained by polishing aluminum with an aluminum oxide compound according to the following schedule. Begin with grits coarser than 320; follow with 320 grit and a wheel speed of 30 m/s (6000 ft/min); complete polishing by buffing with tripoli-based buffing compound at 35 to 41 m/s (7000 to 8000 ft/min). The designation for this finish is AA-M21 (Table 27).

Architectural Building Panel. A matte anodized finish for a building, such as that produced by giving aluminum a matte finish, then chemical cleaning followed by architectural class II natural anodizing, would be designated as AA-M32C12A31:

AA ...... Aluminum Association
M32 ..... Mechanical finish, directional textured, medium satin appearance
C12...... Chemical treatment, inhibited alkaline cleaning
A31...... Anodic coating, architectural class II (10 to 18 μm or 0.4 to 0.7 mil thick), clear (natural)

Architectural Aluminum with Anodized Integral Color. An anodized panel with an integral color for architectural application would be designated as AA-M10C22A42:

AA ...... Aluminum Association
M10 ..... Unspecified as-fabricated finish
C22...... Chemically etched medium matte finish
A42...... Anodic coating, architectural class I (18 μm or 0.7 mil or thicker), integral color

Chromium-Plated Aluminum Panel. The finish for a chromium-plated aluminum panel that is first given a highly specular mechanical finish, then a nonetch chemical cleaning, followed by a thin anodic coating produced in phosphoric acid, and finally

Table 27 Designations for aluminum finishes(a)

Designation	Finish	Designation	Finish
Mechanical finishes — M		Chemical coatings:	
As fabricated:		C40	Unspecified
M10	Unspecified	C41	Acid chromate-fluoride
M11	Specular finish as fabricated	C42	Acid chromate-fluoride-phosphate
M12	Nonspecular finish as fabricated	C43	Alkaline chromate
M1X	Other (to be specified)	C44	Nonchromate
Buffed:		C45	Nonrinsed chromate
M20	Unspecified	C4X	Other (to be specified)
M21	Smooth specular	**Anodic coatings — A**	
M22	Specular	General:	
M2X	Other (to be specified)	A10	Unspecified
Directional textured:		A11	Preparation for other applied coatings
M30	Unspecified	A12	Chromic acid coatings
M31	Fine satin	A13	Hard, wear- and abrasion-resistant coatings
M32	Medium satin	A1X	Other (to be specified)
M33	Coarse satin	Protective and decorative: (Less than 10 μm or 0.4 mil)	
M34	Hand rubbed	A21	Clear (natural)
M35	Brushed	A22	Integral color
M3X	Other (to be specified)	A23	Impregnated color
Nondirectional textured:		A24	Electrolytically deposited color
M40	Unspecified	A2X	Other (to be specified)
M41	Extra fine matte	Architectural class II(c): (10 to 18 μm or 0.4 to 0.7 mil)	
M42	Fine matte	A31	Clear
M43	Medium matte	A32	Integral color
M44	Coarse matte	A33	Impregnated color
M45	Fine shot blast	A34	Electrolytically deposited color
M46	Medium shot blast	A3X	Other (to be specified)
M47	Coarse shot blast	Architectural Class I(c): (18 μm or 0.7 mil)	
M4X	Other (to be specified)	A41	Clear (natural)
Chemical finishes(b) — C		A42	Integral color
Nonetched cleaned:		A43	Impregnated color
C10	Unspecified	A44	Electrolytically deposited color
C11	Degreased	A4X	Other (to be specified)
C12	Inhibited chemical cleaned	**Resinous and other organic coatings(d) — R**	
C1X	Other (to be specified)	R10	Unspecified
Etched:		R1X	Other (to be specified)
C20	Unspecified	**Vitreous (porcelain and ceramic) coatings(d) — V**	
C21	Fine matte	V10	Unspecified
C22	Medium matte	V1X	Other (to be specified)
C23	Coarse matte	**Electroplated and other metal coatings(d) — E**	
C2X	Other (to be specified)	E10	Unspecified
Brightened:		E1X	Other (to be specified)
C30	Unspecified	**Laminated coatings(d) — L**	
C31	Highly specular	L10	Unspecified
C32	Diffuse bright	L1X	Other (to be specified)
C3X	Other (to be specified)		

(a) All designations are preceded by the letters "AA", to identify them as Aluminum Association designations. Examples of methods of finishing are given in the Aluminum Association publication from which the presentation here is derived. (b) Includes chemical conversion coatings, chemical or electrochemical brightening and cleaning treatments. (c) Classification established in Aluminum Association Standards for Anodically Coated Aluminum Alloys for Architectural Applications, October 1978. (d) These designations may be used until more complete series of designations are developed for these coatings

direct chromium plating, would be designated as AA-M21C12A1XE1X:

AA ...... Aluminum Association

M21 ..... Mechanical finish, polished, smooth specular (see smooth specular finish, (above)

C12 ...... Inhibited alkaline cleaned

A1X ..... Specify exact anodizing process

E1X ..... Specify exact chromium plating method

Cleaning and Finishing of Copper and Copper Alloys

By Robert M. Paine
Senior Chemist
Brush Wellman, Inc.
and
Bob Srinivasan
Research Manager
Diversey Wyandotte, Corp.

COATINGS on copper alloys vary according to treatment, appearance, and corrosion resistance. The applications range from simple low cost chemical treatments that provide a uniform surface appearance, to expensive electroplates that provide maximum corrosion resistance. Prior to the application of a protective or decorative coating, the metal surface must be prepared by suitable cleaning procedures. The severity of these cleaning procedures will depend on the past processing history. Copper and copper alloys obtained from major materials suppliers will be received reasonably clean (free from heat treat scale or tarnish) and should only require minimal cleaning prior to further processing. Only in those cases where the processor heat treats will it be necessary to use the full range of cleaning discussed below.

Pickling and Bright Dipping

Pickling in solutions containing 4 to 15 vol % sulfuric acid or 40 to 90 vol % hydrochloric acid is used for the removal of oxides formed on the surface of copper-base materials during mill processing and fabricating operations. The sulfuric acid solution is used to remove (*a*) black copper oxide scale on brass extrusions, forgings, and machined parts; (*b*) oxide on copper tubing, forgings, and machined parts; and (*c*) light annealing scale or tarnish. The hydrochloric acid solution is primarily used for finishing, but is also used to remove scale and tarnish from brass forgings and machined parts, and oxide on copper forgings and machined parts. Conditions for pickling copper-base metals with sulfuric acid and hydrochloric acid are shown in Table 1. Sometimes no additional surface preparation is necessary to produce the uniformity of appearance required for further finishing of copper; however, heavily scaled material may need a bright dip or color dip after pickling.

Except for bright annealed material, copper alloys must be (*a*) pickled after each annealing treatment; (*b*) completely descaled; and (*c*) bright dipped to produce a natural surface color and luster suitable for other finishing treatments such as electroplating or painting. Scale dip and bright dip solutions are given in Table 2. Bright dips for copper-base materials consist of sulfuric and nitric acids in varying proportions with a small amount of water and hydrochloric acid. Proprietary pickle and bright dips, using sulfuric acid and stabilized hydrogen peroxide, are also available. Whereas the conventional bright dips use highly concentrated acids, this sulfuric acid-hydrogen peroxide bath is dilute and reduces worker safety hazards.

After bright dipping and thorough rinsing in cold running water, stain or tarnish may be removed by dipping in a cyanide solution. Some proprietary bright dips, incorporated with corrosion inhibitors, eliminate the use of toxic cyanides and chromates and are easy to waste treat.

When a semibright finish is satisfactory, a dichromate color dip is less expensive and more convenient to use than the conventional acid dip. Color dip should not be used if parts are to be plated or soldered. The following conditions for color dipping include a solution that removes red copper oxide and imparts a film that resists discoloration during storage and work:

- Sodium dichromate . . . 30 to 90 g/L (4 to 12 oz/gal)
- Sulfuric acid, 1.83 sp gr . . . . .5 to 10 vol %
- Water. . . . . rem
- Immersion time. . . . . 30 s
- Temperature of solution. . . . .Room temperature

Extruded yellow brass rod may be pickled, usually in dilute sulfuric acid, to remove light oxide prior to drawing, forging, or machining. Yellow brass forgings are similarly treated for the removal of oxide scale and forging lubricant. Hydrochloric acid solutions may be used instead of sulfuric acid in some applications. Bright dipping follows pickling to complete the removal of all oxide. The yellow brass is then given a color dip to produce a uniform cartridge-brass color. Although brass sand castings are seldom pickled, a solution similar to that used for forgings may be used when pickling is indicated.

A typical acid treatment cycle for copper-base materials is given in the following list. The cycle may be terminated after any water rinse if the desired finish and color have been obtained.

- Pickling
- Cold water rinse
- Scale dip or bright dip
- Cold water rinse
- Cold water rinse
- Color dip
- Cold water rinse
- Cold water rinse
- Hot water rinse
- Air-blast dry

Aluminum bronzes form a tough, adherent aluminum oxide film during hot fabrication. This film can be removed or loosened by the following strong alkaline solution:

- Sodium hydroxide, 10 wt %
- Water, rem
- Temperature of solution, 75 °C (170 °F)
- Immersion time, 2 to 6 min

After this treatment the material can be treated in acid solutions by some of the same cycles used for other copper-base materials.

Alloys containing silicon may form oxides of silicon that are removable only by hydrofluoric acid. Proprietary fluorine-bearing compounds are also available for this purpose. If a dull brown-to-gray appearance is not objectionable, the material may be pickled in the conventional sulfuric acid solution to remove the copper oxides. If a brighter finish is required, one of the solutions in Table 3 may be used.

Alloys containing beryllium that have been heat treated at relatively low temperatures (below 400 °C or 750 °F) can be pickled or bright dipped as any other copper alloy; however, alloys containing beryllium that have been heat treated at temperatures above 400 °C (750 °F), in operations such as solution annealing, and those which have been bright annealed, will generally possess a surface oxide that contains beryllium oxide as a major constituent. This oxide can be difficult to remove if present in a thickness greater than 0.04 to 0.05 μm (1.6 to 2.0 μin.). A 1- or 2-min soak in a solution of 50% sodium hydroxide at 130 °C (265 °F), before acid pickling, facilitates removal of any beryllium oxide.

Nickel silvers and cupro-nickel alloys do not respond readily to the pickling solution usually used for brasses, because nickel oxide has a limited solubility in sulfuric acid. Heavy scaling of these alloys should be avoided during annealing by using a reducing atmosphere. For example, the annealing of 18% nickel silver in a rich reducing atmosphere results in a slight tarnish that is easily removed in the sulfuric acid pickle and dichromate solutions ordinarily used for descaling brass. Controlling the atmosphere during annealing produces a bright metal finish.

Tubing made of 30% cupro-nickel may be annealed in a reducing atmosphere, but not in a brightening atmosphere, to produce a clean surface that does not require acid treatment; 18% nickel silver wire is pretreated in a proprietary hot alkaline cleaning solution and then annealed in a controlled atmosphere furnace to produce a clean and bright surface. The wire is subsequently pickled in sulfuric acid and treated in a dichromate solution to remove zinc sweat. The wire is then finish pickled in 10 to 15% sulfuric acid solution at 60 °C (140 °F).

Table 1 Pickling conditions for copper-based materials

Acid condition	Amount, vol %	35% hydrogen peroxide	Water	Temperature of solution	Immersion time, min
Sulfuric acid(a)	15-20	3-5 vol %	rem	Room temperature to 60 °C (140 °F)	¼-5
Hydrochloric acid(b)	40-90	. . .	rem	Room temperature	1-3

(a) 1.83 sp gr. The bath needs additives to stabilize peroxide and accelerators to maintain etch rate. Proprietary products are available from metal finishing suppliers. (b) 1.16 sp gr

Table 2 Scale dip and bright dip conditions for copper-base metals

These solutions remove scale that is not removed by sulfuric or hydrochloric acid solutions; lower concentrations of nitric acid and higher concentrations of sulfuric acid produce a bright lustrous finish; these solutions can remove 0.0255 mm (0.001 in.) of metal and should not be used when close dimensional tolerances must be maintained.

Solution	Sulfuric acid, vol %(a)	Nitric acid, vol %(b)	Hydrochloric acid(c) g/L	Hydrochloric acid(c) oz/gal	Water, vol %	Temperature of solution	Immersion time, s
Scale dip							
Solution A	0	50	4	½	50	Room temperature	15-60
Solution B	25-35	35-50	4	½	35-40	Room temperature	15-60
Bright dip solution	50-60	15-25	4	½	rem	Room temperature	5-45

(a) 1.83 sp gr. (b) 1.41 sp gr. (c) 1.16 sp gr, excess hydrochloric acid spots brass. Wood soot and activated charcoal are added to the solution to prevent this condition

Tarnish Removal. Tarnish, the surface discoloration formed on copper-base materials during exposure to the atmosphere or to alkaline cleaning, usually consists of a thin film of oxide or sulfide. One of the most commonly used dips for removing tarnish is an aqueous solution of 8 to 60 g/L (1 to 8 oz/gal) of sodium cyanide. The metal to be cleaned is immersed for 1 to 2 min in the cyanide solution at room temperature. Steel tanks are used to contain the solution. Thorough rinsing is required after the dip treatment.

Extreme safety precautions must be followed when cyanide solutions are being used, because cyanide is highly poisonous. To prevent the formation of ammonia, metal previously cleaned in an alkaline solution must be thoroughly rinsed before it is brought into contact with the cyanide solution. Cyanide must not come in contact with acid because lethal hydrocyanic acid is produced. Despite the strict safety precautions that are necessary, cyanide solutions are in common use.

A solution containing 5 to 10 vol % hydrochloric or sulfuric acid may also be used to remove tarnish from copper-base materials. Immersion time is a few seconds in either of these solutions at room temperature. Thorough rinsing in water is required after the treatment. The hydrochloric acid solution may be contained in a vitrified crock, or in rubber-lined or glass-lined tanks. A vitrified crock or lead-lined tank may be used as a container for the sulfuric acid solution. Plastics such as polypropylene or polyvinyl chloride are also suitable containers for hydrochloric and sulfuric acid solutions. These materials are resistant to 50% sulfuric acid and 37% hydrochloric acid up to 65 to 70 °C (150 to 160 °F).

Another noncyanide solution sometimes used for removing tarnish consists of 10% citric acid at 70 to 80 °C (160 to 175 °F). This solution removes most tarnish stains and poses no health or ecological hazards.

Pickling Equipment. Equipment requirements for the automatic pickling of brass forgings of various sizes are given in Table 4. The automatic pickling machine is similar to automatic plating equipment. Solution conditions for automatic pickling of brass forgings are as follows:

- Hydrochloric acid solution at room temperature
- Nitric-sulfuric acid solution, 38 °C (100 °F) maximum
- Sodium dichromate solution, room temperature
- Cold water rinse, 2300 L/h (600 gal/h)
- Hot water rinse, 66 to 82 °C (150 to 180 °F), 4 L/h (1 gal/h)

Compositions of pickling tanks are determined by the type of solution used, such as:

- Hydrochloric acid solution, plastic-lined steel
- Nitric-sulfuric acid solution, type 316 stainless steel
- Color dip, dichromate solution, plastic-lined steel
- Cold water rinse, four tanks, steel
- Hot water rinse, steel or rubber-lined steel

Work baskets require type 316 stainless steel or plastic composition with dimensions of 330 mm (13 in.) by 635 mm (25 in.) by 155 mm (6 in.) and 610 mm (24 in.) by 205 mm (8 in.) by 180 mm (7 in.). Equipment requirements for pickling brass tubes or rods and the equipment for pickling brass tubing are indicated in Table 5.

Tanks for sulfuric acid may be lined with natural rubber. Tanks intended to contain nitric and hydrofluoric acids may be lined with polyvinyl chloride or polypropylene. Additional information on materials and construction of acid tanks is given in the article on mineral acid cleaning in this Volume.

Defects from Pickling. By far the most common defect encountered in pickling brass is the presence of red stain on the metal. Caused by cuprous oxide, this defect may be removed in a dichromate or ferric sulfate pickle or in a sulfuric acid-hydrogen peroxide solution. Red stains may result from the presence of metallic iron in the pickling solution, causing copper to plate out on the parts when the parts come in contact with iron or steel. This is because iron is anodic to copper. Care must be taken to prevent any stray pieces of iron, such as nails or tools, from entering the pickling solution. Red stains can also result from the reaction of sulfuric acid on cuprous oxide (Cu_2O), forming cupric oxide (CuO) and copper. This can be removed in any of the above oxidizing acids.

Oil and lubricants remaining on formed metal must be removed before pickling and dipping. If formed material is to be bright dipped, it should first be degreased, or the lubricant remaining will prevent the bright dip from being effective. If the work is to be annealed and pickled, it should be degreased, usually in a hot water rinse, because the oil may cause excess cuprous oxide and unwanted red staining to form during annealing. In extreme instances, the more volatile constituents of the oil burn off readily, leaving a carbonaceous deposit that must be removed mechanically.

Acid stains will appear on the metal if it is not thoroughly rinsed and cleansed of all remaining acid after pickling or dipping. At least two separate washings in water or one thorough running water wash should be used before drying. Dipping in a soap solution or a buffered salt solution will neutralize traces of acid.

The metal is etched or pitted if it is immersed for too long in a pickling solution or dip, or if the solution is too strong or too hot. The proper time, temperature, and concentration of solu-

Table 3 Pickling conditions for copper alloys containing silicon

Constituent or condition	Solution A(a)	Solution B(a)
Sulfuric acid(b)	5-15 vol %	40-50 vol %
Hydrofluoric acid(c)	½-15 vol %	½-5 vol %
Nitric acid(d)	...	15-20 vol %
Water	rem	rem
Immersion time	½-10 min	5-45 s

(a) Solution at room temperature. (b) 1.83 sp gr. (c) 52%. (d) 1.41 sp gr

Table 4 Production and operating requirements for automatic pickling of brass forgings

Item or condition	Size of forgings Small	Medium	Large
Number of pieces per hour	2500	500	250
Pieces per basket	250	50	25
Pounds pickled per hour	2500	2500	2500
Configuration of forgings	Irregular	Irregular(a)	Irregular(a)
Immersion time	40 s	40 s	40 s
Total cycle time	6 min	6 min	6 min

(a) May contain cavities; require hand loading into baskets

Table 5 Equipment requirements for pickling brass tubes or rods

Oxide scale is removed from 25-mm (1-in.) diam brass tubing and rods which measure 1.9 m (6 ft) to 21 m (70 ft) in length; immersion time varies from 5 to 30 min; the source of heat is steam, 105 000 kJ (100 000 Btu)

Tanks(a)	Material	Solution	Temperature °C	Temperature °F	Overflow(b)
Acid tank	Stainless steel	10% sulfuric acid	38-60	100-140	...
Cold water rinse tank	Wood	...	...	...	380 L/h (100 gal/h)
Hot water rinse tank	Wood	...	54-71	130-160	38 L/h (10 gal/h)

(a) All tanks measure 1.1 m (3.5 ft) by 0.9 m (3 ft) by 21 m (70 ft) and hold 6738 L (1780 gal) for a 0.6 m (2 ft) depth of solution. (b) Connected to bottom of tank

tion may be determined quickly during trial runs.

Safety Practices. Acids, even in dilute solution, can cause serious injuries to the eyes and other portions of the body. Operators should wear face shields, rubber boots, and rubber aprons for protection. Eye fountains and showers adjacent to acid tanks, for use in the event of an accident, should be provided. Adequate ventilation and suitable hoods for the tanks are recommended. Additional information concerning hazards in the use of acids is given in the article on pickling of iron and steel found in this Volume.

Extreme caution should be exercised when adding acid to a water solution, especially if sulfuric acid is used. Sulfuric acid should be added slowly and only while stirring the solution.

Abrasive Blast Cleaning

Abrasive cleaning is used to remove molding, core sand, and investment material from the exterior and interior surfaces of copper-base castings. Selection of the proper kind and particle size of grit determines the type and color of the finish. The coarser grits clean faster but give a rougher finish.

Dry abrasive cleaning of beryllium copper is usually confined to castings. Steel shot is used for general cleaning to remove sand and slight surface imperfections from the casting after mold shakeout. Sands are used to blend in surface areas, to remove heat treat scale and to produce a uniform surface texture. Graded bronze chips, together with the regular commercial abrasives, are used in some applications to impart a better color and finish. Abrasive blast cleaning is seldom used to produce decorative finishes on copper alloys.

Wet blasting offers a means of cleaning previously blasted and machined surfaces without damaging the finished or threaded areas. Wet blasting produces various degrees of satin finish. The process is ideal for removing oxide film acquired during brazing, soldering, welding, or heat treating and for removing smudges, stains, and finger marks. An example is bronze castings that have been machined and brazed. These are wet blasted with quartz (140-grit) for ½ to 5 min to remove braze discoloration and shop dirt. The parts are normally degreased before wet blasting. After the castings are wet blasted, they are cleaned ultrasonically, inspected, and assembled.

Surfaces cleaned by wet blasting are uniform in appearance, although their color is not the same as original grit-blasted surfaces because of the abrasive used. Cleaning action is gentle but effective because water is the carrier.

Mass Finishing

Mass finishing is best suited for stamped, formed, or machined parts. Castings with remnants of gates and parting lines, forgings with heavy scale, flash lines or die marks, and heavily burred, pitted, or dented parts are not well suited for mass finishing. Light burrs can sometimes be removed by a prior bright dip, after which tumbling may be used for radius blending, polishing, and burnishing. High thin burrs of soft alloys are likely to peen over. Mass finishing of soft alloys at excessive speeds with insufficient amounts of solution can result in roughened and indented surfaces. Dry tumbling is generally restricted to small parts of simple shape and maximum dimension less than 50 mm (2 in.).

Abrasives. Aluminum oxide, silicon carbide, limestone, and flintstone are the abrasive materials most often used in mass finishing of copper and copper alloys. Combinations of these abrasives may be used for specific applications. For example, a blend of aluminum oxide and silicon carbide of mesh size 46 to 150 produces a reasonably fine matte surface on parts with heavy burrs. Aluminum oxide has a cutting action, and silicon carbide has a planing action.

Preshaped abrasives of various sizes and shapes also may be used. Although more expensive than material in its natural form, preshaped abrasives permit a more constant and uniform polishing action, are more effective in holes and recesses, and retain their effective cutting shape for a longer period of use.

Compounds. Parts heavily coated with grease or oil or contaminated with dirt or chips should be degreased before mass finishing, preferably in a separate barrel, dip tank, or degreaser. A better practice is to clean parts in the barrel when progressing from rough to finishing cycles and to bright dip before burnishing. A sulfuric-nitric acid or sulfuric acid-hydrogen peroxide bright dip should be used if plating follows burnishing.

Soft water and neutral compounds are preferred for mass finishing copper and copper alloys. The use of liquid soap-free alkaline compounds for mass finishing highly leaded, free-cutting brasses prevents the formation of lead soaps that impair the effectiveness of the operation.

Compounds are classified according to the following categories and are purchased as proprietary materials:

- *Cleaning compounds:* possessing high detergency and buffering action, to remove oils, greases and residues
- *Descaling compounds:* used to remove tarnish from copper alloys. Neutralizing cycles usually follow the use of these compounds
- *Grinding compounds:* used with abrasive mediums for softening the water, saponifying oils, and keeping chips clean. These compounds inhibit tarnish and improve the color of parts
- *Abrasive compounds:* containing grits such as aluminum oxide, silicon carbide, emery, quartz sand, or pumice

Table 6 Operating conditions for mass finishing

Material	Medium	Size of abrasive particles mm	in.	Tumbling time, h	Finish
Heavy cutting					
Brass or bronze castings	Aluminum oxide	6.4-19	0.25-0.75	6-16	Matte
Moderate cutting					
Brass stampings	Aluminum oxide	6.4-19	0.25-0.75	1-6	Light matte
Brass screw-machine parts	Aluminum oxide or granite	6.4-19	0.25-0.75	½-6	Light matte or bright
Light cutting(a)					
Brass stampings or screws(b)	Limestone	3.2-13	0.13-0.50	2-6	Bright

(a) Submerged tumbling is used for fragile and precision parts. (b) Screw-machine parts

Table 7 Operating conditions for bright rolling or water rolling

Parts	Weight of load kg	lb	Additive(a)	Water L	gal	Speed, rev/min	Temperature of solution	Time, min	Finish
Blanks or buttons, copper(b)	363-408	800-900	Cream of tartar	95	25	32	Room temperature	20-60	Bright
Gripper post, nickel over brass	136	300	Proprietary(c)	95	25	32	Room temperature	20-60	Bright
Fragile tubular part, brass(d)	18	40	Proprietary	20	5	7	Room temperature	20-60	Bright

(a) 9 g/L (1 oz/gal). (b) Removal of burrs. (c) Additive containing sodium bicarbonate, cream of tartar, sodium acid pyrohosphate, sodium sulfite, and a wetting agent. (d) Wall thickness is 0.150 mm (0.006 in.)

Surface Finishes. Although mass finishing produces the final finish for many parts, it is used more extensively for cleaning prior to plating and painting or for deburring and polishing before a final finish is applied. Examples of mass finishing applications are given in Table 6.

Bright rolling or water rolling in a barrel is an economical bulk method of finishing small parts. The finishes may be dull, semibright, or of high luster. This operation consists of tumbling the parts in water containing a suitable additive. The water acts as the carrier for the fine burnishing materials. Selection of additive and cycle time controls the surface roughness, stock removal, color, and reflective luster. Examples of water rolling applications are given in Table 7.

A prime consideration for successful water rolling is the cleanness of the parts. Oil, grease, scale, and dirt should be removed before rolling for luster. Basic cleaning and pickling operating conditions used before water rolling are given in Table 8.

For economy, cleaning and pickling should be done in the same barrel by the same operator. For example, oily gripper parts from eyelet machines are placed in an oblique stainless steel barrel and cleaned by a suitable alkaline cleaner. After rinsing, the parts are pickled in a 1 to 2% sulfuric acid solution and rinsed. Clear water and additives, burnishing compounds, are added, and the parts are rolled until the desired finish is obtained. If the work appears to darken during water rolling, the medium should be dumped, the barrel and parts rinsed and then charged with a fresh burnishing compound. Parts suitable for water rolling to a bright finish include posts, sockets, studs, tack buttons, and zipper parts.

Table 8 Operating conditions for cleaning and pickling in water rolling barrels

Barrel is made of Type 304 stainless steel

Solutions	Concentration	Temperature °C	°F	Speed of rotation, rev/min	Time cycle(a), min
Alkaline cleaner(b)	15-20 g/L (2-3 oz/gal)	71	160	7-32	15-20
Pickling solution(c)	1-2%	54	130	7-32	15-20

(a) Rinsed thoroughly in hot water after cycle is finished. (b) Sodium hydroxide or proprietary compound. (c) H_2SO_4

Vibratory finishing is particularly effective for deburring, forming radii, descaling, and removing flash from castings and molded parts. It may also be used for burnishing. Vibratory finishers deburr parts 50 to 75% faster and are more versatile than rotary tumbling barrels. There is no cascading of parts, with the attendant possibility of damage by impact. The process is adaptable to both light castings and formed parts. Vibratory finishing is also effective on internal surfaces and recesses that are not usually worked by rotary tumbling.

The mediums for vibratory finishing of copper and copper alloys are similar to those used for rotary tumbling. Dry mediums are used occasionally, but usually a liquid is added to provide lubrication, suspension of worn-off particles, and a more gentle cleaning action. The selection of medium is frequently by trial and error. Conditions for vibratory finishing are given in Table 9.

Polishing and Buffing

Copper alloy parts are polished after scale removal and dressing or rough cutting, but before final finishing operations, which include buffing, burnishing, or honing. Rough castings normally require two polishing operations before buffing. Forgings and stampings require one polishing operation before buffing. Pipe, tubing, and some stampings can be buffed without previous polishing. Buffing is not required when a brushed or satin finish is desired as the final finish.

Because copper-based materials are softer than steel, fewer stages of successively finer polishing are required to achieve a uniformly fine surface finish.

Table 9 Operating conditions for vibratory finishing

Part sizes	Material	Medium	Parts per load	Ratio of medium to parts(a)	Vibrations per min	Time, min	Purpose
25-mm (1-in.) diam, machined	Nickel, silver	Steel pins and quartz sand	650	4 to 1	1500	40	Deburr
38-mm (1½-in.) diam, machined	Brass	Steel wire brads	500	2 to 1	1300	60	Deburr and form radii
75-mm (3-in.) diam, machined	Brass	Arkansas stone	150	10 to 1	1500	35	Deburr and finish

(a) 0.03 m^3 (1 ft^3) bowl

Table 10 Offhand belt polishing and wheel buffing operations for sand cast red brass parts

Sand cast lavatory fittings made of red brass are finished in a sequence of six operations in preparation for decorative chromium plating; sequential finishing of spout with flat surfaces

Operation	Type of contact wheel	Wheel speed, rev/min	Pieces per hour	Type of abrasive belt	Belt life, pieces	Polishing lubricant or buffing compound
Rough polishing	Cloth(a)	2100	23	80-grit silicon carbide	29	None or light application of grease stick
Final polishing	Cloth(a)	2100	30	220-grit Al_2O_3	49	Grease stick
Spot polishing	Cloth(a)	2100	46	220-grit A_2O_3	77	Grease stick
General buffing	Spiral-sewn, treated cloth sections with intermediate airway(b)	2400	32	...	...	Tripoli
Spot buffing	Spiral-sewn, treated cloth sections with intermediate airway(b)	2400	115	...	...	Tripoli
Color buffing	Spiral-sewn, treated cloth sections with intermediate airway(b)	1700	75	...	...	Silica compound

(a) 355-mm (14-in.) diam, 45-mm (1¾-in.) width, 90 density. (b) 355-mm (14-in.) diam, 60.3-mm (2⅜-in.) width, 18 ply, 86/93.

For many parts, especially those having machined surfaces or those free of defects, a single-stage polishing operation using 180- to 200-grit abrasive on a lubricated belt or setup wheel may be all that is required before buffing. Poor quality surfaces require preliminary rough polishing on a dry belt or wheel with 80- to 120-grit abrasive. Surfaces of intermediate quality may be given a first-stage polishing with 120- to 160-grit abrasive, either dry or lubricated. Belt polishing is generally advantageous for high-production finishing except when special shapes are processed. These are best handled by the contoured faces of setup polishing wheels.

Buffing of copper and copper alloys is usually accomplished with standard sectional cloth wheels operating at moderate speeds of 1200 to 1800 rev/min. Typical wheel speeds for various finishes are as follows.

- 915 to 1675 m/min (3000 to 5500 ft/min) for a dull finish, using 120- to 200-grit aluminum oxide
- 1220 to 1830 m/min (4000 to 6000 ft/min) for a satin finish
- 1675 to 2135 m/min (5500 to 7000 ft/min) for cutting and coloring
- 2135 to 2440 m/min (7000 to 8000 ft/min) for a high luster, using tripoli, lime, and silica with no free grease binder

When it is necessary to mush a buff to the contour of a complicated part, buffing speeds may range between 200 and 1000 rev/min.

Neutral compounds that are free of sulfur must be used to avoid staining in the plating operation when polishing and buffing precede electroplating. Excessively high temperatures during polishing and buffing may cause difficulties in subsequent cleaning and plating operations. When flawless chromium-plated surfaces are required, it is necessary both to buff and color buff the polished copper alloy surfaces before plating. Chromium reproduces all imperfections in the underlying plating or base metal and, because chromium is hard and has a high melting point, it is more resistant to flow and is not readily buffed by normal methods. A good chromium-plated surface can be obtained without the color buff operation, by only polishing and cut-down buffing. An example of offhand belt polishing and wheel buffing operations is given in Table 10.

Scratch brushing is used (*a*) to produce a contrasting surface adjacent to a bright reflective surface; (*b*) to produce an uneven surface for better paint adherence; (*c*) to remove metal during final finishing of parts with intricate recesses that are inaccessible to polishing and buffing wheels; and (*d*) to remove impacted soil and buffing compounds from previous finishing operations, prior to subsequent finishing in some applications.

Various types of scratching mediums are used to produce different finishes, such as butler, satin, directional, sunburst or circular, and matte. These mediums are as follows:

- Wire wheels are used on copper or brass grill work to clean intricate recesses, holes, or ribbed areas and to produce a decorative noncontinuous scratch pattern on ornamental parts such as vases, lamps, and kitchenware
- Emery cloth or paper is a common medium for producing a series of linear or circular parallel lines on flat objects with no sudden changes in contour. This type of decorative finish is applied to fireplace accessories, automotive hub caps, and kitchenware and appliances
- Polishing wheels headed with greaseless compounds produce scratch-brush finishes with parallelism of

Table 11 Suggested sequence of operations for scratch brushing copper alloy parts

Part	Finish desired	Abrasive	Type of wheel	Size of wheel mm	Size of wheel in.	Speed, rev/min
Black fuse body, yellow brass(a)	Dull, smooth, black	None	Tampico	80-mm diam by 75 mm thick, 5 rows wide	7-in. diam by 3 in. thick, 5 rows wide	1200
Silver-plated red brass lipstick case(b)	Semibright	Solution of soap bark and cream of tartar	Nickel-silver wire, 0.100-mm (0.004-in.) diam	150-mm diam by 75 mm thick, 6 rows wide	6-in. diam by 3 in. thick, 6 rows wide	850
Black-on-bronze bookends, highlights relieved(b)	Black background, colored copper highlights, sulfurated potash	Pumice in water	Cloth, sewn sections	180-mm diam by 13 mm wide	7-in. diam by ½ in. wide	850
Silver-plated lipstick cap(c)	Satin	Greaseless rouge, proprietary	Loose cloth wheel	150-mm diam by 51 mm wide	6-in. diam by 2 in. wide	1800
Nickel-plated refrigerator panels or stove parts(d)	Satin	Greaseless compound	Loose cloth wheel	305-mm diam by 50-510 mm wide	12-in. diam by 2-20 in. wide	1800

(a) Clean brush often by running pumice stone across face of wheel. (b) Lacquer after scratch brushing. (c) Lacquer after finishing. (d) Chromium plate after finishing

the directional pattern. Decorative items such as jewelry, building paneling, and built-in refrigerator and stove parts can be finished in this manner

- Soft tampico and manila brushes remove soil from scrollwork and embossed areas on ornate tableware serving sets and jewelry prior to final processing

Table 11 gives the sequence of operations and mediums for scratch-brush finishing several copper alloy products.

Although scratch-brush finishing is useful for producing eye-appealing finishes and as a mechanical means for preparing surfaces for subsequent processing, certain hazards must be recognized. Extreme care and control are required when the part being worked contains patterns with sharp corners or embossments, because the sharpness of detail may be destroyed. In salvage or rework operations, it is difficult and sometimes impossible to blend the original brush pattern into a repaired area from which a defect has been removed by grinding.

Chemical and Electrochemical Cleaning

During fabrication, copper alloys may become coated with lubricating oils, drawing compounds, greases, oxides, dirt, metallic particles, or abrasives. These must be removed by cleaning. The type of cleaning depends on the type of lubricant and other materials to be removed, the equipment available, the environmental restrictions, and the degree of cleanness required. The nature and size of the pieces also influence the selection of equipment or process.

Where permissible, mechanical scrubbing, accomplished by turbulent boiling, pressure spraying, or agitating, aids in the removal of any substances that are not exceptionally adherent. Occasionally, hand brushing may be used for small production quantities.

Lubricants made from animal or vegetable oils or greases, such as tallow, lard oil, palm oil, and olive oil, can usually be removed by saponification. In this process, the parts are immersed in an alkaline solution where the oil reacts with the alkali to form water-soluble soap compounds. Mineral oils that are not saponifiable, such as kerosine, machine oil, cylinder oil, and general lubricating oils, are usually removed from the metal by emulsion cleaning.

Dirt particles, abrasives, metal dust, and inert materials are removed by one or both of these processes. To remove undesirable materials by saponification, emulsification, or similar means, it is necessary to use particular chemicals or combinations of chemicals.

Solvent and Vapor Degreasing. Solvent cleaning of copper alloys involves immersion in special naphthas, such as Stoddard solvent, with flash points over 38 °C (100 °F), for the removal of light grease and light oil. Solvents of this type are preferred to kerosine and to the naphthas used in paints, because less residue remains on the work after the special naphthas have evaporated.

Chlorinated solvents such as trichlorethylene, boiling point 90 °C (189 °F), and perchlorethylene, boiling point 120 °C (250 °F), are sometimes used instead of naphtha. Although chlorinated solvents are less of a fire hazard, they are much more toxic than straight petroleum solvents; however, 1,1,1-trichloroethane has the same range of toxicity as Stoddard solvent or mineral spirits. The use of chlorinated solvents for cleaning brass before annealing may cause staining of the brass.

The straight-chain naphthas are not effective for complete removal of heavy grease, burned-on hydrocarbons, pigmented drawing compounds, and oils containing solid contaminants. Buffing compounds containing tallow, stearic acid, and metallic soaps require cyclic hydrocarbons such as toluol and xylol for effective cleaning. Table 12 gives cycles for solvent cleaning of copper alloys.

Vapor degreasing effectively removes many soils from copper alloys. Stabilized trichlorethylene is used extensively in vapor degreasing because it does not attack copper alloys during degreasing and because it has high solvency for the oils, greases, waxes, tars, lubricants, and coolants in general use in the copper and brass industry. Perchlorethylene is used especially for removing high-melting pitches and

Table 12 Cycles for cleaning copper alloy parts with a solvent cleaner

Part	Solvent cleaner	Temperature of solvent °C	°F	Immersion time, min	Soil removed
Dose cap(a)	Stoddard solvent or mineral spirits	Room temperature to 49	Room temperature to 120	2	Heavy drawing compound
Brass retainer ring(a)	Stoddard solvent or mineral spirits	Room temperature to 49	Room temperature to 120	2	Eyelet machine lubricant
Brass rods	Sawdust, dampened with Stoddard solvent or mineral spirits	Room temperature	Room temperature	5(b)	Mill lubricant

(a) Hand cleaning necessary with fragile parts. (b) Tumbled in barrel

waxes, for drying parts by vaporizing entrapped moisture, and for degreasing thin-gage materials.

Vapor degreasing or solvent cleaning is not effective for removing inert materials and inorganic soils such as metal salts, oxides, or compounds that are not generally soluble in chlorinated solvents. Similarly, vapor degreasing for removal of hard and dry buffing compounds could leave behind the insoluble and hard to clean abrasives. Solvent emulsion cleaners, which although slow, are highly effective for complete removal of buffing compounds.

Solid particles of metal dust or chips that are held on the surface by organic soil can be removed mechanically by the washing action of the solvent. Removal of these particles is accelerated and accomplished better with solvent sprays or by immersion in boiling solvent with the vapor phase of the degreaser.

Emulsion and Alkaline Cleaning. Parts with heavy soils such as machine oils, grease, and buffing compounds are treated first with emulsion cleaners to remove most of the soil. After the parts have been rinsed, the remaining soil is removed by alkaline soak or electrolytic cleaning. Precleaning reduces the contamination of the alkaline solution, extending the life of the solution. Thorough alkaline cleaning must follow the emulsion cleaning cycle before the subsequent acid cycles. Extreme caution must be exercised to avoid dragging emulsifiers through the rinses and into a plating solution, especially an acid solution.

Emulsion cleaning may be accomplished by soaking the work for 3 min or less in a mildly agitated solution. Spraying is helpful only when all surfaces being cleaned can be thoroughly contacted. The thin film of oil remaining after emulsion cleaning acts as a temporary tarnish preventive. Some parts may be stored after drying, depending on the composition of the solution and the metal being cleaned; however, brass may become pitted by prolonged exposure to certain emulsifier films. When emulsion-cleaned zinc-bearing brass parts are cleaned electrolytically by alkalines, sufficient time should be allowed for the dispersal of the emulsifier in the alkaline solution, to avoid pitting. Because of the high cost for solvents and the severe restrictions on their waste disposal, the industry is now using alkaline cleaners wherever possible.

Alkaline cleaners for copper-based materials contain one or more of the compounds listed in Table 13. Solutions for soak cleaning usually contain 30 to 60 g/L (4 to 8 oz/gal) of cleaner and are operated at 60 to 88 °C (140 to 190 °F). For every 6 °C (10 °F) rise in temperature above 60 °C (140 °F), the cleaning time is reduced by about 25%. Regardless of the concentration of the cleaning solution, there is a practical limit to the amount of contamination a given volume can accommodate without redepositing soil on the metal. In this condition, the solution should be discarded even though analysis reveals unused cleaner.

Dissolved air in fresh cleaning solutions is frequently the cause of tarnishing of copper alloys. The air can be eliminated by heating the solution to the boiling point for about ½ h before use. This procedure may not be advisable with every proprietary cleaner because of constituents that break down at the boiling temperature. Some uninhibited alkaline cleaners may also cause a slight darkening or tarnishing of the work surface. The darkening may be removed by dipping in dilute hydrochloric acid or in cyanide solution.

Electrolytic alkaline cleaning is the most reliable method for cleaning parts for plating. The work is the cathode, and steel electrodes are the anodes. Reverse-current anodic cleaning cannot be used for more than a few seconds because copper dissolves in the solution. Copper alloys will tarnish readily during exposure to the oxygen that is released at the anode, but this may be minimized by the addition of inhibitors. In many electroplating operations, anodic cleaning for a few seconds is used to develop tarnish, because it indicates that all soil has been removed. The small amount of metal dissolved by anodic cleaning exposes a more active surface for electroplating, and the light tarnish formed is readily dissolved by a mild hydrochloric or sulfuric acid solution. Current density during anodic electrocleaning of brass should be about 3 A/dm^2 (30 A/ft^2) at 3 to 4 V. Use of high currents will etch the brass and cause dezincification.

Copper-based materials are electrolytically cleaned by cathodic cleaning followed by short time anodic cleaning, or by soak cleaning followed by anodic cleaning. Positively charged particles plated onto the work during cathodic cleaning cause smut, which may cause blistering and poor adhesion of the plated metal if not completely removed. To avoid this condition, a short period of anodic cleaning should follow the cathodic electrocleaning.

Racks and baskets that are used during the application of chromate chemical conversion coatings should not be used in alkaline cleaning solutions. Contamination of the cleaning solution with as little as 10 ppm of hexavalent chromium can cause poor adhesion and blistering of subsequent metal deposits. To avoid the effects of accidental contamination when chromate treatments are used, sodium hydrosulfite should be added to the cleaning solution at the rate of 14 g (½ oz) per 380 L (100 gal). The sodium hydrosulfite reduces hexavalent chromium to trivalent chromium, which does not affect subsequent plating. Periodic additions are required because trivalent chromium is reoxidized to hexavalent by the oxygen released at the anode.

Ultrasonic cleaning of copper-based materials is used when the size of particles remaining on the surface is less than 5 to 10 μm (0.2 to 0.4 mils) in any one direction and for removing dirt and chips that cause smudge, as indicated by a white cloth wipe test. Remove much of the surface contamination by other more economical cleaning procedures before ultrasonic cleaning. This increases the life of the ultrasonic cleaning solution and maintains production efficiency. After ultrasonic cleaning, the parts may be rinsed in deionized or distilled water and dried with warm filtered air or in a vacuum oven.

Ultrasonic cleaning is used with alkaline cleaning solutions, solvents, vapor degreasing solutions, or acid pickling solutions, to increase the rate of cleaning or to complete the removal of soil from areas not completely cleaned by soak or spray procedures. An example of the need for ultrasonic assistance is the cleaning of fine internal threads on brass parts. An alkaline solution containing 15 to 30 g/L (2 to 4 oz/gal) of cleaner may be used for ultrasonic cleaning of such parts.

Preparation for Plating

Before copper alloys are plated, surface oxides are removed and the surfaces chemically activated for adherence of the plate. Heavy oxide scales are usually removed at the mill, so descaling is not normally a part of the preplate treatment. Figures 1 and 2 show sequences of operations required for preparing the surfaces of copper alloys. These operations apply to both lead-free and leaded materials and to soft-soldered assemblies. The subsequent plating operation is assumed to be deposited copper. For many plating operations, where the copper alloy is relatively clean, free of oils, and uncontaminated by buffing compounds, the process can be simplified by eliminating the alkaline soak. If the surface contains buffing compounds, it may require precleaning by soaking in emulsion cleaners. Many plating processes do not use the anodic alkaline clean, especially when an acid dip is used after the cathodic alkaline clean. The acid dip suffices to remove any smut or stain. An acid dip may be simply 15% sulfuric acid or, if the surface needs additional pickling, a bright dip may be used. There are a number of bright dips for copper alloys on the market based on the sulfuric acid-hydrogen peroxide system, (see Table 1), in addition to the example in Fig. 2. A chromate bright dip should not be used before plating because of possible chromium contamination of the plating bath which could lead to blistering, pitting, and peeling problems of the electrodeposit.

Table 13 Compounds used for formulating alkaline cleaners

Component	Soak cleaners, %	Electrolytic cleaners, %
Sodium hydroxide	10-20	10-15
Sodium polyphosphates	5-20	5-20
Sodium orthosilicate, sesquisilicate, metasilicate	30-50	30-50
Sodium carbonate, bicarbonate	0-25	0-25
Resin-type soaps	5-10	None
Organic emulsifiers, wetting agents, chelating agents	2-10	1-3

Fig. 1 Surface preparation required for removal of buffing compounds and tarnish before plating of copper alloys

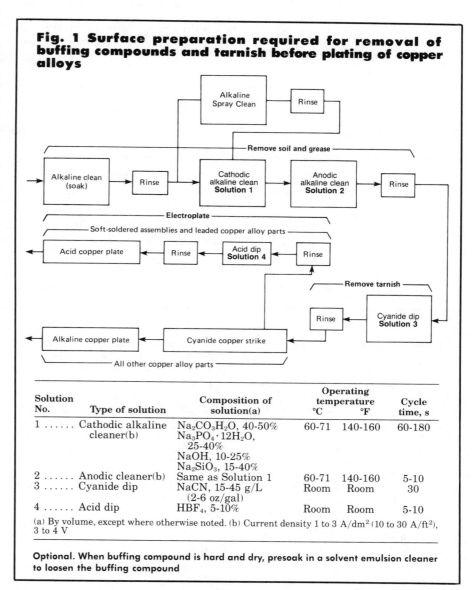

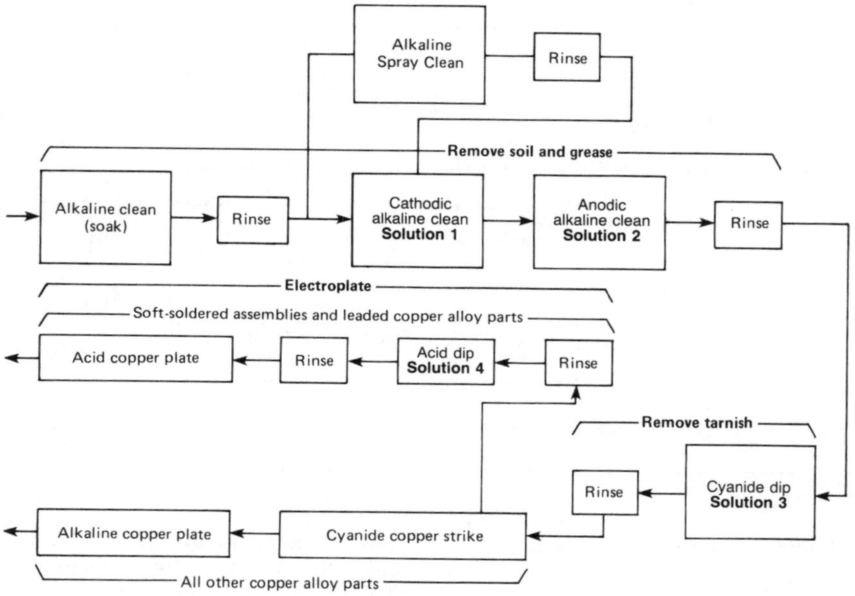

Solution No.	Type of solution	Composition of solution(a)	Operating temperature °C	°F	Cycle time, s
1	Cathodic alkaline cleaner(b)	$Na_2CO_3H_2O$, 40-50% $Na_3PO_4 \cdot 12H_2O$, 25-40% NaOH, 10-25% Na_2SiO_3, 15-40%	60-71	140-160	60-180
2	Anodic cleaner(b)	Same as Solution 1	60-71	140-160	5-10
3	Cyanide dip	NaCN, 15-45 g/L (2-6 oz/gal)	Room	Room	30
4	Acid dip	HBF_4, 5-10%	Room	Room	5-10

(a) By volume, except where otherwise noted. (b) Current density 1 to 3 A/dm^2 (10 to 30 A/ft^2), 3 to 4 V

Optional. When buffing compound is hard and dry, presoak in a solvent emulsion cleaner to loosen the buffing compound

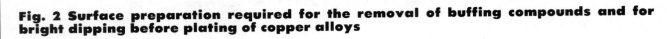

Fig. 2 Surface preparation required for the removal of buffing compounds and for bright dipping before plating of copper alloys

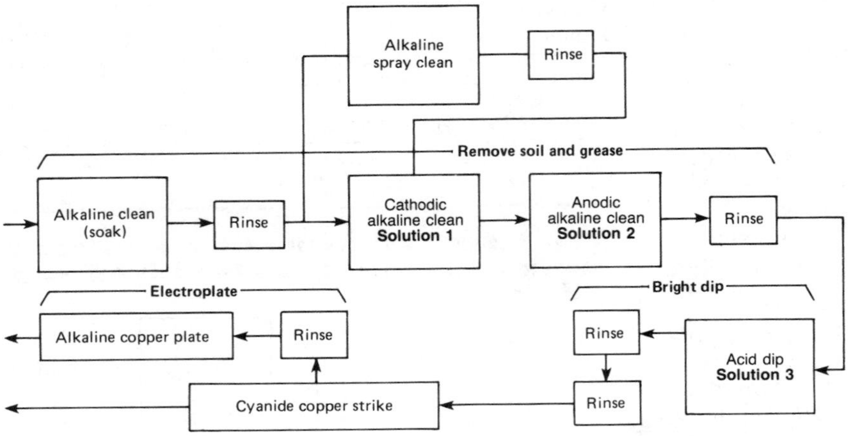

Solution No.	Type of solution	Composition of solution(a)	Operating temperature °C	°F	Cycle time, s
1	Cathodic alkaline cleaner(b)	$Na_2CO_3 \cdot H_2O$, 40-50% $Na_3PO_4 \cdot 12H_2O$, 25-40% NaOH, 10-25% Surface active agent, 1%	60-71	140-160	60-180
2	Anodic cleaner(b)	Same as solution 1	60-71	140-160	5-10
3	Bright dip(c)	H_2SO_4, 65-75% HNO_3, 20-35% HCl, 1.1 g/L (1/8 oz/gal) H_2O, 5-10%	Cool	Cool	5-10

(a) By volume, except where noted. (b) Current density, 1 to 3 A/dm^2 (10 to 30 A/ft^2), 3 to 4 V. (c) Constant agitation of work

Oxides can be removed from cast or heat treated copper alloys by either abrasive or chemical cleaning procedures. Castings always require abrasive cleaning; however, they do not necessarily require subsequent pickling or bright dipping for the removal of oxides. If acid treatment is required, thorough rinsing is mandatory to prevent spotting of the plating by entrapped acid. Stampings and drawn parts can be treated in the same manner as castings; however, it may be less expensive to pickle the parts before abrasive finishing if heavy oxides are present. Screw-machine parts made of leaded copper alloys usually require no treatment for oxide removal before plating, but abrasive cleaning can be used when required.

Beryllium Copper. Surface preparation procedures for beryllium copper are the same as for other copper alloys, unless the beryllium copper has been heat treated and still contains surface oxides resulting from that heat treatment. Such oxides on beryllium copper are more difficult to remove than normal oxides are on other copper alloys. They can be best removed in strong, hot caustic solutions followed by a good bright dip. The procedure outlined in Fig. 2 should be sufficient if the bright dip cycle is increased to 10 to 15 s. The sulfuric-peroxide bright dips also work well on beryllium copper. The alkaline soak clean and the anodic alkaline clean may or may not be required, depending on the degree of surface contamination resulting from heat treatment. In rare cases, the plater is required to remove scale resulting from high temperature, >400 °C (750 °F), heat treatments such as in solution annealing. The parts must be soaked in a 50% solution of sodium hydroxide at 130 °C (265 °F) for 1 or 2 min before acid pickling in a good bright dip.

Powder Metallurgy Parts. In the 1970's, technology has advanced to a level where parts sintered from metal powders are not as porous and can be easily electroplated. A sequence of operations for surface preparation before electroplating is shown in Fig. 3. Because powder metallurgy (P/M) parts are made from different alloys and because porosity varies, a suitable surface preparation and electroplating technique must be developed for each part.

The methods applicable to porous cast parts can be used for low-porosity P/M parts with a density more than 95% of theoretical. P/M strip stock bonded to steel is an example of low-porosity material. The surface preparation and plating techniques vary for parts with higher porosity with a density 85 to 90% of theoretical.

During plating of P/M parts, the pores act as thermal pumps. Solutions are released from or absorbed by the pores, depending on whether the solution is hotter or colder than the part. Deep and interconnecting pores are likely to entrap cleaning or rinsing solutions, which are released slowly during plating. Porosity must be determined before cleaning, and suitable techniques must be used to prevent

Fig. 3 Surface preparation for brass P/M parts to be plated

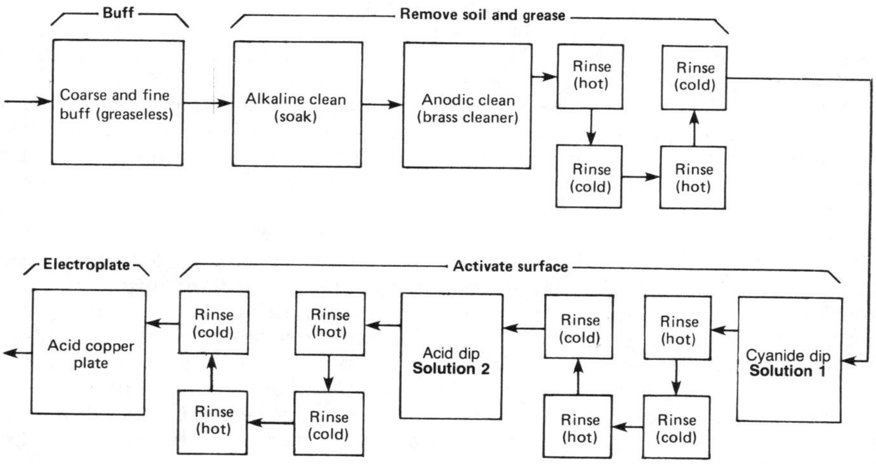

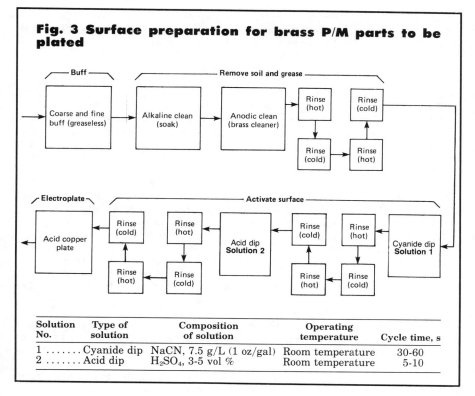

Solution No.	Type of solution	Composition of solution	Operating temperature	Cycle time, s
1	Cyanide dip	NaCN, 7.5 g/L (1 oz/gal)	Room temperature	30-60
2	Acid dip	H_2SO_4, 3-5 vol %	Room temperature	5-10

solution entrapment. Entrapped solutions may cause precipitation of metal compounds on the work surface during plating. The porous nature of P/M parts causes contamination and depletion of all solutions and necessitates that the solutions be closely controlled.

Spotty plating is caused by entrapped solutions that seep out of the pores and stain the plated surface as it dries. Spots may not develop for several days after plating or may appear to be insignificant immediately after plating and subsequently enlarge with time.

A continuous plated coating can seldom be attained on P/M parts unless the pores are closed. Methods for closing the pores include buffing, rolling, burnishing, heat treating, and impregnation. Buffing, rolling, or burnishing are used when appearance is important and dimensional tolerances are not critical. These techniques can be applied either before or after plating. Heat treating is one of the most satisfactory methods for closing pores when surface quality and size are more important than appearance. Occasionally, green compacts are electroplated and then sintered. Another technique is to sinter fully, then to plate and heat treat. This procedure requires washing and acid dipping the parts to neutralize alkalinity prior to heat treating.

Impregnation with copper, lead, tin, waxes, resins, or oils is another method for closing the pores of P/M parts before plating. Thermosetting polyester styrene and silicone resins are suitable impregnating materials. When the polyester-styrene resin is used, the parts are degreased, baked to remove moisture, then impregnated under vacuum and pressure. An emulsion cleaner can be used to remove excess resin from blind holes and threads prior to curing the resin at 120 to 150 °C (250 to 300 °F). Tumbling or polishing can be used to remove excess cured resin.

When silicone resin is used, the surfaces of the powder compact are coated with a thin water-repellent film. Then the compact is heated in air to 205 °C (400 °F), quenched in a solution of 4% silicone (type 200) and 96% perchlorethylene, and baked again for ½ to 2 h. After being pickled in acid, the part can be plated in the conventional manner, preferably in an alkaline bath.

Before being impregnated with oil, the parts are heated in a vapor degreaser and, while hot, immersed in cold oil. After cooling in the oil, the parts are removed, drained of oil, and then rinsed in cold solvent. Parts impregnated with oil have been coated with copper by conventional electroplating processes with resultant satisfactory bonding of the coating.

Electroless Plating

Electroless plating involves the reduction of a metal salt to its metallic state by electrons supplied from the simultaneous oxidation of a reducing agent. The electroless plating solutions are compounded so that the concentration of the reducing salts, metal salts, buffering salts, and the pH controls the rate of reduction of the metal salt and of the oxidation of the reducing agent. If these reactions are not controlled, the metal deposit is reduced spontaneously to a fine powder. The metal to be plated acts as a catalyst and localizes the deposit of the plate to the part itself. Electroless plating completely plates uniform deposits to any thickness desired over intricately shaped parts or assemblies and into deep recesses and long tubular holes. Fine and close-fitting threads can be plated uniformly over the entire threaded area.

On copper-based materials, the process is limited to the plating of nickel, tin, gold, and silver deposits. Solutions for plating these deposits are affected adversely by contaminants, such as cyanides, lead, zinc, manganese, and cadmium. Tin may render the nickel plating solution inoperative; therefore, tin-containing copper alloys must be plated with copper or gold before a final nickel coating is applied.

Because copper is more noble than nickel, it will not act as a catalyst to start the deposition of nickel. This condition is alleviated by contacting the copper workpiece with an active metal such as iron or aluminum that causes the deposition of a thin nickel coating on the workpiece. The nickel coating is of sufficient thickness to continue the oxidation-reduction reaction for further nickel deposition after the activating metal is removed. Another procedure for starting the deposition of nickel is to make the part cathodic and apply a current briefly to the part as it is held in the electroless bath. Electroless nickel plating is particularly useful for coating deep recesses and holes that cannot be adequately electroplated. The reader may refer to the article on electroless nickel plating in this Volume, for further information on the operating conditions of electroless nickel baths.

Electroless plating processes are used to plate copper with gold and silver for decorative purposes. Nickel, gold, and silver are frequently applied to copper electronic components to prevent tarnishing during subsequent processing and to aid in the soldering of semiconductors. Electroless tin plating has been used for copper tubes to prevent corrosion by carbonated waters.

Immersion Plating

Immersion plating, sometimes called galvanic plating, depends on the position that the base metal occupies in the electromotive series with respect to the metal to be deposited from solution. Plating occurs when the metal from a dissolved metal salt is displaced by a more active, less noble metal that is immersed in the solution. As the depositing metal is displaced from the bath, metal from the workpiece dissolves in the solution and becomes a contaminant in the bath. Depleted baths are never replenished, but are replaced with fresh solution.

When immersion plating copper-based materials, only those metals more noble than copper can be displaced from solution, thereby limiting this process to metals such as gold and silver. Immersion-plated deposits are thin, usually in the range of 0.050 to 0.50 μm (0.002 to 0.02 mil).

Compositions and operating temperatures of solutions for immersion plating of gold and silver are:

Gold plating

- Potassium gold cyanide ...... 4 g/L (½ oz/gal)
- Potassium or sodium cyanide ...... 25 g/L (3½ oz/gal)
- Sodium carbonate ...... 30 g/L (4 oz/gal)
- Temperature of solution .... 60 to 80 °C (140 to 180 °F)

Silver plating

- Silver cyanide .... 7.5 g/L (1 oz/gal)
- Sodium or potassium cyanide .... 15 g/L (2 oz/gal)
- Temperature of solution .... 18 to 38 °C (65 to 100 °F)

Electroplating

Metals such as gold, silver, rhodium, nickel, chromium, tin, and cadmium are electrodeposited on copper and copper alloys. Copper electroplating with these metals is used primarily for decorative purposes and for preventing tarnish or corrosion.

Chromium is occasionally plated directly onto copper and copper alloys as a low-cost decorative coating and color match. A plated nickel undercoat is applied when quality plating of chromium is required, to produce color depth and resistance to corrosion, abrasion, and dezincification.

Decorative chromium deposits are characteristically thin and porous. Galvanic action between the chromium and base metal may occur during exposure to aggressive environments and result in accelerated corrosion. With copper alloys of high zinc content, dezincification and eventual lifting and flaking of the deposit may result. For added protection, chromium deposits plated directly on the base metal are coated with clear lacquers or with the more durable combination coatings of silicone and acrylic thermosetting organic resins.

Decorative chromium deposits are usually 0.25 to 0.50 μm (0.01 to 0.02 mil) thick. When chromium is deposited directly onto the base metal, the thickness of the deposit is seldom less than 38 μm (1.5 mils), and for many applications a minimum of 50 μm (2 mils) is specified. Additional information can be found in the article on decorative chromium plating in this Volume.

Hard chromium deposits are used primarily to improve wear resistance and friction characteristics. Because most of the copper-based materials are soft and ductile, the basis metal lacks sufficient hardness to support thick deposits of hard chromium against heavy pressures without the risk of scoring or flaking. Hard chromium is electroplated from solutions under conditions similar to those used for decorative chromium plating; however, the deposits are much thicker, frequently dull in appearance, and generally deposited directly on the basis metal. To conform to exact dimensional tolerances and to avoid buildup of deposit at corners, an excess of deposit is plated and the final dimension is attained by grinding or lapping. Further information can be found in the article on hard chromium plating in this Volume.

Cadmium. Copper-based materials are easily electroplated with cadmium from conventional plating baths. A bath has the following composition:

- Cadmium oxide ...... 22 to 33 g/L (3.0 to 4.5 oz/gal)
- Sodium cyanide ..... 85 to 110 g/L (11.5 to 15.0 oz/gal)
- Sodium hydroxide .... 16 to 24 g/L (2.1 to 3.2 oz/gal)
- Brightening agent ..... as required

Usually, applying an undercoat prior to the deposition of cadmium is unnecessary. The principal reason for cadmium plating is to minimize galvanic corrosion between copper-based materials and other cadmium-plated metals in the same assembly. Cadmium-plated copper parts are used in aircraft, marine, and military applications involving atmospheric exposure.

Gold is usually applied over an electroplated undercoating of nickel or silver, but may be electroplated directly on the copper alloy. In the electronics industry, nickel is plated to copper before gold plating. The nickel barrier stops migration of copper into the gold layer. The gold thickness is about 2.5 μm (0.1 mil). The nickel barrier is normally a minimum of 1.3 μm (0.05 mil) thick; however, exposure of the gold-plated part to elevated temperatures for extended periods of time may require a nickel barrier 127 μm (5 mils) thick. If silver is the barrier, government specifications for electronic applications usually require a minimum thickness of 5 μm (0.2 mil) of silver and 1.3 μm (0.05 mil) of gold.

When gold is plated directly on copper alloy, a deposit of 3 μm (0.1 mil) or more is needed because of rapid diffusion, especially at elevated temperatures, which destroys the electrical and decorative properties of the deposit. Total coverage and freedom from porosity are necessary for corrosion protection.

For greatest adhesion, the part should be plated with a gold strike before the final deposit of gold. One satisfactory strike solution consists of potassium gold cyanide 0.8 g/L (0.1 troy oz/gal as gold) and potassium cyanide 45 g/L (6 oz/gal). To avoid tarnishing, the struck part should be transferred immediately to the plating solution with the current flowing as the work enters the solution. An alternate procedure is to rinse the struck work, then acid dip, rinse, and plate with a suitable current density.

Current density for racked parts varies from 0.2 to 1 A/dm^2 (2 to 10 A/ft^2), depending on the solution used. The rate of deposition depends on the solution and varies from 100 mg/A · min (0.3 A/dm^2 or 3 A/ft^2 for 14.4 min deposits 3 μm or 0.1 mil of gold) in most cyanide and neutral solutions to one third of this rate in most acid solutions. For barrel plating, the limiting current density for all solutions is about one third of that used in rack plating.

Cyanide solutions are the most efficient and have the best throwing power, whereas deposits from acid and neutral solutions are less porous and are not stained by incompletely removed solution.

Patented solutions for plating gold contain 0.03 kg (1 troy oz) of gold per gallon, usually added as potassium gold cyanide, and proprietary brightening agents. Insoluble anodes are used with these solutions. The patent-free hot gold cyanide solution is extremely sensitive to impurities, and the deposits lack brightness. The solution is limited in application. Gold metal anodes are used with this solution.

The proprietary solutions are brightened or colored by codeposited base metals, especially silver, nickel, cobalt, and copper. These base metals may be classed as impurities for some applications. When gold deposits of less than 0.33 μm (0.013 mil) are used to avoid the federal excise tax on a plated item, the deposit should be protected by lacquer or a chromate passivation.

Nickel is plated onto copper-based materials for decorative purposes and as an undercoating for increased corrosion resistance of nickel-chromium coating systems. Because nickel deposits have a yellow cast and tarnish easily, many nickel-plated parts are finished with electroplated chromium. Costume jewelry, lipstick cases, hardware for doors and windows, automotive hub caps, air valves, and plumbing fixtures are examples of copper-based parts regularly plated with nickel, either as a final coat or an undercoat.

Nickel electrodeposits from some solutions reproduce the irregularities in the basis metal surface to a marked degree. The basis metal must be polished and buffed before plating if a smooth uniform plated surface is required. Polishing and buffing are not required when nickel is to be plated on a reasonably smooth surface, from solutions that result in a high degree of leveling of the deposit as it builds up in thickness. Deposits from these solutions are smooth and bright. Nickel plating with high leveling and brightness can provide an attractive mirror-like finish. Proprietary leveler brighteners are added to the nickel bath to achieve high leveling with nickel deposits as thin as 7.6 to 10 μm (0.3 to 0.4 mil).

Numerous types of nickel electroplating solutions are available. The solutions differ in basic composition, preferred operating conditions, and quality of the deposit. These are described in the article on nickel plating in this Volume.

Most decorative plating is done in solutions that yield bright or semi-bright deposits. Although dull nickel deposits may be buffed to a high luster, the cost of buffing limits dull nickel to nondecorative or functional uses.

Black nickel is a decorative finish only. It should be plated over a deposit or basis metal that is corrosion resistant. Very thin deposits of black nickel are recommended because thick deposits lack the ductility and adhesiveness required to withstand shock.

Rhodium. An undercoat of nickel is used when copper alloys are to be plated with rhodium. For decorative purposes, the thickness of rhodium is usually about 0.25 μm (0.01 mil). Deposits of 25 μm (1 mil) or more are used for functional purposes, but these heavier deposits do not have the brightness or color characteristics of the decorative finishes. Rhodium plating solutions are of the sulfuric acid or phosphoric acid type. Formulas and operating conditions are given in the article on rhodium plating in this Volume.

Silver plating is used for decorative and functional purposes. The important functional applications include:

- High skin conductivity for radio frequencies. A minimum of 3 μm (0.1 mil) of silver is required
- Low resistance for electrical contacts. The thickness of silver ranges from very thin deposits, such as those produced by immersion coating to electroplates 13 μm (0.5 mil) thick
- Antiseizing or antigalling surfaces

Silver can be deposited as a dull plate from cyanide solutions or as a bright plate from baths containing suitable brighteners. Details of electrolyte compositions and operating conditions will be found in the article on silver plating in this Volume. Plating times for the deposition of 25 μm (1 mil) of silver, as a function of current density, are as follows:

- 1 A/dm^2 (10 A/ft^2) .... 38 min
- 2 A/dm^2 (20 A/ft^2) .... 18 min
- 3 A/dm^2 (30 A/ft^2) .... 13 min
- 4 A/dm^2 (40 A/ft^2) .... 9 min

Pollution free, proprietary acid silver sulfate plating baths are now also available.

Tin. As a protective coating, tin is not necessarily anodic to copper alloys. A copper alloy may not always be fully protected against a corrosive environment at the expense of tin. Some corrosion products of tin are more noble than copper and can create the type of galvanic conditions that lead to pitting corrosion. Tin may be plated from alkaline stannate baths or from sulfate or fluoborate acid baths. For compositions and operating conditions of tin electrolytes, see the article on tin plating in this Volume.

Tin-Copper Alloys. The 40 to 50% tin alloy, speculum metal, has been used as a decorative coating because of its silver-like color. The bronze alloy coatings containing 10 to 20% tin are producible to match the color of gold and have been used as decorative deposits for this reason. Information on plating bronze alloys is given in the article on bronze plating in this Volume.

Tin-lead alloys can be plated on copper alloys in almost any proportion from a fluoborate bath. An alloy of 7% tin and 93% lead has been used for corrosion resistance, especially to sulfuric and chromic acids. The 60% tin, 40% lead eutectic alloy has excellent solderability and good electrical properties and corrosion resistance. The eutectic alloy may be applied also by hot dipping, but control of thickness is as difficult as with hot dipping in pure tin.

For compositions of solutions for plating 7% tin, 93% lead alloy and 60% tin, 40% lead alloy (solder), see Table 14.

Chemical and Electrolytic Polishing

Chemical or electrolytic polishing imparts a smooth bright finish and removes burrs selectively from certain areas. Usually the same solution will accomplish both purposes, although variations in operating conditions may be desirable, depending on which feature is required.

For electropolishing, the parts are racked and made the anode in an acid solution. Direct current is applied and the process selectively removes more metal from the high points of the surface than from the low. This results in a high polish and the removal of burrs.

Chemical polishing imparts a finish similar to that obtained by electropolishing, but electropolished surfaces are usually smoother and brighter.

Table 14 Bath compositions and operating conditions for plating tin-lead alloys

Conditions for operating include a cathode current density of 3 A/dm² (30 A/ft²); anode to cathode ratio is 2 to 1; solution has maximum pH of 0.2 and is mildly agitated at a temperature of 16 to 38 °C (60 to 100 °F); air agitation cannot be used because it will oxidize the tin

Constituent	Total tin g/L	Total tin oz/gal	Stannous tin g/L	Stannous tin oz/gal	Lead g/L	Lead oz/gal	Free fluoboric acid g/L	Free fluoboric acid oz/gal	Free boric acid g/L	Free boric acid oz/gal	Peptone g/L	Peptone oz/gal	Anode composition
7 Sn, 93 Pb	7.0	0.94	6.0	0.80	88.4	11.8	100.0	13.4	25	3.4	0.50	0.067	7 Sn, 93 Pb
60 Sn, 40 Pb	60.0	8.0	55.0	7.4	25.0	3.4	100.0	13.4	25	3.4	5.0	0.67	63 Sn, 40 Pb

Electropolishing for 4 to 6 min will reduce the surface roughness to between one third and one half the original micron (micro-inch) value. Electropolishing and chemical polishing will remove minor scratches and defects, as well as burrs.

Selection of these processes is based on the following considerations:

- Part configuration, complex enough so that mechanical polishing would be expensive and difficult
- Difficulty in removing entrapped buffing or polishing compounds. Copper-based materials that are polished either electrolytically or chemically do not require any additional cleaning operations other than rinsing
- Parts racked for electropolishing may be electroplated in the same racks, which results in considerable savings in handling costs
- Electrolytically or chemically polished surfaces tarnish less readily than mechanically polished surfaces. Some chemical polishing baths can offer passivation and prevent copper from tarnishing under storage
- Parts that might deform during mechanical polishing or tumbling will not be deformed by electrolytic or chemical polishing
- Mechanical polishing may result in a smeared skin that differs from the base metal and does not accept plating uniformly. Electroplate on electropolished base metal has adhesion superior to that on mechanically finished surfaces

Parts that are electropolished include fishing lures, plumbing fixtures, furniture arms and legs, lamps and lighting fixtures, wire goods, brass appliance parts, jewelry, nameplates, and bezels. Copper-plated parts, such as automobile tail pipes, die castings that have not been buffed, and high-altitude oxygen bottles are electropolished effectively for an improved finish. The thickness of copper plate should be in excess of 20 μm (0.8 mil) before electropolishing. Most of the commercial solutions used for electropolishing copper alloys contain phosphoric acid with additional agents. For further information, see the article on electroplating and electropolishing in this Volume.

A sequence of operations for electropolishing consists of :

- Racking the parts
- Cleaning to remove oil, grease, dirt, and oxide
- Double rinsing in water at room temperature
- Electroplating for 2 to 4 min at about 65 °C (150 °F), using 30 A/dm² (300 A/ft²)
- Double rinsing in water at room temperature and drying

If the parts are to be electroplated after electropolishing, the drying operation in the above sequence is omitted and the process continues with anodic alkaline cleaning until the surface is light brown in color. Then the parts are rinsed in water at room temperature, dipped in 5% sulfuric acid solution, double rinsed in water at room temperature, immersed in the cyanide copper strike solution, and plated.

The quality of an electrolytically or chemically polished surface is affected by the following factors:

- Composition of material: most copper alloys are suitable except those containing appreciable amounts of lead, such as the free-machining alloys. Materials successfully treated include copper, beryllium copper, single-phase bronzes, and single-phase nonleaded brasses
- Condition of the initial surface
- Grain size of the metal: the smaller the grain size the better the finish. A grain size of 0.004 mm (0.0002 in.) yields excellent results; satisfactory results are produced with grain sizes up to about 0.01 mm (0.0004 in.).

Passivation

Passivation refers to the process of forming a protective film on metal. The blue-green patina of copper developed during atmospheric exposure of copper alloys is a protective coating that is aesthetically pleasing. The patina may be artificially produced or accelerated by a solution, developed by the Copper and Brass Research Association, having the formulation:

- Ammonium sulfate ....2.7 kg (6 lb)
- Copper sulfate.......... 85 g (3 oz)
- Ammonia (technical grade, 0.90 sp gr)... 39.6 cm³ (1.34 fl oz)
- Water............... 25 L (6.5 gal)
- Total solution.........27 L (7.3 gal)

A fine spray of the solution should be applied to a chemically clean surface. The film should be permitted to dry before the part is sprayed a second time. Five or six repetitions of the spraying and drying sequence are required. The color begins to develop in about 6 h and at first is somewhat bluer than natural patina. A more attractive color develops as the surface is exposed to natural weathering.

Small copper parts may be coated with an imitation patina by dipping or brushing a solution consisting of the following proportions by weight:

- Copper........................30
- Nitric acid, concentrated 60
- Acetic acid, 6% 600
- Ammonium chloride 11
- Ammonium hydroxide, (technical grade, 0.90 sp gr)....20

When preparing the solution, copper is dissolved in the nitric acid and as soon as the action ceases, the remaining three constituents are added. The solution stands several days before use. Parts treated with this solution are coated with linseed oil.

Coloring

Copper-based materials may be surface treated to produce a variety of colors, ranging from dark reds to black. The final color depends on base metal composition, solution composition, immersion time, and operator skill. Coloring is primarily an art, and practical experience is necessary to develop the skill required to produce uniform finishes consistently. Copper alloys are colored chemically to enhance the appearance of a product, to provide an undercoating for subsequent organic finishes as with brass, and to reduce light reflection in optical systems. Chemical coloring produces a thin layer of a compound on the surface of the basis metal. This layer retains some of the characteristics of the metal surface prior to coloring, such as smooth and lustrous or dull.

The procedures for artificial coloring of metals utilize many of the reactions that occur more slowly under natural conditions. Some colored films not found in nature can be produced artificially. Additional color combinations, such as oxidized and highlighted finishes, can be produced by successive chemical and mechanical operations, to emphasize or to remove partially the chemically colored film.

Coloring copper alloys is essentially a process for coloring copper, because zinc and tin compounds are colorless. These constituents and their concentrations greatly affect many of the chemical reactions and color tones of the coatings formed. A copper content of less than 85% is required to produce a good blue-black finish on brass by an ammoniacal copper sulfate or ammoniacal copper carbonate blackening or blue dip solution. Other solutions are more suitable for coloring high-copper alloys.

After machining and mechanical surface preparation have been completed, the parts should be thoroughly cleaned to remove dirt, oil, grease, and oxide films. Cleaning is important for the development of a uniform film in chemical coloring. The cleaning and deoxidizing procedures should be selected so that the structure of the metal at the surface undergoes a minimum of undesirable change. Acid dipping or bright dipping using nitric-sulfuric acid solution may be necessary to remove oxides and to activate the surface for chemical coloring. A certain amount of trial and error is usually required to establish the most suitable techniques for surface preparation.

Coloring Solutions

Many types of chemical solutions are used for coloring copper alloys. The following formulations and conditions are commonly used in commercial applications to produce the colors indicated.

Alloys with 85% or More Copper

Dark red

- Molten potassium nitrate, 650 to 700 °C (1200 to 1300 °F)
- Immersion time, up to 20 s
- Hot water quench
- Parts must be lacquered

Black

- Solution A
  Liquid sulfur .... 28 g (1 oz)
  or
  Sulfurated potash .... 57 g (2 oz)
  Ammonium hydroxide, sp gr 0.89 .... 7 g (¼ oz)
  Water .... 4 L (1 gal)
  Temperature of solution ..... Room temperature
  Produces a dull black finish
  Reddish bronze to dark brown finishes are obtainable by dry scratch brushing with a fine wire or cloth wheel
- Solution B
  Sulfurated potash .... 1.9 g/L (¼ oz/gal) of water
  Solution strength should be adjusted to blacken the part approximately 1 min
  Too rapid formation of coloring film can result in a nonadherent and brittle film
- Solution C
  Potassium sulfide .... 3.7 to 7.5 g/L (½ to 1 oz/gal) of water
  Immersion time .... Up to 10 s
  Lacquer protection required
- Other solutions
  Several proprietary processes are available for producing a satisfactory black finish
  Alloys blackened by these materials include silicon bronzes, beryllium coppers, bronzes containing up to 8% tin, phosphor bronzes of all types, and brasses, leaded or unleaded, with zinc contents up to 35%

Steel black

- Arsenious oxide, white arsenic .... 113 g (4 oz)
- Hydrochloric acid, sp gr 1.16 .... 240 cm^3 (8 fl oz)
- Water .... 4 L (1 gal)
- Temperature of solution .... approx 82 °C (180 °F)
- Immerse parts in the solution until a uniform color is obtained
- Scratch brush while wet, then dry and lacquer

Black anodizing

- Sodium hydroxide ..... 45 g (16 oz)
- Water .... 4 L (1 gal)
- Temperature .... 82 to 99 °C (180 to 210 °F)
- Current density .... 0.2 to 1 A/dm^2 (2 to 10 A/ft^2)
- Anode-to-cathode ratio .... 1 to 1
- Voltage .... 6 V
- Cathodes .... Steel, carbon, or graphite
- Anodizing time .... 45 s to 3.75 min
- Tank material .... Steel
- Adequate ventilation is required
- After anodizing, the parts are washed in hot and cold water, rinsed in hot water, dried, buffed lightly with a soft cloth wheel, and lacquered, if desired

Reddish bronze to dark brown, statuary bronze

- Sulfurated potash ...... 57 g (2 oz)
- Sodium hydroxide ...... 85 g (3 oz)
- Water .... 4 L (1 gal)
- Temperature .... 77 °C (170 °F)
- Immersion time depends on final color desired
- Parts are usually scratch brushed with a fine wire wheel
- Lacquering is required

Verde antique

- Solution A
  Copper nitrate .... 113 g (4 oz)
  Ammonium chloride .. 113 g (40 oz)
  Calcium chloride ...... 113 g (4 oz)
  Water .... 4 L (1 gal)
- Solution B
  Acetic acid .... 2 L (½ gal)
  Ammonium chloride ... 570 g (20 oz)
  Sodium chloride .... 200 g (7 oz)
  Cream of tartar .... 200 g (7 oz)
  Copper acetate .... 200 g (7 oz)
  Water .... 2 L (½ gal)

Verde antique finishes are known also as patina. They are stippled on brass or copper and dried. Parts are made of copper or copper plated and are usually treated in a sulfide solution to produce a black base color, which results in a dark background. The use of sodium salts in the verde antique solution results in a yellowish color, while ammonium salts impart a bluish cast.

Stippling can be repeated, and when the antique green color appears, immersion in boiling water will produce several different color effects. Other color effects are obtained by using some dry colors such as light and dark chrome green, burnt and raw sienna, burnt and raw umber, ivory drop white, and drop black, or Indian red. After coloring, the surface should be lacquered or waxed. A semiglossy appearance of the lacquered surface may be produced by brushing with paraffin, beeswax, or carnuba wax on a goats-hair brush rotated at about 750 rev/min.

Light brown

- Barium sulfide ....16 g (½ oz)
- Ammonium carbonate....7 g (¼ oz)
- Water.................. 4 L (1 gal)
- Temperature ... Room temperature

Brown

- Potassium chlorate....155 g (5½ oz)
- Nickel sulfate ....78 g (2¾ oz)
- Copper sulfate........ 680 g (24 oz)
- Water.................. 4 L (1 gal)
- Temperature .... 90 to 100 °C (195 to 212 °F)

Alloys with Less than 85% Copper

Black

- Solution A
  A suitable number of brass parts is placed in oblique tumbling barrel made of stainless steel
  Parts are covered with 11 to 19 L (3 to 5 gal) water. 85 g (3 oz) copper sulfate and 170 g (6 oz) sodium thiosulfate are dissolved in warm water and added to the contents of the barrel
  Parts are tumbled for 15 to 30 min to obtain finish
  Solution is drained from barrel
  Parts are washed thoroughly in clean water
  Parts are removed from barrel, dried in sawdust or air blasted
  Parts may be lacquered
- Solution B
  Copper carbonate ....16 g (½ oz)
  Ammonium hydroxide, sp gr 0.89 .... 113 g (4 oz)
  Sodium carbonate ....7 g (¼ oz)
  Water.................. 4 L (1 gal)
  Temperature ....88 to 93 °C (190 to 200 °F)

Statuary bronze

- Copper carbonate ....16 g (½ oz)
- Ammonium hydroxide, sp gr 0.89 .... 113 g (4 oz)
- Sodium carbonate ...... 7 g (¼ oz)
- Water.................. 4 L (1 gal)
- Temperature .... 88 to 93 °C (190 to 200 °F)
- Immerse the parts into the hot solution for 10 s
- Rinse in cold water and dip in solution of dilute sulfuric acid
- Rinse in hot and then cold water
- Clean with soft cloth or sawdust
- Coat with clear lacquer

Blue black

- Copper carbonate .... 0.45 kg (1 lb)
- Ammonium hydroxide, sp gr 0.89 .... 1 L (1 qt)
- Water.................... 3 L (3 qt)
- Temperature .... 54 to 79 °C (130 to 175 °F)
- Excess copper carbonate should be present
- Proper color should be obtained in 1 min

Brown

- Solution A
  Copper sulfate......... 113 g (4 oz)
  Potassium chlorate .... 227 g (8 oz)
  Water.................. 4 L (1 gal)
- Solution B
  Liquid sulfur .... 28 g (1 oz)
  or
  Sulfurated potash....... 57 g (2 oz)
  Water.................. 4 L (1 gal)
- Immerse parts in Solution A for 1 min
- Without rinsing, immerse parts in Solution B for a short time
- Rinse in cold water
- Repeat dipping operation in both solutions until desired color is obtained
- Rinse work in hot water
- Dry in hot sawdust or with an air blast
- Scratch brush with fine wire wheel and lacquer

Old English finish, light brown

- Solution A
  Liquid sulfur ....14 g (½ oz)
  or
  Sulfurated potash....... 28 g (1 oz)
  Water.................. 4 L (1 gal)
- Solution B
  Copper sulfate.......... 57 g (2 oz)
  Water.................. 4 L (1 gal)
- Immerse parts in Solution A
- Without rinsing, immerse in Solution B
- Rinse in cold water
- Repeat dipping operations until light color is obtained
- For uniform finish, scratch brush and repeat dipping operations until desired color is obtained
- Rinse parts in cold and hot water
- Dry in sawdust, scratch brush on a fine wire wheel, and lacquer

Antique green on brass

- Nickel ammonium sulfate ....227 g (8 oz)
- Sodium thiosulfate .... 227 g (8 oz)
- Water.................. 4 L (1 gal)
- Temperature .... 71 °C (160 °F)

Hardware green on brass

- Ferric nitrate .... 28 g (1 oz)
- Sodium thiosulfate .... 170 g (6 oz)
- Water.................. 4 L (1 gal)
- Temperature .... 71 °C (160 °F)

Brown on brass or copper

- Potassium chlorate .. 155 g (5½ oz)
- Nickel sulfate .... 78 g (2¾ oz)
- Copper sulfate........ 680 g (24 oz)
- Water.................. 4 L (1 gal)
- Temperature .... 91 to 100 °C (195 to 212 °F)

Light brown on brass or copper

- Barium sulfide ....14 g (½ oz)
- Ammonium carbonate....7 g (¼ oz)
- Water .... 4 L (1 gal)
- Color is made more clear by wet scratch brushing and redipping

Post-Treatment. Many chemical films must be scratch brushed to remove excess or loose deposits. In addition, contrast in colors may be obtained by (*a*) relieving by scratch brushing with a slurry of fine pumice, (*b*) hand rubbing with an abrasive paste, (*c*) mass finishing, or (*d*) buffing to remove some or all of the colored film from the highlights. Clear lacquers are usually necessary for adequate life and service of chemical films used as outdoor decorative finishes. Finishes for exposure indoors are often used without additional protection to the conversion coating.

Organic Coatings

Information on the materials and methods commonly used for organic finishing of metal surfaces with clear or pigmented coatings is given in the article on painting in this Volume. Tarnishing or discoloration of copper alloys may be retarded or delayed indefinitely by the application of a lacquer. Lacquers should be selected and applied on the basis of the intended service environment of a given product. For exterior service, a dry film thickness of 38 to 50 μm (1.5 to 2 mils) is recommended. In less severe indoor service, a dry film thickness of about 13 to 18 μm (0.5 to 0.7 mil) performs satisfactorily.

Air-Drying and Thermosetting Lacquers. The performance of thermosetting or heat-cured lacquers is superior to that of air-drying lacquers. The use of the thermosetting types is preferred if ovens are available. Distinction should be made, however, between true thermosetting and forced drying. All lacquers can be force dried after a suitable air-flash period to facilitate handling, but thermosetting materials must be heated to an appropriate temperature, 120 to 205 °C (250 to 400 °F) or higher for 5 to 60 min, to cross-link the polymers present in them and to develop their inherent characteristics.

The catalytic activity of copper is such that essentially complete curing of thermosetting lacquers is obtained at temperatures lower than those required with inert substrates. Consequently, many thermosetting lacquers discolor copper alloys severely when heated to temperatures recommended by their suppliers. Such discoloration can be minimized by curing at lower temperatures.

Resins are chosen on the basis of service requirements and include the following materials:

- Alkyds
- Acrylics
- Cellulosics such as cellulose nitrates, ethyl cellulose, and cellulose acetate butyrates
- Epoxies
- Phenolics
- Polyesters
- Silicones
- Urea and melamine formaldehydes
- Vinyls
- Urethanes

An epoxy, cellulose nitrate, or alkyd resin provides satisfactory protection for an inexpensive container, such as a lipstick container. The use of a heat-resisting melamine provides satisfactory service at moderately elevated temperatures. Silicone formulated for high temperature service provides the best heat resistance, but is a more costly material. If resistance to degradation by weather is important, high quality acrylics provide the best results. General characteristics of these resins are given in the article on painting in this Volume.

Cleaning and Finishing of Magnesium Alloys

By the ASM Committee on Cleaning and Finishing of Magnesium Alloys*

FINISHING systems are applied to magnesium alloys primarily for corrosion resistance, although solderability, wear resistance, and other properties sometimes are also desired.

The selection of a suitable finishing system depends on the service environment, particularly with respect to oxygen (air), moisture, chlorides, and temperature. The following environments are listed in the order of increasing corrosiveness: (*a*) oil immersion, (*b*) indoor, (*c*) rural, (*d*) industrial, and (*e*) marine. Magnesium alloys are afforded increasing protection in the following order: (*a*) bare, (*b*) pickled, (*c*) chromate conversion coated, (*d*) anodized, (*e*) electroplated, and (*f*) coated with organic finishes. Anodic coatings are porous and provide no corrosion protection unless sealed with a paint. Table 1 relates applications of various degrees of severity to finishing systems used.

Because of its high position in the electromotive series, magnesium is more susceptible than other common metals to galvanic corrosion. However, galvanic corrosion of magnesium is possible only if bare magnesium is in intimate contact with another bare metal and if a film of moisture covers the joint. Because the presence of moisture is difficult to control, galvanic corrosion is avoided by separating the two metals with a film of paint, plastic tape, sealant, or a metal more compatible with magnesium (see the article on protection of assemblies in *Metals Handbook*, 9th ed., Vol 2).

Proper and adequate cleaning is an important prerequisite to the successful application of finishing systems to magnesium alloy products. Failure to follow suitable cleaning procedures causes future problems of corrosion and product degradation. Mechanical and chemical cleaning methods are used singly or in combination, depending on the specific application and product involved. Cleaning processes of either type require suitable control to ensure repetitive reliability.

Mechanical Cleaning

Mechanical cleaning of magnesium alloy products is accomplished by grinding and rough polishing, dry or wet abrasive blast cleaning, and wire brushing.

Grinding and Rough Polishing. Grinding with belts, wheels, and rotary files is used for cleaning sand castings. Belt grinding, however, usually is used as a finishing operation for the removal of flash and surface imperfections from die castings, or of die marks and scratches from significant surfaces of extrusions. No great danger of surface contamination exists, and no special abrasive or belt backings are necessary.

The possibility of fire resulting from the polishing of magnesium may be virtually eliminated by following the recommendations set forth by the producers of magnesium. These recommendations are summarized in the section of this article on health and safety.

Dry Abrasive Blast Cleaning. Sand blasting is the method of dry abrasive blast cleaning most frequently used on magnesium alloys. Many foundries use flint silica sand with a fineness of 25 or 35 AFS. Occasionally, however, steel grit is used. Usually, castings are blasted immediately after shakeout to reveal any major surface defects. After

*Keith Ball, Manufacturing Engineer Manager, Brooks & Perkins; Donald J. Levy, Senior Staff Scientist, Lockheed Palo Alto Research Laboratory; John F. Pashak, Project Leader, Dow Chemical Company

gates, sprues, and risers are sawed off and other operations are performed to prepare the castings for shipment, a final abrasive blast with sand or steel grit is given just before pickling. Under conventional operating conditions, the two types of abrasives yield equally satisfactory results; but if excessive blast pressure is used, steel grit is more likely to cause surface corrosion, because of the embedding of steel particles in the magnesium surface.

Dry blasting with any abrasive is followed by acid pickling to remove the harmful effects of blasting. Table 2 describes two pickling baths, aqueous solutions of sulfuric and of nitric and sulfuric acids, frequently used for this purpose, and indicates conditions for their use.

Wet abrasive blast cleaning of magnesium alloys is used for (*a*) final finishing before electroplating, (*b*) producing a matte surface before chemical treatments, (*c*) removing carbonaceous matter or heavy corrosion products, and (*d*) removing residual paint after stripping operations. Equipment, abrasives, and process controls are the same as those used for many other materials (see the article on abrasive blast cleaning in this Volume). Remove any contamination by acid pickling, using the sulfuric or nitric-sulfuric baths described in Table 2.

Wire Brushing. Magnesium alloy sheet is wire brushed for in-process cleaning and for the removal of oxides before arc or resistance welding. The wire brushing machine should be adjusted as to the amount of pressure exerted on the sheet. The machine should be designed so that it does not gouge the sheet. The final surface smoothness depends also on the coarseness and composition of the wire. Safety precautions should be observed when wire brushing magnesium as indicated in the section of this article on health and safety.

Table 1 Finishing systems applied to magnesium for various applications

Systems for specific parts	
Cameras, die cast	Dichromate, perchlorethylene dryer, bake 2 h at 150 °C (300 °F), zinc chromate air-dry primer, baked enamel(a)
Chain saws	Dichromate, zinc chromate primer, baked alkyd paint
Dockboards	No finish applied
Drive couplings	No finish applied
Foundry flasks	No finish applied
Furniture	Bright buff with steel wool, bright finish, clear enamel
Garden shears, die cast	Buff, clear baked enamel
Hand trucks	No finish applied
Luggage exteriors	Chrome pickle, thermally bonded vinyl sheet
Luggage frames	Nitric acid prepickle, 13-25 μm (0.5-1 mil), ferric nitrate pickle, 8 μm (0.3 mil), sodium carbonate bleach, clear baked enamel
Missile exteriors	System A: dichromate, vinyl butyrate zinc chromate wash primer, baked phenolic epoxy varnish, 2 coats, enamel, 2 System B: chrome pickle or anodizing depending on alloy, phenolic epoxy paint, 2 coats
Photoengraving plates(b)	System A: zinc immersion coating, flash copper plate, 9.4 μm (0.36 mil), matte-finish hard chromium plate, 3.8 μm (0.15 mil) System B: zinc immersion coating, copper plate, 75 μm (3 mil), chromium plate, 50 μm (2 mil)
Precision inspection equipment	Wax
Stator blades(c)	No finish applied
Tape recorder parts	Chrome pickle, baked epoxy primer, aluminum-pigmented baked epoxy, clear baked epoxy
Satellites	System A: dichromate System B: anodize System C: zinc immersion coating, flash copper plate, 9.1 μm (0.36 mil), silver, 0.0025 μm (0.0001 mil), gold, 0.0025 μm (0.0001 mil)
Systems for indoor exposure	
System A	Vinyl zinc chromate or epoxy calcium chromate primer, lacquer or enamel
System B	Light or heavy anodic coating, epoxy primer, 8-20 μm (0.3-0.8 mil)
System C	Electroplate, 25 μm (1 mil)
Systems for marine or other corrosive exposures	
System A	Light or heavy anodic coating, epoxy primer, 25 μm (1 mil), baked enamel, 1 or 2 coats
System B	Anodize, vinyl wash primer, vinyl or epoxy chromate primer, 2 coats, lacquer or synthetic enamel, 2 or more coats
Systems for outdoor inland exposure	
System A	Vinyl zinc chromate or epoxy chromate primer, 2 coats, compatible synthetic enamel or lacquer, 2 coats
System B	Light anodic coating, epoxy primer, 20 μm (0.7 mil), baked enamel, 8 μm (0.3 mil)
System C	Heavy anodic coating, epoxy primer, 20 μm (0.7 mil)
System D	Electroplate, 50-64 μm (2-2.5 mil)

(a) No primers used under vinyl paints. (b) For wear resistance. (c) For automatic transmissions

Chemical Cleaning

Chemical methods for cleaning magnesium alloys are (*a*) vapor degreasing, (*b*) solvent cleaning, (*c*) emulsion cleaning, (*d*) alkaline cleaning, and (*e*) acid pickling.

Solvent cleaning and vapor degreasing are used to remove oils, forming lubricants, waxes, quenching oils, corrosion-protective oils, polishing and buffing compounds, and other soluble soils and contaminants. Solid particles such as machining dust or chips are removed by the washing action of the solvent as it dissolves the oil or grease that holds the metal fines to the part. These processes must be used before painting, plating, and chemical treatments, as well as before and after machining and forming.

The same methods, equipment, and solvents are used for magnesium as for other metals (see the articles on vapor degreasing and solvent cleaning in this Volume). Trichlorethylene and perchlorethylene are the solvents most often used. Methylene chloride is effective in removing the excess organic-resin compound from the surface of the castings without removing the compound from the pores in the metal.

Emulsion cleaning may be used for the removal of oils and buffing compounds (see the article on emulsion

Table 2 Acid pickling treatments for magnesium alloys

Treatment	Principal applications	Metal removed, μm	Metal removed, mils	Solution: Constituents	Solution: Amount, g/L	Solution: Amount, oz/gal	Solution: Operating temperature, °C	Solution: Operating temperature, °F	Immersion time, min	Tank material or lining
For cast or wrought alloys										
Chromic acid	Remove oxide, flux, corrosion products	None		CrO_3	180	24	21-100 (a)	70-212 (a)	1-15	Stainless steel, 1100 aluminum, lead
Ferric nitrate (b)	Bright finish; maximum corrosion resistance of bare metal; finishing of die castings	8	0.3	CrO_3 $Fe(NO_3)_3 \cdot 9H_2O$ NaF	180 40.0 3.5	24 5.3 0.47	16-38	60-100	1/4-3	Type 316 stainless steel, vinyl, polyethylene
Hydrofluoric acid	Active surface for chemical treatment	3	0.1	50% HF	230	31	21-32	70-90	1/2-5	Type 316 stainless steel, lead, rubber
Nitric acid	Prepickle for ferric nitrate treatment(c)	13-25	0.5-1.0	70% HNO_3	50	6.7	21-32	70-90	1/5-1/2	Stainless steel
For wrought alloys only										
Acetic-nitrate	Remove mill scale; improve corrosion resistance of bare metal	13-25	0.5-1.0	CH_3COOH $NaNO_3$	192 50.0	25.6 6.7	21-27	70-80	1/2-1	3003 aluminum, ceramic, lead
Glycolic-nitrate	Remove mill scale or surface oxides; improve corrosion resistance(d)	12-25	0.5-1.0	70% $CH_2OHCOOH$ 70% HNO_3 $NaNO_3$	230 40 40	31 5.3 5.3	16-49	60-120	1/2-1	Rubber
Chromic-nitrate	Remove mill scale, burned-on graphite; preclean for welding	13	0.5	CrO_3 $NaNO_3$	180 30	24 4	21-32	70-90	3	Stainless steel, lead, rubber, vinyl
Chromic sulfuric	Preclean for spot welding	8	0.3	CrO_3 96% H_2SO_4	180 0.4	24 0.05	21-32	70-90	3	Stainless steel, 1100 aluminum, ceramic, rubber
For cast alloys only										
Nitric sulfuric	Remove effects of blasting from sand castings	50	2.0	70% HNO_3 96% H_2SO_4	77.0 20	10.3 2.7	21-32	70-90	1/6-1/4	Ceramic, rubber, glass
Phosphoric acid	Remove surface segregation from die castings; maximum corrosion resistance of bare metal	13	0.5	85% H_3PO_4	866	116	21-27	70-80	1/2-1	Lead, glass, ceramic, rubber
Sulfuric acid	Remove effects of blasting from sand castings	50	2.0	96% H_2SO_4	30	4	21-32	70-90	1/6-1/4	Ceramic, rubber, lead, glass

(a) For removal of flux, solution must be 88 to 100 °C (190 to 212 °F). (b) For most uniform appearance, die castings must be mechanically finished before being pickled, because the ferric nitrate solution accentuates flow marks and segregation on die-cast surfaces. (c) Use of nitric acid prepickle increases solution life and decreases treatment time in ferric nitrate pickling. (d) Nonvolatile glycolic acid reduces costs compared to acetic acid

cleaning in this Volume). The emulsion cleaner should be neutral or alkaline, with a pH of 7.0 or above, so as not to etch magnesium surfaces. Emulsion cleaners incorporating water with the solvent should be tested before use to avoid possible attack or pitting of the metal.

Alkaline cleaning is the most frequently used method of cleaning magnesium alloys preparatory to painting, chemical treatments, or plating. Alkaline cleaners are also used to remove chromate films from magnesium.

Most magnesium alloys are not attacked by common alkalis except pyrophosphates and some polyphosphates, and even these alkalis do not appreciably attack magnesium above a pH of 12.0. Nearly any heavy-duty alkaline cleaner suitable for low-carbon steel performs satisfactorily on magnesium alloys. The pH of alkaline cleaners for magnesium alloys should be 11.0 or higher.

Soak cleaners are usually based on alkali hydroxides, carbonates, phosphates, and silicates, preferably in combinations of two or more, and also contain natural resinates or synthetic surfactants as emulsifying agents. Soak cleaners are used in concentrations of 30 to 75 g/L (4 to 10 oz/gal) and at 71 to 100 °C (160 to 212 °F). Alkaline cleaners used for spray cleaning cannot use a surface-active wetting agent, because the foaming problem would be too great. In this case, the mechanical force of the spray helps dislodge soils.

Cathodic cleaning uses the work as the cathode in the cleaning solution at approximately 6 V dc. Anodic cleaning is not recommended because of the formation of undesirable oxide or hydroxide films. Pitting of the surface of the magnesium also may result from prolonged anodic cleaning.

A simple aqueous bath for soak or electrolytic cleaning of magnesium alloys is made and used as follows:

Trisodium phosphate ($Na_3PO_4 \cdot 12H_2O$) .... 30 g/L (4 oz/gal)
Sodium carbonate ($Na_2CO_3 \cdot 10H_2O$) .... 30 g/L (4 oz/gal)
Wetting agent .....0.7 g/L (0.1 oz/gal)
Operating temperature ...82 to 100 °C (180 to 212 °F)
Immersion time .... 3 to 10 min

Another formula for soak or electrolytic cleaning, which can be used before electroplating, consists of the following:

Sodium carbonate ($Na_2CO_3 \cdot 10H_2O$) .... 22.5 g/L (3 oz/gal)
Sodium hydroxide (NaOH) ..... 15 g/L (2 oz/gal)
Wetting agent .....0.7 g/L (0.1 oz/gal)
Operating temperature ...82 to 100 °C (180 to 212 °F)
Immersion time .... 3 to 10 min

When either of these baths is used as an electrolytic cleaner, parts are made the cathode at a current density of 1 to 5 A/dm^2 (10 to 50 A/ft^2) at 6 V dc. Many proprietary compounds are better cleaners than the above formulas.

Hard-to-remove soils, such as graphitic lubricants used in the hot forming of magnesium alloy sheet products or in the fabrication of impact extrusions, can best be removed by soaking in this heavy-duty caustic cleaner:

Sodium hydroxide (NaOH) ..... 98 g/L (13 oz/gal)
Wetting agent .....0.7 g/L (0.1 oz/gal)
Operating temperature ...87 to 100 °C (190 to 212 °F)
Immersion time .... 10 to 20 min

A chromic acid pickle, as shown in Table 2, usually is used after cleaning in the above solution. Cleaners containing more than 2% NaOH attack ZK60A; therefore, the above solution should not be used on this alloy.

Acid pickling is required for removal of contamination that is tightly bound to the surface or insoluble in solvents or alkalis. These contaminants include natural oxide tarnish, embedded sand or iron, chromate coatings, welding residues, and burned-on lubricants.

In selecting an acid pickling treatment, consider the type of surface contamination to be removed, the type of magnesium alloy to be treated, and the dimensional loss allowable, as well as the desired surface appearance. Table 2 gives details of acid pickling treatments used for magnesium alloys. Confining these treatments to alloy types indicated avoids the formation of a powdery black smut on pickled surfaces.

When a magnesium alloy product is used bare or with a clear finish, it is desirable to improve the corrosion resistance of the metal surface in addition to providing an attractive appearance. This can be accomplished by the use of ferric nitrate, acetic-nitrate, or phosphoric acid pickle (Table 2). The ferric nitrate pickle deposits an invisible chromium oxide film that passivates the surface and improves corrosion resistance. The acetic-nitrate and phosphoric acid pickles act as sequestrants that effectively remove even invisible traces of other metals from the magnesium surface and thus prevent localized galvanic corrosion. Although all of these methods are effective in improving corrosion resistance, ferric nitrate and phosphoric acid pickles give best results.

Mechanical Finishing

Depending on required appearance, the most frequently used methods of mechanical finishing are barrel tumbling, polishing and buffing, vibratory finishing, wire or fiber brushing, and shot blasting. Shot blasting with 0.10- to 0.25-mm (0.003- to 0.009-in.) glass spheres produces a satin finish. Sand blasting is not recommended because contaminants can be embedded into the magnesium surface.

Mass (Barrel) Finishing. The use of dry barrel finishing is usually limited to the production of a final finish on magnesium alloy parts that are small and relatively thin and for which metal removal must be held to a minimum. Wet barrel finishing is used for deburring, grinding, polishing, burnishing, and coloring. Procedures, materials, and equipment used in wet and dry barrel finishings are discussed in the article on mass finishing in this Volume.

Vibratory finishing combines smoothing and brightening with cleaning to remove dirt, oxides, thin flash, and casting skin. Ceramic and other media are used depending on the surface condition of the part and the finish required. Acid additives etch magnesium surfaces and should never be used. Proper rinsing and drying are important.

Polishing and buffing are not used for final finishing, but for preparing magnesium alloy surfaces for other finishes, such as chemical treatments, anodizing, plating, or painting. As shown in Table 3, the procedures used to produce a highly polished or buffed surface on magnesium alloys are similar to those used for aluminum-based alloys, as is discussed in the article on polishing and buffing in this Volume. However, because most magnesium alloys are appreciably harder than aluminum alloys and tend to drag or tear to a lesser degree, it is unnecessary to use as much lubrication as is usually required with aluminum alloys. When polishing die castings, the metal removal is kept

Table 3 Procedures for polishing and buffing magnesium alloys

Operation	Abrasive grit size	Wheel: Type	Wheel: Diam, mm	Wheel: Diam, in.	Wheel: Speed, m/s	Wheel: Speed, sfm
Rough polishing(a)	60-100(b)	Canvas, felt, sheepskin	150-360	6-14	15.3-25.5	3000-5000
Medium polishing(c)	100-320(b)	Built-up cloth	150-360	6-14	20.4-30.6	4000-6000
Fine polishing	240-400	Cloth or sheepskin	250-360	10-14	22.9-38.2	4500-7500
Satin finishing(d)	50-320(e)	Full-disk buffs(f)	150-300	6-12	15.3-25.5	3000-5000
Buffing(g)	(h)	Cotton buffs	150-410	6-16	20.4-40.8	4000-8000

(a) For rough surfaces only, such as sand castings. (b) Aluminum oxide or silicon carbide; grease stick may be used. (c) For less severe surface condition and before buffing. (d) For final finish or before plating. (e) Greaseless compound. (f) Loose or folded cotton cloth. (g) A cutdown operation to produce a smooth finish with an intermediate degree of brightness. (h) Tripoli or aluminum oxide buffing compound

at a minimum to preserve the relatively thin nonporous outer layer of the casting.

Standard polishing wheels and abrasive belts are used to remove rough surfaces, parting lines, and other surface imperfections from magnesium alloy pieces. Aluminum oxide or silicon carbide abrasives are used, in grit sizes from 60 to 320, depending on the surface roughness and the final finish desired.

Particles of free iron or other heavy metals must not be used in abrasives for the polishing of magnesium alloys, because these metals, when embedded in the surface, can cause cell corrosion effects while parts are in storage or cause a pitting condition in pickling processes used before chemical treatments or electroplating.

Large castings require polishing of significant surfaces before buffing. Fine polishing belts are used on parts with little contour. Parts with more intricate contour are polished on setup wheels or on cloth wheels, using a greaseless compound. Grit sizes from 220 to 320 are most commonly used in polishing operations prior to buffing.

Satin finishing or flexible-wheel polishing is also used on magnesium alloys. This finish is produced by applying a greaseless compound to the surface of a cloth wheel as it is slowly revolved. No lubricant is required.

Magnesium alloys can be buffed to a smooth, bright finish using aluminum oxide or tripoli compounds on a light-count sewed wheel. For color buffing, a dry lime compound may be used. Buffing compounds containing free iron or other heavy metal abrasives should not be used.

A ferric nitrate bright pickle, as shown in Table 2, is used on polished or buffed surfaces prior to the application of a clear organic coating. Using a ferric nitrate bright pickle produces a bright, passive surface that increases the adhesion and corrosion protection of the coating; it also serves to prevent blemishes such as spotting, blooming, and worm tracking in the coating after it has dried.

Precautions that must be exercised in the polishing and buffing of magnesium alloys are discussed in the section of this article on health and safety.

Chemical Finishing Treatments

Chemical finishing treatments for magnesium alloys are used either: (*a*) to provide short-time protection against corrosion and abrasion during shipment or storage, or (*b*) as pretreatments to additional finishing systems.

Figures 1 through 4 present flow charts of processing steps in four frequently used chemical treatments; tabular data accompanying each of these charts indicate operating conditions and applicability of the respective treatments. Note that the chrome pickle and sealed chrome pickle treatments cannot be used on close-tolerance machined parts unless tolerances will permit or allowances for metal removal have been made. The dichromate treatment (Fig. 3) involves no appreciable dimensional changes. Moreover, when a hydrofluoric-sulfuric acid pretreatment and brief exposure to the chrome pickling solution are used, the modified chrome pickle treatment (Fig. 4) involves no appreciable dimensional changes.

Anodic Treatments

The chemical conversion coatings produced on magnesium alloys by anodic treatments provide protection against corrosion, an adhesive base for paint, and a decorative finish. Of the four anodic treating processes described in this article, Chemical Treatment No. 9, Chemical Treatment No. 17, HAE, and Cr-22, all but Chemical Treatment No. 9 provide a hard coating that possesses good abrasion resistance. When anodically treated magnesium alloys are painted, the combination coating offers maximum protection against corrosion under severe conditions, including immersion in fresh water and exposure to outdoor industrial and marine atmospheres.

Surface Preparation. Before conversion coating, all surfaces must be scrupulously clean and free of contamination. Remove oil, grease, and other organic contaminants with suitable solvent cleaners, or with hot alkaline cleaning solutions such as those previously described in the section on chemical cleaning. Scales, oxides, burned-on drawing lubricants, and inorganic corrosion products should be removed in acid pickling solutions of prescribed composition and in accordance with the recommendations given in Table 2.

Parts that have been fabricated by stamping should be deburred to ensure that a coating is not applied to feathered edges that might break off and expose bare metal. Sharp edges and round sharp corners should be made smooth before anodizing. Wherever close dimensional tolerances are entailed, adequate provision for buildup that results from coating should be made.

Chemical Treatment No. 9. A galvanic anodizing treatment in which a source of electric power is not required, Chemical Treatment No. 9 is applied to all forms and alloys of magnesium to produce a protective black coating with good paint-base characteristics. Parts with attachments of other metals may also be treated. Because this process does not result in appreciable dimensional change, the parts are machined to close tolerances before treatment.

Proper galvanic action requires the use of racks, made of stainless steel, Monel, or phosphor bronze. When the workpieces are immersed in the anodizing solution, they are made the anodes, and the tank, if made of low-carbon steel, acts as cathode. If the tank is equipped with a nonmetallic lining, separate steel cathodes must be used.

A processing diagram, together with details of solution compositions and operating conditions for Chemical Treatment No. 9, is presented in Fig. 5.

Chemical Treatment No. 17, which can be applied to all forms and alloys of magnesium, produces a two-phase, two-layer coating. A light green or greenish tan undercoating, about 5.0 μm (0.2 mil) thick, forms at lower voltages. This is covered by a much heavier, second-phase coating about 30.4 μm (1.2 mils) thick, vitreous and dark green. The second-phase coating is relatively brittle and highly abrasion resistant. Formed at higher voltages, its corrosion resistance and paint-based properties are excellent. However, according to ASTM D1732, the dark green, thick coating is preferable to the light green thin coating only when:

- Preliminary removal of surface contamination is not convenient
- The highest degree of abrasion resistance is required from the coating
- A dimensional increase of about 0.02 mm (0.001 in.) can be tolerated
- The part is not to be subjected in service to severe impact, deformation, or flexing, which can cause spalling of the coating

Details of solution compositions and operating conditions for this treatment are given in Fig. 6, which also shows a flow diagram of the processing steps.

Bath Preparation. Half the amount of water required should be heated to

Fig. 1 Chrome pickle treatment

Solution No.	Type of solution	Constituents	Amount g/L	Amount oz/gal	Operating temperature °C	Operating temperature °F	Cycle time, min	Tank material
1	Alkaline cleaner	(a)	(a)	(a)	88-100	190-212	3-10	Low-carbon steel
2	Cold rinse	Water	...	...	Ambient		(b)	Low-carbon steel
3	Chrome pickle(c)	$Na_2Cr_2O_7 \cdot 2H_2O$	180	24	21-32	70-90	½-2(e)	(f)
		HNO_3 (sp gr 1.42)	180	24	...	...		
		Water(d)	rem	rem	...	...	...	...
4	Chrome pickle(g)	$Na_2Cr_2O_7 \cdot 2H_2O$	180	24	21-32	70-90	½-2(j)	Type 316 stainless steel(k)
		HNO_3 (sp gr 1.42)	120-180	16-24	49-60(h) for die castings	120-140(h) for die castings		
		Na_2HF_2, KHF_2, or NH_4HF_2	15	2.0				
		Water(d)	rem	rem	...	...	...	...
5	Hot rinse	Water	...	...	71-82	160-180(m)	...	Low-carbon steel

(a) Type and strength of solution used governed by degree of surface contamination. (b) Rinse thoroughly with adequate inflow of fresh water; agitate. (c) For wrought materials, all alloys, not to be hot formed, and for wrought materials, all alloys, for which tolerances permit removal of approximately 15 μm (0.6 mil) of stock. When treatment is used on alloys M1A and ZK60A to provide paint base, use fresh solutions. (d) Use water from steam condensate or water treated by ion exchange, when available, instead of well or hard tap water. (e) For well-controlled bath, 1 min is standard. (f) Tank material can be of stainless steel or low-carbon steel lined with glass, ceramic, synthetic rubber, or vinyl-based materials. Racks and baskets for use with Solution 3 may be stainless steel or Monel. (g) For sand and permanent mold castings of all alloys for which tolerances permit stock removal of approximately 15 μm (0.6 mil); die castings, all alloys, of 3 to 8 μm (0.1 to 0.3 mil). Dip die castings for 15 to 30 s in water 71 to 82 °C (160 to 180 °F) before immersing in solution. (h) For die castings. (j) For well-controlled bath, 1 min is standard on all products except die castings. Immerse die castings only 10 s. (k) Racks and baskets for use with Solution 4 can be type 316 stainless steel or phosphor bronze. (m) Immerse long enough to heat parts sufficiently to facilitate rapid drying; keep rinse clean with inflowing fresh water

Applicable to all alloys and product forms. Used to provide a base for paint or short-time protection for shipment or storage

71 °C (160 °F), then slowly add ammonium acid fluoride. Add other chemicals and the remainder of the water. The solution should be heated to 82 °C (180 °F), and stirred vigorously for 5 to 10 min. Reheating to 82 °C (180 °F) and stirring should be repeated each time the solution cools to room temperature.

Bath Replenishment. Rise of terminating voltage indicates depletion of bath. Usually, an adjustment in bath composition is required after about 0.5 m^2/L (20 ft^2/gal) of bath has been treated. The amounts of ingredients added are determined by standard methods of chemical analysis of the depleted solution.

Equipment. The anodizing tank and heating coils usually are made of low-carbon steel. However, tanks lined with synthetic rubber or a vinyl-based material may be used, provided the cathodes are of low-carbon steel and a sufficient distance is allowed between the work and the sides of the tank to prevent contact and burning.

Most racks are made of magnesium alloys, and the coating formed on these racks during anodizing can be stripped in an aqueous solution containing 20% chromic acid. Aluminum alloys 5052, 5056, and 220 are also acceptable rack materials, but to avoid chemical attack, these alloys must not be immersed in the solution on open circuit.

Motor generators may be used to supply power for either alternating-current or direct-current installations. Rectifiers are suitable for supplying direct current only. A saturable reactor may be used to generate alternating current.

HAE treatment also produces a two-phase coating and is applied to all forms and alloys of magnesium, provided they contain no attachments or inserts of other metals. A light tan subcoating, about 5.0 μm (0.2 mil) thick, is produced at lower voltages. The thicker phase, formed at higher voltages, is normally dark brown in color and usually about 30 μm (1.2 mils) thick. Both phases have excellent paint-based characteristics.

The dark brown coating is hard and exhibits exceptionally good abrasion resistance, but it spalls under compressive deformation and can adversely affect the fatigue strength of a basis metal less than about 2.5 mm (0.1 in.) thick. The subcoating has no effect on fatigue strength and does not spall.

Solution compositions and operating conditions for the HAE treatment are given in the table that accompanies the flow chart for the process in Fig. 7. The dichromate-bifluoride post-treatment that is part of the HAE treatment, as shown in Fig. 7, provides the coating with high resistance to corrosion.

Bath Preparation. As the first step in preparing the HAE solution, the anhydrous potassium fluoride and trisodium phosphate are dissolved in about half the amount of water required. To this is added an aqueous solution of potassium hydroxide in which aluminum hydroxide has been dissolved. It is difficult to dissolve commercially obtained aluminum hydroxide in a potassium hydroxide solution at room temperature, but it can be dissolved by boiling it in the potassium hydroxide solution for 15 to 20 min.

Metallic aluminum may be dissolved in potassium hydroxide in place of aluminum hydroxide. The reaction, which should be performed under an exhaust hood or in a well-ventilated location, forms a clear liquid and black residue; only the clear liquid is added to the main solution.

Fig. 2 Sealed chrome pickle treatment

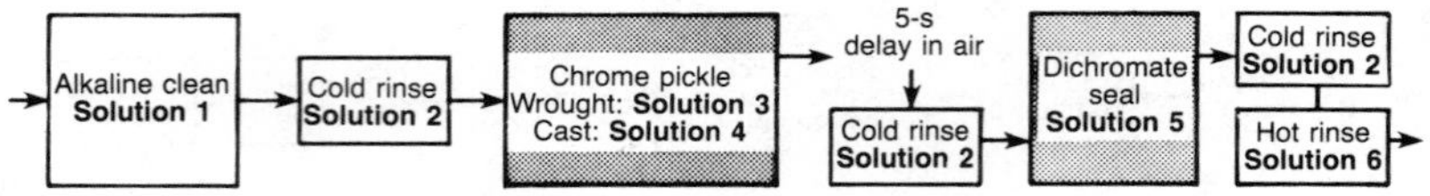

Solution No.	Type of solution	Constituents	Amount g/L	Amount oz/gal	Operating temperature °C	Operating temperature °F	Cycle time, min	Tank material
1	Alkaline cleaner	(a)	(a)	(a)	88-100	190-212	3-10	Low-carbon steel
2	Cold rinse	Water	...	...	Ambient		(b)	Low-carbon steel
3	Chrome pickle(c)	$Na_2Cr_2O_7 \cdot 2H_2O$	180	24	21-32	70-90	½-2(e)	(f)
		HNO_3 (sp gr 1.42)	180	24				
		Water(d)	rem	rem				
4	Chrome pickle(g)	$Na_2Cr_2O_7 \cdot 2H_2O$	180	24	21-32	70-90	½-2(j)	Type 316 stainless steel(k)
		HNO_3 (sp gr 1.42)	120-180	16-24	49-60(h)	120-140(h)		
		$NaHF_2$, KHF_2, or NH_4HF_2	15	2	...	...	...	...
		Water(d)	rem	rem	...	...	...	...
5	Dichromate seal	$Na_2Cr_2O_7 \cdot 2H_2O$	120-180	16-24	99-100	210-212	30	Low-carbon steel
		CaF_2 or MgF_2	2.47	0.33	(boiling	(boiling)	...	...
		Water(d)	rem	rem	...	...	...	...
6	Hot rinse	Water	...	...	71-82	160-180	(m)	Low-carbon steel

(a) Type and strength of solution used governed by degree of surface contamination. (b) Rinse thoroughly with adequate inflow of fresh water; agitate. (c) For wrought materials, all alloys, not to be hot formed, and for wrought materials, all alloys, for which tolerances permit removal of approximately 15 μm (0.6 mil) of stock. When treatment is used on alloys M1A and ZK60A to provide paint base, use fresh solutions. (d) Use water from steam condensate or water treated by ion exchange, when available, instead of well or hard tap water. (e) For well-controlled bath, 1 min is standard. (f) Tank material can be of stainless steel or low-carbon steel lined with glass, ceramic, synthetic rubber, or vinyl-based materials. Racks and baskets for use with Solution 3 may be stainless steel or Monel. (g) For sand and permanent mold castings of all alloys for which tolerances permit stock removal of approximately 15 μm (0.6 mil); die castings, all alloys, of 3 to 8 μm (0.1 to 0.3 mil). Dip die castings for 15 to 30 s in water 71 to 82 °C (160 to 180 °F) before immersing in solution. (h) For die castings. (j) For well-controlled bath, 1 min is standard on all products except die castings. Immerse die castings only 10 s. (k) Racks and baskets for use with Solution 4 can be type 316 stainless steel or phosphor bronze. (m) Immerse long enough to heat parts sufficiently to facilitate rapid drying; keep rinse clean with inflowing fresh water

Applicable to all alloys and product forms. Used as a substitute for dichromate coating or for more protection than is provided by the chrome pickle treatment

Either potassium manganate or potassium permanganate, which is unstable in alkaline solutions and converts to the manganate, is added last. Conversion of the permanganate is evidenced by a color change from purple to green. Conversion can be accelerated by dissolving the permanganate in boiling water and then adding this solution to the bath, or by adding it to the potassium hydroxide solution and boiling. Finally, water is added to make the required volume of solution; continuous agitation is necessary to put all ingredients into solution.

Bath Replenishment. Expensive analyses are not required for maintaining the bath, because depletion of the manganese compound is indicated by a lightening of the characteristic dark brown color of the coating and serves as a bath control. It should be noted, however, that coatings of the lighter brown color have the same properties as those of dark brown color. To maintain the dark brown color of the coating, bath adjustments should be made after about 0.4 m^2/L (15 ft^2/gal) of bath has been treated.

To replenish, 3.7 to 7.5 g/L (0.5 to 1 oz/gal) of potassium manganate or permanganate is added while the bath is being constantly agitated. Additions of aluminum are made at 15 to 20% of the original quantities or aluminum hydroxide (11 g/L or 1½ oz/gal to solution) as previously described. The fluoride and phosphate ingredients deplete at a very slow rate and may be added at six-month intervals if the bath is used continually at the rate of 15 to 20% of original quantities.

Post-Treatment. After anodizing, parts are rinsed thoroughly in water and immersed for 1 min in an aqueous solution containing 77 g (2.7 oz) of sodium dichromate and 377 g (13.3 oz) of ammonium bifluoride per 3.7 L (1 gal) of water. To dry parts, hot air may be used, without rinsing, and parts may be aged in an oven for 7 to 15 h. Aging at about 79 °C (175 °F) and about 85% relative humidity greatly improves the corrosion-protective quality of the coating. Heating of parts, particularly those with thick wall sections, should be done at low humidity until the surface is hot, after which aging is continued at higher humidity.

Equipment. The anodizing tank of double-welded construction and cooling plates or coils are made of black iron. Racks should be of magnesium alloy and should be protected by a suitable vinyl tape at solution level. Steel clamps can be used to clamp the racks to busbars. Galvanized iron, brass, bronze, tin, zinc, and rubber should not be used for equipment that comes in contact with the bath.

Alternating current at 60 Hz is supplied from a constant current regulator capable of delivering the required current over a range of 0 to 110 V.

The post-treatment dip tank must be lined with polyethylene or similar inert material. An oven or humidity cabinet, capable of maintaining temperatures to 86 °C (185 °F) and humidity to 85%, is used for aging.

Cr-22 Treatment. A flow chart of processing steps for the Cr-22 treatment is presented in Fig. 8, together with a table listing bath compositions and operating conditions. The green and black coatings obtainable by vary-

Fig. 3 Dichromate treatment

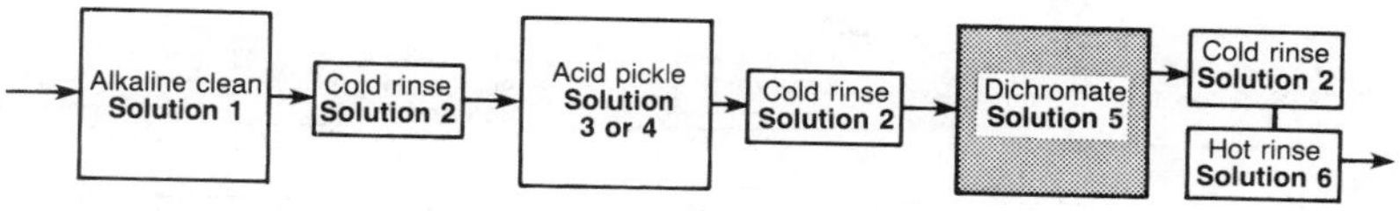

Solution No.	Type of solution	Constituents	Amount g/L	Amount oz/gal	Operating temperature °C	Operating temperature °F	Cycle time, min	Tank material
1	Alkaline cleaner	(a)	(a)	(a)	88-100	190-212	3-10	Low-carbon steel
2	Cold rinse	Water	...	...	Ambient		(b)	Low-carbon steel
3	Acid pickle(c)	60% HF Water(d)	180 rem	24	21-32	70-90	5(e)	Low-carbon steel(f)
4	Acid pickle(g)	$NaHF_2$, KHF_2 or Water(d)	50 rem	6.7	21-32	70-90	5	Low-carbon steel(f)
5	Dichromate	$Na_2Cr_2O_7 \cdot 2H_2O$ CaF_2 or MgF_2 Water(d)	120-180 2.48 rem	16-24 0.33	99-100 (boiling)	210-212 (boiling)	30(h)	Low-carbon steel
6	Hot rinse	Water	...		71-82	160-180	(j)	Low-carbon steel

Note: Racks and baskets for use with all tank materials should be Monel, type 316 stainless steel, or phosphor bronze
(a) Type and strength of solution used governed by degree of surface contamination. (b) Rinse thoroughly with adequate inflow of fresh water; agitate. (c) This bath may be used for all treatable alloys in all forms. This bath must be used for castings that have not been pickled after being sand blasted. (d) Water from steam condensate or water treated by ion exchange should be used when available instead of well or hard tap water. (e) For AZ31A and B, ½ to 1 min. (f) Line tanks with lead or with natural or synthetic rubber. (g) This pickle is preferred for AZ31B and C; an alternate pickle for wrought products and for castings that have been pickled after being sand blasted. (h) For ZK60A, 15 min. (j) Immerse long enough to heat parts sufficiently to facilitate rapid drying; keep rinse clean with inflowing fresh water

Used for maximum corrosion protection and to provide a paint base. Results in no appreciable change in dimensions and can be used on finish-machined wrought and cast products of all alloys except EK30A, EK41A, EZ33A, HK31A, and M1A

ing solution composition, operating temperature, and current density may be applied to all alloys, and although intended primarily as bases for organic finishes, these coatings provide excellent corrosion resistance to unpainted parts when sealed as described in Fig. 8. Inclusions or residual stresses in the magnesium surface occasionally result in the appearance of small nodules in the Cr-22 coating, but the presence of these nodules has no apparent effect on the corrosion protection provided by the coating. Heavier coatings produced by the Cr-22 treatment, like those produced by Chemical Treatment No. 17 and HAE treatments, spall under compressive deformation.

Bath Preparation. To half the required volume of water, the constituents of the bath are added in the order listed in Fig. 8, stirring continuously. When the chemicals are completely dissolved, water is added to bring the bath to the required volume, then sufficient ammonium hydroxide is added to raise the pH to 8.

Bath replenishment can be affected by adding the initial preparation chemicals in quantities determined by conventional chemical analyses for ammonia, phosphorus, fluorine, and hexavalent chromium. Replenishment after the anodizing of 0.6 m^2/L (30 ft^2/gal) of electrolyte, at 1.5 A/dm^2 (15 A/ft^2), has been found satisfactory. To minimize vapor losses at the high operating temperature of Cr-22 baths, polyethylene floats are used to cover bath surfaces.

Equipment. Tanks and heating coils should be made of low-carbon steel. Do not use copper, brass, or zinc for equipment that will come into contact with the bath. Racks should be made of magnesium alloys, and as in all anodizing treatments, electrical contact points should be clean and firm.

The power supply should be capable of delivering a current of 1.5 A/dm^2 (15 A/ft^2) of work area at a maximum of 350 V (60 cycles/s, alternating current).

Repair of Damaged Anodic Coatings. Touch up procedures depend on the nature and extent of damage. Minor damage, scratches, or very small areas where magnesium metal is exposed can be treated as follows:

- Clean damaged area with organic solvent
- Brush with chrome pickle solution (Fig. 1, Solution 4); or with a solution containing 9.7 g/L (1.3 oz/gal) of chromic acid with water and 28 g (1 oz) of calcium sulfate, added in that order and stirred vigorously for 15 min
- Apply suitable primer and finish paint coat after thoroughly rinsing and drying

Larger areas, several square inches, can be reanodized without removing the old coating, or the old coating can be stripped in a 20% solution of chromic acid and the entire part reanodized.

Process Control of Chemical and Anodic Treatments

To achieve desired results with any acid pickle, chemical, or anodic treatment, proper control of temperature and concentration must be maintained. In these treatments, solution temperature governs the rate of metal removal or of film formation. Acid pickles and chemical treatments are conventionally formulated to be used at room temperature to avoid the expense of heating or cooling. Elevated temperatures are required only when chemical action is too slow at ambient temperature, as in alkaline cleaning, chromic acid pickling, and anodizing.

The rate of metal removal by acid pickling also increases with acid concentration. Solutions are usually formulated so that the pickling rate is fast enough to be economical, but not so

Fig. 4 Modified chrome pickle treatment

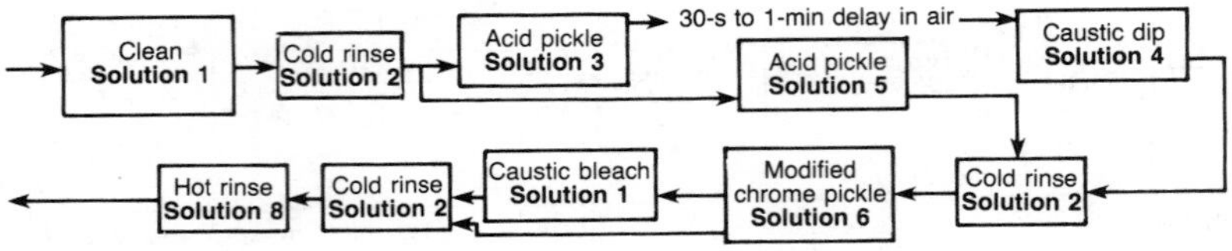

Solution No.	Type of solution	Constituents	Amount g/L	Amount oz/gal	Operating temperature °C	Operating temperature °F	Cycle time, min	Tank material
1	Cleaner(a)	$Na_4P_2O_7$	30	4	77-82	170-180	2-5	Low-carbon steel
		$Na_2B_4O_7 \cdot 10H_2O$	68	9	...	...	...	...
		NaF	8	1	...	...	...	...
		Water(b)	rem	rem	...	...	...	...
2	Cold rinse	Water	...	...	Ambient		(c)	Low-carbon steel
3	Acid pickle(d)	100% H_3PO_4	481-820	65-110	21-32	70-90	1/6-1/4	Low-carbon steel(e)
		Water(b)	rem	rem	...	...	...	...
4	Caustic dip	NaOH	50	6.6	21-82	70-180	1/2	Low-carbon steel
5	Acid pickle(f)	100% HF	143-196	20-25	21-32	70-90	2-5	Low-carbon steel(e)
		H_2SO_4	48	6.4				
		Water(b)	rem	rem	...	...	...	...
6	Modified chrome pickle(g)	$NaHF_2$	15	2	21-32	70-90	1/12-2	Type 316 stainless steel or low-carbon steel with vinyl
		$NaCr_2O_7 \cdot 2H_2O$	180	24	...	...	...	
		$Al_2(SO_4) \cdot 14H_2O$	10	1.3	...	...	...	
		HNO_3 (sp gr 1.42)	120	16	...	...	...	...
		Water(b)	rem	rem	...	...	...	...
7	Caustic bleach(h)	NaOH, 5-10%	...	...	82-100(j)	180-212(j)	1/2	Low-carbon steel
8	Hot rinse(k)	Water	...	...	71-82	160-180	(m)	Low-carbon steel

(a) Vapor degreasing, solvent washing, or conventional alkaline cleaning may be used instead of this special mild-etching cleaner. Metal removal in 5 min, 2.5 to 5.0 μm (0.1 to 0.2 mil). (b) Use water from steam condensate or water treated by ion exchange, when available, instead of well or hard tap water. (c) Rinse thoroughly with adequate inflow of fresh water; agitate. (d) Acid pickle using this solution is preferred pretreatment if metal loss of 13 μm (0.5 mil) can be tolerated. (e) Lined with rubber or vinyl. (f) Alternative treatment; for use if metal loss of 13 μm (0.5 mil) is unacceptable. (g) Metal removal rate, 1.2 μm (0.05 mil) per min. (h) Optional; provides bright appearance. (j) For castings; wrought products are best treated in a room temperature solution. (k) Oven drying, at temperatures not exceeding 150 °C (300 °F), may be substituted for hot water rinse. (m) Immerse long enough to heat parts sufficiently to facilitate rapid drying; keep rinse clean

Used to provide a base for paint or for protection during shipment or storage. Provides a more uniform coating than chrome pickle treatment. Applicable to all alloys and product forms, particularly die castings, for which slight metal losses are acceptable

fast that metal removal is difficult to control or that localized pitting results. As the bath is used, magnesium is dissolved and acid is consumed. Concentrated acid must be added periodically to replenish what is consumed and to maintain the solution within recommended composition limits. As the magnesium content of the solution increases, the pickling rate becomes slower, and eventually the rate becomes too slow, no matter how high the acid content. Chemical treatments cause staining when the magnesium content of solutions used is too high. These conditions can be avoided by discarding solution when its magnesium content reaches about 30 g/L (4 oz/gal).

When only one acid is present in a pickling solution, its concentration may be determined analytically by titration with standard sodium hydroxide solution. Magnesium content can be determined rapidly with atomic absorption spectroscopy or a flame photometer. Other components of the chemical baths can be determined by standard analytical procedures. In the absence of analytical facilities, it is necessary to fortify pickling solutions periodically to maintain the desired reaction rate; solutions for chemical treatment may be maintained by periodically replacing one fourth of the volume with fresh solution.

Table 4 summarizes process control for various acid pickling, chemical, and anodic treatments and indicates the unfavorable effects that result when operating variables are above or below recommended ranges.

In-Process Corrosion. During periods of high humidity, magnesium parts in process often exhibit corrosion products, even though protected with a chrome pickle. This is especially true of sand castings. The corrosion products are readily removed by treating the parts in a boiling dichromate solution as shown in Fig. 3. The parts are then protected from further corrosion by slushing oil.

Difficulties with Chemical and Anodic Treatments

Difficulties experienced with chemical and anodic treatments result from the lack of control of processing variables. One difficulty common to all treatments, chemical or anodic, is nonuniformity of coating or actual failure to coat on certain areas of a part. This is most likely to occur on complex parts and is caused by the entrapment of air in pockets or blind holes. The remedy is to agitate parts when first placing them in the bath and to reposition them periodically to make sure the bath is brought into contact with all areas.

Other difficulties and their corrections vary for each chemical or anodic treatment. Tables 5, 6, and 7 describe the usual appearance of coatings produced by the chrome pickle, sealed chrome pickle, dichromate, Chemical Treatment No. 9, Chemical Treatment No. 17, and HAE treatments, and list

Fig. 5 Chemical Treatment No. 9 (MIL-M-3171A) galvanic anodizing

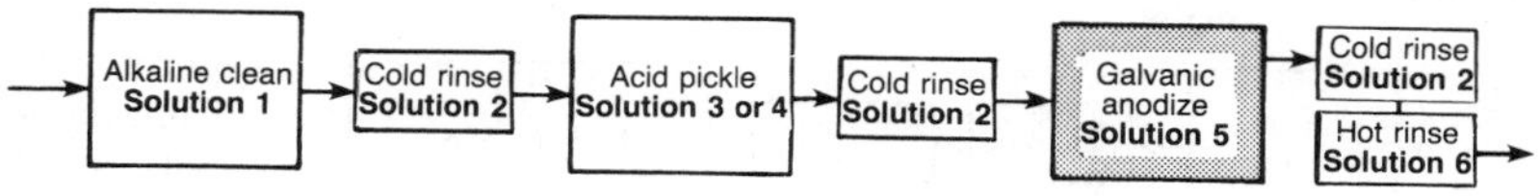

Solution No.	Type of solution	Constituents	Amount g/L	Amount oz/gal	Operating temperature °C	Operating temperature °F	Cycle time, min	Tank material
1	Alkaline cleaner	(a)	(a)	(a)	88-100	190-212	3-10	Low-carbon steel
2	Cold rinse	Water	...	...	Ambient	Ambient	(b)	Low-carbon steel
3	Acid pickle(c)	60% HF	180	24	21-32	70-90	5(e)	Low-carbon steel(f)
		Water(d)	rem	rem				
4	Acid pickle(g)	$NaHF_2$, KHF_2 or NH_4HF_2	50	6.6	21-32	70-90	5	Low-carbon steel(f)
		Water(d)	rem	rem				
5	Galvanic anodize(h)	$(NH_4)_2SO_4$	30	4	49-60	120-140	10-30(j)	(k)
		$Na_2Cr_2O_7 \cdot 2H_2O$	30	4				
		NH_4OH (sp gr 0.880)	2.2	0.3				
		Water(d)	rem	rem				
6	Hot rinse	Water	...	...	71-82	160-180	(m)	Low-carbon steel

Note: Racks and baskets for use with all tank materials may be stainless steel, Monel, or phosphor bronze

(a) Type and strength of solution governed by degree of surface contamination. (b) Rinse thoroughly; agitate. (c) May be used for all alloys in all forms; must be used for castings that have not been pickled after being sand blasted. (d) Water from steam condensate or water treated by ion exchange should be used, when available, instead of well or hard tap water. (e) For AZ31A and B, ½ to 1 min. (f) Lined with lead or with natural or synthetic rubber. (g) An alternative pickle for wrought products and for castings that have been pickled after sand blasting; preferred for AZ31B and C. (h) Current density should not exceed 1 A/dm^2 (10 A/ft^2); at least 753 $A \cdot min/m^2$ (70 $A \cdot min/ft^2$) is required for uniform coating. Maintain solution at pH 5.6 to 6.0 by adding solution containing 5% CrO_3 and concentrated H_2SO_4. (j) Treat parts until a uniform black coating is obtained. (k) If made of low-carbon steel, the tank acts as a cathode. If tank is made of or lined with nonmetallic material, use steel cathode plates. (m) Immerse long enough to sufficiently heat parts to facilitate rapid drying; keep rinse clean with adequate flow of fresh water

common difficulties, as evidenced by deviations from usual coating appearance, together with causes of these difficulties and procedures to be followed for their correction.

For example, the adhesion of paint films applied to dichromated parts can be adversely affected by the presence of a dusty material on the parts after final rinsing and drying. Chemical analysis might indicate this material to be a sulfate precipitated from the domestic water that was being used to replace evaporation losses in the dichromating solution. The problem can be solved by recycling: storing the steam condensate obtained from heating the solution and installing an automatic level-control system to feed this condensate into the dichromating tank.

Although dichromating usually treats the metal satisfactorily, another example of special difficulties might involve a residue of dye-penetrant developer that can remain on the surfaces of castings and interfere with the adhesion of paint films. In this situation, vapor degreasing, either before or after the dichromating treatment, would only set the developer, rather than effect its removal. Thorough cleaning of parts to remove all developer residue immediately after penetrant inspection, and before vapor degreasing, can solve the problem. This cleaning would entail the use of an alkaline cleaner and, if necessary, scrubbing with a bristle brush.

Another example involves corrosion products which were observed surrounding stainless steel hardware that had been inserted into sand castings of alloy AZ91A before dichromating. The parts were free from corrosion immediately after application of the finish; corrosion occurred while parts were in storage.

The corrosion of the magnesium was found to be the result of a galvanic couple set up between the magnesium and stainless steel, caused by the entrapment of acid solution during dichromating. This condition was eliminated by applying the dichromate treatment to the castings before inserting the stainless steel parts, and by applying a coating of zinc chromate primer to the stainless steel parts just before inserting them into the castings.

Precision machined parts that were given a chromic acid pickle might present another example of difficulty if the parts were found to have an etched surface. Analysis of the bath might indicate the presence of a high concentration of chloride and sulfate ions, 0.5 g/L (0.07 oz/gal), of Cl, as NaCl, and 1.7 g/L (0.2 oz/gal) of SO_4, as H_2SO_4, the source of which might be traced to the prepared water. Because chloride and sulfate ions cannot readily be removed, the bath would probably be discarded. To prevent further loss of parts, baths must be checked frequently and test panels treated for 15 to 30 s and examined; an evolution of gas during treatment and a dimensional change indicate that a bath is contaminated and must be discarded.

As another example, spacecraft parts were only given a Chemical Treatment No. 17 anodize for corrosion protection. Although this treatment was satisfactory during indoor inplant processing, white corrosion appeared during standby in the salt air and fog atmosphere at the launch pad. This attractive anodic coating is porous and not corrosion-protective. Parts were sealed with a paint system to correct the problem.

Inspection Methods. Examine coatings on treated parts visually for complete and uniform coverage and for color peculiar to the type of treatment being checked. Buildup caused by the coating is determined most simply and reliably by measurement of the parts, before and after treatment, with a micrometer accurate to 2.5 μm (0.1 mil). If this is not possible, measure test panels that have been treated simulta-

Fig. 6 Chemical Treatment No. 17 (MIL-M-45202) anodizing treatment

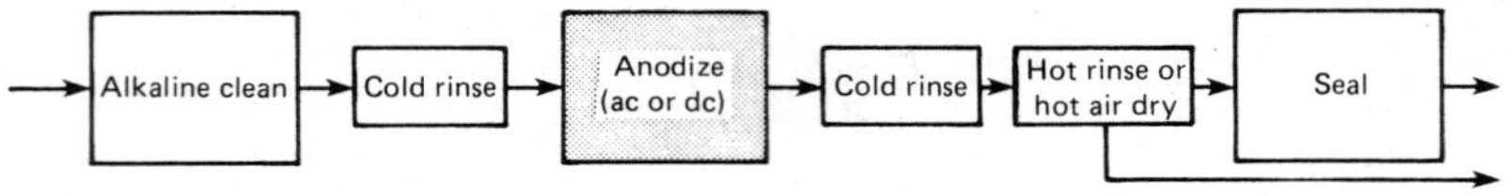

Operating conditions	Alternating current treatment	Direct current treatment
For coatings of both colors and thicknesses		
Composition of aqueous anodizing solution(a):		
NH_4HF_2	240 g/L (32 oz/gal)	360 g/L (48 oz/gal)
$Na_2Cr_2O_7 \cdot 2H_2O$	100 g/L (13.3 oz/gal)	100 g/L (13.3 oz/gal)
85% H_3PO_4	90 mL/L (12 oz/gal)	90 mL/L (12 oz/gal)
Operating temperature(b)	71-82 °C (160-180 °F)	71-82 °C (160-180 °F)
Current density	0.5-5 A/dm^2 (5-50 A/ft^2)	0.5-5 A/dm^2 (5-50 A/ft^2)
For light green coating, thin(c)		
Current consumption(d)	860-1075 A·min/m^2 (80-100 A·min/ft^2)	540-645 A·min/m^2 (50-60 A·min/ft^2)
Terminating voltage(d)	75 V	75 V
Treatment time(d)	4-5 min	2½-3 min
For dark green coating, thick(e)		
Current consumption(d)	4950 A·min/m^2 min (460 A·min/ft^2) min	3230 A·min/m^2 min (300 A·min/ft^2) min
Terminating voltage(d)	100 V	100 V
Treatment time(d)	23 min(f)	15 min(f)
Sealing post-treatment(g)		
Aqueous solution	Sodium tetrasilicate ($Na_2Si_4O_9$), 53 g/L (7 oz/gal)	
Temperature	93-100 °C (200-212 °F)	
Immersion time	15 min	

(a) Use water from steam condensate or water treated by ion exchange, when available, instead of well or hard tap water. (b) Solution will not operate below 60 °C (140 °F). It may be operated up to the boiling point without adversely affecting results. (c) Formed in first phase of treatment. Usual thickness, 5.0 μm (0.2 mil); range, 2.5 to 7.5 μm (0.1 to 0.3 mil). (d) Terminating voltage and treatment time vary, but the number of ampere minutes per unit area remains constant for any one alloy. (e) Formed in second phase of treatment. Usual thickness 30.4 μm (1.2 mils); range, 23 to 38.1 μm (0.9 to 1.5 mils). (f) A denser coating is produced by maintaining the voltage, allowing the current to decrease for 20 min (ac) or 15 min (dc) after the indicated minimum amperage per square metre (foot) has been applied. (g) Optional; to increase corrosion resistance, apply to parts not to receive organic finishes. Immerse in solution, rinse parts in cold, then in hot, water. Air dry

Conditions for cleaning and rinsing same as for galvanic anodizing treatment

neously with the parts before and after treatment.

Process test panels together with parts through all the steps of cleaning, pretreatment coating, and where applicable, post-treatment. Post-treatment should not consist of an additional film of coating, organic or inorganic, applied either by spraying or dipping. After processing for the time required to produce the coating, test panels should be subjected to the following tests:

- *Abrasion resistance*: the test panels should be rubbed by the tip of a 4.7-mm (3⁄16-in.) diam Pyrex rod, the end of which has been heated in a blast flame until it is fire polished and hemispherical. Press the rod against the coating and rub back and forth by hand with just enough pressure to abrade the surface of the rod. The coating should not break or show scratches that penetrate to the basis metal
- *Salt spray (fog) exposure*: the treated test panels, when exposed to salt spray in accordance with Federal Test Method 151 for a specified number of hours, rinsed in tap water, and air dried, should show an average of no more than eight corrosion spots, none larger than 0.8 mm (1⁄32 in.) in greatest dimension. Corrosion appearing at the electrical-contact area or at the edges of the test panels may be disregarded

Plating

Because they are known to provide satisfactory adhesion, nickel and zinc are the only metals that are plated directly on magnesium. However, either of these metals serves as a satisfactory undercoating on which other commonly plated metals may be deposited.

The preparation of magnesium for electroplating is similar in principle to that prescribed for many other metals. First, magnesium must be cleaned to remove all surface contaminants. Then, depending on the surface finish desired, it must be conditioned. This may be accomplished by one or more of the following finishing methods: (*a*) glass bead, (*b*) shot or sand blasting, (*c*) pickling, (*d*) barrel finishing, and (*e*) polishing and buffing. Surface contaminants resulting from these treatments must be removed by additional cleaning. Depending on the type of soil, suitable cleaning methods include vapor degreasing, emulsion cleaning, alkaline soak cleaning, and water rinsing. Finally, the magnesium surface must be activated chemically before it can accept either of the recommended metal undercoatings.

Nickel Undercoating. The deposition of nickel on magnesium must be preceded by two acid etching treatments, one in a chromic-nitric acid solution and another in a solution of hydrofluoric acid. These pretreatments of magnesium, in combination with the deposition of a nickel undercoating, are generally referred to as the chemical-etching process.

Figure 9 illustrates the successive steps involved in alkaline cleaning, etching, and nickel plating. Operating conditions and pertinent data are given in the accompanying table. The nickel undercoating may be deposited from either an electrolytic, acid-fluoride, plating bath or an electroless immersion bath as shown:

Constituent or condition	Value or range
Electroless nickel	
Basic nickel carbonate	9.7 g/L (1.3 oz/gal)
Hydrofluoric acid as 70% HF	6.4 mL/L (0.85 fl oz/gal)
Citric acid	5.2 g/L (0.7 oz/gal)
Ammonium acid fluoride	15 g/L (2.0 oz/gal)
Sodium hypophosphite	20.3 g/L (2.7 oz/gal)
Ammonium hydroxide	30.4 mL/L (4.06 fl oz/gal)
Water	rem
pH (colorimetric)	4.5-6.8(a)
Operating temperature	76-82 °C (170-180 °F)
Agitation required	Mild mechanical
Rate of deposition (approx)	20 μm/h (0.8 mil/h)
Nickel electroplate	
Basic nickel carbonate	120 g/L (16.0 oz/gal)
Hydrofluoric acid as 100% HF	43 mL/L (5.7 fl oz/gal)
Citric acid	40.6 g/L (5.4 oz/gal)
Wetting agent(b)	0.7 g/L(c) (0.1 oz/gal)
Water	rem
pH (colorimetric)	3.0
Operating temperature	48-60 °C (120-140 °F)
Current density	3-10 A/dm^2 (30-100 A/ft^2)
Cathode-rod agitation	0.06-0.08 m/s (12-16 ft/min)

(a) Preferred, 6.5. (b) Example, sodium lauryl sulfate. (c) 1 g/L (0.134 oz/gal)

Fig. 7 HAE anodizing treatment (MIL-C-13335, Amendment 1; MIL-M-45202)

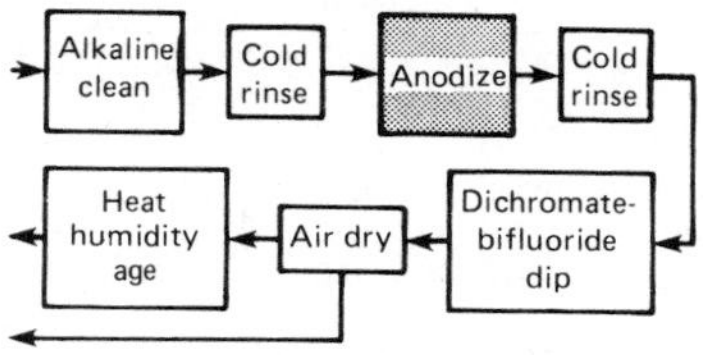

Anodizing(a)	
Composition of solution:	
Potassium hydroxide, KOH	165 g/L (22 oz/gal)
Aluminum hydroxide, $Al(OH)_3$	34 g/L (4.5 oz/gal)(b)
Potassium fluoride, K_2F_2	34 g/L (4.5 oz/gal)
Trisodium phosphate, Na_3PO_4	34 g/L (4.5 oz/gal)
Potassium manganate, K_2MnO_4(c)	19 g/L (2.5 oz/gal)
Water	rem to 3.7 L (to 1 gal)
Operating temperature	Room temperature to 25 °C (to 77 °F)
Current density	1.5-2.5 A/dm^2 (15-25 A/ft^2)
Terminating voltage:	
First-phase coating	65-70 V
Second-phase coating	80-90 V
Treatment time(d):	
First-phase coating	7-10 min
Second-phase coating	60 min
Dichromate-bifluoride dip	
Composition of solution:	
Sodium dichromate, $Na_2Cr_2O_7 \cdot 2H_2O$	20 g/L (2.7 oz/gal)
Ammonium bifluoride, NH_4HF_2	100 g/L (13.3 oz/gal)
Water	rem to 3.7 L (to 1 gal)
Operating temperature	21-32 °C (70-90 °F)
Immersion time	1 min
Heat-humidity aging(e)	
Temperature	77-85 °C (170-185 °F)
Relative humidity	85% approx
Time	7-15 h

(a) Acid pickling is not required before anodizing because of the chemical action of the anodizing bath on the alkaline-cleaned surface. Use alternating current for anodizing. Coatings are deposited in two phases: first a light tan coating 5.0 μm (0.2 mil) thick, then a dark brown coating usually 30.4 μm (1.2 mils) thick. (b) For an exceptionally hard, abrasion resistant, dark brown coating, use 45 to 53 g/L (6 to 7 oz/gal) $Al(OH)_3$. (c) An equal weight of potassium permanganate ($KMnO_4$) may be used instead of potassium manganate; dissolve the permanganate completely in water before adding to the solution; an additional 43 g (1.5 oz) of potassium hydroxide must also be added. (d) Treatment time varies with current density. Increasing current density decreases treatment time; for uniform coating, 2 A/dm^2 (20 A/ft^2) is recommended. (e) Improves corrosion resistance of coated parts

Conditions for cleaning and rinsing same as for galvanic anodizing treatment

To some extent, the two etching treatments undercut surface pits to provide mechanical anchorage for the plated deposits. After a nickel undercoating is applied to this roughened surface, the surface of the coating is not as bright or as smooth as that obtained with the zinc immersion process. However, in common with zinc coating, nickel coating can accept any of the standard electrodeposits, including cadmium, zinc, copper, brass, nickel, black nickel, chromium, gold, silver, and rhodium. Electroless deposits, such as gold or copper, also may be applied over the initial nickel plate.

Zinc Undercoating. After being alkaline cleaned, acid pickled, and surface-activated, magnesium can be prepared for plating by being immersed in a nonelectrolytic chemical bath that produces a thin deposit of zinc directly on the basis metal. The basic steps, operating conditions, and tank materials required for the zinc immersion process are summarized in the flow chart and table of Fig. 10. After the immersion treatment, the very light zinc coating about 2.5 μm (0.1 mil) thick must be protected by a strike copper plate before any other standard electrodeposit may be applied. The following are the various steps that comprise the zinc immersion process, as well as the various solutions used.

Alkaline cleaning, as used in the electroplating of magnesium, consists of soak cleaning or cathodic electrocleaning, or a combination of both. These processes are described in detail in the

Table 4 Effects of lack of control of process variables in acid pickling, chemical, and anodic treatments

Treatment	Solution constituent or condition	Recommended range	Effect of operating out of recommended range: Too low	Too high
Acid pickling treatments				
Acetic-nitrate	Glacial acetic acid	115-300 g/L (15-40 oz/gal)	Slow rate	Uneconomical
	Sodium nitrate	30.0-75 g/L (4-10 oz/gal)	Smut	Slow rate
Chromic acid	Chromic acid	115-375 g/L (15-50 oz/gal)	Slow rate	Uneconomical
Chromic-sulfuric	Chromic acid	150.4-240.6 g/L (20-32 oz/gal)	Unknown(a)	Unknown(a)
	96% sulfuric acid	0.37-0.60 g/L (0.05-0.08 oz/gal)	Unknown(a)	Unknown(a)
	pH	0.4-1.0	Unknown(a)	Unknown(a)
Ferric nitrate(b)	Chromic acid	150.4-225 g/L (20-30 oz/gal)	Rapid rate; yellow stain	Slow rate
	Ferric nitrate	22.5-45.1 g/L (3-6 oz/gal)	Slow rate; rough or frosted	Rapid rate; yellow stain
	Sodium fluoride	2.2-6.7 g/L (0.3-0.9 oz/gal)	Slow rate; smut; rough or frosted appearance	Rapid rate; yellow or gray film
Glycolic-nitrate	70% glycolic acid	115-375 g/L (15-50 oz/gal)	Poor corrosion resistance	Uneconomical
	70% nitric acid	22.5-90.2 g/L (3-12 oz/gal)	Slow rate	Rapid rate
	Sodium nitrate	30.0-75.2 g/L (4-10 oz/gal)	Smut	Slow rate
Hydrofluoric acid	50% hydrofluoric acid	190-565 g/L (25-75 oz/gal)	Powdery dichromate treatment; incomplete MgF_2 film	Uneconomical
Nitric acid	70% nitric acid	30.0-75.2 g/L (4-10 oz/gal)	Slow rate	Rapid rate; rough surface
Nitric-sulfuric	70% nitric acid	30.0-90.2 g/L (4-12 oz/gal)	Smut	Rapid rate
	96% sulfuric acid	9.5-30.0 g/L (1.3-4 oz/gal)	Slow rate	Rapid rate
Phosphoric acid(c)	85% phosphoric acid	865-965 g/L (115-128 oz/gal)	Rapid rate	...
Sulfuric acid	96% sulfuric acid	15.0-45.1 g/L (2-6 oz/gal)	Slow rate	Rapid rate; rough surface
Chemical treatments				
Chrome pickle and sealed chrome pickle	Sodium dichromate	135.3-205 g/L (18-27 oz/gal)	Powdery coating	Shallow etch
	70% nitric acid	60.1-180.4 g/L (8-24 oz/gal)	Shallow etch	Powdery coating
	Temperature	21-49 °C (70-120 °F)	Shallow etch	Powdery coating
	Magnesium	30.0 g/L max (4 oz/gal max)	...	Shallow etch
	Prerinse delay in air	5 s max(d)	...	Powdery coating
Dichromate	Sodium dichromate	120-205 g/L (16-27 oz/gal)(e)	Nonuniform or no coating	Uneconomical
	Calcium fluoride		Powdery coating	No problem
	Free fluoride	0.2% max	...	No coating(f)
	pH(g)	4.1-5.5	Heavy or powdery coating	Nonuniform or no coating
	Temperature	93-100 °C (200-212 °F)	No coating	...

(continued)

(a) Effects of operation outside recommended ranges are not known. Depletion of solution is indicated by the formation of a brassy film and by a pH in excess of 1.0; at this point, add chromic and sulfuric acids, or discard solution. (b) Discard solution when depleted or periodically add one fourth of volume of solution. (c) Dragout loss prevents excessive salt buildup. (d) Do not allow surface to dry out before rinsing. (e) Limited solubility regulates fluoride. (f) Add 0.2% calcium chromate if fluoride is too high. (g) Correct pH with additions of sodium hydroxide or chromic acid. (h) Fortify bath by additions of sodium dichromate to restore original concentration of 180.4 g/L (24 oz/gal) of nitric acid to maintain concentration of used bath at 60 g/L (8 oz/gal) and of 28 g (1 oz) of sodium acid fluoride for every 310 g (11 oz) of nitric acid added. Bath is considered depleted when concentration of nitric acid is below 22.5 g/L (3 oz/gal) and is indicated by poor color and nonuniformity of coating. (j) For fresh bath; no additions required when fortifying. (k) As H_2SO_4. (m) As CrO_3. (n) Control of pH achieved by additions of a solution containing, by weight, 5% CrO_3 and 5% H_2SO_4 (concentrated). (p) Fortify after preparing 0.4 m^2/L (20 ft^2/gal) of solution. (q) Local pitting begins at 1020 g (36 oz), particularly with thorium-containing alloys. (r) Anhydrous. (s) Fortify after treating 0.3 m^2/L (15 ft^2/gal) of solution to maintain dark brown color of coating. (t) 2.2 to 15.0 g/L (0.3 to 2 oz/gal) as Al. (u) Use 1.5 A/dm^2 (15 A/ft^2) for uniformity of coating in recesses. (v) Ideal temperature, 27 °C (80 °F); 38 °C (100 °F) is satisfactory for alloys containing aluminum. (w) Fortify after treating 0.7 m^2/L (36 ft^2/gal) to maintain deposition rate. (x) Control of pH achieved by adding 28% solution of ammonium hydroxide

Table 4 (continued)

Treatment	Solution constituent or condition	Recommended range	Effect of operating out of recommended range: Too low	Too high
Modified chrome pickle(h)	Sodium acid fluoride	11.2-15.0 g/L (1.5-2 oz/gal)	Powdery smut	Powdery coating
	Sodium dichromate	135.3-205.0 g/L (18-27 oz/gal)	Pale color; too reactive	Slow rate
	Aluminum sulfate	7.5-11.2 g/L (1.0-1.5 oz/gal)(j)	Poor coating color on castings	Rapid rate; poor etch
	70% nitric acid	22.5-120.3 g/L (3-16 oz/gal)	Poor etch; slow rate	Pale color; poor etch
	Temperature	21-38 °C (70-100 °F)	Slow rate	Too reactive
Anodic treatments				
Chemical Treatment No. 9	Ammonium sulfate	16.5-24.0 g/L(k) (2.2-3.2 oz/gal)	Slow rate	Powdery coating
	Sodium dichromate	16.5-24.0 g/L(m) (2.2-3.2 oz/gal)	Thin coating	Slow rate
	pH(n)	5.6-6.0	Powdery coating	Slow rate
	Temperature	49-60 °C (120-140 °F)	Slow rate	Rapid rate
Chemical Treatment No. 17	Ammonium acid fluoride(p)	300.8-451.2 g/L(q) (40-60 oz/gal)	Local pitting	Insoluble
	Sodium dichromate(p)	53-120 g/L (7-16 oz/gal)	Thin coating	Uneconomical
	85% phosphoric acid(p)	53-105.2 g/L (7-14 oz/gal)	Soft coating	Uneconomical
	Chloride contamination	3% max	...	Soft coating
	Current density	0.5-5 A/dm^2 (5-50 A/ft^2)	Attack at liquid level	Pitting or burning
	Temperature	71-82 °C (160-180 °F)	No coating	No problem
HAE	Potassium hydroxide	105-190 g/L (14-25 oz/gal)	Pitting; burning; nonuniformity	Pitting; burning; roughness
	Potassium fluoride(r)	15.0-150.4 g/L (2-20 oz/gal)	Spottiness; nonuniformity	Uneconomical
	Trisodium phosphate(s)	15.0-225.6 g/L (2-30 oz/gal)	Light and dark areas	Uneconomical
	Aluminum hydroxide(s)	7.5-53 g/L (1-7 oz/gal)(t)	Coating hardness lowered	Hard, rough coating
	Potassium manganate	3.7-22.5 g/L (0.5-3 oz/gal)	Light brown coating	Dark brown, rough coating
	Current density	0.5-5 A/dm^2 (5-50 A/ft^2)	Slow rate	Nonuniformity in recesses(u)
	Temperature	25-38 °C(v) (77-100 °F)	...	Lower corrosion resistance
Cr-22	Chromic acid(w)	22.5-30.0 g/L (3-4 oz/gal)	Nonuniform or pale coating	Dark coating
	50% hydrofluoric acid(w)	30.0-53 g/L (4-7 oz/gal)	Low voltage, nonuniform coating	Pale coating
	85% phosphoric acid(w)	45.1-90.2 g/L (6-12 oz/gal)	Nonuniform coating	Uneconomical; rough coating
	pH(x)	6.0-6.5	...	Uneconomical
	Current density	1.0-3.0 A/dm^2 (10-30 A/ft^2)	Slow rate	Dark coating
	Temperature	74-93 °C (165-200 °F)	Slower corrosion resistance	...

(a) Effects of operation outside recommended ranges are not known. Depletion of solution is indicated by the formation of a brassy film and by a pH in excess of 1.0; at this point, add chromic and sulfuric acids, or discard solution. (b) Discard solution when depleted or periodically add one fourth of volume of solution. (c) Dragout loss prevents excessive salt buildup. (d) Do not allow surface to dry out before rinsing. (e) Limited solubility regulates fluoride. (f) Add 0.2% calcium chromate if fluoride is too high. (g) Correct pH with additions of sodium hydroxide or chromic acid. (h) Fortify bath by additions of sodium dichromate to restore original concentration of 180.4 g/L (24 oz/gal) of nitric acid to maintain concentration of used bath at 60 g/L (8 oz/gal) and of 28 g (1 oz) of sodium acid fluoride for every 310 g (11 oz) of nitric acid added. Bath is considered depleted when concentration of nitric acid is below 22.5 g/L (3 oz/gal) and is indicated by poor color and nonuniformity of coating. (j) For fresh bath; no additions required when fortifying. (k) As H_2SO_4. (m) As CrO_3. (n) Control of pH achieved by additions of a solution containing, by weight, 5% CrO_3 and 5% H_2SO_4 (concentrated). (p) Fortify after preparing 0.4 m^2/L (20 ft^2/gal) of solution. (q) Local pitting begins at 1020 g (36 oz), particularly with thorium-containing alloys. (r) Anhydrous. (s) Fortify after treating 0.3 m^2/L (15 ft^2/gal) of solution to maintain dark brown color of coating. (t) 2.2 to 15.0 g/L (0.3 to 2 oz/gal) as Al. (u) Use 1.5 A/dm^2 (15 A/ft^2) for uniformity of coating in recesses. (v) Ideal temperature, 27 °C (80 °F); 38 °C (100 °F) is satisfactory for alloys containing aluminum. (w) Fortify after treating 0.7 m^2/L (36 ft^2/gal) to maintain deposition rate. (x) Control of pH achieved by adding 28% solution of ammonium hydroxide

Table 5 Causes and corrections of difficulties with coatings produced by the chrome pickle and sealed chrome pickle treatments

Normal appearance of coating, matte gray to yellow red iridescent(a)

Difficulty	Cause	Corrective procedure
Pale color, shallow etch, slow reaction of solution with metal	Depletion of the chrome pickle bath. Paleness of color should not be confused with the lack of color from insufficient exposure to air during transfer to rinse	Revivify bath by additions of nitric acid and sodium dichromate to restore to proper operating level. Control bath by periodic chemical analysis of ingredients
Bright, brassy—smooth surface with occasional pits(b)	Excess of nitric acid in the bath	Adjust with additions of sodium dichromate to restore proper ratio of dichromate to acid
	Buildup of excessive amount of magnesium nitrate in the bath	Bath has been revivified too many times; discard and use fresh bath
Brown, nonadherent, powdery coatings	Work held in air too long before rinsing	Delay in air; 5 s max; surface should not dry before rinsing
	Ratio of acid to sodium dichromate concentration too high	Adjust concentrations to restore proper ratio
	Solution too hot because of small volume in relation to work	Cool solution or increase volume of solution used
	Metal not properly degreased	Use recommended cleaning method
	Oil film on solution	Prevent by using recommended cleaning procedures(c)
	Solution revivified too often	Discard; use fresh solution

(a) When viewed under magnification, coating exhibits a network of pebbled etching. The appearance of a properly applied chrome pickle coating is influenced to some degree by the age of the solution and by the composition and temper of the alloy treated. (b) When examined under magnification. (c) Correct by skimming off oil and adding small amount, 0.05%, of fluorinated-hydrocarbon wetting agent

article on alkaline cleaning in this Volume. Strong caustic solutions are used to remove segregated metal, graphite, and other surface contaminants. The composition and operating conditions for one water solution of this type are:

Sodium hydroxide,
  NaOH .... 120 to 720 g/L (1 to 6 lb/gal)
Temperature .... 99 to 170 °C (210 to 340 °F)
Time .... 10 to 20 min

Follow this treatment by an acid dip to remove the heavy film of hydroxide.

Acid Pickling. Depending on magnesium alloy composition and product form, the following acid pickling solutions can be used prior to plating. All standard alloys and forms of magnesium can be pickled and chemically brightened by the use of the aqueous solution which follows:

Chromic trioxide,
  CrO_3 .... 180 g/L (24.0 oz/gal)
Ferric nitrate,
  $Fe(NO_3)_3 \cdot 9H_2O$ .... 40 g/L (5.3 oz/gal)
Potassium fluoride,
  KF .... 3.5 to 7 g/L (0.05 to 0.9 oz/gal)
Temperature .... Room temperature, 21 to 32 °C (70 to 90 °F)
Time .... ¼ to 2 min

All magnesium alloys and forms can be pickled under the following conditions:

Phosphoric acid,
  85% H_3PO_4 .... No dilution
Temperature .... Room temperature, 21 to 32 °C (70 to 90 °F)
Time .... ¼ to 5 min

Sheet, extrusions, and forgings can be pickled in the following aqueous solution:

Glacial acetic acid,
  CH_3COOH .... 280 mL/L (37.4 fl oz/gal)
Sodium nitrate .... 80 g/L (10.7 oz/gal)
Temperature .... Room temperature, 21 to 32 °C (70 to 90 °F)
Time .... ½ to 2 min

Where little or no dimensional change is permitted, parts may be pickled in an aqueous solution of chromic acid, 180 g/L (24.0 oz/gal) of chromic trioxide of water at any temperature from room to boiling temperatures. This dip removes graphite, oxides, and other soils without removing an appreciable amount of metal.

Surface Activation. To obtain an adherent and uniform coating of zinc on magnesium, the surface of the magnesium must be activated by immersion in a special acid or alkaline solution. The acid activating bath removes thin oxide or chromate films left by prior pickling or cleaning operations with a minimum of etching of the basis metal. It also produces an equipotentialized surface on which the zinc immersion coating can deposit uniformly. The composition and operating characteristics of an aqueous acid activating solution in common use are:

Phosphoric acid,
  85% H_3PO_4 .... 200 mL/L (26.7 oz/gal)
Ammonium acid fluoride,
  NH_4HF_2 .... 100 g/L (13.3 oz/gal)
Temperature .... 21 to 32 °C (70 to 90 °F)
Time .... 25 s to 2 min

Operation of acid fluoride activating baths at appreciably below 21 °C (70 °F) may result in poor adhesion of the electroplated deposit.

An alkaline pyrophosphate activating solution also removes oxides and other films; by virtue of its alkalinity, it does not dissolve the metals employed for rack contacts. Thus, there is little or no danger that metals such as iron or copper can plate out by immersion at locations adjacent to rack contacts. This form of stray plating prevents the proper deposition of the subsequent zinc immersion coating. The composition

Table 6 Causes and corrections of difficulties with dichromate coatings

Normal appearance varies from light to dark brown, depending on alloy composition

Difficulty	Cause	Corrective procedure
Abnormally heavy coatings	pH of dichromate bath too low, below 4.1	Raise pH by addition of NaOH
	Contact between work and tank	Insulate parts and rack from tank
Loose, powdery coatings	Acid fluoride or HF bath too dilute	Adjust fluoride content
	pH of dichromate bath <4.1	Raise pH by addition of NaOH
	Parts oxidized, corroded, or flux-contaminated	Chromic acid clean parts prior to dichromating
	Contact between work and tank	Insulate parts and rack from tank
Failure to coat, or nonuniform coatings	pH of dichromate bath >5.5	Lower pH by addition of CrO_3
	Dichromate concentration too low	Do not allow dichromate to fall below 120 g/L (16 oz/gal); maintain as high as economical
	Oily matter not removed; previous chrome pickle coating not completely removed	Careful rinsing after alkaline cleaning; use chromic acid pickle to supplement alkaline cleaning for complete removal of chrome pickle coating
	Part not fluoride treated	Give part acid fluoride or HF pickle before dichromating
	Part is made of EK30A, EK41A, EZ33A, HK31A, or M1A	Dichromate treatment does not coat these alloys; use another treatment
	Dichromate bath not kept boiling during treatment period	Solution should actually boil; minimum bath temperature, 93 °C (200 °F)
	High carryover of acid fluoride or HF into dichromate bath	Thoroughly rinse parts after acid fluoride or HF pickle
	Hydrofluoric acid dip too long	Use ½ to 1 min for AZ31A and B; if 5-min dip is used, extend cold water rinse to 5 to 7 min
Pitting or other attack on aluminum inserts or attachments	Hydrofluoric acid dip reacts on aluminum	Use sodium, potassium, or ammonium acid fluoride solution in place of hydrofluoric acid dip

Table 7 Causes and corrections of difficulties with coatings produced by anodizing treatments

Difficulty	Cause	Corrective procedure
Chemical Treatment No. 9(a)		
Normal appearance of thin coating: light tan; thick coating: dark brown		
Grayish, nonuniform appearance	Insufficient or improper cleaning of work before treatment	Use proper methods for adequate cleaning
	Treating solutions depleted	Check pickling bath, HF or acid fluoride, for proper pH(b)
Failure to coat on certain alloys(c)	Lack of good external electrical contact of work with tank or cathode plates	Assure good external electrical contact between work and cathode
	Work touching tank walls or bottom (or cathode plates)	Prevent work from contacting cathode in electrolyte
Chemical Treatment No. 17(d)		
Normal appearance of thin coating: light green; thick coating: dark green		
Deviation from characteristic appearance	Improper balance in content of bath ingredients(e)	Adjust bath ingredients to proper operating range; use chemical analysis to determine need
Spalling of dark green coating	Unusually thick coatings may spall when substrate is bent or otherwise deformed	Spalling can be minimized by sealing coating with an organic finish having low surface tension, low viscosity, and slow drying rate
HAE anodizing treatment(f)		
Normal appearance of thin coating: light tan; thick coating: dark brown		
Significant color lightening	Depletion of the chemical ingredients of the bath(e)	Adjust aluminum and manganese contents to proper level(g)

(a) Usual appearance of thin coating: light tan; thick coating: dark brown. (b) Galvanic anodizing bath is maintained at a pH of 5.6 to 6.0 with additions of a solution containing, by weight, 5% H_2SO_4 (concentrated). (c) Especially alloys without aluminum content such as EK30A, EK41A, EZ33A, HK31A, and HZ32A. (d) Usual appearance of a thin coating: light tan; thick coating: dark brown. A clear, colorless coating can also be deposited by Chemical Treatment No. 17 by use of 40 V as terminal voltage. (e) Solution unbalance may occur after treatment of as little as 0.4 m^2/L (20 ft^2/gal) of solution; however, when applying clear or thin coatings, 2 to 2.6 m^2/L (100 to 130 ft^2/gal) of solution usually can be treated before bath adjustment is indicated. (f) Usual appearance of thin coating: light tan; thick coating: dark brown. (g) Fluoride and phosphate become depleted at an extremely low rate and require adjustment only every few months in production and operating characteristics of the alkaline solution are:

Tetrasodium pyrophosphate ($Na_4P_2O_7$) .... 40 g/L (5.3 oz/gal)
Sodium tetraborate ($Na_2B_4O_7$) .... 70 g/L (9.3 oz/gal)
Sodium fluoride (NaF) .... 20 g/L (2.7 oz/gal)
Temperature.................. 76 °C (170 °F)
Time..................... 2 to 5 min

The bath should not be operated much above 76 °C (170 °F), as the alkaline solution has a tendency to dry on the parts prior to rinsing. For this reason, the transfer from hot activator to water rinse should be as rapid as possible.

Zinc Immersion Coating. Standard zinc immersion baths (Table 8) are based on aqueous solutions of pyrophosphate, a zinc salt, a fluoride salt, and, if required, a small amount of carbonate for adjusting alkalinity to the proper range. The pyrophosphate dissolves oxide and hydroxide films to form water-soluble complexes. By effecting film removal under specific conditions of pH, temperature, and bath concentration, a thin, adherent coating of metallic zinc is deposited. Fluoride is added to control the rate of zinc deposition. To a considerable extent, the quality and adhesion of the zinc deposit depends on the rate of deposition; too rapid a rate usually produces less adherent coatings.

The most commonly used zinc immersion solution has composition and pH as follows:

Zinc sulfate monohydrate ($ZnSO_4 \cdot H_2O$) .... 30 g/L (4.0 oz/gal)
Tetrasodium pyrophosphate ($Na_4P_2O_7$) .... 120 g/L (16.0 oz/gal)
Sodium fluoride (NaF) .... 5 g/L (0.7 oz/gal)
or
Lithium fluoride (LiF) .... 3 g/L (0.4 oz/gal)
Sodium carbonate (Na_2CO_3)................... 5 g/L (0.7 oz/gal)
pH..................... 10.0 to 10.6 (10.2 to 10.4 preferred)

Within limits, the bath constituents can vary from the above formula provided the ratio of pyrophosphate to zinc is kept to the proper range and fluoride concentration is maintained at the level indicated. Water for preparing the bath should be reasonably free from iron and other heavy-metal salts. Ordinary tap water may be used, but deionized water is preferred.

The bath is prepared by first dissolving the zinc sulfate monohydrate in water at room temperature. The solution is then heated to 60 to 82 °C (140 to 180 °F), and tetrasodium pyrophosphate is added slowly, while stirring. The white, fluffy precipitate (sodium zinc pyrophosphate) that first forms dissolves after further stirring for 5 to 10 min or longer, depending on the degree of agitation and temperature. Next, fluoride is added and then carbonate. The amount of carbonate added is the average required to produce the proper pH and should be determined by the actual pH of the solution. Sulfuric or phosphoric acid can be added to reduce pH, when necessary. The pH values given above are those determined

Fig. 8 Cr-22 (MIL-M-45202) anodizing treatment

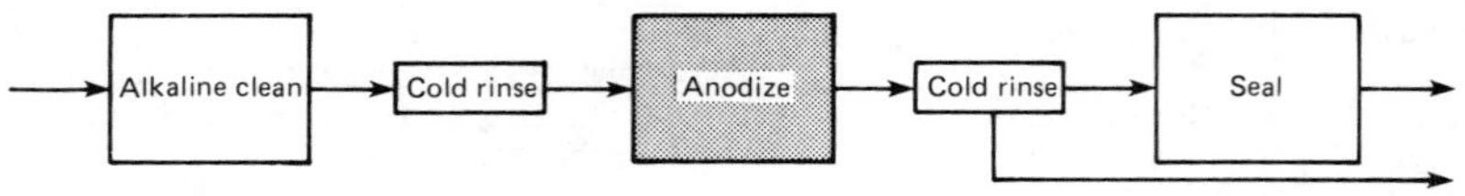

Operating conditions	Type of coating: Green(a)	Black(b)
Composition of anodizing solution		
CrO_3	25 g/L (3.3 oz/gal)	50 g/L (6.7 oz/gal)
50% HF (sp gr, 1.20)	36 mL/L (4.7 fl oz/gal)	35 mL/L (4.7 fl oz/gal)
85% H_3PO_4 (sp gr, 1.70)	50 mL/L (6,7 fl oz/gal)	50 mL/L (6.7 fl oz/gal)
NH_4OH (sp gr, 0.90)	220 mL/L (29 fl oz/gal)	220 mL/L (28 fl oz/gal)
Water(c)	rem	rem
Operating temperature	88-91 °C (190-195 °F)(d)	85 °C (185 °F)(d)
Current density	1.5 A/dm² (15 A/ft²)(e)	2.5 A/dm² (25 A/ft²)(f)
Terminating voltage		
For normal thickness	320 V	320 V
For heavy coatings	350-380 V	350-380 V
Treatment time	12-15 min	12-15 min
Sealing post-treatment(g)		
Solution	10 vol % sodium silicate (42° Bé)	
Temperature	85-100 °C (185-212 °F)	
Immersion time	2 min	

(a) Green coating, conforming to MIL-M-45202, type II, class B, ranges in thicknesses from 10 to 28 μm (0.4 to 1.1 mil); usual thickness, 20 μm (0.8 mil). (b) Black coating, conforming to MIL-M-45202, type II, class C, ranges in thickness from 5 to 25 μm (0.2 to 1.0 mil); usual thickness, 13 μm (0.5 mil). (c) Use water from steam condensate or water treated by ion exchange, when available, instead of well or hard tap water. (d) Recommended; effective range, 74 to 93 °C (165 to 200 °F). (e) Recommended; effective range, 1.0 to 3.0 A/dm² (10 to 30 A/ft²). (f) Recommended; effective range, 2.0 to 3.0 A/dm² (20 to 30 A/ft²). (g) Optional for both types; to increase corrosion resistance, apply to parts not to receive organic finishes

Conditions for cleaning and rinsing same as for galvanic anodizing treatment

Table 8 Composition of zinc immersion bath

Activator	Constituent	Nominal g/L	Nominal oz/gal	Range g/L	Range oz/gal
Acid	Oxalic acid $(CO_2H)_2$	5	0.7	2-10	0.3-1.3
Alkaline	Potassium pyrophosphate ($K_4P_2O_7$)	60	8	50-150	6.5-10.0
	Sodium carbonate (Na_2CO_3)	15	2	10-20	1.3-2.6

Fig. 9 Preparing magnesium alloys for nickel plating

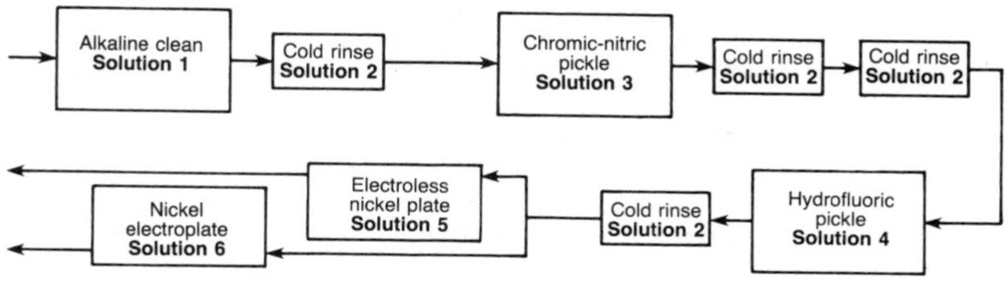

Solution No.	Type of solution	Composition	Amount	Operating temperature °C	Operating temperature °F	Cycle time, min	Tank material
1	Alkaline cleaner(a)	(b)	(b)	82-100	180-212	3-10	Low-carbon steel
2	Cold rinse	Water	...	Ambient		½-1	Low-carbon steel(e)
3	Chrome-nitric pickle(c)	CrO_3	125 mL/L(d) (16 fl oz/gal)	21-32	70-90	½-2	
		70% HNO_3	110 mL/L(d) (14 fl oz/gal)				
4	Hydrofluoric pickle	70% HF	220 mL/L(f) (28.2 fl oz/gal)	21-32	70-90	10	Low-carbon steel(g)
5	Electroless nickel(h)	Basic nickle carbonate	9.7 g/L (1.3 oz/gal)	77-82	170-180	(j)	Low-carbon steel(k)
		Hydrofluoric acid (as 70% HF)	6.2 mL/L (0.8 fl oz/gal)				
		Citric acid	5.2 g/L (0.7 oz/gal)				
		Ammonium acid fluoride	15 g/L (2.0 oz/gal)				
		Sodium hypophosphite	20.3 g/L (2.7 oz/gal)				
		Ammonium hydroxide	30.4 mL/L (4.0 fl oz/gal)				
		Water	rem				
		pH (colorimetric)	4.5-6.8(m)				
6	Nickel electroplate(n)	Basic nickel carbonate	120 g/L (16.0 oz/gal)	49-60	120-140	(j)	Low-carbon steel(p)
		Hydrofluoric acid (as 100% HF)	43 mL/L (5.7 fl oz/gal)				
		Citric acid	40 g/L (5.4 oz/gal)				
		Wetting agent(q)	0.7 g/L(r) (.01 oz/gal)				
		pH (colorimetric)	3.0				

(a) Alkaline cleaning may be preceded by vapor degreasing, emulsion cleaning, or mechanical cleaning. (b) Type and strength of solution governed by degree of surface contamination. (c) Solution must be ventilated. (d) For aluminum containing alloys. For alloys containing no aluminum, concentrations are 226 g (8 oz) CrO_3 and 326 g (11.5 oz) of 70% HNO_3 and water to make 3.8 L (1 gal). (e) Stainless steel or low-carbon steel lined with polyethylene. (f) For alloys containing more than 5% Al. For all other alloys, solution contains 195 g (6.9 oz) 70% HF and water to make 3.8 L (1 gal). (g) Lined with polyethylene, synthetic rubber, or vinyl-based material. (h) Agitation of work is recommended. (j) Time determined by required thickness of nickel deposit. (k) Lined with polyethylene, saran, or Lucite. (m) Preferred, 6.5. (n) Direct current required; agitation of cathode bar recommended. (p) Lined with synthetic rubber or vinyl-based material. (q) Example, sodium lauryl sulfate. (r) 1 g/L (0.13 oz/gal)

by colorimetric methods; electrometric values, using a standard glass electrode, are 0.5 pH lower.

The bath is operated at 79 to 85 °C (175 to 185 °F). Mild agitation is used to prevent stratification, particularly when water is added to replace evaporation losses. Immersion time depends on alloy composition, bath temperature, surface preparation, and other variables; usually it ranges from 1 to 3 min, but it may be as long as 10 min for aluminum-containing alloys.

As indicated in Fig. 10, the tank should be made of stainless steel. Low-carbon steel, unless it is heavily nickel plated, will cause excessive iron contamination of the solution.

The zinc immersion bath has a fairly long life and can be maintained by analysis and regular additions. Eventually, it becomes contaminated with dissolved magnesium and other impurities and must be discarded.

Copper strike plating in a cyanide bath should follow immediately after application of the zinc coating. The work should make electrical contact with the cathode bar before being immersed in the plating bath. If the work is to be subsequently plated in an alkaline bath or in a modified noncorrosive fluoride nickel bath, flash copper deposits of 3 μm (0.1 mil) or less are adequate. However, if an acid bath is to be used, copper deposits of 8 μm (0.3 mil) minimum are required to protect the basis metal from chemical attack, and the work must be thoroughly neutralized and rinsed after the flash plating cycle.

In some applications, because of the complexity of the parts or porosity of the base metal, increasing the minimum thickness of the copper strike plate could be necessary. Castings, because of porosity, generally require a slightly higher minimum thickness of copper when the subsequent plating bath is to be an acid type.

A high current density 3 to 4 A/dm^2 (30 to 40 A/ft^2) is applied for ½ to 1 min, and then reduced to 1.5 to 2.5 A/dm^2 (15 to 25 A/ft^2). Prolonged strike plating at high current density causes poor adhesion and blistering. Any other metal usually electrodeposited can be applied after flash copper plating is completed.

Zinc Undercoating for Die Castings. Whereas AZ91 with 9% aluminum is frequently used for die castings, any surface areas with aluminum-rich segregates are difficult to zincate. The best compromise for castability and plating characteristics is the AZ71 composition with a lower aluminum content of 7%. The section of this article on surface activation outlines normal practice of either an acid or alkaline activation after acid pickling and before zinc immersion plating (Fig. 10). Hot chamber magnesium die castings can be successfully plated using an acid activation followed by an alkaline activation. The patented process follows:

- Acid activation, 0.3 to 1 min, 21 to 29 °C (70 to 85 °F)
- Cold water rinse, minimum 1 min
- Alkaline activation, 0.3 to 1 min, 60 to 65 °C (140 to 150 °F)
- Cold water rinse
- Zinc immersion bath, 60 to 65 °C (140 to 150 °F)

Acid copper is more efficient than semibright nickel for leveling and filling pits in the castings.

Uses of Plated Magnesium. Copper-nickel-chromium systems are being used for both bright and satin decorative finishes on parts for cameras and portable tools. Copper-nickel systems are used for protection against wear and corrosion, as well as to aid in improving the appearance of parts for portable tools. Copper-tin systems, in which the tin plate is subsequently heat flowed in hot oil, are used for corrosion protection and solderability on electronic components.

Corrosion Resistance. Magnesium, being anodic to other metals that are used as plated deposits, would be expected to corrode at a more rapid rate when coatings such as copper and nickel are deposited on it, unless the deposits are pore-free. This has been observed, particularly in salt or marine environments, when these deposits are applied directly on the magnesium surface. However, when zinc is applied as the initial layer, to separate the more noble deposits from the magnesium surfaces, greatly improved results are obtained with regard to the protective value of the subsequent plated coatings. The presence of the zinc layer reduces the galvanic cell action usually encountered once corrosion begins in porous areas of the deposit.

Best corrosion results, especially in marine exposures or with salt-spray testing, are obtained when the plated deposit is pore-free. Porosity in the basis metal promotes porosity in the plated deposits; consequently, corrosion resistance of plated magnesium alloy parts having appreciable surface porosity is less than that of parts that have a minimum of surface porosity. This is evident from the results of salt-spray corrosion studies of plated wrought and sand cast alloys. Less total plate thickness is required to give comparable corrosion resistance over wrought alloys than over cast alloys. The results with wrought alloys are much more consistent over a number of tests than are those with cast alloys, undoubtedly because of the amount of variation in surface porosity or surface quality of the sand castings tested.

In general, on the same type of basis metal, either wrought or cast, corrosion resistance is a function of total plate thickness. Although corrosion tests conducted and reported generally indicate satisfactory performance, service tests should be conducted to determine the best systems and plate thicknesses to give satisfactory corrosion resistance. Suggested minimum total plate thicknesses for decorative chromium plating and copper-nickel-chromium are shown in Table 9.

In addition to the copper-nickel-chromium system, the following systems are being used for finishing magnesium alloy parts in a number of military applications:

- Copper plus tin, with the tin plate being subsequently heat flowed in hot oil
- Copper plus electroless nickel
- Copper plus electroless nickel plus tin, with the tin plate being subsequently heat flowed in hot oil

Field results and accelerated corrosion tests indicate that satisfactory corrosion protection is obtained by use of these systems, which offer about equal protection when deposited to the same total plate thickness.

Table 9 Minimum total plate thicknesses for decorative chromium plating

Service environment	Wrought alloys μm	Wrought alloys mils	Cast alloys μm	Cast alloys mils
Interior	13	0.5	19	0.75
Mild exterior	25	1.0	38	1.5
Average exterior	32	1.25	50	2.0
Severe exterior	38	1.5	64	2.5

Quality Control. The most widely used methods for determining the quality of plated deposits on magnesium alloys are adhesion, relative humidity, and salt spray tests. Relative humidity and salt spray tests are considered standard acceptance tests where corrosion resistance is required.

Adhesion tests are conducted by either (*a*) baking the plated parts at 175 to 260 °C (350 to 500 °F) for 1 h followed by air cooling or (*b*) immersing the plated parts in an oil bath at about 260 °C (500 °F) for a sufficient time for the parts to reach the temperature of the bath followed by quenching in a room-temperature kerosine bath, and then (*c*) inspecting the parts for evidence of blistering. The absence of blistering of the plate generally indicates that proper adhesion of the plate has been obtained. These heat tests are simple, effective, and inexpensive nondestructive methods for determining adhesion and are particularly convenient where 100% testing of parts is required.

Stripping of Plated Deposits. Plated deposits can be satisfactorily stripped from magnesium alloys with little if any attack of the basis metal.

Zinc immersion deposits can be stripped in a room-temperature bath containing 10 to 15 vol % of 70% hydrofluoric acid (HF). The standard acid activator as used prior to zinc immersion coating can also be used.

Chromium can be stripped by reverse current in a hot alkaline solution, such as one containing 60 to 75 g/L (8 to 10 oz/gal) of sodium hydroxide, or combinations of sodium hydroxide and sodium carbonate, in water.

Copper can be stripped by immersion in hot alkaline polysulfide followed by a dip in cyanide, as in stripping copper from steel or die cast zinc.

Nickel can be stripped in a bath containing 15 to 25% hydrofluoric acid and 2% sodium nitrate, using 4 to 6 V. Work

Fig. 10 Preparing magnesium alloys for copper plating

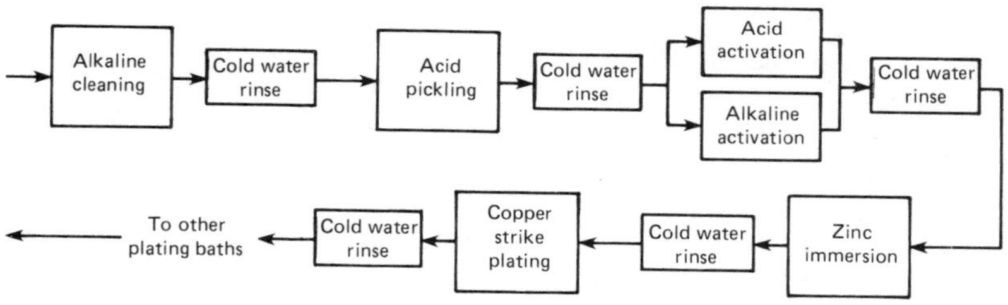

Operation	Cycle time, min	Solution temperature °C	°F	Tank material
Alkaline cleaning(a)	3-10	87-100	190-212	Low-carbon steel
Acid pickling	½-1	21-32	70-90	Ceramic or stainless steel(b)
Acid activation	1-5	21-32	70-90	Rubber, polyethylene, or vinyl plastisol
Alkaline activation	2-5	76	170	Stainless steel
Zinc immersion	1-3	79-85(c)	175-185(c)	Stainless steel
Copper strike plating(d)	6-10	65-71	150-160	Low-carbon steel; rubber
Cold water rinses	½-1	Ambient		Low-carbon steel

(a) May be preceded by solvent cleaning, mechanical cleaning, or both. (b) Mixtures containing fluorides require polyethylene, vinyl-based or synthetic rubber linings. (c) Automatic temperature control preferred. (d) Cathode agitation, constant filtration, and automatic temperature control preferred

See text for details of solution composition and other operating conditions

Table 10 Organic finishing systems for magnesium alloys

Appearance effect	Typical finishing system for Interior service(a)	Exterior service(b)
Bright metal	Buff + ferric nitrate bright pickle + clear epoxy or acrylic	Buff + ferric nitrate bright pickle + clear epoxy or acrylic
Satin finish	Wire brush + ferric nitrate bright pickle + clear epoxy or acrylic	Wire brush + ferric nitrate bright pickle + clear epoxy or acrylic
Tinted clear	Ferric nitrate bright pickle + tinted epoxy or acrylic	Not recommended
Dyed clear	Ferric nitrate bright pickle + clear epoxy or acrylic + dye dip	Not recommended
Metallic	Chrome pickle or dilute chromic acid + epoxy, acrylic, polyvinyl butyral or vinyl pigmented with metal powder or paste	Chrome pickle or dilute chromic acid + 1 coat polyvinyl butyral primer + epoxy or acrylic pigmented with metal powder(c)
Wrinkle	Chrome pickle or dilute chromic acid + standard wrinkle finish	Not generally used in outdoor service
High-gloss enameled	Chrome pickle or dilute chromic acid + epoxy, acrylic, polyurethane or alkyd enamel	Chrome pickle or dilute chromic acid + 1 coat polyvinyl butyral primer + acrylic, alkyd or polyurethane enamel
Leatherette (smooth or textured)	Chrome pickle or dilute chromic acid + vinyl cladding	Chrome pickle or dilute chromic acid + vinyl cladding
	Chrome pickle or dilute chromic acid + vinyl organosol	Chrome pickle or dilute chromic acid + polyvinyl butyral or vinyl primer + vinyl organosol

(a) Suggested minimum total film thickness, 25 μm (1 mil). (b) Suggested minimum total film thickness, 50 μm (2 mil); marine environments may require total film thickness of 75 μm (3 mil). (c) Or metal paste

to be stripped is made anodic; cathodes can be carbon, lead, or magnesium.

Copper, nickel, tin, cadmium, and zinc can be stripped by immersion in the following solution, operated at room temperature: nitric acid (sp gr, 1.42), 1 part by volume; hydrofluoric acid, 70%, 2 parts; water, 2 parts. The tank for containing the solution should be made of polyethylene or some similar material.

Gold, silver, copper, and nickel can be stripped by immersion in proprietary alkaline chemical strippers containing cyanide. Add 15 g/L (2 oz/gal) of sodium hydroxide per gallon of prepared solution to aid in preventing possible etching of the magnesium basis metal. Operating temperature of the bath may range from 21 to 60 °C (70 to 140 °F). Low-carbon steel tanks and heating coils can be used.

Organic Finishing

Organic finishes are used on magnesium alloys to attain various effects in appearance and to provide durability. Organic finishing systems require careful adherence to manufacturers' recommendations on solvents, baking schedules, and compatibility of prime and finish coats to achieve desired film properties and adhesion. A chromate or anodizing treatment before painting roughens the surface and improves adhesion. Bleed-out of dichromate from porous castings or around core-hole plugs is prevented by baking the castings after the dichromate treatment and before the application of an organic finish, particularly one of the baking type.

Some form of coating produced by chemical or anodic treatment is almost always used under an organic finish, either clear or pigmented; an exception would be an item, such as a hand tool, given one coat of aluminized lacquer to provide short-time shelf life or sales appeal.

Table 10 contains information on various appearance effects that are attainable to the mechanical or chemical pretreatment and organic finishing systems required for interior and exterior applications. Military finishing is controlled by specifications; Table 11 lists the various finishing steps involved and indicates the appropriate Government specifications.

Selection of Paint. Both air-drying and baking paints are currently in use. Baking paints are harder and more resistant to attack by solvents. Vinyl alkyds have been used for their resistance to alkali, acrylics for resistance to salt spray, alkyd enamels for exterior durability, and epoxies for good abrasion resistance. Vinyls can withstand temperatures up to about 150 °C (300 °F). Other finishes have increasing temperature resistance in this order: (*a*) modified vinyls, (*b*) epoxies, (*c*) modified epoxies, (*d*) epoxy-silicones, and (*e*) silicones.

Primers for magnesium should be based on such alkali-resistant vehicles

Table 11 Military finishing systems for magnesium alloys

Operation	Appearance	Applicable specifications
Cleaning	...	MIL-M-3171
Chemical treatment	Varies with treatment	MIL-M-3171 or MIL-M-45202
Prime coat	Varies with protective pigment used	MIL-P-15930, MIL-P-52192
Second coat	Matte	TT-E-529
	Semigloss	TT-E-485, TT-E-529 or MIL-L-52043
	Gloss	TT-E-489
Finish coat	Matte	TT-E-529; MIL-L-14486
	Semigloss	TT-E-485, TT-E-529 or MIL-L-52043
	Gloss	TT-E-489

as polyvinyl butyral, acrylic, polyurethane, vinyl epoxy, and baked phenolic. Zinc chromate or titanium dioxide pigments are commonly used in these vehicles for corrosion inhibition. Finish coats are chosen to have:

- Compatibility with primer
- Adaptability to shop conditions and to finishing equipment available
- Ability to perform in service
- Economy consistent to compliance with service requirements

For a detailed discussion of types of paints and methods of application, see the article on painting in this Volume.

Examples of problems that might arise in production painting of magnesium alloy products are given below, together with details of the methods used to solve them.

The first problem, masking multiple threaded inserts in castings before painting, was solved by this method:

- A chemical-resistant finish was applied before drilling and tapping the castings for inserts
- Holes were then tapped and drilled
- Castings were reprocessed through the dichromate to treat threaded holes
- Inserts were set in zinc chromate sealer

Another example involves bubbles and fisheyes appearing in the baked wrinkle-finish paint on dichromated magnesium alloy die castings. This condition was eliminated by lowering the paint-baking temperature from 150 °C (300 °F) to 120 °C (250 °F) and adding a baking cycle of 2 h at 175 °C (350 °F) after the dichromating treatment.

Die castings given a chrome pickle treatment followed by baked coats of primer and wrinkle-finish enamel present another example of problems that might arise. When the primer was applied and baked, small bubbles formed on the film surface; these would later break, leaving an unsightly ring on the surface and affecting uniformity and appearance of the wrinkle finish. Preheating the castings to drive out entrapped air and painting them after they had cooled, greatly reduced the number of bubbles but did not completely eliminate their formation. Painting the castings while they were still hot from the preheating solved the problem.

Stripping of Paint. The ability of magnesium alloys to withstand strong alkalis lessens the problem of paint removal. After painted parts are soaked in a hot, 82 °C (180 °F), solution containing 360 g/L (3 lb/gal) of sodium hydroxide in water, most paints dissolve or lose adhesion to the extent that they can be removed by flushing with water, hand brushing, or other mechanical means. Paint strippers with a phenol or methylene chloride base may be used, but extended soaking of over 24 h causes slight etching of the magnesium alloy, which is undesirable for close-tolerance machined areas.

Stripping vinyl-based enamels from dichromate treated magnesium is more difficult. One method entails softening the coating and weakening the bond by soaking in methylene chloride for ½ to 1 h, then resetting the paint by soaking in hot alkalis for 5 to 30 min. The loose paint is then brushed off, and any remaining paint is removed by wet blasting with 150-mesh aluminum oxide. This stripping operation should be followed as soon as possible by a dichromate treatment.

Chrome pickle and dichromate coatings can usually be removed in hot sodium hydroxide. Old chromate coatings are usually more difficult to remove and work to be stripped of them may require soaking in sodium hydroxide solution followed by immersion in a chromic acid pickling solution.

Special baked primers and other finishes that are difficult to remove in a sodium hydroxide solution may be stripped by being immersed for 5 min or longer in the following solution, at 82 to 100 °C (180 to 212 °F):

Sodium hydroxide	10 to 20 vol %
Diethylene glycol	10 vol %
or ethylene glycol	20 vol %
Water	rem

Health and Safety

Magnesium and magnesium oxide dusts are considered nontoxic. The usual precautions taken to prevent gross inhalation of dirt, dust, and other foreign matter are required. No unusual health hazard has been found associated with the handling, mechanical finishing, or pickling of magnesium alloys, including those containing thorium or beryllium.

Some of the chemical baths used contain chromate or fluorides. Because of the toxic nature of these materials, adequate ventilation is required when baths containing chromate or fluorides are used.

Although bulk magnesium is difficult to ignite unless temperatures are near the melting point, finer particles, such as chips, sawings, and especially sanding dust, are more easily ignited. Proper handling of these materials is required to prevent ignition.

Polishing and Buffing. Special precautions must be followed in the polishing and buffing of magnesium alloys. Magnesium dust or powder must not be allowed to accumulate because it will create extreme fire and explosion hazards. The fire hazards produced in buffing are somewhat less than those produced in polishing. Belt polishing is more potentially dangerous than wheel polishing because the resulting dust is more difficult to collect.

Under no circumstances should steel be polished in the same setup that is used for polishing magnesium, because the polishing of steel causes sparks which can ignite magnesium dust. Because of strong oxidizing properties, heavy chromate coatings also cause sparking when magnesium alloy parts to which they have been applied are being polished.

A system considered safe for the collection of magnesium grinding and polishing dust is one in which the dust is precipitated by a heavy spray of water or low-viscosity mineral oil to form a sludge with a great excess of liquid. The equipment must be designed so that the sludge does not accumulate and dry out and thus become flammable. Inspect the dust-collecting system periodically to ensure that the dust is kept moist.

The best collectors for grinding or polishing stands are small units that serve one or two stands and have the power supply for the grinder or polisher, water circulation, and air blower interconnected so that the grinder or polisher cannot operate unless the collector system is operating. Connect the grinding or polishing machine to the collector by short, straight ducts. Do not allow dry dust to accumulate in the ducts. Clean all polishing machines each day, dust collector and suction pipes at least once a week.

Hydrogen is formed in the reaction of magnesium with water. Ventilate sludge pits adequately to prevent the accumulation of hydrogen. The sludge that collects under the hood frequently must be removed and spread out in thin layers on a noncombustible surface in a safe, isolated area. The sludge may be burned in small quantities. Sludge from dust collectors may be rendered noncombustible by reacting it with a 5% solution of ferrous chloride ($FeCl_2 \cdot 2H_2O$) in an open container placed outdoors.

Central collecting systems in which the dust is passed through long dry ducts, or conventional dust collectors using dry filters, are definitely not recommended, because accident records have indicated dry systems of these types to be extremely dangerous.

Wire Brushing. The dust generated during wire brushing also must be trapped in a water-wash dust collector to minimize fire hazards and prevent the formation of explosive concentrations. This same method of collection has proved suitable for removing the dust during the wire brushing of sheet containing thorium (HK31A and HM21A) and preventing contamination of the air and work area with low-level radioactive dust.

Cleaning and Finishing of Reactive and Refractory Alloys

CLEANING AND FINISHING processes for titanium, tungsten, molybdenum, tantalum, niobium (columbium), zirconium, hafnium, and their alloys are similar to those for other metals. The differences in processes, methods, and cleaning solutions are of major importance to procure maximum use of the finished metal and to maintain safe procedure during the working of these metals.

Generally, a heavy oxide layer resulting from hot working of a metal is removed by grit blasting or other mechanical means, followed by acid pickling in order to achieve the metallic luster of the material. Heavy grease, oil, and black lubricant coatings resulting from cold working of a metal are removed by alkaline or caustic soaking followed by acid pickling. To remove soil or light oil from a metal, cleaning by detergents, solvent washing, or vapor degreasing is generally applied.

Electroplating on refractory and reactive metals is not generally practiced due to difficulty of the process. However, plating by other means for refractory metals, in order to protect the metal from speedy oxidation at elevated service temperatures, is generally mandatory.

The metals are subdivided to cover the following procedures:

- Cleaning and finishing of titanium and titanium alloys
- Cleaning and finishing of tungsten and molybdenum
- Cleaning and finishing of tantalum and niobium
- Cleaning and finishing of zirconium and hafnium alloys

Cleaning and Finishing of Titanium and Titanium Alloys*

Cleaning procedures serve to remove scale, tarnish films, and other contaminants which form or are deposited on the surface of titanium and titanium alloys during hot working, heat treating, and other processing operations. Because their resistance to many corrosive environments is excellent, titanium alloys do not require special surface treatments to improve corrosion resistance; however, several different coatings may be applied to the surface of these materials to develop or to enhance specific properties, such as lubricity and emissivity.

Cleaning and Descaling Problems

The metallurgical and chemical properties of titanium create a number of very special cleaning problems. These include:

- Affinity of titanium to common gases
- Galvanic effects caused by discontinuities in scaled surfaces
- Metallurgical restrictions on the temperature of the descaling media
- Variety of scales encountered in titanium descaling
- Protective coatings used in titanium manufacturing

*By Douglas H. Wilson, Manager, Customer Technical Services, RMI Company, and William G. Wood, Vice President, Technology Research & Development, Kolene Corp.

Fig. 1 Ti-6Al-4V alpha case

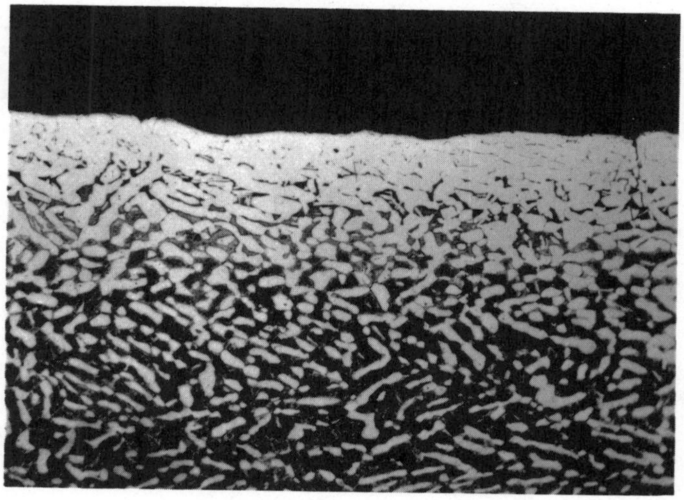

Magnification: 250×

Gas Absorption. The property that causes the most difficulty is the capacity of titanium to absorb common gases including oxygen, hydrogen, and nitrogen, all of which tend to embrittle the product. Because tightly packed hot rolling scale acts as partial protection against additional gas absorption, some mills perform two or three heat treatments over the scale. Each additional heat treatment toughens the scale and compounds descaling difficulties. An additional problem is that heat treating furnace atmospheres are maintained on the oxidizing side, because a diffusion rate of oxygen in titanium is so much lower than that of hydrogen. A layer of oxygen-rich metal, as shown in Fig. 1, develops beneath the resulting scale formation, varying in thickness from 0.05 to 0.07 mm (0.002 to 0.003 in.) in the heat treated condition to 0.15 to 0.20 mm (0.006 to 0.008 in.) in the hot rolled condition. This brittle surface is usually removed by acid or electrolytic chemical pickling.

Galvanic effects and discontinuities in the surface scale are encountered in all types of metal descaling, but appear to be more pronounced in titanium. Although the exact cause of small pits or cells formed in descaled material is debatable, possibilities include alloy or nonmetallic segregations, scale porosity, and surface contamination. A more severe galvanic attack problem is created by patch scale conditions on titanium surfaces, when areas of heavy scale flake away from an apparently uniform surface. The same problem has been observed with superimposed oxides, even though the surface layer may be quite thin and powderlike. Surface contamination with oil, grease, or fingerprints can also create a patch scale condition. All of these factors promote severe localized attack when areas of the basis metal are exposed selectively during descaling. Some producers have considered reoxidation of the product during processing as a possible solution.

Metallurgical Restrictions on Descaling. Solution treated, age-hardenable alloys of titanium are sensitive to time-temperature reactions, and the temperatures of descaling media. The metastable or high beta alloys, which are solution treated and aged at temperatures ranging from 370 to 540 °C (700 to 1000 °F) collective for times of ½ h or more, may induce a subsequent aging effect and cause a change in mechanical properties. This is particularly true in thin-gage sheet materials and may cause property changes of as much as 70 MPa (10 ksi) in tensile strength. The alloys normally should not be allowed to exceed 260 °C (500°F).

Variety of Scale. Another factor that contributes appreciably to titanium descaling problems is the wide variety of scale encountered, including scale formed by annealing, forging, solution treating, stress relieving, extruding, rolling, aging, hot forming, or a combination of several of these operations. With processing temperatures ranging from 425 to 1150 °C (800 to 2100 °F), the scale spectrum for titanium is far broader than for most other difficult-to-descale materials.

Coatings. Protective coatings are often used in titanium manufacturing operations. These coatings, which are an asset and a necessity in manufacturing operations, become a liability and a contaminant in cleaning operations. They are soluble and removable if the proper techniques are used. Protective coatings are applied to titanium surfaces during manufacturing operations to:

- Lubricate and aid in metal flow, good die contouring, and forming operations
- Act as barrier films, reducing gas contamination during high temperature forming and heat treating cycles
- Reduce surface flaws caused by nicking and scratching during manufacturing operations

The gas protective films are usually applied directly to the titanium surface. They are silicate-based materials that deposit uniform fusable films through solvent evaporation. These films form glassy barriers at treatment temperatures up to 815 °C (1500 °F) and are quite effective in reducing oxygen, hydrogen, and nitrogen contamination. Above 815 °C (1500 °F), most of these films are less effective, because spheroidizing creates voids in the film. Lubricant films or abrasion protective films are applied over a silica-based coating. This process has the advantage of providing double protection against scratching and scoring. During hot forming operations and metal surface stretching, some voiding and penetration occurs, creating a titanium oxide on the surface. The contaminant then consists of organic bond or residues, graphite, molybdenum disulfide, silicates, and titanium oxides.

Removal of Scale

Scale is removed from titanium products by several mechanical methods. Abrasive methods, such as grinding and grit blasting, are preferred for removing heavy scale from large sections. Centerless grinding is used for finishing round bars, and wide-belt grinding is used for finishing sheet and strip. Grinding is usually most efficient when it is performed at low wheel and belt speeds.

Most alloy sheet materials with a high aluminum content such as Ti-5Al-2.5Sn are ground to eliminate pits and a rippled condition that develops in hot rolling as a result of discontinuous slip during plastic deformation. Grinding is frequently used to eliminate surface defects before cold rolling. Originally, strip was ground on standard strip grinders, using various oil lubricants; however, oils contributed to fire hazard and several grinding machines were partially or wholly destroyed when the oil ignited. When titanium was ground with aluminum oxide belts, a water lubricant was less effective than air. The water reacted with the aluminum oxide to form a weak hydroxide that proved ineffective as a grinding lubricant.

Fig. 2 Cleaning and belt grinding sequences for Ti-6Al-4V sheet

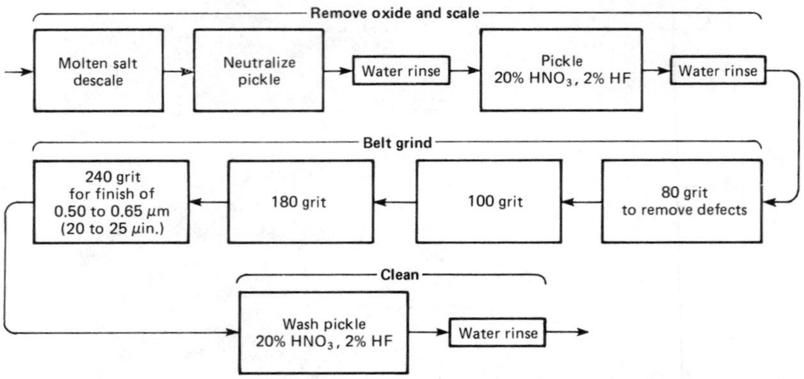

Belt Grinding. Dry belt grinding is dangerous because of the hazards of explosion and fire. It is uneconomical because of poor belt life. When stock is removed during dry grinding, small globules of molten metal and oxide roll along the sheet, causing a type of pitting by burning that is not removed by the grinding. Weld grit scratches and embedded grit result when titanium welds to the dry grit.

A 5% aqueous solution of potassium orthophosphate, K_3PO_4, is widely used as a grinding lubricant. It is applied as a flood at both the entrance and exit side of the contact line. Water-soluble oils, particularly highly chlorinated and sulfochlorinated oils, have also been successful as lubricants. These compounds should be used with care because of the possibility of chloride residues remaining as an integral part of the surface. Both types of lubricants improve grinding efficiency when the belts are coated with aluminum oxide or silicon carbide.

Flooding the work with lubricant is recommended; however, machines built for flooding are equipped with a recirculating and filtering system and waterproof cloth belts, and they are expensive. An alternative is to spray a water-soluble wax fog through atomizing nozzles on the line of contact at both the entrance and exit sides of the belt. The solution should not be sprayed through an ejector that mixes it with air and increases the fire hazard. Instead, it should be sprayed as an atomized liquid. Application of the spray can be controlled to volatilize the lubricant during grinding. This eliminates the need for waterproof belts. Care must be exercised to avoid a buildup of titanium chips that may cause a fire. Chips that would wash away in a flood are not removed by the spray.

Titanium should be ground at belt speeds not exceeding rates of 8 m/s (1500 ft/min). Using a 5% solution of potassium orthophosphate as a lubricant, maximum efficiency is achieved at about 6 m/s (1100 ft/min). Both billy roll and flat table grinding machines have been successful in grinding titanium. Sheet grinding machines, equipped with feed rolls, sometimes leave a ground line on the sheet where the feed rolls, on the exit side of the machine, and interrupt uniform travel when they grip the sheet. A high degree of grinding uniformity is obtained on machines equipped with a flat table and vacuum chuck. On these machines, the table holding the sheet usually oscillates. Traveling-head machines are available also.

The belt grinding sequence is usually begun with an 80-grit belt, when it is necessary to remove more than 0.07 mm (0.003 in.) of stock from the surface of the sheet. Descaling and pickling the sheet before grinding prolongs belt life. Following the initial grit, each successive grit must remove enough stock to eliminate the scratches caused by the previous grit. The alpha alloys, such as Ti-5Al-2.5Sn, are less sensitive to surface condition than the alpha-beta alloys, such as Ti-6Al-4V. Surface pits on Ti-6Al-4V sheet, caused by weld grit scratches, seriously detract from bend ductility. A flow chart for the belt grinding of Ti-6Al-4V sheet is shown in Fig. 2.

Abrasive blast cleaning techniques, either wet or dry, are convenient for removing scale from a variety of titanium products, ranging from massive ingots to small parts. Because it can be used at lower velocities and is less likely to be embedded in the surface, alumina sand is preferred to silica sand.

Sheet in thicknesses to about 0.50 mm (0.020 in.) can be descaled without distortion if fine sand and low velocities are used. Mill scale on titanium products can be removed with coarse high-carbon steel shot or grit, while finished compressor blades can be cleaned with zircon sand of 150 to 200 mesh. The type of product to be cleaned, the cleaning rate, and the cost of the abrasive must be balanced in the selection of a specific blast cleaning method. For a discussion of selection factors and equipment, see the article on abrasive blast cleaning in this Volume.

Mineral abrasive particles, such as silica, zircon, or alumina sands, are used more commonly than metal abrasives for blasting finished or semifinished products. Although these abrasives are more expensive, they produce the finer finish that is required in final processing or service. Adequate safety precautions must be observed to avoid inhalation of fine sand particles. Air-circulating and dust-collecting systems must be cleaned frequently and equipped to cope with the fire hazard associated with titanium dust.

A fine dust remains on the titanium from the blasting operation, particularly when mineral abrasives have been used. This is not considered detrimental, although a washing or pickling cycle following the blast is desirable if the part is to be welded subsequently. The following describes a wet blasting procedure and a dry blasting procedure used for descaling titanium parts:

- *Wet blasting*: parts are wet blast cleaned, using a slurry that consists of 400-mesh aluminum oxide, 40 vol %, and water. Air pressure of 655 kPa (95 psi) is used to pump the slurry in a steady stream with a pressure of about 34 kPa (5 psi). The descaling rate, normally about 50 min/m^2 (5 min/ft^2), depends on the complexity of the part. Distortion and the need for planishing are held to a minimum by placing the blast nozzle at a distance of approximately 50 mm (2 in.) from the workpiece, and by using an angle of impingement of 60°
- *Dry blasting*: Rocket motor case assemblies are dry blasted after final stress relieving at 480 to 540 °C (895 to 1005 °F). Blasting is accomplished with 100- to 150-mesh zircon sand at an air pressure of 275 kPa (40 psi). Each assembly is rotated at 2½ rev/min and is passed at a speed of 65 mm/min (2.5 in./min) between two diametrically opposed fixed position blasting nozzles. The nozzles blast the inside and outside surfaces simultaneously at the same wall location. To prevent distortion, each nozzle is placed at the same distance, 300 mm (12 in.), from the metal surface.

Fig. 3 Effect of surface condition on etched metal surface

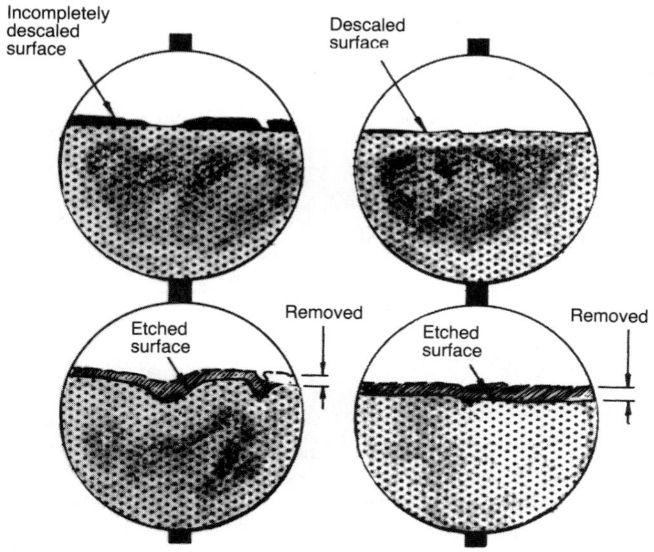

Molten Salt Descaling Baths

Molten salt descaling baths are primarily used for descaling, bar, sheet products, and tubing. With the most effective barrier films available today, some gas penetration of titanium surfaces can be expected at the elevated temperatures required for working and heat treatment. The alpha case or oxygen-enriched layer resulting from this gas reaction is extremely hard and brittle and must be removed. Bar products used for machining finished parts must have this hard scale and oxide removed because these are very abrasive and cause rapid tool wear. Welding or forming must have these scales removed, or poor and small welds are made and forming (hot or cold) is virtually impossible without surface rupture or failure of parts. Removal presents no serious problems since chemical milling techniques have been perfected by the aircraft industry to effect weight savings. In the case of titanium, the purpose is to improve the structural soundness of metal, and the solvent materials applied are of a different chemical composition.

One specific problem encountered in alpha case removal is that the titanium oxide formed is substantially more insoluble in the nitric hydrofluoric etchant than the base metal. Residues of oxide on the surface develop areas resembling craters on the finished product. Examination of the sketch of the artist's conception shown in Fig. 3 indicates the surface as a result of a nonuniform cleaning operation.

Where alpha case removal is a required part of a manufacturing operation, salt bath cleaning is specified because proper cycling practically guarantees a chemically clean surface. Conditioning salt baths fall into two basic categories of high-temperature salt baths and low-temperature salt baths.

High-temperature salt baths may vary in chemical reaction and effectiveness depending on composition. All types operate at a range of 370 to 480 °C (700 to 895 °F). The temperature range is sufficiently high to produce the most rapid reaction possible for soiled and oxide films. High-temperature oxidizing salt baths are also capable of reacting chemically with organic films to destroy them. These baths are also excellent solvents for silicate barrier films. They do require special fixturing to reduce the strong galvanic effects present at these temperatures, and for this reason, they are used in cleaning primary forming operation products such as forgings, extrusions, rolled plate, and sheet. The major advantage of high-temperature oxidizing or reducing salt baths for titanium descaling is their great speed in removing extremely tenacious scale. Although reducing baths have the inherent disadvantage of promoting hydrogen absorption, this can be overcome or minimized by chemical additions. Vacuum degassing is another solution.

A primary producer of titanium sheet uses an oxidizing salt bath for removing the hot work scale in the following sequence of operations:

- Immerse in oxidizing salt for 5 to 20 min at 400 to 480 °C (750 to 895 °F)
- Quench with water 1 min
- Immerse in sulfuric acid, 10 to 40 vol %, for 2 to 5 min at 50 to 60 °C (120 to 140 °F)
- Rinse with water 1 min
- Recycle if necessary

- Pickle in nitric-hydrofluoric acid solution, time and concentration as required

The same producer also uses a sodium hydride reducing salt bath for descaling high beta or metastable beta alloys. A typical cycle using this type of salt is:

- Immerse in reducing salt for 1 to 3 min at 370 °C (700 °F)
- Quench in water 1 min
- Immerse in sulfuric acid, 10 to 40 vol %, for 2 to 5 min at 50 to 60 °C (120 to 140 °F)
- Rinse in water
- Pickle in nitric-hydrofluoric acid solution, time and concentration as required
- Vacuum degas or decontaminate titanium beta alloys that absorb hydrogen in reducing baths

These baths are used by one of the major aerospace contractors for cleaning titanium blades for jet engines. Blade materials are Ti-6Al-4V and Ti-8Al-1Mo-1V. Descaling cycles for removing oxides and proprietary glass-like compounds from these blades are:

- Immerse in oxidizing salt for 15 min at 455 °C (850 °F)
- Rinse in cold water
- Pickle in solution of 35% nitric acid and 3.5% hydrofluoric acid for 1 min max at 20 °C (70 °F)
- Rinse in hot water

Low-Temperature Baths. The temperature range used for cleaning fabricated parts is 200 to 220 °C (390 to 430 °F). Descaling systems based on salts in this temperature range eliminate some of the possible problems associated with higher temperature baths including:

- Age hardening
- Dissimilar metal reactions
- Chemical attack
- Metal distortion
- Hydrogen embrittlement

Salts in this range have a very limited composition because of the effect of various compounds on the melting point. Although they contain oxidizing agents, the effect of these materials is not as aggressive as it is in the high-temperature fused salts. Consequently, organic materials are not destroyed, but are saponified and absorbed. Silicate barrier films and molybdenum disulfide are soluble in these low-temperature salts. The temperature range permits cycling between salt and acid to reduce cleaning times and costs. Examples of salt bath and acid cycle times are given in Table 1.

Table 1 Low-temperature salt bath and acid bath conditions

Sample composition	Scale formation temperature °C	Scale formation temperature °F	Salt bath immersion time(a), min	Acid cleaning bath time(b), min	Acid cleaning bath time(c), s
Ti-6Al-4V	650	1200	2	2	30
Ti-8Al-1Mo-1V	650	1200	2	2	30
Ti-8Al-1Mo-1V	820	1510	5	2	30
Ti-6Al-4V(d)	820	1510	5	5	30
Ti-6Al-4V(e)	950	1745	5	5	60
Ti-8Al-1Mo-1V(f)	950	1745	5	5	60

(a) Salt bath temperature 205 °C (400 °F). (b) Bath composition, 30% sulfuric acid. (c) Bath composition, 30% nitric acid, 3% hydrofluoric acid. (d) Sample recycled in salt bath for 5 min, in sulfuric acid bath for 5 min, in nitric acid-hydrofluoric acid bath for 30 s. (e) Sample recycled in salt bath for 5 min, in sulfuric acid bath for 5 min, in nitric acid-hydrofluoric acid bath for 60 s. (f) Sample recycled in salt bath for 5 min, in sulfuric acid bath for 5 min, in nitric acid-hydrofluoric acid bath for 60 s

Aqueous caustic descaling baths have been developed to remove light scale and tarnish from titanium alloys. Aqueous caustic solutions containing 40 to 50% sodium hydroxide have been used successfully to descale many titanium alloys. One bath containing 40 to 43% sodium hydroxide operates at a temperature near its boiling point 125 °C (260 °F). Descaling normally requires from 5 to 30 min. Immersion time is not critical because little weight loss is encountered after the first 5 min. Caustic descaling conditions the scale so that it is removed readily during subsequent acid pickling.

A more effective aqueous solution contains either copper sulfate or sodium sulfate in addition to sodium hydroxide. This bath operates at a lower temperature, 105 °C (220 °F). A composition of this solution by weight is as follows: 50% sodium hydroxide, 10% copper sulfate pentahydrate ($CuSO_4 5H_2O$), and 40% water. Using immersion times of 10 to 20 min, this bath has proved effective in descaling Ti-6Al-4V and Ti-2.5Al-16V alloys.

Pickling Procedures Following Descaling

All advantages gained through proper conditioning and handling of titanium parts during cleaning can be lost if the composition of the final pickling acid is not controlled. Cold spent acid solutions have increased appreciably the time requirements for pickling and the possible quality problems experienced with hydrogen pickup. Highly concentrated hot acids can be overly aggressive, resulting in surface finish problems, such as a rough and pitted surface caused by preferential acid attack. Sulfuric acid, 35 vol % at 65 °C (150 °F), is recommended for pickling immediately following salt bath conditioning and rinsing to remove molten salt and residual softened scales. An acid of this formula has very little effect on titanium metal. Metal salts in the original and additional acid solutions further minimize these base metal attacks. Table 2 gives conditions for corrosion of titanium in various sulfuric acid pickle baths.

Removal of Tarnish Films

Tarnish films are thin oxide films that form on titanium in air temperatures between 315 and 650 °C (600 and 1200 °F), after exposure at 315 °C (600 °F). The film is barely perceptible, but with increasing temperature and time at temperature, it becomes thicker and darker. The film acquires a distinct straw yellow color at about 370 °C (700 °F), and a blue color at 480 °C (900 °F). At about 650 °C (1200 °F), it assumes the dull gray appearance of a light scale. Alloying elements and surface contaminants also influence the color and characteristics.

Tarnish films are readily removed by abrasive methods, and all but the heaviest films can be removed by acid pickling. Prolonged exposures at temperatures above about 595 °C (1105 °F), in combination with surface contaminants, result in heavier surface films that are not removed satisfactorily by acid pickling, but require descaling treatments for their removal.

Acid Pickling. Acid pickling removes a light amount of metal, usually a few tenths of a mil. It is used to remove smeared metal, which could affect penetrant inspection. Titanium and titanium alloys can be satisfactorily pickled by the following procedure:

Table 2 Corrosion of titanium in sulfuric acid pickle baths

A nitric-hydrofluoric acid solution, which is the final stage brightening in most alloy cleaning lines, should be maintained at a minimum ratio of 15 parts nitric acid to 1 part hydrofluoric acid to reduce hydrogen pickup effects; the concentration of hydrofluoric acid may vary from 1 to 5%, or even higher as long as the ratio is not exceeded; the activity of these pickle solutions is effected by titanium content, and the acids are frequently discarded at a level of 26 g/L (3 oz/gal); the solution used for final brightening can be used for the required alpha case removal also, with careful watch of titanium content

Sulfuric acid concentration, %	Acid addition	Bath temperature °C	Bath temperature °F	Corrosion rate μm/yr	Corrosion rate mils/yr
30	0.5% copper sulfate	38	100	100	4.0
30	1% copper sulfate	38	100	20	0.8
30	10% copper sulfate	38	100	400	16.0
30	0.25% copper sulfate	95	205	76	3.0
10	2% ferrous sulfate	Boiling point	Boiling point	125	5.0
17	7-8% ferrous sulfate	60	140	125	5.0

- Clean thoroughly in alkaline solution to remove all shop soils, soap drawing compounds, and identification inks. If coated with heavy oil, grease, or other petroleum-based compounds, parts may be degreased in trichlorethylene before alkaline cleaning. Degreasing will not be harmful to the part in subsequent processing
- Rinse thoroughly in clean running water after alkaline immersion cleaning
- Pickle for 1 to 5 min in an aqueous nitric-hydrofluoric acid solution containing 15 to 40% nitric acid and 1.0 to 2.0% hydrofluoric acid by weight, and operated at a temperature of 24 to 60 °C (75 to 140 °F). The ratio of nitric acid to hydrofluoric acid should be at least 15 to 1. The preferred acid content of the pickling solution, particularly for alpha-beta and beta alloys, is usually near the middle of the above ranges. A solution of 33.2% nitric acid and 1.6% hydrofluoric acid has been found effective. When the buildup of titanium in the solution reaches 12 g/L (2 oz/gal), discard the solution
- Rinse the parts thoroughly in clean water
- High-pressure spray wash thoroughly with clean water at 55 ± 6 °C (130 ± 10 °F)
- Rinse in hot water to aid in drying. Allow to dry

To avoid excessive stock removal, the recommended immersion times for pickling solutions should not be exceeded. It is equally important to maintain the composition and operating temperature of the bath within the limits prescribed to prevent an excessive amount of hydrogen pickup. Gage loss from all acid pickling after descaling is estimated to be less than 0.025 mm/min (0.001 in./min), depending on the combination of variables used.

Hydrogen contamination is estimated to be 0 to 15 ppm/0.025 mm (0.001 in.) of metal removed, depending on alloy composition and gage material pickled. Data on hydrogen pickup for an alpha, an alpha-beta, and a beta alloy pickled in a 15% nitric acid, 1% hydrofluoric acid bath at 49 to 60 °C (120 to 140 °F) are given in Table 3. Hydrogen contamination can be held to a minimum by maintaining an acid ratio of 10 to 1 or greater of nitric acid to hydrofluoric acid. Hydrogen diffuses more rapidly into the beta phase. Alpha-beta alloys that have $\alpha + \beta$ microstructures, which have been heat treated to complete equilibrium, pick up less hydrogen than microstructures of transformed beta and/or simple mill annealed structures.

Mass Finishing (Barrel Finishing). Oxide films formed by heating to temperatures as high as 650 °C (1200 °F) for 30 min have been effectively removed from Ti-8Mn alloy parts by wet mass finishing. At barrel speeds of 43 000 to 51 000 mm/min (1700 to 2000 in./min), parts have been cleaned satisfactorily in about 1 h. Complete barrel loading procedures for three barrels, ranging from 0.02 to 0.25 m^3 (0.75 to 8.85 ft^3) capacity, are given in Table 3. For more information on mass finishing equipment and procedures, see the article on mass finishing in this Volume.

In mass finishing titanium parts, the ratio of medium to parts should be between 10 and 15 to 1, depending on the size of the parts. Proportionately more medium is required as part size increases. Water is used to cover parts and medium. Surface finish is improved when more water is added, but cycle time required to obtain a given finish is

Table 3 Effect of titanium alloy composition on hydrogen pickup in acid pickling

Pickling bath is an aqueous solution containing 15% nitric acid and 1% hydrofluoric acid by weight; operating temperature is 49 to 60 °C (120 to 140 °F)

Alloy	Thickness mm	Thickness in.	Hydrogen pickup (gage removed) ppm/0.0250 mm (ppm/0.001 in.)
Alpha alloy			
Ti-5Al-2.5Sn	0.50	0.020	0-4
Ti-5Al-2.5Sn	1.00	0.040	0-3
Alpha-beta alloy			
Ti-6Al-4V	0.50	0.020	4-7
Ti-6Al-4V	1.00	0.040	3-5
Beta alloy			
Ti-13V-11Cr-3Al	0.50	0.020	10-15
Ti-13V-11Cr-3Al	1.00	0.040	5-8

increased. The rate of descaling increases directly with barrel speed but is limited by the fragility of the parts being processed. Parts are randomly loaded in the barrel, and rotated at relatively low barrel speeds to minimize distortion and nicking. Conditions for mass finishing of titanium parts are given in Table 4.

Aluminum oxide mediums are the most satisfactory. They do not contaminate the work and have a long useful life. For oxide removal, small well-worn mediums produce the highest finish. To avoid possible metallic contamination, the medium used for titanium should not be used in processing other metals. Strong acid-forming compounds are avoided, principally because they are corrosive and contribute to hydrogen embrittlement. Because of the fire hazard created by fine, dry titanium particles, dry mass finishing of titanium parts is not recommended.

Polishing and Buffing

The polishing and buffing of titanium is accomplished with the same equipment used for other metals. Polishing is frequently done wet, using mineral oil lubricants and coolants. Silicon carbide abrasive cloth belts have been effective. It is common to polish in two or more steps, using a coarser grit initially, such as 60 or 80, to remove gross surface roughness, followed by polishing with 120 or 150 grit to pro-

Table 4 Mass finishing conditions for titanium parts

Capacity, m^3	Capacity, ft^3	Barrel size Diameter, mm	Barrel size Diameter, in.	Barrel size Width, mm	Barrel size Width, in.	Speed, rev/min	Part load, kg	Part load, lb	Medium(a), kg	Medium(a), lb	Water, L	Water, qt	Abrasive compound(b), kg	Abrasive compound(b), lb	Alkaline cleaner(c), kg	Alkaline cleaner(c), lb
0.02	0.75	381	15	178	7	36	1-2	3-4	18	40	1.2	1.25	0.2	0.5	0.2	0.5
0.07	2.33	559	22	240	10	28	4-5	8-12	54	120	4	4	0.7	1.5	0.34	0.75
0.25	8.85	813	32	457	18	20	14-18	30-40	209	460	14	15	2.3	5	0.5	1

(a) Aluminum oxide nuggets 6.4 to 38 mm (0.25 to 1.5 in.) or preformed vitrified chips 4.8 by 9.5 to 7.9 by 28.6 mm 3/16 by 3/8 to 5/16 by 1 1/8 in.). (b) Dry, mildly alkaline compound. (c) Mild cleaner with high soap content

vide a smooth finish. Titanium tends to wear the sharp edges of the abrasive particles, and to load the belts more rapidly than steel. Frequent belt changes are required for effective cutting. A good flow of coolant improves polishing and extends the life of the abrasives.

Dry polishing is more appropriate than wet for some applications. For these operations, belts or cloth wheels with silicon carbide abrasive may be used. Soaps and proprietary compounds may be applied to the belts to improve polishing and to extend belt life. Abrasive belt materials that incorporate solid stearate lubricants offer improved results for dry polishing operations.

Fine polishing of titanium articles for extremely smooth finishes requires several progressive polishing steps with finer abrasives until pumice or rouge types of abrasive are applied. With the softer grades of titanium, such as unalloyed material, fine polishing requires more time and care to prevent scratching. The harder alloy grades can be polished more readily to a surface of high reflectivity. If a matte finish is desired, wet blasting with a fine slurry may be used after initial polishing.

Titanium alloys can be buffed safely. The purpose of buffing is to improve the surface appearance of the metal and to produce a smooth tight surface. Buffing is used as a final finishing operation and is particularly adaptable to finishing a localized area of a part. Parts such as body prosthesis, pacemakers, and heart valves require a highly buffed tight surface to prevent entrapment of particles. Close fitting parts for equipment, such as the modern guidance systems, and electronics applications require highly polished surfaces obtained by buffing. In addition, sheet sizes too large to be processed by other abrasive finishing methods, such as mass finishing or wet blasting, can be economically processed by buffing.

The principal limitations of buffing are (*a*) distortion, caused by the inducement of localized stress, (*b*) surface burning, resulting from prolonged dwell of the buff, (*c*) an inability to process inner or restricted surfaces, and (*d*) the feathering of holes and edges. Proper care of the buffing wheel is essential. Buffing with insufficient compound or a loaded wheel produces burning or distortion of the part. After buffing, no further cleaning of parts is required except degreasing to remove the buffing compound. Further information may be found in the article on polishing and buffing in this Volume.

Wire Brushing

Wire brushing of titanium alloys is not recommended when other finishing methods, such as buffing, can accomplish the objective. Wire brushing used on titanium, in an attempt to remove surface scratches or oxide films, resulted in serious defects. A stiff-bristled wire brush removed surface scratches and oxide films, but the surface was pitted by the wire tips. To avoid pitting, softer wire bristles were tried. The surface of the titanium acquired a burnished appearance; surface layers were cold worked; and grinding scratches, instead of being removed, were filled with smeared metal. Wire brushing with a silicon carbide abrasive grease has been used successfully to remove burrs, break sharp edges from edge radii, and blend chamfers.

Removal of Grease and Other Soils

Removal of grease, oil, and other shop soils from titanium parts is normally accomplished with the same type of equipment and the same cleaning procedures used for stainless steel and high-temperature alloy components. Descriptions of the equipment and procedures used in all of the methods, including vapor degreasing, emulsion and solvent cleaning, and alkaline cleaning, are presented in individual articles in this Volume. Certain aspects of conventional processes must be modified or omitted when titanium alloys are being cleaned.

Vapor degreasing normally employs either trichlorethylene or perchlorethylene. Under certain conditions, these solvents are known to be a cause of stress-corrosion cracking in titanium alloys. Methylethyl ketone is used as a cleaner in situations where chlorinated solutions are not desired. All titanium parts should be acid pickled after vapor degreasing.

Other cleaning methods use chemicals which, if they are left to dry on the part, may have a harmful effect on the properties of titanium. Among these are (*a*) soda ash, borates, silicates, and wetting agents commonly used in alkaline cleaners; (*b*) kerosine and other hydrocarbon solvents used in emulsion cleaners; and (*c*) mineral spirits employed in hand wiping operations. Residues of all these cleaning agents must be completely removed by thorough rinsing. To ensure a surface that is free of contaminants, rinsing is frequently followed by acid pickling.

Chemical Conversion Coatings

Chemical conversion coatings are used on titanium to improve lubricity by acting as a base for the retention of lubricants. Titanium has a severe tendency to gall. Lack of lubricity creates serious problems in applications involving the contact of moving parts in various forming operations.

Conversion coatings are applied by immersing the material in a tank containing the coating solution. Spraying and brushing are alternate methods of application. One coating bath consists of an aqueous solution of sodium orthophosphate, potassium fluoride, and hydrofluoric acid, and can be used with various constituent amounts, immersion times, and bath temperatures. The resultant coatings are composed primarily of titanium and potassium fluorides and phosphates. Several solutions are listed in Table 5. The flow chart of

Table 5 Conversion coating baths for titanium alloys

Bath No.	Bath solution	Composition	Amount g/L	Amount oz/gal	Temperature °C	Temperature °F	pH	Immersion time, min
1	Degreasing solution	$Na_3PO_4 \cdot 12H_2O$	50	6.5	85	185	5.1-5.2	10
		$KF \cdot 2H_2O$	20	2.6				
		HF solution(a)	11.5	1.5				
2	Pickling solution	$Na_3PO_4 \cdot 12H_2O$	50	6.5	27	81	<1.0	1-2
		$KF \cdot 2H_2O$	20	2.6				
		HF solution(a)	26	3.4				
3	Chemical immersion solution	$Na_2B_4O_7 \cdot 10H_2O$	40	5.2	85	185	6.3-6.6	20
		$KF \cdot 2H_2O$	18	2.3				
		HF solution(a)	16	2.1				

(a) Hydrofluoric acid, 50.3% by weight

Fig. 4 Processing sequence used in the chemical coating of titanium alloys

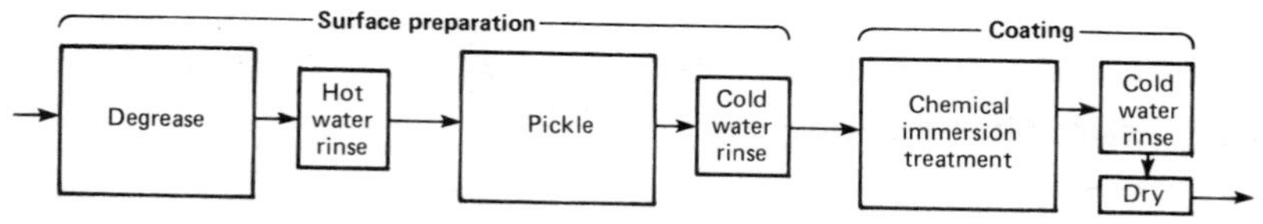

Type of solution	Composition	Operating temperature °C	Operating temperature °F	Cycle time, s
Degreasing	Alkaline cleaner	Boiling point		60-120
Pickling	15% HNO_3, 1% HF in water	23	75	10-60
Chemical immersion	(a)	...	...	...

(a) See Table 5

Fig. 5 Weight of conversion coating as a function of immersion time

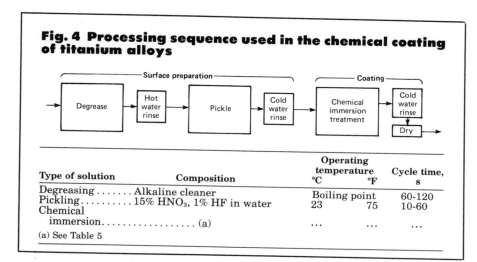

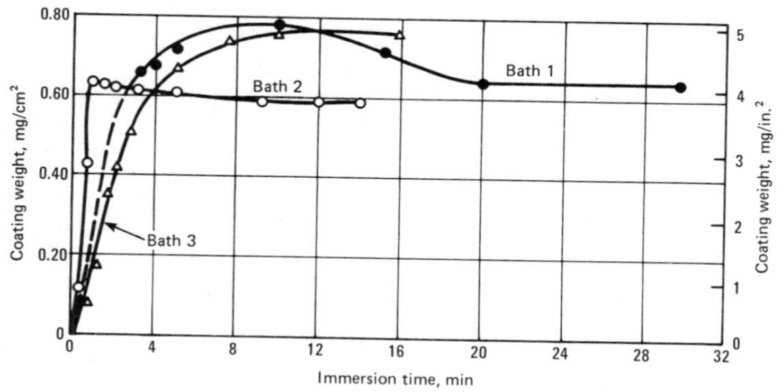

See Table 5

Fig. 4 indicates a processing sequence for coating of titanium alloys.

Cleanness of the part before immersion is critical, and all preliminary cleaning and handling operations must be closely controlled for good results. Finger marks or residual grease on the surface of a part will interfere seriously with the coating process.

The appearance of the baths varies widely during the coating reaction, ranging from rapid bubbling to relative dormancy. Some coatings rub off when still wet; others are adherent. The solutions produce coatings of approximately the same dark gray or black appearance.

The control of pH and immersion time is important. Dissolved titanium and the active fluoride ion make it impossible to use glass electrodes for pH measurements. Indicator paper and colorimetry are the most satisfactory methods for measuring in the degreasing and chemical immersion baths, which are held in the pH range from 5 to 7. The pickling bath is quite acid, and titrametric analysis offers the most practical method of control. When the bath is in the proper coating range, a 20-mL (0.70-fluid oz) sample in 100 mL (3.4 fluid oz) of water will neutralize 11.8 to 12.0 mL (0.4 to 0.41 fluid oz) of normal sodium hydroxide, using a phenolphthalein indicator.

Coating thickness depends on immersion time. In all three baths, a specific time is reached after which the coating weight remains essentially constant. In the fluoride-phosphate baths, a maximum coating weight is reached at some time before this equilibrium point. The maximum coating weight is obtained in about 2 min in the low-temperature bath and in about 10 min in the two other baths. The relation between immersion time and coating weight is shown in Fig. 5.

Results of extensive wiredrawing experiments, given in Table 6, illustrate the effectiveness of conversion coatings when used with various lubricants. Reciprocating wear tests showed that conversion coatings and oxidized surfaces provided some improvement in wear characteristics, but when conversion coated samples were also oxidized, a marked improvement was noted. The conversion coating increases the oxidation rate of titanium at about 425 °C (800 °F) and may increase oxidation rates at temperatures up to 595 °C (1100 °F). The original coating is re-

Table 6 Comparison of conversion coatings in wiredrawing of titanium

Coating	Drawing compound	Total reduction, %	No. of passes	No. of coats	Final condition
Bare	Molybdenum disulfide with grease	...	0	...	Galled
Bare	Soapy wax	...	0	...	Galled
Degreasing bath	Molybdenum disulfide with grease	85	8	2	Smooth
Pickling bath	Molybdenum disulfide with grease	94	17	7	Smooth
Pickling bath	Soapy wax	68	7	3	Galled
Pickling bath(a)	Molybdenum disulfide with grease	70	7	1	Smooth
Chemical immersion bath	Lacquer molybdenum disulfide	63	8	2	Smooth
Chemical immersion bath	Molybdenum disulfide with grease	63	8	3	Smooth

(a) Coating heated for 1 h at 425 °C (795 °F)

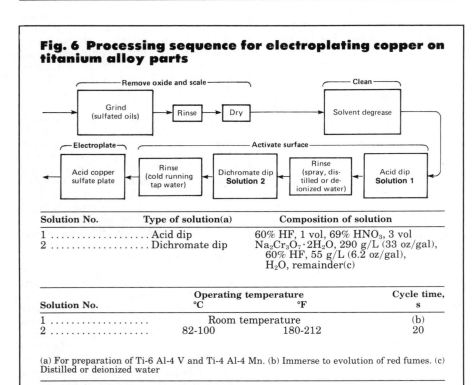

Fig. 6 Processing sequence for electroplating copper on titanium alloy parts

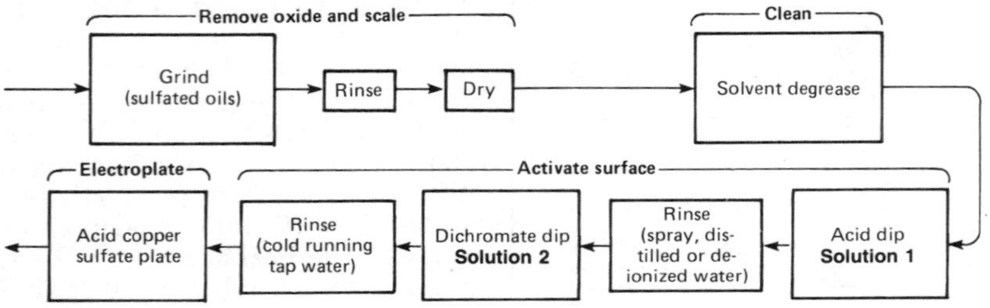

Solution No.	Type of solution(a)	Composition of solution
1	Acid dip	60% HF, 1 vol, 69% HNO_3, 3 vol
2	Dichromate dip	$Na_2Cr_2O_7 \cdot 2H_2O$, 290 g/L (33 oz/gal), 60% HF, 55 g/L (6.2 oz/gal), H_2O, remainder(c)

Solution No.	Operating temperature °C	Operating temperature °F	Cycle time, s
1	Room temperature		(b)
2	82-100	180-212	20

(a) For preparation of Ti-6 Al-4 V and Ti-4 Al-4 Mn. (b) Immerse to evolution of red fumes. (c) Distilled or deionized water

tained above the titanium oxide layer. High-speed rotary tests have indicated marked improvement in the wear characteristics of the metal after conversion coating and lubricating with one part of molybdenum disulfide and two parts of thermosetting eponphenolic resin.

Coatings are easily removable without excessive loss of metal by pickling in an aqueous solution containing 20% nitric acid and 2% hydrofluoric acid by weight.

Electroplating on Titanium

The electrodeposition of copper on titanium and titanium alloys provides a basis for subsequent plating.

Copper Plating. A flow chart outlining the processing sequence for copper plating titanium is shown in Fig. 6. After cleaning and before plating, the surface of the titanium must be chemically activated by immersion in an acid dip and a dichromate dip to obtain adequate adhesion of the plated coating. The compositions and operating temperatures of these activating solutions are in the table with Fig. 6.

Water purity is critical in the composition of activating solutions, although chemicals of technical grade are as effective as, and may be substituted for chemicals of the chemically pure grade. In both the acid and dichromate baths, hydrofluoric acid content is most critical and must be carefully controlled.

After proper activation, titanium may be plated in a standard acid copper sulfate bath. The adhesion of the deposited copper is better than that of 60-40 solder to copper, and the deposit successfully withstands the heat of a soldering iron. The normal thickness of the plated deposit is about 25 μm (1 mil).

Copper-plated titanium wire is available commercially. The outstanding property of this material is the lubricity of its copper-plated surface. The wire can be drawn easily and can be threaded on rolls. Such wire has been used in applications that require electrical surface conductivity.

The titanium wire is plated continuously at a speed of about 60 m/min (200 ft/min) in a copper fluoborate acid bath at a current density of 7.5 to 12.5 A/dm^2 (75 to 125 A/ft^2). The final copper deposit is a thin flash coating. Higher current densities up to 150 A/dm^2 (1500 A/ft^2) have been tried, but if the copper coating is too thick, adhesion is poor.

Platinum Plating. Although titanium is not satisfactory as an anode material because of an electrically resistant oxide film that forms on its surface, application of a thin film of platinum to titanium results in a material with excellent electrochemical properties. Theoretically, the thinnest possible film is sufficient to give the highly desirable low overvoltage characteristics of platinum; furthermore, the film need not be continuous or free of defects to be effective.

The greatest immediate use for platinum-coated titanium is for anodes in the chlorine-caustic industry. Some horizontal-type chlorine cells use expanded metal anodes. From 1.3 to 2.5 μm (0.05 to 0.1 mil) of platinum is applied to the anode surface. Replating of the anodes may be required after about 2 years, depending on the operating conditions. The attrition rate for platinum appears to be about 0.6 g/tonne (0.5 g/ton) of chlorine.

Several platinum and electrode suppliers have developed reliable methods for platinum plating titanium; most use proprietary solutions. A platinum diamino nitrite bath has been used successfully to apply platinum plates to titanium. In this and other procedures, certain precautionary steps are required to achieve adherent, uniform plates. The surface must be cleaned thoroughly, and etched in hydrochloric or hydrofluoric acid to produce a roughened surface. Some procedures also involve a surface activating treatment just before plating. Immersion for 4 min in a solution of glacial acetic acid 895 mL (30 fluid oz) containing hydrofluoric acid 125 mL (4 fluid oz) of 52% hydrofluoric acid, followed by a prompt rinse, appears to be an effective activating treatment if performed immediately before plating. A postplating treatment, consisting of heating to 400 to 540 °C (750 to 1000 °F) for a period of 10 to 60 min, stress relieves the plate, and improves adhesion. This treatment can be done in an air atmosphere, and a light oxide film forms on unplated areas.

Coatings for Emissivity. Electrodeposits and sprayed coatings of gold on titanium are being used to provide a heat-reflecting surface that reduces the temperature of the base metal. Gold-coated titanium has been used for jet engine components.

The gold coating is applied by spraying a gold-containing liquid on chemically clean titanium sheet. This is followed by a baking treatment. Normal coating thickness is about 25 μm (0.1 mil).

Cleaning and Finishing of Tungsten and Molybdenum*

The processes and equipment used for cleaning and finishing tungsten, molybdenum, and their alloys are similar to those for steel and heat-resistant metals with some exceptions and modifications.

Abrasive Blasting

The oxides of tungsten and molybdenum become volatile at relatively low temperatures. Oxides formed on the metal surface during hot working are porous. To remove these oxides, abrasive blasting is not generally required. When this process is used, special caution must be taken. When a plate of these metals is subjected to abrasive blasting on one side for an extended length of time without turning over, the thermal stress caused by the temperature difference across the thickness of the plate may result in cracking of the metal. For abrasive blasting, a silicon carbide of grit sizes ranging from 36 to 120 mesh is used.

Molten Caustic Process

To remove the heavy oxide scale from tungsten, molybdenum, and their alloys, the molten caustic process is used. The caustic may be straight sodium hydroxide, or with an addition of 10% sodium nitrate or sodium nitrite, or with 0.5 to 2.5% sodium hydroxide. The operating temperature ranges from 340 to 400 °C (645 to 750 °F), and the immersion time is from 5 to 20 min or until the bubbling reaction stops. Caution should be taken to avoid any water getting into the bath, or a violent reaction will occur. When the workpiece is removed from the caustic, it is rinsed immediately with a jet of hot water in order to blast off the dissolved material and attached salt.

In the molten salt bath room, a suction hood or other ventilation system should be installed. The operator should use a control in a separate room or be separated from the bath container by a partition. In the vicinity of the bath container, the operator should wear a respirator, a face shield, rubber gloves, and an apron. In case the skin or eyes come in contact with the salt, one should flush the eye or skin area with plenty of water for at least 15 min.

Acid Cleaning and Pickling

When tungsten and molybdenum are slightly oxidized on the surface or after the heavily oxidized workpiece is cleaned with molten caustic, acid cleaning is used. The acid solution consists of 50 to 70 vol % of concentrated nitric acid, 10 to 20% of concentrated hydrofluoric acid, and the remainder water. The cleaning solution is best when maintained at temperatures of 50 to 65 °C (120 to 150 °F). The process of acid pickling is twofold. First the nitrogen dioxide generated by nitric acid oxidizes the metal surface. Then the oxide is dissolved in hydrofluoric acid.

In the pickling area, ventilation and drainage systems should be installed. For the pickling acid, a recycling system to remove the residues and to refresh the acid is recommended. In small operations, without recycling, the acid solution should be neutralized and diluted with plenty of water before pouring into the drainage system.

Electrolytic Cleaning

Electrolytic etching may be used for the preparation and activation of the metals before electroplating. The electrolytes may either be acid or alkaline as the following:

- *Acid electrolyte*: aqueous solution of 5 to 50 vol % concentrated hydrofluoric acid, using 5 V, 60 Hz alternating current
- *Alkaline electrolyte*: aqueous solution of 2 to 10% sodium hydroxide or potassium hydroxide, by weight, with a nickel cathode, using a current density of 230 A/dm^2 (2300 A/ft^2)

Solvent Cleaning

To remove oil, grease, and other contaminants from the surface of tungsten or molybdenum, a common organic solvent such as trichlorethylene, acetone, or isopropanol may be used. Cleaning is usually performed at room temperature by immersing and soaking the part in the solvent. Ultrasonic vibration is sometimes used to loosen soils from deep recesses. For large parts, spray or wiping with the solvent may serve the same cleaning purpose. A proper ventilation system should be installed in the room for solvent cleaning.

Mechanical Grinding and Finishing

To remove oxide scale from tungsten, molybdenum, and their alloys, mechanical grinding is rarely used; however, mechanical grinding is applied to remove surface or edge cracks from the material. Used for this purpose is a coarse grit vitrified-bond grinding wheel coated with a 46 to 60 mesh of aluminum oxide. This type of grinding

*By Chun T. Wang, Senior Research Metallurgist, Teledyne Wah Chang Albany, and R. Terrence Webster, Principal Metallurgical Engineer, Teledyne Wah Chang Albany

wheel is also used for roughing a flat or cylindrical surface. Used for finishing is a finer grit vitrified-bond grinding wheel with a 60 to 120 mesh of silicon carbide. For the cutting of tungsten or molybdenum, an abrasive wheel should also be applied. In this case, a fine-grit 60 to 120 mesh of silicon carbide resinoid bond wheel is preferred. Except for cutting, the grinding of tungsten and molybdenum is usually performed without water or other coolants. A good ventilation system should be installed in a grinding or polishing room. The operator should wear a face shield, mask, and gloves during working. Emery cloth and paper are ineffective for polishing tungsten and molybdenum and are seldom used for finishing.

Electroplating

Molybdenum and tungsten are difficult to plate because the natural oxides formed on the surface of these metals interfere with the adherence of coatings. Chromium has been a favored strike coating because its thermal coefficient of expansion is similar to that of molybdenum. It has good as-plated adherence, and the two metals diffuse to form a solid solution at elevated temperatures. Activation of molybdenum involves anodic etching in various acid media. Four solutions and etching conditions are shown in Table 7. The anodic etching process is as follows:

Table 7 Etching solutions and conditions for molybdenum

Solution composition	Amount g/L	Amount oz/gal	Anodic current density A/dm²	Anodic current density A/ft²	Time, min	Temperature °C	Temperature °F
Sulfuric acid, H_2SO_4	880	1140	2-10	20-100	2-3	21-32	70-90
Phosphoric acid, H_3PO_4	718	93					
Chromate trioxide, CrO_3	100	153	38	380	2	21-32	70-90
Sodium chromate, Na_2CrO_4	10	1					
Sulfuric acid, H_2SO_4	1180	153	22	220	½	21-32	70-90
Chromate trioxide, CrO_3	250	33	15.5	155	⅓	56	135
Sulfuric acid, H_2SO_4	2.5	0.3					

- Chromium is deposited in a chromic-sulfuric acid bath, 250 g/L (33 oz/gal) chromium trioxide and 2.5 g/L (0.3 oz/gal) sulfuric acid, at 90 ± 5 °C (195 ± 9 °F), using a current density of 100 ± 20 A/dm² (1000 ± 200 A/ft²)
- Nickel is deposited in a Watts-type bath over an acid chloride Wood's nickel strike plate, 240 g/L (31 oz/gal) nickel chloride·hexahydrate, 80 g/L (10 oz/gal) hydrochloric acid at room temperature, using a current density of 3 to 10 A/dm² (30 to 100 A/ft²), on the chromium-plated molybdenum
- Molybdenum plated with 25 μm (1 mil) of chromium and 175 μm (7 mils) of nickel is protected in air for 1200 h at 980 °C (1795 °F), and for 100 h at 1200 °C (2190 °F)
- Chromium-strike plated tungsten is carried out in the same bath as for molybdenum at 50 °C (120 °F), for 1 min with a current density of 5 to 15 A/dm² (50 to 150 A/ft²). Watts nickel can be deposited over an acid sulfate strike on chromium-plated tungsten. This is carried out in a bath containing 240 g/L (31 oz/gal) nickel sulfate·hexahydrate and 40 g/L (5 oz/gal) sulfuric acid at room temperature for 2 to 5 min with a current density of 5 to 10 A/dm² (50 to 100 A/ft²)

Another process intended to produce a hydride film as a basis for nickel plating is as follows:

- Tungsten is treated cathodically at 1 to 11 A/dm² (10 to 110 A/ft²) in approximately 2% acid or alkali, by weight, for 5 to 15 s
- Tungsten is rinsed, and nickel plated for 2 min in a Watts nickel bath, pH 4.0, at 65 °C (150 °F), with a current density of 5 to 11 A/dm² (50 to 110 A/ft²)
- The nickel-coated tungsten, after heating under vacuum at 450 to 750 °C (840 to 1380 °F) until gas evolution ceases, is cleaned and activated by strike plating in a Wood's nickel bath and then plated as desired

Recent development shows that gold can also be used as a preplate for the subsequent deposition of other metal coatings. A process for molybdenum is detailed as follows:

- Alkaline ferricyanide etch for 10 s at room temperature in a solution of 200 to 250 g/L (26 to 33 oz/gal) ferricyanide and 75 to 85 g/L (10 to 11 oz/gal) potassium hydroxide
- Water rinse
- Hydrochloric acid, 50 vol %, dip for 10 s at room temperature
- Water rinse
- Acid gold strike, alkaline gold-cyanide-tartrate bath, 1 to 3 min at 0.1 A/dm² (1 A/ft²), deposit 0.12 to 0.72 mg/cm² (0.78 to 4.7 mg/in.²), 0.06 to 0.37 μm (2.4 to 15 μin.)
- Water rinse and dry
- Diffuse in dry hydrogen at 830 °C (1525 °F) for 30 min
- Acid gold strike 4 min at 0.1 to 0.2 A/dm² (1 to 2 A/ft²) to deposit additional 0.2 to 0.5 μm (8 to 20 μin.)
- Water rinse and dry. Diffusion in dry hydrogen at 830 °C (1525 °F) for 30 min
- Acid gold strike for 45 to 75 s at 0.1 to 0.2 A/dm² (1 to 2 A/ft²)
- Water rinse
- Plate desired metal coating. A thermal treatment in dry hydrogen at 830 °C (1525 °F) reduces the natural oxide to form a metal-metal interface and a diffusion bond

Tungsten is first plated with about 0.2 to 2.0 μm (8 to 80 μin.) gold, heated at 1200 to 1370 °C (2190 to 2500 °F) under vacuum for 5 min, and then plated with gold or other metals to the desired thickness.

The TZM, 0.5% titanium, 0.08% zirconium, molybdenum alloy has good strength properties at high temperature, but is subject to galling in threaded fasteners. Rhodium plating on molybdenum tends to eliminate galling. The process is detailed as follows:

- Solvent degrease
- Alkaline clean at 80 °C (175 °F) for 5 min
- Water rinse
- Alkaline ferricyanide etch with agitation for 30 s at room temperature
- Water rinse
- Sulfuric acid, 5 to 10 vol %, dip for 10 s
- Rhodium strike at 1 to 2 A/dm² (10 to 20 A/ft²) at 50 °C (120 °F) to deposit a 0.06 to 0.10 mg/cm² (0.40 to 0.65 mg/in.²) 0.05 to 0.08 μm (2.0 to 3.2 μin.) in an acid rhodium sulfate bath
- Water rinse and dry
- Diffusion heat treat at 1000 °C (1830 °F) for 30 min, 10^{-4} torr (10^{-2} Pa) maximum, purge with helium
- Sulfuric acid dip for 30 s minimum at room temperature
- Rhodium plate at 1 to 2 A/dm² (10 to 20 A/ft²) at 50 °C (120 °F) to thickness of 0.5 μm (2.0 μin.)
- Water rinse and dry

Application of the rhodium strike directly over the natural oxide to TZM is somewhat difficult to accomplish initially because of the usual high tensile stress in rhodium plate. Reducing this stress is more effective in achieving adherence of the rhodium strike than are efforts to reduce the thickness of the natural oxide on TZM.

Iridium coatings about 12 μm (475 μin.) thick, plated in a fused cyanide bath, over chromium on nickel over gold strike-coated molybdenum, offered oxidation protection for about ½ h in air at 1000 °C (1830 °F).

Anodizing

For the anodizing of tungsten and molybdenum, an acetic-based electrolyte for vanadium may be used. It consists of (*a*) acetic acid, (*b*) 0.02 *M* sodium tetraborate·decahydrate, and (*c*) 1.0 *M* additional water. When the water content is less than 1.0 *M*, the conductivity is reduced to an inconvenient extent. While the content is more than 2 *M*, the film formed on molybdenum becomes excessively unstable. In the process, a platinum cathode is preferred, and the temperature is maintained around 25 °C (77 °F), with the presence of air.

The anodic film formed on molybdenum is so unstable that it will change interference colors and eventually disappear when exposed to air. It can be stabilized either by dipping in glacial acetic acid, followed by drying with filter paper, or by using a jet of compressed air immediately after its withdrawal from the electrolyte. Once stabilized, it can be stored in a desiccator almost indefinitely.

Table 8 Metallic coatings for molybdenum

Process type	Material	Thickness range μm	Thickness range μin.
Electrodeposition	Chromium, nickel, gold, iridium, palladium, platinum, rhodium	12.7-76.2	500-3002
Flame sprayed	Nickel-chromium-boron, nickel-silicon-boron, nickel-chromium, nickel-molybdenum	127-254	5004-10 007
Clad or bonded	Platinum, nickel, nickel-chromium, platinum-rhodium	50.8-508	2002-20 015
Molten bath	Chromium	12.7-25.4	500-1001

Table 9 Aluminide coatings for molybdenum

Type	Composition	Deposition process	Thickness range μm	Thickness range μin.	Developer
Aluminum-chromium	20% Al + 80% (55Cr-40Si-3Fe-1Al)	Flame spray	178-254	7013-10 008	Climax
Aluminum-silicon	88% Al-12% Si	Flame spray, hot dipped	12.7-178	500-7013	NRC
Tin-aluminum	90% (Sn-25Al)-10% $MoAl_3$	Slurry dip or spray	50.8-203	2002-7998	Sylcor, G. T. & E.

Table 10 Silicide coatings for molybdenum

Type	Trade name	Developed by	Deposition process
Molybdenum silicide	Disil	Boeing	Fluidized bed
	PFR-6	Pfaudler	Pack cementation
	L-7	McDonnell-Douglass	Slip pack
	LM-5	Linde	Plasma spray
	Molybdenum silicide	Vitro	Electrophoresis
Molybdenum silicide and chromium	W-2	Chromalloy	Pack cementation
	Durak-MG	Chromizing	Pack cementation
	PFR-5	Pfaudler	Pack cementation
Molybdenum silicide, and chromium, boron	Durak-B	Chromizing	Pack cementation
	W-3	Chromalloy	Pack cementation
Molybdenum silicide and chromium, aluminum, boron	Vought II, IX	Chance Vought	Slip pack
Molybdenum silicide and tin-aluminum	...	Battelle-Geneva	Cementation and impregnation

High-Temperature Oxidation-Resistant Coatings for Molybdenum

Ductile metallic overlays are among the first materials investigated as coatings for molybdenum. The materials and processes used are summarized in Table 8. The electrodeposition processes have been discussed in the electroplating section. Chromium is the most promising of the metallic coatings for molybdenum. It offers excellent oxidation protection and is compatible with substrate; however, chromium is embrittled by nitrogen and will crack and spall on repeated thermal cycling. A nickel or nickel alloy overlay will protect chromium from nitriding. A duplex coating will give good service for molybdenum up to 1200 °C (2195 °F).

Noble metal coatings, such as platinum and iridium, can be used on molybdenum at temperatures to about 1425 °C (2595 °F). They provide a significantly longer useful life at all temperatures than the duplex chromium-nickel coatings.

Coatings based on compounds with aluminum, in initial attempts, were to develop molybdenum parts for aircraft gas turbines. Coatings of aluminum-chromium-silicon, aluminum-silicon, and aluminum-tin systems are indicated in Table 9. The aluminide coatings are applied as thick overlay, using a variety of spray or dip processes. The aluminide coatings provide good oxidation protection to molybdenum at temperatures up to 1540 °C (2805 °F). Life at higher temperatures is very short, less than 1 h, probably as a result of rapid interdiffusion.

Interest in coatings for molybdenum shifted from an aluminide to a silicide base in the mid-1950's. A list of the basic silicide coatings is shown in Table 10. Most of the silicide coatings are deposited by pack-cementation diffusion processes. A major deficiency in the performance of silicide-base coatings appears when the system is used

Fig. 7 The effect of air pressure on the maximum temperature for a 4 h life of silicide-coated refractory materials

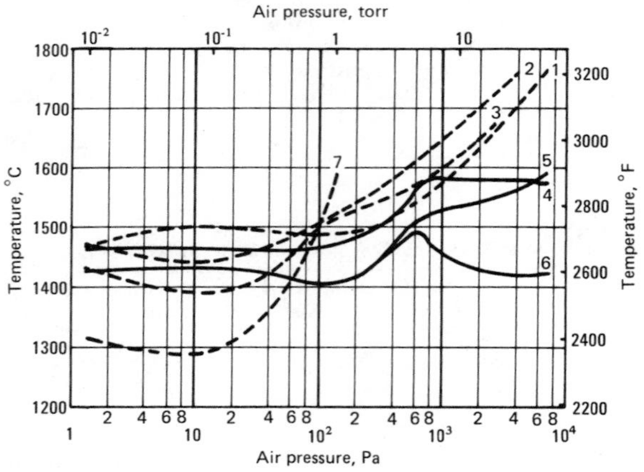

1, TZM/PFR-6; TZM/Disil; 3, TAM/Durak-B; 4, Cb-752/PFR-32; 5, Cb-752/CrTiSi; 6, B-66/CrTiSi; 7, Ta-10W/Sn-Al

Table 11 Oxide coatings for molybdenum

Type	Designation	Deposition process	Thickness range µm	Thickness range mils	Developer
Zirconium oxide-glass	...	Frit, enamel	130-760	5-30	National Bureau of Standards
Chromium-glass	...	Frit, enamel	130-250	5-10	National Bureau of Standards
Chromium-alumina oxide	GE 300	Flame spray over chromium plate	200-380	8-15	General Electric
Alumina oxide	Rockside-A	Flame spray	25-2500	1-100	Norton
Zirconium oxide	Rockside-Z	Flame spray	25-2500	1-100	Norton
Zirconium oxide	ZP-74	Troweling	2500-7600	100-300	Marquardt

in low-pressure environments. As shown in Fig. 7, silicide coatings that protect TZM substrate for 4 h at 1650 to 1760 °C (3000 to 3200 °F) in air at 1 atm cannot be used above 1480 °C (2695 °F) in air at pressure of 0.1 to 1.0 torr (13 to 130 Pa).

Refractory oxides or ceramics are the only materials suitable for the oxidation protection of molybdenum above 1650 °C (3000 °F) for any length of time. The types of oxide coatings that have been used on molybdenum are summarized in Table 11. Ceramic coatings suffer from one common problem. They crack on thermal cycling and tend to spall from the substrate. For only minutes, during single-cycle uses, such as for rocket motors, ceramic coatings provide useful protection from oxidation to 1930 °C (3505 °F) with Al_2O_3 and to 2200 °C (3990 °F) with ZrO_2.

High-Temperature Protective Coatings for Tungsten

A more restrictive situation exists with tungsten. Only the pure disilicide has emerged as useful, and it is protective for a short time to 1925 °C (3500 °F). In the 1090 to 1650 °C (1995 to 3000 °F) range, lives of 10 to 50 h are attained.

To modify the silicides with tungsten, zirconium, and titanium may be a more successful approach. A tungsten-modified silicide pack coating with HfB_2 additives is available commercially. A multi-element umbrella type, consisting of iridium electrodeposited from a fused salt bath, followed by a plasma-sprayed ZrO_2 layer prevents volatilization losses during very high-temperature oxidation. This may be serviceable at 1800 °C (3270 °F) for several hours. In the ceramic approach, a coating of ThO_2 over a tungsten wire mesh can be used at 2925 °C (5300 °F), and a coating of HfO_2 at 2700 °C (4890 °F).

SELECTED REFERENCES

- J. G. Beach and C. L. Faust, Plating on Less Common Metals, *Modern Electroplating,* edited by F. A. Lowenheim, New York: John Wiley & Sons, 1974, p 618-635
- D. J. Levy, C. R. Arnold, and D. H. Ma, Electroplating of Gold and Rhodium Directly on Molybdenum, *AES Symposium on Plating on Difficult-to-Plate Metals,* American Electroplater's Society, 1980
- National Materials Advisory Board, *High-Temperature Oxidation Resistant Coatings,* Washington D. C.: National Academy of Science, 1970
- S. W. H. Yih and C. T. Wang, *Tungsten, Sources, Metallurgy, Properties and Applications,* New York: Plenum Press, 1979

Cleaning and Finishing of Tantalum and Niobium*

The process and equipment used for the cleaning and finishing of tantalum, niobium, and their alloys are essentially similar to those used for steel and heat-resistant alloys with some exceptions and modifications.

Mechanical Grinding and Finishing

Although commercially pure tantalum or niobium is worked at room temperature, their alloys are usually worked at elevated temperatures. Heavy oxide scale is formed. To remove such heavy oxide scale, mechanical

*By Chun T. Wang, Senior Research Metallurgist, Teledyne Wah Chang Albany, and R. Terrence Webster, Principal Metallurgical Engineer, Teledyne Wah Chang Albany

grinding is the most effective method. For a coarse grinding, a vitrified-bond grinding wheel having a 46 to 60 mesh of aluminum oxide is used. For finishing, a finer grit, vitrified or resinoid-bond grinding wheel 60 to 120 mesh of silicon carbide is used. For the cutting of tantalum or niobium, an abrasive wheel is also preferred. The grinding operation usually is performed dry, but the abrasive cutting is used with water cooling.

A good ventilation system should be installed in a grinding or polishing room. The operator should wear a face shield, mask, and gloves during working. Emery cloth and paper are ineffective for polishing tantalum and niobium. They are seldom used for finishing the product of these metals.

Abrasive Blasting

To remove intermediate thickness of the oxide scale of tantalum or niobium, abrasive blasting is applied. The abrasives to be used are usually silicon carbide of grit sizes ranging from 36 to 120 mesh. A good ventilation system is required in the abrasive blasting room, and the operator should wear a mask during working.

Alkaline Cleaning Process

When oxide scale is combined with grease, graphite, molybdenum disulfide, and other lubricants on the workpieces of tantalum and niobium, an alkaline cleaning process is usually used. The starting product for the solution is a solid alkaline material that usually consists of 50 to 80% sodium hydroxide, with the remainder being sodium metasilicate and sodium carbonate. The solids are dissolved into water at a concentration of 0.6 to 1.2 kg/L (5 to 10 lb/gal), and the solution is kept at 65 to 80 °C (150 to 180 °F). The soaking time for the workpiece ranges from a few minutes to a few hours. After soaking, the workpiece should be immediately rinsed with a jet of water in order to blast off the loosened scale and attached salt. The operator should wear a face shield, rubber gloves, and an apron during working.

Acid Cleaning and Pickling

After mechanical grinding, abrasive blasting, or alkaline cleaning, tantalum and niobium are cleaned further with an acid solution. This consists of 40 to 60 vol % of concentrated nitric acid to 10 to 30% of concentrated hydrofluoric acid, and the remainder water. The cleaning solution is best when maintained at temperatures of 50 to 65 °C (120 to 150 °F). After acid pickling, the workpiece should be washed with water or rinsed thoroughly with a jet of water to remove any trace of acids.

Good ventilation and drainage systems should be installed in the acid pickling room. A recycling system to remove the residues and to refresh the acid is preferred both for economical and ecological reasons. For small operations, without recycling, the acid solution should be neutralized and diluted with plenty of water before pouring into the drainage.

Electrolytic Cleaning

Electrolytic etching may be used for the preparation and activation of tantalum and niobium before electroplating. The electrolytes usually used are as follows:

- For niobium, a concentrated hydrofluoric acid, 49% solution, with alternating current at 1 to 5 V and a current density of 22 to 108 A/dm^2 (220 to 1080 A/ft^2) for 1 to 3 min at room temperature is used
- For tantalum, a solution consisting of 90% concentrated sulfuric acid and 10% concentrated hydrofluoric acid, used in the 25 to 40 °C (75 to 105 °F) range with a platinum cathode at a current density of 10 to 50 A/dm^2 (100 to 500 A/ft^2), gives excellent results

Solvent Cleaning

To remove oil, grease, and other contaminants from the surface of tantalum or niobium, a common organic solvent, such as trichlorethylene, acetone, or isopropanol may be used. Cleaning is performed at room temperature by immersing workpieces in solvent. Ultrasonic vibration is sometimes used to loosen soils from deep recesses. For large parts, spraying or wiping with the solvent may serve the same purpose. A proper ventilation system should be installed in the room for solvent cleaning.

Anodizing

Anodizing is a process in which a current passes through an electrolyte from an inert cathode to the anode, on which a thin layer of the oxide forms during the process. The thin but dense oxide layer thus formed usually protects the metal from further oxidation in air and from abrasion. Anodic oxidation for tantalum and niobium is carried out at constant current density, about 1 to 15 A/dm^2 (10 to 150 A/ft^2), at room temperature either in 0.5 to 2%, by weight, ammonium citrate or in 0.5%, by weight, ammonium borate solution, with the pH kept at 9 by adding ammonium hydroxide.

Electroplating

Before electroplating, the substrate to be plated should be electrolytically cleaned or activated. The electrolytic cleaning processes for niobium and tantalum have been mentioned in a previous section. After this, the article may be dipped in an aqueous solution of 50% concentrated nitric acid and 2% concentrated hydrofluoric acid, at 20 to 30 °C (68 to 86 °F).

Copper, chromium, gold, iron, nickel, and platinum are plated on niobium or tantalum. Conditions and solutions for these are given in Table 12.

Iron and nickel deposits are fairly adherent. Other deposits easily peel off the niobium. The presence of a hydride layer between the plate and the niobium limits the strength of the bond, and heat treating is required to decompose the hydride. This involves heating in a vacuum for 1 h at 700 to 1100 °C (1290 to 2010 °F). Improved adhesion of the coating can also be obtained if the surface to be plated is first shot blasted and then flash coated with an undercoat of copper, deposited by chemical displacement from a specified electrolyte.

For plating gold on tantalum, heat treatment before and after the electrolytic plating will enhance the bonding. The temperature for preheating is 1955 °C (3550 °F), and the after treatment is carried out at 1100 °C (2010 °F). Both treatments occur under vacuum.

For plating nickel, cathodic treatment at 2 A/dm^2 (20 A/ft^2) for 20 to 30 min in a solution of 2.5 vol % of 48% hydrofluoric acid, 2.5 vol % of hydrochloric acid, sp gr 1.16, and 95% absolute methanol at 40 to 50 °C (105 to 120 °F), gives a more uniform surface texture on tantalum. After this treatment, 0.3 μm (12 μin.) of nickel is deposited in a Watts bath, pH 4.0, vacuum heat treated at 500 to 600 °C (930 to 1110 °F), then nickel plated

Table 12 Solutions for plating on columbium or tantalum

Solution	Constituent	Amount of constituent g/L	oz/gal	Current density A/dm²	A/ft²	Temperature °C	°F
Copper	$CuSO_4 \cdot 5H_2O$	210	27	2.69	26.9	25	77
	H_2SO_4	80	10				
	Molasses	2.5	0.3				
Chromium	CrO_3	400	52	5.38	53.8	75	165
	H_2SO_4	4	0.5				
Gold	$KAu(CN)_2$	44	5.7	2.15	21.5	54	130
	$K_2C_4H_4O_5$	48	6.2				
	KOH	3	0.9				
	K_2CO_3	10	1.3				
	KCN(a)	30	4				
Iron	$Fe_2SO_4 \cdot 7H_2O$	300	39	3.23	32.3	60	140
	$FeCl_2 \cdot 4H_2O$	42	5.5				
	$(NH_4)_2SO_4$	15	2				
	NaCOOH	15	2				
	H_3BO_3	30	4				
	Wetting agent	1	0.13				
Nickel	$NiSO_4 \cdot 7H_2O$	300	39	1(b)	10	60	140
	$NiCl_2 \cdot 6H_2O$	50	6.5				
	H_3BO_3	30	4				
	Wetting agent	1	0.13				
Platinum	Bakers platinum solution No. 209	(c)	(c)	8.07	80.7	88	190

(a) pH, 12.1. (b) 4.31 A/dm² (43.1 A/ft²). (c) Diluted to 4 g/L (0.52 oz/gal) of contained platinum made slightly ammoniacal

to 10 μm (390 μin.) thick, and again heat treated at 450 °C (840 °F).

Another procedure intended to produce a hydride film as a basis for electroplating with nickel on tantalum or niobium is as follows:

- The metal is treated cathodically at 1 to 11 A/dm² (10 to 110 A/ft²) in 2%, by weight, acid or alkali for 5 to 15 s
- The metal is then rinsed and nickel plated for 2 min in a Watts nickel bath, pH 4.0, at 65 °C (150 °F), with a current density of 5 to 11 A/dm² (50 to 110 A/ft²)
- After vacuum heating at 450 to 750 °C (840 to 1380 °F) until gas evolution ceases, the nickel-coated article is cleaned and activated by strike plating in Wood's nickel bath, and then plated as desired

Oxidation-Resistant Coatings for Niobium

The use of refractory metals at elevated temperatures is being pursued intensively by the aerospace industry. One of the major limitations to progress in this field is the need for coatings to protect the alloys from rapid oxidation and embrittlement at elevated temperatures. These coatings may be classified as (*a*) intermetallic compounds, including silicides and aluminides that form compact or glassy oxide layers; (*b*) alloys that form compact oxide layers; (*c*) noble metals that resist oxidation, and (*d*) stable oxides that provide a physical barrier to the penetration of oxygen.

Many techniques have been used to form protective coatings on niobium alloys. They are as follows:

- *Pack cementation*: immersion of the material to be coated in a pack consisting of inert filler, metal coating elements, and halide activator. Heat treatment is conducted in hydrogen or an inert atmosphere. Coating is formed by vapor transport and diffusion
- *Vacuum pack*: immersion of material to be coated in an all-metal pack of coating elements plus halide activator. Heat treatment is performed in dynamic vacuum or under a partial atmosphere of inert gas. Coating is formed by vapor transport and diffusion
- *Slip pack*: similar to the two processes above, substitution of thin disposable bisque for massive pack. Bisque is sprayed, dipped, or applied to the substrate surface in other ways
- *High-pressure pack cementation*: similar to pack cementation and vacuum pack, with utilization of inert gas pressure in excess of 1 atm (100 kPa)
- *Fused slurry*: uniform application of a metal or alloy slurry to the substrate surface, followed by fusion of the slurry in an inert gas or vacuum environment. Coating forms by liquid-solid diffusion
- *Slurry-sinter*: spray or dip application of metal particles plus binder to substrate surface, followed by solid- or liquid-phase sintering in vacuum or inert environment
- *Fused salt*: electrolytic or nonelectrolytic deposition of metal ions from a fused metal salt solution, such as fluorides, chlorides, bromides, on the substrate material. Pure or alloy coatings are formed
- *Electrophoresis*: deposition of charged metal particles from a liquid suspension onto a substrate surface of opposite electrical potential. Synthesis of particulate deposit generally involves isostatic compaction followed by vacuum or inert atmosphere sintering
- *Aqueous electroplating*: electroplating of metal ions from aqueous solution. Process can involve suspension of particulate materials in electrolyte and occlusion of these particles in metal deposit
- *Fugitive vehicle*: involves spray or dip application of metal slurry onto the substrate surface, followed by vacuum heat treatment. Slurry contains coating elements and fugitive vehicle that is molten at the firing temperature. Fugitive vehicle has high solubility for coating elements and very low solubility for substrate. Coating forms by transfer of elements from the liquid solution to the substrate surface, with subsequent diffusion growth of the coating phase. Fugitive vehicle is eventually removed by evaporation
- *Fluidized bed*: immersion of the material to be coated in a heated fluidized bed of coating elements, using as the fluidizing gases mixtures of reactive halogen gases and hydrogen or argon. Coating forms by disproportionation of gaseous metal halides on substrate surface, followed by diffusion alloying
- *Chemical vapor deposition (CVD)*: entirely gaseous process. Metal halide gases, plus argon or hydrogen, pass over surface of metal to be coated. Coating forms by hydrogen reduction or thermal decomposition of metal halide at substrate surface, followed by diffusion-controlled coating growth
- *Vacuum vapor deposition*: evaporation of a metal or alloy from a filament, liquid bath, or other source, followed by condensation of the metal vapor onto a cold or hot substrate surface. Post diffusion treatment is optional

Table 13 Coating system and processes for niobium

Basic type	Designation	System concept	Composition	Process	Developer
Silicide	Cr-Ti-Si	Complex silicide multilayered	Cr, Ti, Si	Vacuum pack and vacuum slip-pack Fused slurry and pack Fluidized bed Electrophoresis Chemical vapor deposition Electrolytic fused salt	TRW Sylcor Boeing and Pfaudler Vitro Texas Instruments Solar and Pfaudler
Silicide	Disil(a) Disil 3	Modified silicide	Si + V, Cr, Ti Si	Fluidized bed Pack cementation, iodine	Boeing
Silicide	Vought I	Modified silicide	Si-B-Cr	Pack cementation, multicycle	Vought
	Vought II	Modified silicide	Si-Cr-Al	Pack cementation, multicycle	
Silicide	Durak(a) KA	Modified silicide	Si + additives	Pack cementation, single cycle	Chromizing
Silicide	N-2	Modified silicide	Si + Cr, Al, B	Pack cementation, single cycle	Chromalloy
Silicide	S-2	Silicide	Si	Chemical vapor deposition	Fansteel
	M-2	Molybdenum disilicide	Si + Mo	MoO_3 reduction and chemical vapor deposition	
Silicide	PFR(a)	Modified silicide	Si + additives	Pack cementation, fluidized bed, fused salt, slurry dip	Pfaudler
Silicide	AMFKOTE(a)	Modified silicide	Si + additives	Pack cementation, single cycle	American Machine & Foundry
Silicide	...	Liquid phase–solid matrix	Si, Sn, Al	Porous silicide applied by pack or CVD-impregnated with Sn-Al	Battelle Memorial Institute, Geneva
Silicide	LM-5	Multilayered complex silicide	40Mo-40Si, 10CrB-10Al	Plasma spray-diffuse	Linde-Union Carbide
Silicide	...	Modified silicide	Si + additives	Pack cementation	Pratt & Whitney Aircraft–Canel
Silicide	R(a)	Complex silicide	Si-20Fe-20Cr Si-20Cr-5Ti Si + (Cr, Ti, V, Al, Mo, W, B, Fe, Mn)	Fused silicides Fusion of eutectic mixtures	Sylvania
Silicide	V-Cr-Ti-Si	Complex silicide	V-Cr-Ti-Si	Vacuum and high-pressure pack	Solar
Silicide	...	Complex silicide	Mo-Cr-Ti-Si V-Cr-Ti-Si V-Al-Cr-Ti-Si Mo-Cr-Si	Multicycle vacuum pack	TRW

(a) Denotes several trade designations

(continued)

- *Hot dipping*: immersion of material to be coated in a hot liquid bath of the molten coating elements. Coating forms by liquid-solid diffusion
- *Hydride and oxide reduction*: spray or dip application of metal hydrides or oxides on substrate surface, followed by vacuum or hydrogen reduction, respectively
- *Plasma arc*: spray deposition of particulate metal or oxide particles using conventional plasma arc facility
- *Detonation gun*: patented gun detonation metal spraying process
- *Gun metallizing*: utilization of conventional wire or powder metallizing equipment
- *Cladding*: bonding of thin sheet metal cladding to substrate surface by diffusion bonding, forging, rolling, extrusion, and other methods

Coating systems and processes for niobium are shown in Table 13.

Oxidation-Resistant Coatings for Tantalum

Much of the work on tantalum alloys is the outgrowth of approaches taken with niobium alloys. The development of coatings has come generally in the 1970's. Most of the emphasis is on the diffusional growth of intermetallic layers. A few commercial or semicommercial coatings are as follows:

- A series of tin-aluminum-molybdenum-base coatings are used by slurry process. These are limited primarily by poor resistance to reduced pressure and erosion at very high temperatures
- Electrophoretical deposition of binary disilicides combined with

Table 13 (continued)

Basic type	Designation	System concept	Composition	Process	Developer
Silicide	...	Glass-sealed silicide	Si + glass	Silicide by pack cementation or CVD + glass slip overcoat	NGTE
Silicide	TNV 12	Multilayered	Mo, Ti + Si and glass	Slurry sinter application of Mo + Ti powder, pack silicide, glass slurry seal	Solar
Aluminide	AMFKOTE(a)	Modified aluminide	Al + B	Pack cementation	American Machine and Foundry
Aluminide	LB-2	Modified aluminide	88Al-10Cr-2Si	Fused slurry	General Electric, McDonnell Aircraft
Aluminide	...	Modified aluminide	Al-Si-Cr	Fused slurry	Sylvania
			Al-Si-Cr Ag-Si-Al	Hot dip	
Aluminide	...	Multilayered systems	Fe, Cr, Al, Ni, Mo Si, VSi_2, $TiCr_2$, $CrSi_2$, B + Al	Powder metallize + aluminum hot dip	Horizons
Aluminide	...	Simple aluminide	Al	Pack cementation	Chromalloy
Aluminide	...	Multilayered systems	Cr, FeB, NiB, Si, Al_2O_3, SiO_2, ThO_2 + Al	Electroplate dispersions in Ni + Al hot dip	Horizon
Aluminide	...	Modified aluminide	Al + (Si, Ag, Cr)	Silver plate + Al, Cr, Si hot dip	General Electric
Aluminide	...	Multilayered systems	Al_2O_3 + Ti + Al	(Al_2O_3 + TiH), Spray-sinter + Al hot dip	General Electric
Aluminide	Slurry ceramic	Oxide-metal composite	Al_2O_3 + Al	Slurry fusion of Al_2O_3-Al mixture	North American Rockwell
Aluminide	Lunite 2	Aluminide	Al + additives	Fused slurry	Vac Hyd
Aluminide	...	Modified aluminide	Al + Sn	Hot dip	Pfaudler, Sylcor
Zinc	...	Self-healing intermetallic	Zn and Zn + Al Ti, Co, Cu Cr, FeMg, Zr, Cu, Si	Vacuum distillation and hot dip	Naval Research Lab
Oxide	System 400	Glass-sealed oxide	Al_2O_3 + glass (baria, alumina, silica)	Flame spray Al_2O_3 + glass slurry	General Electric
Nichrome	...	Oxidation-resistant alloy	Ni-Cr	Flame spray, detonation gun, plasma arc	Linde--Union Carbide
Chromium carbide	...	Carbide	Cr-C	Plasma spray	Linde-Union Carbide
Noble metal	...	Clad	Pt, Rh	Roll bonding and hermetic sealing	Metals and Controls
Noble metal	...	Barrier-layer-clad	Pt, Rh + Re, Be, Al_2O_3, W, ZrO_2 MgO, SiC, Hf	Noble metal clad over barrier layer-diffusion couple study	Metals and Controls
Noble metal	...	Pure metal	Ir	Fused-salt deposition	Union Carbide

(a) Denotes several trade designations

a molybdenum-vanadium system shows no pest phenomenon during service. This process can be considered commercial for small parts

- A fluidized-bed, three-step silicon-vanadium-silicon process may be used for niobium as well as Ta-10W
- Complex titanium, molybdenum, tungsten, and vanadium modified silicide coatings can be applied by a two-step method: a slurry plus high-temperature sinter of an alloy layer followed by a straight silicide pack. Protection for hundreds of hours at 870 °C (1600 °F) and 1320 °C (2410 °F) in furnace tests has been obtained
- A fused-silicide coating system, particularly Si-20Ti-10Mo, appears practical for coating large, complex aerospace sheet metal components
- A duplex coating consisting of a sintered hafnium boride-molybdenum silicide layer overlaid with a hafnium-tantalum slurry is serviceable at 1820 to 1870 °C (3310 to 3400 °F)

SELECTED REFERENCES

- J. G. Beach and C. L. Faust, *Modern Electroplating,* edited by F. A. Lowenheim, New York: John Wiley & Sons, 1974, p 618-635
- G. L. Miller, *Tantalum and Niobium,* Butterworths Scientific Publications, London, 1959
- National Research Council, NMAB, *High-Temperature Oxidation-Resistant Coatings,* Washington, D. C.: National Academy of Science, 1970
- E. B. Saubestre, Electroplating on Certain Transition Metals, *Journal of Electrochemical Soc.,* Vol 106, 1959, p 305-309
- M. S. Tsirlin, A. V. Kasatkin, and A. V. Byalobzheskii, *Poroshkovaya Metallurgiya, Kiev,* Dec 1978, (12), p 31-34

Cleaning and Finishing of Zirconium and Hafnium Alloys*

Zirconium and hafnium surfaces may require cleaning and finishing for reasons such as preparation for joining, heat treatment, plating, forming, and producing final surface finishes. Special surface preparation and cleaning is generally not required for corrosion resistance because the naturally formed surface oxide protects the metal regardless of surface condition.

Surface Soil Removal

Grease, oil, and lubricants used in machine forming and other fabricating operations may be removed by a number of techniques. Alkaline or emulsion cleaners used in simple soak tanks or with ultrasonic cleaning, acetone or trichlorethylene solvent washing, or vapor degreasing and detergents are all widely used. Hand wiping with solvents such as acetone, alcohol, or trichloroethylene is used for light soil removal. Electrolytic alkaline cleaning is also used. In the electrolytic system, the work can be either anodic or cathodic polarity provided the voltage and current can be controlled to avoid anodizing or spark discharge, and subsequent pitting. Removing these soils is essential before acid etching to provide uniform acid attack. The soils must be removed before heat treatment and joining to prevent contamination and consequent loss of ductility.

Blast Cleaning

Mechanical descaling methods such as sandblasting, shot blasting, and vapor blasting are used to remove hot work scales and hard lubricants from zirconium and hafnium surfaces. Aluminum oxide, silicon carbide, silica sand, and steel grit are satisfactory media for mechanical descaling. Periodic replacement of used media may be required to avoid excessive working of the surfaces by dull particulates. Roughening of exposed surface areas occurs from grit or shot impingement, depending on the grit size used. Any abrasive or shot blast cleaning may induce residual compressive stresses and warpage in the surface of the material, particularly thin sheet. Warpage also may occur in sections which are subsequently chemical milled or contour machined.

Blast cleaning is not intended to eliminate pickling procedures. Abrasive blasting does not remove surface layers contaminated with interstitial elements such as carbon, oxygen, and nitrogen. Generally blast cleaning is followed by a pickling step to ensure complete removal of surface contamination and cold worked layer and to produce a smooth bright finish.

Chemical Descaling

Some scale, as well as forming lubricants, can be removed by proprietary water solutions of strong caustic compounds, or by the use of molten alkaline-based salt baths. The salt baths operate at temperatures of 650 to 705 °C (1200 to 1300 °F) and must be used carefully according to the manufacturer's instructions. Salt bath descaling is accomplished by a series of cycles through (*a*) the salt, (*b*) a water rinse, and (*c*) a sulfuric acid bath to remove scale before final pickling.

Pickling or Etching

Metal removal by a chemical bath of nitric-hydrofluoric acid is used most commonly, although other baths have been used. The usual bath for zirconium, Zircaloys, and hafnium is composed of (*a*) 25 to 50% nitric acid, 70 vol %; (*b*) 2 to 5% hydrofluoric acid, 49 vol %; and (*c*) the remainder water. The acid bath for zirconium-niobium alloys consists of (*a*) 28 to 32% sulfuric acid, sp gr 1.84; (*b*) 28 to 32% nitric acid; (c) 5 to 10% hydrofluoric acid; and (*d*) the remainder water.

The hydrofluoric acid attacks the zirconium and hafnium, and the nitric acid oxidizes the hydrogen formed by the reaction and prevents its absorption by the metal. The ratio of nitric to hydrofluoric acid should not be less than 10 to 1. Except for the zirconium-niobium alloys, the rate of metal removal is linear with hydrofluoric acid concentration and doubles as the bath temperature rises from 43 to 71 °C (110 to 160 °F).

Material etched in the nitric-hydrofluoric acid bath must be rinsed quickly and completely with flowing water, to prevent an insoluble fluoride surface stain from forming. This stain is extremely detrimental to corrosion resistance in hot water and steam environments of nuclear reactors. Extreme precautions to ensure a rapid and effective rinse are required for this service.

Anodizing and Autoclaving

Zirconium and hafnium oxidize readily in air and aqueous solution, but the tenacious oxide layer generally protects the metal from further oxidation. Methods to produce a dense oxide layer to serve as a protective coating for zirconium and hafnium are anodizing and autoclaving. Before anodizing, the article should be cleaned in a 50% nitric acid-50% water solution and then in water with the assistance of ultrasonic vibration.

The anodizing solution used for zirconium and hafnium consists of:

- 45.4 vol % absolute ethanol
- 26.5 vol % water
- 15.2 vol % glycerine
- 7.6 vol % lactic acid, 85% concentration
- 3.8 vol % phosphoric acid, 85% concentration
- 1.5 vol % citric acid

The voltage during anodizing usually reaches 200 to 300 V. The anodized surface becomes a golden color. Caution should be exercised during the anodizing process, especially when the high voltage is reached. Operators should wear rubber gloves. Rubber floor mats are also advisable.

Autoclaving is a standard test for zirconium and hafnium for nuclear industry. The test for zirconium and Zircaloys is carried out at 400 °C (750 °F) under a steam pressure of 10 MPa (1.5 ksi) for 1 to 14 days. The test for hafnium is carried out at 360 °C (680 °F) under a water pressure of 18.6 MPa (2.7 ksi). This procedure may be used to produce a dense oxide coating for zirconium and hafnium. The water used should have a pH value of 6 to 8, with a resistivity of 1×10^4 Ω·m (1×10^6 Ω·cm) minimum when corrected to 22 °C (72 °F). The autoclave should be constructed of the type 300 series stainless steel, and fitted with a pressure gage, pressure records, thermocouples,

*By Chun T. Wang, Senior Research Metallurgist, Teledyne Wah Chang Albany, and R. Terrence Webster, Principal Metallurgical Engineer, Teledyne Wah Chung Albany

Table 14 Electrolyte compositions and conditions for plating on zirconium

Electrolyte	Composition	Amount g/L	Amount oz/gal	Temperature °C	Temperature °F	Current density A/dm²	Current density A/ft²	Heat treatment(a) °C	Heat treatment(a) °F
Copper	$CuSO_4 \cdot 5H_2O$	150-250	20-33	32-42	90-110	16-22	160-220	149-204(b)	300-400(b)
	H_2SO_4	45-110	6-14						
	$NH_2 \cdot CS \cdot NH_2O$	0.002-0.005	0.0003-0.0007						
	H_2O	rem	rem						
Chromium	CrO_3	200-300	26-39	30-50	86-120	10-20	100-200	...	...
	H_2SO_4	2-3	0.3-0.4						
	H_2O	rem	rem						
Nickel	$NiSO_4 \cdot 6H_2O$	320-340	42-44	45-65	115-150	2.1-10	21-100	700(c)	1290(c)
	$NiCl_2 \cdot 6H_2O$	40-50	5-7						
	H_3BO_3	36-40	4.6-5.2						
	H_2O	rem	rem						

(a) After plating. (b) 3 h. (c) 1 h

safety blowout assembly, and a high pressure venting valve. To prepare the article for anodizing, it should be detergent cleaned followed by solvent cleaning with reagent grade acetone, trichlorethylene, or perchlorethylene. The article is cleaned further by acid pickling as described in a previous section. Upon completion of pickling, the article shall be immediately transferred to the first water rinse tank with a temperature maintained at 27 °C (81 °F), and a flow rate of at least two bath changes per minute. It is then transferred to the second water rinse tank at the same temperature for at least 15 min. After this, the article is transferred to the third rinse tank containing water at 80 °C (175 °F) minimum. All rinse water, especially in the third rinse tank should have a pH of 7 (±1) and a resistivity of 5000 Ω·m (500 000 Ω·cm) minimum. After rinsing, the article should be dried by a method which prevents water marks, stains, or contamination on the surfaces of the article, by wiping with a clean, lint-free cloth, and blowing with dry air. Autoclaved zirconium and its alloys show a dense shiny black coating of oxide, while hafnium shows a dense shiny coating of oxide, the color of which ranges from purple or blue to gold.

Electroplating

Hafnium, zirconium, and Zircaloys (zirconium-tin alloys), are difficult to coat with an adherent electroplate because they form a tenacious oxide film in air and aqueous solutions. There are alternative processes for plating nickel on zirconium.

In the first process, the zirconium article is etched in either an ammonium bifluoride-sulfuric acid, 45 g/L (6 oz/gal) of ammonium bifluoride + 0.5 mL (0.02 fluid oz) of sulfuric acid aqueous solution for 1 min at 22 °C (72 °F), or an ammonium bifluoride solution, 45 g/L (6 oz/gal) of ammonium bifluoride aqueous solution for 3 min at 22 °C (72 °F). For nickel plating, the electrolyte consists of 81 g/L (11 oz/gal) of nickel as in $Ni(NH_2SO_3)_2$, 1.0 g/L (0.13 oz/gal) of nickel chloride and 40 g/L (5.2 oz/gal) of boric acid in an aqueous solution, with an ideal surface tension of 0.038 N/m (0.003 lb/ft). The pH value should be kept at 3.8 to 4.0, with the current density of 2.15 A/dm^2 (21.5 A/ft^2), and the temperature at 50 °C (120 °F). Preheating is required to improve bond strength. Carry the treatment at 705 °C (1300 °F) in vacuum for 1 h with the plated article placed in a molybdenum alloy, TZM ring, or constrained case. Because the coefficient of thermal expansion for molybdenum is lower than that of zirconium and nickel, it provides a stress on the plating as the assembly is heated.

In the alternate process, the zirconium article is etched with an aged aqueous activating solution of 10 to 20 g/L (1.3 to 2.6 oz/gal) of ammonium bifluoride and 0.75 to 2.0 g/L (0.1 to 0.26 oz/gal) of sulfuric acid, for about 1 min at ambient temperature. The solution is aged by immersion of a piece of pickled zirconium for at least 10 min at ambient temperature. The next step is to remove any loosely adhering film or smut formed on the article in the activating step. This can be accomplished in one of the following methods:

- Using a chemical solution which is either an aqueous solution of 2 to 10 vol % fluoboric acid or a hydrofluosilic acid of a similar concentration. The solution is maintained at about 25 °C (77 °F), and the article is immersed in the solution for about 1 min
- Submerging the article in water and applying ultrasonic energy of 20 000 to 30 000 cycles/s for 1 to 2 min
- Swabbing the loosely adhering film from the article by rubbing the surface with cotton, paper, or a brush

An optional step of rinsing the article in water can be practiced by using deionized water to free the article of any residual traces of the material used in previous steps.

Preferred metals to be deposited on the article include copper, nickel, and chromium. The chemical compositions and conditions used for the different electrolytes are shown in Table 14. Similar procedures can be used for electroplating metals on hafnium, although information in this respect is sparse.

SELECTED REFERENCES

- J. G. Beach and C. L. Faust, Plating on Less Common Metals, *Modern Electroplating,* edited by F. A. Lowenheim, New York: John Wiley & Sons, 1974, p 618-635
- J. W. Dini, H. R. Johnson, and A. Jones, Plating on Zircaloy-2, *Journal of Less Common Metals,* 79, 1981, p 261-270
- R. E. Donaghy, Process for Electrolytic Deposition of Metals on Zirconium Materials, U. S. Patent 4,137,131
- J. H. Schemel, *ASTM STP639 Manual on Zirconium and Hafnium*

Cleaning and Finishing of Nickel and Nickel Alloys

By J. D. VanDevender
Chemist
Huntington Alloys, Inc.

NICKEL ALLOYS do not require special techniques or precautions for removing shop soils such as soap, drawing compound, oil, grease, cutting fluid, and polishing compound. Oxide, scale, tarnish, or discoloration can be removed from nickel and nickel alloys by mechanical methods such as grinding or abrasive blasting or by chemical methods such as pickling. Conventional methods of cleaning with alkaline compounds, emulsions, or solvents or vapor degreasing may be employed. These processes are described in articles on alkaline cleaning, emulsion cleaning, solvent cleaning, vapor degreasing, and abrasive blast cleaning in this Volume.

Pickling

Pickling is a standard method for producing bright, clean surfaces on nickel alloys, either as an intermediate step during fabrication or as a last step on finished parts. Procedures used for pickling nickel alloys are governed by both material composition and prior thermal treatment. The necessity of pickling can be avoided by using bright-heating practices, that is, the use of inert furnace atmospheres to prevent oxidation. Pickling should not be used to overhaul material by dissolving away appreciable amounts of metal. This practice can cause severe damage by pitting and intergranular attack.

Oxidizing furnace atmospheres, high sulfur content fuels, and air leakage in furnaces cause heavy scale to form on nickel alloys. The metal has a dull, spongy appearance. Hairline cracks are sometimes present, and patches of scale may break away from the surface. In such cases, the underlying metal is rough, and an attractive finish cannot be attained by any pickling method. Table 1 provides selected formulas for pickling nickel alloys. To aid in preparing these solutions, the acids, and respective specific gravities and concentrations are given below:

Acid	°Baumé	Specific gravity	Concentration, wt %
HNO_3	42	1.41	67
H_2SO_4	66	1.84	93
HCl	20	1.16	32
HF	30	1.26	70

Abrasive blasting or grinding, followed by flash pickling, is usually the best method for removing heavy scale. Abrasive blasting requires low capital investment and eliminates the use and disposal of acids. An alternative method is to soak the work in the hydrochloric acid pickle (Table 1, Formula 11) followed by flash pickling, if brightening is necessary.

Precautions must be taken in handling pickling solutions. Noxious and sometimes toxic fumes are liberated during pickling. Positive ventilation, either by using a ventilating hood over the bath or by providing a controlled draft, is required to remove the fumes. Acids must be handled with care, particularly hydrofluoric acid. Use protective clothing, face shields, and rubber gloves. In preparing solutions, always add the acid to the water. This is especially important when diluting sulfuric acid. In preparing the solutions shown in Table 1, add the ingredients to the water in the order listed.

Bath Life. Pickling baths should be analyzed periodically to determine acidity. Only total acidity can be determined in baths composed of more than one acid; it is impossible to distinguish acidity contributed by one acid source from that contributed by another. However, in maintaining a bath, add ingredients in the proportions used for the original solution. Sufficient additions should be made to restore the bath to the initial acidity level.

With the exception of flash pickling solutions, dispose of bath solution when the total metallic content reaches 150 g/L (20 oz/gal). If the bath is used for only one type of alloy, only one element need be determined in the analysis. The remaining metallic elements can be determined from the percentage of each in the composition of the alloy. For example, in analyzing baths used for pickling nickel-chromium alloys the nickel content determines whether the total metallic content is within the 150-g/L limitation.

Flash pickling solutions continue to perform satisfactorily even when nearly saturated with metal salts. Fresh solution should be made up only when salts begin to crystallize on the side of the bath container.

Table 1 Formulas for pickling nickel alloys

Formula No.	Reagents	Weight, %	Amount	Temperature °C	Temperature °F
1	Nitric acid (HNO_3) 1.41 sp gr	20	300 mL 10 oz	70	160
	Water		1000 mL 34 oz		
2(a)	Nitric acid (HNO_3) 1.41 sp gr	10	133 mL 4 oz	75	170
	Sodium chloride (NaCl)	5	63 g 2 oz		
	Water		1000 mL 34 oz		
3	Nitric acid (HNO_3) 1.41 sp gr	20	315 mL 11 oz	50	125
	Hydrofluoric acid (HF) 1.26 sp gr	2	34 mL 1 oz		
	Water		1000 mL 34 oz		
4	Sulfuric acid (H_2SO_4) 1.84 sp gr	25	200 mL 8 oz	80	180
	Water		1000 mL 34 oz		
5	Sulfuric acid (H_2SO_4) 1.84 sp gr	15	111 mL 4 oz	80	180
	Sodium chloride (NaCl)	5	63 mL 2 oz		
	Water		1000 mL 34 oz		
6	Sulfuric acid (H_2SO_4) 1.84 sp gr	15	119 mL 4 oz	20-40	70-100
	Sodium dichromate ($Na_2Cr_2O_7$)	10	135 g 5 oz		
	Water		1000 mL 34 oz		
7	Sulfuric acid (H_2SO_4) 1.84 sp gr	12	82 mL 3 oz	Ambient	
	Sodium fluoride (NaF)	2	23 g 1 oz		
	Water		1000 mL 34 oz		
8	Sulfuric acid (H_2SO_4) 1.84 sp gr	20	171 mL 6 oz	80	180
	Sodium chloride (NaCl)	5	73 g 3 oz		
	Sodium nitrate ($NaNO_3$)	5	73 g 3 oz		
	Water		1000 mL 34 oz		
9	Sulfuric acid (H_2SO_4) 1.84 sp gr	35	1200 mL 41 oz	20-40	70-100
	Nitric acid (HNO_3) 1.41 sp gr	30	1860 mL 63 oz		
	Sodium chloride (NaCl)	0.5	30 g 1 oz		
	Water		1000 mL 34 oz		
10	Hydrochloric acid (HCl) 1.16 sp gr	6	200 mL 7 oz	60	140
	Water		1000 mL 34 oz		
11	Hydrochloric acid (HCl) 1.16 sp gr	12	535 mL 18 oz	80	180
	Cupric chloride ($CuCl_2$)	2	33 g 1 oz		
	Water		1000 mL 34 oz		
12	Hydrochloric acid (HCl) 1.16 sp gr	1	30 mL 1 oz	Ambient	
	Ferric chloride ($FeCl_3$)	1	11 g 0.3 oz		
	Water		1000 mL 34 oz		
13	Sodium hydroxide (NaOH)	15	188 g 7 oz	80	180
	Potassium permanganate ($KMnO_4$)	5	63 g 2 oz		
	Water		1000 mL 34 oz		
14	Ammonium hydroxide (NH_4OH)	2 (b)	20 mL 0.5 oz	Ambient	
	Water		1000 mL 34 oz		
15	Alkaline cleaner		60-75 g/L 7-9 oz/gal	80	180
	Water		1000 mL 34 oz		
16	Agar-agar	1	10 g 0.3 oz	20-65	70-150
	Potassium ferricyanide ($K_3Fe(CN)_6$)	0.1	1 g 0.03 oz		
	Sodium chloride (NaCl)	0.1	1 g 0.03 oz		
	Water		1000 mL 34 oz		

(a) An addition of at least 40 g of nickel per litre to formula 2 will prevent overpickling of chromium-bearing alloys. (b) Volume %

Alloy Groups

For the purpose of discussing pickling procedures, the nickel alloys can be divided into three groups:

- *Nickel alloys:* Nickel 200, Nickel 201, Nickel 270, Duranickel alloy 301, and similar alloys composed primarily of nickel are included in this group
- *Nickel-copper alloys:* Monel alloys 400, K-500 and others are included in this group
- *Nickel-chromium and nickel-iron-chromium alloys:* Inconel alloys 600, 625, 718, and X-750 are examples of nickel-chromium alloys. Incoloy alloys such as alloys 800 and 825 make up the nickel-iron-chromium group

All alloys within any one group have virtually the same pickling characteristics. However, the pickling procedures for alloys within a group must be varied to suit the surface condition of the metal.

Surface Conditions

Alloys within each composition group with similar surface conditions are pickled in the same solutions using the same procedures. Three different surface conditions, primarily depending on the method of prior heating, are generally encountered:

- Bright annealed white metal requiring removal of tarnish by flash pickling
- Bright annealed oxidized metal requiring removal of a layer of reduced oxide, sometimes followed by a flash pickle to brighten
- Black or dark-colored surface requiring removal of adherent oxide film or scale

Tarnish and dullness from bright annealed metal can be removed by flash pickling or bright dipping. Bright annealed white surfaces are generally found on drawn and spun shapes, cold headed rivets, cold drawn wire, and other cold worked products. The white surface is a result of annealing in a reducing, sulfur-free atmosphere and cooling either out of contact with oxygen or by quenching in a 2% (by volume) alcohol solution.

Flash pickling solutions act rapidly, and care must be exercised to prevent overpickling and etching. The solutions are used at room temperature. If the bath is cold, warm it slightly to prevent unduly slow action.

Best results in flash pickling are obtained by first warming the parts by dipping them in hot water, placing them in the acid for a few seconds, and rinsing them again with hot water. Use a second dip in acid if necessary. Badly tarnished metal may require a total of

3 min in acid, but the material should be withdrawn frequently from the bath and inspected to prevent overpickling.

Reduced-oxide surfaces occur when hot worked products, such as forgings and hot rolled material, have been heated after hot working in a reducing, sulfur-free atmosphere and cooled out of contact with air or quenched in an alcohol solution. On such hot worked products, pickling produces a clean surface for further processing by cold forming, or produces the finished surface necessary for items such as rivets.

At annealing temperatures in reducing atmospheres, oxides formed on the high-nickel alloys, except those containing chromium, are readily converted to a spongy, tightly adherent layer. On nickel alloys, the layer consists of metallic nickel; on Monel alloys it is a mixture of metallic nickel and copper.

The oxide film formed on nickel-chromium alloys and nickel-iron-chromium alloys does not undergo complete reduction, making pickling more difficult. The oxide on these alloys is selectively reduced to a mixture of metallic nickel and chromic oxide. The surface color ranges from the characteristic chrome green of chromic oxide to dark brown.

Oxidized or scaled surfaces are present on all hot worked products and heat treated material cooled in air. This type of surface also occurs on the nickel-chromium and nickel-iron-chromium alloys in all conditions except bright heated. The oxide film forms on properly heated nickel and nickel-copper alloys during contact with air after the work is withdrawn from the furnace. The nickel-chromium and nickel-iron-chromium alloys form oxide films even when heated and cooled in atmospheres that keep other alloys bright. Thus, the usual pickling procedure for Inconel and Incoloy alloys is designed to remove oxide and scale.

Nickel-Copper Alloys

Depending on the method of prior heating and cooling, nickel-copper alloys can have any of the surface conditions discussed previously. Accordingly, the appropriate pickling procedure depends on the surface condition of the material.

Tarnish is best removed from bright annealed nickel-copper alloys by pickling in two solutions in sequence. First, the metal should be pickled thoroughly in Formula 2 (Table 1) in brief exposures and rinsed in water at 80 °C (180 °F). After rinsing, the metal should be dipped in Formula 1. Follow the second dip by rapidly rinsing the workpiece and neutralizing it in a 1 to 2% solution (by volume) of ammonia (Formula 14). To dry, dip small workpieces in boiling water and rub them in sawdust or with a clean, dry cloth.

Reduced-Oxide Surfaces. Formula 8 is recommended for reduced-oxide surfaces on nickel-copper alloys if pickling is done on a large-scale basis in fully equipped plants. This acid mixture is more destructive to tanks and racks than solutions used for steel or copper. Steel tanks lined with 4.8-mm (3/16-in.) thick natural rubber and a double layer of yellow acid bricks have proven to be the best and most economical containers for this corrosive solution. After pickling in Formula 8, the metal should be rinsed in hot water and neutralized in a 1 to 2% (by volume) ammonia solution.

The Formula 8 pickling solution works best after a short period of use. Therefore, when preparing new solutions, add about 2% (by volume) of spent solution to the fresh mixture to improve its action.

When a pickling room is not regularly operated, the time required for complete pickling of small lots in Formula 8 is usually a disadvantage. Adequate results in most cases can be obtained by using flash pickling solutions (Formulas 2 and 1). Formula 8 may be used for occasional small jobs, however, if ceramic vessels or wooden barrels are used as containers. The solution can be heated and agitated by injecting live steam either through a rubber hose or a carbon pipe that has a perforated carbon-block end.

Oxidized or Scaled Surfaces. Oxidized nickel-copper alloys, having a thin to moderately thick oxide, are pickled by immersion in Formula 11 followed by brightening in Formula 6. After treatment in the first bath, rinse the work in hot water before transferring it to the brightening dip. Follow the second dip in Formula 6 by rinsing the work in cold water and neutralizing it in a 1 to 2% (by volume) ammonia solution (Formula 14).

Nickel Alloys

Nickel 200, Nickel 201, Nickel 270, Duranickel alloy 301, and similar alloys can have any of the three types of surface conditions. With the exception of flash pickling, nickel alloys are pickled in the same solutions used for Monel alloys. However, maintaining separate baths for the two groups of alloys is usually advisable.

Tarnish. Only one dip, Formula 9, is required to remove tarnish from bright annealed nickel alloys. Dip the metal first in hot water to warm it. Immersion in the acid bath for 5 to 20 s is usually sufficient to produce bright, clean surfaces. After removing from the pickling solution, rinse the metal in hot or cold water and neutralize it in a dilute ammonia solution.

The pickling action of Formula 9 can be retarded, if necessary, by decreasing the amount of acid to as little as one third the volume that is added to the standard formula, as low as 400 mL (13.5 oz) of sulfuric acid and 620 mL (21 oz) of nitric acid. Normally, however, the formula gives the best results in the proportions given in Table 1.

Reduced-Oxide Surface. The solution used to pickle reduced-oxide surfaces on nickel-copper alloys, Formula 8, is also used for nickel alloys. Follow the procedure described for reduced-oxide surfaces under the section of this article on nickel-copper alloys. If both groups of alloys are being pickled, maintain separate baths.

Formula 8 may be used for occasional small lots, but suitable results can usually be obtained in less time by flash pickling in Formula 9.

Oxidized or scaled surfaces on nickel alloys can be pickled with the hydrochloric acid/cupric chloride solution found in Formula 11 used for nickel-copper alloys. A longer time is required, however. Immersion from 1 to 2 h is necessary to obtain a good pickle on nickel alloys.

After removal from the pickling bath, rinse the work with hot water and dip for a few seconds in Formula 9, if brightening is required. Follow the brightening dip with a cold water rinse, and neutralize in a dilute ammonia solution.

Nickel-Chromium and Nickel-Iron-Chromium Alloys

Tarnished or reduced-oxide surfaces are usually not encountered on nickel-chromium or nickel-iron-chromium alloys. These alloys can only be bright annealed in very dry hydrogen or in a vacuum. Oxides formed on their surfaces in other atmospheres do not undergo complete reduction. Oxide or scale is the usual surface pickled on nickel-chromium and nickel-iron-chromium alloys.

Pretreating in a fused-salt bath is strongly recommended to facilitate pickling of oxidized or scaled surfaces. If the metal has been properly heated and cooled, however, it will usually have a surface suitable for direct pickling in Formula 3.

When using a fused-salt bath for pretreating, follow this procedure for pickling nickel-chromium and nickel-iron-chromium alloys:

- Treat in fused-salt bath
- Quench in and spray with water
- Immerse in Formula 4
- Withdraw from bath and rinse with water
- Immerse in Formula 1
- Withdraw from bath and rinse with water
- Pickle in Formula 3 as required

As with the nickel-copper alloys and nickel alloys, an alternative procedure employing a salt bath and Formula 8, with 10 g/L (1 oz/gal) of ferric chloride added to the formula, effectively removes oxide from hot rolled or annealed Inconel alloys 601, 671, and Incoloy alloys 804 and 903. The procedure for the use of a salt bath is described in the section of this article on oxidized or scaled surfaces under nickel-copper alloys.

If the oxide film cannot be readily removed by pickling directly in Formula 3 or if a fused-salt bath is not available, Formula 13 is a useful pretreatment. Soak for 1 to 2 h in Formula 13; remove the work; rinse it to remove all of the caustic solution; and pickle it in Formula 3. On highly refractory oxides, it is sometimes necessary to repeat the cycle.

Adding 7 to 10 g/L (0.9 to 1 oz/gal) of iron to Formula 3 decreases the danger of overpickling. This can be done conveniently by adding the proper weight of scrap iron to the bath when it is prepared.

The nitric acid/hydrofluoric acid bath (Formula 3) must be used with care. Nickel-chromium and nickel-iron-chromium alloys are subject to intergranular attack in this solution if they have been sensitized by heating in or slowly cooling through the 540 to 760 °C (1000 to 1400 °F) temperature range. Time in the bath should be kept to a minimum, and bath temperature must not exceed 50 °C (125 °F). Stress-relieved and age-hardened material can also be sensitive to intergranular attack if the heat treatment involved exposure to sensitizing temperatures.

Salt Baths

Pretreatment baths of fused salts aid in the pickling of many alloys. They are particularly effective in pickling nickel-chromium and nickel-iron-chromium alloys. Several proprietary baths are commercially available; information on their use can be obtained from manufacturers.

Salt baths are of three types: reducing, oxidizing, and electrolytic. Oxidizing baths are usually the least expensive to operate and the easiest to control, while reducing baths are no longer in common use. Electrolytic baths, although more expensive to install and operate, are quite effective for nickel alloys.

Oxidizing salt baths have a base of either sodium hydroxide or potassium hydroxide. Other salts, such as sodium nitrate and sodium chloride, are added to provide controlled oxidizing properties. The sodium hydroxide bath is operated at temperatures of 425 to 540 °C (800 to 1000 °F); a temperature of 480 °C (900 °F) is preferred for descaling nickel alloys. The potassium hydroxide bath operates at lower temperatures, usually 205 to 260 °C (400 to 500 °F).

Treatment time in oxidizing baths is usually 5 to 20 min. In the operation of continuous strand pickling lines, the time may be as short as 15 to 60 s.

The salt bath oxidizes the lower oxides on the surface of the work to form soluble salts and water. Quenching after treatment removes part of the scale and loosens the remainder so that it is easily removed by appropriate acid dips.

Oxidizing salt baths are also effective cleaners. They remove oil, grease, organic materials, and some inorganic substances from metal surfaces.

Electrolytic baths have a sodium hydroxide base and contain other salts, such as sodium chloride and sodium carbonate, which form reducing agents at the cathode and oxidizing agents at the anode when electrically activated. Baths are usually operated at about 480 °C (900 °F), and although electrolytic salt baths can be used as a batch process, they are more suitable for continuous operations such as descaling of strip. Two tanks are normally used for continuous processes. The work is made anodic in the first tank and cathodic in the second.

Specialized Pickling Operations

Although pickling is used most often to remove the oxide or scale formed during heating, it is also used to remove foreign metals and other substances. Several procedures specifically designed for such purposes are applicable to nickel alloys.

Removal of Lead and Zinc. Lead and zinc will embrittle the high-nickel alloys at elevated temperatures. Consequently, when the alloys are formed in dies made of materials containing lead and zinc, all traces of the die material picked up during forming should be removed. When parts will be given intermediate anneals for processing, or will be exposed to high temperatures during service, this becomes especially important.

Formula 10 is used to remove lead and zinc from nickel and nickel-copper alloys. For the nickel-chromium and nickel-iron-chromium alloys, a bath of nitric acid similar to Formula 1 (but with the concentration of nitric acid increased to 30%) is used. After being immersed in the appropriate bath for 15 min, the work is removed, rinsed in water, and dried.

Detection and Removal of Embedded Iron. During mechanical operations such as rolling to shape or hot pressing, small particles of iron may become so firmly embedded in the surfaces of nickel alloys that they cannot be removed by the cleaning methods normally used for dissolving grease or cutting compounds. Under certain corrosive conditions, such iron particles can initiate local attack. For that reason, it is often necessary to test for iron traces and remove them.

For large-scale testing, a solution consisting of about 1% sodium chloride is effective. Use a chemically pure grade of salt to avoid false results from iron that might be present in less pure grades. After 12 to 24 h in the dilute salt solution, iron particles can easily be detected by examining for rust deposits. When compressed air is available, it is usually less expensive to keep the tank full of salt spray with an atomizer than to fill it with the solution.

The ferroxyl test works well for small-scale testing. The test is carried out by applying a potassium ferricyanide solution to the surface of the material. The solution should be made up in approximately the proportions shown in Formula 16. The ingredients are mixed in earthenware, glass, or ceramic vessels and then boiled until all of the agar-agar is dissolved, and a clear liquor is formed. Use chemically pure sodium chloride to avoid iron contamination of the test solution. The warm solution is applied to the surface to be tested and allowed to remain for at least 1 h. The solution jells as it cools, and the presence of iron on the metal surface is indicated by blue spots in the jell.

The ferroxyl test is sensitive enough that minute particles of iron that collect on the surface in the form of shop dust will appear as small blue spots in the jell. Because the iron dust will be washed off with the jell, the small spots should be distinguished from the larger ones caused by embedded iron. When spots of relatively major proportions develop, large iron particles are probably present.

A solution of hydrochloric acid and ferric chloride, Formula 12, is used to remove the embedded iron. This solution should be used cold and should remain in contact with the metal for only the minimum time required for iron removal, not exceeding 1 h. After removing from the solution, rinse the work thoroughly in cold water, rinse again in warm water, and repeat the detection tests to verify the removal of the iron.

Prevention and Removal of Copper Flash. Copper flash sometimes forms on the surface of nickel alloys during pickling. For copper deposits to form, copper ions in the solution must be in the cuprous state, or must pass from the cupric to the cuprous state during the cementing process. Consequently, any agents in the pickling bath that tend to maintain the cupric state will help prevent coppering.

Oxidizing agents such as nitric acid and sodium nitrate promote the action of pickling solutions but become depleted with use. As the pickling bath ages, the concentration of copper ions increases while that of the oxidizing agents decreases. Thus, aging the bath facilitates coppering on areas where the reducing effect of the metal exceeds the oxidizing power of the bath.

When coppering occurs on Monel alloy 400 and other high-nickel alloys containing an appreciable amount of copper, the bath can be restored by adding small amounts of nitric acid or sodium nitrate. Nickel has greater reducing capacity than Monel alloy 400 and requires a greater concentration of oxidizing agents to prevent coppering in solutions containing copper salts. For this reason, Nickel 200 and similar alloys must not be pickled in solutions that have been used for nickel-copper alloys.

Patches of copper will plate out on nickel-copper alloys if steel contacts the alloys while they are wet with acid. Steel tongs or other devices used to handle the work are the usual sources of coppered areas. Coppering is prevented by using tongs, wires, or other handling devices made of Nickel 200 or Monel alloy 400.

Copper flash is readily removed by immersing in an aerated, 4 to 5% ammonia solution at room temperature. Use approximately 125 mL of ammonia to 1 L of water or 1 pt of commercial aqua ammonia to 1 gal of water. The work should only be immersed for about 1 min. Rinse the work in water after dipping it in the ammonia solution.

Electrolytic Pickling. Light oxide films on any of the nickel alloys can be removed by electrolytic pickling in Formula 7. The work should be made anodic by using a current density of 5.4 to 10.75 A/dm^2 (50 to 100 A/ft^2). Electrolytic pickling is also useful for etching ground material to obtain a surface suitable for inspection.

Cleaning of Springs. In general, cleaning springs made of the high-nickel alloys is not recommended after heat treating. The oxide on heat treated springs usually aids in resisting corrosion at high temperatures. When the oxide is removed, resistance to relaxation is often lowered.

If cleaning is necessary for inspection of the springs, treat in a salt bath, quench in water, and rinse to produce a good surface. Table 2 shows the effect of the cleaning method on resistance to relaxation of Inconel alloy X-750 springs age hardened at 730 °C (1350 °F) for 16 h.

Table 2 Effect of cleaning method on relaxation of Inconel alloy X-750 springs(a)

Cleaning method	Relaxation (b), %
No cleaning, as age hardened	10.0
Oxidizing salt bath plus Formula 3	11.5
Tumbling in sand and oil	12.8
Abrasive blasting (120 grit with water	13.3
Shot peening	13.5
Abrasive blasting (standard sand, dry)	14.6

(a) Cold drawn (No. 1 temper) and age hardened at 730 °C (1350 °F) for 16 h. (b) Relaxation after 250 h at 415 MPa (60 ksi) and 540 °C (1000 °F)

Cleaning for Welding. Before maintenance welding is done on high-nickel alloys that have been in service, products of corrosion and other foreign materials must be removed from the area to be welded. Clean bright base metal should extend 50 to 75 mm (2 to 3 in.) from the joint on both sides of the material to prevent the corrosion products from embrittling at welding temperatures. Cleaning can be done mechanically by grinding with a fine wheel or disk, or chemically by pickling.

Flash pickling solutions are effective in cleaning before welding. Apply the solutions by swabbing, brushing, or dipping, if the parts are easily handled. A single dip in Formula 2 is adequate for the nickel-copper alloys. Formula 1 is used for nickel alloys and Formula 3 for nickel-chromium and nickel-iron-chromium alloys.

Finishing

Nickel alloys can be ground, polished, buffed, or brushed by all methods commonly used for other metals.

For high-nickel alloys, a series of operations is required to produce a satisfactory finish. The number and type of operations required depend on the initial finish of the material, the desired final finish, and the type of equipment used. The pressures and speeds of the finishing equipment must be closely controlled. The high-nickel alloys, particularly nickel-chromium and nickel-iron-chromium alloys, do not conduct heat away as rapidly as copper and aluminum. Excessive heat will destroy the true color of the metal and may warp flat, thin articles.

Some general recommendations for finishing operations are given in Table 3. Table 4 lists spindle speeds and corresponding surface speeds for various wheel diameters.

Grinding is often the first operation in a finishing sequence, used to remove large surface imperfections and to rough-down welds prior to polishing and buffing. Rubber-bond grinding wheels are used for nickel and Monel nickel-copper alloys for their cutting effectiveness and for their relative soft-

Table 3 Recommended finishing procedures

Operation	Wheel	Grit No.	Compound	Speed m/s	Speed sfpm
Grinding	Rubber bond	24 or 36	None	40-45	8000-9000
Grinding	Vitrified bond	24 or 36	None	25-30	5000-6000
Roughing	Cotton fabric, sewn sections	60 or 80	None	30-40	6000-7500
Dry fining	Cotton fabric, sewn sections	100 or 120	None	30-40	6000-7500
Greasing	64-68 unbleached sheeting, spirally sewn sections	150 or 180	Polishing tallow or No. 180 emery grease cake	30-40	6000-7500
Grease coloring	88-88 unbleached sheeting, spirally sewn or loose disk; or quilted sheepskin	200 or 220	Polishing tallow or "F" emery grease cake	30-40	6000-7500
Bobbing and sanding	Leather wheel for two bobbing operations, second with medium-density felt wheel	...	Grout	25	5000
Cutting down	88 unbleached sheeting, loose spirally sewn sections or loose-disk wheel	...	Tripoli	40-45	8000-9000
Coloring (bright finish)	88-88 unbleached sheeting, loose spirally sewn sections or loose-disk wheel	...	White aluminum oxide	50	10 000
Coloring (mirror finish)	Loose-disk, 88-88 unbleached sheeting or Canton flannel	...	Green chromium oxide	50	10 000
Brushing	Tampico	...	"F" emery grease cake or grout	5-15	1200-3000

Table 4 Spindle speed for various surface speeds and wheel diameters

Spindle speed was measured in rev/min

Surface speed m/s	Surface speed ft/min	Wheel diameter 150 mm (6 in.)	200 mm (8 in.)	250 mm (10 in.)	300 mm (12 in.)	360 mm (14 in.)	410 mm (16 in.)	460 mm (18 in.)
15	3 000	1930	1450	1150	950	820	710	640
20	4 000	2550	1900	1500	1300	1100	950	850
25	5 000	3200	2400	1900	1600	1375	1200	1050
28	5 500	3500	2600	2100	1750	1500	1300	1175
30	6 000	3800	2850	2300	1900	1650	1425	1275
38	7 500	4800	3550	2850	2400	2100	1800	1600
40	8 000	5100	3800	3100	2550	2200	1900	1700
45	9 000	5750	4300	3450	2850	2450	2150	1900
50	10 000	6400	4750	3800	3200	2750	2400	2100

ness, which reduces the heat generated. Rubber-bond wheels should be operated at a surface speed of 40 to 45 m/s (8000 to 9000 ft/min). Vitrified-bond wheels are preferred for grinding the harder alloys such as the Inconel and Incoloy alloys. These wheels should be operated at a surface speed of 25 to 30 m/s (5000 to 6000 ft/min). Light welds can be ground efficiently with a No. 36 grit wheel, while a No. 24 grit is more practical for heavy welds.

Polishing. The first polishing operation should be done with very fine grit to remove all surface defects and give a base upon which to build the final finish. Wheels of No. 60 to 80 grit are usually required to remove heavy oxide or deep defects.

The first operation should be done dry. After the initial roughing, use tallow on all roll-head wheels of No. 150 grit or finer. Tallow clogs the wheel, giving a smoother finish, and reducing the amount of heat generated.

When polishing flat work, each subsequent operation should be done with a grit 30 to 40 numbers finer than the previous one, until the surface is ready for the brushing or bobbing operations needed to prepare for buffing.

When possible, the scratches produced by the abrasive should cross the scratches left by the preceding operation. When polishing is done in only one direction, the finer abrasive will follow in the grooves made by the coarser abrasive, and the efficiency of the polishing wheels will be impaired.

Wheels for roughing and dry fining should be made of tightly woven unbleached cotton fabric. To prevent chattering, the wheels should be perfectly balanced and should have a soft or cushioned face. Fine grit wheels require more cushions than coarse wheels.

A more resilient and flexible wheel should be used for greasing operations. Greasing wheels should be made of 64 to 68 count unbleached sheeting and should have more cushion than the wheels used for coarser polishing.

Grease coloring may be done on a full-disk, quilted sheepskin wheel or a spirally sewn wheel made of fine-count (88 to 88) heavy sheeting. Compounds of artificial abrasives are preferred for roughing and dry fining. Turkish emery is usually used for greasing and grease coloring.

Bobbing done with emery grout is more like burnishing than polishing and, if performed in one operation, is best done with leather wheels. If done in two operations, the second should be done with a medium-density felt wheel, because felt wheels require less pressure.

The best results are obtained from emery roll-head wheels at surface speeds of 33 to 38 m/s (6500 to 7500 ft/min). The metal drags at slower speeds, and at excessive speeds, the wheel tends to pull up the surface of the metal.

Buffing. For good results in buffing, a high-quality wheel of the proper construction and material is essential. Wheels should be of sturdy, closely woven, high-count sheeting. The close weaves give good cutting, while the heavy threads provide good coloring.

Buffing is usually done in two operations. The first, the cutting-down operation, is done with a sewn buffing wheel operating at a surface speed of 40 to 45 m/s (8000 to 9000 ft/min). The second, the coloring operation, is done with a loose-disk wheel operating at a surface speed of approximately 50 m/s (10 000 ft/min). A loose disk, Canton-flannel wheel is best for buffing to a mirror finish. The high speeds used in buffing create a high heat. Consequently, less pressure must be used than for polishing.

The cutting-down operation is normally performed with tripoli compound, which leaves a haze on high-nickel alloys. Use buffing compound with less grease to promote deep color.

White aluminum oxide and green chromium oxide compounds can be used for the final coloring. Chromium oxide compounds produce less friction and give a truer color.

Brushing. A tampico or wire wheel will produce a brushed finish on the high-nickel alloys. Tampico wheels usually produce a better finish and higher luster and, because of their flexibility, are better for irregular shapes. A tampico wheel is used with emery paste or grout to produce a satin finish, and with pumice and water or pumice and oil to produce a butler finish. For wet brushing, tampico wheels should have wooden hubs.

Wire brushes can be used to produce a satin finish on sheet-metal articles. Brushes should have a wire diameter of 0.10 to 0.20 mm (0.004 to 0.008 in.) and not be made of either steel or brass wire. Small particles from the brush are always embedded in the metal during the process. The steel particles will rust, and the brass particles will discolor the nickel alloys.

Brushing is done at slower speeds than those used for polishing. Brushing speeds are normally 5 to 15 m/s (1200 to 3000 ft/min), depending on the final finish desired. Higher speeds — 20 to 30 m/s (4000 to 6000 ft/min) — are required to produce a bright wire brush finish. Wire brushes operated at too slow a speed produce coarse, undesirable scratches.

Cleaning and Finishing of Zinc Alloys*

ALLOYS used in the manufacture of zinc alloy die castings are made with high-grade zinc with about 4% aluminum, 0.04% magnesium, and either 0.25% (maximum) or 1.0% copper. Die castings are usually dense and fine grained, but do not always have smooth surfaces. Defects sometimes found in the surface layers include cracks, cold shut crevices, skin blisters, and hemispherical pores. Burrs are usually left at parting lines where fins and gates are removed by die trimming. The normal sequence of preparation steps prior to plating includes:

- Smoothing of parting lines
- Smoothing of rough or defective surfaces, if necessary
- Buffing, if necessary
- Precleaning and rinsing
- Alkaline electrocleaning and rinsing
- Acid dipping and rinsing
- Copper striking

Polishing of parting lines is the initial finishing operation on zinc die castings after fins, gates, and risers have been removed. This operation may be carried out by using setup wheels or abrasive belts, tumbling with abrasive media, or vibrating with abrasives. Abrasive sizes from 180 to 220 mesh are commonly used. The abrasive surface may be lubricated to produce a finer texture. Setup wheels should not be operated at more than 30 to 40 m/s (6000 to 8000 sfm). Abrasive belts should not exceed 25 to 35 m/s (5000 to 7000 sfm). Lower speeds from about 18 to 20 m/s (3500 to 4000 sfm) are often used.

Tumbling in horizontal barrels loaded with abrasive stones, such as limestone, preformed and fused aluminum oxide, ceramic shapes or abrasive-loaded plastic chips, and a lubricant such as soap or detergent solution, removes parting line burrs from die castings in 4 to 12 h. Vibration in a bed of resin-bonded abrasive chips removes parting line burrs in 1 to 4 h. Frequencies range from 700 to 2100 cpm and amplitudes from 0.8 to 6.5 mm (1/32 to 1/4 in.).

Polishing Other Surfaces. Although many die castings are smooth enough to require only buffing, spot, or over-all polishing may be required. A 220-mesh abrasive is used on setup wheels or abrasive belts, although finer sizes may suffice at times. The same surface speeds of setup wheels or abrasive belts defined above for polishing parting lines are applicable. The abrasive surface should be lubricated. For further information, refer to the article on polishing and buffing in this Volume.

Buffing of zinc-based die castings should be performed using cloth wheels of suitable stiffness, at speeds not exceeding 35 to 45 m/s (7000 to 9000 sfm). Slower speeds, 18 to 25 m/s (3500 to 5000 sfm), are commonly used. After buffing, or after buffing and coloring, clean the surface by passing it over a relatively clean, dry buffing wheel. In some instances, a separate coloring operation may be needed. The polishing, buffing, and coloring steps must be properly graduated or a poor surface, or the necessity for an additional operation, may result. The buffing compound should be made with a binder that is readily emulsified or saponified during cleaning. For further discussion, see the article on polishing and buffing in this Volume.

The packing of hardened compound into holes and recesses makes cleaning difficult. Extra care in avoiding such packing by wiping to remove excess compound is sometimes desirable and reduces the demand placed on the precleaning operation.

Preparation for Plating. Zinc-based die castings may be prepared for plating using the procedures shown in the flow chart in Fig. 1. Duration of immersion in the alkaline cleaning solution must be kept to a minimum, or the cleaner will attack the zinc. Anodic cleaning is preferred because any film of alkali remaining on the parts after anodic cleaning is removed more easily in the subsequent acid dip than is alkaline film from cathodic cleaning.

Degreasing. Several methods are used to remove greasy soil from zinc-based die castings. These include (*a*) vapor degreasing, (*b*) emulsion cleaning, and (*c*) solvent cleaning. Vapor de-

Reviewed by Roger E. Marce, Manager, Technical Services, Cadmium Council, Inc.

Fig. 1 Surface preparation of zinc-based die castings for electroplating

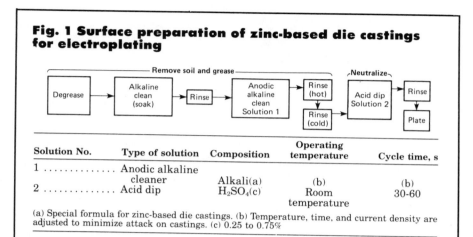

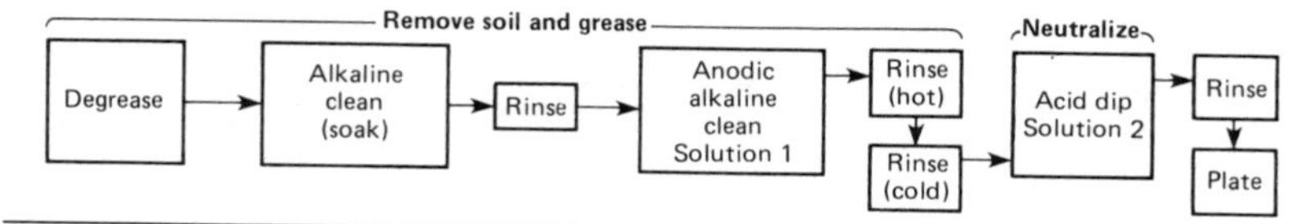

Solution No.	Type of solution	Composition	Operating temperature	Cycle time, s
1	Anodic alkaline cleaner	Alkali(a)	(b)	(b)
2	Acid dip	H_2SO_4(c)	Room temperature	30-60

(a) Special formula for zinc-based die castings. (b) Temperature, time, and current density are adjusted to minimize attack on castings. (c) 0.25 to 0.75%

greasers are often supplemented by spraying with chlorinated solvents. Care must be taken to avoid dragging the solvents into the plating baths, which the solvents would seriously contaminate. Vapor degreasing is not recommended for cleaning buffing compounds, since dry abrasives are left behind which are even harder to clean. Although usually not regarded as heavy-duty cleaners, and despite their relatively high dragout loss, emulsion soaks have their place where carryover of oil or chlorinated solvent cannot be tolerated. Where the dirt load is comparatively light or the volume of work is small, hand scrubbing supplemented by cathodic cleaning may be adequate and most satisfactory. Cathodic cleaning must be followed by anodic cleaning. The mineral solvents commonly used in solvent cleaning are not easily removed during alkaline electrocleaning. When plating is to follow cleaning, these solvents may be dragged over into the plating solutions. Alkaline soak cleaning or thorough washing after solvent cleaning is a usual requirement.

Alkaline Soak Cleaning. Hot mixtures of emulsifiers and wetting agents, combined with mild alkaline phosphates or borates, are used as presoaks to soften and to remove buffing compound. Combining these soaks with ultrasonics is particularly effective on impacted buffing compound. Such detergent soaks are followed by spray cleaning with an alkaline cleaner or at least spray rinsing with warm water before electrocleaning. Sometimes conventional alkaline soak cleaners are used for precleaning die castings with little or no buffing compound on them. These alkaline cleaners must be mild and inhibited because strong alkali attacks the castings.

Alkaline Electrocleaning. In most instances, it is impossible to ensure completely adequate cleaning for electroplating if only degreasing, emulsion cleaning, solvent cleaning, or alkaline soak cleaning is used. Final cleaning should be performed in an electrolytic alkaline cleaner, following the procedures described in the article on alkaline cleaning. Extreme care should be exercised in controlling the current density for electrocleaning zinc die cast. Normally 3 to 4 V and 2.5 to 3.0 A/dm^2 (25 to 30 A/ft^2) is ideal. High current density cleaning attacks zinc severely. This is referred to as burning.

A solution for anodic cleaning contains 26 to 34 g/L (3.4 to 4.5 oz/gal) of mixed alkalis such as trisodium phosphate and sodium metasilicate, 0.5 g/L (0.06 oz/gal) of surfactant, and not more than 0.5 g/L (0.06 oz/gal) of sodium hydroxide. This solution is heated to 70 to 82 °C (158 to 180 °F). Time cycles vary from 25 to 45 s. Lower temperatures may be required if time cycles must be prolonged for more than 45 s, or if the transfer time from the electrocleaner to the first rinse is more than about 30 s. An electrocleaner safe for use after simple trichlorethylene degreasing may be too strong when its alkalinity effect is added to that of a previously used alkaline soak cleaner. Similarly, the relatively long immersion times imposed by some automatic conveyors may necessitate reducing the concentration or temperature of a cleaner or the substitution of a milder cleaner. The more effective the precleaning, the greater is the danger of overcleaning in the electrocleaner. Rinsing after alkaline cleaning must be thorough. Whenever feasible, hot rinsing followed by cold rinsing should be used. Directed sprays may remove solutions from holes and cavities. Good rinsing practice includes both spray and dip rinsing. Agitation of dip rinses is helpful.

Diphase Cleaning. Because extended treatment of zinc-based die castings in alkaline electrocleaners may result in attack on the casting metal, other cleaning methods are often used. Diphase cleaning is widely used. In this method, parts are immersed and soaked in a tank that contains a layer of high-flash-point solvent oil, such as kerosine, that floats on a buffered solution of a weak alkali. This immersion soak loosens the dirt, which drops to the bottom of the tank. The remainder of the gross dirt and the oil is removed by an alkaline power spray and a subsequent water spray rinse. Diphase cleaning is excellent for dirt removal, but some oil may be retained in pores in the die castings and in built-up and cracked rack coatings.

Acid Dipping. When zinc-based die castings are to be electroplated, an acid dip must follow alkaline cleaning. Dilute acid dips should be used after short contacts with dilute mild alkalis and more concentrated acids after prolonged contact with strongly alkaline solutions. Dilute solutions of sulfuric acid with concentrations of 0.25 to 0.75% are commonly used. Similar solutions of hydrofluoric acid or a mixture of sulfuric and hydrofluoric acids are also effective. For very mild acid dipping, citric acid may be used. Because cyanide copper solutions are used as the first plating electrolytes, thorough rinsing is required after acid dipping.

Copper Strike. When copper plate is to be the first deposit in a multiplate system, cyanide copper strike should be used before copper plating. This minimizes blistering of the copper plate. Procedures for copper striking and subsequent plating cycles required for a decorative chromium plated finish are described in the article on decorative chromium plating. The copper strike serves also as a base for electrodeposits of other metals and alloys.

Système Internationale d'Unités (SI)

SI Base Units

Quantity	Unit	Symbol	Quantity	Unit	Symbol	Quantity	Unit	Symbol
Length	metre	m	Thermodynamic temperature	kelvin	K	Luminous intensity	candela	cd
Mass	kilogram	kg	Amount of substance	mole	mol	Plane angle(a)	radian	rad
Time	second	s				Solid angle(a)	steradian	sr
Electric current	ampere	A						

(a) Supplementary unit

SI Derived Units(a)

Quantity	Unit	Symbol	Formula	Quantity	Unit	Symbol	Formula
Frequency (of a periodic phenomenon)	hertz	Hz	s^{-1}	Capacitance	farad	F	C/V
Force	newton	N	$kg \cdot m/s^2$	Electric resistance	ohm	Ω	V/A
Pressure, stress	pascal	Pa	N/m^2	Conductance	siemens	S	A/V
Energy, work, quantity of heat	joule	J	$N \cdot m$	Magnetic flux	weber	Wb	$V \cdot s$
Power, radiant flux	watt	W	J/s	Magnetic flux density	tesla	T	Wb/m^2
Quantity of electricity, electric charge	coulomb	C	$A \cdot s$	Inductance	henry	H	Wb/A
Electric potential, potential difference, electromotive force	volt	V	W/A	Luminous flux	lumen	lm	$cd \cdot sr$
				Illuminance	lux	lx	lm/m^2
				Activity (of radionuclides)	becquerel	Bq	s^{-1}
				Absorbed dose	gray	Gy	J/kg

(a) Derived units in this list include only those units for which special names and symbols have been approved by the General Conference on Weights and Measures (CGPM)

SI Prefixes(a)

Prefix	Multiplication factor	Symbol	Prefix	Multiplication factor	Symbol
exa	1 000 000 000 000 000 000 = 10^{18}	E	deci(b)	0.1 = 10^{-1}	d
peta	1 000 000 000 000 000 = 10^{15}	P	centi(c)	0.01 = 10^{-2}	c
			milli	0.001 = 10^{-3}	m
tera	1 000 000 000 000 = 10^{12}	T			
giga	1 000 000 000 = 10^{9}	G	micro	0.000 001 = 10^{-6}	μ
mega	1 000 000 = 10^{6}	M	nano	0.000 000 001 = 10^{-9}	n
			pico	0.000 000 000 001 = 10^{-12}	p
kilo	1 000 = 10^{3}	k			
hecto(b)	100 = 10^{2}	h	femto	0.000 000 000 000 001 = 10^{-15}	f
deka(b)	10 = 10^{1}	da	atto	0.000 000 000 000 000 001 = 10^{-18}	a

(a) Used to form multiples and decimal fractions of the base and derived SI units. (b) Normally avoided. (c) Use not recommended

Abbreviations and Symbols

A ampere

Å angstrom

AA Aluminum Association

ASCR aluminum conductor, steel reinforced

AEC Atomic Energy Commission

AES American Electroplater's Society

AFS American Foundrymen's Society

A·h ampere hour

AISI American Iron and Steel Institute

AMS Aerospace Material Specification (of SAE)

ANSI American National Standards Institute, Inc.

approx approximately

ASM American Society for Metals

ASME American Society of Mechanical Engineers

ASTM American Society for Testing and Materials

avg average

avoir oz avoirdupois ounce

bal balance

°Bé degrees Baumé

Btu British thermal unit

cal calorie

CASS copper-accelerated acetic acid-salt spray

cm centimetre

cpm cycles per minute

CPVC chlorinated polyvinyl chloride

CVD chemical vapor deposition

d diameter

dc direct current

DEAB N-Diethylamine borane

diam diameter

DMAB N-Dimethylamine borane

DMAC Defense Metals Information Center

DPH diamond pyramid hardness

EB electron beam

ED electrodialysis

EDTA ethylenediamine tetraacetate

EMI electromagnetic interference

EPA Environmental Protection Agency

Eq equation

eV electron volt

EW Erftwerk process

FAS ferrous aluminum sulfate

ft foot

gal gallon

g gram

GPa gigapascal

h hour

HB Brinell hardness

HC hot cathode

HIP hot isostatic pressing

HK Knoop hardness

hp horsepower

HRB Rockwell "B" hardness

HRC Rockwell "C" hardness

HV Vickers hardness

Hz hertz

in. inch

ID inside diameter

in situ in the natural or original position

ISO International Standards Organization

IVD ion vapor deposition

J joule

K kelvin

kg kilogram

kHz kilocycles per second (kilohertz)

ksi kips per square inch

kV kilovolt

kW kilowatt

L litre

lb pound

log common logarithm (base 10)

M molar concentration

m metre

max maximum

MBV modified Bauer-Vogel process

mg milligram

MHC Martin Hard Coat

MHz megahertz (megacycles per second)

min minimum, minute

mL millilitre

mm millimetre

mμ millimicrons

mV millivolt

N newton

N normal concentration

NBS National Bureau of Standards

NF National Formulary

NFPA National Fire Protection Association

nm nanometre

NMAB National Material Advisory Board

No. number

NTA trisodium nitrilotriacetate

OD outside diameter

ORP oxidation-reduction potential

OSHA Occupational Safety and Health Administration

oz ounce

Pa pascal

PACVD plasma-assisted chemical vapor deposition

PEI Porcelain Enamel Institute, Inc.

pH negative logarithm of hydrogen-ion activity

P/M powder metallurgy

POTW Publicly Owned Treatment Works

ppm parts per million

PR periodic reverse

psi pounds per square inch

psig pounds per square inch, gage

PVA polyvinyl alcohol

PVAc polyvinyl acetate

PVC polyvinyl chloride

PVD physical vapor deposition

qt quart

R roentgen

RA activated rosin

Ref reference

rem remainder

rev/min revolutions per minute

RFI radio frequency interference

RMA rosin mildly activated

RO reverse osmosis

s second

SAE Society of Automotive Engineers

SCR silicon controlled rectifier

SEM scanning electron microscope

sfm surface feet per minute

Soc society

sp gr specific gravity

STP standard temperature and pressure

SUS Saybolt universal seconds

t tonne (metric ton or 1000 kg)

TBO time between overhauls

T_m melting point

TR temper rolled

TWA time-weighted average

UI University of Illinois designations

V volt

VM & P varnish makers and painters

vol volume

vol % volume percent

W watt

wt % weight percent

yr year

° degree, angular measure

°C degree Celsius (centigrade)

°F degree Fahrenheit

Δ difference

= equal sign

> greater than

< less than

± maximum deviation

μm micron

μin. micro-inch

− minus, negative ion charge

• multiplied by

× multiplied by, diameters (magnification)

/ per

% percent

+ plus, in addition to, including, positive ion charge

Ω ohm

Index

C

D

F

G

I

K

L

N

Q

R

S

T

U

V

W